Wolf Prize
in Physics

Wolf Prize
in Physics

Edited by

Tsvi Piran
Hebrew University of Jerusalem, Israel

We World Scientific

NEW JERSEY • LONDON • SINGAPORE • BEIJING • SHANGHAI • HONG KONG • TAIPEI • CHENNAI • TOKYO

Published by

World Scientific Publishing Co. Pte. Ltd.

5 Toh Tuck Link, Singapore 596224

USA office: 27 Warren Street, Suite 401-402, Hackensack, NJ 07601

UK office: 57 Shelton Street, Covent Garden, London WC2H 9HE

Library of Congress Cataloging-in-Publication Data

Names: Piran, Tsvi, 1949–   editor.

Title: Wolf prize in physics / edited by Tsvi Piran, Hebrew University of Jerusalem, Israel.

Description: Singapore ; Hackensack, NJ : World Scientific Publishing Co. Pte. Ltd., [2016] |
   Includes bibliographical references.

Identifiers: LCCN 2016024940| ISBN 9789813109858 (hardcover ; alk. paper) |
   ISBN 9813109858 (hardcover ; alk. paper) | ISBN 9789813141025 (pbk. ; alk. paper) |
   ISBN 9813141026 (pbk. ; alk. paper)

Subjects: LCSH: Physics--Awards. | Wolf Foundation Prizes.

Classification: LCC QC28 .W65 2016 | DDC 530.079--dc23

LC record available at https://lccn.loc.gov/2016024940

British Library Cataloguing-in-Publication Data

A catalogue record for this book is available from the British Library.

The editor and the publisher would like to thank the following for their permission to reproduce the articles found in this volume:

American Physical Society, AIP Publishing, Elsevier Science B.V., Springer Publishing, Taylor and Francis Publishing, Royal Society Publishing, EDP Sciences and IOP Publishing

INTRODUCTION

The Wolf Foundation began its activities in 1976, with an initial endowment donated by the Wolf family. The Foundation's founders and major donors were Dr. Ricardo Subirana y Lobo Wolf and his wife Francisca. Since 1978, five or six prizes have been awarded annually in the sciences and arts. The prizes are awarded to "outstanding scientists and artists — irrespective of nationality, race, color, religion, sex or political views — for achievements in the interest of mankind and friendly relations among peoples". Laureates receive their awards from the President of the State of Israel. The prize presentation takes place at a special ceremony at the Knesset Building (Israel´s Parliament), in Jerusalem.

Within a very short time after its initiation, the Wolf prize became one of the major signs for recognition of scientific achievements and excellence. Wolf prize laureates tell the story of the development of science and art in the second half of the twentieth century and the first decades of the twenty-first century. This is true in all fields in which this prize is given, but particularly so in physics.

The first Wolf physics award was given in 1978 to Chien-Shiung Wu for exploring the weak interaction. Since that first prize, awards have been given to fifty-six laureates from all branches of the subject. These include particle physics, condensed matter, general relativity, astrophysics, nuclear physics, atomic physics and quantum information. The scope of the research is demonstrated by the last (2015) prize, which was split between James Bjorken for his contributions to subnuclear physics and Robert Kirshner for forging the path to supernova cosmology. The scope of these two explorations range from a subatomic Fermi scale to the size of the whole Universe, several billion light years, with a factor of 10^{40} between the two. The wide range of topics manifests mankind's quest to explore the deepest secrets of Nature on all scales and in all possible directions.

The different chapters of this book are devoted to different Wolf prize laureates. Each contains two major publications, a short biography and a complete list of publications. More importantly, each also includes a commentary written by the Laureate describing his or her scientific career. These commentaries provide a unique opportunity to have a look at how the scientific process works in physics.

Tsvi Piran
Editor
Schwartzman Chair for Theoretical Physics
Racah Institute for Physics
The Hebrew University of Jerusalem

CONTENTS

Michael E. Fisher

MICHAEL ELLIS FISHER

Birthplace

Born   Fyzabad, Trinidad, West Indies, 3 September 1931

Education

1951   B.Sc. King's College, University of London, 1st Class Honors in Special Physics
1957   Ph.D. King's College, London, Physics (Supervisor: Donald M. MacKay)

Research and Professional Experience

1952–1953	Lecturer in Mathematics, Royal Air Force Technical College
1956–1958	Department of Scientific and Industrial Research, Senior Research Fellow, King's College, London
1958–1962	Lecturer in Theoretical Physics, King's College, London
1962–1964	Reader, King's College, London
1963–1964	Guest Investigator, Rockefeller University, New York
1965	Guest Researcher, Institut des Hautes Etudes Scientifiques, France
1965–1966	Professor of Physics, King's College, London
1966–1973	Professor of Chemistry and Mathematics, Cornell University
1970–1971	Visiting Professor in Applied Physics, Stanford University
1973–1989	Horace White Professor of Chemistry, Physics and Mathematics, Cornell University
1975–1978	Chairman, Department of Chemistry, Cornell University
1979	Morris Loeb Lecturer, Harvard University and Visiting Professor, M.I.T.
1981–1982	Visiting Fellow and Institute Professor, Sackler Institute for Advanced Studies, Tel Aviv University
1983	Guest Researcher, Institut des Hautes Etudes Scientifiques, France
1984	Sherman Fairchild Distinguished Scholar, Caltech
1985	Visiting Professor, Weizmann Institute of Science and in Theoretical Physics at Oxford University
1987–1993	Wilson H. Elkins Professor, Institute for Physical Science and Technology, University of Maryland
1988–2012	Distinguished University Professor, University of Maryland
1993	Lorentz Professor, University of Leiden
1993–	Regent's Professor of the University of Maryland System
1994	Visiting Professor, National Institute of Standards and Technology, Maryland
2012–	Distinguished University Professor Emeritus and Research Professor
2013–	Professor Emeritus of Chemistry and Chemical Biology, Physics, and Mathematics, Cornell University

Membership of Societies

1954	Physical Society, London
1968	Society for Industrial and Applied Mathematics
1968	Mathematical Association of America
1969	Fellow, American Physical Society
1971	Fellow, The Royal Society of London
1972	American Chemical Society
1979	Fellow, American Academy of Arts and Sciences
1983	National Academy of Sciences (Foreign Associate), U.S.A.
1986	Fellow, American Association for the Advancement of Science
1986	Honorary Fellow, Royal Society of Edinburgh
1993	American Philosophical Society
1996	Brasilian Academy of Science (Foreign Member)
1996	Fellow, New York Academy of Sciences (resigned 2002)
1999	Biophysical Society
2000	Indian Academy of Sciencies (Honorary Fellow), Bangalore, India
2001	Society of General Physiologists
2003	Royal Norwegian Society of Sciences and Letters (Foreign Member)
2004	Fellow, Institute of Physics, London

Honours

1971	Sigma Xi RESA Touring Lecturer
1975	Fritz London Memorial Lecturer, Duke University
1977	Exchange Visitor, NSF-Japan Society for Promotion of Science
1977	Walker Ames Professor, University of Washington
1979	Bakerian Lecturer, The Royal Society of London
1981	Fellow, King's College, London
1985	Cherwell-Simon Lecturer, Oxford University
1987	Honorary Doctorate of Science, Yale University
1989	National Science Council Lecturer, Taipei, Taiwan
1991	Conference (Natl. Acad. Sci., Washington) and Festschrift in honor of 60th Birthday
1992	Josiah Willard Gibbs Lecturer, American Mathematical Society
1992	Doctorate of Philosophy Honoris Causa, Tel Aviv University
1993–1995	Vice President, The Royal Society of London
1995	G.N. Lewis Memorial Lecturer, University of California, Berkeley
1997	George Fisher Baker Lecturer in Chemistry, Cornell University
1998	Ferdinand G. Brickwedde Lecturer in Physics, The Johns Hopkins University
2001	Rutgers University Conference and Festschrift in honor of 70th Birthday
2005	Royal Medal, The Royal Society of London
2007	Homi J. Bhabba Lecturer, Tata Institute of Fundamental Research, Mumbai, India
2009	Honorary Doctorate of Philosophy, Weizmann Institute of Science
2010	Life Member, Board of Trustees, Weizmann Institute of Science

2012	Symposium and Banquet celebrating Lifetime Achievements, University of Maryland
2012	Docteur Honoris Causa, École Normale Supérieure de Lyon
2015	Rudranath Capildeo Award for Applied Science and Technology, Gold Medal, State of Trinidad and Tobago

Awards

Irving Langmuir Prize in Chemical Physics, 1970; John Simon Guggenheim Memorial Fellowships, 1970–1971; 1978–1979; New York Academy of Sciences Award in Physical and Math. Sciences, 1978; Guthrie Medal and Prize of the Institute of Physics (U.K.), 1980; Wolf Prize in Physics, 1980; Michelson-Morley Award of Case Western Reserve University, 1982; National Academy of Sciences, 1983 James Murray Luck Award; Boltzmann Medal, IUPAP Commission on Thermodynamics and Statistical Mechanics, 1983; Lars Onsager Medal, Norwegian Institute of Technology, 1993; Hildebrand Award, American Chemical Society, 1995; Hirschfelder Prize in Theoretical Chemistry, University of Wisconsin, 1995; First Lars Onsager Memorial Prize, American Physical Society, 1995; University of Maryland, 1998, Distinguished Scholar-Teacher Award; Michelson Lecturer, Case-Western Reserve University, 1999; Wolfson College Lecturer, Oxford University, 2000; Thomas A. Edison Memorial Lecturer, U.S. Naval Research Laboratory, 2000; Distinguished Lecturer, Institute for Theoretical Physics, The Technion, Israel, 2004; Landau Days-2005, 40th Anniversary of the Founding of the Landau Institute for Theoretical Physics, Chernogolovka, Russia, 2005; J. Robert Oppenheimer Lecturer, University of Berkeley, 2006; C.V. Raman Memorial Lecturer, Indian Institute of Science, Bangalore, 2007; Mark Kac Memorial Lecturer, Los Alamos National Laboratory, 2008; Outstanding Referee, American Physical Society, 2008; 2009 BBVA Foundation, Frontiers of Knowledge Award in Basic Sciences, 2010.

COMMENTARY

My father, Harold Wolf Fisher, worked for Royal Dutch Shell, originally in Venezuela but, on marrying my mother, Jeanne Marie Halter (who came to England as a young girl from Alsace-Lorraine), he moved to Trinidad in the West Indies. Thus, in 1931, I was born in a small town, Fyzabad, near Point Fortin in Trinidad where we lived. We stayed in Trinidad for a year or more before, in the depths of the Great Depression, we returned to live in North West London. My education was, hence, mainly in Great Britain but for two years, 1940–1942, with my mother and sister, Ann Marie, 17 months my junior, we stayed (as evacuees) with kind relatives (the Weinberg's) in Johannesburg, South Africa.

From my earliest age I was filled with curiosity about most things in the world. Asking questions and trying new things became a way of life, always encouraged by my father: it is still innate for me. My mother, the daughter of a school teacher, paid attention to my more formal schooling. Eventually, I gained a part-scholarship to enroll in University College School in Hampstead.

Feeling a need for concrete answers to some of my queries I taught myself calculus from the delightful little book "*Calculus made Easy*" by Silvanus P. Thompson F.R.S. (Second Edition, Macmillan & Co. Ltd., London, 1914) which I bought second-hand in the Charing Cross Road in 1944. I especially took to heart the "*Ancient Simian Proverb*" (printed just after the title page): "*What one fool can do, another can*" (although it was many years before I knew where "*Simia*" was).

Some years later at age 16, under the Labour Government that came to power in the first post-World War II elections in Britain, I won a State Scholarship. Turned down by Cambridge University (on the grounds of youth and not having done military service) I enrolled in King's College, London, in October 1948. To the disappointment of my Chemistry teacher at school, I joined the *Special Physics* (three-year) program; but, by cutting some practical classes, I also sat in on the basic *Analysis* lectures in the Mathematics Department. Ironically, many years later I found myself Chair of the Chemistry Department at Cornell!

Quite soon I met another ex-school boy, Peter W. Higgs (later Wolf Prize and Nobel Prize winner) who was just a year ahead of me. Peter was a great encouragement for me and before long became a close personal friend. He gained the first degree at King's specializing in *Theoretical Physics*; a year later I obtained the second (with *First Class Honours*). Meanwhile, when time allowed, I played, and wrote about the flamenco guitar and, to earn a little cash, taught regularly and performed occasionally.

From age eleven or so, I had a fascination with computing devices, both digital and analog. Following this up by readings in the *Encyclopedia Britannica* I learned about the *hatchet planimeter* (for measuring areas in the plane) and still have the one I made for myself. More ambitious (but not carried to completion) was the construction (in the workshop at school, over-impressing the crafts teacher) of a device, invented by Lord Kelvin, for generating sine waves for Fourier analysis of the tides. But at King's, to my delight, I discovered that Dr. Donald M. MacKay (then a lecturer) had research underway on high speed (for that time!) electronic analog computers. He had worked in radar during the War and in my undergraduate summers I was welcomed to work in his laboratory. In due course (following two years service in the Royal Air Force where, as an officer, I taught mathematics at the Royal Air Force Technical College), my interactions with Donald MacKay led to my Ph.D. thesis with him at King's. In turn, I published my first scientific papers (see [1, 4, 5, 7–10, 12, 13] of *Section B* of the *Publication List*) and eventually a joint book with MacKay [A.2]. Studies of computation also gave rise to my longstanding interest in numerical analysis and related mathematics [16], including my first article to be reprinted [23].

But in 1956, short of immigrating, with wife and two young children, to the United States (where analog computers were actively used in the aircraft industry for some time to come), there seemed no possibilities for going further in analog computation in Britain. So with the aid of the "new" Professor of Theoretical Physics, Cyril Domb [323, 413] (who had already engaged me in some interesting mathematics problems [6, 11]), I shifted fields in the directions he proposed.

From a visit in 1951 Cyril Domb had contacts in Israel, particularly at the Weizmann Institute, that included the distinguished polymer chemist Aharon Katchalsky Katzir.[1] Katchalsky had fascinating experimental results on a polyelectrolyte system in which a fibre underwent contraction in response to chemical change, in striking analogy to a contracting muscle. There seemed a clear need for developing the statistical mechanics of polyelectrolytes; on that basis, with Domb's support I gained a DSIR (Department of Scientific and Industrial Research) postdoctoral Fellowship. Only one paper [14] resulted; but my perspectives were significantly broadened by the study of polymers and I was especially intrigued by the excluded volume problem [15].

Most importantly, by interacting with Cyril's star student from Oxford, Martin F. Sykes, I was led to statistical mechanics on lattices. He introduced me especially to the already infamous *Ising model* of a ferromagnet exactly solved in 1944 in two spatial dimensions by Lars Onsager (at Yale). This system and its behavior in many guises was soon a "leading star" for my researches[2] and remained so for many subsequent decades.[3] Indeed, with a by-then-former postdoctoral associate (or "postdoc"), Helen Au-Yang Perk, a paper on layered Ising models [417] was published 56 years later by virtue of the direct analogy with the *proximity* and *connectivity* ordering effects observed by Gasparini and coworkers in micron-sized boxes of *superfluid helium* [408].

Very soon Martin Sykes and I collaborated and published together. Our initial article [17] appeared in the very first volume of the *American Physical Society's* then novel journal *Physical Review Letters*, soon to be regarded by many as the world's leading journal in Physics! The article explored what became a common theme, in my work, namely, the frequently misleading predictions for the critical point singularities of the traditional van der Waals, Curie–Weiss, mean field, or Landau–Ginzburg theories. Specifically, we predicted that as

---

[1]Katchalsky Katzir became the first President of the Israeli Academy of Science but was later murdered by terrorists at Ben Gurion airport.

[2]Among the earliest articles are [18, 20, 21, 24] from the Publication List.

[3]See, e.g., articles [64,7, 78, 99, 146, 156, 180,6, 197, 242, 273,8, 285, 320,8, 401].

a function of the temperature, T, the susceptibility of an antiferromagnet, say $\chi(T)$, had a *vertical slope* at T_c (where the specific heat singularity occurs) and, as a result, a smooth rounded maximum *above* T_c.

Over subsequent years Martin and I worked with each other, with senior and junior visitors, and with many students in Cyril Domb's lively theory group at King's College.[4] Indeed, Cyril through his broad interests in physics, mathematics, and statistics and his wide contacts brought many new and stimulating problems to us: beyond the Ising model and magnetism in many forms,[5] polymers and excluded volume issues,[6] low-temperature challenges [22], the Heisenberg and quantum-mechanical models,[7] I must especially mention percolation, cluster-size, and related combinatoric problems[8] and last, but far from least, *critical phenomena* and their crucial singularities in fluids and alloys, along with Potts' models and more.[9] In the practical analysis of many of these problems, I also learned from Cyril Domb the previously under-appreciated value of exact series expansions and how to intelligently extrapolate them in order to extract estimates for the location and, more important, the actual nature of their singularities embodied in *critical exponents*.[10] And this then led to the exploitation of Padé approximants[11] and, later, to extensions,[11] including *partial differential approximants* for many-variable series.[12]

Among many visiting speakers, R.B. Potts and C.A. Hurst impressed me in different ways; the latter told of his work with H.S. Green which introduced me to the beauties of Pfaffians. Very quickly I found, as did H.N.V. Temperley and P.W. Kasteleijn [30], that here was a route to the exact solution of the *dimer problem* on planar lattices! Many exciting results followed [31, 41, 63, 277] and decades later, from what became known as the "*FKT algorithm*," combinatorialists pushed much further.

In 1957, needing a more permanent position, I applied for and gained an interview at Queen Mary College for a *Lectureship in Statistics*. The first question was: "Are you, by any chance, a relative of *R.A. Fisher*?" (R.A. Fisher was, indeed, the world's most famous statistician!) Jaws dropped when I replied: "I regret not; but I have read his books!" Although no offer was made to me, I gained some consolation on learning that the position had remained unfilled!

---

[4]Articles with Martin Sykes: [19, 34, 35, 52]; with John Zucker [22] and Basil Hiley [27], and with my own first students, John Essam [28, 40, 52, 89], John Stephenson [41], and David Gaunt [42, 52, 54], Robert Burford [64] and Arthur Ferdinand [69, 78, 156], and with Jill Bonner [36, 44].

[5]See [21,4,5, 36,7, 43, 320].

[6]Excluded volume: [27, 59, 61].

[7]See [38, 43,4, 62].

[8]See [26,8,9, 30,2, 46] and related mathematics [89, 94, 102, 267, 275].

[9]Critical behavior features in [39, 45,8,9, 51,2,8, 64–66, 102, 124,7, 131, 159, 184, 192].

[10]Critical exponents, usually labeled α, β, γ, δ, ν, η, ... : [19, 27, 34,5, 42, 52, 64, 141, 279].

[11]Following G.A. Baker Jr. and J.L. Gammel, Padé approximants were applied in [40, 118, 122] and extended to *differential approximants* in [177, 202].

[12]*PDA's* for two- or more-variable series were introduced and studied in [160,4,7, 204, 211] and, with student Daniel Styer and postdoc Jing-Huei Chen, in [212,3, 237, 247, 257].

Fortunately, a *Lectureship in Theoretical Physics* opened at King's. Repairing a disappointment of some years previously, I was interviewed — along with three others, one a close friend, another a close colleague. At the end of the afternoon I was offered and accepted what was, in actuality, a life-time position. A few years later came a promotion to *Reader*. In 1960, thanks, likely, to an introduction by Cyril Domb to the notable theorist and expositor, Elliott Montroll, I made my first visit to the United States: a summer month at Brookhaven National Laboratory. There I met Mark Kac, the famous probabilist, who invited me, with my family, to spend 1963–1964 with him and George Uhlenbeck, both recently moved to Rockefeller University in New York City. That visit led to offers of professorships at various North American Universities. Thus, following a clap on the shoulder by Mark Kac who said: "Today, Michael, I have perjured myself on your behalf!" I eventually accepted a position in Chemistry and Mathematics at Mark's former home, Cornell University. My wife Sorrel (née Castillejo Claremont) and our four children, Caricia Jane, Daniel Sebastian, Martin John, and Mathew Paul Alejandro, took up residence in Ithaca, N.Y., on 1 July 1966.

In due course, our children went to university, became U.S. citizens, made careers, left home, found partners, and presented us with five delightful grand-daughters, Natalia, Yvette, Paulina, Isabelle, and Maya, and two fine grandsons, Andrew and Benjamin.

In 1987, twenty one years later, for a combination of personal and professional reasons, Sorrel and I moved to the suburbs of Washington, D.C., initially College Park but soon to our own house in Silver Spring, while I took up a professorship at the University of Maryland. Two and half decades further on, in August 2012, I followed the example of my good friend, the Cornell chemist, Ben Widom, and formally retired just before my 81st birthday.

At Rockefeller University in 1963–1964, following Uhlenbeck's insistence on the value of mathematical rigor and the significance of the infinite-volume or *thermodynamic limit*, especially for phase transitions, I developed a longstanding interest in the foundations of statistical mechanics [47]. That led to visits with David Ruelle in IHES near Paris [56] and further mathematically rigorous researches over the following decades.[13]

Mark Kac and his challenge: "Can you hear the shape of a drum?" brought complementary stimulation [60]. From Mark, I also learnt about Toeplitz matrices, and the asymptotic behavior of their determinants. A few years later, Robert Hartwig, a postdoctoral associate, and I obtained some theorems [74] and, for general singular cases, advanced significant conjectures [75] the validation of which kept leading experts busy for thirty or more years!

By contrast, at a 1964 Gordon Conference on Magnetism I met Jim Kouvel, then at General Electric Laboratories in Schenectady. Jointly we established the *non-classical* critical behavior of nickel [50] — using observations made 38 years previously! This unambiguously ended the era of mean-field theories in magnetism.

Two years later in Cornell new problems and new colleagues abounded; close to me in Chemistry were Ben Widom [76, 223], Harold Scheraga [61], and Roald Hoffmann;

---

[13]See: [53,5,7, 67,8, 77, 82, 90, 110]; in 1979 Professor Masuo Suzuki — sent to me at Cornell from Japan by Ryogo Kubo, the international pioneer — was involved [96], and later the Cornell students Gunduz Caginalp and Graeme Milton, and, in 1986, postdoc Vladimir Privman; see [166, 172, 226, 245, 256, 286, 331].

in Condensed Matter Theory, Geoffrey Chester [82, 153], Vinay Ambegaokar, N. David Mermin [82], Neil Ashcroft, John Wilkins, and, visiting, Jim Langer [70, 71]; in experiment: John Reppy — who was a key to my longstanding devotion to helium at low temperatures [70][14] — Watt Webb and, later, Dave Lee and Bob Richardson.

A special role was played by Kenneth G. Wilson, in High Energy theory, since he attended and spoke at the weekly statistical mechanics seminar that Ben Widom and I ran. Ken emphasized the close analogies between field theories and statistical mechanics. His devotion to computation meant he readily appreciated the numerical techniques brought from King's [119]; indeed, he played a central role in computing 3D-Ising-model series expansions for graduate student Howard Tarko [127, 131, 141]. Meanwhile Ken developed what he named the "*renormalization group*" as a powerful and general approach to studying singularities in critical phenomena, condensed matter theory (specifically solving the Kondo problem), and field theories. With a little aid from me [105, 333] the "*epsilon expansion*" was launched (taking *epsilon* as the deviation of the spatial dimension d below $d = 4$ so that $\varepsilon = 4 - d$). That opened the way for many reviews,[15] for calculations by talented students[16] and postdoctoral associates,[17] and further insights[18] including Yang–Lee Edge singularities [174, 350].

The epsilon expansion came to mind, in part no doubt, by how dimensionality enters the excluded volume problem for polymers[19] and also *via* the infinite range-of-interaction Baker–Kac ideas (inspired, in turn, by Robert Brout[20]) which, in my hands, led to $1/d$ expansions.[21]

A particular benefit of renormalization group theory was its justification for *scaling* or *homogeneity* arguments[22] and the path it opened to the systematic study of *crossover*[23] and multicritical phenomena. Specifically, it helped unravel *tricritical points*, which proved subtle[24] when $d = 3$, and the novel *bicritical* and *tetracritical points*[25] arising especially in magnetic materials. It also provided a fresh incentive for disposing of much loved *integral equations* for fluid criticality.[26]

---

[14]In 1972 concerning *supersolids* [116] with Geoffrey Chester's student Kao-Shien Liu; in 1984–1986 [239, 250, 259] with student Peter Weichman and Dr. Michael Stephen at Rutgers; and in 2010–2013 [404,8, 410,4,6,7].

[15]Renormalization group theory reviews: [129, 135, 143,8,9, 184,8, 209, 333].

[16]With students, especially David Nelson and Eytan Domany: [133,7-9, 145,7, 158,9, 161–3, 254].

[17]With postdocs, initially Pierre Pfeuty, especially Ammon Aharony, and later Reinhard Lipowsky: [109, 115, 121,3, 132,6,9, 217, 261,3].

[18]See [125, 152, 171,4,8,9, 321].

[19]See, e.g. [19, 27, 59, 80(b)].

[20]Sadly Robert Brout died only a year or two before his work with Englert was recognized for the Nobel Prize along with Peter Higgs' independent researches.

[21]Inverse dimensionality: [42, 142, 150, 269, 273] with, in 1974–1975, Swiss postdoc Paul Gerber.

[22]Scaling: [66, 104, 154, 209, 217]: for *corrections-to-scaling*, considered with Mustansir Barma from Bombay, see [240,2,9].

[23]Crossover: [109, 119, 129(b), 130,4, 169, 216].

[24]For tricriticality, see: [137, 145, 170,3,5, 181, 198,9] with work by postdoc Stéphane Sarbach from Switzerland.

[25]For bicriticality, with postdoc David Mukamel and others: [138, 147,8, 151,7–9, 161,3, 187,9, 231, 268].

[26]At small rewards for postdoc Shmuel Fishman: [201,3,6, 219].

In another direction, starting with Onsager's exact solutions for 2D Ising models, I set out to study with Arthur Ferdinand, a Trinidadian by birth, *finite-size effects* in the vicinity of critical points [69, 78, 156]. All critical singularities are rounded on a scale found to be set [112–4] by the ratio of the finite linear dimension(s), L (or L_1, L_2, …) to the *correlation length*, $\xi(T)$; this crucial state-dependent parameter, first identified in 1914–1926 by Ornstein and Zernike [45, 76], *diverges* when the critical point itself is approached in a bulk ($L = \infty$) system. In the 1970 Varenna Summer School [104] I was able to expound the theory; but further developments soon ensued.[27] Later, the ideas were extended to finite-size effects on *first-order transitions*.[28] The concepts remain vital tools for computation [376, 384].

In 1978, Pierre-Gilles de Gennes (whom I had met in France in the early 1960s) visited Cornell as a *Professor-at-Large*.[29] A resulting short joint paper in French [176] opened another aspect of finite domain size, namely, a "Casimir-like" force between near-by plates induced by critical fluctuations. This attracted attention, led to experiments, and to further studies of walls and the properties they induce, especially *wetting* and *surface phase transitions*,[30] including with randomness [251, 277]. A short step involved the interactions of *interfaces* or *domain walls* separating phases[31] and to striking phase diagrams with *multiphase points* generated by the so-called *ANNNI* models[32] and their analogs.

Moving to Maryland in 1987 with Dr. Rajiv Singh and students Andrea Liu and Martin Gelfand, kept research and reviews moving steadily [266–276, 278–282]. Before long, however, an unanticipated challenge came from my friend J.M.H. (Anneke) Levelt Sengers, an outstanding experimentalist at NIST.[33] The noted physical chemist K.S. Pitzer had recently studied liquid-liquid coexistence in a special *ionic fluid* or *electrolyte* and claimed a critical exponent *beta* of 0.5, the van der Waals value, in strong disagreement with Anneke's own experiments yielding a by-then-expected *nonionic* value, close to 1/3. Anneke's reason for contacting me, however, was a theory paper that supported Pitzer's value for an ionic

---

[27]Finite size effects studied with colleagues, with postdoc David Jasnow, and with student Michael Barber (later Vice Chancellor of Flinders University, Adelaide, South Australia): [82,7, 92,3, 100,3, 112–4,7, 120, 146, 153, 183, 235, 253, 288].

[28]First-order transitions, initially with Nihat Berker (later President of Sabanci University, Istanbul, Turkey); then with Vladimir Privman: [215, 228,9, 241,4–6,8].

[29]When Mark Kac was asked: "As a Professor-at-Large what are you supposed to do?" He replied: "Just tell the Professors-at-Small what they should do!" In fact, at-large visitors came for a few weeks each year for 3 or 4 years, gave a public lecture, a colloquium, and a seminar or two, and interacted.

[30]Wetting, etc., with postdocs Helen Au-Yang and Hisao Nakanishi: [183,5, 207, 220–2,5, 231,2,4, 282, 305, 332]; and with student Albert Jin: [299, 301,7, 310].

[31]Domain walls and interfaces: [210, 255,8, 262,4,5, 274,6, 281,3,8, 291–4, 314, 322, 346, 385] and exact results with postdoc Lev Mikheev from Russia: [302,8,9, 311].

[32]Note "ANNNI" stands for "Axial" or "Anisotropic Next-Nearest-Neighbor Ising"; the model is described and analyzed in [182,6, 190,4, 200,5,8] with Walter Selke, from Germany, and Julia Yeomans, from England. Extensions, especially for *commensurate melting*, were studied with students David Huse [196, 214,6,8, 224, 230] an Tony Szpilka [262,4,5].

[33]National Institutes for Science and Technology formerly, when I first visited in Washington, NBS, the National Bureau of Standards.

fluid; the false arguments were readily dismissed![34] But the long-range $1/r^{d-2}$ character of ionic or charge-charge interactions posed deep theoretical challenges [313]. These were addressed over two subsequent decades.

A primary task was to extend the 1923 theory of Debye and Hűckel to a full mean-field description of criticality for the simplest hard-sphere ionic models [312] while allowing for neutral dipolar pairs, etc., more general dimensionalities, d, and, especially, fluctuations.[35] Improved theories for chemical association[36] and exact low-density expansions were called for [336, 340,2]. Nonetheless, it became clear that only a focused program of reliable, precise simulations[37] could confirm the growing belief that, even for ionic fluids, Ising criticality triumphs: as it does![38] Eventually exploiting spherical models [380], *exactly soluble*, albeit unrealistic, ionic models were devised and studied.[39]

Witnessing the structure of DNA uncovered in the 50's at King's and close friendships with Eric Barnard and Lewis Wolpert (both becoming distinguished biologists) left me with strong inclinations towards modern biology. Indeed, from 1969–1982, I served on the Editorial Board of *Journal of Theoretical Biology*. But it was only in 1985 when Stanislas Leibler brought to Cornell from Paris his special perspectives on matters biological that my research turned towards "living matter" via the shapes of cells in the guise of planar vesicles and their surprising transformations [270].[40] However, it was the 1998 arrival of A.B. (Tolya) Kolomeisky in Maryland that really brought me to *Biophysics*. As a student of Ben Widom in Chemistry at Cornell, Tolya — following Ben's visit to Jacques Prost in Paris — had published a theory paper on *molecular motors*; I challenged a basic assumption! That led us to an extensive study of motor proteins, microtubules, and general cellular biophysics.[41,42] Eventually, we were invited to publish a review [393] and I joined the Editorial Board of the *Biophysical Journal*!

Other fascinating problems arose: in thermodynamics, the phase diagrams of *critical endpoints*[43] and associated matters of convexity [331, 341]; considering, with postdoc G. (Makis) Orkoulas, the Yang–Yang anomaly in near-critical fluids [356,8] led us to *complete scaling*;[44] colleagues Kurt Binder (in Munich), Jan Sengers, and postdoc Subir Das tempted

---

[34]See [303]; this was neither the first nor the last time that my personal entry into a new field entailed disproving published claims (or hopes) [15, 22, 57, 110, 226, 392, 416].

[35]A vital role was played by postdocs Yan Levin (in part with student Xiaojun Li), Ben Lee, and Stefan Bekiranov; see: [315–9, 324–6,9, 334–7, 340, 362,8, 386].

[36]With student Daniel Zuckerman: [338,9].

[37]*Nonionic simulation*: [352,4, 364,5, 373,4,8, 383,8], with crucial aid from my colleague A.Z. (Thanos) Panagiotopoulos.

[38]Subsequent ionic studies: [361,3,6,7,9] with postdoc Eric Luijten and, mostly with student Young Kim, from Korea: [372,6,7, 384,8, 390,5, 412,5].

[39]With, especially, postdoc Jean-Noël Aqua: [379, 380,2, 402,3].

[40]Picked up with enthusiasm by my first Maryland student, Carlos Camacho from Chile: [284,7,9, 291,7,8, 300,6].

[41]With Kolomeisky: [347, 351,3,5,9, 360, 370,1].

[42]With Young Chan Kim, Denis Tsygankov, and, from China, Yunxin Zhang: [381,9, 394,6,8, 406,9].

[43]See: [292-6, 331, 346,4, 387] with postdocs Paul Upton and Marcia Barbosa and student Sonny Zinn.

[44]Described in [356–8, 373,8, 384^] but partially anticipated in extremely flexible but exactly soluble 1D models [84,5,6,8] explored early on at Cornell with Ubbo Felderhof, from the Netherlands.

me with dynamic critical behavior [391,2,7]; while colleagues K.R. Sreenivasan and Daniel Lathrop (with students Matthew Paoletti and Cecilia Rorai) brought me to fluid dynamics and turbulence in, yet once again, superfluid helium [400,4, 410,6]. Outlets for personal views [343, 399, 405, 411] and even artwork [407] came my way.

As a parting gift on my transfer to Emeritus status, Dean Jayanth Banavar, a one-time student of my first U.S. postdoc at Cornell, David Jasnow, arranged a two-day Celebratory Symposium and Banquet in my honor at Maryland on 26–27 October, 2012. It was a great pleasure for me to see so many friends, colleagues and former collaborators as speakers and listeners and, finally, to express my thanks for the role they had played in my life and science.

JOURNAL OF MATHEMATICAL PHYSICS   VOLUME 5, NUMBER 7   JULY 1964

Correlation Functions and the Critical Region of Simple Fluids[†]

MICHAEL E. FISHER*
The Rockefeller Institute, New York, New York
(Received 4 February 1964)

The "classical" (e.g. van der Waals) theories of the gas–liquid critical point are reviewed briefly and the predictions concerning the nature of the singularities of the coexistence curve, the specific heat, and the compressibilities are compared critically with experiment and with the analytical and numerical results for lattice gas models.

The critical singularities are related to the behavior of the pair correlation function $G(r) = g(r) - 1$ and the Ornstein–Zernike theory of critical scattering is reviewed. Alternative derivations of the theory are discussed and its validity is assessed in relation to experiment and to more detailed theoretical calculations. The nature and magnitude of the expected deviations from the "classical" theory are described. The analogies with critical magnetic phenomena are mentioned briefly.

THIS is a review article about theories of the critical point and their experimental and theoretical validity. Recent work has revealed the shortcomings of the well-known approximate theories and it is hoped that this review, although in the main nonmathematical in character, will provide some stimulus to further theoretical (and experimental) work on this old, but still imperfectly understood problem.

1. THE CLASSICAL THEORY OF THE CRITICAL POINT

At temperatures below its critical temperature T_c a gas can be condensed by isothermal compression. At temperatures above T_c the transition from dense gas to liquid takes place without discontinuity of density or, as far as has been determined experimentally, without any higher-order singularities in the density or other variables. As the temperature increases to T_c the difference between the densities ρ_L and ρ_G of coexisting liquid and gas tends continuously to zero.[1] The limiting density ρ_c, and corresponding pressure p_c, define the *critical point* (see Fig. 1). In this article we shall be concerned with the properties of a simple fluid in the region of its critical point and in particular on the critical isochore $\rho = \rho_c$.

In many respects the behavior of binary fluid mixtures which undergo phase separation is closely analogous to the condensation of simple fluids and most of our remarks can be translated directly into

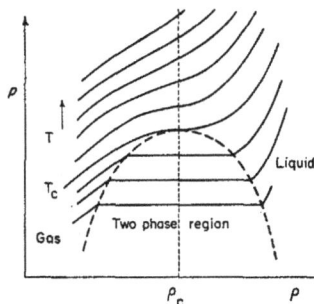

FIG. 1. Schematic isotherms for a simple fluid in the critical region.

such terms.[2] For simplicity, however, we refer in the main only to single-component systems.

As is well known, a qualitative account of condensation phenomena and the critical point is given by the classical equation of van der Waals

$$p/kT = \rho/(1 - b\rho) - a\rho^2/kT, \quad (1.1)$$

provided this is supplemented by the "equal area rule" of Maxwell which ensures that the density ρ is a single-valued function of the pressure p.[3] The general appearance of the isotherms of a van der Waals gas in the critical region is similar to that of a real gas as shown in Fig. 1. In particular the p, ρ isotherms become flatter and flatter as T approaches T_c from above at the critical density and correspondingly the isothermal compressibility

$$K_T = -\frac{1}{V}\left(\frac{\partial V}{\partial p}\right)_T = \frac{1}{\rho}\left(\frac{\partial \rho}{\partial p}\right)_T \quad (1.2)$$

---

* On leave of absence from The Wheatstone Physics Laboratory, King's College, London, England.

[1] It now seems quite well established experimentally that for simple fluids, such as the noble gases, the coexistence curve does *not* have a significant "flat top." For a discussion of this suggestion see O. K. Rice, J. Phys. & Colloid Chem. 54, 1293 (1950); and Ref. 60 below. See also D. R. Thompson and O. K. Rice, quoted in Ref. 11 below.

[2] J. S. Rowlinson, *Liquids and Liquid Mixtures* (Butterworths Scientific Publications, Ltd., London, 1959), Chap. 5.

[3] Maxwell's thermodynamic derivation of the rule is unsatisfactory in that it is necessary to give thermodynamic significance to "unstable" (and "metastable") states on the original van der Waals isotherm. A formally equivalent but theoretically somewhat more convincing argument is to use the minimal properties of the Gibbs free energy (or chemical potential) to eliminate the unwanted parts of the van der Waals isotherm.

which is essentially the reciprocal of this slope, diverges to infinity at the critical point.

The derivation of the van der Waals equation given by Ornstein[4] suggests that it should be a reasonably good approximation when the pair interaction potential $\phi(r)$ of the molecules of the fluid has a short-range strongly repulsive core and a very long-range weakly attractive tail. Indeed Kac, Uhlenbeck, and Hemmer[5] have recently shown that the van der Waals isotherm (*with* flat part) follows rigorously for a one-dimensional gas of hard rods interacting with an attractive exponential potential *in the limit* that the exponential becomes infinitely long-ranged *and* infinitely weak [holding $\int_b^\infty \phi(r)\, dr$ constant]. It seems probable, however, that the behavior of a real gas *in its critical region* is crucially dependent on the finite or relatively *short* range of the attractive parts of more realistic potentials. This is supported by the comparison with real systems and exactly soluble models we present in the next section.

Three principal predictions concerning the critical region which follow from the van der Waals equation are:

(a) that the *coexistence curve* follows a square-root law, i.e., the difference between liquid and gaseous densities vanishes as

$$\rho_L - \rho_G \approx A(T_o - T)^{\frac{1}{2}} \qquad (T \to T_o-); \quad (1.3)$$

(b) that the *compressibility* along the critical isochore diverges as a simple pole,

$$K_T \approx B/|T - T_o| \qquad (\rho = \rho_o,\, T \to T_o+); \quad (1.4)$$

and

(c) that the *specific heat* (at constant volume) along the critical isochore rises to a maximum and then falls discontinuously as T increases through T_o, i.e.,

$$C_V(T) \approx C_o^{\pm} - D^{\pm}\,|T - T_o|, \qquad T \gtrless T_o, \quad (1.5)$$

with $C_o^- - C_o^+ = \Delta C > 0$.

The compressibility of the gas and of the liquid

along the coexistence curve (i.e. at condensation) also diverges as a simple pole as $T \to T_o-$ according to the van der Waals equation but the amplitude corresponding to B in (1.4) is smaller. (The constants A, B, $C_o^{\pm}$ and $D^{\pm}$ can of course be written explicitly in terms of the van der Waals parameters a and b.)

It is important to note that these predictions are not peculiar to the van der Waals equation but follow from almost *all* approximate equations of state. Indeed they are essentially a direct consequence of the implicit or explicit assumption that the free energy and the pressure can be expanded in a Taylor series in density and temperature *at* the critical point: in other words that the critical point is not a singular point of the free energy expressed as a function of ρ and T (except in as far as Maxwell's rule is utilized *below* $T = T_o$).[6]

To demonstrate this put

$$\Delta p = p - p_o, \quad \Delta \rho = \rho - \rho_o, \quad \Delta T = T - T_o, \quad (1.6)$$

and assume that

$$\Delta p = a(T) + b(T)\Delta\rho + c(T)\Delta\rho^2 + d(T)\Delta\rho^3 + \cdots . \quad (1.7)$$

By definition $a(T_o) = 0$ so we assume similarly that

$$a(T) = a_1\Delta T + a_2\Delta T^2 + \cdots . \quad (1.8)$$

Since the compressibility is infinite at the critical point, $b(T_o)$ must vanish so again we assume

$$b(T) = b_1\Delta T + b_2\Delta T^2 + \cdots . \quad (1.9)$$

Finally since $b(T_o) = 0$ and the pressure above T_o must be a monotonic increasing function of ρ we have $C(T_o) = 0$ and, presumably, $C(T)$ is small in the critical region. We thus obtain for small $\Delta\rho$ and ΔT the isotherms

$$\Delta p = a_1\Delta T + b_1\Delta T\Delta\rho + d_0\Delta\rho^3 + \cdots . \quad (1.10)$$

Near the critical point the compressibility is hence given by

$$K_T \simeq \frac{(b_1\rho c)^{-1}}{\Delta T + (3d_0/b_1)\Delta\rho^2}, \quad (1.11)$$

from which the prediction (b) follows. Application of the equal area rule to the isotherms (1.10) with negative ΔT yields the coexistence curve

$$\Delta\rho^2 \simeq (b_1/d_0)\,|\Delta T| \qquad (T \leq T_o), \quad (1.12)$$

which implies the square-root law (a). The divergence

[4] L. S. Ornstein, Dissertation, Leiden 1908; see also Ref. 5.

[5] M. Kac, G. E. Uhlenbeck, and P. C. Hemmer, J. Math. Phys. 4, 216 (1963). It should be noted that the correction to ideal gas behavior, arising from the hard core and represented by the parameter b, is exact in one dimension but only approximate in two or three dimensions. On the other hand the accuracy of the correction represented by the parameter a depends mainly on the long-range nature of the attractive tail of the potential and not so directly on dimensionality. Consequently, although the behavior of a three-dimensional model in the corresponding long-range limit should be van der Waals-*like* [in that Eqs. (1.3) to (1.5) should hold], one should still not expect van der Waals' equation (1.1) to hold precisely.

[6] See, for example, Landau's theory of the critical point and second-order phase transitions [L. D. Landau and E. M. Lifshitz, *Statistical Physics* (Pergamon Press, Ltd., London, 1958), pp. 259–268 and pp. 434–439].

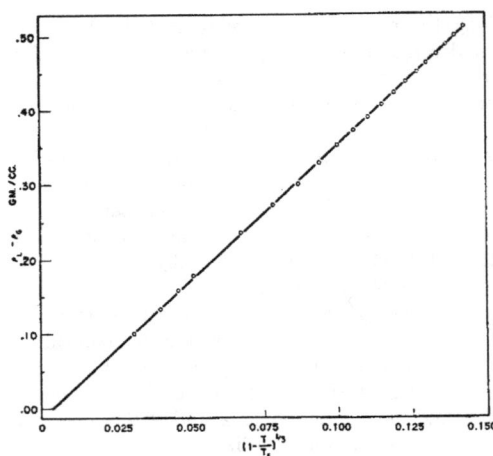

FIG. 2. Coexistence curve for xenon: plot of $(\rho_L - \rho_G)$ vs $[1 - (T/T_c)]^{\frac{1}{3}}$ (after Weinberger and Schneider[9]).

of the compressibility along the coexistence curve as $(T_c - T)^{-1}$ then follows from (1.10).

By integrating (1.9) with respect to Δp one finds the free energy $F(T, \rho)$. If one assumes that the additive constant of integration is also a nonsingular function of T and imposes the continuity of F on the coexistence curve (1.12) one finally derives the prediction (c) of the discontinuity in $C_V(T)$.

2. THE NATURE OF THE CRITICAL SINGULARITIES

Perhaps the most striking test of the predictions of the classical theories is provided by the data on the coexistence curves of simple gases. Some time ago Guggenheim[7] showed that the gases Ne, Ar, Kr, Xe, N_2, and O_2 obey closely a law of corresponding states of the form

$$(\rho_L - \rho_G)/2\rho_c = A(1 - T/T_c)^\beta, \quad (T \to T_c), \quad (2.1)$$

with $\beta = \frac{1}{3}$. This relation was observed to hold with an accuracy of 0.5% (or better) in $\Delta\rho/\rho_c$ and in $\Delta T/T_c$ and over a range of temperatures from $T/T_c \simeq 0.6$ up to within $\frac{1}{2}\%$ of the critical temperature by when $(\rho_L - \rho_G)/2\rho_c$ had fallen to about 0.30. In earlier work on CO_2, Michels, Blaisse, and Michels[8] found that the coexistence curve could be fitted over a similar range but with somewhat greater accuracy by (2.1) with the index $\beta = 0.357$.

These results seem to be in clear disagreement

with the classical prediction $\beta = \frac{1}{2}$ which should be applicable in range of $(T - T_c)$ and $(\rho - \rho_c)$ observed. They certainly show that the experimental coexistence curve is much flatter than the van der Waals curve. However, experiments near the critical region are very difficult to perform; a long time is needed to establish equilibrium and hysteresis phenomena are difficult to avoid; the system is very susceptible to minute amounts of impurities and, due to the large compressibility, highly sensitive to gravitational fields. Indeed as shown by Weinberger and Schneider[9] it is important to take special precautions to reduce the effects of gravity if the true shape of the coexistence curve is to be measured close to T_c. In a very careful study of xenon (see Fig. 2) they extended the density measurements down to temperatures differing from T_c by only 1 part in 30 000 (i.e., $\Delta T/T_c \simeq 0.003\%$) and down to corresponding density differences of $(\rho - \rho_c)/2\rho_c \simeq 0.04$. [The temperature was controlled to within $\pm 0.001\,°C$.] Their data accurately obey the relation (2.1) with a nonclassical value of the index β over about three decades in $(T - T_c)$. Analysis of their measurements indicates[10] that

$$\beta = 0.345 \pm 0.015 \quad (2.2)$$

which is not inconsistent with a value of exactly $\frac{1}{3}$ as may be seen in Fig. 2.

While it is always possible that measurements taken much closer still to the critical point might yet yield the value $\beta = \frac{1}{2}$ it seems reasonable to conclude that the classical theory does *not* provide the correct description of reality. Furthermore measurements of the phase boundaries of binary fluid mixtures near both their upper and lower critical points are also fitted well by the same cube-root law.[11] This suggests that the behavior close to a critical point is insensitive to the detailed nature of the intermolecular forces. To check how far β is really independent of the interaction potentials it would be desirable to have measurements on other systems of an accuracy matching the experiments on xenon. One should note, however,

[7] E. A. Guggenheim, J. Chem. Phys. 13, 253 (1945).

[8] A. Michels, B. Blaisse, and C. Michels, Proc. Roy. Soc. (London) A160, 358 (1937). More accurate measurements on CO_2 have since been made by H. L. Lorentzen, Acta Chem. Scand. 7, 1335 (1953) and later work.

[9] M. A. Weinberger and W. G. Schneider, Can. J. Chem. 30, 422 (1952).

[10] Note that the plot of $\rho_L - \rho_G$ versus $[1 - (T/T_c)]^{1/3}$ in Fig. 2 is a good straight line down to $(\rho_L - \rho_G)/2\rho_c = 0.04$ but does not extrapolate exactly to the origin $[\rho_L - \rho_G$ at $T = T_c]$ as it should. This suggests that the index β is not precisely $\frac{1}{3}$ and a log-log plot leads to the value quoted. The uncertainty reflects the spread of the experimental points about the best straight line.

[11] See Ref. 2, pp. 165-166 and especially the work of O. K. Rice referred to therein. See, also, D. R. Thompson and O. K. Rice, "Shape of the coexistence curve in the perfluoromethylcyclohexane-carbon tetrachloride system, II." J. Am. Chem. Soc. (1964) in press.

CRITICAL REGION OF SIMPLE FLUIDS　　　**947**

TABLE I. Critical indices.

Index	below T_c			above T_c $(\rho = \rho_c)$			
	α'	β	γ'	α	γ	ν	η
Defined in equations	(2.7, 2.12)	(2.1)	(2.8)	(2.7, 2.12)	(2.8)	(3.10, 5.6)	(4.7, 5.3)
Classical theory	$0_{\text{discon.}}$	$\frac{1}{2}$	1	$0_{\text{discon.}}$	1	$\frac{1}{2}$	0
Lattice gases $d = 2$	$0_{\log}$	$\frac{1}{8}$	$1\frac{3}{4}$	$0_{\log}$	$1\frac{3}{4}$	1	$\frac{1}{4}$
Lattice gases $d = 3$	≥ 0	$\simeq\frac{5}{16}$	$\geq 1\frac{1}{4}$	$\geq 0, \leq 0.2$	$1\frac{1}{4}$	$\simeq 0.64$	$\simeq\frac{1}{18}$
Experiment	$\gtrsim 0_{\log}$	0.33–0.36	$\geq 1.27(?)$	≥ 0.1 ?	> 1.1 ?	> 0.55 ?	$> 0(?)$

Note: α', β, and γ' are related by (2.20) and γ, ν, and η by (5.7). The queries, ? and (?), indicate greater and lesser degrees of experimental doubt.

that any value of β between $\frac{1}{2}$ and $\frac{1}{4}$ is inconsistent with the assumption that the critical point is a nonsingular point of the free energy in the sense discussed in the previous section.

The shape of the coexistence curve may also be studied theoretically for more-or-less idealized models of a fluid. The only model so far sufficiently tractable to yield significant predictions in the critical region is the very simplest lattice gas in which each molecule occupies a site of a lattice to the exclusion of other molecules and interacts, attractively, only with nearest-neighboring molecules. This model is equivalent to the well known Ising model of ferromagnetism which has been studied intensively.[12,13]

The exact calculation of the free energy of the plane square lattice gas along its critical isochore (corresponding to zero magnetic field) was first achieved by Onsager.[14] In addition, however, Onsager[15] and Yang[16] were able to find the coexistence curve (corresponding to the spontaneous magnetization). They found the relation (2.1) but with the index

$$\beta = \tfrac{1}{8}. \qquad (2.3)$$

This result, which implies a very flat coexistence curve, is, of course, quite inconsistent with classical theory.[17] It is important, furthermore, to note that the index β is *independent of lattice structure* for all soluble plane Ising lattices (including the triangular, honeycomb, kagomé, and checkerboard lattices[12,13]).

For three-dimensional lattice gases no rigorous theoretical results are available. Nonetheless on the

basis of sufficiently long power-series expansions[18] it has proved possible to draw quite accurate conclusions concerning the shape of the corresponding coexistence curves. [The coefficients are analyzed numerically with the aid of the recently introduced technique of Padé approximants.[19]]

The behavior again appears to be independent of lattice structure. [The simple, body-centered and face-centered cubic lattices[20,21] and the tetrahedral (diamond) lattice[22] have been studied; the latter on the basis of the more sensitive ratio method.[23]] Dimensionality, however, is important since, in contrast to (2.3), the index β is found to lie in the range[21,22]

$$0.303 \leq \beta \leq 0.318 \qquad (2.4)$$

which is consistent with the conjecture $\beta = \frac{5}{16} = 0.31250$.

It is remarkable, and perhaps unexpected, that a model as simple as a lattice gas with only nearest-neighbor interactions should yield a result for the shape of the coexistence curve so close to the experimental results (2.1) and (2.2). The agreement suggests that in the critical region the lattice gas represents rather adequately the pertinent features of a real gas. It appears that only the grosser features of the model—in particular the dimensionality and the short range of the forces—are really essential for obtaining a good description of critical behavior.

It seems probable, nonetheless, that the difference of about 0.025 between the experimental and theoretical values of β is a real discrepancy due, presumably, to the more artificial aspects of the Ising Hamiltonian which, in particular, restricts the molecules to the lattice positions. There remains

[12] A. comprehensive review of the Ising model is C. Domb, Advan. Phys. **9**, Nos. 34, 35 (1960), while Ref. 13 is a brief review of more recent results.

[13] M. E. Fisher, J. Math. Phys. **4**, 278 (1963).

[14] L. Onsager, Phys. Rev. **65**, 117 (1944).

[15] L. Onsager, Nuovo Cimento Suppl. **6**, 261 (1949); see also E. W. Montroll, R. B. Potts, and J. C. Ward, J. Math. Phys. **4**, 308 (1963).

[16] C. N. Yang, Phys. Rev. **85**, 808 (1952); T. D. Lee and C. N. Yang, Phys. Rev. **87**, 410 (1952).

[17] Since β is the inverse of an integer it could still be possible for the free energy to be an analytic function of ρ at the critical point. In view of the logarithmic singularity in the specific heat, however (Ref. 14 and the discussion below), this possibility seems rather remote.

[18] The expansion variable is $\exp[-V_0/kT]$ where V_0 is the depth of the well in the pair interaction potential. Details of the expansions are given in Ref. 11.

[19] G. A. Baker, Jr., J. L. Gammel, and J. G. Wills, J. Math. Anal. Appl. **2**, 405 (1961).

[20] G. A. Baker, Jr., Phys. Rev. **124**, 768 (1961).

[21] J. W. Essam and M. E. Fisher, J. Chem. Phys. **38**, 802 (1963).

[22] J. W. Essam and M. F. Sykes, Physica **29**, 378 (1963).

[23] C. Domb and M. F. Sykes, J. Math. Phys. **2**, 63 (1961).

948   M I C H A E L   E.   F I S H E R

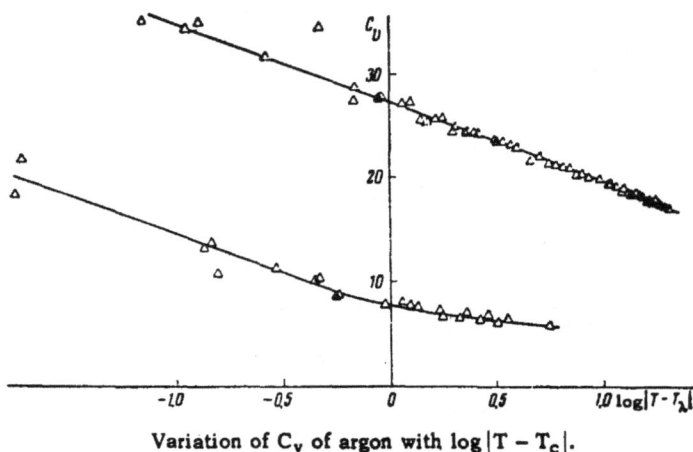

FIG. 3. Variation of the constant-volume specific heat of argon along the critical isochore (after Bagastskii *et al.*[25]). (Note that the logarithm to the base ten of $|T - T_c|$ in °K is plotted.)

Variation of C_V of argon with $\log |T - T_c|$.

($T_c = 150.5°$ K).

the theoretical problem of calculating β for more realistic continuum models. (For convenience the various results for β are collected in Table I.)

Turning now to the question of the specific heats, it has long been known that real gases exhibit a large "anomalous" specific-heat maximum above T_c which lies near the critical isochore and which is not expected on classical theory.[24] Similarly in the two-phase region below the critical point specific heats rise much more rapidly than expected as T approaches T_c. From the earlier measurements one could not conclude with certainty that the specific heat actually became infinite at T_c but recent measurements by Bagatskii, Voronel', and Gusak[25] of $C_V(T)$ for argon along the critical isochore (see Fig. 3) suggest strongly that

$$C_V(T) \to \infty \quad \text{as} \quad T \to T_c\pm. \qquad (2.5)$$

Such a result is again inconsistent with classical theory.

The measurements covered a range from 15% below to 5% above T_c at temperature intervals of 0.04 to 0.05°C (corresponding to $\Delta T/T_c \simeq 0.03\%$). Over a range of one or two decades in $|T - T_c|$ the specific heat could be fitted quite well[25] by a logarithmic singularity of the form

$$C_V(T) \simeq -A^* \log |1 - (T/T_c)| + B^* $$
$$(T \gtrless T_c), \qquad (2.6)$$

where the marked asymmetry of the curve (see

Fig. 3) indicates that $B^+ \ll B^-$ and possibly that $A^+ < A^-$. The specific-heat curve in fact resembles quite closely the famous lambda anomaly displayed by liquid helium at its transition to the superfluid state.[26] [In this case the formula (2.6) is followed very accurately over four or more decades.[26] It should be noted, however, that the lambda point of helium has a quantum-mechanical origin and is not a critical point in the usual sense.]

The data for argon are not at present, accurate enough to confirm (2.6) as closely as might be wished.[26a] To avoid prejudicing the conclusions one should preferably consider a singularity of a form such as

$$C_V(T) \simeq (A^*/\alpha)\{|1 - (T/T_c)|^{-\alpha} - 1\} + B^* $$
$$(T \gtrless T_c), \qquad (2.7)$$

and ask for the experimental value (and uncertainty) of the index α. [When $\alpha \to 0$ this expression reduces to the logarithmic singularity (2.6).] In particular the definite curvature of the plot of $C_V(T)$ versus $\log |T - T_c|$ for $T > T_c$ (see Fig. 3) suggests that the true value of α might be greater than say 0.1. If this is so it seems probable that below T_c the index has a somewhat different value, α' which is probably less than 0.1. It would be valuable to have similar and more extensive measurements on the other noble gases in order to test the relation

[24] See Ref. 2, pp. 100–101; Ref. 8; and A. Michels, J. M. H. Levelt, and W. de Graaff, Physica **24**, 769 (1958).
[25] M. I. Bagatskii, A. V. Voronel', and B. G. Gusak, Zh. Eksperim. i Teor. Fiz. **43**, 728 (1962) [English transl.: Soviet Phys.—JETP **16**, 517 (1963)].
[26] M. J. Buckingham and W. M. Fairbank, in *Progress in Low Temperature Physics III*, edited by C. J. Gorter (North-Holland Publishing Company, Amsterdam, 1961), Chap. 3.
[26a] More recent experiments on oxygen have given very similar results: A. V. Voronel', Yu. R. Chasshkin, V. A. Popov, and V. G. Simkin, Zh. Eksperim. i Teor. Fiz. **45**, 828 (1963) [English transl.: Soviet Phys.—JETP **18**, 568 (1964)].

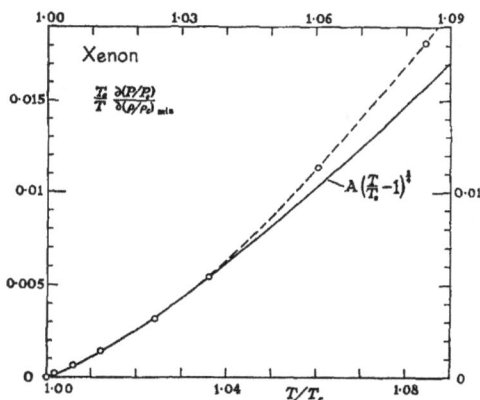

FIG. 4. Variation of the inverse of the maximum isothermal compressibility of xenon (based on the experiments of Habgood and Schneider[30]).

(2.7) more critically and to decide how far a logarithmic singularity might be truly "universal."[26a]

Significant theoretical predictions are again available only for the simple lattice gas models. Onsager's rigorous solution[14] for the plane square lattice (and subsequent results for *all* other soluble plane lattices[12,13]) yielded a *symmetric* logarithmic specific heat singularity, i.e., $\alpha = \alpha' = 0$ and $A^+ = A^-$, $B^+ = B^-$. Although this famous result demonstrates conclusively the weakness of the classical theory and is very suggestive in view of the experimental results it is, unfortunately, restricted to two-dimensional systems.

For three-dimensional Ising lattices the specific-heat series expansions have been calculated both above and below T_c. Numerical analysis of these series indicates that $C_V(T)$ is almost certainly infinite at T_c but the precise nature of the divergence is more difficult to ascertain. Below the critical point the series can be fitted well by a logarithmic singularity (i.e., $\alpha' = 0$).[13,22,27,28] As with the experimental results, however, it is not easy to exclude the possibility of a slightly sharper singularity corresponding in (2.7) to, say, $\alpha' = 0.06$. On the high-temperature side of the transition the series may be fitted moderately well by a logarithmic singularity if $A^+/A^- \simeq \frac{1}{3}$.[27,13] However, careful analysis of the ratios of coefficients definitely suggests a sharper singularity of the form (2.7) with $\alpha \simeq 0.2$.[29,13] This would lead to a shallow parabolic curve for $C_V(T)$ versus $\log|T - T_c|$ not inconsistent with the corresponding (lower) experimental curve in Fig. 3.

One may hope that with further work based on longer series or on more rigorous arguments the theoretical conclusions will be drawn more firmly. The present results are summarized in Table I.

The accurate experimental measurement of the isothermal compressibility of a gas near its critical point is not easy and the classical prediction (1.4) that K_T should diverge as $(T - T_c)^{-1}$ along the isochore does not seem to have been properly tested. In practice plots of $1/K_T$ versus temperature are distinctly concave upwards in the critical region which suggests that the compressibility might diverge more sharply than a simple pole, i.e., as

$$K_T(T) \approx \frac{B}{|(T/T_c) - 1|^\gamma} \quad (\rho = \rho_c, \; T \to T_c) \quad (2.8)$$

with $\gamma > 1$. By way of illustration a plot of $1/K_T$ (suitably normalized) versus T/T_c for xenon is shown in Fig. 4.[30] The experimental results indicated by circles and the broken line, were obtained by differentiation of the experimental isotherms, a procedure which is necessarily subject to appreciable uncertainty when K_T is large. The solid curve in Fig. 4 corresponds to the nonclassical prediction (2.8) with $\gamma = 1.25$ (see below) and is evidently quite consistent with the experimental points near T_c. This fit cannot be considered very significant, however, since estimates of γ based on the data alone are rather indefinite, although they do seem to indicate that γ is greater than 1.1.

More extensive and accurate experimental data would be extremely valuable since the compressibility is an important theoretical parameter and one which is usually somewhat easier to calculate than the specific heat or the coexistence curve. Indeed the theoretical situation for the simple lattice gases is quite unequivocal (see Table I). On the basis of Onsager and Kaufman's exact calculations[31] of the correlation functions it can be shown[32] that the compressibility of the plane square lattice gas (which is isomorphic to the magnetic susceptibility) should diverge as (2.8) with

$$\gamma = 1\tfrac{3}{4}. \quad (2.9)$$

This represents a very large deviation from the classical prediction $\gamma = 1$. Numerical examination of the corresponding series expansions confirms (2.9) in two dimensions for all other lattices and leads,

[27] M. E. Fisher and M. F. Sykes, Physica **28**, 939 (1962).
[28] G. A. Baker, Jr., Phys. Rev. **129**, 99 (1963).
[29] C. Domb and M. F. Sykes, Phys. Rev. **108**, 1415 (1957).

[30] Figure 4 is based on the measurements of H. W. Habgood and W. G. Schneider, Can. J. Chem. **32**, 98 (1954) as presented in their Fig. 4.
[31] B. Kaufman and L. Onsager, Phys. Rev. **76**, 1244 (1949).
[32] See Sec. 5 and M. E. Fisher, Physica **25**, 521 (1959).

in three dimensions, to the estimate

$$\gamma = 1.250, \qquad (2.10)$$

which is again independent of lattice structure and accurate to ± 0.001 or better.[12,20,23]

The compressibility of the simple lattice gases can also be studied on the coexistence curve below T_c.[21] In two dimensions it is found that (2.8) holds with the index $\gamma' = \frac{7}{4}$ so that $\gamma' = \gamma$ although the amplitude B^- is much smaller than B^+. In three dimensions the analysis yields $\gamma' \simeq 1.25 = \gamma$ but cannot at present exclude the possibility that γ' exceeds γ (by perhaps 0.05).

On the basis of heuristic arguments related to the Frenkel–Bijl–Band picture of condensation[33] Essam and Fisher[21,34] conjectured that the indices for the specific heat, coexistence curve and compressibility below T_c were in general related by

$$\alpha' + 2\beta + \gamma' = 2, \qquad (2.11)$$

where the index α' is defined, more precisely, by

$$\alpha' = \lim_{T \to T_c^-} [\log C_V(T) / |\log (T_c - T)|], \qquad (2.12)$$

and similarly for β and γ'. A logarithmic specific heat still corresponds to $\alpha' = 0$ so that the relation (2.11) is certainly verified for the two-dimensional lattice gases (see Table I). The formula remains true even for a van der Waals gas since a discontinuity in specific heat is now also equivalent to $\alpha' = 0$ [assuming only $C_V(T_c-) > 0$].

Rushbrooke[35] has shown that the relation (2.11) can be proved as an inequality (with $\geq$ replacing =) by purely thermodynamic reasoning. His argument as presented applies only to a ferromagnetic system but it may be adapted for a fluid as follows.

Firstly recall that in the two-phase region the specific heat at constant *total* volume is related to the properties of the system in its liquid and gaseous phases separately by[36]

$$C_V = x_L C_\sigma^L + x_G C_\sigma^G$$
$$- T\left(\frac{\partial p}{\partial T}\right)_\sigma \left[x_L \left(\frac{\partial V_L}{\partial T}\right)_\sigma + x_G \left(\frac{\partial V_G}{\partial T}\right)_\sigma \right], \qquad (2.13)$$

where the subscript σ denotes properties along the coexistence curve and where the mole fractions are given by

$$x_L = \frac{V_G - V}{V_G - V_L}, \qquad x_G = \frac{V - V_L}{V_G - V_L}. \qquad (2.14)$$

Now by the standard argument used to relate C_p and C_V one may show that in a single phase

$$C_\sigma = C_V - T\left(\frac{\partial p}{\partial V}\right)_T \left(\frac{\partial V}{\partial T}\right)_p \left(\frac{\partial V}{\partial T}\right)_\sigma. \qquad (2.15)$$

On eliminating the factor $(\partial V/\partial T)_p$ through

$$\left(\frac{\partial V}{\partial T}\right)_\sigma = \left(\frac{\partial V}{\partial T}\right)_p + \left(\frac{\partial V}{\partial p}\right)_T \left(\frac{\partial p}{\partial T}\right)_\sigma, \qquad (2.16)$$

and substituting for C_σ^L and C_σ^G separately in (2.12) one expresses C_V in terms of C_V^L and C_V^G. Finally on introducing the coexisting densities ρ_L and ρ_G and the corresponding isothermal compressibilities K_T^L and K_T^G, one gets

$$C_V(T) = x_L C_V^L + x_G C_V^G$$
$$+ \frac{x_L T}{\rho_L^3 K_T^L} \left(\frac{\partial \rho_L}{\partial T}\right)^2 + \frac{x_G T}{\rho_G^3 K_T^G} \left(\frac{\partial \rho_G}{\partial T}\right)^2. \qquad (2.17)$$

Now C_V^L and C_V^G are necessarily positive since they are essentially mean square energy fluctuations. Consequently all terms on the right of (2.17) are positive and by dropping the first three we obtain the inequality

$$C_V(T) \geq \frac{x_G T}{\rho_G^3 K_T^G} \left(\frac{\partial \rho_G}{\partial T}\right)^2. \qquad (2.18)$$

As the critical point is approached at constant density x_G (and x_L) approaches the value $\frac{1}{2}$, ρ_G (and ρ_L) tends to ρ_c and $(\partial \rho_G/\partial T)$ diverges as $(T_c - T)^{-1+\beta}$.[37] If K_T^G, the compressibility at condensation, diverges as $(T_c - T)^{-\gamma'}$ we obtain

$$\log C_V(T) \geq (2 - 2\beta - \gamma') |\log (T_c - T)|$$
$$+ \cdots . \qquad (2.19)$$

The higher-order terms vanish on dividing by $|\log (T_c - T)|$ and taking the limit $T \to T_c-$ which, by (2.12), yields the index α'. We have thus proved quite generally

$$\alpha' + 2\beta + \gamma' \geq 2. \qquad (2.20)$$

Various consequences follow from this inequality. For the two-dimensional lattice gases the rigorous results[14-16] $\alpha' = 0$ and $\beta = \frac{1}{8}$ show that $\gamma' \geq \frac{7}{4}$ thereby confirming the numerical estimates. [As before, the van der Waals gas corresponds to the case of equality.] If, for a three-dimensional lattice gas, the values $\beta = 0.3125$ and $\gamma' = 1.25$ are

[33] J. Frenkel, *Kinetic Theory of Liquids* (Oxford University Press, London, 1946), Chap. VII; J. Chem. Phys. **7**, 200, 538 (1939); A. Bijl, Doctoral Dissertation, Leiden, 1938; W. Band, J. Chem. Phys. **7**, 324, 927 (1939).

[34] Other conjectures relating the indices γ and β have been made by B. Widom [J. Chem. Phys. **37**, 2703 (1962)].

[35] G. S. Rushbrooke, J. Chem. Phys. **39**, 842 (1963).

[36] See, for example, Ref. 2, p. 41.

[37] We assume (as is true in reality and for the models considered) that as $T \to T_c$ [$\frac{1}{2}(\rho_L + \rho_G) - \rho_c$] does not vanish as rapidly as does $(\rho_L - \rho_G)$.

accepted one must conclude that $\alpha' \simeq 0.125$. Conversely for a logarithmic specific heat singularity the compressibility index γ' would have to exceed 1.25 by about 0.1. At present it is difficult to judge between these alternatives.[21,35]

For a real gas the evidence of Fig. 3 suggests $0.1 > \alpha' \geq 0$ and the coexistence data indicate $\beta < 0.36$. Consequently we should certainly have

$$\gamma' \geq 1.27, \qquad (2.21)$$

and probably $\gamma' \geq 2 - 2(0.345) = 1.31$. It would be most interesting to have an experimental test of this prediction since it differs appreciably from the classical result.

It should be mentioned that Widom and Rice[38] have observed that the critical isotherms of real gases also deviate significantly from the classical prediction being much flatter than the cubic (p, ρ) curve which follows from the van der Waals equation. As yet however, this feature has not been investigated theoretically for lattice gases.

3. PAIR CORRELATION FUNCTION AND CRITICAL SCATTERING

To obtain insight into the microscopic nature of a fluid in the critical region it is natural to consider the many-particle distribution functions $n_s(\mathbf{r}_1, \cdots \mathbf{r}_s)$ which describe the correlations between the constituent molecules. In particular the pair correlation function defined, for a uniform system, by

$$g(\mathbf{r}_{12}) = n_2(\mathbf{r}_1, \mathbf{r}_2)/n_1(\mathbf{r}_1)n_1(\mathbf{r}_2) = n_2(\mathbf{r}_{12})/\rho^2 \qquad (3.1)$$

is of central importance. When the system is in one phase $g(r) \to 1$ as $r \to \infty$ and one may introduce the net correlation function

$$G(\mathbf{r}) = g(\mathbf{r}) - 1 \qquad (3.2)$$

which decays to zero as $r \to \infty$. The deviation of $G(\mathbf{r})$ from zero is a direct measure of the influence of one molecule on another.

As is well known $G(\mathbf{r})$ is rather directly related to the thermodynamic variables of the system. For our purposes the most important result is the so called fluctuation theorem for the isothermal compressibility,

$$k_{\mathrm{B}}T\left(\frac{\partial \rho}{\partial p}\right)_T = k_{\mathrm{B}}T\rho K_T = 1 + \rho \int G(\mathbf{r}) \, d\mathbf{r}, \qquad (3.3)$$

which is a quite general consequence of the laws of statistical mechanics. [For a two-dimensional system the integral in (3.3) is restricted appro-

[38] B. Widom and O. K. Rice, J. Chem. Phys. **23**, 1250 (1955).

priately while for a lattice system it is replaced by a sum.]

Now $G(\mathbf{r})$ is essentially a bounded function (more precisely its integral over a finite region is bounded in virtue of the existence of a maximum density arising from the incompressibility of real molecules.) Hence the fact that K_T becomes infinite at the critical point can only be understood if the integral over $G(\mathbf{r})$ diverges at its upper limits. This means that at the critical point the net correlation function becomes *long-range* in the sense that it decays to zero more slowly than $1/r^3$ (or in d-dimensions than $1/r^d$). It is clearly of interest to know the precise nature of this critical decay and to understand the rate of approach of $G(\mathbf{r})$ to its long-range behavior as $T \to T_c$.

Fortunately the pair correlation function can also be studied directly by scattering waves off the system. In practice experiments are usually performed with light or with x rays but thermal neutrons may also be used.[39] The observed angular dissymmetry is then a direct measure of the degree of correlation. To the extent that multiple scattering may be neglected (first Born approximation) we have for the scattering intensity

$$I(\mathbf{k})/I_0(\mathbf{k}) = \chi(\mathbf{k}) = 1 + \rho \hat{G}(\mathbf{k}), \qquad (3.4)$$

where $I_0(\mathbf{k})$ is the scattering intensity in the absence of correlation (the molecular form factor), $\mathbf{k}$ is the wave vector, $k = (4\pi/\lambda) \sin \frac{1}{2}\theta$, and where

$$\hat{G}(\mathbf{k}) = \int e^{i\mathbf{k}\cdot\mathbf{r}} G(\mathbf{r}) \, d\mathbf{r} \qquad (3.5)$$

is the Fourier transform of $G(\mathbf{r})$; for an isotropic three-dimensional system

$$\hat{G}(k) = 4\pi \int_0^\infty \frac{\sin kr}{kr} G(r) \, r^2 \, dr. \qquad (3.6)$$

The ratio $\chi(k)$ may be regarded as a generalized "susceptibility" since it measures the response of the fluid to an impressed periodic potential of wave number k.[40]

Comparison of (3.5) and (3.4) with the fluctuation relation (3.3) shows that

$$\chi(0) = \lim_{k \to 0} I(k)/I_0(k) = 1 + \rho \hat{G}(0) = k_{\mathrm{B}}T\rho K_T, \qquad (3.7)$$

so that the scattering intensity extrapolated to zero angle is proportional to the isothermal compress-

[39] L. Van Hove, Phys. Rev. **95**, 249 (1954). To discuss neutron scattering fully one must also consider the time dependence of the pair correlation function. Near the critical point, however, the decay of fluctuations probably becomes slower and it is reasonable to neglect this aspect of the problem in first approximation.

[40] P. G. de Gennes, Nuovo Cimento **9**, Suppl. 1, 240 (1958).

ibility. By virtue of the divergence of K_T at the critical point the low-angle scattering must thus become very large as the critical point is approached in the one-phase region. This is the "anomalous" critical scattering long known with visible light as *critical opalescence.* In physical terms one may say that the large compressibility near the critical point allows long-wavelength density fluctuations to grow to large amplitude and these produce visible diffraction.[39]

The classical theory of critical scattering is that developed by Ornstein and Zernike.[41,42] Their results have since been rederived many times and in various ways. In order to discuss the validity of their conclusions we will outline two approaches which characterize most of the derivations: on the one hand the original method of Ornstein and Zernike[41,43] which is perhaps the more mathematical, and on the other hand a semithermodynamic method which concentrates attention on the fluctuations of the free energy and their relation to the gradients of the density deviations. This latter approach was initiated by Rocard[44-46] in the spirit of Einstein's semiphenomenological ideas, but we will follow the presentation of Landau.[46]

Ornstein and Zernike[41] argue on a heuristic basis that the correlation $G(\mathbf{r}_1 - \mathbf{r}_2)$ between molecules 1 and 2 can be regarded as caused by (i) a *direct* influence of 1 on 2 described by the so-called "direct correlation function" $C(\mathbf{r}_1 - \mathbf{r}_2)$ which should be short-ranged [essentially having the range of the pair potential $\phi(r)$], and (ii) an indirect influence propagated directly from 1 to a third molecule at $\mathbf{r}_3$ which in turn exerts its total influence on molecule 2. Integrating over $\mathbf{r}_3$ they thus write the relation

$$G(\mathbf{r}_1 - \mathbf{r}_2) = C(\mathbf{r}_1 - \mathbf{r}_2) + \rho \int C(\mathbf{r}_1 - \mathbf{r}_3) G(\mathbf{r}_3 - \mathbf{r}_2) \, d\mathbf{r}_3. \quad (3.8)$$

In the absence of an independent theory enabling one to calculate $C(\mathbf{r})$ in terms of the molecular parameters this relation is really only a *definition* of the direct correlation function: we will, in the main, adopt this attitude. However, Ornstein and Zernike regarded $C(\mathbf{r})$ as the more basic function

(in its, presumably, closer relation to the intermolecular forces) and they contemplated the possibility of calculating $C(\mathbf{r})$ directly.

On introducing the Fourier transform of $C(\mathbf{r})$ the relation (3.8) (which states that $G(\mathbf{r}_1 - \mathbf{r}_2)$ and $C(\mathbf{r}_1 - \mathbf{r}_2)$ are reciprocal kernels in the sense of the theory of integral equations) can be solved to yield

$$1 + \rho\hat{G}(\mathbf{k}) = 1/[1 - \rho\hat{C}(\mathbf{k})]. \quad (3.9)$$

On substituting in (3.4) one finds for the inverse scattering intensity

$$1/\chi(\mathbf{k}) = 1 - \rho\hat{C}(\mathbf{k}). \quad (3.10)$$

Consequently the divergence of the compressibility $K_T = \chi(0)/k_B T\rho$ at the critical point is associated with the equation

$$1 - \rho\hat{C}(0) = 1 - \rho \int C(\mathbf{r}) \, d\mathbf{r} = 0. \quad (3.11)$$

This shows that the integral of $C(\mathbf{r})$ (i.e., its zeroth moment) remains finite at the cirtical point. Thus $C(\mathbf{r})$ certainly decays to zero more rapidly than $G(\mathbf{r})$ thereby confirming the expectation that it should be relatively short-ranged. To develop the theory further, however, one makes the central assumption that $C(\mathbf{r})$ is strictly short-ranged *at* (and near) the critical point in the sense that its transform $\hat{C}(\mathbf{k})$ has a Taylor series expansion in powers of k^2. In particular, one assumes that the second moment

$$R^2 = \tfrac{1}{2}\rho\langle\cos^2\theta\rangle \int r^2 C(\mathbf{r}) \, d\mathbf{r}, \quad (3.12)$$

exists at the critical point and does not vary rapidly in the vicinity. [In three dimensions the average over angles yields $\langle\cos^2\theta\rangle = \tfrac{1}{3}$.]

We note again that unless $C(\mathbf{r})$ can be calculated in an independent way R^2 will have the status only of a semiphenomenological parameter. The short-range character of $C(\mathbf{r})$ and the existence of R^2 at the critical point constitute a major problem of the theory and we will return to it. We remark here, however, that the virial expansion of $C(\mathbf{r})$ may be obtained quite easily from that for $G(\mathbf{r})$[43,42] and indicates for low densities at least, that $C(\mathbf{r})$ is, in a definite sense, shorter ranged than $G(\mathbf{r})$.

This may be seen in terms of the graphical representations of the respective expansions. All the graphs required[47,48] are connected and have two

[41] L. S. Ornstein and F. Zernike, Proc. Acad. Sci. Amsterdam 17, 793 (1914); Physik. Z. 19, 134 (1918); *ibid.* 27, 761 (1926).

[42] F. Zernike, Proc. Acad. Sci. Amsterdam 18, 1520 (1916).

[43] See also J. Yvon, Nuovo Cimento 9, Suppl. 1, 144 (1958) and Ref. 40.

[44] Y. Rocard, J. Phys. Radium 4, 165 (1933).

[45] See also: M. Fierz in *Theoretical Physics in the Twentieth Century,* edited by M. Fierz and V. F. Weisskopf (Interscience Publishers, Inc., New York, 1960), pp. 175 *et seq.*; M. J. Klein and L. Tisza, Phys. Rev. 76, 1861 (1949).

[46] L. D. Landau and E. M. Lifshitz, *Statistical Physics* (Pergamon Press, Ltd., London, 1958), Sec. 116.

[47] See, for example, E. Meeron, J. Math. Phys. 1, 192 (1960), J. M. J. Van Leeuwen, J. Groeneveld, and J. De Boer, Physica 25, 792 (1959); T. Morita and K. Hiroike, Progr. Theoret. Phys. (Kyoto) 23, 1003 (1960).

[48] G. E. Uhlenbeck and G. W. Ford, in *Studies in Statistical Mechanics,* I, edited by J. De Boer and G. E. Uhlenbeck, (North-Holland Publishing Company, Amsterdam, 1962), Chap. B.III 4.

fixed points (corresponding to the two molecules fixed at distance **r** apart) and $n = 0, 1, 2, \cdots$ field points. Each point is associated with a power of the density ρ while the bonds are, as usual, associated with the Mayer f factors $f_{ij} = \exp[-\phi(r_{ij})/kT] - 1$. If the potential is of strictly finite range b [in the sense that $\phi(r) = 0$ for $r > b$] the factor f_{ij} vanishes unless $r_{ij} \leq b$. This fact restricts the distance up to which a graph of given type can "stretch." Thus in the expansion of $G(r)$ the longest-ranged term in a given order, say ρ^{2m+1}, comes from the open chain of $2m$ bonds which contributes up to distances $r = 2mb$ but not beyond. The graphs entering the expansion of $C(r)$, however, are restricted to be "nonnodal", i.e., none of the field points may be cutting points whose removal would separate the graph into two parts.[47] The open chain of bonds is thus excluded and the longest-ranged contribution to $C(r)$ in order ρ^{2m+1} comes from the graph consisting of two parallel chains of m bonds each. This graph, however, will make no contribution for $r > mb$.

We see therefore that in a given order the range of $C(r)$ is only half that of $G(r)$. It is clear, nevertheless, that one should not expect the virial series to converge well in the critical region (if it converges at all) so that the assumed short-range nature of $C(r)$ cannot be established in this way. For the present, however, we follow Ornstein and Zernike and accept the existence of the second moment R^2 and the possibility of a Taylor series expansion of $\hat{C}(k)$. From (3.10) and (3.12) we then obtain on neglecting terms of order k^4, the scattering formula

$$\chi(k) = 1 + \rho\hat{G}(k) \simeq R^{-2}/(\kappa^2 + k^2), \quad (k^2 \to 0), \quad (3.13)$$

where κ, which has the dimensions of an inverse length, is defined by

$$\kappa^2 = [1 - \rho\hat{C}(0)]/R^2. \quad (3.14)$$

Fourier inversion of the simple Lorentzian scattering curve (3.13) shows that the behavior of $G(r)$ for large r is given in *three* dimensions by the famous result

$$G(r) \simeq \frac{1}{4\pi\rho R^2} \frac{e^{-\kappa r}}{r} \quad (r \to \infty). \quad (3.15)$$

From this one concludes that the correlations decay exponentially with an inverse range κ which, via the fluctuation theorem (3.3), should be related to the compressibility by

$$K_T = A/\kappa^2 \quad (T \to T_c). \quad (3.16)$$

The constant of proportionality is $A = 1/k_B T \rho R^2$

and is expected to be only slowly varying in the critical region. The divergence of K_T at the critical point thus implies

$$\kappa(T) \to 0 \quad \text{as} \quad T \to T_c, \quad (3.17)$$

so that the critical point correlation function is no longer exponentially damped but is predicted to follow the law

$$G_c(r) \simeq D/r \quad (r \to \infty, \ T = T_c). \quad (3.18)$$

In as far as the relation $K_T \sim 1/\kappa^2$ is valid and the classical variation of $K_T(T)$ is accepted, the inverse range will go to zero along the critical isochore as

$$\kappa(T) \simeq \kappa^0 |1 - (T/T_c)|^\nu \quad (T \to T_c), \quad (3.19)$$

with $\nu = \frac{1}{2}$. This conclusion was mentioned by Zernike[42] and is accepted by other authors.[49-52] More generally, however, if one recognizes deviations of the isothermal compressibility from van der Waals behavior one would get the "nonclassical" result $\nu = \frac{1}{2}\gamma > \frac{1}{2}$ [see Eq. (2.8)].

The principal alternative approach to the Ornstein–Zernike theory is based on considering the thermodynamic work, or change in free energy, required to establish a density fluctuation in the system,[44-46,49-52] i.e., a local inhomogeneity. One supposes that a local free-energy density $F(\mathbf{r})$ can be defined for an inhomogeneous system and considers the expansion of F (or its integral over a small but macroscopic volume) about its homogeneous mean value $\bar{F}$ in terms of the local deviation $\delta\rho(\mathbf{r})$ of the density from its mean value ρ. In the spirit of the classical theory of the equation of state one assumes that a Taylor series exists even at the critical point. The first power of $\delta\rho$ may be dropped by virtue of the conservation of particles. The coefficient of $\delta\rho^2$ is, thermodynamically, proportional to $1/K_T$ and so this term must be retained. Since the state of the system will be inhomogeneous, however, one must also expect terms dependent on $\nabla\rho$ the gradient of the density deviation, and on higher derivatives. The necessity for such terms can indeed be seen rather generally from the existence of surface tension which represents, of course, an additive contribution to the free energy directly associated with the density inhomogeneities at an interface. On the grounds of symmetry the leading term will be proportional to $(\nabla\rho)^2$. [The terms $\nabla^2\rho$ and $\rho\nabla^2\rho$ add nothing further after

[49] P. Debye, J. Chem. Phys. **31**, 680 (1959).
[50] E. W. Hart, J. Chem. Phys. **34**, 1471 (1961).
[51] M. Fixman, J. Chem. Phys. **33**, 1357 (1960).
[52] M. Fixman, J. Chem. Phys. **36**, 1965 (1962).

integrating over a small volume.[46]] Consequently, one writes the expansion

$$\Delta F(\mathbf{r}) = \tfrac{1}{2}c\,\delta\rho^2 + \tfrac{1}{2}d(\nabla\rho)^2 + \cdots \qquad (3.20)$$

and assumes that at least for small slowly varying deviations (i.e., small k), the higher-order terms may be neglected to a good approximation even at the critical point. By stability considerations c and d must be positive. Although c will vanish at the critical point d remains nonzero.

On introducing the Fourier components of the density deviation

$$\delta\hat\rho_\mathbf{k} = V^{-1} \int e^{i\mathbf{k}\cdot\mathbf{r}}\,\delta\rho(\mathbf{r})\,d\mathbf{r}, \qquad (3.21)$$

substituting in (3.20) and integrating over the volume of the system one obtains for the total free energy fluctuation

$$\Delta F_{\text{total}} = \tfrac{1}{2}V \sum_\mathbf{k} (c + dk^2)\,|\delta\hat\rho_\mathbf{k}|^2. \qquad (3.22)$$

We notice that each density mode contributes additively to the free energy so that the modes are statistically independent or, in other words, effectively noninteracting. This conclusion is, of course, a direct consequence of the truncation of (3.20) although it also has an immediate physical appeal for long wavelength modes.

Now the Boltzmann factor for a fluctuation $\delta\hat\rho_\mathbf{k}$ is $\exp[-\Delta F_\mathbf{k}/k_\mathrm{B}T]$ and consequently the mean square fluctuation is predicted, at least for small k, to be

$$\langle|\delta\hat\rho_\mathbf{k}|^2\rangle = k_\mathrm{B}T/V(c + dk^2). \qquad (3.23)$$

Now by the definition (3.21) of the Fourier coefficients we have

$$\langle|\delta\hat\rho_\mathbf{k}|^2\rangle = \langle\delta\hat\rho_\mathbf{k}\,\delta\hat\rho_{-\mathbf{k}}\rangle = V^{-1}\int e^{i\mathbf{k}\cdot\mathbf{r}}\langle\delta\rho(0)\,\delta\rho(\mathbf{r})\rangle\,d\mathbf{r},$$

where one integration over $\mathbf{r}$ space has been performed using the (approximate) translational invariance of the system. Since the mean-density fluctuation is zero

$$\langle\delta\rho(0)\,\delta\rho(\mathbf{r})\rangle/\rho = \langle[\rho + \delta\rho(0)][\rho + \delta\rho(\mathbf{r})]\rangle/\rho - \rho,$$
$$= \rho g(\mathbf{r}) + \delta(\mathbf{r}) - \rho,$$

where the second line follows from the definition (3.1) of $g(\mathbf{r})$, the delta function being added to allow for the identity of the correlated particles [which was implicitly excluded in (3.1)]. Substitution shows that

$$(V/\rho)\langle|\delta\hat\rho_\mathbf{k}|^2\rangle = 1 + \rho\hat G(\mathbf{k})$$
$$\simeq \frac{(k_\mathrm{B}T/\rho)}{c + dk^2}, \qquad (3.24)$$

which is clearly equivalent to the Ornstein–Zernike result (3.13).[53]

The relationship between the two approaches to critical scattering theory is revealed by the more recent development of a complete formal theory of the statistical mechanics of nonuniform systems which shows the significance of the direct correlation function in constructing expansions of the thermodynamic variables of inhomogeneous systems.[54-56] In particular the existence of local thermodynamic variables and the convergence of the corresponding expansions turns out to be dependent on the short range nature of $C(r)$.

A fruitful method developed by Lebowitz and Percus[56,57] is to consider the deviations in density produced by an externally imposed potential $U(\mathbf{r})$ when this is chosen to be the potential $\phi(\mathbf{r})$ which would just correspond to a molecule of the fluid fixed at the origin. The induced singlet-density deviation is then related to the pair correlation function in the homogeneous fluid as follows:

$$n_1(\mathbf{r}|\phi) = \rho + \delta n_1(\mathbf{r}) = n_2^0(\mathbf{r})/n_1^0,$$
$$= \rho + \rho G(\mathbf{r}), \qquad (3.25)$$

where the superscripts zero denote the uniform system.[58]

To obtain an equation for $n_1(\mathbf{r}|\phi)$ and hence for $G(r)$ one assumes the inhomogeneous system can be represented by a grand canonical ensemble and one asks for the relation between the external potential $U(r) = \phi(r)$ and the induced-density deviation. To this end it might be natural to try to expand $n_1(\mathbf{r}|\phi)$ as a functional Taylor series in $\phi(r)$ but Lebowitz and Percus show, on the contrary, that it is possible, and more useful, to expand $\phi(r)$ (in combination with the chemical potential) in terms of the density deviation it produces.[55-57]

[53] J. L. Lebowitz (private communication) has pointed out that by the methods of Refs. 55–57 one may show that in an appropriate grand canonical ensemble the fluctuation of the free energy is given to *second order* in $\delta\hat\rho_\mathbf{k}/\rho$ exactly by
$$\Delta F_{\text{total}}/Nk_\mathrm{B}T = \tfrac{1}{2}\sum_\mathbf{k}[1 - \rho\hat C(\mathbf{k})]\,|\delta\hat\rho_\mathbf{k}/\rho|^2.$$
Comparison with (3.22) shows even more directly the equivalence to the Ornstein–Zernike theory.

[54] F. H. Stillinger, Jr. and F. P. Buff, J. Chem. Phys. **37**, 1 (1962).

[55] J. L. Lebowitz and J. K. Percus, Phys. Rev. **122**, 1675 (1961); J. Math. Phys. **4**, 116 (1963).

[56] J. K. Percus, Phys. Rev. Letters **8**, 462 (1962).

[57] J. L. Lebowitz and J. K. Percus, J. Math. Phys. **4**, 248 (1963).

[58] We use the notation $\delta n_1(\mathbf{r})$ rather than $\delta\rho(\mathbf{r})$ as previously, since in the thermodynamic arguments one really is considering a macroscopic or coarse-grained density fluctuation whereas here one refers directly to the microscopic distribution functions. By the same token one should replace $\delta\hat\rho_\mathbf{k}$ in Footnote 53 by $\delta\hat n_1(\mathbf{k})$ whereas in Eq. (3.24) and the preceding steps $g(\mathbf{r})$ and $\hat G(\mathbf{k})$ represent coarse-grained or macroscopic correlation functions.

One anticipates that such an expansion would converge most rapidly when taken at each point, about a uniform system with the same *local* density $n_1(\mathbf{r}|\phi)$. The successive terms of the expansion, which is conveniently derived by the technique of functional differentiation, are then found to be multiple integrals over the density deviation with kernels which involve $C(\mathbf{r}_1 - \mathbf{r}_2)$ and certain higher-order correlation functions.

Now when $C(\mathbf{r})$ is a short-ranged function one may expand these integrals as a "local" series in the spatial derivatives of the density deviation. The justification for this step must rest, of course, on showing that $C(r)$ decays rapidly to zero for large r [and that $n_1(\mathbf{r}|\phi)$ is not too rapidly varying]. If this is the case one finds

$$\phi(\mathbf{r}) = \mu - \mu^0[n_1(\mathbf{r})] + (R^2 k_B T \rho^2/n_1^2)\nabla^2 G(\mathbf{r})$$
$$+ \tfrac{1}{6}(l^2\rho^2/k_B T n_1^4 K_T^2)[\nabla G(\mathbf{r})]^2 + \cdots, \quad (3.26)$$

where μ is the chemical potential and R^2, defined already in Eq. (3.12), is the second moment of the direct correlation function. The length l is defined similarly in terms of $C(r)$ and the three-particle distribution function. Further terms of (3.26) can be written down explicitly and involve higher derivatives of $G(r)$ with coefficients depending on higher moments of $C(r)$ and the further many-particle distribution functions.[55-57]

If we now drop the terms in $(\nabla G)^2$ and higher-order derivatives and expand $n_1(r) = \rho[1 + G(r)]$ to first order in $G(r)$ we obtain

$$R^2 k_B T \nabla^2 G(\mathbf{r}) - \rho(\partial\mu^0/\partial\rho)G(\mathbf{r}) = \phi(\mathbf{r}), \quad (3.27)$$

which should be valid in the asymptotic region where $G(r)$ is small. The derivative of μ^0 arises from expanding $\mu^0[n_1(\mathbf{r})]$ and may be eliminated through the thermodynamic relation $\rho(\partial\mu^0/\partial\rho)_T = (\partial p/\partial\rho)_T = 1/\rho K_T$. On introducing the length Λ by

$$\Lambda^2 = R^2 k_B T \rho K_T$$

$$= \tfrac{1}{2}\langle\cos^2\theta\rangle\rho \int r^2 G(r)\, dr/[1 + \hat{G}(0)] \quad (3.28)$$

[where the last formula follows from (3.12), (3.10) and (3.9)] we obtain the equation

$$\nabla^2 G(r) - \Lambda^{-2}G(r) = \phi(r)/k_B T R^2 \quad (3.29)$$

first derived by Zernike.[42]

If the potential $\phi(r)$ is negligible for large r the asymptotic solution of (3.29) is, in three dimensions,

$$G(r) \approx D e^{-(r/\Lambda)}/r. \quad (3.30)$$

Comparison with (3.15) shows the equivalence to

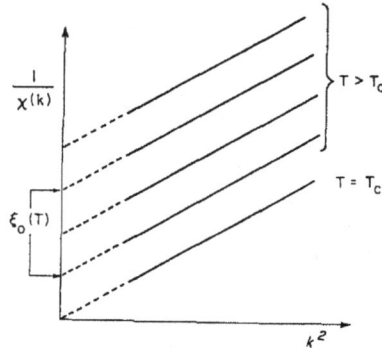

FIG. 5. Dependence of the inverse critical scattering on $k = (4\pi/\lambda)\sin\tfrac{1}{2}\theta$ according to the Ornstein-Zernike theory. The linear intercept $\xi_0(T)$ is proportional to $1/K_T$.

the previous theory and identifies $\Lambda = 1/\kappa$ as the range of correlation. Fourier transformation of Eq. (3.29) leads directly to the previous expression (3.13) for the scattering intensity for small κ^2 and k^2.

The advantage of this last derivation is that the neglected terms are explicitly displayed and that the central role of the direct correlation function is apparent. It does not, however, enable us to decide if the assumption that $C(r)$ is short-ranged at the critical point is justified. On the other hand *away* from the critical point, but in a region where K_T is moderately large the analysis indicates that the Ornstein-Zernike theory should be correct. Indeed the exponential part of the asymptotic decay law for $G(r)$ appears to have a rather wide range of validity in a one-phase fluid system since it depends essentially only on the short-range nature of the interactions.[59]

4. VALIDITY OF THE ORNSTEIN–ZERNIKE THEORY

The principal experimental predictions following from the classical theory of critical scattering [i.e., from Eq. (3.13)] are (a) that $1/\chi(k, T)$, the reciprocal of the relative scattering intensity, should vary *linearly* with k^2 with a temperature-independent coefficient of proportionality, and (b) that the extra-

[59] This conclusion follows from a formulation of statistical mechanics in which the system is taken to be in a cylinder of length L and cross section A. For forces with a hard core and strictly finite range b, a nonsingular integral kernel describes the addition of a layer of thickness b to the cylinder. The thermodynamics and correlation functions are related to the resolvent of this kernel: M. E. Fisher, abstract in Proc. Second Eastern Theoretical Physics Conf., University of North Carolina (October, 1963). D. Ruelle and, independently, J. Groeneveld (to be published) have also shown that in the region where the activity expansion can be proved to converge, $G(r)$ [suitably smoothed] decays to zero at least exponentially fast for strictly short-range potentials.

MICHAEL E. FISHER

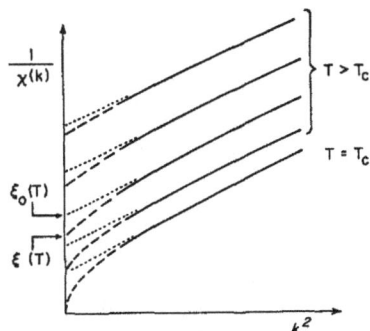

FIG. 6. Schematic variation of the inverse critical scattering expected in view of the limitations of the classical theory. (Compare with Fig. 5.) The apparent linear intercept $\xi_0(T)$ differs from the true intercept $\xi(T)$ which is proportional to $1/K_T$.

polated intercept with the $k^2=0$ axis, $\xi_0(T)$, should be proportional to $1/K_T(T)$ and hence go to zero as $T \rightarrow T_c$. The predicted behavior is shown schematically in Fig. 5.

Most modern tests of the theory have been made by light or x-ray scattering measurements on binary fluid mixtures of relatively complex organic molecules.[60-65] For example, Zimm[60] studied a mixture of perfluoromethylcyclohexane in carbon tetrachloride, Brady and Frisch,[62] perfluoroheptane in iso-octane, while Debye and co-workers[63] and McIntyre, Wims and Green[64] have investigated polystryrene–cyclohexane solutions.

Earlier measurements on binary fluids and other systems have been reviewed by Rice.[65] Recently Thomas and Schmidt[66] have made an extensive series of x-ray measurements on argon at various constant pressures in the critical region. They also give references to other more recent work on single-component systems such as carbon dioxide, ethylene and neon.

Accurate experiments are not easy to perform near the critical point and the interpretation can be confused by multiple scattering. Qualitatively,

however, the theory seems quite well confirmed by the binary fluid measurements. Plots of the reciprocal scattering intensity versus k^2 in the experimentally accessible range are well represented by sets of parallel straight lines whose intercepts $\xi_0(T)$ fall roughly linearly with $(T - T_c)$ as indicated in Fig. 5. Closer inspection, nevertheless, reveals certain "anomalies", although at present these are not much larger than the experimental uncertainties.[64,67,68] In particular one observes (a) a tendency for the scattering curves taken near T_c to be slightly curved and to dip downwards somewhat at the lowest values of k^2 and (b) the intercepts $\xi_0(T)$ obtained by extrapolation of the best straight line fits to the data (all necessarily lying above some k^2_{min}) do not seem to approach zero as T goes to T_c: rather plots of $\xi_0(T)$ versus T along the isochore are slightly concave upwards and tend to level off or to extrapolate to a nonzero value at T_c. These deviations are indicated in Fig. 6.

Green[68] reports that significant deviations from the Ornstein–Zernike theory were also found for the, presumably physically simpler, system of nitrogen at its critical point.[69] No deviations were observed by Thomas and Schmidt[66] in argon but they did not measure along the critical isochore and their lowest values of k^2 were relatively large.

As we show below these deviations are in the direction to be expected theoretically on the basis of an analysis of the limitations of the classical theory. Needless to say, however, it would still be desirable to have more accurate and extensive experimental data, especially for low values of k^2 and for simple systems like the noble gases, in order to elucidate fully the true nature of the critical scattering.

As we have seen the main theoretical problem in justifying the derivation of the Ornstein–Zernike result is to establish the short-range nature of the direct correlation function $C(r)$, or, what is equivalent, to show that its Fourier transform $\hat{C}(k)$ has a Taylor series expansion in powers of k^2 *at* the critical point. Our previous discussion of the thermodynamic variables at the critical point, has shown that the hypothesis of a Taylor expansion in temperature or density is probably not tenable. By analogy we should be prepared for a similar failure for the correlation functions.

An obvious defect of the theory can be seen by

[60] B. H. Zimm, J. Phys. & Colloid Chem. 54, 1306 (1950).
[61] Chow Quantie, Proc. Roy. Soc. (London) A224, 90 (1954); R. Fürth and C. L. Williams, Proc. Roy. Soc. (London) A224, 104 (1954).
[62] G. W. Brady and H. L. Frisch, J. Chem. Phys. 35, 2234 (1961).
[63] P. Debye, H. Coll, and D. Woermann, J. Chem. Phys. 32, 939 (1960); *ibid* 33, 1746 (1960); P. Debye, D. Woermann and B. Chu, J. Chem. Phys. 36, 851 (1962).
[64] D. McIntyre, A. Wims and M. S. Green, J. Chem. Phys. 37, 3019 (1962).
[65] O. K. Rice, "Critical Phenomena" Sec. E, in *Thermodynamics and Physics of Matter*, edited by F. D. Rossini (Princeton University Press, Princeton, New Jersey, 1955).
[66] J. E. Thomas and P. W. Schmidt, J. Chem. Phys. 39, 2506 (1963).

[67] H. L. Frisch and G. W. Brady, J. Chem. Phys. 37, 1514 (1962).
[68] M. S. Green, J. Chem. Phys. 33, 1403 (1960).
[69] R. L. Wild, J. Chem. Phys. 18, 1627 (1950).

considering its application to model systems of dimensionality d different from three. In any number of dimensions the classical expression for the (appropriate) Fourier transform is formally the same, namely

$$\chi(\mathbf{k}) = 1 + \rho\hat{G}(\mathbf{k}) \simeq A/(\kappa^2 + k^2). \quad (4.1)$$

For fixed $\kappa(T) > 0$, that is *away* from the critical point, we find[70] by inverting (4.1) that as r becomes very large

$$G(r) \simeq B_d(e^{-\kappa r}/r^{\frac{1}{2}(d-1)})[1 + O(1/\kappa r)]$$
$$(r \to \infty, \ \kappa \text{ fixed} > 0). \quad (4.2)$$

As we argued at the end of the previous section it seems probable that this result is generally valid for *fixed* $T > T_c$ and large enough r. However, as $\kappa \to 0$ for *fixed* large r we find a different result,[70] namely, for $d \geq 3$

$$G(r) \simeq D_d(e^{-\kappa r}/r^{d-2})[1 + O(\kappa r)]$$
$$(\kappa \to 0, \ r \text{ fixed}) \quad (4.3)$$

while for $d = 2$

$$G(r) \simeq D_2(\log r)e^{-\kappa r}[1 + O(1/\log \kappa r)]$$
$$(\kappa \to 0, \ r \text{ fixed}). \quad (4.4)$$

One notices that $d = 3$ is a rather special case in which both limits (4.2) and (4.3) agree and [by comparison with Eq. (3.15)] the higher-order terms vanish identically!

Now (4.4) implies that at the critical point of a two-dimensional system the correlation function will vary as $D_2 \log r$. This is clearly nonphysical for large r and shows that the assumptions of the theory are certainly not to be trusted for two-dimensional systems.

This defect of the theory can be repaired in an *ad hoc* fashion by retaining further (nonlinear) terms in the density expansions of the free energy [in Eq. (3.20)] or in the Taylor series expansion of $n_1(\mathbf{r})$ [in Eq. (3.26)].[71] It would seem difficult, however, to justify keeping nonlinear terms in $G(r)$ rather than, say nonlinear terms in ∇G or higher-order derivatives of $G(r)$: the more so as the whole question of the convergence of such an expansion is in doubt at the critical point.

The first author to question the theory in respect of the prediction $G(r) \sim 1/r$ at $T = T_c$, for three dimensions was Green[68] who based his arguments on an integral relation for the pair correlation function derived by cluster diagram summation tech-

niques.[47] This so called, hypernetted-chain integral equation may be written

$$1 + G(r) = \exp[-\beta\phi(r) + G(r) - C(r) + E(r)], \quad (4.5)$$

where

$$E(r) = \mathcal{E}\{\rho; G(r)\} \quad (4.6)$$

is a nonlinear integral functional of $G(r)$ known only as an expansion in powers of ρ representable in terms of certain, so called, "basic graphs."[47] [The leading term is of order ρ^2]. The same relations hold for a lattice system (with appropriate definition of the functional $\mathcal{E}$).

Green[68] considered the consequences of assuming that the term $E(r)$ involving the basic graphs, might be neglected at the critical point. Stillinger and Frisch extended his analysis to two-dimensional systems.[72]

To follow the argument let us suppose in greater generality that at the critical point of a d-dimensional system

$$G(r) \simeq D/r^{d-2+\eta} \quad (r \to \infty), \quad (4.7)$$

where the index η $(0 \leq \eta \leq 2)$ measures the departure from the Ornstein–Zernike prediction. [As in Eq. (2.7) $1/r^0$ corresponds to $\log r$.] It follows that

$$\rho\hat{G}(k) \simeq \hat{D}/k^{2-\eta}, \quad (k \to 0) \quad (4.8)$$

so that at fixed density one has, through the relation (3.9),

$$\hat{C}(k) \simeq C(0)[1 - c_0 k^{2-\eta} + \cdots], \quad (k \to 0). \quad (4.9)$$

If $\eta > 0$ asymptotic inversion yields

$$C(r) \simeq F/r^{d+2-\eta} \quad (r \to \infty). \quad (4.10)$$

Thus when $\eta > 0$, the direct correlation function is *also* "long ranged" in the sense that its second moment does not exist, although it certainly decays to zero more rapidly than $G(r)$, in fact by a factor $1/r^{4-2\eta}$. Notice that if $\eta = 0$ we find instead

$$C(r) \simeq F'e^{-\kappa' r}/r, \quad (\kappa'^2 = 1/c_0) \quad (4.11)$$

so that only in this special case is $C(r)$ short ranged in the sense that $\hat{C}(k)$ has a Taylor series expansion at $k^2 = 0$.

To analyze the hypernetted integral relation it is convenient to define

$$S(r) = G(r) - C(r). \quad (4.12)$$

Clearly $S(r)$ must have the same asymptotic be-

[70] M. E. Fisher, Physica **28**, 172 (1962).
[71] M. Fixman, J. Chem. Phys. **36**, 1965 (1962).

[72] F. H. Stillinger, Jr. and H. L. Frisch, Physica **27**, 751 (1961).

havior as $G(r)$. If the potential $\phi(r)$ is short ranged we may expand the exponential in (4.5) for large r to get

$$C(r) = E(r) + \tfrac{1}{2}[S(r) + E(r)]^2 + \cdots . \qquad (4.13)$$

If we now suppose $E(r)$ can be neglected or, more weakly, that $E(r)$ decays faster than $[S(r)]^2$ it follows that

$$C(r) = \tfrac{1}{2}[S(r)]^2 + \cdots . \qquad (4.14)$$

On substituting (4.7) and (4.10) we obtain the consistency relation

$$\eta = 2 - \tfrac{1}{3}d, \qquad (4.15)$$

which fixes the asymptotic form of the correlation functions.

On this basis we would predict that at the critical point in a three-dimensional system $G(r) \sim 1/r^2$ rather than $1/r$.[68] Correspondingly $\hat{G}(k) \sim 1/k = 1/(k^2)^{\frac{1}{2}}$ so that a plot of inverse scattering versus k^2 at, and near $T = T_c$, should be significantly curved downwards for small k^2 (see Fig. 6).

In two dimensions (4.15) leads to $G(r) \sim 1/r^{4/3}$ ($T = T_c$) which is more reasonable than the Ornstein–Zernike result $\log r$. However as pointed out by Stillinger and Frisch,[72] this prediction can be tested against the rigorous result obtained by Onsager and Kaufman[31] for the correlation function of the nearest-neighbor plane square lattice gas at its critical point. [Note that $G(\mathbf{r})$ at $\rho = \rho_a$ is proportional to the spin pair correlation function $\langle S_0^z S_r^z \rangle$ of the Ising ferromagnet in zero field.] This exact result is[31,32,72,73]

$$G(r) \approx D/r^{\frac{1}{4}} \qquad (T = T_c,\ d = 2) \qquad (4.16)$$

so that the true value of the index η is $\tfrac{1}{4}$ rather than $\tfrac{2}{3}$ or zero.

We conclude that both the Ornstein–Zernike theory and Green's argument are incorrect for a two-dimensional lattice. Consequently both must be suspect for three-dimensional lattice systems.

It is interesting to note that if, as seems to be the case, the true value of η is less than $2 - \tfrac{1}{3}d$ one really has, as $r \to \infty$

$$E(r) \approx \tfrac{1}{2}[S(r)]^2 \approx \tfrac{1}{2}[G(r)]^2 \approx D^2/r^{2d-4+2\eta} \qquad (4.17)$$

so that the contribution of the basic graphs is also long ranged although it decays faster than $G(r)$. Since after all, $E(r)$ is a functional of $G(r)$ this is not really surprising.

Of course, the rigorous result (4.16) is known only for nearest-neighbor interactions and one should ask to what extent the behavior of lattices with interactions reaching to further neighbors would be similar. It seems plausible that the correct answer is that the behavior sufficiently near the critical point is qualitatively unchanged provided the range of the potential is finite [for example, if $\phi(r) = 0$ for $r > b$]. The reason for this surmise is that the range of correlation near the critical point becomes, as we have seen, very large compared to the lattice spacing and, indeed, very large compared to the potential range b. The asymptotic correlations are then determined by long chains of interactions and should thus be insensitive to the detailed variation of $\phi(r)$. [The same conclusion is really implicit in Ornstein and Zernike's and in Green's approach.]

The independence of the indices α, β, γ and η of the lattice structure is evidence for this conclusion. Further evidence comes from numerical studies of the compressibility of lattice gases with first and *second* neighbor interactions which indicate unchanged values for the index γ in two and three dimensions.[74]

It is more difficult to assess the relationship of the lattices gases to more realistic *continuum* models. At low temperatures and high densities the properties of a lattice gas will always deviate from continuum behavior but in the critical region the long range of the correlations again suggests an insensitivity to the details of the potential and hence, for a lattice subdivision sufficiently fine relative to the range of the potential, one would expect qualitatively very similar behavior.[75] This conclusion is supported by the apparently quite close resemblance between the critical singularities of real gases and of even the simplest nearest neighbor three-dimensional lattice gases discussed already in Sec. 2.

5. MORE GENERAL ANALYSIS OF CRITICAL SCATTERING

It is clear from the foregoing that the Ornstein–Zernike theory is probably not valid at the critical point of a three-dimensional system. On the contrary one should evidently expect that $G(r)$ behaves asymptotically as $1/r^{1+\eta}$ with $0 < \eta < 1$.[70,72]

A natural way to extend the Ornstein–Zernike theory within the same framework is to consider further powers of k^2 in the expansion of the Fourier

---

[73] See also Ref. 12, pp. 200–201 but notice that an exponent $\tfrac{1}{2}$ is missing on the left of Eq. (108).

[74] M. F. Sykes and N. Dalton (to be published).

[75] In one dimension one may verify explicitly that the behavior of the lattice gas approaches that of the continuum gas as the lattice spacing is made smaller relative to the scale of the potential.

transform of the direct correlation function $\hat{C}(k)$. Inclusion of such terms leads to a representation of $G(r)$ as a sum of increasingly more rapidly damped exponentials of the form (for $d = 3$)

$$G(r) \approx \frac{De^{-\kappa r}}{r} + \frac{D_1 e^{-\kappa_1 r}}{r} + \frac{D_2 e^{-\kappa_2 r}}{r} + \cdots \quad (5.1)$$

with $\kappa < \kappa_1 < \kappa_2 \cdots$. At a fixed temperature, however, such an expansion leads to the same asymptotic behavior for $G(r)$ and to similar scattering at small k^2 as does the original theory. (For smaller enough r and large enough k^2 one must, of course, expect derivations from any general theory since the detailed nature of the potential must eventually make itself felt.)

As one considers T approaching T_c along $\rho = \rho_c$, however, it is possible that more and more exponentials in (5.1) become "excited," so that the first exponential is no longer a good approximation except for extremely large r.[76] This certainly represents essentially what happens for the two-dimensional lattice gas. Here the higher-order range parameters $\kappa_1, \kappa_2, \cdots$ obey the relation[12,14,77]

$$\kappa_n(T) - \kappa(T) \sim n\kappa_0 \, |1 - (T/T_c)|, \quad (5.2)$$

so that the "spectrum" of exponentials closes up as $T \to T_c$ and in fact becomes dense at the critical point.

A similar behavior for a continuum model is suggested by the recent calculations of Hemmer, Kac, and Uhlenbeck[76] for a potential with a strongly repulsive core and a weakly attractive long-range exponential tail. A series similar to (5.1) can be derived but the amplitudes of successive terms are proportional to higher powers of the compressibility K_T and hence the expansion breaks down near the critical point.

These considerations [which really amount to a restatement of our previous conclusion that $\hat{C}(k)$ probably does not have a Taylor series expansion at the critical point although it does for $T > T_c$] indicate that (5.1) is not the best basis for analyzing the deviations from the Ornstein–Zernike theory. Indeed the asymptotic form $1/r^{1+\eta}$ at $T = T_c$ with $\eta \neq 0$ could only arise from (5.1) if the expansion broke down in some way. To investigate the possibilities more generally let us, therefore, extend (4.7)

by writing [70] for $T \geq T_c$ and $\rho = \rho_c$

$$G(r) \simeq (De^{-\kappa r}/r^{d-2+\eta})[1 + Q(\kappa r)] \quad (r \to \infty) \quad (5.3a)$$

where $D = D(T)$ is a relatively slowly varying function of temperature and where $Q(x) \to 0$ as $x \to 1$ and $Q(x)$ does not grow exponentially fast as $x \to \infty$. This expression is still in the spirit of the Ornstein–Zernike theory in as far as the main assumption implicit in (5.3a) is that in the critical region the correlation functions for large r can be described in terms of only *two* lengths: (i) the *range of correlation* $1/\kappa(T)$ which becomes infinite at the critical point, and (ii) an *effective range of direct interaction* $r_0(T)$ which remains finite at the critical point. In (5.3a) r_0 has been absorbed into the coefficient D. [Compare with R of the Ornstein–Zernike theory: Eqs. (3.12) and (3.15).] We could write in analogy with (3.15)

$$D = \bar{d}/\rho r_0^{2+\eta}, \quad (5.3b)$$

where $\bar{d}$ is a dimensionless constant.

If the classical theory is valid away from the critical point, as concluded in Sec. 3, the formula (5.3a) should reduce to (4.2) when $\kappa > 0$ and r is very large, i.e. only the first term in (5.1) should remain. [For the plane square nearest-neighbor lattice gas at $\rho = \rho_c$ one may verify that (4.2) is indeed valid above T_c.[77]] This would imply that for large x

$$1 + Q(x) \approx qx^{\frac{1}{2}(d-3)+\eta} \quad (x \to \infty). \quad (5.3c)$$

Sufficiently close to the critical point the nature of this behavior will not matter. It is clear, however, that if $Q(x)$ becomes of order unity for small x the region of significant deviation from classical theory may be rather small. [Stated alternatively, the further exponentials in (5.1) would be significantly excited only very close to the critical point.]

Accepting (5.3) we may calculate the fluctuation integral (3.3) (making the substitution $x = \kappa r$). For the divergence of the compressibility along the critical isochore this yields

$$\chi(0) = k_B T \rho K_T \simeq \hat{D}_0/\kappa^{2-\eta} \quad (T \to T_c), \quad (5.4)$$

where $\hat{D}_0$ is a slowly varying function of T [of magnitude dependent on $Q(x)$]. Similarly a calculation of the Fourier transform of (5.3) for small k^2 yields, near T_c, the non-Lorentzian critical scattering formula

$$\chi(k) \simeq \hat{D}/(\kappa^2 + k^2)^{1-\frac{1}{2}\eta} \quad (k^2 \to 0), \quad (5.5)$$

where $\hat{D}$ is a slowly varying function of T and k^2.[78]

---

[76] P. C. Hemmer, M. Kac and G. E. Uhlenbeck, J. Math. Phys. 5, 60 (1964), P. C. Hemmer, J. Math. Phys. 5, 75 (1964).

[77] The correlation functions may be expressed as a sum of integrals over the complete set of eigenvalues of the basic transition matrix for an Ising lattice (see Ref. 31). The exact limiting density of these eigenvalues is known from Onsager's work (Ref. 14).

[78] If $Q(x)$ is neglected one has for small η
$$\hat{D}(k^2) = D_0\{1 - \eta\kappa^2/(\kappa^2 + k^2) + \cdots\}.$$

As in our discussion of the Ornstein–Zernike theory the way in which the inverse range of correlation $\kappa(T)$ vanishes as $T \to T_c$ for $\rho = \rho_c$ is related to the nature of the corresponding divergence of the compressibility as $1/(T - T_c)^\gamma$. Assuming, as in (3.19) that

$$\kappa(T) \simeq \kappa^0 \, |1 - (T/T_c)|^\nu \qquad (T \to T_c) \qquad (5.6)$$

and substituting in (5.4) shows that the critical indices are related by

$$\gamma = (2 - \eta)\nu. \qquad (5.7)$$

This relation may be checked for the plane square lattice gas since, as we have seen,[31,32,72,73] $\eta = \frac{1}{4}$ and $\gamma = 1\frac{3}{4}$ (see Table I). Consequently we should have

$$\nu = 1, \qquad (d = 2) \qquad (5.8)$$

in contrast to the classical result $\nu = \frac{1}{2}$. Now the result $\nu = 1$ was in fact derived rigorously by Onsager in his original paper on the Ising model.[14] His derivation is based on the relation

$$e^{-\kappa a} = \lambda_1/\lambda_0, \qquad (5.9)$$

where a is the lattice spacing, and where $\lambda_0(T)$ and $\lambda_1(T)$ are the largest and next largest eigenvalues of the basic matrix which adds a row to the lattice at temperature T.[77]

If we *assume*[79] that ν is still unity for the three-dimensional lattice gas and utilize the result[80] $\gamma = 1\frac{1}{4}$ we would predict $\eta = \frac{3}{4}$. Thence the critical point decay law would be $G_c(r) \sim 1/r^{7/4}$ in closer agreement with Green's result. However the assumption $\nu = 1$ in three dimensions is not much better justified *a priori* than the classical assumption $\nu = \frac{1}{2}$. Furthermore, examination of the other critical indices (see Table I) indicates that the classically predicted behavior is more closely approached the larger the dimensionality of the system.[81] Consequently one might anticipate that for the three-dimensional nearest-neighbor lattice gas ν lies between $\frac{1}{2}$ and 1 and η is *less* than $\frac{1}{4}$. [As mentioned previously the *assumption* $\eta = 0$ leads to $\nu = \frac{1}{2}\gamma = \frac{5}{8}$ when $d = 3$.]

To decide between these various speculations it is necessary to calculate $\kappa(T)$ or some other feature of the correlation functions. Fortunately the Ising model is again sufficiently tractable to allow some progress. It is possible at the critical density, to derive a diagrammatic expansion for the decay factor $e^{-\kappa a}$ in powers of $1/T$ via Eq. (5.9).[82] The required graphs consist of an infinite chain of connected bonds stretching right across the lattice together with nonoverlapping closed polygons, as occur in the expansion of the partition function.[12]

Evaluation of the series[83] for $\kappa(T)$ (with the aid of Padé approximants) reveals a behavior near T_c clearly intermediate between the two-dimensional and classical results in a region $T = T_c$ to $2T_c$. Direct estimation of the index ν, however, proves to be not very accurate but indicates a range $\nu = 0.6$ to 0.7. [The coefficients of the series are difficult to calculate and not very smooth.]

An alternative approach is to study the temperature dependence of the higher moments of the correlation function, namely

$$\mu_s(T) = \rho \int r^s G(\mathbf{r}) \, d\mathbf{r}. \qquad (5.10)$$

The zeroth moment is essentially the compressibility but the second moment is also a direct measure of the range of correlation. It is evident furthermore, that μ_2 is proportional to the curvature of $\rho\hat{G}(k)$ for small k and hence to the true limiting slope of the curve of $1/\chi(k)$ versus k^2 as $k^2 \to 0$.[84] From (5.3) we see that when $T \to T_c$, $\mu_2(T)$ diverges as

$$\mu_2(T) \simeq M_2/\kappa^{4-\eta} \qquad (T \to T_c),$$
$$\simeq M_2'/|1 - (T/T_c)|^\delta, \qquad (5.11)$$

with

$$\delta = (4 - \eta)\nu. \qquad (5.12)$$

The coefficients M_2 and M_2' are slowly varying functions of T. By comparing (5.12) with the index relation (5.7) we see that η and ν can both be determined if γ and δ are known. [In particular only if $\delta = 2\gamma$ would we have $\eta = 0$.]

For the two- and three-dimensional lattice gases expansions for $\mu_2(T)$ is series of powers of $1/T$ at $\rho = \rho_c$ are not too difficult to calculate.[83] [The labor and graphical analysis is similar to that for the compressibility.] The coefficients prove to be rather smoothly varying and for the plane lattices numerical analysis confirms quite accurately the relation (5.12)

---

[79] This assumption was made tentatively in Ref. 70 and tested experimentally by Frisch and Brady (Ref. 67). A good fit was obtained but although the data revealed definite nonclassical behavior, they were not sufficiently accurate to distinguish between $\nu = 1$ and some appreciably lower value.

[80] See the discussion in Sec. 2 and Refs. 23 and 28.

[81] From Table I we see $\gamma = 1\frac{3}{4}$, $1\frac{1}{4}$ for $d = 2$, $d = 3$ to which we may add $\gamma = 1.094$ ($d = 4$); see M. E. Fisher and D. S. Gaunt, Phys. Rev. **133**, A224 (1964). The classical result $\gamma = 1$ corresponds to $d \to \infty$.

[82] M. E. Fisher (to be published).

[83] M. E. Fisher and R. J. Burford (to be published).

[84] The second moment μ_2 is closely related to the length Λ defined in (3.28) and to the "persistence length" L defined by Debye (Ref. 44) explicitly: $\Lambda^2 = \frac{1}{2} \langle \cos^2 \theta \rangle \mu_2/(1 + \mu_0)$ and $L^2 = \mu_2/\mu_0$. From (3.28) one sees that $1/\chi(k) = 1/\chi(0)\{1 + \Lambda^2 k^2 + O(k^4)\}$.

[which predicts $\delta = 3.75$]. This is an important result since it provides support for the original hypothesis (5.3).

Initial estimates for the simple cubic lattice gas[83] yield $\delta = 2.538 \pm 0.003$ and hence

$$\nu = 0.644 \pm 0.003, \quad \eta = 0.060 \pm 0.007. \quad (5.13)$$

Results for other three-dimensional lattices confirm these estimates with lower accuracy. [One might mention that (5.14) is not inconsistent with the conjecture $\eta = \frac{1}{16} = 0.0625$ which is rather natural in view of the estimate $\beta \simeq \frac{5}{16}$ discussed in Sec. 2.]

The values (5.13) are in accordance with our expectation that the three-dimensional results should be closer to the classical predictions.[81] The magnitude of η for the lattice gas is indeed rather close to zero but it is not unlikely that a more realistic continuum model would lead to somewhat larger value, say $\eta \simeq 0.1$.[85] Until more rigorous theories are developed and more realistic models become soluble we must content ourselves with these rough estimates. Accordingly let us review the nature of the critical scattering to be expected on the basis of our analysis.

The peak in the critical scattering is described by (5.5) and (for $\eta > 0$) should be narrower at a fixed temperature near T_c than the Lorentzian curve of same peak height. Correspondingly a plot of the inverse scattering intensity versus k^2 should be somewhat convex although, as suggested in Fig. 6, for larger values of $k^2 > k_{\min}^2$ the curves might *appear* to be reasonably linear. For small values of η, of the magnitude (5.13), it might indeed be rather difficult experimentally to detect the increasing curvature of the scattering plots as $k^2 \to 0$ even though the curve for $T = T_c$ will theoretically have an infinite slope at $k^2 = 0$. In practice a nonzero value of η can probably best be detected by the observation that the apparent linear intercepts $\xi_0(T)$ [see Fig. 6] would not approach zero when $T \to T_c$ as must the true intercepts $\xi(T) = 1/\chi(0)$. In particular the scattering plot taken *at* $T = T_c$ would tend to extrapolate to a small positive value at $k^2 = 0$. [Of course to detect this behavior it is important to have an independent measurement of T_c and *not* to judge T_c by extrapolating $\xi_0(T)$ to zero with T as would otherwise be tempting!] This behavior is reminiscent of the experimental "anomalies" described in the previous Section although, as yet, these can probably not be regarded as fully established.

The likely fact that γ exceeds unity for a real

gas and the consequent curvature of $1/K_T$ versus T would indicate that a plot of the true intercepts, $\xi(T)$, versus T should flatten out as T approaches T_c (and theoretically have zero slope at $T = T_c$). This effect might well be less obvious in a plot of the apparent intercepts $\xi_0(T)$ although it does seem to have been observed.[86]

6. SUMMARY AND CONCLUSIONS

We have shown that the classical theories of the gas–liquid critical point are unsatisfactory both on experimental and theoretical grounds. Thus the coexistence curve must be described experimentally by

$$\rho_L - \rho_G \sim (T_c - T)^\beta, \quad (6.1)$$

with $\beta \simeq 0.33$ to 0.36 (rather than with $\beta = \frac{1}{2}$) while for three-dimensional lattice gas models one finds $\beta \simeq 0.31 \simeq \frac{5}{16}$. The specific heat $C_V(T)$ measured along the critical isochore of a fluid becomes infinite at T_c, diverging approximately as $\log|T - T_c|$. A similar result holds for the lattice gas. Theoretically one also expects that the compressibility above and below T_c should diverge as

$$K_T \sim 1/|T - T_c|^\gamma \quad (6.2)$$

with $\gamma > 1$. This prediction while qualitatively correct awaits quantitative experimental verification. (Values of these indices are given in Table I.)

On theoretical grounds the classical (Ornstein–Zernike) theory of critical scattering has been shown to be unsatisfactory close to the critical point (although it probably is valid away from the critical region when the compressibility is still moderately large). More generally one should expect the scattering intensity to vary as

$$I(k) \sim 1/(\kappa^2 + k^2)^{1-\frac{1}{2}\eta}, \quad (6.3)$$

where $\eta > 0$ and where the range parameter vanishes along the critical isotherm as

$$\kappa(T) \sim (T - T_c)^\nu \quad (6.4)$$

with $\nu(2 - \eta) = \gamma$. Theoretical analysis suggests that the index η might be no larger than 0.1 (see Table I). Consequently it is probably not easy experimentally to detect deviations from the classical theory (for which $\eta = 0$). Nevertheless there are experimental indications of the failure of the classical predictions for small k^2 near T_c and these are consistent with $\eta > 0$. Final confirmation of the theory must, however, rest on further, more extensive and accurate measurements on sufficiently simple fluid systems.

---

[85] For example the effectively more "continuum-like" Heisenberg model of ferromagnetism yields $\gamma \simeq 1\frac{1}{3}$ compared with the Ising value $\gamma = 1\frac{1}{4}$ $(d = 3)$. (See Refs. 90, 91).

[86] D. McIntyre, Ref. 64 and a private communication.

962   MICHAEL E. FISHER

In conclusion one should mention the rather close analogy between the gas–liquid critical point and the Curie point of a ferromagnetic crystal.[87] Indeed almost all our analysis and conclusions apply directly to ferromagnetic systems if appropriately translated. The density deviation $\rho - \rho_c$ should be identified with the magnetization M while the magnetic field H is isomorphic to the chemical potential of the fluid. The critical isochore, $\rho = \rho_c$, corresponds to zero magnetic field (since the mean magnetization then vanishes) and the coexistence curve corresponds to the curve of spontaneous magnetization $M_0(T)$. The specific heat $C_V(T)$ along the critical isochore of a fluid is isomorphic to the specific heat $C_H(T)$ of a ferromagnet in zero field and the compressibilities at condensation and above T_c for $\rho = \rho_c$ correspond essentially to the initial susceptibilities $\chi_0(T) = (\partial M/\partial H)_T$, $(H \to 0)$.

The van der Waals and equivalent classical theories find their precise parallel in the Weiss molecular field theory and its extensions.[88] Similarly the Ornstein–Zernike theory and its developments have been adapted to describe the critical scattering of neutrons by ferromagnets.[89,90] The net pair correlation function $G(\mathbf{r})$ is replaced by the spin–spin correlation functions $\Gamma_{\alpha\beta}(\mathbf{r}) = \langle S_0^\alpha S_r^\beta \rangle$. Theoretically one is able to estimate the susceptibility index γ for the nearest neighbor Heisenberg model above T_c with the approximate result[91,92] $\gamma = \frac{4}{3}$ (independent of spin and lattice structure in three-dimensions).

It is interesting that this nonclassical prediction has been quite accurately confirmed recently.[92-95] Furthermore modern neutron scattering experiments[95] have also given a definite suggestion of deviations from the Ornstein–Zernike theory consistent with a small positive value of the index η.

The spontaneous ferromagnetic moment $M_0(T)$ is not easy to measure near T_c but nuclear magnetic resonance experiments by Benedek and Heller[96] have shown that the somewhat analogous sublattice magnetization (or long range order) of an *anti*ferromagnet (actually MnF_2) varies as $(T - T_c)^\beta$ with $\beta = 0.335 \pm 0.010$. The relation holds with remarkable accuracy up to within millidegrees of the critical or Néel point [$\Delta T/T_c = 0.007\%$].[97] Experiments on antiferromagnets also reveal specific heat infinities at T_c which are approached in approximately logarithmic fashion.[98] The close similarity of these results to the corresponding behavior of fluid systems presents a striking challenge to our theoretical understanding.

ACKNOWLEDGMENTS

I am most grateful to Professor G. E. Uhlenbeck for his criticisms of a first draft of this article and would like to thank Professor Mark Kac and Professor Joel Lebowitz for their comments.

[87] As mentioned in Sec. 2, this relationship is formally exact for a lattice gas and an Ising model ferromagnet. See, for example, T. D. Lee and C. N. Yang, Phys. Rev. **87**, 410 (1952).

[88] See, for example, J. H. Van Vleck, Rev. Mod. Phys. **17**, 27 (1945); P. W. Kasteleijn and J. Van Kranendonk, Physica **22**, 317 (1956).

[89] L. Van Hove, Phys. Rev. **95**, 1374 (1954).

[90] R. J. Elliott and W. Marshall, Rev. Mod. Phys. **30**, 75 (1958).

[91] C. Domb and M. F. Sykes, Phys. Rev. **128**, 168 (1962).

[92] J. L. Gammel, W. Marshall, and L. Morgan, Proc. Roy. Soc. (London) **A275**, 257 (1963).

[93] J. E. Noakes and A. Arrott, J. Appl. Phys. Suppl. **35**, 931 (1964).

[94] J. S. Kouvel [private communication based on an analysis of the measurements by P. Weiss and R. Forrer, Ann. Phys. 5, 153 (1926).]

[95] L. Passell, K. Blinowski, T. Brun, and P. Nielsen, J. Appl. Phys. (Suppl.) **35**, 933 (1964). Note that the linear intercepts with the $k^2 = 0$ axis of the inverse scattering appear to approach a positeve value at $T = T_c$.

[96] P. Heller and G. B. Benedek, Phys. Rev. Letters **8**, 428 (1962).

[97] More recent NMR experiments by Heller and Benedek (private communication) reveal a similar behavior for the spontaneous magnetization of the ferromagnet EuS.

[98] See, for example, W. K. Robinson, and S. A. Friedberg, Phys. Rev. **117**, 402 (1960).

For Computer Physics Communications: Proceedings of CCP 2004

Fluid Coexistence close to Criticality: Scaling Algorithms for Precise Simulation

Young C. Kim and Michael E. Fisher *

Institute for Physical Science and Technology, University of Maryland, College Park, Maryland 20742 USA

Abstract

A novel algorithm is presented that yields precise estimates of coexisting liquid and gas densities, $\rho^{\pm}(T)$, from grand canonical Monte Carlo simulations of model fluids near criticality. The algorithm utilizes data for the isothermal minima of the moment ratio $Q_L(T; \langle \rho \rangle_L) \equiv \langle m^2 \rangle_L^2 / \langle m^4 \rangle_L$ in $L \times \cdots \times L$ boxes, where $m = \rho - \langle \rho \rangle_L$. When $L \to \infty$ the minima, $Q_m^{\pm}(T; L)$, tend to zero while their locations, $\rho_m^{\pm}(T; L)$, approach $\rho^+(T)$ and $\rho^-(T)$. Finite-size scaling relates the ratio $\mathcal{Y} = (\rho_m^+ - \rho_m^-)/\Delta\rho_\infty(T)$ *universally* to $\frac{1}{2}(Q_m^+ + Q_m^-)$, where $\Delta\rho_\infty = \rho^+(T) - \rho^-(T)$ is the desired width of the coexistence curve. Utilizing the exact limiting ($L \to \infty$) form, the corresponding scaling function can be generated in recursive steps by fitting overlapping data for three or more box sizes, $L_1, L_2, \cdots, L_n$. Starting at a T_0 sufficiently far below T_c and suitably choosing intervals $\Delta T_j = T_{j+1} - T_j > 0$ yields $\Delta\rho_\infty(T_j)$ and precisely locates T_c.

The algorithm has been applied to simulation data for a hard-core square-well fluid and the restricted primitive model electrolyte for sizes up to $L/a = 8\text{-}12$ (where a is the hard-core diameter): the coexistence curves can be computed to a precision of $\pm 1\text{-}2\%$ of ρ_c up to $|T - T_c|/T_c = 10^{-4}$ and 10^{-3}, respectively. Universality of the scaling functions and the exponent β is verified and the (T_c, ρ_c) estimates confirm previous values based on data from above T_c. The algorithm extends directly to calculating the diameter, $\rho_{\text{diam}}(T) \equiv \frac{1}{2}(\rho^+ + \rho^-)$, and can lead to estimates of the Yang-Yang ratio. Furthermore, a new, explicit approximant for the basic scaling function $\mathcal{Y}$ permits straightforward estimates of $\Delta\rho_\infty(T)$ from limited Q-data when Ising-type criticality may be assumed.

1. Introduction

In recent years computer simulation has been an important tool to understand the critical behavior of fluids [1]. Various programming algorithms and techniques have been developed to enhance calculations with large-scale computers. However, determining phase boundaries, critical points and the universality classes of complex fluids, such as electrolytes, polymers, colloids, etc., has been and still is a great challenge. Here we present in detail, together with

a new, 'economical' extension, a powerful method developed recently [2] to estimate precisely coexisting liquid and gas densities, $\rho^+(T)$ and $\rho^-(T)$, very close to the critical temperature, T_c. Precise values of $\rho^{\pm}(T)$ can then provide critical parameters and reveal critical exponents, via

$$\Delta\rho_\infty(T) \equiv \rho^+ - \rho^- \approx B|t|^\beta, \ t \equiv (T - T_c)/T_c. \ (1)$$

To determine $\rho^{\pm}(T)$ in simulations it has been customary to observe the grand canonical equilibrium distribution function, $\mathcal{P}_L(\rho; T)$, of the density, $\rho \equiv N/V$, at constant $T < T_c$, where N and $V \equiv L^d$ are the particle number and volume of the system,

---

* Corresponding Author: xpectnil@ipst.umd.edu

respectively. For large L below T_c, the distribution $\mathcal{P}_L(\rho; T)$ exhibits two well separated peaks located near $\rho^{\pm}(T)$. Examining these peaks with aid of the equal-weight prescription [3] provides reasonable estimates for $\rho^{\pm}(T)$. However, when T approaches T_c, the peaks broaden, overlap strongly, and can no longer be separated uniquely thereby precluding reliable estimation of $\rho^{\pm}(T)$ [4]. As well known, the underlying reason is the divergence of the bulk correlation length at criticality as $|t|^{-\nu}$.

To obtain better estimates of $\rho^{\pm}(T)$ valid closer to T_c, we examine instead the Q-parameter defined by [5]

$$Q_L(T; \langle\rho\rangle_L) \equiv \langle m^2\rangle_L^2/\langle m^4\rangle_L, \quad m = \rho - \langle\rho\rangle_L, \quad (2)$$

where $\langle\cdot\rangle_L$ denotes a finite-size grand canonical expectation value at fixed T and chemical potential, μ, in a cubic box of dimensions $L \times L \times \cdots \times L$ with periodic boundary conditions. Below T_c one finds that $Q_L(T; \langle\rho\rangle_L)$ exhibits two minima, say $Q_m^{\pm}(T; L)$, at densities $\rho_m^{\pm}(T; L)$ near $\rho^{\pm}(T)$ [6]: see Fig. 1. When

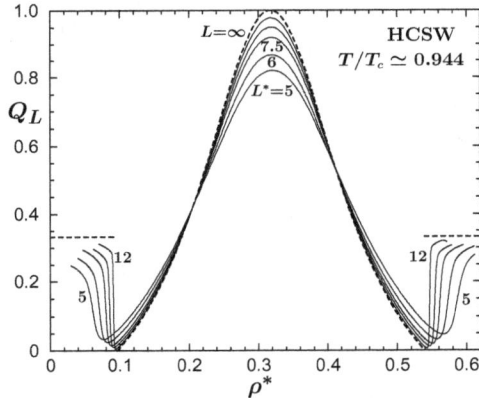

Fig. 1. Plots of simulation data for $Q_L(T; \langle\rho\rangle_L)$ vs. $\rho^* \equiv \langle\rho\rangle a^3$ for the HCSW fluid at $T/T_c \simeq 0.944$ showing minima that approach the limiting coexistence values $\rho^+(T)$ and $\rho^-(T)$. The solid curves are for $L^* \equiv L/a = 12, 9, 7.5, 6$ and 5 (where a is the hard-core diameter); the dashed lines represent the *exact* limiting form (for the estimated values of ρ^+ and ρ^-).

$L \to \infty$ the heights, $Q_m^{\pm}(T; L)$, of these minima decay to zero while their locations, $\rho_m^{\pm}(T; L)$, approach the desired coexistence values $\rho^{\pm}(T)$. Thus understanding the behavior of the minima is potentially rewarding.

To that end, this article explains an *unbiased* scaling algorithm which utilizes calculated values of $Q_m^{\pm}(T; L)$ and $\rho_m^{\pm}(T; L)$ to obtain precise estimates

of the coexistence-curve width or density discontinuity, $\Delta\rho_\infty(T)$. By "unbiased" we specifically mean that not only are the critical parameters T_c and ρ_c left open but, also, *no assumptions* regarding the value of the exponent β in (1) *or* regarding the universality class of the critical point are made (in contrast to earlier approaches [7,8]). The algorithm has been applied to a hard-core square-well (HCSW) fluid and to the restricted primitive model (RPM) electrolyte. Although not demonstrated here, the algorithm extends directly [2,9] to accurately estimate the diameter,

$$\rho_{\text{diam}}(T) \equiv \tfrac{1}{2}[\rho^+(T) + \rho^-(T)]. \quad (3)$$

Furthermore, studying $(Q_m^+ - Q_m^-)$ provides an effective way [2,9] of estimating the strength, $\mathcal{R}_\mu$, of the Yang-Yang anomaly, namely, the relative divergence at criticality of the second derivative of the chemical potential on the phase boundary, $d^2\mu_\sigma/dT^2$ [10].

Finally, on the basis of an explicit expression, approximate but accurate, for a crucial scaling function relating $\Delta\rho_\infty(T)$ to the difference $\rho_m^+(T) - \rho_m^-(T)$, we demonstrate a simple, albeit biased algorithm that requires only limited data for the Q-minima: this should be valuable when, as usual, it may be safely assumed that the criticality is of Ising character [7,8].

2. Theoretical Background

For sufficiently large L at fixed $T < T_c$ it is well established [3,6] that the density distribution, $\mathcal{P}_L(\rho; T)$, asymptotically approaches a sum of two Gaussian peaks which can be written as

$$\mathcal{P}_L(\rho; T) \approx C_L \left\{ \chi_-^{-1/2} \exp[-\beta(\rho - \rho^-)^2 L^d/2\chi_-] + \chi_+^{-1/2} \exp[-\beta(\rho - \rho^+)^2 L^d/2\chi_+] \right\} \times \exp[\beta\rho(\mu - \mu_\sigma)L^d], \quad (4)$$

where $\beta = 1/k_B T$ while $\chi_{\pm}(T)$ are the infinite-volume susceptibilities [defined via $\chi = (\partial\rho/\partial\mu)_T$] evaluated at $\rho = \rho^{\pm}(T)\pm$, and $C_L(\mu, T)$ is a normalization factor. From this expression the parameter $Q_L(T; \langle\rho\rangle_L)$ can be calculated readily and, in particular, the limiting form $Q_\infty(T; \langle\rho\rangle_\infty)$, shown by the dashed lines in Fig. 1, can be derived. However, when criticality is approached at fixed L, the two-Gaussian representation of $\mathcal{P}_L(\rho;T)$ becomes less accurate and it fails badly near criticality.

In the critical region, on the other hand, the behavior of $Q_L(T;\langle\rho\rangle_L)$ can be understood via finite-size scaling theory[11], recently extended to incorporate pressure-mixing in the scaling fields $\tilde{t}$ and $\tilde{\mu}$ [12] (which is essential for describing the Yang-Yang anomaly [10]). For the Q-parameter, which depends on the three variables L, T, and $\langle\rho\rangle_L$, finite-size scaling provides the asymptotic, $t\to 0$, reduced, two-variable representation

$$Q_L(T;\langle\rho\rangle_L) \approx \mathfrak{Q}(tL^{1/\nu};\ \Delta\rho/|t|^\beta), \qquad (5)$$

where $\mathfrak{Q}(x,y)$ is the scaling function, while $\Delta\rho = \langle\rho\rangle_L - \rho_c$ and, as above, ν is the correlation length exponent.

It then follows that the minima, $Q_m^+(T;L)$ and $Q_m^-(T;L)$, and their corresponding displacements, $[\rho_m^+(T;L) - \rho_c]$ and $[\rho_c - \rho_m^-(T;L)]$, should, on approach to criticality, all reduce to functions of the scaled variable $x = tL^{1/\nu}$ alone. Accordingly, the average of Q_m^+ and Q_m^- and, using (1) and (3), the normalized density deviation [1]

$$y \equiv [\langle\rho\rangle_L - \rho_{\text{diam}}(T)]/\Delta\rho_\infty(T), \qquad (6)$$

should scale likewise. Thus we may anticipate (but should plan to *check* in applications) the scaling expressions

$$\bar{Q}_{\min}(T;L) \equiv \tfrac{1}{2}(Q_m^+ + Q_m^-) \approx \mathcal{M}(tL^{1/\nu}), \qquad (7)$$

$$\Delta y_{\min}(T;L) \equiv (y_m^+ - y_m^-) = \frac{\rho_m^+ - \rho_m^-}{\Delta\rho_\infty(T)}$$
$$\approx \mathcal{N}(tL^{1/\nu}), \qquad (8)$$

where $\mathcal{M}(\cdot)$ and $\mathcal{N}(\cdot)$ are appropriate scaling functions, which when properly normalized should be universal.

Before proceeding further, notice that the normalizing divisor $\Delta\rho_\infty(T)$ in (6) and (8) is just the true width of the coexistence curve that we wish to estimate!

Now, at least in principle, one may eliminate the scaling variable $x = tL^{1/\nu}$ between (7) and (8), e.g., by solving (7) for x and substituting in (8), to obtain $\Delta y_{\min}$ as a universal function of $\bar{Q}_{\min}$, say, $\mathcal{Y}(q)$. Of

---

1 The definition of y adopted here differs from that used in [2] by a factor $\tfrac{1}{2}$.

course, this function is not known *a priori*. However, the two-Gaussian limiting form (4) for the density distribution $\mathcal{P}_L(\rho;T)$, which is easily seen to obey scaling close to T_c when x is large, can be used straightforwardly [12] to study the minima of $Q_L(T;\langle\rho\rangle_L)$. In this limit we thence obtain the *exact* and *universal* expansion

$$\Delta y_{\min} = 1 + \tfrac{1}{2}q + O(q^2), \qquad (9)$$

in terms of the auxiliary variable

$$q \equiv \bar{Q}_{\min}\ln(4/e\bar{Q}_{\min}). \qquad (10)$$

As we will explain, this result provides a route to the recursive, numerical construction of the full universal scaling function $\mathcal{Y}(q)$ and, furthermore, to the evaluation of $\Delta\rho_\infty(T)$ and T_c.

3. Scaling Algorithm

The basic idea of our algorithm is to fit data for the Q-minima to the formula (9), starting at some temperature T_0 far enough below T_c that the two-Gaussian form (4) is a good approximation, and then to extend the fits progressively to higher temperatures checking consistency with scaling, i.e., the uniqueness of $\mathcal{Y}(q)$, as the calculations proceed. To make the fits, it is just the sought-for values of $\Delta\rho_\infty(T)$ that must be selected: and in order to vary q in (9) and check the scaling, it is crucial to obtain simulation data for three or more fixed box sizes, say $L_i = L_1, L_2, \cdots, L_n$ ($n \geq 3$), at the same temperatures T_j ($j = 0, 1, 2, \cdots$).

It must also be stressed that high quality, precise data are essential. These can be obtained, as previous studies of the HCSW fluid [4] and the RPM [13] demonstrate, by careful simulation and the use of multiple histogram reweighting [14].

More formally, the initial step is to collect grand canonical Monte Carlo data sets for the Q-minima, $\{Q_m^\pm(T,L), \rho_m^\pm(T,L)\}$, generated at a sufficiently low T_0 as is to be verified by the ease of fitting to (9). This is illustrated in Fig. 2(a) using data for the HCSW fluid; but note, in particular, that the magnitude of the (positive) exponent ψ is arbitrary and may be assigned any graphically convenient value (such as, e.g., $\psi = 2$ or 5: see [2]). However, the reason for the choice made will be explained below.

3

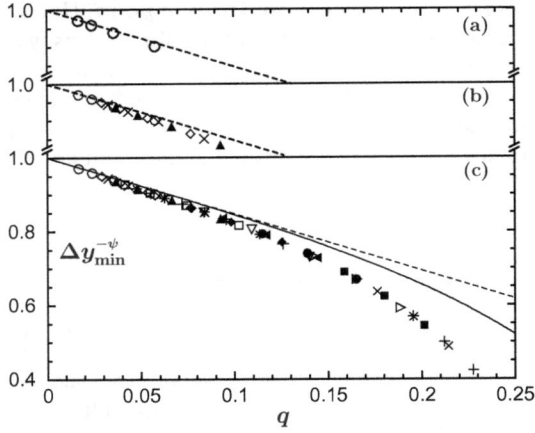

Fig. 2. Scaling plot of $\Delta y_{\min}^{-\psi}$ vs. $q \equiv \bar{Q}_{\min} \ln(4/e\bar{Q}_{\min})$ with $1/\psi = 0.326$ for the HCSW fluid; (a) at $T_0 \simeq 0.903\,T_c$, (b) at T_0 and three higher temperatures up to $T \simeq 0.952\,T_c$, and (c) up to $T \simeq 0.985\,T_c$. The dashed lines represent the exact two-Gaussian limiting form to linear order in q; the solid curve in (c) portrays the full two-Gaussian approximation which evidently deviates significantly from the HCSW results even for $q \simeq 0.05$.

To generate the scaling function successfully, $n = 3$ distinct box sizes may well suffice although $n = 4$ has been used in our calculations. Furthermore, in order to avoid effects arising from *corrections* to the leading scaling behavior, the $L_i^* \equiv L_i/a$ should not be too small (where a measures the particle size). For the HCSW fluid $L^* \gtrsim 7$ sufficed.

At an appropriate T_0 the value of $\bar{Q}_{\min}$ will be small: for our choice of T_0 for the HCSW fluid we had $\bar{Q}_{\min} \lesssim 0.03$; but a somewhat larger value might provide acceptable accuracy. Following the prescription, one density-discontinuity value, say $\Delta\rho_{T_0}$ is then chosen for all the L_i to provide the best fit of $\Delta y_{\min}^{(i)} \equiv [\rho_m^+(T_0, L_i) - \rho_m^-(T_0, L_i)]/\Delta\rho_{T_0}$ to the relation (9) with $q_0^{(i)} \equiv q(T_0, L_i)$. For the HCSW fluid this fit could be achieved accurately to within $\pm 0.2\%$ of $\Delta\rho_{T_0}$. One may also check that the assignment of ψ in any reasonable range has negligible effect on the fitting (which should be weighted more heavily on the smaller values of $q_0^{(i)}$). The optimal value $\Delta\rho_{T_0}$ is then identified as an estimate for $\Delta\rho_\infty(T_0)$.

The next step is to increase T_0 to $T_1 = T_0 + \Delta T_0$ and to compute the new data sets $\{\Delta y_{\min}(T_1; L_i)\}_{i=1}^n$ and $\{q_1^{(i)} \equiv q(T_1; L_i)\}_{i=1}^n$. In doing so, however, the crucial point is to select ΔT_0 small enough that the

new set $\{q_1^{(i)}\}_{i=1}^n$ *overlaps* the previous one $\{q_0^{(i)}\}_{i=1}^n$. When the new data set is in place, one must find, as before, a new value, $\Delta\rho_{T_1}$, such that the new data when plotted overlap and smoothly extend the previous data: see Fig. 2(b). The procedure thereby extends and numerically validates the scaling function up to larger values of q. The new value $\Delta\rho_{T_1}$ can then be taken as an optimal estimate for $\Delta\rho_\infty(T_1)$.

Subsequently, repeating these steps by increasing the temperature to $T_{j+1} = T_j + \Delta T_j$ will extend the scaling function further and generate successive estimates $\Delta\rho_\infty(T_j)$ for $j = 2, 3, \cdots$: see Fig. 2(c). As T_c is approached, one will observe that smaller increments ΔT_j are needed and high quality data become increasingly important.

Since, via (1), $\Delta\rho_\infty(T) \to 0$ when $T \to T_c-$ whereas the interval $\rho_m^+(T; L) - \rho_m^-(T; L)$ does not then vanish, the plot of $\Delta y_{\min}^{-\psi}(q)$ must approach zero at criticality. This behavior is clear in Fig. 3 which presents the full scaling function, $[\mathcal{Y}(q)]^{-\psi}$, as con-

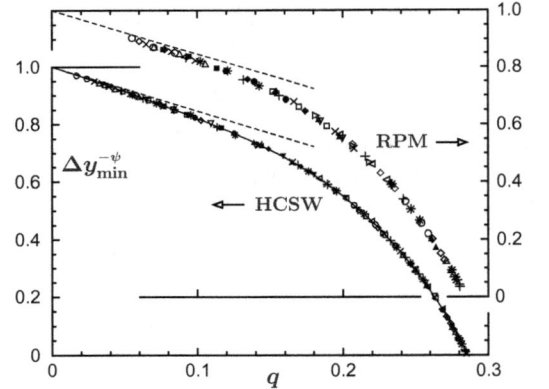

Fig. 3. Scaling plots of $\Delta y_{\min}^{-\psi}$ vs. q for the HCSW fluid and the RPM with $1/\psi$ identified with the Ising exponent $\beta \simeq 0.326$. The dashed lines again show the exact two-Gaussian limit to linear order while the solid curve (passing through the HCSW data) represents the approximant (12).

structed via the algorithm both for the HCSW fluid and for the RPM.

The vanishing of $\Delta y_{\min}^{-\psi}$ at $q = q_c \simeq 0.2860$ generates unequivocal estimates for T_c. For the HCSW fluid, a precision of ± 2 parts in 10^5 is realized: furthermore, the value for T_c agrees well with less precise (± 3 parts in 10^4) estimates obtained by studying the model only *above* T_c [4].

The $Q_L(T; \langle\rho\rangle_L)$ data for the RPM are harder to generate accurately and, moreover, as seen in Fig. 4, the variation of Q_L with $\langle\rho\rangle_L$ turns out to be highly

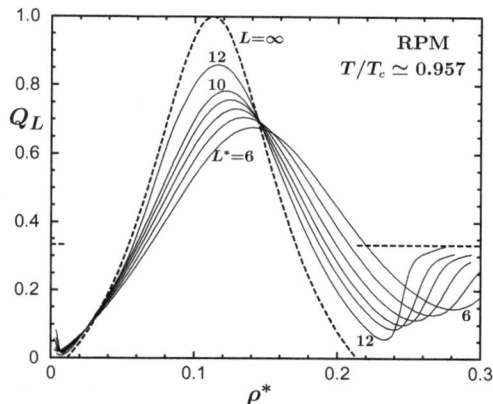

Fig. 4. Simulation data for the restricted primitive model (RPM) electrolyte as in Fig. 1; but notice the significantly greater asymmetry.

asymmetric, in strong contrast to the HCSW data in Fig. 1. (It might be remarked that the marked asymmetry of the RPM seems to be associated with a large, $\mathcal{R}_\mu \simeq 0.26$, value of the Yang-Yang ratio [2].) Nevertheless, the algorithm continues to work surprisingly well, yielding $\Delta\rho_\infty(T)$ to within ± 1-2% down to $|t| \simeq 10^{-3}$ and generating estimates for T_c with a precision of 4 parts in 10^4: these in turn agree completely with previous, $T > T_c$ estimates [13]. It is important to note, furthermore, that to within the uncertainties, the RPM data in Fig. 3 fully confirm the expected universality of the scaling function $\mathcal{Y}(q)$.

The analysis presented above demonstrates that $\Delta y_{\min}^{-\psi}$ must vanish *linearly* with q if one sets $1/\psi = \beta$; but note again that this choice is *not* needed in order to estimate T_c reliably. However, the HCSW fluid is certainly expected to exhibit Ising-type criticality with $\beta_{\mathrm{Is}} \simeq 0.326$; and this is convincingly confirmed by the resulting plot of the coexistence curve shown in Fig. 5. Accordingly, $\psi = 1/\beta_{\mathrm{Is}}$ was selected for use in Figs. 2 and 3. A bonus of the RPM calculations also displayed in Fig. 5 is that β close to β_{Is} again fits well. This result is of value since serious doubts have been raised regarding the universality class of ionic systems [2,13,15,16].

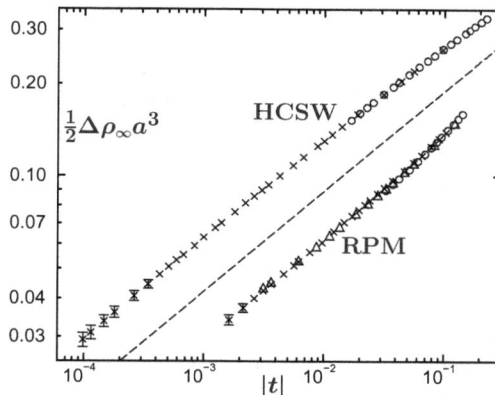

Fig. 5. A log-log plot of the coexistence-curve half-width, $\frac{1}{2}\Delta\rho_\infty a^3$ vs. $t \equiv (T - T_c)/T_c$ (where a is the hard-sphere diameter) for the HCSW fluid and for the RPM (at a $\zeta = 5$ fine-discretization level [2,13]). The crosses follow from the full unbiased scaling algorithm while the circles for $|t| > 10^{-2}$ report the best previous estimates (employing an equal-weight prescription). The triangles for the RPM are estimates obtained from the simple, economical (but biased) algorithm embodied in (11) and (12). The dashed line has a slope corresponding to $\beta = \beta_{\mathrm{Is}} \simeq 0.326$: see (1).

4. An Economical Biased Algorithm

In practice, the full, unbiased scaling algorithm may be inconvenient for some applications since it requires a significant amount of precise data narrowly spaced in temperature, especially when T_c is approached. Furthermore, one needs to calculate reliably an initial set of Q minima at sufficiently low T that the two-Gaussian structure of $\mathcal{P}_L(\rho; T)$ is well obeyed. On the other hand, if one is prepared to accept that a model of interest exhibits Ising-type criticality, one can, in fact, utilize the HCSW scaling plot for $y_{\min}(T; L)$ in Fig. 3 to estimate $\Delta\rho_\infty$ at *any* given T!

To see this most clearly, recall that $\Delta y_{\min}(T; L)$ is (for large enough L and, say, $|t| = (T_c - T)/T_c \lesssim 0.1$) described by a universal scaling function, $\mathcal{Y}(q)$, as Fig. 3 demonstrates. Then we may rewrite (8) in the direct form

$$\Delta\rho_\infty(T) \approx [\rho_{\mathrm{m}}^+(T; L) - \rho_{\mathrm{m}}^-(T; L)]/\mathcal{Y}[q(T; L)], \quad (11)$$

where $q(T; L)$ follows from (7) and (10). In words this simply says that $\mathcal{Y}(q)$ acts as a *correction factor* that transforms the first approximation to the coexistence-curve width derived from the locations of the Q-minima, into the desired answer. Thus, for a selected temperature T one need only locate the minima of $Q_L(T; \langle \rho \rangle_L)$ (for a reasonable value of L), determine the difference $(\rho_m^+ - \rho_m^-)$, calculate q, and substitute in (11). As a wise precaution, using a second value of L will enable one to check that corrections to scaling are negligible.

To facilitate this very simple, albeit biased procedure, we have fitted the HCSW data in Fig. 3 to an expression for $\mathcal{Y}(q)$ that, with $\beta = \beta_{\text{Is}} \simeq 0.326$, embodies the linear vanishing when $q \to q_c \simeq 0.2860$ and the exact small-q behavior (9). Indeed, with $\tilde{q} \equiv q/q_c$, the approximant

$$[\mathcal{Y}(q)]^{-1/\beta}$$
$$\simeq \left(1 - \frac{q}{2\beta}\right) \frac{(1-\tilde{q})(1 + a_2\tilde{q}^2 + a_3\tilde{q}^3)}{1 - \tilde{q} + b_2\tilde{q}^2 + b_3]\tilde{q}^3}, \quad (12)$$

provides an excellent fit (see Fig. 3) for the coefficient values $a_2 = 1.829$, $a_3 = 1.955$, $b_2 = 2.340$, and $b_3 = -1.388$.

We have tested this approach on the RPM (where Ising-type criticality is now well established [13,16]): it yields the triangular data points shown in Fig. 5. These evidently agree well with the results of the full, unbiased algorithm. Thus we believe that the approximant (12) provides a convenient practical tool for accurately estimating the coexistence curves for a wide range of systems of Ising character.

5. Summary

In conclusion we have presented a novel scaling algorithm developed to estimate precisely the coexistence curves of asymmetric fluids near criticality. Both the fuller unbiased and a simpler biased approach have been illustrated using simulation data for the HCSW fluid and the RPM: precise and reliable estimates for $\Delta\rho_\infty \equiv \rho^+(T) - \rho^-(T)$ can be obtained even very close to T_c. The biased approach, using (11) and the accurate scaling function representation (12), should prove especially useful in exploratory investigations. On the other hand, the full algorithm extends to yield estimates of the coexistence diameter, (3), and the Yang-Yang ratio [2,10].

The support of the National Science Foundation (through Grant No. CHE 03-01101) is gratefully acknowledged. Jean-Noël Aqua and Sarvin Moghaddam kindly commented on a draft.

References

[1] See, e.g., A. Z. Panagiotopoulos, Mol. Simul. 9 (1992) 1.

[2] Y. C. Kim, M. E. Fisher, E. Luijten, Phys. Rev. Lett. 91 (2003) 065701.

[3] C. Borgs, S. Kappler, Phys. Lett. A 171 (1992) 37.

[4] G. Orkoulas, M. E. Fisher, A. Z. Panagiotopoulos, Phys. Rev. E 63 (2001) 051507.

[5] K. Binder, Z. Phys. B 43 (1981) 119.

[6] K. Binder, D. P. Landau, Phys. Rev. B 30 (1984) 1477; B. Dünweg, D. P. Landau, Phys. Rev. B 48 (1993) 14182.

[7] A. D. Bruce, N. Wilding, Phys. Rev. Lett. 68 (1992) 193.

[8] Y. C. Kim, M. E. Fisher, J. Phys. Chem. B 108 (2004) 6750.

[9] Y. C. Kim, M. E. Fisher, to be published.

[10] M. E. Fisher, G. Orkoulas, Phys. Rev. Lett. 85 (2000) 696.

[11] M. E. Fisher, M. N. Barber, Phys. Rev. Lett. 28 (1972) 1516.

[12] Y. C. Kim, M. E. Fisher, Phys. Rev. E 68 (2003) 041506.

[13] E. Luijten, M. E. Fisher, A. Z. Panagiotopoulos, Phys. Rev. Lett. 88 (2002) 185701.

[14] A. M. Ferrenberg, R. H. Swendsen, Phys. Rev. Lett. 63 (2002) 1195.

[15] H. Weingärtner, W. Schröer, Adv. Chem. Phys. 116 (2001) 1.

[16] Y. C. Kim, M. E. Fisher, Phys. Rev. Lett. 92 (2004) 185703.

LIST OF PUBLICATIONS

A. Books, Lecture Series, Theses

1. "The Solution of Problems in Theoretical Physics by Electronic Analogue Methods," Ph.D. Thesis (London, 1957), pp. 314.

2. "Analogue Computing at Ultra-High Speed: an Experimental and Theoretical Study" (with D.M. MacKay) Chapman and Hall (London, 1962); John Wiley (New York, 1962) pp. xv.+ 395.

3. "The Nature of Critical Points," lectures delivered at the Summer Institute for Theoretical Physics, University of Colorado, Boulder, 1964; "Lectures in Theoretical Physics," Vol. VIIC, pp. 1–159 (Ed. W.E. Brittin, The University of Colorado Press, 1965).

4. "Природа Критических Точек," М.Ш.Гитерман 1968, Translation of "The Nature of Critical Points" by M.Sh. Giterman, pp. 1–221 (M.I.R., Moscow, 1968).

5. "The Theory of Critical Point Singularities." See [104] (1971) below.

6. "Scaling, Universality, and Renormalization Group Theory." See [209] (1983) below.

7. "Interfaces: Fluctuations, Interactions and Related Transitions." See [281] (1989) below; reprinted (with corrections and additions) in 2004: see [385].

B. Scientific Papers

1. Slide rule versus bead frame, J. Royal Air Force Technical College, 3, 53–60 (Oct. 1954).

2. The matrix approach to filters and transmission lines, Part I, Electronic Engineering 27, 198–204 (1955).

3. The matrix approach to filters and transmission lines, Part II, Electronic Engineering 27, 258–263 (1955).

4. Higher order differences in the analogue solution of partial differential equations, Journées Internationales de Calcul Analogique — Sept. 55 — Proc. Internat. Conf. Analog. Computation (Brussels, Sept. 1955) pp. 208–213.

5. Higher order differences in the analogue solution of partial differential equations, J. Assoc. Comp. Mach. 3, 325–347 (1956).

6. On iterative processes and functional equations (C. Domb and M.E.F.) Proc. Camb. Phil. Soc. 52, 652–662 (1956).

7. On the continuous solution of integral equations by an electronic analogue, Proc. Camb. Phil. Soc. 53, 162–174 (1957).

8.   Optimum design of quarter-squares multipliers with segmented characteristics, J. Sci. Insts. 34, 312–316 (1957).

9.   Avoiding the need for dividing units in setting up differential analyzers, J. Sci. Insts. 34, 334–335 (1957).

10.   A wide band analogue multiplier using crystal diodes and its application to the study of a non-linear differential equation, Electronic Engineering 29, 580–585 (1957).

11.   On random walks with restricted reversals (C. Domb and M.E.F.) Proc. Camb. Phil. Soc. 54, 48–59 (1958).

12.   Proposed methods for the analog solution of Fredholm's integral equation, J. Assoc. Comp. Mach. 5, 357–369 (1958).

13.   Stability and convergence limitations on the use of analogue computers with resistance network analogues, Brit. J. Appl. Phys. 9, 288–291 (1958).

14.   Chain configurations of polymers and polyelectrolytes, J. Chem. Phys. 28, 756–761 (1958).

15.   The excluded volume problem — Remarks on a paper by H.N.V. Temperley, Discussions of the Faraday Soc. No. 25, 1958, "Configurations and Interactions of Macromolecules and Liquid Crystals" (Aberdeen Univ. Press, Aberdeen) 200–2.

16.   On the stabilization of matrices and the convergence of linear iterative processes (M.E.F. and A.T. Fuller) Proc. Camb. Phil. Soc. 54, 417–425 (1958).

17.   The susceptibility of the Ising model of an antiferromagnet (M.E.F. and M.F. Sykes) Phys. Rev. Lett. 1, 321–2 (1958).

18.   Transformations of Ising models, Phys. Rev. 113, 969–981 (1959).

19.   Excluded volume problem and the Ising model of ferromagnetism (M.E.F. and M.F. Sykes) Phys. Rev. 114, 45–58 (1959).

20.   The susceptibility of the plane Ising model, Physica 25, 521–524 (1959).

21.   Lattice statistics in a magnetic field I. A two-dimensional super-exchange antiferromagnet, Proc. Roy. Soc. A 254, 66–85 (1960). Phys. Rev. 114, 45–58 (1959).

22.   On a non-linear differential equation for the zero-point energies of the rare gas solids (M.E.F. and I.J. Zucker) Proc. Camb. Phil. Soc. 57, 107–114 (1961).

23.   Limitations due to noise, stability and component tolerance on the solution of partial differential equations by differential analyzers, J. Elec. and Control 8, 113–126 (1960). REPRINTED in Simulation 3, No. 5, 45–52 (1964).

24.   Lattice statistics in a magnetic field II. Order and correlations of a two-dimensional super-exchange antiferromagnet, Proc. Roy. Soc. A 256, 502–513 (1960).

25. Perpendicular susceptibility of an anisotropic antiferromagnet, Physica 26, 618–622, 1028(1960).

26. The association problem in statistical mechanics — A critique of the treatment of H.S. Green and R. Leipnik (M.E.F. and H.N.V. Temperley) Rev. Mod. Phys. 32, 1029–1031 (1960).

27. Configuration and free energy of a polymer molecule with solvent interactions (M.E.F. and B.J. Hiley) J. Chem. Phys. 34, 1253–67 (1961).

28. Some cluster size and percolation problems (M.E.F. and J.W. Essam) J. Math. Phys. 2, 609–619 (1961).

29. Critical probabilities for cluster size and percolation problems, J. Math. Phys. 2, 620–627 (1961).

30. Dimer problem in statistical mechanics — An exact result (H.N.V. Temperley and M.E.F.) Phil. Mag. 6, 1061–63 (1961).

31. Statistical mechanics of dimers on a plane lattice, Phys. Rev. 124, 1664–1672 (1961).

32. Solution of a combinatorial problem — Intermediate statistics, Amer. J. Phys. 30, 49–51 (1962).

33. On the theory of critical point density fluctuations, Physica 28, 172–180 (1962).

34. Antiferromagnetic susceptibilities of the plane square and honeycomb Ising lattices (M.F. Sykes and M.E.F.) Physica 28, 919–938 (1962).

35. Antiferromagnetic susceptibilities of the simple cubic and body-centered cubic Ising lattices (M.E.F. and M.F. Sykes) Physica 28, 939–956 (1962).

36. The entropy of an antiferromagnet in a magnetic field (J.C. Bonner and M.E.F.) Proc. Phys. Soc. 80, 508–515 (1962).

37. Relation between the specific heat and susceptibility of an antiferromagnet, Phil. Mag. 7, 1731–1743 (1962).

38. Perpendicular susceptibility of the Ising model, J. Math Phys. 4, 124–135 (1963).

39. Lattice statistics — A review and an exact isotherm for a plane lattice gas, J. Math. Phys. 4, 278–286 (1963).

40. Padé approximant studies of the lattice gas and Ising ferromagnet below the critical point (J.W. Essam and M.E.F.) J. Chem. Phys. 38, 802–812 (1963).

40A. Free energy, correlations and phase transitions — A new approach to statistical mechanics [abstract] Proc. 2nd Eastern Theoret. Physics Conf., Chapel Hill, North Carolina, October 1963.

41.   Statistical mechanics of dimers on a plane lattice II. Dimer correlations and monomers (M.E.F. and J. Stephenson) Phys. Rev. 132, 1411–1431 (1963).

42.   Ising model and self-avoiding walks on hypercubical lattices and "high-density" expansions (M.E.F. and D.S. Gaunt) Phys. Rev. 133, A224–A239 (1964).

43.   Magnetism in one-dimensional systems — The Heisenberg model for infinite spin, Amer. J. Phys. 32, 343–346 (1964).

44.   Linear magnetic chains with anisotropic coupling (J.C. Bonner and M.E.F.) Phys. Rev. 135, A640–A658 (1964).
      REPRINTED in:
      (i)   "Mathematical Physics in One-Dimension," Eds. E.H. Lieb and D.C. Mattis (Academic Press, New York 1966) pp. 487–505;
      (ii) "The Many-Body Problem: An encyclopedia of exactly solved models in one dimension," Ed. D.C. Mattis (World Scientific, Singapore, 1993) pp. 717–735.

45.   Correlation functions and the critical region of simple fluids, J. Math Phys. 5, 944–962 (1964).
      REPRINTED in:
      (i)   "The Equilibrium Theory of Classical Fluids," Eds. H.L. Frisch and J.L. Lebowitz (W.B. Benjamin Inc., N.Y., 1964) pp. III–75–93;
      (ii) "Series of selected Papers in Physics," No. 49 (Phys. Soc. Japan, Tokyo, 1972) pp. 66–84.

46.   Cluster size and percolation theory, Proc. I.B.M. Scientific Computing Symposium on Combinatorial Problems, held March 16–18, 1964 (I.B.M., N.Y., 1966) Chap. 11, pp. 179–198.

47.   The free energy of a macroscopic system, Arch. Ratl. Mech. Anal. 17, 377–410 (1964).

48.   Deviations from van der Waals behaviour on the critical isobar, J. Chem. Phys. 41, 1877–1878 (1964).

49.   Specific heat of a gas near the critical point, Phys. Rev. 136, A1599–A1604 (1964).

50.   Detailed magnetic behaviour of nickel near its Curie point (J.S. Kouvel and M.E.F.) Phys. Rev. 136, A1626–A1632 (1964).

51.   The critical behaviour of ferromagnets, Proc. Internat. Conf. Magnetism, Nottingham, 7–11 September 1964 (Phys. Soc., London, 1965) pp. 79–84.

52.   Critical isotherm of a ferromagnet and of a fluid (D.S. Gaunt, M.E.F., M.F. Sykes, and J.W. Essam) Phys. Rev. Lett. 13, 713–715 (1964).

53.   Bounds for the derivatives of the free energy and the pressure of a hard-core system near close packing, J. Chem. Phys. 42, 3852–3856 (1965).

54.   Hard-sphere lattice gases I. Plane-square lattice (D.S. Gaunt and M.E.F.) J. Chem. Phys. 43, 2840–2863 (1965).

55.   Correlation functions and the coexistence of phases, J. Math. Phys. 6, 1643–1653 (1965).

56.   The stability of many-particle systems (M.E.F. and D. Ruelle) J. Math. Phys. 7, 260–270 (1966).

57.   The Theory of Condensation and the Critical Point, Physics 3, 255–283 (1967). Expanded version of The Theory of Condensation, Proc. Centennial Conf. on Phase Transformation, University of Kentucky, 18–20 March 1965 (Univ. of Kentucky, unpublished).

58(a)   Theory of critical fluctuations and singularities, in "Critical Phenomena," Proc. Conf. "Phenomena in the Neighborhood of Critical Points," N.B.S., Washington D.C., 5–8 April, 1965, Eds. M.S. Green and J.V. Sengers (N.B.S. Misc. Publ. 273, 1966) pp. 108–115.
   (b) Notes, definitions and formulas for critical point singularities, *ibid*, pp. 21–26.

59.   The shape of a self-avoiding walk or polymer chain, J. Chem. Phys. 44, 616–622 (1966).

60.   On hearing the shape of a drum, J. Comb. Theory 1, 105–125 (1966).

61.   Effect of excluded volume on phase transitions in bio-polymers, J. Chem. Phys. 45, 1469–1473 (1966).
        REPRINTED in "Theory of Helix-Coil Transitions in Biopolymers," Eds. D. Poland and H.A. Scheraga (Academic Press, N.Y. 1970) pp. 739–743.

62.   Quantum corrections to critical point behavior, Phys. Rev. Lett. 16, 11–14 (1966).

63.   On the dimer solution of planar Ising models, J. Math. Phys. 7, 1776–1781 (1966).

64.   Theory of critical point scattering and correlations I. The Ising model (M.E.F. and R.J. Burford) Phys. Rev. 156, 583–622 (1967).

65.   Magnetic critical point exponents — Their interrelations and meaning, J. Appl. Phys. 38, 981–990 (1967).

66.   The theory of equilibrium critical phenomena, Rep. Prog. Phys. 30, 615–731 (1967).

67.   Critical temperatures of anisotropic Ising lattices I. Lower bounds (C.-Y. Weng, R.B. Griffiths and M.E.F.) Phys. Rev. 162, 475–479 (1967).

68.   Critical temperatures of anisotropic Ising lattices II. Upper bounds, Phys. Rev. 162, 480–485 (1967).

69.   Interfacial, boundary and size effects at critical points (M.E.F. and A.E. Ferdinand) Phys. Rev. Lett. 19, 169–172 (1967).

70.   Intrinsic critical velocity of a superfluid (J.S. Langer and M.E.F.) Phys. Rev. Lett. 19, 560–563 (1967).
        REPRODUCED (on CD-ROM) in "The Physical Review: The first hundred years," Ed. H.H. Stroke (Amer. Inst. Phys. Press, New York, 1995).

71. Resistive anomalies at magnetic critical points (M.E.F. and J.S. Langer) Phys. Rev. Lett. 20, 665–668 (1968).

72. Renormalization of critical exponents by hidden variables, Phys. Rev. 176, 257–272 (1968).

73. The decay of superflow in helium, Proc. Conf. on "Fluctuations in Superconductors," Asilomar, California, March 13–15, 1968, Eds. W.S. Goree and F. Chilton (Stanford Research Institute, Menlo Park, 1968) pp. 357–380.

74. Asymptotic behavior of Toeplitz matrices and determinants (R.E. Hartwig and M.E.F.) Arch. Ratl. Mech. Anal. 32, 190–225 (1969).

75. Toeplitz determinants: Some applications, theorems and conjectures (M.E.F. and R.E. Hartwig) Adv. Chem. Phys. 15, 333–353 (1968).

76. Decay of correlation in linear systems (M.E.F. and B. Widom) J. Chem. Phys. 50, 3756–3772 (1969).

77(a) Rigorous inequalities for critical point correlation exponents, Phys. Rev. 180, 594–600 (1969).
   (b) Exponent inequalities for critical point spin correlation functions [abstract] J. Appl. Phys. 40, 1278 (1969).

78. Bounded and inhomogeneous Ising models I. Specific heat anomaly of a finite lattice (A.E. Ferdinand and M.E.F.) Phys. Rev. 185, 832–846 (1969).
   REPRINTED in "Conformal Invariance and Applications to Statistical Mechanics," C. Itzykson, H. Salem and J.-B. Zuber (World Scientific, Singapore, 1988) pp. 289–303.

79. Mobile-electron Ising ferromagnets [abstract] (M.E.F. and P.E. Scesney) J. Appl. Phys. 40, 1554 (1969).

80(a) Aspects of equilibrium critical phenomena, J. Phys. Soc. Japan, Suppl. 26, 87–93 (1969) [Proc. I.U.P.A.P. Conf. on Statistical Mechanics, Kyoto, September 1968].
   (b) A restatement of Flory's theory of excluded volume in polymers, *ibid* pp. 44–55 [discussion remarks].

81. Phase transitions and critical phenomena, in "Contemporary Physics," Vol. I, Trieste Symposium 1968 [Proc. International Symposium, 7–28 June 1968] (International Atomic Energy Agency, Vienna, 1969) pp. 19–46.

82. Absence of anomalous averages in systems of finite non-zero thickness or cross section (G.V. Chester, M.E.F. and N.D. Mermin) Phys. Rev. 185, 760–762 (1969).

83. Light scattering and pseudospinodal curves: The isobutyric-acid-water system in the critical region (B. Chu, F.J. Schoenes and M.E.F.) Phys. Rev. 185, 219–226 (1969).

84. Phase transitions in one-dimensional cluster-interaction fluids IA. Thermodynamics (M.E.F. and B.U. Felderhof) Ann. Phys.(N.Y.) 58, 176–216 (1970).

85. *idem* IB. Critical behavior (M.E.F. and B.U. Felderhof) Ann. Phys.(N.Y.) 58, 217–267 (1970).

86. Phase transitions in one-dimensional classical fluids with many-body interactions, "Systèmes a un Nombre Infini de Degrés de Liberté," Gif-sur-Yvette, France, 7–10 May 1969, Colloq. Internat. du C.N.R.S. No. 181 (Editions du C.N.R.S., Paris 1970) pp. 87–103.

87. Broken symmetry and decay of order in restricted dimensionality (D. Jasnow and M.E.F.) Phys. Rev. Lett. 23, 286–288 (1969).

88. Phase transitions in one-dimensional cluster-interaction fluids II. Simple logarithmic model (B.U. Felderhof and M.E.F.) Ann. Phys. (N.Y.) 58, 268–280 (1970).

88^ *idem* III. Correlation functions (B.U. Felderhof) Ann. Phys.(N.Y.) 58, 281–300 (1970).

89. Some basic definitions in graph theory (J.W. Essam and M.E.F.) Rev. Mod. Phys. 42, 271–288 (1970).

90. Asymptotic free energy of a system with periodic boundary conditions (M.E.F. and J.L. Lebowitz) Commun. Math. Phys. 19, 251–272 (1970).

91. Visibility of critical-exponent renormalization (M.E.F. and P.E. Scesney) Phys. Rev. A 2, 825–835 (1970).

92. Decay of order in isotropic systems of restricted dimensionality I. Bose superfluids (D. Jasnow and M.E.F.) Phys. Rev. B 3, 895–907 (1971).

93. *idem* II. Spin systems (M.E.F. and D. Jasnow) Phys. Rev. B 3, 907–924 (1971).

94. Sums of inverse powers of sines: Solution to Problem 69–14 [SIAM Review 11, 621 (1969)] SIAM Review 13, 116–119 (1971) [Note corrections].
     REPRINTED in "Problems in Applied Mathematics: Selections from SIAM Review," Ed. M.S. Klamkin (SIAM, Philadelphia, 1990) pp. 157–160.

95. Critical behavior in the one-dimensional cluster-interaction models, Proc. Midwest Conf. Theoret. Phys., University of Notre Dame, 3–4 April 1970 (National Science Foundation, Washington, 1970) pp. 50–60.

96. Zeros of the partition function for the Heisenberg, ferro-electric and general Ising models (M. Suzuki and M.E.F.) J. Math Phys. 12, 235–246 (1971).

97. Phase transitions, symmetry and dimensionality [Chicago Solid State Colloq. Series] Essays in Physics, Vol. 4 (Academic Press, London, 1972) pp. 43–89.

98. Behavior of two-point correlation functions at high temperatures (W.J. Camp and M.E.F.) Phys. Rev. Lett. 26, 73–77 (1971).

99. Behavior of two-point correlation functions near and on a phase boundary (M.E.F. and W.J. Camp) Phys. Rev. Lett. 26, 565–568 (1971).

100(a) Critical behavior in spherical models of finite thickness (M.E.F., G.A.T. Allan and M.N. Barber) [summary] I.U.P.A.P. Conference on Statistical Mechanics, Chicago, March–April 1971.

   (b) Critical behavior of an ideal Bose gas of finite thickness (M.E.F. and M.N. Barber) [abstract] I.U.P.A.P. Conference on Statistical Mechanics, Chicago, March–April 1971.

   (c) Finite size effects in the spherical model and ideal Bose gas (M.E.F., M.N. Barber, and G.A.T. Allan) [abstract] Bull. Am. Phys. Soc. 16, 394 (1971).

100^ Critical temperatures of Ising Lattice films (G.A.T. Allan) Phys. Rev. B 1, 352–356 (1970).

101.   Excitons in Mott-Hubbard insulators (S. Doniach, B.J. Roulet and M.E.F.) Phys. Rev. Lett. 27, 262–265 (1971).

102(a) Theory of critical-point scattering and correlations II. Heisenberg models (D.S. Ritchie and M.E.F.) Phys. Rev. B 5, 2668–2692 (1972).

   (b) Critical scattering in the Heisenberg model (D.S. Ritchie and M.E.F.) [abstract] Amer. Inst. Phys. Conf. Proc. No. 5, p. 1250 (Amer. Inst. Phys., New York, 1972).

103.   Remnant functions (M.E.F. and M.N. Barber) Arch. Ratl. Mech. Anal. 47, 205–236 (1972).

104.   The Theory of Critical Point Singularities, pp. 1–99 in "Critical Phenomena," Proc. 1970 Enrico Fermi International School of Physics, Course No. 51, Varenna, Italy, Ed. M.S. Green (Academic Press for the Italian Physical Society, New York, 1971).
   I.   Phase transitions and the order parameter
   II. Critical exponents and homogeneity
   III. Scaling, power laws and analyticity
   IV. The droplet model: its extensions and implications
   V.   Finite size and boundary effects
   TRANSLATED by M.Sh. Giterman and published by MIR (Moscow, 1973) in Устойчивость и фазовые переходы ("Stability and Phase Transitions") pp. 245–369: "Theory of Critical Point Singularities" from "Critical Phenomena".

105.   Critical exponents in 3.99 dimensions (K.G. Wilson and M.E.F.) Phys. Rev. Lett. 28, 240–243 (1972).
   REPRINTED in:
   (i) "Series of Selected Papers in Physics," No. 78, "Critical Phenomena II," Eds. S. Hikami and Y. Iwasaki (Phys. Soc. Japan, Tokyo, 1982) pp. 17–20;
   (ii) "The Physical Review: The first hundred years," Ed. H.H. Stroke (Amer. Inst. Phys. Press, New York, 1995).

106(a) Estimation of spectra from moments — Application to the Hubbard model (M.E.F. and W.J. Camp) Phys. Rev. B 5, 3730–3737 (1972).

   (b) *idem* [abstract] Bull. Amer. Phys. Soc. 17, 79 (1972).

107.   Spectra, moments, and lattice walks for Hubbard magnetic insulators (B.J. Roulet, M.E.F. and S. Doniach) Phys. Rev. B 7, 403–420 (1973).

108. Ferromagnetic Heisenberg lattice films (D.S. Ritchie and M.E.F.) Amer. Inst. Phys. Conf. Proc. No. 5, Magnetism and Magnetic Materials, 1971 (Amer. Inst. Phys., 1972) pp. 1245–1249.

109. Critical behavior of the anisotropic n-vector model (M.E.F. and P. Pfeuty) Phys. Rev. B 6, 1889–1891 (1972).

110. On discontinuity of the pressure, Commun. Math. Phys. 26, 6–16 (1972).

111. Decay of order in classical many–body systems I. Introduction and formal theory (W.J. Camp and M.E.F.) Phys. Rev. B 6, 946–959 (1972).

111^ *idem* II. (W.J. Camp) Ising model at high temperatures, Phys. Rev. B 6, 960–979 (1972); III. Ising model at low temperatures, Phys. Rev. B 7, 3187–3203 (1973).

112. Scaling theory for finite-size effects in the critical region (M.E.F. and M.N. Barber) Phys. Rev. Lett. 28, 1516–1519 (1972).
REPRINTED in "Finite-Size Scaling," J.L. Cardy, Ed.(North Holland, Amsterdam, 1988) Paper 1.1, pp. 8–11.
REPRODUCED (on CD-ROM) in "The Physical Review: The first hundred years," Ed. H.H. Stroke (Amer. Inst. Phys. Press, New York, 1995).

113. Critical phenomena in systems of finite thickness I. The spherical model (M.N. Barber and M.E.F.) Ann. Phys. (N.Y.) 77, 1–78 (1973).

114. Finite-size and surface effects in Heisenberg films (D.S. Ritchie and M.E.F.) Phys. Rev. B 7, 480–494 (1973).

115. Critical exponents for long-range interactions (M.E.F., S.-K. Ma and B.G. Nickel) Phys. Rev. Lett. 29, 917–920 (1972).

116. Quantum lattice gas and the existence of a supersolid (K.-S. Liu and M.E.F.) J. Low Temp. Phys. 10, 655–683 (1973).

117. Critical phenomena in systems of finite thickness III. Specific heat of an ideal boson film (M.N. Barber and M.E.F.) Phys. Rev. A 8, 1124–35 (1973).

118. Critical phenomena — Series expansions and their analysis [summary] in "Padé Approximants and their Applications," Ed. P.R. Graves-Morris (Academic Press, 1973) pp. 159–162.

119. Crossover exponent — for spin systems (P. Pfeuty, M.E.F., and D. Jasnow) Amer. Inst. Phys. Conf. Proc. No. 10, "Magnetism and Magnetic Materials," 1972 (Amer. Inst. Phys., 1973), pp. 817–821.

120. Helicity Modulus, superfluidity and scaling in isotropic systems (M.E.F., M.N. Barber, and D.M. Jasnow) Phys. Rev. A 8, 1111–1124 (1973).

121.   Critical behavior of magnets with dipolar interactions I. Renormalization group near four dimensions (A. Aharony and M.E.F.) Phys. Rev. B 8, 3323–3341 (1973).

121^   *idem* II–V. (A. Aharony) Phys. Rev. B 8, 3342–48, 3349–57, 3358–62, 3363–70 (1973).

122.   Critical point phenomena — The role of series expansions, Proc. Int. Conf. Padé Approximants, "Continued Fractions and Related Topics" (Univ. Colorado, Boulder, June 1972) Rocky Mountain J. Math. 4, 181–201 (1974).

123.   Dipolar interactions at ferromagnetic critical points (M.E.F. and A. Aharony) Phys. Rev. Lett. 30, 559–562 (1973).

124.   Three-state Potts model and anomalous tricritical points (J.P. Straley and M.E.F.) J. Phys. A 6, 1310–1326 (1973).

125.   Classical, n-component spin systems or fields with negative even integral n, Phys. Rev. Lett. 30, 679–681 (1973).

126.   Discussion remarks in Proc. Symp. "Statistical and Probilistic Problems in Metallurgy," Ed. W.L. Nicholson, Suppl. Adv. Appl. Prob. (Dec. 1972) pp. 46, 67–8, 144–5, 216–220, etc.

127.   Theory of critical point scattering and correlations III. The Ising model below T_c and in a field (H.B. Tarko and M.E.F.) Phys. Rev. B 11, 1217–1253 (1975).

128.   Critical phenomena in films and surfaces [Proc. Conf. Thin Film Phenomena, 15–16 March 1973, IBM San Jose] J. Vacuum Sci. Techn. 10, 665–673 (1973).

129(a) Renormalization group, exponents, and scaling [summary with tables] "Renormalization Group in Critical Phenomena and Quantum Field Theory: Proceedings of a Conference," May 1973, Eds. J.D. Gunton and M.S. Green (Temple University, 1974). pp. 10–20.
   (b) Crossover effects and operator expansions, *ibid*, pp. 65–72.

130.   General scaling theory for critical points, Proc. Nobel Symp. XXIV held in Aspenäsgården, Lerum, Sweden, 12–16 June 1973, "Collective Properties of Physical Systems," Eds. B. Lundqvist and S. Lundqvist (Academic Press, New York, 1974) pp. 16–37.

131.   Critical scattering in a field and below T_c (H.B. Tarko and M.E.F.) Phys. Rev. Lett. 31, 926–930 (1973).

132(a) Scaling function for critical scattering (M.E.F. and A. Aharony) Phys. Rev. Lett. 31, 1238–1241, 1537 (1973).
   (b) *idem* [abstract] I.U.P.A.P. van der Waals Centennial Conference on Statistical Mechanics, Amsterdam, 27–31 August 1973, Eds. C. Prins and M. Blumendal (North-Holland Publ. Co. 1973).

133(a) Soluble renormalization groups and scaling fields for low-dimensional spin systems (D.R. Nelson and M.E.F.) Ann. Phys. (N.Y.) 91, 226–274 (1975).
   (b) Exact renormalization groups for one-dimensional spin systems [summary] Amer. Inst. Phys. Conf. Proc. No. 18, Magnetism and Magnetic Materials, 1973 (Amer. Inst. Phys., N.Y. 1974) pp. 888–890.

134.   Crossover scaling functions for exchange anisotropy (P. Pfeuty, D. Jasnow and M.E.F.) Phys. Rev. B 10, 2088–2112 (1974).

135.   The renormalization group in the theory of magnetism, Proc. Internat. Conf. Magnetism, Moscow, 22–28 August 1973, Vol. I, pp. 51–79 (NAUKA, Moscow, 1974).

136.   Scaling function for two-point correlations I. Expansion near four dimensions (M.E.F. and A. Aharony) Phys. Rev. B 10, 2818–2833 (1974).

136^   *idem* II. Expansion to order $1/n$ (A. Aharony) Phys. Rev. B 10, 2834–44 (1974).

137.   Renormalization-group analysis of metamagnetic tricritical behavior (D.R. Nelson and M.E.F.) Phys. Rev. B 11, 1030–1039 (1975).

138.   Spin flop, supersolids, and bicritical and tetracritical points (M.E.F. and D.R. Nelson) Phys. Rev. Lett. 32, 1350–53 (1974).

139.   Renormalization-group analysis of bicritical and tetracritical points (D.R. Nelson, J.M. Kosterlitz and M.E.F.) Phys. Rev. Lett. 33, 813–817 (1974).
       REPRINTED in "Selected Papers in Physics," No. 78, "Critical Phenomena II," Eds. S. Hikami and Y. Iwasaki (Phys. Soc. Japan, Tokyo, 1982) pp. 51–55.

140.   Gonzalo's scaling function for the equation of state (T.W. Capehart and M.E.F.) Phys. Rev. B 11, 1262–1263 (1975).

141.   Tests of strong scaling in the three-dimensional Ising model (M.E.F. and H.B. Tarko) Phys. Rev. B. 11, 1131–1133 (1975).

142.   Critical temperatures of classical n-vector models on hypercubic lattices (P.R. Gerber and M.E.F.) Phys. Rev. B 10, 4697–4703 (1974).

143.   The renormalization group in the theory of critical behavior, Rev. Mod. Phys. 46, 597–616 (1974).

144.   Susceptibility scaling functions for ferromagnetic Ising films (T.W. Capehart and M.E.F.) Phys. Rev. B 13, 5021–5038 (1976).

145.   Universality of magnetic tricritical points (M.E.F. and D.R. Nelson) Phys. Rev. B 12, 263–266 (1975).

146.   Bounded and inhomogeneous Ising models II. Specific heat scaling function for a strip (H. Au-Yang and M.E.F.) Phys. Rev. B 11, 3469–3487 (1975).

147.   Bicritical and tetracritical points in anisotropic antiferromagnetic systems (J.M. Kosterlitz, D.R. Nelson and M.E.F.) Phys. Rev. B 13, 412–432 (1976).

148.   Theory of multicritical transitions and the spin-flop bicritical point, Amer. Inst. Phys. Conf. Proc. No 24, "Magnetism and Magnetic Materials," 1974 (Amer. Inst. Phys., N.Y., 1975) pp. 273–280.

149.   The renormalization group and its application to critical phenomena [summary] "Statistical Physics," Bose Memorial Symposium, Bangalore, July 1974, Eds. N. Mukunda, A.K. Rajagopal, and K.P. Sinha, Suppl. J. Ind. Inst. Sci., Bangalore, June 1975, pp. 51–55.

150.   Critical temperatures of continuous spin models and the free energy of a polymer (P.R. Gerber and M.E.F.) J. Chem. Phys. 63, 4941–4946 (1975).

151.   Scaling axes and the spin-flop bicritical phase boundaries, Phys. Rev. Lett. 34, 1634–1638 (1975).

152.   Self-interacting walks, random spin systems and the zero-component limit (D. Jasnow and M.E.F.) Phys. Rev. B 13, 1112–1118 (1976).

153.   Monte Carlo study of multicriticality in finite Baxter models (E. Domany, K.K. Mon, G.V. Chester and M.E.F.) Phys. Rev. B 12, 5025–5033 (1975).

154.   Scaling theory of nonlinear critical relaxation (M.E.F. and Z. Rácz) Phys. Rev. B 13, 5039–5041 (1976).

155(a) Regularly spaced defects in Ising models (M.E.F. and H. Au-Yang) J. Phys. C: Solid St. Phys., 8, L418–L421 (1975).
   (b) Critical effects of regularly spaced defects (M.E.F. and H. Au-Yang) [abstract] Amer. Inst. Phys. Conf. Proc. No. 29, Magnetism and Magnetic Materials 1975 (Amer. Inst. Phys., 1976) p. 490.

156.   Bounded and inhomogeneous Ising models III. Regularly spaced defects (H. Au-Yang, M.E.F. and A.E. Ferdinand) Phys. Rev. B 13, 1238–1265 (1976).

156^   *idem* IV. Specific-heat amplitude for regular defects (H. Au-Yang) Phys. Rev. B 13, 1266–1271 (1976).

157.   Crossover behavior of the specific heat and nonordering susceptibility of the anisotropic Heisenberg ferromagnet (P.R. Gerber and M.E.F.) Phys. Rev. B 13, 5042–5053 (1976).

158.   Equations of state for bicritical points II. Ising-like ordered phase (E. Domany, D.R. Nelson and M.E.F.) Phys. Rev. B 15, 3493–3509 (1977).

159.   Magnetization of cubic ferromagnets and the three-component Potts model (D. Mukamel, M.E.F. and E. Domany) Phys. Rev. Lett. 37, 565–568 (1976).

160.   Novel two-variable approximants for studying magnetic multicritical behavior, Proc. Int. Conf. on Magnetism, Amsterdam, 1976, 5E–1; Physica 86–88, 590–592 (1977).

161.   Critical behavior of cubic and tetragonal ferromagnets in a field (D. Mukamel, E. Domany and M.E.F.) Proc. Int. Conf. on Magnetism, Amsterdam, 1976, 7E–1; Physica 86–88, 572–574 (1977).

162(a) Destruction of first-order transitions by symmetry-breaking fields (E. Domany, D. Mukamel and M.E.F.) Phys. Rev. B 15, 5432–5441 (1977).

(b) Symmetry-breaking fields and destruction of first-order transitions (D. Mukamel, E. Domany and M.E.F.) Proc. Int. Conf. Magnetism, Ann. Israel Phys. Soc. 2, 451–454 (1978).

163. Equations of state for bicritical points III. Cubic anisotropy and tetracriticality (E. Domany and M.E.F.) Phys. Rev. B 15, 3510–3521 (1977).

164. Series expansion approximants for singular functions of many variables, in "Statistical Mechanics and Statistical Methods in Theory and Application" – Proc. Symposium in honor of E.W. Montroll, Ed. U. Landman (Plenum Press, N.Y., 1977) pp. 3–32.

165. Critical points and multicritical points [abstract] in "Statistical Mechanics and Statistical Methods in Theory and Application" – Proc. Symposium in honor of E.W. Montroll, Ed. U. Landman (Plenum Press, N.Y., 1977) pp. 1–2.

166. Wall and boundary free energies I. Ferromagnetic scalar spin systems (M.E.F. and G. Caginalp) Commun. Math. Phys. 56, 11–56 (1977).

167 . Partial differential approximants for multicritical singularities (M.E.F. and R.M. Kerr) Phys. Rev. Lett. 39, 667–670 (1977).

168. Lars Onsager 1903–1976 (H.C. Longuet-Higgins and M.E.F.) Bio. Mem. Fell. Roy. Soc. 24, 445–471 (1978).
    REPRINTED (with revisions) in:
    (i) Bio. Mem. N.A.S. (U.S.A.), Vol.60, pp. 182–232(1991);
    (ii) J. Stat. Phys. 78, 605–640 (1995);
    (iii) "The Collected Works of Lars Onsager," Eds. P.C.Hemmer, H. Holden, and S. Kjelstrup Ratke (World Scientific, Singapore, 1996) pp. 9–34.

169. Scaling functions for quantum crossover (I.D. Lawrie and M.E.F.) J. Appl. Phys. 49, 1353–55 (1978).

170. Tricritical scaling in the spherical model limit (S. Sarbach and M.E.F.) J. Appl. Phys. 49, 1350–52 1978).

171. The Yang-Lee edge singularity in spherical models (D.A. Kurtze and M.E.F.) J. Stat. Phys. 19, 205–218 (1978).

172. Wall and boundary free energies II. General domains and complete boundaries (G. Caginalp and M.E.F.) Commun. Math. Phys. 65, 247–280 (1979).

172^ *idem* III. Correlation decay and vector spin systems (G. Caginalp), Commun. Math. Phys, 76, 149–163 (1980).

173. Tricriticality and the failure of scaling in the many-component limit (S. Sarbach and M.E.F.) Phys. Rev. B 18, 2350–2363 (1978).

174. The Yang-Lee edge singularity and φ^3 field theory, Phys. Rev. Lett. 40, 1610–1613 (1978).

175(a)   Nonuniversality of tricritical behavior (M.E.F. and S. Sarbach) Phys. Rev. Lett. 41, 1127–1130 (1978).

(b)   Nonuniversality in magnetic tricriticality (S. Sarbach and M.E.F.) J. Appl. Phys. 50, 1802–1803 (1979).

176.   Phénomenes aux parois dans un mélange binaire critique (M.E.F. and P.-G. de Gennes) C.R. Acad. Sc. Paris, Ser. B 287, 207–209 (1978).

177.   Inhomogeneous differential approximants for power series (M.E. Fisher and H. Au-Yang) J. Phys. A 12, 1677–1692 (1979); 13, 1517 (1980).

178.   Yang-Lee edge singularity in the hierarchial model (G.A. Baker, Jr., M.E.F. and P. Moussa) Phys. Rev. Lett. 42, 615–618 (1979).

179.   Yang-Lee edge singularities at high temperatures (D.A. Kurtze and M.E.F.) Phys. Rev. B 20, 2785–2796 (1979).

180.   Monte-Carlo study of the spatially modulated phase in an Ising model (W. Selke and M.E.F.) Phys. Rev. B 20, 257–265 (1979).

181.   Tricritical coexistence in three dimensions: The multi-component limit (S. Sarbach and M.E.F.) Phys. Rev. B 20, 2797–2817 (1979).

182.   Spatially modulated phases in Ising models with competing interactions (W. Selke and M.E.F.) Proc. Int. Conf. Magnetism, München, Sept. 1979: J. Mag. Magn. Matls. 15–18, 403–404 (1980).

183.   Wall effects in critical systems: Scaling in Ising model strips (H. Au-Yang and M.E.F.) Phys. Rev. B 21, 3956–3970 (1980).

184.   Critical phenomena in statistical mechanics — Aspects of renormalization group theory [Summary of 1979 Cargèse Lectures], "Bifurcation Phenomena in Mathematical Physics and Related Topics," Eds. C. Bardos and D. Bessis (Reidel Publ. Co., Dordrecht, 1980) pp. 61–68.

185(a)   Critical wall perturbations and a local free energy functional (M.E.F. and H. Au-Yang) Physica 101A, 255–264 (1980).

(b)   *idem* [abstract] Bull.Amer.Phys.Soc. 25, No.3, 277(1980).

186.   Infinitely many commensurate phases in a simple Ising model (M.E.F. and W. Selke) Phys. Rev. Lett. 44, 1502–1505; 45, E148 (1980). REPRODUCED (on CD-ROM) in "The Physical Review: The first hundred years," Ed. H.H. Stroke (Amer. Inst. Phys. Press, New York, 1995).

187.   Universality tests at Heisenberg bicritical points (M.E.F., J.-H. Chen, and H. Au-Yang) J. Phys. C 13, L459–464 (1980); corrig. location unknown.

188. The states of matter — A theoretical perspective, Proc. Welch Foundation Conf. XXIII, "Modern Structural Methods," Houston, 1979, Ed. W.O. Milligan (R.A. Welch Foundation, Houston, 1980) pp. 74–145; and discussion contributions, pp. 146–175.

189. Cubic fields, bicritical crossover, the spherical and van der Waals limits (P. Seglar and M.E.F.) J. Phys. C 13, 6613–6625 (1980).

190. Two-dimensional Ising models with competing interactions — A Monte Carlo study (W. Selke and M.E.F.) Z. Physik. B — Condensed Matter 40, 71–77 (1980).

191. Discussion remarks in "Order and Fluctuations in Equilibrium and Nonequilibrium Statistical Mechanics:" XVII International Solvay Conference on Physics, Eds. G. Nicolis, G. Dewel and J.W. Turner (John Wiley and Sons, Inc., New York 1981) pp. 14–22; see also pp. 32, 99–100, 105, and 362.

192. This Week's Citation Classic: Fisher, M.E., The theory of equilibrium critical phenomena, Current Contents, P.C. & E.S., 20(46) 18 (1980).

193. Universality in analytic corrections to scaling for planar Ising models (A. Aharony and M.E.F.) Phys. Rev. Lett. 45, 679–682, 1044 (1980).

194. Low-temperature analysis of the axial next-nearest neighbour Ising model near its multiphase point (M.E.F. and W. Selke) Phil. Trans. Roy. Soc. 302, 1–44 (1981).

195. Yang-Lee edge behavior in one-dimensional systems, Prog. Theoret. Phys. Suppl. 69, 14–29 (1980).

196(a) Multiphase behavior and modulated ordering in soluble Ising models (D.A. Huse, M.E.F. and J.M. Yeomans) Phys. Rev. B 23, 180–185 (1981).
   (b) Exactly soluble Ising models exhibiting multiphase points (D.A. Huse, J.M. Yeomans and M.E.F.) J. Appl. Phys. 52, 2028–30 (1981).

197. Simple Ising models still thrive! A review of some recent progress, Physica 106A, 28–47 (1981).

198. Three-component model and tricritical points: A renormalization group study I. Two dimensions (M. Kaufman, R.B. Griffiths, J.M. Yeomans and M.E.F.) Phys. Rev. B 23, 3448–3459 (1981).

199. Three-component model and tricritical points: A renormalization group study II. General dimensions and the three-phase monohedron (J.M. Yeomans and M.E.F.) Phys. Rev. B 24, 2825–2840 (1981).

200. An infinity of commensurate phases in a simple Ising system: The ANNNI model, J. Appl. Phys. 52, 2014–18 (1981).

201. Critical point correlations of the Yvon-Born-Green equation (G.L. Jones, J.J. Kozak, E. Lee, S. Fishman and M.E.F.) Phys. Rev. Lett. 46, 795–798, 1350 (1981).

202.   Specific heats of classical spin systems and inhomogeneous differential approximants (J.-H. Chen and M.E.F.) J. Phys. A 14, 2553–2566 (1981).

203.   Critical point scaling in the Percus-Yevick equation (S. Fishman and M.E.F.) Physica 106A, 1–13 (1981).

204.   Bicriticality and partial differential approximants (M.E.F. and J.-H. Chen) in "Phase Transitions: Cargèse 1980," Eds. M. Levy, J.C. Le Guillou, and J. Zinn-Justin (Plenum Publ. Corp., New York, 1982) pp. 169–216.

205.   Many commensurate phases in the chiral Potts model or asymmetric clock models (J.M. Yeomans and M.E.F.) J. Phys. C: Solid State Phys. 14, L835–L839 (1981).

206.   Critical scattering and integral equations for fluids (M.E.F. and S. Fishman) Phys. Rev. Lett. 47, 421–423 (1981).

207(a) Scaling theory for the criticality of fluids between plates (M.E.F. and H. Nakanishi) J. Chem. Phys. 75, 5857–5863 (1981).
   (b) Fluid criticality near walls and between plates [abstract and summary] Int. J. Quant. Chem.: Quantum Chem. Symp. 16, 237–240 (1982).

208.   Analysis of the multiphase region in the three-state chiral clock model (J.M. Yeomans and M.E.F.) Physica 127A, 1–37 (1984).

209.   Scaling, universality, and renormalization group theory (Lecture notes prepared with the assistance of A.G. Every) in Lect. Notes in Physics, Vol. 186, "Critical Phenomena," Proc. Summer School at the University of Stellenbosch, January 1982, Ed. F.J.W. Hahne (Springer-Verlag, Berlin, 1983) pp. 1–140.

210.   Wall wandering and the dimensionality dependence of the commensurate-incommensurate transition (M.E.F. and D.S. Fisher) Phys. Rev. B 25, 3192–3198 (1982).

211.   Unbiased estimation of corrections to scaling by partial differential approximants (J.-H. Chen, M.E.F. and B.G. Nickel) Phys. Rev. Lett 48, 630–634 (1982).

212.   Partial differential approximants for multivariable power series I. Definitions and faithfulness (M.E.F. and D.F. Styer) Proc. Roy. Soc. A 384, 259–298 (1982).

213.   Partial differential approximants for multivariable power series II. Invariance properties (D.F. Styer and M.E.F.) Proc. Roy. Soc. A 388, 75–102 (1983).

213^   *idem* III. Enumeration of invariance transformations (D.F. Styer) Proc. Roy. Soc. A 390, 321–339 (1983).

214.   Melting, order, flows, mappings and chaos (M.E.F. and D.A. Huse) "Melting, Localization and Chaos," Proc. Ninth Midwest Solid State Theory Symposium, Argonne, Nov. 1981, Eds. R. Kalia and P. Vashishta (Elsevier Sci. Publ. Co., New York, 1982) pp. 259–293.

215. Scaling for first-order transitions in thermodynamic and finite systems (M.E.F. and A.N. Berker) Phys. Rev. B 26, 2507–2513 (1982).

216. The decoupling point of the axial next-nearest-neighbour Ising model and marginal crossover (D.A. Huse and M.E.F.) J. Phys. C: Solid State Phys. 15, L585–L595 (1982).

217. Nonlinear scaling fields and corrections to scaling near criticality (A. Aharony and M.E.F.) Phys. Rev. B 27, 4394–4400 (1983).

218. Domain walls and the melting of commensurate surface phases (D.A. Huse and M.E.F.) Phys. Rev. Lett. 49, 793–796 (1982).

219. Criticality in the Yvon-Born-Green and similar integral equations (M.E.F. and S. Fishman) J. Chem. Phys. 78, 4227–4244 (1983).

220. Critical phenomena in fluid films: Critical-temperature-shifts (H. Nakanishi and M.E.F.) J. Phys. C, 16, L95–L97 (1983).

221. Multicriticality of wetting, prewetting and surface transitions (H. Nakanishi and M.E.F.) Phys. Rev. Lett. 49, 1565–1568 (1982).

222. Critical point shifts in films (H. Nakanishi and M.E.F.) J. Chem. Phys. 78, 3279–3293 (1983).

223. Surface tension variation in multiphase fluid systems (D.J. Klinger, M.E.F. and B. Widom) J. Phys. Chem. 87, 2841–2845 (1983).

224. Commensurate melting, domain walls, and dislocations (D.A. Huse and M.E.F.) Phys. Rev. B 29, 239–270 (1984).

225. Melting and wetting transitions in the three-state chiral clock model (D.A. Huse, A.M. Szpilka, and M.E.F.) Physica 121A, 363–398 (1983).

226. Continuum fluids with a discontinuity in the pressure (G.W. Milton and M.E.F.) J. Stat. Phys. 32, 413–438 (1983).

227. Critical behavior of a three-dimensional dimer model (S.M. Bhattacharjee, J.F. Nagle, D.A. Huse and M.E.F.) J. Stat. Phys. 32, 361–374 (1983).

228. Convergence of finite-size scaling renormalization techniques (V. Privman and M.E.F.) J. Phys. A: Math. Gen. 16, L295–L301 (1983).
REPRINTED in "Finite-Size Scaling," Ed., J.L. Cardy (North Holland, Amsterdam, 1988) Paper 1.11, pp. 149–181.

229. Finite-size effects at first order transitions (V. Privman and M.E.F.) J. Stat. Phys. 33, 385–417 (1983).
REPRINTED in "Finite-Size Scaling," Ed., J.L. Cardy (North Holland, Amsterdam, 1988) Paper 2.8, pp. 292–298.

230(a)   The melting of commensurate phases and new universality classes [abstract] (M.E.F. and D.A. Huse) J. Phys. Soc. Japan 52, Suppl. (1983) p. 31.

  (b)   Transition to "New Type of Ordered Phase": Four illustrations From the summary talk by M.E.F., *ibid*, pp.00,94,172,234.

231.   Lectures on multicritical behavior in "Multicritical Phenomena," Proc. NATO Advanced Study Institute, Geilo, Norway, 10–21 April 1983, Eds. R. Pynn and A. Skjeltorp (Plenum Publ. Corp., New York, 1984) [summaries]:-
       (a) Multicriticality: A theoretical introduction, pp. 1–5;
       (b) A plenitude of commensurate phases in simple models, pp. 233–235;
       (c) Commensurate melting and domain walls in surface phases, pp. 289–291.

232.   Wetting transitions near bulk triple points (R. Pandit and M.E.F.) Phys. Rev. Lett. 51, 1772–5 (1983).

233.   Nonlinear extensions of square gradient theory for fluid pair correlations (D.J. Klinger, M.E.F. and S. Fishman) J. Chem. Phys. 80, 3392–98 (1984).

234.   Walks, walls, wetting and melting, J. Stat. Phys. 34, 667–729 (1984).

235.   Universal critical amplitudes in finite-size scaling (V. Privman and M.E.F.) Phys. Rev. B 30, 322–327 (1984).

236.   Nonlinearities in differential equations for near-critical pair correlation functions (D.J. Klinger and M.E.F.) Phys. Rev. Lett. 52, 400 (1984).

237.   Partial differential approximants and the elucidation of multisingularities (D.F. Styer and M.E.F.) in Lect. Notes in Math. Vol. 1105, "Rational Approximation and Interpolation," [Proc. Conf., Tampa, 1983] Eds. P.R. Graves-Morris, E.B. Saff and R.S. Varga (Springer-Verlag, Heidelberg, 1984) pp. 313–330.

238.   The winding angle of planar self-avoiding walks (M.E.F., V. Privman and S. Redner) J. Phys. A: Math. Gen. 17, L569–L578 (1984).

239.   Critical behavior of a dilute interacting Bose fluid (M. Rasolt, M.J. Stephen, M.E.F. and P.B. Weichman) Phys. Rev. Lett. 53, 798–801 (1984).

240.   Corrections to scaling and crossover in two-dimensional Ising-like systems (M. Barma and M.E.F.) Phys. Rev. Lett. 53, 1935–38 (1984).

241.   Finite-size rounding of first-order transitions (V. Privman and M.E.F.) J. Appl. Phys. 57, 3327–28 (1985).

242.   Two-dimensional Ising-like systems: Corrections to scaling in the Klauder and double Gaussian models (M. Barma and M.E.F.) Phys. Rev. B 31, 5954–75 (1985).

243.   Walks, walls and ordering in low dimensions, in "Fundamental Problems in Statistical Mechanics VI," Proc. Summer School, Trondheim, Norway, June 1984, Ed. E.G.D. Cohen (Elsevier Science Publ., B.V., Amsterdam, 1985) pp. 1–50.

244. First-order transitions breaking $O(n)$ symmetry: Finite-size scaling (M.E.F. and V. Privman) Phys. Rev. B 32, 447–464 (1985).

245. First-order transitions in spherical models: Finite-size scaling (M.E.F. and V. Privman) Commun. Math. Phys. 103, 527–548 (1986).

246(a) Critical Phenomena: Some recent theoretical developments, J. Appl. Phys. 57, 3265–67 (1985).
   (b) Phase transitions and criticality: Finite-size effects, conformal covariance, and other current problems, Proc. ICM'85, J. Mag. Magn. Matls. 54–57, 646–648 (1986).

247. The validity of hyperscaling in three dimensions for scalar spin systems (M.E.F. and J.-H. Chen) J. Phys. (Paris) 46, 1645–54 (1985).

248. Spin-wave corrections to single-domain behavior in finite ferromagnetic systems (V. Privman and M.E.F.) Proc. ICM'85, J. Mag. Magn. Matls. 54–57, 663–4 (1986).

249. Barma and Fisher Respond [to Comment "Universality among scalar spin systems" by G.A. Baker, Jr. and J.D. Johnson] (M.E.F. and M. Barma) Phys. Rev. Lett. 54, 2462 (1985).

250. Criticality and superfluidity in a dilute Bose fluid (P.B. Weichman, M. Rasolt, M.E.F. and M.J. Stephen) Phys. Rev. B 33, 4632–4663 (1986).

251. Wetting in random systems (R. Lipowsky and M.E.F.) Phys. Rev. Lett. 56, 472–475 (1986).

252. Spatial symmetries and critical phenomena, Proc. Fifth Phillip Morris Science Symposium, 1985, "Natural Products Research: The Impact of Scientific Advances," Ed. C.G. Lunsford (Phillip Morris, Inc., New York, 1987) pp. 17–60.

253. Finite-size effects in the spherical model of ferro-magnetism: Zero-field susceptibility under antiperiodic boundary conditions (S. Singh, R.K. Pathria and M.E.F.) Phys. Rev. B 33, 6415–6422 (1986).

254. The location of renormalization-group fixed points (M.E.F. and M. Randeria) Phys. Rev. Lett. 56, 2332 (1986).

255(a) Interface wandering in adsorbed and bulk phases, pure and impure, J. Chem. Soc., Faraday Trans. 2, 82, 1569–1603 (1986) (Faraday Symp. 20).
   (b) General discussions [Pure-to-random crossover] *loc. cit.* pp. 1818–19.

256. Classifying first-order phase transitions (M.E.F. and G.W. Milton) Physica 138A, 22–54 (1986).

257. Multisingularity and scaling in partial differential approximants. I. (M. Randeria and M.E.F.) Proc. Roy. Soc., A 419, 181–203 (1988).

258(a) Domain-wall interactions and spatially modulated phases (A.M. Szpilka and M.E.F.) Phys. Rev. Lett. 57, 1044–47 (1986).

(b)   Domain-wall interactions and high-order commensurate phases [abstract] (A.M. Szpilka and M.E.F.) Proc. STATPHYS-16, Boston University, August 1986.

259 (a)   Helium in Vycor, constrained randomness and the Harris criterion (P.B. Weichman and M.E.F.) Phys. Rev. B 34, 7652–7665 (1986).
   (b)   Dilute Bose fluids, helium in Vycor and constrained randomness [abstract] (P.B. Weichman and M.E.F.) Bull. Amer. Phys. Soc. 31 (4) 782 (1986).

260.   Long-range crossover and "nonuniversal" exponents in micellar solutions, Phys. Rev. Lett. 57, 1911–14 (1986).

261.   Unusual bifurcation of renormalization group fixed points for interfacial transitions (R. Lipowsky and M.E.F.) Phys. Rev. Lett 57, 2411–14 (1986).

262.   Domain wall interactions I. General features and phase diagrams for spatially modulated phases (M.E.F. and A.M. Szpilka) Phys. Rev. B 36, 644–666 (1987).

263.   Scaling regimes and functional renormalization for wetting transitions (R. Lipowsky and M.E.F.) Phys. Rev. B 36, 2126–2141 (1987).

264.   Domain wall interactions II. High-order phases in the axial next-nearest-neighbor Ising model (M.E.F. and A.M. Szpilka) Phys. Rev. B 36, 5343–5362 (1987).

265.   Domain wall interactions III. High-order phases in the three-state chiral clock model (A.M. Szpilka and M.E.F.) Phys. Rev. B 36, 5363–5376 (1987).

266 (a)   Random coupling crossover in Ising ferromagnets (R.R.P. Singh and M.E.F.) Phys. Rev. B 37, 1980–85 (1988).
   (b)   Bond randomness at Ising criticality (R.R.P. Singh and M.E.F.) Proc. MMM Conf. 1987, J. Appl. Phys. 63, 3082 (1988).

267.   Diffusion from an entrance to an exit, IBM J. Res. Develop. 32, 76–81 (1988).

268.   Disordered systems which escape the bound $\nu \geq 2/d$ (R.R.P. Singh and M.E.F.) Phys. Rev. Lett 60, 548 (1988).

269.   Short-range Ising spin-glasses in general dimensions (R.R.P. Singh and M.E.F.) Proc. MMM Conf. 1987, J. Appl. Phys. 63, 3994–96 (1988).

270 (a)   Thermodynamic behavior of two-dimensional vesicles (S. Leibler, R.R.P. Singh and M.E.F.) Phys. Rev. Lett. 59, 1989–92 (1987).
      REPRINTED (with revisions) in:
      (i)   J.A. Blackman and J. Tagüeña, Eds., "Disorder in Condensed Matter Physics" (Oxford Univ. Press, 1990) Chap. 30, pp. 389–396;
      (ii)   F. Ramos-Gómez, Ed., "Proc. Fifth Mexican School on Statistical Physics," Oaxtepec, 1989 (World Scientific Pub., Singapore, 1991) pp. 41–56.

(b) Membranes and vesicles: Their statistical mechanics in two dimensions, Proc. 3rd Univ. California Conf. Stat. Mech., Ed. C. Garrod (North Holland Publ. Co., 1988) Nucl. Phys. B (Proc. Suppl.) 5A, 165–167 (1988).

(c) The statistical mechanics of two-dimensional vesicles, J. Math. Chem. 4, 395–399 (1990).

271. Condensed matter physics: Does quantum mechanics matter? in "Niels Bohr: Physics and the World," Eds. H. Feshbach, T. Matsui and A. Oleson (Harwood Academic Publ., Chur, 1988) pp. 65–115.

272(a) Phases and phase transitions in less than three dimensions [summary], in "Physics in a Technological World," Proc. XIX General Assembly, I.U.P.A.P., Ed. A.P. French (Amer. Inst. Phys., New York, 1988), pp. 177–183.

(b) Phase transitions and fluctuations in less than three dimension, in "Frontiers of Physics," Proc. Landau Memorial Conf., Tel Aviv, June 1988, Eds. E. Gotsman, Y. Ne'eman and A. Voronel (Pergamon Press, Oxford, 1990) pp. 195–205.

273. Critical points, large-dimensionality expansions, and the Ising spin glass (M.E.F. and R.R.P. Singh) in "Disorder in Physical Systems," Eds. G. Grimmett and D.J.A. Welsh (Oxford Univ. Press, 1990) pp. 87–111.

274. Finite-size effects in surface tension: Thermodynamics and the Gaussian interface model (M.P. Gelfand and M.E.F.) Proc. 10th Symp. Thermophysical Props. 1988, Eds. A. Cezairliyan and J.V. Sengers, Int. J. Thermophys. 9, 713–727 (1988).

275. Double Summations (Solution to Problem 87–109 by D. Sommers) SIAM Review 30, 320–322 (1988).

276. The reunions of three dissimilar vicious walkers (M.E.F. and M.P. Gelfand) J. Stat. Phys. 53, 175–189 (1988); 55, 472 (1989)[E].

277. Wetting in a two-dimensional random-bond Ising model (M. Huang, M.E.F. and R. Lipowsky) Phys. Rev. B 39, 2632–2639 (1989).

278. The three-dimensional Ising model revisited numerically (A.J. Liu and M.E.F.) Physica A 156, 35–76 (1989).

279. Surface tension of the three-dimensional Ising model: A low-temperature series analysis (L.J. Shaw and M.E.F.) Phys. Rev. A 39, 2189–93 (1989).

280. The spin-½ antiferromagnetic XXZ chain: New results and insights (R.R.P. Singh, M.E.F. and R. Shankar) Phys. Rev. B 39, 2562–2567 (1989).

281. Interfaces: Fluctuations, interactions and related transitions, in "Statistical Mechanics of Membranes and Surfaces," Eds. D.R. Nelson, T. Piran, and S. Weinberg (World Scientific Publ., Singapore, 1989) pp. 19–44.
REPRINTED (with corrections and additions) in an Extended Second Edition (World Scientific, Singapore, 2004) pp. 19–47.

282(a) The universal critical adsorption profile from optical experiments (A.J. Liu and M.E.F.) Phys. Rev. A 40, 7202–7221 (1989).

(b) Theoretical analysis of critical adsorption experiments [abstract] (A.J. Liu and M.E.F.) Bull. Amer. Phys. Soc., 34 (3) 650 (1989).

283. The shapes of bowed interfaces in the two-dimensional Ising model (L.-F. Ko and M.E.F.) J. Stat. Phys. 58, 249–264 (1990).

284. Fractal and nonfractal shapes in two-dimensional vesicles, Physica D 38, 112–118 (1989); Proc. Conf. "Fractals in Physics," 1–4 October 1989, Vence, France, Eds. A. Aharony and J. Feder (North Holland Publ. Co., 1989).

285. On the corrections to scaling in three-dimensional Ising models (A.J. Liu and M.E.F.) J. Stat. Phys. 58, 431–442 (1990).

286(a) Phases and phase diagrams: Gibbs' legacy today, in "Proceedings of the Gibbs Symposium, Yale University, May 1989," Eds. D.G. Caldi and G.D. Mostow (Amer. Math. Soc., Rhode Island, 1990) pp. 39–72.

(b) Phases and phase diagrams: Gibbs' legacy today [summary] Physica A 163, 15–16 (1990).

287. Tunable fractal shapes in self-avoiding polygons and planar vesicles (C.J. Camacho and M.E.F.) Phys. Rev. Lett. 65, 9–12 (1990).

288. Finite-size effects in fluid interfaces (M.P. Gelfand and M.E.F.) Physica A 166, 1–74 (1990).

289. Size of an inflated vesicle in two dimensions (A.C. Maggs, S. Leibler, M.E.F. and C.J. Camacho) Phys. Rev. A 42, 691–695 (1990).

290. Low-dimensional quantum antiferromagnets: Criticality and series expansions at zero temperature [extended abstract] Proc. Conf. "Frontiers of Condensed Matter Physics," Physica A 168, 22 (1990).

291. Interfaces, membranes: rough, smooth and interacting [summary] Proc. Conf. "Nonlinear Science: The Next Decade," Physica D 51, 498–500 (1991).

292. Universality and interfaces at critical end points (M.E.F. and P.J. Upton) Phys. Rev. Lett. 65, 2402–2405 (1990).

293. Critical endpoints, interfaces, and walls, Proc. Amer. Chem. Soc. "Symposium on Surfaces and Interfaces," Physica A 172, 77–86 (1991).

294. Fluid interface tensions near critical end points (M.E.F. and P.J. Upton) Phys. Rev. Lett. 65, 3405–8 (1990).

295. Phase boundaries near critical end points I. Thermo-dynamics and universality (M.E.F. and M.C. Barbosa) Phys. Rev. B 43, 11177–184 (1991).

296. Phase boundaries near critical end points II. General spherical models (M.C. Barbosa and M.E.F.) Phys. Rev. B 43, 10635–646 (1991).

297. Semiflexible planar polymeric loops (C.J. Camacho, M.E.F. and R.R.P. Singh) J. Chem. Phys. 94, 5693–5700 (1991).

298. Two-dimensional lattice vesicles and polygons (M.E.F., A.J. Guttmann and S.G. Whittington) J. Phys. A: Math. Gen. 24, 3095–3106 (1991).

299. Effective potentials, constraints, and critical wetting theory (M.E.F. and A.J. Jin) Phys. Rev. B 44, 1430–33 (1991).

300. Simulations of planar vesicles and their transitions (C.J. Camacho and M.E.F.) in "Computer Simulation Studies in Condensed Matter Physics IV," Eds. D.P. Landau, K.K. Mon and H.B. Schuttler (Springer Verlag, Berlin, 1993) pp. 189–193.

301. Effective interface Hamiltonians for short-range critical wetting (A.J. Jin and M.E.F.) Phys. Rev. B 47, 7365–7388 (1993).

302. Microcanonical density functionals for critical systems: An exact one-dimensional example (L.V. Mikheev and M.E.F.) J. Stat. Phys. 66, 1225–1244 (1992).

303. The critical behavior of model electrolytes, Comment on: *J. Chem. Phys. 93, 8405 (1990)*, J. Chem. Phys. 96, 3352–54 (1992); corrected version has an abstract: pp. 3352–55.

304. The isothermal binodal curves near a critical endpoint (Y.C. Kim, M.E.F. and M.C. Barbosa) J. Chem. Phys. 115, 933–950 (2001).

305. Interfacial stiffness and the wetting parameter: The simple cubic Ising model (M.E.F. and H. Wen) Phys. Rev. Lett. 68, 3654 (1992).

306. Rods to self-avoiding walks to trees in two dimensions (C.J. Camacho, M.E.F. and J.P. Straley) Phys. Rev. A 46, 6300–6310 (1992).

307(a) Is short-range "critical" wetting a first-order transition? (M.E.F. and A.J. Jin) Phys. Rev. Lett. 69, 792–5 (1992).
   (b) An interfacial-stiffness instability in critical wetting [abstract] (A.J. Jin and M.E.F.) Bull. Amer. Phys. Soc. 37 (2) 952 (1992) G10 8.

308(a) Surface tension of helium at the superfluid critical end-point (L.V. Mikheev and M.E.F.) J. Low Temp. Phys. 90, 119–138 (1993).
   (b) The λ-point singularity of the liquid helium surface tension [abstract] (M.E.F. and L.V. Mikheev) Bull. Amer. Phys. Soc. 37 (2) 952 (1992) G10 7.

309. Exact variational analysis of layered planar Ising models (L.V. Mikheev and M.E.F.) Phys. Rev. Lett. 70, 186–189 (1993).

310. Stiffness instability in short-range critical wetting (A.J. Jin and M.E.F.) Phys. Rev. B 48, 2642–2658 (1993).

311. Two-dimensional layered Ising models: Exact variational formulation and analysis (L.V. Mikheev and M.E.F.) Phys. Rev. B 49, 378–402 (1994).

312.   Criticality in ionic fluids: Debye-Hückel, Bjerrum and beyond (M.E.F. and Y. Levin) Phys. Rev. Lett. 71, 3826–3829 (1993).

313.   The story of Coulombic criticality, J. Stat. Phys. 75, 1–36 (1994).

314.   On the stiffness of an interface near a wall (M.E.F., A.J. Jin and A.O. Parry) Ber. Bunsenges. Phys. Chem. 98, 357–361 (1994).

315.   Cavity forces and criticality in electrolytes (X.-J. Li, Y. Levin and M.E.F.) Europhys. Lett. 26, 683–688 (1994).

316.   The interaction of ions in an ionic medium (M.E.F., Y. Levin and X.-J. Li) J. Chem. Phys. 101, 2273–82 (1994).

317.   Coulombic criticality in general dimensions (Y. Levin, X.-J. Li and M.E.F.) Phys. Rev. Lett. 73, 2716–19 (1994); 75 [E] 3374 (1995).

318.   On the absence of intermediate phases in the two-dimensional Coulomb gas (M.E.F., X.-J. Li and Y. Levin) J. Stat. Phys. 79, 1–11; 81 [E] 865 (1995).

319.   Criticality in the hard-sphere ionic fluid (Y. Levin and M.E.F.) Physica A 225, 164–220 (1996).

320.   On the critical polynomial of the simple cubic Ising model, J. Phys. A: Math. Gen. 28, 6323–33 (1995).

321.   The universal repulsive-core singularity and Yang-Lee edge criticality (S.-N. Lai and M.E.F.) J. Chem. Phys. 103, 8144–55 (1995).

322.   Universal surface-tension and critical isotherm amplitude ratios in three dimensions (S.-Y. Zinn and M.E.F.) Physica A 226, 168–180 (1996).

323.   Foreword: About the author and the subject, to "The Critical Point" by C. Domb (Taylor and Francis, London, 1996), pp. xiii–xviii.

324(a)   Density fluctuations in an electrolyte from generalized Debye-Hückel theory (B.P. Lee and M.E.F.) Phys. Rev. Lett. 76, 2906–10 (1996).
   (b)   Inhomogeneous Debye-Hückel theory and density fluctuations near Coulombic criticality [abstract] (B.P. Lee and M.E.F.) Bull. Amer. Phys. Soc. 41 (1) 377 1996) I20 1.

325(a)   Ginzburg criterion for Coulombic criticality (M.E.F. and B.P. Lee) Phys. Rev. Lett. 77, 3561–64 (1996).

   (b)   The Ginzburg criterion for ionic criticality according to Debye-Hückel based theories [abstract] (M.E.F. and B.P. Lee) Bull. Amer. Phys. Soc. 41 (1) 377 (1996) I20 2.

326(a)   Dipolar-ion pairs and Debye-Hückel theory: The charging process and reciprocity [abstract] (B.P. Lee and M.E.F.) Bull. Amer. Phys. Soc. 41 (1) 356 (1996) I7 11.
   (b)   Generalized Debye-Hückel theory and ionic criticality (B.P. Lee and M.E.F.) [in preparation].

327.   Criticality in Gaussian-molecule mixtures (S.-N. Lai and M.E.F.) Molec. Phys. 88, 1373–1397 (1996).

328.   The renormalized coupling constants and related amplitude ratios for Ising systems (S.-Y. Zinn, S.-N. Lai, and M.E.F.) Phys. Rev. E 54, 1176–1182 (1996).

329(a) Coulombic criticality in general dimensions: Progress and challenges [extended abstract] Proc. Hayashibara Forum '95: "Coherent Approaches to Fluctuations," Eds. M. Suzuki and N. Kawashima (World Scientific Publ. Co., Singapore, 1996) pp. 2–5.

   (b) The nature of criticality in ionic fluids, Proc. Third Liquid Matter Conference, J. Phys.: Condens. Matter 8, 9103–9109 (1996).

   (c) Criticality in hard-core ionic systems: Progress and challenges [extended abstract] Proc. Inauguration Conf. Asia Pacific Center for Theoretical Physics, "Current Topics in Physics," Eds. Y.M. Cho, J.B. Hong and C.N. Yang (World Scientific Publ. Co., Singapore, 1998) pp. 100–105.

330.   The superfluid transition in a dilute Bose gas: Experiments and theory [abstract] Proc. Inauguration Conf. Asia Pacific Center for Theoretical Physics, "Current Topics in Physics," Eds. Y.M. Cho, J.B. Hong and C.N. Yang (World Scientific Publ. Co., Singapore, 1998) p. 175.

331.   Right and wrong near critical endpoints (M.E.F. and Y.C. Kim) J. Chem. Phys. 117, 779–787 (2002).

332(a) Prewetting transitions in a near-critical metallic vapor (V.F. Kozhevnikov, D.I. Arnold, S.P. Naurzakov and M.E.F.) Phys. Rev. Lett. 78, 1735–1738 (1997).

   (b) Prewetting phenomena in mercury vapor (V.F. Kozhevnikov, D.I. Arnold, S.P. Naurzakov, and M.E.F.) Fluid Phase Equilibria, 150–151, 625–632 (1998).

333.   Renormalization group theory: Its basis and formulation in statistical physics, Rev. Mod. Phys. 70, 653–681 (1998).
        REPRINTED (with revisions) in "Conceptual Foundations of Quantum Field Theory," Ed., T.Y. Cao (Cambridge Univ. Press, 1998) Part IV, Chap. 8, pp. 89–135.

334(a) Critique of primitive model electrolyte theories (D.M. Zuckerman, M.E.F. and B.P. Lee) Phys. Rev. E 56, 6569–6580 (1997).

   (b) Critique of electrolyte theories using thermodynamic bounds (M.E.F., D.M. Zuckerman and B.P. Lee) in "Strongly Coupled Coulomb Systems," Proc. Conf. held in Boston College, August 1997, Eds. G.J. Kalman, J.M. Rommel and K.B. Blagoev (Plenum Press, New York 1998) pp. 415–418.

335(a) Charge oscillations in Debye-Hückel theory (B.P. Lee and M.E.F.) Europhys. Lett. 39, 611–616 (1997).

   (b) Charge oscillations in electrolytes from generalized Debye-Hückel theory [abstract] (B.P. Lee and M.E.F.) Bull. Amer. Phys. Soc. 42 (1) 725 (1997) Q16 9.

336(a) Diverging correlation lengths in electrolytes: Exact results at low densities (S. Bekiranov and M.E.F) Phys. Rev. E 59, 492–511 (1999).

(b) Exact density correlation length for electrolytes at low densities [abstract] (S. Bekiranov and M.E.F.) Bull. Amer. Phys. Soc. 42 (1) 725 (1997) Q16 8.

337. Electrolyte criticality and generalized Debye-Hückel theory (M.E.F., B.P. Lee and S. Bekiranov) in "Strongly Coupled Coulomb Systems," Proc. Conf. held in Boston College, August 1997, Eds. G.J. Kalman, J.M. Rommel and K. B. Blagoev (Plenum Publ. Corp., 1998) pp. 33–41.

338. Exact thermodynamic formulation of chemical association (M.E.F. and D.M. Zuckerman) J. Chem. Phys. 109, 7961–81 (1998).

339. Chemical association via exact thermodynamic formulations (M.E.F. and D.M. Zuckerman) Chem. Phys. Lett. 293, 461–468 (1998).

340(a) Fluctuations in electrolytes: The Lebowitz and other correlation lengths (S. Bekiranov and M.E.F.) Phys. Rev. Lett. 81, 5836–39 (1998).
(b) Charge and density fluctuations in electrolytes: The Lebowitz and other correlation lengths (M.E.F. and S. Bekiranov) Proc. STATPHYS 20, Physica A 263, 466–476 (1999).
(c) Correlation lengths in electrolytes: Exact results and sensible approximations [abstract] Proc. 1999 Int. Conf. on "Strongly Coupled Coulomb Systems," Saint-Malo, France, 4–10 September 1999, Eds. C. Deutsch, B. Jancovici, and M.-M. Gombert (EDP Sciences, Les Ulis, France, 2000) J. Physique IV 10, Pr5–17 (2000).

341. The shape of the van der Waals loop and universal critical amplitude ratios (M.E.F. and S.-Y. Zinn) J. Phys. A: Math. Gen. 31, L629–L635 (1998).

342. Understanding criticality: Simple fluids and ionic fluids [summary] Proc. NATO Advanced Study Institute, "New Approaches to New and Old Problems in Liquid State Theory," Messina, Sicily, Eds. C. Caccamo et al. (Kluwer Academic Publ., Dordrecht, 1999) pp. 3–8.

343. Some views from forty years as a statistical mechanician [retrospective outline] Proc. STATPHYS 20, Physica A 263, 554–556 (1999).

344. Trigonometric models for scaling behavior near criticality (M.E.F., S.-Y. Zinn and P.J. Upton) Phys. Rev. B 59, 14533–545 (1999); Erratum: *ibid* 64, 149901(E) (2001).

345. Crossover scales at the critical points of fluids with electrostatic interactions (A.G. Moreira, M.M. Telo da Gama and M.E.F.) J. Chem. Phys. 110, 10058–66 (1999).

346. Scaling for interfacial tensions near critical endpoints (S.-Y. Zinn and M.E.F.) [arXiv: cond-mat/0410673] Phys. Rev. E 71, 011601:1–17 (2005).

347. The force exerted by a molecular motor (M.E.F. and A.B. Kolomeisky) Proc. Natl. Acad. Sci. USA, 96, 6597–6602 (1999).

348. Understanding criticality: Magnetism as the key [abstract] invited talk in "The History of Magnetism" session, American Physical Society Centennial Meeting, Bull. Amer. Phys. Soc. 44(1) Part II, 1400 (1999) UA01 4.

349. Discussion remarks on quantum field theory, invited contributions in "Conceptual Foundations of Quantum Field Theory," Ed., T.Y. Cao (Cambridge University Press, 1999) pp. 72–73, 86–88, 135, 165, 231, 250, 266–267, 269–272, 274–275, 278–286, 377, 383.

350. Identity of the universal repulsive-core singularity with Yang-Lee edge criticality (Y. Park and M.E.F.) Phys. Rev. E 60, 6323–28 (1999).

351. Molecular motors and the forces they exert (M.E.F. and A.B. Kolomeisky) Physica A 274, 241–266 (1999) and in "Applications of Statistical Physics," Proc. NATO Adv. Res. Workshop, Budapest, May 1999, Eds. A. Gadomski, J. Kertész, H.E. Stanley and N. Vandewalle (Elsevier, Amsterdam, 1999) pp. 241–266.

352. Criticality and crossover in accessible regimes (G. Orkoulas, A.Z. Panagiotopoulos, and M.E.F.) Phys. Rev. E 61 5930–5939 (2000).

353. Periodic sequential kinetic models with jumping, branching, and deaths (A.B. Kolomeisky and M.E.F.) Physica A 279, 1–20 (2000); Erratum: *ibid* 284, 496 (2000).

354. High resolution study of fluid criticality (G. Orkoulas, M.E.F. and A.Z. Panagiotopoulos) in "Computer Simulation Studies in Condensed Matter Physics XIII," Eds. D.P. Landau, S.P. Lewis and H.B. Schüttler (Springer Verlag, Heidelberg, Berlin, 2000) pp. 167–171.

355(a) Simple mechanochemistry describes the dynamics of single kinesin molecules (M.E.F. and A.B. Kolomeisky) Proc. Natl. Acad. Sci. USA 98, 7748–53 (2001).

  (b) Describing the dynamics of single kinesin molecules [abstract] (M.E.F. and A.B. Kolomeisky) Biophys. J. 80, 513a (2001) 2202-Plat.

  (c) Motor proteins: Observations and theory (M.E.F. and A.B. Kolomeisky) [abstract for an invited talk] Bull. Amer. Phys. Soc. 46 (1) 1104 (2001) X21 1.

356. The Yang-Yang anomaly in fluid criticality: Experiment and scaling theory (M.E.F. and G. Orkoulas) Phys. Rev. Lett. 85, 696–699 (2000).

357(a) The Yang-Yang anomaly in fluid criticality: An exactly soluble model [abstract] (C.A. Cerdeiriña, G. Orkoulas and M.E.F.) XV Congress of Statistical Physics, Royal Spanish Physical Society, Salamanca (27–29 March 2008).

  (b) Exactly soluble fluid models exhibiting Yang-Yang anomalies at criticality [abstract and poster] (C.A. Cerdeiriña, G. Orkoulas and M.E.F.) Proc. 7th Liquid Matter Conf. (Lund, Sweden, June 2008).

  (c) [in preparation].

358. The Yang-Yang relation and the specific heats of propane and carbon dioxide (G. Orkoulas, M.E.F. and C. Üstün) J. Chem Phys. 113, 7530–45 (2000).

359(a) Force-velocity relation for growing microtubules (A.B. Kolomeisky and M.E.F.) Biophys. J. 80, 149–154 (2001).

  (b) The growth of microtubules against an external force [abstract] (A.B. Kolomeisky and M.E.F.) Biophys. J. 80, 514a (2001) 2203-Plat.

360.   Extended kinetic models with waiting-time distributions: Exact results (A.B. Kolomeisky and M.E.F.) J. Chem. Phys. 113, 10867–877 (2000).

361.   Coexistence and criticality in size-asymmetric hard-core electrolytes (J.M. Romero-Enrique, G. Orkoulas, A.Z. Panagiotopoulos and M.E.F.) Phys. Rev. Lett. 85, 455861 (2000).

362.   Asymmetric primitive-model electrolytes: Debye-Hückel theory, criticality and energy bounds (D.M. Zuckerman, M.E.F. and S. Bekiranov) Phys. Rev. E. 64, 011206:1–13 (2001).

363(a) The heat capacity of the restricted primitive model (E. Luijten, M.E.F. and A.Z. Panagiotopoulos) J. Chem. Phys. 114, 5468–71 (2001).
   (b) Criticality and charge fluctuations in the restricted primitive model [abstract] (E. Luijten, M.E.F. and A.Z. Panagiotopoulos) Bull. Amer. Phys. Soc. 46 (1) 71 (2001) A11 4.

364.   Precise simulation of criticality in asymmetric fluids (G. Orkoulas, M.E.F. and A.Z. Panagiotopoulos) Phys. Rev. E 63, 051507:1–14 (2001).

365.   The critical locus of a simple fluid with added salt (Y.C. Kim and M.E.F.) J. Phys. Chem. B 105, 11785–95 (2001).

366.   Phase transitions in 2:1 and 3:1 hard-core model electrolytes (A.Z. Panagiotopoulos and M.E.F.) Phys. Rev. Lett. 88, 045701:1–4 (2002).

367(a) Universality class of criticality in the restricted primitive model electrolyte (E. Luijten, M.E.F. and A. Z. Panagiotopoulos) Phys. Rev. Lett. 88, 185701:1–4 (2002).
   (b) How to simulate fluid criticality: The simplest ionic model has Ising behavior but the proof is not so obvious! Proc. Internat. Conf. Theoretical Physics: TH-2002, UNESCO, Paris, 22–27 July 2002, Ann. Henri Poincaré 4, 413–416 (2002).
   (c) Charge fluctuations and criticality in the restricted primitive model electrolyte [abstract] (E. Luijten, M.E.F. and A.Z. Panagiotopoulos) in "STATPHYS 21 Conference Abstracts," Eds, D. López, M. Barbosa and A. Robledo (IUPAP, Cancun, Mexico, 2001) p. 66.

368.   Lattice models of ionic systems (V. Kobolev, A.B. Kolomeisky and M.E.F.) J. Chem. Phys. 116, 7589–7598 (2002).

369.   Criticality in charge asymmetric ionic fluids (J.-N. Aqua, S. Banerjee and M.E.F.) [arXiv:cond-mat/0410692] Phys. Rev. E 72, 041501:1–25(2005).

370.   Motion of kinesin on a microtubule [abstract] (M.E.F. and A.B. Kolomeisky) Biophys. J. 82, 62a (2002) 305-Pos.

371(a) A simple kinetic model describes the processivity myosin-V (A.B. Kolomeisky and M.E.F.) Biophys. J. 84, 1642–1650 (2003).
   (b) A simple stochastic model can explain the motility of myosin V molecules [abstract] (A.B. Kolomeisky and M.E.F.) Biophys. J. 82, 15a (2002) 71-Plat.

372.   Screening in ionic systems: Simulations for the Lebowitz length (Y.C. Kim, E. Luijten and M.E.F.) Phys. Rev. Lett. 95, 145701:1–4 (2005).

373. Asymmetric fluid criticality I. Scaling with pressure mixing (Y.C. Kim, M.E.F. and G. Orkoulas) Phys. Rev. E 67, 061506:1–21 (2003).

374. Asymmetric fluid criticality II. Finite-size scaling for simulations (Y.C. Kim and M.E.F.) [arXiv:condmat/0306331] Phys. Rev. E 68, 041506:1–23 (2003).

375. First-order transitions from singly peaked distributions, Special issue: Physica A 389, 2873–79 (2010).

376(a) Precise simulation of near-critical fluid coexistence (Y.C. Kim, M.E.F. and E. Luijten) [arXiv:cond-mat/0304032] Phys. Rev. Lett. 91, 065701:1–4 (2003).
    (b) Criticality, coexistence and screening in electrolytes: High-resolution simulations [abstract] (M.E.F. and Y.C. Kim) program book for "15th Symposium on Thermophysical Properties," NIST, Boulder, Colorado, June 2003, p. 133.

377. Discretization dependence of criticality in model fluids: a hard-core electrolyte (Y.C. Kim and M.E.F.) [arXiv:cond-mat/0402275] Phys. Rev. Lett. 92, 185703:1–4 (2004).

378. Fluid critical points from simulations: the Bruce-Wilding method and Yang-Yang anomalies (Y.C. Kim and M.E.F.) [arXiv:cond-mat/0310247] J. Phys. Chem. B 108, 6750–6759 (2004).

379. Ionic criticality: an exactly soluble model (J.-N. Aqua and M.E.F.) [arXiv:cond-mat/0311491] Phys. Rev. Lett. 92, 135702:1–4 (2004).

380(a) Reflections on the beguiling but wayward spherical model [Spanish title: "Reflexiones sobre el Modelo Esférico, Hechicero pero Caprichoso"] in "Current Topics in Physics," Proc. Symposium at the Universidad Nacional Autónoma de Mexico, 17 June 2003, in honor of Sir Roger J. Elliott, Eds. R.A. Barrio and K.K. Kaski (Imperial College Press, London, 2005) Chap. 1, pp. 3–31.
    (b) The beguiling but wayward spherical model: Criticality, fluctuations and screening (M.E.F. and J.-N. Aqua) Rev. Mod. Phys. [in preparation].

381(a) Vectorial loading of processive motor proteins: Implementing a landscape picture (Y.C. Kim and M.E.F.) [arXiv:cond-mat/0506185] J. Phys.: Condens. Matt. 17, S3821S3838 (2005).
    (b) Vectorial loading of processive motor proteins: Understanding kinesin [abstract] (M.E.F. and Y.C. Kim) Biophys. J. 86, 527a–528a (2004) 2738-Plat.

382. Charge and density fluctuations lock horns: ionic criticality with power-law forces (J.-N. Aqua and M.E.F.) J. Phys. A: Math. Gen. 37, L241–L248 (2004).

383. Convergence of fine-lattice discretization for near-critical fluids (S. Moghaddam, Y.C. Kim and M.E.F.) [arXiv:cond-mat/0502169] J. Phys. Chem. B 109, 6824–6837(2005).

384. Fluid coexistence close to criticality: Scaling algorithms for precise simulation (Y.C. Kim and M.E.F.) [arXiv: cond-mat/0411736] Comp. Phys. Commun. 169, 295–300 (2005).

384^ Yang-Yang anomalies and coexistence curve diameters: Simulation of asymmetric fluids (Y.C. Kim) [arXiv:cond-mat/0503480] Phys. Rev. E 71, 051501:1–16 (2005).

Wolf Prize in Physics

385. Interfaces: Fluctuations, interactions and related transitions, Chap. 2 in "Statistical Mechanics of Membranes and Surfaces" (Second Edition, enlarged and revised) Eds. D. Nelson, T. Piran and S. Weinberg (World Scientific Publ., Singapore, 2004) pp. 19–47: see [281].

386. How multivalency controls ionic criticality (M.E.F., J.N. Aqua and S. Banerjee) [arXiv:cond-mat/0507077] Phys. Rev. Lett. 95, 135701:1–4 (2005).

387. Interfacial tensions near critical endpoints: Experimental checks of EdGF theory (S.-Y. Zinn and M.E.F.) [arXiv:cond-mat/05043857] Molec. Phys. 103, 2927–2942 (2005): special issue in honor of B. Widom.

388. Singular coexistence-curve diameters: Experiments and simulations (Y.C. Kim and M.E.F.) [arXiv:cond-mat/0507369] Chem. Phys. Lett. 414, 185–192 (2005).

389(a) Kinesin crouches to sprint but resists pushing (M.E.F. and Y.C. Kim) Proc. Natl. Acad. Sci. USA 102, 16209–214 (2005).
 (b) Kinesin binds ATP, crouches and swings 8 nm; But rebuffs help [abstract] (Y.C. Kim and M.E.F.) Biophys. J. 88, 649a (2005) 3189-Pos.
 (c) Kinesin crouches before sprinting and resists forward and leftward loading [abstract] (Y.C. Kim and M.E.F.) Proc. 228th ACS National Meeting, "Biophysical Chemistry and Novel Imaging of Single Molecules and Single Cells," Division of Physical Chemistry, Philadelphia, PA (25 August 2004.)

390. Universality of ionic criticality: Size- and charge-asymmetric electrolytes (Y.C. Kim, M.E.F. and A.Z. Panagiotopoulos) Phys. Rev. Lett. 95, 195703:1–4 (2005).

391(a) Static and dynamic critical behavior of a symmetrical binary fluid: A computer simulation (S.K. Das, J. Horbach, K. Binder, M.E.F. and J.V. Sengers) [arXiv: cond-mat/0603587 22 Mar 2006] J. Chem. Phys. 125, 024506:1–11 (2006).
 (b) Interdiffusion in Critical Binary Mixtures by Molecular Dynamics Simulation (K. Binder, S.K. Das, M.E.F., J. Horbach, and J.V. Sengers) in "Diffusion Fundamentals II; L'Aquila 2007," Eds. S. Brandani, C. Chmelik, J. Kärger and R. Volpe (Leipziger Universitätsverlag, Leipzig, 2007) pp. 120–131.

392. Critical dynamics in a binary fluid: Simulations and finite-size scaling (S.K. Das, M.E.F., J.V. Sengers, J. Horbach and K. Binder) Phys. Rev. Lett. 97, 025702:1–4 (2006).

393. Molecular Motors: A theorist's perspective (A.B. Kolomeisky and M.E.F.) Ann. Rev. Phys. Chem. 58, 675–695 (2007).

394(a) Backstepping, hidden substeps, and conditional dwell times in molecular motors (D. Tsygankov, M. Lindén and M.E.F.) [arXiv:q-bio BM/0611051] Phys. Rev. E 75, 021909:1–16 (2007).
 (b) Back-stepping, dwell times, and hidden substeps in molecular motors [abstract] (D. Tsygankov, M. Lindén and M.E.F.) Biophys. J. 90 (Jan. 2007) 496a: 2369-Pos.

395. Charge fluctuations and correlation lengths in finite electrolytes (Y.C. Kim and M.E.F.) Phys. Rev. E 77, 051502:1–7 (2008).

396(a) Mechanoenzymes under superstall and large assisting loads reveal structural features (D. Tsygankov and M.E.F.) Proc. Natl. Acad. Sci. USA 104, 19321–326 (2007).

(b) Superstall and assisting-load velocities of motor proteins can reveal mechano-chemical structure [abstract] (D. Tsygankov and M.E.F.) Biophys. J. 90 (Jan. 2007) 497a: 2371-Pos.

397. Simulating critical dynamics in liquid mixtures: Short-range and long-range contributions (S.K. Das, J.V. Sengers and M.E.F) J. Chem. Phys. 127, 144506:1–5 (2007).

398. Kinetic models for mechanoenzymes: Structural aspects under large loads (D. Tsygankov and M.E.F) J. Chem. Phys. 128, 015102:1–12 (2008).

399. Boltzmann Award Session; Remarks on the Boltzmann Medalists (*Kurt Binder* and *Giovanni Gallavotti*) Proc. STATPHYS 23, Euro. Phys. J. B 64, 301, 303–5 (2008).

400(a) Velocity statistics distinguish quantum turbulence from classical turbulence (M.S. Paoletti, M.E.F., K.R. Sreenivasan and D.P. Lathrop) Phys. Rev. Lett. 101, 154501:1–4 (2008); designated an Editor's Selection.

(b) Quantum turbulence [abstract] (D.P. Lathrop, M.S. Paoletti, M.E.F., K.R. Sreenivasan) Bull. Amer. Phys. Soc. 54 (19) ET 1, 132 (2009).

(c) Velocity statistics in superfluid and classical turbulence [abstract] (K.R. Sreenivasan, D.A. Donzis, M.E.F., D.P. Lathrop, M.S. Paoletti, and P.K. Young) Bull. Amer. Phys. Soc. 54 (19) ET 2, 132 (2009).

401(a) Comment on a recent conjectured solution of the three-dimensional Ising Model (F.Y. Wu, B.M. McCoy, M.E.F. and L. Chayes) Phil. Mag. 88, 3093–95 (2008).

(b) Rejoinder to the Response to 'Comment on a recent conjectured solution of the three-dimensional Ising Model' (F.Y. Wu, B.M. McCoy, M.E.F. and L. Chayes) Phil. Mag. 88, 3103 (2008).

(c) Erratum for 'Comment' and 'Rejoinder,' (F.Y.Wu *et al.*) Phil. Mag. 89, 195 (2009).

402. Criticality in multicomponent spherical models: Results and cautions (J.-N. Aqua and M.E.F) Phys. Rev. E 79, 011118:1–13(2009).

403. Critical charge and density coupling in ionic spherical models (J.-N. Aqua and M.E.F.) [in preparation].

404. Reconnection dynamics for quantized vortices (M.S. Paoletti, M.E.F. and D.P. Lathrop) Physica D 239, 1367–77 (2010).

404^ 'Search & Discovery': "*Filming vortex lines reconnecting in a turbulent superfluid*," B. Schwarzschild, Physics Today (July 2010) pp. 12–14.

405(a) Atoms and Ions; Universality, singularity and particularity: On Boltzmann's vision a century later [abstract] p. 351,in "*Statistical Physics, High Energy, Condensed Matter and Mathematical Physics*," Proc. Conf. in Honor of C.N. Yang's 85th Birthday, Eds. M.L. Ge, C.H. Oh and K.K. Phua (World Scientific, Singapore (2008): PHOTO p. *xxiv*.

(b) Discussion Remarks following lecture by Kerson Huang and a question by C.N. Yang, *ibid*, pp. 350–351.

406.   Dynamics of the tug-of-war model for cellular transport (Y.-X. Zhang and M.E.F.) Phys. Rev. E 82, 011923:1–14 (2010).

407.   For Professor Thomas Erber in recognition of his 80th Birthday, in *"Doing Physics: Festschrift for Tom Erber,"* Ed. P.W. Johnson (Illinois Inst. Tech. Press, Chicago, IL, 2010) pp. 169–172. [Also pattern on the book's cover.]

408.   Superfluid transitions: Proximity eases confinement, News & Views, Nature Phys. 6, 483–4 (2010). Comment on:

408^   *"Coupling and proximity effects in the superfluid transition in ^{4}He dots,"* J.K. Perron, *et al.* Nature Phys. 6, 499–502 (2010).

409 (a) Measuring the limping of processive motor proteins (Y. Zhang and M.E.F.) J. Stat. Phys. 142, 1218–1251 (2011).

    (b) Limping factors for motor proteins [abstract] (Y. Zhang and M.E.F.) Biophys. J. 100, 120a, 652-Pos (2011).

410 (a) Initial conditions for reconnection calculations of quantized vortices [abstract] (C. Rorai, D.P. Lathrop, M.E.F. and K.R. Sreenivasan) Bull. Amer. Phys. Soc. 55 (16) 327 (2010) MN 5.

    (b) Numerical investigations of reconnection of quantized vortices (C. Rorai, M.E.F., D.P. Lathrop, K.R. Sreenivasan and R.M. Kerr) Bull. Amer. Phys. Soc. 56 (18)DFD (2011) D.21.2.

411 (a) Entrevista con Michael E. Fisher (interview conducted in English by J.M. Ortiz de Zárate and translated by him into Spanish and edited by M.E.F. in both languages) Revista Iberoamericana de Física 6(1), 60–63 (2010).

    (b) Interview with Michael E. Fisher [conducted by José M. Ortiz de Zárate] Europhys. News 42(1), 14–16 (2010).

412.   When is a conductor not perfect? Sum rules fail under critical fluctuations (S.K. Das, Y.C. Kim and M.E.F.) Phys. Rev. Lett. 107, 215701:1–4(2011).

413.   Cyril Domb: A personal view and appreciation [invited preface for special issue of the journal] J. Stat. Phys. 145, 510–517 (2011); 146, 883 (2012).

414.   Quantized vortex reconnection: Fixed points and initial conditions (D.P. Meichle, C. Rorai, M.E.F. and D.P. Lathrop) Phys. Rev. B 86, 014509:1–4 (2012).

415.   Near critical electrolytes: Are the charge-charge sum rules obeyed? (S.K. Das, Y.C. Kim and M.E.F) J. Chem. Phys. 137, 074902:1–12 (2012).

416.   Propagating and annihilating vortex dipoles in the Gross-Pitaevskii equation (C.R. Rorai, K.R. Sreenivasan and M.E.F.) Phys. Rev. B 88, 134522:1–10 (2013).

417.   Criticality in alternating layered Ising models. I. Effects of connectivity and proximity (H. Au-Yang and M.E.F.) Phys. Rev. E 88, 032147:1–12 (2013).

417^   Criticality in alternating layered Ising models. II. Exact scaling theory (H. Au-Yang) Phys. Rev. E 88, 032148:1–8 (2013).

418 (a) Renormalization group theory, the epsilon expansion, and Ken Wilson as I knew him, in *K. G. Wilson Memorial Volume: Renormalization, Lattice Gauge Theory, the Operator Product Expansion and Quantum Fields*, Eds. B.E, Baaquie, K. Huang, M.E. Peskin and K.K. Phua (World Scientific Publishers, Singapore, 2015). Chap. 3, pp. 43–82.

    (b) Renormalization group theory, the epsilon expansion, and Ken Wilson as I knew him, Int. J. Mod. Phys. B 29, 1530006: 1–39 (2015).

419.   Comment concerning the Ising model and two letters by N.H. March (M.E.F. and J.H.H. Perk) Phys. Lett. A 380, 1339–40 (2016).

420.   Soluble model fluids with complete scaling and Yang-Yang features (C.A. Cerdeiriña, G. Orkoulas, and M.E.F) Phys. Rev. Lett . 116, 040601: 1–5 (2016).

421 (a) Statistical Physics in the Oeuvre of Chen Ning Yang, Proc. Conf. 60 Years of Yang-Mills Gauge Field Theories: *CN Yang's Contributions to Physics*, Eds. L. Brink and K.K. Phua (World Scientific Publ., Singapore, 2015) Chap. 10, pp. 167–198.

    (b) Statistical Physics in the Oeuvre of Chen Ning Yang [Invited review] Int. J. Modern Phys. B 29, 1530013: 1–31 (2015).

C. Book Reviews, etc.
(omitting contributions to Math. Reviews, 1966–67)

1.   "Journal of Electronics," Vol. 1, No. 1, in Endeavour, February 1956.

2.   L. Eisenbud and E.P. Wigner, "Nuclear Structure," in Atomics, January 1959.

3.   L.D. Landau and E.M. Lifschitz, "Quantum Mechanics," in Science Progress, 1959.

4.   E. Bodewig, "Matrix Calculus," 2nd Edition, in Proc. Phys. Soc. 74, 1959.

5.   P. and T. Ehrenfest, "The Conceptual Foundations of the Statistical Approach in Mechanics," in Bull. Inst. Phys., 1960.

6.   J.L. Powell and B. Craseman, "Quantum Mechanics," in Science Progress, 1961.

7.   G.G. Hall, "Matrices and Tensors," in Proc. Phys. Soc., November 1963.

8.   H.S.W. Massey and H. Kestelman, "Ancillary Mathematics," 2nd Edition, in Proc. Phys. Soc. 85, 1304 (1965).

9.   J. de Boer and G.E. Uhlenbeck, Eds., "Studies in Statistical Mechanics. III," in Phil. Mag. 10 (1965).

10.   D. Mattis, "The Theory of Magnetism," in Proc. Phys. Soc. 85 (1965).

11.   "A consideration on the solution of partial differential equations by analog computer," T. Miura and J. Iwata, Simulation 6 (2) 105–108 (February 1966) in IEEE Trans. Electronic Computers (1966).

12. "Simulation of Field Problems using General Purpose Resistance Network Analogues," Jacques J. Vidal, Simulation 7, 190–203, October 1966, in IEEE Trans. on Computers c-17, 98 (1968).

13. E.G.D. Cohen, Ed., "Fundamental Problems in Statistical Mechanics. II," in Science, 21 February 1969.

14. B.W. Roos, "Analytic Functions and Distributions in Physics and Engineering," in Physics Today 23, December(12) 52 (1970).

15. C. Domb and M.S. Green, Eds., "Phase Transitions and Critical Phenomena," Vols. 1 and 2, in Science 182, 576 (9 November 1973).

16. H. Haken and M. Wagner, Eds., "Cooperative Phenomena," in American Scientist (1974).

17. Remarks on M.S. Green in "Perspectives in Statistical Physics," Vol. 9 in Studies in Statistical Mechanics, Ed. H.J. Raveche (North Holland, Amsterdam, 1980).

18. "What's Mathematical Physics to Physics? Some examples from past, present, and future," Extract in Appendix to A. Bohm, Physica 124A, 103–114 (1984).

19. J. Mennier, D. Langevin and N. Boccara, Eds., "Physics of Amphiphilic Layers," in Applied Optics 28 (12) 2180 (15 June 1989).

20. Yu A. Izyumov and Yu N. Skryabin, "Statistical Mechanics of Magnetically Ordered Systems" (Consultants Bureau, N.Y., 1988) in Physics Today (January 1990) pp. 75–76.

21. "Mathematical Rigor in Physics: To What End?" Report on a lecture by M.E.F. in Summary from the Symposium "*Modern Perspectives of Mathematics: Mathematics as a Consumer Good, Mathematics in Academia,*" Math. Sciences Institute, Cornell University, 29–31 March (1990) (M.S.I., Cornell Univ., Ithaca, N.Y., 1990) pp. 24–25.

22. J. Willard Gibbs and His Legacy: A Double Centennial, APS March Meeting, Austin, 3 March 2003 (Report by S.G. Brush and M.E.F.) Newsletter of the Forum on the History of Physics of The American Physical Society 9(1) 6–7 (Fall 2003).

D. Miscellaneous Reports, Posters, etc.

1. The dimer problem and related lattice Combinatorics. Notes prepared (with aid) for Summer School on Statistics, Univ. North Carolina, June 1969.

2. MIT Summer School Notes: Renormalization Group and Critical Phenomena (June 1972).

3. MIT Notes as above [revised] (June 1974).

4. Concerning research on tricritical and multicritical points at Cornell University [for MSC and NSF] (June 1974).

5. Theory of Critical Correlations in the Critical Region (with D. Jasnow) Chaps. 1–7: intended as Vol. IV in "Phase Transitions and Critical Phenomena," Eds. C. Domb and M.S. Green (Academic Press, Inc.).

6. Spatially Modulated Phases in Ising Models with Competing Interactions (with Walter Selke) for Workshop on "The Contribution of Neutron Scattering to the Study of Incommensurable Modulations," Inst. Laue-Langevin, Grenoble, 13–14 October 1980.

7. Proof that Scaling cannot be derived from Thermodynamics (intended as an Appendix in H.E. Stanley's "Critical Phenomena," 2nd Edition) February 1981.

8. Some Progress in the Study of Wetting Phenomena (April 1984) for Condensed Matter Theory Program, N.S.F.

9. Helium in Vycor: Criticality in a Dilute Bose Fluid (P.B. Weichman and M.E.F.) Informal Proceedings of the Fifth Oregon Conference on Liquid Helium, 1–3 February 1987.

10. Critical Adsorption Profiles from Reflectivity Data (with Andrea J. Liu) September 1987. Circulated informally; formed a basis for the first part of a paper published in 1989: see [282].

11. Comment on work by A.L. Kholodenko and A.L. Beyerlein concerning the relation between electrolyte models and the spherical model (12 June 1991) [circulated informally].

12. On the Possibility of Convection in the 3M-PVTOS Experiment (by Sheng-Nan Lai, under supervision of M.E.F.) 26 March 1993. [Circulated informally and included in report to A.F.O.S.R. for NASA.]

13. Beyond the Ising Model: Understanding Criticality in Continuum Systems, extended abstract for the International Symposium on *New Developments in Statistical Physics* to celebrate the 60th birthday of Professor Masuo Suzuki, 25–26 March 1997, Univ. of Tokyo. [Circulated in book of Abstracts for Meeting.]

14. Charge Fluctuations in Electrolytes: Exact results at low densities (S. Bekiranov and M.E.F.) Abstract for a (late) contributed poster at the American Physical Society 1998 March Meeting.

15. Criticality in simple fluids: The Yang-Yang anomaly and other novelties [summary] Howard Reiss Symposium at the International Conference on Nucleation and Atmospheric Aerosols, University of Missouri-Rolla, August 2000.

16. The Yang-Yang Anomaly in Fluid Criticality: Exactly Soluble Models (R.T. Willis and M.E.F.) Poster for "Thermo-2005" Mid-Atlantic Meeting on Thermodynamics, University of Maryland, April 2005.

17. ARCC Workshop Statement for "*Phase Transitions in Physics, Computer Science, Combinatoriecs and Probability Theory*," American Institute of Mathematics, Palo Alto, California, 15 July 2006.

18. Definition of Thermodynamic Phases and Phase Transitions (M.E.F. with Charles Radin) set on the website of the American Institute of Mathematics (AIM), December 2006.

19. Backstepping, Dwell Times, and Hidden Substeps in Molecular Motors (D. Tsygankov, M. Lindén, and M.E.F.) Poster (with Abstract) for the 51st Annual Meeting of the Biophysical Society in Baltimore, March 2007.

20. Superstall and Assisting-Load Velocities of Motor Proteins can reveal Mechano-Chemical Structure (D. Tsygankov and M.E.F.) Poster (with Abstract) for the 51st Annual Meeting of the Biophysical Society in Baltimore, March 2007.

21. Limping Factors for Motor Proteins (Y.-X. Zhang and M.E.F.) Poster (with Abstract) for the 55th Annual Meeting of the Biophysical Society in Baltimore, March 2011.

22. Vortex Dipole Dynamics in the Gross-Pitaevskii Equation (C. Rorai, D. P. Lathrop and M.E.F.) [abstract with references] "Dynamics Days 2012," Baltimore, 4–7 January 2012.

Leo P. Kadanoff

LEO P. KADANOFF

Birthplace

Born	New York, 14 January 1937

Education

1957	A.B. Harvard
1958	M.A. Harvard
1960	Ph.D Harvard

Research and Professional Experience

1960–1962	Postdoctoral Research at the Bohr Institute for Theoretical Studies in Copenhagen
1962	Assistant Professor (Physics), University of Illinois
1963–1964	Associate Professor (Physics), University of Illinois
1965–1969	Professor (Physics), University of Illinois
1965	Visiting Professor, Cambridge University, Cambridge, England
1969–1978	University Professor and Professor of Physics and Engineering, Brown University
1978–2015	Professor of Physics, University of Chicago
1981–1984	Director, University of Chicago Material Research Laboratory
1994–1997	Director, University of Chicago Material Research Laboratory
1982–2003	John D. and Catherine T. MacArthur Distinguished Service Professor of Physics and Mathematics

Honor

1962–1967	Alfred P. Sloan Foundation Fellow
	Fellow American Physical Society
1977	Buckley Prize American Physical Society
	Member National Academy of Science (US)
1980	The Wolf Foundation Award
	Fellow American Academy of Arts and Sciences
	Fellow American Association for the Advancement of Science
1986	Elliott Cresson Medal, The Franklin Institute
1989	Boltzmann Medal IUPAP
1990	Quantrell Award, The University of Chicago
1990	Centennial Medal, Harvard University
1997	Member American Philosophical Society
1998	Onsager Prize, American Physical Society
1998	Grande Medaille d'Or, Académie des Sciences de l'Institut de France
1999	The National Medal of Science (US)
2000	The Ryerson Lectureship, University of Chicago
2000	Honorary Doctor in Science, University of Copenhagen

2003	Lorentz Professorship, University of Leiden
2004	Onsager Medal, University of Trondheim, Trondheim Norway
2005	Onsager lectureship, Trondheim Technological Institute
2006	Lorentz Medal, the Royal Dutch Academy of Science
2007	Presidency, The American Physical Society
2011	Newton Prize, Institute of Physics

Community Service

1963–1965	Vice-President NAACP, Urbana Chapter
1973–1978	Member of Technical Committee, Statewide Planning Program, State of Rhode Island
1973–1978	Chairman, Human Resources Sub-Committee, Statewide Planning Program, Rhode Island
1974–1978	President, Urban Observatory (Rhode Island)
1997–1999	Vice President, The Shakespeare Project

Professional Service

1978–1981	Advisory Committee, Institute for Theoretical Physics (Santa Barbara)
1981–1986	Advisory Committee, Schlumberger Doll Research Laboratory, Statistical Physics Program
1983–1986	Board of Physics and Astronomy, National Research Council
1983–1984	Member, Board of Governors, Argonne National Laboratory
1986–1990	Advisory Committee Institute for Theoretical Physics, University of Minnesota
1988–1993	Advisory Board of The Miller Institute for Basic Research in Science, Berkeley
1994–1998	Scientific Advisory Board, Niels Bohr Institute, Copenhagen
1999–2004	Visiting Committee, Harvard Physics Department
1999–2003	Scientific Advising Board, NEC Research Laboratory, New Jersey

COMMENTARY

Words, Words, Words[1]

Words are very important to science. Newton's word, *momentum*, encapsulates his new concept in mechanics. When we use it, we are pushed away from the Aristotelian view of mechanics and into the more modern view.

The statistical physics of the nineteenth century and early twentieth century has given us *molecule, path, entropy, ensemble,* and *phase space.*

In a similar vein, words and phrases have been important to the physics I have used and to the new ideas that I helped to create. Mostly these have been standard, non-scientific words, called into more technical service. Thus, in working with hydrodynamics and with the thermodynamic Green's functions of Martin and Schwinger,[2] I focused on *correlations, response, transport,* and *conservation.* On their technical meaning, these words reflected a focus upon complex physical systems in their dynamical behavior within space and time. Of course, these words were not anyone's property, they were part of a school of thought that was developing space-time studies, most particularly, Bogolyubov, Landau, Schwinger, and Feynman.

In 1965, I applied this approach to the behavior of systems near phase transitions. I made use of the field's crucial words including *order, singularity, homogeneity,* and *free energy.* These words were very effectively used by my predecessors in the field, most notably Michael Fisher and Benjamin Widom. I take pride in the concepts that I helped add to the field, as reflected in such words as *universality, relevant/irrelevant/marginal, block variable and scaling.* Later on, the field reached a culmination with Kenneth Wilson's[3] *renormalization* leading to the recognition of *universality classes.* This work formed the basis for the Wolf prize shared by Fisher, Wilson, and myself.

In subsequent years, I applied the concepts reflected in the abovementioned words to the analysis of urban economics and public policy, to the onset of chaos in dynamical systems, and to turbulent fluid flows.

More recently, I have tried to position myself as an observer and student of efforts being directed to the fundamentals of quantum mechanics. Here too, new concepts are reflected in new word uses as in *entanglement, information,* and new meanings for old ones *spooky, glassy, localization, measurement.*

I have also watched with interest the builders of large computer models some of whom were trying in, I think, an unfortunate way to give new meanings to *predict, verify,* and *validate.* In parallel, I have followed my old interest, public policy, by looking at the economists' love of things *disruptive,* and their consequent disdain of the *sustainable.*

---

[1]W. Shakespeare, Hamlet, act 2, scene 2.

[2]Paul Martin and Julian Schwinger (1959). "Theory of many-particle systems. I.," *Phys. Rev.* **115**, 1342.

[3]Wilson, K. G. (1971). "Renormalization group and critical phenomena. I. Renormalization group and the Kadanoff scaling picture," *Phys. Rev. B* **4**(9), 3174.

Physics Vol. 2, No. 6, pp. 263-272, 1966.   Physics Publishing Co.   Printed in Great Britain.

SCALING LAWS FOR ISING MODELS NEAR T_c*

LEO P. KADANOFF [†]

Department of Physics, University of Illinois
Urbana, Illinois

(Received 3 February 1966)

Abstract

A model for describing the behavior of Ising models very near T_c is introduced. The description is based upon dividing the Ising model into cells which are microscopically large but much smaller than the coherence length and then using the total magnetization within each cell as a collective variable. The resulting calculation serves as a partial justification for Widom's conjecture about the homogeneity of the free energy and at the same time gives his result $s\nu' = \gamma' + 2\beta$.

1. Introduction

IN a recent paper [1] Widom has discussed the consequences of the assumption that the free energy in a system near a phase transition of second order is a homogeneous function of parameters which describe the deviation from the critical point and has shown that this assumption leads to consequences which roughly agree with our present numerical information [2] about the behavior of such systems. Another paper by Widom [3] written at the same time explores the consequences of an apparently quite independent idea: that the behavior of the interface separating droplets of the "wrong phase" within a system just below a phase transition should be quite similar to the behavior of an interface separating a region of fluctuation in the order parameter from the surrounding medium [4]. Here again information is derived which agrees with numerical calculations and experiment [2].

Widom's ideas about interfaces are based upon physical plausibility arguments; his idea about the homogeneity of the singular part of the free energy is not given any very strong justification beyond the fact that it appears to work. In the present paper, the Ising model is analyzed in a manner which is designed to throw light upon how correlations between the order parameter in different regions of the lattice scale when the parameters describing the deviation from the critical point – in this case $T - T_c$ and the applied magnetic field – are changed. Widom's homogeneity condition upon the singular part of the free energy and some of his results for interfaces are then derived as a consequence of these scaling arguments based upon our model.

---

*   This research was supported in part by the National Science Foundation under grant
    NSF-GP 4937.

[†]   A.P. Sloan Foundation Fellow.

Although the argument is carried out in Ising model language, it is clear that the arguments could be generalized to other cases of second order phase transitions.

2. Description of Model

Consider an Ising Model in a weak magnetic field near T_c. This model can be described in terms of two parameters

$$\epsilon = (T - T_c)/T_c, \tag{1}$$

which measures the deviation from the critical temperature, and h, a dimensionless magnetic field defined so that flipping a single spin gives a change in magnetic energy $2h/kT$. A full solution of the Ising model would be obtained if we knew $f(\epsilon, h)$, the free energy per site in the presence of the magnetic field.

To get some feeling for the behavior of the Ising model, imagine that we divided the entire lattice into cells of L lattice sites on a side. Then each cell in this s-dimensional lattice contains L^s lattice sites. We take L to be large, but much smaller than the coherence length, ξ, which describes the range of spin correlations, measured in lattice constants. Since ξ goes to infinity at $\epsilon = 0$ and $h = 0$, it is easy to find an L which satisfies this criterion.

To zeroth order, each cell can be considered to be isolated from the others and from the external magnetic field. Then to zeroth order $f(\epsilon, h) = f_L(\epsilon)$ where $f_L(\epsilon)$ is a free energy per site of a lattice of side L in no magnetic field. Since the small size of the cell tends to eliminate the singularities from the phase transition, $f_L(\epsilon)$ should be an analytic function of ϵ but not of L, i.e.

$$f_L(\epsilon) = f_L^{(0)} + \epsilon f_L^{(1)} + \epsilon^2 f_L^{(2)} + \ldots \tag{2}$$

where f_L^0, f'_L and f_L^2 should not all be analytic functions of L [4].

Next consider interactions of the cell α with the magnetic field. This gives a term in $-\beta\mathcal{H}$ of the form

$$h \sum_{\mathbf{r} \in \alpha} \sigma_{\mathbf{r}} \tag{3}$$

The basic assumption of our analysis is that within each cell, the spins tend to line up so that they mostly point either up or down. That is,

$$\sum_{\mathbf{r} \in \alpha} \sigma_{\mathbf{r}} = \mu_\alpha < \sigma >_L L^s \tag{4}$$

where μ_α is either plus one or minus one. The average spin, $<\sigma>_L$, defined in (4) is given by

$$< \sigma >_L^2 = \sum_{\mathbf{r} \in \alpha} \sum_{\mathbf{r}' \in \alpha} \frac{<\sigma_{\mathbf{r}} \sigma_{\mathbf{r}'}>}{L^{2s}} \tag{5}$$

Because of the "small" size of the cell, this depends strongly on only L, not upon ϵ or h. Thus the interaction with the magnetic field takes the form of a term in $\exp[-\beta\mathcal{H}]$ of

$$\exp\left(\sum_\alpha h \ \mu_\alpha \ L^s < \sigma >_L \right) \tag{6}$$

Next, consider the interaction among cells. The free energy will tend to be larger if the spins on neighboring cells are lined up. There will tend to be a smaller contribution if they are anti-parallel. Then, in net, this tends to make a contribution to the exponential $\exp[-\beta F]$

$$\exp \sum_{\alpha, \beta} \left\{ \mu_\alpha \mu_\beta \, \widetilde{K}(\epsilon, L) + f_{int}(\epsilon, L) \right\}. \tag{7}$$

Here the sum extends over nearest neighbor cells, $\widetilde{K} + f_{int}$ gives the contribution to the free energy when neighboring cells are aligned and, $-\widetilde{K} + f_{int}$ gives the contribution to the free energy when they are out of step. Since the direct interactions between cells which produce f_{int} occur within a distance which is very short compared to the coherence length, we assume that f_{int} is, like $f_L(\epsilon)$, a regular function of ϵ, but not necessarily a regular function of L. On the other hand, $\widetilde{K}$ is perhaps a somewhat more subtle beast. This describes the extra free energy that it costs to put two cells out of step. This involves, then, the rather delicate difference between the ways cells can match up when they are in step and when they are out of step. Nonetheless, it seems reasonable to assume that $\widetilde{K}(\epsilon, L)$ is also a regular function of ϵ; but, we assert this with somewhat less confidence than our other statements relating to this model. In writing (7) we are asserting that the correlations among cells can be totally represented by these interactions among near neighbors and that there are no less direct interactions that we need include in (7) as long ranged interactions. This statement, together with the assertion that the cell can be represented by the double valued variable, μ_α, are the two basic assumptions of this model.

To find the correction to the zeroth order result (1), we sum the product of (6) and (7) over both possible orientations of each μ_α. This sum is, of course, just an Ising model calculation with coupling constant $\widetilde{K}$ and effective magnetic field $\widetilde{h}$. This then gives a contribution to the total free energy

$$\sum_\alpha s \, f_{int}(\epsilon, L) + f(\widetilde{\epsilon}, \widetilde{h}) \tag{8}$$

where $\widetilde{\epsilon}$ measures the deviation of the new coupling constant from the critical value, and where, from (6) the effective magnetic field in the cell problem is

$$\widetilde{h}(h, L) = h \langle \sigma \rangle_L L^s \tag{9}$$

Since the original Ising model problem has within it correlations over many cells, we assert that $\widetilde{\epsilon}$ and $\widetilde{h}$ must be sufficiently small so that the new Ising model problem retains long-ranged correlations in $\langle \mu_\alpha \mu_\beta \rangle$. This requires $\widetilde{\epsilon} \ll 1$, $\widetilde{h} \ll 1$.

Since there are L^s sites per cell, equations (1) and (9) give the free energy as

$$f(\epsilon, h) = f_L(\epsilon) + sL^{-s} f_{int}(\epsilon, L) + L^{-s} f(\widetilde{\epsilon}, \widetilde{h}). \tag{10}$$

Equation (10) is one of the two basic relations in our analysis. The other is obtained by discussing the spin-spin correlation function $\langle \sigma_r \sigma_{r'} \rangle$ for the case in which the distance between spins $|\mathbf{r} - \mathbf{r}'|$, as measured in lattice constants, is much larger than L. Then each σ_r can be replaced by the average of σ_r over the cell in which it lies so that (4) may be used to rewrite the correlation function as

$$\langle \sigma_r \sigma_{r'} \rangle = g(\epsilon, h, |\mathbf{r} - \mathbf{r}'|) = \langle \sigma \rangle_L^2 \langle \mu_\alpha \mu_{\alpha'} \rangle = \langle \sigma \rangle_L^2 g(\widetilde{\epsilon}, \widetilde{h}, |\mathbf{r} - \mathbf{r}'|/L). \tag{11}$$

The last line of (11) follows because the μ's are described by an Ising model with effective distance from T_c, $\tilde{\varepsilon}$, effective magnetic field $\tilde{h}$ and cell L lattice constants long.

3. Analysis of the Model

The free energy is a singular function of ε. For example if the specific heat diverges as $\varepsilon^{-\alpha}$ $[(-\varepsilon)^{-\alpha'}]$ above [below] T_c, for small ε and $h = 0$,

$$
f(\varepsilon, 0) = \begin{cases} f_0 + f_1\varepsilon + f_2\varepsilon^{2-\alpha} + \dots & \text{for } T > T_c \\[2em] f_0 + f_1\varepsilon + f'_2(-\varepsilon)^{2-\alpha'} + \dots & \text{for } T < T_c \end{cases} \tag{12a}
$$

In three dimensions (2) $0 \leqslant \alpha \leqslant 1$, $0 \leqslant \alpha' \leqslant 1$. Another possible behavior has the specific heat above and/or below T_c diverge logarithmically so that α and/or α' are zero but

$$
f(\varepsilon, 0) = f_0 + f_1\varepsilon + 1/2\ \varepsilon^2\ [B - A \log \varepsilon] \text{ for } T > T_c \tag{12b}
$$

and/or

$$
f(\varepsilon, 0) = f_0 + f_1\varepsilon + 1/2\ \varepsilon^2\ [B' - A' \log|\varepsilon|] \text{ for } T < T_c \tag{12c}
$$

We first attack the case in which the α's differ from zero and later consider the possibility of behaviors (12b) and/or (12c).

Consider first, at $h = 0$, the terms in equation (10) which are singular in ε and $\tilde{\varepsilon}$. If we equate these singular terms above T_c, we find

$$
\varepsilon^{2-\alpha} = [\tilde{\varepsilon}]^{2-\alpha}/L^s.
$$

It follows then that

$$
\tilde{\varepsilon} = \varepsilon L^{1/\nu} \text{ above } T_c
$$
$$
\tilde{\varepsilon} = \varepsilon L^{1/\nu'} \text{ below } T_c \tag{13}
$$

and

$$
s\nu = (2 - \alpha) \tag{14a}
$$

$$
s\nu' = (2 - \alpha') \tag{14b}
$$

Notice that if we assert that $\tilde{K}$ is a regular function of ε, $\tilde{\varepsilon}$, which is $\tilde{K} - K_c$, cannot have a discontinuous first derivative with respect to ε. Therefore, our assertion about the regularity of K implies

$$\nu = \nu' \tag{15}$$

which in turn implies $\alpha = \alpha'$. It should be recognized that equation (15) is more dubious than the other assertions of this paper. Consequently, we hold this statement aside and do not use it in simplifying our further analysis.

More generally, when $h \neq 0$, we can equate the singular parts in ϵ and h on both sides of equation (10) and find

$$f_{sing}(\epsilon, h) = \frac{1}{L^s} f_{sing}(\tilde{\epsilon}, \tilde{h}). \tag{16}$$

This can only be true if h is proportional to some power of L, i.e. if

$$<\sigma> = L^{-\psi} \tag{17}$$

and then (16) and (14) can hold true if the singular part of $f(\epsilon, h)$ obeys

$$f_{sing}(\epsilon, h) = \begin{cases} \epsilon^{2-\alpha} F(\epsilon^{(s-\psi)\nu}/h) & \text{for } T > T_c \\[2ex] |\epsilon|^{2-\alpha'} F'(|\epsilon|^{(s-\psi)\nu'}/h) & \text{for } T < T_c \end{cases} \tag{18}$$

This is Widom's homogeneity assertion [1] which we have now derived from our model.

If $\alpha = 0$, and f has the form (12b), then the last term on the right-hand side of equation (10) has a singular part

$$\frac{\tilde{\epsilon}^2}{L^s}[B - A \log \tilde{\epsilon}] = \epsilon^2[B - A \log \epsilon - A \log L^{1/\nu}] \tag{19}$$

if we again employ (13) for this case. The extra term in $\log L^{1/\nu}$ must be cancelled out by some other term on the right-hand side of (10). But since this term is regular in ϵ, the compensating term can be obtained by taking $f_L^{(2)}$ in equation (2) to be

$$f_L^{(2)} = A \log L^{1/\nu} + \text{regular terms} \tag{20}$$

or by a similar structure in $sL^{-s}f_{int}(\epsilon, L)$. When $\alpha' > 0$, this term in (20) does not contribute to the leading singularity in $f(\epsilon, h)$ for $T < T_c$ and consequently the analysis which led to (18) can go through just as before except that a term in $\epsilon^2 \log |\epsilon|$ must be added to F and F'.

But, when α' is also zero, the same term, (20), must cancel out $\log L$ terms both above and below T_c. The coefficient in front of this term cannot change at T_c since f_L is, by hypothesis, regular in ϵ. Then when $\alpha = \alpha' = 0$, and hence $\nu = \nu'$ by (14), we conclude with Widom that

$$A = A' \tag{21}$$

It is relevant to notice that this equality holds in the Onsager solution of the two dimensional

Ising model [4]. It is interesting, but perhaps quite besides the point to notice that it is also true for the Λ-transition [5] of He^4 and in at least one anti-ferromagnetic transition [6].

The arguments of Widom are designed to use the homogeneity of the singular part of $f(\varepsilon, h)$, as exhibited in equation (18), to derive relations between the average magnetization,

$$M \sim <\sigma> \sim \left.\frac{\partial f(\varepsilon, h)}{\partial h}\right|_\varepsilon \tag{22}$$

the spin susceptibility

$$\chi \sim \left.\frac{\partial <\sigma>}{\partial h}\right|_\varepsilon \tag{23}$$

and the parameters, α, α', ν and ν' which we have already defined. If we define β, γ, γ' and δ in the conventional manner [1,2]

$$<\sigma> = |\varepsilon|^\beta \quad \text{for } T < T_c, \ h = 0$$

$$\chi = \varepsilon^{-\gamma} \quad \text{for } T > T_c, \ h = 0$$

$$\chi = |\varepsilon|^{-\gamma'} \text{ for } T < T_c, \ h = 0 \tag{24}$$

$$<\sigma> = |h|^{1/\delta} \text{ for } \varepsilon = 0$$

it follows from (18) that

$$\alpha' + \beta(1 + \delta) = 2 \tag{25a}$$

$$\gamma' + 2\beta + \alpha' = 2 \tag{25b}$$

$$\gamma'/\gamma = (2 - \alpha')/(2 - \alpha). \tag{25c}$$

Since the parameters in (25) are all experimental quantities, these relations could be checked if we could find an Ising model in nature.

In the course of this analysis, we also find that ψ can be written in terms of experimental quantities as

$$2\psi = s - \gamma/\nu. \tag{26}$$

This result is useful in the analysis in the spin-spin correlation function of equation (11), which equation can now be written as

$$g(\varepsilon, h, |r - r'|) = L^{-2\psi} g(\varepsilon L^{1/\nu}, hL^{s-\psi}, |r - r'|/L) \tag{27}$$

for $T > T_c$. The right-hand side of (27) will only be independent of L if g is a homogeneous function of its arguments of the form

$$g(\epsilon, h, |r - r'|) = \begin{cases} \dfrac{1}{|r - r'|^{2\psi}} \, G(\epsilon^{\nu(s-\psi)}/h, |r - r'| \epsilon^{\nu}) & \text{for } T > T_c \\[20pt] \dfrac{1}{|r - r'|^{2\psi}} \, G'(\epsilon^{\nu'(s-\psi)}/h, |r - r'| \epsilon^{\nu'}) & \text{for } T < T_c. \end{cases} \qquad (28)$$

Consequently, the coherence length that we discussed at the beginning of this paper must be at $h = 0$

$$\xi = \begin{cases} \epsilon^{-\nu} & \text{for } T > T_c \\[12pt] |\epsilon|^{-\nu'} & \text{for } T < T_c \end{cases} \qquad (29)$$

and at $\epsilon = 0$

$$\xi = |h|^{-(s-\psi)}. \qquad (30)$$

The result (29) makes contact with the notation of Widom [3] and Fisher [2] and permits us to identify ν and ν' as experimental quantities. Therefore, equations (14a) and (14b) may be viewed as experimental relations.

There is one more experimental relation to be obtained. From (28) as ϵ and h go to zero, g is proportional to $|\mathbf{r} - \mathbf{r'}|^{-2\psi}$. This relation is conventionally [2] written as

$$g(\epsilon = 0, \; h = 0, \; |r - r'|) = \frac{1}{|r - r'|^{s-2+\eta}} \qquad (31)$$

so that, from (26), we can write

$$\gamma = (2 - \eta)\nu \qquad (32)$$

which is our final experimental relation.

4. Discussion and Comparison with Other Theoretical Results

One can view the preceding work as developing relations among the nine "experimental" quantities α, α', β, γ, γ', δ, ν, ν' and η. Equations (14a), (14b), (25a), (25b), (25c) and (31) are the six such relations that we consider to be direct consequences of our model. The relation (15), $\nu = \nu'$, which implies $\alpha = \alpha'$, $\gamma = \gamma'$ is a somewhat more tenuous conclusion. With (15) included, we have seven relations among our nine parameters; if (15) is rejected there are only six such relations. In the first case, then, there are three free parameters; in the second two.

None of these relations are new. Essam and Fisher [7] argued for (25b) on the basis of homogeneity considerations not totally different from those of Widom [1] who made full use of the homogeneity conjecture to get (25a), (25b), (25c) and (31) as well as (15). Besides, (31) can be viewed as a tautology: a definition of the coherence length. In this sense, the relation

$$\gamma' = (2 - \eta)\nu'$$

which follows from (31), (25c), and (14) can also be viewed as trivial.

There exists an alternative derivation of equation (14a) due to Pippard [8] and Ginsberg [9]. Imagine that we were in a situation of zero magnetic field in which the temperature lay just above T_c, i.e. $\epsilon > 0$. Then imagine that a temperature fluctuation occurred which took a region of side X of the material into the ordered state. This would cost a free energy of the order of the volume of the region times the difference in free energy density between that at T and that at T_c, i.e.

$$X^s \epsilon^{(2-\alpha)}. \tag{33}$$

But, the probability that such a fluctuation will occur is proportional to the exponential of the cost in free energy, divided by kT_c. Therefore, the maximum possible value of X, which is the coherence length, is given by setting (33) to be of order kT_c or

$$X_{max} \sim \xi \sim \epsilon^{-(2-\alpha)/s}. \tag{34}$$

Therefore, we find at once $\nu = (2 - \alpha)/s$.

Despite this simple derivation equations (14) are not unassailable. One could argue that, as $s \to \infty$, we know that the molecular field approximation is valid. This gives $\nu = 1/2$, $\alpha = 0$, and (14) fails. But, it may well be that as $s \to \infty$, the molecular field approximation becomes valid closer and closer to T_c; and for all finite s there remains a region within a very small neighborhood of T_c for which the molecular field approximation fails. Then (14) can still be true for all s, in the strict limit as $\epsilon \to 0$.

Precisely this effect is expected to occur in the superconductor [8-10]. Here, the molecular field approximation (Landau-Ginzberg theory) is known to be valid until you get very close to T_c. This theory introduces a coherence length

$$\xi_{L.G.} = |\epsilon|^{-1/2}\xi_0 \tag{35}$$

where ξ_0 is a very large number $\sim 10^4$ lattice constants for a pure superconductor. Over most of the temperature range $\xi_{L.G.}$ is larger than the coherence length defined by (34). Then the molecular field theory remains valid because fluctuations in the order parameter are not important. But very close to T_c, the coherence length of (34) will surpass that of (35), and then the molecular field theory will fail. In this tiny temperature region about T_c, equations (14) are expected to be true if the analysis of reference [10] is correct.

To check the seven relations among physical parameters we compare with known results for two and three dimensional Ising models. A variety of numerical results have been obtained in both two and three dimensions through evaluations of the singularities of power series. These are reviewed in references [2, 11-13]. In the two dimensional case there are also a variety of analytical results available [14-16] all based upon Onsager's solution [4]. For the two dimensional case, equations (28) would imply that the spin-spin correlation function is of the form

$$|r - r'|^{-1/4} H(\varepsilon |r - r'|)$$

when the magnetic field vanishes and $|r - r'| \gg 1$. This result agrees with the conclusions of reference [16].

The second column of Table 1 lists the known values of the nine experimental parameters for the two dimensional Ising model. These all check exactly with Widom's relations.

TABLE 1

Parameter	Two dimensional case		Three dimensional case		
	Value	Reference	Previously calculated value	Reference	Fit to data
α	0	4	$0.0 \leqslant \alpha \lesssim 0.2$	2	0.085
α'	0	4	$0.0 \leqslant \alpha' \lesssim 0.06$	2	0.085
β	1/8	13	$0.303 \leqslant \beta \leqslant 0.318$	2	0.332
γ	7/4	2, 11, 16	1.250 ± 0.001	2	1.250
γ'	7/4	2, 16	$1.23 \leqslant \gamma' \leqslant 1.32$	2	1.250
δ	15	13	5.2 ± 0.15	13	4.78
ν	1	2	0.644 ± 0.003	2	0.638
ν'	1	16		2	0.638
η	1/4	14	0.060 ± 0.007	2	0.045

The first column listed under the three dimensional case gives information which has been obtained via Pade and other numerical methods. The last column, labeled "Fit to data", gives a set of values for these parameters which agrees with all seven of our relations including the more doubtful statement $\nu = \nu'$. This fit agrees with all the known numerical results to within about two and one-half standard deviations. These results indicate at least that the conclusions drawn from our model are not grossly inaccurate.

Acknowledgment

I would like to thank Dr. P.C. Martin and Dr. Brian Josephson for very helpful comments.

References

1. B. WIDOM, *J. Chem. Phys.* 43, 3898 (1965).

2. M. FISHER, *J. Math. Phys.* 5, 944 (1964).

3. B. WIDOM, *J. Chem. Phys.* 43, 3892 (1965).

4. For the two dimensional Ising model, the Onsager solution (L. ONSAGER, *Phys. Rev.* 65, 117, 1944) bears out equation (2).

5. M.J. BUCKINGHAM and W.M. FAIRBANK, in *Progress in Low Temperature Physics III*, (ed. C.J. Gorter), Ch. 3. North Holland, Amsterdam (1961).

6. J. SKALGO, Jr. and S.A. FRIEDBERG, *Phys. Rev. Letters* 13, 133 (1964).

7. J.W. ESSAM and M.E. FISHER, *J. Chem. Phys.* 39, 842 (1963).

8. A.B. PIPPARD, *Proc. Roy. Soc. Lond.* A216, 547 (1953).

9. D.M. GINSBERG and J.S. SHIER, to be published.

10. E.G. BATYEV, A.Z. PATASHINSKII and V.L. POKROVSKII, *Zh. Eksp. Teor. Fiz.* 46, 2093 (1964). English translation in *Sov. Phys.-JETP* 19, 1412 (1964).

11. C. DOMB, *Advanc. Phys.* 9, 34, 35 (1960).

12. M.E. FISHER, *J. Math. Phys.* 4, 278 (1963).

13. M.E. FISHER, in *Proceedings of the International Conference on Phenomena Near Phase Transitions*, to be published.

14. B. KAUFMAN, *Phys. Rev.* 76, 1232 (1949); B. KAUFMAN and L. ONSAGER, *Phys. Rev.* 76, 1244 (1949).

15. G.F. NEWELL and E.W. MONTROLL, *Rev. Mod. Phys.* 25, 353 (1953).

16. L. KADANOFF, to be published in *Nuovo Cimento*.

Slippery Wave Functions V2.01

Leo P. Kadanoff[*]

The James Franck Institute
The University of Chicago
Chicago, IL USA
and
The Perimeter Institute,
Waterloo, Ontario, Canada

March 5, 2013

Abstract

Superfluids and superconductors are ordinary matter that show a very surprising behavior at low temperatures. As their temperature is reduced, materials of both kinds can abruptly fall into a state in which they will support a persistent, essentially immortal, flow of particles. Unlike anything in classical physics, these flows engender neither friction nor resistance. A major accomplishment of Twentieth Century physics was the development of an understanding of this very surprising behavior via the construction of partially microscopic and partially macroscopic quantum theories of superfluid helium and superconducting metals. Such theories come in two parts: a theory of the motion of particle-like excitations, called quasiparticles, and of the persistent flows itself via a huge coherent excitation, called a condensate. Two people, above all others, were responsible for the construction of the quasiparticle side of the theories of these very special low-temperature

[*]e-mail: lkadanoff@gmail.com

1

behaviors: Lev Landau and John Bardeen. Curiously enough they both partially ignored and partially downplayed the importance of the condensate. In both cases, this neglect of the actual superfluid or superconducting flow interfered with their ability to understand the implications of the theory they had created. They then had difficulty assessing the important advances that occurred immediately after their own great work.

Some speculations are offered about the source of this unevenness in the judgments of these two leading scientists.

Contents

1   Introduction

In physics publications, one relatively rarely sees direct statements saying that contemporary authors are wrong. So it is particularly striking when the authors of two of the most important papers in condensed matter physics go out of their way to criticize other physics papers, saying flatly that they are far from the mark. We should look even more carefully when the two criticisms appear to be quite similar, and also somewhat problematic from the perspective of today's understanding of physics.[1]

The two authors are Lev Landau in the 1941 paper entitled[2] *The Theory of Superfluidity in Helium II*[1] and the collective authorship of John Bardeen, Leon Cooper, and Robert Schrieffer, abbreviated BCS, in their blockbuster papers on the theory of superconductivity[2, 3].

1.1   Background

Landau's paper[1] was inspired by the 1938 discovery of superfluid motion in the natural form of helium[8, 9] when that material is taken to sufficiently low temperature. In this surprising motion, helium can move through small cracks and in thin sheets without friction. If put into a circular channel(See Figure (1)), it can flow around apparently forever without slowing down.

Superconductivity, discovered by Heike Kamerlingh Onnes[10] in 1911, exhibits frictionless flow, like that of superfluid helium, but in this case occurring in metals like mercury and aluminum. However, in superconductors electrons are in motion. These charged particles engender an electromagnetic field that tends to push[11] magnetic fields out of the superconducting body.

As the temperature is lowered both superfluidity and superconductivity appear quite abruptly at low temperatures in a change of material behavior called a phase transition[15, 16, 17].

By 1941 some progress had been reached in understanding superconductivity. A good phenomenological theory had been developed by Gorter and

---

[1]A good general reference to the development of this part of science is[6]. A predecessor that carefully assesses the helium work is Sébastien Balibar's *The Discovery of Superfluidity*[7]. When he and I overlap, our conclusions are much the same and were reached independently. The outline for much of this paper was suggested by the late Allan Griffin[5] in his introduction to a Varenna volume on Bose-Einstein condensation. Mistakes and errors of judgment are my own rather than Allan's, of course.

[2]*Helium II* is the name for the state of helium below the temperature for its transition to superfluidity.

Casimir[12] and had been extended to include electromagnetic effects by Fritz and Heinz London[13]. The Gorter-Casimir work included a "two-fluid" analysis that considered the superconductors to be composed of two interpenetrating fluids: a normal fluid and a superfluid. The latter was given all the anomalous properties, including the ability to flow without friction. These fluids were not in any obvious sense quantum. Fritz London, on the other hand, suggested that the superconductor was behaving like a huge atom in which the existence of quantization forced a special rigidity upon the wave function for the low temperature state[14]. London argued that this rigidity tended to reduce the magnetic field within the atom in the same way as the current flow in the superconductor reduced its internal magnetic field[3].

By 1941, some work had also been done on the superfluid behavior of helium. There was an obvious analogy between the eternal currents produced by superconductivity and superfluidity. Tisza had suggested that a helium could also be described by a two-fluid model[26], rather like the one that applied to superconductors. The work of Fritz London and Tisza suggested that perhaps superfluid helium could be likened to the anomalous state of a non-interacting boson system, as had previously been described by Albert Einstein[19].[4]

In 1924, Einstein[19] had developed a theory of the behavior of a noninteracting gas of particles obeying the kind of quantum symmetry first posited by Satyendra Nath Bose[18]. In Einstein's theory, the gas is composed of two parts:

- A spectrum of independently moving particles, with energy-momentum relation $\epsilon = p^2/(2m)$, as is appropriate for noninteracting, nonrelativistic particles.

- A condensate, a single-particle quantum state with wave function $\Psi(\mathbf{r})$, occupied by a finite fraction of all the particles in the system[5]

Einstein's results imply that his phase transition occurs because a finite fraction of all the particles in the fluid fall into a single quantum state, which

[3]The technical name for the magnetic response with this kind of rigidity is *diamagnetism*. So London is saying that a superconductor is like a big diamagnetic atom.

[4]The interaction between Tisza and Fritz London in the period 1937-1939 is described in detail in [7][pages 454-456].

[5]In this paper, I discuss the condensate as if it were always one and only one mode of oscillation, macroscopically occupied. This picture applies in three or higher dimensions. In two dimensions, however, the condensate is spread out over a whole collection of modes.

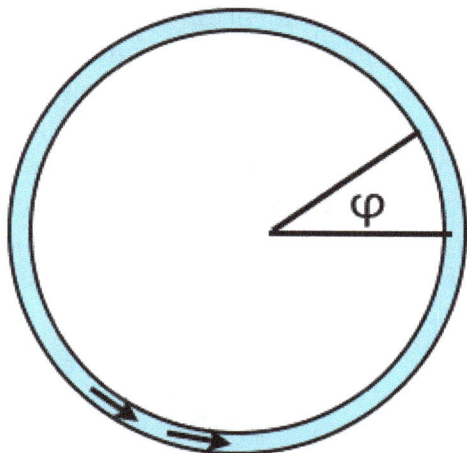

Figure 1: Superflow. Helium is contained in a narrow circular channel. It is set in circular motion at a temperature below its T_c for the transition to superfluidity. The flow will continue essentially indefinitely. In describing its quantum behavior we use a angular coordinate, ϕ.

then might be described by a single wave function $\Psi(\mathbf{r})$. The Bose (or Bose-Einstein) character of the particles permit many different particles to be described by a single wave function. That would be impossible for the other kind of quantum particles, ones that obey Fermi-Dirac system.

However, a very high barrier separated physicists from an acceptable theory of these superflows. Felix Bloch had proven that the lowest energy state of *any* system, including multiply connected material like that in Figure (1) had to have current equal to zero. But superflows could be seen to persist down to very low temperatures so that we might expect their low-temperature configurations to be very much like those in the ground state. There seemed to be no way out of this dilemma.

2 Landau: 1941, 1948

2.1 Quasiparticle analysis

Landau's 1941 paper[1] grapples with many ideas that would subsequently form the center of condensed matter physics. For example, the paper extends concepts of elementary excitations to include excitations in low temperatures helium. Previous work on elementary excitations in normal metals, done by Felix Bloch[20] and others[21], had the excitations given properties closely based upon the properties of a non-interacting gas of particles. As in the gas, the excitations would each have well-defined momentum, $\mathbf{p}$, and the excitation energy was then taken to be a function of the momentum, $\epsilon(\mathbf{p})$. Excitations like these would later[22] be called "quasiparticles."[6] Their behavior was to be analyzed in terms of the Hamiltonian mechanics of their independent motion. Also included were the occasional collisions of the quasiparticles. This point of view became quite celebrated through Landau's later application[27] of quasiparticle concepts to produce a theory of the low temperature behavior of the fluid composed of the helium isotope with atomic weight three, ^{3}He.

It is important to note that Kapitsa's superfluid and Onnes' superconductors are formed from very different constituents. The material studied by Kapitza[8] and Landau-1941[1, 4] is natural helium, mostly composed of the isotope ^{4}He, obeying Bose-Einstein statistics. On the other hand the electrons in metals and the isotope ^{3}He, later studied by Landau[27], obey Fermi-Dirac statistics. In the Bose case, the wave function for the system will be symmetrical under an interchange of the coordinates of any two particles. However, for Fermi particles the wave function for the system will change sign under such an interchange. This difference between the two kinds of particles is likely to manifest itself quite strongly at low temperatures, where quantum properties will be important. As mentioned by Landau, the statistics will determine the detailed scattering and correlation properties of the quasiparticles. But since Landau does not focus upon a condensate, he cannot make use of the fact that condensates are directly possible for Bose particles but can only occur for Fermi particles if they first group together to form "molecules" with even numbers of fermions.

A major portion of the 1941 Landau paper[1] is devoted to a discussion of what we would now call the quasiparticle spectrum of superfluid helium and how that spectrum is connected with the observed superfluidity. An important element of Landau's analysis is his suggestion of a new kind of quasiparticle excitation, given the name, "roton", which has the important property that its energy, $\epsilon(p)$, has a non-zero minimum value, Δ, at a particular value of p.

Most important, Landau here gives three different arguments for superfluidity. The first of these is the famous *Landau criterion*. A material moving

---

[6]An *quasiparticle* is a long-lived, particle-like, excitation in a many-body system. Some people would reserve the word for a fermionic mode and call a bosonic mode a collective excitation. Since the theory of the two kinds of excitations are quite similar, I here use quasiparticle to refer to both.

through a tube can lose momentum to the walls containing it if it is moving fast enough. According to this paper, momentum can be lost if the velocity is greater than

$$v* = \text{minimum } [\epsilon(\mathbf{p})/p] \tag{1}$$

where the minimum is taken with respect to $\mathbf{p}$. If the minimum is zero, as it is for example in the case of non-interacting particles, then there will be no superfluidity.

Since Landau does not really tell us why superfluidity exists, it is hard to tell from his arguments the extent to which his criterion is true.[7] However, given the many mechanisms for broadening the distributions of both energy and momentum, it seems very implausible that any condition like that implied by Eq.(1) can begin to account for the very long-lived nature of the flow of superfluid helium.

Secondly, Landau points out that if the superfluid flow velocity can be constructed as the gradient of some scalar quantity, then the superflow cannot produce a net force on any solid body. However, the quasiparticle microscopics discussed by him does not translate into a gradient superflow.

2.2   Superconductivity

The last requirement for superfluidity is given in Landau's final section, titled "The Problem of Superconductivity." This section starts out by saying that superconductivity and superfluidity are closely related phenomena. It is thus not quite clear why Landau has shifted his focus from helium to metals. Landau mentions the previous suggestion, "advanced several times" that superconductivity arises from a energy gap in its spectrum. This paper previously argued that just this kind of gap, in the roton spectrum, was the basic cause of superfluidity in helium. But then he avers that the criterion of Eq.(1) is a necessary but not sufficient condition for superflow. One needs a special behavior of the entire system, a behavior that Landau says, incorrectly, is quite different in superconductor and superfluid. That special behavior includes motions[1, page 202] "in which the liquid moves as a whole,

---

[7]Recently G. Baym and C. Pethick[61] have argued that the Landau criterion is neither necessary nor sufficient for superfluidity by pointing to counterexamples of both types. They assert that the criterion does work to describe the possible reduction of superfluidity via the loss of momentum in collisions, but that this loss can have the modest result of converting a fraction of the superfluid component of the liquid into the normal fluid rather than the dramatic result of destroying the superfluidity.

in a macroscopic manner." Landau only defines that special behavior at zero temperature, in which he says that for weak magnetic fields the ground state wave function should be multiplied by a factor of the form

$$\prod_{\alpha} \exp[i\chi(\mathbf{r}_a)/\hbar]. \tag{2}$$

where $\chi(\mathbf{r})$ is proportional to the vector potential, $\mathbf{A}(\mathbf{r})$, describing the applied magnetic field at the spatial point $\mathbf{r}$. See [1, equation 9.1], expressed in slightly different form. This is exactly what we would say today. But it is also the rigidity statement of Fritz London[14]. For some reason, Landau fails to mention that this step is old hat. He does not, in fact, refer to London at all.

Landau avoids the next step which we, today, might want him to take: to define a wave function describing the common flow properties of a superconductor in a magnetic field as

$$\Psi(\mathbf{r}) = \exp[i\chi(\mathbf{r})/\hbar], \tag{3}$$

where $\chi(\mathbf{r})$ might be complex, We might then wish Landau to say that the existence of such a wave function is the essential requirement for superfluidity.[8]

The issue in this note is not only these criteria for superfluidity, but also why Landau does not touch upon using a wave function in his requirements for superfluidity. Some insight is perhaps derivable from asides in which Landau goes out of his way to argue that a paper by Laszlo Tisza[26] is incorrect. Tisza is responsible for introducing a two-fluid model of superfluid helium[24] as distinct from the two-fluid model of superconductivity. In a later paper[26], he makes the insightful but not-very-audacious claim that the observed superfluidity of ^4He, a boson fluid, is related to the Bose-Einstein theory for condensation of non-interacting bosons. Tisza further identifies the perfect superfluid with the condensed part of the Bose-Einstein construction. A natural generalization of the theories previously used for electronic excitations[20, 21] would be to use the non-interacting Bose gas as a basis for constructing a theory of superfluid helium. In that case, both a

---

[8]To complete the argument by describing the magnetic perturbation that might exist at non-zero temperatures, one would envision multiplying both sides of a quantum density matrix by factors like that in Eq.(2). This step is taken in a later paper by Ginzburg and Landau[29] described in Sec. (4.1) below.

9

quasiparticle spectrum and a condensate wave function would appear in the helium theory.[9] Indeed, the condensate wave function is regarded today as a crucial part of the superfluid behavior of helium. But, back to Landau, who in 1941 took a different view.

2.3 Landau and Tisza

Landau mentioned Tisza's ideas twice in this paper. On the very first page of his article, he notes that Tisza argued that superfluid helium should be considered to be a Bose gas with a finite fraction of their particles in a state of lowest energy. Further,

> ...[Tisza] suggested that the atoms found in the normal state (a state of zero energy) move through the fluid without friction. This point of view, however, cannot be considered as satisfactory. Apart from the fact that liquid helium has nothing to do with an ideal gas, atoms in the normal state would not behave as "superfluid". On the contrary, nothing would prevent atoms in the normal state from colliding with excited atoms, i.e. when moving through the liquid they would experience a friction and there would be no superfluidity at all. In this way the explanation advanced by Tisza has no foundation in his suggestions but is in direct contradiction to them.[10]

After he discusses the roton excitation and its energy-minimum at energy equal to Δ, Landau presents a footnote[1, page 192]

> It must be mentioned that for an ideal gas $\Delta = 0$ and, therefore, even at absolute zero it would not disclose the phenomenon of superfluidity at any velocities of the flow contrary to Tisza's suggestion.

---

[9]This strategy of taking the noninteracting system to be the template for the construction of a quasiparticle theory, the latter being modified to include a more general energy momentum relation, is exactly the one later followed by Landau in his later construction of a quasiparticle theory of ^{3}He[27].

[10]Part of Landau's trouble with Tisza seems to be contained in the word "normal" as in "normal fluid". They appear to use the word differently so that what Tisza is saying is hard for Landau to interpret.

These words suggest an absolute rejection of Tisza's thesis that what we now call the condensate, having a huge number of particles described by the lowest energy wave function, has anything to do with superfluidity. Landau's position is somewhat delicate in that he certainly accepts the two-fluid model of Tisza and H. London[28] as a descriptor of superfluidity. But the two-fluid model is phenomenological and macroscopic in content. Landau would like to reject out of hand the connection of Einstein's microscopic theory with any description of superfluid helium. In the Twenty-First Century, we might consider this point of view to be quite remarkable since we now consider the superfluid motion to be describable by this condensate wave function. Further, we might remark that this identification of superfluid motion is in reasonable measure based upon the work of Landau himself, specifically but not exclusively in his Ginzburg-Landau[29] paper that will form a later focus of this paper.

Landau returned to Helium II at least twice more. In 1947[34] he wrote a two-page paper in which he used experimental data to correct an error he had made in placing the roton minimum at zero momentum. As a result Landau's revised picture of the quasiparticle spectrum became quite similar to the Feynman picture[33] mostly held today.

In addition, Landau has a 1948 short paper[4] that is mostly devoted to assessing the progress made since 1941 in understanding superfluidity of helium. It does not have the magnificent new ideas of the 1941 work. Landau does point out the recent experimental and theoretical work that has supported his previous predictions while he continues to disagree with Tisza about the form and substance of the latter's work. Once more Landau suggests that the quasiparticle spectrum determines the entire behavior of the helium. He hardly mentions the condensate. Here, Landau devotes six footnotes to Tisza. The last sentence of the first is

> However, his entire quantitative theory (microscopic as well and thermodynamic-hydrodynamic) is in my opinion, entirely incorrect.

The paper continues in the same vein in the text.

> The hydrodynamic equations given by Tisza are, in my opinion, quite unsatisfactory. It is easy to see that in their exact form they even violate the conservation laws!.

11

Landau also mentions N.N. Bogoliubov's important work[22] on weakly coupled Bose gases. Landau mentions that Bogoliubov has determined the "general form" of the quasiparticle theory, while saying that Bogoliubov's work "does not have any direct bearing on the actual liquid Helium II." Further Landau repeats his old assertion that the Landau criterion is sufficient for the deduction of superfluid flow behavior, while ignoring Bogoliubov's deduction of the existence of a condensate. It is, of course, the last point that will prove crucial for understanding superflow. Landau's rhetoric about both Tisza and Bogoliubov has quite a strong tone, perhaps indicative of some extra-scientific reasons for his argument.

I further discuss Landau's motivation in Sec. (5.2) below.

3 BCS: 1957

One and a half decades later, the blockbuster BCS work was published, first as a letter[2] and then in full form[3]. As we shall see, these BCS publications were similar to Landau's work in showing an unwillingness to admit the possibility that a condensate wave function might play an essential role in superflow.

These BCS papers do contain the first satisfactory quasiparticle theory of superconductivity. Over the decades between the 1911 discovery of superconductivity by Heike Kamerlingh Onnes[10] and the BCS work, theorists had put together a phenomenological theory of superconductivity, but there had never been anything close to a successful microscopic theory. The phenomenological theory[13] included equations describing the current flow that described the most important macroscopic facts about the electromagnetism of superconductivity:

- Electrical currents could flow without ohmic resistance.

- They could flow apparently forever around a ring.

- These flow properties appeared abruptly at a critical temperature, T_c and persisted below that temperature.

- Current flow in a pure superconductor eliminated the magnetic field from within the material. This is called the Meissner[11] effect.

In addition to this phenomenological work, there were several attempts at constructing a microscopic theory. None of these described in detail the behavior of real superconducting materials. For this note, the most important previous work was a series of studies by M. R. Schafroth, S, T, Butler and J. M. Blatt, particularly [39], which pointed out a possible analogy between superfluid helium and superconductors, including the existence of a condensate which might drive the flow.

3.1   From Cooper pairs to quasiparticles

The BCS papers were based upon an earlier work in which Leon Cooper[31] worked with an attractive interaction between electrons produced by phonons that had previously been derived by Bardeen and David Pines[30]. Using this interaction, Cooper showed that the interaction produced a tendency for pairs of electrons to form bound pairs. These pairs had opposite spins directions and momenta roughly equal in magnitude and and opposite in direction.

The existence of a possible boson interpretation of superconductivity was evident in the Cooper letter which stated

> It has been suggested that superconducting properties would result if electrons could combine in even groups so that the resulting aggregates would obey Bose statistics[a,b].

- a. V.L. Ginzburg *Uspekhi Fiz. Nauk.* **48**, 25b (1957),
- b. M.R. Schafroth *Phys. Rev.* **100**, 463 (1955)

Indeed it is well known that pairs of identical fermions, viewed from afar, would appear to obey Bose-Einstein statistics. The argument is much older than 1955, and goes back at least to Richard Ogg's 1946 pointer[32] to Bose-Einstein condensation as a source of superconductivity.

Building upon Cooper's work, BCS showed that when there was a multiplicity of Cooper pairs in a metallic situation, the electrons in the metal fell into a state substantially different from that of the usual band theory[20, 21, 23]. In that theory, electrons in a metal are taken to behave rather similarly to a non-interacting gas of particles that obey Fermi-Dirac statistics. They form quasiparticles labeled by particle momentum and spin.

In the BCS paper, the excitations are once more quasiparticles labeled by spin and momentum. The major difference is that, as in the Bogoliubov quasiparticles for Helium II[22], the quasiparticle is a linear combination of a particle with momentum $\mathbf{p}$ and a hole[11] with momentum $-\mathbf{p}$. In addition the energy spectrum of the excitations is strongly modified from the structure of the band theory. The BCS quasiparticles have an energy gap, given the value Δ. The paper then briefly argued that this modified gas of excitations could display the frictionless flow of current that was the first described property of superconductors. In part, the argument was that the gapped excitation spectrum satisfied a form of the Landau criterion, appropriately modified for Fermi-Dirac particles.

3.2   Looking away from Bose-Einstein behavior

One can see the viewpoint of BCS in the introduction to the main paper. They start by listing the facts that should be explained by a theory of superconductivity. The first fact is that superconductivity is produced by a second order phase transition[12]. They continue with the "evidence for an energy gap", and the expulsion of the magnetic field from the interior of the bulk superconductor. Only in fourth place is "effects associated with infinite conductivity". Last comes the dependence of T_c on the isotopic mass of the ions. Thus, the actual infinite conductivity is not by any means of first importance to this work.

The BCS work shies away from discussing a condensate wave function and the previously suggested connection of superconductivity with bose behavior[32, 37, 38, 39]. The first, and only, reference to the behavior of a Bose-Einstein gas is in footnote 18, elicited by a mention of [quantum] coherence of a large number of electrons. The footnote reads

> Our picture differs from that of Schafroth, Butler, and Blatt, *Helv. Phys. Acta* **30**, 93 (1957) who suggest that pseudomolecules of pairs of electrons of opposite spin are formed. They show that if the size of the pseudomolecules is less than the average distance between them, and if other conditions are fulfilled, the system has properties similar to a charged Bose-Einstein gas, including a

---

[11]BCS could have reminded the reader that the 1946 paper of N. N. Bogoliubov[22] contained a very similar form of excitation. They did not do so. This failure pushed the reader away from looking at the analogy between superfluids and superconductors

[12]It is important to the BCS arguments that the transition not be first order in nature. First order phase transitions permit and entail jumps in behavior. Second order ones introduce new behavior at the phase transition, but do so gradually. BCS expect continuity at the phase transition.

> Meissner effect and a critical temperature of condensation. Our pairs are not localized in that sense, and our transition is not analogous to a Bose-Einstein transition.

This statement points the reader away from any analogy with Bose-Einstein condensation. However, to be fair, in footnote 24 BCS agree with the Blatt, Butler, Schafroth group[36, 37] in pointing out that the quantum coherence of the pairs over the entire sample is crucial to the superfluid properties. Nonetheless, the net effect of the arguments of both BCS and Landau is to reject the perfectly reasonable position that understanding the condensate in the Bose-Einstein situation will illuminate the behavior of the superflows in both helium and also superconductors.

The obvious question is whether in downplaying the effect of the condensate wave function, BCS leave anything out. In one sense they do not. They do point out that pairing of electrons with total momentum equal to $\mathbf{q}$ rather than zero will lead to current carrying states. (See Sec. (5) below.) So they have explained, almost in passing, the existence of supercurrents. However, later on, this explanation seems to have disappeared. On page 177 they say "There is thus a one to one correspondence between excited states of the normal and the superconducting phases." This statement does not work for the states carrying supercurrents around a ring. These can also be viewed as excited states of the superconducting phase, and they have no direct analog in the normal phase of an electron gas.

Further BCS do not seem to carry their discussion of the current-carrying states very far. Specifically they do not point out that in the presence of the current, the gap function, Δ, becomes complex and can vary in space as $\exp(i\mathbf{q} \cdot \mathbf{r}/\hbar)$. The reader would be helpfully guided by such a statement. It might even have been helpful to BCS themselves. It seems very strange that neither the contemporary audience nor BCS asked for strong and convincing arguments for the amazing observed stability of the flow of supercurrents.

The omission of the metastable current carrying states makes the gauge invariance of the BCS paper problematical. Gauge invariance, a basic symmetry of quantum physics, involves the fact that changing the phase angle of quantum wave functions as $\Psi(\mathbf{r}) \rightarrow \exp(i\alpha(\mathbf{r}))\Psi(\mathbf{r})$ leaves the quantum system unchanged if one also changes the electromagnetic vector potential, $A(\mathbf{r})$, in a suitable manner. The vector potential, $A(\mathbf{r})$, appears in BCS; the phase angle of the wave function, Δ, does not. This omission leaves the question of whether BCS obeys this basic invariance of quantum physics in limbo. At the time, theorists argued that BCS must be wrong because of its lack of gauge invariance.

However, soon after BCS, in a very important paper, P.W. Anderson showed how the gauge invariance works[40]. Oscillations of the gap parameter, or equivalently of the condensate wave function, produces extra states of the system, states which rescue the gauge invariance of the BCS theory. This new type of excitation is the theoretical grandfather of the Higgs boson of particle physics.

4 Condensate Studies

4.1 Ginzburg-Landau: 1950

Well before BCS, two Soviet authors, Landau and Vitale Ginzburg[29], had an excellent insight into the possible macroscopic behavior of superconductors. They applied the Landau theory of phase transitions[48, 49] to superconductors. Two conditions are required for this application. One condition was that the "order parameter" that described the amount of ordering produced by the superconducting phase transition had to be relatively small. The other was that mean field theory was applicable. In superconductors, the latter condition will ensue when the bound pairs are large enough so that many electrons interact at once. This overlapping pair criterion was very well satisfied in the kind of superconductor known in and before the 1950s. The small order parameter criterion was well satisfied for temperatures near the critical temperature for this phase transition.[13] The only other requirement for the application of the Landau phase transition theory was to get the right symmetry for the order parameter.

These authors assume and assert that the order parameter has the symmetry and behavior of a quantum wave function. Accordingly they take the order parameter to be a complex number, Ψ. They borrow a piece of the Landau theory of second order phase transitions which would give for temperatures, T, near the critical temperature, T_c an equation for the order

---

[13]In unpublished work, Kurt Gottfried and I showed that the region of applicability of mean field theory does not include a small range of temperatures near the critical temperature for onset of superconductivity. However, the range of non-applicability is so narrow as to be irrelevant in almost all studies of simple superconducting metals.

16

parameter of the form:

$$[a(T - T_c) + b|\Psi|^2]\Psi = 0 \tag{4}$$

with a and b being undetermined constants. The appearance of the $|\Psi|^2$ is appropriate for a situation in which the order parameter, Ψ, is a complex number and has a physically undetermined phase angle. Note that the form of Eq.(4) like the form of the 1937 Landau phase transition theory[48, 49] has no accommodation for any spatial dependence of the order parameter. It would, I believe, be natural to make provision for a spatial dependence of the order by including a term in $\nabla^2\Psi$ in Eq.(4)[14]. This extra term would bring this Landau work into line with the classical work of Ornstein and Zernike[35] on the liquid gas phase transition. However, in their 1950 work Ginzburg and Landau do include the possibility of spatial dependence.

They do this by combining Eq.(4) with the Schrödinger equation for the quantum theory of a particle with energy, E, in the presence of a electromagnetic vector potential, $A(\mathbf{r})$,

$$\frac{[-i\hbar\nabla - qA(\mathbf{r})/c]^2}{2m}\Psi(\mathbf{r}) = E\Psi(\mathbf{r}) \tag{5}$$

Here c is the speed of light, $\hbar$ is Planck's constant while q and m are the charge and mass of the particle. When combined the equations become the result now known as the Ginzburg-Landau equation:

$$\{ a(T - T_c) + b|\Psi(\mathbf{r})|^2 + [-i\hbar\nabla - qA(\mathbf{r})/c]^2/(2m)]\}\Psi(\mathbf{r}) = 0 \tag{6a}$$

We need also an equation for the electromagnetic current. which then takes the form

$$J = -i\hbar(\Psi^*\nabla\Psi - \Psi\nabla\Psi^*) - \frac{2q}{c}|\Psi(\mathbf{r})|^2 A(\mathbf{r}) \tag{6b}$$

Since the work is purely phenomenological, a, b, q, and m are all unknown, but the form of the equation looked promising.

Ginzburg and Landau also described a microscopic definition of the condensate wave function in terms of the quantum mechanical density matrix. This definition agrees with the one introduced in parallel by Oliver

---

[14]In my papers[15, 16], I incorrectly assumed that the 1935-7 Landau work did make provision for the spatial dependence of the order parameter by including a term in ∇^2 applied to that parameter. It did not.

Penrose[62], which then, came to be named off-diagonal-long-range-order (ODLRO). ODLRO is considered today to be a defining characteristic of superfluids. The paragraph[29][page 550] in which this definition is given is quite remarkable. It is marred by a crucial misprint.[15] In contrast with the usual confidant language of the authors, it is filled with tentative words: "consider", "suppose", "it might be thought", "it is reasonable to suppose". It is as if the authors could not agree on simple declarative statements. Perhaps this tone is part of the reason why BCS did not pick up on the Ginzburg-Landau paper as a definition of important characteristics of a superconductor. Further the tone partially explains why Ginzburg-Landau is not generally cited as a source of ODLRO.

In 1962, Lev Gor'kov[54] was able to derive the Ginzburg-Landau equation from the BCS theory, obtaining a Ψ proportional to Δ. This derivation then gives the values of the various constants in Eq.(6). Specifically, he found that the charge in question, q, is twice the electron charge. This is the value we might have expected since the wave function describes bound pairs of electrons. At this point, we finally have a conceptually complete theory of the kind of superconductivity that was known in the 1960s and before.[16] However, a complete theory is not yet a complete understanding. It is not clear that in 1962 anyone yet had a good intuitive understanding of the implications of the condensate wave function.

4.2 Abrikosov: 1952, 1957

The first use of the new work was A. A. Abrikosov's application of the Ginzburg-Landau equation to the behavior of superconductors in a magnetic field[41, 42]. The first of these papers described the behavior of films and introduced the idea that there were two kinds of superconductors, Type I and Type II, with the latter being a new kind introduced by Abrikosov. This kind of superconductor has novel properties because it tends to break up into normal and superconducting regions. The second of the papers showed that in a magnetic field type II materials formed vortices, swirls of super-

---

[15]This paragraph and this misprint was pointed out to me by Pierre Hohenberg.

[16]Specifically, BCS had constructed a theory of weak-coupling electronic superconductivity in situations with time-reversal invariance. This theory must be modified for other cases. For example, it fails in the presence of magnetic impurities and also for the subsequently discovered high temperature superconductors. Nobody knows how to do the analog of BCS for these high T_c materials.

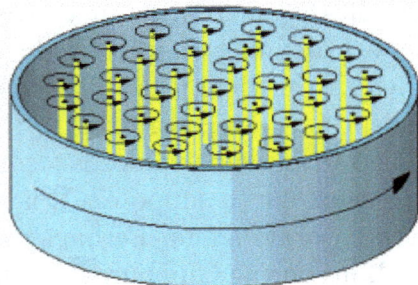

Figure 2: Type II superconductor. A magnetic field produces vortices each containing a quantized unit of magnetic flux.

current surrounding normal regions. (See Figure (2).) Landau termed this behavior "exotic" and at his suggestion this work remained unpublished for several years[43]. After Feynman had published his vortex work[33], Landau relented, Abrikosov finally published[42], and the work became a major contribution to superconductivity theory.[17] Once again, it would appear, Landau undervalued work involving the condensate wave function.

In this context it is interesting to note that the 1957 BCS article does not quote either the Bogoliubov work of 1947[22], the Ginzburg-Landau paper from 1950[29] or the 1952 Abrikosov work[41]. In fact, no Soviet article is quoted. This is perhaps an indication of the bad Soviet-American relations in that period, or more directly of the bad circulation of Soviet work in the U.S.. Or perhaps it is one more indication of the BCS downplaying of the importance of condensate effects.

But the most important condensate effect is yet to come.

4.3 Josephson: 1962

In 1960, Ivar Giaevar found that he could see the structure of the quasiparticle spectrum of superconductors. He used an electrical voltage to pull electrons from the superconductor, through a narrow gap of insulating materials, and thence into another metal[50]. The quasiparticle spectrum could be inferred from a measurement of the current as a function of the voltage across the gap. The first theories of this effect[58, 52] left out the lowest order effect of having the condensate wave function tunnel across a barrier separating

---

[17]Feynman's vortex predictions[33] were preceded by Lars Onsager's[46] and followed by estimates of the size of vortex effects in helium by Penrose and Onsager[47].

Figure 3: Brian Josephson was a graduate student when he described the result of wave-function tunneling between two superconductors. In a *Josephson junction*, the pair wave function tunnels from one side of the barrier through to the other side and interferes with the wave function on the other side. This interference produces a current.

two superconducting materials. A further paper on the subject by Brian Josephson[56] argued that the tunneling of the condensate wave function would produce an additional measurable current, and predicted the size of the current. The specific predictions were very surprising:

- Time Independent Effect: In a situation in which the voltage is the same on the both sides of the barrier, there is nonetheless a possibility for a supercurrent to tunnel across the barrier. The supercurrent is driven by difference in the phase angles, ϕ, between the wave functions, Ψ on the different sides of the gap. The current then has the value $J = J_0 \sin(\delta\phi)$, where $\delta\phi$ is that difference in phase angle. In this way, the wave functions on the two sides of the gap are made directly visible.

- Time Dependent Effect: Correspondingly when there is a voltage, V, across the gap, there is a time dependent current with a value $J = J_0 \sin(2eVt/\hbar)$. This dramatic effect offers the possibility of a measurement of the charge on the electron, e, via a simple measurement of a voltage and a frequency.

Bardeen did not believe that Josephson's calculations were correct[60], most probably because he did not deeply understand the implications of the existence of the condensate wave function and of the Ginzburg-Landau theory.

In a famous debate at the Eighth International Low Temperature Physics Conference at Queen Mary College London in September 1962, Josephson described his work while Bardeen argued that Josephson was wrong. As the experimental facts came in, it became apparent that Josephson's predictions

20

were, in fact, correct. At the next low temperature conference, Bardeen acknowledged that Josephson had indeed been right.

In a series of papers written just before[55] and just after BCS[57, 58, 59, 60], one can see Bardeen working to absorb and use the new concepts[57, 58, 59]. Part of this is an attempt to define a useful tunneling theory[58, 60]. Part is an attempt to use concepts related to condensate wave functions[55, 57, 59]. By the time Bardeen recognized flux quantization[59], he also recognized the existence of condensate wave functions. He even followed Ginzburg-Landau and wrote down[59] a condensate wave function himself. However, his disbelief[60] of Josephson's work is sufficient evidence that, in 1962, he had not yet absorbed all the implications of the physical argument that stretched forward from Einstein's theory.

5   Understanding superflow

5.1   Understanding gained

Like Josephson, we have all the ingredients in front of us so we can describe why superfluids and superconductors exhibit flow without frictional slowing. This understanding arises from a consideration of the superfluid wave function, $\Psi(\mathbf{r})$.

5.1.1   Phase angles

Let us imagine for a moment that there were some wave function $\Psi(\mathbf{r})$ describing a finite fraction of all the particles in a superfluid. Such a wave function would, of course, be complex and have the form

$$\Psi(\mathbf{r}) = \sqrt{\rho(\mathbf{r})} \exp(i\alpha(\mathbf{r})) \tag{7}$$

Here $\rho(\mathbf{r})$ can be interpreted as being proportional to the probability of finding a particle at $\mathbf{r}$ or of the density of particles at $\mathbf{r}$ depending upon whether we interpret this wave function as describing one particle or the entire condensate. In a non-relativistic situation, the phase angle, $\alpha(\mathbf{r})$, is related to the velocity and the momentum of the particle(s) by

$$\mathbf{p}(\mathbf{r}) = m\mathbf{v}(\mathbf{r}) = \hbar\nabla\alpha(\mathbf{r}) \tag{8}$$

We can then give a physical interpretation to the variation of phase angle from place to place, but we cannot give a physical interpretation to the actual value of $\alpha(\mathbf{r})$. In the presence of a vector potential, $m\mathbf{v}(\mathbf{r})$ is replaced by $m\mathbf{v}(\mathbf{r}) - q\mathbf{A}(\mathbf{r})|\Psi(\mathbf{r})|^2/c$. The velocity is then gauge invariant while the momentum, $\mathbf{p}$, the vector potential A, and the phase angle, α, are not gauge invariant. To explain the Meissner effect, one has to describe the behavior of the vector potential and get into electromagnetic theory. Instead of doing this, I shall concentrate on the behavior of superfluid helium.

In helium these is no magnetism and no vector potential. We identify the velocity of Eq.(8) as the superfluid velocity and note that is it automatically the gradient of a scalar, the scalar being phase angle. Landau[1][page 314, where Landau refers to the "Euler paradox"] has pointed us to fluid mechanics and D'Alembert's "paradox" which says that such a fluid velocity distribution cannot produce a net drag force on a body[66, 67]. So the condensate approach automatically gives frictionless flow. However in not admitting a condensate wave function, both BCS and Landau have avoided Eq.(8) and its direct connection with superfluidity.

5.1.2 Toroidal geometry

Let us consider a thin tube of helium twisted into a torus as in Figure (1). We can describe this situation with a wave function with a single coordinate, ϕ, describing angular position in the tube. Since the wave functions must be single valued the possible static wave functions are of the form

$$\Psi(\phi) = \sqrt{N_0}\,\exp(iM\phi). \tag{9}$$

Here M is a quantum number describing the angular momentum of the flow. Because the wave function must be single-valued, M must be an integer. If $M = 0$, there is no flow; any other value gives a quantized flow. Note that this description satisfies the Bloch's theorem requirement (See Sec. (1.1)) that the ground state ($M = 0$) carry no current.

Because this flow encompasses a finite fraction of all the particles in the system it is very hard to stop. Indeed quasiparticle scattering mechanisms like the ones envisioned in the Landau criterion can only produce a finite rate of scattering out of the condensate. However, according Einstein's description of the mechanisms for boson scattering[51], this very same scattering mechanism will return, on the average, an equal number of particles per unit time into the condensed state. Because the condensate contains a huge number of particles, the fluctuations engendered by this process remain small compared to the occupation of the condensate so that these scatterings do not threaten the constancy of the values of M or of the superflow.

If the circular channel holding the helium is relatively broad, the superflow can continue almost indefinitely. But not quite. An Abrikosov-Feynman vortex can move through the channel from inside to out or vice versa. This vortex will carry one unit of angular momentum and change the value of M by ± 1. If one waits long enough, such tunneling of vortices will make the superflow decay to zero[64, 65]. But, this tunneling is essentially a macroscopic process and can be quite slow. In realistic cases, the superflow may well be stable for periods longer than the age of the universe.

5.2   Understanding Postponed

We can certainly conclude that both Landau and BCS are playing down the importance of the condensate dynamics in comparison to the quasiparticles dynamics. We should ask why they engaged in this apparently counterproductive behavior. To some extent this outcome may be manifestations of the "not invented here" syndrome in which each scientist and group emphasizes what it has created and deemphasizes the creations of others. There may well have been an unwillingness to admit something as radically different as a wave function extending through the whole body of the material. Further, these world-leading authors may have been unwilling to admit that relatively uncelebrated scientists may have led the production of a quite radical solution to these very important physics problems.

A more historical argument is suggested in a thesis by Edward Jurkowitz[25] who looked back at the development of quantum theory starting with the old guard: Einstein, Schrödinger, Weyl and moving on to new voices: Heisenberg, Pauli, Dirac, Jordan. The old guard spent considerable effort in constructing unified field theories which often contained an allover wave function describing space-time. They were unable to make these theories work. The newer group preferred to work with operator commutation relations. They were then very suspicious of any wave function which pretended to describe an entire physical system. Einstein's condensate wave function did just that. So perhaps people who had contact with the new voices in the history of quantum theory, like Landau and Bardeen, felt similar suspicions.

One can easily imagine midcentury authors trying to make sense of their research by enforcing an almost complete separation between the microscopic and the macroscopic. They might be willing to admit of the familiar connections given by the microscopic determination of thermodynamic functions and transport processes, but being unwilling to admit of new and exotic connections, like spatially varying order parameters or wave function extended through the whole body of the material.

A fourth reason is recounted by Alan Griffin[5]. After the Einstein paper[19], George Uhlenbeck[68] wrote a thesis in which he said that the Bose-Einstein transition was nonsense because it could only occur in an infinite system. Uhlenbeck retracted his criticism[69] in 1938. Nonetheless Uhlenbeck's objection perhaps turned people away from the Einstein work.

I believe that the noise and bombast of the critical comments in Landau and BCS's work indicated that at some level they were aware that they were making a mistake. It has long been considered suspicious when people "doth protest too much."

6   Fritz London: 1900-1954

In all the years following his first work on superconductivity in 1935, Fritz London was pushing in the right direction. Working with his brother Heinz London, the outcome was an early and correct description of the electrodynamics of superconductors[13, 63]. Fritz London recognized the close analogy between superconductors and superfluids as well as the connection between these problems and Bose-Einstein condensation. His accomplishments can be found in his books on superconductors and superfluids[70, 71]

By no means did London get the sort of recognition that was tendered to Landau, BCS or the other leading lights. London was a Jew and a German in difficult times. Fleeing Germany, he could not get a long-term job in England and had to work hard to finally find a job at Duke University[63].

But recognition did finally come. In 1953, the Royal Netherlands Academy of Sciences gave him the highest recognition of Netherlands Science, its Lorentz medal. Previous winners were Planck, Debye, Sommerfeld, and Kramers. In the same year, his University recognized him with the title of "James Duke Professor".

The most significant recognition, however, came from John Bardeen. In 1972, John contributed monies to be used for the Triennial Award in Low Temperature Physics, which would thenceforth be the Fritz London Award. He also set up a lectureship in London's name at Duke University. In 1990, John wrote a piece, an afterward, for Kostas Gavroglu's *Fritz London, a scientific biography*[63][pp. 267-272], saying, among other things,

> By far the most important step in understanding the phenomena [of superconductivity] was the recognition by Fritz London that both superconductors and superfluid helium are macroscopic quantum systems [...] The key to understanding superfluidity is macroscopic occupation of a quantum state.

6.1 Afterward

However, despite this recognition for Fritz London no similar recognition has been forthcoming for the objects of Landau's and BCS's comments: Tisza and Blatt, Butler, Schafroth.

Physicists became comfortable in working with coherent states of many bosons through experimental maser and laser studies of the late 1950s and the coherent state analyses of Julian Schwinger[72] and Roy Glauber[73]. The Bose-Einstein condensation, abbreviated BEC, has been directly studied in the behavior of cold atomic gases. This condensation is accepted as the basis of all superfluid behavior.

Acknowledgments

This work was partially supported by the University of Chicago NSF-MRSEC under grant number DMR-0820054. I have had instructive conversations on the topics of this paper with David Pines, Silvan Schweber, Gloria Lubkin, Gordon Baym, Edward Jurkowitz, Margaret Morrison, Joel Lebowitz, Roy Glauber, Sébastien Balibar, and Paul Martin.

References

[1] L. Landau, "On the Theory of Superfluidity of Helium II" in L. N. Khlat-nikov, *An Introduction to the Theory of Superfluidity*, translator Pierre C. Hohenberg, pp. 185-204, W.A. Benjamin, New York, (1965).

In Russian in *JETP* **11** 592 (1941) and in English in *J. Phys. U.S.S.R.* **5** 71-90 (1941) also in *Phys. Rev.* **60** (4): 356–358 (1941).

[2] John Bardeen, Leon Cooper, and J. Robert Schrieffer, "Microscopic Theory of Superconductivity," *Phys. Rev.* **106**, pp 162-165 (1957).

[3] John Bardeen, Leon Cooper, and J. Robert Schrieffer, "Theory of Superconductivity", *Phys. Rev.* **108**, pp 1175-1204 (1957). Also [44][page 350].

[4] L. Landau, "On the Theory of Superfluidity," *Phys. Rev.* **75** 884 (1949), also Doklady **61** 253 (1948). also pp 474-477 in *Collected Papers of L.D. Landau*, ed. D. ter Haar, Gordon and Breech Science Publisher, New York London Paris, (1965).

[5] Allan Griffin, "A Brief History of our Understanding of BEC: From Bose to Belieav" , pp. 1-13 in *Bose Einstein Condensation in Atomic Gases*, Varenna meeting, Societa Italiana di Fisica, (1999).

[6] *Out of the Crystal Maze* , ed. Lillian Hoddeson, Ernest Braun, Jörgen Teichmann, Spencer Wert, Oxford University Press, Oxford, (1992), Lillian Hoddeson, Hulmut Schubert, Steve J. Heims,Gordon Baym "Collective Phenomena", pp 489–617.

[7] Sébastien Balibar, "The Discovery of Superfluidity", *Journal of Low Temperature Physics*, **146**, 441-470 (2007).

[8] P. Kapitza, "Viscosity of liquid helium below the λ-point," *Nature* **141**, (3558): 74 (1938).

[9] J.F. Allen, A.D. Misener "Flow of Liquid Helium II," *Nature* **142**, (3597): 643 (1938).

[10] H. K. Onnes, "The resistance of pure mercury at helium temperatures". *Commun. Phys. Lab. Univ. Leiden* **12**: 120 (1911).

[11] W. Meissner, R. Ochsenfeld "Ein neuer Effekt bei Eintritt der Supraleit-fähigkei". *Naturwissenschaften* **21** (44): 787–788 (1933).

[12] C.J. Gorter and H.G.B.Casimir, *Physik Z.* **35** 963 (1934), Z. techn Physik. **15** 539 (1934).

[13] F. London and H. London, *Proc. Roy Soc. London*, Ser. A **149**, 71 (1935).

[14] F. London, "Macroscopic Interpretation of Supraconductivity", *Proc. Roy. Soc.* **152** 24-34, (1935).

[15] Leo Kadanoff, "More is the Same; Mean Field Theory and Phase Transitions," *Journal of Statistical Physics* **137**, 777-797, (2009).

[16] Leo P. Kadanoff, "Theories of Matter: Infinities and Renormalization," *The Oxford Handbook of the Philosophy of Physics*, ed. Robert Batterman, Oxford University Press (2013).

[17] Leo Kadanoff, "Relating Theories via Renormalization," Leo P. Kadanoff, to be published in *Studies in History and Philosophy of Modern Physics* (2013).

[18] S. N. Bose, "Plancks Gesetz und Lichtquantenhypothese," *Zeitschrift für Physik* **26**, 178 (1924).

[19] Albert Einstein, "Quantentheorie des einatomigen idealen Gases," *Sitzungsberichte der Preussischen Akademie der Wissenschaften* **1**, 3 (1925).

[20] Felix Bloch "Über die Quantenmechanik der Elektronen in Kristallgittern," *Z. Physik* **52**, 555–600 (1928).

[21] Paul Hoch, "The Development of the Band Theory of Solids," pp. 182-235 in *Out of the Crystal Maze* , ed. Lillian Hoddeson, Ernest Braun, Jörgen Teichmann, Spencer Wert, Oxford University Press, Oxford, 1992 York: Wiley (1975).

[22] N.N. Bogoliubov, "On the Theory of Superfluidity", *Journal of Physics* **40**, pp. 23-32 (1947). Also [44][page 202].

[23] David Bohm and David Pines "A Collective Description of Electron Interactions," *Phys. Rev* **92** 609-625 (1953).

[24] L Tisza, *Comptes Rendu* **207**, 1035, 1186 (1938). *J. Phys. rad* **1** 350 (1940).

[25] Edward Jurkowitz, thesis, *Interpreting Superconductivity: The History of Quantum Theory and the Theory of Superconductivity and Superfluidity*, University of Toronto, 1995.

[26] L Tisza, *Nature* **141**, 913 (1938).

[27] L. D. Landau, Soviet Phys. *JETP.* 3:920 (1957); *Soviet Phys. JETP.* **5**,101 (1957).

[28] H. London, *Proc. Roy. Soc.* **A 171**, 484 (1939).

[29] Lev Landau and Vitale Ginzburg, "On the Theory of Superconductivity", pp. 546-568 in *Collected Papers of L.D. Landau*, ed. D. ter Haar, Gordon and Breech Science Publisher, New York London Paris, (1965).

[30] John Bardeen and David Pines, *Phys. Rev.* **99**, 1140-1150 (1955). See[44][page 367].

[31] Leon Cooper, *Phys. Rev.* 104, 1189-1190 (1956). See[44][page 350].

[32] Richard A. Ogg, Jr., "Bose-Einstein Condensation of Trapped Electron Pairs. Phase Separation and Super- conductivity of Metal-Ammonia Solutions," *Phys. Rev.* **69** 243-244 (1946).

[33] R. P. Feynman, "Application of quantum mechanics to liquid helium", *Progress in Low Temperature Physics* **1**, 17–53 (1955).

[34] L. D. Landau, "On the theory of supefluidity of Helium II," *J. Phys. (USSR)*, **11**: 91, (1947).

[35] L.S. Ornstein and F. Zernike, *Proc. Acad. Sci. Amsterdam*, **17** , 793(1914); **18** , 1520 (1916);

[36] J. M. Blatt,S, T, Butler, M. R. Schafroth *Phys. Rev.* **100**, 481 (1955)

[37] M. R. Schafroth *Phys. Rev.* **100**, 502 (1955)

[38] M.R. Schafroth *Phys. Rev.* **96** 1149,1442 (1954).

[39] M. R. Schafroth,S, T, Butler,J. M. Blatt *Helv Physica Acta.* **30**, 93 (1957)

[40] P.W. Anderson *Phys. Rev.* **112**, 1900-1916 (1958).

[41] A. A. Abrikosov, *Doklady Akademii Nauk SSSR* **86**, 489 (1952).

[42] A. A. Abrikosov,*Zh. Eksp. Teor. Fiz.* **32**, 1442 (1957); *Sov. Phys.ÑJETP* **5**, 1174(1957).

[43] A. A. Abrikosov, "My years with Landau," *Physics Today,* **26**, 56-60 (1973).

[44] David Pines *The Many-Body Problem,* W.A.Benjamin, New York (1961) includes many important reprints.

[45] Donald G. McDonald "The Nobel Laureate vs the graduate student," *Physics Today* pp. 46-50 July (2001)

[46] L. Onsager, "Statistical hydrodynamics," *Nuovo Cimento,* **6**, 279–287, (1949).

[47] O. Penrose and L. Onsager, "Bose-Einstein condensation and liquid Helium", *Phys. Rev.* **104** 576-584 (1956).

[48] Landau L.D., Phys. Z. Sowjet.,**8**, 113 (1935). Translation in *Collected papers of L. D. Laudau* (ed. D. ter Haar) Gordon and Breach, New York pp. 96-100 (1965).

[49] Landau L.D.,*Zh. Eksp. Teor. Fiz.* **7**, pp. 19–32 (1937). Translation in *Collected papers of L. D. Laudau* (ed. D. ter Haar) Gordon and Breach, New York pp. 193-216 (1965).

[50] Ivar Giaever "Energy Gap in Superconductors Measured by Electron Tunneling," *Physical Review Letters* **5**, (4): 147 (1960). "Electron Tunneling Between Two Superconductors" *Physical Review Letters* **5** (10): 464 (1960); "Electron tunneling and superconductivity" *Reviews of Modern Physics* **46** (2): 245 (1974).

[51] Einstein, A . "Strahlungs-emission und -absorption nach der Quanten-theorie," *Verhandlungen der Deutschen Physikalischen Gesellschaft* **18**: 318-323 (1916).

[52] M.H. Cohen, Leo Falicov, J.C. Phillips, *Phys. Rev. Lett.,* **8** 316 (1962)

[53] Lev Gor'kov, *J. Exptl. Theoret. Phys. U.S.S.R.,* **34** 505 (1958) [translation: Soviet Phys. *–JETP* 7, 505 (1958).

[54] Lev Gor'kov, *J. Exptl. Theoret. Phys. U.S.S.R.,* **36**, 1918 (1959) [translation: *Soviet Phys. –JETP* **9**, 1364 (1959).

[55] John Bardeen, "Theory of Meissner Effect in Superconductors," *Phys. Rev.* **97** 1724-1725 (1955).

[56] Brian Josephson, *Physics Letters,* **1**, 251 (1962).

[57] John Bardeen, "Two Fluid Model of Superconductivity", *Phys. Rev. Lett.* **1** 399-400 (1958)

[58] John Bardeen, "Tunneling from A Many Particle Point of View", *Phys. Rev. Lett.* **6**, 57-59 (1961)

[59] John Bardeen, "Quantization of Flux in a Superconducting Cylinder", *Phys. Rev. Lett.* **7** 162-163 (1961)

[60] John Bardeen, "Tunneling Into Superconductors," *Phys. Rev. Lett.* **9**, 147-149 (1962).

[61] Gordon Baym and C. J. Pethick, "Landau critical velocity in weakly interacting Bose gases," *Phys. Rev. A* **86**, 023602 (2012).

[62] Oliver Penrose, *Phil. Mag.* **42** 1373 (1951).

[63] Kostas Gavroglu's *Fritz London, a scientific biography*, Cambridge University Press, Cambridge (1995)

[64] James Langer and Michael Fisher, *Phys. Rev. Letts.* **19** 560 (1967)

[65] James Langer and Vinay Ambegaokar, *Phys. Rev.* **164**, 498 (1967); D.E. McCumber and B.I. Halperin, *Phys. Rev.* **B1**, 1054 (1970).

30

[66] Jean le Rond d'Alembert, "Memoir XXXIV", *Opuscules Mathématiques,* **5** (first ed.), pp. 132Ð138. (1768).

[67] G. Grimberg, W. Pauls, U. Frisch , "Genesis of dÕAlembertÕs paradox and analytical elaboration of the drag problem", *Physica D* **237** (issue 14Ð17), pp. 1878Ð1886 (2008).

[68] G. E. Uhlenbeck, *thesis*, Leiden (1927).

[69] B. Kahn and G. E. Uhlenbeck, "On the Theory of Condensation," *Physica* **4** 1155-1156, (1937); **5** 399-416, (1938).

[70] Fritz London, *Superfluids,* Volume 1, Dover (1960).

[71] Fritz London, *Superfluids,* Volume II, Dover (1964).

[72] J. Schwinger, private communication, (1959); J. Schwinger, "Theory of quantized fields. III," *Phys. Rev.* **91** 728-740 (1953).

[73] R.J. Glauber, "Coherent and incoherent states of radiation field," *Phys. Rev.* **131** 2766-2788, (1963).

LIST OF PUBLICATIONS

1. Knight Shift in Superconductors, L.P. Kadanoff and P.C. Martin, Phys. Rev. Letts., 3 322 (1959).

2. Transport Properties of Superconductors, L.P. Kadanoff and P.C. Martin, Bull. Am. Phys. Soc. 5 13 (1960).

3. Radiative Transport Within an Ablating Body, L.P. Kadanoff, Trans. ASME: Series C 3 215 (1961).

4. Electromagnetic Properties of Superconductors, V. Ambegaokar and L.P. Kadanoff, Nuovo Cimento 22 914 (1961).

5. Conservation Laws and Correlation Functions, G. Baym and L.P. Kadanoff, Phys. Rev. 124 237 (1961).

6. Theory of Many-Particle Systems II: Superconductivity, L.P. Kadanoff and P.C. Martin, Phys. Rev. 124 670 (1961).

7. Quantum Statistical Mechanics, L.P. Kadanoff and G. Baym, (New York, W.A. Benjamin, 1962), p. 203. Also translated into Russian, Japanese, and Chinese.

8. Comparison Between Theory and Flight Ablation Data, H. Hidalgo and L.P. Kadanoff, Amer. Inst. of Aeronautics and Astronautics, J. 1 41 (1963).

9. Boltzmann Equation for Polarons, L.P. Kadanoff, Phys. Rev. 130 1364 (1963).

10. Effect of Impurities upon Critical Temperature of Anisotropic Superconductors, L.P. Kadanoff and D. Markowitz, Phys. Rev. 131 563 (1963).

11. Hydrodynamic Equations and Correlation Functions, L.P. Kadanoff and P.C. Martin, Ann. Phys. 24 419 (1963).

12. Failure of the Electronic Quasiparticle Picture for Nuclear Spin Relaxation in Metals, L.P. Kadanoff, Phys. Rev. 132 2073 (1963).

13. Perturbation Theoretic Calculation of Polaron Mobility, D.C. Langreth and L.P. Kadanoff, Phys. Rev. 133 A1070 (1964).

14. Transport Theory for Electron-Phonon Interactions in Metals, R.E. Prange and L.P. Kadanoff, Phys. Rev. 134 A566 (1964).

15. Green's Function Formulation of the Feynman Model of the Polaron, L.P. Kadanoff and M. Revzen, Nuovo Cimento 33 397 (1964).

16. Ultrasonic Attenuation in Superconductors Containing Magnetic Impurities, L.P. Kadanoff and I.I. Falko, Phys. Rev. 136 A1170 (1964).

17. The Electron-Phonon Interaction in Normal and Superconducting Metals. Lectures on the Many-Body Problem, Vol. 2 (New York and London, Academic Press, 1964), p. 787.

18. On the Problem of Ultrasonic Attenuation in Superconductors Containing Magnetic Impurities, Proc. 9th Int. Conf. Low Temp. Phys., Columbus, 1964 (New York, Plenum Press, 1965), Part A. pp. 378–80.

19. Green's Functions and Superfluid Hydrodynamics, J.W. Kane, L.P. Kadanoff, J. Math. Phys. 6 1902 (1965).

20. Ultrasonic Attenuation in Superconductors, L.P. Kadanoff and A.B. Pippard, Proc. Roy. Soc. A 292 299 (1966).

21. Scaling Laws for Ising Models Near Critical Points, L.P. Kadanoff, Proc. 966 Midwest Conf. on Theor. Phys., (Bloomington, Ind.).

22. Scaling Laws for Ising Models Near Tc, L.P. Kadanoff, Physics 2 263 (1966).

23. Spin-Spin Correlation in the Two-Dimensional Ising Model, L.P. Kadanoff, Nuovo Cimento 44 276 (1966).

24. Basic Principles of Physics: Electricity, Magnetism, and Heat, (New York, W. A. Benjamin, 1967).

25. Long Range Order in Superfluid Helium, J.W. Kane and L.P. Kadanoff, Phys. Rev. 155 80 (1967).

26. Static Phenomena Near Critical Points: Theory and Experiment, L.P. Kadanoff, W. Gotze, D. Hamblen, R. Hecht, E.A.S. Lewis, V.V. Palciauskas, M. Rayl, J. Swift, D. Aspnes, and J.W. Kane, Rev. Mod Phys. 39 395 (1967).

27. Transport Coefficients Near the Critical Point: A Master-Equation Approach, J. Swift and L.P. Kadanoff, Phys, Rev. 165 310 (1968).

28. Transport Coefficients Near the Liquid-Gas Critical Point, J. Swift and L.P. Kadanoff, Phys. Rev. 166 89 (1968).

29. Wave Function Fluctuations in Finite Superconductors, Comments on Solid State Phys. 1 No. 2 (1968).

30. Transport Coefficients Near Critical Points, Comments on Solid State Phys. 1 5 (1968).

31. Anomalous Electrical Conductivity Above the Superconducting Transition, G. Laramore and L.P. Kadanoff, Phys. Rev. 175 579 (1968).

32. Transport Coefficients Near the Transition of Helium, J. Swift and L.P. Kadanoff, Ann. Phys. 50 312 (1968).

33. Operator Algebra and the Determination of Critical Indices, L.P. Kadanoff, Phys. Rev. Letts. 23 1430 (1969).

34. Correlations Along a Line in the Two-Dimensional Ising Model, Phys. Rev. 188 859 (1969).

35. A City Grows Before Your Eyes, J.R. Voss, L.P. Kadanoff and W.J. Bouknight, Computer Decisions, Dec., 1969.

36. Anomalous Ultrasonic Attenuation Above the Magnetic Critical Point, G.E. Laramore and L.P. Kadanoff, Phys. Rev. 187 619 (1969).

37. Computer Display and Analysis Through Time and Space, L.P. Kadanoff, J.R. Voss, and W.J. Bouknight, Technological Forecasting and Social Change, 2 77 (1970).

38. The Droplet Model and Scaling, L.P. Kadanoff, In Critical Phenomena, Proceedings of the Int. School of Physics, "Enrico Fermi", Course LI, ed., M.S. Green, (New York, Academic Press, 1971), p. 118.

39. Critical Behavior Universality and Scaling. L.P. Kadanoff, In Critical Phenomena, Proceedings of the Int. School of Physics, "Enrico Fermi", Course LI, ed. M.S. Green, (New York, Academic Press, 1971), p. 101.

40. Studying and Displaying Urban Growth Patterns, L.P. Kadanoff, Aerospace and Electronic Systems 7 No. 3 (1971).

41. Determination of an Operator Algebra for the Two-Dimensional Ising Model, L.P. Kadanoff and H. Ceva, Phys. Rev. B. 3 3918 (1971).

42. From Simulation Model to Public Policy: An Examination of Forrester's Urban Dynamics, L.P. Kadanoff, Simulation 16 261 (1971). Reprinted in Best Computer Papers of 1971, ed., O. Petrocelli (New York, Anerbach, 1971). American Scientist, 60 1, 74 (1972).

43. Some Critical Properties of the Eight-Vertex Model, L.P. Kadanoff and F.J. Wegner, Phys. Rev. B 4 2909 (1971).

44. A Modified Forrester Model of the United States as a Group of Metropolitan Areas, L.P. Kadanoff, Urban Dynamics: Extensions and Reflections (San Francisco Press, 1972).

45. Public Policy Conclusions from Urban Growth Models, L.P. Kadanoff and H. Weinblatt, Urban Dynamics: Extensions and Reflections (San Francisco Press, (1972).

46. Public Policy Conclusions from Urban Growth Models, L.P. Kadanoff and H. Weinblatt, IEEE Trans. On Systems, Man, and Cybernetics, SMC-2 159 (1972).

47. Critical Exponents for the Heisenberg Model, M.K. Grover, L.P. Kadanoff and F.J. Wegner, Phys. Rev. B 6 311 (1972).

48. The Brown University National Metropolitan Models, B. Chinitz, G. Crampton, S. Jacobs, L.P. Kadanoff, J. Tucker and H. Weinblatt, In Socio-Economic Systems and Principles, ed., W. Vogt, et al. (University of Pittsburgh, Pennsylvania, 1973). Also published in Synergetics, ed., A. Haken (Stuttgart, B. G. Taubner, 1973).

49. Renormalization Equations: Conceptual Basis and a Simple Example. L.P. Kadanoff, In Renormalization Group in Critical Phenomena and Quantum Field Theory, ed., J.D. Gunton and M.S. Green (Philadelphia, Temple University Press, 1973), p. 21.

50. Renormalization Group Techniques on a Lattice, L.P. Kadanoff, In Cooperative Phenomena, ed., H. Haken, 139, North Holland (1974).

51. A Simulation Model of Urban Labor Markets and Development Policy, L.P. Kadanoff, B. Harrison and B. Chinitz, In Urban Development Models, ed., R. Baxter, et al., The Construction Press (1975).

52. Numerical Evaluations of the Critical Properties of the Two- Dimensional Ising Model, L.P. Kadanoff and A. Houghton, Phys. Rev. B 11 377 (1975).

53. Variational Principles and Approximate Renormalization Group Calculations, L.P. Kadanoff, Phys. Rev. Letts. 16 1005 (1975).

54. Scaling, Universality, and Operator Algebras. In Phase Transitions and Critical Phenomena, ed., C. Domb and M.S. Green, Vol. 5A (New York, Academic Press, p. 1 (1976).

55. From Simulation Model to Public Policy, L.P. Kadanoff, The American Scientist 60 74 (1972).

56. Variational Approximations for Renormalization Group Transformations, L.P. Kadanoff, A. Houghton, and M.C. Yalabik, J. Stat. Phys. 14 171(1976). Also Published in Proc. IUPAP Statistical Mechanics Conference. Hungarian Academy of Sciences, (1976).

57. Notes on Migdal's Recursion Formulas, L.P. Kadanoff, Ann. Phys. 100 359 (1976).

58. The Application of Renormalization Group Techniques to Quarks and Strings, L.P. Kadanoff, Rev. Mod. Phys. 49, 267 (1977). Also published in Lecture Notes in Physics, (Springer-Verlag) 54 276 (1976).

59. Renormalization, Vortices, and Symmetry Breaking Perturbations in the two-dimensional Planar Model, J.V. Jose, L.P. Kadanoff, S. Kirkpatrick, and D. Nelson, Phys. Rev. B 16 1217 (1977).

60. Connections Between the Critical Behavior of the Planar Model and That of the Eight-Vertex Model, L.P. Kadanoff, Phys. Rev. Letts. 39 903 (1977).

61. A Model for Interacting Quarks and Strings, L.P. Kadanoff, Proc. IUPAP Statistical Mechanics Conference, Haifa (1977).

62. Teaching the Renormalization Group, H. Maris and L.P. Kadanoff, Amer. J. Phys. 46 625 (1978).

63. Lattice Coulomb Gas Representation of Two-Dimensional Problems, L.P. Kadanoff, J. Phys. A. 11 1399 (1978).

64. A Variational Real Space Renormalization Group Transformation Based on the Cumulant Expansion, S.J. Shenker, L.P. Kadanoff, and A. Pruisken, J. Phys. A 12 91 (1979).

65. Multicritical Behavior at the Kosterlitz-Thouless Critical Point, L.P. Kadanoff, Ann Phys. 120 39 (1979).

66. High Temperature Series Expansion Methods for Ising Systems with Quenched Impurities, R. Ditzian and L.P. Kadanoff, Phys. Rev. 19 4531 (1979).

67. Correlation Functions on the Critical Lines of the Baxter and Ashkin-Teller Models, L.P. Kadanoff and A. Brown, Ann. Phys. 121 319 (1979).

68. Series Studies of the Four State Potts Model, R. Ditzian and L.P. Kadanoff, J. Phys. A 12 L229 (1979).

69. SMJ's Analysis of Ising Model Correlation Functions, L.P. Kadanoff and M. Kohmoto, Ann. Phys. A 126 371 (1980).

70. Phase Diagram for the Ashkin-Teller Model in Three Dimensions, R. Ditzian, J.R. Banavar G.S. Grest, and L.P. Kadanoff, Phys. Rev. 22 2542 (1980).

71. Singularities Near the Bifurcation Point of the Ashkin-Teller Model, L.P. Kadanoff, Phys. Rev. B 2 1405 (1980).

72. Marginality, Universality and Expansion Techniques for Critical Lines in Two Dimension, A.M.M. Pruisken and L.P. Kadanoff, Phys. Rev. B 22 5154 (1980).

73. Lower Bound RSRG Approximation for a Large η System, M. Kohmoto and L.P. Kadanoff, J. Phys. A 13 3339 (1980).

74. Disorder Variables and Para-Fermions in Two-Dimensional Statistical Mechanics, E. Fradkin and L.P. Kadanoff, Nucl. Phys. B 170 1 (1980).

75. Ground State Entropy and Algebraic Order at Low Temperatures, A.N. Berker and L.P. Kadanoff, J. Phys. A 13 L259 (1980).

76. Planar Model Correlation Functions, L.P. Kadanoff and A. Zisook, J. Phys. A 13 L379 (1980).

77. Correlation Functions on the Critical Line of the Two-Dimensional Planar Model: Logarithms and Correlations to Scaling, A. Zisook and L.P. Kadanoff, Nucl. Phys. B 180 61 (1981).

78. Quantum Mechanical Ground States, Non-Linear Schrodinger Equations, and Linked Cluster Expansions, M. Kohmoto and L.P. Kadanoff, J. Phys. A 14 1291 (1981).

79. Band to Band Hopping in One-Dimensional Maps, L.P. Kadanoff and S.J. Shenker, J. Phys. A 14 L23 (1981).

80. Hamiltonian Studies of the d = 2 Ashkin-Teller Model, M. Kohmoto, M. den Nijs, and L.P. Kadanoff, Phys. Rev. B 24 5229 (1981).

81. Connections among different phase transition problems in two dimensions, M. Kohmoto, L.P. Kadanoff and M.P.M. den Nijs, Physica 106A 122 (1981).

82. Disorder Variables for a Non-Abelian Symmetry Group, L.P. Kadanoff and M. Kohmoto, Nucl. Phys. B 190 671 (1981).

83. Many Point Correlation Functions in a Modified Ising Model, L.P. Kadanoff, Phys. Rev. B 24 5382 (1981).

84. Critical Behavior of a KAM Surface: I. Empirical Results, L.P. Kadanoff and S.J. Shenker, J. Stat. Phys. 27 631 (1982).

85. Scaling for a Critical Kolmogorov-Arnold-Moser, (KAM) Trajectory, L.P. Kadanoff, Phys. Rev. Letts. 47 1641 (1981).

86. Critical Behavior of a KAM Surface: II. Renormalization Approach, L.P. Kadanoff, In Melting, Localization, and Chaos, ed., R.K. Kalia and P. Vashista, North Holland, New York (1982).

87. Renormalization Group Analysis of Bifurcations in Area Preserving Maps, M. Widom and L.P. Kadanoff, Physica 5 D 287 (1982).

88. Quasiperiodicity in Dissipative Systems: A Renormalization Group Analysis, M. Feigenbaum, L.P. Kadanoff, and S.J. Shenker, Physica 5 D 370 (1982).

89. Quantum Model for Commensurate-Incommensurate Transitions, S. Howes, L.P. Kadanoff, and M. den Nijs, Nucl. Phys. B 215 169 (1983).

90. Solutions to the Schrodinger Equation on Some Fractal Lattices, E. Domany, S. Alexander, D. Bensimon and L.P. Kadanoff, Phys. Rev. B 28 3110 (1983).

91. Supercritical Behavior of an Ordered Trajectory, L.P. Kadanoff, J. Stat. Phys. 31 1 (1983).

92. Localization Problem in One Dimension: Mapping and Escape, M. Kohmoto, L.P. Kadanoff, and C. Tang, Phys. Rev. Letts. 50 1870 (1983).

93. Strange Objects in the Complex Plane, M. Widom, D. Bensimon, L.P. Kadanoff, and S.J. Shenker, J. Stat. Phys. 32 443 (1983).

94. Roads to Chaos, L.P. Kadanoff, Physics Today 30 46 (1983).

95. Analysis of Cycles for a Volume Preserving Map, L.P. Kadanoff, unpublished.

96. Extended Chaos and Disappearance of KAM Trajectories, L.P. Kadanoff and D. Bensimon, Physica D 13 82 (1984).

97. Escape from Strange Repellers, L.P. Kadanoff and C. Tang, Proc. Nat. Acad. Sci. 81 1276 (1984).

98. Applications of Scaling Ideas to Dynamics, L.P. Kadanoff, Proc. Erice Summer School on Dynamics, Springer-Verlag (1984).

99. Mean-Field Theories for a Ballistic Model of Aggregation, D. Bensimon, B. Shraiman, and L.P. Kadanoff, Proc. Kinetics, Aggregation and Gellation Conf., Athens, Georgia (1984).

100. From Periodic Motion to Unbounded Chaos: Investigations of the Simple Pendulum, L.P. Kadanoff, Physica Scripta T9 5 (1985).

101. Energy Spectrum for a Fractal Lattice in a Magnetic Field, J.R. Banavar, L.P. Kadanoff, and A.A.M. Pruisken, Phys. Rev. B 31 1388 (1985).

102. Simulating Hydrodynamics: A Pedestrian Model, L.P. Kadanoff, J. Stat. Phys. 39 267 (1985).

103. Scaling in a Ballistic Aggregation Model, L.P. Kadanoff and S. Liang, Phys. Rev. A 31 2628 (1985).

104. Renormalization, Unstable Manifolds and the Fractal Structure of Mode Locking, P. Cvitanovic, M.H. Jensen, L.P. Kadanoff and I. Procaccia, Phys. Rev. Letts. 55 343 (1985).

105. Fractal Singularities in a Measure and How to Measure Singularities on Fractal, L.P. Kadanoff, Cargese School (1985).

106. Circle Maps in the Complex Plane, P. Cvitanovic, M.H. Jensen, L.P. Kadanoff and I. Procaccia, Proc. 6th Int. Symposium on 'Fractals in Physics' eds., L. Pietronero and E. Tosatti, North Holland (1985).

107. Global Universality at the Onset of Chaos: Results of a Forced Rayleigh-Bénard Experiment, M.H. Jensen, L.P. Kadanoff, A. Libchaber, I. Procaccia and J. Stavans, Phys. Rev. Letts. 55 2798 (1985).

108. Complex Analytic Methods for Viscous Flows in Two Dimensions, D. Bensimon, L.P. Kadanoff, S. Liang, B.I. Shraiman and C. Tang, Proc. Mexican Summer School, 1985. Also published in Directions in Condensed Matter Physics, ed. by G. Grinstein and G. Mazenko, World Scientific, p. 51 (1986).

109. More on Microcanonical Paradigms, D. Bensimon, T.C. Halsey, M.H. Jensen, L.P. Kadanoff, A. Libchaber, B.I. Shraiman and J. Stavans, Suppl. to J. Irrep. Events A 17 1597 (1985).

110. Fractal Measures and Their Singularities: The Characterization of Strange Sets, T.C. Halsey, M.H. Jensen, L.P. Kadanoff, I. Procaccia and B.I. Shraiman, Phys. Rev. A 33 1141 (1986). See also erratum: Phys. Rev A34 1601 E (1986).

111. Fractals: Where's the Physics, L.P. Kadanoff, Phys. Today 39 6 (1986).

112. Viscous Flows in Two Dimensions, D. Bensimon, L.P. Kadanoff, S. Liang, B.I. Shraiman and C. Tang, Rev. Mex. Fisica 32 S101 (1986) also published as Rev. Mod. Phys. 58 977–999 (1986).

113. Renormalization-Group Analysis of the Global Structure of the Period-Doubling Attractor, D. Bensimon, L.P. Kadanoff, and M.H. Jensen, Phys. Rev. A 33 3622 (1986).

114. Chaos: A View of Complexity in the Physical Sciences, The Great Ideas Today, Encyclopedia Brittanica, Inc., Chicago, p. 86 (1986).

115. On Two Levels, Phys. Today 39 7 (1986).

116. Computational Physics: Pluses and Minuses, Phys. Today, 39 7 (1986).

117. Saffman-Taylor Bubble Problem, Unpublished.

118. Cathedrals and Other Edifices, Phys. Today 39 (1986).

119. Renormalization Group Analysis of the Global Properties of a Strange Attractor, J. Stat. Phys. 43 395 (1986).

120. On Complexity, Phys. Today, March (1987).

121. Dimensional Calculations for Julia Sets in Proceedings of European Condensed Matter Physics Congress., Physica Scripta T19 19–22 (1987).

122. The Breakdown of KAM Trajectories. D. Bensimon and L.P. Kadanoff, Published in Chaotic Phenomena in Astrophysics, Annals of the New York Academy of Sciences, 497 pp 110–16 (1987).

123. Scaling Structure and Thermodynamics of Strange Sets, M.H. Jensen, L.P. Kadanoff, and I. Procaccia, Phys. Rev. A 36 1409 (1987).

124. Finger Narrowing Under Local Perturbations in the Saffman-Taylor Problem, G. Zocchi, B. Shaw, L.P. Kadanoff, and A. Libchaber, Phys. Rev. A 36 1894 (1987).

125. From Neutrinos to Quasiparticles. Physics Today, pp. 7–9 (1987).

126. A Poiseuille Viscometer for Lattice Gas Automata, L.P. Kadanoff, G. McNamara and G. Zanetti, Complex Systems 1 791–803 (1987).

127. Diffusion of Passive Scalars in Fluid Flows: Maps in Three Dimensions, M. Feingold, L.P. Kadanoff and O. Piro, A Way to Connect Fluid Dynamics to Dynamical Systems: Passive Scalars, in Fractal Aspects of Materials: Disordered Systems, p. 203. Eds. A.J. Hurd, D.A. Weitz and B.B. Mandelbrot (Materials Research Society, Pittsburgh) (1987). Also appears in Universalities in Condensed Matter, Eds. R. Jullian, L. Peliti, R. Rammal and N. Boccara. Springer Proc. Phys. (Springer, Berlin 1988). Transport of Passive Scalars: KAM Surfaces and Diffusion in Three-Dimensional Liouvillean Maps. Instabilities and Nonequilibrium Structures, Eds. E. Tirapequi and D. Villarroel (D. Reidel, Dordrecht) p. 37 (1989).

128. The Big, The Bad and The Beautiful, Physics Today, 9 (1988).

129. Passive Scalars, 3D Volume Preserving Maps and Chaos, M. Feingold, O. Piro and L.P. Kadanoff, J. Stat. Phys. 50 529 (1988).

130. Interactive Computation for Undergraduates, Phys. Today 41 9–11 [Reference Frame] (1988).

131. From Automata to Fluid Flow: Comparisons of Simulation and Theory, L.P. Kadanoff, G. McNamara and G. Zanetti, Phys. Rev. A 40 4527 (1989).

132. Scaling of Hard Thermal Turbulence in Rayleigh Bénard Convection, B. Castaing, G. Gunaratne, F. Heslot, L.P. Kadanoff, A. Libchaber, S. Thomae, X.-Z. Wu, S. Zaleski and G. Zanetti, J. Fluid Mechanics 209 1–30 (1989).

133. Scaling and Universality in Avalanches, L.P. Kadanoff, S.R. Nagel, L. Wu and S-M. Zhou, Phys. Rev A 39 6524–6537 (1989).

134. Fractals and Multifractals in Avalanche Models, Physica D 38 213–214 (1989).

135. Scaling and Structures in the Hard Turbulence Region of Rayleigh Benard Convection, proceedings of Newport Conference on Turbulence (1989), proceedings of Europhysics Conference on Turbulence, Moscow (1989).

136. Scaling and Universality in Statistical Physics, Physica A 163 1–14 (1990).

137. Singularities in Complex Interface Dynamics, W.-S. Dai, L.P. Kadanoff, and S.-M. Zhou, Proceedings of Cargese Conference on Nonlinear Flow Problems (1990).

138. Frequency Power Spectrum of Temperature Fluctuations in Free Convection, X.-Z. Wu, L.P. Kadanoff, A. Libchaber and M. Sano, Phys. Rev. Letts. 64 2140–2144 (1990).

139. Singularities in Complex Interfaces, P. Constantin and L.P. Kadanoff, Phil Trans R Soc London A 333 379–389 (1990).

140. Exact Solutions for the Saffman-Taylor Problem with Surface Tension. Phys. Rev. Letts. 65 2986–2988 (1990).

141. Scaling and Multiscaling (Fractals and Multifractals), Proceedings of Second Latin-American Workshop on Nonlinear Phenomena (1990).

142. Chaos and Complexity: The results of non-linear processes in the physical world — Springer Proceedings in Physics, Volume 57, Evolutionary Trends in the Physical Sciences (eds. M. Suzuki and R. Kubo) Springer-Verlag (1991).

143. Dynamics of a Complex Interface, P. Constantin and L.P. Kadanoff, Physica D 47 450–460 (1991).

144. Interface Dynamics and the Motion of Complex Singularities, W.-S. Dai, L.P. Kadanoff, and S.-M. Zhou, Phys Rev A 43 6672–6682 (1991).

145. Chaos, Computers, and Physics, Sue Copppersmith, L.P. Kadanoff, M.J. Vinson, and A.J. Kolan, an d Scott Wunsch(Laboratory Notes for Physics 251) (1987, 1990, 1991,1997).

146. Transitions in Convective Turbulence: the Role of Thermal Plumes, I. Procaccia, E. Ching, P. Constantin, L.P. Kadanoff, A. Libchaber and X-Z Wu, Phys Rev A, 44 8091 (1991).

147. Turbulence in a Box, L.P. Kadanoff, A. Libchaber, E. Moses and G. Zocchi, La Recherche, 22 628–638 (1991).

148. Stationary Solutions for the Saffman-Taylor Problem with Surface Tension, L.P. Kadanoff and G. Vasconcelos, Phys Rev A, 44 6490–6495 (1991).

149. Complex Structures from Simple Systems, Physics Today, 9 March (1991).

150. Scaling and Multiscaling: Fractals and Multifractals (Review). Chinese Journal of Physics, 29 613–635 (1991).

151. Scaling in Hydrodynamics, to be published in Proceedings of Cargese Summer School, Plenum (1991).

152. Beyond all Orders: Singular Perturbations in a Mapping, C. Amick, S.C.E. Ching L.P. Kadanoff, and V. Rom-Kedar, J. Nonlinear Science, 2 9–68 (1992).

153. Turbulent convection in helium gas, E.S.C. Ching, L.P. Kadanoff, A. Libchaber, and X-Z Wu, Physica D 58 414–422 (1992).

154. Hard Times, Physics Today (October, 1992).

155. Critical Indices for Singular Diffusion, L.P. Kadanoff, A.B. Chhabra, A.J. Kolan, M.J. Feigenbaum, and I. Procaccia, Phys. Rev. A 45 6095–6098 (1992).

156. Droplet Breakup in a Model of the Hele Shaw Cell, P. Constantin, T.F. Dupont, R.E. Goldstein, L.P. Kadanoff, M. Shelley, and S-M Zhou, Phys. Rev. E 47 4169–4181 (1993).

157. Sandpiles, Avalanches and the Statistical Mechanics of Non-Equilibrium Stationary States, A.B. Chhabra, M.J. Feigenbaum, L.P. Kadanoff, A.J. Kolan, and I. Procaccia, Phys. Rev. E 47 3099–3121 (1993).

158. Chaos, Computers, and Physics, L.P. Kadanoff, M.J. Vinson, and A.J. Kolan, (Laboratory Notes for Physical Science 114) (1993).

159. From Order to Chaos: Essays Critical Chaotic and Otherwise, World Scientific (1993).

160. Finite-time singularity formation in Hele-Shaw systems, T.F. Dupont, R.E. Goldstein, L.P. Kadanoff, S-M Zhou, Phys. Rev. E 47 (6) 4182–4196 (1993).

161. The break-up of a heteroclinic connection in a volume preserving mapping, V. Rom-Kedar, L.P. Kadanoff, E.S.C. Ching and C. Amick, Physica D 62 51–65 (1993).

162. Traveling-wave solutions to thin-film equations, S. Boatto, L.P. Kadanoff, and P. Olla, Phys. Rev. E, 48 (6) 4423–4431 (1993).

163. Singularities and Similarities in Interface Flows, A. Bertozzi, M. Brenner, T.F. Dupont and L.P. Kadanoff, in L. Sirovitch Editor, Trends and Perspectives in Applied Mathatematics, Springer Verlag Applied Math Series, Volume 100 (1994), page 155–208.

164. Conditional Averages in Convective Turbulence, L.P. Kadanoff, Physica A 204 341–345 (1994).

165. Bubble, Bubble, Boil, and Trouble, D.H. Rothman and L.P. Kadanoff, Computations in Physics, 8 199–204 (1994).

166. Greats, Reference Frame, L.P. Kadanoff, Physics Today, April (1994).

167. Breakdown of Hydrodynamics in a One Dimensional System of Inelastic Particles, Y. Du, H. Li, L.P. Kadanoff, Phys. Rev. Letts. 74 (8) 1268–1271 (1995).

168. Scaling and Dissipation in the GOY Shell Model, L. Kadanoff, D. Lohse, J. Wang, and R. Benzi, Phys. Fluids 7 (3) 617–629 (1995).

169. A Model of Turbulence, Reference Frame, L.P. Kadanoff, Physics Today, September (1995).

170. How the viscous subrange determines inertial range properties in turbulence shell models, N. Schorghofer, L.P. Kadanoff, and D. Lohse, Physica D 88 40–54 (1995).

171. Scaling and Linear Response in the GOY Turbulence Model, L.P. Kadanoff, D. Lohse, and N. Schörghofer, Physica D 100 165–186 (1997).

172. Inelastic Collapse of Three Particles, T. Zhou and L.P. Kadanoff, Phys. Rev. E, 54 623 (1996).

173. Transitions and Probes in Helium Turbulence, V. Emsellem, L.P. Kadanoff, D. Lohse, P. Tabeling, and J. Wang, Phys. Rev. E, 55 (3) 2672–2681 (1997).

174. Turbulent excursions, L.P. Kadanoff, Fluid Motion, News and Views, Nature, Vol. 382, July 11 (1996).

175. Cascade Models of Turbulence and Mixing, L.P. Kadanoff, Tr. J. of Physics, 21 1–14 (1997).

176. Self-Organized Short-Term Memories, S.N. Coppersmith, T.C. Jones, L.P. Kadanoff, A. Levine, J.P. McCarten, S.R. Nagel, S.C. Venkataramani, and X. Wu, Phys. Rev. Letts. 78 (21) 3983–3986 (1997).

177. Scaling properties of passive scalars in one dimension, L.P. Kadanoff, S. Wunsch and T. Zhou, Physica A 244 190–212 (1997).

178. Can you tell the difference between a stock market crash and a black hole?, L.P. Kadanoff and N. Goldenfeld, for NumeriX www site, unpublished (1997).

179. Built Upon Sand: Theoretical Ideas Inspired by the Flow of Granular Materials, L.P. Kadanoff, Rev. Mod. Phys., 71 (1) 435–444 (1999).

180. Blowups and Singularities, L.P. Kadanoff, Reference Frames, Physics Today, September, pp. 11–12 (1997).

181. Diffusion, Attraction, and Collapse, M. Brenner, P. Constantin, L.P. Kadanoff, A. Schenkel, and S.C. Venkataramani, Nonlinearity 12 1071–1098 (1999).

182 Analysis of a Population Genetics Model With Mutation, Selection, and Pleitropy, S.N. Coppersmith, R.D. Blank, and L.P. Kadanoff, Journal of Statistical Physics **97** 3/4 429–459 (1999).

183. Simple Lessons from Complexity, N. Goldenfeld and L.P. Kadanoff, Science **284** 87–90 (1999).

184. Noise stabilization of self-organized memories, M.L. Povinelli, S.N. Coppersmith, L.P. Kadanoff, S.R. Nagel, and S.C. Venkataramani, Phys. Rev. E. 59 (5) 4970–4982 (1999).

185. From Order to Chaos II, Essays: Critical Chaotic and Otherwise, World Scientific Series on Nonlinear Science, Series A, Vol. 32 (1999).

186. Changing, L.P. Kadanoff, Reference Frame, Physics Today, December, pp. 11–13 (1999).

187. Reversible Boolean Networks I: Distribution of Cycle Lengths, S.N. Coppersmith, Leo P. Kadanoff, Zhitong Zhang, Physica D: Nonlinear Phenomena, Elsenier Science B.V. **149** 11–29 (2001).

188. STATISTICAL PHYSICS Statics, Dynamics and Renormalization, Leo P. Kadanoff, World Scientific (2000).

189. Reversible Boolean Networks II: Phase Transitions, Oscillations, and Local Structures, S.N. Coppersmith, Leo P. Kadanoff, Zhitong Zhang, Physica D: Nonlinear Phenomena, Elsenier Science B.V. **157** 54–74 (2001).

190. The Unreasonable Effectiveness of ..., Leo P. Kadanoff, Reference Frame, Physics Today, November pp. 11–12 (2000).

191. Cascade Models of Turbulence and Mixing, Statistical Methods and Mixing, Leo P. Kadanoff, Mixing: Chaos and Turbulence, edited by Chate et al. Kluwer Academic/Plenum Publishers, New York pp. 385–394 (1999).

192. Class Notes, Phys. 251/CS 279/Math 292, Chaos, Complexity, and Computers, B. Blander, S. Coppersmith, L.P. Kadanoff, A.J. Kolan, M. Magnasco, M.J. Vinson, and S. Wunsch (1999).

193. Book Review: Boltzmann's Science, Irony and Achievement, L.P. Kadanoff, Science **291**, 2553–2554 (March 2001).

194. The 2000 Nora and Edward Ryerson Lecture, "Making a Splash, Breaking a Neck: The Development of Complexity in Physical Systems", L.P. Kadanoff, The University of Chicago Record, **35** 4, April (2001).

195. Book Review by L.P. Kadanoff of Noah's Flood: by William Ryan and Walter Pitman, Perspectives in Biology and Medicine, The Johns Hopkins University Press, Spring, 307–309 (2001).

196. Turbulent Heat Flow: Structures and Scaling, L.P. Kadanoff, Physics Today, pp. 34–39, August (2001).

197. Weak long-ranged Casimir attraction in colloidal crystals, A. Gopinathan, T. Zhou, S.N. Coppersmith, L.P. Kadanoff, and D.G. Grier, Europhysics Letters **57** 3, 451–457 (2002).

198. Sue's Several Heads, the evolution of the natural history museum, L.P. Kadanoff, Perspectives in Biology and Medicine **45** 2, 272–280, The Johns Hopkins University Press (2002).

199. Boolean Dynamics with Random Couplings, M. Aldana, S. Coppersmith and L.P. Kadanoff. In *Perspectives and Problems in Nonlinear Science. A celebratory volule in honor of Lawrence Sirovich.* Springer Applied Mathematical Sciences Series. Ehud Kaplan, Jerrold E. Marsden, and Katepalli R. Sreenivasan Eds. May 2003. ISBN: 0-387-00312-6.

200. Viscous Flow: Approach to Singularity in an Electric Field, C. Yang and L.P. Kadanoff, in process (2002).

201. Models, Morals, and Metaphors, L.P. Kadanoff, Reference Frame, Physics Today, pp. 10–11, February (2002).

202. Book Review: Chemistry, the elements of an Education, L.P. Kadanoff, *Science* **295**, p.448, January (2002). Review of the Book: Uncle Tungsten Memories of a Chemical Boyhood, by Oliver Sacks.

203. The Trouble with Cycles in the *N-K* model, M. Aldana, S. Coppersmith, L.P. Kadanoff and X. Qu, (2002), on Web site: http://discuss.santafe.edu/robustness/stories/

204. Book Review: Wolfram on Celluar Automata, L.P. Kadanoff, Physics Today, 55–56, July (2002). Review of the Book: A New Kind of Science by Stephen Wolfram.

205. Numerical and Theoretical Studies of Noise Effects in the Kauffman Model, Xiaohui Qu, Maximino Aldana, Leo P. Kadanoff. *Journal of Statistical Physics*, **109** 516, December (2002).

206. Geometry, Physics, and Social Commentary, L.P. Kadanoff. Review of Flatterland, by Ian Stewart, *Perspectives in Biology and Medicine*, The Johns Hopkins University Press, **45** 3, 472–474 Summer (2002).

207. Construction of Scientific Knowledge, L.P. Kadanoff. 'unpublished' (2002).

208. Intelligent Design and Complexity Research, L.P. Kadanoff. An Essay and Book Review of the Book: *No Free Lunch: Why Specified Complexity Cannot be Purchased without Intelligence,* by William A. Dembski, 429pp., Roman and Littlefield, 2002. *Journal of Statistical Physics*, **110** (1–2), 451–454 (2003).

209. Bursting Apart, L.P. Kadanoff, *Nature*, Vol. **421** (9), 124 (January 2003).

210. Imports and Exports, L.P. Kadanoff. *Journal of Statistical Physics*, Vol. **111** (5–6), 1391–1396 (June 2003).

211. Excellence in Computer Simulation, L.P. Kadanoff. *Computing in Science and Engineering*, Vol. 6, Issue 2, 57–67 (March/April 2004).

212. Exact solutions for Loewner Evolutions, Wouter Kager, Bernard Nienhuis and L.P. Kadanoff. *Journal of Statistical Physics*, **115** 3/4, 805–822 (May 2004).

213. Discrete Charges on a Two Dimensional Conductor, M.Kl. Berkenbusch, I. Claus, C. Dunn, L.P. Kadanoff, M. Nicewicz, and S.C. Venkataramani. *Journal of Statistical Physics*, **116**, 5/6, (September 2004).

214. Trace for the Loewner Equation with Singular Forcing, L. P. Kadanoff and M. Kleine Berkenbusch. *Nonlinearity*, **17** 4, R41–R54 (2004).

215. The Loewner Equation: Maps and Shapes. Ilya A. Gruzberg and Leo P. Kadanoff. *Journal of Statistical Physics*, **114** 5, 1183–1198 (2004).

216. "Scientific Truth", Leo Kadanoff, ONE HUNDRED REASONS TO BE A SCIENTIST, pages 118–119, ICTP Publications, Trieste (2004).

217. Computational Scenarios. Leo. P. Kadanoff, Physics Today, pp.10–11 (November 2004).

218. Erratum: Trace for the Loewner equation with singular forcing [Nonlinearity, 17 4, R41–R54 (2004)], P. Oikonomou, Leo P. Kadanoff and Marko Kleine Berkenbusch. *Nonlinearity*, **18** 4, 937 (2005).

219. Hip Bone is Connected to…, Leo P. Kadanoff, Physics Today (September 2005).

220. Stochastic Loewner evolution driven by Lévy processes, I. Rushkin, P. Oikonomou, L.P. Kadanoff and I.A. Gruzberg. *J. Stat. Mech.* P01001 (January 2006).

221. Pulled Fronts and the Reproductive Individual, Leo P. Kadanoff, *Journal of Statistical Physics*, p. 1–4 (April 2006).

222. An educational moment, Leo P. Kadanoff, *Physics Today*, (September 2006).

223. 3D Large-Scale DNS of Weakly-Compressible Homogeneous Isotropic Turbulence with Lagrangian Tracer Particles, Robert Fisher, F. Catteneo, P. Constantin, L. Kadanoff, D. Lamb, and T. Plewa, unpublished.

224. Global properties of stochastic Loewner evolution driven by Levy processes, P. Oikonomou, I. Rushkin, I. A. Gruzberg and L. P. Kadanoff, *Journal of Statistical Mechanics*, P01019 (2008).

225. APS, Physics: Aspirations and Goals, Leo Kadanoff, *published in APS News, July 2008.*

226. Asymptotics of eigenvalues and eigenvectors of Toeplitz matrices, H. Dai, Z. Geary and L. P. Kadanoff, *Journal of Statistical Mechanics*, P05012 (2009).

227. Hip Bone is connected to…II, Leo P. Kadanoff, *Physics Today*, March 2009.

228. R. Benzi, etal. Intermittancy and Universality in Turbulence, Phys. Rev. Letts. **100** 234503 (2008).

229. Leo Kadanoff, More is the Same; Mean Field Theory and Phase Transitions, *J Stat Phys.* 137, 777–797 (2009).

230. Theories of Matter: Infinities and Renormalization, Leo Kadanoff, *The Oxford Handbook of the Philosophy of Physics* editor Robert Batterman, Oxford University Press (2011).

231. Relating Theories via Renormalization, Leo Kadanoff, studies in History and Philosophy of Science Part B: Studies in History and Philosophy of Modern Physics **44** (1):22–39 (2013).

232. Expansions for eigenfunction and eigenvalues of large-n Toepliz matrices, Papers in Physics, 2 020003 (2010).

233. "Kenneth Geddes Wilson, 1936–2013, An Appreciation", Leo P. Kadanoff, *Proceeding of StatPhys25 J. Stat. Mech.* (2013) P10016.

234. "Kenneth Geddes Wilson (1936–2013)" Leo P. Kadanoff, *Nature Magazine.* (August 1, 2013).

235. "Real-space renormalization in statistical mechanics" Efi Efrati, Zhe Wang, Amy Kolan, and Leo P. Kadanoff Rev. Mod. Phys. accepted January 2014. arXiv:1301.6323.

236. "Reflections on Gibbs: From Statistical Physics to the Amistad", Leo P. Kadanoff, to be published J. Stat, Phys. Xiv:1403.6162 [physics,.hist-ph] 5 Apr 2014.

237. "The Early Years of Condensed Matter Physics at Illinois — in Celebration of the 80th Birth Year of Charles P. Slichter" — Charlie Slichter & the gang at Urbana arXiv:1403.2458 (March 2014).

238. "Slippery Wave Functions" V2.01 J. Stat. Phys. 152 805–823 (2013) arXiv:1303.0585.

239. "Innovations in Statistical Physics" to be published in Annual Reviws of Condensed Matter Physics arXiv: 1403.6464 (April 2014).

240. Entropy is in Flux arXiv: 1403.6162 (2014).

241. Reflections on Gibbs: From Statistical Physics to the Amistad J Stat Phys 156:1–9 (2014). arXiv:1403.6162 [physics,.hist-ph].

Martin L. Perl

MARTIN L. PERL

Present Position

Professor Emeritus
Kavli Institute for Particle Astrophysics and Cosmology, Stanford University
SLAC National Accelerator Laboratory, Stanford University

Profession

Physicist, engineer, educator

Basic Research Interest

Experimental astrophysics and cosmology
Experimental elementary particle physics
Atom optics
Atom interferometry

Honors

Award: 1995 Nobel Prize in Physics
Award: 1982 Wolf Prize in Physics for discovery of tau lepton
Member: U.S. National Academy of Sciences
Member: American Academy of Arts and Sciences
Honorary Degree: Dr. Sc. Univ. of Belgrade, 2009
Honorary Degree: Dr. Sc. Union College, 2009
Honorary Degree: Dr. Sc. Polytechnic University, 1996
Honorary Degree: Dr. Sc. University of Chicago, 1990
In: Who's Who in America. Who's Who in the World

Birthplace

Born   New York City, 24 June 1927

Education

1942	Graduated from James Madison High School in Brooklyn, New York
1948	Bachelor in Chemical Engineering from Polytechnic Institute of Brooklyn
1955	Ph.D. in Physics from Columbia University

Military Service

1944–1945	U.S. Merchant Marine
1945–1946	U.S. Army

Professional Employment

1948–1950	Chemical engineer with General Electric Company

1950–1955	Graduate research assistant in atomic physics at Columbia University
1955–1963	Instructor, assistant professor, and associate professor of physics at University of Michigan
1963–2004	Professor and Group Leader at SLAC National Accelerator Laboratory, Stanford University
2004-Present	Professor and Professor Emeritus, SLAC National Accelerator Laboratory and Kavli Institute for Particle Astrophysics and Cosmology , Stanford University

Books

Reflections on Experimental Science (1996)

The Tau-Charm Factory (1994), editor

The Search for New Elementary Particles (1992), editor

Physics Careers, Employment and Education (1977), editor

High Energy Hadron Physics (1974)

COMMENTARY

Early Years

In spite of very good school marks, a love of books (particularly in science and mathematics), and a great love of mechanics and chemistry, and winning the physics medal in high school, I never thought of becoming a scientist. As the child of immigrants, I was taught the importance of "making a good living"; a boy should become a doctor, a dentist, a lawyer, an accountant or go into business; a girl should become a teacher or a nurse. I did not think about going into business because the difficulties of the Depression years did not make business a good way to earn a living. My parents and I knew about a few scientists, certainly Pasteur and Einstein, but we did not know that it was possible for a man to earn a living as a scientist.

The War Years, Polytechnic Institute of Brooklyn, and Chemical Engineering

We did know that a man could earn a living as engineer, particularly as a chemical engineer; my father sold printing to local chemical companies. This was an unusual choice for a Jewish boy in the early 1940's because there was still plenty of anti-Semitism in engineering companies. In 1942, I enrolled in the Polytechnic Institute of Brooklyn, and began studying chemical engineering

My studies were interrupted by the war. I wanted to join the United States Army, but I was not yet eighteen and my parents would not give me permission. However, they agreed to me joining the United States Merchant Marine, I was allowed to leave college and become an engineering cadet in the program at the Kings Point Merchant Marine Academy. The training ship was wonderful — it had a main reciprocating steam engine, and direct steam driven pumps and auxiliary machinery; a paradise of mechanics. But when I went to sea for six months as part of the training, I was on a Victory ship with a sealed steam turbine. Very boring. When the war ended with the atom bomb I left the merchant marine. I knew so little about physics that I didn't know why the bomb was so powerful.

I didn't get right back to college. The draft was still in force in the United States. I was drafted, and spent a pleasant year as sergeant at a laboratory in Walter Reed Hospital in Washington, DC, working on encephalitis vaccine; a result of my being a chemical engineering student. I found I was definitely not interested in biology. Finally, I returned to the Polytechnic Institute and received a summa cum laude bachelor degree in Chemical Engineering in 1948.

The skills and knowledge I acquired at the Polytechnic Institute have been crucial in all my experimental work: the use of strength of materials principles in equipment design, machine shop practice, engineering drawing, practical fluid mechanics, inorganic and organic chemistry, chemical laboratory techniques, manufacturing processes, metallurgy, basic concepts in mechanical engineering, basic concepts in electrical engineering, dimensional analysis, speed and power in mental arithmetic, numerical estimation (crucial when depending on a slide rule for calculations), and much more.

Working at General Electric and Union College

Upon graduation, I joined the General Electric Company. After a year in an advanced engineering training program, I settled in Schenectady, New York, working as a chemical engineer in the Electron Tube Division. I worked in an engineering office in the electron tube production factory. Our job was to troubleshoot production problems, to improve production processes, and occasionally to do a little development work. We were not a fancy R&D office. I worked on speeding-up the production of television picture tubes, and then on problems of grid emission in industrial power tubes. These tasks led to a turning point in my life.

I had to learn a little about how electron vacuum tubes worked, so I took a few courses in Union College in Schenectady specifically, atomic physics and advanced calculus. I got to know a wonderful physics professor, Vladimir Rojansky. One day he said to me "Martin, what you are interested in is called physics not chemistry!" At the age of 23, I finally decided to begin the study of physics.

Graduate Study in Physics, I.I. Rabi, and Learning the Physicist's Trade

I entered the physics doctoral program in Columbia University in the autumn of 1950. Looking back, it seems amazing that I was admitted. True, I had a summa cum laude bachelor degree, but I had taken only two courses in physics: one year of elementary physics and a half-year of atomic physics. There were several reasons I could do this in 1950; it could not be done today. Graduate study in physics was primitive in 1950, compared to today's standards. We did not study quantum mechanics until the second year, the first year was devoted completely to classical physics. The most advanced quantum mechanics we ever studied was a little bit in Heitler, and we were not expected to be able to do calculations in quantum electrodynamics.

There was no advising or course guidance by the Columbia Physics Department faculty; students were on their own. I was arrogant about my ability to learn anything fast. Soon I realized I was in trouble, I realized that many of my fellow students were smarter than me and better trained; but it was too late to quit. I had explained the return to school to my astonished parents by telling them that physics was what Einstein did. They thought if Einstein, why not Martin. I survived the Columbia Physics Department, never the best student, but an ambitious and hard-working student. I was married and had one child. I had to get my Ph.D. and once more earn a living.

Just as the Polytechnic Institute was crucial in my learning how to do engineering; just as Union College and Vladimir Rojansky were crucial in my choosing physics; so Columbia University and my thesis advisor, I.I. Rabi, were crucial in my learning how to do experimental physics. I undertook for my doctoral research the problem of using the atomic beam resonance method to measure the quadrupole moment of the sodium nucleus. This measurement had to be made using an excited atomic state, and Rabi had found a way to do this.

As is well known, Rabi was not a "hands-on" experimenter. He never used tools or manipulated the apparatus when I knew him. I learned experimental techniques from older graduate students and by occasionally going to ask for help or advice from Rabi's colleague, Polykarp Kusch. I was on my own in learning the experimenter's trade. I learned quickly, as I tell my graduate students now, there are no answers in the back of the book when the equipment doesn't work or the measurements look strange.

I learned things more precious than experimental techniques from Rabi. I learned the deep importance of choosing one's own research problems. Rabi once told me that he would worry when talking to Leo Szilard that Szilard would propose some idea to Rabi. This was because Rabi wouldn't carry out an idea suggested by someone else, even though he had already been thinking about that same idea.

I learned from Rabi the importance of getting the right answer and checking it thoroughly. When I finished my measurement of the quadrupole moment, I was eager to publish and to get on with earning a living. But Rabi had heard that a similar measurement had been made by an optical resonance method in France. He wrote to the French physicists to see if they had a similar answer. He didn't telephone or cable; he calmly wrote. I waited nervously. Six or eight weeks later he received the answer that they had a similar answer; then, I was allowed to publish. It is far better to be delayed, it is better to be second in publishing a result, than to publish first with the wrong answer.

It was Rabi who always emphasized the importance of working on a fundamental problem, and it was Rabi who sent me into elementary particle physics. It would have been natural for me to continue in atomic physics, but he preached particle physics to me — particularly when his colleagues in atomic physics were in the room. I think that most of that public preaching may have been Rabi's way of deliberately irritating his colleagues.

Michigan, Bubble Chambers, and Working with L.W. Jones on Pion-Proton Elastic Scattering

When I received my Ph.D. in 1955, I had job offers from the Physics Departments at Yale, the University of Illinois, and the University of Michigan. At that time, the first two Physics Departments had better reputations in elementary particle physics, and so I deliberately went to Michigan. I followed a two-part theorem that I always pass on to my graduate students and post-doctoral research associates. Part 1: don't choose the most powerful experimental group or department — choose the group or department where you will have the most freedom. Part 2: there is an advantage in working in a small or new group — then you will get the credit for what you accomplish.

At Michigan, I first worked in bubble chamber physics with Donald Glaser using the Brookhaven COSMOTRON. But I wanted to be on my own. When the Russians flew SPUTNIK in 1957, I saw the opportunity, and jointly with my colleague, Lawrence W. Jones, we wrote to Washington for research money. We began our own research program, using first the now-forgotten luminescent chamber and then spark chambers.

Jones and I measured the differential cross-section high energy pion-proton elastic scattering using the Berkeley BEVATRON. Our purpose was to look for verification of the Regge theory of elastic scattering. Our spark chamber measurements were beautiful, much more precision than bubble chamber measurements. But Regge theory turned out to be a sham.

I Move to the Stanford Linear Accelerator Center

In 1963, I moved to the Stanford Linear Accelerator Center, now the SLAC National Accelerator Laboratory. I formed an experimental group, called Group E. We carried out an early search for new types of particles working with Robert Hofstadter using his large sodium iodide crystals. No new particles found.

My group then carried out our most beautiful optical spark chamber experiment, a measurement of high energy, muon-proton inelastic scattering. Our object was to compare muon-proton inelastic scattering with electron-proton inelastic scattering, the latter being measured at SLAC. We were searching for some hint as to why the intrinsic nature of the muon is different from the intrinsic nature of the electron. By 1971, we found that the two inelastic cross-sections were the same within 15%; the major inaccuracy came from our experiment. There was no hint as to the nature of the muon-electron difference. This led me to think about other ways to search for the nature of the muon-electron difference and ultimately to the discovery of the tau lepton.

The Discovery of the Tau Lepton

The history of the discovery of the tau lepton begins in the late 1960's when my colleagues and I and other experimenters worked on the problem of "how does the muon differ from the electron". I began to speculate that if we could not find the origin of the electron-muon difference, perhaps we could find another charged lepton. This new lepton I called a sequential lepton because it would be in the sequence electron, muon, new heavier lepton. The new heavy lepton would decay as follows:

new heavy lepton → electron or muon + neutrino + neutrino

I realized that a new heavy lepton might be found using electron-+ positron annihilation by looking for events of the form

electron + positron → electron + muon + undetected neutrinos carrying off energy

This search method had many attractive features. If the new lepton was a point particle, we could search up to almost equal to the beam energy, given enough luminosity. The appearance of an (electron + muon + undetected neutrinos carrying off energy) event would be dramatic.

My thoughts in the late 1960's and 1970–1971 about heavy lepton searches using electron + positron annihilation coincided with the beginning of the building of the SPEAR electron + positron storage ring by a SLAC group led by Burton Richter. Gary Feldman and I, and our SLAC Group E, joined Richter's Group C and a Lawrence Berkeley Laboratory Group led by William Chinowsky, Gerson Goldhaber, and George Trilling.

The SPEAR collider began operation in 1973. Eventually SPEAR obtained a total energy of about 8 GeV, but in the first few years the maximum energy with useful luminosity was 4.8 GeV.

In 1974, I began to find electron + muon events with missing energy, that is, events with an electron, an opposite sign muon, no other charged particles, and no visible photons. By early 1975, we had seen dozens of electron + muon events, but those of us who believed we had found a heavy lepton faced two problems: how to convince the rest of our collaboration and how to convince the physics world. The main focus of this early skepticism was the electron, muon and pion identification systems: Had we underestimated hadron misidentification into leptons? Since our identification system only covered about half of 4π, what about undetected photons? What about inefficiencies and cracks in these systems?

Finally in December 1975, the SLAC-LBL experimenters published a paper by Perl *et al.* entitled " Evidence for Anomalous Lepton Production in e + −e Annihilation". The final paragraph read:

"We conclude that the signature e − μ events cannot be explained either by the production and decay of any presently known particles or as coming from any of the well-understood interactions which can conventionally lead to an e and a μ in the final state. A possible explanation for these is the production and decay of a pair of new particles, each having a mass in the range of 1.6 to 2.0 GeV/c^2 ."

This was the first evidence of the existence of the tau lepton.

There is much more to tell about the decade long path from this first evidence to the acceptance of the reality of the tau lepton and the measurement of its properties. And my Nobel Prize in 1995. I do not have the space here.

Search For Fractional Charge Elementary Particles

All known elementary particles that can be isolated such as electrons, photons, and protons, have an electric charge given by nq where n is an integer, including 0, and q is the magnitude of the electron charge. (Quarks have fractional charge such as $q/3$ and $2q/3$, but it is believed that quarks are always bound in other particles such as protons, and cannot be isolated.) There is no understanding of the observed rule that the charges of elementary particles that can be isolated are restricted to nq, hence this rule is simply built into modern elementary particle theory.

The idea for using a modern, computerized, Millikan Oil Drop Method came from Klaus Lackner and others in the early 1990s. At the SLAC National Accelerator Laboratory, a group gathered to develop searches for isolatable, fractional charge particle: Irwin T. Lee, Sewan Fan, Valerie Halyo, Eric R. Lee, Peter C. Kim, Dinesh Loomba, Howard Rogers, Gordon Shaw and myself. The drop ejector was a glass and silicon micromachined drop on demand ejector producing 25 micron diameter target drops.

There are geochemical and astrophysical reasons why fractional charge particles could exist in much greater abundance in some types of asteroidal material compared to accessible terrestrial materials. Therefore, we carried out our final search in 2004–2007 using material from the Allende meteorite.

No fractional charges were detected.

Present Research Idea: Using Atom Interferometry to Search Terrestrially for the Presence of Dark Energy

1. Dark energy is the rather poetic name for the astronomical observations made about fifteen years ago that the expansion rate of the visible universe is accelerating. This phenomena is not explained by the conventional Einstein theory of general relativity and by the measured total matter in the universe

2. Although it is less than two decades since this astonishing discovery, the experimenter or observer desirous of investigating dark energy already faces difficulties. Astronomical observations of the dark energy phenomenon are becoming increasingly precise. But these observations are on the cosmological scale and are unlikely to teach us anything of the essence of dark energy, is it an energy field, is it related to matter, is it just a term in the equations that describe general relativity, is it a fluid? Or is it some other phenomenon that we have never thought about?

3. During my doctoral research with the laureate Isadore Rabi, he repeatedly reminded me "Physics is an experimental science". This is the spirit behind this experiment.

4. About three years ago, I began to think about the possibility of using non-astronomical means to investigate dark energy, and I conceived the idea of searching on earth for the presence of dark energy or other dark contents of the vacuum using atom interferometry. I had never worked in atom interferometry but I have a basic understanding of the technology of atom interferometry.

5. Atom interferometry uses the quantum mechanical interference properties of falling neutral atoms to make very precise measurements of the gravitational field, present precision is 10^{-9}. This precision is achieved through the small de Broglie wavelength of slowly moving atoms: the phase difference can be millions of radians, whereas the phase difference can be measured to 10^{-4} radians. Accuracies of 10^{-15} are expected in future atom interferometry laboratory experiments on the weak equivalence principle. Moreover, atoms have few and well characterized internal degrees of freedom, and couple in a well controlled way to their environment. When the great precision of atomic beam interferometry is combined with improvements that have recently been

demonstrated such as large momentum transfer or common-mode rejection between conjugate interferometers, new applications will become possible. Furthermore, the possibility of future operation in the nearly gravity-free environment of a spacecraft promises truly impressive possibilities.

Evidence for Anomalous Lepton Production in e^+-e^- Annihilation*

M. L. Perl, G. S. Abrams, A. M. Boyarski, M. Breidenbach, D. D. Briggs, F. Bulos, W. Chinowsky,
J. T. Dakin,† G. J. Feldman, C. E. Friedberg, D. Fryberger, G. Goldhaber, G. Hanson,
F. B. Heile, B. Jean-Marie, J. A. Kadyk, R. R. Larsen, A. M. Litke, D. Lüke,‡
B. A. Lulu, V. Lüth, D. Lyon, C. C. Morehouse, J. M. Paterson,
F. M. Pierre,§ T. P. Pun, P. A. Rapidis, B. Richter,
B. Sadoulet, R. F. Schwitters, W. Tanenbaum,
G. H. Trilling, F. Vannucci,‖ J. S. Whitaker,
F. C. Winkelmann, and J. E. Wiss

*Lawrence Berkeley Laboratory and Department of Physics, University of California, Berkeley, California 94720,
and Stanford Linear Accelerator Center, Stanford University, Stanford, California 94305*
(Received 18 August 1975)

We have found events of the form $e^+ + e^- \rightarrow e^{\pm} + \mu^{\mp} +$ missing energy, in which no other charged particles or photons are detected. Most of these events are detected at or above a center-of-mass energy of 4 GeV. The missing-energy and missing-momentum spectra require that at least two additional particles be produced in each event. We have no conventional explanation for these events.

We have found 64 events of the form

$$e^+ + e^- \rightarrow e^{\pm} + \mu^{\mp} + \geq 2 \text{ undetected particles} \quad (1)$$

for which we have no conventional explanation. The undetected particles are charged particles or photons which escape the 2.6π sr solid angle of the detector, or particles very difficult to detect such as neutrons, K_L^0 mesons, or neutrinos. Most of these events are observed at center-of-mass energies at, or above, 4 GeV. These events were found using the Stanford Linear Accelerator Center–Lawrence Berkeley Laboratory (SLAC-

VOLUME 35, NUMBER 22   PHYSICAL REVIEW LETTERS   1 DECEMBER 1975

LBL) magnetic detector at the SLAC colliding-beams facility SPEAR.

Events corresponding to (1) are the signature for new types of particles or interactions. For example, pair production of heavy charged leptons[1-4] having the decay modes $l^- \to \nu_l + e^- + \bar{\nu}_e$, $l^+ \to \bar{\nu}_l + e^+ + \nu_e$, $l^- \to \nu_l + \mu^- + \bar{\nu}_\mu$, and $l^+ \to \bar{\nu}_l + \mu^+ + \nu_\mu$ would appear as such events. Another possibility is the pair production of charged bosons with decays $B^- \to e^- + \bar{\nu}_e$, $B^+ \to e^+ + \nu_e$, $B^- \to \mu^- + \bar{\nu}_\mu$, and $B^+ \to \mu^+ + \nu_\mu$. Charmed-quark theories[5,6] predict such bosons. Intermediate vector bosons which mediate the weak interactions would have similar decay modes, but the mass of such particles (if they exist at all) is probably too large[7] for the energies of this experiment.

The momentum-analysis and particle-identifier systems of the SLAC-LBL magnetic detector[8] cover the polar angles $50° \leq \theta \leq 130°$ and the full 2π azimuthal angle. Electrons, muons, and hadrons are identified using a cylindrical array of 24 lead-scintillator shower counters, the 20-cm-thick iron flux return of the magnet, and an array of magnetostrictive wire spark chambers situated outside the iron. Electrons are identified solely by requiring that the shower-counter pulse height be greater than that of a 0.5-GeV e. Incidently, the e's in the e-μ events thus selected give no signal in the muon chambers; and their shower-counter pulse-height distribution is that expected of electrons. Also the positions of the e's in the shower counters as determined from the relative pulse heights in the photomultiplier tubes at each end of the counters agree within measurement errors with the positions of the e tracks. Hence the e's in the e-μ events are not misidentified combinations of $\mu + \gamma$ or $\pi + \gamma$ in a single shower counter, except possibly for a few events already contained in the background estimates. Muons are identified by two requirements. The μ must be detected in one of the muon chambers after passing through the iron flux return and other material totaling 1.67 absorption lengths for pions. And the shower-counter pulse height of the μ must be small. All other charged particles are called hadrons. The shower counters also detect photons (γ). For γ energies above 200 MeV, the γ detection efficiency is about 95%.

To illustrate the method of searching for events corresponding to Reaction (1), we consider our data taken at a total energy ($\sqrt{s}$) of 4.8 GeV. This sample contains 9550 three-or-more-prong events and 25 300 two-prong events which include $e^+ + e^- \to e^+ + e^-$ events, $e^+ + e^- \to \mu^+ + \mu^-$ events, two-prong hadronic events, and the e-μ events described here. To study two-prong events we define a coplanarity angle

$$\cos\theta_{\text{copl}} = -(\vec{n}_1 \times \vec{n}_{e^+}) \cdot (\vec{n}_2 \times \vec{n}_{e^+}) / |\vec{n}_1 \times \vec{n}_{e^+}| |\vec{n}_2 \times \vec{n}_{e^+}|, \quad (2)$$

where $\vec{n}_1$, $\vec{n}_2$, and $\vec{n}_{e^+}$ are unit vectors along the directions of particles 1, 2, and the e^+ beam. The contamination of events from the reactions $e^+ + e^- \to e^+ + e^-$ and $e^+ + e^- \to \mu^+ + \mu^-$ is greatly reduced if we require $\theta_{\text{copl}} > 20°$. Making this cut leaves 2493 two-prong events in the 4.8-GeV sample.

To obtain the most reliable e and μ identification[9] we require that each particle have a momentum greater than 0.65 GeV/c. This reduces the 2493 events to the 513 in Table I. The 24 e-μ events with no associated photons, called the signature events, are candidates for Reaction (1). The e-μ events can come conventionally from the two-virtual-photon process[10] $e^+ + e^- \to e^+ + e^- + \mu^+ + \mu^-$. Calculations indicate that this source is negligible, and the absence of e-μ events with charge 2 proves this point since the number of charge-2 e-μ events should equal the number of charge-0 e-μ events from this source.

We determine the background from hadron misidentification or decay by using the 9550 three-or-more-prong events and assuming that every particle called an e or a μ by the detector either was a misidentified hadron or came from the decay of a hadron. We use $P_{h \to l}$ to designate the sum of the probabilities for misidentification or decay causing a hadron h to be called a lepton l. Since the P's are momentum dependent[9] we use all the

TABLE I. Distribution of 513 two-prong events, obtained at $E_{\text{c.m.}} = 4.8$ GeV, which meet the criteria $|\vec{p}_1| > 0.65$ GeV/c, $|\vec{p}_2| > 0.65$ GeV/c, and $\theta_{\text{copl}} > 20°$. Events are classified according to the number N_γ of photons detected, the total charge, and the nature of the particles. All particles not identified as e or μ are called h for hadron.

Particles \ N_γ	0	1	>1	0	1	>1
	Total charge = 0			Total charge = ±2		
e-e	40	111	55	0	1	0
e-μ	24	8	8	0	0	3
μ-μ	16	15	6	0	0	0
e-h	20	21	32	2	3	3
μ-h	17	14	31	4	0	5
h-h	14	10	30	10	4	6

VOLUME 35, NUMBER 22   PHYSICAL REVIEW LETTERS   1 DECEMBER 1975

e-h, μ-h, and h-h events in column 1 of Table I to determine a "hadron" momentum spectrum, and weight the P's accordingly. We obtain the momentum-averaged probabilities $P_{h \to e} = 0.183 \pm 0.007$ and $P_{h \to \mu} = 0.198 \pm 0.007$. Collinear e-e and μ-μ events are used to determine $P_{e \to h} = 0.056 \pm 0.02$, $P_{e \to \mu} = 0.011 \pm 0.01$, $P_{\mu \to h} = 0.08 \pm 0.02$, and $P_{\mu \to e} < 0.01$.

Using these probabilities and assuming that all e-h and μ-h events in Table I result from particle misidentifications or particle decays, we calculate for column 1 the contamination of the e-μ sample to be 1.0 ± 1.0 event from misidentified e-e,[11] < 0.3 event from misidentified μ-μ,[11] and 3.7 ± 0.6 events from h-h in which the hadrons were misidentified or decayed. The total e-μ background is then 4.7 ± 1.2 events.[12,13] The sta-

tistical probability of such a number yielding the 24 signature e-μ events is very small. The same analysis applied to columns 2 and 3 of Table I yields 5.6 ± 1.5 e-μ background events for column 2 and 8.6 ± 2.0 e-μ background events for column 3, both consistent with the observed number of e-μ events.

Figure 1(a) shows the momentum of the μ versus the momentum of the e for signature events.[14] Both p_μ and p_e extend up to 1.8 GeV/c, their average values being 1.2 and 1.3 GeV/c, respectively. Figure 1(b) shows the square of the invariant e-μ mass (M_i^2) versus the square of the missing mass (M_m^2) recoiling against the e-μ system. To explain Fig. 1(b) at least two particles must escape detection. Figure 1(c) shows the distribution in collinearity angle between the e and μ ($\cos\theta_{\text{coll}} = -\vec{p}_e \cdot \vec{p}_\mu / |\vec{p}_e||\vec{p}_\mu|$). The dip near $\cos\theta_{\text{coll}} = 1$ is a consequence of the coplanarity cut; however, the absence of events with large θ_{coll} has dynamical significance.

Figure 2 shows the *observed* cross section in the range of detector acceptance for signature e-μ events versus center-of-mass energy with the background subtracted at each energy as described above.[9] There are a total of 86 e-μ events summed over all energies, with a calculated background of 22 events.[12] The corrections to obtain the true cross section for the angle and momentum cuts used here depend on the hypothesis as to the origin of these e-μ events, and the corrected cross section can be many times larger than the observed cross section. While Fig. 2 shows an apparent threshold at around 4 GeV, the statistics are small and the correction fac-

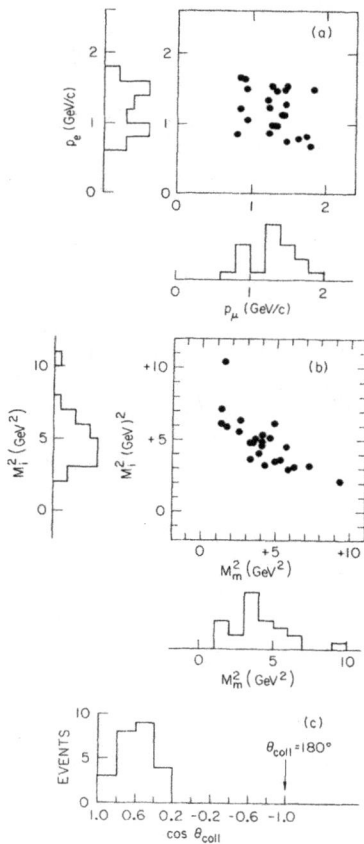

FIG. 1. Distribution for the 4.8-GeV e-μ signature events of (a) momenta of the e (p_e) and μ (p_μ); (b) square of the invariant mass (M_i^2) and square of the missing mass (M_m^2); and (c) $\cos\theta_{\text{coll}}$.

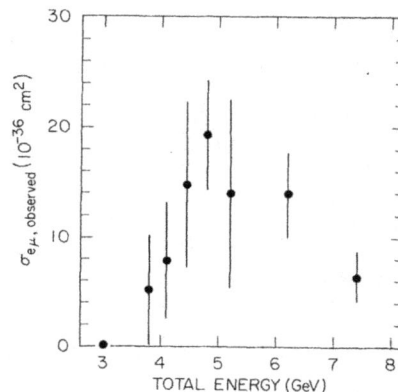

FIG. 2. The *observed* cross section for the signature e-μ events.

VOLUME 35, NUMBER 22   PHYSICAL REVIEW LETTERS   1 DECEMBER 1975

tors are largest for low $\sqrt{s}$. Thus, the apparent threshold may not be real.

We conclude that the signature e-μ events cannot be explained either by the production and decay of any presently known particles or as coming from any of the well-understood interactions which can conventionally lead to an e and a μ in the final state. A possible explanation for these events is the production and decay of a pair of new particles, each having a mass in the range of 1.6 to 2.0 GeV/c^2.

---

*Work supported by the U. S. Energy Research and Development Administration.

†Present address: Department of Physics and Astronomy, University of Massachusetts, Amherst, Mass. 01002.
‡Fellow of Deutsche Forschungsgemeinschaft.
§Centre d'Etudes Nucléaires de Saclay, Saclay, France.
∥Institut de Physique Nucléaire, Orsay, France.

[1] M. L. Perl and P. A. Rapidis, SLAC Report No. SLAC-PUB-1496, 1974 (unpublished).
[2] J. D. Bjorken and C. H. Llewellyn Smith, Phys. Rev. D **7**, 887 (1973).
[3] Y. S. Tsai, Phys. Rev. D **4**, 2821 (1971).
[4] M. A. B. Beg and A. Sirlin, Annu. Rev. Nucl. Sci. **24**, 379 (1974).
[5] M. K. Gaillard, B. W. Lee, and J. L. Rosner, Rev. Mod. Phys. **47**, 277 (1975).
[6] M. B. Einhorn and C. Quigg, Phys. Rev. D (to be published).
[7] B. C. Barish *et al.*, Phys. Rev. Lett. **31**, 180 (1973).
[8] J.-E. Augustin *et al.*, Phys. Rev. Lett. **34**, 233 (1975); G. J. Feldman and M. L. Perl, Phys. Rep. **19C**, 233 (1975).
[9] See M. L. Perl, in Proceedings of the Canadian Institute of Particle Physics Summer School, Montreal, Quebec, Canada, 16–21 June 1975 (to be published).
[10] V. M. Budnev *et al.*, Phys. Rep. **15C**, 182 (1975); H. Terazawa, Rev. Mod. Phys. **45**, 615 (1973).
[11] These contamination calculations do not depend upon the source of the e or μ; anomalous sources lead to overestimates of the contamination.
[12] Using *only* events in column 1 of Table I we find at 4.8 GeV $P_{h \to e} = 0.27 \pm 0.10$, $P_{h \to \mu} = 0.23 \pm 0.09$, and a total e-μ background of 7.9 ± 3.2 events. The same method yields a total e-μ background of 30 ± 6 events summed over all energies. This method of background calculation (Ref. 9) allows the hadron background in the two-prong, zero-photon events to be different from that in other types of events.
[13] Our studies of the two-prong and multiprong events show that there is *no* correlation between the misidentification or decay probabilities; hence the background is calculated using independent probabilities.
[14] Of the 24 events, thirteen are $e^+ + \mu^-$ and eleven are $e^- + \mu^+$.

SLAC-PUB-9191

Large bulk matter search for fractional charge particles*

Irwin T. Lee, Sewan Fan, Valerie Halyo, Eric R. Lee,
Peter C. Kim, Martin L. Perl and Howard Rogers
*Stanford Linear Accelerator Center, Stanford University,
Stanford, California 94309*

Dinesh Loomba
*Department of Physics, University of New Mexico,
Albuquerque, New Mexico 87131*

Klaus S. Lackner
*Department of Earth and Environmental Engineering,
Columbia University,
New York, New York 10027*

Gordon Shaw
*Department of Physics, University of California–Irvine,
Irvine, California 92717*

Submitted to Physical Review D
(Dated: April 2, 2002)

We have carried out the largest search for stable particles with fractional electric charge, based on an oil drop method that incorporates a horizontal electric field and upward air flow. No evidence for such particles was found, giving a 95% confidence level upper limit of 1.17×10^{-22} particles per nucleon on the abundance of fractional charge particles in silicone oil for $0.18e \leq |Q_{residual}| \leq 0.82e$. Since this is the first use of this new method we describe the advantages and limitations of the method.

PACS numbers: 14.80.-j, 13.40.Em, 14.65.-q, 47.60.+i

I. INTRODUCTION

We have carried out the largest search for fractional electric charge elementary particles in bulk matter using 70.1 mg of silicone oil. That is, we looked for stable particles whose charge Q deviates from Ne where N is an integer, including zero, and e is the magnitude of the charge on the electron. No evidence for such particles was found in this amount of silicone oil. We used our new version [1] of the Millikan oil drop method containing two innovations compared to the classical method that we used in Halyo *et al.*[2]. One innovation is that the drop charge is obtained by observing the drop motion in a *horizontal*, alternating electric field compared to the classical use of a vertical electric field [2, 3, 4]. The other innovation is the use of an upward flow of air to reduce the vertical terminal velocity of the drop, which enabled us to use larger drops, about 20.6 μm in diameter compared to the 10 μm drops used in our previous experiments.

We define the residual drop charge, $Q_r = Q - N_l e$ where N_l is the largest integer less than Q/e. We find

the 95% confidence level upper limit on the abundance of fractional charge particles in silicone oil for $0.18e \leq Q_r \leq 0.82e$ is 1.17×10^{-22} particles per nucleon. This experiment was a follow up on our previous search in silicone oil, Halyo *et al.*[2], based on 17.4 mg. In that search we found one drop with anomalous charge, but no such charge was found in the present experiment.

In this paper we describe the experimental method and apparatus in Sec. II. In Sec. III we discuss the measurement precision resulting from the various measurement errors and the calibration methods. The data analysis method, including the criteria used to accept drop charge measurements, is discussed in Sec. IV. Here we pay particular attention to the drop spacing criterion necessitated by interactions between adjacent drops. This is the primary limitation on the rate at which drops can be measured and we had to acquire considerable experience to understand this limitation. We conclude with Sec. V, giving our results, comparing our results with other fractional charge searches, and discussing the applicability and extension of this new Millikan oil drop technique to other searches.

*Work supported by Department of Energy contract DE-AC03-76SF00515.

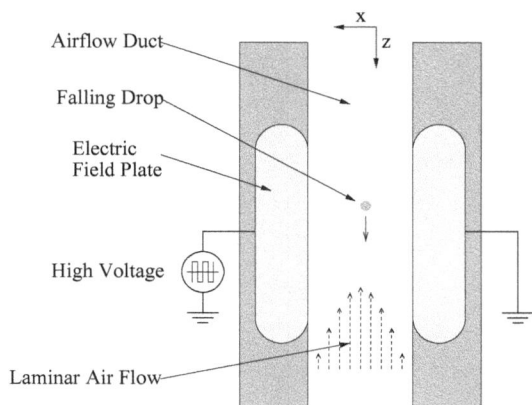

FIG. 1: Basic principles of the experimental method.

II. EXPERIMENTAL METHOD AND APPARATUS

A. Experimental method

The principle of the experimental method is simple. Consider a drop of radius r, density ρ, and charge Q falling in air through a horizontal electric field of strength E, as shown in Fig. 1. Applying Stokes' law the horizontal terminal velocity, v_x is

$$v_x = \frac{QE}{6\pi\eta r} \qquad (1)$$

where η is the viscosity of air. Hence measuring v_x gives Q providing r is known. As explained in Sec. III A, the drop radius is determined from the v_x of integer charge particles. Note that the measurement of Q does not depend on the density of the drop and is also independent of the gravitational force on the drop. The electric field alternates in the $+x$ and $-x$ direction so that the drop is moved back and forth along the x axis. This cancels some sources of error and allows the drop motion to be viewed in a relatively narrow horizontal area. Previous uses of a static horizontal electric field in the Millikan oil drop method were in 1941 by Hopper and Laby [5] who measured the electron charge and by Kunkel [6] in 1950 who measured the charge on dust particles.

If the drop were falling in still air, the vertical terminal velocity would be given by

$$v_{z,term} = \frac{2r^2\rho g}{9\eta} \qquad (2)$$

where g is the acceleration of gravity. However we use an upward flow of air of velocity v_{air} in the $-z$ direction.

Hence the net downward velocity of the drop is

$$v_z = \frac{2r^2\rho g}{9\eta} - v_{air}. \qquad (3)$$

As explained in the next section we want v_z to be small hence we set v_{air} to be close to $v_{z,term}$ but slightly smaller.

B. General description of experiment

Figure 2 is a schematic picture of the apparatus. Drops averaging 20.6 μm in diameter are produced at a rate of 1 Hz using a piezoelectrically actuated drop-on-demand microdrop ejector. The drops fall through the upward moving air in the measurement chamber passing through a horizontal, uniform, alternating electric field. In this figure the electric field is perpendicular to the paper. The electric field alternates as a square wave with a frequency of 2.5 Hz and has an amplitude of about 1.8×10^6 V/m.

A rectangular measurement region 2.29 mm in the x direction by 3.05 mm in the z direction is projected by a lens onto the charge-coupled device (CCD) sensor of a monochrome, digital video camera. A light source consisting of a bank of light emitting diodes (LEDs) provides 10 Hz stroboscopic illumination. As the motion of the drop carries it through the measurement region, its image appears on the surface of the CCD. Thus the camera collects 10 frames per second, the drop appearing as a dark image on a bright background.

In addition to the v_x motion there is also the v_z motion, Eq. (3). Since the camera has a field of view of $Z = 3.05$ mm in the vertical direction, the 10 Hz stroboscopic illumination leads to acquisition of

$$N_{images} = 10Z/v_z = 30.5/v_z \qquad (4)$$

images of any given drop, before the drop moves below the viewing area of the camera. Here v_z is in mm/s. Hence we get a larger number of images per drop, leading to better charge measurement precision, when v_z is small. Of the order of $N_{images} = 15$ are required.

We give an example of the importance of the upward airflow in obtaining this many images. Consider a typical drop of diameter 20.6 μm with a density of 0.913 g/cm^3. From Eq. (2), $v_{z,term} = 11.3$ mm/s. If there were no upward airflow there would be an average of 2.6 images per drop. To obtain $N_{images} = 15$, v_z must be about 2.0 mm/s. Therefore from Eq. (3), v_{air} must be 11.3 − 2.0 = 9.3 mm/s.

Each image from the CCD camera is processed though a framegrabber in a conventional desktop computer, the signal in each pixel being recorded. An analysis program then finds the drop images and calculates the x and z coordinates of the centroid of the drop image. Using all the images of the drop and knowing the time spacing of the images, v_x and v_z are then calculated.

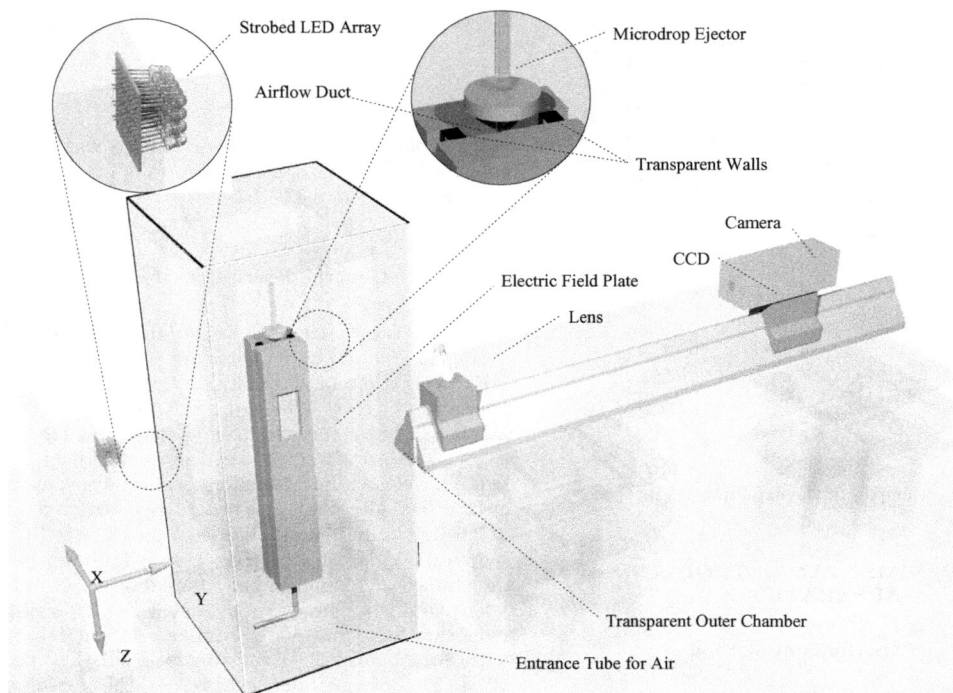

FIG. 2: Diagram of the apparatus. Diagram is to scale, except for the lens and CCD which are shown at 2× scale. Support structures are drawn transparent for clarity.

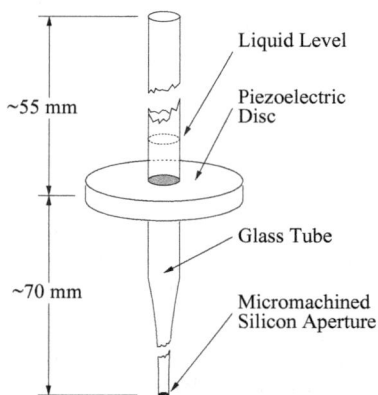

FIG. 3: Drop generator.

C.   Drop generator

The drop generator, Fig. 3, is based on the general design principles used in piezoelectrically actuated, drop-on-demand, inkjet print heads. Our generator is designed for flexibility, allowing a variety of liquids to be used and providing ease of control and maintenance. The body of the generator is glass so as to preserve the purity of the liquid, but the ejection aperture at the bottom is micro-machined silicon [7]. The diameter of the aperture sets approximately the diameter of the drop. Upon application of a short voltage pulse, usually 3 to 20 μs, across the surfaces of the piezoelectric disk, the central hole in the disk contracts in diameter, squeezing the glass tube and sending a pressure pulse down the liquid, ejecting a jet of fluid from the aperture. The forming of a discrete fluid drop from the high speed jet is a complex process with the repeatability of the process and the final diameter of the drop being highly dependent on the properties of the fluid, and on how the fluid is driven. For this reason, the shape and amplitude of the voltage pulse applied to the piezoelectric disk must be specifically tuned for stability and the desired drop size. In addition, it is necessary to experiment with single and double pulsing, varying both the pulse width and the separation between the pulses.

The pressure in the drop generator is maintained slightly below atmospheric pressure by 10 to 30 mm Hg. This helps to retract the excess liquid outside the ejection aperture after the drop has been produced and also prevents leaking of the liquid between pulses.

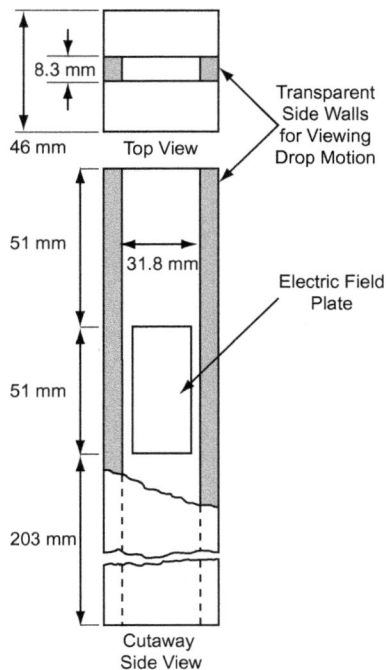

FIG. 4: Airflow tube and measurement chamber.

It is important that the drop generator produce drops of constant radius, the primary reason being that the size of the drops determines the ratio between Q and the measured quantity v_x, from Eq. (1). A secondary reason that we did not initially appreciate has to do with Eqs. 2 and 3. We set v_{air} close to $v_{z,term}$ so that v_z is much smaller than $v_{z,term}$. Thus a small change in r leads to a relatively large change in v_z. This dispersion disrupts the consistent spacing between adjacent drops, and a decreased separation between drops is undersirable for reasons discussed in Sec. IV A 6.

With a clean and newly tuned drop generator we get remarkably uniform drop radii, constant to about $\pm 0.2\%$. The drop generator also ejects in a consistent downward vertical direction along the centerline of the airflow tube. At 1 Hz operation, the drop ejector exhibits slow drifts in its characteristics with time scales of the order of a week. These drifts appear as changes in drop size and as destabilization that manifests as the appearance of satellite drops or inconsistent drop production. Typically, these effects can be compensated for by small changes in the drive parameters or adjustments of the air velocity or both. By the end of the first data set, set 1, the drop ejector had destabilized to the point where it had to be removed from the apparatus for cleaning and refilling. Similarly during the taking of data set 2, the drop ejector and air velocity required periodic small adjustments. The end of set 2 was caused by increasing instability in

drop production which could not be compensated for. We do not know the reason for this behavior.

In our drop generator the silicone oil drops are produced with a spread of charges, $|Q|$ ranging mostly from 0 to about 10 e. A few percent of the drops have larger $|Q|$. As described in Sec. IV A 1 we used drops with $|Q| < 9.5e$ to maintain good precision in the charge measurement. We do not know what sets the charge distribution for a particular drop generator. But we have the general observation that silicone oil gives narrow charge distributions, whereas water, mineral oil and most other organic liquids give broad charge distributions, with $|Q|$ values as large as several 1000 e or even larger.

D. Optical system

Referring to Fig. 2, the stroboscopic light source consists of a rectangular bank of 20 LED's emitting at 660 nm. The pulse length was about 100 μs. The lens, a 135 mm focal length, $f/11$, photographic enlarging lens, images the measurement region onto the face of the CCD camera with a magnification of 2.1.

The rectangular active image area of the CCD camera [8] is 4.8 mm in the horizontal, that is, x, direction, and 6.4 mm in the vertical, that is, z, direction. Hence the viewing area in physical space is 2.29 mm horizontally by 3.05 mm vertically. We remind the reader that the electric field is horizontal. The active imaging area is an array of 240 horizontal picture elements (pixels) and 736 vertical picture elements (pixels). We chose this orientation of the array to maximize the vertical distance, maximizing the number of images per drop.

Given the magnification and pixel density of the CCD, one would expect from geometric optics that the shadow of a 20 μm diameter drop would cover 2 pixels horizontally and 5 pixels vertically. The actual observed shadow typically covered 3 pixels horizontally and 7 pixels vertically, and had an intensity variation that was approximately a two dimensional Gaussian. This can be quantitatively described as the convolution of the simple shadow predicted by geometric optics with a point spread function that is a result of the diffractive effects due to the finite aperture of the lens. We do not and should not observe diffractive effects caused by the small size of the drops.

E. Airflow tube and measurement chamber

Figure 4 shows a slightly simplified, dimensioned drawing of the airflow tube and the measurement chamber. A rectangular duct contains the upward flowing air. It is 8.3 mm wide in the direction of the electric field and 31.8 mm wide in the direction perpendicular to the electric field. The field plates that define the measurement chamber are 51 mm high and 28.6 mm wide. The inner surfaces of the plates are machined flat and are in the

CCD Camera — Image Data → Computer — Sparse Data

Field Index

Frequency Divider

Data Storage

10 Hz Clock

HV Power Supply

Pulse Generator

+V   −V

LED Driver

High Voltage Switcher   HV Monitor

HV Pulse Amplifier

LED Light Source

Electric Field Plates

Drop Generator

LED Pulses ⟶

E Field ⟶

1 s   1 s

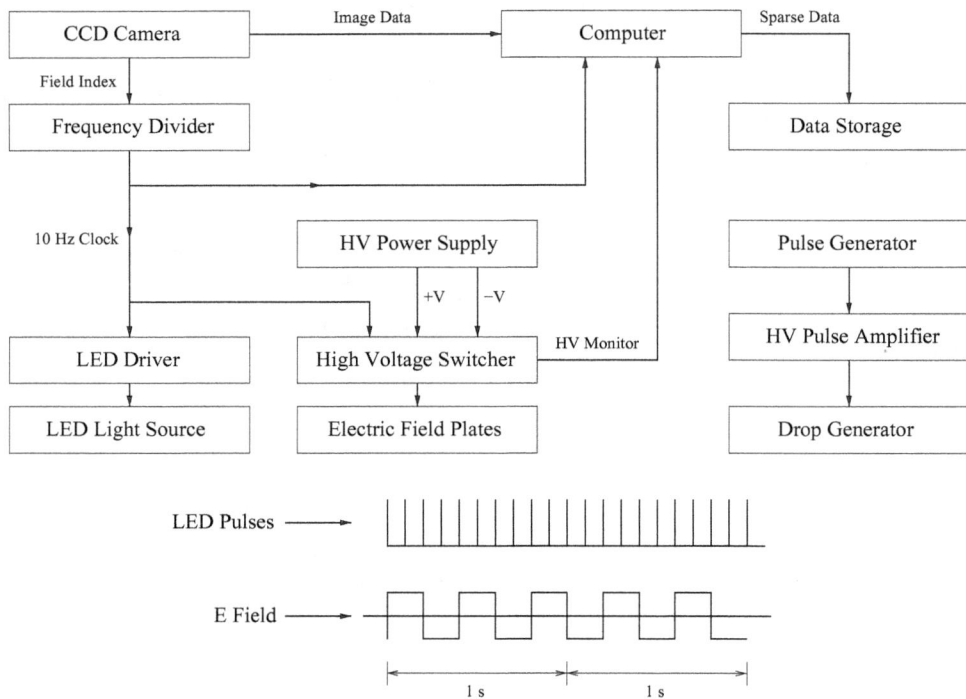

FIG. 5: Schematic of electronic system. The LED and high voltage is synchronized to the 10 Hz clock, while the drop generator runs asynchronously.

same plane as the inner surfaces of the walls of the air-flow tube. The optic axis of the optical system passes through the transparent side walls of the airflow tube.

The air velocity is sufficiently small, with a Reynolds number on the order of $R_e = 50$ so that the flow is laminar. The 203 mm length of air flow tube between the measurement region and the air inlet allows the air to settle into its equilibrium flow pattern. At equilibrium, the velocity profile of the air is approximately parabolic across the narrow direction of the channel (x axis). Across the long axis, the flow has a roughly constant central region and falls to zero at the boundaries [9].

F.   Electronics

All the electronics of the apparatus, Fig. 5, are hard-wired to give reliable timing, independent of the operation of the computer. A 30 Hz handshaking signal from the CCD is divided down to provide a 10 Hz clock that synchronizes the LED strobe, the electric field switcher and the computer image acquisition. The switching of the electric field, which is driven by the clock signal divided by 4, operates at 2.5 Hz. This results in a cycle where two images are acquired with the electric field in one direction, and then two images with the electric field

in the other direction. These relationships between the signals is depicted in the timing diagram of Fig. 5. The drop generator is driven asynchronously at 1 Hz.

G.   Data acquisition and storage

Data acquisition was performed by a single desktop computer running Linux. The computer was equipped with two special components: a digital framegrabber that allowed the capture of image data from the camera and a general purpose input/output interface board with digital I/O and A/D conversion capability. The additional inputs allowed the computer to monitor the state of the experiment as well as a variety of environmental variables.

All software was custom written in C. Hardware dependent code was encapsulated into drivers at the kernel level, which allowed a guarantee of synchronization of the software with apparatus by a combination of hardware and software buffering of the data.

The overall strategy was to acquire data from the apparatus and write them to files in *raw* form for later processing off line. Recall that each image frame contains 736×240 pixels, each digitized to 8 bit accuracy. Therefore acquiring data at 10 Hz produces a data rate from

6

the CCD camera of about 2 MB/s, much too large to be stored. Since an image contains just a few drops, most of the pixels in an image have just the background signal, which allows the information to be stored using sparse storage. As mentioned in Sec. II D, the typical image of drop extends over an area of 3 pixels by 7 pixels. In each frame the regions containing drops were isolated using a thresholding operation. The position of each drop was then measured using a simple center of mass algorithm, and for each drop only a surrounding region containing 13 horizontal pixels by 21 vertical pixels of the image is written to the output file.

H. Data collection

The search was carried out in two sets described in Table I. At the beginning of the Set 1 the drop ejector was operated at 0.5 Hz, then at 1 Hz for the remainder of Set 1 and all of Set 2.

TABLE I: Data collection

Data set	Weeks	Number of drops	Total mass (mg)
1	13	3 377 477	12.1
2	17	13 430 167	58.0

III. CALIBRATION, ERRORS, AND MEASUREMENT PRECISION

A. Electric field and drop radius

Rewrite Eq. (1) in the form

$$v_x = \left(\frac{Q}{6\pi\eta}\right)\left(\frac{E}{r}\right) \qquad (5)$$

Consider nonfractional values of $Q = ne$, $n = 0, \pm 1, \pm 2, \ldots$. Then, as shown in Sec. IV B, the measured values of v_x sharply peak at $n = 0, \pm 1, \pm 2, \ldots$. From a fit to the center of these peaks, Eq. (5) gives the fitted E/r ratio.

The electric field strength, E, is calculated from the measured voltage across the electric field plates, known to 3%, and the plate separation, 8.25 ± 0.01 mm. The plates are parallel to within 0.1 mrad. Inserting the calculated value of E into the fitted E/r ratio, we obtain the drop radius r.

We have two additional checks of the drop radius, one from the measurement of the error caused by the Brownian motion, Sec. III B, and the other from the measurement of the net downward velocity of the drops, v_z in Eq. (3).

The drop radius depends to a moderate extent upon the size and shape of the voltage pulse applied to the drop

generator and to a slight extent upon the age and history of the drop generator. However over periods of hours the average drop radius could be taken to be constant, with fluctuations of $\pm 0.2\%$ for individual drops. Since the data were analyzed in one-hour-long blocks, the average E/r ratio for any given block was known to much better accuracy than this.

B. Brownian motion and drop position measurement errors

The precision of the determination of the drop charge depends upon the precision of the measurement of v_x. Consider the sequence of position measurements x_i of the trajectory of a drop. For two consecutive frames, j and $j-1$, the velocity measurement $v_{x,j}$ is given by

$$v_{x,j} \equiv \frac{x_j - x_{j-1}}{\Delta t}. \qquad (6)$$

Here, Δt is the time between successive frames, 0.1 s in our case. Since Δt is known with very good precision, the error in measuring v_x comes from the error in determining the x_i of the drop centers, and from Brownian motion. Take the error in centroiding to be normally distributed with a standard deviation of σ_c.

During the time Δt between any two successive measurements of the x_i positions, Brownian motion adds a random contribution with standard deviation given by

$$\sigma_b = \sqrt{\frac{kT\Delta t}{3\pi\eta r}}. \qquad (7)$$

Here k is the Boltzmann constant, T is the absolute temperature, η is the viscosity of air and r is the drop radius.

The trajectory of uncharged drops, which have no contribution to their trajectories due to the electric field, can thus be written as

$$x_j = x_0 + \sigma_{c,j} + \sum_{i=1}^{j} \sigma_{b,i} \qquad (8)$$
$$j = 1, 2, \ldots, N_{images}$$

with x_0 set by the initial position of the drop, and where the $\sigma_{c,i}$ ($\sigma_{b,i}$) are normally distributed with a std. dev. of σ_c (σ_b). The analysis for charged drops is similar if the effect of the electric field is first subtracted from the observed data points. It then follows that

$$v_{x,j}\Delta t = \sigma_{b,j} + \sigma_{c,j} - \sigma_{c,j-1} \qquad (9)$$

and

$$\langle v_{x,j}v_{x,k}\rangle \Delta t^2 = \begin{cases} 2\sigma_c^2 + \sigma_b^2, & j = k \\ -\sigma_c^2, & |j-k| = 1 \\ 0, & \text{otherwise.} \end{cases} \qquad (10)$$

Therefore the total error on any given velocity measurement, σ_v, is given by $\sigma_v^2 = 2\sigma_c^2 + \sigma_b^2$, and the centroiding error introduces an anticorrelation with magnitude $-\sigma_c^2$ in two consecutive velocity measurements, due to the shared position measurement. We use the concept summarized in Eq. (10) and the observed distributions of the $v_{x,i}$ (after removal of the contribution due to the alternating electric field) to separate σ_c from σ_b.

We find that averaged over this experiment

$$\sigma_c = 0.31 \ \mu\text{m}, \ \sigma_b = 0.47 \ \mu\text{m}, \quad (11)$$

in the measurement region. Compared to the size of an individual pixel on the CCD, the centroiding error is small, approximately $1/30$ of a pixel. The value of σ_b obtained provides an independent check on the size of the drops, and is consistent with the size determined from the terminal velocity and the electric field drift velocity.

Equation 11 shows that the Brownian motion error, σ_b, is about the same magnitude as the error involved in finding the drop position, σ_c. Therefore substantially reducing σ_c through the use of smaller pixels will not by itself substantially reduce the error on the charge measurement, since the Brownian motion error can only be reduced by increasing N_{images}.

The final charge measurement of a drop is made using a single, detailed best fit to the entire observed trajectory of the drop, and the final error on the charge measurement σ_q is a result of propagating the errors σ_c and σ_b through this calculation.

C. Other sources of errors

We looked for other sources of errors, but all are negligible compared to those in Eq. (10). When we developed the upward air flow method we thought about the possibility that there might be some horizontal air velocity, $v_{x,air}$ in the measurement chamber, contributing an error to σ_v of order $v_{x,air} \times \Delta t$. By studying a large amount of data we found that the distribution of $v_{x,air} \times \Delta t$ had an rms value of 100 nm, and was a fixed property of the measurement region. For comparison $v_e \times \Delta t$ was of the order of 8 μm. These irregularities in $v_{x,air}$ are probably due to residual surface imperfections in the electric field plates. Since the irregularities are constant over long periods of time, they can be accurately measured and corrected for. For this analysis, that was not necessary.

Another possible source of error would be a nonuniformity in the electric field in the measurement region giving a horizontal gradient, dE/dx. This would produce a horizontal force on the drop's induced electric dipole moment. This dipole force acts in addition to the QE force. We found such a dipole force to be negligible compared to the QE force.

A small, vertical deceleration of the drops as they fall through the measurement chamber was observed. This amounted to a change of 30 μm/s in the apparent terminal velocity of the drops as they fell through the measure-

ment region, or a systematic uncertainty in the radius of the drop of the order of 0.3%. We believe that the deceleration is due to the evaporation of the drop as it falls. The magnitude of this effect is small enough such that it can be neglected in the calculation of v_x. As a side note, any systematic uncertainties in the radius of the drops are absorbed by the calibration process described in Sec. III A, and do not affect the final charge measurement.

Similarly, other possible sources of measurement error such as apparatus vibrations, optical distortions and CCD array distortions, and patch nonuniformities on the electric field plates, were negligible.

IV. DATA ANALYSIS AND RESULTS

A. Drop selection criteria

In this section we use $q = Q/e$, a measure of the drop charge in units of the electron charge. We required that all drops used in the analysis meet the criteria in Table II. The criteria are designed to maintain a charge measurement accuracy of approximately 0.03 e and to reject irregular drops caused by inconsistent operation of the drop generator.

1. $q < 9.5$ criterion

For any given drop there is an uncertainty in the radius of approximately 0.2% which contributes to the relative error on q. The *absolute* error on q thus increases linearly with q. Since the absolute error on q must be kept to the order of 0.03 e, restricting the data sample to drops with $q < 9.5$ keeps this contribution to less than 0.02 e. The overall charge distribution is such that only a few percent of the drops have q values outside this range.

2. $\sigma_q < 0.03$ criterion

Primarily, this criterion is a measure of N_{images} of the drop. Brownian motion and centroiding accuracy, characterized by σ_c and σ_b as described in Sec. III B, limit the accuracy of the charge measurement, σ_q. For any given drop, the number of position measurements, N_{images}, and the state of the electric field during those measurements, in addition to σ_b and σ_c, determines this accuracy. As noted earlier, N_{images} was of the order of 15. If a drop has an exceptionally large radius or is falling too far from the centerline of the airflow tube, v_z will be too large and N_{images} will be too small.

3. χ^2 criterion

As mentioned earlier, the final calculation of the charge on a drop is done using a fit to the trajectory of the drop. It was required that the χ^2 probability of the fit to the drop's trajectory be better than 10^{-3}. This rejects a large class of rare artifacts based on the statistical likelihood that the observed deviations from the fitted trajectory could be attributed to the Brownian motion and centroiding errors. For example, a drop would be rejected if it had an anomalous trajectory due to vibrations in the apparatus or due to its charge having been changed by collision with an ion during measurement.

4. v_z criterion

The net downward velocity of the drop, v_z, depends upon the drop radius and the upward air velocity, v_{air}, Eq. (3). This criterion insures consistent drop radii within the hour long data blocks by requiring

$$|v_{z,drop} - v_{z,block}| < 0.124 \text{ mm/s} \qquad (12)$$

where v_z is the measured value for one drop and $v_{z,block}$ is the average value of v_z for all the drops in the one hour data block. Using Eqs. 2 and 3, taking v_{air} as fixed and using an average value for r of 10 μm, this eliminates any drops with r different from the nominal value by more than about $\pm 0.5\%$.

Recall that the air velocity is approximately parabolic and that close to the centerline it is given by

$$v_{air}(x) = v_{air,0} \left[1 - (x/x_w)^2 \right] \qquad (13)$$

where x is the distance along the x axis from the centerline of the airflow tube, $x_w = 4.15$ mm is the distance to the wall of the tube, and $v_{air,0}$ is the air velocity along the centerline. Therefore this criterion indirectly restricts how far the drop can be from the centerline.

5. x deviation criterion

This criterion

$$|x - x_{block}| < 0.19 \text{ mm} \qquad (14)$$

provides a direct constraint on how far a drop may deviate from the centerline in the x direction. Here x_{block} is the average value of x for all the drops in the one hour data block. The purpose of this criterion is to eliminate drops that were produced irregularly. The 0.19 mm upper limit in Eq. (14) was determined by examining the distribution of x positions of drops produced during normal operation of the drop generation and setting the upper limit to eliminate the tails.

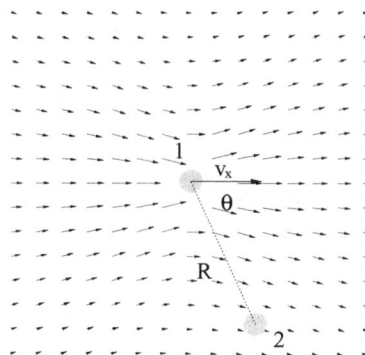

FIG. 6: The viscous coupling between a moving drop 1 on a neighboring drop 2 in still air. The small arrows show the vector velocity of the disturbed air. Note that there is a slight disturbance at the position of drop 2 that will affect the trajectory of drop 2.

6. Minimum distance R between any two drops criterion

The drops interact with one another through their induced electric dipole moments and viscous coupling through the air. Consider two drops of radius r, drop 1 moving with a velocity v_x due to the force of the electric field on its charge, as shown in Fig. 6. This motion will move the surrounding air. At the position of drop 2, the velocity of the air in the x direction, $V_{x,\text{disturbed air}}$, is given by

$$V_{x,\text{disturbed air}} = \frac{3}{4} \frac{v_x r}{R} (1 + \cos^2\theta). \qquad (15)$$

Since drop 2 sits in this disturbed air, its v_x due to the force of the electric field on its charge will have superimposed upon it $V_{x,\text{disturbed air}}$. This will distort the charge measurement. Therefore $V_{x,\text{disturbed air}}$ must be kept small by keeping R, the distance between the drops, much larger than r, the radius of the drop. A large separation also serves to minimize the interaction between the induced electric dipole moments of the drops, which increases as the inverse fourth power of the separation. We require

$$R > 0.62 \text{ mm} \qquad (16)$$

separation between any two drops, which limits these forces to a small fraction of QE.

7. Summary and magnitude of drop selection criteria

Table II gives the percent of drops removed by each criterion averaged over each of the two data sets. The total percent of drops removed is also given. Since the same drop may be removed by several criteria, the total

TABLE II: Drop selection criteria. The entries are the percent removed by each criterion separately. The bottom row gives the total percent of drops removed by all criteria. Since the same drop may be removed by several criteria, the total percent removed is *not* the sum of the percent removed by the individual criteria.

Criterion	Set 1	Set 2
q	0.4	0.7
σ_q	6.0	0.3
χ^2	4.7	2.4
v_z	4.6	1.2
x	9.4	5.1
R	12.5	3.8
Total	22.3	8.7

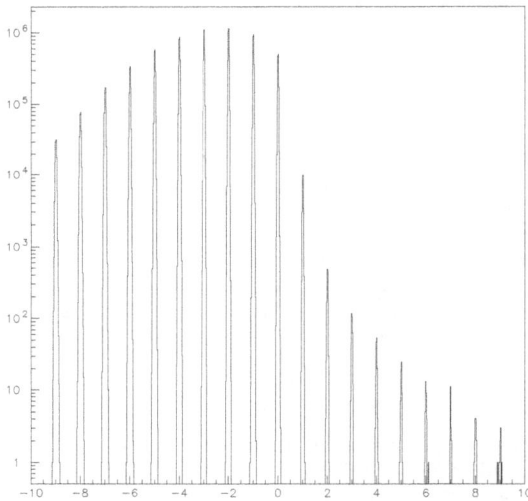

FIG. 8: The q_c charge distribution in units of e.

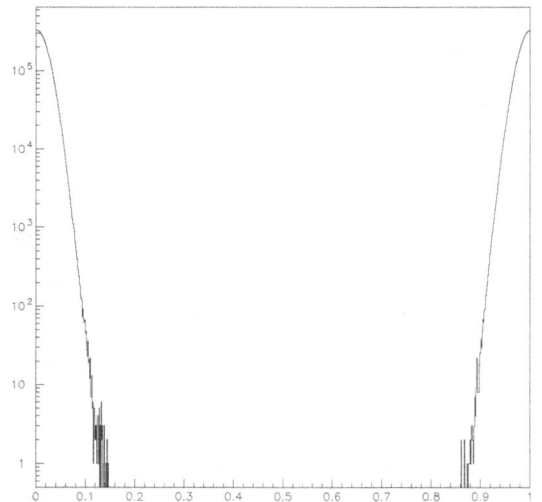

FIG. 7: The q charge distribution in units of e.

percent removed is *not* the sum of the percent removed by the individual criteria.

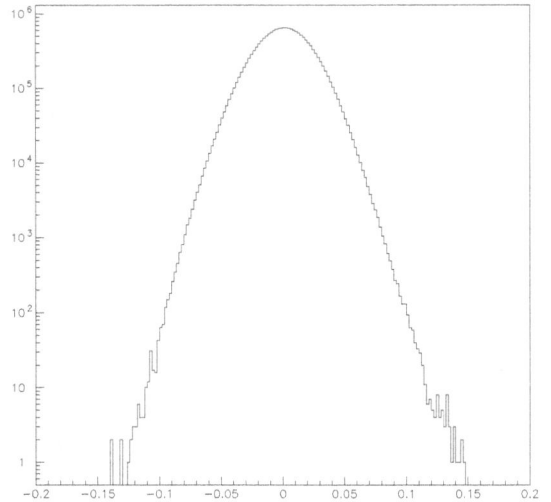

FIG. 9: The q_c residual charge distribution in units of e.

B. Results

After the application of these criteria we had a final data sample of 1.7×10^7 drops of average diameter 20.6 μm. The total mass of the sample was 70.1 mg. Figure 7 shows the charge distribution in units of e. (The asymmetry of the charge distribution is a property of the drop generator as discussed in Sec. II C.) We see sharp peaks at integer numbers of charges and no drops further than 0.15 e from the nearest integer. We emphasize that there is no background subtraction here, this is all

the data after the application of the criteria previously discussed.

To show the shape of the peaks at integer values of q we superimpose them in Fig. 8 using the charge distribution, q_c, defined by $q_c = q - N_c$ where N_c is the signed integer closest to q. The peaks have a Gaussian distribution with a standard deviation of 0.021 e. *The absence of non-Gaussian tails is what allows this search method to be so powerful.*

In Fig. 9 we superimpose the valleys between the peaks using the residual charge distribution, q_r, defined by $q_r = q - N_l$ where N_l is the largest integer less than q. We did not find any drops with residual charge between 0.15 e and 0.85 e. In this residual charge range there are fewer than 1.17×10^{-22} fractional charge particles per nucleon in silicone oil with 95% confidence.

Thus this 70.1 mg search did not confirm the one unusual aspect of our previous 17 mg search, Halyo *et al.*[2], where we found 1 drop with a q_r of about 0.29 e. No such charge was found in this search. While it is of course still possible that the fractional charge found in the 17 mg experiment was real, we are inclined to believe that the 17 mg experiment had a very small background that has been eliminated by the improved method of this experiment.

V. CONCLUSIONS

A. Comparison with other fractional charge searches in bulk material

In Table III we compare this search with previous, larger sample, searches for fractional charge particles in bulk matter. No evidence for fractional charge particles was found in the searches by Marinelli *et al.*[11], Smith *et al.*[12], and Jones *et al.*[13], similar to the null result in the present search.

In their superconducting levitometer search in niobium, LaRue *et al.*[10] claimed to have fractional charge particles with $e/3$ and $2e/3$. But Smith *et al.*[12] who also searched in niobium using a ferromagnetic levitometer method did not find any such evidence in a four times larger sample. At present the results of LaRue *et al.*[10] are not understood and are generally not accepted.

Our search is by far the largest to date and has the smallest upper limit of any search on the concentration of fractional charge particles in bulk matter. But it is important not to generalize our limit to other kinds of bulk matter for several reasons. First, we do not know what happens to fractional charge particles that are in natural matter when that matter is processed. Note that except for the search in meteoritic material by Jones *et al.*[13], all the material in Table III is processed.

Second, if we assume the existence of stable, fractional charge particles, we do not know what natural materials are most likely to have a detectable concentration. Our own thoughts are that the most promising natural material is that found in carbonaceous chondrite asteroids, since they are representative of the primordial composition of the solar system, having not undergone any geochemical or biochemical processes. Hence similar to the motivation of Jones *et al.*[13], our next search will be in meteoritic material from an asteroid.

TABLE III: Searches for fractional charge particles in ordinary matter. All experimenters reported null results except LaRue *et al.*[10]. There are 6.4×10^{20} nucleons in a milligram.

Method	Experiment	Material	Mass(mg)
levitometer	LaRue *et al.*[10]	niobium	1.1
levitometer	Marinelli *et al.*[11]	iron	3.7
levitometer	Smith *et al.*[12]	niobium	4.9
levitometer	Jones *et al.*[13]	meteorite	2.8
liquid drop	Halyo *et al.*[2]	silicone oil	17.4
liquid drop	this search	silicone oil	70.1

B. Remarks on further use of this new method

The purpose of the new method [1] used in this experiment was to allow large drops to be used compared to the classical method, thus increasing the rate at which we could search through a sample and also enabling the use of suspensions of more interesting materials. We have succeeded in doing this, using drops of about 20 μm diameter compared to the approximately 10 μm diameter used in Halyo *et al.*[2]. In the Appendix we discuss further increasing the search rate by using still larger drops and by using multiple columns of drops to increase the total rate of drop production. We find that with this new method the mass per second search rate can be further increased by a factor of the order of 10, but probably not by a factor of a 100.

*

APPENDIX A: INCREASING THE SEARCH RATE

Three are several ways in which the mass per second search rate can be increased in this experimental method.

1. Use of larger drops

The first way to increase the search rate is to use larger drops. Maintenance of the precision of the charge measurement requires that N_{images} increase in proportion to the drop radius. An increase in N_{images} can be accomplished by some combination of a decrease in v_z and an increase in the vertical length Z, Eq. (4). However a significant decrease in v_z requires too fine a balance between v_{air} and r^2, Eq. (3). If we keep v_z constant, an increase of Z can be attained by an increase in the number of vertical direction pixels in the CCD array of the camera. Existing CCD cameras with 10 Hz frame reading rates have twice the 736 vertical pixels used in the present camera and larger arrays will probably be available in the future. Therefore based on this consideration alone, drop diameters of several times 20 μm are feasible.

However, there are two problems that must be considered for drop diameters larger than 30 to 40 μm. The dipole force on a drop in a non-uniform electric field is proportional to the third power of the drop diameter. This force was negligible in this experiment, Sec. III C, but would have to be considered for much larger drops. The other problem is that the maintenance of a small and constant v_z, Eq. (3), requires v_{air} to increase as the square of the drop diameter, possibly leading to non-laminar flow. Therefore without more design and experimental studies, our conservative conclusion is that the drop diameter is limited to about 30 μm. This would lead to an increase of the mass search rate by a factor of 3 compared to the 20 μm drops used in this experiment.

2. Increase of drop production rate per column of drops

Let the drop production rate for a column of drops be n per second. Then the vertical separation between drops in a column is $R = v_{air}/n$. The criteria in Sec. IV A 6 require $R > 0.62$ mm. Using $v_{air} = 2.0$ mm/s, this gives an upper limit on n of about 3 Hz. However our experience in this experiment, Sec. II C, strongly suggests that a maximum 1 Hz rate is conservative practice, because of irregularities in drop production.

3. Increase in the number of drop columns

In this experiment we used one column of drops, however the extension to many columns of drops is straightforward. Of course the horizontal separation between adjacent columns must meet the requirements of Eq. (16), a nominal separation of 1 mm is useful for design purposes. The use of multiple columns requires two changes in the experimental design, namely the number of pixels in the horizontal direction in the CCD array must be increased and the space between the electric field plates must be increased. The latter requirement means the alternating potential applied across the plates must also be increased to keep the electric field constant.

Existing 10 Hz frame rate CCD cameras limit the number of columns to three but improvements in these cameras would probably allow five columns. The corresponding increase in the electric plate spacing and the potential difference is straightforward.

4. Correction for drop to drop interactions

It is clear that the primary constraint limiting the density of drops achievable in the measurement chamber is that of Eq. (16). To reiterate, interaction between the drops due to their induced electric dipole moment and viscous coupling requires that there be a minimum separation allowable between drops. In the limit that these interactions are small, both of these effects can be calculated from first principles, for example as in Eq. (15). In principle then, it should be possible to subtract the effect of these perturbing forces from the measured trajectory of each drop. Given this, it would be possible to relax the constraint on R. Since this possibility requires further study, it is not clear to what extent R can be reduced and throughput increased.

5. Summary

Putting these estimates together we can see how to achieve an improvement on the order of 10 times the present mass per second search rate using existing CCD cameras. Future cameras will probably allow a factor of 15 improvement.

[1] D. Loomba *et al.*, Rev. Sci. Instrum. **71**, 3409 (2000).
[2] V. Halyo *et al.*, Phys. Rev. Lett. **84**, 2576 (2000).
[3] N. M. Mar *et al.*, Phys. Rev. D **53**, 6017 (1996).
[4] M. L. Savage *et al.*, Phys. Lett. **167B**, 481 (1986).
[5] V. D. Hopper and T. H. Laby, Proc. R. Soc. London **A178**, 243 (1941).
[6] W. B. Kunkel, J. Appl. Phys. **21**, 820 (1950); **21**, 833 (1950).
[7] E. R. Lee and M. L. Perl, U.S. Patent 5,943,075, Aug. 24, 1999, Assigned to Stanford University.
[8] The CCD camera, a Cohu 4110, uses the frame transfer method and has digital output.
[9] W. E. Langlois, *Slow Viscous Flow* (Macmillan Company, New York, 1964).
[10] G. S. LaRue *et al.*, Phys. Rev. Lett. **46**, 967 (1981).
[11] M. Marinelli and G. Morpurgo, Phys. Lett. **137B**, 439 (1984).
[12] P. F. Smith *et al.*, Phys. Lett. **153B**, 188 (1985).
[13] W. G. Jones *et al.*, Z. Phys. C **43**, 349 (1989).

LIST OF PUBLICATIONS

1. Large-Angle Elastic Scattering of Negative Pions by Protons at 1.51, 2.01, and 2.53 BeV/c
   By Kwan Wu Lai.

2. Elastic Scattering of Negative Pions by Protons at 2 BeV/c
   By David E. Damouth, L.W. Jones, M.L. Perl.

3. Experimental Study of REACTIONS: $P + P \rightleftarrows \pi + d$ (Combined Report)
   By K.R. Chapman *et al.*

4. Search for a Light Higgs Resonance in Radiative Decays of the $\Upsilon(1S)$ with a Charm Tag
   By BaBar Collaboration (J.P. Lees *et al.*).
   arXiv:1502.06019 [hep-ex].
   10.1103/PhysRevD.91.071102.
   Phys.Rev. D91 (2015) 7, 071102.

5. Search for Long-Lived Particles in e^+e^- Collisions
   By BaBar Collaboration (J.P. Lees *et al.*).
   arXiv:1502.02580 [hep-ex].
   10.1103/PhysRevLett.114.171801.
   Phys.Rev.Lett. 114 (2015) 17, 171801.

6. Measurement of the Branching Fractions of the Radiative Leptonic τ Decays τ to $\tau \rightarrow e\gamma\nu\bar{\nu}$
   and $\tau \rightarrow \mu\gamma\nu\bar{\nu}$ at BABAR
   By BaBar Collaboration (J.P. Lees *et al.*).
   arXiv:1502.01784 [hep-ex].
   10.1103/PhysRevD.91.051103.
   Phys.Rev. D91 (2015) 051103.

7. High-Energy Elastic Pion-Proton Scattering Between 1.5 and 5 GeV/c (*)
   By L.W. Jones, Kwan W. Lai, M.L. Perl, S. Ting, V. Cook, B. Cork, W. Hoiley.

8. Evidence for CP Violation in $B^+ \rightarrow K^*(892)^+\pi^0$ from a Dalitz Plot Analysis of $B^+ \rightarrow K_S^0\pi^+\pi^0$ Decays
   By BaBar Collaboration (J.P. Lees *et al.*).
   arXiv:1501.00705 [hep-ex].

9. Dalitz Plot Analyses of $B^0 \rightarrow D^-D^0K^+$ and $B^+ \rightarrow \bar{D}^0D^0K^+$ Decays
   By BaBar Collaboration (J.P. Lees *et al.*).
   arXiv:1412.6751 [hep-ex].
   10.1103/PhysRevD.91.052002.
   Phys.Rev. D91 (2015) 5, 052002.

10. Measurement of the $D^0 \to \pi^- e^+ \nu_e$ Differential Decay Branching Fraction as a Function of q^2 and Study of Form Factor Parameterizations
     By BaBar Collaboration (J.P. Lees *et al.*).
     arXiv:1412.5502 [hep-ex].
     10.1103/PhysRevD.91.052022.
     Phys.Rev. D91 (2015) 5, 052022.

11. Study of *CP* Asymmetry in $B^0 - \overline{B}^0$ Mixing with Inclusive Dilepton Events
     By BaBar Collaboration (J.P. Lees *et al.*).
     arXiv:1411.1842 [hep-ex].
     10.1103/PhysRevLett.114.081801.
     Phys.Rev.Lett. 114 (2015) 8, 081801.

12. Search for New π^0-Like Particles Produced in Association with a τ-Lepton Pair
     By BaBar Collaboration (J.P. Lees *et al.*).
     arXiv:1411.1806 [hep-ex].
     10.1103/PhysRevD.90.112011.
     Phys.Rev. D90 (2014) 11, 112011.

13. Bottomonium Spectroscopy and Radiative Transitions Involving the $\chi_{bJ}(1P, 2P)$ States at BABAR
     By BaBar Collaboration (J.P. Lees *et al.*).
     arXiv:1410.3902 [hep-ex].
     10.1103/PhysRevD.90.112010.
     Phys.Rev. D90 (2014) 11, 112010.

14. Observation of the Baryonic Decay $\overline{B}^0 \to \Lambda_c^+ \overline{p} K^- K^+$
     By BaBar Collaboration (J.P. Lees *et al.*).
     arXiv:1410.3644 [hep-ex].
     10.1103/PhysRevD.91.031102.
     Phys.Rev. D91 (2015) 3, 031102.

15. Study of $B^{\pm,0} \to J/\psi K^+ K^- K^{\pm,0}$ and Search for $B^0 \to J/\psi \phi$ at BABAR
     By BaBar Collaboration (J.P. Lees *et al.*).
     arXiv:1407.7244 [hep-ex].
     10.1103/PhysRevD.91.012003.
     Phys.Rev. D91 (2015) 1, 012003.

16. Developing Sreativity and Innovation in Engineering and Science
     By Martin L. Perl.
     10.1142/9789812794185_0001.

17. Review of Inclusive Pion Electroproduction
     By M.L. Perl.

18. The Physics of the *B* Factories
     By BaBar and Belle Collaborations (A.J. Bevan *et al.*).
     arXiv:1406.6311 [hep-ex].
     10.1140/epjc/s10052-014-3026-9.
     Eur.Phys.J. C74 (2014) 3026.

19. Monitoring System for the GRID Monte Carlo Mass Production in the H1 Experiment at DESY
    By Elena Bystritskaya, Alexander Fomenko, Nelly Gogitidze, Bogdan Lobodzinski.
    10.1088/1742-6596/513/3/032060.
    J.Phys.Conf.Ser. 513 (2014) 032060.

20. Search for a Dark Photon in e^+e^- Collisions at BaBar
    By BaBar Collaboration (J.P. Lees *et al.*).
    arXiv:1406.2980 [hep-ex].
    10.1103/PhysRevLett.113.201801.
    Phys.Rev.Lett. 113 (2014) 20, 201801.

21. Measurements of Direct *CP* Asymmetries in $B \to X_{sy}$ Decays Using Sum of Exclusive Decays
    By BaBar Collaboration (J.P. Lees *et al.*).
    arXiv:1406.0534 [hep-ex].
    10.1103/PhysRevD.90.092001.
    Phys.Rev. D90 (2014) 9, 092001.

22. Measurements of Branching Fractions, Polarizations, and Direct *CP*-Violation Asymmetries in $B \to \rho K^*$ and $B \to f_0(980)K^*$ decays
    By BaBar Collaboration (Bernard Aubert *et al.*).

23. Cross Sections for the Reactions $e^+e^- \to K_S^0 K_L^0$, $K_S^0 K_L^0 \pi^+\pi^-$, $K_S^0 K_S^0 \pi^+\pi^-$, and $K_S^0 K_S^0 K^+K^-$ from Events with Initial-State Radiation
    By BaBar Collaboration (J.P. Lees *et al.*).
    arXiv:1403.7593 [hep-ex].
    10.1103/PhysRevD.89.092002.
    Phys.Rev. D89 (2014) 9, 092002.

24. Dalitz Plot Analysis of $\eta_c \to K^+K^-\eta$ and $\eta_c \to K^+K^-\pi^0$ in Two-Photon Interactions
    By BaBar Collaboration (J.P. Lees *et al.*).
    arXiv:1403.7051 [hep-ex].
    10.1103/PhysRevD.89.112004.
    Phys.Rev. D89 (2014) 11, 112004.

25. Antideuteron Production in $\Upsilon(nS)$ Decays and in $e^+e^- \to q\bar{q}$ at $\sqrt{s} \approx 10.58$ GeV
    By BaBar Collaboration (J.P. Lees *et al.*).
    arXiv:1403.4409 [hep-ex].
    10.1103/PhysRevD.89.111102.
    Phys.Rev. D89 (2014) 11, 111102.

26. Strange Particle Interactions
    By C.C. Butler *et al.*.

27. Evidence for the Baryonic Decay $\bar{B}^0 \to D^0 \Lambda \bar{\Lambda}$
    By BaBar Collaboration (J.P. Lees *et al.*).
    arXiv:1401.5990 [hep-ex].
    10.1103/PhysRevD.89.112002.
    Phys.Rev. D89 (2014) 11, 112002.

28. CANDELS Visual Classifications: Scheme, Data Release, and First Results
    By Jeyhan S. Kartaltepe *et al.*.
    arXiv:1401.2455 [astro-ph.GA].
    10.1088/0067-0049/221/1/11.
    Astrophys.J.Suppl. 221 (2015) 1, 11.

29. Search for the Decay $\bar{B}^0 \to \Lambda_c^+ \bar{p} p \bar{p}$
    By BaBar Collaboration (J.P. Lees *et al.*).
    arXiv:1312.6800 [hep-ex].
    10.1103/PhysRevD.89.071102.
    Phys.Rev. D89 (2014) 7, 071102.

30. Measurement of the $B \to X_s l^+ l^-$ Branching Fraction and Search for Direct *CP* Violation from a Sum of Exclusive Final States
    By BaBar Collaboration (J.P. Lees *et al.*).
    arXiv:1312.5364 [hep-ex].
    10.1103/PhysRevLett.112.211802.
    Phys.Rev.Lett. 112 (2014) 211802.

31. Evidence for the Decay $B^0 \to \omega\omega$ and Search for $B^0 \to \omega\phi$
    By BaBar Collaboration (J.P. Lees *et al.*).
    arXiv:1312.0056 [hep-ex].
    10.1103/PhysRevD.89.051101.
    Phys.Rev. D89 (2014) 5, 051101.

32. Search for Lepton-Number Violating $B^+ \to X^- \ell^+ \ell'^+$ Decays
    By BaBar Collaboration (J.P. Lees *et al.*).
    arXiv:1310.8238 [hep-ex].
    10.1103/PhysRevD.89.011102.
    Phys.Rev. D89 (2014) 1, 011102.

33. Analysis of Diffractive Jets
    By CMS Collaboration.

34. Measurement of Collins Asymmetries in Inclusive Production of Charged Pion Pairs in e^+e^- Annihilation at BABAR
    By BaBar Collaboration (J.P. Lees *et al.*).
    arXiv:1309.5278 [hep-ex].
    10.1103/PhysRevD.90.052003.
    Phys.Rev. D90 (2014) 5, 052003.

35. Measurement of the $B^+ \to \omega \ell^+ \nu$ Branching Fraction with Semileptonically Tagged *B* Mesons
    By BaBar Collaboration (J.P. Lees *et al.*).
    arXiv:1308.2589 [hep-ex].
    10.1103/PhysRevD.88.072006.
    Phys.Rev. D88 (2013) 7, 072006.

36. Measurement of the $e^+e^- \to p\bar{p}$ Cross Section in the Energy Range from 3.0 to 6.5 GeV
    By BaBar Collaboration (J.P. Lees *et al.*).
    arXiv:1308.1795 [hep-ex].
    10.1103/PhysRevD.88.072009.
    Phys.Rev. D88 (2013) 7, 072009.

37. Measurement of the Mass of the D^0 Meson
    By BaBar Collaboration (J.P. Lees *et al.*).
    arXiv:1308.1151 [hep-ex].
    10.1103/PhysRevD.88.071104.
    Phys.Rev. D88 (2013) 071104.

38. Search for a Light Higgs Boson Decaying to Two Gluons or $s\bar{s}$ in the Radiative Decays of $\Upsilon(1S)$
    By BaBar Collaboration (J.P. Lees *et al.*).
    arXiv:1307.5306 [hep-ex].
    10.1103/PhysRevD.88.031701.
    Phys.Rev. D88 (2013) 3, 031701.

39. FAMA: An Automatic Code for Stellar Parameter and Abundance Determination
    By Laura Magrini *et al.*.
    arXiv:1307.2367 [astro-ph.IM].
    10.1051/0004-6361/201321844.
    Astron.Astrophys. 558 (2013) A38.

40. Precision Measurement of the $e^+e^- \to K^+K^-(\gamma)$ Cross Section with the Initial-State Radiation Method at BABAR
    By BaBar Collaboration (J.P. Lees *et al.*).
    arXiv:1306.3600 [hep-ex].
    10.1103/PhysRevD.88.032013.
    Phys.Rev. D88 (2013) 3, 032013.

41. Production of Charged Pions, Kaons, and Protons in e^+e^- Annihilations into Hadrons at $\sqrt{s} = 10.54$ GeV
    By BaBar Collaboration (J.P. Lees *et al.*).
    arXiv:1306.2895 [hep-ex].
    10.1103/PhysRevD.88.032011.
    Phys.Rev. D88 (2013) 032011.

42. Study of the K^+K^- Invariant-Mass Dependence of CP Asymmetry in $B^+ \to K^+K^-K^+$ Decays
    By BaBar Collaboration (J.P. Lees *et al.*).
    arXiv:1305.4218 [hep-ex].

43. The BABAR Detector: Upgrades, Operation and Performance
    By BaBar Collaboration (B. Aubert *et al.*).
    arXiv:1305.3560 [physics.ins-det].
    10.1016/j.nima.2013.05.107.
    Nucl.Instrum.Meth. A729 (2013) 615-701.

44. Search for CP Violation in $B^0\overline{B}^0$ Mixing using Partial Reconstruction of $B^0 \to D^{*-} X\ell^+\nu_\ell$ and a Kaon Tag
    By BaBar Collaboration (J.P. Lees *et al.*).
    arXiv:1305.1575 [hep-ex].
    10.1103/PhysRevLett.111.101802, 10.1103/PhysRevLett.111.159901.
    Phys.Rev.Lett. 111 (2013) 10, 101802, Phys.Rev.Lett. 111 (2013) 15, 159901.

45. Measurement of the $D^*(2010)^+$ Meson Width and the $D^*(2010)^+ - D^0$ Mass Difference
    By BaBar Collaboration (J.P. Lees *et al.*).
    arXiv:1304.5657 [hep-ex].
    10.1103/PhysRevLett.111.169902, 10.1103/PhysRevLett.111.111801.
    Phys.Rev.Lett. 111 (2013) 11, 111801.

46. Measurement of the $D^*(2010)^+$ Natural Line Width and the $D^*(2010)^+ - D^0$ Mass Difference
    By BaBar Collaboration (J.P. Lees *et al.*).
    arXiv:1304.5009 [hep-ex].
    10.1103/PhysRevD.88.052003, 10.1103/PhysRevD.88.079902.
    Phys.Rev. D88 (2013) 5, 052003, Phys.Rev. D88 (2013) 7, 079902.

47. Measurement of CP-Violating Asymmetries in $B^0 \to (\rho\pi)^0$ Decays Using a Time-Dependent Dalitz Plot Analysis
    By BaBar Collaboration (J.P. Lees *et al.*).
    arXiv:1304.3503 [hep-ex].
    10.1103/PhysRevD.88.012003.
    Phys.Rev. D88 (2013) 1, 012003.

48. Search for $B \to K^{(*)}\nu\overline{\nu}$ and Invisible Quarkonium Decays
    By BaBar Collaboration (J.P. Lees *et al.*).
    arXiv:1303.7465 [hep-ex].
    10.1103/PhysRevD.87.112005.
    Phys.Rev. D87 (2013) 11, 112005.

49. Search for the Rare Decays $B \to \pi\ell^+\ell^-$ and $B^0 \to \eta\ell^+\ell^-$
    By BaBar Collaboration (J.P. Lees *et al.*).
    arXiv:1303.6010 [hep-ex].
    10.1103/PhysRevD.88.032012.
    Phys.Rev. D88 (2013) 3, 032012.

50. Massive Non-Planar Two-Loop Four-Point Integrals with SecDec 2.1
    By Sophia Borowka, Gudrun Heinrich.
    arXiv:1303.1157 [hep-ph].
    10.1016/j.cpc.2013.05.022.
    Comput.Phys.Commun. 184 (2013) 2552-2561.

51. Measurement of an Excess of $\overline{B} \to D^{(*)}\tau^-\overline{\nu}_\tau$ Decays and Implications for Charged Higgs Bosons
    By BaBar Collaboration (J.P. Lees *et al.*).
    arXiv:1303.0571 [hep-ex].
    10.1103/PhysRevD.88.072012.
    Phys.Rev. D88 (2013) 7, 072012.

52. Study of the Decay $\bar{B}^0 \to \Lambda_c^+ \bar{p} \pi^+ \pi^-$ and its Intermediate States
   By BaBar Collaboration (J.P. Lees *et al.*).
   arXiv:1302.0191 [hep-ex].
   10.1103/PhysRevD.87.092004.
   Phys.Rev. D87 (2013) 9, 092004.

53. Study of $e^+e^- \to p\bar{p}$ via Initial-State Radiation at BABAR
   By BaBar Collaboration (J.P. Lees *et al.*).
   arXiv:1302.0055 [hep-ex].
   10.1103/PhysRevD.87.092005.
   Phys.Rev. D87 (2013) 9, 092005.

54. A New Information Architecture, Website and Services for the CMS Experiment
   By Lucas Taylor, Eleanor Rusack, Vidmantas Zemleris.
   10.1088/1742-6596/396/6/062021.
   J.Phys.Conf.Ser. 396 (2012) 062021.

55. The PhEDEx Next-Gen Website
   By R. Egeland, C.H. Huang, P. Rossman, P. Sundarrajan, T. Wildish.
   10.1088/1742-6596/396/3/032117.
   J.Phys.Conf.Ser. 396 (2012) 032117.

56. Time-Integrated Luminosity Recorded by the BABAR Detector at the PEP-II e^+e^- Collider
   By BaBar Collaboration (J.P. Lees *et al.*).
   arXiv:1301.2703 [hep-ex].
   10.1016/j.nima.2013.04.029.
   Nucl.Instrum.Meth. A726 (2013) 203-213.

57. Observation of Direct *CP* Violation in the Measurement of the Cabibbo-Kobayashi-Maskawa Angle γ with $B^\pm \to D^{(*)} K^{(*)\pm}$ Decays
   By BaBar Collaboration (J.P. Lees *et al.*).
   arXiv:1301.1029 [hep-ex].
   10.1103/PhysRevD.87.052015.
   Phys.Rev. D87 (2013) 5, 052015.

58. The Geant4 Visualisation System — A Multi-Driver Graphics System
   By John Allison, Laurent Garnier, Akinori Kimura, Joseph Perl.
   arXiv:1212.6923 [cs.GR].

59. Search for *CP* Violation in the Decays $D^\pm \to K_S^0 K^\pm$, $D_s^\pm \to K_S^0 K^\pm$, and $D_s^\pm \to K_S^0 \pi^\pm$
   By BaBar Collaboration (J.P. Lees *et al.*).
   arXiv:1212.3003 [hep-ex].
   10.1103/PhysRevD.87.052012.
   Phys.Rev. D87 (2013) 5, 052012.

60. Search for Direct *CP* Violation in Singly Cabibbo-Suppressed $D^\pm \to K^+ K^- \pi^\pm$ Decays
   By BaBar Collaboration (J.P. Lees *et al.*).
   arXiv:1212.1856 [hep-ex].
   10.1103/PhysRevD.87.052010.
   Phys.Rev. D87 (2013) 5, 052010.

61. Study of the Reaction $e^+e^- \to \psi(2S)\pi^-\pi^-$ via Initial-State Radiation at BaBar
    By BaBar Collaboration (J.P. Lees *et al.*).
    arXiv:1211.6271 [hep-ex].
    10.1103/PhysRevD.89.111103.
    Phys.Rev. D89 (2014) 11, 111103.

62. Search for a Low-Mass Scalar Higgs Boson Decaying to a τ Pair in Single-Photon Decays of $\Upsilon(1S)$
    By BaBar Collaboration (J.P. Lees *et al.*).
    arXiv:1210.5669 [hep-ex].
    10.1103/PhysRevD.88.071102.
    Phys.Rev. D88 (2013) 7, 071102.

63. Search for Di-Muon Decays of a Low-Mass Higgs Boson in Radiative Decays of the $\Upsilon(1S)$
    By BaBar Collaboration (J.P. Lees *et al.*).
    arXiv:1210.0287 [hep-ex].
    10.1103/PhysRevD.87.031102, 10.1103/PhysRevD.87.059903.
    Phys.Rev. D87 (2013) 3, 031102, Phys.Rev. D87 (2013) 5, 059903.

64. The CEBAF Element Database
    By M.E. Joyce, T.L. Larrieu, C.J. Slominski.
    Conf.Proc. C110328 (2011) 594-596.

65. Measurement of $D^0 - \bar{D}^0$ Mixing and *CP* Violation in Two-Body D^0 Decays
    By BaBar Collaboration (J.P. Lees *et al.*).
    arXiv:1209.3896 [hep-ex].
    10.1103/PhysRevD.87.012004.
    Phys.Rev. D87 (2013) 1, 012004.

66. Study of High-Multiplicity 3-Prong and 5-Prong τ Decays at BABAR
    By BaBar Collaboration (J.P. Lees *et al.*).
    arXiv:1209.2734 [hep-ex].
    10.1103/PhysRevD.86.092010.
    Phys.Rev. D86 (2012) 092010.

67. Study of the Baryonic *B* Decay $B^- \to \Sigma_c^{++}\bar{p}\pi^-\pi^-$
    By BaBar Collaboration (J.P. Lees *et al.*).
    arXiv:1208.3086 [hep-ex].
    10.1103/PhysRevD.86.091102.
    Phys.Rev. D86 (2012) 091102.

68. Measurement of the Time-Dependent *CP* Asymmetry of Partially Reconstructed $B^0 \to D^{*+}D^{*-}$ Decays
    By BaBar Collaboration (J.P. Lees *et al.*).
    arXiv:1208.1282 [hep-ex].
    10.1103/PhysRevD.86.112006.
    Phys.Rev. D86 (2012) 112006.

69. Branching Fraction and Form-Factor Shape Measurements of Exclusive Charmless Semileptonic
    B Decays, and Determination of $|V_{ub}|$
    By BaBar Collaboration (J.P. Lees *et al.*).
    arXiv:1208.1253 [hep-ex].
    10.1103/PhysRevD.86.092004.
    Phys.Rev. D86 (2012) 092004.

70. The Branching Fraction of $\tau^- \to \pi^- K_S^0 K_S^0 (\pi^0) \nu_\tau$ Decays
    By BaBar Collaboration (J.P. Lees *et al.*).
    arXiv:1208.0376 [hep-ex].
    10.1103/PhysRevD.86.092013.
    Phys.Rev. D86 (2012) 092013.

71. A Search for the Decay Modes $B^{+-} \to h^{+-} \tau^{+-} l$
    By BaBar Collaboration (J.P. Lees *et al.*).
    arXiv:1204.2852 [hep-ex].
    10.1103/PhysRevD.86.012004.
    Phys.Rev. D86 (2012) 012004.

72. Observation of Time Reversal Violation in the B^0 Meson System
    By BaBar Collaboration (J.P. Lees *et al.*).
    arXiv:1207.5832 [hep-ex].
    10.1103/PhysRevLett.109.211801.
    Phys.Rev.Lett. 109 (2012) 211801.

73. Measurement of $B(B \to X_s \gamma)$, the $B \to X_s \gamma$ Photon Energy Spectrum, and the Direct *CP*
    Asymmetry in $B \to X_{s+d} \gamma$ Decays
    By BaBar Collaboration (J.P. Lees *et al.*).
    arXiv:1207.5772 [hep-ex].
    10.1103/PhysRevD.86.112008.
    Phys.Rev. D86 (2012) 112008.

74. Precision Measurement of the $B \to X_s \gamma$ Photon Energy Spectrum, Branching Fraction, and
    Direct *CP* Asymmetry $A_{CP}(B \to X_s \gamma)$
    By BaBar Collaboration (J.P. Lees *et al.*).
    arXiv:1207.2690 [hep-ex].
    10.1103/PhysRevLett.109.191801.
    Phys.Rev.Lett. 109 (2012) 191801.

75. Study of $X(3915) \to J/\psi \omega$ in Two-Photon Collisions
    By BaBar Collaboration (J.P. Lees *et al.*).
    arXiv:1207.2651 [hep-ex].
    10.1103/PhysRevD.86.072002.
    Phys.Rev. D86 (2012) 072002.

76. Exclusive Measurements of $b \to s \gamma$ Transition Rate and Photon Energy Spectrum
    By BaBar Collaboration (J.P. Lees *et al.*).
    arXiv:1207.2520 [hep-ex].
    10.1103/PhysRevD.86.052012.
    Phys.Rev. D86 (2012) 052012.

77. Evidence of $B^+ \to X_s \gamma \, \tau^+ \nu$ Decays with Hadronic B Tags
    By BaBar Collaboration (J.P. Lees *et al.*).
    arXiv:1207.0698 [hep-ex].
    10.1103/PhysRevD.88.031102.
    Phys.Rev. D88 (2013) 3, 031102.

78. Search for the Decay Modes $D^0 \to e^+e^-$, $D^0 \to \mu^+\mu^-$, and $D^0 \to e\mu$
    By BaBar Collaboration (J.P. Lees *et al.*).
    arXiv:1206.5419 [hep-ex].
    10.1103/PhysRevD.86.032001.
    Phys.Rev. D86 (2012) 032001.

79. Measurement of CP Asymmetries and Branching Fractions in Charmless Two-Body B-Meson Decays to Pions and Kaons
    By BaBar Collaboration (J.P. Lees *et al.*).
    arXiv:1206.3525 [hep-ex].
    10.1103/PhysRevD.87.052009.
    Phys.Rev. D87 (2013) 5, 052009.

80. Improved Limits on B^0 Decays to Invisible Final States and to $\nu\bar{\nu}\gamma$
    By BaBar Collaboration (J.P. Lees *et al.*).
    arXiv:1206.2543 [hep-ex].
    10.1103/PhysRevD.86.051105.
    Phys.Rev. D86 (2012) 051105.

81. Search for Resonances Decaying to $\eta_c \pi^+ \pi^-$ in Two-Photon Interactions
    By BaBar Collaboration (J.P. Lees *et al.*).
    arXiv:1206.2008 [hep-ex].
    10.1103/PhysRevD.86.092005.
    Phys.Rev. D86 (2012) 092005.

82. Branching Fraction Measurement of $B^+ \to \omega \ell^+ \nu$ Decays
    By BaBar Collaboration (J.P. Lees *et al.*).
    arXiv:1205.6245 [hep-ex].
    10.1103/PhysRevD.87.099904, 10.1103/PhysRevD.87.032004.
    Phys.Rev. D87 (2013) 3, 032004, Phys.Rev. D87 (2013) 9, 099904.

83. Evidence for an Excess of Decays $\bar{B} \to D^{(*)} \tau^- \bar{\nu}_\tau$
    By BaBar Collaboration (J.P. Lees *et al.*).
    arXiv:1205.5442 [hep-ex].
    10.1103/PhysRevLett.109.101802.
    Phys.Rev.Lett. 109 (2012) 101802.

84. Precise Measurement of the $e^+e^- \to \pi^+\pi^-(\gamma)$ Cross Section with the Initial-State Radiation Method at BABAR
    By BaBar Collaboration (J.P. Lees *et al.*).
    arXiv:1205.2228 [hep-ex].
    10.1103/PhysRevD.86.032013.
    Phys.Rev. D86 (2012) 032013.

85. Measurement of Branching Fractions and Rate Asymmetries in the Rare Decays $B \to K^{(*)}\ell^+\ell^-$
By BaBar Collaboration (J.P. Lees *et al.*).
arXiv:1204.3933 [hep-ex].
10.1103/PhysRevD.86.032012.
Phys.Rev. D86 (2012) 032012.

86. Study of the Reaction $e^+e^- \to J/\psi\pi^+\pi^-$ via Initial-State Radiation at BaBar
By BaBar Collaboration (J.P. Lees *et al.*).
arXiv:1204.2158 [hep-ex].
10.1103/PhysRevD.86.051102.
Phys.Rev. D86 (2012) 051102.

87. High-Energy $\pi - p$ Elastic Scattering for Small Momentum Transfers and Forward Dispersion Calculations
By Howard I. Saxer.

88. Rare Decay Modes of $N^*(1950)$
By W. Chinowsky *et al.*.
Submitted to: Phys.Rev..

89. Elastic Scattering of Negative Pions by Protons at 2 BeV/c
By D.E. Damouth, L.W. Jones, M.L. Perl.

90. Further Work on an Image for Determination of Profiles and Images of Weak Beams of High-energy Particles
By Dale Dickinson, Martin Perl.

91. The Status of the Scintillation Chamber
By Martin L. Perl.

92. Measurement of the Hadronic Form-Factor in $D^0 \to K^-e^+\nu_e$
By BaBar Collaboration (Bernard Aubert *et al.*).
arXiv:0704.0020 [hep-ex].
10.1103/PhysRevD.76.052005.
Phys.Rev. D76 (2007) 052005.

93. Search for Lepton-Number Violating Processes in $B^+ \to h^-\ell^+\ell^+$ Decays
By BaBar Collaboration (J.P. Lees *et al.*).
arXiv:1202.3650 [hep-ex].
10.1103/PhysRevD.85.071103.
Phys.Rev. D85 (2012) 071103.

94. Search for Low-Mass Dark-Sector Higgs Bosons
By BaBar Collaboration (J.P. Lees *et al.*).
arXiv:1202.1313 [hep-ex].
10.1103/PhysRevLett.108.211801.
Phys.Rev.Lett. 108 (2012) 211801.

95. Study of *CP* Violation in Dalitz-Plot Analyses of $B^0 \to K^+K^-K_S^0$, $B^+ \to K^+K^-K^+$, and $B^+ \to K_S^0 K_S^0 K^+$
By BaBar Collaboration (J.P. Lees *et al.*).
arXiv:1201.5897 [hep-ex].
10.1103/PhysRevD.85.112010.
Phys.Rev. D85 (2012) 112010.

96. Examples for Exceptional Sequences of Invertible Sheaves on Rational Surfaces
By Markus Perling.
arXiv:0904.0529 [math.AG].
Seminaires et Congres 25 (2010) 369-389.

97. Exceptional Sequences of Invertible Sheaves on Rational Surfaces
By Lutz Hille, Markus Perling.
arXiv:0810.1936 [math.AG].
10.1112/S0010437X10005208.
Compos.Math. 147 (2011) 1230-1280.

98. Initial-State Radiation Measurement of the $e^+e^- \to \pi^+\pi^-\pi^+\pi^-$ Cross Section
By BaBar Collaboration (J.P. Lees *et al.*).
arXiv:1201.5677 [hep-ex].
10.1103/PhysRevD.85.112009.
Phys.Rev. D85 (2012) 112009.

99. B^0 Meson Decays to $\rho^0 K^{*0}$, $f_0 K^{*0}$, and $\rho^- K^{*+}$, Including Higher K^* Resonances
By BaBar Collaboration (J.P. Lees *et al.*).
arXiv:1112.3896 [hep-ex].
10.1103/PhysRevD.85.072005.
Phys.Rev. D85 (2012) 072005.

100. Study of $\overline{B} \to X_u \ell \overline{\nu}$ Decays in $B\overline{B}$ Events Tagged by a Fully Reconstructed B-Meson Decay and Determination of $\|V_{ub}\|$
By BaBar Collaboration (J.P. Lees *et al.*).
arXiv:1112.0702 [hep-ex].
10.1103/PhysRevD.86.032004.
Phys.Rev. D86 (2012) 032004.

101. Search for the $Z_1(4050)^+$ and $Z_2(4250)^+$ States in $\overline{B}^0 \to \chi_{c1} K^- \pi^+$ and $B^+ \to \chi_{c1} K_S^0 \pi^+$
By BaBar Collaboration (J.P. Lees *et al.*).
arXiv:1111.5919 [hep-ex].
10.1103/PhysRevD.85.052003.
Phys.Rev. D85 (2012) 052003.

102. Observation and Study of the Baryonic B-Meson Decays $B \to D^{(*)} p\overline{p} (\pi)(\pi)$
By BaBar Collaboration (P. del Amo Sanchez *et al.*).
arXiv:1111.4387 [hep-ex].
10.1103/PhysRevD.85.092017.
Phys.Rev. D85 (2012) 092017.

103. Amplitude Analysis and Measurement of the Time-Dependent CP Asymmetry of $B^0 \to K_S^0 K_S^0 K_S^0$ Decays
By BaBar Collaboration (J.P. Lees *et al.*).
arXiv:1111.3636 [hep-ex].
10.1103/PhysRevD.85.054023.
Phys.Rev. D85 (2012) 054023.

104. Elastic-Differential Cross Section of $\pi^+ + p$ at 1.5, 2.0, and 2.5 BeV/c
By Victor Cook, Bruce Cork, William R. Holley, Martin L. Perl.
10.1103/PhysRev.130.762.
Phys.Rev. 130 (1963) 762–765.

105. Search for the Decay $D^0 \to \gamma\gamma$ and Measurement of the Branching Fraction for $D^0 \to \pi^0\pi^0$
By BaBar Collaboration (J.P. Lees *et al.*).
arXiv:1110.6480 [hep-ex].
10.1103/PhysRevD.85.091107.
Phys.Rev. D85 (2012) 091107.

106. Search for $\bar{B} \to \Lambda_c^+ X \ell^- \bar{\nu}_\ell$ Decays in Events with a Fully Reconstructed B Meson
By BaBar Collaboration (J.P. Lees *et al.*).
arXiv:1110.6005 [hep-ex].
10.1103/PhysRevD.85.011102.
Phys.Rev. D85 (2012) 011102.

107. A Measurement of the Semileptonic Branching Fraction of the B_s Meson
By BaBar Collaboration (J.P. Lees *et al.*).
arXiv:1110.5600 [hep-ex].
10.1103/PhysRevD.85.011101.
Phys.Rev. D85 (2012) 011101.

108. A Tool for the LHCb MWPC Production Monitoring: The LNF On-Line Database
By A. Sarti.

109. Search for CP Violation in the Decay $\tau^- \to \pi^- K_S^0 \, (\geq 0\pi^0)\nu_\tau$
By BaBar Collaboration (J.P. Lees *et al.*).
arXiv:1109.1527 [hep-ex].
10.1103/PhysRevD.85.099904, 10.1103/PhysRevD.85.031102.
Phys.Rev. D85 (2012) 031102, Phys.Rev. D85 (2012) 099904.

110. Observation of the Rare Decay $B^+ \to K^+\pi^0\pi^0$ and Measurement of the Quasi-Two Body Contributions $B^+ \to K^*(892)^+\pi^0$, $B^+ \to f_0(980)K^+$ and $B^+ \to \chi_{c0}K^+$
By BaBar Collaboration (J.P. Lees *et al.*).
arXiv:1109.0143 [hep-ex].
10.1103/PhysRevD.84.092007.
Phys.Rev. D84 (2011) 092007.

111. Study of $\Upsilon(3S, 2S) \to \eta\Upsilon(1S)$ and $\Upsilon(3S, 2S) \to \pi^+\pi^-\Upsilon(1S)$ Hadronic Transitions
By BaBar Collaboration (J.P. Lees *et al.*).
arXiv:1108.5874 [hep-ex].
10.1103/PhysRevD.84.092003.
Phys.Rev. D84 (2011) 092003.

112. Search for Hadronic Decays of a Light Higgs Boson in the Radiative Decay $\Upsilon \to \gamma A^0$
By BaBar Collaboration (J.P. Lees *et al.*).
arXiv:1108.3549 [hep-ex].
10.1103/PhysRevLett.107.221803.
Phys.Rev.Lett. 107 (2011) 221803.

113. Observation of the Baryonic B Decay $\overline{B}^0 \to \Lambda_c^+\overline{\Lambda}K^-$
By BaBar Collaboration (J.P. Lees *et al.*).
arXiv:1108.3211 [hep-ex].
10.1103/PhysRevD.85.039903, 10.1103/PhysRevD.84.071102.
Phys.Rev. D84 (2011) 071102.

114. Branching Fraction Measurements of the Color-Suppressed Decays $\overline{B}^0 \to D^{(*)0}\pi^0$, $D^{(*)0}\eta$, $D^{(*)0}\omega$, and $D^{(*)0}\eta'$ and Measurement of the Polarization in the Decay $\overline{B}^0 \to D^{*0}\omega$
By BaBar Collaboration (J.P. Lees *et al.*).
arXiv:1107.5751 [hep-ex].
10.1103/PhysRevD.87.039901, 10.1103/PhysRevD.84.112007.
Phys.Rev. D84 (2011) 112007, Phys.Rev. D87 (2013) 3, 039901.

115. The Luminescent Chamber and Other New Detectors
By L.W. Jones, M.L. Perl.

116. Searches for Rare or Forbidden Semileptonic Charm Decays
By BaBar Collaboration (J.P. Lees *et al.*).
arXiv:1107.4465 [hep-ex].
10.1103/PhysRevD.84.072006.
Phys.Rev. D84 (2011) 072006.

117. The Influence of Accelerator Science on Physics Research
By Enzo F. Haussecker, Alexander W. Chao.
10.1007/s00016-010-0049-y.
Phys.Perspect. 13 (2011) 146-160.

118. Measurements of Branching Fractions and *CP* Asymmetries and Studies of Angular Distributions for $B \to \phi\phi K$ Decays
By BaBar Collaboration (J.P. Lees *et al.*).
arXiv:1105.5159 [hep-ex].
10.1103/PhysRevD.84.012001.
Phys.Rev. D84 (2011) 012001.

119. Study of Di-Pion Bottomonium Transitions and Search for the $h_b(1P)$ State
     By BaBar Collaboration (J.P. Lees *et al.*).
     arXiv:1105.4234 [hep-ex].
     10.1103/PhysRevD.84.011104.
     Phys.Rev. D84 (2011) 011104.

120. Search for *CP* Violation Using *T*-Odd Correlations in $D^+ \to K^+ K^0_S \pi^+ \pi^-$ and $D^+_s \to K^+ K^0_S \pi^+ \pi^-$ Decays
     By BaBar Collaboration (J.P. Lees *et al.*).
     arXiv:1105.4410 [hep-ex].
     10.1103/PhysRevD.84.031103.
     Phys.Rev. D84 (2011) 031103.

121. Search for *CP* Violation in the Decay $D^\pm \to K^0_S \pi^\pm$
     By BaBar Collaboration (P. del Amo Sanchez *et al.*).
     arXiv:1011.5477 [hep-ex].
     10.1103/PhysRevD.83.071103.
     Phys.Rev. D83 (2011) 071103.

122. Amplitude Analysis of $B^0 \to K^+ \pi^- \pi^0$ and Evidence of Direct *CP* Violation in $B \to K^* \pi$ Decays
     By BaBar Collaboration (J.P. Lees *et al.*).
     arXiv:1105.0125 [hep-ex].
     10.1103/PhysRevD.83.112010.
     Phys.Rev. D83 (2011) 112010.

123. PACCE: Perl Algorithm to Compute Continuum and Equivalent Widths
     By Rogerio Riffel, Tiberio Borges Vale.
     arXiv:1105.0318 [astro-ph.IM].
     10.1007/s10509-011-0731-9.
     Astrophys.Space Sci. 334 (2011) 351–356.

124. Study of Radiative Bottomonium Transitions Using Converted Photons
     By BaBar Collaboration (J.P. Lees *et al.*).
     arXiv:1104.5254 [hep-ex].
     10.1103/PhysRevD.84.099901, 10.1103/PhysRevD.84.072002.
     Phys.Rev. D84 (2011) 072002.

125. Search for $b \to u$ Transitions in $B^\pm \to [K^\mp \pi^\pm \pi^0]_D K^\pm$ Decays
     By BaBar Collaboration (J.P. Lees *et al.*).
     arXiv:1104.4472 [hep-ex].
     10.1103/PhysRevD.84.012002.
     Phys.Rev. D84 (2011) 012002.

126. Observation of $\eta_c(1S)$ and $\eta_c(2S)$ Decays to $K^+ K^- \pi^+ \pi^- \pi^0$ in Two-Photon Interactions
     By BaBar Collaboration (P. del Amo Sanchez *et al.*).
     arXiv:1103.3971 [hep-ex].
     10.1103/PhysRevD.84.012004.
     Phys.Rev. D84 (2011) 012004.

127. Cross Sections for the Reactions $e^+e^- \to K^+K^-\pi^+\pi^-$, $K^+K^-\pi^0\pi^0$, and $K^+K^-K^+K^-$ Measured Using Initial-State Radiation Events
By BaBar Collaboration (J.P. Lees *et al.*).
arXiv:1103.3001 [hep-ex].
10.1103/PhysRevD.86.012008.
Phys.Rev. D86 (2012) 012008.

128. Measurement of the Mass and Width of the $D_{s1}(2536)^+$ Meson
By BaBar Collaboration (J.P. Lees *et al.*).
arXiv:1103.2675 [hep-ex].
10.1103/PhysRevD.83.072003.
Phys.Rev. D83 (2011) 072003.

129. Evidence for the $h_b(1P)$ Meson in the Decay $\Upsilon(3S) \to \pi^0 h_b(1P)$
By BaBar Collaboration (J.P. Lees *et al.*).
arXiv:1102.4565 [hep-ex].
10.1103/PhysRevD.84.091101.
Phys.Rev. D84 (2011) 091101.

130. A Terrestrial Search for Dark Contents of the Vacuum, such as Dark Energy, Using Atom Interferometry
By Ronald J. Adler, Holger Muller, Martin L. Perl.
arXiv:1101.5626 [astro-ph.CO].
10.1142/S0217751X11054814.
Int.J.Mod.Phys. A26 (2011) 4959-4979.

131. Searches for the Baryon- and Lepton-Number Violating Decays $B^0 \to \Lambda_c^+ \ell^-$, $B^- \to \Lambda \ell^-$, and $B^- \to \bar{\Lambda} \ell^-$
By BaBar Collaboration (P. del Amo Sanchez *et al.*).
arXiv:1101.3830 [hep-ex].
10.1103/PhysRevD.83.091101.
Phys.Rev. D83 (2011) 091101.

132. Measurement of the $\gamma\gamma^* \to \eta$ and $\gamma\gamma^* \to \eta'$ Transition Form Factors
By BaBar Collaboration (P. del Amo Sanchez *et al.*).
arXiv:1101.1142 [hep-ex].
10.1103/PhysRevD.84.052001.
Phys.Rev. D84 (2011) 052001.

133. Measurement of Partial Branching Fractions of Inclusive Charmless B Meson Decays to K^+, K^0, and π^+
By BaBar Collaboration (P. del Amo Sanchez *et al.*).
arXiv:1012.5031 [hep-ex].
10.1103/PhysRevD.83.031103.
Phys.Rev. D83 (2011) 031103.

134. Measurements of Branching Fractions, Polarizations, and Direct *CP*-Violation Asymmetries in $B^+ \to \rho^0 K^{*+}$ and $B^+ \to f_0(980)K^{*+}$ Decays
By BaBar Collaboration (P. del Amo Sanchez *et al.*).
arXiv:1012.4044 [hep-ex].
10.1103/PhysRevD.83.051101.
Phys.Rev. D83 (2011) 051101.

135. Observation of the Decay $B^- \to D_s^{(*)+} K^- \ell^- \overline{\nu}_\ell$
By BaBar Collaboration (P. del Amo Sanchez *et al.*).
arXiv:1012.4158 [hep-ex].
10.1103/PhysRevLett.107.041804.
Phys.Rev.Lett. 107 (2011) 041804.

136. Analysis of the $D^+ \to K^- \pi^+ e^+ \nu_e$ Decay Channel
By BaBar Collaboration (P. del Amo Sanchez *et al.*).
arXiv:1012.1810 [hep-ex].
10.1103/PhysRevD.83.072001.
Phys.Rev. D83 (2011) 072001.

137. Dalitz Plot Analysis of $D_s^+ \to K^+ K^- \pi^+$
By BaBar Collaboration (P. del Amo Sanchez *et al.*).
arXiv:1011.4190 [hep-ex].
10.1103/PhysRevD.83.052001.
Phys.Rev. D83 (2011) 052001.

138. Studies of $\tau^- \to \eta K^- \nu_\tau$ and $\tau^- \to \eta \pi^- \nu_\tau$ at BaBar and a Search for a Second-Class Current
By BaBar Collaboration (P. del Amo Sanchez *et al.*).
arXiv:1011.3917 [hep-ex].
10.1103/PhysRevD.83.032002.
Phys.Rev. D83 (2011) 032002.

139. Measurement of the $B \to \overline{D}^{(*)} D^{(*)} K$ Branching Fractions
By BaBar Collaboration (P. del Amo Sanchez *et al.*).
arXiv:1011.3929 [hep-ex].
10.1103/PhysRevD.83.032004.
Phys.Rev. D83 (2011) 032004.

140. Search for the Decay $B^0 \to \gamma\gamma$
By BaBar Collaboration (P. del Amo Sanchez *et al.*).
arXiv:1010.2229 [hep-ex].
10.1103/PhysRevD.83.032006.
Phys.Rev. D83 (2011) 032006.

141. Measurement of the $B^0 \to \pi^\ell \ell^+ \nu$ and $B^+ \to \eta^{(\prime)} \ell^+ \nu$ Branching Fractions, the $B^0 \to \pi^- \ell^+ \nu$ and $B^+ \to \eta \ell^+ \nu$ Form-Factor Shapes, and Determination of $\left| V_{ub} \right|$
By BaBar Collaboration (P. del Amo Sanchez *et al.*).
arXiv:1010.0987 [hep-ex].
10.1103/PhysRevD.83.052011.
Phys.Rev. D83 (2011) 052011.

142. Search for the Rare Decay $B \to K$ Nu Nubar
     By BaBar Collaboration (P. del Amo Sanchez *et al.*).
     arXiv:1009.1529 [hep-ex].
     10.1103/PhysRevD.82.112002.
     Phys.Rev. D82 (2010) 112002.

143. Observation of new Resonances Decaying to $D\pi$ and $D^*\pi$ in Inclusive e^+e^- Collisions Near $\sqrt{s} = 10.58$ GeV
     By BaBar Collaboration (P. del Amo Sanchez *et al.*).
     arXiv:1009.2076 [hep-ex].
     10.1103/PhysRevD.82.111101.
     Phys.Rev. D82 (2010) 111101.

144. Measurement of the Absolute Branching Fractions for $D_s^- \to \ell^- \bar{\nu}_\ell$ and Extraction of the Decay Constant f_{D_s}
     By BaBar Collaboration (P. del Amo Sanchez *et al.*).
     arXiv:1008.4080 [hep-ex].
     10.1103/PhysRevD.91.019901, 10.1103/PhysRevD.82.091103.
     Phys.Rev. D82 (2010) 091103, Phys.Rev. D91 (2015) 1, 019901.

145. Evidence for $B^+ \to \tau^+ \nu_\tau$ Decays Using Hadronic B Tags
     By BaBar Collaboration (P. del Amo Sanchez *et al.*).
     arXiv:1008.0104 [hep-ex].

146. Exclusive Production of $D_s^+ D_s^-$, $D_s^{*+} D_s^-$, $D_s^{*+} D_s^{*-}$ and via e^+e^- Annihilation with Initial-State-Radiation
     By BaBar Collaboration (P. del Amo Sanchez *et al.*).
     arXiv:1008.0338 [hep-ex].
     10.1103/PhysRevD.82.052004.
     Phys.Rev. D82 (2010) 052004.

147. Search for Production of Invisible Final States in Single-Photon Decays of $\Upsilon(1S)$
     By BaBar Collaboration (P. del Amo Sanchez *et al.*).
     arXiv:1007.4646 [hep-ex].
     10.1103/PhysRevLett.107.021804.
     Phys.Rev.Lett. 107 (2011) 021804.

148. Dalitz-plot Analysis of $B^0 \to \bar{D}^0 \pi^+ \pi^-$
     By BaBar Collaboration (P. del Amo Sanchez *et al.*).
     arXiv:1007.4464 [hep-ex].
     PoS ICHEP2010 (2010) 250.

149. Search for $f_J(2220)$ in Radiative J/ψ Decays
     By BaBar Collaboration (P. del Amo Sanchez *et al.*).
     arXiv:1007.3526 [hep-ex].
     10.1103/PhysRevLett.105.172001.
     Phys.Rev.Lett. 105 (2010) 172001.

150. Search for B^+ Meson Decay to $a_1^+ (1260) K^{*0} (892)$
     By BaBar Collaboration (P. del Amo Sanchez *et al.*).
     arXiv:1007.2732 [hep-ex].
     10.1103/PhysRevD.82.091101.
     Phys.Rev. D82 (2010) 091101.

151. The Possible Detection of Dark Energy on Earth Using Atom Interferometry
     By Martin L. Perl.
     arXiv:1007.1622 [astro-ph.IM].
     10.1393/ncc/i2011-10738-8.
     Nuovo Cim. C033N5 (2010) 375-381.

152. A High Performance Hierarchical Storage Management System for the Canadian Tier-1 Centre at TRIUMF
     By D.C. Deatrich, S.X. Liu, R. Tafirout.
     10.1088/1742-6596/219/5/052008.
     J.Phys.Conf.Ser. 219 (2010) 052008.

153. Observation of the Decay $\overline{B}^0 \rightarrow \Lambda_c^+ \overline{p} p i^0$
     By BaBar Collaboration (B. Aubert *et al.*).
     arXiv:1007.1370 [hep-ex].
     10.1103/PhysRevD.82.031102.
     Phys.Rev. D82 (2010) 031102.

154. Measurement of *CP* Observables in $B^\pm \rightarrow D_{CP} K^\pm$ Decays and Constraints on the CKM Angle γ
     By BaBar Collaboration (P. del Amo Sanchez *et al.*).
     arXiv:1007.0504 [hep-ex].
     10.1103/PhysRevD.82.072004.
     Phys.Rev. D82 (2010) 072004.

155. TSkim: A Tool for Skimming ROOT Trees
     By David Chamont.
     10.1088/1742-6596/219/4/042004.
     J.Phys.Conf.Ser. 219 (2010) 042004.

156. Search for $b \rightarrow u$ Transitions in $B^- \rightarrow DK^-$ and $D^* K^-$ Decays
     By BaBar Collaboration (P. del Amo Sanchez *et al.*).
     arXiv:1006.4241 [hep-ex].
     10.1103/PhysRevD.82.072006.
     Phys.Rev. D82 (2010) 072006.

157. Correlated Leading Baryon-Antibaryon Production in $e^+ e^- \rightarrow c\bar{c} \rightarrow \Lambda_c^+ \overline{\Lambda}_c^- X$
     By BaBar Collaboration (B. Aubert *et al.*).
     arXiv:1006.2216 [hep-ex].
     10.1103/PhysRevD.82.091102.
     Phys.Rev. D82 (2010) 091102.

158. Evidence for the Decay $X(3872) \to J/\psi\omega$
By BaBar Collaboration (P. del Amo Sanchez *et al.*).
arXiv:1005.5190 [hep-ex].
10.1103/PhysRevD.82.011101.
Phys.Rev. D82 (2010) 011101.

159. Study of $B \to X_\gamma$ Decays and Determination of $|V_{td}/V_{ts}|$
By BaBar Collaboration (P. del Amo Sanchez *et al.*).
arXiv:1005.4087 [hep-ex].
10.1103/PhysRevD.82.051101.
Phys.Rev. D82 (2010) 051101.

160. Study of $B \to \pi\ell\nu$ and $B \to \rho\ell\nu$ Decays and Determination of $|V_{ub}|$
By BaBar Collaboration (P. del Amo Sanchez *et al.*).
arXiv:1005.3288 [hep-ex].
10.1103/PhysRevD.83.032007.
Phys.Rev. D83 (2011) 032007.

161. Observation of the Rare Decay $B^+ \to K^+\pi^0\pi^0$
By BaBar Collaboration (P. del Amo Sanchez *et al.*).
arXiv:1005.3717 [hep-ex].

162. Evidence for Direct *CP* Violation in the Measurement of the Cabibbo-Kobayashi-Maskawa Angle γ with $B^\mp \to D^{(*)}K^{(*)\mp}$ Decays
By BaBar Collaboration (P. del Amo Sanchez *et al.*).
arXiv:1005.1096 [hep-ex].
10.1103/PhysRevLett.105.121801.
Phys.Rev.Lett. 105 (2010) 121801.

163. Search for $B^+ \to D^+K^0$ and $B^+ \to D^+K^{*0}$ Decays
By BaBar Collaboration (P. del Amo Sanchez *et al.*).
arXiv:1005.0068 [hep-ex].
10.1103/PhysRevD.82.092006.
Phys.Rev. D82 (2010) 092006.

164. Measurement of D^0-$\bar{D}^0$ Mixing Parameters Using $D^0 \to K_S^0\pi^+\pi^-$ and $D^0 \to K_S^0 K^+K^-$ Decays
By BaBar Collaboration (P. del Amo Sanchez *et al.*).
arXiv:1004.5053 [hep-ex].
10.1103/PhysRevLett.105.081803.
Phys.Rev.Lett. 105 (2010) 081803.

165. *B*-Meson Decays to $\eta'\rho$, $\eta'f_0$, and $\eta'K^*$
By BaBar Collaboration (P. del Amo Sanchez *et al.*).
arXiv:1004.0240 [hep-ex].
10.1103/PhysRevD.82.011502.
Phys.Rev. D82 (2010) 011502.

166. Observation of the $\Upsilon 1^3 D_J$ Bottomonium State Through Decays to $\pi^+\pi^-\Upsilon 1S$
By BaBar Collaboration (P. del Amo Sanchez *et al.*).
arXiv:1004.0175 [hep-ex].
10.1103/PhysRevD.82.111102.
Phys.Rev. D82 (2010) 111102.

167. Search for *CP* Violation Using *T*-Odd Correlations in $D^0 \to K^+K^-\pi^+\pi^-$ Decays
By BaBar Collaboration (P. del Amo Sanchez *et al.*).
arXiv:1003.3397 [hep-ex].
10.1103/PhysRevD.81.111103.
Phys.Rev. D81 (2010) 111103.

168. Measurement of the Branching Fraction for $D_s^+ \to \tau^+\nu_\tau$ and Extraction of the Decay Constant f_{D_s}
By BaBar Collaboration (J.P. Lees *et al.*).
arXiv:1003.3063 [hep-ex].

169. Observation of the Rare Decay $B^0 \to K_s^0 K^\pm \pi^\mp$
By BaBar Collaboration (P. del Amo Sanchez *et al.*).
arXiv:1003.0640 [hep-ex].
10.1103/PhysRevD.82.031101.
Phys.Rev. D82 (2010) 031101.

170. Limits on τ Lepton-Flavor Violating Decays in Three Charged Leptons
By BaBar Collaboration (J.P. Lees *et al.*).
arXiv:1002.4550 [hep-ex].
10.1103/PhysRevD.81.111101.
Phys.Rev. D81 (2010) 111101.

171. Test of Lepton Universality in $\Upsilon(1S)$ Decays at BaBar
By BaBar Collaboration (P. del Amo Sanchez *et al.*).
arXiv:1002.4358 [hep-ex].
10.1103/PhysRevLett.104.191801.
Phys.Rev.Lett. 104 (2010) 191801.

172 Measurement of the $\gamma\gamma^* \to \eta_c$ Transition Form Factor
By BaBar Collaboration (J.P. Lees *et al.*).
arXiv:1002.3000 [hep-ex].
10.1103/PhysRevD.81.052010.
Phys.Rev. D81 (2010) 052010.

173. Observation of the $\chi_{c2}(2p)$ Meson in the Reaction $\gamma\gamma \to D\bar{D}$ at BaBar
By BaBar Collaboration (B. Aubert *et al.*).
arXiv:1002.0281 [hep-ex].
10.1103/PhysRevD.81.092003.
Phys.Rev. D81 (2010) 092003.

174. Searches for Fractionally Charged Particles
By Martin L. Perl, Eric R. Lee, Dinesh Loomba.
10.1146/annurev-nucl-121908-122035.
Ann.Rev.Nucl.Part.Sci. 59 (2009) 47–65.

175. Exploring the Possibility of Detecting Dark Energy in a Terrestrial Experiment Using Atom Interferometry
By Martin L. Perl, Holger Mueller.
arXiv:1001.4061 [physics.ins-det].

176. Search for Charged Lepton Flavor Violation in Narrow Upsilon Decays
By BaBar Collaboration (J.P. Lees *et al.*).
arXiv:1001.1883 [hep-ex].
10.1103/PhysRevLett.104.151802.
Phys.Rev.Lett. 104 (2010) 151802.

177. A Search for $B^+ \to \ell^+ \nu_\ell$ Recoiling Against $\bar{B} \to D^0 \ell^- \bar{\nu} X$
By BaBar Collaboration (Bernard Aubert *et al.*).
arXiv:0912.2453 [hep-ex].
10.1103/PhysRevD.81.051101.
Phys.Rev. D81 (2010) 051101.

178. Measurements of Charged Current Lepton Universality and $|V_{us}|$ Using τ Lepton Decays to $e^- \bar{\nu}_e \nu_\tau$, $\mu^- \bar{\nu}_\mu \nu_\tau$, $\pi^- \nu_\tau$ and $K^- \nu_\tau$
By BaBar Collaboration (Bernard Aubert *et al.*).
arXiv:0912.0242 [hep-ex].
10.1103/PhysRevLett.105.051602.
Phys.Rev.Lett. 105 (2010) 051602.

179. Observation of Inclusive $D^{*\pm}$ Production in the Decay of $\Upsilon(1S)$
By BaBar Collaboration (B. Aubert *et al.*).
arXiv:0911.2024 [hep-ex].
10.1103/PhysRevD.81.011102.
Phys.Rev. D81 (2010) 011102.

180. Search for the $B^+ \to K^+ \nu \bar{\nu}$ Decay Using Semi-Leptonic Tags
By BaBar Collaboration (B. Aubert *et al.*).
arXiv:0911.1988 [hep-ex].

181. Measurement of *CP* Violation Observables and Parameters for the Decays $B^\pm \to DK^{*\pm}$
By BaBar Collaboration (Bernard Aubert *et al.*).
arXiv:0909.3981 [hep-ex].
10.1103/PhysRevD.80.092001.
Phys.Rev. D80 (2009) 092001.

182. Measurements of the τ Mass and the Mass Difference of the τ^+ and τ^- at BABAR
By BaBar Collaboration (Bernard Aubert *et al.*).
arXiv:0909.3562 [hep-ex].
10.1103/PhysRevD.80.092005.
Phys.Rev. D80 (2009) 092005.

183. Measurement of Branching Fractions of B Decays to $K_1(1270)\pi$ and $K_1(1400)\pi$ and Determination of the CKM Angle α from $B^0 \to a_1(1260)^{\pm}\pi^{\mp}$
By BaBar Collaboration (Bernard Aubert *et al.*).
arXiv:0909.2171 [hep-ex].
10.1103/PhysRevD.81.052009.
Phys.Rev. D81 (2010) 052009.

184. Precise Measurement of the $e^+e^- \to \pi^+\pi^-\,(\gamma)$ Cross Section with the Initial State Radiation Method at BABAR
By BaBar Collaboration (Bernard Aubert *et al.*).
arXiv:0908.3589 [hep-ex].
10.1103/PhysRevLett.103.231801.
Phys.Rev.Lett. 103 (2009) 231801.

185. A Search for Invisible Decays of the $\Upsilon(1S)$
By BaBar Collaboration (Bernard Aubert *et al.*).
arXiv:0908.2840 [hep-ex].
10.1103/PhysRevLett.103.251801.
Phys.Rev.Lett. 103 (2009) 251801.

186. Search for a Narrow Resonance in e^+e^- to Four Lepton Final States
By BaBar Collaboration (Bernard Aubert *et al.*).
arXiv:0908.2821 [hep-ex].

187. Searches for Lepton Flavor Violation in the Decays $\tau^{\pm} \to e^{\pm}\gamma$ and $\tau^{\pm} \to \mu^{\pm}\gamma$
By BaBar Collaboration (Bernard Aubert *et al.*).
arXiv:0908.2381 [hep-ex].
10.1103/PhysRevLett.104.021802.
Phys.Rev.Lett. 104 (2010) 021802.

188. Observation and Study of Baryonic B Decays: $B \to D^{(*)}p\bar{p}$, $D^{(*)}p\bar{p}\pi$, and $D^{(*)}p\bar{p}\pi\pi$
By BaBar Collaboration (Bernard Aubert *et al.*).
arXiv:0908.2202 [hep-ex].

189. Study of D_{sJ} Decays to D^*K in Inclusive e^+e^- Interactions
By BaBar Collaboration (Bernard Aubert *et al.*).
arXiv:0908.0806 [hep-ex].
10.1103/PhysRevD.80.092003.
Phys.Rev. D80 (2009) 092003.

190. Measurement of $D^0 \to \bar{D}^0$ Mixing Using the Ratio of Lifetimes for the Decays $D^0 \to K^-\pi^+$ and K^+K^-
By BaBar Collaboration (Bernard Aubert *et al.*).
arXiv:0908.0761 [hep-ex].
10.1103/PhysRevD.80.071103.
Phys.Rev. D80 (2009) 071103.

191. Measurement and Interpretation of Moments in Inclusive Semileptonic Decays $\bar{B} \to X_c \ell^- \bar{\nu}$
By BaBar Collaboration (Bernard Aubert *et al.*).
arXiv:0908.0415 [hep-ex].
10.1103/PhysRevD.81.032003.
Phys.Rev. D81 (2010) 032003.

192. Observation of the Baryonic B-Decay $\bar{B}^0 \to \Lambda_c^+ \bar{p} K^- \pi^+$
By BaBar Collaboration (Bernard Aubert *et al.*).
arXiv:0907.4566 [hep-ex].
10.1103/PhysRevD.80.051105.
Phys.Rev. D80 (2009) 051105.

193. Search for B-Meson Decays to $b_1 \rho$ and $b_1 K^*$
By BaBar Collaboration (Bernard Aubert *et al.*).
arXiv:0907.3485 [hep-ex].
10.1103/PhysRevD.80.051101.
Phys.Rev. D80 (2009) 051101.

194. Observation and Polarization Measurement of $B^0 \to a_1(1260)^+ a_1(1260)^-$ Decay
By BaBar Collaboration (Bernard Aubert *et al.*).
arXiv:0907.1776 [hep-ex].
10.1103/PhysRevD.80.092007.
Phys.Rev. D80 (2009) 092007.

195. B-Meson Decays to Charmless Meson Pairs Containing η or η' Mesons
By BaBar Collaboration (Bernard Aubert *et al.*).
arXiv:0907.1743 [hep-ex].
10.1103/PhysRevD.80.112002.
Phys.Rev. D80 (2009) 112002.

196. Model-Independent Search for the Decay $B^+ \to \ell^+ \nu_\ell \gamma$
By BaBar Collaboration (Bernard Aubert *et al.*).
arXiv:0907.1681 [hep-ex].
10.1103/PhysRevD.80.111105.
Phys.Rev. D80 (2009) 111105.

197. SCTE: An Open-Source Perl Framework for Testing Equipment Control and Data Acquisition
By Luiz C.B. Mostaco-Guidolin, Leonid Ruchko, Ricardo M.O. Galvao.
arXiv:0906.4833 [physics.ins-det].
10.1016/j.cpc.2012.02.013.
Comput.Phys.Commun. 183 (2012) 1511-1518.

198. Search for a Low-Mass Higgs Boson in $\Upsilon(3S) \to \gamma A^0$, $A^0 \to \tau^+ \tau^-$ at BABAR
By BaBar Collaboration (Bernard Aubert *et al.*).
arXiv:0906.2219 [hep-ex].
10.1103/PhysRevLett.103.181801.
Phys.Rev.Lett. 103 (2009) 181801.

199. Measurement of Branching Fractions and CP and Isospin Asymmetries in $B \to K^*(892)\gamma$ Decays
By BaBar Collaboration (Bernard Aubert *et al.*).
arXiv:0906.2177 [hep-ex].
10.1103/PhysRevLett.103.211802.
Phys.Rev.Lett. 103 (2009) 211802.

200. Measurement of the $\gamma\gamma^* \to \pi^0$ Transition Form Factor
By BaBar Collaboration (Bernard Aubert *et al.*).
arXiv:0905.4778 [hep-ex].
10.1103/PhysRevD.80.052002.
Phys.Rev. D80 (2009) 052002.

201. Search for Dimuon Decays of a Light Scalar Boson in Radiative Transitions $\Upsilon \to \gamma A^0$
By BaBar Collaboration (Bernard Aubert *et al.*).
arXiv:0905.4539 [hep-ex].
10.1103/PhysRevLett.103.081803.
Phys.Rev.Lett. 103 (2009) 081803.

202. Time-Dependent Amplitude Analysis of $B^0 \to K_S^0\pi^+\pi^-$
By BaBar Collaboration (Bernard Aubert *et al.*).
arXiv:0905.3615 [hep-ex].
10.1103/PhysRevD.80.112001.
Phys.Rev. D80 (2009) 112001.

203. Search for B^0 Meson Decays to $\pi^0 K_S^0 K_S^0$, $\eta K_S^0 K_S^0$, and $\eta' K_S^0 K_S^0$
By BaBar Collaboration (Bernard Aubert *et al.*).
arXiv:0905.0868 [hep-ex].
10.1103/PhysRevD.80.011101.
Phys.Rev. D80 (2009) 011101.

204. Measurement of the Branching Fraction and $\overline{\Lambda}$ Polarization in $B^0 \to \overline{\Lambda} p \pi^-$
By BaBar Collaboration (Bernard Aubert *et al.*).
arXiv:0904.4724 [hep-ex].
10.1103/PhysRevD.79.112009.
Phys.Rev. D79 (2009) 112009.

205. Measurement of $|V_{cb}|$ and the Form-Factor Slope in $\overline{B} \to D\ell^-\overline{\nu}_\ell$ Decays in Events Tagged by a Fully Reconstructed B Meson
By BaBar Collaboration (Bernard Aubert *et al.*).
arXiv:0904.4063 [hep-ex].
10.1103/PhysRevLett.104.011802.
Phys.Rev.Lett. 104 (2010) 011802.

206. Search for Second-Class Currents in $\tau^- \to \omega\pi^-\nu_\tau$
By BaBar Collaboration (Bernard Aubert *et al.*).
arXiv:0904.3080 [hep-ex].
10.1103/PhysRevLett.103.041802, 10.1016/j.nuclphysbps.2009.03.021.
Phys.Rev.Lett. 103 (2009) 041802.

207. Search for $b \to u$ Transitions in $B^0 \to D^0 K^{*0}$ Decays
By BaBar Collaboration (Bernard Aubert *et al.*).
arXiv:0904.2112 [hep-ex].
10.1103/PhysRevD.80.031102.
Phys.Rev. D80 (2009) 031102.

208. Improved Limits on Lepton Flavor Violating τ Decays to $\ell\phi$, $\ell\rho$, ℓK^* and $\ell\bar{K}^*$
By BaBar Collaboration (Bernard Aubert *et al.*).
arXiv:0904.0339 [hep-ex].
10.1103/PhysRevLett.103.021801.
Phys.Rev.Lett. 103 (2009) 021801.

209. Exclusive Initial-State-Radiation Production of the $D\bar{D}$, $D^*\bar{D}$, and $D^*\bar{D}^*$ Systems
By BaBar Collaboration (Bernard Aubert *et al.*).
arXiv:0903.1597 [hep-ex].
10.1103/PhysRevD.79.092001.
Phys.Rev. D79 (2009) 092001.

210. Search for the Rare Leptonic Decays $B^+ \to \ell^+ \nu_\ell$ ($\ell = e, \mu$)
By BaBar Collaboration (Bernard Aubert *et al.*).
arXiv:0903.1220 [hep-ex].
10.1103/PhysRevD.79.091101.
Phys.Rev. D79 (2009) 091101.

211. Evidence for the $\eta_b(1S)$ Meson in Radiative $\Upsilon(2S)$ Decay
By BaBar Collaboration (Bernard Aubert *et al.*).
arXiv:0903.1124 [hep-ex].
10.1103/PhysRevLett.103.161801.
Phys.Rev.Lett. 103 (2009) 161801.

212. Measurement of the Semileptonic Decays $B \to D\tau^-\bar{\nu}_\tau$ and $B \to D^*\tau^-\bar{\nu}_\tau$
By BaBar Collaboration (Bernard Aubert *et al.*).
arXiv:0902.2660 [hep-ex].
10.1103/PhysRevD.79.092002.
Phys.Rev. D79 (2009) 092002.

213. Search for Dimuon Decays of a Light Scalar in Radiative Transitions $\Upsilon(3S) \to \gamma A^0$
By BaBar Collaboration (Bernard Aubert *et al.*).
arXiv:0902.2176 [hep-ex].

214. Dalitz Plot Analysis of $B^\pm \to \pi^\pm\pi^\pm\pi^\mp$ Decays
By BaBar Collaboration (Bernard Aubert *et al.*).
arXiv:0902.2051 [hep-ex].
10.1103/PhysRevD.79.072006.
Phys.Rev. D79 (2009) 072006.

215. Measurement of Time-Dependent CP Asymmetry in $B^0 \to c\bar{c}K^{(*)0}$ Decays
By BaBar Collaboration (Bernard Aubert *et al.*).
arXiv:0902.1708 [hep-ex].
10.1103/PhysRevD.79.072009.
Phys.Rev. D79 (2009) 072009.

216. Observation of B Meson Decays to ωK^* and Improved Measurements for $\omega\rho$ and ωf_0
By BaBar Collaboration (Bernard Aubert *et al.*).
arXiv:0901.3703 [hep-ex].
10.1103/PhysRevD.79.052005.
Phys.Rev. D79 (2009) 052005.

217. Improved Measurement of $B^+ \to \rho^+\rho^0$ and Determination of the Quark-Mixing Phase Angle α
By BaBar Collaboration (B. Aubert *et al.*).
arXiv:0901.3522 [hep-ex].
10.1103/PhysRevLett.102.141802.
Phys.Rev.Lett. 102 (2009) 141802.

218. Searches for Fractionally Charged Particles: What Should Be Done Next?
By Martin L. Perl.

219. Dalitz Plot Analysis of $B^- \to D^+\pi^-\pi^-$
By BaBar Collaboration (Bernard Aubert *et al.*).
arXiv:0901.1291 [hep-ex].
10.1103/PhysRevD.79.112004.
Phys.Rev. D79 (2009) 112004.

220. Evidence for $B^+ \to \bar{K}^{*0}K^{*+}$
By BaBar Collaboration (Bernard Aubert *et al.*).
arXiv:0901.1223 [hep-ex].
10.1103/PhysRevD.79.051102.
Phys.Rev. D79 (2009) 051102.

221. Search for Lepton Flavor Violating Decays $\tau^- \to l^- K_S^0$ with the BABAR Experiment
By BaBar Collaboration (Bernard Aubert *et al.*).
arXiv:0812.3804 [hep-ex].
10.1103/PhysRevD.79.012004.
Phys.Rev. D79 (2009) 012004.

222. Developing Creativity and Innovation in Engineering and Science
By Martin L. Perl.
10.1142/S0217751X08042754.
Int.J.Mod.Phys. A23 (2008) 4401–4413.

223. Search for the Lepton-Flavor Violating Decays $\Upsilon(3S) \to e^\pm\tau^\mp$ and $\Upsilon(3S) \to \mu^\pm\tau^\mp$
By BaBar Collaboration (Bernard Aubert *et al.*).
arXiv:0812.1021 [hep-ex].

224. Search for the Decay $B^+ \to K_S^0 K_S^0 \pi^+$
By BaBar Collaboration (Bernard Aubert *et al.*).
arXiv:0811.1979 [hep-ex].
10.1103/PhysRevD.79.051101.
Phys.Rev. D79 (2009) 051101.

225. Search for the $Z(4430)^-$ at BABAR
By BaBar Collaboration (Bernard Aubert *et al.*).
arXiv:0811.0564 [hep-ex].
10.1103/PhysRevD.79.112001.
Phys.Rev. D79 (2009) 112001.

226. Can the Existence of Dark Energy Be Directly Detected?
By Martin L. Perl.
arXiv:0809.5083 [hep-ex].
10.1142/S0217751X09047028.
Int.J.Mod.Phys. A24 (2009) 3426-3436.

227. Measurement of the $e^+e^- \to b\bar{b}$ Cross Section Between $\sqrt{s} = 10.54$ and 11.20-GeV
By BaBar Collaboration (Bernard Aubert *et al.*).
arXiv:0809.4120 [hep-ex].
10.1103/PhysRevLett.102.012001.
Phys.Rev.Lett. 102 (2009) 012001.

228. A Search for $B^+ \to \ell^+ \nu_\ell$ Recoiling Against $B^- \to D^0 \ell^- \bar{\nu} X$
By BaBar Collaboration (Bernard Aubert *et al.*).
arXiv:0809.4027 [hep-ex].

229. Measurement of Time Dependent *CP* Asymmetry Parameters in B^0 Meson Decays to ωK_S^0, $\eta' K^0$, and $\pi^0 K_S^0$
By BaBar Collaboration (Bernard Aubert *et al.*).
arXiv:0809.1174 [hep-ex].
10.1103/PhysRevD.79.052003.
Phys.Rev. D79 (2009) 052003.

230. Measurements of the Semileptonic Decays $\bar{B} \to D\ell\bar{\nu}$ and $\bar{B} \to D^*\ell\bar{\nu}$ Using a Global Fit to $DX\ell\bar{\nu}$ Final States
By BaBar Collaboration (Bernard Aubert *et al.*).
arXiv:0809.0828 [hep-ex].
10.1103/PhysRevD.79.012002.
Phys.Rev. D79 (2009) 012002.

231. Evidence for $X(3872) \to \psi(2S)\gamma$ in $B^\pm \to X(3872)K^\pm$ Decays, and a Study of $B \to c\bar{c}\gamma K$
By BaBar Collaboration (Bernard Aubert *et al.*).
arXiv:0809.0042 [hep-ex].
10.1103/PhysRevLett.102.132001.
Phys.Rev.Lett. 102 (2009) 132001.

232. Measurement of the $B^+ \to \omega \ell^+ \nu$ and $B^+ \to \eta \ell^+ \nu$ Branching Fractions
     By BaBar Collaboration (Bernard Aubert *et al.*).
     arXiv:0808.3524 [hep-ex].
     10.1103/PhysRevD.79.052011.
     Phys.Rev. D79 (2009) 052011.

233. Measurement of Branching Fractions and *CP* and Isospin Asymmetries in $B \to K^* \gamma$
     By BaBar Collaboration (Bernard Aubert *et al.*).
     arXiv:0808.1915 [hep-ex].

234. Update of Time-Dependent *CP* Asymmetry Measurements in $b \to c\bar{c}s$ Decays
     By BaBar Collaboration (Bernard Aubert *et al.*).
     arXiv:0808.1903 [hep-ex].

235. Measurements of Time-Dependent *CP* Asymmetries in $B^0 \to D^{(*)+}D^{(*)-}$ Decays
     By BaBar Collaboration (Bernard Aubert *et al.*).
     arXiv:0808.1866 [hep-ex].
     10.1103/PhysRevD.79.032002.
     Phys.Rev. D79 (2009) 032002.

236. Measurement of the Branching Fractions of the Radiative Charm Decays $D^0 \to \bar{K}^{*0}\gamma$ and $D^0 \to \phi\gamma$
     By BaBar Collaboration (Bernard Aubert *et al.*).
     arXiv:0808.1838 [hep-ex].
     10.1103/PhysRevD.78.071101.
     Phys.Rev. D78 (2008) 071101.

237. Study of the $\pi^+\pi^- J/\psi$ Mass Spectrum via Initial-State Radiation at BABAR
     By BaBar Collaboration (Bernard Aubert *et al.*).
     arXiv:0808.1543 [hep-ex].

238. Observation of $B^0 \to \chi_{c0}K^{*0}$ and Evidence for $B^+ \to \chi_{c0}K^{*+}$
     By BaBar Collaboration (Bernard Aubert *et al.*).
     arXiv:0808.1487 [hep-ex].
     10.1103/PhysRevD.78.091101.
     Phys.Rev. D78 (2008) 091101.

239. Measurements of Branching Fractions for $B^+ \to \rho^+\gamma$, $B^0 \to \rho^0\gamma$, and $B^0 \to \omega\gamma$
     By BaBar Collaboration (Bernard Aubert *et al.*).
     arXiv:0808.1379 [hep-ex].
     10.1103/PhysRevD.78.112001.
     Phys.Rev. D78 (2008) 112001.

240. Search for $B \to K^*\nu\bar{\nu}$ Decays
     By BaBar Collaboration (Bernard Aubert *et al.*).
     arXiv:0808.1338 [hep-ex].
     10.1103/PhysRevD.78.072007.
     Phys.Rev. D78 (2008) 072007.

241. Measurement of $B(\tau^- \to \bar{K}^0 \pi^- \nu_\tau)$ Using the BaBar Detector
     By BaBar Collaboration (Bernard Aubert *et al.*).
     arXiv:0808.1121 [hep-ex].
     10.1016/j.nuclphysbps.2009.03.034.
     Nucl.Phys.Proc.Suppl. 189 (2009) 193-198.

242. Dalitz Plot Analysis of $D_s^+ \to \pi^+ \pi^- \pi^+$
     By BaBar Collaboration (Bernard Aubert *et al.*).
     arXiv:0808.0971 [hep-ex].
     10.1103/PhysRevD.79.032003.
     Phys.Rev. D79 (2009) 032003.

243. Search for the Highly Suppressed Decays $B^- \to K^+ \pi^- \pi^-$ and $B^- \to K^- K^- \pi^+$
     By BaBar Collaboration (Bernard Aubert *et al.*).
     arXiv:0808.0900 [hep-ex].
     10.1103/PhysRevD.78.091102.
     Phys.Rev. D78 (2008) 091102.

244. Time-Dependent and Time-Integrated Angular Analysis of $B \to \varphi K_S^0 \pi^0$ and $\varphi K^\pm \pi^\mp$
     By BaBar Collaboration (Bernard Aubert *et al.*).
     arXiv:0808.3586 [hep-ex].
     10.1103/PhysRevD.78.092008.
     Phys.Rev. D78 (2008) 092008.

245. Measurement of *CP*-Violating Asymmetries in the $B^0 \to K^+ K^- K_S^0$ Dalitz Plot
     By BaBar Collaboration (Bernard Aubert *et al.*).
     arXiv:0808.0700 [hep-ex].

246. Measurement of the Branching Fractions of the Color-Suppressed Decays $\bar{B}^0 \to D^0 \pi^0$, $D^{*0} \pi^0$, $D^0 \eta$, $D^{*0} \eta$, $D^0 \omega$, $D^{*0} \omega$, $D^0 \eta'$, and $D^{*0} \eta'$
     By BaBar Collaboration (Bernard Aubert *et al.*).
     arXiv:0808.0697 [hep-ex].

247. Search for B^+-Meson Decay to $a_1^+ K^{*0}$
     By BaBar Collaboration (Bernard Aubert *et al.*).
     arXiv:0808.0579 [hep-ex].

248. Measurement of the Branching Fractions of $\bar{B} \to D^{**} \ell^- \bar{\nu}_\ell$ Decays in Events Tagged by a Fully Reconstructed B Meson
     By BaBar Collaboration (Bernard Aubert *et al.*).
     arXiv:0808.0528 [hep-ex].
     10.1103/PhysRevLett.101.261802.
     Phys.Rev.Lett. 101 (2008) 261802.

249. Measurement of Semileptonic *B* Decays into Orbitally Excited Charmed Mesons
     By BaBar Collaboration (Bernard Aubert *et al.*).
     arXiv:0808.0333 [hep-ex].
     10.1103/PhysRevLett.103.051803.
     Phys.Rev.Lett. 103 (2009) 051803.

250. Search for Invisible Decays of a Light Scalar in Radiative Transitions $\Upsilon(3S) \to \gamma A^0$
By BaBar Collaboration (Bernard Aubert *et al.*).
arXiv:0808.0017 [hep-ex].

251. Evidence for B Semileptonic Decays into the Charmed Baryon Λ_c^+
By BaBar Collaboration (Bernard Aubert *et al.*).
arXiv:0808.0011 [hep-ex].

252. Measurement of $\left|V_{cb}\right|$ and the Form-Factor Slope for $\overline{B} \to D\ell^- \overline{\nu}_\ell$ Decays on the Recoil of Fully Reconstructed B Mesons
By BaBar Collaboration (Bernard Aubert *et al.*).
arXiv:0807.4978 [hep-ex].

253. Measurement of the Branching Fraction, Polarization, and CP Asymmetries in $B^0 \to \rho^0 \rho^0$ Decay, and Implications for the CKM Angle α
By BaBar Collaboration (Bernard Aubert *et al.*).
arXiv:0807.4977 [hep-ex].
10.1103/PhysRevD.78.071104.
Phys.Rev. D78 (2008) 071104.

254. Measurement of $B \to X\gamma$ Decays and Determination of $\left|V_{td}/V_{ts}\right|$
By BaBar Collaboration (Bernard Aubert *et al.*).
arXiv:0807.4975 [hep-ex].
10.1103/PhysRevLett.102.161803.
Phys.Rev.Lett. 102 (2009) 161803.

255. Measurements of $\mathscr{B}(\overline{B}^0 \to \Lambda_c^+ \overline{p})$ and $\mathscr{B}(B^- \to \Lambda_c^+ \overline{p}\pi^-)$ and Studies of $\Lambda_c^+ \pi^-$ Resonances
By BaBar Collaboration (Bernard Aubert *et al.*).
arXiv:0807.4974 [hep-ex].
10.1103/PhysRevD.78.112003.
Phys.Rev. D78 (2008) 112003.

256. Search for Second-Class Currents in $\tau^- \to \omega\pi^- \nu_\tau$
By BaBar Collaboration (Bernard Aubert *et al.*).
arXiv:0807.4900 [hep-ex].

257. Measurement of Branching Fractions of B^0 Decays to $K_1(1270)^+\pi^-$ and $K_1(1400)^+\pi^-$
By BaBar Collaboration (Bernard Aubert *et al.*).
arXiv:0807.4760 [hep-ex].

258. Amplitude Analysis of the Decay $B^0 \to K^+\pi^-\pi^0$
By BaBar Collaboration (Bernard Aubert *et al.*).
arXiv:0807.4567 [hep-ex].

259. Measurement of D^0-$\overline{D}^0$ Mixing from a Time-Dependent Amplitude Analysis of $D^0 \to K^+\pi^-\pi^0$ Decays
By BaBar Collaboration (Bernard Aubert *et al.*).
arXiv:0807.4544 [hep-ex].
10.1103/PhysRevLett.103.211801.
Phys.Rev.Lett. 103 (2009) 211801.

260. Measurement of CP Asymmetries and Branching Fractions in $B^0 \to \pi^+\pi^-$, $B^0 \to K^+\pi^-$, $B^0 \to \pi^0\pi^0$, $B^0 \to K^0\pi^0$ and Isospin Analysis of $B \to \pi\pi$ Decays
By BaBar Collaboration (Bernard Aubert *et al.*).
arXiv:0807.4226 [hep-ex].

261. Search for $B^+ \to \mu^+\nu_\mu$ with Inclusive Reconstruction at BaBar
By BaBar Collaboration (Bernard Aubert *et al.*).
arXiv:0807.4187 [hep-ex].

262. Direct CP, Lepton Flavor and Isospin Asymmetries in the Decays $B \to K^{(*)}\ell^+\ell^-$
By BaBar Collaboration (Bernard Aubert *et al.*).
arXiv:0807.4119 [hep-ex].
10.1103/PhysRevLett.102.091803.
Phys.Rev.Lett. 102 (2009) 091803.

263. Searches for B Meson Decays to $\phi\phi$, $\phi\rho$, $\phi f_0(980)$, and $f_0(980)f_0(980)$ Final States
By BaBar Collaboration (Bernard Aubert *et al.*).
arXiv:0807.3935 [hep-ex].
10.1103/PhysRevLett.101.201801.
Phys.Rev.Lett. 101 (2008) 201801.

264. Measurement of Time-Dependent CP Asymmetry in $B^0 \to K^0_S\pi^0\gamma$ Decays
By BaBar Collaboration (Bernard Aubert *et al.*).
arXiv:0807.3103 [hep-ex].
10.1103/PhysRevD.78.071102.
Phys.Rev. D78 (2008) 071102.

265. Measurement of Ratios of Branching Fractions and CP-Violating Asymmetries of $B^\pm \to D^*K^\pm$ Decays
By BaBar Collaboration (Bernard Aubert *et al.*).
arXiv:0807.2408 [hep-ex].
10.1103/PhysRevD.78.092002.
Phys.Rev. D78 (2008) 092002.

266. Study of Hadronic Transitions Between Υ States and Observation of $\Upsilon(4S) \to \eta\Upsilon(1S)$ Decay
By BaBar Collaboration (Bernard Aubert *et al.*).
arXiv:0807.2014 [hep-ex].
10.1103/PhysRevD.78.112002.
Phys.Rev. D78 (2008) 112002.

267. Study of the Decay $D_s^+ \to K^+K^-e^+\nu_e$
By BaBar Collaboration (Bernard Aubert *et al.*).
arXiv:0807.1599 [hep-ex].
10.1103/PhysRevD.78.051101.
Phys.Rev. D78 (2008) 051101.

268. Observation of the Bottomonium Ground State in the Decay $\Upsilon(3S) \to \gamma\eta_b$
By BaBar Collaboration (Bernard Aubert *et al.*).
arXiv:0807.1086 [hep-ex].
10.1103/PhysRevLett.101.071801.
Phys.Rev.Lett. 101 (2008) 071801, Phys.Rev.Lett. 102 (2009) 029901.

269. Search for $B^0 \to K^{*+}K^{*-}$
By BaBar Collaboration (Bernard Aubert *et al.*).
arXiv:0806.4467 [hep-ex].
10.1103/PhysRevD.78.051103.
Phys.Rev. D78 (2008) 051103.

270. Observation and Polarization Measurements of $B^\pm \to \varphi K_1^\pm$ and $B^\pm \to \varphi K_2^{*\pm}$
By BaBar Collaboration (Bernard Aubert *et al.*).
arXiv:0806.4419 [hep-ex].
10.1103/PhysRevLett.101.161801.
Phys.Rev.Lett. 101 (2008) 161801.

271. Observation of $e^+e^- \to \rho^+\rho^-$ Near $\sqrt{s} = 10.58$ GeV
By BaBar Collaboration (Bernard Aubert *et al.*).
arXiv:0806.3893 [hep-ex].
10.1103/PhysRevD.78.071103.
Phys.Rev. D78 (2008) 071103.

272. A Measurement of *CP* Asymmetry in $b \to s\gamma$ Using a Sum of Exclusive Final States
By BaBar Collaboration (Bernard Aubert *et al.*).
arXiv:0805.4796 [hep-ex].
10.1103/PhysRevLett.101.171804.
Phys.Rev.Lett. 101 (2008) 171804.

273. Measurements of $B \to \{\pi, \eta, \eta'\}\ell\nu_\ell$ Branching Fractions and Determination of $|V_{ub}|$ with Semileptonically Tagged *B* Mesons
By BaBar Collaboration (Bernard Aubert *et al.*).
arXiv:0805.2408 [hep-ex].
10.1103/PhysRevLett.101.081801.
Phys.Rev.Lett. 101 (2008) 081801.

274. Constraints on the CKM Angle γ in $B^0 \to \overline{D}^0 K^{*0}$ and $B^0 \to D^0 K^{*0}$ from a Dalitz Analysis of D^0 and $\overline{D}^0$ Decays to $K_S\pi^+\pi^-$
By BaBar Collaboration (Bernard Aubert *et al.*).
arXiv:0805.2001 [hep-ex].
10.1103/PhysRevD.79.072003.
Phys.Rev. D79 (2009) 072003.

275. Branching Fractions and *CP*-Violating Asymmetries in Radiative *B* Decays to $\eta K\gamma$
By BaBar Collaboration (Bernard Aubert *et al.*).
arXiv:0805.1317 [hep-ex].
10.1103/PhysRevD.79.011102.
Phys.Rev. D79 (2009) 011102.

276. Observation of $B^+ \to b_1^+ K^0$ and Search for B-Meson Decays to $b_1^0 K^0$ and $b_1 \pi^0$
By BaBar Collaboration (Bernard Aubert *et al.*).
arXiv:0805.1217 [hep-ex].
10.1103/PhysRevD.78.011104.
Phys.Rev. D78 (2008) 011104.

277. Measurement of the Mass Difference $m(B^0) - m(B^+)$
By BaBar Collaboration (Bernard Aubert *et al.*).
arXiv:0805.0497 [hep-ex].
10.1103/PhysRevD.78.011103.
Phys.Rev. D78 (2008) 011103.

278. Angular Distributions in the Decays $B \to K^* \ell^+ \ell^-$
By BaBar Collaboration (Bernard Aubert *et al.*).
arXiv:0804.4412 [hep-ex].
10.1103/PhysRevD.79.031102.
Phys.Rev. D79 (2009) 031102.

279. Observation of $B^+ \to \eta \rho^+$ and Search for B^0 Decays to $\eta' \eta$, $\eta \pi^0$, $\eta' \pi^0$, and $\omega \pi^0$
By BaBar Collaboration (Bernard Aubert *et al.*).
arXiv:0804.2422 [hep-ex].
10.1103/PhysRevD.78.011107.
Phys.Rev. D78 (2008) 011107.

280. Improved Measurement of the CKM Angle γ in $B^\mp \to D^{(*)} K^{(*)\mp}$ Decays with a Dalitz Plot Analysis of D Decays to $K_S^0 \pi^+ \pi^-$ and $K_S^0 K^+ K^-$
By BaBar Collaboration (Bernard Aubert *et al.*).
arXiv:0804.2089 [hep-ex].
10.1103/PhysRevD.78.034023.
Phys.Rev. D78 (2008) 034023.

281. Study of B-Meson Decays to $\eta_c K^{(*)}$, $\eta_c(2S) K^{(*)}$ and $\eta_c \gamma K^{(*)}$
By BaBar Collaboration (Bernard Aubert *et al.*).
arXiv:0804.1208 [hep-ex].
10.1103/PhysRevD.78.012006.
Phys.Rev. D78 (2008) 012006.

282. Evidence for *CP* Violation in $B^0 \to J/\psi \pi^0$ Decays
By BaBar Collaboration (Bernard Aubert *et al.*).
arXiv:0804.0896 [hep-ex].
10.1103/PhysRevLett.101.021801.
Phys.Rev.Lett. 101 (2008) 021801.

283. Study of B Meson Decays with Excited η and η' Mesons
By BaBar Collaboration (Bernard Aubert *et al.*).
arXiv:0804.0411 [hep-ex].
10.1103/PhysRevLett.101.091801.
Phys.Rev.Lett. 101 (2008) 091801.

284. Evidence for Direct CP Violation from Dalitz-Plot Analysis of $B^{\pm} \to K^{\pm} \pi^{\mp} \pi^{\pm}$
By BaBar Collaboration (Bernard Aubert *et al.*).
arXiv:0803.4451 [hep-ex].
10.1103/PhysRevD.78.012004.
Phys.Rev. D78 (2008) 012004.

285. Measurement of the Branching Fractions of the Rare Decays $B^0 \to D_s^{(*)+} \pi^-$, $B^0 \to D_s^{(*)+} \rho^-$, and $B^0 \to D_s^{(*)-} K^{(*)+}$
By BaBar Collaboration (Bernard Aubert *et al.*).
arXiv:0803.4296 [hep-ex].
10.1103/PhysRevD.78.032005.
Phys.Rev. D78 (2008) 032005.

286. Changes in Geant4 Electromagnetics from Release 4.6.1 to 4.9.1
By J. Perl.

287. Standard and Unconventional Experiments in Lepton Physics
By Martin L. Perl.

288. A Study of $B \to X(3872) K$, with $X(3872) \to J/\psi \pi^+ \pi^-$
By BaBar Collaboration (Bernard Aubert *et al.*).
arXiv:0803.2838 [hep-ex].
10.1103/PhysRevD.77.111101.
Phys.Rev. D77 (2008) 111101.

289. Measurement of the Spin of the $\Xi(1530)$ Resonance
By BaBar Collaboration (Bernard Aubert *et al.*).
arXiv:0803.1863 [hep-ex].
10.1103/PhysRevD.78.034008.
Phys.Rev. D78 (2008) 034008.

290. Essay: The τ Lepton and Thirty Years of Changes in Elementary Particle Physics Research
By M.L. Perl.
10.1103/PhysRevLett.100.070001.
Phys.Rev.Lett. 100 (2008) 070001.

291. Measurement of the $\tau^- \to \eta \pi^- \pi^+ \pi^- \nu_\tau$ Branching Fraction and a Search for a Second-Class Current in the $\tau^- \to \eta'(958) \pi^- \nu_\tau$ Decay
By BaBar Collaboration (Bernard Aubert *et al.*).
arXiv:0803.0772 [hep-ex].
10.1103/PhysRevD.77.112002.
Phys.Rev. D77 (2008) 112002.

292. Measurement of *CP* Observables in $B^{\pm} \to D_{CP}^0 K^{\pm}$ Decays
By BaBar Collaboration (Bernard Aubert *et al.*).
arXiv:0802.4052 [hep-ex].
10.1103/PhysRevD.77.111102.
Phys.Rev. D77 (2008) 111102.

293. Search for *CP* Violation in Neutral *D* Meson Cabibbo-Suppressed Three-Body Decays
     By BaBar Collaboration (Bernard Aubert *et al.*).
     arXiv:0802.4035 [hep-ex].
     10.1103/PhysRevD.78.051102.
     Phys.Rev. D78 (2008) 051102.

294. Searches for the Decays $B^0 \to \ell^\pm \tau^\mp$ and $B^+ \to \ell^+ \nu \,(\ell = e, \mu)$ Using Hadronic Tag Reconstruction
     By BaBar Collaboration (Bernard Aubert *et al.*).
     arXiv:0801.0697 [hep-ex].
     10.1103/PhysRevD.77.091104.
     Phys.Rev. D77 (2008) 091104.

295. Study of the Exclusive Initial-State-Radiation Production of the $D\bar{D}$ System
     By BaBar Collaboration (Bernard Aubert *et al.*).
     arXiv:0710.1371 [hep-ex].

296. A Measurement of the Branching Fractions of Exclusive $\bar{B} \to D^{(*)}(\pi)\ell^-\bar{\nu}_\ell$ Decays in Events with
     a Fully Reconstructed *B* Meson
     By BaBar Collaboration (Bernard Aubert *et al.*).
     arXiv:0712.3503 [hep-ex].
     10.1103/PhysRevLett.100.151802.
     Phys.Rev.Lett. 100 (2008) 151802.

297. Measurement of the Decay $B^- \to D^{*0}e^-\bar{\nu}_e$
     By BaBar Collaboration (Bernard Aubert *et al.*).
     arXiv:0712.3493 [hep-ex].
     10.1103/PhysRevLett.100.231803.
     Phys.Rev.Lett. 100 (2008) 231803.

298. Time-Dependent Dalitz Plot Analysis of $B^0 \to D^\mp K^0 \pi^\pm$ Decays
     By BaBar Collaboration (Bernard Aubert *et al.*).
     arXiv:0712.3469 [hep-ex].
     10.1103/PhysRevD.77.071102.
     Phys.Rev. D77 (2008) 071102.

299. Measurement of D^0-$\bar{D}^0$ Mixing Using the Ratio of Lifetimes for the Decays $D^0 \to K^-\pi^+, K^-K^+$,
     and $\pi^-\pi^+$
     By BaBar Collaboration (Bernard Aubert *et al.*).
     arXiv:0712.2249 [hep-ex].
     10.1103/PhysRevD.78.011105.
     Phys.Rev. D78 (2008) 011105.

300. Search for Decays of B^0 Mesons into e^+e^-, $\mu^+\mu^-$, and $e^\pm\mu^\mp$ Final States
     By BaBar Collaboration (Bernard Aubert *et al.*).
     arXiv:0712.1516 [hep-ex].
     10.1103/PhysRevD.77.032007.
     Phys.Rev. D77 (2008) 032007.

301. Measurement of the $B \to X_s\gamma$ Branching Fraction and Photon Energy Spectrum Using the Recoil Method
By BaBar Collaboration (Bernard Aubert *et al.*).
arXiv:0711.4889 [hep-ex].
10.1103/PhysRevD.77.051103.
Phys.Rev. D77 (2008) 051103.

302. Improvements in Monte Carlo Simulation of Large Electron Fields
By Bruce A. Faddegon, Joseph Perl, Makoto Asai.
Submitted to: Phys.Med.Biol..

303. Dalitz Plot Analysis of the Decay $B^0(\bar{B}^0) \to K^{\pm}\pi^{\mp}\pi^0$
By BaBar Collaboration (Bernard Aubert *et al.*).
arXiv:0711.4417 [hep-ex].
10.1103/PhysRevD.78.052005.
Phys.Rev. D78 (2008) 052005.

304. Search for Fractional-Charge Particles in Meteoritic Material
By Peter C. Kim, Eric R. Lee, Irwin T. Lee, Martin L. Perl, Valerie Halyo, Dinesh Loomba.
10.1103/PhysRevLett.99.161804.
Phys.Rev.Lett. 99 (2007) 161804.

305. Search for CPT and Lorentz Violation in $B^0 - \bar{B}^0$ Oscillations with Dilepton Events
By BaBar Collaboration (Bernard Aubert *et al.*).
arXiv:0711.2713 [hep-ex].
10.1103/PhysRevLett.100.131802.
Phys.Rev.Lett. 100 (2008) 131802.

306. A Contrarian View of How to Develop Creativiity in Science and Engineering
By Martin L. Perl.

307. Observation of $Y(3940) \to J/\psi\omega$ in $B \to J/\psi\omega K$ at BABAR
By BaBar Collaboration (Bernard Aubert *et al.*).
arXiv:0711.2047 [hep-ex].
10.1103/PhysRevLett.101.082001.
Phys.Rev.Lett. 101 (2008) 082001.

308. Search for Lepton Flavor Violating Decays $\tau^{\pm} \to \ell^{\pm}\omega$ ($\ell = e, \mu$)
By BaBar Collaboration (Bernard Aubert *et al.*).
arXiv:0711.0980 [hep-ex].
10.1103/PhysRevLett.100.071802.
Phys.Rev.Lett. 100 (2008) 071802.

309. A Study of $\bar{B} \to \Xi_c\bar{\Lambda}_c^-$ and $\bar{B} \to \Lambda_c^+\bar{\Lambda}_c^-\bar{K}$ Decays at BABAR
By BaBar Collaboration (Bernard Aubert *et al.*).
arXiv:0710.5775 [hep-ex].
10.1103/PhysRevD.77.031101.
Phys.Rev. D77 (2008) 031101.

310. A Study of Excited Charm-Strange Baryons with Evidence for New Baryons $\Xi_c(3055)^+$ and $\Xi_c(3123)^+$
By BaBar Collaboration (Bernard Aubert *et al.*).
arXiv:0710.5763 [hep-ex].
10.1103/PhysRevD.77.012002.
Phys.Rev. D77 (2008) 012002.

311. Measurements of $e^+e^- \to K^+K^-\eta$, $K^+K^-\pi^0$ and $K_s^0 K^\pm \pi^\mp$ Cross Sections Using Initial State Radiation Events
By BaBar Collaboration (Bernard Aubert *et al.*).
arXiv:0710.4451 [hep-ex].
10.1103/PhysRevD.77.092002.
Phys.Rev. D77 (2008) 092002.

312. Observation of $B^+ \to a_1^+(1260) K^0$ and $B^0 \to a_1^-(1260) K^+$
By BaBar Collaboration (Bernard Aubert *et al.*).
arXiv:0709.4165 [hep-ex].
10.1103/PhysRevLett.100.051803.
Phys.Rev.Lett. 100 (2008) 051803.

313. Search for *CP* Violation in the Decays $D^0 \to K^-K^+$ and $D^0 \to \pi^-\pi^+$
By BaBar Collaboration (Bernard Aubert *et al.*).
arXiv:0709.2715 [hep-ex].
10.1103/PhysRevLett.100.061803.
Phys.Rev.Lett. 100 (2008) 061803.

314. Study of $e^+e^- \to \Lambda\bar{\Lambda}$, $\Lambda\bar{\Sigma}^0$, $\Sigma^0\bar{\Sigma}^0$ Using Initial State Radiation with BABAR
By BaBar Collaboration (Bernard Aubert *et al.*).
arXiv:0709.1988 [hep-ex].
10.1103/PhysRevD.76.092006.
Phys.Rev. D76 (2007) 092006.

315. Observation of the Semileptonic Decays $B \to D^*\tau^-\bar{\nu}_\tau$ and Evidence for $B \to D\tau^-\bar{\nu}_\tau$
By BaBar Collaboration (Bernard Aubert *et al.*).
arXiv:0709.1698 [hep-ex].
10.1103/PhysRevLett.100.021801.
Phys.Rev.Lett. 100 (2008) 021801.

316. Measurements of Partial Branching Fractions for $\bar{B} \to X_u \ell \bar{\nu}$ and Determination of $\left|V_{ub}\right|$
By BaBar Collaboration (Bernard Aubert *et al.*).
arXiv:0708.3702 [hep-ex].
10.1103/PhysRevLett.100.171802.
Phys.Rev.Lett. 100 (2008) 171802.

317. Improved Limits on the Lepton-Flavor Violating Decays $\tau^- \to \ell^-\ell^+\ell^-$
By BaBar Collaboration (Bernard Aubert *et al.*).
arXiv:0708.3650 [hep-ex].
10.1103/PhysRevLett.99.251803.
Phys.Rev.Lett. 99 (2007) 251803.

318. Measurements of the Branching Fractions of $B^0 \to K^{*0}K^+K^-$, $B^0 \to K^{*0}\pi^+K^-$, $B^0 \to K^{*0}K^+\pi^-$, and $B^0 \to K^{*0}\pi^+\pi^-$
    By BaBar Collaboration (Bernard Aubert *et al.*).
    arXiv:0708.2543 [hep-ex].
    10.1103/PhysRevD.76.071104.
    Phys.Rev. D76 (2007) 071104.

319. The $e^+e^- \to 2(\pi^+\pi^-)\pi^0$, $2(\pi^+\pi^-)\eta$, $K^+K^-\pi^+\pi^-\pi^0$ and $K^+K^-\pi^+\pi^-\eta$ Cross Sections Measured with Initial-State Radiation
    By BaBar Collaboration (Bernard Aubert *et al.*).
    arXiv:0708.2461 [hep-ex].
    10.1103/PhysRevD.77.119902, 10.1103/PhysRevD.76.092005.
    Phys.Rev. D76 (2007) 092005, Phys.Rev. D77 (2008) 119902.

320. A Search for $B^+ \to \tau^+\nu$ Decays with Hadronic B Tags
    By BaBar Collaboration (Bernard Aubert *et al.*).
    arXiv:0708.2260 [hep-ex].
    10.1103/PhysRevD.77.011107.
    Phys.Rev. D77 (2008) 011107.

321. Observation of $B^0 \to K^{*0}\bar{K}^{*0}$ and Search for $B^0 \to K^{*0}K^{*0}$
    By BaBar Collaboration (Bernard Aubert *et al.*).
    arXiv:0708.2248 [hep-ex].
    10.1103/PhysRevLett.100.081801.
    Phys.Rev.Lett. 100 (2008) 081801.

322. Time-Dependent Dalitz Plot Analysis of $B^0 \to K_s^0\pi^+\pi^-$
    By BaBar Collaboration (Bernard Aubert *et al.*).
    arXiv:0708.2097 [hep-ex].

323. Measurement of the $B^0 \to X_u^-\ell^+\nu_\ell$ Decays Near the Kinematic Endpoint of the Lepton Spectrum and Search for Violation of Isospin Symmetry
    By BaBar Collaboration (Bernard Aubert *et al.*).
    arXiv:0708.1753 [hep-ex].

324. Measurement of the Branching Fractions of Exclusive $\bar{B} \to D/D^*/D^{(*)}\pi\ell^-\bar{\nu}_\ell$ Decays in Events Tagged by a Fully Reconstructed B Meson
    By BaBar Collaboration (Bernard Aubert *et al.*).
    arXiv:0708.1738 [hep-ex].

325. Evidence for $b \to d\gamma$ Transitions from a Sum of Exclusive Final States in the Hadronic Final State Mass Range $1.0\ \text{GeV}/c^2 < M(X_d) < 1.8\ \text{GeV}/c^2$
    By BaBar Collaboration (Bernard Aubert *et al.*).
    arXiv:0708.1652 [hep-ex].

326. Study of Resonances in Exclusive B Decays to $\bar{D}^{(*)}D^{(*)}K$
    By BaBar Collaboration (Bernard Aubert *et al.*).
    arXiv:0708.1565 [hep-ex].
    10.1103/PhysRevD.77.011102.
    Phys.Rev. D77 (2008) 011102.

327. Improved Measurement of Time-Dependent *CP* Asymmetries and the *CP*-Odd Fraction in the Decay $B^0 \to D^{*+}D^{*-}$
By BaBar Collaboration (Bernard Aubert *et al.*).
arXiv:0708.1549 [hep-ex].
10.1103/PhysRevD.76.111102.
Phys.Rev. D76 (2007) 111102.

328. Measurement of $\cos 2\beta$ in $B^0 \to D^{(*)}h^0$ Decays with a Time-Dependent Dalitz Plot Analysis of $D \to K_s^0\pi^+\pi^-$
By BaBar Collaboration (Bernard Aubert *et al.*).
arXiv:0708.1544 [hep-ex].
10.1103/PhysRevLett.99.231802.
Phys.Rev.Lett. 99 (2007) 231802.

329. Measurement of the Branching Fraction Ratios and *CP* Asymmetries in $B^- \to D^0_{CP}K^-$ Decays
By BaBar Collaboration (Bernard Aubert *et al.*).
arXiv:0708.1534 [hep-ex].

330. Time-Dependent Analysis of the Decay $B^0 \to \rho^0\rho^0$
By BaBar Collaboration (Bernard Aubert *et al.*).
arXiv:0708.1630 [hep-ex].

331. Measurement of the Time-Dependent *CP* Asymmetry in $B^0 \to K^{*0}\gamma$ Decays
By BaBar Collaboration (Bernard Aubert *et al.*).
arXiv:0708.1614 [hep-ex].

332. Search for the Decay $B^+ \to K^+\tau^\mp\mu^\pm$
By BaBar Collaboration (Bernard Aubert *et al.*).
arXiv:0708.1303 [hep-ex].
10.1103/PhysRevLett.99.201801.
Phys.Rev.Lett. 99 (2007) 201801.

333. On Learning in Engineering and Science When Old
By Martin L. Perl.

334. Search for the Rare Charmless Hadronic Decay $B^+ \to a_0^+\pi_0$
By BaBar Collaboration (Bernard Aubert *et al.*).
arXiv:0708.0963 [hep-ex].
10.1103/PhysRevD.77.039903, 10.1103/PhysRevD.77.011101, 10.1103/PhysRevD.77.019904.
Phys.Rev. D77 (2008) 011101.

335. Comparison of Monte Carlo Simulation Results to an Experimental Thick-Target Bremsstrahlung Benchmark
By B. Faddegon, E. Traneus, J. Perl, J. Tinslay, M. Asai.
10.1118/1.2760781.
Med.Phys. 34 (2007) 2429.

336. Observation of the Decay $B^+ \to K^+K^-\pi^+$
     By BaBar Collaboration (Bernard Aubert *et al.*).
     arXiv:0708.0376 [hep-ex].
     10.1103/PhysRevLett.99.221801.
     Phys.Rev.Lett. 99 (2007) 221801.

337. Verification of Bremsstrahlung Splitting in Geant4 for Radiotherapy Quality Beams
     By J. Tinslay, B.A. Faddegon, J. Perl, M. Asai.
     10.1118/1.2761172.
     Med.Phys. 34 (2007) 2504.

338. Search for $b \to u$ Transitions in $B^- \to [K^+\pi^-\pi^0]_D K^-$
     By BaBar Collaboration (Bernard Aubert *et al.*).
     arXiv:0708.0182 [hep-ex].
     10.1103/PhysRevD.76.111101.
     Phys.Rev. D76 (2007) 111101.

339. Evidence for Charged B Meson Decays to $a_1^{\pm}(1260)\pi^0$ and $a_1^0(1260)\pi^{\pm}$
     By BaBar Collaboration (Bernard Aubert *et al.*).
     arXiv:0708.0050 [hep-ex].
     10.1103/PhysRevLett.99.261801.
     Phys.Rev.Lett. 99 (2007) 261801.

340. Observation of B-Meson Decays to $b_1\pi$ and $b_1 K$
     By BaBar Collaboration (Bernard Aubert *et al.*).
     arXiv:0707.4561 [hep-ex].
     10.1103/PhysRevLett.99.241803.
     Phys.Rev.Lett. 99 (2007) 241803.

341. Exclusive Branching Fraction Measurements of Semileptonic τ Decays into Three Charged Hadrons, $\tau^- \to \phi\pi^-\nu_\tau$ and $\tau^- \to \phi K^-\nu_\tau$
     By BaBar Collaboration (Bernard Aubert *et al.*).
     arXiv:0707.2981 [hep-ex].
     10.1103/PhysRevLett.100.011801.
     Phys.Rev.Lett. 100 (2008) 011801.

342. Measurement of the *CP*-Violating Asymmetries in $B^0 \to K_s^0\pi^0$ and of the branching fraction of $B^0 \to K^0\pi^0$
     By BaBar Collaboration (Bernard Aubert *et al.*).
     arXiv:0707.2980 [hep-ex].
     10.1103/PhysRevD.77.012003.
     Phys.Rev. D77 (2008) 012003.

343. Measurement of the $\tau^- \to K^-\pi^0\nu_\tau$ Branching Fraction
     By BaBar Collaboration (Bernard Aubert *et al.*).
     arXiv:0707.2922 [hep-ex].
     10.1103/PhysRevD.76.051104.
     Phys.Rev. D76 (2007) 051104.

344. A Study of *B*-Meson Decays to $\eta_c K^*$ and $\eta_c \gamma K^{(*)}$
     By BaBar Collaboration (Bernard Aubert *et al.*).
     arXiv:0707.2843 [hep-ex].

345. Study of $B^0 \to \pi^0 \pi^0$, $B^\pm \to \pi^\pm \pi^0$, and $B^\pm \to K^\pm \pi^0$ Decays, and Isospin Analysis of $B \to \pi\pi$
     Decays
     By BaBar Collaboration (Bernard Aubert *et al.*).
     arXiv:0707.2798 [hep-ex].
     10.1103/PhysRevD.76.091102.
     Phys.Rev. D76 (2007) 091102.

346. Measurement of the Semileptonic Decays $B \to D\tau^- \bar{\nu}_\tau$ and $B \to D^* \tau^- \bar{\nu}_\tau$
     By BaBar Collaboration (Bernard Aubert *et al.*).
     arXiv:0707.2758 [hep-ex].

347. Measurement of Moments of the Hadronic-Mass and Energy Spectrum in Inclusive
     Semileptonic $\bar{B} \to X_c \ell^- \bar{\nu}$ Decays
     By BaBar Collaboration (Bernard Aubert *et al.*).
     arXiv:0707.2670 [hep-ex].

348. Measurement of the Decay $B^- \to D^{*0} e^- \bar{\nu}_e$
     By BaBar Collaboration (Bernard Aubert *et al.*).
     arXiv:0707.2655 [hep-ex].

349. Mysteries of Mass: Some Contrarian Views From an Experimenter
     By Martin L. Perl.
     10.1063/1.2775915.
     AIP Conf.Proc. 928 (2007) 207-214.

350. Evidence for the $B^0 \to p\bar{p}K^{*0}$ and $B^+ \to \eta_c K^{*+}$ Decays and Study of the Decay Dynamics of
     B Meson Decays into $p\bar{p}h$ Final States
     By BaBar Collaboration (Bernard Aubert *et al.*).
     arXiv:0707.1648 [hep-ex].
     10.1103/PhysRevD.76.092004.
     Phys.Rev. D76 (2007) 092004.

351. Search for Prompt Production of χ_c and $X(3872)$ in $e^+ e^-$ Annihilations
     By BaBar Collaboration (Bernard Aubert *et al.*).
     arXiv:0707.1633 [hep-ex].
     10.1103/PhysRevD.76.071102.
     Phys.Rev. D76 (2007) 071102.

352. Observation of Tree-Level *B* Decays With $s\bar{s}$ Production from Gluon Radiation
     By BaBar Collaboration (Bernard Aubert   *et al.*).
     arXiv:0707.1043 [hep-ex].
     10.1103/PhysRevLett.100.171803.
     Phys.Rev.Lett. 100 (2008) 171803.

353. Branching Fraction and *CP*-Violation Charge Asymmetry Measurements for *B*-Meson Decays to $\eta K^{\pm}$, $\eta \pi^{\pm}$, $\eta' K$, $\eta' \pi^{\pm}$, ωK, and $\omega \pi^{\pm}$
By BaBar Collaboration (Bernard Aubert *et al.*).
arXiv:0706.3893 [hep-ex].
10.1103/PhysRevD.76.031103.
Phys.Rev. D76 (2007) 031103.

354. Measurements of *CP*-Violating Asymmetries in the Decay $B^0 \to K^+K^-K^0$
By BaBar Collaboration (Bernard Aubert *et al.*).
arXiv:0706.3885 [hep-ex].
10.1103/PhysRevLett.99.161802.
Phys.Rev.Lett. 99 (2007) 161802.

355. Search for the Decays $B^0 \to e^+e^-\gamma$ and $B^0 \to \mu^+\mu^-\gamma$
By BaBar Collaboration (Bernard Aubert *et al.*).
arXiv:0706.2870 [hep-ex].
10.1103/PhysRevD.77.011104.
Phys.Rev. D77 (2008) 011104.

356. Search for the decay $B^+ \to \bar{K}^{*0}(892)K^+$
By BaBar Collaboration (Bernard Aubert *et al.*).
arXiv:0706.1059 [hep-ex].
10.1103/PhysRevD.76.071103.
Phys.Rev. D76 (2007) 071103.

357. Determination of the Form Factors for the Decay $B^0 \to D^{*-}\ell^+\nu_\ell$ and of the CKM Matrix Element $|V_{cb}|$
By BaBar Collaboration (Bernard Aubert *et al.*).
arXiv:0705.4008 [hep-ex].
10.1103/PhysRevD.77.032002.
Phys.Rev. D77 (2008) 032002.

358. A Study of $B^0 \to \rho^+\rho^-$ Decays and Constraints on the CKM Angle α
By BaBar Collaboration (Bernard Aubert *et al.*).
arXiv:0705.2157 [hep-ex].
10.1103/PhysRevD.76.052007.
Phys.Rev. D76 (2007) 052007.

359. A Search for $B^+ \to \tau^+\nu$
By BaBar Collaboration (Bernard Aubert *et al.*).
arXiv:0705.1820 [hep-ex].
10.1103/PhysRevD.76.052002.
Phys.Rev. D76 (2007) 052002.

360. Amplitude Analysis of the $B^\pm \to \varphi K^*(892)^\pm$ Decay
By BaBar Collaboration (B. Aubert *et al.*).
arXiv:0705.1798 [hep-ex].
10.1103/PhysRevLett.99.201802.
Phys.Rev.Lett. 99 (2007) 201802.

361. Measurement of *CP*-Violating Asymmetries in $B^0 \to D^{(*)\pm}D^{\mp}$
      By BaBar Collaboration (Bernard Aubert *et al.*).
      arXiv:0705.1190 [hep-ex].
      10.1103/PhysRevLett.99.071801.
      Phys.Rev.Lett. 99 (2007) 071801.

362. Search for $D^0 - \bar{D}^0$ Mixing Using Doubly Flavor Tagged Semileptonic Decay Modes
      By BaBar Collaboration (Bernard Aubert *et al.*).
      arXiv:0705.0704 [hep-ex].
      10.1103/PhysRevD.76.014018.
      Phys.Rev. D76 (2007) 014018.

363. Search for $B^0 \to \phi(K^+\pi^-)$ Decays with Large $K^+\pi^-$ Invariant Mass
      By BaBar Collaboration (Bernard Aubert *et al.*).
      arXiv:0705.0398 [hep-ex].
      10.1103/PhysRevD.76.051103.
      Phys.Rev. D76 (2007) 051103.

364. Amplitude Analysis of the Decay $D^0 \to K^-K^+\pi^0$
      By BaBar Collaboration (Bernard Aubert *et al.*).
      arXiv:0704.3593 [hep-ex].
      10.1103/PhysRevD.76.011102.
      Phys.Rev. D76 (2007) 011102.

365. Measurement of the Absolute Branching Fraction of $D^0 \to K^-\pi^+$
      By BaBar Collaboration (Bernard Aubert *et al.*).
      arXiv:0704.2080 [hep-ex].
      10.1103/PhysRevLett.100.051802.
      Phys.Rev.Lett. 100 (2008) 051802.

366. Search for the Radiative Leptonic Decay $B^+ \to \gamma\ell^+\nu_\ell$
      By BaBar Collaboration (Bernard Aubert *et al.*).
      arXiv:0704.1478 [hep-ex].

367. Branching Fraction and Charge Asymmetry Measurements in $B \to J/\psi\pi\pi$ Decays
      By BaBar Collaboration (Bernard Aubert *et al.*).
      arXiv:0704.1266 [hep-ex].
      10.1103/PhysRevD.76.031101.
      Phys.Rev. D76 (2007) 031101.

368. Measurement of Decay Amplitudes of $B \to (c\bar{c})K^*$ with an Angular Analysis, for $(c\bar{c}) = J/\psi$, $\psi(2S)$ and χ_{c1}
      By BaBar Collaboration (Bernard Aubert *et al.*).
      arXiv:0704.0522 [hep-ex].
      10.1103/PhysRevD.76.031102.
      Phys.Rev. D76 (2007) 031102.

369. The $e^+e^- \to K^+K^-\pi^+\pi^-$, $K^+K^-\pi^0\pi^0$ and $K^+K^-K^+K^-$ Cross Sections Measured with Initial-State Radiation
By BaBar Collaboration (Bernard Aubert *et al.*).
arXiv:0704.0630 [hep-ex].
10.1103/PhysRevD.76.012008.
Phys.Rev. D76 (2007) 012008.

370. Search for Neutral *B*-Meson Decays to $a_0\pi$, $a_0 K$, $\eta\rho^0$, and ηf_0
By BaBar Collaboration (Bernard Aubert *et al.*).
hep-ex/0703038 [HEP-EX].
10.1103/PhysRevD.75.111102.
Phys.Rev. D75 (2007) 111102.

371. Measurement of *CP* Violation Parameters with a Dalitz Plot Analysis of $B^\pm \to D_{\pi^+\pi^-\pi^0} K^\pm$
By BaBar Collaboration (Bernard Aubert *et al.*).
hep-ex/0703037 [HEP-EX].
10.1103/PhysRevLett.99.251801.
Phys.Rev.Lett. 99 (2007) 251801.

372. Production and Decay of Ω_c^0
By BaBar Collaboration (Bernard Aubert *et al.*).
hep-ex/0703030 [HEP-EX].
10.1103/PhysRevLett.99.062001.
Phys.Rev.Lett. 99 (2007) 062001.

373. Measurement of the Relative Branching Fractions of $\bar{B} \to D/D^*/D^{**}\ell^-\bar{\nu}_\ell$ Decays in Events with a Fully Reconstructed *B* Meson
By BaBar Collaboration (Bernard Aubert *et al.*).
hep-ex/0703027 [HEP-EX].
10.1103/PhysRevD.76.051101.
Phys.Rev. D76 (2007) 051101.

374. Improved Measurement of *CP* Violation in Neutral *B* Decays to $c\bar{c}s$
By BaBar Collaboration (Bernard Aubert *et al.*).
hep-ex/0703021.
10.1103/PhysRevLett.99.171803.
Phys.Rev.Lett. 99 (2007) 171803.

375. Evidence for D^0-$\bar{D}^0$ Mixing
By BaBar Collaboration (Bernard Aubert *et al.*).
hep-ex/0703020 [HEP-EX].
10.1103/PhysRevLett.98.211802.
Phys.Rev.Lett. 98 (2007) 211802.

376. Measurement of the Time-Dependent *CP* Asymmetry in $B^0 \to D_{CP}^{(*)}h^0$ Decays
By BaBar Collaboration (Bernard Aubert *et al.*).
hep-ex/0703019 [HEP-EX].
10.1103/PhysRevLett.99.081801.
Phys.Rev.Lett. 99 (2007) 081801.

377. Search for the Rare Decay $B \rightarrow \pi \ell^+ \ell^-$
      By BaBar Collaboration (Bernard Aubert *et al.*).
      hep-ex/0703018.
      10.1103/PhysRevLett.99.051801.
      Phys.Rev.Lett. 99 (2007) 051801.

378. Observation of *CP* Violation in $B^0 \rightarrow K^+ \pi^-$ and $B^0 \rightarrow \pi^+ \pi^-$
      By BaBar Collaboration (Bernard Aubert *et al.*).
      hep-ex/0703016 [HEP-EX].
      10.1103/PhysRevLett.99.021603.
      Phys.Rev.Lett. 99 (2007) 021603.

379. Measurement of *CP*-Violating Asymmetries in $B^0 \rightarrow (\rho \pi)^0$ Using a Time-Dependent Dalitz
      Plot Analysis
      By BaBar Collaboration (Bernard Aubert *et al.*).
      hep-ex/0703008 [HEP-EX].
      10.1103/PhysRevD.76.012004.
      Phys.Rev. D76 (2007) 012004.

380. Measurement of *CP* Asymmetries in $B^0 \rightarrow K_S^0 K_S^0 K_S^0$ Decays
      By BaBar Collaboration (Bernard Aubert *et al.*).
      hep-ex/0702046.
      10.1103/PhysRevD.76.091101.
      Phys.Rev. D76 (2007) 091101.

381. Observation of $B^+ \rightarrow \rho^+ K^0$ and Measurement of its Branching Fraction and Charge Asymmetry
      By BaBar Collaboration (Bernard Aubert *et al.*).
      hep-ex/0702043 [HEP-EX].
      10.1103/PhysRevD.76.011103.
      Phys.Rev. D76 (2007) 011103.

382. Measurement of *CP* Asymmetry in $B^0 \rightarrow K_s^0 \pi^0 \pi^0$ Decays
      By BaBar Collaboration (Bernard Aubert *et al.*).
      hep-ex/0702010.
      10.1103/PhysRevD.76.071101.
      Phys.Rev. D76 (2007) 071101.

383. What Einstein Did Not Know
      By Martin Perl.
      10.1142/S0218301308010143.
      Int.J.Mod.Phys. E17 (2008) 735-757.

384. Measurement of the $B^\pm \rightarrow \rho^\pm \pi^0$ Branching Fraction and Direct *CP* Asymmetry
      By BaBar Collaboration (Bernard Aubert *et al.*).
      hep-ex/0701035.
      10.1103/PhysRevD.75.091103.
      Phys.Rev. D75 (2007) 091103.

385. Measurements of *CP*-Violating Asymmetries in $B^0 \rightarrow a_1^\pm(1260)\,\pi^\mp$ Decays
By BaBar Collaboration (Bernard Aubert *et al.*).
hep-ex/0612050.
10.1103/PhysRevLett.98.181803.
Phys.Rev.Lett. 98 (2007) 181803.

386. Evidence for $B^0 \rightarrow \rho^0\rho^0$ Decay and Implications for the CKM Angle α
By BaBar Collaboration (Bernard Aubert *et al.*).
hep-ex/0612021.
10.1103/PhysRevLett.98.111801.
Phys.Rev.Lett. 98 (2007) 111801.

387. Measurement of the $B^0 \rightarrow \pi^-\ell^+\nu$ Form-Factor Shape and Branching Fraction, and Determination of $\left|V_{ub}\right|$ with a Loose Neutrino Reconstruction Technique
By BaBar Collaboration (Bernard Aubert *et al.*).
hep-ex/0612020.
10.1103/PhysRevLett.98.091801.
Phys.Rev.Lett. 98 (2007) 091801.

388. Branching Fraction Measurements of $B^+ \rightarrow \rho^+\gamma$, $B^0 \rightarrow \rho^0\gamma$, and $B^0 \rightarrow \omega\gamma$
By BaBar Collaboration (Bernard Aubert *et al.*).
hep-ex/0612017.
10.1103/PhysRevLett.98.151802.
Phys.Rev.Lett. 98 (2007) 151802.

389. Measurement of B Decays to $\phi K\gamma$
By BaBar Collaboration (Bernard Aubert *et al.*).
hep-ex/0611037.
10.1103/PhysRevD.75.051102.
Phys.Rev. D75 (2007) 051102.

390. Evidence for the Rare Decay $B^+ \rightarrow D_s^+\pi^0$
By BaBar Collaboration (Bernard Aubert *et al.*).
hep-ex/0611030.
10.1103/PhysRevLett.98.171801.
Phys.Rev.Lett. 98 (2007) 171801.

391. Observation of the Exclusive Reaction $e^+e^- \rightarrow \phi\eta$ at $\sqrt{s}=10.58$ GeV
By BaBar Collaboration (Bernard Aubert *et al.*).
hep-ex/0611028.
10.1103/PhysRevD.74.111103.
Phys.Rev. D74 (2006) 111103.

392. Vector-Tensor and Vector-Vector Decay Amplitude Analysis of $B^0 \rightarrow \phi K^{*0}$
By BaBar Collaboration (Bernard Aubert *et al.*).
hep-ex/0610073.
10.1103/PhysRevLett.98.051801.
Phys.Rev.Lett. 98 (2007) 051801.

393. Search for Lepton Flavor Violating Decays $\tau^{\pm} \to \ell^{\pm}\pi^0, \ell^{\pm}\eta, \ell^{\pm}\eta'$
By BaBar Collaboration (Bernard Aubert *et al.*).
hep-ex/0610067.
10.1103/PhysRevLett.98.061803.
Phys.Rev.Lett. 98 (2007) 061803.

394. Evidence of a Broad Structure at an Invariant Mass of 4.32 GeV/c^2 in the Reaction $e^+e^- \to \pi^+\pi^-\psi(2S)$ Measured at BaBar
By BaBar Collaboration (Bernard Aubert *et al.*).
hep-ex/0610057.
10.1103/PhysRevLett.98.212001.
Phys.Rev.Lett. 98 (2007) 212001.

395. Branching Fraction Measurement of $\overline{B}^0 \to D^{(*)+}\pi^-$ and $B^- \to D^{(*)0}\pi^-$ and Isospin Analysis of $\overline{B} \to D^{(*)}\pi$ Decays
By BaBar Collaboration (Bernard Aubert *et al.*).
hep-ex/0610027.
10.1103/PhysRevD.75.031101.
Phys.Rev. D75 (2007) 031101.

396. A Structure at 2175 MeV in $e^+e^- \to \phi f_0(980)$ Observed via Initial-State Radiation
By BaBar Collaboration (Bernard Aubert *et al.*).
hep-ex/0610018.
10.1103/PhysRevD.74.091103.
Phys.Rev. D74 (2006) 091103.

397. Observation of CP Violation in $B^0 \to \eta'K^0$ Decays
By BaBar Collaboration (Bernard Aubert *et al.*).
hep-ex/0609052.
10.1103/PhysRevLett.98.031801.
Phys.Rev.Lett. 98 (2007) 031801.

398. Measurement of the Absolute Branching Fractions $B \to D\pi, D^*\pi, D^{**}\pi$ with a Missing Mass Method
By BaBar Collaboration (Bernard Aubert *et al.*).
hep-ex/0609033.
10.1103/PhysRevD.74.111102.
Phys.Rev. D74 (2006) 111102.

399. Observation of $B^+ \to \phi\phi K^+$ and Evidence for $B^0 \to \phi\phi K^0$ below η_c Threshold
By BaBar Collaboration (Bernard Aubert *et al.*).
hep-ex/0609027.
10.1103/PhysRevLett.97.261803.
Phys.Rev.Lett. 97 (2006) 261803.

400. Inclusive Λ_c^+ Production in e^+e^- Annihilations at $\sqrt{s} = 10.54$ GeV and in $\Upsilon(4S)$ Decays
By BaBar Collaboration (Bernard Aubert *et al.*).
hep-ex/0609004.
10.1103/PhysRevD.75.012003.
Phys.Rev. D75 (2007) 012003.

401. Observation of an Excited Charm Baryon Ω_c^* Decaying to $\Omega_c^0 \gamma$
By BaBar Collaboration (Bernard Aubert *et al.*).
hep-ex/0608055.
10.1103/PhysRevLett.97.232001.
Phys.Rev.Lett. 97 (2006) 232001.

402. Measurement of the *CP* Asymmetry and Branching Fraction of $B^0 \to \rho^0 K^0$
By BaBar Collaboration (Bernard Aubert *et al.*).
hep-ex/0608051.
10.1103/PhysRevLett.98.051803.
Phys.Rev.Lett. 98 (2007) 051803.

403. Observation of $B^+ \to \bar{K}^0 K^+$ and $B^0 \to K^0 \bar{K}^0$
By BaBar Collaboration (Bernard Aubert *et al.*).
hep-ex/0608036.
10.1103/PhysRevLett.97.171805.
Phys.Rev.Lett. 97 (2006) 171805.

404. Measurement of the $B^0 \to \bar{\Lambda} p \pi^-$ Branching Fraction and Study of the Decay Dynamics
By BaBar Collaboration (Bernard Aubert *et al.*).
hep-ex/0608020.

405. A Search for $B^+ \to \tau^+ \nu$ Recoiling Against $B^- \to D^0 \ell^- \bar{\nu}_\ell X$
By BaBar Collaboration (Bernard Aubert *et al.*).
hep-ex/0608019.

406. Measurement of the Branching Fraction and Time-Dependent *CP* Asymmetry in the Decay $B^0 \to D^{*+} D^{*-} K_s^0$
By BaBar Collaboration (Bernard Aubert *et al.*).
hep-ex/0608016.
10.1103/PhysRevD.74.091101.
Phys.Rev. D74 (2006) 091101.

407. Precise Branching Ratio Measurements of the Decays $D^0 \to \pi^- \pi^+ \pi^0$ and $D^0 \to K^- K^+ \pi^0$
By BaBar Collaboration (Bernard Aubert *et al.*).
hep-ex/0608009.
10.1103/PhysRevD.74.091102.
Phys.Rev. D74 (2006) 091102.

408. Search for D^0-$\bar{D}^0$ Mixing and Branching-Ratio Measurement in the Decay $D^0 \to K^+ \pi^- \pi^0$
By BaBar Collaboration (Bernard Aubert *et al.*).
hep-ex/0608006.
10.1103/PhysRevLett.97.221803.
Phys.Rev.Lett. 97 (2006) 221803.

409. Measurement of Branching Fractions and Charge Asymmetries in B Decays to an η Meson and a K^* Meson
     By BaBar Collaboration (Bernard Aubert *et al.*).
     hep-ex/0608005.
     10.1103/PhysRevLett.97.201802.
     Phys.Rev.Lett. 97 (2006) 201802.

410. Search for Charmonium States Decaying to $J/\psi\gamma\gamma$ Using Initial-State Radiation Events
     By BaBar Collaboration (Bernard Aubert *et al.*).
     hep-ex/0608004.

411. Improved Measurements of the Branching Fractions for $B^0 \to \pi^+\pi^-$ and $B^0 \to K^+\pi^-$, and a Search for $B^0 \to K^+K^-$
     By BaBar Collaboration (Bernard Aubert *et al.*).
     hep-ex/0608003.
     10.1103/PhysRevD.75.012008.
     Phys.Rev. D75 (2007) 012008.

412. Measurement of CP-Violating Asymmetries in $B^0 \to (\rho\pi)^0$ Using a Time-Dependent Dalitz Plot Analysis
     By BaBar Collaboration (Bernard Aubert *et al.*).
     hep-ex/0608002.

413. Branching Fraction Measurements of Charged B Decays to $K^{*+}K^+K^-$, $K^{*+}\pi^+K^-$, $K^{*+}K^+\pi^-$ and $K^{*+}\pi^+\pi^-$ Final States
     By BaBar Collaboration (Bernard Aubert *et al.*).
     hep-ex/0607113.
     10.1103/PhysRevD.74.051104.
     Phys.Rev. D74 (2006) 051104.

414. Measurement of CP-Violating Asymmetries in the $B^0 \to K^+K^-K^0$ Dalitz plot
     By BaBar Collaboration (Bernard Aubert *et al.*).
     hep-ex/0607112.

415. Measurement of the Ratio $\mathcal{B}(B^+ \to Xe\nu)/\mathcal{B}(B^0 \to Xe\nu)$
     By BaBar Collaboration (Bernard Aubert *et al.*).
     hep-ex/0607111.
     10.1103/PhysRevD.74.091105.
     Phys.Rev. D74 (2006) 091105.

416. A Search for the Decays $B^+ \to e^+\nu_e$ and $B^+ \to \mu^+\nu_\mu$ Using Hadronic-Tag Reconstruction
     By BaBar Collaboration (Bernard Aubert *et al.*).
     hep-ex/0607110.

417. Observation of $B \to \eta'K^*$ and Evidence for $B^+ \to \eta'\rho^+$
     By BaBar Collaboration (Bernard Aubert *et al.*).
     hep-ex/0607109.
     10.1103/PhysRevLett.98.051802.
     Phys.Rev.Lett. 98 (2007) 051802.

418. Measurement of Time-Dependent CP Asymmetries in $B^0 \to K_S^0 K_S^0 K_S^0$ Decay
By BaBar Collaboration (Bernard Aubert *et al.*).
hep-ex/0607108.

419. Improved Measurement of CP Asymmetries in $B^0 \to (c\bar{c})K^{(*)0}$ Decays
By BaBar Collaboration (Bernard Aubert *et al.*).
hep-ex/0607107.

420. Measurement of CP Asymmetries and Branching Fractions in $B \to \pi\pi$ and $B \to K\pi$ Decays
By BaBar Collaboration (Bernard Aubert *et al.*).
hep-ex/0607106.

421. Measurement of $\cos 2\beta$ in B $B^0 \to D^{(*)0}h^0$ Decays with a Time-Dependent Dalitz Plot Analysis of $D^0 \to K_S^0 \pi^+ \pi^-$
By BaBar Collaboration (Bernard Aubert *et al.*).
hep-ex/0607105.

422. Measurement of the CKM Angle γ in $B^\mp \to D^{(*)}K^\mp$ Decays with a Dalitz Analysis of $D^0 \to K_S^0 \pi^- \pi^+$
By BaBar Collaboration (Bernard Aubert *et al.*).
hep-ex/0607104.

423. Search for CPT and Lorentz Violation in B^0-$\bar{B}^0$ Oscillations with Inclusive Dilepton Events
By BaBar Collaboration (Bernard Aubert *et al.*).
hep-ex/0607103.

424. Measurements of CP-Violating Asymmetries in B Decays to ωK_S^0
By BaBar Collaboration (Bernard Aubert *et al.*).
hep-ex/0607101.

425. Time-Dependent CP-Violation Parameters in $B^0 \to \eta' K^0$ Decay
By BaBar Collaboration (Bernard Aubert *et al.*).
hep-ex/0607100.

426. Measurement of the Branching Fractions for the Decays $B^+ \to \rho^+ \gamma$, $B^0 \to \rho^0 \gamma$, and $B^0 \to \omega \gamma$
By BaBar Collaboration (Bernard Aubert *et al.*).
hep-ex/0607099.

427. Updated Measurement of the CKM Angle α Using $B^0 \to \rho^+ \rho^-$ Decays
By BaBar Collaboration (Bernard Aubert *et al.*).
hep-ex/0607098.

428. Evidence for the $B^0 \to \rho^0 \rho^0$ Decay and Implications for the CKM Angle α
By BaBar Collaboration (Bernard Aubert *et al.*).
hep-ex/0607097.

429. Measurement of the *CP*-Violating Asymmetries in $B^0 \to K_s^0 \pi^0$ and of the Branching Fraction of $B^0 \to K^0 \pi^0$
      By BaBar Collaboration (Bernard Aubert *et al.*).
      hep-ex/0607096.

430. Measurement of the Pseudoscalar Decay Constant f_{D_s} Using Charm-Tagged Events in e^+e^- Collisions at $\sqrt{s} = 10.58$ GeV
      By BaBar Collaboration (Bernard Aubert *et al.*).
      hep-ex/0607094.
      10.1103/PhysRevLett.98.141801.
      Phys.Rev.Lett. 98 (2007) 141801.

431. Measurements of Branching Fraction, Polarization, and Charge Asymmetry of $B^\pm \to \rho^\pm \rho^0$ and a Search for $B^\pm \to \rho^\pm f_0(980)$
      By BaBar Collaboration (Bernard Aubert *et al.*).
      hep-ex/0607092.
      10.1103/PhysRevLett.97.261801.
      Phys.Rev.Lett. 97 (2006) 261801.

432. A Measurement of *CP*-Violation Parameters in $B^0\overline{B}^0$ Mixing Using Partially Reconstructed $D^{*-}\ell^+\nu_\ell$ Events at BaBar
      By BaBar Collaboration (Bernard Aubert *et al.*).
      hep-ex/0607091.

433. Search for D^0-$\overline{D}^0$ Mixing in the Decays $D^0 \to K^+\pi^-\pi^+\pi^-$
      By BaBar Collaboration (Bernard Aubert *et al.*).
      hep-ex/0607090.

434. Measurement of the $B \to \pi\ell\nu$ Branching Fraction and Determination of $|V_{ub}|$ with Tagged B Mesons
      By BaBar Collaboration (Bernard Aubert *et al.*).
      hep-ex/0607089.
      10.1103/PhysRevLett.97.211801.
      Phys.Rev.Lett. 97 (2006) 211801.

435. Ξ_c' Production at BABAR
      By BaBar Collaboration (Bernard Aubert *et al.*).
      hep-ex/0607086.

436. Measurement of the Hadronic Form-Factors in $D_s^+ \to \phi e^+\nu_e$ Decays
      By BaBar Collaboration (Bernard Aubert *et al.*).
      hep-ex/0607085.

437. A Precision Measurement of the $D_{s1}(2536)^+$ Meson Mass and Decay Width
      By BaBar Collaboration (Bernard Aubert *et al.*).
      hep-ex/0607084.

438. Study of the Exclusive Initial-State Radiation Production of the $D\bar{D}$ System
By BaBar Collaboration (Bernard Aubert *et al.*).
hep-ex/0607083.
10.1103/PhysRevD.76.111105.
Phys.Rev. D76 (2007) 111105.

439. Observation of a New D_s Meson Decaying to DK at a Mass of 2.86 GeV/c^2
By BaBar Collaboration (Bernard Aubert *et al.*).
hep-ex/0607082.
10.1103/PhysRevLett.97.222001.
Phys.Rev.Lett. 97 (2006) 222001.

440. Measurement of Decay Amplitudes of $B \to (c\bar{c}) K^*$ with an Angular Analysis, for $(c\bar{c}) = J/\psi$, $\psi(2S)$ and χ_{c1}
By BaBar Collaboration (Bernard Aubert *et al.*).
hep-ex/0607081.

441. Measurement of the q^2 Dependence of the Hadronic Form Factor in $D^0 \to K^-e^+\nu_e$ Decays
By BaBar Collaboration (Bernard Aubert *et al.*).
hep-ex/0607077.

442. Determination of the Form Factors for the Decay $B^0 \to D^{*-}\ell^+\nu_\ell$ and of the CKM Matrix Element $|V_{cb}|$
By BaBar Collaboration (Bernard Aubert *et al.*).
hep-ex/0607076.

443. Measurement of the Branching Fraction and Photon Energy Moments of $B \to X_s\gamma$ and $A_{CP}(B \to X_{s+d}\gamma)$
By BaBar Collaboration (Bernard Aubert *et al.*).
hep-ex/0607071.
10.1103/PhysRevLett.97.171803.
Phys.Rev.Lett. 97 (2006) 171803.

444. Measurement of the Relative Branching Fractions for $\bar{B} \to D/D^*/D^{**}(D^{(*)}\pi)\ell^-\bar{\nu}_\ell$ with a Large Sample of Tagged B Mesons
By BaBar Collaboration (Bernard Aubert *et al.*).
hep-ex/0607067.

445. Measurement of the $B^+ \to \eta\ell^+\nu$ and $B^+ \to \eta'\ell^+\nu$ Branching Fractions Using $\Upsilon(4S) \to B\bar{B}$ Events Tagged by a Fully Reconstructed B Meson
By BaBar Collaboration (Bernard Aubert *et al.*).
hep-ex/0607066.

446. Search for $b \to u$ Transitions in $B^- \to [K^+\pi^-\pi^0]_D K^-$
By BaBar Collaboration (Bernard Aubert *et al.*).
hep-ex/0607065.

447. Search for the Rare Decay $B^\pm \to a_0^\pm \pi^0$
By BaBar Collaboration (Bernard Aubert *et al.*).
hep-ex/0607064.

448. Searches for B^0 Decays to ηK^0, $\eta\eta$, $\eta'\eta'$, $\eta\phi$, and $\eta'\phi$
By BaBar Collaboration (Bernard Aubert *et al.*).
hep-ex/0607063.
10.1103/PhysRevD.74.051106.
Phys.Rev. D74 (2006) 051106.

449. Observation of the Decays $B^- \to D_s^{(*)+} K^- \pi^-$
By BaBar Collaboration (Bernard Aubert *et al.*).
hep-ex/0607062.

450. Measurement of the $B^0 \to \pi^- \ell^+ \nu$ Form-Factor Shape and Branching Fraction, and Determination of $|V_{ub}|$ with a Loose Neutrino Reconstruction Technique
By BaBar Collaboration (Bernard Aubert *et al.*).
hep-ex/0607060.

451. A Search for the $B^0 \to e^+ e^- \gamma$ and $B^0 \to \mu^+ \mu^- \gamma$ Decays
By BaBar Collaboration (Bernard Aubert *et al.*).
hep-ex/0607058.

452. Measurements of Branching Fractions, Polarizations, and Direct *CP*-Violation Asymmetries in $B \to \rho K^*$ and $B \to f_0(980) K^*$ Decays
By BaBar Collaboration (Bernard Aubert *et al.*).
hep-ex/0607057.
10.1103/PhysRevLett.97.201801.
Phys.Rev.Lett. 97 (2006) 201801.

453. Measurement of the Branching Fractions of the Decays $\overline{B}^0 \to \Lambda_c^+ \overline{p}$ and $B^- \to \Lambda_c^+ \overline{p} \pi^-$
By BaBar Collaboration (Bernard Aubert *et al.*).
hep-ex/0607055.

454. Search for Inclusive Charmless $B \to K^+ X$ and $B \to K^0 X$ Decays
By BaBar Collaboration (Bernard Aubert *et al.*).
hep-ex/0607053.

455. Search for Flavor-Changing Neutral-Current Charm Decays
By BaBar Collaboration (Bernard Aubert *et al.*).
hep-ex/0607051.

456. Search for $B^+ \to X(3872) K^+$, $X(3872) \to J/\psi\gamma$
By BaBar Collaboration (Bernard Aubert *et al.*).
hep-ex/0607050.
10.1103/PhysRevD.74.071101.
Phys.Rev. D74 (2006) 071101.

457. Search for the Rare Decay $B \to \pi \ell^+ \ell^-$
By BaBar Collaboration (Bernard Aubert *et al.*).
hep-ex/0607048.

458. Study of the Lepton Flavor Violating Decay $\tau^- \to \mu^- \eta$
By BaBar Collaboration (Bernard Aubert *et al.*).
hep-ex/0607045.

459. Search for the Reactions $e^+ e^- \to \mu^+ \tau^-$ and $e^+ e^- \to e^+ \tau^-$
By BaBar Collaboration (Bernard Aubert *et al.*).
hep-ex/0607044.
10.1103/PhysRevD.75.031103.
Phys.Rev. D75 (2007) 031103.

460. Measurement of the Mass and Width and Study of the Spin of the $\Xi(1690)^0$ Resonance from $\Lambda_c^+ \to \Lambda \overline{K}^0 K^+$ Decay at BABAR
By BaBar Collaboration (Bernard Aubert *et al.*).
hep-ex/0607043.

461. A Study of $\Xi_c(2980)^+$ and $\Xi_c(3077)^+$
By BaBar Collaboration (Bernard Aubert *et al.*).
hep-ex/0607042.

462. Search for the Baryon and Lepton Number Violating Decays $\tau \to \Lambda h$
By BaBar Collaboration (Bernard Aubert *et al.*).
hep-ex/0607040.

463. Measurements of the Decays $B^0 \to \overline{D}^0 p \overline{p}$, $B^0 \to \overline{D}^{*0} p \overline{p}$, $B^0 \to D^- p \overline{p} \pi^+$, and $B^0 \to D^{*-} p \overline{p} \pi^+$
By BaBar Collaboration (Bernard Aubert *et al.*).
hep-ex/0607039.
10.1103/PhysRevD.74.051101.
Phys.Rev. D74 (2006) 051101.

464. Measurement of B Decays to $\phi K \gamma$
By BaBar Collaboration (Bernard Aubert *et al.*).
hep-ex/0607037.

465. Observation of $e^+ e^-$ Annihilations into the $C = +1$ Hadronic Final States $\rho^0 \rho^0$ and $\phi \rho^0$
By BaBar Collaboration (Bernard Aubert *et al.*).
hep-ex/0606054.
10.1103/PhysRevLett.97.112002.
Phys.Rev.Lett. 97 (2006) 112002.

466. Search for the Decay of a B^0 or $\overline{B}^0$ meson to $\overline{K}^{*0} K^0$ or $K^{*0} \overline{K}^0$
By BaBar Collaboration (Bernard Aubert *et al.*).
hep-ex/0606050.
10.1103/PhysRevD.74.072008.
Phys.Rev. D74 (2006) 072008.

467. Measurement of the Spin of the Ω^- Hyperon at BABAR
      By BaBar Collaboration (Bernard Aubert *et al.*).
      hep-ex/0606039.
      10.1103/PhysRevLett.97.112001.
      Phys.Rev.Lett. 97 (2006) 112001.

468. Search for the Decay $B^0 \to K_S^0 K_S^0 K_L^0$
      By BaBar Collaboration (Bernard Aubert *et al.*).
      hep-ex/0606031.
      10.1103/PhysRevD.74.032005.
      Phys.Rev. D74 (2006) 032005.

469. Study of Inclusive B^- and $\overline{B}^0$ Decays to Flavor-Tagged D, D_s and Λ_c^+
      By BaBar Collaboration (Bernard Aubert *et al.*).
      hep-ex/0606026.
      10.1103/PhysRevD.75.072002.
      Phys.Rev. D75 (2007) 072002.

470. Search for Doubly Charmed Baryons Ξ_{cc}^+ and Ξ_{cc}^{++} in BABAR
      By BaBar Collaboration (Bernard Aubert *et al.*).
      hep-ex/0605075.
      10.1103/PhysRevD.74.011103.
      Phys.Rev. D74 (2006) 011103.

471. Measurement of the $D^+ \to \pi^+\pi^0$ and $D^+ \to K^+\pi^0$ Branching Fractions
      By BaBar Collaboration (Bernard Aubert *et al.*).
      hep-ex/0605044.
      10.1103/PhysRevD.74.011107.
      Phys.Rev. D74 (2006) 011107.

472. Search for $B^+ \to \phi\pi^+$ and $B^0 \to \phi\pi^0$ Decays
      By BaBar Collaboration (Bernard Aubert *et al.*).
      hep-ex/0605037.
      10.1103/PhysRevD.74.011102.
      Phys.Rev. D74 (2006) 011102.

473. Study of $B \to D^{(*)} D_{s(J)}^{(*)}$ Decays and Measurement of D_s^- and $D_{sJ}(2460)^-$ Branching Fractions
      By BaBar Collaboration (Bernard Aubert *et al.*).
      hep-ex/0605036.
      10.1103/PhysRevD.74.031103.
      Phys.Rev. D74 (2006) 031103.

474. Search for the Decay $B^0 \to a_1^\pm \rho^\mp$
      By BaBar Collaboration (Bernard Aubert *et al.*).
      hep-ex/0605024.
      10.1103/PhysRevD.74.031104.
      Phys.Rev. D74 (2006) 031104.

475. Measurement of the η and η' Transition Form-Factors at $q^2 = 112$ GeV^2
By BaBar Collaboration (Bernard Aubert *et al.*).
hep-ex/0605018.
10.1103/PhysRevD.74.012002.
Phys.Rev. D74 (2006) 012002.

476. B Meson Decays to ωK^*, $\omega\rho$, $\omega\omega$, $\omega\phi$, and ωf_0
By BaBar Collaboration (Bernard Aubert *et al.*).
hep-ex/0605017.
10.1103/PhysRevD.74.051102.
Phys.Rev. D74 (2006) 051102.

477. Focal Plane Metrology for the LSST Camera
By Andrew P. A Rasmussen, Layton Hale, Peter Kim, Eric Lee, Martin Perl, Rafe Schindler, Peter Takacs, Timothy Thurston.
Submitted to: Proc.SPIE Int.Soc.Opt.Eng.6273.

478. Search for B Meson Decays to $\eta'\eta'K$
By BaBar Collaboration (Bernard Aubert *et al.*).
hep-ex/0605008.
10.1103/PhysRevD.74.031105.
Phys.Rev. D74 (2006) 031105.

479. Dalitz Plot Analysis of the Decay $B^\pm \to K^\pm K^\pm K^\mp$
By BaBar Collaboration (Bernard Aubert *et al.*).
hep-ex/0605003.
10.1103/PhysRevD.74.032003.
Phys.Rev. D74 (2006) 032003.

480. Geant4 Developments and Applications
By John Allison *et al.*.
10.1109/TNS.2006.869826.
IEEE Trans.Nucl.Sci. 53 (2006) 270.

481. Measurement of Branching Fractions and *CP*-Violating Charge Asymmetries for B Meson Decays to $D^{(*)}\bar{D}^{(*)}$, and Implications for the CKM Angle γ
By BaBar Collaboration (Bernard Aubert *et al.*).
hep-ex/0604037.
10.1103/PhysRevD.73.112004.
Phys.Rev. D73 (2006) 112004.

482. Observation of $\Upsilon(4S)$ Decays to $\pi^+\pi^-\Upsilon(1S)$ and $\pi^+\pi^-\Upsilon(2S)$
By BaBar Collaboration (Bernard Aubert *et al.*).
hep-ex/0604031.
10.1103/PhysRevLett.96.232001.
Phys.Rev.Lett. 96 (2006) 232001.

483. A Study of the $D_{sJ}^*(2317)^+$ and $D_{sJ}(2460)^+$ Mesons in Inclusive $c\bar{c}$ Production Near $\sqrt{s} = 10.6$ GeV
By BaBar Collaboration (Bernard Aubert *et al.*).
hep-ex/0604030.
10.1103/PhysRevD.74.032007.
Phys.Rev. D74 (2006) 032007.

484. Measurement of the $B^- \to D^0 K^{*-}$ Branching Fraction
By BaBar Collaboration (Bernard Aubert *et al.*).
hep-ex/0604017.
10.1103/PhysRevD.73.111104.
Phys.Rev. D73 (2006) 111104.

485. Measurement of $\bar{B}^0 \to D^{(*)0} \bar{K}^{(*)0}$ Branching Fractions
By BaBar Collaboration (Bernard Aubert *et al.*).
hep-ex/0604016.
10.1103/PhysRevD.74.031101.
Phys.Rev. D74 (2006) 031101.

486. Search for the Decay $\tau^- \to 3\pi^- 2\pi^+ 2\pi^0 \nu_\tau$
By BaBar Collaboration (Bernard Aubert *et al.*).
hep-ex/0604014.
10.1103/PhysRevD.73.112003.
Phys.Rev. D73 (2006) 112003.

487. Observation of Decays $B^0 \to D_s^{(*)+}\pi^-$ and $B^0 \to D_s^{(*)-}K^+$
By BaBar Collaboration (Bernard Aubert *et al.*).
hep-ex/0604012.
10.1103/PhysRevLett.98.081801.
Phys.Rev.Lett. 98 (2007) 081801.

488. Study of the Decay $\bar{B}^0 \to D^{*+}\omega\pi^-$
By BaBar Collaboration (Bernard Aubert *et al.*).
hep-ex/0604009.
10.1103/PhysRevD.74.012001.
Phys.Rev. D74 (2006) 012001.

489. Measurements of Branching Fractions, Rate Asymmetries, and Angular Distributions in the Rare Decays $B \to K\ell^+\ell^-$ and $B \to K^*\ell^+\ell^-$
By BaBar Collaboration (Bernard Aubert *et al.*).
hep-ex/0604007.
10.1103/PhysRevD.73.092001.
Phys.Rev. D73 (2006) 092001.

490. Search for the Charmed Pentaquark Candidate $\Theta_c(3100)^0$ in e^+e^- Annihilations at $\sqrt{s} = 10.58$ GeV
By BaBar Collaboration (Bernard Aubert *et al.*).
hep-ex/0604006.
10.1103/PhysRevD.73.091101.
Phys.Rev. D73 (2006) 091101.

491. Measurement of Branching Fractions in Radiative B Decays to $\eta K\gamma$ and Search for B Decays to $\eta'K\gamma$
By BaBar Collaboration (Bernard Aubert *et al.*).
hep-ex/0603054.
10.1103/PhysRevD.74.031102.
Phys.Rev. D74 (2006) 031102.

492. Observation of a Charmed Baryon Decaying to D^0p at a Mass Near 2.94 GeV/c^2
By BaBar Collaboration (Bernard Aubert *et al.*).
hep-ex/0603052.
10.1103/PhysRevLett.98.012001.
Phys.Rev.Lett. 98 (2007) 012001.

493. Search for *T*, *CP* and *CPT* Violation in $B^0 - \overline{B}^0$ Mixing with Inclusive Dilepton Events
By BaBar Collaboration (Bernard Aubert *et al.*).
hep-ex/0603053.
10.1103/PhysRevLett.96.251802.
Phys.Rev.Lett. 96 (2006) 251802.

494. Observation of B^0 Meson Decay to $a_1^\pm(1260)\,\pi^\mp$
By BaBar Collaboration (Bernard Aubert *et al.*).
hep-ex/0603050.
10.1103/PhysRevLett.97.051802.
Phys.Rev.Lett. 97 (2006) 051802.

495. Measurements of *CP*-Violating Asymmetries and Branching Fractions in B Decays to ωK and $\omega\pi$
By BaBar Collaboration (Bernard Aubert *et al.*).
hep-ex/0603040.
10.1103/PhysRevD.74.011106.
Phys.Rev. D74 (2006) 011106.

496. Branching Fraction Limits for B^0 Decays to $\eta'\eta$, $\eta'\pi^0$ and $\eta\pi^0$
By BaBar Collaboration (Bernard Aubert *et al.*).
hep-ex/0603013.
10.1103/PhysRevD.73.071102.
Phys.Rev. D73 (2006) 071102.

497. Measurements of the Branching Fraction and Time-Dependent *CP* Asymmetries of $B^0 \to J/\psi\pi^0$ Decays
By BaBar Collaboration (Bernard Aubert *et al.*).
hep-ex/0603012.
10.1103/PhysRevD.74.011101.
Phys.Rev. D74 (2006) 011101.

498. Measurement of Time-Dependent *CP* Asymmetries in $B^0 \to D^{(*)\pm}\pi^\mp$ and $B^0 \to D^\pm\rho^\mp$ Decays
By BaBar Collaboration (Bernard Aubert *et al.*).
hep-ex/0602049.
10.1103/PhysRevD.73.111101.
Phys.Rev. D73 (2006) 111101.

499. A Counterexample to King's Conjecture
     By Lutz Hille, Markus Perling.
     math/0602258 [math-ag].

500. Measurements of the $B \to D^*$ Form Factors Using the Decay $\overline{B}^0 \to D^{*+}e^-\overline{\nu}_e$
     By BaBar Collaboration (Bernard Aubert *et al.*).
     hep-ex/0602023.
     10.1103/PhysRevD.74.092004.
     Phys.Rev. D74 (2006) 092004.

501. The $e^+e^- \to 3(\pi^+\pi^-), 2(\pi^+\pi^-\pi^0)$ and $K^+K^-2(\pi^+\pi^-)$ Cross Sections at Center-of-Mass Energies
     from Production Threshold to 4.5 GeV Measured with Initial-State Radiation
     By BaBar Collaboration (Bernard Aubert *et al.*).
     hep-ex/0602006.
     10.1103/PhysRevD.73.052003.
     Phys.Rev. D73 (2006) 052003.

502. Determinations of $|V_{ub}|$ from Inclusive Semileptonic B Decays with Reduced Model
     Dependence
     By BaBar Collaboration (Bernard Aubert *et al.*).
     hep-ex/0601046.
     10.1103/PhysRevLett.96.221801.
     Phys.Rev.Lett. 96 (2006) 221801.

503. Measurements of Λ_c^+ Branching Fractions of Cabibbo-Suppressed Decay Modes Involving
     Λ and Σ^0.
     By BaBar Collaboration (Bernard Aubert *et al.*).
     hep-ex/0601017.
     10.1103/PhysRevD.75.052002.
     Phys.Rev. D75 (2007) 052002.

504. Measurements of the Branching Fractions and CP-Asymmetries of $B \to D^0_{CP}K$ Decays
     By BaBar Collaboration (Bernard Aubert *et al.*).
     hep-ex/0512067.
     10.1103/PhysRevD.73.051105.
     Phys.Rev. D73 (2006) 051105.

505. Search for the Rare Decays $B^0 \to D_s^{(*)+}a_{0(2)}^-$
     By BaBar Collaboration (Bernard Aubert *et al.*).
     hep-ex/0512031.
     10.1103/PhysRevD.73.071103.
     Phys.Rev. D73 (2006) 071103.

506. Search for Rare Quark-Annihilation Decays, $B^- \to D_s^{(*)-}\phi$
     By BaBar Collaboration (Bernard Aubert *et al.*).
     hep-ex/0512028.
     10.1103/PhysRevD.73.011103.
     Phys.Rev. D73 (2006) 011103.

507. A Study of $e^+e^- \to p\bar{p}$ Using Initial State Radiation with BABAR
     By BaBar Collaboration (Bernard Aubert *et al.*).
     hep-ex/0512023.
     10.1103/PhysRevD.73.012005.
     Phys.Rev. D73 (2006) 012005.

508. A Search for the Rare Decay $B^0 \to \tau^+\tau^-$ at BABAR
     By BaBar Collaboration (Bernard Aubert *et al.*).
     hep-ex/0511015.
     10.1103/PhysRevLett.96.241802.
     Phys.Rev.Lett. 96 (2006) 241802.

509. Measurements of the Absolute Branching Fractions of $B^\pm \to K^\pm X_{c\bar{c}}$
     By BaBar Collaboration (Bernard Aubert *et al.*).
     hep-ex/0510070.
     10.1103/PhysRevLett.96.052002.
     Phys.Rev.Lett. 96 (2006) 052002.

510. Search for the W-Exchange Decays $B^0 \to D_s^{(*)-} D_s^{(*)+}$
     By BaBar Collaboration (Bernard Aubert *et al.*).
     hep-ex/0510051.
     10.1103/PhysRevD.72.111101.
     Phys.Rev. D72 (2005) 111101.

511. Measurement of the Inclusive Electron Spectrum in Charmless Semileptonic B Decays Near the Kinematic Endpoint and Determination of $|V_{ub}|$
     By BaBar Collaboration (Bernard Aubert *et al.*).
     hep-ex/0509040.
     10.1103/PhysRevD.73.012006.
     Phys.Rev. D73 (2006) 012006.

512. Measurement of Branching Fractions and Resonance Contributions for $B^0 \to \bar{D}^0 K^+ \pi^-$ and Search for $B^0 \to D^0 K^+ \pi^-$ Decays
     By BaBar Collaboration (Bernard Aubert *et al.*).
     hep-ex/0509036.
     10.1103/PhysRevLett.96.011803.
     Phys.Rev.Lett. 96 (2006) 011803.

513. Improved Measurements of Branching Fractions for $B^0 \to \pi^+\pi^-$, $K^+\pi^-$, and Search for K^+K^- at BABAR
     By BaBar Collaboration (Bernard Aubert *et al.*).
     hep-ex/0508046.

514. Measurement of the Branching Ratios $\Gamma(D_s^{*+} \to D_s^+ \pi^0)/\Gamma(D_s^{*+} \to D_s^+ \gamma)$ and $\Gamma(D^{*0} \to D^0 \pi^0)/\Gamma(D^{*0} \to D^0 \gamma)$
     By BaBar Collaboration (Bernard Aubert *et al.*).
     hep-ex/0508039.
     10.1103/PhysRevD.72.091101.
     Phys.Rev. D72 (2005) 091101.

515. Measurement of CP Asymmetries in $B^0 \rightarrow K_s^0 \pi^0 \pi^0$ Decays
By BaBar Collaboration (Bernard Aubert *et al.*).
hep-ex/0508017.

516. Measurements of Neutral B Decay Branching Fractions to $K_s^0 \pi^+ \pi^-$ final states and the charge Asymmetry of $B^0 \rightarrow K^{*+} \pi^-$
By BaBar Collaboration (Bernard Aubert *et al.*).
hep-ex/0508013.
10.1103/PhysRevD.73.031101.
Phys.Rev. D73 (2006) 031101.

517. Search for Lepton Flavor Violation in the Decay $\tau^\pm \rightarrow e^\pm \gamma$
By BaBar Collaboration (Bernard Aubert *et al.*).
hep-ex/0508012.
10.1103/PhysRevLett.96.041801.
Phys.Rev.Lett. 96 (2006) 041801.

518. Measurements of the $B \rightarrow X_s \gamma$ Branching Fraction and Photon Spectrum from a Sum of Exclusive Final States
By BaBar Collaboration (Bernard Aubert *et al.*).
hep-ex/0508004.
10.1103/PhysRevD.72.052004.
Phys.Rev. D72 (2005) 052004.

519. A Study of $b \rightarrow c$ and $b \rightarrow u$ Interference in the Decay $B^- \rightarrow (K^+ \pi^-) DK^{*-}$
By BaBar Collaboration (Bernard Aubert *et al.*).
hep-ex/0508001.
10.1103/PhysRevD.72.071104.
Phys.Rev. D72 (2005) 071104.

520. Measurement of γ in $B^\mp \rightarrow D^{(*)} K^\mp$ and $B^\mp \rightarrow DK^{*\mp}$ Decays with a Dalitz Analysis of $D \rightarrow K_S^0 \pi^- \pi^+$
By BaBar Collaboration (Bernard Aubert *et al.*).
hep-ex/0507101.

521. Dalitz Plot Study of $B^0 \rightarrow K^+ K^- K_S^0$ Decays
By BaBar Collaboration (Bernard Aubert *et al.*).
hep-ex/0507094.

522. Study of $X(3872)$ and $Y(4260)$ in $B^0 \rightarrow J/\psi \pi^+ \pi^- K^0$ and $B^- \rightarrow J/\psi \pi^+ \pi^- K^-$
By BaBar Collaboration (Bernard Aubert *et al.*).
hep-ex/0507090.
10.1103/PhysRevD.73.011101.
Phys.Rev. D73 (2006) 011101.

523. Measurement of Time-Dependent *CP*-Violating Asymmetries in B^0 Meson Decays to $\eta' K_L^0$
By BaBar Collaboration (Bernard Aubert *et al.*).
hep-ex/0507087.

524. Measurement of the $B^0 \to \pi^- \ell^+ \nu$ and $B^+ \to \pi^0 \ell^+ \nu$ Branching Fractions and Determination of $|V_{ub}|$ in $\Upsilon(4S) \to B\bar{B}$ Events Tagged by a Fully Reconstructed B Meson
By BaBar Collaboration (Bernard Aubert *et al.*).
hep-ex/0507085.

525. Measurement of *CP*-Violating Parameters in Fully Reconstructed $B \to D^{(*)\pm}\pi^{\mp}$ and $B \to D^{\pm}\rho^{\mp}$ Decays
By BaBar Collaboration (Bernard Aubert *et al.*).
hep-ex/0507075.

526. A Search for the Decay $B^+ \to \tau^+ \nu_\tau$
By BaBar Collaboration (Bernard Aubert *et al.*).
hep-ex/0507069.
10.1103/PhysRevD.73.057101.
Phys.Rev. D73 (2006) 057101.

527. Measurement of the $\bar{B}^0$ Lifetime and the $B^0\bar{B}^0$ Oscillation Frequency Using Partially Reconstructed $\bar{B}^0 \to D^{*+}\ell^-\bar{\nu}_\ell$ Decays
By BaBar Collaboration (Bernard Aubert *et al.*).
hep-ex/0507054.
10.1103/PhysRevD.73.012004.
Phys.Rev. D73 (2006) 012004.

528. Measurement of Time-Dependent *CP*-Violating Asymmetries in $B^0 \to K_S^0 K_S^0 K_S^0$ Decays
By BaBar Collaboration (Bernard Aubert *et al.*).
hep-ex/0507052.

529. Measurement of the Time-Dependent *CP*-Violating Asymmetry in $B^0 \to K_S^0 \pi^0 \gamma$ Decays
By BaBar Collaboration (Bernard Aubert *et al.*).
hep-ex/0507038.
10.1103/PhysRevD.72.051103.
Phys.Rev. D72 (2005) 051103.

530. Measurement of Branching Fractions and Mass Spectra of $B \to K\pi\pi\gamma$
By BaBar Collaboration (Bernard Aubert *et al.*).
hep-ex/0507031.
10.1103/PhysRevLett.100.189903.
Phys.Rev.Lett. 98 (2007) 211804, Phys.Rev.Lett. 100 (2008) 189903, Phys.Rev.Lett. 100 (2008) 199905.

531. Measurement of the Branching Fraction of B^0 Meson Decay to $a_1^+(1260)\,\pi^-$
By BaBar Collaboration (Bernard Aubert *et al.*).
hep-ex/0507029.

532. Dalitz Plot Analysis of $D^0 \to \overline{K}^0 K^+ K^-$
     By BaBar Collaboration (Bernard Aubert *et al.*).
     hep-ex/0507026.
     10.1103/PhysRevD.72.052008.
     Phys.Rev. D72 (2005) 052008.

533. Amplitude Analysis of the Decay $B^\pm \to \pi^\pm \pi^\pm \pi^\mp$
     By BaBar Collaboration (Bernard Aubert *et al.*).
     hep-ex/0507025.
     10.1103/PhysRevD.72.052002.
     Phys.Rev. D72 (2005) 052002.

534. Evidence for $B^+ \to \overline{K}^0 K^+$ and $B^0 \to K^0 \overline{K}^0$, and Measurement of the Branching Fraction and
     Search for Direct *CP* Violation in $B^+ \to K^0 \pi^+$
     By BaBar Collaboration (Bernard Aubert *et al.*).
     hep-ex/0507023.
     10.1103/PhysRevLett.95.221801.
     Phys.Rev.Lett. 95 (2005) 221801.

535. Measurement of the Partial Branching Fraction for Inclusive Charmless Semileptonic *B*
     Decays and Extraction of $\left| V_{ub} \right|$
     By BaBar Collaboration (Bernard Aubert *et al.*).
     hep-ex/0507017.

536. Measurement of Time-Dependent *CP*-Violating Asymmetries in $B^0 \to K^+ K^- K^0_L$ Decays
     By BaBar Collaboration (Bernard Aubert *et al.*).
     hep-ex/0507016.

537. Measurement of the $B^+ \to p\bar{p}K^+$ Branching Fraction and Study of the Decay Dynamics
     By BaBar Collaboration (Bernard Aubert *et al.*).
     hep-ex/0507012.
     10.1103/PhysRevD.72.051101.
     Phys.Rev. D72 (2005) 051101.

538. A Study of Production and Decays of Ω_c^0 Baryons at BaBar
     By BaBar Collaboration (Bernard Aubert *et al.*).
     hep-ex/0507011.

539. A Precision Measurement of the Λ_c^+ Baryon Mass
     By BaBar Collaboration (Bernard Aubert *et al.*).
     hep-ex/0507009.
     10.1103/PhysRevD.72.052006.
     Phys.Rev. D72 (2005) 052006.

540. Measurements of the Rare Decays $B \to K \ell^+ \ell^-$ and $B \to K^* \ell^+ \ell^-$
     By BaBar Collaboration (Bernard Aubert *et al.*).
     hep-ex/0507005.

541. Dalitz-Plot Analysis of the Decays $B^\pm \to K^\pm \pi^\mp \pi^\pm$
By BaBar Collaboration (Bernard Aubert *et al.*).
hep-ex/0507004.
10.1103/PhysRevD.74.099903, 10.1103/PhysRevD.72.072003.
Phys.Rev. D72 (2005) 072003, Phys.Rev. D74 (2006) 099903.

542. Study of $B \to \pi \ell \nu$ and $B \to \rho \ell \nu$ Decays and Determination of $\left| V_{ub} \right|$
By BaBar Collaboration (Bernard Aubert *et al.*).
hep-ex/0507003.
10.1103/PhysRevD.72.051102.
Phys.Rev. D72 (2005) 051102.

543. Measurement of *CP* Observables for the Decays $B^\pm \to D_{CP}^0 K^{*\pm}$
By BaBar Collaboration (Bernard Aubert *et al.*).
hep-ex/0507002.
10.1103/PhysRevD.72.071103.
Phys.Rev. D72 (2005) 071103.

544. Results from the BaBar Fully Inclusive Measurement of $B \to X_s \gamma$
By BaBar Collaboration (Bernard Aubert *et al.*).
hep-ex/0507001.

545. Point Source Extraction with MOPEX
By David Makovoz, Francine R. Marleau.
astro-ph/0507007.
10.1086/432977.
Publ.Astron.Soc.Pac. 117 (2005) 1113-1128.

546. Measurement of Time-Dependent *CP* Asymmetries and the *CP*-Odd Fraction in the Decay $B^0 \to D^{*+} D^{*-}$
By BaBar Collaboration (Bernard Aubert *et al.*).
hep-ex/0506082.
10.1103/PhysRevLett.95.151804.
Phys.Rev.Lett. 95 (2005) 151804.

547. Observation of a Broad Structure in the $\pi^+ \pi^- J/\psi$ mass spectrum around 4.26 GeV/c^2
By BaBar Collaboration (Bernard Aubert *et al.*).
hep-ex/0506081.
10.1103/PhysRevLett.95.142001.
Phys.Rev.Lett. 95 (2005) 142001.

548. Search for $B^- \to D_s^{(*)-} \phi$
By BaBar Collaboration (Bernard Aubert *et al.*).
hep-ex/0506073.

549. Search for the rare decay $\overline{B}^0 \to D^{*0} \gamma$
By BaBar Collaboration (Bernard Aubert *et al.*).
hep-ex/0506070.
10.1103/PhysRevD.72.051106.
Phys.Rev. D72 (2005) 051106.

550. Measurement of the $B^{\pm} \to \rho^{\pm}\pi^0$ Branching Fraction and Direct *CP* Asymmetry
By BaBar Collaboration (Bernard Aubert *et al.*).
hep-ex/0506069.

551. Search for Lepton-Flavor and Lepton-Number Violation in the Decay $\tau^- \to \ell^{\mp} h^{\pm} h'^-$
By BaBar Collaboration (Bernard Aubert *et al.*).
hep-ex/0506066.
10.1103/PhysRevLett.95.191801.
Phys.Rev.Lett. 95 (2005) 191801.

552. Branching Fraction for $B^+ \to \pi^0 \ell^+ \nu$, Measured in $\Upsilon(4S) \to B\overline{B}$ Events Tagged by $B^- \to D^0 \ell^- \overline{\nu}(X)$ Decays
By BaBar Collaboration (Bernard Aubert *et al.*).
hep-ex/0506065.

553. Branching Fraction for $B^0 \to \pi^- \ell^+ \nu$ and Determination of $|V_{ub}|$ in $\Upsilon(4S) \to B^0 \overline{B}^0$ Events Tagged by $\overline{B}^0 \to D^{(*)+} \ell^- \overline{\nu}$
By BaBar Collaboration (Bernard Aubert *et al.*).
hep-ex/0506064.

554. Measurement of Double Charmonium Production in e^+e^- Annihilations at $\sqrt{s} = 10.6$ GeV
By BaBar Collaboration (Bernard Aubert *et al.*).
hep-ex/0506062.
10.1103/PhysRevD.72.031101.
Phys.Rev. D72 (2005) 031101.

555. Determination of $|V_{ub}|$ from Measurements of the Electron and Neutrino Momenta in Inclusive Semileptonic *B* Decays
By BaBar Collaboration (Bernard Aubert *et al.*).
hep-ex/0506036.
10.1103/PhysRevLett.95.111801.
Phys.Rev.Lett. 95 (2005) 111801, Phys.Rev.Lett. 97 (2006) 019903.

556. Search for the Decay $\tau^- \to 4\pi^- 3\pi^+ (\pi^0) \nu_{\tau}$
By BaBar Collaboration (B. Aubert *et al.*).
hep-ex/0506007.
10.1103/PhysRevD.72.012003.
Phys.Rev. D72 (2005) 012003.

557. Search for the Rare Decays $B^+ \to D^{(*)+} K_S^0$
By BaBar Collaboration (Bernard Aubert *et al.*).
hep-ex/0505099.
10.1103/PhysRevD.72.011102.
Phys.Rev. D72 (2005) 011102.

558. Measurement of Time-Dependent *CP* Asymmetries in $B^0 \to D^{(*)\pm} D^{\mp}$ Decays
By BaBar Collaboration (Bernard Aubert *et al.*).
hep-ex/0505092.
10.1103/PhysRevLett.95.131802.
Phys.Rev.Lett. 95 (2005) 131802.

559. Measurement of the Branching Fraction and Decay Rate Asymmetry of $B^- \to D_{\pi^+\pi^-\pi^0} K^-$
By BaBar Collaboration (Bernard Aubert *et al.*).
hep-ex/0505084.
10.1103/PhysRevD.72.071102.
Phys.Rev. D72 (2005) 071102.

560. Study of the $\tau^- \to 3h^-2h^+\nu_\tau$ Decay
By BaBar Collaboration (Bernard Aubert *et al.*).
hep-ex/0505004.
10.1103/PhysRevD.72.072001.
Phys.Rev. D72 (2005) 072001.

561. Search for $b \to u$ Transitions in $B^- \to D^0 K^-$ and $B^- \to D^{*0} K^-$
By BaBar Collaboration (Bernard Aubert *et al.*).
hep-ex/0504047.
10.1103/PhysRevD.72.032004.
Phys.Rev. D72 (2005) 032004.

562. Measurement of the Cabibbo-Kobayashi-Maskawa Angle γ in $B^\mp \to D^{(*)} K^\mp$ Decays with a Dalitz Analysis of $D \to K_S^0 \pi^- \pi^+$
By BaBar Collaboration (Bernard Aubert *et al.*).
hep-ex/0504039.
10.1103/PhysRevLett.95.121802.
Phys.Rev.Lett. 95 (2005) 121802.

563. Measurement of Time-Dependent *CP*-Violating Asymmetries and Constraints on $\sin(2\beta+\gamma)$ with Partial Reconstruction of $B \to D^{*\mp}\pi^\pm$ Decays
By BaBar Collaboration (Bernard Aubert *et al.*).
hep-ex/0504035.
10.1103/PhysRevD.71.112003.
Phys.Rev. D71 (2005) 112003.

564. Production and Decay of Ξ_c^0 at BABAR
By BaBar Collaboration (Bernard Aubert *et al.*).
hep-ex/0504014.
10.1103/PhysRevLett.95.142003.
Phys.Rev.Lett. 95 (2005) 142003.

565. Evidence for the Decay $B^\pm \to K^{*\pm}\pi^0$
By BaBar Collaboration (Bernard Aubert *et al.*).
hep-ex/0504009.
10.1103/PhysRevD.71.111101.
Phys.Rev. D71 (2005) 111101.

566. Measurement of the Branching Fraction of $\Upsilon(4S) \to B^0 \bar{B}^0$
By BaBar Collaboration (Bernard Aubert *et al.*).
hep-ex/0504001.
10.1103/PhysRevLett.95.042001.
Phys.Rev.Lett. 95 (2005) 042001.

567. Improved Measurement of the CKM Angle α Using $B^0(\overline{B}^0) \to \rho^+\rho^-$ Decays
By BaBar Collaboration (Bernard Aubert *et al.*).
hep-ex/0503049.
10.1103/PhysRevLett.95.041805.
Phys.Rev.Lett. 95 (2005) 041805.

568. Measurement of Branching Fractions and Charge Asymmetries in B^+ Decays to $\eta\pi^+$, ηK^+, $\eta\rho^+$ and $\eta'\pi^+$, and Search for B^0 Decays to ηK^0 and $\eta\omega$
By BaBar Collaboration (Bernard Aubert *et al.*).
hep-ex/0503035.
10.1103/PhysRevLett.95.131803.
Phys.Rev.Lett. 95 (2005) 131803.

569. Search for $B \to J/\psi D$ Decays
By BaBar Collaboration (Bernard Aubert *et al.*).
hep-ex/0503021.
10.1103/PhysRevD.71.091103.
Phys.Rev. D71 (2005) 091103.

570. Measurement of Branching Fraction and *CP*-Violating Asymmetry for $B^0 \to \omega K_S^0$
By BaBar Collaboration (Bernard Aubert *et al.*).
hep-ex/0503018.

571. Measurement of the Branching Fraction and the *CP*-Violating Asymmetry for the Decay $B^0 \to K_S^0 \pi^0$
By BaBar Collaboration (Bernard Aubert *et al.*).
hep-ex/0503011.
10.1103/PhysRevD.71.111102.
Phys.Rev. D71 (2005) 111102.

572. Measurement of the $B^0 \to D^{*-} D_s^{*+}$ and $D_s^+ \to \phi\pi^+$ Branching Fractions
By BaBar Collaboration (Bernard Aubert *et al.*).
hep-ex/0502041.
10.1103/PhysRevD.71.091104.
Phys.Rev. D71 (2005) 091104.

573. Search for Lepton Flavor Violation in the Decay $\tau^\pm \to \mu^\pm\gamma$
By BaBar Collaboration (Bernard Aubert *et al.*).
hep-ex/0502032.
10.1103/PhysRevLett.95.041802.
Phys.Rev.Lett. 95 (2005) 041802.

574. The $e^+e^- \to \pi^+\pi^-\pi^+\pi^-$, $K^+K^-\pi^+\pi^-$, and $K^+K^-K^+K^-$ Cross Sections at Center-of-Mass Energies 0.5–4.5 GeV Measured with Initial-State Radiation
By BaBar Collaboration (Bernard Aubert *et al.*).
hep-ex/0502025.
10.1103/PhysRevD.71.052001.
Phys.Rev. D71 (2005) 052001.

575. Measurement of *CP* Asymmetries in $B^0 \to \phi K^0$ and $B^0 \to K^+ K^- K_S^0$ Decays
By BaBar Collaboration (Bernard Aubert *et al.*).
hep-ex/0502019.
10.1103/PhysRevD.71.091102.
Phys.Rev. D71 (2005) 091102.

576. Measurements of Branching Fractions and Time-Dependent *CP*-Violating Asymmetries in $B \to \eta' K$ Decays
By BaBar Collaboration (B. Aubert *et al.*).
hep-ex/0502017.
10.1103/PhysRevLett.94.191802.
Phys.Rev.Lett. 94 (2005) 191802.

577. Branching Fraction and *CP* Asymmetries of $B^0 \to K_S^0 K_S^0 K_S^0$
By BaBar Collaboration (Bernard Aubert *et al.*).
hep-ex/0502013.
10.1103/PhysRevLett.95.011801.
Phys.Rev.Lett. 95 (2005) 011801.

578. Search for Strange-Pentaquark Production in $e^+ e^-$ annihilation at $\sqrt{s} = 10.58$ GeV
By BaBar Collaboration (B. Aubert *et al.*).
hep-ex/0502004.
10.1103/PhysRevLett.95.042002.
Phys.Rev.Lett. 95 (2005) 042002.

579. A Search for *CP* Violation and a Measurement of the Relative Branching Fraction in $D^+ \to K^- K^+ \pi^+$ Decays
By BaBar Collaboration (Bernard Aubert *et al.*).
hep-ex/0501075.
10.1103/PhysRevD.71.091101.
Phys.Rev. D71 (2005) 091101.

580. Improved Measurements of *CP*-Violating Asymmetry Amplitudes in $B^0 \to \pi^+ \pi^-$ Decays
By BaBar Collaboration (Bernard Aubert *et al.*).
hep-ex/0501071.
10.1103/PhysRevLett.95.151803.
.Rev.Lett. 95 (2005) 151803.

581. Search for Factorization-Suppressed $B \to \chi_c K^{(*)}$ Decays
By BaBar Collaboration (Bernard Aubert *et al.*).
hep-ex/0501061.
10.1103/PhysRevLett.94.171801.
Phys.Rev.Lett. 94 (2005) 171801.

582. Search for the Radiative Decay $B^0 \to \phi \gamma$
By BaBar Collaboration (Bernard Aubert *et al.*).
hep-ex/0501038.
10.1103/PhysRevD.72.091103.
Phys.Rev. D72 (2005) 091103.

583. Limit on the $B^0 \to \rho^0\rho^0$ Branching Fraction and Implications for the CKM Angle α
By BaBar Collaboration (Bernard Aubert *et al.*).
hep-ex/0412067.
10.1103/PhysRevLett.94.131801.
Phys.Rev.Lett. 94 (2005) 131801.

584. Measurement of Branching Fractions and Charge Asymmetries for Exclusive B Decays to Charmonium
By BaBar Collaboration (Bernard Aubert *et al.*).
hep-ex/0412062.
10.1103/PhysRevLett.94.141801.
Phys.Rev.Lett. 94 (2005) 141801.

585. Search for a Charged Partner of the $X(3872)$ in the B Meson Decay $B \to X^- K$, $X^- \to J/\psi\pi^-\pi^0$
By BaBar Collaboration (Bernard Aubert *et al.*).
hep-ex/0412051.
10.1103/PhysRevD.71.031501.
Phys.Rev. D71 (2005) 031501.

586. Measurement of Branching Fraction and Dalitz Distribution for $B^0 \to D^{(*)\pm}K^0\pi^\mp$ Decays
By BaBar Collaboration (Bernard Aubert *et al.*).
hep-ex/0412040.
10.1103/PhysRevLett.95.171802.
Phys.Rev.Lett. 95 (2005) 171802.

587. Branching Fractions and *CP* Asymmetries in $B^0 \to \pi^0\pi^0$, $B^+ \to \pi^+\pi^0$ and $B^+ \to K^+\pi^0$ Decays and Isospin Analysis of the $B \to \pi\pi$ System
By BaBar Collaboration (Bernard Aubert *et al.*).
hep-ex/0412037.
10.1103/PhysRevLett.94.181802.
Phys.Rev.Lett. 94 (2005) 181802.

588. A Brief Review of the Search for Isolatable Fractional Charge Elementary Particles
By M.L. Perl, E.R. Lee, D. Loomba.
10.1142/S0217732304016019.
Mod.Phys.Lett. A19 (2004) 2595-2610.

589. Measurement of the Ratio $\mathcal{B}(B^- \to D^{*0}K^-)/\mathcal{B}(B^- \to D^{*0}\pi^-)$ and of the *CP* Asymmetry of $B^- \to D^{*0}_{CP+}K^-$ Decays
By BaBar Collaboration (Bernard Aubert *et al.*).
hep-ex/0411091.
10.1103/PhysRevD.71.031102.
Phys.Rev. D71 (2005) 031102.

590. A Search for the Decay $B^+ \to K^+\nu\bar{\nu}$
By BaBar Collaboration (Bernard Aubert *et al.*).
hep-ex/0411061.
10.1103/PhysRevLett.94.101801.
Phys.Rev.Lett. 94 (2005) 101801.

591. Measurements of B Meson Decays to ωK^* and $\omega\rho$
By BaBar Collaboration (Bernard Aubert *et al.*).
hep-ex/0411054.
10.1103/PhysRevD.71.031103.
Phys.Rev. D71 (2005) 031103.

592. Automated Electric Charge Measurements of Fluid Microdrops Using the Millikan Method
By Eric R. Lee, Valerie Halyo, Irwin T. Lee, Martin L. Perl.
10.1088/0026-1394/41/5/S05.
Metrologia 41 (2004) S147-S158.

593. Ambiguity-Free Measurement of cos 2β: Time-Integrated and Time-Dependent Angular Analyses of $B \to J/\psi K\pi$
By BaBar Collaboration (Bernard Aubert *et al.*).
hep-ex/0411016.
10.1103/PhysRevD.71.032005.
Phys.Rev. D71 (2005) 032005.

594. Measurement of $B \to D^*$ Form-Factors in the Semileptonic Decay $\overline{B}^0 \to D^{*+}\ell^-\overline{\nu}$
By BaBar Collaboration (Bernard Aubert *et al.*).
hep-ex/0409047.

595. Search for Lepton-Flavor Violation in the Decay $\tau^- \to \ell^\mp h^\pm h^-$
By BaBar Collaboration (Bernard Aubert *et al.*).
hep-ex/0409036.

596. $D^+ \to K^-K^+\pi^+$ Meson Decays: A Search for CP Violation and a Measurement of the Branching Ratio
By BaBar Collaboration (Bernard Aubert *et al.*).
hep-ex/0408136.

597. Improved Measurement of CP Asymmetries in $B^0 \to (c\overline{c})K^{0(*)}$ Decays
By BaBar Collaboration (Bernard Aubert *et al.*).
hep-ex/0408127.
10.1103/PhysRevLett.94.161803.
Phys.Rev.Lett. 94 (2005) 161803.

598. Measurement of the Branching Fractions for Inclusive B^- and $\overline{B}^0$ Decays to Flavor-Tagged D, D_s and Λ_c
By BaBar Collaboration (Bernard Aubert *et al.*).
hep-ex/0408113.
10.1103/PhysRevD.70.091106.
Phys.Rev. D70 (2004) 091106.

599. Measurement of CP-Violating Asymmetries in $B^0 \to (\rho\pi)^0$ Using a Time-Dependent Dalitz Plot Analysis
By BaBar Collaboration (Bernard Aubert *et al.*).
hep-ex/0408099.

600. Search for Decays of B^0 Mesons into Pairs of Charged Leptons: $B^0 \to e^+e^-$, $B^0 \to \mu^+\mu^-$, and $B^0 \to e^\pm \mu^\mp$
     By BaBar Collaboration (Bernard Aubert *et al.*).
     hep-ex/0408096.
     10.1103/PhysRevLett.94.221803.
     Phys.Rev.Lett. 94 (2005) 221803.

601. A Measurement of *CP* Violating Asymmetries in $B^0 \to f_0(980)K_s^0$ Decays
     By BaBar Collaboration (Bernard Aubert *et al.*).
     hep-ex/0408095.

602. Measurement of the Branching Ratio $\Gamma(D_s^{*+} \to D_s^+\pi^0)/\Gamma(D_s^{*+} \to D_s^+\gamma)$
     By BaBar Collaboration (Bernard Aubert *et al.*).
     hep-ex/0408094.

603. Measurement of the Inclusive Electron Spectrum in Charmless Semileptonic B Decays Near the Kinematic Endpoint and Determination of $|V_{ub}|$
     By BaBar Collaboration (Bernard Aubert *et al.*).
     hep-ex/0408075.

604. $B^0 \to K^+\pi^-\pi^0$ Dalitz Plot Analysis
     By BaBar Collaboration (Bernard Aubert *et al.*).
     hep-ex/0408073.

605. Study of $b \to u\ell\bar{\nu}$ Decays on the Recoil of Fully Reconstructed B Mesons and Determination of $|V_{ub}|$
     By BaBar Collaboration (Bernard Aubert *et al.*).
     hep-ex/0408068.

606. Measurement of the $D_{sJ}^*(2317)^+$ and $D_{sJ}(2460)^+$ Properties in $e^+e^- \to c\bar{c}$ Production
     By BaBar Collaboration (Bernard Aubert *et al.*).
     hep-ex/0408067.

607. Search for D^0-$\bar{D}^0$ Mixing Using Semileptonic Decay Modes
     By BaBar Collaboration (Bernard Aubert *et al.*).
     hep-ex/0408066.
     10.1103/PhysRevD.70.091102.
     Phys.Rev. D70 (2004) 091102.

608. Search for Strange Pentaquark Production in e^+e^- Annihilations at $\sqrt{s} = 10.58$ GeV and in $\Upsilon(4S)$ Decays
     By BaBar Collaboration (Bernard Aubert *et al.*).
     hep-ex/0408064.

609. Search for the Decay $B^0 \to \rho^0\rho^0$
     By BaBar Collaboration (Bernard Aubert *et al.*).
     hep-ex/0408061.

610. Measurement of *CP*-Violating Parameters in Fully Reconstructed $B \to D^{(*)}\pi$ and $B \to D\rho$ Decays
By BaBar Collaboration (Bernard Aubert *et al.*).
hep-ex/0408059.

611. Searches for Charmless Decays $B^0 \to \eta\omega$, $B^0 \to \eta K^0$, $B^+ \to \eta\rho^+$, and $B^+ \to \eta'\pi^+$
By BaBar Collaboration (Bernard Aubert *et al.*).
hep-ex/0408058.

612. Measurement of the Ratio of Branching Fractions of Ξ_c^0 Decays to $\Xi^-\pi^+$ and to $\Omega^- K^+$
By BaBar Collaboration (Bernard Aubert *et al.*).
hep-ex/0408056.

613. A Study of $\overline{B}^0 \to D^{(*)0}\overline{K}^{(*)0}$ Decays
By BaBar Collaboration (Bernard Aubert *et al.*).
hep-ex/0408052.

614. A Search for $B^+ \to \tau^+\nu_\tau$ Recoiling Against $B^- \to D^{*0}\ell^-\overline{\nu}_\ell$
By BaBar Collaboration (Bernard Aubert *et al.*).
hep-ex/0408091.

615. Measurement of Time-Dependent *CP*-Violating Asymmetries in B^0 Meson Decays to $\eta'K^0$
By BaBar Collaboration (Bernard Aubert *et al.*).
hep-ex/0408090.

616. Improved Measurements of *CP*-Violating Asymmetries in $B^0 \to \pi^+\pi^-$ Decays
By BaBar Collaboration (Bernard Aubert *et al.*).
hep-ex/0408089.

617. Constraints on r_B and γ in $B^\pm \to D^{(*)0}K^\pm$ Decays by a Dalitz Analysis of $D^0 \to K_S\pi^-\pi^+$
By BaBar Collaboration (Bernard Aubert *et al.*).
hep-ex/0408088.

618. Search for the $D_{sJ}^*(2632)^+$ at BABAR
By BaBar Collaboration (Bernard Aubert *et al.*).
hep-ex/0408087.

619. A Search for the Decay $B^+ \to K^+\nu\overline{\nu}$
By BaBar Collaboration (Bernard Aubert *et al.*).
hep-ex/0408086.

620. Search for an $X(3872)$ Charged Partner in the Decay Mode $X^- \to J/\psi\pi^-\pi^0$ in the B Meson Decays $B^0 \to X^-K^+$ and $B^- \to X^-K_S^0$
By BaBar Collaboration (Bernard Aubert *et al.*).
hep-ex/0408083.

621. Measurement of the Branching Fractions and CP Asymmetries of $B^- \to D^0_{(CP)} K^-$ Decays with the BABAR Detector
      By BaBar Collaboration (Bernard Aubert *et al.*).
      hep-ex/0408082.

622. Study of $B^0(\overline{B}^0) \to \pi^0\pi^0$, $B^\pm \to \pi^\pm\pi^0$ and $B^\pm \to K^\pm\pi^0$ Decays
      By BaBar Collaboration (Bernard Aubert *et al.*).
      hep-ex/0408081.

623. Measurements of Branching Fractions and CP-Violating Asymmetries in B-Meson Decays to the Charmless Two-Body States $K^0\pi^+$, $\overline{K}^0K^+$, and $K^0\overline{K}^0$
      By BaBar Collaboration (Bernard Aubert *et al.*).
      hep-ex/0408080.

624. Evidence for $B^0 \to \rho^0 K^0_S$
      By BaBar Collaboration (Bernard Aubert *et al.*).
      hep-ex/0408079.

625. Study of $e^+e^- \to \pi^+\pi^-\pi^0$ Process Using Initial State Radiation with BaBar
      By BaBar Collaboration (Bernard Aubert *et al.*).
      hep-ex/0408078.
      10.1103/PhysRevD.70.072004.
      Phys.Rev. D70 (2004) 072004.

626. Measurement of CP Asymmetry in $B^0 \to K^+K^-K^0_S$ Decays
      By BaBar Collaboration (Bernard Aubert *et al.*).
      hep-ex/0408076.

627. Measurements of CP Asymmetries in the Decay $B \to \phi K$
      By BaBar Collaboration (Bernard Aubert *et al.*).
      hep-ex/0408072.

628. Measurement of CP-Asymmetries for the Decays $B^\pm \to D^0_{CP}K^{*\pm}$ with the BABAR Detector
      By BaBar Collaboration (Bernard Aubert *et al.*).
      hep-ex/0408069.

629. Measurement of the $B^0 \to K^0_S K^0_S K^0_S$ Branching Fraction
      By BaBar Collaboration (Bernard Aubert *et al.*).
      hep-ex/0408065.

630. Search for the Decay $B^0 \to K^{*+}\rho^-$
      By BaBar Collaboration (Bernard Aubert *et al.*).
      hep-ex/0408063.

631. Measurements of the Branching Fraction and CP-Violating Asymmetries of $B^0 \to K^0_S\pi^0$ Decays
      By BaBar Collaboration (Bernard Aubert *et al.*).
      hep-ex/0408062.

632. Measurement of the Ratio $\mathcal{B}(B^- \to D^{*0}K^-)/\mathcal{B}(B^- \to D^{*0}\pi^-)$ and the CP-Asymmetry of $B^- \to D^{*0}_{CP+}K^-$ Decays with the BABAR Detector
   By BaBar Collaboration (Bernard Aubert *et al.*).
   hep-ex/0408060.

633. Measurement of Neutral B Decay Branching Fractions to $K^0_S \pi^+ \pi^-$ Final States
   By BaBar Collaboration (Bernard Aubert *et al.*).
   hep-ex/0408054.
   10.1103/PhysRevD.70.091103.
   Phys.Rev. D70 (2004) 091103.

634. Study of the $\tau^- \to 3h^- 2h^+ \nu_\tau$ Decay
   By BaBar Collaboration (Bernard Aubert *et al.*).
   hep-ex/0408050.

635. Determination of $|V_{ub}|$ in Inclusive Semileptonic B Meson Decays
   By BaBar Collaboration (Bernard Aubert *et al.*).
   hep-ex/0408045.

636. Study of $B \to D^{(*)+}_{sJ} \overline{D}^{(*)}$ Decays
   By BaBar Collaboration (Bernard Aubert *et al.*).
   hep-ex/0408041.
   10.1103/PhysRevLett.93.181801.
   Phys.Rev.Lett. 93 (2004) 181801.

637. Partial Reconstruction of $B^0 \to D^{*+}_s D^{*-}$ Decays and Measurement of the $D^+_s \to \phi\pi^+$ Branching Fraction
   By BaBar Collaboration (Bernard Aubert *et al.*).
   hep-ex/0408040.

638. Measurement of the $\overline{B}^0$ Lifetime and of the $B^0 \overline{B}^0$ Oscillation Frequency Using Partially Reconstructed $\overline{B}^0 \to D^{*+}\ell^-\overline{\nu}_\ell$ Decays
   By BaBar Collaboration (Bernard Aubert *et al.*).
   hep-ex/0408039.

639. A Search for the Θ^{*++} Pentaquark in $B^\pm \to p\overline{p}K^\pm$
   By BaBar Collaboration (Bernard Aubert *et al.*).
   hep-ex/0408037.

640. Measurement of Exclusive B Decays to Charmonium and K or K^* Branching Fractions with the BABAR Detector
   By BaBar Collaboration (Bernard Aubert *et al.*).
   hep-ex/0408036.

641. Measurement of the Branching Fractions for the Decays $B^0 \to D^{*-}p\overline{p}\pi^+$, $B^0 \to D^-p\overline{p}\pi^+$, $B^0 \to \overline{D}^{*0}p\overline{p}$, and $B^0 \to \overline{D}^0 p\overline{p}$
   By BaBar Collaboration (Bernard Aubert *et al.*).
   hep-ex/0408035.

642. Search for Radiative Penguin Decays $B^+ \to \rho^+\gamma$, $B^0 \to \rho^0\gamma$, and $B^0 \to \omega\gamma$
By BaBar Collaboration (Bernard Aubert *et al.*).
hep-ex/0408034.
10.1103/PhysRevLett.94.011801.
Phys.Rev.Lett. 94 (2005) 011801.

643. Search for $B \to \chi_c K^{(*)}$ Decays
By BaBar Collaboration (Bernard Aubert *et al.*).
hep-ex/0408033.

644. Amplitude Analysis of $B^\pm \to \pi^\pm\pi^\mp\pi^\pm$ and $B^\pm \to K^\pm\pi^\mp\pi^\pm$
By BaBar Collaboration (Bernard Aubert *et al.*).
hep-ex/0408032.

645. A Search for the Rare Decay $B^0 \to D_s^+\rho^-$
By BaBar Collaboration (Bernard Aubert *et al.*).
hep-ex/0408029.

646. Search for $b \to u$ Transitions in $B^- \to \overset{(-)}{D}{}^0 K^-$ and $B^- \to \overset{(-)}{D}{}^{*0} K^-$
By BaBar Collaboration (Bernard Aubert *et al.*).
hep-ex/0408028.

647. Measurement of the $\bar{B}^0 \to D^{(*)+}\ell^-\bar{\nu}_\ell$ Decay Rate and $|V_{cb}|$
By BaBar Collaboration (Bernard Aubert *et al.*).
hep-ex/0408027.
10.1103/PhysRevD.71.051502.
Phys.Rev. D71 (2005) 051502.

648. Measurements of Λ_c^+ Branching Fractions of Cabibbo-Suppressed Decay Modes
By BaBar Collaboration (Bernard Aubert *et al.*).
hep-ex/0408024.

649. Search for Flavor-Changing Neutral Current and Lepton-Flavor Violating Decays of $D^0 \to \ell^+\ell^-$
By BaBar Collaboration (Bernard Aubert *et al.*).
hep-ex/0408023.
10.1103/PhysRevLett.93.191801.
Phys.Rev.Lett. 93 (2004) 191801.

650. Measurement of the Branching Fraction of $e^+e^- \to B^0\bar{B}^0$ at the $\Upsilon(4S)$ Resonance
By BaBar Collaboration (Bernard Aubert *et al.*).
hep-ex/0408022.

651. Observation of B^0 Meson Decays to $a_1^+(1260)\pi^-$
By BaBar Collaboration (Bernard Aubert *et al.*).
hep-ex/0408021.

652. Search for the Decay $B^0 \to J/\psi\gamma$
By BaBar Collaboration (Bernard Aubert *et al.*).
hep-ex/0408018.
10.1103/PhysRevD.70.091104.
Phys.Rev. D70 (2004) 091104.

653. Measurement of the $B^0 \to \phi K^{*0}$ Decay Amplitudes
By BaBar Collaboration (Bernard Aubert *et al.*).
hep-ex/0408017.
10.1103/PhysRevLett.93.231804.
Phys.Rev.Lett. 93 (2004) 231804.

654. Direct *CP* Asymmetry in $B^0 \to K^+\pi^-$ Decays
By BaBar Collaboration (Bernard Aubert *et al.*).
hep-ex/0407057.
10.1103/PhysRevLett.93.131801.
Phys.Rev.Lett. 93 (2004) 131801.

655. A Search for the Rare Leptonic Decay $B^- \to \tau^-\bar{\nu}_\tau$
By BaBar Collaboration (B. Aubert *et al.*).
hep-ex/0407038.
10.1103/PhysRevLett.95.041804.
Phys.Rev.Lett. 95 (2005) 041804.

656. Search for *B*-Meson Decays to Two-Body Final States with $a_0(980)$ Mesons
By BaBar Collaboration (Bernard Aubert *et al.*).
hep-ex/0407013.
10.1103/PhysRevD.70.111102.
Phys.Rev. D70 (2004) 111102.

657. Measurement of Branching Fractions, and *CP* and Isospin Asymmetries, for $B \to K^*\gamma$
By BaBar Collaboration (Bernard Aubert *et al.*).
hep-ex/0407003.
10.1103/PhysRevD.70.112006.
Phys.Rev. D70 (2004) 112006.

658. Measurements of the Branching Fraction and *CP*-Violation Asymmetries in $B^0 \to f_0(980)K_S^0$
By BaBar Collaboration (Bernard Aubert *et al.*).
hep-ex/0406040.
10.1103/PhysRevLett.94.041802.
Phys.Rev.Lett. 94 (2005) 041802.

659. Study of the $B^- \to J/\psi K^-\pi^+\pi^-$ Decay and Measurement of the $B^- \to X(3872)K^-$ Branching Fraction
By BaBar Collaboration (Bernard Aubert *et al.*).
hep-ex/0406022.
10.1103/PhysRevD.71.071103.
Phys.Rev. D71 (2005) 071103.

660. Branching Fractions and *CP* Asymmetries in $B^0 \rightarrow K^+ K^- K_S^0$ and $B^+ \rightarrow K^+ K_S^0 K_S^0$
      By BaBar Collaboration (Bernard Aubert *et al.*).
      hep-ex/0406005.
      10.1103/PhysRevLett.93.181805.
      Phys.Rev.Lett. 93 (2004) 181805.

661. Measurement of Time-Dependent *CP*-Violating Asymmetries in $B^0 \rightarrow K^{*0}\gamma(K^{*0} \rightarrow K_S^0 \pi^0)$
      Decays
      By BaBar Collaboration (Bernard Aubert *et al.*).
      hep-ex/0405082.
      10.1103/PhysRevLett.93.201801.
      Phys.Rev.Lett. 93 (2004) 201801.

662. Search for B^0 Decays to Invisible Final States and to $\nu\bar{\nu}\gamma$
      By BaBar Collaboration (Bernard Aubert *et al.*).
      hep-ex/0405071.
      10.1103/PhysRevLett.93.091802.
      Phys.Rev.Lett. 93 (2004) 091802.

663. A Measurement of the Total Width, the Electronic Width, and the Mass of the $\Upsilon(10580)$
      Resonance
      By BaBar Collaboration (Bernard Aubert *et al.*).
      hep-ex/0405025.
      10.1103/PhysRevD.72.032005.
      Phys.Rev. D72 (2005) 032005.

664. Study of the Decay $B^0(\bar{B}^0) \rightarrow \rho^+ \rho^-$, and Constraints on the CKM Angle α
      By BaBar Collaboration (Bernard Aubert *et al.*).
      hep-ex/0404029.
      10.1103/PhysRevLett.93.231801.
      Phys.Rev.Lett. 93 (2004) 231801.

665. Determination of the Branching Fraction for $B \rightarrow X_c \ell \nu$ Decays and of $|V_{cb}|$ from Hadronic
      Mass and Lepton Energy Moments
      By BaBar Collaboration (Bernard Aubert *et al.*).
      hep-ex/0404017.
      10.1103/PhysRevLett.93.011803.
      Phys.Rev.Lett. 93 (2004) 011803.

666. Measurement of the $B \rightarrow X_s \ell^+ \ell^-$ Branching Fraction with a Sum Over Exclusive Modes
      By BaBar Collaboration (Bernard Aubert *et al.*).
      hep-ex/0404006.
      10.1103/PhysRevLett.93.081802.
      Phys.Rev.Lett. 93 (2004) 081802.

667. Measurement of the Ratio of Decay Amplitudes for $\overline{B}^0 \to J/\psi K^{*0}$ and $B^0 \to J/\psi K^{*0}$
By BaBar Collaboration (Bernard Aubert *et al.*).
hep-ex/0404005.
10.1103/PhysRevLett.93.081801.
Phys.Rev.Lett. 93 (2004) 081801.

668. Searches for B^0 Decays to Combinations of Charmless Isoscalar Mesons
By BaBar Collaboration (Bernard Aubert *et al.*).
hep-ex/0403046.
10.1103/PhysRevLett.93.181806.
Phys.Rev.Lett. 93 (2004) 181806.

669. Measurement of the Direct *CP* Asymmetry in $b \to s\gamma$ Decays
By BaBar Collaboration (Bernard Aubert *et al.*).
hep-ex/0403035.
10.1103/PhysRevLett.93.021804.
Phys.Rev.Lett. 93 (2004) 021804.

670. Measurements of Moments of the Hadronic Mass Distribution in Semileptonic *B* Decays
By BaBar Collaboration (Bernard Aubert *et al.*).
hep-ex/0403031.
10.1103/PhysRevD.69.111103.
Phys.Rev. D69 (2004) 111103.

671. Measurement of the Electron Energy Spectrum and its Moments in Inclusive $B \to Xe\nu$ Decays
By BaBar Collaboration (Bernard Aubert *et al.*).
hep-ex/0403030.
10.1103/PhysRevD.69.111104.
Phys.Rev. D69 (2004) 111104.

672. Measurement of the Time-Dependent *CP* Asymmetry in the $B^0 \to \phi K^0$ Decay
By BaBar Collaboration (Bernard Aubert *et al.*).
hep-ex/0403026.
10.1103/PhysRevLett.93.071801.
Phys.Rev.Lett. 93 (2004) 071801.

673. *B* Meson Decays to $\eta^{(\prime)} K^*$, $\eta^{(\prime)}\rho$, $\eta^{(\prime)}\pi^0$, $\omega\pi^0$, and $\phi\pi^0$
By BaBar Collaboration (Bernard Aubert *et al.*).
hep-ex/0403025.
10.1103/PhysRevD.70.032006.
Phys.Rev. D70 (2004) 032006.

674. A Study of the Impact of Radiation Exposure on Uniformity of Large CsI(Tl) Crystals for the BABAR Detector
By Tetiana Berger-Hryn'ova, Peter Kim, Martin Kocian, Martin Perl, Howard Rogers, Rafe H. Schindler, William Wisniewski.
10.1016/j.nima.2004.07.268.
Nucl.Instrum.Meth. A535 (2004) 452-456.

675. Search for the Decay $B^0 \to p\bar{p}$
     By BaBar Collaboration (Bernard Aubert *et al.*).
     hep-ex/0403003.
     10.1103/PhysRevD.69.091503.
     Phys.Rev. D69 (2004) 091503.

676. Limits on the Decay-Rate Difference of Neutral-*B* Mesons and on *CP*, *T*, and *CPT* Violation
     in $B^0\bar{B}^0$ Oscillations
     By BaBar Collaboration (Bernard Aubert *et al.*).
     hep-ex/0403002.
     10.1103/PhysRevD.70.012007.
     Phys.Rev. D70 (2004) 012007.

677. Measurements of *CP* Violating Asymmetries in $B^0 \to K_S^0 \pi^0$ Decays
     By BaBar Collaboration (Bernard Aubert *et al.*).
     hep-ex/0403001.
     10.1103/PhysRevLett.93.131805.
     Phys.Rev.Lett. 93 (2004) 131805.

678. Observation of the Decay $B \to J/\psi \eta K$ and Search for $X(3872) \to J/\psi \eta$
     By BaBar Collaboration (Bernard Aubert *et al.*).
     hep-ex/0402025.
     10.1103/PhysRevLett.93.041801.
     Phys.Rev.Lett. 93 (2004) 041801.

679. Search for $B^\pm \to [K^\mp \pi^\pm]_D K^\pm$ and Upper Limit on the $b \to u$ Amplitude in $B^\pm \to DK^\pm$
     By BaBar Collaboration (Bernard Aubert *et al.*).
     hep-ex/0402024.
     10.1103/PhysRevLett.93.131804.
     Phys.Rev.Lett. 93 (2004) 131804.

680. Energy and Centrality Dependence of Deuteron and Proton Production in Pb + Pb Collisions
     at Relativistic Energies
     By NA49 Collaboration (T. Anticic *et al.*).
     10.1103/PhysRevC.69.024902.
     Phys.Rev. C69 (2004) 024902.

681. Study of $B^\pm \to J/\psi \pi^\pm$ and $B^\pm \to J/\psi K^\pm$ Decays: Measurement of the Ratio of Branching
     Fractions and Search for Direct *CP* Violation
     By BaBar Collaboration (Bernard Aubert *et al.*).
     hep-ex/0401035.
     10.1103/PhysRevLett.92.241802.
     Phys.Rev.Lett. 92 (2004) 241802.

682. Measurement of the B^+/B^0 Production Ratio from the $\Upsilon(4S)$ Meson Using $B^+ \to J/\psi K^+$ and
     $B^0 \to J/\psi K_S^0$ Decays
     By BaBar Collaboration (Bernard Aubert *et al.*).
     hep-ex/0401028.
     10.1103/PhysRevD.69.071101.
     Phys.Rev. D69 (2004) 071101.

683. Study of High Momentum η' production in $B \to \eta' X_s$
By BaBar Collaboration (Bernard Aubert *et al.*).
hep-ex/0401006.
10.1103/PhysRevLett.93.061801.
Phys.Rev.Lett. 93 (2004) 061801.

684. Search for the Rare Leptonic Decay $B^+ \to \mu^+ \nu_\mu$
By BaBar Collaboration (Bernard Aubert *et al.*).
hep-ex/0401002.
10.1103/PhysRevLett.93.189902.
Phys.Rev.Lett. 92 (2004) 221803.

685. Recent Results from the NA49 Experiment
By NA49 Collaboration (T. Anticic *et al.*).

686. Multistrange Hyperon Production in Pb + Pb Collisions at 30, 40, 80 and 158 A·GeV
By NA49 Collaboration (C. Alt *et al.*).
nucl-ex/0312022.
10.1016/j.ppnp.2004.02.017.
Prog.Part.Nucl.Phys. 53 (2004) 269-272.

687. Measurements of Branching Fractions and *CP*-Violating Asymmetries in B Meson Decays to Charmless Two-Body States Containing a K^0
By BaBar Collaboration (Bernard Aubert *et al.*).
hep-ex/0312055.
10.1103/PhysRevLett.92.201802.
Phys.Rev.Lett. 92 (2004) 201802.

688. Measurement of the Branching Fraction for $B^- \to D^0 K^{*-}$
By BaBar Collaboration (Bernard Aubert *et al.*).
hep-ex/0312051.
10.1103/PhysRevD.69.051101.
Phys.Rev. D69 (2004) 051101.

689. Search for Lepton Flavor Violation in the Decay $\tau^- \to \ell^- \ell^+ \ell^-$
By BaBar Collaboration (Bernard Aubert *et al.*).
hep-ex/0312027.
10.1103/PhysRevLett.92.121801.
Phys.Rev.Lett. 92 (2004) 121801.

690. Λ and $\bar{\Lambda}$ Production in Central Pb – Pb Collisions at 40, 80, and 158 A·GeV
By NA49 Collaboration (T. Anticic *et al.*).
nucl-ex/0311024.
10.1103/PhysRevLett.93.022302.
Phys.Rev.Lett. 93 (2004) 022302.

691. Measurement of Branching Fractions and Charge Asymmetries in $B^\pm \to \rho^\pm \pi^0$ and $B^\pm \to \rho^0 \pi^\pm$
      Decays, and Search for $B^0 \to \rho^0 \pi^0$
      By BaBar Collaboration (Bernard Aubert *et al.*).
      hep-ex/0311049.
      10.1103/PhysRevLett.93.051802.
      Phys.Rev.Lett. 93 (2004) 051802.

692. Measurements of the Mass and Width of the η_c Meson and of an $\eta_c(2S)$ Candidate
      By BaBar Collaboration (Bernard Aubert *et al.*).
      hep-ex/0311038.
      10.1103/PhysRevLett.92.142002.
      Phys.Rev.Lett. 92 (2004) 142002.

693. Limits on the Decay-Rate Difference of Neutral B Mesons and on *CP*, *T*, and *CPT* Violation
      in $B^0 \bar{B}^0$ Oscillations
      By BaBar Collaboration (Bernard Aubert *et al.*).
      hep-ex/0311037.
      10.1103/PhysRevLett.92.181801.
      Phys.Rev.Lett. 92 (2004) 181801.

694. A Systematic Study of Radiation Damage to Large Crystals of CsI(Tl) for the BaBar Detector
      By Tetiana Berger-Hryn'ova, P. Kim, M. Kocian, Martin L. Perl, H. Rogers, R.H. Schindler,
      William J. Wisniewski.

695. Measurement of the Branching Fractions and *CP*-Asymmetry of $B^- \to D^0_{(CP)} K^-$ Decays with
      the BaBar Detector
      By BaBar Collaboration (Bernard Aubert *et al.*).
      hep-ex/0311032.
      10.1103/PhysRevLett.92.202002.
      Phys.Rev.Lett. 92 (2004) 202002.

696. Observation of the Decay $B^0 \to \rho^+ \rho^-$ and Measurement of the Branching Fraction and
      Polarization
      By BaBar Collaboration (Bernard Aubert *et al.*).
      hep-ex/0311017.
      10.1103/PhysRevD.69.031102.
      Phys.Rev. D69 (2004) 031102.

697. Observation of $B^0 \to \omega K^0$, $B^+ \to \eta \pi^+$, and $B^+ \to \eta K^+$ and Study of Related Decays
      By BaBar Collaboration (Bernard Aubert *et al.*).
      hep-ex/0311016.
      10.1103/PhysRevLett.92.061801.
      Phys.Rev.Lett. 92 (2004) 061801.

698   Transverse Momentum Fluctuations in Nuclear Collisions at 158 AGeV
      By NA49 Collaboration (T. Anticic *et al.*).
      hep-ex/0311009.
      10.1103/PhysRevC.70.034902.
      Phys.Rev. C70 (2004) 034902.

699. Measurement of the Average ϕ Multiplicity in B Meson Decay
By BaBar Collaboration (Bernard Aubert *et al.*).
hep-ex/0311008.
10.1103/PhysRevD.69.052005.
Phys.Rev. D69 (2004) 052005.

700. Observation of a Narrow Meson Decaying to $D_s^+ \pi^0 \gamma$ at a Mass of 2.458 GeV/c^2
By BaBar Collaboration (Bernard Aubert *et al.*).
hep-ex/0310050.
10.1103/PhysRevD.69.031101.
Phys.Rev. D69 (2004) 031101.

701. The Discovery of the τ Lepton and the Changes in Elementary Particle Physics in Forty Years
By Martin L. Perl.
10.1007/s00016-003-0218-3.
Phys.Perspect. 6 (2004) 401-427.

702. Sizing of Microdrops
By Irwin T. Lee, Sewan Fan, Valerie Halyo, Peter C. Kim, Eric R. Lee, Martin L. Perl, Howard Rogers.
10.1063/1.1630831.
Rev.Sci.Instrum. 75 (2004) 891-895.

703. Measurement of Time-Dependent CP Asymmetries and Constraints on $\sin(2\beta + \gamma)$ with Partial Reconstruction of $B^0 \to D^{*\mp}\pi^{\pm}$ Decays
By BaBar Collaboration (Bernard Aubert *et al.*).
hep-ex/0310037.
10.1103/PhysRevLett.92.251802.
Phys.Rev.Lett. 92 (2004) 251802.

704. Measurement of Branching Fractions of Color-Suppressed Decays of the $\bar{B}^0$ Meson to $D^{(*)0}\pi^0$, $D^{(*)0}\eta$, $D^{(*)0}\omega$, and $D^0\eta'$
By BaBar Collaboration (Bernard Aubert *et al.*).
hep-ex/0310028.
10.1103/PhysRevD.69.032004.
Phys.Rev. D69 (2004) 032004.

705. J/ψ Production Via Initial State Radiation in $e^+e^- \to \mu^+\mu^-\gamma$ at an e^+e^- Center-of-Mass Energy Near 10.6 GeV
By BaBar Collaboration (Bernard Aubert *et al.*).
hep-ex/0310027.
10.1103/PhysRevD.69.011103.
Phys.Rev. D69 (2004) 011103.

706. Measurement of the Branching Fraction for $B^{\pm} \to \chi_{c0}K^{\pm}$
By BaBar Collaboration (Bernard Aubert *et al.*).
hep-ex/0310015.
10.1103/PhysRevD.69.071103.
Phys.Rev. D69 (2004) 071103.

707. Observation of an Exotic S = –2, Q = –2 Baryon Resonance in Proton-Proton Collisions at the CERN SPS
By NA49 Collaboration (C. Alt *et al.*).
hep-ex/0310014.
10.1103/PhysRevLett.92.042003.
Phys.Rev.Lett. 92 (2004) 042003.

708. Measurement of sin 2β with Hadronic and Previously Unused Muonic J/ψ Decays
By BaBar Collaboration (Bernard Aubert *et al.*).
hep-ex/0309039.
10.1103/PhysRevD.69.052001.
Phys.Rev. D69 (2004) 052001.

709. Search for the B Meson Decay to $\eta'\phi$
By BaBar Collaboration (Bernard Aubert *et al.*).
hep-ex/0309038.

710. Measurements of Branching Fractions in $B \rightarrow \phi K$ and $B \rightarrow \phi\pi$ and Search for Direct *CP* violation in $B^{\pm} \rightarrow \phi K^{\pm}$
By BaBar Collaboration (Bernard Aubert *et al.*).
hep-ex/0309025.
10.1103/PhysRevD.69.011102.
Phys.Rev. D69 (2004) 011102.

711. Measurements of the Branching Fractions of Charged B Decays to $K^{\pm}\pi^{\mp}\pi^{\pm}$ Final States
By BaBar Collaboration (Bernard Aubert *et al.*).
hep-ex/0308065.
10.1103/PhysRevD.70.092001.
Phys.Rev. D70 (2004) 092001.

712. Measurement of the Branching Fraction and Polarization for the Decay $B^- \rightarrow D^{*0}K^{*-}$
By BaBar Collaboration (Bernard Aubert *et al.*).
hep-ex/0308057.
10.1103/PhysRevLett.92.141801.
Phys.Rev.Lett. 92 (2004) 141801.

713. Evidence for the Rare Decay $B \rightarrow K^*\ell^+\ell^-$ and Measurement of the $B \rightarrow K\ell^+\ell^-$ Branching Fraction
By BaBar Collaboration (Bernard Aubert *et al.*).
hep-ex/0308042.
10.1103/PhysRevLett.91.221802.
Phys.Rev.Lett. 91 (2003) 221802.

714. Measurement of $|V_{cb}|$ Using $\bar{B}^0 \rightarrow D^{*+}\ell^-\bar{\nu}_\ell$ Decays
By BaBar Collaboration (Bernard Aubert *et al.*).
hep-ex/0308027.

715. Study of the Decays $B^- \rightarrow D^{(*)+}\pi^-\pi^-$
By BaBar Collaboration (Bernard Aubert *et al.*).
hep-ex/0308026.

716. Observation of the Decay $B^0 \to \rho^+\rho^-$ and Measurement of the Branching Fraction and Polarization
By BaBar Collaboration (Bernard Aubert *et al.*).
hep-ex/0308024.

717. Measurement of the Electric Form Factor of the Neutron at $Q^2 = 0.5$ and 1.0 GeV$^2/c^2$
By BaBar Collaboration (Bernard Aubert *et al.*).
hep-ex/0409035, hep-ex/0308021.
10.1103/PhysRevD.70.091105.
Phys.Rev. D70 (2004) 091105.

718. A Measurement of the Total Width, the Electronic Width and the Mass of the $\Upsilon(10580)$ Resonance
By BaBar and PEP-II Machine Group Collaborations (Bernard Aubert *et al.*).
hep-ex/0308020.

719. Measurement of Time-Dependent CP Asymmetries in $B^0 \to D^{(*)\pm}\pi^\mp$ Decays and Constraints on $\sin(2\beta + \gamma)$
By BaBar Collaboration (Bernard Aubert *et al.*).
hep-ex/0308018.
10.1103/PhysRevLett.92.251801.
Phys.Rev.Lett. 92 (2004) 251801.

720. Measurement of $\mathcal{B}(B^0 \to D_s^{*+}D^{*-})$ and Determination of the $D_s^+ \to \phi\pi^+$ Branching Fraction with a Partial-Reconstruction Method
By BaBar Collaboration (Bernard Aubert *et al.*).
hep-ex/0308017.

721. Measurement of the $B \to X_s\ell^+\ell^-$ Branching Fraction Using a Sum Over Exclusive Modes
By BaBar Collaboration (Bernard Aubert *et al.*).
hep-ex/0308016.

722. Measurement of Branching Fractions and Charge Asymmetries in B Meson Decays to $\eta^{(\prime)}K^*$, $\eta^{(\prime)}\rho$, and $\eta'\pi$
By BaBar Collaboration (Bernard Aubert *et al.*).
hep-ex/0308015.

723. Observation of the Decay $B^0 \to \pi^0\pi^0$
By BaBar Collaboration (Bernard Aubert *et al.*).
hep-ex/0308012.
10.1103/PhysRevLett.91.241801.
Phys.Rev.Lett. 91 (2003) 241801.

724. Measurement of Branching Fractions and CP-Violating Charge Asymmetries in $B^+ \to \rho^+\pi^0$ and $B^+ \to \rho^0\pi^+$ Decays, and Search for $B^0 \to \rho^0\pi^0$
By BaBar Collaboration (Bernard Aubert *et al.*).
hep-ex/0307087.
Submitted to: Phys.Rev.Lett..

725. Measurement of the Inclusive Charmless Semileptonic Branching Ratio of *B* Mesons and Determination of $\left|V_{ub}\right|$
By BaBar Collaboration (Bernard Aubert *et al.*).
hep-ex/0307062.
10.1103/PhysRevLett.92.071802.
Phys.Rev.Lett. 92 (2004) 071802.

726. A Search for $B^+ \to \mu^+ \nu_\mu$
By BaBar Collaboration (Bernard Aubert *et al.*).
hep-ex/0307047.

727. Measurement of the First and Second Moments of the Hadronic Mass Distribution in Semileptonic *B* Decays
By BaBar Collaboration (Bernard Aubert *et al.*).
hep-ex/0307046.

728. Study of Time-Dependent *CP* Asymmetries with Partial Reconstruction of $B^0 \to D^{*\mp}\pi^\pm$
By BaBar Collaboration (Bernard Aubert *et al.*).
hep-ex/0307036.

729. Evidence for the Rare Decay $B \to J/\psi\eta K$
By BaBar Collaboration (Bernard Aubert *et al.*).
hep-ex/0307032.

730. Rates, Polarizations, and Asymmetries in Charmless Vector-Vector *B* Meson Decays
By BaBar Collaboration (Bernard Aubert *et al.*).
hep-ex/0307026.
10.1103/PhysRevLett.91.171802.
Phys.Rev.Lett. 91 (2003) 171802.

731. Measurement of Time-Dependent *CP* Asymmetries and the *CP*-Odd Fraction in the Decay $B^0 \to D^{*+}D^{*-}$
By BaBar Collaboration (Bernard Aubert *et al.*).
hep-ex/0306052.
10.1103/PhysRevLett.91.131801.
Phys.Rev.Lett. 91 (2003) 131801.

732. Search for the Radiative Decays $B \to \rho\gamma$ and $B^0 \to \omega\gamma$
By BaBar Collaboration (Bernard Aubert *et al.*).
hep-ex/0306038.
10.1103/PhysRevLett.92.111801.
Phys.Rev.Lett. 92 (2004) 111801.

733. Magda: Manager for Grid Based Data
By Wen-sheng Deng, Torre Wenaus.
physics/0306105.
eConf C0303241 (2003) TUCT010.

734. Measurements of Branching Fractions and *CP*-Violating Asymmetries in $B^0 \to \rho^{\pm} h^{\mp}$ Decays
By BaBar Collaboration (Bernard Aubert *et al.*).
hep-ex/0306030.
10.1103/PhysRevLett.91.201802.
Phys.Rev.Lett. 91 (2003) 201802.

735. The Use of HepRep in GLAST
By J. Perl, R. Giannitrapani, M. Frailis.
cs/0306059 [cs-gr].
eConf C0303241 (2003) THLT009.

736. Interfacing Interactive Data Analysis Tools with the GRID: The PPDG CS-11 Activity
By D.L. Olson, J. Perl.
10.1016/S0168-9002(03)00457-1.
Nucl.Instrum.Meth. A502 (2003) 420-422.

737. Limits on D^0-$\overline{D}^0$ Mixing and *CP* Violation from the Ratio of Lifetimes for Decay to $K^-\pi^+$, K^-K^+ and $\pi^-\pi^+$
By BaBar Collaboration (Bernard Aubert *et al.*).
hep-ex/0306003.
10.1103/PhysRevLett.91.121801.
Phys.Rev.Lett. 91 (2003) 121801.

738. Strangeness from 20 *A*GeV to 158 *A*GeV
By NA49 Collaboration (C. Alt *et al.*).
nucl-ex/0305017.
10.1088/0954-3899/30/1/011.
J.Phys. G30 (2004) S119-S128.

739. Parallel Reconstruction of CLEO III Data
By Gregory J. Sharp, Christopher D. Jones.
hep-ex/0305032.
eConf C0303241 (2003) MODT001.

740. Measurement of the Branching Fractions for the Exclusive Decays of B^0 and B^+ to $\overline{D}^{(*)}D^{(*)}K$
By BaBar Collaboration (Bernard Aubert *et al.*).
hep-ex/0305003.
10.1103/PhysRevD.68.092001.
Phys.Rev. D68 (2003) 092001.

741. A Search for $B^- \to \tau^- \overline{\nu}$ Recoiling Against a Fully Reconstructed B
By BaBar Collaboration (Bernard Aubert *et al.*).
hep-ex/0304030.

742. Observation of a Narrow Meson Decaying to $D_s^+ \pi^0$ at a Mass of 2.32 GeV/c^2
By BaBar Collaboration (B. Aubert *et al.*).
hep-ex/0304021.
10.1103/PhysRevLett.90.242001.
Phys.Rev.Lett. 90 (2003) 242001.

743. A Search for the Decay $B^- \to K^- \nu \bar{\nu}$
     By BaBar Collaboration (Bernard Aubert *et al.*).
     hep-ex/0304020.

744. Rare B Decays into States Containing a J/ψ Meson and a Meson with $s\bar{s}$ Quark Content
     By BaBar Collaboration (Bernard Aubert *et al.*).
     hep-ex/0304014.
     10.1103/PhysRevLett.91.071801.
     Phys.Rev.Lett. 91 (2003) 071801.

745. Search for D^0-$\bar{D}^0$ Mixing and a Measurement of the Doubly Cabibbo-Suppressed Decay Rate in $D^0 \to K\pi$ Decays
     By BaBar Collaboration (Bernard Aubert *et al.*).
     hep-ex/0304007.
     10.1103/PhysRevLett.91.171801.
     Phys.Rev.Lett. 91 (2003) 171801.

746. Measurements of the Branching Fractions and Charge Asymmetries of Charmless Three-Body Charged B Decays
     By BaBar Collaboration (Bernard Aubert *et al.*).
     hep-ex/0304006.
     10.1103/PhysRevLett.91.051801.
     Phys.Rev.Lett. 91 (2003) 051801.

747. Measurements of *CP*-Violating Asymmetries and Branching Fractions in B Meson Decays to $\eta' K$
     By BaBar Collaboration (Bernard Aubert *et al.*).
     hep-ex/0303046.
     10.1103/PhysRevLett.91.161801.
     Phys.Rev.Lett. 91 (2003) 161801.

748. Limits on the Lifetime Difference of Neutral B Mesons and on *CP*, *T*, and *CPT* Violation in $B^0 \bar{B}^0$ Mixing
     By BaBar Collaboration (Bernard Aubert *et al.*).
     hep-ex/0303043.

749. Observation of B Meson Decays to $\omega \pi^+$, ωK^+, and ωK^0
     By BaBar Collaboration (Bernard Aubert *et al.*).
     hep-ex/0303040.

750. Observation of B Meson Decays to $\eta \pi$ and ηK
     By BaBar Collaboration (Bernard Aubert *et al.*).
     hep-ex/0303039.

751. Evidence for $B^+ \to J/\psi \, p \bar{\Lambda}$ and Search for $B^0 \to J/\psi \, p \bar{p}$
     By BaBar Collaboration (Bernard Aubert *et al.*).
     hep-ex/0303036.
     10.1103/PhysRevLett.90.231801.
     Phys.Rev.Lett. 90 (2003) 231801.

752. A Search for $B^+ \to \tau^+ \nu_\tau$ Neutrino Recoiling Against $B^- \to D^0 \ell^- \overline{\nu}_\ell X$
By BaBar Collaboration (Bernard Aubert *et al.*).
hep-ex/0303034.

753. Branching Fractions in $B \to \phi h$ and Search for Direct *CP* Violation in $B^\pm \to \phi K^\pm$
By BaBar Collaboration (Bernard Aubert *et al.*).
hep-ex/0303029.

754. Observation of the Decay $B^\pm \to \pi^\pm \pi^0$, Study of $B^\pm \to K^\pm \pi^0$, and Search for $B^0 \to \pi^0 \pi^0$
By BaBar Collaboration (Bernard Aubert *et al.*).
hep-ex/0303028.
10.1103/PhysRevLett.91.021801.
Phys.Rev.Lett. 91 (2003) 021801.

755. A Device for Precision Neutralization of Electric Charge of Small Drops Using Ionized Air
By Sewan Fan, Peter C. Kim, Eric R. Lee, Irwin T. Lee, Martin L. Perl, Howard Rogers, Dinesh Loomba.
10.1063/1.1606117.
Rev.Sci.Instrum. 74 (2003) 4305-4309.

756. Measurements of the Branching Fractions of Charged B Decays to $K^+ \pi^- \pi^+$ Final States
By BaBar Collaboration (B. Aubert *et al.*).
hep-ex/0303022.

757. Rates, Polarizations, and Asymmetries in Charmless Vector-Vector B Decays
By BaBar Collaboration (Bernard Aubert *et al.*).
hep-ex/0303020.

758. Study of Time-Dependent *CP* Asymmetry in Neutral B Decays to $J/\psi \pi^0$
By BaBar Collaboration (Bernard Aubert *et al.*).
hep-ex/0303018.
10.1103/PhysRevLett.91.061802.
Phys.Rev.Lett. 91 (2003) 061802.

759. Measurement of the Branching Fraction and *CP*-Violating Asymmetries in Neutral B Decays to $D^{*\pm} D^\mp$
By BaBar Collaboration (Bernard Aubert *et al.*).
hep-ex/0303004.
10.1103/PhysRevLett.90.221801.
Phys.Rev.Lett. 90 (2003) 221801.

760. Directed and Elliptic Flow of Charged Pions and Protons in Pb + Pb Collisions at 40 and 158 A GeV
By NA49 Collaboration (C. Alt *et al.*).
nucl-ex/0303001.
10.1103/PhysRevC.68.034903.
Phys.Rev. C68 (2003) 034903.

761. Measurement of $B^0 \to D_s^{(*)+} D^{*-}$ Branching Fractions and $B^0 \to D_s^{*+} D^{*-}$ Polarization with a Partial Reconstruction Technique
By BaBar Collaboration (Bernard Aubert *et al.*).
hep-ex/0302015.
10.1103/PhysRevD.67.092003.
Phys.Rev. D67 (2003) 092003.

762. Measurement of the CKM Matrix Element $|V_{ub}|$ with $B \to \rho e\nu$ Decays
By BaBar Collaboration (Bernard Aubert *et al.*).
hep-ex/0301001.
10.1103/PhysRevLett.90.181801.
Phys.Rev.Lett. 90 (2003) 181801, eConf C0304052 (2003) WG117.

763. Production of Dry Powder Clots Using Piezoelectric Drop Generator
By Valeriy V. Yashchuk, Alexander O. Sushkov, Dmitry Budker, Eric R. Lee, Irwin T. Lee, Martin L. Perl.

764. Large Bulk Matter Search for Fractional Charge Particles
By Irwin T. Lee.

765. Simultaneous Measurement of the B^0 Meson Lifetime and Mixing Frequency with $B^0 \to D^{*-} \ell^+ \nu_\ell$ Decays
By BaBar Collaboration (Bernard Aubert *et al.*).
hep-ex/0212017.
10.1103/PhysRevD.67.072002.
Phys.Rev. D67 (2003) 072002.

766. Measurement of the B^0 Meson Lifetime with Partial Reconstruction of $B^0 \to D^{*-}\pi^+$ and $B^0 \to D^{*-}\rho^+$ Decays
By BaBar Collaboration (Bernard Aubert *et al.*).
hep-ex/0212012.
10.1103/PhysRevD.67.091101.
Phys.Rev. D67 (2003) 091101.

767. A Study of the Rare Decays $B^0 \to D_s^{(*)+}\pi^-$ and $B^0 \to D_s^{(*)-}K^+$
By BaBar Collaboration (Bernard Aubert *et al.*).
hep-ex/0211053.
10.1103/PhysRevLett.90.181803.
Phys.Rev.Lett. 90 (2003) 181803.

768. Bose–Einstein Correlations of Charged Kaons in Central Pb + Pb Collisions at $E_{\text{beam}} = 158$ AGeV
By NA49 Collaboration (S.V. Afanasiev *et al.*).
nucl-ex/0210018.
10.1016/S0370-2693(03)00102-3.
Phys.Lett. B557 (2003) 157-166.

769. The Practice of Experimental Physics: Recollections, Reflections, and Interpretations
By Martin L. Perl, Mary A. Meyer.
Theor.Histor.Scient. 6 (2002) 206-239.

770. System Size Dependence of Strangeness Production at 158 AGeV
By NA49 Collaboration (C. Hohne *et al.*).
nucl-ex/0209018.
10.1016/S0375-9474(02)01453-7.
Nucl.Phys. A715 (2003) 474-477.

771. A Measurement of the $B^0 \rightarrow J/\psi \pi^+ \pi^-$ Branching Fraction
By BaBar Collaboration (Bernard Aubert *et al.*).
hep-ex/0209013.
10.1103/PhysRevLett.90.091801.
Phys.Rev.Lett. 90 (2003) 091801.

772. Monitoring the Health and Safety of the Acis Instrument On-Board the Chandra X-Ray Observatory
By Shanil N. Virani, Peter G. Ford, Joseph M. DePasquale, Paul P. Plucinsky.
astro-ph/0209124.
10.1117/12.460672.
Proc.SPIE Int.Soc.Opt.Eng. 4844 (2002) 464.

773. Energy Dependence of Λ and $\overline{\Lambda}$ Production at CERN SPS Energies
By NA49 Collaboration (A. Mischke *et al.*).
nucl-ex/0209002.
10.1016/S0375-9474(02)01433-1.
Nucl.Phys. A715 (2003) 453-457.

774. Results on Correlations and Fluctuations from NA49
By NA49 Collaboration (C. Blume *et al.*).
nucl-ex/0208020.
10.1016/S0375-9474(02)01413-6.
Nucl.Phys. A715 (2003) 55-64.

775. Recent Results on Spectra and Yields from NA49
By S.V. Afanasiev *et al.*.
nucl-ex/0208014.
10.1016/S0375-9474(02)01424-0.
Nucl.Phys. A715 (2003) 161-170.

776. GEANT4: A Simulation Toolkit
By GEANT4 Collaboration (S. Agostinelli *et al.*).
10.1016/S0168-9002(03)01368-8.
Nucl.Instrum.Meth. A506 (2003) 250-303.

777. A Measurement of the $B^0 \rightarrow J/\psi \pi^+ \pi^-$ Branching Fraction
By BaBar Collaboration (Bernard Aubert *et al.*).
hep-ex/0203034.

778. Study of Inclusive Production of Charmonium Mesons in B Decay
     By BaBar Collaboration (Bernard Aubert *et al.*).
     hep-ex/0207097.
     10.1103/PhysRevD.67.032002.
     Phys.Rev. D67 (2003) 032002.

779. Measurement of the Branching Fraction for Inclusive Semileptonic B Meson Decays
     By BaBar Collaboration (Bernard Aubert *et al.*).
     hep-ex/0208018.
     10.1103/PhysRevD.67.031101.
     Phys.Rev. D67 (2003) 031101.

780. Production of Dry Powder Clots Using a Piezoelectric Drop Generator
     By Valeriy Yashchuk, Alexander O. Sushkov, Dmitry Budker, Eric R. Lee, Irwin T. Lee, Martin
     L. Perl.

781. Measurement of Branching Fractions of Color Suppressed Decays of the $\bar{B}^0$ Meson to $D^0\pi^0$,
     $D^0\eta$, and $D^0\omega$
     By BaBar Collaboration (Bernard Aubert *et al.*).
     hep-ex/0207092.

782. Dalitz Plot Analysis of D^0 Hadronic Decays $D^0 \to K^0 K^- \pi^+$, $D^0 \to \bar{K}^0 K^+ \pi^-$ and $D^0 \to \bar{K}^0 K^+ K^-$
     By BaBar Collaboration (Bernard Aubert *et al.*).
     hep-ex/0207089.

783. Measurement of Branching Ratios and CP Asymmetries in $B^- \to D^0_{(CP)} K^-$ Decays
     By BaBar Collaboration (Bernard Aubert *et al.*).
     hep-ex/0207087.

784, Measurement of the Branching Fractions for the Exclusive Decays of B^0 and B^+ to $\bar{D}^{(*)}D^{(*)}K$
     By BaBar Collaboration (Bernard Aubert *et al.*).
     hep-ex/0207086.

785. Measurement of the $B^0 \to D^{*-} a_1^+$ Branching Fraction with Partially Reconstructed D^*
     By BaBar Collaboration (Bernard Aubert *et al.*).
     hep-ex/0207085.

786. Measurement of the First Hadronic Spectral Moment from Semileptonic B Decays
     By BaBar Collaboration (Bernard Aubert *et al.*).
     hep-ex/0207084.

787. Search for Decays of B^0 Mesons into Pairs of Leptons
     By BaBar Collaboration (Bernard Aubert *et al.*).
     hep-ex/0207083.

788. Evidence for the Flavor Changing Neutral Current Decays $B \to K\ell^+\ell^-$ and $B \to K^*\ell^+\ell^-$
     By BaBar Collaboration (Bernard Aubert *et al.*).
     hep-ex/0207082.

789. Measurement of the Inclusive Electron Spectrum in Charmless Semileptonic B Decays Near the Kinematic Endpoint
By BaBar Collaboration (Bernard Aubert *et al.*).
hep-ex/0207081.

790. Measurement of the CKM Matrix Element $|V_{ub}|$ with Charmless Exclusive Semileptonic B Meson Decays at BABAR
By BaBar Collaboration (Bernard Aubert *et al.*).
hep-ex/0207080.

791. Measurement of $B^0 \to D_s^{(*)+} D^{*-}$ Branching Fractions and Polarization in the Decay $B^0 \to D_s^{*+} D^{*-}$ with a Partial Reconstruction Technique
By BaBar Collaboration (Bernard Aubert *et al.*).
hep-ex/0207079.

792. Determination of the Branching Fraction for Inclusive Decays $B \to X_s \gamma$
By BaBar Collaboration (Bernard Aubert *et al.*).
hep-ex/0207076.

793. $b \to s\gamma$ Using a Sum of Exclusive Modes
By BaBar Collaboration (Bernard Aubert *et al.*).
hep-ex/0207074.

794. Search for the Exclusive Radiative Decays $B \to \rho\gamma$ and $B^0 \to \omega\gamma$
By BaBar Collaboration (Bernard Aubert *et al.*).
hep-ex/0207073.

795. Measurement of Time-Dependent CP Asymmetries and the CP-Odd Fraction in the Decay $B^0 \to D^{*+}D^{*-}$
By BaBar Collaboration (Bernard Aubert *et al.*).
hep-ex/0207072.

796. Simultaneous Measurement of the B^0 Meson Lifetime and Mixing Frequency with $B^0 \to D^{*-}\ell^+\nu_\ell$ Decays
By BaBar Collaboration (Bernard Aubert *et al.*).
hep-ex/0207071.

797. Measurement of $\sin 2\beta$ in $B^0 \to \phi K_s^0$
By BaBar Collaboration (Bernard Aubert *et al.*).
hep-ex/0207070.

798. A Search for $B^+ \to K^+ \nu\bar{\nu}$
By BaBar Collaboration (Bernard Aubert *et al.*).
hep-ex/0207069.

799. Search for CP Violation in $B^0/\bar{B}^0$ Decays to $\pi^+\pi^-\pi^0$ and $K^\pm\pi^\mp\pi^0$ in Regions Dominated by the $\rho^\pm$ Resonance
By BaBar Collaboration (Bernard Aubert *et al.*).
hep-ex/0207068.
Submitted to: Phys.Rev.Lett..

800. Measurement of the Branching Fraction for $B^\pm \to \chi_{c0} K^\pm$
By BaBar Collaboration (Bernard Aubert *et al.*).
hep-ex/0207066.

801. Measurements of Branching Fractions and Direct *CP* Asymmetries in $\pi^+\pi^0$, $K^+\pi^0$ and $K^0\pi^0$
B Decays
By BaBar Collaboration (Bernard Aubert *et al.*).
hep-ex/0207065.

802. A Search for the Decay $B^0 \to \pi^0\pi^0$
By BaBar Collaboration (Bernard Aubert *et al.*).
hep-ex/0207063.

803. A Study of Time-Dependent *CP* Asymmetry in $B^0 \to J/\psi\pi^0$ Decays
By BaBar Collaboration (Bernard Aubert *et al.*).
hep-ex/0207058.

804. Measurements of Branching Fractions and *CP*-Violating Asymmetries in $B^0 \to \pi^+\pi^-$, $K^+\pi^-$,
K^+K^- Decays
By BaBar Collaboration (Bernard Aubert *et al.*).
hep-ex/0207055.
10.1103/PhysRevLett.89.281802.
Phys.Rev.Lett. 89 (2002) 281802.

805. A Study of the Rare Decays $B^0 \to D_s^{(*)+}\pi^-$ and $B^0 \to D_s^{(*)-}K^+$
By BaBar Collaboration (Bernard Aubert *et al.*).
hep-ex/0207053.

806. Measurement of the *CP* Asymmetry Amplitude sin 2β
By BaBar Collaboration (Bernard Aubert *et al.*).
hep-ex/0207042.
10.1103/PhysRevLett.89.201802.
Phys.Rev.Lett. 89 (2002) 201802.

807. Measurements of Charmless Two-Body Charged *B* Decays with Neutral Pions and Kaons
By BaBar Collaboration (Bernard Aubert *et al.*).
hep-ex/0206053.

808. Search for Deconfinement in NA49 at the CERN SPS
By Peter Seyboth *et al.*.
hep-ex/0206046.
10.1007/BF02705171.
Heavy Ion Phys. 15 (2002) 257-268, Pramana 60 (2003) 725-738.

809. Measurements of the Branching Fractions of Charmless Three-Body Charged *B* Decays
By BaBar Collaboration (Bernard Aubert *et al.*).
hep-ex/0206004.

810. Evidence for the $b \to u$ Transition $B^0 \to D_s^+ \pi^-$ and a Search for $B^0 \to D_s^{*+} \pi^-$
By BaBar Collaboration (Bernard Aubert *et al.*).
hep-ex/0205102.

811. Measurements of Branching Fractions and *CP*-Violating Asymmetries in $B^0 \to \pi^+\pi^-$, $K^+\pi^-$, K^+K^- Decays
By BaBar Collaboration (Bernard Aubert *et al.*).
hep-ex/0205082.

812. Energy Dependence of Pion and Kaon Production in Central Pb + Pb Collisions
By NA49 Collaboration (S.V. Afanasiev *et al.*).
nucl-ex/0205002.
10.1103/PhysRevC.66.054902.
Phys.Rev. C66 (2002) 054902.

813. Large Bulk Matter Search for Fractional Charge Particles
By Irwin T. Lee *et al.*.
hep-ex/0204003.
10.1103/PhysRevD.66.012002.
Phys.Rev. D66 (2002) 012002.

814. Branching Fraction Measurements of the Decays $B \to \eta_c K$, Where $\eta_c \to K\bar{K}\pi$ and $\eta_c \to 4K$
By BaBar Collaboration (Bernard Aubert *et al.*).
hep-ex/0203040.

815. A Measurement of the Neutral B Meson Lifetime Using Partially Reconstructed $B^0 \to D^{*-}\pi^+$ Decays
By BaBar Collaboration (Bernard Aubert *et al.*).
hep-ex/0203038.

816. Measurement of the B^0 Lifetime with Partial Reconstruction of $\bar{B}^0 \to D^{*+}\rho^-$
By BaBar Collaboration (Bernard Aubert *et al.*).
hep-ex/0203036.

817. Rare B Decays to States Containing a J/ψ Meson
By BaBar Collaboration (Bernard Aubert *et al.*).
hep-ex/0203035.

818. Progress Report and Beam Request for 2002
By NA49 Collaboration (S.V. Afanasev *et al.*).

819. New Results from NA49
By NA49 Collaboration (S.V. Afanasev *et al.*).
10.1016/S0375-9474(01)01353-7.
Nucl.Phys. A698 (2002) 104-111.

820. Measurement of the Branching Fraction and *CP* Content for the Decay $B^0 \to D^{*+}D^{*-}$
      By BaBar Collaboration (B. Aubert *et al.*).
      hep-ex/0203008.
      10.1103/PhysRevLett.89.061801.
      Phys.Rev.Lett. 89 (2002) 061801.

821. Improved Measurement of the *CP*-Violating Asymmetry Amplitude sin 2β
      By BaBar Collaboration (Bernard Aubert *et al.*).
      hep-ex/0203007.

822. Search for *T* and *CP* Violation in B^0-$\bar{B}^0$ Mixing with Inclusive Dilepton Events
      By BaBar Collaboration (Bernard Aubert *et al.*).
      hep-ex/0202041.
      10.1103/PhysRevLett.88.231801.
      Phys.Rev.Lett. 88 (2002) 231801.

823. Ξ and $\bar{\Xi}^+$ Production in Central Pb + Pb Collisions at 158 GeV/c per Nucleon
      By NA49 Collaboration (S.V. Afanasiev *et al.*).
      hep-ex/0202037.
      10.1016/S0370-2693(02)01970-6.
      Phys.Lett. B538 (2002) 275-281.

824. Measurement of the B^0 Lifetime with Partially Reconstructed $B^0 \to D^{*-}\ell^+\nu_\ell$ Decays
      By BaBar Collaboration (Bernard Aubert *et al.*).
      hep-ex/0202005.
      10.1103/PhysRevLett.89.011802.
      Phys.Rev.Lett. 89 (2002) 011802, Phys.Rev.Lett. 89 (2002) 169903.

825. Measurement of D_s^+ and D_s^{*+} Production in *B* Meson Decays and from Continuum e^+e^-
      Annihilation at $\sqrt{s} = 10.6$ GeV
      By BaBar Collaboration (Bernard Aubert *et al.*).
      hep-ex/0201041.
      10.1103/PhysRevD.65.091104.
      Phys.Rev. D65 (2002) 091104.

826. Lambda Production in Central Pb + Pb Collisions at CERN-SPS Energies
      By NA49 Collaboration (A. Mischke *et al.*).
      nucl-ex/0201012.
      10.1088/0954-3899/28/7/330.
      J.Phys. G28 (2002) 1761-1768.

827. A Study of Time Dependent *CP*-Violating Asymmetries and Flavor Oscillations in Neutral *B*
      Decays at the $\Upsilon(4S)$
      By BaBar Collaboration (Bernard Aubert *et al.*).
      hep-ex/0201020.
      10.1103/PhysRevD.66.032003.
      Phys.Rev. D66 (2002) 032003.

828. Search for the Rare Decays $B \to K\ell^+\ell^-$ and $B \to K^*\ell^+\ell^-$
By BaBar Collaboration (Bernard Aubert *et al.*).
hep-ex/0201008.
10.1103/PhysRevLett.88.241801.
Phys.Rev.Lett. 88 (2002) 241801.

829. Strange Beauty: Murray Gell-Mann and the Revolution of Twentieth Century Physics
By M.L. Perl.
Phys.Perspect. 3 (2002) 130-131.

830. Remarks on Search Methods for Stable, Massive, Elementary Particles
By Merlin L. Perl.
10.1063/1.1426799.
AIP Conf.Proc. 596 (2001) 156-168.

831. JavaProcMan: A Java Based Process Management System
By Joseph Perl.

832. The FreeHEP Java Library
By Gary Brower, Mark Donszelmann, Julius Hrivnac, Tony Johnson, Cal Loomis, Joseph Perl

833. Converting Equipment Control Software from Pascal to C/C++
By Ludwig Hechler.
physics/0111056.
eConf C011127 (2001) WEAP034.

834. Beam Dynamics Simulations for a DC Gun Based Injector for PERL
By I. Ben-Zvi, X. Wang, F. Zhou.
Conf.Proc. C0106181 (2001) 2266-2268.

835. Measurement of the B^0-$\overline{B}^0$ Oscillation Frequency with Inclusive Dilepton Events
By BaBar Collaboration (Bernard Aubert *et al.*).
hep-ex/0112045.
10.1103/PhysRevLett.88.221803.
Phys.Rev.Lett. 88 (2002) 221803.

836. Measurement of B^0-$\overline{B}^0$ Flavor Oscillations in Hadronic B^0 Decays
By BaBar Collaboration (Bernard Aubert *et al.*).
hep-ex/0112044.
10.1103/PhysRevLett.88.221802.
Phys.Rev.Lett. 88 (2002) 221802.

837. Direct *CP* Violation Searches in Charmless Kadronic *B* Meson Decays
By BaBar Collaboration (Bernard Aubert *et al.*).
hep-ex/0111087.
10.1103/PhysRevD.65.051101.
Phys.Rev. D65 (2002) 051101.

838. A Prototype of the UAL 2.0 Application Toolkit
      By N. Malitsky, J. Smith, J. Wei.
      physics/0111096.
      eConf C011127 (2001) THAP013.

839. Generating EPICS IOC Databases from a Relational Database — A Different Approach
      By Rolf Keitel.
      physics/0111048.
      eConf C011127 (2001) WEAP071.

840. Measurement of $B \to K^* \gamma$ Branching Fractions and Charge Asymmetries
      By BaBar Collaboration (Bernard Aubert *et al.*).
      hep-ex/0110065.
      10.1103/PhysRevLett.88.101805.
      Phys.Rev.Lett. 88 (2002) 101805.

841. Study of *CP*-Violating Asymmetries in $B^0 \to \pi^+\pi^-$, $K^+\pi^-$ Decays
      By BaBar Collaboration (Bernard Aubert *et al.*).
      hep-ex/0110062.
      10.1103/PhysRevD.65.051502.
      Phys.Rev. D65 (2002) 051502.

842. Study of Semi-Inclusive Production of η' Mesons in B Decays
      By BaBar Collaboration (Bernard Aubert *et al.*).
      hep-ex/0109034.

843. Study of $B \to D^{(*)}\overline{D}^{(*)}$ Decays with the BaBar Detector
      By BaBar Collaboration (Bernard Aubert *et al.*).
      hep-ex/0109009.

844. Search for a Lifetime Difference in D^0 Decays
      By BaBar Collaboration (Bernard Aubert *et al.*).
      hep-ex/0109008.

845. Measurement of the Branching Fraction for $B^+ \to K^{*0}\pi^+$
      By BaBar Collaboration (Bernard Aubert *et al.*).
      hep-ex/0109007.

846. Search for Direct *CP* Violation in Quasi Two-Body Charmless B Decays
      By BaBar Collaboration (Bernard Aubert *et al.*).
      hep-ex/0109006.

847. Search for B Decays into $K^0\overline{K}^0$
      By BaBar Collaboration (Bernard Aubert *et al.*).
      hep-ex/0109005.

848. Measurement of the Branching Fractions for $\psi(2S) \to e^+e^-$ and $\psi(2S) \to \mu^+\mu^-$
By BaBar Collaboration (Bernard Aubert *et al.*).
hep-ex/0109004.
10.1103/PhysRevD.65.031101.
Phys.Rev. D65 (2002) 031101.

849. CLEO-c and CESR-c: A New Frontier of Weak and Strong Interactions
By CLEO Collaboration (Roy A. Briere *et al.*).

850. Measurements of the Branching Fractions of Exclusive Charmless B Meson Decays with η' or ω Mesons
By BaBar Collaboration (Bernard Aubert *et al.*).
hep-ex/0108017.
10.1103/PhysRevLett.87.221802.
Phys.Rev.Lett. 87 (2001) 221802.

851. A Study of $B^\pm \to J/\psi\pi^\pm$ and $B^\pm \to J/\psi K^\pm$ Decays: Measurement of the Ratio of Branching Fractions and Search for Direct CP-Violating Charge Asymmetries
By BaBar Collaboration (Bernard Aubert *et al.*).
hep-ex/0108009.
10.1103/PhysRevD.65.091101.
Phys.Rev. D65 (2002) 091101.

852. Search for $B^0 \to a_0^+(980)\pi^-$
By BaBar Collaboration (Bernard Aubert *et al.*).
hep-ex/0107075.

853. Study of CP-Violating Asymmetries in $B \to \pi^\pm\pi^\mp$, $K^\pm\pi^\mp$ Decays
By BaBar Collaboration (Bernard Aubert *et al.*).
hep-ex/0107074.

854. Search for the Decay $B^0 \to \gamma\gamma$
By BaBar Collaboration (Bernard Aubert *et al.*).
hep-ex/0107068.
10.1103/PhysRevLett.87.241803.
Phys.Rev.Lett. 87 (2001) 241803.

855. Measurement of D_s^+ and D_s^{*+} Production in B Meson Decays and from Continuum e^+e^- Annihilations at $\sqrt{s} = 10.6$ GeV
By Collaboration (B. Aubert *et al.*).
hep-ex/0107060.

856. Study of T and CP Violation in $B^0\bar{B}^0$ Mixing with Inclusive Dilepton Events
By BaBar Collaboration (Bernard Aubert *et al.*).
hep-ex/0107059.

857. Measurements of B^0 Decays to $\pi^+\pi^-\pi^0$
By BaBar Collaboration (Bernard Aubert *et al.*).
hep-ex/0107058.

858. Measurement of the Branching Fraction for the Decay $B^0 \to D^{*+}D^{*-}$
By BaBar Collaboration (Bernard Aubert *et al.*).
hep-ex/0107057.

859. Investigation of $B \to D^{(*)}\overline{D}^{(*)}K$ Decays with the BaBar Detector
By BaBar Collaboration (Bernard Aubert *et al.*).
hep-ex/0107056.

860. Measurement of the $B \to J/\psi K^*(892)$ Decay Amplitudes
By BaBar Collaboration (Bernard Aubert *et al.*).
hep-ex/0107049.
10.1103/PhysRevLett.87.241801.
Phys.Rev.Lett. 87 (2001) 241801.

861. Measurement of the Exclusive Branching Fractions $B^0 \to \eta K^{*0}$ and $B^+ \to \eta K^{*+}$
By BaBar Collaboration (Bernard Aubert *et al.*).
hep-ex/0107037.

862. Measurement of the $B^0\overline{B}^0$ Oscillation Frequency in Hadronic B^0 Decays
By BaBar Collaboration (Bernard Aubert *et al.*).
hep-ex/0107036.

863. Search for the Rare Decays $B \to K\ell^+\ell^-$ and $B \to K^*(892)\ell^+\ell^-$
By BaBar Collaboration (Bernard Aubert *et al.*).
hep-ex/0107026.

864. Measurement of Branching Fractions for Exclusive B Decays to Charmonium Final States
By BaBar Collaboration (Bernard Aubert *et al.*).
hep-ex/0107025.
10.1103/PhysRevD.65.032001.
Phys.Rev. D65 (2002) 032001.

865. Measurement of the B^0 and B^+ Meson Lifetimes with Fully Reconstructed Hadronic Final States
By BaBar Collaboration (Bernard Aubert *et al.*).
hep-ex/0107019.
10.1103/PhysRevLett.87.201803.
Phys.Rev.Lett. 87 (2001) 201803.

866. Observation of *CP* Violation in the B^0 Meson System
By BaBar Collaboration (Bernard Aubert *et al.*).
hep-ex/0107013.
10.1103/PhysRevLett.87.091801.
Phys.Rev.Lett. 87 (2001) 091801.

867. Measurement of J/ψ Production in Continuum e^+e^- Annihilations Near $\sqrt{s} = 10.6$ GeV
By BaBar Collaboration (Bernard Aubert *et al.*).
hep-ex/0106044.
10.1103/PhysRevLett.87.162002.
Phys.Rev.Lett. 87 (2001) 162002.

868. The BaBar Detector
By BaBar Collaboration (Bernard Aubert *et al.*).
hep-ex/0105044.
10.1016/S0168-9002(01)02012-5.
Nucl.Instrum.Meth. A479 (2002) 1-116.

869. Status and Future of the NA49 Program on Nucleus Nucleus Collisions at Low SPS Energies
(Addendum 9 to proposal CERN/SPSLC/P264)
By NA49 Collaboration (S.V. Afanasev *et al.*).

870. The Search for Stable, Massive, Elementary Particles
By Martin L. Perl, Peter C. Kim, Valerie Halyo, Eric R. Lee, Irwin T. Lee, Dinesh Loomba,
Klaus S. Lackner.
hep-ex/0102033.
10.1142/S0217751X01003548.
Int.J.Mod.Phys. A16 (2001) 2137-2164.

871. Measurement of *CP*-Violating Asymmetries in B^0 Decays to *CP* Eigenstates
By BaBar Collaboration (Bernard Aubert *et al.*).
hep-ex/0102030.
10.1103/PhysRevLett.86.2515.
Phys.Rev.Lett. 86 (2001) 2515-2522.

872. Observation of Radiative Leptonic Decay of the Tau Lepton
By Xiao-fan Zhou.

873. Rare Decays of η and η' Mesons at CLEO-II
By Dennis Wayne Ugolini.

874. WIRED — World Wide Web Interactive Remote Event Display
By A. Ballaminut *et al.*.
10.1016/S0010-4655(01)00277-6.
Comput.Phys.Commun. 140 (2001) 1-2, 266-273.

875. HepRep: A Generic Interface Definition for HEP Event Display Representables
By J. Perl.

876. A Search for Free Fractional Electric Charge Elementary Particles
By Valerie Halyo.

877. A Measurement of the Decay Asymmetry Parameters in $\Xi_c^0 \to \Xi^- \pi^+$
By CLEO Collaboration (S. Chan *et al.*).
hep-ex/0011073.
10.1103/PhysRevD.63.111102.
Phys.Rev. D63 (2001) 111102.

878. Measurement of the Λ_c^+ Lifetime
By CLEO Collaboration (A.H. Mahmood *et al.*).
hep-ex/0011049.
10.1103/PhysRevLett.86.2232.
Phys.Rev.Lett. 86 (2001) 2232-2236.

879. Study of $B \to \psi(2S) K$ and $B \to \psi(2S) K^*(892)$ Decays
By CLEO Collaboration (S.J. Richichi *et al.*).
hep-ex/0010036.
10.1103/PhysRevD.63.031103.
Phys.Rev. D63 (2001) 031103.

880. Search for Direct *CP* Violation in Ξ Hyperon Decay
By CLEO Collaboration (D.E. Jaffe *et al.*).
hep-ex/0009037.

881. First Observation of the Decays $B^0 \to D^{*-} p\bar{p}\pi^+$ and $B^0 \to D^{*-} p\bar{n}$
By CLEO Collaboration (S. Anderson *et al.*).
hep-ex/0009011.
10.1103/PhysRevLett.86.2732.
Phys.Rev.Lett. 86 (2001) 2732-2736.

882. Observation of $B \to K^\pm \pi^0$ and $B \to K^0 \pi^0$, and Evidence for $B \to \pi^+\pi^-$
By CLEO Collaboration (D. Cronin-Hennessy *et al.*).
10.1103/PhysRevLett.85.515.
Phys.Rev.Lett. 85 (2000) 515-519.

883. Measurements of $B \to D_S^{(*)+} D^{*(*)}$ Branching Fractions
By CLEO Collaboration (S. Ahmed *et al.*).
hep-ex/0008015.
10.1103/PhysRevD.62.112003.
Phys.Rev. D62 (2000) 112003.

884. A Search for $B \to \tau\nu$
By CLEO Collaboration (T.E. Browder *et al.*).
hep-ex/0007057.
10.1103/PhysRevLett.86.2950.
Phys.Rev.Lett. 86 (2001) 2950-2954.

885. Determination of the $B \to D^* \ell\nu$ Decay Width and $|V_{cb}|$
By CLEO Collaboration (J.P. Alexander *et al.*).
hep-ex/0007052.

886. Evidence of New States Decaying into $\Xi_c' \pi$
By CLEO Collaboration (P. Avery *et al.*).
hep-ex/0007050.

887. Observation of New States Decaying into $\Lambda_c^+\pi^-\pi^+$
By CLEO Collaboration (P. Avery *et al.*).
hep-ex/0007049.

888. Observation of the Ω_c^0 Charmed Baryon at CLEO
By CLEO Collaboration (S. Ahmed *et al.*).
hep-ex/0007047.
Int.J.Mod.Phys. A16S1B (2001) 505-507.

889. Study of χ_{c1} and χ_{c2} Meson Production in B Meson Decays
By CLEO Collaboration (G. Brandenburg *et al.*).
hep-ex/0007046.

890. Search for Decays of B^0 Mesons into Pairs of Leptons: $B^0 \to e^+e^-$, $B^0 \to \mu^+\mu^-$ and $B^0 \to e^\pm\mu^\mp$
By CLEO Collaboration (T. Bergfeld *et al.*).
hep-ex/0007042.
10.1103/PhysRevD.62.091102.
Phys.Rev. D62 (2000) 091102.

891. First Observation of the Σ_c^{*+} Baryon and a New Measurement of the Σ_c^+ Mass
By CLEO Collaboration (R. Ammar *et al.*).
hep-ex/0007041.
10.1103/PhysRevLett.86.1167.
Phys.Rev.Lett. 86 (2001) 1167-1170.

892. Study of B Decays to Charmonium States $B \to \eta_c K$ and $B \to \chi_{c0} K$
By CLEO Collaboration (K.W. Edwards *et al.*).
hep-ex/0007012.
10.1103/PhysRevLett.86.30.
Phys.Rev.Lett. 86 (2001) 30-34.

893. Measurements of the Mass, Total Width and Two Photon Partial Width of the η_c Meson
By CLEO Collaboration (G. Brandenburg *et al.*).
hep-ex/0006026.
10.1103/PhysRevLett.85.3095.
Phys.Rev.Lett. 85 (2000) 3095-3099.

894. $B \to D^*\pi^+\pi^-\pi^-\pi^0$, $D^{(*)}\omega\pi^-$ and the Observation of a Wide 1^- $\omega\pi^-$ Enhancement at 1418 MeV
By CLEO Collaboration (M. Artuso *et al.*).
hep-ex/0006018.

895. Study of Charmless Hadronic B Meson Decays to Pseudoscalar Vector Final States
By CLEO Collaboration (C.P. Jessop *et al.*).
hep-ex/0006008.
10.1103/PhysRevLett.85.2881.
Phys.Rev.Lett. 85 (2000) 2881-2885.

896. Measurement of the Relative Branching Fraction of $\Upsilon(4S)$ to Charged and Neutral B-Meson Pairs
      By CLEO Collaboration (J.P. Alexander *et al.*).
      hep-ex/0006002.
      10.1103/PhysRevLett.86.2737.
      Phys.Rev.Lett. 86 (2001) 2737-2741.

897. Precise Measurement of $B^0 - \bar{B}^0$ Mixing Parameters at the $\Upsilon(4S)$
      By CLEO Collaboration (B.H. Behrens *et al.*).
      hep-ex/0005013.
      10.1016/S0370-2693(00)00990-4.
      Phys.Lett. B490 (2000) 36-44.

898. Measurement of the Product Branching Fraction $\mathcal{B}(c \to \Theta_c X) \cdot \mathcal{B}(\Theta_c \to \Lambda X)$
      By CLEO Collaboration (R. Ammar *et al.*).
      hep-ex/0004033.
      10.1103/PhysRevD.62.092007.
      Phys.Rev. D62 (2000) 092007.

899. Study of Exclusive Two-Body B^0 Meson Decays to Charmonium
      By CLEO Collaboration (P. Avery *et al.*).
      hep-ex/0004032.
      10.1103/PhysRevD.62.051101.
      Phys.Rev. D62 (2000) 051101.

900. Measurements of Charm Fragmentation into D_s^{*+} and D_s^+ in e^+e^- Annihilations at $\sqrt{s} = 10.5$ GeV
      By CLEO Collaboration (Roy A. Briere *et al.*).
      hep-ex/0004028.
      10.1103/PhysRevD.62.072003.
      Phys.Rev. D62 (2000) 072003.

901. Resonance Structure of $\tau^- \to K^-\pi^+\pi^-\nu_\tau$ Decays
      By CLEO Collaboration (D.M. Asner *et al.*).
      hep-ex/0004002.
      10.1103/PhysRevD.62.072006.
      Phys.Rev. D62 (2000) 072006.

902. Measurement of $\mathcal{B}(\Lambda_c^+ \to pK^-\pi^+)$
      By CLEO Collaboration (D.E. Jaffe *et al.*).
      hep-ex/0004001.
      10.1103/PhysRevD.62.072005.
      Phys.Rev. D62 (2000) 072005.

903. Search for *CP* Violation in $B^\pm \to J/\psi K^\pm$ and $B^\pm \to \psi(2S) K^\pm$ Decays
      By CLEO Collaboration (G. Bonvicini *et al.*).
      hep-ex/0003004.
      10.1103/PhysRevLett.84.5940.
      Phys.Rev.Lett. 84 (2000) 5940.

904. Study of the Decays $B^0 \to D^{(*)+}D^{(*)-}$
By CLEO Collaboration (Elliot David Lipeles *et al.*).
hep-ex/0002065.
10.1103/PhysRevD.62.032005.
Phys.Rev. D62 (2000) 032005.

905. From Accelerators to Asteroids
By Martin L. Perl.
hep-ex/0002001.

906. Measurement of the B^0 and B^+ Meson Masses from $B^0 \to \psi^{(\prime)}K_S^0$ and $B^+ \to \psi^{(\prime)}K^+$ Decays
By CLEO Collaboration (S.E. Csorna *et al.*).
hep-ex/0001013.
10.1103/PhysRevD.61.111101.
Phys.Rev. D61 (2000) 111101.

907. Study of Two-Body B Decays to Kaons and Pions: Observation of $B \to \pi^+\pi^-$, $B \to K^\pm\pi^0$, and $B \to K^0\pi^0$ Decays
By CLEO Collaboration (D. Cronin-Hennessy *et al.*).
hep-ex/0001010.

908. Measurement of Charge Asymmetries in Charmless Hadronic in B Meson Decays
By CLEO Collaboration (S. Chen *et al.*).
hep-ex/0001009.
10.1103/PhysRevLett.85.525.
Phys.Rev.Lett. 85 (2000) 525-529.

909. Search for $D^0 - \overline{D}^0$ Mixing
By CLEO Collaboration (R. Godang *et al.*).
hep-ex/0001060.
10.1103/PhysRevLett.84.5038.
Phys.Rev.Lett. 84 (2000) 5038-5042.

910. The Search for Fractional Charge Elementary Particles and Very Massive Particles in Bulk Matter
By Martin L. Perl, Valerie Halyo, Peter C. Kim, Eric R. Lee, Irwin T. Lee, Dinesh Loomba.
hep-ex/0001025.

911. Search for the Decay $\overline{B}^0 \to D^{*0}\gamma$
By CLEO Collaboration (M. Artuso *et al.*).
hep-ex/0001002.
10.1103/PhysRevLett.84.4292.
Phys.Rev.Lett. 84 (2000) 4292-4295.

912. Two-Body B Meson Decays to η and η': Observation of $B \to \eta K^*$
By CLEO Collaboration (S.J. Richichi *et al.*).
hep-ex/9912059.
10.1103/PhysRevLett.85.520.
Phys.Rev.Lett. 85 (2000) 520-524.

913. Study of Exclusive Radiative B Meson Decays
      By CLEO Collaboration (T.E. Coan *et al.*).
      hep-ex/9912057.
      10.1103/PhysRevLett.84.5283.
      Phys.Rev.Lett. 84 (2000) 5283-5287.

914. A New Method for Searching for Free Fractional Charge Particles in Bulk Matter
      By Dinesh Loomba, Valerie Halyo, Eric R. Lee, Irwin T. Lee, Peter C. Kim, Martin L. Perl.
      hep-ex/0001026.
      10.1063/1.1308268.
      Rev.Sci.Instrum. 71 (2000) 3409-3414.

915. Search for Free Fractional Electric Charge Elementary Particles
      By V. Halyo, P. Kim, E.R. Lee, I.T. Lee, D. Loomba, M.L. Perl.
      hep-ex/9910064.
      10.1103/PhysRevLett.84.2576.
      Phys.Rev.Lett. 84 (2000) 2576-2579.

916. Update of the Search for the Neutrinoless Decay $\tau \to \mu\gamma$
      By CLEO Collaboration (S. Ahmed *et al.*).
      hep-ex/9910060.
      10.1103/PhysRevD.61.071101.
      Phys.Rev. D61 (2000) 071101.

917. Hadronic Structure in the Decay $\tau^- \to \pi^-\pi^0\nu_\tau$
      By CLEO Collaboration (S. Anderson *et al.*).
      hep-ex/9910046.
      10.1103/PhysRevD.61.112002.
      Phys.Rev. D61 (2000) 112002.

918. Observation of Radiative Leptonic Decay of the τ Lepton
      By CLEO Collaboration (T. Bergfeld *et al.*).
      hep-ex/9909050.
      10.1103/PhysRevLett.84.830.
      Phys.Rev.Lett. 84 (2000) 830-834.

919. Search for $D^0 - \bar{D}^0$ Mixing
      By CLEO Collaboration (M. Artuso *et al.*).
      hep-ex/9908040.

920. Study of Charmless Hadronic B Decays into the Final States $K\pi$, $\pi\pi$, and KK, with the First
      Observation of $B \to \pi^+\pi^-$ and $B \to K^0\pi^0$
      By CLEO Collaboration (Y. Kwon *et al.*).
      hep-ex/9908039.

921. Structure Functions in the Decay $\tau^\mp \to \pi^\mp\pi^0\pi^0\nu_\tau$
      By CLEO Collaboration (T.E. Browder *et al.*).
      hep-ex/9908030.
      10.1103/PhysRevD.61.052004.
      Phys.Rev. D61 (2000) 052004.

922. Measurement of Charge Asymmetries in Charmless Hadronic B Decay
By CLEO Collaboration (T.E. Coan *et al.*).
hep-ex/9908029.

923. Update of the Search for the Neutrinoless Decay $\tau \to \mu\gamma$
By CLEO Collaboration (A. Anastassov *et al.*).
hep-ex/9908025.

924. Resonant Structure of $\tau \to 3\pi\pi^0 \nu_\tau$ and $\tau \to \omega\pi\nu_\tau$ Decays
By CLEO Collaboration (K.W. Edwards *et al.*).
hep-ex/9908024.
10.1103/PhysRevD.61.072003.
Phys.Rev. D61 (2000) 072003.

925. $b \to s\gamma$ Branching Fraction and *CP* Asymmetry
By CLEO Collaboration (S. Ahmed *et al.*).
hep-ex/9908022.

926. Two-Body B Meson Decays to η and η'–Observation of $B \to \eta K^*$
By CLEO Collaboration (S.J. Richichi *et al.*).
hep-ex/9908019.

927. Charmless Hadronic B Decays to Exclusive Final States with a K^*, ρ, ω, or ϕ Meson
By CLEO Collaboration (C.P. Jessop *et al.*).
hep-ex/9908018.

928. First Observation of the Decay $B \to J/\psi\phi K$
By CLEO Collaboration (C.P. Jessop *et al.*).
hep-ex/9908014.
10.1103/PhysRevLett.84.1393.
Phys.Rev.Lett. 84 (2000) 1393.

929. Observation of a Broad L = 1 $c\bar{q}$ State in $B^- \to D^{*+}\pi^-\pi^-$ at CLEO
By CLEO Collaboration (S. Anderson *et al.*).
hep-ex/9908009.
10.1016/S0375-9474(99)00697-1.
Nucl.Phys. A663 (2000) 647-650.

930, Charged Track Multiplicity in B Meson Decay
By CLEO Collaboration (G. Brandenburg *et al.*).
hep-ex/9907057.
10.1103/PhysRevD.61.072002.
Phys.Rev. D61 (2000) 072002.

931. Observation of Radiative Leptonic Decay of the Tau Lepton
By CLEO Collaboration (T. Bergfeld *et al.*).
hep-ex/9907056.

932. Rare Decays of the η'
     By CLEO Collaboration (Roy A. Briere *et al.*).
     hep-ex/9907046.
     10.1103/PhysRevLett.84.26.
     Phys.Rev.Lett. 84 (2000) 26-30.

933. Limit on Tau Neutrino Mass from $\tau^- \to \pi^-\pi^+\pi^-\pi^0\nu_\tau$
     By CLEO Collaboration (M. Athanas *et al.*).
     hep-ex/9906015.
     10.1103/PhysRevD.61.052002.
     Phys.Rev. D61 (2000) 052002.

934. Evidence of New States Decaying into $\Xi_c^*\pi$
     By CLEO Collaboration (J.P. Alexander *et al.*).
     hep-ex/9906013.
     10.1103/PhysRevLett.83.3390.
     Phys.Rev.Lett. 83 (1999) 3390-3393.

935. Measurement of $B \to \rho\ell\nu$ Decay and $|V_{ub}|$
     By CLEO Collaboration (B.H. Behrens *et al.*).
     hep-ex/9905056.
     10.1103/PhysRevD.61.052001.
     Phys.Rev. D61 (2000) 052001.

936. Report on the Tau-Charm Physics Workshop
     By Martin L. Perl, Peter C. Kim.

937. Hadronic Structure in the Decay $\tau^- \to \nu_\tau\pi^-\pi^0\pi^0$ and the Sign of the Tau Neutrino Helicity
     By CLEO Collaboration (D.M. Asner *et al.*).
     hep-ex/9902022.
     10.1103/PhysRevD.61.012002.
     Phys.Rev. D61 (2000) 012002.

938. Measurement of Charm Meson Lifetimes
     By CLEO Collaboration (G. Bonvicini *et al.*).
     hep-ex/9902011.
     10.1103/PhysRevLett.82.4586.
     Phys.Rev.Lett. 82 (1999) 4586-4590.

939. Search for Baryon and Lepton Number Violating Decays of the τ Lepton
     By CLEO Collaboration (R. Godang *et al.*).
     hep-ex/9902005.
     10.1103/PhysRevD.59.091303.
     Phys.Rev. D59 (1999) 091303.

940. Certainty and Uncertainty in the Practice of Science: Electrons, Muons, and Taus
     By M.L. Perl.
     physics/9901037.

941. Measurement of the $B \rightarrow D\ell\nu$ Branching Fractions and Form-Factor
By CLEO Collaboration (John E. Bartelt *et al.*).
hep-ex/9811042.
10.1103/PhysRevLett.82.3746.
Phys.Rev.Lett. 82 (1999) 3746.

942. The Tau Lepton and the Search for New Elementary Particle Physics
By M.L. Perl.
hep-ph/9812400.

943. First Observation of the Decay $B^0 \rightarrow D^{*+}D^{*-}$
By CLEO Collaboration (M. Artuso *et al.*).
hep-ex/9811027.
10.1103/PhysRevLett.82.3020.
Phys.Rev.Lett. 82 (1999) 3020-3024.

944. A Perspective on Tau Physics
By Martin L. Perl.
10.1016/S0920-5632(99)80001-2.
Nucl.Phys.Proc.Suppl. 76 (1999) 3-19.

945. Search for Exclusive Rare Baryonic Decays of B Mesons
By CLEO Collaboration (T.E. Coan *et al.*).
hep-ex/9810043.
10.1103/PhysRevD.59.111101.
Phys.Rev. D59 (1999) 111101.

946. Observation of Two Narrow States Decaying into $\Xi_c^+\gamma$ and $\Xi_c^0\gamma$
By CLEO Collaboration (C.P. Jessop *et al.*).
hep-ex/9810036.
10.1103/PhysRevLett.82.492.
Phys.Rev.Lett. 82 (1999) 492-496.

947. Study of 3-Prong Hadronic τ Decays with Charged Kaons
By CLEO Collaboration (S.J. Richichi *et al.*).
hep-ex/9810026.
10.1103/PhysRevD.60.112002.
Phys.Rev. D60 (1999) 112002.

948. First Observation of the Decay $\tau^- \rightarrow K^{*-}\eta\nu_\tau$
By CLEO Collaboration (M. Bishai *et al.*).
hep-ex/9809012.
10.1103/PhysRevLett.82.281.
Phys.Rev.Lett. 82 (1999) 281-285.

949. Υ Dipion Transitions at Energies Near the $\Upsilon(4S)$
By CLEO Collaboration (S. Glenn *et al.*).
hep-ex/9808008.
10.1103/PhysRevD.59.052003.
Phys.Rev. D59 (1999) 052003.

950. First Observation of $\Upsilon(1S) \to \gamma\pi\pi$
      By CLEO Collaboration (A. Anastassov *et al.*).
      hep-ex/9807031.
      10.1103/PhysRevLett.82.286.
      Phys.Rev.Lett. 82 (1999) 286-290.

951. Further Search for the Two Photon Production of the Glueball Candidate $f_J(2220)$
      By CLEO Collaboration (M.S. Alam *et al.*).
      hep-ex/9805033.
      10.1103/PhysRevLett.81.3328.
      Phys.Rev.Lett. 81 (1998) 3328-3332.

952. First search for *CP* Violation in Tau Lepton Decay
      By CLEO Collaboration (S. Anderson *et al.*).
      hep-ex/9805027.
      10.1103/PhysRevLett.81.3823.
      Phys.Rev.Lett. 81 (1998) 3823-3827.

953. Observation of High Momentum η' Production in *B* Decay
      By CLEO Collaboration (T.E. Browder *et al.*).
      hep-ex/9804018.
      10.1103/PhysRevLett.81.1786.
      Phys.Rev.Lett. 81 (1998) 1786-1790.

954. A Limit on the Mass of the ν_τ
      By CLEO Collaboration (R. Ammar *et al.*).
      hep-ex/9803031.
      10.1016/S0370-2693(98)00539-5.
      Phys.Lett. B431 (1998) 209-218.

955. Radiative Decay Modes of the D^0 Meson
      By CLEO Collaboration (D.M. Asner *et al.*).
      hep-ex/9803022.
      10.1103/PhysRevD.58.092001.
      Phys.Rev. D58 (1998) 092001.

956. Observation of $B^+ \to \omega K^+$ and Search for Related *B* Decay Modes
      By CLEO Collaboration (T. Bergfeld *et al.*).
      hep-ex/9803018.
      10.1103/PhysRevLett.81.272.
      Phys.Rev.Lett. 81 (1998) 272-276.

957. Measurement of the Mass Splittings Between the $b\bar{b}\,\chi_{b,J}(1P)$ States
      By CLEO Collaboration (K.W. Edwards *et al.*).
      hep-ex/9803010.
      10.1103/PhysRevD.59.032003.
      Phys.Rev. D59 (1999) 032003.

958. The Hadronic Transitions $\Upsilon(2S) \to \Upsilon(1S)$
By CLEO Collaboration (J.P. Alexander *et al.*).
hep-ex/9802024.
10.1103/PhysRevD.58.052004.
Phys.Rev. D58 (1998) 052004.

959. First Observation of the Cabibbo Suppressed decay $B^+ \to \overline{D}^0 K^+$
By CLEO Collaboration (M. Athanas *et al.*).
hep-ex/9802023.
10.1103/PhysRevLett.80.5493.
Phys.Rev.Lett. 80 (1998) 5493-5497.

960. Continuum Charged D^* Spin Alignment at $\sqrt{s} = 10.5$ GeV
By CLEO Collaboration (G. Brandenburg *et al.*).
hep-ex/9802022.
10.1103/PhysRevD.58.052003.
Phys.Rev. D58 (1998) 052003.

961. Two-Body B Meson Decays to η and η'–Observation of $B \to \eta' K$
By CLEO Collaboration (B.H. Behrens *et al.*).
hep-ex/9801012.
10.1103/PhysRevLett.80.3710.
Phys.Rev.Lett. 80 (1998) 3710-3714.

962. Measurement of the Branching Ratios for the Decays of D_s^+ to $\eta\pi^+$, $\eta'\pi^+$, $\eta\rho^+$, and $\eta'\rho^+$
By CLEO Collaboration (C.P. Jessop *et al.*).
hep-ex/9801010.
10.1103/PhysRevD.58.052002.
Phys.Rev. D58 (1998) 052002.

963. Search for the Decay $B \to D_{s1}^+(2536)X$
By CLEO Collaboration (M. Bishai *et al.*).
hep-ex/9710023.
10.1103/PhysRevD.57.3847.
Phys.Rev. D57 (1998) 3847-3853.

964. Measurement of the Branching Fractions of $\Lambda_c^+ \to p\overline{K}n(\pi)$
By CLEO Collaboration (M.S. Alam *et al.*).
hep-ex/9709012.
10.1103/PhysRevD.57.4467.
Phys.Rev. D57 (1998) 4467-4470.

965. The Leptons After 100 Years
By M.L. Perl.
10.1063/1.881960.
Phys.Today 50N10 (1997) 34-40.

966. The Search for Elementary Particles with Fractional Electric Charge and the Philosophy of
     Speculative Experiments
     By M.L. Perl, E.R. Lee.
     10.1119/1.18641.
     Am.J.Phys. 65 (1997) 698-706.

967. Measurement of Branching Ratio $Br(D^0 \to K^-\pi^+)$ Using Partial Reconstruction of $\bar{B} \to$
     $D^{*+}X\ell^-\bar{\nu}$
     By CLEO Collaboration (M. Artuso *et al.*).
     hep-ex/9712023.
     10.1103/PhysRevLett.80.3193.
     Phys.Rev.Lett. 80 (1998) 3193-3197.

968. Improved Measurement of the Pseudoscalar Decay Constant f_{D_s}
     By CLEO Collaboration (M. Chadha *et al.*).
     hep-ex/9712014.
     10.1103/PhysRevD.58.032002.
     Phys.Rev. D58 (1998) 032002.

969. New Limits for Neutrinoless Tau Decays
     By CLEO Collaboration (D.W. Bliss *et al.*).
     hep-ex/9712010.
     10.1103/PhysRevD.57.5903.
     Phys.Rev. D57 (1998) 5903-5907.

970. Study of Semileptonic Decays of *B* Mesons to Charmed Baryons
     By CLEO Collaboration (G. Bonvicini *et al.*).
     hep-ex/9712008.
     10.1103/PhysRevD.57.6604.
     Phys.Rev. D57 (1998) 6604-6608.

971. Observation of the radiative decay $D^{*+} \to D^+\gamma$
     By CLEO Collaboration (John E. Bartelt *et al.*).
     hep-ex/9711011.
     10.1103/PhysRevLett.80.3919.
     Phys.Rev.Lett. 80 (1998) 3919-3923.

972. Observation of Exclusive Two-Body *B* Decays to Kaons and Pions
     By CLEO Collaboration (R. Godang *et al.*).
     hep-ex/9711010.
     10.1103/PhysRevLett.80.3456.
     Phys.Rev.Lett. 80 (1998) 3456-3460.

973. Flavor-Specific Inclusive *B* Decays to Charm
     By CLEO Collaboration (T.E. Coan *et al.*).
     hep-ex/9710028.
     10.1103/PhysRevLett.80.1150.
     Phys.Rev.Lett. 80 (1998) 1150-1155.

974. Search for Inclusive $b \to s l^+ l^-$
     By CLEO Collaboration (S. Glenn *et al.*).
     hep-ex/9710003.
     10.1103/PhysRevLett.80.2289.
     Phys.Rev.Lett. 80 (1998) 2289-2293.

975. A New Method for Searching for Massive, Stable, Charged Elementary Particles
     By Martin L. Perl.
     10.1103/PhysRevD.57.4441.
     Phys.Rev. D57 (1998) 4441-4445.

976. Investigation of Semileptonic B Meson Decay to P-Wave Charm Mesons
     By CLEO Collaboration (A. Anastassov *et al.*).
     hep-ex/9708035.
     10.1103/PhysRevLett.80.4127.
     Phys.Rev.Lett. 80 (1998) 4127-4131.

977. Search for Color-Suppressed B Hadronic Decay Processes with CLEO
     By CLEO Collaboration (B. Nemati *et al.*).
     hep-ex/9708033.
     10.1103/PhysRevD.57.5363.
     Phys.Rev. D57 (1998) 5363-5369.

978. The Higher Derivative Expansion of the Effective Action by the String Inspired Method. Part 2
     By D. Fliegner, P. Haberl, M.G. Schmidt, C. Schubert.
     hep-th/9707189.
     10.1006/aphy.1997.5778.
     Annals Phys. 264 (1998) 51-74.

979. Measurements of the Meson-Photon Transition Form Factors of Light Pseudoscalar Mesons at Large Momentum Transfer
     By CLEO Collaboration (J. Gronberg *et al.*).
     hep-ex/9707031.
     10.1103/PhysRevD.57.33.
     Phys.Rev. D57 (1998) 33-54.

980. Search for the Decay $\tau^- \to 4\pi^- 3\pi^+ (\pi^0) \nu_\tau$
     By CLEO Collaboration (K.W. Edwards *et al.*).
     hep-ex/9707029.
     10.1103/PhysRevD.56.R5297.
     Phys.Rev. D56 (1997) 5297-5300.

981. Study of the Decay $\tau^- \to 2\pi^- \pi^+ 3\pi^0 \nu_\tau$
     By CLEO Collaboration (S. Anderson *et al.*).
     hep-ex/9707027.
     10.1103/PhysRevLett.79.3814.
     Phys.Rev.Lett. 79 (1997) 3814-3818.

982. The Physics of the τ Lepton and τ-Neutrino in 1995
     By M.L. Perl.
     In *Perl, M.L.: Reflections on experimental science* 219-265.

983. Scattering of K^+ Mesons by Protons
     By D.I. Meyer, M.L. Perl, D.A. Glaser.
     10.1103/PhysRev.107.279.
     Phys.Rev. 107 (1957) 279-282.

984. The Use of a Sodium Iodide Luminescent Chamber to Study Elastic and Inelastic Scattering
     of Pions in Hydrogen
     By M.L. Perl, Lawrence W. Jones, K. Lai.
     In *Perl, M.L.: Reflections on experimental science* 357-362. In *Berkeley 1960,
     Instrumentation for high-energy physics* 186-191.

985. Negative Pion-Proton Elastic Scattering at 1.51, 2.01, and 2.53 BeV/c Outside the Diffraction
     Peak
     By K.W. Lai, Lawrence W. Jones, M.L. Perl.
     10.1103/PhysRevLett.7.125.
     Phys.Rev.Lett. 7 (1961) 125-126.

986. Pion-Proton Elastic Scattering at 2.00 GeV/c
     By D.E. Damouth, Lawrence W. Jones, M.L. Perl.
     10.1103/PhysRevLett.11.287.
     Phys.Rev.Lett. 11 (1963) 287-290.

987. A Measurement of the Total Cross Section for $e^+e^- \to$ Hadrons at $\sqrt{s} = 10.52$ GeV
     By CLEO Collaboration (R. Ammar *et al.*).
     hep-ex/9707018.
     10.1103/PhysRevD.57.1350.
     Phys.Rev. D57 (1998) 1350-1358.

988. First Observation of $\tau \to 3\pi\eta\nu_\tau$ and $\tau \to f_1\pi\nu_\tau$ Decays
     By CLEO Collaboration (T. Bergfeld *et al.*).
     hep-ex/9706020.
     10.1103/PhysRevLett.79.2406.
     Phys.Rev.Lett. 79 (1997) 2406-2410.

989. A New Measurement of $B \to D^*\pi$ Branching Fractions
     By CLEO Collaboration (G. Brandenburg *et al.*).
     hep-ex/9706019.
     10.1103/PhysRevLett.80.2762.
     Phys.Rev.Lett. 80 (1998) 2762-2766.

990. A New Upper Limit on the Decay $\eta \to e^+e^-$
     By CLEO Collaboration (T.E. Browder *et al.*).
     hep-ex/9706005.
     10.1103/PhysRevD.56.5359.
     Phys.Rev. D56 (1997) 5359-5365.

991, Measurement of the $\overline{B} \to D\ell\overline{\nu}$ Partial Width and Form-Factor Parameters
By CLEO Collaboration (M. Athanas *et al.*).
hep-ex/9705019.
10.1103/PhysRevLett.79.2208.
Phys.Rev.Lett. 79 (1997) 2208-2212.

992. Determination of the Michel Parameters and the τ Neutrino Helicity in τ Decay
By CLEO Collaboration (J.P. Alexander *et al.*).
hep-ex/9705009.
10.1103/PhysRevD.56.5320.
Phys.Rev. D56 (1997) 5320-5329.

993. First Observation of Inclusive B Decays to the Charmed Strange Baryons Ξ_c^0 and Ξ_c^+
By CLEO Collaboration (B. Barish *et al.*).
hep-ex/9705005.
10.1103/PhysRevLett.79.3599.
Phys.Rev.Lett. 79 (1997) 3599-3603.

994. Observation of the Decay $D_s^+ \to \omega\pi^+$
By CLEO Collaboration (R. Balest *et al.*).
hep-ex/9705006.
10.1103/PhysRevLett.79.1436.
Phys.Rev.Lett. 79 (1997) 1436-1440.

995. Search for the Decays $B^0 \to D^{(*)+}D^{(*)-}$
By CLEO Collaboration (D.M. Asner *et al.*).
hep-ex/9704014.
10.1103/PhysRevLett.79.799.
Phys.Rev.Lett. 79 (1997) 799-803.

996. Search for Neutrinoless τ Decays Involving π^0 or η Mesons
By CLEO Collaboration (G. Bonvicini *et al.*).
hep-ex/9704010.
10.1103/PhysRevLett.79.1221.
Phys.Rev.Lett. 79 (1997) 1221-1224.

997. Proposal to Measure Polarized Open Charm Photo Production
By R.G. Arnold *et al.*.

998. Studies of the Cabibbo Suppressed Decays $D^+ \to \pi^0\ell^+\nu$ and $D^+ \to \eta e^+\nu_e$
By CLEO Collaboration (John E. Bartelt *et al.*).
hep-ex/9703013.
10.1016/S0370-2693(97)00649-7.
Phys.Lett. B405 (1997) 373-378.

999. Limit on the Two Photon Production of the Glueball Candidate $f_J(2220)$ at CLEO
By CLEO Collaboration (R. Godang *et al.*).
hep-ex/9703009.
10.1103/PhysRevLett.79.3829.
Phys.Rev.Lett. 79 (1997) 3829-3833.

1000. Bremsstrahlung Suppression Due to the LPM and Dielectric Effects in a Variety of Materials
By SLAC-E-146 Collaboration (P.L. Anthony *et al.*).
hep-ex/9703016.
10.1103/PhysRevD.56.1373.
Phys.Rev. D56 (1997) 1373-1390.

1001. The Inclusive Decays $B \to DX$ and $B \to D^*X$
By CLEO Collaboration (L. Gibbons *et al.*).
hep-ex/9703006.
10.1103/PhysRevD.56.3783.
Phys.Rev. D56 (1997) 3783-3802.

1002. Measurement of the Decay Amplitudes and Branching Fractions of $B \to J/\psi K^*$ and $B \to J/\psi K$ Decays
By CLEO Collaboration (C.P. Jessop *et al.*).
hep-ex/9702013.
10.1103/PhysRevLett.79.4533.
Phys.Rev.Lett. 79 (1997) 4533-4537.

1003. Study of the B^0 Semileptonic Decay Spectrum at the $\Upsilon(4S)$ Resonance
By CLEO Collaboration (M. Artuso *et al.*).
hep-ex/9702007.
10.1016/S0370-2693(97)00336-5.
Phys.Lett. B399 (1997) 321-328.

1004. Tau Neutrino Helicity from $h^\pm$ Energy Correlations
By CLEO Collaboration (T.E. Coan *et al.*).
hep-ex/9701012.
10.1103/PhysRevD.55.7291.
Phys.Rev. D55 (1997) 7291-7295.

1005. $\Lambda\bar{\Lambda}$ Production in Two-Photon Interactions at CLEO
By CLEO Collaboration (S. Anderson *et al.*).
hep-ex/9701013.
10.1103/PhysRevD.56.R2485.
Phys.Rev. D56 (1997) 2485-2489.

1006. Analyses of $D^+ \to K_S^0 K^+$ and $D^+ \to K_S^0 \pi^+$
By CLEO Collaboration (M. Bishai *et al.*).
hep-ex/9701008.
10.1103/PhysRevLett.78.3261.
Phys.Rev.Lett. 78 (1997) 3261-3265.

1007. Study of Gluon Versus Quark Fragmentation in $\Upsilon \to gg\gamma$ and $e^+e^- \to q\bar{q}\gamma$ Events at $\sqrt{s} = 10$ GeV
By CLEO Collaboration (M.S. Alam *et al.*).
hep-ex/9701006.
10.1103/PhysRevD.56.17.
Phys.Rev. D56 (1997) 17-22.

1008. A New Search for Elementary Particles with Fractional Electric Charge Using an Improved Millikan Technique
By Nancy Marie Mar.

1009. Reflections on Experimental Science
By M.L. Perl.
Singapore, Singapore: World Scientific (1996) 537 p. (World Scientific series in 20th century physics: 14).

1010. A Measurement of the Michel Parameters in Leptonic Decays of the τ
By CLEO Collaboration (R. Ammar *et al.*).
10.1103/PhysRevLett.78.4686.
Phys.Rev.Lett. 78 (1997) 4686-4690.

1011. Experiment and Theory in Particle Physics: Reflections on the Discovery of the Tau Lepton
By Martin L. Perl.
In *Minneapolis 1996, Particles and fields, vol. 1* 10-20.

1012. Search for $B \to \mu\bar{\nu}_\mu\gamma$ and $B \to e\bar{\nu}_e\gamma$
By CLEO Collaboration (T.E. Browder *et al.*).
10.1103/PhysRevD.56.11.
Phys.Rev. D56 (1997) 11-16.

1013. Search for Neutrinoless Tau Decays: $\tau \to e\gamma$ and $\tau \to \mu\gamma$
By CLEO Collaboration (K.W. Edwards *et al.*).
10.1103/PhysRevD.55.3919.
Phys.Rev. D55 (1997) 3919-3923.

1014. Experimental Test of Lepton Universality in Tau Decay
By CLEO Collaboration (A. Anastassov *et al.*).
10.1103/PhysRevD.58.119904, 10.1103/PhysRevD.55.2559.
Phys.Rev. D55 (1997) 2559-2576, Phys.Rev. D58 (1998) 119904.

1015. Measurement of the Direct Photon Spectrum in $\Upsilon(1S)$ Decays
By CLEO Collaboration (B. Nemati *et al.*).
hep-ex/9611020.
10.1103/PhysRevD.55.5273.
Phys.Rev. D55 (1997) 5273-5281.

1016. Search for ψ Mesons in Tau Lepton Decay
By CLEO Collaboration (P. Avery *et al.*).
10.1103/PhysRevD.55.R1119.
Phys.Rev. D55 (1997) 1119-1123.

1017. A Search for Nonresonant $B^+ \to h^+h^-h^+$ Decays
By CLEO Collaboration (T. Bergfeld *et al.*).
10.1103/PhysRevLett.77.4503.
Phys.Rev.Lett. 77 (1996) 4503-4507.

1018. Observation of Two Excited Charmed Baryons Decaying into $\Lambda_c^+ \pi^\pm$
By CLEO Collaboration (G. Brandenburg *et al.*).
10.1103/PhysRevLett.78.2304.
Phys.Rev.Lett. 78 (1997) 2304-2308.

1019. First Measurement of the $B \to \pi\ell\nu$ and $B \to \rho(\omega)\ell\nu$ Branching Fractions
By CLEO Collaboration (J.P. Alexander *et al.*).
10.1103/PhysRevLett.77.5000.
Phys.Rev.Lett. 77 (1996) 5000-5004.

1020. Observation of Exclusive B Decays to Final States Containing a Charmed Baryon
By CLEO Collaboration (X. Fu *et al.*).
10.1103/PhysRevLett.79.3125.
Phys.Rev.Lett. 79 (1997) 3125-3129.

1021. Measurement of the Tau Lepton Lifetime
By CLEO Collaboration (R. Balest *et al.*).
10.1016/S0370-2693(96)01163-X.
Phys.Lett. B388 (1996) 402-408.

1022. Reflections on the Discovery of the Tau Lepton
By Martin L. Perl.
In *Ekspong, G. (ed.): Nobel lectures in physics 1991-1995* 168-195.

1023. An Improved Search for Elementary Particles with Fractional Electric Charge
By N.M. Mar *et al.*.
10.1103/PhysRevD.53.6017.
Phys.Rev. D53 (1996) 6017-6032.

1024. Observation of an Excited Charmed Baryon Decaying into $\Xi_c^0 \pi^+$
By CLEO Collaboration (L. Gibbons *et al.*).
10.1103/PhysRevLett.77.810.
Phys.Rev.Lett. 77 (1996) 810-813.

1025. Analysis of $D^0 \to K\bar{K}X$ Decays
By CLEO Collaboration (D.M. Asner *et al.*).
10.1103/PhysRevD.54.4211.
Phys.Rev. D54 (1996) 4211-4220.

1026. A Measurement of $\beta(D^0 \to K^-\pi^+\pi^0)/\beta(D^0 \to K^-\pi^+)$
By CLEO Collaboration (B. Barish *et al.*).
10.1016/0370-2693(96)00159-1.
Phys.Lett. B373 (1996) 334-338.

1027. Measurement of the Branching Fraction for $D_s^- \to \phi\pi^-$
By CLEO Collaboration (M. Artuso *et al.*).
10.1016/0370-2693(96)00503-5.
Phys.Lett. B378 (1996) 364-372.

1028. First Observation of the Decay $\tau^- \to K^- \eta \nu_\tau$
By CLEO Collaboration (John E. Bartelt *et al.*).
10.1103/PhysRevLett.76.4119.
Phys.Rev.Lett. 76 (1996) 4119-4123.

1029. Decays of τ Leptons to Final States Containing K_S^0 Mesons
By CLEO Collaboration (T.E. Coan *et al.*).
10.1103/PhysRevD.53.6037.
Phys.Rev. D53 (1996) 6037-6053.

1030. Limits on Flavor Changing Neutral Currents in D^0 Meson Decays
By CLEO Collaboration (A. Freyberger *et al.*).
10.1103/PhysRevLett.76.3065.
Phys.Rev.Lett. 76 (1996) 3065-3069, Phys.Rev.Lett. 77 (1996) 2147.

1031. Discovery of the Tau: The Role of Motivation and Technology in Experimental Particle Physics
By Martin L. Perl.
SLAC Beam Line 25N4 (1998) 4-27.

1032. Study of $B \to \psi\rho$
By CLEO Collaboration (M. Bishai *et al.*).
10.1016/0370-2693(95)01585-X.
Phys.Lett. B369 (1996) 186-192.

1033. Nobel Leptons
By K.V.L. Sarma.
hep-ph/9512420.
Submitted to: Curr. Sci..

1034. Measurement of Dielectric Suppression of Bremsstrahlung
By P.L. Anthony *et al.*.
hep-ph/9512381.
10.1103/PhysRevLett.76.3550.
Phys.Rev.Lett. 76 (1996) 3550-3553.

1035. Measurement of the Form-Factors for $\overline{B}^0 \to D^{*+}\ell^-\overline{\nu}$
By CLEO Collaboration (J.E. Duboscq *et al.*).
10.1103/PhysRevLett.76.3898.
Phys.Rev.Lett. 76 (1996) 3898-3902.

1036. Tau Decays into Three Charged Leptons and Two Neutrinos
By CLEO Collaboration (M.S. Alam *et al.*).
10.1103/PhysRevLett.76.2637.
Phys.Rev.Lett. 76 (1996) 2637-2641.

1037. Tau Lepton Physics at a Tau Charm Factory
By M.L. Perl.
In *Stanford 1994, Proceedings, The tau-charm factory in the era of B-factories and CESR*
69-94.

1038. An Accurate Measurement of the Landau-Pomeranchuk-Migdal Effect
       By P.L. Anthony *et al.*.
       10.1103/PhysRevLett.75.1949.
       Phys.Rev.Lett. 75 (1995) 1949-1952.

1039. BaBar Technical Design Report
       By BaBar Collaboration (D. Boutigny *et al.*).

1040. Tau Physics at Future Facilities
       By Martin L. Perl.
       10.1016/0920-5632(95)00178-C.
       Nucl.Phys.Proc.Suppl. 40 (1995) 541-555.

1041. The Discovery of the Tau Lepton
       By Martin L. Perl.
       10.1007/978-1-4613-1147-8_15.
       NATO Sci.Ser.B 352 (1996) 277-302.

1042. Letter of Intent for the Study of *CP* Violation and Heavy Flavor Physics at PEP-II
       By BaBar Collaboration (D. Boutigny *et al.*).

1043. Notes on the Landau, Pomeranchuk, Migdal effect: Experiment and Theory
       By Martin L. Perl.
       In *Perl, M.L.: Reflections on experimental science* 463-477.

1044. The Search for New Elementary Particles: Status and Prospects. Proceedings, Workshop,
       Trieste, Italy, May 20-22, 1992
       By G. Herten, L. Beers, M.L. Perl.
       Singapore, Singapore: World Scientific (1993) 303 p. (Int. J. Mod. Phys. A, Proc. Suppl. 3B
       (1993)).

1045. Maximum Acceptance Detector for the Fermilab Collider (MAX)
       By Y.T. Gao *et al.*.

1046. A Method of Obtaining Parasitic e^+ or e^- Beams During SLAC Linear Collider Operation
       By SLAC-E-146 Collaboration (M. Cavalli-Sforza *et al.*).
       10.1109/23.322915.
       IEEE Trans.Nucl.Sci. 41 (1994) 1371-1373.

1047. Quantum Mechanical Suppression of Bremsstrahlung
         By SLAC-E-146 Collaboration (R. Becker-Szendy *et al.*).
       In *Stanford 1993, Proceedings, Spin structure in high energy processes* 519-529, and In
       Aspen 1993, Proceedings, Multiparticle dynamics 527-541. SLAC Stanford - SLAC-PUB-6400
       (93/12, rec.Jan.94) 15 p.

1048. A Measurement of the LPM Effect
       By S.R. Klein *et al.*.
       In *Ithaca 1993, Proceedings, Lepton and photon interactions* 172-197, and SLAC Stanford -
       SLAC-PUB-6378 (93/11,rec.Jan.94) 16 p.

1049. A Study of LPM Suppression of Bremsstrahlung at 25 GeV
By SLAC-E-146 Collaboration (P. Anthony *et al.*).

1050. Efficient Bulk Search for Fractional Charge with Multiplexed Millikan Chambers
By Charles D. Hendricks, Klaus S. Lackner, Martin L. Perl, Gordon L. Shaw.
In *Perl, M.L.: Reflections on experimental science* 481-490.

1051. Tau Physics
By Martin L. Perl.
In *Stanford 1992, The third family and the physics of flavor* 213-252.

1052. Tau Physics
By Martin L. Perl.

1053. The SLD Single Event Display
By R. Dubois, J. Perl.
In *Annecy 1992, Computing in high energy physics '92* 427-431.

1054. From the Psi to Charmed Mesons: Three Years with the SLAC-LBL Detector at Spear
By Gerson Goldhaber.

1055. Searches for New Particles at a Tau Charm Factory
By Martin L. Perl.
In *Trieste 1992, Proceedings, The search for new elementary particles: Status and prospects* 188-210 and SLAC Stanford - SLAC-PUB-5989 (92/11,rec.Apr.93) 23 p.

1056. Beyond the Tau: Other Directions in Tau Physics
By Martin L. Perl.
In *Perl, M.L.: Reflections on experimental science* 308-326.

1057. A Proposal for an Experiment to Study the Interference Between Multiple Scattering and Bremsstrahlung (the LPM Effect)
By M. Cavalli-Sforza *et al.*.

1058. Inclusive Charged Hadron and K^0 Production in Two Photon Interactions
By D. Cords *et al.*.
10.1016/0370-2693(93)90406-8.
Phys.Lett. B302 (1993) 341-344.

1059. An Experimental Test of the LPM Effect: Bremsstrahlung Suppression at High-Energies
By S. Klein *et al.*.

1060. The Discovery of the Tau Lepton
By Martin L. Perl.
In *Stanford 1992, The rise of the standard model* 79-100.

1061. The Discovery of the Tau, 1975–1977: A Tale of Three Papers
By Gary J. Feldman.
Conf.Proc. C9207131 (1992) 631-646.

1062. Measurement of the Charged Multiplicity of Events Containing Bottom Hadrons at $E_{c.m.} =$ 91 GeV
By B.A. Schumm *et al.*.
10.1103/PhysRevD.46.453.
Phys.Rev. D46 (1992) 453-456.

1063. The Tau Charm Factory: Concept and Construction
By Martin L. Perl, Gary D. Niemi.
10.1109/23.159653.
IEEE Trans.Nucl.Sci. 39 (1992) 486-493.

1064. The Tau Charm Factory: Experimental Perspectives
By Martin L. Perl, R.H. Schindler.

1065. The Tau Lepton
By Martin L. Perl.
10.1088/0034-4885/55/6/001.
Rept.Prog.Phys. 55 (1992) 653-722.

1066. A Search for the Production of the Final States $\tau^+\tau^-e^+e^-$, $\tau^+\tau^-\mu^+\mu^-$, and $\tau^+\tau^-\pi^+\pi^-$ in e^+e^- Collisions at $\sqrt{s} = 29$ GeV
By Mark-II Collaboration (T. Barklow *et al.*).
10.1103/PhysRevLett.68.13.
Phys.Rev.Lett. 68 (1992) 13-16.

1067. Measurement of the $b\bar{b}$ Fraction in Hadronic Z^0 Decays with Precision Vertex Detectors
By Robert Jacobsen *et al.*.
10.1103/PhysRevLett.67.3347.
Phys.Rev.Lett. 67 (1991) 3347-3350.

1068. The Future of Tau Physics and Tau Charm Detector and Factory Design
By Martin L. Perl.
In *Perl, M.L.: Reflections on experimental science* 266-307.

1069. Measurement of the Two-Photon Width of the $\eta'(958)$
By F. Butler *et al.*.
10.1103/PhysRevD.42.1368.
Phys.Rev. D42 (1990) 1368-1384.

1070. Two Photon Production of Pion Pairs
By J. Boyer *et al.*.
10.1103/PhysRevD.42.1350.
Phys.Rev. D42 (1990) 1350-1367.

1071. Elementary Particle Physics
By T. Barklow, M. Perl.
In *Meyers, R.A. (ed.): Encyclopedia of modern physics* 171-204. (see Book Index).

1072. The Discovery of the Tau and its Major Properties: 1970–1985
By Martin L. Perl.
Conf.Proc. C9009241 (1990) 3-28.

1073. Tau Charm Factory Design
By Barry C. Barish *et al.*.

1074. Measurement of the Total Hadronic Cross Section in e^+e^- Annihilation at $\sqrt{s} = 29$ GeV
By MARK-II Collaboration (Christoph Von Zanthier *et al.*).
10.1103/PhysRevD.43.34.
Phys.Rev. D43 (1991) 34-45.

1075. A Search for Pair Production of Heavy Stable Charged Particles in Z Decays
By E. Soderstrom *et al.*.
10.1103/PhysRevLett.64.2980.
Phys.Rev.Lett. 64 (1990) 2980-2983.

1076. Study of Four Lepton Final States in e^+e^- Interactions at $\sqrt{s} = 29$ GeV
By M. Petradza *et al.*.
10.1103/PhysRevD.42.2171.
Phys.Rev. D42 (1990) 2171-2179.

1077. Searches for Supersymmetric Particles Produced in Z Boson Decay
By T. Barklow *et al.*.
10.1103/PhysRevLett.64.2984.
Phys.Rev.Lett. 64 (1990) 2984-2987.

1078. Search for Nonminimal Higgs Bosons from Z Boson Decay
By Sachio Komamiya *et al.*.
10.1103/PhysRevLett.64.2881.
Phys.Rev.Lett. 64 (1990) 2881-2884.

1079. A Search for Decays of the Z to Unstable Neutral Leptons with Mass Between 2.5 GeV and 22 GeV
By P. Burchat *et al.*.
10.1103/PhysRevD.41.3542.
Phys.Rev. D41 (1990) 3542.

1080. A Search for Doubly Charged Higgs Scalars in Z Decay
By Morris L. Swartz *et al.*.
10.1103/PhysRevLett.64.2877.
Phys.Rev.Lett. 64 (1990) 2877-2880.

1081. Properties of Leptons
By Martin L. Perl.

1082. A Reanalysis of B^0-$\bar{B}^0$ Mixing in e^+e^- Annihilation at 29 GeV
By A. Weir *et al.*.
10.1016/0370-2693(90)90447-E.
Phys.Lett. B240 (1990) 289.

1083. Measurements of Charged Particle Inclusive Distributions in Hadronic Decays of the Z Boson
       By G.S. Abrams *et al.*.
       10.1103/PhysRevLett.64.1334.
       Phys.Rev.Lett. 64 (1990) 1334.

1084. Radiative Tau Production and Decay
       By D.Y. Wu *et al.*.
       10.1103/PhysRevD.41.2339.
       Phys.Rev. D41 (1990) 2339.

1085. Determination of α_s from a Differential Jet Multiplicity Distribution at SLC and PEP
       By Sachio Komamiya *et al.*.
       10.1103/PhysRevLett.64.987.
       Phys.Rev.Lett. 64 (1990) 987.

1086. Search for Longlived Massive Neutrinos in Z Decays
       By C.K. Jung *et al.*.
       10.1103/PhysRevLett.64.1091.
       Phys.Rev.Lett. 64 (1990) 1091.

1087. Measurement of Z Decays into Lepton Pairs
       By G.S. Abrams *et al.*.
       10.1103/PhysRevLett.63.2780.
       Phys.Rev.Lett. 63 (1989) 2780.

1088. Measurement of the $b\bar{b}$ Fraction in Hadronic Z Decays
       By J.F. Kral *et al.*.
       10.1103/PhysRevLett.64.1211.
       Phys.Rev.Lett. 64 (1990) 1211.

1089. Searches for New Quarks and Leptons Produced in Z Boson Decay
       By G.S. Abrams *et al.*.
       10.1103/PhysRevLett.63.2447.
       Phys.Rev.Lett. 63 (1989) 2447.

1090. Measurement of the B^0 Meson Lifetime
       By S.R. Wagner *et al.*.
       10.1103/PhysRevLett.64.1095.
       Phys.Rev.Lett. 64 (1990) 1095.

1091. First Measurements of Hadronic Decays of the Z Boson
       By MARK-II Collaboration (G.S. Abrams *et al.*).
       10.1103/PhysRevLett.63.1558.
       Phys.Rev.Lett. 63 (1989) 1558.

1092. Measurements of Z Boson Resonance Parameters in e^+e^- Annihilation
       By G.S. Abrams *et al.*.
       10.1103/PhysRevLett.63.2173.
       Phys.Rev.Lett. 63 (1989) 2173.

1093. Search for a Nearly Degenerate Lepton Doublet (L^-, L^0)
By Keith Riles *et al.*.
10.1103/PhysRevD.42.1.
Phys.Rev. D42 (1990) 1-9.

1094. Initial Measurements of Z Boson Resonance Parameters in e^+e^- Annihilation
By G.S. Abrams *et al.*.
10.1103/PhysRevLett.63.724.
Phys.Rev.Lett. 63 (1989) 724.

1095. Search for a Charged Lepton Specific Force in Electron - Positron Collisions
By C.A. Hawkins, Martin L. Perl.
Submitted to: Phys.Rev.Lett..

1096. Search for B Decay to Higgs Bosons for Higgs Boson Masses Between 50 MeV/c^2 and 210 MeV/c^2
By A. Snyder *et al.*.
10.1016/0370-2693(89)90177-9.
Phys.Lett. B229 (1989) 169.

1097. The Tau Charm Factory and Tau Physics
By Martin L. Perl.

1098. The Search for Charged Lepton Specific Forces and the Pegasys Facility
By Martin L. Perl.
In *Stanford 1989, Proceedings, Electronuclear physics with internal targets* 255-267 and SLAC Stanford - SLAC-PUB-4881 (89,rec.Jun.) 13 p.

1099. The Mark-II Vertex Drift Chamber
By James P. Alexander *et al.*.
10.1016/0168-9002(89)91410-1.
Nucl.Instrum.Meth. A283 (1989) 519-527.

1100. The Mark-II Detector for the SLC
By G.S. Abrams *et al.*.
10.1016/0168-9002(89)91217-5.
Nucl.Instrum.Meth. A281 (1989) 55.

1101. Λ_c^+ Production and Semileptonic Decay in 29 GeV e^+e^- Annihilation
By S. Klein *et al.*.
10.1103/PhysRevLett.62.2444.
Phys.Rev.Lett. 62 (1989) 2444.

1102. Tau Physics and Tau Factories
By Martin L. Perl.

1103. Inclusive Charged Hadron and K^0 Production in Two Photon Interactions
By D. Cords *et al.*.
In *Shoresh 1988, Proceedings, Photon photon collisions* 66-68.

1104. The Tau and Beyond: Future Research on Heavy Leptons
       By Martin L. Perl.

1105. An Exploration of the Limits on Charged Lepton Specific Forces
       By C.A. Hawkins, Martin L. Perl.
       10.1103/PhysRevD.40.823.
       Phys.Rev. D40 (1989) 823-832.

1106. Tutorial Guide to the Tau Lepton and Close Mass Lepton Pairs
       By Martin L. Perl.

1107. Limits on New Lepton Pairs (L^-, L^0) with Arbitrary Neutrino Mass
       By D.P. Stoker *et al.*.
       10.1103/PhysRevD.39.1811.
       Phys.Rev. D39 (1989) 1811.

1108. Electron - Positron Collision Physics: 1 MeV to 2 TeV
       By Martin L. Perl.
       In *Perl, M.L.: Reflections on experimental science* 429-449.

1109. Measurement of Single and Double Radiative Low Q^2 Bhabha Scattering at $E_{\text{c.m.}}$ = 29 GeV
       By D. Karlen *et al.*.
       10.1103/PhysRevD.39.1861.
       Phys.Rev. D39 (1989) 1861.

1110. Bose–Einstein Correlations in e^+e^- Collisions
       By I. Juricic *et al.*.
       10.1103/PhysRevD.39.1.
       Phys.Rev. D39 (1989) 1.

1111. Application of the Bootstrap Statistical Method to the Tau Decay Mode Problem
       By K.G. Hayes, Martin L. Perl, Bradley Efron.
       10.1103/PhysRevD.39.274.
       Phys.Rev. D39 (1989) 274.

1112. The Tau Decay Mode Problem
       By Martin L. Perl.

1113. Graduate Education and Research in the Era of Large Accelerators
       By Martin L. Perl.

1114. Inclusive Lepton Production in e^+e^- Annihilation at 29 GeV
       By R.A. Ong *et al.*.
       10.1103/PhysRevLett.60.2587.
       Phys.Rev.Lett. 60 (1988) 2587.

1115. A Statistical Study of Tau Decay Data
       By K.G. Hayes, Martin L. Perl.
       10.1103/PhysRevD.38.3351.
       Phys.Rev. D38 (1988) 3351.

1116. η and η' Production in e^+e^- Annihilation at 29 GeV: Indications for the $D_s^\pm$ Decays into $\eta\pi^\pm$ and $\eta'\pi^\pm$
By G. Wormser *et al.*.
10.1103/PhysRevLett.61.1057.
Phys.Rev.Lett. 61 (1988) 1057.

1117. The Tau Missing Decay Modes Problem and Limits on a Second ν_τ
By Martin L. Perl.
10.1103/PhysRevD.38.845.
Phys.Rev. D38 (1988) 845.

1118. Photon Photon Physics with the Mark-II at Pep
By D. Cords *et al.*.

1119. The One Charged Particle Decay Modes of the Tau
By Martin L. Perl.
10.1111/j.1749-6632.1988.tb51543.x.
Annals N.Y.Acad.Sci. 535 (1988) 516-526.

1120. Measurement of the Tau Lifetime
By D. Amidei *et al.*.
10.1103/PhysRevD.37.1750.
Phys.Rev. D37 (1988) 1750.

1121. New Leptons: An Experimental Review
By K.K. Gan, Martin L. Perl.
10.1142/S0217751X88000229.
Int.J.Mod.Phys. A3 (1988) 531.

1122. Determination of α_s from Energy-Energy Correlations in e^+e^- Annihilation at 29 GeV
By D.R. Wood *et al.*.
10.1103/PhysRevD.37.3091.
Phys.Rev. D37 (1988) 3091.

1123. Two Current Experimental Problems in Heavy Lepton Physics: Tau Decay Modes and Close Mass Pairs
By Martin L. Perl.
10.1016/0920-5632(88)90351-9.
Nucl.Phys.Proc.Suppl. 1B (1988) 347-368.

1124. Science in the Age of Accelerators
By Martin L. Perl.
In *Perl, M.L.: Reflections on experimental science* 501-522.

1125. Observation of Ω^- Production in e^+e^- Annihilation at 29 GeV
By S. Klein *et al.*.
10.1103/PhysRevLett.59.2412.
Phys.Rev.Lett. 59 (1987) 2412.

1126. Upper Limit on the Branching Ratio for the Decay $\tau^- \to \pi^- \eta \nu_\tau$
By K.K. Gan *et al.*.
10.1016/0370-2693(87)91056-2.
Phys.Lett. B197 (1987) 561.

1127. Search for Close Mass Lepton Pairs (L^-, L^0)
By D.P. Stoker, Martin L. Perl.

1128. Measurement of the D^0 Lifetime from the Upgraded Mark-II Detector at PEP
By S.R. Wagner *et al.*.
10.1103/PhysRevD.36.2850.
Phys.Rev. D36 (1987) 2850.

1129. Multihadronic Events at $E_{c.m.}$ = 29 GeV and Predictions of QCD Models from $E_{c.m.}$ = 29 GeV to $E_{c.m.}$ = 93 GeV
By A. Petersen *et al.*.
10.1103/PhysRevD.37.1.
Phys.Rev. D37 (1988) 1.

1130. A New Lepton
By M. Perl.
In *Sutton, C. (Ed.): Building The Universe*, 121-129.

1131. The Grains in the Proton
By M. Perl.
In *Sutton, C. (Ed.): Building The Universe*, 67-76.

1132. Observation of Spin-1 $f_1(1285)$ in the Reaction $\gamma\gamma^* \to \eta^0 \pi^+ \pi^-$
By G. Gidal *et al.*.
10.1103/PhysRevLett.59.2012.
Phys.Rev.Lett. 59 (1987) 2012.

1133. Observation of a Spin-1 Resonance in the Reaction $\gamma\gamma^* \to K^0 K^\pm \pi^\pm$
By G. Gidal *et al.*.
10.1103/PhysRevLett.59.2016.
Phys.Rev.Lett. 59 (1987) 2016.

1134. Study of τ Decay Modes with Multiple Neutral Mesons in the Final States
By K.K. Gan *et al.*.
10.1103/PhysRevLett.59.411.
Phys.Rev.Lett. 59 (1987) 411.

1135. The Experimental View of Particles and Forces in 1987: A Picture Outline of the Talk
By Martin L. Perl.

1136. Search for Heavy Neutrino Production at PEP
By Christopher Wendt *et al.*.
10.1103/PhysRevLett.58.1810.
Phys.Rev.Lett. 58 (1987) 1810.

1137. The Reactions $\gamma\gamma \to K^0 K^\pm \pi^\mp$ and $\gamma\gamma^* \to K^0 K^\pm \pi^\mp$
By G. Gidal *et al.*.

1138. Popular and Unpopular Ideas in Particle Physics
By Martin L. Perl.
10.1063/1.881045.
Phys.Today 39 (1986) 27-30.

1139. Small Multiplicity Events in $e^+e^- \to Z^0$ and Unconventional Phenomena
By Martin L. Perl.
In *Watsonville 1987, Proceedings, SLC Physics* 332-353 and SLAC Stanford - SLAC-PUB-4165
(86,REC.MAR.87) 21p.

1140. Limit on the Decay $D^0 \to e^\pm \mu^\mp$
By Keith Riles *et al.*.
10.1103/PhysRevD.35.2914.
Phys.Rev. D35 (1987) 2914.

1141. Observation of Ξ^- Production in e^+e^- Annihilation at 29 GeV
By S. Klein *et al.*.
10.1103/PhysRevLett.58.644.
Phys.Rev.Lett. 58 (1987) 644.

1142. Heavy Leptons in 1986
By Martin L. Perl.
In *Tahoe City 1986, Proceedings, SLC Physics*, 212-219 and SLAC Stanford - SLAC-PUB-4092
(86,Rec.Nov.) 6p.

1143. Comparison of the Particle Flow in Three-Jet and Radiative Two-Jet Events from e^+e^-
Annihilation at $E_{\text{c.m.}}$ = 29 GeV
By P.D. Sheldon *et al.*.
10.1103/PhysRevLett.57.1398.
Phys.Rev.Lett. 57 (1986) 1398.

1144. Measurement of the Branching Fractions of the τ Lepton Using a Tagged Sample of τ Decays
By P. Burchat *et al.*.
10.1103/PhysRevD.35.27.
Phys.Rev. D35 (1987) 27.

1145. Study of the Decay $\tau^- \to \pi^- \pi^- \pi^+ \nu_\tau$
By W.B. Schmidke *et al.*.
10.1103/PhysRevLett.57.527.
Phys.Rev.Lett. 57 (1986) 527-530.

1146. Remarks on Heavy Leptons
By Martin L. Perl.

1147. The Rotor Electrometer: A New Instrument for Bulk Matter Quark Search Experiments
       By John C. Price, Walter R. Innes, S. Klein, Martin L. Perl.
       10.1063/1.1139079.
       Rev.Sci.Instrum. 57 (1986) 2691.

1148. Measurement of the D^0 and D^+ Lifetimes
       By Larry D. Gladney *et al.*.
       10.1103/PhysRevD.34.2601.
       Phys.Rev. D34 (1986) 2601.

1149. The Superconducting Super Collider Project
       By Martin L. Perl.

1150. Three Maxims for Particle Physics
       By Martin L. Perl.
       New Sci. 109N1489 (1986) 24-28.

1151. Track Finding with the Mark II/SLC Drift Chamber
       By Joseph Perl, Andreas S. Schwarz, Abraham Seiden, Alan J. Weinstein.
       10.1016/0168-9002(86)91251-9.
       Nucl.Instrum.Meth. A252 (1986) 616-620.

1152. A Rotor Electrometer for Fractional Charge Searches
       By Walter R. Innes, Martin L. Perl, John C. Price.

1153. Proposal for the Mark II at SLC
       By Cal Tech-LBL-Santa Cruz-Hawaii-Michigan-SLAC Collaboration (G. Trilling *et al.*).

1154. A Study of Noncollinear Two Charged Particle Events Produced in 29 GeV Electron - Positron
       Annihilation
       By Martin L. Perl *et al.*.
       10.1103/PhysRevD.34.3321.
       Phys.Rev. D34 (1986) 3321.

1155. Charged Meson Pair Production in $\gamma\gamma$ Interactions
       By J. Boyer *et al.*.
       10.1103/PhysRevLett.56.207.
       Phys.Rev.Lett. 56 (1986) 207.

1156. Measurement of the Branching Fractions $\tau^- \to \rho^- \nu_\tau$ and $\tau^- \to K^- \nu_\tau$
       By J.M. Yelton *et al.*.
       10.1103/PhysRevLett.56.812.
       Phys.Rev.Lett. 56 (1986) 812.

1157. The Rotor Electrometer: A New Instrument for Fractional Charge Searches
       By John C. Price.

1158. Inclusive Charged Particle Distribution in Nearly Threefold Symmetric Three Jet Events at $E_{c.m.}$ = 29 GeV
By A. Petersen *et al.*.
10.1103/PhysRevLett.55.1954.
Phys.Rev.Lett. 55 (1985) 1954.

1159. Upper Limit of B^0-$\bar{B}^0$ Mixing in e^+e^- Annihilation at 29 GeV
By T. Schaad *et al.*.
10.1016/0370-2693(85)91490-X.
Phys.Lett. B160 (1985) 188.

1160. Measurement of the Branching Fraction for $\tau^- \to 5\pi^\pm (\pi^0) \nu_\tau$ and an Upper Limit on the ν_τ Mass
By P. Burchat *et al.*.
10.1103/PhysRevLett.54.2489.
Phys.Rev.Lett. 54 (1985) 2489.

1161. Search for Monojet Production in e^+e^- Annihilation
By G.J. Feldman *et al.*.
10.1103/PhysRevLett.54.2289.
Phys.Rev.Lett. 54 (1985) 2289.

1162. Improved Upper Limit on ν_τ Mass
By C. Matteuzzi *et al.*.
10.1103/PhysRevD.32.800.
Phys.Rev. D32 (1985) 800.

1163. Charged Multiplicity of Hadronic Events Containing Heavy Quark Jets
By P.C. Rowson *et al.*.
10.1103/PhysRevLett.54.2580.
Phys.Rev.Lett. 54 (1985) 2580-2583.

1164. Searches for Unstable Neutral Leptons in Low Multiplicity Events From Electron - Positron Annihilation
By Martin L. Perl *et al.*.
10.1103/PhysRevD.32.2859.
Phys.Rev. D32 (1985) 2859.

1165. Λ Production in e^+e^- Annihilation at 29 GeV
By C. de la Vaissiere *et al.*.
10.1103/PhysRevLett.54.2071.
Phys.Rev.Lett. 54 (1985) 2071-2074, Phys.Rev.Lett. 55 (1985) 263.

1166. Measurement of $K^\pm$ and K^0 Inclusive Rates in e^+e^- Annihilation at 29 GeV
By H. Schellman *et al.*.
10.1103/PhysRevD.31.3013.
Phys.Rev. D31 (1985) 3013.

1167. The Search for Neutral Leptons
       By Martin L. Perl.

1168. Experiments Beyond the Standard Model
       By Martin L. Perl.

1169. Inclusive Production of Vector Mesons in e^+e^- Annihilation at 29 GeV
       By H. Schellman *et al.*.

1170. An Upper Limit on the ν_τ Mass
       By C. Matteuzzi *et al.*.
       10.1103/PhysRevLett.52.1869.
       Phys.Rev.Lett. 52 (1984) 1869.

1171. A Measurement of the D^0 Lifetime
       By J.M. Yelton *et al.*.
       10.1103/PhysRevLett.52.2019.
       Phys.Rev.Lett. 52 (1984) 2019.

1172. Precise Measurement of the Tau Lifetime
       By J. Jaros *et al.*.
       10.1103/PhysRevLett.51.955.
       Phys.Rev.Lett. 51 (1983) 955.

1173. Pion Pair Production from $\gamma\gamma$ Collisions at PEP
       By J.R. Smith *et al.*.
       10.1103/PhysRevD.30.851.
       Phys.Rev. D30 (1984) 851.

1174. Study of Heavy Quark Production with the Mark II Detector at PEP
       By G.S. Abrams *et al.*.
       10.1063/1.34470.
       AIP Conf.Proc. 113 (1984) 369-383.

1175. Decays of the $\psi(3097)$ to Baryon - Anti-baryon Final States
       By M.W. Eaton *et al.*.
       10.1103/PhysRevD.29.804.
       Phys.Rev. D29 (1984) 804.

1176. Weak Neutral Currents in e^+e^- Collisions at $\sqrt{s} = 29$ GeV
       By M.E. Levi *et al.*.
       10.1103/PhysRevLett.51.1941.
       Phys.Rev.Lett. 51 (1983) 1941.

1177. Measurement of the Lifetime of Bottom Hadrons
       By N. Lockyer *et al.*.
       10.1103/PhysRevLett.51.1316.
       Phys.Rev.Lett. 51 (1983) 1316.

1178. Status and Future Plans for the Mark II Detector at SLAC
By Martin L. Perl.

1179. Limits on J/ψ and Υ Production in e^+e^- Interactions at $\sqrt{s} = 29$ GeV
By C. Matteuzzi *et al.*.
10.1016/0370-2693(83)90745-1.
Phys.Lett. B129 (1983) 141.

1180. Measurement of $\psi(3097)$ and $\psi'(3686)$ Decays into Selected Hadronic Modes
By M.E.B. Franklin *et al.*.
10.1103/PhysRevLett.51.963.
Phys.Rev.Lett. 51 (1983) 963-966.

1181. Inclusive electron Production in e^+e^- Annihilation at 29 GeV
By M.E. Nelson *et al.*.
10.1103/PhysRevLett.50.1542.
Phys.Rev.Lett. 50 (1983) 1542.

1182. Proposal for the Mark II at SLC
By Cal Tech-LBL-Santa Cruz-Hawaii-Michigan-SLAC Collaboration (G. Trilling *et al.*).

1183. Physics with Linear Colliders in the Tev Center-of-Mass Energy Region
By F. Bulos, V. Cook, I. Hinchliffe, Kenneth D. Lane, David E. Pellett, Martin L. Perl, A. Seiden, H. Wiedemann.
eConf C8206282 (1982) 71-81.

1184. Beyond the Standard Model
By Gordon L. Kane, M.L. Perl.
eConf C8206282 (1982) 18-49.

1185. Variation of the Strong Coupling Constant from a Measurement of the Jet Energy Spread in e^+e^- Annihilation
By B. Lohr *et al.*.
10.1016/0370-2693(83)91175-9.
Phys.Lett. B122 (1983) 90.

1186. Precise Measurement of τ Decay Charged Particle Multiplicity Distribution
By C.A. Blocker *et al.*.
10.1103/PhysRevLett.49.1369.
Phys.Rev.Lett. 49 (1982) 1369.

1187. Measurement of $sd\sigma/Dx$ for Hadron Production by e^+e^- Annihilation at $\sqrt{s} = 5.2$ GeV, 6.5 GeV and 29.0 GeV
By J.F. Patrick *et al.*.
10.1103/PhysRevLett.49.1232.
Phys.Rev.Lett. 49 (1982) 1232.

1188. *D*-Meson Production in e^+e^- Annihilation at $E_{c.m.}$ Between 3.8 and 6.7 GeV
By M.W. Coles *et al.*.
10.1103/PhysRevD.26.2190.
Phys.Rev. D26 (1982) 2190.

1189. A Fractional Charge Search
By Walter R. Innes, S. Klein, Martin L. Perl, John C. Price.

1190. Observation of the Decay $\psi \to \gamma\rho^0\rho^0$
By D.L. Burke *et al.*.
10.1103/PhysRevLett.49.632.
Phys.Rev.Lett. 49 (1982) 632.

1191. Limits on the Production of Point - Like, Charged, Spin 0 Particles in e^+e^- Annihilations at 29 GeV
By C.A. Blocker *et al.*.
10.1103/PhysRevLett.49.517.
Phys.Rev.Lett. 49 (1982) 517.

1192. D^{*+} Production in e^+e^- Annihilation at 29 GeV
By J.M. Yelton *et al.*.
10.1103/PhysRevLett.49.430.
Phys.Rev.Lett. 49 (1982) 430.

1193. Observation of Semileptonic Decays of Charmed Baryons
By E. Vella *et al.*.
10.1103/PhysRevLett.48.1515.
Phys.Rev.Lett. 48 (1982) 1515.

1194. Measurement of the Branching Fraction for the Cabibbo-Suppressed Decay $\tau^- \to K^-\nu_\tau$
By C.A. Blocker *et al.*.
10.1103/PhysRevLett.48.1586.
Phys.Rev.Lett. 48 (1982) 1586.

1195. Resonance Production by Two Photon Interactions at Spear
By P. Jenni *et al.*.
10.1103/PhysRevD.27.1031.
Phys.Rev. D27 (1983) 1031.

1196. Measurement of Energy Correlations in $e^+e^- \to$ Hadrons
By D. Schlatter *et al.*.
10.1103/PhysRevLett.49.521.
Phys.Rev.Lett. 49 (1982) 521.

1197. Hadron Production by e^+e^- Annihilation at Center-of-Mass Energies Between 2.6 GeV and 7.8 GeV. Part 2. Jet Structure and Related Inclusive Distributions
By G. Hanson *et al.*.
10.1103/PhysRevD.26.991.
Phys.Rev. D26 (1982) 991.

1198. Experimental Upper Limits on Branching Fractions for Unexpected Decay Modes of the Tau Lepton
By K.G. Hayes *et al.*.
10.1103/PhysRevD.25.2869.
Phys.Rev. D25 (1982) 2869.

1199. Hadron Production by e^+e^- Annihilation at Center-of-Mass Energies Between 2.6 GeV and 7.8 GeV. Part 1. Total Cross-Section, Multiplicities and Inclusive Momentum Distributions
By J. Siegrist *et al.*.
10.1103/PhysRevD.26.969.
Phys.Rev. D26 (1982) 969.

1200. Measurement of the τ Lifetime
By G.J. Feldman *et al.*.
10.1103/PhysRevLett.48.66.
Phys.Rev.Lett. 48 (1982) 66.

1201. A Study of the Decay $\tau^- \to \pi^- \nu_\tau$
By C.A. Blocker *et al.*.
10.1016/0370-2693(82)90476-2.
Phys.Lett. B109 (1982) 119.

1202. Pion Pair Production in Photon - Photon Collisions at Spear
By A. Roussarie *et al.*.
10.1016/0370-2693(81)90894-7.
Phys.Lett. B105 (1981) 304.

1203. The Status of Heavy Lepton Searches
By Martin L. Perl.

1204. A Measurement of the Cross-Section for Four Pion Production in $\gamma\gamma$ Collisions at Spear
By D.L. Burke *et al.*.
10.1016/0370-2693(81)90691-2.
Phys.Lett. B103 (1981) 153.

1205. A Search for Stable Particles Heavier than the Proton and for $Q = -2/3$ Quarks Produced in e^+e^- Annihilation
By J. Weiss *et al.*.
10.1016/0370-2693(81)90171-4.
Phys.Lett. B101 (1981) 439.

1206. Review of New Experimental Upper Limits on Forbidden Decay Modes of the Tau Lepton
By Kenneth G. Hayes, Martin L. Perl.
10.1063/1.32985.
AIP Conf.Proc. 72 (1981) 602-613.

1207. Measurements of the Properties of d Meson Decays
By R.H. Schindler *et al.*.
10.1103/PhysRevD.24.78.
Phys.Rev. D24 (1981) 78.

1208. e^+e^- Physics Today and Tomorrow: Four Tutorial Lectures Delivered at the Arctic School of Physics 1980
By Martin L. Perl.
10.1088/0031-8949/25/1B/008.
Phys.Scripta 25 (1982) 172.

1209. Measurement of the Branching Fraction for the Cabibbo-Suppressed Decay $\tau^- \to K^{*-}(892)\nu_\tau$
By J. Dorfan *et al.*.
10.1103/PhysRevLett.46.215.
Phys.Rev.Lett. 46 (1981) 215.

1210. Observation of the $\eta_c(2980)$ Produced in the Radiative Decay of the $\psi'(3684)$
By T. Himel *et al.*.
10.1103/PhysRevLett.45.1146.
Phys.Rev.Lett. 45 (1980) 1146.

1211. Observation of the Radiative Transition $\psi \to \gamma E(1420)$
By D.L. Scharre *et al.*.
10.1016/0370-2693(80)90612-7.
Phys.Lett. B97 (1980) 329.

1212. Direct Photon Production at the ψ
By D.L. Scharre *et al.*.
10.1103/PhysRevD.23.43.
Phys.Rev. D23 (1981) 43.

1213. The Electron, Muon, and Tau Heavy Lepton: Are they the Truly Elementary Particles?
By Martin L. Perl.
Sci.Teacher 47N9 (1980) 16.

1214. Measurement of $\gamma\gamma\psi$ Final States in ψ' Decay
By T. Himel *et al.*.
10.1103/PhysRevLett.44.920.
Phys.Rev.Lett. 44 (1980) 920.

1215. Leptons — What are they?
By M. Perl.
New Sci. 81 (1979) 564-566.

1216. The Tau Lepton
By Martin L. Perl.
10.1146/annurev.ns.30.120180.001503.
Ann.Rev.Nucl.Part.Sci. 30 (1980) 299-335.

1217. Searching for Heavy Leptons
By Martin L. Perl.

1218. Recent results from the Mark II Detector at Spear
By SLAC-LBL Collaboration (Jonathan Dorfan *et al.*).
10.1063/1.32179.
AIP Conf.Proc. 59 (1980) 159-184.

1219. Results on Two Photon Interactions from Mark II at Spear
By P. Jenni *et al.*.

1220. Measurement of High-Energy Direct Photons in ψ Decays
By G.S. Abrams *et al.*.
10.1103/PhysRevLett.44.114.
Phys.Rev.Lett. 44 (1980) 114.

1221. Observation of Charmed Baryon Production in e^+e^- Annihilation
By G.S. Abrams *et al.*.
10.1103/PhysRevLett.44.10.
Phys.Rev.Lett. 44 (1980) 10.

1222. Results from the Mark II Detector at Spear
By SLAC-LBL Collaboration (V. Luth *et al.*).

1223. Measurement of the Parameters of the $\psi''(3770)$ Resonance
By G.S. Abrams *et al.*.
10.1103/PhysRevD.21.2716.
Phys.Rev. D21 (1980) 2716.

1224. Results from the Mark II Detector at Spear
By G.S. Abrams *et al.*.

1225. Measurement of the Branching Fraction for $\tau \to \rho \nu$
By G.S. Abrams *et al.*.
10.1103/PhysRevLett.43.1555.
Phys.Rev.Lett. 43 (1979) 1555.

1226. Leptons: What are they?
By Martin L. Perl.

1227. Some First Results from the Mark II at Spear
By G.S. Abrams *et al.*.

1228. Observation of Cabibbo-Suppressed Decays $D^0 \to \pi^-\pi^+$ and $D^0 \to K^-K^+$
By G.S. Abrams *et al.*.
10.1103/PhysRevLett.43.481.
Phys.Rev.Lett. 43 (1979) 481.

1229. First Results from Mark II at Spear
By G.S. Abrams *et al.*.

1230. Measurement of the Radiative Width of the η' in Two-Photon Interactions at SPEAR
By G.S. Abrams *et al.*.
10.1103/PhysRevLett.43.477.
Phys.Rev.Lett. 43 (1979) 477.

1231. Inclusive Production of D Mesons in e^+e^- Annihilation at 7 GeV
By Petros A. Rapidis *et al.*.
10.1016/0370-2693(79)91249-8.
Phys.Lett. B84 (1979) 507.

1232. Heavy Leptons
By Martin L. Perl, William T. Kirk.
10.1038/scientificamerican0378-50.
Sci.Am. 238 (1978) 50-57.

1233. Heavy Lepton Phenomenology
By Martin L. Perl.

1234. Electroproduction of Hadrons from Nuclei
By L.S. Osborne *et al.*.
10.1103/PhysRevLett.40.1624.
Phys.Rev.Lett. 40 (1978) 1624.

1235. Inclusive Electroproduction from Protons and Deuterons
By J.F. Martin *et al.*.
10.1103/PhysRevD.20.5.
Phys.Rev. D20 (1979) 5.

1236. The Tau Heavy Lepton: A Recently Discovered Elementary Particle
By Martin L. Perl.
10.1038/275273a0.
Nature 275 (1978) 273-278.

1237. Inclusive γ and π^0 Production in e^+e^- Annihilation
By D.L. Scharre *et al.*.
10.1103/PhysRevLett.41.1005.
Phys.Rev.Lett. 41 (1978) 1005.

1238. An Experimental Summary of the 13th Rencontre De Moriond
By Martin L. Perl.

1239. Inclusive Electron Production in Multiprong Events Produced by e^+e^- Annihilation
By J.M. Feller *et al.*.
10.1103/PhysRevLett.40.1677.
Phys.Rev.Lett. 40 (1978) 1677, Phys.Rev.Lett. 41 (1978) 518.

1240. Observation of $\mu 3\pi$ Events in e^+e^- Annihilation
By J. Jaros *et al.*.
10.1103/PhysRevLett.40.1120.
Phys.Rev.Lett. 40 (1978) 1120.

1241. Lectures on the Quark Model, Ordinary Mesons, Charmed Mesons, and Heavy Leptons
By M.L. Perl.

1242. Summary of Charmed Particle Studies
By G.S. Abrams *et al.*.

1243. The Lead Glass Wall Addition to the Spear Mark I Magnetic Detector
By J.M. Feller *et al.*.
10.1109/TNS.1978.4329321.
IEEE Trans.Nucl.Sci. 25 (1978) 304-308.

1244. Anomalous e^+e^- and $\mu^+\mu^-$ Events Produced in e^+e^- Annihilation
By F.B. Heile *et al.*.
10.1016/0550-3213(78)90243-2.
Nucl.Phys. B138 (1978) 189.

1245. Evidence for, and Properties of, the Tau Lepton
By Martin L. Perl.

1246. Inclusive K^+ and K^- Electroproduction
By J.F. Martin *et al.*.
10.1103/PhysRevLett.40.283.
Phys.Rev.Lett. 40 (1978) 283.

1247. Inclusive Baryon Production in e^+e^- Annihilation
By M. Piccolo *et al.*.
10.1103/PhysRevLett.39.1503.
Phys.Rev.Lett. 39 (1977) 1503.

1248. Measurement of Semileptonic Decays of d Mesons to Electrons at the $\psi''(3772)$
By J.M. Feller *et al.*.
10.1103/PhysRevLett.40.274.
Phys.Rev.Lett. 40 (1978) 274.

1249. Review of Heavy Lepton Production in e^+e^- Annihilation
By Martin L. Perl.
In *Perl, M.L.: Reflections on experimental science* 197-216.

1250 Observation of D^0 Meson Decay into $K^-\pi^+\pi^0$
By D.L. Scharre *et al.*.
10.1103/PhysRevLett.40.74.
Phys.Rev.Lett. 40 (1978) 74.

1251. Study of D Mesons Produced in the Decay of the $\psi''(3772)$
By I. Peruzzi *et al.*.
10.1103/PhysRevLett.39.1301.
Phys.Rev.Lett. 39 (1977) 1301.

1252. Properties of the Proposed Tau Charged Lepton
       By Martin L. Perl *et al.*.
       10.1016/0370-2693(77)90421-X.
       Phys.Lett. B70 (1977) 487.

1253. A Search for Neutral Heavy Leptons
       By D.I. Meyer *et al.*.
       10.1016/0370-2693(77)90416-6.
       Phys.Lett. B70 (1977) 469.

1254. Comments on the Tau Heavy Lepton
       By Martin L. Perl.

1255. Radiative Decays of the $\psi'(3684)$ into High Mass States
       By William M. Tanenbaum *et al.*.
       10.1103/PhysRevD.17.1731.
       Phys.Rev. D17 (1978) 1731.

1256. Electron-Muon and Electron-Hadron Production in e^+e^- Collisions
       By Angela Barbaro-Galtieri *et al.*.
       10.1103/PhysRevLett.39.1058.
       Phys.Rev.Lett. 39 (1977) 1058.

1257. D Meson Production and Decay in e^+e^- Annihilation at 4.03 GeV and 4.41 GeV Center-of-Mass
       Energy
       By M. Piccolo *et al.*.
       10.1016/0370-2693(77)90534-2.
       Phys.Lett. B70 (1977) 260.

1258. D and D^* Meson Production Near 4 GeV in e^+e^- Annihilation
       By G. Goldhaber *et al.*.
       10.1016/0370-2693(77)90855-3.
       Phys.Lett. B69 (1977) 503.

1259. Recent Results in Electron-Positron Annihilation Above 2 GeV
       By G.J. Feldman, Martin L. Perl.
       10.1016/0370-1573(77)90024-2.
       Phys.Rept. 33 (1977) 285-365.

1260. Observation of a Resonance in e^+e^- Annihilation Just Above Charm Threshold
       By Petros A. Rapidis *et al.*.
       10.1103/PhysRevLett.39.526.
       Phys.Rev.Lett. 39 (1977) 526, Phys.Rev.Lett. 39 (1977) 974.

1261. K^0 Production in e^+e^- Annihilation
       By V. Luth *et al.*.
       10.1016/0370-2693(77)90359-8.
       Phys.Lett. B70 (1977) 120.

1262. Spin Analysis of Charmed Mesons Produced in e^+e^- Annihilation
By H.K. Nguyen *et al.*.
10.1103/PhysRevLett.39.262.
Phys.Rev.Lett. 39 (1977) 262-265.

1263. Evidence for, and Properties of, the New Charged Heavy Lepton
By Martin L. Perl.

1264. Observation of the Decay $D^{*+} \to D^0 \pi^+$
By G.J. Feldman *et al.*.
10.1103/PhysRevLett.38.1313.
Phys.Rev.Lett. 38 (1977) 1313.

1265. Proposal for Checkout of the Mark II Magnetic Detector at Spear
By M.S. Alam *et al.*.

1266. Studies of the $s^{(1/2)} = 3$ GeV to 8 GeV Region Using the Mark II Detector at Spear
By M.S. Alam *et al.*.

1267. A General Survey of Particle Production at PEP
By M.S. Alam *et al.*.

1268. Evidence for Parity Violation in the Decays of the Narrow States Near 1.87 GeV/c^2
By J. Wiss *et al.*.
10.1103/PhysRevLett.37.1531.
Phys.Rev.Lett. 37 (1976) 1531-1534.

1269. Inclusive Anomalous Muon Production in e^+e^- Annihilation
By G.J. Feldman *et al.*.
10.1103/PhysRevLett.38.117.
Phys.Rev.Lett. 38 (1977) 117, Phys.Rev.Lett. 38 (1977) 576.

1270. The Total Cross-Section in e^+e^- Annihilation and the New Particles
By Martin L. Perl.
Comments Nucl.Part.Phys. 7 (1977) 2, 55-63.

1271. Radiative Decays of $\psi(3095)$ and $\psi'(3684)$
By John Scott Whitaker *et al.*.
10.1103/PhysRevLett.37.1596.
Phys.Rev.Lett. 37 (1976) 1596.

1272. Observation of a Narrow Charged State at 1876 MeV/c^2 Decaying to an Exotic Combination
of $K\pi\pi$
By I. Peruzzi *et al.*.
10.1103/PhysRevLett.37.569.
Phys.Rev.Lett. 37 (1976) 569-571.

1273. Properties of Anomalous $e\mu$ Events Produced in e^+e^- Annihilation
By Martin L. Perl *et al.*.
10.1016/0370-2693(76)90399-3.
Phys.Lett. B63 (1976) 466.

1274. Particle Ratios in Inclusive Electroproduction from Hydrogen and Deuterium
By J.F. Martin *et al.*.
10.1016/0370-2693(76)90448-2.
Phys.Lett. B65 (1976) 483.

1275. Interpretation of Anomalous $e\mu$ Events Produced in e^+e^- Annihilation
By Martin L. Perl.

1276. Threshold and Other Properties of u Particle Production in e^+e^- Annihilation
By Martin L. Perl.

1277. Observation in e^+e^- Annihilation of a Narrow State at 1865 MeV/c^2 Decaying to $K\pi$ and $K\pi\pi$
By G. Goldhaber *et al.*.
10.1103/PhysRevLett.37.255.
Phys.Rev.Lett. 37 (1976) 255-259.

1278. ψ(3095) Decays Involving Kaons
By F. Vannucci *et al.*.
Submitted to: Phys.Rev.Lett..

1279. Observation of a Resonance at 4.4 GeV and Additional Structure Near 4.1 GeV in e^+e^- Annihilation
By J. Siegrist *et al.*.
10.1103/PhysRevLett.36.700.
Phys.Rev.Lett. 36 (1976) 700.

1280. Production of $4\pi^\pm$ and $6\pi^\pm$ by e^+e^- Annihilation Between 2.4 GeV and 7.4 GeV
By B. Jean-Marie *et al.*.
Submitted to: Phys.Rev.Lett..

1281. Study of Anomalous Electron Production and of Photon and π^0 Production in e^+e^- Annihilation
By M. Alston-Garnjost *et al.*.

1282. The $e\mu$ Events Produced in e^+e^- Annihilation
By M.L. Perl.

1283. Study of ψ(3.1) at SPEAR
By B. Jean-Marie *et al.*.
In *Paris 1975, Neutrino Physics At High Energies*, Paris 1975, 1-16.

1284. The New Particles Produced in Electron-Positron Annihilation
By M.L. Perl.

1285. A Survey of Muon Pair Photoproduction at 20.5 GeV
By J.T. Dakin *et al.*.

1286. Observation of the Decay $\psi'(3684) \to \psi(3095)\,\eta$
By William M. Tanenbaum *et al.*.
10.1103/PhysRevLett.36.402.
Phys.Rev.Lett. 36 (1976) 402.

1287. The $\psi(3.1)$ and the Search for Other Narrow Resonances of Spear
By Martin Breidenbach *et al.*.

1288. Determination of the G Parity and Isospin of $\psi(3095)$ by Study of Multi-Pion Decays
By B. Jean-Marie *et al.*.
10.1103/PhysRevLett.36.291.
Phys.Rev.Lett. 36 (1976) 291.

1289. Properties of $e\mu$ Events Produced in e^+e^- Annihilation
By Martin L. Perl.

1290. Radiative Decays of the $\psi'(3684)$ to New High Mass States
By G.S. Abrams *et al.*.

1291. Evidence for Jet Structure in Hadron Production by e^+e^- Annihilation
By G. Hanson *et al.*.
10.1103/PhysRevLett.35.1609.
Phys.Rev.Lett. 35 (1975) 1609-1612.

1292. Lectures on Electron-Positron Annihilation. Part 1 — The Production of Hadrons and ψ Particles
By Martin L. Perl.

1293. Observation of an Intermediate State in $\psi'(3684)$ Radiative Cascade Decay
By William M. Tanenbaum *et al.*.
10.1103/PhysRevLett.35.1323.
Phys.Rev.Lett. 35 (1975) 1323.

1294. Decay into Strange Baryon-Anti-Baryon Pairs and an I Spin Determination of the $\psi(3095)$
By G. Goldhaber *et al.*.

1295. Azimuthal Asymmetry in Inclusive Hadron Production by e^+e^- Annihilation
By R. Schwitters *et al.*.
10.1103/PhysRevLett.35.1320.
Phys.Rev.Lett. 35 (1975) 1320.

1296. Lectures on Electron-Positron Annihilation. 2. Anomalous Lepton Production
By Martin L. Perl.
In *Perl, M.L.: Reflections on experimental science* 139-192.

1297. Limits on Charmed Meson Production in e^+e^- Annihilation at 4.8 GeV Center-of-Mass Energy
By A. Boyarski *et al.*.
10.1103/PhysRevLett.35.196.
Phys.Rev.Lett. 35 (1975) 196.

1298. Some Properties of the $\psi'(3.7)$ Resonance, and Features of the Total Hadronic Cross-Section in e^+e^- Annihilation from 2.4 GeV to 5.0 GeV Center-of-Mass Energy
By J.A. Kadyk *et al.*.

1299. The $e\mu$ Events Produced in e^+e^- Annihilation
By Martin L. Perl.

1300. The Quantum Numbers and Decay Widths of the $\psi(3095)$
By A. Boyarski *et al.*.
10.1103/PhysRevLett.34.1357.
Phys.Rev.Lett. 34 (1975) 1357.

1301. Quantum Numbers and Decay Modes of the Resonances $\psi(3095)$ and $\psi(3684)$
By A. Boyarski *et al.*.

1302. Electron-Positron Annihilation Above 2 GeV and the New Particles
By G.J. Feldman, Martin L. Perl.
10.1016/0370-1573(75)90031-9.
Phys.Rept. 19 (1975) 233-293.

1303. $\psi'(3684)$ Radiative Decays to High Mass States
By G.J. Feldman *et al.*.
10.1103/PhysRevLett.35.821.
Phys.Rev.Lett. 35 (1975) 821, Phys.Rev.Lett. 35 (1975) 1184.

1304. The Quantum Numbers and Decay Widths of the $\psi'(3684)$
By V. Luth *et al.*.
10.1103/PhysRevLett.35.1124.
Phys.Rev.Lett. 35 (1975) 1124-1126.

1305. Evidence for Anomalous Lepton Production in e^+e^- Annihilation
By Martin L. Perl *et al.*.
10.1103/PhysRevLett.35.1489.
Phys.Rev.Lett. 35 (1975) 1489-1492.

1306. The New Particles Produced in Electron-Positron Annihilation
By Martin L. Perl.

1307. The Decay of $\psi(3700)$ into $\psi(3100)$
By G.S. Abrams *et al.*.
10.1103/PhysRevLett.34.1181.
Phys.Rev.Lett. 34 (1975) 1181.

1308. Recent Results for e^+e^- Annihilation at SPEAR
By H.L. Lynch *et al.*.

1309. Search for Narrow Resonances in e^+e^- Annihilation in the Mass Region 3.2 GeV to 5.9 GeV
By A. Boyarski *et al.*.
10.1103/PhysRevLett.34.762.
Phys.Rev.Lett. 34 (1975) 762.

1310. Total Cross-Section for Hadron Production by Electron-Positron Annihilation Between 2.4 GeV and 5.0 GeV Center-of-Mass Energy
By J.E. Augustin *et al.*.
10.1103/PhysRevLett.34.764.
Phys.Rev.Lett. 34 (1975) 764.

1311. Muon Pair Photoproduction at 20.5 GeV
By J.T. Dakin *et al.*.
10.1016/0370-2693(75)90330-5.
Phys.Lett. B56 (1975) 405.

1312. A Proposal to Search for Heavy Leptons at SPEAR
By G.J. Feldman *et al.*.

1313. Preliminary Results on Hadron Production in Electron Positron Collisions at SPEAR
By J.E. Augustin *et al.*.
In *Meribel-les-allues 1974, Vol.2, High Energy Leptonic Interactions*, Orsay 1975, 139-160.

1314. ψ Photoproduction at SLAC
By J.T. Dakin *et al.*.

1315. Experimental Upper Limit on the Photoproduction Cross-Section for the $\psi(3105)$
By J.F. Martin *et al.*.
10.1103/PhysRevLett.34.288.
Phys.Rev.Lett. 34 (1975) 288.

1316. Discovery of a Narrow Resonance in e^+e^- Annihilation
By SLAC-SP-017 Collaboration (J.E. Augustin *et al.*).
10.1103/PhysRevLett.33.1406.
Phys.Rev.Lett. 33 (1974) 1406-1408, Adv.Exp.Phys. 5 (1976) 141.

1317. Measurement of $e^+e^- \to e^+e^-$ and $e^+e^- \to \mu^+\mu^-$ at SPEAR
By J.E. Augustin *et al.*.
10.1103/PhysRevLett.34.233.
Phys.Rev.Lett. 34 (1975) 233.

1318. The Search for Heavy Leptons and Muon - Electron Differences
By Martin L. Perl, Petros A. Rapidis.
In *Perl, M.L.: Reflections on experimental science* 48-138.

1319. Measurement of Inclusive Hadron Electroproduction from Hydrogen and Deuterium
       By J.T. Dakin, G.J. Feldman, F. Martin, Martin L. Perl, W.T. Toner.
       10.1103/PhysRevD.10.1401.
       Phys.Rev. D10 (1974) 1401.

1320. The Discovery of a Second Narrow Resonance in e^+e^- Annihilation
       By G.S. Abrams *et al.*.
       10.1103/PhysRevLett.33.1453.
       Phys.Rev.Lett. 33 (1974) 1453-1455, Adv.Exp.Phys. 5 (1976) 150.

1321. 2nd Generation Electronic Neutrino Detector at NAL
       By David H. Frisch, David Luckey, L. Osborne, B. COx, M. Perl.

1322. An Experimental Survey of Positron - Electron Annihilation into Multiparticle Final States in
       the S Range 27 GeV^2 To 74 GeV^2
       By A. Boyarski *et al.*.

1323. Electroproduction of Hadrons from Deuterium
       By J.T. Dakin, G.J. Feldman, F. Martin, Martin L. Perl, W.T. Toner.
       10.1103/PhysRevLett.31.786.
       Phys.Rev.Lett. 31 (1973) 786.

1324. Measurement of ρ^0 and ψ Meson Electroproduction
       By J.T. Dakin, G.J. Feldman, W.L. Lakin, F. Martin, Martin L. Perl, E.W. Petraske, W.T. Toner.
       10.1103/PhysRevD.8.687.
       Phys.Rev. D8 (1973) 687.

1325. Inclusive Electroproduction of Hadrons at 19.5 GeV
       By J.T. Dakin, G.J. Feldman, W.L. Lakin, F. Martin, M.L. Perl, E.W. Petraske, W.T. Toner.
       eConf C720906V2 (1972) 83-97.

1326. An Experimental Survey of Positron - Electron Annihilation into Multiparticle Final States
       in the Center-of-Mass Energy Range 2 GeV to 5 GeV
       By A. Boyarski *et al.*.

1327. A Proposal to Study the Electroproduction of Hadrons
       By J.T. Dakin, G.J. Feldman, F. Martin, Martin L. Perl, R.L. Lanza, David Luckey, J.F. Martin,
       L.S. Osborne.

1328. Isolation of Isoscalar Exchange in $\pi^\pm P \to \rho^\pm P$ at 15 GeV/c
       By W.T. Kaune, W.L. Lakin, Martin L. Perl, E.W. Petraske, J.C. Pratt, J. Tenenbaum.

1329. Electroproduction of ρ and ψ Mesons
       By J.T. Dakin, G.J. Feldman, W.L. Lakin, F. Martin, Martin L. Perl, E.W. Petraske, W.T. Toner.
       10.1103/PhysRevLett.30.142.
       Phys.Rev.Lett. 30 (1973) 142.

1330. Preliminary Results on the Inclusive Electroproduction of Hadrons
By J.T. Dakin, G.J. Feldman, W.L. Lakin, F. Martin, Martin L. Perl, E.W. Petraske, W.T. Toner.

1331. Measurement of the Inclusive Electroproduction of Hadrons
By J.T. Dakin, G.J. Feldman, W.L. Lakin, F. Martin, Martin L. Perl, E.W. Petraske, W.T. Toner.
10.1103/PhysRevLett.29.746.
Phys.Rev.Lett. 29 (1972) 746-749.

1332. Comparison of Muon - Proton And Electron - Proton Inelastic Scattering
By T. Braunstein, W.L. Lakin, F. Martin, Martin L. Perl, W.T. Toner, T.F. Zipf.
10.1103/PhysRevD.6.106.
Phys.Rev. D6 (1972) 106.

1333. A Spark Chamber Measurement of the Reactions $\pi^\pm P \to \rho^\pm P$ at 15 GeV/c
By J.C. Pratt, W.T. Kaune, W.L. Lakin, Martin L. Perl, E.W. Petraske, J. Tenenbaum, W.T. Toner.
10.1016/0370-2693(72)90601-6.
Phys.Lett. B41 (1972) 383.

1334. Comparison of Muon-Proton and Electron-Proton Deep Inelastic Scattering
By W.T. Toner, T.J. Braunstein, W.L. Lakin, F. Martin, M.L. Perl, T.F. Zipf, H.C. Bryant, B.D. Dieterle.
10.1016/0370-2693(71)90080-3.
Phys.Lett. B36 (1971) 251-256.

1335. Neutron-Proton Elastic Scattering from 5 to 30 GeV/c
By B.G. Gibbard, M.J. Longo, Lawrence W. Jones, J.R. O' Fallon, M.N. Kreisler, M.L. Perl.
10.1016/0550-3213(71)90279-3.
Nucl.Phys. B30 (1971) 77-115.

1336. Measurement of the Neutron Form-Factors at a Timelike Momentum Transfer of 4 GeV/c^2
By J.T. Dakin, G.J. Feldman, F. Martin, Martin L. Perl.

1337. Proposal for a Magnetic Detector for SPEAR
By A. Boyarski *et al.*.

1338. How does the Muon Differ from the Electron?
By Martin L. Perl.
10.1063/1.3022841.
Phys.Today 24N7 (1971) 34-44.

1339. Muon - Proton Inelastic Scattering
By B. Dieterle, W. Lakin, F. Martin, M. Perl, E. Petraske, J. Tenenbaum, W. Toner, T. Zipf.

1340. A Proposal for an Experiment to Study Electroproduced Hadrons
By B. Dieterle, W.L. Lakin, F. Martin, E.W. Petraske, Martin L. Perl, J. Tenenbaum, W.T. Toner.

1341. Experimental Determination of the Inelastic Neutron Form-Factor by the Scattering of 12 GeV Muons on Hydrogen, Carbon and Copper
By W.L. Lakin, T. Braunstein, Jack Cox, B. Dieterle, Martin L. Perl, W.T. Toner, T.F. Zipf, Howard C. Bryant.
10.1103/PhysRevLett.26.34.
Phys.Rev.Lett. 26 (1971) 34.

1342. Neutron - Proton Elastic Scattering from 2 GeV/c to 7 GeV/c
By Martin L. Perl, Jack Cox, Michael J. Longo, M. Kreisler.
10.1103/PhysRevD.1.1857.
Phys.Rev. D1 (1970) 1857.

1343. Muon Proton Inelastic Scattering and Vector Dominance
By Martin L. Perl *et al.*.
10.1103/PhysRevLett.23.1191.
Phys.Rev.Lett. 23 (1969) 1191-1194.

1344. Muon Proton Inelastic Scattering q^2 less than 1.2 GeV/c^2
By B. Dieterle *et al.*.
10.1103/PhysRevLett.23.1187.
Phys.Rev.Lett. 23 (1969) 1187-1190.

1345. Observation of $Y^{*+}(1385)K^+$ and $N^{*++}(1236)\rho^0$ Decays of a Nucleon Resonance
By W. Chinowsky *et al.*.
10.1103/PhysRev.171.1421.
Phys.Rev. 171 (1968) 1421-1428.

1346. Differential Cross Sections for $p^+ p^- \rightarrow d^+ \pi^+$ from 1 to 3 BeV
By R.M. Heinz, O.E. Overseth, David E. Pellett, M.L. Perl.
10.1103/PhysRev.167.1232.
Phys.Rev. 167 (1968) 1232-1239.

1347. Production of K Mesons in Three-Body States in Proton-Proton Interactions at 6 BeV/c
By W. Chinowsky, R.R. Kinsey, S.L. Klein, M. Mandelkern, J. Schultz, F. Martin, M.L. Perl, T.H. Tan.
10.1103/PhysRev.165.1466.
Phys.Rev. 165 (1968) 1466-1478.

1348. High Statistics Study of the Production of Charged $\rho^\pm$ Mesons, Neutral ρ^0 Mesons, F^0 Mesons and Nucleon Isobars by Pions
By J.C. Pratt, W.T. Kaune, Jack Cox, B. Dieterle, Martin L. Perl, J. Tenenbaum, W.T. Toner, T.F. Zipf.

1349. Progress Report on Muon and Electron Beam for NAL
By M.L. Perl, Richard Wilson.

1350. Some Considerations on the Design of a Minimal NAL Target Station, Based on Experience at SLAC
By M.L. Perl.

1351. Inelastic Muon Proton Experiments at NAL
By M.L. Perl.
eConf C680615 (1968) 21-40.

1352. Progress Report on Group B. 4, Neutral Beams
By M.L. Perl.

1353. Production and Decay of A1 and A2 Resonances in 16 GeV/c $\pi^- p$ Interactions
By Joseph Ballam *et al.*.
10.1103/PhysRevLett.21.934.
Phys.Rev.Lett. 21 (1968) 934-937.

1354. Large Angle Neutron - Proton Elastic Scattering from 3 Gev/c to 6.8 GeV/c
By Jack Cox, Martin L. Perl, M. Kreisler, Michael J. Longo, S.T. Powell, III.
10.1103/PhysRevLett.21.641.
Phys.Rev.Lett. 21 (1968) 641-644.

1355. A Search for New Particles Produced by High-Energy Photons
By Arpad Barna, Jack Cox, F. Martin, Martin L. Perl, T.H. Tan, W.T. Toner, T.F. Zipf, E.H. Bellamy.
10.1103/PhysRev.173.1391.
Phys.Rev. 173 (1968) 1391-1402.

1356. Analysis of Muon Hydrogen Bubble Chamber Experiments
By Martin L. Perl.

1357. Neutron - Proton Elastic Scattering 8 GeV/c to 30 GeV/c
By Bruce G. Gibbard, Lawrence W. Jones, Michael J. Longo, John R. O'Fallon, Jack Cox, Martin L. Perl, W.T. Toner, M. Kreisler.
10.1103/PhysRevLett.24.22.
Phys.Rev.Lett. 24 (1970) 22-24.

1358. An Investigation of the 1.4 GeV/c^2 Nucleon Isobar in Proton Proton Interactions
By T.H. Tan *et al.*.
10.1016/0370-2693(68)90014-2.
Phys.Lett. B28 (1969) 195-198.

1359. A High-energy Small Phase Space Volume Muon Beam
By Jack Cox, F. Martin, Martin L. Perl, T.H. Tan, W.T. Toner, T.F. Zipf, W.L. Lakin.
10.1016/0029-554X(69)90575-8.
Nucl.Instrum.Meth. 69 (1969) 77-88.

1360. Two-Body Processes with Large Momentum Transfer
By Martin L. Perl.
Proc. of Topical Conf. on High-Energy Collisions of Hadrons, CERN, Jan 1968. (CERN-68-7, vol. 1) p. 252-89.

1361. Comparison of High-Energy Neutron - Proton and Proton Proton Elastic Scattering
       By Jack Cox, Martin L. Perl, M. Kreisler, Michael J. Longo.
       10.1103/PhysRevLett.21.645.
       Phys.Rev.Lett. 21 (1968) 645-647.

1362. Proposal for an Experiment to Search for Fractionally Charged Particles
       By E.H. Bellamy *et al.*.

1363. Energy Loss and Straggling of High-Energy Muons in NaI(Tl)
       By E.H. Bellamy, R. Hofstadter, W.L. Lakin, Jack Cox, Martin L. Perl, W.T. Toner, T.F. Zipf.
       10.1103/PhysRev.164.417.
       Phys.Rev. 164 (1967) 417-420.

1364. An Experimental Limit on High-Energy Diffraction Photoproduction of the ψ Meson
       By Yung-Su Tsai, Jack Cox, F. Martin, Martin L. Perl, T.H. Tan, W.T. Toner, T.F. Zipf.
       10.1103/PhysRevLett.19.915.
       Phys.Rev.Lett. 19 (1967) 915-917.

1365. A Search for Fractionally Charged Particles
       By E.H. Bellamy, R. Hofstadter, W.L. Lakin, Martin L. Perl, W.T. Toner.
       10.1103/PhysRev.166.1391.
       Phys.Rev. 166 (1968) 1391-1394.

1366. Secondary Particle Yields at 0-degrees from the New Stanford Electron Accelerator
       By Arpad Barna, Jack Cox, F. Martin, Martin L. Perl, T.H. Tan, W.T. Toner, T.F. Zipf,
       E.H. Bellamy.
       10.1103/PhysRevLett.18.360.
       Phys.Rev.Lett. 18 (1967) 360-362.

1367. Proposal to Search for New Particles
       By Arpad Barna, Jack Cox, F. Martin, Martin L. Perl, T.H. Tan, W.T. Toner, T.F. Zipf.

1368. Proposal for a Survey of the Mu P Inelastic Interaction at High-Energy
       By J.L. Brown, Jack Cox, F. Martin, Martin L. Perl, T.H. Tan, W.T. Toner, T.F. Zipf.

1369. Neutron Proton Elastic Scattering from 1 GeV to 6 GeV
       By M. Kreisler, F. Martin, Martin L. Perl, Michael J. Longo, S.T. Powell, III.
       10.1103/PhysRevLett.16.1217.
       Phys.Rev.Lett. 16 (1966) 1217-1220.

1370. Pion-Proton Elastic Scattering from 3 GeV/c to 5 GeV/c
       By M.L. Perl, Lawrence W. Jones, C.C. Ting.
       10.1103/PhysRev.132.1252.
       Phys.Rev. 132 (1963) 1252-1272.

1371. The Tau - Charm Factory in the Era of B Factories and CESR. Proceeding, Conference,
       Stanford, USA, August 15-16, 1994. By L.V. (ed.) Beers, M.L. (ed.) Perl

1372. Demonstration of Parity Nonconservation in Hyperon Decay
By F. Eisler *et al.*.
10.1103/PhysRev.108.1353.
Phys.Rev. 108 (1957) 1353-1355.

1373. Empirical Partial Wave Analysis of $\pi + p$ Elastic Scattering above 1 GeV/c
By Martin L. Perl, Mary C. Corey.
10.1103/PhysRev.136.B787.
Phys.Rev. 136 (1964) B787-803.

1374. Some Aspects of the Prospective Experimental Use of the Stanford 2-mile Accelerator
By William Chinowsky *et al.*.

1375. Large Angle $\pi^- + p$ Elastic Scattering at 3.63 GeV/c
By Martin L. Perl, Yong Yung Lee, Erwin Marquit.
10.1103/PhysRev.138.B707.
Phys.Rev. 138 (1965) B707-711.

Peter B. Hirsch

PETER BERNHARD HIRSCH

Birthplace

Born	Berlin, Germany, 16 January 1925
Married	Mabel Anne Kellar, widow of James Noel Kellar, née Stephens, 1 stepson, 1 stepdaughter

Education

1939–1943	Sloane School, Chelsea, London,
1943–1946	St Catharine's College, University of Cambridge, Natural Sciences Tripos, Part II Physics, 1st Class Honours B.A. 1946
1946–1950	Research Student, Cavendish Laboratory, Cambridge, (Supervisor Dr W.H. Taylor), M.A. 1950, PhD 1951

Career

1959–1964	Lecturer in Physics, University of Cambridge
1960–1966	Fellow of Christ's College, Cambridge,
1964–1966	Reader in Physics, University of Cambridge
1966–1992	Isaac Wolfson Professor of Metallurgy, University of Oxford
1992 to date	Emeritus Professor, Dept. of Materials, University of Oxford
1966 to date	Fellow of St Edmund Hall, Oxford
1970–1973	Chairman: Metallurgy and Materials Committee and Member of Engineering Board of Science

Research Council

1982–1984	Chairman: United Kingdom Atomic Energy Authority
1982–1994	Member of United Kingdom Atomic Energy Authority
1982–1987	Chairman: AEA Light Water Reactor Pressure Vessel Study Group
1985–1989	Director of Cogent Limited
1986 to date	Honorary Professor, Beijing University of Iron and Steel Technology
1988–1996	Chairman: Isis Innovation Limited
1993–2002	Chairman: Technical Advisory Group on Structural Integrity
1994–1998	Director (non-executive) Rolls Royce Associates
1996–2000	Chairman: Materials and Processes Advisory Board, Rolls Royce Plc
2000–2001	Director (non-executive) Oxford Medical Image Analysis

Awards

Rosenhain Medal of Institute of Metals 1961; C.V. Boyes Prize of Institute of Physics and Physical Society 1962; Clamer Medal of Franklin Institute 1970; Wihuri International Prize, Helsinki 1971; Hughes Medal of the Royal Society 1973; Platinum Medal of the Metals Society 1976; Royal Medal of the Royal Society 1977; A.A. Griffith Silver Medal, Materials Science Club

1979; Wolf Prize in Physics 1984 (jointly with E. Hahn and T. Maiman); Arthur Von Hippel Award of the Materials Research Society 1983; Bruce Preller Lecture of the Royal Society of Edinburgh 1985; Hatfield Memorial Lecture 1985; Electron Microscopy Society of America Distinguished Scientist Award 1986; Holweck Prize, Institute of Physics and French Physical Society 1988; Gold Medal, Japan Institute of Metals 1989; Scripta Metallurgica et Materialia Outstanding Paper Award for 1992; Acta Metallurgica Gold Medal 1997; G.I. Taylor Lecture, Cambridge Philosophical Society 1994; Royal Society UK-Canada Rutherford Lecture 1992; Van Horn Distinguished Lecturer, CASE Western Reserve University 1993; Heyn Medal of German Society for Materials Science 2002; Lomonosov Gold Medal of the Russian Academy of Sciences 2005

Honour

1975   Knight Bachelor

Honorary Degrees and Fellowships

Hon. Fellow Christ's College, Cambridge 1978; Hon. D.Sc. Newcastle University 1979; Hon. D.Sc. City University 1979; Hon. D.Sc. Northwestern 1982; Hon. Fellow St Catharine's College, Cambridge 1982; Hon. Sc.D. East Anglia 1983; Fellow of Imperial College, London 1988; Hon. D.Eng. Liverpool 1991; Hon. D.Eng. Birmingham 1993; Hon. D. University of York 2013.

Honorary Membership of Professional Societies

Fellow of the Royal Society, 1963; Life Fellow of the Franklin Institute; Hon. Member of Spanish Electron Microscopy Society, 1974; Hon. Member of French Electron Microscopy Society; Hon. Fellow of the Royal Microscopical Society 1977; Hon. Fellow of the Japan Society of Electron Microscopy 1979; Hon. Fellow of the Japan Institute of Metals 1989; Hon. Member of the Materials Research Society of India 1990; Member of Academia Europaea 1989; Hon. Member of the Chinese Electron Microscope Society, 1992; Foreign Member of the Royal Academy of Science and Fine Arts of Belgium 1995; Foreign Associate of the US National Academy of Engineering 2001; Honorary Fellow of the Institute of Materials 2002; Foreign Honorary Member of American Academy of Arts and Sciences, 2005; Foreign Member of the Russian Academy of Sciences 2006.

COMMENTARY

I was born in Berlin, Germany, on 16th January 1925. Because of the Nazi persecution of the Jews, my family emigrated to England. My brother, mother, I and my stepfather all went separately, in that order, between the summer of 1938 and the Spring of 1939. I arrived in England (almost 14 years old) on 1st January 1939 on a Kindertransport, and stayed with my guarantor's kind family in London. Two weeks later, I enrolled in Sloane School in Chelsea, a London County Council (LCC) Secondary School. At the beginning, of the war the school was evacuated to Addlestone in Surrey. After taking the General School Certificate in 1941, I was awarded a LCC Intermediate Scholarship, which enabled me to stay on in the sixth form at school. I chose the science option partly because I thought it would be easier to get jobs with a scientific training. But, unlike my (older) brother, I had no burning desire to study Physics at this time. In spite of the limited laboratory facilities the science teaching was rather good, and the mathematics teaching excellent.

In the spring of 1943 I took the Entrance Examination (aimed at attracting students from state schools, who could not stay on for a third year in the sixth form) run by St Catharine's College and Selwyn College in Cambridge, and was awarded an Exhibition in Natural Sciences at St Catharine's College. I passed the Higher School Certificate of Education in 1943 in Physics, Chemistry, Pure and Applied Mathematics, gaining distinctions in the last three subjects. The LCC awarded me a Senior County Scholarship, which enabled me to take up my place in Cambridge.

I went up to Cambridge in 1943 to read Natural Sciences, and took Physics, Chemistry, Mathematics and Mineralogy for Part I of the Tripos. In those days one had to take three experimental subjects, thus being introduced to a generally unfamiliar non-school subject. I found Mineralogy and Crystallography in particular, fascinating, and this had an important influence on my future career. By the end of Part I, my interest was in Physics, which was my choice for Part II of the Tripos. I obtained first class Honours in Part II Physics and got my B.A. in 1946.

After my Part II, I joined the Crystallography Department of the Cavendish Laboratory as a research student. I was ineligible for a grant from the Department of Scientific and Industrial Research (D.S.I.R.) because of my nationality, but Dr. W.H. Taylor, the Head of the Department, obtained a maintenance grant for me from the British Iron and Steel Research Association.

The topic for my Ph.D. research was the development of a microbeam X-ray diffraction technique and its application to the study of cold-worked metals. The idea of the project came from Professor Sir Lawrence Bragg, who was Head of the Cavendish Laboratory at that time, but my supervisor was Dr. W. H. Taylor. At the time Bragg was interested in the structure and strength of cold worked metals. When a polycrystalline metal is cold-worked, the X-ray diffraction rings are broadened, and the question was whether the broadening is due to the small size of the subgrains into which the original grains were assumed to break up, or whether it is due to strains in the original grains. The idea was that if the X-ray beam diameter is sufficiently small, spotty diffraction rings would be obtained, and from the number of spots the particle size could be determined. The technique required the construction of a high intensity X-ray tube consisting of a high brightness electron gun and a rotating anode target, and a microbeam X-ray camera, using fine bore glass capillaries as collimators. I joined James Noel Kellar who had already started on the project. He had prime responsibility for the X-ray tube, and I for the camera. Tragically, Kellar was killed in a freak sailing accident in Holland in 1948. (I married his widow in 1959, and acquired two step children). Applying the technique to cold worked Al showed that the original grains had broken up into small subgrains, about 2μm in diameter, but that the X-ray line broadening was due to the strain within the subgrains. About the same time Heidenreich at Bell Telephone Laboratories in the USA published electron microscope images of thin foils of cold worked Al which showed these subgrains directly. This suggested that transmission electron microscopy might be an easier way to study the structure of cold worked metals and that it might become possible to observe individual dislocations, the line defects whose movement was thought to be responsible for plastic deformation.

After my Ph.D. I stayed on at the Cavendish to study the structure of coal by X-ray diffraction, but at the same time kept in touch with my successors in the X-ray microbeam team, who applied the method to other cold worked metals. It became clear that the X-ray technique was limited to particle sizes exceeding about 0.5μm. However it was possible to obtain discrete spots from smaller particles on electron diffraction patterns of thin foils, utilising the small diameter electron beams available in an electron microscope. With Kelly and Menter spotty rings were obtained from beaten gold foil, suggesting distorted particles about 2000 Å in diameter. Menter took some micrographs, which showed complex contrast features consisting of bands parallel to the projection of the close packed (111) planes in the structure. The diffraction patterns showed streaks from faults in the stacking of the close packed planes. This suggested to me that the contrast of the bands in the micrographs could be due to the stacking faults, which would cause a phase change of $\alpha = 2\pi\ g.R$ (where g is the reciprocal lattice vector of the reflection, and R is the relative displacement of the planes above and below the fault) and consequent changes in the diffracted and direct beam intensities. About this time it was suggested by Heidenreich and Shockley that dislocations in face centred cubic (fcc) metals would be dissociated into two partial dislocations bounding a ribbon of stacking fault. I conceived the idea that the dislocations in fcc metals might be visible in transmission electron micrographs because of the contrast from the ribbon of stacking fault.

In 1953, I was awarded an ICI Fellowship which gave me freedom to pursue my interest in cold worked metals and in particular the possibility of seeing individual dislocations in thin foils by transmission electron microscopy. I stopped the work on coal in the mid 1950s. Professor Bragg retired from the Cavendish Laboratory in 1953 and was succeeded by Sir Nevill Mott who was very interested in dislocations and a great supporter of my project. In 1954, I recruited MJ Whelan as a research student and put him on a project to observe dislocations in deformed Al by transmission electron microscopy. We used electron microscopes in other research groups and benefitted from access to a new generation

Siemens Electron Microscope in Dr Cosslett's Electron Microscopy Group. In 1956, we observed individual dislocations in thin foils of deformed Al by this technique. (See paper 39 in list of publications). We were very fortunate, because unexpectedly the dislocations started moving under the influence of the stresses set up in the foils probably due to surface contamination produced under the electron beam. Furthermore, the movement left traces behind showing the paths taken by the dislocations. It also turned out that the image contrast was actually due to the local bending of the lattice planes around the defects, rather than to the stacking fault ribbon, the width of which would be far too small in Al.

The new diffraction contrast electron microscopy technique provided the Material Scientist with a powerful tool to study defects in crystalline materials. It had the effect of dispelling the doubts of some Metallurgists at the time as to the relevance of dislocations to the understanding of mechanical properties. Sir Alan Cottrell observed that "Dislocations ceased to be a theory and became observed features of crystalline materials". From then onwards my research concentrated on the study of dislocations by transmission electron microscopy, and the understanding of mechanical properties in terms of dislocation mechanisms. I built up a research group in Cambridge and was very fortunate in attracting excellent research students and postdoctoral workers, several of whom have become Fellows of the Royal Society. The group developed the theory of image contrast from dislocations and stacking faults, and applied the technique to many different problems. Early applications included the observation and identification of dislocation loops in Al, quenched from a high temperature, (the publication is a citation classic) and stacking fault tetrahedra in quenched gold, revealing the defects responsible for quench hardening, the first observations of small dislocation loop damage in neutron irradiated Cu (relevant to irradiation hardening), which opened up a whole new field of irradiation damage studies widely developed by others, the first direct evidence for dissociated dislocations, their motion and interaction in stainless steel, relevant to work hardening, other observations on the work hardened state, the high densities and importance of so-called forest dislocations, and the plasticity of body centred cubic metals. The years 1956 to 1966 were the most productive period in my research career. The diffraction contrast technique is described in the book "*Transmission Electron Microscopy of Thin Crystals*" by Hirsch, Howie, Nicholson, Pashley and Whelan (1965), which is still the authoritative text today. The technique has wide applications in Materials Science, and is used all over the world, in Universities and Industry Laboratories.

In 1960 at the 5th International Congress of Crystallography in Cambridge, I suggested that the lattice friction stress, and the increase in yield stress of body centred cubic metals at low temperatures is due to the three-fold dissociation of the screw dislocation, which has to be constricted before motion can occur. This proposal and its consequences have stimulated much experimental and theoretical research by my group and others.

In 1966, I was offered and accepted the Isaac Wolfson Chair and Headship of the Department of Metallurgy at Oxford, which I held until my retirement in 1992. (I also became a Fellow of St Edmund Hall.) The challenge was to build up the quite small but distinguished department founded by my predecessor Professor W Hume-Rothery a few years earlier, into a viable medium sized department with a wider remit in Materials Science. The enthusiastic support of my colleagues ensured the success of this venture. Inevitably my research activities suffered during this period.

In Cambridge one of my aims had been to explain work hardening in pure metals in terms of dislocation mechanisms. While I made various contributions to different aspects of the problem, I did not manage to develop a satisfactory comprehensive theory for the hardening rate in single crystals of fcc metals, without making *ad hoc* assumptions. In Oxford I decided to concentrate on other projects. In particular, I studied, often in collaboration with colleagues, the plastic properties of intermetallics and dispersion hardened alloys, the brittle-ductile transition of various crystals, dislocations in semi-conductors, indentation plasticity, defects in diamond, and electron channelling imaging of dislocations near the surface of bulk specimens by scanning electron microscopy.

During my time as Head of Department, the main undergraduate Honour School was broadened into a Materials Science course, and two new joint Honour Schools were introduced, one of which included Economics and Management, the other elements of Materials Engineering. The name of the Department was changed to the Department of Materials to reflect its broad Materials remit.

As to the inevitable committee work, my career was affected by joining in 1973 the Atomic Energy Authority (UKAEA), Light Water Reactor Pressure Vessel Study Group (LWRSG), which investigated in depth the structural integrity of the Nuclear Reactor Pressure Vessel of the Pressurised Water Reactor (PWR), which was to be built at Sizewell in the UK. This gave me a lasting interest in structural integrity problems. I was a member of the Board of the UKAEA from 1982 to 1994, and its Chairman from 1982 to 1984. I chaired the LWRSG from 1982 to 1987, and also from 1993 to 2002 its successor committee, the Technical Advisory Group on Structural Integrity (TAGSI), which gave advice on structural integrity issues to the Nuclear Industry.

In the University, I was instrumental in setting up the University's Technology Transfer Company, "Isis Innovation Limited", and was its founder Chairman from 1988 to 1996. While during my initial period the company was quite small, later, the University invested significantly in the company, and today Isis is a large and highly successful technology transfer business.

During the last ten years I have acted as full-time carer for my disabled wife.

I received a Knighthood in 1975. I was elected a Fellow of the Royal Society in 1963, and my work was recognised by election to other academies, the award of Honorary Degrees, and a number of medals and prizes, including the 1984 Wolf Prize in Physics (jointly with E. Hahn, and T. Maiman).

Preprint of a paper to appear in
THE PHILOSOPHICAL MAGAZINE, July 1956

Direct Observations of the Arrangement and Motion of Dislocations in Aluminium

By P. B. HIRSCH, R. W. HORNE and M. J. WHELAN
Crystallographic Laboratory and Electron Microscopy Group,
Cavendish Laboratory, Cambridge†

[Received June 25, 1956]

ABSTRACT

Electron optical experiments on Al foils have revealed individual dislocations in the interior of the metal. The arrangement and movement of individual dislocations have been observed. Most of the dislocations occur in the boundaries of a substructure, the diameter of the subrains being of the order of $1\,\mu$ or more. Tilt-boundaries, networks and dislocation nodes have been resolved. The results apply to aluminium recovered at 350°c after heavy deformation by beating ; the dislocation density is $10^{10}/cm^2$.

The dislocations can be seen to move along traces of (111) slip planes ; the motion of the dislocations can be either rapid or slow and jerky. Cross-slip by the screw dislocation mechanism has been observed frequently.

---

§ 1. INTRODUCTION

RECENT experiments (Hedges and Mitchell 1953, Amelinckx 1956) have revealed the dislocation structures inside optically transparent crystals of AgBr and NaCl. For metals, however, it has so far been possible to demonstrate the presence and arrangement of dislocations only at the surface, for example by the use of etching and preferential precipitation techniques (Lacombe and Beaujard 1948, Wilsdorf and Kuhlmann-Wilsdorf 1955). Most of the information about dislocation arrangements in the interior of a cold-worked metal has been derived from x-ray diffraction experiments (see for example Hirsch 1956). The present paper gives a preliminary account of electron–optical experiments on Al foils, which have revealed the arrangement and motion of individual dislocations in the interior of the metal.

In previous electron microscope studies of electrolytically thinned specimens of Al foils the substructure was observed, but no evidence was advanced to suggest that individual dislocation lines could be seen (Heidenreich 1949). In these new experiments beaten Al foils were examined, after suitable annealing and etching treatments, by electron

---

† Communicated by the Authors.

2   P. B. Hirsch, R. W. Horne *and* M. J. Whelan *on the*

microscopy and electron diffraction, in a high resolution electron microscope. The results suggest that individual dislocation lines are revealed in the interior of the metal on account of the strain field associated with them, and no ' decoration ' is therefore required. The arrangement of the dislocation lines in the sub-boundaries and within the subgrains can be studied directly, and the motion of individual dislocations followed. Many complex effects have been observed, and it is not possible at present to account for more than a few of these in detail. A number of interesting points have however already emerged, and it has therefore been considered worthwhile to publish this preliminary account.

§ 2. Experimental Technique

Specimens of high purity (99·99+%) and commercially pure (99·8%) aluminium were beaten by Messrs. George M. Whiley Ltd., to a thickness of 0·5 µ. The foils were annealed *in vacuo* at 350°c, and subsequently etched in dilute hydrofluoric acid.

The specimens were examined in the Siemens and Halske ' Elmiskop ' electron microscope operating at 80 kv. The foils were found to be transparent to electrons over large areas. High resolution electron micrographs in bright and dark field, and diffraction patterns from preselected areas were taken. The instrumental magnification was ×40 000 for all the plates, except for fig. 1, which was taken at ×8000. The instrument was accurately corrected for astigmatism, using suitable test objects, prior to examining the specimens, and under these conditions the resolution attained was probably of the order of 10 to 20 Å.

§ 3. Evidence for the Visibility of Dislocation Lines

Figures 1–15 show some typical micrographs obtained. Experiments designed to elucidate the nature of the contrast mechanism have shown that the contrast is due to differences in intensities of Bragg reflections, and that the observed effects are complicated owing to certain interference effects. Details of these experiments will be described in a later paper. The following facts, however, leave little doubt that individual dislocation lines are being observed :

(*a*) The specimens contain a substructure of subgrain diameter about 1 µ or more (fig. 1). The misorientations across the boundaries have been determined from diffraction experiments and are found to be about 1° or 2°. These low-angle boundaries presumably consist of arrays or networks of dislocations.

(*b*) The boundaries of the substructure consist (under higher magnification) of black lines or dots (figs. 2–9). The spacings of the dislocations in the boundaries expected from angles of rotation measured in two cases across particular boundaries agreed within a factor of 1·5 with the observed spacings. In view of the difficulties involved in

Arrangement and Motion of Dislocations in Aluminium   3

counting the individual irregularly spaced dislocations, and owing to the uncertainty in knowing the type of boundary involved, this agreement is considered satisfactory.

(*c*) The average distance between the lines or dots is about 100Å, corresponding to average angular misorientations of about $1\frac{1}{2}°$. This is in excellent agreement with the order of magnitude of the rotation determined from many diffraction patterns, from areas covering several subgrain boundaries.

(*d*) On tilting the illumination or the specimen through small angles, the lines are seen to remain fixed in position, although the contrast changes. Experiments such as these and dark field experiments show that the visibility of the lines is due to Bragg contrast, and that they represent a definite property of the specimen.

(*e*) It is already possible in some cases to explain the detailed geometry of the lines and dots in terms of dislocations. Some examples will be given below.

(*f*) When working with large condenser apertures at high beam currents, the lines are observed to move ; in areas with a [001] direction normal to the foil, the movement occurs along straight lines parallel to the traces of (111) slip planes (figs. 10–13). The behaviour of these moving lines is identical with that expected of dislocation lines.

These observations leave little doubt that the lines represent single dislocation lines of unit Burgers vector. There are, however, some complicating features. Some of the dislocation lines have a 'spotty' appearance (figs. 5, 9) ; this effect may be understood in terms of an interference mechanism which will be discussed in a later paper. Some of the micrographs show a fringe structure at the boundaries which is due to the same cause. 'Ghost' images, similar to the dislocations, but displaced from them, are due to Bragg reflections which are included in the image (fig. 5) but which have suffered spherical aberration in the objective lens. In addition to all these effects 'extinction' contours are also observed (Heidenreich 1949). These are mainly due to misorientations caused by buckling of the foil, although thickness contours may also occur. On tilting the illumination or the specimen through angles of the order of 1° or 2° the extinction contours move, whereas the dislocation lines remain fixed.

§ 4. ARRANGEMENT OF DISLOCATIONS

Figure 1 shows the substructure in Al at low magnification. The average subgrain size is about $1\,\mu$, the average angular misorientation about $1\frac{1}{2}°$. The dislocation density measured from other photographs is of the order of $10^{10}/cm^2$. All these figures are in excellent agreement with the results deduced from x-ray data (Hirsch 1952, Gay, Hirsch and

4   P. B. Hirsch, R. W. Horne *and* M. J. Whelan *on the*

Kelly 1953).   The photographs leave no doubt at all that most of the dislocations are in the sub-boundaries, and relatively few within the subgrains.

The broad bands running across the subgrains are extinction
contours (A).   The difference in contrast from one grain to the next is
due to differences in the intensities of Bragg reflexions caused by small
changes in orientation.

Figure 2 shows dislocations spaced uniformly along a boundary.
Generally the boundaries appear to be less regular.   The boundary
planes in this region are all nearly normal to the foil.

Figure 3 shows an area in which most of the boundaries are not normal
to the plane of the foil.   Some of the boundaries appear to consist of
parallel dislocation lines (A), so that they must be pure tilt boundaries,
others (B) appear to contain cross-grids of dislocations.   The micrograph
shows the three-dimensional nature of the substructure ; in particular
the junction of the boundaries can be observed quite clearly.   In some
cases nodes where three dislocation lines meet at a junction can be
recognized (C).   (For a discussion of networks and boundaries in face-
centred cubic crystals reference should be made to Frank (1955), Ball and
Hirsch (1955) and Amelinckx (1956).)

Figure 4 shows a square cross-grid of dislocations (A) representing
probably a twist boundary on (100).   Other networks can be seen at B ;
dislocation nodes can be recognized clearly in boundary junctions at C.
A few isolated dislocations occur within the grain at D.

Figure 5 shows a dislocation node inside a subgrain (A).   The
dislocations appear to consist of a number of spots.   This appearance
is probably due to an interference effect at dislocations with a screw
component, which will be discussed in a later paper.   At B two
dislocations appear to cross in characteristic manner (Read 1954).
A cross-grid can be seen at C.   At D a boundary appears to terminate
in the middle of a subgrain.   The strain around this region is apparent
from the curvature in the extinction contour.   Single dislocation lines
can be seen at E and F; at E the dislocations are considered to be more
nearly parallel to the foil than at F ; the spots at E are thought to be
due to the interference effect mentioned above.   At F, on the other hand,
each pair of spots corresponds to one dislocation steeply inclined to the
foil ; this follows from the experiments on the motion of dislocations
(see § 5).   The two spots on each dislocation pair are thought to be due
to the increased distortion on the top and bottom surfaces due to the
oxide layer which must be present.   More precisely, in the transition
region between metal and oxide the lattice begins to deviate appreciably
from the perfect face-centred cubic arrangement, and the dislocation
may be considered very ‘ joggy ’ in this region.   These jogs are thought
to be responsible for the additional distortion.   This area shows some
typical extinction contours at G.

Figure 6 shows hexagonal networks of dislocations (A, B).   These
are the only clear hexagonal networks observed so far.   Many of the

P. B. HIRSCH et al.   Phil. Mag. Ser. 8, Vol. I, Pl. 23.

Fig. 1

1 μ Mag. $\times 20\,000$

Substructure in Al annealed at 350°c after beating at room temperature. The average subgrain size is about $1\,\mu$, the average angular misorientation about $1\frac{1}{2}°$. The dislocation density is about 10^{10} per cm². Extinction contours are shown at A.

P. B. HIRSCH *et al.*   Phil. Mag. Ser. 8, Vol. I, Pl. 24.

Fig. 2

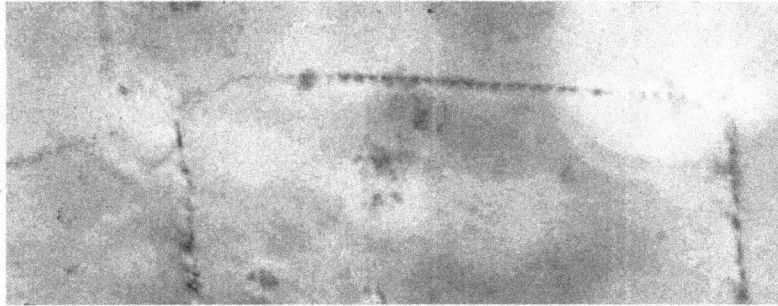

1000 A   Mag. × 100 000

A sub-boundary consisting of uniformly spaced dislocations. The average spacing of the dislocations is about 175 Å.

Fig. 5

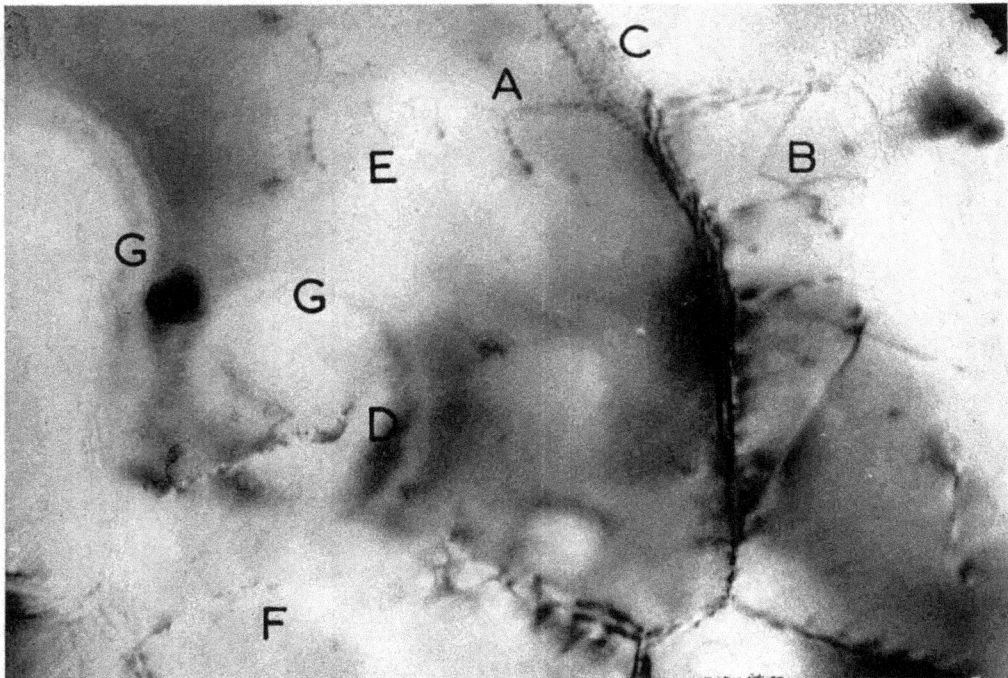

1000 A   Mag. × 100 000

A :—Dislocation node ; the 'spotted' appearance of the dislocation lines is probably due to an interference effect. B :—Crossing of two dislocations. C :—Cross-grid of dislocations. D :—Termination of boundary. E, F :—Single dislocation lines. G :—Extinction contours.

P. B. HIRSCH *et al.*   Phil. Mag. Ser. 8, Vol. I, Pl. 25.

Fig. 3

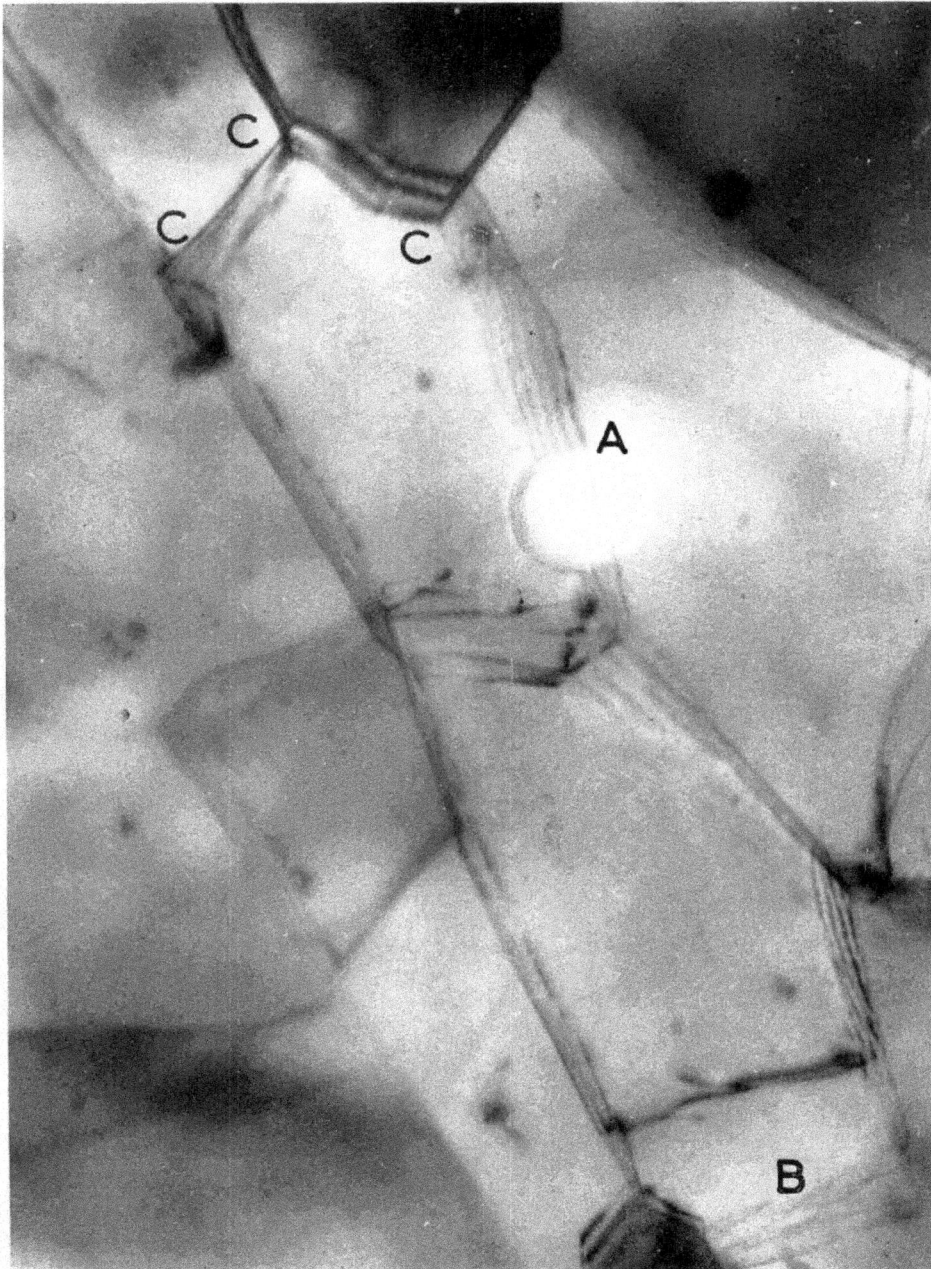

1000 Å

Mag. ×100 000

Boundaries whose planes are not normal to the plane of the foil.   A :—Pure tilt
boundary.   B :—Cross-grid of dislocations.   C :—Dislocation nodes.

P. B. HIRSCH *et al.*   Phil. Mag. Ser. 8, Vol. I, Pl. 26.

Fig. 4

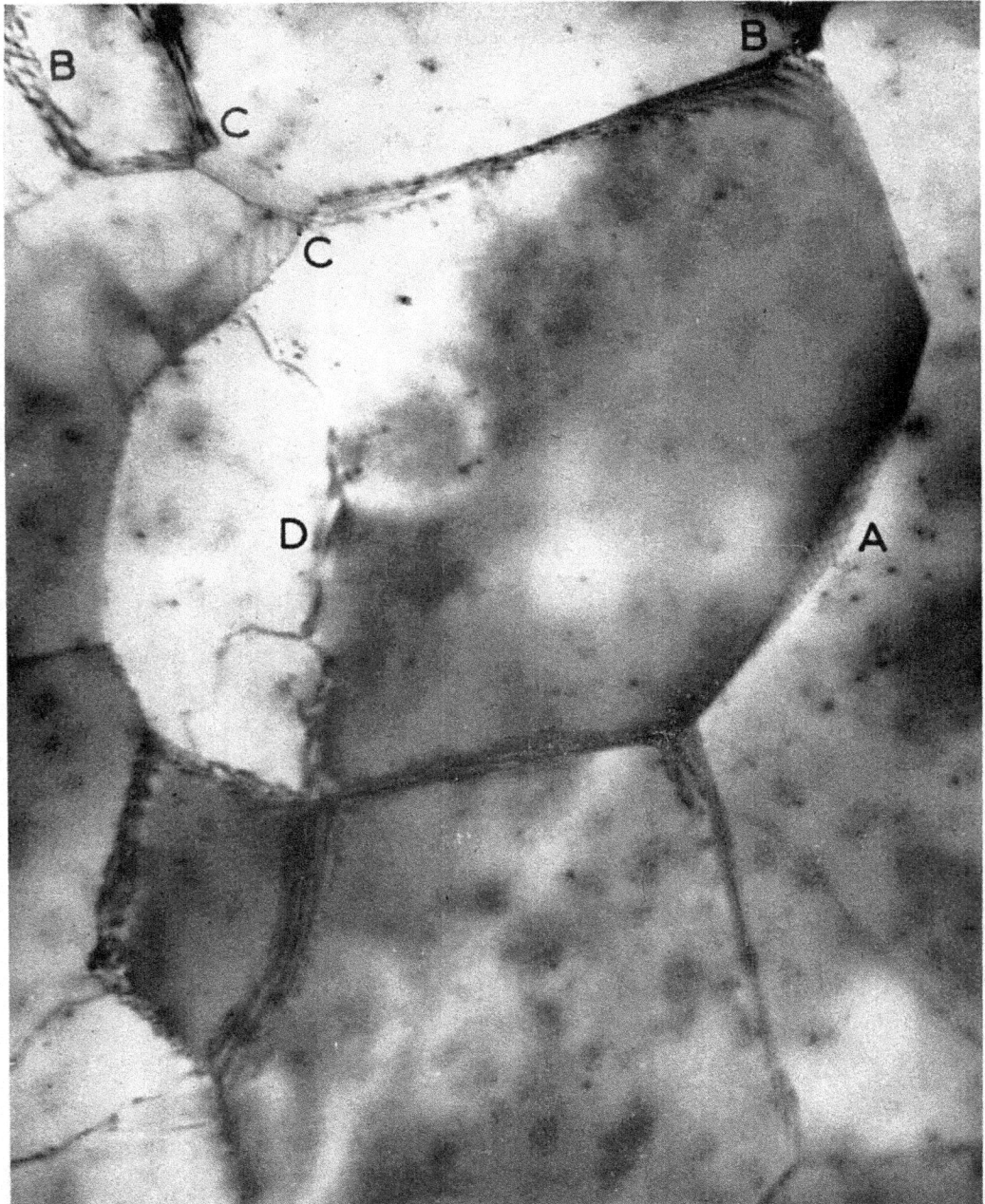

1000 Å   Mag. ×100 000

Boundaries containing dislocation networks.   A :—Square cross-grid.   B :—Other networks
C :—Dislocation nodes.   D :—Isolated dislocations.

P. B. HIRSCH *et al.* Phil. Mag. Ser. 8, Vol. 1, Pl. 27.

Fig. 6

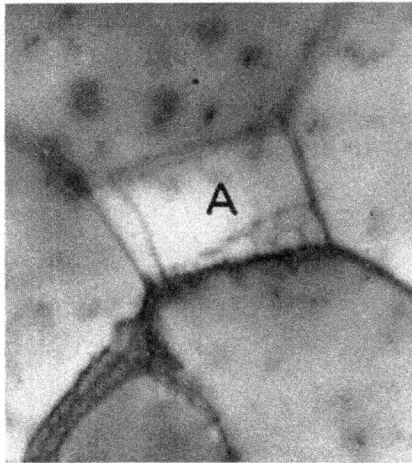

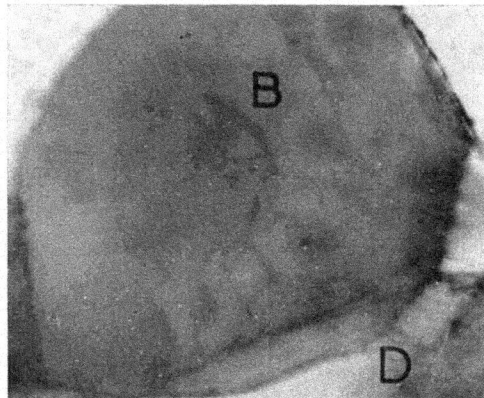

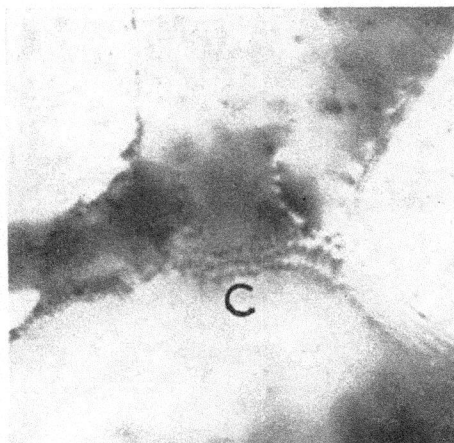

1000 Å   Mag. ×100 000

A,   B :—Hexagonal networks of dislocations.   C :—Indistinct network.
D :—Dislocation nodes.

P. B. HIRSCH *et al.*   Phil. Mag. Ser. 8, Vol. I, Pl. 28.

Fig. 7

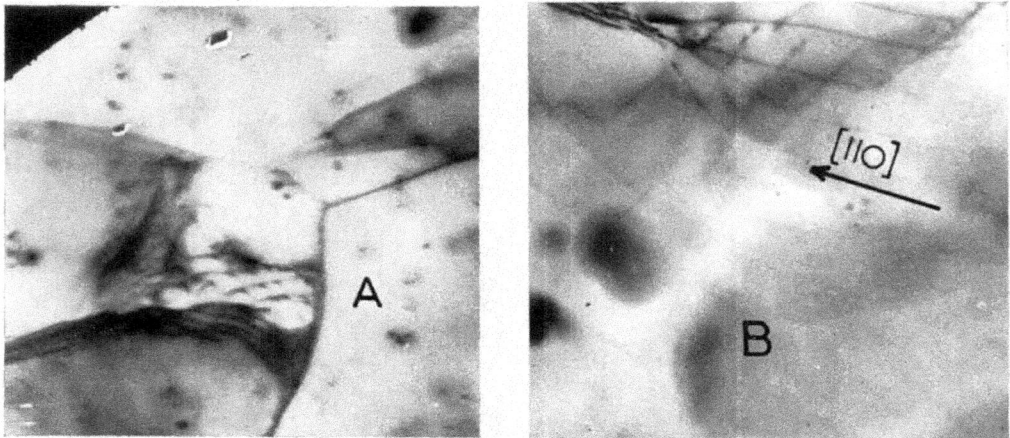

1000 Å   Mag. ×100 000

A, B :—Square cross-grids of dislocations.   The dislocations
in B are approximately parallel to [110] directions.

Fig. 8

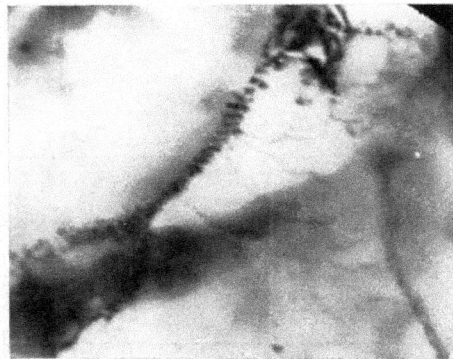

1000 Å   Mag. ×100 000

Irregular network consisting of bowed-out dislocations.

P. B. HIRSCH *et al.* Phil. Mag. Ser. 8, Vol. I, Pl. 29.

Fig. 9

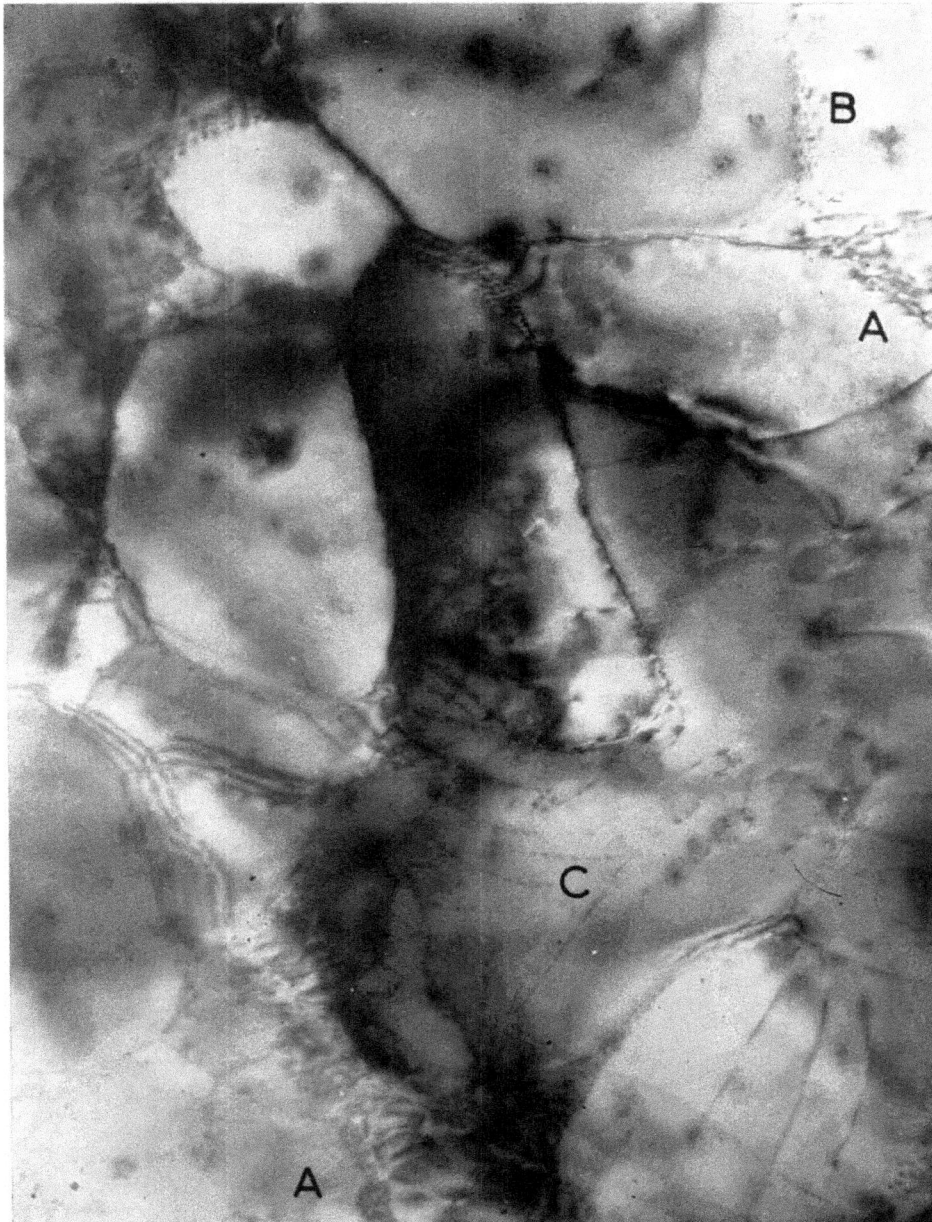

1000 Å Mag. ×100 000

Complex arrangement of dislocations. A:—Irregular
network. B:—Interference effect at a boundary.
C:—Interference effect at single dislocations.

P. B. HIRSCH *et al.*   Phil. Mag. Ser. 8, Vol. I, Pl. 30.

Fig. 10

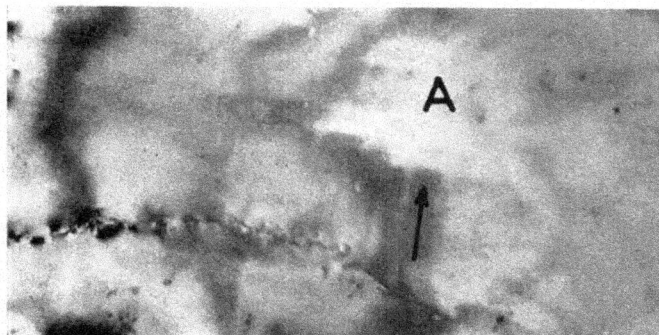

(a)

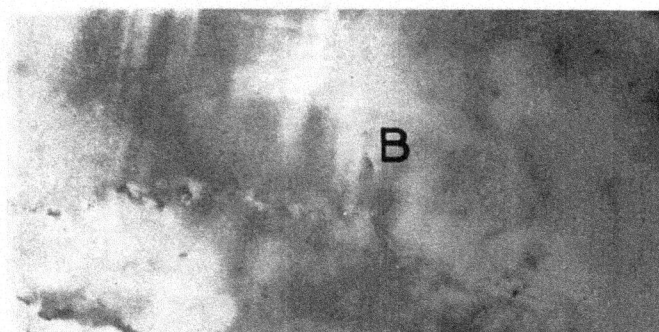

(b)

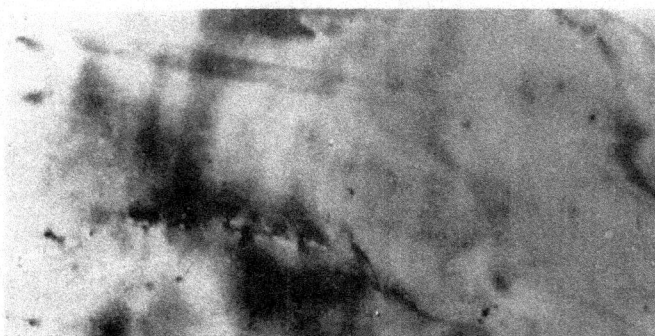

1000 A   (c)   Mag. ×60 000

Sequence showing slip and break-up of a sub-boundary. The length of the
    dislocations (A, B) determines the width of the bands. The bands are
    parallel to traces of (111) planes, and their contrast disappears after
    a few seconds.

P. B. HIRSCH et al.   Phil. Mag. Ser. 8, Vol. I, Pl. 31.

Fig. 11

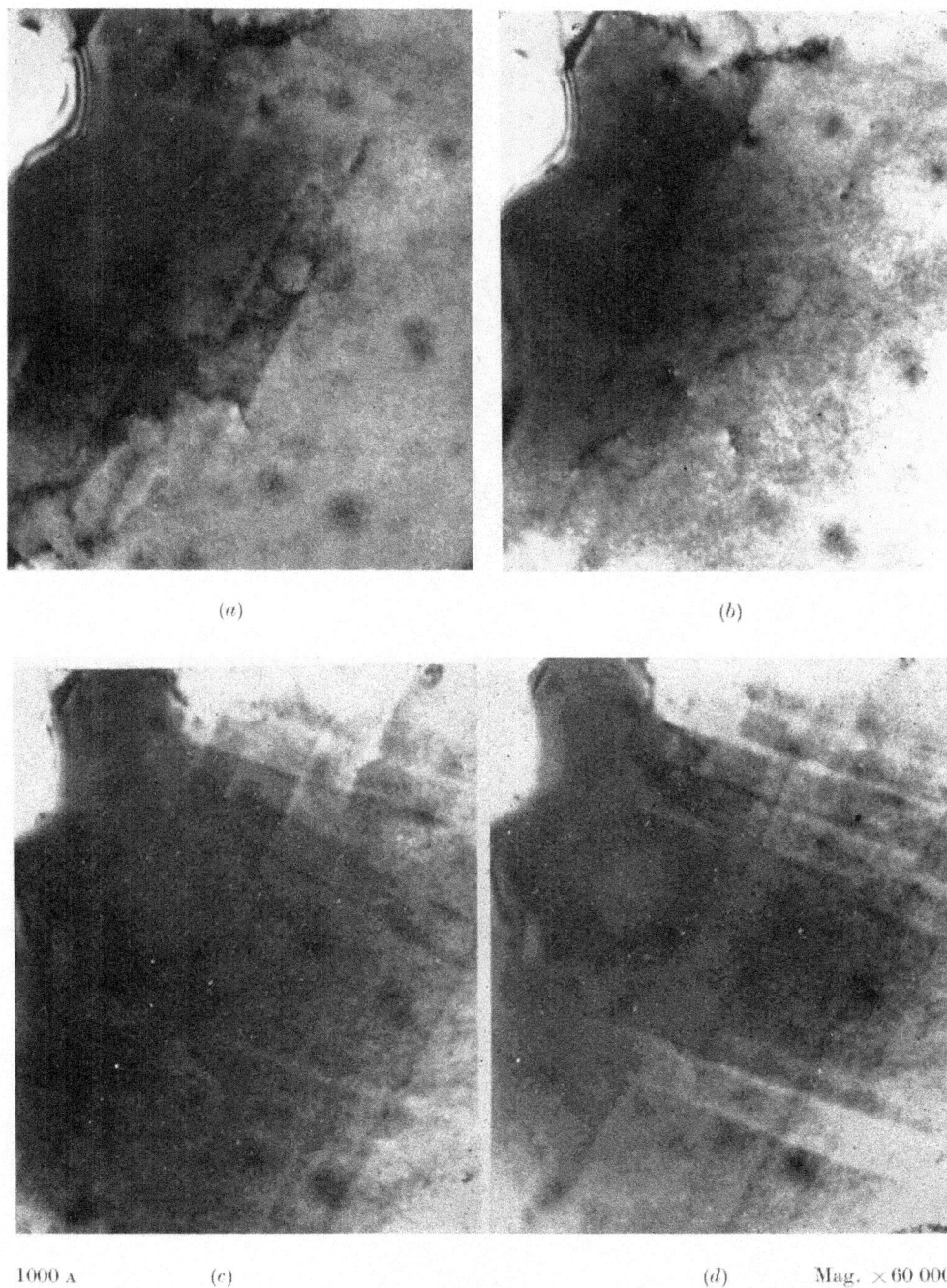

(a)

(b)

1000 Å   (c)   (d)   Mag. ×60 000

Sequence showing slip.   The disappearance of the contrast in the bands can be followed
in the sequence.

P. B. HIRSCH *et al.*　　　　　　　　　　　　　Phil. Mag. Ser. 8, Vol. I, Pl. 32.

Fig. 12

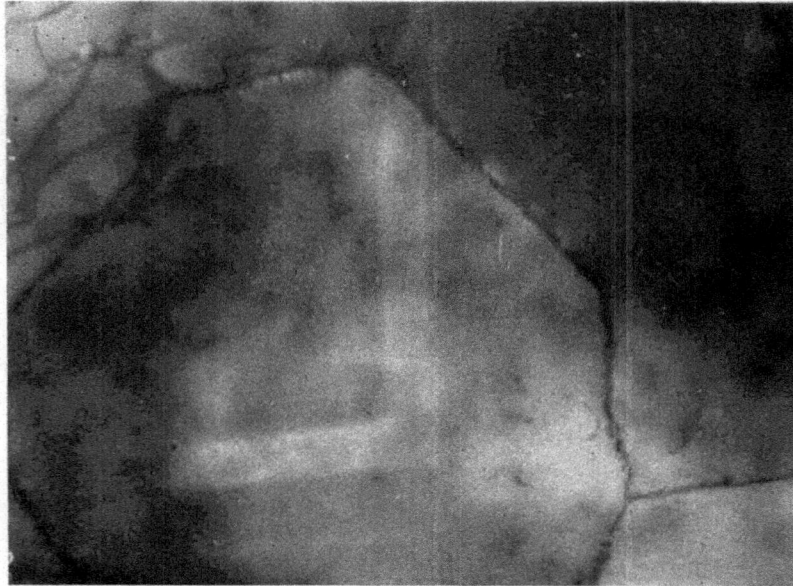

1000 Å　　　　　　　　　　　　　　　　　　Mag. ×60 000

Cross-slip by the screw dislocation mechanism.

Fig. 13

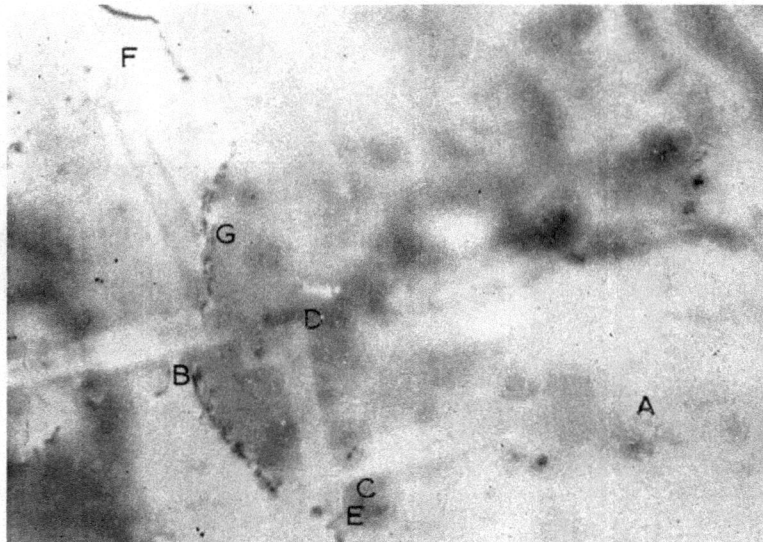

1000 Å　　　　　　　　　　　　　　　　　　Mag. ×60 000

Cross-slip; the dislocation started at A and penetrated the boundary at B.
Note that the original band splits into two branches at C, CD and CE.
Another slip line is seen at FG.

P. B. HIRSCH *et al.*　　　　　　　　　　　Phil. Mag. Ser. 8, Vol. I, Pl. 33.

Fig. 14

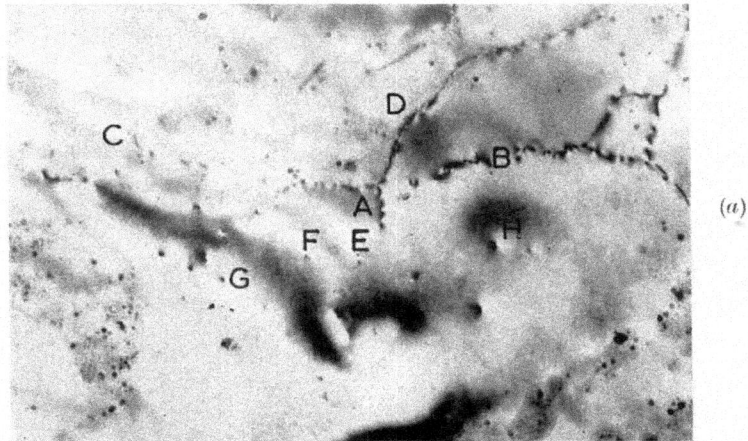

(*a*)

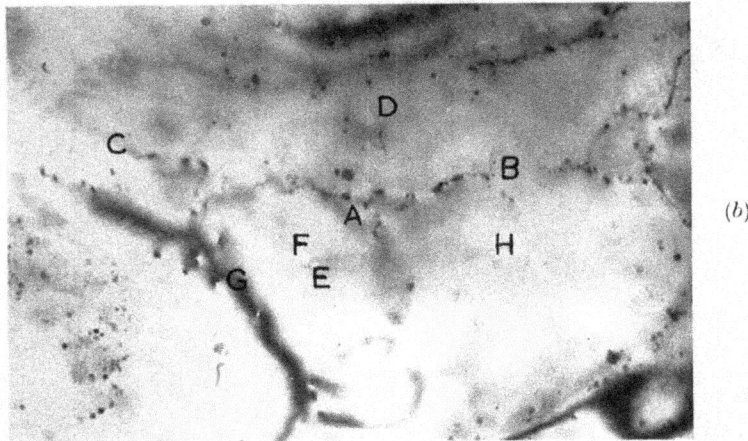

(*b*)

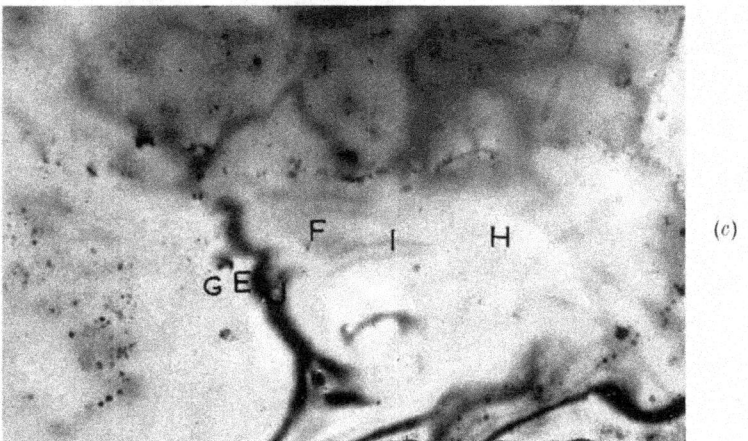

(*c*)

1000 A　　　　　　　　　　　　　　　Mag. ×60 000

Sequence showing slowly-moving dislocations.　Refer to text for detailed description.

Fig. 15

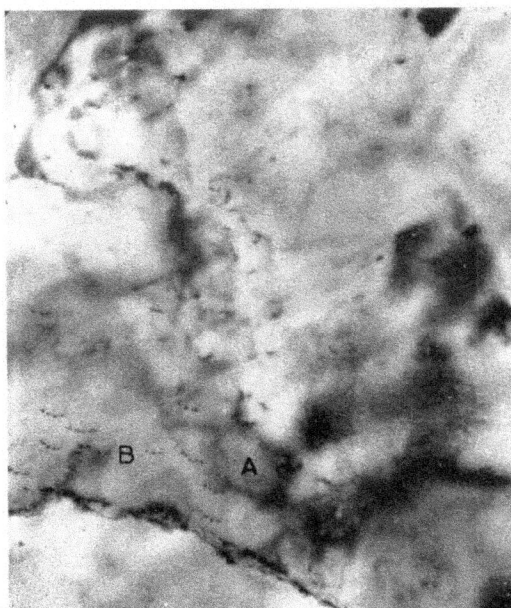
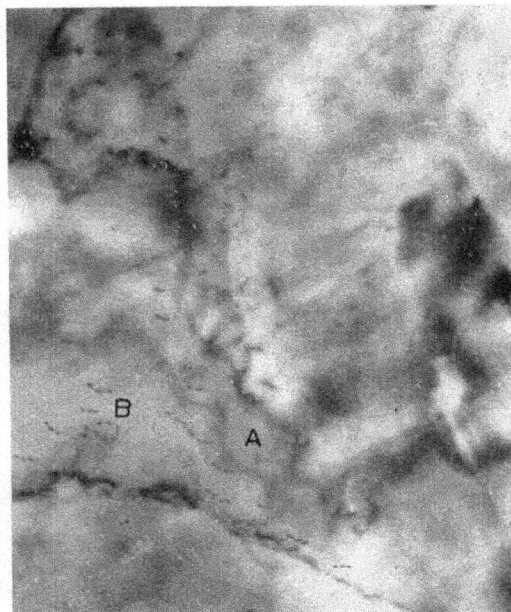

(*a*)   (*b*)

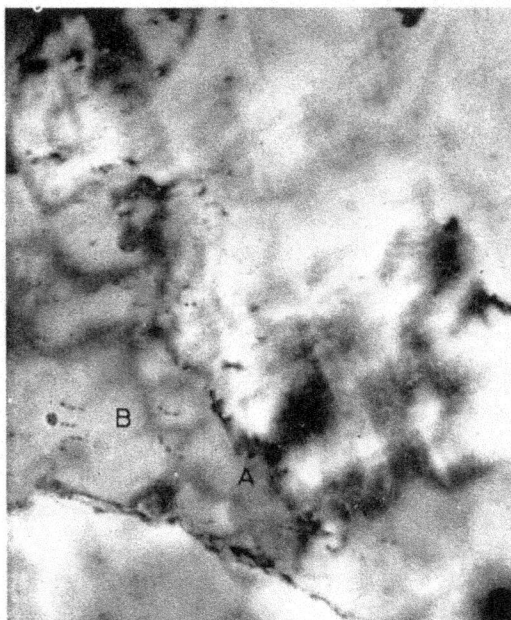
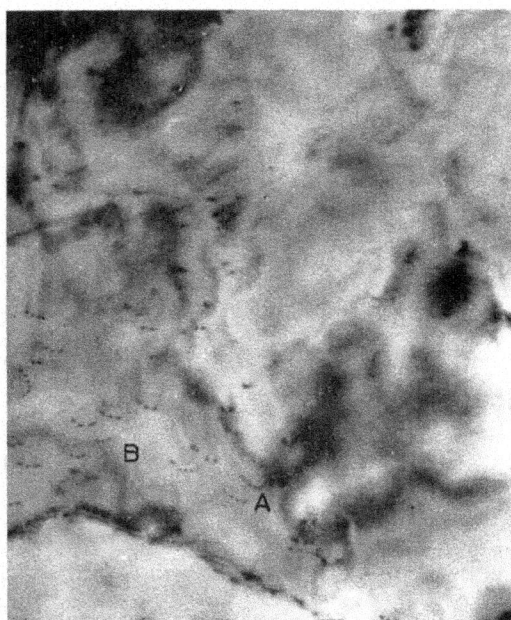

1000 A   (*c*)   (*d*)   Mag. ×60 000

Sequence showing slowly-moving dislocations. Some of the dislocations move
along irregular paths. At A the dislocations may be attempting to
form a boundary. B :—dislocation loops.

Arrangement and Motion of Dislocations in Aluminium   5

boundaries (e.g. C) however appear to consist of more or less distorted hexagonal networks, but the clarity of the networks is spoiled by interference effects. At D the nodes of the dislocations at the boundary junction can again be clearly recognized.

Figure 7 shows square cross-grids of dislocations of large mesh size (A, B). From a diffraction pattern taken from B it was shown that the dislocations in B are parallel to [110] directions and that the normal to the foil is almost [001]. This network, therefore, is likely to be a twist boundary on (001).

Figure 8 shows an irregular network, the dislocations of which are bowed out, presumably owing to some local strain.

In some areas the arrangement of the dislocations is quite complex. Figure 9 shows such a case ; many complex networks can be seen (A). This photograph shows interesting interference effects at boundaries (B) and single dislocations (C).

§ 5. Movement of Dislocations

When working at high beam currents and with large condenser apertures, the dislocations are observed to move. Two types of motion are observed, either rapid or slow and jerky.

First there is a movement of the extinction contours. Subsequently the dislocation lines often bow out and in many cases move. The bowing-out effect is presumably direct confirmation of the mechanism suggested for the decrease in elastic modulus due to dislocations (Mott 1952). The movement of the extinction contours shows that the foil buckles ; the dislocations therefore move presumably under the strain ; their movement may be aided by heat. The temperature rise in the specimen is not known at present.

Figures 10 (*a*), (*b*), (*c*) is a sequence showing the break-up of a sub-boundary. Dislocations leave the boundary and move in straight lines parallel to the direction marked with an arrow. The interval between the exposures is about 15 sec. The moving dislocations leave behind them bands, which have been found by selected area diffraction experiments to be parallel to the traces of (111) slip planes. Thus, the appearance of slip traces at right angles suggests immediately that this area has the usual (100) orientation. The width of the band is governed by the length of the dislocation (e.g. A or B). The bands appear either as white on black, or black on white background ; both types are observed in fig. 10. The edges of the band are always more intense than the middle. After a time interval of the order of several seconds the contrast disappears. It is possible to account for the results tentatively in the following way. The dislocation lines may be ' joggy ', and non-conservative motion of these jogs results in the formation of vacancies and interstitials. Alternatively impurity atoms might be left behind after the dislocation has passed. These point defects cause a reduction in intensity of the Bragg reflexion (equivalent to a temperature factor) and an increase in

the background intensity. The bands on the micrograph appear white
or black according to whether the intensity scattered outside the objective
aperture is greater or smaller than in the surrounding regions. The intense
edges of the bands are then due to the larger numbers of point defects
generated in the surface layer. After the passage of the dislocation,
the point defects are concentrated near a (111) plane. The vanishing
of the slip plane contrast is explained by the diffusion of the defects
away from the slip plane.

In the course of the sequence 10 (a), (b), (c) the sub-boundary is being
depleted of dislocations. The disappearance of the contrast of the bands
can be clearly seen.

If the suggested explanation of the bands is correct, the edges of the
bands correspond to the top and bottom of the foil. Since the orientation
of the foil is known, the thickness of the foil can be claculated from the
width of the bands. The thickness varies from about 500 Å to 1000 Å.

Figures 11 (a), (b), (c), (d) shows slip in another area; the plane of
the foil is apparently not quite parallel to (100). Complicated cross-slip
can be observed. The disappearance of the contrast is again noticeable.

Figure 12 shows a fine example of cross-slip. It is quite clear here that
a single dislocation has transferred from one slip plane to another. This
represents direct proof of the Mott–Frank screw dislocation mechanism
of cross-slip (Mott 1951). The plane of the foil is again approximately
parallel to (100); it should be noted that several of the boundaries in
this region are parallel to (100) and (110), suggesting that they contain
only one or two sets of dislocations which differ in their Burgers vector
or slip plane (Ball and Hirsch 1955).

Cross-slip has been observed very frequently; fig. 13 shows another
example; here the dislocation started at A and eventually penetrated
the boundary at B. An interesting feature about this photograph is
the fact that the original line splits into two branches at C, CD and CE.
The precise process here is uncertain, but it may involve the splitting of
a single dislocation into two. Another slip line is seen at FG.

Figure 14 show a sequence in an area where the dislocations moved
slowly. This type of movement may perhaps correspond to creep.
In addition to sub-boundaries, isolated dislocations can be seen within
the subgrains. The single dislocations are typically more intense at the
ends. Comparing 14 (a) and (b) it is clear that a number of dislocations
have moved, for example at A, B and C. The boundary at D has
disappeared; dislocation E has moved to dislocation F which, judging
from its contrast, lies in another slip plane. This suggests that E is
held up by F because of the difficulty of cutting through a dislocation.
However, fig. 14 (c) shows that E has moved on to G where it is again
stopped. It is also clear that many other dislocations have moved;
in particular H has moved along the path HIJ. This path is quite
irregular and indicates that the dislocation may move by a very intimate
cross-slip mechanism and possibly by climb. Other dislocations also

appear to have moved by similarly irregular paths. Motion of this type tends to be jerky when observed in the microscope.

Figure 15 shows another sequence in which dislocations are moving in very jerky and irregular paths. It appears that the dislocations are attempting to form a boundary at A. Many of the dislocations appear to be bowed at the centre. This indicates clearly that the ends of the dislocations can be moved only with difficulty. Long dislocation loops are seen to move in region B.

Many observations of moving dislocations have been made in the microscope, and many complex effects have been seen. Although the growth of dislocation loops has been observed, so far it has not been possible to locate the dislocation sources. It is clear, however, that sometimes dislocations come out of boundaries, and sometimes they originate in complex regions such as those of fig. 9. On the whole, regions with very well formed polygons are most stable. Many of the dislocations are stopped at the boundaries, others appear to pass through them. The movement of parts of boundaries as a whole has also been observed.

§ 6. Conclusions

The experiments show that individual dislocations can be seen in aluminium foil, and that their arrangement and movement can be studied. While the results reported in this paper are only preliminary, some conclusions can already be drawn. Most of the dislocations are arranged in sub-boundaries; relatively few occur inside the grains. Nevertheless, isolated dislocations are observed in many cases. Pure tilt boundaries, square and hexagonal networks have been observed. Many of the boundaries appear to consist of networks, but the nature of these is confused at present owing to possible interference effects. Dislocation nodes at boundary junctions and inside grains can be observed quite clearly. Many features of the dislocation arrangement are similar to those observed in AgBr (Hedges and Mitchell 1953) and NaCl (Amelinckx 1956), although in this case the dislocations are spaced at distances of only 100Å, compared with the corresponding distance of $1\,\mu$ in the inorganic crystals. It appears now that the theoretical predictions about the geometry of networks (Frank 1955) and of sub-boundaries (Ball and Hirsch 1955) apply for Al as well as for AgBr and NaCl. It follows that the dislocation arrangement in a heavily deformed and recovered metal is similar to that in the well-annealed inorganic crystals; the only difference is one in the scale of the arrangement. Combined diffraction and microscopy experiments are now in progress to study the details of the dislocation arrangements.

Ciné films have been prepared showing the movement of the dislocations, and it is hoped that a detailed study will reveal some of the important features of the motion, and in particular, the nature of dislocation sources and obstacles. So far the experiments have shown

8   *On the Arrangement and Motions of Dislocations in Aluminium*

the bowing out of dislocations, the spreading of dislocation loops and the movement of dislocations along (111) slip planes. The Mott–Frank mechanism of cross-slip has been observed in many cases. The experiments suggest that the dislocations are 'imperfect' (probably very 'joggy') near the surface, and it appears that the oxide film acts as an obstacle.

It is proposed to extend these observations to other metals.

ACKNOWLEDGMENTS

Our thanks are due to Professor N. F. Mott, F.R.S., Dr. W. H. Taylor, and Dr. V. E. Cosslett, for their interest, encouragement and helpful discussions. The micrographs were taken on a Siemens and Halske Elmiskop I electron microscope, purchased through a generous benefaction from the Nuffield Foundation. We are also grateful to Messrs George Whiley and Co. for supplying the beaten foils.

Acknowledgments for grants are due to the Ministry of Supply (P. B. H.), to the Agricultural Research Council (R. W. H.), and to the Department of Scientific and Industrial Research (M. J. W.).

REFERENCES

AMELINCKX, S., 1956, *Phil. Mag.*, **1**, 269.
BALL, C. J., and HIRSCH, P. B., 1955, *Phil. Mag.*, **46**, 1343.
FRANK, F. C., 1955, *Report of the Conference on Defects in Crystalline Solids* (London : Physical Society), p. 159.
GAY, P., HIRSCH, P. B., and KELLY, A., 1953, *Acta Met.*, **1**, 315.
HEDGES, J. M., and MITCHELL, J. W., 1953, *Phil. Mag.*, **44**, 223.
HEIDENREICH, R. D., 1949, *J. Appl. Phys.*, **20**, 993.
HIRSCH, P. B., 1952, *Acta Cryst.*, **5**, 172 ; 1956, *Progress in Metal Physics*, Vol. 6 (Permagon Press), in the press.
LACOMBE, P., and BEAUJARD, L., 1948, *Revue de Metallurgie*, **45**, 317.
MOTT, N. F., 1951, *Proc. Phys. Soc. B*, **64**, 729 ; 1952, *Phil. Mag.*, **43**, 1151.
READ, W. T., 1954, *Dislocations in Crystals* (McGraw-Hill).
WILSDORF, H., and KUHLMANN-WILSDORF, D., 1955, *Report of the Conference on Defects in Crystalline Solids* (London : Physical Society), p. 175.

[499]

A KINEMATICAL THEORY OF DIFFRACTION CONTRAST OF ELECTRON TRANSMISSION MICROSCOPE IMAGES OF DISLOCATIONS AND OTHER DEFECTS

By P. B. HIRSCH, A. HOWIE and M. J. WHELAN*

Crystallographic Laboratory, Cavendish Laboratory, University of Cambridge

(*Communicated by N. F. Mott, F.R.S.—Received 5 August* 1959)

[Plates 3 to 9]

CONTENTS

This paper describes a kinematical theory of electron microscope images of dislocations observed by transmission in thin crystalline foils. The contrast is essentially phase contrast in the Bragg diffracted beams, the phase differences being due to the displacements of the atoms from their positions in the ideally perfect crystal. The theory explains many of the characteristic features of the observed images, such as the dependence of the contrast on orientation, the reversal of contrast on bright and dark field images, the fact that dislocations are generally dark on bright field images, the position and width of the images, the general nature of the profile, the occurrence of dotted dislocations, the invisibility of some dislocations, the dependence of contrast on the inclination of the dislocation, and the occurrence of double images. The theory also accounts satisfactorily for the nature and width of the dislocation images obtained with X-rays.

1. INTRODUCTION

In the last few years the transmission electron microscope technique has become widely used for studying crystal imperfections such as dislocations and stacking faults (Menter 1956; Hirsch, Horne & Whelan 1956; Bollmann 1956, 1957; Whelan, Hirsch, Horne & Bollman 1957; Nicholson & Nutting 1958; for review see Menter 1958; Hirsch 1959; Nutting 1959).

* Now a Mr and Mrs John Jaffé Donation Student.

Essentially three methods have been employed for revealing lattice defects. In the first of these, due to Menter (1956), the lattice planes are resolved directly by ensuring that direct and diffracted waves enter the aperture of the objective lens of the imaging system. When the spacing of the lattice planes is too small to be resolved in present-day electron microscopes, moiré patterns due to overlapping crystals can be used to give images which, under certain conditions, can be considered essentially as magnified images of the crystal lattice (Hashimoto & Uyeda 1957; Bassett, Menter & Pashley 1958). Both these methods reveal the nature of the distortion of the lattice planes near a defect and depend for their success on good resolution and rather stringent specimen requirements. For example, the specimens must be relatively thin to avoid excessive chromatic aberration due to energy losses.

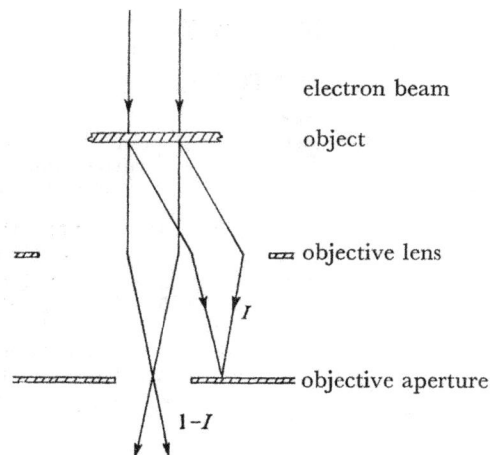

FIGURE 1. Illustrating diffraction contrast. The incident electron beam is diffracted by the specimen and the diffracted beam is removed by the aperture of the objective lens. The intensity of the incident beam is unity, that of the diffracted beam I and that of the transmitted beam $1 - I$. Contrast arises through local variations of the intensities of the diffracted beams.

In the third method, first investigated independently by Bollmann (1956) and by Hirsch *et al.* (1956), image contrast is produced by local differences in the intensities of the Bragg diffracted beams. It is generally arranged that only the direct beam (bright field image) or one diffracted beam (dark field image) enters the aperture of the objective lens (see figure 1). In the bright field image, dark contrast results wherever local conditions in the crystal lead to strong diffraction. This 'diffraction contrast' is sensitive to changes in orientation or thickness, and to displacements of the atoms from their normal positions due to lattice strains. In the perfect crystal this type of contrast is responsible for 'extinction contours', which are bands of constant orientation or of thickness corresponding to conditions of particularly strong diffraction. In the imperfect crystal, dislocations appear as lines because of the atomic displacements near them, and stacking faults give rise to a characteristic fringe pattern. This method of contrast formation has the advantage that the atomic array is not resolved, that resolution is therefore not a limiting factor, and that specimen requirements are less stringent. For example, a projection of the three-dimensional arrangement of dislocations can be obtained for crystals whose thickness would be too great for

THEORY OF DIFFRACTION CONTRAST AT DISLOCATIONS   501

resolution of lattices or of moiré patterns of small spacings. The main application of this method is in the study of the distribution of defects and also of their behaviour, for example, on heating or straining the specimen, while the first two methods are particularly suitable for the study of the nature of the atomic displacements near the defect.

So far no complete theory of contrast for any of these methods is available. Most of the ideas on contrast in the image formed by the electron microscope (Scherzer 1949; Haine 1957; Lenz 1958) seem to be based on theories dealing with the scattering from single atoms (Lenz 1954). In general, of course, the contrast in electron microscope images is determined by the amplitude and phase distribution of the electron waves passing through the objective aperture. Theories in which this distribution is worked out for single atoms apply strictly only to a specimen in which all atoms scatter independently, and therefore can at best only be considered as crude approximations in the case of the most amorphous solids. Even solids such as amorphous carbons give scattering curves which differ considerably, particularly in the all-important low-angle region, from those expected from independently scattering atoms (see, for example, Franklin 1950). This point has been emphasized by Niehrs (1958 a), and is understandable, for if atoms are adjacent to each other, the phase relation between their scattered amplitudes becomes important. The general problem in the case of crystals as for other solids involves the calculation of the distribution of amplitude and phase in the transmitted beam. This has been discussed for the case of lattice resolution by Menter (1956, 1958), Niehrs (1958 b), and Hashimoto, Naiki & Mannami (1958), for moiré patterns by Bassett *et al.* (1958), Dowell, Farrant & Rees (1958), Hashimoto *et al.* (1958), and Kamiya, Nonoyama, Tochigi & Uyeda (1958), and for diffraction contrast for various cases by Heidenreich (1949), Kato (1952 a, b, 1953) and Whelan & Hirsch (1957 a, b). There is, however, as yet no theory of diffraction contrast from dislocations. It is the aim of the present paper to develop an approximate theory which will make it possible to interpret correctly most of the contrast features which are observed on electron micrographs.

To describe the diffraction contrast from dislocations completely, a dynamical theory taking account of the equilibrium between the waves in the imperfect crystal is required, but in spite of the urgent need for such a theory for the correct interpretation of electron micrographs this problem is still unsolved. However, a 'correspondence principle' exists whereby the results of dynamical theory become asymptotic to those of the so-called kinematical theory of diffraction, for large enough deviations from the Bragg angle or for sufficiently thin crystals. Under these conditions the kinematical theory predicts correctly the intensity of the diffracted beam at an infinite distance from the crystal. Most of the interference between the diffracted waves takes place in the crystal itself, and it will be assumed that the diffracted waves at the lower crystal surface (figures 2 and 4) are essentially plane. The diffracted amplitude emerging from any small area of the lower surface of the crystal (at B in figure 2) can then be regarded as coming from a column AB drawn in the direction of the diffracted beam. The kinematical theory gives an expression for the amplitude of this beam, both in the perfect crystal and in the imperfect crystal, when the atoms in the column are displaced from their equilibrium positions. The application of this method to the case of an imperfect crystal containing dislocations is justified by the fact that the diffracted beams from such a crystal are mainly concentrated in directions very close to those corresponding to a perfect crystal (Wilson 1950; Suzuki, reported by Willis 1957),

61-2

and that therefore the waves at the lower surface of the crystal may again be expected to be approximately plane. The calculation of the intensity diffracted from a column as a function of its position in the crystal leads in the first instance to a description of the dark field image, and the bright field image is obtained on the assumption that the two images are complementary. Alternatively, the bright field image can be found directly by calculating the intensity diffracted away from a column in the direction of the direct beam.

In the cases where the dynamical theory has been used, i.e. for the perfect crystal and for the crystal containing a stacking fault, the results predicted by the simple kinematical theory are in good qualitative agreement with those of the more accurate treatment. Some confidence is therefore placed in extending the kinematical theory to the dislocated crystal, and the experimental confirmation of many of its predictions is further justification of the method.

2. Kinematical theory of contrast effects

2·1. *Perfect crystal*

A thin crystal free of defects such as dislocations, faults, etc., but possibly containing long wavelength strains due to bending, is called a perfect crystal. The method of the kinematical theory is illustrated in figure 2, where only one diffracted beam is considered. The intensity

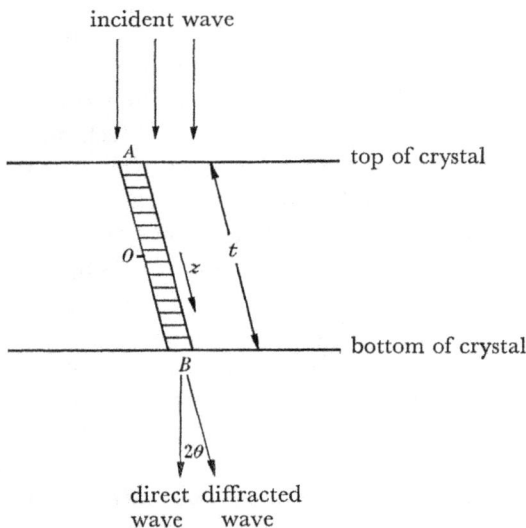

FIGURE 2. Section through a thin crystal showing the column of crystal chosen for calculating the intensity of the dark field image. The column is inclined in the direction of the diffracted beam.

of the dark field image is obtained by calculating the amplitude scattered by a column of crystal in the direction of the diffracted beam shown shaded in figure 2. The amplitude diffracted by the column is

$$A = \sum_j F_j \exp\left(2\pi i(\mathbf{g} + \mathbf{s}) \cdot \mathbf{r}_j\right),\tag{1}$$

where F_j is the scattering factor for electron waves of the contents of the unit cell situated at $\mathbf{r}_j$ in the column, $\mathbf{g}$ is the reciprocal lattice vector corresponding to the Bragg reflexion, and

THEORY OF DIFFRACTION CONTRAST AT DISLOCATIONS   503

$\mathbf{s}$ is a small vector indicating a deviation of the reciprocal lattice point from the reflecting sphere. Since $\mathbf{g}$ is a reciprocal lattice vector, and $\mathbf{r}_j$ is a lattice vector, the product $\mathbf{g}.\mathbf{r}_j$ is an integer.

Hence for a perfect crystal

$$A = \sum_j F_j \exp\left(2\pi i \mathbf{s}.\mathbf{r}_j\right). \qquad (2)$$

If all the unit cells are similar, the F_j's are constant and can be taken outside the summation sign. The summation is then replaced by an integral, and taking the origin O at the middle of the column in figure 2, the amplitude A is found to be proportional to $\{\sin \pi t s\}/\pi s$ and the intensity I proportional to

$$(\sin^2 \pi t s)/(\pi s)^2, \qquad (3)$$

where t is the thickness of the crystal measured along the column and s is the distance of the reciprocal lattice point from the reflecting sphere, also measured along the column. This is the familiar expression for the kinematical intensity distribution around a reciprocal lattice point for a crystal of thickness t, which gives rise to a spike in reciprocal space along the direction of t. Now expression (2) gives the amplitude scattered in a particular direction from a column one unit cell diameter thick. For each reflexion the diffracted rays from such a column are actually spread out over a whole range of directions. However, because of the interference in the crystal between rays diffracted from neighbouring columns the rays at the bottom surface will be concentrated in a small range of angles around that corresponding to the reciprocal lattice point for the operating reflexion. The total intensity diffracted from a column in this small range of angles is given by an expression identical to (3), where the proportionality factor is obtained by calculating the average intensity per unit area of cross-section from a column of finite lateral dimensions and integrating the intensity scattered over the reflecting sphere. In this way the factor is found to be $F^2/k^2V^2\cos^2\theta$ (for the symmetrical Laue case where incident and diffracted beams are inclined at the same angle θ to the normal to the foil surface); $k = \lambda^{-1}$ is the wave vector of the incident electron wave, V is the volume of the unit cell, θ is the Bragg angle, and F is the scattering factor of the contents of a unit cell for electron waves. In most of the subsequent discussion this factor will be omitted. The angle θ is usually of the order of one or two degrees so that the column can be effectively considered as being normal to the crystal surface. Equation (3) shows that the intensity of the diffracted wave oscillates sinusoidally with t for fixed s, a result which is well known from dynamical theory.

The calculation of diffracted amplitudes is conveniently illustrated by means of an amplitude-phase diagram as shown in figure 3. The amplitude scattered by an element of the column lying between depths z and $z+dz$ (measured along the column) is proportional to dz and has a phase angle $2\pi sz$. The amplitude-phase diagram (which represents the sum of all the elements of amplitude scattered by the various parts of the column) is therefore a circle of radius $(2\pi s)^{-1}$, i.e. analogous to the case of diffraction by a slit in physical optics. If we choose the phase zero in the middle of the column, the amplitudes from the top and bottom halves of the crystal in figure 2 are represented by PO and OP' in figure 3, where the arcs PO and OP' are equal to $\frac{1}{2}t$. The resultant amplitude is PP' and as t varies PP' will oscillate. The average diffracted amplitude is proportional to s^{-1} and the average diffracted intensity to s^{-2}. The depth periodicity of the intensity oscillation is $t'_0 = s^{-1}$.

504   P. B. HIRSCH, A. HOWIE AND M. J. WHELAN ON THE

Figure 4 shows schematically the intensity variations of the direct and diffracted waves in the crystal. As s decreases the wavelength of the oscillations t_0' increases, i.e. the radius of the circle in figure 3 increases. According to the kinematical theory t_0' may increase indefinitely. However, for small s dynamical effects become important, and the results of this theory mentioned in §4 show that t_0' has an upper limit t_0 known as the extinction distance for the Bragg reflexion. The depth variation of intensities of direct and diffracted waves gives rise to

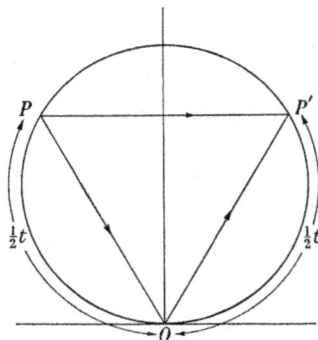

FIGURE 3. Amplitude-phase diagram for a perfect crystal. The circle is of radius $(2\pi s)^{-1}$. PP' is the amplitude diffracted from a crystal of thickness t. The intensity is PP'^2. As s or t varies the value of PP' will oscillate.

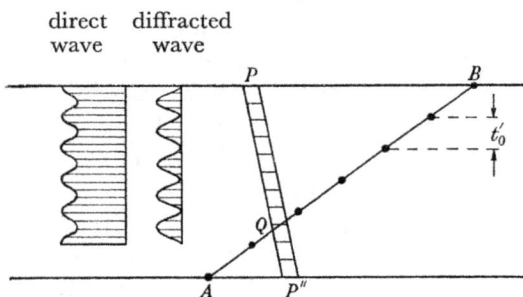

FIGURE 4. Cross-sectional diagram illustrating schematically the intensity oscillations of direct and diffracted waves as a function of depth in the crystal in the kinematical case. The depth periodicity of the intensity oscillations is $t_0' = s^{-1}$. AB is an inclined plane of discontinuity of crystal structure such as a stacking fault, grain boundary, or a crystal surface of a wedge crystal. Fringes will be visible over the region AB. The number of fringes visible is determined essentially by the depth periodicity t_0' and the crystal thickness. The circles represent the points along AB at which the intensity of the diffracted beam is zero.

thickness extinction contours (Heidenreich 1949). For a wedge-shaped crystal (figure 4) the intensity of the directly transmitted wave will clearly vary along the wedge. Figure 5 (a), plate 3, is an example of thickness extinction contours in an aluminium foil containing a depression over a local region; the fringes are contours of equal thickness of the foil. Similar fringes are also observed at grain boundaries running obliquely through the foil (figure 5 (c)), and also, as will be seen later, at stacking faults on oblique planes (figure 5 (d)). A foil of uniform thickness may show extinction contours due to buckling (Heidenreich 1949). In this case t is fixed and s varies across the foil so that the reflecting sphere sweeps through the

THEORY OF DIFFRACTION CONTRAST AT DISLOCATIONS   505

intensity distribution around the reciprocal lattice point. This gives rise to multiple fringes corresponding to the various intensity maxima of equation (3). An example of such bend contours in aluminium is shown in figure 5(*b*). Again, it is found that the kinematical theory gives an adequate qualitative account of the features of the contour, although dynamical theory is required for a detailed interpretation.

2·2. *Imperfect crystal*

A defect generally causes displacements of the atoms from their ideal positions, and if impurity atoms are involved, local changes in the scattering factor F. The latter effect can of course be taken into account, but in this discussion only the atomic displacements which usually have a more important effect on the contrast will be considered.

The amplitude diffracted by a column of imperfect crystal is then given by

$$A = F \sum_j \exp\left(2\pi i (\mathbf{g} + \mathbf{s}) \cdot (\mathbf{r}_j + \mathbf{R})\right), \tag{4}$$

where $\mathbf{R}$ is the displacement of the cell from the ideal position $\mathbf{r}_j$. $\mathbf{R}$ may in general be a function of $\mathbf{r}_j$, as for a dislocation. This reduces to

$$A = F \sum_j \exp\left(2\pi i \mathbf{g} \cdot \mathbf{R}\right) \exp\left(2\pi i \mathbf{s} \cdot \mathbf{r}_j\right), \tag{5}$$

where the product $\mathbf{s} \cdot \mathbf{R}$ in the exponent of (4) has been neglected in comparison with the other terms, since R is usually of atomic dimensions or less, while s is usually the reciprocal of a macroscopic dimension. Thus the displacement of a cell produces a phase angle $\alpha = 2\pi \mathbf{g} \cdot \mathbf{R}$ in the scattered wave, and the resultant amplitude given by equation (5) will differ from that from a perfect crystal. The contrast therefore arises through a phase-contrast mechanism, the phase difference being produced by atomic displacements. Equation (5) may be approximated by an integral. Leaving out the factor F/V

$$A = \int_{\text{column}} \exp\left(i\alpha\right) \exp\left(2\pi i s z\right) \mathrm{d}z, \tag{6}$$

where α is a function of z, depending on both the reflexion $\mathbf{g}$ and the displacement $\mathbf{R}$. The amplitude is essentially the Fourier transform of the phase factor $\exp\left(i\alpha\right)$ over the whole column.

As in the case of a perfect crystal the resultant contrast is obtained by taking into account the interference between neighbouring columns and integrating the diffracted intensity over the range of angles around that corresponding to the reciprocal lattice point. We appeal here to the results of the theories of diffraction from crystals containing stacking faults (Paterson 1952) or dislocations (Wilson 1950; Suzuki, reported by Willis 1957) that in both these cases this range of angles is still small. If the range of angles in the diffracted beam at the lower surface of the crystal is $\mathrm{d}\phi$, the rays contributing to the amplitude at any point on the lower surface will come from a column of diameter $\sim t\,\mathrm{d}\phi$, where t is the length of the column. The range of angles $\mathrm{d}\phi$ is generally quite small; for example, for a crystal of diameter t containing a dislocation, $\mathrm{d}\phi \sim \mathbf{b}/t$ (where $\mathbf{b}$ is the Burgers vector) so that the diameter of the contributing column is of atomic dimensions. This means that the intensity at any point on the lower surface of the crystal will depend mainly on the displacements in the column passing through that point, and will be relatively unaffected by those in neighbouring columns; the total diffracted intensity will be approximately proportional to $|A|^2$, the factor of proportionality being the same as that for the perfect crystal (see §2·1). This is true for crystals with stacking faults, for which the results of the dynamical theory for large deviations from the Bragg angle are found to be identical with those of the kinematical theory (Whelan & Hirsch 1957a, b). The above arguments suggest that the same conclusions apply to crystals with dislocations.

2·3. *Stacking fault*

A simple example which illustrates the use of amplitude-phase diagrams is the case of a stacking fault in a face-centred cubic crystal. Suppose the fault runs obliquely across the foil as shown in figure 4. The two parts of the crystal separated by the fault are displaced

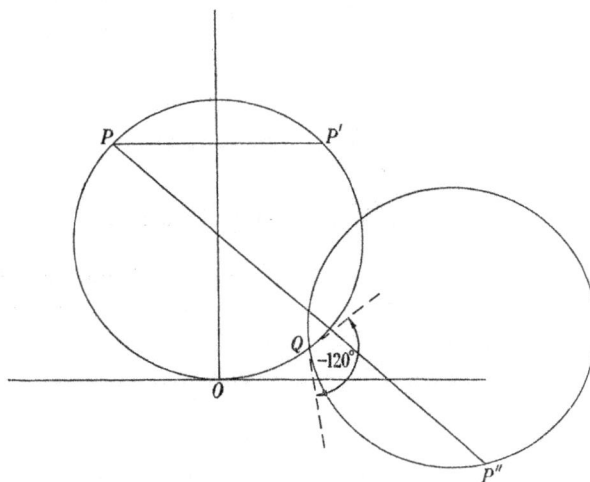

FIGURE 6. Amplitude-phase diagram for a face-centred cubic crystal containing
a stacking fault (figure 4).

relative to each other parallel to the fault plane by the vector $\mathbf{R} = \frac{1}{6}[112]$. This produces a phase difference α in waves diffracted from opposite sides of the fault equal to $\frac{1}{3}\pi[h+k+2l]$, where h, k and l are the indices of the Bragg reflexion. The possible values of α are $\pm 120°$ or $0°$ depending on the indices of the reflexion. The amplitude-phase diagram for $\alpha = -120°$ is shown in figure 6, where the point O corresponds to the middle of the column in figure 4. The circle POP' of radius $(2\pi s)^{-1}$ is called the initial circle; this is the amplitude-phase diagram for the unfaulted crystal. At the point Q on this circle corresponding to the inter-section of the column and the fault plane an abrupt change of phase $-120°$ occurs. The amplitude-phase diagram thereafter follows the final circle, also of radius $(2\pi s)^{-1}$ passing through Q and making an angle of $-120°$ with the initial circle. The amplitude from the part of the column above the fault is PQ, while that from the bottom part is QP''. The resultant amplitude is PP'', which in general is different from PP', the amplitude from the unfaulted crystal. If the fault runs obliquely across the foil, P is fixed (figure 6) but Q and P'' and hence PP'' vary as the position of the column is changed. For any two points on the fault whose depths differ by t'_0, Q occupies identical positions on the initial circle. The corresponding difference in length of the column in the lower crystal will also be t'_0 and hence

THEORY OF DIFFRACTION CONTRAST AT DISLOCATIONS   507

the two positions of P'' are also identical. Thus PP'' will have the same value for these points on the fault. It follows that the contrast is in the form of fringes parallel to the intersection of the fault with the foil surface, the periodicity of the fringes being determined by t_0'. An example of stacking-fault fringes is shown in figure 5 (*d*). The resultant intensity distribution in the fringes can of course be obtained from equation (6).

3. Contrast at dislocations

3·1. *Nature of the problem*

In this section the method of the kinematical theory outlined in the previous section will be applied to the case of a dislocation. If one looks down on a thin crystal containing a dislocation parallel to the surface, the line of vision passing through the centre of the dislocation, it

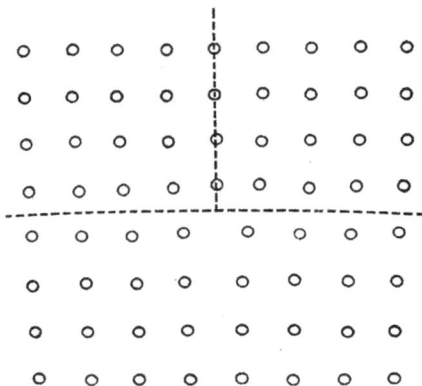

FIGURE 7. Diagram illustrating the local rotation of lattice planes near an edge dislocation in a simple cubic lattice (courtesy Clarendon Press).

is readily seen that the atoms above the dislocation are displaced relative to those below by an amount equal to half the Burgers vector. (This is most easily seen for pure screw or edge dislocations; see, for example, Read (1953).) Thus if the column of figure 2 passes through the centre of the dislocation an abrupt change of phase occurs, and the calculation of contrast is similar to that for a fault. For other positions of the column not passing through the centre of the dislocation, the column is continuously deformed so that the phase angle $\alpha = 2\pi \mathbf{g} . \mathbf{R}$ becomes a continuous function of position. In this simple way it is easy to see how contrast arises.

One significant feature of the contrast from dislocations can be deduced immediately from the fact that the atomic displacements on opposite sides of the dislocation are in opposite directions. (This is shown for an edge dislocation in figure 7.) In equation (6) the phase angle α due to displacement will add to that due to depth in the crystal (i.e. the phase angle $2\pi sz$) on one side of the dislocation, and will subtract from it on the other side. In the former case the effect of the dislocation is to bring the crystal locally further from, and in the latter case nearer to the reflecting position. Generally, therefore, the contrast will lie on one side of the centre of the dislocation. This can also be interpreted in terms of a local rotation of lattice planes, which can be seen for an edge dislocation in figure 7. The conclusions that the

P. B. HIRSCH, A. HOWIE AND M. J. WHELAN ON THE

sense of the phase angle and the rotation of the lattice planes are opposite on opposite sides of a dislocation, and that the contrast appears only on one side, apply to dislocations of all types.

3·2. *Amplitude-phase diagrams*

For the construction of amplitude-phase diagrams and associated calculations for the column of crystal near a dislocation, the simplest case to consider is that of a screw dislocation parallel to the surface of a thin foil and at distances of z_1 and z_2 from the top and bottom surfaces. Figure 8 shows the arrangement and the co-ordinate system used. The atomic displacements around the screw are taken to be the same as in an infinite medium, i.e.

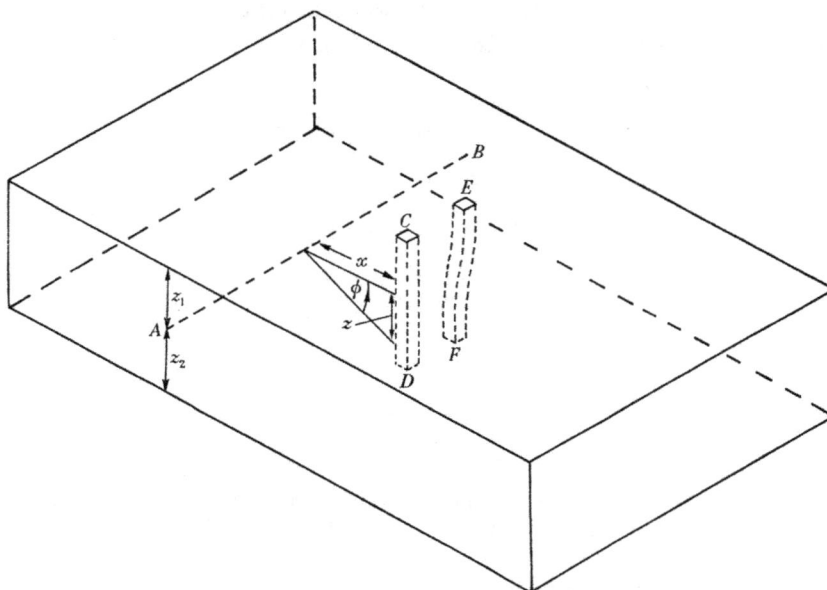

FIGURE 8. Crystal containing a screw dislocation AB parallel to the plane of the foil. A column of crystal CD in the perfect crystal is deformed to shape EF after introduction of the screw dislocation. The diagram illustrates the choice of parameters x, z and ϕ.

the effects of image dislocations in the surfaces are neglected. If these are included additional complications are introduced, but it is thought that the main qualitative results will be unaffected, provided the dislocation is not too close to the surface. Figure 8 shows the nature of the displacements; a column of crystal CD is deformed to a shape EF after introduction of the screw dislocation AB.

The displacement vector $\mathbf{R} = \mathbf{b}\phi/2\pi = (\mathbf{b}/2\pi)\tan^{-1}(z/x)$, where $\mathbf{b}$ is the Burgers vector of the dislocation. Hence

$$\alpha = 2\pi\mathbf{g}.\mathbf{R} = \mathbf{g}.\mathbf{b}\tan^{-1}(z/x) = n\tan^{-1}(z/x). \tag{7}$$

Since $\mathbf{g}$ is a reciprocal lattice vector and $\mathbf{b}$ is an interatomic vector, n is an integer which may take positive and negative values and zero. $n\pi$ is the phase difference between waves scattered immediately above and below the dislocation. Since the atomic scattering factor decreases rapidly with increasing scattering angle the diffracted beams corresponding to small values

THEORY OF DIFFRACTION CONTRAST AT DISLOCATIONS   509

$\mathbf{g}$ are most important in producing contrast, and in this paper the values of n considered are of $n = 0, 1, 2, 3, 4$, (n can always be regarded as positive provided the sign of x is suitably adjusted). The case $n = 0$, i.e. $\alpha = 0$, corresponds to $\mathbf{g}$ and $\mathbf{b}$ being perpendicular to each other. This case in which the dislocation is invisible will be mentioned later (§ 5·2). Now $\mathbf{g}$ is perpendicular to the reflecting planes, and this condition implies that displacements parallel to the reflecting planes produce no contrast. In particular since the Bragg angles are small, displacements parallel to the incident beam, i.e. approximately normal to the foil surface produce no contrast.

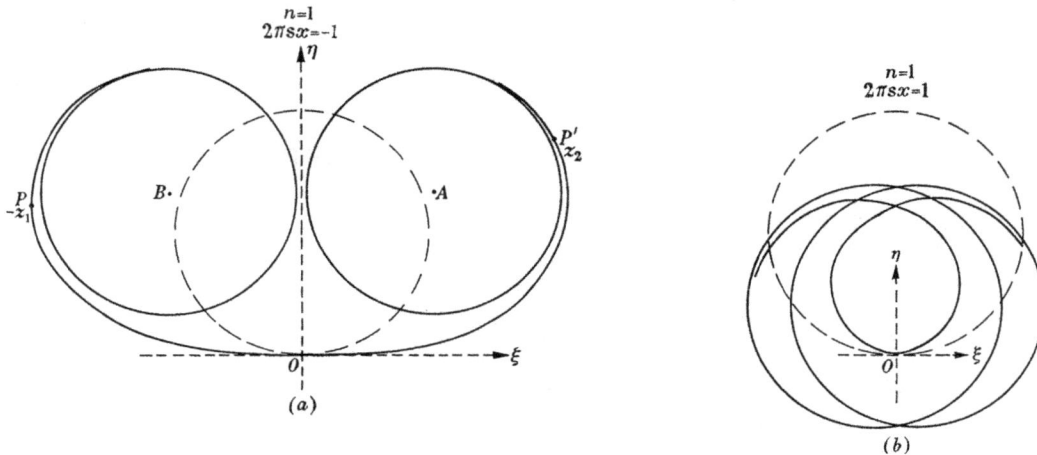

FIGURE 9. Amplitude-phase diagrams for a column of crystal close to a screw dislocation (figure 8). The radius of the broken circle is $(2\pi s)^{-1}$. The diagram is an unwound spiral (a) or wound-up spiral (b). The diagrams (a) and (b) refer to $n = 1$, $2\pi s x = -1$ and $2\pi s x = +1$, respectively. There is a different diagram for each value of $2\pi s x$. The amplitude diffracted by the column of crystal EF in figure 8 is obtained by joining points PP' corresponding to the top and bottom of the column. The separation between the points P and P' measured along the curves is equal to the crystal thickness. The amplitude diffracted from one side of the dislocation (a) is greater than that from the other side (b).

Substitution of (7) in equation (6) gives

$$A = \int_{-z_1}^{+z_2} \exp\left(in\tan^{-1} z/x + 2\pi i s z\right) \mathrm{d}z. \tag{8}$$

Since the integral (8) is an expression for the amplitude-phase diagram (z being distance measured along the curve and $n\tan^{-1} z/x + 2\pi s z$ the slope), diagrams for various situations can be constructed graphically with tolerable accuracy (figures 9, 10 and 14). The following points are to be noted.

(a) The size of the diagram, and therefore the amplitude, are proportional to s^{-1} as in the case of a perfect crystal or a crystal with a stacking fault. The form of the diagram depends on two parameters only, namely, n and $\beta = 2\pi s x$. For fixed s, β will vary with x and $-\infty \leqslant \beta \leqslant \infty$. The centre of the dislocation corresponds to $\beta = 0$.

(b) In figure 9 (a) and (b) the circle corresponding to the perfect crystal is shown by a broken line. This is called the original circle. O corresponds to $z = 0$ in figure 8, i.e. to the

510   P. B. HIRSCH, A. HOWIE AND M. J. WHELAN ON THE

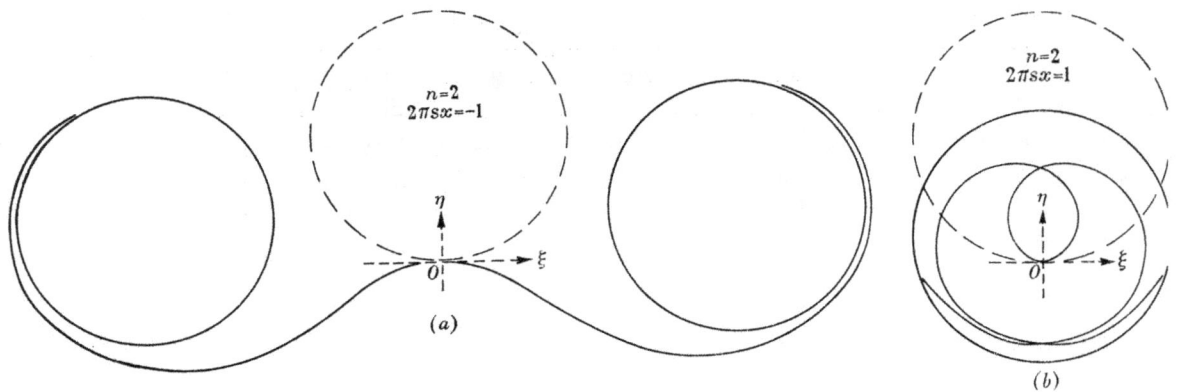

FIGURE 10. Amplitude-phase diagrams for a screw dislocation for $n = 2$, $2\pi s x = \pm 1$.

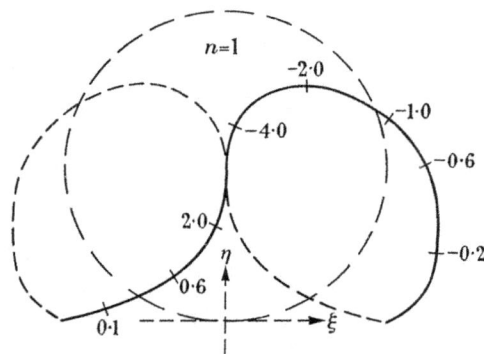

FIGURE 11. Full curve: locus of the centre of the final circle (A in figure 9) as $\beta(= 2\pi s x)$ varies. Broken curve, similar locus of the centre of the initial circle (B in figure 9). The diagram is for $n = 1$, and values of β are marked on the locus. The broken circle represents the original circle of radius $(2\pi s)^{-1}$.

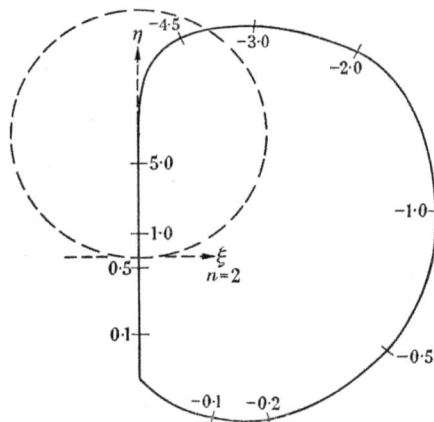

FIGURE 12. Locus of the centre of the final circle for $n = 2$. The broken circle is the original circle of radius $(2\pi s)^{-1}$. Values of $\beta(= 2\pi s x)$ are marked on the locus.

THEORY OF DIFFRACTION CONTRAST AT DISLOCATIONS   511

point on the column nearest the dislocation. At the top of the column CD in figure 8 the crystal is nearly perfect, so that the resultant amplitude scattered from such a region is a chord of a curve which is asymptotic to a circle of radius $(2\pi s)^{-1}$ centred at B in figure 9 (a). Similar considerations hold for the part of the column near the bottom of the foil remote

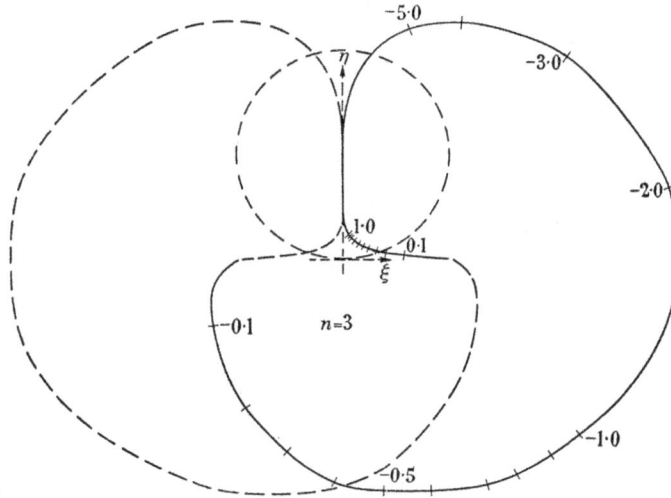

FIGURE 13. Loci of the centres of the initial circle (broken curve) and final circle (full curve) for $n = 3$. The broken circle is the original circle of radius $(2\pi s)^{-1}$. Values of $\beta(= 2\pi sx)$ are marked on the locus.

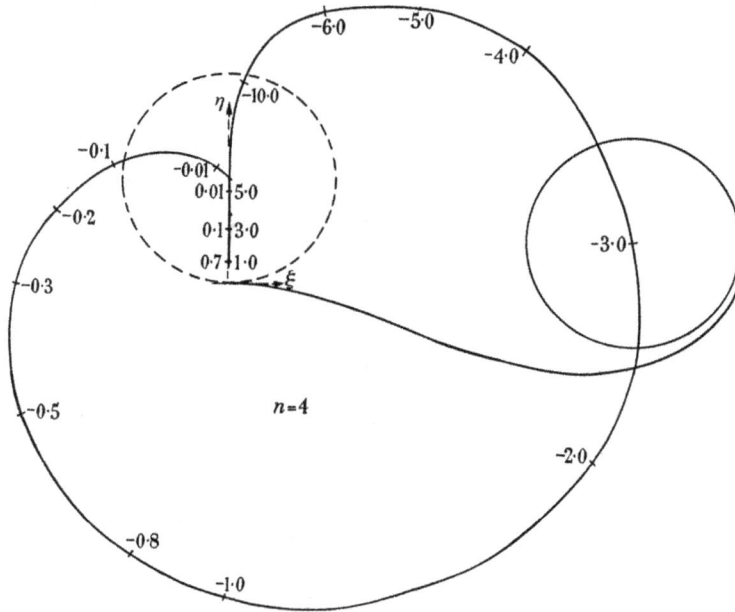

FIGURE 14. Locus of the centre of the final circle for $n = 4$. The broken circle is the original circle of radius $(2\pi s)^{-1}$. Values of $\beta(= 2\pi sx)$ are marked on the locus. The diagram also shows a graphically constructed amplitude-phase diagram for $\beta = 2\pi sx = -3$.

from the dislocation. Here the amplitude-phase diagram is asymptotic to a circle of radius $(2\pi s)^{-1}$ centred at A in figure 9 (a). These two circles are called the initial and final circles, respectively. The complete amplitude-phase diagram is a double spiral curve connecting the initial and final circles. It will be noted that two types of spiral occur; these are shown in figures 9 and 10. For negative β the phase angles in the integral (8) subtract and the resultant spiral can be regarded as generated by unwinding the original circle (figures 9 (a) and 10 (a)). For positive β the phase angles are additive and the spiral winds up (figures 9 (b) and 10 (b)).

(c) The significance of the amplitude-phase diagram is as follows. The amplitude diffracted by a column $-z_1 \leqslant z \leqslant z_2$ is given by the line joining the points PP' corresponding to $-z_1$ and z_2 on figure 9 (a). Usually P and P' will be near the initial and final circles, respectively. The arc length PP' along the double spiral is proportional to $z_1 + z_2$.

(d) For a dislocation in an unbent uniform foil z_1, z_2, s and n are fixed for a given reflexion, and the amplitude diffracted will vary with x, i.e. with $\beta = 2\pi s x$. As β varies the centre of the final circle (with co-ordinates ξ_n and η_n) describes a locus which passes through the centre of the original circle when $\beta = \pm \infty$ (this corresponds to a column far from the dislocation where the crystal is perfect). Figures 11 to 14 show the loci for $n = 1, 2, 3, 4$; values of β are marked on the loci. The co-ordinates ξ_n and η_n have been calculated from the integral (8) with $z_1 = 0$ and $z_2 = \infty$. The calculation is given in the appendix. Knowing the centre of the final circle the amplitude-phase diagram can be constructed graphically. As β varies the initial circle also moves along a locus which is the reflexion of the above locus in the η axis. When $x = \beta = 0$, $\alpha = n \tan^{-1} z/x = \frac{1}{2} n\pi$ (neither s nor $z = 0$). The final circle in this case must pass through the origin and make an angle $\frac{1}{2} n\pi$ with the original circle. These simple considerations are sufficient to fix the initial and final points of the locus.

3·3. *Intensity profiles of dislocation images*

Inspection of a typical amplitude-phase diagram, e.g. figure 9 (a), shows that the diffracted amplitude PP' consists of two parts—a steady component represented by the distance between the centres of the two circles $BA = 2\xi_n$ and a fluctuating part depending on the exact positions of the points P and P' near these circles. As x varies both these contributions to the diffracted amplitude will change, but since the angular position of P' near its circle behaves like $\tan^{-1} z_2/x$ it can be taken as constant for $|x| < \frac{1}{4} z_2$ say. In practice this means that all the variation of the diffracted amplitude within the region where the contrast in the image is appreciable comes from the steady component. Since the amplitude is also proportional to $(2\pi s)^{-1}$, a convenient measure of the diffracted intensity is thus given by the quantity $(4\pi s \xi_n)^2$ which has been calculated (see appendix) and plotted against β for various values of n in figure 15.

The first point of interest is that the dislocation scatters more intensity into the Bragg reflexion than the surrounding regions, and should appear as a dark line on bright field images, and as a bright line on the dark field image. Secondly, the intensity peak occurs for negative values of β, i.e. the intensity scattered is larger on one side of the dislocation than on the other (in fact for $n = 2, 4$, ξ_n is zero for positive β). This of course corresponds to the differences in behaviour of the spiral on opposite sides of the dislocation and to the elementary considerations mentioned in § 3·1. Thus the image of a dislocation should lie to one side

THEORY OF DIFFRACTION CONTRAST AT DISLOCATIONS 513

of the centre of the dislocation; the displacement of the peak from 0 in figure 15 is of the order of the width of the image. Moreover, for $n = 3$ and $n = 4$ two maxima occur— a principal one and a subsidiary one at a smaller value of $|\beta|$. Thus the image should show two peaks on the same side of the dislocation. Dislocations with multiple images have been observed occasionally, but it is not known whether the theory is sufficiently reliable to predict this fine detail. Double images are more readily explained in another way (see §5 (*c*)). All the profiles rise steeply from low values of $|\beta|$ and tail off more gradually.

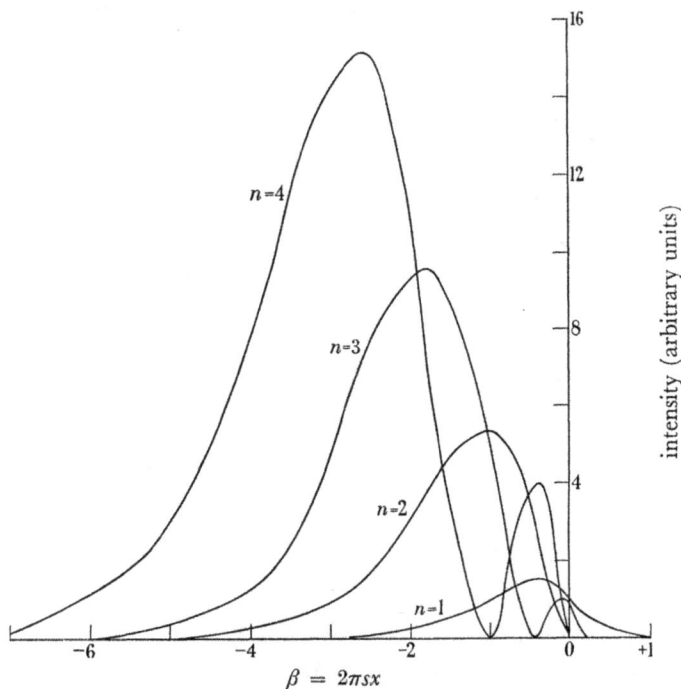

FIGURE 15. Intensity profiles of images of a screw dislocation for various values of n. $(4\pi s \xi_n)^2$ is plotted as a function of $\beta(= 2\pi sx)$. The centre of the dislocation is at $\beta = 0$. Note that the contrast lies to one side of the centre of the dislocation.

It should be noted that the intensity profiles of figure 15 do not take into account the smaller atomic scattering factors for larger values of n. This effect would tend to make the peak heights for different n more nearly equal. The curves of figure 15 can be considered as dislocation profiles for various n for the same value of s. However it will be seen (§4) that for any reflexion the smallest possible effective value of s (corresponding to maximum contrast, since the intensity is proportional to $(2\pi s)^{-2}$) is of the order of the reciprocal of the extinction distance t_0. Reference to table 1 shows that t_0 increases with increasing order of reflexion. Thus, smaller effective values of s are possible for higher values of n. It is thought that the result of these two effects is to give intensity profiles under optimum contrast conditions whose peak heights are approximately independent of n, but whose width increases with increasing values of n so that the profile for $n = 4$ is about five times as wide and as far from the dislocation as that for $n = 1$.

3·4. *Extension of the theory to other dislocation configurations*

(a) *Screw dislocation inclined at an angle ψ to the plane of the foil*

The dislocation AB and the column of crystal are shown in figure 16. The displacement is now $(\mathbf{b}/2\pi) \tan^{-1} y/x = (\mathbf{b}/2\pi) \tan^{-1} (z \cos \psi)/x$. The calculation of the diffracted amplitude proceeds therefore as before x now being replaced by $x/\cos \psi$. This means that the amplitude at $x \cos \psi$ will be the same as that at x for the screw parallel to the surface, i.e. the dislocation will appear narrower by a factor $\cos \psi$ (the amplitude is also less by a factor of $\cos \psi$).

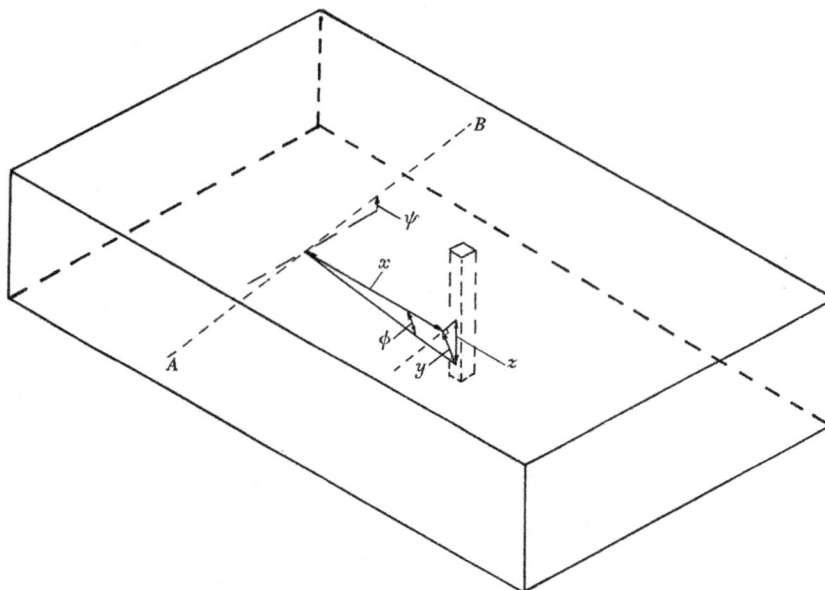

FIGURE 16. Thin foil containing a screw dislocation AB inclined to the plane of the foil. The diagram shows the definition of the distances x, y, z and the angles ϕ, ψ.

Another effect which can occur with an inclined dislocation is an oscillation in the diffracted amplitude along its length because the positions of P and P' (figure 9 (a)) near their respective circles vary for different parts of the dislocation situated at different depths (z). As before, for each z the positions of P and P' are approximately independent of x within the image. Since the origin of z is taken at the level of the dislocation, P and P' will move in such a way as to keep constant the distance PP' measured round the curve and the extremes of the motion will be when either P or P' coincides with O (corresponding to the point where the dislocation meets one or other of the foil surfaces). Figure 17 shows how this may produce large oscillations in some cases and small or no oscillations in others. The points A, B, C and A', B', C' correspond to these different positions of P and P'. Inspection of figure 17 reveals that in all these cases oscillations will always occur when either P or P' is near O and thus the oscillations of the diffracted amplitude which produce a dotted image should be more pronounced when the dislocation is near one of the surfaces of the foil.

THEORY OF DIFFRACTION CONTRAST AT DISLOCATIONS 515

(b) Edge dislocation

The displacements due to an edge dislocation in an infinite medium are (Read 1953):

$$R_1 = \frac{b}{2\pi}\left[\phi + \frac{\sin 2\phi}{4(1-\nu)}\right] \text{parallel to the Burgers vector,}$$

$$R_2 = -\frac{b}{2\pi}\left[\frac{1-2\nu}{2(1-\nu)}\ln|r| + \frac{\cos 2\phi}{4(1-\nu)}\right] \text{normal to the slip plane,}$$

$$R_3 = 0 \text{ parallel to the dislocation line.}$$

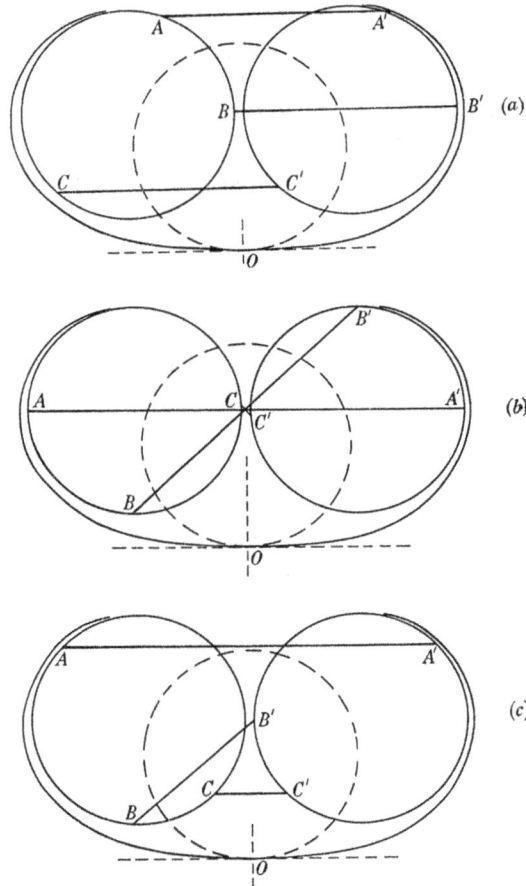

FIGURE 17. Amplitude-phase diagrams for an inclined screw dislocation illustrating how dotted images can occur. In (a), A and A' correspond to the top and bottom of the column in figure 16. For a sufficiently thick foil A and A' will lie close to the initial and final circles and are actually shown on these circles for clarity. As the column is moved along the dislocation, keeping x constant, AA' will move to BB' and CC'. The amplitudes AA', BB' and CC' will remain approximately constant, giving a continuous dislocation image. In (b), AA', BB' and CC' may differ considerably as the position of the column varies giving rise to a dotted image; (c) shows an intermediate case. In all cases the points AA' move in such a way that the total arc length measured along the curves between A and A', B and B', C and C' remains constant and equal to the crystal thickness.

ν is Poisson's ratio; ϕ is an angular co-ordinate in the plane normal to the dislocation line measured from the direction of the Burgers vector.

When the slip plane is parallel to the surface of the foil, R_1 is the displacement which affects the contrast (see §3·2).

For

$$\nu = \tfrac{1}{3}, \quad \alpha = 2\pi\mathbf{g}.\mathbf{R}_1 = n[\phi + \tfrac{3}{8}\sin 2\phi] \approx n\tan^{-1}(2z/x),$$

to a reasonable approximation, where $\tan\phi = z/x$. The amplitude is the same at $2x$ and x for edge and screw dislocations for the same s. Thus the edge dislocation in this orientation will appear about twice as wide as a screw dislocation.

No calculations have been performed for the case of an edge dislocation whose Burgers vector is normal to the foil. In this case $\alpha = 2\pi\mathbf{g}.\mathbf{R}_2$ and the presence of the term in $\ln|r|$ means that the amplitude-phase diagram is no longer asymptotic to a circle for large $|z|$. Two general remarks can be made however.

(1) The phase factor α for this displacement does not change sign with x. Thus any image that results will be symmetrically placed relative to the dislocation.

(2) Most of the effect should come from the $\ln|r|$ term in the phase factor, since the term in $\cos 2\phi$ changes sign twice in any column of crystal. Since the $\ln|r|$ term varies extremely slowly at large distances the image is likely to be very narrow.

3·5. *Effect of divergence of the incident beam*

Under microscopy conditions in the Siemens Elmiskop I electron microscope the maximum divergence of the incident beam is approximately 3×10^{-3} radians ($600\,\mu$ condenser aperture). This corresponds to a fractional variation in s of about $\tfrac{1}{6}$ for a typical reflexion and reasonable value of s. Consequently, all the dislocation image profiles should be averaged over this range of s which corresponds to a variation of approximately $60°$ in the angular position of a point such as P' on the final circle in figure 9(a). Of course the size of the circle also changes with s. Probably this effect would be sufficient to obscure some of the fine details that might otherwise appear on the images and it may be the explanation for the comparative rarity of dotted images. Absorption effects, if important, would also be expected to obscure fine detail except perhaps for dislocations close to the surface.

4. Discussion of the applicability of the theory and its relation to dynamical theory

The basic feature of the kinematical theory is the amplitude-phase diagram and it is necessary to discuss its validity and be aware of its limitation in view of the fact that dynamical effects are known to be important in electron diffraction.

For a perfect crystal the dynamical theory of electron diffraction gives the following expressions for the amplitudes transmitted (T) and diffracted (A) by a crystal of thickness t (see, for example, Whelan & Hirsch 1957 a). The amplitude of the incident wave is taken to be t_0/π.

$$T = (t_0/\pi)\{-\mathrm{i}\cos\pi t\bar{s} - (s/\bar{s})\sin\pi t\bar{s}\}, \tag{9}$$

$$A = (\sin\pi t\bar{s})/(\pi\bar{s}), \tag{10}$$

where

$$\bar{s} = \surd(t_0^{-2} + s^2),$$

THEORY OF DIFFRACTION CONTRAST AT DISLOCATIONS   517

and t_0 is the extinction distance, equal to $(\pi V k \cos\theta)/F$; t_0 depends on the indices of the reflexion mainly through the factor F (§2·1). Since F falls off rapidly with increasing scattering angle the extinction distance increases for higher-order reflexions. Table 1 lists some typical extinction distances for face-centred cubic metals. F has been calculated using the Born approximation (Mott & Massey 1933). It is to be noted that extinction distances for low-order reflexions are about a few hundred ångströms.

TABLE 1. EXTINCTION DISTANCES (Å) OF TYPICAL METALS FOR
100 kV ELECTRONS ($\lambda = 0.037$ Å)

reflexion	111	200	220
Al	646	774	1240
Ni	258	302	468
Cu	268	308	472
Ag	250	285	403
Pt	165	188	262
Au	181	204	281

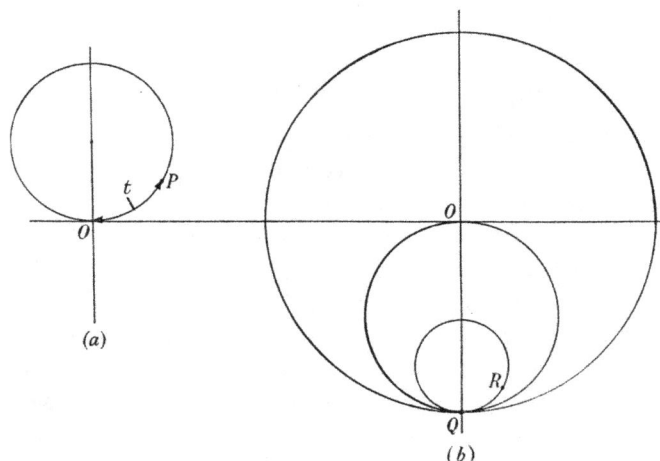

FIGURE 18. Amplitude-phase diagrams representing dynamical scattering from a perfect crystal. (a) Diffracted wave. The radius of the circle is $(2\pi\bar{s})^{-1}$. OP is the dynamical amplitude for a crystal of thickness t. (b) The transmitted wave amplitude is OR, where R lies on a circle QR of radius $t_0(\bar{s}-s)(2\pi\bar{s})^{-1}$. $OQ = t_0/\pi$ and QR is parallel to OP in (a). As the crystal thickness varies both OP and OR oscillate. When s is large compared with t_0^{-1} the circle QR in (b) is much smaller than the circle OP in (a) and approximates to a point. This is the kinematical approximation. For small s both the circles OP and QR expand to a maximum radius $t_0/2\pi$.

Equation (10) is identical with the kinematical expression for the diffracted amplitude with s replaced by $\bar{s}$. The minimum value of $\bar{s}$ is t_0^{-1}.

Amplitude-phase diagrams can be constructed to represent A and T. The diagram for the diffracted wave (figure 18 (a)) is a circle of radius $(2\pi\bar{s})^{-1}$, while that for the transmitted wave is a circle of radius $t_0(\bar{s}-s) \times (2\pi\bar{s})^{-1}$ (figure 18(b)). OP is the diffracted amplitude for a crystal of thickness t; OR is the transmitted amplitude; OP is parallel to QR. As the crystal thickness varies both OP and OR oscillate. For s large compared with t_0^{-1} the circle QR in figure 18 (*b*) has a much smaller radius than the circle in figure 18 (*a*), i.e. OR oscillates only very slightly from OQ. This is the region in which the kinematical approximation is justified. The circle in figure 18 (*a*) is then approximately of radius $(2\pi s)^{-1}$ as assumed in the kinematical theory. However, as $s \to 0$ the oscillations of OP and OR increase in amplitude until eventually when $s = 0$ both OP and OR describe circles of radius $t_0/2\pi$ passing through O. In this case both transmitted and diffracted amplitudes oscillate between zero and t_0/π with $90°$ phase difference. This corresponds to the centre of the dynamical region and illustrates the well-known phenomenon of periodic extinction of the transmitted wave. Thus for a perfect crystal the transition from the kinematical to the dynamical regions is gradual, and no well-defined limits to the applicability of kinematical theory occur. However, it is clear that amplitudes greater than t_0/π have no physical meaning on amplitude-phase diagrams, and that for a dislocated crystal dynamical effects will tend to distort the diagrams to a greater extent for smaller s, so that they are always contained by a circle of radius t_0/π.

5. Comparison of theory with experiment

The theory developed in the previous sections explains a number of features associated with images of dislocations.

5·1. *Nature, width and position of dislocation contrast*

The theory predicts that dislocations should appear as dark lines on bright field micrographs and that the contrast should be reversed in dark field. Figures 19 (*a*) and (*b*), plate 4, provide a striking example of the reversal of contrast in bright and dark field micrographs of dislocations in aluminium. The reflexion producing the contrast is (220). Some cases have been observed where the bright and dark field images are not complementary (dislocations are dark in both), but usually then another high-order reflexion is found to be operating.

The intensity profiles of figure 15 suggest that the width Δx of a dislocation at half maximum intensity for $n = 1$ and 2 should correspond to $\Delta \beta = 2\pi s \Delta x \sim 2$, with possibly a considerably larger width for $n = 3$ and 4. The experimental evidence seems to suggest that $n = 1$, 2 occur frequently while $n = 3$, 4 occur seldom if ever. The width of a dislocation should therefore be $\Delta x \sim (\pi s)^{-1} = t_0'/\pi$. According to the kinematical theory the width should become infinite when s tends to zero. However, we have seen that dynamical conditions effectively limit s to values greater than t_0^{-1}. The maximum width of a dislocation is therefore expected to be about t_0/π or less. Using the extinction distances t_0 of table 1 it is seen that dislocations in aluminium should appear as lines about 200 to 400 Å wide or less depending on the reflexion, while other metals may have smaller maximum dislocation widths (e.g. for Ni about 80 to 150 Å). These considerations are also in general agreement with observations.

The theory predicts that the contrast peak should be to one side of the centre of the dislocation, the displacement of the peak being of the order of the dislocation width. The position of the peak depends on the factors **g**, **b** and **s**; it reverses if any of these parameters changes sign. Figure 20, plate 4, shows a typical example of peak reversal when s changes sign for dislocations in aluminium. The orientation of the normal to the foil is [112] and it is thought that the dislocations lie in (111) with Burgers vectors either $\pm\frac{1}{2}[1\bar{1}0]$ or $\pm\frac{1}{2}[0\bar{1}1]$. Running across the micrograph is a bend extinction contour of the $1\bar{3}1$ reflexion (the

Hirsch et al.   *Phil. Trans. A, volume* 252, *plate* 3

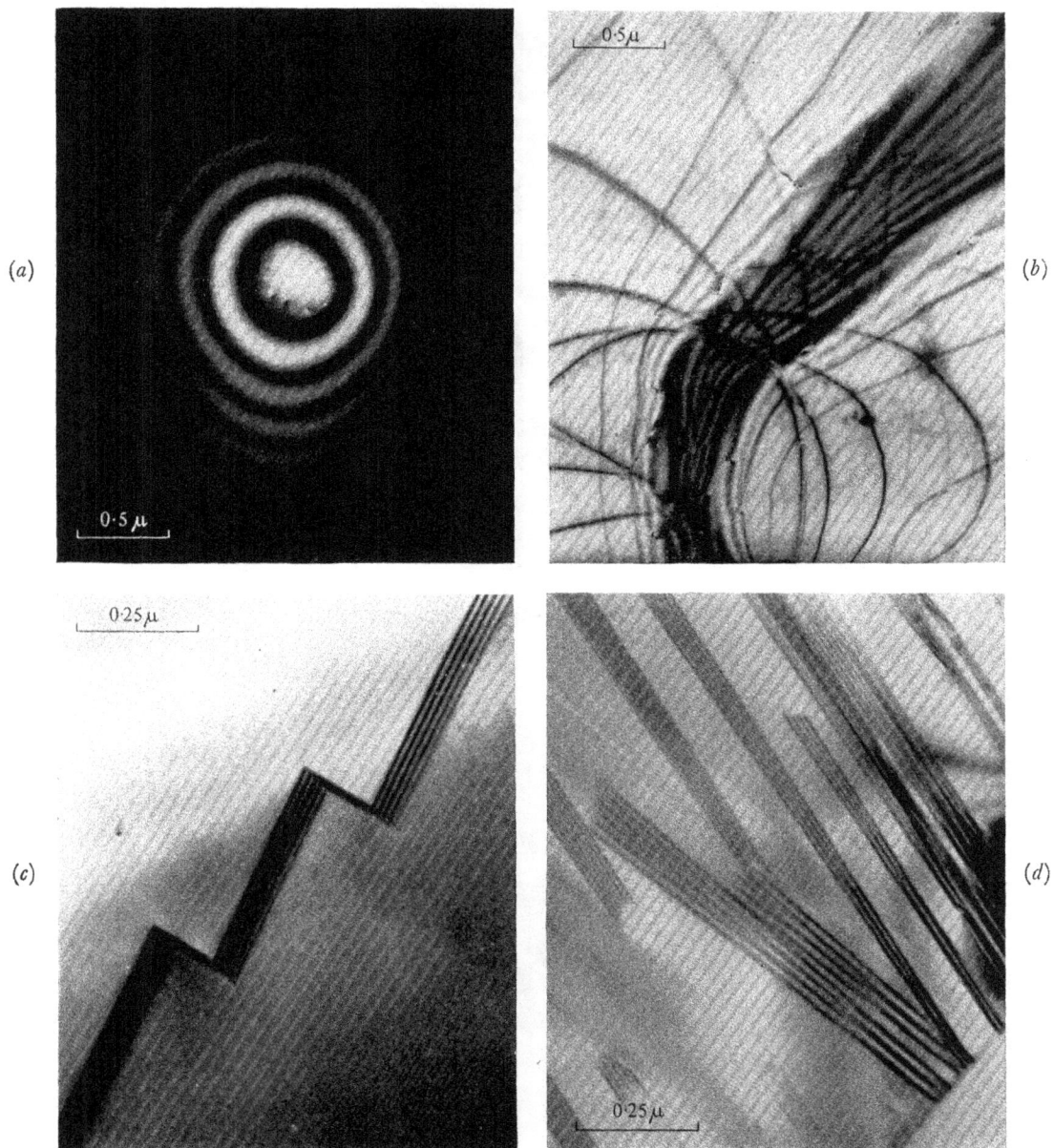

FIGURE 5. Examples of fringes in thin metal foils. (*a*) Thickness extinction contours in aluminium around a nearly circular depression. (Magn. × 30 000.) (*b*) Extinction contours in aluminium due to buckling of the foil. (Magn. × 30 000.) (*c*) Fringes at a twin boundary in stainless steel. (Magn. × 60 000.) (*d*) Fringes at stacking faults in stainless steel. (Magn. × 80 000.)

Hirsch et al.   *Phil. Trans. A, volume* 252, *plate* 4

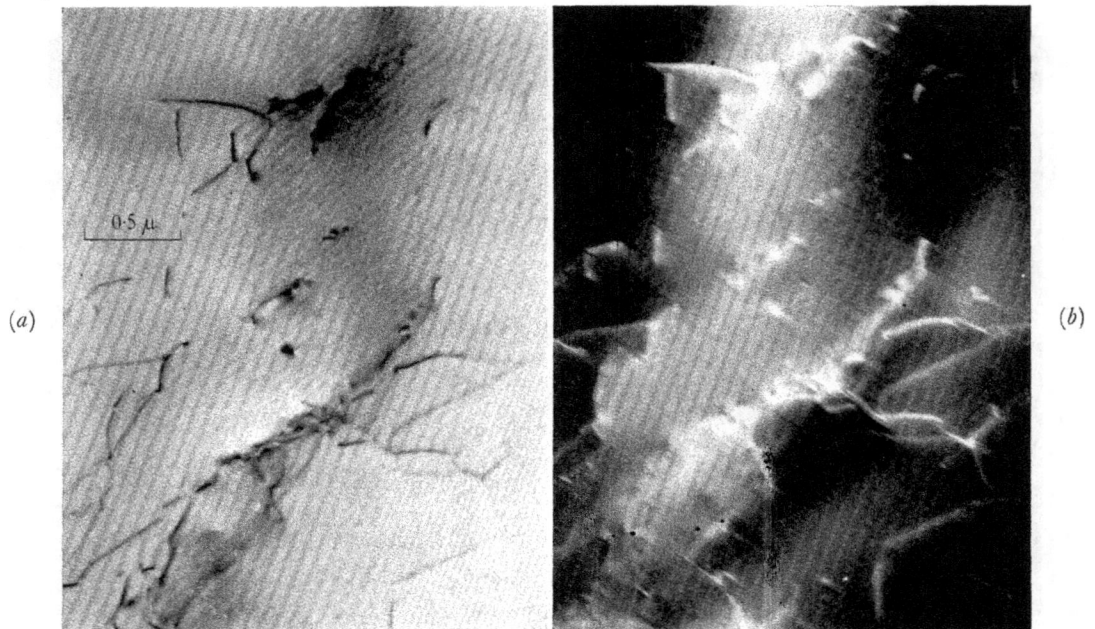

(a)

(b)

FIGURE 19. Bright field and dark field images of dislocations in aluminium. Note that the dislocations appear as light lines on the dark field image and as dark lines on the bright field image. (Magn. × 30 000.)

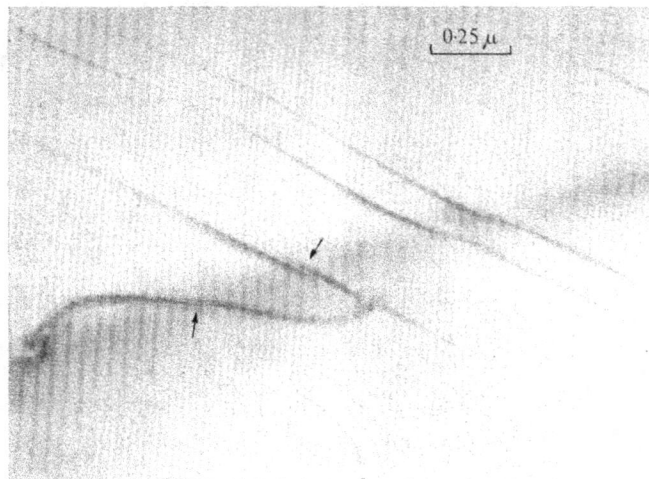

FIGURE 20. Micrograph illustrating the one-sided nature of dislocation contrast in aluminium. Note that the contrast changes from one side of the dislocation to the other where the dislocation crosses the extinction contour. Note also the rapid decrease in contrast on the side facing the dislocation and slower decrease on the other side. The width and total intensity of the image is a maximum on the edge of the extinction contour. (Magn. × 50 000.)

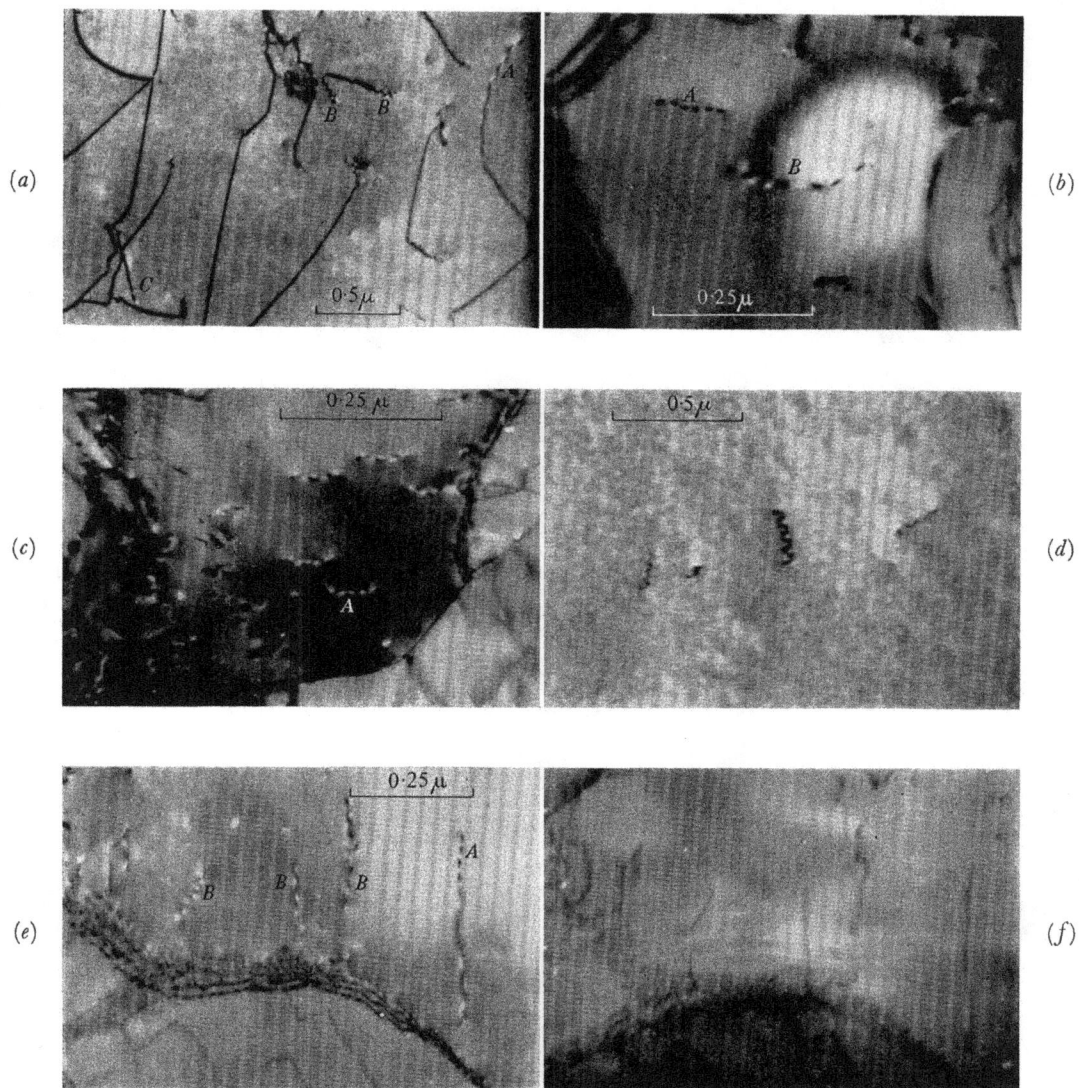

FIGURE 21. Micrographs of aluminium showing special contrast effects at dislocations. (*a*) *A*, alternating black-white appearance, *B*, zigzag contrast, *C*, dotted contrast. (Magn. × 26 000.) (*b*) *A*, dotted contrast; *B*, alternating black and white contrast. (Magn. × 100 000.) (*c*) Dotted, white contrast at dislocations in region of dark extinction contour. (Magn. × 100 000.) (*d*) Zigzag contrast. (Magn. × 40 000.) (*e*) *A*, dotted contrast; *B*, zigzag contrast. (Magn. × 75 000.) (*f*) Same area as (*e*) after tilting the specimen. (Magn. × 75 000.)

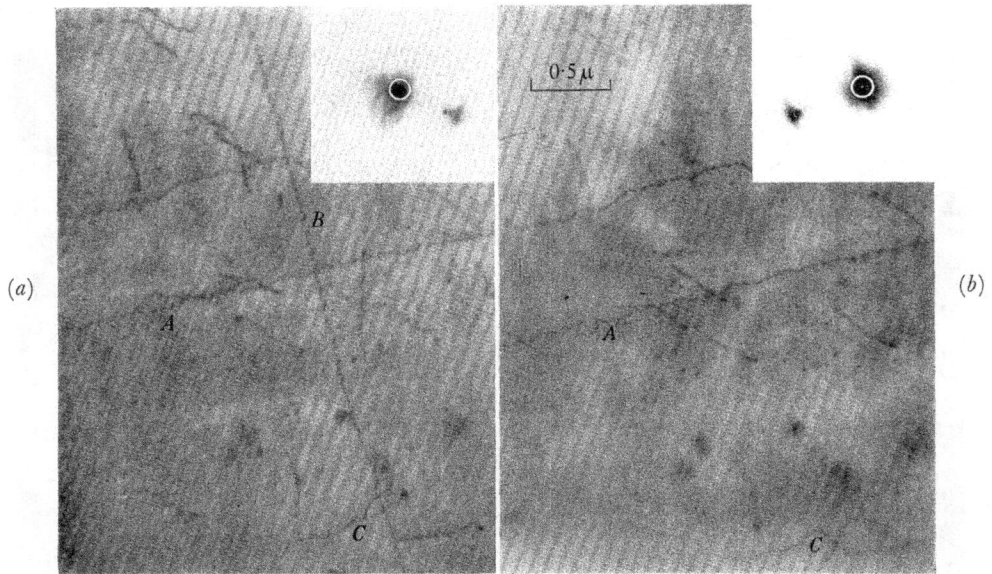

FIGURE 22. Sequence of micrographs (and diffraction patterns) of quenched aluminium + 4% copper alloy showing two helical dislocations A and B. After tilting dislocation B has vanished. Another dislocation has appeared at C. (Magn. × 25 000.)

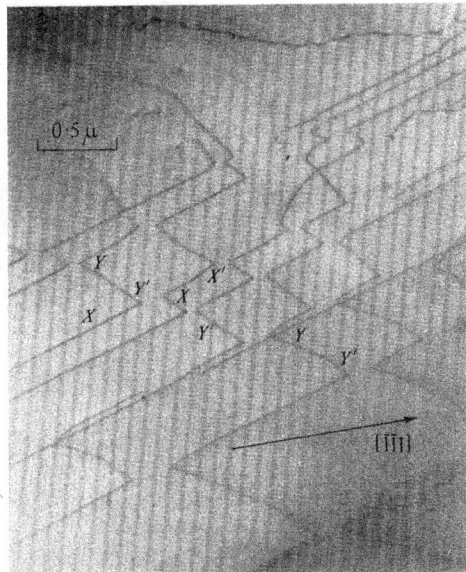

FIGURE 23. Interaction of two sets of dislocations X and Y to produce a resultant dislocation which is invisible. The normal to the foil is [213]. See figure 24 and text for interpretation. (Magn. × 25 000.)

Hirsch et al.   *Phil. Trans.* A, *volume* 252, *plate* 7

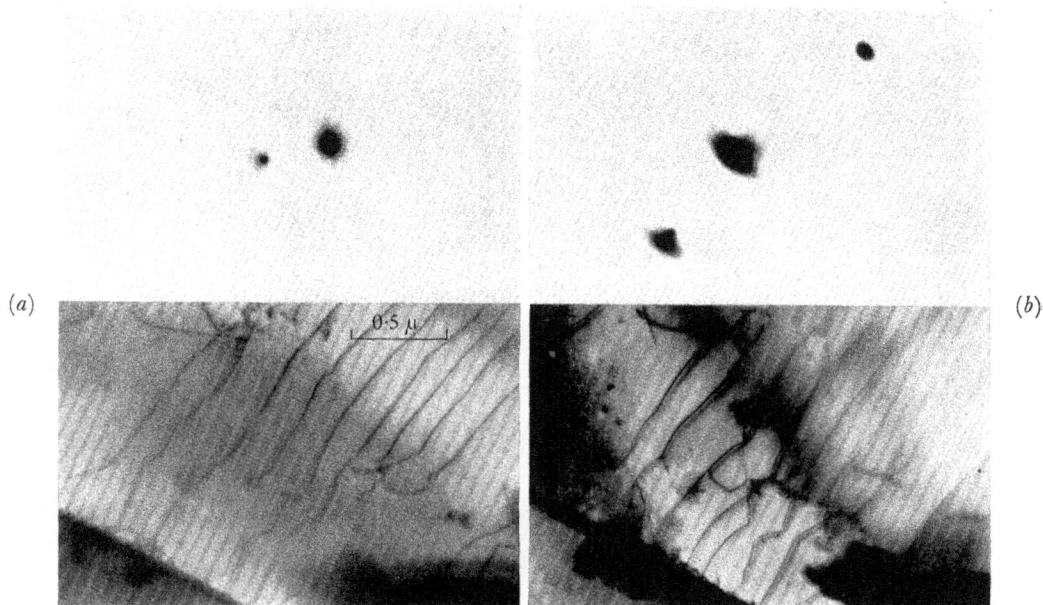

FIGURE 25. Micrographs of nickel showing (*a*) single images and (*b*) double images in the same area. Corresponding diffraction patterns are shown. See text for discussion. (Magn. × 30 000.)

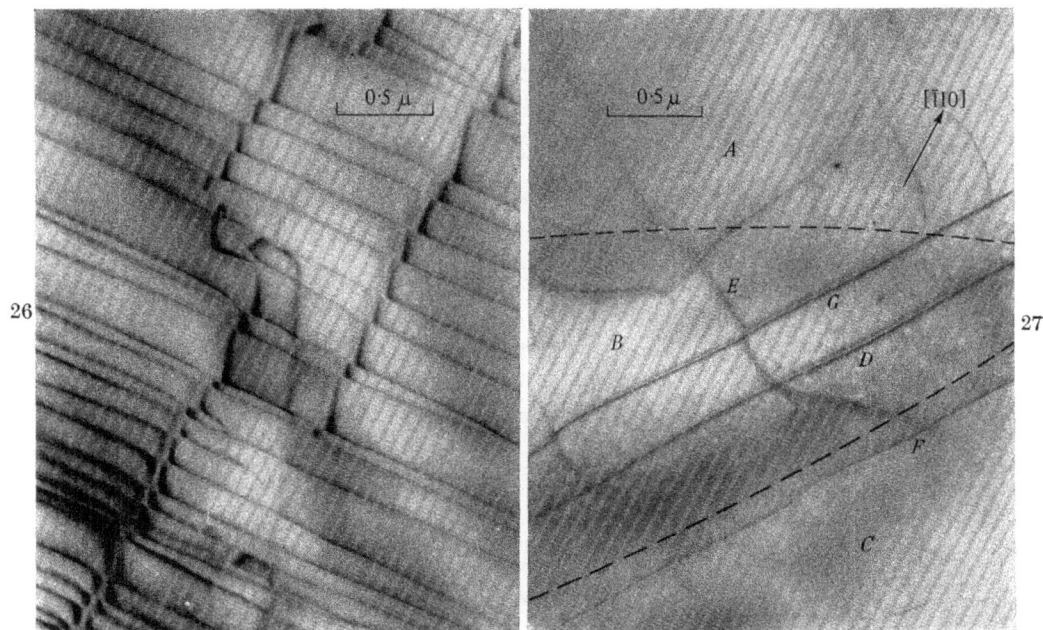

FIGURE 26. Micrograph of interacting dislocations in aluminium. Note the wide double images for one type of dislocation. See text for discussion. (Magn. × 30 000.)

FIGURE 27. Region showing double images of dislocations in aluminium. The approximate centres of high-order extinction contours are marked with broken lines. See text for discussion. (Magn. × 26 000.)

Hirsch et al.   *Phil. Trans. A, volume* 252, *plate* 8

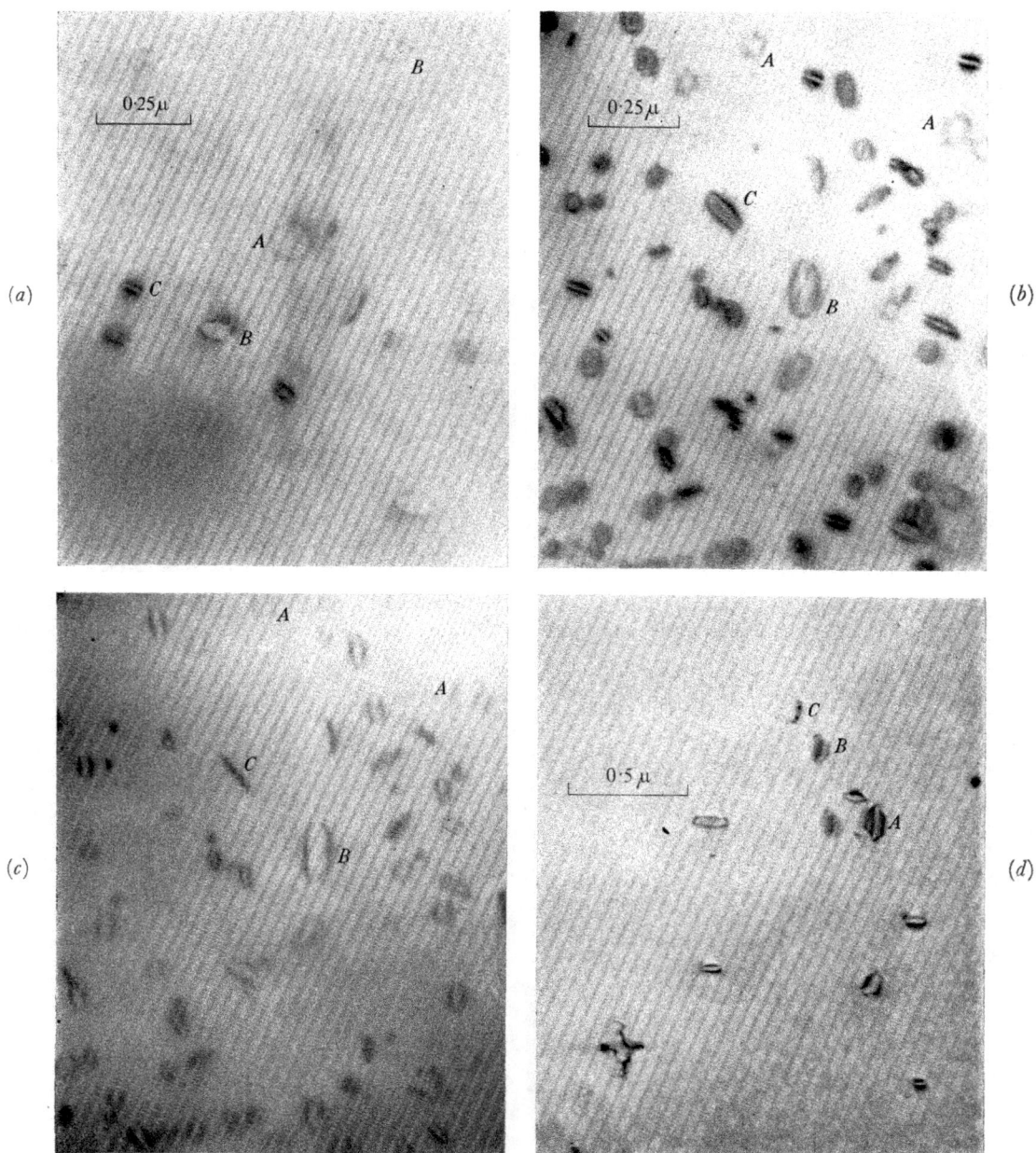

FIGURE 28. Images of dislocation loops in quenched aluminium. (*a*) *A*, hexagonal loop lying in a plane nearly parallel to the foil surface showing uniform contrast. Loops *B* and *C* are on inclined planes and the more steeply inclined parts are less visible. (Magn. × 65 000.) (*b*) *A*, dislocation with nearly uniform contrast. *B*, *C*, double images of dislocation loops. (Magn. × 60 000.) (*c*) Same area as (*b*) after tilting. (Magn. × 60 000.) (*d*) *A*, *B*, hexagonal loops showing double images. *C*, dotted appearance of a loop thought to be lying in a plane normal to that of the foil. (Magn. × 40 000.)

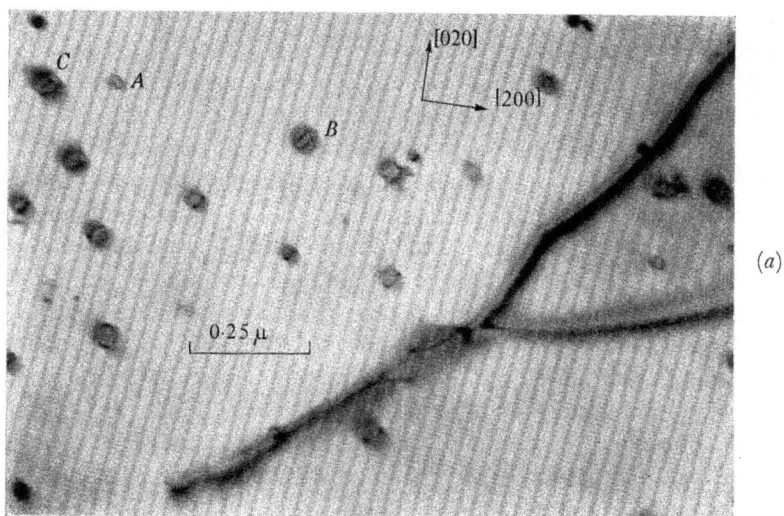

(a)

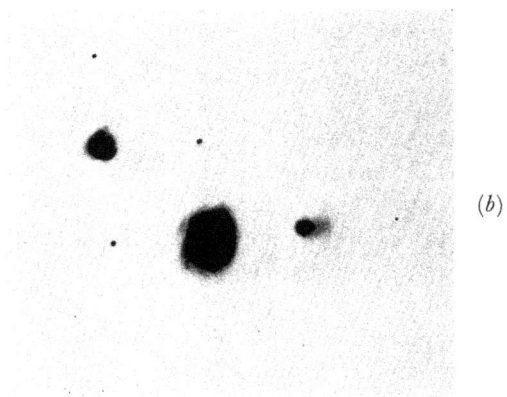

(b)

FIGURE 29. (a) Dislocation loops in quenched aluminium. Note the double images of loops at B and C. See text for discussion. (Magn. ×75000.) (b) Selected area diffraction pattern from (a).

THEORY OF DIFFRACTION CONTRAST AT DISLOCATIONS   519

indices have been determined by selected area diffraction); the values of n are ± 2 for both possible Burgers vectors. At the extinction contour s changes sign and it can be clearly seen that the position of the peak also changes from one side of the dislocation to the other, e.g. at the points marked with arrows. The displacement of the peaks is of the same order as their width, as expected. It will also be noted that the contrast at these dislocations is a maximum just outside the centre of the contour, i.e. in the region where the kinematical theory is expected to be valid. The contrast falls very steeply on the side facing the dislocation and fades out gradually on the other side, just as expected from the profiles of figure 15. The contrast and width of the dislocation also decrease very rapidly with distance from the contour, i.e. with increasing $|s|$. This behaviour of dislocations at bend contours has been observed a number of times. All these facts are in excellent agreement with the predictions of the theory.

A number of more complicated image effects can be seen in figure 21 (a) to (f), plate 5. In figure 21 (a) the dislocations show an alternating black-white appearance (A), zigzag appearance (B), and dotted appearance (C). Figure 21 (b) also shows a dislocation A running through the foil which appears as a series of dark dots along its whole length. Close to B another dislocation is seen in the region of an extinction contour; the change over to the alternating black-white appearance is evident at this dislocation. The theory provides a ready qualitative explanation of the dark dotted contrast (§ 3·4) and also for the frequent occurrence of these phenomena at those parts of inclined dislocations near the foil surface (e.g. figure 21 (a) at C). Sometimes dislocations are observed which appear lighter than their surroundings on bright field micrographs. An example of this is visible in figure 21 (c) where dislocations appear as a series of white dots at A. In these cases the surrounding regions of the foil are usually diffracting strongly. The white dotted appearance is thought to be a dynamical effect not predicted by the present theory. The zigzag effects are also not easily explained on the present theory. Figure 21 (a) at A and B shows an effect which is observed frequently, the contrast is observed to zigzag as the dislocation approaches the top and bottom surfaces of the foil. Sometimes the contrast zigzags along the whole length of the dislocation. Figure 21 (d) shows particularly good examples of zigzag dislocations, which might at first sight be taken to be helical dislocations. However, all these zigzag and dotted contrast effects are extremely sensitive to orientation; by tilting the specimen slightly it is possible to alter the contrast to a continuous dark line. This is shown quite clearly in figure 21 (e) and (f). Figure 21 (e) shows dotted contrast at A and zigzag contrast at B. Figure 21 (f) shows the same area after tilting the specimen slightly; the dislocations now appear as continuous dark lines. It is thought that the detailed explanation of all these effects will require a treatment by dynamical theory (possibly including absorption); the effects are presumably connected with the depth oscillations of the crystal wave fields in the dynamical region. They are mentioned here to emphasize that such effects are not to be interpreted in terms of any structural features of the dislocations.

5·2. *Invisible dislocations and determination of Burgers vectors*

The case $n = 0$, i.e. $\alpha = 0$, for a screw dislocation corresponds to $\mathbf{g}$ and $\mathbf{b}$ being mutually perpendicular and is of considerable interest. In this case the displacements of atoms are parallel to the reflecting planes and no contrast is produced. Similar considerations are expected to hold approximately for an edge dislocation and in general for a mixed dislocation. There is considerable experimental evidence that not all dislocations may be visible on transmission micrographs (Bradley & Phillips 1957). The same contrast mechanism is expected to apply to the new techniques for revealing dislocations by transmission of X-rays (Lang 1958) and probably also to the corresponding X-ray reflexion case (Newkirk 1958; Bonse & Kappler 1958). In the former case (Newkirk 1958) the above condition for invisibility appears to have been discovered experimentally.

Figure 22 (*a*) and (*b*), plate 6, show a striking example of this effect in Al + 4 % Cu alloy quenched from 440 °C. This material contains helical dislocations formed from screw dislocations by vacancy climb (Thomas & Whelan 1959). The axes of the helices are parallel to the Burgers vector. The normal to the foil in figure 22 is near [001] and the helical dislocations A and B have Burgers vectors $\frac{1}{2}[\bar{1}10]$ and $\frac{1}{2}[110]$, respectively. Selected area diffraction patterns (corrected for rotation) are included with the micrographs. In figure 22 (*a*) both helices have a non-vanishing value of $\mathbf{g}.\mathbf{b}$ ($=1$) for the operating 020 reflexion, and both are visible. In figure 22 (*b*) the specimen has been tilted so that $2\bar{2}0$ is the operating reflexion. $\mathbf{g}$ is now perpendicular to B, and this dislocation therefore vanishes. A now has $\mathbf{g}.\mathbf{b} = 2$ and the characteristic rapid decrease of the contrast at one side is apparent. Other dislocations have also vanished, while another has appeared at C showing that a feature which might otherwise have been interpreted as a pinning point is a triple node. Since a helical dislocation is mainly of edge character the theory is seen to be applicable to edges. Moreover, the vanishing of the helix B in figure 22 (*b*) indicates that the displacement normal to the slip plane of an edge dislocation has little effect on the contrast in agreement with the predictions made in § 3·4.

Figure 23, plate 6, shows a rather interesting case of invisible dislocations formed by an interaction process. The normal to the foil is close to [213] and the slip planes involved are $(\bar{1}11)$ (dislocations of type X) and $(1\bar{1}1)$ (dislocations of type Y). The dislocations X lie on parallel but displaced $(\bar{1}11)$ slip planes which cut through the foil at an angle; similar considerations hold for dislocations Y. The directions of X and Y are approximately parallel to the traces of the $(\bar{1}11)$ and $(1\bar{1}1)$ planes on the foil surface suggesting that these dislocations lie in a direction in their slip planes approximately parallel to the surface of the foil. A detailed study of this micrograph shows that an interaction has occurred at the intersection points of dislocations of types X and Y, and that the resultant dislocations (joining the sharp bends) are invisible for contrast reasons. The possibility that the interaction might be an annihilation (Whelan 1959) is ruled out by the fact that very weak contrast is visible on the original photographic plate at some of the resultant dislocations, and by the fact that such an angular dislocation would not be stable in this configuration. The details visible at several of the interactions can be explained satisfactorily in terms of the diagram of figure 24 which shows the arrangement in terms of Thompson's reference tetrahedron (Thompson 1953). Dislocation X lies on the plane ABD (($\bar{1}11$)) nearly parallel to BD with Burgers vector BD ($\frac{1}{2}[01\bar{1}]$), while dislocation Y lies on the plane ACD (($1\bar{1}1$)) nearly parallel to CD with Burgers vector DC ($\frac{1}{2}[\bar{1}01]$). X and Y are nearly in the screw orientation so that at the intersection point it will be energetically favourable for them to cross-slip to the plane BCD ((111)) to form segments X' and Y' which interact to form a resultant dislocation with Burgers vector $BD + DC = BC$ ($\frac{1}{2}[\bar{1}10]$). The segments of dislocation $X - X'$ and $Y - Y'$ therefore lie on

THEORY OF DIFFRACTION CONTRAST AT DISLOCATIONS   521

different planes and the points where they bend from one slip plane to the other give rise to the slight kinks in the dislocations visible on the micrograph (e.g. X, X' and Y, Y').

The Burgers vectors of all the dislocations in figure 23 are therefore determined purely by geometrical considerations. In particular, the Burgers vector of the resultant dislocation is $\frac{1}{2}[\bar{1}10]$ and a selected area diffraction pattern showed that the $\bar{1}\bar{1}1$ reflexion is producing the contrast. This is in agreement with the contrast theory since $n = \mathbf{g} \cdot \mathbf{b}$ is zero for this

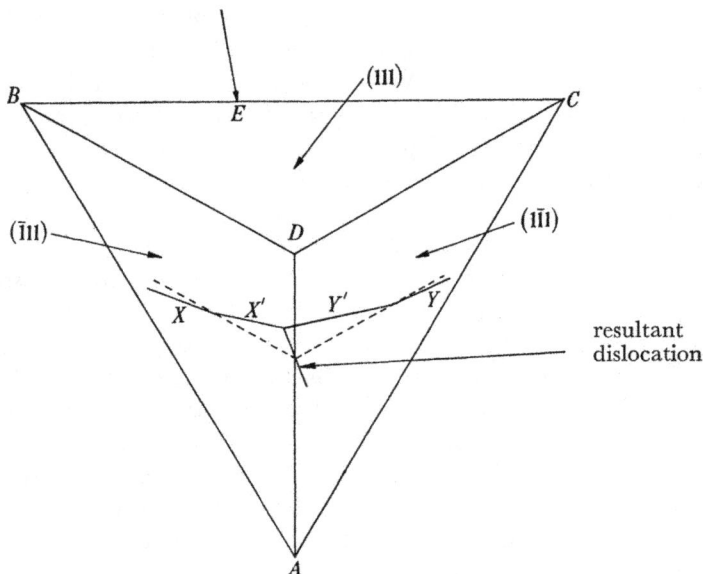

FIGURE 24. Interpretation of the interactions of figure 23. The diagram shows Thompson's tetrahedron representing the slip systems. The dislocation segment X lies in $(\bar{1}11)$ with Burgers vector BD; Y lies in $(1\bar{1}1)$ with Burgers vector DC, while X' and Y' and the resultant dislocation lie in (111). The direction of observation in figure 23 is roughly along EA where E lies on BC. At most of the nodes in figure 23 usually one or other of the segments X' and Y' predominates in length. Also at some interactions all of Y appears to have cross-slipped to Y'. The resultant dislocation is approximately a pure edge dislocation and is invisible in the micrograph of figure 23 because its Burgers vector is perpendicular to the reciprocal lattice vector of the Bragg reflexion producing the contrast.

Burgers vector and reflexion and the resultant dislocation is expected to be invisible. In principle therefore it should be possible to determine the Burgers vector of any dislocation by examining the contrast in several reflexions and by finding reflexions for which the contrast vanishes. However, the facilities for tilting the specimen in the electron microscope are at present not adequate to make this a practical possibility.

It should also be mentioned that the very weak contrast visible at some of the resultant dislocations in figure 23 on the original plate is probably due to the fact that these are very nearly pure edge dislocations orientated so that the displacement R_2 of §3·4 is effective. This displacement will be approximately in the [111] direction, and gives a non-vanishing phase factor with the $\bar{1}\bar{1}1$ reflexion. Thus the experimental observations are in agreement with the conclusions of §3·4 that the contrast due to this displacement is expected to be narrow and weak. It should also be possible to distinguish between screw and edge dislocations in certain orientations by use of this effect.

5·3. *Double images of dislocations*

Double images of dislocations are frequently observed and occur when two sets of reflecting planes which produce contrast on opposite sides of the dislocation are operating simultaneously. If the parameters $\mathbf{g}$ and s describing the two reflexions are labelled with suffixes 1 and 2 we must have either $\mathbf{g}_1.\mathbf{b}$ and $\mathbf{g}_2.\mathbf{b}$ of opposite sign or s_1 and s_2 of opposite sign. The effect is illustrated in figure 25 (*a*) and (*b*), plate 7, for dislocations in nickel. In figure 25 (*a*) the images of the dislocations are single and the diffraction pattern from a selected area shows only one strong diffracted beam. The specimen was then tilted until double images appeared (figure 25 (*b*)). Selected area diffraction then gave the pattern shown with two strong diffracted beams. These observations are insufficient to determine the Burgers vector of the dislocations, but all the possible values for $\mathbf{b}$ have $\mathbf{g}_1.\mathbf{b}$ and $\mathbf{g}_2.\mathbf{b}$ opposite in sign, so that s_1 and s_2 must have the same sign. This example shows again that the displacement of the image is of the same order as its width.

Another example of double images appears in figure 26, plate 7, where two sets of dislocations on different slip planes have intersected and interacted to produce short segments of resultant dislocation. One set of dislocations appears single (either because the two images fall on the same side of the dislocation or because the Burgers vector lies in one set of reflecting planes). The other dislocations have a different Burgers vector and appear as widely separated double images possibly corresponding to $n = 3$. These dislocation images also show clearly the expected gradual fading of the contrast on the side away from the dislocation.

Figure 27, plate 7, shows a situation in aluminium where both sets of intersecting dislocations appear double. In this micrograph two extinction contours *A–B* and *B–C* (marked by broken lines) are visible. These correspond to the $\bar{3}11$ and $2\bar{2}0$ reflexions, respectively. The extinction contours divide the micrograph into three regions *A*, *B* and *C*. Consideration of the reflecting sphere and the two reflexions shows that in regions *A* and *C*, s_1 and s_2 for the two reflexions have opposite sign, while in region *B*, s_1 and s_2 have the same sign. The electron beam passes in the [112] direction and examination of slip traces in the area showed that the normal to the foil was some 6° away from this in the direction of [313]. Consideration of the possible Burgers vectors for the dislocations *G*, *D* and *F* (lying in a (111) slip plane) and *E* (lying in a ($\bar{1}11$) plane) showed that the Burgers vector of *E* must be $\frac{1}{2}[101]$ and that of *G*, *D* and *F* either $\frac{1}{2}[1\bar{1}0]$ or $\frac{1}{2}[10\bar{1}]$. Both these possibilities give opposite values of $\mathbf{g}.\mathbf{b}$ for each Burgers vector in the two reflexions which agrees with the fact that s_1 and s_2 have the same sign in the region *B* where the dislocations appear double. The second of the two possibilities, however, in which the two Burgers vectors are mutually perpendicular provides the more satisfactory explanation of the absence of any attraction or repulsion at the points of intersection. Note the characteristic image of *F* with a sharp rise in contrast on one side and a gradual decrease on the other. The dislocation *G* becomes single after crossing the $\bar{3}11$ contour because the image due to this reflexion crosses to the other side of the dislocation. While the above observations show that double dislocations occur only

THEORY OF DIFFRACTION CONTRAST AT DISLOCATIONS   523

if two or more reflexions are excited simultaneously, Phillips (private communication) has been able to confirm by dark field photographs that each image arises from a different reflexion.

5·4. *Images of dislocation loops*

Figure 28 (*a*) to (*d*), plate 8, show some of the features of the images of dislocation loops observed in quenched aluminium (Hirsch *et al.* 1958). This type of loop lies on a (111) plane and the Burgers vector is not in the plane of the loop. Loops such as those at *A* in figure 28 (*a*) to (*c*) lie on planes nearly parallel to the surfaces of the foil and appear round or hexagonal with uniform contrast. Many of the other loops are characterized by pronounced variation in contrast so that they appear as two arcs. These loops are lying on planes more steeply inclined to the foil; the two arcs correspond to the parts of the loop approximately parallel to the surface while the steeply inclined parts of the loops appear faint and narrow or are altogether invisible, in agreement with the predictions of §3·4 (e.g. at *B* in figure 28 (*a*) and (*c*)).

The Burgers vectors of opposite sides of a loop are of opposite sign, provided the positive sense of the dislocation is taken in the same direction on both sides. The image will therefore lie either completely inside or completely outside the dislocation loop. If two reflexions are operating which give double images, one image lies outside the loop and the other inside. Figure 28 (*b*) and (*c*) show the same area in slightly different orientation. Double images are clearly visible in figure 28 (*b*); the appearance of some of these images, for example at *B*, must not be confused with that expected from loops containing stacking faults. In figure 28 (*c*) only single images are visible. In the case of these double images the actual dislocation loop lies between the two images. If only one image occurs as in figure 28 (*c*), the images of opposite sides of the loop may actually coincide giving the appearance of a single line, such as at *C* in figure 28 (*c*). Comparison with the double image of the same loop on figure 28 (*b*) shows that the single line indeed represents a fairly steeply inclined loop. Figure 28 (*d*) shows further contrast effects observed at hexagonal-shaped loops in aluminium. A dark region is observed inside the hexagonal loop at *B* while the loop at *A* appears to have interference fringes, which might at first sight be attributed to a hexagonal disk of stacking fault contained by a Frank sessile dislocation. However, a detailed study of this contrast effect suggests that the correct explanation depends on double images as outlined above. Two reflexions are operating producing contrast both inside and outside the loop: the contrast inside the loops *A* and *B* in figure 28 (*d*) gives the appearance of fringes. The other loops visible in figure 28 (*d*) show no extra contrast. These loops are on different {111} planes with different Burgers vectors, so that the dislocation contrast is on the same side of the loop for both reflexions.

Other more complicated contrast effects have also been observed at loops, e.g. at *C* in figure 28 (*d*). It is thought that these effects are produced by hexagonal loops on planes almost parallel to the line of sight with one segment of the loops nearly parallel to the foil surface. The double dots at the ends of the line of contrast may be due to the more steeply inclined parts of the hexagonal loop. This interpretation is, however, tentative.

The appearance of the loops and the type of reflexion operating can be used to attempt to determine the Burgers vector. Figure 29 (*a*), plate 9, is a micrograph showing double images; figure 29 (*b*) is the associated diffraction pattern showing strong 200 and $\overline{2}20$

reflexions. The orientation of the normal to the foil is approximately [001]. The loops at A appear single and their elongation suggests that they lie on (111) or (11$\bar{1}$) planes. The double image loops (e.g. at B), however, lie on ($\bar{1}$11) or (1$\bar{1}$1) planes. Examination of the product $\mathbf{g}.\mathbf{b}$ for various Burgers vectors shows that for the single images $\mathbf{b}$ has the possible values $\frac{1}{2}[110]$, $\frac{1}{2}[01\bar{1}]$ and $\frac{1}{2}[011]$. For those with double images $\mathbf{b}$ has the possible values $\frac{1}{2}[\bar{1}10]$, $\frac{1}{2}[\bar{1}01]$ and $\frac{1}{2}[101]$. The first of these values is thought to be most likely as it is the only one which gives rise to $n = 2$ for one of the reflexions, all other values giving ± 1. Such a different value of n would explain the rather different nature of the two images, i.e. one of the images is much wider as at C. All the possibilities for the double loops give values of $\mathbf{g}.\mathbf{b}$ which are of opposite sign for the two reflexions. Therefore s must have the same sign for both reflexions. This has been confirmed directly from the position of the Kikuchi lines for the two reflexions. (The Kikuchi lines are not visible on figure 29 (b).) The double images of the nearby dislocation lines are also apparent.

5·5. *The interpretation of X-ray topographs*

Lang (1958) has recently developed a new technique for revealing dislocations by transmission X-ray diffraction. The method is the X-ray analogue of the dark field technique in electron microscopy. Lang reports that with Ag$K\alpha$ radiation and a crystal of silicon, dislocations are observed as regions of increased diffracting power about 20μ wide. In this case the divergence of the X-ray beam at the specimen ($\sim 2'$ of arc, Lang, private communication) is much greater than the angular range of reflexion of the crystal on dynamical theory ($\sim 1/25'$). Thus an effect averaged over s should be observed, but the same considerations as in the electron case should apply. However, on averaging over s it is possible that the resultant contrast will be in the form of a single peak at the centre of the dislocation. The same conditions for visibility should apply. The X-ray extinction distances t_0 for the 111 and 220 reflexions in silicon are 55 and 46μ, respectively. t_0/π is therefore a good measure of the observed dislocation width in this case too.

Newkirk (1958) has used a reflexion X-ray technique for revealing dislocations in Fe + Si alloys and LiF crystals. The method is essentially an improvement of the Berg–Barrett technique. Although the theory of this paper applies strictly to the transmission technique only, it appears to explain qualitatively Newkirk's results also. Thus with Cr$K\alpha$ radiation ($\lambda = 2\cdot29$ Å), dislocations near the surface of a LiF crystal are visible as regions of enhanced diffracting power a few microns wide (Newkirk, private communication). The extinction distances for the 220 and 200 reflexions of LiF with Cr$K\alpha$ radiation are $\sim 9\mu$. This is smaller than the extinction distances in Lang's experiments. Again t_0/π is seen to be good measure of the dislocation width. Newkirk (private communication) has also studied the dislocation contrast as a function of the indices of the reflexion. The results also agree well with the predictions of the present theory concerning the variation of contrast with $n = \mathbf{g}.\mathbf{b}$. Thus for different reflexions and Burgers vectors the intensity of the image seems to depend only on n, and increases with increasing n. Analysis of the photographs shows that for $n = 0$ the dislocations are invisible, for $n = 1$ and 2 the contrast is intermediate and strong, respectively. The theory (see §3·3) predicts that for optimum contrast conditions the width of the image increases with increasing n. Because of the large divergence of the X-ray beam the resultant image consists of a set of overlapping images corresponding to a range of values of s, and the

THEORY OF DIFFRACTION CONTRAST AT DISLOCATIONS   525

total intensity of this resultant image is expected to increase with increasing width of the dislocation profile and therefore with increasing n, as observed.

It may be noted here that Bonse (1958) has attempted to explain the contrast effects observed with X-rays by estimating the effect of strain and lattice curvature due to dislocations on the width of the reflexions. Such a theory, in which the phase relation between neighbouring diffracting elements is ignored, can at best only lead to very qualitative conclusions.

6. CONCLUSION

The kinematical theory of contrast of electron microscope images of dislocations developed in this paper explains many of the characteristic features of the observed images, such as the orientation dependence of the contrast, the reversal of contrast on bright field and dark field images, the fact that dislocations are generally dark on bright field images, the position and width of the images, the general nature of the profile, the occurrence of dotted dislocations, the invisibility of some dislocations, the dependence of contrast on the inclination of the dislocation, and the occurrence of double images. The main idea of the theory is that the contrast is essentially phase contrast, the phase variations arising from displacements of atoms near the defect. It should be possible to apply the same theory to explain other contrast effects such as the trails left behind by moving dislocations (Hirsch *et al.* 1956), or the contrast near G.P. zones in aluminium+copper alloys (Nicholson & Nutting 1958), provided the nature of the strain field near the defect is known. The theory also accounts satisfactorily for the nature and width of the dislocation images obtained with X-rays, for example, using the techniques of Lang (1958) or Newkirk (1958).

The theory does not, however, explain the 'white' dislocations which have been observed occasionally, or the characteristic black-white or zigzag appearance which is observed more frequently. Both these effects are thought to find their explanation in a full dynamical treatment of the wave propagation in the crystal. There is also some reason to believe that in thicker foils ($t \gtrsim 1000\,\text{Å}$) absorption processes may play some part in the image formation. It is found that in thick regions the bend extinction contours (corresponding to small deviations from the Bragg angle) widen into broad dark bands in which the transmitted and diffracted beams are heavily absorbed. On either side of these bands the foil is particularly transparent and the dislocations are clearly visible. In these regions the absorption appears to be anomalously low; thus dislocations have been seen quite clearly in such areas in foils of aluminium+4% copper alloy about 7500 Å thick (Thomas & Whelan 1959). This anomalous transmission may be analogous to the Borrmann (1941, 1950) effect observed with X-rays and part of the contrast in the image of a dislocation may be due to enhanced absorption near the dislocation. These observations suggest that a full treatment of the dynamical theory including absorption for a dislocated crystal may be required to explain all the details of the observed contrast effects. It might be noted here that while the contrast at dislocations in the X-ray experiments of Lang (1958) can be explained on the basis of the kinematical theory, that observed by Borrmann, Hartwig & Irmler (1958) is also likely to be due to a locally enhanced absorption effect. Most of the evidence presented here has referred to aluminium, for which absorption effects are relatively unimportant even in thicker areas of this material. However, the same contrast effects have also been observed with suitable specimens of other metals, e.g. Cu, Ni, Au, Fe, α-brass, stainless steel, so that the theory is applicable to many metals.

For most practical purposes, however, particularly for transmission electron microscopy, the kinematical theory will probably be sufficient. Its most important application will be in the determination of the Burgers vectors of dislocations. The most satisfactory method for this purpose would be to change the orientation of the specimen until the image disappears; the indices of the operating reflexion are then obtained from the diffraction pattern and **b** is derived from the criterion $\mathbf{g}.\mathbf{b} = 0$. Should there still be some ambiguity this method could be repeated for another reflexion. In order to carry out such experiments satisfactorily electron microscopes must be provided with goniometer stages which allow the specimen to be rotated several degrees about two directions at right angles. Even without such equipment it may be possible to determine **b** in many cases by using the existing stereo-facilities for tilting the specimens about one axis, and observing the nature of the contrast in several reflexions. Thus it should be possible to determine the value of $\mathbf{g}.\mathbf{b}$ from the width of the dislocation images under conditions of optimum contrast; a number of such determinations taken in conjunction with any other information, for example, on the slip plane of the dislocation, should enable **b** to be found. So far, however, no systematic experiments have been carried out to test how these methods can be applied in practice.

The theory can also be useful in other ways; for example, from the variation of the contrast around a dislocation loop, its plane can be determined. And finally the existence of complicated effects such as dotted dislocations or double images shows that care must be taken in interpreting similar contrast effects in terms of segregation or precipitation phenomena.

We are indebted to Professor N. F. Mott, F.R.S., and to Dr W. H. Taylor for their interest and encouragement during the course of this work. Mention should be made of helpful discussions with Dr D. W. Pashley and also with Dr A. Lang and Dr J. B. Newkirk, both of whom showed us unpublished work. Thanks are due to Dr G. Thomas for supplying the specimen from which the micrographs in figure 22 were taken and to Mr J. Silcox for supplying figures 28 and 29. Acknowledgments for financial support are due to Trinity College, Cambridge, for a Research Studentship (A. H.), and to D.S.I.R. and Gonville and Caius College, Cambridge, for Research Fellowships (M. J. W.).

APPENDIX. CALCULATION OF THE CO-ORDINATES ξ_n AND η_n

Consider the integral

$$A = \int_0^\infty \exp\left(\pm in\phi + 2\pi isz\right) dz, \tag{A 1}$$

where $\phi = \tan^{-1}(z/|x|)$ is defined in figure 8. The positive and negative signs are taken for x positive and negative, respectively. The co-ordinates ξ_n and η_n of the centre of the final circle represent the point in the complex plane about which the integral (A 1) oscillates.

Integrate (A 1) by parts and extract the part which does not oscillate at the upper limit:

$$\xi_n + i\eta_n = \frac{1}{2\pi s}\left[i \mp n \int_0^{\frac{1}{2}\pi} \exp\left(\pm in\phi + 2\pi isz\right) d\phi\right]. \tag{A 2}$$

THEORY OF DIFFRACTION CONTRAST AT DISLOCATIONS   527

Put $\tan\phi = u$, then $z = xu$ for positive x and $z = |x|u$ for negative x. Without loss of generality we may take s positive so that $\beta = 2\pi sx$ is positive for positive x and negative for negative x. The integral in (A 2) may then be transformed to

$$\mp n \int_0^\infty \exp\left(\mathrm{i}\,|\beta|\,u\right) \frac{(1\pm \mathrm{i}u)^n}{(1+u^2)^{\frac{1}{2}(n+2)}}\,\mathrm{d}u. \tag{A 3}$$

(A 3) can be written as

$$\mp n \left(1 \pm \frac{\mathrm{d}}{\mathrm{d}\,|\beta|}\right)^n \int_0^\infty \frac{\exp\left(\mathrm{i}\,|\beta|\,u\right)}{(1+u^2)^{\frac{1}{2}(n+2)}}\,\mathrm{d}u. \tag{A 4}$$

Thus from (A 2)

$$\xi_n = \mp \frac{n}{2\pi s}\left(1 \pm \frac{\mathrm{d}}{\mathrm{d}\,|\beta|}\right)^n \int_0^\infty \frac{\cos|\beta|\,u}{(1+u^2)^{\frac{1}{2}(n+2)}}\,\mathrm{d}u, \tag{A 5}$$

$$\eta_n = \frac{1}{2\pi s} \mp \frac{n}{2\pi s}\left(1 \pm \frac{\mathrm{d}}{\mathrm{d}\,|\beta|}\right)^n \int_0^\infty \frac{\sin|\beta|\,u}{(1+u^2)^{\frac{1}{2}(n+2)}}\,\mathrm{d}u. \tag{A 6}$$

ξ_n can be evaluated in terms of modified Bessel functions of the second kind. Watson (1922) gives a formula for the integral in (A 5) which leads to the following expression for ξ_n.

$$\xi_n = \mp \frac{n}{2\pi s}\,\frac{\sqrt{\pi}}{2^{\frac{1}{2}(n+1)}\Gamma(\frac{1}{2}n+1)}\left(1 \pm \frac{\mathrm{d}}{\mathrm{d}\,|\beta|}\right)^n |\beta|^{\frac{1}{2}(n+1)}K_{\frac{1}{2}(n+1)}(|\beta|).$$

The expressions for ξ_n can then be simplified using the recurrence relations for K_ν and its derivatives (Watson 1922).

Watson (1922) also gives a formula for the sine integral in (A 6) in terms of Bessel functions I_ν and the modified Struve function L_ν. Unfortunately, an indeterminancy occurs for even values of n and the values of L_ν are not extensively tabulated.

However, tables of

$$\phi(x) = \int_0^\infty \frac{\sin t}{(x^2+t^2)^{\frac{1}{2}}}\,\mathrm{d}t$$

and of $\phi'(x)$ have been given by Müller (1939). In terms of these,

$$\int_0^\infty \frac{\sin\beta u}{(1+u^2)^{\frac{1}{2}}}\,\mathrm{d}u = \phi(\beta)$$

and

$$\int_0^\infty \frac{\sin\beta u}{(1+u^2)^{\frac{3}{2}}}\,\mathrm{d}u = -\beta\phi'(\beta);$$

also

$$\mathrm{d}/\mathrm{d}\beta\,(\beta\phi') = \beta\phi(\beta)-1.$$

By means of these relations, formulae for η_1 and η_3 can be obtained. To calculate the expressions for η_2 and η_4 we use the relations

$$\int_0^\infty \frac{\sin\beta u}{1+u^2}\,\mathrm{d}u = \tfrac{1}{2}[\exp(-\beta)\,\overline{\mathrm{Ei}}(\beta) - \exp(\beta)\,\mathrm{Ei}(-\beta)],$$

$$\int_0^\infty \frac{\sin\beta u}{(1+u^2)^2}\,\mathrm{d}u = \frac{1}{2}\left[1 - \beta\frac{\mathrm{d}}{\mathrm{d}\beta}\right]\int_0^\infty \frac{\sin\beta u}{1+u^2}\,\mathrm{d}u,$$

$$\int_0^\infty \frac{\sin\beta u}{(1+u^2)^3}\,\mathrm{d}u = \frac{1}{8}\left[3 - \beta\frac{\mathrm{d}}{\mathrm{d}\beta}\right]\left[1 - \beta\frac{\mathrm{d}}{\mathrm{d}\beta}\right]\int_0^\infty \frac{\sin\beta u}{1+u^2}\,\mathrm{d}u,$$

$$\frac{\mathrm{d}}{\mathrm{d}\beta}\,\mathrm{Ei}(-\beta) = \frac{\exp(-\beta)}{\beta},$$

$$\frac{\mathrm{d}}{\mathrm{d}\beta}\,\overline{\mathrm{Ei}}(\beta) = \frac{\exp(\beta)}{\beta},$$

where the functions $\overline{\mathrm{Ei}}(\beta)$ and $\mathrm{Ei}(-\beta)$ are defined and tabulated by Jahnke & Emde (1933). The final expressions for ξ_n and η_n are

$$\xi_1 = \frac{|\beta|}{2\pi s}(K_0(|\beta|) \mp K_1(|\beta|));$$

$$\eta_1 = \frac{|\beta|}{2\pi s}(\phi(|\beta|) \pm \phi'(|\beta|));$$

$$\xi_2 = 0 \quad \text{for} \quad \beta > 0,$$
$$= \frac{1}{2\pi s}2\pi|\beta|\exp(-|\beta|) \quad \text{for} \quad \beta < 0;$$

$$\eta_2 = -\frac{1}{2\pi s}(2\beta\exp(\beta)\mathrm{Ei}(-\beta)+1) \quad \text{for} \quad \beta > 0,$$
$$= \frac{1}{2\pi s}(2|\beta|\exp(-|\beta|)\overline{\mathrm{Ei}}(|\beta|)-1) \quad \text{for} \quad \beta < 0;$$

$$\xi_3 = \frac{|\beta|}{2\pi s}\{4|\beta|[K_1(|\beta|)\mp K_0(|\beta|)]-3K_0(|\beta|)\pm K_1(|\beta|)\};$$

$$\eta_3 = \frac{|\beta|}{2\pi s}\{\pm 4-(3\pm 4|\beta|)\phi(|\beta|)-(4|\beta|\pm 1)\phi'(|\beta|)\};$$

$$\xi_4 = 0 \quad \text{for} \quad \beta > 0,$$
$$= \frac{|\beta|}{2\pi s}4\pi\exp(-|\beta|)(|\beta|-1) \quad \text{for} \quad \beta < 0;$$

$$\eta_4 = \frac{1}{2\pi s}\{4\beta(\beta+1)\exp(\beta)\mathrm{Ei}(-\beta)+4\beta+1\} \quad \text{for} \quad \beta > 0,$$
$$= \frac{1}{2\pi s}\{4|\beta|(|\beta|-1)\exp(-|\beta|)\overline{\mathrm{Ei}}(|\beta|)-4|\beta|+1\} \quad \text{for} \quad \beta < 0.$$

With these expressions for ξ_n and η_n the locus of the centre of the final circle can be drawn for all important values of n as shown in figures 11 to 14.

REFERENCES

Bassett, G. A., Menter, J. W. & Pashley, D. W. 1958 *Proc. Roy. Soc.* A, **246**, 345.
Bollmann, W. 1956 *Phys. Rev.* **103**, 1588.
Bollmann, W. 1957 *Proceedings of the Stockholm Conference on Electron Microscopy*, p.316. Stockholm: Almqvist and Wiksell.
Bonse, U. 1958 *Z. Phys,* **153**, 278.
Bonse, U. & Kappler, E. 1958 *Z. Naturf.* **13**a, 348.
Borrmann, G. 1941 *Phys. Z.* **42**, 157.
Borrmann, G. 1950 *Z. Phys.* **127**, 297.
Borrmann, G., Hartwig, W. & Irmler, H. 1958 *Z. Naturf.* **13**a, 423.
Bradley D. E. & Phillips, R. 1957 *Proc. Phys. Soc.* B, **70**, 533.
Dowell, W. C. T., Farrant, J. L. & Rees, A. L. G. 1958 *Report of Fourth International Conference on Electron Microscopy*. Berlin: Springer-Verlag. (In the Press.)
Franklin, R. E. 1950 *Acta. Cryst.* **3**, 107.
Haine, M. E. 1957 *J. Sci. Instrum.* **34**, 9.
Hashimoto, H. & Uyeda, R. 1957 *Acta Cryst.* **10**, 143.

THEORY OF DIFFRACTION CONTRAST AT DISLOCATIONS   529

Hashimoto, H., Naiki, T. & Mannami, M. 1958 *Report of Fourth International Conference on Electron Microscopy*. Berlin: Springer-Verlag. (In the Press.)

Heidenreich, R. D. 1949 *J. Appl. Phys.* **20**, 993.

Hirsch, P. B. 1959 *Metallurg. Rev.* **4**, 101.

Hirsch, P. B., Horne, R. W. & Whelan, M. J. 1956 *Phil. Mag.* **1**, 677.

Hirsch, P. B., Silcox, J., Smallman, R. E. & Westmacott, K. H. 1958 *Phil. Mag.* **3**, 897.

Jahnke, E. & Emde, F. 1933 *Tables of functions*, p. 78. Leipzig: Teubner.

Kamiya, Y., Nonoyama, M., Tochigi, H. & Uyeda, R. 1958 *Report of Fourth International Conference on Electron Microscopy*. Berlin: Springer-Verlag. (In the Press.)

Kato, N. 1952a *J. Phys. Soc. Japan*, **7**, 397.

Kato, N. 1952b *J. Phys. Soc. Japan*, **7**, 406.

Kato, N. 1953 *J. Phys. Soc. Japan*, **8**, 350.

Lang, A. R. 1958 *J. Appl. Phys.* **29**, 597.

Lenz, F. 1954 *Z. Naturf.* **9**a, 185.

Lenz, F. 1958 *Report of Fourth International Conference on Electron Microscopy*. Berlin: Springer-Verlag. (In the Press.)

Menter, J. W. 1956 *Proc. Roy. Soc.* A, **236**, 119.

Menter, J. W. 1958 *Advanc. Phys.* **7**, 299.

Mott, N. F. & Massey, H. S. W. 1933 *The theory of atomic collisions*, p. 87. Oxford: Clarendon Press.

Müller, R. 1939 *Z. Angew Math. Mech.* **19**, 36.

Newkirk, J. B. 1958 *J. Appl. Phys.* **29**, 995.

Nicholson, R. B. & Nutting, J. 1958 *Phil. Mag.* **3**, 531.

Niehrs, H. 1958a, b *Report of Fourth International Conference on Electron Microscopy*. Berlin: Springer-Verlag. (In the Press.)

Nutting, J. 1959 Article in *The structure of metals*. London: Iliffe.

Paterson, M. S. 1952 *J. Appl. Phys.* **23**, 805.

Read, W. T. 1953 *Dislocations in crystals*, p. 18. New York: McGraw-Hill.

Scherzer, O. 1949 *J. Appl. Phys.* **20**, 20.

Thomas, G. & Whelan, M. J. 1959 *Phil. Mag.* **4**, 511.

Thompson, N. 1953 *Proc. Phys. Soc.* B, **66**, 481.

Watson, G. N. 1922 *Theory of Bessel Functions*, pp. 172, 332. Cambridge University Press.

Whelan, M. J. 1959 *Proc. Roy. Soc.* A, **249**, 114.

Whelan, M. J. & Hirsch, P. B. 1957a *Phil Mag.* **2**, 1121.

Whelan, M. J. & Hirsch, P. B. 1957b *Phil. Mag.* **2**, 1303.

Whelan, M. J., Hirsch, P. B., Horne, R. W. & Bollmann, W. 1957 *Proc. Roy. Soc.* A, **240**, 524.

Willis, B. T. M. 1957 *Proc. Roy. Soc.* A, **239**, 192.

Wilson, A. J. C. 1950 *Research*, **3**, 387.

LIST OF PUBLICATIONS

I. X-Ray Investigations of Cold-Worked Metals

1. An X-ray microbeam examination of a plastically deformed metal. J.N. Kellar, P.B. Hirsch and J.S. Thorp. Nature (1950), 165, 554.

2. An X-ray microbeam technique: I. Collimation. P.B. Hirsch and J.N. Kellar. Proc. Phys. Soc. (1951), B64, 369.

3. An X-ray microbeam technique II. A high intensity X-ray generator. P. Gay, P.B. Hirsch, J.N. Kellar and J.S. Thorp. Proc. Phys. Soc. (1951), B64, 374.

4. An X-ray technique for the study of substructures in materials. P. Gay and P.B. Hirsch. Acta Cryst. (1951), 4, 284.

5. A study of cold-worked aluminium by an X-ray microbeam technique: 1. The measurement of particle size and misorientations. P.B. Hirsch and J.N. Kellar. Acta Cryst. (1952), 5, 162.

6. A study of cold-worked aluminium by an X-ray microbeam technique: 2. Measurement of shapes of spots. P.B. Hirsch. Acta Cryst. (1952), 5,168.

7. A study of cold-worked aluminium by an X-ray microbeam technique: 3. The structure of cold-worked aluminium. P.R. Hirsch. Acta Cryst. (1952), 5, 172.

8. Appendix to paper on low-temperature fractures in tempered alloy steels. A.R. Entwistle. J. Iron & Steel Inst. (1951), 169, 36.

9. An X-ray microbeam investigation of cold-worked aluminium. P.B. Hirsch. Imperfections in nearly perfect crystals. Wiley & Sons, 1952, p. 167.

10. X-ray microbeam techniques. P.B. Hirsch. Chapter 9 of 'X-ray diffraction from polycrystalline materials'. (Peiser, Rooksby and Wilson). Review.

11. X-ray studies of polycrystalline metals deformed by rolling: III. The structure and deformation of metals. P. Gay, P.B. Hirsch and A. Kelly. Acta Cryst. (1954), 7, 41.

12. A method of estimating dislocation densities from X-ray diffraction data. P. Gay, P.B. Hirsch and A. Kelly. Acta Metall. (1953), 1, 315.

13. Étude expérimentale de l'écrouissage des métaux par la différence des microfaisceaux de rayons X. P.B. Hirsch. Rev. de Métallurgie (1953), 50, 333.

14. The determination of grain size from spotty X-ray diffraction rings. P.B. Hirsch. Brit. J. of Appl. Phys. (1954), 5, 257.

II.   Surface Layers

15.   A parallel beam concentrating monochromator for X-rays. R.C. Evans, P.B. Hirsch and J.N. Kellar. Acta Cryst. (1948), 1, 124.

16.   Surface layers on crystals. P.B. Hirsch and J.N. Kellar. Nature (1950), 162, 609.

17.   A non-destructive X-ray method for the determination of the thickness of surface layers. P. Gay and P.B. Hirsch. J. AppI. Phys. (1951), 2, 218.

18.   Asymmetric reflexions from abraded crystals. P. Gay, P.R. Hirsch and J.N. Kellar. Acta Cryst. (1952), 5, 7.

19.   The crystalline character of abraded surfaces. P. Gay and P.B. Hirsch. Symposium on 'Properties of Metallic Surfaces'. (London: Institute of Metals 1952).

III.   Dynamical Theory of X-Ray Reflexions

20.   The intensity of reflexion from perfect and mosaic absorbing crystals. P.B. Hirsch and G.N. Ramachandran. Acta Cryst. (1950), 3, 187.

21.   The transmission and reflexion of X-rays in perfect absorbing crystals. P.B. Hirsch. Acta Cryst. (1952), 5, 176.

IV.   Structure of Coal

22.   X-ray scattering from coals. P.B. Hirsch. Proc. Roy. Soc. (1954), A226, 143.

23.   Recent infra-red and X-ray studies of coal. J.K. Brown and P.B. Hirsch. Nature (1955), 175, 229.

24.   New X-ray data on coals. L. Cartz, R. Diamond and P.B. Hirsch. Nature (1956), 177, 500.

25.   X-ray scattering from carbonised coals. R. Diamond and P.B. Hirsch. Report of London Conference on Industrial Carbon and Graphite' (Society of Chemical Industry 1948), p. 197.

26.   Conclusions from X-ray scattering data on vitrain coals. P.B. Hirsch. Report of Sheffield Conference on 'Science in the use of Coal'. (The Institute of Fuel 1958), p. A29.

27.   A contribution to the structure of coals from X-ray diffraction studies. L. Cartz and P.B. Hirsch. Phil. Trans. Roy. Soc. (1960), 252 A, 557.

28.   Conclusions from X-ray scattering data on vitrain coals. Paper 1. Proc. Conf. on Science in the use of Coal. P.B. Hirsch.

V.   Nuclear Power

29.   Research and nuclear power. P.B. Hirsch, Atom, July 1983, 321, 142.

30.   The fast reactor: perspective and prospects. P.B. Hirsch, Nucl. Energy, 1983, 22 No. 6, 401.

31. The future of UK nuclear power in the European context. P.B. Hirsch and A.A. Farmer, Nuclear Engineering and Design, 1986, 92, 293.

32. An assessment of the integrity of PWR Pressure Vessels. P.B. Hirsch, UKAEA Report, January 1987.

33. '25 Years of TAGSI and LWRSG', P.B. Hirsch. Fracture, Plastic Flow and Structural Integrity (editors P.B. Hirsch and D. Lidbury), I.O.M. Communications Ltd., London p. x, (2000).

34. 'Fracture, Plastic Flow and Structural Integrity', editors P.B. Hirsch and D. Lidbury I.O.M. Communications Ltd., London (2000).

35. "Methods for the Assessment of the Structural Integrity of Components and Structures". Report of 8th TAGSI Symposium, editors D. Lidbury and P.B. Hirsch, Maney, Institute of Materials, Minerals and Mining, London (2003).

VI. Plastic Deformation, Crystal Defects, Electron Microscopy, Materials Science

36. The structure of cold-worked gold: I. A study by electron diffraction. P.B. Hirsch, A. Kelly and J.W. Menter. Proc. Phys. Soc. B (1955), 68, 1132.

37. Surface distributions of dislocations in metals. C.J. Ball and P.B. Hirsch. Phil. Mag. (1955), 46, 1343.

38. Mosaic structure. P.B. Hirsch. Progress in Metal Physics VI (Pergamon Press, London: 1956, p. 236). Review.

39. Direct observations of the arrangement and motion of dislocations in aluminium. P.B. Hirsch, R.W. Horne and M.J. Whelan. Phil. Mag. (1956), 1, 677.

40. Direct observations of the arrangement and motion of dislocations in aluminium. P.B. Hirsch, R.W. Horne and M.J. Whelan. Report of Lake Placid Conference on "Dislocations and Mechanical Properties of Crystals'. (New York: Wiley 1957), p. 92.

41. X-ray studies of tin whiskers. P.B. Hirsch. Report of Lake Placid Conference on 'Dislocations and Mechanical Properties of Crystals'. (New York: Wiley 1957), p. 545.

42. X-ray diffraction from body-centred cubic crystals containing stacking faults. P.B. Hirsch and H.M. Otte. Acta Cryst. (1957), 10, 447.

43. A note on the interpretation of the slip pattern in terms of dislocation movement. P.B. Hirsch. Appendix to paper by J. Nutting and G. Thomas. J. Inst. Metals (1957–58), 86, 7.

44. Dislocations and stacking faults in stainless steel. M.J. Whelan, P.B. Hirsch, R.W. Horne and W. Bollmann. Proc. Roy. Soc. (1957), A240, 524.

45. Electron diffraction from crystals containing stacking faults: I. M.J. Whelan and P.B. Hirsch. Phil. Mag. (1957), 2, 1121.

46. Electron diffraction from crystals containing stacking faults: II. M.J. Whelan and P.B. Hirsch. Phil. Mag. (1957), 2, 1303.

47. The theory of small angle scattering from dislocations. H.H. Atkinson and P.B. Hirsch. Phil. Mag. (1958), 3, 213.

48. The theory of small angle scattering from extended dislocations. H.H. Atkinson and P.B. Hirsch. Phil. Mag. (1958), 3, 862.

49. The effect of stacking fault energy on low temperature creep in pure metals. P.R. Thornton and P.B. Hirsch. Phil. Mag. (1958), 3, 738.

50. Comparison of dislocation arrangements and movements in a number of metals. P.B. Hirsch, P. Partridge and H. Tomlinson. Report of 4th International (Berlin) Conference on Electron Microscopy (1960), p. 536.

51. Kinematical theory of Bragg diffraction phase contrast of electron microscope images of dislocations and stacking faults in crystals. P.B. Hirsch, A. Howie and M.J. Whelan. Report of 4th International (Berlin) conference on Electron Microscopy (1960), p. 527.

52. Dislocation loops in quenched aluminium. P.B. Hirsch, J. Silcox, R.E. Smallman and K.H. Westmacott. Phil. Mag. (1950), 3, 897.

53. An electron optical study of defects in quenched gold. J. Silcox and P.B. Hirsch. Phil. Mag. (1958), 3, 72.

54. Dislocation loops in quenched metals. P.B. Hirsch and J. Silcox. Report of Cooperstown Conference on 'Growth and Perfection of Crystals' (New York: Wiley 1958), p. 262.

55. Observations of dislocations in metals. P.B. Hirsch. Report of Symposium on 'Internal Stresses and Fatigue' (Detroit: General Motors 1958), p. 139.

56. An electron microscope study of stainless steel deformed in fatigue and simple tension. P.B. Hirsch, P.G. Partridge and R.L. Segall. Phil. Mag. (1959), 4, 721.

57. Direct experimental evidence of dislocations. P.B. Hirsch. Metall. Reviews (1959), 4, 101. Review.

58. Observations of dislocations in metals by transmission electron microscopy. P.B. Hirsch. Contribution to London Symposium on 'The Application of Thin-Film Techniques to the Electron Microscopic Examination of Metals'. J. Inst. Met. (1958–59), 87, 406.

59. A kinematical theory of contrast of electron microscope images of dislocations. P.B. Hirsch, A. Howie and M.J. Whelan. Phil. Trans. Roy. Soc. (1960), A252, 499.

60. Direct observations of defects in neutron-irradiated copper. J. Silcox and P.B. Hirsch. Phil. Mag. (1959), 4, 1356.

61. A comparison of stored energy, flow stress and dislocation distribution in cold-worked polycrystalline silver. J. Bailey and P.B. Hirsch. Phil. Mag. (1960), 5, 485.

62. Electron microscope in Solid state physics and metallurgy. P.B. Hirsch. Proc. Eur. Reg. Conf. on Electron Microscopy, Delft 1960, Vol. 1, p. 212.

63. Flow stress of Al and Cu at high temperature. P.B. Hirsch and D.H. Warrington. Phil. Mag. (1961), 6, 735.

64. Techniques and applications of transmission electron microscopy. P.B. Hirsch. Report of Summer School on Radiation Damage in Solids, Ispra 1960.

65. The dislocation distribution in face-centred cubic metals after fatigue. R.L. Segall, P.G. Partridge and P.B. Hirsch. Phil. Mag. (1961), 6, 1493.

66. Electron microscope studies of defects in crystals. P.B. Hirsch. Kyoto (Japan) Symposium on Electron Diffraction, 1961.

67. Seeing atomic defects in metals. P.R. Hirsch. Proc. Roy. Instn. (1962), 39, 176.

68. Recrystallisation processes in some polycrystalline metals. J.E. Bailey and P.B. Hirsch. Proc. Roy. Soc. (1962), A267, 11.

69. International Conference on magnetism and crystallography, Kyoto 1961. W.P. Wolf and P.B. Hirsch. British J. Appl. Phys. (1962), 13, 257.

70. Some observations on radiation damage in metals. P.B. Hirsch, R.M.J. Cotterill and M.W. Jones. Electron Microscope Conference (1962), Philadelphia.

71. Some recent studies on defects in deformed metals. P.B. Hirsch and J.W. Steeds. Electron Microscope Conference (1962), Philadelphia.

72. Technique and application of transmission electron microscopy. P.B. Hirsch 'Radiation damage in solids', Academic Press (1962), 39.

73. Electron Microscope Studies of Defects in Crystals. P.B. Hirsch. Proc. of Intern. Conf. on Magnetism and Crystallography 1961, Vol. 11, J. Phys. Soc. Japan (1962), 17. Supplement B-II.

74. Extended jogs in dislocations in face-centred cubic metals. P.B. Hirsch. Phil. Mag. (1962), 7, 67.

75. The strain-rate dependence of the flow stress of copper single crystals. P.B. Hirsch, P.R. Thornton and T.E. Mitchell. Phil. Mag. (1962), 7, 337.

76. On the production of X-rays in thin metal foils. P.B. Hirsch, A. Howie and M.J. Whelan. Phil. Mag. (1962), 7, 2095.

77. The dependence of cross-slip on stacking-fault energy in face-centred cubic metals and alloys. P.R. Thornton, T.E. Mitchell and P.B. Hirsch. Phil. Mag. (1962), 7, 1349.

78. Dislocation distributions and hardening mechanisms in metals. P.B. Hirsch. Presented at the Conference on the Relation between Structure and Strength in Metals and Alloys, Teddington, 1963.

79. Rearrangements of dislocations in stainless steel during electropolishing. U. Valdre and P.B. Hirsch. Phil. Mag. (1963), 8, 237.

80. Work-hardening in niobium single crystals. P.B. Hirsch, T.E. Mitchell and R.A. Foxall. Phil. Mag. (1963), 8, 1895. Note also P.B. Hirsch in Proc. 5th International Congress of Crystallography, Cambridge (1960), Cambridge University Press, p. 139.

81. A theory of linear strain hardening in crystals. P.B. Hirsch. Offprinted from the Discussions of the Faraday Society, 1964, No. 38.

82. Elastic interaction between prismatic dislocation loops and straight dislocations. F. Kroupa and P.B. Hirsch. Offprinted from the Discussions of the Faraday Society, 1964, No. 38.

83. Effect of surface stress relaxation on the electron microscopy images of dislocations normal to thin metal foils. W.J. Tunstall, P.B. Hirsch and J.W. Steeds. Phil. Mag. (1964), 9, 99.

84. Effect of thermal diffuse scattering on propagation of high energy electrons through crystals. C.R. Hall and P.B. Hirsch. Proc. Roy. Soc. (1965), A286, 158.

85. Effect of weak Bragg reflected beams on absorption of electrons. C.R. Hall and P.B. Hirsch. Phil. Mag. (1965), 12, 539.

86. Transmission Electron Microscopy of Thin Crystals. P.B. Hirsch, A. Howie, R.B. Nicholson, D.W. Pashley, and M.J. Whelan (1965). Butterworths: London.

87. Stacking fault strengthening. P.B. Hirsch and A. Kelly. Phil. Mag. (1965), 12, 881

88. The contributions from thermal diffuse scattering and weak beams to the absorption of high energy electrons. C.R. Hall and P.B. Hirsch. Int. Conf. on Electron Diffraction and Crystal Defects, Melbourne, 1965.

89. Electron microscope investigations of dislocations. P.B. Hirsch. Int. Conf. on Electron Diffraction and Crystal Defects, Melbourne, 1965.

90. The deformation of magnesium single crystals. P.B. Hirsch and J.S. Lally. Phil. Mag. (1965), 12, 595.

91. Crystal defects. Chapter in 'Encyclopaedia Dictionary of Physics'. Pergamon Press. P.B. Hirsch.

92. The impact of transmission electron microscopy in the science of materials. Electron Microscopy in Materials Science, 1971. (New York: Academic Press) p. 3.

93. Dislocation structures in deformed single crystals and work-hardening theories. P.B. Hirsch and T.E. Mitchell. AIME Seminar on Work-hardening, Chicago, U.S.A., 1966. Editors: Hirth and Weertman (New York: Gordon and Breach) 1968, p. 65.

94. The effect of point defects on absorption of high energy electrons passing through crystals. C.R. Hall, P.B. Hirsch and G.R. Booker. Phil. Mag. (1966), 14, 979.

95. The plasticity of pure niobium single crystals. M.S. Duesbery, R.A. Foxall and P.B. Hirsch. J. de Phys. (1966), 27, C3–193.

96. A critical examination of the long-range stress theory of work-hardening. P.M. Hazzledine and P.B. Hirsch. Phil. Mag. (1967), 15, 121.

97. The deformation of niobium single crystals. R.A. Foxall, M.S. Duesbery and P.B. Hirsch. Can. J. Phys. (1967), 45, 607.

98. Some comments on the interpretation of the 'Kikuchi-like reflection patterns' observed by scanning electron microscopy. G.R. Booker, A.M.B. Shaw, M.J. Whelan and P.B. Hirsch. Phil. Mag. (1967) 16, 1185.

99. Effect of core structure on dislocation mobility with special reference to b.c.c. metals. M.S. Duesbery and P.B. Hirsch. Report on Battelle Conference on Dislocation Dynamics, Seattle, 1967. Eds: Rosenfield, Hahn, Bement and Jaffee. (McGraw Hill, New York, 1968) p. 57.

100. Stage II work hardening in crystals. P.B. Hirsch and T.E. Mitchell. Can. J. Phys. (1967), 45, 663.

101. Absorption parameters in electron diffraction theory. C.J. Humphreys and P.B. Hirsch. Phil. Mag. (1968), 18, 115.

102. Absorption parameters in electron microscopy. C.J. Humphreys, P.B. Hirsch and M.J. Whelan. 4th European Regional Conference on Electron Microscopy, Rome, 1968, p. 287.

103. Chromatic aberration and absorption in high voltage electron microscopy. P.B. Hirsch and C.J. Humphreys. 4th European Conference on Electron Microscopy, Rome 1968, p. 99.

104. Some aspects of the deformation of body-centred cubic metals. P.B. Hirsch. Trans. Jap. Inst. Metals SupI. (1968), 9, 30.

105. The movement of a dislocation through random arrays of point and parallel line obstacles. A.J.E. Foreman, P.B. Hirsch and F.J. Humphreys. Conf. on Fundamental Aspects of Dislocation Theory, Washington, 1969. Editors: Simmons, de Wit, and Bullough. Nat. Bureau of Standards, Vol. 2, p. 1083.

106. Plastic deformation of two-phase alloys containing small nondeformable particles. P.B. Hirsch and F.J. Humphreys. (1969) Physics of Strength and Plasticity. Editor: E. Argon (MIT Press) p. 189.

107. On the mechanism of cross-slip of dislocations on particles. M.S. Duesbery and P.B. Hirsch. Conf. on Fundamental Aspects of Dislocation Theory, Washington (1969). Eds.: Simmons, de Wit, and Bullough. Nat. Bureau of Standards, Vol. 2, p. 1115.

108. The deformation of single crystals of copper and copper-zinc alloys containing alumina particles. I — Macroscopic properties and workhardening theory. P.B. Hirsch and F.J. Humphreys. Proc. Roy. Soc. (1970) A318, 45.

109. The deformation of single crystals of copper and copper-zinc alloys containing alumina particles. II — Microstructure and dislocation-particle interactions. F.J. Humphreys and P.B. Hirsch. Proc. Roy. Soc. (1970), A318, 73.

110. The dynamical theory of scanning electron microscope channelling patterns. P.B. Hirsch and C.J. Humphreys. Proc. 3rd Annual Scanning Electron Microscope Symposium, Illinois, 1970, p. 449.

111. The theory of scanning electron microscope channelling patterns for normal and tilted specimens. P.B. Hirsch and C.J. Humphreys. Septième Congres International de Microscopie Électronique, Grenoble, 1969, Vol. 1, p. 229.

112. The orientation dependence of backscattered electrons in scanning electron microscopy. P.B. Hirsch and C.J. Humphreys. 2nd Aust. Conf. on Electron Microscopy, Canberra, 1970.

113. The microstructure and work-hardening in stage I of single crystals of dispersion hardened copper alloys. P.B. Hirsch and F.J. Humphreys. 2nd Int. Conf. on the Strength of Metals and Alloys, Asilomar, 1970. (ASM), Vol. 2, p. 545.

114. The effect of temperature on the mechanical properties and microstructure of single crystals of copper containing dispersed oxide particles. F.J. Humphreys, P.B. Hirsch and D. Gould. 2nd Int. Conf. on the Strength of Metals and Alloys, Asilomar 1970. (ASM), Vol. 2, p. 550.

115. The form of the defect clusters produced in copper by Cu+ Ion Irradiation. M.M.Wilson and P.B. Hirsch. Phil. Mag. 1972, 25, 983.

116. High voltage electron microscopy — Penetration and contrast. P.B. Hirsch and C.J. Humphreys. Proc. 5th Euro. Conf. on Electron Microscopy (1972), p. 520.

117. Some recent trends and application of transmission and scanning electron microscopy of crystalline materials. 5th Int. Symposium on Materials, Berkeley 1971. Editor: G. Thomas, 1972. (Los Angeles: University of California Press), p. 1.

118. A dynamical theory for the contrast of perfect and imperfect crystals in the scanning electron microscope using backscattered electrons. J.P. Spencer, C.J. Humphreys and P.B. Hirsch, Phil. Mag. 1972, 26, 193.

119. Workhardening, recovery and Bauschinger effect in alloys with dispersions of small particles. D. Gould, P.M. Hazzledine, P.B. Hirsch and F.J. Humphreys. Third Intern. Conf. on the Strength of Metals and Alloys, Cambridge 1973. (London: Jnst. of Metals and Iron and Steel Institute), Vol. I, p. 31.

120. Comment on 'Dispersion Hardening in Metals' by E.W. Hart. P.B. Hirsch and F.J. Humphreys, Scripta Met. 1973, 7, 259.

121. The rôle of dislocations in solid state reactions. P.B. Hirsch. Proc. 7th Intern. Symposium on Reactivity of Solids, Bristol 1972, p. 362.

122. Interlamellar slip in polyethylene. G.W. Groves and P.B. Hirsch. J. Mat. Sci. 1969, 4, 929.

123. High voltage electron microscopy in the U.K. P.B. Hirsch. Proc. 3rd Intern. Conf. on High Voltage Electron Microscopy, Oxford 1973 (London: Academic Press). Editors: P.R. Swann, C.J. Humphreys and M.J. Goringe. p.1. Also Research Policy 1974, 3, 78.

124. The Physics of Metals II — Defects (1975). Editor: P.B. Hirsch (Cambridge University Press).

125. Research in material science and technology. P.B. Hirsch. S.R.C. Annual Report 1972–73, p. 32.

126. Some electron microscope studies of lattice defects. P.B. Hirsch. Proc. 8th Intern. Conf. on Electron Microscopy, Canberra 1974. Editors: J.V. Sandars and D.J. Goodchild. (Canberra: Australian Academy of Science), p. 458.

127. Thermal diffuse scattering of electrons. P.B. Hirsch, C.J. Humphreys and M.J.Whelan. Abstracts of papers at Intern. Crystallography Conf., Melbourne 1974 (Melbourne: Australian Academy of Science), p. 333.

128. Dissociation of near-screw dislocations in germanium and silicon. A. Gomez, D.J.H. Cockayne, P.B. Hirsch and V. Vitek. Phil. Mag. 1975, 31, 105.

129. The Bauschinger effect, work-hardening and recovery in dispersion-hardened copper crystals. D. Gould, P.B. Hirsch and F.J. Humphreys. Phil. Mag. 1974, 30, 1353.

130. A coplanar Orowan loops model for dispersion hardening. P.M. Hazzledine and P.B. Hirsch. Phil. Mag. 1974, 30, 1331.

131. A new type of planar defect in slightly reduced MoO3. W. Thoeni and P.B. Hirsch. 4th Intern. Conf. on HVEM, Toulouse, 1975.

132. Formation, structure and movement of jogs in copper alloys. C.B. Carter and P.B. Hirsch. 4th Intern. Conf. on the Strength of Metals and Alloys, Nancy, 1976, 70.

133. Work-hardening and recovery of dispersion hardened alloys. F.J. Humphreys and P.B. Hirsch. Phil. Mag. 1976, 34, 373.

134. Dislocation structures in deformed copper-silica crystals. F.J. Humphreys and P.B. Hirsch. 4th Intern. Conf. on the Strength of Metals and Alloys, Nancy, 1976, 204.

135. The reduction of MoO3 at low temperatures. W. Thoeni and P.B. Hirsch. Phil. Mag. 1976, 33, 639.

136. The formation and glide of jogs in low stacking-fault energy face-centred cubic materials. C.B. Carter and P.B. Hirsch. Phil. Mag. 1977, 35, 1509.

137. On the mobility of dislocations in germanium and silicon. A.M. Gomez and P.B. Hirsch. Phil. Mag. 1977, 36, 169.

138. Dislocations and mechanical properties of metals. Proc. Intern. School of Physics 'Enrico Fermi', 1974, 'Atomic Structure and Mechanical Properties of Metals', Varenna, North-Holland 1976, 133.

139. Point defect cluster hardening. P.B. Hirsch. Proc. Conf. on Point Defect Behaviour and Difftisional Processes. The Metals Society, 1976, 95.

140. Direct observation of the reduction of MoO3 at low temperatures. W. Thoeni, P.L. Gai and P.R. Hirsch. Journal of the Less-Common Metals, 1977, 54, 263.

141. The reduction of MoO3 at low temperatures. II. W. Thoeni, P.L. Gai and P.B. Hirsch. Phil. Mag. 1977, 35, 781.

142. Characterisation of lattice defects. P.B. Hirsch. Proc. 9th Intern. Congr. on Electron Microscopy, Toronto 1978 (Microscopical Society of Canada, Toronto) Vol. III, p. 369.

143. Electron microscope studies of climb of dissociated dislocations. C.B. Carter, D. Cherns, P.B. Hirsch and H. Saka. Proc. 9th Intern. Congr. on Electron Microscopy, Toronto 1978 (Microscopical Society of Canada, Toronto). Vol. I, p. 324.

144. Study of crystal defects by diffraction contrast in the electron microscope. P.B. Hirsch. Electron Diffraction 1927–1977, ed. P.J. Dobson, J.R. Pendry and C.J. Humphreys (Institute of Physics, London) 1978, p. 335.

145. Applications of HVEM to Materials Science. P.B. Hirsch. Proc. 5th Intern. Conf. on HVEM, Kyoto, Japan 1977 (Japan Soc. of Electron Microscopy, Tokyo), Supplement of Journal of Electron Microscopy 1977, 26, p. 21.

146. Crystal gazing with the electron microscope. P.B. Hirsch. The David Martin 42nd Royal Society BAYS Lecture, British Association for the Advancement of Science, Bath, September 1978.

147. Studies of metal oxide catalysts. P.B. Hirsch. Institute of Physics EMAG Conference on Applications of Electron Microscopy in Chemistry, December 1977.

148. Workhardening in Ordered Alloys. P.B. Hirsch and R.C. Crawford. Proc. ICSMA5, Aachen, 1979 (Oxford: Pergamon Press), p. 89.

149. Mechanism of climb of dissociated dislocations. D. Cherns, P.B. Hirsch and H. Saka. Proc. ICSMA5, Aachen, 1979 (Oxford: Pergamon Press) p. 295.

150. The dissociation of dislocations in GaAs. A.M. Gomez and P.B. Hirsch. Phil. Mag. A., 1978, 38, 733.

151. A mechanism for the effect of doping on dislocation mobility. P.B. Hirsch. J. de Physique Supplement 6, 1979, 40, C6–l 17.

152. Recent results on the structure of dislocations in tetrahedrally coordinated semiconductors. J. de Physique Supplement 6, 1979, 40, C6–27.

153. Mechanism of climb of dissociated dislocations. D. Cherns, P.B. Hirsch and H. Saka, Proc. Roy. Soc. Lond. A 371, 1980, 213.

154. Mechanism of climb of dissociated dislocations. D. Cherns, P.B. Hirsch and H. Saka, Proc. 6th Australian Conference on Electron Microscopy, 1980.

155. The structure of dislocations in quartz under electron irradiation. D. Cherns, P.B. Hirsch, J.L. Hutchison, M.L. Jenkins and S. White. Proc. 6th Australian Conference on Electron Microscopy, 1980.

156. Lattice images of naturally deformed quartz. D. Cherns, P.B. Hirsch, J.L. Hutchison, M.L. Jenkins and S. White, Electron Microscopy 1980, 1, 374.

157. Electron irradiation-induced vitrification of dislocation cores in quartz. D. Cherns, P.B. Hirsch, J.L. Hutchison, M.L. Jenkins and S. White, Nature, 1980, 287, 3.

158. Migration of a grain boundary through a dispersion of particles. P.M. Hazzledine, P.B. Hirsch and N. Louat, 1st Risø International symposium on Metallurgy and Materials Science, 1980, 159.

159. The structure and electrical properties of dislocations in semiconductors. P.B. Hirsch. J. Microscopy, 1980, 118(1), 3.

160. High resolution electron microscope studies of defects in crystals. P.B. Hirsch, Micron, 1980, 11, 243.

161. High resolution electron microscopy of defects in crystals. P.B. Hirsch. Electron Microscopy 1980, 1, 146.

162. Structure images of dislocations in silicon. J.L. Hutchison, C.J. Humphreys, A. Ourmazd and P.B. Hirsch, Electron Microscopy 1980, 1, 304.

163. Mechanism of climb of dissociated dislocations. D. Cherns, G.J. Hardy, P.B. Hirsch and H. Saka, Micron, 1980, 11, 277.

164. The structure of dislocations in quartz under electron irradiation. D. Cherns, P.B. Hirsch, J.L. Hutchison, M.L. Jenkins and S. White, Micron, 1980, 11, 291.

165. Electronic and mechanical properties of dislocations in semiconductors. P.B. Hirsch, Proc. Materials Research Society Symposium on Defects in Semiconductors, Boston (North-Holland), Eds. J. Narayan and T.Y. Tan) 1981, 257.

166. Plastic deformation and electronic mechanisms in semiconductors and insulators. P.B. Hirsch, J. de Physique, 1981, 42, 6(suppl), C3, 149.

167. Concluding Remarks. P.B. Hirsch, Royal Society Meeting on Fracture Mechanics in Design and Service 'Living with Defects'. Phil. Trans. R. Soc. Lond., 1981, A299, 237.

168. In-situ straining in the HVEM of neutron-irradiated copper crystals. E. Johnson and P.B. Hirsch, Phil. Mag. A., 1981, 43, 157.

169. The effect of alloying additions on collision cascades in heavy-ion irradiated copper solid solutions. A.Y. Stathopoulos, C.A. English, B.L. Eyre and P.B. Hirsch, Phil. Mag. A., 1981, 44 , 309.

170. Lattice images of the cores of 30o partials in silicon. G.R. Anstis, P.B. Hirsch, C.J. Humphreys, J.L. Hutchison and A. Ourmazd, Institute of Physics Conference Series No. 60, Microscopy of Semiconducting Materials, 1981, (Eds. A.G. Cullis and D.C. Joy), 15.

171. Relaxation of dislocations in deformed silicon. P.B. Hirsch, A. Ourmazd and P. Pirouz, Institute of Physics Conference Series No.60, Microscopy of Semiconducting Materials, 1981 (Eds. A.G. Cullis and D.C. Joy), 29.

172. The climb of dissociated dislocations in semiconductors. A. Ourmazd, D. Cherns and P.B. Hirsch, Institute of Physics Conference Series No.60, Microscopy of Semiconducting Materials, 1981 (Eds. A.G. Cullis and D.C. Joy), 39.

173. Antiphase domain boundary tubes in plastically deformed ordered Fe 30.5 wt. % Al alloy. C.T. Chou and P.B. Hirsch, Phil. Mag., 1981, 44, 1415; and Proceedings of the Institute of Physics Electron Microscopy and Analysis Group, Conference Series No.61, Editor M.J. Goringe, 1981.

174. The climb of dissociated dislocations. D. Cherns, G.J. Hardy, P.B. Hirsch, A. Ourmazd and H. Saka, Proc. Int. Conf. on Dislocation Modelling, Gainesville, Florida (Eds. M.F. Ashby *et al.*, Pergamon) 1981, 564.

175. Determination of the stacking fault energy of diamond. P. Pirouz, N. Sumida, A.R. Lang and P.B. Hirsch, Proc. Int. Diamond Conf., Oxford 1982.

176. Anti-phase domain boundary tubes in Ni3AI. C.T. Chou, P.B. Hirsch, M. McLean and E. Hondros, Nature, 1982, 300, 621.

177. The rôle of electron microscopy in materials science. P.B. Hirsch, Aust. J. Phys., 1982, 35, 727.

178. Effect of doping on mechanical properties, recrystallisation and diffusion in semiconductors. P.B. Hirsch, Microscopy of Semiconducting Materials Conference, Oxford 1983, Inst. Phys. Conf. Ser. No. 67.

179. Dissociation of dislocations in diamond. P. Pirouz, D.H.J. Cockayne, N. Sumida, P.B. Hirsch and A.R. Lang, Proc. Roy. Soc. Lond. 1983, A.386, 241.

180. Electron microscopy of antiphase domain boundary tubes in deformed ordered alloys. C.T. Chou and P.B. Hirsch, Proc. Roy. Soc. Lond. A., 1983, 387, 91.

181. Dark-field electron microscopy of dissociated dislocations and surface steps in silicon using forbidden reflections. A. Ourmazd, G.R. Anstis and P.B. Hirsch, Phil. Mag. A., 1983, 48, No. 1, 139.

182. A simple technique for measuring doping effects on dislocation motion in silicon. S.G. Roberts, P. Pirouz and P.B. Hirsch, Colloque International CNRS Propnétés et Structure des Dislocations dans les Semiconductors, Aussois, 1983.

183. Introductory remarks. P.B. Hirsch, Phil. Trans. Roy. Soc. Lond. 1983, A310, 5.

184. Transformation of (001) platelets to voidites in type Ia diamond. P. Pirouz, J.C. Barry and P.R. Hirsch, 34th Annual Diamond Conference, Bristol, 1984, 8.

185. The effect of doping on the plasticity and fracture around indentations in Ge and GaAs. S.G. Roberts, P. Pirouz, P.D. Warren and P.B. Hirsch, Conference on Dislocations: Core Structure and Physical Properties, Aussois, France, 1984, 8.

186. Dislocations in semiconductors. P.B. Hirsch, Eshelby Memorial Symposium, Sheffield, 1984, Fundamentals of Deformation and Fracture, (Cambridge University Press; editors B.A. Bilby, K.J. Miller, J.R. Willis), p. 325.

187. Fine Structure of antiphase domain boundary tubes in B2 ordered alloys. C.T. Chou, P.B. Hirsch, P.M. Hazzledine and G.R. Anstis, 3rd Asia-Pacific Conference and Workshop on Electron Microscopy, 1984.

188. Dislocations in Semiconductors. P.B. Hirsch, Proc. of Conference on Dislocations and Properties of Real Materials, London 1984, ed. M.H. Loretto, Institute of Metals, London, 1985, p. 333.

189. Dislocations in Semiconductors, P.B. Hirsch, Materials Science and Technology, 1985, 1, 666.

190. Indentation Plasticity and polarity of hardness on {111} faces of GaAs. P.B. Hirsch, P. Pirouz, S.G. Roberts and P.D. Warren, Phil.Mag.B, 1985, 52, 761.

191. Effects of Doping on Mechanical Properties of Semiconductors. P.B. Hirsch, P. Pirouz, S.G. Roberts and P.D. Warren, 2nd International Conference on Science of Hard Materials, Rhodes, 1984, Inst. Phys. Conf. Ser. No.75, p.83, (Adam Hilger, London) 1986, 83.

192. Electronic effects on plasticity of covalently bonded materials. P.B. Hirsch, in Deformation of Ceramics II, 1983, ed. R.E. Tressler and R.C. Bradt.

193. Knoop hardness anisotropy on {00l} faces of Ge and GaAs. S.G. Roberts, P.D. Warren and P.B. Hirsch, J.Materials Research 1986, 1, 162–176.

194. Application of HVEM to materials science: recent results and future prospects. P.B. Hirsch, Proceedings XIth International Congress on Electron Microscopy, Kyoto, 1986, Eds. T. Imura, S. Maruse and T. Suzuki, Japanese Soc. Electron Microscopy, Tokyo, p. 1013.

195. Voidites in diamond: evidence for a crystalline phase containing nitrogen. P.B. Hirsch, J.L. Hutchison and J. Titchmarsh, Proceedings XIth International Congress on Electron Microscopy, Kyoto, 1986, Eds. T. Imura, S. Maruse and T. Suzuki, Japanese Soc. Electron Microscopy, Tokyo, p. 1703.

196. Platelets, dislocation loops and voidites in diamond. P.B. Hirsch, P. Pirouz and J.C. Barry, Proc. Roy.Soc., 1986, A407, 239.

197. Direct observations of moving dislocations: reflections on the thirtieth anniversary of the first recorded observations of moving dislocations by transmission electron microscopy. P.B. Hirsch, Materials Science and Engineering, 84, (1986) 1–10.

198. Citation Classic: Dislocation loops in quenched aluminium. P.B. Hirsch, J. Silcox, R.E. Smallman and K.H. Westmacott, Phil. Mag. 3:897–908: Current Contents Vol. 18, No. 10, p. 18.

199. Dislocation mobility and crack tip plasticity at the ductile-brittle transition. P.B. Hirsch, S.G. Roberts and J. Samuels, Revue de Physique Appliquée, 1988, 23, 409–418.

200. The dynamics of dislocation generation at crack tips and the ductile-brittle transition. P.B. Hirsch, S.G. Roberts and J. Samuels, Scripta Met., 1987, 21, 1523–1528.

201. Antiphase domain boundary tubes in ordered alloys. P.M. Hazzledine and P.B. Hirsch, Materials Research Society 81, 1987, 75. (Eds. N.S. Stoloff, C.C. Koch, C.T. Liu, 0. Izumi.)

202. Voidites in diamond:- evidence for a crystalline phase containing nitrogen. P.B. Hirsch, J.L. Hutchison and J. Titchmarsh, Phil. Mag. A (1986) 54, 2, L49.

203. The dynamics of dislocation generation at crack tips and the brittle-ductile transition. P.B. Hirsch, S.G. Roberts, J. Samuels and P.D. Warren. Strength of Metals and Alloys (editors: P.O. Kettunen, T.K. Lepistö and M.E. Lektonen) (Pergamon Press, Oxford) 1988, Vol. 2, p. 1083.

204. The brittle-ductile transition in silicon. II. Interpretation. P.B. Hirsch, S.G. Roberts and J. Samuels, Proc. Roy. Soc. A, (1989) 421, 25–53.

205. Dislocation dynamics and crack tip plasticity at the brittle-ductile transition. P.B. Hirsch, S.G. Roberts, J. Samuels and P.D. Warren. Proceedings of 7th International Conference on Fracture, Houston 1989, (eds. K. Salama, K. Ravi-Chandar, D.M.R. Taplin and P. Rama Rao) (Advances in Fracture Research, Pergamon Press, Oxford) 1989, 139–158.

206. TEM in materials science — Past, present and future. P.B. Hirsch, J. Microscopy, 1989, 155, pt. 3, 361.

207. Dislocation dynamics and the brittle-ductile transition in precracked silicon. P.B. Hirsch, S.G. Roberts, J. Samuels and P.D. Warren, Int. Symp. on Structural Properties of Dislocations in Semiconductors, 1989. Inst. Phys. Conf. Ser. No. 104, 4.

208. The dynamics of edge dislocation generation along a plane orthogonal to a Mode I crack. P.B. Hirsch and S.G. Roberts, Scripta Metall. 1989, 23, 6, 925–930.

209. Fundamentals of the brittle-ductile transition. P.B. Hirsch, Commemorative Lecture by the 34th Gold Medalist of the Japan Institute of Metals, Materials Transactions, JIM, 1989, 30, 11, 841–855.

210. Formation of antiphase-domain boundary tubes in B2 ordered alloys by cross-slip and annihilation of screw dislocations. C.T. Chou, P.M. Hazzledine, P.B. Hirsch and G.R. Anstis, Phil. Mag. A, 1987, 56, 6, 799–813.

211. Imaging of dislocations using back-scattered electrons in a scanning electron microscope. J.T. Czernuszka, N.J. Long, E.D. Boyes and P.B. Hirsch, Phil. Mag. Lett. (1990) 62, 227.

212. The brittle-ductile transition in silicon. P.B. Hirsch and S.G. Roberts, Phil. Mag. (1991), 64, 55.

213. Electron Channelling Contrast Imaging of Dislocations. J.T. Czernuszka, N.J. Long and P.B. Hirsch, Proc. ICEM XII, Seattle (1990), Vol. I, p. 600.

214. Contribution of Microscopy to Society — Instrumentation and Computation. P.B. Hirsch. Proc. ICEM XII, Seattle (1990), Vol. I, p. 2.

215. Report of the discussion at the 'Round Table on the Yield Stress Anomaly', P.B. Hirsch, J. Phys. III, (1991), 1, 989.

216. Frontiers of electron microscopy in materials science. P.B. Hirsch (1990), Proc. of 3rd Conf. on Frontiers of Electron Microscopy in Materials Science, Argonne, USA, Ultramicroscopy (1991), 37, 1.

217. Nucleation and propagation of misfit dislocations in strained epitaxial layer systems. P.B. Hirsch, Proc. of POLYSE 90, Schwabisch Hall, Germany (1990), Springer-Verlag, Berlin, 470,1990.

218. A new theory of the anomalous yield stress in L12 alloys. P.B. Hirsch, Scripta Met. & Mater., (1991), 25, 1725.

219. Modelling the yield stress anomaly in intermetallics. P.R. Hirsch, ICSMA9 (1991), p. 3.

220. The brittle-ductile transition and dislocation mobility in silicon and sapphire. S.G. Roberts, H.S. Kim and P.B. Hirsch, ICSMA9, (1991), p. 317.

221. A new theory of the anomalous yield stress in Ll2 alloys. P.B. Hirsch, Phil. Mag. A (1992), 65, 569.

222. A model of the anomalous yield stress for (111) slip in L12 alloys. P.B. Hirsch, Progress in Materials Science, (1992), 36, 63.

223. Locking and unlocking of screws and superkinks, and the yield stress anomaly in L12 alloys. P.B. Hirsch, Proceed. of NATO Workshop on High Temperature Intermetallics, Irsee, Germany (1992), p. 197.

224. Are new materials interesting? P.B. Hirsch, 1990 Wolfson Lectures on The Science of New Materials, Blackwell, Oxford, 1992, (ed. G.A.D. Briggs).

225. Modelling the brittle-ductile transition. P.B. Hirsch, and S.G. Roberts, Proc. of Symposium on Materials Modelling: From Theory to Technology, Oxford 1991, lOP Publishing Ltd., p. 181.

226. Imaging defects by TEM and SEM, P.B. Hirsch, 5th Asia Pacific Electron Microscopy Conference, Beijing, 1992, p. 8.

227. Electron channelling contrast imaging in the JSM-6400 scanning electron microscope. A.J. Wilkinson, J.T. Czernuszka, N.J. Long and P.B. Hirsch. JEOL News, 30E, 1, 1–4, 1992.

228. Electron channelling contrast imaging of defects in Si1-xGex epilayers. A.J. Wilkinson, J.T. Czernuszka, N.J. Long and P.B. Hirsch. Micro 92, July 7–10, London, 1992.

229. Imaging defects in SiGe/Si heterostructures using electron channelling contrast. A.J. Wilkinson, C. Anstis, J.T. Czernuszka, N.J. Long and P.B. Hirsch. Xth ECEM (EUREM 92), Granada, Spain, September 7–11, 1992.

230. Dislocation-controlled stable crack growth in Mo and MgO. A.S. Booth, M. Ellis, S.G. Roberts and P.B. Hirsch. Materials Science and Engineering A164, 270–274 (1993).

231. Glide sequences of deformation in the (111) plane of Ni3Ga single crystals in the yield stress anomaly. A. Couret, Y.Q. Sun and P.B. Hirsch. Phil Mag. A67, 29–50 (1993).

232. The yield stress anomaly in L12 alloys. P.B. Hirsch. Mater. Res. Soc. Symp. Proc. Vol. 288, 33–43 (1993).

233. Observations of dislocations relevant to the anomalous yield stress in L12 alloys. P.B. Hirsch and Y.Q. Sun. Materials Science and Engineering A164, 395–400 (1993).

234. Dislocation-driven slow crack growth by microcleavage in semi-brittle crystals. P.B. Hirsch, A.S. Booth, M. Ellis and S.G. Roberts. Scripta Metall et Mater. 27, 1723–1728 (1992).

235. Dislocation dynamics and brittle-to-ductile transitions. S.G. Roberts, M. Ellis and P.B. Hirsch. Materials Science and Engineering A164, 135–140 (1993).

236. Electron channelling contrast imaging of interfacial defects in strained silicon-germanium layers on silicon. A.J. Wilkinson, C.R. Anstis, J.T. Czernuszka, N.J. Long and P.B. Hirsch. Phil. Mag. A68, 59–80 (1993).

237. The strain rate sensitivity of the flow stress and the mechanism of deformation of single crystals of Ni3(AlHf)B. S.S. Ezz and P.B. Hirsch. Phil. Mag. A69, 105–127 (1994).

238. Computer simulation of the motion of screw dislocations in Ni3Al. C.T. Chou and P.B. Hirsch. Phil. Mag. A68, 1097–1 128 (1993).

239. Dislocation activity and brittle-ductile transitions in single crystals. S.G. Roberts, A.S. Booth and P.B. Hirsch. Mater. Science and Engineering A176, 91–98 (1994).

240. Effects of surface relaxation on electron channelling contrast images of misfit dislocations. A.J. Wilkinson, P.B. Hirsch, J.T. Czernuszka and N.J. Long. Proc. 13th Int. Cong. Electron Microsc. 95–96 (1994).

241. Electron channelling contrast imaging of defects in semiconductors. A.J. Wilkinson, P.B. Hirsch, J.T. Czernuszka and N.J. Long. Int. Phys. Conf. Series No. 134, 755–762 (1993).

242. Dislocations, cracks and brittleness in single crystals. S.G. Roberts, P.B. Hirsch, A.S. Booth, M. Ellis and F.C. Serbena. Physica Scripta T49, 420–426 (1993).

243. Crack-tip plasticity and quasi-brittle fracture of single crystals. P.B. Hirsch. Plastic Deformation of Ceramics, p. 1 (Editors: R. Bradt, C. Brooks and J. Routbort; Plenum Press, New York) 1995.

244. The effects of surface stress relaxation on electron channelling contrast images of 2 dislocations. A.J. Wilkinson and P.B. Hirsch. Phil. Mag. A72 , 81 (1995).

245. Strain-rate dependence of the flow stress and work hardening of γ. S.S. Ezz, Y.Q. Sun and P.B. Hirsch. Materials Science and Engineering Al 92/193 (1995) 45–52.

246. The effect of room temperature prestrain on the yield stress of single crystals of γ 'in the yield stress anomaly range of temperatures. S.S. Ezz, and P.B. Hirsch. High Temperature Ordered Intermetallic Alloys VI, MRS Symp. Proc. 364, 719 (1995).

247. Strain-rate dependence of the flow stress and workhardening of single crystals of Ni3(Al,Hf)B. S.S. Ezz, Y.Q. Sun and P.B. Hirsch. High Temperature Ordered Intermetallic Alloys VI, MRS Symp. Proc. 364, 695 (1995).

248. Yield stress reversibility and the operation of Frank-Read sources in L12 alloys in the anomalous yield stress range for octahedral slip. S.S. Ezz and P.B. Hirsch. High Temperature Ordered Intermetallic Alloys VI, MRS Symp. Proc. 364, 35 (1995).

249. The operation of Frank-Read sources, yield stress reversibility and the strain-rate dependence of the flow stress in the anomalous yielding regime of L12 alloys. S.S. Ezz and P.B. Hirsch. Phil. Mag. A72, 383 (1995).

250. The effect of room temperature deformation on the yield stress anomaly in Ni3(AlHf)B. S.S. Ezz and P.B.Hirsch. Phil. Mag. A73, 1069 (1996).

251. The behaviour of macrokinks under stress in the yield stress anomaly of L12 alloys. P.B. Hirsch. Czech Journal of Physics 45, 921 (1995).

252. Comments on the brittle-to-ductile transition — I: Cooperative dislocation generation instability; II: Dislocation dynamics and the strain-rate dependence of the transition temperature. P.B. Hirsch and S.G. Roberts. Acta Mater. 44 , 2361 (1996).

253. Comment on the reply to comments on the brittle-to-ductile transition — I: Cooperative dislocation generation instability; II: Dislocation dynamics and the strain-rate dependence of the transition temperature. P.B. Hirsch and S.G. Roberts. Scripta Materialia 35, 291 (1996).

254. Kear-Wilsdorf locks, jogs and the formation of antiphase-boundary tubes in Ni3Al. P.B. Hirsch. Phil. Mag. A74, 1019 (1996).

255. Modelling the Brittle-Ductile Transition. P.B. Hirsch. Acta Metallurgica Sinica 33, 225 (1997).

256. Modelling plastic zones and the brittle-ductile transition. P.B. Hirsch and S.G. Roberts. Phil. Trans. Roy. Soc. 355, 1991 (1997).

257. Modelling crack-tip plastic zones and brittle-ductile transitions. P.B. Hirsch and S.G. Roberts. In George R. Irwin symposium on Cleavage and Fracture e.g. K.S. Chan (TMS, Warrendale, Pennsylvania 1997) p. 137.

258. Electron Diffraction Based Techniques in Scanning Electron Microscopy of Bulk Materials. A.J. Wilkinson and P.B. Hirsch. Micron 28, 279 (1997).

259. Yield stress anomalies in single crystals of Ti 54.5at% Al: I. Overview and <011] superdislocation slip. S. Jiao, N. Bird, P.B. Hirsch and G. Taylor. Phil. Mag. 78, 777 (1998).

260. A mechanism for "double-half dislocation loop" nucleation in low misfit epitaxial GexSi1-xon Si. P.B. Hirsch. Microscopy of Semicond. Confer. (Oxford) Instit. of Phys. Confer. Series 157, 121 (1997).

261. Comment on the "Simulation of the Brittle-Ductile Transition in Si single crystals using dislocation mechanics". P.B. Hirsch and S.G. Roberts. Scripta Mater. 37, 190 (1997).

262. Modelling the threshold conditions for propagation of stage I fatigue cracks. A.J. Wilkinson, S.G. Roberts and P.B. Hirsch. Acta Mater. 46, 379 (1998).

263. Yield stress anomalies in single crystals of Ti 54.5at%Al: II. 1/2[112](111) slip. S. Jiao, N. Bird, P.B. Hirsch and G. Taylor. Phil. Mag. A79, 609–625 (1999).

264. Electron Microscopy in Materials Science — A historical perspective, P.B. Hirsch. Proc. of Centennial Symp.on the Electron. Ed. A. Kirkland and P.D. Brown (Institute of Materials, London) 64–81 (1998).

265. Decoherence in Electron Backscattering by Kinked Dislocations, S.L. Dudarev, J. Ahmed, P.B. Hirsch and A.J. Wilkinson. Acta. Cryst. A, 55, 234–245, (1999).

266. Characteristics of ordinary ½ <110] slip in single crystals of γ-TiAl, S. Jiao, N. Bird, P.B. Hirsch and G. Taylor. High Temperature Ordered Intermetallic Alloys VIII, Materials Research Society Proc., 552, KK8.11.1–8.11.5, (1999).

267. Nature of slip in γ-TiAl above the yield stress anomaly temperature, S. Jiao, N. Bird, P.B. Hirsch and G. Taylor. High Temperature Ordered Intermetallic Alloys VIII, Materials Research Society Proc. 552, KK8.12.1–8.12.6, (1999).

268. Yield stress anomaly for ½<112]{111} slip in γ-TiAl, S. Jiao, N. Bird, P.B. Hirsch and G. Taylor. High Temperature Ordered Intermetallic Alloys VIII, Materials Research Society Proc. 552, KK1.9.1–1.9.6, (1999).

269. Modelling crack tip plastic zones and brittle-ductile transitions, P.B. Hirsch and S.G. Roberts. Cleavage Fracture, ed. K.S. Chan (The Minerals, Metals and Materials Society, Warrendale Pa.) 137–145 (1997).

270. Climb/glide dislocation sources at low misfit GexSi1-x/Si(001) interfaces, S. Jiao, P.B. Hirsch and D.D. Perovic. Microscopy of Semiconductor Materials, Inst. Phys. Conf. Ser. 164, p. 269, (1999).

271. 'Early days of diffraction contrast transmission electron microscopy', P.B. Hirsch, in Topics in Electron Diffraction and Microscopy of Materials, I.O.P. Microscopy in Materials Science Series, I.O.P., Bristol, p. 1, (1999).

272. 'Topics in Electron Diffraction and Microscopy of Materials', ed. P.B. Hirsch, I.O.P. Microscopy in Materials Science Series, I.O.P. Bristol (1999).

273. Yield stress anomalies in single crystals of Ti-54.5 at % Al. III Ordinary slip, S. Jiao, N. Bird, P.B. Hirsch and G. Taylor, Phil Mag. A, 81, 213 (2001).

274. Climb/glide dislocation sources at low misfit Gex Si1-x Si (001) interfaces. S. Jiao, P.B. Hirsch and D.D. Perovic, Phil. Mag. A, 81, 1041 (2001).

275. Modelling the initiation of cleavage fracture of ferritic steels. S.G. Roberts, S.J. Noronha, A.J. Wilkinson and P.B. Hirsch, Acta Mater., 50, 1229 (2002).

276. Co-operative Nucleation of Shear Dislocation Loops. Y.Q. Sun, P.M. Hazzledine and P.B. Hirsch, Phys. Rev. Lett. 88, 065503–1 (2002).

277. Commentary on Workhardening. P.B. Hirsch, Dislocations in Solids, Vol. 11, Editor F.R.N. Nabarro and M.S. Duesbery, Elsevier (2002) p. XXV.

278. Kear-Wilsdorf Locks and the Formation of Anti-Phase Boundary Tubes in L12 Alloys. P.B. Hirsch Phil. Mag. A 83, 1007 (2003).

279. On the core structure of 112 2 1 edge dislocations in γ-TiAl. C. Lang, P.B. Hirsch and D.J.H. Cockayne, Phil. Mag. Lett., 84, 139 (2004).

280. On the stability of the three-fold symmetrically dissociated screw dislocation in the b.c.c. lattice. P.B. Hirsch and Y.Q. Sun, Transactions of the Royal Society of South Africa, 58, 129 (2003).

281. Comment on "The locking of 35o <110> dislocations in NiAl" by N. Yang and Y.Q. Sun. P.B. Hirsch, Scripta Materialia, 54, 313 (2005).

282. Modelling the upper yield point and the brittle-ductile transition of silicon wafers in threepoint bend tests. S.G. Roberts and P.B. Hirsch, Phil. Mag. 86, 4099 (2006).

283. 50 Years of TEM of Dislocations. Phil. Mag. Special Issue, Editors P.B. Hirsch, D.J.H. Cockayne, J.C.H. Spence and M.J. Whelan, Phil. Mag. 86, No. 29–31 (2006).

284. 50 Years of TEM of Dislocations: Past, Present and Future. P.B. Hirsch, D.J.H. Cockayne, J.C.H. Spence and M.J. Whelan, Phil. Mag. Special Issue, 86, 4519 (2006).

285. Selection rules for Bloch wave scattering for HREM imaging of imperfect crystals along symmetry axes. P.D. Nellist, E.C. Cosgriff, P.B. Hirsch and D.J.H. Cockayne, Phil. Mag. 88, 135 (2008).

286. Bloch wave analysis of depth dependent strain effects in high resolution electron microscopy. P.D. Nellist, E.C. Cosgriff, P.B. Hirsch and D.J.H. Cockayne, Microscopy & Microanalysis, 14, 92 (2008).

287. Bloch wave analysis of depth dependent strain effects in high resolution electron microscopy. P.D. Nellist, E.C. Cosgriff, P. B. Hirsch and D.J.H. Cockayne (2008). In Proceedings EMC 2008, Aachen, Springer-Verlag pp. 139–140 (2008).

288. ADF STEM imaging of screw dislocations viewed end-on. E.C. Cosgriff, P.D. Nellist, P.B. Hirsch, Z. Zhou and D.J.H. Cockayne, Phil. Mag. 90, 4361 (2010).

289. Electron microscopy and diffraction of defects, nanostructures, interfaces and amorphous materials conference to mark the retirement of Professor David Cockayne FRS Oxford, 7 September 2009. P. B. Hirsch, A. I. Kirkland and P. D. Nellist. Phil. Mag. 90 (2010).

290. Effect of Eshelby twist on core structure of screw dislocations in molybdenum: Atomic structure and electron microscope image simulations. R. Groger, K.J. Dudeck, P.D. Nellist, V. Vitek, P.B. Hirsch and D.J.H. Cockayne, Phil. Mag. 91, 2364 (2011).

300. The dissociation of the [a+c] dislocation in GaN. P.B. Hirsch, J.G. Lozano, S. Rhode, M.K. Horton, M.A. Moram, S. Zhang, M.J. Kappers, C.J. Humphreys, A. Yasuhara, E. Okunishi and P.D. Nellist, Phil. Mag. 93, 28–30, 3925 (2013).

301. Conference Proceedings, EMAG, York, 2013. The dissociation of the [a+c] dislocation in GaN. P.D. Nellist, P.B. Hirsch, S. Rhode, M.K. Horton, J.G. Lozano, A. Yasuhara, E. Okunishi, S. Zhang, S.-L. Sahonta, M.J. Kappers, C.J. Humphreys and M.A. Moram. In Press.

302. Conference Proceedings, EMAG, York, 2013. Observation of depth-dependent atomic displacements related to dislocations in GaN by optical sectioning in the STEM. J.G. Lozano, M.P. Guerrero-Lebrero, A. Yasuhara, E. Okinishi, S. Zhang, C.J. Humphreys, P.L. Galindo, P.B. Hirsch and P.D. Nellist. In Press.

VII.   Miscellaneous

303. "Foreword" to M.J. Whelan Festschrift. P.B. Hirsch. Micron 28, 77 (1997).

304. Foreword to Festschrift in Honour of Dr T.W. Mitchell's 60th birthday. P.B. Hirsch. Phil. Mag. 78, 525 (1998).

305. Dr Michael S. Duesbery: an appreciation. P.B. Hirsch, Phil. Mag. A, 81, 1021 (2001).

306. J.W. Christian, F.R.S. (obituary), P.B. Hirsch, Materials Science and Technology 17, 609 (2001).

307. Professor J.W. Christian, F.R.S. (1926–2001), (obituary), P.B. Hirsch, Materials World 9, 46 (2001).

308. Obituary of Professor Angus Hellawell. P.B. Hirsch, Materials World, 14, Number 9 (2006).

309. Obituary of Professor John Hunt FRS. P.B. Hirsch, St Edmund Hall Magazine (2013).

310. Prelude: A brief biography of Sir Alan Cottrell FRS, FREng 17th July 1919–15 February 2012. P.B. Hirsch, 93, 3697 (2013).

311. Isis Innovation — How did it start? P.B. Hirsch and C. Quinn, The Oxford Magazine, 8th Week, Trinity Term, June 2008.

Albert J. Libchaber

ALBERT J. LIBCHABER

Birthplace

Born	Paris, France, 23 October 1934

Education

1956	Bachelor's Degree in Mathematics, University of Paris
1958	Ingénieur des Télécommunications, Ecole Nationale Supérieure des Télécommunications
1959	Master's of Science Degree in Physics, University of Illinois
1965	Ph.D. in Physics, Ecole Normale Supérieure, University of Paris

Research and Professional Experience

1959–1961	Military Service, Atomic Division of the Army
1962	Attaché de Recherche CNRS, Ecole Normale Supérieure
1965–1966	Member of Technical Staff, Bell Telephone Laboratory
1967	Maitre de Recherche CNRS, Ecole Normale Supérieure
1974	Directeur de Recherche CNRS, Ecole Normale Supérieure
1983	Professor, Department of Physics and James Franck and Enrico Fermi Institutes, University of Chicago,
1987	Paul Snowdon Russell Distinguished Service Professorship, University of Chicago
1991	Professor, Department of Physics, Princeton University
1991–2003	Fellow, NEC Research Institute, Princeton University
1993	James S. McDonnell Distinguished University Professor, Princeton University
1994	Professor of Physics, The Rockefeller University
1995	Detlev W. Bronk Professor, The Rockefeller University
2001	Raman Chair, Indian Academy of Science

Membership of Societies

1982	Académie des Sciences, Paris
1986	American Academy of Arts and Sciences
2007	National Academy of Science, Washington

Honours

1980	Palmes Académiques1993 Légion d'honneur
2003	Doctor Honoris Causa, Weizmann Institute of Science
2005	Doctor Honoris Causa, University of Copenhagen

Awards

1968	Médaille d'argent CNRS
1979	Grand Prix de Physique, Prix Ricard

1986	Wolf Prize, Israel
1986–1991	MacArthur Fellow
1999	Prix des Trois Physiciens, Foundation de France

COMMENTARY

I was born in 1934 in Paris, 10th arrondissement, Porte Saint Martin, one of the entrances to the old city fortification. After graduating from one of the French School Systems, École Nationale Supérieure des Telecommunications, I went to work at the University of Illinois with John Bardeen (a two-time Nobel Prize winner) in 1958. It is there that my interests in Physics developed, through the observation and interaction with Bardeen, a brilliant and modest scientist. He taught me the beauty and rigor of theoretical physics and how to always be available to students.

After an obligatory two years in the French Army, as part of the first French atom bomb program in the Sahara desert, I went to work for a Ph.D at Ecole Normale Supérieure in 1962.

For my Ph.D years, with Pierre Aigrain, we discovered a new mode of electromagnetic wave propagation in metals, when a high DC magnetic field is present. We called it Helicon Waves. The wave propagates along the magnetic field, is circularly polarized and dispersive. We later realized that this mode was known in plasma physics as Whistler modes propagating along the earth's magnetic field.

I was then invited to join Bell Laboratories, Murray Hill, for two years, where I pursued the study of plasma effects in metals, Helicon waves, Alfven waves, and radio frequency size effects in collaboration with C. C. Grimes.

I came back to Ecole Normale in 1967 as Maitre de Recherche CNRS (Centre Nationale de la Recherché Scientifique) and built a low temperature laboratory. I quickly became interested in dynamical system theory and realized that the low temperature property of liquid helium, such as its low viscosity, allowed to test nonlinear phenomena. We built, with Jean Maurer, thermal convection cells and discovered many routes to chaos, period doubling, intermittency, Ruelle–Takens scenario. The period doubling cascade may be the most beautiful nonlinear phenomenon with perfect scaling, the Feigenbaum numbers. We then pursued the studies in mercury, a very low Prandtl number fluid. By then I became Directeur de Recherche CNRS. During my stay in France, I was influenced and educated in theoretical physics by Philippe Nozieres, a great theorist.

In 1982, I was invited as professor at the University of Chicago, physics department, and developed a long and fruitful interaction with Leo Kadanoff, another great theoretical physicist. My research then took three directions, all related to nonlinear dynamics.

In the first direction, we looked at more global features of the phase space orbit. We studied experimentally a 2-Torus, a periodically forced Rayleigh–Bénard experiment in mercury with golden mean winding number. We determined two scaling indices, in agreement with circle map theory. Also, for the first time, we measured a continuous spectrum of scaling indices, the multifractal indices.

In a second direction, Francois Heslot and I built an experiment on helium gas at low temperature, again Rayleigh–Bénard convection, but at a very high Rayleigh Number. We measured the Kolmogorov scaling for thermal turbulence, observed that the temperature fluctuations at high turbulence are not Gaussian, but exponential, and developed a model of what we called hard thermal turbulence.

Finally, we studied the nonlinear dynamics of interfaces, being solid/liquid, liquid/liquid, and liquid/gas. The most fundamental one is the penetration of a gas into a viscous fluid in a thin channel, essentially two-dimensional. In this famous Saffman–Taylor problem, the long term solution is a single finger of gas occupying one half of the channel. We showed that the third dimension is important with the formation of a liquid film in the gas finger, that the ½ solution is not really observed and that at high velocity the finger destabilizes by tip-splitting. We also studied the dynamics of the Mullins–Sekerka instability in liquid/liquid crystal interfaces.

This brings me to 1991. I became interested in the dynamical properties of biological systems. We realized then that we could study biological objects in the tens of nanometer scales with advanced optics and software, and that new manipulation techniques were developing like the optical tweezer. I thus accepted to move to the Physics department at Princeton University, and also accepted an offer as Fellow of a new Princeton institution, the NEC research laboratory. We developed four new biophysics studies there.

In the first one, interacting with Stanislas Leibler, we develop a research program on the cell cytoskeleton, with its two main proteins actin and tubulin that polymerize to define the cellular skeleton. We measure, for the first time, the actin filament rigidity, its persistence length, in a single molecule experiment. The study of microtubules led us to look at how microtubules form and deform vesicles, as a model of cellular organization.

Associated to the cytoskeleton there was very active interest in molecular motors moving on those tubules. One explanation for molecular motors motion was Brownian ratchets. Following such models we built the first optical motor working on Brownian principles, *a tour de force* by Luc Faucheaux.

Another very active subject was related to Leon Adleman's proposal that one can solve NP complete problems using DNA and molecular biology, in short a DNA computer. We then solved one NP complete problem using DNA: the maximal clique problem.

With the same philosophy, applying physical ideas to biology, X. L. Wu and I studied the collective modes of bacteria colonies. Bacteria were introduced in a suspended soap film, defining thus a 2D geometry. And there it was easy to analyze the collective mode of *E. coli* bacteria. For all the work at NEC Shumo Liu, a biologist, was an essential member of the laboratory.

This leads us to 1994, where I was invited to join Rockefeller University. We first developed a lot of new techniques, inventing various detectors. Being condensed matter physicists, we knew material science well. Benoit Dubertret and I showed first that quantum dots are toxic as biological materials. We showed that by incorporating the dots in phospholipids micelles they became biologically inert and we were able to inject 10^9 dots in Xenopus embryos and followed up to the tadpole stage the fluorescence of the organism, thus developing experiments in embryogenesis.

Associated to it we showed for the first time that gold nanoparticles are excellent quenchers of fluorescence and could measure Forster transfer distance for fluorescence quenching.

Franck Vollmer and I then showed that optical resonance (whispering gallery mode) in hundreds of micron size silica spheres can be used to measure DNA mismatches, proteins, bacteria, at the single molecule level.

Finally, we introduced molecular beacons, DNA probes that form a stem-and-loop structure and possess an internal quenched fluorophore and we showed that they are ideal for DNA target recognition. It was a beautiful Ph.D. work of Gregoire Bonnet.

During all this time I did not forget hydrodynamics. I performed two experiments with Jun Zhang. In the first one, using a flowing suspended soap film, we studied all the modes of a one-dimensional flag in two-dimensional wind. The second experiment was related to plate tectonics. We showed that in a Rayleigh–Bénard experiment with free top surface, a Floater can oscillate in time confirming Tuzo Wilson model of the Atlantic closing and reopening.

In the last 10 years, Dieter Braun and I have been looking at the origin of life problem, and the possibility that it originated in porous submarine hydrothermal mounds, called thermal vents. Taking into account temperature gradients there, we showed that DNA replication is feasible, caused by laminar thermal convection, as the molecules are continuously cycled between hot and cold regions of a chamber. The so-called PCR amplification is thus a natural phenomena of thermal convection. On the other hand, we also showed that thermophoresis can accumulate charged biopolymers like DNA or RNA and thus reduces the problem of a critical concentration needed for life onset.

Finally, we wanted to understand the steps needed to produce a cell from the early soup.

For that we indulged in producing artificial cells. Vincent Noireaux and I were successful in producing cells that could express genes for about a week, polymerize cytoskeleton proteins on the phospholipid membrane, express pores at the membrane and for that you need to encapsulate in a phospholipid bilayer a cellular extract, the genes and their regulation. This is an approximate model of Von Neumann theory of self-reproduction. Of course, the ultimate aim of the experiment is self-reproduction, one of the essential aspects of biology.

In all of my scientific career I was surrounded by extremely talented students and postdocs. Some of the names appear in the text, but one should understand that experimental scientific research is a collective activity and from this laboratory more than 30 physicists come out and populate the world as active professors in India, China, Japan, Europe, Israel, USA, Brazil, and Canada.

Tome 43   No 7   1er AVRIL 1982

LE JOURNAL DE PHYSIQUE · LETTRES

J. Physique — *LETTRES* **43** (1982) L-211 - L-216   1er AVRIL **1982**, PAGE L-211

Classification
Physics Abstracts
47.20 — 47.65

Period doubling cascade in mercury, a quantitative measurement

A. Libchaber, C. Laroche and S. Fauve

Groupe de Physique des Solides de l'Ecole Normale Supérieure,
24, rue Lhomond, 75231 Paris Cedex 05, France

(*Reçu le 21 décembre 1981, accepté le 12 février 1982*)

Résumé. — Observation de la cascade de doublement de période dans une expérience de Rayleigh-Bénard sur le mercure. La cellule expérimentale a un rapport d'aspect $\Gamma = 4$ et comporte quatre rouleaux convectifs. Un champ magnétique de 270 G est appliqué le long de l'axe des rouleaux. Le nombre de Feigenbaum mesuré est $\delta = 4,4$. Le rapport des sous harmoniques successifs est de l'ordre de 14 dB pour les sous harmoniques les plus bas mesurés.

Abstract. — Observation of the period doubling cascade in a Rayleigh-Bénard experiment in mercury. The experimental cell has an aspect ratio $\Gamma = 4$ and contains four convective rolls. A DC magnetic field of 270 G is applied along the convective roll axis. The measured Feigenbaum number is $\delta = 4.4$. The ratio of the successive subharmonics is of the order of 14 dB for the lowest measured subharmonics.

The period doubling cascade as a route to chaos is now well documented theoretically [1] as well as experimentally [2]. We present here new results on a Rayleigh-Bénard experiment in a cell of liquid mercury with an aspect ratio larger than in previously reported works on helium or water (four convective rolls). The very good signal to noise ratio of the experiment allows a precise determination of the Feigenbaum number and of the power ratio of the successive subharmonics. In this experiment, a DC magnetic field, applied along the convective roll axis, introduces an extra damping of the oscillators which favours the period doubling cascade.

1. **An experimental dynamical system with controlled dissipation.** — At low Prandtl number, the relevant instability of two-dimensional convective rolls, which leads to time dependent convection and chaos, is the oscillatory instability [3]. It was shown [2] in experiments of convection in liquid helium that for a range of Prandtl numbers and wavenumbers of the convective pattern, one route to chaos is a period doubling cascade, which follows the Feigenbaum scenario [1]. It is customary to compare this type of experiments to a one-dimensional mapping. But this mapping relates to strongly dissipative modes. A more realistic mapping is therefore a two-dimensional one with two parameters, a constraint and a viscosity (area contraction in phase space). The

simplest quadratic one is the Henon mapping [4] which reduces for infinite contraction to a one-dimensional mapping. In a way, we want to find a physical system depending on those two parameters. In a Rayleigh-Bénard experiment, the constraint is the Rayleigh number R. At low Prandtl number, the damping of the modes associated with the oscillatory instability depends on the Prandtl number and the wavenumber of the convective pattern [3]. But it is not easy to control such parameters experimentally, and furthermore, they do not affect selectively the oscillatory instability. On the contrary, a horizontal magnetic field, parallel to the convective roll axis, will add a quantitative damping to the oscillatory instability, as we have recently shown [5]. This increased damping of the oscillators will favour the period doubling cascade as a route to chaos [6].

2. **Experimental apparatus.** — Our experimental system has been described in detail elsewhere [5, 7]. We use a parallelepipedic cell of aspect ratio $\Gamma = 4$ (dimensions $7 \times 7 \times 28$ mm) with, in the convective state, four rolls parallel to the shorter side wall (see reference five for visualization). The top and bottom boundaries consist of two thick copper plates. The lateral boundaries are made of plexiglass. In the centre of the bottom plate, a small NTC (negative temperature coefficient) thermistor is located on a 2 mm diameter hole and is adjacent to the convective fluid. The temperature signal, given by the bolometer, is analysed by a 5 420 A H. P. digital analyser. An

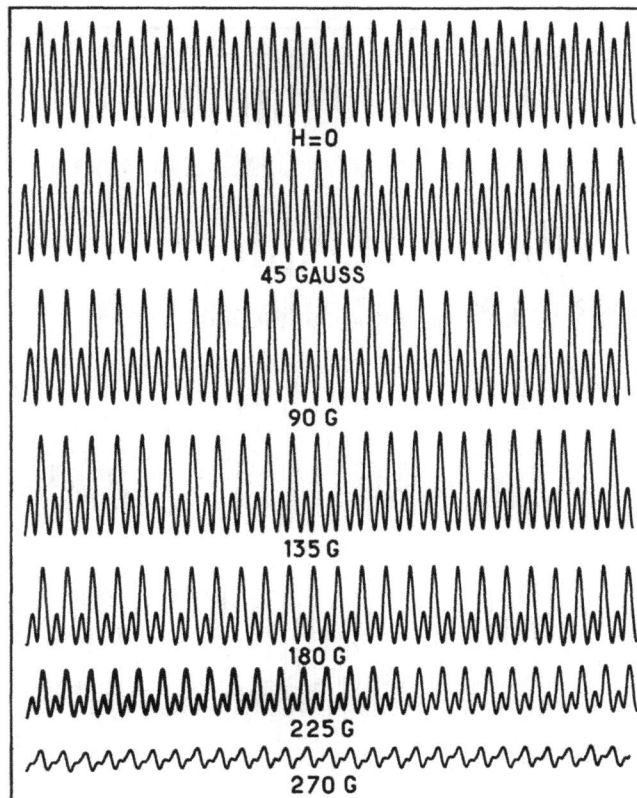

Fig. 1. — Effect of a magnetic field parallel to the roll axis. The Rayleigh number is $R/R_c = 2.9$ at zero field. The two oscillators are locked ($f_1/f_2 = 2$), the amplitude of oscillator f_1 about 20 dB larger than the amplitude of oscillator f_2 for $H = 0$. As H increases, the two oscillators tend to the same amplitude and keep their locking ratio 2.

electromagnet provides a uniform horizontal magnetic field. The experimental convective cell is placed in a vacuum chamber.

3. Effect of a DC magnetic field, parallel to the roll axis, on the oscillating state. — Defining the Rayleigh number for the onset of convection as R_c, the first time dependent instability observed is the oscillatory instability, its onset being at $R \simeq 2\,R_c$. As previously reported [7], the frequency of the oscillatory mode increases linearly with the Rayleigh number. It allows, by measuring the frequency, a precise determination of each bifurcation point. Increasing slightly the Rayleigh number, a second frequency f_2 appears in the temperature spectrum which, within a very small range of R, locks with f_1, the ratio being $f_1 = 2\,f_2$.

We then apply a DC magnetic field parallel to the roll axis up to a field of 270 G. We present, in figure 1, the effect of the magnetic field on the direct local temperature recording. Two distinct regimes appear.

For $H > 100$ G, the evolution is the one described in our previous study [5]. The frequency of the oscillatory instability increases and so does the damping of the mode. This will tend at a higher field to a complete inhibition of the oscillatory instability.

But for $H < 100$ G, and the effect is already noticeable around 10 G, the relative oscillating strength of the two locked oscillators changes. This sensitivity to a very small field is surprising.

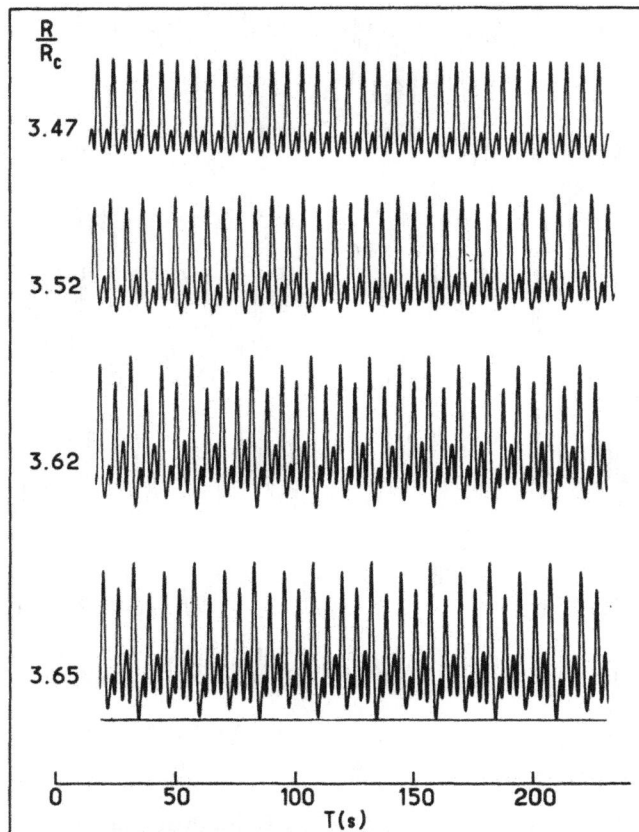

Fig. 2. — Direct time recordings of temperature for various stages of the period doubling cascade showing the onset of $f/4$ $(R/R_c = 3.52)$, $f/8$ $(R/R_c = 3.62)$, $f/16$ $(R/R_c = 3.65)$.

L-214   JOURNAL DE PHYSIQUE — LETTRES   N° 7

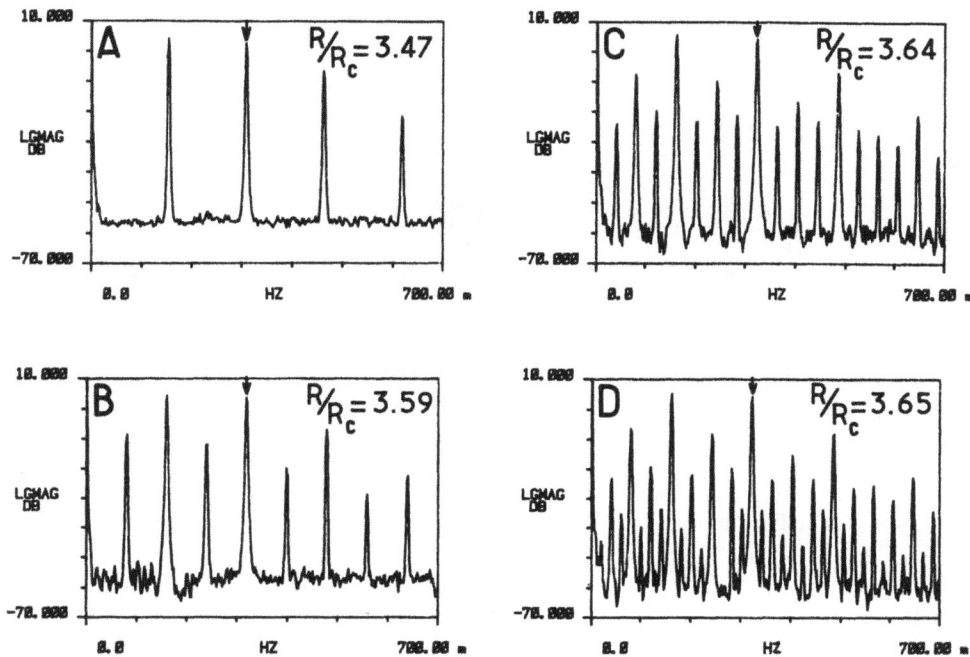

Fig. 3. — The Fourier spectrum. Arrows indicate the peak at the frequency f_1.

The net result of this phenomenon is that, at 270 G, the two modes are highly damped and have about the same oscillating strength, whereas at zero field the mode f_1 has an amplitude about 20 dB larger than the mode f_2.

4. The period doubling cascade. — From now on, we keep a constant magnetic field $H = 270$ G. A typical recording is shown in figure 2 for $R/R_c = 3.47$. Let us note that it shows a striking similarity with a recent Cray machine simulation done by Upson *et al.* [8] on a Rayleigh-Bénard experiment in a small box. The corresponding Fourier spectrum is shown in figure 3A.

As we increase the Rayleigh number, a period doubling cascade develops up to $f/32$, well resolved as far as the onset values up to $f/16$.

In table I, we present the various onset values of the cascade of pitchfork bifurcations. In figure 2, the temperature recordings are presented, showing the development of the subharmonics $f/4$, $f/8$ and $f/16$, for $R/R_c = 3.52$, 3.62 and 3.65. Their respective Fourier spectra are presented in figures 3B, C, D for values of R/R_c close to the preceding ones.

Table I

Onset of bifurcations	R/R_c		
$f/4$	3.485		$\mu_{4/8} = 3.5$ (~ 11 dB)
$f/8$	3.618 3	$\delta = 4.4 \pm 0.1$	
$f/16$	3.648 6		$\mu_{8/16} = 5$ (14 dB)

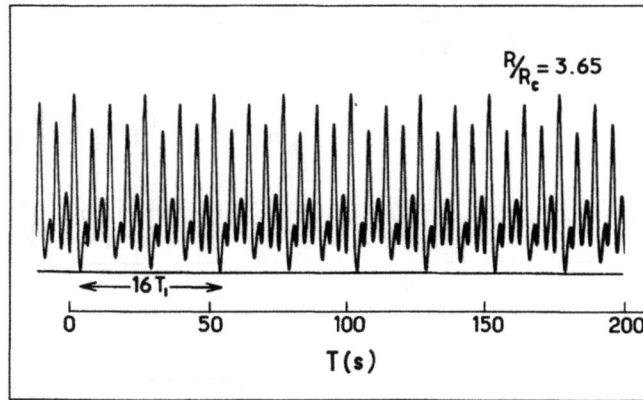

Fig. 4. — Enlargement of the recording showing the development of the cascade.

If we compute the Feigenbaum number δ for the last three bifurcations, we get

$$\delta = \frac{R_8 - R_4}{R_{16} - R_8} = 4.4 \pm 0.1 \, .$$

This is to be compared with the theoretical asymptotic value [1] $\delta = 4.669...$

We can also compute the ratio of the successive subharmonics amplitude called μ.

This ratio is measured directly on the temperature recordings. We show in figure 4 an enlargement of the temperature signal after the $f/16$ bifurcation and in figure 5 an enlargement of its Fourier spectrum. The last value of μ measured is $\mu \sim 5$ to be compared with theoretical values between 4.58 and 6.5 (the first one is given by Nauenberg and Rudnick [9], the second one by Feigenbaum [1]). This measurement of μ led to some confusion in the past, which can be understood if we look at the Fourier spectrum in figures 3D and 5. It is clear there that the odd harmonics of $f/16$ have an amplitude which is modulated and depends on the order of the harmonics.

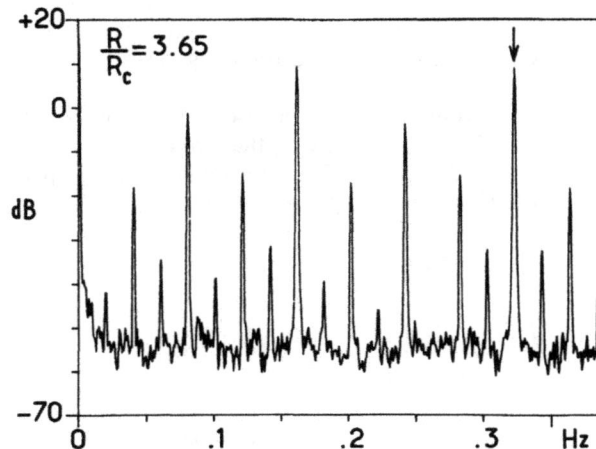

Fig. 5. — Enlargement of the Fourier spectrum corresponding to figure 4.

L-216   JOURNAL DE PHYSIQUE — **LETTRES**   No 7

It is thus not clear how to calculate the ratio from the Fourier spectrum. The numbers we give in table I are derived from the direct recording.

We have also observed beyond the accumulation point the inverse cascade. Its study and the subtle phenomena of locking windows in the chaotic state will be presented elsewhere.

Let us add some final remarks. Recent computer experiments have shown the role of the breakdown of some spatial symmetries on the onset of chaos [10]. In this respect, the presence of a DC magnetic field plays an important role, such as, for example, introducing an anisotropic effective viscosity. Clearly, for other values of the field and especially at very high fields, other routes to chaos will be found. We must also stress the fact that a very small magnetic field affects the oscillating modes which is somewhat surprising.

References

[1] FEIGENBAUM, M. J., *Phys. Lett.* **75A** (1979) 375.
    COLLET, P. and ECKMANN, J. P., *Iterated maps on the interval as dynamical systems* (Birkhaüser, Boston) 1980.
[2] MAURER, J. and LIBCHABER, A., *J. Physique Lett.* **40** (1979) L-419.
    LIBCHABER, A. and MAURER, J., *J. Physique Colloq.* **41** (1980) C3-51.
    GIGLIO, M., MUZZATI, S. and PERINI, U., *Phys. Rev. Lett.* **47** (1981) 243.
    LINSAY, P. S., *Phys. Rev. Lett.* **47** (1981) 1349.
[3] BUSSE, F. H., *Rep. Progr. Phys.* **41** (1978) 1929.
[4] HENON, M., *Commun. Math. Phys.* **77** (1976) 50.
[5] FAUVE, S., LAROCHE, C. and LIBCHABER, A., *J. Physique Lett.* **42** (1981) L-455.
[6] ARNEODO, A., COULLET, P., TRESSER, C., LIBCHABER, A., MAURER, J. and D'HUMIÈRES, D., *About the observation of the uncompleted cascade*, preprint.
[7] FAUVE, S. and LIBCHABER, A., *Chaos and order in nature*, H. Haken Ed., (Springer) 1981.
[8] UPSON, C. D., GRESHO, P. M., SANI, R. L., CHAN, S. T. and LEE, R. L., Lawrence Livermore Lab., Preprint UCRL 85555.
[9] NAUENBERG, M. and RUDNICK, J., UCSC preprint.
[10] LIPPS, F. B., *J. Fluid Mech.* **75** (1976) 113.
    Mc LAUGHLIN, F. B. and ORSZAG, S. A., preprint.

RAPID COMMUNICATIONS

PHYSICAL REVIEW A   VOLUME 36, NUMBER 12   DECEMBER 15, 1987

Transitions to turbulence in helium gas

F. Heslot, B. Castaing, and A. Libchaber

James Franck and Enrico Fermi Institutes, University of Chicago, 5640 S. Ellis Avenue, Chicago, Illinois 60637

(Received 13 July 1987)

Experimental study in gaseous helium at low temperature (4 K) of thermal convection up to a Rayleigh number $R = 10^{11}$. Three regimes are observed, a chaotic state up to $R = 2.5 \times 10^5$, a soft-turbulence state up to $R = 4 \times 10^7$, and then a hard-turbulence state. We associate those three regimes to the boundary-layer formation and dynamics.

We present in this Rapid Communication preliminary results on a Rayleigh-Bénard experiment in gaseous helium at low temperature (4 K), and very high Rayleigh numbers (R). As shown by Threlfall[1] it is possible to cover 11 orders of magnitude in the Rayleigh number, with very small changes in the Prandtl number, by changing the pressure inside the cell.[2] Careful measurements of the Nusselt number (relative effective conductivity of the gas) led Threlfall to discover several transitions in the buoyancy state, as previously reported for other fluids by Malkus and others.[3] In the experiment presented here, our main emphases are local measurements of the temperature at two different points inside the cell in order to characterize the various transitions. The main results are as follows. At low Rayleigh number we observe the onset of the oscillatory instability[4] followed by the now well-known routes to chaos. We label the chaotic state itself in terms of the attractors correlation dimension. We then observe the transition to a spatial disorder through the loss of coherence between detectors for $R = 2.5 \times 10^5$. At higher Rayleigh numbers two successive regimes with slightly different power laws for the Nusselt number versus Rayleigh number are distinguished, the transition in our geometry being at $R = 4 \times 10^7$. The time recordings, power spectra, and histograms of the temperature fluctuations are different for the two regimes. We associate the two regimes to a laminar and a turbulent boundary layer.

A detailed description of the cell can be found and will be published.[5] Let us give only its main spirit. The thin-wall stainless-steel cell is thermally isolated from the main liquid-helium bath by a vacuum jacket. It is a vertical cylinder with equal diameter and height (8.7 cm). The gas filling tube has a two-meter-long thermalization on the upper plate before entering the cell and can be closed near the cell in the helium bath. The upper plate is thermally regulated by a linear research LR-130 regulation system. The bottom plate is heated with a constant dc power, using a four-wire method for the precise measurement of power. Both plates are made out of thick electronic copper with calibrated Ge thermometers in each.[6] Two local arsenic-doped silicon far-infrared detectors,[7] 200 μm in size, are used to measure local temperature fluctuations. The "bottom" one is placed 200 μm above the bottom plate and inside the cell, and half a radius in distance from the cell axis. The "center" bolometer is placed right above the bottom one, at equal distances from both plates (see inset of Fig. 1).

Figure 1 shows the Nusselt number versus Rayleigh number dependence up to $R = 10^{11}$. To span all this range we have used five different helium densities, whose corresponding pressures are 3, 8.5, 34, 138, and 625 Torr. This allowed large overlap in the study. The calibration between upper and lower thermometers was checked for each pressure, with corrections always inferior to 0.4 mK. We could thus be confident in results from temperature differences of $\Delta T \approx 3$ mK to $\Delta T \approx 1$ K, which represents more than two decades in R for each pressure. Four different domains in R number will be presented and are delineated in Fig. 1: The low R number including the chaotic state (R up to 2.5×10^5); the transition region where one loses coherence between bolometers (2.5×10^5 up to 5×10^5); a first turbulent state, which could be called "soft" turbulence and has many of the characteristics of large-aspect ratio-phase turbulence[8] for R up to 4×10^7; finally, above this value a transition to a state showing large intermittency in the center bolometer, and the beginning of a power law for its spectrum. It could be associated with a turbulent boundary layer. We call it a "hard"-turbulence state.

In the first domain, the onset of convection ($R = 5.8 \times 10^3$ for this aspect ratio), the onset of the oscillatory instability (9×10^4), and the onset of a chaotic state (1.5×10^5) can be easily measured. Figure 2 shows typi-

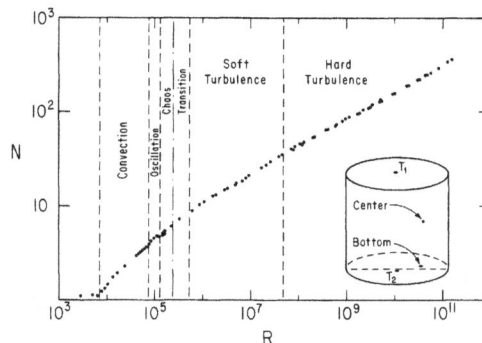

FIG. 1. Nusselt number vs Rayleigh number. The domains for the various transitions are defined. The inset shows a sketch of the experimental cell, whose height and diameter are 8.7 cm.

© 1987 The American Physical Society

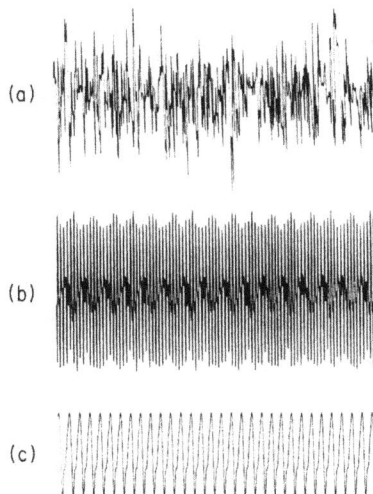

FIG. 2. Time recordings for three different states for the center bolometer. The time scales are different for each recording. (a) Soft turbulence, (b) quasiperiodicity, (c) oscillatory instability.

cal recordings of the local bolometers signal for the oscillatory region (c), and for a quasiperiodic state (b). In all this region, even in the chaotic state, the coherence function[9] between the bottom and center bolometers is close to one for all significant frequencies. The routes to chaos observed included one through a period-doubling cascade, but this is by now well documented[10] and we shall not detail it. Let us just remark that in this quite large cell, the signal to noise is above 80 dB, and thus with good stability and local bolometers, those routes to chaos can be studied in detail. We have measured the correlation[11] dimension D of the chaotic state. At the onset of chaos $D \simeq 2 \pm 0.1$. It increases rapidly, reaching a value of $D = 4$ at $R = 2 \times 10^5$. As the number of points required to calculate the dimension increases exponentially with the dimension, the correlation dimension is a useful measure only for a relatively restricted range of Rayleigh numbers.

At $R = 2.5 \times 10^5$ the coherence function between our two bolometers starts to decrease and at $R = 5 \times 10^5$ it has a negligible value. This defines our second domain. The evolution is quite odd. First the amplitude of the coherence function decreases at low frequency but spreads to a higher frequency range before disappearing slowly. This transition region where the coherence between bolometers disappears, shows that the cell is now stratified with a different dynamical behavior near the center and close to the walls. We thus associate this transition with the formation of a boundary layer. Our main evidence for this is the Nusselt-number (N) evolution in the next domain, the third one, which goes up to $R = 4 \times 10^7$.

The Nusselt-number evolution with R is well charac-

ized there by the following curve-fitting formula

$$N = 1 + 0.096 \times R^{0.333 \pm 0.005} \qquad (1)$$

The linear relation between N and $R^{1/3}$ is characteristic of a boundary-layer control flow,[12] indicating that the vertical length scale of the cell is no more relevant.

The power spectra of the two bolometers in this soft-turbulence regime are similar, with no power law. There are close to already published data[8] on large aspect ratio cells, for low Rayleigh numbers. This suggests that we have a regime of detached boundary layers at the corners of the cell.[13] Thus convective rolls germinate near the corners where an adverse pressure gradient is present, and this regime is similar to large-aspect ratio phase turbulence.[8,14] A typical recording is shown in Fig. 2(a).

For $R = 4 \times 10^7$ a transition to a new state, the fourth one in our nomenclature, occurs. Let us call it a hard-turbulence state. It is visible in the slope of the Nusselt number, which extends on four decades of Rayleigh numbers:

$$N = 1 + 0.2 \times R^{0.282 \pm 0.006} . \qquad (2)$$

Also, the shape of the temperature fluctuations in the center bolometer, its power spectrum, and histogram are different in this hard-turbulence state from the soft-turbulence one. The temperature recordings as a function of time for the two bolometers are shown in Fig. 3.

The histogram[15] for the center probe, the probability density for the temperature fluctuations, is shown in Fig. 4. Its change of shape clearly reveals the transition. It is Gaussian-like for the soft-turbulence case [Fig. 4(a)] and exponential for the hard-turbulence one [Fig. 4(b)]. Also, the width of the distribution is smaller in the hard turbulence regime. This width decreases, as R increases, following a power law, but the shape remains the same for the highest R number studied.

We postulate that this is a regime of a turbulent boundary layer oscillating and with abrupt detachment of thermal plumes,[12] which leads to sharp temperature peaks

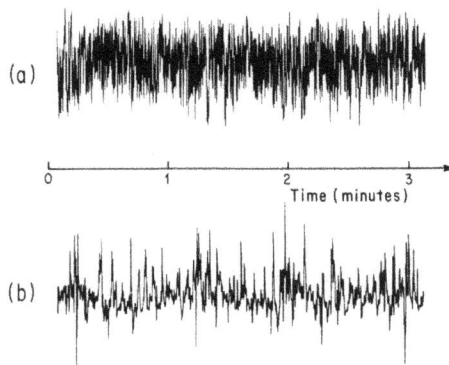

FIG. 3. Time recordings for the hard-turbulence state for $R = 4 \times 10^9$. (a) Bottom bolometer, (b) center bolometer.

5872   F. HESLOT, B. CASTAING, AND A. LIBCHABER   36

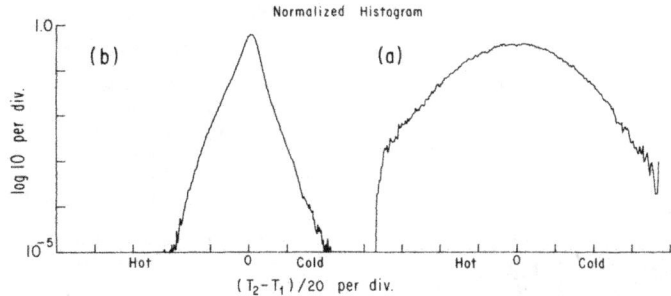

FIG. 4. Histogram of the time recordings: (a) soft-turbulence regime, (b) hard-turbulence regime. $T_2 - T_1$ represents the temperature difference across the cell.

in the center of the cell. Those plumes are localized objects, while in the soft turbulence, we have extended objects, the boundary between convective rolls.

The temperature recordings for the two bolometers is shown in Fig. 3. The center bolometer signal is characterized by a distribution of sharp peaks of variable heights. The bottom signal contains higher frequencies and also a well-defined oscillation at a frequency $1/t_0$, seen all over the cell, as can be measured from the coherence function shown in Fig. 5(c). This reemergence of coherence is one of the striking characteristics of the new state. Figures 5(a) and 5(b) show the Fourier spectra. One sees clearly that, above the frequency $1/t_0$, a power law develops for the center bolometer with a slope -1.4 ± 0.1, and the power law tends to spread to a larger band of frequencies as the R number increases. Below $1/t_0$ the spectrum is flat.

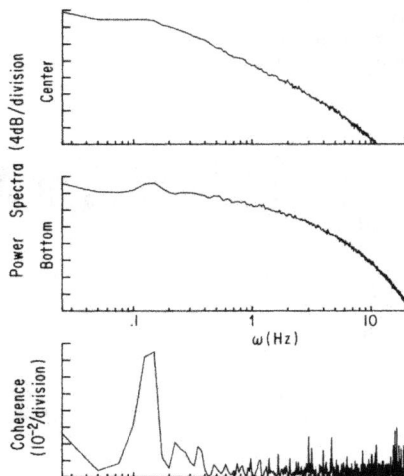

FIG. 5. Power spectra and coherence function for the center and bottom bolometer in the hard-turbulence state.

Looking at the distribution of the peaks in the center bolometer signal, we analyzed the time intervals between peaks, and found two groups in the density of time intervals. In the long-time group a well-developed peak also shows clearly the time t_0. The short-time one could be characteristic of plumes, whereas t_0 seems to be the characteristic time of the convective circulation. We have measured the evolution of t_0 with R number and it follows:

$$1/t_0 = 5.10^{-2} R^{1/2} \text{ in units of } \kappa/L^2 . \tag{3}$$

To sum up, the model proposed is that the distribution of peaks seen in the center bolometer, Fig. 3(b), is caused by thermal plumes emitted from the boundary layer. Through the stretching and folding of those detached plumes, by flow advection, the beginning of a developed turbulent state sets in.

In this experimental study of the transition from chaos to turbulence, we have in some ways brought together the two approaches of turbulence. The dynamical system approach[10] is a very good analysis of the onset of disorder, as one slowly increases the R number starting from the onset of convection. It is followed by a state where space disorder sets in. In this regime, the development of a boundary layer,[16] laminar first and later turbulent,[17] leads to two distinct turbulent states, in agreement with a wealth of observations in fluid mechanics,[18] and the description becomes more statistical in nature.[19] One may suppose, but the experiments have to be performed, that at higher R numbers a fully developed turbulent state is present, but at what R number? By working near the critical point of helium we should be able to increase the control parameter by four orders of magnitude. The hard-turbulence state may be amenable to a statistical description. The soft-turbulence state might be more difficult to understand. We have thus outlined our program.

We would like to thank S. Thomae and L Kadanoff for illuminating discussions. This work was supported by the National Science Foundation under Grant No. DMR-8316204.

RAPID COMMUNICATIONS

[1]D. C. Threlfall, Ph.D. thesis, University of Oxford, 1976 (unpublished); J. Fluid Mech. **67**, 1 (1975); **67**, 17 (1975). R. D. McCarthy, Nat. Bur. Stand. (U.S.) Tech. Note No. 631 (1972).

[2]We define the Rayleigh number as

$$R = g a \Delta T L^3 / \nu \kappa \; ,$$

where g is the gravity acceleration, L is the height of the cell ($L = 8.7$ cm), $\Delta T = T_2 - T_1$, where T_2 is the bottom temperature and T_1 the top one. α is the isobaric thermal expansion, ν is the kinematic viscosity, and κ the heat diffusivity. These three quantities are taken at $T = (T_1 + T_2)/2$.

[3]W. V. R. Malkus, Proc. R. Soc. London Ser. A **225**, 185 (1954); H. Janaka and H. Miyata, Int. J. Heat Mass Transfer **23**, 1973 (1980).

[4]F. H. Busse, Rep. Prog. Phys. **41**, 1928 (1978).

[5]F. Heslot, Thèse de Doctorat d'Etat, Université Paris VI, 1986 (unpublished).

[6]Cryocal, Germainium Thermometers GR1000.

[7]These bolometers are due to the courtesy of Professor R. Hildebrandt, Giles A. Novak, and M. Dragovan. See also, A. E. Lange, E. Kreysa, S. E. McBride, and P. L. Richards, Int. J. Infrared Millimeter Waves **4**, 689 (1983); J. C. Mather, Appl. Opt. **21**, 1125 (1982).

[8]G. Ahlers and R. P. Behringer, Phys. Rev. Lett. **40**, 712 (1978); A. Libchaber and J. Maurer, J. Phys. (Paris) Lett. **39**, 369 (1978).

[9]The coherence $C(w)$ between two signals is defined in the following way: Let $A(w)$ and $B(w)$ be their power spectra and $K(\omega)$ the Fourier transform of their cross correlation. Then

$$C(w) = \frac{K(w)}{[A(w) \times B(w)]^{1/2}} \; .$$

[10]J. P. Eckmann, Rev. Mod. Phys. **53**, 643 (1981); *Universality in Chaos,* edited by P. Cvitanovic (Adam Hilger, Philadelphia, 1985).

[11]P. Grassberger and I. Proccacia, Physica D **9**, 198 (1983); J. P. Eckmann and D. Ruelle, Rev. Mod. Phys. **57**, 617 (1985).

[12]L. M. Howard, in *Proceedings of the 11th International Congress of Applied Mechanics, Munich, 1964,* edited by H. Görtler (Springer, Berlin, 1964); D. J. Tritton, in *Turbulence and Predictability in Geophysical Fluid Dynamics and Climate Dynamics,* Proceedings of the Enrico Fermi School of Physics, Course 88, edited by M. Ghil, R. Benzi, and G. Parisi (North-Holland, Amsterdam, 1985).

[13]E. Reshotko, Am. Rev. Fluid Mech. **8**, 311 (1976).

[14]R. P. Behringer, Rev. Mod. Phys. **57**, 657 (1985); M. C. Cross and A. Newell, Physics D **10**, 299 (1984).

[15]We present, in fact, the probability density $P(T), P(T)dT$ being the probability for the bolometer to be found between the temperatures T and $T + dT$.

[16]T. J. Foster and S. Waller, Phys. Fluids **28**, 455 (1985).

[17]H. Schlichting, *Boundary Layer Theory* (McGraw-Hill, New York, 1979).

[18]D. J. Tritton, *Physical Fluid Dynamics* (Van Nostrand, New York, 1977), Chap. 11.

[19]H. Tennekes and J. L. Lumley, *A First Course in Turbulence* (MIT Press, Cambridge, MA, 1972).

LIST OF PUBLICATIONS

1. Kumar, P., Lehmann, J., & Libchaber, A. (2012). Kinetics of bulge bases in small RNAs and the effect of pressure on it. *PLOS ONE, 7*(8). http://dx.doi.org/10.1371/journal.pone.0042052

2. Maeda, Y., Tlusty, T., & Libchaber, A. (2012). Effects of long DNA folding and small RNA stem-loop in thermophoresis. *Proceedings of the National Academy of Sciences of the United States of America, 109*(44), 17972–17977. http://dx.doi.org/10.1073/pnas.1215764109

3. Iyer K.V., Maharana S., Gupta, S., Libchaber, A., Tlusty, T., & Shivashankar, G. V. (2012). Modeling and experimental methods to probe the link between global transcription and spatial organization of chromosomes. *PLOS ONE, 7*(10). http://dx.doi.org/10.1371/journal.pone.0046628

4. Kumar, P. & Libchaber, A. (2012). Pressure and temperature dependence of growth and morphology of Escherichia coli: Experiments and stochastic model. *Biophysical Journal, 105*(3), 783–793. http://dx.doi.org/10.1016/j.bpj.2013.06.029

5. Salman, H., Brenner, N., Tung, C. K., Elyahu, N., Stolovicki, E., Moore, L., ... Braun, E. (2012). Universal protein fluctuations in populations of microorganisms. *Physical Review Letters, 108*(23), 51–55. http://dx.doi.org/10.1103/PhysRevLett.108.238105

6. Maeda, Y., Nakadai, T., Shin, J., Uryu, K., Noireaux, V., & Libchaber, A. (2012). Assembly of MreB filaments on liposome membranes: A synthetic biology approach. *ACS Synthetic Biology, 1*(2), 53–59. http://dx.doi.org/10.1021/sb200003v

7. Demir, M., Douarche, C., Yoney, A., Libchaber, A., & Salman, H. (2011). Effects of population density and chemical environment on the behavior of Escherichia coli in shallow temperature gradients. *Physical Biology, 8*(6), 6–13. http://dx.doi.org/10.1088/1478-3975/8/6/063001

8. Shimamoto, Y., Maeda, Y., Ishiwata, S., Libchaber, A., & Kapoor, T. M. (2011). Insights into the micromechanical properties of the metaphase spindle. *Cell 145*(7), 1062–1074. http://dx.doi.org/10.1016/j.cell.2011.05.038

9. Maeda, Y., Buguin, A., & Libchaber, A. (2011). Thermal separation: interplay between the Soret Effect and entropic force gradient. *Physical Review Letters, 107*(3), 038301. http://dx.doi.org/10.1103/PhysRevLett.107.038301

10. Noireaux, V., Maeda, Y., & Libchaber, A. (2011). Development of an artificial cell, from self-organization to computation and self-reproduction. *Proceedings of the National Academy of Sciences of the United States of America, 108*(9), 3473–3480. http://dx.doi.org/10.1073/pnas.1017075108

11. Lehmann, J., Cibils, M., & Libchaber, A. (2009). Emergence of a code in the polymerization of amino acids along RNA templates. *PLOS ONE, 4*(6), 97–104. http://dx.doi.org/10.1371/journal.pone.0005773

12. Douarche, C., Buguin, A., Salman, H., & Libchaber, A. (2009). E. coli and oxygen: A motility transition. *Physical Review Letters, 102*(19), 198101. http://dx.doi.org/10.1103/PhysRevLett.102.198101

13. Lehmann J., & Libchaber, A. (2008). Degeneracy of the genetic code and stability of the base pair at the second position of the anticodon. *RNA, 14*(7), 1264–1269. http://dx.doi.org/10.1261/rna.1029808

14. Ren, H. C., Vollmer, F., Arnold, S., & Libchaber, A. (2007). High-Q microsphere biosensor — Analysis for adsorption of rodlike bacteria. *Optics Express, 15*(25), 17410–17423. http://dx.doi.org/10.1364/OE.15.017410

15. Salman, H., & Libchaber, A. (2007). A concentration-dependent switch in the bacterial response to temperature. *Nature Cell Biology, 9*(9), 1098–1100. http://dx.doi.org/10.1038/ncb1632

16. Lehmann, J., Reichel, A., Buguin, A., & Libchaber, A. (2007). Efficiency of a selfaminoacylating ribozyme: Effect of the length and base-composition of its 3' extension. *RNA, 13*(8), 1191–1197. http://dx.doi.org/10.1261/rna.500907

17. Liu, S., & Libchaber, A. (2006). Some aspects of E. coli promoter evolution observed in a molecular evolution experiment. *Journal of Molecular Evolution, 62*(5), 536–550. http://dx.doi.org/10.1007/s00239-005-0128-x

18. Gaetz, J., Gueroui, Z., Libchaber, A., & Kapoor, T. M. (2006). Examining how the spatial organization of chromatin signals influences metaphase spindle assembly. *Nature Cell Biology, 8*(9), 924–932. http://dx.doi.org/10.1038/ncb1455

19. Salman, H., Zilman, A., Loverdo, C., Jeffroy, M., & Libchaber, A. (2006). Solitary modes of bacterial culture in a temperature gradient. *Physical Review Letters, 97*(11), 118101. http://dx.doi.org/10.1103/PhysRevLett.97.118101

20. Noireaux, V., Bar-Ziv, R., Godefroy, J.,Salman, H., & Libchaber, A. (2005). Toward an artificial cell based on gene expression in vesicles. *Physical Biology, 2*(3), P1–P8. http://dx.doi.org/10.1088/1478–3975/2/3/P01

21. Ren, H. C., Goddard, N. L., Altan-Bonnet, G., & Libchaber, A. (2004). Discerning aggregation in homogeneous ensembles: A general description of photo counting spectroscopy in diffusing systems. *Physical Review E, 6905*(5), 542–553. http://dx.doi.org/ 10.1103/PhysRevE.69.051916

22. Braun, D., & Libchaber, A. (2004). Molecular lock-in: Imaging kinetics of a DNA hairpin in frequency space for each pixel of a CCD. *Biophysical Journal, 86*(1), 169A.

23. Noireaux, V., & Libchaber, A. (2004). A vesicle bioreactor as a step toward an artificial cell assembly. *Proceedings of the National Academy of Sciences of the United States of America, 101*(51), 17669–17674. http://dx.doi.org/10.1073/pnas.0408236101

24. Tlusty, T., Bar-Ziv, R., & Libchaber A. (2004). High-fidelity DNA sensing by protein binding fluctuations. *Physical Review Letters, 93*(25), 258103. http://dx.doi.org/10.1103/PhysRevLett.93.258103

25. Gueroui, Z., & Libchaber, A. (2004). Single-molecule measurements of goldquenched quantum dots. *Physical Review Letters, 93*(16), 166108. http://dx.doi.org/10.1103/PhysRevLett.93.166108

26. Braun, D., & Libchaber, A. (2004). Thermal force approach to molecular evolution. *Physical Biology, 1*(1), P1–8. http://dx.doi.org/10.1088/1478–3967/1/1/P01

27. Liu, S. M., & Libchaber, A. (2004). Production and assay of a transcription activator CRP in coupled in vitro transcription and translation. *Biotechniques 36*(4), 596, 598, 600.

28. Braun, D., & Libchaber, A (2003). Lock-in by molecular multiplication. *Applied Physics Letters, 83*(26), 5554–5556.

29. Goddard, N. L., Thaler, D. S., & Libchaber, A. (2003). Why the genetic code? *Biophysical Journal, 84*(2), 307A.

30. Vollmer, F., Arnold, S., Braun, D., Teraoka, I., & Libchaber, A. (2003). DNA detection from the shift of whispering gallery modes in multiple microspheres. *Biophysical Journal, 84*(2), 295A.

31. Braun, D., Goddard, N. L., & Libchaber, A. Thermophoretic trapping of DNA and convective PCR: new physical approaches to the origin of life. *Biophysical Journal, 84*(2), 6A.

32. Altan-Bonnet, G., Libchaber, A., & Krichevsky, O. (2003). Bubble dynamics in double-stranded DNA. *Physical Review Letters, 90*(13), 230–233. http://dx.doi.org/10.1103/PhysRevLett.90.138101

33. Noireaux, V., Bar-Ziv, R., & Libchaber, A. (2003). Principles of cell-free genetic circuit assembly. *Proceedings of the National Academy of Sciences of the United States of America, 100*(22), 12672–12677. http://dx.doi.org/10.1073/pnas.2135496100

34. Braun, D., Goddard, N. L., & Libchaber, A. (2003). Exponential DNA replication by laminar convection. *Physical Review Letters, 91*(15), 158103. http://dx.doi.org/10.1103/PhysRevLett.91.158103

35. Vollmer, F., Arnold, S., Braun, D., Teraoka, I., & Libchaber, A. (2003). Multiplexed DNA quantification by spectroscopic shift of two microsphere cavities. *Biophysical Journal, 85*(3), 1974–1979.

36. Vollmer, F., Arnold, S., & Libchaber, A. (2002). Novel, fiber-optic biosensor based on morphology dependent resonances in dielectric micro-spheres. *Biophysical Journal, 82*(1), 161A-162A.

37. Bar-Ziv, R., Tlusty, T., & Libchaber, A. (2002). Protein-DNA computation by stochastic assembly cascade. *Proceedings of the National Academy of Sciences of the United States of America, 99*(18), 11589–11592.

38. Braun, D., & Libchaber, A. (2002). Computer-based photon-counting lock-in for phase detection at the shot-noise limit. *Optics Letters, 27*(16), 1418–1420.

39. Goddard, N., Bonnet, G., Krichevsky, O., & Libchaber, A. (2002). Misfolded loops decrease the effective rate of DNA hairpin formation — Reply. *Physical Review Letters, 8806*(6), 286.

40. Braun, D., & Libchaber, A. (2002). Trapping of DNA by thermophoretic depletion and convection. *Physical Review Letters, 89*(18), 188103.

41. Dubertret, B., Skourides, P., Norris, D. J., Noireaux, V., Brivanlou, A. H., & Libchaber, A. (2002). In vivo imaging of quantum dots encapsulated in phospholipid micelles. *Science, 298*(5599), 1759–1762.
http://dx.doi.org/10.1126/science.1077194

42. Vollmer, F., Braun, D., Libchaber A., Khoshsima, M., Teraoka, I., & Arnold, S. (2002). Protein detection by optical shift of a resonant microcavity. *Applied Physics Letters, 80*(21), 4057–4059.

43. Dubertret, B., Calame, M., & Libchaber, A. J. (2001). Single-mismatch detection using gold-quenched fluorescent oligonucleotides. *Nature Biotechnology, 19*(7), 365–370.

44. Bar-Ziv, R., & Libchaber, A. (2001). Effects of DNA sequence and structure on binding of RecA to single-stranded DNA. *Proceedings of the National Academy of Sciences of the United States of America, 98*(16), 9068–9073.

45. Wu, X. L., & Libchaber, A. (2001). Comment on 'Particle diffusion in a quasi-twodimensional bacterial bath' — Reply. *Physical Review Letters, 86*(3), 557.

46. Dubertret, B., Liu, S. M., Ouyang, Q., & Libchaber, A. (2001). Dynamics of DNAprotein interaction deduced from in vitro DNA evolution. *Physical Review Letters, 86*(26), 6022–6025.

47. Bonnet, G., Krichevsky, O., & Libchaber, A. (2000). Unusual DNA-breathing modes unravel transcription initiation specificity. *Biophysical Journal, 78*(1), 132A.

48. Goddard, N. L., Bonnet, G., Krichevsky, O., & Libchaber, A. (2000). Sequence dependent rigidity of single stranded DNA. *Physical Review Letters, 85*(11), 2400–2403.

49. Liu, S. M., Shivashankar, G. V., Sano, T., & Libchaber, A. (2000). Method for linking a synthesized protein to it mRNA-DNA complex. *Biotechniques, 29*(4), 792, 794, 796, 798.

50. Shivashankar, G. V., Liu, S., & Libchaber, A. (2000). Control the expression of anchored genes using micron scale heater. *Applied Physics Letters, 76*(24), 3638–3640.

51. Zhang, J., Childress, S., Libchaber, A., & Shelley, M. (2000). Flexible filaments in a flowing soap film as a model for one-dimensional flags in a two-dimensional wind. *Nature, 408*, 835–839. http://dx.doi.org/10.1038/35048530

52. Zhang, J., & Libchaber, A. (2000). Periodic boundary motion in thermal turbulence. *Physical Review Letters, 84*(19), 4361–4364. http://dx.doi.org/10.1103/PhysRevLett.84.4361

53. Wu. X. L., & Libchaber, A. (2000). Particle diffusion in a quasi-two-dimensional bacterial bath. *Physical Review Letters, 84*(13), 3017–3020. http://dx.doi.org/10.1103/PhysRevLett.84.3017

54. Bonnet, G., & Libchaber, A. (1999). Optimal sensitivity in molecular recognition. *Physica A, 263*(1–4), 68–77.

55. Shivashankar, G. V., Liu, S., & Libchaber, A. (1999). Micro-chip patterning with bio-molecules and biological cells using localized light. *Biophysical Journal, 76*(1), A398.

56. Shivashankar, G. V., Feingold, M., Krichevsky, O., & Libchaber, A. (1999). RecA polymerization on double-stranded DNA by using single-molecule manipulation: The role of ATP hydrolysis. *Proceedings of the National Academy of Sciences of the United States of America, 96*(14), 7916–7921.

57. Bonnet, G., Tyagi, S., Libchaber, A., & Kramer, F. R. (1999). Thermodynamic basis of the enhanced specificity of structured DNA probes. *Proceedings of the National Academy of Sciences of the United States of America, 96*(11), 6171–6176.

58. Shivashankar, G. V., Stolovitsky, G., & Libchaber, A. (1998). Backscattering from a tethered bead as a probe of DNA flexibility. *Applied Physics Letters, 73*(3), 291–293.

59. Shivashankar, G. V., & Libchaber, A. (1998). Single molecule observation of duples DNA extension by Rec A. *Biophysical Journal, 74*(2), A242.

60. Goulian, M., Mesquita, O. N., Fygenson, D., K., Nielsen, C., Andersen, O. S., & Libchaber, A. (1998). Gramicidin channel kinetics under tension. *Biophysical Journal, 74*(1), 328–337.

61. Libchaber, A. (1998). Genome stability, cell motility and force generation. *Progress of Theoretical Physics Supplement, 130*, 1–8. http://dx.doi.org/10.1143/PTPS.130.1

62. Zhang, J., Childress, S., & Libchaber, A. (1998). Non-Boussinesq effect: Asymmetric velocity profiles in thermal convection. *Physics of Fluids, 10*(6), 1534–1536. http://dx.doi.org/10.1063/1.869672

63. Shivashankar, G. V., & Libchaber, A. (1998). Biomolecular recognition using submicron laser lithography. *Applied Physics Letters, 73*(3), 417–419.

64. Bonnet, G., Krichevsky, O., & Libchaber, A. (1998). Kinetics of conformational fluctuations in DNA hairpin-loops. *Proceedings of the National Academy of Sciences of the United States of America, 95*(15), 8602–8606.

65. Fygenson, D. K., Elbaum, M., Shraiman, B., & Libchaber, A. (1997). Mcotrubules and vesicles under controlled tension. *Physical Review E, 55*(1), 850–859.

66. Fygenson, D. K., Marko, J. F., & Libchaber, A. (1997). Mechanics of microtubule-based membrane extension. *Physical Review Letters, 79*(22), 4497–4500.

67. Kaplan, P. D., Qi, O. Y., Thaler, D. S., & Libchaber, A. (1997). Parallel overlap assembly for the construction of computational DNA libraries. *Journal of Theoretical Biology, 188*(3), 333–341.

68. Goulian, M., Feygenson, D., Mesquita, O. N., Moses, E., Nielsen, C., Andersen, O. S., & Libchaber, A. (1997). The effect of membrane tension on gramicidin A channel kinetics. *Biophysical Journal, 72*(2), A191.

69. Houchmandzadeh, B., Marko, J. F., Chatenay, D., & Libchaber, A. (1997). Elasticity and structure of eukaryote chromosomes studied by micromanipulation and micropipette aspiration. *The Journal of Cell Biology, 139*(1), 1–12.
http://dx.doi.org/10.1083/jcb.139.1.1

70. Ouyang, Q., Kaplan, P. D., Liu, S. M., & Libchaber, A. (1997). DNA solution of themaximal clique problem. *Science, 278*(5337), 446–449.
http://dx.doi.org/10.1126/science.278.5337.446

71. Libchaber, A. (1996). Is temperature passive or active in hard turbulence? *Physica D, 97*(1–3), 155–157.

72. Goulian, M., & Libchaber, A. (1996). A new technique for probing inter-membrane interactions. *Journal of General Physiology, 107*(3), 311–312.

73. Belmonte, A., & Libchaber, A. (1996). Thermal signature of plumes in turbulent convection: The skewness of the derivative. *Physical Review E, 53*(5), 4893–4898.
http://dx.doi.org/10.1103/PhysRevE.53.4893

74. Elbaum, M., Fygenson, D. K., & Libchaber, A. (1996). Buckling microtubules in vesicles. *Physical Review Letters, 76*(21), 4078–4081 (1996).
http://dx.doi.org/10.1103/PhysRevLett.76.4078

75. Faucheux, L., & Libchaber, A. (1995). Selection of Brownian particles. *Journal of the Chemical Society, Faraday Transactions, 91*, 3163–3166.
http://dx.doi.org/10.1039/FT9959103163

76. Faucheux, L. P., Bourdieu, L. S., Kaplan, P. D., & Libchaber, A. (1995). Optical thermal ratchet. *Physical Review Letters, 74*(9), 1504–1507.
http://dx.doi.org/10.1103/PhysRevLett.74.1504

77. Bourdieu, L., Magnasco, M., Winkelmann, D., & Libchaber, A. (1995). Actin filaments on myosin beds: The velocity distribution, *Physical Review E, 52*(6), 6573–6579.
http://dx.doi.org/10.1103/PhysRevE.52.6573

78. Bourdieu, L., Duke, T., Elowitz, M. B., Winkelmann, D. A., Leibler, S., & Libchaber, A. (1995). Spiral defects in motility assays: A measure of motor protein force. *Physical Review Letters, 75*(1), 176–179. http://dx.doi.org/10.1103/PhysRevLett.75.176

79. Winkelmann, D. A., Bourdieu, L., Ott, A., Kinose, F., & Libchaber, A. (1995). The flexibility of myosin attachment to surfaces influences F-actin motion. *Biophysical Journal, 68*(6), 2444–2453.

80. Faucheaux, L., Stolivitzky, G., & Libchaber, A. (1995). Periodic forcing of a Brownian particle. *Physical Review E, 51*(6), 5239–5250. http://dx.doi.org/10.1103/PhysRevE.51.5239

81. Kuchnir Fygenson, D., Flyvbjerg, H., Sneppen, K., Libchaber, A., & Leibler, S. (1995). Spontaneous nucleation of microtubules. *Physical Review E, 51*(5), 5058–5063.

82. Kaplan, P. D., Faucheux, L., & Libchaber, A. (1994). Direct observation of the entropic potential in a binary suspension. *Physical Review Letters, 73*(21), 2793–2796. http://dx.doi.org/10.1103/PhysRevLett.73.2793

83. Belmonte, A., Tilgner, A., & Libchaber, A. (1995). Turbulence and internal waves in side-heated convection. *Physical Review E, 51*(6), 5681–5687. http://dx.doi.org/10.1103/PhysRevE.51.5681

84. Faucheux, L., & Libchaber, A. (1994). Confined Brownian motion. *Physical Review E, 49*(6), 5158–5163. http://dx.doi.org/10.1103/PhysRevE.49.5158

85. Fygenson, D., Braun, E., & Libchaber, A. (1994). Phase diagram of microtubules. *Physical Review E, 50*(2), 1579–1588. http://dx.doi.org/10.1103/PhysRevE.50.1579

86. Belmonte, A., Tilgner, A., & Libchaber, A. (1994). Temperature and velocity boundary layers in turbulent convection. *Physical Review E, 50*(1), 269–279. http://dx.doi.org/10.1103/PhysRevE.50.269

87. Ott, A., Magnasco, M., Simon, A., & Libchaber, A. (1993). Measurement of the persistance length of filamentous actin using fluorescence microscopy. *Physical Review E, 48*(3), 1642–1645.

88. Braun, E., Faucheux, L., Libchaber, A., McLaughlin, D., Muraki, D., & Shelley, M. (1993). Filamentation and undulation of self focused laser beams in liquid crystals. *Europhysics Letters, 23*(4), 239. http://dx.doi.org/10.1209/0295-5075/23/4/001

89. Braun, E., Faucheux, L., & Libchaber, A. (1993). Strong self-focusing in nematic liquid crystals. *Physical Review A, 48*(1), 611–622. http://dx.doi.org/10.1103/PhysRevA.48.611

90. Moses, E., Zocchi, G., & Libchaber, A. (1993). A localized heat source: An experimental study of laminar plumes. *Journal of Fluid Mechanics, 251*, 581–601. http://dx.doi.org/10.1017/S0022112093003532

91. Belmonte, A., Tilgner, A., & Libchaber, A. (1993). Boundary layer length scales in thermal turbulence. *Physical Review Letters, 70*(26), 4067–4070. http://dx.doi.org/10.1103/PhysRevLett.70.4067

92. Tilgner, A., Belmonte, A., & Libchaber, A. (1993). Temperature and velocity profiles of turbulent convection in water. *Physical Review E, 47*(4), 2253–2257.

93. Ching, E., Kadanoff, L., Libchaber, A., & Wu, X. Z. (1992). Turbulent convection in helium gas. *Physica D, 58*, 414–422.

94. Simon, A. & Libchaber, A. (1992). Escape and synchronization of a Brownian particle. *Physical Review Letters, 68*(23), 3375–3378.

95. Smith, D., Wu, X. Z., Witten, T., & Libchaber, A. (1992). Viscous finger narrowing at the coil-stretch transition in a dilute polymer solution. *Physical Review A, 45*(4), 2165–2168.

96. Wu, X. Z., & Libchaber, A. (1992). Scaling relations in thermal turbulence: The aspect ratio dependence. *Physical Review A, 45*(2), 842–845.

97. Flesselles, J. M., Magnasco, M., & Libchaber, A. (1991). From discs to hexagons and beyond: A study in two dimensions. *Physical Review Letters, 67*(18), 2489–2492. http://dx.doi.org/10.1103/PhysRevLett.67.2489

98. Procaccia, I., Ching, E., Constantin, P., Kadanoff, L., Libchaber, A., & Wu, X. Z. (1991). Transitions in convective turbulence: The role of thermal plumes. *Physical Review A, 44*(12), 8091–8102. http://dx.doi.org/10.1103/PhysRevA.44.8091

99. Flesselles, J. M., Simon, A., & Libchaber, A. (1991). Dynamics of onedimensional interfaces: An experimentalist's view. *Advances in Physics, 40*(1), 1–51. http://dx.doi.org/10.1080/00018739100101462

100. Berge, B., Faucheux, L., Schwab, K., & Libchaber, A. (1991). Experimental evidence for faceted crystal growth in two dimensions. Nature, 350, 322.

101. Wu, X. Z., & Libchaber, A. (1991). Non-Boussinesq effects in free thermal convection. Physical Review A, 43(6), 2833–2839. http://dx.doi.org/10.1103/PhysRevA.43.2833

102. Moses, E., Zocchi, G., Procaccia., I., & Libchaber, A. The dynamics and interaction of laminar thermal plumes. *Europhysics Letters, 14*(1), 55. http://dx.doi.org/10.1209/0295-5075/14/1/010

103. Zocchi, G., Moses, E., & Libchaber, A. (1990). Coherent structures in turbulent convection, an experimental study. *Physica A, 166*(3), 387–407. http://dx.doi.org/10.1016/0378-4371(90)90064-Y

104. Berge, B., Simon, A., & Libchaber, A. (1990). Dynamics of gas bubbles in monolayer. *Physical Review A, 41*(12), 6893–6900. http://dx.doi.org/10.1103/PhysRevA.41.6893

105. Simon, A., & Libchaber, A. (1990). Moving interface: The stability tongue and phenomena within. *Physical Review A, 41*(12), 7090–7093. http://dx.doi.org/10.1103/PhysRevA.41.7090

106. Molho, P., Simon, A., & Libchaber, A. (1990). Peclet number and crystal growth in a channel. *Physical Review A, 42*(2), 904–910. http://dx.doi.org/10.1103/PhysRevA.42.904

107. Wu, X. Z., Kadanoff, L., Sano, M., & Libchaber, A. (1990). Frequency power spectrum of temperature fluctuations in free convection. *Physical Review Letters, 64*(18), 2140–2143. http://dx.doi.org/10.1103/PhysRevLett.64.2140

108. Sano, M., Wu, X. Z., & Libchaber, A. (1989). Turbulence in helium-gas free convection. *Physical Review A, 40*(11), 6421–6430. http://dx.doi.org/10.1103/PhysRevA.40.6421

109. Bechhoefer, J., Simon, A., Oswald, P., & Libchaber, A. (1989). Destabilization of a flat nematic-isotropic interface. *Physical Review A, 40*(4), 2042–2056. http://dx.doi.org/10.1103/PhysRevA.40.2042

110. Simon, A., Bechhoefer, J., & Libchaber, A. (1988). Solitary modes and the Eckhaus instability in directional solidification. *Physical Review Letters, 61*(22), 2574–2577. http://dx.doi.org/10.1103/PhysRevLett.61.2574

111. Wu, X. W., Castaing, B., Heslot, F., & Libchaber, A. (1988). Scaling properties of soft thermal turbulence in Rayleigh-Bénard convection. *Universalities in Condensed Matter, Springer Proceedings in Physics, 32*, 208–212. http://dx.doi.org/10.1007/978-3-642-51005-2_41

112. Gross, S., Zocchi, G., & Libchaber, A. (1988). Ondes et plumes de couche limite thermique. *Comptes Rendus de l'Académie des Sciences Paris, Série 2, 307*(5), 447–452.

113. Castaing, B., Gunaratne, G., Heslot, F., Kadanoff, L., Libchaber, A., Thomae, S., ... Zanetti, G. (1989). Scaling of hard thermal turbulence in Rayleigh–Bénard convection. *Journal of Fluid Mechanics, 204*, 1–30. http://dx.doi.org/10.1017/S0022112089001643

114. Glazier, J., & Libchaber, A. (1988). Quasi-periodicity and dynamical systems: An experimentalist's view. *IEEE Transactions on Circuits and Systems, 35*, 790–809.

115. Bechhoefer, J., Oswald, P., Germain, C., & Libchaber, A. (1988). Observations of cellular and dendritic growth of a smectic-B — Smectic-A interface. *Physical Review A, 37*(5), 1691–1696. http://dx.doi.org/10.1103/PhysRevA.37.1691

116. Glazier, J., Gunaratne, G., & Libchaber, A. (1988). $f(\alpha)$ curves: Experimental results. *Physical Review A, 37*(2), 523–530. http://dx.doi.org/10.1103/PhysRevA.37.523

117. Bechhoefer, J., Guido, H., & Libchaber, A. (1988). Solidification dans un Capillaire Rectangulaire. *Comptes Rendus de l'Académie des Sciences Paris, Série 2, 306*(10), 619–625.

118. Libchaber, A. (1987). From chaos to turbulence in Bénard convection. *Proceedings of the Royal Society of London, 413*(1844), 63–69.

119. Oswald, P., Bechhoefer, J., Lequeux, F., & Libchaber, A. (1987). Pattern formation behind a moving cholesteric — Smectic-A interface. *Physical Review A, 36*(12), 5832–5838. http://dx.doi.org/10.1103/PhysRevA.36.5832

120. Heslot, A., Castaing, B., & Libchaber, A. (1987). Transitions to turbulence in helium gas. *Physical Review A, 36*(12), 5870. http://dx.doi.org/10.1103/PhysRevA.36.5870

121. Zocchi, G., Shaw, B., Kadanoff, L., & Libchaber, A. (1987). Finger narrowing under local perturbations in the Saffman-Taylor problem. *Physical Review A, 36*(4), 1894–1900. (1987). http://dx.doi.org/10.1103/PhysRevA.36.1894

122. Oswald, P., Bechhoefer, J., & Libchaber, A. (1987). Instabilities of a moving nematic-isotropic interface. *Physical Review Letters, 58*(22), 2318–2321. http://dx.doi.org/10.1103/PhysRevLett.58.2318

123. Jensen, M. H., Pelce, P., Zocchi, G., & Libchaber, A. (1987). Effects of gravity on the Saffman-Taylor meniscus: Theory and experiment. *Physical Review A, 35*(5), 2221–2227. http://dx.doi.org/10.1103/PhysRevA.35.2221

124. Bechhoefer, J., & Libchaber, A. (1987). Testing shape selection in directional solidification. *Physical Review B, 35*(3), 1393–1396. http://dx.doi.org/10.1103/PhysRevB.35.1393

125. Tabeling, P., Zocchi, G., & Libchaber, A. (1987). An experimental study of the Saffman–Taylor instability. *Journal of Fluid Mechanics, 177*, 67–82. http://dx.doi.org/10.1017/S0022112087000867

126. Glazier, J.A., Jensen, M. H., Libchaber, A., & Stavans, J. (1986). The structure of Arnold tongues and the $f(\alpha)$ spectrum for period-doubling: Experimental results. *Physical Review A, 34*(2), 1621–1624. http://dx.doi.org/10.1103/PhysRevA.34.1621

127. Jensen, M. H., Kadanoff, L. P., Procaccia, I., Stavans, J., & Libchaber, A. (1985). Global universality at the onset of chaos: Results of a forced Rayleigh–Bénard experiment. *Physical Review Letters, 55*(25), 2798–2801.
     http://dx.doi.org/10.1103/PhysRevLett.55.2798

128. Tabeling, P., & Libchaber, A. (1986). Film draining and the Saffman–Taylor problem. Physical Review A, 33(1), 794–796.
     http://dx.doi.org/10.1103/PhysRevA.33.794

129. Stavans, J., Heslot, F., & Libchaber, A. (1985). Fixed winding number and the quasiperiodic route to chaos in a convective fluid. *Physical Review Letters, 55*(6), 596–599.
     http://dx.doi.org/10.1103/PhysRevLett.55.596

130. Heslot, F., & Libchaber, A. (1985). Unidirectional crystal growth and crystal anisotropy. *Physica Scripta, T9*, 126–129.
     http://dx.doi.org/10.1088/0031-8949/1985/T9/020

131. Fauve, S., Laroche, C., Libchaber, A., & Perrin, B. (1984). Chaotic phases and magnetic order in a convective fluid. *Physical Review Letters, 52*(20), 1774–1777.
     http://dx.doi.org/10.1103/PhysRevLett.52.1774

132. Fauve, S., Laroche, C., & Libchaber, A. (1984). Horizontal magnetic field and the oscillatory instability onset. *Journal de Physique Lettres, 45*(3), 101–105.
     http://dx.doi.org/10.1051/jphyslet:01984004503010100

133. Arneodo, A., Coullet, P., Tresser, C., Libchaber, A., Maurer, J., & d'Humieres, D. (1983). About the observation of an uncompleted cascade in a Rayleigh–Bénard experiment. Physica 6D, 6(3), 385–392.
     http://dx.doi.org/10.1016/0167-2789(83)90020-9

134. D'Humieres, D., Beasley, M. R., Huberman, B. A., & Libchaber, A. (1982). Chaotic states and routes to chaos in the forced pendulum. *Physical Review A, 26*(6), 3483–3496.
     http://dx.doi.org/10.1103/PhysRevA.26.3483

135. Libchaber, A., Fauve, S., & Laroche, C. (1983). Two-parameter study of the routes to chaos. *Physica D, 7*(1–3), 73–84.
     http://dx.doi.org/10.1016/0167-2789(83)90117-3

136. Libchaber, A., Laroche, C., & Fauve, S. (1982). Period doubling cascade in mercury, a quantitive measurement. *Journal de Physique Lettres, 43*(2), 211–216.
     http://dx.doi.org/10.1051/jphyslet:01982004307021100

137. Fauve, S., Laroche, C., & Libchaber, A. (1981). Effect of a horizontal magnetic field on convective instabilities in mercury. *Journal de Physique Lettres, 42*(21), 455–457.
     http://dx.doi.org/10.1051/jphyslet:019810042021045500

138. Maurer, J., & Libchaber, A. (1980). Effect of the Prandtl number on the onset of turbulence in liquid 4He. *Journal de Physique Lettres, 41*(21), 515–18.
     http://dx.doi.org/10.1051/jphyslet:019800041021051500

139. D'Humieres, D., Launay, A., & Libchaber, A. (1980). Pinning of vortices in nucleopores effect on second-sound resonators. *Journal of Low Temperature Physics, 38*(1–2), 207–222. http://dx.doi.org/10.1007/BF00115276

140. Libchaber, A., & Maurer, J. (1979). A reduced-geometry Rayleigh–Bénard experiment: Frequency multiplication, frequency pulling and frequency demultiplication. *Journal de Physique Colloques, 41*, 5.

141. Maurer, J., & Libchaber, A. (1979). Rayleigh–Bénard experiment in liquid helium: Frequency locking and the onset of turbulence. *Journal de Physique Lettres, 40*(16), 419–423. http://dx.doi.org/10.1051/jphyslet:019790040016041900

142. Balibar, S., Buechner, J., Castaing, B., Laroche, C., & Libchaber, A. (1978). Experiments on superfluid 4He evaporation. *Physical Review B, 18*(7), 3096–3104. http://dx.doi.org/10.1103/PhysRevB.18.3096

143. Libchaber, A., & Maurer, J. (1978). Local probe in a Rayleigh–Bénard experiment in liquid helium. *Journal de Physique Lettres, 39*(21), 369–372. http://dx.doi.org/10.1051/jphyslet:019780039021036900

144. Castaing, B., & Libchaber, A. (1978). Roton second sound and the roton-roton interaction potential. *Journal of Low Temperature Physics, 31*(5), 887–896. http://dx.doi.org/10.1007/BF00116056

145. Thome, H., Couder, Y., & Libchaber, A. (1978). Riedel anomaly and nonlinear effects in Josephson point contacts. *Journal of Applied Physics, 49*(3), 1200–1207. http://dx.doi.org/10.1063/1.325007

146. Libchaber, A., & Toulouse, G. (1976). The return of solitons. *La Recherche, 7*, 1027.

147. Balibar, S., Buechner, J., Castaing, B., Laroche, C., & Libchaber, A. (1977). Observation of a maximum roton velocity in superfluid 4He. *Physics Letters A, 60*(2), 135–136. http://dx.doi.org/10.1016/0375-9601(77)90406-6

148. Castaing, B., & Libchaber, A. (1975). Thermal wave in body centered cubic helium. *Journal de Physique Lettres, 36*, L309.

149. Laroche, C., Perrin, B., & Libchaber, A. (1975). Effect of a wire in front of a hole on superfluid helium flow. *Physics Letters A, 55*(1), 29–30. http://dx.doi.org/10.1016/0375-9601(75)90382-5

150. Guthmann, C., Maurer, J., Belin, M., Bok, J., & Libchaber, A. (1975). Dynamic behavior of superconducting microbridges. *Physical Review B, 11*(5), 1909–1913. http://dx.doi.org/10.1103/PhysRevB.11.1909

151. Hulin, J. P., d'Humieres, D., Perrin, B., & Libchaber, A. (1974). Critical velocities for superfluid-helium flow through a small hole. *Physical Review A, 9*(2), 885. http://dx.doi.org/10.1103/PhysRevA.9.885

152. Guthmann, C., d'Haenens, J. P., & Libchaber, A. (1973). Magnetoacoustic wave in an electron-hole gas — Bismuth. II. effect of anisotropy and Landau damping in a geometry where q is not perpendicular to B. *Physical Review B, 8*(2), 561–570.
http://dx.doi.org/10.1103/PhysRevB.8.561

153. Balibar, S., Perrin, B., & Libchaber, A. (1972). Electromagnetic waves and quantum effects in aluminum. *Journal of Physics F, 2*(4), 629.
http://dx.doi.org/10.1088/0305-4608/2/4/008

154. Perrin, B., d'Humieres, D., Laroche, C., Hulin, J. P., & Libchaber, A. (1972). Josephson-like phase-coherence oscillations in superfluid helium. *Physical Review Letters, 28*(24), 1551.
http://dx.doi.org/10.1103/PhysRevLett.28.1551

155. Hulin, J. P., Laroche, C., Perrin, B., & Libchaber, A. (1972). Analog of the d.c. Josephson effect in superfluid helium. Physical Review A, 5(4), 1830.
http://dx.doi.org/10.1103/PhysRevA.5.1830

156. Guthmann, C., d'Haenens, J. P., & Libchaber, A. (1971). Magnetoacoustic wave in an electron-hole gas — Bismuth. *Physical Review B, 4,* 1538.
http://dx.doi.org/10.1103/PhysRevB.4.1538

157. Libchaber, A., Adams, G., & Grimes, C. C. (1970). Radio-frequency size effect in potassium: Relation to cyclotron resonance, helicon waves and sound waves. *Physical Review B, 1*(2), 361–365.
http://dx.doi.org/10.1103/PhysRevB.1.361

158. Libchaber, A., & Grimes, C. C. (1969). Resonant damping of helicon waves in potassium. *Physical Review, 178,* 1145–1154.
http://dx.doi.org/10.1103/PhysRev.178.1145

159. D'Haenens, J. P., Libchaber, A., Laroche, C., & Le Hericy, J. (1968). Gantmakher effects at microwave frequencies. *Physics Letters A, 28*(5), 312–313.
http://dx.doi.org/10.1016/0375-9601(68)90304-6

160. G. Weisbuch and A. Libchaber. (1967). Helicon-Gantmakher wave interactions in copper. *Physical Review Letters, 19*(9), 498.
http://dx.doi.org/10.1103/PhysRevLett.19.498

161. Guthmann, C., & Libchaber, A. (1967). Alfven waves in bismuth at very high magnetic fields. *Comptes Rendus de l'Académie des Sciences B, 265,* 319–22.

162. Hansen, J. W., Grimes, C. C., & Libchaber, A. (1967). High frequency phase sensitive detection applied to helicon wave studies. *Review of Scientific Instruments, 38*(7), 895–897.
http://dx.doi.org/10.1063/1.1720917

163. Nanney, C. A., Libchaber, A., & Garno, J. P. (1966). Helicon drift current interaction in a layered semiconductor structure. *Applied Physics Letters, 9*(11), 395–397.
http://dx.doi.org/10.1063/1.1754626

164. Libchaber, A., & Veilex, R. (1962). Helicon wave propagation in indium antimonide at room temperature. *Proceedings of the International Conference on the Physics of Semiconductors, Exeter, July 1962*, 138–140.

165. Libchaber, A., & Veilex, R. (1962). Wave propagation in a gyromagnetic solid conductor: Helicon waves. *Physical Review Letters, 127*(3), 774–776. http://dx.doi.org/10.1103/PhysRev.127.774

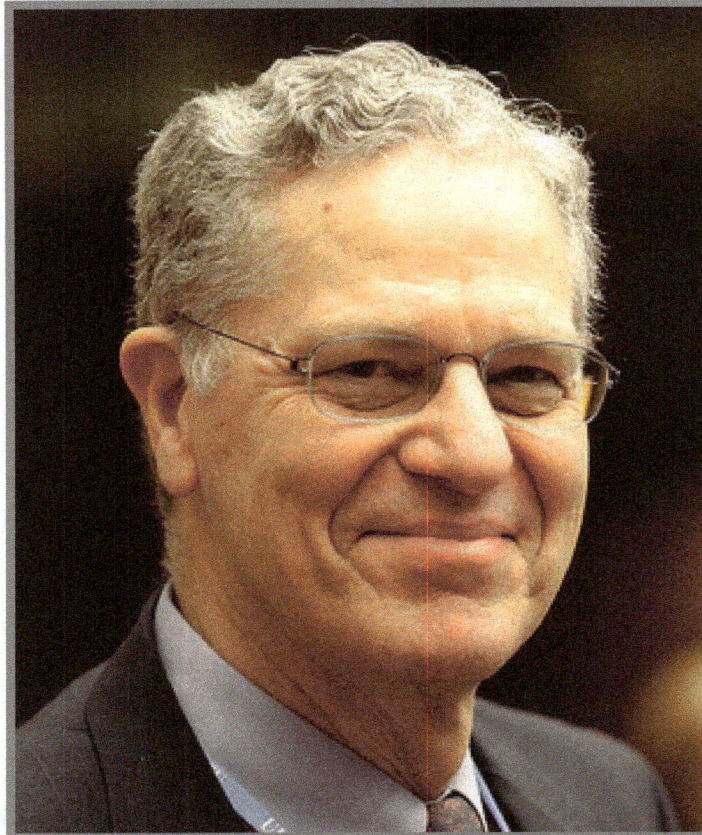

Joseph H. Taylor

JOSEPH HOOTON TAYLOR, JR.

Birthplace

Born	Philadelphia, 29 March 1941
Married	Marietta Bisson Taylor; four children

Education

1963	Haverford College B.A. (Physics, with Honors)
1968	Harvard University Ph.D. (Astronomy)

Research and Professional Experience

1968–1969	Research Fellow & Lecturer, Harvard University
1969–1972	Assistant Professor, University of Massachusetts
1976	Visiting Scientist, CSIRO (Australia)
1973–1977	Associate Professor, University of Massachusetts
1977–1981	Associate Director, Five College Radio Astron. Obs.
1977–1981	Professor of Astronomy, University of Massachusetts
1980–1982	Professor of Physics, Princeton University
1982–1986	Professor of Physics, Princeton University Eugene Higgins
1986–2006	Professor of Physics, Princeton University James S. McDonnell Distinguished
1997–2003	Dean of the Faculty, Princeton University
2006–	Professor of Physics, Emeritus, Princeton University James S. McDonnell Distinguished University

Research Specialization

Radio Astronomy, Pulsars, Experimental Gravitation

Membership of Societies

1981	National Academy of Sciences
1982	American Academy of Arts and Sciences
1992	American Philosophical Society

Honors and Awards

Bart J. Bok Prize, 1975; George Darwin Lecturer (Royal Astronomical Society), 1980; Dannie Heineman Prize (American Astronomical Society and American Institute of Physics), 1980; Chancellor's Medal, University of Massachusetts, 1980; MacArthur Foundation Prize Fellow, 1981–1986; Henry Draper Medal (National Academy of Sciences), 1985; D.Sc. (Honorary), University of Chicago, 1985; Fellow, American Physical Society, 1986; Tomalla Foundation Prize in Gravitation and Cosmology, 1987; The Magellanic Premium (American Philosophical Society), 1990; John J. Carty Award for the Advancement of Science (National Academy of Sciences), 1991; Einstein Prize Laureate (Albert Einstein Society, Bern) 1991; Wolf Prize in

Physics, 1992; Nobel Prize in Physics, 1993; D.Sc. (Honorary), University of Massachusetts, 1994; Docteur Honoris Causa (Universite de Montreal), 1995; John Scott Medal (City of Philadelphia), 1995; Karl Schwarzschild Medal, Astronomische Gessellschaft, 1997; Joseph Priestly Medal (Dickinson College), 2008; D.Sc. (Honorary), Princeton University, 2012.

Publications

Approximately 200 articles and book chapters in the professional research literature of physics and astronomy, and one book.

COMMENTARY

I was born on March 29, 1941, in Philadelphia, Pennsylvania, the second son of Joseph Hooton Taylor and Sylvia Evans Taylor. When I was seven, we moved back to the family farm in Cinnaminson Township, New Jersey, then operated by my paternal grandfather. We were four children, joined later by two more, plus two Evans cousins; like the farm's peaches and tomatoes, the eight of us grew and ripened in a healthy and carefree environment on the eastern bank of the Delaware River. Among my fondest boyhood memories are collecting stone arrowheads left on that land by its much earlier inhabitants, and erecting, together with my brother Hal, numerous large, rotating, ham-radio antennas, high above the roof of the three-story Victorian farmhouse. With one such project, we managed to shear off the brick chimney, flush with the roof, much to the consternation of our parents. That incident was one of many practical lessons of my youth, not all absorbed in the most timely fashion, involving ill-advised shortcuts toward some goal.

Both the Evans and Taylor families have deep Quaker roots going back to the days of William Penn and his Philadelphia experiment. My parents were living examples of frugal Quaker simplicity, twentieth-century style; their very lives taught lessons of tolerance for human diversity and the joys of helping and caring for others. Our house was large, open, and friendly. To my knowledge it has never been (nor indeed can be) locked. In our school years, Hal and I filled most of the third floor with working ham-radio transmitters and receivers. Our rigs were largely built from a mixture of post-war surplus equipment and junk television sets. We learned by experience that when you need high voltage, the power company's 6,000-to-120-volt transformers work admirably in reverse; and that most amplifiers will oscillate, especially if you don't want them to.

I was educated mostly at Quaker institutions, in particular, Moorestown Friends School and Haverford College. In school, mathematics was my first academic love. Somewhat backward high-school introductions to chemistry and physics (I failed to recognize them as such at the time) did not dampen any enthusiasm for science, they just gave me more time for sports, then a greater passion. Soccer, basketball, baseball, golf, and tennis claimed much of my energy through the Haverford years. Concurrently, though, I began discovering the delights of what science is really about. A fascinating senior honors project in physics allowed me to combine a working knowledge of radio-frequency electronics with an awakening appreciation of scientific inquiry, and to build a working radio telescope. My principal references were an old friend, *The Radio Amateur's Handbook*, and an early book on radio astronomy by Pawsey and Bracewell. This thoroughly enjoyable honors project cannot really qualify as research — everything I accomplished had been done by others, years before — but it provided excellent lessons in problem-solving of various kinds. It also delivered a valid reason for selecting something I had been hoping to find: a desirable field of physics in which to pursue graduate studies.

My academic work in the Harvard departments of Astronomy, Physics, and Applied Mathematics was the hardest I ever remember working, at least during my first year there. I suppose every beginning graduate student feels that he or she has something to prove; anyway, I certainly did. But my thesis research in radio astronomy was, once again, thoroughly enjoyable. My mentor, Alan Maxwell, knew the field and its participants well. He gave me plenty of flexibility, provided inroads and introductions when I needed them, and taught me (among many other things) the importance of clear, well-crafted writing in a scientific paper. Ron Bracewell again played an unwitting role; his 1965 book *The Fourier Transform and its Applications* came out just in time to give me some crucial insights necessary for analyzing the data for my thesis. It also prepared me for understanding the signal-processing techniques that later became important in my study of pulsars.

Following a post-doctoral year at Harvard I moved to the University of Massachussets, Amherst, teaching there in the Department of Physics and Astronomy and working with a group of young and enthusiastic colleagues to start a highly successful radio astronomy effort. My first Ph.D. student, Russell A. Hulse, was part of that group. Our discovery together of the first known binary pulsar led to my later work in high-precision pulsar timing, first at the University of Massachusetts and later at Princeton University, to which I moved in 1981. This pioneering work led to the Wolf Prize in 1991 and the Nobel Prize (together with Hulse) in 1993. I taught in the Princeton Physics Department for 25 years, serving also as Princeton's Dean of the Faculty from 1997 to 2003. I retired in 2006.

I have noticed in recent years that many budding scientists worry much more than I ever did about what the future may bring: how to get into the best university, work with the biggest names, find the best post-doctoral fellowship, and secure the ideal university position. My own psychological bent, insofar as it has influenced any professional decisions, is to pursue a path promising enjoyment along the way, without looking too far ahead. Perhaps related to my Quaker upbringing, I've always valued personal involvement in a di cult task over appeals to eminence or authority; I like the challenge of re-examining a problem from fresh perspectives. Ultimately, I believe that in important matters we are mostly self-taught, but in a way that is strongly reinforced by cooperative human relationships. I have worked in two extremely stimulating intellectual environments, first at the University of Massachusetts and more recently at Princeton. I'm fortunate to have associated with some uniquely gifted individuals who have been especially compatible co-seekers of diverse truths and pleasures: among them Dick Manchester, Russell Hulse, Joel Weisberg, Thibault Damour, gifted students too numerous to name, and especially my dearly beloved wife, Marietta Bisson Taylor.

THE ASTROPHYSICAL JOURNAL, **345**:434–450, 1989 October 1
© 1989. The American Astronomical Society. All rights reserved. Printed in U.S.A.

FURTHER EXPERIMENTAL TESTS OF RELATIVISTIC GRAVITY USING THE BINARY PULSAR PSR 1913+16

J. H. TAYLOR
Joseph Henry Laboratories and Physics Department, Princeton University

AND

J. M. WEISBERG
Physics and Astronomy Department, Carleton College
Received 1989 February 13; accepted 1989 March 24

ABSTRACT

Pulse time-of-arrival observations of the binary pulsar PSR 1913+16 now extend over approximately 14 years. The data are consistent with a straightforward model allowing for the motion of the Earth, special and general relativistic effects within the solar system, dispersive propagation in the interstellar medium, relativistic motion of the pulsar in its orbit, and deterministic spin-down behavior of the pulsar itself. The results show that at the present level of precision, the PSR 1913+16 system can be modeled dynamically as a pair of orbiting point masses. A total of five Keplerian and five "post-Keplerian" orbital parameters can now be determined, most of them with remarkably high precision. The masses of the pulsar and its companion are determined (within general relativity) to be $m_1 = 1.442 \pm 0.003$ and $m_2 = 1.386 \pm 0.003$ times the mass of the Sun, respectively, and the orbit is found to be decaying at a rate equal to 1.01 ± 0.01 times the general relativistic prediction for gravitational radiation damping.

Our results represent the first experimental tests of gravitation theory not restricted to the weak-field, slow-motion limit in which nonlinearities and radiation effects are negligible. The excellent agreement of observation with theory shows conclusively that gravitational radiation exists, at the level predicted by general relativity. We also use the results to calculate improved upper limits on the rate of change of the Newtonian gravitational constant, and the fractional energy density (relative to closure density) of a cosmic background of ultra–low-frequency gravitational radiation. These limits are, respectively, $\dot{G}/G = (1.2 \pm 1.3) \times 10^{-11}$ yr^{-1}, and $\Omega_g < 0.04$ at frequencies 10^{-9} to 10^{-12} Hz.

Subject headings: gravitation — pulsars — radiation mechanisms — relativity — stars: binaries

I. INTRODUCTION

The 8 hour binary pulsar PSR 1913+16 has proved to be an outstanding relativity laboratory, perhaps even surpassing the high expectations held shortly after its discovery (Hulse and Taylor 1975). Observations of its pulse arrival times have provided the data necessary (1) to specify the orbital elements and masses of both the pulsar and the companion star; (2) to determine that the orbit is decaying at precisely the rate expected from the emission of gravitational radiation; (3) to rule out several theories of gravitation; and (4), for the first time, to probe experimentally some details of the solution to the general relativistic two-body problem (Taylor *et al.* 1976; Taylor, Fowler, and McCulloch 1979; Taylor and Weisberg 1982; Weisberg and Taylor 1981, 1984). Interesting applications of the data are still being found: some recent published results include the determination of a new upper limit to $\dot{G}/G$ (the fractional rate of change of the Newtonian gravitational constant; Damour, Gibbons, and Taylor 1988) and the first plausible evidence for gravito-magnetic effects, in the form of geodetic precession of the PSR 1913+16 spin axis (Weisberg, Romani, and Taylor 1989).

The pulsar has now been observed regularly for 14 years, with steadily improving data acquisition equipment and techniques. Parameters describing the system have been derived from least-squares fits of arrival-time data to timing models of increasing sophistication and accuracy (Blandford and Teukolsky 1976; Epstein 1977; Haugan 1985, 1988; Damour and Deruelle 1986). As the time span of available data has

lengthened and the quality of the data has improved, we have been able to tighten significantly the constraints on the measured physical parameters. In this paper we provide new determinations of the pulsar and orbital parameters, together with a thorough discussion of errors in the data and their effect on the estimated parameter values.

The plan of our paper is as follows. In § II we describe the techniques used to measure pulse arrival times, along with some details of the data acquisition hardware used for this task since 1974. Section III summarizes the development of the detailed relativistic models used to analyze the data, and § IV describes our implementation of these models in software and presents the results obtained for parameters of the models. Consequences of the parameter values are discussed in § V from a point of view free of any assumptions about the strong-field nature of gravity. Then, in § VI, we discuss the implications in the more restricted circumstance in which general relativity is assumed correct. Section VII summarizes our conclusions and outlines the prospects for further progress.

II. MEASURING TOPOCENTRIC PULSE ARRIVAL TIMES

PSR 1913+16 is a pulsar with short period ($P = 59$ ms), large dispersion measure (DM = 169 cm^{-3} pc), and low flux density ($S = 0.7$ mJy at 1400 MHz). It moves in a binary orbit of short period ($P_b = 7.75$ hr) and high eccentricity ($e = 0.617$), with a maximum speed around 400 km s^{-1}. All of these characteristics make it an unusually difficult pulsar to observe. With a very few exceptions (for example, Lyne and Ritchings

434

1977; Backer *et al.* 1985), useful observations of PSR 1913+16 have been made only with the benefit of the uniquely high sensitivity of the 305 m telescope of the Arecibo Observatory.[1]

Our observations consist of more than 4480 measurements of pulse arrival times at Arecibo between 1974 September and 1988 July. A consistent procedure has been common to nearly all of the measurements. On a given day, the pulsar is observed for the full tracking range of the telescope, about 2.5 hr. The periodic pulsar waveform is averaged for intervals of about 5 minutes, and the resulting integrated profiles are recorded along with time tags corresponding to the first digitized bin of the average profile and occurring near the midpoint of the integration.

Other details of the observations have varied widely over the 14 years, and a total of 13 distinct combinations of data acquisition equipment have been used. The changes were made in order to take advantage of new feed antennas, new receivers, and enhanced signal processing hardware as they were developed. Most changes occurred during the first 6 years of the experiment, during an era of extensive upgrading of the Arecibo telescope's capabilities. Since 1981, we have taken particular care to minimize changes, and to calibrate the resulting instrumental offsets when changes have been deemed desirable. Most of the timing measurements of PSR 1913+16, and indeed all of those with uncertainties less than 40 μs, have been made since 1981 February.

a) Data Acquisition Systems

Important specifications of the first 11 observing systems, including frequency, bandwidth, effective time resolution, and typical measurement uncertainty, were presented and discussed in an earlier paper (Taylor and Weisberg 1982, hereafter TW). The information is summarized and brought up to date in Table 1. The highest quality data discussed in TW are those recorded with "observing system K," which remained in service until late 1984 and has become known as our "Mark I" observing system. It took advantage of a pair of $2 \times 32 \times 250$

---

[1] The Arecibo Observatory is part of the National Astronomy and Ionosphere Center, operated by Cornell University under contract with the National Science Foundation.

kHz filter-bank spectrometers, the detected outputs of which were sent to a pair of 32 channel de-dispersing circuits (Orsten 1970; Boriakoff 1973). The two de-dispersed output signals, each corresponding to the total power in summed orthogonal polarizations from an 8 MHz bandwidth, were routed to a dual 1024 bin digital signal averager clocked in synchronism with the topocentric pulsar period.

As shown in Table 1, a Mark II system was brought into service in 1984 October. This equipment was designed for optimum performance in timing millisecond pulsars (Rawley 1986; Rawley *et al.* 1987; Rawley, Taylor, and Davis 1988). It makes use of one of the $2 \times 32 \times 250$ kHz filter-bank spectrometers, followed by a bank of 32 concurrently operating signal averagers. The total accepted bandwidth is 8 MHz, only half of that used by the Mark I system. Dispersion delays are removed in software before the 32 single-channel profiles are combined to form a grand average for each 5 minute integration.

In 1988 July we began using a Mark III system which combines some of the advantages of both the Mark I and Mark II approaches. Its signal-averaging hardware is multiplexed to allow concurrent averaging of up to 32 signal channels; its speed is adequate for use with the fastest millisecond pulsars; and, with the help of a new $2 \times 32 \times 1.25$ MHz filter bank, it utilizes the full 40 MHz bandpass of the dual-polarization 22 cm feed of the Arecibo telescope. A disadvantage of the present version of the Mark III system is that its frequency resolution is a relatively poor 1.25 MHz per channel. Consequently, the uncorrectable dispersion smearing is a rather large 630 μs.

In each of the 13 observing systems, the detected pulsar signal was sampled continuously at a constant (or nearly constant) rate, and averaged for 5 minutes in synchronism with the apparent pulsar period. In some of the earliest observations, before the PSR 1913+16 parameters were known well enough to allow an adequate ephemeris to be computed in advance of each observing session, raw data samples were recorded on magnetic tape and the Doppler shifting and synchronous averaging accomplished after the observations were completed. Since 1978.4, the average pulsar waveforms have been accumulated in real time, with a computerized ephemeris and associated hardware adjusting the sampling rate to allow

TABLE 1

OBSERVING SYSTEMS USED AND SUMMARY OF AVAILABLE DATA

Dates	Frequency (MHz)	Total Bandwidth (MHz)	Frequency Channels	System Noise Temperature (K)	Time Resolution (μs)	TOA Uncertainty (μs)	Number of Observations
A. 1974 Sep–Dec[a]	430	8.0	32	175	5000	275	524
B. 1975 Apr–1976 Nov[b]	430	0.64, 3.2	32	175	2000	310	112
C. 1975 Jun–1976 Feb	430	0.25	1	175	2000	890	75
D. 1976 Nov–Dec[a,b]	430	0.64	32	175	750	155	73
E. 1977 Jul–Aug[b]	430	0.64	32	175	340	150	52
F. 1978 Jun–1981 Feb	430	3.34	504	175	43	75	573
G. 1977 Jul–Aug	1410	8.0	32	80	125	75	57
H. 1977 Dec	1410	8.0	32	80	125	55	72
I. 1978 Mar–Apr[a]	1410	8.0	32	80	125	50	116
J. 1980 Jul–1981 Feb	1410	8.0	32	80, 40	200	85	312
Mark I. 1981 Feb–1984 Dec	1410	16.0	64	40	125	20	1719
Mark II. 1984 Oct–1988 Jul	1408	8.0	32	40	125	31	638
Mark III. 1988 Jul–	1404	40.0	32	40	640	16	159

[a] Raw data samples were recorded on magnetic tape, with signal averaging done afterward in software. All other observations used real-time signal averaging, synchronized by means of a precomputed ephemeris.
[b] Some or all of these observations were made with only one polarization.

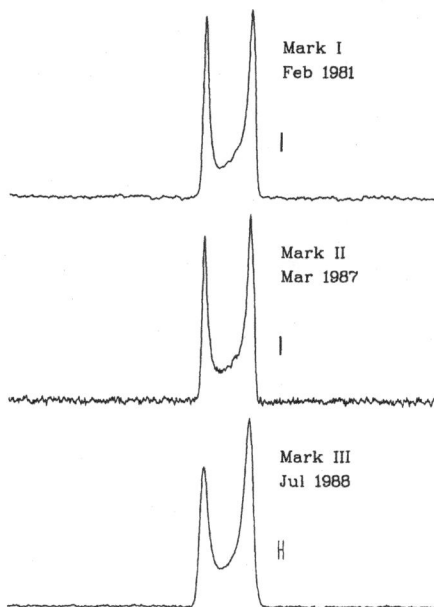

Fig. 1.—Average profiles of PSR 1913+16 as observed with the Mark I, Mark II, and Mark III data acquisition systems at frequencies near 1408 MHz. The effective time resolutions, which are dominated by dispersion smearing, are indicated by bars to the right of each pulse. The full period (59.03 ms) is plotted, and the gradual weakening of component 1 relative to component 2 is a real effect (Weisberg, Romani, and Taylor 1989).

for the small but significant changes in pulsar period during an integration.

Some representative samples of pulse profiles obtained with the pre-1981 observing systems were shown in Figure 1 of TW. These plots illustrated the improvement in time resolution obtained over the first 7 years, as well as the inevitable trade-offs between resolution and signal-to-noise ratio and the significant dependence of pulse shape on frequency. Figure 1 of the present paper shows profiles obtained with the Mark I, Mark II, and Mark III systems at frequencies near 1408 MHz. Each profile is the average of several hundred 5 minute integrations.

b) Determining Times of Arrival and Their Uncertainties

An effective pulse time of arrival (TOA) was calculated for each 5 minute average profile in the following fashion. The observed profile, typically consisting of 1024 numbers representing relative flux density throughout a full pulsar rotation period, was matched with a high signal-to-noise "standard profile" in order to determine the exact phase of the pulsar waveform. This "template matching" was done by least-squares fitting, either in the time domain or in the Fourier transform domain. In either case, a χ^2 "goodness-of-fit" statistic was minimized with respect to three parameters: the level of the baseline, the pulsar signal strength, and the phase offset between observed and standard profiles.

The time delay corresponding to the measured phase was added to the time corresponding to the first bin of the observed profile. In addition, for data acquired since late 1982, a correction (typically amounting to a few microseconds) was applied

to the time reckoned by the observatory's master clock, so that the TOA was ultimately referred to Coordinated Universal Time as kept by the US Naval Observatory or the National Institute of Standards and Technology[2] (Davis *et al.* 1985; Rawley *et al.* 1987). The resulting topocentric TOAs, which consitute our basic data set for all further analysis, are specified in units of proper atomic time on Earth, or terrestrial dynamical time (TDT).

We estimate an uncertainty for each TOA as part of the template-matching procedure. Typical values range from a maximum of nearly 1 ms for a small fraction of the early data to 20 μs or less for most of the measurements made since 1981. Because the estimated uncertainties of individual TOAs are themselves subject to significant errors, we improve statistical reliability by using daily average values of uncertainties scaled by the known dependence of telescope gain and system noise temperature on zenith angle.

III. DEVELOPMENT OF THE TIMING MODEL

The phase of a pulsar waveform at a particular time of observation depends on a number of factors, including (1) the spin-down behavior of the pulsar, (2) dispersive delays in the interstellar medium, (3) motion of the Earth within the solar system, (4) orbital motion of the pulsar, if any, and (5) relevant instrumental delays. Analysis of pulsar timing data requires a detailed model of these effects in order to predict the pulsar phases. Differences between predicted and observed values can then be used to determine corrections to various assumed model parameters, as described in § IV.

a) Single Pulsars

An adequate timing model for single pulsars has been developed in stages of increasing accuracy, beginning soon after pulsars were discovered (see, for example, Hunt 1971; Manchester and Peters 1972; Manchester, Taylor, and Van 1974; Manchester and Taylor 1977). Early observations required accuracies only at the 100 μs level; however, the discovery of millisecond pulsars in 1982 made it essential to achieve accuracy in the submicrosecond range. Our current model, embodies in a FORTRAN program called TEMPO, which has been continually evolving since 1972, meets this goal—as proved by extensive tests with 6 years of data from the millisecond pulsar PSR 1937+21. Some details of the model, particularly those not relevant to the orbital motion of binary pulsars, were presented in a paper by Rawley, Taylor, and Davis (1988).

Our timing model for single pulsars is concisely specified in an equation relating the topocentric pulse arrival times t to the corresponding infinite-frequency relativistic coordinate times t_b of pulse arrival at the solar system barycenter (SSB):

$$t_b = t - D/f^2 + (\mathbf{r} \cdot \hat{\mathbf{n}})/c + \Delta_{E\odot} - \Delta_{S\odot} . \quad (1)$$

In this equation t is measured in the topocentric TDT system, while t_b is in units of barycentric dynamical time (TDB). The observing frequency f is expressed in the rest frame of the SSB; the dispersion constant D, measured in hertz, is conventionally related to the commonly quoted dispersion measure by $D = \mathrm{DM}/(2.41 \times 10^{-16})$. In the remaining terms of equation (1), $\mathbf{r}$ is a vector from the SSB to the phase center of the telescope at the time of observation, $\hat{\mathbf{n}}$ is a unit vector toward the pulsar, c is the speed of light, $\Delta_{E\odot}$ is the solar system "Einstein delay,"

---

[2] Formerly the National Bureau of Standards.

or combined effect of gravitational redshift and time dilation due to motions of the Earth and other bodies, and $\Delta_{S\odot}$ is the "Shapiro delay" caused by propagation of the pulsar signal through curved spacetime in the solar system.

The Einstein delay $\Delta_{E\odot}$ amounts to an integral of the expression

$$\frac{d\Delta_{E\odot}}{dt} = \sum_i \frac{Gm_i}{c^2 r_i} + \frac{v_\oplus^2}{2c^2} - \text{constant} . \quad (2)$$

The sum is taken over all significant masses m_i in the solar system, excluding the Earth; r_i is the distance of m_i from Earth; $v_\oplus$ is the velocity of Earth with respect to the SSB; and the additive constant is chosen so that the average value of the right-hand side over a long time interval is equal to zero (for a related discussion see Backer and Hellings 1986).

The Shapiro delay $\Delta_{S\odot}$ is the time-delay analog of the well-known bending of light at the limb of the Sun (Shapiro 1964). Its magnitude is readily computed from the relation

$$\Delta_{S\odot} = -\frac{2GM_\odot}{c^3} \log (1 + \cos \theta) , \quad (3)$$

where $M_\odot$ is the mass of the Sun and θ is the pulsar-Sun-Earth angle at the time of observation; we have neglected the eccentricity of the Earth's orbit. We note in passing that the product $GM_\odot$ is known with higher precision than either G or $M_\odot$ separately; in equation (3) and elsewhere throughout this paper we adopt the constants used in the JPL ephemeris, $GM_\odot = 1.3271243999 \times 10^{26}$ cm^3 s^{-2}, $GM_\odot/c^3 = 4.925490947$ μs, and $c = 2.99792458 \times 10^{10}$ cm s^{-1}.

Accelerations of nonbinary pulsars relative to the SSB are small enough that barycentric dynamical times t_b are equivalent to proper times T in the rest frame of the pulsar, up to an additive constant and a nearly constant Doppler and gravitational redshift scale factor. Therefore, if a pulsar's rotational energy is gradually being dissipated by some deterministic damping mechanism, and the phase is further affected by a stochastic term $\epsilon(T)$, the phase at proper time $T = t_b - t_0$ is given by

$$\phi(T) = \nu T + \tfrac{1}{2}\dot{\nu}T^2 + \tfrac{1}{6}\ddot{\nu}T^3 + \cdots + \epsilon(T) . \quad (4)$$

In this expression $\nu = 1/P$ is the spin frequency, and dots over symbols represent derivatives with respect to time. With a suitable choice of the reference epoch t_0, the calculated phase for a barycentric arrival time $t_b = t_0$ will vanish, and thus t_0 represents a nominal infinite-frequency pulse arrival time at the SSB.

All reasonable dissipation mechanisms yield braking torques proportional to a moderate power of ν, which guarantees that magnitudes of the frequency derivatives decrease rapidly after the first, with $\dot{\nu}/\nu \approx \ddot{\nu}/\dot{\nu}$, and so forth. For PSR 1913+16, the frequency and first derivative are $\nu = 16.9$ s^{-1} and $\dot{\nu} = -2.5 \times 10^{-15}$ s^{-2}, so the expected second derivative is $\ddot{\nu} = 3 \times 10^{-31}$ s^{-3}. Such a small value would be unobservably small over our data span. Moreover, experience has shown (Cordes and Downs 1985) that pulsars with small $\dot{\nu}/\nu$ usually have little or no observable "timing noise"—that is, $\epsilon(T)$ is also very small. Thus there is good reason to expect only the first two terms on the right-hand side of equation (4) to be significant for PSR 1913+16.

b) Blandford-Teukolsky

Orbital motion of a pulsar involves significant accelerations. Consequently, the analysis of binary pulsar timing data

requires an additional transformation to obtain the proper time T corresponding to each TOA. Blandford and Teukolsky (1976, hereafter BT) derived the first useful formulae for this purpose, soon after the discovery of PSR 1913+16. Their model assumes the orbit to be a slowly precessing Keplerian ellipse in a plane inclined at angle i to the plane of the sky. The pulsar and its companion are assumed to obey essentially Newtonian dynamical laws. The largest short-period relativistic effects—gravitational redshift and time dilation, fully analogous to the Einstein delays within the solar system already discussed—are calculated separately and patched into the model afterward. Any additional effects are accommodated in a phenomenological manner, by allowing for nonzero time derivatives of the orbital elements. Thus, the model is formulated in a way that is free of assumptions about the correct relativistic theory of gravity, and it can be used to detect and measure effects that were not explicitly incorporated at the outset.

In the BT model, the transformation from barycentric arrival time t_b to pulsar proper time T is defined by

$$t_b - t_0 = T + \{x \sin \omega(\cos E - e)$$
$$+ [x \cos \omega(1 - e^2)^{1/2} + \gamma] \sin E\}$$
$$\times \left\{1 - \frac{2\pi}{P_b} [x \cos \omega(1 - e^2)^{1/2} \cos E - x \sin \omega \sin E]\right.$$
$$\left. \times (1 - e \cos E)^{-1}\right\} , \quad (5)$$

where P_b, e, and ω are the binary orbital period, orbital eccentricity, and longitude of periastron; $x \equiv (a_1 \sin i)/c$ is the projected semimajor axis of the pulsar orbit in time units; γ measures the combined effect of gravitational redshift and time dilation; and the eccentric anomaly E is defined by Kepler's equation,

$$E - e \sin E = \frac{2\pi}{P_b} (t_b - T_0) , \quad (6)$$

in which T_0 is a reference time of periastron passage, measured in the TDB system.

Precession of the longitude of periastron is accommodated in the model by setting $\omega = \omega_0 + \dot{\omega}(t_b - T_0)$ in equation (5), and orbital changes caused by gravitational radiation or any other mechanism can be detected and measured by similarly allowing for time derivatives $\dot{P}_b$, $\dot{x}$, and $\dot{e}$. Blandford and Teukolsky (1976) argued that a quasi-Newtonian phenomenological model was an adequate approximation for treating the PSR 1913+16 data available in 1976, and this assertion was borne out in practice (Taylor et al. 1976). It is no longer true, however, as shown in § IV, so a better model has become essential for extracting the maximum possible information from the data.

To facilitate discussion in the remainder of § III, we present in Table 2 a list of parameters used in the BT model and in the others we are about to describe. Each model includes right ascension α, declination δ, proper-motion components μ_α and μ_δ, a reference epoch t_0, and the pulsar frequency ν and its derivatives $\dot{\nu}$ and $\ddot{\nu}$. In addition, the models for binary pulsars include the five Keplerian parameters, x, e, T_0, P_b, and ω_0, plus at least two "post-Keplerian" (PK) parameters. Seven orbital parameters are sufficient to specify a binary system completely, up to arbitrary rotations about the line of sight (see, for

TABLE 2

PARAMETERS OF THE TIMING MODELS

Parameter	Single Pulsar	BT Model	EH Model	DD Model	DDGR Model
α	Y	Y	Y	Y	Y
δ	Y	Y	Y	Y	Y
μ_α	Y	Y	Y	Y	Y
μ_δ	Y	Y	Y	Y	Y
t_0	Y	Y	Y	Y	Y
$v = 1/P$	Y	Y	Y	Y	Y
$\dot{v}$	Y	Y	Y	Y	Y
$\ddot{v}$	Y	Y	Y	Y	Y
$x = (a_1 \sin i)/c$	...	Y	Y	Y	Y
e	...	Y	Y	Y	Y
T_0	...	Y	Y	Y	Y
P_b	...	Y	Y	Y	Y
ω_0	...	Y	Y	Y	Y
$\dot{\omega} = 2\pi k/P_b$	...	Y	Y	Y	...
γ	...	Y	Y	Y	...
$\dot{P}_b$	...	Y	Y	Y	...
$s = \sin i$	...	...	Y	Y	...
r	...	...	...	Y	...
$M = m_1 + m_2$	...	...	...	...	Y
m_2	...	...	...	...	Y
$\dot{x}$	...	Y	Y	Y	...
$\dot{e}$	...	Y	Y	Y	...

NOTE.—"Y" denotes inclusion of the parameter in the model.

example, Smarr and Blandford 1976). Additional measurables can therefore yield tests of the nature of gravity or of the "cleanliness" of the binary system. In the BT model, the most significant PK parameters are $\dot{\omega}$, γ, and $\dot{P}_b$, the latter yielding a quantitative test for the existence of gravitational radiation. Secular derivatives $\dot{x}$ and $\dot{e}$ may also be included if desired.

c) Epstein-Haugan

More precise observations demand a more accurate timing model, and Epstein (1977, 1979) developed an improved formula that includes the Shapiro time delay in the binary system and the leading short-term periodic terms in the orbital motion. Instead of adding relativistic corrections and phenomenological time derivatives to an otherwise Newtonian treatment of orbital motion, Epstein used the two-body solution found by Wagoner and Will (1976) for the Einstein-Infeld-Hoffmann equation of motion in general relativity. Since the solution closely approximates a precessing ellipse, a description can be made in terms of an osculating orbit whose most significant parameters have the same names as their Keplerian counterparts. Epstein's (1977) equations (A22) and (A16), which we will not reproduce here, define the transformation from t_0 to T in analogy with equations (5) and (6) for the BT model. As shown in Table 2, the gravitational redshift and time dilation parameter γ, and the phenomenological time derivatives, $\dot{\omega}$, $\dot{P}_b$, etc., are also included as in the BT model. An additional free parameter, the orbital projection factor $\sin i$, is added to the model to quantify the post-Newtonian effects.

By 1984, having acquired several years of data with accuracies in the 20 μs range, we discovered that fitting them to the Epstein model produced recognizably nonrandom residuals. At about the same time, Haugan (1985) derived and integrated post-Newtonian equations of motion for a two-body system and compared his results with those of Epstein. This process revealed a flaw in the Epstein model, amounting to the fact that in an eccentric orbit the rate of periastron advance is not constant but rather is modulated by an orbital phase-dependent term. More specifically, the longitude of periastron

advances in proportion to the true anomaly rather than the mean anomaly. Haugan derived a corrected version of Epstein's equation (A22) which we used to carry out a satisfactory solution yielding the first direct measurement of $\sin i$ (Weisberg and Taylor 1984).

More recently, Haugan (1988, hereafter H88) has recognized (and we have independently proved) that the parameterization used in Epstein (1977) and Haugan (1985) yields a measurement of "$\sin i$" that is principally a test for the presence of the expected periodic modulation of the rate of apsidal advance (which is identical in form, whether caused by relativistic or Newtonian effects) rather than of relativistic orbital-shape and Shapiro-delay effects (for further discussion see Damour and Taylor 1989). Haugan's 1988 paper contains a suggested re-parameterization which separates these effects cleanly, and we illustrate the effect of these changes in § IVc.

d) Damour-Deruelle

Damour and Deruelle (1986) called attention to another distinguishing aspect of the original Epstein-Haugan (EH) model: it departs in a fundamental way from the theory-independent phenomenological approach of BT. Under the assumption that general relativity correctly describes gravitation in the nonlinear (strong-field) regime, the EH model yields a meaningful estimate of a fourth PK parameter, $\sin i$—and thus a consistency check of the cleanliness of the binary pulsar system as well as a test for gravitational radiation (Weisberg and Taylor 1984). However, the EH parameterization folds several independent effects, the largest of which is not necessarily even relativistic, into the single parameter $\sin i$. If general relativity is not valid under strong-field conditions, the meaning of $\sin i$ measured by the EH prescription is unclear. Since no other experimental tests of gravity in strong-field conditions exist, Damour and Deruelle argue persuasively that a more theory-independent approach to the data analysis is highly desirable (see also Damour 1988).

With these ideas in mind, they devised an elegant new method for solving the relativistic two-body problem to post-Newtonian order (Damour and Deruelle 1985). From the solution they derive a new timing formula for binary pulsars (Damour and Deruelle 1986, hereafter DD). This model is valid under very general assumptions about the nature of gravity in strong-field conditions (Damour 1988). The essential transformation relating solar system barycentric time t_b to pulsar proper time T is summarized by the expression

$$t_b - t_0 = T + \Delta_R + \Delta_E + \Delta_S + \Delta_A \, . \qquad (7)$$

Here Δ_R, the "Roemer time delay," is the propagation time across the binary orbit, analogous to the solar system term $(\mathbf{r} \cdot \hat{\mathbf{n}})/c$ in equation (1); Δ_E and Δ_S are the orbital Einstein and Shapiro delays, analogous to $\Delta_{E\odot}$ and $\Delta_{S\odot}$ in the solar system; and Δ_A is a time delay associated with aberration caused by rotation of the pulsar. The Δ's are defined by

$$\Delta_R = x \sin \omega[\cos u - e(1 + \delta_r)]$$
$$+ x[1 - e^2(1 + \delta_\theta)^2]^{1/2} \cos \omega \sin u \, , \qquad (8)$$

$$\Delta_E = \gamma \sin u \, , \qquad (9)$$

$$\Delta_S = -2r \log \{1 - e \cos u - s[\sin \omega(\cos u - e)$$
$$+ (1 - e^2)^{1/2} \cos \omega \sin u]\} \, , \qquad (10)$$

$$\Delta_A = A\{\sin [\omega + A_e(u)] + e \sin \omega\}$$
$$+ B\{\cos [\omega + A_e(u)] + e \cos \omega\} \, , \qquad (11)$$

for which one also needs Kepler's equation, the relation between eccentric anomaly u and true anomaly $A_e(u)$, and the time dependence of ω:

$$u - e \sin u = 2\pi \left[\left(\frac{T - T_0}{P_b} \right) - \frac{\dot{P}_b}{2} \left(\frac{T - T_0}{P_b} \right)^2 \right], \quad (12)$$

$$A_e(u) = 2 \arctan \left[\left(\frac{1 + e}{1 - e} \right)^{1/2} \tan \frac{u}{2} \right], \quad (13)$$

$$\omega = \omega_0 + k A_e(u) . \quad (14)$$

As shown in Table 2, the PK parameters of the DD model include $\dot{\omega} \equiv 2\pi k / P_b$, γ, and $\dot{P}_b$, which have essentially the same meanings as in the BT and EH models.[3] In addition, the equation for Δ_S contains two measurable parameters, $s \equiv \sin i$ and r, which characterize the "shape" and "range" of the Shapiro delay. In principle, the small quantities A, B, δ_r, and δ_θ might also be measured. However, as DD point out, these quantities are nearly degenerate with other parameters, and prospects for their measurements appear poor over any reasonable time scale. Their effects are discussed further in the Appendix.

e) Damour-Deruelle (General Relativity)

For the sake of completeness, we present one final relativistic timing model for binary pulsars, which we call the DDGR model (Taylor 1987). It amounts to a variation of the DD model in which, contrary to their theory-independent approach, general relativity is explicitly assumed to be the correct theory of gravity. The minimum possible parameter set is used, with two PK parameters—and no ad hoc parameters or extra secular derivatives are introduced. The chosen PK parameters are the total system mass, $M \equiv m_1 + m_2$, and the companion star's mass, m_2 (m_1 is the mass of the pulsar). Equations (8)–(14) remain applicable, but the variables k, γ, $\dot{P}_b$, s, r, A, B, δ_r, and δ_θ, instead of being free (or potentially free) parameters, are determined by the following relations:

$$\frac{a_R^3}{P_b^2} = \frac{GM}{4\pi^2} \left[1 + \left(\frac{m_1 m_2}{M^2} - 9 \right) \frac{GM}{2a_R c^2} \right]^2, \quad (15)$$

$$k = \frac{3GM}{c^2 a_R (1 - e^2)}, \quad (16)$$

$$\gamma = \frac{e P_b G m_2 (m_1 + 2m_2)}{2\pi c^2 a_R M}, \quad (17)$$

$$\dot{P}_b = -\frac{192\pi}{5c^5} \left(\frac{2\pi G}{P_b} \right)^{5/3} f(e) m_1 m_2 M^{-1/3}, \quad (18)$$

$$f(e) = (1 + \tfrac{73}{24} e^2 + \tfrac{37}{96} e^4)(1 - e^2)^{-7/2}, \quad (19)$$

$$s = \frac{cxM}{a_R m_2}, \quad (20)$$

$$r = G m_2 / c^3, \quad (21)$$

$$A = \left[\frac{P}{P_b x (1 - e^2)^{1/2}} \right] \left(\frac{a_R m_2}{cM} \right)^2, \quad (22)$$

$$B = 0, \quad (23)$$

---

[3] The parameter $\dot{\omega}$ is strictly not the instantaneous rate of change of ω, but rather its average over an orbital period.

$$\delta_r = \frac{G}{c^2 a_R M} (3m_1^2 + 6m_1 m_2 + 2m_2^2), \quad (24)$$

$$\delta_\theta = \frac{G}{c^2 a_R M} \left(\frac{7}{2} m_1^2 + 6m_1 m_2 + 2m_2^2 \right). \quad (25)$$

We reemphasize that these equations are specific to general relativity, and for the sake of brevity we have given equations defining the aberration constants A and B that are valid only if the pulsar spin axis is closely aligned with the orbital angular momentum, Further details and references can be found in Damour and Deruelle (1985), DD, and TW.

IV. DETERMINING THE SYSTEM PARAMETERS

a) Basic Procedure

Implementation of the four timing models in our computer program TEMPO is a complex but reasonably straightforward process. The program transforms each observed TOA from its topocentric (TDT) value t to a barycentric (TDB) value t_b, using equation (1). The position and velocity of the telescope at the time of an observation are determined by interpolating a solar system ephemeris and correcting, after the fact, for measured irregularities in the rotation rate of the Earth. Earth-rotation data are obtained from the bulletin published by the International Earth Rotation Service, and ephemeris data are taken from either the Center for Astrophysics PEP740R or the Jet Propulsion Laboratory DE200 ephemeris. Both the CfA and the JPL ephemerides are based on numerical integrations of a solar system model whose parameters have been adjusted for best fit to an extensive data base of solar system observations; comparisons between the two (e.g., Prózsyński 1984), as well as our experience in observing millisecond pulsars (Rawley, Taylor, and Davis 1988), confirm that either ephemeris has more than sufficient accuracy for use in analyzing the data from PSR 1913+16. The solar system Einstein delay $\Delta_{E\odot}$ is evaluated by interpolating a table of numerically integrated values included with the CfA ephemeris, or, when the JPL ephemeris is being used, by means of a semi-analytical model developed at the Bureau des Longitudes (BDL) in Paris (Fairhead, Bretagnon, and Lestrade 1988).

The pulsar proper time T is computed from t_b by a process that depends on the binary model being used: equations (5) and (6) are used for the BT model, Haugan's (1985) equations (69) and (71) for the EH model, equations (7)–(14) for the DD model, and equations (7)–(25) for the DDGR model. We call attention to the fact that some of these equations have been presented in the inverse of the form needed. For example, equations (7)–(14) define t_b in terms of T, whereas the opposite is actually required. Some useful short-cut procedures for the necessary inversions are presented in DD.

The pulsar rotational phase $\phi(T_i)$ is determined from equation (4) with $\epsilon = 0$, and a χ^2 statistic measuring goodness of fit of the model to the data is computed from

$$\chi^2 = \sum_{i=1}^N \left(\frac{\phi(T_i) - n_i}{\sigma_i / P} \right)^2. \quad (26)$$

In this equation, n_i is the closest integer to each computed $\phi(T_i)$, and σ_i is the estimated uncertainty of T_i (see § IIb). Along with the phases, TEMPO calculates partial derivatives $\partial \phi / \partial \xi |_{T_i}$ with respect to each of the model parameters ξ. These include, in addition to the physical parameters listed in Table 2, up to five instrumental parameters to account for imperfectly calibrated differences between observing systems. Standard linearized techniques (see, for example, Bevington 1969; Press *et al.* 1986) are then used to minimize χ^2 with respect to as many of the model parameters as desired, while others are held fixed. This process requires the evaluation and inversion of a covariance matrix, from which corrections to the estimated parameters and their formal uncertainties are derived.

The integers n_i in equation (26) represent the number of pulsar periods (or rotations of the spinning neutron star) elapsed between each T_i and the reference epoch at which $\phi = 0$. Since imperfect parameter values can lead to ambiguities of whole cycles across gaps in the data, newly discovered binary pulsars require a "bootstrap" procedure in which ambiguities are detected and systematically resolved while the estimated parameter values are being improved (Taylor 1989). For PSR 1913 + 16 the system parameters have been known well enough since 1975 that no ambiguities have existed. The measured phases $\phi(T_i)$ are typically within ± 0.001 of the nearest integer, even after intervals of a year or more with no observations. It is worth pointing out that the remarkably high precisions for pulsar parameters obtainable from timing observations are largely a result of this ability to connect phase throughout data sets spanning many years.

If the model fitted to the data is an adequate one, and if the data are free of systematic errors, the global minimum of χ^2 should be close to the number of degrees of freedom $N - J$, where J is the number of fitted parameters. In addition, the postfit normalized residuals should have the character of random white noise, with no significant correlations and a Gaussian amplitude distribution, and the formal parameter errors should be realistic estimates of the true uncertainties.

Real data at best approximate such ideals, so it is essential to make quantitative estimates of the size of systematic contributions to the errors. One useful tactic is to study carefully the statistical character of the postfit residuals; another scheme particularly helpful in cases like the present, where subsets of the data have nonuniform quality and may be influenced by different instrumental effects, is to perform solutions in which various subsets of data are intentionally given zero weight. The effect of the unweighted data on the fitted parameter values can then be directly assessed, and their corresponding residuals can give immediate indications of possible systematic errors. We have carried out numerous tests of this kind, and have used them to determine the uncertainty estimates quoted in the next section. Additional details are given in the Appendix.

b) Astrometric and Spin Parameters

The first eight items listed in Table 2, the astrometric and spin parameters of the pulsar, are only peripherally related to the main topic of this paper. These parameters depend on annual and secular terms in the timing model, so their values are sensitive to TOA errors that change over long time scales. Consequently, for high-precision measurement of α, δ, v, $\dot{v}$, etc., the frequent changes of data acquisition equipment during 1974–1977 reduce our earliest data to questionable value. The parameter values quoted in Table 3 are based on solutions carried out for data acquired since 1981. For convenience we quote the pulsar spin parameters in terms of both frequency v and period P.

For several reasons, we list two full sets of results in Table 3. The solution labeled B1950.0 (CfA) is based on the PEP740R ephemeris, including its numerically integrated values for the solar system relativistic clock corrections $\Delta_{E\odot}$. The solution labeled J2000.0 (JPL) is based on the DE200 ephemeris and the

TABLE 3

ASTROMETRIC AND SPIN PARAMETERS OF PSR 1913 + 16

	COORDINATE SYSTEM	
PARAMETER	B1950.0(CfA)	J2000.0(JPL)
α	$19^h13^m12^s46549(15)$	$19^h15^m28^s00018(15)$
δ	$16°01'08''189(3)$	$16°06'27''4043(3)$
μ_α (mas yr^{-1})	-3.2 ± 1.8	-3.2 ± 1.8
μ_δ (mas yr^{-1})	1.2 ± 2.0	1.2 ± 2.0
v(s^{-1})	16.940539303217(2)	16.940539303295(2)
$\dot{v}$(10^{-15} s^{-2})	$-2.47559(2)$	$-2.47583(2)$
$|\ddot{v}|$(10^{-27} s^{-3})	<6	<6
t_0(JED 2,445,888 +)	0.745517962	0.745517886
P(ms)	59.029997929883(7)	59.029997929613(7)
$\dot{P}$(10^{-18} s s^{-1})	8.62629(8)	8.62713(8)
$|\ddot{P}|$(10^{-29} s s^{-1})	<2	<2

NOTE.—Figures in parentheses are uncertainties in the last digits quoted.

BDL semianalytical model for $\Delta_{E\odot}$. For the reference epoch t_0 we list a TDB Julian ephemeris date, roughly centered within our highest quality data, for which $\phi[T(t_0)] = 0$. Thus, t_0 represents a nominal infinite-frequency barycentric pulse arrival time, as well as a reference epoch for the remaining astrometric and spin parameters.

It is known that the coordinate systems of the CfA and JPL ephemerides are not oriented identically (Prószyński 1984; Bartel *et al.* 1985; Backer *et al.* 1985). For this reason, applying any standard recipe (e.g., Standish 1982) to convert the B1950.0 position of PSR 1913 + 16 to J2000.0 coordinates will not yield exactly the J2000.0 position given in Table 3. Furthermore, our two solutions incorporate different definitions of the unit of time, because they implicitly use different constants for the last term in equation (2). The measured values of v, $\dot{v}$, P, and $\dot{P}$ depend on the value of this constant, and therefore on the time interval over which it was determined. Because the BDL semianalytical model includes cyclic terms with periods up to 1000 yr, while the CfA numerical integration extends over only 50 yr, the BDL model is probably the preferable system to use. We hasten to point put that while these subtleties will be important to others observing PSR 1913 + 16, both now and in the future, they have no bearing on the remaining results of this paper. Both the CfA and the JPL/BDL systems are good enough descriptions of solar system effects that they have negligible impact on measurement of the orbital parameters.

c) Orbital Parameters

We now turn to the orbital parameters of the PSR 1913 + 16 system, whose measurement and interpretation form the central theme of our paper. A representative set of solutions for the optical parameters is summarized in Tables 4 and 5. For purposes of discussion we include in these tables one solution each for the EH model, the reparameterized Haugan model (H88), and the DDGR model; two for the BT model; and three for the DD model. The first seven solutions are based on the Mark I, Mark II, and Mark III data alone—that is, all of the high-quality data acquired since 1981 February. The last solution, labeled DD(3), includes all data back to 1974 with estimated uncertainties less than about 500 μs. An example of the postfit residuals for solution DD(1) is presented in Figure 2, both as a function of date and as a function of orbital phase.

Table 4 lists values obtained for the five Keplerian orbit parameters. Small but significant differences exist between the BT and other values for x, and between the BT, EH, H88, and

TESTS OF RELATIVISTIC GRAVITY WITH PSR 1913+16

TABLE 4

KEPLERIAN ORBITAL PARAMETERS

Solution	$x = (a_1 \sin i)/c$ (s)	e	T_0 (JED 2,445,888+)	P_b (s)	ω_0 (degrees)
BT(1)	2.341774(9)	0.6171472(10)	0.61724862(8)	27906.980894(2)	220.1426(2)
BT(2)	2.34178(10)	0.6171467(11)	0.61724862(8)	27906.980894(2)	220.143(2)
EH	2.341749(12)	0.617127(5)	0.61724861(8)	27906.980895(2)	220.1428(2)
H88	2.341752(19)	0.617128(6)	0.61724859(10)	27906.980895(2)	220.1428(3)
DDGR	2.341754(9)	0.6171314(10)	0.61724857(7)	27906.980891(2)	220.1428(2)
DD(1)[a]	2.341754(9)	0.6171313(10)	0.61724860(8)	27906.980895(2)	220.1428(2)
DD(2)[a]	2.34176(10)	0.617132(3)	0.6172486(2)	27906.980895(2)	220.143(2)
DD(3)	2.341761(9)	0.6171304(10)	0.61724861(8)	27906.980894(2)	220.1426(2)

NOTE.—Figures in parentheses are uncertainties in the last digits quoted.

[a] Preferred solutions.

remaining values of e. Such differences are to be expected, because the precise definitions of parameters in the three basic models are not identical. Note that we have chosen a reference periastron passage time, T_0, in 1984 July—near the middle of our span of highest quality data—whereas most previously published solutions used a T_0 in 1974 September. The freedom of choice for T_0 arises, of course, from the discrete freedom (modulo 2π) in choosing an origin for the eccentric anomaly in Kepler's equation.

Table 5 lists all of the significantly nonzero PK parameters. Two such parameters exist for the DDGR model, while the BT model has three, EH and H88 each have four, and DD has as many as five [two of which were held fixed in solutions DD(1) and DD(3)]. The remaining columns of Table 5 contain upper limits for $|\dot{x}|$ and $|\dot{e}|$ obtained from the BT(2) and DD(2) solutions, as well as the number of instrumental free parameters, the value of χ^2, and the number of degrees of freedom for all eight solutions.

With the understandable exceptions already mentioned, the solutions yield consistent parameter estimates within the measurement uncertainties. All of the Keplerian parameters are determined to six or more significant figures, as are $\dot{\omega}$ and M, the most important of the PK group.[4] The next most significant PK parameter, γ (or m_2 in the DDGR model), is determined to a fractional accuracy of about 0.2%. The orbital period decay rate, $\dot{P}_b$, is known to 1%, and $\sin i$ in the EH model is determined to about 5%. All of these parameters, and

the limits for $\dot{x}$ and $\dot{e}$ as well, are readily determined by the linearized least-squares procedure outlined in § IV*a*.

A striking demonstration of the significance of gravitational redshift and time dilation effects in the orbit is produced by setting $\gamma = 0$, minimizing χ^2 with respect to everything else, and comparing the postfit residuals with those produced in a "good" solution in which γ was allowed to vary. Figure 3 shows the results of such a test, directly comparable to the bottom portion of Figure 2. The rapid precession of periastron causes the shape of the orbit delay curve (eqs. [7]–[14]) to change substantially over a few years, yielding large and highly correlated residuals if γ is constrained to zero.

Precession of the periastron and decay of the orbital period also give rise to large effects readily amenable to graphical display. For this purpose, we carried out solutions for each of 19 localized blocks of data, solving for just two parameters, T_0 and ω_0, in each block. The periastron time T_0 was always chosen to corespond to an orbit near the middle of the data segment. All other parameters were held fixed at the values found in the J2000.0 (JPL) solution in Table 3 and the DD(1) solution in Tables 4 and 5. The localized values of T_0 and ω_0 are listed in Table 6 and plotted in Figure 4. To allow error bars to be seen in the figure, we include an expanded-scale version at the bottom in which the expected ω corresponding to the DD(1) parameters has been subtracted. It is obvious that apsidal motion is a huge effect in the PSR 1913+16 system, confirming its strongly relativistic character. The line of apsides of this binary orbit has rotated through nearly 60° during our 14 years of observations.

Figure 5 is a similar graph showing the observable effects of

---

[4] At the present level of accuracy, it is important to specify that $\dot{\omega}$ is measured in units of degrees per Julian year (365.25 days).

TABLE 5

POST-KEPLERIAN PARAMETERS

Solution	$\dot{\omega}$ (degrees yr^{-1})	γ (ms)	$\dot{P}_b$ (10^{-12})	$s = \sin i$	r (μs)	M ($M_\odot$)	m_2 ($M_\odot$)	$\|\dot{x}\|$ (10^{-13})	$\|\dot{e}\|$ (10^{-14} s^{-1})	Instrumental Parameters	χ^2	dof
BT(1)	4.22660(4)	4.288(10)	−2.428(26)	...	...	...	...	0[a]	0[a]	1	2617.8	2500
BT(2)	4.22660(18)	4.29(11)	−2.427(30)	...	...	...	...	<2.7	<4.3	1	2613.3	2498
EH	4.22661(4)	4.281(10)	−2.429(26)	0.73(4)	...	...	...	0[a]	0[a]	1	2551.8	2499
H88	4.22660(4)	4.281(10)	−2.429(27)	$0.71^{+0.06}_{-0.10}$	...	...	...	0[a]	0[a]	1	2551.6	2499
DDGR	...	...	...	...	...	2.82837(4)	1.386(3)	0[a]	0[a]	1	2556.6	2501
DD(1)[b]	4.22659(4)	4.289(10)	−2.427(26)	0.734[a]	6.83[a]	...	...	0[a]	0[a]	1	2552.0	2500
DD(2)[b]	4.22659(18)	4.29(11)	−2.428(34)	[c]	[c]	...	...	<2.4	<1.9	1	2551.9	2496
DD(3)	4.22656(4)	4.296(11)	−2.435(34)	0.734[a]	6.83[a]	...	...	0[a]	0[a]	5	4878.7	4001

NOTE.—Figures in parentheses are uncertainties in the last digits quoted.

[a] Parameter held fixed at this value.

[b] Preferred solutions.

[c] See text and Fig. 7.

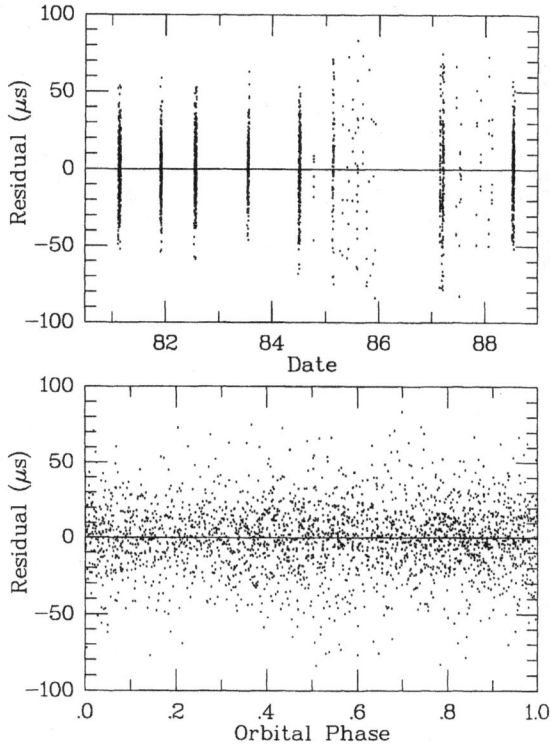

FIG. 2.—Postfit residuals from the DD(1) solution in Tables 4 and 5, plotted separately against date and orbital phase.

TABLE 6

TIMES AND LONGITUDES OF PERIASTRON PASSAGE

Year	T_0 (JED 2,440,000+)	ω_0 (degrees)
1974.78.....	2331.446132(3)	178.983(4)
1974.94.....	2389.585675(2)	179.654(3)
1976.94.....	3118.5909664(19)	188.087(2)
1977.59.....	3356.6401066(18)	190.846(2)
1977.96.....	3493.5910292(11)	192.426(2)
1978.24.....	3593.3972500(6)	193.5845(8)
1978.43.....	3663.487700(2)	194.393(4)
1978.82.....	3807.5445721(9)	196.0622(11)
1979.32.....	3988.4231539(8)	198.1547(9)
1980.11.....	4276.5368946(7)	201.4867(8)
1980.59.....	4455.4774927(7)	203.5581(8)
1981.15.....	4656.38191738(14)	205.88350(15)
1981.92.....	4938.35870602(12)	209.14664(14)
1982.56.....	5172.53186888(11)	211.85639(14)
1983.56.....	5536.55001230(11)	216.06875(14)
1984.52.....	5888.61724869(17)	220.1428(2)
1985.51.....	6249.7284133(4)	224.3214(5)
1987.19.....	6864.0695866(3)	231.4306(5)
1988.54.....	7358.57869939(14)	237.1527(2)

NOTE.—Figure in parentheses are uncertainties in the last digits quoted.

orbital decay. The ordinate, measured in seconds and labeled "orbital phase shift," corresponds to the difference between the measured values of T_0 listed in Table 6 and the values that would be expected if there were no dissipative effects in the orbit. Again, as an aid to visualizing the measurement uncertainties, the figure includes an expanded-scale section showing differences between observed and expected periastron times. In the expanded section we also illustrate, with a time origin at

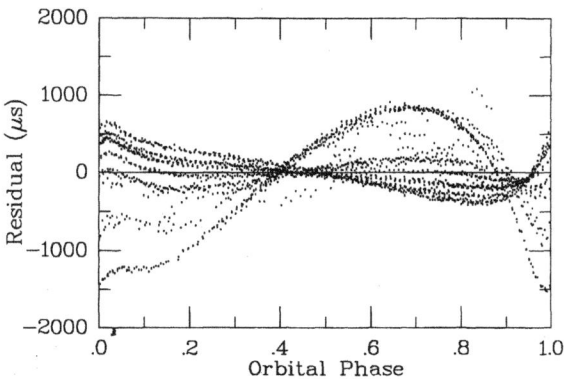

FIG. 3

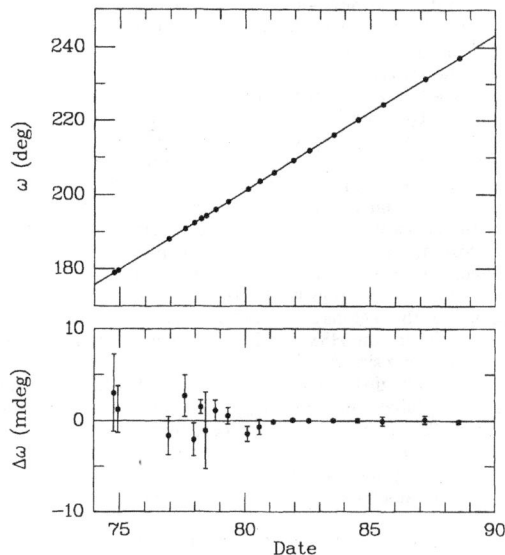

FIG. 4

FIG. 3.—Postfit residuals from a solution similar to DD(1), but constrained by setting $\gamma \equiv 0$ to illustrate the significance of orbital time dilation and gravitational redshift effects. Notice that the vertical scale is 20 times larger than that of Fig. 2.

FIG. 4.—*Top:* Values of the longitude of periastron, ω, measured within 19 localized blocks of data throughout our 14 yr data span. *Bottom:* Differences $\Delta\omega$ between the locally measured values of ω and the values expected according to the DD(1) parameter set.

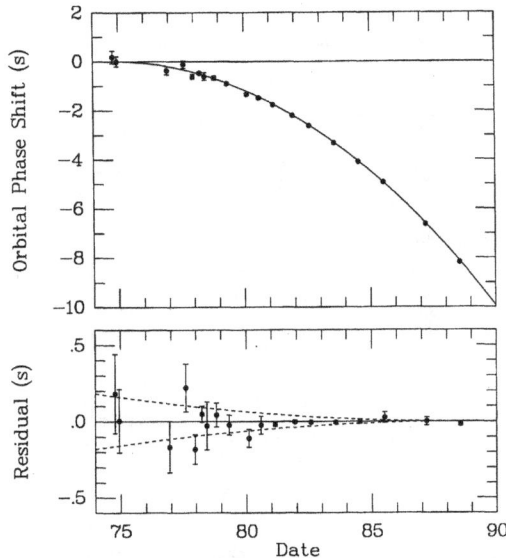

FIG. 5.—*Top:* Cumulative shift of the times of periastron passage relative to a nondissipative model in which the orbital period remains fixed at its 1974.78 value. *Bottom:* Differences between the locally measured periastron times and those expected according to the DD(1) parameter set. Dashed curves illustrate differential trends that would be expected (relative to epoch 1988.54) if the rate of orbital decay $\dot{P}_b$ were 2% larger or 2% smaller.

1988.5, the differential trends that would correspond to 2% larger and 2% smaller values of $\dot{P}_b$. These curves make it clear that the rate of orbital decay can be measured to approximately 1% accuracy with the existing observations.

With the quality of data that we have so far been able to achieve, the values and uncertainties of $\sin i$ in the H88 model, and especially r in the DD model, cannot be reliably determined by linearized minimization of χ^2. This circumstance occurs because the curvature of the χ^2 hypersurface with respect to these parameters is small enough that, with the present measurement uncertainties, the acceptable zone around the global minimum is larger than the region over which the curvature can be treated as constant. Some important information on the parameters in question can still be obtained, however. A useful strategy is to plot values of $\Delta\chi^2$ with respect to the poorly determined parameters, thereby mapping out the regions around their most likely values.

Such a plot for the H88 solution is shown in Figure 6. It was made by holding $\sin i$ fixed at each of the values 0.05, 0.10,..., 0.95, and minimizing χ^2 with respect to all other model parameters. The global minimum χ^2_{min} was subtracted from each resulting $\chi^2(\sin i)$, and the differences interpolated to produce the smooth curve shown. Intervals with $\Delta\chi^2 < 1$ and $\Delta\chi^2 < 4$ formally correspond to 68% and 95% confidence ranges for the single parameter $\sin i$. In Table 5 we quote the 68% or "1 σ" limits.

A map of $\Delta\chi^2(s, r)$ for the DD(2) solution was prepared in a similar way and is reproduced in Figure 7. In this case the list of assumed parameter values is two-dimensional, and the $\Delta\chi^2$ differences were interpolated to facilitate drawing contours by hand. In the figure we use an abscissa linear in $\cos i$ rather than $\sin i$, because it is $\cos i$ that has a uniform *a priori* distribution,

and this helps to illustrate the volume of parameter space excluded by the data. The highest plotted contours, 2.3 and 6.2 units above the minimum, represent the boundaries of nominal 68% and 95% confidence regions for s and r collectively. The contours for $\Delta\chi^2 = 1.0$ and 0.5 have low significance, but help to illustrate where significant contours might fall if higher quality data were available. In the regions to the right of the contours for $\Delta\chi^2 = 6.2$, the value of χ^2 rises very steeply, so that larger values of s and r are quite incompatible with the data. Figure 7 shows that the most likely values of s and r are $s \approx 0.6$ and $r \approx 7$ μs, with uncertainties that are moderately large and interdependent in a complicated way.

V. CONCLUSIONS INDEPENDENT OF A SPECIFIC THEORY OF GRAVITY

a) Astrometry and Rotational Stability

Astrometry of PSR 1913+16 and its surrounding optical field has already been the subject of a number of papers. Taylor, Fowler, and McCulloch (1979) published a pulsar timing position accurate to about 0″.1, from which a candidate optical identification was suggested by Crane, Nelson, and Tyson (1979). More recently, Backer *et al.* (1985) measured the pulsar's position with the VLA and presented an updated timing position, based on our data through 1984 July. The two radio positions were shown to be consistent with each other but inconsistent with that of the optical candidate as remeasured by Elliott *et al.* (1980). The position listed in Table 3 is consistent with the earlier timing position (to within 1.8 σ in right ascension and 1.4 σ in declination), and it has considerably smaller uncertainties. It is clear that the optical candidate is an unrelated foreground or background star which happens to lie approximately 0″.5 north of the pulsar position. This conclusion is consistent with recent three-color CCD photometry by Boeshaar *et al.* (1988), who conclude that the star is most likely an unassociated dA-K main-sequence field star.

Our solution for the proper motion of PSR 1913+16 is formally quite significant, but for reasons described in the

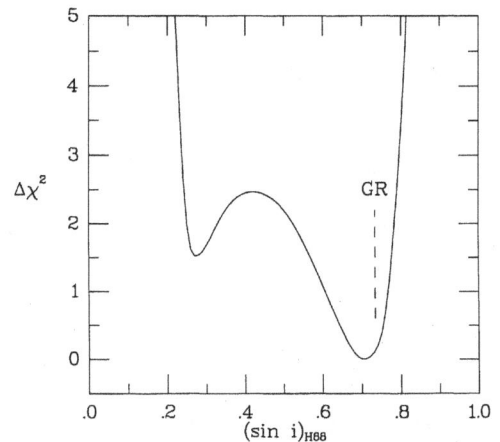

FIG. 6.—The smooth curve represents values of $\Delta\chi^2$ obtained from the H88 solution when $\sin i$ was held fixed at a number of values throughout its range. The expected value of $\sin i$ according to the DDGR solution—the simplest interpretation of the PSR 1913+16 system within general relativity—is marked by the vertical dashed line.

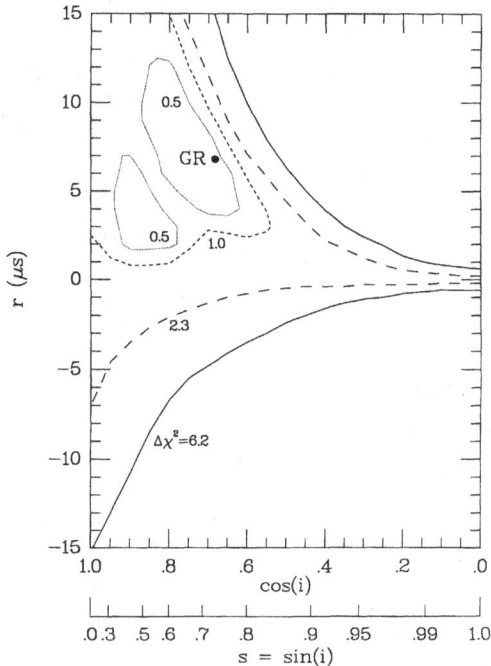

FIG. 7.—Contours of constant $\Delta\chi^2$ in the (s, r)-plane for solution DD(2). Values of s and r corresponding to the DDGR solution are marked by a filled circle.

exceeding ~ 30 μs over at least 7 years (see Fig. 10 and the Appendix) implies an "activity parameter" $A < -5.7$, making PSR 1913 + 16 one of the most stable of all known pulsars (Cordes and Downs 1985).

b) Orbital Dynamics

It was recognized at the time of discovery of PSR 1913 + 16 that its orbital elements imply a high-mass, high-velocity binary system containing regions with relativistically strong gravitational fields, and that periastron precession, gravitational redshift, and time dilation should all produce readily measurable effects in its pulse timing data (Hulse and Taylor 1975). Within a few months, many other experimental possibilities had also been suggested, including the long-predicted, but never before observed, gravitational radiation-reaction effects (Esposito and Harrison 1975; Wagoner 1975) and "magnetic" aspects of gravity as experienced by the rapidly spinning pulsar (Damour and Ruffini 1974; Esposito and Harrison 1975; Barker and O'Connell 1975). Most important, the PSR 1913 + 16 system was seen to offer a first-ever experimental opportunity to probe the nature of gravity in strong-field conditions (Damour 1988; see this paper also for additional references).

As described in §§ III and IV, analysis of PSR 1913 + 16 timing data according to the DD model yields estimates of five phenomenological PK parameters, three of which are determined to 1% precision or better. Two PK parameters suffice to complete a dynamical specification of the binary system within a particular relativistic theory of gravity. Additional parameters can then be used to provide information on the adequacy of the physical and mathematical models, and—if all else is well—on the gravitational theory in question.

A theory of gravity successfully passing the tests posed by this experiment must account for the well-determined values of $\dot{\omega}$, γ, and $\dot{P}_b$, as well as the limits placed on s, r, $\dot{x}$, and $\dot{e}$. A theory would be in serious trouble if descriptions of the binary system based on different parameter subsets were to lead to conflicting conclusions, such as incompatible estimates of the masses of the two stars. Although a theory might be technically "rescued" by introducing astrophysical complications accompanied by additional free parameters, such ad hoc additions are unlikely to be persuasive unless *all* viable theories of gravity require them.

We call attention to the fact that values of χ^2 for the two BT solutions (see Table 5) are significantly higher than for any of the other fits based on post-1981 data. The relatively poor fit of the BT model to the data is illustrated in Figure 8, in which we have plotted residuals from the BT(2) and DD(1) solutions, in each case averaged into 30 equally spaced bins of orbital phase. The inescapable conclusion is that a Newtonian solution to the gravitational two-body problem, even when retrofitted with Einstein delays and phenomenological time derivatives of the orbital elements, is inadequate to describe the PSR 1913 + 16 system to the accuracy demanded by our data.

The DD model is the most theory-independent way of characterizing the observed deviations. Our constraints on parameters s and r (Fig. 7) rule out values of s greater than about 0.8 (unless $|r|$ is very small) and negative values of r. Negative values of r are not meaningful, so far as we know, within any theory of gravity. However, they are mathematically permitted by equation (10), so it is reassuring to find that, according to our data, almost certainly $r > 0$. The value of s must lie

Appendix we quote it in Table 3 with with conservative estimates of the uncertainties. In Galactic coordinates, the angular motion amounts to to $\mu_l = -0.4 \pm 1.5$, $\mu_b = 3.4 \pm 2.3$ mas y^{-1}. The dispersion measure implies a distance of about 5 kpc (Taylor and Manchester 1981), so the pulsar appears to be moving out of the Galactic plane with a transverse velocity of roughly 80 km s^{-1} in the solar system reference frame. There is not much evidence of the expected effect of Galactic rotation, which should contribute a proper motion toward the Galactic center with $\mu_l \approx -5$ mas yr^{-1} (see Rawley, Taylor, and Davis 1988 and references therein). If the Galactic rotation effect is being canceled by peculiar motion of the pulsar relative to its local neighborhood, the pulsar's total velocity in that frame is at least 140 km s^{-1}, and probably higher when the unknown radial component is included. Models of circumstances surrounding the pulsar's birth strongly suggest velocities of ~ 150–250 km s^{-1} (Cordes and Wasserman 1984; Burrows and Woosley 1986; Bailes 1988), which would then be consistent with our observations.

The characteristic timing age of PSR 1913 + 16, $P/\dot{P} = 2 \times 10^8$ yr, is much larger than those for most pulsars, and the surface magnetic field strength, typically estimated from $B = 3 \times 10^{19}(P\dot{P})^{1/2} = 2 \times 10^{10}$ G, is much smaller. We find only marginal evidence for nonzero values of the second derivative and stochastic noise terms of equation (4). As described in the Appendix, attempts to measure such effects over a decade or more, at the extremely low levels of a few milliperiods of accumulated phase, are fraught with difficulties. However, we believe the conservatively stated upper limits for $\ddot{\nu}$ and $\dot{P}$ quoted in Table 3 to be reliable. The lack of timing noise

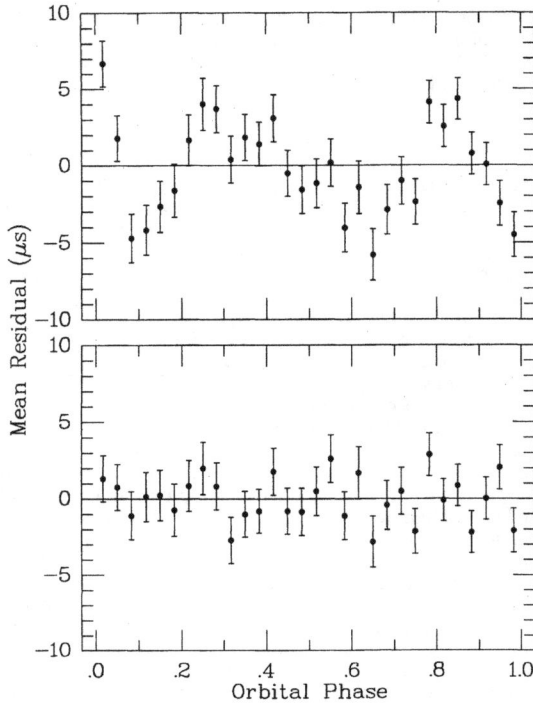

FIG. 8.—Postfit residuals for the BT(2) (*top*) and DD(1) (*bottom*) solutions, averaged into 30 equally spaced bins of orbital phase. The BT model (quasi-Newtonian motion, with "add-on" relativistic effects and time derivatives) is obviously not an adequate approximation.

between zero and one (with suitable definition of ω) for equation (10) to make sense.

VI. CONCLUSIONS BASED ON GENERAL RELATIVITY

We now use the measured orbital parameters to analyze the PSR 1913+16 system under the assumption that general relativity is the correct theory of gravity, at least in the non-quantum regime. Similar treatments within the framework of other theories will be pursued in another paper (Damour and Taylor 1989).

a) Masses of the Two Stars

Table 5 shows that the DDGR solution, with exactly two PK parameters—the minimum number required to specify all of the astrophysical unknowns—has a χ^2 scarcely larger than those of the DD(1), EH, H88, and DD(2) solutions, which have 1, 2, 2, and 5 additional free parameters, respectively. Thus, general relativity successfully accounts for all aspects of our data on the PSR 1913+16 system while using the simplest possible dynamical model: a pair of point masses moving under their mutual gravitational interaction. According to the DDGR solution, the mass of the pulsar is $m_1 = M - m_2 = 1.442 \pm 0.003 \ M_\odot$, and the mass of the companion is $m_2 = 1.386 \pm 0.003 \ M_\odot$.

The success of general relativity in accounting for our observations is further strengthened by the values of the PK parameters obtained in the H88 and DD solutions. (The EH

parameter $\sin i$ is not particularly useful in this regard, as mentioned in § IIIc.) As shown for Haugan's model in TW equations (2)–(8), and for the DD model in equations (15)–(25) of § III, a specific value for each PK parameter—together with the well-determined Keplerian orbital elements—specifies a parametric relationship between m_1 and m_2 in general relativity. If Einstein's theory is valid, and the binary system as uncomplicated as our model assumes (see § VIb), then the parametric curves should all meet at one point in the (m_1, m_2)-plane.

In practice, experimental uncertainties transform the parametric curves for well-determined parameters into strips of finite width. Contours of $\Delta\chi^2$ for poorly determined parameters like s and r can be mapped into the (m_1, m_2)-plane, as well. Strips and contours for the PSR 1913+16 system in general relativity are plotted in Figure 9. We also plot the curve corresponding to $\sin i = 1$. The region below this curve corresponds to $\sin i > 1$, and is therefore forbidden.

Notice that the values of m_1 and m_2 determined from the DDGR model, marked with a black dot near the center of the figure, are consistent with *all* of the well-determined parameters, and with the most likely values of $(\sin i)_{H88}$, s, and r as well. For reference, the values of $(\sin i)_{H88}$, s, and r corresponding to the DDGR values of m_1 and m_2 are also marked in Figures 6 and 7. In both cases the nominal general relativity values are close to location of the χ^2 minima. We mention in passing that Damour and Schäfer (1988) have shown that higher order general relativistic effects modify slightly the dependence of M on $\dot\omega = 2\pi k/P_b$ (eq. [16]). The corrected value for the total system mass M is $2.82827 + 0.00004 \ M_\odot$, smaller than the value given in Table 5 by $0.00010 \ M_\odot$. This correction has a negligible effect on the individual masses that

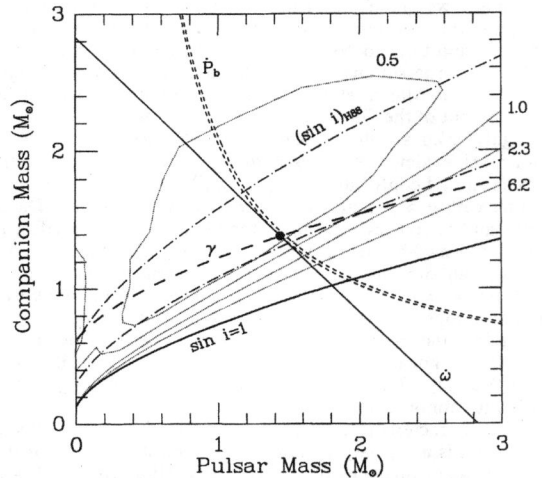

FIG. 9.—Restrictions on the pulsar mass, m_1, and companion mass, m_2, imposed by general relativity are indicated by curves labeled $\dot\omega$, γ, $\dot P_b$, and $(\sin i)_{H88}$ (the Haugan 1988 $\sin i$ parameter). Uncertainties in $\dot\omega$ and γ are smaller than the widths of their plotted curves; two curves are plotted for $\dot P_b$, and $(\sin i)_{H88}$, bracketing the uncertainty range. Numerically labeled dotted curves represent a mapping of $\Delta\chi^2$ contours for parameters r and s from Fig. 7. Companion masses below the curve labeled $\sin i = 1$ are incompatible with the mass function. The point marked with a filled circle corresponds to the mass values given for the DDGR solution in Table 5.

we quote. Note, however, that it illustrates how the high-precision timing of PSR 1913 + 16 constitutes a new laboratory for relativistic gravity.

b) Astrophysical Considerations

From its discovery, the PSR 1913 + 16 system has been known to consist most likely of two neutron stars (Hulse and Taylor 1975; Webbink 1975; Smarr and Blandford 1976). Alternative possibilities have been thoroughly explored, and the results have always led to either theoretical or observational difficulties. For example, a white dwarf companion with $m_2 = 1.386\ M_\odot$ would be barely stable, since its mass is precariously close to the Chandrasekhar limit; in any event, it seems impossible for close binary evolution to leave a white dwarf in the system without also circularizing the orbit (Srinivasan and van den Heuvel 1982; van den Heuvel 1987, and references therein). The stripped helium core of a post-main-sequence secondary would probably be visible (Crane, Nelson, and Tyson 1979) but, as we described in § V*a*, no visible counterpart has been found. In short, all models of the system involving white dwarf or helium star companions appear to be highly contrived and therefore unlikely.

Almost certainly, then, both components of the PSR 1913 + 16 system are neutron stars. In that case, Smarr and Blandford (1976) and others have shown that nonrelativistic contributions to $\dot\omega$ and $\dot P_b$ are utterly negligible. Burrows and Woosley (1986) have emphasized that the "baryonic masses" of the neutron stars are 10%–15% larger than the gravitational masses we measure, the difference having been carried off by neutrinos emitted during final collapse of the supernova cores. The measured parameters leave only a moderate amount of freedom in many details of the system's evolution (Cordes and Wasserman 1984; Burrows and Woosley 1986). Both the pulsar and the companion, for example, are too heavy to have been formed by accretion of matter onto a white dwarf. The system probably evolved from two massive stars with main-sequence masses between 16 and 18 $M_\odot$, and a small degree of asymmetry in the second supernova explosion is needed to keep the orbit bound with the observed eccentricity.

c) Gravitational Radiation

Substituting $M = 2.82827 \pm 0.00004\ M_\odot$, $m_2 = 1.442 \pm 0.003\ M_\odot$, and the Keplerian parameter in equations (18) and (19) yields the prediction of general relativity for the rate of orbital period change due to gravitational radiation. The result is $(-2.40216 \pm 0.00021) \times 10^{-12}$, in excellent accord with our measured value $(-2.427 \pm 0.026) \times 10^{-12}$. The uncertainty in the theoretical value is dominated by the uncertainty in m_2. We hasten to point out that the six different values of $\dot P_b$ listed in Table 5 are in no sense independent estimates, since they all depend on the same observations. It is *not* significant that all six measurements are approximately 1 σ lower than the theoretical value of $\dot P_b$.

Equations (18) and (19) were first derived by Peters and Mathews (1963), heuristically starting from a "quadrupole formula" dating back to Einstein (1918; see also Landau and Lifshitz 1962). More rigorous and more complete derivations of the gravitational radiation energy flux, or directly of the observable rate of orbital period change in binary systems of strongly self-gravitating bodies, have been actively pursued in recent years, especially after the announcement of the measurement of $\dot P_b$ in PSR 1913 + 16 (Taylor, Fowler, and McCulloch 1979). The results of these studies have confirmed the validity

of equations (18) and (19); for reviews of recent work and references see Will (1986) and Damour (1987). Blanchet and Damour (1989) and Blanchet and Schäfer (1989) have investigated the gravitational radiation loss contributions from terms of higher order in (v/c). Blanchet and Schäfer (1989) find that the fractional error in equation (18), caused by ignoring higher order terms, is only 2.15×10^{-5}, which is negligible at present levels of accuracy.

Our test for gravitational radiation in general relativity can be summarized by the ratio of observed to expected orbital decay rates,

$$\frac{\dot P_b(\text{observed})}{\dot P_b(\text{theory})} = 1.010 \pm 0.011 \ . \qquad (27)$$

To the best of our knowledge, there are no other plausible candidates for contributions to $\dot P_b$ as large as 1% of the gravitational radiation value. Three of the closest possibilities—transverse motion of the system, Galactic acceleration, and mass-energy loss caused by pulsar spin-down—each probably contribute no more than a few tenths of 1% of the measured effect (Shapiro and Terzian 1976; Will 1981). Thus the 1% agreement is an impressive confirmation of Einstein's theory and, more specifically, a verification of its ability to predict effects involving strong and rapidly varying gravitational fields.

d) The Newtonian Constant G

Damour, Gibbons, and Taylor (1988) have shown that the good agreement between the measured value $\dot P_b$ and the general relativistic prediction can be used to place a stringent limit on the rate of change of the Newtonian gravitation constant. Recent interest in Kaluza-Klein and superstring theories has brought renewed interest in possible variation of fundamental coupling constants, because these theories predict changes on the time scale of the Hubble expansion. Our improved precision for $\dot P_b$ now allows us to tighten the limit even further. If $\delta \dot P_b = \dot P_b(\text{observed}) - \dot P_b(\text{theory})$, the relevant equation and the new limit can be written as

$$\frac{\dot G}{G} = -\frac{\delta \dot P_b}{2P_b} = (1.2 \pm 1.3) \times 10^{-11}\ \text{yr}^{-1} \ . \qquad (28)$$

This limit is comparable to those obtained from active radar ranging data between Earth and the Viking landers on Mars (Hellings *et al.* 1983; Reasenberg 1983). Further details may be found in the paper by Damour, Gibbons, and Taylor (1988).

e) Ultra–Low-Frequency Gravitational Radiation

Alternatively, the difference between observed and theoretical values of $\dot P_b$ can be used to place an upper limit on Ω_g, the energy density of ultra–low-frequency gravitational radiation expressed as a fraction of closure density. Bertotti, Carr, and Rees (1983) show that gravitational waves with periods greater than the span of the observations, but less than the light-travel time to PSR 1913 + 16, would manifest themselves as an extra contribution to $\delta \dot P_b$. The resulting limit on Ω_g can be written as

$$\Omega_g < \frac{1}{2}\left(\frac{\delta \dot P_b}{P_b H_0}\right)^2 = 0.04 h^{-2} \ , \qquad (29)$$

where $H_0 = 100h$ km s^{-1} Mpc^{-1} is the Hubble constant. At the ultra-low frequencies of 10^{-9} to 10^{-12} Hz, this limit is the best available constraint on a stochastic gravitational wave background.

VII. SUMMARY AND PROSPECTS

Accurate time-of-arrival measurements of pulses from the binary pulsar PSR 1913+16 over the last 14 years have enabled us to measure five Keplerian and five post-Keplerian orbital elements, all but two of them with precisions of 1% or better. The evidence indicates that both stellar components are collapsed objects and may be considered point masses in the orbital analysis. Within general relativity, the component masses have been determined with an accuracy of about 0.2%. The orbit is losing energy within 1% of the rate predicted for gravitational radiation damping in general relativity—which we interpret as inconvertible evidence for the existence of gravitational waves. Moreover, the excellent agreement between predicted and observed orbital decay rates provides a limit comparable to the best available for the rate of change of the Newtonian gravitation constant G, and the best available limit on the energy density of a cosmic gravitational wave background at ultra-low frequencies below 10^{-9} Hz.

In another paper (Damour and Taylor 1989), we will explore the fates of several other theories of gravity when their strong-field consequences are used to interpret the parameters of the PSR 1913+16 system. Future observations with 10 times better precision would enable measurements of the post-Keplerian parameters s and r with accuracies around 5%. This is a very exciting prospect because of the considerably tightened constraints that they would place on gravitational theories. Such observations would be feasible with the Arecibo telescope if it were upgraded according to current plans.

The data for this experiment could not have been obtained without the dedication and skills of many individuals at the Arecibo Observatory. We are also most grateful for early work by R. A. Hulse, L. A. Fowler, and P. M. McCulloch, and for essential contributions to the Mark II and Mark III data acquisition systems by L. A. Rawley, D. R. Stinebring, and T. H. Hankins. We have benefited enormously from correspondence and conversations with M. P. Haugan and especially T. Damour, who also provided a critical reading of this manuscript. R. N. Manchester and D. Nice have contributed importantly to the development of TEMPO, and J. F. Chandler and E. M. Standish furnished the CfA and JPL ephemeris data. Our work has been supported financially by a number of grants from the National Science Foundation.

APPENDIX

ANALYSIS OF EXPERIMENTAL ERRORS

Uncertainties for the parameters listed in Tables 3–6 are a combination of our best estimates of both random and systematic errors, and are intended to represent (probably conservative) 1 σ confidence intervals. Random errors were calculated by the standard methods accompanying least-squares fitting procedures (e.g., Bevington 1969; Press *et al.* 1986). In this appendix we describe the methods used to estimate bounds on the possible systematic errors.

The best kind of systematic measurement errors are those large enough to be easily recognized, and stable enough that they can be retroactively removed or allowed for. One example of such errors is a class which affects all of our TOAs to some degree, and which is particularly important when nonoverlapping data obtained with dissimilar observing systems are analyzed together. Because of widely different instrumental resolutions, a variety of receiver bandwidths, and the frequency-dependent pulse shape of PSR 1913+16, data obtained with each distinct observing system require a standard profile unique to that system. Unless simultaneous (or nearly simultaneous) observations are made, there will be unknown offsets between TOAs obtained with a given system relative to others.

Such uncalibrated offsets exist for all of our data acquired before 1978, with observing systems A–E, G, and H (see Table 1). Therefore, in solutions including these data, such as solution DD(3) discussed in § IV, additional instrumental parameters must be introduced and their values estimated along with the physically interesting parameters. One consequence is that postfit residuals and χ^2 values for these solutions will be artificially smaller than they might otherwise have been; another consequence is that the early data do not help much in determining high-precision values of the astrometric and spin parameters of the pulsar.

Residuals for the DD(3) solution are plotted as a function of date in Figure 10. Five instrumental parameters were estimated in this fit, respectively characterizing the offsets of observing systems A, B + C, D, E + G, and H. Since 1978, and especially since 1981, we have taken care to measure the offsets between TOAs obtained with different observing systems, so that additional empirical parameters would not be required. A minor exception to this rule applies to the 1983 July observations, which were carried out after some supposedly innocuous changes had been made in the equipment to permit observations of the millisecond pulsar PSR 1937+21. The single instrumental parameter included in the first seven solutions listed in Tables 4 and 5 accounts for the uncalibrated offset caused by these modifications.

More insidious systematic errors are those small enough to be individually undetectable, or with a time dependence that approximates some linear combination of terms in the model. We have taken two approaches toward allowing for possible errors of this kind. First, as described in TW, we have looked for telltale correlations among nearly adjacent postfit residuals by averaging groups of n consecutive values, and testing to see whether the standard deviations of the resulting averages decrease as $n^{-1/2}$. The results of such tests for the Mark I, Mark II, and Mark III observing systems are illustrated in Figure 11. (Similar results for the earlier observing systems were presented in Fig. 3 of TW.) With the exception of the offset in the 1983 July data mentioned earlier, we find no evidence for correlated errors in the data acquired since 1981, at least down to the 1–3 μs range.

Our second means of identifying possibly significant systematic errors concentrates on those parameters which come closest to making our "experimental design matrix" singular. For purposes of discussion, we reproduce in Table 7 the most significant portions of the normalized covariance matrix for solution DD(1). The submatrix at the top of the table includes all covariances among the astrometric and spin parameters and the single instrumental offset, O_1, while the bottom portion contains covariances among the orbital parameters and between each of them and the pulsar frequency v. Together, these two sections include all elements of the full (15×15) matrix with absolute values greater than 0.3. The largest off-diagonal term is the covariance of 0.974

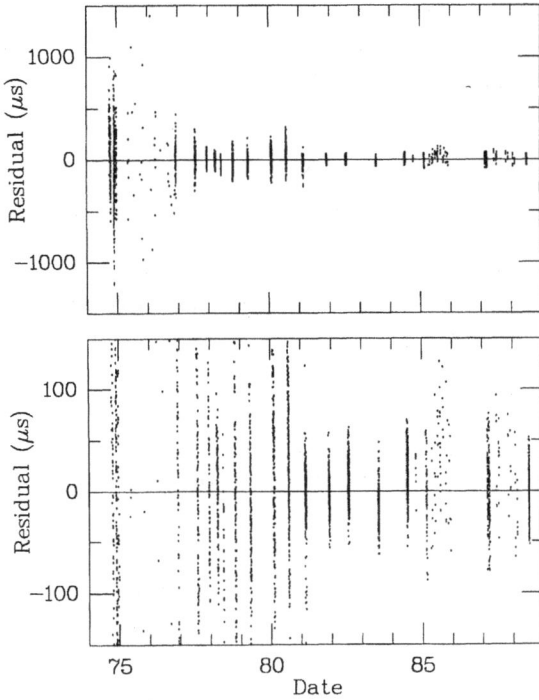

FIG. 10

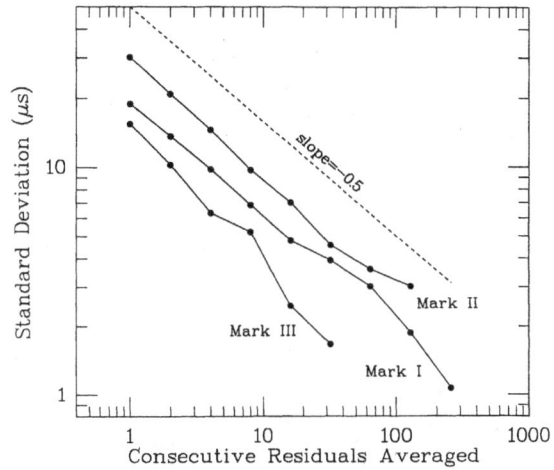

FIG. 11

FIG. 10.—Postfit residuals from the DD(3) solution, which includes all data back to 1974 with estimated uncertainties less than 500 μs. The lower panel presents the same data on a 10 times expanded vertical scale; the slight negative offset around 1983.5 illustrates the consequence of not allowing for a small instrumental offset in the 1983 July data.

FIG. 11.—Standard deviations among averages of postfit residuals obtained from sequences of consecutive measurements with the Mark I, Mark II, and Mark III systems. If the measurement errors are uncorrelated, the standard deviations should decrease as $n^{-1/2}$, as indicated by the dashed line.

TABLE 7

COVARIANCE MATRIX FOR SOLUTION DD(1)

	α	δ	μ_α	μ_δ	ν	$\dot\nu$	O_1
α	1.000						
δ	0.652	1.000					
μ_α	0.812	0.726	1.000				
μ_δ	0.590	0.662	0.474	1.000			
ν	−0.379	−0.505	−04.72	−0.329	1.000		
$\dot\nu$	0.008	0.162	0.328	−0.420	−0.346	1.000	
O_1	−0.080	0.047	−0.195	0.085	0.090	−0.420	1.000
ρ	0.863	0.918	0.905	0.916	0.969	0.905	0.610

	x	e	T_0	P_b	ω_0	$\dot\omega$	γ	$\dot P_b$
x	1.000							
e	−0.025	1.000						
T_0	−0.104	−0.009	1.000					
P_b	−0.215	−0.006	0.449	1.000				
ω_0	−0.903	0.107	0.431	0.355	1.000			
$\dot\omega$	−0.580	0.080	0.366	0.859	0.684	1.000		
γ	0.974	−0.207	−0.073	−0.211	−0.897	−0.584	1.000	
$\dot P_b$	−0.019	−0.027	−0.393	−0.186	0.035	−0.037	−0.013	1.000
ν	−0.058	−0.210	−0.206	−0.670	−0.072	−0.594	−0.013	0.075
ρ	0.991	0.839	0.957	0.952	0.992	0.983	0.994	0.819

between the parameters x and γ; these parameters become fully decoupled only on the time scale of apsidal motion, approximately 85 years.

The last row in each submatrix of Table 7 contains the "global correlation coefficient," ρ, defined as the maximum correlation between a given parameter and all possible linear combinations of the other parameters (see, for example, Eadie *et al.* 1971). The global correlation provides a relative indication of the measurability of a parameter, given the times at which measurements have been made. Values of ρ extremely close to unity indicate probable difficulties. Not surprisingly, γ has the largest global correlation, 0.994, followed by ω_0 (0.992) and x (0.991). If significant orbital phase-dependent systematic errors were present in the data, it would not be surprising to find these parameters affected by larger amounts (relative to their formal errors) than the other parameters.

The nearly random nature of the residuals in Figure 10 shows that our efforts directed toward long-term stability of the measurements have been largely successful, and that our model of the nonbinary aspects of the PSR 1913+16 system is reasonably good. On the other hand, there are some long-term (i.e., not orbit phase–dependent) systematic trends still visible in the residuals, partucularly in the data from observing system J and, to a lesser extent, system F. For this reason the most trustworthy results from the astrometric and spin parameters, and especially their time derivatives, come from the post-1981 data.

The parameter values quoted in Table 3 were taken from solution DD(1) and a similar solution based on the CfA ephemeris. The quoted uncertainties for all parameters except $\ddot{v}$, $\dot{P}$, μ_α, and μ_δ are 2–3 times the formal errors, and reflect our estimates of the largest systematic errors that might have escaped notice through the tests outlined above. The quoted uncertainties for the remaining parameters in Table 3 are 6–10 times the formal errors, and allow for the contingencies that (1) some of the uncalibrated offsets could be real and (2) timing noise and improperly calibrated offsets may have affected the proper-motion values.

For the orbital parameters as well, the most reliable solutions are based on the post-1981 data. As can be seen in the top panel of Figure 2, these observations were obtained in seven concentrated sessions each of about 2 weeks duration, plus some scattered measurements made with the Mark II system from late 1984 through early 1988. To test for the presence of session-dependent errors and their effect on the measured parameters, we carried out eight solutions similar to DD(1) except that the data from each block, in turn, was given zero weight. Except for the 1983 July offset already mentioned, this procedure brought to light no further evidence of systematic errors—and, in particular, no evidence for orbital phase-dependent errors. In these solutions even the values of γ, ω_0, and x, which have the largest global correlation coefficients, varied by no more than the uncertainties quoted in Tables 4 and 5, which are twice the formal standard errors.

Solutions carried out using the DD model require explicit assumptions for the values of δ_r, δ_θ, A, and B in equations (8) and (11). In the theory-independent phenomenological approach of §§ IV and V, the values of these parameters cannot be specified *a priori*, and yet they are too small to be measured from the existing data. In order to test the sensitivity of other parameters to the values assumed for the four small quantities, we carried out a series of solutions in which each of the first three, in turn, was first taken to have its value in general relativity given by equations (22)–(25), and then 50% larger and 50% smaller values. Similarly, B was taken to have the values 0 and ±50% smaller values. Similarly, B was taken to have the values 0 and ±50% of the nominal value of A. The χ^2 values of these fits were nearly identical (the extremes differing by less than 0.5 out of 2552), and variations in the fitted parameters were never as much as twice the uncertainties quoted in Tables 4 and 5.

REFERENCES

Backer, D. C., Fomalont, E. B., Goss, W. M., Taylor, J. H., and Weisberg, J. M. 1985, *A.J.*, **90**, 2275.
Backer, D. C., and Hellings, R. W. 1986, *Ann. Rev. Astr. Ap.*, **24**, 537.
Bailes, M. 1988, *Astr. Ap.*, **202**, 109.
Barker, B. M., and O'Connell, R. R. 1975, *Ap. J. (Letters)*, **199**, L25.
Bartel, N., Capallo, R. J., Ratner, M. I., Rogers, A. E. E., Shapiro, I. I., and Whitney, A. R. 1985, *A.J.*, **90**, 318.
Bertotti, B., Carr, B. J., and Rees, M. J. 1983, *M.N.R.A.S.*, **203**, 945.
Bevington, P. R. 1969, *Data Reduction and Error Analysis for the Physical Sciences* (New York: McGraw-Hill).
Blanchet, L., and Damour, T. 1989, *Phys. Rev.*, in press.
Blanchet, L., and Schäfer, G. 1989, *M.N.R.A.S.*, in press.
Blandford, R. D., and Teukolsky, S. A. 1976, *Ap. J.*, **205**, 580 (BT).
Boeshaar, P. C., Tyson, J. A., Pildis, R. A., and Wenk, R. A. 1988, *Bull. AAS*, **20**, 1048.
Boriakoff, V. 1973, Ph.D. thesis, Cornell University.
Burrows, A., and Woosley, S. P. 1986, *Ap. J.*, **308**, 680.
Cordes, J. M., and Downs, G. S. 1985, *Ap. J. Suppl.*, **59**, 343.
Cordes, J. M., and Wasserman, I. 1984, *Ap. J.*, **279**, 798.
Crane, P., Nelson, J. E., and Tyson, J. A. 1979, *Nature*, **280**, 367.
Damour, T. 1987, in *Gravitation in Astrophysics*, ed. B. Carter and J. B. Hartle (New York: Plenum), p. 3.
————. 1988, in *Proc. Second Canadian Conf. on General Relativity and Relativistic Astrophysics*, ed. A. Coley, C. Dyer, and T. Tupper (Singapore: World Scientific), p. 315.
Damour, T., and Deruelle, N. 1985, *Ann. Inst. H. Poincaré (Phys. Théorique)*, **43**, 107.
————. 1986, *Ann. Inst. H. Poincaré (Phys. Théorique)*, **44**, 263 (DD).
Damour, T., Gibbons, G. W., and Taylor, J. H. 1988, *Phys. Rev. Letters*, **61**, 1151.
Damour, T., and Ruffini, R. 1974, *C.R. Acad. Sci., Paris*, **279**, A971.
Damour, T., and Schäfer, G. 1988, *Nuovo Cimento*, **B101**, 127.
Damour, T., and Taylor, J. H. 1989, in preparation.
Davis, M. M., Taylor, J. H., Weisberg, J. M., and Backer, D. C. 1985, *Nature*, **315**, 547.
Eadie, W. T., Drijard, D., James, F. E., Roos, M., and Sadoulet, B. 1971, *Statistical Methods in Experimental Physics* (Amsterdam: North-Holland).

Einstein, A. 1918, *Preuss. Akad. Wiss. Sitzber. Berlin*, 154.
Elliott, K. H., Peterson, B. A., Wallace, P. T., Jones, D. H. P., Clements, E. D., Hartley, K. F., and Manchester, R. N. 1980, *M.N.R.A.S.*, **192**, 51P.
Epstein, R. 1977, *Ap. J.*, **216**, 92.
————. 1979, *Ap. J.*, **231**, 644.
Esposito, L. W., and Harrison, E. R. 1975, *Ap. J. (Letters)*, **196**, L1.
Fairhead, L., Bretagnon, P., and Lestrade, J.-F. 1988, in *The Earth's Rotation and Reference Frames for Geodesy and Geodynamics*, ed. A. K. Babcock and G. A. Wilkins (Dordrecht: Kluwer), p. 419.
Haugan, M. P. 1985, *Ap. J.*, **296**, 1.
————. 1988, preprint (H88).
Hellings, R. W., Adams, P. J., Anderson, J. D., Keesey, M. S., Lau, E. L., and Standish, E. M. 1983, *Phys. Rev. Letters*, **51**, 1609.
Hulse, R. A., and Taylor, J. H. 1975, *Ap. J. (Letters)*, **195**, L51.
Hunt, G. C., 1971, *M.N.R.A.S.*, **153**, 119.
Landau, L. D., and Lifshitz, E. M. 1962, *The Classical Theory of Fields* (Reading, MA: Addison-Wesley), p. 366.
Lyne, A. G., and Ritchings, R. T. 1977, *Nature*, **268**, 606.
Manchester, R. N., and Peters, W. 1972, *Ap. J.*, **173**, 221.
Manchester, R. N., and Taylor, J. H. 1977, *Pulsars* (San Francisco: Freeman).
Manchester, R. N., Taylor, J. H., and Van, Y. Y. 1974, *Ap. J. (Letters)*, **189**, L119.
Orsten, G. S. F. 1970, *Rev. Sci. Instr.*, **41**, 957.
Peters, P. C., and Mathews, J. 1963, *Phys. Rev.*, **131**, 435.
Press, W. H., Flannery, B. P., Teukolsky, S. A., and Vetterling, W. T. 1986, *Numerical Recipes* (Cambridge: Cambridge University Press).
Prózyński, M. 1984, in *Millisecond Pulsars*, ed. S. P. Reynolds and D. R. Stinebring (Green Bank: NRAO), p. 287.
Rawley, L. A. 1986, Ph.D. thesis, Princeton University.
Rawley, L. A., Taylor, J. H., and Davis, M. M. 1988, *Ap. J.*, **326**, 947.
Rawley, L. A., Taylor, J. H., Davis, M. M., and Allan, D. W. 1987, *Science*. **238**, 761.
Reasenberg, R. D. 1983, *Phil. Trans. Roy. Soc. London, A*, **310**, 227.
Shapiro, I. I. 1964, *Phys. Rev. Letters*, **13**, 798.
Shapiro, S. L., and Terzian, Y. 1976, *Astr. Ap.*, **52**, 115.
Smarr, L. L., and Blandford, R. 1976, *Ap. J.*, **207**, 574.
Srinivasan, G., and van den Heuvel, E. P. J. 1982, *Astr. Ap.*, **108**, 143.

450   TAYLOR AND WEISBERG

Standish, E. M. 1982, *Astr. Ap.*, **114**, 297.
Taylor, J. H. 1987, in *General Relativity and Gravitation*, ed. M. A. H. MacCallum (Cambridge: Cambridge University Press), p. 209.
————. 1989, in *Proc. NATO Advanced Study Institute, Timing Neutron Stars*, ed. H. Ögelman and E. P. J. van den Heuvel, (Dordrecht: Kluwer), p. 17.
Taylor, J. H., Fowler, L. A., and McCulloch, P. M. 1979, *Nature*, **277**, 437.
Taylor, J. H., Hulse, R. A., Fowler, L. A., Gullahorn, G. E., and Rankin, J. M. 1976, *Ap. J.* (*Letters*), **206**, L53.
Taylor, J. H., and Manchester, R. N. 1981, *A.J.*, **86**, 1953.
Taylor, J. H., and Weisberg, J. M. 1982, *Ap. J.*, **253**, 908 (TW).
van den Heuvel, E. P. J. 1987, in *The Origin and Evolution of Neutron Stars*, ed. D. J. Helfand and J.-H. Huang (Dordrecht: Reidel), p. 393.

Wagoner, R. V. 1975, *Ap. J.* (*Letters*), **196**, L63.
Wagoner, R. V., and Will, C. M. 1976, *Ap. J.*, **210**, 764.
Webbink, R. F. 1975, *Astr. Ap.*, **41**, 1.
Weisberg, J. M., Romani, R. W., and Taylor, J. H. 1989, *Ap. J.*, in press.
Weisberg, J. M., and Taylor, J. H. 1981, *Gen. Rel. Grav.*, **13**, 1.
————. 1984, *Phys. Rev. Letters*, **52**, 1348.
Will, C. M. 1981, *Theory and Experiment in Gravitational Physics* (Cambridge: Cambridge University Press).
————. 1986, *Canadian J. Phys.*, **64**, 140.

J. H. TAYLOR: Physics Department, Princeton University, Princeton, NJ   08544

J. M. WEISBERG: Department of Physics and Astronomy, Carleton College, Northfield, MN   55057

Binary pulsars and relativistic gravity[*]

Joseph H. Taylor, Jr.

Princeton University, Princeton, New Jersey 08544

I. SEARCH AND DISCOVERY

Work leading to the discovery of the first pulsar in a binary system began more than twenty years ago, so it seems reasonable to begin with a bit of history. Pulsars burst onto the scene (Hewish *et al.*, 1968) in February 1968, about a month after I completed my Ph.D. at Harvard University. Having accepted an offer to remain there on a post-doctoral fellowship, I was looking for an interesting new project in radio astronomy. When *Nature* announced the discovery of a strange new rapidly pulsating radio source, I immediately drafted a proposal, together with Harvard colleagues, to observe it with the 92 m radio telescope of the National Radio Astronomy Observatory. By late spring we had detected and studied all four of the pulsars which by then had been discovered by the Cambridge group, and I began thinking about how to find further examples of these fascinating objects, which were already thought likely to be neutron stars. Pulsar signals are generally quite weak, but have some unique characteristics that suggest effective search strategies. Their otherwise noise-like signals are modulated by periodic, impulsive waveforms; as a consequence, dispersive propagation through the interstellar medium makes the narrow pulses appear to sweep rapidly downward in frequency. I devised a computer algorithm for recognizing such periodic, dispersed signals in the inevitable background noise, and in June 1968 we used it to discover the fifth known pulsar (Huguenin *et al.*, 1968).

Since pulsar emissions exhibited a wide variety of new and unexpected phenomena, we observers put considerable effort into recording and studying their details and peculiarities. A pulsar model based on strongly magnetized, rapidly spinning neutron stars was soon established as consistent with most of the known facts (Gold, 1968). The model was strongly supported by the discovery of pulsars inside the glowing, gaseous remnants of two supernova explosions, where neutron stars should be created (Large, *et al.*, 1968; Staelin and Reifenstein, 1968), and also by an observed gradual lengthening of pulsar periods (Richards and Comella, 1969) and polarization measurements that clearly suggested a rotating source (Radhakrishnan and Cooke, 1969). The electrodynamical properties of a spinning, magnetized neutron star were studied theoretically (Goldreich and Julian, 1969) and shown to be plausibly capable of generating broad-

---

[*]Nobel Lecture, presented to the Royal Swedish Academy of Sciences on 8 December 1993.

band ratio noise detectable over interstellar distances. However, the rich diversity of the observed radio pulses suggested magnetospheric complexities far beyond those readily incorporated in theoretical models. Many of us suspected that detailed understanding of the pulsar emission mechanism might be a long time coming—and that, in any case, the details might not turn out to be fundamentally illuminating.

In September 1969 I joined the faculty at the University of Massachusetts, where a small group of us planned to build a large, cheap radio telescope especially for observing pulsars. Our telescope took several years to build, and during this time it became clear that whatever the significance of their magnetospheric physics, pulsars were interesting and potentially important to study for quite different reasons. As the collapsed remnants of supernova explosions, they could provide unique experimental data on the final stages of stellar evolution, as well as an opportunity to study the properties of nuclear matter in bulk. Moreover, many pulsars had been shown to be remarkably stable natural clocks (Manchester and Peters, 1972), thus providing an alluring challenge to the experimenter, with consequences and applications about which we could only speculate at the time. For such reasons as these, by the summer of 1972 I was devoting a large portion of my research time to the pursuit of accurate timing measurements of known pulsars, using our new telescope in western Massachusetts, and to planning a large-scale pulsar search that would use bigger telescopes at the national facilities.

I suspect it is not unusual for an experiment's motivation to depend, at least in part, on private thoughts quite unrelated to avowed scientific goals. The challenge of a good intellectual puzzle, and the quiet satisfaction of finding a clever solution, must certainly rank highly among my own incentives and rewards. If an experiment seems difficult to do, but plausibly has interesting consequences, one feels compelled to give it a try. Pulsar searching is the perfect example: it's clear that there must be lots of pulsars out there, and, once identified, they are not so very hard to observe. But finding each one for the first time is a formidable task, one that can become a sort of detective game. To play the game you invent an efficient way of gathering clues, sorting, and assessing them, hoping to discover the identities and celestial locations of all the guilty parties.

Most of the several dozen pulsars known in early 1972 were discovered by examination of strip-chart records, without benefit of further signal processing. Nevertheless, it was clear that digital computer techniques would

Reviews of Modern Physics, Vol. 66, No. 3, July 1994   0034-6861/94/66(3)/711(9)/$05.90

be essential parts of more sensitive surveys. Detecting new pulsars is necessarily a multidimensional process; in addition to the usual variables of two spatial coordinates, one must also search thoroughly over wide ranges of period and dispersion measure. Our first pulsar survey, in 1968, sought evidence of pulsar signals by computing the discrete Fourier transforms of long sequences of intensity samples, allowing for the expected narrow pulse shapes by summing the amplitudes of a dozen or more harmonically related frequency components. I first described this basic algorithm (Burns and Clark, 1969) as part of a discussion of pulsar search techniques, in 1969. An efficient dispersion-compensating algorithm was conceived and implemented soon afterward (Manchester *et al.*, 1972; Taylor, 1974), permitting extension of the method to two dimensions. Computerized searches over period and dispersion measure, using these basic algorithms, have by now accounted for discovery of the vast majority of nearly 600 known pulsars, including forty in binary systems (Taylor *et al.*, 1993; Camilo, 1994).

In addition to private stimuli related to "the thrill of the chase," my outwardly expressed scientific motivation for planning an extensive pulsar survey in 1972 was a desire to double or triple the number of known pulsars. I had in mind the need for a more solid statistical basis for drawing conclusions about the total number of pulsars in the Galaxy, their spatial distribution, how they fit into the scheme of stellar evolution, and so on. I also realized (Taylor, 1972) that it would be highly desirable ". . . to find even *one* example of a pulsar in a binary system, for measurement of its parameters could yield the pulsar mass, an extremely important number." Little did I suspect that just such a discovery would be made, or that it would have much greater significance that anyone had foreseen! In addition to its own importance, the binary pulsar PSR 1913+16 is now recognized as the harbinger of a new class of unusually short-period pulsars with numerous important applications.

An up-to-date map of known pulsars on the celestial sphere is shown in Fig. 1. The binary pulsar PSR 1913+16 is found in a clump of objects close to the Galactic plane around longitude 50°, a part of the sky that passes directly overhead at the latitude of the Arecibo Observatory in Puerto Rico. Forty of these pulsars, including PSR 1913+16, were discovered in the survey that Russell Hulse and I carried out with the 305 m Arecibo telescope (Hulse and Taylor, 1974, 1975a, 1975b). Figure 2 illustrates the periods and spin-down rates of known pulsars, with those in binary systems marked by larger circles around the dots. All radio pulsars slow down gradually in their own rest frames, but the slow-down rates vary over nine orders of magnitude. Figure 2 makes it clear that binary pulsars are special in this regard. With few exceptions, they have unusually small values of both period and period derivative—an important factor which helps to make them especially suitable for high-precision timing measurements.

Much of the detailed implementation and execution of

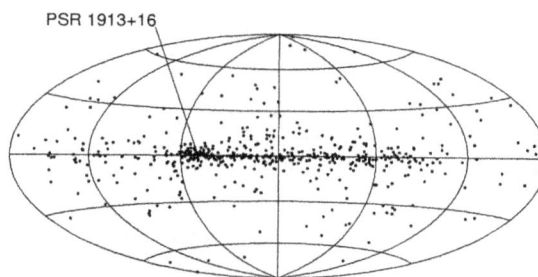

FIG. 1. Distribution of 558 pulsars in Galactic coordinates. The Galactic center is in the middle, and longitude increases to the left.

our 1973–74 Arecibo survey was carried out by Russell Hulse. He describes that work, and particularly the discovery of PSR 1913+16, in his accompanying lecture (Hulse, 1994). The significant consequences of our discovery have required accurate timing measurements extending over many years, and since 1974–76 I have pursued these with a number of other collaborators. I shall now turn to a description of these observations.

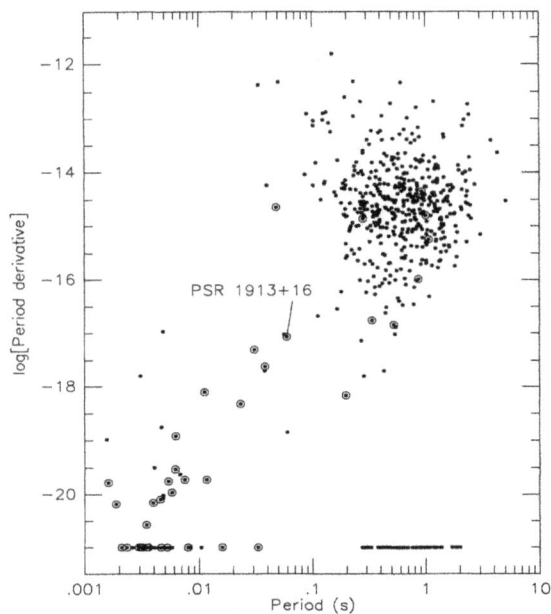

FIG. 2. Periods and period derivatives of known pulsars. Binary pulsars, denoted by larger circles around the dots, generally have short periods and small derivatives. Symbols aligned near the bottom represent pulsars for which the slow-down rate has not yet been measured.

II. CLOCK-COMPARISON EXPERIMENTS

Pulsar timing experiments are straightforward in concept: one measures pulse times of arrival (TOAs) at the telescope, and compares them with time kept by a stable reference clock. A remarkable wealth of information about a pulsar's spin, location in space, and orbital motion can be obtained from such simple measurements. For binary pulsars, especially, the task of analyzing a sequence of TOAs often assumes the guise of another intricate detective game. Principal clues in this game are the recorded TOAs. The first and most difficult objective is the assignment of unambiguous pulse numbers to each TOA, despite the fact that some of the observations may be separated by months or even years from their nearest neighbors. During such inevitable gaps in the data, a pulsar may have rotated through as many as 10^7–10^{10} turns, and in order to extract the maximum information content from the data, these integers must be recovered *exactly*. Fortunately, the correct sequence of pulse numbers is easily recognized, once attained, so you can tell when the game has been "won."

A block diagram of equipment used for recent pulsar timing observations (Taylor, 1991) at Arecibo is shown in Fig. 3. Incoming radio-frequency signals from the antenna are amplified, converted to intermediate frequency, and passed through a multichannel spectrometer equipped with square-law detectors. A bank of digital signal averagers accumulates estimates of a pulsar's periodic wave form in each spectral channel, using a precomputed digital ephemeris and circuitry synchronized with the observatory's master clock. A programmable synthesizer, its output frequency adjusted once a second in a phase-continuous manner, compensates for changing Doppler shifts caused by accelerations of the pulsar and the telescope. Average profiles are recorded once every few minutes, together with appropriate time tags. A log is kept of small measured offsets (typically of order 1 μs) between the observatory clock and the best

available standards at national time-keeping laboratories, with time transfer accomplished via satellites in the Global Positioning System.

An example of pulse profiles recorded during timing observations of PSR 1913+16 is presented in Fig. 4, which shows intensity profiles for 32 spectral channels spanning the frequency range 1383–1423 MHz, followed by a "de-dispersed" profile at the bottom. In a five-minute observation such as this, the signal-to-noise ratio is just high enough for the double-peaked pulse shape of PSR 1913+16 to be evident in the individual channels. Pulse arrival times are determined by measuring the phase offset between each observed profile and a long-term average with much higher signal-to-noise ratio. Differential dispersive delays are removed, the adjusted offsets are averaged over all channels, and the resulting mean value is added to the time tag to obtain an equivalent TOA. Nearly 5000 such five-minute measurements have been obtained for PSR 1913+16 since 1974, suing essentially this technique. Through a number of improvements in the data-taking systems (Taylor *et al.*, 1976; McCulloch *et al.*, 1979; Taylor *et al.*, 1979; Taylor and Weisberg, 1982, 1989; Stinebring *et al.*, 1992), the typical uncertainties have been reduced from around 300 μs in 1974 to 15–20 μs since 1981.

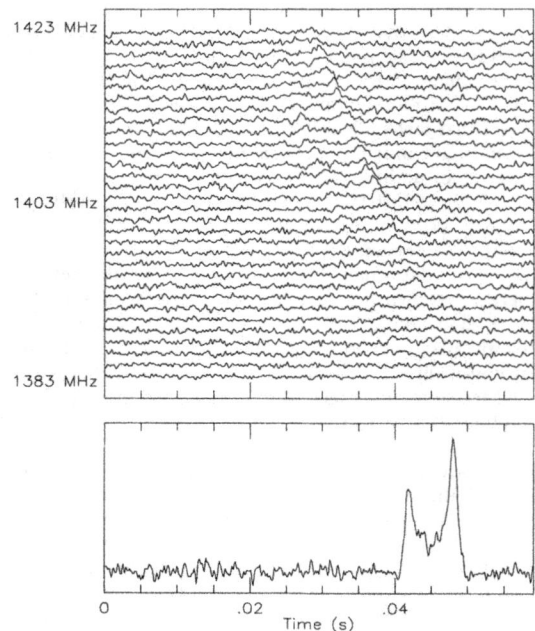

FIG. 4. Pulse profiles obtained on April 24, 1992 during a five-minute observation of PSR 1913+16. The characteristic double-peaked shape, clearly seen in the de-dispersed profile at the bottom, is also discernible in the 32 individual spectral channels.

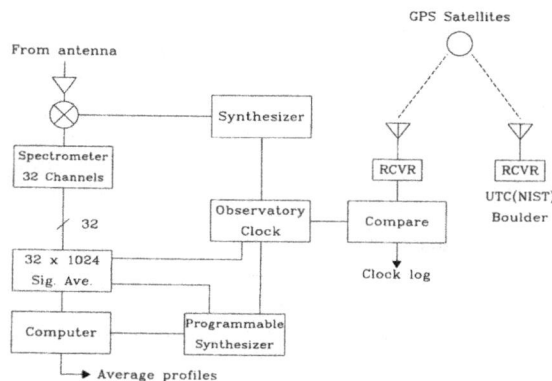

FIG. 3. Simplified block diagram of equipment using for timing pulsars at Arecibo.

III. MODEL FITTING

In the process of data analysis, each measured topocentric TOA, say t_{obs}, must be transformed to a corresponding proper time of emission T in the pulsar frame. Under the assumption of a deterministic spin-down law, the rotational phase of the pulsar is given by

$$\phi(T) = \nu T + \frac{1}{2}\dot{\nu}T^2 , \qquad (1)$$

where ϕ is measured in cycles, $\nu \equiv 1/P$ is the rotation frequency, P the period, and $\dot{\nu}$ the slowdown rate. Since a topocentric TOA is a relativistic space-time event, it must be transformed as a four-vector. The telescope's location at the time of a measurement is obtained from a numerically integrated solar-system model, together with published data on the Earth's unpredictable rotational variations. As a first step one normally transforms to the solar-system barycenter, using the weak-field, slow-motion limit of general relativity. The necessary equations include terms depending on the positions, velocities, and masses of all significant solar-system bodies. Next, one accounts for propagation effects in the interstellar medium; and finally, for the orbital motion of the pulsar itself.

With presently achievable accuracies, all significant terms in the relativistic transformation can be summarized in the single equation

$$T = t_{\text{obs}} - t_0 + \Delta_C - D/f^2 + \Delta_{R\odot}(\alpha,\delta,\mu_\alpha,\mu_\delta,\pi)$$
$$+ \Delta_{E\odot} - \Delta_{S\odot}(\alpha,\delta)$$
$$- \Delta_R(x,e,P_b,T_0,\omega,\dot{\omega},\dot{P}_b) - \Delta_E(\gamma) - \Delta_S(r,s) . \qquad (2)$$

Here t_0 is a nominal equivalent TOA at the solar-system barycenter; Δ_C represents measured clock offsets; D/f^2 is the dispersive delay for propagation at frequency f through the interstellar medium; $\Delta_{R\odot}$, $\Delta_{E\odot}$, and $\Delta_{S\odot}$ are propagation delays and relativistic time adjustments within the solar system; and Δ_R, Δ_E, and Δ_S are similar terms for effects within a binary pulsar's orbit. Subscripts on the various Δ's indicate the nature of the time-dependent delays, which include "Römer," "Einstein," and "Shapiro" delays in the solar system and in the pulsar orbit. The Römer terms have amplitudes comparable to the orbital periods times $v/2\pi c$, where v is the orbital velocity and c the speed of light. The Einstein terms, representing the integrated effects of gravitational redshift and time dilation, are smaller by another factor ev/c, where e is the orbital eccentricity. The Shapiro time delay is a result of reduced velocities that accompany the well-known bending of light rays propagating close to a massive object. The delay amounts to about 120 μs for one-way lines of sight grazing the Sun, and the magnitude depends logarithmically on the angular impact parameter. The corresponding delay within a binary pulsar orbit depends on the companion star's mass, the orbital phase, and the inclination i between the orbital angular momentum and the line of sight.

Figure 5 illustrates the combined orbital delay $\Delta_R + \Delta_E + \Delta_S$ for PSR 1913+16, plotted as a function of orbital phase. Despite the fact that the Einstein and Shapiro effects are orders of magnitude smaller than the Römer delay, they can still be measured separately if the precision of available TOAs is high enough. In fact, the available precision is very high indeed, as one can see from the lone data point shown in Fig. 5 with $50\,000\sigma$ error bars.

Equations (1) and (2) have been written to show explicitly the most significant dependences of pulsar phase on as many as nineteen *a priori* unknowns. In addition to the rotational frequency ν and spin-down rate $\dot{\nu}$, these phenomenological parameters include a reference arrival time t_0, the dispersion constant D, celestial coordinates α and δ, proper-motion terms μ_α and μ_δ, and annual parallax π. For binary pulsars the terms on the third line of Eq. (2), with as many as ten significant orbital parameters, are also required. The additional parameters include five that would be necessary even in a purely Keplerian analysis of orbital motion: the projected semimajor axis $x \equiv a_1 \sin i/c$, eccentricity e, binary period P_b, longitude of periastron ω, and time of periastron T_0. If the experimental precision is high enough, relativistic effects can yield the values of five further "post-Keplerian" parameters: the secular derivatives $\dot{\omega}$ and $\dot{P}_b$, the Einstein parameter γ, and the range and shape of the orbital Shapiro delay, r and $s \equiv \sin i$. Several earlier versions of this formalism for treating timing measurements of binary pulsars exist (Blandford and Teukolsky, 1976; Epstein, 1977; Haugan, 1985), and have been historically important to our progress with the PSR 1913+16 experiment. The elegant framework outlined here was derived during 1985–86 by Damour and Deruelle (1985, 1986).

Model parameters are extracted from a set of TOAs by calculating the pulsar phases $\phi(T)$ from Eq. (1) and minimizing the weighted sum of squared residuals,

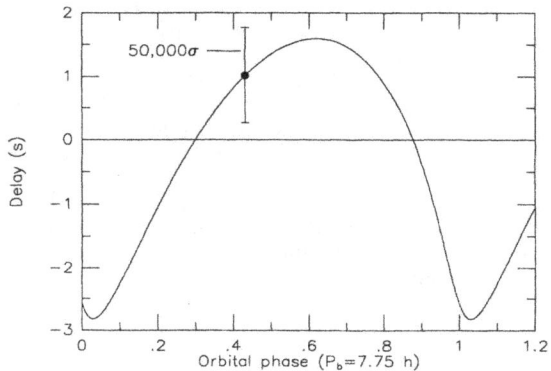

FIG. 5. Orbital delays observed for PSR 1913+16 during July, 1988. The uncertainty of an individual five-minute measurement is typically 50 000 times smaller than the error bar shown.

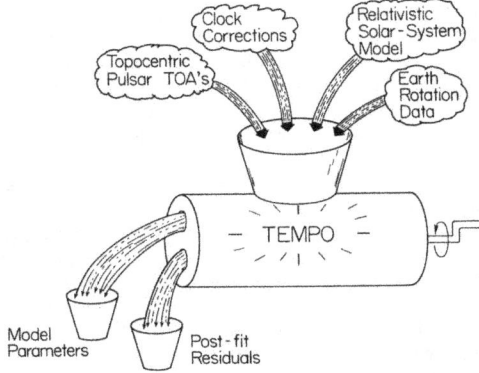

FIG. 6. Schematic diagram of the analysis of pulsar timing measurements carried out by the computer program TEMPO. The essential functions are all described in the text.

$$\chi^2 = \sum_{i=1}^{N} \left[\frac{\phi(T_i) - n_i}{\sigma_i / P} \right]^2 , \qquad (3)$$

with respect to each parameter to be determined. In this equation, n_i is the closest integer to $\phi(T_i)$, and σ_i is the estimated uncertainty of the ith TOA. In a valid and reliable solution the value of χ^2 will be close to the number of degrees of freedom, i.e., the number of measurements N minus the number of adjustable parameters. Parameter errors so large that the closest integer to $\phi(T_i)$ may not be the correct pulse number are invariably accompanied by huge increases in χ^2; this is the reason for my earlier statement that correct pulse numbering is easily recognizable, once attained. In addition to providing a list of fitted parameter values and their estimated uncertainties, the least-squares solution produces a set of post-fit residuals, or differences between measured TOAs and those predicted by the model (see Fig. 6). The post-fit residuals are carefully examined for evidence of systematic trends that might suggest experimental errors, or some inadequacy in the astrophysical model, or perhaps deep physical truths about the nature of gravity.

Necessarily some model parameters will be easier to measure than others. When many TOAs are available, spaced over many months or years, it generally follows that at least the pulsar's celestial coordinates, spin parameters, and Keplerian orbital elements will be measurable with high precision, often as many as 6–14 significant digits. As we will see, the relativistic parameters of binary pulsar orbits are generally much more difficult to measure—but the potential rewards for doing so are substantial.

IV. THE NEWTONIAN LIMIT

Thirty-five binary pulsar systems have now been studied well enough to determine their basic parameters, including the Keplerian orbital elements, with good accu-

racy. For each system the orbital period P_b and projected semimajor axis x can be combined to give the mass function.

$$f_1(m_1, m_2, s) = \frac{(m_2 s)^3}{(m_1 + m_2)^2} = \frac{x^3}{T_\odot (P_b / 2\pi)^2} . \qquad (4)$$

Here m_1 and m_2 are the masses of the pulsar and companion in units of the Sun's mass, $M_\odot$; I use the shorthand notations $s \equiv \sin i$, $T_\odot \equiv GM_\odot / c^3 = 4.925\,490\,947 \times 10^{-6}$ s, where G is the Newtonian constant of gravity. In the absence of other information, the mass function cannot provide unique solutions for m_1, m_2, or s. Nevertheless, likely values of m_2 can be estimated by assuming a pulsar mass close to $1.4 M_\odot$ (the Chandrasekhar limit for white dwarfs) and the median value $\cos i = 0.5$, which implies $s = 0.87$. With this approach one can distinguish three categories of binary pulsars, which I shall discuss by reference to Fig. 7: a plot of binary pulsar companion masses versus orbital eccentricities.

Twenty-eight of the binary systems in Fig. 7 have orbital eccentricities $e < 0.25$ and low-mass companions likely to be degenerate dwarfs. Most of these have nearly circular orbits; indeed, the only ones with eccentricities more than a few percent are located in globular clusters, and their orbits have probably been perturbed by near collisions with other stars. Five of the binaries have much larger eccentricities and likely companion masses of $0.8 M_\odot$ or more; these systems are thought to be pairs of neutron stars, one of which is the detectable pulsar. The large orbital eccentricities are almost certainly the result of rapid ejection of mass in the supernova explosion creating the second neutron star. Finally, at the upper right of Fig. 7 we find two binary pulsars that

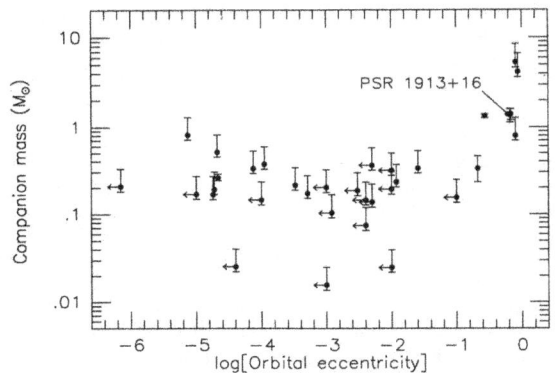

FIG. 7. Masses of the companions of binary pulsars, plotted as a function or orbital eccentricity. Near the marked location of PSR 1913+16, three distinct symbols have merged into one; these three binary systems, as well as their two nearest neighbors in the graph, are thought to be pairs of neutron stars. The two pulsars at the upper right are accompanied by high-mass main-sequence stars, while the remainder are believed to have white-dwarf companions.

move in eccentric orbits around high-mass main-sequence stars. These systems have not yet evolved to the stage of a second supernova explosion. Unlike the binary pulsars with compact companions, these two systems have orbits that could be significantly modified by complications such as tidal forces or mass loss.

V. GENERAL RELATIVITY AS A TOOL

As Russell Hulse and I suggested in the discovery paper for PSR 1913+16 (Hulse and Taylor, 1975a), it should be possible to combine measurements of relativistic orbital parameters with the mass function, thereby determining masses of both stars and the orbital inclination. In the post-Keplerian (PK) framework outlined above, each measured PK parameter defines a unique curve in the (m_1, m_2) plane, valid within a specified theory of gravity. Experimental values for any two PK parameters (say, $\dot{\omega}$ and γ, or perhaps r and s) establish the values of m_1, m_2, and s unambiguously. In general relativity the equations for the five most significant PK parameters are as follows (Damour and Deruelle, 1986; Taylor and Weisberg, 1989; Damour and Taylor, 1992):

$$\dot{\omega} = 3 \left[\frac{P_b}{2\pi} \right]^{-5/3} (T_\odot M)^{2/3} (1-e^2)^{-1} , \quad (5)$$

$$\gamma = e \left[\frac{P_b}{2\pi} \right]^{1/3} T_\odot^{2/3} M^{-4/3} m_2 (m_1 + 2m_2) , \quad (6)$$

$$\dot{P}_b = -\frac{192\pi}{5} \left[\frac{P_b}{2\pi} \right]^{-5/3} \left[1 + \frac{73}{24} e^2 + \frac{37}{96} e^4 \right]$$
$$\times (1-e^2)^{-7/2} T_\odot^{5/3} m_1 m_2 M^{-1/3} , \quad (7)$$

$$r = T_\odot m_2 , \quad (8)$$

$$s = x \left[\frac{P_b}{2\pi} \right]^{-2/3} T_\odot^{-1/3} M^{2/3} m_2^{-1} . \quad (9)$$

Again the masses m_1, m_2, and $M \equiv m_1 + m_2$ are expressed in solar units. I emphasize that the left-hand sides of Eqs. (5) through (9) represent directly measurable quantities, at least in principle. Any two such measurements, together with the well-determined values of e and P_b, will yield solutions for m_1 and m_2, as well as explicit predictions for the remaining PK parameters.

The binary systems most likely to yield measurable PK parameters are those with large masses and high eccentricities and which are astrophysically "clean," so that their orbits are overwhelmingly dominated by the gravitational interactions between two compact masses. The five pulsars clustered near PSR 1913+16 in Fig. 7 would seem to be especially good candidates, and this has been borne out in practice. In the most favorable circumstances, even binary pulsars with low-mass companions and nearly circular orbitals can yield significant post-Keplerian measurements. The best present example is PSR 1855+09: its orbital plane is nearly parallel to

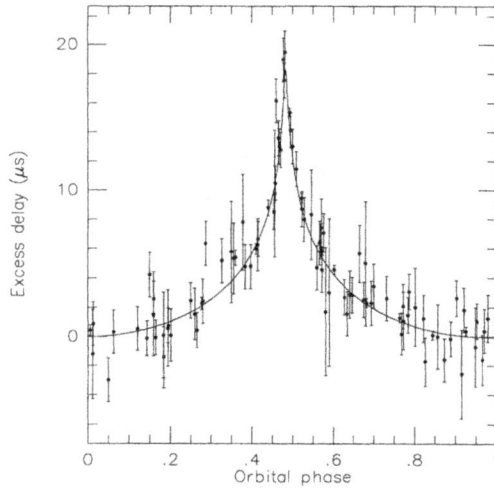

FIG. 8. Measurements of the Shapiro time delay in the PSR 1855+09 system. The theoretical curve corresponds to Eq. (10), and the fitted values of r and s can be used to determine the masses of the pulsar and companion star.

the line of sight, greatly magnifying the orbital Shapiro delay. The relevant measurements (Rawley et al., 1988; Ryba and Taylor, 1991; Kaspi et al., 1994) are illustrated in Fig. 8, together with the fitted function $\Delta_S(r,s)$, in this case closely approximated by

$$\Delta_S = -2r \log(1 - s \cos[2\pi(\phi - \phi_0)]) , \quad (10)$$

where ϕ is the orbital phase in cycles and $\phi_0 = 0.4823$ the phase of superior conjunction. The fitted values of r and s yield the masses $m_1 = 1.50^{+0.26}_{-0.14}$, $m_2 = 0.258^{+0.028}_{-0.016}$. In a

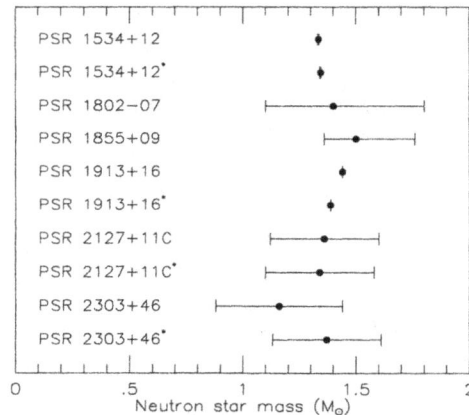

FIG. 9. The masses of ten neutron stars, measured by observing relativistic effects in binary pulsar orbits. Asterisks after pulsar names denote companions to the observed pulsars.

similar way, all binary pulsars with two measurable **PK** parameters yield solutions for their component masses. At present, most of the experimental data on the masses of neutron stars (see Fig. 9) come from such timing analyses of binary pulsar systems (Taylor and Dewey, 1988; Thorsett *et al.*, 1993, and references therein).

VI. TESTING FOR GRAVITATIONAL WAVES

If three or more post-Keplerian parameters can be measured for a particular pulsar, the system becomes over-determined, and the extra experimental degrees of freedom transform it into a calibrated laboratory for testing relativistic gravity. Each measurable **PK** parameter beyond the first two provides an explicit, quantitative test. Because the velocities and gravitational energies in a high-mass binary pulsar system can be significantly relativistic, strong-field and radiative effects come into play. Two binary pulsars, PSRs 1913+16 and 1534+12, have now been timed well enough and long enough to yield three or more **PK** parameters. Each one provides significant tests of gravitation beyond the weak-field, slow-motion limit (Damour and Taylor, 1982; Taylor *et al.*, 1992).

PSR 1913+16 has an orbital period $P_b \approx 7.8$ h, eccentricity $e \approx 0.62$, and mass function $f_1 \approx 0.13 M_\odot$. With the available data quality and time span, the Keplerian orbital parameters are actually determined with fractional accuracies of a few parts per million, or better. In addition, the **PK** parameters $\dot{\omega}$, γ, and $\dot{P}_b$ are determined with fractional accuracies better than 3×10^{-6}, 5×10^{-4}, and 4×10^{-3}, respectively (Taylor and Weisberg, 1989; Taylor, 1993). Within any viable relativistic theory of gravity, the values of $\dot{\omega}$ and γ yield the values of m_1 and m_2 and a corresponding prediction for $\dot{P}_b$ arising from the damping effects of gravitational radiation. At present levels of accuracy, a small kinematic correction (approximately 0.5% of the observed $\dot{P}_b$) must be included to account for accelerations of the solar system and the binary pulsar system in the Galactic gravitational field (Damour and Taylor, 1991). After doing so, we find that Einstein's theory passes this extraordinarily stringent test with a fractional accuracy better than 0.4% (see Figs. 10 and 11). The clock-comparison experiment for PSR 1913+16 thus provides direct experimental proof that changes in gravity propagate at the speed of light, thereby creating a dissipative mechanism in an orbiting system. It necessarily follows that gravitational radiation exists and has a quadrupolar nature.

PSR 1534+12 was discovered just three years ago, in a survey by Aleksander Wolszczan (1991) that again used the huge Arecibo telescope to good advantage. This pulsar promises eventually to surpass the results now available from PSR 1913+16. It has orbital period $P_b \approx 10.1$ h, eccentricity $e \approx 0.27$, and mass function $f_1 \approx 0.31 M_\odot$. Moreover, with a stronger signal and narrower pulse than PSR 1913+16, its TOAs have considerably smaller measurement uncertainties, around 3 μs for five-minute

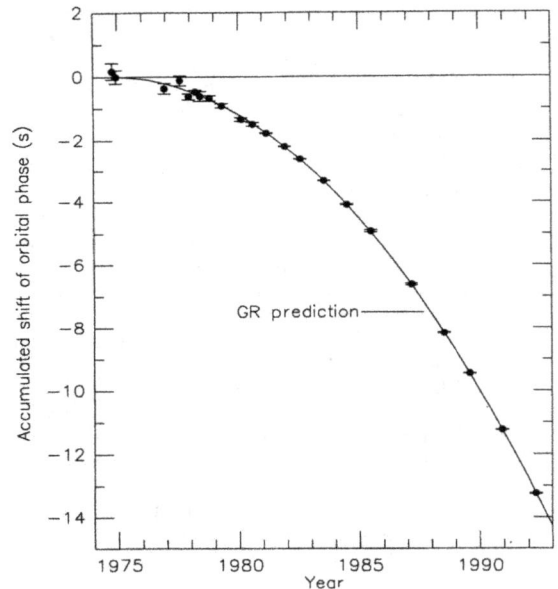

FIG. 10. Accumulated shift of the times of periastron in the PSR 1913+16 system, relative to an assumed orbit with constant period. The parabolic curve represents the general relativistic prediction for energy losses from gravitational radiation.

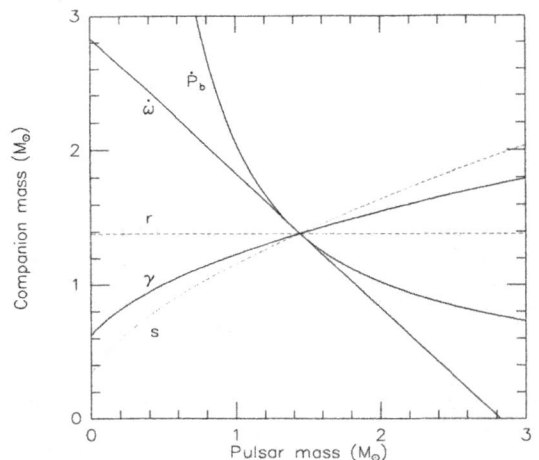

FIG. 11. Solid curves correspond to Eqs. (5)–(7) together with the measured values of $\dot{\omega}$, γ, and $\dot{P}_b$. Their intersection at a single point (within the experimental uncertainty of about 0.35% in $\dot{P}_b$), establishes the existence of gravitational waves. Dashed curves correspond to the *predicted* values of parameters r and s; these quantities should become measurable with a modest improvement in data quality.

718   Joseph H. Taylor, Jr.: Binary pulsars and relativistic gravity

observations. Results based on 15 months of data (Taylor *et al.*, 1992) have already produced significant measurements of four PK parameters: $\dot{\omega}$, γ, r, and s. In recent work not yet published, Wolszczan and I have measured the orbital decay rate, $\dot{P}_b$, and found it to be in accord with general relativity at about 20% level. In fact, *all* measured parameters of the PSR 1534+12 system are consistent within general relativity, and it appears that when the full experimental analysis is complete, Einstein's theory will have passed three more very stringent tests under strong-field and radiative conditions.

I do not believe that general relativity necessarily contains the last valid words to be written about the nature of gravity. The theory is not, of course, a quantum theory, and at its most fundamental level the universe appears to obey quantum-mechanical rules. Nevertheless, our experiments with binary pulsars show that, whatever the precise directions of future theoretical work may be, the correct theory of gravity must make predictions that are asymptotically close to those of general relativity over a vast range of classical circumstances.

ACKNOWLEDGMENTS

Russell Hulse and I have many individuals to thank for their important work, both experimental and theoretical, without which our discovery of PSR 1913+16 could not have borne fruit so quickly or so fully. Most notable among these are Roger Blandford, Thibault Damour, Lee Fowler, Peter McCulloch, Joel Weisberg, and the skilled and dedicated technical staff of the Arecibo Observatory.

REFERENCES

Blandford, R., and S. A. Teukolsky, 1976, "Arrival-time analysis for a pulsar in a binary system," Astrophys. J. **205**, 580–591.
Burns, W. R. and B. G. Clark, 1969, "Pulsar search techniques," Astron. Astrophys. **2**, 280–287.
Camilo, F., 1994, "Millisecond pulsar searches," in *Lives of the Neutron Stars, NATO ASI Series,* edited by A. Alpar (Kluwer, Dordrecht).
Damour, T., and N. Deruelle, 1985, "General relativistic celestial mechanics of binary systems. I. The post-Newtonian motion," Ann. Inst. Henri Poincaré Phys. Théor. **43**, 107–132.
Damour, T., and N. Deruelle, 1986, "General relativistic celestial mechanics of binary systems. II. The post-Newtonian timing formula," Ann. Inst. Henri Poincaré Phys. Théor. **44**, 263–292.
Damour, T., and J. H. Taylor, 1991, "On the orbital period change of the binary pulsar PSR 1913+16," Astrophys. J. **366**, 501–511.
Damour, T., and J. H. Taylor, 1992, "Strong-field tests of relativistic gravity and binary pulsars," Phys. Rev. D **45**, 1840–1868.
Epstein, R., 1977, "The binary pulsar: Post Newtonian timing effects," Astrophys. J. **216**, 92–100.

Gold, T., 1968, "Rotating neutron stars as the origin of the pulsating radio sources," Nature **218**, 731–732.
Goldreich, P., and W. H. Julian, 1969, "Pulsar electrodynamics," Astrophys. J. **157**, 869–880.
Haugan, M. P., 1985, "Post-Newtonian arrival-time analysis for a pulsar in a binary system," Astrophys. J. **296**, 1–12.
Hewish, A., S. J. Bell, J. D. H. Pilkington, P. F. Scott, and R. A. Collins, 1968, "Observation of a rapidly pulsating radio source," Nature **217**, 709–713.
Huguenin, G. R., J. H. Taylor, L. E. Goad, A. Hartai, G. S. F. Orsten, and A. K. Rodman, 1968, "New pulsating radio source," Nature **219**, 576.
Hulse, R. A., 1994, "The discovery of the binary pulsar," in Les Prix Nobel (The Nobel Foundation).
Hulse, R. A., and J. H. Taylor, 1974, "A high sensitivity pulsar survey," Astrophys. J. **191**, L59–L61.
Hulse, R. A. and J. H. Taylor, 1975a, "Discovery of a pulsar in a binary system," Astrophys. J. **195**, L51–L53.
Hulse, R. A., and J. H. Taylor, 1975b, "A deep sample of new pulsars and their spatial extent in the galaxy," Astrophys. J. **201**, L55-L59.
Kaspi, V. M., J. H. Taylor, and M. Ryba, 1994, "High-precision timing of millisecond pulsars. III. Long-term monitoring of PSRs B1855+09 and B1937+21," Astrophys. J. (in press).
Large M. I., A. E. Vaughan, and B. Y. Mills, 1968, "A pulsar supernova association," Nature **220**, 340–341.
Manchester, R. N., and W. L. Peters, 1972, "Pulsar parameters from timing observations," Astrophys. J. **173**, 221–226.
Manchester, R. N., J. H. Taylor, and G. R. Huguenin, 1972, "New and improved parameters for twenty-two pulsars," Nature Phys. Sci. **240**, 74.
McCulloch, P. M., J. H. Taylor, and J. M. Weisberg, 1979, "Tests of a new dispersion-removing radiometer on binary pulsar PSR 1913+16," Astrophys. J. **227**, L133-L137.
Radhakrishnan, V., and D. J. Cooke, 1969, "Magnetic poles and the polarization structure of pulsar radiation," Astrophys. Lett. **3**, 225–229.
Rawley, L. A., J. H. Taylor, and M. M. Davis, 1988, "Fundamental astrometry and millisecond pulsars," Astrophys. J. **326**, 947–953.
Richards, D. W., and J. M. Comella, 1969, "The period of pulsar NP 0532," Nature **222**, 551–552.
Ryba, M. F., and J. H. Taylor, 1991, "High precision timing of millisecond pulsars. I. Astrometry and masses of the PSR 1855+09 system," Astrophys. J. **371**, 739–748.
Staelin, D. H., and E. C. Reifenstein, III, 1968, "Pulsating radio sources near the Crab Nebula," Science **162**, 1481–1483.
Stinebring, D. R., V. M. Kaspi, D. J. Nice, M. F. Ryba, J. H. Taylor, S. E. Thorsett, and T. H. Hankins, 1992, "A flexible data acquisition system for timing pulsars," Rev. Sci. Instrum. **63**, 3551–3555.
Taylor, J. H., 1972, "A high sensitivity survey to detect new pulsars," research proposal submitted to the US National Science Foundation, September, 1972.
Taylor, J. H., 1974, "A sensitive method for detecting dispersed radio emission," Astron. Astrophys. Suppl. Ser. **15**, 367.
Taylor, J. H., 1991, "Millisecond pulsars: Nature's most stable clocks," Proc. IEEE **79**, 1054–1062.
Taylor, J. H., 1993, "Testing relativistic gravity with binary and millisecond pulsars," in *General Relativity and Gravitation 1992*, edited by R. J. Gleiser, C. N. Kozameh, and O. M. Moreschi (Institute of Physics, Bristol), pp. 287–294.
Taylor, J. H., and R. J. Dewey, 1988, "Improved parameters for four binary pulsars," Astrophys. J. **332**, 770–776.

Taylor, J. H., L. A. Fowler, and P. M. McCulloch, 1979, "Measurements of general relativistic effects in the binary pulsar PSR 1913+16," Nature **277**, 437.

Taylor, J. H., R. A. Hulse, L. A. Fowler, G. W. Gullahorn, and J. M. Rankin, 1976, "Further observations of the binary pulsar PSR 1913+16," Astrophys. J. **206**, L53–L58.

Taylor, J. H., R. N. Manchester, and A. G. Lyne, 1993, "Catalog of 558 pulsars," Astrophys. J. Suppl. Ser. **88**, 529–568.

Taylor, J. H., and J. M. Weisberg, 1982, "A new test of general relativity: Gravitational radiation and the binary pulsar PSR 1913+16," Astrophys. J. **253**, 908–920.

Taylor, J. H., and J. M. Weiberg, 1989, "Further experimental tests of relativistic gravity using the binary pulsar PSR 1913+16," Astrophys. J. **345**, 434–450.

Taylor, J. H., A. Wolszczan, T. Damour, and J. M. Weisberg, 1992, "Experimental constraints on strong-field relativistic gravity," Nature **355**, 132–136.

Thorsett, S. E., Z. Arzoumanian, M. M. McKinnon, and J. H. Taylor, 1993, "The masses of two binary neutron star systems," Astrophys. J. **405**, L29–L32.

A. Wolszczan, 1991, "A nearby 37.9 ms radio pulsar in a relativistic binary system," Nature **350**, 688–690.

LIST OF PUBLICATIONS

1. Taylor, J. H., "Low Cost Radio Telescope," Am. J. Physics, 32, 546 (1964).

2. Taylor, J. H., "Brightness Distribution of Antares from Lunar Occultations," Nature, 210, 105 (1966).

3. Taylor, J. H., "Lunar Occultations of Five Radio Sources," Astrophys. J., 146, 646 (1966).

4. Taylor, J. H., "Two Dimensional Brightness Distributions of Radio Sources from Lunar Occultation Observations," Astrophys. J., 150, 421 (1967).

5. Taylor, J. H., and De Jong, M. L., "Models of Nine Radio Sources from Lunar Occultation Observations," Astrophys. J., 151, 33 (1968).

6. Taylor, J. H., "Lunar Occultations of Radio Sources," Ph.D. Dissertation, Harvard University (1968).

7. Huguenin, G. R., Taylor, J. H., Goad, L. E., Hartai, A., Orsten, G. S. F., and Rodman, A. K., "New Pulsating Radio Source," Nature, 219, 576 (1968).

8. Maxwell, A., and Taylor, J. H., "Lunar Occultations of the Radio Source Sagittarius A," Astrophys. Lett., 2, 191 (1968).

9. Huguenin, G. R., and Taylor, J. H., "Dynamic Spectra of Pulsars," Astrophys. Lett., 3, 107 (1969).

10. Taylor, J. H., and Huguenin, G. R., "Two New Pulsating Radio Sources," Nature, 221, 816 (1969).

11. Taylor, J. H., "Catalog of 37 Pulsars," Astrophys. Lett., 3, 205 (1969).

12. Huguenin, G. R., Taylor, J. H., and Jura, M., "Dynamic Spectra of Pulsars in the Frequency Range 110–420 MHz," Astrophys. Lett., 4, 71 (1969).

13. Taylor, J. H., Jura, M., and Huguenin, G. R., "Periodic Intensity Fluctuations in Pulsars," Nature, 223, 797 (1969).

14. Huguenin, G. R., Taylor, J. H., and Troland, T. H., "The Radio Emission from Pulsar MP 0031+07," Astrophys. J., 162, 727 (1970).

15. Taylor, J. H., and Huguenin, G. R., "Observations of Rapid Fluctuations of Intensity and Phase in Pulsar Emissions," Astrophys. J., 167, 273 (1971).

16. Pollen, D. A., Lee, J. R., and Taylor, J. H., "How Does the Striate Cortex Begin the Reconstruction of the Visual World?" Science, 173, 74 (1971).

17. Huguenin, G. R., Manchester, R. N., and Taylor, J. H., "Properties of Pulsars," Astrophys. J., 169, 97 (1971).

18. Huguenin, G. R., Taylor, J. H., Hjellming, R. M., and Wade, C. M., "Interferometric Observations of Pulsars at 2.7 and 8.1 GHz," Nature Phys. Sci., 234, 50 (1971).

19. Taylor, J. H., Huguenin, G. R., Hirsch, R. M., and Manchester, R. N., "Polarization of the Drifting Subpulses of Pulsar 0809+74," Astrophys. Lett, 9, 205 (1971).

20. Taylor, J. H., and Huguenin, G. R., "Periodic Modulation of Pulsar Signals," in Pulsars and High Energy Activity in Supernova Remnants, Rome: Academia Nazionale dei Lincei, p. 65 (1972).

21. Pollen, D. A., Taylor, J. H, and Lee, J. R., Reply to "Does the Striate Cortex Begin the Reconstruction of the Visual World?" Science, 176, 317 (1972).

22. Manchester, R. N. and Taylor, J. H., "Parameters of 61 Pulsars," Astrophys. Lett., 10, 67 (1972).

23. Taylor, J. H., Huguenin, G. R., and Hirsch, R. M., "Search for Pulsed Radio Emission from Scorpius X-1 and Cygnus X-1," Astrophys. J. (Letters), 172, L17 (1972).

24. Manchester, R. N., Huguenin, G. R., and Taylor, J. H., "Polarization of the Crab Pulsar Radiation at Low Radio Frequencies," Astrophys. J. (Letters), 174, L19 (1972).

25. Manchester, R. N., Taylor, J. H., and Huguenin, G. R., "New and Improved Parameters for Twenty-two Pulsars," Nature Phys. Sci., 240, 74 (1972).

26. Manchester, R. N., Taylor, J. H., and Huguenin, G. R., "Frequency Dependence of Pulsar Polarization," Astrophys. J. (Letters), 179, L7 (1973).

27. Huguenin, G. R., Taylor, J. H., and Helfand, D. J., "Slow Variations of Pulsar Intensities," Astrophys. J. (Letters), 181, L139 (1973).

28. Pollen, D. A., and Taylor, J. H., "The Striate Cortex and the Spatial Analysis of Visual Space," in Third Study Program, ed. F. Schmidt, F. Woerden, and T. Melnechuk, Cambridge: M.I.T. Press, p. 239 (1973).

29. Taylor, J. H., "Pulsar Receivers and Data Processing," Proc. I.E.E.E., 61, 1295 (1973).

30. Manchester, R. N., Tademaru, E., Taylor, J. H., and Huguenin, G. R., "Pulse Emission Mechanisms in Pulsars," Astrophys. J., 185, 951 (1973).

31. Manchester, R. N., Taylor, J. H., and Van, Y.-Y., "Detection of Pulsar Proper Motion," Astrophys. J. (Letters), 189, L119 (1974).

32. Taylor, J. H., "A Sensitive Method for Detecting Dispersed Radio Emission," Astron. Astrophys. Suppl., 15, 367 (1974).

33. Hulse, R. A., and Taylor, J. H., "A High Sensitivity Pulsar Survey," Astrophys. J. (Letters), 191, L59 (1974).

34. Manchester, R. N., and Taylor, J. H., "Period Irregularities in Pulsars," Astrophys. J. (Letters), 191, L63 (1974).

35. Taylor, J. H., Manchester, R. N., and Huguenin, G. R., "Observations of Pulsar Radio Emission: I. Total-Intensity Measurements of Individual Pulses," Astrophys. J., 195, 513 (1975).

36. Hulse, R. A., and Taylor, J. H., "Discovery of a Pulsar in a Binary System," Astrophys. J. (Letters), 195, L51 (1975).

37. Manchester, R. N., Taylor, J. H., and Huguenin, G. R., "Observations of Pulsar Radio Emission: II. Polarization of Individual Pulses," Astrophys. J., 196, 83 (1975).

38. Taylor, J. H., "Automated Pulsar Observations at the Arecibo Observatory and the Five College Radio Astronomy Observatory," in Automation of Optical Telescopes, ed. T. Mc-Cord, Cambridge: M.I.T. (1975).

39. Taylor, J. H., "Discovery of a Pulsar in a Binary System," Annals. N. Y. Acad. Sci., 262, 490 (1975).

40. Helfand, D. J., Manchester, R. N., and Taylor, J. H., "Observations of Pulsar Radio Emission: III. Stability of Integrated Pro_les," Astrophys. J., 198, 661 (1975).

41. Taylor, J. H., and Manchester, R. N., "Observed Properties of 147 Pulsars," Astron. J., 80, 794 (1975).

42. Hulse, R. A., and Taylor, J. H., "A Deep Sample of New Pulsars and their Spatial Extent in the Galaxy," Astrophys. J. (Letters), 201, L55 (1975).

43. Taylor, J. H., "Discovery of the First Binary Pulsar," in Yearbook of Science and Technology, New York: McGraw-Hill, p. 335 (1976).

44. Taylor, J. H., Hulse, R. A., Fowler, L. A., Gullahorn, G. E., and Rankin, J. M., "Further Observations of the Binary Pulsar PSR 1913+16," Astrophys. J. (Letters), 206, L53 (1976).

45. Helfand, D. J., Taylor, J. H., and Manchester, R. N., "Pulsar Proper Motions," Astrophys. J. (Letters), 213, L1 (1977).

46. Taylor, J. H., and Manchester, R. N., "Galactic Distribution and Evolution of Pulsars," Astrophys. J., 215, 885 (1977).

47. Manchester, R. N., and Taylor, J. H., Pulsars, San Francisco: Freeman (1977).

48. Taylor, J. H., and Manchester, R. N., "Recent Observations of Pulsars," Ann. Rev. Astron. Astrophys., 15, 19 (1977).

49. Taylor, J. H., "The Relationship of Pulsars to Supernovae," Annals N. Y. Acad. Sci., 302, 101 (1977).

50. Taylor, J. H., "High Velocity Pulsars," in Yearbook of Science and Technology, New York: McGraw-Hill, p. 309 (1977).

51. Manchester, R. N., Lyne, A. G., Taylor, J. H., Durdin, J. M., Large, M. I., and Little, A. G., "The Second Molonglo Pulsar Survey|Discovery of 155 Pulsars," Mon. Not. Roy. Astron. Soc., 185, 409 (1978).

52. Damashek, M., Taylor, J. H., and Hulse, R. A., "Parameters of 17 Newly Discovered Pulsars in the Northern Sky," Astrophys. J. (Letters), 225, L31 (1978).

53. Taylor, J. H., "Pulsars as a Galactic Population," in The Large Scale Characteristics of the Galaxy, IAU Symposium No. 84, ed. W. B. Burton, Dordrecht: Reidel, p. 119 (1979).

54. Fowler, L. A., Taylor, J. H., and Cordes, J. M., "Progress Report on the Binary Pulsar 1913+16," Australian J. Phys., 32, 35 (1979).

55. Phinney, S., and Taylor, J. H., "A Sensitive Search for Radio Pulses from Primordial Black Holes and Distant Supernovae," Nature, 277, 117 (1979).

56. McCulloch, P. M., Taylor, J. H., and Weisberg, J. M., "Tests of a New Dispersion-Removing Radiometer on Binary Pulsar PSR 1913+16," Astrophys. J. (Letters), 227, L133 (1979).

57. Taylor, J. H., Fowler, L. A., and McCulloch, P. M., "Measurements of General Relativistic Effects in the Binary Pulsar PSR 1913+16," Nature, 277, 437 (1979).

58. Durdin, J. M., Large, M. I., Little, A. G., Lyne, A. G., Manchester, R. N., and Taylor, J. H., "An Unusual Pulsar PSR 0826+34," Mon. Not. Roy. Astron. Soc., 186, 39 (1979).

59. Taylor, J. H., and McCulloch, P. M., "Evidence for the Existence of Gravitational Radiation from the Binary Pulsar PSR 1913+16," Annals N. Y. Acad. Sci., 336, 442 (1980).

60. Taylor, J. H., "Gravitation," in Yearbook of Science and Technology, New York: McGraw-Hill, p. 171 (1980).

61. Helfand, D. J., Taylor, J. H., Backus, P. R., and Cordes, J. M., "Pulsar Timing. I. Observations from 1970 to 1978," Astrophys. J., 237, 206 (1980).

62. McCulloch, P. M., Taylor, J. H., and Fowler, L. A., "Gravitational Radiation and the Binary Pulsar," in Gravitational Radiation, Collapsed Objects and Exact Solutions, ed. C. Edwards, Berlin: Springer-Verlag, p. 5 (1980).

63. Weisberg, J. M., and Taylor, J. H., "Gravitational Radiation from an Orbiting Pulsar," Gen. Relativ. Gravitation, 13, 1 (1981).

64. Backus, P. R., Taylor, J. H., and Damashek, M., "No Radio Pulses from M87," Astrophys. J. (Letters), 244, L65 (1981).

65. Taylor, J. H., "Binary Pulsars," in Pulsars, IAU Symposium No. 95, ed. W. Sieber and R. Wielebinski, Dordrecht: Reidel, p. 361 (1981).

66. Weisberg, J. M., Taylor, J. H., and Fowler, L. A., "Gravitational Waves from an Orbiting Pulsar," Scientific American, 245, No. 4, p. 74 (October 1981).

67. Manchester, R. N., and Taylor, J. H., "Observed and Derived Parameters for 330 Pulsars," Astron. J., 86, 1953 (1981).

68. Taylor, J. H., "Pulsar," in Encyclopedia of Science and Technology, New York: McGraw-Hill (1982).

69. Damashek, M., Taylor, J. H., Backus, P. R., and Burkhardt, R. K., "Northern Hemisphere Pulsar Survey: A Third Radio Pulsar in a Binary System," Astrophys. J. (Letters), 253, L57 (1982).

70. Taylor, J. H., and Weisberg, J. M., "A New Test of General Relativity: Gravitational Radiation and the Binary Pulsar PSR 1913+16," Astrophys. J., 253, 908 (1982).

71. Backus, P. R., Taylor, J. H., and Damashek, M., "Improved Parameters for 67 Pulsars from Timing Observations," Astrophys. J. (Letters), 255, L63 (1982).

72. Taylor, J. H., "The Binary Pulsars," in Cosmology and Astrophysics: Essays in Honor of Thomas Gold, ed. Y. Terzian and E. M. Bilson, Ithaca, NY: Cornell University Press, p. 99 (1982).

73. Taylor, J. H., "Gravitational Radiation and the Binary Pulsar," in Proceedings of the Second Marcel Grossman Meeting on Relativity, ed. R. Ru_ni, Amsterdam: North Holland, p. 15 (1983).

74. Backer, D. C., Kulkarni, S., and Taylor, J. H., "Timing Observations of the Millisecond Pulsar," Nature, 301, 314 (1983).

75. Romani, R. W., and Taylor, J. H., "An Upper Limit on the Stochastic Background of Ultralow-Frequency Gravitational Waves," Astrophys. J. (Letters), 265, L35 (1983).

76. Manchester, R. N., Newton, L. M., Cooke, D. J., Backus, P. R., Taylor, J. H., Damashek, M., and Condon, J. J., "Further Observations of the Long-Period Binary Pulsar PSR 0820+02," Astrophys. J., 268, 832 (1983).

77. Weisberg, J. M., and Taylor, J. H., "Observations of Post-Newtonian Timing Effects in the Binary Pulsar PSR 1913+16," Phys. Rev. Letters, 52, 1348 (1984).

78. Taylor, J. H., Gwinn, C. R., Weisberg, J. M., and Rawley, L. A., "Pulsar Astrometry," in VLBI and Compact Radio Sources, IAU Symposium No. 110, ed. R. Fanti et al., Dordrecht: Reidel, p. 347 (1984).

79. Gwinn, C. R., Taylor, J. H., Weisberg, J. M., and Rawley, L. A., "The Parallax of Pulsar 0950+08 and the Local Free Electron Density," Local Interstellar Medium, IAU Colloq. No. 81, ed. Y. Kondo, F. C. Brahweiler, and B. D. Savage, NASA Conf. Publ. 2345, p. 281 (1984).

80. Davis, M. M., Taylor, J. H., Weisberg, J. M., and Backer, D. C., "Arrival Time Observations of the Millisecond Pulsar 1937+21," in Millisecond Pulsars, ed. S. P. Reynolds and D. R. Stinebring, Green Bank, WV: National Radio Astronomy Observatory, p. 12 (1984).

81. Kulkarni, S. R., Davis, M. M., and Taylor, J. H., "No Detectable Millisecond Pulsar in the PSR 1913+16 System," in Millisecond Pulsars, ed. S. P. Reynolds and D. R. Stinebring, Green Bank, WV: National Radio Astronomy Observatory, p. 59 (1984).

82. Dewey, R. J., Stokes, G. H., Segelstein, D. J., Taylor, J. H., and Weisberg, J. M., "The Period Distribution of Pulsars," in Millisecond Pulsars, ed. S. P. Reynolds and D. R. Stinebring, Green Bank, WV: National Radio Astronomy Observatory, p. 234 (1984).

83. Weisberg, J. M., and Taylor, J. H., "Further Observations of the Eight Hour Binary Pulsar PSR 1913+16," in Millisecond Pulsars, ed. S. P. Reynolds and D. R. Stinebring, Green Bank, WV: National Radio Astronomy Observatory, p. 317 (1984).

84. Davis, M. M., Taylor, J. H., Weisberg, J. M., and Backer, D. C., "High Precision Timing Observations of the Millisecond Pulsar PSR 1937+21," Nature, 315, 547 (1985).

85. Lyne, A. G., Manchester, R. N., and Taylor, J. H., "The Galactic Population of Pulsars," Mon. Not. Roy. Astron. Soc., 213, 613 (1985).

86. Stokes, G. H., Taylor, J. H., and Dewey, R. J., "A New Binary Pulsar in a Highly Eccentric Orbit," Astrophys. J. (Letters), 294, L21 (1985).

87. Dewey, R. J., Taylor, J. H., Weisberg, J. M., and Stokes, G. H., "A Search for Low Luminosity Pulsars," Astrophys. J. (Letters), 294, L25 (1985).

88. Taylor, J. H., "Fast Rotating Pulsars," in Compact Galactic and Extragalactic X-Ray Sources, ed. Y. Tanaka and W. Lewin, Tokyo: Institute of Space and Astronautical Science, p. 1 (1985).

89. Stokes, G. H., Taylor, J. H.,Weisberg, J. M., and Dewey, R. J., "A Survey for Short-Period Pulsars," Nature, 317, 787 (1985).

90. Backer, D. C., Fomalont, E. B., Goss, W. M., Taylor, J. H., andWeisberg, J. M., "Accurate Timing and Interferometer Positions for the Millisecond Pulsar 1937+21 and the Binary Pulsar 1913+16," Astron. J., 90, 2275 (1986).

91. Gwinn, C. R., Taylor, J. H., Weisberg, J. M., and Rawley, L. A., "Measurement of Pulsar Parallaxes by VLBI," Astron. J., 91, 338 (1986).

92. Rawley, L. A., Taylor, J. H., and Davis, M. M., "Period Derivative and Orbital Eccentricity of Binary Pulsar 1953+29," Nature, 319, 383 (1986).

93. Taylor J. H., and Stinebring, D. R., "Recent Progress in the Understanding of Pulsars," Ann. Rev. Astron. Astrophys., 24, 285 (1986).

94. Dewey, R. J., Maguire, C. M., Rawley, L. A., Stokes, G. H., and Taylor, J. H., "Binary Pulsar with a Very Small Mass Function," Nature, 322, 712 (1986).

95. Segelstein, D. J., Rawley, L. A., Stinebring, D. R., Fruchter, A. S., and Taylor, J. H., "New Millisecond Pulsar in a Binary System," Nature, 322, 714 (1986).

96. Stokes, G. H., Segelstein, D. J., Taylor, J. H., and Dewey, R. J., "Results of Two Surveys for Fast Pulsars," Astrophys. J., 311, 694 (1986).

97. Taylor, J. H., "Binary Pulsars: Observations and Implications," in The Origin and Evolution of Neutron Stars, ed. D. J. Helfand and J.-H. Huang, Dordrecht: Reidel, p. 383 (1987).

98. Taylor, J. H., "Astronomical and Space Experiments to Test Relativity," in General Relativity and Gravitation, ed. M. A. H. MacCallum, Cambridge: Cambridge University Press, p. 209 (1987).

99. Taylor, J. H., "Pulsars: An Overview of Recent Developments," in 13th Texas Symposium on Relativistic Astrophysics, ed. M. P. Ulmer, Singapore: World Scientific, pp. 467–477 (1987).

100. Rawley, L. A., Taylor, J. H., Davis, M. M., and Allan, D. W., "Millisecond Pulsar PSR 1937+21: A Highly Stable Clock," Science, 238, 761 (1987).

101. Kulkarni, S. R., Clifton, T. R., Backer, D. C., Foster, R. S., Fruchter, A. S., and Taylor, J. H., "A Fast Pulsar in Radio Nebula CTB 80," Nature, 331, 50 (1988).

102. Fruchter, A. S., Taylor, J. H., Backer, D. C., Clifton, T. R., Foster, R. S., and Wolszczan, A., "Timing and Scintillation Observations of the Fast Pulsar in CTB 80," Nature, 331, 53 (1988).

103. Rawley, L. A., Taylor, J. H., and Davis, M. M., "Fundamental Astrometry and Millisecond Pulsars," Astrophys. J., 326, 947 (1988).

104. Foster, R. S., Backer, D. C., Taylor, J. H., and Goss, W. M., "Period Derivative of the Millisecond Pulsar in Globular Cluster M28," Astrophys. J. (Letters), 326, L13 (1988).

105. Dewey, R. J., Taylor, J. H., Maguire, C. M., and Stokes, G. H., "Period Derivatives and Improved Parameters for 66 Pulsars," Astrophys. J., 332, 762 (1988).

106. Taylor, J. H., and Dewey, R. J., "Improved Parameters for Four Binary Pulsars," Astrophys. J., 332, 770 (1988).

107. D'Amico, N., Manchester, R. N., Durdin, J. M., Stinebring, D. R., Stokes, G. H., Taylor, J. H., and Brissenden, R. J. V., "A Survey for Millisecond Pulsars at Molonglo," Mon. Not. Roy. Astron. Soc., 234, 437 (1988).

108 Fruchter, A. S., Stinebring, D. R., and Taylor, J. H., "An Eclipsing Binary Millisecond Pulsar," Nature, 333, 237 (1988).

109. Damour, T., Gibbons, G., and Taylor, J. H., "Upper Limit on Rate of Change of the Gravitational Constant," Phys. Rev. Letters, 61, 1151 (1988).

110. Taylor, J. H., "Timing Binary and Millisecond Pulsars," in Timing Neutron Stars, ed. H. Ogelman and E. van den Heuvel, Dordrecht: Kluwer Academic Publishers, p. 17 (1989).

111. Taylor, J. H. "Relativity, Pulsar Time, and Atomic Time," in Frequency Standards and Metrology, ed. A. De Marchi, Berlin: Springer-Verlag, pp. 332–337 (1989).

112. Taylor, J. H., and J. M. Weisberg, "Further Experimental Tests of Relativistic Gravity Using the Binary Pulsar PSR 1913+16," Astrophys. J., 345, 434 (1989).

113. Weisberg, J. M., Romani, R. W., and Taylor, J. H., "Evidence for Geodetic Precession in the Binary Pulsar PSR 1913+16," Astrophys. J., 347, 1030 (1989).

114. Stinebring, D. R., Bower, G., Fruchter, A. S., Klein, J. R., Ryba, M. A., Taylor, J. H., Thorsett, S. E., and Weisberg, J. M., "Eclipse Duration and Post-Eclipse Delay in PSR 1957+20," Fourteenth Symposium on Relativistic Astrophysics, ed. E. J. Fenyves (New York: N. Y. Acad. Sci.), p. 414 (1989).

115. Thorsett, S. E., Fruchter, A. S., Stinebring, D. R., and Taylor, J. H., "PSR 1957+20: Polarization and the Nature of the Eclipsing Medium," Fourteenth Symposium on Relativistic Astrophysics, ed. E. J. Fenyves (New York: N. Y. Acad. Sci.), p. 420 (1989).

116. Taylor, J. H., "Radio Pulsars at Gamma-Ray Energies," in Proceedings of the Gamma Ray Observatory Science Workshop, ed. W. N. Johnson, Greenbelt, MD: NASA, p. 4–143 (1990).

117. Taylor, J. H., "Radio Pulsar Timing Observations for GRO," in The Energetic Gamma-Ray Experiment Telescope (EGRET) Science Symposium, ed. C. Fichtel, S. Hunter, P. Sreekumar, and F. Stecker, Greenbelt, MD: NASA Conference Publication No. 3071, p. 81 (1990).

118. Fruchter, A. S., Berman, G., Bower, G., Convery, M., Goss, W. M., Hankins, T. H., Klein, J. R., Nice, D. J., Ryba, M. F., Stinebring, D. R., Taylor, J. H., Thorsett, S. E., and Weisberg, J. M., "The Eclipsing Millisecond Pulsar PSR 1957+20," Astrophys. J., 341, 642–650 (1990).

119. Stinebring, D. R., Ryba, M. F., Taylor, J. H., and Romani, R. W., "The Cosmic Gravitational Wave Background: Limits from Millisecond Pulsar Timing," Phys. Rev. Lett., 65, 285 (1990).

120. Nice, D. J., Thorsett, S. E., Taylor, J. H., and Fruchter, A. S., "Observations of the Eclipsing Binary Pulsar in Terzan 5," Astrophys. J. (Letters), 361, L61 (1990).

121. Taylor, J. H., "Basic Physics and Cosmology from Pulsar Timing Data," in High Energy Astrophysics: American and Soviet Perspectives ed. W. H. G. Lewin, G. W. Clark, and R. Sunyayev, Washington: National Academy Press, p. 385 (1991).

122. Taylor, J. H. "Pulsar Timing and Relativistic Gravity," in Atomic Physics 12, AIP Conference Proceedings vol. 233, ed. J. C. Zorn and R. R. Lewis, New York: American Institute of Physics, pp. 540–548.

123. Damour, T., and Taylor, J. H., "On the Orbital Period Change of the Binary Pulsar PSR 1913+16," Astrophys. J., 366, 501 (1991).

124. Ryba, M. F., and Taylor, J. H., "High Precision Timing of Millisecond Pulsars. I. Astrometry and Masses of the PSR 1855+09 System," Astrophys. J., 371, 739 (1991).

125. Ryba, M. F., and Taylor, J. H., "High Precision Timing of Millisecond Pulsars. II. Astrometry, Orbital Evolution, and Eclipses of PSR 1957+20," Astrophys. J., 380, 557 (1991).

126. Taylor, J. H., "Millisecond Pulsars: Nature's Most Stable Clocks," Proc. IEEE, 79, 1054 (1991).

127. Taylor, J. H. "Recent Observations of Recycled Pulsars," in X-Ray Binaries and Recycled Pulsars, ed. E. P. J. van den Heuvel and S. A. Rappaport, Dordrecht: Kluwer Academic Publishers, pp. 87–92 (1992).

128. Taylor, J. H., Wolszczan, A., Damour, T., and Weisberg, J. M., "Experimental Constraints on Strong-Field Relativistic Gravity," Nature, 355, 132 (1992).

129. Damour, T., and Taylor, J. H., "Strong-Field Tests of Relativistic Gravity and Binary Pulsars," Phys. Rev. D, 45, 1840–1868 (1992).

130. Stinebring, D. R., Kaspi, V. M., Nice, D. J., Ryba, M. F., Taylor, J. H., Thorsett, S. E., and Hankins, T. H., "A Flexible Data Acquisition System for Timing Pulsars," Rev. Sci. Instrum., 63, 3551–3555, (1992).

131. Thompson, D. J., Arzoumanian, Z., Bertsch, D. L., Brazier, K. T. S., D'Amico, N., Fichtel, C. E., Fierro, J. M., Hartman, R. C., Hunter, S. D., Johnston, S., Kanbach, G., Kaspi, V., Kniffen. D. A., Lin, Y. C., Lyne, A. G., Manchester, R. N., Mattox, J. R., Mayer-Hasselwander, H. A., Michelson, P. F., Montigny, C. v., Nel, H. I., Nice, D., Nolan, P. L., Pinkau, K., Rothermel, H., Schneid, E. J., Sommer, M., Sreekumar, P., and Taylor, J. H., "Pulsed High Energy -rays from the Radio Pulsar PSR 1706+44," Nature, 359, 615–616 (1992).

132. Taylor, J. H., "Testing Gravity with Binary Pulsars," Proc. Am. Phil. Soc. 00, 000–000 (1992).

133. Taylor, J. H., "Pulsar Timing and Relativistic Gravity," Phil. Trans. Roy. Soc. London, 341, 117–134 (1992).

134. Nice, D. J., Taylor, J. H., and Fruchter, A. S., "Two Newly Discovered Millisecond Pulsars," Astrophys. J. (Letters), 402, L49–L52 (1993).

135. Nolan, P. L., Arzoumanian, Z., Bertsch, D. L., Chiang, J., Fichtel, C. E., Fierro, J. M., Hartman, R. C., Hunter, S. D., Kanbach, G., Kni_en, D. A., Kwok, P. W., Lin, Y. C., Mattox, J. R., Mayer-Hasselwander, H. A., Michelson, P. F., von Montigny, C., Nel, H. I., Nice, D., Pinkau, K., Rothermel, H., Schneid, E., Sommer, M., Sreekumar, P., Taylor, J. H., and Thompson, D. J., "Observations of the Crab Pulsar and Nebula by the EGRET Telescope on the Compton Gamma Ray Observatory," Astrophys. J., 409, 697–704 (1993).

136. Taylor, J. H., and Cordes, J. M., "Pulsar Distances and the Galactic Distribution of Free Electrons," Astrophys. J., 411, 674–684 (1993).

137. Taylor, J. H., "Pulsar Timing and Relativistic Gravity," Class. Quant. Grav., 10, S167–S174 (1993).

138. Taylor, J. H., "Probing Relativistic Gravity with Pulsar Timing Experiments," in Particle Astrophysics, ed. J. Tr^an Thanh V^an, Gif-sur-Yvette, France: Editions Frontieres, pp. 367–374 (1993).

139. Thorsett, S. E., Arzoumanian, Z., McKinnon, M. M., and Taylor, J. H., "The Masses of Two Binary Neutron Star Systems," Astrophys. J. (Letters), 405, L29–L32 (1993).

140. Taylor, J. H., Manchester, R. N., and Lyne, A. G., "Catalog of 558 Pulsars," Astrophys. J. Suppl. Series, 88, 529–568 (1993).

141. Taylor, J. H., "Testing Relativistic Gravity with Binary and Millisecond Pulsars," in General Relativity and Gravitation 1992, ed. R. J. Gleiser, C. N. Kozameh, and O. M. Moreschi, Bristol: Institute of Physics Publishing, pp. 287–294 (1993).

142. Thorsett, S. E., Arzoumanian, Z., and Taylor, J. H., "PSR 1620+26: A Binary Radio Pulsar with a Planetary Companion?" Astrophys. J. (Letters), 412, L33–L36 (1993).

143. Camilo, F., Nice, D. J., and Taylor, J. H., "Discovery of Two Fast-Rotating Pulsars," Astrophys J. (Letters), 412, L37–L40 (1993).

144. Arzoumanian, Z., Nice, D. J., Taylor, J. H., and Thorsett, S. E., "Timing Behavior of 96 Radio Pulsars," Astrophys. J., 422, 671–680 (1994).

145. Arzoumanian, Z., Fruchter, A. S. and Taylor, J. H., "Orbital Variability in the Eclipsing Pulsar Binary PSR B1957+20," Astrophys. J., 426, L85–L88 (1994).

146. Kaspi, V. M., Taylor, J. H., and Ryba, M. F., "High-Precision Timing of Millisecond Pulsars. III. Long-Term Monitoring of PSRs B1855+09 and B1937+21," Astrophys. J., 428, 713–728 (1994).

147. Taylor, J. H., "Binary Pulsars and Relativistic Gravity," Rev. Mod. Phys., 711–719, (1994).

148. Taylor, J. H., "Binary Pulsars and Relativistic Gravity," in Les Prix Nobel, Stockholm: Norstedts Tryckeri, pp. 80–101 (1994).

149. Kanbach, G., Arzoumanian, Z., Bertsch, D. L., Brazier, K. T. S., Chiang, J., Fichtel, C. E., Fierro, J. M., Hartman, R. C., Hunter, S. D., Kni_en, D. A., Lin, Y. C., Mattox, J. R., Mayer-Hasselwander, H. A., Michelson, P. F., von Montigny, C., Nel, H. I., Nice, D., Nolan, P. L., Pinkau, K., Rothermel, H., Schneid, E., Sommer, M., Sreekumar, P., Taylor, J. H., and Thompson, D. J., "EGRET Observations of the Vela Pulsar, PSR 0833+45," Astron. Astrophys., 289, 855–867 (1994).

150. Nice, D. J., Sayer, R. W., and Taylor, J. H., "A Search for Radio Pulsars in the Direction of Egret High-Latitude Point Sources," in The Second Compton Symposium," ed. C. E. Fichtel, N. Gehrels, and J. P. Norris, New York: American Institute of Physics, pp. 82–83 (1994).

151. Thompson, D. J.. Arzoumanian, Z., Bertsch, D. L., Brazier, K. T. S., Chiang, J., D'Amico, N., Dingus, B. L., Esposito, J. A., Fierro, J. M., Fichtel, C. E., Hartman, R. C., Hunter, S. D., Johnston, S., Kanback, G., Kaspi, V. M., Kni_en, D. A., Lin, Y. C., Lyne, A. G., Manchester, R. N., Mattox, J. R., Mayer-Hasselwander, H. A., Michelson, P. F., von Montigny, C., Nel, H. I., Nice, D. J., Nolan, P. L., Ramanamurthy, P. V., Shemar, S. L., Schneid, E. J., Sreekumar, P, and Taylor, J. H., "EGRET High-Energy Gamma-Ray Pulsar Studies. I. Young Spin-Powered Pulsars," Astrophys. J., 436, 229–238 (1994).

152. Nice, D. J., and Taylor, J. H., "PSR J2019+2425 and PSR J2322+2057 and the Proper Motions of Millisecond Pulsars," Astrophys. J., 441, 429–435 (1995).

153. Taylor, J. H., "Binary Pulsars and Relativistic Gravity," J. Astrophys. Astron., 16, 307–325 (1995).

154. Fierro, J. M., Arzoumanian, Z., Bailes, M., Bell, J. F., Bertsch, D. L., Brazier, K. T. S., Chiang, J., D'Amico, N., Dingus, B. L., Esposito, J. A., Fichtel, C. E., Hartman, R. C., Hunter, S. D., Johnston, S., Kanback, G., Kaspi, V. M., Kni_en, D. A., Lin, Y. C., Lyne, A. G., Manchester, R. N., Mattox, J. R., Mayer-Hasselwander, H. A., Michelson, P. F., von Montigny, C., Nel, H. I., Nice, D. J., Nolan, P. L., Schneid, E. J., Sreekumar, P, and Taylor, J. H., Thompson, D. J., and Willis, T. D., "EGRET High-Energy Gamma-Ray Pulsar Studies. II. Individual Millisecond Pulsars," Astrophys. J., 447, 807–812 (1995).

155. Nice, D. J., Fruchter, A. S., and Taylor, J. H., "A Search for Fast Pulsars along the Galactic Plane," Astrophys. J., 449, 156–163 (1995).

156. Cognard, I., Shrauner, J. A., Taylor, J. H., and Thorsett, S. E., "Giant Radio Pulses from a Millisecond Pulsar," Astrophys. J., 457, L81–L84 (1996).

157. Camilo, F., Nice, D. J., and Taylor, J. H., "A Search for Millisecond Pulsars at Galactic Latitudes 50_ < b < 20_," Astrophys. J., 461, 812–819 (1996).

158. Nel, H. I., Arzoumanian, Z., Bailes, M., Brazier, K. T. S., Chiang, J., D'Amico, N., Esposito, J. A., Fichtel, C. E., Fierro, J. M., Hunter, S. D., Johnston, S., Kanback, G., Kaspi, V. M., Kni_en, D. A., Lin, Y. C., Lyne, A. G., Manchester, R. N., Mattox, J. R., Mayer-Hasselwander, H. A., Merck, M., Michelson, P. F., Nice, D. J., Nolan, P. L., Ramanamurthy, P. V., Taylor, J. H., Thompson, D. J., and Westbrook, C., "EGRET High-Energy Gamma-Ray Pulsar Studies. III. A Survey," Astrophys. J., 465, 898–906 (1996).

159. Nice, D. J., Sayer, R. W., and Taylor, J. H., "PSR J1518+4904: A Mildly Relativistic Binary Pulsar System," Astrophys. J., 466, L87–L90 (1996).

160. Camilo, F., Nice, D. J., Shrauner, J. A., and Taylor, J. H., "Princeton-Arecibo Declination-Strip Survey for Millisecond Pulsars. I.," Astrophys. J., 469, 819–829 (1996).

161. Arzoumanian, Z., Phillips, J. A., Taylor, J. H., and Wolszczan, A., "Radio Beam of the Relativistic Binary Pulsar B1534+12," Astrophys. J., 470, 1111–1117 (1996).

162. Nel, H. I., Arzoumanian, Z., Bailes, M., Brazier, K. T. S., D'Amico, N., Esposito, J. A., Fichtel, C. E., Fierro, J. M., Hunter, S. D., Johnston, S., Kanback, G., Kaspi, V. M., Kniffen, D. A., Lin, Y. C., Lyne, A. G., Manchester, R. N., Mattox, J. R., Mayer-Hasselwander, H. A., Merck, M., Michelson, P. F., Nice, D. J., Nolan, P. L., Ramanamurthy, P. V., Taylor, J. H., Thompson, D. J., and Westbrook, C., "EGRET Pulsar Upper Limits," Astron. Astrophys. Supp., 120, 89–93 (1996).

163. Nice, D. J., Sayer, R. W., and Taylor, J. H., "The Green Bank Northern Sky Survey: Discovery of a New Neutron Star–Neutron Star Binary," in Pulsars: Problems and Progress, IAU Colloq. 160, ed. S. Johnston, M. A. Walker, and M. Bailes, Kramer, San Francisco: Astron. Soc. Pacific, 11–12 (1996).

164. Shrauner, J. A., Stairs, I. H., Dewey, R. J., Krumholz, M., Taylor, H. E., Taylor, J. H., and Thorsett, S. E., "Mark IV: A Phase Coherent Observing System for Pulsars," in Pulsars: Problems and Progress, IAU Colloq. 160, ed. S. Johnston, M. A. Walker, and M. Bailes, Kramer, San Francisco: Astron. Soc. Pacific, 23–24 (1996).

165. Taylor, J. H., "The Next Five Years of Pulsar Timing," in Pulsars: Problems and Progress, IAU Colloq. 160, ed. S. Johnston, M. A. Walker, and M. Bailes, Kramer, San Francisco: Astron. Soc. Pacific, 65–71 (1996).

166. Kramer, M., Doroshenko, O., Jessner, A., Wielebinski, R., Wolszczan, A., Camilo, F., Taylor, J. H., and Xilouris, K. M., "Millisecond Pulsar Timing in Effelsberg," in Pulsars: Problems and Progress, IAU Colloq. 160, ed. S. Johnston, M. A. Walker, and M. Bailes, Kramer, San Francisco: Astron. Soc. Pacific, 95–96 (1996).

167. Thorsett, S. E., Shrauner, J. A., Cognard, I., and Taylor, J. H., "Observations of Giant Pulses from Pulsar PSR B1937+21," in Pulsars: Problems and Progress, IAU Colloq. 160, ed. S. Johnston, M. A. Walker, and M. Bailes, Kramer, San Francisco: Astron. Soc. Pacific, 209 (1996).

168. Sayer, R. W., Nice, D. J., and Taylor, J. H., "The Green Bank Northern Sky Survey for Fast Pulsars," Astrophys. J., 474, 426–432 (1997).

169. Matsakis, D. N., Taylor, J. H., and Eubanks, T. M., "A Statistic for Describing Pulsar and Clock Stabilities," Astron. Astrophys., 326, 924–928 (1997).

170. Stairs, I. H., Arzoumanian, Z., Camilo, F., Lyne, A. G., Nice, D. J., Taylor, J. H., Thorsett, S. E., and Wolszczan, A., "Measurement of Relativistic Orbital Decay in the PSR B1534+12 Binary System," Astrophys. J., 505, 352–357 (1998).

171. Shrauner, J. A., Taylor, J. H., and Woan, G., "The Second Cambridge Pulsar Survey at 81.5 MHz," Astrophys. J., 509, 785–792 (1998).

172. Taylor, J. H., "Binary Pulsars and Relativistic Gravity," in The Universe Unfolding, ed. H. Bondi and M. Weston-Smith, Oxford: Clarendon, pp. 315–333 (1998).

173. Thompson, D. J., Bailes, M., Bertsch, D. L., Cordes, J., D'Amico, N., Esposito, J. A., Finley, J., Hartman, R. C., Hermsen, W., Kanbach, G., Kaspi, V. M., Kniffen, D. A., Kuiper, L., Lin, Y. C., Lyne, A., Manchester, R. N., Matz, S. M., Mayer-Hasselwander, H. A., Michelson, P. F., Nolan, P. L., Ogelman. H., Pohl, M., Ramanamurthy, P. V., Sreekumar, P., Reimer, O., Taylor, J. H., and Ulmer, M., "Gamma Radiation from PSR B1055+52," Astrophys. J., 516, 297–306 (1999).

174. Nice, D. J., Taylor, J. H., and Sayer, R. W., "Timing Observations of the J1518+4904 Double Neutron Star System," in Pulsar Timing General Relativity and the Internal Structure of Neutron Stars, ed. Z. Arzoumanian, F. van der Hooft, and E. P. J. van den Heuvel, Amsterdam: Royal Netherlands Academy of Arts and Sciences, pp. 79–83 (1999).

175. Arzoumanian, Z., Taylor, J. H., and Wolszczan, A., "Evidence for Relativistic Precession in the Pulsar Binary B1534+12," in Pulsar Timing General Relativity and the Internal Structure of Neutron Stars, ed. Z. Arzoumanian, F. van der Hooft, and E. P. J. van den Heuvel, Amsterdam: Royal Netherlands Academy of Arts and Sciences, pp. 79–83 (1999).

176. Stairs, I. H., Nice, D. J., Thorsett, S. E., and Taylor, J. H., "Recent Arecibo Timing of the Relativistic Binary PSR B1534+12," in Gravitational Waves and Experimental Gravity, XXXIV Rencontres de Moriond, ed. ... (2000).

177. Wolszczan, A., Doroshenko, O., Konaki, M., Kramer, M., Jessner, A., Wielebinski, R., Camilo, F., Nice, D. J., and Taylor, J. H., "Timing Observations of Four Millisecond Pulsars with the Arecibo and Effelsberg Radiotelescopes," Astrophys. J., 528, 907–912 (2000).

178. Stairs, I. H., Splaver, E. M., Thorsett, S. E., Nice, D. J., and Taylor, J. H., "A Baseband Recorder for Radio Pulsar Observations," Mon. Not. Roy. Astron. Soc., 314, 459–467 (2000).

179. Stairs, I. H., Thorsett, S. E., Taylor, J. H., and Arzoumanian, Z., "Geodetic Precession in PSR B1534+12," in Pulsar Astronomy|2000 and Beyond, IAU Colloq. 177, ed. M. Kramer, N. Wex, and R. Wielebinski, San Francisco: Astron. Soc. Pacific, 121–124 (2000).

180. Weisberg, J. M., and Taylor, J. H., "General Relativistic Precession of the Spin Axis of Binary Pulsar B1913+16: First Two Dimensional Maps of the Emission Beam," in Pulsar Astronomy|2000 and Beyond, IAU Colloq. 177, ed. M. Kramer, N. Wex, and R. Wielebinski, San Francisco: Astron. Soc. Pacific, 127–130 (2000).

181. Kalogera, V., Narayan, R., Spergel, D. N., and Taylor, J. H., "The Coalescence Rate of Double Neutron Star Systems," Astrophys. J., 556, 340-356 (2001).

182. McKee, C. F., and Taylor, J. H., "A Blueprint for Astronomy's Next Ten Years," Sky and Telescope, 101, 38-42 (2001).

183. McKee, C. F., and Taylor, J. H., "How we Reached Consensus," Sky and Telescope, 101, 42 (2001).

184. McKee, C. F., and Taylor, J. H., "Education and Outreach," Sky and Telescope, 101, 44 (2001).

185. McKee, C. F., and Taylor, J. H., *et al.*, "Astronomy and Astrophysics in the New Millennium," Washington, DC: National Academy Press, 246 pages (2001).

186. Stairs, I. H., Thorsett, S. E., Taylor, J. H., and Wolszczan, A., "Studies of the Relativistic Binary Pulsar PSR B1534+12: I. Timing Analysis," Astrophys. J., 581, 501–508 (2002).

187. Taylor, J. H., "Stellar Timekeepers," in Knowledge from Nature: Twenty-one Discoveries that Changed the World, ed. L. Garwin and T. Lincoln, Tokyo: Nature Japan K. K., pp. 128–138 (2002) (in Japanese).

188. Weisberg, J. M, and Taylor, J. H., "General Relativistic Geodetic Spin Precession in Binary Pulsar B1913+16: Mapping the Emission Beam in Two Dimensions," Astrophys. J., 576, 942–949 (2002).

189. Taylor, J. H., "Stelar Timekeepers," in A Century of Nature, ed. L. Garwin and T. Lincoln, Chicago: University of Chicago Press, pp. 147–153 (2003).

190. Taylor, J. H., "Archaeological Reections on the Dawn of Pulsar Civilization," in Radio Pulsars,, ed. M. Bailes, D. J. Nice, and S. E. Thorsett, San Francisco: Astronomical Society of the Pacific, Conference Series Volume 302, pp. 37–43 (2003).

191. Weisberg, J. M, and Taylor, J. H., "The Relativistic Binary Pulsar B1913+16," in Radio Pulsars,, ed. M. Bailes, D. J. Nice, and S. E. Thorsett, San Francisco: Astronomical Society of the Pacific, Conference Proceedings Volume 302, pp. 93–98 (2003).

192. McKee, C. F., and Taylor, J. H., "The Future of Small Telescopes In The New Millennium, Volume I — Perceptions, Productivities, and Policies," ed. T. D. Oswalt., Astrophysics and Space Science Library, Volume 287, Dordrecht: Kluwer Academic Publishers, p. 7 (2003).

193. Champion, D. J., Lorimer, D. R., McLaughlin, M. A., Cordes, J. M., Arzoumanian, Z., Weisberg, J. M., and Taylor, J. H., "PSR J1829+2456: A Relativistic Binary Pulsar," Mon. Not. Roy. Astron. Soc., 350, L61–L65, (2004).

194. Weisberg, J. M., and Taylor, J. H., "Relativistic Binary Pulsar B1913+16: Thirty Years of Observations and Analysis," San Francisco: Astronomical Society of the Pacific, Conference Series, 328, pp. 25–31 (2005).

195. Weisberg, J. M., Nice, D. J., and Taylor, J. H., "Timing Measurements of the Relativistic Binary Pulsar PSR B1913+16," Astrophys. J., 722, 1030–1034 (2010).

Michael V. Berry

MICHAEL VICTOR BERRY

Academic History

1959–1962	Undergraduate at Exeter University, leaving with honours degree in physics
1962–1965	Postgraduate research in St Andrews University, leaving with PhD in theoretical physics
1965–1967	Postdoctoral research in theoretical physics at Bristol University, as Department of Scientific and Industrial Research Fellow
1967–1974	Lecturer in Physics at Bristol University
1974–1979	Reader in Physics at Bristol University
1979–1988	Professor of Physics at Bristol University
1988–2006	Royal Society Research Professor at Bristol University
2006–present	Melville Wills Professor of Physics at Bristol University (Emeritus)

Awards

1959	London Academy of Music and Dramatic Art, Verse and prose speaking medal
1978	Maxwell Medal and Prize of the Institute of Physics
1982	Elected to the Royal Society of London
1983	Elected Fellow of the Royal Society of Arts
1983	Elected Fellow of the Royal Institution
1986	Elected Member of the Royal Society of Sciences, Uppsala
1989	Elected member of the European Academy
1990	Julius Edgar Lilienfeld prize of the American Physical Society
1990	Paul Dirac Medal and prize of the Institute of Physics
1990	Elected to Indian Academy of Sciences
1990	Royal Medal of the Royal Society
1993	Naylor Prize, London Mathematical Society
1994	Louis-Vuitton Moët-Hennessey 'Science for Art' prize (Paris)
1995	Hewlett-Packard Europhysics Prize
1995	Elected Foreign Member of the National Academy of Science of the USA
1995	Elected Member of the London Mathematical Society
1996	Dirac Medal and Prize of the International Centre for Theoretical Physics, Trieste
1996	Knight Bachelor Queen's Birthday Honours, 15 June
1997	Kapitsa Medal of the Russian Academy of Sciences
1997	LEGO Master Builder Badge
1998	Wolf Prize in Physics
1998	Elected a Governor of the Weizmann Institute
1999	Honorary Fellow, Institute of Physics
2000	Elected Foreign Member of Royal Netherlands Academy of Arts and Sciences

2000	Ig Nobel prize in physics
2001	Onsager Medal (Norwegian Technical University, Trondheim)
2002	Novartis/Daily Telegraph 'Visions of Science' competition, 1st prize (Science as Art), 3rd prize (Science Concepts)
2005	Elected to Royal Society of Edinburgh
2005	Polya Prize, London Mathematical Society
2005	Chancellor's Medal, University of Bristol
2008–2010	Leverhulme Emeritus Fellow
2010	Elected first Honorary Member of the Mexican Mathematical Society
2012–2014	Leverhulme Emeritus Fellow
2012	Distinguished Visiting Professor, Marymount University, Virginia
2014	Richtmyer Memorial Lecture Award, American Association of Physics Teachers
2015	Lorentz Medal, Royal Netherlands Academy of Arts and Sciences
2015–2017	Leverhulme Emeritus Fellow

COMMENTARY
(Numbers refer to items on my publications list)

Many incompletely understood phenomena lurk in the borderlands between physical theories — between classical and quantum, between rays and waves... These borderlands — the domain of physical asymptotics — are my intellectual habitat, with an emphasis on geometrical aspects of waves (especially phase) and chaos.

A source of delight is uncovering down-to-earth or dramatic and sometimes beautiful [352, 363] examples of abstract mathematical ideas: the arcane in the mundane [354, 374]. Examples are:

mathematical singularities in rainbows [213, 266] and the patterns on the bottom of swimming-pools [056];

a laser pointer shone through irregular bathroom-window glass, illustrating abstract aspects of wave interference;

optics with transparent overhead-projector plastic sheets, illustrating polarization singularities, matrix degeneracies, and quantum measurement [303, 361], and Anderson localization [281];

a levitating spinning-top, illustrating adiabatic stability and geometric phases [271, 285];

twists and turns with a belt, illustrating the behaviour of identical particles in quantum mechanics [286], responsible for the impenetrability of matter, lasers, superconductivity...

Oriental magic mirrors [383], directly displaying the Laplacian.

Tsunamis, which are caustics in spacetime [376], and, when focused, spacetime caustics on a cusped caustic [399]

My contribution to particle physics

What is the elementary particle of sudden understanding? It is the clariton. Any scientist will recognise the "Aha!" moment when this particle is created. But there are snags: all too frequently, today's clariton is annihilated by tomorrow's anticlariton. So many of our scribblings disappear beneath a rubble of anticlaritons. It is not always easy to detect my particle: the observation might at best be a quasiclariton. And how sad that a clariton is sometimes less than astonishing — well known to those who know well, a mere claritino.

Some recent and current areas of interest

1. Quantum mechanics, chaos and the primes (with Jon Keating). This concerns the relation between the Riemann zeros — harmonies in the music of the primes — and energy levels of classically chaotic quantum systems, whose detailed understanding began in the mid-1980s [154, 163, 227, 307, 339]. We study the statistics of the zeros [175], the asymptotics of the Riemann–Siegel formula for calculating them [223, 265], and efforts to find the operator whose eigenvalues are the imaginary parts of the zeros [306, 440]. It is possible that the Riemann zeros could be seen [451] and they can be heard [454, 456].

2. Quantum chaology for systems with mixed chaology (with Jon Keating and Henning Schomerus). Bifurcations of closed orbits cause the moments of the fluctuations of the level counting function (spectral staircase) to diverge semiclassically [294] according to power laws, whose 'twinkling exponents' depend on a competition between bifurcations [320], related to the competition between catastrophes to dominate the fluctuations of twinkling starlight [058, 114, 318].

3. Spin-statistics connection, e.g. Pauli principle (with Jonathan Robbins). Our non-relativistic theory [286, 322] incorporates the indistinguishability of identical spinning particles into quantum geometry in a new way. The Pauli sign is an unusual kind of geometric phase, though there is an analogue [253] for single spins. There are many physical and mathematical ramifications [319].

4. Singularities of bright light (caustics). With Christopher Howls, I have written an account of the mathematical properties of diffraction catastrophes, a new class of integrals [049, 086] with many applications to physics, including spectacular patterns of wave interference associated with focusing [079, 089, 105, 106, 256, 270]. This is part of the Digital Library of Mathematical Functions project (of which I am also a member of the editorial board), which we hope will go online on 2005. See [326] and http://dlmf.nist.gov/

Conical diffraction is a type of extreme singularity, now elucidated in detail after nearly two centuries [360, 381, 386, 387, 392, A]. Regarding tsunamis as caustics in spacetime facilitates an accurate calculation of the wave profile [376], and their focusing onto a cusped caustic by a submerged island is a singularity on a singularity [399]. Looking at caustics and the associated coalescing images [398] involves Husimi functions and complexified diffraction catastrophes.

5. Singularities of faint light ('Optical vorticulture'). In scalar waves, these are phase singularities — that is wavefront dislocations [034, 312] or optical vortices [296] — zeros of the field; in vector waves, they are lines of purely circular or purely linear polarization, related to singularities in local expectation values of photon spin. In white light, phase singularities are decorated with universal coloured interference patterns [346, 347]. A helicoidal wavefront, whose phase pitch can be fractional, evolves into an elaborate pattern of singularities [359]. I am studying the general properties of these singularities [362, 365, 368, 395], and, with Mark Dennis, their statistics in random waves (e.g. black-body radiation) [321, 324, 340], and their knottedness [328, 332, 333,]. and their interactions [394]. In crystal optics, singularities appear in the space of directions, along wth a new type of degeneracy singularity; this viewpoint leads to clarification and simplification of this old subject, especially for crystals that are absorbing and optically active (chiral) as well as anisotropic [355] and, bianosotropic [379], and the polarization pattern of the blue sky [373]. The phase singularities of classical optics are windows, through which can be glimpsed the faint glimmering of the quantum vacuum [364].

6. Asymptotics and relations between theories. Less general theories in physics are limits of more general ones, but the limits are usually singular [260, 341]. An example is the classical limit of quantum mechanics [433], and classicalization through decoherence, where tiny uncontrolled external influences suppress the quantum suppression of classical chaos [337]. Refined studies of these singular limits must involve divergent series, Stokes phenomenon [181, 190, 201], and (work with Christopher Howls) resurgence and hyperasymptotics [208, 223, 244, 249, 261], and band-limited functions can oscillate arbitrarily faster than their fastest Fourier components [252, 262, 388]. Surprisingly, the exponentially small contributions whose appearance and disappearance are governed by the Stokes phenomenon [181] can sometimes dominate [370]. Another universal asymptotic phenomenon is oscillations generated by infinitely repeated differentiation [377]. Asymptotics is a unifying idea: the same mathematics that governs rainbows also describes tsunamis [376]. These studies are part of humanity's long struggle to get to grips with infinity [241].

7. Nonhermitian operators. When there is absorption or loss, physical operators can be strongly nonhermitian, and much of the resulting new physics is associated with energy-level degeneracies [372], as in near-resonant atom optics [293, 295], and unstable lasers [332, 334, 350]. Crystal optics singularities [355, 379] (see section 5 above) exhibit a variety of nonhermitian phenomena. When the nonhermitian operators have antiunitary symmetry (e.g. PT), spectra can be real [325, 345]. The evolution of systems slowly driven by nonhermitian operators is strikingly different from hermitian evolution [441], with implications for optical polarization [442].

8. Extreme coherence. The addition of many discrete waves with quadratically-varying phases, as in the near field of diffraction gratings or in time-dependent quantum mechanics of periodic waves, striking interference phenomena occur, dominated by the Gauss sums of number theory. Such hypercoherent waves generate fantastic spirals in their complex planes [171, 179], and elaborate carpets in space (Talbot effect) or spacetime (quantum revivals)) [274, 275, 283, 304, 329].

9. Superoscillations. Functions can oscillate arbitrarily faster than their fastest Fourier component [262, 252]. This mathematical phenomenon is unexpectedly common [412], and has physical implications [388, 431, 443, 449, 461]. More generally, it is a central feature of Aharonov's weak measurement scheme [429, 437, 445].

Proc. R. Soc. Lond. A **392**, 45–57 (1984)
Printed in Great Britain

Quantal phase factors accompanying adiabatic changes

By M. V. Berry, F.R.S.

*H. H. Wills Physics Laboratory, University of Bristol,
Tyndall Avenue, Bristol BS8 1TL, U.K.*

(*Received* 13 *June* 1983)

A quantal system in an eigenstate, slowly transported round a circuit C by varying parameters $\mathbf{R}$ in its Hamiltonian $\hat{H}(\mathbf{R})$, will acquire a geometrical phase factor $\exp\{i\gamma(C)\}$ in addition to the familiar dynamical phase factor. An explicit general formula for $\gamma(C)$ is derived in terms of the spectrum and eigenstates of $\hat{H}(\mathbf{R})$ over a surface spanning C. If C lies near a degeneracy of $\hat{H}$, $\gamma(C)$ takes a simple form which includes as a special case the sign change of eigenfunctions of real symmetric matrices round a degeneracy. As an illustration $\gamma(C)$ is calculated for spinning particles in slowly-changing magnetic fields; although the sign reversal of spinors on rotation is a special case, the effect is predicted to occur for bosons as well as fermions, and a method for observing it is proposed. It is shown that the Aharonov–Bohm effect can be interpreted as a geometrical phase factor.

1. Introduction

Imagine a quantal system whose Hamiltonian $\hat{H}$ describes the effects of an unchanging environment, and let the system be in a stationary state. If the environment, and hence $\hat{H}$, is slowly altered, it follows from the adiabatic theorem (Messiah 1962) that at any instant the system will be in an eigenstate of the instantaneous $\hat{H}$. In particular, if the Hamiltonian is returned to its original form the system will return to its original state, apart from a phase factor. This phase factor is observable by interference if the cycled system is recombined with another that was separated from it at an earlier time and whose Hamiltonian was kept constant.

My purpose here is to explain how the phase factor contains a circuit-dependent component $\exp(i\gamma)$ in addition to the familiar dynamical component $\exp(-iEt/\hbar)$ which accompanies the evolution of any stationary state. A general formula for γ in terms of the eigenstates of $\hat{H}$ will be obtained in §2. If the circuit is close to a degeneracy in the spectrum of $\hat{H}$, γ takes a particularly simple form which will be derived in §3; this contains, as a special case, the sign change around a degeneracy of the eigenstates of a system whose Hamiltonian is real as well as Hermitian (Herzberg & Longuet-Higgins 1963; Longuet-Higgins 1975; Mead 1979; Mead & Truhlar 1979; Mead 1980 *a*, *b*; Berry & Wilkinson 1984).

A particle of any spin in an eigenstate of a slowly-rotated magnetic field is another case where γ can be calculated explicitly (§4), and gives predictions that could be

[45]

46 M. V. Berry

tested experimentally. This phase factor exists for bosons as well as fermions. A special case is the sign change of spinors slowly rotated by 2π, predicted by Aharonov & Susskind (1967); this will be shown to be different from the dynamical phase factors measured in experiments on precessing neutrons (reviewed by Silverman 1980).

Finally, it is shown in § 5 that physical effects of magnetic vector potentials in the absence of fields, predicted by Aharonov & Bohm (1959) and observed by Chambers (1960), can be understood as special cases of the geometrical phase factor.

2. General formula for phase factor

Let the Hamiltonian $\hat{H}$ be changed by varying parameters $\boldsymbol{R} = (X, Y, \dots)$ on which it depends. Then the excursion of the system between times $t = 0$ and $t = T$ can be pictured as transport round a closed path $\boldsymbol{R}(t)$ in parameter space, with Hamiltonian $\hat{H}(\boldsymbol{R}(t))$ and such that $\boldsymbol{R}(T) = \boldsymbol{R}(0)$. The path will henceforth be called a circuit and denoted by C. For the adiabatic approximation to apply, T must be large.

The state $|\psi(t)\rangle$ of the system evolves according to Schrödinger's equation

$$\hat{H}(\boldsymbol{R}(t))\,|\psi(t)\rangle = i\hbar\,|\dot{\psi}(t)\rangle. \tag{1}$$

At any instant, the natural basis consists of the eigenstates $|n(\boldsymbol{R})\rangle$ (assumed discrete) of $\hat{H}(\boldsymbol{R})$ for $\boldsymbol{R} = \boldsymbol{R}(t)$, that satisfy

$$\hat{H}(\boldsymbol{R})\,|n(\boldsymbol{R})\rangle = E_n(\boldsymbol{R})\,|n(\boldsymbol{R})\rangle, \tag{2}$$

with energies $E_n(\boldsymbol{R})$. This eigenvalue equation implies no relation between the phases of the eigenstates $|n(\boldsymbol{R})\rangle$ at different $\boldsymbol{R}$. For present purposes any (differentiable) choice of phases can be made, provided $|n(\boldsymbol{R})\rangle$ is single-valued in a parameter domain that includes the circuit C.

Adiabatically, a system prepared in one of these states $|n(\boldsymbol{R}(0))\rangle$ will evolve with $\hat{H}$ and so be in the state $|n(\boldsymbol{R}(t))\rangle$ at t.

Thus $|\psi\rangle$ can be written as

$$|\psi(t)\rangle = \exp\left\{\frac{-i}{\hbar}\int_0^t dt'\, E_n(\boldsymbol{R}(t'))\right\} \exp\left(i\gamma_n(t)\right)|n(\boldsymbol{R}(t))\rangle. \tag{3}$$

The first exponential is the familiar dynamical phase factor. In this paper the object of attention is the second exponential. The crucial point will be that its phase $\gamma_n(t)$ is *non-integrable*; γ_n cannot be written as a function of $\boldsymbol{R}$ and in particular is not single-valued under continuation around a circuit, i.e. $\gamma_n(T) \neq \gamma_n(0)$.

The function $\gamma_n(t)$ is determined by the requirement that $|\psi(t)\rangle$ satisfy Schrödinger's equation, and direct substitution of (3) into (1) leads to

$$\dot{\gamma}_n(t) = i\langle n(\boldsymbol{R}(t))\,|\,\nabla_{\boldsymbol{R}} n(\boldsymbol{R}(t))\rangle \cdot \dot{\boldsymbol{R}}(t). \tag{4}$$

Phase factors accompanying adiabatic changes 47

The total phase change of $|\psi\rangle$ round C is given by

$$|\psi(T)\rangle = \exp\left(\mathrm{i}\gamma_n(\mathrm{C})\right)\exp\left\{\frac{-\mathrm{i}}{\hbar}\int_0^T \mathrm{d}t\, E_n(\boldsymbol{R}(t))\right\} |\psi(0)\rangle, \tag{5}$$

where the *geometrical phase change* is

$$\gamma_n(\mathrm{C}) = \mathrm{i}\oint_\mathrm{C} \langle n(\boldsymbol{R})|\, \nabla_R\, n(\boldsymbol{R})\rangle \cdot \mathrm{d}\boldsymbol{R}. \tag{6}$$

Thus $\gamma_n(\mathrm{C})$ is given by a circuit integral in parameter space and is independent of how the circuit is traversed (provided of course that this is slow enough for the adiabatic approximation to hold). The normalization of $|n\rangle$ implies that $\langle n|\nabla_R n\rangle$ is imaginary, which guarantees that γ_n is real.

Direct evaluation of $|\nabla_R n\rangle$ requires a locally single-valued basis for $|n\rangle$ and can be awkward. Such difficulties are avoided by transforming the circuit integral (6) into a surface integral over any surface in parameter space whose boundary is C. In order to employ familiar vector calculus, parameter space will be considered as three-dimensional, and this will turn out to be the important case in applications; the generalization to higher dimensions will be outlined at the end of this section.

Stokes's theorem applied to (6) gives, in an obvious abbreviated notation.

$$\gamma_n(\mathrm{C}) = -\operatorname{Im}\iint_\mathrm{C} \mathrm{d}\boldsymbol{S}\cdot\nabla\times\langle n|\nabla n\rangle, \tag{7a}$$

$$= -\operatorname{Im}\iint_\mathrm{C} \mathrm{d}\boldsymbol{S}\cdot\langle\nabla n|\times|\nabla n\rangle, \tag{7b}$$

$$= -\operatorname{Im}\iint_\mathrm{C} \mathrm{d}\boldsymbol{S}\cdot\sum_{m\neq n}\langle\nabla n|m\rangle\times\langle m|\nabla n\rangle, \tag{7c}$$

where $\mathrm{d}\boldsymbol{S}$ denotes area element in $\boldsymbol{R}$ space and the exclusion in the summation is justified by $\langle n|\nabla n\rangle$ being imaginary. The off-diagonal elements are obtained from (2) as

$$\langle m|\nabla n\rangle = \langle m|\nabla \hat{H}|n\rangle/(E_n - E_m), \quad m\neq n. \tag{8}$$

Thus γ_n can be expressed as

$$\gamma_n(\mathrm{C}) = -\iint_\mathrm{C} \mathrm{d}\boldsymbol{S}\cdot\boldsymbol{V}_n(\boldsymbol{R}), \tag{9}$$

where

$$\boldsymbol{V}_n(\boldsymbol{R}) \equiv \operatorname{Im}\sum_{m\neq n} \frac{\langle n(\boldsymbol{R})|\nabla_R \hat{H}(\boldsymbol{R})|m(\boldsymbol{R})\rangle \times \langle m(\boldsymbol{R})|\nabla_R \hat{H}(\boldsymbol{R})|n(\boldsymbol{R})\rangle}{(E_m(\boldsymbol{R}) - E_n(\boldsymbol{R}))^2}. \tag{10}$$

Obviously $\gamma_n(\mathrm{C})$ is zero for a circuit which retraces itself and so encloses no area.

Equations (9) and (10) embody the central results of this paper. Because the dependence on $|\nabla n\rangle$ has been eliminated, phase relations between eigenstates with different parameters are now immaterial, and (as is evident from the form of (10)), it is no longer necessary to choose $|m\rangle$ and $|n\rangle$ to be single-valued in $\boldsymbol{R}$: any solutions of (2) may be employed without affecting the value of V_n. This is a surprising conclusion, as can be seen by comparing (9) with (7a) which show that V_n is the curl of a vector, $\langle n|\nabla n\rangle$, and $\langle n|\nabla n\rangle$ certainly does depend on the choice of phase of the (single-valued) eigenstate $|n(\boldsymbol{R})\rangle$. The dependence on phase is of the following kind: if $|n\rangle \to \exp\{i\mu(\boldsymbol{R})\}|n\rangle$ then $\langle n|\nabla n\rangle \to \langle n|\nabla n\rangle + i\nabla\mu$ (in another context the importance of such gauge transformations has been emphasized by Wu & Yang (1975)). Thus the vector is not unique but its curl is. The quantity V_n is analogous to a 'magnetic field' (in parameter space) whose 'vector potential' is $\operatorname{Im}\langle n|\nabla n\rangle$. In Appendix A it is shown directly from (10) that $\nabla \cdot V_n$ vanishes, thus confirming that (9) gives a unique value for $\gamma_n(C)$.

Using perturbation theory, Mead & Truhlar (1979) obtained essentially the formulae (9) and (10) for an infinitesimal circuit, in a study of molecular electronic states which (in the Born–Oppenheimer approximation) depend parametrically on nuclear coordinates. Their phase factor was not intended to apply to a $|\psi\rangle$ that evolves slowly under the time-dependent Schrödinger equation, but to the variation of eigenstates $|n\rangle$ under a particular phase-continuation rule in $\boldsymbol{R}$-space which can be shown to give the same result.

In parameter spaces of higher dimension, Stokes's theorem cannot be employed to transform the circuit integral (6). The appropriate generalization, provided by the theory of differential forms (see, for example, Arnold 1978, chap. 7), transforms (6) into the integral of a 2-form over a surface bounded by C. The surprising result (10) can now be expressed as follows: independently of the choice of phases of the eigenstates, there exists in parameter space a *phase 2-form*, which gives $\gamma(C)$ when integrated over any surface spanning C. This 2-form is obtained from (10) by replacing ∇ by the exterior derivative d and $\times$ by the wedge product $\wedge$. The validity of this generalization is consistent with the observation that in the three-dimensional version there are infinitely many choices of interpolating Hamiltonian (and hence of parameter spaces) on the surfaces bounded by C, and the geometrical phase factor is independent of the choice.

Professor Barry Simon (1983), commenting on the original version of this paper, points out that the geometrical phase factor has a mathematical interpretation in terms of holonomy, with the phase two-form emerging naturally (in the form (7b)) as the curvature (first Chern class) of a Hermitian line bundle.

3. Degeneracies

The energy denominators in (10) show that if the circuit C lies close to a point $\boldsymbol{R}*$ in parameter space at which the state n is involved in a degeneracy, then $V_n(\boldsymbol{R})$, and hence $\gamma_n(C)$, is dominated by the terms m corresponding to the other states involved. We shall consider the commonest situation, where the degeneracy involves only two states, to be denoted $+$ and $-$, where $E_+(\boldsymbol{R}) \geqslant E_-(\boldsymbol{R})$. For $\boldsymbol{R}$ near $\boldsymbol{R}*$, $\hat{H}(\boldsymbol{R})$ can be expanded to first order in $\boldsymbol{R} - \boldsymbol{R}*$, and

$$V_+(\boldsymbol{R}) = \operatorname{Im} \frac{\langle +(\boldsymbol{R})| \nabla \hat{H}(\boldsymbol{R}*)|-(\boldsymbol{R})\rangle \times \langle -(\boldsymbol{R})| \nabla \hat{H}(\boldsymbol{R}*)|+(\boldsymbol{R})\rangle}{(E_+(\boldsymbol{R}) - E_-(\boldsymbol{R}))^2}. \tag{11}$$

Obviously $V_-(\boldsymbol{R}) = -V_+(\boldsymbol{R})$, so that $\gamma_-(C) = -\gamma_+(C)$.

Phase factors accompanying adiabatic changes 49

Without essential loss of generality we can take $E_\pm(\boldsymbol{R}^*) = 0$ and $\boldsymbol{R}^* = 0$. $H(\boldsymbol{R})$ can be represented by a 2×2 Hermitian matrix coupling the two states. The most general such matrix satisfying the given conditions depends on three parameters X, Y, Z which will be taken as components of $\boldsymbol{R}$, and by linear transformation in $\boldsymbol{R}$-space can be brought into the following standard form

$$\hat{H}(\boldsymbol{R}) = \frac{1}{2}\begin{bmatrix} Z & X - \mathrm{i}Y \\ X + \mathrm{i}Y & -Z \end{bmatrix}. \tag{12}$$

The eigenvalues are

$$E_+(\boldsymbol{R}) = -E_-(\boldsymbol{R}) = \tfrac{1}{2}(X^2 + Y^2 + Z^2)^{\frac{1}{2}} = \tfrac{1}{2}R. \tag{13}$$

Thus the degeneracy is an isolated point at which all three parameters vanish. This illustrates an old result of Von Neumann & Wigner (1929): for generic Hamiltonians (Hermitian matrices), it is necessary to vary three parameters in order to make a degeneracy occur accidentally, that is, not on account of symmetry. Alternatively stated, degeneracies have co-dimension three.

The form (12) was chosen to exploit the fact that

$$\nabla \hat{H} = \tfrac{1}{2}\hat{\boldsymbol{\sigma}}, \tag{14}$$

where the components $\hat{\sigma}_X$, $\hat{\sigma}_Y$, $\hat{\sigma}_Z$ of the vector operator $\hat{\boldsymbol{\sigma}}$ are the Pauli spin matrices. When evaluating the matrix elements in (11) it greatly simplifies the calculations to take advantage of the isotropy of spin and temporarily rotate axes so that the Z-axis points along $\boldsymbol{R}$, and to employ the following relations, which come from the commutation laws between the components of $\hat{\boldsymbol{\sigma}}$:

$$\hat{\sigma}_X | \pm \rangle = | \mp \rangle, \quad \hat{\sigma}_Y | \pm \rangle = \pm \mathrm{i} | \mp \rangle, \quad \hat{\sigma}_Z | \pm \rangle = \pm | \pm \rangle. \tag{15}$$

With these rotated axes, (11) gives

$$\left.\begin{aligned}
V_{X+} &= (\mathrm{Im} \langle + | \hat{\sigma}_Y | - \rangle \langle - | \hat{\sigma}_Z | + \rangle)/2R^2 = 0, \\
V_{Y+} &= (\mathrm{Im} \langle + | \hat{\sigma}_Z | - \rangle \langle - | \hat{\sigma}_X | + \rangle)/2R^2 = 0, \\
V_{Z+} &= \mathrm{Im} \langle + | \hat{\sigma}_X | - \rangle \langle - | \hat{\sigma}_Y | + \rangle = 1/2R^2.
\end{aligned}\right\} \tag{16}$$

Reverting to unrotated axes, we obtain

$$V_+(\boldsymbol{R}) = \boldsymbol{R}/2R^3. \tag{17}$$

Now use of (9) shows that the phase change $\gamma_+(\mathrm{C})$ is the flux through C of the magnetic field of a monopole with strength $-\tfrac{1}{2}$ located at the degeneracy. Thus we obtain the pleasant result, valid for the natural choice (12) of standard form for $\hat{H}$, that the geometrical phase factor associated with C is

$$\exp\{\mathrm{i}\gamma_\pm(\mathrm{C})\} = \exp\{\mp \tfrac{1}{2}\mathrm{i}\Omega(\mathrm{C})\}, \tag{18}$$

where $\Omega(\mathrm{C})$ is the *solid angle* that C subtends at the degeneracy; Ω is, in a sense, a measure of the *view* of the circuit as seen from the degeneracy. The phase factor is

50   M. V. Berry

independent of the choice of surface spanning C, because Ω can change only in multiples of 4π (when the surface is deformed to pass through the degeneracy).

An important special case of (18) occurs when C consists entirely of *real* Hamiltonians and so is confined to the plane $Y = 0$ (cf. (12)). The energy levels $E_\pm$ intersect conically in the space E, X, Z, whose origin, where the degeneracy occurs, is a 'diabolical point' of the type recently studied by Berry & Wilkinson (1984) in the spectra of triangles. This illustrates the result that for real symmetric matrices, degeneracies have co-dimension two: see Appendix 10 of Arnold 1978. If C encloses the degeneracy, $\Omega = \pm 2\pi$; if not, $\Omega = 0$. Thus the phase factor (18) is

$$\exp\{i\gamma_\pm(C)\} = -1, \quad \text{if C encircles the degeneracy,}$$
$$= +1, \quad \text{otherwise,} \tag{19}$$

which expresses the sign changes of real wavefunctions as a degeneracy involving them is encircled, a phenomenon first described by Herzberg & Longuet-Higgins (1963). (Sign changes are not restricted to circuits involving real Hamiltonians: (18) shows that the phase factor is -1 if C lies in *any* plane through the degeneracy and encircles it.)

Confirmation of the correctness of (17) can be obtained without the rotation-of-axes trick, by a lengthy calculation of (11) involving explicit formulae for the eigenvectors $|\pm(\boldsymbol{R})\rangle$ of the matrix (12). Alternatively, direct continuation of the eigenvectors may be attempted. This cannot be accomplished for all circuits by means of (6) because it is not possible to construct eigenvectors that are globally single-valued continuous functions of $\boldsymbol{R}$; multivaluedness can be reduced to singular lines connecting the degeneracy with infinity, and in the analogue $\boldsymbol{V}(\boldsymbol{R})$ these appear as Dirac strings attached to the monopole. Such approaches obscure the simplicity and essential isotropy of the solid-angle result (17).

Using topological arguments not involving explicit formulae for $\gamma_n(C)$, Stone (1976) proved that if C is expanded from one point $\boldsymbol{R}$ and contracted on to another so as to sweep out a surface enclosing a degeneracy, then the geometrical phase factor traverses a circle in its Argand plane. This property (which follows easily from (18)), is the Hermitian generalization of the sign-reversal test for degeneracy.

4. Spins in magnetic fields

A particle with spin s (integer or half-integer) interacts with a magnetic field $\boldsymbol{B}$ via the Hamiltonian

$$\hat{H}(\boldsymbol{B}) = \kappa\hbar\boldsymbol{B}\cdot\hat{\boldsymbol{s}}, \tag{20}$$

where κ is a constant involving the gyromagnetic ratio and $\hat{\boldsymbol{s}}$ is the vector spin operator with $2s+1$ eigenvalues n with integer spacing and that lie between $-s$ and $+s$. The eigenvalues are

$$E_n(\boldsymbol{B}) = \kappa\hbar Bn, \tag{21}$$

and so there is a $(2s+1)$-fold degeneracy when $\boldsymbol{B} = 0$. (The special case $s = \frac{1}{2}$ reproduces the two-fold degeneracy considered in the last section.) We consider the components of B as the parameters R in our previous analysis, and calculate the phase change $\gamma_n(\mathrm{C})$ of an eigenstate $|n, s(\boldsymbol{B})\rangle$ of $\hat{s}$ in the direction along $\boldsymbol{B}$, as $\boldsymbol{B}$ is slowly varied (and hence the spin rotated) round a circuit C.

The vector $\boldsymbol{V}_n(\boldsymbol{B})$ as given by (10) can be expressed by using (20) and (21) as

$$\boldsymbol{V}_n(B) = \frac{\mathrm{Im}}{B^2} \sum_{m \neq n} \frac{\langle n, s(\boldsymbol{B}) | \hat{s} | m, s(\boldsymbol{B}) \rangle \times \langle m, s(\boldsymbol{B}) | \hat{s} | n, s(\boldsymbol{B}) \rangle}{(m-n)^2}. \tag{22}$$

To evaluate the matrix elements we again temporarily rotate axes so that the Z-axis points along $\boldsymbol{B}$, and employ the following generalization of (15):

$$\left.\begin{aligned}
(\hat{s}_X + \mathrm{i}\hat{s}_Y) |n, s\rangle &= [s(s+1) - n(n+1)]^{\frac{1}{2}} |n+1, s\rangle, \\
(\hat{s}_X - \mathrm{i}\hat{s}_Y) |n, s\rangle &= [s(s+1) - n(n-1)]^{\frac{1}{2}} |n-1, s\rangle, \\
s_Z |n, s\rangle &= n |n, s\rangle.
\end{aligned}\right\} \tag{23}$$

It is clear that only states with $m = n \pm 1$ are coupled with $|n\rangle$ in (22), and that V_x and V_y are zero because they involve off-diagonal elements of $\hat{s}_Z$. To find V_Z, we make use of (23) to obtain

$$\left.\begin{aligned}
\langle n \pm 1, s| s_X |n, s\rangle &= \tfrac{1}{2}[s(s+1) - n(n \pm 1)]^{\frac{1}{2}}, \\
\langle n \pm 1, s| s_Y |n, s\rangle &= \mp \tfrac{1}{2}\mathrm{i}[s(s+1) - n(n \pm 1)]^{\frac{1}{2}},
\end{aligned}\right\} \tag{24}$$

then (22) gives

$$\begin{aligned}
\boldsymbol{V}_{Zn} &= \frac{\mathrm{Im}}{B^2} \{ \langle n| s_X |n+1\rangle \langle n+1 |s_Y |n\rangle - \langle n| s_Y |n+1\rangle \langle n+1| s_X |n\rangle \\
&\quad + \langle n| s_X |n-1\rangle \langle n-1 |s_Y |n\rangle - \langle n| s_Y |n-1|\rangle \langle n-1| s_X |n\rangle \} \\
&= \frac{n}{B^2}.
\end{aligned} \tag{25}$$

Reverting to unrotated axes, we obtain

$$V_n(\boldsymbol{B}) = n\boldsymbol{B}/B^3. \tag{26}$$

Now, use of (9) shows that $\gamma_n(\mathrm{C})$ is the flux through C of the 'magnetic field' of a monopole $-n$ located at the origin of magnetic field space. Thus the geometrical phase factor is

$$\exp\{\mathrm{i}\gamma_n(\mathrm{C})\} = \exp\{-\mathrm{i}n\Omega(\mathrm{C})\}, \tag{27}$$

where $\Omega(\mathrm{C})$ is the solid angle that C subtends at $\boldsymbol{B} = 0$. Note that γ_n depends only on the eigenvalue n of the spin component along $\boldsymbol{B}$ and not on the spin s of the particle, so that γ_n is insensitive to the strength $2s + 1$ of the degeneracy at $\boldsymbol{B} = 0$.

It follows from (27) that any phase change can be produced by varying $\boldsymbol{B}$ round a suitable circuit. For fermions (half-integer n), a whole turn of $\boldsymbol{B}$ (rotation through 2π in a plane, giving $\Omega = 2\pi$) produces a phase factor -1. In the special case $n = \tfrac{1}{2}$ this shows that the sign change of spinors on rotation and the sign change of wavefunctions round a degeneracy have the same mathematical origin. For bosons (integer n), a whole turn of $\boldsymbol{B}$ produces a phase factor $+1$. To produce a sign change, different circuits are required; if $n = 1$, for example, varying $\boldsymbol{B}$ round a cone of semiangle $60°$ will give $\Omega = \gamma = \pi$ and hence a phase factor -1.

The following experiment could be carried out to test the predictions embodied in (27). A polarized monoenergetic beam of particles in spin state n along a magnetic field $\boldsymbol{B}$ is split into two. Along the path of one beam $\boldsymbol{B}$ is kept constant. Along the path of the other beam, $\boldsymbol{B}$ is kept constant in magnitude but its direction is varied slowly (in comparison with the dynamical precession frequency) round a circuit C subtending a solid angle Ω, the field being generated by an arrangement enabling Ω to be changed. The beams are then combined and the count rate at a detector is measured as a function of Ω. The dynamical phase factor (the second exponential in (5) is the same for both beams because the energy $E_n(\boldsymbol{B})$ (21) is insensitive to the direction of $\boldsymbol{B}$. There will in addition be a propagation phase factor which can be made unity by adjusting the path-length of one of the beams when $\Omega = 0$. The resulting fringes occur as a consequence of the geometrical phase factor. If C is a circuit round a cone of semiangle θ, the predicted intensity contrast is

$$I(\theta) = \cos^2(n\pi(1 - \cos\theta)). \tag{28}$$

I wish to emphasize that this proposed experiment is different from those carried out by Rauch *et al.* (1975, 1978) and Werner *et al.* (1975) (see Silverman 1980) with *unpolarized* neutrons in a *constant* magnetic field. Those neutrons were not in an eigenstate, and their phase changed dynamically, rather than geometrically, under the Hamiltonian (20) (with $\boldsymbol{B}$ along Z and $\hat{\boldsymbol{\sigma}}$ replacing $\hat{\boldsymbol{s}}$) according to the evolution operator

$$\exp(-\mathrm{i}\hat{H}t/\hbar) = \exp(-B\kappa t\hat{\sigma}_Z) = \cos\tfrac{1}{2}\kappa Bt \begin{bmatrix} 1 & 0 \\ 0 & 1 \end{bmatrix} + \mathrm{i}\sin\tfrac{1}{2}\kappa Bt \begin{bmatrix} 1 & 0 \\ 0 & -1 \end{bmatrix}. \tag{29}$$

The sign changed whenever $\tfrac{1}{2}\kappa Bt$ was an odd multiple of π, and this was interpreted on the basis of precession theory as corresponding to odd numbers of complete rotations about $\boldsymbol{B}$.

5. AHARONOV–BOHM EFFECT

Consider a magnetic field consisting of a single line with flux Φ. For positions $\boldsymbol{R}$ not on the flux line, the field is zero but there must be a vector potential $\boldsymbol{A}(\boldsymbol{R})$ satisfying

$$\oint_C \boldsymbol{A}(\boldsymbol{R})\cdot\mathrm{d}\boldsymbol{R} = \Phi, \tag{30}$$

for circuits C threaded by the flux line. Aharonov & Bohm (1959) showed that in quantum mechanics such vector potentials have physical significance even though they correspond to zero field. I shall now show how their effect can be interpreted as a geometrical phase change of the type described in § 2.

Let the quantal system consist of particles with charge q confined to a box situated at $\boldsymbol{R}$ and not penetrated by the flux line (figure 1). In the absence of flux

Phase factors accompanying adiabatic changes   **53**

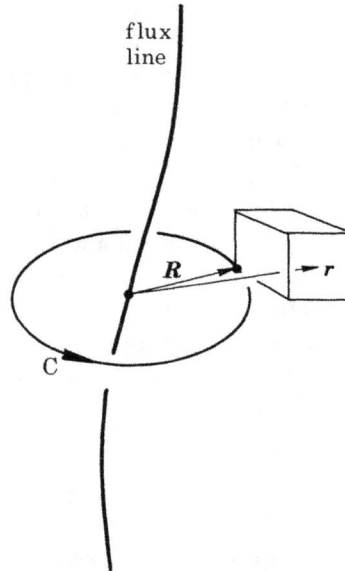

FIGURE 1. Aharonov–Bohm effect in a box transported round a flux line.

$(A = 0)$, the particle Hamiltonian depends on position $\hat{r}$ and conjugate momentum $\hat{p}$ as follows:

$$\hat{H} = H(\hat{p}, \hat{r} - R), \tag{31}$$

and the wavefunctions have the form $\psi_n(r - R)$ with energies E_n independent of R. With non-zero flux, the states $|n(R)\rangle$ satisfy

$$H(\hat{p} - qA(\hat{r}), \hat{r} - R)|\,n(R)\rangle = E_n\,|n(R)\rangle, \tag{32}$$

an equation whose exact solutions are obtained by multiplying ψ_n by an appropriate Dirac phase factor, giving

$$\langle r|\,n(R)\rangle = \exp\left\{\frac{iq}{\hbar}\int_R^r dr' \cdot A(r')\right\}\psi_n(r - R). \tag{33}$$

These solutions are single-valued in r and (locally) in R. The energies are unaffected the vector potential.

Now let the box be transported round a circuit C threaded by the flux line; in this particular case it is not necessary that the transport be adiabatic. After completion of the circuit there will be a geometrical phase change that can be calculated from (6) and (33) by using

$$\langle n(R)|\,\nabla_R n(R)\rangle = \iiint d^3r\,\psi_n^*(r - R)\left\{\frac{-iq}{\hbar}A(R)\,\psi_n(r - R) + \nabla_R\psi_n(r - R)\right\}$$

$$= -\,iqA(R)/\hbar. \tag{34}$$

54   M. V. Berry

(The vanishing of the second term in braces follows from the normalization of ψ_n.) Evidently in this example the analogy between $\mathrm{Im}\,\langle n|\nabla n\rangle$ and a magnetic vector potential becomes a reality. Thus

$$\gamma_n(\mathrm{C}) = \frac{q}{\hbar}\oint_{\mathrm{C}} A(R)\cdot \mathrm{d}R = q\Phi/\hbar, \tag{35}$$

which shows that the phase factor is independent of n, and also of C if this winds once round the flux line. The phase factor could be observed by interference between the particles in the transported box and those in a box which was not transported round the circuit.

In elementary presentations of the Aharonov–Bohm effect (including its anticipation by Ehrenburg & Siday 1949), the Dirac phase factor is often invoked in comparing systems passing opposite sides of a flux line. Such invocations are subject to the objection that the wavefunction thus obtained is not single-valued. One way to avoid this objection is by summation over all contributions (whirling waves) representing different windings round the flux line (Schulman 1981; Berry 1980; Morandi & Menossi 1984). Another way, adopted in the original paper by Aharonov & Bohm, is to solve Schrödinger's equation exactly for scattering in the flux line's vector potential. The argument of the preceding paragraphs, which employs the geometrical phase factor, is a third way of obtaining the Aharonov–Bohm effect by using only single-valued wavefunctions.

Mead (1980a, b), employs the term 'molecular Aharonov–Bohm effect' in a different context, to describe how degeneracies in electron energy levels affect the spectrum of nuclear vibrations. He explains two options, both leading to the same vibration spectrum. The first option is to continue the electronic states round degeneracies (in the space of nuclear coordinates) in the manner described in this paper, thus causing the electronic wavefunctions to be multi-valued, with a compensating multi-valuedness in the nuclear states, which must be incorporated into their boundary conditions. The alternative is to enforce single-valuedness on the electronic (and hence also the nuclear) states, and this introduces a vector potential into the Schrödinger equation for nuclear motion. In general one may expect such effects whenever an isolated system is considered as being divided into two interacting parts, each slaved to a different aspect of the other (in the molecular case, electron states are slaved to nuclear coordinates, and nuclear states are slaved to the electronic states and wavefunctions). The systems considered in this paper might be regarded as a special case, in which the coupling is with 'the rest of the Universe' (including us as observers). The only role of the rest of the Universe is to provide a Hamiltonian with slowly-varying parameters, thus forcing the system to evolve adiabatically with phase continuation governed by the time-dependent Schrödinger equation.

6. DISCUSSION

A system slowly transported round a circuit will return in its original state; this is the content of the adiabatic theorem. Moreover its internal clocks will register the passage of time; this can be regarded as the meaning of the dynamical phase factor. The remarkable and rather mysterious result of this paper is that in addition the system records its history in a deeply geometrical way, whose natural formulation (9) and (10) involves phase functions hidden in parameter-space regions which the system has not visited.

The total phase of the transported state (5) is dominated by the dynamical part, because $T \to \infty$ in the adiabatic limit, and it might be thought that this must overwhelm the geometrical phase γ_n and make its physical effects difficult to detect. This objection can be met by observing that the strengths of non-adiabatic transitions are exponentially small in T if $\hat{H}$ changes smoothly (Hwang & Pechukas 1977), so that essentially adiabatic evolution can occur even when the dynamical phase is only a few times greater than 2π.

As we saw in §3, degeneracies in the spectrum of $\hat{H}(\mathbf{R})$ are the singularities of the vector $\mathbf{V}(\mathbf{R})$ (equation (10)) in parameter space, and so have an important effect on the geometrical phase factor. This is reminiscent of the part played by singularities of an analytic function, but the analogy is imperfect: if $\gamma(\mathrm{C})$ were completely singularity-determined, $\mathbf{V}(\mathbf{R})$ would be the sum of the 'magnetic fields' of 'monopoles' situated at the degeneracies (cf. (17)) and so would have zero curl, which is not the case (zero curl, unlike zero divergence, is not a property which is invariant under deformations of $\mathbf{R}$ space, and in the general case the sources of $\mathbf{V}$ are not just monopoles but also 'currents' distributed continuously in parameter space). A closer analogy is with wavefront dislocation lines, which are phase singularities of complex wavefunctions in three-dimensional position space (Nye & Berry 1974; Nye 1981; Berry 1981), that dominate the geometry of wavefronts without completely determining them.

In view of the emphasis on degeneracies as organizing centres for phase changes, it is worth remarking that close approach of energy levels is not a necessary condition for the existence of nontrivial geometrical phase factors. Indeed, our examples have shown that $\gamma(\mathrm{C})$ can be non-zero even if C involves isospectral deformations of $\hat{H}(\mathbf{R})$ (in the Aharonov–Bohm illustration, the levels E_n do not depend on $\mathbf{R}$ at all).

The results obtained here are not restricted to quantum mechanics, but apply more generally, to the phase of eigenvectors of any Hermitian matrices under a natural continuation in parameter space. Therefore they have implications throughout wave physics. For example, the electromagnetic field of a single mode travelling along an optical fibre will change sign if the cross section of the fibre is slowly altered so that its path (in the space of shapes) surrounds a shape for which the spectrum of the Helmholtz equation is degenerate (such as one of the diabolical triangles discovered by Berry & Wilkinson 1984).

56   M. V. Berry

I thank Dr J. H. Hannay, Dr E. J. Heller and Dr B. R. Pollard for several suggestions, and Professor Barry Simon for showing me, before publication, his paper which comments on this one. This work was not supported by any military agency.

REFERENCES

Aharonov, Y. & Bohm, D. 1959 *Phys. Rev.* **115**, 485–491.
Aharonov, Y. & Susskind, L. 1967 *Phys. Rev.* **158**, 1237–1238.
Arnold, V. I. 1978 *Mathematical methods of classical dynamics.* New York: Springer.
Berry, M. V. 1980 *Eur. J. Phys.* **1**, 240–244.
Berry, M. V. 1981 Singularities in waves and rays. In *Les Houches Lecture Notes for session XXXV* (ed. R. Balian, M. Kléman & J.-P. Poirier), pp. 453–543. Amsterdam: North-Holland.
Berry, M. V. & Wilkinson, M. 1984 *Proc. R. Soc. Lond.* A **392**, 15–43.
Chambers, R. G. 1960 *Phys. Rev. Lett* **5**, 3–5.
Ehrenburg, W. & Siday, R. E. 1949 *Proc. phys. Soc.* B **62**, 8–21.
Herzberg, G. & Longuet-Higgins, H. C. 1963 *Discuss. Faraday Soc.* **35**, 77–82.
Hwang, J.-T. & Pechukas, P. 1977 *J. chem. Phys.* **67**, 4640–4653.
Longuet-Higgins, H. C. 1975 *Proc. R. Soc. Lond.* A **344**, 147–156.
Mead, C. A. 1979 *J. chem. Phys.* **70**, 2276–2283.
Mead, C. A. 1980a *Chem. Phys.* **49**, 23–32.
Mead, C. A. 1980b *Chem. Phys.* **49**, 33–38.
Mead, C. A. & Truhlar, D. G. 1979 *J. chem. Phys.* **70**, 2284–2296.
Messiah, A. 1962 *Quantum mechanics*, vol. 2. Amsterdam: North-Holland.
Morandi, G. & Menossi, E. 1984 *Nuovo Cim.* B (Submitted.)
Nye, J. F. 1981 *Proc. R. Soc. Lond.* A **378**, 219–239.
Nye, J. F. & Berry, M. V. 1974 *Proc. R. Soc. Lond.* A **336**, 165–190.
Rauch, H., Wilfing, A., Bauspiess, W. & Bonse, U. 1978 *Z. Phys.* B **29**, 281–284.
Rauch, H., Zeilinger, A., Badurek, G., Wilfing, A., Bauspiess, W. & Bonse, U. 1975 *Physics Lett.* A **54**, 425–427.
Schulman, L. S. 1981 Techniques and Applications of Path Integration. New York: John Wiley.
Silverman, M. P. 1980 *Eur. J. Phys.* **1**, 116–123.
Simon, B. 1983 *Phys. Rev. Lett.* (In the press.)
Stone, A. J. 1976 *Proc. R. Soc. Lond.* A **351**, 141–150.
Von Neumann, J. & Wigner, E. P. 1929 *Phys. Z.* **30**, 467–470.
Werner, S. A., Colella, R., Overhauser, A. W. & Eagen, C. F. 1975 *Phys. Rev. Lett.* **35**, 1053–1055.
Wu, T. T. & Yang, C. N. 1975 *Phys. Rev.* D **12**, 3845–3857.

APPENDIX A

To show that $\gamma(C)$ is independent of the surface spanning C, it is necessary to prove that $V(\boldsymbol{R})$ (equation (10)) has zero divergence. This can be accomplished by expressing V in terms of the vector Hermitian operator $\hat{\boldsymbol{B}}$ defined by

$$\hat{\boldsymbol{B}} \equiv -\mathrm{i} \sum_n |\nabla n\rangle \langle n|. \tag{A 1}$$

From (8), the off-diagonal elements of $\hat{\boldsymbol{B}}$ are

$$\langle n| \hat{\boldsymbol{B}} |m\rangle = -\mathrm{i} \langle m| \nabla H |n\rangle/(E_n - E_m), \quad m \neq n. \tag{A 2}$$

Phase factors accompanying adiabatic changes 57

Thus (10) becomes

$$V = \text{Im} \langle n | \hat{\boldsymbol{B}} \times \hat{\boldsymbol{B}} | n \rangle. \tag{A 3}$$

Now we can calculate the divergence:

$$\nabla \cdot \boldsymbol{V} = \text{Im} \{ \langle \nabla n | \cdot \hat{\boldsymbol{B}} \times \hat{\boldsymbol{B}} \, | n \rangle + \langle n | \, \boldsymbol{B} \times \boldsymbol{B} \cdot | \nabla n \rangle + \langle n | \, \nabla \cdot (\hat{\boldsymbol{B}} \times \hat{\boldsymbol{B}}) \, | n \rangle \}, \tag{A 4}$$

Use of a consequence of (A 1), namely

$$| \nabla n \rangle = \text{i} \hat{\boldsymbol{B}} \, | n \rangle \tag{A 5}$$

gives

$$\nabla \cdot \boldsymbol{V} = n (- \hat{\boldsymbol{B}} \cdot \hat{\boldsymbol{B}} \times \hat{\boldsymbol{B}} + \hat{\boldsymbol{B}} \times \hat{\boldsymbol{B}} \cdot \hat{\boldsymbol{B}}) | n \rangle + \text{Im} \langle n | \, (\nabla \times \hat{\boldsymbol{B}} \cdot \hat{\boldsymbol{B}} - \hat{\boldsymbol{B}} \cdot \nabla \times \boldsymbol{B} \, | n \rangle. \tag{A 6}$$

For the curl of $\boldsymbol{B}$, (A 1) and (A 5) give

$$\nabla \times \hat{\boldsymbol{B}} = + \text{i} \sum_n | \nabla n \rangle \times \langle \nabla n | = \text{i} \sum_n \hat{\boldsymbol{B}} \, | n \rangle \times \langle n | \hat{\boldsymbol{B}} = \text{i} \hat{\boldsymbol{B}} \times \hat{\boldsymbol{B}} \rangle, \tag{A 7}$$

whence $\nabla \cdot \boldsymbol{V}$ vanishes by the dot-cross rule for triple products.

This result is valid everywhere except at the 'monopole' singularities arising from degeneracies.

Proc. R. Soc. Lond. A **422**, 7–21 (1989)
Printed in Great Britain

Uniform asymptotic smoothing of Stokes's discontinuities

By M. V. Berry, F.R.S.

H. H. Wills Physics Laboratory, Tyndall Avenue, Bristol BS8 1TL, U.K.

(*Received* 14 *June* 1988)

Across a Stokes line, where one exponential in an asymptotic expansion maximally dominates another, the multiplier of the small exponential changes rapidly. If the expansion is truncated near its least term the change is not discontinuous but smooth and moreover universal in form. In terms of the singulant F – the difference between the larger and smaller exponents, and real on the Stokes line – the change in the multiplier is the error function

$$\pi^{-\frac{1}{2}} \int_{-\infty}^{\sigma} \mathrm{d}t \exp\left(-t^2\right) \quad \text{where} \quad \sigma = \operatorname{Im} F / (2 \operatorname{Re} F)^{\frac{1}{2}}.$$

The derivation requires control of exponentially small terms in the dominant series; this is achieved with Dingle's method of Borel summation of late terms, starting with the least term. In numerical illustrations the multiplier is extracted from Dawson's integral (erfi) and the Airy function of the second kind (Bi): the small exponential emerges in the predicted universal manner from the dominant one, which can be 10^{10} times larger.

1. Introduction

Stokes's phenomenon concerns the behaviour of small exponentials while hidden by large ones (Stokes 1864, 1871, 1889, 1902). Such exponentials occur commonly in the asymptotic approximation of functions $y(X; k)$ defined by integrals or differential equations and dependent on a large parameter k and variables $X = (X_1, X_2, \ldots)$. In the simplest case there are just two exponentials, and the lowest-order approximation incorporating both can be written

$$y(X; k) \approx M_+(X; k) \exp\{k\phi_+(X)\} + \mathrm{i}S(X; k) M_-(X; k) \exp\{k\phi_-(X)\}$$

$$(\operatorname{Re}\phi_+(X) > \operatorname{Re}\phi_-(X)), \quad (1)$$

where the dominant and subdominant contributions are labelled $+$ and $-$, and the factor i is inserted for later convenience. The prefactors $M_\pm$ vary slowly with X and their k-dependences are simple powers. Attention will here focus on the *Stokes multiplier function* $S(X; k)$ weighting the subdominant exponential; this varies rapidly when X is near the *Stokes line* of y, where

$$\operatorname{Im}\left[\phi_+(X) - \phi_-(X)\right] = 0. \quad (2)$$

On the Stokes line (a set of codimension 1) it can be said that the dominance of $+$ over $-$ is maximal.

[7]

8   M. V. Berry

The need to retain the subdominant term, even though it is numerically insignificant in (1) as $k \to \infty$, and the need for Stokes's multiplier, both spring from a common cause; maintaining the validity of (1) when X crosses *anti*-Stokes lines (far from Stokes's lines) on which $\mathrm{Re}\,[\phi_+ - \phi_-] = 0$ and the exponential previously called $-$ becomes the dominant one. (Readers are warned that some authors employ the term Stokes line to denote what we here call an anti-Stokes line, and vice versa.)

A simple illustrative example, to which we shall later return, due to Stokes (1864) and well described by Dingle (1973, hereinafter called I) is the complex error function

$$y(X;k) = \int_{-i\infty}^{Z} \mathrm{d}t \exp(kt^2), \left.\right\}$$
$$(Z = X_1 + iX_2). \qquad\qquad (3)$$

Near the positive real Z axis the dominant contribution to y as $k \to \infty$ comes from the end-point of integration $t = Z$;

$$y \sim (2kZ)^{-1} \exp(kZ^2) \quad (Z \text{ positive real}). \qquad (4)$$

Thus $\phi_+ = Z^2$ and $M_+ = (2kZ)^{-1}$. Near the positive imaginary axis this would predict that y is exponentially small, which is false because the integral is then dominated by the stationary point at $t = 0$, giving

$$y \sim i(\pi/k)^{\frac{1}{2}} \quad (Z \text{ positive imaginary}). \qquad (5)$$

This would suggest the asymptotics

$$y \sim (2kZ)^{-1} \exp(kz^2) + i(\pi/k)^{\frac{1}{2}} \qquad (6)$$

involving $\phi_- = 0$ and $M_- = (\pi/k)^{\frac{1}{2}}$. Thus the Stokes line (2) is the real axis $X_2 = 0$, and the anti-Stokes lines, where dominance is exchanged, are the diagonals $X_1 = \pm X_2$. But (6) fails near the *negative* imaginary axis because it would predict $y \sim i(\pi/k)^{\frac{1}{2}}$ whereas it is obvious from (3) that y is exponentially small and given by the continuation of (4). To encompass the three regions discussed, we must write

$$y \sim (2kZ)^{-1} \exp(kZ^2) + iS(Z;k)(\pi/k)^{\frac{1}{2}}, \qquad (7)$$

incorporating Stokes's multiplier S which must change from 0 to 1 between the anti-Stokes lines $X_2 = -X_1$ and $X_2 = +X_1$. This change in S is Stokes's phenomenon.

The conventional view (Stokes 1864) is powerfully (and unconventionally) argued by Dingle in I. It asserts that the change in S is discontinuous and localized at the Stokes line: on one side, S takes a value, S_- say; on the other, $S = S_- + 1$; on the line itself, $S = S_- + \frac{1}{2}$. For the example (3) the intuition behind this view is illustrated by figure 1, which shows how the steepest-descent contours of the integral $(\mathrm{Im}\,t^2 = \mathrm{Im}\,Z^2)$ change discontinuously across the Stokes line, suddenly bringing in the subdominant contribution from the stationary point at $t = 0$

Smoothing Stokes's discontinuities 9

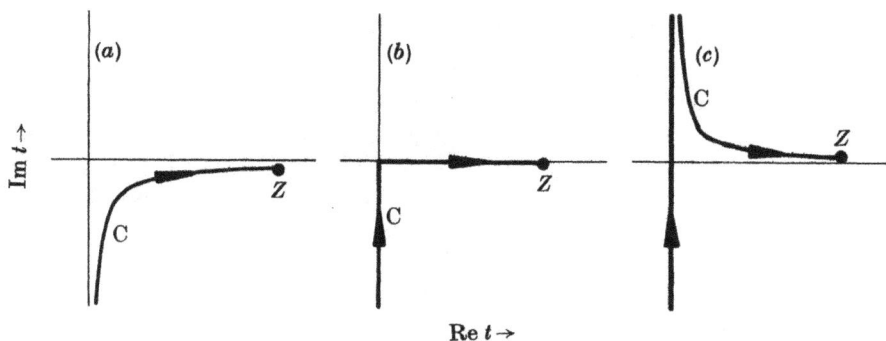

FIGURE 1. Steepest-descent contours of (3) from $t = -i\infty$ to $t = Z = X_1 + iX_2$, for three values of Z near the Stokes line $X_2 = 0$.

($S_- = 0$ in this case). It is worth repeating Stokes's description of the asymptotic emergence of his discontinuity. As a Stokes line is crossed,

> ...the inferior term enters as it were into a mist, is hidden for a little from view, and comes out with its coefficient changed. The range during which the inferior term remains in a mist decreases indefinitely as [the asymptotic parameter] increases indefinitely. (Stokes 1902)

From the context it is clear that Stokes is referring to asymptotic series interpreted by truncation near their least term. My aim here is to dispel Stokes's mist and show that his discontinuity is an artefact of poor resolution: with the appropriate magnification, S changes smoothly. Moreover, with the appropriate variable to describe the crossing of Stokes's line the change in the function $S(X; k)$ is *universal*, that is, the same for all problems in a wide class.

Obtaining this result requires control of the magnitude of the dominant exponential contribution with error small compared to the size of the subdominant exponential. Such control will be achieved by analysing the dominant asymptotic series (in descending powers of k) of which only the first term is included in (1). The series is

$$y(X; k) = M_+ \exp(k\phi_+) \sum_{r=1}^{\infty} a_r, \left.\vphantom{\sum_{r=1}^{\infty}}\right\}$$
$$(a_0 = 1; \quad a_r \propto k^{-r}). \qquad (8)$$

This diverges and so is numerically meaningless, but Dingle (I) explains how it can nevertheless be regarded as a coded representation of y, which can be reconstructed exactly (in principle and sometimes in practice) by proper interpretation of the late terms $r \gg 1$. The interpretation reveals how the subdominant exponential (together with S) originates from the divergence of the late terms. My derivation of the leading-order functional form of S across a Stokes line is a simple development within Dingle's interpretative scheme.

10   M. V. Berry

2. Derivation of the Stokes multiplier function

The obstruction preventing the series in (8) from converging is the existence of the subdominant exponential. (In the example (3) the series appended to (4) is obtained by an expansion about the integration limit $t = Z$, and divergence originates in the stationary point at $t = 0$.) This subdominant exponential engenders in the late terms a_r a remarkable universality, best expressed in terms of the complex quantity

$$F \equiv k(\phi_+ - \phi_-). \tag{9}$$

Dingle calls this the *singulant*; on a Stokes line, it is real and positive. He shows that

$$a_r \to \frac{M_-(r-\beta)!}{2\pi M_+ F^{r-\beta+1}} \quad \text{as} \quad r \to \infty. \tag{10}$$

For example, if y is defined by an integral for which (as in the example (3)) the dominant exponential comes from a limit of integration and the subdominant one from a stationary point, then $\beta = \frac{1}{2}$ (I, pp. 111, 145). If y is defined by an integral for which both exponentials are associated with stationary points, then $\beta = 1$ (I, pp. 135, 145). If y is a solution of a second-order differential equation, say

$$\partial_z^2 y = k^2 Q(Z)\, y \tag{11}$$

with $\operatorname{Re} Q > 0$ in the region of $Z = X_1 + iX_2$ being studied, for which the exponentials come from the two primitive phase-integral (JWKB) approximations (giving

$$M_\pm = Q^{-\frac{1}{4}} \quad \text{and} \quad \phi_\pm = \pm \int_a^Z Q^{\frac{1}{2}}\, dz'$$

where a is a simple zero of $Q^2(Z)$), then again $\beta = 1$ (I, p. 299) (for an nth-order zero, (10) is multiplied by $2\cos(\pi/(n+2))$).

To interpret (8) Dingle employs Borel summation, not for the whole series as is customary, but for the nth term and beyond, where n is close to the value $r \sim |F|$ for which a_r is least. Assuming this formal procedure is valid, we obtain, by using (10),

$$y \approx M_+ \exp(k\phi_+) \sum_{r=0}^{n-1} a_r + i M_- S_n(F) \exp(k\phi_-), \tag{12}$$

where

$$S_n(F) \equiv \frac{-i}{2\pi} \exp(F) \sum_{r=n}^{\infty} \frac{(r-\beta)!}{F^{r-\beta+1}}. \tag{13}$$

The interpretation is obtained by writing the factorial as an integral and performing the summation

$$S_n(F) = \frac{-i \exp(F)}{2\pi F^{1-\beta}} \int_0^\infty dS \exp(-S)\, S^{-\beta} \sum_{r=n}^{\infty} \left(\frac{S}{F}\right)^r$$

$$= \frac{-i}{2\pi} \int_0^\infty dt\, \frac{t^{n-\beta} \exp\{F(1-t)\}}{1-t}. \tag{14}$$

To complete the interpretation it is necessary to specify the t-contour relative to the pole at $t = 1$. This corresponds to specifying the contour of integration (as

Smoothing Stokes's discontinuities 11

will be illustrated later), or the desired solution of a differential equation, when defining y. Different choices differ by real constants, corresponding to the value of the quantity S_- in section 1. We specify that the contour passes above $t = 1$, so that

$$S_n(F) = \tfrac{1}{2} - \frac{\mathrm{i}}{2\pi} \fint_0^\infty \mathrm{d}t \frac{t^{n-\beta} \exp\{F(1-t)\}}{1-t}, \tag{15}$$

where now the principal value of the integral is taken. This choice corresponds to the situation in example (3), where the Stokes multiplier switches on from zero ($S_- = 0$) as $\mathrm{Im}\,F$ increases through zero.

Now we identify S_n as the Stokes multiplier and determine its dominant asymptotics when F is large and nearly real. A crucial simplification (corresponding to the evaluative interpretation of asymptotic series adopted by Stokes) occurs if we truncate the r-sum near its least term, that is at

$$n - 1 = \mathrm{Int}\,(|F| + \alpha), \tag{16}$$

where α is of order unity (as is β). With this truncation, the stationary point of (15) almost coincides with its pole $t = 1$, whose neighbourhood therefore dominates the integral. Let

$$F \equiv A + \mathrm{i}B \tag{17}$$

(where A and B are real with $A \gg 1$ and $B \ll A$) and

$$n - \beta \equiv A + \mu, \quad \text{i.e.} \quad \mu = \mathrm{Int}\,(|F| + \alpha) - \beta - A + 1, \tag{18}$$

(so μ is of order unity) and change variables to $x \equiv t - 1$. Then expanding the integrand in (15) to third order about $x = 0$ gives

$$S_n(F) = \tfrac{1}{2} + \frac{\mathrm{i}}{2\pi} \fint_{-1}^\infty \frac{\mathrm{d}x}{x} \exp\{(A+\mu)\ln(1+x) - Ax - \mathrm{i}Bx\}$$

$$\approx \tfrac{1}{2} + \frac{1}{2\pi} \int_{-\infty}^\infty \mathrm{d}x(1 + \mu x + \tfrac{1}{3}Ax^3) \exp\left(-\tfrac{1}{2}Ax^2\right)(\sin Bx + \mathrm{i}\cos Bx)/x$$

$$= \frac{1}{2}\left(1 + \frac{1}{\pi}\int_{-\infty}^\infty \frac{\mathrm{d}x}{x} \exp\left(-\tfrac{1}{2}Ax^2\right) \sin Bx\right)$$

$$+ \frac{\mathrm{i}}{2\pi}\int_{-\infty}^\infty \mathrm{d}x \exp\left(-\tfrac{1}{2}Ax^2\right)(\mu + \tfrac{1}{3}Ax^2)\cos Bx$$

$$= \frac{1}{\sqrt{\pi}}\int_{-\infty}^{B/(2A)^{\frac{1}{2}}} \mathrm{d}t \exp\left(-t^2\right) - \mathrm{i}(2\pi A)^{-\frac{1}{2}}(\mathrm{Fract}\,\{|F| + \alpha\}$$

$$+ \beta - \alpha - \tfrac{4}{3} - B^2/6A) \times \exp\left(-B^2/2A\right). \tag{19}$$

The real part dominates, and comparison with (1) and (12) yields the change in the Stokes multiplier as

$$S(\sigma) = \frac{1}{\sqrt{\pi}}\int_{-\infty}^\sigma \mathrm{d}t \exp\left(-t^2\right) \tag{20}$$

12   M. V. Berry

involving the *Stokes variable*

$$\sigma(X;k) = B/(2A)^{\frac{1}{2}} = \operatorname{Im} F/(2\operatorname{Re} F)^{\frac{1}{2}}$$
$$= k^{\frac{1}{2}}\operatorname{Im}(\phi_+ - \phi_-)/\{2\operatorname{Re}(\phi_+ - \phi_-)\}^{\frac{1}{2}}. \tag{21}$$

Equations (20) and (21) carry the following implication: under a magnification of order $k^{\frac{1}{2}}$, the multiplier varies smoothly from S_- to $S_- + 1$ across the Stokes line, the functional dependence on the natural variable σ being that of the error function.

According to (19), the imaginary part of $S_n(F)$ is smaller than the subdominant exponential in (12) by a factor $A^{-\frac{1}{2}} \approx |F|^{-\frac{1}{2}} \sim k^{-\frac{1}{2}}$. This gives the assurance that the dominant series has been controlled to better-than-exponential accuracy. Thus to detect the multiplier numerically, as we will do in the next section, it is sufficient to subtract from the exact function y the dominant series taken up to its nearly least term, i.e.

$$\left[y\exp(-k\phi_-) - M_+ \exp(F) \sum_{r=0}^{\operatorname{Int}(|F|+\alpha)} a_r\right]\Big/ M_- \to \mathrm{i}\,(S(\sigma) + S_-) \quad \text{as} \quad |F| \to \infty \tag{22}$$

independently of α provided α is of order unity. This equation embodies our main result.

Any alteration in α changes the number of terms included in the sum (cf. 16), and is compensated by a change in the imaginary part of (19) which is small compared with $S(\sigma)$. Two natural choices for the 'best' α are (i) that which minimizes the imaginary part of (19) on the average, which gives (because the average of Fract$\{x\}$ is $\frac{1}{2}$) $\alpha = \beta - \frac{5}{6}$; (ii) that for which the smallest a_r has $r = n$ or $r = n+1$, which is the case if $\alpha = \beta - \frac{1}{2}$.

The imaginary terms in (19) constitute the lowest-order approximation in the technique of 'terminants' or 'converging factors', which has been employed (see I and Olver (1974)) to correct the dominant series representing y on the Stokes line. Here, of course, we are studying the variation of y *across* a Stokes line.

3. Numerical illustrations

According to (22), the Stokes multiplier function, which is of order unity, is the difference of two large quantities (of order $\exp(F)$). To detect S numerically, two conditions must be satisfied: y must be calculable to better-than-exponential precision, and the coefficients a_r in the asymptotic expansion must be known (or calculable) for large r. There follow two examples for which these conditions are met.

The first is *Dawson's integral* (Abramowitz & Stegun 1964)

$$y \equiv \int_0^Z \mathrm{d}t \exp(kt^2) = k^{-\frac{1}{2}}\operatorname{erfi}(k^{\frac{1}{2}}Z) \tag{23}$$

which differs by $\mathrm{i}\,(\pi/k)^{\frac{1}{2}}$ from the example (3). We are interested in $Z = X_1 + \mathrm{i}X_2$ near the positive real axis. Obviously y is real on the Stokes line, so that in (7) S must vanish when $X_2 = 0$, implying in turn that in the general expression (14) the principal value must be taken and that $S_- = -\frac{1}{2}$ in (22). Figure 2 shows the

Smoothing Stokes's discontinuities 13

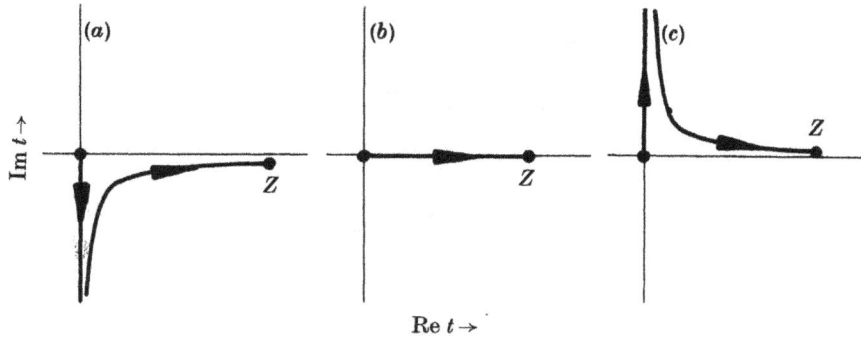

FIGURE 2. Steepest-descent contours of Dawson's integral (23) from $t = 0$ to $t = Z = X_1 + \mathrm{i}X_2$, for three values of Z near the Stokes line $X_2 = 0$.

steepest-descent contours, for comparison with those in figure 1. The singulant is given by (9) and (7) as

$$F = kZ^2 \tag{24}$$

so that the Stokes variable is

$$\sigma = (2k)^{\frac{1}{2}}X_1 X_2 / (X_1^2 - X_2^2)^{\frac{1}{2}} \approx (2k)^{\frac{1}{2}}X_2. \tag{25}$$

In the expansion (8) the coefficients a_r can be found by elementary asymptotics, for example (I, p. 5) changing the integration variable in (23) by $kt^2 = kZ^2 + u$ and expanding the jacobian about $u = 0$, with the result

$$a_r = (r - \tfrac{1}{2})! \, F^{-r} / \sqrt{\pi}. \tag{26}$$

Thus the limiting form (10) is here exact for all r, and $\beta = \tfrac{1}{2}$.

Incorporating (24)–(26) into the general result (22) with $S_- = -\tfrac{1}{2}$ we obtain, as the formula for the asymptotic emergence of the multiplier,

$$\mathrm{Im}\left\{ \mathrm{erfi}\,(F^{\frac{1}{2}})/\sqrt{\pi} - (2\pi)^{-1}\exp{(F)} \sum_{r=0}^{\mathrm{Int}\,(|F|+\alpha)} (r - \tfrac{1}{2})! / F^{r + \frac{1}{2}} \right\}$$

$$\to \frac{1}{\sqrt{\pi}} \int_0^{\sigma} \mathrm{d}t \exp{(-t^2)} \quad \text{as} \quad |F| \to \infty. \tag{27}$$

We shall here regard σ and $|F|$ as given, and obtain X_1 and X_2 by inversion of (24) and (25):

$$k^{\frac{1}{2}}X_1 = |F|^{\frac{1}{2}}\cos\theta; \quad k^{\frac{1}{2}}X_2 = |F|^{\frac{1}{2}}\sin\theta$$

where

$$\theta = \tfrac{1}{2}\arccos\{[1 + (\sigma^2/|F|)^2]^{\frac{1}{2}} - \sigma^2/|F|\}. \tag{28}$$

Tables 1–4 show the results of a numerical test of (27) over a range of values of σ, for singulants $|F| = 5$ and $|F| = 25$ and truncation variables $\alpha = -\tfrac{1}{3}$ and $\alpha = 0$ (these are the two 'best' choices of α described at the end of §2). The integral for erfi was evaluated along a straight-line contour from X_1 to $X_1 + \mathrm{i}X_2$ (the integral from 0 to X_1 being real) by the extended Simpson's rule; sufficient accuracy was achieved with 200 steps for $|F| = 5$ and 2000 steps for $|F| = 25$, and checked by evaluating the convergent series for erfi.

14   M. V. Berry

Table 1. Comparison of theory and 'experiment' for the Stokes multiplier function for Dawson's integral (23)

The singulant modulus is $|F| = 5$ and the truncation variable is $\alpha = -\frac{1}{3}$. σ is the Stokes variable (21) and $F = k(X_1 + iX_2)^2$ with X_1 and X_2 given by (28). (lhs and rhs are left-hand and right-hand sides respectively.)

σ	$2\,\mathrm{Im}\,\mathrm{erfi}\,(F^{\frac{1}{2}})/\sqrt{\pi}$	$2 \times$ lhs of (27)	$2 \times$ (lhs $-$ rhs) of (27)
0.2	21.532 335 635 1	0.220 432 726 808	$-0.002\,277\,090\,460\,5$
0.4	32.406 989 074	0.421 905 998 99	$-0.006\,499\,177\,409\,1$
0.6	30.369 785 607 3	0.591 079 840 706	$-0.012\,791\,995\,848$
0.8	21.029 192 239 2	0.722 780 152 631	$-0.019\,336\,679\,001\,1$
1.0	11.224 112 560 1	0.818 898 722 383	$-0.023\,815\,907\,806\,3$
1.2	4.455 199 927 81	0.885 393 233 85	$-0.024\,931\,438\,183\,8$
1.4	0.939 754 805 283	0.929 422 218 948	$-0.022\,870\,317\,935\,5$
1.6	$-0.434\,973\,606\,01$	0.957 562 519 129	$-0.018\,790\,516\,202$
1.8	$-743\,819\,764\,526$	0.975 062 006 02	$-0.014\,031\,146\,904$
2.0	$-0.634\,984\,356\,598$	0.985 737 130 522	$-0.009\,586\,512\,135\,48$
2.2	$-0.411\,962\,286\,068$	0.992 177 544 661	$-0.005\,960\,263\,230\,16$
2.4	$-0.190\,731\,569\,089$	0.996 048 680 562	$-0.003\,263\,089\,936\,87$

Table 2. As table 1 with $\alpha = 0$

σ	$2\,\mathrm{Im}\,\mathrm{erfi}\,(F^{\frac{1}{2}})/\sqrt{\pi}$	$2 \times$ lhs of (27)	$2 \times$ (lhs $-$ rhs) of (27)
0.2	21.532 335 635 1	0.242 465 868 574	0.019 756 051 305 7
0.4	32.406 989 074	0.463 805 332 974	0.035 400 156 575
0.6	30.369 785 607 3	0.648 329 058 716	0.044 457 222 162 2
0.8	21.029 192 239 2	0.788 860 592 493	0.046 743 760 860 5
1.0	11.224 112 560 1	0.886 578 386 719	0.043 863 756 53
1.2	4.455 199 927 81	0.948 451 996 877	0.038 127 324 843 3
1.4	0.939 754 805 283	0.983 881 887 394	0.031 589 350 511 1
1.6	$-0.434\,973\,606\,01$	1.001 915 070 31	0.025 562 034 975 8
1.8	$-0.743\,819\,764\,526$	1.009 685 586 97	0.020 592 434 041 3
2.0	$-0.634\,984\,356\,598$	1.012 031 282 78	0.016 707 640 125
2.2	$-0.411\,962\,286\,068$	1.011 844 149 78	0.013 706 341 889 7
2.4	$-0.190\,731\,569\,089$	1.010 670 520 05	0.011 358 749 547 4

Table 3. As table 1 with $|F| = 25$

σ	$2\,\mathrm{Im}\,\mathrm{erfi}\,(F^{\frac{1}{2}})/\sqrt{\pi}$	$2 \times$ lhs of (27)	$2 \times$ (lhs $-$ rhs) of (27)
0.2	7 833 454 975.39	0.222 258 567 81	$-0.000\,451\,249\,458\,156$
0.4	2 631 901 621.35	0.427 119 255 066	$-0.001\,285\,921\,333\,19$
0.6	$-4\,826\,024\,257.27$	0.600 670 814 514	$-0.003\,201\,022\,040\,09$
0.8	$-3\,211\,113\,836.27$	0.737 760 543 823	$-0.004\,356\,287\,808\,85$
1.0	1 495 138 180.67	0.840 577 602 386	$-0.002\,137\,027\,802\,68$
1.2	1 999 214 434.28	0.908 263 206 482	$-0.002\,061\,465\,551\,5$
1.4	137 285 256.727	0.946 264 266 968	$-0.006\,028\,269\,915\,31$
1.6	$-647\,807\,326.949$	0.970 328 092 575	$-0.006\,024\,942\,755\,88$
1.8	$-311\,385\,089.924$	0.987 966 120 243	$-0.001\,127\,032\,680\,75$
2.0	42 695 984.0144	0.997 133 061 29	0.001 809 418 632 56
2.2	98 642 118.0045	0.999 141 067 266	0.001 003 259 375 05
2.4	39 463 275.4458	0.998 938 336 968	$-0.000\,373\,433\,530\,283$

Smoothing Stokes's discontinuities **15**

TABLE 4. AS TABLE 1 WITH $|F| = 25$ AND $\alpha = 0$

σ	$2\operatorname{Im}\operatorname{erfi}(F^{\frac{1}{2}})/\sqrt{\pi}$	$2 \times$ LHS of (27)	$2 \times$ (LHS $-$ RHS) of (27)
0.2	7 833 454 975.39	0.226 700 782 776	0.003 990 965 507 66
0.4	2 631 901 621.35	0.435 605 049 133	0.007 199 872 734 19
0.6	$-4 826 024 257.27$	0.612 331 390 381	0.008 459 553 826 61
0.8	$-3 211 113 836.27$	0.751 268 863 678	0.009 152 032 045 89
1.0	1 495 138 180.67	0.854 352 474 213	0.011 637 844 023 5
1.2	1 999 214 434.28	0.920 836 210 251	0.010 511 538 217 4
1.4	137 285 256.727	0.956 623 911 858	0.004 331 374 974 52
1.6	$-647 807 326.949$	0.978 073 358 536	0.001 720 323 204 81
1.8	$-311 385 089.924$	0.993 239 462 376	0.004 146 309 451 82
2.0	42 695 984.0144	1.000 409 543 51	0.005 085 900 857 02
2.2	98 642 118.0045	1.000 999 227 17	0.002 861 419 274 76
2.4	39 463 275.4458	0.999 897 457 659	0.000 585 687 160 54

Successive columns clearly show the decreasing orders of magnitude of the first term on the left-hand side of (27) (i.e. the integral), whose order is $\exp(|F|)/|F|^{\frac{1}{2}}$ (at least for small σ); the whole left-hand side of (27) (i.e. the 'experimental' Stokes multiplier), whose order is unity; and the difference between the two sides of (27), whose order can be shown (by easy extension of the argument leading to (19)) to be $|F|^{-1}$. The two sides of (27) are shown plotted against σ for $|F| = 5$ and 25 in figures $3a, b$, for the two values of α. Evidently the agreement between 'experimental' and 'theoretical' multipliers improves with increasing $|F|$, as it should. The two curves for the 'best' α bracket the theoretical curve.

For the second example we take the *Airy function of the second kind* (Abramowitz & Stegun 1964),

$$y = \int_C dt \exp\{k(-\tfrac{1}{3}t^3 + tZ)\} = 2\pi k^{-\frac{1}{3}} \operatorname{Bi}(Zk^{\frac{2}{3}}) \tag{29}$$

where C is the contour shown in figure $4a$. We are interested in $Z = X_1 + iX_2$ near the positive real axis. The integrand has stationary points at $t_{\pm} = \pm Z^{\frac{1}{2}}$, so that the dominant and subdominant exponents in (1), and the singulant, are

$$\phi_{\pm} = \pm\tfrac{2}{3}Z^{\frac{3}{2}}, \quad F = \tfrac{4}{3}kZ^{\frac{3}{2}}, \tag{30}$$

where the roots are positive real when Z is positive real, that is on the Stokes line. Thus the Stokes variable (21) is

$$\sigma = (\tfrac{2}{3}k)^{\frac{1}{2}}\operatorname{Im}(X_1 + iX_2)^{\frac{3}{2}}/[\operatorname{Re}(X_1 + iX_2)^{\frac{3}{2}}]^{\frac{1}{2}} \approx (\tfrac{3}{2}k)^{\frac{1}{2}}X_2/X_1^{\frac{1}{4}}. \tag{31}$$

This shows that for fixed k the 'width' of the Stokes zone increases slowly (as $X^{\frac{1}{4}}$) away from the origin.

Figures $4b$–d show the steepest-descent contours for Z nearly positive real. Evidently the contribution of the subdominant exponential reverses across the Stokes line and vanishes on it, as for Dawson's integral, so again $S_- = -\tfrac{1}{2}$. After taking into account the double contour through the dominant stationary point, the prefactors $M_{\pm}$ can be obtained from the simplest steepest-descent argument, giving

$$M_+ = 2M_- = 2(\pi^2/k^2Z)^{\frac{1}{4}}. \tag{32}$$

16   M. V. Berry

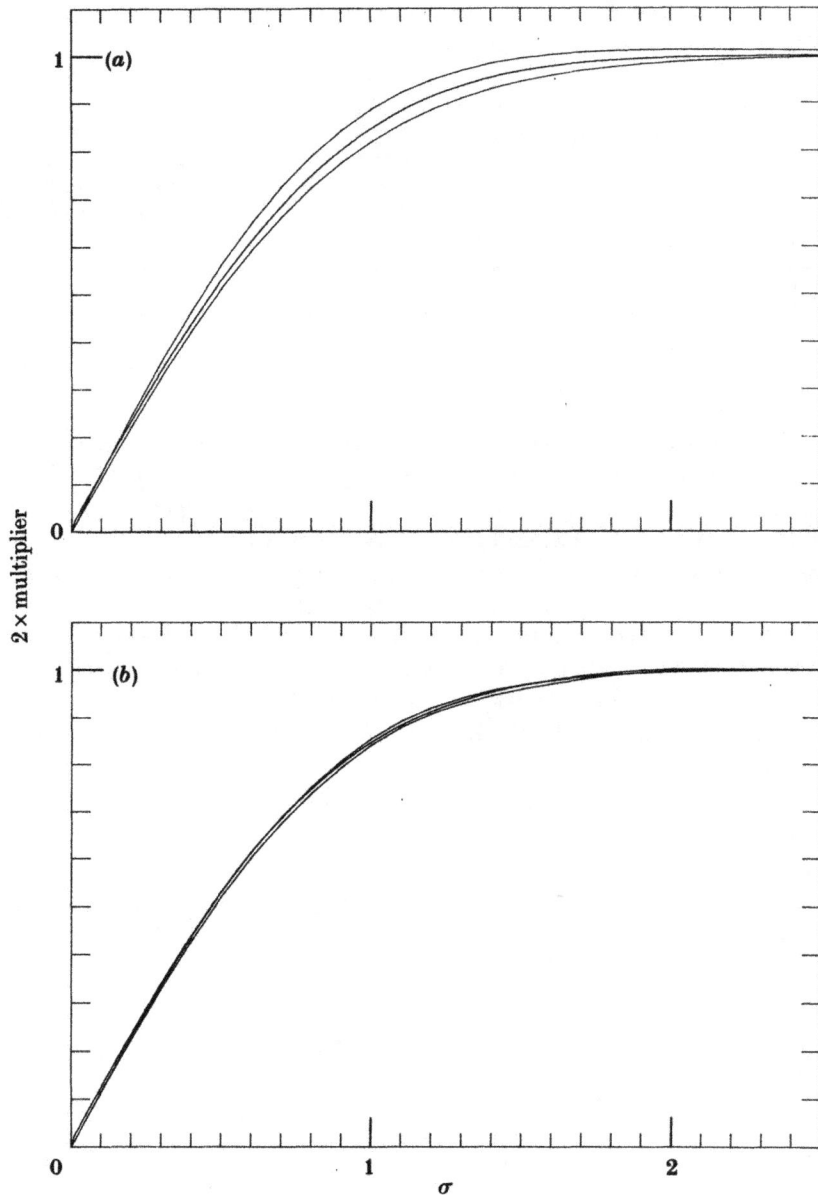

FIGURE 3. Stokes multiplier function for Dawson's integral erfi (equation (23)) for (*a*) singulant $|F| = 5$; (*b*) singulant $|F| = 25$. The middle curve is the 'theoretical' multiplier (RHS of (27)) and the upper and lower curves are the 'experimental' multipliers (LHS of (27)) for $\alpha = 0$ and $\alpha = -\frac{1}{2}$ respectively.

Smoothing Stokes's discontinuities 17

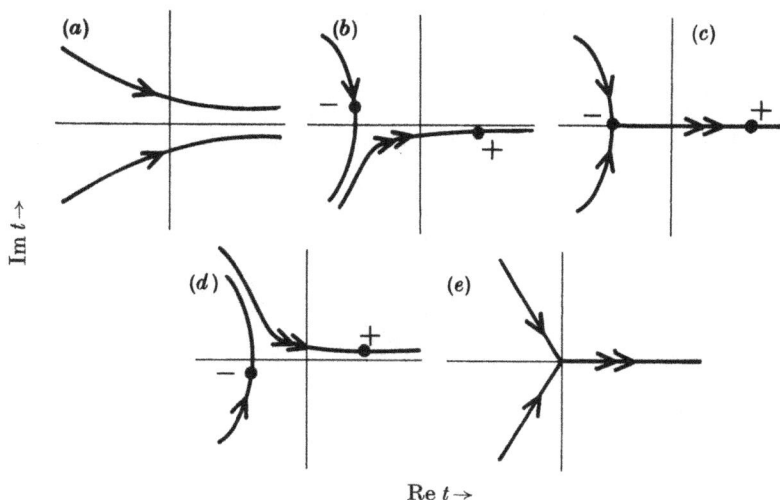

FIGURE 4. Integration contours for the Airy function Bi (equation (29)). (a) generic contour; (b) steepest-descent contour for $\operatorname{Im} Z < 0$; (c) steepest-descent contour for $\operatorname{Im} Z = 0$; (d) steepest-descent contour for $\operatorname{Im} Z > 0$; (e) contour for numerical integration of Bi. In b–d the subdominant and dominant stationary points are labelled + and −. Double arrows mean contours counted twice.

Further steepest-descent analysis (or phase-integral solution of Airy's equation $\partial_Z^2 y = k^2 Z y$) gives the expansion coefficients as

$$a_r = (r - \tfrac{1}{6})!\,(r - \tfrac{5}{6})!/(2\pi r!\,F^r). \qquad (33)$$

This satisfies the initial condition $a_0 = 1$ (because $(-\tfrac{1}{6})!\,(-\tfrac{5}{6})! = 2\pi$), and conforms to the limit (10) with $\beta = 1$ (because $(r-\mu)!\,(r-\nu)!/r! \to (r-\mu-\nu)!$ as $r \to \infty$).

Incorporating (29)–(33) into the general result (22) with $S_- = -\tfrac{1}{2}$, including the contour doubling, we obtain, as the formula for the asymptotic emergence of the multiplier,

$$2\operatorname{Im}\left\{ \pi^{\frac{1}{2}} (\tfrac{3}{4}F)^{\frac{1}{6}} \exp{(\tfrac{1}{2}F)}\,\mathrm{Bi}\,[(\tfrac{3}{4}F)^{\frac{2}{3}}] - \exp{(F)} \sum_{r=0}^{\operatorname{Int}[|F|+\alpha]} (r - \tfrac{1}{6})!\,(r - \tfrac{5}{6})!/(2\pi r!\,F^r) \right\}$$

$$\to \frac{2}{\sqrt{\pi}} \int_0^\sigma \mathrm{d}t\,\exp{(-t^2)} \quad \text{as} \quad |F| \to \infty. \quad (34)$$

Again we shall regard σ and $|F|$ as given, obtaining F by solving (30) and (31) for X_1 and X_2;

$$\left. \begin{array}{l} k^{\frac{2}{3}} X_1 = (\tfrac{3}{4}|F|)^{\frac{2}{3}} \cos\phi; \quad k^{\frac{2}{3}} X_2 = (\tfrac{3}{4}|F|)^{\frac{2}{3}} \sin\phi, \\[2mm] \text{where} \qquad \phi = \tfrac{2}{3} \arccos\{[1 + (\sigma^2/|F|)^2]^{\frac{1}{2}} - \sigma^2/|F|\}. \end{array} \right\} \quad (35)$$

Tables 5 and 6 show the results of a numerical test of (34) over a range of values of σ, for singulants $|F| = 5$ and $|F| = 25$. For these integer $|F|$ the two 'best' truncation variables, $\alpha = \tfrac{1}{6}$, and $\alpha = \tfrac{1}{2}$, give identical sums in (34). The function Bi was computed from the extended Simpson's rule, after deforming the contour C

18 M. V. Berry

TABLE 5. COMPARISON OF THEORY AND 'EXPERIMENT' FOR THE STOKES
MULTIPLIER FUNCTION FOR THE MODIFIED AIRY FUNCTION (29)

The singulant modulus is $|F| = 5$ and the truncation variable is $\alpha = \frac{1}{6}$. σ is the Stokes variable (21) and $F = \frac{4}{3}k(X_1 + iX_2)^{\frac{3}{2}}$ with X_1 and X_2 given by (35).

σ	Im $\{\pi^{\frac{1}{2}}(\frac{3}{4}F)^{\frac{1}{6}} e^{\frac{1}{2}F} \mathrm{Bi}\,((\frac{3}{4}F)^{\frac{2}{3}})\}$	LHS of (34)	(LHS−RHS) of (34)
0.2	86.460 499 672 4	0.221 345 661 477	−0.012 178 151 010 6
0.4	123.976 073 85	0.424 404 297 191	−0.013 568 051 778
0.6	105.332 800 481	0.595 782 204 695	−0.015 902 967 923 4
0.8	59.829 175 322 5	0.729 510 930 471	−0.018 496 302 177 4
1.0	18.783 679 441 2	0.826 574 068 164	−0.020 239 852 379 9
1.2	−5.020 034 878 88	0.892 523 281 855	−0.020 434 857 650 3
1.4	−13.954 827 399 5	0.934 750 647 189	−0.019 103 607 977 4
1.6	−14.786 787 214	0.960 427 535 479	−0.016 780 434 441 5
1.8	−12.617 998 357 4	0.975 396 073 945	−0.014 129 113 588 7
2.0	−9.963 092 930 78	0.983 860 233 76	−0.011 664 948 150 8
2.2	−7.698 262 589 58	0.988 566 623 58	−0.009 657 971 616 79
2.4	−5.977 844 120 23	0.991 179 082 126	−0.008 167 187 509 41

TABLE 6. AS TABLE 5 WITH $|F| = 25$

σ	Im $\{\pi^{\frac{1}{2}}(\frac{3}{4}F)^{\frac{1}{6}} e^{\frac{1}{2}F} \mathrm{Bi}\,((\frac{3}{4}F)^{\frac{2}{3}})\}$	LHS of (34)	(LHS−RHS) of (34)
0.2	68 717 981 224.8	0.218 231 201 172	−0.015 292 611 315 5
0.4	19 591 748 977.8	0.422 904 968 262	−0.015 067 380 707 3
0.6	−44 424 281 459.3	0.599 472 045 898	−0.012 213 126 72
0.8	−24 788 677 708.8	0.739 608 764 648	−0.008 398 467 999 97
1.0	16 363 698 359.3	0.840 934 753 418	−0.005 879 167 126 17
1.2	16 545 841 811.1	0.908 313 751 221	−0.004 644 388 284 93
1.4	−1 014 224 788.75	0.949 258 327 484	−0.004 595 927 682 19
1.6	−6 181 168 808.05	0.972 358 703 613	−0.004 849 266 307 28
1.8	−2 073 873 258.79	0.984 602 928 162	−0.004 922 259 371 89
2.0	856 749 591.829	0.990 872 144 699	−0.004 653 037 211 44
2.2	909 857 385.36	0.993 932 247 162	−0.004 292 348 035 32
2.4	237 699 764.863	0.995 340 585 709	−0.004 005 683 926 31

(figure 4*a*) of the integral (29) to that in figure 4*e*, and checked by evaluating the convergent series for Bi.

As for Dawson's integral, successive columns clearly show the expected decreasing orders of magnitude. In the second column, containing the values of the first term on the left-hand side of (34), the order is now $\exp(|F|)$, which is larger by $|F|^{\frac{1}{2}}$ than the corresponding order for Dawson's integral. The left and right sides of (34) are plotted against σ for $|F| = 5$ and $|F| = 25$ in figures 5*a*, *b*.

It is again evident that the agreement between 'experimental' and 'theoretical' multipliers is excellent and improves with increasing $|F|$. A feature of this example, not present for Dawson's integral, is that it illustrates the general case in which the limit (10), forming the basis of our theory of the multiplier, is an approximation valid for the late terms a_r, rather than being exact for all r.

Stokes (1964) himself illustrated the change in the multiplier, by computing an Airy function at two complex arguments with phases $\pm 30°$ from a Stokes line, and a common modulus for which $|F| = \sqrt{128}$. Then (30), (21) and (20) give $\sigma = -2$ and $S(\sigma) = 0.005$, for which the change has barely begun, and $\sigma = +2$ and $S(\sigma) = 0.995$, for which it is virtually complete.

Smoothing Stokes's discontinuities 19

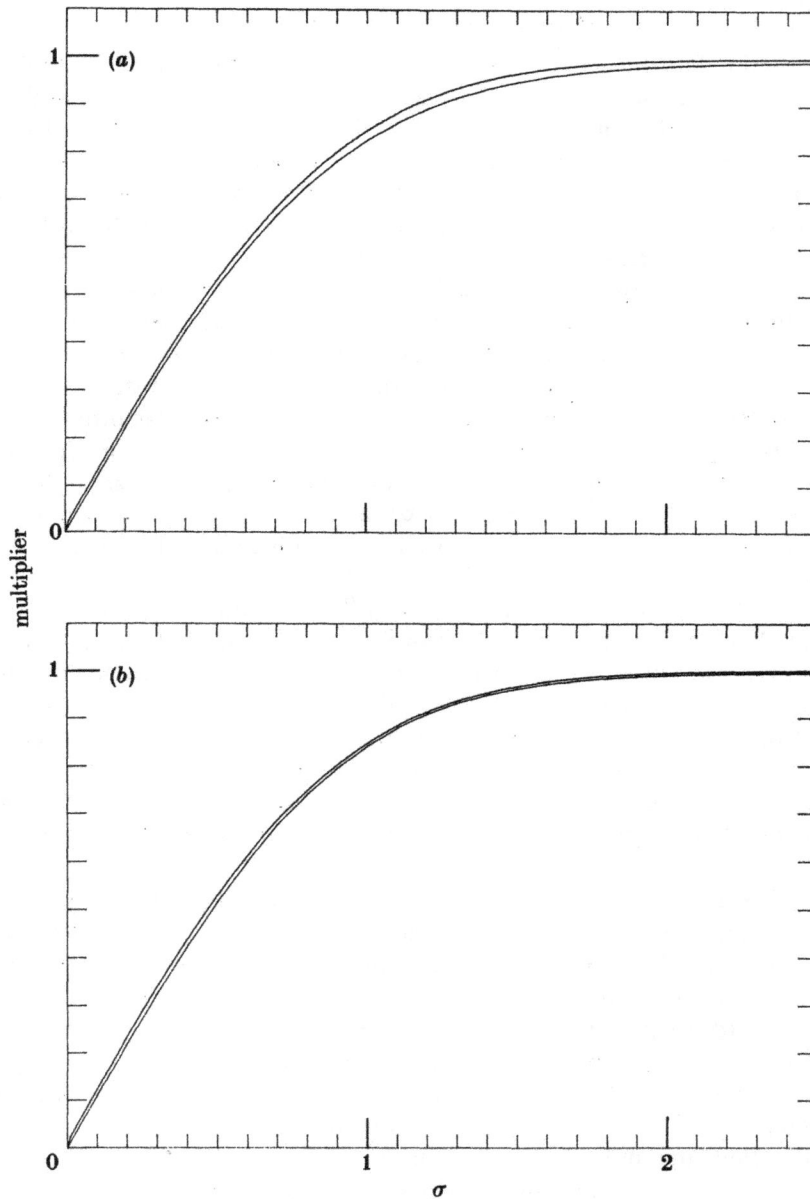

FIGURE 5. Stokes multiplier function for Airy function of the second kind (Bi) (equation (29)) for (a) singulant $|F| = 5$; (b) singulant $|F| = 25$. The upper curves are the 'theoretical' multipliers (RHS of (34)) and the lower curves are the 'experimental' multipliers (LHS of (34)) for $\alpha = \frac{1}{6}$.

20   M. V. Berry

4. Concluding remarks

We have found that across the Stokes line, on which the large exponential is maximally dominant, the multiplier of the small exponential varies rapidly but smoothly in a universal manner that is almost the simplest imaginable; the error function of a natural variable (equation (21)) depending on the singulant (difference between exponents). In numerical computations the Stokes multiplier (of order unity) emerged stably, and in agreement with theory, as the difference between two large quantities (of order up to 10^{10}, cf. table 6).

The universality class appears to be large. Our derivation depended on Dingle's (I) factorial formula (10) for the late terms a_r of the dominant asymptotic series. This type of asymptotics applies (at least) to integrals of functions involving $\exp(k\phi)$ as $k \to \infty$, whether dominated by an end-point or a stationary point; and to the solutions of ordinary differential equations with first-order turning points (for a turning point of order n the multiplier (20) is simply magnified by $2\cos(\pi/(n+2))$). But the results remain valid for some kinds of superfactorial divergence, for example $a_r \to (r!)^m/k^r$ as $r \to \infty$ (as can be shown by multiple Borel summation or, more easily, with multiplication formulae for factorials). It would be interesting to establish the limits of universality.

A trivial way to lose both the universality and simplicity of the main result is truncate the dominant series far from the least term, so that in (16) $|\alpha| \gg 1$. In these circumstances a general theory is still possible provided the limit (10) remains valid for all the Borel-summed terms, that is for all $r > n$, but is unsatisfactory in two respects. First, the multiplier S acquires an imaginary part comparable to its real part. Second, the multiplier now depends on $D \equiv (|F|-n)/(2\,\mathrm{Re}\,F)^{\frac{1}{2}}$ as well as σ, and when $|D|$ is large S rises to values of order $\exp(D^2)$ and oscillates with period of order D^{-1} over a total σ-range of order $|D|$ (rather than unity).

Our treatment is limited by being restricted to the situation where there is just one dominant and one subdominant exponential (equation (1)). What happens when there are more? The following conjecture is natural. Let the exponents be $k\phi_1, k\phi_2, \ldots$, ordered so that $\mathrm{Re}\,\phi_i > \mathrm{Re}\,\phi_j$ if $i < j$, and define for each pair the singulant $F_{ij} \equiv k(\phi_i - \phi_j)$ with $i < j$. Then across the line $\mathrm{Im}\,F_{ij} = 0$ the jth exponential is maximally dominated by the ith and its switching-on is described by the universal Stokes multiplier function that we have obtained, the appropriate singulant being $F = F_{ij}$. A proof (or disproof) is desirable.

It is worth pointing out that although in our examples (erfi and Bi) the parameters X were components of a complex variable $X_1 + iX_2$, the parameter space need not possess a complex structure. To illustrate this, consider the oscillatory integral describing the cusp diffraction catastrophe (Pearcey's integral)

$$y(X_1, X_2; k) = \int_{-\infty}^{\infty} dt \exp\{ik(\tfrac{1}{4}t^4 + \tfrac{1}{2}X_1 t^2 + X_2 t)\} \qquad (36)$$

as studied by Wright (1980). The asymptotics are dominated by real stationary points giving oscillatory contributions (waves). In the (real) parameter space X_1, X_2 there are three real stationary points inside the cusp $27X_2^2 + 4X_1^3 = 0$, and

Smoothing Stokes's discontinuities 21

one real and two complex stationary points outside. In this outside region the real stationary point always contributes to (36), and one of the complex ones never does. Wright discovered that the (exponentially small) contribution of the second complex stationary point switches on across the *Stokes set*, which is the different cusp $27X_2^2 - (5 + 3\sqrt{3})X_1^3 = 0$.

An immediate application of the main result is to the *birth of reflected waves* as described by $y''(x) + k^2 n^2(x)\,y(x) = 0$, where $n(x)$ is a real non-zero refractive index profile. The reflected wave is exponentially small and (if the incident wave is defined by the dominant W.K.B. expansion truncated at its least term) switches on where the Stokes line from the nearest complex turning point crosses the x-axis. The switch occurs over $|F|^{\frac{1}{2}}$ wavelengths, where the singulant $|F|$ is the exponent in the reflection amplitude. (This interpretation originated in conversation with Professor R. G. Littlejohn.)

Finally, there ought to be connections between this work (and, more generally, Dingle's interpretative theory of asymptotic series (I)) and Écalle's recent doctrine of *résurgence* (Écalle 1981, 1984; see also Voros 1983; Pham *et al.* 1989).

I thank Professor F. Pham for the hospitality of the mathematics department of the University of Nice, where this research was carried out.

References

Abramowitz, M. & Stegun, I. A. 1964 *Handbook of Mathematical Functions.* Washington: National Bureau of Standards.

Dingle, R. B. 1973 *Asymptotic expansions: their derivation and interpretation.* New York & London: Academic Press. (Referred to as I in the text.)

Écalle, J. 1984 'Cinq applications des fonctions résurgentes'. (Preprint 84T62 Orsay.)

Écalle, J. 1981 'Les fonctions résurgentes'. (In 3 volumes.) *Publ. Math. Université de Paris-Sud.*

Olver, F. W. J. 1974 *Asymptotics and special functions*, ch. 14. New York & London: Academic Press.

Pham, F., Nosmas, C. & Candelpergher, B. 1989 'Résurgence, quantized canonical transformations and multi-instanton expansions'. In *Prospect in Algebraic Analysis*, dedicated to Professor M. Sato on his sixtieth birthday. (In the press.)

Stokes, G. G. 1864 *Trans. Camb. phil. Soc.* **10**, 106–128. Reprinted in *Mathematical and Physical papers by the late Sir George Gabriel Stokes.* Cambridge University Press 1904, vol. IV, 77–109.

Stokes, G. G. 1871 *Trans. Camb. phil. Soc.* **11**, 412–425. Reprinted in *Mathematical and Physical papers by the late Sir George Gabriel Stokes.* Cambridge University Press 1904, vol. IV, 283–298.

Stokes, G. G. 1889 *Proc. Camb. phil. Soc.* **6** (6). Reprinted in *Mathematical and Physical papers by the late Sir George Gabriel Stokes.* Cambridge University Press 1905, vol. V, 221–225.

Stokes, G. G. 1902 *Acta math., Stockh.* **26**, 393–397. Reprinted in *Mathematical and Physical papers by the late Sir George Gabriel Stokes.* Cambridge University Press 1905, vol. V, 283–287.

Voros, A. 1983 *Ann. Inst. H. Poincaré* **39**, 211–338.

Wright, F. J. 1980 *J. Phys.* A **13**, 2913–2928.

LIST OF PUBLICATIONS

1. Berry, M V, Clark, R C and Rijnierse, P J, 1965, *Proc. Phys. Soc.* **86**, 242–4 'Note on the invariance of the phase difference between two waves'.

2. Berry, M V, 1965, 'The diffraction of light by ultrasound', Ph.D. Thesis, St Andrews University.

3. Berry, M V, 1966, *The diffraction of light by ultrasound* (Academic Press).

4. Berry, M V, 1966, *Physica* **32**, 1582–90, 'Solution of the Raman-Nath Equation for light diffracted by ultrasound at normal incidence'.

5. Berry, M V, 1966, *Proc. Phys. Soc.* **88**, 285–92, 'Semiclassical scattering phase shifts in the presence of metastable states'.

6. Berry, M V, 1966, *Proc. Phys. Soc.* **89**, 479–90, 'Uniform approximation for potential scattering involving a rainbow'.

7. Lloyd, P and Berry, M V, 1967, *Proc. Phys. Soc.* **91**, 678–88, 'Wave propagation through an assembly of spheres IV. Relations between different multiple scattering theories'.

8. Berry, M V, 1967, *Uniformly approximate solutions for short-wave problems.* (SERC research report).

9. Berry, M V, 1968, *Engineering Outline* **152**, 779–82 'Liquid Structure Theory'.

10. Berry, M V, 1969, *Science Progress* (Oxford) **57**, 43–64 'Uniform approximation: A new concept in wave theory'.

11. Berry, M V, 1969, *J. Phys. B*, **2**, 381–92 'Uniform approximations for glory scattering and diffraction peaks'.

12. Berry, M V, 1969, *Engineering Outline*, **187**, 121–4 'Rheology'.

13. Berry, M V, 1970, *Engineering Outline*, **230**, 581–4 'Liquid surface physics'.

14. Berry, M V and Gibbs, D F, 1970, *Proc. R. Soc. A,* **314**, 143–52, 'The interpretation of optical projections'.

15. Berry, M V and Reznek, S R, 1971, *J. Phys. A*, **4**, 77–84, 'A simple theory for the densities of coexistent liquid and vapour through the transition region'.

16. Berry, M V, 1971, *J. Phys. C*, **4**, 697–722, 'Diffraction in crystals at high energies'.

17. Berry, M V, 1971, 'Transition from quantum to classical theory for HEED', (in *Proc. 25th Anniv. meeting Emag Inst. Phys (UK)*) 122–4.

18. Berry, M V, 1971, *Physics Education*, **6**, 79–84, 'The molecular mechanism of surface tension'.

19. Berry, M V, 1972, *Science Progress*, **60**, 125–8, review of *Introduction to meterological optics*, by R A R Tricker.

20. Berry, M V, Durrans, R F, Evans, R, 1972, *J. Phys. A*, **5**, 166–70, 'The calculation of surface tension for simple liquids'.

21. Berry, M V, 1972, *J. Phys. A*, **5**, 272–91, 'On deducing the form of surfaces from their diffracted echoes'.

22. Berry, M V, 1972, *Physics Education*, **7**, 1–6, 'Reflections on a Christmas-tree bauble'.

23. Berry, M V and Mount, K E, 1972, *Reps. Prog. Phys.* **35**, 315–97, 'Semiclassical approximations in wave mechanics'.

24. Doyle, P A and Berry, M V, 1972, 'Semiclassical prediction of increased penetration near certain voltages' in *Proc. EMCON, 72*, pp. 452–3.

25. Berry, M V and Buxton, B F, 1972, 'A new interpretation of bend contours in terms of semiclassical mechanics'. in *Proc. EMCON 72*, pp. 454–5.

26. Berry, M V and Ozorio de Almeda, A, 1972, 'Towards a manageable theory for cross-grating HEED' in *Proc. EMCON 72*, pp. 456–7.

27. Nye, J F, Berry, M V and Walford, M E R, 1972, *Nature*, **240**, No. 97, 7–9, 'Measuring the change in thickness of the Antarctic ice sheet'.

28. Berry, M V, 1972, *Times Higher Education Supplement 24.11.1972*, Review of *Relativistic Quantum Theory*, by Lifshitz *et al.*

29. Berry, M V, 1973, *Phil. Trans. R. Soc. A*, **273**, 611–54, 'The statistical properties of echoes diffracted from rough surfaces'.

30. Berry, M V and Doyle, P A, 1973 *J. Phys. C*, **6**, L6–9, 'Interpreting electron micrographs of amorphous solids'.

31. Doyle, P A and Berry, M V, 1973, *Z. fur Naturforsch*, **28a**, 571–6, 'Absorption and penetration in the semiclassical theory of high-energy electron diffraction'.

32. Berry, M V and Ozorio de Almeida, A M, 1973, *J. Phys. A*, **6**, 1451–60, 'Semiclassical approximation of the radial equation with two-dimensional potentials'.

33. Berry, M V, Buxton, B F and Ozorio de Almeida, A M, 1973, *Radiation effects*, **20**, 1–24, 'Between wave and particle — The semiclassical method for interpreting high-energy electron micrographs in crystals'.

34. Nye, J F and Berry, M V, 1974, *Proc. R. Soc. A*, **336**, 165–90, 'Dislocations in wave trains'.

35. Berry, M V, 1974, *J. Phys. A*, **7**, 231–45, 'Simple fluids near rigid solids — Statistical mechanics of density and contact angle'.

36. Atkinson, P and Berry, M V, 1974, *J. Phys. A,* **7**, 1293–302, 'Random noise in ultrasonic echoes diffracted by blood'.

37. Buxton, B F and Berry, M V, 1974, 'A General theory of the critical voltage effect' in *High-voltage electron microscopy,* ed. Swann *et al.*, Academic press, 60–3.

38. Berry, M V and Greenwood, D A, 1975, *Am. J. Phys.* **43**, 91, 'On the ubiquity of the sine wave'.

39. Berry, M V, 1974, *Science Progress,* **61**, 595–7, review of *The optics of rays, wavefronts and caustic,* by O N Stavroudis.

40. Berry, M V, 1975, *Nature,* **254**, 465, review of *Introductory eigenphysics* by C A Croxton.

41. Berry, M V, 1975, *J. Phys. A,* **8**, 566–84, 'Cusped rainbows and incoherence effects in the rippling-mirror model for particle scattering from surfaces'.

42. Berry, M V, 1975, *Science Progress,* **62**, 356–60, review of *Gravitation,* by G W Misner, K Thorne and J A Wheeler.

43. Berry, M V, 1975, *Times Higher Educational Supplement* review of 'Quantum Physics and Ordinary Language' by T Bergstein.

44. Berry, M V, 1975, *Journal of Glaciology,* **15**, 65–74, 'Theory of radio echoes from glacier beds'.

45. Berry, M V, 1975, *Surface Science,* Vol. 1, 291–327, 'Liquid Surfaces', IAEA-SMR-15/9.

46. Berry, M V, 1975, *Science Progress,* **62**, 638–40, review of *System Identification: Method and applications* by H H Kagiwada.

47. Berry, M V, 1975, *J. Phys. A,* **8**, 1952–71, 'Attenuation and focussing of electromagnetic surface waves rounding gentle bends'.

48. Berry, M V, 1976, *Physics Bulletin,* March, 107–8, 'Waves as catastrophes'.

49. Berry, M V, 1976, *Advances in Physics,* **25**, 1–26, 'Waves and Thom's theorem'.

50. Berry, M V, 1976, *Nature,* **261**, 174 review of *Twentieth Century Physics,* by Joseph Norwood.

51. Berry, M V and Tabor, M, 1976, *Proc. R. Soc. A,* **349**, 101–23, 'Closed orbits and the regular bound spectrum'.

52. Buxton, B F and Berry, M V, 1976, *Phil. Trans. R. Soc. A,* **282**, (No. 1308), 485–525. 'Bloch wave degeneracies in systematic high energy electron diffraction'.

53. Berry, M V, 1976, in *The greatest thinkers* (ed. de Bono) Weidenfeld and Nicolson. Essays on Maxwell and Einstein.

54. Berry, M V, 1976, *Principles of cosmology and gravitation,* (Cambridge University Press).

55. Berry, M V and Tabor, M, 1977, *J. Phys. A*, **10**, 371–9, 'Calculating the bound spectrum by path summation in action-angle variables'.

56. Berry, M V and Nye, J F, 1977, *Nature*, **267**, 34–6, 'Fine structure in caustic junctions'.

57. Berry, M V and Mackley, M R, 1977, *Phil. Trans. R. Soc. A*, **287**, 1–16 (No.1337). 'The six roll mill: Unfolding an unstable persistently extensional flow'.

58. Berry, M V, 1977, *J. Phys. A*, **10**, 2061–81, 'Focusing and twinkling: Critical exponents from catastrophes in non-Gaussian random short waves'.

59. Navascues, G and Berry, M V, 1977, *Mol. Phys*, **34**, 649–64. 'The statistical mechanics of wetting'.

60. Berry, M V and Hannay, J H, 1977, *J. Phys. A*, **10**, 1809–21, 'Umbilic points on Gaussian random surfaces'.

61. Berry, M V and Tabor, M, 1977, *Proc. R. Soc. A*, **356**, 375–94, 'Level clustering in the regular spectrum'.

62. Berry, M V, 1977, *Phil. Trans. R. Soc. A*, **287**, 237–71, 'Semi-classical mechanics in phase space: Astudy of Wigner's function'.

63. Berry, M V, 1977, *J. Phys. A*, **10**, L193–4, 'Remarks on degeneracies of semiclassical energy levels'.

64. Berry, M V, 1977, *J. Phys. A*, **10**, 2083–91. 'Regular and irregular semiclassical wave functions'.

65. Berry, M V, 1977, in *The Fontana dictionary of modern thought*, (Collins/Fontana: A Bullock and O Stallybrass, eds.) about 200 entries on physics.

66. Berry, M V, 1977, *J. Sci. Ind. Res.* (New Delhi) **36**, No. 3 103–5, 'Catastrophe theory: A new mathematical tool for scientists'.

66a. Berry, M V, 1977 *Nature*, **270** 382–383.

67. Berry, M V, 1978, *J. Phys. A*, **11**, 27–37. 'Disruption of wavefronts: Statistics of dislocations in incoherent Gaussian random waves'.

68. Berry, M V, 1978, *Nature*, **271**, 486, Review of '*Catastrophe theory: Selected papers 972–1977*' by E C Zeeman.

69. Berry, M V, *Physics Bulletin*, **29**, 177, Review of *Elements of wave propagation in random media* by B J Uscinski.

70. Berry, M V and Hannay, J H, 1978, *Nature*, **273**, 573 Comment on 'Topography of Random Surfaces' by R S Sayles and T R Thomas, *Nature*, **271**, 431–4, 1978.

71. Navascues, G and Berry, M V, 1978, 'A statistical — Mechanical_theory for the solid-liquid interface' in *Wetting, spreading and adhesion* , ed. Padday, Academic Press, 83–92.

72. Berry, M V, 1978, *Nature,* **274**, 930 Review of *Catastrophe theory: The revolutionary new way of understanding how things change* by A Woodcock and M Davis.

73. Berry, M V, 1978 *Nature,* **275**, 75–6. Review of *Catastrophe theory and its applications* by T Poston and I Stewart.

74. Berry, M V, 1978, *La Recherche* **92**, 760–8 'Les Jeux de lumière dans l'eau'.

75. Berry, M V, 1978, 'Catastrophes in semiclassical mechanics' in *Rencontre de CARGESE sur les singularites et leurs applications'* ed. F Pham, 1975, p. 133–5.

76. Berry, M V, 1978, 'Regular and Irregular Motion' in *Topics in Nonlinear Mechanics,* ed. S Jorna, *Am.Inst.Ph.Conf.Proc* No. **46**, 16–120.

77. Berry, M V and Balazs, N L, 1979, *J. Phys. A,* **12**, 625–42 'Evolution of semiclassical quantum states in phase space'.

78. Berry, M V and Balazs, N L, 1979, *Am. J. Phys.* **4**, 264–7 'Nonspreading wave packets'.

79. Berry, M V, Nye, J F and Wright, F J, 1979, *Phil. Trans. R. Soc. A,* **291**, 453–84, 'The elliptic umbilic diffraction catastrophe'.

80. Berry, M V, 1979, *J. Phys. A,* **12**, 781–97, 'Diffractals'.

81. Berry, M V, 1979, 'Catastrophe and fractal regimes in random waves' in *Structural stability in physics'*, eds.W Güttinger and H Eikemeier, Springer, 43–50.

82. Berry, M V, 1979, 'Distribution of modes in fractal resonators', in *Structural stability in physics,* eds W Güttinger and H Eikemeier, Springer, 51–3.

83. Berry, M V, 1979, 'Catastrophe and stochasticity in semiclassical quantum mechanics', in *Structural stability in physics,* eds. W Güttinger and H Eikemeier, Springer, 122–5.

84. Berry, M V, Balazs, N L, Tabor, M and Voros, A, 1979, *Ann. Phys. N.Y.* **122**, 26–63, 'Quantum maps'.

85. Berry, M V, 1979, *The Sciences,* **19**, No. 8, 18–20, 'Forms of Light'.

86. Berry, M V and Wright, F J, 1980, *J. Phys. A,* **13**, 149–60, 'Phase-space projection identities for diffraction catastrophes'.

87. Berry, M V, 1980, 'Quantization of mappings and other simple classical models', in *Nonlinear Dynamics,* ed. R H G Helleman, *Ann. N.Y. Acad. Sci.,* Vol. **357**, 183–202.

88. Berry, M V, 1980, 'Some geometric aspects of wave motion: Wavefront dislocations, diffraction catastrophes, diffractals', in *Geometry of the Laplace operator,* eds. R Osserman and A Weinstein, *Proc. Symp. App. Maths* **36**, AMS, 13–28.

89. Berry, M V and Upstill, C, 1980 *Progress in Optics XVIII*, 257–346, 'Catastrophe optics: Morphologies of caustics and their diffraction patterns'.

90. Berry, M V and Lewis, Z V, 1980, *Proc. R. Soc. A*, **370**, 459–484, 'On the Weierstrass-Mandedlbrot fractal function'.

91. Berry, M V, 1980, *L'Astronomia*, **5**, 35–38, 'L'Atmosfera come laboratorio di ottica'.

92. Berry, M V, 1980, *Nature*, **285**, 597. Review of *Lattice Path Counting and Applications*, by Sri Gopal Mohanty.

93. Berry, M V, 1980, *Nature*, **286**, 191. Review of *Bifurcation Theory and Applications in Scientific Disciplines*, eds. O Gürel and O E Rössler.

94. Berry, M V, 1980, *Nature*, **286**, 542. Review of *Works on the Foundations of Statistical Physics*, by N S Krylov.

95. Hannay, J H and Berry, M V, 1980, *Physica* **1D**, 267–290, 'Quantization of linear maps on a torus — Fresnel diffraction by a periodic grating'.

96. Berry, M V, Chambers, R G, Large, M D, Upstill, C and Walmsley, J C, 1980, *Eur. J. Phys.* **1**, 154–162, 'Wavefront dislocations in the Aharonov-Bohm effect and its water wave analogue'.

97. Berry, M V, 1980, *Eur. J. Phys.* **1**, 240–244, 'Exact Aharonov-Bohm wave function obtained by applying Dirac's magnetic phase factor'.

98. Berry, M V, 1981, *Ann. Phys.* **131**, 163–216, 'Quantizing a classically ergodic system: Sinai's billiard and the KKR method'.

99. Richens, P J and Berry, M V, 1981, *Physica* **1D**, 495–512, 'Pseudo-integrable systems in classical and quantum mechanics'.

100. Berry, M V and Shepherd, P J, 1981, *Physics Bulletin*, 238, 'Physics and Weapons'.

101. Berry, M V, 1981, *Eur. J. Phys.* **2**, 22–28, 'A curious multifoliate caustic in the magnetic Green function'.

102. Berry, M V, 1981, *Eur. J. Phys.* **2**, 91–102, 'Regularity and chaos in classical mechanics, illustrated by three deformations of a circular billiard'.

103. Berry, M V and Blackwell, T M, 1981 *J. Phys. A*, **14**, 3101–3110, 'Diffractal echoes'.

104. Korsch, H J and Berry, M V, 1981, *Physica* **3D**, 627–636, 'Evolution of Wigner's phase-space density under a non-integrable quantum map'.

105. Berry, M V, 1981, 'Singularities in Waves' in *Les Houches Lecture Series Session XXXV*, eds. R Balian, M Kléman and J-P Poirier, North-Holland: Amsterdam, 453–543.

106. Berry, M V, 1982, *J. Phys. A*, **15**, L385–388, 'Wavelength-independent fringe spacing in rainbows from falling neutrons'.

107. Berry, M V, 1982, Review of *Semiclassical approximation in quantum mechanics*', by V P Maslov and M V Fedoriuk in *Phys. Bull.* **33**, 241.

108. Berry, M V, 1982, *J. Phys. A*, **15**, 2735–2749 'Universal power-law tails for singularity-dominated strong fluctuations'.

109. Berry, M V, 1982, Review of '*The Accidental Universe*' by Paul Davies in *Nature*, **300**, 133–4.

110. Berry, M V, 1982, *J. Phys. A*, **15**, 3693–3704, 'Semiclassical weak reflections above analytic and nonanalytic potential barriers'.

111. Berry, M V and Hajnal, J V, 1983, *Optica Acta* **30**, 23–40, 'The shadows of floating objects and dissipating vortices'.

112. Berry, M V, 1983, Review of *The fractal geometry of nature*, by B B Mandelbrot, in *New Scientist* **97**, No. 1342, 250.

113. Berry, M V, 1983, Review of *Image analysis and mathematical morphology* by J Serra, in *Physics Bulletin* **34**, 252.

114. Walker, J G, Berry, M V and Upstill, C, 1983, *Optica Acta* **30**, 1001–1010, 'Measurement of twinkling exponents of light focused by randomly rippling water'.

115. Berry, M V, 1983, 'Semiclassical Mechanics of regular and irregular motion' in *Les Houches Lecture Series Session XXXVI*, eds. G Iooss, R H G Helleman and R Stora, North Holland, Amsterdam, 171–271.

116. Berry, M V, 1983 Review of *Regular and stochastic motion* by A J Lichtenberg and M A Lieberman, in *Nature* **305**, 456.

117. Berry, M V, Hannay, J H & Ozorio de Almeida, A M, 1983, *Physica* **8D**, 229–242, 'Intensity moments of semiclassical wavefunctions'.

118. Berry, M V, 1984, 'Structures in semiclassical spectra: A question of scale' in *The Wave-Particle Dualism*, 231–252, eds. S Diner, D Fargue, G Lochak, F Selleri, (D Reidel)

119. Berry, M V and Wilkinson, M, 1984, *Proc. R. Soc. A*, **392**, 15–43, 'Diabolical points in the spectra of triangles'.

120. Berry, M V, 1984, *Proc. R. Soc. A*, **392**, 45–57, 'Quantal phase factors accompanying adiabatic changes'.

121. Wright, F J and Berry, M V, 1984, *J. Acoust. Soc. Amer.* **75**, 733–748, 'Wave-front dislocations in the sound-field of a pulsed circular piston radiator'.

122. Berry, M V, 1984, *J. Phys. A*, **17**, 1225–1233, 'The adiabatic limit and the semiclassical limit'.

123. Berry, M V and Klein, G, 1984, *J. Phys. A*, **17**, 1805–1815, 'Newtonian trajectories and quantum waves in expanding force fields'.

124. Berry, M V, 1984, *Physica*, **10D**, 369–378, 'Incommensurability in an exactly-soluble quantal and classical model for a kicked rotator'.

125. Berry, M V, Indekeu, J O, Tabor, M and Balazs, N L, 1984, *Physica*, **11D**, 1–24, 'Nonlocal maps'.

126. Berry, M V and Robnik, M, 1984, *J. Phys. A*, **17**, 2413–2421, 'Semiclassical level spacings when regular and chaotic orbits coexist'.

127. Berry, M V, 1984, *Physics Letters*, **104A**, 306–9, 'Comment on "New Representation of Quantum Chaos"'.

128. Berry, M V, 1984, *Phys. Bull.* **35**, 338, Review of *Sunsets, twilights and evening skies,* by Aden and Marjorie Meinel.

129. Berry, M V, 1984, *Phys. Bull.* **35**, 437 Review of *Universality in chaos,* ed. by Predrag Cvitanovic.

130. Berry, M V, 1984, *Bristol University Newsletter* ,15 November, p. 2, Obituary of P A M Dirac.

131. Berry, M V, 1984, *Input,* **37**, 1164–1171, 'Patterns from nature'.

132. Berry, M V, 1985, *J. Phys. A*, **18**, 15–27, 'Classical adiabatic angles and quantal adiabatic phase'.

133. Berry, M V, 1985, *Phys. Bull.* **36**, 177 Review of *Symplectic techniques in physics* by Victor Guillemin and Shlomo Sternberg.

134. Berry, M V, 1985, 'Aspects of Degeneracy' in *Chaotic behavior in quantum systems,* ed. Giulio Casatil, Plenum, New York, 123–140.

135. Berry, M V, 1985, 'Quantum, classical and semiclassical adiabaticity' in *Theoretical and Applied Mechanics,* eds. F I Niordson and N Olhoff, Elsevier, North-Holland, Amsterdam, pp. 83–96.

136. Robnik, M and Berry, M V, 1985, *J. Phys. A*, **18**, 1361–1378, 'Classical billiards in magnetic fields'.

137. Berry, M V, 1985, *Nature* **315**, 779, Review of *An idiot's fugitive essays on science: Methods, criticism, training, circumstances,* by C Truesdell.

138. Tanner, L H and Berry, M V, 1985, *J. Phys. D*, **18**, 1037–1061, 'Dynamics and optics of oil hills and oilscapes'.

139. Berry, M V, 1985, 'A problem in semiclassical adiabatic theory', in *Mthodes Semiclassiques en Mchanique Quantique,* eds. B Helffer et D Robert, Publications de L'Université de Nantes 23–27.

140. Berry, M V, 1985, *Proc. R. Soc. A*, **400**, 229–251, 'Semiclassical theory of spectral rigidity'.

141. Berry, M V, 1985, Interview in *Corriere della Provincia (Como)*, September 16.

142. Berry, M V, 1985, *Nature* **318**, 241. Review of *Deterministic chaos*, by H G Schuster.

143. Berry, M V, 1985, *Prometheus* **1**, 41–79, 'Ipotesi di scala e fluttuazioni non gaussiane nella teoria catastrofica della onde', eds. Paolo Bisogno, Augusto Forti, (Italian translation of 'Scaling and nongaussian fluctuations in the catastrophe theory of waves').

144. Berry, M V, 1986, 'Twinkling exponents in the catastrophe theory of random short waves' in *Wave propagation and scattering*, ed. B J Uscinsci, Oxford, Clarendon Press, 11–35.

145. Berry, M V and Robnik, M, 1986, *J. Phys. A*, **19**, 649–668, 'Statistics of energy levels without time-reversal symmetry: Aharonov-Bohm chaotic billiards'.

146. Robnik, M and Berry, M V, 1986, *J. Phys. A*, **19**, 669–682, 'False time-reversal violation and energy level statistics: The role of anti-unitary symmetry'.

147. Berry, M V and Mondragon, R J, 1986, *J. Phys. A*, **19**, 873–885, 'Diabolical points in one-dimensional Hamiltonians quartic in the momentum'.

148. Berry, M V and Robnik, M, 1986, *J. Phys. A*, **19**, 1365–1372, 'Quantum states without time-reversal symmetry: Wavefront dislocations in a non-integrable Aharonov-Bohm billiard'.

149. Berry, M V and Percival, I. C, 1986, *Optica Acta*, **33**, 577–591, 'Optics of fractal clusters such as smoke'.

150. Berry, M V, 1986, *J. Phys. A*, **19**, 2281–2296, 'Spectral zeta functions for Aharonov-Bohm quantum billiards'.

151. Berry, M V, 1986, *Nature* **323**, 590. Review of *The beauty of fractals: Images of complex dynamical systems,* by H-O Peitgen and P H Richter, Springer.

152. Berry, M V, 1986, *New Scientist* 16 October, 60. Review of *Symmetry: Unifying Human Understanding*, ed. by Istvan Hargittai.

153. Berry, M V, 1986, 'The unpredictable bouncing rotator: A chaology tutorial machine' in *Dynamical systems: A renewal of mechanism*, eds. S Diner, D Fargue, G Lochak, World Scientific 3–12.

154. Berry, M V, 1986, 'Riemann's zeta function: A model for quantum chaos?' in *Quantum chaos and statistical nuclear physics*, eds. T H Seligman and H Nishioka, Springer Lecture Notes in Physics No. 263, 1–17.

155. Berry, M V, 1986, 'Fluctuations in numbers of energy levels', in *Stochastic processes in classical and quantum systems*, eds. S Albeverio, G Casati and D Merlini, Springer Lecture Notes in Physics, No. 262, 47–53.

156. Berry, M V, 1986, 'Adiabatic phase shifts for neutrons and photons', in *Fundamental aspects of quantum theory*, eds. V Gorini and A Frigerio, Plenum, NATO ASI series vol. 144, 267–278.

157. Berry, M V, 1986, 'The Aharonov-Bohm effect is real physics not ideal physics' in *Fundamental aspects of quantum theory*, eds. V Gorini and A Frigerio, Plenum, NATO ASI Series Vol. 144, 319–320.

158. Berry, M V, *J. Opt. Soc. Amer.* **A4**, 561–569, 'Disruption of images: The caustic-touching theorem'.

159. Berry, M V, 1987, *Nature*, **326**, 277–278, 'Interpreting the anholonomy of coiled light'.

160. Berry, Michael and Berry, Monica, 1987, *New Scientist*, 2 April p. 56. Review of *The problems of mathematics* by Ian Stewart, Oxford Univ. Press.

161. Berry, M V and Mondragon, R J, l 1987, *Proc. R. Soc. A*, **412**, 53–74 'Neutrino billiards: Time-reversal symmetry-breaking without magnetic fields'.

162. Berry, M V, 1987 *J. Phys. A*, **20**, 2389–2403, 'Improved eigenvalue sums for inferring quantum billiard geometry'.

163. Berry, M V, 1987, *Proc. R. Soc. A*, **413**, 183–198, 'Quantum chaology', (The Bakerian Lecture).

164. Berry, M V, 1987, *Proc. R. Soc. A*, **414**, 31–46, 'Quantum phase corrections from adiabatic iteration'.

165. Berry, M V, 1987, *New Scientist*, 19 November, 44–47, 'Quantum physics on the edge of chaos'.

166. Berry, M V, 1987, *Nature*, **330**, 293–294 Review of *Chaos: Making a new science*, by James Gleick.

167. Berry, M V, 1987, *J. Mod. Optics*, **34**, 1401–1407, 'The adiabatic phase and Pancharatnam's phase for polarized light'.

168. Keating, J P and Berry, M V, 1987, *J. Phys. A*, **20**, L 1139–1141, 'False singularities in partial sums over closed orbits'.

169. Berry, M V, 1987, *Semiclassical chaology* in *Quantum measurement and chaos*, eds. E R Pike and Sarben Sarkar, Plenum, NATO ASI series B, vol. 161, pp. 81–87.

170. Berry, M V, 1988, *Cogito*, Vol. 2, No. 1, 1–5 Interview 'Chaos and Order'.

171. Berry, M V and Goldberg, J 1988, *Nonlinearity*, **1**, 1–26, 'Renormalization of curlicues'.

172. Berry, M V and Hannay, J, 1988, *J. Phys. A*, **21**, L325–331, 'Classical non-adiabatic angles'.

173. Scharf, R, Dietz, B, Kus, M, Haake, F and Berry, M V, 1988, *Europhys. Lett.* **5**, 383–389, 'Kramers' degeneracy and quartic level repulsion'.

174. Berry, M V, 1988, 'The electron at the end of the universe' in *A passion for science*, eds. L Wolpert and A Richards; Oxford University Press, 38–51.

175. Berry, M V, 1988, *Nonlinearity* **1**, 399–407, 'Semiclassical formula for the number variance of the Riemann zeros'.

176. Berry, M V, 1988, *Nature* **335**, 22 Review of *Mathematics and the unexpected* by Ivar Ekeland.

177. Berry, M V, 1988, 'Classical chaos and quantum eigenvalues' in *Order and Chaos in nonlinear physical systems,* eds. Stig Lundqvist, Norman H March and Mario P Tosi; Plenum Press, New York and London) 341–348.

178. Berry, M V, 1988, 'The geometric phase', *Scientific American,* **259** (6), 26–34.

179. Berry, M V, 1988, 'Random renormalization in the semiclassical long-time limit of a precessing spin', *Physica D,* **33**, 26–33.

180. Berry, M V, 1988, 'Le dé passement interne des paradignes de la physique classique' (translation — 'Breaking the paradigms of classical physics from within'), in 'Logos et Théorie des catastrophes', ed: J Petitot, (Editions Patino, Geneva) 106–117.

181. Berry, M V, 1989, 'Uniform asymptotic smoothing of Stokes's discontinuities', *Proc. R. Soc. A,* **422** 7–21.

182. Berry, M V, 1989 *New Scientist,* 11 March, p. 66, review of 'The science of Fractal Images', eds: H-O Peitgen and D Saupe.

183. Berry, M V, 1989 *Nature* **338** 685–6, review of *Journey into Light: Life and Science of C V Raman* by G Venkataraman.

184. Berry, M V, 1989, 'Quantum scars of classical closed orbits in phase space', *Proc. R. Soc. A,* **423** 219–231.

185. Berry, M V, 1989, *Principles of Cosmology and Gravitation* (Adam Hilger) (corrected reprint of item 54).

186. Berry, M V, 1989, 'Studies of Semiclassical Spectra: Where next?' in *Atomic Physics II,* eds: S Haroche, J C Gay, S Grynberg World Scientific 277–279.

187. Berry, M V, 1989 'The quantum phase, five years after' in 'Geometric Phases in Physics', eds: A Shapere, F Wilczek. (World Scientific) 7–28.

188. Mondragon, R J and Berry, M V, 1989, 'The quantum phase 2-form near degeneracies: Two numerical studies' *Proc. R. Soc. A,* **424** 263–278.

189. Berry, M V, 1989, 'Fringes decorating anticaustics in ergodic wavefunctions', *Proc. R. Soc. A,* **424** 279–288.

190. Berry, M V, 1989, 'Stokes' phenomenon; smoothing a Victorian discontinuity', *Publ. Math. of the Institut des Hautes Études scientifique,* **68** 211–221.

191. Berry, M V, 1989, 'Quantum chaology, not quantum chaos', *Physica Scripta* **40** 335–336.

192. Berry, M V, 1989, 'Falling fractal flakes', *Physica D*, **38** 29–31.

193. Berry, M V, 1989, *Times Higher Ed. suppl.* review of 'Does God play dice: The mathematics of chaos' by Ian Stewart.

194. Berry, M V, 1989, *Bull. Lond. Math. Soc.*, review of 'Multiphase Averaging for classical systems', by P Lochak and C Meunier.

195. Berry, M V, 1989, *Physics World* November, p. 45, review of 'Buckminster Fuller's Universe: An appreciation', by Lloyd Steven Seiden.

196. Berry, M V, 1990, *Proc.Roy. Institution of Gt Britain*, **61**, 189–204, 'Chaology; the emerging science of unpredictability'.

197. Berry, M V, 1990 *Proc. R. Soc. A*, **427**, 265–280, 'Waves near Stokes lines'.

198. Berry, M V, 1990, 'Quantum adiabatic anholonomy' in *Anomalies, phases, defects'* eds: U M Bregola, G Marmo and G Morandi (Naples: Bibliopolis) 125–181.

199. Berry, M V, 1990, 'Generalized rainbows in wave physics: How catastrophe theory has helped' in *Rainbows and catastrophes* (ed. N Neskovic, Boris Kidric Institute, Belgrade) 19–23.

200. Berry, M V and Howls, C J, 1990, *J. Phys. A*, **23**, L243–L246, 'Fake Airy functions and the asymptotics of reflectionlessness'.

201. Berry, M V, 1990, *Proc. R. Soc. A*, **429**, 61–72, 'Histories of adiabatic quantum transitions'.

202. Berry, M V, 1990, *Current Contents*, **21** (No. 19), 14, 'Catastrophes and waves' (account of [49] as 'Citation Classic').

203. Berry, M V, 1990, *Highlights in Physics* (SERC) 50–51, 'Quantum theory near the classical limit'.

204. Berry, M V and Howls, C J, 1990, *Nonlinearity* **3** 281–291, 'Stokes surfaces of diffraction catastrophes with codimension three'.

205. Berry, M V and Lim, R, 1990, *J. Phys. A*, **23**, L655–L657, 'The Born-Oppenheimer electric gauge force is repulsive near degeneracies'.

206. Berry, M V, 1990, *Proc. R. Soc. A,* **430** 405–411, 'Geometric amplitude factors in adiabatic quantum transitions'.

207. Berry, M V, 1990, *Current Contents*, **21** (No. 33), 16, 'Quantum asymptotics of rainbows' (account of [6] as 'Citation Classic').

208. Berry, M V and Howls, C J, 1990, *Proc. R. Soc. A*, **430**, 653–668, 'Hyperasymptotics'.

209. Berry, M V, 1990, *Physics World*, October, 21–22, 'Minds, quantum measurement, and gravity'.

210.   Berry, M V and Keating, J P, 1990, *J. Phys. A*, **23** 4839–4849, 'A rule for quantizing chaos?'.

211.   Berry, M V, 1990, *Proc. R. Soc. A*, **431**, 531–537 'Budden and Smith's 'additional memory' and the geometric phase'.

212.   Berry, M V, 1990, *Physics Today*, **43** (No. 12) 34–40, 'Anticipations of the geometric phase'.

213.   Berry, M V, 1990, *Current Science*, **59**, 1175–1191, 'Beyond Rainbows'.

214.   Goldberg, J, Smilansky, U, Berry, M V, Schweitzer, W, Wunner, G and Zeller, G, 1991, *Nonlinearity*, **4**, 1–14, 'The parametric number variance'.

215.   Berry, M V, 1991, *Physics World March*, p. 52, review of 'The correspondence between Sir George Gabriel Stokes and Sir William Thomson, Baron Kelvin of Largs', ed. David B Wilson.

216.   Berry, M V, 1991, *Bristol University Newsletter*, March 21, p. 8, 'Heisenberg's Sofa'.

217.   Berry, M V, 1991, In 'Sir Charles Frank OBE FRS, an eightieth birthday tribute' Eds. R G Chambers, J E Enderby, A Keller, A R Lang and J W Steeds (Adam Hilger, Bristol), 'Bristol Anholonomy Calendar', pp. 207–219.

218.   Berry, M V, 1991, in 'Quantum Coherence', ed. Jeeva S Anandan (World Scientific) 'Wave Geometry: A plurality of singularities' pp. 92–98.

219.   Berry, M V, 1991, *Physics World July*, pp.12–13, 'What's wrong with these conference Proceedings?'

220.   Berry, M V, 1991, in 'Quantum Chaos', Eds. H A Cerdeira, R Ramaswamy, M C Gutzwiller and G Casati, 'Introductory remarks', pp. vii-viii.

221.   Lim, R and Berry, M V, 1991, *J. Phys. A*, **24**, 3255–3264, 'Superadiabatic Tracking of Quantum Evolution'.

222.   Berry, M V, 1991, *Proc. R. Soc. A*, **434**,465–472 'Infinitely many Stokes smoothings in the Gamma function'.

223.   Berry, M.V. and Howls, C.J. 1991, *Proc. R. Soc. A*, **434** 657–675. 'Hyperasymptotics for integrals with saddles'.

224.   Berry, M V, 1991, In 'Nonlinear and chaotic Phenomena in Plasmas, Solids and Fluids', Eds: W Rozmus and J A Tuszynski (World Scientific) 'Quantal reflections of classical chaos' (Abstract), p. 2.

225.   Berry, M V, 1991, *Proc. R. Soc. A*, **435**, 437–444, 'Stokes phenomenon for superfactorial asymptotic series'.

226.   Berry, M V, 1991, *New Scientist*, 30 November, Letter about the green flash.

227. Berry, M V, 1991, 'Some quantum-to-classical asymptotics', in *Les Houches Lecture Series* LII (1989), eds. M-J Giannoni, A Voros and J Zinn-Justin, North-Holland, Amsterdam, 251–304.

228. Berry, M V, 1992, in 'Huygens' principle 1690–1990; theory and applications', eds: H Bok, H A Ferwerda, North-Holland: Amsterdam), 'Rays, wavefronts and phase: A picture book of cusps', pp. 97–111.

229. Berry, M V, 1992, in 'The New Scientist guide to chaos', ed: Nina Hall (Penguin Books: London), 'Quantum physics on the edge of chaos', pp. 184–195 (reprint of item 165).

230. Berry, M V, 1992, in 'New Trends in Nuclear Collective Dynamics', eds: Y Abe, H Horiuchi and K Matsuyanagi (*Springer proceedings in Physics* 58), 'quantum chaology: Our knowledge and ignorance', pp. 177–181.

231. Berry, M V, 1992, in 'New Trends in Nuclear Collective Dynamics', eds: Y Abe, H Horiuchi and K Matsuyanagi (*Springer proceedings in Physics* 58), 'True Quantum Chaos?' An Instructive Example', pp. 183–186.

232. Robbins, J M and Berry, M V, 1992, *Proc. R. Soc. A*, **436**, 631–661, 'The geometric phase for chaotic systems'.

233. Berry, M V and Keating, J P, 1992, *Proc. R. Soc. A*, **437** 151–173, 'A new approximation for $z(1/2 + it)$ and quantum spectral determinants'.

234. Berry, M V, 1992, in 'Asymptotics beyond all orders', ed: H Segur and S Tanveer (Plenum, New York 1991) 'Asymptotics, superasymptotics, hyperasymptotics', pp. 1–14.

235. Berry, M V, 1992, *Physics World* June, p. 62, review of 'Understanding the present: Science and the soul of modern man' by Bryan Appleyard.

236. Berry, M V, 1992, *Encyclopedia Britannica* (15th ed.) Vol. 2, p. 948 'Catastrophe Theory' (original considerably edited).

237. Berry, M V, 1992, *Encyclopedia Britannica* (15th ed.) Vol. 3, p. 92 'Chaos' (original considerably edited).

238. Robbins, J M and Berry, M V, 1992, *J. Phys. A*, **25** p. L961–L965 'Discordance between quantum and classical correlation moments for chaotic systems'.

239. Berry, M V, 1992, *Physics World* December, p. 40, Review of 'Genius: Richard Feynman and Modern Physics', by James Gleick.

240. Berry, M V, 1992, *Nature*, **360**, 376–377, Review of 'Pi in the sky: Counting, Thinking, and Being', by John Barrow.

241. Berry, M V and Howls, C J, 1993, *Physics World*, June, 35–39, 'Infinity interpreted'.

242. Berry, M V and Robbins, J M, 1993, *Proc. R. Soc. A*, **442** 641–658, 'Classical geometric forces of reaction: An exactly solvable model'.

243. Berry, M V and Robbins, J M, 1993, *Proc. R. Soc. A*, **442** 659–672, 'Chaotic classical and half-classical adiabatic reactions: Geometric magnetism and deterministic friction'.

244. Berry, M V and Howls, C J, 1993, *Proc. R. Soc. A*, **443**, 107–126, 'Unfolding the high orders of asymptotic expansions with coalescing saddles: Singularity theory, crossover and duality'.

245. Berry, M V and Lim, R, 1993, *J. Phys. A*, **26**, 4737–4747, 'Universal transition prefactors derived by superadiabatic renormalisation'.

246. Berry, M V, 1993, *Inst. Phys. Conf. Ser.* **133**, 133–134, 'Quantum chaology, prime numbers and Riemann's zeta function' (abstract and bibliography).

247. Berry, M V, 1993, *Physics World* December, pp. 46–7, review of 'Light and color in the Outdoors' by M G J Minnaert.

248. Berry, M V, 1993, *Bull. Lond. Math. Soc.* **25**, 411–412, review of 'Lectures on Mechanics' by J E Marsden.

248a. Berry, M V, 1993, 'Visiting Nablus', Bristol University Physics Department occasional Newsletter.

249. Berry, M V and Howls, C J, 1994, *Proc. R. Soc. A*, **444,** 201–216, 'Overlapping Stokes smoothings: Survival of the error function and canonical catastrophe integrals'.

250. Berry, M V, 1994, in 'Selected papers on geometrical aspects of scattering' P L Marston, ed. (SPIE Optical Engineering Press: Washington). Reprints of papers 202, 49, 79, 114.

251. Berry, M V, 1994, *Nature*, **364**, 529, review of 'The quark and the jaguar' by Murray Gell-Mann.

252. Berry, M V, 1994, *J. Phys. A*, **27**, L391–L398, 'Evanescent and real waves in quantum billiards, and Gaussian Beams'.

253. Robbins, J M and Berry, M V, 1994, *J. Phys. A*, **27**, L435–L438, 'A geometric phase for $m = 0$ spins'.

254. Berry, M V, 1994, *Physics World*, August, p. 49, review of 'The beat of a different drum: The life and science of Richard Feynman', by Jagdish Mehra.

255. Berry, M V, 1994, *Applied Optics*, **33,** 4563–4568, 'Supernumerary ice-crystal halos?'.

256. Berry, M V and Wilson, A N, 1994, *Applied Optics*, **33**, 4714–4718, 'Black-and-white fringes and the colours of caustics'.

257. Berry, M V, 1994, *Current Science*, **67**, 220–223, 'Pancharatnam, virtuoso of the Poincaré sphere: An appreciation'.

258. Berry, M V and Keating, J P, 1994, *J. Phys. A*, **27**, 6167–6176, 'Persistent current flux correlations calculated by quantum chaology'.

259. Berry, M V, 1994, *La Recherche* **269**, 1066–1067 'Michael Berry, un géomètre des ondes' (interview).

260. Berry, M V, 1994, 'Asymptotics, singularities and the reduction of theories', for Proc. 9th Int. Cong. Logic, Method., and Phil. of Sci. IX, 597–607, eds. D Prawitz, B Skyrms and D Westerståhl.

261. Berry, M V and Howls, C J, 1994, *Proc. R. Soc. A*, **447**, 527–555, 'High orders of the Weyl expansion for quantum billiards: Resurgence of periodic orbits, and the Stokes phenomenon'.

262. Berry, M V, 1994, 'Faster than Fourier', in 'Quantum Coherence and Reality; in celebration of the 60th Birthday of Yakir Aharonov' (J S Anandan and J L Safko, eds.) World Scientific, Singapore, pp. 55–65.

263. Berry, M V, 1995, in *Fundamental Problems of quantum theory*, eds: D M Greenberger and A Zeilinger. *Ann.N.Y.Acad.Sci.* **755**, 303–317, 'Some two-state quantum asymptotics'.

264. Berry, M V, 1995, in *Quantum systems: New trends and methods* (eds. A O Barut, I D Feranchuk, Ya M Shnir and L M Tomil 'chik) World Scientifc, 387–392, 'Quantum mechanics, chaos and the Riemann zeros'.

265. Berry, M V, 1995, *Proc. R. Soc. A*, **450,** 439–462, 'The Riemann-Siegel formula for the zeta function: High orders and remainders'.

266. Berry, M V, 1995, 'Natural Focusing', in 'The Artful Eye' (Richard Gregory, John Harris, Priscilla Heard and David Rose, eds) Oxford University Press, pp. 311–323.

267. Berry, M V, 1995, *Physics World*, December, pp. 52–3, review of 'Nature's Numbers: Discovering Order and Pattern in the Universe' by Ian Stewart.

268. Berry, M V, in 'Advances in Quantum Phenomena', eds: E S Beltrametti and J M Levy-Leblond (Plenum: New York and London) NATO ASI Series B: Physics Vol. 347, 'Three comments on the Aharonov-Bohm effect', 353–354.

269. Berry, M V and Klein, S, 1996, *J. Mod. Opt.* **43**, 165–180, 'Geometric phases from stacks of crystal plates'.

270. Berry, M V and Klein, S, 1996 ,*Proc. Natl. Acad. Sci. USA* **93** pp. 2614–2619, 'Colored diffraction catastrophes'.

271. Berry, M V, 1996, *Proc. R. Soc. A*, **452**, 1207–1220, 'The Levitron^TM: An adiabatic trap for spins'.

272. Berry, M V, 1996, 'Levitron Physics', explanatory leaflet distributed with the Levitron by *Fascinations Toys and Gifts* (Seattle).

273. Berry, M V and Morgan, M A, 1996, *Nonlinearity* **9** 787–799, 'Geometric angle for rotated rotators, and the Hannay angle of the world'.

274. Berry, M V and Klein, S, 1996, *J. Mod. Opt.* **43**, 2139–2164, 'Integer, fractional and fractal Talbot effects'.

275. Berry, M V, *J. Phys. A*, **29** 6617–6629, 'Quantum fractals in boxes'.

276. Berry, M V, 1997, *Nature* **385** 33, review of 'Would-be worlds: How simulation is changing the frontiers of science', by John Casti.

277. Berry, M V, 1997, *Nexus News* February, pp. 4–5, interview by Nina Hall: 'Caustics, catastrophes' and quantum chaos'.

278. Berry, M V and Sinclair, E C, *J. Phys. A*, **30** 2853–2861, 'Geometric magnetism in massive chaotic billiards'.

279. Berry, M V, 1997, *Afleidung* (Amsterdam) **3**, April, 11–13, Interview with M Postma and M Sallé.

280. Berry, M V, 1997, *Lego Review* January, p. 10, report 'A professorial riddle' on Möbius spring ring.

281. Berry M V and Klein, S, 1997, *Eur. J. Phys.* **18** 222–228, 'Transparent mirrors: Rays, waves and localization'.

282. Berry, M V, 1997, *Bristol University Newsletter* **27** (22 May), 3, Oration for Professor Yakir Aharonov.

283. Berry, M V, 1997, 'Quantum and optical arithmetic and fractals', in *The Mathematical beauty of Physics* (eds: J-M Drouffe and J-B Zuber) (Singapore: World Scientific) pp. 281–294.

284. Berry, M V and Klein, S 1997, 'Diffraction near fake caustics', *Eur.J.Phys*, **18** 303–306.

285. Berry, M V and Geim, A K, 1997, 'Of flying frogs and levitrons' *Eur. J. Phys.* **18** 307–313.

286. Berry, M V and Robbins, J M, 1997, *Proc. R. Soc. A*, **453** 1771–1790 'Indistinguishability for quantum particles: Spin, statistics and the geometric phase'.

287. Berry, M V and Klein, S, 1997 'Die Farben von Kaustiken: Katastrophen in Regentropfen und Strukturglas' *Phys.Blätt* **53** 1095 – 1098.

288. Berry, M V, 1997, *New Scientist* **2109** (22 November), 50, Desert island book review (M Abramowitz and I A Stegun: Handbook of Mathematical Functions).

289. Berry, M V, 1997, *Physics World* **10** (December) 41–42, 'Slippery as an eel', review of 'The fire in the eye', by David Park.

290. Berry, M V, 1997, 'Aharonov-Bohm geometric phases for rotated rotators' *J. Phys. A*, **30** 8355–8362.

291. Berry, M V, 1998, 'Paul Dirac: The purest soul in physics', *Physics World* **11** (February) 36–40.

292. Berry, M V, Foley, J T, Gbur, G and Wolf, E, 1998, 'Non-propagating string excitations' *Am. J. Phys.* **66** 121–3.

293. Berry, M V and O'Dell, D H J, 1998, 'Diffraction by volume gratings with imaginary potentials' *J. Phys. A*, **31** 2093–2101.

294. Berry, M V, Keating, J P and Prado, S, 1998, 'Orbit bifurcations and spectral statistics' *J. Phys. A*, **31** L245–L254.

295. Berry, M V, 1998, 'Lop-sided diffraction by absorbing crystals' *J. Phys. A*, **31** 3493–3502.

296. Berry, M V, 1998, 'Much ado about nothing: Optical dislocation lines (phase singularities, zeros, vortices...)', In *Singular optics*, **3487** (Ed, Soskin, M S) Frunzenskoe, Crimea, *SPIE*, 3487, pp. 1–5.

297. Berry, M V, 1998, 'Paraxial beams of spinning light', In *Singular optics*, 3487 (Ed, Soskin, M S) Frunzenskoe, Crimea, *SPIE*, **3487**, pp. 6–11.

298. Berry, M V, 1998, Review of 'Copenhagen' (play by Michael Frayn) *Nature*, **394**, 735 20 August 1998.

299. Berry, M V, 1998, 'Wave dislocations in non-paraxial Gaussian beams' *J. Mod. Optics*, **45**, 1845–1858.

300. Berry, M V, 1998, 'Extreme twinkling, and its opposite' (summary), in 'Gravitation and Relativity: At the turn of the millennium' (Proc GR15) (Eds: N Dadhich and J Narlikar), (IUCAA, Pune), 23–24.

301. Berry, M V, 1998, Foreword to 'Global properties of simple quantum systems — Berry's phase and others' by Hua-Zhong Li (Shanghai Scientific and Technical Publishers).

302. Berry, M V, 1998, Foreword to 'Introduction to quantum computation', eds: Hoi Kwong Lo, Sandu Popescu and Tim Spiller (World Scientific).

303. Berry, M V, Bhandari, R and Klein, S, 1999, 'Black plastic sandwiches demonstrating biaxial anisotropy', *Eur. J. Phys.* **20** 1–14.

303a. Berry, M V, 1999, 'A week in Beirut', Bristol University Newsletter, 28 January (Vol. 29, No. 8).

304. Berry, M V and Bodenschatz, E, 1999, 'Caustics, multiply-reconstructed by Talbot interference', *J. Mod. Opt.* **46** 349–365.

305. Berry, M V and O'Dell, D, 1999, 'Ergodicity in wave-wave diffraction', *J. Phys. A*, **32** 3571–3582.

306. Berry, M V and Keating, J P, 1999, '$H = xp$ and the Riemann zeros' in 'Supersymmetry and trace formulae: Chaos and disorder' (eds I V Lerner, J P Keating) Plenum (New York) 355–367.

307. Berry, M V and Keating, J P, 1999, *SIAM Review* **41** 236–266, 'The Riemann zeros and eigenvalue asymptotics'.

308. Berry, M V, 1999, 'A theta-like sum from diffraction physics', *J. Phys. A*, **32** L329–L336.

309. Berry, M V, 1999, 'Aharonov-Bohm beam deflection: Shelankov's formula, exact solution, asymptotics and an optical analogue', *J. Phys. A*, **32** 5627–5641.

310. Berry, M V and Shelankov, A, 1999, 'The Aharonov-Bohm wave and the Cornu spiral', *J. Phys. A*, **32** L447–L455.

311. Berry, M V, 1999, 'Darkness behind the curtain' Bristol University Newsletter Vol. 30, No. 1 P12.

312. Berry, M V, 1999, Foreword to 'Optical Vortices', eds: M Vasnetsov and K Staliunas (New York: Nova Science Publishers).

313. Berry, M V, 1999, in 'On Minnaert's Shoulders: Twenty years of the "Light and Color" Conferences, Optical Society of America CD-ROM (C L Adler, ed) Reprints of papers 158, 255 and 256.

314. Berry, M V, 2000, 'Making waves in physics. 'Three wave singularities from the miraculous 1830s', *Nature* **403** 21–6 January 2000.

315. Berry, M V, 2000, 'Connections' *Impact* (Leiden University Physics Student Magazine) **11**, 12.

316. Berry, M V, 2000, Review of 'Natural focusing and fine structure of light', by J F Nye, *Contemp. Phys* **41** 118–119.

317. Berry, M V, 2000, Review of 'The Genius of Science: A Portrait Gallery of Twentieth-Century Physicists', by A Pais, *Physics World* June p56.

318. Berry, M V, 2000, 'Spectral twinkling' in *Proc.Int School of Physics "Enrico Fermi"*, course **CXLIII** (eds: G Casati, I Guarneri and U Smilansky), IOS Press, Amsterdam, 45–63.

319. Berry, M V & Robbins, J M, 2000, 'Quantum Indistinguishability: Alternative constructions of the transported basis', *J. Phys. A*, **33**, L207–L214.

320. Berry, M V, Keating, J P, Schomerus, H, 2000 'Universal twinkling exponents for spectral fluctuations associated with mixed chaology', *Proc. R. Soc. A*, **456** 1659–1668.

321. Berry, M V and Dennis, M R, 2000, 'Phase singularities in isotropic random waves', *Proc. R. Soc. A*, **456** 2059–2079.

322. Berry, M V and Robbins, J M, 2000, 'Quantum indistinguishability: Spin-statistics without relativity or field theory?', in *Spin-Statistics Connection and Commutation Relations*, (eds: R C Hilborn and G M Tino), American Institute of Physics **CP545**.

323. Berry, M V, 2000, 'Odessa, little and large' *Bristol University Newsletter* 6 December, p. 8.

324. Berry, M V and Dennis, M R, 2001 ,'Polarization singularities in isotropic random waves' *Proc. R. Soc. A*, **457** 141–155.

325. Bender, CM, Berry, M V, Meisinger, P M, Savage, V M and Simsek, M, 2001, 'Complex WKB analysis of energy-level degeneracies of non-Hermitian Hamiltonians', *J. Phys. A*, **34** L31–L36.

326. Berry, M V, 2001, 'Why are special functions special?', *Physics Today*, April, pp. 11–12.

327. Berry, M V, 2001, 'Spectral twinkling: A new example of singularity-dominated strong fluctuations (summary)', *Physica Scripta*, **T90** 15.

328. Berry, M V, 2001, 'Knotted zeros in the quantum states of hydrogen' *Found. Phys.* **31** 659–667.

329. Berry, M V, Marzoli, I and Schleich, W, 2001, 'Quantum carpets, carpets of light' *Physics World* (June), 39–44.

330. Berry, M V, 2001, 'Geometry of phase and polarization singularities, illustrated by edge diffraction and the tides', in 'Second international conference on Singular Optics (Optical Vortices): Fundamentals and applications' *SPIE* **4403** (Bellingham Washington), 1–12.

331. Berry, M V, 2001, 'Asymptotics of Evanescence', *J. Modern Optics* **48** 1535–1541.

332. Berry, M V and Dennis, M R, 2001, 'Knotted and linked phase singularities in monochromatic waves', *Proc. R. Soc. A*, **457** 2251–2263.

333. Berry, M V and Dennis, M R, 2001, 'Knotting and unknotting of phase singularities: Helmholtz waves, paraxial waves and waves in 2 + 1 spacetime', *J. Phys. A*, **34** 8877–8888.

334. Berry, M V, Storm, C, and van Saarloos, W, 2001 'Theory of unstable laser modes: Edge waves and fractality' *Optics Commun.* **197**, 393–402.

335. Berry, M V, 2001, 'Spectral twinkling: A new example of singularity-dominated strong fluctuations (summary)' *Physica Scripta* **T90** 15.

336. Berry, M V, 2001, 'Fractal modes of unstable lasers with polygonal and circular mirrors' *Optics Communications* **200** 321–330.

337. Berry, M V, 2001, 'Chaos and the semiclassical limit of quantum mechanics (is the moon there when somebody looks?)', in *Quantum Mechanics: Scientific perspectives on divine action* (eds: Robert John Russell, Philip Clayton, Kirk Wegter-McNelly and John Polkinghorne), Vatican Observatory CTNS publications, pp. 41–54.

338. Berry, M V, 2001, 'Indistinguishable spinning particles' in *XIII*[th] *International Congress of Mathematical Physics* (Eds: A Fokas, A Grigoryan, T Kibble and B Zegarlinski), International Press of Boston, pp. 29–30.

339. Berry, M V and Keating, J P, 2002, 'Clusters of near-degenerate levels dominate negative moments of spectral determinants' *J. Phys. A*, **35** L1–L6.

340.   Berry, M V, 2002, 'Statistics of nodal lines and points in chaotic quantum billiards: Perimeter corrections, fluctuations, curvature' *J. Phys. A*, **35** 3025–3038.

341.   Berry, M V, 2002, 'Singular Limits' *Physics Today* May pp. 10–11.

342.   Berry, M V, 2002, 'Exuberant interference: Rainbows, tides, edges, (de)coherence....' *Phil. Trans. R. Soc. A*, **360** 1023–1037.

343.   Berry, M V, 2002, Comments on Stephen Wolfram's 'A new kind of science', *The Daily Telegraph* 15 May, page 22.

344.   Berry, M V and Ishio, H, 2002, 'Nodal densities of Gaussian random waves satisfying mixed boundary conditions' *J. Phys. A*, **35** 5961–5972.

345.   Bender, C M, Berry, M V and Mandilara, A, 2002, 'Generalized PT symmetry and real spectra', *J. Phys. A*, **35** L467–L471.

346.   Berry, M V, 2002, 'Coloured phase singularities' *New Journal of Physics*, **4** 66.1–66.14.

347.   Berry, M V, 'Exploring the colours of dark light' *New Journal of Physics*, **4** 74.1–74.14.

348.   Berry, M V, 2002, 'Paul Dirac: The purest soul in physics', *Nonesuch* (*University of Bristol*) Autumn, 22–25 (shortened version of paper 291).

349.   Berry, M V, 2002, Review of 'The Rainbow Bridge: Rainbows in Art, Myth and Science' by Raymond L Lee and Alastair B Fraser, *Physics World* **15** (No. 11) 49.

350.   Berry, M V, 2003, 'Mode degeneracies and the Peterman excess-noise factor for unstable lasers', *Journal of Modern Optics* **50**, No. 1, 63–81.

351.   Berry, M V, 2003, 'Outstanding visions of science' *Excellence in Science* (The Royal Society of London), February, p. 9.

352.   Berry, M V, 2003, 'The Art of Physics: Snapshots from Recent Research', for research (Bristol University Magazine) March, pp. 3, 5, 7, 9, 11, 13 & 15.

353.   Berry M V, 2003, 'Paraxial beams of spinning light', in *Optical Angular Momentum* (eds: L Alten, Stephen M Barnett and Miles J Padgett) Institute of Physics Publishing) pp. 65–74 (reprint of paper 297).

354.   Berry, M V, 2003, 'Making Light of Mathematics', *Bull. Amer. Math. Soc.* **40**, 229–237.

355.   Berry, M V and Dennis, M R, 2003, 'The optical singularities of birefringent dichroic chiral crystals', *Proc. R. Soc. A*, **459**, 1261–1292.

356.   Berry, M V and Dennis, M R, 2003, 'The singularities of crystal optics', in proceedings of ICO conference on Polarization Optics, University of Joensuu press (Physics), pp. 18–19 (summary of paper 355).

357. Dennis, M R and Berry, M V, 2003, 'Polarization singularities in paraxial and nonparaxial fields', in proceedings of ICO conference on Polarization Optics, University of Joensuu press (Physics), p. 20.

358. Berry, M V, 2003, 'Quantum Chaology', in *Quantum: A guide for the perplexed*, by Jim Al-Khalili (Weidenfeld and Nicolson), pp. 104–5.

359. Berry, M V, 2004, 'Optical vortices evolving from helicoidal integer and fractional phase steps', *J. Optics. A*, **6**, 259–268.

360. Berry, M V, 2004, 'Conical diffraction asymptotics: Fine structure of Poggendorff rings and axial spike', *J. Optics. A*, **6** 289–300.

361. Berry, M V and Dennis, M R, 2004, 'Black polarization sandwiches are square roots of zero', *J. Optics. A*, **6**, S24–S25.

362. Berry, M V, 2004, 'The electric and magnetic polarization singularities of paraxial waves', *J. Optics.A*, **6** 475–481.

363. Berry, M V, 2004, 'Sightings' (interview with Felice Frankel), *Amer. Sci.* **92** (No. 3) 268–269.

364. Berry, M V and Dennis, M R, 2004, 'Quantum cores of optical phase singularities' *J. Optics. A*, **6** S178–S180.

365. Berry, M V, 2004, 'Riemann-Silberstein vortices for paraxial waves', *J. Optics. A*, **6** S175–S177.

366. Berry, M V, Dennis, M R and Soskin, M S, 2004, 'The plurality of optical singularities' *J. Optics. A*, **6** (Editorial Introduction to special issue).

367. Zafra, R, Bergeman, T, Berry, M, Balian, R and Voros, A, 2004, 'Nandor Balazs (obituary)', *Physics Today* May, p. 74, and Michael Berry's extended version.

368. Berry, M V, 2004, 'Index formulae for singular lines of polarization', *J. Optics. A*, **6**, 675–678.

369. Berry, M V, 2004, 'The study of empirical laws that determine unpredictable events' (reprint of paper #196), in History and Philosophy of Science for African Undergraduates (ed: Helen Lauer) (Hope Publications, Ibadan, Nigeria) Chapter 26, pp. 383–395.

370. Berry, M V, 2004, 'Asymptotic dominance by subdominant exponentials', *Proc. R. Soc. A*, **460**, 2629–2636.

371. Berry, M V, 2004, Transcript of TV interview, in 'Talking science' by Adam Hart-Davis (John Wiley & Sons, Chichester 2004) pp. 24–40.

372. Berry, M V, 2004, 'Physics of nonhermitian degeneracies', *Czech. J. Phys.* **54** 1039–1047.

373. Berry, M V, Dennis, M R and Lee, R L Jr, 2004, 'Polarization singularities in the clear sky', *New Journal of Physics* **6** 162 (doi: 10.1099/1367-2630/1/162) includes press release from the journal).

374. Berry, M V, 2004, 'Physics for taxi-drivers', *Physics World* December, p. 15.

375. Berry, M V, 2004, 'Benefiting from fractals' (A tribute to Benoit Mandelbrot), *Proc. Symp. Pure Mathematics* **72,1** 31–33.

375a. Berry, M V, 2004, 'Living with Physics', in 'One Hundred Reasons to be a Scientist' (ed: K. Sreenivasan) (Abdus Salam Centre for Theoretical Physics, Trieste) pp. 47–49.

376. Berry, M V, 2005, 'Tsunami asymptotics', *New Journal of Physics* **7** 129.

377. Berry, M V, 2005, 'Universal oscillations of high derivatives', *Proc. R. Soc. A*, **461**, 1735–1751.

378. Berry, M V and Ishio, H, 2005, 'Nodal-line densities of chaotic quantum billiard modes satisfying mixed boundary conditions', *J. Phys. A*, **38** L513–L518.

379. Berry MV 2005 'The optical singularities of bianisotropic crystals', *Proc. R. Soc. A*, **461** 2071–2098.

380. Ahmed, Zafar, Bender, Carl, M and Berry, M V, 2005, 'Reflectionless potentials and PT Symmetry', *J. Phys. A*, **38** L627–L630.

381. Berry, M V, Jeffrey, M and Mansuripur, M R, 2005, 'Orbital and spin angular momentum in conical diffraction', *J. Optics. A: Pure Appl. Opt.* **7**, 685–690.

382. Berry, M V, 2005, 'Phase vortex spirals', *J. Phys. A*, **38**, L745–L751.

383. Berry, M V, 2006, 'Oriental magic mirrors and the Laplacian image', *Eur. J. Phys.* **27** 109–118.

384. Berry, M V, 2006, review of 'The equations', by Sandor Bais, *Nature* **2** 65.

385. Berry, M V, 2006, 'Inaugural editorial', *Proc. R. Soc. A*, **462** 1.

386. Berry, M V and Jeffrey, M R, 2006, 'Chiral conical diffraction', *J. Opt. A: Pure Appl. Opt.* **8** 363–372.

387. Berry, M V, Jeffrey, M R and Lunney, J G, 2006, 'Conical diffraction: Observations and theory', *Proc. R. Soc. A*, **462** 1629–1642.

388. Berry, M V and Popescu, S, 2006, 'Evolution of quantum superoscillations, and optical superresolution without evanescent waves', *J. Phys. A*, **39** 6965–6977.

389. Berry, M V, 'Proximity of degeneracies and chiral points' *J. Phys. A*, **39** 10013–10018.

390. Berry, M V, 2006 ,review of 'Fantastic realities' by Frank Wilczek *Nature* **442** 870.

391. Berry, M V, 2006, Inaugural podcast for Institute of Physics, posted 1 August 2006 http:// podcasts.iop.org/index.php?post_id=114098%22 .

392. Berry, M V and Jeffrey, M R, 2006, 'Conical diffraction complexified: Dichroism and the transition to double refraction', *J. Optics A*, **8**, 1043–1051.

393. Berry, M V, 2006, 'Optical vorticulture', in 'Topology in Ordered Phases' (eds: Satoshi Tanda, Toyoki Matsuyama, Migaku Oda, Yasuhiro Asano & Kousuke Yakubo) World Scientific, pp. 3–4. Book includes a cd-rom containing all Bristol papers on phase and polarization singularities up to 2005.

394. Berry, M V and Nye, J F, 'John Michael Ziman' *Biographical Memoirs of the Royal Society*, **52**, 479–491.

394a. Berry, M V, 2006, 'The interface between mathematics and physics' (Panel discussion) *Irish. Math. Soc. Bulletin* **58**, 33–54.

395. Berry, M V and Dennis, M R, 2007, 'Topological events on wave dislocation lines: Birth and death of loops, and reconnection', *J. Phys. A*, **40** 65–74.

396. Berry, M V, 2007, 'Vortex-free complex landscapes and umbilic-free real landscapes', *J. Phys. A*, **40**, F185–F192.

397. Berry, M V, 2007, 'Wave dislocations threading interferometers' *Proc. R. Soc. A* (online reference doi:10.1098/rspa.2007.1842).

398. Berry, M V, 'Looking at coalescing images and poorly resolved caustics', *J. Optics A*, **9** 649–657.

399. Berry, M V, 2007, 'Focused tsunami waves', *Proc. R. Soc. A*, **463**, 3055–3071

400. Berry, M V and Jeffrey, M R, 2007, 'Conical diffraction: Hamilton's diabolical point at the heart of crystal optics', *Progress in Optics* **50**, 13–50.

401. Berry, M V, 2008, 'Three quantum obsessions', *Nonlinearity*, **21**, T19–T26.

402. Berry, M V and McDonald, K T, 2008, 'Exact and geometrical-optics energy trajectories in twisted beams', *J. Optics A*, **10** 035005.

403. Berry, M V and Dennis, M R, 2008, 'Boundary-condition-varying circle billiards and gratings: the Dirichlet singularity', *J. Phys. A*, **41**135203.

404. Berry, M V, 'Waves near zeros' *Coherence and Quantum Optics IX* (The Optical Society of America, Washington DC), Eds: N P Bigelow, J H Eberly & C R Stroud Jr, pp. 37–41.

405. Berry, M V, 'The arcane in the mundane', in Les Déchiffreurs: Voyages en mathématiques, Eds: Jan-Francois Dars, Annick Lesne & Anne Papillaut (Éditions Belin, France), pp. 134–135.

406. Berry, M V, 2008, 'Optical lattices with PT symmetry are not transparent.' *J. Phys. A*, **41**, 244007.

407. Berry, M V, 2008, 'Divagações nocturnas de um físico teórico', Gazeta de Fisica, May 2008, http://tektix.serveftp.com:8080/gfisica/index.jsp?page=articles&id=52&lang=pt (translation of unpublished article aa).

408. Berry, M V and Shukla, P, 2008, 'Tuck's incompressibility function: statistics of zeta zeros and eigenvalues', *J. Phys. A*, **41** 385202.

409. Berry, M V, 2008, Report and speech at John Ziman plaque unveiling, Physics South-West, November, pp. 5–6.

410. Berry, M V and Pollard, B, 2008, 'Physics in Bristol', *Phys. Perspect.* **10** 468–480.

411. Berry, M V, 2008, 'My (nearly) half-century in Bristol', in '100: A collection of words and images celebrating the centenary of the University of Bristol' (Ed; Barry Taylor, University of Bristol).

412. Berry, M V and Dennis, M R, 2009, 'Natural superoscillations in monochromatic waves in D dimension', *J. Phys. A*, **42** 022003.

413. Berry, M V, 2009, 'Hermitian boundary conditions at a Dirichlet singularity: the Marletta-Rozenblum model', *J. Phys. A*, **42** 165208.

414. Berry, M V, 2009, 'Optical currents', *J. Optics. A*, **11**, 004001.

415. Berry, M V, 2009, 'Transitionless quantum driving', *J. Phys. A*, **42** 365303 (9pp).

416. Berry, M V, 2009, 'John Michael Ziman', in the *Oxford Dictionary of National Biography*.

417. Berry, M V and Shukla, P, 2009, 'Spacings distributions for real symmetric 2 x 2 generalized gaussian ensembles', *J. Phys. A*, **42** 485102 (13pp).

418. Berry, M V, 2010, 'Editorial', *Proc. R. Soc. A*, **466** 1–2.

419. Berry, M V and Shukla, P, 2010, 'High-order classical adiabatic reaction forces: Slow manifold for a spin model', *J. Phys. A*, **43**, 045102 (27pp).

420. Berry, M V, 2010 'Geometric Phase memories' *Nature Physics* **6** 148–150.

421. Berry, M V and Howls, C J, 2010, 'Integrals with coalescing saddles', chapter 36 of the NIST Digital Library of Mathematical Functions (Eds: Frank W J Olver, Daniel W Lozier, Ronald F Boisvert & Charles W Clark), Cambridge University Press. Available online at http://dlmf.nist.gov/

422. Berry, M V, 2010, 'in Introductory section of Special Issue on Spin-Statistics' *Foundations of Physics* **40**, 681–683.

423. Berry, M V, 2010, 'Conical diffraction from an N-Crystal cascade', *J. Opt.* **12** 075704 (8pp).

424. Berry, M V, 2010, 'Horse calculus', *Annals of improbable research*, **16** (No. 4 pp.10–11) (corrigendum J. Opt. 12 (2010) 089801).

425. Berry, M V and Howls, C J, 2010, 'Axial and focal-plane diffraction catastrophe integrals', *J. Phys. A*, **43**, 375206 (13pp).

426. Berry, M V, 2010, 'Aptly named Aharonov-Bohm effect has classical analogue, long history', Letter in *Physics Today*, August, p. 8.

427. Berry, M V and Popescu, S, 2010, 'Semifluxon degeneracy choreography in Aharonov-Bohm billiards', *J. Phys. A*, **43** 354005 (11pp).

428. Berry, M V, 2010, 'Asymptotics of the many-whirls representation for Aharonov-Bohm scattering', *J. Phys. A*, **43** 354002 (9pp).

429. Berry, M V and Shukla, P, 2010, 'Typical weak and superweak values', *J. Phys. A*, **43** 354024 (9pp).

430. Berry, M V, 2010, 'After-dinner remarks at the 60th birthday celebration for Celso Grebogi', in Nonlinear Dynamics and Chaos: Advances and Perspectives (Eds: Marco Thiel, Jürgen Kurths, Carmen Romero, Alessandro Moura & György Karoly), Springer 2010, pp. 7–9.

431. Berry, M V, 2010, 'Quantum backflow, negative kinetic energy, and optical retro-propagation', *J. Phys. A*, **43** 415302 (15pp).

432. Berry, M V, 2010, Foreword to 'New Directions in Linear Acoustics and Vibration: Quantum Chaos, Random matrix Theory, and Complexity' (eds: Matthew Wright & Richard Weaver) Cambridge: University Press.

433. Berry, M V, 2010, review of 'Reexamining the Quantum-Classical Relation: Beyond Reductionism and Pluralism', by Alisa Bokulich, in *Brit. J. Philos. Sci*; doi: 10.1093/bjps/axq022.

434. Berry, M V and Shukla, P, 2011, 'Slow manifold and Hannay angle for the spinning top', *Eur. J. Phys.* **32** 115–127.

435. Berry, M V, 2011, 'Lateral and transverse shifts in reflected dipole radiation', *Proc. R. Soc. A*, **467**, doi:10.1098/rspa.2011.0081.

436. Berry, M V, 2011, review of 'The Beginning of Infinity: Explanations that Transform the World' by David Deutsch, in *Times Higher Education*, 31 March, 50–51.

437. Berry, M V, Dennis, M R, McRoberts, B and Shukla, P, 2011, 'Weak value distributions for spin ½', *J. Phys. A*, **44**, 205301.

438. Berry, M V and Dennis, M R, 2011, 'Stream function for optical energy flow', *J. Opt.* **13**, 064004.

439. Berry, M V and Cornwell, J, 2011, Obituary of Robert Balson Dingle, Published online by the Royal Society of Edinburgh: http://www.royalsoced.org.uk/cms/files/fellows/obits_alpha/dingle_robert.pdf.

440. Berry, M V and Keating, J P, 2011, 'A compact hamiltonian with the same asymptotic mean spectral density as the Riemann zeros', *J Phys A* **44** 285203 (14pp).

441. Berry, M V and Uzdin, R, 2011, 'Slow nonhermitian cycling: Exact solutions and the Stokes phenomenon', *J. Phys. A*, **44** 435303 (26pp).

442. Berry, M V, 2011, 'Optical polarization evolution near a non-Hermitian degeneracy', *J. Optics* **13**, 115701 (15pp).

443. Berry, M V, Brunner, N, Popescu, S and Shukla, P, 'Can apparent neutrino superluminal speeds be explained as a quantum weak measurement?", *J. Phys. A*, **44** 492001 (5pp) (http://arxiv.org/abs/1110.2832).

444. Berry, M V, 2012, 'Editorial: Papers we reject' *Proc. R. Soc. A*, **468** 1 (doi:10.1098/rspa.2011.0564).

445. Berry, M V and Shukla, Pragya, 2012 'Pointer supershifts and superoscillations in weak measurements', *J. Phys. A*, **45** 015301 (14pp).

446. Berry, M V, 2012, 'Causal wave propagation for relativistic massive particles: physical asymptotics in action' *Eur. J. Phys.* **33** 279–294.

447. Berry, M V, 2012, 'Martin Gutzwiller and his periodic orbits', in *The legacy of Martin Gutzwiller, Communs. Swiss Phys. Soc.* **37**, 27–30.

448. Berry, M V and Dennis, M R, 2012, 'Reconnections of wave vortex lines', *Eur. J. Phys.* **33** 723–731.

449. Berry, M V, 2012, 'Superluminal speeds for relativistic random waves', *J. Phys. A*, **45**, 185308 (14pp).

450. Berry, M V and Shukla, Pragya, 2012, 'Classical dynamics with curl forces, and motion driven by time-dependent flux', *J. Phy. A*, **45** 305201 (18pp).

451. Berry, M V, 2012, 'Riemann zeros in radiation patterns', *J. Phy. A*, **45** 302001 (9pp).

452. Berry, M V, 2012, Contribution to 'Tribute to Vladimir Arnold' (eds: B Khesin & S Tabachnikov) *Not. AMS.* **59**, 378–399.

453. Berry, M V, 2012, Contribution to 'Glimpses of Beniot B Mandelbrot (1924–2010) (eds: M Barnsley & M Frame) *Not. AMS* **59**, 1056–1063.

454. Berry, M V, 2012, 'Hearing the music of the primes: auditory complementarity and the siren song of zeta', *J. Phys. A*, **45**, 382001 (7pp).

454a. Berry, M V, 2012, 'Beware the double colon', *Ann. Improb. Res.* **18** (No. 6), 2.

455. Berry, M V, 2013, 'Impact and influence: valedictory editorial' *Proc. R. Soc. A* **469** 20120698.

456. Berry, M V & Shukla, Pragya, 2013, 'Hearing random matrices and random waves', *New. J. Phys.* **15** 013026 (11pp).

457. Berry, M V, 2013, 'A note on superoscillations associated with Bessel beams', *J. Opt.* **15** 044006 (5pp).

458. Berry, M V, 2013, 'Circular lines of circular polarization in three dimensions, and their transverse-field counterparts', *J. Opt.* **15** 044024 (5pp).

459. Berry, M V, 2013, 'Much ado about rather little', *Learned Publishing* **26**, 77.

460. Berry, M V, 2013, 'Classical limits' in *The Theory of the Quantum World (Proceedings of the 25th Solvay Physics Conference on Physics)* (eds: David Gross, Marc Henneaux & Alexander Sevrin) Singapore: World Scientific, pp. 52–54.

461. Berry, M V, 2013, 'Exact nonparaxial transmission of subwavelength detail using super-oscillations', *J. Phys. A*, **46**, 205203 (15pp).

462. Berry, M V, 2013, Review of 'Time Reborn: From the Crisis of Physics to the Future of the Universe', by Lee Smolin, in *Times Higher Education*, 27 June, p. 50.

463. Berry, M V, 2013, 'Five momenta', *Eur. J. Phys.* **34**,1337–1348.

464. Berry, M V, 2013, 'Curvature of wave streamlines' *J. Phys. A*, **46**, 395202 (6pp)

465. Berry, M V, 2013, 'Raman and the mirage revisited: confusions and a rediscovery', *Eur. J. Phys.* **34**, 1423–1437

466. Berry, M V and Shukla, Pragya, 2013, 'Physical curl forces: dipole dynamics near optical vortices', *J. Phys. A*, **46** 422001 (9pp).

467. Berry, M V, 2013, 'Superoscillations, endfire and supergain', in *Quantum Theory: A Two-Time Success Story; Yakir Aharonov Festschrift* (editors: D Struppa & J Tollaksen) (Springer) pp. 327–336.

468. Barnett, S M and Berry, M V, 2013, 'Superweak momentum transfer near optical vortices', *J. Opt.* **15** 125701 (6pp).

469. Berry, M V and Shukla, Pragya, 2014, 'Superadiabatic forces on a dipole: exactly soluble model for a vortex field', *J Phys A*, **47**, 125201 (16pp).

470. Berry, M V, 2014, 'Remembering Akira Tonomura', in *In memory of Akira Tonomura: Physicist and Electron Microscopist*, (editors: K Fujikawa & Y A Ono) (World Scientific), 30–32.

471. Baeriswyl, D, Berry, M V and Vollhardt, D, 2014, 'Martin Charles Gutzwiller', *Physics Today*, June, 60.

472. Berry, M V, 2014, 'A tribute to Frank Olver (1924–2013)' *Analysis and Applications* **12** No. 4 ix–x.

473. Berry, M V and Moiseyev, N, 2014, 'Superoscillations and supershifts in phase space: Wigner and Husimi function interpretations', *J Phys A*, **47**, 315203 (14pp).

474. Berry, M V, 2014, Foreword to *Reductionism, Emergence and Levels of Reality: the Importance of being Borderline*, Sergio Chibarro, Angelo Vulpiani and Lamberto Rondoni, Springer: Heidelberg, New York, Dordrecht & London, pp. vii–viii.

475. Berry, M V and Shukla, Pragya, 2015, 'Hamiltonian curl forces', *Proc. R. Soc. A*, **471**, 20150002 (13pp).

476. Berry, M V, 2015, 'Nature's optics and our understanding of light', *Contemporary Physics* (celebrating the International Year of Light) **56** 2–16.

477. Aiello, Andrea and Berry, M V, 2015, 'Note on the helicity decomposition of spin and orbital optical currents', *J. Optics* **17** 062001 (4pp).

478. Berry, Michael, 2015, 'Chasing the Silver Dragon', *Physics World*, July, 45–47.

479. Berry, M V, 2015, 'The squint Moon and the witch ball' *New J. Phys.* **17** 060201 (11pp).

480. Berry, M V, 2015, review of 'Einstein, his space and times', by Steven Gimbel (Yale University Press), in *Jewish Renaissance*, July 2015, p. 60.

481. Berry, M V 2015, 'Riemann zeros in radiation patterns: II. Fourier transforms of zeta', *J. Phys. A*, **48** 385203 (8pp).

482. Berry, M V and Howls, C J, 2015, 'Divergent series: taming the tails', In *The Princeton Companion to Applied Mathematics*, ed. N. J. Higham, pp. 634–640. Princeton, NJ: Princeton University Press.

Anthony J. Leggett

ANTHONY J. LEGGETT

Anthony J. Leggett was born in London, England in March 1938. He attended Balliol College, Oxford where he majored in Literae Humaniores (classical languages and literature, philosophy and Greco-Roman history), and thereafter Merton College, Oxford where he took a second undergraduate degree in Physics. He completed a D. Phil. (Ph.D.) degree in theoretical physics under the supervision of D. ter Haar. After postdoctoral research in Urbana, Kyoto and elsewhere he joined the faculty of the University of Sussex (UK) in 1967, being promoted to Reader in 1971 and to Professor in 1978. In 1983 he became John D. and Catherine T. Macarthur Professor at the University of Illinois at Urbana-Champaign, a position he currently holds. His principal research interests lie in the areas of condensed matter physics, particularly high-temperature superconductivity, and the foundations of quantum mechanics.

Memberships of Learned Societies

1980	Royal Society
1991	American Philosophical Society
1996	American Academy of Arts and Sciences
1997[*]	Foreign Associate of the National Academy of Sciences
1999	Foreign Member, Russian Academy of Sciences
	Fellow of the Institute of Physics, American Physical Society
1999	Honorary Fellow of the Institute of Physics (UK)

[*]now regular member following receipt of US citizenship in Aug. 2001 (dual with UK).

Awards

1975	Maxwell Medal and Prize (Institute of Physics, UK)
1981	Fritz London Memorial Award
1981	Simon Memorial Prize
1992	Paul Dirac Medal and Prize (Institute of Physics, UK)
1994	John Bardeen Prize (with G. M. Eliashberg) (M^2S)
1999	Eugene Feenberg Memorial Medal
2003	Wolf Prize in Physics (with B. I. Halperin)
2003	Nobel Prize in Physics (with A. A. Abrikosov and V. L. Ginzburg)

Main Visiting Appointments Since 1973

April 1973	Visiting Professor, Cornell University
July 1974	Visiting Professor, Cornell University
April 1975	Visiting Professor, University of Illinois
1973–4	Royal Society Japan Fellow, University of Tokyo
Sept.–Dec.1976	Visiting Lecturer at UST, Kumasi, Ghana
1977	Visiting Lecturer at UST, Kumasi, Ghana
April 1980	Bethe Lecturer, Cornell University

Jan.–Aug. 1983	Visiting Scientist, Cornell University
Feb. 1985	Morris Loeb Lecturer, Harvard University
	Visiting Professor, Inst. of Theoretical Physics
Mar.–Jun. 1990	University of Minnesota
June 1996	BBV Foundation Professor, Universidad Autonoma de Madrid
June–July 1997	Visiting Professor, ENS, Paris
June 1998	Astor Lecturer, Oxford University
Nov. 2001	Mueller Lecturer, Penn State University
March 2002	Cave Memorial Lecturer, Queens University, Kingston, Ontario
Jan–Feb. 2003	Visiting Professor, Univ. of Florida
Aug. 2005–	Visiting Professor, National University of Singapore,
Oct. 2006–	Mike and Ophelia Lazaridis Distinguished Visiting Professor, University of Waterloo, Ontario, Canada
Jan. 2013–	Visiting Professor and Director, Shanghai Center for Complex Physics, Shanghai Jiaotong University

Community Roles

Spring 1984	Co-organizer, ITP Program on Quantum Noise in Macroscopic Systems
1987–93	Divisional Associate Editor, Physical Review Letters
Feb.–June 1998	Co-organizer, ITP Program on Bose Einstein Condensation,
1997–8	External examiner, National University of Singapore
June 2002	Co-organizer, Benasque Workshop on Physics of Ultracold Dilute Atomic Gases
2002(?)–9	Editorial Board, New Journal of Physics
2003(?)–7	Board of Reviewing Editors, Science
2005(?)–	Editorial Board, Proc. Nat. Acad. Sci.

Books

The Problems of Physics, OPUS series, Oxford University Press, 1987 (reissued 2006)
Quantum Liquids: Bose Condensation and Cooper Pairing in Condensed Matter Systems, Oxford University Press, Oxford, 2006

Papers

About 125 full papers in refereed

Journals

About 65 in conference proceedings etc.

COMMENTARY

My original academic interests lay on the humanities side, and I completed an undergraduate degree in Literae Humaniores ("Greats") at Balliol College, Oxford. However, by the time of my graduation, my interests had shifted towards physics, and thanks to a fortunate conjunction of the stars, I was able to take a second undergraduate degree in that subject and continue to a D. Phil. (Ph.D.) degree under the supervision of the late Dr. Dirk ter Haar.

The title of my thesis was not very informative. "Some problems in the theory of many-body systems"; in the first half, I discussed some aspects of the interaction of phonons with a nearly linear dispersion relation, and in the second, I treated the properties of a dilute low-temperature solution of 4-He in liquid 3-He (a system which we now know — but did not know then — is essentially nonexistent in nature).

Following the completion of my D. Phil. degree in the summer of 1964, I obtained a one-year postdoctoral fellowship with David Pines at the University of Illinois at Urbana-Champaign. This was a turning point in my career: I decided to try to put together the output of two important recent theories, the Landau theory of a normal Fermi liquid and the BCS theory of superconductivity, and came up with a phenomenological way of doing so which I found rather pleasing. (Many years later, I discovered that some of my results had been obtained earlier by the late Anatoly Larkin by a more microscopic approach, but, perhaps fortunately, I did not know this at the time). I think it is likely that had I not had the good luck to make this discovery, I would have quit academic physics at this point to go into high-school teaching.

During the period 1965–1967 I spent a year in the group of Professor Takeo Matsubara at Kyoto University and then a "roving" year at Oxford, Harvard and (again) Illinois, before taking up a more permanent position at the University of Sussex, one of a batch of new "plateglass" universities set up in Britain in the early 60's. I eventually spent fifteen years at Sussex; and overall have rather warm memories of them. (Another Sussex alumnus, the well-known author Ian McEwan, who was a student there in the late 60's and set his novel Sweet Tooth there in the 70's, seems to have rather more jaundiced recollections). One aspect of the Sussex environment which I especially liked was the implicit assumption (at least as I perceived it) that if one did an effective teaching job, one had earned one's salary, and research was then an optional extra pleasure, not a duty. For the first few years I worked on various topics in quantum fluids and solids, including the conjectured phenomenon which later became known as "supersolidity", but was gradually seduced, in part no doubt because of my philosophical background, by the perennial conceptual problems in the foundations of quantum mechanics, and was just about to make a move into that area, when I heard the news from Cornell about their NMR experiments on liquid helium-3 below 3 mK.

My first thought was that these results were so bizarre that they might be the first indication of a breakdown of quantum mechanics, and I originally set out to demonstrate rigorously that they could have no explanation within accepted theory. Of course, that was not how things turned out, and after a few months I was able to account for the existing NMR data, and predict new ones, using standard quantum and statistical mechanics. For the next ten years or so I worked in the new field of superfluid 3-He; one piece of work of which I am particularly proud was the conjecture that the unexpectedly low degree of metastability of the A phase is due to nucleation of the more stable B phase by cosmic-ray muons — a conjecture which seems likely to be true in at least some sample cells.

However, I could not suppress for ever my curiosity about the foundations of quantum mechanics and in particular about whether the Schroedinger's Cat paradox is for real, and towards the end of the 70's, in part because of interactions with an experimental colleague, Terry Clark, began to wonder whether superconducting devices based on the Josephson effect (nowadays known as "flux qubits") would allow one to "build Schroedinger's Cat in the laboratory". Having been steeped in the formal "quantum measurement" literature, I was initially sceptical that a meaningful experimental test of the validity of the quantum-mechanical superposition principle even at the relatively modest level of flux qubits could survive the effects of decoherence; but I eventually convinced myself that this was a matter for quantitative calculation rather than prejudice, and over the next few years I devoted myself to the analysis of the effects of dissipation and decoherence on the quantum-mechanical behavior of a macroscopic variable, trying in particular to devise a scheme by which one could make predictions of quantities such as the rate of escape by tunnelling from a metastable well purely from a knowledge of classically measurable parameters such as the resistance shunting a flux qubit; the earlier part of this work was done in collaboration with my Brazilian graduate student Amir Caldeira, with whom I wrote a number of papers.

One rather unusual episode during my Sussex period was a couple of semesters spent, in the fall semesters of 1976 and 1977, at the University of Science and Technology in Kumasi, Ghana, with which we had an exchange arrangement. This was interesting experience in many ways, and in particular, gave me some feeling for the difficulties under which scientists working in the third world labor–difficulties which have perhaps been somewhat mitigated, but by no means obviated, by the information revolution.

In the spring of 1982, I received a rather attractive offer of a named chair from the University of Illinois at Urbana-Champaign, where I had spent one of my postdoctoral years, and eventually moved there in the fall of 1983. One consequence of the move was a drastic reduction in my formal teaching load; at Sussex I had spent 12–15 contact hours per week on undergraduate teaching, while at Illinois I had one (usually) graduate course in the fall semester and the spring completely free. However, I am not clear that my research output increased proportionately! Anyway, I spent much of my first few years at Illinois continuing the work on the quantum mechanics of (possibly dissipative) macroscopic systems such as flux qubits, and in particular published, with Anupam Garg, a scheme for confronting the predictions of quantum mechanics with those of a class of world-views we called "macrorealistic". It is ironic to recall, in 2015, that in 1985 and even up to 1999 the whole idea of trying to generate quantum superpositions at the level of flux qubits seemed to many so outrageous that (I am told) at least one "expert" proclaimed that such experiments would be a waste of the taxpayers' money!

Since 1983, my career at the University of Illinois has been pretty much in the standard North American mould; I eventually got interested in other areas such as ultracold gases and high-temperature superconductivity, and supervised a number of graduate students in both these areas. Highlights of this period include my co-authorship of the 1993 experiment by my colleague Dale van Harlingen which firmly established the now accepted symmetry of the cuprate order parameter, and (of course!) my receipt of both the Wolf and the Nobel prizes in the year 2003. At 77, I try to keep going — there are too many interesting questions out there to quit just yet!

PHYSICAL REVIEW

LETTERS

| VOLUME 54 | 4 MARCH 1985 | NUMBER 9 |

Quantum Mechanics versus Macroscopic Realism: Is the Flux There when Nobody Looks?

A. J. Leggett

Department of Physics,[a] *University of Illinois at Urbana-Champaign, Urbana, Illinois 61801, and Department of Physics, Harvard University, Cambridge, Massachusetts 02138*

and

Anupam Garg
University of Illinois at Urbana-Champaign, Urbana, Illinois 61801
(Received 19 November 1984)

It is shown that, in the context of an idealized "macroscopic quantum coherence" experiment, the predictions of quantum mechanics are incompatible with the conjunction of two general assumptions which are designated "macroscopic realism" and "noninvasive measurability at the macroscopic level." The conditions under which quantum mechanics can be tested against these assumptions in a realistic experiment are discussed.

PACS numbers: 03.65.Bz, 05.30.−d, 74.50.+r, 85.25.+k

Despite sixty years of schooling in quantum mechanics, most[1] physicists have a very non-quantum-mechanical notion of reality at the macroscopic level, which implicitly makes two assumptions. (A1) Macroscopic realism: A macroscopic system with two or more macroscopically distinct[2,3] states available to it will at all times *be* in one or the other of these states. (A2) Noninvasive measurability at the macroscopic level: It is possible, in principle, to determine the state of the system with arbitrarily small perturbation on its subsequent dynamics. A direct extrapolation of quantum mechanics to the macroscopic level denies this. The aim of this Letter is (1) to point out that under certain conditions the experimental predictions of the conjunction of (A1) and (A2) are incompatible with those of quantum mechanics extrapolated to the macroscopic level, and (2) to investigate how far these conditions may be met in a realistic experiment.

To this end, let us consider the (as yet unobserved) phenomenon of "macroscopic quantum coherence" (MQC) in an rf SQUID.[4] We take the potential $V(q)$ for the trapped magnetic flux q to be reflection symmetric (see Fig. 1) with minima at $\pm q_0$ far enough apart that states in which q is close to $+q_0$ and $-q_0$ can be regarded as macroscopically distinct. For an

isolated SQUID, quantum mechanics predicts that if the flux is initially in one well, it will oscillate back and forth with some frequency Δ_0. A more realistic quantum mechanical calculation[5] which includes the irremovable environmental effects shows that for low enough temperature and weak enough coupling to the environment, the oscillations are not entirely des-

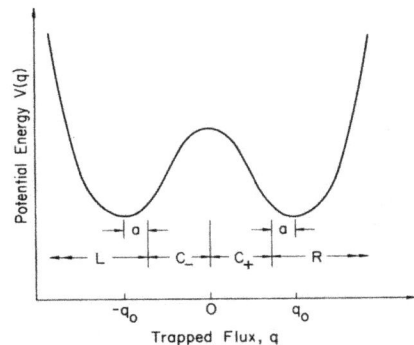

FIG. 1. The potential $V(q)$ for the trapped flux q. The various notations are explained in the text.

VOLUME 54, NUMBER 9   PHYSICAL REVIEW LETTERS   4 MARCH 1985

troyed, but are merely underdamped. Since it is under these conditions that our argument is most pertinent, we shall assume that the experimental constraints on achieving them, while stringent,[6] can, in fact, be met.

Let us divide the possible values of q into four regions L, C_-, C_+, and R, as shown in Fig. 1, where $x_0 << a << q_0$, x_0 being the zero-point width that a wave packet would have in either well if the other were absent. We define a quantity Q, which equals $+1$ (-1) if the system is observed to be in region R (L). If we temporarily ignore the minuscule probability of finding the system in $C_\pm$, quantum mechanics predicts (and we assume that experiment will find) that any observation of Q will find only the values ± 1.

It immediately follows from (A1) that for an ensemble of systems prepared in some way at time t_0,[7] we can define (i) joint probability densities $\rho(Q_1,Q_2)$, $\rho(Q_1,Q_2,Q_3)$, etc. for Q to have the values Q_i at times t_i (we take $t_0 < t_1 < t_2 \ldots$), (ii) correlation functions $K_{ij} \equiv \langle Q_i Q_j \rangle$. The probability densities must be consistent with one another, which implies, e.g.,

$$\sum_{Q_2 = \pm 1} \rho(Q_1,Q_2,Q_3) = \rho(Q_1,Q_3). \tag{1}$$

From this, we can derive inequalities similar to those of Bell[8] or of Clauser *et al.*[9] for the Einstein-Podolsky-Rosen (EPR) experiment[10] with the times t_i playing the role of the polarizer settings. For example, we have

$$1 + K_{12} + K_{23} + K_{13} \geq 0, \tag{2a}$$

$$|K_{12} + K_{23}| + K_{14} - K_{24}| \leq 2. \tag{2b}$$

If we assume that (A2) can be realized in an actual experiment (we shall discuss this below), then these correlations and probabilities can be measured, and we can test whether (1) and (2) hold.

We can also test (1) and (2) against the predictions of quantum mechanics. For definiteness, we consider the case of "Ohmic" dissipation, which has been studied in detail by Chakravarty and Leggett.[5] The behavior of the system can be parametrized by Δ_r, a renormalized tunneling frequency, by ω_c, the highest frequency scale at which the environment can respond, and by α, a dimensionless dissipation coefficient. Typically $\Delta_r << \Delta_0 << \omega_c$. If, as in Chakravarty and Leggett, we ignore the so-called "interblip effects" (which is a good approximation both for very low T and α, when the flux executes underdamped oscillations, and for high values of T and α, when we have overdamped relaxation), then although we cannot rigorously prove, we can very plausibly argue that for t_i, $|t_j - t_i| >> \omega_c^{-1}$, K_{ij} is essentially independent of the choice of the initial ensemble and equals $P(t_j - t_i)$ as defined there.[5] One can further argue

that

$$\rho(Q_1,Q_2,Q_3) \simeq \rho(Q_1,Q_2)\rho(Q_2,Q_3). \tag{3}$$

It is now clear from experience with Bell-type inequalities that if $P(t)$ is not too heavily damped, then quantum mechanics will violate conditions (1) and (2). Consider, for example, the expression (24) of Chakravarty and Leggett for $P(t)$ at $T=0$, and set $\Delta P(t) = 0$.[5] Since the "incoherent" part, $P_{inc}(t)$, is always negative, we will overestimate the left-hand side of (2a) if we neglect this altogether. Any value of α for which a violation of (2a) is thus obtained will be less than the critical value of α beyond which (2a) is always satisfied. The reader can verify that for $t_2 - t_1 = t_3 - t_2 = 2.3\Delta_{eff}^{-1}(\approx \frac{2}{3}\pi\Delta_{eff}^{-1})$, and $\alpha \leq 0.11$, Eq. (2a) is indeed violated. A similar underestimate of the critical α value can be obtained from (2b) but with $P_{inc}(t)$ replaced with its asymptotic long-time form (which overestimates its magnitude).[11] Doing this we find that Eq. (2b) is violated for $t_2 - t_1 = t_3 - t_2 = t_4 - t_3 = 0.84/\Delta_{eff} (\approx \frac{1}{4}\pi\Delta_{eff}^{-1})$, and $\alpha \leq 0.08$.[12] Note, however, that quantum mechanics and macroscopic realism continue to differ even in the overdamped regime. Using the methods of Ref. 5, we can show that for $Q_1 = Q_3 = 1$, quantum mechanics would have the left-hand side of Eq. (1) exceed the right-hand side by

$$[\hbar \cot(\pi\alpha)/4\tau k_B T]\exp[-(t_3 - t_2)/\tau],$$

a quantity that can assume negative values.[13]

There is a slight difficulty in this argument arising from the nonzero (but exponentially small) probability of finding the system in regions $C_\pm$ (see Fig. 1), which is that once the system can have nearby q values, the concept of "macroscopically distinct states" becomes somewhat blurred. The easiest solution to this problem is to modify the macroscopic realism postulate (A1) to allow the system to be in a superposition of only two *neighboring* states (R and C_+, C_+ and C_-, etc.). We now assign to Q the value $+1$ (-1) if the system is in R (L) alone, in C_+ (C_-) alone, or in a superposition of R and C_+ (L and C_-). The only combination which can affect Eqs. (1) and (2) is C_+ and C_-. Its contribution, however, cannot be more than a few times the total probability for finding the system in either C_+ or C_-, which is vanishingly small, and the incompatibility of quantum mechanics and macroscopic realism is not affected.

We now turn to the vexing question of whether the assumption (A2) of noninvasive measurability is likely to hold in practice. Indeed, ever since Heisenberg's "invention" of the "γ-ray microscope," we have all learned *not* to make such assumptions when dealing with microsystems, and at first sight there is no reason to treat macrosystems differently. We can, nevertheless, make (A2) seem extremely natural and plausible

VOLUME 54, NUMBER 9   PHYSICAL REVIEW LETTERS   4 MARCH 1985

by introducing the idea of an *ideal negative result* experiment. This is defined to be an experiment in which the measuring apparatus interacts with the system (and then very strongly) *only* if the latter has one value of $Q(t)$ (say $+1$), and does not interact at all otherwise. We can then confidently infer that $Q(t)$ has the value -1, if at time t the system does not elicit a response from the apparatus. Conjoined with the assumption of macroscopic reality, this strongly suggests that the system also had $Q(t') = -1$ for t' immediately prior to the measurement at time t, and therefore that (at least in the limit of an arbitrarily short measurement) the apparatus could not have affected the dynamics of the system, i.e., that (A2) holds. Unlike the two-slit experiment where such a measurement can be made by shining light on one slit only, it is highly doubtful whether the analogous measurement could be made for an rf SQUID, but the difficulty seems to be technical and not conceptual. Under the assumption that an ideal negative-result experiment can be conducted, it is plain that all the quantities in Eqs. (1) and (2) can be measured. Suppose, for example, we wish to measure $\rho(Q_1 = 1, Q_3 = 1)$. Since the dynamics after t_3 are not of interest, we can use an ordinary measurement at t_3, and an ideal negative-result setup at t_1, which responds only if $Q(t_1) = -1$. We then simply discard those members of our ensemble which produce a response at t_1. Of the remainder, we count the number which have $Q(t_3) = 1$ and divide by the total number of members of the ensemble to obtain $\rho(Q_1 = 1, Q_3 = 1)$. By using a different ideal negative-result setup on another large and identical ensemble, we can obtain $\rho(Q_1 = -1, Q_3 = 1)$. We can thus calculate a value of K_{13}, and assumption (A2) allows us to assert that this is the K_{13} characteristic of the original ensemble.

An alternative to making ideal negative-result measurements is to couple the system to a microscopic probe. For example, in principle one could fire a neutron through the SQUID ring with its spin transverse to the magnetic field with a velocity such that it would precess precisely through an angle $\pm\pi/2$ if $q = \pm q_0$, and with a Larmor frequency much larger than Δ_{eff} but much less than the small oscillation frequency in either well. Let us consider how this method could be used to measure $\rho(Q_1, Q_2, Q_3)$, for example. For simplicity, let us prepare the system in a definite state (say $Q_1 = +1$) at time t_1 itself. We then fire our neutron to pass through the ring at t_2, and measure the flux at t_3 directly. Since the SQUID-neutron interaction is effectively instantaneous on the scale of Δ_{eff}^{-1}, we can infer the value of Q at time t_2 by measuring the neutron spin *at any time after* t_2, or even t_3![14] A little thought shows that the quantum mechanical prediction (3) still holds with extra (small) corrections due to the finite duration of the measurement at t_2. Similar small

corrections enter into the macroscopic-realistic predictions (1) and (2), so that once again, the conflict between quantum mechanics and assumptions (A1) and (A2) is not affected.

In conclusion it should be emphasized that, should the quantum mechanically predicted results be obtained in a situation where they conflict with postulates (A1) and (A2), this would, of course, not be formally in conflict with the arguments so often given in discussions of the quantum theory of measurement to the effect that once a microsystem has interacted with a realistic measuring device, the device (and, if necessary, the microsystem) behave *as if* it were in a definite (and noninvasively measurable) macroscopic state: The macroscopic systems suitable for a macroscopic quantum coherence experiment are certainly *not* suitable to be measuring devices, at least under the conditions specified. But such a result might cause us to think a great deal harder about the significance of the "as if"!

This research was supported through the MacArthur Professorship endowed by the John D. and Catherine T. MacArthur Foundation. One of us (A.J.L.) is grateful to the Department of Physics at Harvard University for hospitality during the period where some of this work was performed.

(a)Present and permanent address.

[1]One must, of course, exclude here the genuine adherents of the relative-state ("many worlds") and mentalistic ("reduction-by-consciousness") interpretations of quantum mechanics. We strongly suspect that the number of physicists who in fact genuinely adhere to either of these interpretations (in the sense that it really makes a difference to the way they think about the macroscopic world) is considerably less than the number who claim to!

[2]One can, of course, argue *ad nauseam* about the precise meaning of the phrase "macroscopically distinct." One specific objection which is sometimes raised with respect to a hypothetical experiment on a SQUID ring is that the difference in flux values between the two potential minima can be at most a fraction of the flux quantum $\phi_0 \equiv \pi\hbar/e$ [A. J. Leggett, in *Proceedings of NATO Advanced Study Institute on Percolation, Localization, and Superconductivity*, edited by A. Goldman and S. Wolf (Plenum, New York, 1984)]; it is therefore (it is argued) "only of order $\hbar$" and therefore still in the quantum domain. We would regard this particular objection as merely verbal, since it is a historical accident that we treat the constants e and $\hbar$ as independent "fundamental constants" rather than say $\hbar$ and ϕ_0. A more sweeping objection is that any phenomenon which involves quantum interference effects can *by definition* not occur "at the macroscopic level." One can no more argue with this view than with the claim that the mere fact that a certain kind of behavior can be programmed into a computer *ipso facto* disqualifies it from being "intelligent." For our present purpose it is adequate that the "disconnectivity" (as defined as Ref. 3) of the superposition of "left" and "right" states is

VOLUME 54, NUMBER 9   PHYSICAL REVIEW LETTERS   4 MARCH 1985

of the order of the total number of electrons in the device ($\sim 10^{15}$–10^{23}).

[3]A. J. Leggett, Prog. Theor. Phys. **69**, 80 (1980).

[4]Leggett, Ref. 2.

[5]S. Chakravarty and A. J. Leggett, Phys. Rev. Lett. **52**, 5 (1984). A much more detailed treatment of the argument leading to the results quoted in this reference, and of the corrections $\Delta P(t)$ in Eq. (24) due to "interblip" effects, is contained in A. J. Leggett *et al.*, to be published. In particular, it is shown that for any finite t a rigorous upper bound, which tends to zero as α^2 for small α, can be placed on the magnitude of the deviation of $P(t)$ from the expression given by the first two terms of (24) with $A(\alpha) = 1$, $q(\alpha) = 0$. Further, inspection of the formalism used by Chakravarty and Leggett makes it extremely plausible (though we have not yet succeeded in giving a rigorous proof) that the difference between the K_{ij} defined below and $P(t_j - t_i)$ is itself at most of the order of this deviation. If this is so, the effect of all these corrections would be at most a (probably very small) correction to the "critical" values of α estimated in the text.

The work of Chakravarty and Leggett applies to the case of "Ohmic" dissipation (the case almost certainly realized in a SQUID), for which the spectral function $J(\omega)$ defined there is proportional to ω for small ω. One can show (see Leggett *et al.*) that for environments with $J(\omega) \sim \omega^p$, $p > 1$, quantum effects are less severely suppressed than in the Ohmic case. While environments with spectra corresponding to $0 < p < 1$ do not appear to be excluded by any *a priori* consideration, no mechanism which would produce such a state of affairs is known for SQUIDS, and it would presumably have dramatic (and, so far at least, unobserved) effects on the dynamics in the classically accessible regime. However, it is not known at present whether "$1/f$ noise" can be treated within the general framework of these references, and so our results may not be applicable to systems where such noise is appreciable.

[6]See, e.g., R. de Bruyn Ouboter, in *Proceedings of the International Symposium on the Foundation of Quantum Mechanics in the Light of New Technology*, edited by S. Kamefuchi (Physical Society of Japan, Tokyo, 1984).

[7]In practice this is more likely to be a time ensemble.

[8]J. S. Bell, Physics (N.Y.) **1**, 195 (1964). The inequality (15) of this paper is essentially the same as (2a) of our text.

[9]J. F. Clauser, M. A. Horne, A. Shimony, and R. A. Holt, Phys. Rev. Lett. **23**, 880 (1969). The inequality (1a) of this paper is (2b) of our text. See J. F. Clauser and A. Shimony, Rep. Prog. Phys. **41**, 1881 (1978), for the various interpretations of this inequality, and also for discussions pertaining to imperfect/inefficient detectors, many of which are applicable to the problem at hand.

[10]A. Einstein, B. Podolsky, and N. Rosen, Phys. Rev. **47**, 777 (1935).

[11]This argument exploits the additional fact that the error made in replacing $P_{\text{inc}}(t)$ by its asymptotic form decreases with increasing t. The replacement, therefore, adds three negative quantities and one positive quantity (whose magnitude is less than that of any of the negative quantities) to the left-hand side of (2a). The net effect is to underestimate the left-hand side.

[12]It is easy to show that irrespective of the form of $P(t)$, (2a) and (2b) are maximally violated (if at all) for a given value of α for equally spaced times t_i.

[13]If α is too close to an integer or half-integer, the discrepancy is not accurately given by the expression in the text.

[14]We note in passing that since the neutron can be quite far from the SQUID at t_3, the situation has many of the seemingly paradoxical aspects of the EPR experiment. For example, suppose that the neutron spin was measured before the flux was measured at t_3, and that the two measurements were separated by a timelike interval. A local realist could argue that a measurement on the microsystem (neutron) was affecting the macrosystem (SQUID)!

VOLUME 57, NUMBER 3   PHYSICAL REVIEW LETTERS   21 JULY 1986

Dynamics of the ^{3}He A-B Phase Boundary

S. Yip and A. J. Leggett

Department of Physics, University of Illinois at Urbana-Champaign, Urbana, Illinois 61801
(Received 10 February 1986)

We investigate the friction coefficient of the moving boundary. At not too low temperatures the friction is ascribed primarily to transmission and Andreev reflection of quasiparticles; the resulting theoretical value of the terminal velocity v_{AB} is in reasonable agreement with the data of Buchanan, Swift, and Wheatley. At lower temperatures we predict a saturation of v_{AB} and underdamped oscillations of the pinned boundary.

PACS numbers: 67.50.Fi, 64.70.Ja

The phase transition between the A and B phases[1,2] of liquid ^{3}He is first order, and its equilibrium temperature T_{AB} is a strong function of pressure P and magnetic field H. Since, moreover, the A phase supercools very appreciably, it should be possible to study the structure and behavior of the A-B phase boundary under a wide variety of conditions. Indeed, its static properties have been investigated both experimentally[3] and theoretically.[3–6]

In this Letter we outline the general physical principles which we believe govern the *dynamics*[7] of the phase boundary, and apply them to discuss quantitatively its terminal velocity in the regime investigated by Buchanan, Swift, and Wheatley.[8] We also discuss more briefly and qualitatively the behavior at low temperatures. We hope to supply more details[9] elsewhere.

Essential to our argument are the following order-of-magnitude inequalities, which hold over almost all of the region of interest[10]: (1) The width of the boundary, $d(T)$, which is[4,5] a few times the temperature-dependent coherence length[2] $\xi(T)$, is much shorter than the mean free path $l(T)$ of an excitation in either bulk phase (cf. Greaves and Leggett[11]). (2) The velocity of the boundary, v, is small compared to the Fermi velocity v_F (cf. Ref. 8, and below). (3) The characteristic time associated with the motion of the boundary, which is of order d/v for the experiment of Ref. 8 and of order of the reciprocal vibrational frequency for the pinned boundary, may be either the reciprocal vibrational frequency for the pinned boundary, may be either large or small compared to the quasiparticle collision time $\tau(T) \sim l(T)/v_F$, but is always long compared to the characteristic adjustment time $\hbar/\Delta(T)$ of the Cooper pairs, where $\Delta(T)$ is the rms gap in either bulk phase.

In general, if we consider the displacement $\mathbf{u}$ of a small area of the boundary under conditions (2) and (3), we should expect it to satisfy an equation of the general form

$$M^*\ddot{\mathbf{u}} + \Gamma\dot{\mathbf{u}} = \mathbf{F} \tag{1}$$

where M^*, Γ, and $\mathbf{F}$ are respectively the inertial mass,

friction coefficient, and external (conservative) force per unit area of the boundary. The precise form of $\mathbf{F}$ depends on the geometry and thermodynamic conditions: In the case of vibrations of the pinned boundary the main contribution to it comes from the surface tension, while for free motion it should usually be adequate to take it as the difference ΔG_{AB} of the Gibbs free energy per unit volume at the pressure and temperature of the metastable A liquid.[12]

In applying formula (1) let us first consider the experiments of Buchanan, Swift, and Wheatley.[8] We will verify below that at the relevant temperatures the relaxation time M^*/Γ should be of order 10^{-9} s, so that we would expect that in those experiments (time scale $\gtrsim 10^{-2}$ s) the boundary should move at a terminal velocity v_{AB} given by $v_{AB} = \Delta G_{AB}/\Gamma$. Thus it remains only to calculate the friction coefficient Γ.

Let us assume that the boundary is moving (relative to the cell walls) with velocity $\mathbf{v}$ ($\equiv\dot{\mathbf{u}}$) while the A phase some distance ahead of it, say a few mean free paths, is characterized by two-fluid flow with normal and superfluid velocities $\mathbf{v}_n$ and $\mathbf{v}_s$, respectively. We shall initially assume that $\mathbf{v}_n$ and $\mathbf{v}_s$ are both zero (on both A and B sides), and return later to examine this assumption. To calculate the friction coefficient Γ, we assume that the condensed Cooper pairs transform their wave functions adiabatically as the boundary passes and therefore contribute no frictional force (though see below). Then the only mechanism which can produce such a friction is the change of momentum suffered by a normal quasiparticle when reflected from, or transmitted across, the moving boundary. In considering this effect it is essential to note that the time spent by a typical quasiparticle in the boundary region, which is of order d/v_F, is always short compared to a typical collision time [condition (1) above], and hence the relaxation of the transmitted or reflected quasiparticles to equilibrium takes place overwhelmingly in the bulk A or B phase, where the order parameter is not a function of position. Thus in strong contrast to the case of (say) a moving A-phase texture whose characteristic dimension is large compared to

VOLUME 57, NUMBER 3   PHYSICAL REVIEW LETTERS   21 JULY 1986

the mean free path $l(T)$, the concept of orbital viscosity[13] is irrelevant, and the correct calculation simply consists in working out the extra change in the total momentum of the quasiparticle system per unit time due to reflection from, or transmission across, the moving boundary and setting it equal to $\Gamma\mathbf{v}$.

Consider for definiteness a quasiparticle incident on the boundary from the A side with energy E and a wave vector $\mathbf{k}$ of magnitude greater than the Fermi wave vector k_F (E and $\mathbf{k}$ are measured in the rest frame of the superfluid, i.e., that of the cell walls), and in a direction specified by azimuthal and polar angles θ, ϕ, where θ is measured from the inward normal $\hat{\mathbf{n}}$ to the boundary and ϕ from the common plane of $\hat{\mathbf{n}}$ and the characteristic orbital vector[2] $\mathbf{l}$. Since the A-phase gap Δ_A is a function of θ and ϕ, the quasiparticle "kinetic energy" $\epsilon_k \cong \hbar v_F(k - k_F)$ (i.e., the normal-state energy of the state $\mathbf{k}$ relative to the Fermi energy) may be regarded as a function of E, θ, and ϕ:

$$\epsilon_A(E,\theta,\phi) \equiv + (E^2 - |\Delta_A(\theta,\phi)|^2)^{1/2}. \quad (2)$$

Suppose for definiteness that E is less than the isotropic B-phase gap Δ_B. Then the quasiparticle will, of course, be reflected from the boundary, not by ordinary but by Andreev[14] reflection. Consider first the case $\mathbf{v} = 0$. Then the reflection process must conserve both the component of $\mathbf{k}$ parallel to the boundary and the energy E, so that the quasiparticle simply emerges as a quasihole with kinetic energy $-\epsilon_A$; moreover, provided $\cos\theta$ is not too small ($\lesssim \Delta/\epsilon_F$) the resulting change in its momentum, which we label $\Delta p^{(0)}$, is simply $-(2\epsilon_A/v_F \cos\theta)\hat{\mathbf{n}}$. Such a process, and the obviously related ones (cf. below), give rise to a finite force on the boundary even when $\mathbf{v} = 0$; it is straightforward to show[9] that this force is just the quasiparticle contribution to the change in free energy when the A-B boundary is adiabatically displaced, and it should

therefore be counted as part of the conservative force $\mathbf{F}$ in Eq. (1).

Now suppose that the boundary is moving with finite velocity $\mathbf{v}$. Now what must be conserved, apart from the parallel component of $\mathbf{k}$, is the energy *in the frame of the moving boundary*, that is the quantity $E' \equiv E - \hbar\mathbf{k}\cdot\mathbf{v}$ ($+$const). Expanding up to first order in $v/(v_F \cos\theta)$, we find that the *extra* momentum change $\Delta p^{(1)}$ over and above that for the stationary boundary is given by the expression

$$\Delta p^{(1)} = [(2E/v_F^2)\sec^2\theta]\mathbf{v}. \quad (3)$$

To find the frictional force due to reflection of A-phase quasiparticles, we must multiply the total momentum change, $\Delta p^{(0)} + \Delta p^{(1)}$, by the flux of quasiparticles incident on the boundary with energy in the range dE and wave vectors in the solid angle $d\Omega$, integrate over energy and (relevant) angle, and keep the term linear in $\mathbf{v}$ [the resulting expression is clearly, for each value of θ, the leading term in an expansion in $v/v_F \cos\theta$]. The relevant flux is given by the expression

$$(4\pi)^{-1}\frac{dn}{d\epsilon}\frac{E}{\epsilon_A}f(E)\left\{\frac{v_F\epsilon_A}{E}\cos\theta + v\right\},$$

where $dn/d\epsilon$ is the (normal state) density of states (of both spins) at the Fermi surface, $\epsilon_A(E,\theta,\phi)$ is given by Eq. (2), and $f(E)$ is the Fermi function. Note that we have explicitly assumed, here, that the incident quasiparticle distribution is the thermal equilibrium one in the frame of the walls (cf. below).

To obtain the total friction coefficient Γ we clearly have to generalize the calculation to include (a) quasiholes, (b) quasiparticles and quasiholes incident from the B phase, and (c) the possibility of transmission across the boundary. Omitting the straightforward algebra involved, we quote the final result for Γ:

$$\Gamma = (2/v_F)(dn/d\epsilon)\int(d\Omega/4\pi)|\sec\theta|\int_0^\infty dE\, Ef(E)\{(2-\overline{T})[\theta(E - |\Delta_A(\theta,\phi)|) + \theta(E - \Delta_B)]$$

$$-\theta(E - \Delta_{\max}(\theta,\phi))\overline{T}[\epsilon_A^2(E,\theta,\phi) + \epsilon_B^2(E)]/\epsilon_A\epsilon_B\}. \quad (4)$$

Here $\theta(x)$ is the usual Heaviside step function, ϵ_B is $+(E^2 - \Delta_B^2)^{1/2}$, $\Delta_{\max}(\theta,\phi)$ is the larger of $|\Delta_A(\theta,\phi)|$ and Δ_B, $\overline{T}(E,\theta,\phi)$ is the spin-averaged transmission coefficient, and the angular integral goes over *all* solid angle. Equation (4) is the principal quantitative result of this paper. The right-hand side is clearly logarithmically divergent unless some lower cutoff is put on $|\cos\theta|$. Since the approximations implicit in the calculation fail whenever $|\cos\theta|$ is small compared to any of the quantities Δ/ϵ_F, l/ξ_0, or v/v_F, we should presumably take the cutoff to be of the order of the largest of these. In the numerical calculations quoted below we have taken it to be equal to v/v_F; since the depen-

dence on the cutoff is only logarithmic and the other two quantities are at most of order 5×10^{-3}, while v/v_F is never less[8] than about 5×10^{-4} except possibly for the last data point, the error so incurred is never more than a factor of about 1.4 and is usually much less.

The general order of magnitude of the quantity Γ, for $T \sim T_c$, is easily seen to be $\Delta^2(T)(dn/d\epsilon)v_F^{-1}$; note that this is a factor of order $(\Delta/\epsilon_F)^2$ smaller than the "friction coefficient" we should expect if all the atoms in the A liquid were reflected in the usual (not Andreev) way from the moving boundary. This fact is

VOLUME 57, NUMBER 3   PHYSICAL REVIEW LETTERS   21 JULY 1986

crucial in the discussion of a possible finite v_n and v_s, to which we now turn.

If v_n is finite, then, since on the relevant time scale the mass current is zero, $v_n - v_s$ is also finite; moreover, in general v_n and v_s may be different on the two sides of the boundary. Under these conditions the analysis is quite delicate even[9] for $v_{nA} = v_{nB}$, and we shall not go into it here, merely remarking that we expect (this contribution to) the friction to vanish when $v_n = v$ on both sides of the boundary (even if $v_{SA} \neq v_{SB} \neq v$) and in general to be a function of $v_n - v_s$ (as is also the conservative force F) on a scale which is likely to be not the Fermi velocity v_F but the much smaller "pair-breaking critical velocity" $v_c \equiv 2\Delta/p_F \sim (\Delta/\epsilon_F)v_F$. (Indeed, when $|v_n - v_s|$ exceeds v_c the very concepts of the two-fluid model may break down.) Thus, *prima facie*, the necessary conditions to apply the results of the above calculation, done for $v_n = v_s = 0$, should be (a) $|v_n| \ll |v|$, and (b) $|v_n - v_s| \ll v_c$. At first sight it seems very unlikely that these conditions, particularly (b), would be fulfilled in a realistic experiment: Even if we exclude the case of any substantial preexisting counterflow in the metastable A liquid (as we shall for present purposes) it appears *a priori* likely that the very motion of the boundary would itself induce a v_n (hence a $v_n - v_s$) of order at least v_c, if not greater, in the liquid ahead of it.

That this argument fails is entirely due to the anomalously small value of the coefficient Γ [Eq. (4)]. Suppose we define a momentum-transfer coefficient K between the bulk phases and the walls by $\dot{P} \sim K v_n$, where $\dot{P}$ is the momentum transferred per unit time between the bulk A and B liquids and the walls and v_n is a typical value of the normal velocity in the bulk. A straightforward hydrodynamic calculation[9] shows that in a tube of radius R, K is at least of order[15] ηR if $R \lesssim l(T)$ and of order $\eta R^2 / l(T)$ in the opposite case, where η is the ordinary normal-fluid viscosity. Thus the ratio of ΓR^2 to K is at most of order $(\Delta/\epsilon_F)^2$, and since these two quantities play the role of two conductances in series it follows that the smaller, ΓR^2, totally dominates and so the bulk normal velocity is only of order $(\Delta/\epsilon_F)^2 v \lesssim (\Delta/\epsilon_F)v_c \ll v_c$. We conclude that provided there was no substantial two-fluid counterflow initially in the A phase, the expression (4) is an excellent approximation to the actual friction on the boundary.

In the limit of low temperature ($k_B T \ll \Delta_B$) we can set $\bar{T} = 0$ in Eq. (4) without appreciable error, and the right-hand side is then also independent of the cutoff provided that θ_0, the angle made by the l vector with $\hat{n}$, is not too close to $\pi/2$. In fact, in this limit Γ is just equal to $[2E_A(T)/v_F]|\sec\theta_0|$, where $E_A(T)$ is the thermal energy of the A phase, and thus is proportional to T^4 (though see below). In the region of the

experiments of Buchanan, Swift, and Wheatley[8] ($T/T_c \gtrsim 0.6$), explicit evaluation of Eq. (4) requires a knowledge of the transmission coefficient $\bar{T}$. For present purposes we shall make the simple *Ansatz* $\bar{T} = \theta(E - \Delta_{max}(\theta, \phi))$. With this approximation (which will if anything underestimate Γ and hence overestimate v_{AB}, the error decreasing with distance from T_{AB}), and the experimental values[16] of $\Delta G_{AB}(P,T)$ (with $H = 0$), we obtain the terminal velocity v_{AB} of the boundary as a function of temperature; the comparison with the (lower field) data of Buchanan, Swift, and Wheatley[8] is shown in Fig. 1.[17] The apparently impressive quantitative agreement of the melting-pressure "perpendicular" curve with the 33.6-bar data may be an accident, since under the conditions of the experiment one might *prima facie* expect l to lie predominantly *across* the tube, i.e., parallel to the boundary (cf. Ref. 2, Sec. X): on the time scale of the experiment there is, of course, no time for l to adjust (cf. Ref. 13). However, it is clear that both the order of magnitude and the general trend of v_{AB} are in good agreement with the theory. (On the rapid upturn of the 27-bar theoretical curve, see footnote 16.) Note that from the point of view of the theory, the pressure independence of the limiting slope of v_{AB} as a function of $\Delta T/T_{AB}$ may be a numerical accident: In fact for P close to the polycritical pressure P_c, we expect this slope to vary as $P - P_c$.

The theory developed above is appropriate to the temperature regime of the experiment ($T/T_c \gtrsim 0.6$). However, we should expect that at the lowest temperatures the terminal velocity would be limited by a quite different mechanism, namely "pair breaking" (excitation of the Cooper pairs from the "ground pair"[2] to

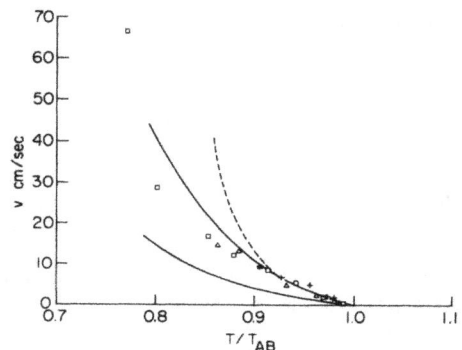

FIG. 1. Velocity of the phase boundary $v \equiv v_{AB}$ as a function of the reduced temperature $T/T_{AB}(P)$. Data points are from Ref. 8. Full curves, theoretical results at melting pressure (upper and lower curves are for $\hat{l}$ respectively normal and parallel to the plane of the phase boundary). Broken curve, theoretical result at 27 atm with $\hat{l}$ normal to boundary. All curves are for $H = 0$.

VOLUME 57, NUMBER 3   PHYSICAL REVIEW LETTERS   21 JULY 1986

the "excited pair" state) by the time-varying self-consistent pair field of the moving boundary. A straightforward order-of-magnitude calculation[9] using adiabatic perturbation theory gives at $T = 0$ (for $v \ll v_F$ and $\hbar\omega \ll \Delta$) a damping $\Gamma(v{:}\omega)v$, where

$$\Gamma(v{:}\omega) \sim (dn/d\epsilon)(\Delta^2/v_F)(\xi/d)^4(v/v_F)^3$$

for the freely moving boundary (provided $v \gg v_c$ and the superfluid is at rest with respect to the walls) and $\Gamma(v{:}\omega) \sim (dn/d\epsilon)(\Delta^2/v_F)(\hbar\omega/\Delta)^3$ for the pinned oscillating case. Thus, for example, if we define the quantity $\lambda \equiv \Delta G_{AB}(P,0)/(\Delta_B^2\, dn/d\epsilon)$ ($\sim 10^{-2}$ at the melting pressure in zero field), we find that the limiting low-temperature value of the terminal velocity should be of order $\lambda^{1/4}(d/\xi)v_F$ (i.e., possibly comparable to v_F itself), and should be attained at a temperature of order $\lambda^{3/16}(\xi/d)^{1/4}T_c$; we hope to give a more quantitative estimate elsewhere.

We have also estimated the inertial mass of the boundary by assuming that the "kinetic energy" associated with the time variation of the order parameter which its motion induces is of the same order as in a homogeneous situation. In this way we find[9] $M^* \sim \sigma/v_F^2 \sim 10^{-12}$ g/cm², where σ is the surface tension. Note that this is only a fraction of order $(\Delta/\epsilon_F)^2$ of the total mass of the liquid in the boundary region; this is not surprising since, when the boundary moves, only a very small fraction of the atoms actually change their state.

With the above estimates for M^* and Γ, it is straightforward to obtain a qualitative idea of the behavior of the boundary in many situations of experimental interest.[9] For example, if the boundary is pinned at low temperatures in an aperture of radius R ($\gg d$), it is easy to see that the fundamental mode will have a frequency of order $v_F R^{-1}$, and will be underdamped provided $(T/T_c)^4 \ll \xi/R$, a condition which for reasonable R, say 100 μm, is probably just within the range of existing cryogenic technology.

We are very grateful to Scott Buchanan, Greg Swift, and the late John Wheatley for many helpful discussions and for keeping us continuously informed of the progress of their experiment. We also enjoyed helpful discussions with Gordon Baym and Chris Pethick. This work was supported by the National Science Foundation under Grant No. DMR 83-15550.

[1]J. C. Wheatley, Rev. Mod. Phys. **47**, 415 (1975) (experimental review).

[2]A. J. Leggett, Rev. Mod. Phys. **47**, 331 (1975) (theoretical review).

[3]D. D. Osheroff and M. C. Cross, Phys. Rev. Lett. **38**, 905 (1977).

[4]M. C. Cross, in *Quantum Fluids and Solids 1977*, edited by S. B. Trickey, E. D. Adams, and J. W. Dufty (Plenum, New York, 1977).

[5]R. Kaul and H. Kleinert, J. Low Temp. Phys. **38**, 539 (1980). Note that this calculation is valid in the Ginzburg-Landau region; the resulting value of d/ξ (~ 2.5) may not be applicable at low temperature.

[6]S. Yip, Phys. Rev. B **32**, 2915 (1985).

[7]The oscillations of the boundary at $T = 0$ are discussed by A. V. Markelov, Pis'ma Zh. Eksp. Teor. Fiz. **42**, 151 (1985) [JETP Lett. **42**, 186 (1985)] using concepts apparently quite different from ours. We also note the interesting similarities and differences between the present problem and that of the ⁴He solid-superfluid phase-boundary dynamics [see, e.g., R. M. Bowley and D. O. Edwards, J. Phys. (Paris) **44**, 723 (1983)].

[8]D. S. Buchanan, G. W. Swift, and J. C. Wheatley, preceding Letter [Phys. Rev. Lett. **57**, 341 (1985)].

[9]S. Yip and A. J. Leggett, unpublished.

[10]Condition (1) fails very close to T_c ($1 - T/T_c \lesssim 10^{-4}$). Condition (2) may be marginal for free expansion of the B phase at very low temperatures; cf. below. Condition (3) would fail, e.g., for pinning in ultrasmall orifices, or high-frequency modes in larger ones.

[11]N. A. Greaves and A. J. Leggett, J. Phys. C **16**, 4383 (1983).

[12]A simple estimate based on the thermal-impedance results of Ref. 6 shows that the boundary acts as an almost perfectly diathermal barrier, except at the lowest temperatures where the temperature dependence of G is negligible anyway. Also, the heating effect due to the latent heat is negligible except for (say) $1 - T/T_{AB} \lesssim 10^{-2}$; see Ref. 8.

[13]D. N. Paulson, R. L. Kleinert, and J. C. Wheatley, in *Proceedings of the Fourteenth International Conference on Low Temperature Physics*, edited by M. Krusius and M. Vuorio (North-Holland, Amsterdam, 1975), Vol. 5, p. 417; M. C. Cross and P. W. Anderson, *ibid.*, Vol. 1, p. 29.

[14]A. F. Andreev, Zh. Eksp. Teor. Fiz. **46**, 1823 (1964) [Sov. Phys. JETP **19**, 1228 (1964)].

[15]This estimate fails at very low temperatures in very wide tubes, but by that time other mechanisms dominate the friction anyway; cf. below.

[16]At melting pressure we used the data of W. P. Halperin *et al.*, Phys. Rev. B **13**, 2124 (1976); at 27 atm, the data given in Fig. 16 of Ref. 1. Note that extrapolation of the latter almost certainly overestimates ΔG_{AB} and hence v_{AB} for T appreciably below T_{AB}.

[17]Note that at general temperatures there is no reason for Γ to be a monotonic function of θ, even in the indicated approximation; indeed, an analytic calculation (Ref. 9) in the limit $T \to T_{AB} \to T_c$ shows that in this limit the value of Γ at $\theta_0 = \pi/2$ is nearly equal to that at $\theta_0 = 0$, but the value at $\sin^{-1}3^{-1/2}$ is about half the latter. This feature, and more generally the strong anisotropy of Γ, raise the intriguing possibility that the boundary may spontaneously roughen and then propagate by a "tacking" mechanism.

LIST OF PUBLICATIONS

(a) Papers

1. "On the Theory of He4 Impurities in He3" (with D. ter Haar). Physics Letters 11, 129 (1964).

2. "Spin Susceptibility of a Superfluid Fermi Liquid". Phys. Rev. Lett. 14, 536 (1965).

3. "Finite Linewidths and 'Forbidden' Three-Phonon Interactions" (with D. ter Haar). Phys. Rev. 139A, 779 (1965).

4. "Theory of a Superfluid Fermi Liquid. I. General Formalism and Static Properties". Phys. Rev. 140A, 1869 (1965).

5. "Theory of a Superfluid Fermi Liquid. II. Collective Oscillations". Phys. Rev. 147, 1, 119 (1966).

6. "Persistence of Zero Sound in the Possible Superfluid Phase of He3". Prog. Theor. Phys. (Kyoto) 36, 417 (1966).

7. "Number-Phase Fluctuations in Two-Band Superconductors". Prog. Theor. Phys. (Kyoto) 36, 901 (1966).

8. "Electron-Phonon Interactions in the Transition Metals". Prog. Theor. Phys. (Kyoto) 36, 931 (1966).

9. "Inequalities, Instabilities and Renormalization in Metals and Other Fermi Liquids" Ann. Phys. (NY) 46,76 (1968).

10. "Spin Echoes in Liquid Helium-3 and Mixtures: A Predicted New Effect" (with M. J. Rice). Phys. Rev. Lett. 20, 586 (1968).

11. "Spin Diffusion and Spin Echoes in Liquid 3He at Low Temperatures". J. Phys. C 3, 448 (1970).

12. "On the Anomalous CMN-He3 Thermal Boundary Resistance" (with M. Vuorio). J. Low Temp. Phys. 3, 359 (1970).

13. "Mass and Spin Diffusion Near the Lambda Line in Dilute 3He-4He Mixtures" (with M. A. Eggington). J. Low Temp. Phys. 5, 275 (1971).

14. "Can a Solid be 'Superfluid'?" Phys. Rev. Lett. 25, 1543 (1970).

15. "Interpretation of Recent Results on He3 below 3 mK: A New Liquid Phase?". Phys. Rev. Lett. 29, 1227 (1972).

16. "On the Minimum Entropy of a Large System at Low Temperatures". Ann. Phys. (NY) 72, 1, 80 (1972).

17. "Microscopic Theory of NMR in an Anisotropic Superfluid (3HeA)". Phys. Rev. Lett. 31, 352 (1973).

18. "Topics in the Theory of Helium". Physica Fennica 8, 125 (1973).

19. "NMR Lineshifts and Spontaneously Broken Spin-Orbit Symmetry. I. General Concepts". J. Phys. C 6, 3187 (1973).

20. "Is ODLRO a Necessary Condition for Superfluidity?" (with M. A. Eggington). Collective Phenomena 2,81 (1975).

21. "Implications of the 3He Phase Diagram Below 3 mK". Prog. Theor. Phys. (Kyoto) 51, 1275 (1974).

22. "Spin Dynamics of an Anisotropic Fermi Superfluid (3He)". Ann. Phys. (NY) 85, 11 (1974).

23. "A Theoretical Description of the New Phases of Liquid 3He". Revs. Mod. Phys. 47, 331 (1975).

24. "NMR in 3He-A and 3He-B: The Intrinsic Relaxation Mechanism" (with S. Takagi). Phys. Rev. Lett. 34, 1424 (1975).

25. "Ringing-Down of the 'Wall-Pinned' Mode in 3He-B as a Test of Theories of Spin Relaxation". Phys. Rev. Lett. 35, 1178 (1975).

26. "The New Phases of Liquid 3He". Comments on Solid State Physics VII,45 (1976).

27. "Orbital Dynamics of 3He-A" (with S. Takagi). Phys. Rev. Lett. 36, 1379 (1976).

28. "Orientational Dynamics of Superfluid 3He: A 'Two-Fluid' Model. I. Spin Dynamics with Relaxation" (with S. Takagi). Ann. Phys. (NY) 106, 79 (1977).

29. "Macroscopic Parity Violation Due to Neutral Currents?" Phys. Rev. Lett. 39, 587 (1977).

30. "Superfluid 3He-A is a Liquid Ferromagnet". Nature 270, 585 (1977).

31. "Orientational Dynamics of Superfluid 3He: A 'Two-Fluid' Model. II.Orbital Dynamics" (with S. Takagi). Ann. Phys. (NY) 110, 353 (1978).

32. "Macroscopic Effect of P- and T-Nonconserving Interactions in Ferroelectrics: A Possible Experiment?" Phys. Rev. Lett. 41, 586 (1978).

33. "Macroscopic Quantum Systems and the Quantum Theory of Measurement". Prog. Theor. Phys. (Kyoto), Supplement No. 69, 80 (1980).

34. "Influence of Dissipation on Quantum Tunnelling in Macroscopic Systems" (with A. O. Caldeira). Phys. Rev. Lett. 46, 211 (1981).

35. "Atomic Hydrogen in an Inhomogeneous Magnetic Field: Density Profile and Bose–Einstein Condensation) (with V. V. Goldman and I. F. Silvera). Phys. Rev. B 24, 2870 (1981).

36. "Comment on `Probabilities for Quantum Tunnelling Through a Barrier with Linear Passive Dissipation'" (with A. O. Caldeira). Phys. Rev. Lett. 48, 1571 (1982).

37. "Quasiparticle Ballistics, Textural Andreev Reflection and Low Temperature Transport In 3He-A" (with N. A. Greaves). J. Phys. C 16, 4383 (1983).

38. "Quantuum Tunnelling in a Dissipative System" (with A. O. Caldeira). Ann. Phys. (NY) 149, 374 (1983). Erratum, ibid. 153, 445 (1984).

39. "Path Integral Approach to Quantum Brownian Motion" (with A. O. Caldeira). Physica 121A, 587 (1983). Erratum, ibid. 130A, 374(1985).

40. "Dynamics of the Two-State System with Ohmic Dissipation" (with S. Chakravarty). Phys. Rev. Lett. 52, 5 (1984).

41. "Macroscopic Quantum Tunnelling and All That" in Essays in Theoretical Physics in Honor of Dirk ter Haar, Pergamon, Oxford (1984), p. 95.

42. "Quantum Tunnelling in the Presence of an Arbitrary Linear Dissipation Mechanism". Phys. Rev. B 30, 1208 (1984).

43. "Comment on `Bell's Theorem: Does the Clauser-Horne Inequality Hold for All Local Theories?'" (with A. Garg). Phys. Rev. Lett. 53, 1019 (1984).

44. "Nucleation of 3He-B from the A Phase: A Cosmic-Ray Effect?" Phys. Rev. Lett. 53, 1096 (1984).

45. "Response to Comment of Hakonen *et al.*". Phys. Rev. Lett. 54, 246 (1985).

46. "Influence of Damping on Quantum Interference: An Exactly Soluble Model" (with A. O. Caldeira). Phys. Rev. B3, 1059 (1985).

47. "Some Recent Applications of Instanton and Related Techniques in Condensed-Matter Physics". Prog. Theor. Phys. Suppl. 80, 10 (1985).

48. "Quantum Mechanics Versus Macro-Realism: Is the Flux there = 0B when Nobody Looks?" (with A. Garg). Phys. Rev. Lett. 54, 857 (1985).

49. "Response to Comment of Kraichnan" (with A. Garg). Phys. Rev. Lett. 54, 2724 (1985).

50. "Dissipative Quantum Tunnelling at Finite Temperatures" (with D. Waxman). Phys. Rev. B32, 4450 (1985).

51. "Quantum Mechanics at the Macroscopic Level" in Directions in Condensed Matter Physics, (Memorial Volume for Shang-Keng Ma)(1986), ed. G. Grinstein and G. Mazenko, World Scientific, Singapore, 1986, p. 87.

52. "Dynamics of the 3He A-B Phase Boundary" (with S. Yip). Phys. Rev. Lett. 57, 345 (1986).

53. "Dynamics of the Dissipative Two-State System" (with S. Chakravarty, A. T. Dorsey, M. P. A. Fisher, A. Garg and W. Zwerger). Revs. Mod. Phys. 59, 1 (1987).

54. "Reflections on the Quantum Measurement Paradox" in Quantum Implications: Essays in Honor of David Bohm, ed. B. J. Hiley and F. D. Peat (Routledge and Kegan Paul, London and New York (1987), p. 85.

55. "Comment on `Realism and Quantum Flux Tunnelling" (with A. Garg). Phys. Rev. Lett.

56. "Experimental Approaches to the Quantum Measurement Paradox". Foundations of Physics, 18, 939 (1988).

57. "Return of A Hysteretic Josephson Junction to the Zero-Voltage State: I-V Characteristics and Quantum Retrappoing" (with Y. C. Chen and M. P. A. Fisher). J. Appl. Phys., 64, 3199, 1988.

58. "Comment on `How the Result of a Measurement of a Component of the Spin of a Spin-1/2 Particle Can Turn out to be 100' ". Phys. Rev. Lett. 62, 2325 (1989).

59. "Can Solid-State Effects Enhance the Cold Fusion Rate?"(with G. Baym) Nature 340, 45 (1989).

60. "Exact Upper Limit on Barrier Penetration Probabilities in Many-body Systems: Application to 'Cold Fusion'" (with G. Baym), Phys. Rev. Lett. 63, 191 (1989).

61. "Nucleation and Growth of 3He-B in the Supercooled A Phase" (with S. Yip), in Superfluid 3-He, ed. L. P. Pitaevskii and W. P. Halperin (North-Holland, Amsterdam) (1990).

62. "Low Temperature Properties of Amorphous Materials: Through a Glass Darkly" (with C. C. Yu), Comments on Condensed Matter Physics 14, 231 (1988).

63. "Quantum Mechanics and Macroscopic Realism" in Trends in Theoretical Physics, ed. P. J. Ellis and Y. C. Tang (1990).

64. "On the Concept of Spontaneously Broken Gauge Symmetry in Condensed Matter Physics" (with F. Sols), Foundations of Physics, 21, 353 (1991).

65. "Comment on 'Quantum Limitations on the Measurement of Magnetic Flux" Phys. Rev. Lett. 63, 2159 (1989).

66. "Sign of the Coupling between T-violating Groundstates in Second-order Perturbation Theory" (with A. G. Rojo), Phys. Rev. Lett. 67, 3614 (1991).

67. "On the Nature of Research in Condensed Matter Physics", Foundations of Physics, 22, 221 (1992).

68. "Inertial Mass of a Moving Singularity in a Fermi Superfluid" (with J. M. Duan), Phys. Rev. Lett. 68, 1216 (1992).

69. "Some Aspects of c-axis Coupling and Transport in the Copper Oxide Superconductors", Brazil. Journal of Physics 22, 129 (1992).

70. "Nucleation of the AB Transition in Superfluid 3He: Experimental and Theoretical Considerations" (with P. Schiffer and D. D. Osheroff), Prog. Low Temp. Phys. Vol. XIV, ed. W. P. Halperin and L. P. Pitaevskii, (Elsevier, Amsterdam, 1995), pp. 159–211.

71. "Experimental Determination of the Superconducting Pairing State In YBCO from the Phase Coherence of YBCO-Pb SQUIDS" (with D. A. Wollman, D. J. Van Harlingen, W. C. Lee and D. M. Ginsberg), Phys. Rev. Lett. 71, 2134 (1993).

72. "Tensor Magnetothermal Resistance in YBa2Cu307-x via Andreev Scattering of Quasiparticles" (with F. Yu, M. B. Salamon, W. C. Lee and D. M. Ginsberg), Phys. Rev. Lett., 74, 5136 (1995).

73. "Is 'Relative Quantum Phase' Transitive?", Foundations of Physics 25, 113 (1995).

74. "Time's Arrow and the Quantum Measurement Problem", in Time's Arrows Today, ed. S. Savitt, Cambridge University Press, 1995.

75. "Response to Klemm" (with D. A. Wollman and D. J. Van Harlingen), Phys. Rev. Letters 73, 1871 (1994).

76. "Response to Klemm et al." (with F. Yu and M. B. Salamon), Phys. Rev. Letters 77, 3058 (1995).

77. "Josephson experiments on the High-Temperature Superconductors", Phil. Mag. B 74, 509 (1996).

78. "Experimental Constraints on the Pairing State of the Cuprate Superconductors: An Emerging Consensus" (with J. F. Annett and N. D. Goldenfeld), in D. M. Ginsberg, ed., Physical Properties of high Temperature Suprerconductors, Vol. V, World Scientific, Singapore, 1995.

79. "Interlayer Tunnelling Models of Cuprate Superconductivity: Implications of a Recent Experiment", Science 274, 587–589 (1996).

80. "Some Properties of a Spin-1 Fermi Superfluid with Attractive Interaction" (with A. G. K. Modawi), J. Low Temp. Phys. 109, 625–639 (1997).

81. "Nonlocal Effects on the Magnetic Penetration Depth in d-wave Superconductors" (with I. Kosztin) Phys. Rev. Letters 79, 135–138 (1997).

82. "Free Energy of an Inhomogeneous Superconductor: A Wave Function Approach" (with I. Kosztin, S. Kos and M. Stone), Phys. Rev.B 58, 9365–9384 (1998).

83. "Josephson Effect between Trapped Bose–Einstein Condensates" (with I. Zapata and F. Sols) Phys. Rev.A 57, R 28-31 (1998).

84. "Macroscopic Quantum Tunnelling of a Bose Condensate with Attractive Interaction" (with M. Ueda), Phys. Rev. Letters 82, 1576–1579 (1998).

85. "On the Superfluid Density of an Arbitrary Many-Body System at T=0", J. Stat. Phys. 93, 927–941 (1998).

86. "Comment on `Phase and Phase Diffusion of a Split Bose–Einstein Condensate" (with F. Sols), Phys. Rev. Letters 81, 1344 (1998)

87. "Ueda and Leggett Reply" (with M. Ueda), Phys. Rev. Letters 81, 1342 (1998)

88. "Cuprate Superconductivity:Dependence of T_c on the c-axis Layering Structure" Phys. Rev. Letters 83, 392–395 (1999).

89. "A 'Mid-Infrared' Scenario for Cuprate Superconductivity", Proc. Nat. Acad. Sciences 96, 8365–8372 (1999).

90. "Superfluidity", Revs. Mod. Phys. 71, S318-323 (1999).

91. "The Physical Basis of 3-He A-B Nucleation" (with P. Schiffer and D. D. Osheroff), Phys. Rev. Letters 82, 395 (1999)

92. "Some Thought-Experiments involving Macrosystems as Illustrations of Various Interpretations of Quantum Mechanics", Found. Phys. 29, 445 (1999).

93. "Ground-state properties of a rotating Bose–Einstein condensate with attractive interaction" (with M. Ueda), Phys. Rev. Letters 83, 1489 (1999).

94. "Leggett replies", Phys. Rev. Letters 85, 3984 (2000).

95. "Interlayer c-axis transport in the normal state of cuprates" (with M. Turlakov), Phys. Rev. B 63, 064518/1-8 (2001).

96. "Bose Einstein Condensation in the Alkali Gases: some Fundamental Concepts" Revs. Mod. Phys. 73, 307–356 (2001).

97. "An upper bound on the condensate fraction in a Bose gas with weak repulsion", New J. Phys. 3, 23 (2001).

98. "The Josephson plasmon as a Bogoliubov quasiparticle" (with G. S. Paraoanu, S. Kohler and F. Sols, J. Phys, B 34,4689 (2001).

99. "Probing quantum mechanics towards the macroscopic world: Where do we stand?", Physica Scripta T102.69 (2002).

100. "Condensation energy and high-temperature superconductivity" (with D. van der Marel, J. W. Loram and J. R. Kirtley), Phys. Rev.B 66, 140501 (2002).

101. "Measurement theory and interference of spinor Bose–Einstein condensates" (with S. Ashhab), Phys. Rev.A 65, 023604-1-10 (2002).

102. "High energy low temperature physics: Production of phase transitions and topological defects by energetic particles in superfluid 3-He", J. Low Temp. Phys. 126, 775 (2002).

103. "Testing the limits of quantum mechanics: Motivation, state ofplay, prospects", J. Phys: Cond. Mat. 14, R415 (2002).

104. "Philippe Nozieres: Feenberg medallist 2001: Microscopic and phenomenological foundations of the theory of quantum many-bony systems" (with E. Krotscheck and J. W. Negele), Int. J. Mod. Phys. B 17, 4947 (2003).

105. "Anomalous diffusion near the Fermi surface" (with G. L. Warner), Phys. Rev. B 68, 174516-1-7 (2003).

106. "Phase dynamics after connection of two separated Bose–Einstein condensates" (with I. Zapata and F. Sols), Phys. Rev. A 67, 21603-1-4 (2003).

107. "Bose–Einstein condensation in a harmonic trap: Effect of interactions on T_c", J. Stat. Phys. 110, 903 (2003).

108. "Sum rule analysis of umklapp processes and Coulomb energy: Application to cuprate superconductivity" (with M.Turlakov), Phys. Rev. B 67, 94517-1-8 (2003).

109. "The relation between the Gross-Pitaevskii and Bogoliubov descriptions of a dilute Bose gas", New J. of Physics 5,103 (2003).

110. "Quench dynamics of a superfluid Fermi gas" (with G. L. Warner), Phys. Rev. B 71,134514 (2003).

111. "Nonlocal hidden variables and quantum mechanics: An incompatibility theorem", Found. Phys. 33, 1469 (2003); erratum, ibid.

112. "Nuclear magnetic esonance in ultra-small samples of superfluid 3-He", Synthetic Metals 141, 51 (2004).

113. "Some thoughts about two-dimensionality and superconductivity", J. of Super-conductivity and Novel Magnetism 19, 187 (2006).

114. "Probing quantum mechanics towards the macroscopic world: How far have we come?", Prog. Theor. Phys.supplement 170,100 (2007).

115. "How far do EPR-Bell experiments constrain physical collapse theories?", J. Phys. A 40,12 (2007).

116. "Bose–Einstein condensation of spin-1/2 atoms with conserved total spin" (with S. Ashhab), Phys. Rev. A 68, 63612-1-7 (2008).

117. "Realism and the physical world", Reps. Prog. Phys. 71, 022001-1-6 (2008).

118. "Sum-rule analysis of radio-frequency spectroscopy of ultra-cold Fermi gas" (with Shizhong Zhang), Phys. Rev. A 73033614 (2008).

119. "Universal properties of the unitary Fermi gas" (with Shizhong Zhang), Phys. Rev. A 79, 023601 (2009)

120. "Spin polarization of half-quantum vortices in systems with equal spin pairing" (with V. Vakaryuk), Phys. Rev. Letters 103, 057003 (2009).

121. "Comment on "Possible experience: From Boole to Bell" by K. Hess *et al.*" (with A. Garg), Europhysics Letters 81, 40001 (2010).

122. "The superfluid phases of 3-He: BCS theory", Mod. Phys. Letters 24, 525 (2010).

123. "Universal sound absorption in amorphous solids: A theory of elastically coupled generic blocks" (with D. C. Vural), J. Noncrystalline Solids 357, 3528 (2011).

124. "Absence of spontaneous magnetic order of lattice spins coupled to itinerant electrons in one or two dimensions" (with D. Loss and F. L. Pedrocchi), Phys. Rev. Letters 107, 107201 (2011).

125. "BEC-BCS crossover with Feshbach resonance for a three-hyperfine-species model" (with Guojun Zhu), Phys. Rev. A 87, 023627 (2013).

126. "Tunnelling two-level systems" model of the low-temperature properties of glasses: Are "smoking-gun" experiments possible?" (with D. C. Vural), J. Phys. Chem. 117, 12766 (2013).

127. "Andreev bound states:some semiclassical reflections" (with Yiruo Lin), JETP (2014).

(b) Conference Contributions, Published Lecture Notes, Journalism, etc.

1. "Fermi-Liquid Effects in the Superfluid Phase". Quantum Fluids, ed. D. F. Brewer, North-Holland, Amsterdam (1966).

2. "Spin Echoes in Very Degenerate Fermi Systems". Proc. 11th Int. Conf. on Low Temperature Physics, St. Andrews (1966).

3. "Application of the Fermi-Liquid Theory to Metals" in Theory of Metals and the Many-Body Problem, Proc. of the IXth Winter School of Theoretical Physics in Karpacz, Acta Universitatis Wratislaviensis 181, Wroclaw (1972).

4. Critical Review of I. Lakatos and A. Musgrave "Criticism and the Growth of Knowledge", Second Order (Ife, Nigeria) I, 80 (1972).

5. "NMR in an Anisotropic Superfluid (3He)". Proc. 24th Nobel Symposium, Aspen=E4sgarden, Sweden (1973).

6. "Fermi-Ekitai, Hitoohooteki Tyooryuudootai, Ekitai 3He no Atarasii So ni tuite" (with S. Takagi and Y. Ono) (in Japanese) Bussei Kenkyuu 23, 275 and 339 (1974).

7. "The New Phases of 3He". Proc. EPS Conf. on Liquid and solid Helium, Haifa, Israel (1975).

8. "New Phases of Liquid 3He". Physics Bulletin 225, 92 (1974).

9. "NMR in the New Phases of 3He: A Unique Situation?". Proc. 18th Ampere Congress, Nottingham (1974) p. 31.

10. Low Temperature Phases of Helium". Physics Bulletin 26, 311 (1975).

11. "The New Phases of Liquid 3He" (invited plenary talk). Proc. 14th Int. Conf. on Low Temperature Physics, ed. M. Krusius and M. Vuorio, North-Holland (1975) Vol. 5, p. 52.

12. "The New Phases of 3He". Endeavor, May 1975.

13. "Friendly Fermions of Helium-Three". New Scientist, 7 Oct 1976, p. 38.

14. "The 'Arrow of Time' and Quantum Mechanics" in The Encyclopaedia of Ignorance, ed. M. Weston-Smith and R. Duncan, Pergamon Press, Oxford (1977).

15. "Prospects in Ultralow Temperature Physics" (invited plenary talk). Proc. 15th Int. Conf. on Low Temperature Physics, J. de Physique Colloque C6, 1264 (1978).

16. "Diatomic Molecules and Cooper Pairs". Proc. XVI Karpacz Winter School, published in Modern Trends in the Theory of Condensed Matter, 115, 13, Springer-Verlag (1979).

17. "Cooper Pairing in Spin Polarized Fermi Systems". Proc. Aussois Conf. on Spin Polarized Quantum Systems. J. de Physique Colloque C7, 19 (1980).

18. "Quantum Tunnelling and Noise in SQUIDS". Proc. VI Int. Conf. on Noise in Physical Systems, Gaithersburg, MD (1981), NBS Special Publication No. 614, p. 355.

19. "The Unique Liquid, 3He" (11th Fritz London Memorial Award Lecture) Proc. 16th Int. Conf. on Low Temperature Physics, Physica 109 B&C, 1393,(1982).

20. "Quasiparticle Ballistics in 3He-A" (with N. A. Greaves) in Quantum Fluids and Solids — 1983 (AIP Conf. Proc. No. 103), 254 (1983).

21. "Macroscopic Quantum Tunnelling and Related Effects in Josephson Systems" in Proc. NATO ASI on Percolation, Localization and Superconductivity, Les Arcs, France, 1983, Pergamon (1984).

22. "The Superposition Principle in Macroscopic Systems" in Proc. Int. Symp. on the Foundations of Quantum Mechanics in the Light of New Technology, ed. S. Kamefuchi *et al.,* Physical Society of Japan (1984).

23. "Schroedinger's Cat and Her Laboratory Cousins". Contemp. Phys. 25, 583 (1984). (Japanese translation: "Schroedinger no neko to doozokutati", Kagaku 54, 693 and 761 (1984).)

24. "Quantum Mechanics at the Macroscopic Level" in The Lesson of Quantum Theory, ed. J. de Boer, E. Dal and O. Ulfbeck, (Proc. Niels Bohr Centennial Symposium, Copenhagen, October, 1985), North-Holland, Amsterdam (1986), p. 24.

25. "Quantum Mechanics and Realism at the Macroscopic Level: Is an Experimental Discrimination Feasible?" in New Techniques and Ideas in Quantum Measurement Theory, Annals of the New York Academy of Sciences, New York, Vol. 480 (1986), p. 21.

26. "Superfluidity of 3He and Heavy Fermions". J. Magn. Magn. Mat. 63 & 64,406 (1987).

27. "The Current Status of Quantum Mechanics at the Macroscopic Level", Proc. 2nd Int. Symposium on the Foundations of Quantum Mechanics in the Light of New Technology, ed. M. Namiki, Y. Ohnuki, Y. Murayama and S. Nomura, Phys. Soc. Japan, Tokyo (1987).

28. "Velocity of Propagation of the 3He A-B Transition" (with S. Yip). Can. J. Phys. 65, 1514 (1987).

29. "Quantum Mechanics at the Macroscopic Level" in Chance and Matter, ed. J. Souletie, J. Vannimenus and R. Stora, North-Holland, Amsterdam (1987).

30. "Macroscopic Quantum Tunnelling and Related Matters" (invited plenary talk), Proc. 18th Int. Conf. on Low Temperature Physics, Jpn. J. Appl. Phys. 26, Suppl. 26-3, 1986 (1987).

31. "The Quantum Mechanics of a Macroscopic Variable: Some Recent Results and Current Issues" in Frontiers and Borderlines in Many-particle Physics. Soc. Italiana di Fisica, Bologna (1988).

32. "Superconductivity and Superfluidity" in The New Physics, ed. P. C. W. Davies,Cambridge University Press (1989).

33. "Quantum and Classical Concepts at the One-electron Level", in Nano-structure Physics and Fabrication, ed. Mark A. Reed and Wiley P. Kirk, Academic Press, San Diego, 1990.

34. "Quantum mechanics of Complex Systems, I and II" in Proc. 1989 NATO Summ. School, Evora, Portugal (Plenum Press, 1990).

35. "The 'Cold Fusion' Problem", ibid.

36. "Dephasing and Non-dephasing Collisions in Nanostructures" in Physics of Granular Nanoelectronics", ed. D. K. Ferry, J. Barber and C. Jacoboni NATO ASI Series, 1990.

37. "Some Considerations Related to the Quantization of Charge in Mesoscopic Systems", ibid.

38. "Amorphous Materials at Low Temperatures: Why are they so similar?" in Proc. LT19, Physics B 169, 322 (1991).

39. "Superfluidity and Superconductivity", in Low Temperature Physics, ed. M. J. Hoch and R. H. Lemmer, Springer-Verlag, Berlin (1991).

40. "High Magnetic Fields and Ultralow Temperatures in Superfluid 3He", Proc. Conf. on Physical Phenomena in High Magnetic Fields, Tallahassee, May 1991, ed. E. Manousakis *et al.*, Addison-Wesley (1991).

41. "The 3He A-B Interface", J. Low Temp. Phys. 87, 571 (1992).

42. "The Effect of Dissipation on Tunnelling", Proc. 4th Int'l. Symposium on Foundations of Quantum Mechanics in the Light of New Technology, ed. M.Tsukada *et al.*, JJAP, Tokyo (1992).

43. "D-wave Superconductivity: the Lifetime Problem", Physica B&C 199–200, 291 (1994).

44. "Broken Symmetry in a Bose Condensate", in Bose–Einstein Condensation, ed. A. Griffin, D. W. Snoke and S. Stringari, Cambridge University Press, New York (1994).

45. "Phase-Sensitive Tests of the Symmetry of the Pairing State in the High-Temperature Superconductors-Evidence for dx2-y2 Symmetry in YBCO" (with D. J. Van Harlingen, D. A. Wollman and D. M. Ginsberg), Physica C 235, 122 (1994).

46. "Present and Future of Macroscopic Quantum Coherence", in Phenomenology of Unification from Present to Future, ed. G. Diambrini-Palazzi *et al.*, World Scientific, Singapore (1994).

47. "Macroscopic Quantum Effects in Magnetic Systems: an Overview", Proc. NATO Workshop on Quantum Tunnelling in Magnetic Systems, ed. L. Gunther (in press).

48. "As a Martian Might See Us: Subversive Reflections on the Practice of Physics", Current Science (India).

49. "Superfluids and Superconductors", in History of Twentieth Century Physics, ed. L. Brown, A. Pais and A. B. Pippard, American Physical Society and Institute of Physics, London (in press).

50. "Phase coherence in YBCO-Pb SQUIDs: direct experimental determination of the symmetry of the pairing state in high temperature superconductors" (with D. A. Wollman, D. J. Van Harlingen, W. C. Lee and D. M. Ginsberg), Physica B 194, 1669 (1994).

51. "Constraints on the pairing state of the cuprate superconductors" (with J. F. Annett and N. D. Goldenfeld), J. Low Temp.Phys.105, 473 (1996).

52. "The Paired Electron", in Electron, ed. M. Springford, Cambridge University Press, Cambridge, UK 1997, pp. 148–181.

53. "The Symmetry of the Order Parameter in the Cuprate Superconductors" (with J. F. Annett and N. D. Goldenfeld), Proc. Tenth Anniversary HTS Workshop on Physics, Materials and Applications, ed. B. Batlogg *et al.*, World Scientific, Singapore 1996, pp. 63–67.

54. "How can we use low-temperature systems to throw light on questions of more general interest?", J. Low Temp. Phys.110, 719–728 (1998).

55. "Bose-condensed alkali gases as testbeds for fundamental quantum mechanics", Physica Scripta T76, 199–202 (1998).

56. "Coherent atomic beams from Bose condensates",in Quantum Aspects of Beam Physics, ed. Pisin Chen, World Scientific, Singapore, 1998, pp. 190–199.

56. "BEC: The alkali gases from the perspective of research on liquid helium", in Proc. Atomic Physics 16 (AIP Conf. Proceedings 477), ed. W. E. Baylis and G. W. F. Drake, American Institute of Physics, Woodbury, NY, 1999, pp. 154–169.

57. "The significance of the MQC experiment", J. Supercond. 12, 683–7 (2000).

58. "Topics in the theory of the ultracold dilute alkali gases", Mod. Phys. Letters B, 14, suppl. issue, 1-42 (2000).

59. "Qubits,cbits,decoherence,quantum measurement and environment", in Foundations of quantum information, quantum computation, communication, decoherence and all that", p. 3 (2002).

60. "Superconducting units-a major roadblock dissolved?", Science 296, 86 (2003).

61. "Nobel lecture: Superfluid 3-He: The early days as seen by a theorist", Revs. Mod. Phys. 76, 999 (2004).

62. "The quantum measurement problem", Science 307, 871 (2003).

63. "What DO we know about high T_c?" Nature Physics 2,134 (2006).

64. "Bose–Einstein condensation",in Compendium of Quantum Physics, 2006.

65. "Quantum liquids", Science 319, 5867 (2008).

66. "Majorana fermions:a pedagogical essay" in Doing Physics: A Festschrift for Tom Erber (2011).

(c) Books

1. The Problems of Physics, Oxford University Press, Oxford, 1987.

2. Quantum Tunnelling in Condensed Media, (ed., with Yu. Kagan), Elsevier, Amsterdam (1992).

3. Quantum Liquids:Boase condensation and Cooper pairing in condensed matter systems,Oxford 2006.

(d) Translations

1. "Nuclear Theory: The Quasiparticle Method" by A. B. Migdal (W. A. Benjamin, New York, 1968).

2. "Approximation Methods in Quantum Mechanics" by A. B. Migdal and V. Krainov (W. A. Benjamin, New York, 1969).

3. "Qualitative Methods in Quantum Theory" by A. B. Migdal (W. A. Benjamin Inc., Reading, Mass., 1977).

(e) Other

"Notes on the Writing of Scientific English for Japanese Physicists" Nihon Buturi Gakkai Si 21, 790 (1966). (Japanese translation: Denki Tuusin Bassatu, Feb/Mar 1973).

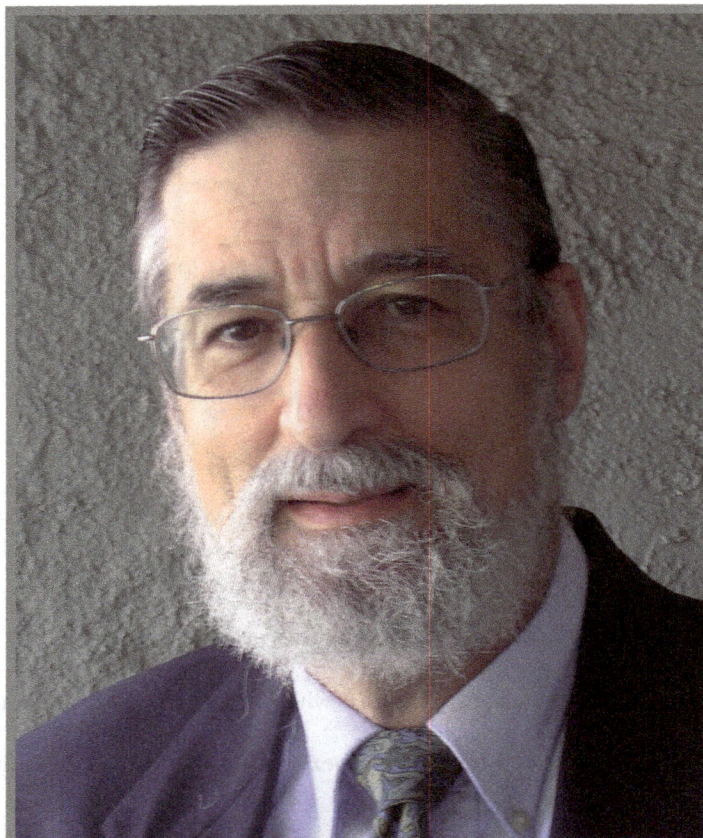

Bertrand I. Halperin

BERTRAND I. HALPERIN

Birthplace

Born   Brooklyn, N.Y., 6 December 1941

Education

1961   A.B. Physics, Harvard University
1963   M.A. Physics, University of California, Berkeley
1965   Ph.D. Physics, University of California, Berkeley

Positions Held

1965–1966   National Science Foundation Postdoctoral Fellow, Ecole Normale Supérieure, Paris
1966–1976   Member of Technical Staff, Bell Laboratories, Murray Hill, NJ
1969–1970   Lecturer, Harvard University (on leave from Bell Laboratories)
1976–present   Professor of Physics, Harvard University
1988–1991   Chairman of the Harvard Physics Department
1999–2004   Scientific Director, Harvard Center for Imaging and Mesoscale Structures
1992–present   Hollis Professor of Mathematicks and Natural Philosophy, Harvard University
2004–present   Senior Fellow in the Harvard Society of Fellows

Organizations and Awards

Member. National Academy of Sciences; American Philosophical Society
Fellow. American Physical Society; American Academy of Arts and Sciences

1982   Oliver E. Buckley Condensed Matter Physics Prize, American Physical Society
2001   Lars Onsager Prize, American Physical Society
2003   Wolf Prize in Physics
2004   Teaching Award, Phi Beta Kappa, Harvard Chapter
2007   Dannie Heineman Prize of the Göttingen Akademie der Wissenschaften
2009   Lars Onsager Medal and Lecture, Norwegian University of Science and Technology
2013   Doctor of Philosophy, Honoris Causa, Weizmann Institute of Science, 2013

COMMENTARY

Bertrand Halperin was born in Brooklyn, New York, on December 6, 1941. His parents, Morris and Eva Halperin had both immigrated to the United States from the Ukraine when they were children. Bertrand attended public schools in Brooklyn through the twelfth grade, and entered Harvard College with sophomore standing in September 1958, where he received an AB in physics, summa cum laude, in June 1961. He then stayed at Harvard as a graduate student for one additional year, before transferring to the University of California, Berkeley, in September 1962.

In 1964, when his Berkeley thesis advisor, John Hopfield moved to Princeton University, Halperin went with him as a visiting graduate student. After receiving his Ph.D. from Berkeley in 1965, he spent one year as a postdoctoral fellow at the Ecole Normale Supérieure in Paris, and then was employed for ten years as a Member of the Technical Staff at Bell Laboratories in Murray Hill, New Jersey. He joined the faculty of Harvard University, as a Professor of Physics, in 1976.

Halperin was drawn towards science and mathematics from an early age. In large part he was influenced by his father, who had done some graduate work in mathematics, and had at one time hoped to pursue an academic career in the field, but had been forced by economic realities of the Great Depression to abandon this goal in favor of a civil service job with the U. S. Customs Service. His father maintained his interest in math throughout his life, however, and he loved to solve puzzles as well as to tutor the children of friends and relatives in mathematics at all levels.

Halperin was supported in his studies by several outstanding teachers in junior high school and high school, who also encouraged him to study subjects on his own. By the end of his freshman year at Harvard, Halperin was convinced that he wanted to pursue a career in physics. His particular interest in condensed matter physics, however, developed during a summer internship at Los Alamos National Laboratory, where he worked with an experimental group using inelastic neutron scattering to measure the vibrational spectrum of aluminum. He was particularly inspired by Brillouin's book on phonons and electrons in crystals, which he read in preparation for this work.

At Berkeley, the thesis project that John Hopfield gave to Halperin concerned the low-energy tail of optical absorption by excitons in alkali halides. It turned out that this problem was related mathematically to questions about the low-energy tail of an electron band in a semiconductor, in the presence of random impurities, which was being studied by other theorists including Melvin Lax at Bell Laboratories. Halperin spent the summer of 1964 at Bell, working with Lax, and the results of this collaboration became a central part of Halperin's doctoral thesis.

During the period 1966–1976, when Halperin was a staff member at Bell Laboratories, he found the environment there to be exceptionally stimulating. An important outcome of this period was a close collaboration with a colleague, Pierre Hohenberg, on the theory of the dynamic behavior of systems at, or close to, a critical-point phase transition. Examples included the superfluid phase transition in liquid helium, the gas–liquid critical point, and magnetic transitions of various types. The joint publications of Halperin and Hohenberg began with a *Physical Review Letter*, in 1967, which proposed a generalization of *scaling laws* to dynamic critical phenomena. Later papers applied the newly developed *renormalization group* methods to dynamic critical properties, and delineated the various *universality classes* for dynamic critical phenomena, determined by symmetries and conservation laws. (Several of these articles also involved Shang-Keng Ma and Eric Siggia as additional collaborators.) The 1977 *Reviews of Modern Physics* article by Hohenberg and Halperin, which summarized these results, has become a basic reference for the field.

During his years at Bell Laboratories, Halperin also worked on a variety of other problems in condensed matter physics, often concerning quantum mechanical phenomena in metals, semiconductors and semimetals. Important collaborators in and outside of Bell included T. M. Rice, James Langer and Vinay Ambegaokar.

In 1972, Halperin was a coauthor with P. W. Anderson and C. M. Varma of a highly-cited paper that argued that the anomalous thermal conductivity as well as the excess specific heat observed in glasses at low temperatures might be caused by localized defects with two nearly-degenerate quantum levels. These ideas were later extended by Halperin and his student James Black to include observed nonlinear phenomena such as phonon echoes and spectral diffusion.

Soon after moving to Harvard, in 1976, Halperin developed a close collaboration with David R. Nelson on problems related to the theory of phase transitions in two-dimensional systems. Most notably, Halperin and Nelson argued that melting of a two-dimensional crystal could occur via a two-step process, through an intermediate phase that had only short range positional order but long-range correlations in the orientation of the bonds between neighboring atoms. The new "hexatic" order proposed for this intermediate phase was later observed in layered three-dimensional liquid crystals. In other work, together with a number of other collaborators, Halperin and Nelson examined dynamic properties of two-dimensional systems, including transport in thin superconductors and superfluid films.

At Harvard, Halperin maintained an interest in the transport properties of highly disordered systems. Among other results, Halperin and collaborators Pabitra Sen, Shechao Feng and Christopher Lobb analyzed the behavior of an elastic network near a percolation, and demonstrated the differences between transport properties of various continuum percolation problems and those of previously studied discrete lattice models.

In 1988, a paper by Sudip Chakravarty, Halperin and Nelson discussed the possibility of a zero-temperature quantum phase transition in a two-dimensional Heisenberg antiferomagnet. The paper introduced the notion of a quantum critical regime at finite temperatures, where scaling laws for the static and dynamical properties would be dominated by the zero-temperature quantum critical point. The paper also argued that a layer of La_2CuO_4, the parent compound of the first of the newly-discovered high-temperature cuprate superconductors, should be on the *ordered* side of this critical point, contrary to earlier speculation.

Since 1981, a major focus of Halperin's interest has been on a set of remarkable phenomena, known as the *quantum Hall effects*, which take place in two-dimensional electron systems at low temperatures in strong magnetic fields. In a paper published in 1982, Halperin showed that observation of the quantized Hall effect implies the existence of unidirectional current-carrying *edge states*, which are impervious to localization by impurities. This edge transport description formed the basis for subsequent developments, including extensions to *fractional quantized Hall states* and to the description of interference effects in devices containing quantized Hall samples with narrow constrictions. It also foreshadowed recent developments in the theory of *topological insulators*, which are insulators with conducting edge or surface states.

In 1983, Halperin published a paper that showed how Robert Laughlin's explanation of the newly discovered fractional quantized Hall effect could be generalized to include arbitrary fractions with odd denominator. The paper demonstrated the theoretical possibility of fractions with even denominator, which were later observed in experiments. The paper also noted that electron spins need not be fully aligned in fractional quantized Hall states, which implied that there could be phase transitions between states with different degrees of spin alignment, as was later observed in a variety of experiments.

In a 1984 *Physical Review Letter*, Halperin showed that elementary charged quasiparticles in fractional quantized Hall states could be described as particles that obey *fractional statistics*, a concept that had been previously introduced as a mathematical abstraction, but had never been shown to occur in a real physical system. The existence of quasiparticles obeying fractional statistics, and more complicated extensions to non-Abelian statistics, have emerged as intriguing possibilities in a number of other two-dimensional systems.

In a 1993 paper with Patrick Lee and Nicholas Read, Halperin showed how one could understand experiments on systems with a Landau level at or near half filling, where a quantized Hall conductance is not observed. The theory employed a type of analysis, previously used for quantized Hall systems, in which electrons are mathematically transformed into another type of particle interacting with a *gauge field* of the Chern–Simons type. This analysis allowed predictions for dynamic properties, including electrical resistance and anomalous interactions with surface acoustic waves, which were later confirmed by experiments.

Halperin has continued to work on many aspects of quantum Hall systems in recent years. This work has benefited from collaborations with a large number of theorists, including Rudolf Morf, Steve Simon, Leonid Levitov, Bernd Rosenow, and especially, Ady Stern. Other portions of Halperin's work have been directed to understanding experiments on a variety of other systems at low temperatures, including one-dimensional metals and spin systems, electrons in very small semiconductor devices, and structures that combine superconductors and semiconductors. Much of this work has involved collaborations with experimenter colleagues, especially Charles Marcus and Amir Yacoby.

VOLUME 41, NUMBER 2   PHYSICAL REVIEW LETTERS   10 JULY 1978

Theory of Two-Dimensional Melting

B. I. Halperin and David R. Nelson

Department of Physics, Harvard University, Cambridge, Massachusetts 02138

(Received 17 May 1978)

The consequences of a theory of dislocation-mediated two-dimensional melting are worked out for triangular lattices. Dissociation of dislocation pairs first drives a transition into a "liquid crystal" phase with exponential decay of translational order, but power-law decay of sixfold orientational order. A subsequent dissociation of *disclination* pairs at a higher temperature then produces an isotropic fluid. The critical behavior, as well as the effect of a periodic substrate, is discussed.

Kosterlitz and Thouless[1] have proposed a model of two-dimensional melting, in which the "topological order" of a solid phase is destroyed by the dissociation of dislocation pairs. Similar ideas,[1] with vortices taking the place of dislocations, have led to a rather detailed theory[2] of the superfluid transition in two dimensions. In this Letter, we summarize the consequences of dislocation-mediated melting of triangular lattices, on both smooth and periodic substrates. A more detailed derivation will be given elsewhere.[3]

Consider the properties of a two-dimensional triangular solid on a smooth substrate. By definition, the solid has nonzero long-wavelength elastic constants. The structure factor exhibits[4] power-law singularities, $S(\vec{q}) \sim |\vec{q} - \vec{G}|^{-2 + \eta_{\vec{G}}}$, near a set of reciprocal lattice vectors $\{\vec{G}\}$, with exponents $\eta_{\vec{G}}$ related to the Lamé elastic constants $\mu_R(T)$ and $\lambda_R(T)$ by $\eta_{\vec{G}} = k_B T |\vec{G}|^2 (3\mu_R + \lambda_R)/4\pi\mu_R(2\mu_R + \lambda_R)$. These singularities, which replace the δ-function Bragg peaks found in three-dimensional solids, reflect power-law decay at large distances of the correlation function $\langle \exp\{i\vec{G} \cdot [\vec{u}(\vec{r}) - \vec{u}(\vec{0})]\} \rangle$, where $\vec{u}(\vec{r})$ is the lattice displacement at point $\vec{r}$. One can also define an order parameter (analogous to $e^{i\vec{G} \cdot \vec{u}}$) for bond orientations, namely $\psi \equiv e^{6i\theta}$, where $\theta(\vec{r})$ is the orientation, relative to the x axis, of a bond between two nearest-neighbor atoms at $\vec{r}$. In a solid, θ is given in terms of the displacement field, $\theta = \frac{1}{2}(\partial_y u_x - \partial_x u_y)$. The solid phase exhibits long-range orientational order, since $\langle \psi^*(\vec{r})\psi(\vec{0}) \rangle$ approaches a nonzero constant at large $\vec{r}$.[5]

If melting is indeed characterized by an unbinding of dislocation pairs at a temperature T_m, one expects that a density $n_f(T)$ of free dislocations above T_m will lead to exponential decay of the translational order parameter $e^{i\vec{G} \cdot \vec{u}}$, with a correlation length $\xi_+(T) \approx n_f^{-2}$. This length diverges as $T \to T_m^+$ [see (6) below]. The structure factor $S(\vec{q})$ is now finite at all Bragg points, and the Lamé coefficients vanish at long wavelengths. We shall see, however, that orientational order persists, in the sense that bond-angle correlations now decay algebraically, $\langle \psi^*(\vec{r})\psi(\vec{0}) \rangle \sim 1/r^{\eta_6(T)}$. This phase can be described as a liquid crystal, similar to a two-dimensional nematic, but with a sixfold rather than twofold anisotropy. The exponent $\eta_6(T)$ is related to the Franck constant $K_A(T)$, which is the coefficient of $\frac{1}{2}|\nabla\theta|^2$ in the free-energy density: $\eta_6(T) = 18k_B T/\pi K_A(T)$. We find that K_A is infinite just above T_m, but decreases with increasing temperatures, until a temperature T_i, where dissociation of *disclination* pairs drives a transition into an isotropic phase in which both the translational and orientational order decay exponentially.

The liquid-crystal phase is isomorphic to a two-dimensional superfluid, except that $\pm 60°$ disclinations play the role of vortices. The transition at T_i should belong to the same universality class as the superfluid transition, and we expect, in particular, that[2] $\eta_6(T_i) = \frac{1}{4}$. Although disclination pairs are very tightly bound in the solid phase, screening by a gas of free dislocations produces a weaker logarithmic binding for $T_m < T < T_i$. It is interesting to note that an isolated dislocation can itself be regarded as a tightly bound disclination pair,[6] separated by one lattice constant.

To see the origin of these results, let us decompose the displacement field of a solid into a smoothly varying phonon field $\vec{\varphi}(\vec{r})$, and a part due to dislocations.[1] The Hamiltonian $\mathcal{H}_E$ for the solid, within continuum elasticity theory,[6] then breaks into two parts, $\mathcal{H}_E = \mathcal{H}_0 + \mathcal{H}_D$, with

$$\frac{\mathcal{H}_0}{k_B T} = \frac{1}{2} \int \frac{d^2 r}{a_0^2} [2\bar{\mu}\,\varphi_{ij} + \bar{\lambda}\varphi_{ii}^2], \tag{1}$$

$$\frac{\mathcal{H}_D}{k_B T} = -\frac{1}{8\pi} \sum_{\vec{R} \neq \vec{R}'} \left[K_1 \vec{b}(\vec{R}) \cdot \vec{b}(\vec{R}') \ln\left(\frac{|\vec{R} - \vec{R}'|}{a}\right) + K_2 \frac{\vec{b}(\vec{R}) \cdot (\vec{R} - \vec{R}')\vec{b}(\vec{R}') \cdot (\vec{R} - \vec{R}')}{|\vec{R} - \vec{R}'|^2} \right] + \frac{E_c}{k_B T} \sum_{\vec{R}} |\vec{b}(\vec{R})|^2. \tag{2}$$

In (1), φ_{ij} is related to the smooth part of the displacement field, $\varphi_{ij} = \frac{1}{2}(\partial_i \varphi_j + \partial_j \varphi_i)$, and $\overline{\mu}$ and $\overline{\lambda}$ are "reduced" elastic constants, given by the usual Lamé coefficients μ and λ multiplied by the square of the lattice spacing, a_0^2, and divided by $k_B T$. In (2), $\vec{b}(\vec{R})$ is a dimensionless dislocation Burgers vector of the form $\vec{b}(\vec{R}) = m(\vec{R})\vec{e}_1 + n(\vec{R})\vec{e}_2$, where $m(\vec{R})$ and $n(\vec{R})$ are integers, and $\vec{e}_1$ and $\vec{e}_2$ are unit vectors spanning the underlying Bravais lattice. The sums in (2) are over, say, a square mesh with spacing a of sites in physical space, and the $\vec{b}(\vec{R})$ must satisfy a vector charge neutrality condition, $\sum_{\vec{R}} \vec{b}(\vec{R}) = 0$. The quantities K_1 and K_2 are equal, $K_1 = K_2 \equiv K = 4\overline{\mu}(\overline{\mu} + \overline{\lambda})/(2\overline{\mu} + \overline{\lambda})$, and E_c is the core energy of a dislocation.

If dislocations only exist in bound pairs at low temperatures, one expects that they can be ignored, and that the long-wavelength properties of the solid will simply be given by (1), with suitably renormalized elastic constants. The properties of the solid phase quoted above follow directly from this observation.

One of us[7] has studied the properties of (2) in the absence of the dot product or angular terms ($K_2 = 0$). It was found that dislocations are indeed unimportant at low temperatures (large K_1), and that a dislocation-unbinding transition was controlled by the terminus of a fixed *surface*, parametrized by K_1 and $\vec{e}_1 \cdot \vec{e}_2$. Here, we restrict ourselves to the triangular lattice ($\vec{e}_1 \cdot \vec{e}_2 = \frac{1}{2}$), and extend this treatment to the full dislocation Hamiltonian $\mathcal{K}_D$, taking into account the neglected angular terms. Recursion relations for K and $y \equiv \exp(-E_c/k_B T)$ can in fact be obtained rather straightforwardly, by considering the renormalization of elastic constants due to dislocation pairs, in analogy to calculations of the effect of vortices on the superfluid density in a ⁴He film.[8] Integrating over mesh sizes between a and ae^l, we obtain partially dressed parameters $\overline{\mu}(l)$, $\overline{\lambda}(l)$, $y(l)$, and $K(l)$, which satisfy, to $O(y^2(l))$,

$$\frac{d\overline{\mu}^{-1}}{dl} = 3\pi y^2 e^{-K/8\pi} I_0\left(\frac{K}{8\pi}\right), \tag{3}$$

$$\frac{d[\overline{\mu} + \overline{\lambda}]^{-1}}{dl} = 3\pi y^2 e^{-K/8\pi}\left[I_0\left(\frac{K}{8\pi}\right) + I_1\left(\frac{K}{8\pi}\right)\right], \tag{4}$$

$$\frac{dy}{dl} = \left(2 - \frac{K}{8\pi}\right)y + 2\pi y^2 e^{-K/16\pi} I_0\left(\frac{K}{16\pi}\right), \tag{5}$$

where $I_0(x)$ and $I_1(x)$ are modified Bessel functions of the first and second kind. We find that

$$K^{-1}(l) = \frac{1}{4}\left\{\overline{\mu}^{-1}(l) + [\overline{\mu}(l) + \overline{\lambda}(l)]^{-1}\right\}$$

for all l, so that its recursion relation can be obtained trivially from (3) and (4).

As in Ref. 7, $y(l)$ is driven to zero at large l, for all temperatures below a critical value T_m. Above T_m, $y(l)$ is ultimately driven toward large values and $K(l)$ is driven towards zero, an instability we associate with dislocation-pair unbinding.

Following Kosterlitz[2] and Ref. 7, we determine the behavior near T_m by studying Eqs. (3)–(5) near the critical value $K_c = 16\pi$. We identify the correlation length $\xi_+(T)$ with ae^{l*}, with $l*$ chosen such that $K(l*) \approx \frac{1}{2}K_c$. In this way, we find that

$$\xi_+(T) \approx a\exp[b/(T/T_m - 1)^{0.44817\cdots}], \tag{6}$$

as $T \to T_m^+$, where b is a constant, and $0.44817\ldots$ can be expressed in terms of the roots of a quadratic equation with Bessel-function coefficients. The specific heat exhibits only an essential singularity, $C_p \sim \xi_+^{-2}$, while the structure factor at the Bragg points is given by $S(\vec{G}) \sim \xi_+^{2-\eta_{\vec{G}}}$. Taking over the discussion for the superfluid density in Ref. 8, we find that the reduced shear modulus in the solid phase is

$$\overline{\mu}_R(T) = \lim_{l \to \infty} \overline{\mu}(l).$$

It follows from Eqs. (3)–(5) that $\mu_R(T)$ approaches a finite limiting value as $T \to T_m^-$. Just below T_m we find $\mu_R(T) = \mu_R(T_m)[1 + \text{const}(T_m - T)^{0.44817\cdots}]$, with a similar result for $\lambda_R(T)$. There is a universal relationship involving the elastic constants at T_m,

$$\lim_{T \to T_m^-}\left\{\frac{1}{\mu_R(T)} + \frac{1}{\mu_R(T) + \lambda_R(T)}\right\} = \frac{a_0^2}{4\pi k_B T_m}. \tag{7}$$

This corresponds to the critical value $K_c = 16\pi$, and is also suggested by the "entropy argument" of Kosterlitz and Thouless.[1]

The results for orientational correlations above T_m follow from a calculation of the Franck constant K_A:

$$\frac{k_B T}{K_A} = \lim_{q \to 0} q^2 \langle \hat{\theta}(q)\hat{\theta}(-q) \rangle$$

$$= \lim_{q \to 0} \frac{q_i q_j}{q^2} \langle \hat{b}_i(q)\hat{b}_j(-g) \rangle, \tag{8}$$

where $\hat{\theta}(q)$ and $\hat{b}_i(q)$ are the Fourier-transformed orientational and Burgers-vector fields, respectively. The second line of (8) follows because the contribution of $\vec{\varphi}(\vec{r})$ to K_A^{-1} is zero and because the dislocation part of $\hat{\theta}(q)$ is just[3,6] $\hat{\theta}(q) = -iq_j\hat{b}_j(q)/q^2$. To estimate K_A just above T_m, we use its transformation properties under the renormal-

VOLUME 41, NUMBER 2   **PHYSICAL REVIEW LETTERS**   10 JULY 1978

ization group, $K_A(K(0), y(0)) = e^{2l} K_A(K(l), y(l))$. Choosing $l = l^* = \ln(\xi_+/a)$, we can evaluate K_A using Debye-Hückel theory, which amounts to treating $\vec{b}(\vec{R})$ as a continuous vector field, rather than restricting it to discrete points on a Bravais lattice. Upon Fourier transformation, $\mathcal{H}_D$ becomes

$$\frac{\mathcal{H}_D}{k_B T} = \frac{1}{2V} \sum_{\vec{q}} \left[\frac{K}{2q^2} \left(\delta_{ij} - \frac{q_i q_j}{q^2} \right) + \frac{2 E_c a^2}{k_B T} \delta_{ij} \right] \hat{b}_i(\vec{q}) \hat{b}_j(-\vec{q}), \tag{9}$$

where V is the volume. Since the term proportional to the transverse projection operator in (9) does not contribute to (8), one obtains $K_A(K(l^*), y(l^*)) \approx 2 E_c(l^*) a^2 = O(k_B T_m)$. It follows that the physical Franck constant is $K_A \sim \xi_+^2(T)$. The algebraic decay of orientational order above T_m and the relationship between $\eta_6(T)$ and $K_A(T)$ are straightforward consequences of this result.

Many experimental investigations of two-dimensional melting are carried out on films adsorbed onto a regular substrate,[9] and so it is important to determine the effect of a periodic potential on our results. One must now distinguish between a "floating solid," characterized by power-law Bragg singularities at reciprocal lattice vectors $\{\vec{G}\}$ which vary continuously with coverage and temperature, and an "epitaxial solid," having δ-function Bragg peaks at a lattice of vectors including the substrate reciprocal lattice $\{\vec{K}\}$ as a proper subset. The floating solid should be rather similar to the solid on a smooth substrate discussed in this paper. Figure 1 shows a schematic phase diagram with fluid, floating solid, and epitaxial phases. A region of two-phase coexistence is also shown, separating epitaxial phase I from a dilute fluid or "vapor."[8,10] We expect an increasing multiplicity and complexity of epitaxial phases with decreasing temperatures.

To understand Fig. 1, consider first the effect of a weak substrate potential commensurate with the lattice of the adsorbed film. Let $\{\vec{M}\}$ be the set of vectors common to $\{\vec{G}\}$ and $\{\vec{K}\}$, and let M_0 be the minimum length of nonzero vectors in $\{\vec{M}\}$. Let us expand the potential on the reciprocal lattice $\{\vec{K}\}$, and focus our attention on

$$\mathcal{H} = \mathcal{H}_E + \sum_{|\vec{M}| = M_0} h_{\vec{M}} \sum_{\vec{r}} e^{i\vec{M} \cdot \vec{u}(\vec{r})}, \tag{10}$$

where $h_{\vec{M}}$ is the potential strength at $\vec{M}$; the terms displayed in (10) are the most important ones for weak potentials. The renormalization-group eigenvalue for $h_{\vec{M}}$ is easily shown to be $\lambda_{\vec{M}} = 2 - \frac{1}{2}\eta_{\vec{G}}|_{\vec{G} = \vec{M}}$, so that any commensurate perturbation becomes relevant at sufficiently low temperatures. If M_0 is sufficiently small ($M_0 \lesssim 8\pi/a_0$), λ_{M_0} remains relevant out to quite high temperatures and a floating solid can never exist; there is then a transition directly from a low-temperature expitaxial phase into a fluid, as shown for epitaxial phases I and III. For large enough M_0 ($M_0 \gtrsim 8\pi/a_0$), however, there is a temperature window where $\lambda_{M_0} < 0$ and $K_R > 16\pi$, indicating that a floating solid is stable to both substrate perturbations and dislocation unbinding. The dotted line in Fig. 1 shows a locus of such points, where the floating solid has the same periodicity as epitaxial phase II, which exists at lower temperatures.

It can be shown[3] that the transitions from floating solid to fluid and from floating solid to epitaxial phase II (marked A and B in Fig. 1) are both describable at long wavelengths by a Hamiltonian of the form (2) with $K_1 \neq K_2$. Indeed, these transitions are essentially dual to each other.[3] The situation is very similar to the "$\cos p\theta$" perturbations discussed by José et al.[11] The transition from epitaxial phase II to a floating solid at points other than B may connect two phases with different periodicities; its nature is not yet known. We expect that the transition from floating solid to fluid will be everywhere qualitatively similar to our discussion of dislocation unbinding on a smooth substrate. The orientational bias imposed by the substrate, however, should alter or eliminate the liquid-crystal isotropic transition discussed above.

We wish to emphasize in closing that we have only explored consequences of the dislocation model of melting perturbatively in $y = \exp(-E_c/$

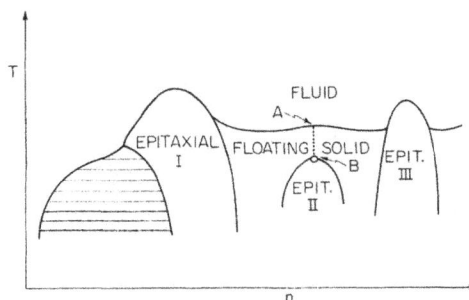

FIG. 1. Hypothetical phase diagram for a submonolayer adsorbed film on a periodic substrate as a function of density n and T.

VOLUME 41, NUMBER 2   PHYSICAL REVIEW LETTERS   10 JULY 1978

$k_B T$). Although the theory is stable and self-consistent, we cannot rule out other mechanisms for melting, perhaps leading to a first-order transition. A "premature" unbinding of disclinations (before dislocations dissociate) might constitute such a mechanism.

We have benefitted from discussions with S. Aubry, A. N. Berker, M. Kléman, and R. Peierls. This work was supported by the National Science Foundation, under Grant No. DMR 77-10210, and by the Harvard Materials Research Laboratory program. One of us (D.R.N.) received a Junior Fellowship from the Harvard Society of Fellows.

[1]J. M. Kosterlitz and D. J. Thouless, J. Phys. C **6**, 118 (1973), and to be published.

[2]J. M. Kosterlitz, J. Phys. C **7**, 1046 (1974); see also J. José, L. P. Kadanoff, S. Kirkpatrick, and D. R. Nelson, Phys. Rev. B **16**, 1217 (1977).

[3]D. R. Nelson and B. I. Halperin, to be published.

[4]See, e.g., Y. Imry and L. Gunther, Phys. Rev. B **3**, 3939 (1971).

[5]N. D. Mermin, Phys. Rev. **176**, 250 (1968).

[6]F. R. N. Nabarro, *Theory of Dislocations* (Clarendon, New York, 1967).

[7]D. R. Nelson, Phys. Rev. B (to be published).

[8]D. R. Nelson and J. M. Kosterlitz, Phys. Rev. Lett. **39**, 1201 (1977).

[9]J. G. Dash, *Films on Solid Surface* (Academic, New York, 1975).

[10]See, e.g., A. N. Berker, S. Ostlund, and F. A. Putnam, Phys. Rev. B **17**, 3650 (1978).

[11]José *et al.*, Ref. 2.

VOLUME 41, NUMBER 7   PHYSICAL REVIEW LETTERS   7 AUGUST 1978

THEORY OF TWO-DIMENSIONAL MELTING. B. I. Halperin and David R. Nelson [Phys. Rev. Lett. **41**, 121 (1978)].

Dr. A. P. Young has kindly called our attention to a sign error in the second term on the right-hand side of Eq. (2), which led to incorrect coefficients in the recursion relations (3)–(5). Consequently, the numerical value of the exponent in Eq. (6), for the correlation length, and in the singular part of the elastic constants is wrong. The correct value of this exponent (quoted as $0.448\,17\ldots$ in our Letter) is $0.369\,63\ldots$. Dr. Young had independently obtained the correct behavior of the correlation length.

The corrected forms of Eqs. (3)–(5) are

$$\frac{d\bar{\mu}^{-1}}{dl} = 3\pi y^2 e^{K/8\pi} I_0\left(\frac{K}{8\pi}\right), \tag{3}$$

$$\frac{d[\bar{\mu}+\bar{\lambda}]^{-1}}{dl} = 3\pi y^2 e^{K/8\pi}\left[I_0\left(\frac{K}{8\pi}\right) - I_1\left(\frac{K}{8\pi}\right)\right]. \tag{4}$$

$$\frac{dy}{dl} = \left(2 - \frac{K}{8\pi}\right)y + 2\pi y^2 e^{K/16\pi} I_0\left(\frac{K}{8\pi}\right). \tag{5}$$

Two additional misprints should be corrected: Equations (8) and (9) should read

$$\frac{k_B T}{K_A} = \lim_{q\to 0} q^2\langle\hat{\theta}(\vec{q})\hat{\theta}(-\vec{q})\rangle = \lim_{q\to 0}\frac{q_i q_j}{q^2}\langle\hat{b}_i(\vec{q})\hat{b}_j(-\vec{q})\rangle a_0^2, \tag{8}$$

$$\frac{\mathcal{H}_D}{k_B T} = \frac{1}{2V}\sum_{\vec{q}}{}'\left[\frac{K}{q^2}\left(\delta_{ij} - \frac{q_i q_j}{q^2}\right) + \frac{2E_c a_0^2}{k_B T}\delta_{ij}\right]\hat{b}_i(\vec{q})\hat{b}_j(-\vec{q}). \tag{9}$$

VOLUME 52, NUMBER 18   PHYSICAL REVIEW LETTERS   30 APRIL 1984

Statistics of Quasiparticles and the Hierarchy of Fractional Quantized Hall States

B. I. Halperin

Physics Department, Harvard University, Cambridge, Massachusetts 02138
(Received 9 November 1983)

Quasiparticles at the fractional quantized Hall states obey quantization rules appropriate to particles of fractional statistics. Stable states at various rational filling factors may be constructed iteratively by adding quasiparticles or holes to lower-order states, and the corresponding energies have been estimated.

PACS numbers: 05.30.-d, 03.65.Ca, 71.45.Nt, 73.40.Lq

Observations of the fractional quantized Hall effect[1] show that there exist special stable states of a two-dimensional electron gas, in strong perpendicular magnetic field B, occurring at a set of rational values of ν, the filling factor of the Landau level. Laughlin[2] has constructed an explicit trial wave function (product wave function) to explain the states at $\nu = 1/m$, with m an odd integer, and has argued that the elementary excitations from the stable states are quasiparticles with fractional electric charge. Among the proposals to explain the other observed fractional Hall steps are hierarchical schemes, in which higher-order stable states ν_{s+1} are built up by adding quasiparticles to a stable state ν_s of smaller numerator and denominator.[3-5]

In the present note, we observe that the quantization rules which determine the allowed quasiparticle spacings are just those that would be expected for a set of identical charged particles that obey *fractional statistics*—i.e., such that the wave function changes by a complex phase factor when two particles are interchanged. Moreover, by assuming that the dominant interaction between quasiparticles is just the Coulomb interaction between the quasiparticle charges, we are led to a natural set of approximations for the ground-state energies and energy gaps at all levels of the hierarchy.

The appearance of fractional statistics in the present context is strongly reminiscent of the fractional statistics introduced by Wilczek to describe charged particles tied to "magnetic flux tubes" in two dimensions.[6] As in Ref. 6, the quasiparticles can *also* be described by wave functions obeying Bose or Fermi statistics, the various representations being related by a "singular gauge transformation." The boson description was, in fact, used in Refs. 3 and 4 and the fermion description in Ref. 5. However, the boson or fermion descriptions require, in effect, a long-range interaction between quasiparticles which alters the usual quantization rules. The transformation between representations is analogous to the well-known transformation between im-penetrable bosons and fermions in one dimension.

As in previous discussions of the fractional quantized Hall effect, we consider a two-dimensional system of electrons in the lowest Landau level, with a uniform positive background. The filling factor ν is defined by $\nu = n/2\pi l_0^2$, where n is the density of electrons, and $l_0 \equiv |Be/\hbar c|^{1/2}$ is the magnetic length; hence ν is the number of electrons per quantum of flux.

Let ν_s be a stable rational filling factor obtained at level s of the hierarchy. I assert that the low-lying energy states for filling factors near to ν_s can be described by the addition of a small density of quasiparticle excitations to the ground state at ν_s. The elementary quasiparticle excitations are of two types—particlelike "p excitations" and holelike "h excitations"—having charges $q_s e$ and $-q_s e$, respectively, according to a sign convention described below. For the present purposes we need only consider states with one type of excitation present. We shall describe these states by a pseudo wave function Ψ, which is a function of the coordinates $\vec{R}_k$ of the N_s quasiparticles present. I assert that the the allowed pseudo wave functions can be written in the form

$$\Psi[\vec{R}_k] = P[Z_k]Q_s[Z_k]\prod_{k=1}^{N_s}\exp(-|q_s||Z_k|^2/4l_0^2),$$

$$(1)$$

where $Z_k = X_k \mp iY_k$ is the position in complex notation, with the sign depending on the sign of the charge of the quasiparticle, $P[Z_k]$ is a *symmetric polynomial* in the variables Z_k, and

$$Q_s = \prod_{k<l}|Z_k - Z_l|^{-\alpha/m_s}.$$

$$(2)$$

In Eq. (2), $\alpha = \pm 1$, according to whether we are dealing with particle- or hole-type excitations, and m_s is a rational ≥ 1, to be specified by an iterative equation below. We may interpret $|\Psi[\vec{R}_k]|^2$ as the probability density for finding a quasiparticle at each of the positions $\vec{R}_1, \ldots, \vec{R}_{N_s}$, at least in the case

VOLUME 52, NUMBER 18   PHYSICAL REVIEW LETTERS   30 APRIL 1984

that the $\vec{R}_k$ are not too close to each other. Since the quasiparticles have a finite size (of order l_0), however, there is no direct significance to the behavior of $|\Psi|^2$ when two positions R_k and R_l come very close together. The wave function is normalized if $\int |\Psi|^2 = 1$, and two wave functions Ψ and Ψ' are orthogonal if $\int \Psi^* \Psi' = 0$.

The pseudo wave function (1)–(2) can be derived in different ways, starting from various microscopic descriptions that have been proposed[2-5] for the electronic state with quasiparticle or quasihole excitations. I shall give below a derivation for p excitations using the *pair model* proposed in Ref. 3.

Because there is no direct physical significance to the phase of the pseudo wave function, it is permissible to *redefine* the factor Q in Eq. (1) by *removing the absolute value sign* in Eq. (2). (This operation may be described as a singular gauge transformation.)[6] If $m_s \neq 1$, the new wave function is a multivalued function of the positions $\{R_k\}$, and one should consider it as a function defined on the appropriate Riemann surface for $\{Z_k\}$. [Alternatively one could use a single-valued definition and specify discontinuities along cuts in the variable $(Z_k - Z_l)$.] Now if we continuously interchange the positions of two quasiparticles, the wave function will change by a complex phase factor $(-1)^{\pm 1/m_s}$, with the sign depending on the sense of rotation as the quasiparticles pass by each other. Although the extra phase factor is perhaps a complication, the pseudo wave function now has the esthetically pleasing property that it is an eigenstate of the differential operator $[\nabla_k \mp i q_s e \vec{A}(\vec{R}_k)/\hbar c]^2$ with special boundary conditions at the points $Z_k = Z_l$, where $\vec{A}$ is the vector potential in the symmetric gauge. Then Eq. (1) may be described as a general wave function appropriate to a collection of particles of charge $\pm q_s e$ obeying fractional statistics, all in the lowest Landau level. Of course, in the special case $m_s = 1$, the quasiparticles are ordinary fermions.

In order to find the ground-state configuration for a given density n_s of quasiparticles, we must find the symmetric polynomial $P[Z]$ which leads to the minimum expectation value of the repulsive interaction between the quasiparticles. Using the same reasoning as Laughlin in Ref. 2, we expect that certain choices of P can lead to specially low energies, namely,

$$P[Z_k] = \prod_{k<l} (Z_k - Z_l)^{2p_s+1}, \qquad (3)$$

where p_{s+1} is a positive integer. The probability

distribution $|\Psi|^2$ is then that of a classical one-component plasma[2] with dimensionless inverse temperature $\Gamma = 2m_{2+1}$, where

$$m_{s+1} = 2p_{s+1} - \alpha_{s+1}/m_s, \qquad (4)$$

and $\alpha_{s+1} = 1$ or -1 as particlelike or holelike quasiparticles are involved. The density of the plasma is fixed by a charge neutrality condition,[2] so that the number of quasiparticles in an area $2\pi l_0^2$ is just $n_s = |q_s|/m_{s+1}$. Since each quasiparticle has charge $\alpha_{s+1} q_s$, we may readily calculate the electron density in the new stable state, and we find the filling factor

$$\nu_{s+1} = \nu_s + \alpha_{s+1} q_s |q_s|/m_{s+1}. \qquad (5)$$

If we multiply the pseudo wave function described above by the factor $\prod_k Z_k$, for $k = 1, \ldots, N_s$, we find a deficiency near the origin of $1/m_{s+1}$ quasiparticles of level s. We identify this state as a hole excitation at level $s+1$. Similarly, we may construct a p excitation having an *excess* of $1/m_{s+1}$ quasiparticles at the origin. The iterative equation for q_s is thus

$$q_{s+1} = \alpha_{s+1} q_s/m_{s+1}. \qquad (6)$$

Together with the starting conditions $\nu_0 = 0$, $q_0 = m_0 = \alpha_1 = 1$, the iterative equations (4)–(6) give a sequence of rational filling factors ν_s for any choice of the sequence $\{\alpha_s, p_s\}$. At the level $s = 1$, we recover Laughlin's states with $\nu_1 = 1/m_1 = 1, \frac{1}{3}, \frac{1}{5}, \ldots$ for various choices of p_1. If we add holes to the state $\nu_1 = 1$, we find at level $s = 2$, the complements to the Laughlin states, $\nu_2 = \frac{2}{3}, \frac{4}{5}, \frac{6}{7} \ldots$. (In order to stay in the lowest Landau level, we impose the restriction $\alpha_2 = -1$, if $\nu_1 = 1$.) From the state $\nu_1 = \frac{1}{3}$, we achieve such states as $\nu_2 = \frac{2}{5}$ or $\frac{4}{11}$, with p excitations, and $\nu_2 = \frac{2}{7}$ or $\frac{4}{13}$, with h excitations.

It can be shown, after some algebra, that the allowed values of ν_s may be expressed as continued fractions in terms of the finite sequences $\{\alpha_s, p_s\}$ and that they are identical to those of Haldane.[4] (I have used the opposite sign for α, however, and here p is one-half of Haldane's.) As noted by Haldane, every rational value of ν with odd denominator, with $0 < \nu \leq 1$, is obtained once in this way. There will *not* be a quantized Hall step at *every* such rational ν, however. We know that there exists a maximum allowed value m_c for the parameter m_s, such that if at any stage of the hierarchy the calculated m_s is greater than m_c, then the quasiparticles at the density n_s will form a Wigner crystal rather than a quantum-liquid state.[2] There is then no stabilization of the electron density at the corresponding ν_s, and there will be no meaning to any further states in the hierarchy constructed from this ν_s.

The pseudo wave function (1)–(3) leads to a natural estimate of the potential energy of the system, if we assume that the dependence on the positions of the quasiparticles can be approximated by the pairwise Coulomb interaction between point particles of charge $q_s e$, in the background dielectric constant ϵ. If $E(\nu)$ is the energy per quantum of magnetic flux, we have

$$E(\nu_{s+1}) \cong E(\nu_s) + n_s \epsilon_s^{\pm} + n_s |q_s|^{5/2} u_{pl}(m_s), \qquad (7)$$

where $\epsilon_s^{\pm}$ is the energy to add one particlelike excitation or one holelike excitation, together with neutralizing uniform background, to the state ν_s, and u_{pl} is a smooth function of m_s, given (approximately) by Laughlin's interpolation formula[5]

$$u_{pl}(m) = \frac{-0.814}{m^{1/2}} \left[1 - \frac{0.230}{m^{0.64}}\right] \left[\frac{e^2}{\epsilon l_0}\right]. \qquad (8)$$

We recall that $u_{pl}(m)$ is the potential energy per particle that one would find for a system of electrons at filling factor $\nu = 1/m$ if one approximates the pair correlation function $g(r)$ for the electrons by the pair correlation function $g_{pl}(r)$ for a one-component plasma at inverse temperature $\Gamma = 2m$; the factor $|q_s|^{5/2}$ in the last term of (7) reflects the smaller charge and larger magnetic length for our quasiparticles.

In order to use Eq. (7), we need an iterative formula for the quasiparticle energies $\epsilon_s^{\pm}$. It is convenient to write

$$\epsilon_s^{\pm} = \tilde{\epsilon}_s^{\pm} \pm m_s^{-1}[\epsilon_{s-1} + \tfrac{3}{2}|q_{s-1}|^{5/2} u_{pl}(m_s)]. \qquad (9)$$

The quantity in square brackets is the energy it would take to add one quasiparticle or quasihole of level $s-1$, if one could keep the Laughlin product form (3) for the polynomial P, and simply increase the density n_{s-1} by means of a reduction, of order $1/N$, in the magnetic length l_0 which controls the distance scale in Eq. (1).[7] The term $\tilde{\epsilon}_s^{\pm}$ in (9) may be called the *proper* excitation energy; it is relatively small, but is presumably positive for both quasiparticles and holes. For the proper hole energy, we use the approximate formula

$$\tilde{\epsilon}_s^- = 0.313|q_{s-1}|^{5/2} m_s^{-9/4}(e^2/\epsilon l_0). \qquad (10)$$

This form has the correct dependence on the charge q_{s-1}; it passes through the exact value $0.313(e^2/\epsilon l_0)$, for $q_0 = 1$, $m_1 = 1$, and it yields $\tilde{\epsilon}_1^- = 0.264$, $\tilde{\epsilon}_1^- = 0.0837$, for $m_1 = 3$ and $m_1 = 5$, in close agreement with the values obtained by Laughlin.[5,7]

Unfortunately, there does not exist at the present time any reliable calculation of the quasiparticle excitation energy. Therefore, *for purposes of illustration,* I have made the arbitrary approximation $\tilde{\epsilon}_s^+ = \lambda \tilde{\epsilon}_s^-$, where λ is a constant independent of m_s. The resulting curve for $E(\nu)$ is plotted in Fig. 1, for the choice $\lambda = 3$, after subtraction of the "plasma approximation" $E_{pl} = \nu u_{pl}(\nu^{-1})$, which is a smooth function of ν. We can see that there are downward pointing cusps in the energy visible at the low-order rational ν with odd denominators. The approximation also gives *upward*-pointing cusps at all rational ν with even denominators; in fact, I

find small discontinuities in E, not visible on the scale of the figure, at all these even points except for $\nu = \tfrac{1}{2}$, where continuity is guaranteed by the particle-hole symmetry of the cohesive energy, which is respected exactly by the present approximation.[7] Clearly the upward-pointing cusps are unphysical; the system could always lower its energy by breaking up into small regions of larger and smaller density; alternatively there may be a different type of ground state with still lower energy at these values of ν. The behavior of the approximate energy curve near the low-order rationals of odd denominator should be qualitatively and semiquan-

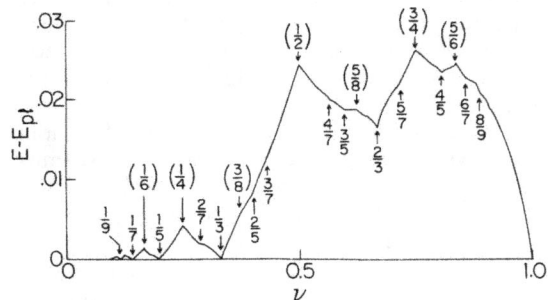

FIG. 1. Potential energy per quantum of magnetic flux, in units of $e^2/\epsilon l_0$, as a function of filling factor ν of the first Landau level, from approximate formulas (7)–(10). Smooth function $E_{pl}(\nu) = \nu u_{pl}(\nu^{-1})$ has been subtracted off.

VOLUME 52, NUMBER 18   PHYSICAL REVIEW LETTERS   30 APRIL 1984

titatively correct, however. More reliable estimates will be possible when p-excitation energies have been properly calculated, and when corrections are included such as the finite quasiparticle size and effects of virtual excitations of particle-hole pairs.

With the approximation described above, the energy gap $\tilde{\epsilon}_s^+ + \tilde{\epsilon}_s^-$ is equal to

$$0.313(1+\lambda)|q_s|^{5/2}m_s^{1/4}(e^2/\epsilon l_0)$$

[cf. (6) and (10)]. Except for the rather weak factor $m_s^{1/4}$, the gap is determined by the value of

$|q_s|^{-1}$, which is the *denominator* of the fraction ν_s. This is in qualitative agreement with reported experimental observations on GaAs samples.[1]

Finally, we derive by induction the starting equation (1). For $s=0$, the Z_k are positions of bare electrons, and Eqs. (1) and (2) are correct, with $q_0 = m_0 = \alpha_1 = 1$. We assume that the p excitations of level $s=1$ can be formed out of *pairs* of electrons, by a generalization of Eq. (23) of Ref. 3. A system containing N_1 pairs of electrons, together with $N_0 - 2N_1$ unpaired electrons, is then described by choosing the polynomial in (1) to have the (schematic) form

$$P[Z_k] = \mathscr{S}\,\bar{P}[z_i]\prod_{i<j}(z_i-z_j)^{8p_1-4}\prod_{i,\gamma}(z_i-\tilde{Z}_\gamma)^{4p_1-1}\prod_{\gamma<\delta}(\tilde{Z}_\gamma-\tilde{Z}_\delta)^{2p_1}, \tag{11}$$

where z_i are the positions of the centers of gravity of the bound pairs, $\tilde{Z}_\gamma$ are the positions of the *unpaired* electrons, $\bar{P}$ is a symmetric polynomial, and $\mathscr{S}$ is an operator which symmetrizes with respect to the positions of all N_0 electrons. I have assumed that the separation between two members of a pair is small, and have dropped the variables describing these separations. To calculate the probability distribution of the pairs, we ignore the symmetrizer $\mathscr{S}$, and take the trace of $|\Psi[Z_k]|^2$ over the unpaired electron positions $\tilde{Z}_\gamma$. The result can be expressed in the form $|\bar{\Psi}[z_i]|^2\Phi[z_i]$, where $\bar{\Psi}$ has again the form of (1) and (2), with P replaced by $\bar{P}$, and with $m_1 = 2p_1 - 1/m_0$, $\alpha = 1$, and $q_1 = q_0/m_1$, while the remaining factor Φ is the partition function of a classical one-component plasma with sources of strength $2 - m_1^{-1}$, located at the positions z_i. Now Φ will be independent of the positions z_i, provided that the sources are sufficiently separated so that their screening clouds do not overlap. Thus it is consistent to interpret $\bar{\Psi}$ as a pseudo wave function for the positions of the pairs. Higher levels may be obtained iteratively.

Derivation of the pseudo wave function for hole excitations is more complicated because of the necessity to use an integral representation, such as Eq. (25) of Ref. 3.[7]

The author is grateful for helpful discussions with R. B. Laughlin, R. Morf, H. Stormer, P. A. Lee, D. Yoshioka, S. Girvin, and P. Ginsparg. This work was supported in part by the National Science Foundation under Grant No. DMR-82-07431.

[1]See H. L. Stormer *et al.*, Phys. Rev. Lett. **50**, 1953 (1983).

[2]R. B. Laughlin, Phys. Rev. Lett. **50**, 1395 (1983).

[3]B. I. Halperin, Helv. Phys. Acta **56**, 75 (1983).

[4]F. D. M. Haldane, Phys. Rev. Lett. **51**, 605 (1983).

[5]R. B. Laughlin, in Proceedings of the Conference on Electronic Properties of Two-Dimensional Systems, Oxford, 1983 (to be published).

[6]F. Wilczek, Phys. Rev. Lett. **49**, 957 (1982).

[7]Details will be given elsewhere.

LIST OF PUBLICATIONS

1. B. I. Halperin, Green's Functions for a Particle in a One-Dimensional Random Potential, Phys. Rev. 139, A104 (1965).

2. B. I. Halperin and M. Lax, Impurity Band Tails in the High Density Limit. I. Minimum Counting Methods, Phys. Rev. 148, 722 (1966).

3. M. Lax and B. I. Halperin, Impurity Band Tails in Degenerate Semiconductors, Journal of the Physical Soc. of Japan, Vol. 21, Supplement (1966), p. 218.

4. B. I. Halperin and M. Lax, Impurity Band Tails in the High Density Limit. II. Higher Order Corrections, Phys. Rev. 155, 802 (1967).

5. B. I. Halperin and P. C. Hohenberg, Generalization of Scaling Laws to Dynamical Properties of a System Near its Critical Point, Phys. Rev. Lett. 19, 700 (1967).

6. B. I. Halperin, Properties of a Particle in a One Dimensional Random Potential, in Advances in Chemical Physics, edited by I. Prigogine (J. Wiley and Sons, London 1968) Vol. XIII, p. 123.

7. B. I. Halperin and T. M. Rice, The Excitonic State at the Semiconductor-Semimetal Transition, in Solid State Physics, edited by F. Seitz, D. Turnbull, and H. Ehrenreich (Academic Press, N.Y. 1968) Vol. 21, p. 115.

8. B. I. Halperin and T. M. Rice, Possible Anomalies at the Semimetal-Semiconductor Transition, Rev. Mod. Phys. 40, 755 (1968).

9. A. S. Barker, Jr., B. I. Halperin, and T. M. Rice, Antiferromagnetic Energy Gap in Chromium, Phys. Rev. Lett. 20, 384 (1968).

10. T. M. Rice, A. S. Barker, Jr., B. I. Halperin and D. B. McWhan, Antiferromagnetism in and its Alloys, J. Appl. Phys. 40, 1337 (1969).

11. B. I. Halperin and P. C. Hohenberg, Scaling Laws for Dynamic Phenomena Near a Critical Point, Journal of the Physical Society of Japan, Vol. 26, Supplement, p. 131 (1969).

12. B. I. Halperin and P. C. Hohenberg, Scaling Laws for Dynamic Critical Phenomena, Phys. Rev. 177, 952 (1969).

13. V. Ambegaokar and B. I. Halperin, Voltage Due to Thermal Noise in the D. C. Josephson Effect, Phys. Rev. Lett. 22, 1364 (1969).

14. B. I. Halperin and P. C. Hohenberg, A Hydrodynamic Theory of Spin Waves, Phys. Rev. 188, 899 (1969).

15. T. M. Rice and B. I. Halperin, Kohn Anomalies in Tungsten and Other Chromium-Group Metals, Phys. Rev. B 1, 509 (1970).

16. D. E. McCumber and B. I. Halperin, Time Scale for Intrinsic Resistive Fluctuation in Thin Superconducting Wires, Phys. Rev. B $\underline{1}$, 1054 (1970).

17. A. B. Harris, D. Kumar, B. I. Halperin and P. C. Hohenberg, Spin-Wave Damping and Hydrodynamics in the Heisenberg Antiferromagnet, J. Appl. Phys. $\underline{41}$, 1361 (1970).

18. A. B. Harris D. Kumar, B. I. Halperin and P. C. Hohenberg, Dynamics of an Antiferromagnets at Low temperatures: Spin-Wave Damping and Hydrodynamics. Phys. Rev. B $\underline{3}$, 961 (1971).

19. B. I. Halperin, Comment on the Hydrodynamics of Unsaturated Helium Films, J. Low Temp. Phys. $\underline{3}$, 255(1970).

20. J. R. Tucker and B. I. Halperin, Onset of Superconductivity in One-Dimensional Systems, Phys. Rev. B $\underline{3}$, 3768 (1971).

21. V. Ambegaokar, B. I. Halperin and J. S. Langer, Hopping Conductivity in Disordered Systems, Phys. Rev. B $\underline{4}$, 2612 (1971).

22. V. Ambegaokar, B. I. Halperin and J. S. Langer, Theory of Hopping Conductivity is Disordered Systems, J. Non. Cryst. Sol. $\underline{8\text{--}10}$, 492 (1972).

23. P. W. Anderson, B. I. Halperin and C. M. Varma, Anomalous Low Temperature Thermal Properties of Glasses and Spin-Glasses, Phil. Mag. $\underline{25}$, 1 (1972).

24. B. I. Halperin, P. C. Hohenberg and S. Ma, Calculation of Dynamic Critical Properties Using Wilson's Expansion Methods, Phys. Rev. Lett. $\underline{29}$, 1548 (1972).

25. B. I. Halperin, Electronic States in Disordered Systems, Proc. of Summer School on the Theory of Condensed Matter, Kiljava, Finland, Aug. 1971; Phys. Fennica $\underline{8}$, 215 (1973).

26. B. Golding, J. E. Graebner, B. I. Halperin and R. J. Schutz, Nonlinear Phonon Propagation in Fused Silica Below 1 K, Phys. Rev. Lett. $\underline{30}$, 223 (1973).

27. M. Pollak and B. I. Halperin, Comments on D. C. Conductivity in Amorphous Semiconductors, Solid State Communications $\underline{13}$, 869 (1973).

28. B. I. Halperin, Rigorous Inequalitites for the Spin-Relaxation Function in the Kinetic Ising Model, Phys. Rev. B $\underline{8}$, 4437 (1973).

29. B. I. Halperin, Aspects of Time-Dependent Critical Phenomena, in Proceedings of the Twenty-Fourth Nobel Symposium, Edited by B. Lundqvist and S. Lundqvist, Academica Press, New York (1974) p. 54.

30. B. I. Halperin, T. C. Lubensky and S. Ma, First-Order Phase Transitions in Superconductors and Smectic-A Liquid Crystals, Phys. Rev. Lett. $\underline{32}$, 292 (1974).

31. B. I. Halperin and T. C. Lubensky, On the Analogy Between Smectic-A Liquid Crystals and Superconductors, Solid State Communications, $\underline{14}$, 997 (1974).

32. B. I. Halperin, P. C. Hohenberg and S. Ma, Renormalization Group Methods for Critical Dynamics; I. Recursion Relations and Effects of Energy Conservation, Phys. Rev. B $\underline{10}$, 139 (1974).

33. B. I. Halperin, P. C. Hohenberg, and E. D. Siggia, Renormalization Group Calculations of Divergent Transport Coefficients of Critical Points, Phys. Rev. Lett. $\underline{32}$, 1289 (1974).

34. H. Fukuyama, T. M. Rice, C. M. Varma and B. I. Halperin, Some Properties of the One-Dimensional Fermi Model, Phys. Rev. B $\underline{10}$, 3775 (1974).

35. B. I. Halperin, Dynamic Properties of the Multi-Component Bose Fluid, Phys. Rev. B $\underline{11}$, 178 (1975).

36. D. J. Bergman and B. I. Halperin, Hydrodynamic Theory Applied to Fourth Sound in a Moving Superfluid, Phys. Rev. B $\underline{11}$, 4253 (1975).

37. A. Aharony and B. I. Halperin, Exact Relations among Amplitudes at Critical Points of Marginal Dimensionality, Phys. Rev. Lett. $\underline{35}$, 1308 (1975).

38. D. J. Bergman and B. I. Halperin, Critical Behavior of an Ising Model on a Cubic Compressible Lattice, Phys. Rev. B $\underline{13}$, 2145 (1976).

39. B. I. Halperin, P. C. Hohenberg and E. D. Siggia, Renormalization Group Treatment of the Critical Dynamics of Superfluid Helium, the Isotropic Antiferromagnet, and the Easy-Plane Ferromagnet, Phys. Rev. B $\underline{13}$, 1299 (1976).

40. E. D. Siggia, B. I. Halperin and P. C. Hohenberg, Renormalization-Group Treatment of the Critical Dynamics of the Binary-Fluid and Gas-Liquid Transitions, Phys. Rev. B $\underline{13}$, 2110 (1976).

41. B. I. Halperin, P. C. Hohenberg and S. K. Ma, Renormalization-Group and Methods for Critical Dynamics: II Detailed Analysis of the Relaxational Models, Phys. Rev. B $\underline{13}$, 4119 (1976).

42. B. I. Halperin, Theory of Dynamic Critical Properties, Proceeding of the International Conference on Statistical Physics, Budapest, August 1975.

43. B. I. Halperin and C. M. Varma, Defects and the Central Peak near Structural Phase Transitions, Phys. Rev. B $\underline{14}$, 4030 (1976).

44. P. C. Hohenberg, A. Aharony, B. I. Halperin and E. D. Siggia, Two Scale Factor Universality and the Renormalization Group, Phys. Rev. B $\underline{13}$, 2986 (1976).

45. B. I. Halperin, Can Tunneling Levels Explain the Anomalous Properties of Glasses at Very Low Temperature?, in The Glass Transition and the Nature of the Glassy State, edited by M. Goldstein and R. Simha, Annals of the New York Academy of Sciences, 279, 173 (1976).

46. P. C. Hohenberg, E. D. Siggia and B. I. Halperin, Density-Correlation Function for liquid Helium near T_λ in the Symmetric Planar-Spin Model, Phys. Rev. B $\underline{14}$, 2865 (1976).

47. P. C. Hohenberg and B. I. Halperin, Theory of Dynamic Critical Phenomena, Rev. Mod. Phys. <u>49</u>, 435 (1977).

48. B. I. Halperin and W. M. Saslow, Hydrodynamic Theory of Spin Waves in Spin Glasses and other Systems with Non-Collinear Spin Orientations, Phys. Rev. B <u>16</u>, 2154 (1977).

49. J. L. Black and B. I. Halperin, Spectral Diffusion, Phonon Echoes, and Saturation Recovery in Glasses at Low Temperatures, Phys. Rev. B <u>16</u>, 2879 (1977).

50. V. Ambegaokar, B. I. Halperin, D. R. Nelson and E. Siggia, Dissipation in Two-Dimensional Superfluids, Phys. Rev. Lett. <u>40</u>, 783 (1978).

51. P. M. Platzman and B. I. Halperin, The Effect of Parallel Magnetic Field on Interband Transition in Two-Dimensional Systems, Phys. Rev. B <u>18</u>, 226 (1978).

52. R. Bausch and B. I. Halperin, Renormalization Group Analysis of the Critical Dynamics of a Hamiltonian Model with a Scalar Displacive Transition, Phys. Rev. B <u>18</u>, 190 (1978).

53. B. I. Halperin and D. R. Nelson, Theory of Two-Dimensional Melting, Phys. Rev. Lett. <u>41</u>, 121 (1978). Erratum: Phys.Rev. Lett. <u>41</u>, 519 (1978).

54. R. G. Petschek and B. I. Halperin, Proton-Spin-Resonance Relaxation Times Near the Ordering Transition in $NH_4 Cl$, Phys. Rev. B <u>19</u>, 166 (1979).

55. R. A. Pelcovits and B. I. Halperin, Two-Dimensional Ferro-electric Crystals, Phys. Rev. B <u>19</u>, 4614 (1979).

56. D. R. Nelson and B. I. Halperin, Dislocation Mediated Melting in Two Dimensions, Phys. Rev. B <u>19</u>, 2457 (1979).

57. B. I. Halperin and D. R. Nelson, Resistive Transition in Superconducting Films, J. Low Temp. Phys. <u>36</u>, 599 (1979).

58. B. I. Halperin, Superfluidity, Melting and Liquid Crystal Phases in Two-Dimensions, in Proceeding of Kyoto Summer Institute 1979 — Physics of Low Dimensional Systems, edited by Y. Nagaoka and S. Hikami (Publications Office, Progress of Theoretical Physics, Kyoto, 1979), p. 53.

59. D. S. Fisher, P. M. Platzman and B. I. Halperin, Phonon-Ripplon Coupling and the Two-Dimensional Electron Solid on a Liquid Helium Surface, Phys. Rev. Lett. <u>42</u>, 798 (1979).

60. D. S. Fisher, B. I. Halperin and R. Morf, Defects in the Two-Dimensional Electron Solid and Implications for Melting, Phys. Rev. B <u>20</u>, 4692 (1979).

61. D. R. Nelson and B. I. Halperin, Solid and Fluid Phases in Smectic Layers with Tilted Molecules, Phys. Rev. B <u>21</u>, 5312 (1980).

62. V. Ambegaokar, B. I. Halperin, D. R. Nelson and E. D. Siggia, Dynamics of Superfluid Films, Phys. Rev. B <u>21</u>, 1806 (1980).

63. A. Zippelius, B. I. Halperin and D. R. Nelson, Dynamics of Two-Dimensional Melting, Phys. Rev. B 22, 2514 (1980).

64. P. C. Hohenberg, B. I. Halperin and D. R. Nelson, Deviation from Dynamic Scaling at the Superfluid Transition in Two and Three Dimensions, Phys. Rev. B 22, 2373 (1980).

65. S. N. Coppersmith, D. S. Fisher, B. I. Halperin, P. A. Lee and W. F. Brinkman, Dislocations and the Commensurate-Incommensurate Transition in Two Dimensions, Phys. Rev. Lett. 46, 549 (1981); 46, 869 (E) (1981); and Phys. Rev. B25, 349 (1982).

66. S. Ostlund and B. I. Halperin, Dislocation-Mediated Melting of Anisotropic Layers, Phys. Rev. B 23, 335 (1981).

67. B. I. Halperin, Theory of Melting and Liquid-Crystal Phases in Two Dimensions, Proceeding of the Colloque Pierre Curie, Paris, September 1-5, 1980 (in press).

68. B. I. Halperin, Statistical Mechanics of Topological Defects, in Physics of Defects, Les Houches 1980 Session XXXV, ed. by R. Balian, M. Kléman and J.-P. Poirier (North Holland, 1981) p. 813.

69. C. Dasgupta and B. I. Halperin, Phase Transition in a Lattice Model of Superconductivity, Phys. Rev. Lett. 47, 1556 (1981).

70. R. Bruinsma, B. I. Halperin and A. Zippelius, Motion of Defects and Stress Relaxation in Two-Dimensional Crystals, Phys. Rev. B 25, 579 (1982).

71. B. I. Halperin, Quantized Hall Conductance, Current-Carrying Edge States and the Existence of Extended States in a Two-Dimensional Disordered Potential, Phys. Rev. B 25, 2185 (1982).

72. C. Henley, H. Sompolinsky and B. I. Halperin, Spin-Resonance Frequencies in Spin-Glass with Random Anisotropies, Phys. Rev. B 25, 5489 (1982).

73. A. Weinrib and B. I. Halperin, Distribution of Maxima, Minima, and Saddle Points of the Intensity of Laser Speckle Patterns, Phys. Rev. B 27, 413 (1983).

74. H. J. Schulz and B. I. Halperin and C. L. Henley, Dislocation Interaction in a Adsorbate Solid near the Commensurate-Incommensurate Transition, Phys. Rev. B 26, 3797 (1982).

75. A. Weinrib and B. I. Halperin, Critical Phenomena in Systems with Long-Range-Correlated Quenched Disorder, Phys. Rev. B 27, 413 (1983).

76. E. Brézin, B. I. Halperin and S. Leibler, Critical Wetting: The Domain of Validity of Mean Field Theory, J. Physique 44, 775 (1983).

77. E. Brézin, B. I. Halperin and S. Leibler, Critical Wetting in Three Dimensions, Phys. Rev. Lett. 50, 1387 (1983).

78. R. Rammal, G. Toulouse, M. T. Jaekel and B. I. Halperin, Quantized Hall Conductance and Edge States: Two-dimensional Strips with a Periodic Potential, Phys. Rev. B 27, 5142 (1983).

79. D. Yoshioka, B. I. Halperin and P. A. Lee, Ground State of Two-Dimensional Electrons in Strong Magnetic Fields and 1/3 Quantized Hall Effect, Phys. Rev. Lett. $\underline{50}$, 1219 (1983).

80. B. I. Halperin, Theory of the Quantized Hall Conductance, Helvetica Physica Acta, $\underline{56}$, 75 (1983).

81. C. Kallin and B. I. Halperin, Surface-Induced Charge Disturbances and Piezoelectricity in Insulating Crystals, Phys. Rev. B $\underline{29}$, 2175 (1984).

82. J. C. Hensel, B. I. Halperin and R. C. Dynes, A Dynamical Model for the Absorption and Scattering of Ballistic Phonons by the Electron Inversion Layer in Silicon, Phys. Rev. Lett. $\underline{51}$, 2302 (1983).

83. J. C. Hensel, R. C. Dynes, B. I. Halperin and D. C. Tsui, Scattering and Absorption of Ballistic Phonons by the Electron Inversion Layer in Silicon: Theory and Experiment, Surface Science $\underline{142}$, (1984).

84. D. Yoshioka, B. I. Halperin and P. A. Lee, The Ground State of the 2D Electrons in a Strong Magnetic Field and the Anomalous Quantized Hall Effect, Surface Science $\underline{142}$, 155 (1984).

85. B. I. Halperin, Statistics of Quasiparticles and the Hierarchy of Fractional Quantized Hall States, Phys. Rev. Lett. $\underline{52}$, 1583 (1984). Erratum: Phys. Rev. Lett. $\underline{52}$. 2390(1984).

86. C. Kallin and B. I. Halperin, Excitations from a Filled Landau Level in the Two-Dimensional Electron Gas, Phys. Rev. B $\underline{30}$, 5655 (1984).

87. C. Kallin and B. I. Halperin, Many-Body Effects on the Cyclotron Resonance in a Two-Dimensional Electron Gas, Phys. Rev. B $\underline{31}$, 3635 (1985).

88. S. Feng, P. N. Sen, B. I. Halperin and C. J. Lobb, Percolation on 2-D Elastic Networks with Rotationally Invariant Bond-Bending Forces, Phys. Rev. B $\underline{30}$, 5386 (1984).

89. B. I. Halperin, S. Feng and P. N. Sen, Differences Between Lattice and Continuum Percolation Transport Exponents, Phys. Rev. Lett. $\underline{54}$, 2391 (1985).

90. B. I. Halperin, S. Feng and P. N. Sen, Transport Properties near the Percolation Threshold of Continuum Systems, in <u>Localization and Metal Insulator Transitions</u>, ed. by H. Fritzsche and D. Adler (Plenum, N.Y. 1985).

91. D. R. Nelson and B. I. Halperin, Pentagonal and Icosahedral Order in Rapidly Cooled Metals, Science $\underline{229}$, 4710 (1985).

92. P. N. Sen, J. N. Roberts and B. I. Halperin, Non-Universal Critical Exponents for Transport in Percolating Systems with a Distribution of Bond Strengths, Phys. Rev. B $\underline{32}$, 3306 (1985).

93. B. I. Halperin, Calculations of the Fractional Quantized Hall Effect, Proceedings of Sixth Conf. On Electronic Properties of Two-Dimensional Systems (Kyoto, Sept. 1985). Surface Science 170, 115 (1986).

94. R. Morf and B. I. Halperin, Monte-Carlo Evaluation of Trial Wavefunctions for the Fractional Quantized Hall Effect: I. Disc Geometry, Phys. Rev. B $\underline{33}$, 221 (1986).

95. B. I. Halperin, Condensed Matter Physics and Materials Research, in <u>Advancing Materials Research</u>, edited by P. A. Psaras and H. D. Langford (National Academy Press 1987) p. 131.

96. M. Ma, B. I. Halperin and P. A. Lee, Strongly Disordered Superfluids: Quantum Fluctuations and Critical Behavior. Phys. Rev. B $\underline{34}$, 3136 (1986).

97. P. N. Sen, S. Feng, B. I. Halperin and M. F. Thorpe, Elastic Properties of Depleted Networks and Continua. To appear in Proceedings of the Les Houches Workshop April 1985.

98. B. I. Halperin, The 1985 Nobel Prize in Physics, Science $\underline{231}$, 820 (1986).

99. B. I. Halperin, The Quantized Hall Effect, Scientific American $\underline{254}$, No. 4, p. 52 (1986).

100. J. Eisinger and B. I. Halperin, Effects of Spatial Variation in Membrane Diffusability and Solubility on the Lateral Transport of Membrane Components. Biophys. J. $\underline{50}$, 513 (1986).

101. S. Feng, B. I. Halperin and P. N. Sen, Transport Properties of Continuum Systems Near the Percolation Treshold. Phys. Rev. B $\underline{35}$, 197 (1987).

102. R. Morf, N. d'Ambrumenil and B. I. Halperin, Microscopic Wavefunctions for the Fractional Quantized Hall States at v = 2/5 and 2/7. Phys. Rev. B $\underline{34}$, 3037 (1986).

103. M. Rasolt, B. I. Halperin and D. Vanderbilt, Dissipation in the Quantum Hall Effect of a Multivalley Semiconductor due to a "Valley Wave" Channel, Phys. Rev. Lett. $\underline{57}$, 126 (1986).

104. B. I. Halperin, Z. Tesanovic and F. Axel. Compatibiltiy of Crystalline Order and the Quantized Hall Effect. Phys. Rev. Lett. $\underline{57}$, 922 (1986) (Comment).

105. S. Chakravarty, S. Kivelson, G. T. Zimanyi and B. I. Halperin, Effect of Quasiparticle Tunneling on Quantum Phase Fluctuations and the Onset of Superconductivity in Granular Films, Phys. Rev. B $\underline{35}$, 7256 (1987).

106. B. I. Halperin, Degenerately-Doped Semiconductors in Strong Magnetic Fields, in *Condensed Matter Theories, Vol. 2* (Proceedings of the International Workshop at Argonne, Ilinois, July 21–25 1986) edited by P. Vashista, R. K. Kalia and R. F. Bishop (Plenum Press, N.Y., 1987). p. 259.

107. R. Morf and B. I. Halperin, Monte Carlo Evaluation of Trial Wavefunctions for the Fractional Quantized Hall Effect: Spherical Geometry, Z. Phys. Rev. B $\underline{68}$, 391 (1987).

108. H. A. Fertig and B. I. Halperin, Transmission Coefficient of an Electron Through a Saddle Point Potential in a Magnetic Field, Phys. Rev. B $\underline{36}$, 7969 (1987).

109. H. A. Fertig and B. I. Halperin, The Hypernetted Chain Approximation and Quasi-particle Energies in the 1/3 Fractional Quantized Hall Effect, Phys. Rev. B $\underline{36}$, 6302 (1987).

110. B. I. Halperin, Possible States of a Three-Dimensional Electron Gas in a Strong Magnetic Field, in Proceedings of the 18th International Conference on Low Temperature Physics, Kyoto, 1987, Japanese Journal of Applied Physics 26, Suppl. 26–3, 1913 (1987).

111. Z. Tesanovic and B. I. Halperin, Multivalley Electron Gas in a Strong Magnetic Field, Phys. Rev. B 36, 4888 (1987).

112. S. Chakravarty, B. I. Halperin and D.R. Nelson, Low-Temperature Behavior of Two-Dimensional Quantum Antiferromagnets, Phys. Rev. Lett. 60, 1057 (1988).

113. S. Tyc and B. I. Halperin, Random Resistor Network with an Exponentially Wide Distribution of Bond Conductances, Phys. Rev. B 39, 877 (1989).

114. S. Chakravarty, B. I. Halperin and D.R. Nelson, Two-Dimensional Quantum Heisenberg Antiferromagnet at Low Temperatures, Phys. Rev. B 39, 2344 (1989).

115. S. Tyc, B. I. Halperin and S. Chakravarty, Dynamic Properties of a Two-Dimensional Heisenberg Antiferromagnet at Low-Temperatures, Phys. Rev. Lett. 62, 835 (1989).

116. Z. Tesanovic, F. Axel and B. I. Halperin, Hall Crystal Versus Wigner Crystal, Phys. Rev. B 39, 8535 (1989).

117. Y. Chen, F. Wilczek, E. Witten and B. I. Halperin, On Anyon Superconductivity, Internat'l Journal Modern Physics, 3, 1001 (1989).

118. L. Brey and B. I. Halperin, Spin-Density-Wave Instability in Wide Parabolic Quantum Wells, Phys. Rev. B 40, 11634 (1989).

119. L. Brey, N. F. Johnson and B. I. Halperin, Optical and Magneto-Optical Absorption in Parabolic Quantum Wells, Phys. Rev. B 40, 10647 (1989).

120. B. I. Halperin, J. March-Russell and F. Wilczek, Consequences of Time-Symmetry Reversal-Violation in Models of High-T_c Superconductors, Phys. Rev. B 40, 8726 (1989).

121. L. Schwartz, J. R. Banavar and B. I. Halperin. Biased-Diffusion Calculations of Electrical Transport in Inhomogeneous Continuum Systems, Phys. Rev. B 40, 9155 (1989).

122. S. Tyc and B. I. Halperin, Damping of Spin Waves in a Two-Dimensional Heisenberg Antiferromagnet at Low Temperatures, Phys. Rev. B 42, 2096 (1990).

123. L. Brey, J. Dempsey, N. F. Johnson and B. I. Halperin, Infrared Optical Absorption in Imperfect Parabolic Quantum Wells, Phys. Rev. B 42, 1240 (1990).

124. B. I. Halperin, The Search for Anyon Superconductivity, Proceedings of the International Seminar on Theory of High Temperature Superconductivity, Dubna USSR, 3–6 July, 1990.

125. J. Dempsey, N. F. Johnson, L.Brey and B. I. Halperin, Collective Modes in Quantum Dot Arrays in Magnetic Fields, Phys. Rev. B 42, 11708 (1990).

126. M. Hagiwara, K. Katsumata, I. Affleck, B. I. Halperin, and J. P. Renard, Observation of S = 1/2 Degrees of Freedom in an S = 1 Linear-Chain Heisenberg Antiferromagnet, Phys. Rev. Lett. 65, 3181 (1990).

127. B. I. Halperin, The Hunt for Anyon Superconductivity, in <u>The Physics and Chemistry of</u> <u>Oxide</u> <u>superconductors</u>, edited by Y. Iye and H.Fukuyama, (Springer-Verlag, Berlin 1992) p. 439.

128. B. I. Halperin, The Electron Gas Between Two and Three Dimensions, in <u>Physical Phenomena</u> <u>in High Magnetic Fields</u>, E. Manousakis, ed.,(Addison Wesley, 1991).

129. P. P. Mitra, B. I. Halperin, and I. Affleck, Temperature-Dependence of the ESR Spectrum of the Chain-End S = 1/2 Degrees ofFreedom in an S = 1 Linear Chain, Phys. Rev. B 45, 5299 (1992).

130. M. Hagiwara, K Katsumata, J. P. Renard, I. Affleck and B. I.Halperin, Hyperfine Structure due to the S = 1/2 Degrees of Freedom in an S = 1 Linear-Chain Antiferromagnet, J. Magnetism and Magnetic Mat., 104, 839 (1992).

131. B. I. Halperin, Quantum Antiferromagnets in One and Two Dimensions, J. Magnetism and Magnetic Mat., 104–107, 761 (1992).

132. J. Dempsey and B. I. Halperin, Magnetoplasma Excitations in Parabolic Quantum Wells: Hydrodynamic Model, Phys. Rev. B 45, 1719 (1992).

133. J. Dempsey and B. I. Halperin, Tilted-Field Magneto-Optical Absorption in an Imperfect Parabolic Quantum Well, Phys. Rev. B 45, 3902 (1992).

134. B. I. Halperin, Local Magnetic Fields in an Anyon Superconductor, Phys. Rev. B 45, 5504 (1992).

135. B. Y. Gelfand and B. I. Halperin, Hartree Calculation of Local Magnetic Fields in an Anyon Superconductor, Phys. Rev. B 45, 5517 (1992).

136. M. Kohmoto, B. I. Halperin and Y.-S. Wu, Diophantine Equation for the 3D Quantum Hall Effect, Phys. Rev. B 45, 13488 (1992).

137. J. Dempsey and B. I. Halperin, Magneto-optical Absorption in an Overfilled Parabolic Quantum Well. I. Tilted Field, Phys. Rev. B 47, 4662 (1993).

138. J. Dempsey and B. I. Halperin, Magneto-optical Absorption in an Overfilled Parabolic Quantum Well. II In-plane field, Phys. Rev. B 47, 4674 (1993).

139. B. I. Halperin, P. A. Lee and Nicholas Read, Theory of the Half-filled Landau Level, Phys. Rev. B 47, 7312 (1993).

140. B. I. Halperin, Theories for n = 1/2 in Single- and Double-Layer Systems, Proceedings of Tenth Int'l Conf. on Electronic Properties of Two-Dimensional Systems, Newport, R.I. May 31–June 4, 1993, Surface Science 305, 1 (1994).

141. B. I. Halperin, The Quantized Hall Effect at Even-Denominator Fractions and What Happens When It Does Not Occur, in Physics News in 1992, P.F. Schewe, Ed., American Institute of Physics (1992).

142. J. Dempsey, B. Y. Gelfand and B. I. Halperin, Electron-Electron Interactions and Spontaneous Spin Polarization in Quantum Hall Edge States, Phy. Rev. Lett. 70, 3639 (1993).

143. S. He, P. M. Platzman and B. I. Halperin, Tunneling into a Two-Dimensional Electron System in a Strong Magnetic Field, Phys. Rev. Letters 71, 777 (1993).

144. M. Kohmoto, B. I. Halperin and Y.-S. Wu, Quantized Hall Effect in 3D Periodic Systems, Physica B 184, 30 (1993).

145. S. He, P. M. Platzman and B. I. Halperin, One-Electron Green's Function and Electron Tunneling in a Two Dimensional Electron System in a Strong Magnetic Field, Surf. Sci. 305, 398, (1994).

146. P. N. Sen, L. M. Schwartz, P. P. Mitra and B. I. Halperin, SurfaceRelaxation and the Long Time Diffusion Coefficient in Porous Media: Periodic Geometries, Phys. Rev. B 49, 215 (1994).

147. B. Y. Gelfand, J. Dempsey and B. I. Halperin, Effects of Electron-Electron Interactions on the Ground State of Quantum Edge States, Surface Science 305, 166 (1994).

148. P. P. Mitra and B. I. Halperin, Understanding the Far-Infrared Absorption in the S=1 Antiferromagnet Chain Compound NENP, Phys. Rev. Letters 72, 912 (1994).

149. B. Y. Gelfand and B. I. Halperin, Edge Electrostatics of Mesa-Etched Sample and Edge-to-Bulk Scattering Rate in the Fractional Quantum Hall Regime, Phys. Rev. B 49, 1862 (1994).

150. S. Simon and B. I. Halperin, Finite Wave-vector Electromagnetic Response of Fractional Quantized Hall States, Phys. Rev. B 48, 17368 (1993).

151. S. Simon, B. I. Halperin, Response Function of the Fractional Quantized Hall States on a Sphere I: Fermion Chern-Simons Theory, Phys. Rev. B 24, 49, 50 (1994).

152. S. He, S. Simon and B. I. Halperin, Response Function of the Fractional Quantized Hall States on a Sphere II: Exact Diagonalization, Phys. Rev. B 50, 1823 (1994).

153. B. I. Halperin, Developments in the Theory of the Quantum Hall Effect in Proceedings of the 22nd International Conference on the Physics of Semiconductors, Vancouver, Canada 1994, edited by D.J. Lockwood, World Scientific, 1995) Vol. 2, p. 975.

154. P. P. Mitra and B. I. Halperin, Effects of Finite Gradient Pulse Widths in Pulsed Field Gradient Diffusion Measurements, J. Mag. Res. 113, 94 (1995).

155. B.Y. Gelfand and B. I. Halperin, Orbital Magnetization of a Finite Anyon System: A Hartree Calculation on a Cylinder, Phys. Rev. B 50, 13449 (1994).

156. B. Y. Gelfand and B. I. Halperin, Hartree Calculation of Local Magnetic Fields in an Anyon Superconductor: Dependence on the Background Dielectric Constant, Phys. Rev. B $\underline{50}$, 16745 (1994).

157. B. I. Halperin, Recent Developments in the Theory of the Quantum Hall Effect, Chinese J. Physics $\underline{33}$, 1(1995).

158. S. H. Simon and B. I. Halperin, Explanation for the Resistivity Law in Quantum Hall Systems, Phys. Rev. Letters $\underline{73}$, 3278 (1994).

159. A. Stern and B .I. Halperin, Singularities in the Fermi Liquid Description of a Partially Filled Landau Level and the Energy Gaps of Fractional Quantum Hall Liquids, Phys. Rev. B $\underline{52}$, 5890 (1995).

160. J. M. Golden and B. I. Halperin, Relation Between Barrier Conductance and Coulomb Blockade Peaks Splitting for Tunnel-Coupled Quantum Dots, Phys. Rev. B $\underline{53}$, 3893 (1996).

161. B. I. Halperin, Fermion Chern-Simons Theory and the Unquantized QuantumHall Effect, chapter to appear in New Perspectives in Quantum Hall Effects, edited by S. Das Sarma and A. Pinczuk, (John Wiley & Sons, 1997) pp. 225–264.

162. M. Ya. Azbel and B. I. Halperin, Landau Levels in the Presence of Dilute Short-Range Scatterers, Phys. Rev. B $\underline{52}$, 14098 (1995).

163. I. M. Ruzin, N. R. Cooper and B. I. Halperin, Nonuniversal Behavior of Finite Quantum Hall Systems as a Result of Weak Macroscopic Inhomogenities, Phys. Rev. B $\underline{53}$, 1558 (1996).

164. D. B. Chklovskii and B. I. Halperin, Understanding the Dynamics of Fractional Edge States with Composite Fermions, Surface Science $\underline{361}/\underline{362}$, 79 (1996).

165. A. Stern and B. I. Halperin, The Physical Significance of Singularities in the Chern-Simons Fermi Liquid Description of a Partially Filled Landau Level, Surface Science 361/362, 42 (1996).

166. I. Affleck and B. I. Halperin, On a Renormalization Group Approach to Dimensional Crossover, J. Phys. A $\underline{29}$, 2627 (1996).

167. J. M. Golden and B. I. Halperin, Higher-Order Results for the Relation between Channel Conductance and the Coulomb Blockade for two Tunnel-Coupled Quantum Dots, Phys. Rev. B $\underline{54}$, 16757 (1996).

168. S. H. Simon, A. Stern and B. I. Halperin, Composite Fermions with Orbital Magnetization, Phys. Rev. B $\underline{54}$, 11114 (1996).

169. N. R. Cooper, B. I. Halperin and I. M. Ruzin, Thermoelectric Response of an Interacting Two-Dimensional Electron Gas in a Quantizing Magnetic Field, Phys. Rev B $\underline{55}$, 2344 (1997).

170. C.-K. Hu and B. I. Halperin, Scaling Function for the Number of Alternating Percolation Clusters on Self-Dual Finite Square Lattices, Phys. Rev. B $\underline{55}$, 2705 (1997).

171. N. R. Cooper, B. I. Halperin, C.-K. Hu, and I. M. Ruzin, Statistical Properties of the Low-Temperature Conductance Peak-Heights for Corbino Discs in the Quantum Hall Regime, Phys. Rev. B 55, 4551 (1997).

172. J. M. Golden and B. I. Halperin, Correction to the Universal Behavior of the Coulomb-Blockade Peak-Splitting for Quantum Dots Separated by a Finite Barrier, Phys. Rev. B 56, 4716 (1997).

173. D. B. Chklovskii and B. I. Halperin, Consequences of a Possible Adiabatic Transition Between n = 1/3 and n = 1 quantum Hall states in a Narrow Wire, Phys. Rev. B 57, 3781 (1998).

174. D.B. Chklovskii and B. I. Halperin, Possibility of an Adiabatic Point Contact Between n = 1/3 and n = 1 Quantum Hall States, Physica E1, 75 (1997).

175. A. V. Shytov, L. S. Levitov and B. I. Halperin, Tunneling into the Edge of a Compressible Quantum Hall State, Phys. Rev. Letters, 80, 141 (1998).

176. F. von Oppen, A. Stern and B. I. Halperin, Composite Fermions in Modulated Structures: Transport and Surface Acoustic Waves, Phys. Rev. Letters 80, 4494 (1998).

177. S. D. Berger and B. I. Halperin, Parity Effect in a Small Superconducting Particle, Phys. Rev. B 58, 5213 (1998).

178. L. S. Levitov, A. V. Shytov and B. I. Halperin, Effective Action and Green's Function for aCompressible Quantum Hall Edge State, Physics-Uspekhi 41, 141 (1998).

179. B. I. Halperin and A. Stern, Comment on Towards a Field Theory of Fractional Quantum Hall States, Phys. Rev. Letters 80, 5457 (1998).

180. C. Mudry, P. W. Brouwer, B. I. Halperin, V. Gurarie and A. Zee, Density of States in a Non-Hermitian Lloyd Model, Phys. Rev. B 58, 13539 (1998).

181. A. Stern, B. I. Halperin, F. von Oppen and S. H. Simon, Half-Filled Landau Level as a Fermi Liquid of Dipolar Quasiparticles, Phys. Rev. B 58, 13539 (1998).

182. F. von Oppen, B. I. Halperin, S. H. Simon and A. Stern, The Half-Filled Landau Level-Composite Fermions and Dipoles, to appear in Festkorperprobleme: Advances in Solid State Physics 39 (Vieweg, Braunschweig, 1999) p. 203 (cond-mat/9903430).

183. Y. Oreg and B. I. Halperin, Longitudinal Drag and Hall Drag in a Bilayer System with Pinholes, Phys. Rev B 60, 5679 (1999).

184. P. W. Brouwer, Y. Oreg, and B. I. Halperin, Mesoscopic Fluctuations of the Ground-State Spin of a Small Metal Particle, Phys. Rev. B 60, R13977 (1999).

185. Y. Oreg, K. Byczuk, and B. I. Halperin, Spin Configurations of a Carbon Nanotube in a Nonuniform External Potential, Phys. Rev. Letters 84, 365 (2000).

186. F. von Oppen, B. I. Halperin and A. Stern, Conductivity Tensor of Striped Quantum Hall Phases, Phys. Rev. Letters 84, 2937 (2000).

187. P. W. Brouwer, X. Waintal and B. I. Halperin, Fluctuating Spin g-Tensor in Small Metal Grains, Phys. Rev. Letters <u>85</u>, 369 (2000).

188. L. Marinelli, B. I. Halperin and S. H. Simon, Quasiparticle Spectrum of d-Wave Superconductors in the Mixed State, Phys. Rev. B <u>62</u>, 3488 (2000).

189. L. S. Levitov, A. V. Shytov and B. I. Halperin, Effective Action of a Compressible QH State Edge: Application toTunneling, Phys. Rev. B <u>64</u>, 075322 (2001)

190. B. I. Halperin, A. Stern, Y. Oreg, J. N. H. J. Cremers, J. A. Folk and C. M Marcus, Spin-Orbit Effects in a GaAs Quantum Dot in a Parallel Magnetic Field, Phys. Rev. Letters <u>86</u>, 2106 (2001).

191. A. Lopatnikova, S. H. Simon, B. I. Halperin, and X. G. Wen, Striped States in Quantum Hall Effect: Deriving a Low-Energy Theory from Hartree-Fock, Phys. Rev. B <u>64</u>, 155301 (2001).

192. B. Rosenow and B. I. Halperin, Non-Universal Behavior of Scattering Between Fractional Quantum Hall Edges, Phys. Rev. Lett. <u>88</u>, 096404 (2002).

193. L. Marinelli and B. I. Halperin, Quasiparticle Spectrum of d-Wave Superconductors in the Mixed State: A Large Fermi-Velocity-Anisotropy Study, Phys. Rev. B <u>65</u>, 014516 (2002).

194. J. M. Golden and B. I. Halperin, Coulomb Blockade of Strongly Coupled Quantum Dots Studied via Bosonization of a Channel with a Finite Barrier, Phys. Rev. B <u>65</u>, 115326 (2002).

195. Y. Oreg, P. W. Brouwer, X. Waintal and B. I. Halperin, Spin, Spin-Orbit, and Electron-Electron Interactions in Mesoscopic Systems, "Nano-Physics and Bio-Electronics," edited by T. Chakraborty, F. Peeters, and U. Sivan (Elsevier, 2002.) pp. 147-186.

196. P. W. Brouwer, J. N. H. J. Cremers and B. I. Halperin, Weak Localization and Conductance Fluctuations of a Chaotic Quantum Dot with Tunable Spin-Orbit Coupling, Phys. Rev. B <u>65</u>, 081302 (2002).

197. E. Demler, D-W. Wang, S. Das Sarma and B. I. Halperin, Quantum Hall Stripe Phases at Integer Filling Factors, Solid State Comm. <u>123</u>, 243 (2002).

198. A. Stern and B. I. Halperin, Strong Enhancement of Drag and Dissipation at the Weak- to Strong- Coupling Phase Transition in a Bi-Layer System at a Total Landau Level Filling n = 1, Phys. Rev. Lett. <u>88</u>, 106801 (2002).

199. Gregory A. Fiete, Gergely Zarand, Bertrand I. Halperin, and Yuval Oreg, Kondo Effect and STM Spectra through Ferromagnetic Nanoclusters, Phys. Rev. B <u>66</u>, 024431, (2002).

200. A. Brataas, Y. Tserkovnyak, G. E. W. Bauer and B. I. Halperin, Spin Battery Operated by Ferromagnetic Resonance, Phys. Rev. B <u>66</u>, 060404 (2002).

201. Mikito Koshino, Hideo Aoki and B. I. Halperin, Wrapping Current Versus Bulk Integer Quantum Hall Effect in Three Dimension, Phys. Rev. B <u>66</u>, 081301 (2002).

202. A. Aharony, O. Entin-Wohlman, B. I. Halperin and Y. Imry, Phase Measurement in the Mesoscopic Aharonov-Bohm Interferometer, Phys. Rev. B $\underline{66}$, 115311 (2002).

203. Y. Tserkovnyak, B. I. Halperin, O. M. Auslaender and A Yacoby, Finite-Size Effects in Tunneling Between Parallel Quantum Wires, Phys. Rev. Lett. $\underline{89}$, 136805 (2002).

204. D. W. Wang, S. Das Sarma, E. Demler and B. I. Halperin, Magnetoplasmon Excitations and Spin Density Instabilities in an Integer Quantum Hall System with a Tilted Magnetic Field, Phys. Rev. B $\underline{66}$, 195334 (2002).

205. L. Borda, G. Zarand, W. Hofstetter, B. I. Halperin and J. von Delft, SU(4) Fermi Liquid State and Spin Filtering in a Double Quantum Dot System, Phys. Rev. Lett. $\underline{90}$, 026602 (2003).

206. B. I. Halperin, Ady Stern and S. M. Girvin, *dc* Voltage Step-up Transformer based on a Bi-layer n = 1 Quantum Hall System, Phys. Rev B $\underline{67}$, 235313 (2003).

207. Y. Tserkovnyak, B. I. Halperin, O. M. Auslaender and A. Yacoby, Interference and Zero-biasAnomaly in Tunneling Between Luttinger-Liquid Wires, Phys. Rev. B $\underline{68}$, 125312 (2003).

208. E.G. Mishchenko and B. I. Halperin, Transport Equations for a Two-Dimensional Electron Gas with Spin-Orbit Interaction, Phys. Rev. B 68, 045317 (2003).

209. B. I. Halperin, Composite Fermions and the Fermion-Chern-Simons Theory, Physica E $\underline{20}$, 71 (2003).

210. B. I. Halperin, Tunneling between parallel quantum wires, Ann. Henri Poincare, 4, Suppl. 2, S633 (2003).

211. Y. Tserkovnyak, B. I. Halperin, O.M. Auslaender, and A. Yacoby, Signatures of Spin-Charge Separation in Double-Quantum-Wire Tunneling, in Proceedings of the NATO Workshop on Theory of Quantum Transport in Metallic and Hybrid Nanostructures, St. Petersburg, August 2003, edited by V. Kozub and V. Vinokur. (Kluwer Academic Publishers, Dordecht, The Netherlands, in press)

212. Gerrit E.W. Bauer, Arne Brataas, Yaroslav Yserkovnyak, Bertrand I. Halperin, Maciej Zwierzycki, and Paul J. Kelly, Dynamic Ferromagnetic Proximity Effect in Photoexcited Semiconductors, Phys. Rev. Lett. $\underline{92}$, 145301 (2004).

213. E. G. Mishchenko, A. V. Shytov and B. I. Halperin, Spin Current and Polarization in Impure Two-Dimensional Electron Systems with Spin-Orbit Coupling, Phys. Rev. Lett. $\underline{93}$, 226601 (2004).

214. A. Polkovnikov, E. Altman, E. Demler, B. I. Halperin and M. D. Lukin, Decay of Super-Currents in Condensates in Optical Lattices, Journal of Superconductivity $\underline{17}$, 577 (2004).

215. Y. Tserkovnyak, G. A. Fiete and B. I. Halperin, Mean-Field Magnetization Relaxation in Conducting Ferromagnets, Appl Phys. Lett. $\underline{84}$, 5234 (2004).

216. O. M. Auslaender, H. Steinberg, A. Yacoby, Y. Tserkovnyak, B. I. Halperin, R. de Picciotto, K. W. Baldwin, L. N. Pfeiffer and K. W. West, Many-Body Dispersions in Interacting Ballistic Quantum Wires, Solid State Commun. <u>131</u>, 657 (2004).

217. Y. Tserkovnyak, A. Brataas, G. E. W. Bauer and B. I. Halperin, Nonlocal magnetization dynamics in ferromagnetic heterostructures, Rev. Mod. Phys. <u>77</u>, 1375 (2005).

218. E. Altman, A. Polkovnikov, E. Demler, B. I. Halperin and M. D. Lukin, Superfluid-insulator transition in a moving system of interacting bosons, Phys. Rev. Lett. <u>95</u>, 020402.

219. A. Polkovnikov, E Altman, E. Demler, B. I. Halperin and M. D. Lukin, Decay of a superfluid currents in a moving system of strongly interacting bosons, Phys. Rev. A <u>71</u>, 06313 (2005).

220. A. Auerbach, I. Finkler, B. I. Halperin and A. Yacoby, Steady states of a microwave irradiated quantum hall gas, Phys. Rev. Lett. <u>94</u>, 196801 (2005).

221. G.A. Fiete, J. Qian, Y. Tserkovnyak and B. I. Halperin, Theory of momentum resolved tunneling into short quantum wire, Phys. Rev. B <u>72</u>, 045315 (2005).

222. O. M. Auslaender, H. Steinberg, A. Yacoby, Y. Tserkovnyak, B. I. Halperin, K. W. Baldwin, L. N. Pfeiffer and K. W. West, Spin-Charge separation and localization in one dimension, Science 1 April 2005 308: 88–92.

223. H.-A. Engel, B. I. Halperin and E. Rashba, Theory of spin Hall conductivity in n-Doped GaAs, Phys. Rev. Lett. <u>95</u>, 166605 (2005).

224. A. V. Shytov, E. G. Mishchenko, H.-A.Engel and B. I. Halperin, Small-angle impurity scattering and the spin Hall conductivity in two-dimensional semiconductor systems, Phys. Rev. B <u>73</u>, 075316 (2006).

225. A. Stern and B. I. Halperin, Proposed experiments to probe the non-abelian v = 5/2 quantum Hall state, Phys. Rev. Lett. <u>96</u>, 016802 (2006).

226. H. Steinberg, O. M. Auslaender, A. Yacoby, J. Qian, G. A. Fiete, Y. Tserkovnyak, B. I. Halperin, K. W. Baldwin, L. N. Pfeiffer and K. W. West, Localization transition in a ballistic quantum wire, Phys. Rev. B <u>73</u>, 113307 (2006).

227. I. Finkler, B. I. Halperin, A. Auerbach and A. Yacoby, Domain patterns in the microwave-induced zero-resistance state, Journal of Statistical Physics <u>125</u>, 1093 (2006).

228. H.-A. Engel, E. I. Rashba and B. I. Halperin, Theory of Spin Hall Effects in Semiconductors, in Handbook of Magnetism and Advanced Magnetic Materials, H. Kronmüller and S. Parkin, eds. (John Wiley & Sons, Chichester, UK, 2007) Vol. 5, pp. 2858–2877. Cond-mat/0603306.

229. I. Adagideli, G. E. W. Bauer and B. I. Halperin, Detection of current-induced spins by ferromagnetic contact, Phys. Rev. Lett. <u>97</u>, 256601 (2006).

230. Y. Tserkovnyak and B. I. Halperin, Magnetoconductance Oscillations in Quasiballistic Multimode Nanowires, Phys. Rev. B <u>74</u>, 245327 (2006).

231. H.-A. Engel, E. I. Rashba and B. I. Halperin, Out-of-plane Spin Polarization from in-plane Electric and Magnetic Fields, Phys. Rev. Lett. <u>98</u>, 036602 (2007).

232. Y. Tserkovnyak, B. I. Halperin, A. A. Kovalev and A. Brataas, Boundary spin Hall effect in a two-dimensional semiconductor system with Rashba spin-orbit coupling, Phys. Rev. B <u>76</u>, 085319 (2007).

233. B. Rosenow and B. I. Halperin, Influence of Interactions on Flux and Back-Gate Periods of Quantum Hall Interfereometers, Phys. Rev. Lett. <u>98</u>, 106801 (2007).

234. B. I. Halperin, Spin-charge separation, tunneling, and spin transport in one-dimensional metals, Journal of Applied Physics <u>101</u>, 081607 (2007).

235. J. J. Krich and B. I. Halperin, Cubic Dresselhaus Spin-Orbit Coupling in 2D Electron Quantum Dots, Phys. Rev. Lett <u>98</u>, 226802 (2007).

236. I. G. Finkler, H. A. Engel, E. I. Rashba and B. I. Halperin, Spin generation away from boundaries by nonlinear transport, Phys. Rev. B <u>75</u>, 241202 (2007).

237. M. Levin, B. I. Halperin, and B. Rosenow, Particle-hole symmetry and the Pfaffian state, Phys. Rev. Lett. <u>99</u>, 236806 (2007).

238. J. Qian and B. I. Halperin, Hartree-Fock calculations of finite inhomogeneous quantum wire, Phys. Rev. B <u>77</u>, 085314 (2008).

239. B. Rosenow, B. I. Halperin, S. H. Simon, and A. Stern, Bulk-edge coupling in the non-abelian n = 5/2 quantum Hall interferometer, Phys. Rev. Lett. <u>100</u>, 226803 (2008).

240. D. A. Abanin, A. V. Shytov, L. S. Levitov and B. I. Halperin, Nonlocal charge transport mediated by spin diffusion in the Spin-Hall Effect regime, Phys. Rev. B <u>79</u>, 035304 (2009).

241. I. Dimov, B. I. Halperin, and C. Nayak, Spin order in paired Quantum Hall states, Phys. Rev. Lett. <u>100</u>, 126804 (2008).

242. R. G. Pereira, N. Laflorencie, I. Affleck and B. I. Halperin, Kondo Screening Cloud and Charge Quantization in Mesoscopic Devices, Phys. Rev. B <u>77</u>, 125327 (2008).

243. H. Steinberg, G. Barak, A. Yacoby, L. N. Pfeiffer, K. N. West, B. I. Halperin and K. Le Hur, Charge fractionalization in quantum wires, Nature Physics <u>4</u>, 116 (2008).

244. B. I. Halperin, The Peculiar Properties of Quantum Hall Systems, in *Jahrbuch der Akademie der Wissenschaften zu Göttingen, 2007* (Walter de Gruyter, Berlin, 2008) p. 237. (Manuscript of acceptance talk for the 2007 Dannie Heineman Prize).

245. J. Krich and B. I. Halperin, Spin-polarized current generation from quantum dots without magnetic fields, Phys. Rev. B <u>78</u>, 035338 (2008).

246. K. Le Hur, B. I. Halperin, and A. Yacoby, Charge fractionalization in non-chiral Luttinger liquid systems, Annals of Physics <u>323</u>, 3037 (2008). (arXiv:0803.0744).

247. B, Rosenow and B. I. Halperin, Signatures of neutral quantum Hall modes through low-density constrictions, Phys. Rev. B <u>81</u>, 165313 (2010).

248. I. G. Finkler and B. I. Halperin, Are Microwave-Induced Zero-Resistance States Necessarily Static? Phys. Rev. B <u>79</u>, 085315 (2009) (Editors' Suggestion).

249. J. Qian, B. I. Halperin and E. J. Heller, Imaging and manipulating electrons in a 1D quantum dot with Coulomb blockade microscopy, Phys. Rev. B <u>81</u>, 125823 (2010).

250. M. Levin and B. I. Halperin, Collective states of non-abelian quasiparticles in a magnetic field, Phys. Rev. B <u>79</u>, 205301 (2009).

251. K. Yang and B. I. Halperin, Thermopower as a possible probe of non-abelian quasiparticle statistics in fractional quantum Hall liquids, Phys. Rev. B <u>79</u>, 115317 (2009). (Editors' Suggestion).

252. B. Rosenow, B. I. Halperin, S. H. Simon and A. Stern, Exact Solution for Bulk-Edge Coupling in the Non-Abelian n = 5/2 Quantum Hall Interferometer, Phys. Rev. B <u>80</u>, 155309 (2009).

253. K. Esaki, M. Sato, M. Kohmoto and B. I. Halperin, Zero-modes, energy gaps and edge states of anisotropic honeycomb lattice in a magnetic field, Phys. Rev. B <u>80</u>, 125405 (2009).

254. M. S. Rudner, I. Neder, L. S. Levitov and B. I. Halperin, Phase-sensitive probes of nuclear polarization in spin-blockaded transport, Phys. Rev. B <u>82</u>, 041311(R) (2010).

255. B. I. Halperin and D. J. Bergman, Heterogeneity and Disorder: Contributions of Rolf Landauer, Physica B <u>405</u>, 2908 (2010); arXiv:0910.0993.

256. A. Stern, B. Rosenow, R. Ilan and B. I. Halperin, Interference, Coulomb blockade, and the identification of non-abelian quantum Hall states, Phys. Rev. B <u>82</u>, 085321 (2010). (Editors' Suggestion.)

257. M. Gullans, J. J. Krich, J. M. Taylor, H. Bluhm, B. I. Halperin, C. M. Marcus, M. Stopa, A. Yacoby and M. D. Lukin, Dynamic Nuclear Polarization in Double Quantum Dots, Phys. Rev. Lett. <u>104</u>, 226807 (2010). (Selected for Synopsis in "Physics")

258. Bertrand I. Halperin, Gil Refael and Eugene Demler, Resistance in Superconductors, in *BCS: 50 Years*, edited by Leon N. Cooper and Dmitri Feldman, (World Scientific Publishing Co., 2010), pp. 185-226; arXiv:1005.3347. Reproduced in Int. J. Mod. Phys. <u>24</u>, 4039 (2010).

259. Andrew C. Potter, Erez Berg, Daw-Wei Wang and Bertrand I. Halperin, Eugene Demler, Superfluidity and Dimerization in a Multilayered System of Fermionic Polar Molecules, Phys. Rev. Lett. <u>105</u>, 220406 (2010).

260. Bertrand I. Halperin, On the Quantum Theory of Condensed Matter, in *Quantum Theory Of Condensed Matter — Proceedings of The 24th Solvay Conference On Physics,* edited by Bertrand I. Halperin and Alexander Sevrin, (World Scientific Publishing Company, Singapore 2010).

261. N. d'Ambrumenil, B. I. Halperin and R.H. Morf, Model for the Dissipative Conductance in Fractional Quantum Hall States, Phys. Rev. Lett. <u>106</u>, 126804 (2011).

262. Bertrand I. Halperin, Ady Stern, Izhar Neder and Bernd Rosenow, Theory of the Fabry-Pérot Quantum Hall Interferometer. Phys. Rev. B <u>83</u>, 155440 (2011)

263. Gilad Barak, Loren N. Pfeiffer, Ken W. West, Bertrand I. Halperin and Amir Yacoby, Spin reconstruction in quantum wires subject to a perpendicular magnetic field, arXiv:1012.1845.

264. Izhar Neder, Mark S. Rudner, Hendrik Bluhm, Sandra Foletti, Bertrand I. Halperin and Amir Yacoby, Semiclassical model for the dephasing of a two-electron spin qubit coupled to a coherently evolving nuclear spin bath, Phys. Rev. B <u>84</u>, 035441 (2011). (Editors' Suggestion)

265. Jay D. Sau, B. I. Halperin, K. Flensberg and S. Das Sarma, Number conserving theory for topologically protected degeneracy in one-dimensional fermions, Phys. Rev. B <u>84</u>, 144509 (2011).

266. Bertrand I. Halperin, Yuval Oreg, Ady Stern, Gil Refael, Jason Alicea and Felix von Oppen, Adiabatic manipulations of Majorana fermions in a three-dimensional network of quantum wires, Phys. Rev B <u>85</u>, 144501 (2012). (Editors' Suggestion)

267. Dimitrije Stepanenko, Mark Rudner, Bertrand I. Halperin and Daniel Loss, Singlet-triplet splitting in double quantum dots due to spin orbit and hyperfine interactions, Phys. Rev. B <u>85</u>, 075416 (2012).

268. Jacob J. Krich, Bertrand I. Halperin and Alán Aspuru-Guzik, Nonradiative lifetimes in intermediate band materials - absence of lifetime recovery, J. Appl. Phys. <u>112</u>, 013707 (2012).

269. Jay D. Sau, Takuya Kitagawa and Bertrand I. Halperin, Conductance Beyond the quantum limit and charge pumping in quantum wires, Phys. Rev. B <u>85</u>, 155425 (2012). (Editors' Suggestion)

270. Jelena Klinovaja, Dimitrije Stepanenko, Bertrand I. Halperin and Daniel Loss, Exchange-based CNOT gates for singlet-triplet qubits with spin orbit interaction, Phys. Rev. B <u>86</u>, 085423 (2012).

271. Jay D. Sau, Erez Berg and Bertrand I. Halperin, On the possibility of the fractional ac Josephson effect in non-topological conventional superconductor-normal-superconductor junctions, arXiv:1206.4596

272. Gilad Ben-Shach, Chris R. Laumann, Izhar Neder, Amir Yacoby and Bertrand I. Halperin, Detecting Non-Abelian Anyons by Charging Spectroscopy, Phys. Rev. Lett. 110, 106805 (2013).

273. M. Gullans, J. J. Krich, J. M. Taylor, B. I. Halperin and M. D. Lukin, Preparation of Non-equilibrium Nuclear Spin States in Double Quantum Dots, Phys. Rev. B **88**, 035309 (2013).

274. Benjamin E. Feldman, Andrei J. Levin, Benjamin Krauss, Dmitry Abanin, Bertrand. I. Halperin, Jurgen H. Smet and Amir Yacoby, Fractional Quantum Hall Phase Transitions and Four-flux Composite Fermions in Graphene, Phys. Rev. Lett. 111, 076802 (2013).

275. Dmitry A. Abanin, Benjamin E. Feldman, Amir Yacoby, Bertrand I. Halperin, Fractional and integer quantum Hall effects in the zeroth Landau level in graphene Phys. Rev. B $\underline{88}$, 115407 (2013).

276. Arijeet Pal, Emmanuel I. Rashba and Bertrand I. Halperin, Driven nonlinear dynamics of two coupled exchange-only qubits. Phys. Rev. X $\underline{4}$, 011012 (2014).

277. Izhar Neder, Mark S. Rudner and Bertrand I. Halperin, The theory of coherent dynamic nuclear polarization in quantum dots. Phys. Rev. B $\underline{89}$, 085403 (2014).

278. Angela Kou, Benjamin E. Feldman, Andrei J. Levin, Bertrand I. Halperin, Kenji Watanabe, Takashi Taniguchi and Amir Yacoby, Electron-hole asymmetric integer and fractional quantum Hall effect in bilayer graphene. Science $\underline{345}$, 6192 (2014).

Edited Book

Quantum Theory Of Condensed Matter — Proceedings of The 24th Solvay Conference On Physics
edited by Bertrand I. Halperin and Alexander Sevrin, (World Scientific Publishing Company, Singapore, 2010.)

François Englert

ENGLERT, FRANÇOIS

Birthplace

Born   Belgian, 6 November 1932
Married Five children

Studies

Ingénieur Civil Electricien-Mécanicien (1955) (Université Libre de Bruxelles [U.L.B.]); Docteur en Sciences Physiques (1959) (U.L.B.).

Academic Status

Assistant at the U.L.B. (1956–1959); Research Associate at Cornell University (Ithaca) USA, (1959–1960); Assistant Professor at Cornell University (Ithaca) USA, (1960–1961); Lecturer at the U.L.B., (1961–1964); Professor at the U.L.B. since 1964; Director (with Robert Brout) of the Theoretical Physics Group at the U.L.B. (1980–1998); Emeritus at the U.L.B. since 1998; Permanent Sackler Fellow since 1988 and Senior Professor at Tel-Aviv University since 1992.

Scientific Prizes

A. WETREMS Prize in Mathematical and Physical Sciences (1977); First award of the International Gravity Contest, (with R. Brout and E. Gunzig) (1978); FRANCQUI Prize (1982); High Energy and Particle Physics EPS Prize (with R. Brout and P.W. Higgs) (1997); WOLF Prize in Physics (with R. Brout and P.W. Higgs) (2004); SAKURAI APS Prize (with R.Brout, Guralnik, Hagen and P.W. Higgs) (2010); Prince of Asturias Prize (with P.W. Higgs and CERN) (2013); NOBEL Prize in Physics (with P.W. Higgs) (2013).

Honorary Titles

Hereditary nobility with title BARON.
Honorary member of the European Physical Society.
Honorary member of the Solvay Institutes.
Honorary President of the "Jeunesses Scientifiques of Belgium".
Honorary doctorates from VUB and U. Mons (Belgium); from U. Blaise Pascal (France); from U. Edinburg (U.K.); From U. Athens (Greece); from U. Beijing and U. Shangai (China).

COMMENTARY

I first studied Engineering at the "Université Libre de Bruxelles" and got my degree as electrical-mechanical engineer in 1955. I realized that my interest was less in detailed applications of scientific advances than in the understanding of the underlying structures and I decided to learn physics. As an assistant in the engineering department, I could finance new studies and got my physics masters degree in 1958 and my PhD in 1959.

I had discovered a passion for research and I was thrilled when the same year, based on recommendations and on previous publications in *Condensed Matter Physics*, I was offered a two-year position in the United States at Cornell University, Ithaca (NY), as a Research Associate to the young Professor Robert Brout. I immediately accepted and left for Ithaca.

Our first contact was unexpectedly warm. During my stay at Cornell, the convergence of our vision of science and life laid the groundwork for lasting collaboration and a lifelong friendship. In Ithaca, we worked together in condensed matter physics and in the statistical theory of phase transitions, mainly on ferromagnetism and superconductivity. We realized the importance of spontaneous symmetry breaking in phase transitions and we were extremely impressed when Nambu showed how this notion could be transferred to elementary particle physics to explain the small pion mass on the hadron scale. This work and his beautiful analysis of superconductivity in field theoretic terms drove us later to study the fate of spontaneous symmetry breaking in the context of gauge theory.

In fall 1961, I was scheduled to return to Belgium. By that time, our collaboration and our friendship had become deeply rooted. I was offered a University Professorship at Cornell but I was missing Europe. I decided not to accept it and to return to Belgium. Robert and his wife Martine had a similar attraction for the Old Continent; Robert got a Guggenheim fellowship and they joined me with their children in Belgium. After a few months, the social life there and our personal relationship persuaded Robert to resign from his professorship at Cornell University and to settle permanently at the Université Libre de Bruxelles in Brussels. He eventually acquired Belgian nationality and together we directed the Theoretical Physics Group at the ULB.

In Brussels, we resumed our analysis on spontaneous symmetry breaking, both in the statistical theory of phase transition and, inspired mainly by Nambu's work, in field theory. This is how we discovered the mass generating mechanism that may now indeed be viewed as a phase transition from a high temperature phase in the early Universe, where elementary particle were massless, to the present low temperature phase where their mass arises from a generalization of spontaneous symmetry breaking to Yang–Mills fields, namely the BEH mechanism.

At the ULB, Brout and I initiated a research group in fundamental interactions, that is in the search for the general laws of nature. Joined by brilliant students, many of them becoming world renowned physicists, our group contributed to the many fields at the frontier of the challenges facing contemporary physics. While the mechanism discovered in 1964 was developed all over the world to encode the nature of weak interactions in a

"Standard Model," our group contributed to the understanding of strong interactions and quark confinement, general relativity and cosmology. There we introduced the idea of a primordial exponential expansion of the universe, later called inflation, which we related to the origin of the universe as a quantum fluctuation of gravity and matter, a scenario which I still think is conceptually the correct one. During these developments, our group extended our contacts with other Belgian universities and got involved in many international collaborations.

With our group and many other collaborators, I analyzed fractal structures, supergravity, string theory, infinite Kac–Moody algebras and more generally, all tentative approaches to what I consider as the most important problem in fundamental interactions: the solution to the conflict between the classical Einsteinian theory of gravitation, namely general relativity, and the framework of our present understanding of the world, quantum theory. Although this conflict appears experimentally to affect known results only at very tiny scales of the order of 10^{-32} cm, transcending it would amount to overcoming a conceptual mistake. As such, a solution of this conflict might affect our understanding of the laws of nature at all scales and is crucial for attempting to reach a rational understanding of the origin of the Universe.

Robert was less interested in these new developments and concentrated more on cosmology. Our collaboration had become less frequent but our friendship was unaffected. He passed away on May 3, 2011 after a prolonged illness and missed the remarkable discovery of the BEH scalar boson at CERN and the awarding of the Nobel Prize.

BROKEN SYMMETRY AND THE MASS OF GAUGE VECTOR MESONS*

F. Englert and R. Brout

Faculté des Sciences, Université Libre de Bruxelles, Bruxelles, Belgium

(Received 26 June 1964)

It is of interest to inquire whether gauge vector mesons acquire mass through interaction[1]; by a gauge vector meson we mean a Yang-Mills field[2] associated with the extension of a Lie group from global to local symmetry. The importance of this problem resides in the possibility that strong-interaction physics originates from massive gauge fields related to a system of conserved currents.[3] In this note, we shall show that in certain cases vector mesons do indeed acquire mass when the vacuum is degenerate with respect to a compact Lie group.

Theories with degenerate vacuum (broken symmetry) have been the subject of intensive study since their inception by Nambu.[4-6] A characteristic feature of such theories is the possible existence of zero-mass bosons which tend to restore the symmetry.[7,8] We shall show that it is precisely these singularities which maintain the gauge invariance of the theory, despite the fact that the vector meson acquires mass.

We shall first treat the case where the original fields are a set of bosons ψ_A which transform as a basis for a representation of a compact Lie group. This example should be considered as a rather general phenomenological model. As such, we shall not study the particular mechanism by which the symmetry is broken but simply assume that such a mechanism exists. A calculation performed in lowest order perturbation theory indicates that those vector mesons which are coupled to currents that "rotate" the original vacuum are the ones which acquire mass [see Eq. (6)].

We shall then examine a particular model based on chirality invariance which may have a more fundamental significance. Here we begin with a chirality-invariant Lagrangian and introduce both vector and pseudovector gauge fields, thereby guaranteeing invariance under both local phase and local γ_5-phase transformations. In this model the gauge fields themselves may break the γ_5 invariance leading to a mass for the original Fermi field. We shall show in this case that the pseudovector field acquires mass.

In the last paragraph we sketch a simple argument which renders these results reasonable.

(1) Lest the simplicity of the argument be shrouded in a cloud of indices, we first consider a one-parameter Abelian group, representing, for example, the phase transformation of a charged boson; we then present the generalization to an arbitrary compact Lie group. The interaction between the φ and the A_μ fields is

$$H_{\text{int}} = ieA_\mu \varphi^* \overleftrightarrow{\partial}_\mu \varphi - e^2 \varphi^* \varphi A_\mu A_\mu, \tag{1}$$

where $\varphi = (\varphi_1 + i\varphi_2)/\sqrt{2}$. We shall break the symmetry by fixing $\langle \varphi \rangle \neq 0$ in the vacuum, with the phase chosen for convenience such that $\langle \varphi \rangle = \langle \varphi^* \rangle = \langle \varphi_1 \rangle / \sqrt{2}$.

We shall assume that the application of the

theorem of Goldstone, Salam, and Weinberg[7] is straightforward and thus that the propagator of the field φ_2, which is "orthogonal" to φ_1, has a pole at $q = 0$ which is not isolated.

We calculate the vacuum polarization loop $\Pi_{\mu\nu}$ for the field A_μ in lowest order perturbation theory about the self-consistent vacuum. We take into consideration only the broken-symmetry diagrams (Fig. 1). The conventional terms do not lead to a mass in this approximation if gauge invariance is carefully maintained. One evaluates directly

$$\Pi_{\mu\nu}(q) = (2\pi)^4 i e^2 [g_{\mu\nu}\langle\varphi_1\rangle^2 - (q_\mu q_\nu/q^2)\langle\varphi_1\rangle^2]. \quad (2)$$

Here we have used for the propagator of φ_2 the value $[i/(2\pi)^4]/q^2$; the fact that the renormalization constant is 1 is consistent with our approximation.[9] We then note that Eq. (2) both maintains gauge invariance ($\Pi_{\mu\nu}q_\nu = 0$) and causes the A_μ field to acquire a mass

$$\mu^2 = e^2\langle\varphi_1\rangle^2. \quad (3)$$

We have not yet constructed a proof in arbitrary order; however, the similar appearance of higher order graphs leads one to surmise the general truth of the theorem.

Consider now, in general, a set of boson-field operators φ_A (which we may always choose to be Hermitian) and the associated Yang-Mills field $A_{a,\mu}$. The Lagrangian is invariant under the transformation[10]

$$\delta\varphi_A = \sum_{a,A'}\epsilon_a(x)T_{a,AB}\varphi_{B'}$$

$$\delta A_{a,\mu} = \sum_{c,b}\epsilon_c(x)c_{acb}A_{b,\mu} + \partial_\mu\epsilon_a(x), \quad (4)$$

where c_{abc} are the structure constants of a compact Lie group and $T_{a,AB}$ the antisymmetric generators of the group in the representation defined by the φ_B.

Suppose that in the vacuum $\langle\varphi_{B'}\rangle \neq 0$ for some B'. Then the propagator of $\sum_{A,B'}T_{a,AB'}\varphi_A$

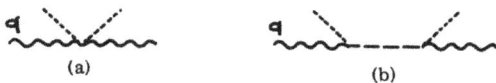

(a)   (b)

FIG. 1. Broken-symmetry diagram leading to a mass for the gauge field. Short-dashed line, $\langle\varphi_1\rangle$; long-dashed line, φ_2 propagator; wavy line, A_μ propagator. (a) → $(2\pi)^4 i e^2 g_{\mu\nu}\langle\varphi_1\rangle^2$, (b) → $-(2\pi)^4 i e^2 (q_\mu q_\nu/q^2) \times \langle\varphi_1\rangle^2$.

$\times\langle\varphi_{B'}\rangle$ is, in the lowest order,

$$\left[\frac{i}{(2\pi)^4}\right] \sum_{A,B',C'} \frac{T_{a,AB'}\langle\varphi_{B'}\rangle T_{a,AC'}\langle\varphi_{C'}\rangle}{q^2}$$

$$\equiv \left[\frac{-i}{(2\pi)^4}\right]\frac{(\langle\varphi\rangle T_a T_a\langle\varphi\rangle)}{q^2}.$$

With λ the coupling constant of the Yang-Mills field, the same calculation as before yields

$$\Pi_{\mu\nu}{}^a(q) = -i(2\pi)^4\lambda^2(\langle\varphi\rangle T_a T_a\langle\varphi\rangle)$$

$$\times[g_{\mu\nu} - q_\mu q_\nu/q^2],$$

giving a value for the mass

$$\mu_a{}^2 = -(\langle\varphi\rangle T_a T_a\langle\varphi\rangle). \quad (6)$$

(2) Consider the interaction Hamiltonian

$$H_{int} = -\eta\bar{\psi}\gamma_\mu\gamma_5\psi B_\mu - \epsilon\bar{\psi}\gamma_\mu\psi A_\mu, \quad (7)$$

where A_μ and B_μ are vector and pseudovector gauge fields. The vector field causes attraction whereas the pseudovector leads to repulsion between particle and antiparticle. For a suitable choice of ϵ and η there exists, as in Johnson's model,[11] a broken-symmetry solution corresponding to an arbitrary mass m for the ψ field fixing the scale of the problem. Thus the fermion propagator $S(p)$ is

$$S^{-1}(p) = \gamma p - \Sigma(p) = \gamma p[1 - \Sigma_2(p^2)] - \Sigma_1(p^2), \quad (8)$$

with

$$\Sigma_1(p^2) \neq 0$$

and

$$m[1 - \Sigma_2(m^2)] - \Sigma_1(m^2) = 0.$$

We define the gauge-invariant current $J_\mu{}^5$ by using Johnson's method[12]:

$$J_\mu{}^5 = -\eta\lim_{\xi\to 0}\bar{\psi}'(x+\xi)\gamma_\mu\gamma_5\psi'(x),$$

$$\psi'(x) = \exp[-i\int_{-\infty}^x \eta B_\mu(y)dy^\mu\gamma_5]\psi(x). \quad (9)$$

This gives for the polarization tensor of the

Volume 13, Number 9   PHYSICAL REVIEW LETTERS   31 August 1964

pseudovector field

$$\Pi_{\mu\nu}^{5}(q) = \eta^2 \frac{i}{(2\pi)^4} \int \mathrm{Tr}\{S(p-\tfrac{1}{2}q)\Gamma_{\nu 5}(p-\tfrac{1}{2}q;p+\tfrac{1}{2}q)$$

$$\times S(p+\tfrac{1}{2}q)\gamma_\mu\gamma_5$$

$$-S(p)[\partial S^{-1}(p)/\partial p_\nu]S(p)\gamma_\mu\} d^4p, \quad (10)$$

where the vertex function $\Gamma_{\nu 5} = \gamma_\nu\gamma_5 + \Lambda_{\nu 5}$ satisfies the Ward identity[5]

$$q_\nu\Lambda_{\nu 5}(p-\tfrac{1}{2}q;p+\tfrac{1}{2}q) = \Sigma(p-\tfrac{1}{2}q)\gamma_5 + \gamma_5\Sigma(p+\tfrac{1}{2}q), \quad (11)$$

which for low q reads

$$q_\nu\Gamma_{\nu 5} = q_\nu\gamma_\nu\gamma_5[1-\Sigma_2] + 2\Sigma_1\gamma_5$$

$$-2(q_\nu p_\nu)(\gamma_\lambda p_\lambda)(\partial\Sigma_2/\partial p^2)\gamma_5. \quad (12)$$

The singularity in the longitudinal $\Gamma_{\nu 5}$ vertex due to the broken-symmetry term $2\Sigma_1\gamma_5$ in the Ward identity leads to a nonvanishing gauge-invariant $\Pi_{\mu\nu}^{5}(q)$ in the limit $q \to 0$, while the usual spurious "photon mass" drops because of the second term in (10). The mass of the pseudovector field is roughly $\eta^2 m^2$ as can be checked by inserting into (10) the lowest approximation for $\Gamma_{\nu 5}$ consistant with the Ward identity.

Thus, in this case the general feature of the phenomenological boson system survives. We would like to emphasize that here the symmetry is broken through the gauge fields themselves. One might hope that such a feature is quite general and is possibly instrumental in the realization of Sakurai's program.[3]

(3) We present below a simple argument which indicates why the gauge vector field need not have zero mass in the presence of broken symmetry. Let us recall that these fields were in-

troduced in the first place in order to extend the symmetry group to transformations which were different at various space-time points. Thus one expects that when the group transformations become homogeneous in space-time, that is $q \to 0$, no dynamical manifestation of these fields should appear. This means that it should cost no energy to create a Yang-Mills quantum at $q = 0$ and thus the mass is zero. However, if we break gauge invariance of the first kind and still maintain gauge invariance of the second kind this reasoning is obviously incorrect. Indeed, in Fig. 1, one sees that the A_μ propagator connects to intermediate states, which are "rotated" vacua. This is seen most clearly by writing $\langle\varphi_1\rangle = \langle[Q\varphi_2]\rangle$ where Q is the group generator. This effect cannot vanish in the limit $q \to 0$.

*This work has been supported in part by the U. S. Air Force under grant No. AFEOAR 63-51 and monitored by the European Office of Aerospace Research.

[1] J. Schwinger, Phys. Rev. 125, 397 (1962).

[2] C. N. Yang and R. L. Mills, Phys. Rev. 96, 191 (1954).

[3] J. J. Sakurai, Ann. Phys. (N.Y.) 11, 1 (1960).

[4] Y. Nambu, Phys. Rev. Letters 4, 380 (1960).

[5] Y. Nambu and G. Jona-Lasinio, Phys. Rev. 122, 345 (1961).

[6] "Broken symmetry" has been extensively discussed by various authors in the Proceedings of the Seminar on Unified Theories of Elementary Particles, University of Rochester, Rochester, New York, 1963 (unpublished).

[7] J. Goldstone, A. Salam, and S. Weinberg, Phys. Rev. 127, 965 (1962).

[8] S. A. Bludman and A. Klein, Phys. Rev. 131, 2364 (1963).

[9] A. Klein, reference 6.

[10] R. Utiyama, Phys. Rev. 101, 1597 (1956).

[11] K. A. Johnson, reference 6.

[12] K. A. Johnson, reference 6.

Volume 119B, number 4,5,6   PHYSICS LETTERS   23/30 December 1982

SPONTANEOUS COMPACTIFICATION OF ELEVEN-DIMENSIONAL SUPERGRAVITY *

F. ENGLERT [1]
CERN, Geneva, Switzerland
and Faculté des Sciences, Université Libre de Bruxelles, Belgium

Received 31 August 1982

Three distinct geometries on S_7 arise as solutions of the classical equations of motion in eleven dimensions. In addition to the conventional riemannian geometry, one can also obtain the two exceptional Cartan–Schouten compact flat geometries with torsion. Possible implications of these results are mentioned.

Simple supergravity in eleven dimensions [1] may turn out to be of more fundamental significance than the ungauged version of the maximal ($N = 8$) extension of four-dimensional supergravity [2]; in the trivial dimensional reduction process leading from the former theory to the latter, relevant information may well have been lost. Following this line of thought, it is of interest to look for possible non-trivial classical solutions in eleven dimensions which would describe a compactification of seven dimensions into a sphere S_7 and which would keep four space-time dimensions non-compact. Indeed a Kaluza–Klein dimensional reduction [3] around $S_7 \equiv SO(8)/SO(7)$ would lead to an $SO(8)$ gauge theory which could be compared and perhaps identified with the gauged version of the $N = 8$ theory [4]: in this way one would learn whether or not this gauged version is also contained in the eleven-dimensional theory [+1].

In this paper we shall derive the surprising result that there exists not only one, but three geometries on S_7 which emerge as solutions of the equations of motion in eleven dimensions. In addition to the "trivial" riemannian geometry on S_7, one also obtains the two

"non-trivial" geometries for which the linear connection in S_7 contains just the torsion terms required to define an absolute parallelism on the seven sphere. This property of the eleven-dimensional action is particularly remarkable because, apart from group spaces, S_7 is the only parallelizable compact manifold, a feature linked to the fact that it is the space of unit octonions [7].

We now present and briefly discuss these solutions. We defer to a separate publication the detailed analysis of the supersymmetry and gauge symmetries they would induce in the four-dimensional space–time background.

The bosonic action of the theory is

$$S = \int \sqrt{g^{(11)}}\, \mathrm{d}^{(11)}x\, \{ -\tfrac{1}{2}R - \tfrac{1}{48}F_{MNPQ}F^{MNPQ}$$

$$+ [\sqrt{2}/6 \cdot (4!)^2]\,(1/\sqrt{g^{(11)}})\,\epsilon^{M_1 M_2 \cdots M_{11}}$$

$$\times F_{M_1 M_2 M_3 M_4} F_{M_5 M_6 M_7 M_8} A_{M_9 M_{10} M_{11}} \}. \quad (1)$$

We use the metric $(+, - - ..)$, $\epsilon^{0123\cdots} = +1$ and $F_{MNPQ} = 4!\,\partial_{[M}A_{NPQ]}$. The field equations are

$$R_{MN} - \tfrac{1}{2}g_{MN}R =$$

$$= -\tfrac{1}{48}(8F_{MPQR}F_N{}^{PQR} - g_{MN}F_{SPQR}F^{SPQR}), \quad (2)$$

$$F^{MNPQ}{}_{;M} = -[\sqrt{2}/2 \cdot (4!)^2]\,(1/\sqrt{g^{(11)}})\,\epsilon^{M_1 \cdots M_8 NPQ}$$

$$\times F_{M_1 M_2 M_3 M_4} F_{M_5 M_6 M_7 M_8}. \quad (3)$$

---

* Supported in part by the Belgian State under the contract ARC 79/83-12.
[1] Permanent address: Pool de Physique, Université Libre de Bruxelles, Campus Plaine, CP 225 Bd. du Triomphe, 1050 Bruxelles, Belgium
[+1] This possibility was already suggested by Duff and Pope [5] (see Duff and Toms [6]).

Volume 119B, number 4,5,6	PHYSICS LETTERS	23/30 December 1982

We look for solutions of (2) and (3) which describe a direct product of S_7 with a four-dimensional Einstein space to cope with an eventual cosmological constant. Hence we must have

$$R_{mn} = \gamma g_{mn} \quad (m, n = 4, 5, 6, 7, 8, 9, 10), \tag{4}$$

$$R_{\mu\nu} = \gamma' g_{\mu\nu} \quad (\mu, \nu = 0, 1, 2, 3), \tag{5}$$

$$R_{m\mu} = 0, \tag{6}$$

where γ and γ' are constants and $\gamma < 0$.

Eq. (3) is consistent with the existence of a non-vanishing expectation value for the auxiliary field $F^{\mu\nu\sigma\lambda}$, namely [8]

$$F^{\mu\nu\sigma\lambda} = (1/\sqrt{-g^{(4)}})\epsilon^{\mu\nu\sigma\lambda} f/(4!)^{1/2},$$

$$f = \text{constant}. \tag{7}$$

If we assume that no other components of F^{MNPQ} acquire a non-vanishing expectation value, we see that (4), (5) and (6) solve eq. (2) provided

$$f^2 = -F_{\mu\nu\sigma\lambda}F^{\mu\nu\sigma\lambda} = -72\gamma, \quad \gamma' = -2\gamma. \tag{8, 9}$$

Thus the "trivial" S_7 is indeed a solution for arbitrary γ, or equivalently arbitrary radius.

Let us now admit non-vanishing expectation values for the four-scalars F^{mnpq}. The solution (7) remains valid, but (3) now yields an additional equation:

$$F^{mnpq}{}_{;m} = [\sqrt{2}/(4!)^{3/2}] f(1/\sqrt{-g^{(7)}})\epsilon^{npqrstu}F_{rstu}. \tag{10}$$

It will turn out that the solutions of eq. (10) are related to the torsion tensors required to parallelize S_7. We shall therefore first rederive the Cartan–Schouten equations which these tension tensors must satisfy.

A Riemann space is parallelizable if, by adding to the riemannian symmetric connection $\Gamma^m{}_{np}$ a suitable totally antisymmetric torsion tensor $S^m{}_{np}$ ($S_{mnp} = S_{[mnp]}$), one can "flatten" the space in the sense that

$$R^m{}_{npq}(\{\Gamma^s{}_{tu} + S^s{}_{tu}\}) = 0. \tag{11}$$

Using the cyclic identities for $R^m{}_{npq}(\{\Gamma^s{}_{tu}\})$ one easily obtains the necessary and sufficient conditions for such a torsion, and hence for a parallelism, to exist:

$$\overset{0}{R}_{mnpq} = S_{tmn}S^t{}_{pq} - S_{t[mn}S^t{}_{p]q}, \tag{12}$$

$$S_{mnp;q} = S_{t[mn}S^t{}_{p]q} = S_{[mnp,q]}. \tag{13}$$

Here and in what follows $\overset{0}{R}_{mnpq} \equiv R_{mnpq}(\{\Gamma^s{}_{tu}\})$ and the covariant derivatives are taken with respect to the Riemann connection $\Gamma^s{}_{tu}$; the last equality in (13) follows from the antisymmetry of $S_{t[mn}S^t{}_{p]q}$ in the *four* indices m, n, p and q. Contracting any two indices in (12) with g^{rs} and $S^{rs}{}_t$ and using the explicit form of $\overset{0}{R}_{mnpq}$ for S_7, namely

$$\overset{0}{R}_{mnpq} = \tfrac{1}{6}\gamma(g_{mp}g_{nq} - g_{mq}g_{np}), \tag{14}$$

one gets

$$S^{tr}{}_m S_{trn} = \gamma g_{mn}, \tag{15}$$

$$S_{tm}{}^r S_{rn}{}^s S_{sp}{}^t = \tfrac{1}{2}\gamma S_{mnp}. \tag{16}$$

Eqs. (15) and (16) were first obtained by Cartan and Schouten [7] who showed that their solutions define two types of parallelisms on S_7.

We shall show that each parallelism is described by a torsion $\underset{+}{S}_{mnp}$ or $\underset{-}{S}_{mnp}$ which satisfy different linear partial differential equations. To this effect, consider the algebraic equations

$$X^{mnp} \equiv X^{[mnp]} = \pm(6/-\gamma)^{1/2}(1/4!)(1/\sqrt{-g^{(7)}})$$

$$\times \epsilon^{mnpqrst}X_{u[rs}X^u{}_{t]q}. \tag{17}$$

Performing the ϵ algebra, one may verify that both eqs. (17) admit solutions satisfying (12) and which of course also satisfy (15) and (16). These solutions determine the torsions $\underset{+}{S}$ and $\underset{-}{S}$ up to a sign which can be fixed by (13). Using the latter equation in the right-hand side of (17) one gets

$$\underset{+}{S}^{mnp} = +(6/-\gamma)^{1/2}(1/4!)(1/\sqrt{-g^{(7)}})$$

$$\times \epsilon^{mnpqrst}\underset{+}{S}_{[rst,q]}, \tag{18}$$

$$\underset{-}{S}^{mnp} = -(6/-\gamma)^{1/2}(1/4!)1/(\sqrt{-g^{(7)}})$$

$$\times \epsilon^{mnpqrst}\underset{-}{S}_{[rst,q]}. \tag{19}$$

These "fields" satisfy $\underset{\pm}{S}^{mnp}{}_{;m} = 0$; they may be interpreted as "potentials" which, in this "Landau gauge", are dual (or anti-dual) to their field strength

Volume 119B, number 4,5,6   PHYSICS LETTERS   23/30 December 1982

$S_{\pm[mnp,q]}$ ($S_{-[mnp,q]}$). These potentials are eigenfunctions of the Laplace operator on S_7:

$$S_{\pm}^{mnp;q}{}_{;q} = -\tfrac{2}{3}\gamma S_{\pm}^{mnp}. \tag{20}$$

Eq. (20) follows from (13) using (15) and (16); it may also be obtained directly from (18) or (19) using

$$S_{\pm[rst,q]} = S_{\pm rst;q}.$$

We now solve eq. (10) by postulating

$$F_{mnpq} = \lambda S_{+[npq,m]}, \tag{21}$$

or

$$F_{mnpq} = \lambda S_{-[npq,m]}, \tag{22}$$

where λ is a constant. Indeed, using (18), (19) and (20) we can satisfy (10) provided

$$\pm f = \sqrt{-\gamma}\, 2^{5/2}. \tag{23}$$

As the right-hand side of (23) is positive, one must choose (21) or (22) according to the sign of f.

The expressions (7), (21) [or (22)] and (23) define solutions of eq. (3) if g_{mn} is the metric tensor of S_7. One must still verify (4), (5) and (6). From (2) we see that these equations can be satisfied only if $F_{mpqr}F_n{}^{pqr}$ is proportional to g_{mn}. This is indeed the case as follows from (21) [or (22)] and (18) [or (19)]: inverting (18) [or (19)] and performing the ϵ algebra, we get, using (15)

$$F_{mpqr}F_n{}^{pqr} = \tfrac{2}{3}\lambda^2\gamma^2 g_{mn}. \tag{24}$$

Eq. (2) now yields (4), (5) and (6) and determines λ^2:

$$\lambda^2 = -12/\gamma. \tag{25}$$

Thus we have obtained, in addition to the "trivial" S_7 solution characterized by $F_{mnpq} = 0$, new solutions where the four-scalars take the values (21) or (22) with λ^2 determined by (25). The results corresponding to (8) and (9) are, for these new solutions,

$$f^2 = -F_{\mu\nu\sigma\lambda}F^{\mu\nu\sigma\lambda} = -32\gamma, \tag{26}$$

$$F^2 = F_{mpqr}F^{mpqr} = -56\gamma, \tag{27}$$

$$\gamma' = -\tfrac{5}{3}\gamma, \tag{28}$$

and γ is, of course, still arbitrary. The geometrical significance of (21) and (22) is clear from eqs. (18) and

(19): the admissible expectation values of the four-scalars F_{mnpq} are just those for which the potentials A_{npq} in the "Landau gauge" are, up to a scale factor whose absolute value is determined by (25), equal to one of the torsion tensors $S_{\pm}^{npq}$ which "flattens" the seven-sphere.

A few concluding remarks are in order:

(a) the non-trivial solutions [(26), (27), (28)], as well as the trivial one [(8), (9)], will necessarily lead, after a Kaluza–Klein reduction around $S_7 \equiv SO(8)/SO(7)$, to four-dimensional actions containing $SO(8)$ gauge fields with a coupling constant proportional to $|\gamma|$ in Planck units. What about supersymmetry? It can be shown [5] that, provided four-dimensional space-time is taken to be the anti-de Sitter space satisfying eq. (9), the trivial solution on S_7 preserves the full supersymmetry of the original action (1) with fermions included, but that supersymmetry is broken by the non-trivial solution. Hence the four-dimensional action deducible from the reduction around the trivial S_7 geometry will be fully supersymmetric and this suggests of course that it might describe the gauged version of $N = 8$ supergravity [4]. On the other hand, the reduction around the non-trivial S_7 geometries will reveal a spontaneous breakdown of supersymmetry [+2].

(b) The appealing mathematical properties of the non-trivial S_7 solutions are marred by the reappearance of a cosmological constant [eq. (28)] which was already there in the trivial S_7 solution: thus the "flattening" of S_7 is not accompanied by a vanishing of the cosmological constant. However, the value of the measurable cosmological constant can anyhow not be predicted without understanding the effect of the quantum fluctuations which are not considered here.

(c) Finally, we have not at all discussed the relative stability of the different classical solutions. There may be no sensible way of attacking this problem as long as the physical implications of the "wrong" sign induced by the conformal part of gravity are not mastered. This may require the link between field theory and cosmology provided by the conjecture that the path integral over metrics is stabilized by the entropy stored in a "foam of universes" [9]. In this respect, it is perhaps encouraging to notice that if eleven-dimensional

---

[+2] I thank M. Duff for pointing out to me that the spontaneous character of this breaking is preserved under the Kaluza–Klein dimensional reduction process.

Volume 119B, number 4,5,6   PHYSICS LETTERS   23/30 December 1982

supergravity is physically relevant, and if spontaneous compactification does take place, a multiplicity of "universes" will already be unavoidably generated at the level of "multi-soliton" solutions.

I thank the members of the Department of Physics and Astronomy of the Tel-Aviv University where part of this work was completed, for their warm hospitality and active interest. In particular I am grateful to T. Banks and A. Casher for illuminating discussions. I am also greatly indebted to R. Brout, M. Duff, B. de Wit, M. Green, H. Nicolai and P. Windey for encouragement and clarifying remarks.

References

[1] E. Cremmer, B. Julia and J. Scherk, Phys. Lett. 76B (1978) 409.

[2] E. Cremmer and B. Julia, Phys. Lett. 80B (1978) 48; Nucl. Phys. B159 (1979) 141.

[3] B. de Wit, Dynamical theory of groups and fields (Gordon and Breach, New York, 1965).

[4] B. de Wit and H. Nicolai, Phys. Lett. 108B (1981) 285; CERN preprint TH-3291 (1982).

[5] M.J. Duff and C.N. Pope, to be published; B. de Wit, private commnuication.

[6] M.J. Duff, CERN preprint TH 3232 (1982); M.J. Duff and D.J. Toms, CERN preprint TH 3259 (1982).

[7] E. Cartan and J.A. Schouten, Proc. K. Akad. Wet. Amsterdam 29 (1926) 933.

[8] P.G.O. Freund and M.A. Rubin, Phys. Lett. 97B (1980) 233.

[9] A. Casher and F. Englert, Phys. Lett. 104B (1981) 117; F. Englert, Quantum field theory and cosmology, in Fundamental interactions: Cargèse 1981 (Plenum, New York, 1982).

LIST OF PUBLICATIONS

1. F. Englert, C. Jauquet, Théorie du dipôle replié, Revue H.F. III, 157 (1955).

2. F. Englert, Application de la théorie des groupes au calcul du couplage spin-orbite dans les cristaux, Académie royale de Belgique, Bulletin de la Classe des Sciences, 5e serie, **43**, 273 (1957).

3. F. Englert, Interaction entre un petit et un grand systeme, Il Nuovo Cimento **10**, 560 (1958).

4. P. Aigrain, F. Englert, "Les semiconducteurs", Monographie Dunod, 203 pp — Dunod, Paris (1958) — Version espagnole, Eudeba, Buenos Aires (1966).

5. F. Englert, Comportement d'un petit système quantique dans un milieu faiblement dissipatif, Journal of Physics and Chemistry of Solids **11**, 78 (1959).

6. F. Englert, Renormalisation de masse d'un électron dans un corps noir, Académie royale de Belgique, Bulletin de la Classe des Sciences, 5e série, **45**, 782 (1959).

7. R. Brout, F. Englert, Linked cluster expansion in quantum statistics, Phys. Rev. **120**, 1519 (1960).

8. F. Englert, R. Brout, Dielectric formulation of quantum statistics, Phys. Rev. **120**, 1085 (1960).

9. F. Englert, Effective hamiltonian formalism in the exciton problem and in superconduc-tivity, published in the "Proceedings of the International Conference on Semiconductor Physics" (Prague), 34 (1960).

10. F. Englert, Theory of a Heisenberg ferromagnet in the random phase approximation, Phys. Rev. Lett. **5**, 102 (1960).

11. F. Englert, M. Antono , Band theory of ferromagnetism, Physica **30**, 429 (1964).

12. F. Englert, Linked cluster expansion in the statistical theory of ferromagnetism, Phys. Rev. **129**, 567 (1963).

13. R. Stinchombe, G. Horwitz, F. Englert, R. Brout, Thermodynamic behavior of the Heisenberg ferromagnet, Phys. Rev. **130**, 155 (1963).

14. G. Horwitz, F. Englert, R. Brout, Zero temperature properties of many fermion systems, Phys. Rev. **130**, 404 (1963).

15. J. De Coen, F. Englert, R. Brout, Linked cluster expansion in N-boson problem, Physica **30**, 1293 (1964).

16. F. Englert, Fonctions de réponse et leurs relations avec les phénomènes d'équilibre, published in "Etudes des phénomènes irréversibles, Morgins 1963, 207–250", Bureau — Lausanne, (1963).

17. F. Englert, R. Brout, Broken symmetry and the mass of gauge vector mesons, Phys. Rev. Lett. **13**, 321 (1964).

18. F. Englert, R. Brout, A Possible Argument which eliminates the hitherto unrealized representations of SU_3, Phys. Rev. Lett. **12**, 682 (1964).

19. F. Englert, R. Brout, Remark on the misuse of the active interpretation of general relativity, Physics Letters **16**, 250 (1965).

20. F. Englert, R. Brout, M.F. Thiry, Vector mesons in the presence of broken symmetry, Il Nuovo Cimento **43**, 244 (1966).

21. R. Brout, F. Englert, Gravitational Ward identity and the principle of equivalence, Phys. Rev. **141**, 1231 (1966).

22. C. De Dominicis, F. Englert, Potential correlation function duality in statistical mechanics, Journal of Math. Phys. **8**, 2143 (1967).

23. F. Englert, C. De Dominicis, Lagrangian field theory in terms of Green's functions, Il Nuovo Cimento **53A**, 1007 (1968).

24. F. Englert, C. De Dominicis, Self generating interactions, Il Nuovo Cimento **53A**, 1021 (1968).

25. F. Englert, R. Brout, H. Stern, Field theoretic formulation of bootstrap theory, Il Nuovo Cimento **58A**, 601 (1968).

26. F. Englert, R. Brout, Current algebra and partial symmetry, Il Nuovo Cimento **55A**, 543 (1968).

27. R. Brout, F. Englert, Asymptotic form factors and scattering amplitudes, Physics Letters **27B**, 647 (1968).

28. F. Englert, R. Brout, P. Nicoletopoulos and C. Truffin, The optical limit of relativistic scattering, Il Nuovo Cimento **64A**, 561 (1969).

29. R. Brout, F. Englert, C. Truffin, Chiral symmetry and linear trajectories, Physics Letters **29B**, 590 (1969).

30. F. Englert, R. Brout, C. Truffin, Meson mass spectrum, Physics Letters 29B, 686 (1969).

31. F. Englert, R. Brout, H. Stern, Mass quantization conditions from infinite supercon-vergence and chiral symmetry, Il Nuovo Cimento **66A**, 845 (1970).

32. F. Englert, Self consistent approach to the fine structure constant, Il Nuovo Cimento **16A**, 557 (1973).

33. F. Englert, J.-M. Frère, P. Nicoletopoulos, Vertex bootstrap for the ne structure constant, Il Nuovo Cimento **19A**, 395 (1974).

34. R. Brout, F. Englert, C. Truffin, $\beta\omega\pi$ decay and SU_3 breaking, Phys. Rev. **D9**, 2694 (1974).

35. F. Englert, J.-M. Frère, P. Nicoletopoulos, Low energy eigenvalue condition from dynamically broken gauge symmetry, Physics Letters **52B**, 433 (1974).

36. F. Englert, R. Brout, Dynamical theory of weak and electromagnetic interactions, Physics Letters **49B**, 77 (1974).

37. F. Englert, J.-M. Frère, P. Nicoletopoulos, Infrared versus ultraviolet driven dynamical mass generation in ladder approximation, Physics Letters **59B**, 346 (1975).

38. F. Englert, J.-M. Frère, P. Nicoletopoulos, Dynamical symmetry breakdown in pure Yang-Mills theory, Nuclear Physics **B95**, 269 (1975).

39. F. Englert, Dynamical symmetry breakdown, published in "Weak and electromagnetic interactions at high energies, Cargèse 1975, Part A 265–190", Plenum Press, New York and London (1976).

40. F. Englert, E. Gunzig, C. Truffin, P. Windey, Conformal invariant general relativity with dynamical symmetry breakdown, Physics Letters **57B**, 73 (1975).

41. F. Englert, R. Gastmans, C. Truffin, Conformal invariance in quantum gravity, Nuclear Physics **B117**, 407 (1976).

42. F. Englert, Mechanisms of dynamical symmetry breaking, published in "International Center for Theoretical Physics reports" (Trieste) IC/76 (1976).

43. F. Englert, P. Windey, Quantization conditions for 't Hooft monopoles, Phys. Rev. D, 2728 (1976).

44. R. Brout, E. Englert, W. Fischler, Magnetic confinement in non abelian gauge theory, Phys. Rev. Lett. **36**, 649 (1976).

45. R. Brout, F. Englert, J.-M. Frère, On the origin of the pion in confinement schemes, Nuclear Physics **B134**, 327 (1978).

46. F. Englert, P. Windey, Electric confinement and magnetic superconductors, Nuclear Physics **B135**, 529 (1978).

47. R. Brout, F. Englert, E. Gunzig, The creation of the universe as a quantum phenomenon, Annals of Physics **115**, 78 (1978).

48. F. Englert, Electric and magnetic confinement schemes, published in "Hadron structure and lepton-hadron interactions, Cargèse 1977, 503–560", Plenum Press, New York and London (1979).

49. F. Englert, P. Windey, Electric confinement and magnetic superconductors, Nuclear Physics **B135**, 529 (1978).

50. R. Brout, F. Englert, E. Gunzig, The causal universe (First Prize for 1978 awarded by Gravity Research Foundation), Gen. Rel. and Grav. Journal **10**, 1 (1979).

51. R. Brout, F. Englert, P. Spindel, Cosmological origin of the grand-uni cation mass scale, Phys. Rev. Lett. **43**, 417 (1979).

52. R. Brout, F. Englert, J.-M. Frère, E. Gunzig, P. Nardone, C. Truffin, P. Spindel, Cosmogenesis and the origin of the fundamental length scale, Nuclear Physics **B170**, 228 (1980).

53. F. Englert, The creation of the universe as the breakdown of a grand uni ed symmetry, published in "Physical Cosmology, Les Houches, session XXXII, 1979, 515–532", North Holland Publishing Company, Amsterdam — New York — Oxford (1980).

54. R. Brout, F. Englert, TCP conservation in cosmology, Nuclear Physics **B180**, 181 (1981).

55. J.-M. Blairon, R. Brout, F. Englert, J. Greensite, Chiral symmetry breaking in the action formulation of lattice gauge theory, Nuclear Physics **B180**, 439 (1981).

56. F. Englert, J.-M. Frère, Time-dependent perturbation theory with permanent effects, Nuclear Physics **B180**, 468 (1981).

57. A. Casher and F. Englert, The quantum era, Physics Letters **104B**, 117 (1981).

58. F. Englert, Quantum eld theory and cosmology, published in "Theory of fundamental interaction, Cargèse 1981", Plenum Press, New York and London (1982).

59. F. Englert, Spontaneous compactification of eleven dimensional supergravity, CERN Preprint TH 3394 (1982), Physics Letters **119B**, 339 (1982).

60. B. Biran, B. De Wit, F. Englert, H. Nicolaï, Gauged N=8 supergravity and its break-ing from spontaneous compactification, CERN preprint TH 3489 (1982), Physics Letters **124B**, 45 (1983).

61. F. Englert, From quantum cosmology to quantum gravity published in "Unification of the fundamental particle interactions II", edited by J. Ellis and S. Ferrara, Plenum Press, New York and London (1983).

62. F. Englert, M. Rooman, P. Spindel, Supersymmetry breaking by torsion and Ricci-flat squashed seven-spheres, Physics Letters **127B**, 47 (1983).

63. F. Englert, M. Rooman, P. Spindel, Symmetries in eleven-dimensional supergravity compactified on a parallelized seven-sphere, Physics Letters **130B**, 50 (1983).

64. B. Biran, A. Casher, F. Englert, M. Rooman, P. Spindel, The fluctuating seven-sphere in eleven-dimensional supergravity, Physics Letters **134B**, 179 (1984).

65. A. Casher, F. Englert, H. Nicolaï, M. Rooman, The mass spectrum of supergravity on the sound seven-sphere, Nuclear Physics **B243**, 173 (1984).

66. F. Englert, The quest for unification, published in "Recent developments in quantum field theory", eds. J. Ambjorn, B.J. Durhuus and J.L. Petersen, Elsevier Sc. Pub. North Holland, 39 (1985).

67. A. Casher, F. Englert, H. Nicolaï, A. Taormina, Consistent superstrings as solutions of the d=26 bosonic string theory, CERN preprint TH 4220/85, Physics Letters **162B**, 121 (1985).

68. F. Englert, A. Neveu, Non-abelian compactification of the interacting bosonic string, CERN preprint TH 4168/85, Physics Letters **163B**, 349 (1985).

69. F. Englert, Metric space-time from field propagation on fractal structures, CERN preprint TH 4091/85, published in "From SU(3) to gravity", ed. E. Gotsman and G. Tauber, 35 (1985).

70. F. Englert, H. Nicolaï, A. Schellekens, Superstrings from 26 dimensions, Nuclear Physics **B264**, 514 (1986).

71. F. Englert, Hidden superstrings, published in "Festschrift in honor of Y. Nambu", (1986).

72. F. Englert, J.-M. Frère, M. Rooman and P. Spindel, Metric space-time as fixed point of the renormalization group equations on fractals structures, Nuclear Physics **B280**, 147 (1987).

73. F.A. Bais, F. Englert, A. Taormina, P. Zizzi, Torus compactification for non simply laced groups, Nuclear Physics **B279**, 529 (1987).

74. F. Englert, From disorder to space-time geometry, Foundations of Physics **17**, 621 (1987).

75. R. Brout, F. Englert, Cosmologie quantique, published in "La nouvelle encyclopédie Diderot, Vol.: Aux con ns de l'univers", ed. J. Schneider, Fayard, Fondation Diderot, Paris (1987).

76. F. Englert, From Classical to Quantum Cosmology published in "XVII G.I.F.T. International seminar on cosmology and particle physics", eds. E. Alvarez, C.A. Dominguez, L.E. Ibanes and M. Quiros, World Scientific, Singapore, (1987).

77. Y. Aharonov, F. Englert, J. Orlo , Macroscopic fundamental strings in cosmology, Physics Letters **B199**, 366 (1987).

78. F. Englert, A. Sevrin, W. Troost, A. Van Proeyen, P. Spindel, Loop algebras and superalgebras based on S^7, Journal of Math. Phys. **29**, 281 (1988).

79. F. Englert, J. Orloff, T. Piran, Fundamental strings and large scale structure formation, Physics Letters **B212**, 423 (1988).

80. F. Englert, String thermodynamics and cosmology, published in the "Proceedings of the Boulder NATO workshop on superstrings", Plenum Press (1988).

81. F. Englert, Quantum physics without time, Physics Letters **B228**, 111 (1989).

82. F. Englert, J. Orloff, Universality of the closed string phase transition, Nuclear Physics **B334**, 472 (1990).

83. F. Englert, From quantum correlations to time and entropy, published in "The Gardener of Eden, edited by P. Nicoletopoulos and J. Orloff, in Physicalia Magazine, Vol. 12 special issue", in honour of R. Brout's 60th birthday (1990).

84. A. Casher, F. Englert, Entropy generation in quantum gravity, Classical and Quantum Gravity **9**, 2231 (1992).

85. A. Casher, F. Englert, Black hole tunneling entropy and the spectrum of gravity, gr-qc/9212010; Classical and Quantum Gravity **10**, 2479 (1993).

86. A. Casher, F. Englert, Entropy generation in quantum gravity and black hole remnants, gr-qc/9404025; published in "String theory, quantum gravity and the unification of the fundamental interactions" Ed. by M. Bianchi, F. Fucito, E. Marinari, A. Sagnotti, World Scientific and dedicated to F. Englert on the occasion of his sixtieth birthday (1993).

87. F. Englert, S. Massar, R. Parentani, Source vacuum fluctuations of black hole radiance, gr-qc/9404026; Classical and Quantum Gravity **11**, 2919 (1994).

88. F. Englert, B. Reznik, Entropy generation by tunneling in 2+1-Gravity, gr-qc/9401010; Phys. Rev. **D 50**, 2692 (1994).

89. F. Englert, The black hole history in tamed vacuum, gr-qc/9408005; (1994).

90. F. Englert, Operator weak values and black hole complementarity, gr-qc/9502039; published in "Proceedings of The Oskar Klein Centenary (19–21 September 1944)" Ed. by U. Lindström, World Scientific (1995).

91. F. Englert, L. Houart, P. Windey, The black hole entropy can be smaller than A/4, hep-th/9503202; Physics Letters **B372**, 111 (1996).

92. F. Englert, L. Houart, P. Windey, Black hole entropy and string instantons, hep-th/9507061; Nuclear Physics **B458**, 231 (1996).

93. F. Englert, On the black hole unitarity issue, hep-th/9705115, published in the proceedings of the "Workshop on frontiers in field theory, quantum gravity and string theory, December 1996, Puri (India)" (1996).

94. F. Englert, L. Houart, P. Windey, Thermodynamics of black hole in presence of string instantons, hep-th/9606179; published in the "Proceedings of the 2nd International A.D. Sakharov conference on physics, Moscow, (20–23 May 1996)" Ed. by I.M. Dremlin and A.M. Semikhatov, World Scientific (1997).

95. A. Casher, F. Englert, N. Itzhaki, S. Massar, R. Parentani, Black hole horizon fluctuations, hep-th/9606106; Nuclear Physics **B484**, 419 (1997).

96. Y. Aharonov, R. Brout, F. Englert, La Notion de Temps; published in "Le vieillissement", Laus Medicinae, ULB University Press (1997).

97. R. Argurio, F. Englert, L. Houart, Intersection rules for p-branes, hep-th/9701042; Physics Letters **B398**, 61 (1997).

98. R. Argurio, F. Englert, L. Houart, P. Windey, On the opening of branes, hep-th/9704190; Physics Letters **B408**, 151 (1997).

99. F. Englert, E. Rabinovici, Statistical entropy of Schwarzschild black holes, hep-th/9801048; Physics Letters **B426**, 264 (1998).

100. R. Argurio, F. Englert, L. Houart, Statistical entropy of the four dimensional Schwarzschild black hole, hep-th/9801053; Physics Letters **B426**, 275 (1998).

101. R. Brout and F. Englert, Spontaneous symmetry breaking in gauge theories: A historical survey, hep-th/9802142; published in the "Proceedings of the international europhysics conference on high energy physics, Jerusalem, Israel, 19–25 Aug. 1997" Ed. by Lellouch, Daniel; Mikenberg, Giora and Rabinovici, Eliezer — Springer, Berlin, (1999).

102. F. Englert, Primordial in ation, hep-th/9911185; published in "Lectures on basics and highlights of fundamental physics, Proceedings of the international school of subnuclear physics **Vol 37** 516", Ed. by A. Zichichi, World Scientific (1999).

103. K. Bautier, F. Englert, M. Rooman, P. Spindel The Fefferman-Graham ambiguity and ADS black holes, hep-th/0002156; Physics Letters **B479**, 291 (2000).

104. F. Englert, L. Houart, A. Taormina, Brane fusion and the emergence of fermionic strings, hep-th/0106235; JHEP **0108**, 013, (2001).

105. F. Englert, A brief course in spontaneous symmetry breaking. 2- Modern Times: The BEH mechanism. Presented at the "Corfu summer institute on elementary particle physics (Corfu 2001), Corfu, Greece, 31 Aug–20 Sep 2001", hep-th/0203097, (2002).

106. F. Englert, L. Houart, A. Taormina, The bosonic ancestor of closed and open fermionic strings. "Contributed to meeting on strings and gravity: Tying the forces together, Brussels, Belgium, 19–21 Oct 2001", hep-th/0203098, (2002).

107. Auttakit Chattaraputi, F. Englert, L. Houart, A. Taormina, The bosonic mother of fermionic D-branes, hep-th/0207238; JHEP **0209**, 037, (2002).

108. Auttakit Chattaraputi, F. Englert, L. Houart, A. Taormina, Fermionic subspaces of the bosonic string. Published in the "Proceedings of the workshop on the quantum structure of space-time and the geometrical nature of the fundamental interactions, Leuven, Belgium, 13–19 Sep 2002", and of the "6th International workshop on conformal field theory and integrable systems, Chernogolovka, Russia, 15–21 Sep 2002.", hep-th/0212085, (2002).

109. F. Englert, L. Houart, A. Taormina, P. West. The symmetry of M-theories, hep-th/0304206; JHEP **0309**, 020, (2003).

110. F. Englert, L. Houart, P. West. Intersection rules, dynamics and symmetries, hep-th/0307024; JHEP **0308**, 025, (2003).

111. F. Englert, L. Houart. G^{+++} Invariant formulation of gravity and M-theories: Exact BPS solutions, hep-th/0311255; JHEP **0401**, 002, (2004).

112. F. Englert, L. Houart. From brane dynamics to a Kac-Moody invariant formulation of M-theories; published in the "Proceedings of 27th Johns Hopkins workshop on current problems in

particle theory: Symmetries and mysteries of M-Theory, Goteborg, Sweden, 24–26 Aug 2003", hep-th/0402076 (2004).

113. F. Englert, L. Houart. G^{+++} Invariant formulation of gravity and M-theories: Exact intersecting brane solutions; hep-th/0405082; JHEP **0405**, 059, (2004).

114. F. Englert, M. Henneaux and L. Houart. From very-extended to overextended gravity and M-theories, hep-th/0412184; JHEP **0502**, 070, (2005).

115. F. Englert, Broken symmetry and Yang-Mills theory, published in "50 years of Yang-Mills theory", editor G.'t Hooft, World Scientific, hep-th/0406162, (2005).

116. F. Englert, L. Houart, Axel Kleinschmidt, Hermann Nicolai, Nassiba Tabti. An E9 multiplet of BPS states, hep-th/0703285; JHEP **0705**, 065, (2007).

117. R. Brout, F. Englert, Spontaneous broken symmetry, published in "Comptes Rendus Physique", **8** (2007).

118. F. Englert, L. Houart. The emergence of fermions and the E11 content, Presented at the "Workshop on quantum mechanics of fundamental systems: The quest for beauty and simplicity: Dedicated to Claudio Bunster on the occasion of his 60th birthday, Valdivia, Chile, 10–11 Jan 2008", arXiv:0806.4780 [hep-th], (2008).

119. F. Englert, K. Peeters, A. Taormina, Twenty-four near-instabilities of Caspar-Klug viruses, arXiv:0804.4275 [q-bio.BM]; Phys. Rev. **E 78**, 031908 (2008). [Selected by the Virtual Journal of Biological Physics Research (September 15, 2008), http://www.vjbio.org]

120. F. Englert and Ph. Spindel, The hidden horizon and black hole unitarity, arXiv:1009.6190 [hep-th]; JHEP **1012**, 065, (2010).

121. Symmetry breaking and the Scalar boson: Evolving perspectives, published in "Proceedings of the 47th Rencontres de Moriond: 2012 Electroweak interactions and unified theories", editors Etienne Augé, Jacques Dumarchez, Jean-Marie Frère, Lydia Iconomidou-Fayard, Jean Tran Thanh Van, ARISF, arXiv:1204.5382 [hep-th], (2012).

122. F. Englert, The BEH mechanism and its scalar boson, Nobel Lecture (2013), Rev. Mod. Phys. **86**, 843, (2014).

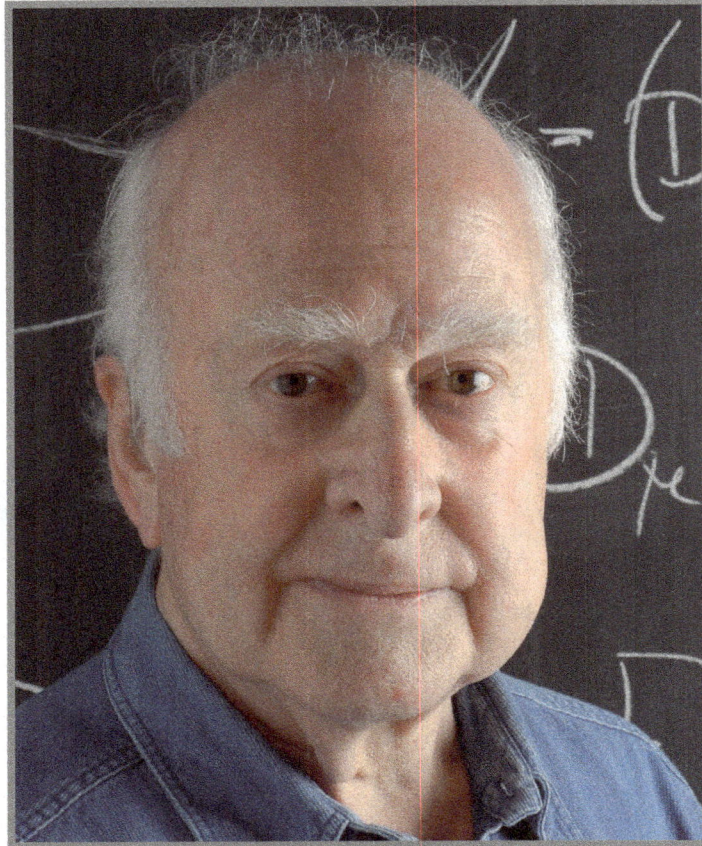

Peter W. Higgs

PETER WARE HIGGS

Birthplace

    Born   29 May 1929 at Elswick, Newcastle upon Tyne, Northumbria, United Kingdom

Childhood

    1930–1941   Birmingham
    1941–1946   Bristol

Secondary Education

    1940–1941   Halesowen Grammar School, Worcestershire
    1941–1946   Cotham Grammar School, Bristol
    1946–1947   City of London School

University Education

    1947–1954   King's College, University of London

Degrees

    1950   BSc (First Class Honours) in Physics
    1951   MSc
    1954   PhD

Professional Career

Royal Commission for the Exhibition of 1851 Senior Student

    1953–1954   King's College London
    1954–1955   University of Edinburgh
    1955–1956   Senior Research Fellow, University of Edinburgh

ICI Research Fellow, University of London

    1956–1957   University College
    1957–Dec 1958   Imperial College
    Jan 1959–1960   Temporary Lectureship in Mathematics, University College
    1960–1970   Lecturer in Mathematical Physics, University of Edinburgh
    1965–1966   (On leave University of North Carolina, Chapel Hill)
    1970–1980   Reader in Mathematical Physics, University of Edinburgh
    Oct–Dec 1976   (On leave at CERN, Geneva)
    1980–1996   Professor of Theoretical Physics, University of Edinburgh
    1996–   Professor Emeritus, University of Edinburgh

Fellowships

1974   Fellow of the Royal Society of Edinburgh (FRSE)
1983   Fellow of the Royal Society, London (FRS)
1991   Fellow of the Institute Of Physics (FInstP)
1998   Fellow of King's College London
1999   Honorary Fellow of the Institute Of Physics
2008   Fellow of the University of Swansea
2013   Honorary Fellow of the Royal Scottish Society of Arts
2013   Honorary Member of the Saltire Society
2013   Honorary Fellow of the Science Museum London

Prizes

1981   Hughes Medal, Royal Society (with T W B Kibble)
1984   Rutherford Medal, Institute of Physics (with T W B Kibble)
1990   Scottish Science Award, Saltire Society and Royal Bank of Scotland
1993   James Scott Prize Lectureship, Royal Society of Edinburgh (delivered April 1995)
1997   Paul Dirac Medal and Prize, Institute of Physics
1997   High Energy and Particle Physics Prize, European Physical Society (with R Brout, F Englert)
2000   Royal Medal, Royal Society of Edinburgh
2004   Wolf Prize in Physics (with R Brout and F Englert)
2009   Oskar Klein Memorial Lecture and Medal, Stockholm Academy of Sciences
2010   J J Sakurai Prize, American Physical Society (with R Brout, F Englert, G S Guralnik, C R Hagen and T W B Kibble)
2012   Higgs Medal, Royal Society of Edinburgh
2013   Nonino Prize 'Man of Our Time'
2013   Edinburgh Medal of the Edinburgh International Science Festival (with CERN)
2013   Prince of Asturias Award for Technical and Scientific Research (with F Englert and CERN)
2013   Nobel Prize in Physics (with F Englert)
2014   Galileo Galilei
2013   Coply Medal, Royal Society

Honorary Degrees

1997   DSc University of Bristol
1998   DSc University of Edinburgh
2002   DSc University of Glasgow
2009   DSc King's College London
2010   DSc University College London
2012   DSc University of Cambridge
2012   DSc Heriot-Watt University
2013   PhD SISSA, Trieste
2013   DSc University of Durham
2013   DSc University of Manchester
2014   DSc University of St Andrews

2014   DSc Free University of Brussells (ULB)
2015   DSc University of North Carolina at Chapel Hill
2015   DSc Queen's Univeristy Belfast 2015

Other Awards

2012/2013   Companion of Honour in the New Year Honours List
2013   Edinburgh Award, City of Edinburgh
2013   Freedom of the City of Bristol
2014   Freedom of the City of Newcastle
2014   Freedom of the City of Edinburgh

COMMENTARY

Peter Higgs was born on 29 May 1929 in the Elswick district of Newcastle upon Tyne, UK. He graduated with First Class Honours in Physics from King's College, University of London, in 1950. A year later, he was awarded an MSc and started research, initially under the supervision of Charles Coulson and, subsequently, Christopher Longuet-Higgins. In 1954, he was awarded a PhD for a thesis entitled 'Some Problems in the Theory of Molecular Vibrations', work which signalled the start of his life-long interest in the application of the ideas of symmetry to physical systems.

In 1954, Peter Higgs moved to the University of Edinburgh for his second year as a Royal Commission for the Exhibition of 1851 Senior Student, and remained for a further year as a Senior Research Fellow. He returned to London in 1956 to take up an ICI Research Fellowship, spending a year at University College and a little over a year at Imperial College, before taking up an appointment as Temporary Lecturer in Mathematics at University College. In October 1960, Peter Higgs returned to Edinburgh, taking up a lectureship in Mathematical Physics at the Tait Institute. He was promoted to Reader in 1970, became a Fellow of the Royal Society of Edinburgh in 1974 and was promoted to a Personal Chair of Theoretical Physics in 1980. He was elected Fellow of the Royal Society in 1983 and Fellow of the Institute of Physics in 1991. He retired in 1996, becoming Professor Emeritus at the University of Edinburgh. He was awarded a Fellowship of the University of Swansea in 2008. He was awarded an Honorary Membership of the Saltire Society and a Fellowship of the Royal Scottish Society of the Arts in 2013.

Peter Higgs' contribution to physics has been recognised by numerous academic honours: the Hughes Medal of the Royal Society (1981, shared with Tom Kibble), the Rutherford Medal of the Institute of Physics (1984, also shared with Tom Kibble), the Saltire Society & Royal Bank of Scotland Scottish Science Award (1990), the Royal Society of Edinburgh James Scott Prize Lectureship (1993), the Paul Dirac Medal and Prize of the Institute of Physics (1997), and the High Energy and Particle Physics Prize of the European Physical Society (1997, shared with Robert Brout and François Englert), Royal Medal of the Royal Society of Edinburgh (2000), Wolf Prize in Physics (2004, shared with Robert Brout and François Englert), the Stockholm Academy of Sciences Oskar Klein Memorial Lecture and Medal (2009) and the American Physical Society J J Sakurai Prize (2010), shared with Robert Brout, François Englert, Gerry Guralnik, Carl Hagen and Tom Kibble). He received a unique personal Higgs medal from the Royal Society of Edinburgh on 1 October 2012 and the 2013 Nonino Prize 'Man of Our Time'. He shared the award of the 2013 Edinburgh International Science Festival Edinburgh Medal with CERN and the 2013 Prince of Asturias Award for Technical and Scientific Research with François Englert and CERN.

He has received honorary degrees from the Universities of Bristol (1997), Edinburgh (1998), Glasgow (2002), King's College London (2009), University College London (2010), Cambridge (2012), Heriot-Watt (2012), Scuola Internazionale Superiore di Studi Avanzati

Trieste (2013), Durham (2013), Manchester (2013), St Andrews (2014), ULB Brussels (2014), University of North Carolina at Chapel Hill (2015) and Queen's University Belfast (2015).

In 2011, he was awarded the Edinburgh Award for his outstanding contribution to the city.

He was granted the Freedom of the Cities of Bristol (2013), Newcastle (2014) and Edinburgh (2014).

In the 2012/2013 New Year Honours List, he was appointed a Companion of Honour.

On 10 December 2013, he received the 2013 Nobel Prize in Physics along with Francois Englert.

On 20 July 2015, he was awarded the Copley Medal of the Royal Society.

VOLUME 13, NUMBER 16   PHYSICAL REVIEW LETTERS   19 OCTOBER 1964

BROKEN SYMMETRIES AND THE MASSES OF GAUGE BOSONS

Peter W. Higgs

Tait Institute of Mathematical Physics, University of Edinburgh, Edinburgh, Scotland

(Received 31 August 1964)

In a recent note[1] it was shown that the Goldstone theorem,[2] that Lorentz-covariant field theories in which spontaneous breakdown of symmetry under an internal Lie group occurs contain zero-mass particles, fails if and only if the conserved currents associated with the internal group are coupled to gauge fields. The purpose of the present note is to report that, as a consequence of this coupling, the spin-one quanta of some of the gauge fields acquire mass; the longitudinal degrees of freedom of these particles (which would be absent if their mass were zero) go over into the Goldstone bosons when the coupling tends to zero. This phenomenon is just the relativistic analog of the plasmon phenomenon to which Anderson[3] has drawn attention: that the scalar zero-mass excitations of a superconducting neutral Fermi gas become longitudinal plasmon modes of finite mass when the gas is charged.

The simplest theory which exhibits this behavior is a gauge-invariant version of a model used by Goldstone[2] himself: Two real[4] scalar fields φ_1, φ_2 and a real vector field A_μ interact through the Lagrangian density

$$L = -\tfrac{1}{2}(\nabla \varphi_1)^2 - \tfrac{1}{2}(\nabla \varphi_2)^2$$
$$- V(\varphi_1{}^2 + \varphi_2{}^2) - \tfrac{1}{4} F_{\mu\nu} F^{\mu\nu}, \quad (1)$$

where

$$\nabla_\mu \varphi_1 = \partial_\mu \varphi_1 - e A_\mu \varphi_2,$$

$$\nabla_\mu \varphi_2 = \partial_\mu \varphi_2 + e A_\mu \varphi_1,$$

$$F_{\mu\nu} = \partial_\mu A_\nu - \partial_\nu A_\mu,$$

e is a dimensionless coupling constant, and the metric is taken as −+++. L is invariant under simultaneous gauge transformations of the first kind on $\varphi_1 \pm i\varphi_2$ and of the second kind on A_μ. Let us suppose that $V'(\varphi_0{}^2) = 0$, $V''(\varphi_0{}^2) > 0$; then spontaneous breakdown of U(1) symmetry occurs. Consider the equations [derived from (1) by treating $\Delta\varphi_1$, $\Delta\varphi_2$, and A_μ as small quantities] governing the propagation of small oscillations

about the "vacuum" solution $\varphi_1(x) = 0$, $\varphi_2(x) = \varphi_0$:

$$\partial^\mu \{\partial_\mu(\Delta\varphi_1) - e\varphi_0 A_\mu\} = 0, \quad (2a)$$

$$\{\partial^2 - 4\varphi_0{}^2 V''(\varphi_0{}^2)\}(\Delta\varphi_2) = 0, \quad (2b)$$

$$\partial_\nu F^{\mu\nu} = e\varphi_0\{\partial^\mu(\Delta\varphi_1) - e\varphi_0 A_\mu\}. \quad (2c)$$

Equation (2b) describes waves whose quanta have (bare) mass $2\varphi_0\{V''(\varphi_0{}^2)\}^{1/2}$; Eqs. (2a) and (2c) may be transformed, by the introduction of new variables

$$B_\mu = A_\mu - (e\varphi_0)^{-1}\partial_\mu(\Delta\varphi_1),$$

$$G_{\mu\nu} = \partial_\mu B_\nu - \partial_\nu B_\mu = F_{\mu\nu}, \quad (3)$$

into the form

$$\partial_\mu B^\mu = 0, \quad \partial_\nu G^{\mu\nu} + e^2\varphi_0{}^2 B^\mu = 0. \quad (4)$$

Equation (4) describes vector waves whose quanta have (bare) mass $e\varphi_0$. In the absence of the gauge field coupling ($e = 0$) the situation is quite different: Equations (2a) and (2c) describe zero-mass scalar and vector bosons, respectively. In passing, we note that the right-hand side of (2c) is just the linear approximation to the conserved current: It is linear in the vector potential, gauge invariance being maintained by the presence of the gradient term.[5]

When one considers theoretical models in which spontaneous breakdown of symmetry under a semisimple group occurs, one encounters a variety of possible situations corresponding to the various distinct irreducible representations to which the scalar fields may belong; the gauge field always belongs to the adjoint representation.[6] The model of the most immediate interest is that in which the scalar fields form an octet under SU(3): Here one finds the possibility of two nonvanishing vacuum expectation values, which may be chosen to be the two $Y = 0$, $I_3 = 0$ members of the octet.[7] There are two massive scalar bosons with just these quantum numbers; the remaining six components of the scalar octet combine with the corresponding components of the gauge-field octet to describe

508

VOLUME 13, NUMBER 16   PHYSICAL REVIEW LETTERS   19 OCTOBER 1964

massive vector bosons. There are two $I = \frac{1}{2}$ vector doublets, degenerate in mass between $Y = \pm 1$ but with an electromagnetic mass splitting between $I_3 = \pm\frac{1}{2}$, and the $I_3 = \pm 1$ components of a $Y = 0$, $I = 1$ triplet whose mass is entirely electromagnetic. The two $Y = 0$, $I = 0$ gauge fields remain massless: This is associated with the residual unbroken symmetry under the Abelian group generated by Y and I_3. It may be expected that when a further mechanism (presumably related to the weak interactions) is introduced in order to break Y conservation, one of these gauge fields will acquire mass, leaving the photon as the only massless vector particle. A detailed discussion of these questions will be presented elsewhere.

It is worth noting that an essential feature of the type of theory which has been described in this note is the prediction of incomplete multiplets of scalar and vector bosons.[8] It is to be expected that this feature will appear also in theories in which the symmetry-breaking scalar fields are not elementary dynamic variables but bilinear combinations of Fermi fields.[9]

[1]P. W. Higgs, to be published.

[2]J. Goldstone, Nuovo Cimento 19, 154 (1961); J. Goldstone, A. Salam, and S. Weinberg, Phys. Rev. 127, 965 (1962).

[3]P. W. Anderson, Phys. Rev. 130, 439 (1963).

[4]In the present note the model is discussed mainly in classical terms; nothing is proved about the quantized theory. It should be understood, therefore, that the conclusions which are presented concerning the masses of particles are conjectures based on the quantization of linearized classical field equations. However, essentially the same conclusions have been reached independently by F. Englert and R. Brout, Phys. Rev. Letters 13, 321 (1964): These authors discuss the same model quantum mechanically in lowest order perturbation theory about the self-consistent vacuum.

[5]In the theory of superconductivity such a term arises from collective excitations of the Fermi gas.

[6]See, for example, S. L. Glashow and M. Gell-Mann, Ann. Phys. (N.Y.) 15, 437 (1961).

[7]These are just the parameters which, if the scalar octet interacts with baryons and mesons, lead to the Gell-Mann–Okubo and electromagnetic mass splittings: See S. Coleman and S. L. Glashow, Phys. Rev. 134, B671 (1964).

[8]Tentative proposals that incomplete SU(3) octets of scalar particles exist have been made by a number of people. Such a rôle, as an isolated $Y = \pm 1$, $I = \frac{1}{2}$ state, was proposed for the κ meson (725 MeV) by Y. Nambu and J. J. Sakurai, Phys. Rev. Letters 11, 42 (1963). More recently the possibility that the σ meson (385 MeV) may be the $Y = I = 0$ member of an incomplete octet has been considered by L. M. Brown, Phys. Rev. Letters 13, 42 (1964).

[9]In the theory of superconductivity the scalar fields are associated with fermion pairs; the doubly charged excitation responsible for the quantization of magnetic flux is then the surviving member of a U(1) doublet.

PHYSICAL REVIEW   VOLUME 145, NUMBER 4   27 MAY 1966

Spontaneous Symmetry Breakdown without Massless Bosons*

PETER W. HIGGS†

Department of Physics, University of North Carolina, Chapel Hill, North Carolina

(Received 27 December 1965)

We examine a simple relativistic theory of two scalar fields, first discussed by Goldstone, in which as a result of spontaneous breakdown of $U(1)$ symmetry one of the scalar bosons is massless, in conformity with the Goldstone theorem. When the symmetry group of the Lagrangian is extended from global to local $U(1)$ transformations by the introduction of coupling with a vector gauge field, the Goldstone boson becomes the longitudinal state of a massive vector boson whose transverse states are the quanta of the transverse gauge field. A perturbative treatment of the model is developed in which the major features of these phenomena are present in zero order. Transition amplitudes for decay and scattering processes are evaluated in lowest order, and it is shown that they may be obtained more directly from an equivalent Lagrangian in which the original symmetry is no longer manifest. When the system is coupled to other systems in a $U(1)$ invariant Lagrangian, the other systems display an induced symmetry breakdown, associated with a partially conserved current which interacts with itself via the massive vector boson.

I. INTRODUCTION

THE idea that the apparently approximate nature of the internal symmetries of elementary-particle physics is the result of asymmetries in the stable solutions of exactly symmetric dynamical equations, rather than an indication of asymmetry in the dynamical equations themselves, is an attractive one. Within the framework of quantum field theory such a "spontaneous" breakdown of symmetry occurs if a Lagrangian, fully invariant under the internal symmetry group, has such a structure that the physical vacuum is a member of a set of (physically equivalent) states which transform according to a nontrivial representation of the group. This degeneracy of the vacuum permits nontrivial multiplets of scalar fields (which may be either fundamental dynamic variables or polynomials constructed from them) to have nonzero vacuum expectation values, whose appearance in Feynman diagrams leads to symmetry-breaking terms in propagators and vertices. That vacuum expectation values of scalar fields, or "vacuons," might play such a role in the breaking of symmetries was first noted by Schwinger[1] and by Salam and Ward.[2] Under the alternative name, "tadpole" diagrams, the graphs in which vacuons

appear have been used by Coleman and Glashow[3] to account for the observed pattern of deviations from $SU(3)$ symmetry.

The study of field theoretical models which display spontaneous breakdown of symmetry under an internal Lie group was initiated by Nambu,[4] who had noticed[5] that the BCS theory of superconductivity[6] is of this type, and was continued by Glashow[7] and others.[8] All these authors encountered the difficulty that their theories predicted, *inter alia*, the existence of a number of massless scalar or pseudoscalar bosons, named "zerons" by Freund and Nambu.[9] Since the models which they discussed, being inspired by the BCS theory, used an attractive interaction between massless fermions and antifermions as the mechanism of symmetry breakdown, it was at first unclear whether zerons occurred as a result of the approximations (including the usual cutoff for divergent integrals) involved in handling the models or whether they would still be there in an exact solution. Some authors,

* This work was partially supported by the U. S. Air Force Office of Scientific Research under grant No. AF-AFOSR-153-64.

† On leave from the Tait Institute of Mathematical Physics, University of Edinburgh, Scotland.

[1] J. Schwinger, Phys. Rev. **104**, 1164 (1954); Ann. Phys. (N. Y.) **2**, 407 (1957).

[2] A. Salam and J. C. Ward, Phys. Rev. Letters **5**, 390 (1960); Nuovo Cimento **19**, 167 (1961).

[3] S. Coleman and S. L. Glashow, Phys. Rev. **134**, B671 (1964).

[4] Y. Nambu and G. Jona-Lasinio, Phys. Rev. **122**, 345 (1961); **124**, 246 (1961); Y. Nambu and P. Pascual, Nuovo Cimento **30**, 354 (1963).

[5] Y. Nambu, Phys. Rev. **117**, 648 (1960).

[6] J. Bardeen, L. N. Cooper, and J. R. Schrieffer, Phys. Rev. **106**, 162 (1957).

[7] M. Baker and S. L. Glashow, Phys. Rev. **128**, 2462 (1962); S. L. Glashow, *ibid.* **130**, 2132 (1962).

[8] M. Suzuki, Progr. Theoret. Phys. (Kyoto) **30**, 138 (1963); **30**, 627 (1963); N. Byrne, C. Iddings, and E. Shrauner, Phys. Rev. **139**, B918 (1965); **139**, B933 (1965).

[9] P. G. O. Freund and Y. Nambu, Phys. Rev. Letters **13**, 221 (1964).

wishing to identify their zerons with known massive scalar or pseudoscalar mesons, were prepared to spoil the elegance of their theories by adding symmetry-breaking terms to the Lagrangian in order to generate masses.

That zerons must indeed be present in Lorentz invariant field theories in which an internal symmetry breaks down spontaneously was first shown by Goldstone.[10] He clarified the nature of the phenomenon considerably by exhibiting it in a model of a self-interacting scalar field, where the vacuon is the vacuum expectation value of the field itself, rather than that of a bilinear combination of fermion operators which occurs in the BCS model and its progeny. In a theory of this type the breakdown of symmetry occurs already at the level of classical field theory, where vacuons are just nontrivial translationally invariant solutions of the field equations, and zerons, whose existence is readily demonstrated, are small-amplitude waves (superimposed on a "vacuon" solution) whose frequency tends to zero as their wavelength tends to infinity. In a later paper[11] the proof of the Goldstone theorem, as it is now known, was generalized to allow for the possibility that vacuons might be formed from polynomials of any degree in the fundamental field variables of a dynamical system.

During the last few years the problem of how to avoid massless Goldstone bosons has received much attention. Attempts in this direction have been encouraged by the observation that the BCS model does not contain such excitations, provided that Coulomb interactions are taken into account.[12] Klein and Lee[13] showed that in a nonrelativistic theory the spectral representations upon which the more sophisticated proofs of Goldstone's theorem are based are not so restricted in form as to allow the proof to go through, and they conjectured that this might remain true in some relativistic theories. But Gilbert[14] pointed out that their extra terms are ruled out in relativistic theories by the requirement of manifest Lorentz covariance. The present writer restored the status quo to a limited extent by remarking[15] that radiation gauge formulations of gauge field theories, of which electrodynamics is the simplest and best known example, can describe Lorentz-invariant dynamical systems despite the lack of manifest covariance of some of the equations. The freedom which Klein and Lee hoped for in the spectral representations is thereby restored sufficiently to invalidate the Goldstone theorem. From a more physical standpoint one may regard this as an effect of Coulomb interactions, treated now as part of a relativistic field theory.

More recently Guralnik, Hagen, and Kibble[16] and Lange[17] have studied how the failure of global (as distinct from local) conservation laws in spontaneous breakdown theories is related to the existence of Goldstone bosons. Meanwhile, proofs of the theorem have reached new levels of sophistication within the language (or languages) of axiomatic field theory.[18] It has been pointed out that theories of the type proposed in Ref. 15 do not contradict the Goldstone theorem, but rather represent a departure from the assumptions, such as *manifest* covariance and *manifest* causality, upon which it is based. Such considerations do not seem relevant to the possible usefulness of such theories in generating zeron-free models of spontaneous symmetry breakdown, a point which seems to have been overlooked by those[19] who proclaim the failure of the Nambu-Glashow program.

In parallel with the development of "superconductor" models a program has emerged for describing the weak,[20] and possibly also the strong,[21] interactions by an extension of the gauge principle[22] which operates in electrodynamics. According to this principle the symmetry group of the Lagrangian is to be enlarged from global to local (i.e., coordinate-dependent) transformations: To maintain the invariance of derivative terms it is necessary to couple the currents of the group generators to a multiplet of vector fields belonging to the adjoint representation.[23] Like the "superconductor" theories, these gauge theories have suffered from a zero-mass difficulty: The gauge principle appears to guarantee that the associated vector field quanta are massless, in

[10] J. Goldstone, Nuovo Cimento 19, 154 (1961).
[11] J. Goldstone, A. Salam, and S. Weinberg, Phys. Rev. 127, 965 (1962).
[12] P. W. Anderson, Phys. Rev. 112, 1900 (1958).
[13] A. Klein and B. W. Lee, Phys. Rev. Letters 12, 266 (1964).
[14] W. Gilbert, Phys. Rev. Letters 12, 713 (1964).
[15] P. W. Higgs, Phys. Letters 12, 132 (1964).

[16] G. S. Guralnik, C. R. Hagen, and T. W. B. Kibble, Phys. Rev. Letters 13, 585 (1964). These authors appear to attribute the failure of a local conservation law to yield a global conservation law to the lack of manifest covariance of a theory. In fact this happens even in manifestly covariant models of spontaneous breakdown.
[17] R. V. Lange, Phys. Rev. Letters 14, 3 (1965).
[18] R. F. Streater, Proc. Roy. Soc. (London) A287, 510 (1965). The proof of the Goldstone theorem given here is based on axioms which include manifest causality. Radiation gauge theories escape *this* version of the theorem by virtue of their (quite innocuous) acausality. The question of the extent to which one may give up manifest covariance and causality in a theory without losing covariance and causality of the physics which it describes deserves further study. If there are contexts in which it is possible other than the gauge theories which we are discussing, then there are probably other escape routes from the Goldstone theorem. See also D. Kastler, D. W. Robinson, and A. Swieca, Commun. Math. Phys. (to be published).
[19] R. F. Streater, Phys. Rev. Letters 15, 475 (1965). The generalized Goldstone theorem proved by this author and extended by N. Fuchs, Phys. Rev. Letters 15, 911 (1965) also relies on manifest causality and is therefore inapplicable to gauge theories.
[20] S. A. Bludman, Phys. Rev. 100, 372 (1955); Nuovo Cimento 9, 433 (1958); S. L. Glashow, Nucl. Phys. 10, 107 (1959); 22, 579 (1961).
[21] A. Salam and J. C. Ward, Nuovo Cimento 11, 568 (1959); Phys. Rev. 136, B763 (1964); J. J. Sakurai, Ann. Phys. (N. Y.) 11, 1 (1960).
[22] C. N. Yang and R. L. Mills, Phys. Rev. 96, 191 (1954); R. Shaw, dissertation, Cambridge University, 1954 (unpublished); R. Utiyama, Phys. Rev. 101, 1597 (1956).
[23] M. Gell-Mann and S. L. Glashow, Ann. Phys. (N. Y.) 15, 437 (1961).

perturbation theory at least. But the only known mass-less vector boson is the photon; the existing evidence suggests[24] that all other vector bosons must be massive. In particular, the hypothetical intermediate vector bosons of weak interactions, which in a gauge theory belong to the gauge field multiplet, must be much heavier than the known hadrons. For the most part, advocates of gauge theories have met this difficulty either by spoiling the gauge invariance of their theories with explicit mass terms or by taking comfort from the argument of Schwinger[25] that a sufficiently strong gauge-field coupling might generate mass. Recently, however, it was shown by Englert and Brout[26] and by the present writer[27] that gauge vector mesons acquire mass if the symmetry to whose generators they are coupled breaks down spontaneously, however weak their coupling may be. In Ref. 27 this phenomenon was exhibited in a classical field theory, and it was pointed out that in such a theory the longitudinal polarization of the massive vector excitation replaces the massless scalar excitation which would occur in the absence of coupling to the gauge field. Thus it now appears that the spontaneous breakdown program of Nambu *et al.* and the gauge field program of Salam *et al.* stand or fall together. Each saves the other from its zero-mass difficulty.

The purpose of the present paper is to amplify and substantiate the assertions made in Refs. 15 and 27 by displaying the behavior of the simplest possible rela-tivistic field theory which combines spontaneous break-down of symmetry under a compact Lie group with the gauge principle. That is to say, we take the sym-metry group to be the trivial Abelian group $U(1)$, we take as the fundamental dynamic variables a pair of Hermitian scalar fields $\Phi_1(x)$, $\Phi_2(x)$ together with the Hermitian vector gauge field $A_\mu(x)$, and we induce spontaneous breakdown by means of the simplest $U(1)$-invariant self-interaction of $\Phi_a(x)$ which will do the trick, namely, a combination of quadratic and quartic terms. In the absence of the gauge field coupling, the model is just one which Goldstone[10] first discussed.[28]

In Sec. II the behavior of the small-amplitude wave solutions of the classical field equations is used as a guide in formulating a perturbation theory in which the major effects of spontaneous breakdown are already taken into account in zero order. The radiation gauge commutators of the zero-order approximation are ob-tained and used to provide an explicit realization of the spectral representation which was proposed in Ref. 15. In Sec. III decay and scattering amplitudes are cal-culated in lowest order and it is verified that they are

gauge-invariant, Lorentz-invariant, and causal despite the lack of manifest covariance and causality of the radiation gauge. In Sec. IV it is shown that the same amplitudes may be derived by a manifestly covariant and causal method from an equivalent Lagrangian which lacks the original symmetry. Finally, our con-clusions are summarized in Sec. V, and the way in which coupling between a system of the kind described here and other symmetric dynamical systems may lead to partially conserved currents is sketched.

In subsequent papers we propose to elaborate these considerations, both with regard to symmetry groups and with regard to mechanisms of symmetry break-down, so as to make closer contact with particle physics.

II. THE MODEL

The Lagrangian density from which we shall work is given by[29]

$$\mathcal{L} = -\tfrac{1}{4}g^{\kappa\mu}g^{\lambda\nu}F_{\kappa\lambda}F_{\mu\nu} - \tfrac{1}{2}g^{\mu\nu}\nabla_\mu\Phi_a\nabla_\nu\Phi_a + \tfrac{1}{4}m_0{}^2\Phi_a\Phi_a - \tfrac{1}{8}f^2(\Phi_a\Phi_a)^2. \quad (1)$$

In Eq. (1) the metric tensor $g^{\mu\nu} = -1$ ($\mu = \nu = 0$), $+1$ ($\mu = \nu \neq 0$) or 0 ($\mu \neq \nu$), Greek indices run from 0 to 3 and Latin indices from 1 to 2. The $U(1)$-covariant derivatives $F_{\mu\nu}$ and $\nabla_\mu\Phi_a$ are given by

$$F_{\mu\nu} = \partial_\mu A_\nu - \partial_\nu A_\mu,$$
$$\nabla_\mu\Phi_1 = \partial_\mu\Phi_1 - eA_\mu\Phi_2,$$
$$\nabla_\mu\Phi_2 = \partial_\mu\Phi_2 + eA_\mu\Phi_1.$$

At first sight this theory appears to be scalar electro-dynamics augmented by a quartic self-interaction. However, what appears to be the bare-mass term has the wrong sign. In conjunction with the quartic term this feature has the consequence that the field equations

$$\partial_\nu F^{\mu\nu} = ej^\mu, \quad j_\mu = \Phi_2\nabla_\mu\Phi_1 - \Phi_1\nabla_\mu\Phi_2,$$
$$\nabla_\mu\nabla^\mu\Phi_a + \tfrac{1}{2}(m_0{}^2 - f^2\Phi_b\Phi_b)\Phi_a = 0, \quad (2)$$

in the classical theory possess a coordinate-independent solution

$$A_\mu = 0, \quad \Phi_b\Phi_b = m_0{}^2/f^2. \quad (3)$$

The invariance of the Lagrangian (1) under the local $U(1)$ transformations

$$A_\mu(x) \rightarrow A_\mu(x) + e^{-1}\partial_\mu\Lambda(x),$$
$$\Phi_1(x) \rightarrow \Phi_1(x)\cos\Lambda(x) + \Phi_2(x)\sin\Lambda(x), \quad (4)$$
$$\Phi_2(x) \rightarrow -\Phi_1(x)\sin\Lambda(x) + \Phi_2(x)\cos\Lambda(x),$$

is reflected in the existence of a one-parameter family of static solutions defined by Eq. (3).[30] In the classical

---

[24] See, for example, S. Weinberg, Phys. Rev. Letters **13**, 495 (1964).
[25] J. Schwinger, Phys. Rev. **125**, 397 (1962); **128**, 2425 (1962).
[26] F. Englert and R. Brout, Phys. Rev. Letters **13**, 321 (1964).
[27] P. W. Higgs, Phys. Rev. Letters **13**, 508 (1964).
[28] I understand from Dr. Goldstone (private communication) that he and W. Gilbert at one time considered adding a gauge field to the model.

[29] We do not explicitly perform the symmetrizations which a correct quantum-mechanical treatment would demand. They are not necessary for the purposes of the present paper and would, in any case, be dealt with more satisfactorily in a first-order formalism.
[30] Strictly speaking, global $U(1)$ invariance suffices to guarantee this result.

theory, any one of these solutions,

$$\Phi_1 = \eta\cos\alpha, \quad \Phi_2 = \eta\sin\alpha,$$

where $\eta = m_0/f$, defines a possible asymmetric configuration of stable equilibrium, stability being ensured by the sign of the quartic term in Eq. (1). Quantum mechanically each solution, regarded as the "bare" value of the vacuon $\langle\Phi_a\rangle$, corresponds to a different possible vacuum state.[31]

Let us choose $\alpha = \pi/2$ and linearize the classical field equations (2) by treating A_μ, Φ_1, and $\Phi_2 - \eta$ as small quantities. We obtain

$$\partial_\nu F^{\mu\nu} = -e^2\eta^2 B^\mu, \quad \partial_\mu B^\mu = 0,$$
$$(\Box - m_0^2)\chi = 0,$$

in which we have introduced the notation

$$B_\mu = A_\mu - (e\eta)^{-1}\partial_\mu\Phi,$$
$$\Phi = \Phi_1, \quad \chi = \Phi_2 - \eta. \qquad (5)$$

As we remarked in Ref. 27, these are the linear field equations which, after quantization, describe free vector bosons of mass $e\eta$ and free scalar bosons of mass m_0. The longitudinal vector excitation becomes the Goldstone scalar excitation in the limit $e \to 0$. The Lagrangian to which these field equations belong is given by

$$\mathcal{L}_0 = -\tfrac{1}{4}F_{\mu\nu}F^{\mu\nu} - \tfrac{1}{2}g^{\mu\nu}(\partial_\mu\Phi - m_1 A_\mu)(\partial_\nu\Phi - m_1 A_\nu)$$
$$- \tfrac{1}{2}g^{\mu\nu}\partial_\mu\chi\partial_\nu\chi - \tfrac{1}{2}m_0^2\chi^2, \quad (6)$$

where we have written m_1 for the vector boson mass $e\eta$.

The foregoing analysis of classical small-amplitude wave propagation suggests the following perturbation theoretic treatment of the quantized theory. We rewrite Eq. (1) in the form $\mathcal{L} = \mathcal{L}_0 + \mathcal{L}_{\text{int}}$, where $\mathcal{L}_0$ is given by Eq. (6) apart from a trivial additive constant and

$$\mathcal{L}_{\text{int}} = eA^\mu(\chi\partial_\mu\Phi - \Phi\partial_\mu\chi) - em_1\chi A_\mu A^\mu - \tfrac{1}{2}fm_0\chi(\Phi^2 + \chi^2)$$
$$- \tfrac{1}{2}e^2 A_\mu A^\mu(\Phi^2 + \chi^2) - \tfrac{1}{8}f^2(\Phi^2 + \chi^2)^2. \quad (7)$$

Note that the ancestry of (6) plus (7) in the symmetric Lagrangian (1) is embodied in the relation $m_1/m_0 = e/f$ between the bare masses and the bare coupling constants. Our perturbation theory now consists in developing transition amplitudes in powers of e and f

(or, equivalently, in inverse powers of η), the masses m_0 and m_1 being treated as of order zero.[32] Thus in Eq. (7) all five cubic vertices are of the first degree and all five quartic vertices are of the second in the expansion parameter. It will be found that, with few exceptions, gauge-invariant results are obtained only when all Feynman graphs of the same degree are summed.

We first write down the commutators and propagators corresponding to the bare Lagrangian $\mathcal{L}_0$. Apart from the terms in χ, this is just the second-order Lagrangian of a model proposed by Boulware and Gilbert[33] as an illustration of the possibility of a gauge-invariant theory describing a massive vector boson. We shall study it in a radiation gauge; the Lorentz gauge formulation, which even in quantum electrodynamics leads to unnecessary complications such as redundant states, is here inconsistent with the canonical com, mutation rules, as was pointed out by Guralnik, Hagen and Kibble.[16] In a radiation gauge defined by the condition

$$(\partial A) + (n\partial)(nA) = 0,$$

where n^μ is a constant timelike unit vector and (ab) denotes $a_\mu b^\mu$, the variables A_μ and Φ may be expressed in terms of the massive vector field B_μ which was introduced in Eq. (5):

$$A_\mu = B_\mu + m_1^{-1}\partial_\mu\Phi,$$
$$\Phi = -m_1[(\partial^2) + (n\partial)^2]^{-1}[(\partial B) + (n\partial)(nB)]. \quad (8)$$

Since $\mathcal{L}_0$, when expressed in terms of B_μ and χ, is just the usual second order Lagrangian for free vector and scalar bosons, we may immediately write down the covariant commutators:

$$[B_\mu(x), B_\nu(y)] = -i(g_{\mu\nu} - m_1^{-2}\partial_\mu\partial_\nu)\Delta(x - y, m_1^2),$$
$$[\chi(x), \chi(y)] = -i\Delta(x - y, m_0^2), \qquad (9)$$

where $\Delta(x, m^2) = i(2\pi)^{-3}\int d^4 k\, e^{i(kx)}\epsilon(k^0)\delta(k^2 + m^2)$. Then Eq. (8) enables us to deduce the nonvanishing commutators of A_μ, Φ, and χ:

$$[A_\mu(x), A_\nu(y)] = -i\{g_{\mu\nu} - [(n_\mu\partial_\nu + n_\nu\partial_\mu)(n\partial) + \partial_\mu\partial_\nu]$$
$$\times [(\partial^2) + (n\partial)^2]^{-1}\}\Delta(x - y, m_1^2),$$
$$[A_\mu(x), \Phi(y)] = -im_1 n_\mu(n\partial)$$
$$\times [(\partial^2) + (n\partial)^2]^{-1}\Delta(x - y, m_1^2), \qquad (10)$$
$$[\Phi(x), \Phi(y)] = -i(n\partial)^2[(\partial^2) + (n\partial)^2]^{-1}\Delta(x - y, m_1^2),$$
$$[\chi(x), \chi(y)] = -i\Delta(x - y, m_0^2).$$

We also note the commutator relation

$$[B_\mu(x), \Phi(y)] = -im_1^{-1}[(\partial^2)n_\mu - (n\partial)\partial_\mu](n\partial)$$
$$\times [(\partial^2) + (n\partial)^2]^{-1}\Delta(x - y, m_1^2). \quad (11)$$

---

[31] The orthogonality of the worlds built upon different vacua may be understood as a consequence of the impossibility in the classical theory of a displacement of the system from one static configuration to another, on account of the infinite inertia associated with such a motion. To see this, imagine a one-dimensional model consisting of an infinite uniform elastic string subjected to a force field of cylindrical symmetry about the axis from which transverse displacements Φ_1, Φ_2 are measured. (We omit the gauge field from this do-it-yourself model.) If the force is such that stable equilibrium occurs when the whole string is at a distance η from the axis in any direction, then the system exhibits broken rotational symmetry. Displacement of the string from equilibrium at one orientation to equilibrium at another is impossible, since the moment of inertia about the axis is infinite. Waves on the string do not conserve angular momentum about the axis, since the string as a whole can emit or absorb angular momentum without recoiling.

[32] When one comes to consider radiative corrections, it becomes necessary to make these statements about the renormalized rather than the bare masses and coupling constants.

[33] D. C. Boulware and W. Gilbert, Phys. Rev. **126**, 1563 (1962).

The generator $Q(t)$ of infinitesimal *global* $U(1)$ transformations (that is, transformations (4) with Λ constant and infinitesimal) on the hypersurface $(nx)+t=0$ is $\int d\sigma_\mu j^\mu$, where $d\sigma_\mu$ is the volume element of the hypersurface and $j_\mu(x)$ is given by Eq. (2). The invariance of the Lagrangian (1) under these transformations leads to the local conservation law, $\partial_\mu j^\mu=0$. However, even in the absence of the gauge field coupling, the four-dimensional integral of this equation fails to yield the usual global conservation law, $Q(t)$ = constant, since the flux of j_μ across the surface of a large sphere bounding the hypersurface does not tend to zero as the radius tends to infinity. That this is so can be seen by noting that the lowest order approximation to j_μ is $-e\eta^2 B_\mu$ (or $\eta\partial_\mu\Phi$ in the absence of the gauge field): Matrix elements of this operator do not decrease sufficiently rapidly at large spatial distances for the flux term to vanish. (In normal theories the lowest order term in j_μ is quadratic, giving a better asymptotic behavior of the matrix elements.) Strictly speaking, the "operator" $Q(t)$ is now not merely time-dependent but nonexistent, since $\int d\sigma_\mu j^\mu$ diverges as a result of the same bad asymptotic behavior of j^μ. However, certain commutators, such as $[Q(t),\Phi_a(y)]$, do still exist.[34]

The zero-order approximation to the commutator vacuum expectation value $\langle i[j_\mu(x),\Phi_1(y)]\rangle$, upon which so much of the discussion of the Goldstone theorem has centered,[35] is found by replacing j_μ by $-e\eta^2 B_\mu$ and using Eq. (11): It is

$$\eta[(n\partial)\partial_\mu-(\partial^2)n_\mu](n\partial)[(\partial^2)+(n\partial)^2]^{-1}\Delta(x-y, m_1^2).$$

Its Fourier transform,

$$-2\pi\eta[(nk)k_\mu-(k^2)n_\mu](nk)$$
$$\times[(k^2)+(nk)^2]^{-1}\epsilon(k^0)\delta(k^2+m_1^2), \quad (12)$$

provides an explicit realization of a spectral representation of the form obtained in Ref. 15, the Lorentz invariance of the spectrum of intermediate states now being made clear. We are led to conjecture that the vacuum expectation value of the exact commutator may be of the form

$$\langle\Phi_2\rangle[(n\partial)\partial_\mu-(\partial^2)n_\mu](n\partial)[(\partial^2)+(n\partial)^2]^{-1}$$
$$\times\int_0^\infty dm^2\rho(m^2)\Delta(x-y, m^2), \quad (13)$$

where $\rho(m^2)$ is a nonnegative spectral function satisfy-ing the sum rule

$$\int_0^\infty dm^2\rho(m^2)=1.$$

It may be noted that when $m_1=0$ in Eq. (12), corresponding to $e=0$, we recover the manifestly covariant spectral representation $-2\pi\eta k_\mu\epsilon(k^0)\delta(k^2)$ and with it the Goldstone theorem.

We define the propagators of the system described by $\mathcal{L}_0$ to be the quantities $\langle T^*A_\mu(x)A_\nu(y)\rangle$, etc., obtained from the corresponding commutators in Eq. (10) by substituting for Δ the scalar propagator Δ_F given by

$$\Delta_F(x,m^2)=(2\pi)^{-4}\int d^4k\, e^{i(kx)}(k^2+m^2-i\epsilon)^{-1}.$$

Then we may calculate S-matrix elements by using the simple Feynman rules based on the Nishijima-Wick expansion of the expression $T^*\exp(i\int d^4x\,\mathcal{L}_{\text{int}})$ for the S-operator in the interaction picture.[36] We thereby avoid the n_μ-dependent terms, in addition to those already introduced by the radiation gauge, which the use of simple chronological ordering and the Dyson-Wick expansion would entail.

III. DECAY AND SCATTERING AMPLITUDES

As an illustration of the physical content of the model we now calculate in lowest order the matrix elements for the simplest processes which it describes. We shall verify that, despite the unpromising appearance of the radiation gauge propagators, these matrix elements are gauge invariant and Lorentz invariant.

In applying the Feynman rules we shall need, in addition to the propagators, the wave functions $a_\mu(k,\sigma)$ and $\phi(k,\sigma)$ which correspond to the annihilation by the operators A_μ and Φ, respectively, of a vector meson from an incoming state of momentum k and spin component σ. They are related by the Fourier transform of Eq. (8) to the usual vector meson wave function $b_\mu(k,\sigma)$, which has the explicit form

$$b^\mu(k,0)=(\omega/m_1)(|\mathbf{k}|/\omega,\mathbf{k}/|\mathbf{k}|),$$
$$b^\mu(k,\pm 1)=2^{-1/2}(0, \mathbf{e}_1\pm i\mathbf{e}_2), \quad (14)$$

where $\omega=(|\mathbf{k}|^2+m_1^2)^{1/2}$ and $\mathbf{e}_1$, $\mathbf{e}_2$, $\mathbf{k}/|\mathbf{k}|$ form a right-handed orthonormal triad. Actually, all that we shall need is the relation

$$a_\mu=b_\mu+(ik_\mu/m_1)\phi, \quad (15)$$

by which matrix elements may be expressed in terms of wave functions b_μ and ϕ, the desired invariance being achieved by the cancellation of all terms containing factors ϕ. Similar considerations apply to out-

[34] In Ref. 18 it is proved that in a manifestly causal theory this commutator (or at least certain of its matrix elements) is independent of t, despite the breakdown of the global conservation law. The gauge field coupling destroys manifest causality and induces a time dependence in this commutator: In the zero-order approximation it oscillates at a frequency m_1.

[35] See Refs. 11, 13, 14, and 15. In Refs. 14 and 15 it is implied erroneously that the commutator $[Q(t),\Phi_a(y)]$ is independent of t. Fortunately, the discussion of the Goldstone theorem in these papers does not depend on this assumption.

[36] P. T. Matthews, Phys. Rev. 76, 684 (1949); K. Nishijima, Progr. Theoret. Phys. (Kyoto) 5, 405 (1950). The most general conditions for the validity of this expression have been stated by C. S. Lam, Nuovo Cimento 38, 1755 (1965).

going states and associated complex conjugate wave functions.

i. Decay of a Scalar Boson into Two Vector Bosons

The process occurs in first order (four of the five cubic vertices contribute), provided that $m_0 > 2m_1$. Let p be the incoming and k_1, k_2 the outgoing momenta. Then

$$M = i\{e[a^{*\mu}(k_1)(-ik_{2\mu})\phi^*(k_2) + a^{*\mu}(k_2)(-ik_{1\mu})\phi^*(k_1)] - e(ip_\mu)[a^{*\mu}(k_1)\phi^*(k_2) + a^{*\mu}(k_2)\phi^*(k_1)] - 2em_1 a_\mu^*(k_1)a^{*\mu}(k_2) - fm_0\phi^*(k_1)\phi^*(k_2)\}.$$

By using Eq. (15), conservation of momentum, and the transversality $(k_\mu b^\mu(k)=0)$ of the vector wave functions we reduce this to the form

$$M = -2iem_1 b^{*\mu}(k_1)b_\mu^*(k_2) - iem_1^{-1}(p^2 + m_0^2)\phi^*(k_1)\phi^*(k_2). \quad (16)$$

We have retained the last term, which we shall need in calculating scattering amplitudes; when the incident particle is on the mass shell it vanishes and we are left with the invariant expression

$$M = -2iem_1 b^{*\mu}(k_1)b_\mu^*(k_2). \quad (17)$$

Conservation of angular momentum allows three possibilities for the spin states of the decay products: They may be both right-handed, both left-handed, or both longitudinal $(\sigma_1 = \sigma_2 = +1, -1, \text{ or } 0)$. With the help of the explicit vectors (14), we find

$$M(+1,+1) = M(-1,-1) = 2iem_1,$$
$$M(0,0) = ifm_0(1 - 2e^2/f^2).$$

We note that as $e \to 0$ the amplitudes for decay to transverse states tend to zero, but the amplitude $M(0,0)$ tends to the value ifm_0 which we would calculate from the vertex $-\frac{1}{2}fm_0\Phi^2\chi$ for the decay of one massive into two massless scalar bosons in the original Goldstone model. (The sign change arises from the factor i which is associated with the term ϕ in each b_μ.)

ii. Vector Boson-Vector Boson Scattering

Let k_1, k_2 be the incoming and k_1', k_2' the outgoing momenta. The process occurs as a second-order effect of the cubic vertices, by exchange of a scalar boson in the s, t, or u channel, where $s = -(p_1 + p_2)^2$, $t = -(p_1 - p_1')^2$, $u = -(p_1 - p_2')^2$. It also occurs as a direct effect of two of the quartic vertices. Equation (16) enables us to write down

$$M_s = i^2\{-2em_1 b_\mu^*(k_1')b^{*\mu}(k_2') + em_1^{-1}(s - m_0^2)\phi^*(k_1')\phi^*(k_2')\} \times i(s - m_0^2)^{-1}\{-2em_1 b_\nu(k_1)b^\nu(k_2) + em_1^{-1}(s - m_0^2)\phi(k_1)\phi(k_2)\}$$

and similar expressions for M_t and M_u. The quartic vertices yield a contribution given by

$$M_{\text{direct}} = i(-2e^2)\{a_\mu^*(k_1')a^{*\mu}(k_2')\phi(k_1)\phi(k_2) + 5 \text{ similar terms}\} + i(-3f^2)\phi^*(k_1')\phi^*(k_2')\phi(k_1)\phi(k_2)$$
$$= -2ie^2\{b_\mu^*(k_1')b^{*\mu}(k_2')\phi(k_1)\phi(k_2) + 5 \text{ similar terms}\} + i(4e^2 - 3f^2)\phi^*(k_1')\phi^*(k_2')\phi(k_1)\phi(k_2).$$

It is only when we combine these four contributions that we obtain (after some algebra) the invariant expression

$$M_{\text{total}} = M_s + M_t + M_u + M_{\text{direct}} = -4ie^2 m_1^2\{(s - m_0^2)^{-1}b^{*\mu}(k_1')b^{*\mu}(k_2')b_\nu(k_1)b^\nu(k_2) + (t - m_0^2)^{-1}b_\mu^*(k_1')b^\mu(k_1)b_\nu^*(k_2')b^\nu(k_2) + (u - m_0^2)^{-1}b_\mu^*(k_1')b^\mu(k_2)b_\nu^*(k_2')b^\nu(k_1)\}. \quad (18)$$

iii. Vector Boson-Scalar Boson Scattering

Let k, p be the momenta of the incoming vector and scalar boson, respectively, and k', p' be their outgoing momenta. Again there are four contributions, M_s, M_t, M_u, and M_{direct}. In the s and u channels a vector boson is exchanged and it turns out that the various propagators, $\langle T^*A_\mu A_\nu\rangle$, $\langle T^*A_\mu\Phi\rangle$, and $\langle T^*\Phi\Phi\rangle$, occur only in the combination $\langle T^*B_\mu B_\nu\rangle$. We obtain the expression

$$M_s = i^2\{-2em_1 b^{*\mu}(k') + ieq^\mu\phi^*(k')\}i(g_{\mu\nu} + m_1^{-2}q_\mu q_\nu) \times (s - m_1^2)^{-1}\{-2em_1 b^\nu(k) - ieq^\nu\phi(k)\},$$

where $q = k + p$ and $s = -q^2$, and a similar expression for M_u. In the t channel a scalar boson is exchanged, and we find that

$$M_t = i^2\{-3fm_0\}i(t - m_0^2)^{-1}\{-2em_1 b_\mu^*(k')b^\mu(k) + em_1^{-1}(t - m_0^2)\phi^*(k')\phi(k)\},$$

where $t = -(k - k')^2$. Finally, the contribution of the quartic vertices is given by

$$M_{\text{direct}} = i\{-2e^2[b_\mu^*(k') - im_1^{-1}k_\mu'\phi^*(k')] \times [b^\mu(k) + im_1^{-1}k^\mu\phi(k)] - f^2\phi^*(k')\phi(k)\}.$$

Again the four contributions sum to the invariant expression

$$M_{\text{total}} = -2im_1^2\{2e^2(s - m_1^2)^{-1}[b_\mu^*(k')b^\mu(k) + m_1^{-2}p_\mu'b^{*\mu}(k')p_\nu b^\nu(k)] + 3f^2(t - m_0^2)^{-1}b_\mu^*(k')b^\mu(k) + 2e^2(u - m_1^2)^{-1}[b_\mu^*(k')b^\mu(k) + m_1^{-2}p_\mu b^{*\mu}(k')p_\nu'b^\nu(k)]\} - 2ie^2 b_\mu^*(k')b^\mu(k). \quad (19)$$

A similar matrix element may be written down for the process, vector pair $\leftrightarrow$ scalar pair, by making appropriate interchanges of incoming and outgoing momenta and wave functions.

iv. Scalar Boson-Scalar Boson Scattering

This is the only simple process in which no invariance problems arise in lowest order: The vertices which are involved contain no vector boson factors. We find that

$$M_{\text{total}} = M_s + M_t + M_u + M_{\text{direct}}$$
$$= -9if^2m_0^2\{(s-m_0^2)^{-1} + (t-m_0^2)^{-1}$$
$$+ (u-m_0^2)^{-1} + (3m_0^2)^{-1}\}. \quad (20)$$

IV. EQUIVALENT LAGRANGIAN

In the previous section we have illustrated the Lorentz and gauge invariance of the model by the somewhat unsophisticated device of performing a few lowest order calculations. From a more sophisticated point of view we remark that Lorentz invariance may be proved by constructing the generators of the Lorentz group and verifying their commutation relations. Provided that the Lagrangian (1) is first properly symmetrized, the proof goes through as in quantum electrodynamics[37]; spontaneous breakdown of the internal symmetry is irrelevant to the argument, which depends only on the equal-time commutators of products of field operators.

Concerning gauge invariance we remark that our result, that (in lowest order at least) decay and scattering amplitudes depend only on the gauge-invariant vector wave functions $b_\mu(k,\sigma)$, suggests that it must be possible to rewrite the theory in a form in which only gauge-invariant variables appear. Indeed, if one were shown only the expressions (17)–(20), he would guess that they had been derived from an interaction Lagrangian given by

$$\mathcal{L}_{\text{int}}' = -em_1B_\mu B^\mu \chi - \tfrac{1}{2}fm_0\chi^3 - \tfrac{1}{2}e^2B_\mu B^\mu \chi^2 - \tfrac{1}{8}f^2\chi^4. \quad (21)$$

We shall now show that the expressions (7) and (21) are equivalent by finding a transformation of the total Lagrangian (1) which takes the one into the other.

We note that the gauge dependence of the classical dynamic variables may be expressed in the form

$$\Phi_1(x) = R(x)\cos\Theta(x),$$
$$\Phi_2(x) = R(x)\sin\Theta(x), \quad (22)$$
$$A_\mu(x) = B_\mu(x) - e^{-1}\partial_\mu\Theta(x),$$

where $R(x)$ and $B_\mu(x)$ are gauge invariant variables and the transformations (4) take the simple form, $\Theta(x) \rightarrow \Theta(x) - \Lambda(x)$. The classical Lagrangian (1), expressed in terms of the new variables, takes the form

$$\mathcal{L}' = -\tfrac{1}{4}F_{\mu\nu}F^{\mu\nu} - \tfrac{1}{2}g^{\mu\nu}\partial_\mu R\partial_\nu R$$
$$- \tfrac{1}{2}e^2B_\mu B^\mu R^2 + \tfrac{1}{4}m_0^2R^2 - \tfrac{1}{8}f^2R^4. \quad (23)$$

Gauge invariance here has ensured the disappearance

of the variable Θ from the scene. [What we have done is to exploit the freedom which *local* $U(1)$ invariance gives us to "rotate" the entire two-component field $\Phi_a(x)$ onto one of the "axes." In a theory with only *global* $U(1)$ invariance this rotation cannot be performed on the entire field but only on the static solution (3).] The existence of the solution $B_\mu = 0$, $R^2 = m_0^2/f^2$ suggests the substitution $R(x) = \eta + \chi(x)$. In this way, we find immediately that, apart from an additive constant, $\mathcal{L}' = \mathcal{L}_0' + \mathcal{L}_{\text{int}}'$, where

$$\mathcal{L}_0' = -\tfrac{1}{4}F_{\mu\nu}F^{\mu\nu} - \tfrac{1}{2}m_1^2B_\mu B^\mu - \tfrac{1}{2}g^{\mu\nu}\partial_\mu\chi\partial_\nu\chi - \tfrac{1}{2}m_0^2\chi^2. \quad (24)$$

The expression (24) is the same as (6), except that the exactly gauge-invariant variables B_μ and χ which we have just defined replace their interaction picture counterparts.

We conjecture that the equivalence demonstrated here between the classical Lagrangians (1) and (23) may be extended to the corresponding quantum mechanical operators, provided that careful attention is given to the ordering problems which may arise, for example, in the definition of the current j_μ.[38]

V. DISCUSSION

The foregoing considerations illustrate our contention in Refs. 15 and 27 that the extension of a spontaneously broken internal symmetry of a Lorentz-invariant Lagrangian from global to local transformations not only may but actually does change zerons into the longitudinal states of massive vector bosons. Since we believe the value of this simple model to lie in the insight which it may give into this phenomenon when one looks at it as simple-mindedly as we have here, we shall not go into more difficult questions, such as radiative corrections and renormalization, in the present paper.

We note that in this model the original symmetry is almost unrecognizable in the physical states. Even without the gauge field coupling, the invalidation of the usual argument leading to the conservation of $Q(t)$ [39] by the asymptotic behavior of the term $\langle\Phi_2\rangle\partial_\mu\Phi_1$ in j_μ destroys the commutativity of Q with the Hamiltonian: Consequently, the one-particle states are not eigenstates of Q and the masses within the Φ_a multiplet are not degenerate, but at least the multiplet structure remains. The gauge field coupling obscures even the multiplet structure: The scalar doublet is now incomplete, having lost its formerly massless member to form the longitudinal polarization of the vector singlet.

---

[37] See B. Zumino, J. Math. Phys. **1**, 1 (1960).

[38] J. Schwinger, Phys. Rev. Letters **3**, 296 (1959); K. Johnson, Nucl. Phys. **25**, 431 (1961).

[39] In passing, we remark that this feature of spontaneous breakdown theories seems to call into question the validity of the results obtained on the basis of chirality conservation by Nambu and his collaborators. See Y. Nambu and D. Lurié, Phys. Rev. **125**, 1429 (1962); Y. Nambu and E. Shrauner, *ibid.* **128**, 862 (1962); E. Shrauner, *ibid.* **131**, 1847 (1963).

In view of the rather drastic nature of the symmetry breakdown which we have just summarized, it is of interest to inquire what happens when this system is coupled to a second in a Lagrangian which retains local $U(1)$ invariance and contains no additional mechanisms for spontaneous breakdown. To be specific, let us take the second system to be a pair of "baryons" of "charges" $\pm\frac{1}{2}$, together with their antiparticles, and let us assume that the Φ-baryon interaction is of the Yukawa type. The total Lagrangian is then given by

$$\mathcal{L}_{\text{total}} = \mathcal{L}(A, \Phi) - \bar{\psi}_a(\gamma^\mu \nabla_\mu + M)\psi_a + g[\Phi_1(\bar{\psi}_1\psi_2 + \bar{\psi}_2\psi_1) + \Phi_2(\bar{\psi}_2\psi_2 - \bar{\psi}_1\psi_1)], \quad (25)$$

in which $\mathcal{L}(A, \Phi)$ is the expression (1), $\nabla_\mu\psi_1 = \partial_\mu\psi_1 - \frac{1}{2}eA_\mu\psi_2$, $\nabla_\mu\psi_2 = \partial_\mu\psi_2 + \frac{1}{2}eA_\mu\psi_1$ and we have, without loss of generality, made a choice of a phase angle in writing down the invariant Yukawa term. But for the presence of this last term, the Lagrangian would be invariant under global $U(1)$ transformations on Φ and ψ *independently*; that is, the symmetry would be $U(1) \times U(1)$ and the currents $j_\mu(\Phi)$ and $j_\mu(\psi)$ would be *separately* conserved. Thus, despite the nonconservation of $Q(\Phi)$ brought about by the structure of the first term in (25), there would still be conservation of $Q(\psi)$.

The Yukawa term reduces the symmetry to $U(1)$, the divergences of the individual currents now being given by

$$\partial_\mu j^\mu(\psi) = g[\Phi_1(\bar{\psi}_1\psi_1 - \bar{\psi}_2\psi_2) + \Phi_2(\bar{\psi}_1\psi_2 + \bar{\psi}_2\psi_1)]$$
$$= -\partial_\mu j^\mu(\Phi). \quad (26)$$

We observe that spontaneous breakdown of the symmetry in the Φ system breaks the symmetry of the ψ system to an extent which depends on the coupling constant g. In the spirit of the perturbative approach which we have been using, we may evaluate the major part of the effects on the ψ system by replacing Φ by its vacuum expectation value. Then in Eq. (25) the

term $g\eta(\bar{\psi}_2\psi_2 - \bar{\psi}_1\psi_1)$ removes the baryon mass degeneracy, and Eq. (26) becomes

$$\partial_\mu j^\mu(\psi) = g\eta(\bar{\psi}_1\psi_2 + \bar{\psi}_2\psi_1) + \text{higher order terms.} \quad (27)$$

If we were to modify the Lagrangian (25) by adding to it such $U(1)$ invariant baryon-antibaryon interactions as would produce a doublet of low-mass scalar bound states with wave functions transforming as $\bar{\psi}_1\psi_2 + \bar{\psi}_2\psi_1$ and $\bar{\psi}_2\psi_2 - \bar{\psi}_1\psi_1$, then Eq. (27) would be approximately a partial conservation law of the type which has proved so successful for the axial currents of the weak interactions.[40] Moreover, the current $j_\mu(\psi)$ interacts with itself via the massive intermediate vector boson which the $\Phi - A_\mu$ coupling produces.

There appears to be some hope that the basic ingredients of our model, namely the combination of spontaneous symmetry breakdown with the gauge principle, may provide the basis for an understanding of the broken symmetries of high-energy physics. In a subsequent paper we shall discuss models in which the breakdown of higher symmetries such as $SU(3)$ is treated in the same fashion.

ACKNOWLEDGMENTS

I have learned much about the Goldstone theorem from discussions with Dr. G. S. Guralnik, Dr. C. R. Hagen, Dr. T. W. B. Kibble, and Dr. R. F. Streater. I wish to thank Professor Bryce and Professor Cécile DeWitt for the hospitality of the Institute of Field Physics at the University of North Carolina, where this work was completed.

[40] It would be the type of partial conservation law proposed by Y. Nambu, Phys. Rev. Letters **4**, 380 (1960) and by M. Gell-Mann, Phys. Rev. **125**, 1067 (1962), rather than that proposed by M. Gell-Mann and M. Lévy, Nuovo Cimento **16**, 705 (1960), in which the low-lying scalar or pseudoscalar states are treated as elementary rather than composite in the context of a field theory.

LIST OF PUBLICATIONS

1. "Theoretical Determination of Electron Density in Organic Molecules" (with C A Coulson, S L Altmann and N H March) Nature **168** 1039 (1951).

2. "Perturbation Method for the Calculation of Molecular Vibration Frequencies I" J. Chem. Phys. **21** 1131 (1953).

3. "A Method for Computing Zero-Point Energies" J. Chem. Phys. **21** 1330 (1953).

4. "Vibration Spectra of Helical Molecules" Proc. Roy. Soc. **A220** 472 (1953).

5. "Vibrational Modifications of the Electron Density in Molecular Crystals I" Acta. Cryst. **6** 232 (1953).

6. "Perturbation Method for the Calculation of Molecular Vibration Frequencies II" J. Chem. Phys. **23** 1448 (1955).

7. "Perturbation Method for the Calculation of Molecular Vibration Frequencies III" J. Chem. Phys. **23** 1450 (1955).

8. "Vibrational Modifications of the Electron Density in Molecular Crystals II" Acta. Cryst. **8** 99 (1955).

9. "A Method for Calculating Thermal Vibration Amplitudes from Spectroscopic Data" Acta. Cryst. **8** 619 (1955).

10. "Vacuum Expectation Values as Sums over Histories" Nuovo Cimento (10) **4** 1262 (1956).

11. "On Four-Dimensional Isobaric Spin Formalisms" Nuclear Physics **4** 1262 (1957).

12. "Integration of Secondary Constraints in Quantized General Relativity" Phys. Rev. Lett. **1** 373 (1958).

13. "Integration of Secondary Constraints in Quantized General Relativity" Phys. Rev. Lett. **3** 66 (1959).

14. "Quadratic Lagrangians and General Relativity" Nuovo Cimento (10) **11** 816 (1959).

15. "Broken Symmetries, Massless Particles and Gauge Fields" Physics Letters **12** 132 (1964).

16. "Broken Symmetries and the Masses of Gauge Bosons" Phys. Rev. Letters. **13** 508 (1964).

17. "Spontaneous Symmetry Breakdown without Massless Bosons" Phys. Rev. **145** 1156 (1966).

18. "Spontaneous Symmetry Breaking" two lectures at the 14th Scottish Universities Summer School in Physics (1973). Published in "Phenomenology of Particles at High Energy" R L Crawford, R Jennings (eds.) Academic Press (1974) ISBN 9780121971502.

19. "Dynamical Symmetries in a Spherical Geometry I" J. Phys. **A12** 309 (1979).

20. "SBGT and All That", International Conference "50 Years of Weak Interactions from the Fermi Theory to the W" Wingspread, Racine, Wisconsin (29 May–1 June 1984). Published in the conference proceedings by University of Wisconsin at Madison and reproduced in AIP. Conf. Proc. **300**:159-163 (1994).

21. "Inventing an Elementary Particle", INFN Eloisatron Project 9th Workshop "Higgs Particles — Physics Issues and Experimental Searches in High-energy Collisions", Erice, Italy (15–26 Jul 1989). Published in "Higg(s) Particle(s): Physics Issues and Experimental Searches in High-Energy Collisions" A Ali (ed.) Ettore Majorana International Science Series **50** 1-5 Plenum Press (1990) ISBN 9780306435898.

22. "Spontaneous Symmetry Breaking 25 Years Ago", 26th International Conference on Subnuclear Physics "Physics up to 200 TeV", Erice, Italy (16–24 Jul 1990). Published in "Physics up to 200TeV" A Zichichi (ed.) The Subnuclear Series **28** 439–444 Plenum Press (1991) ISBN 9780306439353.

23. Panel Session "Spontaneous Breaking of Symmetry" (with L M Brown, R Brout, T Y Cao, Y Nambu) 3rd International Symposium on the History of Particle Physics "The Rise of the Standard Model" (1992): published in "The Rise of the Standard Model", L Hoddesdon, L M Brown, M Riordan, M Dresden (eds.) Cambridge University Press, (1997) ISBN 978052157165.

24. "My Life as a Boson: The Story of 'The Higgs" Inaugural Conference of the Michigan Center for Theoretical Physics "2001 A Spacetime Odyssey" Ann Arbor, Michigan (21–25 May 2002). Published in "2001 A Spacetime Odyssey" M J Duff, J T Liu (eds.) World Scientific (2002) ISBN 9789810248062 and reproduced in Int. J. Mod. Phys. **A17S1** 86–88 (2002).

25. "Prehistory of the Higgs Boson" Comptes Rendus Physique **8** 970–972 (2007).

26. "Evading the Goldstone Theorem" Rev. Mod. Phys. **86** 851 (2014).

27. "Evading the Goldstone Theorem" Annalen der Physik **526** 211 (2014).

Daniel Kleppner

DANIEL KLEPPNER

Birthplace

Born	New York City, New York, U.S., 16 December, 1932

Education

1953	B.Sc. Williams College, Williamstown Massachusetts, U.S.
1955	B.A. Cambridge University, Cambridge, England
1959	Ph.D. Harvard University, Cambridge, Massachusetts, U.S. research supervisor: Norman F. Ramsey, Jr.

Research and Professional Experience

1959–1962	Postdoctoral research, Harvard University
1962–1966	Assistant Professor of Physics, Harvard University
1966–1973	Associate Professor of Physics, Massachusetts Institute of Technology (M.I.T.)
1974–1985	Professor of Physics, M.I.T.
1985–2003	Lester Wolfe Professor of Physics M.I.T.
2003–	Lester Wolfe Professor of Physics, Emeritus, M.I.T.
1987–2003	Associate Director, Research Laboratory of Electronics, M.I.T.
2000–2008	Director, MIT/Harvard Center for Ultracold Atoms

Honors

1986	American Academy of Arts and Science, member
1988	National Academy of Sciences, member
2002	Academies of Science, Paris, foreign associate
2007	American Philosophical Society, member

Awards

1985	Davisson-Germer Prize, American Physical Society
1990	James Edgar Lilienfeld Prize, American Physical Society
1991	William F. Meggers Award, Optical Society of America
1996	James Rhyne Killian Faculty Achievement Award, M.I.T.
1997	Oersted Medal, American Association of Physics Teachers
2005	Leo Szilard Award, American Physical Society
2006	Wolf Foundation Prize in Physics
2007	Frederick Ives Medal, Optical Society of America
2008	National Medal of Science
2014	Benjamin Franklin Medal, Franklin Institute

COMMENTARY

Like many scientists, my career stemmed from childhood enthusiasms. Although my parents had no connections with the world of science, they were supportive of my older brother and me as we built things from rowboats to radios, and pursued hobbies such as photography, electric trains, furniture-making and sailing. We lived in New Rochelle, a suburb of New York City, and I attended New Rochelle high school. In my senior year I had a charismatic teacher, Arthur B. Hussey, who cemented my interest in physics. This was during the period following World War II when physicists were heroes because of the atom bomb and radar. At Williams College, the physics department provided me with laboratory space for a personal research project and several faculty members were generous in giving advice about how to do things in the laboratory and how to understand vacuum tube electronics. By graduating a year early, I managed to attend Cambridge University for two years as a Fulbright scholar before starting graduate work. At Cambridge, I had the good fortune to have as my tutor Kenneth A. Smith. There was a solid wall between undergraduate studies and research at Cambridge, but Smith illustrated his tutorials with ideas from his research in the new area of molecular beam magnetic resonance. One afternoon he described a proposal by I. I. Rabi, the inventor of molecular beams magnetic resonance, to use this technique to create a clock based on a natural atomic frequency — an atomic clock — that might be sufficiently precise to reveal how gravity affects time, a phenomenon predicted by Einstein's theory of general relativity. In the following year, when an opportunity arose for actually creating such a clock, I jumped at it.

At Harvard University, where I pursued my graduate education, I joined Norman F. Ramsey's research group, arriving at an opportune moment: Ramsey had just conceived the idea for a new type of atomic clock — the hydrogen maser — which might be precise enough to demonstrate Einstein's prediction of the effect of gravity on time. My doctoral research showed that the maser was feasible and I built the first such device as a post-doctoral researcher. The hydrogen maser was subsequently used to confirm Einstein's prediction of the effect of gravity on the rate of clocks by R. F. C. Vessot. (The first observation of the effect of gravity on time was made by R. V. Pound and G. A. Rebka, colleagues of ours at Harvard, using a different approach.) Over the years the hydrogen maser turned out to be useful for probing basic physics. Furthermore, it had a number of unanticipated applications: The hydrogen maser enabled the development of very long baseline radio interferometry for radio astronomers and, along with other types of atomic clocks, enabled the creation of the Global Positioning System (GPS). In 1989, Ramsey received the Nobel Prize for developing methods for high precision measurements, in particular, the hydrogen maser.

In 1966, I joined the faculty at the Massachusetts Institute of Technology where I remained until becoming Professor Emeritus, in 2003. My early research at M.I.T. involved studies of fundamental theory using the hydrogen maser. Subsequently my research interest expanded to include Rydberg atom physics, cavity quantum electrodynamics and the search for Bose–Einstein condensation.

Rydberg Atoms. In the mid 1960s radio astronomers discovered signals emitted when protons in stellar clouds capture free electrons, forming hydrogen atoms in high-lying states. The electrons can be pictured as moving in a gigantic Bohr orbit. As they cascade to lower energy levels they emit radiation in the microwave regime. In the 1970s the advent of lasers and laser spectroscopy made it possible to create and study these atoms, called *Rydberg* atoms, in the laboratory. Our first studies were devoted to understanding the structure of the atoms in electric fields and how atoms ionize in sufficiently large fields. The studies revealed an unexpected sensitivity in the dynamics to perturbations of basic symmetries in the system and inspired a number of experimental tools that have been useful to our group and others.

We next turned our attention to the problem of Rydberg atoms in magnetic fields. Our initial goal was to understand the behavior of the system using the tools of quantum mechanics, as we had done for the electric field problem. However, when Karl Welge at the University of Bielefeld showed that there were surprising connections between quantum behavior and classical mechanics of an atom, the focus of our research changed to understanding the connections between quantum and classical physics. The problem turned out to be unexpectedly rich because the classical motion became chaotic in certain combinations of energy and field strength. The connection between quantum mechanics and classical chaos, is the subject of what came to be called *quantum chaos*. John Delos of the University of William and Mary invented a number of powerful theoretical tools for interpreting our experiments and we collaborated on the problem for some time. A number of other groups entered the field but the subject of quantum chaos remains elusive.

Cavity Quantum Electrodynamics. In the 1960s an important goal in atomic physics was to measure the Lamb shift more precisely. The Lamb shift is an energy shift that is predicted by the theory of quantum electrodynamics. It is difficult to measure precisely because of the short lifetime of one of the states involved. The lifetime is limited by spontaneous emission, the process by which an excited atom radiates its energy as it descends to a lower energy state. It was generally believed that spontaneous emission is a universal property of atoms but I pointed out in a summer school lecture that by putting the atom in a cavity — a small enclosure with conducting walls — that spontaneous emission could be suppressed. At that time the idea was not technically feasible but a decade later, Rydberg atoms made it possible to demonstrate the phenomenon. Our demonstration of inhibited spontaneous emission helped to trigger a new area of study. If the cavity is tuned to the radiation frequency, the opposite effect occurs: spontaneous emission is enhanced, a phenomenon predicted by Edward M. Purcell in 1946. With a sufficiently high quality cavity that atom does not decay to a lower state but the atom-cavity system oscillates as the energy is transferred between the two systems. Herbert Walther in Munich and Serge Haroche in Paris were the first to witness this effect. The study of such atom-cavity systems, dubbed cavity QED, turned out to open a new area of quantum physics. Advances in cavity QED culminated in the award of the Nobel Prize to Haroche in 2013. Cavity QED effects have now been achieved at optical frequencies, notably by Jeffrey Kimble in Pasadena. Furthermore, they have been exploited in a totally different energy regime using not atoms but electrical circuit components, in what is being called circuit QED, and there is a general expectation that useful applications will follow.

Bose–Einstein Condensation. The observation of Bose–Einstein condensation in an atomic gas in 1995 gave rise to a new field at the interface of condensed matter and atomic physics. The discovery was honored in 2001 by the Nobel Prize in Physics to Eric Cornell, Wolfgang Ketterle and Carl Wieman. Previous to that, in 2006, the technology critical for the achievement of BEC — laser cooling — was honored by award of the Nobel Prize to Steven Chu, Claude Cohen-Tannoudji and William D. Phillips. Today, research spawned by BEC is centered on many-body physics, quantum information theory, quantum measurements and quantum simulations. The scope of this new field continues to expand but when the search for atomic BEC started in the mid 1970s, the subject was obscure and regarded, if at all, as a novelty.

My colleague Thomas J. Greytak and I started to search for atomic BEC in 1976, stimulated by a Letter by Lewis H. Nosanow and William C. Stwalley. They pointed out that under suitable conditions atomic hydrogen could remain in the gaseous state as the temperature is reduced to zero. As a result, it should be capable of exhibiting a quantum phase transition — BEC — that had been predicted by Einstein. The goal required achieving a combination of low temperature and high density that was so different from anything yet conceived that the experiments were most kindly described as "visionary. Nevertheless, we set out. Isaac F. Silvera and J. T. M. (Jook) Walraven in Amsterdam started out at the same time, and other groups soon joined the hunt.

Our approach was based on an idea from the hydrogen maser: confining an atomic gas in a container with a noninteracting surface. The hope was that if the container were coated with superfluid helium the atoms would survive wall collisions yet come into thermal equilibrium with the surface. The atoms would be confined in the container by applying a high magnetic field. Initial experiments were encouraging and we pushed our method to achieve higher and higher densities. However, in the mid 1980s, when we were within a factor of ten of the required density, progress was halted by the process by which two atoms join to form a molecule when a third atom is nearby to carry off the energy released — three-body recombination. The density was limited by three-body recombination of atoms while they were adsorbed on the surface. Given this limit to the density, the only route to BEC required going to much lower temperature. At extremely low temperatures, however, the density of atoms adsorbed on the surface would increase and the atoms would rapidly be lost. We needed to eliminate the surfaces. By confining the cold hydrogen in space by magnetic fields — using what is called a magnetic trap —, the problem of surface recombination is avoided. We built a magnetic trap following a design proposed by David E. Pritchard. To demonstrate its operation, we let the atoms escape into a hyperfine resonator, creating a cryogenic hydrogen maser. Two problems remained: The first was to cool the trapped atoms to much lower temperatures, and the second was to detect the atoms *in situ*.

At about this time, attempts at laser cooling of alkaline metal atoms were starting to succeed in a number of laboratories. Unfortunately, laser cooling was not well suited to hydrogen. At this time Harald F. Hess, then a postdoctoral researcher in our group, proposed evaporative cooling. Slowly reducing the depth of the trap and allowing fast atoms to escape while the remaining atoms maintained thermal equilibrium through collisions would force the temperature forced downward. Evaporative cooling worked for us and the technique rapidly became ubiquitous to other studies of BEC and ultra cold atoms. The remaining task was to develop a non-destructive way for observing the trapped gas. Laser spectroscopy offered the possibility, but executing it on the trapped gas was a major challenge. In 1995 Eric A. Cornell and Carl Wieman at JILA, and Wolfgang Ketterle at MIT observed BEC in an alkaline atom gas. In 1996 our group finally observed BEC in hydrogen. Although hydrogen occupies a special place in atomic physics, the atom proved itself to be poorly suited to studies of BEC. Nevertheless, hydrogen helped to spawn interest the field of atomic quantum fluids. The discovery of BEC launched a new field of research and nineteen years later, interest continues to grow.

People

Many students and postdoctoral researchers spent long hours in the laboratory carrying out the research sketched here, and they often contributed important ideas. In addition, a variety of collaborators worked with me over the years. I cannot cite every one in the space available and rather than to cite only a few, I have opted simply to acknowledge my indebtedness to all with whom I have worked over the years and to thank them for all that they have taught me.

VOLUME 47, NUMBER 4   PHYSICAL REVIEW LETTERS   27 JULY 1981

Inhibited Spontaneous Emission

Daniel Kleppner

*Research Laboratory of Electronics and Department of Physics, Massachusetts Institute of Technology,
Cambridge, Massachusetts 02139*

(Received 28 April 1981)

The radiative properties of an atom in a cavity differ fundamentally from the atom's radiative properties in free space. Spontaneous emission is inhibited if the cavity has characteristic dimensions which are small compared to the radiation wavelength, and enhanced if the cavity is resonant. The cavity causes slight shifts in the energies of the atom, analogous to radiative shifts. Experiments are proposed for observing these effects.

PACS numbers: 32.80.-t, 03.70.+k

Spontaneous emission is the most visible manifestation of the dynamical interaction between matter and the vacuum; the primal role of vacuum fields in driving every excited atom to its ground state is often regarded as a basic fact of nature. The purpose of this Letter is to point out that vacuum states can be radically altered by a cavity and that it should be possible to realize experimental conditions in which an atom is effectively decoupled from the vacuum and cannot radiate. In such a situation the atom's excited state is, to first approximation, a real eigenstate which is devoid of the usual radiative energy width. The energy levels are slightly altered by the cavity, and the Lamb shift is modified. By changing the experimental conditions, the rate of spontaneous emission can be vastly increased compared to the free-space value.

Our starting point is the observation made many years ago by Purcell[1] that the spontaneous emission rate A for a two-state system is increased if the atom is surrounded by a cavity tuned to the transition frequency, ν. If the quality factor of the cavity is Q, then the spontaneous rate in the cavity is $A_c \simeq QA$. Physically, the cavity enhances the strength of the vacuum fluctuations at ν, increasing the transition rate. Conversely, the decay rate decreases when the cavity is mistuned.[2] If ν lies below the fundamental frequency of the cavity, spontaneous emission is significantly inhibited. In the case of an ideal cavity, no mode is available for the photon and spontaneous emission cannot occur.

The rate A for spontaneous transitions from initial state $|i\rangle$ to final state $|f\rangle$ can be written by the "golden rule"[3]

$$A = [\,|\langle f|H|i\rangle|^2/\hbar^2\,]\rho(\nu).$$

Here $|i\rangle$ is the excited state of the atom in the absence of any photons, $|f\rangle$ is the final state of the atom with a single photon, H is the Hamiltonian for the atom-field interaction, $\rho(\nu)$ is the density of photon states, and the matrix element is volume normalized. In free space $\rho_s = 2(4\pi\nu^2/c^3)$; the factor of 2 is due to the two allowed polarization states. In a cavity the density of modes is essentially $(1 \text{ mode})\Delta\nu^{-1}V^{-1}$, where $\Delta\nu = \nu/Q$ is the cavity's resonance width and V is the cavity's volume. Taking $V \simeq (\lambda/2)^3 = (c/2\nu)^3$, we have $\rho_c \simeq 8\nu^2Q/c^3$. Under the assumption that the cavity mode is doubly degenerate, then $\rho_c/\rho_s \simeq (2/\pi)Q$ and $A_c \simeq QA$ as previously described.

The presence of any conducting surfaces near the radiator affects the mode density and the radiation rate. Parallel conducting planes can somewhat alter the emission rate but can only reduce the rate by a factor of 2 because of the existence of TEM modes[4] which are independent of the separation.[2] In order to eliminate spontaneous emission every propagating mode must be suppressed. A completely enclosed perfectly conducting cavity would accomplish this, but a waveguide below cutoff could also serve the purpose. The waveguide, which can be viewed as a cavity with ends removed to infinity, is experimentally attractive because atoms can pass through it freely.

The wave vector for the fundamental mode of a waveguide is $k = (2\pi/c)(\nu^2 - \nu_0^2)^{1/2}$, where ν_0 is the fundamental guide cutoff frequency. Imposing periodic boundary conditions, $kL = 2\pi n$, where L is a length long compared to the diameter of the waveguide, yields for the fundamental mode

$$\rho_0(\nu) = \frac{2}{V}\frac{dn}{d\nu} = \frac{4}{cA_g}\frac{\nu}{(\nu^2 - \nu_0^2)^{1/2}}.$$

A_g is the area of the guide; $A_g \simeq \zeta(c/2\nu_0)^2$, where ζ is a numerical constant of order unity. The factor of 4 results from the twofold degeneracy

VOLUME 47, NUMBER 4   PHYSICAL REVIEW LETTERS   27 JULY 1981

between positive and negative values of n, and the twofold degeneracy in polarization which is appropriate for a circular waveguide. Near cutoff the waveguide causes a resonancelike enhancement of the spontaneous emission rate. If $\nu < \nu_0$, $\rho_0(\nu) = 0$ and spontaneous emission cannot occur.

Higher modes in the waveguide contribute to the spontaneous rate for frequencies above their cutoff values. If ν_j is the cutoff frequency for the jth mode, then the total mode density $\rho_g(\nu)$ is

$$\rho_g(\nu) = \frac{16\nu_0^2}{\xi c^3} \sum_j \frac{\nu}{(\nu^2 - \nu_j^2)^{1/2}} .$$

The sum is over all modes for which $\nu_j < \nu$. As the frequency increases and more and more modes are included, $\rho_g(\nu)$ approaches the free-space mode density, as Fig. 1(a) shows. The cavity changes the spontaneous emission rate by the ratio $R(\nu) = \rho_g(\nu)/\rho_s(\nu)$, which is plotted in Fig. 1(b).

A realistic treatment of spontaneous emission in a waveguide must take into account resistive losses. (Radiative losses are expected to be small because of the guide's large length-to-diameter ratio.) The quantum mechanical theory of radiation in a damped system is beyond the scope of this Letter, but the important features can be understood from simple qualitative arguments. In the vicinity of cutoff a waveguide can behave like a cavity with $Q \simeq a/\delta$, where a is the characteristic dimension of the waveguide and δ is the skin depth. As a result the singularities in $R(\nu)$ become finite with a height of approximately Q. There is a nonzero response in the cutoff region, $\nu < \nu_0$, where $R(\nu) \sim 1/Q$. Thus the spontaneous decay rate is decreased compared to free space by a factor of order Q in the region $\nu < \nu_0$, and increased by a factor of Q in the regions $\nu \simeq \nu_0$.

The experimental challenge in demonstrating inhibited spontaneous emission is to find a system which has a spontaneous transition with a wavelength long enough to allow construction of a practical waveguide, but a lifetime short enough for the atom to radiate before escaping from the apparatus. Electric dipole transitions between Rydberg states can satisfy these requirements. In order to avoid radiative decay by optical or other short-wavelength transitions, it is essential to employ states which have but one allowed radiative decay channel. This suggests employing states of the maximum angular momentum, $l = n - 1$, where n is the principal quantum number. To make the discussion concrete, we shall con-

sider the transition $|n, n-1, n-1\rangle \rightarrow |n-1, n-2, n-2\rangle$, where the quantum numbers are $|n, l, m\rangle$. This transition corresponds classically to dipole radiation by a charge in a circular orbit. The wavelength and spontaneous lifetimes are approximately $\lambda = 4.6 \times 10^{-6} n^3$ cm and $\tau = 9.3 \times 10^{-11} n^5$ s. For example, for the transition $n = 25 \rightarrow 24$, $\lambda = 0.07$ cm and $\tau = 900$ μs. A typical thermal atom travels 40 cm in this lifetime. The cutoff diameter of the waveguide is 0.04 cm, and a Q in excess of 10^3 should be attainable.

Spontaneous emission for $\Delta m = -1$ transitions by atoms near the axis of a circular waveguide occurs via coupling to the $TE_{1,j}$ and $TM_{1,j}$ modes.

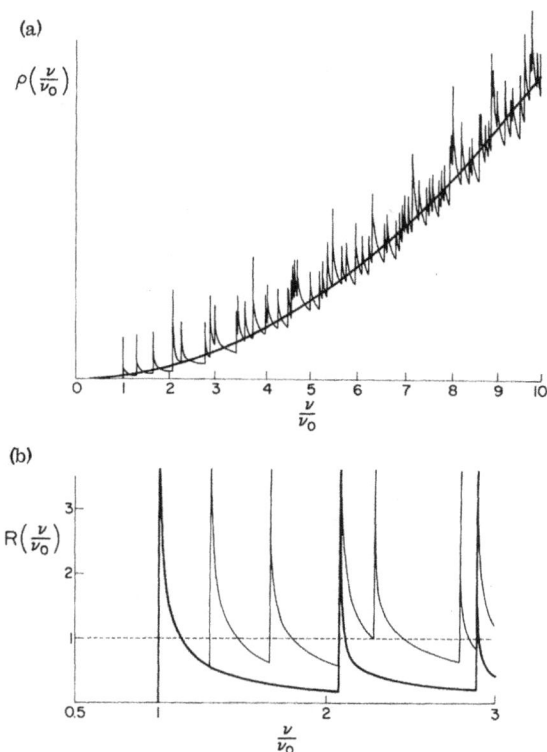

FIG. 1. (a) Mode density in a perfectly conducting cylindrical waveguide. Frequency is in units of the lowest cutoff frequency of the waveguide, $\nu_0 = 0.29c/a$, where a is the radius. For clarity, the singularities have been truncated. The smooth curve represents the mode density in free space. (b) Ratio of waveguide mode density to free-space mode density. The heavy line is for the modes which couple to a $|\Delta m| = 1$ transition for an atom on axis: $TE_{1,j}$ and $TM_{1,j}$. The change in the spontaneous emission rate from the free-space value is proportional to the ratio $R(\nu)$.

VOLUME 47, NUMBER 4   PHYSICAL REVIEW LETTERS   27 JULY 1981

The density ratio $R(\nu)$ for these modes is shown in Fig. 1(b). (It is interesting to note that the spontaneous emission rate into these modes can also be inhibited by a significant factor, greater than 4, in several frequency ranges above ν_0.) In principle, spontaneous emission due to quadrupole and higher-order processes can occur even when the dipole radiation is inhibited by the cavity, but their rates are so low that they should not cause difficulty. Another possible source of confusion, stimulated emission by blackbody radiation, can be avoided by operating the cavity at cryogenic temperatures.

Observing enhanced spontaneous emission is a much simpler experimental task than observing inhibited emission; one can emply high-n transitions whose natural lifetimes are so long that spontaneous emission would normally be unobservable. This allows a general scaling up of the waveguide diameter and relieves the need to employ the highest angular momentum states. Such an experiment is similar in concept to the "one-atom maser" proposed by Haroche.[5] The one-atom maser consists of a Rydberg atom in a very-high-Q etalonlike resonator. The distinction between the two proposals lies in the choice of cavity mode. In a fundamental mode cavity the spontaneous rate is completely determined by the cavity; there is no threshold for enhancement. For a high-mode optical resonator, enhanced emission into a single mode must compete with spontaneous emission in all the other modes of space, and it can only be observed if the cavity Q is sufficiently large.

Numerous experimental schemes are possible for studying the effect of a waveguide on the spontaneous emission rate. The initial- and final-state distribution can be determined by selective field ionization,[6] a sensitive technique which works well at the single-atom level. The primary problem in observing inhibited spontaneous emission is to produce Rydberg atoms in the "circular" state. A number of schemes are possible and experimental work is in progress.

An atom surrounded by a cavity forms a coupled quantum system, and the atom's energy levels are inevitably shifted by the cavity. The leading interaction is essentially a dipolar Van der Waals attraction between the atom and the conducting walls. As Casimir and Poulder pointed out in a well-known paper,[7] Van der Waals interactions arise from alterations in the zero-point energy of a system. An atom in a cavity has many features in common with the hypothetical system Casimir and Poulder employed to motivate their discussion: an atom interacting with an infinite perfectly conducting plane. The behavior of an atom in a cylindrical waveguide can be understood qualitatively by considering an atom at the center of a spherical cavity. Such a model gives the correct order of magnitude of the level shifts, though the perturbation does not have the proper angular dependence.

Consider a one-electron atom with an instantaneous dipole moment $-e\vec{r}$ centered in a cavity of radius a. The induced field is $\vec{E} = -e\vec{r}/a^3$, and the atom-field interaction is $V = -e^2 r^2/a^3$. For a "circular" state $r^2 \simeq n^4 a_0^2$, where a_0 is the Bohr radius, and $V = -(e^2/a_0)(a_0/a)^3 n^4$. The energy shift between level n and $n-1$ is $\Delta V \simeq -4(e^2/a_0)(a_0/a)^3 n^3$. If the cavity is near cutoff, then $a \simeq \lambda/2\pi \simeq n^3 a_0/\alpha$ and $\Delta V \simeq -4(e^2/a_0)\alpha^3 n^{-6}$. The fractional change in energy is $4\alpha^3 n^{-3}$. For $n = 25$, the fractional change is 10^{-10}. By operating on a high-n transition in a small-diameter cavity, the fractional shift can be enhanced. It is interesting to note that retardation corrections to the Van der Waals interaction are negligible since the radiative modes are suppressed. In principle the cavity also alters the Lamb shift, though the effect is very small.

Experimental studies of inhibited spontaneous emission seem worthwhile in view of the novelty of the situation and the ongoing controversies about the physical nature of the vacuum fields.[8] Any scheme for suppressing spontaneous emission is potentially useful to the art of precision measurements, though perturbations due to the atom-cavity interaction may limit such applications. In any case, observations of inhibited or enhanced spontaneous emission would provide an interesting bridge between atomic physics and macroscopic electrodynamics.

I thank Subir Sachdev for carrying out the computations of the mode densities under the Massachusetts Institute of Technology Undergraduate Research Opportunity Program, and Michael M. Kash for helpful comments. This work was supported by the Joint Services Electronics Program, the National Science Foundation, and the U. S. Office of Naval Research.

[1]E. M. Purcell, Phys. Rev. <u>69</u>, 681 (1946).

[2]D. Kleppner, in *Atomic Physics and Astrophysics*, edited by M. Chretien and E. Lipworth (Gordon and Breach, New York, 1971), Sec. 6.3, p. 5.

[3]A lucid discussion of spontaneous emission and other

VOLUME 47, NUMBER 4   PHYSICAL REVIEW LETTERS   27 JULY 1981

vacuum interactions is given by E. A. Power, *Introductory Quantum Electrodynamics* (American Elsevier, New York, 1964).

[4]Standard microwave notation is used. Cf. J. D. Jackson, *Classical Electrodynamics* (Wiley, New York, 1975), 2nd ed.

[5]S. Haroche, in *Atomic Physics 7*, edited by D. Kleppner and F. M. Pipkin (Plenum, New York, 1981) p. 181.

[6]D. Kleppner, in Proceedings of the Les Houches Summer School, edited by J. C. Adams and R. Ballian (Gordon and Breach, New York, to be published).

[7]H. B. G. Casimir and D. Poulder, Phys. Rev. 73, 360 (1948); see also Ref. 3.

[8]*Radiative Theory and Quantum Electrodynamics*, edited by A. O. Barut (Plenum, New York, 1980).

Bose-Einstein condensation in atomic hydrogen

T. J. Greytak [1], D. Kleppner, D. G. Fried, T. C. Killian [2], L. Willmann, D. Landhuis,
S. C. Moss

*Physics Department and Center for Materials Science and Engineering, Massachusetts Institute of Technology,
Cambridge, MA 02139, USA*

Abstract

The addition of atomic hydrogen to the set of gases in which Bose-Einstein condensation can be observed expands the range of parameters over which this remarkable phenomenon can be studied. Hydrogen, with the lowest atomic mass, has the highest transition temperature, 50 μK in our experiments. The very weak interaction between the atoms results in a high ratio of the condensate to normal gas densities, even at modest condensate fractions. Using cryogenic rather than laser precooling generates large condensates. Finally, two-photon spectroscopy is introduced as a versatile probe of the phase transition: condensation in real space is manifested by the appearance of a high density component in the gas, condensation in momentum space is readily apparent in the momentum distribution, and the phase transition line can be delineated by following the evolution of the density of the normal component.

Keywords: spin-polarized hydrogen; Bose-Einstein condensation; two-photon spectroscopy; cold collision frequency shift

1. Introduction

When our MIT group began studying spin-polarized atomic hydrogen in 1976 it was thought to be the only gas in which there was even a remote chance of obtaining Bose-Einstein condensation. Using a quantum theory of corresponding states Hecht [1] had argued in 1959 that, due to its very light mass and weak interactions, spin-polarized atomic hydrogen would remain a gas down to absolute zero. Interest in hydrogen was revived in 1976 with the publication of many-body cal-

culations [2] confirming that the ground state of spin-polarized atomic hydrogen was indeed a gas. It was also noted that even if there were a way to keep other elements in the gaseous state, hydrogen, being the lightest, would have the highest transition temperature at a given atomic density.

Spin-polarized hydrogen was first stabilized by Silvera and Walraven [3] in 1980 at Amsterdam. In subsequent years groups at British Columbia, Cornell, Harvard, Moscow, and Turku, as well as our own, used spin-polarized atomic hydrogen to study many interesting physical phenomena from collision processes to spin-waves [4]. The observation of BEC remained a long range and elusive goal.

When BEC in a gas was finally observed in 1995 it was in alkali-metal vapors [5–7]. The reasoning

---
[1] Corresponding author. E-mail: greytak@mit.edu
[2] Present address: Physics Laboratory, National Institute of Standards and Technology, Gaithersburg, MD 20899-8424

Article published in Physica B 280 (2000) 20-26

of 1976, although true, proved to be irrelevant. The ground state of the alkali-metals is certainly a solid. Yet the density in the alkali experiments was so low that the three-body collisions leading to nucleation did not seriously limit the lifetime of the gas. At a gas density of 10^{14} atoms/cm^3 the transition temperature in sodium is 1.5 μK, 23 times lower than it would be in atomic hydrogen. However, the development of laser cooling of atoms allowed the alkali vapors to be pre-cooled to temperatures low enough that evaporative cooling could be used to take them into the sub-μK realm.

Since 1995, many fruitful experiments have been performed using alkali gases, greatly expanding the field of BEC physics. The study of Bose-condensed hydrogen may lead to unique contributions. Hydrogen possesses special attractions in that it affords relatively weak interactions between atoms, the possibility of condensing large numbers of atoms for long periods of time, and atomic states whose energies can be calculated with exquisite accuracy.

2. Trapping and Cooling

The hyperfine diagram of the $1S$ state of atomic hydrogen is shown in Fig. 1. The lowest two states, **a** and **b**, are high field seeking states. They can not be confined in a static magnetic trap because it is not possible to have a maximum in the magnitude of the magnetic field in a source free region. Most of the early experiments on spin-polarized hydrogen were carried out on these states confined by a combination of magnetic fields and liquid helium covered walls. Ultimately, however, recombination into molecules — due to adsorption on the walls at low temperatures or three-body recombination in the gas at high densities — precluded achieving conditions necessary for BEC. Attention then turned to confining the low field seeking states, **c** and **d**, in pure magnetic traps having a local minimum in the field. (Recently Safonov et al. [8] used changes in the surface three-body recombination rate to detect the formation of a quasicondensate

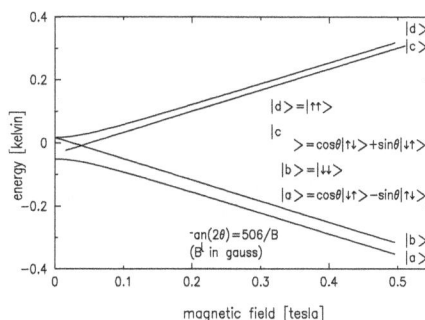

Fig. 1. Hyperfine diagram for the ground state of atomic hydrogen.

in a two-dimensional layer of **b** state atoms.)

The experiments described here were carried out on the doubly polarized **d** state confined in a trap of the Ioffe-Pritchard form. Near the bottom of such a trap the potential rises parabolically from a finite minimum (a bias field is necessary to prevent the spin of the atom from flipping when passing through a zero in the field); thus surfaces of constant magnetic field are ellipsoids of revolution about the vertical axis. A unique feature of our trap is its high aspect ratio of 400:1, the ratio of the major (vertical) axis to the minor axis. The loading of the trap from a low temperature atomic discharge source is described elsewhere [9]. Immediately after loading, the trap contains about 10^{14} atoms at a temperature of 40 mK.

The subsequent cooling of the atoms by three orders of magnitude is carried out by evaporation [10]. The most energetic atoms are allowed to escape from the trap at a rate slow enough that the remaining atoms can continuously readjust their energy distribution to lower temperatures through collisions. A competing mechanism causes heating by removing some of the least energetic atoms from the high density region at the bottom of the trap by two-body spin relaxation. The temperature is determined by a balance between these two mechanisms and comes to steady state at a fraction of the trap depth varying from 1/12 at 40 mK to about 1/7 at 40 μK.

Initially we allow the atoms to escape by lowering the confining potential at one end of the trap

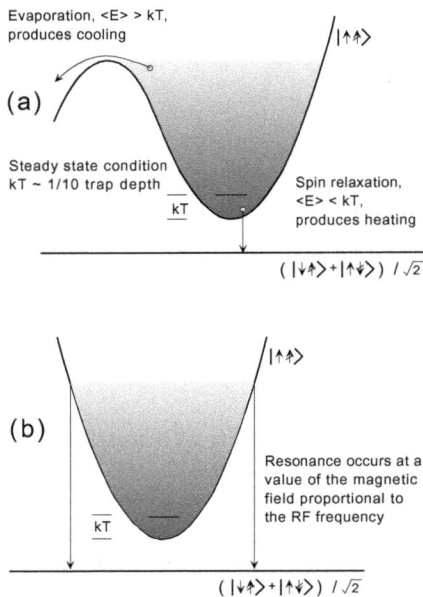

Fig. 2. Sketches of potential energy versus position demonstrating evaporative cooling of hydrogen by (a) lowering a trapping field, or (b) by RF ejection of atoms.

as shown in Fig. 2a. This method becomes inefficient below about 100 μK when atoms promoted to high energy states near the center of the trap have a high probability of losing that energy in a collision before being able to reach the end of the trap and escape [11]. To solve this problem we use RF ejection of the atoms [12] below a temperature of 120 μK. RF ejection was first applied to the evaporative cooling of alkali-metal vapors [13]. An RF magnetic field flips the spins of atoms in that region of the trap having a particular value of the trapping magnetic field (see Fig. 2b). Atoms whose spins are reversed are no longer confined. By starting with an RF field resonant with the highest fields in the trap, then slowly lowering the frequency, successively lower energy atoms can be expelled. Since all the atoms on a specific energy surface in the trap are affected, the process is more efficient than evaporation through one end, particularly in a trap as long and thin as ours. It was the application of RF ejection to the hydrogen that finally allowed us to achieve the conditions necessary for BEC.

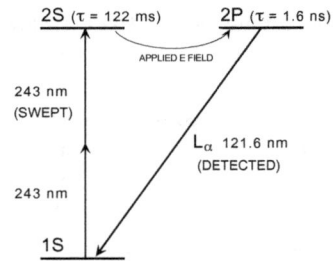

Fig. 3. Two-photon spectroscopy of the 1S-2S transition in hydrogen.

We monitor the temperature by three different techniques: measuring the shape of the energy distribution as the atoms are rapidly dumped from the trap [14], using the known relation between the temperature and the trap depth, and (at low densities) measuring the time of flight broadening of the Doppler-free part of the two-photon spectrum [15]. The density is determined by measuring the two-body dipole decay rate. The densities quoted in this paper are the maximum densities, those at the bottom of the trap.

3. Two-photon Spectroscopy

We use two-photon spectroscopy of the 1S-2S transition to study the trapped hydrogen atoms [15]. When illuminated with photons of energy exactly half the 1S to 2S level spacing, the atoms are promoted to the 2S state by the absorption of two photons (see Fig. 3). This differs from the one photon processes used to manipulate and study the alkali-metals in two important ways. First, the resonance is extremely narrow allowing it to be used for very high resolution spectroscopy. Second, the absorption is so weak that we can not detect the resonance by a decrease in the amplitude of the transmitted beam. Instead, we apply an electric field to the atoms which mixes the long lived 2S state with the short lived 2P state. The atom then returns to the ground state by the emission of a Lyman-α photon. We detect the resonance by recording the Lyman-α production as a function of the frequency of the illuminating beam.

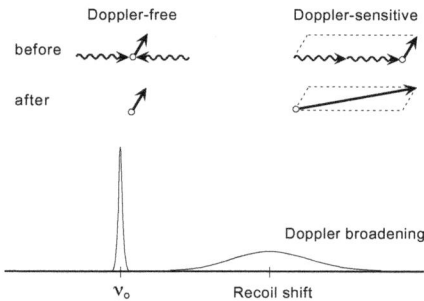

Fig. 4. Features of the two-photon spectrum in a standing wave. For hydrogen the recoil shift referenced to the 243 nm excitation radiation is 6.7 MHz. The Doppler width in this schematic would be appropriate to a temperature of about 50 μK, but the Doppler-free line would be much taller and narrower than indicated.

In our experiment a laser beam at 243 nm is reflected back on itself by a mirror at the bottom of the cell creating a standing wave in the trap. Atoms that absorb two co-propagating photons produce a recoil shifted and Doppler broadened feature in the spectrum (see Fig. 4). The shape of the Doppler line gives the momentum distribution in the gas and thus provides another measure of the temperature. Atoms that absorb two counter-propagating photons transfer no momentum to the atoms. The width of the resulting feature in the spectrum could, in principle, be determined only by the natural lifetime of the 2S state. In our experiments, however, the width of this feature is limited to 1 kHz (or one part in 10^{12}) by the finite coherence time of our laser source.

In 1996 Jamieson, Dalgarno and Doyle [16] pointed out that the interactions between the atoms cause a density dependent mean field shift of each of the hydrogen energy levels. They calculated that the resulting shift in the 1S-2S transition frequency would be negative, proportional to the density, and of a magnitude which would be observable in our experiments. We measured the shift [17] by studying the Doppler-free component in normal hydrogen and found $\Delta\nu_{1S-2S} = 2\Delta\nu_{243nm} = \chi n$, where n is the atomic density and $\chi = -3.8 \pm 0.8 \times 10^{-16}$ MHz cm^3. Since there is a distribution of densities in the trap, the Doppler-

free component is broadened as well as shifted by the interaction. The Doppler-sensitive component is broadened and shifted by similar amounts, but these effects are small compared to the Doppler broadening and recoil shift.

4. Bose-Einstein Condensation

Two-photon spectroscopy allowed us to identify three characteristic features of Bose-Einstein condensation: condensation in real space, condensation in momentum space, and the phase diagram [18].

If a non-interacting Bose gas were to be cooled below the transition temperature, a finite fraction of the atoms would fall into the lowest energy single particle quantum state. For a harmonic trap such as that used here, the lowest energy eigenstate is that of a three dimensional harmonic oscillator. This gives rise to a condensation in real space since the spatial extent of the ground state wavefunction is much smaller that that of the normal gas. Consequently the density of atoms would become extremely high over a narrow region at the bottom of the trap. This effect is evident in Fig. 5 which shows the Doppler-free portion of the two-photon spectrum of a gas cooled just below its transition at a temperature of 50 μK and a density of 2×10^{14} cm^{-3}. The strong sharp peak on the right is due to the non-condensed (normal) component and is shifted to the red of the free atom resonance by an amount determined by its density. The weak broad feature, which appears only below the transition, is due to the condensate. Note that a red shift of 0.8 MHz corresponds to a density of 4×10^{15} cm^{-3}, 20 times higher than the density of the normal gas.

The total number of atoms represented in Fig. 5 is 2×10^{10}, determined from the density and the known effective volume of the trap. The fraction of the atoms in the condensate can be determined by examining the ratio of the areas under the condensate and normal features of the spectrum, taking into account the fact that the laser beam illumi-

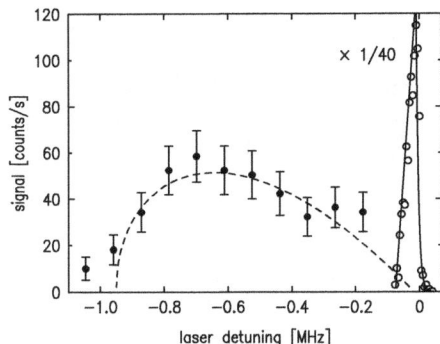

Fig. 5. Doppler-free spectrum of the condensate (broad feature) and the normal component (narrow feature).

Atom	m (amu)	T_c (μK)	a (nm)	a/λ_T	N_c
H	1	50	0.06	2.7×10^{-4}	10^9
Li	7	0.3	-1.5	-1.2×10^{-3}	10^3
Na	23	2.0	2.8	1.1×10^{-2}	10^7
Rb	87	0.7	5.4	2.4×10^{-2}	10^6

Table 1

Typical values of mass, transition temperature, scattering length, perturbation parameter and number of atoms in the condensate for the atomic species that have been Bose condensed.

nates all of the condensate but only a fraction of the normal gas. We estimate the condensate fraction to be about 5%, corresponding to a total of 10^9 condensate atoms.

If 10^9 atoms were to be put in the single particle ground state of this particular trap, the condensate density would be about 200 times greater than the measured value. The observed density is the result of a mean field repulsion or spreading pressure due to the interactions between the atoms. Since the condensate density is substantially higher than that of the coexisting normal gas, we can treat the condensate using a zero-temperature Thomas-Fermi approximation (a mean field model which neglects the kinetic energy terms). The dashed curve in Fig. 5 shows the Thomas-Fermi density profile for an interacting gas in a parabolic trap with a maximum condensate density of 4.8×10^{15} cm^{-3}. The resulting condensate is 15 μm in diameter and 5 mm in length. This should be compared with the 3 μm diameter which would result from the single particle ground state if there were no interactions. As a self-consistency check one can calculate the condensate fraction from the Thomas-Fermi model using as input parameters the s-wave scattering length, $a_{1S-1S} = 0.0648$ nm [19], the measured maximum condensate density, the temperature of the normal component, and the trap geometry. The calculated condensate fraction is 6%, in good agreement with the value determined by spectral weights [20].

A quantitative measure of the interactions in low temperature gases is the ratio of the scattering length to the deBroglie wavelength, a/λ_T. As shown in Table 1 this ratio is substantially smaller in hydrogen than in the other BEC gases. It is the weakness of the interactions in hydrogen which allows the ratio of the condensate density to the normal density to be very large, even at modest condensate fractions. Note that the density ratio of 24:1 for the data in Fig. 5 would, for a homogeneous gas, only be reached at a reduced temperature $T/T_c = 0.12$. Thus for many purposes our condensates are nearly pure and can be treated by zero-temperature models.

Textbook treatments of Bose-Einstein condensation in a homogeneous gas emphasize condensation in momentum space. A drastic reduction in momentum also occurs for condensate atoms in a trap since the momentum spread is determined only by the uncertainty principle and the spatial extent of the condensate. Therefore a very narrow feature should appear in the Doppler-sensitive portion of the two-photon spectrum accompanying the formation of the condensate. In our case, where the condensate has a length of about 5 mm along the propagation direction of the photons, the Doppler width would be about 100 Hz. Note that this is narrower than would be the case for non-interacting atoms in the single particle ground state. Although this intrinsic width is obscured in our experiments by the distribution of density shifts in the condensate, the resulting width is still much narrower than the Doppler width of the normal component. (Stenger et al. [21] have recently

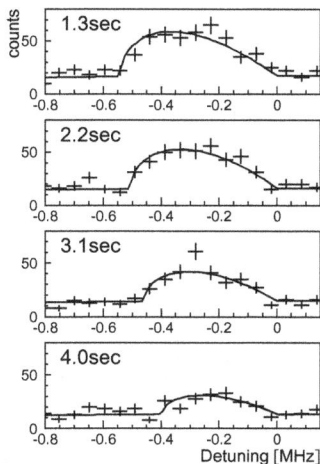

Fig. 6. Condensate feature sitting on top of the much broader contribution from the normal component in the Doppler-sensitive portion of the two-photon spectrum. Times are measured from the end of forced evaporative cooling. The number of atoms in the condensate is about the same as in Fig. 5, but the trap volume is larger resulting in a lower condensate density.

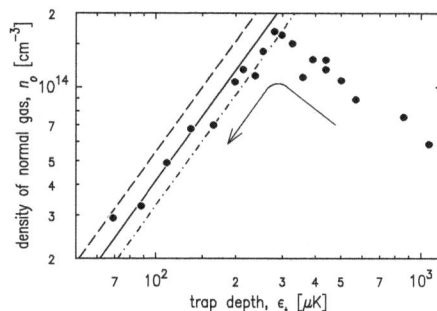

Fig. 7. Density of the normal component of the gas as the trap depth is reduced. The lines (dashed, solid, dot-dashed) indicate the BEC transition line assuming sample temperatures of ($\frac{1}{5}$th, $\frac{1}{6}$th, $\frac{1}{7}$th) the trap depth.

used two-photon Bragg scattering to measure the intrinsic momentum distribution in a Na condensate.) Figure 6 shows the condensate feature in the Doppler-sensitive part of the two-photon spectrum, plotted relative to the recoil shifted resonance and fit to a Thomas-Fermi density profile.

The atoms giving rise to the feature displayed in Fig. 6, those excited from the condensate by the absorption of two co-propagating photons, pick up enough momentum to be ejected from the trap with a divergence angle of the order of 10^{-3}. They are expected to form a narrow, intense beam of coherent atoms. Although we did not have the instrumentation to detect this beam in our initial experiments, we believe that much interesting physics can be done with this beam in future experiments. Kozuma et al. [22] have used a related process, optically induced Bragg diffraction, to couple sodium atoms out of a Bose-Einstein condensate into coherent wave packets with sharply defined momentum.

If one were to add atoms to a non-interacting

Bose gas held at a fixed temperature the density would increase until it reached a critical value $n_c(T) = 2.612(2\pi m k_B T)^{3/2}/h^3$. If more atoms were added they would condense into the lowest single particle energy eigenstate. The density of the normal component would remain constant. Thus $n_c(T)$, which is virtually unchanged by the presence of a weak interaction, can be regarded as a phase transition line in a plot of the density of the normal component versus temperature. Since the two-photon spectroscopy allows us to determine the densities of the condensate and normal gas separately, we are able to plot our cooling results in exactly this manner.

During cooling by RF evaporation the shape of the magnetic trap is held constant. Although the total number of atoms in the trap decreases, the density of remaining atoms increases since they lose energy and settle deeper into the potential well (see Fig. 7). Once we cool below a certain temperature — determined by the number of atoms in the trap — the density of the normal component decreases along a line consistent with $n_c(T)$.

A plot such as Fig. 7 shows the temperature and the density of the normal component. It does not indicate the size of the condensate under those conditions. That could vary, depending on the number of atoms remaining in the trap. In our experiments the size of the condensate is determined by a balance between the rate at which the atoms leave the

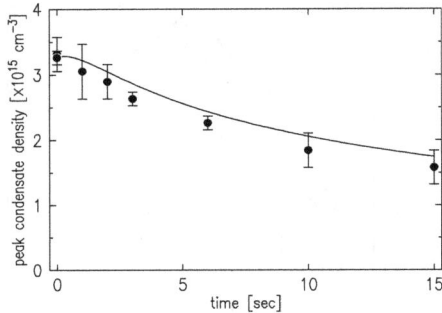

Fig. 8. Time evolution of the peak density in the condensate. The solid curve was generated from a model of the
loss mechanisms.

trap due to two-body spin relaxation (primarily in
the condensate) and the rate at which the condensate is replenished by atoms from the normal component. As Hijmans et al. have pointed out [23], the
weakness of the interaction between the hydrogen
atoms limits how far below the transition temperature hydrogen can be taken in an equilibrium situation. The small s-wave scattering length leads to
an elastic scattering rate in the thermal cloud well
below that in the alkali-metals, thus limiting the
evaporative cooling rate. In addition, the weak interaction leads to a high density in the condensate
which greatly increases the loss rate. Consequently
the condensate fractions that can be obtained in
hydrogen are limited to a few percent.

Since the condensate contains only a small fraction of the atoms, the normal gas acts as a large
reservoir that continually replenishes the condensate as atoms are lost through dipolar relaxation.
This replenishment dramatically increases the apparent lifetime of the condensate. We studied the
process by obtaining spectra at various hold times
after the end of the forced evaporative cooling cycle. The time evolution of the condensate is displayed in Fig. 8 by plotting the peak condensate
density, n_p, as a function of the hold time (here
the sample is held without exposure to the probe
laser). The peak condensate density decreases by
a factor of two (corresponding to a population decrease of $2^{5/2} \simeq 6$) in 15 s. This should be compared to the characteristic condensate decay time

due to dipolar relaxation, $\tau_{dip,c} = 7/(2gn_p)$, where
$g = 1.2 \times 10^{-15}$ cm^3/s is the dipolar decay rate
constant. For $n_p = 2 \times 10^{15}$ cm^{-3}, $\tau_{dip,c} \simeq 1.5$ s.
The 15 s lifetime of the condensate thus indicates
feeding of the condensate from the thermal cloud.

The number of atoms fed into the condensate
roughly matches the number lost during the 15 s
shown in Fig. 8. Estimates of the total trap population at $t = 0$ then set an upper limit on the
number lost, which scales as $n_p^{7/2}$ [24]. This strong
dependence provides a useful consistency check on
the determination of the peak condensate density
from the observed spectral shift. For example, if
n_p were actually a factor of two larger than stated
here [20], then the entire normal gas would be ex-
pended after the 15 s observation. This is clearly
not the case since a large condensate is still present
after 15 s.

A simple model [24] of the trapped gas has been
used to quantitatively test our understanding of
the system. The dynamics are dictated by losses
due to dipolar relaxation in the condensate and
normal gas, and evaporation from the normal gas.
Equilibrium between the normal gas and condensate is assumed. The expected behavior of the condensate is shown by the solid line in Fig. 8. The
agreement with experiment indicates a good understanding of the system.

The large reservoir of normal atoms could be
useful in the creation of a bright, sustained, CW
atom laser. Apparently, in our system 10^9 atoms
per second are being condensed out of the normal component. Efficient coupling of these atoms
into a coherent beam (instead of losing them from
the trap) could have broad applications ranging
from fundamental measurements enhanced by entangled quantum states to holographic nanolithog-
raphy.

The achievement of BEC in hydrogen has attracted wide attention for several reasons. First,
it represents the first new Bose-condensed species
since the initial alkali experiments. Also, hydrogen
condensates are of considerable interest because
the interatomic interactions can be calculated to
a high degree of accuracy. This may allow preci-

sion tests of not only many-body theories of the condensate but also the theory of ultra-cold collisions. Furthermore, our experiment was the first to exploit high resolution spectroscopy for detection of a condensate. This is noteworthy both for the demonstration of the technique and because the spectroscopy of hydrogen has long been a cornerstone for precision measurements in physics. Great potential exists for further increases in precision using cold, trapped hydrogen.

Acknowledgement

This research is supported by the National Science Foundation and the Office of Naval Research. The Air Force Office of Scientific Research contributed to the early stages. L. W. acknowledges support from Deutsche Forschungsgemeinschaft. D. L. and S. M. are grateful for support from the National Defense Science and Engineering Graduate Fellowship Program.

References

[1] C. E. Hecht, Physica **25**, 1159 (1959)

[2] W. C. Stwalley and L. H. Nosanow, Phys. Rev. Lett. **36**, 910 (1976) and references therein

[3] I. F. Silvera and J. T. M. Walraven, Phys. Rev. Lett. **44**, 164 (1980)

[4] For reviews of the early work on spin-polarized hydrogen see T. J. Greytak and D. Kleppner, in *New Trends in Atomic Physics*, G. Grynberg and R. Stora, eds. (North-Holland, Amsterdam, 1984); I. F. Silvera and J. T. M. Walraven, in *Progress in Low Temperature Physics*, D. F. Brewer, ed. (North-Holland, Amsterdam, 1986), Vol. X; J. T. M. Walraven in *Quantum Dynamics of Simple Systems*, G.-L. Oppo, S. M. Barnett, E. Riis, and M. Wilkinson, eds. (Institute of Physics Publishing, Bristol, 1994)

[5] M. H. Anderson, J. R. Ensher, M. R. Matthews, C. E. Wieman, and E. A. Cornell, Science **269**, 198 (1995)

[6] K. B. Davis, M.-O. Mewes, M. R. Andrews, N. J. van Druten, D. S. Durfee, D. M. Kern, and W. Ketterle, Phys. Rev. Lett. **75**, 3969 (1995)

[7] C. C. Bradley, C. A. Sackett, J. J. Tollett, and R. G. Hulet, Phys. Rev. Lett. **75**, 1687 (1995); C. C. Bradley,

C. A. Sackett, and R. G. Hulet, Phys. Rev. Lett. **78**, 985 (1997)

[8] A. I. Safonov, S. A. Vasilyev, I. S. Yasnikov, I. I. Lukashevich, and S. Jaakkola, Phys. Rev. Lett. **81**, 4545 (1998)

[9] T. J. Greytak, in *Bose-Einstein Condensation*, edited by A. Griffin, W. W. Snoke, and S. Stringari (Cambridge University Press, Cambridge, England, 1995). p. 131

[10] H. Hess, Phys. Rev. B **34**, 3476 (1986); T. J. Tommila, Europhys. Lett.**2**, 789 (1986); N. Masuhara, J. M. Doyle, J. C. Sandberg, D. Kleppner, T. J. Greytak, H. F. Hess, and G. P. Kochanski, Phys. Rev. Lett. **61**, 935 (1988)

[11] E. L. Sukov, J. T. M. Walraven, and G. V. Shlyapnikov, Phys. Rev. A **49**, 4778 (1994); Phys. Rev. A **53**, 3403 (1996)

[12] D. E. Pritchard, K. Helmerson, and A. G. Martin, in *Atomic Physics 11*, S. Haroche, J. C. Gay, and G. Grynberg eds. (World Scientific, Singapore, 1989), p. 179

[13] W. Petrich, M. H. Anderson, J. R. Ensher, and E. A. Cornell, Phys. Rev. Lett. **74**, 3352 (1995); K. B. Davis, M.-O. Mewes, M. A. Joffe, M. R. Andrews, and W. Ketterle, Phys. Rev. Lett. **74**, 5202 (1995)

[14] J. M. Doyle, J. C. Sandberg, N. Masuhara, I. A. Yu, D. Kleppner, and T. J. Greytak, J. Opt. Soc. Am. B **6**, 2244 (1989)

[15] C. L. Cesar, D. G. Fried, T. C. Killian, A. D. Polcyn, J. C. Sandberg, I. A. Yu, T. J. Greytak, D. Kleppner, and J. M. Doyle, Phys. Rev. Lett. **77**, 255 (1996)

[16] M. J. Jamieson, A. Dalgarno, and J. M. Doyle, Mol. Phys., **87**, 817 (1996)

[17] T. C. Killian, D. G. Fried, L. Willmann, D. Landhuis, S. C. Moss, T. J. Greytak, and D. Kleppner, Phys. Rev. Lett. **81**, 3807 (1998)

[18] D. G. Fried, T. C. Killian, L. Willmann, D. Landhuis, S. C. Moss, D. Kleppner, and T. J. Greytak, Phys. Rev. Lett. **81**, 3811 (1998)

[19] M. J. Jamieson, A. Dalgarno, and M. Kimura, Phys. Rev. A **51**, 2626 (1995)

[20] In the spectroscopic determination of the condensate density we have used the same constant of proportionality χ between the frequency shift and the density that we found for the normal gas. For excitation out of a Bose condensate the absence of exchange effects should reduce χ by a factor of 2. If this were the case the maximum density in the condensate would be twice as high and the condensate fraction calculated from the Thomas-Fermi analysis would be 25%, inconsistent with the value obtained from the relative spectral weights. We do not as yet understand the origin of this discrepancy.

[21] J. Stenger, S. Inouye, A. P. Chikkatur, D. M. Stamper-Kurn, D. E. Pritchard, and W. Ketterle, Phys. Rev. Lett. **82**, 4569 (1999)

[22] M. Kozuma, L. Deng, E. W. Hagley, J. Wen, R. Lutwak, K. Helmerson, S. L. Rolston, and W. D. Phillips, Phys. Rev. Lett. **82**, 871 (1999)

[23] T. W. Hijmans, Y. Kagan, G. V. Shlyapnikov, and J. T. M. Walraven, Phys. Rev. B **48**, 12886 (1993)

[24] D. G. Fried, Ph.D. thesis, Massachusetts Institute of Technology, 1999

LIST OF PUBLICATIONS

1. Broken Atomic Beam Resonance Experiment, *Phys. Rev. Lett.* **1**, 323 (1958), Kleppner, D; Ramsey, NF; Fjelstadt, P; Kleppner, D; Lerner, R; Trigg, G eds., 1995, 11.

2. Atomic Hydrogen Maser, *Phys. Rev. Lett.* **5**, 361 (1960), Goldenberg, HM; Kleppner, D; Ramsey, NF.

3. Theory of the Hydrogen Maser, *Phys. Rev.* **126**, 603–615 (1962), Kleppner, D; Goldenberg, HM; Ramsey, NF.

4. Properties of the Hydrogen Maser, *Applied Optics*, **1**, 55–60 (1962), Kleppner, D; Goldenberg, HM; Ramsey, NF.

5. Storage Technique for Atomic Hydrogen, *Rev. Sci. Instr.*, **33**, p. 248 (1962), Berg, HC; Kleppner, D.

6. Hyperfine Separation of Ground State Atomic Hydrogen, *Phys. Rev. Lett.* **11**, 338 (1963), Crampon, SB; Ramsey, NF; Kleppner, D.

7. Stimulated Emission and the Maser, *Science Teacher*, **31**, 13 (1964), Kleppner, D.

8. Stark Shift of Hydrogen Hyperfine Separation, *Phys. Rev. Lett.* **13**, 22 (1964), Fortson, EN; Kleppner, D; Ramsey, NF.

9. Hydrogen Maser Principles and Techniques, *Phys. Rev.* **138**, A972 (1965), Kleppner, D; Crampton, SB; Berg, HD; Ramsey, NF; Vessot, RFC; Peters, HF; Vanier, J.

10. Hyperfine Separation of Deuterium, *Phys. Rev.*, **141**, 55 (1966), Crampton, SB; Robinson, HG; Kleppner, D; Ramsey, NF.

11. Absolute Value of the Proton g-Factor, *Phys. Rev. Lett.* **17**, 405 (1966), Myint, T; Kleppner, D; Ramsey, NF; Robinson, HG.

12. Hyperfine Separation of Tritium, *Phys. Rev* **158**, 14 (1967), Mathur, BS; Crampton, SB; Kleppner, D; Ramsey, NF.

13. Atomic Differential Spin-Exchange and Scattering, *Phys. Rev. Lett.* **19**, 1363 (1967), Pritchard, DE; Burnham, DC: Kleppner, D.

14. Avoiding Lagrange Multiplier in Introductory Statistical Mechanics, *Amer. J. of Phys.* **36**, 843 (1968).

15. Alkali-Alkali Differential Spin-Exchange Scattering I, *Phys. Rev. A* **2**, 1922 (1970), Pritchard, DE; Carter, GM; Chu, FY; Kleppner, D.

16. The Effect of Nuclear Mass on the Bound-Electron g-Factor, *Phys. Rev. Lett.* **28**, 1159 (1972), Walther, FG; Phillips, WD; Kleppner, D. Myint, T; Walther, FG.

17. Stark Mixing Spectroscopy in Cesium, *Optic Communications,* **12**, 198 (1974), Zimmerman, M; Ducas, TW; Littman, MG; Kleppner, D.

18. Spin-Rotation Coupling in the Alkali-Rare van der Walls Molecule KAr, *Phys. Rev. Lett.* **32**, 507 (1974), Mattison, EM; Pritchard, DE; Kleppner, D.

19. Alkali-Metal Hyperfine Shift in the van der Waals Molecule KAr, *Phys. Rev, Lett.* **33**, 397 (1974), Freeman, RR; Mattison EM; Pritchard DE; Kleppner D.

20. Stark Ionization of High-Lying States of Sodium, *Phys. Rev. Lett.* **35**, 366 (1975), Ducas, TW; Littman, MG; Freeman, RR; Kleppner, D.

21. Magnetic Moment of the Proton in $H2O$ in Terms of the Bohr Magneton, *Phys. Rev. Lett.* **35**, 1619 (1975), Phillips, WD; Cooke, WE; Kleppner, D.

22. Argon-Induced Hyperfine Shift in Potassium, *Phys. Rev. A* **13**, 907 (1976), Freeman, RR; Pritchard, DE; Kleppner, D.

23. Spin-Rotation Interaction in the van der Waals Molecule KAr, *J. of Chem. Phys.* **64**, 1194 (1976), Freeman, RR; Mattison, E; Pritchard, DE; Kleppner, D.

24. Radiative Lifetimes of Selected Vibrational Levels in the A1u State of Na22–2, *J. of Chem. Phys.* **65**, 842 (1976), Ducas, TW; Littman, MG; Zimmerman, ML; Kleppner, D.

25. Structure of Sodium Rydberg States in Weak to Strong Electric Field, *Phys. Rev. Lett.* **36**, m788 (1976), Littman, MG; Zimmerman, ML; Ducas, TW; Freeman, RR; Kleppner, D.

26. Tunneling Rates for Excited States of Na in Static Electric Field, *Phys. Rev. Lett.* **37**, 486 (1976), Littman, MG; Zimmerman, ML; Kleppner, D.

27. Core Polarization and Quantum Defects in high-angular-momentum states of alkali atoms, *Phys. Rev. A* **14**, 1614 (1976), Freeman, R; Kleppner, D.

28. Magnetic Moment of the Proton in H2O in Bohr Magnetons, *Metrologia,* **13**, 179 (1977), Phillips, WD; Cooke, WE; Kleppner, D.

29. Crossed Beam Determination of Na-Rare Gas Fine Structure Changing Cross Sections, *Phys. Rev. Lett.* **38**, 1018 (1977), Phillips, WD; Glaser, CL; Kleppner, D.

30. Stark Structure of Barium Rydberg States, *J. de Phys.* **11**, L11 (1978), Zimmerman ML; Ducas, TW; Littman, MG; Kleppner, D.

31. Diamagnetic Structure of Na Rydberg States, *Phys. Rev. Lett.* **40**, 1083 (1978), Zimmerman, ML; Castro, JC; Kleppner, D.

32. Field Ionization Processes in Excited Atoms, *Phys. Rev. Lett.* **41**, 103 (1978), Littman, MG; Kash, MM; Kleppner, D.

33. Detection of Far-Infrared Radiation Using Rydberg Atoms, *Ap. Phys. Lett.* **35**, 382 (1979), Ducas, TW; Spencer, WP; Vaidyanathan, AG; Hamilton, WH; Kleppner, D.

34. The Stark Structure of the Rydberg States of Alkali-Metal Atoms, *Phys. Rev A* **20**, 2251–2275 (1979), Zimmerman, ML, Littman, MG; Kash, MM; Kleppner, D.

35. Hyperfine Resonance of Gaseous Atomic Hydrogen at 4.2K, *Phys. Rev. Lett.* **42**, 1039 (1979), Crampton, SB; Greytak, TJ; Kleppner, D.

36. Production and Confinement of Spin Polarized Hydrogen, *J. de Phys.* **41**, Colloque C7–151 (1980), Cline, RW; Greytak, TJ; Kleppner, D; Smith-Cline, DA.

37. Approximate Symmetry for Hydrogen in a Uniform Magnetic Field, *Phys. Rev. Lett.* **45**, 1092 (1980), Zimmerman, ML; Kash, MM; Kleppner, D.

38. Origin and Structure of the Quasi-Landau Resonances, *Phys. Rev. Lett.* **45**, 1780 (1980), Castro, JC; Zimmerman, M, Hulet, RG, Kleppner, D.

39. Magnetic Confinement of Spin-Polarized Atomic Hydrogen, *Phys. Rev. Lett.* **45**, 2117 (1980), Cline, RW; Smith, DA; Greytak, TJ; Kleppner, D.

40. Dynamical Effects at Level Crossings: A Study of the Landau-Zener Effect Using Rydberg Atoms, *Phys. Rev. A* **23**, 2978 (1981), Rubbmark, JR; Kash, MM; Littman, MG; Kleppner, D.

41. Measurements of Lifetimes of Sodium Rydberg States in a Cooled Environment, *Phys. Rev. A* 978 (1981), Spencer, WP; Vaidyanathan, AG; Kleppner, D; Ducas, TW.

42. Inhibited Spontaneous Emission, *Phys. Rev. Lett.* **47**, 233 (1981), Kleppner, D.

43. Nuclear Polarization of Spin-Polarized Hydrogen, *Phys. Rev. Lett.* **47**, 1195 (1981), Cline, R; Greytak, TJ; Kleppner, D.

44. Temperature Dependence of Blackbody-Radiation-Induced Transfer Among Highly Excited States of Sodium, *Phys. Rev. A* **25**, 380 (1982), Spencer, WP, Vaidyanathan, AG; Kleppner, D; Ducas, TW.

45. Photoionization by Blackbody Radiation, *Phys. Rev. A* **26**, 1490 (1982), Spencer, WP; Vaidyanathan, AG; Kleppner, D; Ducas, TW.

46. Experimental Study of Nonadiabatic Core Interactions in Rydberg States of Calcium, *Phys. Rev. A* **26**, 3346 (1982), Vaidyanathan, AG; Spencer, WP, Rubbmark, JR; Kuiper, H; Fabre, C; Kleppner, D.

47. Highly Excited Atoms, ML *Uspekekhi Fisicheskikh Nauk.* **137**, 339–360 (1982), Kleppner, D; Littman, MG; Zimmerman, ML; Kleppner D.

48. Comparisons of Approximate Bases for Hydrogen in a Magnetic Field, *Phys. Rev. A* **27**, 2731 (1983), Zimmerman, ML; Hulet, RG; Kleppner, D.

49. Observation of Three-Body Recombination in Spin-Polarized Hydrogen, *Phys. Rev. Lett.* **51**, 1430 (1983), Hess, HF; Bell, DA; Kochanski, GP; Cline, RW; Kleppner, D; Greytak, TJ.

50. Rydberg Atoms in "Circular" States, *Phys. Rev. Lett.* **51** 1430, 1983, Hulet, RG; Kleppner, D.

51. Temperature and Magnetic Field Dependence of Three-Body, Recombination in Spin-Polarized Hydrogen, *Phys. Rev. Lett.* **52** 1520 (1984), Hess, HF; Bell, DA; Kochanski, GP; Kleppner, D; Greytak, TJ.

52. Progress in Applying Stable Atomic Hydrogen Methods to Polarized Proton Sourcs and Jets, *J. de Phys.* **46**, C2–665 (1985), Kleppner, D.

53. Atomic Stabilization through Electronic and Nuclear Polarization, *Ann. Phys. Fr.* **10**, 877–882 (1985), Kleppner, D.

54. Inhibited Spontaneous Emission by a Rydberg Atom, *Phys. Rev. Lett.* **55**, 2137 (1985), Hulet, RG; Hilfer, ES; Kleppner, D.

55. Relaxation and Recombination in Spin-Polarized Atomic Hydrogen, *Phys. Rev. B* **34**, 7670 (1986), Bell, DA; Hess, HF; Kochanski, GP; Buchman, S; Pollack, L; Xiao, YM; Kleppner, D; Greytak, TJ.

56. Manipulating Atoms with Light, *Helvetica Physica Acta* **59**, 717 (1986), Klpener D.

57. Spin-Polarized Hydrogen Maser, *Phys. Rev. A* **34**, 1902 (1986), Hess, HF; Kochanski, GP, Doyle, JM; Greytak, TJ; Kleppner, D.

58. Bose-Einstein Condensation in an External Potential, *Phys. Rev. A* **35**, 4354 (1987), Bagnato, V; Pritchard, DE; Kleppner, D.

59. Magnetic Trapping of Spin-Polarized Atomic Hydrogen, *Phys. Rev. Lett.* **59**, 672 (1987), Hess, HF; Kochanski, GP, Doyle, JM, Masuhara, N; Kleppner, D; Greytak, TJ.

60. Evaporative Cooling of Spin-Polarized Atomic Hydrogen, *Phys. Rev. Lett.* **61**, 935 (1988), Masuhara, N; Doyle, JM; Sandberg, JC; Kleppner, D; Greytak, TJ; Hess, HF; Kochanski, GP.

61. A Study of One- and Two-Photon Rabi Oscillations, *Phys. Rev. A* **40**, 5103 (1989), Hughey, BJ; Gentile, T; Kleppner, D; Ducas, T.

62. Energy Distributions of Trapped Atomic Hydrogen, *J. of the Opt. Society of Amer.*, B-6, 2244 (1989) Doyle, JM, Sandberg, JC; Masuhara, N; Kleppner, D; Greytak, TJ.

63. Orderly Structure in the Positive-Energy Spectrum of a Diamagnetic Rydberg Atom, *Phys. Rev. Lett.* **63**, 1133–1136 (1989). Iu, CH; Wekch, GR;, Kash, MM; Kleppner, D.

64. iamagnetic Structure of Lithium n~21, *Spect. Acta,* **45**A, 57 (1989), Welch, GR; Kash, M; Iu, CH; Hsu, L; Kleppner, D.

65. Experimental Study of Energy Level Statistics in a Regime of Regular Classical Motion, *Phys. Rev. Lett.* **62**, 893 (1989), Welch, GR; Kash, M; Iu, CH; Hsu, L; Kleppner, D.

66. Positive Energy Structure of the Ryberg Diamagnetic Spectrum, *Phys. Rev. Lett.* **62**, 1975 (1989), Welch, GR; Kash, M; Iu, CH; Hsu, L; Kleppner, D.

67. A Split High Q Superconducting Cavity, *Rev. Sci. Inst.* **61**, 1940 (1990), Hughey, B; Gentile, T; Ducas, TW; Kleppner, D.

68. Experimental Study of Small Ensembles of Atoms in a Microwave Cavity, *Phys. Rev. A* **41**, 6245 (1990), Hughey, B; Gentile, T. Ducas, TW; Kleppner, D.

69. The Recombination of Atomic Hydrogen Below 1K, *Physica B* **165**, 745 (1990), Buchman, S; Greytak, TJ; Pollack, L; Xiao, Y; Kleppner, D.

70. Microwave Spectroscopy of Calcium Rydberg States, *Phys. Rev. A* **42**, 440 (1990), Gentile, TR; Hughey, BJ; Kleppner, D.

71. Positive Energy Spectroscopy of the Diamagnetic Lithium System, *Comments on Atomic and Molecular Physics,* **25**, 301 (1991).

72. The Diamagnetic Rydberg Atom: Confrontation of Calculated and Observed Spectra, *Phys. Rev. Lett.* **66**, 1456 (1991), Iu, CH; Welch, GR; Kash, MM; Kleppner, D; Delande, D; Gay, JC.

73. Hydrogen in the Submillikelvin Regime: Sticking Probability on Superfluid 4HE, *Phys. Rev. Lett.* **67**, 603 (1991), Doyle, JM; Sandberg, JS; Yu, IA; Cesar, CL; Kleppner, D; Greytak, TJ.

74. Observation of Nonstatistical Ortho-para Ratio in Hydrogen Recombination at Low Temperatures, *J. Chem. Phys.* **96**, 4032 (1992), Xiao, YM; Buchman, S; Pollack, L; Kleppner, D; Greytak, TJ.

75. Nuclear Relaxation During the Formation of H_2 from Spin-Polarized Hydrogen, *Phys. Rev. B* **48**, 744 (1993), Xiao, YM; Buchman, S; Pollack, L; Kleppner, D; Greytak, TJ.

76. New Class of Universal Correlations in the Spectra of Hydrogen in a Magnetic Field, *Phys. Rev. Lett.* **71** (2899) 1993, Simmons, BD; Hashimoto, A; Courtney, M; Kleppner, D; Altshuler, BL.

77. Evidence for Universal Quantum Reflection of Hydrogen from Liquid 4He, *Phys. Rev. Lett.* **71**, 1589 (1993), Yu, IA; Doyle, JM; Sandberg, JC, Cesar, CL; Kleppner, D; Greytak, TJ.

78. Comments on the 1992 Antihydrogen Workshop, *Hyperfine Interactions,* **76**, 389–391 (1993), Kleppner. D.

79. Surface Reflection of Submillikelvin Atomic Hydrogen from Thin He Films: Substrate Effects Superfluid, *Physica B* **194**, 17 (1994), Yu, IA; Doyle, JM; Sandberg, JC, Cesar, CL, Kleppner, D.

80. Quantum Reflection of Submillikelvin Atomic Hydrogen from Bulk Superfluid He, *Physica B* **194**, 15 (1994), Yu, IA; Doyle, JM; Sandberg, JC; Cesar, CL; Kleppner, D; Greytak, TJ.

81. Evaporative Cooling of Atomic Hydrogen, *Physica B* **194**, 13 (1994), Doyle, JM; Sandberg, J; Cesar, CL, Kleppner, D, Greytak, TJ.

82. Long-Period Orbits in the Stark Spectrum of Lithium, *Phys. Rev. Lett.* **73**, 1340 (1994), Courtney, M; Jiao, H; Spellmeyer, N; Kleppner, D.

83. Recurrences Associated with a Classical Orbit in the Node of a Quantum Wavefunction, *Phys. Rev. A* **52**, 3695 (1995), Shaw, JA; Delos, JB; Courtney, M; Kleppner, D.

84. Classical, Semiclassical, and Quantum Dynamics in the Lithium Start System, *Phys. Rev. A* **51**, 3604 (1995) Courtney, M; Spellmeyer, N; Jiao, H; Kleppner, D.

85. Closed Orbit Bifurcations in Continuum Stark Spectra, *Phys. Rev. Lett.* **74**, 1538 (1995), Courtney, M; Jiao, H; Spellmeyer, N; Kleppner, D; Gao, J; Delos, JB.

86. Core Induced Chaos in Diamagnetic Lithium, *Phys. Rev. A* **53**, 178 (1996), Courtney, M; Kleppner, D.

87. Two-photon Spectroscopy of Trapped Atomic Hydrogen, *Phys. Rev. Lett.* **77**, 255 (1996), Cesar, CL; Fried, DG; Killian, TC, Polcyn, AD; Sandberg, JC; Yu, IA; Greytak, TJ; Kleppner, D; Doyle, J.

88. Circular States of Atomic Hydrogen, *Phys. Rev. A* **56**, 143 (1997), Lutwak, R; Holley, J; Chang, P; Paine, S; Kleppner, D Ducas, T.

89. The Cat and the Moon: Chaos and High Precision, 1997 Oersted Medal, *Amer. J. of Phys.* **65**, 816 (1997). Recurrence Spectroscopy of a Time-Dependent System: A Rydberg Atom in an Oscillating Field, *Phys. Rev. Lett.* **79**, 1650 (1997), Spellmeyer, N; Kleppner, D; Haggerty, MR; Kondratovich, V; Delos, JB; Gao, J.

90. Extracting Classical Trajectories from Atomic Spectra, *Phys. Rev. Lett.* **81**, 1592–1595 (1998), Haggerty, MR; Delos,JB; Spellmeyer, N; Kleppner, D; Greytak, TJ.

91. Cold Collision Frequency Shift of the 1S-2S Transition in Hydrogen, *Phys. Rev. Lett.* **81**, 3807 (1998), agaxinexsKillian, TC; Fried, DG; Landhuis, D; Moss, SC; Kleppner, D; Greytak, TJ.

92. Bose-Einstein Condensation of Atomic Hydrogen, *Phys. Rev. Lett.* **81**, 3811 (1998), Fried, DG; Killian, TC; Willmann, L; Landhuis, D; Moss, SC; Greytak, TJ; Kleppner, D.

93. A Short History of Atomic Physics in the Twentieth Century, *Rev. Mod. Physics* **71–2** (Special Issue), 584–587 (1999), Kleppner, D.

94. Two-Photon Doppler-free Spectroscopy of Trapped Atoms, *Phys. Rev. A* **59**–6, 4564–4570 (1999), Cesar, CL; Kleppner, D.

95. Classically Forbidden Recurrences in the Photoabsorption Spectrum of Lithium, *Phys. Rev A* **62**, 043409 (2000), Kondratovich, V; Delos, JB; Spellmeyer, N; Kleppner, D.

96. Beyond Quantum Mechanics: Insights from the Work of Martin Gutzwiller, *Found. of Physics*, **31**(94) 593–612 (2001, Kleppner, D; Delos, JB.

97. Classically Forbidden Processes in Photoabsorption Spectra, *Physica Scripta* **T90**, 189–105 (2001), Delos, JB; Kondratovich, V; Wang, DM; *et al.*

98. Sum Rule for the Optical Spectrum of a Trapped Gas, *Phys. Rev.* **65**, 033617 (2002), Oktel, MO; Killian, TC, Kleppner, D; Levitov, L.

99. Inelastic Collision Rates of Trapped Metastable Hydrogen, *Phys. Rev. A* **67**, 022718 (2003) Landhuis, D; Matos, L; Moss, SC; Steinberger, JK; Vant, K; Willmann, L; Greytak, TJ; Kleppner, D.

100. Report of the American Physical Society Study Group on Boost-phase Intercept System for National Missile Defense: Scientific and Technical Issues Issues, *Reviews of Modern Physics* **76**, S1–S424 (2004), Barton, DK; Falcone, R; Kleppner, D; Lamb, FK; Lynch, HL; Moncton, D; Montague, D; Mosher, DE, Priedhorsky, W; Tigner, M; Vaughan, DR.

101. Direct Frequency Comb Generation from an Octave-spanning Prismless Ti: Sapphire Laser, *Optics Lett.* **29**, 1683–1685 (2004), Matos, L; Kleppner, D; Kuzucu, O; Schibli, TR; Kim, J; Ippen, EP; Kaertner, FX.

102. A New Path to Ultracold Hydrogen, *Canadian J. of Phys.* **83**, 293–300 (2005) deCarvalho, R; Brahms, N; Newman, B; Doyle, JM, Kleppner, D; Greytak, T.

103. Magnetic Trapping of Silver and Copper, and Anomalous Spin Relaxation in the Ag-He System, *Phys. Rev. Lett.* **101**, 103002 (2008), Brahms, N; Newman, B; Johnson, C; Greytak, T; Kleppner, D; Doyle, J.

104. Research Data in thes Digital Age, *Science*, **325** (5939), 368 (2009), Kleppner, D; Sharp, PA.

105. Zeeman Relaxation of Cold Atomic Iron and Nickel in Collisions with He-3, *Phys. Rev. A* **81**, 062706 (2010), Johnson, C; Newman, B; Brahms, N; Doyle, JM; Kleppner, D; Treytak, TJ.

106. Magnetic Relaxation in Dysprosium-Dysprosium Collisions, *Phys. Rev. A* **83**, 012713 (2011), Newman, BK; Brahms, N; Au, YS; Johnson, C; Connolly, CB; Doyle, JM; Kleppner, D; Greytak, TJ.

107. On the Radial Dependence of the Herschbach Effect, *Mol. Physics*, **110**, 1591–1592 (2012).

Invited Papers and Published Lectures

1. The Atomic Hydrogen Maser, D. Kleppner; *Advances in Quantum Electronics,* p. 543; Columbia University Press, NY, Singer, JR, ed., 1961.

2. Electron-Proton g-Factor Ratio, D. Kleppner; Winkler, PF; *Physics of One-and Two-Electron Atoms*, North-Holland, Bopp; Kleinpoppen, eds., 1969.

3. A Determination of g_i/g_p in Atomic Hydrogens, *Physics of One- and Two-Electron Atoms*, p. 146, North-Holland, Bopp; Kleinpoppen, eds., 1969.

4. An Orbiting Clock Experiment to Determine the Gravitational Red Shift, *Astrophysics and Space Science* **3**, Reiden Pub, Holland, D. Kleppner; Vessot, RFC; Ramsey, NF; Kopal, Z ed., 1970.

5. Experiments with Atomic Hydrogen, D. Kleppner; *Atomic Physics and Astrophysics*, p. 3, Gordon and Breach, NY, Cretien, M; Lipworth, E eds., 1971.

6. Excited Atoms,, Kleppner, *Atomic Physics 5*, p. 269, Plenum Pub, NY, Marrus, R; Prior, M; Shugard, H eds., 1977.

XX Some recent studies of Rydberg states of one- and two-electron atoms, Kleppner, D., *États Atomique et Moléculaires couples a un continiuum atomes et molecules les Hautement Excités, Colloques Internationauz du CNRS, NO. 273, Paris (1977), eds. S. Feneuille and J. C. Lehman, pp. 227–235.*

7. The Spectroscopy of Highly Excited Atoms, D. Kleppner; *Progress in Atomic Spectroscopy*, p. 713, Plenum Press, NY, 1978.

8. Laboratory Studies of Rydberg Atoms, D. Kleppner; *Radio Recombination Lines*, Reidel Pub., Shaver, P ed., 1980.

9. Turning off the Vacuum, D. Kleppner; *Laser Spectroscopy V*, Vol. 80, p. 111, Springer-Verlag, McKellar, ARW; Oka, T; Stoicheff, BP eds., 1981.

10. Atoms in Very Strong Fields, D. Kleppner; Les Houches Session XXXIV, North-Holland, 1982,

11. Cold Hydrogen Techniques for Polarized Proton Production, Kleppner D., *Polarized Proton Ion Sources, AIP Conference Proceedings*, Vol. 80, p. 111, Krisch, AD; Lin, ATM eds., 1982.

12. Spin-Polarized Hydrogen: New Possibilities for Polarized Sources and Targets, D. Kleppner and T.J. Greytak, *AIP Conference Proceedings*, Vol. 95, pp. 546, Bunce, GM, ed. 1983.

13. Optics for a Spin-Polarized Hydrogen Atomic Beam, D. Kleppner, international symposium on the production and neutralization of negative ions and beams; Upton, NY (USA), 1983, LBL–1686; CONF 831180–12.

14. Lectures on Spin-Polarized Hydrogen, T. J.Greytak , D. Kleppner *New Trends in Atomic Physics*, Les Houches Summer School, North-Holland, 1984, G. Grynberg and R. Stora, eds.

15. A Future for Hydrogen, D. Kleppner, *Atomic Physics 11*, p 665, World Scientific, Singapore, Singapore, v(1989), S. Haroche, J.C. Gay, and G. Grynberg, eds.

16. An Introduction to Cavity Quantum Electrodynamics, D. Kleppner, *Proceedings of the OJI International Seminar on Highly Excited Atoms and Molecules* (Fuji-Yoshida, Japan, Kano, SS and Matsuzawa, M eds. 1986.

17. Trapped Atom Physics, D. Kleppner; *The Hydrogen Atom*, Berlin, p. 68; Bassani, GF; Inguscio, M; Haensch,TW eds., 1989.

18. Experimental Study of Two-Photon Rabi Oscillations, T. Gentile; BJ Hughey; TW Ducas; D. Kleppner; *Coherence and Quantum Optics VI*, Plenum Press, New York, p. 521–525, Eberly, JH *et al* eds, 1990.

19. An Introduction to Rydberg, Atoms, D. Kleppner; Atoms in Unusual Situations, 1986, p. 57, Plenum Press, New York, Briand, JP ed., 1990.

20. Experimental Studies of Chaos in an Atomic System, D. Kleppner, Les Houche Lectures, July 1990, Dalibard, J; Raimond, JM. JM, Zinn-Justin J, eds, Fundamental Systems in Quantum Optics Elsevier Science BV, 1992, D. Kleppner.

21. Experimental Studies of Chaos in an Atomic System, Kleppner D; *Fundamental Systems in Quantum Optics*, Elsevier Science BV, Dalibard, J; Raimond, JM, Zinn-Justin J eds., 1992.

22. Quantum Chaos and Laser Spectroscopy, D. Kleppner, Atomic Physics 13; AIP Conference Proc. Vol. 275, 417–424, 199, H. Walther, T.W. Haänsch, and B. Neizert, eds.

23. Quantum Chaos and Laser Spectroscopy, D. Kleppner, *Proceedings of the International School of Physics Enrico Fermi, 1992*, Course CXX, North- Holland, NY, 1994, Zeller, M ed.

24. Bose-Einstein Condensation of Atomic Hydrogen, M. Inguscio, S. Stringari and C. E. Wieman (1998), eds. Kleppner; Greytack, TJ; Killian, TC; Fried, DG; Willmann, L; Landhuis, D; Moss, SC; *Proceedings of the International School of Physics Enrico Fermi*, Vol. 140, pp. 177–199, 1996.

25. Millimeter-Wave Measurement of the Rydberg Frequency,Lutwak; R; Holley, J; DeVries, J; D. Kleppner; *Proceedings of the Fifth Symposium on Frequency Sta ndards and Metrology; AIP Conference Proceedings*, **549**, p. 259, 1996 L, Bergquist, JC ed., 2000.

26. Quantum Chaos and Rydberg Atoms in Strong Fields, Courtney M; Jiao H; Spellmeyer, N; D. Kleppner; *Quantum–Classicl Correspondence, 1997, Course 140*, International Press.

27. Polarized Atomic Hydrogen Beam Studies in the Michigan Ultra-cold Jet, Raymond, RS; Blinov, BB; Borisov, NS; Cheng, J; Davidenko, AM; Fimushkin, W; Gladycheva, SE; Grishin, VN; Kageya, T; Kantsyrev, DY; D. Kleppner; Krisch, AD; Luppov, VG; Morozov, VS; Murray, JR; Newumann, JJ; Yankama, B; *AIP Conference Proceedings*, Vol. 549, pp. 674–675.

28. Michigan Ultra-cold Polarized Atomic Hydrogen Target, Blinov, BB; Gladycheva, SE; Kageya, T; Kantsyrev, DY; Krisch, AD; Luppov, VG; Morozov, VS; Murray, JR; Raymond, RS; Borisov, NS; Fimushkin, W; Grishin, VN; Mysnik, AI; D. Kleppner; *AIP Conference Proceedings* **570**, pp. 856–860, 2001.

29. Introduction to a Nobel Symposium, D. Kleppner, *Atomic Physics 20; AIP Conference Proceedings*, Vol. 770, pp. 3–8, 2005.

30. A Beginner's Guide to ICAP, D. Kleppner, *Atomic Physics 30. SIP Conference Proceedings*, **869**, pp. 3–12 (2006).

Reports

Physics Through the 1990s: Atomic, Molecular and Optical Physics. NRC (1986), Panel on Atomic, Molecular Optical Physis, D. Kleppner (chair); C.L. Cocke Jr; A. Dalgarno; R.W. Field; T.W. Haensch; N.F. Lane; J. Macek, J; F.M. Pipkin; I.A. Sellin.

Boost-phase defense against intercontinental ballistic missiles, *Physics Today* Jan 1957, p.p. 30–35, D. Kleppner; F.K. Lamb, D.E. Mosher.

Ensuring the Integrity, Accessibility, and Stewardship of Research Datain the Digital Age, NAS, 2009, D. Kleppner and P.A. Sharp, co-chairs.

Book Reviews

Einstein and the Quantum: The Quest of the Valiant Swabian, by Douglas Stone, *Physics Today*, April, 2014, p. 48, D. Kleppner.

Fact and Fraud in Science: Cautionary Tales from the Front Lines of Science, by David Goodstein, *Physics in Perspective*, **13**, 244–247 (2011). D. Kleppner.

J. Robert Oppenheimer by Robert P. Crease (book review), *Physics in Perspective*, **B9**, 505–508 (2007), D. Kleppner.

Splitting the Second: The Story of Atomic Time by A. Jones, *Nature*, **410** (6832), 1027 (2001), D. Kleppner.

Rydberg Atoms by T.F. Gallagher, *Nature*, **372**, 332 (1994), D. Kleppner.

Magazine Articles

Rereading Einstein on Radiation, *Physics Today*, **58**, Feb. 2005, pp. 30–33.

Research Data in the Digital Age, Kleppner D; Sharp, PA, *Science* **325**(5939), 78 (2009).

One Hundred Years of Quantum Physics, *Science* **289**(5481), 893–898 (2000); comments *Science* **289**(5487), 1052 (2000), Kleppner D; Jaciw R.

Cavity Quantum Electrodynamics, *Physics Today*, Jan. 1989, pp. 24–30, Haroche, S; Kleppner D.

Highly Excited Atoms, *Scientific American*, **244**(5), 130 (1985), Kleppner D, Littman, MG; Zimmerman, ML.

Research Data in the Digital Age, Kleppner D; Sharp, PA, *Science* **325**(5939), 78 (2009).

Research in Small Groups, *Phys. Today*, March 1985, 78–85, Kleppner D.

Miscellaneous

Hydrogen: The Essential Element, *Nature*, **422** (6928), 119 (2002), D. Kleppner.

Physics and Common Nonsense, *Annals of the New York Academy of Sciences*, 775–126 (The Flight from Science and Reason), Gross, PR; Levitt, N; Lewis, MW eds., 1996.

Comments on the 1992 Antihydrogen Workshop, D. Kleppner; Science Publishers, Baltzer, JC ed., 1993 Vernon Hughes' Worlds of Atoms, *A Festschrift in honor of Vernon W. Hughes,* Yale University, pp. 389–391, 1992.

Fine and Hyperfine Spectra and Interactions, D. Kleppner and FW. Drake, *Encyclopedia of Physics, 1991,* R.G. Lerner and G. TriggL eds, VCH Publishers, Inc, p. 396, D. Kleppner; Drake, GWF. Niels Bohr and Atomic Physics Today, D. Kleppner.

Niels Bohn: Physics and the World, 1988, Harwood, p. 45, H. Feschbach; T. Matsui and A. Oleson, eds. 1988.

Frontiers of Atomic Physics, D. Kleppner; *Physics in a Technological World,* p. 347, AIP, New York; French, AP eds.,1988.

Fine and Hyperfine Spectra and Interactions, *Encyclopedia of Physics* 1981, p. 326, Addison-Wessley, Reading, MA; D. Kleppner; Lerner, R; Trigg, G; eds.

Essays from the Reference Frame column of Physics Today

A Genuine Teaching Experience, *Physics Today,* October 2009, p. 8 (education), comments: April 10, 2010. p. 11.

What Sam Said, *Physics Today,* February 2009, p. 8 (science writing).

Hanbury Brown's Steamroller, *Physics Today,* August 2008, p. 8 (statistical physics), comments: March 2009, p. 10.

Master Michelson's Measurement, *Physics Today,* August 2007, p. 8 (history and meteorology).

Time Too Good to be True, *Physics Today,* March 2006, p. 10 (metrology).

The Master of Dispersion, *Physics Today,* November 2005, p. 10 (history).

Professor Feshbach and His Resonance, *Physics Today,* August 2004, p. 12 (atomic physics).

On the Matter of the Meter, *Physics Today,* March 2001, p. 11 (Metrology), comments: December 12, 2000, p. 12.

The Yin and Yang of Hydrogen, *Physics Today,* April 1999, 11 (history).

Nibbling the Bullet, *Physics Today,* June 1998, p. 13 (science and society) comments: *Physics Today,* October 1998, p. 13.

A Beginner's Guide to the Atomic Laser, *Physics Today*, August 1997, p. 11 (atomic physics).

The Fuss about Bose-Einstein Condensation, *Physics Today*, August 1996, p. 13 (atomic physics) comments: *Physics Today*, January 1998, p. 90.

Memoirs of Schrodinger's Cat, *Physics Today*, November 1996, p. 11 (history) comments: *Physics Today*, April 1997, p. 94.

The Most Tenuous of Molecules, *Physics Today*, October 1995, p. 11 (molecular physics).

Confessions of a Committee Junkie, *Physics Today*, January 1995, p. 9 (behavior modification).

Some Small Big Science, *Physics Today*, October 1994, p. 9 (styles of physics) comments: *Physics Today*, February 1995, p. 11.

Where I Stand, *Physics Today*, January 1994, p. 9 (science and society).

Thoughts on Being Bad, *Physics Today*, August 1993, p. 9 (science and society).

The Gem of General Relativity, *Physics Today*, April 1993, p. 9 (radio astronomy).

About Benjamin Thompson, *Physics Today*, September 1992, p. 9 (history).

Fretting About Statistics, *Physics Today*, July 1992, p. 9 (experimental physics).

MRI for the Third World, Physics Today, March 1992, p. 9 (science and society).

A Lesson in Humility, *Physics Today*, December 1991, p. 9 (predictions in science).

Kudos for a Radical College, *Physics Today*, January 1991, p. 9 (history).

With Apologies to Casimir, *Physics Today*, October 1990, p. 9 (quantum theory).

Night Thoughts on the NSF, *Physics Today*, April 1990, p. 9 (science policy).

A Passion for Precision, November 1989, p. 9 (experimental physics).

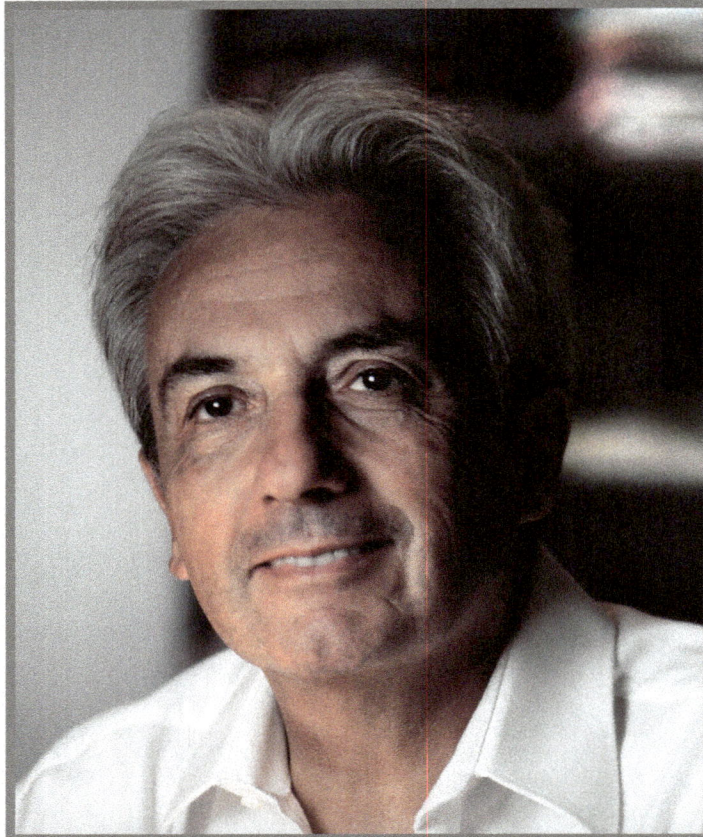

Albert Fert

ALBERT FERT

Birthplace

Born	Carcassonne, France, 7 March 1938
Nationality	French

Education

1957–1962	École Normale Supérieure (Paris): Maîtrise de Mathématiques, Maîtrise de Physique
1970	Université Paris-Sud: Doctorat ès Sciences Physiques (Supervisor: I.A. Campbell)

Main Positions

1962–1964	Teaching assistant at Université de Grenoble (Grenoble, France)
1965–1976	Assistant Professor at Université Paris-Sud (Orsay, France)
1976–	Professor of Physics at Université Paris-Sud (Eméritus Prof. since 2008)
1995–	Scientific Director of Unité Mixte de Physique CNRS/Thales

Main Positions

1964	Société Française de Physique
2004	Académie des Sciences
2008	Académie des Technologies

Awards

1994	International Prize for New Materials (together with P. Grünberg and S.S.P. Parkin), American Physical Society (APS)
1994	Magnetism Prize (together with P. Grünberg), International Union for Pure and Applied Physics (IUPAP)
1994	Jean Ricard Grand Prize for Physics, French Physical Society
1997	Hewlett-Packard Europhysics Prize (together with P. Grünberg and S.S.P. Parkin), European Physical Society (EPS)
2003	Gold Medal, French National Center for Scientific Research
2007	Japan Prize (together with P. Grünberg), Japan Prize Foundation.
2007	Wolf Prize in Physics, Wolf Prize Foundation (together with P. Grünberg)
2007	Nobel Prize in Physics (together with P. Grünberg)

Honors

2008	Grand Officier de l'Ordre du Mérite
2012	Commandeur de la Légion d'Honneur
	Honoris Causa Doctorate of 12 universities

COMMENTARY

The discovery of the Giant Magnetoresistance (GMR) in 1988 separates two different periods of my carrier. Before 1988 was the time of a discreetly active and creative research in which the discovery of the GMR and several of my recent contributions to the development of spintronics have their roots. My second period, after 1988, is characterized by a much more effervescent research activity in the multiple directions of the development of spintronics, an effervescence which continues today in my works on magnetic skyrmions, spin-orbitronics or graphene.

My Ph.D. Thesis (1970) was supervised by Ian Campbell in the Laboratoire de Physique des Solides at Orsay. Its subject was to test the theoretical predictions of Neville Mott on the influence of the orientation of the electron spin on the electrical transport in ferromagnetic metals. By combining several types of experiments, I confirmed this prediction and I also showed that very strong contrasts between the conduction by opposite spin directions can be obtained by doping ferromagnetic metals with selected impurities. Moreover, some of my experiments on ferromagnetic metals doped with two different types of impurity were based on a concept anticipating the concept of GMR in magnetic multilayers. At this time, however, it was not technologically possible to fabricate multilayers with thin enough layers to exploit this concept. It is only in the mid-eighties, with the development in the technologies of microelectronics, that I could come back to the ideas of my thesis and discover the GMR of the magnetic multilayers.

After my Ph.D. and a post.doc. fellowship at Leeds in 1971, I became Assistant Professor at Université Paris-Sud (Orsay) and in 1976, Professor in the same University. Between 1971 and the discovery of the GMR in 1988, as the leader of a small group at the Laboratoire de Physique des Solides in Orsay, I could develop a very active research in many fields of condensed matter physics. I worked on amorphous magnetic systems, weak localization, spin glasses and also, in the continuation of the research of my thesis, on several spin dependent electrical transport phenomena. As my Ph.D. research had been on spin-polarized currents due to the exchange interaction between the spin of conduction electrons and local spins of magnetic materials, I wanted to explore new directions and in particular the poorly known phenomena that can be expected from the spin-orbit interactions (the spin-orbit coupling is a relativistic correction to quantum mechanics). The fashionable topics at this time were the Kondo effect and the spin glasses, and my works on skew scattering, quadrupole scattering, Spin Hall Effect or Dyaloshinskii–Moriya interactions attracted a moderate attention. But I was really happy to pioneer this field. And today, almost thirty years after, I am really happy when I see that the Spin Hall Effect, the Dyaloshinskii–Moriya interactions and other spin-orbit phenomena are becoming important topics in the development of spintronics.

In 1985, at a conference in San Diego, I met my ex-student Alain Friederich who had become the leader of a research group at the company Thomson–CSF. Taking a drink together on the side of the swimming pool, we began to talk about the new technology of Molecular Beam Epitaxy developed at Thomson–CSF to fabricate semiconductor heterostructures. Before the second drink we had decided to collaborate for an extension of this technology to the fabrication of metallic multilayers, one of the objectives being to test some of my ideas on electron transport in magnetic multilayers. The collaboration started in 1986, we fabricated Fe/Cr multilayers in which the relative orientation of magnetizations in successive Fe layers could be switched from antiparallel to parallel and the discovery of the GMR came at the beginning of 1988 (my Wolf and Nobel co-laureate Peter Gruenberg found independently similar results at about the same time). We published our discovery before the end of 1988 in a *Physical Review Letters* presenting both the experimental results and their interpretation in the model of spin-dependent conduction developed during my Ph. D. (my Wolf and Nobel co-laureate Peter Gruenberg found independently similar results also in 1988).

The discovery of the GMR triggered an intense international research activity on spin transport in magnetic nanostructures, a research field that is called today spintronics (often described as a new type of electronics exploiting the magnetism of the electrons). And my activity also took off in the multiple directions of the development of this field. I must say that, until the discovery of the GMR, I had worked with very small teams, a couple of Ph. D. students and a couple of post-docs. A couple of years after the GMR, Agnès Barthélémy and Frederic Petroff got permanent positions to work with me, Frédéric Nguyen Van Dau was recruited by Thomson–CSF, and this small team was the base of the joint laboratory of the CNRS (National Center of Research) and company Thomson–CSF (Thales today) that was created in 1995 before progressively growing until today. I do not want to list all the significant contributions we had on the many axes of the development of spintronics: GMR and TMR (Tunneling Magnetoresistance), spin transport in semiconductors, carbon nanotubes, graphene and organic molecules, spin transfer experiments for the switching of nanomagnets, the current-induced motion of magnetic domain walls or the generation of microwave oscillations, oxitronics (electroresistance of ferroelectric tunnel junctions, multiferroics) and today spin-orbitronics, magnetic skyrmions and neuromorphic components. I was also fortunate in collaborating with excellent theoreticians, Peter Levy at New York University for the development of the first quantum model of the GMR and the theory of the Spin Hall Effect, Thierry Valet at Thomson–CSF for the development of the Valet–Fert model introducing several essential concepts of spintronics, Henri Jaffres, in my team, for our well-known model of spin injection, Jozef Barnas in Posnan for problems of Coulomb blockade and spin transfer, Mairbek Chshief in Grenoble on skew scattering.

In 2007, with successively the Japan Prize, the Wolf Prize, the Nobel Prize and successive ceremonies, speeches, interviews by journalists, I had to slow down my research for a while. It took me two or three years to restore my activity to its usual level. Actually, to restart in a new direction, I came back to some of my ideas in the seventies and eigthies on the role of the relativistic interaction called spin-orbit coupling (SOC) on spin transport and nanomagnetism. This led me, first, in collaboration with Otani's group in Tokyo and with Peter Levy at NYU, to the demonstration of the large spin Hall effects that can be obtained by skew scattering on impurities with large SOC as Bi. I also came back to my work on the Dzyaloshinskii–Moriya interactions (DMI) induced at the interface between a magnetic film and a metal with large SOC. A collaboration with the team of André Thiaville's in Orsay allowed us to show that the DMIs play an essential role in the symmetry of the magnetic domain walls and their current-induced motion. And these DMIs led me also to my recent work on magnetic skyrmions. These new topics, with also the spin/charge conversion by interfacial SOC effects, compose an emerging field of spintronics that begins to be called spin-orbitronics. This research field and particularly studies on magnetic skyrmions will probably be one of my main research topics in the next years.

My final comment is on the (difficult) choice of two favorite papers for this book. The first I have chosen, written a few years after my Ph.D., is a review of all the results of my thesis and it also includes an extended presentation of the two-current model of the conduction in ferromagnetic metals developed by Ian Campbell and me. In particular, it describes the experiments on ternary alloys anticipating the Giant Magnetoresistance (impurities with opposite or similar spin asymmetries corresponding to magnetic layers with opposite or parallel magnetizations) and, more generally, it presents many basic concepts of the exploitation of the spins in electronic transport. The second paper is the *Physical Review Letter* reporting on the discovery of the GMR. If there were three places like on the podium of the Olympic Games, I would have liked to add either the 1993 *Physical Review* article on the Valet–Fert model or my 2013 papers on skyrmions.

J Phys. F: Metal Phys., Vol. 6, No. 5, 1976. Printed in Great Britain. © 1976

Electrical resistivity of ferromagnetic nickel and iron based alloys

A Fert and I A Campbell

Laboratoire de Physique des Solides†, Université de Paris-Sud, 91405 Orsay

Received 1 December 1975

Abstract. We discuss in detail the theoretical basis for the two-band model with spin-mixing which has been widely applied to the analysis of the transport properties of ferromagnetic metals. This model is shown to have much more general validity than the original presentation suggested. The model is then applied to resistivity data in Ni and Fe based alloys to obtain a consistent set of parameters for the scattering within each spin band for various impurities, together with temperature dependent pure metal scattering rates.

1. Introduction

Following Mott's ideas (Mott 1964), we showed some years ago that several transport properties of nickel and iron based ferromagnetic alloys can be explained by assuming conduction in parallel by the spin up and spin down electrons (Campbell *et al* 1967, Fert and Campbell 1968, 1971). The physical basis of this two current model is the dominance of spin-conserving scattering and the weakness of the spin-flip collisions in a ferromagnetic alloy at low temperature. This two current model was taken up by many authors, developed in several directions and sometimes criticized: we refer chiefly to Farrell and Greig (1968), Durand and Gautier (1970), Schwerer and Conroy (1971), Price and Williams (1973), Greig and Rowlands (1974), Dorleijn and Miedema (1975), Jaoul and Campbell (1975) for studies on Ni based alloys and to the work of Loegel and Gautier (1971) on Co based alloys. It seems that the different scattering of the spin ↑ and spin ↓ electrons has been generally confirmed and that the various papers are nearly in agreement as to the ratios $\alpha = \rho_{0\downarrow}/\rho_{0\uparrow}$ of the spin ↑ and spin ↓ resistivities $\rho_{0\uparrow}$ and $\rho_{0\downarrow}$ induced by a given impurity in a given host. On the contrary, opinions are divided as to the occurrence of a temperature-dependent spin-mixing mechanism which would result from scattering by spin waves and would tend to equalize the currents as the temperature is raised, some authors preferring the idea of two independent currents at all temperatures.

In this paper, we report some resistivity measurements on Ni and Fe based alloys and we also review the general pattern of the electrical conduction in these ferromagnetic alloys.

In §2, we place the two current and spin-mixing model on a more solid theoretical basis. In particular, it is shown that the conduction electrons of nickel are far from

---

† Laboratoire associé au CNRS.

850   *A Fert and I A Campbell*

being free electrons (e.g. from the cyclotron resonance performed by Goy and Grimes 1973) and an extension of the previous models to band electrons is needed.

In §3, we describe the experimental procedure used to obtain the results reported in §§4, 5 and 6.

In §4, we report measurements of the residual resistivity of ternary dilute alloys $Ni_{1-x-y}A_xB_y$ and the analysis of the deviations from Matthiessen's rule (MR). This type of study which we introduced for Fe based alloys (Campbell *et al* 1967) has also been done on Ni based alloys by Leonard *et al* (1969) and by Dorleijn and Miedema (1975) and allows a straightforward determination of the impurity resistivities for the spin ↑ and spin ↓ electrons. We interpret the spin ↑ and spin ↓ resistivities of transition impurities on the Friedel model for the ferromagnetic transition alloys.

In §5, we report experimental results on the temperature dependence of the resistivity of $Ni_{1-x}T_x$ alloys (where T is a transition metal, $7 \times 10^{-4} < x < 3 \times 10^{-2}$). The deviations from MR at low temperature are approximately independent of the concentration when the residual resistivity is large enough ($\rho_0 \gtrsim 1 \, \mu\Omega \, cm$) and can be then interpreted fairly well in the two current model. We argue the need of a spin mixing by spin waves to explain the results.

In §6, we report measurements of the temperature dependence of the resistivity of iron based dilute alloys and discuss their interpretation in the two current model.

The two current model has also been used to interpret other transport properties of ferromagnetic alloys: thermo–electric power (Cadeville *et al* 1969, Farrell and Greig 1970), spontaneous anisotropy of resistivity (Campbell *et al* 1970), extraordinary Hall effect (Fert and Jaoul 1972). However, the discussion of these transport properties is not in the scope of this paper and we shall only consider the experimental data concerning the resistivity and its temperature dependence.

2. Model for two current conduction with inter-current scattering

Two current (or two band) models have often been used to describe the electrical conduction in metals. One assumes that two groups of electrons (e.g. belly and neck electrons) carry current *independently* in parallel, which means that one neglects the momentum transfer by scattering from one to the other group.

In ferromagnetic alloys, we separate the electrons into two groups in the same way, but here the groups are spin ↑ and spin ↓ electrons, as scattering with conservation of spin can certainly be assumed to dominate, at least at low temperatures†. We will however improve on the traditional two current models by including the transfer of momentum between the two groups of electrons by spin-flip scattering. We present in this section a model which describes this conduction by two coupled currents. This model provides a generalized justification for a simpler previous model which assumed a conduction band of quasi-free electrons (Campbell *et al* 1967, Fert 1969).

We recall first how the resistivity can be calculated by using the variational method (Ziman 1960). In the notation of Ziman the Boltzmann equation can be written

$$X = P\psi(k\sigma) \tag{1}$$

† As usual, we call spin ↑ the majority spin direction. Note that the separation of the conduction electrons into two groups according to their spin is justified only if the mean free path is much shorter than the size of the magnetic domains. This condition is generally fulfilled except for very pure metals.

Resistivity of ferromagnetic Ni and Fe based alloys 851

where P is the scattering operator and where $\psi(k\sigma)$ is the extra-energy distribution defined by

$$f(k\sigma) = f^0(\epsilon_k) - \psi(k\sigma)(\partial f^0/\partial \epsilon_k). \tag{2}$$

The variational principle (Ziman 1960) leads to the determination of $\psi(k\sigma)$ by minimizing

$$\frac{\langle \psi, P\psi \rangle}{(\langle \psi, X \rangle)^2}. \tag{3}$$

The electrons being divided into two groups according to their spin σ, we choose a trial function $\psi(k\sigma)$ in the form of a linear combination of two functions $\phi_\sigma(k)$ $(\sigma = \pm\frac{1}{2})$:

$$\psi(k\sigma) = \eta_\sigma \phi_\sigma(k) \tag{4}$$

or

$$\psi(k\sigma) = \sum_{\sigma'} \eta_{\sigma'} \phi_{\sigma'}(k)\,\delta_{\sigma\sigma'}. \tag{5}$$

The distributions $\phi_\uparrow(k)$ and $\phi_\downarrow(k)$ are supposed known and independent of the scattering process, while the coefficients $\eta_\uparrow$ and $\eta_\downarrow$ have to be adjusted to minimize (3). This is the implicit hypothesis for almost all two current models: the distribution of relaxation rate within a given electron group is supposed independent of the scatterer type but the relative relaxation of the two groups depends on the scatterer.

The values of η_σ which minimize (3) are given (Ziman 1960) by the equations

$$\mathscr{E}X_\sigma = \sum_{\sigma'} P_{\sigma\sigma'}\,\eta_{\sigma'} \tag{6}$$

with

$$X_\sigma = -\int (v_k \cdot u)\,e\,\frac{\partial f^0}{\partial \epsilon_k}\Phi_\sigma(k)\,\mathrm{d}k \tag{7}$$

$$P_{\sigma\sigma'} = \frac{1}{2k_B T}\sum_{\sigma''\sigma'''}\big[\phi_\sigma(k)\,\delta_{\sigma\sigma''} - \phi_\sigma(k')\,\delta_{\sigma\sigma'''}\big]P(k_{\sigma''}k'_{\sigma'''})$$
$$\times \big[\phi_{\sigma'}(k)\,\delta_{\sigma'\sigma''} - \phi_{\sigma'}(k')\,\delta_{\sigma'\sigma'''}\big]\,\mathrm{d}k\,\mathrm{d}k' \tag{8}$$

where u is the unit vector in the direction of the electric field and where $P(k\sigma'', k'\sigma''')$ is the equilibrium rate between the states $(k\sigma'')$ and $(k'\sigma''')$.

The resistivity ρ is then given (Ziman 1960) by

$$\rho^{-1} = \sum_{\sigma\sigma'} X_\sigma (P^{-1})_{\sigma\sigma'} X_{\sigma'}. \tag{9}$$

If we define

$$\rho_\uparrow = \frac{P_{\uparrow\uparrow}}{X_\uparrow^2} + \frac{P_{\downarrow\downarrow}}{X_\uparrow X_\downarrow} \tag{10}$$

$$\rho_\downarrow = \frac{P_{\downarrow\downarrow}}{X_\downarrow^2} + \frac{P_{\uparrow\downarrow}}{X_\uparrow X_\downarrow} \tag{11}$$

$$\rho_{\uparrow\downarrow} = -\frac{P_{\uparrow\downarrow}}{X_\uparrow X_\downarrow} \tag{12}$$

852   *A Fert and I A Campbell*

one obtains

$$\rho = \frac{\rho_\uparrow \rho_\downarrow + \rho_{\uparrow\downarrow}(\rho_\uparrow \rho_\downarrow)}{\rho_\uparrow + \rho_\downarrow + 4\rho_{\uparrow\downarrow}}. \tag{13}$$

This equation for the resistivity has been already found in the case of two equal groups of free electrons (Campbell *et al* 1967, Fert 1969). The interest of the present derivation is to show that this simple expression for the resistivity is quite general, and that the resistivity in a two-band model including inter-band scattering is always given by an expression of the form (13) with the parameters ρ_σ, $\rho_{\uparrow\downarrow}$ defined by the equations (10) to (12). In fact, although our interest at present is ferromagnetic metals, this expression would apply equally well if the two groups of electrons were for instance the 'neck' and 'belly' electrons of a noble metal. We point out that, after (10), (11) and (12), the resistivities $\rho_\uparrow$, $\rho_\downarrow$ and $\rho_{\uparrow\downarrow}$ depend linearly on the scattering operator, so that different scattering processes provide additive contributions to the three resistivities. This justifies the application of Mathiessen's rule for $\rho_\uparrow$, $\rho_\downarrow$ and $\rho_{\uparrow\downarrow}$. This property results from the assumption that $\phi_\sigma(k)$ is independent of the scatterer type.

In the absence of spin-flip collisions ($P(k_\uparrow, k'_\downarrow) = 0$), $\rho_{\uparrow\downarrow}$ cancels out and one obtains the classical expression of the resistivity in a model with two independent currents

$$\rho = \frac{\rho_\uparrow \rho_\downarrow}{\rho_\uparrow + \rho_\downarrow}. \tag{14}$$

Approximate but practical expressions for $\rho_\uparrow$, $\rho_\downarrow$ and $\rho_{\uparrow\downarrow}$ are obtained if one chooses:

$$\phi_\uparrow(k) = \phi_\downarrow(k) = k \cdot u. \tag{15}$$

($k \cdot u$ is the extra-energy distribution for a spherical Fermi surface and a scattering probability depending only on the scattering angle.) One obtains:

$$\rho_\sigma = \frac{1}{X_\sigma^2 k_B T} \sum_{\sigma'} \int (k \cdot u)[(k - k') \cdot u] P(k\sigma, k'\sigma') \, dk \, dk' \tag{16}$$

$$\rho_{\uparrow\downarrow} = \frac{1}{X_\uparrow X_\downarrow k_B T} \int (k \cdot u)(k' \cdot u) P(k_\uparrow, k'_\downarrow) \, dk \, dk'. \tag{17}$$

The expression (16) for ρ_σ is a classical resistivity expression derived by the variational method (Ziman 1960). Both spin flip and non-spin flip transitions contribute to ρ_σ. It can be remarked that expression (16) for ρ_σ contains the factor $(k - k') \cdot u$ which involves the momentum transferred to the lattice in the transition $k \rightarrow k'$, whereas this factor is replaced in expression (17) for $\rho_{\uparrow\downarrow}$ by $k' \cdot u$ which involves the momentum gained by the final spin direction.

Two current models for ferromagnetic alloys have generally assumed that the two electrons groups are constituted by the spin $\uparrow$ and spin $\downarrow$s electrons and that the d electrons carry a negligible current. However, according to recent data on the band structure and on the s–d hybridization the difference of effective mass between the d and the 's' electrons (with d hybridization) is less marked than it was generally assumed. In our model we suppose that the band of spin σ includes all the electrons with spin σ, with s, d or hybridized character. Stopping at this stage (equation 13), rather than trying to treat the true 'many band' conductor, is meaningful if the additivity rule for the parameters $\rho_\uparrow$, $\rho_\downarrow$ and $\rho_{\uparrow\downarrow}$ is reasonably well obeyed. In practice the rule will remain valid if changes in scatterer produce much weaker effects on the distribution of scattering rates over the Fermi surface than on the ratio of spin ↑ to spin ↓ scattering. If the interband scattering is relatively weak, the dominant mechanism of deviation from MR will then result from the two current conduction.

We can consider briefly the case of Ni. Fermi surface calculations and measurements (Goy and Grimes 1973, Tsui 1967, Wang and Callaway 1974) show that the clear distinction of the simple s–d model between heavy d electron states and light sp states is an over-simplification; there are in reality light $d_↓$ orbits near the point X while some $sp_↓$ band orbits are quite heavy. For the spin ↑ states, the neck orbits are particularly light. In addition, in the presence of impurities we should also consider the different scattering probabilities for different parts of the Fermi surface. In practice, these complicated effects will be hidden inside $\rho_↑$ and $\rho_↓$.

The existence of a spin–orbit effect poses a different problem. Realistic ferromagnetic metal band structures (Wang and Callaway 1974) show Fermi surface regions of hybridized spin ↑ and spin ↓ character. How does this affect the two band approach? We can still include this in the formalism—we can arbitrarily draw the line between spin ↑ and spin ↓ somewhere in the hybridized region, and scatterings across this boundary will contribute both to $\rho_σ$ and to $\rho_{↑↓}$. It is important to remember the existence of this type of contribution to $\rho_{↑↓}$ in the interpretation of experimental results.

We want to conclude this section by a discussion of the several mechanisms of spin-mixing in ferromagnetic Ni or Fe based alloys. Spin-flip occurs first in the collisions with spin waves. This mechanism induces a resistivity $\rho_{↑↓}(T)$ of which the order of magnitude is that indicated by the experimental data (Fert 1969) and is considered as the principal mechanism of spin-mixing. Another contribution to $\rho_{↑↓}(T)$ comes from collisions between spin ↑ and spin ↓ electrons but has been shown to be negligible (Bourguart et al 1968). The spin-mixing term induced by spin wave scattering (and also the electron collisions) vanishes at 0 K, but there is also a residual spin-mixing. First, there is a spin-flip scattering by the impurities due to the spin–orbit coupling, but from EPR data (Monod 1968) the spin-flip cross section of a 3d impurity in Cu is about 100 times smaller than its non-spin flip cross section. The same order of magnitude is expected for the spin-flip cross section of the 3d impurities in Ni or Fe, so that the residual term $\rho_{↑↓}(0)$ induced by this mechanism should be negligible. In fact we have conventionally taken the impurity spin flip scattering to be zero throughout—the effect of a small impurity spin flip term is probably indistinguishable from small changes in the apparent values of $\rho_σ$ for that impurity.

Spin-mixing at 0 K can also result from the combined action of the internal magnetic induction and of the spin–orbit coupling. The resulting spin mixing term is independent of the concentration of impurities and so is only effective in alloys of low residual resistivity. A zero temperature term $\rho_{↑↓}(0)$ introduced phenomenologically explains the behaviour of a number of experimental parameters in dilute alloys in which the residual resistivity is below about 1 μΩ cm (Jaoul and Campbell 1975).

In this paper we concentrate on alloys in the concentration range where the residual spin-mixing can be neglected.

3. Experimental

The alloys were prepared at the CEN Grenoble in a levitation furnace with base metal obtained by zone melting (RRR = 700 and 25 for Ni and Fe respectively). The

854   *A Fert and I A Campbell*

alloys were drawn into wires of diameter 1 mm and annealed at 900°C for the Ni based alloys, 800°C for the iron based alloys. The absence of precipitates was checked by micrography. Most of the binary alloys and all the ternary alloys were analysed by colorimetry or atomic absorption spectrophotometry.

The resistivity was measured between 1·3 K and 77 K and at several temperatures between 77 K and 300 K by a classical DC method (potentiometer + nanovoltmeter). The accuracy of the resistance measurement was generally limited (except for the less concentrated alloys) by the stability of the current ($\pm 2 \times 10^{-5}$). The temperature in the helium range was measured by a germanium resistor with an accuracy of 0·1 K.

4. Residual resistivity of ternary dilute alloys

If the residual spin-mixing is neglected (see discussion at the end of §2), the spin ↑ and spin ↓ electrons of ferromagnetic alloys carry the current independently in the low temperature limit. The residual resistivity of a binary dilute alloy **MA** is then

$$\rho_A = C_A \frac{\rho_{A\uparrow}\rho_{A\downarrow}}{\rho_{A\uparrow} + \rho_{A\downarrow}} \tag{18}$$

where C_A is the concentration of impurities A and where $C_A\rho_{A\uparrow}$ and $C_A\rho_{A\downarrow}$ are the residual resistivities for each spin direction.

We consider now a ternary alloy **MAB** containing a concentration C_A of impurities A and C_B of impurities B. If we suppose that the impurities A and B add their resistivity in each current ($\rho_\sigma = C_A\rho_{A\sigma} + C_B\rho_{B\sigma}$), the model predicts a deviation of the alloy residual resistivity from Mathiessen rule:

$$\Delta\rho = \rho_{AB} - (\rho_A + \rho_B) = \frac{(\alpha_A - \alpha_B)^2 \rho_A \rho_B}{(1 + \alpha_A)^2 \alpha_B \rho_A + (1 + \alpha_B)^2 \alpha_A \rho_B} \tag{19}$$

where $\alpha_A = \rho_{A\downarrow}/\rho_{A\uparrow}$ and $\alpha_B = \rho_{B\downarrow}/\rho_{B\uparrow}$. The analysis of the deviations from MR in alloys **MAB** with several proportions of A and B can be used to determine α_A and α_B.

We have measured the residual resistivity of binary $Ni_{1-x}A_x$ and ternary $Ni_{1-x-y}A_xB_y$ alloys ($7 \times 10^{-4} < x$ or $y < 3 \times 10^{-2}$, A or B = Co, Fe, Mn, Cr, V, Ti). The residual resistivity of most of the alloys was sufficient for the residual spin-mixing to be neglected.

From the measurements on binary alloys we have determined the residual resistivity per at% of various impurities. Then we have measured the residual resistivity of ternary alloys and observed clear deviations from MR which are shown on figure 1 for NiVCo, NiVFe, NiCrMn and NiCrTi alloys. We found a set of parameters α_i to fit the experimental results. The best fit has been obtained with the values α given in table 1† and the agreement between the calculated and experimental points can be seen on figure 1‡. We note that most impurities can be separated into a first group (Co, Fe, Mn) with very high values of α and a second group (Cr, V) for which α is smaller than one. This explains the large deviations from MR for ternary alloys containing an element of each group (e.g. **NiCoV**). On the contrary the

---

† We have chosen the set of parameters given in table 1 rather than the set α^{-1} as being more consistent with the electronic structure of the impurities.

‡ A technique for estimating α values unambiguously from such data is given in the Appendix.

Resistivity of ferromagnetic Ni and Fe based alloys 855

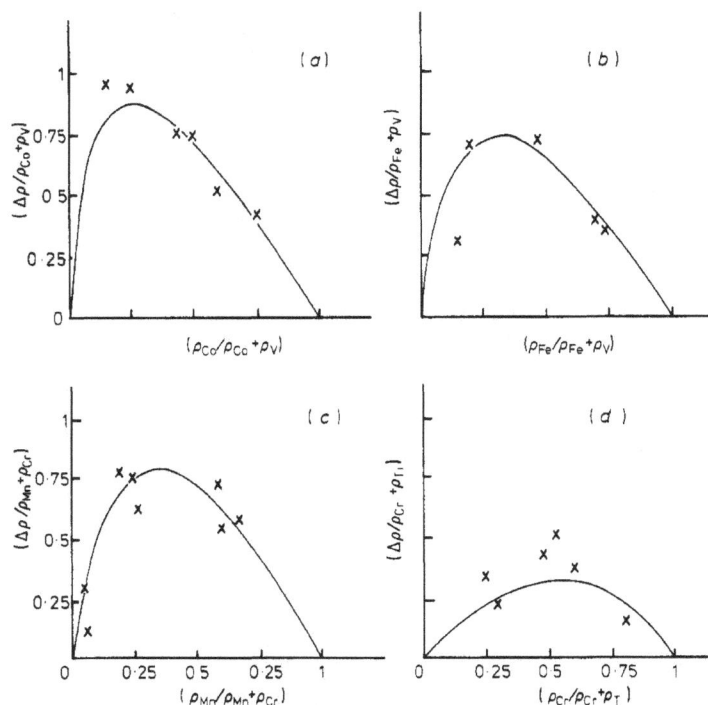

Figure 1. Deviations from Matthiessen's rule in Ni based ternary alloys (relative deviations against relative resistivities). (*a*), **NiCoV**; (*b*), **NiFeV**; (*c*), **NiMnCr**; (*d*), **NiCrTi**. The full curves are calculated from expression (19) with the values of α listed in table 1.

deviations are negligible when both impurities have nearly the same ratio α (this has also been observed by Dorleijn and Miedema (1975)). The latter result shows that additional deviations from MR induced, for instance, by the anisotropy of the relaxation on the Fermi surface are much smaller than the deviations linked with the spin dependent scattering.

The results for the spin ↑ and spin ↓ resistivities of the impurities

$$\rho_{A\uparrow} = \frac{1}{1 + \alpha_A}\rho_A \qquad \rho_{A\downarrow} = \frac{\alpha_A}{1 + \alpha_A}\rho_A$$

Table 1. The residual resistivity per at%, the parameter $\alpha = \rho_{0\downarrow}/\rho_{0\uparrow}$ and the spin ↑ and spin ↓ residual resistivities per at% for 3d impurities in nickel.

Impurity	Ti	V	Cr	Mn	Fe	Co
Resistivity ρ_0 ($\mu\Omega$ cm/at%)	2·9	4·5	5·0	0·61	0·35	0·145
$\alpha = \rho_{0\downarrow}/\rho_{0\uparrow}$	4	0·55	0·45	15	20	30
	$(3 < \alpha < 5)$	$(0.5 < \alpha < 0.6)$	$(0.35 < \alpha < 0.5)$	$(11.5 < \alpha < 17)$	$(15 < \alpha < 23)$	$(23 < \alpha < 33)$
$\rho_{0\uparrow}$ ($\mu\Omega$ cm/at%)	3·6	12·7	16·1	0·65	0·37	0·15
$\rho_{0\downarrow}$ ($\mu\Omega$ cm/at%)	14·5	7·0	7·2	9·8	7·4	4·6

856   *A Fert and I A Campbell*

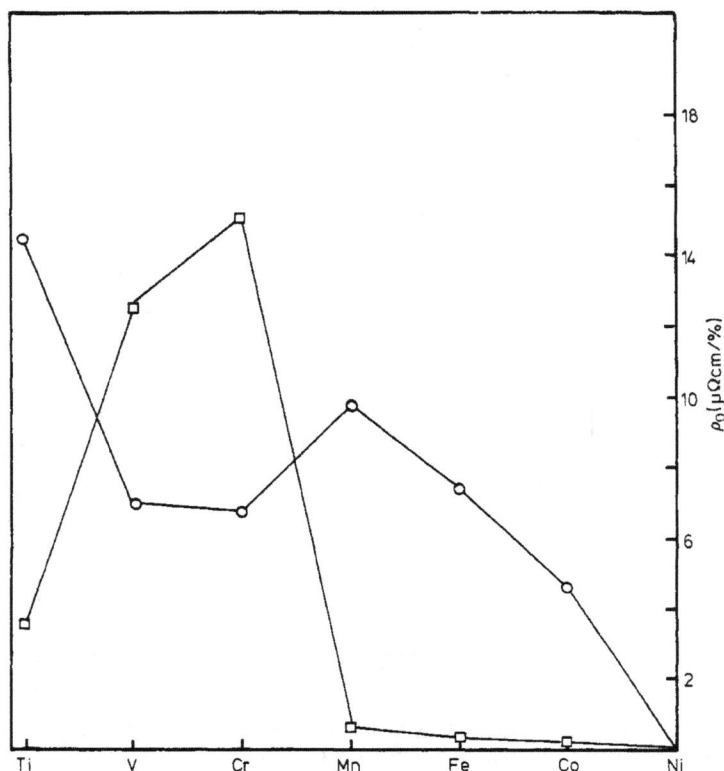

Figure 2. The resistivity of 3d impurities in nickel for each spin direction. □, $\rho_{0\uparrow}$; ○, $\rho_{0\downarrow}$.

are also listed in table 1 and are plotted on figure 2. These results are in approximate agreement with those derived by Leonard *et al* (1969) and by Dorleijn and Miedema (1975) from the residual resistivity of ternary alloys.

The interpretation of the resistivities $\rho_{0\uparrow}$ and $\rho_{0\downarrow}$ of 3d impurities (figure 2) on the basis of the Friedel model (Friedel 1967) is well known. We shall describe it briefly.

When the difference in the number of 3d electrons between the nickel and the impurity is large (Cr, V, Ti) a spin $\uparrow$ d bound state is repelled above the spin $\uparrow$ d band. This explains, for instance, that the magnetic moment of Cr, V or Ti impurities is opposite to the nickel moment. The resonance of the $s_\uparrow$ electrons with the bound state (formation of a virtual bound state) explains the peak of $\rho_{0\uparrow}$ around Cr (figure 2). For **NiCr** one can deduce from the change in magnetic moment $(\mathrm{d}\mu/\mathrm{d}C \simeq -4\,\mu_B$ after Collins and Low 1965) that the centre of the virtual bound state (VBS) is above E_F and that only one of the five d$\uparrow$ states is occupied. The phase shift of the spin $\uparrow$ d partial waves at E_1 is then $\eta_2 = \pi/5$ and for 0·3 conduction electrons per atom and per spin direction, the expected resistivity per at% is

$$\rho_{0\uparrow} = 10^{-2} \times \frac{20\pi\hbar}{nk_\mathrm{F}} \sin^2 \eta_2 \simeq 41\,\mu\Omega\,\mathrm{cm}$$

The experimental value appears somewhat lower: 16·1 $\mu\Omega$ cm.

Resistivity of ferromagnetic Ni and Fe based alloys   857

For V and Ti, $\rho_{0\uparrow}$ remains rather large, which seems to show that, even for Ti, the VBS is not repelled well above the Fermi level, in agreement with the conclusions of the work of Caudron *et al* (1973) on the specific heat of Ni based alloys.

For Co, Fe and Mn impurities $\rho_{0\uparrow}$ is very small. This is due to the presence of only s states at E_{F} for the spin $\uparrow$ direction and, in the absence of resonance effects, to the weakness of s–s scattering.

The resistivities $\rho_{0\downarrow}$ are fairly large for all the impurities, which certainly results from the high density of s and d states at E_1 for the spin $\downarrow$ direction and from the resulting strong scattering of the spin $\downarrow$ electrons in these states.

5. Variation of the resistivity of binary nickel based alloys with temperature

5.1. Dependence on temperature expected from the two current model

At finite temperature T, the resistivity ρ_σ for the current of spin σ electrons can be written as the sum of residual and T dependent terms:

$$\rho_\sigma = \rho_{0\sigma} + \rho_{i\sigma}(T). \tag{20}$$

In addition the occurrence of spin non-conserving scattering can be expressed by a spin-mixing resistivity $\rho_{\uparrow\downarrow}(T)$. We suppose $\rho_{\uparrow\downarrow}(0) = 0$ (we have discussed this assumption at the end of §2). One derives from (13):

$$\rho(T) = \frac{[\rho_{0\uparrow} + \rho_{i\uparrow}(T)][\rho_{0\downarrow} + \rho_{i\downarrow}(T)] + \rho_{\uparrow\downarrow}(T)[\rho_{0\uparrow} + \rho_{i\uparrow}(T) + \rho_{0\downarrow} + \rho_{i\downarrow}(T)]}{\rho_{0\uparrow} + \rho_{i\uparrow}(T) + \rho_{0\downarrow} + \rho_{i\downarrow}(T) + 4\rho_{\uparrow\downarrow}(T)}. \tag{21}$$

In principle, using sets of alloys of different concentrations, the different parameters can be estimated independently at each temperature (see Appendix). We will limit the discussion here to the more qualitative aspects.

In the temperature range where the residual resistivities $\rho_{0\uparrow}$ and $\rho_{0\downarrow}$ are much larger than $\rho_{i\uparrow}(T)$, $\rho_{i\downarrow}(T)$ and $\rho_{\uparrow\downarrow}(T)$ (e.g. $T \lesssim 45\,\mathrm{K}$ for $\rho_0 \simeq 1\,\mu\Omega\,\mathrm{cm}$) one can limit the expression (21) to the terms of first order in $\rho_{i\sigma}(T)$ and $\rho_{\uparrow\downarrow}(T)$. One obtains

$$\rho_T(T) = \rho(T) - \rho_0 = \left(1 + \frac{(\alpha - \mu)^2}{(1 + \alpha)^2\mu}\right)\rho_i(T) + \frac{(\alpha - 1)^2}{(\alpha + 1)^2}\rho_{\uparrow\downarrow}(T) \tag{22}$$

where:
   (i) ρ_0 is the residual resistivity ($\rho_0 = \rho_{0\downarrow}\rho_{0\downarrow}/(\rho_{0\uparrow} + \rho_{0\downarrow})$) and $\alpha = \rho_{0\downarrow}/\rho_{0\uparrow}$.
   (ii) $\rho_i(T)$ is defined by

$$\rho_i(T) = \frac{\rho_{i\uparrow}(T)\rho_{i\downarrow}(T)}{\rho_{i\uparrow}(T) + \rho_{i\downarrow}(T)}$$

and

$$\mu = \rho_{i\downarrow}(T)/\rho_{i\uparrow}(T).$$

The temperature dependent scattering (by phonons and magnons) is then characterized by $\rho_{\uparrow\downarrow}(T)$, $\rho_i(T)$ and μ. The parameter μ should vary slowly with the temperature and, in a first approximation, will be supposed constant in the helium temperature range. We point out that $\rho_i(T)$ is not the resistivity ρ_T of the pure metal ($\rho_T = \rho(T) - \rho_0$). Actually the resistivity ρ_T of a 'pure' ferromagnetic metal is very indirectly linked to $\rho_i(T)$. First, in metals of usual purity, the dominant scattering in the helium temperature range is the scattering by the residual impurities and the resistivity ρ_T should depend on the impurity type in the same way as that of definite alloys. Secondly the resistivity of pure ferromagnetic metals is enhanced (Schwerer and Silcox 1968) by the magnetoresistance effect in the internal induction ($4\pi M$ + demagnetizing field); from Kohler's rule this enhancement vanishes for concentrated alloys. Both mechanisms certainly explain the very wide spread of the experimental data for the resistivity of pure nickel at low temperature: from $14 \times 10^{-12} T^2 \, \Omega$ cm (Schwerer and Silcox 1968) to $26 \times 10^{-12} T^2 \, \Omega$ cm (White and Tainsh 1967). After correction for the enhancement by magnetoresistance, Schwerer and Silcox (1968) find $9.5 \times 10^{-12} T^2 \, \Omega$ cm. It turns out that the resistivity of the pure metal can be only of little help in determining $\rho_i(T)$. We shall rather consider $\rho_i(T)$ as an unknown function to determine from the data on the resistivity of dilute alloys.

Expression (22) predicts that, at low temperature, ρ_T (or the deviation from MR) depends on the type of impurities in solution (*via* α) but not on their concentration. If the residual spin-mixing term $\rho_{\uparrow\downarrow}(0)$ cannot be neglected (low resistivity alloys) it can be shown (Jaoul and Campbell 1975) that the expression (22) for the resistivity at low temperature still holds, provided that α is replaced by a concentration dependent α'

$$\alpha' = \frac{\rho_{0\downarrow} + 2\rho_{\uparrow\downarrow}(0)}{\rho_{0\uparrow} + 2\rho_{\uparrow\downarrow}(0)}.$$

At high temperatures (room temperature and above), if $\rho_{i\uparrow}(T)$, $\rho_{i\downarrow}(T)$ and $\rho_{\uparrow\downarrow}(T)$ dominate $\rho_{0\uparrow}$ and $\rho_{0\downarrow}$, the main contribution of the impurities arises from terms of first order in $\rho_{0\uparrow}$ and $\rho_{0\downarrow}$ in the development of (21), and so the deviations from MR are predicted to be proportional to the impurity concentration.

Finally, as the Curie temperature is approached, other effects will come into play which we do not discuss here.

5.2. General features of the experimental deviations from Matthiessen's rule

5.2.1. A typical behaviour of the deviations is shown on figure 3 for two NiCr alloys; the deviations are observed to be independent of concentration at low temperature and approximately proportional to the concentration at high temperature. The two current model prediction is thus obeyed.

5.2.2. The deviations at low temperature are nearly concentration independent in a wide concentration range but begin to drop for alloys of low residual resistivity (figure 4). This effect has already been observed in NiCu alloys by Greig and Rowlands (1974) and can be explained by a zero temperature spin-mixing $\rho_{\uparrow\downarrow}(0)$ (Jaoul and Campbell 1975). We will consider as far as possible only the concentration range where the deviations are nearly independent of the concentration.

5.2.3. NiMn is a special case: one observes in figure 4 that $\rho_\uparrow$ increases regularly with the concentration. An additional mechanism for deviation from MR certainly occurs (Mills *et al* 1971, Rowlands 1973). But, even for NiMn, there is a range of residual resistivity ($0.3 \, \mu\Omega$ cm $< \rho_0 < 0.7 \, \mu\Omega$ cm) where the deviations from MR depend relatively little on the concentration and can be explained by a two current conduction.

Resistivity of ferromagnetic Ni and Fe based alloys 859

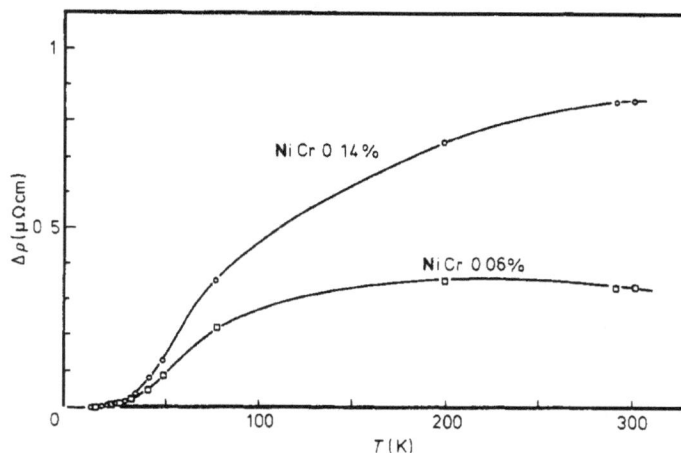

Figure 3. The deviation from Mattheissen's rule against the temperature for two NiCr alloys.

5.3. Resistivity of nickel based alloys at low temperature

We only consider the experimental data in the range of residual resistivity where ρ_T (that is $\rho(T) - \rho_0$) is, at least approximately, independent of the impurity concentration (figure 4). For **NiCo**, **NiFe**, **NiMn**, **NiCr**, **NiV** and **NiCrMn** alloys, we have obtained a good fit of the experimental data with the expression (22) of $\rho_\uparrow$ by taking:

(i) the values of α derived from the measurements on ternary alloys (§4) and listed in table 1 (for **Ni** + 0·07 at% Cr + 0·2 at% Mn the effective value of α calculated from table 1 is 1·5).

(ii) $\mu = 3·6$.

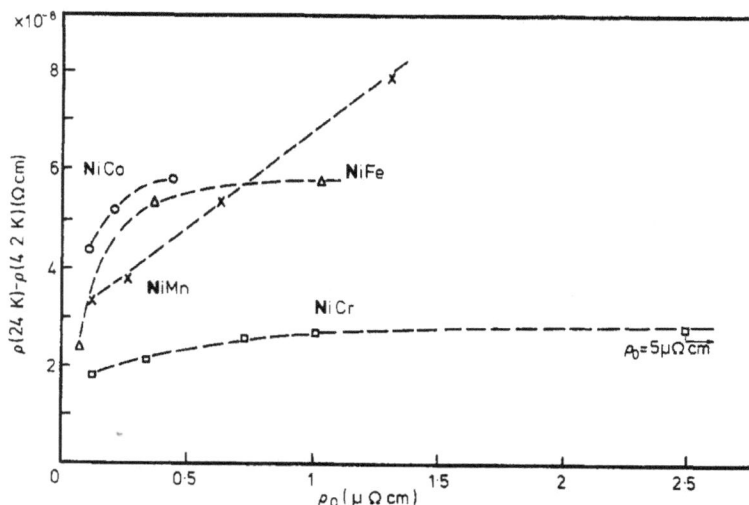

Figure 4. The resistivity increase between 4·2 K and 24 K against the concentration for several types of alloys.

860   *A Fert and I A Campbell*

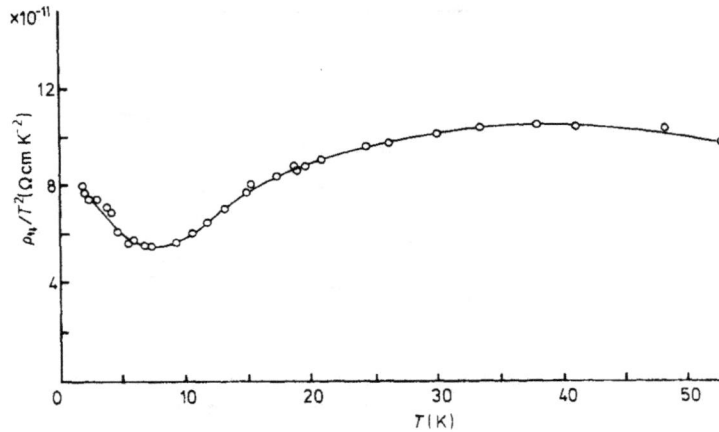

Figure 5. The ratio of $\rho_{\uparrow\downarrow}(T)$ to T^2 against T for Ni. The circles correspond to values obtained using equation (22) and data from Ni + 3 at% Fe. The curve corresponds to the function $\rho_{\uparrow\downarrow}(T)$ which has been adopted for the interpretation of the experimental results for all the nickel based alloys.

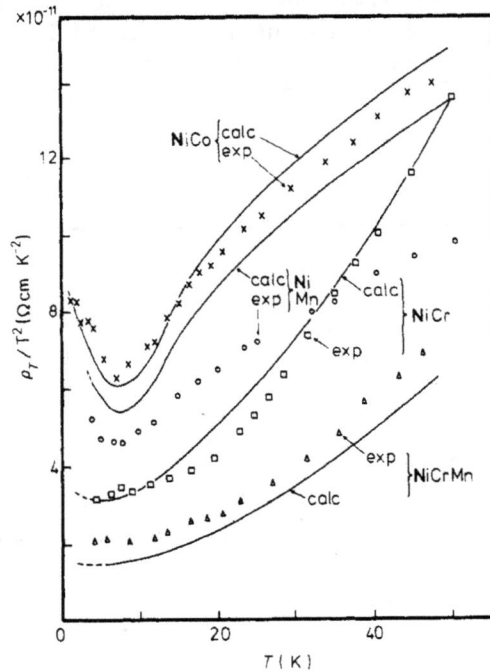

Figure 6. The ratio ρ_T/T^2 is plotted against T for: ×, Ni + 3 at% Co ($\rho_0 = 0.44\ \mu\Omega$ cm); $\bigcirc$, Ni + 0.4 at% Mn ($\rho_0 = 0.25\ \mu\Omega$ cm); $\square$, Ni + 0.14 at% Cr ($\rho_0 = 0.73\ \mu\Omega$ cm) and $\triangle$, Ni + 0.07 at% Cr + 0.2 at% Mn ($\rho_0 = 0.82\ \mu\Omega$ cm). The full lines have been calculated from expression (22) with the parameters indicated in the text.

(iii) $\rho_i(T) = 9.5 \times 10^{-12} T^2 + 1.7 \times 10^{-14} T^4$ (in $\Omega\,\text{cm}$ if T in K).

(iv) the spin-mixing resistivity $\rho_{\uparrow\downarrow}(T)$ which yields an accurate fit for the alloy Ni + 3 at% Fe. The function $\rho_{\uparrow\downarrow}(T)/T^2$ is shown on figure 5.

On figure 6 we have plotted the experimental and calculated values of ρ_T/T^2 for NiCo, NiMn, NiCr and NiCrMn alloys (a plot of ρ_T/T^2 instead of ρ_T appeared more convenient; the figure does not include the plots for NiFe and NiV which are little different from those for NiCo and NiCr respectively). The model explains why ρ_T becomes very large when α is very different from unity, that is for NiCo ($\alpha \simeq 30$), NiFe ($\alpha \simeq 20$), NiMn ($\alpha \simeq 15$); this comes from the term $(\alpha - 1)^2/[(\alpha + 1)^2]\rho_{\uparrow\downarrow}(T)$ in expression (22). The model also explains the very different form of temperature dependence of ρ_T/T^2 according to whether $\alpha \lesssim 1$ or $\alpha \gg 1$. For $\alpha < 1$ or $\alpha \simeq 1$ (NiCr, NiCrMn and also NiV not plotted on figure 6) with $\mu = 3.6$, the dominant term in (22) is $\{1 + [(\alpha - \mu)^2/(1 + \alpha)^2\mu]\}\rho_i(T)$ and the temperature dependence of ρ_T reflects that of $\rho_i(T)$; on the contrary for $\alpha \gg 1$ (NiCo, NiFe, NiMn) the dominant term in (22) is $[(\alpha - 1)^2/(\alpha + 1)^2]\rho_{\uparrow\downarrow}(T)$ and the temperature dependence of ρ_T reflects that of $\rho_{\uparrow\downarrow}(T)$. We note that the calculated curves reproduce fairly well the features of the experimental plot: curvature, crossing of the curve of NiCr with those of NiMn and NiCo etc. The experimental curves for NiMn and NiCrMn are somewhat below the calculated curves but these alloys are less concentrated and we believe that the discrepancy can be mostly ascribed to the $\rho_{\uparrow\downarrow}(0)$ mechanism.

5.4. *Need for spin-mixing*

It has been suggested by Greig and Rowlands (1974) that the temperature dependence of the resistivity of Ni based alloys could as well be explained in a model with two independent currents (without spin-mixing). According to these authors, the experimental data could be explained by the first term of (22) alone

$$\rho_T = \left(1 + \frac{(\alpha - \mu)^2}{(1 + \alpha)^2 \mu}\right)\rho_i(T).$$ (23)

A value of μ much smaller than 1 is then needed to explain the high values of ρ_T for NiCo, NiFe, NiMn ($\alpha \gg 1$) and also its rather high value for NiCr ($\alpha < 1$). Thus, Greig and Rowlands (1974) obtained a reasonable agreement with their experimental data for NiFe and NiCr with $\mu = 0.2$, $\alpha_{Fe} = 8$, $\alpha_{Cr} = 0.8$. We have several objections to this interpretation:

(i) With $\mu = 0.2$ the expression (23) predicts that ρ_T of NiCrMn alloys should progressively increase from NiCr to NiMn (as $0.2 < \alpha_{Cr} < \alpha_{Mn}$). This is in contradiction to our observation for NiCrMn of a resistivity ρ_T definitely smaller than that for NiCr and NiMn.

(ii) Greig and Rowlands (1974) took $\rho_i(T)$ equal to the resistivity of a pure nickel specimen ($15 \times 10^{-12} T^2 \,\Omega\,\text{cm}$ in the low temperature limit). But Schwerer and Silcox (1967) have shown that the resistivity of pure nickel is enhanced by a magnetoresistance effect in the internal induction and they found that the resistivity of nickel in zero induction would be definitely lower ($\rho_{Ni}(B = 0) = 9.5 \times 10^{-12} T^2$ at low temperature). As the effect of internal induction can be neglected in alloys (on account of Kohler's rule), the term $\rho_i(T)$ in expressions (22) or (23) is to be taken in zero induction and should not exceed the resistivity $\rho_{Ni}(B = 0)$ found by Schwerer and

Silcox. A choice of $\rho_i(T)$ smaller than $\rho_{Ni}(B = 0)$ leads to larger values of $\rho_T/\rho_i(T)$ which cannot be explained by expression (23).

(iii) The values of α estimated on the zero spin-mixing assumption, particularly the value for Cr, are incompatible with values obtained from the ternary alloy data (see §4 and the Appendix).

5.5. Interpretation of the spin-mixing term

It has been shown that the spin-mixing could not be due to collisions between electrons with opposite spins but should be ascribed to electron–magnon collisions (Fert 1969, Mills *et al* 1971). The contribution from the electron–magnon collisions has been calculated on an s–d model and can be written

$$\rho_{\uparrow\downarrow}(T) = \frac{3\pi}{8} \frac{mSJ^2 k_B T}{Ne^2 \hbar \epsilon_F D k_F^2} f\left(\frac{\epsilon_m}{k_B T}\right)$$

where

$$f(x) = \left(\frac{x}{1 - e^{-x}} - \lg(e^x - 1)\right) = \begin{cases} -\ln x, & x \ll 1 \\ xe^{-x}, & x \gg 1. \end{cases}$$

S is the spin of the d local moment, J the constant of the s–d exchange, D the spin-wave stiffness constant, $\epsilon_m = \hbar D q_1^2$ where q_1 is the gap between the spin $\uparrow$ and spin $\downarrow$ Fermi surfaces in k space (in the calculation q_1 is considered constant but should actually correspond to a mean value of the gap).

The function $\rho_{\uparrow\downarrow}(T)/T^2$ has a broad maximum (Fert 1969) at $T \simeq \epsilon_m/k_B$ and this maximum could correspond to that observed on the experimental plot of $\rho_{\uparrow\downarrow}/T^2$ (figure 5). Taking D from Shirane *et al* (1968), $S = 0.3$, $SJ = 0.12$ eV, $\epsilon_F = 5.4$ eV, $q_1 = \frac{1}{20} k_F$ one finds:

$$\epsilon_m/k_B = 17 \text{ K}$$

with a value of $\rho_{\uparrow\downarrow}/T^2$ at the maximum

$$(\rho_{\uparrow\downarrow}/T^2)_{max} = 0.38 \times 10^{-14} \ \Omega \text{ cm} \ (T \text{ in K}).$$

In comparison the maximum of $\rho_{\uparrow\downarrow}/T^2$ on the experimental plot of figure 5 is about $10^{-14} \ \Omega$ cm and occurs at about 40 K.

At low temperature the model with two spin $\uparrow$ and spin $\downarrow$ Fermi surfaces separated by q_1 predicts an exponential decrease of $\rho_{\uparrow\downarrow}(T)$ as $\exp[-\epsilon_m/k_B T]$. In contrast, if

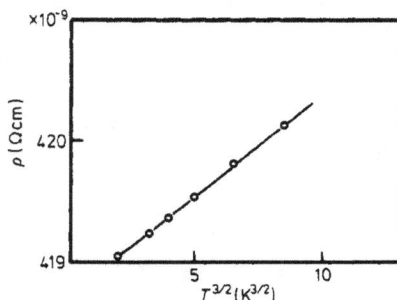

Figure 7. Resistivity of a NiCo 3 at% alloy against $T^{3/2}$.

Resistivity of ferromagnetic Ni and Fe based alloys 863

Table 2. Experimental and calculated deviations from MR in several nickel base alloys at 77 K, 200 K and 300 K [$\Delta\rho = \rho$ (alloy) $-\rho$ (residual) $-\rho$ (pure Ni)].

Alloy	Residual Resistivity $\rho_0(\mu\Omega\,cm)$	77 K		200 K		300 K	
		$(\Delta\rho/\rho_0)_{cal}$	$(\Delta\rho/\rho_0)_{exp}$	$(\Delta\rho/\rho_0)_{cal}$	$(\Delta\rho/\rho_0)_{exp}$	$(\Delta\rho/\rho_0)_{cal}$	$(\Delta\rho/\rho_0)_{exp}$
NiCo3at%	0·44	1·02	1·00	2·27	2·43	3·11	3·38
NiFe3at%	1·04	0·44	0·44	1·19	1·25	1·77	1·70
NiMn1at%	0·63	0·54	0·51			1·24	1·49
NiCr1at%	4·96	0·14	0·138	0·44	0·46	0·58	0·60
NiCr0·14at%	0·73	0·42	0·47	0·50	0·90	0·75	0·95

one supposes that the spin ↑ and spin ↓ Fermi surfaces touch or are very near in some places, a $T^{3/2}$ law is expected. The experimental results support the second hypothesis: the upturn of $\rho_{\uparrow\downarrow}/T^2$ below 10 K corresponds approximately to a $T^{3/2}$ dependence at low temperature. It can be checked on figure 7 that the temperature dependence of ρ_T for **NiFe** is actually very little different from a $T^{3/2}$ dependence.

We conclude that collisions with spin-waves can explain the order of magnitude of $\rho_{\uparrow\downarrow}(T)$ but that a precise interpretation of the variation of $\rho_{\uparrow\downarrow}(T)$ with T is very problematic.

5.6. Deviations from Matthiessen's rule at 77 K, 200 K and 300 K

The deviations from MR at 77 K, 200 K and 300 K for several alloys (table 2) have been compared to the deviations calculated from expression (21). The impurity parameters ($\rho_{0\uparrow}, \rho_{0\downarrow}$) being those derived from the residual resistivity of ternary alloys (table 1), the best fit is obtained with the host parameters [$\mu, \rho_i(T), \rho_{\uparrow\downarrow}(T)$] listed in table 3. The agreement between experimental and calculated deviations (table 2) is reasonably good but the determination of μ, $\rho_i(T)$ and $\rho_{\uparrow\downarrow}(T)$ does not turn out to be very accurate; sets of parameters in which μ is increased somewhat while $\rho_{\uparrow\downarrow}(T)$ is decreased (or vice versa) give almost as good fits. We find μ always larger than 1, which could be due, independently of the scattering mechanisms, to the lighter effective masses of the spin ↑ electrons. It appears also that the spin-mixing term $\rho_{\uparrow\downarrow}(T)$, up to 300 K, does not become much larger than $\rho_i(T)$ and that, at RT, the two currents are not yet completely mixed.

6. Resistivity of iron based alloys

We have measured the resistivity of **FeNi** ($C = 1$ and 2 at%), **FeMn** ($C = 1\cdot4$ and 1·9 at%), **FeCr** ($C = 0\cdot3$ and 0·6 at%) between 1·5 K and 300 K and the resistivity

Table 3. Values of the parameters μ, $\rho_i(T)$, $\rho_{\uparrow\downarrow}(T)$ which have been used to interpret the deviations from MR at 77 K, 200 K, 300 K in nickel based alloys.

Temperature	$\rho_{\uparrow\downarrow}(T)$ ($\mu\Omega\,cm$)	$\rho_i(T)$ ($\mu\Omega\,cm$)	μ	$\rho_{i\uparrow}(T)$ ($\mu\Omega\,cm$)	$\rho_{i\downarrow}(T)$ ($\mu\Omega\,cm$)
77	0·9 ± 0·3	0·32	5 ± 1	0·38	1·9
200	5 ± 2	2·5	5 ± 2	3	15
300	11 ± 4	5·4	4 ± 2	6·7	27

864   *A Fert and I A Campbell*

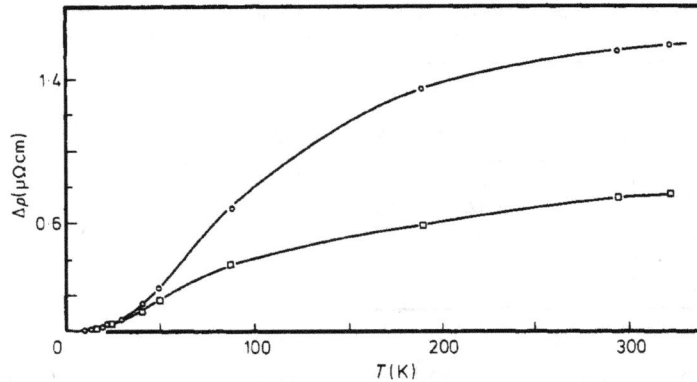

Figure 8. Deviation from Matthiessen's rule against T for two FeCr alloys: O, FeCr 0·6% and ☐, FeCr 0·3%.

of FeTi ($C = 1$ at%), FeV ($C = 1·8$ at%), FeCo ($C = 1$ at%), FeSi ($C = 1$ at%) at 4·2 K and 293 K. The results for the room temperature deviations are in good agreement with those of Arajs *et al* (1969).

The main features of the deviations from MR are similar to those of Ni based alloys. Figure 8 shows the deviations from MR for two FeCr alloys: one observes that the deviations are approximately independent of the concentration C below 30 K and proportional to C at room temperature. The same behaviour is observed for FeMn and FeNi alloys.

Figure 9 shows $\rho_T (\rho_T = \rho(T) - \rho_0)$ at low temperature for two FeCr alloys (0·3 and 0·6 at%) and pure Fe. The resistivity ρ_T of the FeCr alloys turns out to be about five times the resistivity of pure iron and nearly the same for the two alloys. We however have only studied two alloys for each type of impurity and we do not know if ρ_T remains approximately independent of C in a wider concentration range. Figure 10 shows the resistivity ρ_T of FeMn, FeNi alloys and pure Fe: for FeMn ρ_T is larger than the resistivity of pure iron by about a factor of six and for FeNi only a little larger.

These main features of ρ_T for the Fe based alloys suggest that the measurements can be interpreted in a two-current model, as for the Ni based alloys. However we have not measured the residual resistivities of ternary alloys† to determine independently the ratios α of the impurities, and all the parameters of the two current model [$\alpha, \mu, \rho_i(T), \rho_{\cdot}(T)$] must be derived from the data on binary alloys only. We have tentatively chosen $\mu = 1$, which seems a reasonable value as the spin ↑ and spin ↓ densities of states at the Fermi level are not very different in the conduction band of iron. We have also assumed $\rho_i(T) = \rho$ (pure Fe) without correction for the magnetoresistance effect in the internal field. We then obtained the best fit between the experimental and calculated (from expression 22) values of ρ_1 at low temperature with

$$\alpha_{Cr} = \tfrac{1}{11} \text{ (or } 11) \qquad \alpha_{Mn} = \tfrac{1}{6} \text{ (or } 6) \qquad \alpha_{Ni} = 3 \text{ (or } \tfrac{1}{3}).$$

It can be checked on figure 11 that the fit of expression (22) to the experimental data for FeCr and FeMn yield very nearly the same values of $\rho_{\uparrow\downarrow}(T)$. We keep

† Except a few measurements reported in a previous paper (Campbell *et al* 1967).

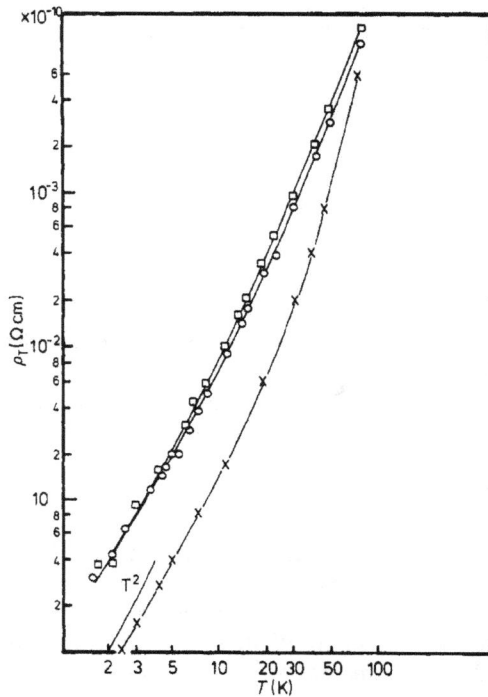

Figure 9. Temperature dependent part of the resistivity against T for: ○, FeCr 0·3% and □, FeCr 0·6% alloys and ×, pure iron at low temperature (logarithmic plot).

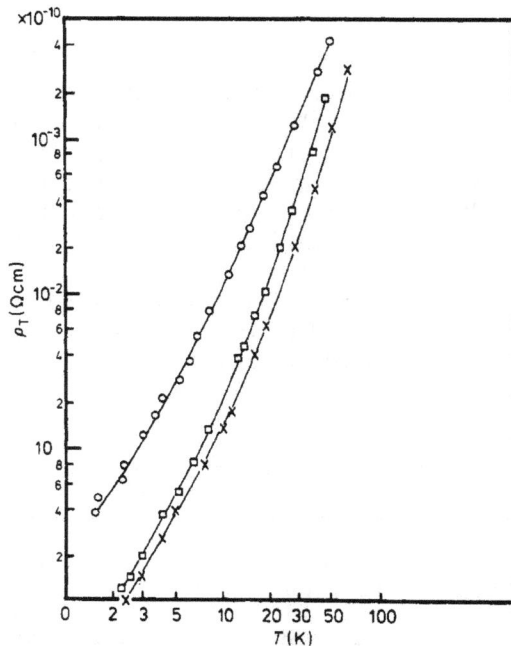

Figure 10. Temperature dependent part of the resistivity against T for: ○, FeMn; □, FeNi alloys and ×, pure Fe (logarithmic plot).

866 *A Fert and I A Campbell*

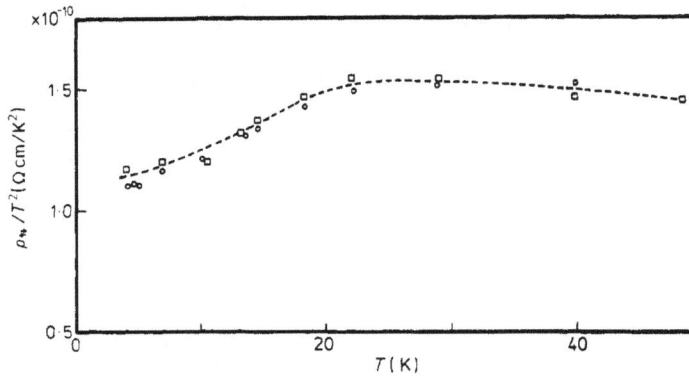

Figure 11. $\rho_{\uparrow}(T)/T^2$ against T for Fe. The circles and the squares correspond to the values of $\rho_{\uparrow\downarrow}(T)$ which fit expression (22) with the experimental data for, respectively, FeMn and FeCr alloys.

the values $\alpha_{Cr} = \frac{1}{11}$, $\alpha_{Mn} = \frac{1}{6}$, $\alpha_{Ni} = 3$ which agree better with the Friedel model for ferromagnetic transition alloys than the inverse values.

The deviations from MR at room temperature for FeCr, FeMn and FeNi alloys are listed in table 4. The impurity resistivity of these alloys being definitely smaller than their thermal resistivity at RT, the deviation from MR at RT is predicted by the two current model (with $\mu = 1$) to be approximately:

$$\frac{\Delta\rho_{RT}}{\rho_0} = \frac{(\alpha - 1)^2}{4\alpha}. \tag{24}$$

The values of $\Delta\rho_{RT}/\rho_0$ have been calculated with $\alpha_{Cr} = \frac{1}{11}$, $\alpha_{Mn} = \frac{1}{6}$, $\alpha_{Ni} = 3$ and are listed in table 4. The agreement with the experimental values is fairly good.

For FeTi, FeV, FeCo, FeSi, the only data are the deviations from MR at RT: $\Delta\rho/\rho_0 = 0.58$ for FeTi (1 at%), $\Delta\rho/\rho_0 = 1.5$ for FeV (1.8 at%) $\Delta\rho/\rho_0 \lesssim 0$ for FeCo (1 at%), $\Delta\rho/\rho_0 \lesssim 0$ for FeSi (1 at%). Assuming $\Delta\rho/\rho_0$ at RT given by expression (24), we have derived the values of α for each impurity. The values of α (after a choice between α and α^{-1}), $\rho_{0\uparrow}$, $\rho_{0\downarrow}$ are listed in table 5. The resistivities $\rho_{0\uparrow}$ and $\rho_{0\downarrow}$ are plotted on figure 11.

The results of table 5 or figure 11 can be interpreted in the Friedel model for the ferromagnetic transition alloys. Friedel (1967) showed that, in the case of an unfilled d band and of a repulsive potential, resonance effects occur if the Fermi

Table 4. Experimental and calculated deviations from MR in several iron based alloys at 77 K, 200 K and 300 K [$\Delta\rho = \rho$ (alloy) $- \rho$ (residual) $- \rho$ (pure Fe)].

Impurity	Cr		Mn		Ni	
Concentration (at%)	0.3	0.6	1.4	1.9	1	2
Residual resistivity ($\mu\Omega$ cm)	0.8	1.5	2.4	3.2	1.9	2.6
$(\Delta\rho/\rho_0)_{exp}$	0.82	0.99	2.14	2.02	0.47	0.38
$(\Delta\rho/\rho_0)_{cal} = (\alpha - 1)^2/4\alpha$	1.04		2.27		0.33	

Resistivity of ferromagnetic Ni and Fe based alloys 867

Table 5. The residual resistivity per at%, the parameter $\alpha = \rho_{0\downarrow}/\rho_{0\uparrow}$ and the spin $\uparrow$ and spin $\downarrow$ residual resistivities per at% for impurities in iron.

Impurity	Ti	V	Cr	Mn	Co	Ni
Resistivity						
$\rho_0 (\mu\Omega\,\text{cm/at}\%)$	2·9	1·4	2·6	1·7	2	1·8
$\alpha = \rho_{0\downarrow}/\rho_{0\uparrow}$	$\frac{1}{4}$	$\frac{1}{8}$	$\frac{1}{6}$	$\frac{1}{11}$	1	3
$\rho_{0\uparrow} (\mu\Omega\,\text{cm/at}\%)$	14·5	13·5	18	21	2·0	2·4
$\rho_{0\downarrow} (\mu\Omega\,\text{cm/at}\%)$	3·6	1·6	3	1·9	2·0	7·2

level is just below the top of the band. For iron the Fermi level is just below the top of the spin $\uparrow$ d band and repulsive potentials (Mn, Cr, V, Ti impurities) push up a spin $\uparrow$ d virtual bound state in the d band through the Fermi level; this explains the progressive lowering of the magnetic moment for Mn, Cr, V and Ti impurities. If we consider the transport properties, a strong s$\uparrow$–d$\uparrow$ scattering should be associated with the high density of spin $\uparrow$ d states at the Fermi level for repulsive potentials. This is the explanation of the high value of $\rho_{0\uparrow}$ for Mn, Cr, V and Ti impurities (figure 12).

A resonance effect at the spin $\uparrow$ Fermi level is not expected for an attractive potential (Friedel 1967), which explains why for Co or Ni impurities, $\rho_{0\uparrow}$ is much lower than for Mn, Cr, V or Ti.

For the spin $\downarrow$ electrons, the Fermi level is in the middle of the d band, so that resonance effects are not expected for the spin $\downarrow$ direction. This is consistent with the low values of $\rho_{0\downarrow}$ for the 3d impurities.

7. Conclusion

We have discussed in detail the two band model for conduction in ferromagnetic metals, and have shown that the formula for the two band model including mixing

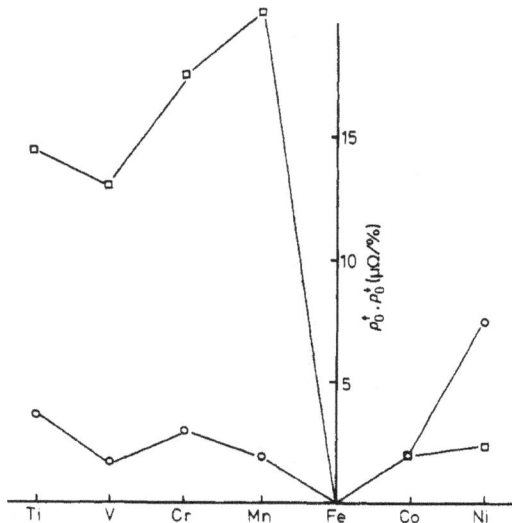

Figure 12. The resistivity of 3d impurities in iron for each spin direction: $\Box$, $\rho_{0\uparrow}$; O, $\rho_{0\downarrow}$. α for Ti $= \frac{1}{4}$, V $= \frac{1}{8}$, Cr $= \frac{1}{6}$, Mn $= \frac{1}{11}$, Co $\simeq 1$ and Ni $\simeq 3$.

868   *A Fert and I A Campbell*

(equation 22) is of rather general validity. This puts the model, which has been widely used in interpreting experimental data, on a firmer theoretical footing.

We then presented experimental data for the temperature dependence of the resistivity of binary Ni and Fe based alloys and for the deviations from Matthiessen's rule at low temperature in ternary Ni based alloys. The results were analysed to give consistent values of the parameters $\alpha = \rho_\uparrow^0/\rho_\downarrow^0$ for each impurity. These values follow the qualitative predictions of the s–d model. We also obtain temperature dependent pure metal intra-spin-band and spin-mixing scattering rates.

Appendix

The parameters of the two current model can be extracted from the experimental results in an unambiguous manner by an appropriate analysis of the data, as long as sufficient carefully determined points exist.

The simplest case is that of ternary alloys at low temperatures assuming zero spin mixing (this is a reasonable approximation when the total impurity resistivity is greater than $1\,\mu\Omega$ cm for all samples). Equation (19) can be rewritten

$$\frac{1}{a}\frac{\Delta\rho}{\rho_A} + \frac{1}{b}\frac{\Delta\rho}{\rho_B} = 1 \tag{A1}$$

with

$$a = (\alpha_A - \alpha_B)^2/\alpha_B(1 + \alpha_A)^2$$
$$b = (\alpha_A - \alpha_B)^2/\alpha_A(1 + \alpha_B)^2. \tag{A2}$$

This is in fact the Kohler–Sondheimer–Wilson relation. A plot of $\Delta\rho/\rho_A$ against $\Delta\rho/\rho_B$ should give a straight line with intercepts a and b. When these values are determined, the simultaneous equations

$$\alpha_A = [(a\alpha_B)^{1/2} + \alpha_B]/[1 - (a\alpha_B)^{1/2}]$$
$$\alpha_B = [\alpha_A - (b\alpha_A)^{1/2}]/[(b\alpha_A)^{1/2} + 1] \tag{A3}$$

can be solved graphically to give the values of α_A, α_B with their associated uncertainties.

As an example, we plot in figure 13 results for NiMnCr alloys in the form of equation (A1). With the values of a and b obtained from this plot, ($a = 6, b = 2$) a graphical solution of equations (A3) is given in figure 14. We obtain unambiguously

$$\alpha_{Mn} = 11 \pm 2 \qquad \alpha_{Cr} = 0.37 \pm 0.05.$$

Our own data together with results in the literature have been analysed in this way to give the values for the different impurities.

A similar approach, including the possibility of spin-mixing can be used for binary alloys at non-zero temperature as pointed out by Schwerer and Conroy (1971). If there is zero spin-mixing at zero temperature and we define as usual

$$\Delta(T) = \rho_{alloy}(T) - [\rho_{alloy}(0) + \rho_{pure}(T)] \tag{A4}$$

then we can write

$$\frac{1}{a}\frac{\Delta(T)}{\rho(0)} + \frac{1}{b}\frac{\Delta(T)}{\rho_{pure}(T)} = 1 \tag{A5}$$

Resistivity of ferromagnetic Ni and Fe based alloys 869

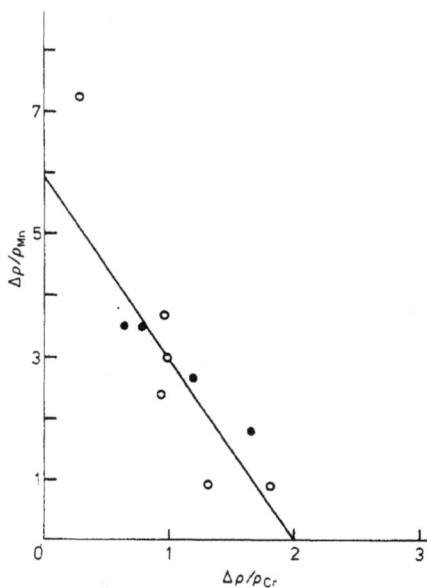

Figure 13. Deviations from Matthiessens rule for **NiCrMn** alloys plotted in the form of equation (A1). ○. present work; ●, Dorleijn and Miedema (1975).

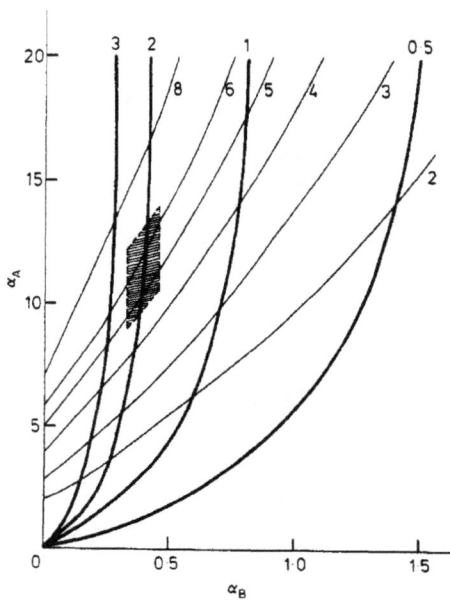

Figure 14. Graphic solutions of equations (A3). Hatched area corresponds to the range of possible values of *a* and *b* for the system NiMnCr estimated from figure 13. Heavy curves: equation (A3), fixed values of *a*. Light curves: equation (A3), fixed values of *b*.

870   *A Fert and I A Campbell*

with

$$a = \frac{[\alpha(1 + 2r) - (\mu + 2r)]^2}{\alpha[\mu + 1 + 4r]^2}$$

$$b = \frac{[\alpha(1 + 2r) - (\mu + 2r)]^2}{[\mu + r + \mu r][\alpha + 1]^2}.$$

(A6)

Here, α and μ are as defined in §5, $r = \rho_{\uparrow\downarrow}(T)/\rho^{\uparrow}_{\text{pure}}(T)$. Results at a fixed temperature for varying concentrations of two sets of binary alloys (impurity A and impurity B) give the four parameters α_A, α_B, μ and r. In fact, Schwerer and Conroy (1971) found that it was necessary to introduce zero temperature spin-mixing which means that the form (A4) is no longer valid. It is possible to return to this form by referring all measurements to a hypothetical 'unmixed' state. The corrections to be applied are

$$\Delta'(T) = \Delta(T) + \Delta_0$$

and

$$\rho'(0) = \rho(0) - \Delta_0$$

with

$$\Delta_0 = \tfrac{1}{2}\{[(\rho(0) - \rho_{\uparrow\downarrow}(0))^2 + 4\rho(0)\rho_{\uparrow\downarrow}(0)(\alpha - 1)^2/(\alpha + 1)^2]^{1/2} - |(\rho(0) - \rho_{\uparrow\downarrow}(0))|\}$$

Then if $\rho_{\uparrow\downarrow}(0)$ is chosen correctly, the relation (A5) should hold for the corrected parameters at each temperature.

References

Arajs S, Schwerer F C and Fischer R M 1969 *Phys. Stat. Solidi* **33** 731–40
Bourquart A, Daniel E and Fert A 1968 *Phys. Lett.* **26A** 260–3
Cadeville M C, Gautier F, Robert C and Roussel J 1969 *Solid St. Commun.* **7** 1701–4
Campbell I A, Fert A and Jaoul O 1970 *J. Phys. C: Solid St. Phys.* **3** 595–601
Campbell I A, Fert A and Pomeroy A R 1967 *Phil. Mag.* **15** 977–83
Caudron R, Caplain R, Meanier J J and Costa P 1973 *Phys. Rev. B* **8** 5247–56
Collins M F and Low G 1965 *Proc. Phys. Soc.* **86** 536–43
Dorleijn J F and Miedema A R 1975 *J. Phys. F: Metal Phys.* **5** 487–96
Durand J and Gautier F 1970 *J. Phys. Chem. Solids* **31** 2773–87
Farrel T and Greig D 1968 *J. Phys. C: Solid St. Phys.* **1** 1359–69
—— 1970 *J. Phys. C: Solid St. Phys.* **3** 138–45
Fert A 1969 *J. Phys. C: Solid St. Phys.* **2** 1784–8
Fert A and Campbell I A 1968 *Phys. Rev. Lett.* **21** 1190–2
—— 1971 *J. Phys. (Paris)* **32** suppl. C1 46–50
Fert A and Jaoul O 1972 *Phys. Rev. Lett.* **28** 303–7
Friedel J 1967 *Rendiconti della Scuala Internazionale di Fisica 'Enrico Fermi'* **XXXVII** Corso (New York: Academic Press)
Goy P and Grimes C C 1973 *Phys. Rev. B* **7** 299–306
Greig D and Rowlands J A 1974 *J. Phys. F: Metal Phys.* **4** 232–46
Jaoul O and Campbell I A 1975 *J. Phys. F: Metal Phys.* **5** L69–73
Leonard P, Cadeville M C, Durand J and Gautier F 1969 *J. Phys. Chem. Solids* **30** 2169–B
Loegel B and Gautier F 1971 *J. Phys. Chem. Solids* **32** 2723–35
Mills D L, Fert A and Campbell I A 1971 *Phys. Rev. B* **4** 196–201
Monod P 1968 *PhD Thesis* (Orsay)
Mott N F 1964 *Adv. Phys.* **13** 325
Price D C and Williams G 1973 *J. Phys. F Metal Phys.* **3** 810–24
Rowlands J A 1973 *J. Phys. F: Metal Phys.* **3** L149–53

Resistivity of ferromagnetic Ni and Fe based alloys   871

Schwerer F C and Conroy J W 1971 *J. Phys. F: Metal Phys.* **1** 877–91
Schwerer F C and Silcox J 1968 *Phys. Rev. Lett.* **20** 101–3
Shirane G, Minkiewicz Y J and Nathans R 1968 *J. Appl. Phys.* **39** 383–8
Tsui D C 1971 *Phys. Rev.* **2** 669–83
Wang C S and Callaway J 1974 *Phys. Rev.* **B 9** 4897–907
White G K and Tainsh R J 1967 *Phys. Rev. Lett.* **19** 105–6
Ziman J M 1960 *Electrons and Phonons* (Oxford: Clarendon Press) p 275

VOLUME 61, NUMBER 21   PHYSICAL REVIEW LETTERS   21 NOVEMBER 1988

Giant Magnetoresistance of (001)Fe/(001)Cr Magnetic Superlattices

M. N. Baibich, [a] J. M. Broto, A. Fert, F. Nguyen Van Dau, and F. Petroff

Laboratoire de Physique des Solides, Université Paris-Sud, F-91405 Orsay, France

P. Eitenne, G. Creuzet, A. Friederich, and J. Chazelas

Laboratoire Central de Recherches, Thomson CSF, B.P. 10, F-91401 Orsay, France

(Received 24 August 1988)

We have studied the magnetoresistance of (001)Fe/(001)Cr superlattices prepared by molecular-beam epitaxy. A huge magnetoresistance is found in superlattices with thin Cr layers: For example, with $t_{Cr} = 9$ Å, at $T = 4.2$ K, the resistivity is lowered by almost a factor of 2 in a magnetic field of 2 T. We ascribe this giant magnetoresistance to spin-dependent transmission of the conduction electrons between Fe layers through Cr layers.

PACS numbers: 75.50.Rr, 72.15.Gd, 75.70.Cn

There is now considerable interest in the study of multilayers composed of magnetic and nonmagnetic metals and great advances have been obtained in the understanding of their magnetic properties.[1-4] Recently the transport properties of magnetic multilayers and thin films have been investigated and have revealed interesting properties resulting from the interplay between electron transport and magnetic behavior.[5-7] In this Letter we present magnetoresistance measurements on (001)Fe/(001)Cr superlattices prepared by molecular-beam epitaxy (MBE). In superlattices with thin Cr layers, the magnetoresistance is very large (a reduction of the resistivity by a factor of about 2 is observed in some samples). This giant magnetoresistance raises exciting questions and moreover is promising for applications.

The (001)Fe/(001)Cr bcc superlattices have been grown by MBE on (001) GaAs substrates under the following conditions: The residual pressure of the MBE chamber was 5×10^{-11} Torr, the substrate temperature was generally around 20°C, the deposition rate was about 0.6 Å/s for Fe and 1 Å/s for Cr. This deposition rate was obtained by use of specially designed evaporation cells in which a crucible of molybdenum is heated by electron bombardment. The individual layer thicknesses range from 9 to 90 Å and the total number of bilayers is generally around 30. The growth of the superlattices and their characterization by reflection high-energy electron diffraction, Auger-electron spectroscopy, x-ray diffraction, and scanning-transmission-electron microscopy have been described elsewhere.[8] Note that the Cr (Fe) Auger line disappears during the growth of a Fe (Cr) layer. This, as well as the main features of the scanning-transmission-electron-microscopy cross sections, rules out a deep intermixing of Fe and Cr.[8] However, the Auger effect, which averages the concentrations over a depth of about 12 Å, cannot probe the interface roughness at the atomic scale. Surface extended x-ray-absorption fine-structure experiments have been started to probe this roughness more precisely.

The magnetic properties of the Fe/Cr superlattices have been investigated by magnetization and torque measurements.[9] The magnetization is in the plane of the layers and an antiferromagnetic (AF) coupling between the adjacent Fe layers is found when the Cr thickness t_{Cr} is smaller than about 30 Å.[9] A signature of this AF interlayer coupling is shown in Fig. 1: As the Cr thickness decreases below 30 Å, the hysteresis loop is progressively tilted. For example, with $t_{Cr} = 9$ Å, a field $H_S \simeq 2$ T is needed to overcome the antiferromagnetic coupling and to saturate the magnetization at about the bulk Fe value. When the applied field is decreased to zero, the AF coupling brings the magnetization back to about zero. As can be seen from the variation of the low-field slopes in Fig. 1, the AF coupling steeply increases when t_{Cr} decreases from 30 to 9 Å. The existence of such AF couplings has already been found in Fe/Cr sandwiches by the light-scattering and magneto-optical measurements

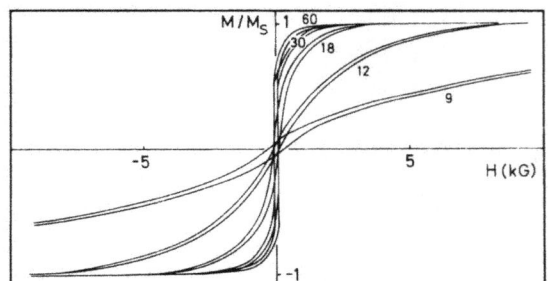

FIG. 1. Hysteresis loops at 4.2 K with an applied field along [110] in the layer plane for several (001)Fe/(001)Cr superlattices: [(Fe 60 Å)/(Cr 60 Å)]₅, [(Fe 30 Å)/(Cr 30 Å)]₁₀, [(Fe 30 Å)/(Cr 18 Å)]₃₀, [(Fe 30 Å)/(Cr 12 Å)]₁₀, [(Fe 30 Å)/(Cr 9 Å)]₄₀, where the subscripts indicate the number of bilayers in each sample. The number beside each curve represents the thickness of the Cr layers.

2472

VOLUME 61, NUMBER 21   PHYSICAL REVIEW LETTERS   21 NOVEMBER 1988

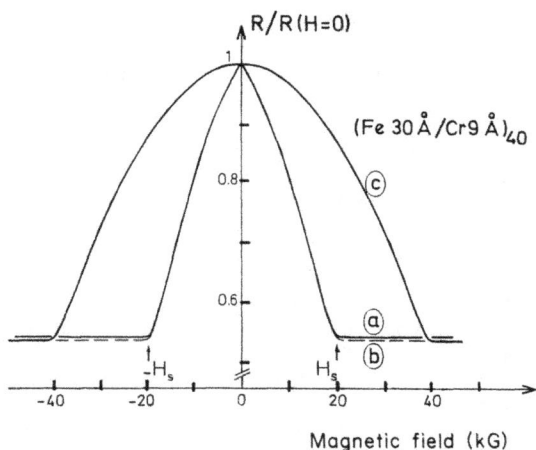

FIG. 2. Magnetoresistance of a [(Fe 30 Å)/(Cr 9 Å)]$_{40}$ superlattice of 4.2 K. The current is along [110] and the field is in the layer plane along the current direction (curve a), in the layer plane perpendicular to the current (curve b), or perpendicular to the layer plane (curve c). The resistivity at zero field is 54 $\mu\Omega$ cm. There is a small difference between the curves in increasing and decreasing field (hysteresis) that we have not represented in the figure. The superlattice is covered by a 100-Å Ag protection layer. This means that the magnetoresistance of the superlattice alone should be slightly higher.

of Grünberg *et al.*[4] and by the spin-polarized low-energy electron-diffraction experiments of Carbone and Alvarado.[10] The AF coupling between the Fe layers has been ascribed to indirect exchange interactions through the Cr layers, but a theoretical model of these interactions is still lacking.[4,9]

The magnetoresistance of the Fe/Cr superlattices has been studied by a classical ac technique on small rectangular samples. Examples of magnetoresistance curves at 4.2 K are shown in Figs. 2 and 3. The resistance decreases during the magnetization process and becomes practically constant when the magnetization is saturated. The curves a and b in Fig. 2 are obtained for applied fields in the plane of layers in the longitudinal and transverse directions, respectively. The field H_S is the field needed to overcome the AF couplings and to saturate the magnetization (compare with Fig. 1). In contrast, fields applied perpendicularly to the layers (curve c) have to overcome not only the AF coupling but also the magnetic anisotropy, so that the magnetoresistance is saturated at a field higher than H_S.

The most remarkable result exhibited in Figs. 2 and 3 is the huge value of the magnetoresistance. For $t_{Cr} = 9$ Å and $T = 4.2$ K, see Fig. 2, there is almost a factor of 2 between the resistivities at zero field and in the saturated state, respectively (in absolute value, the resistivity change is about 23 $\mu\Omega$ cm). By comparison of the results for three different samples in Fig. 3, it can be seen

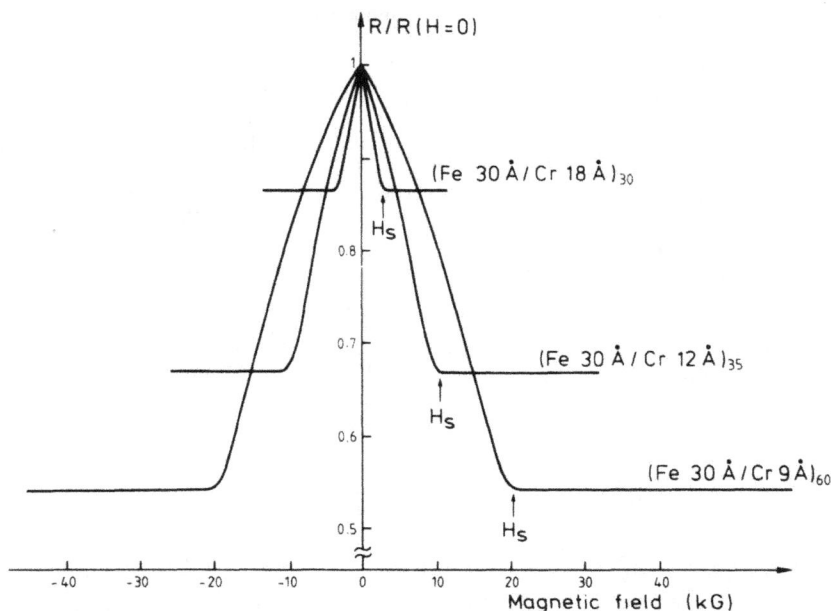

FIG. 3 Magnetoresistance of three Fe/Cr superlattices at 4.2 K. The current and the applied field are along the same [110] axis in the plane of the layers.

VOLUME 61, NUMBER 21   PHYSICAL REVIEW LETTERS   21 NOVEMBER 1988

that the magnetoresistance is lowered when the Cr thickness increases. At the same time, an increase of t_{Cr} weakens the AF coupling and the saturation field H_S decreases. Similarly, we find that both the magnetoresistance and H_S decrease when the temperature increases: typically, from 4.2 K to room temperature, the saturation magnetoresistance is lowered by about a factor of 2, while H_S is reduced by about 30%. We point out that the magnetoresistance is still very significant at room temperature.

Summarizing the data simply, giant magnetoresistance effects are obtained in antiferromagnetically coupled Fe/Cr superlattices by our aligning the magnetizations of adjacent Fe layers with an external field. We propose that this magnetoresistance arises from spin-dependent transmission of the conduction electrons through the thin Cr layers. First, we point out that perfect interfaces would produce only specular reflections and diffractions of the electron waves, without significant change of the longitudinal resistivity. Significant effects on the resistivity are expected only from scattering by interface roughness. Second, we note that the scattering can be strongly spin dependent in a ferromagnetic transition metal.[11] For Cr impurities in bulk Fe, the resistivity cross section for spin-down → spin-down scattering is about 6 times smaller than for spin-up → spin-up scattering (we call up and down the majority and minority spin directions, respectively).[11] This leads us to propose the following explanation for the giant magnetoresistance of the Fe/Cr superlattices. Thin Cr layers (9 Å ≈ 3 lattice constants) between thicker Fe layers having parallel magnetizations should scatter the conduction electrons roughly as Cr impurities in bulk Fe, which implies weak interface scattering and a high coefficient of coherent transmission for the spin-down electrons. Thus, at $H > H_S$, the current is carried by the spin-down electrons with a low resistivity, $\rho \approx \rho_\downarrow \ll \rho^\uparrow$, where $\rho_\uparrow$ and $\rho_\downarrow$ are the resistivities for the spin-up and spin-down currents, respectively. At zero field, with Cr layers between two Fe layers having antiparallel magnetizations, the resistivity is expected to be definitely higher for two reasons. First, for the electrons of one of the Fe layers, not only the Cr atoms but also the Fe atoms of the antiparallel layer represent possible scattering potentials and reduce the coherent transmission. Second the spin-up and spin-down currents are averaged, which suppresses the short-circuit effect by one direction of the first case. An applied field, by aligning the magnetizations, progressively opens the spin-down → spin-down channel and lowers the resistivity. For thicker Cr layers, or at higher temperatures, the probability of spin-flip scattering within the Cr layers increases and, by mixing the spin channels, weakens the magnetoresistance. Similar concepts have been introduced by Cabrera and Falicov[12] for the resistivity of Bloch walls, and more recently by Johnson and Silsbee[13] for the problem of spin injection from a ferromagnet, and their formalisms could probably be adapted to be applied to the magnetoresistance of our multilayers. This is not at all within the scope of this Letter which is aimed only to present experimental results and to suggest an interpretation.

In conclusion, we have found a giant magnetoresistance in (001)Fe/(001)Cr superlattices when, for thin Cr layers (9, 12, and 18 Å), there is an antiparallel coupling of the neighbor Fe layers at zero field. The highest magnetoresistance is observed in [(Fe 30 Å)/(Cr 9 Å)]$_{40}$: The resistivity is reduced by almost a factor of 2 when the magnetization is saturated. We interpret our results in terms of spin-dependent transmission between ferromagnetic layers. The giant magnetoresistance of the Fe/Cr superlattices may result from an interplay of the orientation of the Fe layers by an applied field with the spin-dependent transmission between Fe layers through a Cr layer. If one considers that strongly spin-dependent conduction occurs in many ferromagnetic transition-metal alloys,[11] the type of magnetoresistance found in Fe/Cr should be observed in other transition-metal superlattices. The existence of a giant magnetoresistance in Fe/Cr is promising for applications to magnetoresistance sensors. In the samples we have studied, the saturation fields are obviously too high for applications but a large magnetoresistance at relatively small fields can probably be obtained by thickening of the Fe layers (in a given field, this enhances the torque on the magnetic layers). Alternatively high magnetoresistance effects with weaker AF couplings should probably be observed with other couples of transition metals.

This work is supported in part by the Ministère de la Recherche et de l'Enseignement Supérieur, Grants No. MRES/PMFE RE 86-50-016 and No. RE 86-50-020, and by the Direction des Recherches, Etudes et Techniques (Ministère de la Défense), Grant No. DRET 87/1344. One of us (M.N.B.) wishes to acknowledge financial support from Consello Nacional de Desenvolvimento Cientifico e Tecnologico (Brazil).

---

[a]Permanent address: Instituto de Fisica, Universidade Federal do Rio Grande do Sul, C.P. 15051, 91500 Porto Alegre, Rio Grande do Sul, Brazil.

[1]M. B. Salamon, S. Sinha, J. J. Rhyne, J. E. Cunningham, R. W. Erwin, and C. P. Flynn, Phys. Rev. Lett. **56**, 259 (1986).

[2]C. F. Majkrzak, J. M. Cable, J. Kwo, M. Hong, D. B. McWhan, Y. Yafet, and C. Vettier, Phys. Rev. Lett. **56**, 2700 (1986).

[3]J. R. Dutcher, B. Heinrich, J. F. Cochran, D. A. Steigerwald, and W. F. Egelhoff, J. Appl. Phys. **63**, 3464 (1988).

[4]P. Grünberg, R. Schreiber, Y. Pang, M. B. Brodsky, and H. Sowers, Phys. Rev. Lett. **57**, 2442 (1986); F. Saurenbach, U. Walz, L. Hinchey, P. Grünberg, and W. Zinn, J. Appl. Phys. **63**, 3473 (1988).

[5]E. Velu, C. Dupas, D. Renard, J. P. Renard, and J. Seiden,

VOLUME 61, NUMBER 21   PHYSICAL REVIEW LETTERS   21 NOVEMBER 1988

Phys. Rev. B **37**, 668 (1988).

[6]E. D. Dahlberg, K. Riggs, and G. A. Prinz, J. Appl. Phys. **63**, 4270 (1988).

[7]M. Rubinstein, F. J. Rackford, V. W. Fuller, and G. A. Prinz, Phys. Rev. B **37**, 8699 (1988).

[8]P. Etienne, G. Creuzet, A. Friederich, F. Nguyen Van Dau, A. Fert, and J. Massies, Appl. Phys. Lett. **53**, 162 (1988).

[9]F. Nguyen Van Dau, A. Fert, M. N. Baibich, J. M. Broto, S. Hadjhoudj, H. Hurdequint, J. P. Redoules, P. Etienne, J. Chazelas, G. Creuzet, A. Friederich, and J. Massies, in Proceedings of the International Conference on Magnetism, Paris, France, 1988 (to be published).

[10]C. Carbone and S. F. Alvarado, Phys. Rev. B **36**, 2433 (1987).

[11]A. Fert and I. A. Campbell, J. Phys. F **6**, 849 (1976); I. A. Campbell and A. Fert, in *Ferromagnetic Materials*, edited by E. P. Wohlfarth (North-Holland, Amsterdam, 1982), Vol. 3, p. 769.

[12]G. G. Cabrera and L. M. Falicov, Phys. Status Solidi (b) **61**, 539 (1974), and **62**, 217 (1974).

[13]M. Johnson and R. H. Silsbee, Phys. Rev. B **37**, 3312, 5328 (1988).

LIST OF PUBLICATIONS

1. Fert A., Averbuch P. (1964) Déplacement de Knight de l'hydrogène absorbé par le palladium. *J. de Physique* 25, 297.

2. Campbell I.A., Fert A., Pomer A.R. (1967) Evidence for two current conduction in iron. *Phil. Mag.* 15, 137.

3. Bourquard A., Daniel E., Fert A. (1968) Effet des collisions entre électrons sur la résistivité électrique d'alliages magnétiques. *Physics Letters* 26A, 260.

4. Fert A., Campbell I.A. (1968) Two current conduction in nickel. *Phys. Rev. Letters* 21, 1190.

5. Fert A. (1969) Two current conduction in ferromagnetic metals and spin wave electron collisions. *J. of Physics* C 2, 1784.

6. Fert A., Campbell I.A., Ribault M. (1970) Magnetoresistance of nickel alloys and scattering dependent on spin orientation. *J. Applied Physics* 91, 1428.

7. Fert A., Campbell I.A., Jaoul O. (1970) Thespontaneous resistivity anisotropy of Ni based alloys. *J. of Physics* F (Metal Physics) 1, 595.

8. Fert A., Campbell I.A. (1971) Transport properties of ferromagnetic transition metals. *J. de Physique* 32, C1, 46.

9. Mills D., Fert A., Campbell I.A. (1971) Temperature dependence of the electrical resistivity of dilute ferromagnetic alloys. *Phys. Rev.* B 4, 196.

10. Fert A. (1971) Comments on two band models for the electrical conduction in metals. *J. of Physics* F 1, L42.

11. Campbell I.A., Fert A., Caplin A.D. (1971) A correlation between magnetoresistance and deviations form Matthiessen's rule. *J. of Physics* F 1, L58.

12. Fert A., Jaoul O. (1972) Left right assymetry in the scattering of electrons by magnetic impurities and Hall effect. *Phys. Rev. Letters* 28, 303.

13. Fert A., Jaoul O. (1972) Skew scattering by Ce impurities in La. *Solid State Comm.* 11, 759.

14. Fert. A. (1973) Diffusion "skew" par des impuretés magnétiques. *L'Effet Hall Extraordinaire* (edited by Université de Genève).

15. Friederich A., Fert A., Sierro P. (1973) Skew scattering by rare earth impurities in silver. *Solid State Comm.* 13, 997.

16. Fert A. (1973) Skew scattering in alloys with cerium impurities. *J. of Physics* F 3, 2126.

17. Fert A., Jaoul O. (1974) Hall effect induced by cerium impurities in lanthanum. *Proceedings of 13th Conference on LowTemperature Physics* (edited by Plenum Press) Vol. 3, 488.

18. Jaoul O., Fert A. (1974) The extraordinary Hall effect of nickel based alloys. *Proceedings of the 8th International Conference on Magnetism* Vol. V, 557.

19. Fert A. (1974) On the extraordinary Hall effect of the rare earth metals. *J. de Physique Lettres* 35, L107.

20. Friederich A., Fert A. (1974) Electron scattering by the electronic quadrupole moment of rare earth impurities. *Phys. Rev. Letters* 33, 1214.

21. Fert A., Friederich A. (1975) Skew scattering and quadrupole scattering by rare earth impurities (review) *Conference Proceedings* n°24 (American Institute of Physics), 466.

22. Fert A., Friederich A. (1976) Skew scattering by rare earth impurities in gold, silver and aluminium *Phys. Rev.* B 13, 397.

23. Asomoza R., Creuzet G., Fert A., Reich R. (1976) Quadrupole scattering: The resistivity anisotropy of rare earth impurities in gadolinium, *Solid State Comm.* 18, 905.

24. Fert A., Campbell I.A. (1976) Electric resistivity of ferromagnetic nickel and iron based alloys. *J. of Physics* F, Vol. 6, 849.

25. Jaoul O., Campbell I.A., Fert A. (1977) Spontaneous resistivity anisotropy in Ni alloys. *J. Mag. Mag. Mat.* 5, 23.

26. Fert A. (1977) Transport properties of magnetic alloys: Scattering assymetries. *Physica* 86 88 B, 491.

27. Senoussi S., Campbell I.A., Fert A. (1977) Evidence for local orbital moments on Ni and Co in Pd. *Solid State Comm.* 21, 269.

28. Asomoza R., Fert A., Sanchez D. (1977) Resistivity anisotropy of rare earth impurities in gadolinium. *Physica* 86 88 B, 528.

29. Fert A., Asomoza R., Campbell I.A., Meyer R., Jouve H. (1977) Contribution de l'ordre magnétique à la résistivité d'alliages amorphes. *C.R.A.S.* 285B.

30. Fert A., Asomoza R., Sanchez D., Spanjaard D., Friederich A. (1977) Magneto transport properties of noble metals containing rare earth impurities. I. Quadrupole scattering by rare earth impurities in gold. *Phys. Rev.* B 16, 5040.

31. Fert A., Levy P.M. (1977) Magneto transport properties of noble metals containing rare earth impurities. II. Theory. *Phys. Rev.* B 16, 5052.

32. Asomoza R., Fert A., Campbell I.A., Meyer R. (1977) Resistivity of Dy Ni and Ho Ni amorphous alloys. *J. of Physics* F. 7, L327.

33. Fert A., Levy P.M. (1978) Magnetotransport properties of noble metals containing rare earth impurities. *J. Appl. Phys* 49, 2168.

34. Fert A., Campbell I.A. (1978) Non axial electric field gradients in amorphous rare earth alloys. *J. of Physics* F 8, L57.

35. Fert A. (1978) Transport dans les alliages magnétiques: les asymétries de diffusion. *J. de Physique* C2, 215.

36. Hamzic A., Senoussi S., Campbell I.A., Fert A. (1978) The extraordinary Hall effect in dilute Pd based alloys. *Solid State Comm.* 617 .

37. Hamzic A., Senoussi S., Campbell I.A., Fert A. (1978) The magnetoresistance of Pd based dilute ferromagnetic alloys. *J. of Physics* F. 8, 1947.

38. Asomoza R., Campbell I.A., Fert A., Liénard A., Rebouillat J.P. (1979) Magnetic and transport properties of Ni RE amorphous alloys. *J. of Physics* F9, 349.

39. Asomoza R., Campbell I.A., Fert A. (1979) Transport properties of amorphous rare earth alloys. *J. de Physique* 5, C5 225.

40. Ousset J.C., Creuzet G., Fert A. (1979) High field magnetoresistance of silver containing rare earth impurities. *J. de Physique* 5, C5 40.

41. Fert A., Spanjaard D. (1979) Non axial electric field gradients in amorphous rare earth alloys. *J. de Physique* 5, C5 248.

42. Fert A., Asomoza R. (1979) Transport properties of magnetic amorphous alloys. *J. Appl. Phys.* 50, 1886.

43. Creuzet G., Fert A., Spanjaard D. (1979) Magnetoresistance of silver single crystal containing rare earth impurities. *J. Appl. Phys.* 50, 1901.

44. Boucher B., Friederich A., Fert A. (1979) Electron paramagnetic resonance in Nd Ag amorphous alloys. *Solid State Comm.* 30, 443 (1979).

45. Hamzic A., Senoussi S., Campbell I.A., Fert A. (1980) Orbital magnetism of transition metal impurities in platinum. *J. Mag. Mag. Mat.* 15 18, 921.

46. Garoche P., Fert A., Veyssié J.J., Boucher D. (1980) Specific heat of rare earth amorphous alloys. *J. Mag. Mag. Mat.* 15 18, 1394.

47. Fert A., Hamzic A. (1980) Hall effect from skew scattering in magnetic alloys, *The Hall Effect and its Applications* (Plenum Press), 77.

48. Asomoza R., Biéri J.B., Fert A., Boucher B., Ousset J.C. (1980) Hall effect in silver rare earth amorphous alloys. *J. de Physique* C8, 467.

49. Ousset J.C., Ulmet J.P., Asomoza R., Biéri J.B., Askenazy S., Fert A. (1980) Resistivity and magnetoresistance of silver rare earth amorphous alloys. *J. de Physique* C8 470.

50. Fert A., Asomoza R., Creuzet G., Ousset J.C. (1980) Magneto transport in f electron systems: Quadrupolar and orbital exchange effects". *Crystalline Electric Field and Structural Effects in f Electron Systems* (Plenum Press), 389.

51. Fert A., Garoche P., Boucher B., Durand J. (1980) Electric field gradients in amorphous f electron systems. *Crystalline Electric Field and Structural Effects in f Electron Systems* (Plenum Press), 491.

52. Fert A., Levy P.M. (1980) Role of anisotropic exchange interactions in determining the properties of spin glasses. *Phys. Rev. Letters* 44, 1538.

53. Levy P.M., Fert A. (1981) Anisotropy induced by non magnetic impurities in CuMn spin glass alloys. *Phys. Rev.* B 23, 4667.

54. Ousset J.C., Carrère G., Ulmet J.P., Greuzet G., Fert A. (1981) High field magneto resistance of noble metals containing rare earth impurities. *J. Mag. Mag. Mat.* 24, 7.

55. Levy P.M., Fert A. (1981) Origin of magnetic anisotropy in transition metal sppin glass alloys. *J. Appl. Phys.* 52, 1718.

56. Ousset J.C., Askenazy S., Fert A. (1981) Anisotropic scattering by rare earth impurities: Effect of a high magnetic field. *Physics in High Magnetic Fields* (Springer Verlag, Berlin), 161.

57. Lacueva G., Levy P.M., Creuzet G., Fert A. (1981) Anisotropic k f interactions from spin orbit coupled 5d intermediate states. *Solid State Comm.* 38, 551.

58. Fert A., Friederich A., Hamzic A. (1981) Hall effect in dilute magnetic alloys. *J. Mag. Mag. Mat.* 24, 231.

59. Biéri J.B., Sanchez J., Fert A., Bertrand D., Fert A. R. (1982) Crystal field and local symmetry in rare earth amorphous alloys. *J. Appl. Phys.* 53, 2347.

60. Levy P.M., Morgan Pond G., Fert A. (1982) Anisotropy of spin glasses. *J. Appl. Phys.* 53, 2168.

61. Campbell I.A., Fert A. (1982) Transport properties of ferromagnets. *Ferromagnetic Materials* Vol. 3 (North Holland), 747.

62. Fert A., Hippert F. (1982) Anisotropy of spin glasses from torque measurements. *Phys. Rev. Letters* 49, 15083.

63. Levy P.M., Fert A. Lacueva G. (1982) Unified approach to some transport and EPR properties of noble metals with rare earth impurities. *Phys. Rev.* B 26, 109.4

64. Hippert F., Alloul H., Fert A. (1982) Anisotropy of CuMn spin glasses. *J. Appl. Phys.* 53, 7702.

65. Pureur P., Creuzet G., Fert A. (1982) Magnetostriction of an Yttrium monocrystal doped with terbium impurities. *Crystalline Electric Field in f electron system* (edited by Jack E. Crow, Robert P. Guertin, Ted W. Mihalisin, Plenum Press), 255.

66. Fert A., Pureur P., Hippert F., Baberschke K., Bruss F. (1982) Anisotropic spin glasses. *Phys. Rev.* B26, 5300.

67. Fert A., Senoussi S., Arvanitis D. (1983) How to change or remove the anisotropy of spin glasses with some other reflections on the anisotropy. *J. de Physique Lettres* 44, L345.

68. Fert A. (1983) Spin glasses. *Comptes Rendus de l'Ecole franco-brésilienne de Physique de Rio de Janeiro 1982*.

69. Asomoza R., Reich R., Fert A. (1983) Heavy rare earth alloys: Preparation and extraordinary Hall effect. *J. of Less Common Metals* 90, 177.

70. Sablik M.J., Pureur P., Creuzet G., Fert A., Levy P.M. (1983) Quadrupole scattering in PrAl. *Phys. Rev.* B 28, 7.

71. Campbell I.A., Arvanitis D., Fert A. (1983) Critical line for strong irreversibility in spin glass and ferro spin glass alloys. *Phys. Rev. Lett.* 51, 57.

72. Biéri J.B., Fert A., Creuzet G., Ousset J.C. (1984) Magnetoresistance of amorphous CuZr: Weak localization in three dimensional metallic systems. *Solid State Comm.* 49, 849.

73. Fert A., Arvanitis D., Hippert F. (1984) Triad anisotropy in spin glasses. *J. Appl. Phys.* 55, 1640.

74. Baberschke K., Pureur P., Fert A., Wendler R. (1984) Magnetic properties of spin glasses with uniaxial anisotropy. *Phys. Rev.* B 29, 4999.

75. Campbell I.A., de Courtenay N., Fert A. (1984) Irreversibility onset in spin glasses from torque measurements. *J. de Physique Lettres*, 45, L565.

76. Wendler R., Pureur P., Fert A., Baberschke K. (1984) Magnetic properties of Gd doped Y and Sc single crystals. *J. Mag. Mag. Mat.* 45, 185.

77. Biéri J.B., Fert A., Creuzet G., Ousset J.C. (1984) Weak localization in amorphous alloys. *J. Appl. Phys.* 55, 1948.

78. Fert A., J..B. Biéri. (1984) Localization effects in metallic glasses. *Proceedings of the LITPIM Conference* (edited by L.Schweitzer and B. Kramer).

79. de Courtenay N., Fert A., Campbell I.A. (1984) Role of random anisotropy in determining the phase diagram of spin glasses. *Phys. Rev.* B 30, 6791.

80. Fert A., Levy P.M. (1985) Relaxation of Ytterbium in palladium. *J. de Physique Lettres*, 46, L53.

81. S. Goldberg, P.M. Lévy, Fert A. (1985) Anisotropy in metallic spin glasses arising from Au impurities. *Phys. Rev.* B 31, 3106.

82. Pureur P., Fert A., Baberschke K., Wendler R. (1985) Magnetic ordering in YGd alloys. *J. Appl. Phys.* 57, 1985.

83. Fert A., de Courtenay N., Campbell I.A. (1985) Random anisotropy and phase diagram of spin glasses. *J. Appl. Phys.* 57, 3398.

84. Creuzet G., Fert A., Gaonach G., Gignoux D. (1985) Anisotropic magnetoelastic coupling in CeNi intermediate valence compound. *Physica* 130B, 138.

85. Lacueva G., Levy P.M., Fert A. (1985) Crystal field of Gd in hexagonal metals. *Phys. Rev.* B 31, 6245.

86. Brown P.J., Coudron D.R., Fert A., Pureur P. (1985) Helimagnetic structure in diluted YGd alloys. *J. de Physique Lettres*, 46, L1139.

87. Pureur P., Creuzet G., Fert A. (1985) Magnetostriction of single crystals of yttrium doped with rare earth impurities. *J. Mag. Mag. Mat.* 53, 121.

88. Fert A., Pureur P., Hamzic A., Kappler J.P. (1985) Hall effect in Ce Y Pd mixed valence alloys. *Phys. Rev.* B 32, 7003.

89. S. Goldberg, Levy P.M., Fert A. (1986) Anisotropy in binary metallic spin glass alloys. *Phys. Rev.* B 33, 276.

90. Pureur P., Creuzet G., Fert A. (1986) Anisotropic thermal expansion of yttrium metal and yttrium based alloys. *J. Mag. Mag. Mat.* 60, 161.

91. Hadzic Leroux M., Hamzic A., Fert A., Lapierre F., Laborde J. (1986) Hall effect in heavy fermion systems: UPt, UAl, CeAl, CeRu Si. *Europhys. Letters* 1, 579.

92. de Courtenay N., Bouchiat H., Hurdequint H., Fert A. (1986) Influence of Dzyaloshinsky Moriya interactions on the critical behaviour of metallic spin glasses. *J. de Physique* 47, 1507.

93. Biéri J.B., Fert A., Creuzet G., Schuhl A. (1986) Weak localization in metallic glasses. I. Magnetoresistance. *J. of Physics* F 16, 2099.

94. Fert A., Hamzic A., Levy P.M. (1987) Theory of the Hall effect in heavy fermion and mixed valence systems. *J. Mag. Mag. Mat.* 63 64, 535.

95. Lapierre J., Haen P., Hamzic A., Fert A., Kappler J.P. (1987) Hall effect in heavy fermion systems. *J. Mag. Mag. Mat.* 63 64, 338.

96. Malaurent J.C., Duval H., Fert A. (1987) Etude de la Structure d'une Multicouche Synthétique par la Méthode de Diffraction par Dispersion d'Energie des Rayons X. *J. Appl. Cryst.* 20, 491.

97. de Courtenay N., Bouchiat H., Hurdequint H., Fert A. (1987) Influence of Dzyaloshinsky Moriya Interactions on the Critical Behaviour of Metallic Spin Glasses. *J. Appl. Phys.* 61, 4097.

98. Fert A., Levy P.M. (1987) Theory of the Hall Effect in Heavy Fermion Compounds. *Phys. Rev.* B36, 1987.

99. Fert A. (1987) The Hall effect in magnetic materials. *Physics of Magnetic Materials* (edited by M. Takahashi and S. Maekawa, World Scientific) 503.

100. Fert A., de Courtenay N., Bouchiat H. (1988) Experimental arguments for a non zero transition temperature in RKKY Heisenberg spin glasses without anisotropy. *J. de Physique* 49, 1173.

101. O'Shea M.J., Fert A. (1988) Rigid spin rotation in amorphous rare earth alloys. *Phys. Rev.* B 37, 9824 (1988).

102. Etienne P., Creuzet G., Friederich A., Nguyen Van Dau F., Fert A., Massies J. (1988) Molecular beam epitaxial growth of Fe/Cr multilayers on GaAs(001). *Appl. Phys. Lett.* 53, 162.

103. Sas B., Broto J.M., de Courtenay N., Fert A. (1988) Weak localizatio in the magnetoresistance of Ni TM B amorphous alloys. *Sol. State Comm.* 66, 777.

104. Hamzic A., Fert A., Miljak M., Horn S. (1988) Hall effect in the heavy fermion compound Ce Pt Si.*Phys. Rev.* B 38, 714.

105. Levy P.M., Fert A. (1988) Anomalous velocity contributions to the Hall effect in Kondo systems. *Europhys. Letters* 7, 463.

106. Baibich M.N., Fert A., Nguyen Van Dau F., Broto J.M., Pétroff F., Etienne P., Creuzet G., Friederich A. (1988) Giant magnetoresistancein Fe(001)/Cr(001) superlattices. *Phys. Rev. Letters* 61, 2472.

107. Broto J.M., F. Gonzalez F., Fert A., Sanchez J., Besnus J.M., Kappler J.P. (1988) Magneto transport properties of the Kondo lattice Yb Pd Si: Evidence for a new phase below 1.4 K. *J. Mag. Mag. Mat.* Vol. 76–77, 2 289–290.

108. Hamzic A., Fert A. (1988) Hall effect in heavy fermion systems. *J. Mag. Mag. Mat.* Vol. 76–77, 2 221–222.

109. Djerbi R., Haen P., Lapierre F., Mignot J.-M., Fert A., Hamzic A., Kappler J.P. (1988) Influence of Y and La alloying on the anomalous Hall effect of Ce Ru Si. *J. Mag. Mag. Mat.* Vol. 76–77, 2 267–268.

110. Levy P.M., Fert A. (1988) Crystal field effect on the Hall effect in heavy fermion systems. *J. Mag. Mag. Mat.* 76–77, 238.

111. Hamzic A., Fert A., Broto J.M., Miljak M., Bauer E., Gratz E. (1988) Hall effect in Kondo alloys Ce Cu Al. *J. de Physique* 49, C8, 777.

112. Bonville P., Broto J.M., Fert A., Gonzales Jimenez F., Hamzic A., Imbert P., Jehano G. (1988) Magnetotransport properties of the Kondo lattice Yb Pd Si. *J. de Physique*, 49, C8 727.

113. Nguyen Van Dau F., Fert A., Baibich M.N., Hadjoudj S., Etienne P., Creuzet G., A. Friederich (1988) Magnetic properties of (001)Fe/(001)Cr multilayers. *J. de Physique* 49, C8 1663.

114. Fert A., de Courtenay N., Bouchiat H. (1988) Influence of anisotropy on the critical behaviour of spin glasses. *J. de Physique*, 49, C8 1049.

115. Hamzic A., Fert A., Broto J.M., Miljak M., Horn S. (1988) Hall effect in CePtSi. *J. de Physique*, 49, C8 775.

116. Levy M.P., Guo W., Fert A., Cox D. (1988) The Hall effect in Kondo systems. *J. de Physique*, 49, C8 777.

117. Hamzic A., Fert A., Pureur P., Gignoux D. (1989) Anomalous Hall effect in mixed valence compounds. *J. Mag. Mag.Mat.* 78, 208.

118. Etienne P., Chazelas J., Creuzet G., Friederich A., Massies J., Nguyen Van Dau F., Fert A. (1989) Growth of single crystal Fe/Cr multilayers on GaAs(001) by molecular beam epitaxy. *J. Crystal Growth* 95, 410.

119. O'Shea M.J., Fert A. (1989) Movable uniaxial anisotropy in amorphous Tb Fe. *Europhys. Letters* 9, 283.

120. Levy P.M., Fert A. (1989) Effect of crystal fields on the Hall effect in Kondo type systems. *Phys. Rev.* B 39, 12224.

121. Biéri J.B., Bertrand D., Fert A., Redoules J. (1989) Distribution of crystal electric fields in amorphous rare earth alloys. *J. Mag. Mag. Mat.* 80, 246.

122. Cherradi N., Audouard A., Marchal G., Broto J.M., Fert A. (1989) Electron localization effects in Au/Si multilayers. *Phys. Rev.* B 39, 7424.

123. Etienne P., Massies J., Nguyen Van Dau F., Barthélémy A., Fert A. (1989) Critical steps in the molecular beam epitaxy of high quality Ag/Fe superlattices on GaAs(001). *Appl. Phys. Letters* 55, 2239.

124. Barthélémy A., Baibich M.N., Broto J.M., Cabanel R., Creuzet G., Etienne P., Fert A., Friederich A., Lequien S., Nguyen Van Dau F., Ounadjela K. (1989) Giant Magnetoresistance of Fe/Cr superlattices. *Growth and Characterization and Properties of Ultrathin Magnetic Films and Multilayers* (edited by Jonker B.T., Heremans J.P., Marinero E.E. – MRS) Vol. 151, 43.

125. Audouard A., Cherradi N., Broto J.M., Marchal G., Fert A. (1990) Electron localization and interaction effects in three dimensional Au Si amorphous alloys and Au/Si multilayers. *J. of Physics* (Condens. Matter) 2, 377.

126. Barthélémy A., Fert A., Etienne P., Cabanel R., Lequien S., Nguyen Van Dau F., Creuzet G. (1990) Magnetic and Transport Properties of Fe/Cr Superlattices. *J. Appl. Phys.* 67, 5908.

127. O'Shea M.J., Lee K., Fert A. (1990) The magnetic state and its macroscopic anisotropy in amorphous rare earth alloys films. *J. Appl. Phys.* 67, 5769.

128. Levy P.M., Ounadjela K., Fert A., Sommers C.B., Zhang S. (1990) Theory of magnetic superlattices: Interlayer exchange coupling and magneto resistance of transition metal structures. *J. Appl. Phys.* 67, 5914.

129. Etienne P., Lequien S., Massies J., Cabanel R., Fert A., Barthélémy A. (1990) A comparative study of the MBE growth of Ag/Fe, Ag/Cr and Fe/Cr superlattices on GaAs(001). *J. Appl. Phys.* 67, 5400.

130. Cabanel R., Etienne P., Lequien S., Creuzet G., Barthélémy A., Fert A. (1990) Magnetic properties of Fe bcc(001)/Ag fcc(001) superlattices. *J. Appl. Phys.* 67, 5409 (1990).

131. Fert A. (1990) "Magnetic and Transport Properties of Multilayers". *Metallic Multilayers, Materials Science Forum*, Vols. 59–60 *(Trans. Tech. Publications, Zurich)* 439.

132. Audouard A., Kazoun A., Broto J.M., Marchal G., Fert A. (1990) Temperature and field dependence of the 2D Conductivity in Au Si /Si and Cu/Si multilayers. *Phys. Rev.* B 42, 2728.

133. Levy P.M., Zhang S., Fert A. (1990) Electrical conductivity of magnetic multilayered structures. *Phys. Rev. Letters* 65, 1643.

134. Petroff F., Barthélémy A., Fert A., Etienne P., Lequien S. (1990) Magnetoresistance of Fe/Cr superlattices. *J. Mag.Mag. Mat.* 93, 95.

135. Mosca D.H., Barthélémy A., Petroff F., Fert A., Schroeder P.A. (1990) Magnetoresistance of Co-based multilayered Structures. *J. Mag. Mag. Mat.* 93, 480.

136. Ounadjela K., Sommers C.B., Fert A., D. Stoeffer, F. Gautier, V.L. Moruzzi (1991) Electronic Structure and Interlayer coupling in Fe/Cr Superlattices. *Europhysics Letters* 15, 875 (1991).

137. d'Onofrio L., Hamzic A., Fert A. (1991) Magnetotransport properties of YbNiSn. *Physica* B 171, 266.

138. Hamzic A., Fert A., Bauer E. (1991) Hall Effect in CeCu4Ga and CeCu3Ga2 alloys. *Physica* B 171, 263.

139. Mosca D.H., Petroff F., Fert A., Schroeder P.A. (1991) Oscillatory interlayer coupling in Co/Cu magnetic multilayers. *J. Mag. Mag. Mat.* 94, 1.

140. Pereira P., O'Shea M.J., Fert A. (1991) Magnetic properties and structure ofg DyNi/Mo multilayers. *J. Appl. Phys.* 69, 5992.

141. Barthélémy A., Fert A. (1991) Theory of the magnetoresistance in magnetic multilayers: Analytical expressions from a semiclassical approach. *Phys. Rev.* B 43, 13124.

142. Petroff F., Barthélémy A., Mosca D.H., Lottis D.K., Fert A., Schroeder P.A., Lequien S. (1991) Oscillatory exchange and magnetoresistance in Fe/Cu multilayers. *Phys. Rev.* B 44, 5355.

143. Fert A. (1991) Transport properties of metallic multilayers. *Science and Technology of Nanostructured Magnetic Materials* (Hadjipanayis G.C., Prinz G.A. — NATO ASI Series B Physics, Springer) Vol. 259, 221.

144. Zhang S., Levy P.M., Fert A. (1992) The conductivity and magnetoresistance of magnetic multilayers structures. *Phys. Rev.* B 45, 8689 (1992).

145. Fert A., Lottis D.K. (1992) Magnetotransport phenomena. *Concise Encyclopedia of Magnetic and Superconducting Materials* (edited by J. Evetts, Pergamon Press) 287.

146. Barthélémy A., Fert A., Etienne P., Cabanel R., Lequien S. (1992) Temperature dependence of the interface anisotropy in Fe(001)/Ag(001) superlattices. *J. Mag. Mag. Mat.* 104–107, 1816.

147. Lottis D.K., Petroff F., Fert A., Konczykowski M. (1992) Local Hall probe magnetometry: Application to Fe/Cu multilayers. *J. Mag. Mag. Mat.* 104–107, 1811.

148. Fert A., Barthélémy A., Etienne P., Lequien S., Lottis D.K., Mosca D.H., Petroff F., Schroeder P.A. (1992) Magnetic multilayers: Oscillatory interlayer exchange and giant magnetoresistance. *J. Mag. Mag. Mat.* 104–107, 1712.

149. Thiaville A., Fert A. (1992) Twisted spin configuration in thin magnetic layers with interface anisotropy. *J. Mag. Mag. Mat.* 113, 161.

150. Piraux L., Fert A., Schroeder P.A., Loloee R., Etienne P. (1992) Large magneto-thermoelectric power in Co/Cu, Fe/Cu and Fe/Cr multilayers. *J. Mag. Mag. Mat.* 110, L247.

151. Pizzini S., Baudelet F., Chandesris D., Magnan H. George J.-M., Petroff F., Fert A. (1992) Structural characterization of Fe/Cu multilayers by EXAFS. *Phys. Rev.* B 46, 1253.

152. Valet T., Jacquet J.C., Galtier P., Coutellier J.M. et Pereira L.G., Morel R., Lottis D., Fert A. (1992) Interplay between oscillatory exchange coupling and coercivities in giant magnetoresistive [$Ni_{80}Fe_{20}$/Cu/Co/Cu] multilayers. *Appl. Phys. Lett.* 61, 3187.

153. Pizzini S., Baudelet F., Fontaine A., Chandesris D., Magnan H., Fert A., Petroff F. et C. Marlière (1992) Structure cristallographique de multicouches métalliques et magnétiques étudiés par spectroscopie d'absorption X. *J. de Physique* IV, C3–185.

154. Valet T., Fert A. (1993) Theory of the perpendicular magnetoresistance in magnetic multilayers. *Magnetism and Structure in Systems of Reduced Dimension* (edited by R.F.C. Farrow et al, Plenum Press, New York) 143.

155. Hamzic A., Fert A. (1993) Hall effect in heavy fermion systems. *Frontiers in Solid State Sciences*, Volume 2, 131.

156. Valet T., Fert A. (1993) Classical theory of perpendicular giant magnetoresistance in magnetic multilayers. *J. Mag. Mag. Mat.* 121, 378.

157. Schuhl A., Childress J.R., George J.-M., Galtier P., Durand O., Barthélémy A., Fert A. (1993) Magnetic and magnetotransport properties of Pd-based superlattices. *J. Mag. Mag. Mat.* 121, 275.

158. Durand O., George J.-M., Childress J.R., Lequien S., Schuhl A., Fert A. (1993) Low ferromagnetic moment observed in fcc (111) Fe/Cu multilayers. *J. Mag. Mag. Mat.* 121, 140.

159. Lottis D., Fert A., Morel R., Pereira L.G., Jacquet J.C., Galtier P., Coutellier J.M., Valet T. (1993) Magnetoresistance in rf-sputtered (NiFe/Cu/Co/Cu) spin-valve multilayers. *J. Appl. Phys.* 73, 5515.

160. Pizzini S., Baudelet F., Fontaine A., Chandesris D., Magnan H., Fert A., C. Marlière (1993) Structure of metallic multilayers studied by X-ray absorption spectroscopy. *Applied Surface Science* 69, 7.

161. Valet T., Fert A. (1993) Theory of the perpendicular magnetoresistance in magnetic multilayers. *Phys. Rev.* B 48, 7099.

162. Childress J.R., Schuhl A., George J.-M., Durand O., Galtier P., Cros V., Ounadjela K., Kergoat R., Fert A. (1993) Magnetic and magnetotransport properties of epitaxial (100) Fe/Pd superlattices. *Magnetism and Structure in Systems of Reduced Dimension*, (edited by R.F.C. Farrow *et al.*, Plenum Press, New York) 165.

163. Piraux L., Fert A., Schroeder P.A., Loloee R., Etienne P., Valet T. (1993). Thermoelectric power of magnetic multilayers. *Magnetism and Structure in Systems of Reduced Dimension* (edited by R.F.C. Farrow *et al.*, Plenum Press, New York) 335.

164. George J.-M., Barthélémy A., Durand O., Duvail J.L., Fert A., Galtier P., Heckmann O., Pereira L.G., Petroff F., Valet T. (1993) Structural and magnetoresistance properties of Co/Cu multilayers doped with Fe. *Materials Research Society Symposium Proceedings* Vol. 313, 737.

165. Barnas J., Fert A. (1993) Transport in magnetic layered structures. *Physica Scripta*, Vol. T49, 264.

166. Fert A., Valet T., Barnas J. (1994) Perpendicular magnetoresistance in magnetic multilayers: Theoretical model and discussion. *J. Appl. Phys.* 75, 6693.

167. George J.-M., Pereira L.G., Barthélémy A., Petroff F., Steren L., Duvail J.L., Fert A., Loloee R., Holody P., Schroeder P.A. (1994) Inverse spin-valve-type magnetoresistance in spin engineered multilayered structures. *Phys. Rev. Lett.* 72, 408.

168. Childress J.R., Duvail J.L., Jasmin S., Barthélémy A., Fert A., Schuhl A., Durand O., Galtier P. (1994) Perpendicular magnetic anisotropy in Co_xPd_{1-x} alloy films grown by molecular Beam epitaxy. *J. Appl. Phys.* 75, 6412.

169. Duvail J.L., Fert A., Pereira L.G., Lottis D.K. (1994) Calculation of the temperature dependence of the giant MR and application to Co/Cu multilayers. *J. Appl. Phys.* 75, 7070.

170. Barnas J., Fert A. (1994) Interface resistance for perpendicular transport in layered magnetic structures. *Phys. Rev.* B 49, 12835.

171. Holody P., Steren L.B., Morel R., Fert A., Loloee R., Schroeder P.A. (1994) Giant magnetoresistance in hybrid magnetic nanostructures including both layers and clusters. *Phys. Rev.* B 50, 12 999 (1994).

172. Barnas J., Fert A. (1994) Interfacial scattering and interface resistance for perpendicular transport in multilayers. *J. Mag. Mag. Mat.* 136, 260.

173. Piraux L., George J.-M., Despres J.F., Leroy C., Ferain E., Legras R., Ounadjela K., Fert A. (1994) Giant magnetoresistance in magnetic multilayered nanowires. *Appl. Phys. Lett.* 65, 2484.

174. Fert A., Bruno P. (1994) Interlayer coupling and magnetoresitances in multilayers. *Ultrathin Magnetic Structures* Vol. II, B. Heinrich and J.A.C. Bland (Eds), Springer Verlag, 82.

175. Barthélémy A., Duvail J.L., Etienne P., Fert A., George J.-M., Morel R., Pereira L.G., Petroff F., Schuhl A., Schroeder P.A. (1994) Oscillatory interlayer exchange and giant magnetoresistance in magnetic multilayers. *Journal of Low Dimensional Structures*, 3, 17.

176. Steren L., Barthélémy A., Duvail J.L., Fert A., Morel R., Petroff F., Holody P., Loloee R., Schroeder P.A. (1995) Angular dependence of the giant magnetoresistance effect. *Phys. Rev.* B 51, 292.

177. Loloee R., Schroeder P.A., Pratt W.P. Jr, Bass J., Fert A. (1995) Giant magnetoresistance in Ag/Co and Cu/Co multilayers with very thin Co layers. *Physica* B 204, 274.

178. Fert A., Grünberg P., Barthélémy A., Leng Q., Petroff F., W. Zinn (1995) Layered Magnetic Structures: Interlayer Exchange Coupling and Giant Magnetoresistance. *J. Mag. Mag. Mat.* 140–144, 1.

179. Steren L., Morel R., Barthélémy A., Petroff F., Fert A., Holody P., Loloee R., Schroeder P.A. (1995) Giant Magnetoresistance in Hybrid Magnetic Nanostructures. *J. Mag. Mag. Mat.* 140–144, 495.

180. Cros V., Duvail J.L., Steren L., Barthélémy A., Petroff F., Fert A., Durand O., Schuhl A. (1995) A new multilayer system for GMR: $Pd_{1-x}Co_x/Ag$. *J. Mag. Mag. Mat.* 140–144, 611.

181. Barthélémy A., Cros V., Duvail J.L., Fert A., Morel R., Parent F., Petroff F., Steren L.B. (1995) Giant Magnetoresistance in magnetic nanostructures. *Nanostructured Materials* 6, 217.

182. Barnas J.and Fert A. (1995) Interfacial scattering in layered magnetic structures. *J. Mag. Mag. Mat.* 140–144, 509.

183. Fert A., Barthélémy A., Galtier P., Holody P., Morel R., Petroff F., P. Schroeder, Steren L.B., Valet T. (1995) Giant Magnetoresistance in magnetic nanostructures. Recent developments. *Mater. Sci. Eng.* B31 1–9.

184. Petroff F., Cros V., Fert A., Lamolle S., Wiedmann M., Schuhl A. (1995) Two-dimensional magnetic properties of PdFe layers. *Mat. Res. Soc. Symp. Proc.* Vol. 384, 259.

185. Fert A., Duvail J.L., Valet T. (1995) Spin relaxation effects in the perpendicular magnetoresistance of magnetic multilayers. *Phys. Rev.* B 52, 6513.

186. Duvail J.L., Barthélémy A., Steren L.B., Morel R., Petroff F., Sussiau M., Wiedmann M., Fert A., Holody P., Loloee R., Schroeder P.A. (1995) Giant magnetoresistance in hybrid nanostructures. *J. Mag. Mag. Mat.* 151, 324.

187. Schroeder P.A., Holody P., Loloee R., Duvail J.L., Barthélémy A., Steren L.B., Morel R., Fert A. (1995) Giant magnetoresistance in hybrid magnetic nanostructures including both layers and clusters. *Mat. Res. Proc.* Vol. 384, 415.

188. Fert A., Lee S.F. (1996) Theory of bipolar spin switch. *Phys. Rev.* B 53, 6554.

189. Faini G., Cornette A., Cros V., Lee S.F., Barthélémy A., Petroff F., Fert A. (1996) Fabrication of micro-sensors integrated with single nanometric magnetic particles: Detection of the reversal of the magnetization. *Microelectronic Engineering* 30,483.

190. Piraux L., Dubois R., Fert A. (1996) Perpendicular giant magnetoresistance in multilayered nanowires. *J. Mag. Mag. Mat.* 159, 287.

191. Dauguet P., Gandit P., Chaussy J., Lee S.F., Fert A., Holody P. (1996) Angular dependence of the perpendicular giant magnetoresistance of multilayers. *Phys. Rev.* B 54, 1083.

192. Piraux L., Dubois S., Marchal C., Beuken J.M., Filipozzi L., Desprès J.F., Ounadjela K., Fert A. (1996) Perpendicular magnetoresistance in Co/Cu multilayered nanowires. *J. Mag. Mag. Mat.* 156, 317.

193. Fert A., Morel R., Barthélémy A., Cros V., Duvail J.L., George J.-M., Parent F., Pétroff F., Vouille C. (1996) Oscillatory interlayer exchange and giant magnetoresistance in magnetic multilayers. The 8th Latin American congress on surface science: Surfaces, vacuum, and their applications — *AIP Conference Proceedings* 378, 466.

194. Morel R., Steren L.B., Barthélémy A., Parent F., Fert A. (1996) Giant magnetoresistance and clusters size distribution in Co/Ag granular monolayers. *Surf. Rev. Letters* 3, 1065.

195. Ben Youssef J., Le Gall H., Bouziane K., El Harfaoui M., Koshkina O., Desvignes J.M., Fert A. (1996) Faisabilite et croissance avec haute reproductibilitè de multicouches métalliques à magnétorésistance géante par pulvérisation diode RF. *J. de Phys.* C7, 161.

196. Dubois S., Beuken J.M., Piraux L., George J.-M., Duvail J.L., Maurice J.L., Fert A. (1997) Perpendicular giant magnetoresistance of NiFe/Cu and Co/Cu multilayered nanowires. *J. Mag. Mag. Mat.* 165, 30.

197. Fert A., Lee S.F. (1997) Spin injection: Theory and application to Johnson's spin switch. *J. Mag. Mag. Mat.* 165, 115.

198. Dubois S., Marchal C., Beuken J. M., Piraux L., Duvail J. L., Fert A., George J.-M., Maurice J.L. (1997) Perpendicular giant magnetoresistance in NiFe/Cu multilayered nanowires. *Appl. Phys. Lett.* 70, 396.

199. Cros V., Lee S.F., Faini G., Hamzic A., Fert A. (1997) Detection of the magnetization reversal in submicron Co particles by GMR measurements. *J. Mag. Mag. Mat.* 165, 512.

200. Ben Youssef J., Le Gall H., Bouziane K., El Harfaoui M., Koshkina O., Desvignes J.M., Fert A. (1997) Correlation of GMR with texture and interfacial roughness in optimized rf sputtering deposited Co/Cu multilayers. *J. Mag. Mag. Mat.* 165, 288.

201. Bouziane K., Ben Youssef J., Koshkina O., Le Gall H., Desvignes J.M., Fert A. (1997) Comparative study of giant magnetoresistance in CoZr/Cu and Co/Cu multilayers. *J. Mag. Mag. Mat.* 165, 284.

202. Dupuis V., Tuaillon J., Perez A., Melinon P., Parent F., Steren L.B., Morel R., Fert A., Mangin S., Barbara B. (1997) From superparamagnetism to magnetically ordered structures in systems of transition metal clusters embedded in matrices. *J. Mag. Mag. Mat.* 165, 42.

203. Fettar F., Streren L.B., Barthélémy A., Morel R., Fert A., Barnard J.A., Jarratt J.D. (1997) GMR in CoFe/Ag multilayers with discontinuous magnetic layers. *J. Mag. Mag. Mat.* 165, 316.

204. Piraux L., Dubois S., Ferain E., Legras R., George J.-M., Duvail J.L., Fert A. (1997) Anisotropic transport and magnetic properties of arrays of submicronic wires. *J. Mag. Mag. Mat.* 165, 352.

205. Fert A., Morel R., Barthélémy A., Cros V., Duvail J.L., George J.-M., Parent F., Pétroff F., Vouille C. (1997) Magnétorésistance géante dans les nanostructures magnétiques. *J. de Physique* C 6–151.

206. Parent F., Teraillon J., Steren L.B., Dupuis V., Perez A., Morel R., Fert A. (1997) Giant magnetoresistance in Co-Ag granular films prepared by low energy cluster beam deposition. *Phys. Rev.* B 55, 3683.

207. Vouille C., Fert A., Barthélémy A., Hsu S.Y., Loloee R., Schroeder P.A. (1997) Inverse CPP-GMR in (A/Cu/Co/Cu) multilayers (A = NiCr, FeCr, FeV) and discussion of the spin asymmetry induced by impurities. *J. Appl. Phys.* 81, 4573.

208. Hsu S.Y., Barthélémy A., Holody P., Loloee R., Schroeder P.A., Fert A. (1997) Towards a unified picture of spin dependent transport in longitudinal GMR, perpendicular GMR and bulk alloys. *Phys. Rev. Lett.* 78, 2652.

209. Schelp L.F., Fert A., Fettar F., Holody P., Lee S.F., Maurice J.L., Petroff F., Vaurès A. (1997) Spin dependent tunneling with Coulomb blockade. *Phys. Rev.* B 56, R5747.

210. Viret M., Drouet M., Contour J-P., Nassar J., Fermon C., Fert A. (1997) Low-field colossal magnetoresistance in manganite tunnel spin valves. *Europhysics Letters* 39, 545.

211. Barnas J., Baksalary O., Fert A. (1997) Angular dependence of giant magnetoresistance in magnetic multilayers. *Phys. Rev.* B 56, 6079.

212. Barnas J., Lee S.F., Holody P., Petroff F., Fert A. (1998) Tunneling magnetoresistance in ferromagnetic junctions: bias dependence. *Acta Physica polonica* A 93, 387.

213. Desmicht R., Farini G., Lee S.F., Fert A., Petroff F., Vaures A. (1998) Coulomb blockade in mesoscopic tunneling devices. *Appl. Phys. Letters* 72, 386.

214. Barnas J., Fert A. (1998) Magnetoresistance oscillations due to charging effects in double ferromagnetic tunnel junctions. *Phys. Rev. Letters* 80, 1058.

215. Maurice J.L., Himhoff D., Etienne P., Durand O., Dubois S., Piraux L., George J.-M., Galtier P., Fert A. (1998) Microstructure of magnetic multilayers grown by electrodeposition in membranes nanopores. *J. Mag. Mag. Mat.* 184, 1.

216. Schroeder P.A., Holody P., Loloee R., Vouille C., Barthélémy A., Fert A., Hsu S.Y. (1998) Unified picture of spin-dependent transport in GMR multilayers and ferromagnetic alloys. *J. Mag. Mag. Mat.*177–181, 1464.

217. Nassar J., Viret M., Drouet M., Contour J-P., Fermon C., Fert A. (1998) Low-field colossal magnetoresistance in manganite tunnel junctions. *Mat. Res. Soc. Symp. Proc.* 494, 231.

218. Briatico J., Maurice J.L., Petroff F., Seneor P., Vaurès A., Fert A. (1998) TEM of cobalt clusters embeddedin amorphous alumina. *Electron Microscopy* 417.

219. Nassar J., HehnM., Vaures A., Petroff F., Fert A. (1998) Magnetoresistance of ferromagnetic tunnel junctions with alumina barriers formed by rf sputter-etching in Ar/O_2 plasma. *Appl. Phys. Letters* 73, 698.

220. Montaigne F., Nassar J., Vaurès A., Nguyen Van Dau F., Petroff F., Schuhl A., Fert A. (1998) Enhanced tunnel magnetoresistance at high bias voltage in double barrier planar junctions. *Appl. Phys. Letters* 73, 2829.

221. Fettar F., Maurice J.L., Petroff F., Schelp L.F., Vaurès A., Fert A. (1998) TEM observations of nanometer thick cobalt layers deposits in alumina sandwiches. *Thin Solid Films* 319, 120.

222. Piraux L., Dubois S., Fert A., Belliard L. (1998) Temperature dependence of the perpendicular giant magnetoresistance in Co/Cu multilayered nanowires. *European J. of Physics* B 4, 413.

223. Barnas J., Fert A. (1998) Effects of spin accumulation on single-electron tunneling in a double ferromagnetic microjunction. *Europhys. Letters* 44, 85.

224. Duvail J. L., Dubois S., Piraux L., Vaurès A., Fert A., Rousseaux F., Decanini D. (1998) Electrodeposition of patterned magnetic nanostructures. *J. Appl. Phys.* 84, 6359.

225. Barnas J., Lee S.F., Holody P., Petroff F., Fert A. (1998) Tunneling magnetoresistance in ferromagnetic junctions: bias dependence. *Acta Physica polonica* A 93, 387.

226. Briatico J., Maurice J.L., Petroff F., Seneor P., Vaures A., Fert A. (1998) TEM of Co clusters embedded in amorphous alumina. *Electron Microscopy* 417.

227. Barnas J.and Fert A. (1998) Effects of spin accumulation on single-electron tunneling in a double ferromagnetic microjunction. *Europhys. Letters* 44, 85.

228. Park W., Loloee R., Caballero J., Pratt W.P., Schroeder P.A., Bass J., Fert A., Vouille C. (1999) Test of unified picture of spin dependent transport in perpendicular (CPP) giant magnetoresistance and bulk alloys. *J. Appl. Phys* 85, 4542.

229. Barnas J., Fert A. (1999) Interplay of spin accumulation and Coulomb blockade in double ferromagnetic tunnel junctions. *J. Mag. Mag. Mat.* 192, 391.

230. Dubois S., Piraux L., George J.-M., Ounadjela K., Duvail J.L., Fert A. (1999) Evidence for short spin diffusion length in permalloy from the giant magnetoresistance of multilayered nanowires. *Phys. Rev.* B 60, 477.

231. Seneor P., Fert A., Maurice J.L., Montaigne F., Petroff F., Vaurès A. (1999) Large magnetoresistance in tunnel junctions with an iron oxide electrode. *Appl. Phys. Letters* 74, 4017.

232. Vouille C., Barthélémy A., Fert A., Schroeder P.A., Hsu S. H., Reilly A., Loloee R. (1999) Microscopic mechanisms of the Giant Magnetoresistance. *Phys. Rev.* B 60, 6710.

233. De Teresa J.M., Barthélémy A., Fert A., Contour J-P., Lyonnet R., Montaigne F., Seneor P., Vaurès A. (1999) Inverse tunnel magnetoresistance in $Co/SrTiO_3/La_{0.7}Sr_{0.3}MnO_3$: New ideas on spin polarized tunneling. *Phys. Rev.* Letters. 82, 4288.

234. Fert A., Vouille C. (1999) Magnetoresistance overview: AMR, GMR, TMR, CMR. *Magnetische Schichtsysteme* (edited by Deutcher Zukunftspreis), D1.1.

235. De Teresa J.M., Barthélémy A., Fert A., Contour J-P., Montaigne F., Seneor P. (1999) Role of the metal-oxide interface in determining the spin polarization of magnetic tunnel junctions. *Science* 286, 507.

236. Viret M., Nassar J., Contour J-P., Fermon C., Fert A. (1999) Spin polarized tunneling as a probe of half-metallic ferromagnetism in mixed-valence manganites. *J. Mag. Mag. Mat.* 198–199, 1.

237. Fert A., Piraux L. (1999) Magnetic nanowires. Special Issue "Magnetism Beyond 2000" of *J. Mag. Mag. Mat.*, Vol. 200, 338–358.

238. Barthélémy A., Fert A., Petroff F. (1999) Giant Magnetoresistance of Magnetic Multilayers. *Handbook of Magnetic Materials* (edité par K. Buschow, Elsevier), p. 1–98.

239. Barnas J., Fert A. (1999) Electronic transport in ferromagnetic heterostructures in metallic and tunneling regimes. *Molecular Physics Reports* 24, 11.

240. De Teresa J.M., Barthélémy A., Fert A., Contour J-P., Lyonnet R., Seneor P., Vaurès A., Montaigne F. (1999) Inverse magnetoresistance in Manganite/$SrTiO_3$/Co tunel junctions. *Mater. Res. Soc. Proc.* 574, 293.

241. Martinek J., Barnas J., Michalek G., Bulka B.R., Fert A. (1999) Spin effects in single-electron tunneling in magnetic junctions. *J. Mag. Mag. Mat.* 207, 1.

242. De Teresa J.M., Barthélémy A., Contour J-P., Fert A. (2000) Manganite-based magnetic tunnel junctions: New ideas on spin-polarized tunneling. *J. Mag. Mag. Mat.* 211, 160.

243. Montaigne F., Gogol P., Maurice J.L., Nguyen Van Dau F., Petroff F., Fert A., Schuhl A. (2000) Magnetoresistive tunnel jun ctions deposited on laterally modulated substrates. *Appl. Phys. Lett.* 76, 3286.

244. Barthélémy A., Fert A., Nguyen Van Dau F. (2000) Giant magnetoresistance. *Encyclopedia of Materials: Science and Technology* (edited by Buschow K.H.J., Kahn R.W., Flemings M.C., Ilschner B., Kramer E.J., Mahajan S. — Elsevier Science) 3521.

245. Barnas J., Martinek J., Michalek G., Bulka B.R., Fert A. (2000) Spin effects in ferromagnetic single electron transistor. *Phys. Rev.* B 62, 12363.

246. Fert A., Barthélémy A., Ben Youssef J., Contour J-P., Cros V., De Teresa J.M., Hamzic A., George J.-M., Faini G., Grollier J., Jaffres H., Le Gall H. (2001) Review of recent results on spin polarized tunneling and magnetic switching by spin injection. *Mat. Sci. Eng.* B 84, 1.

247. Grollier J., Cros V., Hanzic A., George J.-M., Jaffres H., Fert A., Faini G., Ben Youssef J., Le Gall H. (2001) Magnetization reversal by spin injection in Co/Cu/Co pillars. *Appl. Phys. Lett.* 78, 3663.

248. Bowen M., Cros V., Petroff F., Fert A., Martinez Boubeta C., Cebollada A., Briones F. (2001) Large magnetoresistance in Fe/MgO/FeCo(001) epitaxial tunnel junctions on GaAs(001). *Appl. Phys. Lett.* 79, 1655.

249. Fert A., Jaffrès H. (2001) Conditions for efficient spin injection from a ferromagnetic metal into a semiconductor. *Phys. Rev.* B 64, 184420.

250. Eddrief M., Wang Y., Etgens V.H., Mosca D.H., George J.-M., Fert A., Bourgognon C. (2001) Epitaxial growth and magnetic properties of Fe(111) films on Si(111) substrates using GaSe(001) as a template. *Phys. Rev.* B 63, 144506.

251. Mosca D.H., George J.-M., Maurice J-L., Fert A., Eddrief M., Etgens V.H. (2001) Magnetoresistance of Fe/ZnSe/Fe planar junctions. *J. Mag. Mag. Mat.* 226–231, 932.

252. Carrey J., Bouzehouane K., George J.-M., Ceneray C., Fert A., Vaurès A. (2001) Resistance measurements of nanocontacts made by electrodeposition into pinholes of alumina layers. *Appl. Phys. Lett.* 79, 3158.

253. Mosca D.H., George J.-M., Maurice J-L., Fert A., Eddrief M., Etgens V.H. (2001) Magnetoresistance of Fe/ZnSe/Fe tunnel junctions. *J. Mag. Mag. Mat.* 226, 932.

254. Zhang S., Levy P.M., Fert A. (2002) Mechanisms of spin-polarized current-driven magnetization switching. *Phys. Rev. Letters* 88, 236601.

255. Jaffrès H., Fert A. (2002) Spin injection from a ferromagnetic metal into a semiconductor. *J. Appl. Phys.* 91, 8111.

256. Fettar F., Lee S.F., Petroff F., Vaurès A., Holody P., Schelp L.F., Fert A. (2002) Temperature and voltage dependence of the resistance and magnetoresistance of discontinuous double tunnel junctions. *Phys. Rev.* B 65, 174415.

257. Fert A., George J.-M., Jaffrès H. (2002) Spin injection and experimental detection of spin accumulation. *J. of Physics D: Appl. Phys.* 35, 2443.

258. Pailloux F., Imhoff D., Sikora T., Barthélémy A. , Maurice J-L., Contour J-P., Colliex C., Fert A. (2002) Nanoscale analysis of $SrTiO_3/La_{2/3}Sr_{1/3}MnO_3$ interfaces. *Phys. Rev.* B 66, 014417.

259. Grollier J., Lacour D., Cros V., Hamzic A., Vaurès A., Fert A. (2002) Switching of the magnetic configuration of a spin valve by current-induced wall motion. *J. Appl. Phys.* 92, 4825.

260. Barthélémy A., Fert A., Contour J-P., Bowen M., Cros V., De Teresa J.M., Hamzic A., Faini G., George J.-M., Grollier J., Montaigne F., Pailloux F., Petroff F., Vouille C. (2002) Giant magnetoresistance and spin electronics. *J. Mag. Mag. Mat.* 242, 68.

261. Besse M., Cros V., Barthélémy A., Vogel J., Petroff F., Mirone A., Decorse P., Berthet P., Szotek Z., Temmerman W.M., Dhesi S.S., Brookes N.B., Rogalev A., Fert A. (2002) Experimental evidence of ferrimagnetic structure in Sr_2FeMoO_6: A requirement for half-metallicity. *Europhys. Lett.* 60, 608.

262. Besse M., Pailloux F., Barthélémy A., Bouzehouane K., Fert A., Olivier J., Durand O., Wyczisk F., Bisarro R., Contour J-P. (2002) Suitable methods to identify epitaxial Sr_2FeMoO_6 thin films. *J. Crystal Growth* 241, 448.

263. Bowen M., Bibes M., Barthélémy A., Contour J-P., Anane A., Lemaitre Y., Fert A. (2003) Nearly total spin polarization in $La_{2/3}Sr_{1/3}MnO_3$ from tunneling experiments. *Appl. Phys. Letters* 82, 233.

264. George J.-M., Fert A., Faini G. (2003) Direct measurement of spin accumulation in a metallic mesoscopic structure. *Phys. Rev.* B 67, 012410.

265. Mattana R., George J.-M., Jaffrès H., Nguyen Van Dau F., Fert A., Lépine B., Guivarch A., Jézéquel G. (2003) Electrical detection of spin accumulation in a p-type GaAs quantum well. *Phys. Rev. Letters* 90, 166601.

266. Maurice J-L., Pailloux F., Imhoff D., Contour J-P., Barthélémy A., Bowen M., Colliex C., Fert A. (2003) Nanoscale analysis of a $Co-SrTiO_3$ interface in a magnetic tunnel junction. *MRS Proceedings* 746, 145.

267. Grollier J., Cros V., Jaffrès H., Faini G., Ben Youssef J., Le Gall H., Fert A. (2003) Field dependence of magnetization reversal by spin transfer. *Phys. Rev.* B 67, 174402.

268. Grollier J., Boulenc P., Cros V., Hamzic A., Vaures A., Fert A., Faini G. (2003) Switching a spin-valve back and forth by current-induced domain wall motion. *Appl. Phys. Lett.* 83, 509.

269. Samet L., Himhoff D., Maurice J-L., Contour J-P., Gloter A., Manoubi T., Fert A., Colliex C. (2003) EELS study of interfaces in magnetoresistive LSMO/STO/LSMO tunnnel junctions. *Europhysics. J.* B 34, 179.

270. Fert A., George J.-M., Jaffrès H., Mattana R., Seneor P. (2003) The new era of spintronics. *Europhysics News* Vol. 34, No. 6.

271. Maurice J-L., Pailloux F., Imhoff D., Bonnet N., Samet L., Barthélémy A., Contour J-P., Colliex C.and Fert A. (2003) Atomic scale analysis of an all-oxide magnetic tunnel junction. *Euro. Phys. J. Appl. Phys.* 24, 215.

272. Fert A., Cros V., George J.-M., Grollier J., Jaffrès H., Hamzic A., Vaurès A., Faini G., Ben Youssef J., Le Gall H. (2004) Magnetization reversal by injection and transfer of spin. *J. Mag. Mag. Mat.* 272, 1706.

273. Grollier J., Boulenc P., Cros V., Hamzic A., Vaures A., Fert A., Faini G. (2004) Spin-transfer-induced domain wall in a spin valve. *J. Appl. Phys.* 95, 6777.

274. Lim C.K., Devolder T., Chappert C., Grollier J., Cros V., Vaurès A., Fert A., and G. Faini (2004) Domain wall displacement induced by subnanosecond pulsed current. *Appl. Phys. Lett.* 84, 2820.

275. Seneor P., Lidgi N., Carrey J., Jaffres H., Nguyen Van Dau F., Friederich A., Fert A. (2004) Principle of a variable capacitor based on Coulomb blockade of nanometric-size cluster *Europhys. Lett.* 65, 699.

276. Carrey J., Seneor P., Lidgi N., Jaffrès H., Nguyen Van Dau F., Fert A., Friederich A., Montaigne F., and Vaurès A. (2004) Capacitance variation of an assembly of clusters in the Coulomb blockade regime. *J. Appl. Phys.* 95, 1265.

277. V. Garcia, Bibes M., Barthélémy A., Bowen M., Jacquet E., Contour J-P., and Fert A. (2004) Temperature dependence of the interfacial spin polarization of La2/3Sr1/3MnO3. *Phys. Rev.* B 69, 052403.

278. AlHajDarwish M., Kurt H., Urazdhin S., Fert A., Loloee R., Pratt W.P., Bass J. (2004) Controlled normal and inverse current-induced magnetization switching and magnetoresistance in magnetic nanopillars. *Phys. Rev. Letters* 93, 157203.

279. George J.-M., Jaffrès H., Mattana R., Elsen M., Nguyen Van Dau F., Fert A., Lépine B., Guivarch A., Jézéquel G. (2004) Electrical spin detection in GaMnAs-based tunnel junctions: Theory and experiments. *Molecular Physics Reports* 40, 23.

280. AlHajDarwish M., Fert A., Pratt W.P. Jr, and Bass J. (2004) Inverted current-driven switching in Fe(Cr)/Cr/Fe(Cr) nanopillars. *J. Appl. Phys.* 95, 6771.

281. Mattana R., Elsen M., George J.-M., Jaffrès H., Nguyen Van Dau F., Fert A., Wyczisk M.F., Olivier J., Galtier P., Lépine B., Guivarch A., and G. Jézéquel (2005) Chemical profile and magnetoresistance of $Ga_{1-x}M_xAs/GaAs/AlAs/GaAs/Ga_{1-x}M_xAs$ tunnel junctions. *Phys. Rev.* B 71, 075206.

282. Gajek M., Bibes M., Barthélémy A., Bouzehouane K., Fusil S., Varela M., Fontcuberta J., Fert A. (2005) Spin filtering through $BiMnO_3$ tunnel barriers. *Phys. Rev.* B 72, 020406.

283. Barnas J., Fert A., Gmitra M., Weymann I., Dugaev V.K. (2005) From giant magnetoresistance to current-induced switching by spin transfer. *Phys. Rev.* B 72, 024426.

284. Bowen M., Barthélémy A., Bibes M., Jacquet E., Contour J-P., Fert A., Cicacci F., Duo L., Bertacco R. (2005) Spin-polarized tunneling spectroscopy in tunnel junctions with half-metallic electrodes. *Phys. Rev. Lett.* 95, 137203.

285. Gajek M., Herranz G., Bibes M., Fusil S., Bouzehouane K., Varela M., Fontcuberta J., Barthélémy A., Fert A. (2005) Perovskite-based heterostructures integrating ferromagnetic-insulating La0.1Bi0.9MnO3. *J. Appl. Phys.* 97, 103909.

286. Bowen M., Barthélémy A., Bibes M., Jacquet E., Contour J-P., Fert A., Wortmann D., Blügel S. (2005) Half-metallicity proven using fully spin-polarized tunneling. *J. of Physics Cond. Mat.* 17, L407.

287. Elsen M., Boulle O., George J.-M., Jaffrès H., Cros V., Fert A., Lemaître A., Giraud R., Faini G. (2006) Spin tranfer experiments on (GaMn)As/(InGa)As/(GaMn)As tunnel junctions. *Phys. Rev.* B 73, 035303.

288. Gimtra M., Barnas J., Fert A., Weygmann I., Dugaev V.K. (2006) Current induced switching due to spin-transfer torque in spin-valves. *Phys. Stat. Sol.* [c] 3, n°1, 97.

289. Barnas J., Fert A., Gimtra M., Weygmann I., Dugaev V.K. (2006) Macroscopic description of spin-transfer torque. *Mat. Sc. Engin.* B126, 271.

290. Theodoropoulou N., Sharma A., Loloee R., Pratt W.P., Bass J., Fert A., Jaffrès H. (2006) Interface-specific resistance and scattering asymmetry of permalloy. *J. Appl. Phys* 99, 1.

291. Grollier J., Cros V., Fert A. (2006) Synchronization of spin-transfer oscillators driven by stimulated microwave currents. *Phys. Rev.* B (Rapid Com.) 73, 060409 [R].

292. Herranz G., Balestic M., Bibes M., Ranchal R., Hamzic A., Tafra E., Bouzeouhane K., Jacquet E., Contour J-P., Barthélémy A., Fert A. (2006) Full oxide heterostructure combining a high-Tc dilute ferromagnet with a high-mobility conductor. *Phys. Rev.* B 73, 064403.

293. Lûders U., Herranz G., Bibes M., Bouzeouhane K., Jacquet E., Contour J-P., Fusil S., Bobo J-F., Fontcuberta J., Barthélémy A., Fert A. (2006) Hybrid perovskite-spinel magnetc tunnel junctions based on conductive ferromagnetic NiFe$_2$O$_3$. *J. Appl. Phys.* 99, 08K301.

294. Herranz G., Ranchal R., Bibes M., Jaffrès H., Jacquet E., Maurice J-L., Bouzeouhane K. Wyczisk F., Tafra E., Balestic M., Hamzic A., Colliex C., Contour J-P., Barthélémy A., Fert A. (2006) A high Curie temperature dilute magnetic system with large spin polarization. *Phys. Rev. Lett;* 96, 027207.

295. Lüders U., Barthélémy A., Bibes M., Bouzehouane K., Fusil S., Jacquet E., Contour J-P., Bobo J-F., Fontcuberta J., Fert A. (2006) NiFe$_2$O$_3$: A versatile spinel material brings nw opportunities for spintronics. *Adv. Mat.* 18, 1733.

296. Bernand-Mantel A., Seneor P., Lidgi N., Munoz M., Fusil S., Bouzehouane K., Deranlot C., Vaurès A., Petroff F., Fert A. (2006) Evidence for spin injection into a single metallic nanoparticle: A step towards nanospintronics. *Appl. Phys. Lett.* 89, 062502.

297. Lüders U., Bibes M., Bouzehouane K., Jacquet E., Contour J-P., Fusil S., Bobo J-F., Fontcuberta J., Barthélémy A., Fert A. (2006) Spin-filtering through ferrimagnetic NiFe$_2$O$_3$ tunnel barriers. *Appl. Phys. Lett.* 88, 082505.

298. Levy P.M., Fert A. (2006) Spin transfer in magnetic tunel junctions with hot electrons. *Phys. Rev. Lett.* 97, 097205.

299. Fert A., Barthélémy A., Petroff F. (2006) Spin transport in magnetic multilayers and tunnel junctions in Nanomagnetism. *Contemporary Concepts of Condensed matter Science, Elsevier* (edited by D.L. Mills and J.A.C. Bland) Vol. 1, 153–225.

300. Levy P.M., Fert A. (2006) Effect of bias-induced spin flips on spin-torque: Enhancement of current and spin-torque in magnetic tunnel junctions. *Phys. Rev.* B 74, 224446.

301. Herranz G., Balestic M., M; Bibes, Hamzic A., Jaffrès H., Bouzehouane K., E. Jacket, Contour J-P., Barthélémy A., Fert A. (2007) High-spin polarized Co-doped (La,SR)TiO_3 thin films on high-mobility Sr TiO_3 substrates. *J. Mag. Mag. Mat.* 310, 2111.

302. Fert A., George J.-M., Jaffrès H., Mattana R. (2007) Semiconductors between spin-polarized sources and drains (review article). *IEEE Transactions on Electron Devices*, Vol. 54, No. 5 (Special Issue on Spintronics), 921.

303. Laribi S., Cros V., Munoz M., Grollier J., Hamzic A., Deranlot C., Fert A., Martinez E., Lopez-Diaz L., Vila L., Faini G., Zoll S., Fournel R. (2007) Reversible and irreversible current induced domain wall motionin CoFeB based spin valve stripes. *Appl. Phys. Lett.* 90, 232505.

304. Bernand-Mantel A., Seneor P., Lidgi N., Munoz M., Cros V., Fusil S., Bouzehouane K., Deranlot C., Vaurès A., Petroff F., Fert A. (2007) Evidence for spin injection into a single nanoparticle: A step towards nanospintronics. *Appl. Phys. Lett.* 89, 062502.

305. Hueso L.H., Pruneda J.M., Ferrari V., Burnell G., Valdés-Herrera J.P., Simons B.D., Littlewood P.B., Atacho E., Fert A., Mathur N.D. (2007) Transformation of spin information into large electrical signals using carbon nanotubes. *Nature* 445, 410.

306. Gajek M., Bibes M., Fusil S., Bouzehouane K., Fontcuberta J., Barthélémy A., Fert A. (2007) Tunnel junctions with multiferroics barriers. *Nature Materials* 6, 296.

307. Boulle O., Cros V., Grollier J., Pereira L.G., Deranlot C., Petroff F., Faini G., Barnas J., Fert A. (2007) Shaped angular dependence of the spin-transfer torque and microwave generation without magnetic field. *Nature Physics*, 3, 492.

308. Herranz G., Balestic M., Bibes M., Carrétéro C., Tafra E., Jacquet E., Bouzehouane K., Deranlot C., Hamzic A., Broto J-M., Barthélémy A., Fert A. (2007) High mobility in $LaAlO_3$/$SrTiO_3$ heterostructures: Origin, dimensionality and perspective. *Phys. Rev. Lett.* 98, 216803.

309. Lüders U., Bibes M., Fusil S., Bouzehouane K., Jacquet E., Sommers C.B., Contour J-P., Bobo J-F., Barthélémy A., Fert A., Levy P.M. (2007) Bias dependence of tunnel magnetoresistance in spin filtering tunnel junctions: Experiments and theory. *Phys. Rev.* B76, 134412.

310. Chappert C., Fert A., Nguyen Van Dau F. (2007) The emergence of spintronics in data storage. *Nature Materials* 6, 813.

311. Fert A. (2008) Nobel Lecture: Origin, development, and future of spintronics. *Rev. Mod. Phys.* 80, 1517.

312. Fert A. (2008) The present and the future of spintronics. *Thin Solid Films* 517, 2.

313. Georges B., Grollier J., Darques M., Cros V., Deranlot C., Marcilhac B., Faini G., Fert A. (2008) Coupling efficiency for phase locking of a spin-transfer oscillator to a microwave current. *Phys. Rev. Lett.* 101, 123714.

314. Boulle O., Cros V., Grollier J., Deranlot C., Petroff F., Barnas J., Fert A. (2008) Microwave excitations associated with a wavy angular dependence of the spin transfer torque: Model and experiments. *Phys. Rev.* B 77, 174403.

315. Georges B., Grollier J., Cros V., Fert A. (2008) Impact of the electrical connection of spin transfer nano-oscillators on their synchronization: An analytical study. *Appl. Phys. Lett.* 92, 232504 (2008).

316. Fert A. (2008) Origin, development and future of spintronics (Nobel lecture). *Angewandte Chemie-International Edition* 47, 5956.

317. Rode K., Mattana R., Anane A., Cros V., Jacquet E., Contour J-P., Petroff F., Fert A., Arrio M.A., Sainctavit P., Rogalev A. (2008) Magnetism of (Zn, Co)O thin films probed by X-ray absorption spectroscopies. *Appl. Phys. Lett.* 92, 012509.

318. Chappert C., Fert A., Nguyen Van Dau F. (2008) The emergence of spintronics in data storage. *Nature Materials* 4, S32.

319. Khvalkovskiy A.V., Zvezdin K.A., Gorbunov Y., Cros V., Grollier J., Fert A., Zvezdin A.K. (2009) High Domain Wall Velocities due to Spin Currents Perpendicular to the Plane. *Phys. Rev. Lett.* 102, 067206.

320. Tran M., Jaffrès H., Deranlot C., George J.-M., Fert A., Miard A., Lemaître A. (2009) Enhancement of the Spin Accumulation at the Interface between a Spin-Polarized Tunnel Junction and a Semiconductor. *Phys. Rev. Lett.* 102, 036601.

321. Pizzini S., Uhlrich V., Vogel J., Rougemaille N., Laribi S., Cros V., Jimenez E., Camarero J., Bonnet E., Bofim M., Mattana R., Deranlot C., Petroff F., Ulysses C., Faini G., Fert A. (2009) High domain wall velocity at zero magnetic field induced by low current densities in spin valve nanotripes. *Appl. Phys. Express* 2023003.

322. Ruotolo A., Cros V., Georges B., Dussaux A., Grollier J., Deranlot C., Guillemet R., Bouzehouane K., Fusil S., Fert A. (2009) Spin transfer induced dynamics of an array of interacting vortices in a periodic potential. *Nature Nanotechnology*, 4, 528.

323. Matsumoto R., Fukushima A., Yakushijin K., Yakata S., Nangahama T., Kubota H., Katayama T., Suzuki Y., Ando K., Yuasa S., Georges B., Cros V., Grollier J., Fert A. (2009) Spin-torque-induced switching and precession in fully epitaxial Fe/MgO/Fe magnetic tunnel junctions. *Phys. Rev.* B 80, 174405.

324. Georges B., Grollier J., Cros V., Fert A., Fukushima A., Kubota H., Yakushijin K.n, Yuasa S., Ando K. (2009) Origin of the spectral linewidth in nonlinear spin-transfer oscillators based on MgO tunnel junctions. *Phys. Rev.* B 80, 060404®.

325. Georges B., Grollier J., Cros V., Marcilhac B., D.G. Crété, Mage J.-C., Fert A., Fukushima A., Kubota H., Yakushijin K., Yuasa S., Ando K. (2009) Frequency converter based on nanoscale MgO Magnetic Tunnel Junctions. *Appl. Phys. Express.* 2, 123003.

326. Cros V., Fert A., Seneor P., Petroff F. (2009) The 2007 Nobel Prize in Physics "The Spin" in the serie "Progress in Mathematical Physics", Vol. 55, Springer, 147.

327. Bernand-Mantel A., Seneor P., Bouzehouane K., Fusil S., Deranlot C., Petroff F., Fert A. (2009) Anisotropic magneto-Coulomb effects and magnetic single-electron-transistor action in asingle nanoparticle. *Nature Physics* 5, 920.

328. Barraud C., Seneor P., Mattana R., Fusil S., Bouzehouane K., Deranlot C., Graciozi P., Hueso L., Bergenti, Dediu V., Petroff F., Fert A. (2010) Unravelling the role of the interface for spin injection into organic semiconductors. *Nature Physics* 6, 615.

329. Dussaux A., Georges B., Grollier J., Cros V., Khvalkovski A.V., Fukushima A., Konoto M., Kubota H., Yakushijin K., Yuasa S., Zvczdi K.A., Ando K., Fert A. (2010) Large microwave generation from current-driven magnetic vortex oscillators in magnetic tunnel junctions. *Nature Communications* 1, 8.

330. Metaxas P.J., Anane A., Cros V., Grollier J., Deranlot C., Lemaitre Y., Xavir S., Ulysse C., Faini G., Petroff F., Fert A. (2010) Current-induced resonan depinning of a transverse magnetic domain in a spin valve nanostrip. *Appl. Phys. Lett.* 97, 182506.

331. Jaffrès H., Fert A., George J.-M. (2010) Spin transport in multiterminal devices: large spin signals in devices with confined geometry. *Phys. Rev.* B, 140408(R).

332. Bernand-Mantel A., Bouzehouane K., Seneor P., Fusil S., Deranlot C., Brenac A., Notin L., Morel R., Petroff F., Fert A. (2010) A versatile nanotechnology to connect individual nano-objects for the fabrication of hybrid singl-electron devices. *Nanotechnology* 21, 445201.

333. Dlubak B., Seneor P., Anane A., Barraud C., Deranlot C., Deneuve D., Servet B., Mattana R., Petroff F., Fert A. (2010) Are Al_2O_3 ansd MgO tunnel barriers suitable for spin injection in grapheme? *Appl. Phys. Lett.* 97, 092502.

334. Barraud C., Deranlot C., Seneor P., Mattana R., Dlubak B., Fusil S., Bouzehouane K., Deneuve D., Petroff F., Fert A. (2010) Magnetoresistance in magnetic tunnel junctions grown on flexible substrates. *Appl. Phys. Lett.* 96, 022504.

335. Fert A. (2010) Giant magnetoresistance. *Scolarpedia* 6(2): 6982.

336. Locatelli N., Naletov V.V., Grollier J., De Loubens G., Cros V., Deranlot C., Ulysse C., Faini G., Klein O., Fert A. (2011) Dynamics of two coupled vortices in a spin valve nanopillar excited by spin transfer torque. *Appl. Phys. Lett.* 98, 062501.

337. Niimi Y., Morota M., Wei D.H., Deranlot C., Basletic M., Hamzic A., Fert A., Otani Y. (2011) Extrinsic Spin Hall Effect induced by Ir impurities in copper. *Phys. Rev. Lett.* 106, 126601.

338. Fert A., Levy P.M. (2011) Spin Hall Effect induced by resonant scattering on impurities in metals. *Phys. Rev. Lett.* 106, 157208.

339. Dussaux A., Khvalkovskiy A. V., Grollier J., Cros V., Fukushima A., Konoto M., Kubota H., Yakushijin K., Yuasa S., Ando K., and Fert A. (2011) Phase locking of vortex-based spin transfer oscillators. *Appl. Phys. Lett.* 98, 132506.

340. Matsumoto R., Chanthbouala A., Grollier J., Cros V., Fert A., Nishimura K., Nagamine Y., Maehara H., Tsunekawa K., Fukushima A., Yuasa S. (2011) Spin-Torque Diode Measurements of MgO-Based Magnetic Tunnel Junctions with Asymmetric Electrodes. *Appl. Phys. Express* 4, 063001.

341. Grollier J., Chanthbouala A., Matsumoto R., Anane A., Cros V., Nguyen van Dau F., Fert A. (2011) Magnetic domain wall motion by spin transfer. *Compte Rendus de Physique*, 12, 309.

342. Chanthbouala A., Matsumoto R., Grollier J., Cros V., Anane A., Fert A., Khvalkovskiy A.V., Zvezdin K.A., Nishimura K., Nagamine Y., Maehara H., Tsunekawa K., Fukushima A., Yuasa S. (2011) Vertical-current-induced domain-wall motion in MgO-based magnetic tunnel junctions with low current densities. *Nature Physics* 7, 626.

343. Bortolotti P., Dussaux A., Grollier J., Cros V., Fikushima A., Kubota H., Yakushijin K., Yuasa S., Ando K., Fert A. (2012) Temperature dependence of microwave voltage emission associated to spin-transfer induced vorte oscillation in magnetic tunnel junction. *Appl. Phys. Lett.* 100, 042408, 2012.

344. Dussaux, Khvalkovskiy A. V., Bortolotti P., Grollier J., Cros V., and Fert A. (2012) Field dependence of spin-transfer-induced vortex dynamics in the nonlinear regime. *Phys. Rev.* B 86, 014402.

345. Fert A. (2012) A Treasure in My Office: Friedel's Publications and Lecture Notes. *Journal of Superconductivity and Novel Magnetism* 25, Issue 3, 557.

346. Metaxas P.J., Sampaio J., Chanthbouala A., Matsumoto R., Lequeux S., Anane A., Cros V., Grollier J., Fert A., Zvezdin K.A., Yakushijin K., Kubota H., Fukushima A., Yuasa S. (2012) High spin torques induced domain wall speeds in magnetic tunnel junctions. *Scientific Reports.*

347. Thiaville A., Rohart S., Jué E., Cros V., Fert A. (2012) Dynamics of Dzyaloshinskii domain walls in ultrathin magnetic films. *Europhys. Letters* 100, 57002.

348. Seneor P., Dlubak B., Martin M.-.B., Anane A., Jaffres H., and Fert A. (2012) "Spintronics with graphene," *MRS Bulletin*, Vol. 37, No. 12, pp. 1245–1254.

349. Dlubak B., Martin M.-.B., Deranlot C., Bouzehouane K., Fusil S., Mattana R., Petroff F., Anane A., Seneor P., and Fert A. (2012) Homogeneous pinhole free 1 nm Al_2O_3 tunnel barriers on grapheme. *Appl. Phys. Lett.*, Vol. 101, No. 20, 203104.

350. Dlubak B., Martin M.-.B., Deranlot C., Servet B., Xavier S., Mattana R., Sprinkle M., Berger C., De Heer W.A., Petroff F., Anane A., Seneor P., Fert A. (2012) "Highly efficient spin transport in epitaxial graphene on SiC," *Nature Physics*, Vol. 8, No. 7, pp. 557–561.

351. Dlubak B., Martin M.-.B., Weatherup R.S., Yang H., Deranlot C., Blume R., Schloegl R., Fert A., Anane A., Hofmann S., Seneor P., Robertson J. (2012) "Graphene-Passivated Nickel as an Oxidation-Resistant Electrode for Spintronics." *ACS Nano* 6(12), 10930.

352. Galbiati M., Barraud C., Tatay S., Bouzehouane K., Deranlot C., Jacquet E., Fert A., Seneor P., Mattana R., and Petroff F. (2012) "Unveiling Self-Assembled Monolayers' Potential for Molecular Spintronics: Spin Transport at High Voltage". *Advanced materials* (Deerfield Beach, Fla.), Vol. 24, Issue 48, 6429–6432.

353. Gradhand M., Fedorov D.V., Zahn P., Mertig I., Otani Y., Niimi Y., Vila L., Fert A. (2012) Perfect alloys for spin hall current-induced magnetization switching. *SPIN* 02, 1250010.

354. Tatay S., Barraud C., Galbiati M., Seneor P., Mattana R., Bouzehouane K., Deranlot C., Jacquet E., Forment-Aliaga A., Jegou P., Fert A., Petroff F. (2012) Self-Assembled Monolayer-Functionalized Half-Metallic Manganite for Molecular Spintronics. *ACS Nano* 6, 8753.

355. Laczkowski P., Vila L., Nguyen V.-D., Marty A., Attané J.-P., Jaffrès H., George J.-M., Fert A. (2012) Enhancement of the spin signal in permalloy/gold multiterminal nanodevices by lateral confinement. *Phys. Rev.* B 85, 220404(R).

356. Seneor P., Dlubak B., M-B. Martin, Anane A., Jaffres H., Fert A. (2012) Sointronics with grapheme. *MRS Bulletin* 37, 1245.

357. Fert A. (2012) Historical overview: from electronic transport in magnetic materials to spintronics. *Handbook of spin transport and nanomagnetism* (edited by E.Y. Tsymbal and I. Zutic, *CRC Press*) Part I.

358. Khvalkovskiy A.V., Cros V., Apalkov D., Nikitin V., Krounbi M., Zvezdin K.A., Anane A., Grollier J., Fert A. (2013) Matching domain wall configuration and spin-orbit torques for very efficient domain-wall motion. *Phys. Rev.* B 87, 020402 R.

359. Fert A., Cros V., Sampaio J. (2013) Skyrmions on the track. *Nature Nanotechnology*, Vol. 8, No. 7, pp. 557–561.

360. Metaxas P.J., Sampaio J., Chanthbouala A., Matsumoto R., Anane A., Fert A.; Zvedin K.A., Yakashiji K., Kubota H., Fukushima A., Yuasa A., Nishimura K., Nagamine Y., Maehara H., Tsunekawa K., Cros V., Grollier J. (2013) High domain wall velocity via spin transfer torque using vertical current injection. *Scientific Reports* 3, 1829.

361. Sampaio J., Cros V., Rohart S., Thiaville A.and Fert A. (2013) Nucleation, stability and current-induced motion of isolated magntic skyrmions in nanostructures. *Nature Nanotechnology* 8, 839.

362. Rojas-Sanchez J.C., Vila L., Desfonds G., Gambarelli S., Attané J.-P., De Teresa J.M., Magen C., Fert A. (2013) Spin-to-charge conversion using Rashba coupling at the interface between nonmagnetic materials. *Nature Communications* 4, 2944.

363. d'Allivy Kelly O., Anane A., Bernard R., Ben Youssef J., Hahn C., Molpeceres A.-H., Jacquet E., Deranlot C., Bortolotti P., Mage J.-C., De Loubens G., Klein O., Cros V., Fert A. (2013) Inverse Spin Hall Effect in nnaometer-thick YIG/Pt system. *Appl. Phys. Lett.* 103, 8, 082408.

364. Levy P.M., Yang H., Chshiev M., Fert A. (2013) Spin Hall effect induced by Bi impurities in Cu: Skew scattering and side-jump. *Phys. Rev.* B 88, 214432.

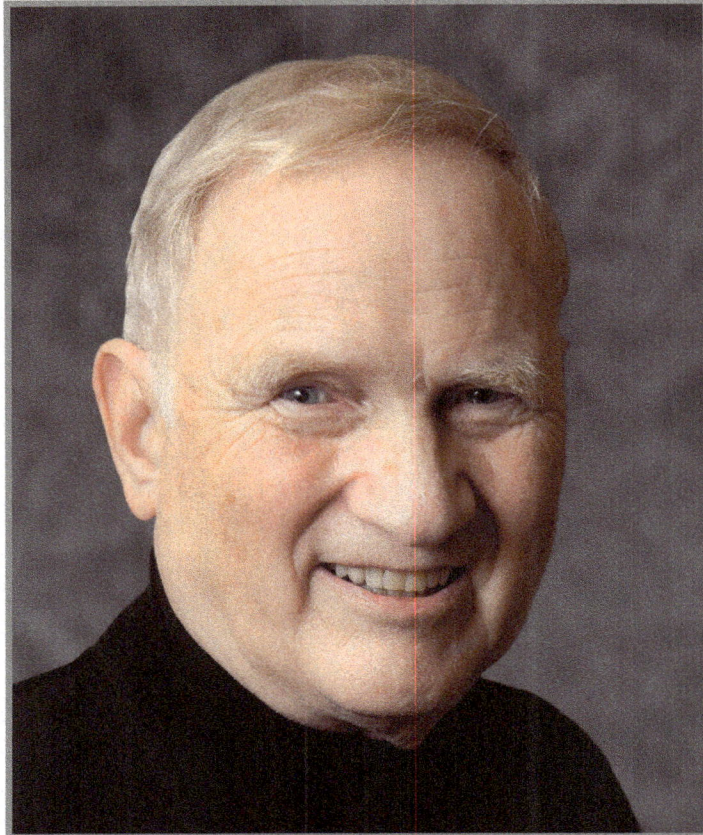

John F. Clauser

JOHN F. CLAUSER

Birthplace

Born	Pasadena, CA, USA, 1 December 1942

Education

1960	Baltimore Polytechnic Institute, High School Diploma, Adv. College Prep.
1964	Caltech, BS, Physics
1966	Columbia University, MA, Physics
1970	Columbia University, PhD, Physics (Research Sponsor, Patrick Thaddeus)

Employment and Research Experience

1969–1975	Postdoctoral Research Associate, Physics Dept., Univ. of Calif. — Berkeley, and Lawrence Berkeley National Laboratory
1975–1986	Research Physicist, Lawrence Livermore National Laboratory
1986–1987	Senior Scientist, Science Applications International Corp., (SAIC, Emeryville, CA)
1988–1989	Consultant, Inventor, (J. F. Clauser & Assoc., Livermore, CA)
1990–1997	Research Physicist, Univ. of Calif. — Berkeley, Physics Dept.
1997–	Research Physicist, Consultant, Inventor (J.F. Clauser & Assoc., Walnut Creek, CA)

Awards and Honors

1982 *The Reality Foundation Prize* ($6000, shared with John S. Bell, CERN): *"The Reality Foundation Prize to John F. Clauser for experimental and theoretical research into quantum foundations, for devising variations of the Bell Inequality which subject local realistic models of nature to experimental tests, for performing experiments which decisively exclude semi-classical models of the radiation field, for carrying out experiments which establish non-locality as a general feature of objective reality."*

2010 The Wolf Foundation Prize in Physics ($100,000, Shared with Alain Aspect and Anton Zeilinger): *"The Prize Committee for Physics has unanimously decided that the 2010 Wolf Prize be jointly awarded to: John F. Clauser, J.F. Clauser & Assoc., Walnut Creek, CA, U.S.A., Alain Aspect, Institut d'Optique, Palaiseau, France, Anton Zeilinger, University of Vienna & Institute for Quantum Optics and Quantum Information, Austrian Academy of Sciences, Vienna, Austria, For their fundamental conceptual and experimental contributions to the foundations of quantum physics, specifically an increasingly sophisticated series of tests of Bell's inequalities or extensions there of using entangled quantum states."*

2011 The Thomson Reuters Citation Laureate — Physics: " for his tests of Bell's Inequalities and quantum entanglement. (for Nobel Prize nomination)

COMMENTARY

1959–1960: As a high-school student, John built the world's first computer-driven "video games" to win awards at the '59 and '60 National High-school Science Fairs, the U.S. Navy "Science Cruiser" Award, and the "J. J. Chapman Memorial" Award ("Most Proficient Student in Electricity").

1965–1969: As a graduate student, with thesis sponsor, Patrick Thaddeus (Columbia), John made the 3rd measurement of the 2.7° cosmic microwave background radiation, and first observed interstellar Carbon-13.

1969: As a graduate student (independently of his thesis work), and with Abner Shimony and Mike Horne (Boston Univ.) and Richard Holt (Harvard), John proposed the first experimental test of Bell's Theorem. This work introduced the Clauser–Horne–Shimony–Holt (CHSH) inequality as a constraint for hidden-variables theories, and later, for "Objective Local Theories and "Local Realism".

1971: John discovered the basic flaw in John vonNeumann's "Informal Hidden-Variable Argument".

1972: At the UC Berkeley Physics Dept., working with PhD student Stuart Freedman, John carried out the first experimental test of the CHSH-Bell's Theorem predictions. This was the world's first observation of quantum entanglement, and was the first experimental observation of a violation of a Bell inequality.

1974: With Mike Horne (Stonehill College, MA), John first showed that a generalization of Bell's Theorem provides severe constraints for all local realistic theories of nature (a.k.a. objective local theories). This work introduced the Clauser–Horne (CH) inequality as the first fully general experimental requirement set by local realism. It also introduced the "CH no-enhancement assumption", whereupon the CH inequality reduces to the CHSH inequality, and whereupon associated experimental tests also constrain local realism. Experimental tests of this inequality have finally been performed in 2013 at Univ. of Vienna and Univ. of Illinois, Urbana-Champaign.

1974: At UC Berkeley, John made the first observation of sub-Poissonian statistics for light (via a violation of the Cauchy–Schwarz inequality for classical electromagnetic fields), and thereby, for the first time, demonstrated an unambiguous particle-like character for photons.

1976: At UC Berkeley, John carried out the second published experimental test of the CHSH-Bell's Theorem prediction, thereby refuting the unpublished experimental results of a similar test by Richard Holt and Francis Pipkin (Harvard).

1976: At UC Berkeley, John carried out the first experimental measurement of the circular-polarization correlation of quantum-mechanically entangled photons.

1976: With Abner Shimony and Michael Horne, pointed out the importance of the observer's freedom of choice (as a criticism of John Bell's '*Theory of Local Beables*').

1978: With Abner Shimony, John published the first comprehensive review of the experimental and theoretical status of the use of Bell's Theorem for testing local realism.

1975–1986: At Lawrence Livermore National Laboratory, with Tom Simonen and Ron Goodman, John designed and built a ruby-laser Thomson scattering plasma diagnostic to measure electron temperature in the 2XIIB magnetic mirror (controlled fusion) experiment. With Grant Logan, he designed and built Langmuir probe arrays to measure the spatial distribution of plasma elecron and ion temperatures and plasma potential in 2XIIB. With Eric Silver and UC Davis PhD student, Bruce Failor, he designed and built x-ray spectrometers and x-ray cameras for the analysis of relativistic *Bremstrahlung* x-ray emission from the TMX-U tandem magnetic mirror experiment.

1975–1986: At Lawrence Livermore National Laboratory, as Group Leader for the 2XIIB experiment, John first experimentally demonstrated the scaling limitations imposed by the drift-cyclotron-loss-cone plasma micro-instability in minimum-B magnetic mirror machines.

1975–1986: At Lawrence Livermore National Laboratory, John discovered strong electron-temperature gradients in the 2XIIB magnetic mirror machine, parallel to a magnetic field. This discovery was totally unexpected, and was a direct refutation of the commonly believed assertion by Lyman Spitzer (Princeton) that a plasma's electron temperature is always constant along magnetic field lines. It was also pivotal for the subsequent invention by Grant Logan and William Fowler of "electron thermal barriers" that, in turn, were used in the TMX-U experiment at LLNL. That invention subsequently earned Logan and Fowler the James Clark Maxwell Prize.

1975–1986: As the leader of a group of computer scientists at Lawrence Livermore National Laboratory, John and his group designed and built the TMX-U experiment's data-acquisition and analysis system.

1984: John invented and built a dramatic new design for racing sailboat keels. Via John's work as a consultant to the St. Francis Yacht Club's America's Cup Challenge Syndicate, this keel design was implemented on their revolutionary 12-meter yacht, USA. Virtually all present-day racing sailboat keels are a direct evolution of this design concept.

1985–1987: At SAIC John helped develop a 3D imaging system using thermal neutrons for counter-terrorist airline checked-baggage inspection, and wrote the tomography software for this system. Then, to diminish its false-positive rate, he invented dual-energy article-gradient material-specific x-ray imaging. SAIC has subsequently adapted this latter invention for use in inspecting carry-on luggage. Both of these systems were put in use at most international airports. Following the Sept. 11, 2001 attack on the US, their use was mandated for all US international airports. They are now manufactured by Perkin Elmer.

1987–1991: John first proposed and patented atom interferometers, as useful ultra-sensitive sensors for inertial and gravity forces, and for petroleum prospecting and well-logging. (US Patents 4,874,942 and 4,992,656, Australian Patent # 637,654, Canadian Patent # 2,033, 341).

1988: John first proposed the use of atom interferometry, in spacecraft for the testing of general relativity.

1992: With UC Berkeley Physics PhD student Matthias Reinsch, John first deduced the number-theoretical properties of the fractional Talbot effect for finite gratings, first calculated this effect's spectral resonance properties, and first combined the fractional

Talbot effect with the Lau effect to provide a new form of lens-free ultra-short wavelength interferometer, which he named the Talbot–Lau interferometer.

1990–1997: At the UC Berkeley, Physics Dept., with postdoc Shifang Li, John first used Talbot–Lau interferometry to build deBroglie-wave atom interferometers.

1990–1997: At the UC Berkeley, Physics Dept., with postdoc Shifang Li, John adapted Talbot–Lau interferometry to provide the first realization of the "Heisenberg-Microscope" experiment.

1994: John first proposed the use of Talbot–Lau matter-wave interferometry with very massive particles (e.g. "small rocks and live viruses") as a probe for limits to the validity of quantum theory, as indicated in various attempts to understand the "quantum measurement problem". Anton Zeilinger's group at Univ. of Vienna has since carried out such experiments.

1996: With J. Dowling (Redstone Arsenal, AL) John first discovered that Young's N-slit interferometer can be used to find the prime factors of an integer N.

1997: John first proposed the possibility of observing a "Temporal Talbot-Effect". This effect has since been observed, and further, has been used for factoring integers.

1998: John was granted US Patent 5,812,629, "*Ultrahigh Resolution Interferometric X-ray Imaging*".

1998: As a consultant to the Jet Propulsion Laboratory's Quantum Computing Technologies Group, John evaluated the use of matter-wave interferometry in earth-orbiting satellites for surveillance and gravity imaging of the earth's surface.

1999–present: With support from the National Cancer Institute, John is presently involved in research at his own private lab to develop his US Patent 5,812,629 to provide a clinically useful device (for mammography). To this end, he has observed Talbot–Lau interference with x-rays, and has used it to obtain actual phase-contrast x-ray images. Recently, other workers have also (re)discovered this technique and experiments using it are in progress at various laboratories in Switzerland, Japan and the USA.

John has published over 300 technical articles in physics journals. He is a member of the American Physical Society, the Optical Society of America, SPIE, the Radiological Society of North America, and United States Sailing Association. He is also an avid sailboat racer.

Rep. Prog. Phys., Vol. 41, 1978. Printed in Great Britain

Bell's theorem: experimental tests and implications

JOHN F CLAUSER† and ABNER SHIMONY‡§

† Lawrence Livermore Laboratory—L-437, Magnetic Fusion Energy Division, Livermore, California 94550, USA
‡ Departments of Physics and Philosophy, Boston University, Boston, Massachusetts 02215, USA

Abstract

Bell's theorem represents a significant advance in understanding the conceptual foundations of quantum mechanics. The theorem shows that essentially all local theories of natural phenomena that are formulated within the framework of realism may be tested using a single experimental arrangement. Moreover, the predictions by these theories must significantly differ from those by quantum mechanics. Experimental results evidently refute the theorem's predictions for these theories and favour those of quantum mechanics. The conclusions are philosophically startling: either one must totally abandon the realistic philosophy of most working scientists, or dramatically revise our concept of space–time.

This review was received in February 1978.

§ Work supported in part by the National Science Foundation.

1882   *J F Clauser and A Shimony*

Contents

Bell's theorem: experimental tests and implications 1883

1. Introduction

Realism is a philosophical view, according to which external reality is assumed to exist and have definite properties, whether or not they are observed by someone. So entrenched is this viewpoint in modern thinking that many scientists and philosophers have sought to devise conceptual foundations for quantum mechanics that are clearly consistent with it. One possibility, it has been hoped, is to reinterpret quantum mechanics in terms of a statistical account of an underlying hidden-variables theory in order to bring it within the general framework of classical physics. However, Bell's theorem has recently shown that this cannot be done. The theorem proves that all realistic theories, satisfying a very simple and natural condition called locality, may be tested with a single experiment against quantum mechanics. These two alternatives necessarily lead to significantly different predictions. The theorem has thus inspired various experiments, most of which have yielded results in excellent agreement with quantum mechanics, but in disagreement with the family of local realistic theories. Consequently, it can now be asserted with reasonable confidence that either the thesis of realism or that of locality must be abandoned. Either choice will drastically change our concepts of reality and of space–time.

The historical background for this result is interesting, and represents an extreme irony for Einstein's steadfastly realistic position, coupled with his desire that physics be expressable solely in simple geometric terms. Within the realistic framework, Einstein *et al* (1935, hereafter referred to as EPR) presented a classic argument. As a starting point, they assumed the non-existence of action-at-a-distance and that some of the statistical predictions of quantum mechanics are correct. They considered a system consisting of two spatially separated but quantum-mechanically correlated particles. For this system, they showed that the results of various experiments are predetermined, but that this fact is not part of the quantum-mechanical description of the associated systems. Hence that description is an incomplete one. To complete the description, it is thus necessary to postulate additional 'hidden variables', which presumably will then restore completeness, determinism and causality to the theory.

Many in the physics community rejected their argument, preferring to follow a counter-argument by Bohr (1935), who believed that the whole realistic viewpoint is inapplicable. Many others, however, felt that since both viewpoints lead to the same observable phenomenology, a commitment to either one is only a matter of taste. Hence, the discussion, for the greater part of the subsequent 30 years, was pursued perhaps more at physicists' cocktail parties than in the mainstream of modern research.

Starting in 1965, however, the situation changed dramatically. Using essentially the same postulates as those of EPR, J S Bell showed for a *Gedankenexperiment* of Bohm (a variant of that of EPR) that no deterministic local hidden-variables theory can reproduce all of the statistical predictions by quantum mechanics. Inspired by that work, Clauser *et al* (1969, hereafter referred to as CHSH) added three contributions. First, they showed that his analysis can be extended to cover actual systems, and that experimental tests of this broad class of theories can be performed. Second, they introduced a very reasonable auxiliary assumption which allows tests to be performed

1884   *J F Clauser and A Shimony*

with existing technology. Third, they specifically proposed performing such a test by examining the polarisations of photons produced by an atomic cascade, and derived the required conditions for such an experiment.

Curiously, the transition to a consideration of real systems introduced new aspects to the problem. EPR had demonstrated that any ideal system which satisfies a locality condition must be deterministic (at least with respect to the correlated properties). Since that argument applies only to ideal systems, CHSH therefore had postulated determinism explicitly. Yet, it eventually became clear that it is not the deterministic character of these theories that is incompatible with quantum mechanics. Although not stressed, this point was contained in Bell's subsequent papers (1971, 1972)—any non-deterministic (stochastic) theory satisfying a more general locality condition is also incompatible with quantum mechanics. Indeed it is the objectivity of the associated systems and their locality which produces the incompatibility. Thus, the whole realistic philosophy is in question! Bell's (1971) result, however, is in a form that is awkward for an experimental test. To facilitate such tests, Clauser and Horne (1974, hereafter referred to as CH) explicitly characterised this broad class of theories. They then gave a new incompatibility theorem that yields an experimentally testable result and derived the requirements for such a test. Although such an experiment is difficult to perform (and in fact has not yet been performed), they showed that an assumption weaker in certain respects than the one of CHSH allowed the experiments proposed earlier by CHSH to be used as a test for these theories also.

The interpretation of all of the existing results requires at least some auxiliary assumptions, although experiments are possible for which this is not the case. Even though some of the assumptions are very reasonable, this fact allows loopholes still to exist. Experiments now in progress or being planned will be able to eliminate most of these loopholes. However, even now one can assert with reasonable confidence that the experimental evidence to date is contrary to the family of local realistic theories. The construction of a quantum-mechanical world view as an alternative to the point of view of the local realistic theories is beyond the scope of this review.

Section 2 of this review summarises the argument of EPR, appendix 1 discusses various critical evaluations of it, and appendix 2 summarises briefly the history of hidden-variables theories. Section 3 describes the versions of Bell's theorem discussed above as well as some others. Section 4 discusses the requirements for a fully general test and shows why such an experiment is a difficult one to perform. Section 5 is devoted to a description of the cascade-photon experiments proposed by CHSH. First, it discusses the auxiliary assumptions by CHSH and CH. Second, calculations of the quantum-mechanical predictions for these experiments are summarised. Third, there is a discussion of the actual cascade-photon experiments performed so far (Freedman and Clauser 1972, Holt and Pipkin 1973, Clauser 1976, Fry and Thompson 1976). All but the second agree very well with the quantum-mechanical predictions, thus providing significant evidence against the entire family of local realistic theories. Section 5 ends with a critique of the CH and CHSH assumptions. Section 6 summarises and discusses related experiments measuring the polarisation correlation of photons produced in positronium annihilation (Kasday *et al* 1975, Faraci *et al* 1974, Wilson *et al* 1976, Bruno *et al* 1977) and an experiment measuring the spin correlation of proton pairs (Lamehi-Rachti and Mittig 1976). Section 7 is devoted to an evaluation of the experimental results obtained so far and to the prospects for future experiments.

Bell's theorem: experimental tests and implications 1885

2. The Einstein–Podolsky–Rosen argument

A profound argument for the thesis that a quantum-mechanical description of a physical system is incomplete was presented by EPR in 1935. Their argument rests upon three premises. (i) Some of the quantum-mechanical predictions concerning observations on a certain type of system, consisting of two spatially separated particles, are correct. (ii) A very reasonable criterion for the existence of 'an element of physical reality' is proposed: 'If, without in any way disturbing a system, we can predict with certainty (i.e., with probability equal to unity) the value of a physical quantity, then there exists an element of physical reality corresponding to this physical quantity' (EPR 1935, p777). (iii) There is no action-at-a-distance in nature.

The system which they study consists of two particles, which are prepared in a state such that the sum of their momenta in a given direction (p_1+p_2) and the difference of their positions (x_1-x_2) are both definite. The wavefunction $\delta(x_1-x_2-a)$ quantum mechanically describes this system, for it is an eigenfunction of the operator x_1-x_2 with eigenvalue a, and of the operator p_1+p_2 with eigenvalue 0. By measuring the position of particle 1 one can predict with certainty, according to quantum mechanics, what value will be found if the position of particle 2 is then measured (immediately). In view of premise (iii) the prediction is made without in any way disturbing particle 2, since the two particles are spatially separated. EPR therefore infer that the position of particle 2 has a definite predetermined value, not included in the description by the wavefunction $\delta(x_1-x_2-a)$. By an analogous argument EPR also infer that the momentum of particle 2 has a definite value, contrary to the uncertainty principle. (Of course, the same argument, starting with measurements made upon particle 2, allows them to infer that particle 1 also has both a definite position and a definite momentum.) Hence EPR reach the conclusion that at least in this particular situation the quantum-mechanical description is incomplete. Although they do not use the term 'hidden variables', this expression can be appropriately used to apply to the parts of the complete state which are not comprised in the quantum-mechanical description, and which suffice to fix the outcomes of measurements that are not fully determined quantum mechanically.

In our opinion the reasoning of EPR is impeccable, once an ambiguity in the phrase 'can predict', which occurs in the second premise, is removed. In a narrow sense, one can predict the value of a quantity only *when an experimental arrangement is chosen for determining the value of that quantity*. In a broad sense, one can predict the value of a quantity *if it is possible to choose an experimental arrangement for determining it*. If the narrow sense is accepted, then the argument of EPR clearly does not go through, since the experimental arrangements for measuring the position and momentum of a particle are incompatible. From the standpoint of physical realism the broad sense of 'can predict' is the appropriate one, since from that viewpoint, one conceives a physical system to have a definite set of properties independently of their being observed, but which may of course be explored at the option of the experimenter. In the situation envisaged by EPR one can predict, in the broad sense, both x_2 and p_2. Hence if this sense of the ambiguous phrase is adopted, their argument does go through. An assumption of physical realism clearly underlies the argument by EPR. Bohr's (1935) answer to EPR, defending the completeness of quantum mechanics, consisted essentially of a critique of the realism which they had taken for granted (see appendix 1).

A variant of EPR's argument was given by Bohm (1951), formulated in terms of discrete states. He considered a pair of spatially separated spin-$\frac{1}{2}$ particles produced

J F Clauser and A Shimony

somehow in a singlet state, for example, by dissociation of the spin-0 system. Various spin components of each of these particles may then be measured independently at the option of the experimenter. The spin part of the state vector is given by:

$$\Psi = \frac{1}{\sqrt{2}} \left[u_{\hat{n}}^{+}(1) \otimes u_{\hat{n}}^{-}(2) - u_{\hat{n}}^{-}(1) \otimes u_{\hat{n}}^{+}(2) \right]. \tag{2.1}$$

Here $\boldsymbol{\sigma}.\hat{n}u_{\hat{n}}^{\pm}(1) = \pm u_{\hat{n}}^{\pm}(1)$, so that $u_{\hat{n}}^{\pm}(1)$ quantum mechanically describes a state in which particle 1 has spin 'up' or 'down', respectively, along the direction $\hat{n}$; $u_{\hat{n}}^{\pm}(2)$ has an analogous meaning concerning particle 2. Since the singlet state Ψ is spherically symmetric, $\hat{n}$ can specify any direction. Suppose that one measures the spin of particle 1 along the $\hat{x}$ axis. The outcome is not predetermined by the description Ψ. But from it, one can predict that if particle 1 is found to have its spin parallel to the $\hat{x}$ axis, then particle 2 will be found to have its spin antiparallel to the $\hat{x}$ axis if the $\hat{x}$ component of its spin is also measured. Thus, the experimenter can arrange his apparatus in such a way that he can predict the value of the $\hat{x}$ component of spin of particle 2 presumably without interacting with it (if there is no action-at-a-distance). Likewise, he can arrange the apparatus so that he can predict any other component of the spin of particle 2. The conclusion of the argument is that all components of spin of each particle are definite, which of course is not so in the quantum-mechanical description. Hence, a hidden-variables theory seems to be required.

Some comments are in order concerning EPR's premises in the light of Bell's theorem. If premise (i) is taken to assert that all of the quantum-mechanical predictions are correct, then Bell's theorem has shown it to be inconsistent with premises (ii) and (iii). Actually, in the body of their argument EPR used only a few predictions with probability one, which are atypical in quantum mechanics, whereas the discrepancies which Bell exhibited between local realistic theories and quantum mechanics involved statistical predictions. If it was EPR's intention to aim at a hidden-variables theory which is local and realistic, and which agrees with all the statistical predictions of quantum mechanics—as many readers have understood them—then, of course, Bell's theorem shows mathematically that this aim cannot be achieved. We shall not try to answer the historical question of their intent. Two statements, however, can be made with confidence. First, the argument from their premises is valid, once the above-mentioned ambiguity is cleared up. Second, the physical situation which they envisaged is of immense value for examining the philosophical implications of quantum mechanics and (via Bell's work) for exploring the limitations of the family of local realistic physical theories.

3. Bell's theorem

There is a vast literature concerning the consistency of hidden-variables theories with the algebraic structure of the observables of quantum mechanics. The major results of this literature are summarised in appendix 2, but they are not indispensable for understanding the content and implications of Bell's theorem. Heuristically, however, this literature was very important for Bell's work. In the course of preparing a review article on 'impossibility' proofs of hidden-variables interpretations of quantum mechanics, Bell studied the theories proposed by de Broglie (1928) and Bohm (1952). He noticed, as Bohm had already realised, that in order to reproduce the quantum-theoretic predictions for a system of EPR type, they postulated the

Bell's theorem: experimental tests and implications 1887

existence of non-local interactions between spatially separated particles. Bell was thus led to ask whether the peculiar non-locality exhibited by these models is a generic characteristic of hidden-variables theories that agree with the statistical predictions by quantum mechanics. He proved (Bell 1965) that the answer is positive for the whole class of deterministic hidden-variables theories in the domain of ideal apparatus and systems. Stronger versions of this theorem, which also constrain actual systems, were later proved by Bell himself and by others. These versions state that essentially all realistic local theories of natural phenomena may be tested in a single experimental arrangement against quantum mechanics, and that these two alternatives necessarily lead to observably different predictions.

In this section we review some of these derivations, which we shall refer to collectively as 'Bell's theorem'. Our purpose here is to arrive at versions of Bell's theorem which satisfy the following criteria. (i) The hypotheses seem to be inescapable for anyone who is committed to physical realism and to the non-existence of action-at-a-distance. (ii) Discrepancies with the predictions by quantum mechanics occur in at least one situation which is experimentally realisable. Criterion (i) is, in our opinion, very close to having been achieved, although the hypotheses are violated by some pathological instances of local realistic theories. Criterion (ii) has essentially been achieved; however, the experiment which it specifies is difficult, and has not yet been performed. Additional assumptions, not implicit in locality and realism, have been relied upon to allow easier experiments to be considered. (The assumptions and experiments are discussed in §§5 and 6.) Unfortunately, this fact leaves open various loopholes (discussed in §§5–7). It must be stressed, however, that the existence of these loopholes in no way diminishes the mathematical validity of the versions of Bell's theorem presented in this section.

3.1. Deterministic local hidden-variables theories and Bell (1965)

In his paper of 1965 Bell considered Bohm's *Gedankenexperiment*, described above in §2. That system consists of two spin-$\frac{1}{2}$ particles, prepared in the quantum-mechanical singlet state Ψ given by equation (2.1). Let $A_{\hat{a}}$ be the result of a measurement of the spin component of particle 1 of the pair along the direction $\hat{a}$, and let $B_{\hat{b}}$ be that of particle 2 along direction $\hat{b}$. We take the unit of spin as $\hbar/2$; hence, $A_{\hat{a}}, B_{\hat{b}} = \pm 1$.

The product $A_{\hat{a}}.B_{\hat{b}}$ is a single observable of the two-particle system (even though two distinct operations are needed in order to measure it). It is represented quantum mechanically by a self-adjoint operator on the Hilbert space associated with the system. For this *Gedankenexperiment* one can readily calculate the quantum-mechanical prediction for the expectation value of this observable†:

$$[E(\hat{a}, \hat{b})]_{\Psi} = \langle \Psi | \boldsymbol{\sigma}_1.\hat{a}\boldsymbol{\sigma}_2.\hat{b} | \Psi \rangle = -\hat{a}.\hat{b}. \tag{3.1}$$

A special case of equation (3.1) contains the determinism implicit in this idealised system. When the analysers are parallel, we have:

$$[E(\hat{a}, \hat{a})]_{\Psi} = -1 \tag{3.2}$$

for all $\hat{a}$. Thus, one can predict with certainty the result B, by previously obtaining

---

† The notation of this review is to use the wavefunction or the letters QM as a subscript to denote the quantum-mechanical prediction. We omit the subscript for predictions by the class of theories included by the postulates of Bell's theorem, when this convention does not cause confusion.

the result A (EPR's premise (ii)). Since the quantum-mechanical state Ψ does not determine the result of an individual measurement, this fact (via EPR's argument) suggests that there exists a more complete specification of the state in which this determinism is manifest. We denote this state by the single symbol λ, although it may well have many dimensions, discrete and/or continuous parts, and different parts of it interacting with either apparatus, etc. Presumably the quantum state Ψ is a related partial specification of this state. We thus define a deterministic hidden-variables theory as any physical theory which postulates the existence of states of a system, for which the observables of quantum mechanics always have definite values.

Let Λ be the space of the states λ for an ensemble comprised of a very large number of the observed systems. We make no restrictions as to what type of space this is, nor to its dimensionality, nor do we require linearity for operations with it, but of course we require that a set of Borel subsets of Λ be defined, so that probability measures can be defined upon it. We represent the distribution function for the states λ on the space Λ by the symbol ρ. For this ensemble we take ρ to have norm one:

$$\int_\Lambda \mathrm{d}\rho = 1. \tag{3.3}$$

In a deterministic hidden-variables theory the observable $A_{\hat{a}}.B_{\hat{b}}$ has a definite value $(A_{\hat{a}}.B_{\hat{b}})(\lambda)$ for the state λ. For these theories Bell defined locality as follows: *a deterministic hidden-variables theory is local if for all $\hat{a}$ and $\hat{b}$ and all $\lambda \in \Lambda$ we have:*

$$(A_{\hat{a}}.B_{\hat{b}})(\lambda) = A_{\hat{a}}(\lambda).B_{\hat{b}}(\lambda). \tag{3.4}$$

That is, once the state λ is specified and the particles have separated, measurements of A can depend only upon λ and $\hat{a}$ but not $\hat{b}$. Likewise measurements of B depend only upon λ and $\hat{b}$. Any reasonable physical theory that is realistic and deterministic and that denies the existence of action-at-a-distance is local in this sense. (More general definitions of 'local' will be considered in §3.3.) For such theories the expectation value of $A_{\hat{a}}.B_{\hat{b}}$ is then given by

$$E(\boldsymbol{a}, \boldsymbol{b}) = \int_\Lambda A_{\hat{a}}(\lambda) B_{\hat{b}}(\lambda)\, \mathrm{d}\rho. \tag{3.5}$$

Bell's (1965) proof of the theorem consists of showing that if the locality condition (3.4) and the condition (3.2) for partial agreement with quantum mechanics are both satisfied, then the expectation values satisfy a simple inequality. This inequality is then an alternative prediction to that by quantum mechanics for the expectation value of $A_{\hat{a}}.B_{\hat{b}}$. The predictions made by this inequality are quantitatively different from those of equation (3.1).

The demonstration is straightforward. Equation (3.2) can hold if and only if

$$A_{\hat{a}}(\lambda) = -B_{\hat{a}}(\lambda) \tag{3.6}$$

holds for all $\lambda \in \Lambda$. Using equation (3.6) we calculate the following function, which involves three different possible orientations of the analysers:

$$E(\hat{\boldsymbol{a}}, \hat{\boldsymbol{b}}) - E(\hat{\boldsymbol{a}}, \hat{\boldsymbol{c}}) = -\int_\Lambda [A_{\hat{a}}(\lambda) A_{\hat{b}}(\lambda) - A_{\hat{a}}(\lambda) A_{\hat{c}}(\lambda)]\, \mathrm{d}\rho$$

$$= -\int_\Lambda A_{\hat{a}}(\lambda) A_{\hat{b}}(\lambda)[1 - A_{\hat{b}}(\lambda) A_{\hat{c}}(\lambda)]\, \mathrm{d}\rho.$$

Since $A, B = \pm 1$, this last expression can be written:

$$|E(\hat{\boldsymbol{a}}, \hat{\boldsymbol{b}}) - E(\hat{\boldsymbol{a}}, \hat{\boldsymbol{c}})| \leqslant \int_\Lambda [1 - A_{\hat{b}}(\lambda) A_{\hat{c}}(\lambda)]\, \mathrm{d}\rho.$$

Bell's theorem: experimental tests and implications 1889

Using equations (3.3), (3.5) and (3.6) we have:

$$|E(\hat{a}, \hat{b}) - E(\hat{a}, \hat{c})| \leqslant 1 + E(\hat{b}, \hat{c}). \tag{3.7}$$

Inequality (3.7) is the first of a family of inequalities which are collectively called 'Bell's inequalities'.

A simple instance of the disagreement between the predictions of equation (3.1) and inequality (3.7) is provided by taking $\hat{a}$, $\hat{b}$ and $\hat{c}$ to be coplanar, with $\hat{c}$ making an angle of $2\pi/3$ with $\hat{a}$, and $\hat{b}$ making an angle of $\pi/3$ with both $\hat{a}$ and $\hat{c}$. Then:

$$\hat{a}.\hat{b} = \hat{b}.\hat{c} = \tfrac{1}{2} \qquad \hat{a}.\hat{c} = -\tfrac{1}{2}.$$

For these directions:

$$|[E(\hat{a}, \hat{b})]_{\Psi} - [E(\hat{a}, \hat{c})]_{\Psi}| = 1 \qquad \text{while} \qquad 1 + [E(\hat{b}, \hat{c})]_{\Psi} = \tfrac{1}{2}.$$

These values do not satisfy inequality (3.7). Hence the quantum-mechanical prediction and that by inequality (3.7) are incompatible, at least for some pairs of analyser orientations.

The version of Bell's theorem just proved can be summarised as follows: no deterministic hidden-variables theory satisfying equation (3.2) and the locality condition (3.1) can agree with all of the predictions by quantum mechanics concerning the spins of a pair of spin-$\tfrac{1}{2}$ particles in the singlet state.

3.2. Foreword to the non-idealised case

Any argument whose scope is strictly limited to a discussion of ideal systems is of little value to working physicists, who endeavour to describe systems that can and do occur in practice. The immense heuristic value of Bell's (1965) argument, outlined in §3.1, is that it leads to formulations that provide direct experimental predictions for systems which can actually be produced in a laboratory. By itself, the derivation given in §3.1 is insufficient to do this, because of its reliance upon the existence of a pair of analyser orientations for which there is a perfect correlation. That is, the above proof hinges strongly upon the condition that equation (3.2) hold exactly. Use is made of this equation in three ways. First, it allows the proof to go through mathematically. Second, determinism is derivable from it and does not have to be postulated separately. Finally, for reasons to be discussed, it assumes that the locality postulate is reasonable.

Unfortunately, equation (3.2) cannot hold exactly in an actual experiment. Any real detector will have an efficiency less than 100%, and any real analyser will have some attenuation as well as some leakage into its orthogonal channel. Since we are attempting to deal with not just one but a whole class of theories, it is quite possible that in some of these theories the above imperfections are inherently correlated with the measurement and detection processes in a way that depends upon the state λ. The problems which arise when these three implications cannot be drawn will be considered in turn.

The problem concerning the derivation's mathematical reliance upon equation (3.2) was first solved by CHSH. They demonstrated that a different proof of the theorem follows from the above formalism, without requiring equation (3.2) to hold. They derived a different inequality that is violated by the quantum-mechanical predictions for systems which never achieve the perfect correlation of equation (3.2), but which do achieve a necessary minimum correlation. The inequality which results from their analysis is:

$$|E(a, b) - E(a', b)| + E(a, b') + E(a', b') \leqslant 2. \qquad (3.8)$$

When equation (3.2) does hold, inequality (3.8) implies inequality (3.7) as a special case. Since essentially this same inequality was subsequently derived by Bell (1971) for the more general non-deterministic case presented in §3.4, we will not present the CHSH derivation here.

The second problem—that determinism is no longer derivable—is not a serious one. One needs merely to assume that determinism holds for the theories under consideration. Indeed this was the approach by CHSH. Thereby, they produced a very powerful result, which constrains deterministic local hidden-variables theories for realisable systems. However, it was subsequently noticed by Bell (1971, 1972) and Clauser and Horne (1974) that this assumption is not needed. On the contrary, a weakening of the locality requirement can be made which still allows inequality (3.8) to be derived, but which significantly increases the scope of the theorem. The theorem thus applies to a class of fundamentally stochastic theories, as well as to deterministic theories in which there are hidden variables in the apparatus.

The third problem is a very delicate one, yet one of great importance. In the idealised situation, whenever a particle is observed at one apparatus an associated particle is *always* observed at the other apparatus. The selection of the sub-ensemble of observed particles from among all of those emitted by the source depends only upon the collimator and source geometry and can have no dependence upon the parameters $\hat{a}$ and $\hat{b}$. Hence ρ was defined for the observed particles, and one can then be confident that it is independent of $\hat{a}$ and $\hat{b}$.

In the actual case, on the other hand, observed particles are paired with particles which, for some reason, are not observed at all, i.e. in neither a spin-up nor a spin-down channel.

The sub-ensemble which we used in the idealised case is then further partitioned into four disjoint sub-ensembles, i.e. those for which (*a*) both particles are observed, (*b*) only particle 1 is observed, (*c*) only particle 2 is observed, and (*d*) neither particle is observed. The distribution ρ of the union of these four sets is clearly independent of $\hat{a}$ and $\hat{b}$. However, the mode of partitioning may well depend upon $\hat{a}$ and $\hat{b}$, since the detection and various attenuation processes occur 'downstream' from the analysers. Hence there is no reason to expect that the composition, and thus the distribution, of each sub-ensemble is independent of $\hat{a}$ and $\hat{b}$. (This fact was noticed by Pearle (1970) and Clauser and Horne (1974). The latter contrived a hidden-variables theory in which ρ becomes dependent upon $\hat{a}$ and $\hat{b}$ when sub-ensemble (*a*) alone is considered and which yields exactly the quantum-mechanical predictions for the system.) Thus if we are to use equation (3.3) for a normalisation condition, and to expect that ρ is independent of $\hat{a}$ and $\hat{b}$, the ensemble for which it is defined must also include the unobserved particles. Since their number is unknown and may be very large, it is no longer obvious how to compare the prediction by inequality (3.8) with experiment.

Three approaches to this problem have been pursued. The approach used by CHSH is to introduce an auxiliary assumption, that if a particle passes through a spin analyser, its probability of detection is independent of the analyser's orientation. Unfortunately, this assumption is not contained in the hypotheses of locality, realism or determinism. Moreover, it also has the undesirable feature that it makes the process of 'passage' or 'non-passage' a primitive one, and thereby excludes from consideration theories for which partial passage is appropriate.

Bell's theorem: experimental tests and implications 1891

A second approach, that used by Bell (1971) (although not specifically stated but clear from the context), is to employ an auxiliary apparatus ('event-ready' detectors) to measure the number of pairs emitted by the source. This possibility is shown schematically in figure 1. For this scheme, one can simply take the ensemble to consist of the particles which actually trigger the 'event-ready' detectors. Whether or not a triggering occurs clearly does not depend upon the analyser orientations. No problem with locality arises from the presence of the signals propagating to the remote apparatuses, since these signals can be simply considered as part of the state λ. Unfortunately, in practice most conceivable 'event-ready' detectors depolarise or destroy the particles. The value of this approach is thus limited.

An altogether different approach was employed by Clauser and Horne (1974). They derived an inequality from the hypotheses of locality and realism in which only ratios of the observed particle detection probabilities appear, and the normalisation condition equation (3.3) is not required for its derivation. The influence of the size

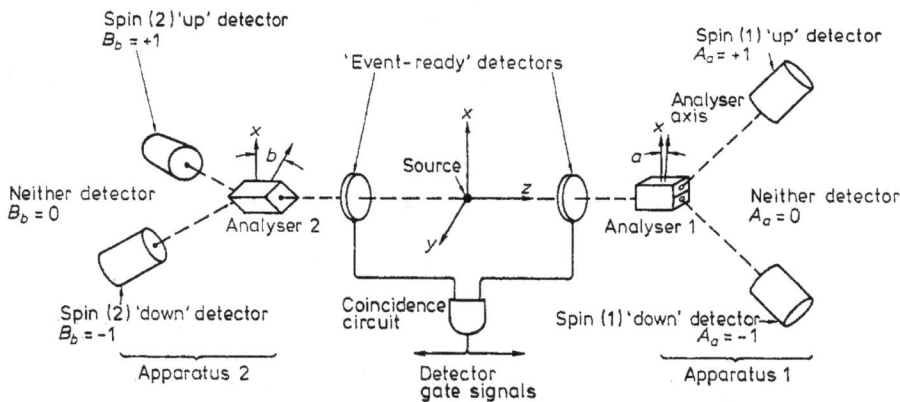

Figure 1. Apparatus configuration used for Bell's 1971 proof. 'Event-ready' detectors signal both arms that a pair of particles has been emitted. For a given gate signal, the result on either arm is assigned the value $+1$ if the corresponding spin-up detector responds, -1 if the spin-down detector responds, and 0 if neither detector responds.

of the ensemble thus vanishes. Their apparatus arrangement does not have the 'event-ready' detectors of figure 1, nor does it have two detectors for each apparatus but only one. It is thus much simpler, and is shown schematically in figure 2.

In the remainder of this section, we will show how these latter two approaches proceed. First, however, we will discuss the aforementioned generalisation of the locality postulate to include inherently stochastic theories.

3.3. Generalisation of the locality concept

Consider either of the experimental configurations for Bohm's *Gedankenexperiment*, described in §3.2. Actually there is nothing in the proof which requires the systems to be spin-$\frac{1}{2}$ particles. They may be any discrete-state quantum-mechanically correlated emissions. (However, not all quantum-mechanically correlated systems are predicted to violate the resulting inequalities. A careful choice is required to find one which is an appropriate test case.) In Bohm's *Gedankenexperiment* the symbols $\hat{a}$ and $\hat{b}$ are taken to represent the orientations of the Stern–Gerlach magnets used for

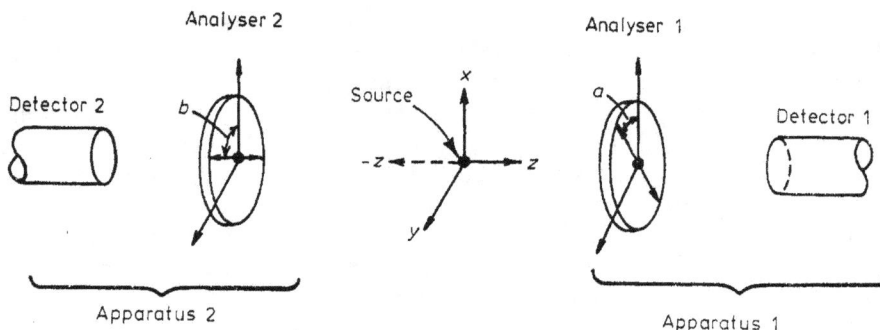

Figure 2. Apparatus configuration used in the proofs by CHSH and by CH. A source emitting particle pairs is viewed by two apparatuses. Each apparatus consists of an analyser and an associated detector. The analysers have parameters a and b respectively, which are externally adjustable. In the above example a and b represent the angles between the analyser axes and a fixed reference axis.

measuring the associated spin components. However, in general, a and b may represent any associated apparatus parameters under control by the experimenter. As before, A_a and B_b represent the measurement outcomes at apparatuses 1 and 2, respectively. Appropriate values will be assigned to these outcomes, as necessary.

The preceding definition of locality will now be generalised to include systems whose evolution is inherently stochastic, as well as to include deterministic systems with additional random variables associated with either apparatus, and that may locally affect their experimental outcomes. Suppose a pair of correlated systems, which have a joint state λ, separate. They then continue to evolve perhaps in an inherently stochastic way, and given λ, a and b, one can define probabilities for any particular outcome at either apparatus. We allow that, given λ, these two probabilities may each depend upon the associated (local) apparatus parameter, a or b respectively, and of course upon λ, but we assume that these probabilities are otherwise independent of each other.

This definition of locality seems very common-sensical. It says that the outcome (or the probability of outcomes) of a measurement performed on one part of a composite system is independent of what aspects of the other component the experimenter chooses to measure. It by no means excludes the possibility of obtaining knowledge concerning system 2 from an examination of system 1. The state λ contains information common to both systems, and a measurement on one of these presumably reveals some of this. Nor does it prevent a measurement performed on one component of a composite system from locally disturbing that component. What it does prescribe, in essence, is that the measured value of a quantity on one system is not causally affected by what one chooses to measure on the other system, since the two systems are well separated (e.g. space-like separated) when the measurements are performed.

3.4. Bell's 1971 proof

We now describe Bell's (1971) proof, using this generalised locality definition. The apparatus configuration appropriate to this proof was discussed in §3.2 and is shown in figure 1. Given that a particle pair was emitted into the associated apparatuses, the results of either measurement can have one of three possible outcomes, to

Bell's theorem: experimental tests and implications 1893

which the following values were assigned by Bell:

$$A_a(\lambda) = \begin{cases} +1, & \text{`spin-up' detector triggered by particle 1} \\ -1, & \text{`spin-down' detector triggered by particle 1} \\ 0, & \text{particle 1 not detected} \end{cases} \quad (3.9(a))$$

and

$$B_b(\lambda) = \begin{cases} +1, & \text{`spin-up' detector triggered by particle 2} \\ -1, & \text{`spin-down' detector triggered by particle 2} \\ 0, & \text{particle 2 not detected.} \end{cases} \quad (3.9(b))$$

For a given state λ of the emitted composite system, we denote the expectation values for these quantities by the symbols $\bar{A}_a(\lambda)$ and $\bar{B}_b(\lambda)$. In the general case these average values will differ from the values assigned by equations (3.9). Since the values for A and B are bounded by 1, it follows that:

$$|\bar{A}_a(\lambda)| \leqslant 1 \qquad |\bar{B}_b(\lambda)| \leqslant 1. \quad (3.10)$$

Using the general definition of locality of §3.3, we can write the expectation value for the product $A_a B_b$ as:

$$E(a, b) = \int_\Lambda \bar{A}_a(\lambda)\bar{B}_b(\lambda) \, \mathrm{d}\rho. \quad (3.11)$$

Since we are including in our ensemble only those particles which have previously triggered the 'event-ready' detectors, we are assured that the distribution ρ and the range of integration Λ are independent of a and b. Now consider the expression:

$$E(a, b) - E(a, b') = \int_\Lambda [\bar{A}_a(\lambda)\bar{B}_b(\lambda) - \bar{A}_a(\lambda)\bar{B}_{b'}(\lambda)] \, \mathrm{d}\rho$$

where we take a' and b' to be alternative settings for analysers 1 and 2, respectively. This can be rewritten as:

$$E(a, b) - E(a, b') = \int_\Lambda \bar{A}_a(\lambda)\bar{B}_b(\lambda)[1 \pm \bar{A}_{a'}(\lambda)\bar{B}_{b'}(\lambda)] \, \mathrm{d}\rho$$
$$- \int_\Lambda \bar{A}_a(\lambda)\bar{B}_{b'}(\lambda)[1 \pm \bar{A}_{a'}(\lambda)\bar{B}_b(\lambda)] \, \mathrm{d}\rho.$$

Using inequalities (3.10), we then have:

$$|E(a, b) - E(a, b')| \leqslant \int_\Lambda [1 \pm \bar{A}_{a'}(\lambda)\bar{B}_{b'}(\lambda)] \, \mathrm{d}\rho + \int_\Lambda [1 \pm \bar{A}_{a'}(\lambda)\bar{B}_b(\lambda)] \, \mathrm{d}\rho$$

or

$$|E(a, b) - E(a, b')| \leqslant \pm [E(a', b') + E(a', b)] + 2 \int_\Lambda \mathrm{d}\rho.$$

Hence:

$$-2 \leqslant E(a, b) - E(a, b') + E(a', b) + E(a', b') \leqslant 2. \quad (3.12)$$

By re-definition of the parameters a, a', b and b' in the central expression of (3.12), the minus sign may be permuted to any one of the four terms. Inequality (3.12) and its permutations are one form of Bell's inequality, and represent a general prediction for the theories covered by the above assumptions.

In order to complete the proof of the theorem, it is sufficient to show that in at least one situation the predictions by quantum mechanics contradict inequality (3.12). The quantum-mechanical prediction $[E(\hat{a}, \hat{b})]_{\text{QM}}$ for the two spin-$\frac{1}{2}$ particle example, when due account is taken of imperfections in the analysers, detectors and state preparation, will be of the form:

$$[E(\hat{a}, \hat{b})]_{\text{QM}} = C\hat{a}.\hat{b} \quad (3.13)$$

where the coefficient C is bounded by one for actual systems, and is equal to plus or minus one only in the idealised case. Suppose we take $\hat{a}$, $\hat{a}'$, $\hat{b}$ and $\hat{b}'$ to be coplanar vectors as shown in figure 3 with $\phi = \pi/4$, and calculate:

$$[E(\hat{a}, \hat{b}) - E(\hat{a}, \hat{b}') + E(\hat{a}', \hat{b}) + E(\hat{a}', \hat{b}')]_{\mathrm{QM}} = 2\sqrt{2}C.$$

There is a wide range of values for C for which the prediction by inequality (3.12) disagrees with that by equation (3.13). Hence the proof is complete.

3.5. The proof by Clauser and Horne

Clauser and Horne (1974) also proved Bell's theorem for general local realistic theories, including inherently stochastic theories. Their proof is noteworthy in that it

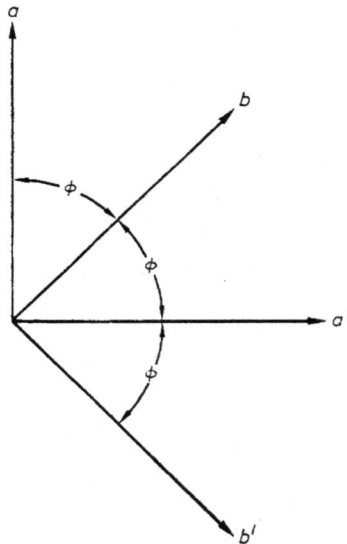

Figure 3. Optimal orientations for a, a', b and b'. If the correlation is of the form $C_1 + C_2 \cos n\phi$, then the maximum violation of the inequalities occurs at $n\phi = \pi/4$.

defines an experiment which might actually be performed and which does not require that auxiliary assumptions be made. The apparatus configuration which they used for the proof was introduced by Clauser *et al* (1969) and is shown schematically in figure 2. In contrast to the configuration of figure 1, theirs has only one detector in each arm and no 'event-ready' detectors. For each analyser/detector assembly there are only two possible outcomes: 'count' and 'no-count'. The results are thus formulated in terms of probabilities for single and coincidence counts, rather than the expectation values considered in §§3.1 and 3.4.

Suppose that during a period of time, while the adjustable parameters have the values a and b, the source emits, say, N of the two-component systems of interest. For this period, denote by $N_1(a)$ and $N_2(b)$ the number of counts at detectors 1 and 2, respectively, and by $N_{12}(a, b)$ the number of simultaneous counts from the two detectors (coincidence counts). When N is sufficiently large, the probabilities for these

Bell's theorem: experimental tests and implications 1895

results for the whole ensemble (with due allowance for random errors) are given by:

$$p_1(a) = N_1(a)/N$$
$$p_2(b) = N_2(b)/N \tag{3.14}$$
$$p_{12}(a, b) = N_{12}(a, b)/N.$$

CH derive an inequality which constrains ratios of the probabilities in equations (3.14) rather than their absolute magnitudes. Thereby, the influence of the quantity N vanishes, so that it does not have to be measured.

Their derivation is straightforward. Following the discussion of §3.3, we expect a well-defined probability $p_1(\lambda, a)$ of detecting component 1, given the state λ of the composite system and the parameter a of the first analyser; a probability $p_2(\lambda, b)$ of detecting component 2, given λ and b; and a probability $p_{12}(\lambda, a, b)$ of detecting both components, given λ, a and b. Following our discussion of §3.3, we assume that, given λ, a and b, the probabilities $p_1(\lambda, a)$ and $p_2(\lambda, b)$ are independent. Thus we write the probability of detecting both components as

$$p_{12}(\lambda, a, b) = p_1(\lambda, a)p_2(\lambda, b). \tag{3.15}$$

The ensemble average probabilities of equations (3.14) are then given by:

$$p_1(a) = \int_\Lambda p_1(\lambda, a)\, \mathrm{d}\rho$$
$$p_1(b) = \int_\Lambda p_2(\lambda, b)\, \mathrm{d}\rho \tag{3.16}$$
$$p_{12}(a, b) = \int_\Lambda p_1(\lambda, a)p_2(\lambda, b)\, \mathrm{d}\rho.$$

To proceed, CH introduce the following lemma, the proof of which may be found in their paper: if x, x', y, y', X, Y are real numbers such that $0 \leqslant x, x' \leqslant X$ and $0 \leqslant y, y' \leqslant Y$, then the inequality:

$$-XY \leqslant xy - xy' + x'y + x'y' - Yx' - Xy \leqslant 0 \tag{3.17}$$

holds. Inequality (3.17) and equation (3.15) yield:

$$-1 \leqslant p_{12}(\lambda, a, b) - p_{12}(\lambda, a, b') + p_{12}(\lambda, a', b) + p_{12}(\lambda, a', b') - p_1(\lambda, a') - p_2(\lambda, b) \leqslant 0. \tag{3.18}$$

Integrating inequality (3.18) over λ with distribution ρ, and using equation (3.16), one obtains the result:

$$-1 \leqslant p_{12}(a, b) - p_{12}(a, b') + p_{12}(a', b) + p_{12}(a', b') - p_1(a') - p_2(b) \leqslant 0. \tag{3.19}$$

(Obtaining the left-hand inequality also required the use of equation (3.3), but the right-hand one did not. Since the left-hand inequality requires a measurement of the absolute magnitude of probabilities, the 'event-ready' detectors of figure 1 will be needed to test it.) The right-hand side of inequality (3.19) can be rewritten in the following form:

$$\frac{p_{12}(a, b) - p_{12}(a, b') + p_{12}(a', b) + p_{12}(a', b')}{p_1(a') + p_2(b)} \leqslant 1. \tag{3.20(a)}$$

As desired, it involves only a quantity that is independent of N. Using equations (3.14), and defining $R(a, b)$ as the rate of coincident detections, and $r_1(a)$ and $r_2(b)$ as the rate of single-particle detections by either apparatus, inequality (3.20(a)) can be rewritten directly in terms of a ratio of observable count rates:

$$\frac{R(a, b) - R(a, b') + R(a', b) + R(a', b')}{r_1(a') + r_2(b)} \leqslant 1. \qquad (3.20(b))$$

Inequalities (3.20) are thus a general prediction for any local realistic theory of natural phenomena.

In order to complete the proof of the theorem it suffices to exhibit an instance in which the quantum-mechanical counterpart to inequalities (3.20) fails. This is done in §4, when we discuss the experimental requirements for a valid test of these theories.

3.6. Symmetry considerations

Almost all of the experiments which have been proposed for testing the predictions by Bell's inequalities involve pairs of polarised particles (either photons or massive particles). In these experiments the parameters a and b, considered abstractly in §§3.4 and 3.5, are taken to be orientation angles relative to some reference axis in a fixed plane. In most of these experiments, the method of preparing the pairs of polarised particles attempts to achieve cylindrical symmetry about a normal to the fixed plane and reflection symmetry with respect to planes through this normal. This symmetry is exhibited in the quantum-mechanical predictions for detection rates and correlations:

$[p_1(a)]_{QM}$ and $[r_1(a)]_{QM}$ are independent of a.
$[p_2(b)]_{QM}$ and $[r_2(b)]_{QM}$ are independent of b.
$[p_{12}(b)]_{QM}$, $[R(a, b)]_{QM}$ and $[E(a, b)]_{QM}$ are functions only of $|a-b|$.

We now assume that the corresponding predictions for local realistic theories exhibit the same symmetries:

$$p_1(a) \equiv p_1 \text{ and } r_1(a) \equiv r_1 \text{ are independent of } b$$

$$p_2(b) \equiv p_2 \text{ and } r_2(b) \equiv r_2 \text{ are independent of } b \qquad (3.21)$$

$$p_{12}(a, b) \equiv p_{12}(|a-b|), \ R(a, b) \equiv R(|a-b|) \text{ and } E(a, b) \equiv E(|a-b|).$$

We must emphasise two points concerning equation (3.21). First, these symmetry relations do not simply follow from the corresponding quantum-mechanical symmetry relations or from the symmetry of the experimental arrangement, for one does not know what symmetry-breaking factors may lurk at the level of the hidden variables. Second, no harm is done in assuming equations (3.21), since they are susceptible to experimental verification.

Now suppose that we take a, a', b and b' so that:

$$|a-b| = |a'-b| = |a'-b'| = \tfrac{1}{3}|a-b'| = \phi$$

as in figure 3. With the use of equation (3.21), inequalities (3.12) and (3.20) simplify to

$$|3E(\phi) - E(3\phi)| \leqslant 2 \qquad (3.22)$$

and

$$S(\phi) \leqslant 1 \qquad (3.23)$$

Bell's theorem: experimental tests and implications 1897

where we have defined:

$$S(\phi) \equiv \frac{3p_{12}(\phi) - p_{12}(3\phi)}{p_1 + p_2} \qquad (3.24(a))$$

in terms of probabilities, or equivalently in terms of count rates:

$$S(\phi) \equiv \frac{3R(\phi) - R(3\phi)}{r_1 + r_2}. \qquad (3.24(b))$$

3.7. The proof by Wigner, Belinfante and Holt

A simple method of proving Bell's theorem for deterministic local hidden-variables theories was invented independently by Wigner (1970) and Belinfante (1973), and extended by Holt (1973). The method consists of subdividing the space Λ of states of a two-component system into subspaces corresponding to various possible values of the observables of interest, and then performing some easy calculations on the measures of these subspaces. Rather than duplicate their proofs, which are readily available, we show how their method can be used to derive the inequality of CH.

Consider the apparatus configuration of figure 2. Assume that the detection or non-detection of component 1 is completely determined by the parameter a of the first analyser and the state of the composite system, but is independent of the parameter b of the other analyser, and so forth for component 2. As such, the discussion is for the restricted situation in which determinism applies. Under this assumption, we can exhaustively subdivide the space Λ into 16 mutually disjoint subspaces Λ $(ij; kl)$, where each letter can take on the value 0 or 1, with 1 denoting detection and 0 non-detection; with i and j referring to the results if the parameter of the first analyser is chosen respectively to be a or a'; and with k and l referring to the results if the parameter of the second analyser is chosen respectively to be b or b'. For example, $\Lambda(10; 01)$ is the subspace in which component 1 will be detected if its associated parameter is chosen to be a but will not if the parameter is chosen to be a', while component 2 will not be detected if its associated parameter is chosen to be b but will be detected if that parameter is chosen to be b'. (Note that there is no question of simultaneously examining detection or non-detection for two different values of a parameter. Indeed, such observations are mutually exclusive. Rather, the subspace is defined in terms of what will happen if any one of the various experiments is performed. Since the theories are assumed to be deterministic, these values are all determined once a, b, λ and the apparatus configuration are specified.) If a probability measure ρ is assumed to be given on Λ (determined presumably by the way in which the composite system is prepared), then $\rho(ij; kl)$ is defined to be the probability that the composite state is in $\Lambda(ij; kl)$. Clearly, all $\rho(ij; kl)$ are non-negative. Because the 16 subspaces are disjoint and exhaustive, we have:

$$\sum_{ijkl} \rho(ij; kl) = 1. \qquad (3.25)$$

We now define $p_1(a)$ to be the probability that component 1 will be detected if its parameter is chosen to be a; $p_2(b)$ to be the probability that component 2 will be detected if its parameter is chosen to be b; and $p_{12}(a, b)$ to be the probability of joint detection of both components if the two parameters are chosen respectively to be a and b. Analogous definitions are given for the other values of the parameters. Then we have:

$$p_{12}(a, b) = \rho(11; 11) + \rho(11; 10) + \rho(10; 11) + \rho(10; 10)$$

$$p_{12}(a, b') = \rho(11; 11) + \rho(11; 01) + \rho(10; 11) + \rho(10; 01)$$

$$p_{12}(a', b) = \rho(11; 11) + \rho(11; 10) + \rho(01; 11) + \rho(01; 10)$$

$$p_{12}(a', b') = \rho(11; 11) + \rho(11; 01) + \rho(01; 11) + \rho(01; 01) \qquad (3.26)$$

$$p_1(a') = \rho(11; 11) + \rho(11; 10) + \rho(11; 01) + \rho(11; 00)$$
$$+ \rho(01; 11) + \rho(01; 10) + \rho(01; 01) + \rho(01; 00)$$

$$p_2(b) = \rho(11; 11) + \rho(11; 10) + \rho(10; 11) + \rho(10; 10)$$
$$+ \rho(01; 11) + \rho(01; 10) + \rho(00; 11) + \rho(00; 10).$$

It follows that:

$$p_{12}(a, b) - p_{12}(a, b') + p_{12}(a', b) + p_{12}(a', b') - p_1(a') - p_2(b)$$

$$= -\rho(11; 01) - \rho(11; 00) - \rho(10; 11) - \rho(10; 01) - \rho(01; 10)$$

$$- \rho(01; 00) - \rho(00; 11) - \rho(00; 10). \quad (3.27)$$

Consequently, we recover inequality (3.19) derived by Clauser and Horne for the more general stochastic case:

$$-1 \leqslant p_{12}(a, b) - p_{12}(a, b') + p_{12}(a', b) + p_{12}(a', b') - p_1(a') - p_2(b) \leqslant 0.$$

The demonstration of the incompatibility between this inequality and quantum mechanics is thus the same as that of § 3.5, and hence the theorem is proved.

3.8. Stapp's proof

Stapp's version of Bell's theorem (1971, 1977) appears to be very general, for it dispenses with all assumptions about the state of the system and about probability measures on the space of states. The proof was generalised by Eberhard (1977) to include realisable systems. Stapp considered a long series of N occurrences of Bohm's *Gedankenexperiment*. In each occurrence a pair of spin-$\frac{1}{2}$ particles is produced in the singlet state in a space–time region S_0. The particles propagate in opposite directions along a given axis. Particle 1 proceeds to a space–time region S_1, where it is deflected 'up' or 'down' by a Stern–Gerlach magnet oriented in either the $\hat{a}$ or the $\hat{a}'$ direction, and particle 2 proceeds to the region S_2 where it is deflected up or down by a magnet oriented in either the $\hat{b}$ or the $\hat{b}'$ direction. S_1 and S_2 are supposed to have space-like separation, and the choice for orienting the first magnet along $\hat{a}$ or $\hat{a}'$ is made when particle 1 is in S_1, and similarly with the choice for orienting the second magnet. Let the number 1 or -1 be recorded for a particle entering the field of a Stern–Gerlach magnet accordingly as it is deflected 'up' or 'down'. Let $r_{\alpha j}(\hat{a}, \hat{b})$ (where $\alpha = 1, 2$ and $j = 1, \ldots, N$) be the number recorded for the αth particle of the jth pair if the two magnets are oriented in the $\hat{a}$ and $\hat{b}$ directions respectively, and let $r_{\alpha j}(\hat{a}, \hat{b}')$, $r_{\alpha j}(\hat{a}', \hat{b})$ and $r_{\alpha j}(\hat{a}', \hat{b}')$ have analogous meanings. Clearly the orientations $\hat{a}$ and $\hat{a}'$ are mutually exclusive, as are $\hat{b}$ and $\hat{b}'$. Although only one of the four possible pairs of orientations $(\hat{a}, \hat{b})$, $(\hat{a}, \hat{b}')$, $(\hat{a}', \hat{b})$, $(\hat{a}', \hat{b}')$ can occur in the real world, Stapp made an assumption

Bell's theorem: experimental tests and implications 1899

of 'counterfactual definiteness', that $r_{\alpha j}(\hat{a}, \hat{b})$, $r_{\alpha j}(\hat{a}, \hat{b}')$, etc, are all definite numbers. In addition, he made an assumption of individual locality, that:

$$r_{1j}(\hat{a}, \hat{b}) = r_{1j}(\hat{a}, \hat{b}') \qquad (3.28(a))$$

$$r_{1j}(\hat{a}', \hat{b}) = r_{1j}(\hat{a}', \hat{b}') \qquad (3.28(b))$$

$$r_{2j}(\hat{a}, \hat{b}) = r_{2j}(\hat{a}', \hat{b}) \qquad (3.28(c))$$

$$r_{2j}(\hat{a}, \hat{b}') = r_{2j}(\hat{a}', \hat{b}'). \qquad (3.28(d))$$

Stapp then showed that the $8N$ numbers $r_{\alpha j}(\hat{a}, \hat{b})$, etc, must disagree with some of the statistical predictions of quantum mechanics. Some critics of Stapp have argued that his assumption of counterfactual definiteness is understandable only from the stand-point of a deterministic local hidden-variables theory. Stapp (1978, §4) has replied, however, that his assumption requires no commitment to determinism, but only to the possibility of speaking (as is commonly done in the sciences) of possible worlds as well as the actual one. He makes the explicit assumption that each of the four choices $(\hat{a}, \hat{b})$, $(\hat{a}, \hat{b}')$, $(\hat{a}', \hat{b})$ and $(\hat{a}', \hat{b}')$ is made in some possible world. It may nevertheless be objected that Stapp has not given a reason for demanding the existence of a quad-ruple of possible worlds which mesh together as in equations $(3.28(a)–(d))$. The combination of no action-at-a-distance with the idea of possible worlds only seems to require four pairs of possible worlds, one pair meshing as in equation $(3.28(a))$, one as in equation $(3.28(b))$, etc. It is not obvious why these four relations need to govern a cluster of four possible worlds unless determinism is supposed. An answer to this objection is provided by Stapp (1978), in which the following equivalence theorem is proved.

Let I be the set of individual outcomes $r_{\alpha j}(c, d)$, where c is $\hat{a}$ or $\hat{a}'$, d is $\hat{b}$ or $\hat{b}'$, α is 1 or 2, and j is $1, \ldots, N$. Let $P(I)$ be the set of probabilities

$$P = (\{r_1 | \hat{a}\}, \{r_2 | \hat{b}\}, \{r_1, r_2 | \hat{a}, \hat{b}\})$$

determined by the appropriate frequencies in I:

$$\{r_1 | \hat{a}\} = N(r_1, \hat{a})/N$$

(where $N(r_1, \hat{a})$ is the number of j such that $r_1 = r_{1j}(\hat{a}, \hat{b}) = r_{1j}(\hat{a}, \hat{b}')$ by individual locality),

$$\{r_1, r_2 | \hat{a}, \hat{b}\} = N(r_1, r_2, \hat{a}, \hat{b})$$

(where $N(r_1, r_2, \hat{a}, \hat{b})$ is the number of j such that $r_1 = r_{1j}(\hat{a}, \hat{b})$ and $r_2 = r_{2j}(\hat{a}, \hat{b})$), etc.

Let L_P be the set of P which satisfy the following probabilistic locality conditions: there exists a discrete set of λ, a probability weight function ρ defined on this set, and probabilities $p_1(\lambda, \hat{a}, r_1)$, $p_2(\lambda, \hat{b}, r_2)$ for the outcomes r_1 and r_2 respectively (given λ and given $\hat{a}$ or $\hat{b}$), such that:

$$\{r_1, r_2 | \hat{a}, \hat{b}\} = \sum_{\lambda} \rho(\lambda) p_1(\lambda, \hat{a}, r_1) p_2(\lambda, \hat{b}, r_2)$$

$$\{r_1 | \hat{a}\} = \sum_{\lambda} \rho(\lambda) p_1(\lambda, \hat{a}, r_1) \qquad (3.29)$$

$$\{r_2 | \hat{b}\} = \sum_{\lambda} \rho(\lambda) p_1(\lambda, \hat{b}, r_2).$$

Finally, let L be the set of I which satisfy the individual locality conditions $(3.28(a)–(d))$. Then the equivalence theorem asserts (i) if $I \in L$, then $P(I) \in L_P$, (ii) if $P \in L_P$, then there is an $I \in L$ such that $P(I)$ is approximately equal to P.

Note that (3.29) is essentially the CH probabilistic locality condition, except that a sum over discrete values of λ is used instead of an integral over the space Λ; but since the integral can always be approximated by a sum, this difference is not crucial. Because of this equivalence theorem, the theorem of CH and of Bell that no P which belongs to L_P can agree statistically with quantum mechanics entails that no I which belongs to L can agree statistically with quantum mechanics and, conversely, the theorem of Stapp that no I belonging to L can agree statistically with quantum mechanics implies the theorem of CH and Bell. Stapp's equivalence theorem, therefore, shows that, contrary to appearances, his proof of Bell's theorem and those of CH and Bell (1971) are of equal strength. J S Bell (personal communication) has remarked that part (ii) of Stapp's equivalence theorem is an example of the possibility of simulating a stochastic process with a deterministic one.

3.9. Other versions of Bell's theorem

Several other versions of Bell's theorem have been discovered. The proofs are mathematically correct, but with hypotheses in some respects problematic, either from a philosophical point of view or from their inherent restriction to idealised systems.

A very general derivation of Bell's theorem has been presented by Bell (1976). It was critically evaluated by Shimony *et al* (1976), who challenged one of the premises. Bell (1977) replied to this criticism. If we retain the notation of §3.8, we can express the essential assumption of Bell (1976) in the following way: the complete state of region S_1 is independent of the choice between $\boldsymbol{b}$ and $\boldsymbol{b}'$ in S_2, and likewise the complete state of S_2 is independent of the choice between $\boldsymbol{a}$ and $\boldsymbol{a}'$ in S_1. Shimony *et al* (1976) criticised this assumption on the ground that the backward light cones of S_1 and S_2 overlap in a region S, and it is possible that a factor in S affecting the choice between $\boldsymbol{b}$ and $\boldsymbol{b}'$ leaves some trace in S_1. Bell's reply (1977) to this objection stresses the spontaneity of the experimenter's choice between $\boldsymbol{b}$ and $\boldsymbol{b}'$ and between $\boldsymbol{a}$ and $\boldsymbol{a}'$, but this answer seems to us to depend upon too strong a commitment to indeterminism for his argument to be fully general (see also Shimony 1978).

The proof by d'Espagnat (1975) has the virtue of staying quite close to the original ideas of EPR by reasoning in terms of the intrinsic properties of the system. We shall not try to summarise his argument, partly because of its length and partly because of a premise which is impossible to be realised experimentally. D'Espagnat assumes (as Bell did in 1965) that one has a system like a pair of spin-$\frac{1}{2}$ particles in the singlet state, such that one can measure an observable of one of the pair and then infer with absolute certainty the value of a corresponding observable of the other pair (equation (3.2)). The same criticism can also be made of the arguments of Gutkowski and Masotto (1974), Selleri (1978) and Schiavulli (1977) but it should be noted that they derive a number of generalisations of Bell's inequalities which have not been obtained elsewhere.

4. Considerations regarding a general experimental test

Following Bell's (1965) results, many readers believed that local realistic theories were *ipso facto* discredited, because quantum mechanics has been so abundantly confirmed in a variety of experimental situations. Indeed, some of the most striking

Bell's theorem: experimental tests and implications   1901

confirmations of quantum mechanics, such as the spectrum of helium, concerned correlated pairs of particles. However, upon careful examination, one finds that situations exhibiting the disagreement discovered by Bell are rather rare, and none had ever been experimentally realised. Moreover, the reasoning of the previous sections indicates that the treatment of correlated but spatially separated systems may well be the point of greatest vulnerability of quantum mechanics. In view of the consequences of Bell's theorem it is thus important to design experiments to test explicitly the predictions made for local realistic theories via Bell's theorem.

Starting with the simple configurations specified in §3, the first problem is to find a suitable system whose quantum-mechanical predictions directly violate the predictions in the theorem, but additionally one that is accessible with available technology. In fact, this has not yet been done! (Possible avenues in this direction are discussed in §7.) In the present section we examine the requirements for a fully general test, and see why the problem is difficult. Since the presence of auxiliary counters is required by the apparatus configuration of figure 1, and usually these depolarise or destroy the emissions, we will confine our discussion to the apparatus configuration of figure 2.

We thus compare the quantum-mechanical predictions for this configuration with those by inequality (3.23). The left-hand side of inequality (3.19) is not considered here, since it cannot be expressed in terms of ratios of observable probabilities. It will, however, become useful for the discussion of §5.

4.1. Requirements for a general experimental test

Consider an experiment, with a configuration similar to that of figure 2, whose quantum-mechanical predictions take the following form:

$$[p_{12}(\phi)]_{\text{QM}} = \tfrac{1}{4} \eta_1 \eta_2 f_1 g [\epsilon_+^1 \epsilon_+^2 + \epsilon_-^1 \epsilon_-^2 \, F \cos(n\phi)]$$

$$[p_1]_{\text{QM}} = \tfrac{1}{2} \eta_1 f_1 \epsilon_+^1 \tag{4.1}$$

$$[p_2]_{\text{QM}} = \tfrac{1}{2} \eta_2 f_2 \epsilon_+^2.$$

This general form is characteristic of the quantum-mechanical predictions for the actual experiments of interest (see, for example, equation (5.15)). In these expressions η_i represents the effective quantum efficiency of detector $i (i = 1, 2)$, and

$$\epsilon_+^i \equiv \epsilon_M^i + \epsilon_m^i \qquad \epsilon_-^i \equiv \epsilon_M^i - \epsilon_m^i. \tag{4.2}$$

The terms ϵ_M^i and ϵ_m^i are the maximum and minimum transmissions of the analysers relative to the pertinent orthogonal basis. The functions f_1 and f_2 are the collimator efficiencies, i.e. the probability that an appropriate emission enters apparatus 1 or 2. Typically, these are simply proportional to the collimator acceptance solid angles. The function g is the conditional probability that, given emission 1 enters apparatus 1, then emission 2 will enter apparatus 2. The function F is a measure of the initial-state purity and the inherent quantum-mechanical correlation of the two emissions. For the actual cascade-photon experiments (see §5), these functions depend on the collimator solid angles. The values of n are 1 or 2 depending upon whether the experiment is performed with fermions or bosons.

Inserting equation (4.1) into the definition of $S(\phi)$, equation (3.24(a)), we find the quantum-mechanical prediction for this function to be given by:

$$S_{\text{QM}}(\phi) = \tfrac{1}{4} \eta g \{ 2\epsilon_+ + [3 \cos(n\phi) - \cos(3n\phi)] \, F(\epsilon_-^2/\epsilon_+) \}. \tag{4.3}$$

Here for simplicity we have taken $\eta \equiv \eta_1 = \eta_2$, $f_1 = f_2$, $\epsilon_+ = \epsilon_+^1 = \epsilon_+^2$ and $\epsilon_- \equiv \epsilon_-^1 = \epsilon_-^2$. Selecting the optimum value $\phi = \pi/4n$, one finds that the condition for a violation of inequality (3.23) is given by:

$$\eta g \epsilon_+ [\sqrt{2}\,(\epsilon_-/\epsilon_+)^2\,F + 1] > 2. \qquad (4.4)$$

Thus, a correlation experiment with values in the domain specified by inequality (4.4) is capable of distinguishing between the prediction, inequality (3.23), and that of quantum theory, equation (4.3). Although such experiments are apparently possible, there is at present no existing experimental result in this domain, and thus none in violation of any inequality which does not require additional assumptions for its derivation.

For a direct test of inequality (4.4) the requirements are stringent, which accounts for the fact that, so far, no such experiment has been attempted.

(i) A source must emit pairs of discrete-state systems, which can be detected with high efficiency.

(ii) Quantum mechanics must predict strong correlations of the relevant observables of each pair (polarisations in the experiments so far). Correspondingly, the ensemble of pairs must have high quantum-mechanical purity.

(iii) The analysers must be capable of allowing systems in certain states to pass with great efficiency, while simultaneously rejecting nearly all of those in orthogonal states.

(iv) The collimators (and filters if these are necessary to remove unwanted emissions, etc) must have very high transmittances and not depolarise the emissions.

(v) The source must produce the systems via a two-body decay. A three- (or more) body decay cannot be used, because the resulting angular correlation will make $g \ll 1$.

(vi) Another requirement should be added in order to achieve an airtight argument against locality: the parameters a and b must be rapidly changed while the emissions are in flight. A detection event should be space-like separated from the corresponding parameter change event at the far apparatus (see §7). This suggestion was first made by Bohm and Aharonov (1957).

For a practical experiment, it is of course also necessary for the counting rate to be sufficiently high to make the required integration time reasonable.

4.2. Three important experimental cases

Let us examine how the failure of any of these parameters to approximate the ideal case prevents a violation of inequality (3.23) from arising. Figure 4 shows the prediction by equation (4.3) for three important cases of interest, along with the prediction by inequality (3.23).

Case I, nearly ideal. In the domain of nearly ideal apparatus, we have $g \approx \epsilon_+ \approx \epsilon_- \approx \eta \approx 1$. For these conditions we find a violation of inequality (3.23) for a wide range of ϕ, with a maximum violation at $n\phi = \pi/4$.

Case II, poor detector efficiencies or co-focusing. When $g \ll 1$ holds, because of imperfect collimator alignment and/or a weak angular correlation inherent in a three-body decay, or when $\eta \ll 1$ holds, because the detector efficiencies are low, then the amplitude of $S(\phi)$ contracts in amplitude about a value close to zero. The quantum-mechanical predictions enter a domain where no violation of inequality (3.23) occurs.

Bell's theorem: experimental tests and implications 1903

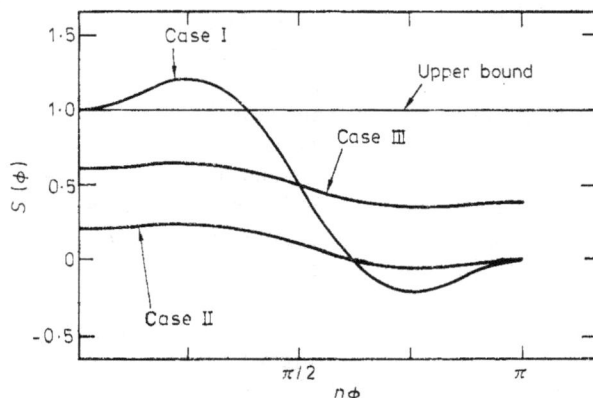

Figure 4. Typical dependence of $S(\phi)$ upon $n\phi$, for cases I–III. Upper bound for $S(\phi)$ set by inequality (3.23) is $+1$. Case I experiments (nearly ideal) have $\eta \approx g \approx F \approx \epsilon_+ \approx \epsilon_- \approx 1$. Case II experiments have nearly ideal parameters $F \approx \epsilon_- \approx \epsilon_+ \approx 1$, but have $\eta \ll 1$ and/or $g \ll 1$. Case III experiments have nearly ideal parameters $\eta \approx g \approx 1$, but have $F \ll 1$ and/or $\epsilon_-/\epsilon_+ \ll 1$.

This case is typical of the low-energy cascade-photon experiments to be described in §5.

Case III, weak correlation. The third case occurs when the predicted correlation is weak. The correlation coefficient F and/or the parameter ϵ_-/ϵ_+ may be much less than unity. This will occur, for example, if the emissions are only weakly correlated, if the initial state is impure, if the emissions suffer significant depolarisation in passing through the apparatus, or if the analyser efficiencies are low. The curve $S(\phi)$ then contracts in amplitude symmetrically about a value slightly less than $+\frac{1}{2}$, and again no violation of inequality (3.23) occurs. This case is typical of the positronium annihilation and the proton–proton S-wave scattering experiments to be described in §6.

The manner in which the amplitude of $S(\phi)$ contracts is of more importance than it may seem. To perform a test of the local realistic theories in the domain of case II and III experiments requires a credible auxiliary assumption that $S(\phi)$ can be rescaled somehow to an amplitude sufficient to violate the inequalities. Case II experiments (discussed in §5) are more favourable in this respect than are those of case III. For the former, a replacement of p_1 and p_2 (singles rates) with carefully selected coincidence rates can provide this rescaling at the small price of accepting only a very mild auxiliary assumption. On the other hand, rescaling case III experiments (see §6) requires one to assume a certain *ad hoc* modification of the basic correlation coefficient F. However, the measurement of this coefficient is in many respects a primary objective of the experiment. Any such assumption must then be scrutinised very carefully, for it inherently becomes the weak point of the experiment.

5. Cascade-photon experiments

The essential problem in testing the predictions in Bell's theorem against those by quantum mechanics is to find experimentally realisable situations in which the

1904   *J F Clauser and A Shimony*

quantum-mechanical predictions directly violate Bell's inequalities. In §4 we showed that to do so with available apparatus is difficult. The situation is not hopeless, however. Clauser *et al* (1969) showed that with a mild supplementary assumption, actual experiments are predicted by quantum mechanics to yield a violation of Bell's inequality, and they proposed such an experiment.

Their suggestion is to measure the correlation in linear polarisation of photon pairs emitted in an atomic cascade. Figure 5 shows a schematic diagram of a typical apparatus for doing this, that of Freedman and Clauser (1972), who reported the first such test. The photons were emitted in a $J=0 \rightarrow J=1 \rightarrow J=0$ atomic cascade. The decaying atoms were viewed by two symmetrically placed optical systems, each consisting of two lenses, a wavelength filter, a rotatable and removable polariser, and a single-photon detector. The following quantities were measured: $R(\phi)$, the coincidence rate for two-photon detection as a function of the angle ϕ between the planes of linear polarisation, defined by the orientations of the inserted polarisers; R_1, the

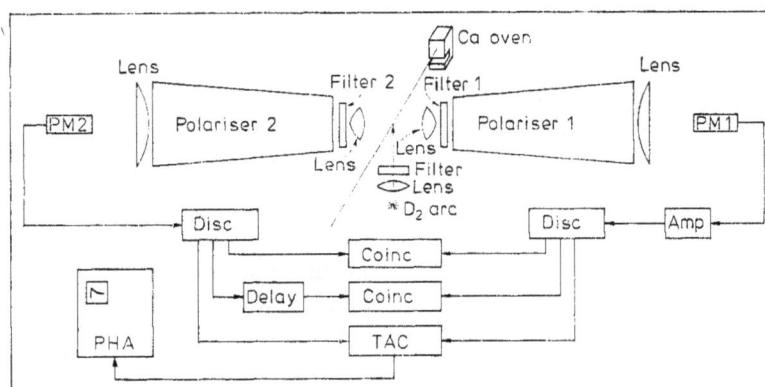

Figure 5. Schematic diagram of apparatus and associated electronics of the experiment by Freedman and Clauser. Scalers (not shown) monitored the outputs of the discriminators and coincidence circuits (figure after Freedman and Clauser).

coincidence rate with polariser 2 removed; R_2, the coincidence rate with polariser 1 removed; R_0, the coincidence rate with both polarisers removed.

The details of this experiment along with other similar ones will be discussed in §5.3. First, however, we describe the auxiliary assumption(s) which render this a reasonable test, and present the resulting inequalities. Then we describe the quantum-mechanical predictions for this and similar arrangements.

5.1. Predictions by local realistic theories

5.1.1. Assumptions for cascade-photon experiments. The initial assumption by CHSH is, given that a pair of photons *emerges* from the polarisers, the probability of their joint detection is independent of the polariser orientations a and b. Clauser and Horne (1974) showed that an alternative assumption leads to the same results. Their assumption is that for every pair of emissions (i.e. for each value of λ), the probability of a count with a polariser in place is less than or equal to the corresponding probability with the polariser removed. The assumption of CH is stronger than that of CHSH in so far as it is stated for each value of λ, whereas CHSH make an assertion only for the total sub-ensemble of photons which pass through the polarisers. On the other hand, the assumption of CH is more general, in that the processes 'passage' and 'non-passage' through a polariser (which are not observable, and which are inappropriate for many possible theories) are not considered primitive. Furthermore, CH only assume an inequality, which is weaker than the equality of CHSH. Both assumptions, in our opinion, are physically plausible, but each gives a certain loophole to those who wish to defend local hidden-variables theories in spite of the experimental evidence which will be presented below.

Let us discuss the consequences of the CH assumption. We denote by the symbol ∞ an apparatus configuration in which the analyser is absent. Let $p_1(\lambda, \infty)$ denote the probability of a count from detector 1 when analyser 1 is absent and the state of the emission is λ. A similar probability $p_2(\lambda, \infty)$ may be defined for apparatus 2. Thus, the assumption is that:

$$0 \leqslant p_1(\lambda, a) \leqslant p_1(\lambda, \infty) \leqslant 1$$
$$0 \leqslant p_2(\lambda, b) \leqslant p_2(\lambda, \infty) \leqslant 1 \tag{5.1}$$

for every λ, and for all values of a and b. Inequalities (5.1) and (3.17) and arguments similar to those which led from (3.16) to (3.19) yield immediately the result:

$$-p_{12}(\infty, \infty) \leqslant p_{12}(a, b) - p_{12}(a, b') + p_{12}(a', b) + p_{12}(a', b') - p_{12}(a', \infty) - p_{12}(\infty, b) \leqslant 0. \tag{5.2}$$

Note that all terms in inequality (5.2) are joint probabilities for coincident counts at the two detectors. Inequality (3.19), in contrast, contains the two terms p_1 and p_2, which are probabilities of a count at a single detector. Furthermore, both the upper and lower limits in inequality (5.2) can be written as a ratio of probabilities, so that both can be tested without the need for the 'event-ready' detectors of figure 1.

5.1.2. Additional symmetries. Again we can invoke a rotational invariance argument similar to that of §3.6; thus we require:

(i) $p_{12}(a, \infty)$ is independent of a, and likewise $R_1(a) \equiv R_1$
(ii) $p_{12}(\infty, b)$ is independent of b, and likewise $R_2(b) \equiv R_2$
(iii) $p_{12}(a, b) \equiv p_{12}(\phi)$, and likewise $R(a, b) \equiv R(\phi)$, where $\phi = |a - b|$.

These conditions are not always satisfied, and they obviously fail when each of the particles has a definite linear polarisation. However, for all of the actual experiments to be described in this section, the conditions are at least satisfied by the quantum-mechanical predictions, and more importantly no experimental deviations from them have been detected. It is noteworthy that this set of conditions is frequently satisfied, even in situations where some of those of §3.6 are not. For example, in many of the cascade-photon experiments, the singles rate r_2 contains an extraneous contribution from excitation to the intermediate state of the cascade by channels not involving the first level of the cascade. Such excitation may result in the emission of polarised light at the wavelength of the second photon of the cascade, but no coincidences.

With these conditions, inequality (5.2) becomes:

$$-p_{12}(\infty, \infty) \leqslant 3p_{12}(\phi) - p_{12}(3\phi) - p_{12}(a', \infty) - p_{12}(\infty, b) \leqslant 0 \tag{5.3}$$

for all a' and b.

Since the emission rates in all of the various experiments were held constant, and in most cases monitored by an auxiliary apparatus, we can write the ratios of probabilities as ratios of count rates:

$$p_{12}(\phi)/p_{12}(\infty, \infty) = R(\phi)/R_0$$
$$p_{12}(a, \infty)/p_{12}(\infty, \infty) = R_1/R_0 \tag{5.4}$$
$$p_{12}(\infty, b)/p_{12}(\infty, \infty) = R_2/R_0.$$

Inserting equations (5.4) into inequality (5.3), we can write this form of Bell's inequality in terms of coincidence rates:

$$-R_0 \leqslant 3R(\phi) - R(3\phi) - R_1 - R_2 \leqslant 0. \tag{5.5}$$

Inequality (5.5) was first derived by Clauser *et al* (1969), but by using their alternative auxiliary assumption.

Freedman (1972) showed that inequality (5.5) can be further contracted to a form which is very convenient for comparison with experimental results. If we take the optimal value for upper-limit violation by cascade-photon experiments $\phi = \pi/8$, then inequality (5.5) becomes:

$$-R_0 \leqslant 3R(\pi/8) - R(3\pi/8) - R_1 - R_2 \leqslant 0.$$

On the other hand, if we take the optimal value for lower-limit violation $\phi = 3\pi/8$, using the fact that $9\pi/8$ represents the same angle as $\pi/8$, it becomes:

$$-R_0 \leqslant 3R(3\pi/8) - R(\pi/8) - R_1 - R_2 \leqslant 0.$$

Dividing both inequalities by R_0, and subtracting the second inequality from the preceding one, we obtain the simple inequality:

$$|R(\pi/8) - R(3\pi/8)|/R_0 \leqslant \tfrac{1}{4}. \tag{5.6}$$

Inequality (5.6) has the advantage that it can be checked by measuring the frequency of joint detection of photons with the polarisers in only two different relative orientations, and it dispenses with the need to measure rates with only one polariser removed.

5.2. Quantum-mechanical predictions for a $J = 0 \rightarrow 1 \rightarrow 0$ two-photon correlation

5.2.1. An idealised case. Even if ideal polarisation analysers and photo-detectors are assumed, the violation or non-violation of inequality (5.6) depends upon the quantum state in which the photon pairs are prepared. It is instructive to demonstrate that a violation does occur with perfect apparatus if the photons are propagating in opposite directions from the source along the $\hat{z}$ axis, with total angular momentum 0 and total parity $+1$. Their state is an ideal limit of ones which can actually be prepared in a laboratory. The polarisation part of the two-photon wavefunction is:

$$\Psi_0 = \frac{1}{\sqrt{2}}\left[\begin{pmatrix} 1 \\ 0 \\ 0 \end{pmatrix} \otimes \begin{pmatrix} 1 \\ 0 \\ 0 \end{pmatrix} + \begin{pmatrix} 0 \\ 1 \\ 0 \end{pmatrix} \otimes \begin{pmatrix} 0 \\ 1 \\ 0 \end{pmatrix} \right] \tag{5.7}$$

where $\begin{pmatrix} 1 \\ 0 \\ 0 \end{pmatrix}$ represents polarisation along the $\hat{x}$ axis and $\begin{pmatrix} 0 \\ 1 \\ 0 \end{pmatrix}$ represents polarisation along the $\hat{y}$ axis, and where the first of two juxtaposed column vectors refers to photon 1 and the second to photon 2. A projection operator for linear polarisation along an

Bell's theorem: experimental tests and implications 1907

axis, lying in the xy plane and making an angle θ with the $\hat{x}$ axis is:

$$Q(\theta) \equiv \begin{bmatrix} \cos^2\theta & \cos\theta\sin\theta & 0 \\ \cos\theta\sin\theta & \sin^2\theta & 0 \\ 0 & 0 & 0 \end{bmatrix} \qquad (5.8)$$

as one can check by noting that the vector $\left(\begin{smallmatrix}\cos\theta\\\sin\theta\\0\end{smallmatrix}\right)$, which represents linear polarisation in this direction, is an eigenvector of $Q(\theta)$ with eigenvalue 1. Similarly the vector $\left(\begin{smallmatrix}-\sin\theta\\\cos\theta\\0\end{smallmatrix}\right)$, representing linear polarisation perpendicular to this direction, is an eigenvector of $Q(\theta)$ with eigenvalue 0, as is $\left(\begin{smallmatrix}0\\0\\1\end{smallmatrix}\right)$ which represents polarisation along the z axis (which of course is excluded by transversality). Consequently, the quantum-mechanical prediction for this case is:

$$[R(\phi)/R_0]_{\Psi_0} = \langle \Psi_0 | Q(a) \otimes Q(b) | \Psi_0 \rangle = \tfrac{1}{4}(1 + \cos 2\phi) \qquad (5.9)$$

where, as before, we have taken $\phi = |a-b|$. From this result we find that the quantum-mechanical predictions

$$[R(\pi/8)/R_0 - R(3\pi/8)/R_0]_{\Psi_0} = \tfrac{1}{4}\sqrt{2} \qquad (5.10)$$

violate inequality (5.5).

5.2.2. Quantum-mechanical predictions for $J=0\to1\to0$ cascade, ideal analysers, and finite solid-angle detectors. Consider a $J=0\to J=1\to J=0$ atomic cascade in which no angular momentum is exchanged with the nucleus, and in which both transitions are electric dipole. Since the atom is both initially and finally in states with zero total angular momentum, and since there is a parity change in each transition, the emitted photon pair has zero total angular momentum and even parity. We can therefore exactly write the angular wavefunction of the photon pair as:

$$\Psi = \frac{1}{\sqrt{3}}\left[Y_{1,1}{}^1(\hat{\eta}_1)Y_{1,-1}{}^1(\hat{\eta}_2) - Y_{1,0}{}^1(\hat{\eta}_1)Y_{1,0}{}^1(\hat{\eta}_2) + Y_{1,-1}{}^1(\hat{\eta}_1)Y_{1,1}{}^1(\hat{\eta}_2) \right] \qquad (5.11)$$

where $\hat{\eta}_1$ and $\hat{\eta}_2$ are variable directions of propagation of the first and second photons, and where $Y_{jm}{}^1$ is the spherical vector function for total angular momentum j, magnetic quantum number m, and parity -1 (see, for example, Akhiezer and Beretstetskii (1965) for notation). Now suppose that the lenses which make the photons impinge normally upon the polarisation analysers collect light in cones of half-angle ξ. The wavefunction of a photon pair which emerges from the pair of lenses can be represented as $D(\xi)\Psi$, where $D(\xi)$ is an operator which is exhibited in the appendix to Shimony (1971). An argument is outlined in that paper that if ξ is infinitesimal, then $D(\xi)\Psi$ is equal (except for normalisation) to the ideal two-photon polarisation vector Ψ_0 of equation (5.7). This is a reasonable result, since there is no orbital angular momentum if the two photons propagate along a straight line. Therefore the fact that the photon pair has total angular momentum 0 implies that it has zero spin angular momentum, as in the state Ψ_0. Of course, a finite value of ξ is essential in an actual experiment in order to obtain a non-vanishing count rate. The quantum-mechanical prediction for the coincidence rates with the polarisation state $D(\xi)\Psi$ is then:

$$[R(\phi)/R_0]_{D(\xi)\Psi} = \langle D(\xi)\Psi | Q(a) \otimes Q(b) | D(\xi)\Psi \rangle = \tfrac{1}{4} + \tfrac{1}{4}F_1(\xi)\cos(2\phi) \qquad (5.12)$$

where $F_1(\xi)$ is a monotonically decreasing function, has the value 1 for $\xi = 0$, and diminishes to 0·9876 at $\xi = 30°$. Equation (5.12) shows a somewhat weaker polarisation correlation than one finds in equation (5.9), as a result of the admixture of orbital angular momentum states when light is collected in a non-zero solid angle. However, the diminution of correlation is small, even for fairly large values of ξ, and it is evident that inequality (5.5) will be violated by the probabilities of equation (5.12), with $\xi = 30°$.

5.2.3. Quantum-mechanical correlation for $J = 0 \to 1 \to 0$ cascade in an actual experiment. In an actual experiment one does not have ideal linear polarisation analysers, and equation (5.12) must be corrected in order to take into account the inefficiency of actual analysers. We let $\epsilon_M{}^j$ be the maximum transmittance of the jth analyser ($j = 1, 2$) and $\epsilon_m{}^j$ be the minimum transmittance. (The former is 1 and the latter is 0 for an ideal analyser, but values for the analysers will be given in the summaries below of the experiments which have actually been performed.) Then equation (5.9) must be replaced by the following:

$$
\begin{aligned}
[R(a, b)/R_0]_{QM} = {} & \epsilon_M{}^1 \epsilon_M{}^2 \langle D(\xi)\Psi | Q(a) \otimes Q(b) | D(\xi)\Psi \rangle \\
& + \epsilon_M{}^1 \epsilon_m{}^2 \langle D(\xi)\Psi | Q(a) \otimes \bar{Q}(b) | D(\xi)\Psi \rangle \\
& + \epsilon_m{}^1 \epsilon_M{}^2 \langle D(\xi)\Psi | \bar{Q}(a) \otimes Q(b) | D(\xi)\Psi \rangle \\
& + \epsilon_m{}^1 \epsilon_m{}^2 \langle D(\xi)\Psi | \bar{Q}(a) \otimes \bar{Q}(b) | D(\xi)\Psi \rangle
\end{aligned}
\tag{5.13}
$$

where

$$
\bar{Q}(a) = 1 - Q(a) \qquad \bar{Q}(b) = 1 - Q(b).
\tag{5.14}
$$

We thus find:

$$
[R(\phi)/R_0]_{QM} = \tfrac{1}{4}(\epsilon_M{}^1 + \epsilon_m{}^1)(\epsilon_M{}^2 + \epsilon_m{}^2) + \tfrac{1}{4}(\epsilon_M{}^1 - \epsilon_m{}^1)(\epsilon_M{}^2 - \epsilon_m{}^2)F_1(\xi)\cos 2\phi
\tag{5.15}
$$

(see Clauser *et al* 1969, Horne 1970, Shimony 1971). Again, the quantum-mechanical counterpart of inequality (5.5) is violated, if suitable values of the transmittances are used.

5.2.4. Other cascades. If the photon pair is obtained from a $J = 1 \to J = 1 \to J = 0$ cascade with equal populations in the initial Zeeman sublevels and no coherence among them (so that the density matrix of the initial level is $\tfrac{1}{3}I$), but the preceding experimental arrangement is otherwise unchanged, then the quantum-mechanical prediction for the probability of joint detection is the same as the right-hand side of equation (5.15), except that $F_1(\xi)$ is replaced by $-F_2(\xi)$, where $F_2(0) = 1$. The function $F_2(\xi)$ decreases monotonically more rapidly than $F_1(\xi)$ (Clauser *et al* 1969, Horne 1970, Holt 1973). A systematic survey of other possible cascades has been made by Fry (1973).

5.3. Description of experiments

So far, there have been four experiments of the type just described. Three of these have agreed with the quantum-mechanical predictions, and one has agreed with the predictions by local realistic theories via Bell's theorem.

5.3.1. Experiment by Freedman and Clauser (1972). Freedman and Clauser (1972, see also Freedman 1972) observed the 5513 Å and 4227 Å pairs produced by the $4p^2\ {}^1S_0 \to 4p4s\ {}^1P_1 \to 4s^2\ {}^1S_0$ cascade in calcium. Their arrangement is shown schematically in figure 5. Calcium atoms in a beam from an oven were excited by resonance absorption to the $3d4p\ {}^1P_1$ level, from which a considerable fraction decayed to the $4p^2\ {}^1S_0$ state at the top of the cascade. No precaution was necessary for eliminating isotopes with non-zero nuclear spin, since 99·855% of naturally occurring calcium has zero nuclear spin. Pile-of-plates polarisation analysers were used, with transmittances $\epsilon_M^1 = 0.97 \pm 0.01$, $\epsilon_m^1 = 0.038 \pm 0.004$, $\epsilon_M^2 = 0.96 \pm 0.01$, $\epsilon_m^2 = 0.037 \pm 0.004$. Each analyser could be rotated by angular increments of $\pi/8$, and the plates could be folded out of the optical path on hinged frames. The half-angle ξ subtended by the primary lenses was $30°$. Coincidence counting was done for 100 s periods; periods during which all plates were removed alternated with periods during which all were inserted. In each run the ratios $R(\pi/8)/R_0$ and $R(3\pi/8)/R_0$ were determined. Corrections were made for accidental coincidences, but even without this correction, the results still significantly violated inequality (5.6). The average ratios for roughly 200 h of running time are:

$$[R(\pi/8)/R_0]_{\text{expt}} = 0.400 \pm 0.007 \qquad [R(3\pi/8)/R_0]_{\text{expt}} = 0.100 \pm 0.003$$

and therefore:

$$[R(\pi/8)/R_0 - R(3\pi/8)/R_0]_{\text{expt}} = 0.300 \pm 0.008$$

in clear disagreement with inequality (5.6). The quantum-mechanical predictions are obtained from equation (5.15) (with allowances for uncertainties in the measurement of the transmittances and the subtended angle):

$$[R(\pi/8)/R_0 - R(3\pi/8)/R_0]_{\text{QM}} = (0.401 \pm 0.005) - (0.100 \pm 0.005) = 0.301 \pm 0.007.$$

The agreement between the experimental results with the quantum-mechanical predictions is excellent. Agreement is also found for other values of the angle ϕ, as well as for measurements made with only one or the other polariser removed.

5.3.2. Experiment by Holt and Pipkin (1973). Holt and Pipkin (1973, see also Holt 1973) observed 5676 Å and 4047 Å photon pairs produced by the $9^1P_1 \to 7^3S_1 \to 6^3P_0$ cascade in the zero nuclear-spin isotope ^{198}Hg (see figure 6 for a partial level diagram of mercury). Atoms were excited to the 9^1P_1 level by a 100 eV electron beam. The density matrix of the 9^1P_1 level was found to be approximately $\frac{1}{3}I$ by measurements of the polarisation of the 5676 Å photons, so that equation (5.15) with $F_1(\xi)$ replaced by $-F_2(\xi)$ is used to calculate the quantum-mechanical predictions for the coincidence counting rates. Calcite prisms were employed as polarisation analysers, with measured transmittances:

$$\epsilon_M^1 = 0.910 \pm 0.001 \qquad \epsilon_M^2 = 0.880 \pm 0.001$$

$$\epsilon_m^1 < 10^{-4} \qquad \epsilon_m^2 < 10^{-4}.$$

The half-angle ξ was taken to be $13°$ ($F_2(13°) = 0.9509$). The quantum-mechanical prediction is:

$$[R(3\pi/8)/R_0 - R(\pi/8)/R_0]_{\text{QM}} = 0.333 - 0.067 = 0.266$$

which only marginally exceeds the value $\frac{1}{4}$ allowed by inequality (5.6). The experimental result in 154·5 h of coincidence counting, however, is:

$$[R(3\pi/8)/R_0 - R(\pi/8)/R_0]_{\text{expt}} = 0.316 \pm 0.011 - 0.099 \pm 0.009 = 0.216 \pm 0.013$$

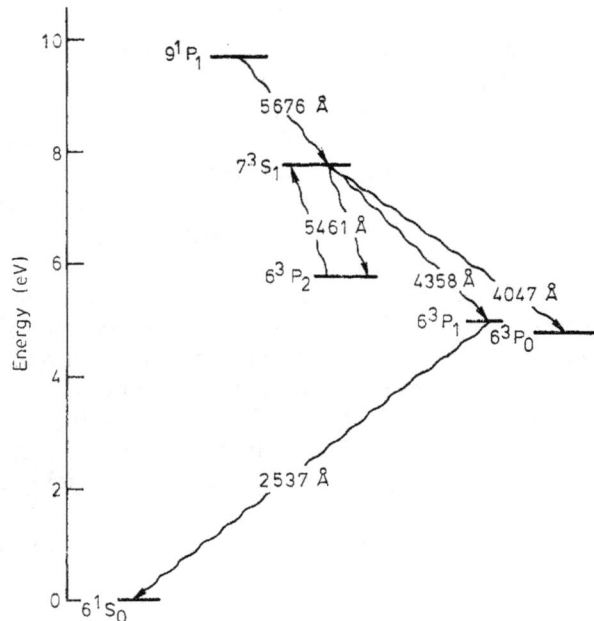

Figure 6. Partial level scheme for atomic mercury. Experiments by Holt and Pipkin, and by Clauser, excited the atoms to the 9^1P_1 level by electron bombardment, and observed photons emitted by the $9^1P_1 \rightarrow 7^3S_1 \rightarrow 6^3P_0$ cascade. The experiment by Fry and Thompson excited atoms to the 6^3P_2 (metastable) level by electron bombardment. Downstream, the atoms were excited by a tunable dye laser to the 7^3S_1 level, and photons were observed from the $7^3S_1 \rightarrow 6^3P_1 \rightarrow 6^1S_0$ cascade.

in good agreement with inequality (5.6) but in sharp disagreement with the quantum-mechanical prediction. Since this result is very surprising, Holt and Pipkin took great care to check possible sources of systematic error: the contamination of the source by isotopes with non-zero nuclear spin, perturbation by external magnetic or electric fields, coherent multiple scattering of the photons (radiation trapping), polarisation sensitivity of the photomultipliers, and spurious counts from residual radioactivity and/or cosmic rays, etc.

One such systematic error was found in the form of stresses in the walls of the Pyrex bulb used to contain the electron gun and mercury vapour. Estimates of the optical activity of these walls were then made, and the results were corrected correspondingly. (The values presented above include this correction.) It is noteworthy, however, that only the retardation sum for both windows was measured, for light entering the cell from one side and exiting through the opposite side. On the other hand, in the present experiment in which light exits from both windows, the relevant quantity is the retardation difference.

It is also noteworthy that in the subsequent experiment by Clauser (§5.3.3), a correlation was first measured which agreed with the results of Holt and Pipkin. Stresses were then found in one lens which were due to an improper mounting. (These were too feeble to be detected by a simple visual check using crossed Polaroids.) The stresses were removed, the experiment was re-performed, and excellent agreement with quantum mechanics was then obtained. On the other hand, Holt and Pipkin did not repeat their experiment when they discovered the stresses in their bulb.

Bell's theorem: experimental tests and implications 1911

A second criticism is that Holt and Pipkin took the solid-angle limit to be that imposed by a field stop placed outside the collimating lenses. It is possible that lens aberrations may have allowed a larger solid angle than they recognised. A ray-tracing calculation was in fact performed to assure that this was not the case. However, a solid stop ahead of the lens would have given one greater confidence that this did not, in fact, occur.

5.3.3. Experiment by Clauser (1976). Clauser (1976) repeated the experiment of Holt and Pipkin, using the same cascade and same excitation mechanism, though with a source consisting mainly of the zero-spin isotope ^{202}Hg. (The depolarisation effect due to some residual non-zero nuclear spin isotopes was calculated, using some results of Fry (1973).) Pile-of-plates polarisers were used with transmittances:

$$\epsilon_M{}^1 = 0.965 \qquad \epsilon_m{}^1 = 0.011 \qquad \epsilon_M{}^2 = 0.972 \qquad \epsilon_m{}^2 = 0.008$$

and the half-angle ξ taken to be $18.6°$. The quantum-mechanical prediction is:

$$[R(3\pi/8)/R_0 - R(\pi/8)/R_0]_{QM} = 0.2841.$$

The experimental result, from 412 h of integration, is:

$$[R(3\pi/8)/R_0 - R(\pi/8)/R_0]_{expt} = 0.2885 \pm 0.0093$$

in excellent agreement with the quantum-mechanical prediction, but in sharp disagreement with inequality (5.6).

5.3.4. Experiment by Fry and Thompson (1976). Fry and Thompson (1976) observed the 4358 Å and 2537 Å photon pairs emitted by the $7^3S_1 \to 6^3P_1 \to 6^1S_0$ cascade in the zero nuclear-spin isotope ^{200}Hg. Their experiment is shown schematically in figure 7.

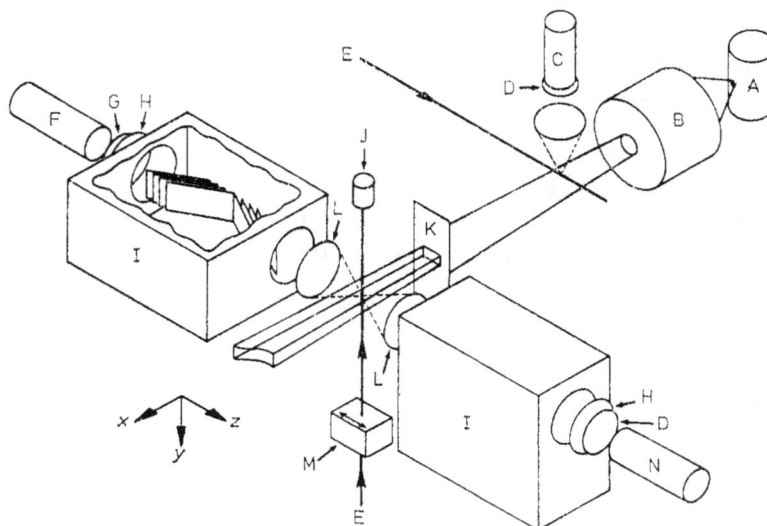

Figure 7. Schematic diagram of the experimental arrangement of Fry and Thompson. Polariser plate arrangement is also indicated. Actual polarisers have 14 plates. A, Hg oven; B, solenoid electron gun; C, RCA 8575; D, 4358 Å filter; E, 5461 Å laser beam; F, Amperex 56 DUVP/03; G, 2537 Å filter; H, focusing lens; I, pile-of-plates polariser; J, laser beam trap; K, atomic beam defining slit; L, light collecting lens; M, crystal polariser; N, RCA 8850 (figure after Fry and Thompson).

An atomic beam consisting of natural mercury was used as a source of ground-state (6^1S_0) atoms. The excitation of these to the 7^3S_1 level occurred in two steps at different locations along the beam. First, the atoms were excited by electron bombardment to the metastable 6^3P_2 level. Downstream, where all rapidly decaying states had vanished, a single isotope was excited to the 7^3S_1 level by resonant absorption of 5461 Å radiation from a narrow-bandwidth tunable dye laser. The technique provided a high data accumulation rate, since only the cascade of interest was excited. Photons were collected over a half-angle ξ of $19\cdot9° \pm 0\cdot3°$, and pile-of-plates analysers were used, with transmittances:

$$\epsilon_M{}^1 = 0\cdot98 \pm 0\cdot01 \qquad \epsilon_m{}^1 = 0\cdot02 \pm 0\cdot005 \qquad \epsilon_M{}^2 = 0\cdot97 \pm 0\cdot01 \qquad \epsilon_m{}^2 = 0\cdot02 \pm 0\cdot005.$$

The density matrix of the 7^3S_1 level was ascertained by polarisation measurements of the 4358 Å photons; it was found to be diagonal even though the Zeeman sublevels were not equally populated. The quantum-mechanical prediction is:

$$[R(3\pi/8)/R_0 - R(\pi/8)/R_0]_{QM} = 0\cdot294 \pm 0\cdot007.$$

The experimental result is:

$$[R(3\pi/8)/R_0 - R(\pi/8)/R_0]_{expt} = 0\cdot296 \pm 0\cdot014$$

in excellent agreement with the quantum-mechanical prediction, but again in sharp disagreement with inequality (5.5). Because of the high pumping rate attainable with the dye laser, it was possible to gather the data in a remarkably short period of 80 min which, of course, diminished the probability of errors due to variations in the operation of the apparatus, and facilitated checking for systematic errors.

5.4. Are the auxiliary assumptions for cascade-photon experiments necessary and reasonable?

We have seen that the data from the cascade-photon experiments are *sufficient* to refute the whole family of local realistic theories, if either the CHSH or the CH auxiliary assumption is accepted. Both assumptions are very reasonable. Yet both are conceivably false. One may ask the question: are the experimental data, by themselves, sufficient to refute the theories? Alternatively, is at least some auxiliary assumption *necessary*? The answer was given by CH, who contrived a local hidden-variables model, the predictions of which agree exactly with those of quantum mechanics.

One may then ask how reasonable are these assumptions. In particular do they disagree with any known experimental data? A similar question may also be asked about the counter-example. It seems highly artificial, but are any of its implications experimentally testable?

5.4.1. Critique of the CH and CHSH assumptions. The CHSH assumption (§5.1) is, given that a pair of photons emerges from the polarisers, the probability of their joint detection is independent of the polariser orientation a and b. It may appear that the assumption can be established experimentally by measuring detection rates when a controlled flux of photons of known polarisation impinges on each detector. From the standpoint of local realistic theories, however, these measurements are irrelevant, since the distribution ρ when the fluxes are thus controlled is almost certain to be different from that governing the ensemble in the correlation experiments. We thus

Bell's theorem: experimental tests and implications   1913

see no way of directly testing this assumption, and thus no experiments with which it disagrees.

It is noteworthy, however, that there exists an important hidden-variables theory— the semiclassical radiation theory—which correctly predicts a large body of atomic physics data, but which denies both the CHSH assumption as well as a presupposition of it. The presupposition is that one can speak unequivocally of a photon's passage or non-passage through the polarisation analysers. In the semiclassical radiation theory, however, a photon partially passes its respective polariser and departs with a reduced (classical) amplitude. Furthermore, this amplitude depends upon the polariser's orientation and thereby determines the probability of the photon's subsequent detection (in violation of the CHSH assumption that all photons have the same detection probability, independent of either polariser's orientation). Nonetheless, the predictions for this theory are consistent with those by inequality (5.2), and the theory is refuted by the cascade-photon experiments (Clauser 1972). Evidently an alternative assumption is possible which allows inequality (5.2) to constrain theories denying this presupposition.

Such an assumption was provided by CH. This assumption (§5.1) is that for every pair of emissions, the probability of a count with the polariser in place is less than or equal to the corresponding probability with the polariser removed. This assumption appears reasonable because the insertion of a polarisation analyser imposes an obstacle between the source of the emissions and the detector, and it is natural to believe that an obstacle cannot increase the probability of detection. To be sure, we know of situations in which the insertion of an additional optical element (apparently an obstacle) does increase the probability of detection, e.g. the insertion of a diagonally oriented linear polariser between two crossed polarisers. However, the situation appropriate to the CH assumption is quite different from the one just mentioned, since no polarising elements follow the inserted polarisers. Moreover, if the third polariser is a two-channel device, such as a Wollaston prism, the increased detection rate observed in one channel occurs at the expense of the detection rate in the ortho-gonal channel. The sum of the rates from both channels actually decreases when the second polariser is inserted. Correspondingly, if the third polariser is replaced by a polarisation-insensitive detector, in a closer parallel to the situation of the cascade-photon experiments, then the detection rate is always reduced when a polariser is inserted ahead of this detector.

These considerations, unfortunately, are by no means sufficient to prove the CH assumption, since these observations concern ensemble-average probabilities. The CH assumption requires that the probability be diminished upon the insertion of a polariser *for all* λ.

5.4.2. The counter-example by Clauser and Horne. Clauser and Horne (1974) produced a local hidden-variables model whose predictions agree exactly with those by quantum mechanics. In their model the rate at which photons jointly pass through the polari-sation analysers is in agreement with Bell's inequalities, but the joint detection rate agrees with the quantum-mechanical predictions. The model requires that the detected photon pairs be selected in a very special manner from among those which pass through the analysers, and that those which have not passed through a polariser have a different detection probability from those which have. Although the selection is done entirely locally, it does have the appearance of being highly artificial and, indeed, almost conspiratorial against the experimenter.

The model applies only as long as the net detector efficiencies are smaller than a certain maximum value p_{max} which depends on the analyser efficiencies ϵ_+ and ϵ_-: that is (using the notation of §4) when:

$$[\eta f]_{expt} \leqslant p_{max}. \tag{5.16}$$

With the conditions holding for the experiment by Freedman and Clauser, these values are $[\eta f]_{expt} \approx 0.004$ and $p_{max} \approx 0.4$. This comparison can be improved somewhat by reference to some experimental results by Clauser†. Inequality (5.16) then becomes $\eta_{expt} \leqslant p_{max}$. For the experiment by Freedman and Clauser the value $\eta_{expt} \approx 0.06$ holds.

Despite our caution concerning the CHSH and CH assumptions, we regard the experimental refutation which relies upon them to be compelling. It is striking that only a highly artificial model has so far been found which is local and yet yields quantum-mechanical detection rates in the cascade-photon experiments, and even this model can be excluded by rather modest improvements in the apparatus. There is also some hope for a theorem to the effect that any model consistent with the experimental data will have anomalous features as does the CH model.

6. Positronium annihilation and proton–proton scattering experiments

6.1. Historical background

Two experiments testing predictions based on Bell's theorem have been performed using the high-energy photons produced by positronium annihilation. The historical background of these experiments is interesting. Wu and Shaknov (1950) determined the parity of the ground state of positronium by a method suggested by Wheeler that consisted of measuring the polarisation correlation of γ rays produced by positronium annihilation. The photons Compton-scattered, and two-photon coincidences were observed as a function of azimuthal scattering angles, a and b. Two relative angles 0 and $\pi/2$ were employed. From the ratio of these two coincidence rates they were able to infer that the parity of the ground state is negative. Bohm and Aharonov (1957), with different motivations, showed that these data are explained by quantum mechanics if the polarisation state of the photon pair is assumed to be:

$$\Psi_1 = \frac{1}{\sqrt{2}} \left[\begin{pmatrix} 1 \\ 0 \\ 0 \end{pmatrix} \otimes \begin{pmatrix} 0 \\ 1 \\ 0 \end{pmatrix} - \begin{pmatrix} 0 \\ 1 \\ 0 \end{pmatrix} \otimes \begin{pmatrix} 1 \\ 0 \\ 0 \end{pmatrix} \right] \tag{6.1}$$

† Clauser (1974) noticed that the parameters of existing experimental results were inappropriate to determine whether or not transmission and reflection of a photon at a dielectric surface (similar to one of the surfaces in a pile-of-plates polariser) are, in fact, mutually exclusive possibilities. He thus performed an experiment which confirmed that they are. This behaviour is in marked contrast to that of the semiclassical radiation theory, in which a photon is simultaneously transmitted and reflected by the surface. He also performed a variation of this experiment (unpublished) in which the dielectric surface was replaced by a fine mesh mirror ($\approx 50\%$ transmission), and again photons were observed to be either transmitted or reflected but not both simultaneously. One can conclude from this result that, at least for the purposes of local realistic theories, the simultaneous emission of a photon into any two different solid-angle elements does not occur. Since the probabilities relevant to the CH counter-example are conditional upon the photons actually entering the collimator, it follows that the solid-angle parameter f can be dropped for the purposes of inequality (5.16).

Bell's theorem: experimental tests and implications 1915

but are incompatible with the assumption that the ensemble of photon pairs can be described by a mixture of states, each of which is a product of two single-photon polarisation states. They therefore concluded that the data of Wu and Shaknov confirm the existence of states of two-particle systems which are 'non-separable', even though the particles are spatially remote from each other (see appendix 1).

Clauser *et al* (1969) investigated the possibility of using the arrangement of Wu and Shaknov, perhaps with some variation, for the further purpose of checking whether the observed frequencies can violate Bell's inequalities. It is, indeed, easy to show that if efficient linear polarisation analysers existed for 0·5 MeV photons (with transmittances $\epsilon_M - \epsilon_m$ greater than $\sim 0·83$), then the quantum-mechanical values for the coincidence rates, with the joint polarisation state given by equation (6.1), can violate the inequalities (3.12) or (3.20). Unfortunately no such analysers exist, and Compton scattering does no more than give a scattering distribution, described by the Klein–Nishina formula, that is dependent upon the direction of linear polarisation. They concluded that no variant of the Wu–Shaknov experiment can provide a test of the predictions based on Bell's theorem (see Horne 1970).

6.2. The experiment by Kasday, Ullman and Wu

It was argued by Kasday *et al* (1970, 1975, hereafter referred to as KUW; see also Kasday 1971) that such photon pairs can be used to test the predictions based on Bell's theorem if one accepts two auxiliary assumptions: (i) in principle, ideal linear polarisers can be constructed for high-energy photons; (ii) the results, which would be obtained in an experiment using ideal analysers, and those obtained in a Compton scattering experiment, are correctly related by quantum theory.

Their experimental arrangement (a variant of that of Wu and Shaknov) is shown schematically in figure 8. Positrons were emitted by a ^{64}Cu source, stopped and

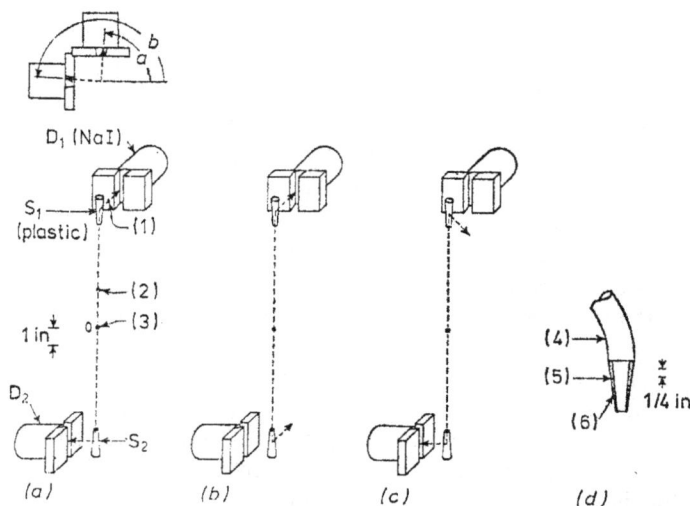

Figure 8. Schematic diagram of the experimental arrangement of KUW. The lead collimator is not shown. (*a*) Four-fold coincidence event; (*b*) (*c*), three-fold coincidence events; (*d*) detail of scatterer. *a*, *b* are the azimuthal angles of the scattered photons. (1) Scattered γ with energy E, absorbed by D_1, (2) annihilation γ, (3) positron source and absorber, (4) light pipe, (5) plastic scatterer, (6) MgO-coated aluminium light reflector (figure after Kasday *et al*).

annihilated in copper at the place labelled by 0. The annihilation γ-rays were emitted in all directions; the vertical direction was selected by a lead collimator. The scatterers were plastic scintillators. Lead slits selected a narrow range of acceptance azimuthal angle about the angles a and b. The top slit-detector assembly was then rotated to vary the relative azimuthal angle. Accepted coincidence events had a fourfold coincidence among the two scatterers and two detectors, as well as a sum-energy requirement that the total energy deposited in each scatterer plus detector equals the annihilation energy. It is noteworthy that this is the only experiment which employs the arrangement of figure 1. (Here the ensemble consists of the pairs jointly scattered by the scintillators.)

KUW applied assumptions (i) and (ii) as follows. Imagine two ideal linear polarisation analysers in the plane perpendicular to the direction of propagation of the selected annihilation photons (the vertical direction in figure 8), which are respectively oriented in the directions a and b of the two slits. If the state λ of the photon pair is given, a deterministic hidden-variables theory will determine whether each photon will pass through its respective analyser. This is their use of assumption (i). If photon 1 will pass its ideal analyser, then linear polarisation in the a direction is assigned to it; and if photon 2 will correspondingly pass its analyser, then it is assigned linear polarisation in the direction b. KUW then use assumption (ii) to assert that the angular scattering distribution of each respective photon is given by the Klein–Nishina formula (a distribution which is dependent upon the photon's initial linear polarisation). With the Klein–Nishina formula one can calculate the probability that the scattered photons will enter the respective acceptance slits. Quantum mechanics makes a definite prediction for this joint probability. Deterministic local hidden-variables theories together with assumptions (i) and (ii) also imply an inequality governing this probability which will disagree with the quantum-mechanical predictions. The experimental data of KUW are in good quantitative agreement with the quantum-mechanical predictions.

This experiment is less decisive, in our opinion, as a refutation of the family of local realistic theories than are the cascade-photon experiments discussed in §5, because it relies upon assumptions which are considerably stronger than the assumption needed by the latter. If assumptions (i) and (ii) are not made, then a local hidden-variables model can be constructed (Horne 1970, Bell 1971 (see Kasday 1971)) which yields the same predictions for the experiment as those by quantum mechanics. This consideration, by itself, is not a fully sufficient reason to prefer the cascade-photon experiments, since a local hidden-variables model, albeit a much more artificial one, also exists which yields quantum-mechanical predictions for those experiments (see §5.4).

The relative strengths of the supplementary assumptions provides a better reason for preference. There is one respect in which assumption (ii) of KUW is quite unconvincing. The only definite polarisation states acknowledged by quantum theory are the various modes of elliptic polarisation (circular and linear polarisation being special cases of these). Since quantum theory can be used to calculate the relationship required for assumption (ii) between ideal and Compton polarimeters only when the state of a photon is one recognised by quantum theory itself, this assumption presupposes that photons which enter the Compton polarimeters are in a quantum-mechanically describable state. Such a supposition is strongly in conflict with the postulates of Bell's theorem. The state λ presumably is not such a state, and moreover there is no prescription within quantum theory for calculating the results of an experiment for these more general states.

Bell's theorem: experimental tests and implications 1917

In a general hidden-variables scheme, the state λ of the photons clearly cannot be represented as one of definite linear polarisation (the special case in which it can is the hypothesis studied by Furry). KUW's decomposition of the state λ of a photon into linear and/or circular polarisation basis states is undoubtedly not possible in general.

Indeed, even in a quantum-mechanical treatment of the problem, the photons are acknowledged not to be in a state of definite polarisation. Quantum mechanically, neither photon's polarisation is in a definite state, but each is in what is known as an 'improper mixture' of such states (see d'Espagnat (1976) for a discussion of improper mixtures). In quantum theory, the only correct procedure for handling such systems is to perform calculations for the composite two-photon state. Thus we see that the 'marriage' between quantum mechanics and a general local realistic theory required by assumption (ii) results in a fatally incorrect handling of both theories.

6.3. The experiments by Faraci et al, Wilson et al and Bruno et al

An experiment very similar to that of KUW (but with ^{22}Na as a source) was performed by Faraci *et al* (1974) with very different results. Their data disagree sharply with the quantum-mechanical predictions based upon the polarisation state of equation (6.1), and are at the extreme limit permitted by Bell's inequalities (given the assumptions of KUW). Their data also showed a variation in correlation strength which depends upon the source-to-scatterer distances. Since their paper is quite condensed, it is difficult to conjecture whether or not a systematic error is responsible for these results. KUW, however, present various criticisms of this work as well as a clarification of various misinterpretations of their own work by these authors.

Wilson *et al* (1976) repeated the experiment using ^{64}Cu as a source. In contrast with the results of Faraci *et al*, they found complete agreement with the quantum-mechanical predictions, and no significant variation of the correlation strength when the scatterer positions were changed.

Bruno *et al* (1977) also repeated the experiment using ^{22}Na as a source, but used alternatively Cu and Plexiglass as the annihilator. To discriminate against multiple scattering events they imposed a sum-energy restriction as did KUW and Wilson *et al* (but not Faraci *et al*), and also varied the scatterer sizes. Residual triplet-positronium contribution was ascertained by the use of the different annihilator materials. Again, no violation of the quantum-mechanical prediction was observed, for any of various source–scatterer distances.

6.4. Proton–proton scattering experiment by Lamehi-Rachti and Mittig

The only test of the predictions in Bell's theorem which has been performed so far not using photons is that of Lamehi-Rachti and Mittig (1976). They measured the spin correlations in proton pairs prepared by low-energy S-wave scattering. The scattering geometry is shown in figure 9. Protons from the Saclay tandem accelerator were scattered by a target containing hydrogen. The incident and recoil protons each entered analysers at $\theta_{lab}=45°$ ($\theta_{cm}=90°$). The protons were scattered by a carbon foil, and detected at positions labelled L_1 or R_1, and L_2 or R_2 in the figure. Coincidences were sought between detectors on opposite arms, as they varied the azimuthal angle of the detector pair of one arm.

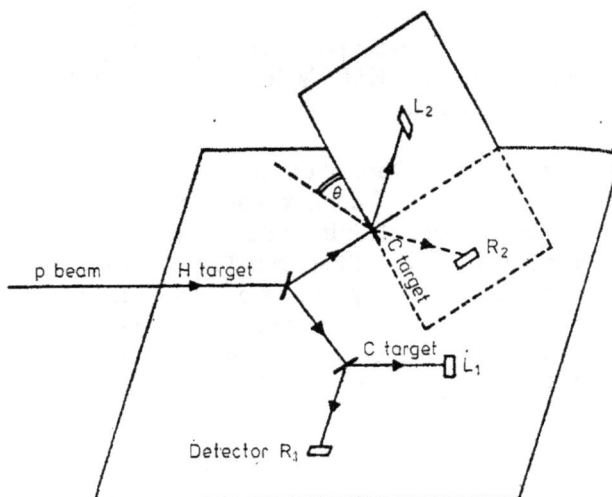

Figure 9. Proton–proton scattering geometry for the experiment by Lamehi-Rachti and Mittig (after Lamehi-Rachti and Mittig 1976).

Auxiliary assumptions similar to those required for the positronium experiments allow them to compare the data with the predictions for local realistic theories. It should be noted that their geometry requires an additional assumption not necessary for the positronium experiments. Since the analysers are only sensitive to the transverse components of the spin, and since $\theta_{cm} = 90°$, the correlation is of the form:

$$E(a, b) = C \cos a \cos b$$

and cannot violate inequalities (3.12) or (3.19) no matter what value $C \leqslant 1$ has. They thus assume that the quantum-mechanically predicted rotational invariance of the S-wave scattering (supposing negligible triplet contribution) allows them to decompose the correlation into a rotationally invariant part (singlet) and a non-rotationally invariant part (triplet). They then extrapolate the results back to a form which violates Bell's inequalities, and rely upon other experimental evidence to set an upper limit to the triplet scattering contribution. An arrangement in which one of the protons is electrostatically deflected through 90°, or magnetically precessed through 90°, would have eliminated the need for this last assumption.

They obtain good agreement with the quantum-mechanical predictions. If one accepts their assumptions, then Bell's inequalities are violated. However, even more reliance on quantum mechanics is needed than for the positronium experiments, and the criticisms of those experiments apply here more acutely.

7. Evaluation of the experimental results and prospects for future experiments

7.1. Two problems

There are two very different problems involved in evaluating the experiments so far performed for testing the predictions in Bell's theorem. The first is to determine the significance of the anomalous results of Holt and Pipkin and Faraci *et al*. The second is to determine what possibilities remain open if only the experiments which

Bell's theorem: experimental tests and implications   1919

favour quantum mechanics are accepted as veridical, and the anomalous results are attributed to spurious effects. In this section we discuss both of these problems.

7.1.1. Significance of the experimental discrepancies. The probability is extremely high, in our opinion, that the results contradicting the predictions by quantum mechanics were due to systematic errors. This opinion is *not* based on a conservative acknowledgment of the great success of quantum mechanics in the atomic domain. Rather, it is based upon the consideration that quantum mechanics predicts strong correlations, whereas Bell's theorem sets a limit upon such correlations. Virtually any conceivable systematic error will wash out a strong correlation so as to produce results in accordance with Bell's theorem, rather than speciously strengthen a weak correlation. We also note that the predictions by quantum mechanics are quantitatively precise. Therefore, in order to maintain that a local realistic theory governs nature, one must invoke experimental errors not only to explain a violation of the inequalities in seven out of nine experiments, but also to explain a very close quantitative agreement with the quantum-mechanical predictions in these seven. In view of the delicacy of these experiments we are not surprised that two anomalous results were obtained among nine. Experience with the experimental techniques and an awareness of the probable systematic errors, one expects, will lead to greater uniformity of results in later repetitions of the experiments. The results of the more recent experiments already indicate this to be so.

7.1.2. Loopholes with auxiliary assumptions. The assumptions for cascade-photon experiments are criticised in §5.4 and those for the positronium and the proton–proton scattering experiments in §§6.2 and 6.4. The opinion is advanced that those for the former are considerably weaker than those for the latter; hence, the cascade-photon experiments are to be preferred. Evidently, none of these assumptions can be directly tested, and thus neither argument is at present fully conclusive.

On the other hand, an indirect test of the assumptions of CH and CHSH may become possible. The counter-example for the cascade-photon experiments (in contrast to that for the positronium and the proton–proton scattering experiments) exploits minor technological imperfections in the apparatus. Indeed, improvements in the polariser efficiencies and/or the photomultiplier quantum efficiencies can make this counter-example obsolete. There is, to our knowledge, nothing fundamentally restricting significant improvements in either of these.

The cascade-photon experiments performed so far were all done on a very small budget (in comparison with modern large-scale experimentation). They were designed simply for testing inequality (5.2), and the various arrangements were sufficient to that end. Now suppose that a theorem (a strengthening of the one conjectured in §5.4) can be proved that the model of CH is essentially the only local hidden-variables model which reproduces quantum-mechanical data in the cascade-photon experiments. Since only a modest improvement in some of these parameters is sufficient to rule out this counter-example, the added expense of a significantly improved apparatus, in our opinion, would be justified.

7.2. Experiments without auxiliary assumptions about detector efficiencies

Even though the experimental results concerning local realistic theories appear highly convincing, it is still desirable to have an experiment for which auxiliary assumptions are not required. It was shown in §4 that the requirements for such a scheme are demanding. Experiments using photons for this purpose do not appear to be feasible in the foreseeable future, since there seems to be no way of resolving the dilemma that highly efficient polarisation analysers can be achieved only for low-energy photons, while highly efficient detectors can be made only for high-energy photons. Furthermore, the two-body decay requirement is problematic with low-energy photons. Charged particles are evidently unusable, since an elegant argument by Bohr (see Mott and Massey 1965) indicates that magnetic state selection of their spin components violates the uncertainty principle. Hence, most schemes under consideration involve using either neutral particles and/or discrete states other than those associated with spin components. For example, Bell (1971) and Clauser *et al* (see Fehrs 1973) were inspired by a paper of Inglis (1961) to consider the charge-conjugation correlations shown in the decay of neutral kaon pairs produced by proton–antiproton annihilation. It was concluded that the exponentially decaying envelope of the correlations precludes the observation of a direct violation of Bell's inequalities in this system.

There is hope that the requirements for efficient analysis and detection can be achieved by observing the dissociation fragments of a metastable molecule, with a pair of Stern–Gerlach magnets as analysers. The latter have virtually 100% transmission, and proper design of the magnetic fields can minimise spurious spin-flips (Majorana transitions) during propagation of the decay fragments. Alkali metal and halogen atoms, if used as the decay fragments, can be detected individually by ionisation or electron attachment at a hot surface with nearly 100% efficiency. The parameters a and b can be taken to be the amplitudes of suitable resonant radio-frequency fields, applied in such a way as to coherently rotate the particle spins. Such an experiment holds promise of testing local realistic theories without any auxiliary assumptions, and with no loopholes other than the possibility of communication between the analysers (see §7.3).

7.3. *Preventing communication between the analysers*

Both the special and general theories of relativity preclude the existence of action-at-a-distance. This fact is, of course, the primary motivation for the various locality postulates considered above. However, in all of the experiments described so far, action-at-a-distance in the relativistic sense is not precluded, since the analysers are always kept at fixed orientations for periods of several seconds. Thus, there is ample time for information about the orientation of one analyser to be transmitted by some unknown mechanism (consistent with relativity theory) to the other apparatus (and/or other particle) thereby influencing its results. It is thus conceivable that such a mechanism is instrumental in producing quantum-mechanical coincidence counting rates in the above experiments. To test this possibility requires an experiment in which the parameters a and b are adjusted with great rapidity while the correlated particles are in flight. If the event consisting of the adjustment of the parameter a of the first analyser is wholly space-like separated from the detection event of particle 2, and similarly concerning adjustment of parameter b and the detection of particle 1, then no signal with subluminal speed can convey information about the orientation of one analyser to the other apparatus in time to affect the probability of detecting the respective particles. In other words, if the parameters a and b are adjusted with sufficient rapidity, then the non-occurrence of action-at-a-distance implies locality. For photons the required rapid adjustment of the analyser orientations can be accomplished, for example, by using modern electro-optical devices such as high-speed Pockell's cells. Aspect (1976) proposed the use of acousto-optical devices for basically the same purpose.

However, even with such devices it is impossible to block the loophole completely. Since the backward light cones of the detection and adjustment events overlap, it may be claimed that events in the overlap region are responsible for determining the choices of the parameters *a* and *b* as well as the observed results. In this way the quantum-mechanical coincidence counting rates can still be accounted for without any direct causal connection between opposite sides of the experiment, and hence without introducing action-at-a-distance. Such an argument, however, seems unacceptable on methodological grounds, for it could be used to justify an *ad hoc* dismissal of any disagreeable data in almost any conceivable scientific experiment.

7.4. *Conclusion*

Although further experimental investigations of the family of theories governed by Bell's theorem are desirable, we are tentatively convinced that no theory of this kind can correctly describe the physical world. Nonetheless, we find this conclusion disturbing, since the philosophical point of view which most working scientists have found natural, at least until quite recently, requires a local realistic theory. Because of the evidence in favour of quantum mechanics from the experiments based upon Bell's theorem, we are forced either to abandon the strong version of EPR's criterion of reality—which is tantamount to abandoning a realistic view of the physical world (perhaps an unheard tree falling in the forest makes no sound after all)—or else to accept some kind of action-at-a-distance. Either option is radical, and a comprehensive study of their philosophical consequences remains to be made.

Appendix 1. Criticism of EPR argument by Bohr, Furry and Schrödinger

The argument of EPR is powerful, since their conclusion surely follows from their plausible premises. Most of the community of physicists rejected EPR's conclusion, however, because of a reply by Bohr (1935), which essentially consisted of a subtle analysis of their premise (ii). His argument is that when the phrase 'without in any way disturbing the system' is properly understood it is incorrect to say that system 2 is not disturbed by the experimentalist's option to measure *a* rather than *a'* on system 1.

'Of course there is, in a case like that just considered, no question of a mechanical disturbance of the system under investigation during the last critical stage of the measuring procedure. But even at this stage there is essentially the question of *an influence on the very conditions which define the possible types of predictions regarding the future behaviour of the system*. Since these conditions constitute an inherent element of the description of any phenomenon to which the term "physical reality" can be properly attached, we see that the argumentation of the mentioned authors does not justify their conclusion that quantum-mechanical description is essentially incomplete.'

It is beyond the scope of the present review to analyse Bohr's claim that the term 'reality' can be used unambiguously in microphysics only when the experimental arrangement is specified. We are not convinced that Bohr ever succeeded in giving a coherent statement of his philosophical position (see, for example, Shimony 1971, Stein 1972, Hooker 1972). We must admit, however, in consideration of the experimental evidence presented in this review against EPR's conclusion, that Bohr's position remains as one of the few feasible options concerning the foundations of quantum mechanics.

An early important reaction to the argument by EPR was to question premise (i). Furry (1936) and Schrödinger (1935) independently considered the possibility that, after systems 1 and 2 become spatially separated and cease effectively to interact, their joint wavefunction no longer has the form (2.1), but rather becomes a mixture of simple product states, each having the form:

$$\Psi'_{\hat{n}} = u_{\hat{n}}^{\pm}(1) \otimes u_{\hat{n}}^{\mp}(2). \tag{A1.1}$$

For each element of the mixture, both 1 and 2 are then in definite quantum states. This possibility is sometimes called 'Furry's hypothesis', but that nomenclature is inappropriate. What Furry did was to show that for any choice of $u_{\hat{n}}(1)$ and $u_{\hat{n}}(2)$, there exist in principle pairs of observables, M of 1 and S of 2, such that the statistical predictions for joint measurements of M and S based upon mixtures of the $\Psi'_{\hat{n}}$ are different from those based upon equation (2.1). Since Furry believed quantum mechanics to be correct, he concluded that a state like that of equation (2.1) does not automatically evolve into a mixture of the $\Psi'_{\hat{n}}$ when 1 and 2 separate from each other (see the conclusion of §4 of his paper). It is more appropriate to call this scheme 'Schrödinger's hypothesis', since he explicitly stated that it may be true. Strong evidence against this hypothesis was presented in 1957 by Bohm and Aharonov (see §6.1). More recent experimental evidence confirming their conclusions has been discussed by Kasday (1971) and Clauser (1972, 1977). It is noteworthy that this hypothesis is such a natural one that many physicists apparently believe it to be a resolution of the EPR 'paradox' without recognising the theoretical and experimental evidence against it.

Appendix 2. Hidden-variables theories

A theory which asserts that the quantum-mechanical description of a physical system is incomplete and requires supplementation in order to specify completely the state of the system is commonly called a *hidden-variables theory*†. Those properties of the system which are proposed as supplements to the quantum-mechanical description are commonly called *hidden variables* or sometimes *hidden parameters*. The history of hidden-variables theories is quite intricate, because the various proponents and opponents have made different assumptions about the conditions of adequacy which a hidden-variables theory should satisfy. We shall review here only as much of this history as is needed to provide the background for Bell's theorem. Other reviews of this subject (some including discussions of Bell's theorem) may be found in Bell (1966), Capasso *et al* (1970), Belinfante (1973), Jammer (1974) and d'Espagnat (1976).

In 1926–7 deBroglie wrote several papers proposing an interpretation of the wavefunction very different from that of Bohr. He supposed that the wavefunction

† The literature is not always consistent on this point, and many authors have included a hypothesis of determinism in their definition of a hidden-variables theory. In this review hidden-variables theories for which determinism holds are referred to as deterministic hidden-variables theories.

associated with a particle is a physically real field propagating in physical space in accordance with the Schrödinger equation. He also supposed that the particle always has a definite position and a definite momentum. Thus his interpretation was actually a hidden-variables theory. Finally, he assumed an intimate coupling between the particle and the field described by the function Ψ, so that the latter can be considered a 'guiding wave' or 'pilot wave' for the particle. This coupling then accounts for interference and diffraction phenomena. Several serious difficulties were found in deBroglie's theory (see deBroglie 1960), especially concerning many-particle systems and the S-wave state of a particle. As a result, deBroglie set aside his investigations of this kind until he was re-encouraged by the work of Bohm in 1952.

We shall not discuss in detail the various other models considered so far (see, for example, Madelüng 1926, deBroglie 1953, Bohm and Vigier 1954, Freistadt 1957, Andrade e Silva and Lochak 1969). Some of these models assume that in addition to the potential recognised in classical mechanics the particle is subject to a 'quantum potential' $h^2\nabla^2 R/2mR$, where R is the amplitude of the wavefunction. Other models assume that the wavefunction describes an averaged or smoothed state of a fluid medium, subject to random fluctuations which are not taken into account by the wavefunction, but which are nevertheless important for understanding the statistical behaviour of particles moving in the medium. For the most part, the advocates of these models do not claim that they are anything but tentative descriptions of the subquantum level of the physical world. Their significance lies in providing existence proofs that a theory can be deterministic in character, and nevertheless agree with many of the statistical predictions by quantum mechanics. We have found, in our discussion of Bell's theorem in §3, that a theory can achieve complete agreement with quantum mechanics only if it is non-local.

Leaving aside the locality problem, we may ask how is it possible that so many mathematically consistent hidden-variables theories have been devised when various theorems have claimed that the structure of the class of quantum-mechanical observables precludes such theories? We now discuss two such theorems.

A2.1. Von Neumann's theorem

The most famous theorem of this type is due to von Neumann (1932). Let $\mathcal{O}$ be the class of observables, and suppose that every self-adjoint operator on a Hilbert space $\mathcal{H}$ of dimension greater than 1 represents a member of $\mathcal{O}$ (but it is not excluded that $\mathcal{O}$ has other members). A state is specified by defining an expectation value $\mathrm{Exp}(A)$ on every $A \in \mathcal{O}$, and it is assumed that Exp satisfies the following conditions.

(i) $\mathrm{Exp}(1) = 1$, where 1 is the observable which, by definition, always has the value unity.

(ii) For each $A \in \mathcal{O}$ and each real number r, $\mathrm{Exp}(rA) = r\mathrm{Exp}(A)$.

(iii) If A is non-negative, then $\mathrm{Exp}(A) \geqslant 0$.

(iv) If $A, B, C, \ldots$, are arbitrary observables, then there is an observable $A + B + C + \ldots$ (which does not depend upon the choice Exp) such that:

$$\mathrm{Exp}(A + B + C + \ldots) = \mathrm{Exp}(A) + \mathrm{Exp}(B) + \mathrm{Exp}(C) + \ldots.$$

The theorem asserts that there exists a self-adjoint operator A on $\mathcal{H}$ such that $\mathrm{Exp}(A^2) \neq [\mathrm{Exp}(A)]^2$, i.e. the state defined by Exp is not dispersion-free over the quantum-mechanical observables.

The intuitive meaning of the conclusion of von Neumann's theorem is that no state of the system—not even a state different from those recognised by quantum mechanics —can assign definite values simultaneously to all quantum-mechanical observables. Von Neumann's theorem is mathematically correct, but its physical significance is doubtful. The fact that it was often cited over three decades as a proof for the completeness of quantum mechanics is a kind of historical aberration, and has wrought much confusion. Its crucial weakness is the supposition that any possible state of the system must satisfy condition (iv), even when $A, B, C, \ldots$, are non-commuting operators and therefore represent observables which, according to quantum mechanics, cannot be simultaneously measured. The actual procedure for measuring $A + B$, when A and B do not commute, is different from the procedures for measuring A and B separately and does not presuppose any information about the value of either A or B. Consequently, the fact that the additivity of condition (iv) is satisfied by quantum-mechanical states is a peculiarity of quantum mechanics, and there is no reason to suppose that it is satisfied by non-quantum-mechanical states. This criticism of the physical significance of von Neumann's theorem was made by Siegel (1966) and Bell (1966). (In the two-dimensional Hilbert space of a spin-$\frac{1}{2}$ particle, Bell also constructed a family of dispersion-free states which are physically reasonable, even though they violate condition (iv). Pearle (1965) and Kochen and Specker (1967) independently constructed similar models.) There is evidence† that Einstein was critical of condition (iv) as early as 1938 and therefore did not consider von Neumann's theorem to be an obstacle to the 'completion' of quantum mechanics as demanded by the argument of EPR. An interesting survey by Jammer (1974, pp272–7) shows that others were critical of condition (iv), but not with complete clarity.

A2.2. Gleason's theorem

In 1957 Gleason proved a theorem which is free from the unphysical condition (iv) of von Neumann's theorem, and which has frequently been considered to be a decisive proof of the impossibility of any consistent hidden-variables theory. We shall not state the theorem itself but rather shall state a corollary‡ which can be compared directly with von Neumann's theorem.

Let $\mathcal{O}$ be a class of observables containing all those represented by the self-adjoint operators on a Hilbert space of dimension greater than 2, and let Exp be a real-valued function over $\mathcal{O}$, which satisfies the following conditions:

1, 2, 3, as in von Neumann's theorem.

4'. If $A, B, C, \ldots$, are commuting self-adjoint operators on $\mathcal{H}$, then

$$\mathrm{Exp}(A + B + C + \ldots) = \mathrm{Exp}(A) + \mathrm{Exp}(B) + \mathrm{Exp}(C) + \ldots .$$

Then Exp is not dispersion-free.

This corollary is weaker than von Neumann's theorem in one respect: it does not apply to a Hilbert space of dimension 2, and therefore it permits the models of Bell, Kochen and Specker, and Pearle. But it is much stronger in one crucial respect: it requires additivity only over commuting operators, for which the values of $A + B + C +$

---

† Professor P G Bergmann was an assistant to Einstein at that time, and he reported Einstein's criticism to one of us. We regret that we have no evidence concerning Einstein's opinion of von Neumann's argument in 1935, when the paper of EPR was written.

‡ There are several direct proofs of essentially this corollary which do not rely upon the main theorem of Gleason: Bell (1966), Kochen and Specker (1967) and Belinfante (1973). The proof given by Jammer (1974, pp298–9) is the same as that of Bell, who is not credited.

Bell's theorem: experimental tests and implications   **1925**

... can in principle be determined by summing the results of simultaneous measurements of $A, B, C, \ldots,$.

The conditions for this corollary are physically plausible, and its conclusion seems to be strong enough to preclude all non-trivial hidden-variables theories (i.e. all which apply to a system which is quantum-mechanically represented by a Hilbert space of dimension greater than 2). However, Bell (1966) pointed out the possibility of a family of non-trivial hidden-variables theories which do not satisfy all the conditions of the corollary and therefore are not bound by its conclusion. Suppose A, B and C are self-adjoint operators such that A commutes with both B and C, but B and C do not commute with each other. Therefore A can in principle be measured simultaneously with B, or it can be measured simultaneously with C, and different experimental arrangements are required for the two measurements. In the corollary it was assumed that when the state of the system is fully specified, the function $\mathrm{Exp}(X)$ has a definite value for each observable X, however X is measured. 'It was tacitly assumed that measurement of an observable must yield the same value independently of what other measurements may be made simultaneously' (Bell 1966, p451). But there is no *a priori* reason that this assumption should be true. 'The result of an observation may reasonably depend not only on the state of the system (including hidden variables) but also on the complete description of the apparatus' (Bell 1966, p451). It is physically reasonable, therefore, to consider hidden-variables theories in which the expectation values have the form $\mathrm{Exp}(X; \mathscr{C})$, where $\mathscr{C}$ indicates the 'context' of the measurement of X, i.e. all the quantities measured simultaneously with X. Gleason's theorem and its corollary do not preclude the possibility that such a 'contextual' hidden-variables theory can be dispersion-free, so that the result of measuring any observable is precisely determined by the state of the system (including hidden variables) together with the 'context' of the measurement.

Bell's proposal of a new family of hidden-variables theories sheds light on models like the one given by Bohm in 1952. This model antedated Gleason's work, but Bohm (1952, p187) defended it against von Neumann's impossibility theorem in the following way:

'the so-called "observables" are ... not properties belonging to the observed system alone, but instead potentialities whose precise development depends just as much on the observing apparatus as on the observed system. In fact, when we measure the momentum "observable", the final result is determined by hidden parameters in the momentum-measuring device as well as by hidden parameters in the observed electron.'

This passage does not propose the consideration of contextual hidden-variables theories as explicitly as Bell does, but it can be construed retrospectively as implicitly agreeing with Bell.

The next step in the history of hidden-variables theories was taken by Bell, once he was convinced that impossibility theorems like that of Gleason do not establish *a priori* the inconsistency of hidden-variables models. By taking these models seriously, he was free to examine whether they shared any physically unreasonable properties in spite of their mathematical consistency, and to inquire whether such properties are inevitable in any hidden-variables theory which agrees with the predictions by quantum mechanics. In this way he was heuristically led to the study of locality.

1926   *J F Clauser and A Shimony*

Acknowledgments

We wish to thank Professor Michael A Horne for many valuable suggestions for this review. One of us (AS) wishes to thank the Department of Theoretical Physics of the University of Geneva for its hospitality.

References

Akhiezer A I and Berestetskii V B 1965 *Quantum Electrodynamics* (New York: Interscience)
Andrade e Silva J L and Lochak G 1969 *Quanta* (New York: McGraw-Hill)
Aspect A 1976 *Phys. Rev.* D **14** 1944–51
Belinfante F J 1973 *A Survey of Hidden-Variables Theories* (Oxford: Pergamon)
Bell J S 1965 *Physics* **1** 195–200
—— 1966 *Rev. Mod. Phys.* **38** 447–52
—— 1971 *Foundations of Quantum Mechanics* ed B d'Espagnat (New York: Academic) pp171–81
—— 1972 *Science* **177** 880–1
—— 1976 *Communication at the 6th Gift Conf., Jaca, June 1975* Res. Th 2053–CERN
—— 1977 *Epistemological Lett.* **15** 79–84
Bohm D 1951 *Quantum Theory* (Englewood Cliffs, NJ: Prentice Hall) pp614–22
—— 1952 *Phys. Rev.* **85** 169–93
Bohm D and Aharonov Y 1957 *Phys. Rev.* **108** 1070–6
Bohm D and Vigier J P 1954 *Phys. Rev.* **96** 208–16
Bohr N 1935 *Phys. Rev.* **48** 696–702
de Broglie L 1926 *C. R. Acad. Sci., Paris* **183** 447–8
—— 1927 *C. R. Acad. Sci., Paris* **184** 273
—— 1928 *J. Phys. Radium* **8** 225–41
—— 1953 *La Physique Quantique: Restera-t-elle Indeterministe?* (Paris: Gauthier Villars)
—— 1960 *Non-Linear Wave Mechanics* (Amsterdam: Elsevier)
Bruno M, d'Agostino M and Maroni C 1977 *Nuobo Cim.* **40** B142–52
Capasso V, Fortunato D and Selleri F 1970 *Riv. Nuovo Cim.* **2** 149–99
Clauser J F 1972 *Phys. Rev.* A **6** 49–54
—— 1974 *Phys. Rev.* D **9** 853–60
—— 1976 *Phys. Rev. Lett.* **36** 1223–6
—— 1977 *Nuovo Cim.* **33** 740–6
Clauser J F and Horne M A 1974 *Phys. Rev.* D **10** 526–35
Clauser J F, Horne M A, Shimony A and Holt R A 1969 *Phys. Rev. Lett.* **23** 880–4
Colodny R G (ed) 1972 *Paradigms and Paradoxes* (Pittsburgh, Pa.: University of Pittsburgh Press)
Eberhard P 1977 *Nuovo Cim.* **38** B75
Einstein A, Podolsky B and Rosen N 1935 *Phys. Rev.* **47** 777–80
d'Espagnat B (ed) 1971 *Foundations of Quantum Mechanics. Proceedings of the International School of Physics 'Enrico Fermi'* Course XLIX (New York: Academic)
—— 1975 *Phys. Rev.* D **11** 1424–35
—— 1976 *Conceptual Foundations of Quantum Mechanics* (Reading, Mass.: Benjamin) 2nd edn
Faraci G, Gutkowski S, Notarrigo S and Pennisi A R 1974 *Lett. Nuovo Cim.* **9** 607–11
Fehrs M H 1973 *PhD Thesis* Boston University
Freedman S J 1972 *Lawrence Berkeley Lab. Rep. No* LBL 391
Freedman S J and Clauser J F 1972 *Phys. Rev. Lett.* **28** 938–41
Freistadt H 1957 *Nuovo Cim. Suppl.* **5** 1–70
Fry E S 1973 *Phys. Rev.* A **8** 1219–32
Fry E S and Thompson R C 1976 *Phys. Rev. Lett.* **37** 465–8
Furry W H 1936 *Phys. Rev.* **49** 393–9
Gleason A M 1957 *J. Math. Mech.* **6** 885–93
Gutkowski D and Masotto G 1974 *Nuovo Cim.* **22** B121–9
Holt R A 1973 *PhD Thesis* Harvard University

Holt R A and Pipkin F M 1973 *Preprint* Harvard University

Hooker C A 1972 *Paradigms and Paradoxes* ed R G Colodny (Pittsburgh, Pa.: University of Pittsburgh Press) pp67–302

Horne M A 1970 *PhD Thesis* Boston University

Inglis D R 1961 *Rev. Mod. Phys.* **33** 1–7

Jammer M 1974 *The Philosophy of Quantum Mechanics* (New York: Wiley)

Kasday L R 1971 *Foundations of Quantum Mechanics* ed B d'Espagnat (New York: Academic) pp195–210

Kasday L R, Ullman J D and Wu C S 1970 *Bull. Am. Phys. Soc.* **15** 586

—— 1975 *Nuovo Cim.* **25** B633–61

Kochen S and Specker E 1967 *J. Math. Mech.* **17** 59–87

Lamehi-Rachti M and Mittig W 1976 *Phys. Rev.* **14** 2543–55

Madelüng E 1926 *Z. Phys.* **40** 322–6

Mott N F and Massey H S W 1965 *The Theory of Atomic Collisions* (Oxford: Oxford University Press) pp214–9

von Neumann J 1932 *Mathematische Grundlagen der Quantenmechanik* (Berlin: Springer-Verlag) (Engl. trans. 1955 *Mathematical Foundations of Quantum Mechanics* (Princeton, NJ: Princeton University Press) pp307–25)

Pearle P 1965 *Preprint* Harvard University

—— 1970 *Phys. Rev.* D **2** 1418–25

Schiavulli L 1977 *Preprint* Universitá di Bari

Schrödinger E 1935 *Proc. Camb. Phil. Soc.* **31** 555–63

Selleri F 1978 *Foundations of Physics* **8** 103–16

Shimony A 1971 *Foundations of Quantum Mechanics* ed B d'Espagnat (New York: Academic) pp182–94, 470–80

—— 1978 *Epistemological Lett.* **18** 1–3

Shimony A, Horne M A and Clauser J F 1976 *Epistemological Lett.* **13** 1–8

Siegel A 1966 *Differential Space, Quantum Systems, and Prediction* ed N Wiener, A Siegel, B Rankin and W T Martin (Cambridge, Mass.: MIT Press)

Stapp H P 1971 *Phys. Rev.* D **3** 1303–20

—— 1978 *Whiteheadian Approach to Quantum Theory and the Generalised Bell's Theorem* Foundations of Physics to be published

Stein H 1972 *Paradigms and Paradoxes* ed R G Colodny (Pittsburgh, Pa.: University of Pittsburgh Press) pp367–438

Wigner E P 1970 *Am. J. Phys.* **38** 1005–9

Wilson A R, Lowe J and Butt D K 1976 *J. Phys. G: Nucl. Phys.* **2** 613–24

Wu C S and Shaknov I 1950 *Phys. Rev.* **77** 136

PHYSICAL REVIEW D   VOLUME 10, NUMBER 2   15 JULY 1974

Experimental consequences of objective local theories*

John F. Clauser

Department of Physics and Lawrence Berkeley Laboratory, University of California, Berkeley, California 94720

Michael A. Horne

Department of Physics, Stonehill College, North Easton, Massachusetts 02356
(Received 10 August 1973; revised manuscript received 8 April 1974)

A broad class of theories, called "objective local theories," is defined, motivation for considering these theories is given, and experimental consequences of the class are investigated. An extension of previous analyses by Bell and by Clauser *et al.* shows that predictions of objective local theories and of quantum mechanics differ, and that an experimental test of the entire family of objective local theories can be performed. The experimental requirements are given. Objective local theories satisfying a plausible but experimentally untestable supplementary assumption are shown to be incompatible with existing experimental data.

I. INTRODUCTION

Two papers by Bell have shown that the statistical predictions of quantum mechanics, for certain spatially separated yet correlated two-particle systems, are incompatible with a broad class of local theories. Bell's earlier paper considers the consequences of a physically reasonable locality condition within the domain of an ideal *Gedankenexperiment*.[1] He demonstrates that any theory which satisfies the locality condition must also be deterministic if certain quantum-mechanical predictions are valid for the idealized case. Bell's further analysis shows, however, that any deterministic local theories are necessarily incompatible with some other quantum-mechanical predictions for the *Gedankenexperiment*.

Upon examining the proof in Bell's earlier paper, one might conjecture that it is essentially the deterministic character of the class of theories that is incompatible with quantum mechanics.[2] That is, if the hypotheses assumed for the *Gedankenexperiment* were slightly relaxed so that determinism is no longer derivable,[3] then the incompatibility with the quantum-mechanical predictions will also be removed. This conjecture is incorrect. Bell shows in his second paper[4] that any stochastic theory satisfying the locality condition is also incompatible with quantum mechanics.

Inspired by Bell's first paper, Clauser *et al.*[5] showed that his analysis can be extended to cover realizable systems and that experimental tests of the broad class of local theories covered by these theorems can be performed. Although existing two-particle sources and/or analyzer-detector apparatuses appear insufficiently close to ideal for the desired experiment, Clauser *et al.* showed that, with a plausible but untestable supplementary assumption concerning detector efficiencies,

deterministic local theories are incompatible with the quantum-mechanical predictions for a realizable experiment. Experimental results obtained by Freedman and Clauser[6] are in excellent agreement with the relevant quantum-mechanical predictions and thereby indicate that any deterministic local theory is untenable if the supplementary assumption is true. Clauser *et al.* in their original proposal for this experiment restricted their discussion to deterministic local theories. Following Bell's second paper, however, it was noted[7] that the experiment also indicates that any stochastic local theory is untenable if the same supplementary assumption concerning the detectors is made. However, the question of the experimental testability of stochastic local theories with either a weaker auxiliary assumption or with no auxiliary assumption has not previously been discussed.

The present paper extends the previous discussions of deterministic and stochastic local theories in several respects: (a) In preparation for the extension, we characterize explicitly a broad class of theories, which we designate as *objective local theories* (OLT), and discuss the fundamental premises which motivate them. Incompatibility of this class of theories with quantum mechanics has essentially been demonstrated by Bell,[4] but the result is in a form that is not practically experimentally testable. (b) We give a new incompatibility theorem that yields an experimentally testable result. (c) We show that, without an auxiliary assumption, neither the existing results of Freedman and Clauser nor any future refinement of their experiment employing more efficient detectors can provide a test of OLT because the angular correlation of the photon pairs is unsuitable. We note, however, that there do exist two-particle sources which are suitable for a test. (d) We state a supplementary assumption, weaker than that previously employed, and prove that it is sufficient to ensure the incompatibility of OLT and the experimental results of Freedman and Clauser. (e) We construct an explicit OLT model which reproduces the results of that experiment. We thereby prove that the Freedman-Clauser results constitute a refutation of only those OLT which satisfy our (or some similar) supplementary assumption.

II. OBJECTIVE LOCAL THEORIES

We will formulate and motivate objective local theories in the context of the experimental arrangement shown schematically in Fig. 1. A source of coincident two-particle emissions is viewed by two analyzer-detector assemblies 1 and 2. Each apparatus has an adjustable parameter; let a denote the value of the parameter at apparatus 1, and b that at apparatus 2. In Fig. 1, a and b are taken to be angles specifying the orientations of the analyzers, e.g., the axes of linear polarizers for photons, or the directions of the field gradients of Stern-Gerlach magnets for spin-$\frac{1}{2}$ particles. However, neither of these particular interpretations of the parameters a and b is essential for the discussion which follows; a and b may denote the values of any adjustable parameter at apparatus 1 and 2, respectively. Finally, in addition to an adjustable component and a detector, each apparatus may (and in practice does) contain various other components, such as additional filters to shield the detectors from unwanted radiations, etc. Since we require that these additional apparatus components remain in place during the experiment, we will ignore them in the discussion. Similarly, we ignore and assume constant any other macroscopic variables, such as those describing the source-apparatus geometry.

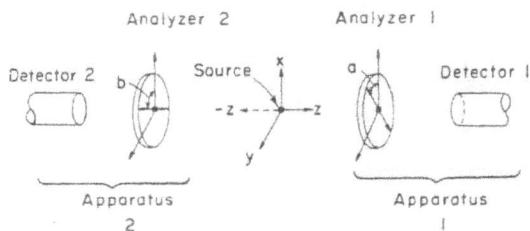

FIG. 1. Scheme considered for a discussion of objective local theories. A source emitting particle pairs is viewed by two apparatuses. Each apparatus consists of an analyzer and an associated detector. The analyzers have parameters, a and b respectively, which are externally adjustable. In the above example, a and b represent the angles between the analyzer axes and a fixed reference axis.

During a period of time, while the adjustable parameters have the values a and b, the source emits, say, N of the two-particle systems of interest.[8] For this period, denote by $N_1(a)$ and $N_2(b)$ the number of counts at detectors 1 and 2, respectively, and by $N_{12}(a, b)$ the number of simultaneous counts from the two detectors (coincident counts).[9] If N is sufficiently large, then the ensemble probabilities of these results are

$$p_1(a) = N_1(a)/N,$$
$$p_2(b) = N_2(b)/N, \qquad (1)$$
$$p_{12}(a, b) = N_{12}(a, b)/N.$$

Consider one of the two-component emissions from the source. Physical theories, classical, quantum-mechanical, and presumably more general ones as well, characterize a physical system with a *state*. Moreover, during the system's existence, its state in general evolves. Consider the state specification of the above system at a time intermediate between its emission and its impingement on either apparatus.[10] Denote this state by λ. Note that we do not necessarily make a commitment to the completeness of this state specification, i.e., it may or may not describe the ultimate essence of the system at the chosen time. But neither do we make any restriction on the possible complexity of λ, nor do we assume it has any special characteristics; in short, we assume no model. As the state described initially by λ subsequently evolves, it may or may not trigger a count at apparatus 1, and similarly it may or may not do so at apparatus 2. The initial state λ, if it serves the same role as in existing theories, will suffice to determine *at least* the probabilities of these events.[11] Let the probabilities of a count being triggered at apparatus 1 and 2 be $p_1(\lambda, a)$ and $p_2(\lambda, b)$, respectively, and let $p_{12}(\lambda, a, b)$ be the probability that both counts are triggered.[12]

Since, in general, every system in the ensemble emitted by the source may not have the same initial state, we allow a mixture of states. Let $\rho(\lambda)$ be the normalized probability density characterizing the ensemble of emissions.[13] In terms of the quantities just defined, the ensemble probabilities given in Eqs. (1) are

$$p_1(a) = \int_\Gamma d\lambda\, \rho(\lambda) p_1(\lambda, a),$$
$$p_2(b) = \int_\Gamma d\lambda\, \rho(\lambda) p_2(\lambda, b), \qquad (2)$$
$$p_{12}(a, b) = \int_\Gamma d\lambda\, \rho(\lambda) p_{12}(\lambda, a, b),$$

where Γ is the space of the states λ. The formulation (2) is quite general. Nothing so far has been assumed that is not satisfied by quantum mechanics. It suffices in Eqs. (2) to let λ define a quantum state and let $\rho(\lambda)$ define a distribution over quantum states implicit in a mixture.

Hereafter, we focus our attention on a special case of formulation (2) in which

$$p_{12}(\lambda, a, b) = p_1(\lambda, a)p_2(\lambda, b) . \qquad (2')$$

What considerations motivate this factored form? Clearly, if each source emission consists of two well-localized subsystems, e.g., a pair of objective particles, and there is no action at a distance, then the factored form is a reasonable locality condition. More generally, the factored form is a natural expression of a field-theoretical point of view, which in turn is an extrapolation from the common-sense view that there is no action at distance.

Fields propagate and can impinge upon different localized objects, among them pieces of apparatus. Since λ describes the field initially,[14] and the parameter a is associated with only one apparatus, it is reasonable that there is a well-defined probability $p_1(\lambda, a)$ that that apparatus will be triggered. The impingement of the field on this apparatus will naturally modify the field. However, if the events marking the action of the field on two pieces of apparatus (triggering or not in each case) have spacelike separation, then the reaction upon the field due to impingement on the first apparatus will not have time to affect the part of the field impinging the second apparatus, and conversely. Hence, the probability p_2 that the second apparatus will be triggered will not depend on whether or not the first has been triggered, or upon the choice of the parameter a at the first apparatus, or even upon the presence of the first apparatus. A similar assertion of independence holds for p_1, whence $(2')$ follows. In view of the foregoing motivation of Eq. $(2')$, we call any theory in which it holds an objective local theory.[15]

It should be recalled that the wave function was introduced by Schrödinger from a field-theoretical point of view, and most physicists have continued to think of it field-theoretically, whatever their precautions when they are on guard about the character of quantum mechanics. Hence, we conjecture that Eq. $(2')$ is implicit in the thinking of many physicists.[16] Whether or not this is correct, it is apparent that quantum mechanics is not of the form $(2')$. For many two-quanta sources, every

emission is described by the same pure state, suggesting that this quantum state be identified with one value of λ, say λ', and that $\rho(\lambda) = \delta(\lambda - \lambda')$ (no mixture). However, when the two-particle quantum state is not a simple product of single-particle states but is instead a superposition of such products, the quantum-mechanical probability p_{12} does not in general admit the factorization $(2')$. The only alternative for saving Eq. $(2')$ is to identify the quantum-mechanical pure state with $\rho(\lambda)$, i.e., with a mixture of other states λ.[17] But, as we shall see, this identification is impossible for some quantum-mechanical pure states.

III. EXPERIMENTAL CONSEQUENCES

Measurement of the probabilities (1) requires not only the numbers $N_1(a)$, $N_2(b)$, and $N_{12}(a, b)$, which are directly observed quantities in a counting experiment, but also the number N, which is generally unobservable in an experiment counting microsystems. In practice, the number of emissions N occurring during any time period is usually deduced from the counting data for that period, since intervening counters will in practice depolarize (if not destroy) the systems. But this deduction always depends upon the currently accepted theoretical description of the whole phenomenon—the source, the apparatus, and their interactions. That is, N is actually deduced from Eqs. (1) *themselves* with the p's supplied by the theory at hand. Clearly, any such method of determining N must not be employed in an experimental test of competing theories. Therefore, in this section, we derive a consequence of Eq. $(2')$ which is experimentally testable without N being known, and which contradicts the quantum-mechanical predictions.

Let a and a' be two orientations of analyzer 1, and let b and b' be two orientations of analyzer 2. The inequalities

$$0 \le p_1(\lambda, a) \le 1 ,$$
$$0 \le p_1(\lambda, a') \le 1 ,$$
$$0 \le p_2(\lambda, b) \le 1 ,$$
$$0 \le p_2(\lambda, b') \le 1 \qquad (3)$$

hold if the probabilities are sensible. These inequalities and the theorem in Appendix A give

$$-1 \le p_1(\lambda, a)p_2(\lambda, b) - p_1(\lambda, a)p_2(\lambda, b') + p_1(\lambda, a')p_2(\lambda, b) + p_1(\lambda, a')p_2(\lambda, b') - p_1(\lambda, a') - p_2(\lambda, b) \le 0$$

for each λ. Multiplication by $\rho(\lambda)$ and integration over λ gives

$$-1 \le p_{12}(a, b) - p_{12}(a, b') + p_{12}(a', b) + p_{12}(a', b') - p_1(a') - p_2(b) \le 0 \qquad (4)$$

as a necessary constraint on the statistical predictions of any OLT. If, for some reason such as rotational invariance, it is found experimentally that $p_1(a)$ and $p_2(b)$ are constant, and that $p_{12}(a,b) = p_{12}(\phi)$ holds, where $\phi = |b-a|$ is the angle between the analyzer axes, then (4) becomes

$$-1 \le 3p_{12}(\phi) - p_{12}(3\phi) - p_1 - p_2 \le 0 . \qquad (4')$$

Here, a, a', b, and b' have been chosen so that

$$|a-b| = |a'-b| = |a'-b| = \tfrac{1}{3}|a-b'| = \phi .$$

The upper limits in (4) and in $(4')$ are experimentally testable without N being known. Inequalities (4) and $(4')$ hold perfectly generally for any systems described by OLT. These are new results not previously presented elsewhere. The relationship between them and the previous inequalities of Bell is discussed in Appendix B.

IV. INCOMPATIBILITY WITH QUANTUM MECHANICS

We now present the conditions under which the predictions (4) and $(4')$ are incompatible with those of quantum mechanics. Consider an experiment with a configuration as described above whose quantum-mechanical predictions take the following form:

$$p_{12}(\phi) = \tfrac{1}{4}\eta_1\eta_2 f_1 g[\epsilon_+^1 \epsilon_+^2 + \epsilon_-^1\epsilon_-^2 F \cos(n\phi)] ,$$
$$p_1 = \tfrac{1}{2}\eta_1 f_1 \epsilon_+^1 , \qquad\qquad (5)$$
$$p_2 = \tfrac{1}{2}\eta_2 f_2 \epsilon_+^2 .$$

This general form is characteristic of many correlation phenomena, e.g., the spin-$\tfrac{1}{2}$–spin-$\tfrac{1}{2}$ correlation *Gedankenexperiment* introduced by Bohm[18] and used by Bell,[1,4] and the actual experiments of Freedman and Clauser,[6] Wu and Shaknov,[19] and Kasday et al.[20] Inserting the predictions (5) into $(4')$ and selecting the optimum value $\phi = 45°/n$, one finds that the condition for a violation of the upper bound is

$$\eta g \epsilon_+ [\sqrt{2}(\epsilon_-/\epsilon_+)^2 F + 1] \ge 2 . \qquad (6)$$

Here for simplicity we have taken $\eta_1 = \eta_2 \equiv \eta$, $f_1 = f_2$, $\epsilon_+^1 = \epsilon_+^2 \equiv \epsilon_+$, and $\epsilon_-^1 = \epsilon_-^2 \equiv \epsilon_-$. Thus, a correlation experiment with values in the domain specified by (6) is capable of distinguishing between OLT and quantum theory. Although experiments are possible for which this is the case, there are at present no existing experimental results satisfying (6), and thus none which are in violation of $(4')$ and/or (4).

Consider next the specific example of the photon pairs emitted by a $J=0 \to J=1 \to J=0$ atomic cascade. A source of such pairs was used in the experiment of Freedman and Clauser. Figure 2 is a diagram of their experiment; it is clearly of the

configuration discussed in the previous section. For this arrangement, the predictions are given by Eqs. (5) with $n=2$ and the following identifications[5]: η_i is the quantum efficiency of detector i ($i=1,2$), and

$$\epsilon_+^i \equiv \epsilon_M^i + \epsilon_m^i \text{ and } \epsilon_-^i \equiv \epsilon_M^i - \epsilon_m^i .$$

Here ϵ_M^i is the efficiency of polarizer i for light polarized parallel to the polarizer axis, and ϵ_m^i is the efficiency for light polarized perpendicular to it. The function $f_1 = f_2 = f(\theta)$ is the probability that the $J=0 \to J=1$ ($J=1 \to J=0$) emission enters apparatus 1 (apparatus 2),

$$f(\theta) = \tfrac{1}{2}(1 - \cos\theta) , \qquad (7)$$

with θ being the half-angle of the cones subtended by each detector aperture. The function $g = g(\theta)$ is the conditional probability, or angular correlation factor, that if the $J=0 \to J=1$ emission enters apparatus 1 then the $J=1 \to J=0$ emission will enter apparatus 2,

$$g(\theta) = \frac{3}{8} \frac{[G_2(\theta)]^2 + \tfrac{1}{2}[G_3(\theta)]^2}{1-\cos\theta} . \qquad (8)$$

Finally,

$$F = F(\theta) = \frac{2[G_1(\theta)]^2}{[G_2(\theta)]^2 + \tfrac{1}{2}[G_3(\theta)]^2} \qquad (9)$$

reflects a depolarization effect due to noncollinearity of the two photons and approaches unity for infinitesimal detector apertures ($\theta \to 0$). The functions G_1, G_2, and G_3 are given in Ref. 5.

Inserting Eqs. (8) and (9) into (6) one finds that because of the relatively small value of $g(\theta)$ condition (6) is *not* satisfied for any value of the detector half-angle θ, even if the analyzer and detector efficiencies are ideal ($\eta = \epsilon_+ = \epsilon_- = 1$).[21] Therefore, for cascade-photon experiments, the quantum-mechanical predictions are compatible with (4) even in the domain of ideal apparatus. The

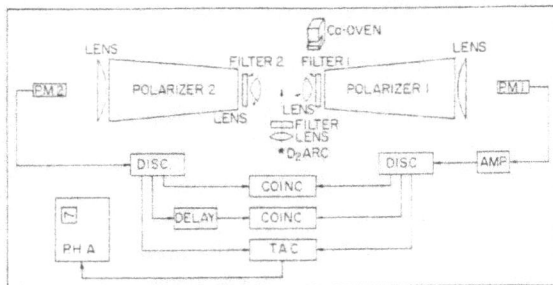

FIG. 2. Schematic diagram of apparatus and associated electronics of the experiment by Freedman and Clauser. Scalers (not shown) monitored the outputs of the discriminators and coincidence circuits. (Figure after Freedman and Clauser.)

insufficient magnitude of the angular correlation factor $g(\theta)$ is a consequence of the fact that an atomic cascade is a three-body decay, the atom being the third body.

However, for correlated particles produced in certain two-body decays, the quantum-mechanical predictions violate (4). The annihilation of ground-state positronium into two γ rays or the dissociation of a spin-0 or a spin-1 molecule into two spin-$\frac{1}{2}$ particles produce correlations of the form (5). Since these are two-body decays, $g(\theta) = 1$ holds even for small θ (provided the center-of-mass velocity of the decaying object is sufficiently small). Even with $g = 1$, inequality (6) imposes rather stringent conditions on the efficiencies of the analyzers and detectors. But there appears to be no *a priori* reason why such conditions cannot be achieved in practice. This question will be the subject of future work.

V. CONSEQUENCES WITH A SUPPLEMENTARY ASSUMPTION

Until a correlation experiment employing highly efficient analyzers and detectors is performed on two-body decays,[22] it is desirable to exhibit a physically plausible supplementary assumption

which makes the existing cascade-photon experiment applicable as a test of OLT. In this section we state an assumption, weaker than the one previously presented by Clauser *et al.*, and prove that it is sufficient to make OLT incompatible with existing experimental results. In the next section we prove, with an explicit OLT model, that this or some other supplementary assumption is necessary for such an application.

The assumption is that, for every emission λ, the probability of a count with a polarizer in place is less than or equal to the probability with the polarizer removed. Let ∞ denote the absence of the polarizer, and let $p_1(\lambda, \infty)$ denote the probability of a count from detector 1 when the polarizer is absent and the emission is λ. A similar probability $p_2(\lambda, \infty)$ may be defined for apparatus 2. Thus, our assumption is

$$0 \le p_1(\lambda, a) \le p_1(\lambda, \infty) \le 1 \, ,$$
$$0 \le p_2(\lambda, b) \le p_2(\lambda, \infty) \le 1 \, , \tag{10}$$

for every λ, and for all values of a and b. We call this the *no-enhancement assumption*.[23] We now make an argument analogous to that which led from (3) to (4). Inequalities (10) and the theorem of Appendix A yield immediately the result

$$-p_{12}(\infty, \infty) \le p_{12}(a, b) - p_{12}(a, b') + p_{12}(a', b) + p_{12}(a',b') - p_{12}(a', \infty) - p_{12}(\infty, b) \le 0 \, , \tag{11}$$

where

$$p_{12}(x, y) \equiv \int_\Gamma d\lambda \, \rho(\lambda) p_1(\lambda, x) p_2(\lambda, y)$$

for all x and y. With the same conditions used in writing (4'), (11) becomes

$$p_{12}(\infty, \infty) \le 3p_{12}(\phi) - p_{12}(3\phi) - p_{12}(a', \infty) - p_{12}(\infty, b) \le 0 \, . \tag{11'}$$

Note that all terms in (11) are joint probabilities for coincident counts at the two detectors. Inequality (4), in contrast, contains the two terms p_1 and p_2 which are probabilities of a count at a single detector. The upper limit of (11), or (11'), is identical to the previous result of Clauser *et al.*, but their derivation was restricted to deterministic local theories and employed an auxiliary assumption stronger than the no-enhancement assumption.

The quantum-mechanical prediction for $p_{12}(\phi)$ in the cascade-photon experiment was given in Eqs. (5) and (7)–(9). The predictions for the other joint probabilities occurring in (11) are[5]

$$p_{12}(a', \infty) = \tfrac{1}{2}\eta_1\eta_2 f(\theta)g(\theta)\epsilon_+^1 \, ,$$
$$p_{12}(\infty, b) = \tfrac{1}{2}\eta_1\eta_2 f(\theta)g(\theta)\epsilon_+^2 \, , \tag{12}$$
$$p_{12}(\infty, \infty) = \eta_1\eta_2 f(\theta)g(\theta) \, .$$

These predictions violate the upper bound in (11') provided

$$\epsilon_+\left[\sqrt{2}\,(\epsilon_-/\epsilon_+)^2 F(\theta) + 1\right] \ge 2 \tag{13}$$

holds. As before, we have used the optimum value $\phi = 22\tfrac{1}{2}°$ and set $\epsilon_+^1 = \epsilon_+^2 \equiv \epsilon_+$ and $\epsilon_-^1 = \epsilon_-^2 \equiv \epsilon_-$. Note that neither the angular correlation factor nor the detector efficiencies appear in (13). The apparatus of Freedman and Clauser satisfied (13), and the experimental results, which confirmed the quantum-mechanical predictions, substantially violated the upper bound of (11'). Consequently, in view of the theorem just given, any OLT that do not admit enhancement are untenable.[24]

VI. NECESSITY OF A SUPPLEMENTARY ASSUMPTION

To prove that an auxiliary assumption is necessary if the cascade-photon experiment is to refute OLT, we exhibit an explicit OLT model which reproduces the results of that experiment. Since the experimental results are in agreement with the quantum-mechanical predictions, it suffices to construct a model which reproduces these predictions. For simplicity, however, we exhibit the model only for the ideal case in which the detectors subtend infinitesimal solid angles $[\theta \to 0$ and $F(\theta) \to 1]$. The extension of the model to the finite solid angles of the actual experiment $[\theta = 30°$ and $F(\theta) = 0.99]$ introduces nothing new, and the additional complexity obscures the point. With these simplifications, the predictions to reproduce are

$$p_{12}(\phi)/p_{12}(\infty, \infty) = \tfrac{1}{4}(\epsilon_+^1 \epsilon_+^2 + \epsilon_-^1 \epsilon_-^2 \cos 2\phi) ,$$
$$p_{12}(a, \infty)/p_{12}(\infty, \infty) = \tfrac{1}{2}\epsilon_+^1 , \qquad (14)$$
$$p_{12}(\infty, b)/p_{12}(\infty, \infty) = \tfrac{1}{2}\epsilon_+^2 .$$

Only ratios could be measured, since, with N unknown, the actual values of the p's were experimentally inaccessible.

The model is as follows: Each emission pair consists of two particles, such that particle 1 travels along the $+z$ axis to apparatus 1, and particle 2 travels along the $-z$ axis to apparatus 2. Both members of the pair possess a common state variable λ which is simply an azimuthal angle; that is, it specifies a direction perpendicular to the flight axis from some reference axis (see Fig. 3). The ensemble of emitted pairs is characterized by a normalized isotropic density

$$\rho(\lambda)d\lambda = \frac{d\lambda}{2\pi} , \quad 0 \le \lambda \le 2\pi . \qquad (15)$$

From the same reference axis, we specify the orientations of polarizers 1 and 2 by the angles a and b. With the polarizer removed, the probability of a count at each detector is a constant, independent of λ:

$$p_1(\lambda, \infty) = c_1 , \qquad (16)$$
$$p_2(\lambda, \infty) = c_2 . \qquad (17)$$

The probability of a count at detector 1, given an emission λ and the setting a of an inserted polarizer, is

$$p_1(\lambda, a) = \tfrac{1}{2}c_1[\epsilon_+^1 + \epsilon_-^1 \cos 2(\lambda - a)] . \qquad (18)$$

At detector 2, the probability is

$$p_2(\lambda, b) = \tfrac{1}{2}c_2 \pi \epsilon_+^2/\delta \qquad (19)$$

for $b - \tfrac{1}{2}\delta \le \lambda \le b + \tfrac{1}{2}\delta$ and $\pi + b - \tfrac{1}{2}\delta \le \lambda \le \pi + b + \tfrac{1}{2}\delta$; $p_2(\lambda, b)$ is zero otherwise. Here δ is a function of the ratio $\epsilon_-^2/\epsilon_+^2$, and is defined by the relation

$$(\sin \delta)/\delta = \epsilon_-^2/\epsilon_+^2 . \qquad (20)$$

Clearly, this model is an OLT provided the probabilities (16) through (19) are less than unity. Evaluation of the ensemble probabilities yields

$$p_{12}(a, b) = \tfrac{1}{4}c_1 c_2[\epsilon_+^1 \epsilon_+^2 + \epsilon_-^1 \epsilon_-^2 \cos 2(a - b)] ,$$
$$p_{12}(a, \infty) = \tfrac{1}{2}c_1 c_2 \epsilon_+^1 ,$$
$$p_{12}(\infty, b) = \tfrac{1}{2}c_1 c_2 \epsilon_+^2 ,$$
$$p_{12}(\infty, \infty) = c_1 c_2 ,$$

which agree with the quantum-mechanical ratios (14).

Note that for some values of the polarizer parameters $\epsilon_\pm^i$ and $\epsilon_-^i/\epsilon_+^i$ the model is enhancement-free. Since the function $p_1(\lambda, a)$ is less than $p_1(\lambda, \infty)$ for all physically sensible values of ϵ_+ and ϵ_-, it is enhancement-free in general. To see this note that, since $0 \le \epsilon_m \le \epsilon_M \le 1$ holds for any polarizer, it follows that $0 \le \epsilon_-^i/\epsilon_+^i \le 1$ and $\epsilon_+^i \le 2[(\epsilon_-^i/\epsilon_+^i) + 1]^{-1}$ also hold. However, $p_2(\lambda, b)$ is enhancement-free if and only if

$$\epsilon_+^2 \le 2\delta/\pi . \qquad (21)$$

The values $\epsilon_+^2 \approx 1.00$ and $\epsilon_-^2/\epsilon_+^2 \approx 0.94$ of the Freedman-Clauser apparatus do not satisfy (21), and therefore, as expected from the theorem of the previous section, the model requires enhancement to reproduce the experimental results. Note finally that, even with enhancement, $p_2(\lambda, b)$ remains sensible provided the constant c_2 of the model is sufficiently small. The value of $p_2(\lambda, b)$ is sensible for all λ and b provided $c_2 \le 2\delta(\pi \epsilon_+^2)^{-1}$. For the actual values of the experiment this condition is $c_2 \le 0.38$.

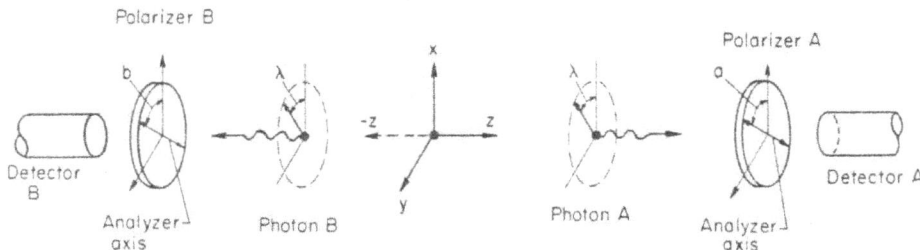

FIG. 3. Coordinate system for OLT model. Photon particles A and B carry the same azimuthal direction λ, which, along with the analyzer orientation a or b, determines the probability of a count at the associated detector.

VII. CONCLUSIONS

Physicists have consistently attempted to model microscopic and macroscopic phenomena in terms of objective entities, preferably with some definable structure. The present paper has addressed the question of whether or not the existing formalism of quantum mechanics can be recast or perhaps reinterpreted in a manner which restores the objectivity of nature, and thus allows such models (deterministic or not) to be made. We have found that it is not possible to do so in a natural way, consistent with locality, without an observable change of the experimental predictions.[26]

ACKNOWLEDGMENT

The authors gratefully acknowledge many valuable discussions with A. Shimony.

APPENDIX A: TWO INEQUALITIES

We prove the following theorem: Given six numbers x_1, x_2, y_1, y_2, X, and Y such that

$$0 \leq x_1 \leq X ,$$
$$0 \leq x_2 \leq X ,$$
$$0 \leq y_1 \leq Y , \quad \text{(A1)}$$
$$0 \leq y_2 \leq Y ,$$

then the function $U = x_1 y_1 - x_1 y_2 + x_2 y_1 + x_2 y_2 - Y x_2 - X y_1$ is constrained by the inequalities

$$-XY \leq U \leq 0 . \quad \text{(A2)}$$

To establish the upper bound, consider two cases. First assume that $x_1 \geq x_2$ and rewrite (A2),

$$U = (x_1 - X)y_1 + (y_1 - Y)x_2 + (x_2 - x_1)y_2 .$$

We have thus assumed the last term to be nonpositive. Inequalities (A1) require the first two terms to be nonpositive, and the validity of the upper bound is demonstrated for this case. Next assume the other alternative, i.e., that $x_1 < x_2$, and use this assumption to bound U, thus:

$$U = x_1(y_1 - y_2) + (x_2 - X)y_1 + x_2(y_2 - Y)$$
$$\leq x_1(y_1 - y_2) + (x_2 - X)y_1 + x_1(y_2 - Y)$$
$$= (x_2 - X)y_1 - x_1(y_1 - Y) \leq 0 .$$

Thus, the upper bound is established in general.

The proof of the lower bound follows from a consideration of three cases. First, assume $x_2 \geq x_1$. The validity of the lower bound is apparent by inspection when written in the form

$$U + XY = (X - x_2)(Y - y_1) + x_1 y_1 + (x_2 - x_1)y_2 ,$$

since (A1) requires all three terms to be non-

negative. Similarly for the case $y_1 \geq y_2$, inspection reveals the correctness of the lower bound when written

$$U + XY = (X - x_2)(Y - y_1) + x_2 y_2 + x_1(y_1 - y_2) .$$

Finally, suppose neither of the two previous cases holds; that is, $x_2 < x_1$ and $y_1 < y_2$. Then write

$$U + XY = (X - x_2)(Y - y_1)$$
$$- (x_1 - x_2)(y_2 - y_1) + x_2 y_1 .$$

The sum of the first two terms in non-negative since now $(X - x_2) \geq (x_1 - x_2) > 0$ and $(Y - y_1) \geq (y_2 - y_1) > 0$. By (A1) the final term is also non-negative. Q.E.D.

APPENDIX B: INEQUALITIES (4) AND BELL'S THEOREM

Inequalities (4) assumed the arrangement of Fig. 1, where each apparatus consists of an analyzer with a single-channel output followed by a photomultiplier. Bell[4] considers a two-channel apparatus, e.g., a birefringent polarizing crystal followed by two photomultipliers, one photomuliplier monitoring the ordinary ray emerging from the crystal and the other monitoring the extraordinary ray.[27] Let $p_1^+(\lambda, a)$ denote the probability of a count in the ordinary channel of apparatus 1, and $p_1^-(\lambda, a)$ denote the probability of a count in the extraordinary channel, for the orientation a of the analyzer and initial state λ of the emission. Let $p_2^+(\lambda, b)$ and $p_2^-(\lambda, b)$ be similarly defined for apparatus 2. Bell considers the correlation function defined by[28]

$$P(a, b) \equiv \int_\Gamma d\lambda \, \rho(\lambda) [\, p_1^+(\lambda, a) - p_1^-(\lambda, a)]$$
$$\times [\, p_2^+(\lambda, b) - p_2^-(\lambda, b)]$$
$$= p_{12}^{++}(a, b) - p_{12}^{+-}(a, b)$$
$$- p_{12}^{-+}(a, b) + p_{12}^{--}(a, b) ,$$
$$p_{12}^{jk}(a, b) \equiv \int_\Gamma d\lambda \, \rho(\lambda) p_1^j(\lambda, a) p_2^k(\lambda, b) , \quad \text{(B1)}$$

$$\text{for } j, k = \pm 1 .$$

Using the fact that each square bracket in (B1) is bounded by ± 1, he proves that P is constrained by the inequalities

$$-2 \leq P(a, b) - P(a, b') + P(a', b) + P(a', b') \leq 2 . \quad \text{(B2)}$$

Note that for a direct experimental test of (B2) the number of emissions N must be known, and that in actual practice this probably cannot be found without either destroying or at least depolarizing the particles. [If N can be found, (B2) suffices for an experimental test.] Here we wish to show the relations between (4) and (B2).

First, (B2) is a corollary of (4). In an experi-

$$-1 \le p_{12}^{jk}(a,\,b) - p_{12}^{jk}(a,\,b') + p_{12}^{jk}(a',\,b) + p_{12}^{jk}(a',\,b') - p_1^j(a') - p_2^k(b) \le 0\,, \qquad (B3)$$

where $j = \pm 1$ and $k = \pm 1$ indicate which detectors are considered. Multiplying the inequalities for which $j \ne k$ by -1 and adding these to the inequalities for which $j = k$, we obtain (B2).

Second, with Bell's formulation, (4) is not an immediate corollary of (B2). For a given analyzer setting a and emission λ, there are three possible results at apparatus 1: a count in the + detector, a count in the − detector, or no count in either detector. Let $p_1^0(\lambda,\,a)$ denote the probability of no count. Clearly

$$p_1^+(\lambda,\,a) + p_1^-(\lambda,\,a) + p_1^0(\lambda,\,a) = 1$$

ment employing two detectors (+ and −) behind each double-channel analyzer, inequalities (4) are still applicable and provide four sets of inequalities,

holds, which implies

$$p_1^+(\lambda,\,a) - p_1^-(\lambda,\,a) = 2p_1^+(\lambda,\,a) + p_1^0(\lambda,\,a) - 1\,. \qquad (B4)$$

With (B4), and a similar expression for apparatus 2, the expression (B1) defining the correlation function becomes

$$\begin{aligned} P(a,\,b) = {}& 4p_{12}^{++}(a,\,b) - 2p_1^+(a) - 2 - p_2^+(b) + 1 \\ & + 2p_{12}^{+0}(a,\,b) + 2p_{12}^{0+}(a,\,b) + p_{12}^{00}(a,\,b) \\ & - p_1^0(a) - p_2^0(b)\,. \end{aligned} \qquad (B5)$$

Insertion of (B5), and similar expressions for $P(a',\,b)$, $P(a,\,b')$ and $P(a',\,b')$, into (B2) yields

$$-1 - Q \le p_{12}^{++}(a,\,b) - p_{12}^{++}(a,\,b') + p_{12}^{++}(a',\,b) + p_{12}^{++}(a',\,b') - p_1^+(a') - p_2^+(b) \le -Q\,, \qquad (B6)$$

where

$$\begin{aligned} Q \equiv \tfrac12 \big[& p_{12}^{+0}(a,\,b) + p_{12}^{0+}(a,\,b) + \tfrac12 p_{12}^{00}(a,\,b) - p_{12}^{+0}(a,\,b') - p_{12}^{0+}(a,\,b') - \tfrac12 p_{12}^{00}(a,\,b') + p_{12}^{+0}(a',\,b) \\ & + p_{12}^{0+}(a',\,b) + \tfrac12 p_{12}^{00}(a',\,b) + p_{12}^{+0}(a',\,b') + p_{12}^{0+}(a',\,b') + \tfrac12 p_{12}^{00}(a',\,b') - p_1^0(a') - p_2^0(b) \big]\,. \end{aligned}$$

If we could establish that $Q \ge 0$, then the upper bound of (B6) would be identical to the experimentally useful upper bound of (4). No reason is immediately apparent that this is so, and unfortunately the terms p_{12}^{00}, p_1^0, and p_2^0 occurring in Q are unobservable, since they are probabilities of nothing happening. Thus auxiliary assumptions, and/or auxiliary experimentation as well, are necessary to test (B2) or, equivalently, (B6).

However, if Bell's formulation is modified *at the beginning*, his method of proof can be employed to obtain (4). Consider, instead of the correlation function (B1), the function

$$\begin{aligned} P'(a,\,b) \equiv {}& \int_\Gamma d\lambda\, \rho(\lambda) [2p_1^+(\lambda,\,a) - 1][2p_2^+(\lambda,\,b) - 1] \\ = {}& 4p_{12}^{++}(a,\,b) - 2p_1^+(a) - 2p_2^+(b) + 1\,. \qquad (B7) \end{aligned}$$

Since each square bracket in P' is bounded by −1 and +1, Bell's method is applicable and yields

$$-2 \le P'(a,\,b) - P'(a,\,b') + P'(a',\,b) + P'(a',\,b') \le 2\,. \qquad (B8)$$

Using Eq. (B7), inequalities (B8) become

$$\begin{aligned} -1 \le {}& p_{12}^{++}(a,\,b) - p_{12}^{++}(a,\,b') + p_{12}^{++}(a',\,b) \\ & + p_{12}^{++}(a',\,b') - p_1^+(a') - p_2^+(b) \le 0\,, \end{aligned}$$

which, suppressing superscripts, is (4). In the context of a double-channel experiment, the three other sets of inequalities given in (B3) can be obtained in similar fashion.[29]

*Work by one of us (J.F.C.) was supported by the U. S. Atomic Energy Commission.

[1] J. S. Bell, Physics (N.Y.) **1**, 195 (1965).

[2] K. Popper, in *Perspectives in Quantum Theory: Essays in Honor of Alfred Landé*, edited by W. Yourgrau and A. Van der Merwe (MIT Press, Cambridge, Mass., 1971), p. 182. For a reply to this conjecture, see J. S.

Bell, Science **177**, 880 (1972).

[3] Clauser *et al*. (Ref. 5) did relax the experimentally unattainable conditions of the *Gedankenexperiment*, but nevertheless retained determinism as a separate hypothesis.

[4] J. S. Bell, in *Foundations of Quantum Mechanics, Proceedings of the International School of Physics "Enrico*

Fermi," Course XLIX, edited by B. d'Espagnat (Academic, New York, 1971), p. 171. For a discussion of the experimental testability of his results, see Appendix B.

[5]J. F. Clauser, M. A. Horne, A. Shimony, and R. A. Holt, Phys. Rev. Lett. <u>23</u>, 880 (1969); J. F. Clauser, Bull. Am. Phys. Soc. <u>14</u>, 578 (1969); A. Shimony, in *Foundations of Quantum Mechanics* (Ref. 4), p. 182; M. A. Horne, Ph.D. thesis, Boston University, 1970 (unpublished).

[6]S. J. Freedman and J. F. Clauser, Phys. Rev. Lett. <u>28</u>, 938 (1972); S. J. Freedman, Ph.D. thesis, University of California, Berkeley, 1972 [Lawrence Berkeley Laboratory Report No. LBL-391 (unpublished)]. See also Ref. 24.

[7]A. Shimony, in *Foundations of Quantum Mechanics* (Ref. 4), p. 191.

[8]We use the word "particle" in the conventional way to identify experimental phenomena exemplifying the general situation studied. Thus we do not assume here the existence of a microscopic entity with a "particle-like" structure.

[9]The practical criterion for "coincident counts" always involves a coincidence time window τ: Pairs of counts separated in time by less than τ are defined to be coincident. This procedure may appear to make the definition of $N_{12}(a,b)$ ambiguous since in general it will depend upon the experimenter's choice for τ. However, this dependence is usually insensitive to variations in τ which satisfy $s \ll \tau \ll 1/r$, where r is the average count rate at either detector, and s is the typical time separation of "true" coincidence pairs. Thus, we tacitly require the experimental arrangement to be such that this condition obtains (suitable source strength, time separation of pairs, etc.). If a sufficiently weak source is used, the ratio of "chance" coincident counts to "true" coincident counts can be made arbitrarily small, and the corresponding dead time can also be minimized.

[10]For the case where the emissions are spatially localized objects in flight from the source to each apparatus, such an intermediate time clearly exists. If the emissions are spatially extended systems (fields), which may take substantial time to exit the source, we consider the system after the emission process has commenced but before the leading edge has impinged on either apparatus.

[11]Even though we have introduced λ as the state of a specific *single* system, the assumed objectivity of the system described by this state allows us to consider an ensemble of these, physically identical to the extent that they are all characterized by the same λ. The probabilities are to be associated with this ensemble. Clearly, this procedure is conceptually sound, even in cases where we cannot in practice prepare the pure λ ensemble.

[12]Previous discussions (Refs. 1, 4, and 5) consider separately the passage of the emission through the analyzer and the detection of the emission after it has passed. Since "passage through analyzer" is not directly observable for microsystems, we consider as a unit the complete causal process from initial emission to final macroscopic effect (count). Moreover, the concept "passage through analyzer" is inappropriate for many field theories (classical, quantum-mechanical,

or others) in which the emission may partially pass the analyzer.

[13]We should emphasize that an assumption is made here. By writing the density as $\rho(\lambda)$, instead of a more general conditional $\rho(\lambda|a,b)$, we deny such objective and local possibilities as these:

(a) Systems of some type originate the source which, reflecting off the analyzers and returning to the source, significantly effect the ensemble in a manner dependent upon a and b.

(b) Systems originate at the analyzers and impinge upon the source, thus effecting the ensemble in a manner dependent upon a and b.

(c) Systems originate within the intersection of the backward light cones of both analyzers and the source. These propagate into the spatial region of the whole apparatus, and simultaneously effect both the experimenters' selections of analyzer orientations and the emissions from the source.

In principle, (a) and (b) can be ruled out experimentally by rapidly and repeatedly changing the analyzer orientations immediately before each data-collection period and then stopping each data-collection period before the new orientations can be communicated (at the speed of light) to the source. The more general type-(c) conspiracies are difficult to rule out experimentally. But, for example, if the orientations a and b are selected at random by two physicists supposedly acting independently, significant cases of (c) require that their selections are *not* independent but are, in fact, strongly correlated with each other as well as with the source emissions.

[14]Because the form of λ is unspecified, we make no commitment to any specific field model, but adopt a quite general view that a field is an objective physical entity extended throughout space.

[15]Recall footnotes 10, 11, and 13 for additional informal justification of the name. The class of OLT includes the deterministic local hidden-variable theories discussed by Bell (Ref. 1) and Clauser *et al.* (Ref. 5) and is essentially the class of stochastic hidden-variable theories "with a certain local character" considered by Bell (Ref. 4). Hence we drop the name "hidden-variable theory," since traditionally the term has been used to identify theories characterized essentially by dispersion-free states, i.e., deterministic theories.

[16]Jauch [in *Foundations of Quantum Mechanics* (Ref. 4), p. 22] makes a similar conjecture when he suggests that "On the whole, and in spite of pretensions to the contrary, physicists are usually more inclined to Realism than to Positivism." However, for a specific example of implicit support for (2'), see M. L. Goldberger and K. M. Watson, Phys. Rev. <u>134</u>, B919 (1964). Considering a system composed of two spin-$\frac{1}{2}$ particles, they comment: "In general therefore the observation of the orientation of spin 1 with respect to the axis has an instantaneous effect on the state of the second spin. With the interpretation of an observation as making a selection among the members of an ensemble, this is in no sense surprising." But a denial of any instantaneous effect for each member of this ensemble is tantamount to assuming (2'). The surprise, then, for an advocate of OLT, is that (2') and quantum mechanics are incompatible.

[17] This identification, for the restricted case that each state of the mixture is a product of single-particle *quantum-mechanical* states, was conjectured by W. H. Furry [Phys. Rev. **49**, 393 (1936); **49**, 476 (1936)] and shown to be in disagreement with experiment by D. Bohm and Y. Aharonov [Phys. Rev. **108**, 1070 (1957)]. An OLT is the generalization to mixtures of *any* type of product states.

[18] D. Bohm, *Quantum Theory* (Prentice-Hall, Englewood Cliffs, N. J., 1951), p. 614.

[19] C. S. Wu and I. Shaknov, Phys. Rev. **77**, 136 (1950).

[20] L. Kasday, J. Ullman, and C. S. Wu, Bull. Am. Phys. Soc. **15**, 586 (1970); see also L. Kasday, in *Foundations of Quantum Mechanics* (Ref. 4), p. 195.

[21] Actually, in atomic cascade experiments, the failure to violate (4′) is much worse than indicated in (6), since the singles probabilities p_1 and p_2 in (5) do not include the large contributions due to unpaired emissions from each transition of the cascade.

[22] The experiment of Kasday, Ullman, and Wu (Ref. 20) has confirmed the quantum-mechanical predictions for coincident Compton scattering of photon pairs produced in positronium annihilation. However, explicit OLT models exist (Refs. 5 and 20) which demonstrate that the analyzers (Compton polarimeter) are too inefficient to reveal a discrepancy between general OLT and quantum mechanics.

[23] Semiclassical radiation theories are OLT which satisfy the no-enhancement assumption. For an earlier discussion of these theories and two-photon correlations, see J. F. Clauser, Phys. Rev. A **6**, 49 (1972). See also J. F. Clauser, Phys. Rev. D **9**, 853 (1974).

[24] Experimental results in conflict with those of Freedman and Clauser have been found by R. A. Holt [Ph.D. thesis, Harvard University, 1973 (unpublished)]. He carried out the design of Ref. 5, as they did, but with a different atomic source of photon pairs and different optical arrangements. Holt's data agree with inequality (11′) and differ significantly from the predicted quantum-mechanical values. Further work is being pursued at various laboratories to explore this discrepancy.

[25] The upper bound on c_2, necessary to keep $p_2(\lambda, b)$ sensible, depends on the analyzer efficiencies ϵ_M^2 and ϵ_m^2 (through the functions ϵ_+^2, ϵ_-^2, and δ) and is independent of the detector efficiency η_2. On the other hand, a physical interpretation of Eq. (16) suggests that $c_2 = \eta_2$. Then if both the detector efficiencies are larger than 0.38, some probability of the model will not be sensible. However, there are undoubtedly OLT models that remain sensible for substantially larger efficiencies. Moreover, the identification $c_2 = \eta_2$ is not a logical necessity. The quantum efficiency definition for a photomultiplier depends on a long chain of experimentation and accepted theory. Recall the related comments concerning N in Sec. III.

[26] A qualitative argument with conclusions similar to ours has been given by B. d'Espagnat, in *Proceedings of the 1972 Trieste Conference on the Physicist's Conception of Nature*, edited by J. Mehra (Reidel, Dordrecht, Holland, 1973). See also B. d'Espagnat, *Conceptional Foundations of Quantum Mechanics* (Benjamin, Menlo Park, California, 1971), Chaps. 8, 9, and 16.

[27] Bell actually considers spin-$\frac{1}{2}$ pairs; we give the photon analogy.

[28] Bell does not write Eqs. (B1), but this definition is implicit in his discussion.

[29] If the no-count outcomes (denoted 0) are considered as well as the + and − ones, there are nine sets of inequalities. But only the four given in (B3) are of immediate experimental utility, since the other five all contain probabilities of nothing happening.

LIST OF PUBLICATIONS

Selected Publications by John F. Clauser on the Foundations of Quantum Mechanics

1. J. F. Clauser, 1969, *Proposed Experiment to Test Local Hidden-Variable Theories*, Bul. Amer. Phys. Soc., **14**, 578.

2. J. F. Clauser, M. A. Horne, A. Shimony, R. A. Holt, 1969, *Proposed Experiment to Test Local Hidden-Variable Theories*, Phys. Rev. Lett., **23**, 880.

3. J. F. Clauser, 1971, *VonNeumann's Informal Hidden-Variable Argument*, Amer. J. Phys., **39**, 1095.

4. J. F. Clauser, 1971, *Reply to Dr. Wigner's Objections*, Amer. J. Phys., **39**, 1098.

5. J. F. Clauser, 1972, *Experimental Limitation to the Validity of Semi-classical Radiation Theories*, Phys. Rev., A **6**, 69.

6. S. J. Freedman and J. F. Clauser, 1972, *Experimental Test of Local Hidden-Variable Theories*, Phys. Rev. Lett., **28**, 938.

7. J. F. Clauser, 1972, *Formalism and Reality*, Science **175**, 871 (1972).

8. J. F. Clauser, 1973, *Experimental Limitations to the Validity of Semiclassical Radiation* Theorie, in *Coherence and Quantum Optics: Proceedings of the Third Rochester Conference on Coherence and Quantum Optics, June 21–23, 1972*, ed. by L. Mandel and E. Wolf (Plenum Press, New York).

9. J. F. Clauser, 1973, *Localization of Photons* in *Coherence and Quantum Optics: Proceedings of the Third Rochester Conference on Coherence and Quantum Optics, June 21–23, 1972*, ed. by L. Mandel and E. Wolf (Plenum Press, New York).

10. J. F. Clauser, 1974, *Experimental Distinction Between the Quantum and Classical Field Theoretic Prediction for the Photoelectric Effect*, Phys. Rev. D, **9**, 853.

11. J. F. Clauser and M. A. Horne, 1974, *Experimental Consequences of Objective Local Theories*, Phys. Rev., D **10**, 526.

12. J. F. Clauser, 1976, *Philosophical Motivations of Bell's Theorem and the Experimenter's Problem*, invited paper presented at *Thinkshops on Physics - Experimental Quantum Mechanics `Ettore Majorana'*, Erice, Sicily, in *Progress in Scientific Culture*, (#4) 1976, ed. by A. Zichichi (Tipographia "Cartograf", Trapani, Italy, 1977).

13. J. F. Clauser, 1976, *Experiments Performed at Lawrence Berkeley Laboratory Bearing Relation to Bell's Inequality*, invited paper presented at *Thinkshops on Physics - Experimental Quantum Mechanics `Ettore Majorana'*, Erice, Sicily, in *Progress in Scientific Culture*, **1** (#4) 1976, ed. by A. Zichichi (Tipographia "Cartograf", Trapani, Italy, 1977).

14. A. Shimony, M. A. Horne, and J. F. Clauser, 1976, *Comment on `The Theory of Local Beables'*, Epistemological Lett. **13**, 1, republished along with comments by J.S. Bell as '*An exchange on local Beables*' in Dialectica, **39**, 97 (1985).

15. J. F. Clauser, 1976, *Measurement of the Circular-Polarization Correlation in Photons from an Atomic Cascade*, Il *Nuovo Cimento*, **33B**, 740.

16. J. F. Clauser, 1976, *Experimental Investigation of a Polarization Correlation Anomaly*, Phys. Rev. Lett., 36, 1223.

17. J. F. Clauser and A. Shimony, 1978, *Bell's Theorem: Experimental Tests and Implications*, Rpts. on Progr. in Phys. 41, 1881.

18. J. F. Clauser, 2002, *Early History of Bell's Theorem*, Chapter 6 in *Quantum [Un]speakables*, ed by R. A. Bertlmann and A. Zeilinger (Springer, Berlin).

19. J. F. Clauser, 2003, *Early History of Bell's Theorem*, in *Coherence and Quantum Optics VIII*, ed. by Bigelow *et al.* (Kluwer Academic/Plenum Publishers), pp. 19-43.

Selected Publications by John F. Clauser on Atom Interferometry and Talbot-Lau Interferometry

1. J. F. Clauser, 1987,88, *Ultra-High Sensitivity Accelerometers and Gyroscopes Using Neutral-Atom Matter-Wave Interferometry*, Physica, **B** 151, 262 (1988), presented at the *International Workshop on Matter-Wave Interferometry in the Light of Schrodinger's Wave Mechanics*, Vienna, 14–16 Sept. 1987.

2. J.F. Clauser, 1989, 1991, US Patents # 4,874,942 (filed 1987, issued 1989) and # 4,992,656 (filed 1989, issued 1991), *Rotation, Acceleration, and Gravity Sensors, Using Quantum-Mechanical Matter-Wave Interferometry with Neutral Atoms and Molecules*.

3. J. F. Clauser, 1988, *Ultra-Sensitive Inertial Sensors via Neutral Atom Interferometry*, in *Relativistic Gravitational Experiments in Space*, NASA Conference Publ. 3046, R. Hellings editor, presented at the *NASA Workshop on Relativistic Gravitational Experiments in Space*, Annapolis MD, 28–30 June 1988.

4. J. F. Clauser, 1993, Australian Patent # 637,654 (filed 1989, issued 1993), *Atom Interferometry Gyroscopes, Accelerometers and Gravity Gradiometers*.

5. J. F. Clauser, 1996, Canadian Patent # 2,033,341 (filed 1991, issued 1996), *Atom Interferometry Gyroscopes, Accelerometers and Gravity Gradiometers*.

6. J. F. Clauser and M. W. Reinsch, 1992, *New theoretical and experimental results in Fresnel optics with applications to matter-wave and x-ray interferometry*, Appl. Phys. B, **54**, 380 (1992), presented at *The WE Heraeus Seminar on Optics and Interferometry with Atoms*, 10–12 June, 1992, Insel Reichenau, Konstanz, Germany, as part of an invited talk,: *Separated - Beam and Talbot - vonLau Interferometry with Potassium*.

7. J. F. Clauser, 1995, *Results of* atom *interferometry experiments with potassium*, in *Proceedings of the International Symposium on Fundamental Problems in Quantum Physics* (ISFPQP — Oviedo, Spain, 29 Aug.–3 Sept., 1993), (Kluwer, Great Britain, 1995).

8. J. F. Clauser and S. Li, 1994, *Talbot-vonLau atom interferometry with slow cold potassium*, Phys. Rev., A **49**, R2213.

9. J. F. Clauser and S. Li, 1994, *"Heisenberg microscope" decoherence atom interferometry*, Phys. Rev., A **50**, 2430.

10. J. F. Clauser and S. Li, 1994, *Matter-Wave/Atom Interferometry*, in *New Techniques and Analysis in Optical Measurements*, presented at *Interferometry '94* (16–20 May, 1994, Warsaw), *Proceedings of SPIE*, Vol. 2340, 2–13.

11. J. F. Clauser and J. Dowling, 1996, *Factoring integers with Young's N-slit interferometer*, Phys. Rev. A 53, 4587.

12. J. F. Clauser, 1997, *deBroglie-wave interference of small rocks & live viruses*, Chapter 1 in *Experimental Metaphysics: Quantum Mechanical Studies for Abner Shimony*, ed. by R. S. Cohen *et al.* (Kluwer Academic Publishers, Great Britain, 1997), pp. 1–11.

13. J. F. Clauser, 1998, *Ultrahigh Resolution Interferometric X-ray Imaging*, U. S. Patent 5,812,629 (filed 1997, issued 1998).

14. J. F. Clauser, and S. Li, 1997, *Generalized Talbot-Lau Atom* Interferometry, Chapter 3 in *Atom Interferometry*, ed. by P. Berman (Academic Press, San Diego, 1997), pp. 121–151.

15. J.F. Clauser, 2000, *Surveilance and Gravity-Imaging of the Earth's Surface, and Satellite Gravity Gradiometry with Atom Interferometers.* (Jet Propulsion Laboratory Quantum Computing Technologies Group report).

List of publications in other areas available on request.

Anton Zeilinger

ANTON ZEILINGER

Birthplace

Born	Ried/Innkreis, Austria, 20 May 1945

Education

1971	Ph.D., University of Vienna

Professional Career

2013–present	President, Austrian Academy of Sciences
2004–2013	Director, IQOQI Vienna, Austrian Academy of Sciences
1999–present	Professor of Experimental Physics, University Vienna
1990–1999	Professor of Experimental Physics, University Innsbruck
1988–1989	Professor of Physics (Lehrstuhlvertretung), Technical University Munich
1983–1990	Associate Professor, Vienna University of Technology
1981–1983	Associate Professor of Physics, M.I.T. (Visiting)
1979–1983	Assistant Professor, Atominstitut Vienna
1977–1978	Fulbright Fellow at M.I.T. under Prof. C.G. Shull (Nobel Laureate 1994), USA
1972–1979	Research Assistant, Atominstitut Vienna, with Professor Helmut Rauch

Visiting Affiliations

2010	Visiting Researcher, Part-Time, Merton College, Oxford
2001–2004	Senior Humboldt Fellow, Humboldt University, Berlin, Germany
1998	Visiting Research Fellow, Merton College, Oxford University, UK
1995	Chaire Internationale, Collège de France, Paris, France
1983–1990	Regular summer research appointments at M.I.T., USA
1974–1989	Guest Researcher (part-time), Institut Laue-Langevin, Grenoble, France

Distinguished Lectures (Selection, Since 2010)

Hofstadter Lecture, Stanford University (2015); Montroll Memorial Lecture, University of Rochester (2014); Herzberg Memorial Lecture, Canadian Association of Physicists (2012); Racah Lecture, Hebrew University, Jerusalem (2012); Cherwell-Simon Lecture, Oxford University (2012); Festkolloquium, 500. WE-Heraeus Seminar (2012); Vice Chancellors Lecture, Univ. of Cape Town (2011); Mark W. Zemansky Lecture, City College of NY (2011); Festvortrag, 75th Annual Meeting, German Physical Society (2011); Van Vleck Lecture, University of Minnesota (2011); Ockham Lecture, Merton College, Oxford (2010); Dvorak Memorial Lecture, University Prague (2010); Celsius Lecture, Uppsala Univ. (2010).

Distinguished Memberships (Selection)

Fellow, The World Academy of Sciences (TWAS) (2014); Foreign Associate, National Academy of Sciences of Belarus (2014); Foreign Associate, U.S. National Academy of Sciences (2014);

Fellow, American Assoc. for the Advancement of Science (AAAS) (2012); Full Member, Academia Europaea (2011); Foreign Member, Académie des Sciences, Institut de France (2009); Foreign Member, Serbian Academy of Sciences and Arts (2006); Honorary Member, Slovak Academy of Sciences (2005); Member, German National Academy of Sciences Leopoldina (2005); Member, Berlin-Brandenburg Academy of Sciences (2002); Fellow, American Physical Society (1999); Member, Austrian Academy of Sciences (1998).

Honorary Professorships and Doctorates

Honorary Doctor, National Academy of Sciences of Ukraine (2014); Honorary Doctor, Gdansk University, Poland (2006); Honorary Doctor, Humboldt University Berlin, Germany (2005); Honorary Professor, University of Science and Technology of China (1996).

Prizes and Awards (Selection)

Grand Decoration of Honour in Gold for Services to the Republic of Austria (2015); Großer Tiroler Adler Orden, Federal Province Tyrol (2013); Urania Medal, Urania Berlin (2013); Ben Gurion Medal (2010); Wolf-Prize in Physics, Wolf Foundation, Israel (2010); Great Cross of Merit with Star of the Federal Republic of Germany (2009); Inaugural Isaac Newton Medal, Institute of Physics, UK (2008); King Faisal Prize, King Faisal Foundation, Saudi Arabia (2005); Descartes Prize, European Commission (2005); Johannes Kepler-Prize, Science Prize of Upper Austria (2002); Order Pour le Mérite for Sciences and Arts, Germany (2001); Senior Humboldt Fellow Prize, Alexander von Humboldt-Stiftung, Germany (2000); Austrian Scientist of the Year (1996).

COMMENTARY

When I was born in May 1945, World War II had just ended. My birthplace was in the American Zone of Austria. Strong connections with America became important in my life. In May 1955, our family moved to Vienna, where my father had become a University professor in a biochemical field. My first memory of Vienna is of sitting by the radio and listening to the live report of the ceremony of signing of the Staatsvertrag, the treaty that finally brought Austria independence again.

In Vienna, I attended a classical gymnasium (high school) which included Ancient Greek and Latin. I enjoyed an excellent, inspiring and exciting teacher in physics and mathematics, Mr. Lederer. This broad spectrum in humanities and in the sciences became important for me for the rest of my life, not only for my career.

When in 1963 I enrolled at the University of Vienna, there was practically no curriculum. The concept of Bachelor and Master did not exist. You had to write a PhD thesis in the end, and pass a "Rigorosum", the final examination where you had to demonstrate your knowledge of physics, mathematics, and philosophy. Before the PhD thesis, one was very free to choose the professors and the courses. For me, this freedom was essential, because I could focus on what really interested me.

A crucial decision was with whom to do my PhD thesis. I was fascinated by Helmut Rauch, a very young lecturer at the Technical University of Vienna. Under his supervision, I did a PhD thesis on neutron scattering for magnetic studies. A new neutron interferometer experiment was built up then at the Vienna reactor by Rauch and colleagues. We expected that the magnetic moment, or spin, of the neutron should play a fundamental role. I found out which, by pulling an all-nighter, calculating what the effect of the magnetic field on the neutron inside an interferometer would be, while keeping an ear out for my one-year old daughter. By dawn, I had the solution. The resulting experiment became my first work in Foundations of Physics. From Helmut Rauch, I learned the basic methodologies of scientific work. I also learned that it is important to follow one's own curiosity, to focus on questions rather than goals, to be open for surprises, and to explore new ideas even if — or maybe especially if — one does not really know what one is talking about.

A very crucial event for me was my attendance at the workshop "Thinkshops in Physics", organized in 1976 in Erice, Sicily, by John Bell and Bernard d'Espagnat. I reported on the recent neutron experiment. The workshop was my first contact with Bell's Inequalities, the Einstein–Podolsky–Rosen Paradox, and the like. At the meeting, I met Valentin Telegdi, an émigré from the Austro-Hungarian Empire. He encouraged me to follow my ambition to work at M.I.T. with the later Nobel Prize winner Cliff Shull, and he helped me to make the connections via Viki Weisskopf. Without his help, I would certainly not have ended up where I am today.

At M.I.T., I essentially built up a neutron interferometry activity. It was fantastic to have an expert basically on any question of physics, right on the faculty. Cliff Shull taught me that it is always worthwhile to do an experiment one order of magnitude better than what one thinks is necessary. He and Helmut Rauch also taught me never to rely on the explanations and interpretations given by others, but to always think things through completely by myself.

In 1983, I accepted a position at the Technical University of Vienna and returned to Austria. For some time, I thought that this was the biggest mistake of my life. But then, in 1987, a new window opened up. Daniel Greenberger from the City College of New York, with whom I had already worked at M.I.T., came to visit Vienna on a Fulbright Fellowship. The first day, when we sat together, we realized that both of us had been observing that so far nobody had investigated entangled systems of more than two particles. What we found, together with Mike Horne, put Bell's Theorem on a new level. Little did we know that our GHZ (Greenberger–Horne–Zeilinger) states were also to become an important concept in the emerging ideas of quantum information, including quantum computation.

It became my goal then to realize such multi-particle entangled states. Nobody knew how to make such states in the laboratory. Most of the tools necessary did not exist. It took us ten years, until 1997, when we finally succeeded. We did many wrong turns, and had to learn slowly the relationship between entanglement and information, and what this means for laboratory realizations. Along the road, we did the quantum teleportation experiment, because inadvertently, the tools we developed for a GHZ experiment were the same as those necessary for quantum teleportation. This became only possible because I switched, in the beginning of the 1990s, from neutron experiments to photons and atoms, starting a significant quantum optics initiative in Austria. It was made possible by a major grant from the Austrian Science Foundation and at the same time by having been appointed Chair of Experimental Physics in Innsbruck, which resulted in seemingly unlimited resources.

Later again, in 1998, I gave up our atom optics work, where we had done some interferometry, and decided to explore the possibility of having quantum interference with macromolecules. This was not easy, because the atom optics experiment still produced significant publications on a regular basis, so some of my young colleagues did not like my decision. But we were more than richly rewarded by observing the first quantum interference of fullerene molecules in 1999.

In the years since then, I always refocused my work on entanglement and its possible applications in quantum communication and quantum computation. That way, we were again able to open new doors. For example, we verified the entanglement of orbital angular momentum states, which opened up the possibility of entanglement for higher-dimensional systems. I wonder what the next steps of my research will be, but I am sure that the question what to give up and what therefore to start new will come up again.

In conclusion, I feel extremely privileged to leading such a rich scientific life.

Experimental Test of Quantum Nonlocality in Three-Photon Greenberger–Home–Zeilinger Entanglement

Jian-Wei Pan[*], Dik Bouwmeester[†], Matthew Daniell[*], Harald Weinfurter[‡] & Anton Zeilinger[*]

[*]*Institut für Experimentalphysik, Universität Wien, Boltzmanngasse 5, 1090 Wien, Austria*
[†]*Clarendon Laboratory, University of Oxford, Parks Road, Oxford OX1 3PU, UK*
[‡]*Sektion Physik, Ludwig-Maximilians-Universität of München, Schellingstrasse 4/III, D-80799 München, Germany*

Bell's theorem[1] states that certain statistical correlations predicted by quantum physics for measurements on two-particle systems cannot be understood within a realistic picture based on local properties of each individual particle—even if the two particles are separated by large distances. Einstein, Podolsky and Rosen first recognized[2] the fundamental significance of these quantum correlations (termed 'entanglement' by Schrödinger[3]) and the

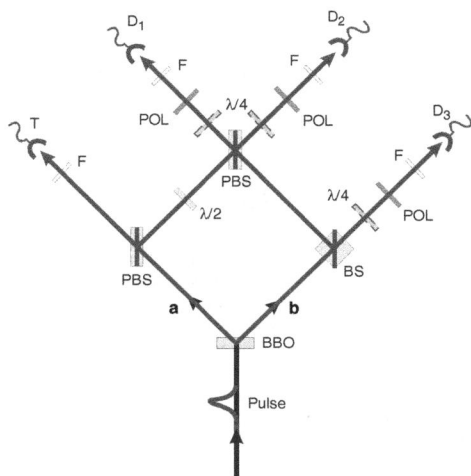

Figure 1 Experimental set-up for Greenberger–Horne–Zeilinger (GHZ) tests of quantum nonlocality. Pairs of polarization-entangled photons[28] (one photon *H* polarized and the other *V*) are generated by a short pulse of ultraviolet light (~ 200 fs, λ = 394 nm). Observation of the desired GHZ correlations requires fourfold coincidence and therefore two pairs[29]. The photon registered at T is always *H* and thus its partner in **b** must be *V*. The photon reflected at the polarizing beam-splitter (PBS) in arm **a** is always *V*, being turned into equal superposition of *V* and *H* by the λ/2 plate, and its partner in arm **b** must be *H*. Thus if all four detectors register at the same time, the two photons in D$_1$ and D$_2$ must either both have been *VV* and reflected by the last PBS or *HH* and transmitted. The photon at D$_3$ was therefore *H* or *V*, respectively. Both possibilities are made indistinguishable by having equal path lengths via **a** and **b** to D$_1$ (D$_2$) and by using narrow bandwidth filters (*F* ≈ 4 nm) to stretch the coherence time to about 500 fs, substantially larger than the pulse length[30]. This effectively erases the prior correlation information and, owing to indistinguishability, the three photons registered at D$_1$, D$_2$ and D$_3$ exhibit the desired GHZ correlations predicted by the state of equation (1), where for simplicity we assume the polarizations at D$_3$ to be defined at right angles relative to the others. Polarizers oriented at 45° and λ/4 plates in front of the detectors allow measurement of linear *H'*/*V'* (circular *R*/*L*) polarization.

two-particle quantum predictions have found ever-increasing experimental support[4]. A more striking conflict between quantum mechanical and local realistic predictions (for perfect correlations) has been discovered[5,6]; but experimental verification has been difficult, as it requires entanglement between at least three particles. Here we report experimental confirmation of this conflict, using our recently developed method[7] to observe three-photon entanglement, or 'Greenberger–Horne–Zeilinger' (GHZ) states. The results of three specific experiments, involving measurements of polarization correlations between three photons, lead to predictions for a fourth experiment; quantum physical predictions are mutually contradictory with expectations based on local realism. We find the results of the fourth experiment to be in agreement with the quantum prediction and in striking conflict with local realism.

We first analyse certain quantum predictions for the entangled three-photon GHZ state:

$$|\Psi\rangle = \frac{1}{\sqrt{2}}(|H\rangle_1|H\rangle_2|H\rangle_3 + |V\rangle_1|V\rangle_2|V\rangle_3) \quad (1)$$

where H and V denote horizontal and vertical linear polarizations respectively. This state indicates that the three photons are in a quantum superposition of the state $|H\rangle_1|H\rangle_2|H\rangle_3$ (all three are horizontally polarized) and the state $|V\rangle_1|V\rangle_2|V\rangle_3$ (all three are vertically polarized) with none of the photons having a well-defined state on its own.

We consider now measurements of linear polarization along directions H'/V' rotated by 45° with respect to the original H/V directions, or of circular polarization L/R (left-handed, right-handed). These new polarizations can be expressed in terms of the original ones as:

$$|H'\rangle = \frac{1}{\sqrt{2}}(|H\rangle + |V\rangle)$$
$$|V'\rangle = \frac{1}{\sqrt{2}}(|H\rangle - |V\rangle) \quad (2)$$
$$|R\rangle = \frac{1}{\sqrt{2}}(|H\rangle + i|V\rangle)$$
$$|L\rangle = \frac{1}{\sqrt{2}}(|H\rangle - i|V\rangle) \quad (3)$$

For convenience we will refer to a measurement of H'/V' linear polarization as an x measurement and one of R/L circular polarization as a y measurement.

Representing the GHZ state (equation (1)) in the new states by using equations (2) and (3), one obtains the quantum predictions for measurements of these new polarizations. For example, for the case of measurement of circular polarization on, say, both photon 1 and 2, and linear polarization H'/V' on photon 3, denoted as a yyx experiment, the state may be expressed as:

$$|\Psi\rangle = \frac{1}{2}(|R\rangle_1|L\rangle_2|H'\rangle_3 + |L\rangle_1|R\rangle_2|H'\rangle_3$$
$$+ |R\rangle_1|R\rangle_2|V'\rangle_3 + |L\rangle_1|L\rangle_2|V'\rangle_3) \quad (4)$$

This expression implies, first, that any specific result obtained in any individual or in any two-photon joint measurement is maximally random. For example, photon 1 will exhibit polarization R or L with the same probability of 50%, or photons 1 and 2 will exhibit polarizations RL, LR, RR or LL with the same probability of 25%. Second, given any two results of measurements on any two photons, we can predict with certainty the result of the corresponding measurement performed on the third photon. For example, suppose photons 1 and 2 both exhibit right-handed R circular polar-

ization. Then by the third term in equation (4), photon 3 will definitely be V' polarized.

By cyclic permutation, we can obtain analogous expressions for any experiment measuring circular polarization on two photons and H'/V' linear polarization on the remaining one. Thus, in every one of the three yyx, yxy, and xyy experiments, any individual measurement result—both for circular polarization and for linear H'/V' polarization—can be predicted with certainty for every photon given the corresponding measurement results of the other two.

Now we will analyse the implications for local realism. As these predictions are independent both of the spatial separation and of the relative time order of the three measurements, we consider them performed simultaneously in a given reference frame—say, for conceptual simplicity, in the reference frame of the source. Then, as Einstein locality implies that no information can travel faster than the speed of light, this requires any specific measurement result obtained for any photon never to depend on which specific measurements are performed simultaneously on the other two nor on their outcome. The only way then for local realism to

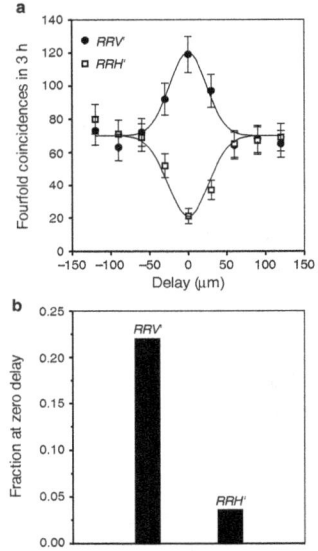

Figure 2 A typical experimental result used in the GHZ argument. This is the yyx experiment measuring circular polarization on photons 1 and 2 and linear polarization on the third. **a**, Fourfold coincidences between the trigger detector T, detectors D_1 and D_2 (both set to measure a right-handed polarized photon), and detector D_3 (set to measure a linearly polarized H' (lower curve) and V' (upper curve) photon as a function of the delay between photon 1 and 2 at the final polarizing beam-splitter). We could adjust the time delay between paths **a** and **b** in Fig. 1 by translating the final polarizing beam-splitter (PBS) and using additional mirrors (not shown in Fig. 1) to ensure overlap of both beams, independent of mirror displacement. At large delay, that is, outside the region of coherent superposition, the two possibilities HHH and VVV are distinguishable and no entanglement results. In agreement with this explanation, it was observed within the experimental accuracy that for large delay the eight possible outcomes in the yyx experiment (and also the other experiments) have the same coincidence rate, whose mean value was chosen as a normalization standard. **b**, At zero delay maximum GHZ entanglement results; the experimentally determined fractions of RRV' and RRH' triples (out of the eight possible outcomes in the yyx experiment) are deduced from the measurements at zero delay. The fractions were obtained by dividing the normalized fourfold coincidences of a specific outcome by the sum of all possible outcomes in each experiment—here, the yyx experiment.

explain the perfect correlations predicted by equation (4) is to assume that each photon carries elements of reality for both x and y measurements that determine the specific individual measurement result[5,6,8].

For photon i we call these elements of reality X_i with values $+1(-1)$ for $H'(V')$ polarizations and Y_i with values $+1(-1)$ for $R(L)$; we thus obtain the relations[8] $Y_1Y_2X_3 = -1$, $Y_1X_2Y_3 = -1$ and $X_1Y_2Y_3 = -1$, in order to be able to reproduce the quantum predictions of equation (4) and its permutations.

We now consider a fourth experiment measuring linear H'/V' polarization on all three photons, that is, an xxx experiment. We investigate the possible outcomes that will be predicted by local realism based on the elements of reality introduced to explain the earlier yyx, yxy and xyy experiments.

Because of Einstein locality any specific measurement for x must be independent of whether an x or y measurement is performed on the other photon. As $Y_iY_i = +1$, we can write $X_1X_2X_3 = (X_1Y_2Y_3)(Y_1X_2Y_3)(Y_1Y_2X_3)$ and obtain $X_1X_2X_3 = -1$. Thus from a local realist point of view the only possible results for an xxx experiment are $V'V'V'$, $H'H'V'$, $H'V'H'$, and $V'H'H'$.

How do these predictions of local realism for an xxx experiment compare with those of quantum mechanics? If we express the state given in equation (1) in terms of H'/V' polarization using equation (2) we obtain:

$$|\Psi\rangle = \frac{1}{2}(|H'\rangle_1|H'\rangle_2|H'\rangle_3 + |H'\rangle_1|V'\rangle_2|V'\rangle_3$$

$$+ |V'\rangle_1|H'\rangle_2|V'\rangle_3 + |V'\rangle_1|V'\rangle_2|H'\rangle_3 \qquad (5)$$

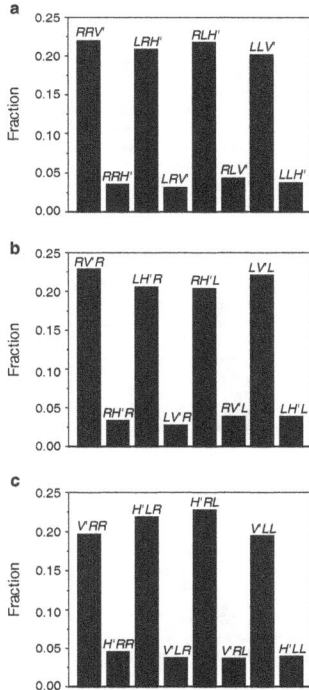

a *yyx*, **b** *yxy*, **c**, *xyy*.

Thus we conclude that the local realistic model predicts none of the terms occurring in the quantum prediction and vice versa. This means that whenever local realism predicts that a specific result will definitely occur for a measurement on one of the photons based on the results for the other two, quantum physics definitely predicts the opposite result. For example, if two photons are both found to be H' polarized, local realism predicts the third photon to carry polarization V' while quantum physics predicts H'. This is the GHZ contradiction between local realism and quantum physics.

In the case of Bell's inequalities for two photons, the conflict between local realism and quantum physics arises for statistical predictions of the theory; but for three entangled particles the conflict arises even for the definite predictions. Statistics now only results from the inevitable experimental limitations occurring in any and every experiment, even in classical physics.

A diagram of our experimental set-up is given in Fig. 1. The method to produce GHZ entanglement for three spatially separated photons is a further development of the techniques that have been used in our previous experiments on quantum teleportation[9] and entanglement swapping[10]. GHZ entanglement has also been inferred for nuclear spins within single molecules from NMR measurements[11], though there a test of nonlocality is impossible.

In the experiment GHZ entanglement is observed under the condition that the trigger detector T and the three GHZ detectors D_1, D_2 and D_3 all actually register a photon. As there are other detection events where fewer detectors fire, this condition might raise doubts about whether such a source can be used to test local realism. The same question arose earlier for certain experiments involving photon pairs[12,13] where a violation of Bell's inequality was only achieved under the condition that both detectors used register a photon. It was often believed[14,15] that such experiments could never, not even in their idealized versions, be genuine tests of local realism. However, this has been disproved[16]. Following the same line of reasoning, it has recently been shown[17] that our procedure permits a valid GHZ test of local realism. In essence, both the Bell and the GHZ arguments exhibit a conflict between detection events and the ideas of local realism.

As explained above, demonstration of the conflict between local realism and quantum mechanics for GHZ entanglement consists of four experiments, each with three spatially separated polarization measurements. First, we perform yyx, yxy and xyy experiments. If the results obtained are in agreement with the predictions for a GHZ state, then for an xxx experiment, our consequent expectations using a local-realist theory are exactly the opposite of our expectations using quantum physics.

For each experiment we have eight possible outcomes of which ideally four should never occur. Obviously, no experiment either in classical physics or in quantum mechanics can ever be perfect; therefore, even the outcomes which should not occur will occur with some small probability in any real experiment. The question is how to deal with such spurious events in view of the fact that the original GHZ argument is based on perfect correlations.

We follow two independent possible strategies. In the first strategy we simply compare our experimental results with the predictions both of quantum mechanics and of a local realist theory for GHZ correlations, and assume that the spurious events are attributable to experimental imperfection that is not correlated to the elements of reality a photon carries. A local realist might argue against that approach and suggest that the non-perfect detection events indicate that the GHZ argument is inapplicable. In our second strategy we therefore accommodate local-realist theories, by assuming that the non-perfect events in the first three experiments indicate a set of elements of reality which are in conflict with quantum mechanics. We then compare the local realist prediction for the xxx experiment obtained under that assumption with the experimental results.

The observed results for two possible outcomes in a yyx experiment

Figure 3 All outcomes observed in the *yyx*, *yxy* and *xyy* experiments, obtained as in Fig. 2. **a**, *yyx*; **b**, *yxy*; **c**, *xyy*. The experimental data show that we observe the GHZ terms predicted by quantum physics (tall bars) in a fraction of 0.85 ± 0.04 of all cases and 0.15 ± 0.02 of the spurious events (short bars) in a fraction of all cases. Within our experimental error we thus confirm the GHZ predictions for the experiments.

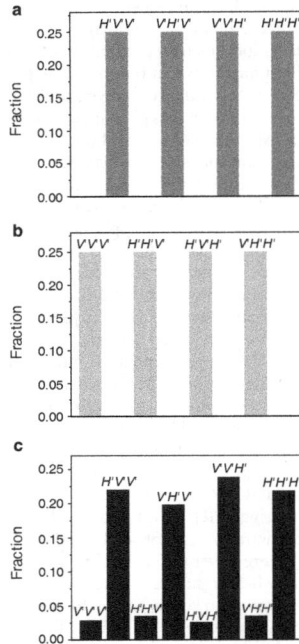

Figure 4 Predictions of quantum mechanics and of local realism, and observed results for the *xxx* experiment. **a, b,** The maximum possible conflict arises between the predictions for quantum mechanics (**a**) and local realism (**b**) because the predicted correlations are exactly opposite. **c,** The experimental results clearly confirm the quantum predictions within experimental error and are in conflict with local realism.

prediction for an event predicted by quantum theory in our *xxx* experiment will use at least one spurious event in the earlier measurements together with two correct ones. Therefore, the fraction of correct events in the *xxx* experiment can at most be equal to the sum of the fractions of all spurious events in the *yyx*, *yxy*, and *xyy* experiments, that is, 0.45 ± 0.03. However, we experimentally observed such terms with a fraction of 0.87 ± 0.04 (Fig. 4c), which violates the local realistic expectation by more than eight standard deviations.

Our latter argument is equivalent to adopting an inequality of the kind first proposed by Mermin[18]. This second analysis succeeds because our average visibility of $71\% \pm 4\%$ clearly surpasses the minimum of 50% necessary for a violation of local realism[18-20]. However, we realise that, as for all existing two-particle tests of local realism, our experiment has rather low detection efficiencies. Therefore we had to invoke the fair sampling hypothesis[21,22], where it is assumed that the registered events are a faithful representative of the whole.

Possible future experiments could include: further study GHZ correlations over large distances with space-like separated randomly switched measurements[23]; extending the techniques used here to the observation of multi-photon entanglement[24]; observation of GHZ-correlations in massive objects like atoms[25]; and investigation of possible applications in quantum computation and quantum communication protocols[26,27].   □

1. Bell, J. S. On the Einstein-Podolsky-Rosen paradox. *Physics* **1,** 195–200 (1964); reprinted Bell, J. S. *Speakable and Unspeakable in Quantum Mechanics* (Cambridge Univ. Press, Cambridge, 1987).
2. Einstein, A., Podolsky, B. & Rosen, N. Can quantum-mechanical description of physical reality be considered complete? *Phys. Rev.* **47,** 777–780 (1935).
3. Schrödinger, E. Die gegenwärtige Situation in der Quantenmechanik. *Naturwissenschaften* **23,** 807–812; 823–828; 844–849 (1935).
4. Aspect, A. Bell's inequality test: more ideal than ever. *Nature* **390,** 189–190 (1999).
5. Greenberger, D. M., Horne, M. A. & Zeilinger, A. in *Bell's Theorem, Quantum Theory, and Conceptions of the Universe* (ed. Kafatos, M.) 73–76 (Kluwer Academic, Dordrecht, 1989).
6. Greenberger, D. M., Horne, M. A., Shimony, A. & Zeilinger, A. Bell's theorem without inequalities. *Am. J. Phys.* **58,** 1131–1143 (1990).
7. Bouwmeester, D., Pan, J.-W., Daniell, M., Weinfurter, H. & Zeilinger, A. Observation of three-photon Greenberger-Horne-Zeilinger entanglement. *Phys. Rev. Lett.* **82,** 1345–1349 (1999).
8. Mermin, N. D. What's wrong with these elements of reality? *Phys. Today* **43,** 9–11 (1990).
9. Bouwmeester, D. *et al.* Experimental quantum teleportation. *Nature* **390,** 575–579 (1997).
10. Pan, J.-W., Bouwmeester, D., Weinfurter, H. & Zeilinger, A. Experimental entanglement swapping: Entangling photons that never interacted. *Phys. Rev. Lett.* **80,** 3891–3894 (1998).
11. Laflamme, R., Knill, E., Zurek, W. H., Catasti, P. & Mariappan, S. V. S. NMR Greenberger-Horne-Zeilinger states. *Phil. Trans. R. Soc. Lond. A* **356,** 1941–1947 (1998).
12. Ou, Z.Y. & Mandel, L. Violation of Bell's inequality and classical probability in a two-photon correlation experiment. *Phys. Rev. Lett.* **61,** 50–53 (1988).
13. Shih, Y. H. & Alley, C. O. New type of Einstein-Podolsky-Rosen-Bohm experiment using pairs of light quanta produced by optical parametric down conversion. *Phys. Rev. Lett.* **61,** 2921–2924 (1988).
14. Kwiat, P., Eberhard, P. E., Steinberger, A. M. & Chiao, R. Y. Proposal for a loophole-free Bell inequality experiment. *Phys. Rev. A* **49,** 3209–3220 (1994).
15. De Caro, L. & Garuccio, A. Reliability of Bell-inequality measurements using polarization correlations in parametric-down-conversion photons. *Phys. Rev. A* **50,** R2803–R2805 (1994).
16. Popescu, S., Hardy, L. & Zukowski, M. Revisiting Bell's theorem for a class of down-conversion experiments. *Phys. Rev. A* **56,** R4353–R4357 (1997).
17. Zukowski, M. Violations of local realism in multiphoton interference experiments. *Phys. Rev. A* **61,** 022109 (2000).
18. Mermin, N. D. Extreme quantum entanglement in a superposition of macroscopically distinct states. *Phys. Rev. Lett.* **65,** 1838–1841 (1990).
19. Roy, S. M. & Singh, V. Tests of signal locality and Einstein-Bell locality for multiparticle systems. *Phys. Rev. Lett.* **67,** 2761–2764 (1991).
20. Zukowski, M. & Kaszlikowski, D. Critical visibility for N-particle Greenberger-Horne-Zeilinger correlations to violate local realism. *Phys. Rev. A* **56,** R1682–1685 (1997).
21. Pearle, P. Hidden-variable example based upon data rejection. *Phys. Rev. D* **2,** 1418–1425 (1970).
22. Clauser, J. & Shimony, A. Bell's theorem: experimental tests and implications. *Rep. Prog. Phys.* **41,** 1881–1927 (1978).
23. Weihs, G., Jennewein, T., Simon, C., Weinfurter, H. & Zeilinger, A. Violation of Bell's inequality under strict Einstein locality conditions. *Phys. Rev. Lett.* **81,** 5039–5043 (1998).
24. Bose, S., Vedral, V. & Knight, P. L. Multiparticle generalization of entanglement swapping. *Phys. Rev. A* **57,** 822–829 (1998).
25. Haroche, S. Atoms and photons in high-Q cavities: next tests of quantum theory. *Ann. NY Acad. Sci.* **755,** 73–86, (1995).
26. Briegel, H.-J., Duer, W., Cirac, J. I. & Zoller, P. Quantum repeaters: The role of imperfect local operations in quantum communication. *Phys. Rev. Lett.* **81,** 5932–5935 (1998).
27. Cleve, R. & Buhrman, H. Substituting quantum entanglement for communication. *Phys. Rev. A* **56,** 1201–1204 (1997).
28. Kwiat, P. G. *et al.* New high intensity source of polarization-entangled photon pairs. *Phys. Rev. Lett.* **75,** 4337–4341 (1995).

are shown in Fig. 2. The six remaining possible outcomes of a *yyx* experiment have also been measured in the same way and likewise in the *yxy* and *xyy* experiments. For all three experiments this results in 24 possible outcomes whose individual fractions thus obtained are shown in Fig. 3.

Adopting our first strategy, we assume that the spurious events are attributable to unavoidable experimental errors; within the experimental accuracy, we conclude that the desired correlations in these experiments confirm the quantum predictions for GHZ entanglement. Thus we compare the predictions of quantum mechanics and local realism with the results of an *xxx* experiment (Fig. 4) and we observe that, again within experimental error, the triple coincidences predicted by quantum mechanics occur and not those predicted by local realism. In this sense, we believe that we have experimentally realized the first three-particle test of local realism following the GHZ argument.

We then investigated whether local realism could reproduce the *xxx* experimental results shown in Fig. 4c, if we assume that the spurious non-GHZ events in the other three experiments (Fig. 3) actually indicate a deviation from quantum physics. To answer this we adopt our second strategy and consider the best prediction a local realistic theory could obtain using these spurious terms. How, for example, could a local realist obtain the quantum prediction $H'H'H'$? One possibility is to assume that triple events producing $H'H'H'$ would be described by a specific set of local hidden variables such that they would give events that are in agreement with quantum theory in both an *xyy* and a *yxy* experiment (for example, the results $H'LR$ and $LH'R$), but give a spurious event for a *yyx* experiment (in this case, LLH'). In this way any local realistic

29. Zeilinger, A., Horne, M. A., Weinfurter, H. & Zukowski, M. Three particle entanglements from two entangled pairs. *Phys. Rev. Lett.* **78**, 3031–3034 (1997).

30. Zukowski, M., Zeilinger, A. & Weinfurter, H. Entangling photons radiated by independent pulsed source. *Ann. NY Acad. Sci.* **755**, 91–102 (1995).

Acknowledgements

We thank D. M. Greenberger, M. A. Horne and M. Zukowski for useful discussions. This work was supported by the Austrian Fonds zur Förderung der Wissenschaftlichen Forschung, the Austrian Academy of Sciences and the Training and Mobility of Researchers programme of the European Union.

Correspondence and requests for materials should be addressed to A.Z. (e-mail: Zeilinger-office@exp.univie.ac.at).

Wave–particle duality of C_{60} molecules

Markus Arndt, Olaf Nairz, Julian Vos-Andreae, Claudia Keller, Gerbrand van der Zouw & Anton Zeilinger

Institut für Experimentalphysik, Universität Wien, Boltzmanngasse 5, A-1090 Wien, Austria

Quantum superposition lies at the heart of quantum mechanics and gives rise to many of its paradoxes. Superposition of de Broglie matter waves[1] has been observed for massive particles such as electrons[2], atoms and dimers[3], small van der Waals clusters[4], and neutrons[5]. But matter wave interferometry with larger objects has remained experimentally challenging, despite the development of powerful atom interferometric techniques for experiments in fundamental quantum mechanics, metrology and lithography[6]. Here we report the observation of de Broglie wave interference of C_{60} molecules by diffraction at a material absorption grating. This molecule is the most massive and complex object in which wave behaviour has been observed. Of particular interest is the fact that C_{60} is almost a classical body, because of its many excited internal degrees of freedom and their possible couplings to the environment. Such couplings are essential for the appearance of decoherence[7,8], suggesting that interference experiments with large molecules should facilitate detailed studies of this process.

When considering de Broglie wave phenomena of larger and more complex objects than atoms, fullerenes come to mind as suitable candidates. After their discovery[9] and the subsequent invention of efficient mass-production methods[10], they became easily available. In our experiment (see Fig. 1) we use commercial, 99.5% pure, C_{60} fullerenes (Dynamic Enterprises Ltd, Twyford, UK) which were sublimated in an oven at temperatures between 900 and 1,000 K. The emerging molecular beam passed through two collimation slits, each about 10 µm wide, separated by a distance of 1.04 m. Then it traversed a free-standing nanofabricated SiN_x grating[11] consisting of nominally 50-nm-wide slits with a 100-nm period.

At a further distance of 1.25 m behind the diffraction grating, the interference pattern was observed using a spatially resolving detector. It consisted of a beam from a visible argon-ion laser (24 W all lines), focused to a gaussian waist of 8 µm width (this is the size required for the light intensity to drop to $1/e^2$ of that in the centre of the beam). The light beam was directed vertically, parallel both to the lines of the diffraction grating and to the collimation slits. By using a suitable mirror assembly, the focus could be scanned with micrometre resolution across the interference pattern. The absorbed light then ionized the C_{60} fullerenes via heating and subsequent thermal emission of electrons[12]. The detection region

was found to be smaller than 1 mm in height, consistent with a full Rayleigh length of 800 µm and the strong power dependence of this ionization process. A significant advantage of the thermionic mechanism is that it does not detect any of the residual gases present in the vacuum chamber. We could thus achieve dark count rates of less than one per second even under moderately high vacuum conditions (5×10^{-7} mbar). The fullerene ions were then focused by an optimized ion lens system, and accelerated to a BeCu conversion electrode at -9 kV where they induced the emission of electrons which were subsequently amplified by a Channeltron detector.

Alignment is a crucial part of this experiment. In order to be able to find the beam in the first place, our collimation apertures are movable piezo slits that can be opened from 0 to 60 µm (in the case of the first slit) and from 0 to 200 µm (for the second slit). The vacuum chamber is rigidly mounted on an optical table together with the ionizing laser, in order to minimize spatial drifts.

The effect of gravity also had to be considered in our set-up. For the most probable velocity (220 m s^{-1}), the fullerenes fall by 0.7 mm while traversing the apparatus. This imposes a constraint on the maximum tilt that the grating may have with respect to gravity. As a typical diffraction angle into the first-order maximum is 25 µrad, one can tolerate a tilt angle of (at most) about one mrad before molecules start falling from one diffraction order into the trajectory of a neighbouring order of a different velocity class. The experimental curves start to become asymmetric as soon as the grating tilt deviates by more than 500 µrad from its optimum vertical orientation.

The interference pattern of Fig. 2a clearly exhibits the central maximum and the first-order diffraction peaks. The minima between zeroth and first orders are well developed, and are due to destructive interference of C_{60} de Broglie waves passing through neighbouring slits of the grating. For comparison, we show in Fig. 2b the profile of the undiffracted collimated beam. The velocity distribution has been measured independently by a time-of-flight method; it can be well fitted by $f(v) = v^3 \exp(-(v-v_0)^2/v_m^2)$, with $v_0 = 166$ m s^{-1} and $v_m = 92$ m s^{-1} as expected for a transition between a maxwellian effusive beam and a supersonic beam[13]. The most probable velocity was $v = 220$ m s^{-1}, corresponding to a de Broglie wavelength of 2.5 pm. The full-width at half-maximum was as broad as 60%, resulting in a longitudinal coherence length of about 5 pm.

The essential features of the interference pattern can be understood using standard Kirchhoff diffraction theory[14] for a grating with a period of 100 nm, by taking into account both the finite width of the collimation and the experimentally determined velocity distribution. The parameters in the fit were the width of the collimation, the gap width s_0 of a single slit opening, the effective beam width of the detection laser and an overall scaling factor. This model, assuming all grating slits to be perfect and identical, reproduces very well the central peak of the interference pattern shown in Fig. 2a, but does not fit the 'wings' of this pattern.

Agreement with the experimental data, including the 'wings' in

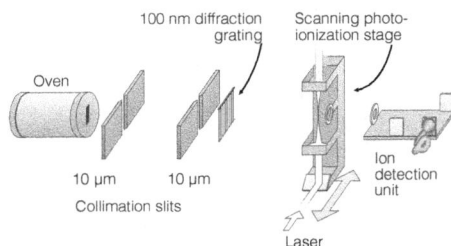

Figure 1 Diagram of the experimental set-up (not to scale). Hot, neutral C_{60} molecules leave the oven through a nozzle of 0.33 mm $\times$ 1.3 mm $\times$ 0.25 mm (width $\times$ height $\times$ depth), pass through two collimating slits of 0.01 mm $\times$ 5 mm (width $\times$ height) separated by 1.04 m, traverse a SiN_x grating (period 100 nm) 0.1 m after the second slit, and are detected via thermal ionization by a laser 1.25 m behind the grating. The ions are then accelerated and directed towards a conversion electrode. The ejected electrons are subsequently counted by a Channeltron electron multiplier. The laser focus can be reproducibly scanned transversely to the beam with 1-µm resolution.

Fig. 2a, can be achieved by allowing for a gaussian variation of the slit widths over the grating, with a mean open gap width centred at $s_0 = 38$ nm with a full-width at half-maximum of 18 nm. That best-fit value for the most probable open gap width s_0 is significantly smaller than the 55 ± 5 nm specified by the manufacturer (T. A. Savas and H. Smith, personal communication). This trend is consistent with results obtained in the diffraction of noble gases and He clusters, where the apparently narrower slit was interpreted as being due to the influence of the van der Waals interaction with the SiN_x grating during the passage of the molecules[15]. This effect is expected to be even more pronounced for C_{60} molecules owing to their larger polarizability. The width of the distribution seems also justified in the light of previous experiments with similar gratings: both the manufacturing process and adsorbents could account for this fact (ref. 16, and T. A. Savras and H. Smith, personal communication). Recently, we also observed interference of C_{70} molecules.

Observation of quantum interference with fullerenes is interesting for various reasons. First, the agreement between our measured and calculated interference contrast suggests that not only the highly symmetric, isotopically pure $^{12}C_{60}$ molecules contribute to the interference pattern but also the less symmetric isotopomeric variants $^{12}C_{59}{}^{13}C$ and $^{12}C_{58}{}^{13}C_2$ which occur with a total natural abundance of about 50%. If only the isotopically pure $^{12}C_{60}$ molecules contributed to the interference, we would observe a much larger background.

Second, we emphasize that for calculating the de Broglie wavelength, $\lambda = h/Mv$, we have to use the complete mass M of the object. Thus, each C_{60} molecule acts as a whole undivided particle during its centre-of-mass propagation.

Last, the rather high temperature of the C_{60} molecules implies broad distributions, both of their kinetic energy and of their internal energies. Our good quantitative agreement between experiment and theory indicates that the latter do not influence the observed coherence. All these observations support the view that each C_{60} molecule interferes with itself only.

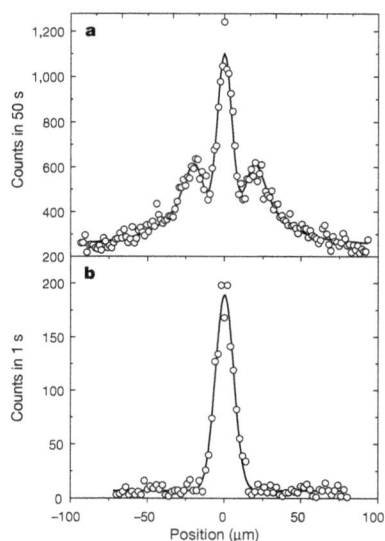

Figure 2 Interference pattern produced by C_{60} molecules. **a**, Experimental recording (open circles) and fit using Kirchhoff diffraction theory (continuous line). The expected zeroth and first-order maxima can be clearly seen. Details of the theory are discussed in the text. **b**, The molecular beam profile without the grating in the path of the molecules.

In quantum interference experiments, coherent superposition only arises if no information whatsoever can be obtained, even in principle, about which path the interfering particle took. Interaction with the environment could therefore lead to decoherence. We now analyse why decoherence has not occurred in our experiment and how modifications of our experiment could allow studies of decoherence using the rich internal structure of fullerenes.

In an experiment of the kind reported here, 'which-path' information could be given by the molecules in scattering or emission processes, resulting in entanglement with the environment and a loss of interference. Among all possible processes, the following are the most relevant: decay of vibrational excitations via emission of infrared radiation, emission or absorption of thermal blackbody radiation over a continuous spectrum, Rayleigh scattering, and collisions.

When considering these effects, one should keep in mind that only those scattering processes which allow us to determine the path of a C_{60} molecule will completely destroy in a single event the interference between paths through neighbouring slits. This requires $\lambda \ll d$; that is, the wavelength λ of the incident or emitted radiation has to be smaller than the distance d between neighbouring slits, which amounts to 100 nm in our experiment. When this condition is not fulfilled decoherence is however also possible via multi-photon scattering[7,8,17].

At $T \approx 900$ K, as in our experiment, each C_{60} molecule has on average a total vibrational energy of $E_v \approx 7$ eV (ref. 18) stored in 174 vibrational modes, four of which may emit infrared radiation at $\lambda_{vib} \approx 7$–$19\,\mu$m (ref. 10) each with an Einstein coefficient of $A_k \approx 100\,s^{-1}$ (ref. 18). During its time of flight from the grating towards the detector ($\tau \approx 6$ ms) a C_{60} molecule may thus emit on average 2–3 such photons.

In addition, hot C_{60} has been observed[19] to emit continuous blackbody radiation, in agreement with Planck's law, with a measured integrated emissivity of $\epsilon \approx 4.5\ (\pm 2.0) \times 10^{-5}$ (ref. 18). For a typical value of $T \approx 900$ K, the average energy emitted during the time of flight can then be estimated as only $E_{bb} \approx 0.1$ eV. This corresponds to the emission of (for example) a single photon at $\lambda \approx 10\,\mu$m. Absorption of blackbody radiation has an even smaller influence as the environment is at a lower temperature than the molecule. Finally, since the mean free path for neutral C_{60} exceeds 100 m in our experiment, collisions with background molecules can be neglected.

As shown above, the wavelengths involved are too large for single photon decoherence. Also, the scattering rates are far too small to induce sufficient phase diffusion. This explains the decoupling of internal and external degrees of freedom, and the persistence of interference in our present experiment.

A variety of unusual decoherence experiments would be possible in a future extension of the experiment, using a large-area interferometer. A three-grating Mach–Zehnder interferometer[6] seems to be a particularly favourable choice, since for a grating separation of up to 1 m we will have a molecular beam separation of up to 30 μm, much larger than the wavelength of a typical thermal photon. In this case, the environment obtains 'which-path' information even through a single thermal photon, and the interference contrast should thus be completely destroyed. The parameters that could be controlled continuously in such an experiment would then be the internal temperature of the fullerenes, the temperature of the environment, the intensity and frequency of external laser radiation, the interferometer size, and the background pressure of various gases.

An improved interferometer could have other applications. For example, in contrast to previous atom-optical experiments[20–22] which were limited to the interaction with only a few lines in the whole spectrum, interferometry with fullerenes would enable us to study these naturally occurring and ubiquitous thermal processes and wavelength-dependent decoherence mechanisms for (we believe) the first time. Another possible application of molecule interferometry is precision metrology; the improved interferometer could be used to measure molecular polarizabilities[6]. Moreover, it might be possible to nanofabricate SiC patterns on Si substrates using C_{60} interferometry.

Furthermore, we note the fundamental difference between isotopically pure C_{60}, which should exhibit bosonic statistics, and the isotopomers containing one ^{13}C nucleus, which should exhibit fermionic statistics. We intend to explore the possibility of observing this feature, for example by showing the different rotational symmetry between the two species in an interferometer[23,24].

In our experiment, the de Broglie wavelength of the interfering fullerenes is already smaller than their diameter by a factor of almost 400. It would certainly be interesting to investigate the interference of objects the size of which is equal to or even bigger than the diffracting structure. Methods analogous to those used for the present work, probably extended to the use of optical diffraction structures, could also be applied to study quantum interference of even larger macromolecules or clusters, up to small viruses[25,26]. □

Received 30 June; accepted 2 September 1999.

1. de Broglie, L. Waves and quanta. *Nature* **112**, 540 (1923).
2. Davisson, C. J. & Germer, L. H. The scattering of electrons by a single crystal of nickel. *Nature* **119**, 558–560 (1927).
3. Estermann, I. & Stern, O. Beugung von Molekularstrahlen. *Z. Phys.* **61**, 95–125 (1930).
4. Schöllkopf, W. & Toennies, J. P. Nondestructive mass selection of small van der Waals clusters. *Science* **266**, 1345–1348 (1994).
5. Halban, H. v. Jr & Preiswerk, P. Preuve expérimentale de la diffraction des neutrons. *C.R. Acad. Sci.* **203**, 73–75 (1936).
6. Berman, P. (ed.) *Atom Interferometry* (Academic, 1997).
7. Zurek, W. H. Decoherence and the transition from quantum to classical. *Phys. Today* 36–44 (October 1991).
8. Giulini, D. *et al. Decoherence and the Appearance of the Classical World in Quantum Theory* (Springer, Berlin, 1996).
9. Kroto, H. W., Heath, J. R., O'Brien, S. C., Curl, R. F. & Smalley, R. E. C_{60}: buckminsterfullerene. *Nature* **318**, 162–166 (1985).
10. Krätschmer, W., Lamb, L. D., Fostiropoulos, K. & Huffman, D. R. A new form of carbon. *Nature* **347**, 354–358 (1990).
11. Savas, T. A., Shah, S. N., Schattenburg, M. L., Carter, J. M. & Smith, H. I. Achromatic interferometric lithography for 100-nm-period gratings and grids. *J. Vac. Sci. Technol. B* **13**, 2732–2735 (1995).
12. Ding, D., Huang, J., Compton, R. N., Klots, C. E. & Haufler, R. E. cw laser ionization of C_{60} and C_{70}. *Phys. Rev. Lett.* **73**, 1084–1087 (1994).
13. Scoles, G. (ed.) *Atomic and Molecular Beam Methods* Vol. 1 (Oxford Univ. Press, 1988).
14. Born, M. & Wolf, E. *Principles of Optics* (Pergamon, Oxford, 1984).
15. Grisenti, R. E., Schöllkopf, W., Toennies, J. P., Hegerfeldt, G. C. & Köhler, T. Determination of atom-surface van der Waals potentials from transmission-grating diffraction intensities. *Phys. Rev. Lett.* **83**, 1755–1758 (1999).
16. Grisenti, R. E. *et al.* He atom diffraction from nanostructure transmission gratings: the role of imperfections. *Phys. Rev. A* (submitted).
17. Joos, E. & Zeh, H. D. The emergence of classical properties through interaction with the environment. *Z. Phys. B* **59**, 223–243 (1985).
18. Kolodney, E., Budrevich, A. & Tsipinyuk, B. Unimolecular rate constants and cooling mechanisms of superhot C_{60} molecules. *Phys. Rev. Lett.* **74**, 510–513 (1995).
19. Mitzner, R. & Campbell, E. E. B. Optical emission studies of laser desorbed C_{60}. *J. Chem. Phys.* **103**, 2445–2453 (1995).
20. Pfau, T., Spälter, S., Kurtsiefer, Ch., Ekstrom, C. R. & Mlynek, J. Loss of spatial coherence by a single spontaneous emission. *Phys. Rev. Lett.* **73**, 1223–1226 (1994).
21. Clauser, J. F. & Li, S. "Heisenberg microscope" decoherence atom interferometry. *Phys. Rev. A* **50**, 2430–2433 (1994).
22. Chapman, M. S. *et al.* Photon scattering from atoms in an atom interferometer: Coherence lost and regained. *Phys. Rev. Lett.* **75**, 3783–3787 (1995).
23. Werner, S. A., Colella, R., Overhauser, A. W. & Eagen, C. F. Observation of the phase shift of a neutron due to precession in a magnetic field. *Phys. Rev. Lett.* **35**, 1053–1055 (1975).
24. Rauch, H. *et al.* Verification of coherent spinor rotation of fermions. *Phys. Lett. A* **54**, 425–427 (1975).
25. Clauser, J. F. in *Experimental Metaphysics* (eds Cohen, R. S., Home, M. & Stachel, J.) 1–11 (Kluwer Academic, Dordrecht, 1997).
26. Arndt, M., Nairz, O., van der Zouw, G. & Zeilinger, A. in *Epistemological and Experimental Perspectives on Quantum, Physics* (eds Greenberger, D., Reiter, W. L. & Zeilinger, A.) 221–224 (IVC Yearbook, Kluwer Academic, Dordrecht, 1999).

Acknowledgements

We thank M. Haluška, H. Kuzmany, R. Penrose, P. Scheier, J. Schmiedmayer and G. Senn for discussions. This work was supported by the Austrian Science Foundation FWF, the Austrian Academy of Sciences, the TMR programme of the European Union, and the US NSF.

Correspondence and requests for materials should be addressed to A.Z.
(e-mail: zeilinger-office@exp.univie.ac.at).

LIST OF PUBLICATIONS

1. H. Rauch, E. Seidl, A. Zeilinger, *Neutron-Depolarisationsmessungen an Dy in der Umgebung des ferromagnetischen Umwandlungspunktes*, Z. f. Angew. Phys. **32**, 109 (1971).

2. H. Rauch, A. Zeilinger, *Neutron Depolarization Measurements on a Dy Single Crystal*, Atomkernenergie **19**, 167 (1972).

3. A. Zeilinger, G. Reitsamer, *Fokussierende Effekte in der Elektrophorese durch variable Potentialgradienten*, Journal of Chromatogr. **93**, 41 (1974).

4. E. Seidl, A. Zeilinger, *A Simple Device for Growing Single Crystals of Reactive Materials*, J. Phys. E, Scientific Instruments **7**, 1030 (1974).

5. A. Zeilinger, *Zur Verwendung einkristalliner Fasern in Neutronenbeugungsexperimenten*, Nucl. Instr. Meth. **120**, 525 (1974).

6. H. Böck, E. Seidl, A. Zeilinger, *Uranium Diffusion into Fission Chamber Electrodes*, J. Nucl. Mat. **54**, 159 (1974).

7. H. Rauch, A. Zeilinger, *Recent Neutron Physical Experiments at the Triga Mark II Reactor Vienna*, 3rd European Conference of Triga Reactor Users, Neuherberg, Published by General Atomic GA-TOC-6, 9–11 (1974).

8. A. Zeilinger, H. Rauch, *Measurement of Hydrogen Distributions by Neutron Radiography*, 3rd European Conference of Triga Reactor Users, Neuherberg, Published by General Atomic GA-TOC-6, 4–25 (1974).

9. M. Mannoussakis, H. Rauch, A. Zeilinger, *Investigation of Hydrogen Motion In Liquids By Neutron Radiography*, In "Radiography with Neutrons", M. R. Hawkesworth Ed, British Nuclear Energy Society, London, 143 (1975).

10. A. Zeilinger, R. Huebner, *Measurement of Moisture Motion Under a Temperature Gradient in a Concrete for SNR-300 Using Thermal Neutrons*, 3rd International Conference on Structural Mechanics in Reactor Technology, Paper H 1/9 in the Transactions. London, (1975).

11. H. Böck, A. Zeilinger, *Radiographic Examination of Irradiated Incore Neutron Detectors*, Nucl. Instr. Meth. **129**, 147 (1975).

12. A. Zeilinger, *Neutron Radiography as a Tool for the Detection and Measurement of Hydrogen Distributions*, International Conference on Peaceful Uses of Atomic Energy for Scientific and Economic Development, Proceedings. Iraqi Atomic Energy Organization, Baghdad, 310 (1975).

13. H. Rauch, A. Zeilinger, G. Badurek, A. Wilfing, W. Bauspiess, U. Bonse, *Verification of Coherent Spinor Rotation of Fermions*, Phys. Lett. A **54**, 425–7 (1975).

14. A. Zeilinger, R. Huebner, *Untersuchung des Feuchtetransports in einem Beton des SNR-300 durch Neutronentransmission*, Kerntechnik **18**, 119 (1976).

15. G. Badurek, H. Rauch, A. Zeilinger, W. Bauspiess, U. Bonse, *Measurements of Neutron Interference and Polarization Effects Caused by Nuclear and Magnetic Interaction*, Phys. Lett. A **56**, 244–6 (1976).

16. G. Badurek, H. Rauch, A. Zeilinger, W. Bauspiess, U. Bonse, *Phase Shift and Spin Rotation Phenomena in Neutron Interferometry*, Phys. Rev. D **14**, 1177 (1976).

17. G. Eder, A. Zeilinger, *Interference Phenomena and Spin Rotation of Neutrons by Magnetic Materials*, Il Nuovo Cimento B **34**, 76 (1976).

18. A. Zeilinger, W. A. Pochman, H. Rauch, M. Suleiman, *Neutronographic Measurements of the Diffusion of Hydrogen and Hydrogeneous Substances in Liquid and Solids*, 4th European Conference of Triga Reactor Users, Vienna, Published by General Atomic, GA-TOC-8, 42856 (1976).

19. A. Zeilinger, M. Suleiman, H. Rauch, *Experimental Diffusion Measurements of Light and Heavy Water Mixing Using Neutron Radiography*, Atomkernenergie **28**, 183 (1976).

20. A. Zeilinger, *General Formulation of Spin Rotations in Neutron Interferometry*, Zeitschrift für Physik B **25**, 97–100 (1976).

21. H. Rauch, G. Badurek, W. Bauspiess, U. Bonse, A. Zeilinger, *Determination of Scattering Lengths and Magnetic Spin Rotations by Neutron Interferometry*, International Conference on the Interaction of Neutrons with Nuclei, Lowell, Mass, Proceedings, published by ERDA (CONF-760715-P2), 1027 (1976).

22. A. Zeilinger, W. A. Pochman, *A New Method for the Measurement of Hydrogen Diffusion in Metals*, J. Appl. Phys. **47**, 5478 (1976).

23. A. A. Harms, A. Zeilinger, *A New Formulation of Total Unsharpness in Radiography*, Phys. Med. Biol. **22**, 70 (1977).

24. A. Zeilinger, W. A. Pochman, *Neutron Radiographic Measurements of the Diffusion of H in b-Ti, V, Nb and Ta*, J. Phys. F, Metal Phys. **7**, 575 (1977).

25. H. Rauch, A. Zeilinger, *Hydrogen Transport Studies Using Neutron Radiography*, Atomic Energy Review **15**, 249 (1977).

26. W. A. Pochman, A. Zeilinger, H. Böck, *Detection of Cracks in Triga Fuel Rods by Neutron Radiography*, Atomkernenergie **29**, 231 (1977).

27. A. A. Harms, A. Zeilinger, *A Comment on the Total Unsharpness in Radiography*, Phys. Med. Biol. **22**, 1207 (1977).

28. H. Rauch, E. Seidl, A. Zeilinger, W. Bauspiess, U. Bonse, *Hydrogen Detection in Metals by Neutron Interferometry*, J. Appl. Phys. **49**, 2731 (1978).

29. A. Zeilinger, C. G. Shull, M. A. Horne, G. L. Squires, *Two-Crystal Neutron Interferometry*, International Workshop on Neutron Interferometry, Grenoble 1978, Proceedings in "Neutron Interferometry" U. Bonse, H. Rauch (eds.), Oxford University Press, 48–59 (1979).

30. A. Zeilinger, *Some Magnetic and Spin Effects in Neutron Interferometry*, International Workshop on Neutron Interferometry, Grenoble 1978, Proceedings, Oxford University Press, 241 (1979).

31. M. A. Horne, A. Zeilinger, *Fizeau Effects for Thermal Neutrons*, International Workshop on Neutron Interferometry, Grenoble 1978, Proceedings, Oxford University Press, 350–4 (1979).

32. A. Zeilinger, *Laue-Case Dynamical Neutron Diffraction With Perfect Nonmagnetic Crystals in Magnetic Fields*, International Workshop on Neutron Interferometry, Grenoble 1978, Proceedings, Oxford University Press, 355 (1979).

33. A. Zeilinger, C. G. Shull, *Magnetic Field Effects on Dynamical Diffraction of Neutrons by Perfect Crystals*, Phys. Rev. B **19**, 3957 (1979).

34. A. Zeilinger, *On the Aharanov-Bohm Effect*, Lett. Nuovo Cimento Serie 2 **25**, 333–6 (1979).

35. A. Zeilinger, *Perfect Crystal Neutron Optics*, International Workshop on Imaging Processes and Coherence in Physics. Les Houches 1979, Proceedings in Vol. 112, Lecture Notes in Physics, Springer Verlag, 267 (1980).

36. A. Zeilinger, *Polarization Effects in Neutron Diffraction at Perfect Non-Magnetic Crystals*, International Conference on Polarized Neutrons in Condensed Matter Research; Zaborow, Poland 1979. Nukleonika **25**, 871 (1980).

37. G. Badurek, H. Rauch, A. Zeilinger, *Neutron Phase Echo Concept and a Proposal for a Dynamical Neutron Polarization Method*, International Workshop on Neutron Spin Echo, Grenoble 1979; in "Neutron Spin Echo" (F. Mezei, Ed.), Lecture Notes in Physics, **128**, Springer Verlag, 136 (1980).

38. H. J. Bernstein, A. Zeilinger, *Exact Spin Rotation by Precession During Neutron Interferometry*, Phys. Lett. A **75**, 169 (1980).

39. G. Badurek, H. Rauch, A. Zeilinger, *Dynamic Concepts in Neutron Polarization*, Zeitschrift für Physik B **38**, 303 (1980).

40. C. G. Shull, A. Zeilinger, G. L. Squires, M. A. Horne, D. K. Atwood, J. Arthur, *Anomalous Flight Time of Neutrons through Diffracting Crystals*, Phys. Rev. Lett. **44**, 1715–8 (1980).

41. R. Gähler, A. G. Klein, A. Zeilinger, *Neutron Optical Tests of Nonlinear Wave Mechanics*, Phys. Rev. A **23**, 1611–7 (1981).

42. H. Rauch, A. Zeilinger, *Demonstration of SU(2)-Symmetry by Neutron Interferometry*, Hadronic Journal **4**, 1280 (1981).

43. A. Zeilinger, *General Properties of Lossless Beam Splitters in Interferometry*, Am. J. Phys. **49**, 882 (1981).

44. A. G. Klein, G. I. Opat, A. Cimmino, A. Zeilinger, W. Treimer, R. Gähler, *Neutron Propagation in Moving Matter: The Fizeau Experiment with Massive Particles,* Phys. Rev. Lett. **46**, 1551–4 (1981).

45. A. Zeilinger, *Spin Directions of Interfering Beams in Quantum Interferometry,* Nature **294**, 544–6 (1981).

46. A. Zeilinger, R. Gähler, C. G. Shull, W. Treimer, *Experimental Status and Recent Results of Neutron Interference Optics,* Symposium on Neutron Scattering, Argonne, AIP Conference Proceedings **89**, 93 (1981).

47. D. Bader, H. Rauch, A. Zeilinger, *An Ultra Small Angle Scattering Study of Hydrogen in Metals,* Z. Naturforsch. **37a**, 512 (1982).

48. G. Badurek, H. Rauch, J. Summhammer, U. Kischko, A. Zeilinger, *Direct Verification of the Quantum Spin-State Superposition Law,* J. Phys. A **16**, 1133–9 (1983).

49. J. Summhammer, G. Badurek, H. Rauch, U. Kischko, A. Zeilinger, *Direct Observation of Fermion Spin Superposition by Neutron Interferometry,* Phys. Rev. A **27**, 2523–32 (1983).

50. M. A. Horne, A. Zeilinger, G. I. Opat, A. G. Klein, *Neutron Phase Shift in Moving Matter,* Phys. Rev. A **28**, 1 (1983).

51. A. Zeilinger, C. G. Shull, J. Arthur, M. A. Horne, *Bragg-Case Neutron Interferometry,* Phys. Rev. A **28**, 487–489 (1983).

52. A. Zeilinger, T. J. Beatty, *Bragg Diffraction and Surface Reflection of Neutrons from Perfect Crystals at Grazing Incidence,* Phys. Rev. B **27**, 7239–7250 (1983).

53. A. Zeilinger, M. A. Horne, C. G. Shull, *Search for Unorthodox Phenomena by Neutron Interference Experiments,* Proceedings of the International Symposium on Foundations of Quantum Mechanics in the Light of New Technology, (S. Kamefuchi *et al.,* Eds.), Phys. Soc. Japan, Tokyo, 289–293 (1983).

54. D. Greenberger, M. A. Horne, C. G. Shull, A. Zeilinger, *Delayed Choice Experiments with the Neutron Interferometer,* Proc. Int. Symp. on Foundations of Quantum Mechanics in the Light of New Technology, (S. Kamefuchi *et al.,* Eds.), Phys. Soc. Japan, Tokyo, 294–300 (1983).

55. A. Zeilinger, *Progress in Physics with Neutrons at Small Reactors,* Use and Development of Low and Medium Flux Research Reactors, Atomkernenergie **44**, Supplement, 3 (1984).

56. A. Zeilinger, M. A. Horne, H. J. Bernstein, *Symmetry Violations and Schwinger Scattering in Neutron Interferometry,* ILL-Workshop on Reactor-Based Fundamental Physics, J. physique **45**, C3-209 (1984).

57. A. Zeilinger, *Generalized Aharanov-Bohm and Wheeler-Type Delayed Choice Experiments with Neutrons,* ILL-Workshop on Reactor Based Fundamental Physics, J. physique **45**, C3-213–216 (1984).

58. A. G. Klein, A. Zeilinger, *Wave Optics with Cold Neutrons,* ILL-Workshop on Reactor-Based Fundamental Physics, J. physique **45**, C3-239 (1984).

59. C. G. Shull, A. Zeilinger, *A One-Axis Flight-Time Neutron Spectrometer for Student Use*, International Union of Crystallography 13th Congress, Hamburg 1984. Acta Cryst. A **30**, Supplement, C-466 (1984).

60. A. Zeilinger, C. G. Shull, M. A. Horne, S. A. Werner, *Measurement of the Effective Mass Enhancement of the Deflection of Neutrons in Perfect Crystals*, International Union of Crystallography 13th Congress, Hamburg 1984. Acta Cryst. A **30**, Supplement, C-345 (1984).

61. A. Zeilinger, K. Svozil, *Measuring the Dimension of Space-Time*, Phys. Rev. Lett. **54**, 2553–5 (1985).

62. J. Arthur, C. G. Shull, A. Zeilinger, *Dynamical Neutron Diffraction in a Thick-Crystal Interferometer*, Phys. Rev. B **32**, 5753 (1985).

63. M. A. Horne, A. Zeilinger, *A Bell-Type EPR Experiment Using Linear Momenta*, Symposium on the Foundations of Modern Physics, Joensuu, P. Lahti, P. Mittelstaedt (eds.), World Scientific Publ. (Singapore), 435–9 (1985).

64. A. Zeilinger, *Testing Quantum Superposition with Cold Neutrons*, Oxford Quantum Gravity Discussion Conference 1984, published in "Quantum Concepts in Space and Time" (C. J. Isham, R. Penrose, Eds.) Oxford University Press, 17 (1986).

65. A. Zeilinger, M. A. Horne, *Neutron Lenses in Interferometry*, Physica B **136**, 141 (1986).

66. K. D. Finkelstein, C. G. Shull, A. Zeilinger, *Magnetic Neutrality of the Neutron*, Physica B **136**, 131–3 (1986).

67. A. Zeilinger, *Long Wavelength Neutron Interferometry*, Proceedings of the Workshop on the Investigation of Fundamental Interactions with Cold Neutrons, NBS Special Publication 711, G. Greene (ed.), Gaithersburg, 112 (1986).

68. A. Zeilinger, *Generalized Aharanov-Bohm Experiments with Neutrons*, Fundamental Aspects of Quantum Theory, Como 1985, V. Gorrini, A. Figuereido (eds.), Plenum press, 311 (1986).

69. A. Zeilinger, *Complementarity in Neutron Interferometry*, Physica B **137**, 235 (1986).

70. A. Zeilinger, C. G. Shull, M. A. Horne, K. D. Finkelstein, *Effective Mass of Neutrons Diffracting in Crystals*, Phys. Rev. Lett. **57**, 3089–92 (1986).

71. A. Zeilinger, *Three Gedanken Experiments on Complementarity in Double-Slit Diffraction*, New Techniques and Ideas in Quantum Measurement Theory D. Greenberger (ed.), Annals of the New York Academy of Sciences, 480 (1986).

72. A. Zeilinger, *Testing Bell's Inequalities with Periodic Switching*, Phys. Lett. A **118**, 1 (1986).

73. M. A. Horne, A. Zeilinger, *Einstein-Podolsky-Rosen Interferometry*, New Techniques and Ideas in Quantum Measurement Theory D. Greenberger (ed.), Annals of the New York Academy of Sciences **480**, 469 (1986).

74. K. Svozil, A. Zeilinger, *Dimension of Space-Time*, Int. J. Mod. Phys. A **1**, 971 (1986).

75. A. Zeilinger, *Das Einstein-Podolsky-Rosen-Paradoxon*, "Ganzheitsphysik", Grazer Gespräche 1986, M. Heindler, F. E. Moser (Hrsg.), TU Graz, 29 (1987).

76. A. Zeilinger, *Interpretationsprobleme und Paradigmensuche in der Quantenmechanik*, "Ganzheitsphysik", Grazer Gespräche 1986, M. Heindler, F. E. Moser (Hrsg.), TU Graz, 212 (1987).

77. M. Horne, A. Zeilinger, *A Possible Spin-Less Experimental Test of Bell's Inequality*, "Microphysical Reality and Quantum Formalism" A. van der Merwe, F. Selleri, G. Tarozzi (eds.), Kluwer (Dordrecht), 401 (1988).

78. G. Grössing, A. Zeilinger, *Quantum Cellular Automata*, Complex Systems **2**, 197 (1988).

79. K. Svozil, A. Zeilinger, *Breakdown of Quantum Electrodynamics in (g-2) Experiments*, Physica Scripta T **21**, 122 (1988).

80. G. Grössing, A. Zeilinger, *A Conservation Law in Quantum Cellular Automata*, Physica D **31**, 70–7 (1988).

81. J. Kamesberger, A. Zeilinger, *Numerical Solution of a Nonlinear Schrödinger Equation for Neutron Optics Experiments*, Physica B **151**, 193 (1988).

82. A. Zeilinger, M. A. Horne, *Neutron Focusing Effects in Perfect-Crystal Systems*, Physica B **151**, 157 (1988).

83. M. A. Horne, K. D. Finkelstein, C. G. Shull, A. Zeilinger, H. J. Bernstein, *Neutron Spin-Pendellösung Resonance*, Physica B **151**, 189 (1988).

84. G. Grössing, A. Zeilinger, *Structures in Quantum Cellular Automata*, Physica B **151**, 366 (1988).

85. A. Zeilinger, R. Gähler, C. G. Shull, W. Treimer, W. Mampe, *Single and Double Slit Diffraction of Neutrons*, Rev. Mod. Phys. **60**, 1067–73 (1988).

86. A. Zeilinger, H. Rauch, *Neutron Interferometry, A Status Report*, In „Nejtronnaja Fizika" Akademie der Wissenschaften der UdSSR, Moskau, 146 (1988).

87. F. Selleri, A. Zeilinger, *Local Deterministic Description of Einstein-Podolski-Rosen Experiments*, Found. Phys. **18**, 1141 (1988).

88. G. Grössing, A. Zeilinger, *Quantum Cellular Automata, A Corrigendum*, Complex Systems **2**, 611 (1988).

89. A. Zeilinger, *Quantum Implications, Essays in Honour of David Bohm. Invited Book Review*, Physics Today **41**, 72 (1988).

90. D. Greenberger, M. A. Horne, A. Zeilinger, *Going beyond Bell's Theorem*, "Bell's Theorem, Quantum Theory, and Conceptions of the Universe", M. Kafatos (ed.), Kluwer, Dordrecht, 69–72 (1989).

91. A. Zeilinger, K. Eder, R. Gähler, M. Gruber, W. Mampe, *The New Very-Cold-Neutron Optics Setup at ILL*, Nucl. Instr. Meth. A **284**, 171 (1989).

92. M. A. Horne, A. Zeilinger, *Speakable and Unspeakable in Quantum Mechanics; Invited Book Review*, Amer. J. Phys. **42**, 630 (1989).

93. M. A. Horne, A. Shimony, A. Zeilinger, *Two-Particle Interferometry*, Phys. Rev. Lett. **62**, 2209 (1989).

94. M. Gruber, K. Eder, A. Zeilinger, R. Gähler, W. Mampe, W. Drexel, *A Phase-Grating Interferometer for Very-Cold Neutrons*, Phys. Lett. A **140**, 363 (1989).

95. A. Zeilinger, M. A. Horne, *Aharonov-Bohm with Neutrons*, Physics World **2**, 23 (1989).

96. A. Zeilinger, *Fundamentale Experimente mit Materiewellen und deren Interpretation*, "Wieviele Leben hat Schrödingers Katze? Zur Physik und Philosophie der Quantenmechanik", J. Audretsch und K. Mainzer (Hrsg.), B. I. Wissenschaftsverlag, Mannheim, Wien, Zürich, (1990).

97. A. Zeilinger, *The Planck Stroll*, Amer. J. Phys. **58**, 103 (1990).

98. T. Chattopadhyay, A. Zeilinger, M. Wacenovsky, H. W. Weber, O. B. Hyun, D. K. Finnemore, *Search for Magnetic Ordering of Tm Moments in TmBa Cu O down to 90 mK*, Solid State Commun. **73**, 721 (1990).

99. D. M. Greenberger, M. A. Horne, A. Zeilinger, *Bell's Theorem without Inequalities*, "Sixty-Two Years of Uncertainty: Historical, Philosophical, and Physical Inquiries into the Foundations of Quantum Mechanics", Arthur I. Miller (ed.), Plenum, N. Y, (1990).

100. M. A. Horne, A. Shimony, A. Zeilinger, *Introduction to Two-Particle Interferometry*, "Sixty-Two Years of Uncertainty: Historical, Philosophical, and Physical Inquiries into the Foundationsof Quantum Mechanics", Arthur I. Miller (ed.), Plenum, N. Y, (1990).

101. A. Zeilinger, *Experiment and Quantum Measurement Theory*, "Quantum Theory Without Reduction", M. Cini, J.-M. Levy-Leblond (eds.), Hilger, Bristol, 9 (1990).

102. A. Zeilinger, *Problemi di interpretazione e ricerca di paradigmi in meccanica quantistica*, "Che cos' e la realta", Franco Selleri (ed.), Jaca Book, Milano, (1990).

103. D. M. Greenberger, M. A. Horne, A. Shimony, A. Zeilinger, *Bell's Theorem Without Inequalities*, Amer. J. Phys. **58**, 1131–43 (1990).

104. M. A. Horne, A. Shimony, A. Zeilinger, *Two-Particle Interferometry*, Nature **347**, 429 (1990).

105. A. Zeilinger, H. J. Bernstein, D. M. Greenberger, M. A. Horne, *Quantum Reality and Higher-Order Correlations: Two Remarks on Entanglement*, Symposium on the Foundations of Modern Physics, P. Lahti, P. Mittelstaedt (eds.), World Scientific Publ, Singapore, 487 (1990).

106. M. Horne, A. Shimony, A. Zeilinger, *Down-conversion Photon Pairs: A New Chapter In the History of Quantum Mechanical Entanglement*, Quantum Coherence, Jeeva Anandan (ed.), World Scientific Publishing Co, Singapore, (1990).

107. G. Grössing, A. Zeilinger, *Zeno's Paradox in Quantum Cellular Automata*, Automata. Physica D **50**, 321–6 (1991).

108. J. Summhammer, A. Zeilinger, *Fundamental and applied neutron interferometry*, Physica B **174**, 396–402 (1991).

109. R. Gähler, A. Zeilinger, *Wave-Optical Experiments with Very-Cold Neutrons*, Amer. J. Phys. **59**, 316 (1991).

110. K. Eder, M. Gruber, A. Zeilinger, R. Gähler, W. Mampe, *Diffraction of Very-Cold Neutrons at Phase Gratings*, Physica B **172**, 329 (1991).

111. A. Zeilinger, R. Gähler, M. A. Horne, *On the Topological Nature of the Aharonov-Casher Effect*, Phys. Lett. **154**, 93 (1991).

112. M. Zukowski, A. Zeilinger, *Test of the Bell's Inequality Based on Phase and linear Momentum as well as Spin*, Phys. Lett. A **155**, 69 (1991).

113. Anton Zeilinger, *Physiker auf der Suche nach der Wirklichkeit*, Naturherrschaft. Wie Mensch und Welt sich in der Wissenschaft begegnen, 99–129 (1991).

114. M. Tschernitz, R. Gähler, W. Mampe, B. Schillinger, A. Zeilinger, *Precision measurement of single slit diffraction with very cold neutrons*, Phys. Lett. A **164**, 365–368 (1992).

115. A. Zeilinger, M. A. Horne, D. M. Greenberger, *Higher-Order Quantum Entanglement*, Proceedings "Squeezed States and Quantum Uncertainty", College Park, D. Han, Y. S. Kim, W. W. Zachary (eds.), NASA Conference Publication 3135, National Aeronautics and Space Administration, (1992).

116. A. Zeilinger, M. A. Horne, D. Greenberger, *Bell's Theorem Without Inequalities and Beyond*, Quantum Measurements in Optics, P. Tombesi, D. F. Walls (eds.), Plenum Press, New York, 369 (1992).

117. Anton Zeilinger, *Physik und Wirklichkeit - Neuere Entwicklungen zum Einstein-Podolsky-Rosen Paradoxon*, Naturwissenschaft und Weltbild. Mathematik und Quantenphysik in unserem Denk- und Wertesystem, 99–121 (1992).

118. A. Zeilinger, H. J. Bernstein, D. M. Greenberger, M. A. Horne, M. Zukowski, *Controlling Entanglement in Quantum Optics*, Quantum Control and Measurement, H. Ezawa, Y. Murayama (eds.), Elsevier Science Publishers, 9 (1993).

119. D. M. Greenberger, H. J. Bernstein, M. A. Horne, A. Shimony, A. Zeilinger, *Proposed GHZ experiments using cascades of down-conversions*, Quantum Control and Measurement, H. Ezawa, Y. Murayama (eds.), Elsevier Science Publishers, 23 (1993).

120. G. Badurek, H. Weinfurter, R. Gähler, A. Kollmar, S. Wehinger, A. Zeilinger, *Nondispersive Phase of the Aharonov-Bohm Effect*, Phys. Rev. Lett. **71**, 307–311 (1993).

121. H. J. Bernstein, D. M. Greenberger, M. A. Horne, A. Zeilinger, *Bell theorem without inequalities for two spinless particles*, Phys. Rev. A **47**, 78–84 (1993).

122. A. Zeilinger, M. Zukowski, M. A. Horne, H. J. Bernstein, D. M. Greenberger, *Einstein-Podolsky-Rosen Correlations in Higher Dimensions*, Fundamental Aspects of Quantum Theory, J. Anandan, J. L. Safko (eds.), World Scientific, Singapore, (1993).

123. M. Zukowski, A. Zeilinger, M. A. Horne, A. K. Ekert, *Event-Ready-Detectors Bell Experiment via Entanglement Swapping*, Phys. Rev. Lett. **71**, 4287–90 (1993).

124. K. Eder, J. Felber, R. Gähler, R. Golub, W. Mampe, E. M. Rasel, A. Zeilinger, *Interferometry with Very Cold Neutrons*, Proceedings "Waves and Particles in Light and Matter", Plenum Publishing Company, London, (1993).

125. Translation into German: R. Gähler, A. Zeilinger, *Wave-optical experiments with very cold neutrons*, Am. J. Phys. 59, 4 (1991). *Wellenoptische Experimente mit sehr kalten Neutronen, trans. G. Theysohn*, PhuD **3**, 217 (1994).

126. D. M. Greenberger, M. A. Horne, A. Zeilinger, *Multiparticle Interferometry and the Superposition Principle*, Physics Today 46 **8**, (1993).

127. M. Reck, A. Zeilinger, *Quantum Phase Tracing of Correlated Photons in Optical Multiports*, Quantum Interferometry, F. DeMartini, A. Zeilinger (eds.), World Scientific, Singapore, (1994).

128. M. Horne, D. Greenberger, A. Zeilinger, *Two-Particle-Fringes Dependent on the sum of the Coordinates*, Quantum Interferometry, F. DeMartini, A. Zeilinger (eds.), World Scientific, Singapore, (1994).

129. T. J. Herzog, J. G. Rarity, H. Weinfurter, A. Zeilinger, *Frustrated Two-Photon Creation via Interference*, Phys. Rev. Lett. **72**, 629 (1994).

130. A. Zeilinger, *Probing Higher Dimensions of Hilbert Space in Experiment*, Acta Physica Polonica A **85**, 717 (1994).

131. A. Zeilinger, H. J. Bernstein, M. A. Horne, *Information Transfer with Two-State, Two-Particle Quantum Systems*, J. Mod. Opt. **41**, 2375 (1994).

132. M. Reck, A. Zeilinger, H. J. Bernstein, P. Bertani, *Experimental Realization of Any Discrete Unitary Operator*, Phys. Rev. Lett. **73**, 58–61 (1994).

133. P. Kwiat, H. Weinfurter, T. Herzog, A. Zeilinger, M. Kasevich, *Experimental Realization of Interaction-Free Measurement*, Proceedings "Symposium on the Foundations of Modern Physics 1994", Helsinki, Finland, 129 - 138 (1994).

134. T. Herzog, J. G. Rarity, H. Weinfurter, A. Zeilinger, *Reply to Senitzky, "Classical Interpretation of "Frustrated Two-Photon Creation via Interference""*, Phys. Rev. Lett. **73**, 3040–3041 (1994).

135. I. Jex, S. Stenholm, A. Zeilinger, *Hamiltonian Theory of a Symmetric Multiport*, Preprint Series in Theoretical Physics HU-TFT-94–43, Research Institute for Theoretical Physics, University of Helsinki, Helsinki (1994), Opt. Comm. **117**, 95 (1995).

136. P. Kwiat, H. Weinfurter, T. Herzog, A. Zeilinger, M. Kasevich, *Experimental Realization of Interaction-Free Measurement*, Fundamental Problems in Quantum Theory, D. M. Greenberger, A. Zeilinger (eds.), Annals of the New York Academy of Sciences **755**, 383 (1995).

137. P. Kwiat, H. Weinfurter, T. Herzog, A. Zeilinger, *Interaction-Free Measurement*, Phys. Rev. Lett. **74**, 4763–6 (1995).

138. K. Raum, M. Koellner, A. Zeilinger, M. Arif, R. Gähler, *Effective-Mass Enhanced Deflection of Neutrons in Noninertial Frames*, Phys. Rev. Lett. **74**, 2859 (1995).

139 H. Weinfurter, T. Herzog, P. G. Kwiat, J. G. Rarity, A. Zeilinger, M. Zukowski, *Frustrated Downconversion: Virtual or Real Photons?*, Fundamental Problems in Quantum Theory, D. M. Greenberger, A. Zeilinger (eds.), Annals of the New York Academy of Sciences **755**, 61 (1995).

140. M. Zukowski, A. Zeilinger, H. Weinfurter, *Entangling Photons Radiated by Independent Pulsed Sources*, Fundamental Problems in Quantum Theory, D. M. Greenberger, A. Zeilinger (eds.), Annals of the New York Academy of Sciences **755**, 91 (1995).

141. K. Raum, M. Koellner, A. Zeilinger, R. Gähler, *Effective Mass-enhanced Deflection of Neutrons in Noninertial Frames*, Fundamental Problems in Quantum Theory, D. M. Greenberger, A. Zeilinger (eds.), Annals of the New York Academy of Sciences **755**, 888 (1995).

142. G. Krenn, A. Zeilinger, *Entangled Entanglement*, Fundamental Problems in Quantum Theory, D. M. Greenberger, A. Zeilinger (eds.), Annals of the New York Academy of Sciences **755**, 873 (1995).

143. D. M. Greenberger, M. A. Horne, A. Zeilinger, *Nonlocality of a Single Photon?*, Phys. Rev. Lett. **75**, 2064 (1995).

144. K. Mattle, M. Michler, H. Weinfurter, A. Zeilinger, M. Zukowski, *Non-Classical Statistics at Multiport Beam Splitters*, Applied Physics B **60**, S111 (1995).

145. P. G. Kwiat, K. Mattle, H. Weinfurter, A. Zeilinger, A. V. Sergienko, Y. H. Shih, *New High-Intensity Source of Polarization-Entangled Photon Pairs*, Phys. Rev. Lett. **75**, 4337–41 (1995).

146. A. Zeilinger, *Einstein-Podolsky-Rosen Interferometry*, Annals of the Israel Phys. Society Vol. 12, 57–72 (1995).

147. A. Zeilinger, *Quantum Correlations Beyond Bell's Inequalities*, Advances in Quantum Phenomena, E. Beltrametti, J. M. Levy-Leblond (eds.), Plenum, New York, 215 (1995).

148. A. Zeilinger, *Experiment, Entanglement and the Foundations of Quantum Mechanics*, The Foundational Debate, W. DePauli-Schimanovich *et al.* (eds.), Kluwer, Dordrecht, 13–19 (1995).

149. E. M. Rasel, M. K. Oberthaler, H. Batelaan, J. Schmiedmayer, A. Zeilinger, *Atom Wave Interferometry with Diffraction Gratings of Light*, Phys. Rev. Lett. **75**, 2633 (1995).

150. B. Dopfer, P. G. Kwiat, H. Weinfurter, A. Zeilinger, *Brillouin Scattering and Dynamical Diffraction of Entangled Photon Pairs*, Phys. Rev. A **52**, R2531 (1995).

151. T. J. Herzog, P. G. Kwiat, H. Weinfurter, A. Zeilinger, *Complementarity and the Quantum Eraser*, Phys. Rev. Lett. **75**, 3034 (1995).

152. M. Zukowski, A. Zeilinger, M. A. Horne und A. Ekert, *Extensions of Bell Theorem: Experiment Involving Independent Sources in "Event-Ready" Configuration*, Fundamental Problems in Quantum Physics, M. Ferrero, A. van der Merwe (eds.), Kluwer, Dordrecht, 363–373 (1995).

153. D. Greenberger, A. Zeilinger, *Quantum Theory: Still Crazy After All These Years*, Physics World **8**, 33 (1995).

154. G. Weihs, M. Reck, H. Weinfurter, A. Zeilinger, *All-fiber three-path Mach-Zehnder interferometer*, Opt. Lett. **21**, 302 (1996).

155. A. Zeilinger, H. Weinfurter, *Informationsübertragung und Informationsverarbeitung in der Quantenwelt*, Physikalische Blätter **52**, 37136 (1996).

156. A. Zeilinger, *On the Interpretation and Philosophical Foundation of Quantum Mechanics*, Vastakohtien todellisuus, Festschrift for K. V. Laurikainen U. Ketvel *et al.* (eds.), Helsinki University Press, (1996).

157. M. Michler, K. Mattle, H. Weinfurter, A. Zeilinger, *Interferometric Bell-State Analysis*, Phys. Rev. A **53**, R1209–12 (1996).

158. P. W. Milonni, H. Fearn, A. Zeilinger, *Theory of Two-Photon Down-Conversion in the Presence of Mirrors*, Phys. Rev. A **53**, 4556 (1996).

159. K. Mattle, H. Weinfurter, P. G. Kwiat, A. Zeilinger, *Dense Coding in Experimental Quantum Communication*, Phys. Rev. Lett. **76**, 4656–59 (1996).

160. G. Weihs, M. Reck, H. Weinfurter, A. Zeilinger, *Two-Photon Interference in Optical Fiber Multiports*, Phys. Rev. A **54**, 893 (1996).

161. S. Bernet, M. Oberthaler, R. Abfalterer, J. Schmiedmayer und A. Zeilinger, *Modulation of atomic de Broglie waves using Bragg diffraction*, Quantum Semiclass. Opt. **8**, 497 (1996).

162. S. N. Chormaic, S. Franke, J. Schmiedmayer, A. Zeilinger, *Concepts of Temporal Mach-Zehnder Interferometry with Atoms*, acta physica slovaca **46**, 463 (1996).

163. P. Domokos, P. Adam, J. Janszky, A. Zeilinger, *Atom de Broglie Wave Deflection by a Single Cavity Mode in the Few-Photon Limit: Quantum Prism*, Phys. Rev. Lett. **77**, 1663 (1996).

164 E. M. Rasel, M. K. Oberthaler, H. Batelaan, S. Bernet, J. Schmiedmayer, A. Zeilinger, *An Interferometer for Atoms with Standing Light Waves*, Proceedings "Coherence and Quantum Optics VII", Eberly, Mandel, Wolf (eds.), Plenum Press, New York, 549.(1996).

165. P. G. Kwiat, H. Weinfurter, A. Zeilinger, *Interaction-Free Measurement of a Quantum Object: On the Breeding of "Schrödinger Cats"*, Coherence and Quantum Optics VII, Eberly, Mandel, Wolf (eds.), Plenum Press, New York, 673 (1996).

166. M. K. Oberthaler, S. Bernet, Ernst M. Rasel, J. Schmiedmayer, A. Zeilinger, *Inertial Sensing with Classical Atomic Beams*, Phys. Rev. A **54**, 3165 (1996).

167. G. Krenn, A. Zeilinger, *Entangled Entanglement*, Phys. Rev. A **54**, 1793 (1996).

168. M. K. Oberthaler, R. Abfalterer, S. Bernet, J. Schmiedmayer, A. Zeilinger, *Atom Waves in Crystals of Light*, Phys. Rev. Lett. **77**, 4980–83 (1996).

169. S. Bernet, M. K. Oberthaler, R. Abfalterer, J. Schmiedmayer, A. Zeilinger, *Coherent Frequency Shift of Atomic Matter Waves*, Phys. Rev. Lett. **77**, 5160 (1996).

170. K. Mattle, M. Eibl, H. Weinfurter, A. Zeilinger, *Experimental Quantum Communication*, Quantum Interferometry, F. DeMartini *et al.* (eds.), VCH Publishers, Weinheim, 119 (1996).

171. D. N. Greenberger, M. A. Horne, A. Zeilinger, *Tangled Concepts about Entangled States*, Quantum Interferometry, F. DeMartini *et al.* (eds.), VCH Publishers, Weinheim, 119 (1996).

172. D. Greenberger, A. Zeilinger, *Teoria kwantowa: wciaz zwariowana po tylu latach (Quantum theory: still crazy after all these years)*, Postepy Fizyki **47**, 339 (1996).

173. A. Zeilinger, T. Herzog, M. A. Horne, P. G. Kwiat, K. Mattle, H. Weinfurter, *Path Information in Quantum Interferometry*, Proceedings "Coherence and Quantum Optics VII", Eberly, Mandel, Wolf (eds.), Plenum Press, New York, 305 (1996).

174. P. G. Kwiat, H. Weinfurter, A. Zeilinger, *Quantum Seeing in the Dark*, Scientific American **275**, 52–58 (1996).

175. P. G. Kwiat, K. Mattle, H. Weinfurter, A. Zeilinger, *Polarization-Entangled Photons and Quantum Dense Coding*, Optics and Photonics News 7 **No. 12**, 14 (1996).

176. K. Raum, M. Weber, A. Zeilinger, *Gravity and inertia in neutron crystal optics and VCN interferometry*, J. Phys. Soc. Japan v. **65 (suppl. A)**, 277–280 (1996).

177. Anton Zeilinger, *The Changing Metaphysics of Science*, Remarks at the Final Panel of the Interdisciplinary Workshop, (1996).

178. Anton Zeilinger, *Jenseits jeder Gewißheit. Das Rätsel der Quantenwelt*, Landesmuseum Joanneum, Graz, (1996).

179. H. Batelaan, S. Bernet, M. K. Oberthaler, E. M. Rasel, J. Schmiedmayer und A. Zeilinger, *Classical and Quantum Atom Fringes*, Atom Interferometry, Paul R. Berman (ed.), Academic Press, San Diego, 85 (1997).

180. A. Zeilinger, *Entanglement and Indistinguishability: Coherence Experiments with Photon Pairs and Triplets*, Proceedings, ICAP 1996, in "Atomic Physics 15", H. B. van Linden van den Heuvell *et al.* (eds.), World Scientific, 47 (1997).

181. G. Weihs, H. Weinfurter, A. Zeilinger, *Towards a Long Distance Bell-Experiment with Independent Observers*, Experimental Metaphysics: "Quantum Mechanical Studies for Abner Shimony", Robert S. Cohen, M. A. Horne, John Stachel (eds.), Kluwer Academic publishers, Dordrecht, Netherlands, 271–280 (1997).

182. M. Zukowski, A. Zeilinger, M. A. Horne, *Realizable Higher-dimensional Two-particle Entanglements via Multiport Beam Splitters*, Phys. Rev. A **55**, 2564 (1997).

183. Č. Brukner, A. Zeilinger, *Diffraction of Matter Waves in Space and in Time*, Phys. Rev. A **56**, 3804–3824 (1997).

184. A. Zeilinger, M. A. Horne, H. Weinfurter, M. Zukowski, *Three-Particle Entanglements from Two Entangled Pairs*, Phys. Rev. Lett. **78**, 3031–4 (1997).

185. A. Zeilinger, *Quantum Teleportation and the Non-Locality of Information*, Phil. Trans. R. Soc. Lond. A **355**, 2401–4 (1997).

186. D. Bruss, A. Ekert, S. F. Huelga, J.-W. Pan, A. Zeilinger, *Quantum Computating with Controlled-NOT and a Few Qubits*, Phil. Trans. R. Soc. Lond. A **355**, 2259 (1997).

187. S. Bernet, R. Abfalterer, C. Keller, M. K. Oberthaler, J. Schmiedmayer, A. Zeilinger, *Atom Holography at Light Structures*, J. Imag. Sci. and Tech. **41**, 324 (1997).

188. R. Abfalterer, C. Keller, S. Bernet, M. K. Oberthaler, J. Schmiedmayer, A. Zeilinger, *Nanometer Definition of Atomic Beams with Masks of Light*, Phys. Rev. A. **56**, R4365 (1997).

189. C. Keller, M. K. Oberthaler, R. Abfalterer, S. Bernet, J. Schmiedmayer, A. Zeilinger, *Tailored Complex Potentials and Friedel's Law in Atom Optics*, Phys. Rev. Lett. **79**, 3327 (1997).

190. R. Abfalterer, S. Bernet, C. Keller, M. K. Oberthaler, J. Schmiedmayer, A. Zeilinger, *Atom Waves in Standing Light Waves*, acta physica slovaca **47**, 165 (1997).

191. H. Batelaan, E. M. Rasel, M. Oberthaler, J. Schmiedmayer, A. Zeilinger, *Anomalous Transmission in Atom Optics*, J. Mod. Opt. **44**, 2629 (1997).

192. H. Weinfurter, M. Reck, A. Zeilinger, *Quantum Cryptography, Communication and Computation: From Application to Utopia*, Proc. 'Second Intern. Austrian-Israeli Technion Symposium', Graz, Austria, 35 (1997).

193. Č. Brukner, A. Zeilinger, *Nonequivalence between Stationary Matter Wave Optics and Stationary Light Optics*, Phys. Rev. Lett. **79**, 2599–2603 (1997).

194. P. Kwiat, H. Weinfurter, A. Zeilinger, *Wechselwirkungsfreie Quantenmessung*, Spektrum der Wissenschaft **42** (1997).

195. G. Weihs, H. Weinfurter, A. Zeilinger, *A Test of Bell's inequalities with independent oberservers*, Acta-Physica-Slovaca v. **47**, 337–340 (1997).

196. D. Bouwmeester, A. Zeilinger, *Atoms that agree to differ*, Nature **388**, 827–829 (1997).

197. A. Zeilinger, *Get set for the quantum revolution*, Physics World **54** (1997).

198. D. Bouwmeester, J.-W. Pan, K. Mattle, M. Eibl, H. Weinfurter, A. Zeilinger, *Experimental Quantum Teleportation*, Nature **390**, 575–579 (1997).

199. G. Krenn, A. Zeilinger, *Reply to "Comment on 'Entangled entanglement'"*, Phys. Rev. A **56**, 4336 (1997).

200. S. Bernet, R. Abfalterer, C. Keller, J. Schmiedmayer, A. Zeilinger, *Diffractive matter wave optics in time*, JOSA-B **15**, 2817 (1998).

201. D. Bouwmeester, J.-W. Pan, K. Mattle, M. Eibl, H. Weinfurter, A. Zeilinger, *Experimental Quantum Teleportation*, Phil. Trans. R. Soc. Lond. A **356**, 1733 (1998).

202. D. Bouwmeester, J.-W. Pan, K. Mattle, M. Eibl, H. Weinfurter, A. Zeilinger, M. Zukowski, *Experimental Quantum Teleportation of Arbitrary Quantum States*, J. Appl. Phys. B **67**, 749 (1998).

203. G. Weihs, T. Jennewein, Ch. Simon, H. Weinfurter, A. Zeilinger, *Violation of Bell's inequality under strict Einstein locality conditions*, Phys. Rev. Lett. **81**, 5039–5043 (1998).

204. M. Zukowski, A. Zeilinger, M. A. Horne, H. Weinfurter, *Quest for GHZ states*, Int. Conference on 'Quantum Future', Wroczlaw, Poland, 1997, Acta Phys. Pol. **93**, 187 (1998).

205. A. Zeilinger, *Quantum Entanglement: A Fundamental Concept Finding its Applications*, Proceedings of the Nobel Symposium 104 "Modern Studies of Basic Quantum Concepts and Phenomena", E. B. Karlsson and E. Brändas (eds.), Physica Scripta **76**, 203 (1998).

206. J.-W. Pan, D. Bouwmeester, H. Weinfurter, A. Zeilinger, *Experimental Entanglement Swapping: entangling photons that never interacted*, Phys. Rev. Lett. **80**, 3891–94 (1998).

207. D. Bouwmeester, J. Schmiedmayer, H. Weinfurter, A. Zeilinger, *Quantum Coherence in Experiment: From Teleportation to Massive Objects*, Proc. of the GR-15 Conference "Gravitation and Relativity: At the turn of the Millenium", IUCAA, Pune, 333 (1998).

208. J.-W. Pan, A. Zeilinger, *Greenberger-Horne-Zeilinger-state analyzer*, Phys. Rev. A **57**, 2208 (1998).

209. C. Keller, R. Abfalterer, St. Bernet, M. K. Oberthaler, J. Schmiedmayer, A. Zeilinger, *Absorptive Masks Of Light: — A Useful Tool For Spatial Probing In Atom Optics*, J. Vac. Sci. Tech. **16**, 3850–3854 (1998).

210. R. Abfalterer, S. Bernet, C. Keller, M. K. Oberthaler, J. Schmiedmayer, A. Zeilinger, *Atomic de Broglie waves in periodic light structures, in Macroscopic quantum coherence*, EMPD '98 Proceedings, International Conference on Energy Management and Power Delivery, Singapore (Singapore) World Scientific 323, 301–315 (1998).

211. M. Horne, I. Jex, A. Zeilinger, *Schroedinger base states in strong periodic media, in macroscopic quantum coherence*, EMPD '98 Proceedings, International Conference on Energy Management and Power Delivery, Singapore. (Singapore) World Scientific 323, 284–300 (1998).

212. H. Weinfurter, D. Bouwmeester, K. Mattle, J.-W. Pan, M. Eibl, A. Zeilinger, J. Brendel, N. Gisin, J. G. Rarity, P. R. Tapster, *Quantum Communication and Entanglement*, Proc. 14th European Meeting on Cybernetics and System Research, Ed. By Trappl *et al.*, 1,95, 346 (1998).

213. D. Bouwmeester, J.-W. Pan, H. Weinfurter, A. Zeilinger, *Experimental Quantum Teleportation and Entanglement Swapping*, Technical Digest, EQUEC'98, Glasgow, 184 (1998).

214. A. G. White, J. R. Mitchell, O. Nairz, P. G. Kwiat, ‚*Interaction-free' imaging*, Phys. Rev. A **58**, 605–613 (1998).

215. A. Zeilinger, *Fundamentals of Quantum Information*, Physics World, 35 (1998).

216. D. Bouwmeester, J.-W. Pan, M. Daniell, H. Weinfurter, M. Zukowski, A. Zeilinger, *Reply to the comment "A posteriori teleportation" by Braunstein and Kimble*, Nature **394**, 841 (1998).

217. K. Raum, J. Felber, M. A. Horne, P. Geltenbort, A. Zeilinger, *The Equivalence Principle in Quantum Mechanics and Neutrons that Fall Upwards*, (1998).

218. C. Keller, S. Bernet, J. Schmiedmayer, A. Zeilinger, *Coherent Channeling of Atomic deBroglie Waves*, (1998).

219. D. Bouwmeester, J.-W. Pan, M. Daniell, H. Weinfurter, A. Zeilinger, *Observation of Three-Photon Greenberger-Horne-Zeilinger Entanglement*, Phys. Rev. Lett. **82**, 1345–49 (1999).

220. J.-W. Pan, A. Zeilinger, *Introduction to Quantum Teleportation*, Physics **28**, 609 (1999).

221. A. Zeilinger, *Experiment and the Foundations of Quantum Physics*, More Things in Heaven and Earth, A Celebration of Physics at the Millenium, American Physical Society, B. Bederson (ed.), Rev. Mod. Phys. **71**, 288–297 (1999).

222. C. Keller, S. Bernet, J. Schmiedmayer, A. Zeilinger, *Coherent Quantum Channeling of Atomic Matter Waves*, (1999).

223. S. Bernet, R. Abfalterer, C. Keller, J. Schmiedmayer, A. Zeilinger, *Matter wave sidebands from a complex potential with temporal helicity in complex space*, Proc. Roy. Soc. (London) 455, 1509–1520 (1999).

224. M. Zukowski, A. Zeilinger, M. A. Horne, H. Weinfurter, *Independent photons and entanglement. A short overview*, Sixth Conference on Conceptual and Philosophical Problems in Physics, Peyresque, 8–12. 09. 1997, Int. J. Theor. Phys. **98**, 501–517 (1999).

225. M. K. Oberthaler, R. Abfalterer, S. Bernet, C. Keller, J. Schmiedmayer, A. Zeilinger, *Dynamical Diffraction of Atomic Matter Waves by Crystals of Light*, Phys. Rev. A **60**, 456–472 (1999).

226. Č. Brukner, A. Zeilinger, *Malus´ Law and Quantum Information*, Acta Phys. Slovaca **89**, 647–652 (1999).

227. C. Simon, G. Weihs, A. Zeilinger, *Quantum Cloning and Signaling*, Acta Phys. Slovaca **49**, 755 (1999).

228. A. Zeilinger, *A Foundational Principle for Quantum Mechanics*, Found. Physics **29**, 631–643 (1999).

229. C. Keller, J. Schmiedmayer, A. Zeilinger, T. Nonn, S. Dürr, G. Rempe, *Adiabatic following in standing-wave diffraction of atoms*, Appl. Phys. B **69**, 303–309 (1999).

230. M. Horne, I. Jex, A. Zeilinger, *Schrödinger wave functions in strong periodic potentials with applications to atom optics*, Phys. Rev. A **59**, 2190–2202 (1999).

231. Č. Brukner, A. Zeilinger, *Operationally Invariant Information in Quantum Measurements*, Phys. Rev. Lett. **83**, 3354–3357 (1999).

232. C. Keller, J. Schmiedmayer, A. Zeilinger, *Matter Wave Diffraction at Standing Light Waves*, Epistemological and Experimental Perspectives on Quantum Physics, Greenberger *et al.* (eds.), Kluwer Academic, Netherlands, 245–247 (1999).

233. Č. Brukner,, A. Zeilinger, *Quantum Complementarity and Information Invariance*, Epistemological and Experimental Perspectives on Quantum Physics, Greenberger *et al.* (eds.), Kluwer Academic, Netherlands, 231–234 (1999).

234. G. Van der Zouw, A. Zeilinger, *Observation of the Nondispersivity of Scalar Aharonov-Bohm Phase Shifts by Neutron Interferometry*, Epistemological and Experimental Perspectives on Quantum Physics, Greenberger *et al.* (eds.), Kluwer Academic, Netherlands, 263–265 (1999).

235. G. Weihs, T. Jennewein, C. Simon, H. Weinfurter, A. Zeilinger, *A Bell Experiment under Strict Einstein Locality Conditions*, Epistemological and Experimental Perspectives on Quantum Physics, Greenberger *et al.* (eds.), Kluwer Academic, Netherlands, 267–269 (1999).

236. M. Daniell, D. Bouwmeester, J.-W. Pan, H. Weinfurter, A. Zeilinger, *Observation of Three-Particle Entanglement*, Epistemological and Experimental Perspectives on Quantum Physics, Greenberger *et al.* (eds.), Kluwer Academic, Netherlands, 239–243 (1999).

237. M. Zukowski, A. Zeilinger, M. A. Horne, H. Weinfurter, *Independent Photons and Entanglement. A Short Overview*, Int. J. Theor. Phys. **38**, 501–517 (1999).

238. M. Arndt, O. Nairz, G. Van-Der-Zouw, A. Zeilinger, *Towards Coherent Matter Wave Optics with Macromolecules*, Epistemological and Experimental Perspectives on Quantum Physics, Greenberger *et al.* (eds.), Kluwer Academic, Netherlands, 221–224 (1999).

239. P. G. Kwiat, A. G. White, I. Appelbaum, J. R. Mitchell, O. Nairz, G. Weihs, H. Weinfurter, A. Zeilinger, *High-Efficiency Quantum Interrogation Measurements via the Quantum Zeno Effect*, Phys. Rev. Lett. **83**, 4725–4728 (1999).

240. A. Zeilinger, *Three- and Four-Photon Correlations and Entanglement: Quantum Teleportation and Beyond*, Quantum Coherence and Decoherence, Y. A. Ono, K. Fujikawa (eds.), (ISQM Tokyo) Elsevier Science, 19–26 (1999).

241. G. van der Zouw, A. Zeilinger, P. Hoghhoj, R. Gähler, P. Geltenbort, J. Butterworth, *Testing the Proportionality of the Neutron's Gravitational and Intertial Mass*, ILL Annual Report 1999, 62–63 (1999).

242. A. Mair, A. Zeilinger, *Entangled States of Orbital Angular Momentum of Photons*, Epistemological and Experimental Perspectives on Quantum Physics, Greenberger *et al.* (eds.), Kluwer Academic, Netherlands, 249–252 (1999).

243. S. Chiangga, P. Zarda, T. Jennewein, H. Weinfurter, *Towards practical quantum cryptography*, Appl. Phys. B **69**, 389–393 (1999).

244. P. Zarda, S. Chingga, T. Jennewein, H. Weinfurter, *Quantum mechanics and Secret Communications*, Epistemological and Experimental Perspectivs of Quantum Physics. Kluwer Academic, Netherlands, 271–273 (1999).

245. A. Zeilinger, *In retrospect: chosen by Anton Zeilinger. Albert Einstein: Philosopher Scientist*, Nature **398**, 210–211 (1999).

246. Č. Brukner, A. Zeilinger, *Information Content of an Elementary System and the Foundations of Quantum Physics*, World Scientific, Proceedings of the 14th International Conference on Laser Spectroscopy, Innsbruck, Austria, June 1999, (1999).

247. M. Arndt, O. Nairz, J. Voss-Andreae, C. Keller, G. Van der Zouw, A. Zeilinger, *Wave-particle duality of C60 molecules*, Nature **401**, 680–682 (1999).

248. D. Bouwmeester, J.-W. Pan, H. Weinfurter, A. Zeilinger, *Experimental Quantum Teleportation of Qubits and Entanglement Swapping*, Epistemological and Experimental Perspectives on Quantum Physics, Greenberger *et al.* (eds.), Kluwer Academic, Netherlands, 127–140 (1999).

249. C. Simon, G. Weihs, A. Zeilinger, *Optimal quantum cloning and universal NOT without quantum gates*, J. Mod. Opt. **47**, 233–246 (2000).

250. D. Bouwmeester, J.-W. Pan, H. Weinfurter, A. Zeilinger, *High Fidelity Teleportation of Independent Qubits*, J. Mod. Opt. **47**, 279–289 (2000).

251. M. Arndt, A. Zeilinger, *Wo ist die Grenze der Quantenwelt?*, Physikalische Blätter **56**, 69–72 (2000).

252. G. van der Zouw, M. Weber, J. Felber, R. Gähler, P. Geltenbort, A. Zeilinger, *Aharonov-Bohm and gravity experiments with the very-cold-neutron interferometer*, Nuclear Instruments and Methods in Physics Research A **440**, 568–574 (2000).

253. T. Jennewein, Christoph Simon, Gregor Weihs, H. Weinfurter, A. Zeilinger, *Quantum Cryptography with Entangled Photons*, Phys. Rev. Lett. **84**, 4729–4732 (2000).

254. C. Simon, G. Weihs, A. Zeilinger, *Optimal Quantum Cloning via Stimulated Emission*, Phys. Rev. Letters **84**, 2993–2996 (2000).

255. D. Greenberger, M. Horne, A. Zeilinger, *Similarities and Differences Between Two-Particle and Three-Particle Interference*, Fortschr. Phys. **48**, 243–252 (2000).

256. C. Keller, J. Schiedmayer, A. Zeilinger, *Requirements for coherent atom channeling*, Optics Communications **179**, 129–135 (2000).

257. S. Bernet, R. Abfalterer, C. Keller, M. K. Oberthaler, J. Schiedmayer, A. Zeilinger, *Matter waves in time-modulated complex light potentials*, Phys. Rev. A. **62**, 023606-1-20 (2000).

258. Č. Brukner,, A. Zeilinger, *Encoding and Decoding in Complementary Bases with Quantum Gates*, J. Mod. Opt. **47**, 2233–2246 (2000).

259. O. Nairz, M. Arndt, J. Voss-Andreae, C. Keller, G. Van-der-Zouw, A. Zeilinger, *Coherence in C60 and C70 interference*, Fundamentals of Quantum Optics V Innsbruck (Austria), F. Ehlotzky (ed.), 104–109 (2000).

260. H. Weinfurter, D. Bouwmeester, T. Jennewein, J.-W. Pan, G. Weihs, A. Zeilinger, *Quantum Communication and Entanglement*, Proc. 2000 IEEE International Symposium on Circuits and Systems, Ed. by Hasler *et al.*, 2, 346 (2000).

261. M. Daniell, J. W. Pan, G. Weihs, A. Zeilinger, *High Fidelity Entanglement Swapping*, Conference Digest, CLEO'2000, Nice, 7 (2000).

262. M. Daniell, J. W. Pan, A. Zeilinger, *The GHZ-experiment*, Fundamentals of Quantum Optics V Innsbruck (Austria), Springer Verlag, F. Ehlotzky (ed.), 89 (2000).

263. M. Arndt, O. Nairz, J. Voss-Andreae, C. Keller, G. Van-der Zouw, J. W. Pan, M. Daniell, A. Zeilinger, *Superposition and entanglement: experimental frontiers of quantum mechanics*, Fundamentals of Quantum Optics V Innsbruck (Austria), Springer Verlag, F. Ehlotzky (ed.), 50 (2000).

264. J. W. Pan, D. Bouwmeester, M. Daniell, H. Weinfurter, A. Zeilinger, *Three-photon GHZ entanglement and quantum information*, Proc. 15th European Meetings on Cybernetics and Systems Research in Vienna, Austria, Carvallo *et al.* (eds.), 2, 247 (2000).

265. A. Carollo, G. M. Palma, C. Simon, A. Zeilinger, *Tensor-product states and local indistinguishability: an optical linear implementation*, AIP-Conference-Proceedings **513**, 79–82 (2000).

266. J.-W. Pan, D. Bouwmeester, M. Daniell, H. Weinfurter, A. Zeilinger, *Experimental test of quantum nonlocality in three-photon Greenberger-Horne-Zeilinger entanglement*, Nature **403**, 515–519 (2000).

267. A. Zeilinger, *Quantum Teleportation*, Scientific American, 32–41 (2000).

268. T. Jennewein, U. Achleitner, G. Weihs, H. Weinfurter, A. Zeilinger, *A Fast and Compact Quantum Random Number Generator*, Rev. Sci. Instr. **71**, 1675–1680 (2000).

269. O. Nairz, M. Arndt, A. Zeilinger, *Experimental Challenges in Fullerene Interferometry*, J. Mod. Opt. **47**, 2811–2821 (2000).

270. A. Zeilinger, *Quantenexperimente zwischen Photon und Fulleren*, Physik in unserer Zeit **5**, 199–202 (2000).

271. A. Zeilinger, *Quanten-Teleportation*, Spektrum der Wissenschaft, 30–40 (2000).

272. A. Zeilinger, *The quantum jungle revisited. Book review of 'The New World of Mr Tompkins' by G. Gamow and R. Stannard*, Nature **405**, 618 (2000).

273. A. Zeilinger, *Quantum Entangled Bits Step Closer to Information Technology*, Science **289**, 405–406 (2000).

274. C. Simon, M. Zukowski, H. Weinfurter, A. Zeilinger, *A feasible "Kochen-Specker" experiment with single particles*, Phys. Rev. Lett. **85**, 1783–6 (2000).

275. M. Arndt, O. Nairz, J. Petschinka, J. Voss-Andreae, G. v. d. Zouw, C. Keller, A. Zeilinger, *Coherence and Decoherence in de Broglie Interference of Fullerenes*, IQEC 2000, Conference Digest, Nice, 115 (2000).

276. D. Kaszlikowski, P. Gnacinski, M. Zukowski, W. Miklaszewski, A. Zeilinger, *Violations of Local Realism by Two Entangled N-Dimensional Systems Are Stronger than for Two Qubits*, Phys. Rev. Lett. **85**, 4418–4421 (2000).

277. A. Zeilinger, *The Quantum Centennial*, Nature **408**, 639–641 (2000).

278. T. Jennewein, C. Simon, G. Weihs, H. Weinfurter, A. Zeilinger, *Quantum Cryptography with Entangled Photons*, Phys. Rev. Lett. **84**, 4729–4732 (2000).

279. Č. Brukner, A. Zeilinger, *Conceptual Inadequacy of the Shannon Information in Quantum Measurements*, Phys. Rev. A **63**, 022113 1–10 (2001).

280. M. Arndt, O. Nairz, J. Petschinka, A. Zeilinger, *High Contrast Interference with C60 and C70*, C. R. Acad. Sci. Paris, t. 2 Série IV, 581–585 (2001).

281. S. Franke-Arnold, M. Arndt, A. Zeilinger, *Magneto-optical effects with cold Lithium atoms*, J. Phys. B. At. Mol. Opt. Phys. **34**, 2527–2536 (2001).

282. J. W. Pan, C. Simon, Č. Brukner, A. Zeilinger, *Entanglement Purification for Quantum Communication*, Nature **410**, 1067–1070 (2001).

283. O. Nairz, A. Zeilinger, *Matter-wave interference of Fullerenes*, SPIE's International Technical Group Newsletter, Special Issue: Hidden Holography **12**, 5 (2001).

284. J.-W. Pan, M. Daniell, S. Gasparoni, G. Weihs, A. Zeilinger, *Experimental Demonstration of Four-Photon Entanglement and High-Fidelity Teleportation*, Phys. Rev. Lett. **86**, 4435–4438 (2001).

285. C. Simon, Č. Brukner, A. Zeilinger, *Hidden-variable theorems for real experiments*, Phys. Rev. Lett. **86**, 4427 (2001).

286. T. Jennewein, G. Weihs, A. Zeilinger, *Schrödingers Geheimnisse — Absolut sichere Kommunikation durch Quantenkryptographie*, Heise, ct magazin für computer technik **6**, 260–268 (2001).

287. O. Nairz, B. Brezger, M. Arndt, A. Zeilinger, *Diffraction of Complex Molecules by Structures Made of Light*, Phys. Rev. Lett **87**, 160401 (2001).

288. A. Mair, A. Vaziri, G. Weihs, A. Zeilinger, *Entanglement of the Orbital Angular Momentum States of Photons*, Nature **412**, 313–316 (2001).

289. Č. Brukner, M. Zukowski, A. Zeilinger, *The essence of entanglement*, quant-ph 0106119 (2001). Translated into Chinese by Qiang Zhang and Yong-de Zhang, New Advances in Physics (Journal of the Chinese Physical Society).

290. G. Weihs, A. Zeilinger, *Photon statistics at beam-splitters: an essential tool in quantum information and teleportation*, "Coherence and Statistics of Photons and Atoms", J. Perina (ed.), Wiley, (2001).

291. O. Nairz, M. Arndt, A. Zeilinger, *Experimental verification of the Heisenberg uncertainty principle for fullerene molecules*, Phys. Rev. A **65**, 32109 (2002).

292. M. Arndt, O. Nairz, A. Zeilinger, *Interferometry with Macromolecules: Quantum Paradigms Tested in the Mesoscopic World*, Quantum [Un]Speakables. From Bell's Theorem to Quantum Information, R. Bertlmann, A. Zeilinger (eds.), Springer (2002).

293. M. Arndt, O. Nairz, A. Zeilinger, *Wave-Particle Duality*, Year Book of Science & Technology, McGraw-Hill **88** (2002).

294. B. Brezger, L. Hackermüller, St. Uttenthaler, J. Petschinka, M. Arndt, A. Zeilinger, *Matter-Wave Interferometer for Large Molecules*, Phys. Rev. Lett. **88**, 100404 (2002).

295. J. Lawrence, Č. Brukner, A. Zeilinger, *Mutually unbiased binary observable sets on N qubits*, Phys. Rev. A **65**, 32320 (2002).

296. T. Jennewein, G. Weihs, J.-W. Pan, A. Zeilinger, *Experimental Nonlocality Proof of Quantum Teleportation and Entanglement Swapping*, Phys. Rev. Lett. **88**, 17903 (2002).

297. A. Vaziri, G. Weihs, A. Zeilinger, *Superposition of the Orbital Angular Momentum for Applications in Quantum Experiments*, J. Opt. B: Quantum Semiclass **4**, 47–51 (2002).

298. A. Vaziri, G. Weihs, A. Zeilinger, *Experimental Two-Photon Three-Dimensional Quantum Entanglement*, Phys. Rev. Lett. **89**, 240401-1 (2002).

299. Č. Brukner, A. Zeilinger, *Young's experiment and the finiteness of information*, Phil. Trans. R. Soc. Lond. A **360**, 1061–1069 (2002).

300. R. D. Gill, G. Weihs, A. Zeilinger, M. Zukowski, *No time loophole in Bell's theorem: The Hess-Philipp model is nonlocal*, P. Natl. Acad. Sci. USA **99**, 14632–14635 (2002).

301. J.-W. Pan, A. Zeilinger, *Multi-Photon Entanglement and Quantum Non-Locality*, Quantum [Un] Speakables. From Bell's Theorem to Quantum Information, R. Bertlmann, A. Zeilinger (eds.), Springer, 225–240 (2002).

302. A. Zeilinger, *Bell's Theorem, Information and Quantum Physics*, Quantum [Un]Speakables. From Bell's Theorem to Quantum Information, R. Bertlmann, A. Zeilinger (eds.), Springer, 241–254 (2002).

303. Č. Brukner, M. Zukowski, A. Zeilinger, *Quantum communication complexity protocol with two entangled qutrits*, Phys. Rev. Lett. **89**, 197901 (2002).

304. A. Vaziri, G. Weihs, A. Zeilinger, *Superpositions of the orbital angular momentum for applications in quantum experiments*, J. Opt. B: Quantum Semiclass. Opt. **4**, 47–51 (2002).

305. A. C. Elitzur, S. Dolev, A. Zeilinger, *Time-Reversed EPR and the Choice of Histories in Quantum Mechanics*, quant-ph 0205182, (2002).

306. Č. Brukner, A. Zeilinger, *Information and fundamental elements of the structure of quantum theory*, in "Time, Quantum, Information", L. Castell, O. Ischebeck (eds.), Springer (2003).

307. A. Zeilinger, *Why the Quantum? It from Bit? A Participatory Universe? Three Far-Reaching, Visionary Challenges from John Archibald Wheeler and How They Inspired a Quantum Experimentalist*, in "Spiritual Information", Charles L. Harper (ed.), Templeton Foundation Press, 201–220 (2003).

308. J.-W. Pan, S. Gasparoni, M. Aspelmeyer, T. Jennewein, A. Zeilinger, *Experimental realization of freely propagating teleported qubits*, Nature **421**, 721–725 (2003).

309. B. Brezger, M. Arndt, A. Zeilinger, *Concepts of Talbot-Lau Interferometry*, J. Opt. B: Quantum Semiclass. Opt. **5**, 82–89 (2003).

310. J.-W. Pan, S. Gasparoni, R. Ursin, G. Weihs, A. Zeilinger, *Experimental entanglement purification of arbitrary unknown states*, Nature **423**, 417–422 (2003).

311. Č. Brukner, J.-W. Pan, Ch. Simon, G. Weihs, A. Zeilinger, *Probabilistic instantaneous quantum computation*, Phys. Rev. A **67**, 034304 (2003).

312. A. Vaziri, J.-W. Pan, G. Weihs, A. Zeilinger, *Concentration of higher dimensional entanglement: Qutrits of Photon Orbital Angular Momentum*, Phys. Rev. Lett. **91**, 227902 (2003).

313. Z.-B. Chen, J.-W. Pan, Y. -D. Zhang, Č. Brukner, A. Zeilinger, *All-versus-nothing violation of local realism for two entangled photons*, Phys. Rev. Lett. **90**, 160408 (2003).

314. K. Hornberger, S. Uttenthaler, B. Brezger, L. Hackermüller, M. Arndt, A. Zeilinger, *Collisional decoherence observed in matter wave interferometry*, Phys. Rev. Lett. **90**, 160401 (2003).

315. M. Aspelmeyer, H. R. Böhm, T. Gyatso, T. Jennewein, R. Kaltenbaek, M. Lindenthal, G. Molina-Terriza, A. Poppe, K. Resch, M. Taraba, R. Ursin, P. Walther, A. Zeilinger, *Long-Distance Free-Space Distribution of Quantum Entanglement*, Science **301**, 621–623 (2003).

316. L. Hackermüller, K. Hornberger, B. Brezger, A. Zeilinger, M. Arndt, *Decoherence in a Talbot-Lau interferometer: The influence of molecular scattering*, Appl. Phys. B **77**, 781–787 (2003).

317. T. Jennewein, G. Weihs, A. Zeilinger, *Photon Statistics and Quantum Teleportation Experiments*, Proc. Waseda Int. Sympo. On Fundamental Physics — New Perspectives in Quantum Physics — J. Phys. Soc. Jpn. **72**, 168–173 (2003).

318. L. Hackermüller, St. Uttenthaler, K. Hornberger, E. Reiger, B. Brezger, A. Zeilinger, M. Arndt, *Wave Nature of Biomolecules and Fluorofullerenes*, Phys. Rev. Lett. **91**, 090408 (2003).

319. O. Nairz, M. Arndt, A. Zeilinger, *Quantum Interference Experiments with Large Molecules*, American Journal of Physics **71**, 319 (2003).

320. Č. Brukner, A. Zeilinger, *Erratum: Conceptual inadequacy of the Shannon information in quantum measurements [Phys. Rev. A 63, 022113 (2001)]*, Phys. Rev. A **67**, 049901(E) (2003).

321. Č. Brukner, M. S. Kim, J.-W. Pan, A. Zeilinger, *Correspondence between continuous-variable and discrete quantum systems of arbitrary dimensions*, Phys. Rev. A **68**, 062105 (2003).

322. R. Kaltenbaek, M. Aspelmeyer, T. Jennewein, Č. Brukner, M. Pfennigbauer, W. R. Leeb, A. Zeilinger, *Proof-of-Concept Experiments for Quantum Physics in Space*, Proceedings of SPIE, Quantum Communications and Quantum Imaging, R. Meyers, Y. Shih (eds.), 252–268 (2003).

323. M. Aspelmeyer, T. Jennewein, M. Pfennigbauer, W. Leeb, A. Zeilinger, *Long-Distance Quantum Communication with Entangled Photons using Satellites*, IEEE Journal of Selected Topics in Quantum Electronics, special issue on "Quantum Internet Technologies" (2003).

324. M. Pfennigbauer, W. Leeb, M. Aspelmeyer, T. Jennewein, A. Zeilinger, *Free-Space Optical Quantum Key Distribution Using Intersatellite Links*, Proccedings of the CNES — Intersatellite Link Workshop (2003).

325. M. Arndt, A. Zeilinger, *Wave-particle experiments with large molecules*, J. S. Al-Khalili, "Quantum: A guide for the perpelexed", Weidenfeld & Nicolson, (2003).

326. A. Zeilinger, *Quantum Teleportation (updated version of the 2001 contribution)*, Scientific American Collection The Edge of Physics (2003).

327. T. Jennewein, G. Weihs, J.-W. Pan, A. Zeilinger, *Reply to Ryff's Comment on "Experimental Nonlocality Proof of Quantum Teleportation and Entanglement Swapping"*, quant-ph/0303104 (2003).

328. B. Brezger, M. Arndt, A. Zeilinger, *Concepts for near-field interferometers with large molecules*, Journal of Optics B: Quantum and Semiclassical Optics **5**, 82–89 (2003).

329. R. D. Gill, G. Weihs, A. Zeilinger, M. Zukowski, *Comment on "Exclusion of time in the theorem of Bell" by K. Hess and W. Philipp*, Europhys. Lett. **61**, 282–283 (2003).

330. L. Hackermüller, K. Hornberger, B. Brezger, A. Zeilinger, M. Arndt, *Decoherence of matter waves by thermal emission of radiation*, Nature **427**, 711–714 (2004).

331. K. Sanaka, T. Jennewein, J.-W. Pan, K. Resch, A. Zeilinger, *Experimental Nonlinear Sign Shift for Linear Optics Quantum Computation*, Phys. Rev. Lett. **92**, 017902 (2004).

332. Č. Brukner, M. Zukowski, J.-W. Pan, A. Zeilinger, *Bell's Inequalities and Quantum Communication Complexity*, Phys. Rev. Lett. **92**, 127901 (2004).

333. G. Molina-Terriza, A. Vaziri, J. Rehacek, Z. Hradil, A. Zeilinger, *Triggered Qutrits for Quantum Communication Protocols*, Phys. Rev. Lett. **92**, 168903 (2004).

334. A. Poppe, A. Fedrizzi, R. Ursin, H. R. Böhm, T. Lörunser, O. Maurhardt, M. Peev, M. Suda, C. Kurtsiefer, H. Weinfurter, T. Jennewein, A. Zeilinger, *Practical quantum key distribution with polarization entangled photons*, Opt. Express **12**, 3865–3871 (2004).

335. Sara Gasparoni, Jian-Wei Pan, Philip Walther, Terry Rudolph, Anton Zeilinger, *Realization of a photonic CNOT gate sufficient for quantum computation*, Phys. Rev. Lett. **93**, 020504 (2004).

336. R. Ursin, T. Jennewein, M. Aspelmeyer, R. Kaltenbaek, M. Lindenthal, P. Walther, A. Zeilinger, *Quantum teleportation across the Danube*, Nature **430**, 849 (2004).

337. M. Aspelmeyer, Č. Brukner, A. Zeilinger, *Entangled photons and quantum communication*, D. Estève, J. -M. Raimand, J. Dalibard, eds, Proceedings of the Les Houches Summer School 2003 (Les Houches, Volume Session LX), D. Estève, J. -M- Raimond, J. Dalibard (eds.), Elsevier Science, 335–353, (2004).

338. P. Walther, J.-W. Pan, M. Aspelmeyer, R. Ursin, S. Gasparoni, A. Zeilinger, *De Broglie Wavelength of a Nonlocal Four-Photon State*, Nature **429**, 158–161 (2004).

339. M. Arndt, L. Hackermüller, K. Hornberger, A. Zeilinger, *Organic molecules and decoherence experiments in a molecule interferometer*, Proceedings of the Workshop on „Multiscale Methods in Quantum Mechanics", Birkhäuser, Boston (2004).

340. A. Zeilinger und A. Zeilinger, *Gesetze der Natur — Natur der Gesetze*, Studien zur Politik und Verwaltung, hrsg. Von Christian Brünner, Wolfgang Mantl, Manfried Welan, Band 90/I, Böhlau Verlag Wien 1217–1222, (2004).

341. T. Jennewein, A. Zeilinger, *Quantum Noise and Quantum Communication*, Fluctuations and Noise in Photonics and Quantum Optics II, edited by Peter Heszler, Derek Abbot, Julio R. Gea-Banacloche, Philip R. Hemmer, Proceedings of SPIE **5468** (SPIE, Bellingham, WA, 2004), 1–9 (2004).

342. M. Aspelmeyer, P. Walther, T. Jennewein, A. Zeilinger, *Nonlocal photon number states for quantum metrology*, Proc. of SPIE Vol. **5551**, 15–20 (2004), Quantum Communications and Quantum Imaging II, R. Meyers, Y. Shih (eds.).

343. Hannes R. Böhm, Paul S. Böhm, Markus Aspelmeyer, Caslav Brukner, Anton Zeilinger, *Exploiting the randomness of the measurement basis in quantum cryptography: Secure Quantum Key Growing without Privacy Amplification*, quant-ph/0408179 (2004).

344. P. Villoresi, F. Tamburini, M. Aspelmeyer, R. Ursin, C. Pachello, G. Bianco, C. Barbieri, T. Jennewein, A. Zeilinger, *Space-to-ground quantum-communication using an optical ground station: A feasibility study*, Proc. of SPIE Vol. **5551**, 113–120 (2004), Quantum Communications and Quantum Imaging II, R. Meyers, Y. Shih (eds.).

345. A. Zeilinger, G. Weihs, T. Jennewein, M. Aspelmeyer, *Happy Centenary, Photon*, Nature **433**, 230–238 (2005).

346. A. Zeilinger, *Time Travel*, "New Scientist's Book of 100 Things to Do Before You Die" (2005).

347. P. Walther, K. J. Resch, T. Rudolph, E. Schenck, H. Weinfurter, V. Vedral, M. Aspelmeyer, A. Zeilinger, *Experimental One-Way Quantum Computing*, Nature **434**, 169–176 (2005).

348. K. J. Resch, M. Lindenthal, B. Blauensteiner, H. R. Böhm, A. Fedrizzi, C. Kurtsiefer, A. Poppe, T. Schmitt-Manderbach, M. Taraba, R. Ursin, P. Walther, H. Weier, H. Weinfurter, A. Zeilinger, *Distributing entanglement and single photons through an intra-city, free-space quantum channel*, Opt. Express **13**, 203 (2005).

349. G. Molina-Terriza, A. Vaziri, R. Ursin, A. Zeilinger, *Experimental Quantum Coin Tossing*, Phys. Rev. Lett. **94**, 40501 (2004).

350. M. Arndt, K. Hornberger, A. Zeilinger, *Probing the limits of the quantum world*, Physics World, March 2005, 35 (2005).

351. T. Jennewein, Č. Brukner, M. Aspelmeyer, A. Zeilinger, *Experimental Proposal of Switched "Delayed-Choice" for Entanglement Swapping*, Int. J. Quantum Inf. **3**, 73–79 (2005).

352. M. Peev, M. Nölle, O. Maurhardt, T. Lorünser, M. Suda, A. Poppe, R. Ursin, A. Fedrizzi, A. Zeilinger, *A Novel Protocol-Authentication Algorithm Ruling Out a Man-in-the Middle Attack in Quantum Cryptography*, Int. J. Quantum Inf. **3**, 225 (2005).

353. K. Resch, P. Walther, A. Zeilinger, *Full characterization of a three-photon GHZ state using quantum state tomography*, Physical Review Letters **94**, 070402 (2005).

354. P. Walther, K. Resch, Č. Brukner, A. Steinberg, J.-W. Pan, A. Zeilinger, *Quantum nonlocality obtained from local states by entanglement purification*, Physical Review Letters **94**, 040504 (2005).

355. M. Arndt, L. Hackermüller, K. Hornberger, A. Zeilinger, *Coherence and Decoherence Experiments with Fullerenes, in:* "Decoherence, Entanglement and Information Protection in Complex Quantum Systems", Springer, 2005.

356. P. Walther, K. J. Resch, A. Zeilinger, *Local conversion of Greenberger-Horne-Zeilinger states to approximate W states*, Physical Review Letters **94**, 240501 (2005).

357. P. Walther, M. Aspelmeyer, K. J. Resch, A. Zeilinger, *Experimental violation of a cluster state Bell inequality*, Physical Review Letters **95**, 020403 (2005).

358. P. Walther, A. Zeilinger, *Decoherence, Entanglement and Information Protection in Complex Quantum Systems*, in: "Quantum Logics Based on Four Photon Entanglement", Springer, 2005.

359. M. Aspelmeyer, T. Jennewein, G. Weihs, A. Zeilinger, *Physik der Photonen*, Spektrum der Wissenschaft **1** (2005).

360. M. Pfennigbauer, M. Aspelmeyer, W. R. Leeb, G. Badurek, G. Baister, T. Dreischer, T. Jennewein, G. Baister, G. Neckamm, J. M. Perdigues, J. Summhammer, H. Weinfurter, A. Zeilinger, *Satellite-based quantum communication terminal employing state-of-the-art technology*, JON **4**, 549–560 (2005).

361. P. Walther, A. Zeilinger, *Experimental realization of a photonic Bell-state analyzer*, Physical Review A **72**, 010302(R) (2005).

362. Č. Brukner, M. Aspelmeyer, A. Zeilinger, *Complementarity and Information in "Delayed-Choice for Entanglement Swapping"*, Foundations of Physics **35**, 1909–1919 (2005).

363. A. Stibor, K. Hornberger, L. Hackermüller, A. Zeilinger, M. Arndt, *Talbot-Lau interferometry with fullerenes: Sensitivity to inertial forces and vibrational dephasing*, Laser Physics **15**, 10–17 (2005).

364. A. Zeilinger, *Verschränkung - ein Quantenrätsel für jedermann*, : in: "Aus den Elfenbeintürmen der Wissenschaft. 1. XLAB Science Festival." Wallstein Verlag, Göttingen (2005).

365. Č. Brukner, A. Zeilinger, *Quantum Physics as a Science of Information*, in: "Quo Vadis Quantum Mechanics?", edited by A. Elitzur, S. Dolev, N. Kolenda, (Springer, 2005).

366. K. Sanaka, K. J. Resch, A. Zeilinger, *Filtering Out Photonic Fock States*, Physical Review Letters **96**, 083601-1-4 (2006).

367. Č. Brukner, V. Vedral, A. Zeilinger, *Crucial Role of Quantum Entanglement in Bulk Properties of Solids*, Phys. Rev. A **73**, 012110 (2006).

368. S. Gröblacher, T. Jennewein, A. Vaziri, G. Weihs, A. Zeilinger, *Experimental Quantum Cryptography with Qutrits*, New J. Phys. **8** (2006).

369. A. Zeilinger, *Essential quantum entanglement*, in: "The New Physics" (2006).

370. M. Aspelmeyer, H. R. Böhm, A. Fedrizzi, S. Gasparoni, T. D. Jennewein, M. Lindenthal, G. Molina-Terriza, A. Poppe, K. Resch, R. Ursin, P. Walther, A. Zeilinger, *Advanced Quantum Communications Experiments with Entangled Photons*, in "Quantum Communications and Cryptography", Editor(s): Alexander V. Sergienko, Taylor and Francis Group, LLC, 45–81 (2006).

371. P. Walther, K. J. Resch, Č. Brukner, A. Zeilinger, Experimental Entangled Entanglement, Physical Review Letters **97**, 020501 (2006).

372. R. Kaltenbaek, B. Blauensteiner, M. Zukowski, M. Aspelmeyer, A. Zeilinger, *Experimental interference of independent photons*, Phys. Rev. Lett. **96**, 240502 (2006).

373. A. Zeilinger, *Quantum Communication and Quantum Computation with Entangled Photons*, in: "Foundations of Quantum Mechanics in the Light of New Technology: ISQM Tokyo '05, Proceedings of the 8th International Symposium, Hatoyama, Saitama, Editor(s): Ishioka S., Fujikawa K. , World Scientific Publishing Co. Pte. Ltd. , 24–28 (2006).

374. C. Bonato, M. Aspelmeyer, T. Jennewein, C. Pernechele, P. Villoresi, A. Zeilinger, *Influence of satellite motion on polarization qubits in a Space-Earth quantum communication link*, Opt. Express **14**, 10050–10059 (2006).

375. H. R. Böhm, S. Gigan, G. Langer, J. Hertzberg, F. Blaser, D. Bäuerle, K. Schwab, A. Zeilinger, M. Aspelmeyer, *A high reflectivity high-Q micromechanical Bragg mirror*, Applied Physics Letters **89**, 223101 (2006).

376. A. Zeilinger, J. Kofler, *La dissolution du paradoxe*, Sciences et Avenir Hors-Série, No. 148, p. 54 (Oct. /Nov. 2006).

377. P. Walther, A. Zeilinger, *Quantum Entanglement, Purification, and linear-optics quantum gates with photonic qubits*, QP-PQ Quantum Probability and White Noise Analysis, Vol. 19, Quantum Information and Computing, L. Accardi, M. Ohya, N. Watanabe (eds.), World Scientific, 360–369 (2006).

378. S. Gigan, H. R. Böhm, M. Paternostro, F. Blaser, G. Langer, J. B. Hertzberg, K. C. Schwab, D. Bäuerle, M. Aspelmeyer, A. Zeilinger, *Self-cooling of a micromirror by radiation pressure*, Nature **444**, 67–70 (2006).

379. M. S. Tame, R. Prevedel, M. Paternostro, P. Böhi, M. S. Kim, A. Zeilinger, *Experimental Realization of Deutsch's Algorithm in a One-way Quantum Computer*, Phys. Rev. Lett. **98**, 140501 (2007).

380. T. Schmitt-Manderbach, H. Weier, M. Fürst, R. Ursin, F. Tiefenbacher, T. Scheidl, J. Perdigues, Z. Sodnik, C. Kurtsiefer, J. G. Rarity, A. Zeilinger, H. Weinfurter, *Experimental Demonstration of Free-Space Decoy-State Quantum Key Distribution over 144 km*, Phys. Rev. Lett. **98**, 010504 (2007).

381. D. Vitali, S. Gigan, A. Ferreira, H. R. Böhm, P. Tombesi, A. Guerreiro, V. Vedral, A. Zeilinger, M. Aspelmeyer, *Optomechanical entanglement between a movable mirror and a cavity field*, Phys. Rev. Lett. **98**, 030405 (2007).

382. R. Prevedel, M. Aspelmeyer, Č. Brukner, T. Jennewein, A. Zeilinger: *Photonic Entanglement as a Resource in Quantum Information Processing*, J. Opt. Soc. Am. B **24**, 241–248 (2007). [also: selected for the Virtual Journal of Quantum Information (Issue 2, Vol. 7, Feb 2007)].

383. P. Walther, M. Aspelmeyer, A. Zeilinger, *Heralded generation of multi-photon entanglement*, Phys. Rev. A **75**, 12313 (2007).

384. R. Prevedel, P. Walther, F. Tiefenbacher, P. Böhi, R. Kaltenbaek, T. Jennewein, A. Zeilinger, *High-speed linear optics quantum computing using active feed-forward*, Nature **445**, 65–69 (2007).

385. S. Gröblacher, T. Paterek, R. Kaltenbaek, C. Brukner, M. Zukowski, M. Aspelmeyer, A. Zeilinger, *An experimental test of non-local realism*, Nature **446**, 871–875 (2007).

386. R. Ursin, F. Tiefenbacher, T. Schmitt-Manderbach, H. Weier, T. Scheidl, M. Lindenthal, B. Blauensteiner, T. Jennewein, J. Perdigues, P. Trojek, B. Ömer, M. Fürst, M. Meyenburg, J. Rarity, Z. Sodnik, C. Barbieri, H. Weinfurter, A. Zeilinger, *Entanglement-based quantum communication over 144 km*, Nat. Phys. **3**, 481–486 (2007); *Nature* Highlight of the Year 2007.

387. R. Prevedel, A. Stefanov, P. Walther, A. Zeilinger, *Experimental realization of a quantum game on a one-way quantum computer*, New J. Phys. **9**, 205 (2007).

388. M. Stütz, S. Gröblacher, T. Jennewein, A. Zeilinger, *How to create and detect N-dimensional entangled photons with an active phase hologram*, Appl. Phys. Lett. **90**, 261114 (2007).

389. H. Hübel, M. R. Vanner, T. Lederer, B. Blauensteiner, T. Lorünser, A. Poppe, A. Zeilinger, *High-fidelity transmission of polarization encoded qubits from an entangled source over 100 km of fiber*, Opt. Express **15**, 7853–7862 (2007).

390. R. Ursin, F. Tiefenbacher, T. Jennewein, A. Zeilinger, *Applications of quantum communication protocols in real world scenarios towards space*, Elektrotechnik & Informationstechnik **5**, 149–153 (2007).

391. T. Paterek, A. Fedrizzi, S. Gröblacher, T. Jennewein, M. Zukowski, M. Aspelmeyer, A. Zeilinger, *Experimental test of non-local realistic theories without the rotational symmetry assumption*, Phys. Rev. Lett. **99**, 210406 (2007).

392. A. Fedrizzi, T. Herbst, A. Poppe, T. Jennewein, A. Zeilinger, *A wavelength-tunable fiber-coupled source of narrowband entangled photons*, Opt. Express **15**, 15377–15386 (2007). arXiv:0706.2877 [quant-ph]

393. J. M. Perdigues Armengol, B. Furch, C. J. de Matos, O. Minster, L. Cacciapuoti, M. Pfennigbauer, M. Aspelmeyer, T. Jennewein, R. Ursin, T. Schmitt-Manderbach, G. Baister, J. Rarity, W. Leeb, C. Barbieri, H. Weinfurter, A. Zeilinger, *Quantum Communications at ESA: Towards a Space Experiment on the ISS*, Proceedings of the 58th International Astronautical Congress, Hyderabad, India, 24–28 September 2007, IAF/IAA (2007).

394. A. Zeilinger, *Long-Distance Quantum Cryptography with Entangled Photons*, in: "Quantum Communications Realized", Editors: Yasuhiko Arakawa, Masahide Sasaki, Hideyuki Sotobayashi, Proceedings of SPIE **6780**, 67800B (2007).

395. A. Zeilinger, *Der Zufall als Notwendigkeit für eine offene Welt*, in: "Der Zufall als Notwendigkeit", by: Anton Zeilinger, Helmut Leder, Elisabeth Lichtenberger, Jürgen Mittelstraß and others, Beiträge aus unterschiedlichen Disziplinen über Einfluss und Wirkung des Zufalls, No. 132, 19–24 (Picus, Vienna, June 2007).

396. R. Prevedel, M. S Tame, A. Stefanov, M. Paternostro, M. S. Kim, A. Zeilinger, *Experimental demonstration of decoherence-free one-way quantum processing*, Phys. Rev. Lett. **99**, 250503 (2007). arXiv:0708.0960 [quant-ph]

397. A. Zeilinger, Foreword to: "Christian Doppler. Life and Work - Principle and Applications", Editors: E. Hiebl, M. Musso, p. 9 (Living Edition, Pöllauberg, 2007).

398. A. Zeilinger, *Quantum teleportation*, McGraw-Hill Encyclopedia of Science & Technology, Vol. 14, pp. 705–706, 10th edition, McGraw-Hill (2007).

399. A. Zeilinger, *Von Einstein zum Quantencomputer. Wirklichkeit und Information in der Quantenwelt*, in: "Vom Urknall zum Bewusstsein — Selbstorganisation der Materie", Editors: K. Sandhoff, W. Donner *et al.* , Verhandlungen der Gesellschaft Deutscher Naturforscher und Ärzte, No. 124, 33–35 (Thieme, Stuttgart, 2007).

400. P. Walther, M. D. Eisaman, A. Nemiroski, A. V. Gorshkov, A. S. Zibrov, A. Zeilinger, M. D. Lukin, *Multi-photon entanglement: From quantum curiosity to quantum computing and quantum repeaters*, Proc. of SPIE, Vol. 6664 (2007).

401. B. Sanders, Y. Yamamoto, A. Zeilinger, *Optical Quantum Information Science. Introduction*, J. Opt. Soc. Am. B **24**, 162–162 (2007).

402. R. Prevedel, A. Zeilinger, *Entanglement and One-Way Quantum Computing*, www. 2physics. com (2007).

403. A. Zeilinger, *Split world* (Book Review of 'Decoherence and the Quantum-to- Classical Transition' by Maximilian Schlosshauer), Nature **451,** 18 (2008).

404. R. Ursin, T. Jennewein, J. Kofler, J. M. Perdigues, L. Cacciapuoti, C. d. Matos, M. Aspelmeyer, A. Valencia, T. Scheidl, A. Acin, C. Barbieri, G. Bianco, S. Cova, D. Giggenbach, W. Leeb, R. H. Hadfield, R. Laflamme, N. Lütkenhaus, G. Milburn, M. Peev, T. Ralph, J. G. Rarity, R. Renner, N. Solomos, W. Tittel, J. P. Torres, M. Toyoshima, P. Villoresi, I. Walmsley, G. Weihs, H. Weinfurter, M. Zukowski, A. Zeilinger, *Space-QUEST. Experiments with quantum entanglement in space*, Proceedings of the 2008 Microgravity Sciences and Process Symposium, arXiv:0806.0945 [quant-ph] (2008). Europhysics News Vol. 40, No. 3, 2009, pp. 26–29, DOI: 10. 1051/ epn/2009503.

405. P. Villoresi, T. Jennewein, F. Tamburini, M. Aspelmeyer, C. Bonato, R. Ursin, C. Pernechele, V. Luceri, G. Bianco, A. Zeilinger, C. Barbieri, *Experimental verification of the feasibility of a quantum channel between space and Earth*, New J. Phys. **10**, 033038 (2008), Highlight of New J. Phys. for 2008.

406. A. Zeilinger, *Wozu Wissenschaft heute?*, Wissenschaftsbericht der Stadt Wien 2007, City of Vienna (2008).

407. D. Abbott, J. Gea-Banacloche, P. W. Davies, S. Hameroff, A. Zeilinger, J. Eisert, H. Wiseman, S. M. Bezrukov, H. Frauenfelder, *Plenary Debate: Quantum Effects in Biology: Trivial or Not?*, World Scientific, Fluctuation and Noise Letters **8**, C5–C26 (2008).

408. A. Zeilinger, *Quantum Computation and Quantum Communication with Entangled Photons*, in: "Coherence and Quantum Optics IX", Editors: N. P. Bigelow, J. H. Eberly, C. R. Stroud, Jr., 299–300 (Optical Society of America, Washington, DC, 2008).

409. J. Armengol, B. Furch, C. J. d. Matos, O. Minster, L. Cacciapuoti, M. Pfennigbauer, M. Aspelmeyer, T. Jennewein, R. Ursin, T. Schmitt-Manderbach, G. Baister, J. G. Rarity, W. Leeb, C. Barbieri, H. Weinfurter, A. Zeilinger, *Quantum communications at ESA: Towards a space experiment on the ISS*, Acta Astronautica **63**, 165–78 (2008).

410. M. Aspelmeyer, A. Zeilinger, *A quantum renaissance*, Physics World, July 2008, 22 (2008).

411. A. Zeilinger, *Die Wirklichkeit der Quanten*, Spektrum der Wissenschaft (November 2008), 54–63.

412. A. Zeilinger, *On the Interpretation and Philosophical Foundation of Quantum Mechanics*, in: "Grenzen menschlicher Existenz", Ed. H. Daub, 184–201, Michael Imhof Verlag (2008).

413. S. Gröblacher, S. Gigan, H. R. Böhm, A. Zeilinger, M. Aspelmeyer, *Radiation-pressure self-cooling of a micromirror in a cryogenic environment*, Europhys. Lett. **81**, 54003 (2008).

414. J. Kofler, R. Ursin, C. Brukner, A. Zeilinger, *Comment on: Testing the speed of 'spooky action at a distance'*, arXiv:0810.4452 [quant-ph] (2008).

415. R. Kaltenbaek, R. Prevedel, M. Aspelmeyer, A. Zeilinger, *High-fidelity entanglement swapping with fully independent sources*, Phys. Rev. A **79**, 040302(R) (2009).

416. P. Villoresi, R. Ursin, A. Zeilinger, *Single photons from a satellite: Quantum communication in space*, SPIE Newsroom (2009), DOI: 10. 1117/2. 1200902. 1398.

417. C. Schmid, N. Kiesel, U. Weber, R. Ursin, A. Zeilinger, H. Weinfurter, *Quantum teleportation and entanglement swapping with linear optics logic gates*, New J. Phys. **11**, 033008 (2009).

418. A. Treiber, A. Poppe, M. Hentschel, D. Ferrini, T. Lorünser, E. Querasser, T. Matyus, H. Hübel, A. Zeilinger, *A fully automated entanglement-based quantum cryptography system for telecom fiber networks*, New J. Phys. **11**, 045013 (2009).

419. A. Fedrizzi, T. Herbst, M. Aspelmeyer, M. Barbieri, T. Jennewein, A. Zeilinger, *Anti-symmetrization reveals hidden entanglement*, New J. Phys. **11**, 103052 (2009).

420. A. Fedrizzi, R. Ursin, T. Herbst, M. Nespoli, R. Prevedel, T. Scheidl, F. Tiefenbacher, T. Jennewein, A. Zeilinger, *High-fidelity transmission of entanglement over a high-loss free-space channel*, Nat. Phys. **5**, 389–392 (2009).

421. A. Fedrizzi, R. Ursin, A. Zeilinger, *Transmission of Entangled Photons over a High-Loss Free-Space Channel*, www. 2Physics. com (May 30, 2009).

422. T. Jennewein, R. Ursin, M. Aspelmeyer, A. Zeilinger, *Performing high-quality multi-photon experiments with parametric down-conversion*, J. Phys. B: At. Mol. Opt. Phys. **42**, 114008 (2009).

423. M. Peev, C. Pacher, R. Alléaume, C. Barreiro, J. Bouda, W. Boxleitner, T. Debuisschert, E. Diamanti, M. Dianati, J. F. Dynes, S. Fasel, S. Fossier, M. Fuerst, J. D. Gautier, O. Gay, N. Gisin, P. Grangier, A. Happe, Y. Hasani, M. Hentschel, H. Hübel, G. Humer, T. Laenger, M. Legré, R. Lieger, J. Lodewyck, T. Lorünser, N. Lütkenhaus, A. Marhold, T. Matyus, O. Maurhardt, L. Monat, S. Nauerth, J. B. Page, A. Poppe, E. Querasser, G. Ribordy, S. Robyr, L. Salvail,

A. W. Sharpe, A. J. Shields, D. Stucki, M. Suda, C. Tamas, T. Themel, R. T. Thew, Y. Thoma, A. Treiber, P. Trinkler, R. Tualle-Brouri, F. Vannel, N. Walenta, H. Weier, H. Weinfurter, I. Wimberger, Z. L. Yuan, H. Zbinden, A. Zeilinger, *The SECOQC quantum key distribution network in Vienna*, New J. Phys. **11**, 075001 (2009).

424. C. Brukner, A. Zeilinger, *Information Invariance and Quantum Probabilities*, Foundations of Physics **39**, 677–689 (2009). arXiv:0905.0653 [quant-ph]

425. T. Scheidl, R. Ursin, A. Fedrizzi, S. Ramelow, X. S. Ma, T. Herbst, R. Prevedel, L. Ratschbacher, J. Kofler, T. Jennewein, A. Zeilinger, *Feasibility of 300 km quantum key distribution with entangled states*, New J. Phys. **11**, 085002 (2009). arXiv:1007.4645 [quant-ph]

426. R. Prevedel, G. Cronenberg, M. S. Tame, M. Paternostro, P. Walther, M. S. Kim, A. Zeilinger, *Experimental Realization of Dicke States of up to Six Qubits for Multiparty Quantum Networking*, Phys. Rev. Lett. **103**, 020503 (2009).

427. M. Arndt, M. Aspelmeyer, A. Zeilinger, *How to extend quantum experiments*, Fortschritte Phys. 1–10 (2009).

428. X. Ma, A. Quarry, J. Kofler, T. Jennewein, A. Zeilinger, *Experimental violation of a Bell inequality with two different degrees of freedom of entangled particle pairs*, Phys. Rev. A **79**, 042101 (2009).

429. M. Peev, C. Pacher, T. Lorünser, M. Noelle, A. Poppe, A. Zeilinger, *Vulnerability of "a novel protocol-authentication ruling out a man-in-the middle attack in quantum cryptography"*, Int. J. Quantum Inf. **7**, 1401–1407 (2009).

430. M. Hentschel, H. Hübel, A. Poppe, A. Zeilinger, *Three-color Sagnac source of polarization-entangled photon pairs*, Opt. Express **17**, 23153–23159 (2009).

431. S. Ramelow, L. Ratschbacher, A. Fedrizzi, N. K. Langford, A. Zeilinger, *Discrete, Tunable Color Entanglement*, Phys. Rev. Lett. **103**, 25360 (2009).

432. T. Paterek, J. Kofler, R. Prevedel, P. Klimek, M. Aspelmeyer, A. Zeilinger, Č. Brukner, *Logical independence and quantum randomness*, New J. Phys. **12**, 013019 (2010). This article was chosen for IOP select. Media coverage: Highlight in Europhys. News 41/2, 10 (2010).

433. S. Barz, G. Cronenberg, A. Zeilinger, P. Walther, *Heralded generation of entangled photon pairs*, Nat. Photonics **4,** 553–556 (2010).

434. T. Scheidl, R. Ursin, J. Kofler, S. Ramelow, X. Ma, T. Herbst, L. Ratschbacher, A. Fedrizzi, N. K. Langford, T. Jennewein, A. Zeilinger, *Violation of local realism with freedom of choice*, P. Natl. Acad. Sci. USA **104**, 19708–19713 (2010).

435. J. Kofler, A. Zeilinger, *Quantum Information and Randomness*, Eur. Rev. **18**, 469–480 (2010).

436. A. Zeilinger, M. Aspelmeyer, *Une incroyable illusion de réalité*, La Recherche **38** (2010).

437. X.-S. Ma, S. Zotter, J. Kofler, T. Jennewein, A. Zeilinger, *Experimental generation of single photons via active multiplexing*, Phys. Rev. A **83**, 043814 (2011).

438. M. Wiesniak, T. Paterek, A. Zeilinger, *Entanglement in mutually unbiased bases*, New J. Phys. **13**, 053047 (2011).

439. X.-S. Ma, B. Dakic, W. Naylor, A. Zeilinger, P. Walther, *Quantum simulation of the wavefunction to probe frustrated Heisenberg spin systems*, Nat. Phys. **7**, 399–405 (2011).

440. R. Lapkiewicz, P. Li, C. Schaeff, N. K. Langford, S. Ramelow, M. Wiesniak, A. Zeilinger, *Experimental non-classicality of an indivisible quantum system*, Nature **474**, 490–493 (2011).

441. M. Sasaki, M. Fujiwara, H. Ishizuka, W. Klaus, K. Wakui, M. Takeoka, S. Miki, T. Yamashita, Z. Wang, A. Tanaka, K. Yoshino, Y. Nambu, S. Takahashi, A. Tajima, A. Tomita, T. Domeki, T. Hasegawa, Y. Sakai, H. Kobayashi, T. Asai, K. Shimizu, T. Tokura, T. Tsurumaru, M. Matsui, T. Honjo, K. Tamaki, H. Takesue, Y. Tokura, J. F. Dynes, A. R. Dixon, A. W. Sharpe, Z. L. Yuan, A. J. Shields, S. Uchikoga, M. Legré, S. Robyr, P. Trinkler, L. Monat, J. -B. Page, G. Ribordy, A. Poppe, A. Allacher, O. Maurhart, T. Länger, M. Peev, A. Zeilinger, *Field test of quantum key distribution in the Tokyo QKD Network*, Opt. Express **19**, 10387–10409 (2011).

442. N. K. Langford, S. Ramelow, R. Prevedel, W. J. Munro, G. J. Milburn, A. Zeilinger, *Efficient quantum computing using coherent photon conversion*, Nature **478**, 360–363 (2011).

443. T. Fujii , S. Matsuo, N. Hatakenaka, S. Kurihara, A. Zeilinger, *Quantum circuit analog of the dynamical Casimir effect*, Phys. Rev. B **84**, 174521 (2011).

444. X. -S. Ma, S. Zotter, N. Tetik, A. Qarry, T. Jennewein, A. Zeilinger, *A high-speed tunable beam splitter for feed-forward photonic quantum information processing*, Opt. Express **19**, 22723–22730 (2011).

445. S. Barz, E. Kashefi, A. Broadbent, J. Fitzsimons, A. Zeilinger, P. Walther, *Demonstration of Blind Quantum Computing*, Science **335**, 303–307 (2012).

446. S. Ramelow, A. Fedrizzi, A. Poppe, N. K. Langford, A. Zeilinger, *Polarization-entanglement-conserving frequency conversion of photons*, Phys. Rev. A **85**, 013845 (2012).

447. M. Aspelmeyer, Č. Brukner, A. Zeilinger, *Festschrift Dedicated to Daniel Greenberger and Helmut Rauch – Editorial*, Found. Phys. **42**, 1–3 (2012).

448. X.-S. Ma, S. Zotter, J. Kofler, R. Ursin, T. Jennewein, C. Brukner, A. Zeilinger, *Experimental delayed-choice entanglement swapping*, Nat. Phys. **8**, 480–485 (2012).

449. C. Schaeff, R. Polster, R. Lapkiewicz, R. Fickler, S. Ramelow, A. Zeilinger, *Scalable fiber integrated source for higher-dimensional path-entangled photonic quNits*, Opt. Express **20**, 16145–16153 (2012).

450. M. A. Hohensee, B. Estey, P. Hamilton, A. Zeilinger, H. Mueller, *Force-Free Gravitational Redshift: Proposed Gravitational Aharonov-Bohm Experiment*, Phys. Rev. Lett. **108**, 230404 (2012).

451. B. Wittmann, S. Ramelow, F. Steinlechner, N. K. Langford, N. Brunner, H. M. Wiseman, R. Ursin, A. Zeilinger, *Loophole-free Einstein-Podolsky-Rosen experiment via quantum steering*, New J. Phys. **14**, 053030 (2012).

452. J.-W. Pan, Z.-B. Chen, C. -Y. Lu, H. Weinfurter, A. Zeilinger, M. Zukowski, *Multiphoton entanglement and interferometry*, Rev. Mod. Phys. **84**, 777–838 (2012).

453. J. Kofler, M. Singh, M. Ebner, M. Keller, M. Kotyrba, A. Zeilinger, *Einstein-Podolsky-Rosen correlations from colliding Bose-Einstein condensates*, Phys. Rev. A **86**, 032115 (2012).

454. X.-S. Ma, T. Herbst, T. Scheidl, D. Wang, S. Kropatschek, W. Naylor, B. Wittmann, A. Mech, J. Kofler, E. Anisimova, V. Makarov, T. Jennewein, R. Ursin, A. Zeilinger, *Quantum teleportation over 143 kilometres using active feed-forward*, Nature **489**, 269–273 (2012).

455. X.-S. Ma, S. Kropatschek, W. Naylor, T. Scheidl, J. Kofler, T. Herbst, A. Zeilinger, R. Ursin, *Experimental quantum teleportation over a high-loss free-space channel*, Opt. Express **20**, 23126–23137 (2012).

456. R. Fickler, R. Lapkiewicz, W. N. Plick, M. Krenn, C. Schaeff, S. Ramelow, A. Zeilinger, *Quantum Entanglement of High Angular Momenta*, Science **338**, 640–643 (2012); selected as one of Physics World magazine's TOP 10 breakthroughs 2012.

457. C. Pacher, A. Abidin, T. Lorünser, M. Peev, R. Ursin, A. Zeilinger, J-A. Larsson, *Attacks on quantum key distribution protocols that employ non-ITS authentication*, arXiv:1209.0365 [quant-ph] (2012).

458. B. Dakić, Y. O. Lipp, X. Ma, M. Ringbauer, S. Kropatschek, S. Barz, T. Paterek, V. Vedral, A. Zeilinger, Č. Brukner, P. Walther, *Quantum discord as resource for remote state preparation*, Nat. Phys. **8**, 666–670 (2012).

459. M. Schlosshauer, J. Kofler, A. Zeilinger, *A Snapshot of Foundational Attitudes Towards Quantum Mechanics*, Stud. Hist. Phil. Mod. Phys. **44**, 222–230 (2013).

460. S. Ramelow, A. Mech, M. Giustina, S. Gröblacher, W. Wieczorek, A. Lita, B. Calkins, T. Gerrits, S. Woo Nam, A. Zeilinger, R. Ursin, *Highly efficient heralding of entangled single photons*, Opt. Express **21**, 6707–6717 (2013).

461. M. Krenn, R. Fickler, M. Huber, R. Lapkiewicz, W. N. Plick, S. Ramelow, A. Zeilinger, *Entangled singularity patterns of photons in Ince-Gauss modes*, Phys. Rev. A **87**, 012326 (2013).

462. J. Kofler, S. Ramelow, M. Giustina, A. Zeilinger, *On 'Bell violation using entangled photons without the fair-sampling assumption'*, arXiv:1307.6475 [quant-ph] (2013).

463. G. A. D. Briggs, J. N. Butterfield, A. Zeilinger, *The Oxford Questions on the foundations of quantum physics*, Proc. R. Soc. A **469**, 20130299 (2013).

464. W. N. Plick, R. Lapkiewicz, S. Ramelow, A. Zeilinger, *The Forgotten Quantum Number: A short note on the radial modes of Laguerre-Gauss beams*, arXiv:1306.6517 [quant-ph] (2013).

465. R. Lapkiewicz, P. Li, C. Schaeff, N. K. Langford, S. Ramelow, M. Wiesniak, A. Zeilinger, *Comment on "Two Fundamental Experimental Tests of Nonclassicality with Qutrits"*, arXiv:1305.5529 [quant-ph] (2013).

466. R. Fickler, M. Krenn, R. Lapkiewicz, S. Ramelow, A. Zeilinger, *Real-Time Imaging of Quantum Entanglement*, Scientific Reports **3**, 1914 (2013).

467. M. Giustina, A. Mech, S. Ramelow, B. Wittmann, J. Kofler, J. Beyer, A. Lita, B. Calkins, T. Gerrits, S. W. Nam, R. Ursin, A. Zeilinger, *Bell violation using entangled photons without the fair-sampling assumption*, Nature **497**, 227–230 (2013).

468. X. Ma, J. Kofler, A. Qarry, N. Tetik, T. Scheidl, R. Ursin, S. Ramelow, T. Herbst, L. Ratschbacher, A. Fedrizzi, T. Jennewein, A. Zeilinger, *Quantum erasure with causally disconnected choice*, P. Natl. Acad. Sci. USA **110**, 1221–1226 (2013).

469. M. Schlosshauer, J. Kofler, A. Zeilinger, *The interpretation of quantum mechanics: from disagreement to consensus?*, Ann. Phys. **525**, A51–A54 (2013).

470. W. N. Plick, M. Krenn, R. Fickler, S. Ramelow, A. Zeilinger, *Quantum orbital angular momentum of elliptically-symmetric light*, Phys. Rev. A **87**, 033806 (2013).

471. T. Scheidl, F. Tiefenbacher, R. Prevedel, F. Steinlechner, R. Ursin, A. Zeilinger, *Crossed-crystal scheme for femtosecond-pulsed entangled photon generation in periodically poled potassium titanyl phosphate*, Phys. Rev. A. **89**, 042324 (2014).

472. X. Ma, B. Dakić, S. Kropatschek, W. Naylor, Y. Chan, Z. Gong, L. Duan, A. Zeilinger, P. Walther, *Towards photonic quantum simulation of ground states of frustrated Heisenberg spin systems*, Scientific Reports **4**, 3583 (2014).

473. M. Krenn, M. Huber, R. Fickler, R. Lapkiewicz, S. Ramelow, A. Zeilinger, *Generation and confirmation of a (100 × 100)-dimensional entangled quantum system*, P. Natl. Acad. Sci. USA **111**, 6122–6123 (2014).

474. R. Fickler, R. Lapkiewicz, S. Ramelow, A. Zeilinger, *Quantum Entanglement of Complex Photon Polarization Patterns in Vector Beams*, Phys. Rev. A **89**, 060301(R) (2014).

475. G. B. Lemos, V. Borish, G. D. Cole, S. Ramelow, R. Lapkiewicz, A. Zeilinger, *Quantum Imaging with Undetected Photons*, Nature **512**, 409–412 (2014).

476. R. Fickler, R. Lapkiewicz, M. Huber, M. P. J. Lavery, M. P. Padgett, A. Zeilinger, *Interface between path and OAM entanglement for high-dimensional photonic quantum information*, Nature Communications **5**, 4502 (2014).

477. M. Krenn, R. Fickler, M. Fink, J. Handsteiner, M. Malik, T. Scheidl, R. Ursin, A. Zeilinger, *Communication with spatially modulated light through turbulent air across Vienna*, New J. Phys. **16**, 113028 (2014).

478. M. Keller, M. Kotyrba, F. Leupold, M. Singh, M. Ebner, A. Zeilinger, *A Bose-Einstein condensate of metastable helium for quantum correlation experiments*, Phys. Rev. A **90**, 063607 (2014).

479. A. Zeilinger, *Eugene Wigner — A Gedanken Pioneer of the Second Quantum Revolution*, EPJ Web of Conferences **78**, 01010 (2014).

480. T. Herbst, T. Scheidl, M. Fink, J. Handsteiner, B. Wittmann, R. Ursin, A. Zeilinger, *Teleportation of entanglement over 143 km*, arXiv:1403.0009 [quant-ph] (2014).

481. M. Krenn, J. Handsteiner, M. Fink, R. Fickler, A. Zeilinger, *Twisted photon entanglement through turbulent air across Vienna*, arXiv:1507.06551 [quant-ph] (2015).

482. M. Malik, M. Erhard, M. Huber, M. Krenn, R. Fickler, A. Zeilinger, *Multi-photon entanglement in high dimensions*, arXiv:1509.02749 [quant-ph] (2015).

483. M. Krenn, M. Malik, R. Fickler, R. Lapkiewicz, A. Zeilinger, *Automated Search for new Quantum Experiments*, arXiv:1509.02561 [quant-ph] (2015).

484. M. Lahiri, R. Lapkiewicz, G. Barreto Lemos, A. Zeilinger, *Theory of Quantum Imaging with Undetected Photons*, Phys. Rev. A **92**, 013832 (2015).

485. M. J. Padgett, F. M. Miatto, M. Lavery, A. Zeilinger, R. W. Boyd, *Divergence of an orbital-angular-momentum-carrying beam upon propagation*, New J. Phys. **17**, 023011 (2015).

486. C. Schaeff, R. Polster, M. Huber, S. Ramelow, A. Zeilinger, *Experimental access to higher-dimensional entangled quantum systems using integrated optics*, Optica **2**, 523–529 (2015).

487. X.-S. Ma, J. Kofler, A. Zeilinger, *Delayed-choice gedanken experiments and their realizations*, arXiv:1407.2930 [quant-ph] (2014), accepted to Rev. Mod. Phys (2015).

Maximilian Haider

MAXIMILIAN HAIDER

M. Haider was born in Austria as a son and a grandson of watchmakers. Already as child of the age of approximately 10 he started to open and to dismantle the somewhat larger wall clocks but he also could assemble them to a working clock too. Hence, his interest in precision mechanics started very early. Later with the age of 14 he left school and started an apprenticeship as an ophthalmic optician where he created his curiosity in physics. After he finished his examination as an optician he attended a school for higher level professional education as optician in Cologne, Germany. At this stage he noticed that his curiosity in Physics was not satisfied and, hence, he started his 2nd career by passing exams to receive the permission to study physics in Kiel and later in Darmstadt, Germany.

In Darmstadt, he joined the group of Prof. H. Rose to carry out his Diploma-Thesis on the development of a new multipole-element for the already progressing correction project which was guided by Prof. O. Scherzer and Prof. H. Rose. During this project, he became convinced that the correction of aberration of an electron microscope is feasible. However, it was also clear to him that for a successful project a new modern transmission electron microscope (TEM) was needed which had to proof beforehand that its resolving power is only aberration limited and not by any other imperfection like stage vibrations or instabilities of power supplies or any disturbance else. After having finished the Diploma in Physics (1982) he wanted to continue with the correction project but O. Scherzer passed away (Nov. 1982) short timer after and the German Research Society (DFG) refused to fund this project under the theoretician Prof. H. Rose afterwards. Hence, in order to be able to stay within the development of electron microscopes and to carry out a PhD thesis he joined the group of Dr. A. Jones (Oct. 1983) who lead the Physical Instrumentation Program at the European Molecular Biology Laboratory (EMBL) in Heidelberg, Germany. His supervisor of his PhD Thesis was still Prof. H. Rose. The instrumental development as part of his thesis was a new high resolution electron energy loss spectrometer for a dedicated Cryo-STEM (a Scanning-TEM with a superconductive objective lens) at the EMBL. Already during his work as a physicist within a biological laboratory he tried to convince biologists to set-up an aberration corrector project at the EMBL. In 1986, he finished successfully the development of the energy spectrometer and received his PhD in Physics from the Technical University of Darmstadt, Germany in January 1987. At about the same time he was successful by starting as a member of the group of A. Jones a project to compensate the chromatic and the spherical aberration of a low voltage scanning electron microscope (LV-SEM) for biological applications.

In 1990, he was appointed as a Group Leader of STEM Development and Maintenance for one year together with A. Jones and afterwards alone after A. Jones left the EMBL for retirement in 1991. He further specialized in electron optics and instrumentation by continuing the development of the Cc/Cs-corrected SEM together with J. Zach. Already at late 80's he started intensive discussions with H. Rose on prospects of a Cs-corrected modern high resolution TEM which could in 1990 be finalized as a joint Grant proposal together with H. Rose and K Urban submitted to the VW-Foundation for the funding of the Cs-corrector project for TEM in 1991. "Cs" means in this context the spherical aberration coefficient. This proposal was accepted in late Summer 1991 and only the project plan had to be revised in such a way to have a milestone incorporated around midterm in order to give the funding Foundation the chance to evaluate the progress of this project. This revised project plan was accepted in Fall 91 and more group members could be hired to start this project beginning of 1992. This Cs-corrector project went well and in 1995 the 2nd phase of this project could be started with the purchase of a new modern 200 kV TEM. In 1996 the compensation of the Cs could be demonstrated but due to instabilities not yet an improvement of resolution. In June 1997 the breakthrough of a substantial improvement of resolution from 0.24 nm down to about 0.12 nm could be achieved. A substantial gain of contrast when imaging non-periodic objects could be demonstrated and many scientists around the world were really impressed and excited.

Almost in parallel in 1995 the success of Cc/Cs-corrector project for a LVSEM could be proven by an improvement of resolution from around 5.6 nm down to about 1.8 nm. This work was mainly carried out by the group member J. Zach and attracted the interest of the Japanese company JEOL. In early 1996 discussions with JEOL were started on a joined project and with this perspective he founded together with J. Zach the company CEOS which concentrates on the development of correctors and other advanced electron optical components. The company CEOS has since its foundation a steadily growth from 3 employees at the beginning now up to around 40. CEOS is involved in several unique developments for high resolution electron microscopy. Among others there are, for example, the TEAM and the SALVE project. With TEAM (Transmission Electron Aberration corrected Microscopy) which aimed for ultra high resolution down to 50 pm in TEM as well as in STEM mode and SALVE (Sub Angstrom Low Voltage Electron microscopy) which will concentrate in the low energy regime for the imaging of beam sensitive materials with high resolution all stated goals could be achieved. Until end of 2015 almost exactly 600 correctors have been installed worldwide with the corrector technology of CEOS.

In 2008, M. Haider was appointed as a Honorary Professor at Karlsruhe Institute of Technology. Since this time he offers lectures in Theoretical Electron Optics and Seminars for undergraduate Students. He received several awards like the Beckurts Award in 2006, the Honda Prize in 2007, the Wolf Prize in 2011 and the BBVA Award in 2014. And recently he was awarded an Honorary Fellowship of the Royal Microscopy Society and the NIMS Award in Tsukuba/Japan.

COMMENTARY

1. K.H. Beckurts Foundation (only in German available)

Kurzfassung

Die wissenschaftliche Leistung liegt auf dem Gebiet der Elektronenoptik. Die Elektronenoptik ist die Grundlage einer internationalen elektronenoptischen Industrie, deren Schwerpunkte bei Elektronenmikroskopen und in der Elektronenstrahllithographie liegen.

Die *Elektronenmikroskopie* stellt in ihren verschiedenen Formen, der Durchstrahlungselektronenmikroskopie (TEM), der Rasterdurchstrahlungsmikroskopie (STEM) und Rasterelektronenmikroskopie (SEM), ein Schlüsselinstrumentarium für die Naturwissenschaften, die Medizin und die Technik bereit, welches mit Bezug auf seine bisherige Leistung und auf seinen zukünftigen Beitrag nicht überschätzt werden kann.

Die Leistung der Elektronenoptik war bislang durch Aberrationen, insbesondere die sphärische Aberration der verwendeten elektromagnetischen Linsen begrenzt. Frühe Versuche, aberrationskorrigierte Elektronenlinsen zu realisieren, waren nicht erfolgreich. Damit galt die Elektronenoptik bis Ende der Achtzigerjahre als technisch ausentwickelt und nicht mehr weiter mit Bezug auf ihr Auflösungsvermögen verbesserbar.

2. Honda-Foundation:

The basic theory of aberration correction for high-resolution imaging was introduced by Dr. Otto Scherzer of Germany in the 1940's. Many researchers attempted, but failed, its implementation as an aberration-corrected electron microscope; and experts had questioned its technical feasibility by the time the laureates, who thought otherwise, were teamed in 1989. The laureates refined the basic theory in light of materials science and combined it with electron optical engineering techniques to attain the mechanical stability required for electron microscopy. In 1997, they succeeded in making an aberration-corrected TEM that is capable of high-resolution imaging of atomic structures.

This aberration-corrected microscopic technology used for the TEM is now made available to microscope manufacturers in Germany, Japan and other countries through CEOS, a project spin-off company headed by Dr. Haider, to be distributed worldwide for its applications. The TEM has become one of the essential instruments for research and development on an atomic level. It is used not only to produce ultrafine particles for advanced, high-integrated semiconductor devices, but to analyze and examine the atomic arrangements, structures, and binding of various materials in industries like metallic engineering, biotechnology, and nanotechnology. Many users expect new materials could be discovered and macroscopic properties could be analyzed at atomic levels by use of this technology.

In the research project, Dr. Rose has been chiefly responsible for the basic design of the corrector and the refinement of the theory of image formation, whereas Dr. Urban has worked the application of the refined theory based on his expertise in materials science, and Dr. Haider has used his knowledge in electron optical engineering for the elaborate design and development of this new aberration-corrected technology. The Honda Foundation recognizes the three physicists for their spirit of challenge as well as substantial contributions to human life through their sophisticated technological achievement that we believe embodies the ethos of Ecotechnology.

3. Wolf-Foundation:
For their development of aberration-corrected electron microscopy, allowing the observation of individual atoms with picometer precision, thus revolutionizing materials science.

4. Royal Microscopy Society:
Dr. Max Haider, *CEOS GmbH* - recognised for his pioneering work in spherical aberration correction in Transmission Electron Microscopes. His work with Professor Harald Rose as well as Professor Ondrej Krivanek developed a technology that overcame the 60 year old problem of spherical aberration in the electron microscope and has revolutionised the performance of the TEM. He then went on to co-found CEOS GmbH, developing spherical and chromatic aberration correctors and is still currently producing the majority of correctors installed in TEMs today.

5. NIMS Award Committee
The annual NIMS Award is awarded to top researchers from all over the world for outstanding scientific achievements and a contribution to the work of the National Institute for Materials Science (NIMS) in Japan.

This year, themed "Materials Innovation driven by Advanced Characterization", three German researchers have been nominated to receive the award: Harald Rose of the University of Ulm, Maximilian Haider of the Karlsruhe Institute of Technology, and Knut Wolf Urban of the Research Centre Juelich will receive the award for their groundbreaking work on improving the electron microscope, which has greatly benefited the scientific community as a whole.

The three researchers developed a device that corrects aberrations in electron microscopy (blur and distortion), which for the first time allowed for clear observation of interfaces and defects. The device also enabled accurate identification of atomic positions and consequently studies on the relationships between the properties of materials and atomic arrangement. The selection committee highly regarded the fact that these events later significantly contributed to the advancement of materials research.

Electron microscopy image enhanced

One of the biggest obstacles in improving the resolution of the electron microscope has always been the blurring of the image caused by lens aberrations. Here we report a solution to this problem for a medium-voltage electron microscope which gives a stunning enhancement of image quality.

Even today, more than 60 years after the invention of the transmission electron microscope[1], the point resolution is still limited by the spherical aberration of its objective lens. Up until now, the only way to improve matters was to impose a numerical correction for this spherical aberration that was based either on a series of images of the object taken under variable objective-focus conditions[2] or on the application of holographic techniques[3].

Fifty years ago, Scherzer[4] suggested that the two principal axial aberrations, chromatic and spherical, could be corrected by electrostatic or magnetic multipole elements, but this proved to be beyond the technology available at that time. Our solution for spherical aberration correction derives from a much more recent suggestion by Rose[5], and depends on two electromagnetic hexapoles and four additional lenses.

The principle by which correction is achieved is based on the fact that the primary aberrations of second order from the first hexapole are compensated by the second hexapole element. However, the two hexapoles additionally induce a residual secondary third-order spherical aberration which is rotationally symmetric[6] and proportional to the square of the hexapole strength.

The appertaining coefficient of spherical aberration is of the opposite sign to that of the objective lens. In this way, the spherical aberration of the entire system can be

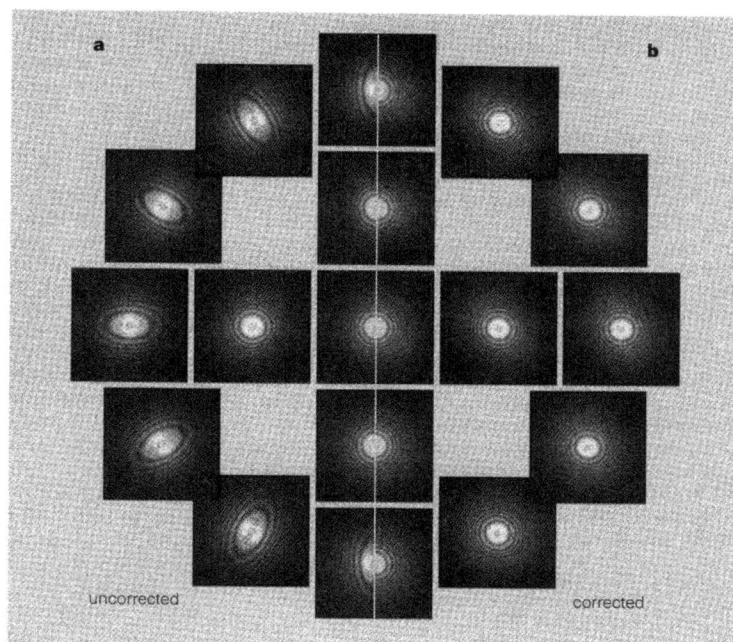

Figure 1 Diffractogram tableau of the microscope. **a,** Uncorrected, and **b,** after correction for spherical aberration and proper alignment. The beam tilt angle is 10.8 mrad in both cases and the azimuthal angles vary between 0 and 2π in steps of $\pi/6$. The essentially identical shape of the diffractograms in **b** indicates the vanishing spherical aberration.

compensated by exciting the hexapoles appropriately[7].

In the transmission electron microscope, off-axial aberrations are as important as axial ones. Therefore, in order to maintain a finite field of view, a semi-aplanatic objective lens system had to be constructed which fulfils Abbe's sine condition. This means that it is not only free of spherical aberration but also of off-axial coma and parasitic axial aberrations resulting from misalignment of the optical system.

Spherical aberration correction is not enough to improve the resolution of the microscope. It is necessary to compensate as well for the parasitic second-order axial aberrations, coma and threefold astigmatism, and adequately to suppress the non-

spherical axial third-order aberrations, star aberration and fourfold astigmatism. For this purpose, the aberration coefficients are determined by means of an extended version of the diffractogram tableau method[8].

With the corrector switched off, a diffractogram tableau of an amorphous sample is recorded digitally (Fig. 1a) and evaluated on-line in terms of induced defocus and twofold astigmatism. Switching on the corrector introduces misalignment aberrations and an appropriate alignment procedure has to be carried out; this semi-automatic alignment takes about 15 minutes.

The diffractogram tableau of the corrected and aligned system is shown in Fig. 1b. All diffractograms are of roughly the same shape, indicating the properties

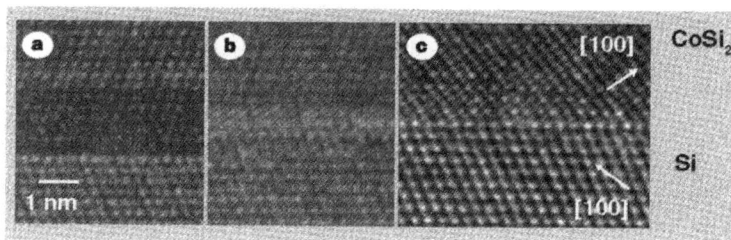

Figure 2 Structure images of an epitaxial Si(111) CoSi$_2$ interface demonstrating the production of image artefacts by the effect of contrast delocalization due to spherical aberration. Images in **a** and **b** were taken in the uncorrected microscope at Scherzer defocus and Lichte defocus, respectively. **c,** Image taken in the aberration-corrected state at Scherzer defocus does not show any delocalization (C_s = 0.05 mm).

of an aplanatic objective lens. Even for beam tilt angles as high as 30 mrad, the values of defocus and two-fold astigmatism induced by the residual axial aberrations remain small.

The basic instrument used for this development was a standard 200 kV Philips CM 200 ST equipped with a field emission gun. The corrector increases the column height by 24 cm but leaves the other operating modes and functions of the microscope unaffected. With this modified microscope, we were able to demonstrate the correction of the spherical aberration and to realize an improvement of the point resolution from 0.24 nm to better than 0.14 nm.

The main advantage of spherical aberration correction is that structure-imaging artefacts due to contrast delocalization can to a great extent be avoided. These artefacts have turned out to be a major obstacle for the application of instruments with field emission guns in defect and interface studies[9]. Contrast delocalization arises from the width of the aberration discs belonging to the individual diffracted electron waves whose diameter increases with the spherical aberration.

Figure 2a shows, in a cross-sectional preparation, the interface of $CoSi_2$ grown epitaxially on Si(111) seen along the <110> direction. The image was taken under Scherzer defocus without correction.

At the interface, an approximately 2-nm-broad region of darker contrast can be seen. The width depends on the defocus value reaching a minimum at the Lichte defocus[10] (Fig. 2b). If the structure is imaged in the aberration-corrected condition (Fig. 2c), the delocalization has essentially disappeared and the interface is atomically sharp.

Maximilian Haider*, Stephan Uhlemann*, Eugen Schwan

European Molecular Biology Laboratory, Postfach 102209, 69012 Heidelberg, Germany
*Present address: CEOS GmbH, Im Neuenheimer Feld 519, 69120 Heidelberg, Germany

Harald Rose

Institut für Angewandte Physik, Technische Hochschule Darmstadt, 64289 Darmstadt, Germany

Bernd Kabius, Knut Urban

Institut für Festkörperforschung, Forschungszentrum Jülich GmbH, 52425 Jülich, Germany

1. Knoll, M. & Ruska, E. *Z. Physik* **78**, 318–339 (1932).
2. Kirkland, E.J. *Ultramicroscopy* **15**, 151–172 (1984).
3. Lichte, H. *Ultramicroscopy* **20**, 293–304 (1986).
4. Scherzer, O. *Optik* **2**, 114–132 (1947).
5. Rose, H. *Optik* **85**, 19–24 (1990).
6. Beck, V.D. *Optik* **53**, 241–255 (1979).
7. Haider, M., Braunshausen, G. & Schwan, E. *Optik* **99**, 167–179 (1995).
8. Zemlin, F., Weiss, K., Schiske, P., Kunath, W. & Herrmann, K.-H. *Ultramicroscopy* **3**, 49–60 (1977).
9. Thust, A., Coene, W.M.J., Op de Beek, M. & Van Dyck, D. *Ultramicroscopy* **64**, 211–230 (1996).
10. Lichte, H. *Ultramicroscopy* **38**, 13–22 (1991).

CHAPTER **2**

Present and Future Hexapole Aberration Correctors for High-Resolution Electron Microscopy

Maximilian Haider, Heiko Müller, and **Stephan Uhlemann***

* Corrected Electron Optical Systems GmbH, Heidelberg, Germany

Advances in Imaging and Electron Physics, Volume 153, ISSN 1076-5670, DOI: 10.1016/S1076-5670(08)01002-1.

44   Maximilian Haider *et al.*

I. INTRODUCTION

Ever since the invention of the electron microscope by Ruska (Knoll and Ruska, 1932), endeavors to improve the resolving power of this new type of microscope were continuously ongoing. At first the attempt was to achieve a resolution superior to that of light microscopes. As soon as this goal was attained, theoreticians started to support the efforts of experimental physicists. The goal was twofold: First, to develop mathematical methods to calculate the electron ray path and the electron optical properties of electron lenses and, second, to understand the image formation in an electron microscope. Very soon after the theoreticians started their investigations it became clear that for fundamental reasons the quality of charged-particle lenses must be rather poor compared to high-quality lenses in light optics. In charged-particle optics the primary lens aberrations cannot be compensated by just a combination of various lenses (Scherzer, 1936). Hence, in the early days of charged-particle optics the search for aberration correctors was already started (Scherzer, 1947). A survey of the early history is provided by Harald Rose in the first chapter of this volume.

A. Aberrations and Information Transfer

An electron microscope transfers spatial information from the illuminated object to the magnified image by means of electromagnetic lenses. This information transfer cannot be perfect. The performance of a charged-particle optics instrument is characterized by two different properties: the optical aberrations and the instrumental information limit. The aberrations of the optical system cause residual phase shifts that deteriorate the phases of the scattered waves in a conventional transmission electron microscope (CTEM) or of the probe-shaped illumination in the scanning transmission electron microscope (STEM). As a result of interference with falsified phase information the image recorded by the detector is fudged or at least artifacts are introduced.

Aberrations are classified in two main categories. The effect of axial aberrations depends only on the aperture and affects the on-axis image point in the same manner as the off-axis image points. Off-axial aberrations are only visible at off-axis image points and do not affect the on-axis image point. At the image plane aberrations can be described quantitatively by aberration coefficients. From the aberration coefficients both the residual phase shift and the delocalization can be calculated. In geometrical optics it is common practice to classify the aberration coefficients according to their Seidel order. The Seidel order of an aberration determines the dependency on the beam parameters. Actually, it is the sum of exponents of the aperture angle and the image coordinate in the delocalization monomial.

For high-resolution imaging primarily axial aberrations and coma-type off-axial aberrations are important. The latter depend linearly on the image coordinate and limit the number of equally well-resolved image points, whereas the axial aberrations determine the achievable optimum point resolution of the instrument.

For example, for phase-contrast imaging in a CTEM the point resolution at Scherzer defocus $C_1 = -1.2 \, (C_3\lambda)^{1/2}$ can be defined as $d = 0.7 \left(C_3\lambda^3\right)^{1/4}$, where C_3 (or C_s) denotes the third-order spherical aberration of the objective lens and λ the wavelength of the electrons. In general, the definition of instrumental resolution is a very difficult matter. What we actually can see depends strongly on the contrast mechanism and also on the specimen and its preparation. Whenever resolution numbers are given or compared, it is prudent to ask what assumptions have been made and which definition of point resolution has been used for the assessment.

For a TEM equipped with an aberration corrector the achievable optimum instrumental resolution is at best determined by the higher-order residual intrinsic aberrations of the optical system. The residual intrinsic aberrations are present for fundamental reasons even for the idealized instrument. They can be assessed very precisely during the optical design of the instrument. For a real microscope unavoidable manufacturing tolerances and misalignments cause additional parasitic aberrations. In order to gain the full benefits from aberration correction all parasitic aberrations must be compensated or at least minimized by an appropriate alignment of the system. For this purpose every aberration corrector has a considerable number of alignment deflectors and stigmators in addition to the principal optical elements. For a well-designed and well-corrected instrument the residual parasitic aberrations should not limit the attainable spatial resolution. Hence, for any aberration-corrected instrument the definition of a complete and efficient set of alignment tools and the implementation of robust and precise methods for aberration measurement are the most crucial ingredients.

In addition to the coherent effects discussed previously, incoherent effects on the optical performance must also be considered. The axial chromatic aberration of a TEM makes the defocus energy-dependent. Since the electron beam is not monochromatic, the recorded image intensity results from an incoherent average over the chromatic focus spread. This is the most prominent incoherent effect that damps the information transfer in an electron microscope for high spatial frequencies. For bright-field imaging in a CTEM the instrumental information limit is governed by the phase-contrast envelope function as a result of chromatic focus spread and lateral incoherence. In a real system, additionally high-voltage ripple, instability of the lens currents, and noise-induced image spread impair the information limit.

The instrumental information limit is the most important parameter for a C_s-corrected TEM. With an aberration corrector the point resolution can be improved up to the information limit but never beyond. Because any multipole C_s corrector consists of focusing elements the corrector slightly increases the total chromatic aberration. On the other hand, because C_s is corrected the delocalization is minimized and, hence, the influence of the lateral incoherence (determined by the effective size of the electron source) on the information limit is largely suppressed. Hence, the information limit after C_s correction should be as good as for the uncorrected instrument.

B. Advent of Hexapole Correctors

An important period for the development of hexapole correctors was in the 1970s when several researchers investigated the optical action of hexapole fields but did not yet notice the advantageous character of hexapoles for the compensation of the third-order spherical aberration. In 1978, Beck for the first time proposed the use of hexapole fields to compensate for the spherical aberration of an objective lens (Beck, 1979). In the following years Albert Crewe in Chicago worked on proposals for STEM hexapole correctors (Crewe, 1980; Crewe and Kopf, 1980; Crewe, 1982), and in Germany Rose analyzed different corrector setups and proposed a new design free of fourth-order aperture aberrations (Rose, 1981). Crewe finally could convince the Department of Energy (DOE) and IBM to fund the development of a hexapole corrector for a dedicated STEM aiming for 0.05 nm resolution. This attempt never did succeed. According to Crewe's own perception funding ran out too early and the project was stopped before the prototype could be finalized (Crewe, 2002).

C. The EMBL Project

When Scherzer died in 1982, the TEM corrector project at Darmstadt, Germany—aiming for simultaneous C_c and C_s correction by means of a quadrupole-octupole corrector—was stopped (Koops, Kuck, and Scherzer, 1977). However, the participants of the Darmstadt project presevered in their strong conviction that aberration correction *is* feasible and its successful realization should only be a matter of better understanding, patience, and improved technologies for machining, corrector control, and alignment. With this attitude a new correction project was started at the European Molecular Biology Laboratory (EMBL) at Heidelberg, Germany. The new goal was to set up a C_c- and C_s-corrected low-voltage scanning electron microscope (LVSEM) for biological applications (Zach, 1989). This project started in 1987 with the design and construction of a quadrupole-octupole corrector (Haider, 1990). During this time discussions with Harald

Rose were ongoing to determine how to stimulate a correction project for high-resolution electron microscopy. At the EUREM meeting in York in 1988, the idea was born to search for a possibility to correct only the spherical aberration in CTEM (or STEM, but less favorable at those days) and to combine it with a field-emission gun to avoid the necessity of a C_c corrector for TEM. At that time field-emission guns were just under development at various places and almost available for commercial instruments.

In 1989, at the *Dreiländertagung* in Salzburg, Austria, the atomic-resolution microscope (ARM) project was introduced. This project at Stuttgart, Germany, aimed for highest resolution by means of a 1.25-MeV CTEM (Phillipp *et al.*, 1994). Rose, who attended this conference, stated during the presentation of this project vituperating that it would be far better to invest a much smaller amount of money in the development of a corrected TEM instead for such a microscope (Rose, 1989). In 1990 he published the outline of a C_s-corrected 300-kV CTEM for sub-angstrom resolution (Rose, 1990). Rose could convince Urban to move forward for such a project and to evaluate the benefits of a C_s-corrected TEM for materials science. Finally, a grant proposal was submitted to the German Volkswagen Foundation at the end of 1990. This was the birth of a successful joint project of three groups: the theory group of Rose at Darmstadt, the physical instrumentation group of Haider at the EMBL, and the materials science group of Urban at Jülich, Germany. By the end of 1991 the funding for this project was ensured and the development of the first C_s-corrected 200-kV CTEM could be started. A Philips CM200F was selected as the base instrument because of the demand for a modern high-resolution microscope with field-emission gun. The assumption was that the performance of this system is limited mainly by the spherical aberration and not by chromatic focus spread. The point resolution of this CTEM, equipped with the so-called super-twin objective lens, was given by 0.24 nm at Scherzer focus and an information limit of at least 0.16 nm was guaranteed by the manufacturer. The phase-contrast transfer function (PCTF) for the uncorrected instrument at Scherzer focus is shown in Figure 1. The point resolution is given by the first zero of the PCTF while the information limit is the spatial frequency for which the PCTF envelope is reduced to $1/e^2$ due to the chromatic aberration and the lateral incoherence. For the uncorrected instrument the effect of the lateral incoherence on the information limit is sensitive to the illumination conditions and the choice of the defocus. The PCTF in Figure 1 represents typical conditions for high-resolution imaging. Shifting from Scherzer to Lichte focus (for minimized delocalization) or reducing the semi-convergence angle of the illumination and, hence, the current density at the specimen further could slightly improve the nominal instrumental information limit.

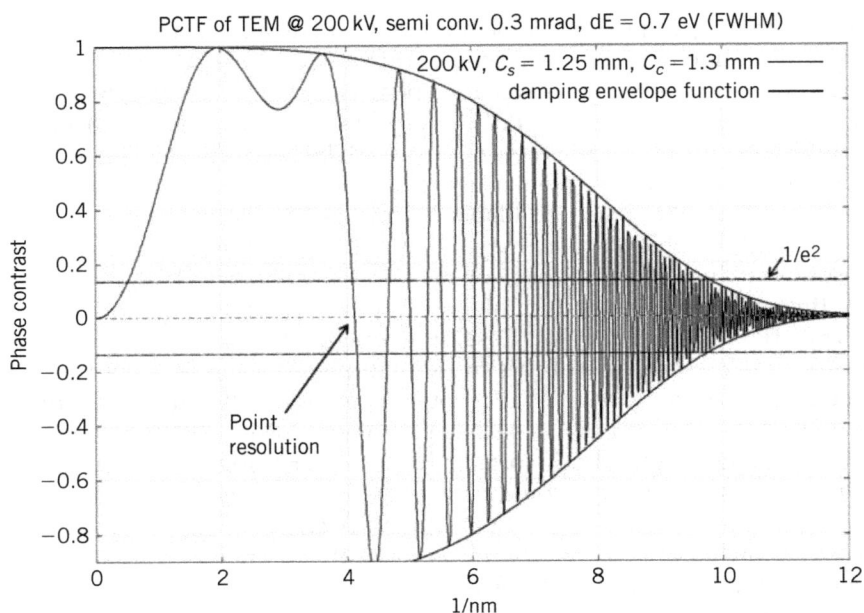

FIGURE 1 The phase contrast transfer function of an uncorrected 200-kV CTEM at Scherzer focus. The point resolution and the contrast level used to define the information limit are indicated.

In 1991 the director general of the EMBL, Lennart Philipson, accepted our project even though the main research of the EMBL is molecular biology and not applied physics. Therefore, it was agreed that the benefits of C_s-corrected CTEM in structural biology would be evaluated during the course of the project. The project was divided into a two-step process: During the first phase, an old 200-kV TEM was used with the goal to achieve a proof of principle. After this first milestone, in a second project phase the improvement of the point resolution by means of a hexapole corrector integrated in a new CM200F needed to be demonstrated. For the new 200-kV TEM we asked for an improved high-tension supply and an increased overall stability of the instrument. Our goal for the final C_s-corrected instrument was to come close to the optimum information limit of $I_L = 9/nm$ due to the total chromatic aberration and the energy width of the field-emission source.

D. The EMBL Corrector

The design of the EMBL hexapole corrector consists of two modules: (1) an upper part with a double transfer lens and additional image deflectors and stigmators and (2) a lower part with two multipole stages to generate the strong hexapole fields, two transfer lenses, and a final adapter lens. The lower part of the original corrector is shown in Figure 4. The transfer lens doublet in the upper part is necessary to match the coma-free plane of the objective lens with the coma-free plane of the hexapole corrector (Rose, 1990). The coma-free plane of the objective lens is situated very close to its back-focal plane; without the transfer system, off-axial third-order coma and spherical aberration of fifth-order C_5 would be too large to achieve a substantial improvement in high-resolution. This matter is discussed in depth in Section II.

It is technologically demanding to insert two transfer lenses into the bore of the lower iron circuit of the objective lens without changing the design of the objective lens itself (see Figure 2). In order to have sufficient room for the additional lenses we used the larger objective lens iron circuit of a 300-kV lens for the 200-kV system and designed a water-cooled double-gap lens with just one coil and a floating inner pole piece instead of two separate lenses (see Figure 3). The original image deflectors and stigmators of the CM200F were removed and replaced by miniaturized versions integrated in the bore of the transfer system.

A few iterations were needed to find a solution that provided the necessary alignment tolerances and the mechanical and thermal stability required for high-resolution imaging. A compact design is essential

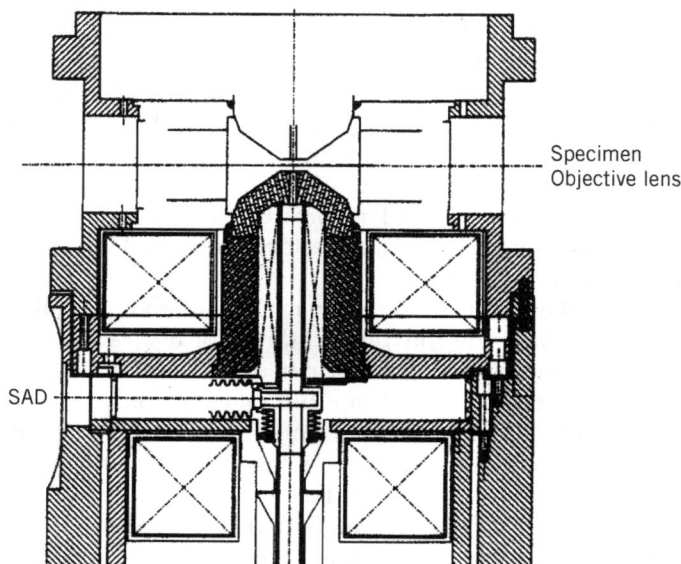

Specimen
Objective lens

SAD

Q3

FIGURE 2 The unmodified objective lens of the CM200F. The first transfer doublet of the corrector must be inserted in the bore of the lower iron circuit.

50   Maximilian Haider *et al.*

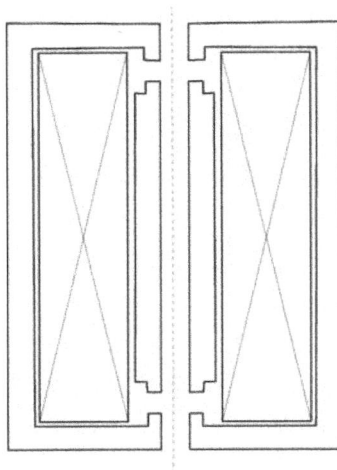

FIGURE 3   Drawing of first transfer system for the EMBL corrector with floating double pole piece, two gaps, and just one coil.

to leave sufficient room between the housing of the coil and the soft iron material to prevent a transfer of heat from the high-current density (finally $\sim20\,\text{A}/\text{mm}^2$) within the coil to the iron circuit of the objective lens and to avoid mechanical vibrations caused by the flow of cooling water very close to the objective lens.

The core module of the hexapole corrector consisting of the lower and the upper multipole element with two transfer lenses in between and additional alignment deflectors was also designed and constructed. We opted for a twelve-pole design to allow for additional weak multipole fields on the principal correction elements for alignment purposes. The twelve-pole elements were made of one cylinder of soft iron to avoid asymmetries of the magnetic flux (see Figure 14). The corrector module with all the wires to excite the various multipole fields, deflectors, and round lenses is depicted in Figure 4 (Haider *et al.*, 1995).

In addition to the corrector hardware itself, we developed all necessary electronics drivers and the software required for aberration assessment, corrector control and alignment. Fast personal computers to calculate a fast Fourier transformation (FFT) of a full image recorded by the charge coupled device (CCD) within a short time were not yet available; therefore, a workstation with an additional image processor board was used. For online calculations new routines have been implemented so that the state of alignment could be measured within about one minute by a fully automatized Zemlin tableau technique. For the acquisition of a Zemlin tableau (Zemlin *et al.*, 1978) the illumination is tilted and guided around a hollow cone. As objects we used amorphous films of strongly scattering

FIGURE 4 The inner part of the EMBL hexapole corrector with the two stages of twelve-pole elements at the top and bottom and between the two lenses of the transfer doublet.

material (e.g., tungsten). In the diffractograms—the modulus of the two-dimensional (2D) discrete Fourier transform of the image—the so-called Thon rings show the phase shift as a function of the scattering angle. The phase shift depends on the aberrations within the imaging system. Depending on the state of alignment (e.g., which aberrations are dominant) an outer tilt angle between 6 and 30 mrads is used and 12 or more images are taken at different tilt positions. If axial aberrations are present, the diffractograms are disturbed by tilt-induced twofold astigmatism and defocus. We developed computer routines that measure defocus and twofold astigmatism very precisely by a new online correlation method with a library of theoretical diffractograms. Given the full set of defocus and astigmatism measurements of the individual diffractograms of the tableau, the full set

52   Maximilian Haider *et al.*

of axial aberrations can be calculated from a linear system of equations. The new methods worked semi-automatically. The only requirements were choosing some parameters and starting the measurement. The result is a list of the axial aberrations that characterize the state of correction (Uhlemann and Haider, 1998).

In 1995 the EMBL group demonstrated the improvement of resolution by means of a quadrupole-octupole C_c/C_s corrector in a low-voltage SEM (Zach and Haider, 1995). This success—for the first time ever the improvement of instrumental resolution by means of an aberration corrector had been shown—also encouraged the hexapole corrector project.

After the successful implementation and an appropriate alignment of the C_s-corrected CM200F in February 1997 an information limit of $\sim$0.18 nm was measured. Since the first transfer lens system was now considered sufficiently stable this indicated the presence of another limitation that had to be detected and eliminated.

Meanwhile, the director general at the EMBL had changed and physical instrumentation was not considered appropriate and was no longer supported at the laboratory. As a consequence, Haider's group (EM development and STEM application) was given notice that their main field of research had to be stopped. With this decision it was clear that only a timely breakthrough with respect to the information limit could turn the project into a success because in case of failure we could hardly expect a second chance outside the EMBL to continue the project until an improvement of resolution could be demonstrated.

From our investigations it was clear that the restriction of the information limit was caused by either a perturbation near the selected area (SA) plane or within the projector system. Since the magnification within the hexapole corrector is adjustable within a certain range by a variation of the first transfer lens doublet, we could obtain a better information limit if the intermediate magnification in the hexapole corrector was increased or if an additional intermediate image was positioned between the last hexapole and the SA plane. Hence, either the projection system or the electron path between the lower hexapole and the SA plane caused the limitation. Unfortunately, with this beam path in the corrector C_s correction was no longer possible.

Q4

The projector system could not be changed; therefore, we decided to design a new much stronger double-gap adapter lens with an additional intermediate image and increased the magnification at the SA plane by a factor of three. This decision was made four weeks before the turn-off switch had to be used at the CM200F because the microscope had to be transferred to the group of Urban at Jülich, Germany. We then needed only three weeks for the design, the construction of the iron circuit (including the necessary annealing process of the soft iron), and for making a new coil. Just 8 days before the C_s-corrected TEM was to be moved we incorporated this

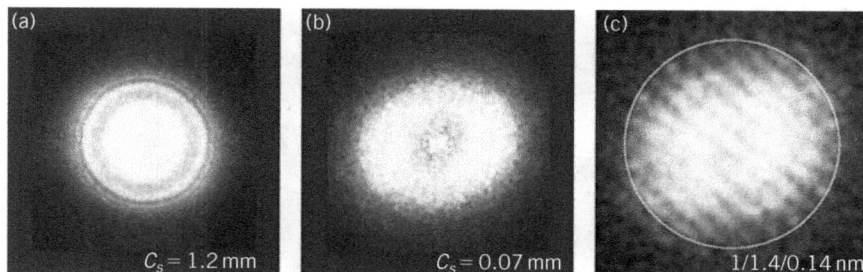

FIGURE 5 First proof of the correction of the third-order spherical aberration C_s without a deterioration of the information limit. Three diffractograms with the same calibration of the spatial frequency are shown. (a) The uncorrected mode at Scherzer focus, (b) the corrected mode near Gauss focus, and (c) the corrected mode superimposed by Young fringes. For the C_s-corrected CM200F the information limit is of 1/(0.12 nm) and 1/(0.14 nm), respectively, for two perpendicular directions. (From Uhlemann and Haider, 1998.)

new lens and one day later we started with the alignment. The first images obtained with this modification were disappointing because the images were very unstable; with this result we lost hope of succeeding with this development. Nevertheless, half a day later the new lens had stabilized and the information limit could be measured in the good direction at 0.12 nm and in the not-so-good at $\sim$0.14 nm. We also took images of amorphous tungsten to illustrate at the diffractogram the modified characteristic of the contrast transfer function (CTF) as shown in Figure 5 (Uhlemann and Haider, 1998).

This result was the hoped-for breakthrough (Haider *et al.*, 1998). The first C_s-corrected CTEM with an improvement of the point resolution by a factor of about two was in place. The exact alignment of the instrument was achieved the following night by means of many Zemlin tableaus. The result of the alignment is shown in Figure 6. In the diffractograms for an outer tilt angle of 18 mrad (shown at the right-hand side) almost no induced astigmatism is visible. This illustrates vividly that the C_s-corrected microscope can focus electrons scattered into large angles without introducing additional disturbing phase shifts.

Within the remaining six days before the CM200F had to be moved to its new site on July 31, 1997, Bernd Kabius demonstrated the superior resolution with seven samples he had prepared for the examination of the first C_s-corrected CTEM. At GaAs in [110] orientation (Figure 7) he could resolve the dumbbell-shaped structure of the pairs of Ga and As atoms with a spacing of 0.14 nm in the raw image. At an interface of $CoSi_2$/Si he demonstrated the vanishing delocalization when working very close to the Gaussian focus with $C_s \approx 0$. He compared the result with a defocus series of images of the same interface (Figure 8) taken without hexapole corrector

54   Maximilian Haider *et al.*

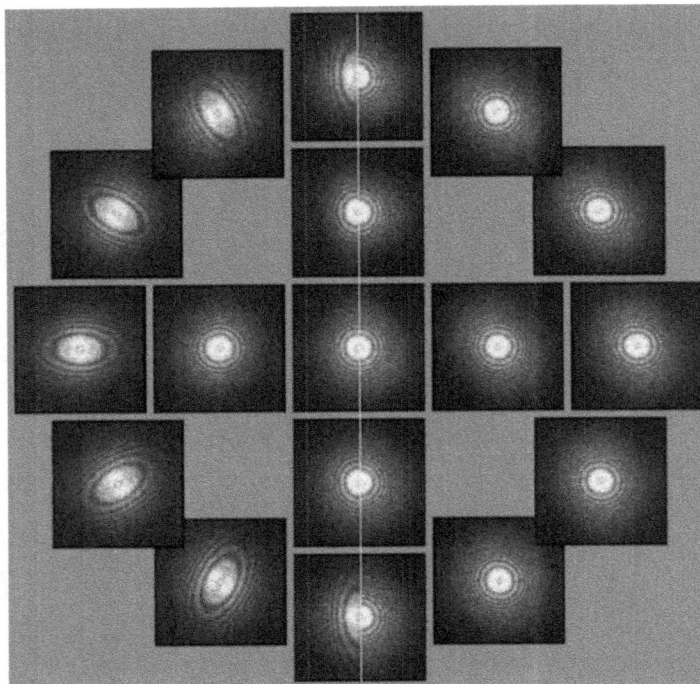

FIGURE 6 Zemlin tableau of the uncorrected (left half) and the corrected CTEM (right half). The outer tilt angle is 10.8 mrad. (From Haider *et al.*, 1998.)

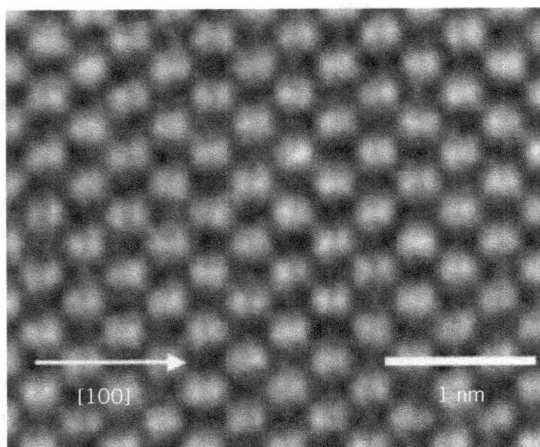

FIGURE 7 Image of GaAs [110] orientation. The distance between the two atom columns in the dumbbell structure in [100] direction is 0.14 nm. (From Rose, Haider, and Urban, 1998.)

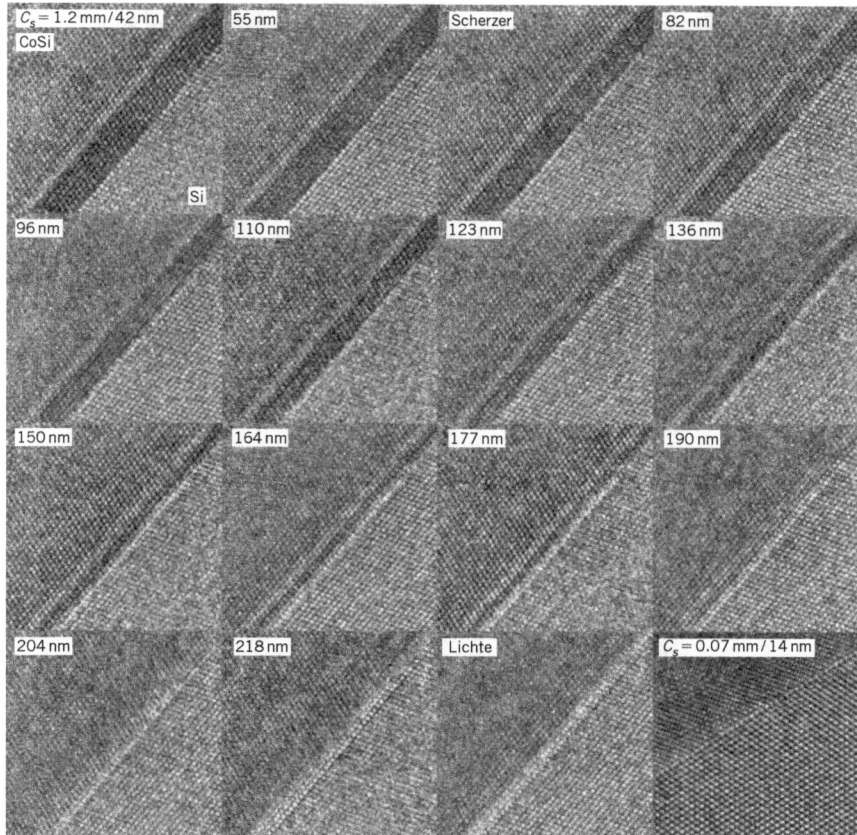

FIGURE 8 Focus series of a $CoSi_2$/Si interface without C_s correction. The last image is taken with C_s corrected. In Scherzer focus the delocalization is very pronounced. At Lichte defocus the delocalization can be reduced but it is still present; only in the C_s-corrected mode is the image of the interface atomically sharp.

where the strong delocalization due to C_s is clearly visible even at Lichte focus. The benefit of vanishing delocalization was also demonstrated at an $YBa_2Cu_3O_{7-x}$/$SrTiO_3$ substrate interface shown in Figure 9.   Q5

The original CM200 equipped with the EMBL hexapole corrector was reinstalled at Jülich in August 1997. Since that time it is—apart from some downtime for service—operating routinely as a user instrument.

E. Commercialization

Urged by the anticipated closure of the electron optics instrumentation group at the EMBL and motivated by the successful realization of the first aberration corrected LVSEM, Haider and Zach founded the Corrected

56   Maximilian Haider *et al.*

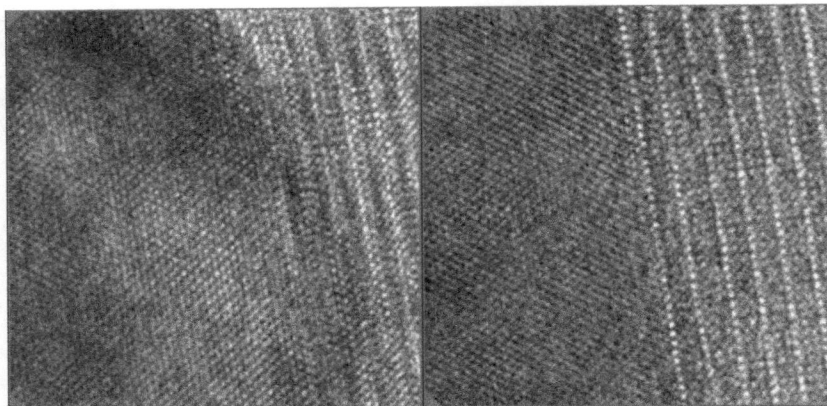

Q6

FIGURE 9   Early attempt to image the substrate interface of the high-temperature superconductor $YBa_2Cu_3O_{7-x}$/$SrTiO_3$. The left side of the figure shows strong delocalization due to C_3 (left), and no delocalization is seen with C_3 corrected (right).

Electron Optical Systems (CEOS) company in Heidelberg in 1996. In the following two years, two more people from the EMBL group joined CEOS. The first project at CEOS was the development of a commercial prototype of the LVSEM C_c/C_s corrector in co-operation with the JEOL company. This development was completed successfully from 1996 to 1999 and the corrector later became commercially available from JEOL (Kazumori *et al.*, 2004).

In 1997 the unique success of the first C_s-corrected CTEM with improved resolution generated a strong demand for commercial availability of TEM C_s correctors in the scientific community. Therefore, CEOS decided to further develop the hexapole corrector and to transform it into a commercial product.

For routine applications several components of the system had to be revised to simplify operation and to improve reliability. In addition, we aimed for an optimization of the electron optical properties and added more alignment flexibility for bright- and dark-field imaging and diffraction mode. Finally we achieved the following:

1. Reduced the number of necessary current drivers from 56 to 29
2. Modified the first transfer lens doublet by implementing two independent coils and iron circuits
3. Added additional stigmators between the objective lens and the first transfer lens for a more precise compensation of twofold and threefold astigmatism without hysteresis
4. Used hexapoles consisting of six separate pole pins and a yoke instead of the monolithic twelve-pole elements and optimized the magnetic flux to avoid saturation and remanence effects,

5. Increased the magnification between objective lens and corrector to reduce the C_c contribution of the transfer lenses from 25–30% down to 15–20% of the C_c of the objective lens.

With this improved design of the hexapole corrector, CEOS started co-operations with the EM manufacturers Carl Zeiss (LEO at that time), JEOL, Philips (now FEI), and Hitachi in the 200-kV TEM market. The hexapole corrector for the conventional TEM has been named CETCOR (*CEOS TEM COR*rector). The first commercial CETCOR was shipped to JEOL in October 2002. At this time the CEOS company had grown to 15 people, several of whom are former members of Rose's theoretical charged-particle optics group at Darmstadt.

Shortly after the first commercial system for CTEM was finished it became obvious that the hexapole corrector is also very suitable as a probe corrector in STEM (Haider, Uhlemann, and Zach, 2000). In order to improve the resolution in STEM the illumination angle has to be increased. This additionally provides a higher beam current for a C_s-corrected STEM. The CEOS hexapole corrector for STEM is called CESCOR (where the *S* stands for STEM). From 2001 to 2003 the first commercial hexapole corrector for STEM was developed for a JEOL 2010F (S)TEM.

This system, finally installed at Oxford University, is shown in Figure 10 (Hutchison *et al.*, 2005). It is the first double-corrected (S)TEM, equipped with two hexapole correctors: one in the illumination system and one in the imaging system. With this system high-resolution TEM imaging can be combined with the analytical capabilities offered by a C_s-corrected STEM (Sawada *et al.*, 2005).

The precise Zemlin tableau technique based on the analysis of diffractograms could not be used for aberration measurement for a STEM C_s corrector. Diffractograms of STEM images do not provide the same information as for CTEM images. We finally implemented a combination of the tableau technique with an analysis of the STEM point-spread function. The latter method was originally developed for the LVSEM corrector by Joachim Zach. To analyze the aberrations of the probe-forming system the illuminating beam is tilted between the gun and the corrector and the point-spread function is deconvoluted from STEM images in overfocus, Gaussian focus, and underfocus. Then defocus and twofold astigmatism are calculated from the shape of the point-spread function. The evaluation of the tableau afterward uses the same methods as used for CTEM.

For the CTEM hexapole corrector two transfer lenses are used in front of the first hexapole element. For STEM a transfer system between the lower hexapole and the objective lens is also necessary to avoid the strong spherical aberration of fifth-order C_5 as a combination aberration. To avoid scan coma in STEM, additionally, the scan unit must be placed between the corrector and the objective lens. This could be achieved by using a

58   Maximilian Haider *et al.*

FIGURE 10   First double-corrected 200-kV (S)TEM. JEOL 2010 F equipped with
CESCOR integrated between condenser and objective lens and CETCOR between
objective lens and projector.

single transfer lens or, preferentially, by using the most often available
condenser mini-lens close to the objective lens in addition to a further
transfer lens to maintain flexibility when adjusting the correction strength
for different high voltages. Over the years hexapole correctors have been
adapted to several objective lenses with different pole pieces. However,
the preferential choice is still a high-resolution small-gap pole piece with
small C_3 and C_c. This system has minimum chromatic focus spread and
small fifth-order residual intrinsic aberrations.

   With the advent of commercially available correctors most TEM manufacturers started a redesign of their columns by concentrating more on the
overall mechanical and electrical stability. Zeiss, for example, introduced

FIGURE 11   CESCOR for dedicated STEM Hitachi HD2700 (Kimoto *et al.*, 2007). The   Q2
filter box for the electronics jacks and the connectors for the cooling water are visible.
The total column extension after integration of the corrector between condenser and
objective lens is ~300 mm.

a hanging column for their high-resolution TEM (Essers *et al.*, 2002), and FEI developed the Titan 80-300 series (Van der Stam *et al.*, 2005).

For the new FEI platform CEOS developed the first 300-kV hexapole correctors for CTEM and STEM. During this time the stability requirements increased, and several design improvements were necessary also for the hexapole correctors. The noise-induced image spread caused by the alignment deflectors has been further reduced. We improved the stability of the current drivers and reduced the maximum strength of the deflectors. The attainable machining tolerances for the electron optical elements has a direct influence on the required relative stability of the power supplies. This is due to the fact that a poor mechanical precision of the electron optical components requires larger excitation ranges for the current drivers of the alignment dipoles in order to counterbalance the parasitic aberrations. The more precise the mechanical alignment is, the lower is the contribution of the alignment elements to the total noise budget. Therefore, the precise construction of the corrector helps to improve the performance of the TEM, although the actual electron ray path can be aligned up to the required precision with the available deflectors and stigmators. The available noise budgets for the various components can be analyzed for a given goal of resolution when designing a new corrector (Haider *et al.*, 2008). The required precision of the mechanics as well as strength and stability for the power supplies must be defined at the same time. With the advanced TEM platforms equipped with gun monochromators (Tiemeijer, 1999; Uhlemann and Haider, 2002; Walther and Stegmann, 2006) in order to reduce the chromatic focus spread information and a CTEM hexapole corrector information transfer in the diffractogram at spatial frequencies above 12/nm at 200 kV was demonstrated (Schlossmacher *et al.*, 2005; Freitag *et al.*, 2005). A Young fringe pattern recorded for an amorphous sample is shown in Figure 12. This indicates the improved stability of the commercial correctors. The precise alignment attainable with the latest hexapole correctors for CTEM can be observed with a Zemlin tableau with an outer tilt angle of 50 mrad (Figure 13).

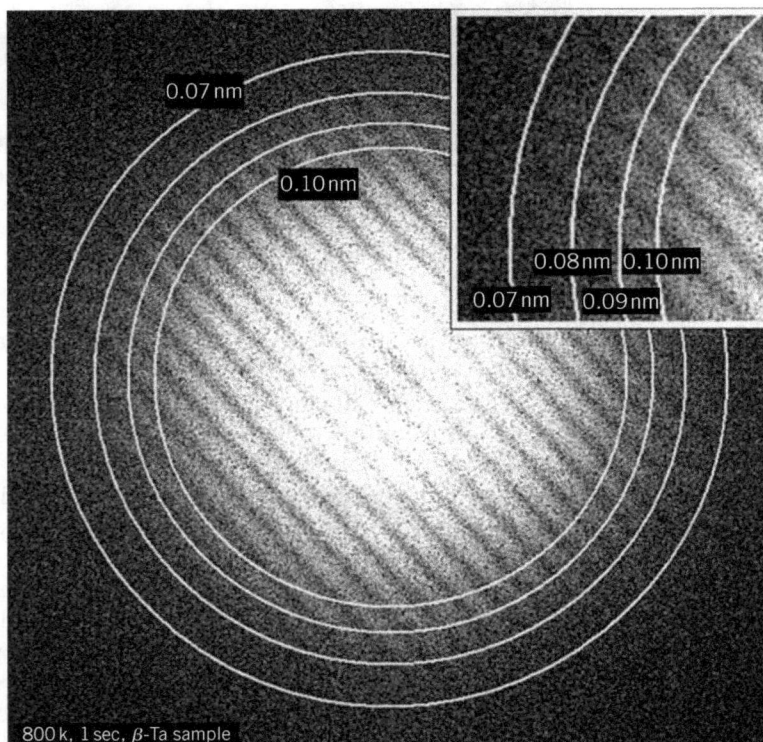

FIGURE 12   Young fringe pattern obtained with a Zeiss UHRTEM consisting of a CETCOR and the CEOS gun monochromator (Uhlemann and Haider, 2002). It demonstrates an information limit of better than 0.08 nm (or 12.5/nm) at 200 kV. The diffractogram has been calculated from the image of an amorphous β-tantalum sample. (From Schlossmacher *et al.*, 2005.)

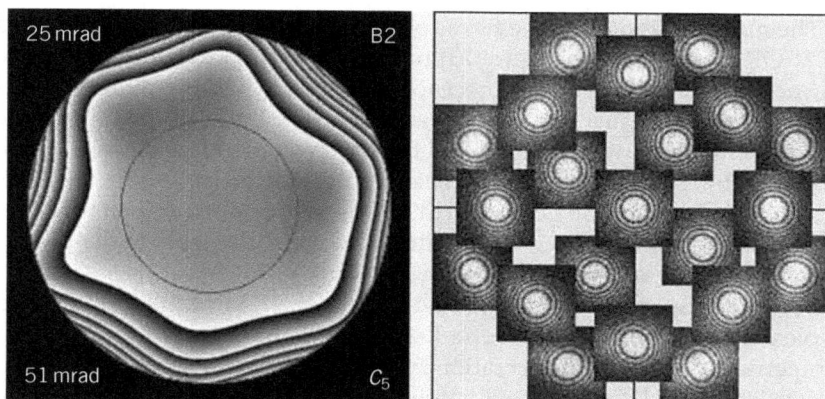

FIGURE 13 Zemlin tableau obtained with a CETCOR for an outer tilt angle of 50 mrad. Although the tilt angle is very large the induced twofold astigmatism and defocus are almost not observable. With this alignment almost aberration-free imaging down to 0.1 nm at 200 kV is possible.

To avoid drift effects after changing thermal load in the corrector we have introduced a constant power design for the transfer lenses. A constant power lens uses a bifilar coil driven by two separate power supplies such that the focal length can be changed while the total power dissipation in the two lens coils is kept constant.

F. Further Progress of Hexapole Correctors

In 2004 the transmission electron aberration-corrected microscope (TEAM) project set the goal of 50 pm resolution in STEM. To achieve this ambitious goal the residual phase shifts for large aperture angles of $\geqslant 40$ mrad must be controlled. This requirement made a partial redesign of the CEOS STEM hexapole corrector necessary.

For the advanced hexapole corrector (D-COR) special care had been taken to reduce the intrinsic sixfold astigmatism A_5 introduced by the corrector and to offer the possibility to compensate for the fourth-order parasitic aberrations by appropriate alignment tools (Müller *et al.*, 2005; Müller *et al.*, 2006b). This was possible because our understanding of the detailed mechanisms of how the intrinsic combination aberrations and residual parasitic aberrations are produced in the system had improved substantially over the years. By design we reduced the coefficient of the intrinsic sixfold astigmatism by more than one order of magnitude from 2—3 mm down to below 200 μm (Müller *et al.*, 2006a). Additionally, by tuning the transfer lenses between objective lens and corrector the coefficient of the intrinsic fifth-order spherical aberration C_5 can be corrected (Hartel *et al.*, 2004). For details, refer to Section II of this chapter.

The magnetic circuit of the hexapole elements as used in CESCOR and CETCOR (compare with Figure 23) has been changed to further improve the mechanical and magnetic precision. For the D-COR the multipole elements are made of just one piece of soft iron. The inner shape of the pole pieces is crafted by wire erosion and the coils were wound up on the pole pins. For this purpose the coil bodies are constructed of two halve-shells and mounted on the pole pins. With this design we came back to the original monolithic solution as used for the first multipole elements at the EMBL. The two designs are compared in Figure 14.

For easier operation additional alignment methods were defined and implemented in software. For the D-COR alignment tools for automatic compensation of all axial aberrations up to and including the fourth-order now exist. In total eighteen real-valued aberration coefficients can be controlled during the user alignment. A screen shot of the D-COR user interface is shown in Figure 15 (Hartel *et al.*, 2007).

The alignment of the first hexapole corrector at the EMBL took several months. Today the factory alignment for a corrector can be done in about one week thanks to more experience and computer-assisted alignment methods. The factory alignment needs to be performed just once during the lifetime of a corrector. The result of the coarse alignment is a fingerprint of the actual manufacturing tolerances and misalignments of an individual system. From our experience we can state that even if a corrector is dismantled and reassembled, most of these data are still valid and can be reused for the precise alignment. Any changes that occur over time due to drift are mainly lower-order aberrations such as twofold and

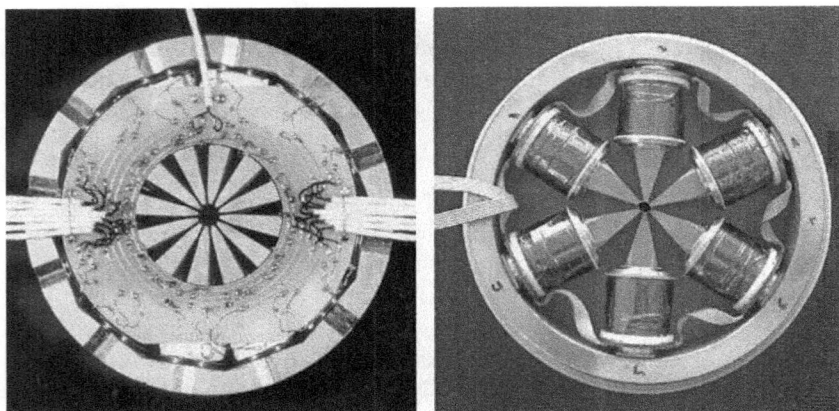

FIGURE 14   Pictures of the multipole elements used for the excitation of the principal hexapole fields. The left-hand photo shows the first multipole constructed at the EMBL as a dodecapole element and the right-hand side shows the design of the hexapole element for the D-COR-type corrector.

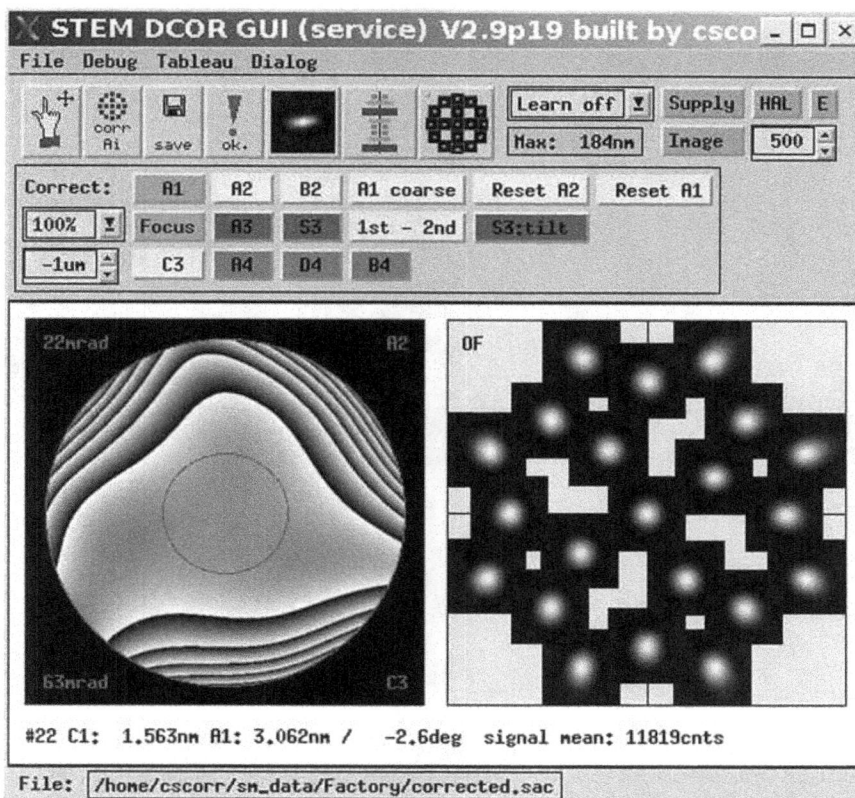

FIGURE 15 Screen shot of the graphical user interface for control and alignment of the CEOS D-COR hexapole corrector for STEM with auto-alignment tools for all fourth-order parasitic aberrations. The measured residual aberrations up to fifth order are visualized by a simulated phase plate image.

threefold astigmatism and axial coma. Residual aberrations in third- and higher-order are stable over a long time, typically days or weeks. Occasionally the auto-alignment tools must be used to maintain the well-corrected state of the instrument.

One impressive example of this stability was the shipping of a high-resolution TEM equipped with a C_s corrector—showing sub-angstrom resolution—from Europe to the annual Microscopy and Microanalysis Conference in Hawaii in the summer of 2005. The TEM arrived in large crates. After installation and electronics test the hexapole corrector was turned on and just slight tuning of second-order coma was necessary to reduce the coma from 100 nm to ~20 nm and sub-angstrom resolution could be observed again.

64   Maximilian Haider *et al.*

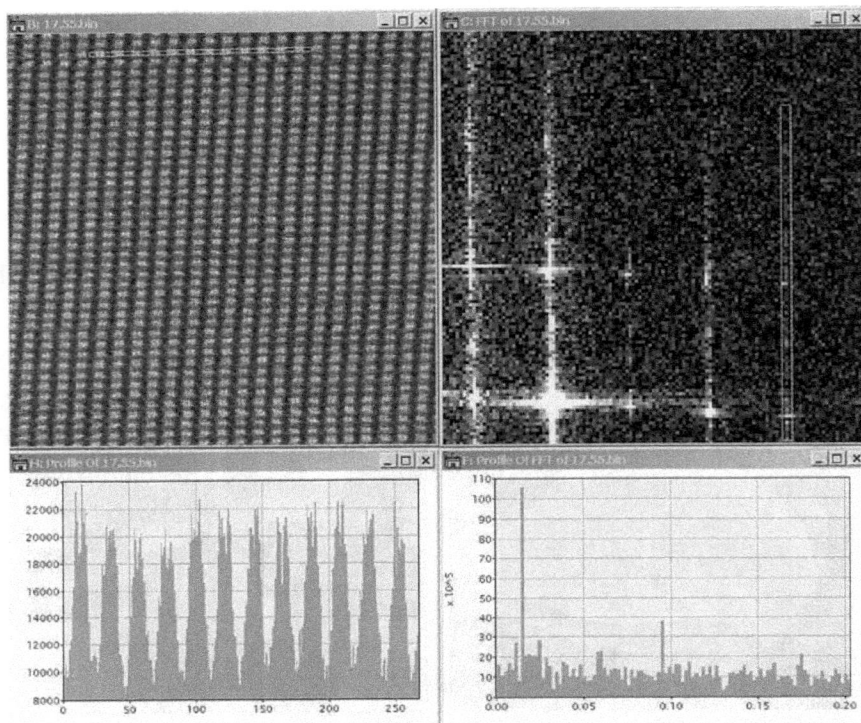

FIGURE 16 STEM image of the dumbbell structure of Ge in 211 orientation (upper left) obtained with an D-COR hexapole corrector incorporated into a FEI Titan TEM operated at 300 kV. The dumbbell structure has a spacing of 0.082 nm and is well resolved (lower left). The reflections for the 444 direction (0.082 nm) and 426 direction (0.076 nm) are clearly visible in the diffractogram (upper right). The line scan shows additionally a weak signal for the 408 direction (0.063 nm) (lower right).

The D-COR design further pushed the limits for STEM resolution. With the first prototype installed at CEOS, Heidelberg in a standard FEI Titan 80-300, we were able to demonstrate 0.082 nm dumbbell resolution in an image of Ge in 211 orientation at 300 kV (Figure 17). In the diffractogram the reflections for the 426 and 408 directions also are visible corresponding to spatial frequencies of 0.128 nm and 0.158 nm, respectively (Hartel, 2007).

Q1

With the final TEAM 0.5 instrument equipped with a D-COR the 0.063 nm Ga dumbbells of GaN in 211 orientation have been resolved with clear reflections in the diffractogram corresponding to 0.049 nm for the 555 direction as shown in (Figure 17). Additionally, with an Au sample 0.048 nm reflection, in the diffractogram were demonstrated. These results (in the summer of 2007) approved the ambitious goal of the TEAM project for STEM at 300 kV and, finally, set a new world record in STEM resolution (Kisielowski *et al.*, 2008).

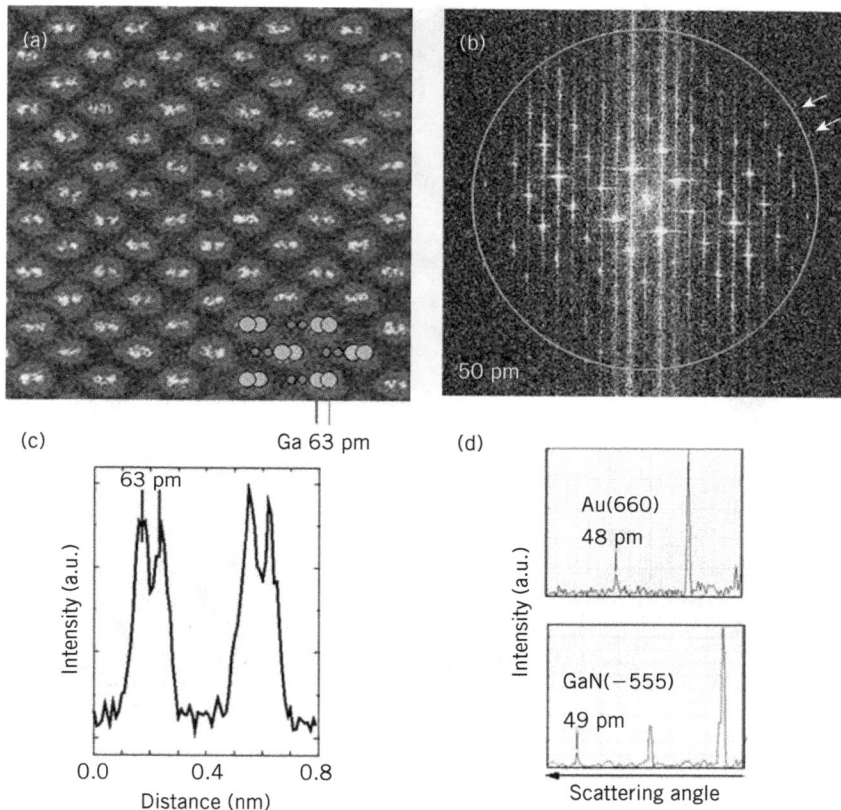

FIGURE 17 STEM image of the dumbbell structure of GaN in 211 orientation (a) obtained with an D-COR hexapole corrector incorporated into the FEI TEAM 0.5 instrument operated at 300 kV. The dumbbell structure has a spacing of 0.063 nm and is well resolved (c). A reflection for the 555 direction corresponding to a spacing of 0.049 nm is clearly visible in the diffractogram (b). At an Au sample a weak signal for the 660 direction (0.048 nm) could be demonstrated (d).

With the D-COR-type design it is possible to use very large STEM apertures. The ronchigram shown in Figure 18 shows a phase-flat area (sweet spot) of ~60 mrad. Nevertheless, the aperture used for these systems is much smaller since the STEM d_{50} probe size (e.g., the diameter of the disc that contains 50% of the total probe current) is primarily governed by the chromatic focus spread. For a standard Schottky-field emission gun (FEG) at 300 kV the optimum aperture amounts to ~25 mrad. To exploit the full benefits of the advanced hexapole STEM corrector it would be advantageous to combine the corrector with a cold or a high-brightness Schottky-FEG equipped with a gun monochromator. With an energy spread reduced down to 0.3 eV full-width half-maximum

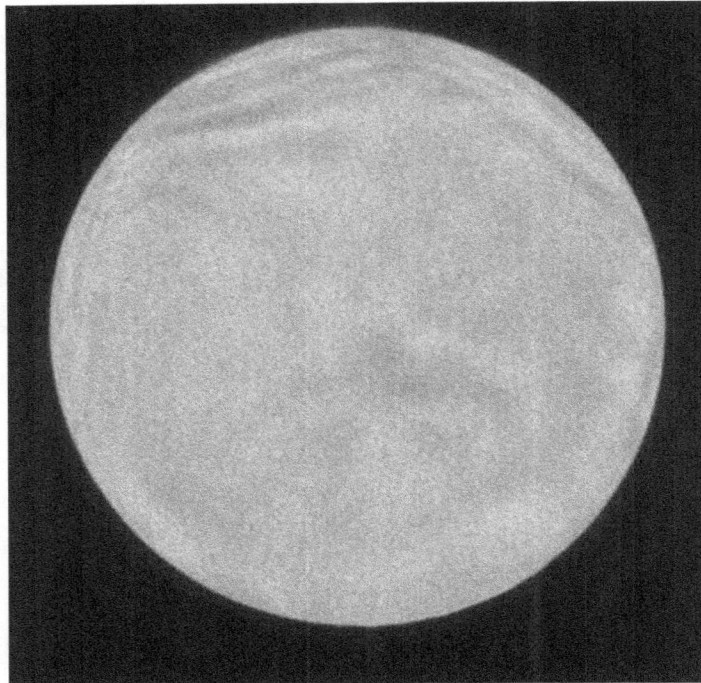

FIGURE 18 In-focus ronchigram with an aperture radius of 80 mrad taken with a CEOS D-COR at 300 kV. The phase-flat area amounts to ∼60 mrad.

(FWHM) the optimum aperture angle (for minimum d_{50}) would increase to ∼40 mrad at 300 kV. The attainable d_{50} STEM probe size in the zero-current limit (zero size virtual source) versus the aperture angle for different C_s-corrected systems is plotted in Figure 19 (Müller *et al.*, 2006a). To obtain the finally achievable STEM probe size the results must be convoluted with the shape of the geometrical image of the virtual source at the specimen plane. The calculations compare two systems: a reduced-gap 200-kV objective lens with small C_3 and small C_c and a standard-gap 300-kV objective lens. With a Schottky-type FEG with $\Delta E = 0.7$ eV (FWHM) both systems are limited by chromatic focus spread and not by fifth-order aberrations. If the energy width is reduced to $\Delta E = 0.3$ eV (FWHM), the aperture angle can be increased accordingly. For a CESCOR-type design the sixfold astigmatism now becomes the limiting aberration for both systems. In this regime the advanced D-COR-type design is most appropriate and can be used to push the STEM probe size to $d_{50} \leqslant 50$ pm and to increase the probe current without loss of spatial resolution for an acceleration voltage of 200 kV or 300 kV.

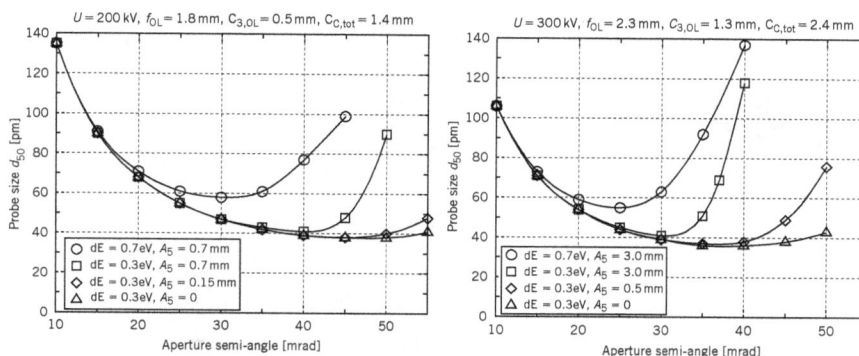

FIGURE 19 Theoretically attainable d_{50} STEM probe size at the zero-current limit for a 200-kV and 300-kV instrument with different energy length $\Delta E \cdot C_c$ and different residual sixfold astigmatism coefficients A_5 (Müller *et al.*, 2006a) and (Hartel *et al.*, 2007).

A larger illumination angle increases the probe current, which is favorable not only for a better signal-to-noise ratio but first of all for analytical work such as electron energy-loss spectroscopy (EELS). For EELS a strong localization of the probe current is important to correctly correlate spatial and analytic information.

By increasing the aperture angle depth resolution can also be improved, because the depth of focus depends quadratically on the aperture angle. Depth resolution will become more important in the future if the expectations with respect to tomography and depth sectioning at atomic scale can be fulfilled (Kisielowski *et al.*, 2008). An illustrative example is show in Figure 21. In STEM individual layers of the specimen can be imaged selectively by focusing the beam at different depth sections.

During the past years the application of hexapole correctors in TEM produced many exciting scientific results. C_s correction improved the visibility of light atoms in bright-field images. Here, imaging with small negative C_3 is a very useful technique to optimize the contrast in the images. This is illustrated by an image of $SrTiO_3$ in Figure 20. In this image the atomic columns are visible as bright spots on dark ground (Jia, Lentzen, and Urban, 2004).

The reduction of the delocalization becomes very obvious in CTEM images taken at lower accelerating voltages. When working at 80 kV, because of the strong interactions of electrons with the objects at this energy the electrons are scattered more often in large angles and, hence, the delocalization is more pronounced. This can be demonstrated by images taken of single carbon nanotubes (SCNTs) in uncorrected and corrected mode (Figure 22).

Q7

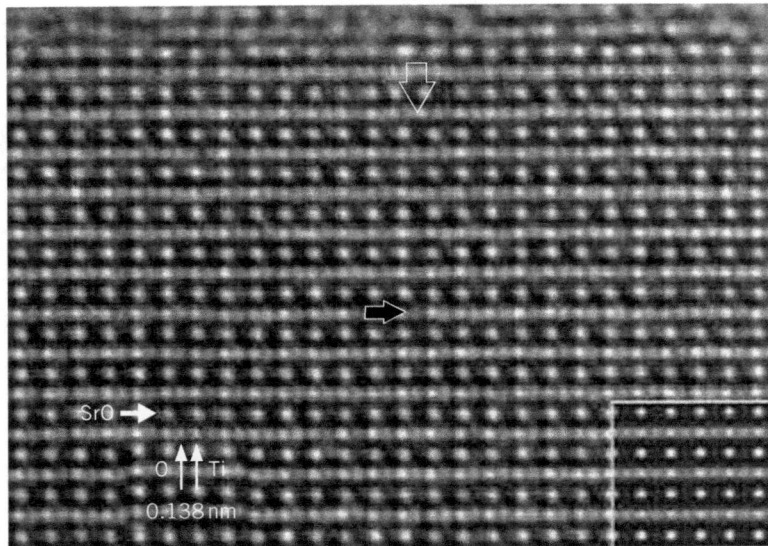

FIGURE 20   SrTiO$_3$ sample in [110] orientation imaged with negative third-order spherical aberration $C_3 \approx -40\,\mu$m and positive defocus $C_1 = 8$ nm. The atomic columns are visible as bright spots on dark ground. The oxygen columns situated between the titanium columns in [001] direction are clearly visible; it is even possible to measure the variable oxygen occupancy quantitatively by comparing image simulations with the recorded contrast. (From Jia, Lentzen, and Urban, 2003.)

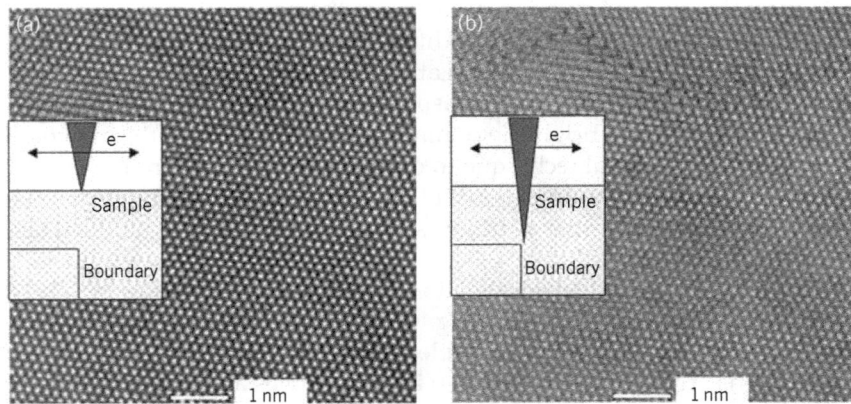

FIGURE 21   STEM depth sectioning with different defocus applied reveals structural boundary in bulk material.

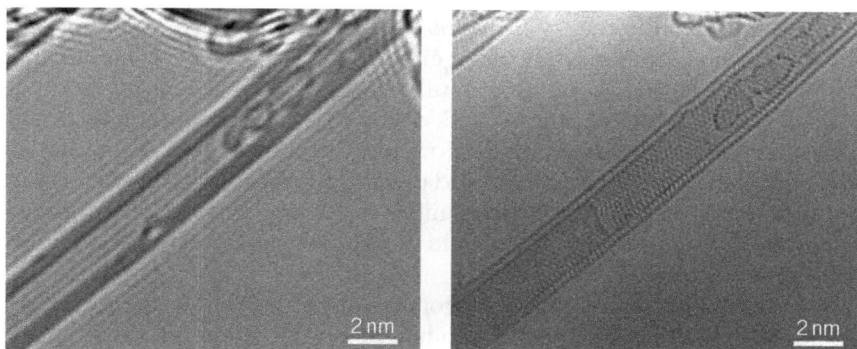

FIGURE 22 Image of a single carbon nanotube taken at 80 kV without (left) and with C_s correction (right). The delocalization due to the spherical aberration is very strong in the case of an uncorrected CTEM.

G. Future Hexapole Correctors

After reviewing the developments of hexapole aberration correctors over the past 10 years the question remains: What will be required in the future and which developments can be anticipated?

The spatial resolution in CTEM and STEM has improved considerably compared with the situation 10 years ago. Therefore, it cannot be expected that the coming 10 years will improve the resolving power by the same factor again. The present limit in high-resolution imaging is no longer due to the coherent axial aberrations but is related more to the chromatic aberration and other incoherent disturbances. Hence, the further development of hexapole correctors will focus more on new fields of applications such as far-field lenses for in situ electron microscopy, Lorentz microscopy for imaging of magnetic domains, and applications such as projection optics in which a very large field of view is mandatory (Rose, 2002). Structural biology with low-dose imaging and single-particle reconstruction also could become an exciting new field for the application of aberration correctors.

The increasing availability of C_s-corrected instruments will initiate the invention of new imaging techniques and applications and these innovations most probably will generate further demands for improved instrumentation. In general, a hexapole corrector is a useful choice whenever for a certain instrument or application the information limit is considerable better than the C_s-limited point resolution. For example, one could imagine designing a fully electrostatic hexapole corrector for low-voltage or ion-beam applications. An electrostatic hexapole corrector would work essentially like a magnetic one but the design could be very compact.

70   Maximilian Haider *et al.*

A reduction of the energy spread is essential for improved resolution in TEM. This can be achieved by using either a cold FEG or a gun monochromator. For the latter a design that is fully dispersion-free seems preferential to keep the brightness of the Schottky-type field emitter as high as possible (Uhlemann and Haider, 2002). For electrons scattered in larger angles in CTEM then higher-order axial and coma-type off-axial aberrations gain importance. The latter limit the number of equally well-resolved image points and, hence, effectively the field of view for a given target resolution (Haider *et al.*, 2008).

The dominant coma-type aberration of a present CTEM is the azimuthal or anisotropic off-axial coma of a magnetic objective lens. This is caused by the Larmor rotation and cannot be avoided for a single-gap magnetic lens. For this reason the present CTEM instruments equipped with a hexapole-corrector are called *semi-aplanatic*. An aplanatic system would be completely free of third-order off-axial coma. In Section III we propose novel designs for hexapole correctors with three and more multipole stages that can be used to compensate for the spherical aberration and the third-order off-axial coma of the objective lens simultaneously.

The correction of the chromatic aberration is the next major step to achieve highest resolution in STEM and CTEM. The TEAM project has recognized this necessity and aims for the development of a C_c-/C_s-corrected CTEM with the goal to allow for an information limit of 0.05 nm at 200 kV. Nevertheless, a hexapole C_s corrector could be combined with a C_c corrector; for this purpose a design based on electric-magnetic quadrupoles and octupoles seems to be the more appropriate choice (Rose, 1971a; Rose 2004). With this type of corrector it is possible to correct simultaneously for the chromatic aberration, the spherical aberration, and the azimuthal coma of the objective lens (Haider *et al.*, 2008). A first prototype quadrupole-octupole C_c-/C_s-corrector using a quadrupole Wien filter for chromatic correction has been developed by CEOS on behalf of Argonne National Lab (Hartel *et al.*, 2008). If this novel design proves successful, it will certainly start an new area in aberration-corrected electron microscopy.

II. PRESENT HEXAPOLE CORRECTORS

A. Hexapole Elements

In a standard electron microscope all optical elements except for alignment deflectors and stigmators are cylinder symmetric. For systems equipped with a multipole aberration corrector this is no longer true. Hexapole C_s correctors use strong magnetic fields with threefold symmetry to correct for the spherical aberration of the objective lens. The optical elements producing these fields are called *hexapoles* or *sextupoles*, since the arrangement of their pole pieces has sixfold symmetry.

FIGURE 23 Magnetic hexapole element as used in a hexapole C_s corrector. The diameter of the outer yoke is 152 mm and the bore is 8 mm. The length of the element in z-direction amounts to $L_{HP} = 30$ mm. The liner tube placed inside the bore and the field clamps are not shown. (From Müller *et al.*, 2006a.)

Figure 23 shows an embodiment of such an element. The hexapole consists of a cylindrical outer yoke with six pole pins pointing toward the optic axis. The tips of the pole pins have an optimized shape to produce a strong and accurate hexapole field in close vicinity to the optic axis. The coils mounted at the pole pins are excited by currents whose directions alternate from pin to pin.

1. Magnetic Field

The magnetic flux density B near the optic axis can be described by the gradient of a magnetic scalar potential ψ_3. Sufficiently far away from the entrance and exit faces of the element it adopts the form of a plane hexapole field

$$\psi_3 = \mathrm{Re}\left\{\Psi_3 \overline{w}^3\right\} = \Psi_{3c}\left(x^3 - 3xy^2\right) + \Psi_{3s}\left(3x^2y - y^3\right), \qquad (1)$$

$$B = -\nabla\Psi = -2\partial_{\overline{w}}\Psi_3 = -3\Psi_3\overline{w}^2. \qquad (2)$$

To simplify the mathematics we use a complex notation where the lateral distance from the optic axis is $w = x + iy$ and its complex conjugate $\overline{w} = x - iy$. The hexapole strength along the optic axis is described by two real-valued components, Ψ_{3c} and Ψ_{3s}, corresponding to a cosine-like and sine-like azimuthal modulation of the potential ψ_3, respectively. In the complex notation we have $\Psi_3 = \Psi_{3c} + i\Psi_{3s}$. The azimuthal orientation of the hexapole is determined by an angle $0° \leq \varphi \leq 30°$. For $\varphi = 0°$ the hexapole strength is real for $\varphi = 30°$ imaginary.

2. Hexapole Strength

A rather simple argument based on Ampere's law and the assumption of infinite permeability $\mu \rightarrow \infty$ of the magnetic material can be used (Haider, Bernhardt, and Rose, 1982) to relate the maximum hexapole strength along the optic axis to the current driving the coils of the element

$$|\Psi_3| = F_{HP} \frac{\mu_0 NI}{R_{HP}^3}, \qquad (3)$$

where R_{HP} denotes the bore radius, NI the current in ampere turns, and F_{HP} is a Fourier factor in the order of one depending on the exact shape of the pole tips. More detailed investigations consider not only the correct Fourier factor F_{HP} but also the exact fringe field at the entrance and exit face of the element.

For this purpose numeric field calculations must be performed. The result of such a calculation (Müller *et al.*, 2006) using the boundary element method for an unsaturated magnetic hexapole element is shown in Figure 24. The course of the hexapole strength along the optic axis can be approximated roughly by a box-shaped function as long as fringing field effects can be neglected. This sharp cut-off fringing field (SCOFF) approximation is a useful tool for initial theoretical investigations. Very often the principal behavior of a corrector system can be described by rather simple analytical relations based on SCOFF calculations. The SCOFF length L of the hexapole field is larger than the physical length of the element L_{HP}; typically the FWHM of the axial multipole strength is used.

B. Aberrations of Hexapole Fields

Before the advent of aberration correctors in electron microscopy hexapole elements—most often simply air coils—have been used as stigmators for threefold astigmatism A_2. Hexapole fields do not influence the paraxial path of rays since paraxial optics considers only constant and linear fields coming from deflectors, quadrupoles, and round lenses. Effectively, a beam transversing a hexapole field is shaped threefold.

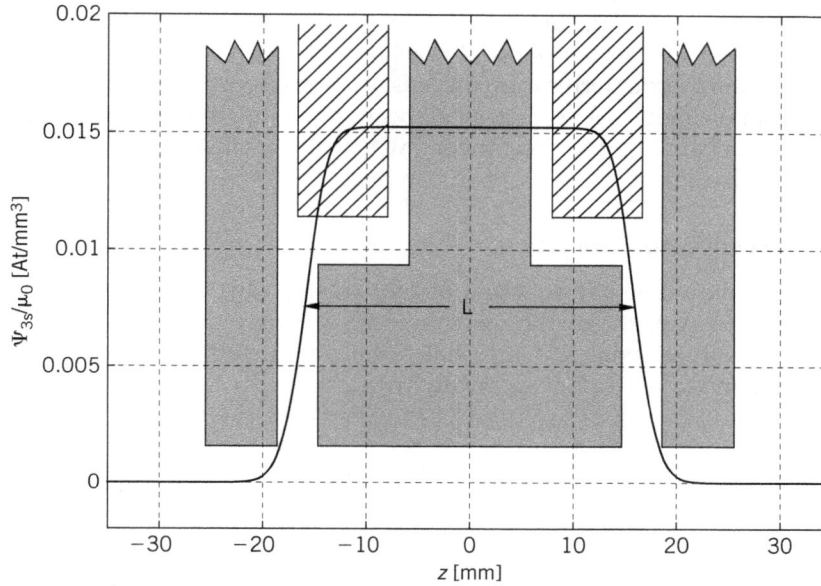

FIGURE 24 Course of the axial hexapole strength for a magnetic hexapole element with length $L_{HP} = 30$ mm and bore radius $R_{HP} = 4.05$ mm enclosed by field clamps in front and behind the element. The hashed area indicates the coil with unit excitation $NI = 1$ At. The data were obtained by a 3D magnetic boundary element method based on the indirect reduced magnetic potential approach for unsaturated magnetic material with permeability $\mu/\mu_0 \approx 10^4$. (From Müller *et al.*, 2006.)

1. Equation of Motion

To describe this quantitatively we add the hexapole field as a nonlinear perturbation at the right-hand side of the linear paraxial ray equation. To simplify the mathematics we neglect the influence of the fringe fields and assume a box-shaped hexapole strength $\Psi_3(z)$ along the optic axis. Electrons traveling through a constant hexapole field experience a Lorentz force perpendicular to the direction of the magnetic field

$$\frac{d^2}{dz^2}u + \frac{\eta^2}{4}B(z)^2 u = 3i\eta\Psi_3(z)\,\overline{u}^2, \tag{4}$$

with $\eta = \sqrt{\frac{|e|}{2m_0 U_0^\star}}$, where e denotes the charge of the electron, m_0 the mass of rest, c the speed of light, and $U_0^\star = U_0\left(1 + \frac{|e|}{2m_0 c^2}U_0\right)$ the relativistically modified acceleration voltage.

The left-hand side of the equation of motion [Eq. (4)] describes the focusing action of the round lenses and the free-space propagation in between.

The hexapole strength causes a small deviation of the electron trajectory u from the course of the paraxial ray. Equation (4) is not exact because it neglects third-order terms form the lens fields and fourth-order terms from the hexapole fields. Nevertheless, all beam plots shown in the subsequent figures are based on the exact theory, which also includes round-lens and fringe-field effects.

2. Successive Approximation

To solve the nonlinear equation of motion [Eq. (4)], we perform a perturbation expansion of the electron path $u = u(z)$ with respect to the complex-valued axial and off-axial beam parameters at the object $\alpha = \vartheta_x + i\vartheta_y$ and $\gamma = x + iy$, respectively

$$u(z) = \sum_{i=1}^{\infty} u^{(i)}(z). \tag{5}$$

Here, $u^{(n)}$ denotes the path deviation of order n. The paraxial trajectory $u^{(1)} = \alpha u_\alpha + \gamma u_\gamma$ is the general solution of the homogeneous linear equation at the left-hand side of Eq. (4). This paraxial solution is cylinder-symmetric, and the two fundamental rays are real-valued with respect to the Larmor frame of reference even if magnetic round lenses are present. By variation of the coefficients we find an integral equation for the general solution of the dynamic problem [Eq. (4)] including the hexapole perturbation

$$u(z) = \left(\alpha + \int_{z_\gamma}^{z} 3i\eta\Psi_3\,\overline{u}^2 u_\gamma dz\right) \cdot u_\alpha + \left(\gamma - \int_{z_\alpha}^{z} 3i\eta\Psi_3\,\overline{u}^2 u_\alpha dz\right) \cdot u_\gamma. \tag{6}$$

For a system without aperture the lower bounds of the integration are the object plane $z_\alpha = z_\gamma = z_0$. In this case $\alpha u'_\alpha(z_0)$ is equal to the slope of the exact trajectory at the object plane $z = z_0$. For the canonical fundamental rays with initial values

$$u_\alpha(z_0) = 0, \qquad u'_\alpha(z_0) = 1,$$
$$u_\gamma(z_0) = 1, \qquad u'_\gamma(z_0) = m_\gamma, \tag{7}$$

the Helmholtz-Lagrange invariant is unity

$$u'_\alpha u_\gamma - u_\alpha u'_\gamma = 1, \quad \text{for all } z \geq z_0. \tag{8}$$

In this mixed representation the axial ray u_α has the dimension of a length while the field ray is dimensionless. The slope parameter m_γ describes the convergence of illumination in CTEM. For parallel illumination we must set $m_\gamma = 0$. Formally convergent or divergent illumination corresponds to a linear transformation of the off-axial ray $u_\gamma \to u_\gamma + m_\gamma u_\alpha$. In STEM this operation corresponds to a change of the position of the aperture plane $z = z_a$ with $u_\gamma(z_a) = 0$. For a system with aperture a constant multiple of the axial ray must be added at the right-hand side of Eq. (6) to make the path deviation $u - u^{(1)}$ vanish at the aperture plane $z = z_a$. This is identical to choosing $z_\gamma = z_a$ for the lower bound of the first integral. In this case, $\alpha u_\alpha(z_a)$ is equal to the aperture coordinate of the exact trajectory.

3. Primary Aberrations

The integral equation [Eq. (6)] can be solved iteratively by successive approximation. Substituting $u \to u^{(1)}$ at the right-hand side yields

$$\bar{u}^2 = \left(\overline{\alpha u_\alpha + \gamma u_\gamma}\right)^2 = \bar{\alpha}^2 u_\alpha^2 + 2\overline{\alpha\gamma}u_\alpha u_\gamma + \bar{\gamma}^2 u_\gamma^2,$$

and the second-order path deviation adopts the form

$$u^{(2)} = \bar{\alpha}^2 u_{\overline{\alpha\alpha}} + \overline{\alpha\gamma}u_{\overline{\alpha\gamma}} + \bar{\gamma}^2 u_{\overline{\gamma\gamma}}. \tag{9}$$

This shows that due to the structure of Eq. (6) only a very limited number of index combinations are allowed by symmetry—in this case 3 of 10.

In the following we chose a fixed orientation for the hexapole field $\Psi_3 = i\Psi_{3s}$. In this case, the right-hand side of Eq. (4) and, therefore, all aberration rays become real-valued.

With this choice the evaluation of the integral [Eq. (6)] results in an explicit representation for the second-order aberration rays

$$u_{\overline{\alpha\alpha}} = -u_\alpha \cdot \int_{-\infty}^{z} 3\eta\Psi_{3s}u_\alpha^2 u_\gamma dz + u_\gamma \cdot \int_{-\infty}^{z} 3\eta\Psi_{3s}u_\alpha^3 dz$$

$$u_{\overline{\alpha\gamma}} = -u_\alpha \cdot \int_{-\infty}^{z} 6\eta\Psi_{3s}u_\alpha u_\gamma^2 dz + u_\gamma \cdot \int_{-\infty}^{z} 6\eta\Psi_{3s}u_\alpha^2 u_\gamma dz \tag{10}$$

$$u_{\overline{\gamma\gamma}} = -u_\alpha \cdot \int_{-\infty}^{z} 3\eta\Psi_3 u_\gamma^3 dz + u_\gamma \cdot \int_{-\infty}^{z} 3\eta\Psi_{3s}u_\alpha u_\gamma^2 dz.$$

By choosing the fundamental rays $u_\alpha = f_0$ and $u_\gamma = -z/f_0$ according to Figure 25 we obtain a piecewise representation of the second-order

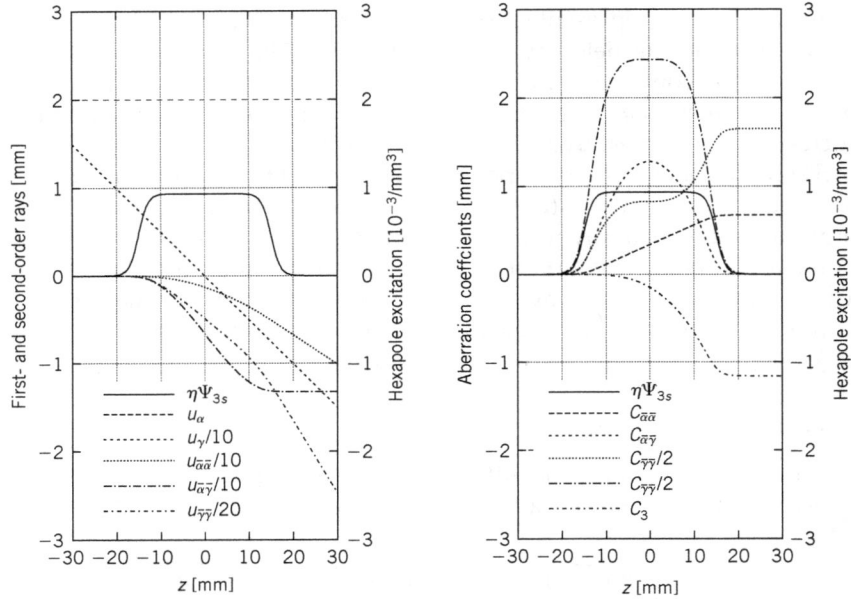

FIGURE 25 Single hexapole element with a certain choice for the paraxial fundamental rays $u_\alpha = f_0$ and $u_\gamma = -\frac{z}{f_0}$. At the left the course of the second-order aberration rays is depicted, and at the right the corresponding aberration coefficients are shown. The parameters are $L = 30$ mm and $f_0 = 2$ mm.

aberration rays in SCOFF approximation

$$
\begin{array}{cccc}
 & z < -\frac{L}{2} & -\frac{L}{2} \leq z \leq \frac{L}{2} & \frac{L}{2} < z \\[2mm]
u_{\overline{\alpha\alpha}} = 0 & & -\frac{3}{2}\eta\Psi_{3s}f_0^2\left(z+\frac{L}{2}\right)^2 & -3\eta\Psi_{3s}f_0^2 L z \\[2mm]
u_{\overline{\alpha\gamma}} = 0 & & -\frac{1}{4}\eta\Psi_{3s}\left(L^3 - 3L^2 z - 4z^3\right) & -\frac{1}{2}\eta\Psi_{3s}L^3 \\[2mm]
u_{\overline{\gamma\gamma}} = 0 & & -\frac{\eta\Psi_{3s}}{64f_0^2}\left(16z^4 + 8L^3 z + 3L^4\right) & -\frac{\eta\Psi_{3s}}{4f_0^2}L^3 z.
\end{array}
\tag{11}
$$

In Figure 25 the course of the fundamental rays and the second-order aberration rays calculated for a realistic shape of the hexapole field are depicted with $L = 30$ mm and $f_0 = 2$ mm. The results are in good agreement with the piecewise approximation in Eq. (11).

The aberration rays are represented by linear combinations of the paraxial rays with two independent integral coefficients. These coefficients are function of z. In general, for each aberration ray u_X in arbitrary order with index combination X, two coefficient functions c_X and C_X exist

$$u_X(z) = u_\alpha(z) c_X(z) + u_\gamma(z) C_X(z),$$
$$u'_X(z) = u'_\alpha(z) c_X(z) + u'_\gamma(z) C_X(z), \qquad (12)$$

which fully describe the course of the aberration ray and of its first derivative. The coefficient function C_X can be interpreted as the projection of the aberration ray u_X on the off-axial fundamental ray u_γ and c_X as that on the axial fundamental ray u_α. Evaluated at some image plane $z = z_i$ the value $C_X(z_i) = \frac{u_X(z_i)}{u_\gamma(z_i)}$ is the image aberration referred back to the object plane $z = z_0$, while the other coefficient c_X governs the aberrations at the diffraction image where $u_\gamma = 0$. For this reason the coefficients C_X and c_X are sometimes called *image* and *slope* coefficients, respectively.

The complex-valued beam parameters allow classification of the aberration rays and coefficients according to their behavior under rotation of the frame of reference about the optic axis. The transformation

$$\alpha \to \alpha e^{i\theta}, \qquad \gamma \to \gamma e^{i\theta} \qquad (13)$$

causes the aberration figure to rotate by an integer multiple $(\nu + 1)$ of the angle θ. Therefore, for a general index combination $\alpha^{N_\alpha} \bar{\alpha}^{N_{\bar\alpha}} \gamma^{N_\gamma} \bar{\gamma}^{N_{\bar\gamma}}$ the azimuthal multiplicity is given by

$$\nu = N_\alpha - N_{\bar\alpha} + N_\gamma - N_{\bar\gamma} - 1. \qquad (14)$$

For an optical system with cylinder symmetry only index combinations with multiplicity $\nu = 0$ are allowed; if in addition hexapole fields are present, the allowed multiplicities are $\nu = 3 \cdot k$, with integer k. Most aberration coefficients are complex-valued, characterized by magnitude and orientation. The real part describes the isotropic and the imaginary part the anisotropic component.

The most important intrinsic aberrations that can occur in a system with threefold symmetry are listed by the common names and the corresponding aberration coefficients in traditional notation and index notation in Table I.

The coefficient $C_{\bar\alpha\bar\alpha}$ in Eq. (10) evaluated at an image plane is the threefold astigmatism A_2. In SCOFF approximation we find the stigmator strength

$$A_2 = C_{\bar\alpha\bar\alpha} = \int_{z_0}^{z_i} 3\eta \Psi_{3s} u_\alpha^3 \, dz = 3\eta \Psi_{3s} L f_0^3. \qquad (15)$$

TABLE I Important aberration coefficients of hexapole-type correctors.

Aberration ray	Aberration coefficient	Multiplicity ν	Order n	Common name
$u_{\overline{\alpha\alpha}}$	$A_2, C_{\overline{\alpha\alpha}}$	-3	2	Threefold astigmatism
$u_{\alpha\alpha\bar\alpha}$	$C_3, C_s, C_{\alpha\alpha\bar\alpha}$	0	3	Third-order spherical aberration
$u_{\alpha\alpha\bar\gamma}, u_{\alpha\bar\alpha\gamma}$	$K_3, C_{\alpha\alpha\bar\gamma}, C_{\alpha\bar\alpha\gamma}$	0	3	Third-order off-axial coma
$u_{\alpha\gamma\bar\gamma}$	$FC_3, C_{\alpha\gamma\bar\gamma}$	0	3	Third-order field curvature
$u_{\bar\alpha\gamma\gamma}$	$FA_3, C_{\bar\alpha\gamma\gamma}$	0	3	Third-order field astigmatism
$u_{\gamma\gamma\bar\gamma}$	$D_3, C_{\gamma\gamma\bar\gamma}$	0	3	Third-order distortion
$u_{\alpha\alpha\alpha\bar\alpha}, u_{\alpha\bar\alpha\bar\alpha\bar\alpha}$	$D_4, C_{\alpha\alpha\alpha\bar\alpha}, C_{\alpha\overline{\alpha\alpha\alpha}}$	± 3	4	Fourth-order three-lobe aberration
$u_{\alpha\alpha\alpha\gamma}, u_{\alpha\overline{\alpha\alpha}\gamma}$	$C_{\alpha\alpha\alpha\gamma}, C_{\alpha\overline{\alpha\alpha}\gamma}$	± 3	4	Off-axial star aberration
$u_{\overline{\alpha\alpha\alpha}\gamma}$	$C_{\overline{\alpha\alpha\alpha}\gamma}$	-3	4	Off-axial fourfold astigmatism
$u_{\alpha\alpha\alpha\bar\alpha\bar\alpha}$	$C_5, C_{\alpha\alpha\alpha\bar\alpha\bar\alpha}$	0	5	Fifth-order spherical aberration
$u_{\overline{\alpha\alpha\alpha\alpha\alpha}}$	$A_5, C_{\overline{\alpha\alpha\alpha\alpha\alpha}}$	-6	5	Fifth-order sixfold astigmatism

These only have rotational, threefold and sixfold symmetry ν.

For the system depicted in Figure 25 with a nodal plane $u_\gamma = 0$ at the center of the hexapole field, second-order threefold distortion

$$C_{\overline{\gamma}\overline{\gamma}} = \int_{z_0}^{z_i} 3\eta \Psi_{3s} u_\gamma^2 u_\alpha dz = \frac{\eta \Psi_{3s}}{4f_0} L^3 \tag{16}$$

also would be visible at the image plane. The course of all four aberration coefficients for this system is depicted in Figure 25.

4. Secondary Aberrations

We now direct our attention to the third-order aberrations of a single hexapole field. For the next step of the iteration procedure, we substitute $u \to u^{(1)} + u^{(2)}$ at the right-hand side of the integral Eq. (6). From Eq. (9) and

$$\left(\overline{u^{(1)} + u^{(2)}}\right)^2 = \left(\overline{\alpha}u_\alpha + \overline{\gamma}u_\gamma + \alpha\alpha\overline{u}_{\overline{\alpha}\overline{\alpha}} + \alpha\gamma\overline{u}_{\overline{\alpha}\overline{\gamma}} + \gamma\gamma\overline{u}_{\overline{\gamma}\overline{\gamma}}\right)^2$$

we find by collecting the third-order terms

$$u^{(3)} = \alpha^2\overline{\alpha}u_{\alpha\alpha\overline{\alpha}} + \alpha^2\overline{\gamma}u_{\alpha\alpha\overline{\gamma}} + \alpha\overline{\alpha}\gamma u_{\alpha\overline{\alpha}\gamma} + \alpha\gamma\overline{\gamma}u_{\alpha\gamma\overline{\gamma}}$$
$$+ \overline{\alpha}\gamma^2 u_{\overline{\alpha}\gamma\gamma} + \gamma^2\overline{\gamma}u_{\gamma\gamma\overline{\gamma}}. \tag{17}$$

According to relation (14) only aberrations with multiplicity $\nu = 0$ occur. This is a very remarkable and important result. The axial and off-axial third-order aberrations caused by a hexapole field have exactly the same structure as those of a round lens.

The first term on the right-hand side of Eq. (17) evaluated at the image corresponds to the spherical aberration C_3. From the integral Eq. (6) we find

$$C_3 = \frac{u_{\alpha\alpha\overline{\alpha}}(z_i)}{u_\gamma(z_i)} = \int_{z_0}^{z_i} 6\eta \Psi_{3s} u_\alpha^2 \overline{u}_{\overline{\alpha}\overline{\alpha}} \, dz. \tag{18}$$

For the system from Figure 25 we now substitute Eq. (11) and perform the integration. For the coefficient of the spherical aberration this results in

$$C_3 = C_{\alpha\alpha\overline{\alpha}} = -3 \left|\eta \Psi_{3s}\right|^2 L^3 f_0^4 < 0, \tag{19}$$

while according to the Scherzer theorem for a round lens the coefficient of the spherical aberration is always positive. It is actually this finding that makes hexapole correctors attractive for correcting for round lens aberrations.

C. Hexapole Corrector

Use of hexapole elements for the correction of the spherical aberration requires a setup for which the strong axial second-order aberrations of the hexapole elements can be avoided while the third-order effect on C_3 is maintained. Two hexapole stages can be combined such that their three-fold astigmatism cancels and the negative contributions of both stages to the third-order spherical aberration add up. This step was taken by (Beck, 1979) and (Crewe, 1980). Unfortunately, these early systems suffered from axial fourth-order three-lobe aberration $D_4 = C_{\alpha\alpha\alpha}$, as pointed out by (Rose, 1981). This aberration prevents a useful application, even in a probe-forming microscope (STEM). Rose proved that the shape of the axial bundle must be cylinder-symmetric at least in second order everywhere outside the corrector and not only at the final image plane in order to obtain a real improvement in resolution with a hexapole C_s corrector. In other words, it is necessary to avoid both the image coefficient $C_{\overline{\alpha}\alpha}$ and the slope coefficient $c_{\overline{\alpha}\alpha}$ of the axial second-order aberration ray $u_{\overline{\alpha}\alpha}$ outside the corrector to eliminate D_4.

1. Basic Setup

Rose (1981, 1990) proposed a minimum setup of two hexapoles and two round lenses to circumvents the aforementioned problem. Today, this system (depicted in Figure 26) forms the basis for all available hexapole correctors. It consists of two symmetric hexapole elements with a telescopic $4f$ round-lens doublet in between. The mid-plane S_1 of the first hexapole field is conjugated to the mid-plane S_2 of the second. This arrangement has the advantage that all second-order aberration rays [Eq. (10)] vanish identically outside the corrector, since the four conditions

$$\int \eta \Psi_3 u_\alpha^3 dz = \int \eta \Psi_3 u_\alpha^2 u_\gamma dz = \int \eta \Psi_3 u_\alpha u_\gamma^2 dz = \int \eta \Psi_3 u_\gamma^3 dz = 0 \quad (20)$$

are simultaneously fulfilled due to a local double-symmetric arrangement of the fields and fundamental rays. The hexapole fields are symmetric with respect to the planes S_0, S_1, and S_2 while the lens fields are anti-symmetric with respect to plane S_0. The axial ray u_α is symmetric with respect to S_1 and S_2 and anti-symmetric with respect to S_0. The field ray u_γ has the opposite symmetry. It is symmetric with respect to S_0 and anti-symmetric with respect to S_1 and S_2. The full symmetry correction of all second-order residual aberrations makes this design attractive for STEM and CTEM. The course of the second-order aberration rays and coefficient functions through the corrector is depicted in Figure 26. A piecewise representation in SCOFF approximation can be obtained directly from Eq. (11) by

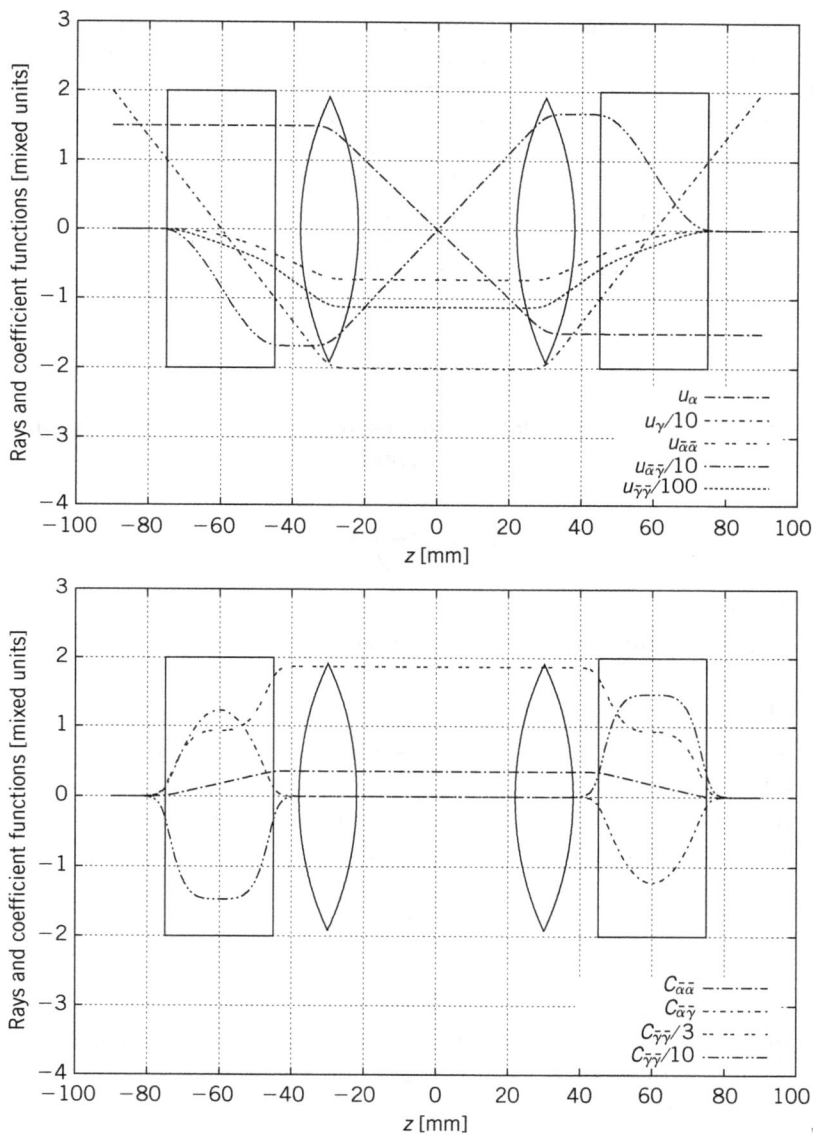

FIGURE 26 Hexapole doublet with local double symmetry with respect to the mid-plane between the transfer lenses S_0 and the mid-planes S_1 and S_2 of the first and second hexapole fields respectively. The top diagram shows the fundamental rays and the second-order aberration rays; the bottom diagram shows the corresponding coefficient functions.

symmetric continuation since the course of the second-order aberration rays $u_{\overline{\alpha}\alpha}$ and $u_{\overline{\gamma}\gamma}$ is symmetric with respect to plane S_0, while that of $u_{\overline{\alpha}\gamma}$ is anti-symmetric. Additionally, the one-to-one imaging between the planes S_1 and S_2 by the anti-symmetric round lens doublet is free of Larmor rotation and third-order off-axial coma and geometric distortion.

According to Hosokawa (2004) the conditions in Eq. (20) can still be fulfilled if the focal length of the first and second transfer lens are different $f_1 \neq f_2$. In this case, the transfer lenses also must be operated as a telescopic doublet with hexapole elements of length L_1 and

$$L_2 = \left(\frac{f_2}{f_1}\right)^2 L_1$$

placed at the front and back focal planes, respectively. For this system the hexapole excitations must fulfill the condition

$$\Psi_3^{(1)} = \left(\frac{f_1}{f_2}\right)^5 \Psi_3^{(2)}.$$

The latter relation shows that the required hexapole excitations rapidly diverge if the ratio f_1/f_2 is substantially different from 1.

2. Third-Order Aberrations

We now derive the third-order aberration coefficients for the hexapole corrector as depicted in Figure 26. At the asymptotic image plane $z \to \infty$ behind the corrector the third-order path deviation [Eq. (17)] can be rearranged in the form

$$u^{(3)}/u_\gamma = \alpha^2\overline{\alpha}\,C_{\alpha\alpha\overline{\alpha}} + \alpha^2\overline{\gamma}\,C_{\alpha\alpha\overline{\gamma}} + \alpha\overline{\alpha}\gamma\,C_{\alpha\overline{\alpha}\gamma} + \alpha\gamma\overline{\gamma}\,C_{\alpha\gamma\overline{\gamma}}$$
$$+ \overline{\alpha}\gamma^2\,C_{\overline{\alpha}\gamma\gamma} + \gamma^2\overline{\gamma}\,C_{\gamma\gamma\overline{\gamma}}. \tag{21}$$

With the SCOFF approximation for the second-order aberration rays and the iteration formula [Eq. (6)] it is straightforward to calculate all third-order aberration coefficients in Eq. (21). The spherical aberration of both hexapoles simply adds up to

$$C_3 = C_{\alpha\alpha\overline{\alpha}} = -6\,|\eta\Psi_{3s}|^2 L^3 f_0^4. \tag{22}$$

The coefficient of the off-axial coma vanishes due to symmetry

$$K_3 = C_{\alpha\alpha\overline{\gamma}} = \frac{1}{2}\overline{C}_{\alpha\overline{\alpha}\gamma} = \int 6\eta\Psi_{3s}\,\overline{u}_{\overline{\alpha}\alpha}u_\alpha u_\gamma dz = \frac{1}{2}\int 6\eta\Psi_{3s}\overline{u}_{\overline{\alpha}\gamma}u_\alpha^2\,dz = 0 \tag{23}$$

since both integrands are anti-symmetric with respect to mid-plane S_0. The relation between the two coma-type coefficients $C_{\alpha\alpha\bar\gamma}$ and $C_{\alpha\bar\alpha\gamma}$ can be derived by partial integration from the explicit representation of the second-order aberrations rays.

The coefficient of field curvature is

$$FC_3 = C_{\alpha\gamma\bar\gamma} = \int 6\eta\Psi_{3s}\,\overline{u}_{\overline{\alpha\gamma}}u_\alpha u_\gamma dz = \frac{3}{5}\,|\eta\Psi_{3s}|^2\,L^5, \tag{24}$$

and for field astigmatism we find

$$FA_3 = C_{\bar\alpha\gamma\gamma} = \int 6\eta\Psi_{3s}\,\overline{u}_{\overline{\gamma\gamma}}u_\alpha^2 dz = -\frac{3}{5}\,|\eta\Psi_{3s}|^2\,L^5 = -FC_3. \tag{25}$$

The coefficient of distortion again vanishes due to symmetry

$$D_3 = C_{\gamma\gamma\bar\gamma} = \int 6\eta\Psi_{3s}\,\overline{u}_{\overline{\gamma\gamma}}u_\alpha u_\gamma dz = 0. \tag{26}$$

Finally, the slope coefficient of the distortion ray $u_{\gamma\gamma\bar\gamma}$, which has no effect on the image but causes spherical aberration at the diffraction image, evaluates to

$$E_3 = c_{\gamma\gamma\bar\gamma} = -\int 6\eta\Psi_{3s}\,\overline{u}_{\overline{\gamma\gamma}}u_\gamma^2 dz = -\frac{3}{56}\,|\eta\Psi_{3s}|^2\,\frac{L^7}{f_0^4}. \tag{27}$$

Formula (22) verifies that the hexapole corrector has negative spherical aberration $C_3 < 0$ that can be used to compensate for the always positive spherical aberration $C_{3,\mathrm{OL}} > 0$ of an objective lens. We further see that the corrector itself is free of off-axial coma $C_{\alpha\alpha\bar\gamma}$ and geometrical distortion $C_{\gamma\gamma\bar\gamma}$. Up to third order the hexapole doublet behaves exactly like a round lens with negative spherical aberration. This means that with this corrector device the rays corresponding to large-aperture angles are refracted less strongly than the paraxial rays. For any round lens the opposite is true.

The correction principle of the hexapole doublet is illustrated vividly by a 3D plot of the shape of the axial beam through the corrector (Figure 27). All path deviations up to third-order have been considered. The three-fold shape of the beam between the hexapole elements is clearly visible. Although at the entrance the beam is cylindrical, it is shaped like a divergent cone at the exit. The angle of divergence increases with the third power of the lateral distance from the optic axis. This is exactly the effect of the negative third-order spherical aberration.

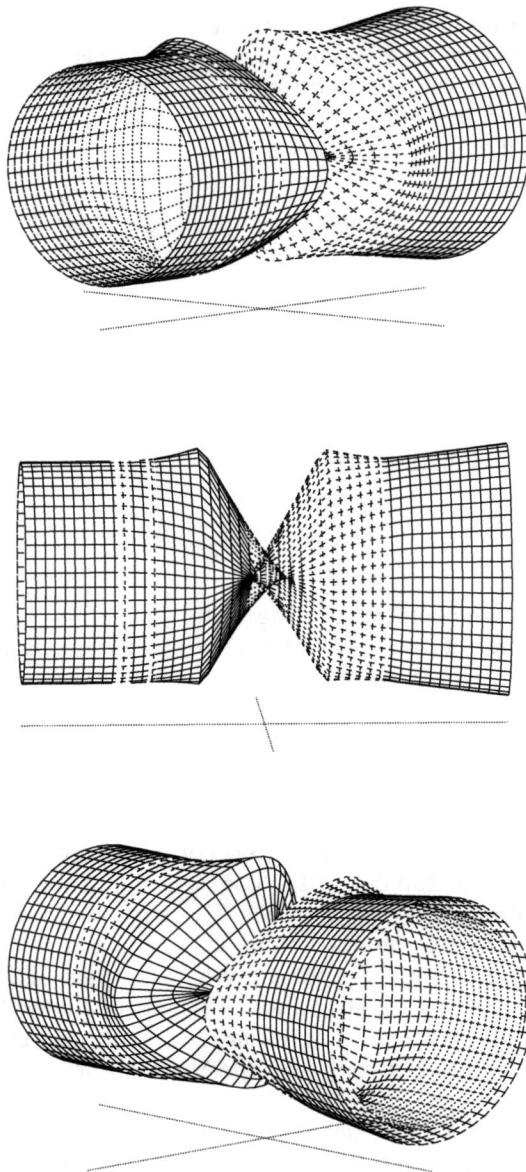

FIGURE 27 Three-dimensional view of the second-order and third-order approximation of a thin sheet of the beam with $|\alpha| =$ constant. Entrance view (top), side view (center), and exit view (bottom) of the corrector.

The intermediate amount of threefold astigmatism A_2 required to correct for a certain C_3 is quantified by

$$A_2 = f_0 \sqrt{\frac{3\,|C_3|}{2L}}. \qquad (28)$$

The derivation of the approximate relations for the third-order coefficients does not consider contributions of the transfer lenses and of the fringe fields. In contrast, all beam plots are based on semi-analytical calculations considering the realistic field shape and the rigorous theory. The results shown in Figure 28 verify that the round-lens contribution to C_3 is negligible. Off-axial coma and distortion at the image are not affected by the symmetric arrangement of the transfer lenses as seen in Figure 29.

Nevertheless, the threefold beam shape interacts with third-order aberrations of the transfer lenses. The exact calculations shown in Figure 30 show that the field astigmatism [Eq. (25)] is not purely real valued. The weak imaginary component is due to the magnetic transfer lenses; hence, for field astigmatism and field curvature we must consider small additional contributions—$FA_{3,TL}$ and $FC_{3,TL}$, respectively—which deteriorate the one-to-one imaging from plane S_1 to plane S_2. Both coefficients depend only on the refraction power and the gap geometry of the transfer lens doublet and not on the intermediate magnification.

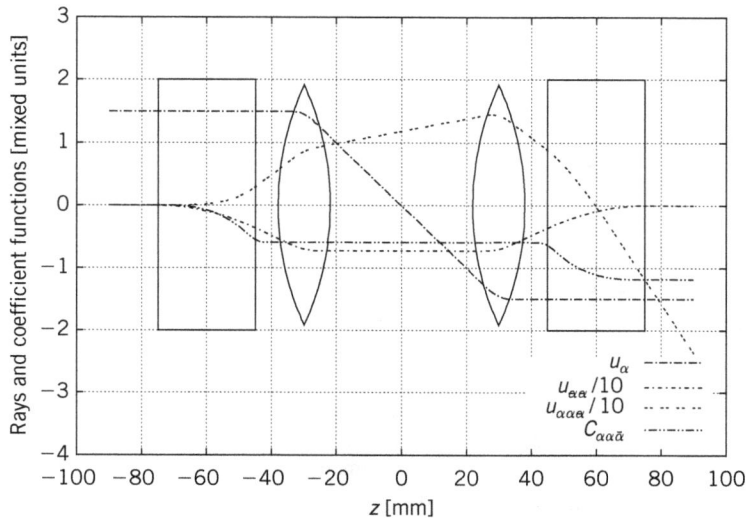

FIGURE 28 Coefficient function of third-order spherical aberration $C_3 = C_{\alpha\alpha\bar{\alpha}}$ with the axial fundamental rays u_α and the axial aberration rays of second and third-order $u_{\bar{\alpha}\alpha}$ and $u_{\alpha\alpha\bar{\alpha}}$, respectively.

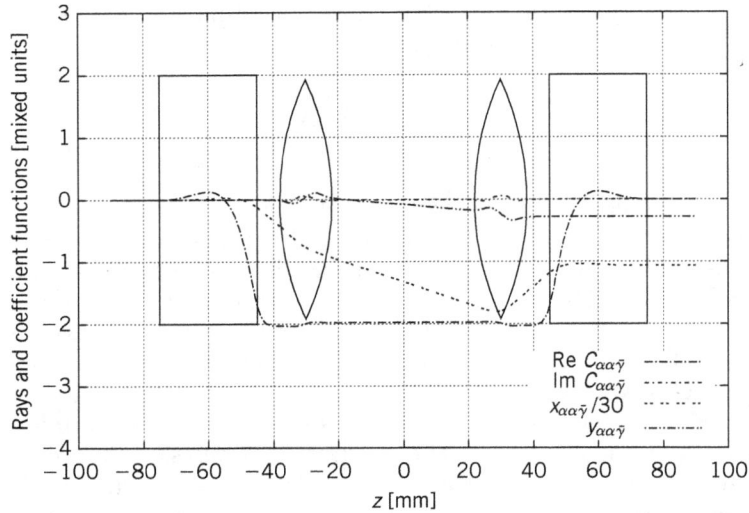

FIGURE 29 Coefficient function of the third-order off-axial coma $K_3 = C_{\alpha\alpha\bar{\gamma}}$ and the corresponding aberration ray $u_{\alpha\alpha\bar{\gamma}}$. Both the real ($x$) and imaginary ($y$) parts are shown.

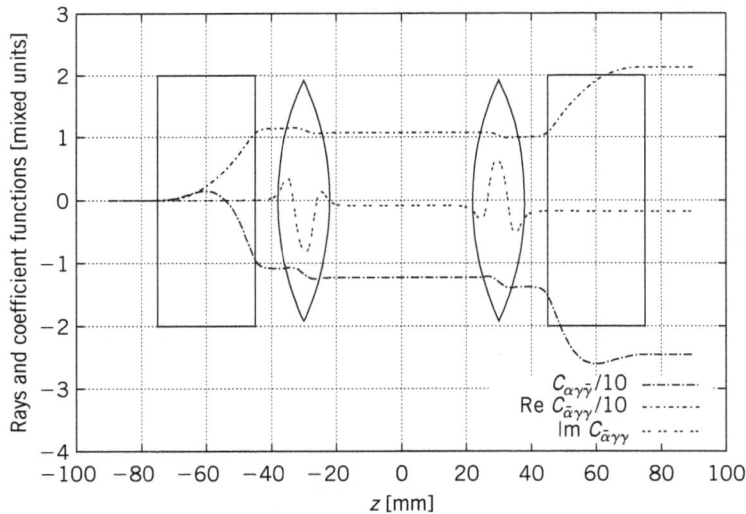

FIGURE 30 Coefficient functions of the third-order field curvature $FC_3 = C_{\alpha\gamma\bar{\gamma}}$ and the field astigmatism $FA_3 = C_{\bar{\alpha}\gamma\gamma}$. For the field-astigmatism the real and imaginary parts are shown.

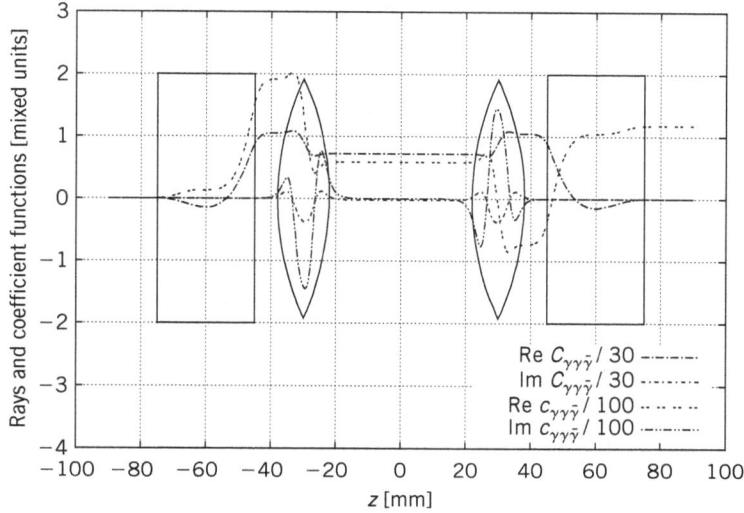

FIGURE 31 Coefficient functions of the third-order distortion $D_3 = C_{\gamma\gamma\bar{\gamma}}$ and the third-order spherical aberration at the diffraction plane $E_3 = c_{\gamma\gamma\bar{\gamma}}$ along the optic axis. Both the real and imaginary parts are shown.   Q2

3. Correction Strength

The expression in Eq. (22) for C_3 derived shows that the correction strength depends on the square of the hexapole excitation and on the fourth power of the beam radius f_0 inside the hexapole elements. Additionally, we observe that C_3 correction demands extended hexapole elements. The limit $L \to 0$ yields $C_3 \to 0$ even if the integral hexapole strength $\int \eta \Psi_{3s}\, dz = \eta \Psi_{3s} L$ is kept constant. This behavior is due to the fact that the negative C_3 is a secondary aberration of the hexapole field (or in other words, a combination aberration).

So far we have discussed the hexapole corrector independent of the objective lens. Using Eq. (3) we can summarize our findings by a useful rule of thumb for the current driving the hexapole coils,

$$\eta \mu_0 NI \approx \frac{R_{\mathrm{HP}}^3}{f_0^2} \sqrt{\frac{C_{3,\mathrm{OL}}}{6L^3}}, \tag{29}$$

where $C_{3,\mathrm{OL}}$ denotes the spherical aberration of the objective lens to be corrected. To estimate the performance of the entire aberration-corrected instrument it is essential to consider the optical coupling between the corrector and the objective lens. At least one additional transfer lens is

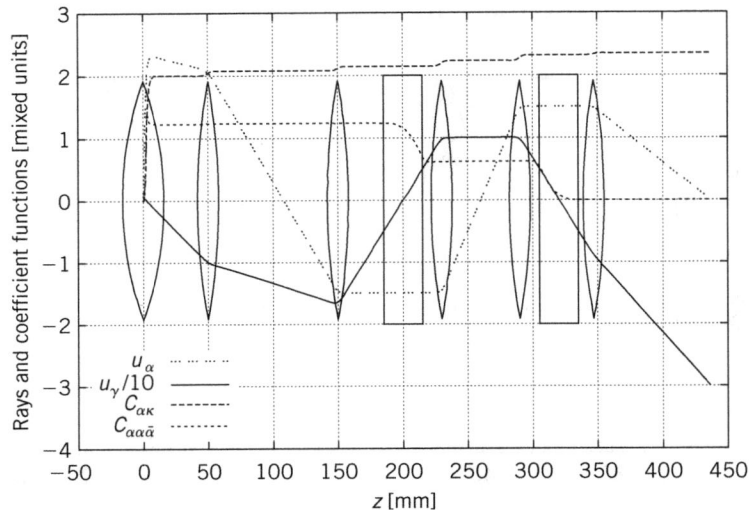

FIGURE 32 Hexapole corrector integrated in CTEM with objective lens, transfer lens system, and adapter lens. The course of the fundamental rays u_α and u_γ and of the coefficient functions of the axial chromatic $C_c = C_{\alpha\kappa}$ and third-order spherical aberration $C_3 = C_{\alpha\alpha\bar{\alpha}}$ are also plotted.

necessary between the aberration corrector and the objective lens. Most systems actually use a transfer lens doublet to gain further flexibility in alignment.

A typical beam path for a CTEM equipped with a hexapole C_s corrector with two transfer lenses between the objective lens and the first hexapole is depicted in Figure 32. Below the corrector an additional adapter lens is used to focus the telescopic beam at an intermediate image plane; typically the selected area diaphragm can be inserted there. In STEM the corrector is integrated between the condenser system and the objective lens. In this case, the STEM aperture is situated above the corrector. Again one or two transfer lenses are required between the corrector and the objective lenses.

By adjusting the objective lens excitation and the focal length of the transfer lenses the coma-free aperture plane of the objective lens field (usually this is roughly the focal plane of the objective lens) can be imaged at or near the center S_1 of the first hexapole element while the magnification between these conjugated planes can be tuned over a certain range (most often between $0.5 \leq M_{TL} \leq 1.0$). We should carefully distinguish this angular magnification from the lateral magnification at the mid-plane S_0 of the corrector, which is actually $\frac{f_{TL}}{M_{TL}f_{OL}} \approx 15 - 30$, where f_{OL} and f_{TL} denote the focal length of the objective lens and of the transfer lenses between the

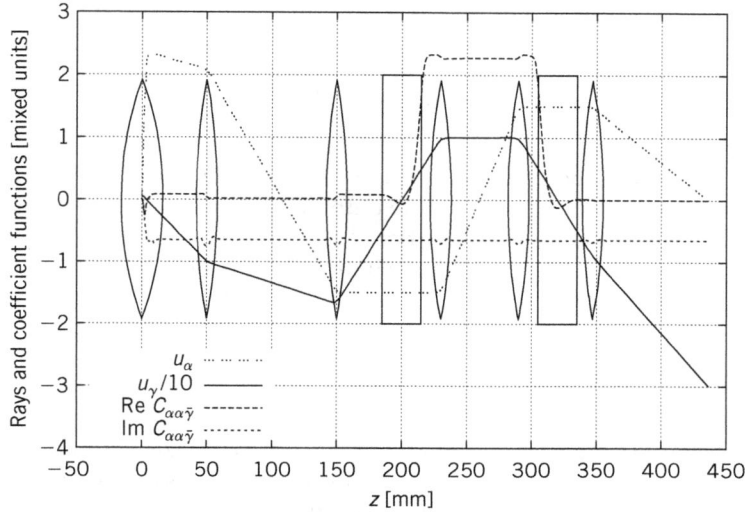

FIGURE 33   Coefficients of radial (isotropic) and azimuthal (anisotropic) off-axial coma $C_{\alpha\alpha\bar{\gamma}}$ for a CTEM equipped with hexapole corrector. The transfer lenses are adjusted such that the semi-aplanatic condition is fulfilled.

Q2

hexapole stages, respectively. Changing the magnification M_{TL} effectively changes the height of the axial ray

$$f_0 = M_{TL} f_{OL} \tag{30}$$

inside the hexapoles. By this means the correction strength of the hexapole corrector is tuned according to Eq. (29). The correction strength stays constant if the hexapole current is changed simultaneously keeping the product $NI \cdot f_0^2$ constant. If the magnification M_{TL} is kept constant and only the hexapole excitation is changed, we can tune the total spherical aberration C_3 between overcompensation and undercompensation

$$\Delta C_3 = -2 C_{3,OL} \frac{\Delta I}{I}. \tag{31}$$

This is very useful in allowing for the optimization of phase contrast in the CTEM, which requires alternating signs (e.g., $C_1 > 0$, $C_3 < 0$, and $C_5 > 0$).

4. Chromatic Aberration

The transfer lenses of the corrector increase the total chromatic aberration of the system. Their contribution to the coefficient C_c is clearly visible in

Figure 32. This agrees with the estimate

$$C_{c,\mathrm{TL}} = \frac{f_0^2}{f_{\mathrm{TL}}} = M_{\mathrm{TL}}^2 \frac{f_{\mathrm{OL}}^2}{f_{\mathrm{TL}}} \tag{32}$$

for the chromatic aberration of a single transfer lens. We observe that the corrector adds about 15–25% to the chromatic aberration of the objective lens. The additional C_c is proportional to the square of the intermediate magnification M_{TL}. Hence, high excitations of the hexapole elements are preferable to minimize the total chromatic aberration.

5. Semi-Aplanatic Condition

For CTEM imaging the course of the field ray between the objective lens and the corrector is essential. Only if the coma-free plane of the objective lens is matched with the center of the hexapoles at S_1 and S_2 does the third-order radial (isotropic) part of the off-axial coma, $\mathrm{Re}\left\{C_{\alpha\alpha\bar\gamma}\right\}$, vanish. If this condition is fulfilled, the corrected system is called *semi-aplanatic* (Rose, 1990). It is not fully aplanatic since the azimuthal off-axial coma of the magnetic objective lens caused by Larmor rotation is still present. The course of the off-axial coma for a semi-aplanatic system is shown in Figure 29.

If the coma-free aperture planes are not matched, residual real-valued off-axial coma is introduced. The amount of off-axial coma is proportional to the distance Δ_{HP} between S_1 and the image of the coma-free aperture plane of the objective lens

$$\Delta C_{\alpha\alpha\bar\gamma} = -\Delta_{\mathrm{HP}} \frac{C_{3,\mathrm{OL}}}{f_0^2}. \tag{33}$$

For any C_3-corrected CTEM the off-axial coma is independent of the convergence of the illumination, since changing the illumination angle changes the course of the rays in the total system. In this case, the total $C_3 = 0$ is relevant in Eq. (33) and, hence, no off-axial coma is induced.

6. Field Aberrations

For a semi-aplanat the only residual intrinsic third-order aberrations at the image are field curvature and field astigmatism. For an objective lens with focal length f_{OL} and spherical aberration $C_{3,\mathrm{OL}}$, we find the relation

$$C_{\alpha\gamma\bar\gamma} = -C_{\bar\alpha\gamma\gamma} = \frac{1}{10} \cdot \frac{C_{3,\mathrm{OL}} L^2}{(M_{\mathrm{TL}} f_{\mathrm{OL}})^4}. \tag{34}$$

This contribution of the hexapole fields dominates the intrinsic contribution of the objective lens. The parameters L and M_{TL} could be used to minimize field curvature and field astigmatism, or a different corrector design with more then two hexapole stages could be used to eliminate this aberration completely (Rose, 2002). Nevertheless, field curvature and field astigmatism are an issue only at low magnification. In this case, the Nyquist frequency of the CCD detector becomes so small that C_s correction is no longer useful; therefore, it seems most appropriate to disable the corrector for low magnification.

D. Higher-Order Aberrations

The next steps in the successive approximation scheme reveal further higher-order aberrations. Here, we restrict our attention to the axial aberrations only.

1. Fourth-Order Aberrations

For the fourth-order path deviation we find

$$u^{(4)} = \alpha^4 u_{\alpha\alpha\alpha\alpha} + \alpha\bar{\alpha}^3 u_{\alpha\overline{\alpha\alpha\alpha}}. \tag{35}$$

This follows by substituting $u^{(3)} = \alpha u_\alpha + \bar{\alpha}^2 u_{\overline{\alpha\alpha}} + \alpha^2 \bar{\alpha} u_{\alpha\alpha\bar{\alpha}}$ in the iteration formula [Eq. (6)] after collecting the fourth-order terms. The image coefficients are

$$C_{\alpha\alpha\alpha\alpha} = \int 3\eta \Psi_{3s}\, \bar{u}_{\overline{\alpha\alpha}}^2 u_\alpha dz, \tag{36}$$

$$C_{\alpha\overline{\alpha\alpha\alpha}} = \int 6\eta \Psi_{3s}\, \bar{u}_{\alpha\alpha\bar{\alpha}} u_\alpha^2 dz. \tag{37}$$

The second coefficient can be evaluated further after inserting the explicit representation of the spherical aberration ray. We obtain by partial integration the relation

$$C_{\alpha\overline{\alpha\alpha\alpha}} = 4\overline{C}_{\alpha\alpha\alpha\alpha}. \tag{38}$$

According to this result the three-lobe aberration at the image has the form

$$u^{(4)}/u_\gamma = \alpha^4 D_4 + 4\alpha\bar{\alpha}^3 \overline{D}_4. \tag{39}$$

Since Ψ_{3s} is a symmetric function and u_α is anti-symmetric, we find from Eq. (36) that the three-lobe aberration $D_4 = C_{\alpha\alpha\alpha\alpha}$ vanishes at the image plane but not the corresponding slope coefficient

$$c_{\alpha\alpha\alpha\alpha} = -\int 3\eta\Psi_{3s}\,\overline{u}_{\alpha\alpha}^2 u_\gamma dz. \qquad (40)$$

As a direct effect the latter is visible as a threefold distortion at the diffraction plane. Residual three-lobe aberration shows up also at the image if the intermediate image plane is moved away from the mid-plane S_0 of the corrector. For this reason focusing through the corrector should be avoided. To shift the position of an image plane (e.g., the SA plane below the corrector in CTEM) always excitation of the adapter lens should be changed; the objective lens should never be tuned. Also in STEM the probe semi-angle should be tuned by changing lenses *above* the corrector only.

From Eq. (40) we can calculate a SCOFF approximation for the slope coefficient of the three-lobe aberration

$$c_{\alpha\alpha\alpha\alpha} = \frac{9}{10}\eta^3\Psi_{3s}^3 f_0^3 L^6. \qquad (41)$$

In Figure 34 the fourth-order aberration rays $u_{\alpha\alpha\alpha\alpha}$ and $u_{\alpha\overline{\alpha}\alpha\alpha}$ are plotted according to an exact calculation. Although the three-lobe aberration becomes complex-valued due to the influence of the transfer lenses, it is still point-corrected at the image. The contribution of the transfer lenses

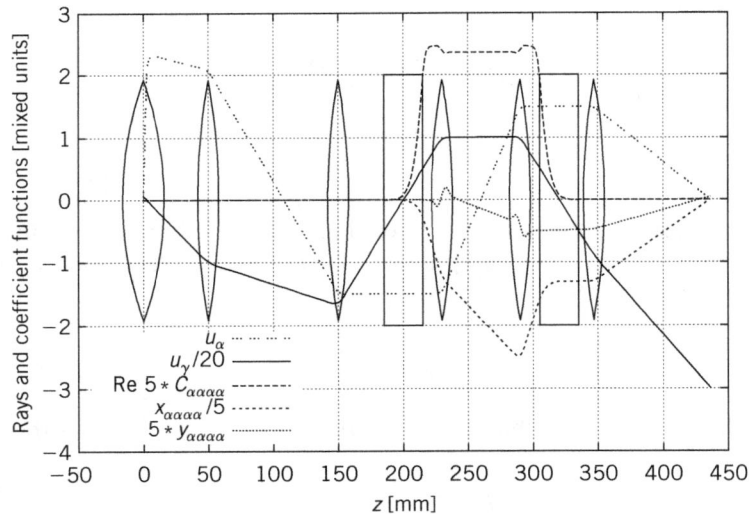

FIGURE 34 Fourth-order axial aberration ray $u_{\alpha\alpha\alpha\alpha} = x_{\alpha\alpha\alpha\alpha} + iy_{\alpha\alpha\alpha\alpha}$ and coefficient function of the three-lobe aberration $D_4 = C_{\alpha\alpha\alpha\alpha}$ for a CTEM with hexapole corrector.

to D_4 cancels by symmetry. It is important to note that inside the second hexapole the aberration ray $u_{\alpha\alpha\alpha} \approx f_0 c_{\alpha\alpha\alpha}$ is almost constant. The slope coefficient $c_{\alpha\alpha\alpha}$ is affected by the third-order aberrations of the transfer lenses. Actually their residual field astigmatism $FA_{3,TL}$ makes the $u_{\alpha\alpha\alpha}$ ray complex-valued. This effect can be estimated by considering an off-axial image point $u = \alpha f_0$ at plane S_1 with slope $u' = \bar{\alpha}^2 A_2/f_0$. At the conjugated plane S_2 a path deviation is induced by the field astigmatism of the transfer lenses. The result adds to the aberration ray of the three-lobe aberration inside the second hexapole field

$$u_{\alpha\alpha\alpha} = -\frac{9}{10}\eta^3 \Psi_{3s}^3 L^6 f_0^4 + 3\eta\Psi_{3s} L f_0^4 FA_{3,TL}. \tag{42}$$

This effect is clearly visible in Figure 34. Although it is rather small, it has important consequence for the residual fifth-order aberrations.

At this stage, we can summarize that for a system equipped with a hexapole C_s corrector, no residual axial aberrations up to and including fourth order occur. The first nonvanishing axial aberrations of the hexapole corrector are of fifth order.

2. Fourth-Order Off-Axial Aberrations

Here we note that the hexapole corrector has coma-type off-axial aberrations only in fourth order, which depend linearly on the radius of the field of view. Nevertheless, the number of equally well-resolved image points for a present C_s-corrected CTEM is limited by the azimuthal third-order coma of the magnetic objective lens and not by the corrector. This matter and how a hexapole corrector can be used to correct for azimuthal off-axial coma are discussed in greater detail in Section III.

3. Fifth-Order Aberrations

The next step of the iteration procedure results in the fifth-order path deviation

$$u^{(5)} = \alpha^3 \bar{\alpha}^2 u_{\alpha\alpha\alpha\bar{\alpha}\bar{\alpha}} + \bar{\alpha}^5 u_{\overline{\alpha\alpha\alpha\alpha\alpha}}. \tag{43}$$

Evaluated at the image we obtain

$$u^{(5)}/u_\gamma = \alpha^3 \bar{\alpha}^2 C_5 + \bar{\alpha}^5 A_5. \tag{44}$$

with the coefficients of the fifth-order spherical aberration C_5 and the six-fold astigmatism A_5. The hexapole corrector has an intrinsic positive C_5, which adds to that of the objective lens. Additionally, a strong combination effect can occur between objective lens and corrector, which then contributes to the total C_5 of the system.

4. Tuning Fifth-Order Spherical Aberration

Since the aberration ray of the spherical aberration between objective lens and corrector has the form

$$u_{\alpha\alpha\bar{\alpha}} = u_\gamma C_{3,\text{OL}}, \tag{45}$$

the combination effect vanishes if the nodal plane of the field ray $u_\gamma = 0$ coincides with the coma-free aperture plane of the corrector. This is the standard situation as depicted in Figure 35. If the transfer lenses between the objective lens and the corrector are changed such that C_3 correction is maintained but the coma-free aperture plane of the corrector is separated by the distance Δ_{HP} from the plane conjugated to the coma-free aperture plane of the objective lens, a combination contribution ΔC_5 is intentionally introduced

$$\Delta C_5 = -3\Delta_{\text{HP}} \left(\frac{C_{3,\text{OL}}}{f_0} \right)^2. \tag{46}$$

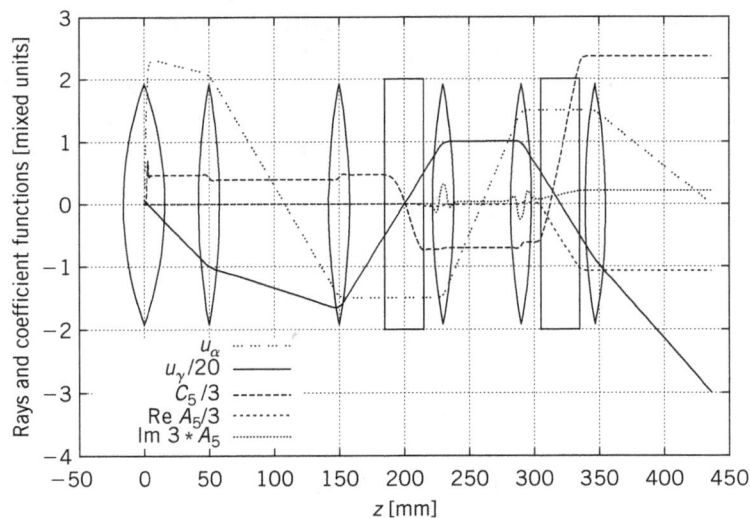

FIGURE 35 Coefficient function of the fifth-order spherical aberration $C_5 = C_{\alpha\alpha\alpha\bar{\alpha}\bar{\alpha}}$ and sixfold astigmatism $A_5 = C_{\bar{\alpha}\bar{\alpha}\bar{\alpha}\bar{\alpha}\bar{\alpha}}$ for a CTEM with hexapole corrector. For the sixfold astigmatism A_5 both the real and imaginary parts are plotted.

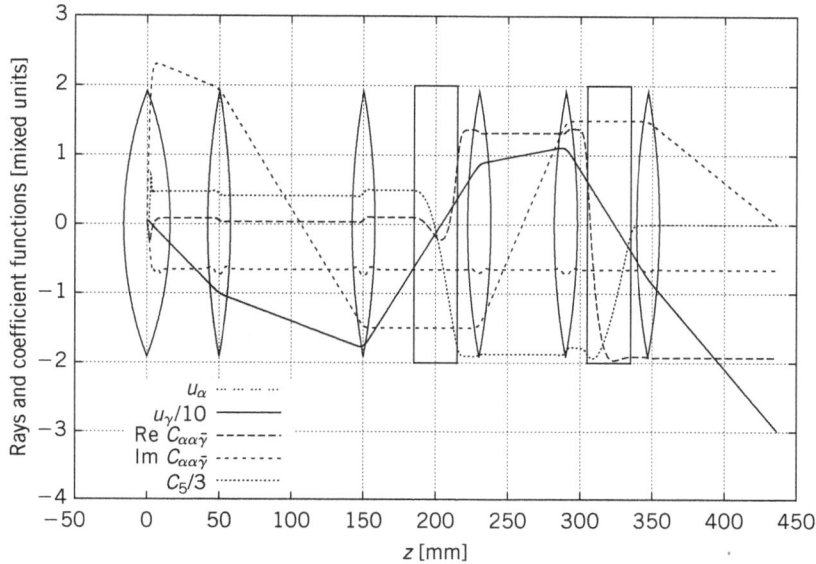

FIGURE 36 Transfer lenses adjusted for $C_5 = 0$. The fifth-order spherical aberration C_5 and both components for the third-order off-axial coma $C_{\alpha\alpha\bar{\gamma}}$ are plotted.

This tunable contribution can be used to counterbalance the intrinsic contribution of the objective lens and of the corrector with the effect that the total fifth-order spherical aberration becomes zero $C_5 = 0$. Figure 36 shows a system in which the transfer lenses have been tuned for C_5-free imaging. It is immediately clear that in this case some off-axial coma must be accepted. By combining Eqs. (33) and (46) the residual amount of off-axial radial coma can be estimated as

$$\Delta C_{\alpha\alpha\bar{\gamma}} = \frac{1}{3} \cdot \frac{\Delta C_5}{C_{3,\text{OL}}} \in \mathcal{R}, \qquad (47)$$

where ΔC_5 is the sum of the primary fifth-order spherical aberration of the objective lens and of the corrector. Unfortunately, Eq. (47) shows that the off-axial coma introduced for C_5-free alignment is rather pronounced, since ΔC_5 is typically considerably larger than C_3. Therefore, C_5 correction is most useful in STEM, where the residual off-axial coma does not matter since the scan is performed between the corrector and the objective lens.

The possibility of tuning the fifth-order spherical aberration in a system with C_3 corrector has been discussed by (Rose, 1971b). He already states Eq. (46). For a STEM equipped with hexapole corrector this relation is also mentioned by Crewe (1981). Later Hartel *et al.* (2004) demonstrated the possibility to tune C_5 experimentally (see Figure 36).

96 Maximilian Haider *et al.*

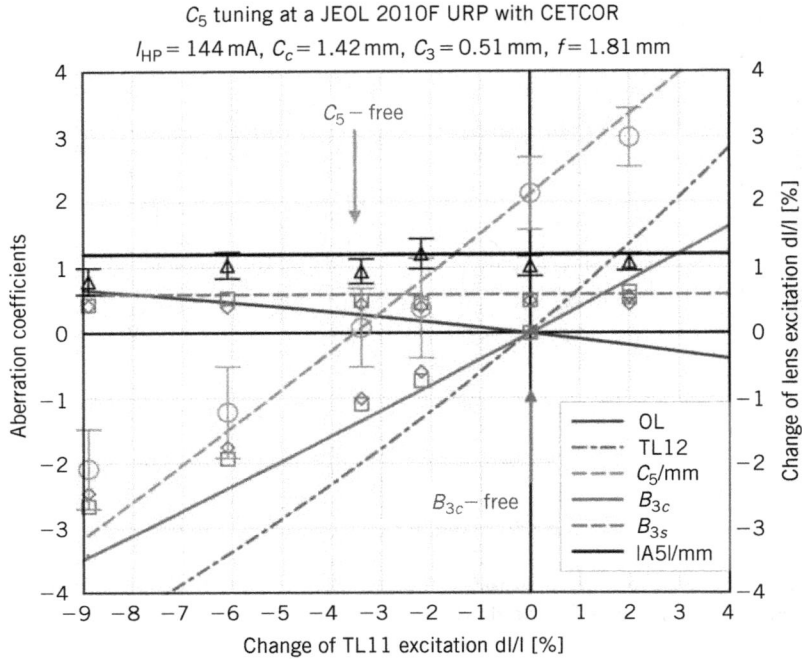

FIGURE 37 Experimental verification of semi-aplanatic (here $K_3 = B_{3c} + i\,B_{3s}$) and C_5-free alignment for a CTEM hexapole corrector with small gap objective lens taken. (From Müller *et al.*, 2007.)

After it had been demonstrated that a system with $C_3 = C_5 = 0$ is feasible, the sixfold astigmatism remained the only residual axial aberration in fifth-order and the question arose how A_5 could be corrected or at least minimized.

5. Minimizing Sixfold Astigmatism

The dominant contribution to the sixfold astigmatism is caused by the combination of the fourth-order aberration ray $u_{\alpha\alpha\alpha\alpha}$ with the hexapole field. With the approximation from Eq. (42) we can estimate the sixfold astigmatism by

$$A_5 = -\frac{27}{5}\,|\eta\Psi_{3s}|^4 f_0^6 L^7 + 18\,|\eta\Psi_{3s}|^2 f_0^6 L^2 \cdot \overline{FA}_{3,\text{TL}}. \tag{48}$$

This result is in good agreement with more accurate calculations. The influence of the transfer lenses causing the second term in Eq. (48) typically

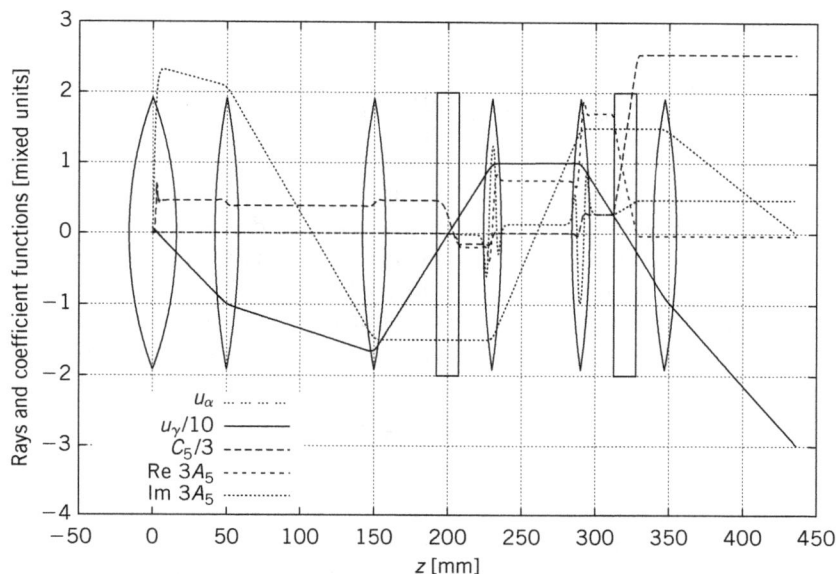

FIGURE 38 Axial fifth-order coefficient functions C_5 and A_5 for a hexapole corrector with optimally chosen length L_{opt} of the hexapole elements and C_5-free alignment of the transfer lenses.

is rather small. By using the expressions from Eq. (22) we can recast this expression

$$A_5 = -\frac{3}{20}L\left(\frac{C_{3,OL}}{M_{TL}f_{OL}}\right)^2 + 3\frac{(M_{TL}f_{OL})^2 C_{3,OL}}{L} \cdot \overline{FA}_{3,TL}. \qquad (49)$$

For most corrector systems the first term strongly dominates (see Figure 38). Therefore, the most obvious way to reduce A_5 is to use an objective lens with a narrow gap. For such a system we have $C_{3,OL} = 0.5$–$0.6\,$mm and, hence, the residual A_5 after C_3 correction can be well below $1\,$mm, and C_c is sufficiently small for optimum high-resolution performance. The alternative measure to increase M_{TL} is not attractive since this also increases the C_c contribution of the corrector.

The second complex-valued term in Eq. (49) due to $\overline{FA}_{3,TL}$ becomes important if the beam inside the hexapoles is large or if the hexapoles are short. Since the real part of field astigmatism for the anti-symmetric transfer doublet is positive a certain length L_{opt} exists such that the real part A_{5x} of the sixfold astigmatism vanishes. By setting $A_{5x} = 0$ in Eq. (49)

Wolf Prize in Physics

98   Maximilian Haider *et al.*

we derive

$$L_{\mathrm{opt}} = \sqrt{20\frac{FA_{3x,\mathrm{TL}}}{C_{3,\mathrm{OL}}}\left(M_{\mathrm{TL}}f_{\mathrm{OL}}\right)^2} > 0. \tag{50}$$

This shows that the most efficient way to reduce A_5 is to reduce the length of the hexapole elements if feasible down to the optimum where $A_{5x} = 0$. Additionally, it can be advantageous to maximize the field astigmatism $FA_{3x,\mathrm{TL}}$ of the transfer doublet by reducing the gap and bore dimensions to avoid infeasibly small values for L_{opt}. Since the imaginary part $\overline{FA}_{3y,\mathrm{TL}}$ of the field astigmatism is non-zero, this procedure introduces a residual

$$A_{5y} = -3\frac{\left(M_{\mathrm{TL}}f_{\mathrm{OL}}\right)^2 C_{3,\mathrm{OL}}}{L}FA_{3y,\mathrm{TL}}. \tag{51}$$

This effect sets an upper limit for the tolerable $FA_{3y,\mathrm{TL}}$. Fortunately, for a magnetic transfer doublet the real part $FA_{3x,\mathrm{TL}}$ strongly dominates the imaginary part $FA_{3y,\mathrm{TL}}$ of the field astigmatism. Hence, the unwanted increase of A_{5y} is small Eq. (50) shows that the magnitude of the sixfold astigmatism $|A_5|$ adopts a minimum if L is varied for constant M_{TL} or if M_{TL} is varied for constant L.

Figure 38 plots the course of the fifth-order aberration coefficients through the corrector for a system with reduced L and optimized transfer lens geometry. For a certain hexapole excitation both coefficients $A_{5x} = 0$ and $C_5 = 0$ are corrected. The data are based on an exact calculation. The residual sixfold astigmatism in this system A_{5y} is in good agreement with Eq. (51). Figure 38 shows that the transfer lenses additionally contribute directly to A_{5x}. This minor effect is due to the third-order field curvature of the transfer lenses and has been neglected in the derivation of Eq. (48). For this reason the exact calculation results in a slightly larger optimum length L_{opt} than predicted from Eq. (50).

Figure 39 shows the result of the minimization of $|A_5|$ performed for the design of the CEOS advanced hexapole STEM corrector D-COR (Müller *et al.*, 2007). For the D-COR systems the aim was to reduce the residual A_5 to such an extent that STEM resolution better than 50 pm with a semi-aperture angle of $\vartheta = 38$ mrad at 300 kV can be achieved. For the first and second prototype of this corrector the residual A_{5x} and A_{5y} have been measured. Figure 40 shows the behavior of both coefficients during a tuning of the intermediate magnification M_{TL}. To keep C_3 corrected the hexapole excitation must be adjusted accordingly. With increasing excitation we find a linear decrease of A_{5x} and finally a change of sign while A_{5y} stays almost constant and close to zero. The slightly different results

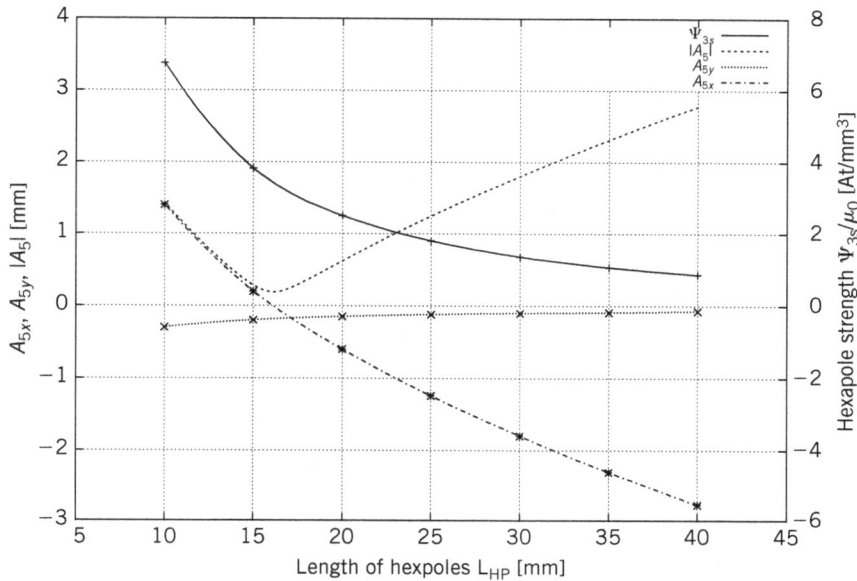

FIGURE 39 Variation of A_{5x} and A_{5y} with the length of the hexapole elements for a typical hexapole corrector according to Müller *et al.* (2006a).

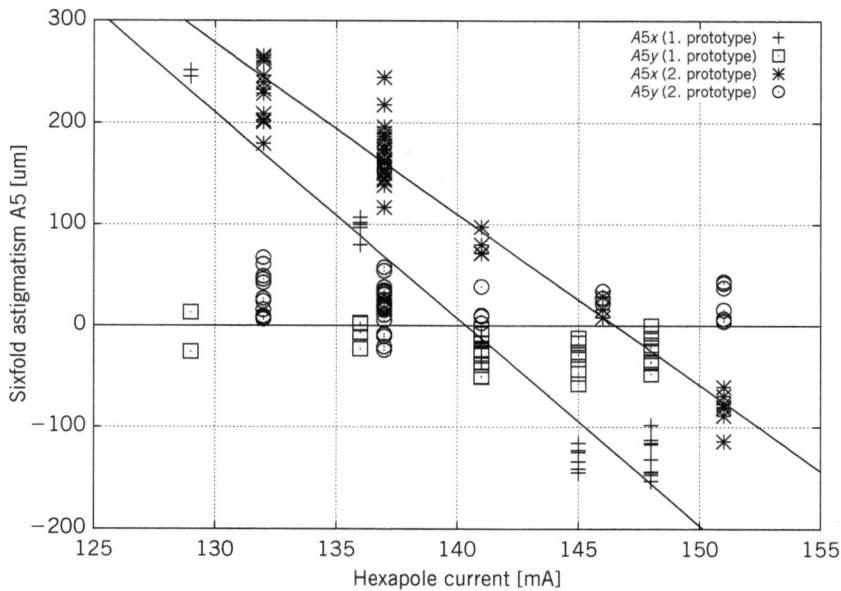

FIGURE 40 Measured values of the sixfold astigmatism for the first two prototypes of the advanced CEOS STEM hexapole corrector (D-COR). (From Hartel *et al.*, 2007.)

100 Maximilian Haider *et al.*

for the two prototypes can be attributed to manufacturing tolerances and non-ideal alignment. Additionally, the true quantitative assessment of fifth-order aberrations on the level of only a few hundred micrometers is questionable with the currently available methods for aberration measurement in STEM. Within the TEAM project the advanced hexapole STEM corrector has been used to set the new milestone of STEM resolution better than 50 pm as mentioned in Section I.

6. Elimination of Scan Coma

The D-COR-type hexapole STEM corrector with minimized A_5 and C_5-free alignment as depicted in Figure 38 is the optimum system for a high-resolution STEM equipped with a cold-FEG or gun monochromator. To obtain a large scan field the scan-induced axial coma in second order must be carefully considered. The usual practice of aligning the pivot point of the scan with the front focal plane or with the center of the coma-free aperture plane of the objective lens is not sufficient in this case. The scan coils situated between the corrector and the objective lens must be operated such that both components of the scan coma, the isotropic and the anisotropic (or azimuthal) part, are compensated simultaneously. Effectively, this makes the scan skew with respect to the optic axis.

7. Further Axial Aberrations

The optical performance of TEM instruments equipped with hexapole C_s correctors typically is not limited by residual intrinsic aberrations but by incoherent effects such as focus spread or image spread. The sum of all incoherent effects finally determines the information limit of the instrument. An advanced hexapole design with C_3/C_5 correction and strongly suppressed sixfold astigmatism A_5 is necessary if the chromatic focus spread in a TEM is strongly reduced by using a cold FEG or a gun monochromator. For every aberration-corrected system it is also mandatory to correct for the parasitic aberrations caused by manufacturing tolerances of the optical elements and misalignments. This requires precise methods for aberration assessment and highly sophisticated alignment strategies. Under optimum conditions even the residual parasitic aberrations should not limit the attainable optical resolution.

The next higher-order intrinsic aberrations of a hexapole corrector occur in order six and seven. By our iteration method we find for the path deviation at the image

$$u^{(6)} + u^{(7)} = 2\alpha^5 \overline{\alpha} D_6 + 5\alpha^2 \overline{\alpha}^4 \overline{D}_6 + \alpha^4 \overline{\alpha}^3 C_7 + \alpha^7 G_7 + 7\alpha \overline{\alpha}^6 \overline{G}_7. \quad (52)$$

The coefficients are the three-lobe aberration in sixth order D_6, the seventh-order spherical aberration C_7, and the sixfold chaplet aberration G_7. For a carefully designed corrector system, none of these aberrations are limiting.

III. HEXAPOLE APLANATS

A. Semi-Aplanat Versus Aplanat

About 30 years ago Vernon Beck discovered the negative spherical aberration of a hexapole doublet (Beck, 1979). Ten years later, Rose pointed out that it is essential to adapt the hexapole C_s corrector to the objective lens by a transfer lens system, in order to make it applicable to the conventional TEM (Rose, 1990). The reason for the transfer system is twofold. First, it avoids a large fifth-order spherical aberration that would otherwise arise as a strong combination between the objective lens and the corrector. On the other hand, the transfer lens system allows imaging of the coma-free plane of the objective lens into the coma-free plane of the corrector. The isotropic off-axial coma of the system vanishes only if this condition is met. We would call an optical system that is free of spherical aberration and off-axial coma *aplantic*. However, unfortunately, the magnetic objective lens also introduces anisotropic off-axial coma that still remains even if the coma-free planes are matched. Rose suggested the term *semi-aplanat* for a spherical aberration-corrected system that is free of isotropic coma and still has anisotropic off-axial coma (Rose, 1990). The delocalization area of both the isotropic (radial) and the anisotropic (azimuthal) off-axial coma are illustrated in Figure 41.

Today hexapole *semi-aplanats* consisting of a magnetic round lens and a hexapole doublet are commercially available options for both the CTEM and the STEM operation mode of the microscope. Optimized and stable current drivers and high-voltage supplies and the increasing availability of monochromators have dramatically improved the information limit of the microscopes. Sub-Ångstroem resolution is now readily available, and hence, apertures in the range of 25 mrad to even 50 mrad become usable.

Considering these large apertures, the remaining anisotropic part of the off-axial coma seriously limits the field of view that can be imaged with same, nearly constant phase shift. We can draw a circle of radius R around a perfect corrected axial point, which indicates that an additional $\pi/4-$phase shift due to the off-axial coma is surpassed:

$$\frac{2\pi}{\lambda} \cdot R \cdot \left| C_{\alpha\alpha\bar{\gamma}} \right| \cdot \vartheta^3 \left(1 - \frac{3}{4} \right) < \frac{\pi}{4}. \tag{53}$$

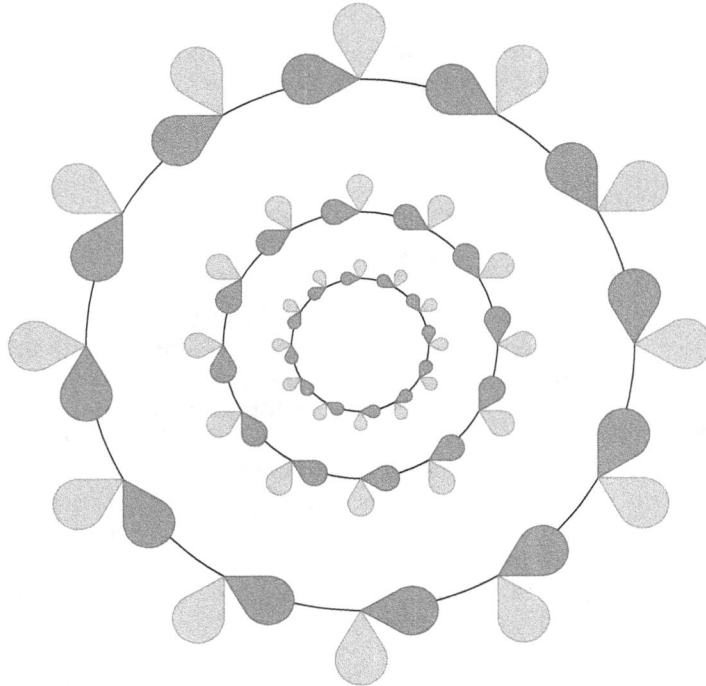

FIGURE 41   The delocalization area of both the isotropic/radial (light grey) and the anisotropic/azimuthal (dark grey) off-axial coma. For increasing circles in the image plane, the shape of the comet-tail–like delocalization figure is depicted. The diameter of the shape increases linearly with the distance to the aberration-free axial point. If a *semi-aplanat* is perfectly aligned, only the azimuthal component remains.

Here ϑ denotes the aperture semi-angle that corresponds to the highest spatial frequency in consideration $g_{\max} = \vartheta/\lambda$ (e.g., the information limit). The factor $(1 - 3/4) = 1/4$ accounts for the compensation of the third-order phase shift of the coma by a first-order phase shift induced by a change of the image magnification and/or rotation that one would immediately recognize.

Assuming a typical dimensionless coefficient of $|C_{\alpha\alpha\bar{\gamma}}| = 2/3$ for the anisotropic off-axial coma the $\pi/4$–radius can be illustrated as shown in Table II. According to this criterion, only $600\ldots2400$ points of size $1/g_{\max}$ can be resolved. In this chapter, we discuss advanced hexapole correctors that are also capable of eliminating the anisotropic part of the off-axial coma. Combining these correctors with a magnetic objective lens results in a true *aplanatic* optical system, which we will consequently call an *aplanat*.

TABLE II Field of view limited by the off-axial coma. The radius $R_{\frac{\pi}{4}} = \frac{\lambda}{2} / (\left|C_{\alpha\alpha\bar{\gamma}}\right| \cdot \vartheta^3)$ of a circle in the image plane around an axial point for which the additional phase shift of an anisotropic off-axial coma $\left|C_{\alpha\alpha\bar{\gamma}}\right| = 2/3$ is less than or equal $\pi/4$. Here N denotes the number of pixels of size $1/g_{max}$ in the diameter $2R_{\frac{\pi}{4}}$.

λ[pm]	ϑ[mrad]	$g_{max} = \frac{\vartheta}{\lambda}$ [1/nm]	$R_{\frac{\pi}{4}}$ [nm]	$N = 2R_{\frac{\pi}{4}} \cdot g_{max}$[1000 px]
2.5	25	10	120	2.4
2.5	50	20	15	0.6
2	25	12.5	96	2.4
2	40	20	23	0.9

B. Advancing the CTEM Hexapole C_s Corrector

Various options are available to correct the off-axial coma in quadrupole-octupole correctors (and eventually the coma correction there is even mandatory). However, in this chapter we want to adhere to the simplicity and the advantages of the hexapole corrector with rotational symmetric Gaussian rays. Hence, the only focusing elements are (transfer-)round lenses. We tailor an appropriate set of combination aberrations induced by the strong threefold astigmatism, which in turn travels through hexapole fields.

1. Requirements

Since we do not want to compromise any of the advantageous properties of the original semi-aplanat, the advanced hexapole corrector must meet a set of constraints. It must

1. Be completely free of second-order geometrical aberrations at any plane outside the corrector
2. Produce a negative C_s to compensate for that of the objective lens
3. Produce a negative Im $\left\{C_{\alpha\alpha\bar{\gamma}}\right\}$ to compensate for that of the objective lens
4. Be free of fourth-order axial aberrations at the image plane (at least point correction).

Moreover, a preferable *aplanat* would additionally

1. Have a C_c that is not (much) higher than that of the *semi-aplanat*
2. Have a small fifth-order sixfold astigmatism A_5, at least below the $\pi/4-$limit
3. Have an optimally chosen spherical aberration C_5 that enables phase contrast.

2. Integral Formulation of Essential Requirements for Aplanats

As discussed for the basic setup of any hexapole corrector, all second-order aberration coefficients in [Eq. (20)] must vanish behind the corrector; the aberration rays are zero in every plane. However, within the corrector the second-order aberration rays are present and intentionally chosen. Again, the aberration ray of the threefold astigmatism

$$u_{\overline{\alpha\alpha}}(z) = u_\alpha(z) \cdot \int_{-\infty}^{z} 3i\eta\, u_\alpha^2 u_\gamma \Psi_3 dz - u_\gamma(z) \cdot \int_{-\infty}^{z} 3i\eta\, u_\alpha^3 \Psi_3 dz$$

$$= u_\alpha(z) \cdot a_2(z) + u_\gamma(z) \cdot A_2(z)$$

plays an important role, because it carries two $\overline{\alpha}$ parameters.

Since the corrector should correct mainly axial aberrations, we further restrict the selection of suitable correctors to systems that fulfill "locally"

$$\int_{local} u_\alpha^2 u_\gamma \Psi_3 dz = 0. \tag{54}$$

This means that hexapole elements are placed at aperture planes or anti-symmetric around intermediate image planes. Hence, the coefficient function $a_2(z)$ of the aberration ray $u_{\overline{\alpha\alpha}}$ does not arise outside the hexapole element and will not produce combination aberrations with other elements. The aberration ray of the threefold astigmatism then can be written as

$$u_{\overline{\alpha\alpha}}(z) = -u_\gamma(z) \cdot \int_{-\infty}^{z} 3i\eta u_\alpha^3 \Psi_3 dz = +u_\gamma(z) \cdot A_2(z) \tag{55}$$

We now generalize the results from Sections II.C.2 and II.D for complex Ψ_3 and A_2 with varying orientation (arguments) along the z-axis. The orientation is always measured with respect to the rotating Larmor frame of reference. The results are written as:

$$C_{\alpha\alpha\overline{\gamma}} = -\int 6i\eta u_\alpha u_\gamma^2 \overline{A}_2 \Psi_3 dz \quad \text{and} \tag{56}$$

$$D_4 = C_{\alpha\alpha\alpha\alpha} = -\int 3i\eta u_\alpha u_\gamma^2 \overline{A}_2^2 \Psi_3 dz. \tag{57}$$

The existing *semi-aplanat* (consisting of a magnetic objective lens and the two-stage hexapole corrector) suffers from the remaining anisotropic part $\mathrm{Im}\left\{C_{\alpha\alpha\bar{\gamma}}\right\}$ of the off-axial coma. Hence, the corrector should produce only

$$\mathrm{Im}\left\{C_{\alpha\alpha\bar{\gamma}}\right\} = -\int 6\eta u_{\alpha}u_{\gamma}^{2}\cdot\mathrm{Re}\left\{\overline{A}_{2}\Psi_{3}\right\}dz, \tag{58}$$

and no real part. Even harder to accomplish is the requirement that the coma correction must not introduce an accompanying axial fourth-order aberration

$$\mathrm{Im}\{D_{4}\} = -\int 3\eta u_{\alpha}u_{\gamma}^{2}\cdot\mathrm{Re}\left\{\overline{A}_{2}^{2}\Psi_{3}\right\}dz. \tag{59}$$

As can be seen from Eqs. (58) and (59), the integral kernels for the coma and the three-lobe aberration are nearly identical. They have the same signature in the fundamental rays. Hence, most of the conceivable hexapole correctors will produce neither $\mathrm{Im}\left\{C_{\alpha\alpha\bar{\gamma}}\right\}$ nor $\mathrm{Im}\{D_{4}\}$ or, even worse, both.

C. Aplanats Without Axial Fourth-Order Aberrations

The first finding from the last section is that the hexapole elements of the corrector must be *rotated* with respect to each other. Otherwise it would not be possible to obtain a non-zero contribution to the azimuthal coma [Eq. (58)] since we can deduce $\mathrm{Re}\left\{\overline{A}_{2}\Psi_{3}\right\} \sim \mathrm{Re}\left\{i\,\overline{\Psi}_{3}\Psi_{3}\right\} = 0$. A coma correction essentially needs the following two steps:

1. It must produce A_{2} in a first hexapole element of the corrector and,
2. Let this aberration combine with a second hexapole element, which is rotated against the first.

This immediately results in the second finding, that we need *at least three* separate hexapole elements. This is obvious since the total A_{2} must be compensated behind the corrector, and astigmatism A_{2} of *two* rotated hexapole fields cannot cancel each other.

We will now further reduce the variety of still possible hexapole-type correctors by requiring them to have a *mid-plane symmetry*. This means that the fields (and hence the geometry of the field producing optical elements), the fundamental rays, and the primary aberration functions of the corrector shall be symmetric or anti-symmetric functions with respect to the mid-plane z_{m} of the corrector. In detail, these requirements are stated as

$$
\begin{aligned}
u_{\alpha}(z_{m}+z) &= \pm u_{\alpha}(z_{m}-z), & u_{\gamma}(z_{m}+z) &= \pm u_{\gamma}(z_{m}-z),\\
\Psi_{3c}(z_{m}+z) &= \pm\Psi_{3c}(z_{m}-z), & A_{2x}(z_{m}+z) &= \pm A_{2x}(z_{m}-z),\\
\Psi_{3s}(z_{m}+z) &= \pm\Psi_{3s}(z_{m}-z), & A_{2y}(z_{m}+z) &= \pm A_{2y}(z_{m}-z).
\end{aligned}
$$

These symmetry requirements are the most obvious way to tailor the aberration integral kernels such that contributions of the first and the second half of the corrector either can compensate for each other or can add up to twice the contribution of one half. (As mentioned previously, Hosokawa suggested (Mitsuishi, 2006) that it is also possible to apply scaling rules between both halves of the corrector, which to some extent results in the same cancellation of unwanted integrals. Such generalized, scaled symmetry rules eventually also could be applied here to some extent.) We indicate the chosen symmetry of functions by attaching a sign top-right to the function (e.g., the function u_α^- would have the symmetry $u_\alpha(z_m + z) = -u_\alpha(z_m - z)$, and u_α^+ the symmetry $u_\alpha(z_m + z) = +u_\alpha(z_m - z)$, respectively).

The integral kernel of the imaginary coma [Eq. (58)] now is written as a symmetric function

$$u_\alpha u_\gamma^2 \cdot (A_{2y}\Psi_{3s} + A_{2x}\Psi_{3c})\Big|_{z_m+z} = +u_\alpha u_\gamma^2 \cdot (A_{2y}\Psi_{3s} + A_{2x}\Psi_{3c})\Big|_{z_m-z},$$

while the kernel of the accompanying three-lobe aberration [Eq. (59)] should be anti-symmetric to cancel

$$u_\alpha u_\gamma^2 \cdot \left[2A_{2x}A_{2y}\Psi_{3s} + \left(A_{2x}^2 - A_{2y}^2\right)\Psi_{3c}\right]\Big|_{z_m+z}$$
$$= -u_\alpha u_\gamma^2 \cdot \left[2A_{2x}A_{2y}\Psi_{3s} + \left(A_{2x}^2 - A_{2y}^2\right)\Psi_{3c}\right]\Big|_{z_m-z}.$$

Here we again have set $A_2(z) = A_{2x}(z) + iA_{2y}(z)$ and $\Psi_3(z) = \Psi_{3c}(z) + i\Psi_{3s}(z)$, assuming four different real functions for the respective real and imaginary parts. Now we are able to find all possible symmetries, which provide correction of the anisotropic coma without introduction of Im $\{D_4\}$. The solutions are listed in Table III. Note that the symmetry type of u_γ does not matter due to the square. However, it is always opposite to that symmetry type chosen for u_α.

1. Type (a) corrector with u_α^+: three hexapoles only

This type of corrector seems to allow the shortest systems with a minimum of three hexapole elements, since one can be placed at the symmetry plane z_m, which is an aperture plane for type (a) correctors (see Figure 42). The anti-symmetry of A_{2x} is achieved by exactly overcompensating that induced by the first hexapole by a factor of two. The other component A_{2y} is introduced in the first and compensated in the last hexapole. In other terms, the hexapoles have the orientations $\varphi - \Delta\varphi, \varphi, \varphi + \Delta\varphi$. In the first and third hexapole Ψ_3 and A_2 are perpendicular; no coma correction

TABLE III   All possible choices for the symmetry of fundamental rays, hexapole field, and threefold astigmatism, if a corrector is to produce anisotropic coma and no three-lobe aberration.

Type	u_α	A_2		Ψ_3	
(a)	u_α^+	A_{2x}^-	A_{2y}^+	Ψ_{3c}^-	Ψ_{3s}^+
(b)	u_α^-	A_{2x}^-	A_{2y}^+	Ψ_{3c}^+	Ψ_{3s}^-
(c)	u_α^-	A_{2x}^-	A_{2y}^+	Ψ_{3c}^+	Ψ_{3s}^+
(d)	u_α^+	A_{2x}^-	A_{2y}^-	Ψ_{3c}^-	Ψ_{3s}^-

occurs. The center hexapole, however, is rotated by $\pm\Delta\varphi$ against both the entrance and exit A_2, and therefore produces coma with

$$\mathrm{Im}\{\Delta C_{\alpha\alpha\bar\gamma}\} = - \int_{center\ hexapole} 6\eta\, u_\alpha^+ u_\gamma^2 A_{2y}^+ \Psi_{3s}^+ dz.$$

The summand with $A_{2x}^- \Psi_{3c}^-$ is zero in the center hexapole. All three hexapoles produce negative contributions to C_s.

A disadvantage of the simple setup in Figure 42 is the lack of flexibility. First, if coma and C_s correction should be adjusted separately, the first and third hexapole field would have to be rotated. A separate adjustment is mandatory, however, if the magnification between the objective lens and the corrector is changed for some reason and preservation of both corrections is desired. Second, even if rotatable hexapole fields are provided, a magnification change and the subsequent adjustment of the C_s correction will destroy the elimination of the second-order threefold distortion, since this can only be done once for one excitation by choosing the length of the central hexapole element. The other second-order integral, which is not compensated locally within each hexapole, is A_{2x}, see Eq. (10). This compensation, however, can always be achieved by balancing the strength of center hexapole against the two outer, which is required to make A_{2x} anti-symmetric.

However, this lack of flexibility might not be too bothersome, since the optimization for minimum A_5 requires a fixed (optimum) excitation of the hexapoles. Moreover, this optimized operation mode also fixes the magnification between objective lens and corrector and, hence, the fundamental rays. Thus, this corrector combined with an objective lens can be considered as an optimized aplanat for a given high-resolution mode of the microscope, and as a semi-aplanat for all other operation conditions.

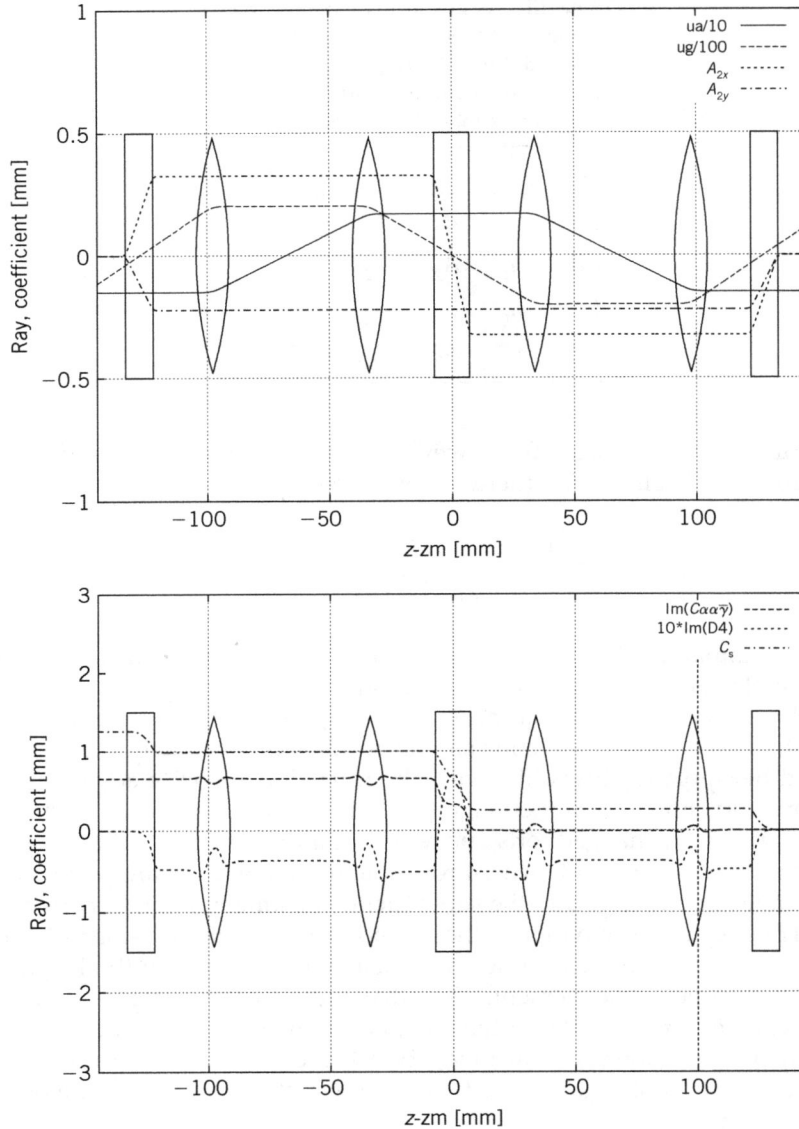

FIGURE 42 Minimum setup for a hexapole corrector that corrects C_s and $\text{Im}\{C_{\alpha\alpha\bar{\gamma}}\}$ simultaneously without introducing fourth-order axial aberrations. Both coefficients to be corrected start with the initial values of the objective lens on the left-hand side (bottom). The choice of the symmetric fundamental and primary aberration rays is also shown (top).

2. Type (a) corrector with u_α^+: three hexapoles plus two weak hexapole pairs

More flexibility provides a second solution of the same symmetry type (a) as shown in Figure 43. Here the coma correction is provided by two hexapole pairs positioned around the intermediate images between the strong hexapoles. The coma integral is

$$\mathrm{Im}\{C_{\alpha\alpha\bar\gamma}\} = -\int 6\eta\, u_\alpha^+ u_\gamma^2 A_{2x}^- \Psi_{3c}^- dz,$$

which is much more efficient here, since u_y is about maximal and constant and A_{2x} is the larger astigmatism component. Complementary to the minimal setup we find here $A_{2y}^+ \Psi_{3s}^+ \approx 0$, since A_{2y} is about three orders of magnitude smaller than A_{2x}.

Rose suggested using the original two-stage hexapole corrector, completed by two pairs of weak hexapole elements centered around the images before and after the first hexapole (Rose, 2005). This is approximately the first half of the corrector in Figure 43. For the situation depicted there, one would expect about $D_{4y} = -60\,\mu\mathrm{m}$, which would vary, depending on the C_s to be corrected and on the intermediate magnification. However, as derived above, this unwanted D_{4y} cannot be avoided if a two-stage corrector produces imaginary off-axial coma. The solution of Figure 43 also adds mid-plane symmetry to A_{2y} (and a third hexapole plus its lens doublet) and enables the cancellation of D_{4y}.

3. Type (b) corrector with u_α^-: a minimum of four hexapoles

Figure 44 shows the simplest solution with an anti-symmetric axial fundamental ray. As shown in Table III, the threefold astigmatism has the same symmetry as in the above solutions, while the hexapole components exchange their symmetries with Ψ_{3c}^+ and Ψ_{3s}^-. In contrast to the solution of Section III.C.1, this solution fulfills all constraints for the primary second-order aberrations by symmetry relations; therefore, it is more flexible. The intermediate magnification can be adjusted, and hence the $C_s/C_{\alpha\alpha\bar\gamma}$ correction without violating the second-order constraints of Eq. (20). Only four hexapole fields are needed; however, the inner two must be rotatable to allow for separate adjustments of the spherical aberration and the anisotropic coma. The second disadvantage of this system is the necessary third round lens doublet and the resulting additional length of $4f$.

The coma correction this time is given by the integral

$$\mathrm{Im}\{\Delta C_{\alpha\alpha\bar\gamma}\} = -\int 6\eta\, u_\alpha^- u_\gamma^2 \left(A_{2x}^- \Psi_{3c}^+ + A_{2y}^+ \Psi_{3s}^- \right) dz,$$

where both terms contribute equally.

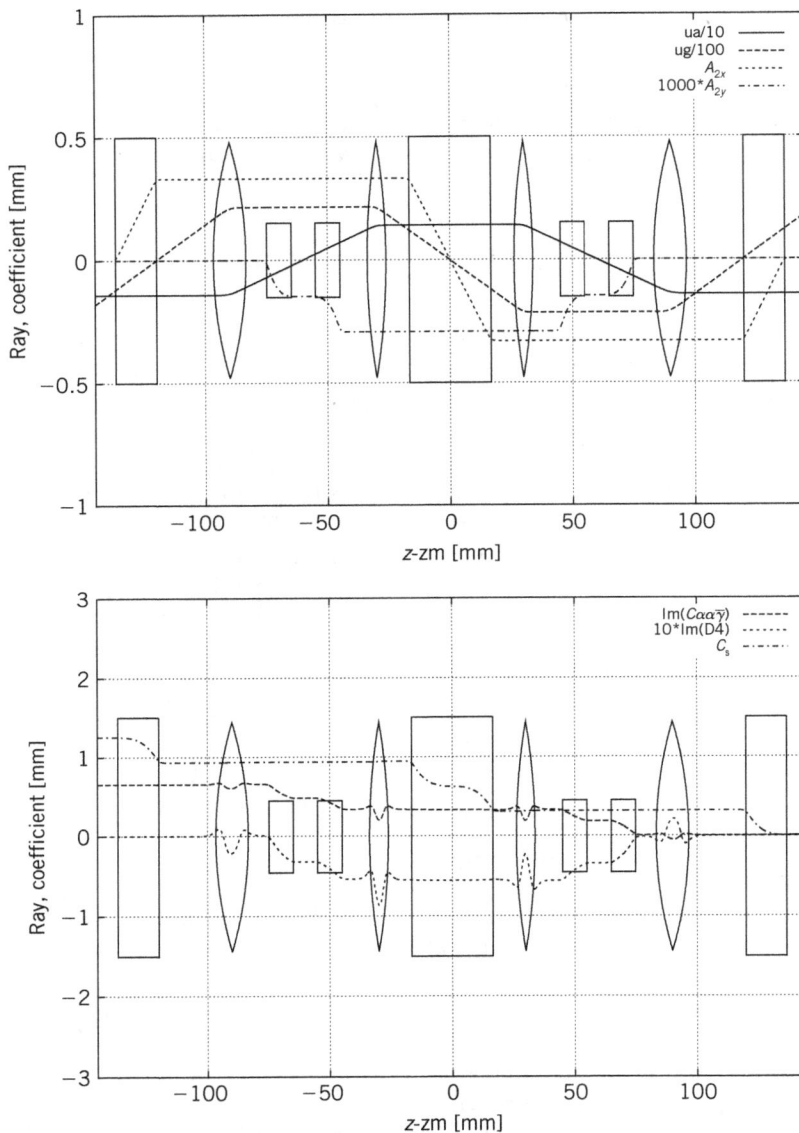

FIGURE 43 More flexible setup for a hexapole corrector of symmetry type (a) that corrects C_s and $\mathrm{Im}\left\{C_{\alpha\alpha\bar{\gamma}}\right\}$ simultaneously without introducing fourth-order axial aberrations. It consists of three strong hexapoles with $\Psi_3 \equiv i\Psi_{3s}$ (large boxes) and two locally anti-symmetric pairs of weak hexapole elements with $\Psi_3 \equiv \Psi_{3c}$ (small boxes). The choice of the symmetric fundamental and primary aberration rays is depicted (top). The correction of the coma and the spherical aberration can be assigned to the weak and the strong hexapole fields, respectively (bottom).

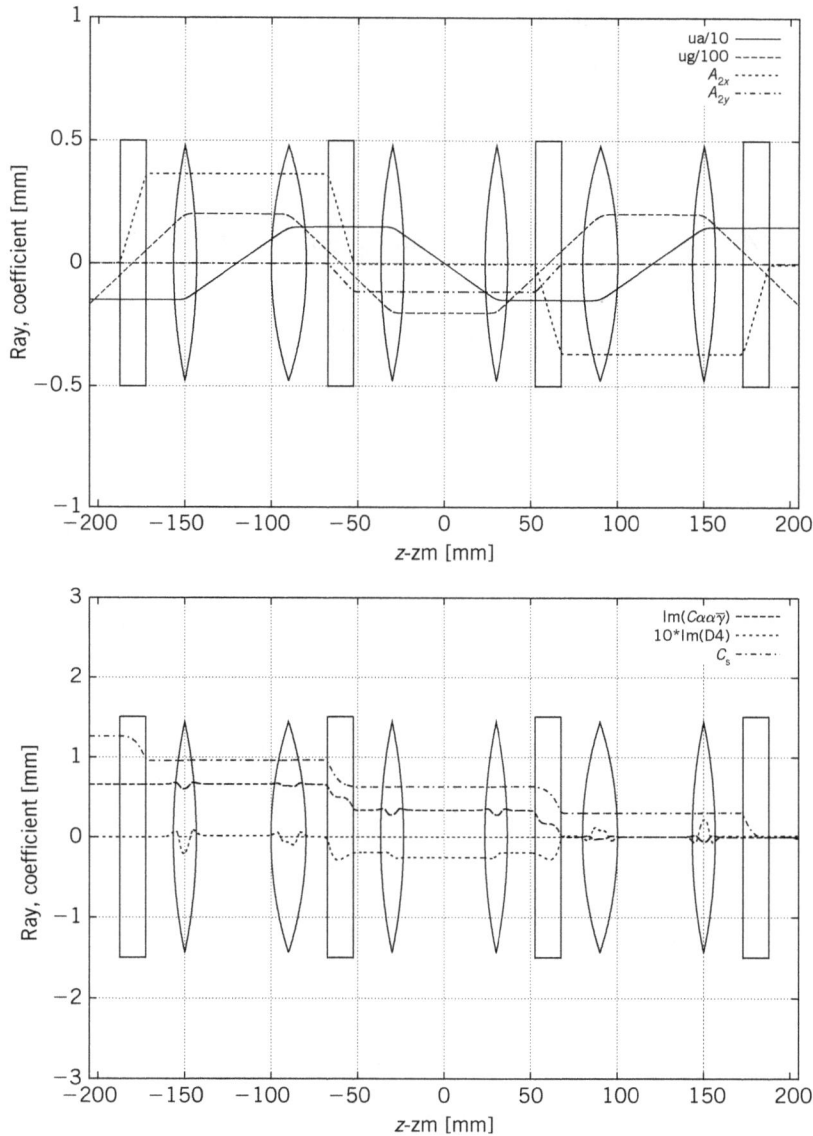

FIGURE 44 Minimum setup for a hexapole corrector with anti-symmetric axial ray u_α^- that corrects C_s and $\mathrm{Im}\left\{C_{\alpha\alpha\bar\gamma}\right\}$ simultaneously without introducing fourth-order axial aberrations. The choice of the symmetric fundamental and primary aberration rays is depicted (top). The orientation of the hexapole fields is φ, $-\varphi + \Delta\varphi$, $\varphi - \Delta\varphi$, $-\varphi$. The correction of the coma can be assigned to the second and third hexapole, while all four contribute to the C_s correction (bottom).

4. Type (c) and (d) correctors

Finally the type (c) and (d) correctors of Table III remain. They do *not* exist. This is due to the following contradiction: From Eq. (55) we again deduce the integral for A_{2x} with an arbitrary orientation of the hexapoles along the z-axis

$$A_{2x}(z) = \int_{-\infty}^{z} 3\eta \, u_\alpha^3 \Psi_{3s} \, d\tilde{z}, \quad \text{and} \quad \frac{d}{dz} A_{2x}(z) = 3\eta \, u_\alpha^3 \Psi_{3s} \Big|_z.$$

According to Table III, type (c) correctors shall have

$$A_{2x}^- \to \left(\frac{d}{dz} A_{2x}\right)^+ \to \left(u_\alpha^3 \Psi_{3s}\right)^+,$$

which is in contradiction to u_α^- and Ψ_{3s}^+. The same argument holds for A_{2y}, and hence type (c) correctors cannot be constructed. If we exchange the symmetries of u_α and Ψ_{3s} for type (d), we find the same contradiction.

D. Feasibility and Prediction of Properties

After the elimination of the anisotropic coma of the objective lens, the *isoplanar field of view* is considerably enlarged. The area that can be imaged without an additional phase shift is now limited by higher-order, generalized coma-type aberrations, which are generated by the corrector itself—namely, $C_{\alpha\alpha\alpha\gamma}$ and $C_{\overline{\alpha}\overline{\alpha}\overline{\alpha}\gamma}$, the off-axial variants of the third-order axial star aberration and the third-order fourfold astigmatism. These coefficients can be calculated. Here we again restrict our considerations to the combination aberrations of the hexapole. If we insert two second-order aberration rays and the off-axial coma ray in the iteration formula [Eq. (6)], we eventually find

$$\bar{u}^2 = \left(\ldots \bar{\alpha} u_\alpha + \alpha^2 \bar{u}_{\overline{\alpha}\overline{\alpha}} + \alpha\gamma \bar{u}_{\overline{\alpha}\gamma} + \bar{\alpha}^2 \gamma \bar{u}_{\alpha\alpha\gamma}\right)^2$$

$$= \ldots + 2\alpha^3 \gamma \bar{u}_{\overline{\alpha}\overline{\alpha}} \bar{u}_{\overline{\alpha}\gamma} + 2\bar{\alpha}^3 \gamma u_\alpha \bar{u}_{\alpha\alpha\gamma}.$$

Considering again Eq. (54) we can write

$$C_{\alpha\alpha\alpha\gamma}^{\text{HP}} = \int 6\eta u_\alpha \text{Im}\left\{\overline{A}_2 \bar{u}_{\overline{\alpha}\gamma} \Psi_3\right\} dz - i \int 6\eta u_\alpha \text{Re}\left\{\overline{A}_2 \bar{u}_{\overline{\alpha}\gamma} \Psi_3\right\} dz$$

$$C_{\overline{\alpha}\overline{\alpha}\overline{\alpha}\gamma}^{\text{HP}} = \int 6\eta u_\alpha^2 \text{Im}\left\{\bar{u}_{\alpha\alpha\gamma} \Psi_3\right\} dz - i \int 6\eta u_\alpha^2 \text{Re}\left\{\bar{u}_{\alpha\alpha\gamma} \Psi_3\right\} dz.$$

Interestingly, only the two-stage *semi-aplanat* with $\Psi_3 \equiv i\Psi_{3s}$ suffers from strong real hexapole contributions and the imaginary parts vanish, since it uses the symmetries u_α^-, A_2^+, $u_{\overline{\alpha\gamma}}^-$, and Ψ_3^+. These real components vanish for all aplanats in Section III.C since they use the symmetries in Table III. However, the aplanats also have these off-axial generalized coma coefficients, but only their imaginary part. Here we note without proof that these imaginary aberrations are caused mainly by combination aberrations of the second-order aberrations with the transfer lenses. However, they can be calculated to be typically one order of magnitude smaller than the strong real coefficients. We use them to determine the size of the enlarged *isoplanatic field of view*. It is given by the radius of a circle, for which an additional phase shift resulting from the fourth-order off-axial aberrations is still below $\pi/4$:

$$R_{\frac{\pi}{4}} = \frac{\lambda}{8} \cdot \left(\left(3 - \sqrt{8}\right) |C_{\alpha\alpha\alpha\gamma}| \cdot \vartheta^4 + \frac{1}{4} |C_{\overline{\alpha\alpha}\gamma}| \cdot \vartheta^4 \right)^{-1}. \qquad (60)$$

Here, by the factor $3 - \sqrt{8} \approx 1/5.8$ we take into account that the phase shift of a twofold star aberration can be counterbalanced by a twofold astigmatism A_1, which has the off-axial variant $C_{\overline{\alpha}\gamma}$. In the same manner, as the phase-shift of C_s ($\sim \vartheta^4$) can be counterbalanced by a defocus C_1 ($\sim \vartheta^2$) at Lichte focus, here we find for the resulting phase shift of the two coefficients at some off-axial point γ

$$S(\alpha, \gamma) = \frac{2\pi}{\lambda} \cdot \chi(\alpha, \gamma) = \frac{2\pi}{\lambda} \cdot \mathrm{Re} \left\{ \overline{\gamma}\alpha^2 \cdot \left(\alpha\overline{\alpha} \cdot \overline{C}_{\alpha\alpha\alpha\gamma} + \frac{1}{2} \cdot C_{\overline{\alpha}\gamma} \right) \right\}.$$

As discussed in Section II.C, the necessary aberration $C_{\overline{\alpha}\gamma}^{\mathrm{opt}} = 2\vartheta^2 (2 - \sqrt{8}) \overline{C}_{\alpha\alpha\alpha\gamma}$ can be easily produced in a hexapole corrector, if we slightly tune the transfer lenses to adjust the integral for $C_{\overline{\alpha}\gamma}$ rf. to Eq. (10). For the off-axial four-fold astigmatism no such compensation can be achieved.

Additionally, the optimization of the fifth-order sixfold astigmatism A_5 by choosing the appropriate length of the hexapole fields can be done as for two-stage STEM corrector (see details in Section II.D.5). Here we note that a (sufficiently) small rest will remain (Table IV). Also the fifth-order spherical aberration eventually can be chosen optimally for phase contrast, since the eigen aberration value of the corrector depends on the setup of the corrector and on the geometry of hexapole fields. The results for the examples in Section III.C.3 are listed in Table IV for illustration.[1]

---

[1]The values marked with * of the conventional two-stage hexapole corrector include the anisotropic coma of the objective lens.

114   Maximilian Haider *et al.*

TABLE IV Properties of the correctors described in section III.C, assuming the parameters $\lambda = 2$ pm and an aperture of $\alpha = 40$ mrad.

Type	$\|C_{\alpha\alpha\alpha\gamma}\|$	$\|C_{\overline{\alpha\alpha\alpha}\gamma}\|$	$R_{\frac{\pi}{4}}$	$N = 2R_{\frac{\pi}{4}} \cdot g_{max}$	$\|A_5\|$	C_5^{tot}	C_c^{tot}
	[1]	[1]	[μm]	[1000px]	[mm]	[mm]	[mm]
CETCOR	13	16	0.02 (0.01*)	0.37*	3.0	6.7	2.3
3 hexapoles	3.0	0.3	0.17	6.6	0.6	+4.5	2.5
3 + 4 hexapoles	1.3	3.6	0.09	3.5	0.15	− 5.0	2.4
4 hexapoles	0.81	0.14	0.56	22	0.13	+4.0	2.7

The objective lens has the typical coefficients $C_s = 1.3$ mm, $C_{\alpha\alpha\overline{\gamma}} = 0.66$, and $C_c = 2$ mm. Note that the necessary information limit of $g_{max} = 20$/nm in the example would require a cold field emission gun or a monochromator.

IV. CONCLUSION

Starting any new endeavor is always difficult. Several circumstances must "come together" to create the opportunity to take off. First, someone must be convinced that the time for an idea has come. In the case of the hexapole corrector, the applicants for funding believed strongly in the success of their endeavor—although a first attempt by Crewe and co-workers had failed at Chicago. A second vital requirement is an organization that is willing to provide funding, is sufficiently patient, and will even provides additional funding if the project runs out of time. The hexapole corrector project received such support from the Volkswagen-Stiftung, especially in the person of Herbert Steinhardt. Third, the project needs a physical home. The open-minded environment of the EMBL at Heidelberg was the optimum breeding ground. It attracted people from all over the world and it provided excellent resources, including technical support and workshops for constructing hardware. Thus the "physical instrumentation group" with its group leader Max. Haider had very good starting conditions for the realization of the hexapole corrector.

However, even this favorable situation did not ensure success. A complex system such as a "high-resolution electron microscope plus corrector" has a large number of parameters. There is always a good chance to get lost, to optimize the wrong thing, or to do an experiment that provides more questions than answers. From today's point of view, the crucial point was identifying the most important parameters. Even then we often found ambiguous answers, both reasonable, but mutually contradictory, as listed below:

- Compact design ⇔ no thermal drift
- Low upper frequency limit ⇔ no remanence
- Low sensitivity and chromatic aberration ⇔ no saturation

- Relaxed mechanical tolerances ⇔ small parasitic aberrations
- Large alignment ranges ⇔ low dipole noise
- Small number of coils ⇔ small residual parasitic aberrations

A deeper understanding of the non-ideal hexapole corrector with manufacturing tolerances and alignment elements attained by computer-algebraic perturbation methods was helpful to trim the tree of possibilities. Of equal importance was the increasing availability of hardware for numerical image processing to characterize the state of alignment in an efficient and reliable manner and to generate the required feedback for the corrector control. After several failures and disappointments, the group at the EMBL finally succeeded as reviewed in section I.

The consolidation phase of the hexapole corrector in the years after 1997 at CEOS led to robust and reproducible C_s correctors in different flavors. Again theoretical methods were used for optimization, simplification, and adaption. A completely new mechanical design and new construction methods were developed and we experienced the difference between academic and industrial research. In the following years, the hexapole corrector became a usable tool also for nonexpert users. Reliable recipes for alignment "from scratch" have been developed and refined. Today, these recipes allow trained service engineers to perform the factory alignment within a few days after a new corrector has been assembled and integrated into the microscope column.

Over the years, the theoretical understanding of the hexapole corrector constantly grew. We developed efficient and compact algebraic models for both the *ideal* and the *non-ideal* perturbed corrector. For the former, many relations among the geometry, the excitation of the elements, and the intrinsic aberrations were derived. These rules of thumb (discussed in section II) provide insight into how the corrector works. They allow fairly good estimations of what happens if one or more parameters of the corrected microscope are changed.

The most surprising results of this further theoretical work are the following two: First, the strong fifth-order sixfold astigmatism of the hexapole corrector, which was considered to be its handicap and ultimate limit, can almost be *eliminated*. The hexapole corrector, therefore, can be used also for very large apertures of ≥50 mrad, provided the electron source is sufficiently monochromatic. These large apertures were infeasible for the EMBL corrector, and an advanced hexapole corrector based on a new design of the principal correction elements has been developed for the TEAM project.

Second, it is very surprising that rather small modifications of the correction concepts additionally allow for the correction of the anisotropic coma of the objective lens. The rotation of the hexapoles with respect to each other enables a true *aplanat* with no linear off-axial aberrations in

116   Maximilian Haider *et al.*

third order. The resulting aplanatic system maintains all other favorable properties of the hexapole semi-aplanat.

The most flexible solution for an aplanat is that proposed in Section III.C. Although it consists of seven hexapoles, probably only three of them require the support of a ferromagnetic circuit. For the weak pairs near the intermediate images planes, wire coils should be sufficient. The correction of the spherical aberration and the anisotropic coma can be adjusted independently and subsequently. The strong hexapoles require six poles only and have a fixed orientation.

The most promising solution, however, seems to be the four hexapole solution of Section 3.3. It immediately enlarges the *isoplanatic field of view* by a factor of more than 20, compared to the field of view, which is restricted by the anisotropic coma of the objective lens alone. The minimization of A_5 (see Section II.D.4) requires the operation at a given hexapole excitation. Therefore, the required rotation of the inner two hexapole elements might not matter too much since it can be fixed by design. However, this version of an aplanat has two more transfer lenses and therefore is longer than the other solutions.

It seems very likely that an aplanatic system will be the next step in the future development of the hexapole corrector for applications that demand a large isoplanatic field of view.

ACKNOWLEDGMENTS

The funding of the C_s-correction project by the Volkswagen-Stiftung, additional financial and technical support by Philips, Eindhoven, NL, and the indirect support of the development by the DFG SATEM-project for the Research Center Jülich are gratefully acknowledged.

Especially for the important C_s-correction project, the authors would like to thank the former reviewers of this project who took the risk and supported the idea by recommending this project to the Volkswagen-Stiftung. At the Volkswagen-Stiftung we would like to emphasize the support in the background by the former officer of the Volkswagen-Stiftung, Dr. H. Steinhardt. At the EMBL we would like to thank the former director Prof. L. Philipson, who opened the opportunity to perform the C_s correction project at the EMBL. We also would like to thank the former members of the EMBL who contributed to the project, including H. Wittman (drawing office), P. Raynor (electronics), Dr. E. Schwan (setup of instrument), D. Mills (maintenance), Dr. R. Wepf (samples), and for the critical and indispensable discussions Dr. J. Zach, and at the CEOS company we would like to thank K. Hessenauer for the fast design and construction of the new adapter lens.

The provision of samples and the continuous support of the project by B. Kabius (former FZ Jülich and now at Argonne National Laboratory) is gratefully acknowledged.

We also would like to thank Dr. G. Benner, Zeiss, and Prof. Dr. U. Kaiser for the provision of images.

The development of the advanced hexapole corrector for STEM was conducted as part of the TEAM project funded by the U.S. Department of Energy, Office of Science.

The support for high-resolution STEM imaging by Rolf Ernie was indispensable and is gratefully acknowledged.

The authors are very grateful to their colleagues at CEOS company for their contributions to research and development of the hexapole correctors and to the user community and the manufacturers of the many C_s-corrected instruments installed worldwide who stimulate and support the improvement of the existing and the development of new corrector systems.

Finally, we would like to acknowledge our esteemed teacher, Harald Rose, who laid the basis for all our work and who constantly supports us with his helpful advice.

REFERENCES

Beck, V. D. (1979). A hexapole spherical aberration corrector. *Optik* **53**, 241–255.

Carl Zeiss AG. (2005). Press release.

Crewe, A. V. (1980). Studies on sextupole correctors. *Optik* **57**, 313–327.

Crewe, A. V. (1982). A system for the correction of axial aperture aberrations in electron lenses. *Optik* **60**, 271–281.

Crewe, A. V. (2002). Some Chicago aberrations. *Microsc. Microanal.* **8**, 4–5.

Crewe, A. V., and Kopf D. (1980). Limitation of sextupole correctors. *Optik* **56**, 391–399.

Essers, E, Benner G., Orchowski, A., Kappel, R., and Trunz, M. (2002). New approach for ultra-stable TEM column support frame. Proceedings of 15th International Congress on Electron Microscopy, Durban, South Africa, pp. 355–356.

Freitag, B., Kujawa, S., Mul, P. M., and Ringnalda, J. (2005). Breaking the spherical and chromatic aberration barrier in TEM. *Ultramicroscopy* **102**, 209–214.

Haider, M. (1990). Electron microscope development and STEM application. *EMBL Research Reports*, 142–146.

Haider, M., Bernhardt, W., and Rose, H. (1982). Design and test of an electric and magnetic dodecapole lens. *Optik* **63**, 9–23.

Haider, M., Braunshausen, G., and Schwan, E. (1995). Correction of the spherical aberration of a 200 kV TEM by means of a hexapole corrector. *Optik* **99**, 167–179.

Haider, M., Müller, H., Uhlemann, S., Zach, J., Löbau, U., and Höschen, R. (2008). Prerequisites for a C_c/C_s-corrected ultrahigh-resolution TEM. *Ultramicroscopy* **108**, 167–178.

Haider, M., Uhlemann, S., and Zach, J. (2000). Upper limits for the residual aberrations of a high-resolution aberration-corrected STEM. *Ultramicroscopy* **81**, 163–175.

Haider, M., Uhlemann, S., Schwan, E., Rose, H., Kabius, B., and Urban, K. (1998). Electron microscopy image enhanced. *Nature* **392**, 768–769.

Hartel, P., Müller, H., Uhlemann, S., and Haider, M. (2004). Residual aberrations of hexapole-type C_s-correctors. In Proceedings 13th European Microscopy Congress, vol. I (D. Van Dyck & P. Oostveldt, eds.), pp. 41–42. Antwerp: Belgian Society for Microscopy.

Hartel, P., Müller, H., Uhlemann, S., and Haider, M. (2007). Experimental set-up of an advanced hexapole C_s-corrector. *Micros. Microanal.* **13**, 1148–1149.

Hartel, P., Müller, H., Uhlemann, S., Zach, J., Löbau, U., Höschen, R., and Haider, M. (2008). Demonstration of C_c/C_s correction in HRTEM. Proceeding of 14th European Microscopy Congress, Aachen, Germany.

Hosokawa, F. (2004). Spherical aberration corrector for electron microscope. United States Patent, US 6836373B2.

Hutchison, J. L., Titchmarsh, J. M., Cockayne, D. J. H., Doole, R. C., Hetherington, C. J. D., Kirkland, A. I., and Sawada, H. (2005). A versatile double aberration-corrected energy filtered HREM/STEM for materials science. *Ultramicroscopy* **103**, 7–15.

Jia, C. L., Lentzen, M., and Urban, K. (2003). Atomic-resolution imaging of oxygen in perovskite ceracmics. *Science* **299**, 870–873.

Jia, C. L., Lentzen, M., and Urban, K. (2004). High-resolution transmission electron microscopy using negative spherical aberration. *Microsc. Microanal.* **10**, 174–184.

118   Maximilian Haider *et al.*

Kazumori, H., Honda, K., Matsuya, M., Date, M., and Nielsen, C. (2004). Field emission SEM with a spherical and chromatic aberration corrector. *Microsc. Microanal.* **10**, 1370–1371.

Kimoto, K., Nakamura, K., Aizawa, S., Isakozawa, S., Matsui, Y. (2007). Development of dedicated STEM with high stability. *J. Electron Microsc.* **56** (1), 17–20.

Kisielowski C., Freitag B., Bischoff M., van Lin H., Lazar S., Knippels G., Tiemeijer P., van der Stam M., von Harrach S., Stekelenburg M., Haider M., Uhlemann S., Müller H., Hartel P., Kabius B., Miller D., Petrov I., Olson E. A., Donchev T., Kenik E. A., Lupini A. R., Bentley J., Pennycook S. J., Anderson I. M., Minor A. M., Schmid A. K., Duden T., Radmilovic V., Ramasse Q. M., Watanabe M., Erni R., Stach E. A., Denes P., and Dahmen U. (2008). Detection of single atoms and buried defects in three dimensions by aberration-corrected electron microscope with 0.5 Åinformation limit. *Microsc. Microanal.* **14**, 454–462.

Knoll, M., and Ruska, E. (1932). Das Elektronenmikroskop. *Z. Physik* **78**, 318–339.

Koops, H., Kuck, G., and Scherzer, O. (1977). Erprobung eines elektronenoptischen Achromators. *Optik* **48**, 225–236.

Mitsuishi, K., Takeguchi, M., Kondo, Y., Hosokawa, F., Kimiharu, O., Sannomiya, T., Hori, M., Iwama, T., Kawazoe, M., and Furuya, K. (2006). Ultrahigh-vacuum third-order spherical aberration corrector for a scanning transmission electron microscope. *Micros. Microanal.* **12**, 456–460.

Müller, H., Uhlemann, S., Hartel, P., and Haider, M. (2005). Optical design, simulation and alignment of present-day and future aberration correctors for the TEM. Proceedings of Microscopy Conference, Davos, Switzerland.

Müller, H., Uhlemann, S., Hartel, P., and Haider, M. (2006a). Advancing the hexapole C_s-corrector for the the STEM. *Microsc. Microanal.* **12**, 442–455.

Müller, H., Uhlemann, S., Hartel P., and Haider M. (2006b). Aberration corrected optics: from an idea to a device. Proceedings of 7th International Conference on Charged Particle Optics, Cambridge, UK.

Phillipp, F., Höschen, R., Osaki, M., Möbus, G., and Rühle, M. (1994). New high-voltage atomic resolution microscope approaching 0.1 nm point resolution installed in Stuttgart. *Ultramicroscopy* **56**, 1–10.

Rose, H. (1971a). Abbildungseigenschaften sphärisch korrigierter elektrononoptischer Achromate. *Optik* **33**, 1–24.

Rose, H. (1971b). Elektronenoptische Aplanate. *Optik* **34**, 285–311.

Rose, H. (1981). Correction of aperture aberrations in magnetic systems with threefold symmetry. *Nucl. Instr. Meth.* **187**, 187–199.

Rose, H. (1989). Private communication.

Rose, H. (1990). Outline of a spherically corrected semi-aplanatic medium-voltage transmission electron microscope. *Optik* **85**, 19–24.

Rose, H. (2002). Advances in electron optics. *In* "High-Resolution Imaging and Spectroscopy of Materials" (F. Ernst and M. Rühle, Eds.), pp. 189–269. Springer Verlag, Berlin.

Rose, H. (2004). Outline of an ultracorrector compensating for all primary chromatic and geometrical aberrations of charged-particle lenses. *Nucl. Instr. Meth.* A **519**, 12–27.

Rose, H. (2005). Private communication.

Rose, H., Haider, M., and Urban, K. (1998). Elektronenmikroskopie mit atomarer Auflösung. *Phys. Bl.* **54** (5), 411–416.

Sawada, H. (2008). Private communication.

Sawada, H., Tomita, T., Naruse, M., Honda, T., Hambridge, P., Hartel, P., Haider, M., Hetherington, C., Doole, R., Kirkland, A., Hutchison, J., Titchmarsh, J., and Cockayne, D. (2005). Experimental evaluation of a spherical aberration-corrected TEM and STEM. *J. Electron Microsc.* **54**, 119–121.

Scherzer, O. (1936). Über einige Fehler von Elektronenlinsen. *Z. Phys.* **101**, 593–603.

Scherzer, O. (1947). Sphärische und chromatische Korrektur von Elektronenlinsen. *Optik* **2**, 114–132.

Schlossmacher P., Matjevic M., Thesen A., and Benner G. (2005). Breaking through the Barrier, Imaging & Microscopy **2/2005**, 50–52.

Shao, Z. (1988). On the fifth-order aberration in a sextupole corrected probe forming system. *Rev. Sci. Instrum.* **59**, 2429–2437.

Tiemeijer, P. C. (1999). Measurement of Coulomb interactions in an electron beam monochromator. *Ultramicroscopy* **78**, 53–62.

Uhlemann, S., and Haider, M. (1998). Residual wave aberrations in the first spherical aberration corrected transmission electron microscope. *Ultramicroscopy* **72**, 109–119.

Uhlemann, S., and Haider, M. (2002). Experimental set-up of a fully electrostatic monochromator for a 200 kV TEM. *In* Proceedings 15th International Congress on Electron Microscopy, vol. I. (J. Engelbrecht, T. Sewell, M. Witcomb, R. Cross, and P. Richards, eds.), pp. 327–328. Durban: Microscopy Society of Southern Africa.

Van der Stam, M., Stekelenburg, M., Freitag, B., Hubert, D., and Ringnalda, J. (2005). A new aberration-corrected transmission electron microscope for a new era. Microscopy and Analysis **19** (4), 9–11.

Walther, T., and Stegmann, H. (2006). Preliminary results from the first monochromated and aberration corrected 200 kV field-emission scanning transmission electron microscope. *Microsc. Microanal.* **12**, 498–505.

Zach, J. (1989). Design of a high-resolution low-voltage scanning electron microscope. *Optik* **83**, 30–40.

Zach, J., and Haider, M. (1995). Correction of spherical and chromatic aberration in a low voltage SEM. *Optik* **98**, 112–118.

Zemlin, F., Weiss, K., Schiske, P., Kunath, W., and Herrmann K.-H. (1978). Coma-free alignment of high-resolution electron microscopes with the aid of optical diffractograms. *Ultramicroscopy* **3**, 49–60.

LIST OF PUBLICATIONS

1. M. Haider, W. Bernhardt and H. Rose, Design and test of an electric and magnetic dodecapole lens, *Optik* **63** (1982) 9.

2. M. Haider, Pure Z-contrast: Can it be achieved with a double deflection electron spectrometer?, *Inst. Phys. Conf. Ser.* **93** (1988) 361.

3. M. Haider, C. Boulin and A. Epstein, A versatile multichannel STEM phase-contrast detector, *Inst. Phys. Conf. Ser.* **93** (1988) 123.

4. N. Webster, M. Haider and H. Houf, Design and construction of a multipole element control unit, *Rev. Sci. Instrum.* **59/6** (1988) 999.

5. M. Haider, A corrected double-deflection electron spectrometer equipped with a parallel recording system, *Ultramicroscopy* **28** (1989) 190.

6. M. Haider, Filtered dark-field and pure Z-contrast: Two novel imaging modes in a STEM, *Ultramicroscopy* **28** (1989) 240.

7. A.V. Jones and M. Haider, Modular detector system for a STEM, *Scanning Microsc.* **3** (1989) 33.

8. M. Haider and B. Bormann, Concentration determination of chromatin in unstained resin embedded sections by means of Z-contrast in STEM, *Proc. XIIth Int. Congr. for EM*, Toronto, Canada Vol. 2. (1990) 186.

9. P. Siedle, H.-J. Butt, E. Bamberg, D.N. Wang, W. Kühlbrandt, J. Zach and M. Haider, Determining the form of atomic force microscope tips, *Inst. Phys. Conf. Ser.* **130** (1992) 361.

10. M. Haider, J. Zach and G. Schäfer, A high resolution low-voltage scanning electron microscope for biological applications, *Proc. EUREM* Vol. 3 (1992) 729.

11. J. Zach and M. Haider, A high-resolution low voltage scanning electron microscope, *Proc. EUREM* Vol. 1 (1992) 49.

12. S. Weinstein, W. Jahn, M. Laschever, T. Arad, W. Tichelaar, M. Haider, C. Glotz, T. Boeckh, Z. Berkovitch-Yellin, F. Franceschi and A. Yonath, Derivatization of ribosomes and of tRNA with an undecagold cluster: crystallographic and functional studies, *J. Crystal Growth* **122** (1992) 286.

13. B. Bohrmann, M. Haider and Kellenberger, Concentration evaluation of chromatin in unstained resin-embedded sections by means of low-dose ratio-contrast imaging in STEM, *Ultramicroscopy* **49** (1993) 235.

14. M. Haider, Instrumental developments for electron microscopes, *Proc. 51st. Ann. Meeting of MSA*, San Francisco Press (1993) 624.

15. M. Haider and J. Zach, State of the development of multipole correctors for a probe-forming system and a high resolution 200 kV TEM, *51st. Ann. Meeting of MSA*, San Francisco Press (1993) 624.

16. T. Bastian, M. Akke, A. Epstein, M. Haider, J. Khazaie and C. Boulin, A user friendly acquisition system for Cryo-Electron Microscopy, *Open Bus Systems*, Zürich (1993) 179.

17. M. Haider, G. Braunshausen and E. Schwan, Correction of the spherical aberration in a 200 kV TEM by means of Hexapoles, *Optik* **94**, Supt. 5 (1993) 18.

18. G. Braunshausen and M. Haider, A comfortable semi-automatic, quasi-online comafree-alignment procedure for TEMs with remote-control capability and equipped with a CCD-based fast DAP-enhanced image processing facility, *Optik* **94**, Supt. 5 (1993) 82.

19. G. Braunshausen and M. Haider, Development of a versatile, general-purpose digital image acquisition & analysis system for accurate measurement of the resolution limiting electron-optical imaging parameters as well as remote-tuning of high-resolution TEM performance, *Optik* **94**, Supt. 5 (1993) 82.

20. M. Haider, A. Epstein, P. Jarron and C. Boulin, A versatile, software configurable multichannel STEM detector for angle-resolved imaging, *Ultramicroscopy* **54** (1994) 41.

21. W. Tichelaar, C. Ferguson, J.-C. Olivo, K. Leonard and M. Haider, A novel method of Z-contrast imaging in STEM applied to double-labeling, *J. Microscopy* **175** (1994) 10.

22. R. Wepf, U. Aebi, A. bremer, M. Haider, P. Tittmann, J. Zach and H. Gross, High resolution SEM of biological macromolecular complexes, *52nd. Ann. Meeting of MSA*, San Francisco Press (1994) 1026.

23. M. Haider, J. Zach and R. Wepf, Physical aspects of corrected high resolution low voltage scanning electron microscope, *Proc. ICEM* **13** (Paris 1994) 55.

24. S. Uhlemann, M. Haider and H. Rose, Procedures for adjusting and controlling the alignment of a spherically corrected electron microscope, *Proc. ICEM* **13** (Paris 1994) 193.

25. M. Haider, G. Braunshausen and E. Schwan, State of the development of a Cs corrected high resolution 200 kV TEM, *ICEM* **13** (Paris 1994) 195.

26. G. Braunshausen and M. Haider, Towards online assessment of all parameters controlling HRTEM image quality, *ICEM* **13** (Paris 1994) 201.

27. M. Haider, State of the development of spherically corrected 200 kV TEMM, *Optik* **100**, Supt. 6 (1995) 4.

28. E. Schwan and M. Haider, Achtpolelement zur korrektur des dreizähligen astigmatismus, *Optik* **100**, Supt. 6 (1995) 4.

29. G. Braunshausen, M. Haider and J. Zach, Estimating the amount of 3-fold astigmatism present in high-resolution TEM images, *Optik* **100**, Supt. 6 (1995) 57.

30. J. Zach and M. Haider, Correction of spherical and chromatic aberration in a low voltage SEM, *Optik* **98** (1995) 112.

31. M. Haider, G. Braunshausen and E. Schwan, Correction of the spherical aberration of a 200 kV TEM by means of a hexapole-corrector, *Optik* **99** (1995) 167.

32. J. Zach and M. Haider, Aberration correction in a low voltage SEM by a multipole corrector, *Nucl. Instr. Method A* **363** (1995) 316.

33. M. Haider and J. Zach, State of the development of a multipole corrector for a probe-forming system and a high resolution 200 kV TEM, 53. *Proc. MSA-Meeting* (Kansas City 1995) 596.

34. S. Uhlemann, M. Haider, E. Schwan and H. Rose, Towards a resolution enhancement in the corrected TEM, *Proc. EUREM* (Dublin 1996).

35. M. Haider, Correctors for electron microscopes: Tools or toys for scientists?, *Proc. EUREM* (Dublin 1996).

36. P. Koeck, R. Schroeder, M. Haider and K. Leonard, Unconventional immuno double labeling by energy filtered transmission electron microscopy, *Ultramicroscopy* **62** (1996) 65.

37. M. Haider and S. Uhlemann, Seeing is not always believing: Reduction of artefacts by an improved point resolution with a spherical aberration corrected 200 kV transmission electron microscope, 55. *MSA-Meeting* (1997) 1179.

38. M. Haider, S. Uhlemann, E. Schwan and B. Kabius, Development of a spherical corrected 200 kV TEM: Current state of this project and results obtained so far, *Optik* Supt. 7 (106), (1997) 7.

39. S. Uhlemann and M. Haider, Residual wave aberrations and point resolution of the first corrected TEM, *Optik* Supl. 7 (106) (1997) 7.

40. S. Uhlemann and M. Haider, Residual wave aberrations in the first spherical aberration corrected transmission electron microscope, *Ultramicroscopy* **72** (1998) 109.

41. M. Haider and S. Uhlemann, Towards sub-angstrom resolution with a 200 kV TEM by means of a Cs-corrector and a computer controlled alignment procedure, 56. *MSA-Meeting* (Atlanta 1998), 50.

42. B. Kabius, M. Haider, S. Uhlemann, E. Schwan, K. Urban and H. Rose, First applications of a spherical aberration corrected microscope in material science, *Microsc. Microanal.* **4** (Suppl. 2), (1998) 384.

43. H. Rose, M. Haider und K. Urban, Sichtbarmachung von Atomen im Elektronenmikroskop durch korrigieren der begrenzenden Linsenfehler, *Physikalische Blätter* **54** (1998) 411.

44. M. Haider, H. Rose, S. Uhlemann, E. Schwan, B. Kabius and K. Urban, Electron microscopy image enhanced, *Nature*, **392**, (1998) 768.

45. M. Haider, H. Rose, S. Uhlemann, E. Schwan, B. Kabius and K. Urban, A spherical-aberration-corrected 200 kV transmission electron microscope, *Ultramicroscopy* **75** (1998) 53.

46. M. Haider and J. Zach, Resolution improvement of electron microscopes by means of correctors to compensate for axial aberrations, *Proc. ICEM-14*, Cancun, Mexico Vol. 1 (1998), 53.

47. M. Haider, and S. Uhlemann, Computer control of the first Cs-corrected 200 keV TEM, *Proc. ICEM-14*, Cancun, Mexico Vol. 1 (1998) 265.

48. M. Haider, H. Rose, S. Uhlemann, B. Kabius and K. Urban, Towards 0.1 nm resolution with the first spherically corrected transmission electron microscope, *J. Electron. Microsc.* **47/5**, (1998) 395.

49. K. Urban, B. Kabius, M. Haider and H. Rose, A way to higher resolution: Spherical-aberration correction in a 200 kV Transmission Electron Microscope, *J. Electron. Microsc.* **48/6**, (1999) 821.

50. M. Haider, S. Uhlemann and J. Zach, Upper limits for the residual aberrations of a high resolution aberration-corrected STEM, *Ultramicroscopy* **81** (2000) 163.

51. M. Haider, Towards sub-Angstrom point resolution by correction of spherical aberration, *EUREM* **12**, Brno, Czech Republic, 2000 I 145.

52. B. Kabius, M. Haider, S. Uhlemann, E. Schwan, K. Urban and H. Rose, First application of a spherical-aberration-corrected transmission electron microscope in material science, *J. Electron. Microsc.* **51** (Suppl. 1), (2002) 51.

53. J.L. Hutchison, J.M. Titchmarsh, D.J.H. Cockayne, G. Möbus, C.J. Hetherington, R.C. Doole, F. Hosokawa, P. Hartel and M. Haider, Applications of a Cs corrected HRTEM in materials science, *Microsc. Microanal.* **8** (Suppl. 2), (2002) 10.

54. H. Müller, S. Uhlemann and M. Haider, Benefits and possibilities of Cc-correction for TEM/STEM, *Microsc. Microanal.* **8** (Suppl. 2), (2002) 12.

55. G. Benner, G. Lang, A. Orchowski, W.D. Rau and M. Haider, Realization of a field emission gun with advanced köhler illumination, *Microsc. Microanal.* **8** (Suppl. 2), (2002) 486CD.

56. H. Müller S. Uhlemann and M. Haider, Benefits and possibilities of Cc-correction for TEM/STEM, *Micr. and Microanal.* Vol. 8, Suppl. S02 12.

57. S. Uhlemann and M. Haider, Experimental set-up of a purely electrostatic monochromator for high-resolution and analytical purposes of a 200 kV TEM, *Microsc. Microanal.* **8** (Suppl. 2) 584.

58. J.L. Hutchison, J.M. Titchmarsh, D.J.H. Cockayne, G. Möbus, C.J. Hetherington, R.C. Doole, F. Hosokawa, P. Hartel and M. Haider, A double Cs corrected TEM /STEM, *Proc. 15th ICEM*, Durban, S. Africa Vol. 3 (2002) 900.

59. M. Haider, P. Hartel, F. Kahl and S. Uhlemann, Correction of spherical aberration for high resolution imaging, *Proc. 15th ICEM*, Durban, S. Africa Vol. 3 (2002) 27.

60. S. Uhlemann and M. Haider, Experimental set-up of a fully electrostatic monochromator for a 200 kV TEM, *Proc. 15th ICEM*, Durban, S. Africa Vol. 3 (2002) 327

61. G. Benner, G. Lang, A. Orchowski, W.D. Rau and M. Haider, Advanced köhler illumination with field emission gun, *Proc. 15th ICEM*, Durban, S. Africa Vol. 3 (2002) 187.

62. J.L. Hutchison, J.M. Titchmarsh, D.J.H. Cockayne, G. Möbus, C.J.D. Hetherington, R.C. Doole, F. Hosokawa, P. Hartel and M. Haider, A Cs corrected HRTEM: Initial applications in materials science, *JEOL news*, Vol. 37E (2002) 2.

63. F. Hosokawa, T. Tomita, M. Naruse, T. Honda, P. Hartel and M. Haider, A spherical aberration-corrected 200 kV TEM, *J. Electr. Microsc.* **52**, (2003) 3.

64. J.L. Hutchison, J.M. Titchmarsh, D.J.H. Cockayne, G. Mobus, C.J. Hetherington, R.C. Doole, F. Hosokawa, P. Hartel and M. Haider, Applications of a Cs corrected HRTEM in materials science, *Microsc. Microanal.* **9** (Suppl. 2) 10.

65. G. Benner, A. Orchowski, M. Haider and P. Hartel, State of the first aberration-corrected, monochromized 200 kV FEG-TEM, *Microsc. Microanal.* **9** (Suppl. 2), (2003) 938.

66. G. Benner, M. Matjievic, A. Orchowski, B. Schindler, M. Haider and P. Hartel, State of the first aberration corrected, monochromatized 200 kV FEG-TEM, *Microsc. and Microanal.* Vol. 9, (Suppl. 3) (2003) 38.

67. F. Hosokawa , T. Tomita, M. Naruse, T. Honda, P. Hartel and M. Haider, A spherical aberration-corrected 200 kV TEM, *J. Electron Microscopy* **52/1** (2003), 3–10.

68. M. Haider, Current and future developments in order to approach a point resolution of dpr ~ 0.5 Å with a TEM, *Microsc. and Microanal.* **9** (Suppl. 2) 930.

69. A.I. Kirkland, J.M. Titchmarsh, J.L. Hutchison, D.J.H. Cockayne, C.J.D. Hetherington, R.C. Doole, H. Sawada, M. Haider and P. Hartel, A double aberration corrected, energy filtered HREM/STEM, *JEOL news*, Vol. 39–1 (2004) 2.

70. S. Bals, B. Kabius, M. Haider, V. Radmilovic and C. Kisielowski, Annular aark field imaging in a TEM, *Solid State Communications* **130** (2004) 675.

71. M. Haider, H. Mueller and P. Hartel, Present state and future trends of aberration correction, *Proc. 8APEM*, Kanazawa, Japan, (2004) 16.

72. H. Sawada, T. Tomita, M. Naruse, T. Honda, P. Hartel, M. Haider, C. Hetherington, R. Doole, A. Kirkland, J. Hutchison, J. Titchmarsh and D. Cockayne, A 200 kV TEM with Cs Correctors for Illumination and Imaging, *Proc. 8APEM*, Kanazawa, Japan, (2004) 20.

73. H. Sawada, T. Tomita, M. Naruse, T. Honda, P. Hartel, M. Haider, C. Hetherington, R. Doole, A. Kirkland, J. Hutchison, J. Titchmarsh and D. Cockayne, Cs corrector for illumination, *Micros. and Microanal.* Vol. 10 Suppl. S02 1004.

74. H. Sawada, T. Tomita, T. Kaneyama, F. Hosokawa, M. Naruse, T. Honda, P. Hartel, M. Haider, N. Tanaka, C. Hetherington, R. Doole, A. Kirkland, J. Hutchison, J. Titchmarsh and D. Cockayne, Cs corrector for imaging, *Micros. and Microanal.* Vol. 10 Suppl. S02 976.

75. H. Lichte, D. Geiger, M. Lehmann, M. Haider and B. Freitag, Electron holography with Cs-corrected TEM journal, *Micr. Microanal.* **10** Suppl. S04 40.

76. H. Lichte, D. Geiger, M. Lehmann, M. Haider and B. Freitag, Electron holography with Cs-corrected TEM journal, *Micr. Microanal.* **10** Suppl. S02 112.

77. G. Benner, M. Matijevic, A. Orchowski, P. Schlossmacher, A. Thesen, M. Haider and P. Hartel, Sub-angstrom and sub-eV resolution with the analytical SATEM, *Micr. and Microanal.* **10** Suppl. S03 6.

78. M. Haider and H. Müller, Design of an electron optical system for the correction of the chromatic aberration Cc of a TEM objective lens, *Micr. Microanal.* **10** (2004) Suppl. S03.

79. M. Haider, H. Müller and P. Hartel, HRTEM vs HRSTEM: Are these instruments complementary or competitive?, *Proc. EMC 2004*, Antwerpen, Vol. 1 (2004) 3.

80. G. Benner, E. Essers, M. Matijevic, A. Orchowski, P. Schlossmacher, A. Thesen, M. Haider and P. Hartel, Performance of monochromized and aberration-corrected TEMs, *Proc. EMC 2004*, Antwerpen, Vol. 1 (2004) 51.

81. P. Hartel, H. Müller, S. Uhlemann and M. Haider, Residual aberrations of hexapole-type Cs-correctors, *Proc. EMC 2004*, Antwerpen, Vol. 1 (2004) 41.

82. D. Geiger, H. Lichte, M. Lehmann, M. Haider and B. Freitag, Electron holography with Cs-corrected TEM, *Proc. EMC 2004*, Antwerpen, Vol. 1 (2004) 187.

83. S. Bals, B. Kabius M. Haider, V. Radmilovic and C. Kisielowski, Annular dark field imaging in a transmission electron microscope, *Proc. EMC 2004*, Antwerpen, Vol. 1 (2004) 385.

84. M. Haider and H. Müller, Is there a road map of aberration correction towards ultra-high resolution in TEM and STEM?, *Proc. MSA, Microsc Microanal* **11** (Suppl 2), (2005) 546.

85. H. Müller, S. Uhlemann, P. Hartel and M. Haider, Optical design, simulation, and alignment of present-day and future aberration correctors for the transmission electron microscope, *Proc. Dreiländertagung MC 2005* Davos, (2005).

86. H. Sawada, T. Tomita, M. Naruse, T. Honda, P. Hambridge, P. Hartel, M. Haider, C. Hetherington, R. Doole, A. Kirkland, J. Hutchison, J. Titchmarsh and D. Cockayne, Experimental evaluation of a spherical aberration-corrected TEM and STEM, *J. Electr. Microscopy* **54**/**2** (2005) 119.

87. H. Müller, S. Uhlemann, P. Hartel and M. Haider, Aberration corrected optics: from an ideato a device, *Proceed. International CPO, Physics Procedia* **1**, pp. 167–178 (Eds. Munro E. & Rouse J.)

88. H. Müller, S. Uhlemann, P. Hartel and M. Haider, Advancing the hexapole Cs-corrector for the scanning transmission electron microscope, *Micr. Microanal.* **12** (2006) 442–455.

89. M. Haider, H. Mueller and S. Uhlemann, Improvement path for the hexapole Cs-corrector towards 0.5 a resolution, *Micr. and Microanal.*, Vol. 12, Supplement S02, Aug 2006.

90. G. Benner, A. Orchowski, M. Haider and P. Hartel, State of the first aberration-corrected, monochromized 200kV FEG-TEM, *Micr. and Microanal.*, Vol. 12, Supplement S02, Aug 2006.

91. H. Lichte, D. Geiger, M. Lehmann, M. Haider and B. Freitag, Electron holography with Cs-corrected TEM, *Micr. and Microanal.*, Vol. 12, Supplement S02, Aug 2006.

92. C. Dwyer, A. I. Kirkland, P. Hartel, H. Müller, and M. Haider, Electron nanodiffraction using sharply-focused parallel probes, *Applied Physics Letters* (Vol. 90, No. 15): 151104.

93. M. Haider, U. Loebau, R. Hoeschen, H. Müller, S. Uhlemann and J. Zach, State of the development of a Cc & Cs corrector for TEAM, *Micr. and Microanal.*, Vol. 13, Supplement S02, Aug 2007.

94. C. Dwyer, P. Hartel, H. Muller, M. Haider and A.I. Kirkland, Uses of a STEM Cs-corrector in electron diffractive imaging journal, *Micr. and Microanal.* Vol. 13 S02 958.

95. P. Hartel, H. Müller, S. Uhlemann and M. Haider, Experimental set-up of an advanced hexapole Cs-corrector, *Micr. and Microanal.*, Vol. 13, Supplement S02, Aug 2007.

96. D. Geiger, H. Lichte, M. Linck, M. Lehmann, B. Freitag and M. Haider, Improved performance of electron holography with tecnai F20 Cs-corr, *Microsc. Microanal.* 13 (Suppl. 3), (2007) 36.

97. P. Hartel, H. Müller, S. Uhlemann and M. Haider, First experimental results with advanced hexapole Cs-correctors, *Microsc. Microanal.* 13 (Suppl. 3), (2007) 4.

98. J. Zach and M. Haider, Next generation of electron microscopes, *NanoS Guide 2008* (2007) 27–31, Wiley-Verlag VCH.

99. M. Haider, H. Müller, S. Uhlemann, J. Zach, U. Loebau and R. Hoeschen, Prerequisites for a Cc/Cs-corrected ultra high resolution TEM, *Ultramicroscopy* **108** (2007) 167–178.

100. M. Haider, U. Loebau, R. Hoeschen, H. Müller, S. Uhlemann and J. Zach, Progress of the development of a Cc & Cs Corrector for TEAM, M*icr. and Microanal.*, Vol. 14, Supplement S02, Aug 2008.

101. M. Haider, H. Müller, S. Uhlemann, P. Hartel and J. Zach, Developments of aberration correction systems for current and future requirements, *Proceedings EMC*, Vol. 1 (2008) pp. 9–10.

102. P. Hartel, H. Müller, S. Uhlemann, J. Zach, U. Löbau, R. Höschen and M. Haider, Demonstration of CC/CS-correction in HRTEM, *Proceedings EMC*, Vol. 1 (2008) pp. 27–28.

103. U. Kaiser, A. Chuvilin, R.R. Schröder, M. Haider and H. Rose, Sub-Ångstrøm Low-Voltage Electron Microscopy — Future reality for deciphering the structure of beam-sensitive nanoobjects?, *Proceedings EMC*, Vol. 1 (2008) pp. 35–36.

104. D. Geiger, A. Rother, M. Linck, H. Lichte, M. Lehmann, M. Haider and B. Freitag, Electron holography with Cs-corrected Tecnai F20 — Elimination of the incoherent damping introduced by the biprism in conventional electron microscopes, *Proceedings EMC*, Vol. 1 (2008) pp. 255–256.

105. C. Kisielowski, B. Fereitag, M. Bischoff, H. van Lin, S. Lazar, G. Knippels, P. Tiemeijer, M. van der Stam, S. von Harrach, M. Stekelenburg, M. Haider, S. Uhlemann, H. Müller, P. Hartel, B. Kabius, D. Miller, I. Petrov, E. A. Olson, T. Donchev, E.A. Kenik, A. Lupini, J. Bentley, S. Pennycook, I.M. Anderson, A.M. Minor, A.K. Schmid, T. Duden, V. Radmilovic, Q. Ramasse, M. Watanabe, R. Erni, E.A. Stach, P. Denes and U. Dahmen, Detection of single atoms and burified defects in three dimesnions by aberration-corrected electron microscope with 0.5 A information limit, *Micr. and Microanal.* **14** (2008) 469–477.

106. M. Haider, H. Müller and S. Uhlemann, Present and future hexapole aberration correctors for high-resolution electron microscopy, *Adv. in Imaging and Electron Physics*, Vol. 153, (2008) 43–121 Elsevier.

107. M. Haider, H. Müller, S. Uhlemann, P. Hartel and J. Zach, Recent corrector developments for high-resolution electron microscopy, *Proceedings APMC*, (2008) p. 14, Korean *Jour. of Micr.*

108. M. Haider, P. Hartel, H. Müller, S. Uhlemann and J. Zach, Current and future aberration correctors for the improvement of resolution in electron microscopy, *Phil. Trans. R. Soc. A* 2009 367 3665–3682.

109. M. Haider, P. Hartel, H. Müller, S. Uhlemann and J. Zach, Development of correctors: From O. Scherzer to TEAM, *Microsc. Microanal.* 15 (Suppl. 2), (2009) 150.

110. B. Kabius, P. Hartel, M. Haider, H. Müller, S. Uhlemann, U. Löbau and J. Zach, First Application of Cc corrected imaging for high-resolution and energy filtered TEM, *Microsc. Microanal.* 15 (Suppl. 2), (2009) 1456,

111. B. Kabius, P. Hartel, M. Haider, H. Müller, S. Uhlemann, U. Loebau, J. Zach and H. Rose, First application of Cc-corrected imaging for high-resolution and energy-filtered TEM, *J. Electron. Microsc.* (Tokyo) (2009) **58**(3) pp. 147–155.

112. M. Haider, To new dimensions in electron microscopy with an advanced aberration corrector, *Proceedings BCEIA* (2009) A1.

113. M. Haider, P. Hartel, R. Hoeschen, U. Loebau, H. Müller, S. Uhlemann and J. Zach, Aberration Correctors in Electron Microscopy: From the first ideas of O. Scherzer to sophisticated correction systems, *Proceedings MC2009*, Vol. I, p. 49.

114. P. Hartel, H. Müller, S. Uhlemann, J. Zach and M. Haider, Benefits of simultaneous Cc- and Cs-correction, *Microsc. Microanal.* 16 (Suppl. 2), (2010) 114.

115. M. Haider, P. Hartel, H. Müller, S. Uhlemann and J. Zach, Information transfer in a TEM corrected for spherical and chromatic aberration, *Microsc. Microanal.* **16**, 393–408.

116. D.C. Bell, C. Russo, S. Meyer, G. Benner and M. Haider, Advantages of monochromated low voltage aberration corrected TEM, *Microsc. Microanal.* 16 (Suppl. 2), (2010) 118.

117. U. Kaiser, J. Meyer, J. Biskupek, J. Leschner, A.N. Khlobystov, H. Müller, P. Hartel, M. Haider, S. Eyhusen and G. Benner, High resolution 20 kV TEM of nanosystems: First results towards sub-angstrom low voltage EM (SALVE-microscopy), *Microsc. Microanal.* 16 (Suppl. 2), (2010) 1702.

118. I. Maßmann, S. Uhlemann, H. Müller, P. Hartel, J. Zach, M. Haider, Y. Taniguchi, D. Hoyle and R. Herring, Realization of the first aplanatic transmission electron microscope journal, *Micr. & Microanal.* Vol. 17 Suppl. S2 1270.

119. H. Müller, I. Massmann, S. Uhlemann, P. Hartel, J. Zach and M. Haider, Aplanatic imaging systems for the transmission electron microscope, *Nucl. Instr. Methods in Phys. Research A* **645** (2011) pp. 20–27.

120. U. Kaiser, J. Biskupek, J.C. Meyer, J. Leschner, L. Lechner, H. Rose, M. Stöger-Pollach, A.N. Khlobystov, P. Hartel, H. Müller, M. Haider, S. Eyhusen and G. Benner, Transmission electron microscopy at 20 kV for imaging and spectroscopy, *Ultramicroscopy* **111** (2011) 1239–1246.

121. B. Barton, D. Rhinow, A. Walter, R. Schröder, G. Benner, E. Majorovits, M. Matijevic, H. Niebel, H. Müller, M. Haider, M. Lacher, S. Schmitz, P. Holik, W. Kühlbrandt, In-focus electron microscopy of frozen hydrated biological samples with a Boersch phase plate, *Ultramicroscopy* **111**(12) (2011) pp. 1696–1705.

122. H. Müller, M. Haider, V. Gerheim and J. Zach, A quadrupole optics with large aspect ratio for an anamorphotic electrostatic phase plate without beam blocking, *Micr. & Micranal.* Vol. 18 Suppl.S2 494.

123. J. Biskupek, P. Hartel. M. Haider and U. Kaiser, Effects of residual aberrations explored on single-walled carbon nanotubes, *Ultramicroscopy* **116** (2012) 1–7.

124. S. Uhlemann, H. Müller, P. Hartel, J. Zach and M. Haider, Thermal magnetic field noise limits resolution in transmission electron microscopy, *Phys. Rev. Lett.* **111**, 046101.

125. M. Haider, H. Mueller and P. Hartel, High resolution TEM/STEM by means of advanced Instrumentation, *Micr. & Micranal.* Vol. 19 Suppl.S2 304.

126. R.A. Herring, D. Hoyle, Y. Taniguchi and M. Haider, The ultra-stable scanning transmission electron holography microscope, *Micr. & Micranal.* Vol. 19 Suppl.S2 320.

127. S. Uhlemann, H. Muller, P. Hartel, J. Zach and M. Haider, Instrumental resolution limit by thermal noise from conductiveparts, *Micr. & Micranal.* Vol. 19 Suppl.S2 598.

128. H. Mueller, S. Uhlemann, J. Zach, P. Hartel and M. Haider, Assessment of the TEM information limit by means of tilted Illumination, *Micr. & Micranal.* Vol. 19 Suppl.S2 612.

129. S. Uhlemann, H. Müller, P. Hartel, J. Zach and M. Haider, Thermal magnetic field noise limits resolution in transmission electron microscopy, *Phys. Rev. Lett.* **111**, 046101.

130. H. Müller, S. Uhlemann, P. Hartel, J. Zach and M. Haider, Overview of commercially available CEOS hexapole-type aberration, *Microscopy and Microanalysis* Vol. 20 Sup. S3 pp. 934–935.

131. P. Hartel, M. Linck, F. Kahl, H. Müller and M. Haider, On proper phase contrast imaging in aberration corrected TEM, *Microscopy and Microanalysis* Vol. 20 Sup. S3 p.p 926–927.

132. M. Haider, S. Uhlemann, P. Hartel and H. Müller, Towards high resolution in TEM and STEM: What are the limitations and achievements, *Microscopy and Microanalysis* Vol. 20 Sup. S3 pp. 378–379.

133. F. Börrnert, H. Müller, T. Riedel, M. Linck, A.I. Kirkland, M. Haider, B. Büchner, H. Lichte, A flexible multi-stimuli *in situ* (S)TEM: Concept, optical performance, and outlook, *Ultramicroscopy*, Volume **151**, pp. 31–36.

134. S. Uhlemann, H. Müller, J. Zach, M. Haider, Thermal magnetic field noise: Electron optics and decoherence, *Ultramicroscopy*, Volume **151**, 199–210.

135. T. Akashi, Y. Takahashi, T. Tanigaki, T. Shimakura, T. Kawasaki, T. Furutsu, H. Shinada, H. Müller, M. Haider, N. Osakabe and A. Tonomura, Aberration corrected 1.2-MV cold field-emission transmission electron microscope with a sub-50-pm resolution, *Appl. Phys. Lett.* **106**.

Harald H. Rose

HARALD H. ROSE

Birthplace

Born	Bremen, Germany, 14 February 1935

Education

1955–1961	Studies in Physics and Mathematics at the Darmstadt University of Technology, (TU Darmstadt), Germany
1961	Master Degree in Physics, TU Darmstadt
1965	Ph.D. in Physics, TU Darmstadt

Employment and Scientific Career

1961–1962	Research Consultant, Air Force Cambridge Research Laboratories, Bedford, Mass
1962–1965	Research Associate, Institute of Theoretical Physics, TU Darmstadt
1965–1970	Post Doc. Position at the Institute of Theoretical Physics, TU Darmstadt
1970–1971	Assistant Professor, Faculty of Physics and Mathematics, TU Darmstadt
1971–1975	Associate Professor, Institute of Applied Physics, TU Darmstadt
1972–1973	Sabbatical year at the Enrico Fermi Institute of the University of Chicago
1976–1980	Principal Research Scientist, New York State Department of Health, Division of Laboratories and Research, Albany, NY
1977–1980	Adjunct Professor, Faculty of Physics, Rensselaer Polytechnic Institute, Troy, NY
1980–2000	Professor of Physics, Institute of Applied Physics, TU Darmstadt, Germany
1987	Visiting Professor at the Department of Physics of the Jiaotong University Xian, Xian, China
1995/96	Sabbatical year at the Institute of Applied Physics, Cornell University (6 months) and the Institute of Plasma Research of the University of Maryland, College Park, MD (5 months).
2000/1	Research Fellow at the Department of Materials Science, Oak Ridge National Laboratory, Oak Ridge and Visiting Professor at the University of Tennessee, Knoxville, TN
2001/2	Research Fellow at the Department of Materials Science, Argonne National Laboratory, Argonne IL
2003–2005	Research Fellow at the Advanced Light Source, Lawrence Berkeley National Laboratory, Berkeley, CA
2010–	Carl Zeiss Senior Professor at the University of Ulm, Germany

Scientific Achievements and Honors

Publications:
More than 200 reviewed articles in scientific journals
9 major review articles in scientific books, 1 book entitled "Geometrical Charged-Particle Optics", Springer, 2009, second edition 2013

Patents:
Inventor of 110 patents on scientific instruments and electron-optical components, partly manufactured by various companies

Honorary Membership in Scientific Societies

Honorary member of the Microscopy Society of America, the German Society of Electron Microscopy, and of the 141 Committee of the Japanese Society for the Promotion of Sciences
Honorary Fellowship of the Royal Microscopy Society

Awards

Distinguished Scientist Award 2003 of the Microscopy Society of America
Consulting Professor of the Jiaotong University, Xi'an, China (since 1987)
2005 Award of the 141 Committee of the Japanese Society for the Promotion of Sciences (JSPS)
Karl Heinz Beckurts Award 2006 together with Professor Maximilian Haider (Univ. Karlsruhe) and Professor Knut Urban (FZ Juelich)
Honda-Prize 2008 together with Professor Maximilian Haider and Professor Knut Urban
Robert-Wichard-Pohl-Preis 2009 of the Deutsche Physikalische Gesellschaft
Wolf-Prize 2011 in Physics together with Professor Maximilian Haider and Professor Knut Urban
Honorary Professorship of the Jiaotong University, Xi'an, China (2011)
Frontiers of Knowledge Award 2013 of the BBVA Foundation together with Prof. M. Haider und Professor K. Urban
NIMS 2015 Award of the National Institute of Materials Science of Japan together with Professor M. Haider und Professor K. Urban

COMMENTARY[*]

1. The Early Years

Born in Bremen on 14 February 1935, a Thursday, Harald Hermann Rose was the second child of Anna-Louise and Hermann Rose. Both mother and father were very mathematically talented. Hermann had actually started to study mathematics at the university when his father lost his fortune during the hyperinflation years in Germany in the early 1920s. Forced to look for a way of making a living, Hermann left university and joined Kaffee-Handels-Aktien-Gesellschaft (Kaffee HAG) in Bremen. In 1907 Kaffee HAG had become the world's first producer of decaffeinated coffee. The company, after ceasing operations during the First World War, resumed production in 1923 and Hermann became a travelling salesman selling the product in the northern part of Germany. He was so successful that he was appointed the company's sales representative for the German state of Hesse where he opened an office in Darmstadt. His family followed him when their son Harald was 2 years old. And from that time on up to the present day Darmstadt remains the home of the Rose family. In 1938, Hermann was lucky to be able to acquire from the state the last free plot of land, Prinz-Christians-Weg 5, just below the Mathildenhöhe, the focal point of the German Jugendstil (Art Nouveau) with world-famous pieces of extraordinary architecture built by the stars of the Art-Nouveau movement between 1899 and 1914. Having grown up in this extraordinary environment, Harald is an expert on this epoch and enjoys taking private visitors on a tour, preferably before dinner.

Hermann Rose's life was tragically overshadowed by the rise of the Nazi regime and the Second World War. The house was finished in 1939, and Hermann was drafted in early 1940. Harald last saw his father when he came home for a few day's leave for his son's birthday in 1944. Shortly afterwards, he was reported missing in action on the eastern front. The house burned down the same year during an RAF air raid on the city in the night of 11 to 12 September 1944 with a loss of more than 11,000 lives. His mother and brother miraculously survived. Harald, together with his school class, had been evacuated to the country, to Winterkasten, a small village in the Odenwald. Here he saw the end of the war in 1945, at the age of just 10 years. At the end of 1945 he passed the examination for the *Realgymnasium* in Darmstadt. In the following nine years he learned English, Latin and French. While he liked the first two languages, in the third he did not learn more than "what is necessary to read the menu in a restaurant". Looking back, Harald remembers the particularly hard years between 1945 and the German currency reform in 1948 when, like many other families in the badly hit city, his mother and the two boys had to live under primitive conditions. In their case, this meant that they lived in the cold and damp basement

[*]In quest of perfection in electron optics: A biographical sketch of Harald Rose on the occasion of his 80th birthday.
Knut W. Urban, Peter Grünberg Institute, Germany. E-mail address: k.urban@fz-juelich.de.

of ruins of their house in Prinz-Christians-Weg. Even with rationing, food was very hard to get.

2. Student at Darmstadt University

Harald was a good pupil. His favourite subject was mathematics. He recalls that he "never did anything for mathematics but was always by far the best in class" much to the frustration of his classmates. After the obligatory nine years at the *Gymnasium* he passed his final examination (*Abitur*) with excellent marks. When he was about to start his studies at the Technische Hochschule (today Technische Universität) of Darmstadt in 1955, the economic situation of the family was still critical. Following the advice of his grandfather ("Physik ist eine brotlose Kunst"; "there is no money in physics") he enrolled in electrical engineering. Before he was allowed to do so he had to take a 6-month apprenticeship at the Hessische Elektrizitäts-AG, the local supplier of electricity, to learn the essentials of craftsmanship in electrical engineering. After only one semester in electrical engineering, however, Harald decided to follow his own inclinations and changed to physics and mathematics. Although he was able to live at his parents' house, which was gradually being rebuilt, he had to earn his own living as a construction worker during the vacations.

After the extraordinarily short time of only 3 semesters, he passed the examination for the 'Vordiplom' (roughly equivalent to a bachelor's degree) in 1957. In 1958, during his 5th semester, Otto Scherzer, professor of theoretical physics at the university, recognized Harald's talent and offered him a paid position as an assistant for practical exercises in theoretical physics. And the following year he received a state scholarship. This relieved some of the financial pressure. Harald remembers that, like many of the war generation in Germany, he was a very hard-working student. Nevertheless, lacking at that time the self-confidence which later became part of his character, he was after discussing the written mathematics pre-diploma's test with his fellow students so uncertain about his own solutions to the ten problems posed that he did not dare to go and see his mathematics professor to ask about the results. When he eventually did so, it turned out that of the 360 students participating he had the best test results. For his 'Diplomarbeit' (roughly master's thesis) he joined Otto Scherzer, whose institute he already knew quite well. Harald dived into quantum mechanics calculating the electron scattering cross sec-tions of elements with the idea of deriving the type of atoms present in a sample from quantitative measurements of scattering intensity in images taken in the transmission electron microscope (TEM). As we know today, he was far ahead of his time. It was not only that with the very limited computation facilities available at the time, the required calculations were, if possible at all by making far-reaching approximations, extremely time-consuming. Furthermore, with the technical state of the instruments such studies were far out of reach. Nevertheless his in-depth studies in quantum-mechanical scattering theory prepared the ground for one of the major aspects of Harald's later work. He completed his university studies after 11 semesters in 1961 with a diploma in physics.

3. In Scherzer's Institute

During the war Otto Scherzer was a radar specialist with the German navy. Since in the immediate post-war period German science hardly existed anymore and, in addition, Otto Scherzer's university career had been interrupted he, like many German scientists at that time, went to the United States in 1947 joining Fort Monmouth, New Jersey, a US Army installation conducting, among other things, research on communications electronics for the US Signal Corps. There he started a collaboration with Heinz Fischer, another German-born scientist. This continued after Scherzer left the USA in late 1948 when his professorship in Darmstadt had been re-established. It was Fischer who in 1961 invited Harald Rose, the newly graduated theoretical physicist, to spend a year as a research consultant at the Air Force Cambridge Research Laboratories, Bedford, Massachusetts. Heinz Fischer became co-author of Harald's first published paper reporting on the nanosecond time response of indium antimonide photodetectors [1].

Returning to Darmstadt in 1962, Harald Rose joined the staff at Otto Scherzer's Institute of Theoretical Physics. In his seminal paper of 1936, Scherzer had shown that the spherical aberration of round electron lenses is unavoidable in principle (Scherzer's theorem) [2]. Furthermore, unlike the case of light-optical lenses [3], it is not possible to construct round lenses with negative spherical aberration which would enable this aberration to be compensated. Nevertheless, as already expressed in his 1947 paper [4], Scherzer was convinced that departure from rotational symmetry offered a promising approach for correcting the axial aberrations of electron lenses. During the following 35 years, it was Scherzer's chief concern to prove this theoretically and experimentally. He had an agreement with the university that — unusually for a theorist — he was allowed to operate his own electron optics mechanical and electronic workshop.

His first attempt in the direction of realizing spherical aberration correction was also the world's first step towards the realization of aberration-corrected electron optics. It became known as the Scherzer–Seeliger corrector, a system built and tested by Scherzer's doctor student Robert Seeliger [5] between 1949 and 1954 for the TEM. After Seeliger had completed his doctor degree, the system was transferred by Gottfried Möllenstedt to the University of Tübingen where Möllenstedt conducted a number of experiments with it [6]. The system consisted of two electrostatic cylinder lenses, a round lens and three octupoles. It turned out that the corrector did not improve the resolution of the microscope because this was limited by mechanical and electromagnetic instabilities rather than by optical aberrations. Nevertheless, Seeliger demonstrated that the corrector provided a negative spherical aberration, which could be adjusted to compensate for the spherical aberration of the objective lens. To make this proof-of-principle experiment even clearer and more convincing, Möllenstedt employed illumination with a large cone angle of 2×10^{-2} rad. In this way, he increased the effective spherical aberration to such an extent that it became resolution-limiting thus resulting in a strongly blurred image. Upon compensation of this spherical aberration by means of the octupoles, the resolution improved by a factor of about seven accompanied by a striking increase in contrast.

4. Doctoral Thesis and *Habilitation*

Scherzer and his young co-worker Harald Rose were convinced that although the first step had been successfully taken much more systematic work was necessary. The two agreed that for his doctoral thesis Harald should explore the imaging conditions of unround electron lenses in a wider perspective [7]. Harald received his 'doctor rerum naturalium' (Dr. rer. nat.) in 1965 from the Physics Faculty of the University of Darmstadt (Fig. 1). Of the numerous publications following his thesis, he considers the paper entitled "On Correctability of Lenses for Fast Electrons" [8] to be particularly relevant. In this work, he expanded Scherzer's theorem to include relativistic electrons. Furthermore, generalizing earlier work by Hawkes [9] and by himself [10], he proved that for chromatic correction unround magnetic elements are not sufficient but electric quadrupoles are indispensable. A principle that remains the basis of every chromatically corrected system even today, apart from electron mirrors [11]. Harald reports on the difficult time he had when he discovered that in his calculations he could not return from the relativistic case to the (non-relativistic) Scherzer formulation. After an extended search for errors in his manuscript he discovered that there are many correct non-relativistic forms and that the form he obtained from his relativistic formula differed from that obtained by Scherzer.

Fig. 1. Harald Rose after his doctor examination at Technische Hochschule Darm-stadt in 1965 (courtesy H. Rose).

In the meantime, Harald Rose had started to work for his *Habilitation*, a process including a postdoctoral thesis that is required in Germany to be granted the 'venia legendi', permission to teach at a university, and to become a university professor. The written part, the *Habilitationsschrift*, was published in 1971 under the title "Properties of spherically corrected achromatic lenses" [12]. Harald proved that all correctors known at the time, based on quadrupoles, introduced large off-axial coma and were not suitable for TEM [13]. In order to compensate for spherical aberration, chromatic aberration and off-axial coma, he proposed a novel aplanatic corrector utilizing symmetry properties. This corrector was built and tested successfully within the framework of the Darmstadt Project. This project demonstrated for the first time the simultaneous correction of chromatic and spherical aberration by a quadrupole–octupole system [14, 15]. Unfortunately, the system still suffered from instabilities, and, since computer control was not feasible at the time, the optical adjustment was awkward (Rose: "it took more time to adjust a specific electron optical state than this would have time to last"). The Darmstadt Project was suspended after Scherzer's retirement in 1977, and it was abandoned when he died in 1982.

5. The Professor

Following his *Habilitation* in 1970, Harald Rose was appointed '*Privatdozent*' (the equivalent of university lecturer, but without appropriate payment) and, one year later, Professor of Theoretical Physics (H2) at the University of Darmstadt. In 1972 he was invited by Albert Crewe to join his group at the University of Chicago for a year. During this time Harald worked on phase contrast in scanning transmission electron microscopy [16]. He proposed an innovative detector for the scanning transmission electron microscope (STEM) built up from a number of annular zones. Adding up the signals in the proper way (analogous to the case of a Fresnel zone plate) he achieved a substantial enhancement of phase contrast. In addition, together with Mike Isaacson Harald worked on inelastic electron scattering. Their joint paper became the first in which a key issue of inelastic scattering that is relevant for electron energy-loss spectroscopy (EELS) in STEM and energy-filtered transmission electron microscopy (EFTEM) in TEM, the phenomenon of 'delocalization', was experimentally proven and treated theoretically [17]. Following Harald's calculations, they designed an experiment in which they moved a scanning probe slowly towards a carbon (foil) sample edge. They indeed received an energy-loss signal in an attached spectrometer although the beam was still four nanometres away from the sample. Harald pursued this topic, later together with Helmut Kohl, for more than 10 years [18–20]. He returned to Darmstadt at the end of 1973. Together with Erich Plies he started to work on energy filter systems [21, 22], a topic that he did not leave for more than 25 years, in which he laid the foundation for all modern in-column energy filtering systems [23, 24].

However, like many German physicists at the time, Harald was attracted by the flourishing though highly competitive science scene in the US. In 1976 he left his position at Darmstadt and moved to the US. He was appointed Principal Research Scientist at the New York Department of Health, Division of Laboratories and Research at Albany, NY, and adjunct professor in the Faculty of Physics at Rensselaer Polytechnic Institute at Troy, NY. In the following years (Figs. 2, 3), Harald worked among other things on the optimization of detector geometries in STEM [25]. In a remarkable paper of pioneering character, he discussed what is known today as preservation of elastic contrast in EFTEM images and, by extending the original split detector concept of Dekkers and de Lang [26] for differential phase contrast he introduced the four-quadrant split electron detector for STEM [27]. In another important paper, together with his co-worker Jürgen Fertig, he investigated for the first time the propagation of electrons in STEM through crystalline objects employing a coherent superposition of Bloch waves. The results showed that strong crosstalk between neighbouring atomic columns occurs if the cone angle of the incident wave exceeds the Bragg angle [28]. Their results were verified later by image simulations employing multi-slice simulation programs which did not exist at that time. Although Harald maintained links to Albany till 1986 on the basis of third-party funding projects, he returned to Darmstadt University in 1980 to become full Professor for Theoretical Physics (C3) there (Figs. 4, 5). He spent another year (1995–1996) as visiting professor in the Institute of Applied Physics at Cornell University, Ithaca, NY, (Fig. 6) and the Institute of Plasma Research, University

Fig. 2. After the doctor examination of Rose's student Erich Plies in 1978 (courtesy E. Plies).

Fig. 3. With Otto Scherzer at Darmstadt in 1978 (courtesy E. Plies).

Fig. 4. Harald Rose's group at Darmstadt University in 1991. From left, upper row: A. Berger, C. Dinges, R. Spehr, H. Rose, Y. Taniguchy. Middle row: D. Preikszas, V. Gerheim, P. Stallknecht, S. Uhlemann. Front row: A. Weickenmeyer, M. Hammel (courtesy R. Spehr).

Fig. 5. At 'Dreiländertagung für Elektronenmikroskopie' (German, Austrian and Swiss Societies for Electron Microscopy), 1989, in Salzburg/Austria, (courtesy E. Plies).

Fig. 6. At Cornell University, Ithaca, NY, in 1995 (courtesy H. Rose).

of Maryland, College Park, MD. He retired from his position in Darmstadt in 2000 at the age of 65.

6. After Retirement: The Research Fellow

In the following years (Figs. 7–9), Harald Rose was a Research Fellow at the Department of Materials Science at Oak Ridge National Laboratory and a Visiting Professor at the University of Tennessee, Knoxville, TN (2000–2001), at the Department of Materials

Science, Argonne National Laboratory, Argonne, IL, (2001–2002) and at the Advanced Light Source, Lawrence Berkeley Laboratory, Berkeley, CA (2003–2005). Here he became a consultant for the TEAM project of the Department of Energy [29]. After many years of preparation, Harald Rose's comprehensive text book "Geometrical Charged-Particle Optics" appeared in the Springer Series in Optical Sciences. This remarkable book, the summary of a lifetime's work in science and in university teaching is now in its second edition [30]. In 2009 Harald was appointed ZEISS Senior Professor at the University of Ulm and is still working (see below) in Ute Kaiser's group as a consultant in the SALVE project on a low-energy transmission electron microscope [31].

Harald Rose has published about 220 original papers and holds more than 110 patents.

Fig. 7. During Harald Rose's 70th anniversary symposium at NCEM Berkeley 2005 (courtesy M. Haider).

Fig. 8. At European Electron Microscopy Congress (MC2011), 2011, in Kiel/Germany (courtesy W. Jäger).

Fig. 9. Together with Maximilian Haider (left) and Knut Urban (right) during the BBVA Frontiers of Knowledge Award Ceremony in Madrid/Spain in 2014 (courtesy BBVA Foundation).

7. A Few Additional Aspects of the Private Life of an Eminent Scientist

During his time at university, Harald did not only work very hard, he was also a very active sportsman. Together with his team, Harald, in his position as captain, won the German university hockey championship three times. He continued to enjoy hockey up to the age of fifty, playing for his local club in Darmstadt. He remarks: "Hockey is a very demanding type of sport, but the risk of injury increases with age. I loved it, but I had to leave it. It was high time." Under the influence of his wife Dorothee, a talented tennis player (she won the club championships many times), he discovered his own passion for the "white" sport. Even today he spends many hours a week on court with his tennis partners. He recalls with pleasure long hiking tours in the German and Italian Alps and in the Rocky Mountains. For decades downhill skiing in the Dolomites with Dorothee, both were excellent skiers, was a fixture in his personal schedule. He adds: "South Tyrol is superb; impressive mountains, wonderful slopes, and in the evening you can enjoy a bottle of the local Lagrein wine with 'Schüttelbrot (a flat unleavened rye bread speciality), bacon and roasted chestnuts." For many years, with some of his former hockey teammates, he has undertaken hiking tours in early autumn lasting several days, preferably in German wine-growing areas. The one for next year is already at the planning stage. Actually with respect to wine he is a real expert, a connoisseur par excellence. Collecting rare wines and the produce of famous wineries in his deep old-fashioned sandstone cellar is his passion. Enjoying a glass of wine with friends in the evening, he always says: "Life is too short to waste time drinking bad wine".

8. The Change of Paradigm: The Double-Hexapole Corrector for TEM and STEM

Harald Rose is internationally recognized as a pioneer in many fields of modern electron optics. However, it is no exaggeration to consider one of his projects as outstanding, in particular with respect to its extraordinary success and widespread application in the modern electron optical industry and in materials science. This is the realization of the double-hexapole system for the correction of spherical aberration in both TEM and STEM in collaboration with Max Haider at EMBL Heidelberg and Knut Urban at Forschungszentrum Jülich between about 1990 and 1998.

At quite an early stage, hexapoles (sextupoles) were treated theoretically with respect to their possible use in spherical-aberration correction systems. Hawkes, in 1965 [32, 33], was the first to point out that in a second approximation extended hexapoles exhibit third-order aberrations equivalent to those of round lenses, and Rose and Plies [34, 35] showed in 1973 that the spherical aberration can have a negative sign. At the end of the 1970s, much of the work on both sides of the Atlantic concentrated on STEM. In TEM, the laterally extended image information has to pass the objective lens. Therefore, correction has to be achieved over the whole image plane including the off-axis aberrations. In STEM, it suffices to correct the axial beam. In 1979, Beck showed in Chicago that a combination of a round lens and two hexapoles has a negative spherical aberration and can be made free of axial second-order aperture aberration [36]. However, his system and that suggested later by Crewe [37] introduced large second-order off-axial aberrations and a large threefold fourth-order aperture aberration which prevented any appreciable improvement in resolution. This fourth-order aberration results from the combination of the large non-corrected second-order field aberrations with the third-order axial aberrations. The problem was solved by Rose in 1981 [38], who proposed, for STEM, a hexapole corrector consisting of a telescopic round-lens doublet and two identical sextupoles, one of them centred at the front focal point of the first round lens and the other at the back focal point of the second lens. A round-lens transfer-doublet images the first hexapole inversely onto the second hexapole, whereupon all second-order path deviations cancel out. On the other hand, the rotationally symmetric third-order path deviations add up. The resulting negative spherical aberration is proportional to the square of the hexapole strength, which can be adjusted to eliminate or adjust [39, 40] the combined spherical aberration of the corrector and the objective lens.

At the end of the 1980s, the situation of aberration correction in electron optics was desperate. Scherzer's TEM experiments had already been discontinued many years before, and Crewe's STEM developments had to be stopped due to a lack of funding [41]. Though much progress had been achieved in Darmstadt, in Tübingen, in Cambridge [42] and in Chicago, it was obvious that funding was not suspended without reason. The basic instruments were not sufficiently stable, neither mechanically nor electronically, and the means of quantitatively diagnosing the current state of the instrument with respect to its optical parameters were not accurate enough and were also too slow. The same held for the techniques and devices developed to eliminate the determined aberrations. Once measured or adjusted the optical parameters were not stable enough for a sufficiently long time to justify hopes that it would ever be possible to carry out realistic scientific work with these instruments in the foreseeable future. Materials science, traditionally a strong supporter of electron microscopy, had entirely lost interest in aberration correction with the fatal consequence that the electron optics community became more and more isolated. From the present perspective, where electron microscopy — on the basis of aberration correction — is a booming field, those dark years of electron microscopy instrumentation can hardly be imagined. However, those who were active in those days recall that it was difficult to attract good people to the field and, very important, to get any funding proposal through since, last but not least, the materials scientists among the referees were, to say the least, difficult to convince.

Furthermore, starting in the 1970s a strong lobby for high-voltage electron microscopy had become established in Europe, Japan and the United States. The goal was to avoid the 'obviously unrealistic' hardware aberration correction and to increase resolution via a reduction of the electron wavelength. The high-voltage electron microscopes, the 'dinosaurs' of electron microscopy, at a height of typically 10–15 m, operating at acceleration voltages of 1 million volts (up to 3 MV) required huge installations and an investment of typically 15 million dollars (up to as much as 30 million). Although some of these instruments reached ångström resolution and, due to reduced electron inelastic scattering, enabled researchers to work with thicker specimens, electron radiation damage (primarily by atom displacement) rapidly changed or even destroyed the specimen structure within minutes. During the 1980s, this and the extraordinary investment (the sums invested in high-voltage electron microscopy were orders of magnitude higher than those ever put into the development of aberration correction) fed serious doubts about the value of high-voltage electron microscopy. Nevertheless, in retrospect it is obvious that the high-voltage movement had a deleterious effect on the field: aberration correction lost support both in science and in the funding agencies. In a sense it was only logical that at the end of the 1980s a panel of experts advised the American National Science Foundation to stop funding the apparently fruitless correction projects. As a result, experimental work on aberration correction was abandoned worldwide.

At about the same time, in September 1989, during the 'Dreiländertagung' (the traditional quadrennial meeting of the electron microscopy societies of Austria, Germany and Switzerland) at Salzburg, Austria, Max. Haider (at that time at EMBL Heidelberg), Harald Rose and Knut Urban (at Research Centre Jülich) had intensive discussions on the prospects for aberration correction. Indeed, there was something new: Harald had arrived at Salzburg with the outline of a new aberration corrector in his bag which, according to a conservative estimate, had a chance of being technically realized with the current state of the art of electronics technology. So far all the more recent advanced concepts published in the literature had referred to STEM. This one, however, since it also addressed the problems of off-axis aberrations, seemed to be capable of bringing about aberration-corrected TEM. Since this was optically the most demanding case, correction for STEM could be treated later on the basis of the same concept, concentrating on axial aberrations only. The new semi-aplanatic corrector, known today as the 'Rose corrector', involves two hexapoles and two round-lens transfer doublets [43]. The system is corrected for spherical aberration and isotropic coma. Since the corrected system is rather simple and because hexapoles do not affect the paraxial path of rays, the three scientists were convinced that the successful correction of spherical aberration and radial off-axial coma in a modern 200 kV TEM equipped with a field emission gun should be feasible, contrary to the belief of all the 'experts'. Urban, who had spent two decades of his professional life with high-voltage microscopy and had just moved from Stuttgart to Jülich to escape once and for all this kind of 'big-machine physics', was enthusiastic about the prospects for materials science, in particular about the chance to reach atomic resolution at last.

Harald calls himself 'stubborn', and those who know him are aware of his occasional impatience and his unwillingness to suffer fools gladly. However, it was their patience and pragmatism that was put to the test when the three tried to obtain funding. The unfortunate attitude of the funding agencies to call for innovation, but to withdraw as soon as somebody comes up with a really innovative project, put a strain on their nerves. Here they had an (as they were convinced) excellent idea and, as far as one could reasonably judge, the prospects were good that it could be realized. Nevertheless, the fact that in the US funding for aberration correction had been suspended also appeared reason enough for the German and the European funding agencies to turn down all their approaches. When at last the Volkswagen Foundation showed interest in the project asking them to submit a proposal, they realized that first of all Harald's publication in 'Optik' was only familiar to the (few) specialists. In addition, in materials science there was either no interest in atomic resolution or, even more disappointingly, the lattice-fringe images published as results of uncorrected lower-resolution microscopy were erroneously taken as atomically resolved since people were not aware of the non-locality and the errors of this type of imaging mode (a striking example of this misinterpretation, which is not uncommon even today among non-specialists, is critically discussed in [44]). This gave rise to the danger that, during refereeing, the proposal would be rejected out of hand due to simple but fatal ignorance: 'no need' or 'nothing new'. On the other hand, it was obvious that the three would only have one single opportunity. In this situation Urban, as a recognized member of the materials science community, decided to offer and give numerous talks in university and science institution seminars and colloquia in Germany and abroad and asked congress organizers (Seattle [45] and Darmstadt [46]) for the opportunity to hold special sessions in order to advertise the need for and the benefit of atomic resolution in materials science. The goal was to make the project known, and not only this, the three aimed to win over as many (potential) reviewers as possible to their side before they could even have seen the actual proposal. The fact that they were well advised to do so became evident when it turned out much later that the proposal was accepted at the final reviewers' meeting with a knife-edge majority of a single vote.

As always after a real change of paradigm, it is today, now that the problem of aberration correction in electron optics is solved and atom-by-atom materials science studies are part of our everyday life, hardly possible to take oneself back in time, to a time when science was quite obviously not prepared for atomic-resolution electron microscopy [47]. Actually the lack of confidence in the Volkswagen proposal could hardly be better demonstrated than by the fact that the reviewers only granted a small fraction of the requested funding, and they set up a critical milestone to be reached before any further money would flow. The project had to provide unambiguous experimental proof on a conventional electron microscope, essentially working as an electron optical test bench, that the new corrector concept worked perfectly and proved to provide the negative C_3 necessary to compensate the spherical aberration of the objective lens to zero. The Haider, Rose and Urban project started under the Volkswagen Foundation number I/67181 in autumn 1991. The outright experimental proof of a functioning corrector compensating the spherical aberration of the objective lens to zero was delivered to the Volkswagen Foundation in summer 1994 and submitted for publication on 10 October 1994 [48]. This was three and four years, respectively, before an alternative project, a quadrupole–octupole corrector for a STEM instrument (started in 1995) reached a comparable stage [49, 50]. About 60 years after the invention of the electron microscope, at last it was proven that aberration correction worked in high-quality electron optics, and a new era had dawned.

After having delivered the requested proof that their corrector worked in practice, the three received funding in order to purchase a CM200 with a field emission gun. After installation of the double-hexapole corrector in this instrument, they were able to demonstrate in summer 1997 that they could almost halve the point resolution of the original (uncorrected) instrument of 0.24 nm by reaching a resolution better than 0.14 nm. And in the night of 24–25 June 1997, Bernd Kabius, a member of Urban's group at Jülich, succeeded, in Haider's laboratory at Heidelberg (shortly before the instrument had to be shipped to Jülich to continue developments there after the management of the institute at Heidelberg had decided to shut down electron microscopy development at EMBL), in reaching genuine atomic resolution in (110) oriented GaAs samples (resolving the 0.14 nm dumbbell atom pairs), a record for a 200 kV instrument. Furthermore, he demonstrated by imaging a Si(111)/CoSi$_2$ interface that for the first time interfaces could be imaged in a field-emission gun microscope free of contrast delocalization, a problem that, in uncorrected instruments, impeded high-resolution investigations due to a spherical-aberration-induced overlap of information from both lattices at the interface. Kabius developed his film negatives during the night in Heidelberg, transported them by car to Jülich for printing and wrapped them up so that he could rush to Eindhoven to join Urban, who had tried for a long time in vain ("we realize that direct atomic resolution is Urban's passion but we don't see a market for this") to persuade Philips Electron Optics to engage in commercial construction of aberration-corrected electron microscopes. Although Philips, represented by Karel van der Mast, convinced by the images, held out the prospect of engaging in aberration-corrected electron microscopy on the very same day, the frustrating experience with an ignorant but complacent scientific public was not over yet. The paper sent on 2 October 1997 to Nature was declined due to, quoting the editors, "a lack of general scientific interest". It was only after some 'clarifying' correspondence that Nature agreed to provide some limited space for a very condensed version [51] of the original paper which was submitted on 24 November 1997 (published in 23 April 1998). This was four years before the already mentioned project of a quadrupole–octupole corrector for a STEM instrument managed to likewise demonstrate an improvement of the resolution compared to that of the basic instrument by installing the corrector [52].

The extraordinary vision expressed in Harald's 1990 Optik paper [43] and the success of the double-hexapole corrector concept can hardly be better demonstrated than by the fact that of the estimated 400–450 aberration-corrected electron microscopes installed worldwide today, of which approximately 150 are TEMs and 250 are STEMs, all TEMs and all but about (estimated) 10% of all STEMs are equipped with the double-hexapole corrector in the imaging or the beam-formation system. In materials science, where till the end of the 1990s, one could observe a dramatic shrinking of the participation of electron microscopy in high-quality research work, atomic-resolution images are ubiquitous today [53].

9. The TEAM, SMART and SALVE Projects

Spherical aberration correction was accomplished, and the results of application of the corrected CM200 TEM to materials science as reported by Haider, Rose and Urban were very promising indeed. What next? In July 2000 Murray Gibson at Argonne National Laboratory organized a workshop as a forum for the discussion of the project of a versatile materials laboratory accommodated in the objective lens of a sub-angstrom resolution TEM [54]. In principle this idea was not new. In-situ studies had for many years been one of the strongholds of high-voltage electron microscopy. However, in order to avoid or at least reduce the risk of radiation damage, Gibson was aiming at a medium-voltage electron microscope (about 300 kV). Conventional instruments at that electron energy would never be able to produce sub-angstrom resolution and at the same time provide sufficiently wide objective lens pole-piece dimensions to allow in-situ facilities to be accommodated. Here aberration correction might be able to help. The aberrations introduced upon enlarging the pole-piece dimensions could, according to Gibson's idea, be corrected by a suitably designed spherically and chromatically corrected objective lens. During the workshop, Gibson was appointed co-ordinator of a project proposal to the DOE to be worked out together by Argonne National Laboratory, Brookhaven National Laboratory, Lawrence Berkeley National Laboratory, Oak Ridge National Laboratory, and Frederick Seitz Materials Research Laboratory. In late 2000, Gibson invited Harald Rose to Argonne to work out the concept of a suitable aberration-corrected objective lens.

The correction of chromatic aberration necessitated, as Rose had already shown in 1967 [8], the incorporation of electric and magnetic quadrupoles. He showed that by imposing symmetry conditions on the multipole fields and the course of the fundamental paraxial rays a large number of aberrations cancel out. The largest reduction is achieved by imposing symmetry conditions on the system as a whole and on each half of it. The optimum system is then obtained by replacing each sextupole element of the corrector by a telescopic quadrupole–octupole quintuplet. By optimizing Rose's design, CEOS [55] obtained an achromatic aplanatic system theoretically satisfying all requirements [56]. The most stringent constraints are the extremely high stabilities of about 2×10^{-8} required for the electric and magnetic quadrupole fields correcting for the chromatic aberration. The set target of also adding chromatic correction to spherical correction was maintained even after the project was re-designed, and the idea of the in-situ experimental facility was pushed more into the background. In 2002, upon Gibson's change as director to the Argonne Advanced Photon Source, Uli Dahmen became project coordinator. The project was approved by the DOE and initiated in 2004. The companies FEI and CEOS won the contract. The instruments (TEAM 0.5, C_S corrected TEM and STEM, and TEAM 1, C_S corrected STEM and C_S/C_C corrected TEM) were installed at the National Center for Electron Microscopy (NCEM) at Lawrence Berkeley National Laboratory in 2008. A resolution of 65 pm was demonstrated in TEM mode at 300 kV and of 55 pm at 200 kV [57–59]. In 2009 a record resolution of 47 pm resolving the dumbbell spacing in a [114] or-iented Germanium crystal was demonstrated by ADF STEM at 300 kV [60].

No description of the highlights of Harald Rose's work should omit the OMEGA in-column electron-energy filter and, in particular, the high-performance MANDOLINE filter [24, 61, 62]. This filter was developed within the framework of the SESAM project funded by the German Research Foundation (DFG). It was installed in the SESAM microscope operating in Manfred Rühle's group at the Max Planck Institute in Stuttgart and has by far the highest dispersion and transmissivity of all imaging energy filters to date. This filter allows isochromatic energy filtering of large object areas with an energy resolution below 0.05 eV.

Harald Rose was also involved in the SMART (Spectro-Microscope for All Relevant Techniques) project, one of the most ambitious projects in surface science. [63, 64]. When the Berlin electron synchrotron BESSY II was launched in 1997, funds were made available for novel projects exploiting the capabilities of the new photon source. Harald was asked at that time by Alex M. Bradshaw and Eberhard Umbach, the initial organizers of the SMART project, to become a member of a group of scientists who were engaged in developing an aberration-corrected low-energy electron microscope which could operate as a low-energy electron microscope (LEEM) and as a photo-emission electron microscope (PEEM). Harald's task involved the development of an electric and magnetic immersion objective lens with minimum aberrations [64], the design and construction of an aberration-free beam separator [65] and a mirror corrector compensating for the chromatic and spherical aberrations of the microscope [66, 67]. These elements were designed, built and successfully tested in Darmstadt by many of Harald's co-workers who later joined the CEOS or ZEISS companies. The components were subsequently incorporated in the SMART microscope, which was assembled in Berlin. This microscope is still operating today at BESSY II, and in the LEEM mode it has reached a resolution of 3 nm at an electron energy as low as 15 eV [68].

One of the most important ideas that Harald promulgated for correcting aberrations is the importance of symmetry requirements. For example, by imposing several symmetry conditions on the arrangement of imaging energy filters he showed that all primary aberrations cancel out. Although this procedure simultaneously eliminates the dispersion outside the filter, the resulting achromatic system is well suited as a monochromator with internal energy selection. The dispersion-free electrostatic CEOS monochromator designed by Kahl and Rose [69] utilizes double symmetry for eliminating all second-order aberrations outside the monochromator. The recently developed magnetic NION monochromator employs the same principles [70, 71].

At present Harald is a member the SALVE (Sub-Angstrom Low-Voltage Electron microscopy) project initiated by Ute Kaiser at the University of Ulm [31]. The object of this project is to build a dedicated chromatically and spherically corrected low-voltage transmission electron microscope for the investigation of low-nuclear charge, in particular carbon-based materials. The paradigmatic type of specimen is graphene. This material suffers from knock-on atom displacement damage at electron energies higher than about 80 keV. At lower values it is fairly stable but the resolution decreases with voltage [72] making chromatic correction indispensable. At lower voltages, however, ionization damage by electron–electron interaction (radiolysis) increases in materials with molecular bonds [73]. Rose's most recent paper deals with the maximum resolution that can be achieved under such circumstances as a function of signal-to-noise-ratio and of critical dose [74]. Now that ZEISS has withdrawn from the project, SALVE is being continued on a new basis by FEI together with CEOS. The corrector is based on an original design by Harald Rose and his former student M. Kuhn and compensates for chromatic as well as for spherical aberration.

10. Summary

Many of the theoretical concepts developed by Harald Rose over the last 50 years have been fundamental for electron optics. Indeed, some of them have changed the whole complexion of this field and are fundamental to modern electron microscopy, both in TEM and in STEM mode. In addition many of the prime players in electron optics like Volker Gerheim, Maximilian Haider, Peter Hartel, Georg Hoffstaetter, Frank Kahl, Helmut Kohl, Stefan Lanio, Heiko Müller, Erich Plies, Dirk Preikszas, Gerald Schönecker, Stefan Uhlemann and Joachim Zach are Harald's former students. And at 80 years of age Harald is still active, still teaching, still enjoying the contact with young scientists and the involved discussion of tricky problems with his colleagues.

With this dedicated issue of Ultramicroscopy, the members of the electron microscopy community would like to thank Harald Rose for dedicating his professional life to their field and thereby enriching the life of those active in it.

Acknowledgements

These biographical notes are based in part on a recorded interview with Harald Rose held on 22–25 October 2014 in Jülich. In the text, excerpts of a historical review written by Rose for Journal of Electron Microscopy [75] are used with the author's permission. The author is also grateful for papers, photographs, preprints and statistical data supplied by Wolfgang Jäger, Max. Haider, Peter Hartel, Ute Kaiser, Erich Plies and Stefan Uhlemann. He also would like to acknowledge valuable comments and suggestions by the referees. An unpublished detailed account of an investigation into the time sequence and priority in bringing about aberration correction in electron optics after 1989 (used in Section 8) can be found on and downloaded from the ER-C web site [76].

References

[1] H. Fischer, H. Rose, Nanosecond Time Responses of Different Indium Antimonide Photodetectors: Developments in Applied Spectroscopy, Plenum Press, New York (1963) 29.

[2] O. Scherzer, Über einige Fehler von Elektronenlinsen, Z. Physik 101 (1936) 593.

[3] E. Abbe, On new methods for improving spherical correction, applied to the construction of wide-angled object glasses, J. R. Microsc. Soc. 1 (1879) 812.

[4] O. Scherzer, Sphärische und chromatische Korrektur von Elektronenlinsen, Optik 2 (1947) 114.

[5] R. Seliger, Die sphärische Korrektur von Elektronenlinsen mittels nicht rotationssymmetrischer Abbildungselemente, Optik 8 (1951) 311.

[6] R. Möllenstedt, Elektronenmikroskopische Bilder mit einem nach O. Scherzer sphärisch korrigierten Objektiv, Optik 13 (1956) 209.

[7] H. Rose, Allgemeine Abbildungseigenschaften unrunder Elektronenlinsen mit gerader Achse, Optik 24 (1966) 36.

[8] H. Rose, On correctability of lenses for fast electrons, Optik 26 (1967) 289.

[9] P.W. Hawkes, The paraxial chromatic aberrations of rectlinear orthogonal systems, Optik 22 (1965) 543.

[10] H. Rose, Über den sphärischen und chromatischen Fehler unrunder Elektronenlinsen, Optik 25 (1967) 587.

[11] J. Zach, M. Haider, Correction of spherical and chromatic aberration in a low-voltage SEM, Optik 99 (1995) 112.

[12] H. Rose, Properties of spherically corrected achromatic lenses, Optik 33 (1971) 1.

[13] H. Rose, Aplanatic electron lenses, Optik 34 (1971) 286.

[14] H. Koops, G. Kuck, O. Scherzer, Erprobung eines elektronenoptischen Achromators, Optik 48 (1977) 225.

[15] W. Bernhard, Erprobung eines sphärisch korrigierten Elektronenmikroskops, Optik 57 (1980) 73.

[16] H. Rose, Phase contrast in scanning transmission electron microscopy, Optik 39 (1974) 416.

[17] M. Isaacson, J.P. Langmore, H. Rose, Determination of nonlocalization of inelastic scattering of electrons by electron microscopy, Optik 41 (1974) 92.

[18] H. Rose, Image formation by inelastically scattered electrons in electron microscopy, Optik 45 (1976) 139.

[19] H. Rose, Image formation by inelastically scattered electrons in electron microscopy 2, Optik 45 (1976) 187.

[20] H. Kohl, H. Rose, Theory of image formation by inelastically scattered electrons in the electron microscope, in: P.W. Hawkes (Ed.), Advances in Imaging and Electron Physics, 65, Academic Press, London, UK, Orlando, FL, 1985, p. 173.

[21] E. Plies, H. Rose, Axial aberrations of magnetic deflection systems with a curved axis, Optik 34 (1971) 171.

[22] H. Rose, E. Plies, Design of a magnetic energy analyzer with small aberrations, Optik 40 (1974).

[23] H. Rose, Aberration correction of homogeneous magnetic deflection systems, Optik 51 (1978) 15.

[24] S. Uhlemann, H. Rose, The Mandoline filter-a new high-performance imaging filter for sub-ev EFTEM, Optik 96 (1994) 163.

[25] H. Rose, J. Fertig, Influence of detector geometry on image properties of STEM for thick objects, Ultramicroscopy 2 (1976) 77.

[26] N.H. Dekkers, H. de Lang, Differential phase contrast in a STEM, Optik 41 (1974) 452.

[27] H. Rose, Nonstandard imaging methods in electron microscopy, Ultramicroscopy 2 (1977) 251.

[28] J. Fertig, H. Rose, Resolution and contrast of crystalline objects in high-resolution scanning transmission electron microscopy, Optik 59 (1981) 407.

[29] U. Dahmen, R. Erni, V. Radmilowic, C. Kisielowski, M.D. Rossell, P. Denes, Background, status and future of the Transmission Electron Aberration-Corrected Microscope project, Phil. Trans. R. Soc. A 367 (2009) 3795.

[30] H. Rose, Geometrical Charged-Particle Optics (Springer Series in Optical Sciences) vol. 142, 2nd ed., Springer, Heidelberg, 2012.

[31] U. Kaiser, J. Biskupek, J.C. Meyer, J. Leschner, L. Lechner, H. Rose, M. Stöger-Pollach, A.N. Khlobystov, P. Hartel, H. Müller, M. Haider, S. Eyhusen, G. Benner, Transmission electron microscopy at 20 kV for imaging and spectroscopy, Ultramicroscopy 111 (2011) 1239.

[32] P.W. Hawkes, The geometrical aberrations of general electron optical systems I, Philos. Trans. R. Soc. A 257 (1965) 479.

[33] P.W. Hawkes, The geometrical aberrations of general electron optical systems II, Philos. Trans. R. Soc. A 257 (1965) 523.

[34] H. Rose, E. Plies, Correction of aberrations in electron optical systems with curved axis, in: P.W. Hawkes (Ed.), Image Processing and Computer-aided Design in Electron Optics, Academic Press, London, 1973, p. 344.

[35] E. Plies, Korrektur der Öffnungsfehler elektronenoptischer Systeme mit krummer Achse und durchgehend astigmatismusfreien Gausschen Bahnen, Optik 38 (1973) 502.

[36] V. Beck, A hexapole spherical aberration corrector, Optik 53 (1979) 241.

[37] A.V. Crewe, Studies on sextupole correctors, Optik 57 (1980) 313.

[38] H. Rose, Correction of aperture aberrations in magnetic systems with threefold symmetry, Nucl. Instrum. Methods Phys. Res. 187 (1981) 187.

[39] C.L. Jia, M. Lentzen, K. Urban, Atomic-resolution imaging of oxygen in perovskite-ceramics, Science 299 (2003) 870.

[40] M. Lentzen, The tuning of a Zernike phase plate with defocus and variable spherical aberration and its use in HRTEM imaging, Ultramicroscopy 99 (2004) 211.

[41] A.V. Crewe, Some Chicago aberrations, Microsc. Microanal. 10 (2004) 1.

[42] J.H.M. Deltrap, Ph.D. thesis, University of Cambridge (1964).

[43] H. Rose, Outline of a spherically corrected semiaplantic medium-voltage transmission electron microscope, Optik 85 (1990) 19.

[44] D.B. Williams, C.B. Carter, Transmission Electron Microscopy, Part 3: Imaging, 2nd ed., Springer, New York, NY (2003) 391.

[45] 12th International Congress on Electron Microscopy, Seattle, WA, August 1990.

[46] 25. Jahrestagung der Deutschen Gesellschaft für Elektronenmikroskopie, September 1991.

[47] K. Urban, Is science prepared for atomic-resolution electron microscopy? Nat. Mater. 8 (2009) 260.

[48] M. Haider, G. Braunshausen, E. Schwan, Correction of the spherical aberration of a 200 kV TEM by means of a hexapole-corrector, Optik 99 (1995) 167 (Submitted 10 October 1994).

[49] O.L. Krivanek et al., Aberration correction in the STEM, in: J. M. Rodenburg (Ed.), Proceedings of the EMAG meeting (IOP Conference Series), 1997, p. 35, vol. 153.

[50] O.L. Krivanek, N. Dellby, A.R. Lupini, Towards sub-Angstrom electron beams, Ultramicroscopy 78 (1999) 1.

[51] M. Haider, S. Uhlemann, E. Schwan, H. Rose, B. Kabius, K. Urban, Electron microscopy image enhanced, Nature 392 (1998) 768 (Submitted 24 November 1997. Published 23 April 1998).

[52] N. Dellby, O.L. Krivanek, P.D. Nellist, P.E. Batson, A.R. Lupini, Progress in aberration corrected scanning transmission electron microscopy, J. Electron Microsc. 50 (2001) 177 (Submitted on 18 January 2001).

[53] K. Urban, Studying atomic structures by aberration-corrected transmission electron microscopy, Science 321 (2008) 506.

[54] ⟨http://mrl.illinois.edu/facilities/center-microanalysis-materials/team-project⟩; TEAM Report 2000.pdf (accessed 2014).

[55] CEOS is a company founded in 1996 by the two former Rose students Maximilian Haider and Joachim Zach in Heidelberg/Germany.

[56] M. Haider, H. Müller, S. Uhlemann, J. Zach, U. Loebau, R. Höschen, Prerequisites for a Cc/Cs-corrected ultrahigh-resolution TEM, Ultramicroscopy 108 (2008) 167.

[57] Ch. Kisielowski, B. Freitag, M. Bischoff, H. van Lin, S. Lazar, G. Knippels, P. Tiemeijer, M. van der Stam, S. von Harrach, M. Stekelenburg, M. Haider, S. Uhlemann, H. Müller, P. Hartel, B. Kabius, D. Miller, I. Petrov, E.A. Olson, T. Donchev, E.A. Kenik, A.R. Lupini, J. Bentley, S.J. Pennycook, I.M. Anderson, A.M. Minor, A.K. Schmid, T. Duden, V. Radmilovic, Q.M. Ramasse, M. Watanabe, R. Erni, E.A. Stach, P. Denes, U. Dahmen, Detection of single atoms and buried defects in three dimensions by aberration-corrected electron microscope with 0.5-Å information limit, Microsc. Microanal. 14 (2008) 469.

[58] M. Haider, P. Hartel, H. Müller, S. Uhlemann, J. Zach, Information transfer in a TEM corrected for spherical and chromatic aberration, Microsc. Microanal. 16 (2010) 393.

[59] P. Eccius, M. Boese, T. Duden, U. Dahmen, Operation of TEAM I in a user environment at NCEM, Microsc. Microanal. 18 (2012) 676.

[60] R. Erni, M.D. Rossell, Ch. Kisielowski, U. Dahmen, Atomic-resolution imaging with sub-50-pm electron probe, Phys. Rev. Lett. 102 (2009) 096101.

[61] S. Lanio, H. Rose, D. Krahl, Test and improved design of a corrected imaging magnetic energy filter, Optik 73 (1986) 56.

[62] H. Rose, D. Krahl, Electron optics of imaging energy filters, in: R. Reimer (Ed.), Energy-Filtering Electron Microscopy, Springer, Berlin, 1995, p. 43.

[63] R. Fink, M.R. Weiss, E. Umbach, D. Preikszas, H. Rose, R. Spehr, P. Hartel, W. Engel, R. Degenhardt, R. Wichtendahl, H. Kuhlenbeck, W. Erlebach, K. Ihrmann, R. Schlögl, H.-J. Freund, A.M. Bradshaw, G. Lilienkamp, T. Schmidt, E. Bauer, G. Benner, SMART: a planned ultrahigh-resolution spectromicroscope for BESSY II, J. Electron Spectrosc. Relat. Phenom. 84 (1997) 231.

[64] D. Preikszas, H. Rose, Procedures for minimizing the aberrations of electromagnetic compound lenses, Optik 100 (1995) 179.

[65] H. Müller, D. Preikszas, H. Rose, A beam separator with small aberrations, J. Electron Microsc. 48 (1999) 191.

[66] D. Preikszas, H. Rose, Correction properties of electron mirrors, J. Electron Microsc. 46 (1997) 1.

[67] P. Hartel, D. Preikszas, R. Spehr, H. Müller, H. Rose, Mirror corrector for low-voltage electron microscopes, in: P.W. Hawkes (Ed.), Advances in Imaging and Electron Physics, 120, Academic Press, San Diego, CA, London, UK, 2001, p. 41.

[68] Th Schmidt, U. Groh, R. Fink, E. Umbach, O. Schaff, W. Engel, B. Richter, H. Kuhlenbeck, R. Schlögl, H.-J. Freund, A.M. Bradshaw, D. Preikszas, P. Hartel, R. Spehr, H. Rose, G. Lilienkamp, E. Bauer, G. Benner, XPEEM with energy filtering: Advantages and first results from the SMART project, Surf. Rev. Lett., 9, 223.

[69] F. Kahl, H. Rose, Outline of an electron monochromator with small Börsch effect. in: H.A. Calderon, M.J. Yacaman (Eds.), Proceedings of the 14th International Conference on Electron Microscopy, Institute of Physics, Cancun, Bristol and Philadelphia, 1998, p. 71.

[70] O.L. Krivanek, T.C. Lovejoy, M.F. Murfitt, G. Skone, P.E. Batson, N. Dellby, Towards sub-10 meV ernergy resolution STEM-EELS, J. Phys.: Conf. Ser. 522 (2014) 012023.

[71] O.L. Krivanek, T.C. Lovejoy, N. Dellby, T. Aoki, R.W. Carpenter, P. Rez, E. Soignard, J. Zhu, P.E. Batson, M.J. Lagos, R.F. Egerton, P.A. Crozier, Vibrational spectroscopy in the electron microscope, Nature 514 (2014) 209.

[72] J. Barthel, A. Thust, Quantification of the information limit of transmission electron microscopes, Phys. Rev. Lett. 101 (2008) 200801.

[73] R. Egerton, Mechanisms of radiation damage in beam-sensitive specimens, for TEM accelerating voltages between 10 and 300 kV, Microsc. Res. Tech. 75 (2012) 1550.

[74] Z. Lee, H. Rose, O. Lethinen, J. Biskupek, U. Kaiser, Electron dose dependence of signal-to-noise ratio, atom contrast and resolution in transmission electron microscope images, Ultramicroscopy 145 (2014) 3.

[75] H. Rose, Historical aspects of aberration correction, J. Elect. Microsc. 58 (2009) 77.

[76] ⟨http://www.er-c.org/news/publications/piaceo.pdf⟩ (accessed 2014).

Optik   85, No. 1 (1990) 19–24   © Wissenschaftliche Verlagsgesellschaft mbH, Stuttgart

Outline of a spherically corrected semiaplanatic medium-voltage transmission electron microscope

H. Rose

Insitut für Angewandte Physik Technische Hochschule Darmstadt, FRG

Outline of a spherically corrected semiaplanatic medium-voltage transmission electron microscope. A spherically corrected semiaplanatic objective lens for a subangstrom medium-voltage transmission electron microscope (TEM) is outlined. The aplanatic corrector consists of two telescopic round-lens doublets and two sextupoles centered about the nodal points of the second doublet. If the corrector is incorporated into a 300 kV TEM equipped with a field emission gun a resolution limit of 0.6 Å and 10^4 equally-well-resolved image points per diameter can be obtained. For achieving this performance the magnetic field of the objective lens must be stabilized with a relative accuracy of 1 ppm, while the fields of the corrector elements require at most a stability of 10 ppm.

Entwurf eines sphärisch korrigierten semiaplanatischen Mittelspannungs-Elektronenmikroskops. Eine in dritter Ordnung sphärisch korrigierte rein magnetische Objektivlinse, deren isotrope Koma beseitigt ist, wird vorgeschlagen. Das korrigierte Objektiv besteht aus einer Objektivlinse, zwei teleskopischen Rundlinsen-Dubletts und zwei Sextupolen, deren Mitten in den Knotenebenen des zweiten Dubletts liegen. Falls der Korrektor in ein 300 kV Transmissions-Elektronenmikroskop eingebaut wird, das mit einer Feldemissionskathode ausgestattet ist, können 10^4 Bildpunkte pro Durchmesser mit einer Auflösungsgrenze von 0.6 Å gleich gut aufgelöst werden. Um eine solche Auflösung zu erzielen, müssen die Beschleunigungsspannung und das Magnetfeld der zu korrigierenden Objektivlinse auf 1 ppm stabil gehalten werden. Für die Felder der Korrektorelemente genügt dagegen eine Stabilität von 10 ppm.

1. Introduction

To realize a transmission electron microscope with a resolution limit below 1 Å at voltages of about 100 kV it is necessary to correct both the spherical aberration and the chromatic aberration of the objective lens. Unfortunately, the correction of the chromatic aberration is rather complicated because it requires the incorporation of electric and magnetic quadrupoles for systems with a straight optic axis. Moreover, the quadrupole fields must be stabilized with a relative accuracy of better than about 0.5 ppm to achieve subangstrom resolution.

On the other hand the spherical aberration can be corrected solely with magnetic elements. This possibility simplifies the mechanical set-up considerably because all magnetic poles can be arranged outside the vacuum tube. So far the extreme difficulties to stabilize and to align the quadrupole-octupole correctors with the required accu-

Received March 22, 1990.

Harald Rose, Institut für Angewandte Physik, Technische Hochschule Darmstadt, Hochschulstrasse 6, D-6100 Darmstadt, FRG.

racy has prevented a successful improvement of the resolution by these correctors.

The amount of expenditure necessary for correction reduces considerably if the chromatic aberration can be kept below 1 Å. In this case only the spherical aberration must be compensated. Contrary to the chromatic aberration the spherical aberration can be corrected by sextupole elements which do not affect the paraxial path of rays. This possibility is of great advantage because the fields of the sextupole elements need only to be stabilized with a relative accuracy of about 10^{-4} in the case of atomic resolution.

According to these facts it would be extremely helpful if the correction of the chromatic aberration could be avoided. The axial chromatic aberration

$$\Delta_c = \alpha \frac{\Delta E}{E} C_c \tag{1}$$

depends on the limiting aperture angle α, the relative energy width $\Delta E/E$ of the beam and the constant C_c. Hence, to suppress the influence of the chromatic aberration these quantities must be made as small as possible. Unfortunately, the limiting aperture angle must be increased with decreasing resolution limit d according to the Rayleigh criterion

$$d \approx 0.61 \frac{\lambda}{\alpha}. \tag{2}$$

The wavelength λ of the electrons is connected with the accelerating voltage ϕ via the relation

$$\lambda \approx \sqrt{\frac{150 \, V}{\phi^*}} \, \text{Å}, \tag{3}$$

where $\phi^* \approx \phi(1 + \phi/MV)$ is the so-called relativistically modified acceleration potential. Assuming a Gaussian energy distribution the minimum achievable resolution limit determined by the combined effect of diffraction and chromatic aberration is found as

$$d_{\min} \approx 1.15(\lambda \, C_c |\Delta E|/E)^{1/2}. \tag{4}$$

By taking further into account the relation (3) it follows from (4), that for fixed C_c and ΔE the resolution limit decreases in proportion to $E^{-3/4}$ with increasing energy $E = e\phi$. To appreciably suppress the influence of the chromatic aberration the full-width half maximum $2|\Delta E|$ of the energy distribution must also be made small. This can be achieved by employing a field emission gun with

$$\Delta E \approx \pm \, 0.25 \, \text{eV}. \tag{5}$$

If, in addition, the voltage is increased from 100 kV to about 300 kV the resolution can be increased by another factor of about two provided that the constant of the chromatic aberration can be kept fixed. To obtain a rough estimate of the obtainable resolution limit of a spherically corrected 300 kV TEM we assume a chromatic aberration constant $C_c \approx 2$ mm. Taking into account this value, a field emission gun with $|\Delta E| \approx 0.25$ eV and considering $\lambda \approx 2 \cdot 10^{-2}$ Å we derive from (4) as the minimum achievable resolution limit

$$d_{min} \approx 0.6 \text{ Å}. \qquad (6)$$

To guarantee this resolution the objective lens must be free of spherical aberration up to the limiting aperture angle $\alpha \approx 2 \cdot 10^{-2}$. Such a large aperture can only be utilized if the third-order spherical aberration is eliminated. Since the chromatic aberration can be tolerated, the correction of the spherical aberration by a purely magnetic sextupole corrector is most promising.

2. Aplanatic sextupole corrector

Sextupoles can only be used for correcting the spherical aberration of a TEM if they are arranged and excited in such a way that they do not introduce any second-order aberration in the final image. This condition is met by a system consisting of two identical sextupoles which are centered about the nodal points of a telescopic rotation-free magnetic round-lens doublet. The geometry of the doublet is shown in fig. 1. The nodal points, which are located outside of the magnetic field, coincide with the outer focal points of the two constituent lenses.

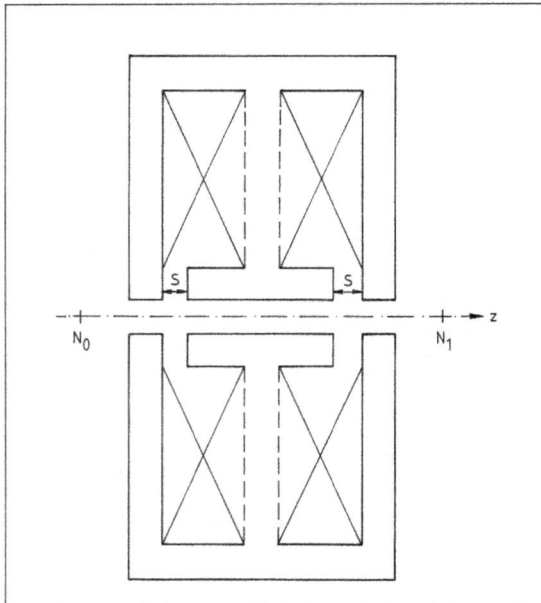

Fig. 1.   Geometry of the iron casing of the telescopic round-lens doublet.

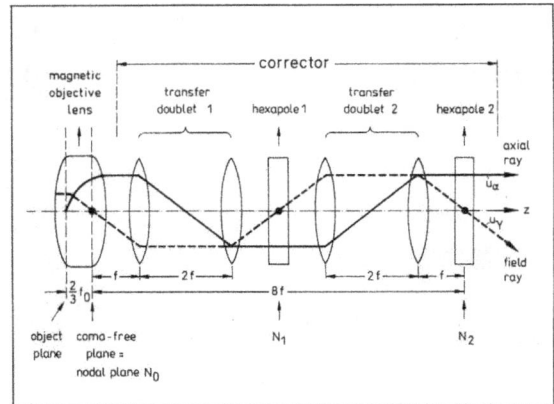

Fig. 2.   Schematic arrangement of the elements of the spherically corrected semiaplanatic objective lens.

It should be noted that the sextupole correctors proposed by Crewe and coworkers [1], [2], [3], [4] for the correction of the spherical aberration of a scanning transmission electron microscope (STEM) are only free of *axial* second-order aberrations. Owing to the non-vanishing second-order off-axial aberrations these correctors cannot be used in a TEM because they result in an untolerably small field of view. Moreover, since these systems possess only a single magnetic round lens between the sextupoles they can also cause a slight fourth-order aberration which prevents an appreciable improvement of the resolution. A non-aplanatic spherically corrected system using sextupoles has been proposed by Shao [5] for the TEM.

The coma-free point of the objective lens is located at a distance of about $(2/3) f_0$ from the object plane if this plane is placed at the field maximum. Here f_0 is the focal length of the objective lens. The coma-free points of the new corrector depicted in fig. 2 are identical with its nodal points. Since the sextupoles are centered about these points, they cannot be placed inside or close to the magnetic field of the objective lens. Therefore, if the corrector is placed behind a conventional objective lens, a large third-order coma is induced preventing a sufficiently large number of equally-well-resolved object points. To obtain a semiaplanatic (free of isotropic coma) system two additional measures are necessary. Firstly, the objective lens must be designed as a probe forming lens such that the coils are located in the region facing the condenser. In this case the distance between the object plane and the lower pole piece can be made rather short. As a result the coma-free point of the condenser-objective lens is located rather close to the exit face of the iron body of the lens. Secondly, the coma-free point of the objective lens must coincide with the coma-free point of the corrector. Since a sextupole element is centered about this point it is not possible to bring the coma-free points directly together. Therefore we must look for an useful alternative. A successful way is to image the coma-free point N_0

of the objective lens by a coma-free round lens into the nodal point N_1 of the corrector. This is indeed possible by means of the telescopic round-lens doublet shown in fig. 1.

Since the coma-free nodal points of this doublet are located well outside its iron body, we can arrange the doublet in such a way that its nodal point on the condenser side is placed into the coma-free plane of the objective lens while the other nodal point is placed at the center of the first sextupole. The resulting "semiaplanatic" system shown in fig. 2 forms an "8-f system", where f is the focal length of each of its constituent round lenses. This quadruplet forms a thick telescopic round lens. The sextupoles for the correction of the spherical aberration are centered about the focal points on the image side of the second and fourth lens, respectively. The coils of the first doublet adjacent to the objective lens are connected in series such that the direction of the current is opposite to that of the objective lens. Owing to this measure the anisotropic coma of the doublet counterbalances to some extent the deleterious anisotropic coma of the objective lens. On the other hand the coils of the second doublet are connected in series opposition so that the excitations of its two lenses are equal and opposite whatever the strength of the current. Accordingly, this doublet is rotation-free and does not introduce an anisotropic coma.

The four round lenses of the corrector contribute to both the chromatic aberration and the spherical aberration. While the contribution to the spherical aberration is of no concern, since this aberration is eliminated by the sextupoles, care must be taken for the chromatic aberration which cannot be compensated. To guarantee that the axial chromatic aberration introduced by the four round lenses of the corrector is small compared with that of the objective lens the focal length f of these lenses must be chosen at least a factor of 10 larger than the focal length f_0 of the objective lens. Since these corrector lenses are weak lenses their contribution C_{cC} to the constant of chromatic aberration of the entire system can be readily calculated by employing the weak-lens approximation. The result

$$C_{cC} \approx 4\frac{f_0^2}{f} \tag{7}$$

indicates that this constant is inversely proportional to the focal length f.

Hence, the constant of the axial chromatic aberration of the entire system consisting of the objective lens and the corrector is

$$C_c = C_{c0} + C_{cC} \approx C_{c0} + 4\frac{f_0^2}{f}, \tag{8}$$

where C_{c0} is the chromatic aberration constant of the objective lens.

As an example we consider a super twin-lens with $f_0 = 2$ mm and $C_{c0} \approx 1.5$ mm for a voltage of 300 kV. Assuming further a focal length $f = 30$ mm for each lens of the corrector we obtain from (8) for the constant C_c of

the entire system

$$C_c \approx 2 \text{ mm}. \tag{9}$$

The total length of the corrector is about 24 cm. This length can easily be tolerated because it does not significantly lengthen the column of the microscope. As has been shown in the introduction this value of the chromatic aberration constant limits the resolution to about 0.6 Å if a field emission gun with $\Delta E = \pm 0.25$ eV is used. This resolution limit is approximately a factor 4 smaller than that of the uncorrected microscope.

Owing to these considerations the correction of only the spherical aberration seems to be very advantageous in the case of medium voltage provided that a field emission gun is used.

3. Performance of the spherically corrected semiaplanatic system

If the sextupoles are exactly centered about the nodal points of the second telescopic doublet, the corrector does not introduce second-order aberrations at any plane behind the last sextupole. The residual aberrations of the sextupoles are primarily determined by the so-called combination aberrations. Their contribution to the axial third-order aberrations is entirely rotationally symmetric [6].

For the corrector shown in fig. 2 the constant of the third-order spherical aberration is found as

$$C_{3c} = 4\frac{f_0^4}{f^4}C_{3t} - 3\frac{e}{m}\frac{\psi_3^2}{\phi^*}l^3 f_0^4. \tag{10}$$

Here C_{3t} is the constant of the third-order spherical aberration of each transfer lens. The sextupole strength

$$\psi_3 \approx \frac{\mu_0 nI}{a^3} \tag{11}$$

depends on the number of ampere-turns nI about each of the poles of the sextupole element, on its bore radius a and on the permeability μ_0 of the vacuum. The length l denotes the extension of the effective sextupole field. The expression (10) has been derived by employing the sharp cut-off fringing field (*SCOFF*) approximation for the sextupoles.

Since the transfer lenses of the corrector are weak-lenses, their focal length f is large compared with the gap width S between the iron poles. In this case the constant C_{3t} is roughly given by

$$C_{3t} \approx \frac{f^3}{S^2}. \tag{12}$$

It follows from (10) that the contribution of the sextupoles to the constant

$$C_3 = C_{30} + C_{3c} \tag{13}$$

of the entire system is negative. Hence, it is always possible by properly choosing the excitation of the sextupoles to eliminate the spherical aberration. Correction of this aberration is obtained if the sextupole strength ψ_3 fulfills the condition

$$3 \frac{e}{m} \frac{\psi_3^2}{\phi^*} l^3 f_0^4 = C_{30} + 4 \frac{f_0^4}{f^4} C_{3t} = C_3. \tag{14}$$

To obtain a rough survey over the number of ampère-turns which are required for the correction of the spherical aberration, we consider again a 300 kV TEM equipped with a super twin-lens. For this lens we have $C_{30} \approx 1.3$ mm, $f_0 = 2$ mm. The data of the corrector are assumed as follows:

$$l = 15 \text{ mm}, \ a = 4 \text{ mm}, \ f = 30 \text{ mm}, \ S = 8 \text{ mm}.$$

Using these values it follows from (12) and (13) that the contribution of the four transfer lenses to the third-order spherical aberration of the round-lenses is negligibly small. The number of ampere-turns for the coils of the sextupoles is found from (11) and (14) as

$$nI \approx 170 \text{ A}.$$

Such ampère-turns can be realized in practice without any difficulties. It is advantageous to excite the sextupole fields within a twelvepole element, because this element also allows one to superimpose a magnetic dodecapole field for compensating the non-rotationally symmetric component of the fifth-order aperture aberration produced by the sextupole fields.

The maximum magnetic induction between any two adjacent poles of the twelvepole element stays smaller than 0.1 Tesla. This value together with the number of ampere-turns can be reduced further by decreasing the bore radius a of the multipole element.

3.1 Number of equally-well-resolved image points

A measure for the size of the image field, whose elements are resolved with the same resolution, is the number N of equally-well-resolved image points per diameter. This number is determined by the off-axial aberrations.

Since the anisotropic third-order coma of the objective lens is only partly counterbalanced by the anisotropic coma of the first doublet, this aberration limits primarily the number of image points. In this case N is given by the relation

$$N \approx \frac{2}{K_a a^2}. \tag{15}$$

The coefficient K_a of the anisotropic coma for the super twin-lens is somewhat smaller than about 0.5. Using this value together with $\alpha = 2 \cdot 10^{-2}$ we obtain at least 10^4 equally-well-resolved image points per diameter. Since this value is sufficiently large for routine work, there is no need to put much effort on the correction of the anisotropic coma.

If the third-order coma is largely eliminated, the number of image points is limited by the fourth-order off-axial aberration which depends linearly on the position of the object point. With the assumption $f \gg f_0$ the following

relation is found for the number of image points:

$$N \approx 3 \left(\frac{f_0}{\alpha l}\right)^3 \left(\frac{l}{C_{30}}\right)^{3/2}. \tag{16}$$

Using the same values as in the other example we obtain $N \approx 3.5 \cdot 10^4$. This value holds true as long as the coefficient K_a is smaller than 0.1.

It follows from these considerations that the corrected objective lens resolves at least 10^4 image points per diameter, regardless if the anisotropic third-order coma is eliminated or not.

4. Alignment and stability requirements

To resolve the theoretical number of object points in an actual corrected microscope the sextupoles must (a) be centered very accurately about the optic axis and (b) be precisely aligned in axial direction. If the central plane of the sextupole field is shifted by a distance ε away from the nodal point of the second doublet an off-axial second-order aberration arises which is found as

$$\Delta = \sqrt{6} \, \alpha \frac{\varepsilon}{f_0} \sqrt{\frac{C_3}{l}} \, \varrho_0. \tag{17}$$

Here ϱ_0 is the off-axial distance of the object point. For the object points located at the border of the field of view this distance is

$$\varrho_0 = \frac{N}{2} d. \tag{18}$$

If we require that the aberration Δ for these points must be smaller than the resolution limit d, we obtain for the tolerable displacement ε the condition

$$\varepsilon < \frac{f_0}{\alpha N} \sqrt{\frac{2l}{3 C_3}}. \tag{19}$$

Using the stated values, for our example, it follows from (8) that 10^4 image points can only be resolved if the displacement ε is smaller than 28 µm. To fulfil this requirement it may be necessary to precisely determine the positions of the nodal points of the telescopic doublet before the corrector is assembled. If this accuracy cannot be achieved in practice an electric adjustment is necessary. For this purpose each of the two multipole elements must be split up into two elements which can be excited separately.

By varying the excitations of these elements the center of gravity of the total sextupole field is shifted along the axis. According to this behaviour the center of gravity of the sextupole field can exactly be placed at the nodal point of the doublet.

Owing to mechanical inaccuracies and the unavoidable inhomogeneities of the magnetic material one can hardly produce a pure hexapole field in practice with a single current. In addition the symmetry axis of the hexapole field does generally not exactly coincide with the optic axis of the system. Therefore the two multipole elements of the corrector must be designed in such a way that within each of these elements additional dipole and quadrupole fields can be excited independently beside the hexapole field. By means of these trimming fields it is possible (a) to eliminate the most disturbing parasitic field components and (b) to shift the hexapole field electrically in lateral direction. This shift, obtained by exciting a proper quadrupole field within the multipole element, enables one to accurately center the hexapole field about the optic axis of the system.

To obtain a rough estimate on the required accuracy of the alignment we assume that the symmetry axis of the hexapole field is laterally displaced by a small distance δ. Such a misplacement causes a quadrupole field about the optic axis. The resulting quadrupole strength

$$\Psi_2 = 3\bar{\delta}\Psi_3 \qquad (20)$$

is proportional to the hexapole strength Ψ_3 and to the conjugate complex value $\bar{\delta}$ of the displacement δ. The phase of δ determines the azimuthal orientation of the displacement. Since the parasitic quadrupole field produces an axial astigmatism in the final image, it can be compensated by the stigmator. However, to keep the influence of residual combination aberrations as small as possible it is advantageous to eliminate the deviations from the ideal trajectory as close as possible near the positions where they originate.

Considering that the sextupole strength is connected by the correction condition (14) with the spherical aberration constant C_3 we obtain for the radius $\Delta\varrho$ of the axial astigmatism, refered back to the object plane, the relation

$$\Delta\varrho = \alpha|\delta|(6C_3/l)^{1/2} \qquad (21)$$

Using the stated values for the 300 kV TEM we derive at

$$\Delta\varrho = 2 \cdot 10^{-2}|\delta|.$$

Accordingly correction of the misplacement error is necessary if the displacement of the axis of the hexapole field from the optic axis is larger than about 50 d.

Because the sextupole fields do not affect the paraxial path of rays the sextupole currents need only to be stabilized with a relative accuracy of somewhat better than 10^{-4}.

The stability condition for the magnetic field B_t of the transfer lenses is

$$\frac{\Delta B_t}{B_t} < \frac{1}{2}\frac{f}{\alpha}\frac{d}{f_0^2}. \qquad (22)$$

For a system with $d = 0.6$ Å, $\alpha = 2 \cdot 10^{-2}$, $f_0 = 2$ mm and $f = 30$ mm these fields must be stabilized slightly better than 10^{-5}.

The stability condition for the magnetic field of the objective lens is given by

$$\frac{\Delta B_0}{B_0} < \frac{d}{2C_{c0}\alpha}. \qquad (23)$$

For our example ($C_{c0} = 1.5$ mm, $\alpha = 2 \cdot 10^{-2}$, $d = 0.6$ Å) we obtain the condition $\Delta B_0/B_0 < 10^{-6}$.

It follows from the values for the required stability of the individual elements that only the magnetic field of the objective lens must be stabilized with a very high accuracy. As a result the alignment and the operation of the sextupole corrector should not be connected with serious difficulties as it has been the case in the past for the quadrupole-octupole correctors.

5. Conclusion

The incorporation of a highly stabilized field emission gun into a medium-voltage electron microscope offers the possibility for obtaining subangstrom resolution by only correcting the spherical aberration of the objective lens. Since the sextupole corrector is entirely magnetic no electric elements are required for correction as it would be the case if the chromatic aberration must be eliminated too. To guarantee that the theoretical resolution limit of about 0.6 Å at 300 kV can be achieved in practice both the magnetic field of the objective lens and the acceleration voltage must be stabilized with an accuracy of 1 ppm.

Owing to the entirely magnetic components of the semiaplanatic objective lens and the very moderate stability requirements for the corrector elements, it is very likely that the proposed system will be successful in practice. Because for many objects the acceleration voltage can be kept below the critical voltage for the occurance of knock-on processes, the proposed subangstrom electron microscope offers a very advantageous alternative to the presently used high-voltage (MV) electron microscopes. To exploit the spherically corrected medium-voltage electron microscope to its fullest capability an imaging magnetic energy filter should be incorporated. Such an instrument would be especially suited for obtaining high-resolution structural and chemical information about radiation-insensitive solid state objects like those investigated in material science.

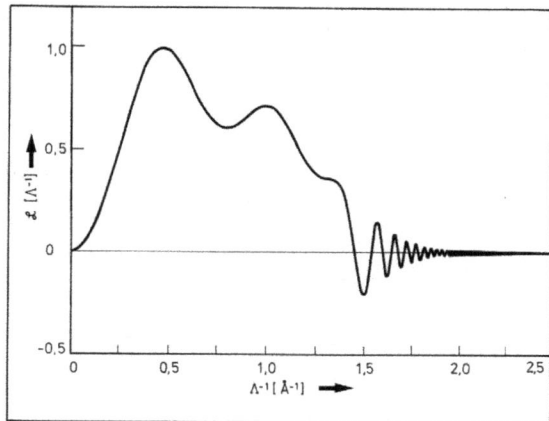

Fig. 3. **Phase contrast transfer function of a spherically corrected 300 kV TEM equipped with a field emission cathode. The PCTF has been optimized for high-resolution imaging by properly adjusting the defocus and the third-order spherical aberration.**

24   H. Rose, Outline of a spherically corrected semiaplanatic medium-voltage transmission electron microscope

To elucidate the capability of a spherically corrected 300 kV TEM with $\Delta E = \pm 0.25$ eV and $C_c = 2.1$ mm we have depicted in fig. 3 the corresponding phase contrast transfer function $(PCTF)$ $L(\Lambda^{-1})$ as a function of the spatial frequency Λ^{-1}. This transfer function has been adjusted for high-resolution imaging by choosing the free parameters $C_1 = -\Delta f = 147$ Å and $C_3 = -86$ μm in such a way that the defocus and the third-order spherical aberration optimally counterbalance the fifth-order spherical aberration whose constant has been assumed as $C_5 = 10$ cm. This value can be considered as an upper limit of the factual fifth-order spherical aberration constant of the instrument. As can be seen from the figure object spacings in the region between 0.7 Å and 7 Å are well transferred without contrast reversal. Such a transfer of high spatial frequencies far exceeds the possibilites of any presently available uncorrected electron microscope.

Summarizing, we can state that the proposed method of correction is a very promising means for improving the resolution of electron microscopes without introducing untolerable knock-on processes. Since the resolution is limited by the chromatic aberration the information which can directly be obtained about the object is approximately the same as that obtainable by indirect holographic methods with a large amount of expenditure.

References

[1] A. V. Crewe and D. Kopf, Optik **55** (1980) 1.
[2] A. V. Crewe and D. Kopf, Optik **56** (1980) 391.
[3] A. V. Crewe, Optik **57** (1980) 313.
[4] Ximen Jiye, Z. Shao, A. V. Crewe, Optik **70** (1985) 37.
[5] Z. Shao, Optik **80** (1988) 61.
[6] H. Rose, Nucl. Instr. Methods **187** (1981) 187.

Ultramicroscopy 2 (1977) 251–267
© North-Holland Publishing Company

NONSTANDARD IMAGING METHODS IN ELECTRON MICROSCOPY *

H. ROSE

Division of Laboratories and Research, New York State Department of Health, Albany, NY 12201, U.S.A.

Received 7 January 1977

A rigorous treatment of image formation, taking into account inelastic scattering and partially coherent illumination, is given for the CTEM. The results also hold for a STEM if the momentum vectors of the incident and outgoing waves are interchanged and the flight direction of the electrons is reversed. A generalized optical theorem is found which connects the anti-Friedel part of the elastic scattering amplitude with the squares of the elastic *and* inelastic scattering amplitudes. This theorem has important consequences for all approximation methods. For example, it shows that inelastic scattering influences the elastic scattering amplitude.

It is further shown that the observed wings in optical diffractograms, obtained from micrographs of amorphous material taken using tilted beam illumination, can only be explained if double scattering (elastic and inelastic) is assumed.

Hollow-cone illumination strongly suppresses the structural noise in the phase contrast image. The corresponding phase contrast transfer function (PCTF) is calculated for two different cone angles, assuming that no beam-limiting aperture is present. In addition, the PCTFs of a corrected microscope are calculated for two different illumination modes, one yielding negative and the other positive phase contrast. Finally, the contrast transfer has been investigated for a STEM operating with a split detector.

1. Introduction

Objects which do not significantly change the intensity of incident radiation, but only shift the phase of the scattered wave, are commonly called phase objects. In contrast, amplitude objects change the intensity of the transmitted beam. This change does not necessarily require a real absorption of electrons within the specimen. This occurs only in very thick objects, in which some electrons are slowed down by multiple inelastic scattering processes and are absorbed within the object. In electron microscopy change in the intensity of the transmitted beam is caused mainly by scattering. All scattered electrons which are either backscattered or removed from the beam by the objective aperture can be considered absorbed because these processes have the same effect as a real absorption of particles within the object.

Thin biological objects scatter only a very small fraction of the incident electrons, and the scattering is strongly forward-directed. Thus many of the scattered electrons pass through the hole of the aperture. The fraction which is removed by the aperture is the smaller the higher the resolution. These objects can be treated to very good approximation as pure phase objects, since the absorption is so small that no detectable contrast is created in the image.

In light optics, phase objects are made visible by a phase-shifting plate first introduced by Zernike [1]. The phase plate is placed in the back focal plane of the objective lens and shifts either the phase of the diffracted light or that of the direct beam. Positive phase contrast is obtained for a phase difference of odd multiples of π (destructive interference), while negative phase contrast (constructive interference) occurs for even multiples of π. Ideal phase plates which shift only the phase do not exist in electron microscopy, since each foil consists of atoms which also scatter the electrons. This scattering causes a frosted-glass effect and blurs the image. Furthermore, the foils must be homogeneous and of precise thickness over the entire aperture. Only recently has it become possible to produce phase plates with the required accuracy.

As early as 1949, Otto Scherzer [2] found a phase-

---

* This paper is dedicated to my esteemed teacher and colleague, Professor Dr. Otto Scherzer, on the occasion of his 68th birthday.

contrast method which avoids the use of foils entirely by producing the necessary phase shift in the lens itself. Scherzer took advantage of the fact that all round electron lenses are poor compared with the glass lenses used in light optics. In a famous paper [3] he showed that the spherical and chromatic aberrations of rotationally symmetric electron lenses are unavoidable. In part, Scherzer's theorem states that the outer zones of an electron lens always have a higher refractive power than the inner ones. This spherical aberration shifts the phase of the electrons which pass through the outer zones of the lens; the phase shift increases rapidly with increasing radius of the lens zone.

From light optics it was known that the effect of this aberration can be partially counterbalanced by a properly chosen defocus. Scherzer demonstrated that it is possible to choose the defocus in such a way that an almost constant phase shift of $\frac{1}{2}\pi$ is obtained over a wide aperture range. As a result, the phase of the diffracted beam is shifted by an amount of approximately $\frac{1}{2}\pi$ while the phase of the unscattered beam remains unaffected. It has become customary to call the optimum defocused object plane the Scherzer focus.

Scherzer also found an ingenious method to circumvent his own theorem [4]. He showed that it is possible to correct the third-order spherical aberration using a corrector which consists of nonrotationally symmetric elements. One can consider the corrector as "glasses" to eliminate the aberrations of the poor objective lenses. Quadrupole-octopole correctors are extremely complicated to align and to stabilize. This is one of the reasons that no success has yet been achieved. At present two correctors are being tested, one in Scherzer's own laboratory and the other at the University of Chicago. If correction works (in fact, we know of no physical law which prevents it from doing so), it will be possible to improve resolution and phase contrast further, since then the defocus *and* the third-order spherical aberration are free parameters which can be utilized to increase the angular range over which the phase shift γ stays close to its ideal value of $\frac{1}{2}\pi$. Due to the importance of subatomic resolution for biology and material science the entire scientific community hopes for a success of the correction efforts which undoubtedly would crown the work of Otto Scherzer.

The standard phase-contrast method in high-resolution electron microscopy uses axial coherent illumination, which guarantees that the envelope of the con-

trast transfer function stays constant over the transmitted frequency spectrum. However, it also introduces severe structural noise and artifacts, created by an overlap of Fresnel fringes into the image. The substrate noise, which appears as granularity, is a frequently encountered effect in the imaging of statistically distributed objects by highly coherent radiation. The observed patterns are known as *speckles*. The artifacts and speckles can be strongly suppressed by using partially coherent illumination. Unfortunately, this also reduces the object contrast, but the signal-to-noise ratio (S/N), which determines the visibility of object details, does improve in many cases. The S/N is defined as the quotient of object and noise contrasts. The total noise results from the statistical spatial fluctuations of the electrons and by the structural noise. Since quantum noise contrast and structural noise contrast are statistical quantities, they sum quadratically to produce total noise contrast.

The reduction of the speckles will improve as the phase-contrast transfer function (PCTF) approaches the triangular shape of the corresponding function of a self-luminous object. It was suggested [5–7] that hollow-cone illumination, which yields higher resolution and suppresses artifacts and structural noise considerably, would improve the quality of phase contrast images in many cases. However, experimentalists did not take notice of these suggestions until recently [8].

The impetus to explore the possibilities of phase contrast for nonstandard partially coherent illumination (hollow-cone illumination) originated from earlier work on phase contrast in the scanning transmission electron microscope (STEM). In the following we shall investigate the most promising arrangements for both the conventional transmission electron microscope (CTEM) and the STEM.

The theory of image formation in electron microscopy is not yet fully developed; as a result confusing statements are quite often made. For example, the winged diffractograms, first obtained by Parsons and Hoelke [9] from micrographs of various amorphous specimens taken with tilted illumination, have been interpreted implicitly [10] or explicitly [11] as an effect of interference of the inelastically scattered wave with the "corresponding" wave of the unscattered beam. Such an explanation is rather dubious for the following reasons. First, no corresponding unscattered beam exists because the energy width of the incident beam was much smaller than the energy loss suffered by the inelastically scattered electrons. Second, the inelastic scattering distribution (especially in the case of plasmon excitation) is strongly peaked in the forward direction. Therefore no electrons that are only inelastically scattered should be found in the high spatial frequency region. Third, the object was supposed to be so thin that plural scattering could be neglected.

A more rigorous theoretical treatment in fact shows that the observed wings are due to double scattering. Strictly speaking, they result from interference between the elastically *and* inelastically scattered part of the wave and the part that is only inelastically scattered. Therefore, the experiments of Parsons and Hoelke can be considered proof for the assumption that additional inelastic scattering processes do not significantly destroy spatial coherence. To my knowledge, nobody seems to have pointed out this important fact. There remains of course the question of why the calculations [10, 11] satisfactorily describe the observed phenomenon. An explanation will be given in this paper.

2. Image formation in CTEM and STEM

The influence of tilted illumination on contrast and resolution was investigated by Scherzer in his famous treatise [2] although he did not outline all the mathematical details. This may be the reason that the theory of image formation under tilted illumination has recently been reinvented by many authors. If we consider the reciprocity theorem [12,13], most papers on the theory of image formation in the STEM also completely describe the CTEM images for tilted illumination, since the detector angle in the STEM corresponds to the illumination angle in the CTEM. Differences occur only if inelastic scattering is incorporated in the theory. The inelastically scattered electrons pass through the objective lens in the CTEM but not in the STEM.

The wave function in the image plane of a CTEM, $\Psi_I(\boldsymbol{\rho})$, is obtained from the wave function in the object plane by a sequence of Fourier transforms. The wave function in the detector plane of a STEM, $\Psi_D(\boldsymbol{\rho})$, is obtained by an analogous procedure. The vector $\boldsymbol{\rho}$ is perpendicular to the optical axis and indicates the location of the image point referred to the object plane. To simplify our calculations we assume for the CTEM

that all incident beams from different directions are completely incoherent with each other. This assumption implies an infinitely extended effective source. It is not valid in the case of illumination of only a small area of the object. No post-specimen lens is assumed for the STEM. With these assumptions in mind, the detector angle of the STEM and the illumination angle of the CTEM can be considered conjugate quantities.

Let us assume an illuminating plane wave whose direction is given by the wave vector

$$K = e_z k \cos \Theta + K_\perp,$$
$$K_\perp = k(e_x \cos \Phi + e_y \sin \Phi) \sin \Theta, \tag{1}$$

where $\lambda = 2\pi/k$ is the wavelength of the incident electron. The angles Θ and Φ determine the direction of incidence; e_x, e_y and e_z are the unit vectors in x, y and z direction. The direction of the outgoing elastically ($n = 0$) or inelastically ($n \neq 0$) scattered electron is determined by the wave vector

$$k_n = k_n e_z \cos \theta + k_{n\perp},$$
$$k_{n\perp} = k_n(e_x \cos \phi + e_y \sin \phi) \sin \theta, \tag{2}$$

where θ and ϕ are its angular coordinates. In the small angle approximation the perpendicular momentum components

$$K_\perp \approx k\Theta, \qquad k_{n\perp} = k_n\theta \tag{3}$$

are proportional to the angles Θ and θ which subtend the optical axis in the direction of the incident and outgoing electron, respectively. After the inelastic process the electron has suffered an energy loss

$$\Delta E_n = \hbar(\omega_n - \omega_0), \tag{4}$$

where $\hbar\omega_0$ and $\hbar\omega_n$ are the energies of the incident and outgoing electrons, respectively, resulting in a smaller wave number $k_n = 2\pi/\lambda_n$.

When these definitions are used, the wave function in the image plane $z = z_I$ has the form

$$\Psi_I = A(\Theta, \Phi)\, e^{-i\gamma(K)}\, e^{i(kz_I - \omega_0 t)}$$

$$+ i \sum_{n=0}^{\infty} e^{i(k_n z_I - \omega_n t)} \int \frac{A(\Theta, \phi)}{\lambda_n}$$

$$\times e^{-i\gamma(k_n)} F_n(k_n, K)\, e^{i(k_n - K)r}\, d\Omega_\theta, \tag{5}$$

where $A(\Theta, \phi)$ is the aperture function. If no beam stop is present in the back focal plane, the aperture function has the form

$$A(\Theta, \phi) = A(\Theta) = \begin{cases} 1 & \text{for} \quad \Theta < \theta_0 \\ 0 & \text{for} \quad \Theta > \theta_0 \end{cases} \tag{6}$$

which guarantees that the direct beam is only transmitted if Θ is smaller than the limiting aperture angle θ_0. The three-dimensional position vector $r = \rho + e_z z_0$ indicates a point in the object plane $z = z_0$. The transition of the object state from its ground state $n = 0$ to the final state n is *not* necessarily the effect of only a single scattering process; it can be the result of successive scattering events. The scattering amplitudes $F_n(k_n, K)$ describe the angular distribution of the scattered electron wave; $F_0(k_0, K) = F(k, K)$ is the elastic scattering amplitude of the entire object. The angular integration $d\Omega_\theta$ has to be taken over the entire solid angle of the aperture; $\gamma(k_n) = \gamma(\theta, \phi, \Delta E_n)$ is the phase shift of the objective lens for an electron which has suffered an energy loss ΔE_n. For an astigmatism-free round lens γ has the well-known form [2]

$$\gamma(\theta, \phi, \Delta E_n) = \gamma(\theta, \Delta E_n)$$

$$= k_n(\tfrac{1}{4}C_3\theta^4 - \tfrac{1}{2}\Delta f\theta^2 + C_c(\Delta E_n/2E_0)\theta^2), \tag{7}$$

where E_0 is the mean kinetic energy of the incident electrons, and C_3, C_c and Δf are the constants of spherical aberration, chromatic aberration, and defocus.

The wave function in the detector plane of a STEM is obtained from eq. (5) if we replace the index I by D and make substitutions in the integrand:

$$k_n \rightarrow -k, \quad K \rightarrow -K_n, \quad \lambda_n \rightarrow \lambda. \tag{8}$$

By taking into account the reciprocity relation

$$F_n(-k, -K_n) = F_n(K_n, k) \tag{9}$$

of the scattering amplitudes we finally arrive at

$$\Psi_D = A(\Theta, \Phi) \, e^{-i\gamma(-K)} \, e^{i(kz_D - \omega_0 t)}$$

$$+ i \sum_{n=0}^{\infty} e^{i(k_n z_D - \omega_n t)} \int \frac{A(\theta, \phi)}{\lambda} \, e^{-i\gamma(-k)}$$

$$\times F_n(K_n, k) \, e^{i(K_n - k)r} \, d\Omega_\theta . \tag{10}$$

The described procedure can be considered a reversal of the propagation direction of the electron wave (see fig. 1). The succession of object plane $z = z_0$ and focal

plane $z = z_F$ changes when the flight direction of the electron is reversed. As a result the phase shift γ given by eq. (7) is only invariant with respect to inversion ($\gamma(k) = \gamma(-k)$) if we define the defocus differently for the STEM and the CTEM. The invariance is guaranteed if we define

$$\Delta f_{\text{CTEM}} = z_0 - z_F, \quad \Delta f_{\text{STEM}} = z_F - z_0. \tag{11}$$

With these definitions a positive defocus always represents an underfocused lens for the CTEM as well as the STEM. The number of incident electrons per image element in the CTEM is proportional to

$$\bar{j} = (1/k)|\text{Im} \, \overline{\Psi^* \, \partial\Psi/\partial z} \tag{12}$$

the current density averaged over the illumination time t_0; Im indicates that the imaginary part has to be taken. During the illumination time the interference terms of waves with different frequencies ($\omega_m \neq \omega_n$) oscillate many times and do not contribute to the recorded image intensity. Therefore only the elastically scattered wave can interfere with the unscattered beam. For purely elastic scattering the current density j is time-independent and takes its well-known form $j = \bar{j} = \Psi\Psi^*$.

Using (5) and (12) we obtain for the time-averaged current density in the image plane the expression

$$\bar{j}_I(\rho, K) = A(\Theta, \Phi) \left\{ 1 - \frac{2}{\lambda} \, \text{Im} \int e^{-i[\gamma(k) - \gamma(K)]} \right.$$

$$\left. \times e^{i(k-K)r} \, F(k, K) \, d\Omega_\theta \right\}$$

$$+ \frac{1}{k} \sum_{n=0}^{\infty} \frac{k_n}{\lambda_n^2} \iint e^{-i[\gamma(k_n) - \gamma(k_n')]}$$

$$\times A(\theta, \phi) A(\theta', \phi') F_n(k_n, K) F_n^*(k_n', K)$$

$$e^{i(k_n - k_n')r} \, d\Omega_\theta \, d\Omega_{\theta'} . \tag{13}$$

This formula is valid for arbitrary objects. Of course a great deal of work is necessary to calculate the scattering amplitudes F_n for a specific object. The first term in eq. (13) describes the constant background intensity. The second term results from the interference of the *elastically* scattered wave with the unscattered wave. The last term, which contains the quadratic terms of the scattering amplitudes, entirely describes the image intensity in the case of dark-field imaging, while in the case of bright-field imaging it *partly* describes the scattering absorption contrast. It also contains the contribution of the inelastically scattered

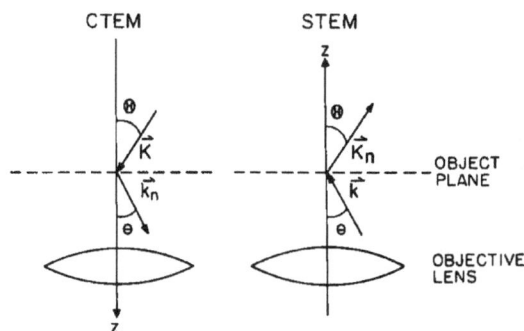

Fig. 1. The momentum vectors of an incident and outgoing electron for the STEM and CTEM and the definition of the corresponding angles of incidence and deflection. Note that in going from the STEM to the CTEM the direction of the z-axis is reversed.

electrons to the image formation. It should be noted that the relations (5), (10) and (13) are valid only as long as the imaged object remains within the isoplanatic region. Only in this case is the phase shift γ independent of the location of the specimen in the object plane.

2.1. Preservation of phase contrast by additional inelastic scattering

McFarlane [10] and Krakow [11] have treated the influence of inelastic scattering on the image contrast by assuming the existence of linear phase contrast terms for both elastic and inelastic scattering. In addition, they neglected the quadratic terms.

A contrast term which is linear in the inelastic scattering amplitudes F_n does not occur in the rigorous theoretical treatment outlined in section 2 since the unscattered and inelastically scattered parts of the electron wave cannot interfere with each other. Therefore, inelastic scattering is automatically disregarded if the quadratic terms are neglected. Nevertheless, a careful investigation of the quadratic terms indicates that an interference between *inelastically* scattered waves is possible. For example, the elastically scattered and the unscattered parts of the electron wave remain coherent if both undergo the same subsequent inelastic scattering process. Indeed, their phases are changed due to the energy loss, but their phase difference remains the same.

Inelastic scattering occurs predominantly in forward

direction. In particular, the excitation of nonlocalized electrons such as free electrons in metals (plasmon excitation) or π-electrons in macromolecules causes almost no deflection at all. Taking this into account, we can arrive at the results of [10] under the following assumptions: (a) the foil is so thick that many electrons have undergone *two* scattering processes; and (b) inelastic scattering causes an energy loss but no deflection.

Using these assumptions the third term of eq. (13) was evaluated by means of the Glauber formalism [14]. If we expand this term in a power series with respect to the object potential, it becomes apparent that a third-order term is responsible for the observed wings in the diffractogram. More precisely, this term is a convolution of the elastic scattering amplitude with the Fourier transform of the nonlocal potential. The calculations for an amorphous object of uniform thickness show that inelastic scattering reduces the elastic scattering amplitude by a factor

$$B = \exp[-N_i], \quad \text{where} \quad N_i = \Sigma_i/S \qquad (14)$$

is the number of inelastic mean free path lengths; Σ_i is the total inelastic scattering cross section, and S is the area of the amorphous object. It should be noted that $1 - B$ is identical with the free parameter P in [10].

Almost all electrons are scattered elastically at least once in objects which are significantly thicker than the mean free path length. In this case no distinct reference beam exists and, as a result the wings of the diffractogram start to blur with increasing foil thickness. They are completely washed out in the limiting case of very thick specimens.

Our calculations have shown that for $N_e \lesssim 1$ additional inelastic scattering can be described to a very good approximation by a properly chosen energy distribution of the elastically scattered electrons.

The results of our calculations can be summarized as follows: (a) additional inelastic scattering does not significantly destroy spatial coherence; (b) the wings observed in optical diffractograms of micrographs of amorphous material taken under tilted illumination are due to double scattering (elastic *and* inelastic); and (c) no wings should be observed for very thin ($N_i \ll 1$) or very thick objects ($N_e \gg 1$).

One way to check these statements experimentally would be to take micrographs of different foil thicknesses under identical imaging conditions.

3. The generalized optical theorem

From eq. (13) an important relation can be obtained for the scattering amplitudes. However, instead of starting from eq. (13), it is more convenient to start from the corresponding expression for the current density in the detector plane of a STEM, $\bar{j}_D$, derived from eq. (13) if we perform the substitutions presented in (8) in the integrands of the second and third terms and replace k'_n by $-k'$. We further assume that no particles have been absorbed within the object. Thus the current density $\bar{j}_D$ must be divergenceless. In other words, the total incident current $I_0 = \pi\theta_0^2$ must be identical to the total outgoing current:

$$\int \bar{j}_D \, d\Omega_\Theta = I_0 \ . \tag{15}$$

This condition leads after some algebraic manipulations to the identity:

$$\frac{1}{2i}\{F(k',k) - F^*(k,k')\}$$
$$= \frac{1}{4\pi}\sum_{n=0}^{\infty} k_n \int F_n(K_n,k)F_n^*(K_n,k') \, d\Omega_\Theta \ , \tag{16}$$

where the integration has to be taken over the entire solid angle 4π. This relation is an extended version of Glauber's "generalized optical theorem" [14] as it explicitly takes into account elastic *and* inelastic scattering.

It is convenient to separate the elastic scattering amplitude

$$F(k',k) = F_s(k',k) + iF_a(k',k) \tag{17}$$

into two parts, where

$$F_s(k',k) = [F(k',k) + F^*(k,k')]/2 \tag{18a}$$

is the "Friedel" term and

$$F_a(k',k) = [F(k',k) - F^*(k,k')]/(2i) \tag{18b}$$

the "anti-Friedel" term. Comparing eq. (18b) with the left side of eq. (16) immediately shows that the generalized optical theorem connects only the anti-Friedel part of the *elastic* scattering amplitude with the quadratic terms of the elastic *and* inelastic scattering amplitudes. The identity (16) is necessary to preserve the number of electrons. It should be noted that all reliable approximation methods to describe the image contrast have to fulfill this condition. For example, neglecting the quadratic terms in eq. (13) requires that the anti-Friedel part of the elastic scattering amplitude must

also be neglected; otherwise, all calculations are *not* consistent within the order of the approximation.

Most authors who have investigated the significance of the anti-Friedel term in expressions of bright-field image contrast have not considered the quadratic terms. Their results must be viewed cautiously (e.g., [15]), because the optical theorem has been violated. The same holds true if the squared terms are taken into account and the anti-Friedel part of the elastic scattering amplitude is dropped in the phase contrast term of eq. (13).

In the case of $k = k'$, eq. (16) reduces to the familiar form of the optical theorem:

$$\text{Im } F(k,k) = k(\Sigma_e + \Sigma_i)/(4\pi) = k\Sigma_{tot}/(4\pi) \ . \tag{19}$$

This identity is a requirement of particle conservation and connects the imaginary part of the elastic scattering amplitude in the forward direction with the total scattering cross section Σ_{tot}. The latter is the sum of the elastic scattering cross section Σ_e and the inelastic scattering cross-section Σ_i of the entire object.

It should be mentioned that we have not made use of the small-angle approximation in the derivation of the generalized optical theorem (16). Therefore, its validity is not limited to small angles. If we neglect the quadratic terms in eq. (16) we restrict ourselves to the kinematic approximation and arrive immediately at Friedel's law:

$$F(k',k) = F^*(k,k') = F_B(k'-k) \ . \tag{20}$$

The second identity becomes obvious if we remember that the first order Born approximation of the scattering amplitude F_B depends only on the scattering vector $k'-k$.

The generalized optical theorem can be utilized to derive the second-order Born approximation of the elastic scattering amplitude by inserting the first-order scattering amplitudes in the right side of eq. (16). This procedure immediately shows that the imaginary part of the elastic scattering amplitude of a single atom also depends on the inelastic processes within the atom. If we neglect inelastic scattering, the first-order approximation of the anti-Friedel term $F_a^{(1)}$ is proportional to the contribution $F_B^{(2)}$ of the second-order Born approximation to the elastic scattering amplitude:

$$F_a^{(1)}(k',k) = \frac{k}{4\pi}\int F_B(K-k) F_B^*(K-k') \, d\Omega_\Theta$$

$$= F_a^{(1)}(k'-k) = -iF_B^{(2)} \ . \tag{21}$$

The second identity becomes obvious if we choose the direction of the scattering vector $K - k$ as the axis of the spherical coordinate system. This can always be done, since the integration has to be taken over the entire sphere. It should be mentioned that the range of the first-order Born approximation is greater than that of the weak-phase object approximation because the curvature of the Ewald sphere has been neglected in the latter. Eq. (21) is valid as long as the object is so thin that plural electron scattering can be disregarded. Due to the fact that single scattering is concentrated in the forward direction we can replace the integration over the sphere by an integration over the plane which is tangent to the sphere at $K - k$. In this case, the anti-Friedel term $F_a^{(1)}$ is proportional to the autocorrelation function of the Friedel term.

4. Contrast in the CTEM

The aperture and tilt angles commonly used in electron microscopy are very small. Furthermore, the scattering is concentrated in the forward direction as long as the objects are thin enough to rule out significant beam broadening by plural scattering. Under these conditions, in eq. (13) and eq. (16) we can replace the integrations over the solid angles by integrations over the tangential planes that are perpendicular to the optical axis:

$$d\Omega_\theta \approx d^2\boldsymbol{\theta}, \qquad d\Omega_\Theta \approx d^2\boldsymbol{\Theta}. \qquad (22)$$

A vast number of different definitions for the image contrast can be found in the electron optics literature. For example, an inexperienced reader may easily become confused when the same contrast is positive in one paper but negative in another. In the following we use the standard notation of light optics, which we hope will become the standard definition in electron microscopy too. We define the image contrast as

$$C(\boldsymbol{\rho}) = 1 - I(\boldsymbol{\rho})/I_0, \qquad (23)$$

where I_0 denotes the average background intensity. We assume partially coherent illumination, which means that the illumination·has a certain angular divergence. Furthermore, the intensity may vary within the solid angle Ω_0 of the incident beam. For the STEM this angle is equal to the solid angle $\pi\theta_0^2$ of the objective aperture. We describe the variation in angular intensity by an illumination function

$$D(\Theta)/\Omega_0 . \qquad (24)$$

In the case of uniform illumination $D = 1$. For the STEM, D can be considered the detector response function. Negative values of D cannot occur in the CTEM but are possible in the STEM [6, 16], where they indicate a signal subtraction. By using the illumination function, eq. (24), the intensity in the image of a CTEM for partially coherent illumination may be written as

$$I(\boldsymbol{\rho}) = \frac{1}{\Omega_0}\int D(\Theta)\, j(\boldsymbol{\rho},\Theta)\, d^2\Theta, \qquad (25)$$

where j is the image intensity for a single tilt angle. The integration has to be taken over the entire Θ plane The same equation holds true for the image intensity of a STEM if we interpret $j = j_D$ as the current density distribution in the detector plane. Replacing the anti-Friedel term in the intensity distribution (13) by the right side of the generalized optical theorem (16) allows us to unambiguously discriminate the phase contrast created by the electrons that are only elastically scattered from the scattering absorption contrast and the contrast produced by the energy loss electrons. If we insert the modified expression eq. (13) for the intensity in eq. (25) and also use eq. (23) we find for the contrast

$$C(\boldsymbol{\rho}) = \frac{2}{\lambda\Omega_0}\, \mathrm{Im}\iint A(\boldsymbol{\theta})\, D(\Theta)\, A(\Theta)$$

$$\times\, e^{i[\gamma(\Theta)-\gamma(\theta)+k(\boldsymbol{\theta}-\Theta)\boldsymbol{\rho}+kz_0(\theta^2-\Theta^2)/2]}$$

$$\times\, F_s(\boldsymbol{\theta},\Theta)\, d^2\boldsymbol{\theta}\, d^2\Theta + \frac{1}{4\pi^2\Omega_0}\, \mathrm{Re}\sum_{n=0}^{\infty}\frac{k_n}{k}$$

$$\times\iiint A(\boldsymbol{\theta})\, D(\Theta)\, \{k^2 A(\Theta)$$

$$\times\, e^{i[\gamma(\Theta)-\gamma(\theta)+k(\boldsymbol{\theta}-\Theta)\boldsymbol{\rho}+kz_0(\theta^2-\Theta^2)/2]}$$

$$\times\, F_n(\boldsymbol{\theta}',\Theta)\, F_n^*(\boldsymbol{\theta}',\boldsymbol{\theta}) - k_n^2 A(\boldsymbol{\theta}')$$

$$\times\, e^{i[\gamma_n(\boldsymbol{\theta}')-\gamma_n(\theta)+k_n(\boldsymbol{\theta}-\boldsymbol{\theta}')\boldsymbol{\rho}+k_n z_0(\theta^2-\theta'^2)/2]}$$

$$\times\, F_n(\boldsymbol{\theta},\Theta)\, F_n^*(\boldsymbol{\theta}',\Theta)\}\, d^2\boldsymbol{\theta}\, d^2\boldsymbol{\theta}'\, d^2\Theta, \qquad (26)$$

where $\gamma_n(\theta) = \gamma(\theta, \Delta E_n)$ is the phase shift of the in-elastically scattered electrons attributable to the lens aberrations. The first term of eq. (26), which describes the phase contrast of the elastically scattered electrons, is proportional to the Friedel term of the elastic scattering amplitude. The quadratic terms yield the scattering absorption contrast as well as the phase contrast of the inelastically scattered electrons (plural scattering effects). We can easily demonstrate the different behavior of the two terms by considering the mean contrast $\bar{C} = (1/S) \int C \, d^2 \rho$ of a single object. The integration over the object area S can be extended over the entire image plane since the contrast is zero outside of S. The imaginary part of the integral over the linear term is zero. Thus the first term in eq. (26) represents a pure phase contrast. On the other hand, integration over the quadratic terms yields in the case $D = 1$, $\Theta < \theta_0$ the expected result

$$\bar{C} = \Sigma_A / S , \qquad (27)$$

where the scattering cross section Σ_A of the specimen determines the fraction of the elastically *and* inelastically scattered electrons that are removed by the aperture diaphragm. If the anti-Friedel term is neglected, we obtain the unrealistic result that the scattering absorption contrast is negative and proportional to the number of scattered electrons which pass through the hole of the aperture diaphragm. In the bright-field mode, scattering absorption always produces positive contrast; the object appears as a dark spot on a bright background. On the other hand, the phase contrast can have a positive or negative sign depending on the phase shift.

Both phase contrast and scattering absorption contrast are simultaneously present in the bright-field image. Of course the magnitude of each contribution strongly depends on the specific object under investigation and on the illumination and imaging modes. For thin biological objects the phase contrast dominates, while for thicker objects the scattering absorption contrast becomes the larger term. If the absolute values of the two components become similar, negative phase contrast should be avoided because it will be largely compensated by the positive scattering absorption contrast. Yet the contrast will be greatly improved if both terms have the same sign.

5. Phase contrast for partially coherent illumination

The curvature of the Ewald sphere can be neglected for objects which are not thicker than the depth of view

$$\Delta = \lambda/(2\theta_0^2) . \qquad (28)$$

In this case the phase-grating approximation [17] can be used to calculate the phase contrast. The phase-grating approximation considers all atoms as projected onto the object plane and neglects the influence of inelastic scattering. Inelastic scattering can be incorporated using the high-energy approximation of Glauber [14]. A plane Ewald sphere means that the scattering amplitude depends only on the projection

$$\omega = \theta - \Theta = (k_\perp - K_\perp)/k \qquad (29)$$

of the scattering vector $k - K$. The scattering angle ω should not be confused with the frequencies ω_n defined in eq. (4).

Most biological objects are weak scatterers and can be considered weak phase objects provided they are thinner than the mean free path length l. The scattering amplitudes of these objects fulfill Friedel's law, eq. (20). The range of validity of the weak phase object approximation also depends on the imaging mode; e.g., in the case of hollow-cone dark-field illumination the curvature of the Ewald sphere can only be neglected if the object is thinner than $\lambda/(2\Theta^2)$.

As an example of this condition let us consider two atoms located on the optical axis. The two atoms cannot be resolved by an observer looking along the axis. On the other hand, an off-axis observer can separate the atoms and do this more easily in proportion to his distance from the axis. Yet the observer has to remain within the angular range of the scattered electrons in order to see the atoms at all. Only a few electrons are scattered in angles larger than the characteristic atomic scattering angle

$$\theta_A = 1/(ka) , \text{ where } a \approx 0.56 \, \text{Å} Z^{-1/4} \qquad (30)$$

is the screening radius of an atom of atomic number Z. To obtain a reasonable collection efficiency, tilt angles larger than θ_A should be avoided at present day resolutions.

When these facts are taken into consideration, the STEM dark-field images can only be described correctly by the weak phase object approximation if the object thickness h fulfills the conditions

$$h \leqslant l \quad \text{and} \quad h \leqslant \lambda/(2\theta_A^2) \approx 2.5 Z^{-1/2} \text{Å}^2/\lambda. \tag{31}$$

As an example, let us assume an accelerating voltage of 60 kV and a carbonaceous object ($Z = 6$). For these values the standard phase object approximation [17] correctly describes the STEM dark-field images only if the object and the supporting film together are not thicker than 25 Å. This thickness is an order of magnitude smaller than the corresponding mean free path length l. The same condition for the specimen thickness must be fulfilled in a CTEM which uses hollow-cone dark-field illumination with a cone angle $\Theta \approx \theta_A$. In the case of point scatterers ($a = 0$) eq. (30) and eq. (31b) show that the weak phase object approximation holds only for infinitely thin objects, while the first-order Born approximation, which contains information about the z location of different atoms, remains valid as long as plural scattering can be ignored. It should be noted that the range in which the phase object approximation is valid can be significantly enlarged if the direction of the particle path inside the object is chosen so that it lies half-way between the directions of the incident and outgoing wave vectors.

A detailed discussion of the influence of the tilt angle on the degree of coherence can be found in the article by Fertig and Rose in this issue.

6. Nonstandard phase contrast of thin objects

Most biological objects are so thin that the quadratic terms in eq. (26) can be neglected. In these cases a linear contrast transfer does exist. Because of the small aperture angles the Ewald sphere can be approximated by a plane. Under these conditions the Fourier transform

$$\widetilde{C}(\omega) = \mathcal{B}(\omega) F_p(\omega)/\lambda \tag{32}$$

of the image contrast can be written as the product of the elastic scattering amplitude F_p of the projected object and a phase contrast transfer function (PCTF)

$$\mathcal{B}(\omega) = i/\Omega_0 \int A(\Theta) D(\Theta) \{ A(\omega - \Theta) e^{-i[\gamma(\omega - \Theta) - \gamma(\Theta)]}$$

$$- A(\omega + \Theta) e^{i[\gamma(\omega + \Theta) - \gamma(\Theta)]} \} d^2\Theta. \tag{33}$$

In the special case of axial illumination ($\Theta = 0$) this expression reduces to

$$\mathcal{B}(\omega) = 2\mathcal{L}(\omega) = -2 \sin \gamma(\omega). \tag{34}$$

The notation $\mathcal{B}$ for the PCTF was introduced by Hanssen [18]; the notation $\mathcal{L}(\omega) = \mathcal{B}(\omega)/2$ is widely used in optics.

The PCTF is generally complex. In some special cases it may become either real or entirely imaginary depending on the chosen illumination. Eq. (33) also describes the phase contrast in the STEM if one interprets Θ as the angle of detection.

McFarlane [10] emphasized that the nonrotationally symmetric terms of the phase shift $\gamma(\omega \pm \Theta) - \gamma(\Theta)$ could not be predicted on general grounds of symmetry. However, according to aberration theory such terms do occur if the lens is tilted and shifted with respect to the optical axis of the microscope. This occurs in the case of tilted illumination, where the central ray which intersects the center of the object plane can be considered the new optical axis. This axis is generally curved and is *not* identical with the symmetry axis of the microscope. Replacing the angle θ by the scattering angle ω ($\theta = \omega + \Theta$) defines a new optical axis because the scattering angle is measured from the direction of incidence. Tilting the axis introduces not only first-order terms (defocus and axial astigmatism) but also a zero-order term (displacement) and second-order terms (axial coma and threefold astigmatism). The small threefold axial astigmatism does not usually have to be taken into account because it is caused by the fifth-order spherical aberration of the objective lens.

All terms in the phase shift that contain uneven powers of the scattering angle ω do not show up in the diffractogram. Indeed the third-order coma and the parasitic second-order axial aberrations cannot be determined from a Thon [19] diffractogram. Although these aberrations do not affect the diffractogram, they may nevertheless limit the actual resolution of the microscope.

In the case of hollow-cone illumination there is no specific direction of incidence. The corresponding PCTF is real and depends only on the absolute value $|\omega| = \omega$ of the scattering angle.

6.1. CTEM phase contrast with hollow-cone illumination

Phase contrast under partially coherent illumination was first investigated by Hanssen and Trepte [18]. These authors assumed hollow-cone illumination of infinitely small width, and the cone angle Θ was chosen to coincide with the beam-limiting aperture angle θ_0. However, a slight misalignment of the illumination angle causes a very strong reduction in the PCTF. Furthermore, it is extremely difficult to adjust the beam-limiting aperture with the required accuracy.

Therefore two questions remain: first, can a smooth PCTF be obtained in the case of hollow-cone illumination of finite angular width; and second, how does the beam-limiting aperture affect the shape of the PCTF?

We have investigated the extreme case, in which no beam-limiting aperture is present. Two different optimal hollow-cone illumination modes were found; one yields positive, the other negative phase contrast. In the latter mode the objects appear as bright spots surrounded by dark rings.

It is convenient to express the defocus and the tilt angles by means of normalized parameters

$$\tau = \Delta f (k/C_3)^{1/2}, \qquad \nu = \Theta (C_3/\lambda)^{1/4}. \qquad (35)$$

The parameter τ was defined by Scherzer [2]. At present the notation $\Delta = \tau/\sqrt{2\pi}$ [18] is widely used.

In the absence of a beam-limiting aperture the optimum PCTF for negative phase contrast is obtained for $\tau \approx 3.7$, $\nu_1 \approx 1.04$ and $\nu_2 \approx 1.3$, where ν_1 and ν_2 are the limiting normalized inner and outer angles of the hollow-cone illumination. In fig. 2 the PCTF is shown for these parameters as a function of the normalized scattering angle. For comparison we have plotted the PCTF in the case of an infinitely small angular width of the hollow cone and a cone angle $\nu_1 = \nu_2 \approx 1.2$. With increasing scattering angle, this curve oscillates very rapidly. The oscillations result from the contributions of the unfavorable lens zones; they do not occur in the presence of a beam-limiting aperture. The locations of the maxima and minima of the oscillating PCTF are functions of the cone angle. When the cone has a small angular width the maxima and minima of different cone angles start to overlap in the high spatial frequency region, and the slope of the PCTF smoothes. With increasing angular width the maxima and minima of the wider oscillations in the low spatial frequency region also start to overlap. Indeed, a very broad angular width completely washes out the oscillations; unfortunately, it also strongly reduces the PCTF. Therefore a compromise has to be made between the magnitude of the PCTF and the suppression of the oscillations. The number of oscillations increases with the size of the cone angle.

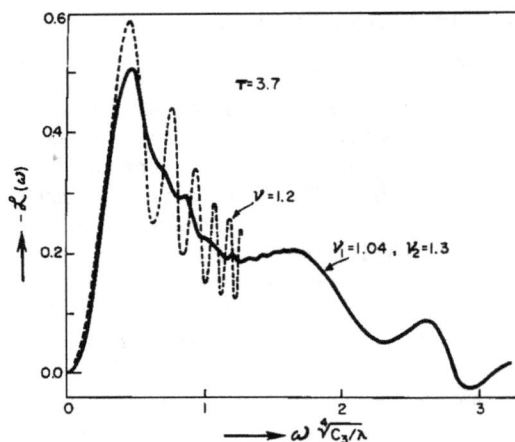

Fig. 2. Phase contrast transfer function of a CTEM used with hollow-cone illumination in the absence of a beam-limiting aperture plotted as a function of the normalized scattering angle; —— extended angular width of the hollow-cone illumination, – – – infinitely small angular width. The illumination parameters were chosen to give an optimum negative phase contrast.

Negative phase contrast is of little use, since most thin biological objects have to be stained with heavy-atom compounds to become detectable. Heavy-atom stains also produce a positive amplitude contrast, thus diminishing the negative phase contrast. Therefore a positive phase contrast is always desirable. If the illumination angle is only slightly smaller than the limiting aperture a large fraction (at least half) of the scattered electrons is removed by the diaphragm and a relatively large scattering absorption contrast is produced.

This contrast is further enhanced by a positive phase contrast if the illumination conditions of Hanssen and Trepte [18] are applied. For a finite width of the hollow cone a reduction of the PCTF can be avoided if we apply a defocus $\tau = \sqrt{\pi}$ (which is smaller than that of Hanssen and Trepte by a factor of $\sqrt{2}$) and an objective aperture $\theta_0 \approx 1.39 (\lambda/C_3)^{1/4}$. Almost no reduction in the PCTF occurs if the limiting illumination angles stay within the range $1.31 \leqslant \nu_1 \leqslant \nu_2 \leqslant 1.39$.

To demonstrate the influence of the aperture, we have plotted in fig. 3 the PCTF for $\nu_1 = 1.31$, $\nu_2 = 1.39$, assuming that no beam-limiting aperture is present. The PCTF has almost the same shape that it would have if an aperture were present [18]. The only dif-

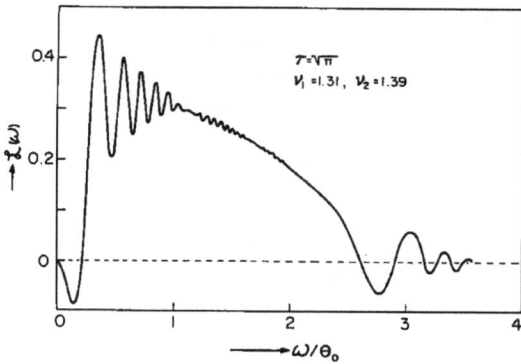

Fig. 3. Phase contrast transfer function of a CTEM which uses hollow-cone illumination of finite angular width yielding positive phase contrast; the absence of a beam-limiting aperture was assumed.

ferences are the "ripples" in the mid-frequency region and the weak negative contrast transfer in the low and high spatial frequency regions. Due to the triangular shape of the PCTF, artifacts and structural noise will be greatly suppressed in the image. For very high spatial frequencies the PCTF is a strongly damped, slowly oscillating function. Therefore the high spatial frequencies will not be transferred whether an aperture is present or not.

Instead of using a condensor aperture it is more advantageous to illuminate the object using a tilted beam which rotates on a cone around the optical axis. Then the illumination angles can easily be adjusted with the necessary accuracy.

If the correction works, the resolution is limited by the fifth-order spherical aberration, and the third-order spherical aberration becomes an additional free parameter which can be used to improve the PCTF further. To suppress the structural noise of nonresolved object layers and to obtain three-dimensional images using minimum doses, Hoppe [20] and Scherzer [5] suggested that low-dose micrographs be made with tilted illumination from different directions. The superposition of images taken with a given tilt angle Θ at different azimuths Φ represents an image obtained with hollow-cone illumination. Scherzer demonstrated that the phase shift of a corrected microscope can be properly adjusted for tilted beam illumination at large tilt angles. There are two different optimum illumination modes that allow a relatively large angular width of the hollow cone. These modes yield phase contrasts opposite

in sign. Assuming a relatively broad angular width of the hollow-cone illumination, we have calculated the corresponding PCTFs in the case of an optimal chosen aperture angle θ_0. The resulting PCTFs of the two illumination modes are plotted in fig. 4a and b for different values,

$$\tau = (3\pi^2)^{1/3}\Delta f(C_5\lambda^2)^{-1/3} \tag{36}$$

of the normalized defocus. The chosen illumination and aperture angles and the third-order aberration are indicated in each figure. The sign of the fifth-order spherical aberration constant C_5 determines which mode gives the positive phase contrast. The optimum PCTFs have a triangular shape, similar to that produced by self-luminous objects, and do not differ very much in magnitude from those of an uncorrected microscope (see figs. 2, 3). The PCTF is strongly reduced for object planes z_0 which lie at distances

$$|z_0 - z_{opt}| \gtrsim \tfrac{1}{4}(C_5\lambda^2)^{1/3} = \Delta_c \tag{37}$$

from the optimal focused plane $z = z_{opt}$. Therefore

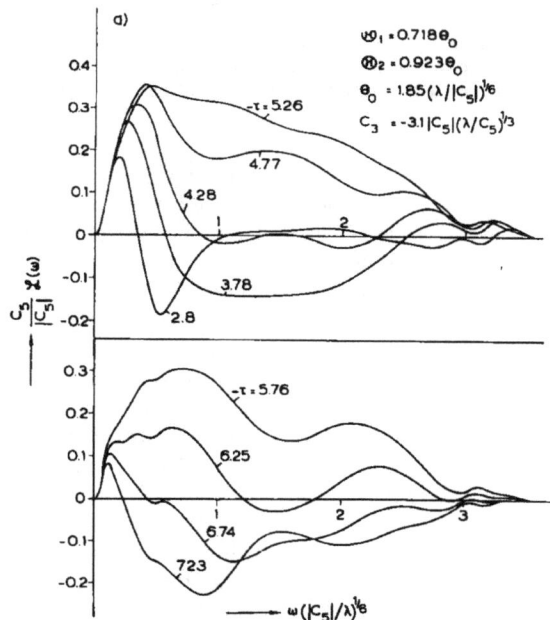

Fig. 4a. Phase contrast transfer functions of a spherically corrected CTEM used with hollow-cone illumination yielding positive phase contrast for $C_5 > 0$. The contrast is reversed if C_5 changes sign; Θ_1 and Θ_2 indicate the inner and outer illumination angles, θ_0 is the beam-limiting aperture angle and τ the normalized defocus (36).

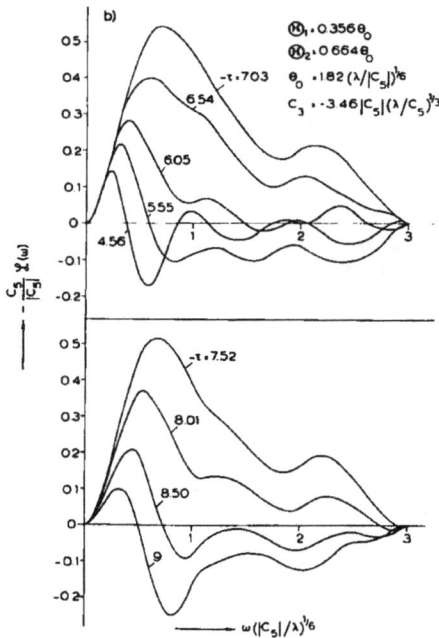

Fig. 4b. Phase contrast transfer functions of a corrected CTEM used with hollow-cone illumination, yielding negative phase contrast for $C_5 > 0$. (See 4a legend for explanation of symbols.)

only objects located within the region $|z_0 - z_{opt}| \lesssim \Delta_c$ will be imaged with relatively high phase contrast. For all frequencies the PCTFs are more than twofold smaller than those for coherent axial illumination, but the useful frequency range is almost doubled. Therefore the contrast of objects such as single atoms, which have a nearly constant frequency spectrum over the entire transmitted frequency region, may become only slightly reduced. In addition, the structural noise of the nonresolved object layers is strongly suppressed by hollow-cone illumination. Thus the S/N determining the amount of information can in some cases be significantly improved by going from axial to partially coherent illumination. With partially coherent illumination the contrast of relatively thick objects or heavy atoms can be further enhanced when a beam-limiting aperture is used to create a high scattering absorption contrast, provided the tilt angle lies close to the limiting aperture angle.

Finally, it should be mentioned that for the large illumination angles considered in this section, the PCTF

cannot be represented as a product of an envelope function times the PCTF for coherent axial illumination. This is only possible if the illumination angles are very small compared to the objective aperture angle [21].

6.2. Differential phase contrast

The STEM allows the use of difference signal techniques for image formation. Nonstandard phase contrast methods applying these techniques were first suggested by the author [6] and later by Dekkers and DeLang [22] and Veneklasen [16].

In a STEM the elastically scattered and unscattered parts of the beam interfere with each other and modulate the intensity distribution in the illumination cone beneath the object. The corresponding intensity distribution in the detector plane can be interpreted as a time-varying Gabor hologram [23]. The information contained in the interference pattern can be extracted by a properly designed arrangement of detectors. The greatest signal is obtained if the regions of constructive and destructive interference are covered by separate detectors.

The interference pattern depends strongly on the aberrations of the lens, yet it does not vanish for an aberration-free lens used in in-focus operation ($\gamma = 0$).

Dekkers and DeLang suggested that the difference signal of a split detector could be used to form the image. This detector consists of two half discs with the bisection perpendicular to the scanning direction.

To avoid the necessity of mechanically rotating the split detector when the direction of the scan is switched from x to y, it is advantageous to split the detector into four sections (see fig. 5) and combine their signals properly. In the following treatment we choose the x-axis as the direction of the scan.

It was first pointed out by the author [24] on theoretical grounds that differential phase contrast images represent the gradient of the object potential taken in the direction of the scan. The light optics experiments of Dekkers and DeLang [22] and of Stewart [25] confirmed this suggestion. Stewart also calculated differential phase contrast images of a one-dimensional phase grating. Unfortunately, these authors considered only the PCTF for one-dimensional objects. To obtain complete information on the contrast transfer properties of the split detector system the two-dimensional PCTF $\mathcal{L}(\omega) = \mathcal{L}(\omega, \psi)$ must be known. The angle ψ

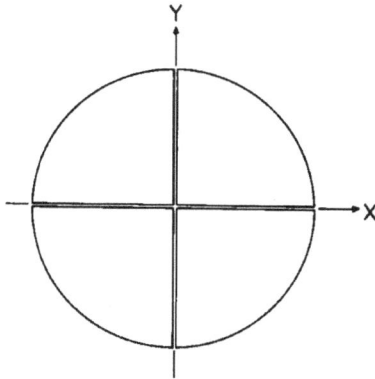

Fig. 5. Scheme of a STEM split detector.

subtends the direction of the projected scattering vector and the direction of the scan (see fig. 6).

The STEM allows us to manipulate the signal of the recorded electrons. In addition to producing the image directly from the difference signal, it is also possible to use the signal integrated with respect to time for image formation. This is equivalent to an integration in the x direction, since the length of scan is proportional to time. Therefore the time-integrated signal should give an image of the object potential itself. Such images can also be obtained directly from the difference signal of a ring detector system [6].

To decide which of the two methods is more appropriate, we must compare the S/Ns of the corresponding images using the same doses of incident electrons. In the absence of noise it is sufficient to compare the PCTFs of the two imaging modes. The PCTF of the ring detector system has an almost triangular shape and depends only on the absolute value ω/λ of the spatial frequency vector. A STEM operating in the phase contrast mode utilizes half of the elastically scattered electrons that remain within the illumination cone [6].

The PCTF of the image produced by the split detector system is not rotationally symmetric and depends on both the absolute value and the direction of the scattering vector. The detector response function $D(\Theta)$ of the split detector is antisymmetric with respect to an inversion through the origin:

$$D(\Theta) = -D(-\Theta). \tag{38}$$

By using this relation we derive from eqs. (33) and (34) the following expression for the differential phase con-

trast image of a STEM:

$$\mathcal{L}(\omega) = i/(\pi\theta_0^2) \int_0^\pi \int_0^{\theta_0} \{A(\omega - \Theta) \cos[\gamma(\omega - \Theta) - \gamma(\Theta)]$$

$$- A(\omega + \Theta) \cos[\gamma(\omega + \Theta) - \gamma(\Theta)]\} \Theta d\Theta \, d\Phi . \tag{39}$$

This PCTF is a pure imaginary quantity, and it is antisymmetric under inversion through the origin:

$$\mathcal{L}(\omega) = -\mathcal{L}(-\omega) = -\mathcal{L}^*(\omega) . \tag{40}$$

It should be noted that exactly the same PCTF (39) is obtained for a CTEM in the case of half-cone illumination if the limiting angle of the illumination and objective aperture coincide. To prove this statement we consider the PCTF of the images recorded independently by each of the two halves of the split detector. From eq. (33) we obtain

$$\mathcal{L}_1 = \tfrac{1}{2}(\mathcal{L}_r + i\mathcal{L}_i) , \qquad \mathcal{L}_2 = \tfrac{1}{2}(\mathcal{L}_r - i\mathcal{L}_i) = \mathcal{L}_1^* , \tag{41}$$

where the imaginary part $i\mathcal{L}_i$ is identical with (39).

The phase contrast vanishes for a disc detector covering the entire illumination cone. This behavior can only occur if the condition $\mathcal{L}_1 + \mathcal{L}_2 = 0$ is fulfilled. By inserting the relations presented in (41) into this equation it follows immediately that the real part $\mathcal{L}_r$ must be zero. From the reciprocity theorem it follows that the CTEM image taken with half-cone illumination is identical to the STEM image obtained with a half-disc detector. Yet the PCTF of the CTEM is twofold higher than the corresponding PCTF of a STEM. They become identical if the difference signal of the split detector forms the STEM image. It should be noted that our result does not agree with the calculations of Stewart [25], who obtained different PCTFs for the CTEM applying half-cone illumination and the STEM using the difference signals of a split detector. Our investigations showed that a difference between the PCTFs occurs only for thick objects, in which case the weak phase object approximation is no longer valid and the contributions of the nonlinear terms lead to different contrast transfers for the two imaging modes.

The split detector yields the greatest signal in the case of a perfect lens and in-focus operation ($\gamma = 0$). The corresponding PCTF is identical with the ratio of the shaded area to the illumination area $\pi\theta_0^2$ shown in fig. 6. The shaded area is that common to two half

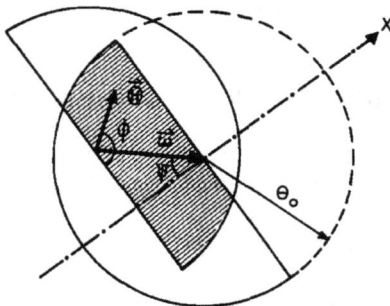

Fig. 6. The area of integration (shaded) of eq. (39) for the phase contrast transfer function of a STEM using a split detector and an ideal lens ($\gamma = 0$). The scan is in the x direction.

circles of radius θ_0 whose origins are separated by a distance ω. The bounding diameters are perpendicular to the direction of the scan and form an angle $\pi/2 - \psi$ with the direction of the scattering vector. An analytical expression has been found for the PCTF and is given in the Appendix.

We have evaluated this formula and plotted the lines $\mathcal{L}(\omega_x, \omega_y) = \mathcal{L}(\omega, \psi) = $ constant in fig. 7. To demonstrate the dependence of the PCTF on the direction of the scattering angle, we have also plotted $\mathcal{L}(\omega, \psi)$ as a function of ω for different angles ψ (fig. 8). No spatial frequencies perpendicular to the direction of the scan ($\psi = \frac{1}{2}\pi$) are transferred. As can be

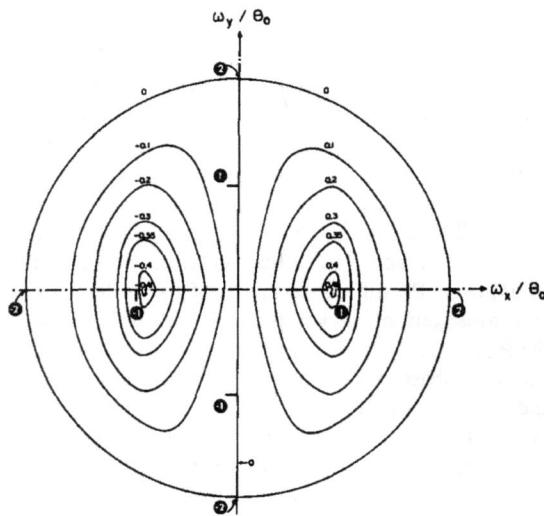

Fig. 7. The two-dimensional plot of $-i\mathcal{L}(\omega)$ of the phase contrast transfer function $\mathcal{L}(\omega)$ of the differential phase contrast STEM image in the case of an ideal lens ($\gamma = 0$).

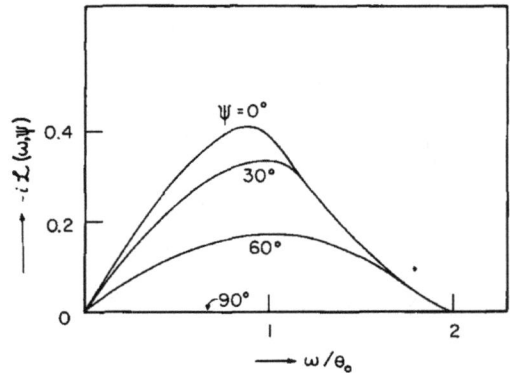

Fig. 8. The phase contrast transfer function $\mathcal{L}(\omega, \psi)$ of the differential phase contrast image for different angles ψ between the directions of the scan and the spatial frequency ω/λ. An ideal lens ($\gamma = 0$) is assumed.

seen from fig. 7, the PCTF is symmetric with respect to ω_y/θ_0 and antisymmetric with respect to ω_x/θ_0. Therefore, we can write

$$\mathcal{L}(\omega) = i\omega_x \mathcal{L}'(\omega)/\theta_0 , \qquad (42)$$

where $\mathcal{L}'(\omega) = \mathcal{L}'(\omega_x^2, \omega_y^2)$ is real and an even function with regard to both the ω_x and ω_y coordinate. This function also represents the PCTF of an image formed by the integrated difference signal of the split detector. This is due to the fact that an integration in real space with respect to the x coordinate is equivalent to a multiplication by $(i\omega_x k)^{-1}$ in reciprocal space.

The function $\mathcal{L}'(\omega)$ can be used to check the degree of accuracy with which the differential phase contrast image represents the gradient of the object structure. An exact gradient will be obtained when $\mathcal{L}'(\omega)$ is rotationally symmetric: $\mathcal{L}'(\omega) = \mathcal{L}'(\omega)$. Fig. 9 demonstrates that $\mathcal{L}'(\omega)$ is not completely rotationally symmetric. Nevertheless, the deviations are small, as can be seen in fig. 10, which shows the PCTF for the two sections $\psi = 0$ and $\psi = \frac{1}{2}\pi$ in which the deviations are the strongest.

Therefore, the image obtained from the direct signal of the split detector in fact represents to a very good approximation an image of the object gradient in the direction of scan. Differential phase contrast images are especially suited to visualizing edges and density variations.

In the case of weak phase objects (31) and $\gamma \approx$ constant over the aperture the anti-Friedel term and the quadratic terms of the current density distribution in the detector plane are invariant under inversion through the origin. Thus these terms are not present in the expression for the difference signal of the split detector. This signal remains unchanged if the split detector partly covers the dark-field region. The dark-field intensity distribution in the detector plane loses its invariance under inversion: for strong scatterers, $\gamma \neq$ constant, and when the curvature of the Ewald sphere must be taken into account. For example, asymmetry becomes extremely strong if we consider a three-dimensional crystal whose axes are tilted with respect to the optical axis.

It can be seen from figs. 9 and 10 that the PCTF $\mathcal{L}'(\omega)$ has no zeros and resembles closely the well-known triangular contrast transfer function of a self-luminous object in the case of an ideal lens. Thus image formation using the time-integrated signal of a split detector could be considered the optimum imaging method. Unfortunately, this holds true only in the absence of noise. In reality, a parasitic noise is always present. The influence of the noise on the image quality is revealed by the power spectrum

$$P(\omega) = 4|\mathcal{L}'(\omega)|^2 |F(\omega)/\lambda|^2 + \theta_0^2 |N(\omega)|^2/\omega_x^2 , \quad (43)$$

where $N(\omega)$ is the noise spectrum of the direct signal. It follows from eq. (43) that the integration suppresses

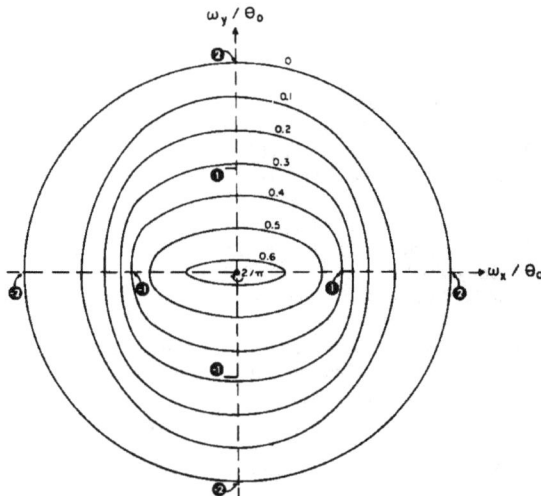

Fig. 9. The two-dimensional phase contrast transfer function $\mathcal{L}'(\omega)$ of an ideal STEM ($\gamma = 0$) produced using the time-integrated signal of a split detector for the image formation.

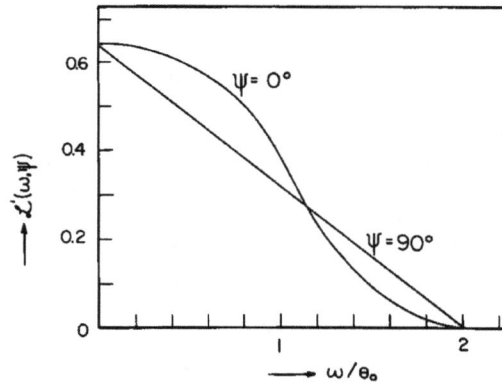

Fig. 10. The phase contrast transfer function $\mathcal{L}'(\omega,\psi)$ of an ideal STEM which operates in the mode described in fig. 9 for spatial frequencies parallel ($\psi = 0$) and perpendicular ($\psi = 90°$) to the direction of the scan.

the noise in the high spatial frequency region for frequencies with components $\omega_x = \omega \cos \psi > \theta_0$ and it strongly enhances the noise when the frequency components ω_x are small compared to θ_0. Therefore it is not possible to arbitrarily enhance spatial information which has been suppressed in the direct signal. To avoid artifacts the low spatial frequencies have to be filtered out from the integrated signal. Then the resulting PCTF becomes very similar to that of the ring detector system [26]. A more careful investigation shows that when a high dose is used (low noise) the integrated signal of a split detector gives better images, while the ring detector is better for low-dose imaging. Yet the differences are too small to recommend use of only one of the two modes. Moreover, the split detector using the integrated signal has two distinct disadvantages: first, the image resolution always has a slight directional dependence; and second, unless the exact intensity at the beginning of each line scan is known, the image will be severely distorted in the direction perpendicular to the scan. The intensity at the start can be obtained, at least in principle, from a single scan along the edge of the image in the y direction. The integrated signal of this scan can be stored to serve as the standard for the intensity at the boundary. The better the starting points of each scan in the x direction coincide with the corresponding points of the y scan, the smaller will be the distortions in the final image.

7. Conclusions

Inelastic scattering does not necessarily result in a loss of coherence. Indeed, the inelastically scattered wave cannot interfere with the unscattered wave, but it can interfere with the elastically *and* inelastically scattered wave if the excitation process has been the same in both cases. Therefore the phase contrast does not vanish when all electrons have been scattered inelastically. Since inelastic scattering occurs predominantly in the forward direction it has approximately the same effect on the phase contrast image as a broadening of the energy width of the incident electrons. Due to the chromatic aberration of the objective lens, the phase contrast diminishes in the CTEM with increasing energy width. This does not happen in the STEM because no imaging lenses are beneath the object. Hence the phase contrast in the STEM will not be significantly deteriorated by additional inelastic scattering. This fact is especially important for the imaging of relatively thick biological objects.

In biological specimens approximately two inelastic scattering processes occur on the average for each elastic scattering event. Consequently, when the object thickness becomes comparable to the inelastic mean free path length the CTEM phase contrast image will be strongly blurred while the corresponding STEM image remains sharp. The STEM image will only deteriorate when the object becomes so thick that most electrons undergo more than one *elastic* scattering process.

Hollow-cone bright-field illumination may become a useful imaging mode in the CTEM. If the cone angles are similar in size to the limiting angle of the aperture diaphragm, the structural noise of the supporting film as well as artifacts will be strongly suppressed. Furthermore, the scattering absorption contrast is very high for this imaging mode because at least half of the scattered electrons are removed by the aperture diaphragm independent of the resolution. This behavior is quite different from that in the case of axial illumination, in which the scattering absorption contrast decreases with increasing resolution. On the other hand, the phase contrast is significantly higher for this imaging mode but, unfortunately, so is the noise contrast. The total contrast for hollow-cone illumination is strongly enhanced if the contributions of phase and scattering absorption contrast are equal in sign and magnitude. The latter condition is fulfilled in the case of strong scatterers and low voltages.

Our investigations of the STEM have shown that the use of a split detector can be quite advantageous for the imaging of weak phase objects. The difference signal of the two half-disc detectors forms a differential phase contrast image which represents to a very good approximation the gradient of the object structure. If instead of this signal the time-integrated signal is displayed on the viewing screen, an image is obtained which fairly represents the object structure. The corresponding PCTF has no zeros. Because no spatial frequencies will be suppressed, the large as well as the small object details should be equally well imaged. Yet this is only true in the absence of noise. In practice, noise is always present and the low spatial frequencies have to be filtered out to avoid severe artifacts. The resulting images are almost the same as those obtained with a ring detector arrangement.

The imaging modes we have outlined here have some advantages over the standard imaging methods. It is up to the experimentalists to show to what extent the theoretical predictions hold true in practice.

Acknowledgement

I wish to thank Dr. W. Pöhner and J. Fertig for performing the numerical calculations.

References

[1] F. Zernike, Z. Tech. Phys. 16 (1935) 454.
[2] O. Scherzer, J. Appl. Phys. 20 (1949) 20.
[3] O. Scherzer, Z. Phys. 101 (1936) 539.
[4] O. Scherzer, Optik 2 (1947) 114.
[5] O. Scherzer, Optik 38 (1973) 387.
[6] H. Rose, Optik 39 (1973/74) 416.
[7] H. Rose, in: Proc. Eighth Int. Congr. Electron Microscopy, Canberra, Vol. I (1974) 212.
[8] W. Krakow, in: Proc. Thirty-Fourth Annual EMSA Conf., Miami Beach, (1976) 566.
[9] J.R. Parsons and C.W. Hoelke, Phil. Mag. 30 (1974) 135. Also in: Proc. Eighth Int. Congr. Electron Microscopy, Canberra, Vol. I (1974) 212.
[10] S.C. McFarlane, J. Phys. C.: Solid State Phys. 8 (1975) 2814.
[11] W. Krakow, D.G. Ast, W. Goldfarb and B.M. Siegel, Phil. Mag. 33 (1976) 985.
[12] J.M. Cowley, Appl. Phys. Letters 15 (1969) 58.
[13] E. Zeitler and M.G.R. Thomson, Optik 31 (1970) 258.
[14] R.J. Glauber, in: Lectures in Theoretical Physics, Boul-

der (Interscience Publishers, New York, 1958) Vol. I, 315.

[15] J. Frank, in: Image Processing and Computer-Aided Design in Electron Optics, ed. P.W. Hawkes (Academic Press, New York, 1973) 196.

[16] L.H. Veneklasen, Optik 44 (1975) 447.

[17] J.M. Cowley and L.F. Moodie, Acta Cryst. 10 (1957) 609.

[18] K.J. Hanssen and L. Trepte, Optik 33 (1971) 766.

[19] F. Thon, Z. Naturf. 21a (1966) 476.

[20] W. Hoppe, Z. Naturf. 27a (1972) 919.

[21] J. Frank, Optik 38 (1973) 519.

[22] N.H. Dekkers and H. DeLang, Optik 41 (1974) 452.

[23] D. Gabor, Proc. Roy. Soc. A197 (1949) 454.

[24] H. Rose, Optik 45 (1976) 139.

[25] W.C. Stewart, J. Opt. Soc. Am. 66 (1976) 813.

[26] H. Rose, Optik 42 (1975) 217.

Appendix

The PCTF for the differential phase contrast image can be expressed analytically in the case of a perfect STEM ($\gamma = 0$). Different formulas were obtained depending on the angle ψ and the absolute value ω of the scattering vector $\boldsymbol{\omega}$. Using the abbreviations

$$u^2 = 1 - (\omega/\theta_0)^2 \cos^2\psi , \qquad v = (\omega/\theta_0)\sin\psi - u,$$

$$\mathcal{L}_i(x) = \begin{cases} \dfrac{2}{\pi}\left[\arccos\tfrac{1}{2}x - \tfrac{1}{2}x(1 - \tfrac{1}{4}x^2)^{1/2}\right] & \text{for } 0 \leqslant x \leqslant 2 \\[2mm] 0 & \text{for } x \geqslant 2 \end{cases}$$

The PCTF can be written

(a) $0 \leqslant \psi \leqslant \psi_c$

$$-i\mathcal{L}(\omega,\psi) = \tfrac{1}{2}\mathcal{L}_i(2u) + (1/\pi)(u-v)(1-u^2)^{1/2} \quad \text{for } \omega \leqslant 2\theta_0\sin\psi.$$

$$-i\mathcal{L}(\omega,\psi) = \mathcal{L}_i(\omega/\theta_0) - \mathcal{L}_i[2(1-u^2)^{1/2}] \quad \text{for } \omega \geqslant 2\theta_0\sin\psi.$$

(b) $\psi > \psi_c$

$$-i\mathcal{L}(\omega,\psi) = \tfrac{1}{2}\mathcal{L}_i(2u) + (1/\pi)(u-v)(1-u^2)^{1/2}$$

for $\omega \leqslant 2\theta_0(1 + 3\cos^2\psi)^{-1/2}$,

$$-i\mathcal{L}(\omega,\psi) = \tfrac{1}{2}\mathcal{L}_i(2v) + \tfrac{1}{2}\mathcal{L}_i[2(1-u^2)^{1/2}]$$
$$+ (2/\pi)v(1-v^2)^{1/2} - (1/\pi)(u+v)(1-u^2)^{1/2}$$

for $2\theta_0(1 + 3\cos^2\psi)^{-1/2} \leqslant \omega \leqslant 2\theta_0\sin\psi$,

$$-i\mathcal{L}(\omega,\psi) = \mathcal{L}_i(\omega/\theta_0) \qquad \text{for } \omega > 2\theta_0\sin\psi.$$

The critical angle ψ_c which separates the two cases is given by

$$\psi_c = \arcsin(3)^{-1/2} \approx 35.2°.$$

These formulas have been evaluated numerically; the results are shown in figs. 7 and 8.

LIST OF PUBLICATIONS

1. Fischer, H. und Rose, H., Nanosecond Time Responses of Different Indium Antimonide Photodetectors; Developments in Applied Spectroscopy, Volume 2, (Plenum Press New York 1963) 29–33.

2. Rose, H., Allgemeine Abbildungseigenschaften unrunder Elektronenlinsen mit gerader Achse; Optik **24** (1966/67) 36–59.

3. Rose, H., Gaußsche Dioptrik begrenzter unrunder Elektronenlinsen; Optik **24** (1966/67) 108–121.

4. Rose, H., Über den sphärischen und den chromatischen Fehler unrunder Elektronenlinsen; Optik **25** (1967) 587–597.

5. Rose, H., Über die Korrigierbarkeit von Linsen für schnelle Elektronen; Optik **26** (1967) 284–298.

6. Rose, H., ber die Berechnung der Bildfehler elektronenoptischer Systeme mit gerader Achse; Teil 1: Optik **27** (1968) 466–474. Teil 2: Optik **27** (1968) 497–514.

7. Rose, H., Der Zusammenhang der Bildfehler-Koeffizienten mit den EntwicklungsKoeffizienten des Eikonals; Optik **28** (1968/69) 462–474.

8. Reichenbach, M. und Rose, H., Entwurf eines korrigierten magnetischen Objektivs; Optik **28** (1968/69) 475–487.

9. Rose, H., Berechnung eines elektronenoptischen Apochromaten; Optik **32** (1970) 144–164.

10. Rose, H., Abbildungseigenschaften sphärische korrigierter elektronenoptischer Achromate; Optik **33** (1971) 1–24.

11. Rose, H. und Petri, U., Zur systematischen Berechnung elektronenoptischer Bildfehler; Optik **33** (1971) 151–165.

12. Plies, E. und Rose, H., Über die axialen Bildfehler magnetischer Ablenksysteme mit krummer Achse; Optik **34** (1971) 171–190.

13. Rose, H., Elektronenoptische Aplanate; Optik **34** (1971) 285–311.

14. Rose, H., Zur Gaußschen Dioptrik elektrisch-magnetischer Zylinderlinsen; Optik **36** (1972) 19–36.

15. Rose, H. und Plies, E., BRD Patent 222 11 22 (1972).

16. Rose, H. und Moses, R. W., Jr., Minimaler Öffnungsfehler magnetischer Rund- und Zylinderlinsen bei feldfreiem Objektraum; Optik **37** (1973) 316–336.

17. Rose, H. und Plies, E., Correction of Aberrations in Electron Optical Systems with Curved Axes, in P.W. Hawkes, Image Processing and Computer-Aided Design in Electron Optics. (Academic Press, New York, 1973) 344–369.

18. Rose, H., To what extend are Inelastically Scattered Electrons Useful in the STEM? Proc. 31st Ann. EMSA Conf. New Orleans (1973) 286–287.

19. Rose, H., Phase Contrast in Scanning Transmission Electron Microscopy, Optik **39** (1973/74) 416–436.

20. Rose, H. und Plies, E., Entwurf eines fehlerarmen magnetischen Energie-Analysators; Optik **40** (1974) 336–341.

21. Rose, H. und Pöhner, W., Die Auflösungsgrenze korrigierter Elektronenmikroskope bei verbeulter Wellenfläche, Optik **41** (1974) 69–90.

22. Rose, H., Resolution and Contrast Transfer in fixed Beam and Scanning Transmission Electron Microscopy, Proc. of the 18th International Congress on electron Microscopy, Canberra (1974) 212–213.

23. Rose, H., Isaacson, M. and Langmore, J.P., Determination of the non-Localization of the Inelastic Scattering of Electrons by Electron Microscopy, Optik **41** (1974) 92–97.

24. Isaacson, M., Langmaore, J. and Rose, H., Measurements of the Non-Local Nature of Inelastic Scattering of Electrons in Carbon, Proc. 32nd EMSA Conference St. Louis (1974) 374–375.

25. Rose, H., Theorie der Bildentstehung im Elektronenmikroskop I, Optik **42** (1975) 217–244.

26. Rose, H., Phase Contrast in High Resolution Scanning Transmission Electron Microscopy, U.S. Patent Nr. 3908124, 1975

39. Fertig, J. and Rose, H., Calculation of STEM Dark-Field Images, Proc. of the 35th Annual Meeting of the Electron Microscopy Society of America, Boston, August 22–26, 1977, 192–193.

40. Rose, H. and Fertig, J., Improvement of TEM Bright Field Images Using HollowCone Illumination, Proc. of the 35th Annual Meeting of the Electron Microscopy Society of America, Boston, August 22–26, 1977, 200–201.

41. Rose, H., Aberration Correction of Homogeneous Magnetic Deflection Systems, Optik **51** (1978) 15–38.

42. Fertig, J. and Rose, H., Image of a Sulfur Atom: Fact or Artifact? Optic **51** (1978) 213–220.

43. Pejas, W. and Rose, H., Outline of an Imaging Magnetic Energy Filter Free of Second-Order Aberrations, Proc. of the 9th Int. Congr. Electr. Microscopy, Toronto, 1978, Vol. 1, 44–45.

44. Fertig, J. and Rose, H., Computer Simulation of Dark-Field Images as a Tool for Image Interpretation. Proc. of the 9th Int. Congr. Electr. Microscopy, Toronto, 1978, Vol. 1, 238–239.

45. Rose, H., Influence of Plural Scattering on the Image Quality of thick Amorphous objects in Transmision Electron Microscopy, Proc. of the 9th Int. Congr. Electr. Microscopy, Toronto, Vol. 3, 230–243.

46. Fertig, J. and Rose, H., On the Theory of Image Formation in the Electron Microscope II: Signal-to-Noise Consideration of Specimen Resolution and Quality of Phase Contrast Images Obtained by Hollow Cone Illumination, Optik . **54** (1979) 165–191.

47. Rose, H. and Pejas, W., Optimization of Imaging Magnetic Energy Filters Free of Second-Order Aberrations. Optik **54** (1979) 235–250.

48. Spehr, R. and Rose, H., Studies of the Boersch Effect in Stigmatic and Astigmatic Focusing, Proc. of the 37th Annual Meeting of the Electron Microscopy Society of America, San Antonio, Texas, 1979 Claitors Publ. Baton Rouge (1979) p. 570–571.

49. Eusemann, R., Fertig, J. and Rose, H., Probability Distribution of Contrast in Electron Micrographs of Amorphous Foils. Proc. of the 7th European Congress on Electron Microscopy, The Hague, 1980, Vol. 1, 128–129.

50. Kohl, H., Rose, H. and Schnabl, H., Dose Rate Effect at Low Temperatures in FBEM and STEM Due to Object Heating. Proc. of the 7th European Congress on Electron Microscopy, The Hague, 1980, Vol. 1, 130–131.

51. Rose, H. and Spehr, R., On the Theory of the Boersch Effect, Optik **57** (1980) 339–364.

52. Kohl, H., Rose, H. and Schnabl, H., Dose-rate Effect at Low Temperatures in FBEM and STEM Due to Object Heating, Optik **58** (1981), 11–25.

53. Rose, H., Correction of Aperture Aberrations in Magnetic Systems with Threefold Symmetry. Nucl. Instr. and Methods 187 (1981) 187–199.

54. Fertig, J. and Rose, H., Resolution and Contrasts of Crystalline Objects in High-Resoltuion Scanning Transmision Electron Microscopy. Optik **59** (1981) 407–429.

55. Bichsel, M., Rose, H. and Spehr, R., Comparison of low-voltage high-current electron microprobes using different types of cathodes. - In: Intern Congr. EM, Hamburg, Aug. 17–24, 82, Congr. Proceedings Vol. 1 (1982) 253–254.

56. Eusemann, R., Rose, H., Dubochet,, F., Electron Scattering in ice and organic materials. - In: Journal of Microscopy, 128, 3 (1982), 239–249.

57. Eusemann, R., Rose, H., Quality of STEM bright field images in the presence of strong scatterers. — In: 10. Intern. Congr. on Electron Microscopy, Hamburg, August 17–24, 1982, Congr. Propceedings, Vol. 1 (1982), 201–202.

58. Haider, M., Bernhardt, W. und Rose, H., Design and test of an electric and magnetic dodecapole lens. Optik **63** (1982), 9–23.

59. Rose, H. and Spehr, R., Energy Broadening in High-Density Electron and Ion Beams: The Boersch Effect. Advances in Electronics and Electron Physics, Supplement 13 C, Academic Press, New York 1983, 475–530.

60. Rose, H., Otto Scherzer gestorben, Optik **63**, (1983) 185–188.

61. Kohl, H. and Rose, H., Delocalization of Electronic Excitations as a Function of Spectrometer Acceptance Angle, Proc. 8th European Congress on Electron Microscopy, Vol. 1, Budapest (1984), 445–446.

62. Hugo, von, D., Kohl, H. and Rose, H., Semianalytical Treatment of Beam Broadening in Crystals, Proc. 8th European Congress on Electron Microscopy, Vol. 1, Budapest (1984), 135–136.

63. Zach, J. and Rose, H., Aberrations of Time-Dependent Deflection Fields, Proc. 8th European Congress on Electron Microscopy, Vol. 1, Budapest (1984), 49–50.

64. Eusemann, R. and Rose, H., Structural Noise and Image Quality in Dark-Field Micrographs of Thin Amorphous Objects, Proc. 8th European Congress on Electron Microscopy, Vol. 1, Budapest (1984), 179–180.

65. Rose, H., Information Transfer in Transmission Electron Microscopy, Ultramicroscopy **15** (1984) 173–192, North-Holland, Amsterdam.

66. Rose, H. and Zach, J., Optimization of Stroboscobic Electron Deflection Systems, In: Electron Optical Systems for Microscopy, SEM Inc. (1984) 127–135.

67. Weidenhausen, A., Spehr, R. and Rose, H., Stochastic ray deflections in focused charged particle beams, Optik **69**, No. 3 (1985) 126–134.

68. Kohl, H. and Rose, H., Theory of Image Formation by Inelastically Scattered Electrons in the Electron Microscope, Adv,. Electronics and Electron Physics, Vol. 65 (Acad. Press, New York, 1985) pp. 173–227.

69. Lanio, S., Rose, H. and Krahl, D., Test and improved design of a corected imaging magnetiv energy filter, Optik **73** (1986) 56–68.

70. Schnecker, G., Rose, H. and Spehr, R., Fast charge-simulation procedure for determining the electron-optical properties of electrostatic systems, Ultramicroscopy **20** (1986) 203–216.

71. Zach, J. and Rose, H., Efficient Detection of Secondary Electrons in Low-Voltage Scanning Electron Microscopy, Scanning 8 (1986) 285–293.

72. Rose, H., Hamiltonian Magnetic Optics, Nucl. Instr. and Methods (1987 in press).

73. Rose, H., Optimization of imaging energy filters for high-resolution analytical electron microscopy, Proc. 1, Beijing conference on ibnstrumental analysis, Beijing 1985.

74. Rose, H., Hamiltonian Magneticv Optics, Nuclear Instruments in Physics Research A 258 (1987) 374–401, North-Holland, Amsterdam.

75. Hugo, von, D., Kohl, H. und Rose, H., Delokalisierungseffekte bei der positionsbestimmung von Fremdatomen im Kristall nach der ALCHEMI-Methode, Optik **77**, Supplement 3 1987) 27.

76. Rose, H., Anwendungen und Grenzen der hochauflösenden Elektronenmikroskopie, Optik **77**, Supplement 3 (1987) 48.

77. Weickenmeier, A., Kohl, H., Hugo, von, D. und Rose, H., Bestimmung der Atompositionen mit EELS, Optik **77**, Supplement 3 (1987) 61.

78. Wolf, B., Zach, J. und Rose, H., Detektorobjektiv fr Elektronenstrahlmetechnik, Optik **77**, Supplement 3 (1987) 62.

79. Zach, J. und Rose, H., Entwurf einer korrigierten Elektronensonde fr die Niederspannungs-Rasterelektronenmikroskopie, Optik **77**, Supplement 3 (1987) 63.

80. Rose, H., The retarding Wien filter as a high-performance imaging filter, Optik **77**, No. 1 (1977) 26–34.

81. Degenhardt, R. and Rose, H., Correction of second-rank aberrations of magnetic deflection systems by imposing symmetry conditions. Inst. Phys.Conf.Ser. No: 93, Volume 1, Chapter 1, 113–114, EUREM 88, York, England 1988.

82. Zach, J. und Rose, H., High-resoltion low-voltage electron microprobe with large SE detection efficiency, Int.Phys.Conf.Ser. No.93: Volume 1, Chapter 1, 81–82, EUREM 88, York, England, 1988.

83. Hugo, von, Dirk, Kohl, H. und Rose, H., Effect of delocalized excitation in channeling enhanced microanalysis, Optik **77** (1988) 19–24.

84. Rose, H., Optimization of imaging energy filters for high resolution analytical electron microscopy, Tagungsband des franzsich-amerikanischen Symposiums ber "Spectroscopies ä tres haute resolution spatiale sous faisceau d'electrons primaires" in Aussois (1988), 261–269.

85. Schecker, G., Spehr, R. and Rose, H., Design of an Electron Line Cathode, Microelectronic Engineering 9 (1989) 259–262, North-Holland.

86. Wolf, B., Zach, J. und Rose, H., Analysis of the local field effects on voltage measurements with an in-lens spectrometer, J. Phys.E: Sci Instrum. **22** (1989) 720–725.

87. Hammel, M., Kohl, H. and Rose, H., Information Enhancement in the STEM by Employing a Multiple-Signals Imaging Procedure, Proceedings of the XIIth International Congress for Electron Microscopy, Seattle, 1990, Vol. 1, pp. 120–121.

88. Rose, H., New Feasible Concepts of Aberration Correction for Realizing HighResolution Electron Microscopes, Proceedings of the XIIth International Congress for Electron.

89. Hammel, M., Colliex, C., Mory, C., Kohl, H. und Rose, H., Defocus determination in the STEM by phase contrast methods, Ultramicroscopy **34** (1990) 257–269.

90. Rose, H., Inhomogeneous Wien filter compensating for the chromatic and spherical aberration of low-voltage electron microscopes, Optik **84** (1990) 91–107.

91. Rose, H., Outline of a spherically corrected semiaplanatic medium-voltage transmission electron microscope, Optik **85** (1990) 19–24.

92. Rose, H., Electrostatic energy filter as monochromator of a highly coherent electron source, Optik **85** (1990) 95–98.

93. Schönecker, G., Spehr, R. and Rose, H., Fast charge-simulation procedure for planar and simple three-dimensional electrostatic fields, Nuclear Instruments and Methods in Physics Research A **298** (1990) 360–376.

94. Degenhardt, R. and Rose, H., A compact aberration-free imaging filter with inside energy selection, Nuclear Instruments and Methods in Physics Research A **298** (1990) 171–178.

95. Rose, H., New Feasible Concepts of Aberration Correction for Realizing HighResolution Electron Microscopes, Proceedings of the XIIth International Congress for Electron Microscopy Vol. I, Seattle (1990) 202–203.

96. Berger, A., Spehr, R., Rose, H., Computer simulations of trajectory displacement with consideration of the energy width of the beam, Optik **86** (1990) 77–80.

97. Hammel, M., Kohl, H. und Rose, H., Information enhancement in the scanning transmission electron microscope by employing a multiple signals imaging procedure, Proc. 12th Intern. Congress on Electron Microscopy, Se

98. Rose, H., Elektronenspiegel mit fehlerfreiem Strahlteiler als Korrektor fr den Farbund Öffnungsfehler, Optik **88**, Suppl. 4 (1991) 53.

99. Hoffstätter, G. und Rose, H., Theoretische Auflösungsgrenze sphärisch korrigierter Elektronenmikroskope, Optik **88**, Suppl. 4 (1991) 54.

100. Hammel, M., Kohl, H. und Rose, H., Lorentzmikroskopie im STEM mit Unterdrckung des Kontrats nicht-magnetischen Ursprungs, Optik **88**, Suppl. 4 (1991) 94.

101. Preikszas, D. und Rose, H., Optimierung rotationssymmetrischer elektromagnetischer Elektronenlinsen, Optik **88**, Suppl. 4 (1991) 99.

102. Rose, H., Electron-optical fundamentals for high-resolution electron microscopy, Proceedings of the Autumn School Halle on "High-Resolution Electron MicroscopyFundamentals and Applications, pp. 6–15, Halle 1991.

103. Rose, H., Preikszas, D., Outline of a versatile corrected LEEM, Optik **92**, No. 1 (1992) 31–44.

104. Rose, H., Mirror-Hexapole corrector with compact beam splitter for eliminating the chromatic and spherical aberration of low-voltage electron microcopes, in: Proc. 50th Meet. EMSA, Boston, San Francisco Press, Inc. 1992, Vol. 1, pp. 94–95.

105. Rose, H., Correction of aberrations, a promising method for improving the performance of electron microscopes, in: Proc. EUREM 92, Granada, Spain, 1992, Vol. 1, pp. 47–48.

106. Dinges, C., Kohl, H. and Rose, H., Image simulation for hollow-cone illumination, in: Proceedings of the Autumn School Halle on "Image Interpretation and Image Processing in Electron Microscopy", pp. 164–173, Halle 1992.

107. Hammel, M. und Rose, H., Resolution and optimum conditions for dark-field STEM and CTEM imaging. Ultamicroscopy **49** (1993) 81–86.

108. Weickenmeier, A., Quandt, E., Kohl, H., Rose, H. und Niedrig, H., Computation and measurement of chracteristic energy-loss large-angle convergent-beam patterns of molybdenum selenide, Ultramicroscopy **49** (1993) 210–219.

109. Hoffstätter, G. H. and Rose, H., Gauge invariance in the eikonal method, Nuclear Instruments and Methods in Physics Research, **A328** (1993) North Holland.

110. Kahl, D. und Rose, H., Möglichkeiten und Grenzen der off-axis STEM Holographie, Optik **94**, Suppl. 5 (1993) 78.

111. Gerheim, V., Uhlemann, S. und Rose, H., Berechnung eines variablen Projektivsystems für ein energiefilterndes Mittelspannungs-Transmissions-Elektronenmikroskop, Optik **94**, Suppl. 5 (1993) 100.

112. Preikszas, D. und Rose, H., Chromatische und sphärische Korrektur mittels eines Elektronenspiegels, Optik **94**, Suppl. 5 (1993) 101.

113. Dinges, C., Rose, H. und Kohl, H., Bildsimulation fr Hohlkegel-Beleuchtung, Optik **94**, Suppl. 5 (1993) 52.

114. Stallknecht, P., Weickenmeier, A., Rose, H. und Kohl, H., Störungstheoretischer Ansatz zur Reduzierung der Reflexanzahl bei der Rechnersimulation elektronenmikroskopischer Beugungsbilder, Optik **94**, Suppl. 5 (1993) 55.

115. Uhlemann, and Rose, H., Comparison of the Performance of Existing and Proposed Imaging Energy Filters, in: Proc. 13th Int. Cong. on Electron Microscopy, Paris 1994, Vol. 1, 163–164.

116. Uhlemann, S., Haider, M. and Rose, H., Procedures for Adjusting and Controlling the Alignment of a Spherically Corrected Electron Microscope, in: Proc. 13th Int. Cong. on Electron Microscopy, Paris 1994, Vol. 1, 193–194.

117. Preikszas, D. and Rose, H., Corrected Low-Energy Electron Microscope for MultiMode Operation, in: Proc. 13th Int. Cong. on Electron Microscopy, Paris 1994, Vol. 1, 197–198.

118. Kahl, F. and Rose, H., Feasibility of Off-Axis STEM Holography, in: Proc. 13th Int. Cong. on Electron Microscopy, Paris 1994, Vol. 1, 285–286.

119. Hartel, P., Berger, A. and Rose, H., Influence of Thermal Vibrations on High-Angle Dark-Field Images in STEM, in: Proc. 13th Int. Cong. on Electron Microscopy, Paris 1994, Vol. 1, 349–350.

120. Dinges, C. and Rose, H., Computer Simulation of High-Resolution Images of Biological Objects, in: Proc. 13th Int. Cong. on Electron Microscopy, Paris 1994, Vol. 1, 493–494.

121. Uhlemann, S. and Rose, H., The MANDOLINE filter — A new high-performance imaging filter for sub-eV EFTEM, Optik **96** (1994) 163–178.

122. Dinges, C., Kohl, H., Rose, H., High-resolution imaging of crystalline objects by hollow-cone illumination, Ulramicroscopy **55** (1994) 91–100.

123. Rose, H., Correction of aberrations, a promising means for improving the spatial and energy resolution of energy-filtering electron microscopes, Ultramicroscopy **56** (1994) 11–25.

124. Rose, H. and Krahl, D., Electron Optics of Imaging Energy Filters in Energy Filtering Transmission Electron Microscopy (L. Reimer ed.), Springer Series in Optical Sciences, Vol.71 (Springer, 1995).

125. Dinges, C. and Rose, H., Simulation of Filtered and Unfilered TEM Images and Diffraction Patterns, phys.stat.sol(a) 150, 23, (1995).

126. Hammel, M. and Rose, H., Optimum rotationally symmetric detector configurations for phase-contrast imaging in scanning transmission electron microscopy, Ultramicroscopy **58** 403–415 (1995).

127. Rose, H. and Preikszas, D., Time-dependent perturbation for calculating the aberrations of systems with large ray gradients, Nuclear Instruments and Methods in Physics Research **A 363** 301–315 (1995).

128. Dinges, D., Berger, A. and Rose, H., Simualtion of TEM images considering phonon and electrons excitations, Ultramicroscopy **60** 49–70 (1995).

129. Preikszas, D. and Rose, H., Procedures for minimizing the aberrations of electromagnetic compound lenses, Optik Reprint **100** No. 4, 179–187 (1995).

130. Kahl, F. and Rose, H., Theoretical Concepts of Electron Holography, in: Adv. Imag. Electr. Phys., Vol. 94, Acad. Press 1995, 197–256.

131. Rose, H., Preikszas, D. and Müller, H., Outline of a corrected PEEM-LEEM, Proceedings of the 5th Seminar on Recent Particle Optics and Surface Physics Instrumentation, Brno, 1996, Institute of Scientific Instruments Academy of Sciences of the Czech Republik.

132. Rose, H., Hartel, P. and Dinges, C.. Simulation of high-angle annular dark field images considering elastic scattering and phonon excitations, Proceedings of the 5th Seminar on recent Particle Optics and Surface Physics Instrumentation, Brno 1996, Institute of Scientific Instruments.

133. Rose, H., Kahl, F., Design of electron monochromator with small Boersch effect, Proceedings of the 5th Seiar on Recent Particle Optics and Surface Physics Instrumentation, Brno 1996, Institute of Scientific Instruments Academy of Sciences of the Czech Republik.

134. Hartel, P., Dinges, C. and Rose, H., Simulation of high annular dark field images considering elastic scattering and phonon excitations, XI. European Congress on Microscopy, Dublin, Ireland, 1996.

135. Müller, H., Preikszas, D. and Rose, H., Precise design of a corrected beam separator for a high-resolution PEEM-LEEM, XI. European Congress on Microscopy, Dublin, Ireland, 1996.

136. Kahl, F. and Rose, H. Design of electron monochromator with small Boersch effect, XI. European Congress on Microscopy, Dublin, Ireland, 1996.

137. Angert, I., Burmester, C., Dinges C, Rose, H. and Schröder, R., Elastic and inelastic scattering cross-sections of amorphous layers of cabron and vitrified ice, Ultramicrocopy **63** (1996)181.

138. Preikszas, D. and Rose, H., Correction properties of electron mirrors, Journal of Electron Microscopy (1997) 1–9.

139. Rose, H., State and Prospects of Corrected High-Resolution Energy-Filtering Electron Microscopes, Proceedings of the 7th International BCEI Conference, Shanghai, Vol. A, (1997).

140. Müller, H., Rose, H. and Schorsch, P., A coherence function approach to image simulation, Journal of Microscopy **190** (1997) 73–88.

141. Preikszas, D. and Rose, H., Correction properties of electron mirrors, Journal of Electron Microscopy **1**(1997) 1–9.

142. Fink, R., Weiss, M. R., Umbach, E., Preikszas, D., Rose, H., Spehr, R., Hartel, P., Engel, W., Degenhardt, R., Wichtendahl, R., Kuhlenbeck, H., Erlebach, W., Ihmann, K., Schlögl, R., Freund, H.-J., Bradshaw, A.M., Lilienkamp, G. Schmidt, Th., Bauer, E., Benner, G., SMART: a planned ultrahigh-resolution spectromicroscope for BESSY II, Journal of Electron Spectroscopy and Related Phenomena **84** (1997) 231–250.

143. Rose, H., Haider, M. Uhlemann, S., Kabius, B., Urban, K., Design, Performance, and First Results of the Spherically Corrected 200 kV TEM, Proceedings of the International Symposium on Atomic Level Characterization for New Materials and Devices, Maui, Hawaii (1997) 12–17.

144. Rose, H., Haider, M. and Urban, K. Elekronenmikroskopie mit atomarer Auflösung, Phys. B. **54** (1998) 411–416.

145. Haider, M., Uhlemann, S., Schwan, E., Rose, H., Kabius, B. and Urban, K., Electron microscopy image enhanced, Nature **392** (1998) 768–769.

146. Kahl, F. and Rose, H., Outline of an Electron Monochromator with small Boersch effect, Proceedings of the 6th International Seminar on Recent Trends in Charged Particle Optics and Surface Physics Instrumentation, Brno, Czech. Republic (1998) 37–38.

147. Gerheim, V. and Rose, H., Quadrupole projector system with variable magnification for energy-filtering transmission electron microscopes, Proceedings of the 6th International Seminar on Recent Trends in Charged Particle Optics and Surface Physics Instrumentation, Brno, Czech. Republic (1998) 22–23.

148. Hartel, P., Preikszas, D., Spehr, R. and Rose, H., Test of a beam separator for a corrected PEEM/LEEM, Proceedings of the 6th International Seminar on Recent Trends in Charged Particle Optics and Surface Physics Instrumentation, Brno, Czech. Republic (1998) 24–25.

149. Müller, H. and Rose, H., A coherence function approach to multislice theory, Proceedings of the 6th International Seminar on Recent Trends in Charged Particle Optics and Surface Physics Instrumentation, Brno, Czech. Republic (1998) 49–50.

150. Hartel, P., Preikszas, D. Spehr, R. and Rose, H., Design and test of the mirror corrector for an ultrahigh-resolution spectromicroscope, Proc. of the 14th International Congress on Electron Microscopy, Cancun, Mexiko (1998), Vol. 1, 57–58.

151. Kahl, F. and Rose, H., Outline of an electron monochromator with small Boersch effect, Proc. of the 14th International Congress on Electron Microscopy, Cancun, Mexiko (1998), Vol. 1, 71–72.

152. Müller, H., Rose, H. and Schorsch, P., Coherence function approach to image simulation, Journal of Micoscopy, Vol. 190 (1998) 73–88.

153. Kabius, B., Urban, K. Haider, M., Uhlemann, S. Schwan, E. and Rose, H., First applications of a spherical-aberration corrected transmission electron microscope in materials science, Proc. of the 14th International Congress on Electron Microscopy, Cancun, Mexiko (1998), Vol. 1, 609–610.

154. Haider, M., Rose, H., Uhlemann, S., Schwan, E., Kabius, B. and Urban, K., A spherical-aberration-corrected 200 kV transmission electron microscope, Ultramicroscopy **75** (1998) 53–60.

155. Haider, M., Rose, H. Uhlemann, S., Kabius, B. and Urban, K., Towards 0.1 nm resolution with the first spherically corrected transmission electron microscope, Journal of Electron Microscopy **47** (1998) 395–405.

156. Müller, H., Rose, H. and Schorsch, P., A coherence function approach to image simulation, Journal of Microscopy, **190** (1998) 73–88.

157. Wichtendahl, R., Fink, R., Kuhlenbeck, H., Preikszas, D., Rose, H., Spehr, R., Hartel, P., Engel, W., Schlögl, R., Freund, H.-J., Bradshaw, A.M., Lilienkamp, G., Schmidt, Th. Bauer, E., Benner, G. and Umbach, E., SMART: An aberrationcorrected XPEEM/LEEM with energy filter, Surface Review and Letters, 5 (1998) 1249–1256.

158. Rose, H., Prospects for realizing a sub-A sub-eV resolution EFTEM, Ultramicroscopy **78** (1999) 13–25.

159. Müller, H., Preikszas, D. and Rose, H., A beam separator with small aberrations, Journal of Electron Microscopy **48** (1999) 191–204.

160. Müller, H., Schorsch, P. and Rose, H., Berücksichtigung der thermisch-diffusen und der unelastischen Elektroenstreuung bei der HRTEM Bildsimuolatin, Optik, Supplement 8, Vol. 10 (1999) 905.

161. Rose, H., High-performance electron microscopes of the future, Proc. 12th Europ. Congr. on Electr. Microscopy, Brno 2000, Vol. III, 21–24.

162. Müller, H., Schorsch, P. and Rose, H., Partial Coherence and HREM Simulation, Proc. 12th Europ. Congr. on Electr. Microscopy, Brno 2000, Vol. III, 55–56.

163. Preikszas, D., Hartel, P., Spehr, R. and Rose, H., SMART electron optics, Proc. 12th Europ. Congr. on Electr. Microscopy, Brno 2000, Vol. III, 81–82.

164. Spehr, R., Hartel, P., Müller, H., Peikszas, D. and Rose, H., Design and manufacturing of magnetic electron optical systems with midsection symmetry, Proc. 12th Europ. Congr. on Electr. Microscopy, Brno 2000, Vol. III, 85–86.

165. Gerheim, V. and Rose, H., Quadrupole projector system with variable magnification for energy-filtering transmission electron microscopes, Proc. 12th Europ. Congr. on Electr. Microscopy, Brno 2000, Vol. III, 95–96.

166. Schmid, P., Janzen, R. and Rose, H., Outline of a variable-axis lens with arbitrary shift of the axis in one direction, Proc. 12th Europ. Congr. on Electr. Microscopy, Brno 2000, Vol. III, 97–99.

167. Hartel, P., Preikszas, D., Spehr, R. and Rose, H., Performance of the mirror corrector for an ultrahigh-resolution spectromicroscope, Proc. 12th Europ. Congr. on Electr. Microscopy, Brno 2000, Vol. III, 153–154.

168. Weißbäcker, Ch. and Rose, H., Electrostatic correction of the chromatic and spherical aberration of charged particle lenses, Proc. 12th Europ. Congr. on Electr. Microscopy, Brno 2000, Vol. III, 157–158.

169. Schorsch, P., Müller, H. and Rose, H., An implementation of the coherence function multislice method, Proc. 12th Europ. Congr. on Electr. Microscopy, Brno 2000, Vol. III, 403–404.

170. Kahl, F. and Rose, H., Design of a monochromator for electron sources, Proc. 12th Europ. Congr. on Electr. Microscopy, Brno 2000, Vol. III, 459–460.

171. Schmid, P. and Rose, H., Outline of a variable-axis lens with arbitrary shift of the axis in one direction to be published in J. Vac. Soc. B. **19**(6) 2001 2555–2565.

172. Weißbäcker, Ch. and Rose, H., Electrostatic correction of the chromatic and of the spherical aberration of charged-particle lenses I, J. Electron Microscopy **50** (2001) 383–390.

173. Weißbäcker, Ch. and Rose, H., Electrostatic correction of the chromatic and of the spherical aberration of charged-particle lenses II, J. Electron Microscopy **51** (2002) 45–51.

174. Hartel, P., Preikszas, D., Spehr, R., Müller, H. and Rose, H., Mirror Corrector for low-voltage in "Advances in Imaging and Electron Physics", ed. P. Hawkes, Acad. Press. **120** (2002) 41–133.

175. Rose, H., Theory of electron-optical achromats and apochromats, Ultramicroscopy **93** (2002) 293–303.

176. Rose, H., Advances in Electron Optics, in: "High-Resolution Imaging and Spectrometry of Materials" eds F. Ernst and M. Rühle, Berlin Springer Verlag, (2003) 183–270.

177. Müller, H. and Rose, H., Electron Scattering , in "High-Resolution Imaging and Spectroscopy of Materials" eds. F. Ernst and M. Rühle, Berlin Springer Verlag, (2003) - .

178. Rose, H., Five-Dimensional Hamilton-Jacobi Approach to Relativistic Quantum Mechanics, in: Advances in Imaging and Electron Physics, Vol.**102** (2004) pp. 247–285, Acad. Press.

179. Rose, H., Outline of an ultracorrector compensating for all primary chromatic and geometrical aberrations of charged-particle lenses, Nucl. Instr. Meth. in Phys Res. **A 519** (2004) 12–27.

180. Rose, H., Prospects for aberration-free electron microscopy, Ultramicroscopy **103** (2005) 1–6.

181. Rose, H. and C. Kisielowski, On the Reciprocity of TEM and STEM, Microsc Microanal **11** (Suppl 2), (2005).

182. Rose, H. and R. Spehr, Effect of Coulomb Interactions on Resolution in Ultra-fast Electron Microscopy, Microsc Microanal **12** (Suppl 2), (2006).

183. Rose, H., Aberration correction in Electron Microscopy, Int. J. Mat. Res. **97** (2006) 885–889.

184. Rose, H., Criteria and Prospects for Realizing Optimum Electron Microscopes, Microsc Microanal **13** (Suppl 2), (2007) 134–135.

185. Rose, H., Theory of electron-optical achromats and apochromats, Ultramicroscopy **93** (2002) 293–303.

186. Rose, H., Outline of an ultracorrector compensating for all primary chromatic and geometrical aberrations of charged-particle lenses, Nucl. Instr. Meth. in Phys Res. **A 519** (2004) 12–27.

187. Rose, H., Five-Dimensional Hamilton-Jacobi Approach to Relativistic Quantum Mechanics, in: Advances in Imaging and Electron Physics, Vol. 102 (2004) pp. 247–285, Acad. Press.

188. Rose, H., Prospects for aberration-free electron microscopy, Ultramicroscopy **103** (2005) 1–6.

189. Rose, H., Aberration Correction in Electron Microscopy, Int. Journal of Materials Research **97** (2006) 885 - 889.

190. Rose, H. and R. Spehr, R., Effect of Coulomb Interactions on Resolution in Ultrafast Electron Microscopy, Proc. Microscopy and Microanalysis 2006, Chicago, 1412 CD.

191. Schröder, R.R., Barton, B., Rose, H. and Benner, G., Contrast Enhancement by Anamorphotic Phase Plates in an Aberration Corrected TEM, Microsc. Microanal. **13** (Suppl 2), (2007) 136–137.

192. Rose, H., Optimum Imaging Methods in Electron Microscopy, Proc. 8th Multinational Congress on Microscopy, Prague 2007, pp. 21–22.

193. Rose, H., Optics of high-performance electron microscopes, Science and Technology of Advanced Materials **9** (2008) 1–30.

194. Rose, H., The route towards sub-Å low-voltage electron microscopy, Proc. 11th Intern. Seminar on Recent Trends in Charged Particle Optics, Brno (2008), pp. 99–100.

195. Kaiser, U., Chuvilin, A., Schröder, R.R., Haider, M. and Rose, H., Sub-Ångstroem Low-Voltage Electron Microscopy — Future reality for deciphering the structure of beam-sensitive nanoobjects? Proc. 14th European Microscopy Congress EMC 2008, Vol. 1, pp. 35–36.

196. Rose, H., History of Direct Aberration Correction, in: Adv. in Imaging and Electron Physics, Vol. 153, pp. 261–280 (Elsevier, 2008).

197. Kabius, B. and Rose, H., Novel Aberration Correction Concepts, in: Adv. in Imaging and Electron Physics, Vol. 153, pp. 1–36 (Elsevier, 2008).

198. Rose, H., Geometrical Charged-Particle Optics, Springer Series in Optical Sciences **142** (Springer, Heidelberg, 2009).

199. Rose, H., Historical aspects of aberration correction, Journal of Electron Microscopy **58** (2009) 77–85.

200. Rose, H., Future trends in aberration-corrected electron microscopy, Phil. Tans. R. Soc. A (2009) **367**, 3809–3823.

201. Rose, H., Eine Brille für Elektronen, Physik Journal **8** (2009) Nr. 8/9, 61–66.

202. Rose, H., In Remembrance of Otto Scherzer, the Eminent Pioneer of Electron Optics, Microscopy and Microanalysis **15**, 2009, Supplement 2, 1452CD, (Cambridge, University Press).

203. Rose, H., Theoretical aspects of image formation in the aberration-corrected electron microscope, Ultramicroscopy **110** (2010) 488–499.

204. Kabius, B., Hartel, P., Haider, M., Müller, H., Uhlemann, S., Loebau, U., Zach, J. and Rose, H., First application of C_C-corrected imaging for high-resolution energy-filtered TEM, J. Electron Microscopy (Tokyo) (2009) **58**(3) pp. 147–155.

205. Kaiser, U., Biskupek, J., Meyer, J.C., Leschner, J., Lechner, L., Rose, H., Stöger-Pollach, M., Khlobystov, A.N. Hartel, P., Müller, H., Haider, M., Eyhusen, S. and Benner, G., Transmission electron microscopy at 20 kV for imaging and spectroscopy, Ultramicroscopy, **111**, 8, 1239–1246 (2011).

206. Lee, Z., Rose, H., Lehtinen, O., Biskupek, J. and Kaiser, U., Initiates file download Electron dose dependence of signal-to-noise ratio, atom contrast and resolution in transmission electron microscope images Ultramicroscopy (2014).
DOI 10.1016/j,Ultramicroscopy 2014.01.010

207. Lee, Z., Rose, H., Hambach, R., Wachsmuth, P. and Kaiser, U., Initiates file download The influence of inelastic scattering on EFTEM images — Exemplified at 20kV for graphene and silicon Ultramicroscopy **134** (2013), 102–112.

208. Lee, Z., Meyer, J.C., Rose, H. and Kaiser, U., Optimum HRTEM image contrast at 20 kV and 80 kV — Exemplified by grapheme, Ultramicroscopy **112**, 39–46 (2012).

Jacob D. Bekenstein

JACOB D. BEKENSTEIN

Birthplace

Born   Mexico City, Mexico, 1 May 1947

Education

1969   B. S. & M. S., Polytechnic Institute of Brooklyn
1972   Ph.D., Princeton University (Supervisor: John Archibald Wheeler) Research and Professional Experience
1972–1974   Post-doctoral Fellow, University of Texas at Austin
1974–1976   Senior Lecturer, Ben-Gurion University, Israel
1976–1978   Associate Professor, Ben-Gurion University, Israel
1978–1990   Professor of Physics, Ben-Gurion University, Israel
1990–   Professor of Physics, The Hebrew University of Jerusalem, Givat Ram, Jerusalem, Israel

Membership of Societies

1974–   International Society for General Relativity and Gravitation
1979–   International Astronomical Union
1997–   Israel Academy of Sciences and Humanities, Israel
2003–   World Jewish Academy of Sciences

Awards

1977   Ernst David Bergmann Prize for Science (Israel)
1981   First Prize, Gravity Research Foundation (USA)
1981   Landau Prize for Research in Physics (Israel)
1988   Rothschild Prize in the Physical Sciences (Israel)
2005   Israel Prize in Physics
2011   Weizmann Prize in the Exact Sciences (Israel)
2012   Wolf Prize in Physics

COMMENTARY

I was born in 1947 in Mexico City, Mexico, obtaining my early education through junior high school in the "Yavneh" Jewish school. I became interested early in chemistry, and invested much spare time with two schoolmates in devising small rockets, which sometimes actually flew. My family moved to the United States in 1962. There I acquired my lifetime interest in astronomy through long sessions observing observing with a small telescope. I graduated from S. J. Tilden senior high school in Brooklyn in 1965, and then went to the Polytechnic Institute of Brooklyn (nowadays the Polytechnic Institute of New York University) to study chemistry. Discovering that the chemical subjects I found most interesting (thermodynamics, kinetics, atomic structure ...) were actually physics, I transferred to the physics program at the end of the first year. Within Polytechnic's unified honors program I completed the requirements for the BS and MS degrees simultaneously in 1969.

My MS thesis adviser, Joseph Krieger, guided me through the first steps in research. We jointly published two papers which used the venerable WKB approximation to obtain highly accurate energy eigenvalues for hydrogen-like atoms; these competed favorably with fourth order perturbation theory results. Krieger convinced me to go on to Ph.D. studies, which I did at Princeton University.

Arriving in Princeton during the summer of 1969 to do the experimental project required for the Ph.D. under the supervision of Constantino Goulianos and David Bartlett, I learned a lot of elementary particle theory while I puttered around the lab. My project was to analyze data from an ongoing accelerator experiment in which fast deuterons were fired at iron nuclei to check the possibility that time reversal symmetry is violated by the electromagnetic interaction. I used my results to check a popular model for the internal wave function of the deuteron. I discovered then that it is very pleasurable to learn by experimenting, even though it seems one gets few facts out of a lot of work, but I found out that I am not too good with my hands. I made up my mind to be a theorist.

I passed the general physics graduate exam after one year at Princeton instead of the allotted two, and joined John A. Wheeler's gravitation theory research group. Wheeler was an important formative influence on my career. He was rightly famous for his joint work with Niels Bohr on nuclear fission, for the invention of the ubiquitous S-matrix formalism, and for his influential activity in resurrecting research into Einstein's general relativity during the mid 50's, as well as for half a dozen other accomplishments. At the time, he had just invented the name "black hole" for the object which attracted much of the attention of his group. Black holes then were hardly the buzzwords they are today, and many respectable scientists did not even believe they existed.

Early on in the group I contributed to the evidence supporting Wheeler's conjecture "black holes have no hair", that is, the black hole solutions of general relativity can have only three parameters: the mass, the charge and the angular momentum of the black hole. For example a black hole, even though formed by the collapse of a mass of baryons, was not supposed to have baryon number as a describing parameter, even though baryon number is ostensibly conserved. I became worried that maybe a black hole could become "dressed in a field", for example, a scalar or vector field which was sourced by baryon number, and that such an example would overthrow Wheeler's conjecture. After much work I invented an analytic method which showed clearly that such solutions of the equations must be self-contradictory, and so I came by a set of "no hair" theorems, later published in three papers in 1972. I also used the method to prove Kip Thorne and John Dykla's conjecture that the black holes in the Brans–Dicke theory of gravity are identical to those in Einstein's general relativity. But I lost the discovery; Stephen Hawking disclosed a similar result at the 1971 general relativity conference in Copenhagen. Methods similar to mine have meanwhile been employed by many to rule out this or that type of black hole hair. It is true that some types of black hole hair are allowed, but the spirit of "no hair" is maintained, and the subject has remained active and important.

After the said relativity conference (which was also a formative event for me), I took off from a hint in the work by Demetrious Christodoulou, Roger Penrose and Hawking proposing that a black hole must be endowed with an entropy proportional to the area of its horizon or boundary. Information theory notions were important in the arguments. Wheeler encouraged me to go forward and suggested the use of Planck's length squared to convert area into a dimensionless quantity (entropy is dimensionless unless you speak to chemists). From black hole entropy, it immediately followed from thermodynamics that a black hole has associated with it a temperature inversely proportional to its mass. The Planck constant appears in both the entropy and temperature, making it clear that black hole thermodynamics is a quantum subject, a window into the mysterious world of quantum gravity.

My doctoral thesis, which I defended on May 1, 1972 (my 25th birthday), was based mostly on the "no hair" theorems and black hole thermodynamics. It included a discussion of the generalized second law of thermodynamics, a new principle that maintains that in every process whereby ordinary entropy is swallowed by a black hole, the sum of the black hole entropy and the ordinary entropy remaining outside the black hole is an increasing quantity. I first published this law with examples in a 1972 paper in the *Lettere al Nuovo Cimento*.

That summer I travelled to the Les Houches summer school in physics which was to be another formative event in my scientific career. There I made reacquaintance with Bryce DeWitt, one of the school organizers and a prominent member of the physics department at Austin where I had just obtained a postdoctoral position. DeWitt was one of the few physicists who did not, at the time, laugh at the black hole thermodynamics. James Bardeen, Brandon Carter and Hawking, all lecturers at the school, ridiculed the thermodynamic viewpoint of black holes as being no more than a poor analogy. Instead they proposed a mechanical viewpoint of transformations of black holes which they developed into a 1973 paper which is often hailed as one of the sources of black hole thermodynamics (in fact it actually asserts there is no such thermodynamics). In between lectures and hikes in the Mont Blanc country around Les Houches, I wrote my long 1973 *Physical Review D* paper "Black holes and entropy" (to date my most cited paper). Simultaneously I wrote a paper showing that the phenomenon of rotating black hole superradiance, first put in evidence by Charles Misner and by Yakov Zel'dovich, follows from a simple argument using the law of black hole horizon growth discovered earlier by Hawking. My argument showed that a type of superradiance will also arise for a charged nonrotating black hole.

As a postdoc in Austin I diversified. I wrote a highly cited paper in support of the generalized second law. I invented a method — still in use — to construct solutions of the gravitational equations whose source is a conformal scalar field; I then used this method to discover a hairy black hole solution (unfortunately not stable). I found cosmological solutions with matter and a conformal field which sidestep the big bang singularity. I made my first foray into astrophysics by showing that a forming black hole recoils at high speed due to gravitational wave emission. I followed this with a joint paper with Richard Bowers, a fellow postdoc, in which we made a list of early type high velocity stars suspect of having a companion black hole formed in a supernova explosion in accordance with an hypothesis of Adrian Blaauw. Finally, I published another paper in *Lettere al Nuovo Cimento* explaining why a black hole should have a discrete and uniformly spaced horizon area spectrum. The large output reflected my anxiety then: that the postdoctoral years would be the last period when I could do research unencumbered. The fear proved right; in later years, teaching and family duty took a toll of my research time.

Towards the end of my postdoctoral stint, Hawking published a *Nature* paper announcing that a black hole radiates spontaneously (the renowned Hawking radiation), and has a temperature of the form predicted by black hole thermodynamics. Thus, he did reverse his previous criticism of my work. Suddenly people stopped sneering at black hole thermodynamics and it became reputable and important. About this time, in 1974, I took up my first faculty job, as a senior lecturer at the Ben-Gurion University in Israel.

Hawking's calculation, published in 1975, provided the proportionality constant in the law for black hole entropy: it is a quarter. Forty years later, there is still much research activity in re-deriving the entropy–area formula, including the proportionality constant; the methods are legion. Hawking radiation also supplied a further example of the generalized second law at work. For when a black hole emits radiation, it losses mass, so its area and entropy must decrease. On the other hand, the radiation being thermal, carries entropy away. In a 1975 paper I demonstrated that, mode for mode, the generalized second law is respected, not only for pure Hawking emission, but also when the black hole is immersed in a radiation reservoir with some different temperature. The generalized second law always works, as certified repeatedly by many workers over the intervening forty years.

The just mentioned paper employed information theory to compute the probabilities and entropies. Starting from it, my student Amnon Meisels and I calculated the full conditional probabilities for black hole emission and absorption under external radiative excitation. At the time, information theory methods were widely regarded with suspicion by the gravitational theory community, but various people verified our results by means of quantum field theory (which entails laborious work). Nowadays, quantum information theory has come to the fore in black hole studies and an approach such as ours would be a choice one. Many years later, my former student Marcelo Schiffer and I showed that the mentioned conditional probabilities for black hole emission is a universal one, applying also to all hot grey bodies.

Another type of research that has occupied me over the years has to do with magnetohydrodynamics, the flow of magnetized fluids and plasma. I first got interested in the subject during my Ph. D. days, but first wrote on it with my first doctoral student, Eli Oron, in 1978. Much of this research has concerned conservation laws for magneto flows, with the latest papers on this written with my student Asaf Oron (Eli's son) in 2001, and my postdoctoral fellow Gerold Betschart in 2006.

In 1980 I went on sabbatical to the University of Californian at Santa Barbara. There I developed an observation which I had recorded in the appendix to my 1973 paper on black hole entropy. I thus came up with the universal entropy bound, which maintains that the entropy capacity of a physical system is limited in terms of its total mass-energy and its linear size. This was the first example of an entropy bound; in this category, Gerard 't Hooft's much later holographic bound is most famous. Because of the close relation between entropy and information, entropy bounds are also bounds on the information capacity of systems, and as such are of great interest in fundamental physics. In the years since then I have often written on various aspects of the universal and the holographic entropy bounds. In the heyday of string theory, a large fraction of theoretical physics research was focused on the holographic correspondence which, like the universal and the holographic entropy bounds, is rooted in the black hole entropy.

I have long been interested in the possibility of variation of the physical constants, particularly the fine-structure constant. My 1982 paper on the subject introduced a paradigm — fine structure constant regarded as a scalar field — which first made it possible to connect its evolution with cosmology without recurring to conjectural ideas. This approach, which covers temporal and spatial variations, has been used by many ever since. My latest paper on the subject was with Schiffer in 2009. Philosophically related to that issue is whether masses of particles can vary; this is equivalent to a variation of the gravitational constant. In 1976, I had already devised VMT, a theory of such variation, a variant of scalar-tensor theory. For long, it was considered a viable theory of gravity. I last wrote about it with Meisels in 1980. This venture got me interested in alternatives to general relativity in general.

After the said sabbatical, a conversation with my colleague Mordehai Milgrom got me interested in his MOND scheme for obviating the need for dark matter. MOND presupposes that when the gravitational field is very weak, as it is within or between galaxies, Newton's law of gravitation is modified in a definite way. We collaborated in a 1984 paper which cast MOND as a (nonrelativistic) gravitational field theory equipped with a Lagrangian. That work set me thinking on how to make the MOND idea covariant. For a long period of time, the absence of a relativistic form of MOND was widely regarded as proof that it should not be taken seriously. After several false starts over the years — all record in publications — I hit upon the right recipe in TeVeS, the first consistent relativistic MOND. The *Physical Review* paper on TeVeS (one of my most cited ones) came out in 2004, and formed the basis for numerous imitations. For a long while, I continued to explore its consequences;

significant were a 2006 paper with my colleague Joao Magueijo, and two papers with my student Eva Sagi in 2008.

I again went on sabbatical in 1986–1987 to the University of Toronto, where the focus of my research was galactic dynamics, the theory describing the motions of the stars in galaxies and the sources responsible for them. I continued work with MOND, but also considered independent problems involving dynamical friction (the friction suffered by a star moving through a cloud of other stars). I wrote significant papers on this with my students Eyal Maoz and Ran Zamir. In the midst of these events, I moved, in 1990, to the Hebrew University in Jerusalem where I have worked ever since.

The 1990's brought a change in the theorist's perception of gravity and black holes. As I spent a second sabbatical in Santa Barbara, the "information paradox" that rose out of the Hawking radiation phenomenon was being investigated by increasing number of workers, and it continues to intrigue in view of the imperfect understanding as yet of the black hole entropy. Indeed, there has been a switch in emphasis: one tends to focus on the entropy instead of on the radiation, thus reversing the evolution in thinking that occurred in the mid-1970's. The holographic principle, arising right out of the black hole entropy formula, has become a central signpost of research in fundamental physics, and has served to link the subject of quantum gravity with that of condensed matter physics, a veteran field, and quantum information, a nascent field. Though I have contributed some papers on the information problem and the holographic principle, the riddles that originated with the introduction of black hole entropy will probably be cleared only by new ideas.

PHYSICAL REVIEW D   VOLUME 7, NUMBER 8   15 APRIL 1973

Black Holes and Entropy*

Jacob D. Bekenstein†

Joseph Henry Laboratories, Princeton University, Princeton, New Jersey 08540
and Center for Relativity Theory, The University of Texas at Austin, Austin, Texas 78712‡
(Received 2 November 1972)

There are a number of similarities between black-hole physics and thermodynamics. Most striking is the similarity in the behaviors of black-hole area and of entropy: Both quantities tend to increase irreversibly. In this paper we make this similarity the basis of a thermodynamic approach to black-hole physics. After a brief review of the elements of the theory of information, we discuss black-hole physics from the point of view of information theory. We show that it is natural to introduce the concept of black-hole entropy as the measure of information about a black-hole interior which is inaccessible to an exterior observer. Considerations of simplicity and consistency, and dimensional arguments indicate that the black-hole entropy is equal to the ratio of the black-hole area to the square of the Planck length times a dimensionless constant of order unity. A different approach making use of the specific properties of Kerr black holes and of concepts from information theory leads to the same conclusion, and suggests a definite value for the constant. The physical content of the concept of black-hole entropy derives from the following generalized version of the second law: When common entropy goes down a black hole, the common entropy in the black-hole exterior plus the black-hole entropy never decreases. The validity of this version of the second law is supported by an argument from information theory as well as by several examples.

I. INTRODUCTION

A black hole[1] exhibits a remarkable tendency to increase its horizon surface area when undergoing any transformation. This was first noticed by Floyd and Penrose[2] in an example of the extraction of energy from a Kerr black hole by means of what has come to be known as a Penrose process.[3] They suggested that an increase in area might be a general feature of black-hole transformations. Independently, Christodoulou[4,5] had shown that no process whose ultimate outcome is the capture of a particle by a Kerr black hole can result in the decrease of a certain quantity which he named the irreducible mass of the black hole, $M_{\rm ir}$. In fact, most processes result in an increase in $M_{\rm ir}$ with the exception of a very special class of limiting processes, called reversible processes, which leave $M_{\rm ir}$ unchanged. It turns out that $M_{\rm ir}$ is proportional to the square root of the black hole's area[5,6] [see (1)]. Thus Christodoulou's result implies that the area increases in most processes, and thus it supports the conjecture of Floyd and Penrose. Christodoulou's conclusion is also valid for charged Kerr black holes.[4,6]

By an approach radically different from Christodoulou's, Hawking[7] has given a general proof that the black-hole surface area cannot decrease in *any* process. For a system of several black holes Hawking's theorem implies that the area of each individual black hole cannot decrease, and more-

over that when two black holes merge, the area of the resulting black hole (provided, of course, that one forms) cannot be smaller than the sum of initial areas.

It is clear that changes of a black hole generally take place in the direction of increasing area. This is reminiscent of the second law of thermodynamics which states that changes of a closed thermodynamic system take place in the direction of increasing entropy. The above comparison suggests that it might be useful to consider black-hole physics from a thermodynamic viewpoint: We already have the concept of energy in black-hole physics, and the above observation suggests that something like entropy may also play a role in it. Thus, one can hope to develop a thermodynamics for black holes – at least a rudimentary one. In this paper we show that it is possible to give a precise definition of black-hole entropy. Based on it we construct some elements of a thermodynamics for black holes.

There are some precedents to our considerations. The idea of making use of thermodynamic methods in black-hole physics appears to have been first considered by Greif.[8] He examined the possibility of defining the entropy of a black hole, but lacking many of the recent results in black-hole physics, he did not make a concrete proposal. More recently, Carter[9] has rederived the result of Christodoulou[4,5] that the irreducible mass of a Kerr black hole is unchanged in a reversible trans-

formation by applying to the black hole the criterion for a thermodynamically reversible transformation of a rigidly rotating star.[10] Carter's
example shows the possibilities inherent in the
use of thermodynamic arguments in black-hole
physics.

In this paper we attempt a unification of black-
hole physics with thermodynamics. In Sec. II we
point out a number of analogies between black-hole
physics and thermodynamics, all of which bring
out the parallelism between black-hole area and
entropy. In Sec. III, after a short review of elements of the theory of information, we discuss
some features of black-hole physics from the point
of view of information theory. We take the area of
a black hole as a measure of its entropy – entropy
in the sense of inaccessibility of information about
its internal configuration. We go further in Sec. IV
and propose a specific expression for black-hole
entropy in terms of black-hole area. Earlier[11,12]
we had proposed this same expression on different
grounds; here we find the value of a previously unknown constant by means of an argument based on
information theory. In Sec. V we propose a generalization of the second law of thermodynamics
applicable to black-hole physics: When some common entropy goes down a black hole, *the black-
hole entropy plus the common entropy in the black-
hole exterior never decreases.*[11,12]

In Secs. VI and VII we construct several examples
which provide support for the generalized second
law. In addition, we analyze in Sec. VII a thought
experiment proposed by Geroch[13] in which, with
the help of a black hole, heat is apparently converted entirely into work in violation of the second
law. We show that, in fact, due to fundamental
physical limitations the conversion efficiency is
somewhat smaller than unity. Moreover, the efficiency is no greater than the maximum efficiency
allowed by thermodynamics for the heat engine
which is equivalent to the Geroch process, so that
this process cannot be regarded as violating the
second law.

II. ANALOGIES BETWEEN BLACK-HOLE PHYSICS AND THERMODYNAMICS

We have already mentioned the resemblance between the tendency of black-hole area to increase,
and the tendency of entropy to increase. Changes
of a black hole or of a system of black holes select
a preferred direction in time: that in which the
black-hole area increases. Likewise, changes of
a closed thermodynamic system select a preferred
direction in time: that in which the entropy
increases. This parallelism between black-hole
area and entropy goes even deeper.

Black-hole area turns out to be as intimately
related to the degradation of energy as is entropy.
In thermodynamics the statement "the entropy has
increased" implies that a certain quantity of energy has been degraded, i.e., that it can no longer
be transformed into work. Now, as Christodoulou
has emphasized,[4,5] the irreducible mass M_{ir} of a
Kerr black hole, which is related to the surface
area A of the black hole by[14]

$$M_{ir} = (A/16\pi)^{1/2} , \qquad (1)$$

represents energy which cannot be extracted by
means of Penrose processes.[3] In this sense it is
inert energy which cannot be transformed into
work. Thus, an increase in A, and hence in M_{ir},
corresponds to the degradation (in the thermodynamic sense) of some of the energy of the black
hole.

The irreducible mass of a Schwarzschild black
hole is just equal to its total mass. Thus, no energy can be extracted from such a black hole by
means of Penrose processes. However, the merger of two Schwarzschild black holes can yield energy in the form of gravitational waves.[7] The only
restriction on the process is that the total black-
hole area must not decrease as a result of the
merger.[7] However, the sum of the irreducible
masses of individual black holes may (in fact,
does) decrease. We see that for a system of several black holes the degraded energy E_d is more
appropriately given by

$$E_d = (\sum A/16\pi)^{1/2} = (\sum M_{ir}^2)^{1/2} \qquad (2)$$

than by $\sum M_{ir}$. According to this formula the degraded energy of a system of black holes is smaller than the sum of degraded energies of the black
holes considered separately. Thus by combining
Schwarzschild black holes which are already
"dead," one can still obtain energy.[7] Analogously,
by allowing two thermodynamic systems which are
separately in equilibrium to interact, one can
obtain work, whereas each system by itself could
have done no work. From the above observations
the parallelism between black-hole area and
entropy is again evident.

We shall now construct the black-hole analog of
the thermodynamic expression

$$dE = TdS - PdV . \qquad (3)$$

For convenience we shall from now on write all
our equations in terms of the "rationalized area"
of a black hole α defined by

$$\alpha = A/4\pi . \qquad (4)$$

Consider a Kerr black hole of mass M, charge Q, and angular momentum $\vec{L}$. (3-vectors here refer to components with respect to the Euclidean frame at infinity.) Its rationalized area is given by[5,7]

$$\alpha = r_+{}^2 + a^2$$

$$= 2Mr_+ - Q^2 , \qquad (5)$$

where

$$\bar{a} = \vec{L}/M , \qquad (6)$$

$$r_\pm = M \pm (M^2 - Q^2 - a^2)^{1/2} . \qquad (7)$$

Differentiating (5) and solving for dM we obtain

$$dM = \Theta d\alpha + \vec{\Omega} \cdot d\vec{L} + \Phi dQ , \qquad (8)$$

where

$$\Theta \equiv \tfrac{1}{4} (r_+ - r_-)/\alpha , \qquad (9a)$$

$$\vec{\Omega} \equiv \vec{a}/\alpha , \qquad (9b)$$

$$\Phi \equiv Q r_+/\alpha . \qquad (9c)$$

In (8) we have the black-hole analog of the thermodynamic expression (3): The terms $\vec{\Omega} \cdot d\vec{L}$ and ΦdQ clearly represent the work done *on* the black hole by an external agent who increases the black hole's angular momentum and charge by $d\vec{L}$ and dQ, respectively. Thus $\vec{\Omega} \cdot d\vec{L} + \Phi dQ$ is the analog of $-PdV$, the work done *on* a thermodynamic system. Comparing our expression for work with the expressions for work done on rotating[15] and charged[16] bodies, we see that $\vec{\Omega}$ and Φ play the roles of rotational angular frequency and electric potential of the black hole, respectively.[5,9] The α in (8) resembles the entropy S in (3) as we have noted before: For any change of the black hole $d\alpha \geqslant 0$,[5-7] while for any change of a closed thermodynamic system $dS \geqslant 0$. Moreover, it is clear from (7) and (9a) that Θ, the black-hole analog of temperature T, is non-negative just as T is. From the above observations the formal correspondence between (3) and (8) is evident.

All the analogies we have mentioned are suggestive of a connection between thermodynamics and black-hole physics in general, and between entropy and black-hole area in particular. But so far the analogies have been of a purely formal nature, primarily because entropy and area have different dimensions. We shall remedy this deficiency in Sec. IV by constructing out of black-hole area an expression for black-hole entropy with the correct dimensions. Preparatory to this we shall now look

at black-hole physics from the point of view of the theory of information.

III. INFORMATION AND BLACK-HOLE ENTROPY

The connection between entropy and information is well known.[17,18] The entropy of a system measures one's uncertainty or lack of information about the actual internal configuration of the system. Suppose that all that is known about the internal configuration of a system is that it may be found in any of a number of states with probability p_n for the nth state. Then the entropy associated with the system is given by Shannon's formula[17,18]

$$S = -\sum_n p_n \ln p_n . \qquad (10)$$

This formula is uniquely determined by a few very general requirements which are imposed in order that S have the properties expected of a measure of uncertainty.[17]

It should be noticed that the above entropy is dimensionless. This simply means that we choose to measure temperature in units of energy. Boltzmann's constant is then dimensionless.

Whenever new information about the system becomes available, it may be regarded as imposing some constraints on the probabilities p_n. For example, the information may be that several of the p_n are, in fact, zero. Such constraints on the p_n always result in a decrease in the entropy function.[18] This property is formalized by the relation[17,18]

$$\Delta I = -\Delta S , \qquad (11)$$

where ΔI is the new information which corresponds to a decrease ΔS in one's uncertainty about the internal state of the system. Equation (11) is the basis for Brillouin's identification of information with negative entropy.[18] All the above comments apply to such divers systems as a quantity of gas in a box or a telegram. A familiar example of the relation between a gain of information and a decrease in entropy is the following. Some ideal gas in a container is compressed isothermally. It is well known that its entropy decreases. On the other hand, one's information about the internal configuration of the gas increases: After the compression the molecules of the gas are more localized, so that their positions are known with more accuracy than before the compression.

The second law of thermodynamics is easily understood in the context of information theory. The entropy of a thermodynamic system which is not in equilibrium increases because information about the internal configuration of the system is being lost during its evolution as a result of the washing out of the effects of the initial conditions. It is possible for an exterior agent to cause a decrease in the entropy of a system by first acquiring information about the internal configuration of the system. The classic example of this is that of Maxwell's demon.[18] But information is never free. In acquiring information ΔI about the system, the agent inevitably causes an increase in the entropy of the rest of the universe which exceeds ΔI.[18] Thus, even though the entropy of the system decreases in accordance with (11), the over-all entropy of the universe increases in the process.

The conventional unit of information is the "bit" which may be defined as the information available when the answer to a yes-or-no question is precisely known (zero entropy). According to the scheme (11) a bit is also numerically equal to the maximum entropy that can be associated with a yes-or-no question, i.e., the entropy when no information whatsoever is available about the answer. One easily finds that the entropy function (10) is maximized when $p_{yes} = p_{no} = \frac{1}{2}$. Thus, in our units, one bit is equal to ln2 of information.

Let us now return to our original subject, black holes. In the context of information a black hole is very much like a thermodynamic system. The entropy of a thermodynamic system in equilibrium measures the uncertainty as to which of all its internal configurations compatible with its macroscopic thermodynamic parameters (temperature, pressure, etc.) is actually realized. Now, just as a thermodynamic system in equilibrium can be completely described macroscopically by a few thermodynamic parameters, so a bare black hole in equilibrium (Kerr black hole) can be completely described (insofar as an exterior observer is concerned) by just three parameters: mass, charge, and angular momentum.[1] Black holes in equilibrium having the same set of three parameters may still have different "internal configurations." For example, one black hole may have been formed by the collapse of a normal star, a second by the collapse of a neutron star, a third by the collapse of a geon. These various alternatives may be regarded as different possible internal configurations of one and the same black hole characterized by their (common) mass, charge, and angular momentum. It is then natural to introduce the concept of black-hole entropy as the measure of the *inaccessibility* of information (to an exterior observer) as to which particular internal configuration of the black hole is actually realized in a given case.

At the outset it should be clear that the black-hole entropy we are speaking of is *not* the thermal entropy inside the black hole. In fact, our black-hole entropy refers to the equivalence class of all black holes which have the same mass, charge, and angular momentum, not to one particular black hole. What are we to take as a measure of this black-hole entropy? The discussion of Sec. II predisposes us to single out black-hole area. But to be more general we shall only assume that the entropy of a black hole, S_{bh}, is some *monotonically increasing* function of its rationalized area:

$$S_{bh} = f(\alpha). \tag{12}$$

Although our motivating discussion for the introduction of the concept of the black-hole entropy made use of the specific properties of stationary black holes, we shall take (12) to be valid for any black hole, including a dynamically evolving one, since the surface area is well defined for any black hole. This choice is supported by the following observations.

As mentioned earlier, the entropy of an evolving thermodynamic system increases due to the gradual loss of information which is a consequence of the washing out of the effects of the initial conditions. Now, as a black hole approaches equilibrium, the effects of the initial conditions are also washed out (the black hole loses its hair)[1]; only mass, charge, and angular momentum are left as determinants of the black hole at late times. We would thus expect that the loss of information about initial peculiarities of the hole will be reflected in a gradual increase in S_{bh}. And indeed Eq. (12) predicts just this; by Hawking's theorem S_{bh} increases monotonically as the black hole evolves. This agreement is evidence in favor of the choice (12).

We mentioned earlier that the possibility of causing a decrease in the entropy of a thermodynamic system goes hand in hand with the possibility of obtaining information about its internal configuration. By contrast, an exterior agent cannot acquire any information about the interior configuration of a black hole. The one-way membrane nature of the event horizon prevents him from doing so.[1] Therefore, we do not expect an exterior agent to be able to cause a decrease in the black hole's entropy. Equation (12) is in agreement with this expectation; by Hawking's theorem S_{bh} never decreases. Here we have a new piece of evidence in favor of the choice (12).

One possible choice for f in (12), $f(\alpha) \propto \alpha^{1/2}$, is untenable on various grounds. Consider two black holes which start off very distant from each other. Since they interact weakly we can take the total black-hole entropy to be the sum of the S_{bh} of each black hole. The black holes now fall together, merge, and form a black hole which settles down to equilibrium. In the process no information about the black-hole interior can become available; on the contrary, much information is lost as the final black hole "loses its hair." Thus, we expect the final black-hole entropy to exceed the initial one. By our assumption that $f(\alpha) \propto \alpha^{1/2}$, this implies that the irreducible mass [see (1)] of the final black hole exceeds the sum of irreducible masses of the initial black holes. Now suppose that all three black holes are Schwarzschild ($M = M_{ir}$). We are then confronted with the prediction that the final black-hole mass exceeds the initial one. But this is nonsense since the total black-hole mass can only decrease due to gravitational radiation losses. We thus see that the choice $f(\alpha) \propto \alpha^{1/2}$ is untenable.

The next simplest choice for f is

$$f(\alpha) = \gamma \alpha, \tag{13}$$

where γ is a constant. Repetition of the above argument for this new f leads to the conclusion that the final black-hole area must exceed the total initial black-hole area. But we know this to be true from Hawking's theorem.[7] Thus the choice (13) leads to no contradiction. Therefore, we shall adopt (13) for the moment; later on we shall exhibit some more positive evidence in its favor.

Comparison of (12) and (13) shows that γ must have the units of (length)$^{-2}$. But there is no constant with such units in classical general relativity. If in desperation we appeal to quantum physics we find only one truly universal constant with the correct units[14]: $\hbar^{-1}$, that is, the reciprocal of the Planck length squared. (Compton wavelengths are not universal, but peculiar to this or that particle; they clearly have no bearing on the problem.) We are thus compelled to write (12) as

$$S_{bh} = \eta \hbar^{-1} \alpha, \tag{14}$$

where η is a dimensionless constant which we may expect to be of order unity. This expression was first proposed by us earlier[11,12] from a different point of view.

We need not be alarmed at the appearance of $\hbar$ in the expression for black-hole entropy. It is well known[15] that $\hbar$ also appears in the formulas for entropy of many thermodynamic systems that are conventionally regarded as classical, for example, the Boltzmann ideal gas. This is a reflection of the fact that entropy is, in a sense, a count of states of the system, and the underlying states of any system are always quantum in nature. It is thus not totally unexpected that $\hbar$ appears in (14). These observations also suggest that it would be

somewhat pretentious to attempt to calculate the precise value of the constant $\eta \hbar^{-1}$ without a full understanding of the quantum reality which underlies a "classical" black hole. Since there is no hope at present of obtaining such an understanding, we bypass the issue, and in the next section we use a semiclassical argument to arrive at a value for $\eta \hbar^{-1}$ which should be quite close to the correct one.

IV. EXPRESSION FOR BLACK-HOLE ENTROPY

In our attempt to obtain a value for $\eta \hbar^{-1}$ we shall employ the following argument. We imagine that a particle goes down a Kerr black hole. As it disappears some information is lost with it. According to the discussion of Sec. III we expect the black-hole entropy, as the measure of inaccessible information, to reflect the loss of the information associated with the particle by increasing by an appropriate amount. How much information is lost together with the particle? The amount clearly depends on how much is known about the internal state of the particle, on the precise way in which the particle falls in, etc. But we can be sure that the absolute minimum of information lost is that contained in the answer to the question "does the particle exist or not?" To start with, the answer is known to be yes. But after the particle falls in, one has no information whatever about the answer. This is because from the point of view of this paper, one knows nothing about the physical conditions inside the black hole, and thus one cannot assess the likelihood of the particle continuing to exist or being destroyed. One must, therefore, admit to the loss of one bit of information (see Sec. III) at the very least.

Our plan, therefore, is to compute the minimum possible increase in the black hole's area which results from the disappearance of a particle down the black hole, then to compute the corresponding minimum possible increase of black-hole entropy by means of our original formula (12), and finally to identify this increase in entropy with the loss of one bit of information in accordance with the scheme (11). If our procedure is reasonable we should then recover the functional form of f given by (13), together with a definite value for γ.

There are many ways in which a particle can go down a black hole, all leading to varying increases in black-hole area. We are interested in that method for inserting the particle which results in the smallest increase. This method has already been discussed by Christodoulou[4-6] in connection with his introduction of the concept of irreducible mass. The essence of Christodoulou's method is that if a freely falling point particle is captured by a Kerr black hole from a turning point in its orbit, then the irreducible mass and, consequently, the area of the hole are left unchanged. For reasons that will become clear presently we wish to allow the particle to have a nonzero radius. As shown in Appendix A, Christodoulou's method can be generalized easily so as to allow for this, as well as for the possibility that the particle is brought to the horizon by some method other than by free fall. We find in Appendix A that the increase in area for the generalized Christodoulou process is no longer precisely zero. But interestingly enough, the minimum increase in rationalized area, $(\Delta\alpha)_{min}$, turns out to be independent of the parameters of the black hole. For a spherical particle of rest mass μ, and proper radius b,

$$(\Delta\alpha)_{min} = 2\mu b . \tag{15}$$

For a point particle $(\Delta\alpha)_{min} = 0$; this is Christodoulou's result.

Expression (15) gives the minimum possible increase in black-hole area that results if a given particle is added to a Kerr black hole. We can try to make $(\Delta\alpha)_{min}$ smaller by making b smaller. However, we must remember that b can be no smaller than the particle's Compton wavelength $\hbar\mu^{-1}$, or than its gravitational radius 2μ, whichever is the larger. The Compton wavelength is the larger for $\mu \lesssim 2^{-1/2}\hbar^{1/2}$, and the gravitational radius is the larger for $\mu > 2^{-1/2}\hbar^{1/2}$ ($2^{-1/2}\hbar^{1/2} \approx 10^{-5}g$). Thus, if $\mu \lesssim 2^{-1/2}\hbar^{1/2}$, then $2\mu b$ can be as small as $2\mu\hbar\mu^{-1} = 2\hbar$. But if $\mu > 2^{-1/2}\hbar^{1/2}$, then $2\mu b$ can be no smaller than $4\mu^2 > 2\hbar$. We conclude that quantum effects set a lower bound of $2\hbar$ on the increase of the rationalized area of a Kerr black hole when it captures a particle. Moreover, this limit can be reached only for a particle whose dimension is given by its Compton wavelength. Of course, only such an "elementary particle" can be regarded as having no internal structure. Therefore, the loss of information associated with the loss of such a particle should be minimum. And indeed we find that the increase in black-hole entropy is smallest for just such a particle. This supports our view that $2\hbar$ is the increase in rationalized area associated with the loss of one bit of information.

Following our program we shall equate the minimum increase in black-hole entropy, $(\Delta S_{bh})_{min} = 2\hbar\, df/d\alpha$, with $\ln2$, the entropy increase associated with the loss of one bit of information. Integration of the resulting equation gives $f(\alpha) = (\frac{1}{2}\ln2)\hbar^{-1}\alpha$. Thus, we have arrived again at (13) by an alternate route, and have obtained the value of γ into the bargain. We now have

$$S_{bh} = (\tfrac{1}{2}\ln2)\hbar^{-1}\alpha , \tag{16}$$

which is of the same form as (14). Our argument has determined the dependence of S_{bh} on α in a straightforward manner. However, our value $\eta = \frac{1}{2}\ln2$ might presumably be challenged on the grounds that it follows from the assumption that the smallest possible radius of a particle is precisely equal to its Compton wavelength whereas the actual radius is not so sharply defined. Nevertheless, it should be clear that if η is not exactly $\frac{1}{2}\ln2$, then it must be very close to this, probably within a factor of two. This slight uncertainty in the value of η is the price we pay for not giving our problem a full quantum treatment. However, in what follows we shall suppose that $\eta = \frac{1}{2}\ln2$. Examples to be given later will show that this value leads to no contradictions.

How is the entropy of a system of several black holes defined? It is natural to define it as the sum of individual black-hole entropies. Then Hawking's theorem tells us that the total black-hole entropy of the system cannot decrease. But this is just what we would expect since the information lost down the black holes is unrecoverable. This observation lends support to our choice.

In conventional units (16) takes the form

$$S_{bh} = (\tfrac{1}{2}\ln2/4\pi)\,kc^3\hbar^{-1}G^{-1}A$$

$$= (1.46 \times 10^{48}\ \text{erg}\ {}^\circ\text{K}^{-1}\,\text{cm}^{-2})A , \tag{17}$$

where k is Boltzmann's constant. We see that the entropy of a black hole is enormous. For example, a black hole of one solar mass would have $S_{bh} \approx 10^{60}$ erg ${}^\circ$K^{-1}. By comparison the entropy of the sun is $S \approx 10^{42}$ erg ${}^\circ$K^{-1}; those of a white dwarf or a neutron star of one solar mass even smaller. The large numerical value of black-hole entropy serves to dramatize the highly irreversible character of the process of black-hole formation. We may define a characteristic temperature for a Kerr hole by the relation $T_{bh}^{-1} = (\partial S_{bh}/\partial M)_{L,Q}$ which is the analog of the thermodynamic relation $T^{-1} = (\partial S/\partial E)_V$. By using (8) and (16) we find

$$T_{bh} = 2\hbar\,(\ln2)^{-1}\Theta$$

$$= (0.165\ {}^\circ\text{K cm})\,(r_+ - r_-)\,(r_+^2 + a^2)^{-1} , \tag{18}$$

where $r_\pm$ and a are to be given in centimeters. We introduce this T_{bh} in anticipation of our discussion of an example in Sec. VII. But we emphasize that one should not regard T_{bh} as *the* temperature of the black hole; such an identification can easily lead to all sorts of paradoxes, and is thus not useful.

V. THE GENERALIZED SECOND LAW

Suppose that a body containing some common entropy goes down a black hole. The entropy of the visible universe decreases in the process. It would seem that the second law of thermodynamics is transcended here in the sense that an exterior observer can never verify by direct measurement that the total entropy of the whole universe does not decrease in the process.[19] However, we know that the black-hole area "compensates" for the disappearance of the body by increasing irreversibly. It is thus natural to conjecture that the second law is not really transcended provided that it is expressed in a generalized form: *The common entropy in the black-hole exterior plus the black-hole entropy never decreases.* This statement means that we must regard black-hole entropy as a genuine contribution to the entropy content of the universe.

Support for the above version of the second law comes from the following argument. Suppose that a body carrying entropy S goes down a black hole (which may have existed previously or may be formed by the collapse of the body). The S is the uncertainty in one's knowledge of the internal configuration of the body. So long as the body was still outside the black hole, one had the option of removing this uncertainty by carrying out measurements and obtaining information up to the amount S. But once the body has fallen in, this option is lost; the information about the internal configuration of the body becomes truly inaccessible. We thus expect the black-hole entropy, as the measure of inaccessible information, to increase by an amount S. Actually, the increase in S_{bh} may be even larger because any information that was available about the body to start with will also be lost down the black hole. Therefore, if we denote by ΔS_c the change in common entropy in the black-hole exterior ($\Delta S_c \equiv -S$), then we expect that

$$\Delta S_{bh} + \Delta S_c = \Delta(S_{bh} + S_c) > 0. \qquad (19)$$

This is just the generalized second law which we proposed above: The generalized entropy $S_{bh} + S_c$ never decreases. Examples supporting this law will be given in Sec. VI-VII.

This is a good place to mention the question of fluctuations. We know that the common entropy of a closed thermodynamic system can decrease spontaneously as a result of statistical fluctuations, i.e., the second law, being a statistical law, is meaningful only if statistical fluctuations are small. Is black-hole entropy also subject to decreases of a statistical nature? Not classically – Hawking's theorem guarantees that. Quantum mechanically

there are two ways by which the black-hole entropy can undergo statistical decreases. One of them depends on the quantum fluctuations of the metric of the black hole which one has reasons to expect.[1] Such fluctuations would be reflected in small random fluctuations in the area, and thus in the entropy of the black hole, and some of these fluctuations would be expected to be decreases in entropy. However, even if one regards a black hole as a purely classical object, it is still possible for its area and entropy to undergo small decreases when the black hole absorbs a single quantum under certain conditions.[20] However, the probability of such an event occurring in any given trial is very small. Therefore, the decrease in area and entropy is of a statistical nature, and is quite analogous to the decrease in entropy of a thermodynamic system due to statistical fluctuations. This discussion serves us warning that the law (19) is expected to hold only insofar as statistical fluctuations are negligible.

We noticed earlier (Sec. IV) the very large magnitude of black-hole entropy. In fact, one can say that the black-hole state is the maximum entropy state of a given amount of matter. The point is that in the gravitational collapse of a body into a black hole, the loss of information down the black hole is the maximum allowed by the laws of physics. Thus if the body collapses to form a Kerr black hole, all information about it is lost with the exception of mass, charge, and angular momentum.[1] These quantities are given in terms of Gaussian integrals,[1] and so information about them cannot be lost. But all other information about the body is eventually lost. Therefore, the resulting black hole must correspond to the maximum (generalized) entropy which can be associated with the given body.

VI. EXAMPLES OF THE GENERALIZED SECOND LAW AT WORK

In the examples which follow we endeavor to subject the generalized second law to the most stringent test possible in each case by maximizing the entropy going down the black hole with a given body while minimizing the associated increase in black-hole entropy.

A. Harmonic Oscillator

As a first example we take an harmonic oscillator composed of two particles of rest mass $\frac{1}{2}m$ each connected by a nearly massless spring of spring constant K. We imagine the oscillator to be enclosed in a spherical box and to be maintained at temperature T. We assume for simplicity that conditions are such that the oscillator is nonrelativistic ($T \ll m$). Let ω be the vibrational frequency of the oscillator. Then the (normalized) probability that the oscillator is in its nth quantum state is given by the canonical distribution

$$p_n = (1 - e^{-x})^{-1} e^{-nx} , \quad x = \hbar\omega / T . \tag{20}$$

The entropy of the oscillator as computed from (10) is

$$S = x(e^x - 1)^{-1} - \ln(1 - e^{-x}) , \tag{21}$$

and the mean vibrational energy,

$$\langle E \rangle \equiv \sum p_n (n + \tfrac{1}{2})\hbar\omega,$$

is

$$\langle E \rangle = [(e^x - 1)^{-1} + \tfrac{1}{2}] \hbar\omega . \tag{22}$$

We remark that the thermal distribution (20) maximizes the entropy of the oscillator for given $\langle E \rangle$, and is thus ideally suited to our plan for subjecting the generalized second law to the most stringent test possible.

Suppose that the box goes down a Kerr black hole. The corresponding increase in black-hole entropy cannot be smaller than the lowest limit derived by the method of Appendix A. From (15) and (16) we have $\Delta S_{bh} \geqslant \mu b \hbar^{-1} \ln 2$, where b is the outer radius of the box and μ is its total rest mass. Clearly b must be at least as large as half of the mean value $\langle y \rangle$ of the separation of the two masses y. And $\langle y \rangle$ in turn must clearly be larger than Δy, the root mean square of the thermal oscillation of y [$(\Delta y)^2 \equiv \langle (y - \langle y \rangle) \rangle$], so that y will always be positive. Now according to the (quantum) virial theorem $\tfrac{1}{2}\langle E \rangle$ is equal to the mean potential energy of the oscillator $\tfrac{1}{2}K(\Delta y)^2$. Since the reduced mass of the oscillator is $\tfrac{1}{4}m$, we have $K = \tfrac{1}{4}m\omega^2$. We thus find from all the above that

$$b > \langle E \rangle^{1/2} m^{-1/2} \omega^{-1} .$$

Remembering that $\mu > m + \langle E \rangle$ (because the box itself must have some mass) we obtain

$$\Delta S_{bh} > \langle E \rangle^{1/2} m^{-1/2} (\hbar\omega)^{-1} (m + \langle E \rangle) \ln 2 . \tag{23}$$

We assume that the entropy given by (21) is the only contribution to the entropy in the box. This amounts to neglecting the contribution of the black body radiation in the box, etc., a sensible procedure if T is not very high. Then $\Delta S_c = -S$ and we have

$$\Delta(S_{bh} + S_c) > \xi^{-1/2}(1 + \xi)\left[\tfrac{1}{2} + (e^x - 1)^{-1}\right] \ln 2$$
$$- x(e^x - 1)^{-1} + \ln(1 - e^{-x}) , \tag{24}$$

where we have introduced the notation $\xi \equiv m\langle E \rangle^{-1}$ and used Eqs. (21) and (22) for S and $\langle E \rangle$. We now show that $\Delta(S_{bh} + S_c) > 0$ as required by the generalized second law. The expression in (24) regarded as a function of x for given ξ has a single minimum at

$$x = x_m$$
$$\equiv \xi^{-1/2}(1 + \xi)\ln 2$$

which has the value

$$\tfrac{1}{2} x_m + \ln[1 - \exp(-x_m)] .$$

Our assumption that the oscillator is nonrelativistic means that $\xi \gg 1$, and hence that $x_m \gg 2\ln 2$. Under these conditions the minimum is positive (in fact, it is positive for $\xi \gtrsim 1$). It follows immediately that $\Delta(S_{bh} + S_c)$ is positive for all x and all ξ which are compatible with the requirement of a nonrelativistic oscillator. The generalized second law is obeyed over the entire regime for which our treatment is valid.

B. Beam of Light

As a second example we consider a beam of light which is aimed at a Kerr black hole. This example is particularly interesting because it shall bring us face to face with the issue of fluctuations as a limitation on the applicability of the second law. We shall restrict our attention only to those cases for which geometrical optics is a valid approximation. We shall thus represent the path of the beam by a null geodesic in the Kerr background.

We shall take it that the beam is thermalized at a certain temperature T. This implies that its entropy is a maximum for given energy. The entropy is easily calculated; in fact, the entropy and energy for each mode in the beam are given by the same expressions (21) and (22) which apply to a harmonic oscillator, except that one must omit the zero-point energy term $\tfrac{1}{2}\hbar\omega$. The total entropy S and mean energy $\langle E \rangle$ of the beam are obtained by integrating these expressions weighed by the conventional density of states

$$\rho = 2\omega^2 (2\pi)^{-3} V d\Omega \tag{25}$$

over all ω. In (25) V is the volume of the beam and $d\Omega$ is the solid angle it subtends. Integrating by parts the expression for S, one easily obtains the relation

$$S = \tfrac{4}{3}\langle E \rangle T^{-1} , \qquad (26)$$

which, not surprisingly, is identical to that for radiation inside a black-body cavity of temperature T.[15] [In Ref. 12 (26) was given with an incorrect numerical factor.]

As the beam nears the black hole, it is deflected by the gravitational field. Insofar as its effects on electromagnetic radiation are concerned, a stationary gravitational field can be mocked up by an appropriate nonabsorbing refractive medium in flat spacetime.[21] But the propagation of a beam of light through such a medium is a reversible process.[22] We infer from this that the entropy of the beam will remain unchanged as the beam nears the black hole. Thus the entropy change of the visible universe when the beam goes down the hole is just

$$\Delta S_c = -\tfrac{4}{3}\langle E \rangle T^{-1} . \qquad (27)$$

What is the increase in black-hole entropy associated with the process? From (8) we see that the increase in α is minimized when the angular momentum that the hole gains from the beam is maximized for given $\langle E \rangle$. Now, the gain in angular momentum is limited because the beam will not be captured if it carries too much angular momentum. In Appendix B we take this into account in calculating (in the geometrical optics limit) the minimum possible increase in α for given $\langle E \rangle$ of the beam. We find that $\Delta\alpha \geq \beta M \langle E \rangle$ where β ranges from 8 for the case of a Schwarzschild hole to $4(1-\sqrt{3}/2)$ for the case of an extreme Kerr hole, this last value being the smallest possible β. From (16) it follows that

$$\Delta S_{\rm bh} \geq (\tfrac{1}{2}\beta \ln2) M \hbar^{-1} \langle E \rangle . \qquad (28)$$

Our assumption (Appendix B) that geometrical optics is always applicable means that the bulk of wavelengths in the beam are much shorter than the characteristic dimension of the hole $\approx M$. Thus, if ω_c is some characteristic frequency in the beam, then we require that $\omega_c \gg M^{-1}$. From the form of the Planck spectrum (22) we see that $\hbar\omega_c \approx T$; therefore (27) tells us that

$$|\Delta S_c| \ll \tfrac{4}{3} M \hbar^{-1} \langle E \rangle . \qquad (29)$$

Comparison of (28) and (29) shows that a violation of the generalized second law (19) cannot arise in the regime under consideration.

In the above discussion the condition that geometrical optics be applicable prevented us from taking T to be arbitrarily small. As a result it

turned out to be impossible for $|\Delta S_c|$ to exceed $\Delta S_{\rm bh}$, and so a violation of the second law was ruled out. But there is a way to circumvent the restriction on T. One simply selects the temperature T (arbitrarily) to be as small as one pleases, and arranges for all frequencies $\omega < \omega_c'$ to be filtered out of the beam. Here $\omega_c' \gg M^{-1}$ is a definite frequency unrelated to T. It should be clear that geometrical optics will be a valid approximation for this case also, so that we may take over the result (28). But the result (27) must be modified since we are here dealing with a truncated frequency spectrum. We are mostly interested in the regime $T \ll \hbar \omega_c'$. Then for all frequencies in the beam $x = \hbar\omega/T \gg 1$. It follows from (21) and (22) that for each mode the entropy to energy ratio is T^{-1} ($S \approx x e^{-x}$, $\langle E \rangle \approx \hbar\omega e^{-x}$). Therefore instead of (27) we have

$$\Delta S_c = -\langle E \rangle T^{-1} . \qquad (30)$$

It now appears that if

$$T < T_c \equiv \hbar(\tfrac{1}{2}\beta M \ln2)^{-1} ,$$

then $\Delta S_{\rm bh} + \Delta S_c$ will be negative in contradiction with the generalized second law.

The resolution of the above paradox is that in the regime $T < T_c$ statistical fluctuations are already dominant so that our entire picture of the process is invalid. To verify the importance of fluctuations we calculate the mean number of quanta N in the beam by integrating the mean number of quanta per mode, $(e^x - 1)^{-1}$, weighed by the density of states (25) over all $\omega > \omega_c'$. For $T = T_c$ we get (recall that $\hbar\omega_c'/T_c \gg 1$ by our assumptions)

$$N \approx \frac{V}{M^3} \frac{d\Omega}{4\pi} \, \delta^{-3}(\delta M \omega_c')^2 \exp(-\delta M \omega_c') , \qquad (31)$$

where $\delta \equiv \tfrac{1}{2}\beta \ln2$ ($0.2 \leq \delta \leq 2.8$). It is clear that for any beam aimed at the black hole $d\Omega/4\pi \ll 1$. Recalling that $M\omega_c' \gg 1$ by assumption, we see from (31) that each quantum occupies a mean volume much larger than M^3. But the cross section of the beam must be smaller than $\sim M^2$ if the beam is to go down the black hole. Thus the mean separation between quanta is much larger than M, the characteristic dimension of the black hole. In case $T < T_c$ the above effect is even more accentuated.

We conclude that in the regime $T \lesssim T_c$ for which the second law (19) appears to break down, our description of the process as a continuous beam going down the black hole is invalidated by the large fluctuations in the concentration of energy in the beam (or equivalently, the large fluctuations in the energy of each section of the beam). In this regime $\langle E \rangle$ is no longer a good measure of the actual energy. It appears, therefore, that statistical fluctuations are responsible for the breakdown of the second law in the context in which we have applied it here. But we can demonstrate that the law has not lost all its meaning by adopting a point of view more suitable to the circumstances at hand than the one used above.

We take the point of view that quanta are going down the black hole one at a time, rather than in a continuous stream. Thus we must check the validity of the generalized second law for the infall of each quantum. The analysis of Appendix B still leads to formula (28) for the increase in black-hole entropy except that $\langle E \rangle$ is replaced by $\hbar\omega$, the energy of the quantum. To compute the common entropy going down the black hole we reason as follows. From our point of view a quantum of definite frequency is going down the black hole. Thus we are no longer dealing with the probability distribution (20); instead we shall ascribe probability $\frac{1}{2}$ to each of the two possible polarizations of the quantum. Then according to (10) the entropy associated with the quantum is $\ln 2$. Therefore,

$$\Delta S_{bh} + \Delta S_c \geq (\tfrac{1}{2}\beta \ln 2)M\omega - \ln 2. \qquad (32)$$

Since $\frac{1}{2}\beta \geq 0.268$, and since we are assuming that $M\omega \gg 1$, we see that $\Delta S_{bh} + \Delta S_c$ is in fact positive: The generalized second law is upheld for the infall of each quantum.

VII. A PERPETUAL MOTION MACHINE USING A BLACK HOLE?

Geroch[13] has described a procedure using a black hole which appears to violate the second law of thermodynamics by converting heat into work with unit efficiency. He envisages a box filled with black-body radiation which is slowly lowered by means of a string from far away down to the horizon of a black hole, at which point its energy as measured from infinity vanishes. Therefore, if the box's rest mass is μ, then the agent lowering the string obtains work equal to μ out of the process. The box is then allowed to emit into the black-hole radiation of (proper) energy $\Delta\mu$. Finally, the agent retrieves the box; since its rest mass is now $\mu - \Delta\mu$, he must do work $\mu - \Delta\mu$ to accomplish this. Therefore, in the whole process the agent obtains net work $\Delta\mu$ at the expense of heat $\Delta\mu$ — conversion with unit efficiency. We shall now show that, in fact, due to fundamental physical limitations, the efficiency of the Geroch process is slightly smaller than unity, so that no violation of the second law is entailed here.

The box under consideration must have a nonzero radius (see below). Because of this its energy as measured from infinity is never quite zero when it is as close to the horizon as it can possibly be. We shall assume that the box is in the shape of a sphere of radius b. Then according to the analysis of Appendix C the minimum value of the energy is

$$E = 2\mu b\Theta , \qquad (33)$$

where Θ is defined by (9a). It follows that in lowering the box from infinity to the horizon, the agent obtains only work $\mu(1 - 2b\Theta)$ rather than μ. After the box has radiated into the black hole, its rest mass becomes $\mu - \Delta\mu$ and according to (33) its energy at the horizon is just $2(\mu - \Delta\mu)b\Theta$. Thus the agent must do work $(\mu - \Delta\mu)(1 - 2b\Theta)$ to retrieve the box to infinity where its energy is $\mu - \Delta\mu$. Therefore, in the over-all process the agent obtains net work $\Delta\mu(1 - 2b\Theta)$ in exchange for the expenditure of heat $\Delta\mu$. The efficiency of conversion is

$$\epsilon = 1 - 2b\Theta , \qquad (34)$$

which is smaller than unity. In practical situations $b \ll r_+$ so that $b\Theta \ll 1$ and the efficiency can be quite near to unity. But the departure of ϵ from unity, albeit small, serves to resolve the problem raised by Geroch's example: There is no violation of the Kelvin statement of the second law.[23]

We must now explain why b cannot be arbitrarily small. Physically the reason is that the box must be large enough for the wavelengths characteristic of radiation of some temperature T to fit into it. More formally we can argue as follows. The frequency of the photon ground state associated with the box, ω_0, cannot exceed that frequency ω_p at which the Planck photon-number spectrum

$$\propto \omega^2 [\exp(\hbar\omega/T) - 1]^{-1}$$

peaks. Otherwise the frequencies of all photon states would lie in the exponential tail of the spectrum, the occupation number of each state would be small, and the resulting large fluctuations would make the concept of temperature meaningless. We have the conventional relation $\omega_0 b' = \pi$, where b' is the interior radius of the box $(b' < b)$, and we easily find that $\hbar\omega_p < 2T$. Therefore, $\omega_0 < \omega_p$ implies that $b > \pi\hbar/2T$. It is thus clear that there is a lower limit for b.

We may write the efficiency (34) in a more transparent form by recalling that $\Theta = \frac{1}{2} T_{bh} \ln 2/\hbar$ [see (18)], where T_{bh} is the characteristic temperature associated with the black hole. Since $b > \pi\hbar/2T$ we find that

$$\epsilon < 1 - T_{bh}/T . \tag{35}$$

We now recall that the efficiency of a heat engine operating between two reservoirs, one at temperature T and the second at temperature $T_{bh} < T$, is restricted by $\epsilon \leqslant 1 - T_{bh}/T$. We thus see that the Geroch process is no more efficient than its "equivalent reversible heat engine." This observation makes it evident again that Geroch's process is not in violation of the second law of thermodynamics. Finally, we wish to remark that since our primary formula (33) is valid only when the box is small compared to the black hole ($b\Theta \ll 1$), we can vouch for the validity of (35) only when $T_{bh} \ll T$. However, due to the smallness of T_{bh} this condition will be satisfied in all cases of practical interest.

We now verify that the Geroch process is in accord with the generalized second law (19). We mentioned earlier that the agent obtains work $\Delta\mu(1 - 2b\Theta)$ for a decrease $\Delta\mu$ in the rest mass of the box. This means that the black hole's mass must increase by $2\Delta\mu b\Theta$ in the complete process. According to (8) and (16) the corresponding increase in black hole entropy is $\Delta S_{bh} = \Delta\mu b\hbar^{-1}\ln 2$ (angular momentum is not added to the hole; see Appendix C). But since $b > \pi\hbar/2T$ we have that

$$\Delta S_{bh} > (\tfrac{1}{2}\pi \ln 2)\Delta\mu/T . \tag{36}$$

On the other hand, the decrease in entropy of the box is clearly $\Delta\mu/T$ (heat/temperature). Thus

$$\Delta S_c = -\Delta\mu/T . \tag{37}$$

From (36) and (37) it follows that $\Delta(S_{bh} + S_c) > 0$ as required by the generalized second law.

ACKNOWLEDGMENTS

The author is grateful to Professor J. A. Wheeler for his interest in this work, his encouragement and his penetrating criticism. He also thanks Professor Karel Kuchař and Professor Brandon Carter, and Dr. Bahram Mashhoon for valuable suggestions and comments.

APPENDIX A

Here we shall calculate the minimum possible increase in black-hole area which must result when a spherical particle of rest mass μ and proper radius b is captured by a Kerr black hole. We are interested in the increase in area ascribable to the particle itself, as contrasted with any increase incidental to the process of bringing the particle to black-hole horizon. For example, there is some *circumstantial* evidence for believing that when the particle is lowered into the black hole by a string, there occurs an increase in black-hole area even as the particle is being lowered.[11] Furthermore, the area will experience an additional increase due to the gravitational waves radiated into the black hole by the string as it relaxes when the particle is dropped.[11] Similarly, if the particle falls freely to the horizon it emits gravitational waves into the hole even before it falls in; the amount of radiation may even be significant.[24] This radiation will also result in an increase in area. Here we shall ignore all these incidental effects and concentrate on the increase in area caused by the particle all by itself.

We assume that the particle is neutral so that it follows a geodesic of the Kerr geometry when falling freely. We shall employ Boyer-Lindquist coordinates for the charged Kerr metric[25]

$$ds^2 = g_{tt}dt^2 + 2g_{t\phi}dt\,d\phi + g_{\phi\phi}d\phi^2 + g_{rr}dr^2 + g_{\theta\theta}d\theta^2 . \tag{A1}$$

For later reference we give g_{rr}:

$$g_{rr} = (r^2 + a^2\cos^2\theta)\Delta^{-1} , \tag{A2}$$

where

$$\Delta \equiv r^2 - 2Mr + a^2 + Q^2 . \tag{A3}$$

The event horizon is located at $r = r_+$ where $r_\pm$ are defined by (7). We have

$$\Delta = (r - r_-)(r - r_+) . \tag{A4}$$

First integrals for geodesic motion in the Kerr background have been given by Carter.[25] Christodoulou[5] uses the first integral

$$E^2[r^4 + a^2(r^2 + 2Mr - Q^2)] - 2E(2Mr - Q^2)ap_\phi$$

$$-(r^2 - 2Mr + Q^2)p_\phi^2 - (\mu^2 r^2 + q)\Delta = (p_r\Delta)^2 \tag{A5}$$

as a starting point of his analysis. In (A5) $E = -p_t$ is the conserved energy, p_ϕ is the conserved component of angular momentum in the direction of the axis of symmetry, q is Carter's fourth constant of the motion,[25] μ is the rest mass of the particle, and p_r is its covariant radial momentum. Following Christodoulou we solve (A5) for E:

$$E = Bap_\phi + \{[B^2 a^2 + A^{-1}(r^2 - 2Mr + Q^2)]p_\phi{}^2$$

$$+ A^{-1}[(\mu^2 r^2 + q)\Delta + (p_r \Delta)^2]\}^{1/2},$$

$$\tag{A6}$$

where

$$A \equiv r^4 + a^2(r^2 + 2Mr - Q^2), \tag{A7}$$

$$B \equiv (2Mr - Q^2)A^{-1}. \tag{A8}$$

At the event horizon $\Delta = 0$ [see (A4)] so that there

$$A = A_+ = (r_+{}^2 + a^2)^2,$$

$$\tag{A9}$$

$$B = B_+ = (r_+{}^2 + a^2)^{-1}.$$

Furthermore, at the horizon $Ba = \Omega$ [see (9b)], and the coefficients of $p_\phi{}^2$ and $\mu^2 r^2 + q$ in (A6) vanish. However,

$$p_r \Delta = (r^2 + a^2 \cos^2\theta)p^r$$

does not vanish at the horizon in general. If the particle's orbit intersects the horizon, then we have from (A6) that

$$E = \Omega p_\phi + A_+{}^{-1/2} |p_r \Delta|_+.$$

As a result of the capture, the black hole's mass increases by E and its component of angular momentum in the direction of the symmetry axis increases by p_ϕ. Therefore, according to (8) the black hole's rationalized area will increase by $\Theta^{-1} A_+{}^{-1/2} |p_r \Delta|_+$. As pointed out by Christodoulou this increase vanishes *only if* the particle is captured from a turning point in its orbit in which case $|p_r \Delta|_+ = 0$. In this case we have

$$E = \Omega p_\phi. \tag{A10}$$

The above analysis shows that it is possible for a black hole to capture a *point* particle without increasing its area. How is this conclusion changed if the particle has a nonzero proper radius b? First we note that regardless of the manner in which the particle arrives at the horizon (being lowered by a string, splitting off from a second particle which then escapes, etc.), it must clearly acquire its parameters E, p_ϕ, and q while every part of it is still outside the horizon, i.e., while it is not yet part of the black hole. Moreover, as the particle is captured, it must already be detached from whatever system brought it to the horizon, so that it may be regarded as falling freely. Therefore, Eq. (A6) should always de-

scribe the motion of the particle's center of mass at the moment of capture.

It should be clear that to generalize Christodoulou's result to the present case one should evaluate (A6) not at $r = r_+$, but at $r = r_+ + \delta$, where δ is determined by

$$\int_{r_+}^{r_+ + \delta} (g_{rr})^{1/2} dr = b$$

($r = r_+ + \delta$ is a point a proper distance b outside the horizon). Using (A2) we find

$$b = 2\delta^{1/2}(r_+{}^2 + a^2 \cos^2\theta)^{1/2}(r_+ - r_-)^{-1/2}. \tag{A11}$$

To obtain this we have assumed that $r_+ - r_- \gg \delta$ (black hole not nearly extreme). Expanding the argument of the square root in (A6) in powers of δ, replacing δ by its value given by (A11), and keeping only terms to $O(b)$ we get

$$E = \Omega p_\phi + [(r_+{}^2 - a^2)(r_+{}^2 + a^2)^{-1}p_\phi{}^2 + \mu^2 r_+{}^2 + q]^{1/2}$$

$$\times \tfrac{1}{2} b (r_+ - r_-)(r_+{}^2 + a^2)^{-1}(r_+{}^2 + a^2 \cos^2\theta)^{-1/2}$$

$$\tag{A12}$$

Here we have already assumed that the particle reaches a turning point as it is captured since we know that this minimizes the increase in black-hole area. Equation (A12) is the generalization to $O(b)$ of the Christodoulou condition (A10).

What is q in (A12)? We can obtain a lower bound for it as follows. From the requirement that the θ momentum p_θ be real it follows that [25]

$$q \geq \cos^2\theta [a^2(\mu^2 - E^2) + p_\phi{}^2/\sin^2\theta]; \tag{A13}$$

the equality holds when $p_\theta = 0$ at the point in question. If we replace E in (A13) by Ωp_ϕ [see (A12)] we obtain

$$q \geq \cos^2\theta [a^2\mu^2 + p_\phi{}^2(1/\sin^2\theta - a^2\Omega^2)],$$

which is correct to zeroth order in b. We know that $1/\sin^2\theta \geq 1$; it is easily shown that $a^2\Omega^2 \leq \tfrac{1}{4}$ for a charged Kerr black hole. Therefore $q \geq a^2\mu^2 \cos^2\theta$. Substituting this into (A12) we find

$$E \geq \Omega p_\phi + \tfrac{1}{2}\mu b(r_+ - r_-)(r_+{}^2 + a^2)^{-1} \tag{A14}$$

which is correct to $O(b)$. By retracing our steps we see that the equality sign in (A14) corresponds to the case $p_\phi = p_\theta = p^r = 0$ at the point of capture. The increase in black-hole area, computed by means of (8), (9a), and (A14), is

$$\Delta\alpha \geq 2\mu b \ . \tag{A15}$$

This gives the fundamental lower bound on the increase in black-hole area. We note that it is independent of M, Q, and L.

APPENDIX B

Here we shall calculate the minimum possible increase in black-hole area which must result when a light beam of energy $E > 0$ coming from infinity is captured by a Kerr black hole. If the black hole is nonrotating the increase is simply obtained by setting $dM = E$ in (8):

$$\Delta\alpha = 8ME \quad \text{for} \quad a = 0. \tag{B1}$$

If the black hole is rotating, $\Delta\alpha$ can be minimized by maximizing the angular momentum p_ϕ which is brought in by the beam [see (8)]. To accomplish this we consider the effective potential V for the motion of a massless particle in a Kerr background.

This V is just the value of E given by (A6) regarded as a function of r for $\mu = 0$ and $p_r = 0$ (E equals V at a turning point). This potential starts off at a value Ωp_ϕ at $r = r_+$ (see Appendix A), increases with r, reaches a maximum, and then falls off to zero as $r \to \infty$. For the beam to be captured by the hole it is necessary that p_ϕ be small enough for the peak of the potential barrier to be smaller than E of the beam. The optimum case we seek corresponds to the peak being just equal to E so that p_ϕ has its largest possible value.

It is clear that we must take q in (A6) as small as possible in order to have the lowest possible potential peak for given p_ϕ. Let us first take $q < 0$. Then according to Carter[25] there are solutions to the geodesic equation only if $|p_\phi| < aE$. From (8) it follows that

$$\Delta\alpha = \Theta^{-1}(E - \Omega p_\phi)$$

$$> \Theta^{-1}E(1 - \Omega a).$$

But since $\Omega a \leq \tfrac{1}{2}$ and $\Theta^{-1} \geq 8M$ it follows that

$$\Delta\alpha > 4ME \quad \text{for} \quad q < 0. \tag{B2}$$

Next we take $q = 0$. Two cases are possible[25]: Either $|p_\phi| < aE$ as above so that (B2) is again applicable, or else the orbit is purely equatorial. In the second case one may calculate the peak of the barrier numerically and then find the optimum increase in α. It turns out that $(\Delta\alpha)_{min}$ decreases

monotonically with a for fixed M. The limit of $(\Delta\alpha)_{min}$ as $a \to M$ may be computed analytically because in this limit one can find the height of the potential analytically with sufficient accuracy. One finds

$$(\Delta\alpha)_{min} \to 4(1 - \tfrac{1}{2}\sqrt{3})ME \quad \text{as} \quad a \to M. \tag{B3}$$

It is clear that for $q > 0$ the potential peak will be higher and the increase in area will be larger than the one given by (B3). Thus we find that the minimum increase in area results when the beam is captured by an extreme Kerr black hole from a purely equatorial orbit.

APPENDIX C

Here we compute the value of the energy (as measured from infinity) of a particle of rest mass μ and proper radius b which is hanging from a string just outside the horizon of a Kerr black hole. It is clear that the particle will not be moving in the r or θ directions; hence $p^r = p_\theta = 0$ for it. We cannot claim that the particle does not move in the ϕ direction. In fact, since it will be within the ergosphere in general, it cannot avoid moving in the ϕ direction.[1] Our intuitive notion that the particle is "not moving" must be applied only in a locally nonrotating (Bardeen) frame.[26] The particle is at rest in such a frame if for it

$$\frac{d\phi}{dt} = -\frac{g_{t\phi}}{g_{\phi\phi}}.$$

It follows that

$$p_\phi = \mu\left(g_{t\phi}\frac{dt}{d\tau} + g_{\phi\phi}\frac{d\phi}{d\tau}\right) = 0.$$

If the particle were to be dropped, it would clearly keep its energy E and it would still have $p_\phi = p^r = p_\theta = 0$, at least momentarily. We may thus compute E for the particle hanging in the string at a proper distance b from the horizon by setting $p_\phi = 0$ in (A12). For q we take the value given by (A13) with the equality sign ($p_\theta = 0$) $p_\phi = 0$, and $E = 0$ [since for $p_\phi = 0$, E is of $O(b)$]. Thus

$$q = \mu^2 a^2 \cos^2\theta$$

and

$$E = \tfrac{1}{2}\mu b(r_+ - r_-)(r_+^2 + a^2)^{-1}$$

$$= 2\mu b\Theta \ . \tag{C1}$$

*Based in part on the author's Ph.D. thesis, Princeton University, 1972; work supported in part by the National Science Foundation Grants No. GP 30799X to Princeton University and No. GP 32039 to the University of Texas at Austin.

†National Science Foundation Predoctoral Fellow when this work was initiated.

‡Present address.

[1]C. W. Misner, K. S. Thorne, and J. A. Wheeler, *Gravitation* (Freeman, San Francisco, 1973).

[2]R. Penrose and R. M. Floyd, Nature <u>229</u>, 177 (1971).

[3]R. Penrose, Riv. Nuovo Cimento <u>1</u>, 252 (1969).

[4]D. Christodoulou, Phys. Rev. Letters <u>25</u>, 1596 (1970).

[5]D.Christodoulou, Ph.D. thesis, Princeton University, 1971 (unpublished).

[6]D. Christodoulou and R. Ruffini, Phys. Rev. D <u>4</u>, 3552 (1971).

[7]S. W. Hawking, Phys. Rev. Letters <u>26</u>, 1344 (1971); contribution to *Black Holes*, edited by B. DeWitt and C. DeWitt (Gordon and Breach, New York, 1973).

[8]J. M. Greif, Junior thesis, Princeton University, 1969 (unpublished).

[9]B. Carter, Nature <u>238</u>, 71 (1972).

[10]Ya. B. Zel'dovich and I. D. Novikov, *Stars and Relativity* (University of Chicago Press, Chicago, 1971), p. 268.

[11]J. D. Bekenstein, Ph.D. thesis, Princeton University, 1972 (unpublished).

[12]J. D. Bekenstein, Lett. Nuovo Cimento <u>4</u>, 737 (1972).

[13]R. Geroch, Colloquium at Princeton University, December 1971.

[14]We use units with $G = c = 1$ unless otherwise specified.

[15]L. D. Landau and E. M. Lifshitz, *Statistical Physics* (Addison-Wesley, Reading, Mass., 1969).

[16]L. D. Landau and E. M. Lifshitz, *Electrodynamics of Continuous Media* (Addison-Wesley, Reading, Mass., 1960), p. 5.

[17]The mathematical definition of information was first given by C. E. Shannon; the relevant papers are re-printed in C. E. Shannon and W. Weaver, *The Mathematical Theory of Communications* (University of Illinois Press, Urbana, 1949). An elementary introduction to information theory is given by M. Tribus and E. C. McIrvine, Sci. Amer. <u>225</u> (No. 3), 179 (1971). The derivation of statistical mechanics from information theory was first carried out by E. T. Jaynes, Phys. Rev. <u>106</u>, 620 (1957); <u>108</u>, 171 (1957).

[18]The relation between information theory and thermodynamics is discussed in detail by L. Brillouin, *Science and Information Theory* (Academic, New York, 1956), especially Chaps. 1, 9, 12—14.

[19]This was first pointed out to the author by J. A. Wheeler (private communication).

[20]J. D. Bekenstein, Phys. Rev. D <u>7</u>, 949 (1973).

[21]See for example, A. M. Volkov, A. A. Izmest'ev, and G. V. Skrotskii, Zh. Eksp. Teor. Fiz. <u>59</u>, 1254 (1970) [Sov. Phys. JETP <u>32</u>, 686 (1971)].

[22]See E. Hisdal, Phys. Norv. <u>5</u>, 1 (1971), and references cited therein.

[23]In Ref. 12 we gave an alternate resolution of the paradox posed by Geroch based on the apparent tendency of the black-hole area to increase as the box is being lowered.[11] The present approach is preferable in that it is independent of the validity of the interpretation given in Ref. 11, and in that it fits very well into the thermodynamic approach of this paper as will be evident presently.

[24]See M. Davis, R. Ruffini, and J. Tiomno, Phys. Rev. D <u>5</u>, 2932 (1972) for the radiation of a particle falling radially into a Schwarzschild black hole. It is not clear whether the large amount of radiation found to go down the black hole is a device of the linearized approximation, or whether the effect will persist for other types of orbits or for Kerr black holes. Therefore we do not base any of our arguments here on this effect as we did in Refs. 11 and 12.

[25]B. Carter, Phys. Rev. <u>174</u>, 1559 (1968).

[26]J. M. Bardeen, Astrophys. J. <u>161</u>, 103 (1970).

Relativistic gravitation theory for the MOND paradigm

Jacob D. Bekenstein*

Racah Institute of Physics, Hebrew University of Jerusalem
Givat Ram, Jerusalem 91904 ISRAEL

(Dated: February 2, 2008)

The modified newtonian dynamics (MOND) paradigm of Milgrom can boast of a number of successful predictions regarding galactic dynamics; these are made without the assumption that dark matter plays a significant role. MOND requires gravitation to depart from Newtonian theory in the extragalactic regime where dynamical accelerations are small. So far relativistic gravitation theories proposed to underpin MOND have either clashed with the post-Newtonian tests of general relativity, or failed to provide significant gravitational lensing, or violated hallowed principles by exhibiting superluminal scalar waves or an *a priori* vector field. We develop a relativistic MOND inspired theory which resolves these problems. In it gravitation is mediated by metric, a scalar field and a 4-vector field, all three dynamical. For a simple choice of its free function, the theory has a Newtonian limit for nonrelativistic dynamics with significant acceleration, but a MOND limit when accelerations are small. We calculate the β and γ PPN coefficients showing them to agree with solar system measurements. The gravitational light deflection by nonrelativistic systems is governed by the same potential responsible for dynamics of particles. To the extent that MOND successfully describes dynamics of a system, the new theory's predictions for lensing by that system's visible matter will agree as well with observations as general relativity's predictions made with a dynamically successful dark halo model. Cosmological models based on the theory are quite similar to those based on general relativity; they predict slow evolution of the scalar field. For a range of initial conditions, this last result makes it easy to rule out superluminal propagation of metric, scalar and vector waves.

PACS numbers: 95.35.+d,95.30.Sf, 98.62.Sb, 04.80.Cc

I. INTRODUCTION

In the extragalactic regime, where Newtonian gravitational theory would have been expected to be an excellent description, accelerations of stars and gas, as estimated from Doppler velocities and geometric considerations, are as a rule much larger than those due to the Newtonian field generated by the visible matter in the system [1, 2]. This is the "missing mass" problem [3] or "acceleration discrepancy" [4]. It is fashionable to infer from it the existence of much dark matter in systems ranging from dwarf spheroidal galaxies with masses $\sim 10^6 M_\odot$ to great clusters of galaxies in the $10^{13} M_\odot$ regime [3, 5]. And again, galaxies and clusters of galaxies are found to gravitationally lense background sources. When interpreted within general relativity (GR), this lensing is anomalously large unless one assumes the presence of dark matter in quantities and with distribution similar to those required to explain the accelerations of stars and gas. Thus extragalactic lensing has naturally been regarded as confirming the presence of the dark matter suggested by the dynamics.

But the putative dark matter has never been identified despite much experimental and observational effort [6]. This raises the possibility that the acceleration discrepancy as well as the gravitational lensing anomaly may reflect departures from Newtonian gravity and GR on galactic and larger scales. Now alternatives to GR are traditionally required to possess a Newtonian limit for small velocities and potentials; thus the acceleration discrepancy also raises the possibility that the correct relativistic gravitational theory may be of a kind not generally considered hitherto.

In the last two decades Milgrom's modified Newtonian dynamics (MOND) paradigm [7, 8, 9] has gained recognition as a successful scheme for unifying much of extragalactic dynamics phenomenology without invoking "dark matter". In contrast with earlier suggested modifications of Newton's law of universal gravitation [10, 11, 12, 13], MOND is characterized by an acceleration scale $\mathfrak{a}_0$, not a distance scale, and its departure from Newtonian predictions is acceleration dependent:

$$\tilde{\mu}(|\mathbf{a}|/\mathfrak{a}_0)\mathbf{a} = -\boldsymbol{\nabla}\Phi_{\mathrm{N}}. \tag{1}$$

Here Φ_{N} is the usual Newtonian potential of the visible matter, while $\tilde{\mu}(x) \approx x$ for $x \ll 1$ and $\tilde{\mu}(x) \to 1$ for $x \gg 1$.

---

*Electronic address: bekenste@vms.huji.ac.il; URL: http://www.phys.huji.ac.il/~bekenste/

Milgrom estimated $\mathfrak{a}_0 \approx 1 \times 10^{-8}$ cm s^{-2} from the empirical data. In the laboratory and the solar system where accelerations are strong compared to $\mathfrak{a}_0$, formula (1) goes over to the Newtonian law $\mathbf{a} = -\boldsymbol{\nabla}\Phi_N$.

Milgrom constructed formula (1) to agree with the fact that rotation curves of disk galaxies become flat outside their central parts. That far out a galaxy of mass M exhibits an approximately spherical Newtonian potential. The scales are such that $|\boldsymbol{\nabla}\Phi_N| \approx GMr^{-2} \ll \mathfrak{a}_0$ in this region, and so Eq. (1) with $\tilde{\mu}(x) \approx x$ gives $|\mathbf{a}| \approx (GM\mathfrak{a}_0)^{1/2}r^{-1}$ which has the r dependence appropriate to the centripetal acceleration v_c^2/r of a radius independent rotational velocity v_c—an asymptotically flat rotation curve. In fact one obtains the relation $M = (G\mathfrak{a}_0)^{-1}v_c^4$ which leads to the *prediction* that for any class of galaxies with a constant mass to luminosity ratio Υ in a specified spectral band, the luminosity in that band should scale as v_c^4. And indeed, there exists an empirical law of just this form: the Tully–Fisher law [14] (TFL) relating near infrared (H–band) luminosity L_H of a spiral disk galaxy to its rotation velocity, $L_H \propto v_c^4$, with the proportionality factor being constant within each galactic morphology class.

This version of the TFL was established only after MOND was enunciated [15]. It is in harmony with the MOND prediction in two ways. First, the infrared light of a galaxy comes mostly from cool dwarf stars which make up most of its mass (hence giving a tight correlation between M of the predicted relation and L_H of the empirical law). Second, the proportionality coefficient varies from class to class as would be expected from the observed correlation between Υ of a galaxy and its morphology.

In the alternative dark matter paradigm (which casts no doubt on standard gravity theory), flat rotation curves are explained by assuming that every disk galaxy is nested inside a roundish spherical halo of dark matter [16] whose mass density drops approximately like r^{-2}. The halo is supposed to dominate the gravitational field in the outer parts of the galaxy. This makes the Newtonian potential approximately logarithmic with radius in those regions, thereby leading to an asymptotically flat rotation curve. In practice the dark halo resolution works only after fine tuning. It is an observational fact that for bright spiral galaxies the rotation curve in the optically bright region is well explained in Newtonian gravity by the observed matter [17]. But, as mentioned, in the outer regions the visible matter's contribution must be dwarfed by the halo's. So fine tuning is needed between the dark halo parameters (velocity dispersion and core radius) and the visible disk ones [18, 19].

This fine tuning problem is exacerbated by the TFL $L_H \propto v_c^4$. Because the infrared luminosity comes from the visible matter in the galaxy, but the rotation velocity is mostly set by the halo, the TFL also requires fine tuning between halo and disk parameters. The standard dark matter explanation of the r^{-2} profile of an halo is that it arises naturally from primordial cosmological perturbations [20]. The visible galaxy is regarded as forming by dissipational collapse of gas into the potential well of the halo. The fine tuning mentioned is then viewed as resulting from the adjustment of the halo to the gravitation of the incipient disk [19, 21]. But the TFL is observationally a very sharp correlation; in fact, it is the basis for one of the most reliable methods for gauging distances to spiral galaxies. Such sharpness is hardly to be expected from statistical processes of the kind envisaged in galaxy formation, a point emphasized by Sanders [22]. So in the dark matter picture the TFL is something of a mystery.

There are other MOND successes. Milgrom predicted early that in galaxies with surface mass density well below $\mathfrak{a}_0 G^{-1}$, the acceleration discrepancy should be especially large [8]. In dwarf spirals this property was established empirically years later [23], and it is now known to be exhibited by a large number of low surface brightness galaxies [24]. Another example: MOND successfully predicts the detailed shape of a rotation curve from the observed matter (stars and gas) distribution on the basis of a single free parameter, Υ, down to correlating features in the velocity field with those seen in the light distribution [25, 26, 27, 28]. This is especially true in the case of low surface mass density disk galaxies for which MOND's predictions are independent of the specific choice of $\tilde{\mu}(x)$ [29], and these MOND theoretical rotation curves fit the observed curves of a number of low surface brightness dwarf galaxies [27, 30, 31] very well. By contrast, the dark halo paradigm requires one or two free parameters apart from Υ to approximate the success of the MOND predictions [32]. In fact, even when the empirical data is analyzed within the dark halo paradigm, it displays the preferred acceleration scale $\mathfrak{a}_0$ of MOND [33].

Occasionally doubt has been cast on MOND's ability to describe clusters of galaxies properly [34]. Many of these exhibit accelerations not small on scale $\mathfrak{a}_0$, yet conventional analysis suggests they contain much dark matter in opposition to what MOND would suggest. Sanders has recently reanalyzed the problem [35] with the conclusion that these clusters may contain much as yet undiscovered baryonic matter in the core which should be classed as "visible" in connection with MOND. Other MOND successes, outside the province of disk galaxies, have been reviewed elsewhere [22, 32, 36].

So the simple MOND formula (1) is very successful. But it is not a theory. Literally taken the MOND recipe for acceleration violates the conservation of momentum (and of energy and of angular momentum) [7]. And MOND entails a paradox: why does the center of mass of a star orbit in its galaxy with anomalously large acceleration given by Eq. (1) with $\tilde{\mu} \ll 1$, while each parcel of gas composing it is subject to such high acceleration that is should, by the same formula, be accelerated Newtonially ? [7]. In short, the MOND formula is not a consistent theoretical scheme. Neither is MOND, as initially stated, complete. For example, it does not specify how to calculate gravitational lensing by galaxies and clusters of galaxies. As is well known, in standard gravity theory light deflection is well described only by relativistic theory (GR). And whereas Newtonian cosmological models work well for part of the cosmological evolution, MOND cosmological models built in analogy with their Newtonian counterparts, though sometimes agreeing with phenomenology [34], can yield peculiar predictions [37] (but see Ref. 38). In short, a complete, consistent theoretical underpinning of the MOND paradigm which accords with observed facts, and is also relativistic, has been lacking.

This lack is being resolved in measured steps. A first step was the lagrangian reformulation of MOND [39] called AQUAL (see Sec. II A). AQUAL cures the nonconservation problems and resolves the paradox of the galactic motion of an object whose parts accelerate strongly relative to one another; it does so in accordance with a conjecture of Milgrom [7]. And for systems with high symmetry AQUAL reduces exactly to the MOND formula (1).

A relativistic generalization of AQUAL is easy to construct with help of a scalar field which together with the metric describes gravity [39] (see Sec. II C 1 below). It reduces to MOND approximately in the weak acceleration regime, to Newtonian gravity for strong accelerations, and can be made consistent with the post-Newtonian solar system tests for GR. But relativistic AQUAL is acausal: waves of the scalar field can propagate superluminally in the MOND regime (see the appendix of Ref. 39 or Appendix A here). The problem can be traced to the aquadratic kinetic part of the lagrangian of the theory which mimics that in the original AQUAL. A theory involving a second scalar field, PCG, was thus developed to bypass the problem [4, 40, 41] (see Sec. II C 2 below). PCG may be better behaved causally than relativistic AQUAL [42], but it brings woes of its own. It is marginally in conflict with the observed perihelion precession of Mercury [4], and in common with relativistic AQUAL, PCG predicts extragalactic gravitational lensing which is too weak if there is indeed no dark matter. This last problem is traceable to a feature common to PCG and relativistic AQUAL: the physical metric is conformal to the metric appearing in the Einstein-Hilbert action [43].

One way to sidestep this problem without discarding the MOND features is to exploit the direction defined by the gradient of the first scalar field to relate the physical metric to the Einstein metric by a disformal transformation (see Ref. [43] or Sec. II C 3 below). But it turns out that with this relation the requirement of causal propagation acts to depress gravitational lensing [44], rather than enhancing it as is observationally required. The persistence of the lensing problem in modified gravitational theories has engendered a folk theorem to the effect that it is impossible for a relativistic theory to simultaneously incorporate the MOND dynamics, observed gravitational lensing and correct post-Newtonian behavior without calling on dark matter [45, 46, 47, 48].

Needless to say, this theorem cannot be proved [49]. Indeed, by the simple device of relating the physical and Einstein metrics via a disformal transformation based on a *constant* time directed 4-vector, Sanders [50] has constructed an AQUAL like "stratified" relativistic theory which gives the correct lensing while ostensibly retaining the MOND phenomenology and consistency with the post-Newtonian tests. Admittedly Sanders' stratified theory is a preferred frame theory, and as such outside the traditional framework for gravitational theories. But it does point out a trail to further progress.

The present paper introduces TeVeS, a new relativistic gravitational theory devoid of *a priori* fields, whose nonrelativistic weak acceleration limit accords with MOND while its nonrelativistic strong acceleration regime is Newtonian. TeVeS is based on a metric, and dynamic scalar and 4-vector fields (one each); it naturally involves one free function, a length scale, and two positive dimensionless parameters, k and K. TeVeS passes the usual solar system tests of GR, predicts gravitational lensing in agreement with the observations (without requiring dark matter), does not exhibit superluminal propagation, and provides a specific formalism for constructing cosmological models.

In Sec. II we summarize the foundations on which a workable relativistic formulation of MOND must stand. We follow this with a brief critical review of relativistic AQUAL, PCG and disformal metric theories, some of whose elements we borrow. Sec. III A builds the action for TeVeS while Sec. III B derives the equations for the metric, scalar and vector fields. In Sec. III C we demonstrate that TeVeS has a GR limit for a range of small k and K. This is shown explicitly for cosmology (Sec. III C 1) and for quasistatic situations like galaxies (Sec. III C 2). All the above applies for any choice of the free function; in Sec. III E we make a simple choice for it which facilitates further elaboration. For spherically symmetric systems the nonrelativistic MOND limit is derived in Sec. IV B, while the Newtonian limit is recovered for modestly small k in Sec. IV C. The above conclusions are extended to nonspherical systems in Sec. IV D. Sec. V shows that the theory passes the usual post-Newtonian solar system tests. Sec. VI demonstrates that for given dynamics, TeVeS gives the same gravitational lensing as does a dynamically successful dark halo model within GR. In Sec. VII we discuss TeVeS cosmological models with flat spaces showing that they are very similar to the corresponding GR models (apart from the question of cosmological dark matter which is left open), and demonstrating that the scalar field evolves little, and so can be taken to be small and positive. As discussed next in Sec. VIII, this last conclusion serves to rule out superluminal propagation in TeVeS.

II. THEORETICAL FOUNDATIONS FOR THE MOND PARADIGM

A. AQUAL: nonrelativistic field reformulation of MOND

However successful empirically when describing motions of test particles e.g. stars in the collective field of a galaxy, formula (1) is not fully correct. It is easily checked that a pair of particles accelerating one in the field of the other according to (1) does *not* conserve momentum. Thus the MOND formula by itself is not a theory. It is, however, a simple matter to construct a fully satisfactory *nonrelativistic* theory for MOND ([39]). Suppose we retain the Galilean and rotational invariance of the Lagrangian density which gives Poisson's equation, but drop the requirement of linearity of the equation. Then we come up with

$$\mathcal{L} = -\frac{\mathfrak{a}_0^2}{8\pi G} \, f\!\left(\frac{|\boldsymbol{\nabla}\Phi|^2}{\mathfrak{a}_0^2}\right) - \rho\Phi. \tag{2}$$

Here ρ is the mass density, $\mathfrak{a}_0$ is a scale of acceleration introduced for dimensional consistency, and f is some function. Newtonian theory (Poisson's equation) corresponds to the choice $f(y) = y$. From Eq. (2) follows the gravitational field equation

$$\boldsymbol{\nabla} \cdot [\tilde{\mu}(|\boldsymbol{\nabla}\Phi|/\mathfrak{a}_0) \, \boldsymbol{\nabla}\Phi] = 4\pi G\rho, \tag{3}$$

where $\tilde{\mu}(\sqrt{y}) \equiv df(y)/dy$. Because of its AQUAdratic Lagrangian, the theory has been called AQUAL [4]. The form of f and the value of $\mathfrak{a}_0$ must be supplied by phenomenology. We assume

$$f(y) \longrightarrow \begin{cases} y & y \gg 1; \\ \frac{2}{3}y^{3/2} & y \ll 1. \end{cases} \tag{4}$$

For systems with spherical, cylindrical or planar geometry, Eq. (3) can be integrated once immediately. With the usual prescription for the acceleration,

$$\mathbf{a} = -\boldsymbol{\nabla}\Phi, \tag{5}$$

the solution corresponds precisely to the MOND formula (1). This is no longer true for lower symmetry. However, numerical integration reveals that (1) is approximately true, in most cases to respectable accuracy [51].

The mentioned inexactness of Eq. (1) for systems such as a discrete collection of particles is at the root of the mentioned violation of the conservation laws. Because AQUAL starts from a Lagrangian, it respects all the usual conservation laws (energy, momentum and angular momentum), as can be checked directly [39]. This supplies the appropriate perspective for the mentioned failings of MOND. AQUAL also supplies the tools for showing that Newtonian behavior of the constituents of a large body, e.g. a star, is consistent with non-Newtonian dynamics of the latter's center of mass in the weak collective field of a larger system, e.g. a galaxy.

To summarize, whenever parts of a system devoid of high symmetry move with accelerations weak on scale $\mathfrak{a}_0$, the field $\boldsymbol{\nabla}\Phi$ which defines their accelerations is to be calculated by solving the AQUAL equation (3). AQUAL then becomes the nonrelativistic field theory on which to model the relativistic formulation of the MOND paradigm.

B. Principles for relativistic MOND

A relativistic MOND theory seems essential if gravitational lensing by extragalactic systems and cosmology are to be understood without reliance on dark matter. What principles should the relativistic embodiment of the MOND paradigm adhere to ? The following list is culled from those suggested by Bekenstein [4, 43], Sanders [52] and Romatka [53].

1. Principles

- *Action principle*   The theory must be derivable from an action principle. This is the only way known to guarantee that the necessary conservation laws of energy, linear and angular momentum are incorporated automatically. It is simplest to take the action as an integral over a local lagrangian density. A nonlocal action has been tried [47], but the resulting theory fails on account of gravitational lensing.

5

- *Relativistic invariance* Innumerable elementary particle experiments provide direct evidence for the universal validity of special relativity. The action should thus be a relativistic scalar so that all equations of the theory are relativistically invariant. Implied in this is the correspondence of the theory with special relativity when gravitation is negligible. This proviso rules out preferred frame theories.
- *Equivalence principle* As demonstrated with great accuracy (1 part in 10^{12}) by the Eötvös–Dicke experiments [54], free particles with negligible self–gravity fall in a gravitational field along universal trajectories (weak equivalence principle). For slow motion (the case tested by the experiments), the equation $\mathbf{a} = -\boldsymbol{\nabla}\Phi$, which encapsulates the universality, is equivalent to the geodesic equation in a (curved) metric $\tilde{g}_{\alpha\beta}$ with $\tilde{g}_{tt} \approx -1 - 2\Phi$. For light propagating in a static gravitational field, such a metric would predict that all frequencies as measured with respect to (w.r.t.) observers at rest in the field undergo a redshift measured by Φ. This is experimentally verified [55] to 1 part in 10^4. It thus appears that a curved metric $\tilde{g}_{\alpha\beta}$ describes those properties of spacetime in the presence of gravitation that are sensed by material objects. According to Schiff's conjecture [54, 56], this implies that the theory must be a metric theory, i.e., that in order to account sfor the effects of gravitation, all *nongravitational* laws of physics, e.g. electromagnetism, weak interactions, etc. must be expressed in their usual laboratory forms but with the metric $\tilde{g}_{\alpha\beta}$ replacing the Lorentz metric. This is the Einstein equivalence principle [54].
- *Causality* So as not to violate causality and thereby compromise the logical consistency of the theory, the equations deriving from the action should not permit superluminal propagation of any measurable field or of energy and linear and angular momenta. Superluminal here means exceeding the speed which is invariant under the Lorentz transformations. By Lorentz invariance of Maxwell's equations this is also the speed of light. In curved space, where curvature can cause waves to develop tails, the maximal speed is that of wavefronts, typically that of the high frequency components.
- *Positivity of energy* Fields in the theory should never carry negative energy. From the quantum point of view this is a precaution against instability of the vacuum. This principle is usually taken to mean that the energy density of each field should be nonnegative at each event (local positivity). The fact that the gravitational field itself cannot be generically assigned an energy density shows that this popular conception is overly stringent. A more useful statement of positivity of energy is that any bounded system must have positive energy (global positivity instead of the stronger local positivity). For example, the gravitational field can carry negative energy density locally (at least in the Newtonian conception), yet for pure gravity and in some cases in the presence of matter, a complete gravitating system is subject to the positive energy theorems [57]. Also, there are examples of scalar fields whose local energy density is of indefinite sign, yet a complete stationary system of such fields with sources has positive mass [58]. Of course, local positivity implies global positivity.
- *Departures from Newtonian gravity* The theory should exhibit a preferred scale of *acceleration* below which departures from Newtonian gravity should set in, even at low velocities.

2. Requirements

The relativistic embodiment of MOND should predict a number of well established phenomena. For example, we expect the following.

⋆ *Agreement with the extragalactic phenomenology*: The nonrelativistic limit of the theory should make predictions in agreement with those of the AQUAL equation, which is known to subsume much extragalactic phenomenology. This is checked for TeVeS in Sec. IV B.

⋆ *Agreement with phenomenology of gravitational lenses*: The theory should predict correctly the lensing of electromagnetic radiation by extragalactic structures which is responsible for gravitational lenses and arcs. In particular, it should give predictions similar to those of GR within the dark matter paradigm. This point is established for TeVeS in Sec. VI.

⋆ *Concordance with the solar system*: The theory should make predictions in agreement with the various solar system tests of relativity [54]: deflection of light rays, time delay of radar signals, precessions of the perihelia of the inner planets, the absence of the Nordtvedt effect in the lunar orbit, the nullness of aether drift, etc. TeVeS is confronted with the first three tests in Sec. V.

⋆ *Concordance with binary pulsar tests* : The theory should make predictions in harmony with the observed pulse times of arrival from the various binary pulsars. These contain information about relativistic time delay, periastron precession and the orbit's decay due to gravitational radiation. They thus constitute a test of the strong *potential* limit of the theory.

⋆ *Harmony with cosmological facts*: The theory should give a picture of cosmology in harmony with basic empirical facts such as the Hubble expansion, its timescales for various eras, existence of the microwave background, light element abundances from primordial nucleosynthesis, etc. The similarity of cosmological evolution in GR and in

TeVeS is established in Sec. VII, though the problem of how to eliminate cosmological dark matter with TeVeS is left open.

C.   Some antecedent relativistic theories

It is now in order to briefly review *some* of the previous attempts to give a relativistic theory of MOND. This will introduce the concepts to be borrowed by TeVeS, and help to establish the notation and conventions that we shall follow. A metric signature +2, and units with $c = 1$ are used throughout this paper. Greek indeces run over four coordinates while Latin ones run over the spatial coordinates alone.

1.   Relativistic AQUAL

It is well known that theories constructed, for example, by using a local function of the scalar curvature as Lagrangian density, have a purely Newtonian limit for weak potentials. So if we steer away from nonlocal actions, then AQUAL behavior cannot arise from merely modifying the gravitational action. The theory one seeks has to involve degrees of freedom other than the metric.

In the first relativistic theory with MOND aspirations, relativistic AQUAL [39], the physical metric $\tilde{g}_{\alpha\beta}$ was taken as conformal to a primitive (Einstein) metric $g_{\alpha\beta}$, i.e., $\tilde{g}_{\alpha\beta} = e^{2\psi} g_{\alpha\beta}$ with ψ a real scalar field. In order not to break violently with GR, which is well tested in the solar system (and to some extent in cosmology), the gravitational action was taken as the Einstein-Hilbert's one built out of $g_{\alpha\beta}$. The MOND phenomenology was implanted by taking for the Lagrangian density for ψ

$$\mathcal{L}_\psi = -\frac{1}{8\pi G L^2} \tilde{f}\left(L^2 g^{\alpha\beta} \psi_{,\alpha} \psi_{,\beta}\right),\tag{6}$$

where $\tilde{f}$ is some function (not known *a priori*), and L is a constant with dimensions of length introduced for dimensional consistency. Note that when $\tilde{f}(y) = y$, $\mathcal{L}_\psi$ is just the lagrangian density for a linear scalar field, but in general $\mathcal{L}_\psi$ is aquadratic.

To implement the universality of free fall, one must write all lagrangians of matter fields using a single metric, which is taken as $\tilde{g}_{\alpha\beta}$ (not $g_{\alpha\beta}$ which choice would make the theory GR). Thus for example, the action for a particle of mass m is taken as

$$S_m = -m \int e^\psi \left(-g_{\alpha\beta} dx^\alpha dx^\beta\right)^{1/2}.\tag{7}$$

Hence test particle motion is nongeodesic w.r.t. $g_{\alpha\beta}$ but, of course, geodesic w.r.t. $\tilde{g}_{\alpha\beta}$. Evidently this last is the metric measured by clocks and rods, hence the physical metric. Addition of a constant to ψ merely multiplies all masses by a constant (irrelevant global redefinition of units), so that the theory is insensitive to the choice of zero of ψ.

For slow motion in a quasistatic situation with nearly flat metric $g_{\alpha\beta}$, and in a weak field ψ, $e^\psi(-g_{\alpha\beta}dx^\alpha dx^\beta)^{1/2} \approx (1 + \Phi_N + \psi - \mathbf{v}^2/2)dt$, were $\Phi_N = -(g_{tt} + 1)/2$ is the Newtonian potential determined by the mass density ρ through the linearized Einstein equations for $g_{\alpha\beta}$, and $\mathbf{v}$ is the velocity defined w.r.t. the Minkowski metric which is close to $g_{\alpha\beta}$. Thus the particle's lagrangian is $m(\mathbf{v}^2/2 - \Phi_N - \psi)$; this leads to the equation of motion

$$\mathbf{a} \approx -\boldsymbol{\nabla}(\Phi_N + \psi).\tag{8}$$

How is ψ determined ? For stationary weak fields the Lagrangian *density* for ψ, including a point source of physical mass M at $\mathbf{r} = 0$, is from the above discussion and Eqs. (6)-(7),

$$\mathcal{L}_\psi = -\frac{1}{8\pi G L^2} \tilde{f}\left(L^2 (\boldsymbol{\nabla}\psi)^2\right) - \psi M \delta(\mathbf{r}).\tag{9}$$

Comparing Eqs. (9) and (2) we conclude that ψ here corresponds to Φ of mass M as computed from AQUAL's Eq. (3), provided we take $\tilde{f} = f$ and $L = 1/\mathfrak{a}_0$. Whenever $|\boldsymbol{\nabla}\psi| \gg |\boldsymbol{\nabla}\Phi_N|$ (Φ_N is the Newtonian potential of the same mass distribution), the equation of motion (8) reduces to (5), and we obtain MOND like dynamics. For the choice of MOND function (4) the said strong inequality is automatic in the deep MOND regime, $|\boldsymbol{\nabla}\psi| \ll \mathfrak{a}_0$, because $\tilde{\mu} \ll 1$ there.

In the regime $|\boldsymbol{\nabla}\psi| \gg \mathfrak{a}_0$, $\tilde{\mu} \approx 1$ and $f(y) \approx y$ so that ψ reduces to Φ_N. It would seem from Eq. (8) that a particle's acceleration is then twice the correct Newtonian value. However, this just means that the measurable Newton's constant G_N is twice the bare G appearing in $\mathcal{L}_\psi$ or in Einstein's equations. It is thus clear, regarding dynamics, that the relativistic AQUAL theory has the appropriate MOND and Newtonian limits depending on the strength of $\boldsymbol{\nabla}\psi$.

But relativistic AQUAL has problems. Early on [4, 39, 42] it was realized that ψ waves can propagate faster than light. This acausal behavior can be traced to the aquadratic form of the lagrangian, as explained in Appendix A. A second problem [43, 53] issues from the conformal relation $\tilde{g}_{\alpha\beta} = e^{2\psi} g_{\alpha\beta}$. Light propagates on the null cones of the physical metric; by the conformal relation these coincide with the lightcones of the Einstein metric. This last is calculated from Einstein's equations with the visible matter and ψ field as sources. Thus so long as the ψ field contributes comparatively little to the energy momentum tensor, it cannot affect light deflection, which will thus be that due to the visible matter alone. But in reality galaxies and clusters of galaxies are observed to deflect light stronger than the visible mass in them would suggest. Thus relativistic AQUAL fails to accurately describe light deflection in situations in which GR requires dark matter. It is thus empirically falsified.

Relativistic AQUAL bequeaths to TeVeS the use of a scalar field to connect Einstein and physical metrics, a field which satisfies an equation reminiscent of the nonrelativistic AQUAL Eq. (3).

2. *Phase coupling gravitation*

The Phase Coupled Gravity (PCG) theory was proposed [4, 40, 42] in order to resolve relativistic AQUAL's acausality problem. It retains the two metrics related by $\tilde{g}_{\alpha\beta} = e^{2\psi} g_{\alpha\beta}$, but envisages ψ as one of a pair of mutually coupled real scalar fields with the Lagrangian density (our definitions here differ slightly from those in Ref. 4)

$$\mathcal{L}_{\psi,A} = -\frac{1}{2} \left[g^{\alpha\beta}(A_{,\alpha}A_{,\beta} + \eta^{-2}A^2\psi_{,\alpha}\psi_{,\beta}) + \mathcal{V}(A^2) \right] \tag{10}$$

Here η is a real parameter and $\mathcal{V}$ a real valued function. The coupling between A and ψ is designed to bring about AQUAL-like features for small $|\eta|$. The theory receives its name because matter is coupled to ψ, which is proportional to the phase of the self-interacting complex field $\chi = A e^{i\psi/\eta}$.

Variation of $\mathcal{L}_{\psi,A}$ w.r.t. A leads to (all covariant derivatives and index raising w.r.t. $g_{\alpha\beta}$)

$$A^{,\alpha}{}_{;\alpha} - \eta^{-2}A\,\psi_{,\alpha}\psi^{,\alpha} - A\mathcal{V}'(A^2) = 0 \tag{11}$$

In the variation w.r.t. ψ we must include the Lagrangian density of a source, say a point mass M at rest at $\mathbf{r} = 0$ [c.f. S_m in Eq. (7)]:

$$\left(A^2 g^{\alpha\beta}\psi_{,\beta}\right)_{;\alpha} = \eta^2 e^\psi M \delta(\mathbf{r}) \tag{12}$$

The connection with AQUAL is now clear. For sufficiently small $|\eta|$ the $A^{,\alpha}{}_{;\alpha}$ term in Eq. (11) becomes negligible, and the other two establish an algebraic relation between $\psi_{,\alpha}\psi^{,\alpha}$ and A^2. Substituting this in Eq. (12) gives the AQUAL type of equation for ψ that would derive from $\mathcal{L}_\psi$ in Eq. (6).

The PCG Lagrangian's advantage over that of the relativistic AQUAL's is precisely in that it involves first derivatives only in quadratic form. This would seem to rule out the superluminality generating X^α dependent terms discussed in Appendix A. In practice things are more complicated. A detailed local analysis employing the eikonal approximation [42] shows that there are superluminal ψ perturbations, for example when $\mathcal{V}'' < 0$. However, the same analysis shows that such superluminality occurs only when the background solution is itself locally unstable. This makes the said causality violation moot.

One way to obtain MOND phenomenology from PCG is to choose $\mathcal{V}(A^2) = -\frac{1}{3}\varepsilon^{-2}A^6$ with ε a constant with dimension of energy. Although with this choice $\mathcal{V}'' < 0$ which makes for unstable backgrounds, we only need this form for small A; $\mathcal{V}$ can take different form for large argument. Then in a static situation with nearly flat $g_{\alpha\beta}$ and weak ψ, Eqs. (11)-(12) reduce to

$$\nabla^2 A - \eta^{-2}A(\boldsymbol{\nabla}\psi)^2 + \varepsilon^{-2}A^5 = 0, \tag{13}$$
$$\boldsymbol{\nabla} \cdot (A^2 \boldsymbol{\nabla}\psi) = \eta^2 M \delta(\mathbf{r}). \tag{14}$$

The spherically symmetric solution of Eqs. (13)-(14) is

$$A = (\kappa\varepsilon/r)^{1/2}; \quad d\psi/dr = (\eta\varpi/4\kappa r) \tag{15}$$
$$\varpi \equiv (\eta M/\pi\varepsilon); \quad \kappa \equiv 2^{-3/2}\left(1 + \sqrt{1+4\varpi^2}\right)^{1/2} \tag{16}$$

One may evidently still use Eq. (8):

$$a_r = -GM/r^2 - (\eta^2 M/4\pi\varepsilon\kappa r) \tag{17}$$

Thus a $1/r$ force competes with the Newtonian one. For small M it starts to dominate at a fixed radius scale r_c, just as in Tohline's [59] and Kuhn-Kruglyak's [60] non-Newtonian gravity theories. Here $r_c = 2\pi G\varepsilon/\eta^2$. By contrast for $M \gg M_c \equiv \frac{1}{2}\pi\varepsilon/\eta$, $\kappa \approx \frac{1}{2}\sqrt{\varpi}$ and the $1/r$ force scales as $M^{1/2}$ and begins to dominate when the Newtonian acceleration drops below the fixed acceleration scale

$$\mathfrak{a}_0 \equiv \eta^3/(4\pi G\varepsilon). \tag{18}$$

For $a_r \ll \mathfrak{a}_0$ the circular velocity whose centripetal acceleration balances the $1/r$ force is $v_c = (G\mathfrak{a}_0 M)^{1/4}$, precisely as in MOND. Thus $\mathfrak{a}_0$ here is to be identified with Milgrom's constant $\mathfrak{a}_0$. We conclude that, with a suitable choice of parameters, PCG with a sextic potential recovers the main features of MOND: asymptotically flat rotation curves and the TFL for disk galaxies. Specifically, the choice $\eta = 10^{-8}$ and $\varepsilon = 10^{53}$ erg gives $\mathfrak{a}_0 = 8.7 \times 10^{-9}$cm s^{-2}, $M_c = 8.7 \times 10^6 M_\odot$ and $r_c = 5.2 \times 10^{19}$ cm. Now since r_c is larger than the Hubble scale, the Tohline-Kuhn-Kruglyak $1/r$ force is comparatively unimportant. Hence for $M \gg 10^7 M_\odot$ we should have MOND, and for low masses almost Newtonian behavior. This is about right: globular star clusters at $10^4 - 10^5 M_\odot$ show no missing mass problem.

However, the above parameters are bad from the point of view of the solar system tests of gravity, as summarized in Appendix B. But the gravest problem with PCG is that it, just as AQUAL, provides insufficient light deflection [43]. Here again, the conformal relation between Einstein and physical metric is to blame. TeVeS incorporates PCG's Lagrangian density (10) in the limit of small η in which A becomes nondynamical.

3. Theories with disformally related metrics

The light deflection problem can be solved only by giving up the relation $\tilde{g}_{\alpha\beta} = e^{-2\psi} g_{\alpha\beta}$. It was thus suggested [43] to replace this conformal relation by a disformal one, namely

$$\tilde{g}_{\alpha\beta} = e^{-2\psi}(Ag_{\alpha\beta} + \mathcal{B}L^2\psi_{,\alpha}\psi_{,\beta}), \tag{19}$$

with A and B functions of the invariant $g^{\mu\nu}\psi_{,\mu}\psi_{,\nu}$ and L a constant length unrelated, of course, to that in Eq. (6). This relation already allows ψ to deflect light via the $\psi_{,\alpha}\psi_{,\beta}$ term in the physical metric. However, it was found [44] that if one insists on causal propagation of both light and gravitational waves w.r.t. the light cones of the physical metric, then the sign required of $\mathcal{B}$ is opposite that required to enhance the light deflection coming from the metric $g_{\alpha\beta}$ alone. Thus the cited disformal relation between metrics, if respecting causality, will give weaker light deflection than would $g_{\alpha\beta}$ were it the physical metric.

This last observation of Ref. 44 has given rise to a folk belief that relativistic gravity theories which attempt to supplant dark matter's dynamical effects necessarily reduce light deflection rather than enhancing it [34, 46, 47, 48]. However, as remarked by Sanders, the mentioned problem disappears if the term $\psi_{,\alpha}\psi_{,\beta}$ is replaced by $\mathfrak{U}_\alpha\mathfrak{U}_\beta$, where $\mathfrak{U}_\alpha$ is a constant 4-vector which, at least in the solar system and within galaxies, points in the time direction [50]. Specifically Sanders takes $\tilde{g}_{\alpha\beta} = e^{-2\psi} g_{\alpha\beta} - 2\mathfrak{U}_\alpha\mathfrak{U}_\beta \sinh(2\psi)$.

This "stratified" gravitation theory is reported to do well in the confrontation with the solar system tests, and to possess the right properties to explain the coincidence between $\mathfrak{a}_0$ of MOND and the Hubble scale [7]. But its vector $\mathfrak{U}_\alpha$ is an *a priori* nondynamical element whose direction is selected in an unspecified way by the cosmological background. This means the theory is a preferred frame theory (although it is reported to be protected on this account against falsification in the solar system and other strong acceleration systems by its AQUAL behavior [50]). This is obviously a conceptual shortcoming which TeVeS removes, but the latter's debt to the stratified theory should be underlined.

III. FUNDAMENTALS OF TeVeS

A. Fields and actions

TeVeS is based on three dynamical gravitational fields: an Einstein metric $g_{\mu\nu}$ with a well defined inverse $g^{\mu\nu}$, a timelike 4-vector field $\mathfrak{U}_\mu$ such that

$$g^{\alpha\beta}\mathfrak{U}_\alpha\mathfrak{U}_\beta = -1, \tag{20}$$

and a scalar field ϕ; there is also a nondynamical scalar field σ (the acronym TeVeS recalls the theory's Tensor-Vector-Scalar content). The physical metric in TeVeS, just as in Sanders' stratified theory, is obtained by stretching the Einstein metric in the spacetime directions orthogonal to $\mathfrak{U}^\alpha \equiv g^{\alpha\beta}\mathfrak{U}_\beta$ by a factor $e^{-2\phi}$, while shrinking it by the same factor in the direction parallel to $\mathfrak{U}^\alpha$:

$$\tilde{g}_{\alpha\beta} = e^{-2\phi}(g_{\alpha\beta} + \mathfrak{U}_\alpha\mathfrak{U}_\beta) - e^{2\phi}\mathfrak{U}_\alpha\mathfrak{U}_\beta \tag{21}$$

$$= e^{-2\phi}g_{\alpha\beta} - 2\mathfrak{U}_\alpha\mathfrak{U}_\beta \sinh(2\phi) \tag{22}$$

It is easy to verify that the inverse physical metric is

$$\tilde{g}^{\alpha\beta} = e^{2\phi}g^{\alpha\beta} + 2\mathfrak{U}^\alpha\mathfrak{U}^\beta \sinh(2\phi) \tag{23}$$

where $\mathfrak{U}^\alpha$ will *always* mean $g^{\alpha\beta}\mathfrak{U}_\beta$.

The geometric part of the action, S_g, is formed from the Ricci tensor $R_{\alpha\beta}$ of $g_{\mu\nu}$ just as in GR:

$$S_g = (16\pi G)^{-1} \int g^{\alpha\beta}R_{\alpha\beta}(-g)^{1/2}d^4x. \tag{24}$$

Here g means the determinant of metric $g_{\alpha\beta}$. This choice is made in order to keep TeVeS close to GR in some sense to be clarified below.

In terms of two constant positive parameters, k and ℓ, the action for the pair of scalar fields is taken to be of roughly PCG form,

$$S_s = -\frac{1}{2}\int \left[\sigma^2 h^{\alpha\beta}\phi_{,\alpha}\phi_{,\beta} + \frac{1}{2}G\ell^{-2}\sigma^4 F(kG\sigma^2)\right](-g)^{1/2}d^4x, \tag{25}$$

where $h^{\alpha\beta} \equiv g^{\alpha\beta} - \mathfrak{U}^\alpha\mathfrak{U}^\beta$ and F is a free dimensionless function (it is related to PCG's potential $\mathcal{V}$). No overall coefficient is required for the kinetic term; were it included, it could be absorbed into a redefinition of σ and thereby in k and ℓ. Because ϕ is obviously dimensionless, the dimensions of σ^2 are those of G^{-1}. Thus k is a dimensionless constant (it could be absorbed into the definition of F, but we choose to exhibit it), while ℓ is a constant length.

Because no kinetic σ terms appear, the "equation of motion" of σ takes the form of an algebraic relation between it and the invariant $h^{\alpha\beta}\phi_{,\alpha}\phi_{,\beta}$, and when this is substituted for σ in S_s, the phenomenologically successful AQUAL type action for ϕ appears. We could, of course, have written this last action directly. The present route is more suggestive of the possible origin of the action; for example, S_s resembles the action for a complex self-interacting scalar field $\eta\sigma\exp(\imath\phi/\eta)$ in the limit of small η. The term $-\sigma^2\mathfrak{U}^\alpha\mathfrak{U}^\beta\phi_{,\alpha}\phi_{,\beta}$ here included in the scalar's action is new; its role is to eliminate superluminal propagation of the ϕ field, a recalcitrant problem in AQUAL type theories.

The action of the vector $\mathfrak{U}^\alpha$ is taken to have the form

$$S_v = -\frac{K}{32\pi G}\int \left[g^{\alpha\beta}g^{\mu\nu}\mathfrak{U}_{[\alpha,\mu]}\mathfrak{U}_{[\beta,\nu]} - 2(\lambda/K)(g^{\mu\nu}\mathfrak{U}_\mu\mathfrak{U}_\nu + 1)\right](-g)^{1/2}d^4x, \tag{26}$$

where antisymmetrization in a pair of indeces is indicated by surrounding them by square brackets, e.g. $A_{[\mu}B_{\nu]} = A_\mu B_\nu - A_\nu B_\mu$. In Eq. (26) λ is a spacetime dependent Lagrange multiplier enforcing the normalization Eq. (20) (we shall calculate λ later), while K is a dimensionless constant since $\mathfrak{U}^\alpha$ is dimensionless. Thus TeVeS has two dimensionless parameters, k and K, in addition to the dimensional constants G and ℓ. The kinetic terms in Eq. (26) are chosen antisymmetric not because of any desire for gauge symmetry, which is broken by the form of the physical metric anyway, but because this choice precludes appearance of second derivatives of $\mathfrak{U}_\alpha$ in the energy–momentum tensor of TeVeS (see next subsection). The action S_v is a special case of that in Jacobson and Mattingly's generalization of GR with a preferred frame [61].

In accordance with the equivalence principle, the matter action in TeVeS is obtained by transcribing the flat spacetime lagrangian $\mathcal{L}(\eta_{\mu\nu}, f^\alpha, \partial_\mu f^\alpha, \cdots)$ for fields written schematically f^α as

$$S_m = \int \mathcal{L}(\tilde{g}_{\mu\nu}, f^\alpha, f^\alpha{}_{|\mu}, \cdots)(-\tilde{g})^{1/2}d^4x, \tag{27}$$

where the covariant derivatives denoted by $|$ are taken w.r.t. $\tilde{g}_{\mu\nu}$. This has the effect that the spacetime delineated by matter dynamics has the metric $\tilde{g}_{\mu\nu}$. The appearance of $(-\tilde{g})^{1/2}$ here requires us to specify its relation to $(-g)^{1/2}$. In Appendix C we show that

$$(-\tilde{g})^{1/2} = e^{-2\phi}(-g)^{1/2} \tag{28}$$

By coupling to matter only through $\tilde{g}_{\alpha\beta}$, the field $\mathfrak{U}_\alpha$ is totally different from the Lee-Yang 4-vector field with gravitation strength interaction [66], whose existence is ruled out by the equivalence principle tests as well as by cosmological symmetry arguments [66, 67].

B. Basic equations

We shall obtain the basic equations by varying the total action $S = S_g + S_s + S_v + S_m$ wth respect to the basic fields $g^{\alpha\beta}$, ϕ, σ and $\mathfrak{U}_\alpha$. To this end we must be explicit about how $\tilde{g}_{\alpha\beta}$, which enters into S_m, varies with the basic fields. Taking increments of Eq. (23) we get

$$\delta\tilde{g}^{\alpha\beta} = e^{2\phi}\delta g^{\alpha\beta} + 2\sinh(2\phi)\mathfrak{U}_\mu\delta g^{\mu(\alpha}\mathfrak{U}^{\beta)} + 2\big[e^{2\phi}g^{\alpha\beta} + 2\mathfrak{U}^\alpha\mathfrak{U}^\beta\cosh(2\phi)\big]\delta\phi + 2\sinh(2\phi)\mathfrak{U}^{(\alpha}g^{\beta)\mu}\delta\mathfrak{U}_\mu \quad (29)$$

where symmetrization in a pair of indeces is indicated by surrounding them by round brackets, e.g. $A^{(\mu}B^{\nu)} = A^\mu B^\nu + A^\nu B^\mu$.

1. Equations for the metric

When varying S w.r.t. $g^{\alpha\beta}$ we recall that $\delta S_g = (16\pi G)^{-1}G_{\alpha\beta}(-g)^{1/2}\delta g^{\alpha\beta}$ ($G_{\alpha\beta}$ denotes the Einstein tensor of $g_{\alpha\beta}$) while

$$\delta S_m = -\frac{1}{2}\tilde{T}_{\alpha\beta}(-\tilde{g})^{1/2}\delta\tilde{g}^{\alpha\beta} + \ldots \quad (30)$$

where the ellipsis denotes variations of the f^α fields, and $\tilde{T}_{\alpha\beta}$ stands for the physical energy–momentum tensor defined with the metric $\tilde{g}_{\alpha\beta}$. We get

$$G_{\alpha\beta} = 8\pi G\Big[\tilde{T}_{\alpha\beta} + (1 - e^{-4\phi})\mathfrak{U}^\mu\tilde{T}_{\mu(\alpha}\mathfrak{U}_{\beta)} + \tau_{\alpha\beta}\Big] + \Theta_{\alpha\beta} \quad (31)$$

where

$$\tau_{\alpha\beta} \equiv \sigma^2\Big[\phi_{,\alpha}\phi_{,\beta} - \frac{1}{2}g^{\mu\nu}\phi_{,\mu}\phi_{,\nu}\,g_{\alpha\beta} - \mathfrak{U}^\mu\phi_{,\mu}\big(\mathfrak{U}_{(\alpha}\phi_{,\beta)} - \frac{1}{2}\mathfrak{U}^\nu\phi_{,\nu}\,g_{\alpha\beta}\big)\Big]$$
$$- \frac{1}{4}G\ell^{-2}\sigma^4 F(kG\sigma^2)g_{\alpha\beta} \quad (32)$$

$$\Theta_{\alpha\beta} \equiv K\Big(g^{\mu\nu}\mathfrak{U}_{[\mu,\alpha]}\mathfrak{U}_{[\nu,\beta]} - \frac{1}{4}g^{\sigma\tau}g^{\mu\nu}\mathfrak{U}_{[\sigma,\mu]}\mathfrak{U}_{[\tau,\nu]}\,g_{\alpha\beta}\Big) - \lambda\mathfrak{U}_\alpha\mathfrak{U}_\beta \quad (33)$$

When varying $g^{\alpha\beta}$ in S_v we have used Eq. (20) to drop a term proportional to $g_{\alpha\beta}$.

2. Scalar equation

Variation of σ in S_s gives the relation between σ and $\phi_{,\alpha}$ ($F' \equiv dF(\mu)/d\mu$),

$$-kG\sigma^2 F - 1/2\,(kG\sigma^2)^2 F' = k\ell^2 h^{\alpha\beta}\phi_{,\alpha}\phi_{,\beta} \quad (34)$$

In carrying out the variation w.r.t. ϕ it must be remembered that this quantity enters in S_m exclusively through $\tilde{g}^{\alpha\beta}$, so that use must be made of Eqs. (29)-(30):

$$\big[\sigma^2 h^{\alpha\beta}\phi_{,\alpha}\big]_{;\beta} = \big[g^{\alpha\beta} + (1 + e^{-4\phi})\mathfrak{U}^\alpha\mathfrak{U}^\beta\big]\tilde{T}_{\alpha\beta} \quad (35)$$

In view of Eq. (34) this is an equation for ϕ only, with $\tilde{T}_{\alpha\beta}$ as source.

Suppose we define a function $\mu(y)$ by

$$-\mu F(\mu) - \frac{1}{2}\mu^2 F'(\mu) = y. \quad (36)$$

so that $kG\sigma^2 = \mu(k\ell^2 h^{\alpha\beta}\phi_{,\alpha}\phi_{,\beta})$. We may now recast Eq. (35) as

$$\big[\mu\,(k\ell^2 h^{\mu\nu}\phi_{,\mu}\phi_{,\nu})\,h^{\alpha\beta}\phi_{,\alpha}\big]_{;\beta} = kG\big[g^{\alpha\beta} + (1 + e^{-4\phi})\mathfrak{U}^\alpha\mathfrak{U}^\beta\big]\tilde{T}_{\alpha\beta}. \quad (37)$$

This equation is reminiscent of the relativistic AQUAL scalar equation [see Appendix A, Eq. (A1)], albeit with the replacement $g^{\alpha\beta} \mapsto h^{\alpha\beta}$ in the l.h.s. In quasistatic situations we may replace $h^{\alpha\beta}$ by $g^{\alpha\beta}$ so that Eq. (37) has the same structure as the AQUAL equation.

3.   Vector equation

Variation of S w.r.t. $\mathfrak{U}_\alpha$ and use of Eq. (29) gives the vector equation

$$K\mathfrak{U}^{[\alpha;\beta]}{}_{;\beta} + \lambda\mathfrak{U}^\alpha + 8\pi G\sigma^2\mathfrak{U}^\beta\phi_{,\beta}g^{\alpha\gamma}\phi_{,\gamma} = 8\pi G(1-e^{-4\phi})g^{\alpha\mu}\mathfrak{U}^\beta\tilde{T}_{\mu\beta} \tag{38}$$

As mentioned, λ here is a Lagrange multiplier. It can be solved for by contracting the previous equation with $\mathfrak{U}_\alpha$. Substituting it back gives

$$K\left(\mathfrak{U}^{[\alpha;\beta]}{}_{;\beta} + \mathfrak{U}^\alpha\mathfrak{U}_\gamma\mathfrak{U}^{[\gamma;\beta]}{}_{;\beta}\right) + 8\pi G\sigma^2\left[\mathfrak{U}^\beta\phi_{,\beta}\,g^{\alpha\gamma}\phi_{,\gamma} + \mathfrak{U}^\alpha(\mathfrak{U}^\beta\phi_{,\beta})^2\right]$$
$$= 8\pi G(1-e^{-4\phi})\left[g^{\alpha\mu}\mathfrak{U}^\beta\tilde{T}_{\mu\beta} + \mathfrak{U}^\alpha\mathfrak{U}^\beta\mathfrak{U}^\gamma\tilde{T}_{\gamma\beta}\right] \tag{39}$$

This equation has only three independent components since both sides of it are orthogonal to $\mathfrak{U}_\alpha$. It thus determines three components of $\mathfrak{U}^\alpha$ with the fourth being determined by the normalization (20). Like any other partial differential equation, the vector equation does not by itself determine $\mathfrak{U}_\alpha$ uniquely.

C.   General relativity limit

TeVeS has three parameters: k, ℓ and K. Here we show first that in several familiar contexts the limit $k \to 0$, $\ell \propto k^{-3/2}$, $K \propto k$ of it corresponds to standard GR for any form of the function F. Many of the intermediate results will beuseful in Sec. V and VII. We then expand on a remark by Milgrom that the GR limit actually follows under more general circumstances: $K \to 0$ and $\ell \to \infty$.

Whenever a specific matter content is needed, we shall assume the matter to be an ideal fluid. Its energy-momentum tensor has the familiar form

$$\tilde{T}_{\alpha\beta} = \tilde{\rho}\tilde{u}_\alpha\tilde{u}_\beta + \tilde{p}(\tilde{g}_{\alpha\beta} + \tilde{u}_\alpha\tilde{u}_\beta), \tag{40}$$

where $\tilde{\rho}$ is the proper energy density, $\tilde{p}$ the pressure and $\tilde{u}_\alpha$ the 4-velocity, all three expressed in the physical metric. We may profitably simplify Eq. (37) in any case when for symmetry reasons $\tilde{u}_\alpha$ is collinear with $\mathfrak{U}_\alpha$. In order that the velocity be normalized w.r.t. $\tilde{g}_{\alpha\beta}$, we must take in that case $\tilde{u}_\alpha = e^\phi \mathfrak{U}_\alpha$ from which follows

$$\tilde{g}_{\alpha\beta} + \tilde{u}_\alpha\tilde{u}_\beta = e^{-2\phi}(g_{\alpha\beta} + \mathfrak{U}_\alpha\mathfrak{U}_\beta). \tag{41}$$

Substituting this in $\tilde{T}_{\alpha\beta}$ allows us to rewrite Eq. (37) as

$$\left[\mu\left(k\ell^2h^{\mu\nu}\phi_{,\mu}\phi_{,\nu}\right)h^{\alpha\beta}\phi_{,\alpha}\right]_{;\beta} = kG(\tilde{\rho} + 3\tilde{p})\,e^{-2\phi}. \tag{42}$$

This form is suitable for the analysis of cosmology as well as static systems.

1.   Cosmology

Not only important in itself, cosmology is relevant for setting boundary conditions in the study of TeVeS in the solar system and other localized weak gravity situations. We shall confine our remarks to Friedmann-Robertson–Walker (FRW) cosmologies, for which the metric can be given the form

$$g_{\alpha\beta}\,dx^\alpha dx^\beta = -dt^2 + a(t)^2[d\chi^2 + f(\chi)^2(d\theta^2 + \sin^2\theta\,d\varphi^2)]. \tag{43}$$

Here $f(\chi) \equiv \sin\chi, \chi, \sinh\chi$ for closed, flat and open spaces, respectively.

In applying Eq. (37) we shall assume that the fields ϕ, σ and $\mathfrak{U}^\alpha$ partake of the symmetries of the FRW spacetime. Thus we take these fields to depend solely on t. Also since there are no preferred spatial directions, $\mathfrak{U}^\alpha$ must point in the cosmological time direction: $\mathfrak{U}^\alpha = \delta_t{}^\alpha$ (that this is possible distinguishes $\mathfrak{U}^\alpha$ from the Lee-Yang field which is ruled out in FRW cosmology [67]). Obviously this is a case where $\tilde{u}_\alpha = e^\phi \mathfrak{U}_\alpha$; the scalar equation then takes the form

$$a^{-3}\partial_t[a^3\mu(-2k\ell^2\dot{\phi}^2)\dot{\phi}] = -\tfrac{1}{2}kG(\tilde{\rho} + 3\tilde{p})e^{-2\phi}, \tag{44}$$

where an overdot signifies $\partial/\partial t$. The first integral is

$$\mu(-2k\ell^2\dot{\phi}^2)\dot{\phi} = \frac{-k}{2a^3}\int_0^t G(\tilde{\rho} + 3\tilde{p})e^{-2\phi}a^3 dt. \tag{45}$$

As is customary in scalar–tensor theories, we have dropped an additive integration constant; this has the effect of ameliorating any divergence of $\dot\phi$ as $a \to 0$. In fact we can see that the r.h.s. of the equation behaves there as $k(\tilde\rho+3\tilde p)e^{-2\phi}t$. We observe that as $k \to 0$ with $\ell \propto k^{-3/2}$, $\dot\phi$ will behave as k with the argument of μ staying constant. Thus regardless of the form of μ, we have $\dot\phi \sim k$. It is thus consistent to assume that ϕ itself is of $\mathcal{O}(k)$ throughout cosmological history. This despite the possible divergence of $\dot\phi$ at the cosmological singularity, since the rate of that divergence is also proportional to k, as we have just seen. Recalling that $kG\sigma^2 = \mu$, we conclude that σ^2 is of $\mathcal{O}(k^{-1})$ in the cosmological solutions (otherwise μ would vary with k whereas its argument stayed constant).

Let us check whether our assumption that $\mathfrak{U}^\alpha = \delta_t{}^\alpha$ is consistent with the vector equation (38). The choice $\mathfrak{U}^\alpha = \delta_t{}^\alpha$ makes $\mathfrak{U}^{[\alpha;\beta]} = 0$. For a comoving ideal fluid $\mathfrak{U}^\beta \tilde T_{\alpha\beta} = -e^{2\phi}\tilde\rho\,\mathfrak{U}_\alpha$. Thus the spatial components of the vector equation (38) vanish identically, while the temporal one informs us that

$$\lambda = 8\pi G\big[\sigma^2\dot\phi^2 - 2\tilde\rho\sinh(2\phi)\big]. \tag{46}$$

Our previous comments make it clear that λ is of $\mathcal{O}(k)$.

Turning to the gravitational equations (31)-(33) we first note that in the limit $\{k \to 0,\, \ell \propto k^{-3/2},\, K \propto k\}$, $\tau_{\alpha\beta}$ and $\Theta_{\alpha\beta}$ are both $\mathcal{O}(k)$. It follows that $G_{\alpha\beta} = 8\pi G\tilde T_{\alpha\beta} + \mathcal{O}(k)$. Since the difference between $\tilde g_{\alpha\beta}$ and $g_{\alpha\beta}$ is also of $\mathcal{O}(k)$, it is obvious that $\tilde G_{\alpha\beta} = 8\pi G\tilde T_{\alpha\beta} + \mathcal{O}(k)$ so that any cosmological model based on TeVeS differs from the corresponding one in GR only by terms of $\mathcal{O}(k)$. In FRW cosmology TeVeS has GR as its limit when $k \to 0$ with $\ell \propto k^{-3/2}$ and $K \propto k$.

2. *Quasistatic localized system*

We now turn to systems such as the solar system, or a neutron star, which may be thought of as quasistatic situations in asymptotically flat spacetime (at least up to sub–cosmological distances). We shall idealize them as truly static systems with time independent metrics of the form

$$g_{\alpha\beta}\,dx^\alpha dx^\beta = g_{tt}(x^k)\,dt^2 + g_{ij}(x^k)\,dx^i dx^j \tag{47}$$

and no energy flow. The scalar and vector equations have a variety of joint solutions. We shall single out the physical one by requiring the boundary condition that $\phi \to$ const. at spatial infinity, the constant being just the value of ϕ from the cosmological model in which our localized system is embedded. Likewise, we shall require that $\mathfrak{U}^\alpha \to \delta_t{}^\alpha$ so that the vector field matches the cosmological field at "spatial infinity".

We first show that $\mathfrak{U}^\alpha = N\xi^\alpha$, with $\xi^\alpha = \delta_t{}^\alpha$ the Killing vector associated with the static character of the spacetime, is an acceptable solution (with $N \equiv (-g_{\alpha\beta}\xi^\alpha\xi^\beta)^{-1/2}$, $\mathfrak{U}^\alpha$ is properly normalized). Let us consider the expression $g^{\alpha\mu}\mathfrak{U}^\beta\tilde T_{\mu\beta} + \mathfrak{U}^\alpha\mathfrak{U}^\beta\mathfrak{U}^\gamma\tilde T_{\gamma\beta}$ appearing in the source of the vector equation (39) for this choice of $\mathfrak{U}^\alpha$. Its $\alpha = t$ component is $N\left(\tilde T^t{}_t + \mathfrak{U}_t\mathfrak{U}^t\tilde T^t{}_t\right) = 0$, while the $\alpha = i$ component is $N\left(g^{ij}\tilde T_{jt} + \mathfrak{U}^i(\mathfrak{U}^t)^2\tilde T_{tt}\right)$ which also vanishes because $\tilde T_{jt} = 0$ (no energy flow). Turn now to the l.h.s. of Eq. (39). Because $\mathfrak{U}^\alpha$ has only a (time–independent) temporal component, $\mathfrak{U}^\alpha\phi_{,\alpha} = 0$, and the only nonvanishing components of $\mathfrak{U}^{[\alpha,\beta]}$ are the jt ones, and they depend only on the x^j. Hence $\mathfrak{U}^{[i,\beta]}{}_{;\beta} = 0$ so that the $\alpha = i$ components of the l.h.s. of the equation vanish. What is left of the $\alpha = t$ component is $K(\mathfrak{U}^{[t,\beta]}{}_{;\beta} + \mathfrak{U}^t\mathfrak{U}_t\mathfrak{U}^{[t,\beta]}{}_{;\beta})$ which vanishes by the normalization of $\mathfrak{U}^\alpha$. Hence $\mathfrak{U}^\alpha = N\xi^\alpha$ satisfies the vector equation for any k and K. We have not succeeded in proving that this is the unique solution, but this seems to be a reasonable supposition.

Now, as $k \to 0$, the scalar equation (37) reduces to $(\mu h^{\alpha\beta}\phi_{,\alpha})_{;\beta} = 0$. Multiplying this by $\phi(-g)^{1/2}$, discarding all time derivatives, and integrating over space gives, after an integration by parts and application of the boundary condition at infinity, that $\int \mu\, g^{\alpha\beta}\phi_{,\alpha}\phi_{,\beta}(-g)^{1/2}d^3x = 0$. Because for any static metric, g^{ij} is positive definite and, when defined, $\mu > 0$, this equation is satisfied only by $\phi = $ const. throughout. But for $k \to 0$, the cosmological model has $\phi \to 0$. Hence as $k \to 0$, $\phi \to 0$ in all the space.

Returning to the full scalar equation (37) and recalling that $\ell \propto k^{-3/2}$, it is easy to see that for small but finite k the *gradient* of ϕ scales as k. From the last paragraph it then follows that $\phi = \mathcal{O}(k)$. These last conclusions are actually independent of the form of μ because its argument goes to a nonzero constant in the limit $k \to 0$. We recall [see Eq. (34)] that as $k \to 0$, $\sigma^2 \propto k^{-1}$. Thus the scalars' energy-momentum tensor $\tau_{\alpha\beta}$ is of $\mathcal{O}(k)$ (recall $\ell \propto k^{-3/2}$). From the $\alpha = t$ component of Eq. (38) we see that $\lambda = \mathcal{O}(k) + \mathcal{O}(K)$. Hence $\Theta_{\alpha\beta} = \mathcal{O}(k) + \mathcal{O}(K)$. In addition, the term in the gravitational equations (31) proportional to $1 - e^{-4\phi}$ is itself of $\mathcal{O}(k)$; hence we have $G_{\alpha\beta} = 8\pi G\tilde T_{\alpha\beta} + \mathcal{O}(k) + \mathcal{O}(K)$. Since the difference between $\tilde g_{\alpha\beta}$ and $g_{\alpha\beta}$ is of $\mathcal{O}(\phi)$, namely $\mathcal{O}(k)$, it is obvious that $\tilde G_{\alpha\beta} = 8\pi G\tilde T_{\alpha\beta} + \mathcal{O}(k) + \mathcal{O}(K)$. Thus for quasistatic situations also, TeVeS has GR as its limit when $k \to 0$ with $\ell \propto k^{-3/2}$ and $K \propto k$.

13

In conclusion, the limit $\{k \to 0, \ell \propto k^{-3/2}, K \propto k\}$ of TeVeS is GR, both in cosmology and in quasistatic localized systems.

D. Generic general relativity limit

Milgrom (private communication) has remarked that GR actually follows from TeVeS in the more general limit $K \to 0$ and $\ell \to \infty$ with k arbitrary. This is easily seen after the change of variables $\phi \mapsto \phi_* \equiv \ell\phi$, $\sigma \mapsto \sigma_* \equiv \sqrt{k}\sigma$, whereby only $\tilde{g}_{\alpha\beta}$ and S_s are changed:

$$\tilde{g}_{\alpha\beta} = e^{-2\phi_*/\ell}g_{\alpha\beta} - 2\mathfrak{U}_\alpha\mathfrak{U}_\beta \sinh(2\phi_*/\ell) \tag{48}$$

$$S_s = -\frac{1}{2k^2\ell^2} \int \left[k\sigma_*{}^2 h^{\alpha\beta}\phi_{*,\alpha}\phi_{*,\beta} + \frac{1}{2}G\sigma_*{}^4 F(G\sigma_*{}^2) \right](-g)^{1/2}d^4x, \tag{49}$$

Thus as $\ell \to \infty$ the scalar action disappears and ϕ_* decouples from the theory. In addition, with $K \to 0$, the vector's action S_v disappears apart from the term with λ. All this means that the r.h.s. of the Einstein equations (31) retains only the $\tilde{T}_{\alpha\beta}$ and $\lambda\mathfrak{U}_\alpha\mathfrak{U}_\beta$ terms. But according to the vector equation (38), from which the terms with differentiated ϕ_* and $\mathfrak{U}_\alpha$ have dropped out, $\lambda \to 0$ because $(1 - e^{-4\phi_*/\ell}) \to 0$. Accordingly, we get the usual Einstein equations. Since $g_{\alpha\beta}$ and $\tilde{g}_{\alpha\beta}$ coincide as $\ell \to \infty$, we get exact GR.

In this paper we shall assume that $k \ll 1$ and $K \ll 1$ without restricting ℓ. Empirical bounds on k and K are discussed in Secs. IV C and V.

E. The choice of F

Because we have no theory for the functions $F(\mu)$ or $y(\mu)$, there is great freedom in choosing them. In this paper we shall adopt, as an example, the form

$$y = \frac{3}{4}\frac{\mu^2(\mu - 2)^2}{1 - \mu} \tag{50}$$

plotted in Fig. 1. As y ranges from 0 to ∞, $\mu(y)$ increases monotonically from 0 to unity; for small y, $\mu(y) \approx \sqrt{y/3}$. For negative y the function $\mu(y)$ is double-valued. As y decreases from 0, one branch decreases monotonically from $\mu = 2$ and tends to unity as $y \to -\infty$, while the second increases monotonically from $\mu = 2$ and diverges as $y \to -\infty$. We adopt the second (far right) branch as the physical one.

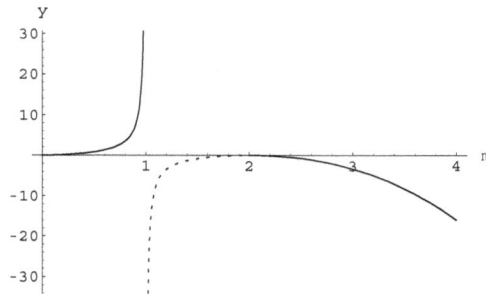

FIG. 1: **The function $y(\mu)$ as relevant for quasistationary systems, $0 < \mu < 1$, and for cosmology, $2 < \mu < \infty$.**

What features of the above $y(\mu)$ are essential for the following sections ? The denominator in Eq. (50) is included so that μ shall asymptote to unity for $y \to \infty$ (the Newtonian limit, c.f. Sec. IV C). The factor μ^2 ensures that the MOND limit is contained in the theory (see Sec. IV B), while the factor $(\mu - 2)^2$ ensures there exists a monotonically decreasing branch of $\mu(y)$ which covers the whole of the range $y \in [0, -\infty)$ (relevant to cosmology, c.f. Sec. VII) and only it.

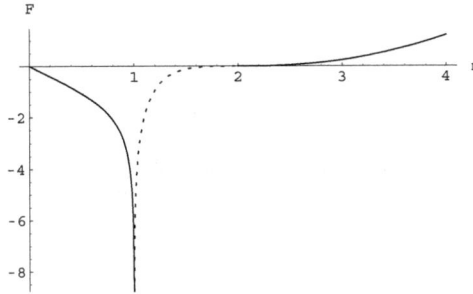

FIG. 2: **The function $F(\mu)$ as relevant for quasistationary systems, $0 < \mu < 1$, and for cosmology, $2 < \mu < \infty$.**

Integrating Eq. (36) with $y(\mu)$ we obtain (see Fig. 2)

$$F = \frac{3}{8} \frac{\mu \left(4 + 2\mu - 4\mu^2 + \mu^3\right) + 2\ln[(1-\mu)^2]}{\mu^2}, \tag{51}$$

where we ignore a possible integration constant (which will, however, be useful in Sec. VII F below). Obviously $F < 0$ in the range $\mu \in (0, 1)$ (relevant for quasistationary systems) but $F > 0$ for $\mu > 2$ (the cosmological range). Where negative, F contributes negative energy density in the energy momentum tensor (32). Despite this there seems to be no collision with the requirement of positive overall energy density (see Secs. V and VII A).

IV. NONRELATIVISTIC LIMIT OF TeVeS

Sec. III C 2 shows that in quasistatic systems TeVeS approaches GR in the limit $\{k \to 0,\ \ell \sim k^{-3/2},\ K \sim k\}$. But in what limit do we recover standard Newtonian gravity ? And where is MOND, which is antagonistic to Newtonian gravity, in all this ? This section shows that with our choice of F, both Newtonian and MOND limits emerge from TeVeS for small gravitational potentials, but that MOND requires in addition small gravitational *fields*, just as expected from Milgrom's original scheme.

A. Quasistatic systems

We are here concerned with a quasistatic, *weak* potential and *slow* motion situation, such as a galaxy or the solar system. As in Sec. III C 2, quasistatic means we can neglect time derivatives in comparison with spatial ones. Let us assume that the metric $g_{\alpha\beta}$ is nearly flat and that $|\phi| \ll 1$. Then linearization of Eq. (31) in terms of the Newtonian potential V generated by the energy content on its r.h.s. gives $g_{tt} = -(1 + 2V) + \mathcal{O}(V^2)$. From the prescription given in Sec. III C 2, $\mathfrak{U}_\alpha = -[1 + V + \mathcal{O}(V^2)]\delta_{t\alpha}$. It follows from Eq. (22) that to $\mathcal{O}(\phi)$ and $\mathcal{O}(V)$, $\tilde{g}_{tt} = -(1 + 2V + 2\phi)$. Thus in TeVeS the total potential governing all nonrelativistic motion is $\Phi = V + \phi$. We should remark that if asymptotically $\phi \to \phi_c \neq 0$, the $\tilde{g}_{tt}$ does not there correspond to a Minkowski metric. This is remedied by rescaling the time (or spatial) coordinates by factors e^{ϕ_c} (or $e^{-\phi_c}$). With respect to the new coordinates the metric is then asymptotically Minkowskian. In this paper we assume throughout that $|\phi_c| \ll 1$; Sec. VII shows this is consistent with cosmological evolution of ϕ.

How is Φ related to Φ_N, the Newtonian gravitational potential generated by the mass density $\tilde{\rho}$ according to Poisson's equation with gravitational constant G ? To relate ϕ to Φ_N we first set temporal derivatives in Eq. (42) to zero which means replacing $h^{\alpha\beta}\phi_{,\alpha} \to g^{\alpha\beta}\phi_{,\alpha}$:

$$\left[\mu\left(k\ell^2 g^{\mu\nu}\phi_{,\mu}\phi_{,\nu}\right) g^{\alpha\beta}\phi_{,\alpha}\right]_{;\beta} = kG(\tilde{\rho} + 3\tilde{p})\,e^{-2\phi}. \tag{52}$$

This equation is still exact. Next we replace $g^{\alpha\beta} \to \eta^{\alpha\beta}$ as well as $e^{-2\phi} \to 1$. This is the nonrelativistic approximation. Further, to be consistent we must neglect $\tilde{p}$ compared to $\tilde{\rho}$; keeping the former would be tantamount to accepting that V is not small. Thus

$$\boldsymbol{\nabla} \cdot \left[\mu\left(k\ell^2(\boldsymbol{\nabla}\phi)^2\right)\boldsymbol{\nabla}\phi\right] = kG\tilde{\rho}. \tag{53}$$

This is just the AQUAL equation (3) with a suitable reinterpretation of the function μ. Now comparing Eq. (53) with Poisson's equation we see that

$$k^{-1}\mu|\nabla\phi| = \mathcal{O}(|\nabla\Phi_N|) \tag{54}$$

This will be made more precise below in situations with symmetry.

We now show that it is consistent to take $V = C\Phi_N$, with C a constant close to unity (to be determined). The starting point are the modified Einstein equations (31). With F as in (51), $F < 0$ simultaneously with $F' < 0$ for $0 < \mu < 1$; it follows from Eq. (36) that $\mu|F| < y$. Now the F term on the r.h.s. of Eq. (31) is $-2\pi G^2\ell^{-2}\sigma^4 F(kG\sigma^2)g_{\alpha\beta} = -2\pi k^{-2}\ell^{-2}\mu^2 F(\mu)g_{\alpha\beta}$. Similarly, since $\phi_{,t} = 0$ here, the terms on the r.h.s. involving $\phi_{,\alpha}$ are of order $8\pi G\sigma^2 h^{\gamma\delta}\phi_{,\gamma}\phi_{,\delta}\,g_{\alpha\beta} = 8\pi k^{-2}\ell^{-2}\mu y(\mu)g_{\alpha\beta}$. Thus by our earlier remark the ϕ derivative terms in $\tau_{\alpha\beta}$ dominate the F term, and by Eq. (54) they are of order $8\pi k\mu^{-1}(\nabla\Phi_N)^2$. But $(\nabla\Phi_N)^2$ is precisely the type of source (Newtonian gravitational energy or stress density) needed to compute the first nonlinear or $\mathcal{O}(\Phi_N{}^2)$ contributions to the metric. As we shall see in Sec. VII, we need $k \sim 10^{-2}$, so that if all we desire is to compute the metric to $\mathcal{O}(\Phi_N)$, and μ is not very small, then all of $\tau_{\alpha\beta}$ may be neglected.

Further, since $\mathfrak{U}_\alpha = -[1 + V + \mathcal{O}(V^2)]\delta_{t\alpha}$, the $\mathfrak{U}_{[\alpha,\beta]}{}^2$ terms in $\Theta_{\alpha\beta}$ have the form $(C\nabla\Phi_N)^2$; we drop them for the same reason that we dropped the $\mathcal{O}(\Phi_N{}^2)$ term in $\tau_{\alpha\beta}$. It follows that in the weak potential approximation the spatio-temporal and spatial-spatial components of Einstein's equations are exactly the same as in GR because the term proportional to $1 - e^{-4\phi}$ can be dropped by virtue of the slow motion condition which suppresses the spatio-temporal components of $T_{\alpha\beta}$. The temporal-temporal component of Einstein's equations depends on λ, and is thus another story. From Eqs. (38) and (40) and the observation that $\mathfrak{U}^\alpha\phi_{,\alpha} = 0$,

$$\lambda = K\mathfrak{U}_\alpha\mathfrak{U}^{[\alpha;\beta]}{}_{;\beta} - 16\pi G\tilde{\rho}\sinh(2\phi). \tag{55}$$

With our $\mathfrak{U}_\alpha$ the first term is $K\mathfrak{U}_t\mathfrak{U}^{[t;\beta]}{}_{;\beta} = -KC\nabla^2\Phi_N + KC^2\mathcal{O}(\nabla\Phi_N{}^2)$, where by Poisson's equation $\nabla^2\Phi_N = 4\pi G\tilde{\rho}$. Further, as we shall see in Sec. V, ϕ is always very close to its aforementioned asymptotic value ϕ_c (which is just ϕ's very slowly varying cosmological value). Dropping the $C^2\mathcal{O}(\nabla\Phi_N{}^2)$ contribution for the same reason as above gives

$$\lambda \approx -8\pi G[KC/2 + 2\sinh(2\phi_c)]\tilde{\rho}. \tag{56}$$

Substituting this in Eq. (33) and combining the result with the $(1 - e^{-4\phi_c})$ term in the G_{tt} equation Eq. (31), we see that $(e^{-2\phi_c} + KC/2)\tilde{\rho}$ replaces the source $\tilde{\rho}$ appropriate in the weak potential approximation to GR. By linearizing the G_{tt} equation as done in GR, we conclude that

$$V = (e^{-2\phi_c} + KC/2)\Phi_N \tag{57}$$

which verifies the claim that V is proportional to Φ_N. Indeed, since the proportionality constant here must be identical with C, we have $C = (1 - K/2)^{-1}e^{-2\phi_c}$. Since we shall show in Sec. VII that it is consistent to assume $|\phi_c| \ll 1$, and assume that $K \ll 1$, we shall replace C everywhere by $\Xi \equiv 1 + K/2 - 2\phi_c$. In particular

$$\Phi = \Xi\Phi_N + \phi. \tag{58}$$

In summary, Eq. (58), which is subject to corrections of $\mathcal{O}(\Phi_N{}^2)$, quantifies the difference at the nonrelativistic level between TeVeS and GR, a difference which is in harmony with our conclusion in Sec. III C 2. We shall use it until we turn to post-Newtonian corrections. The condition "μ is not very small" which we imposed above to be able to neglect the $\tau_{\alpha\beta}$ contribution to the gravitational equations is not restrictive. For the Newtonian limit we shall see that $\mu \approx 1$. And when $\mu \ll 1$ (extreme MOND limit relevant for extragalactic phenomena), the consequent corrections of $\mathcal{O}(\Phi_N{}^2)$ (with large coefficient) to V are entirely ignorable because this potential is then dominated by ϕ in the expression for Φ, c.f. Eq. (59).

B. The MOND limit: spherical symmetry

First for orientation we assume a spherically symmetric situation. Then from Eq. (53) together with Gauss' theorem we infer that

$$\nabla\phi = (k/4\pi\mu)\nabla\Phi_N. \tag{59}$$

In view of Eq. (58) we have

$$\tilde{\mu}\nabla\Phi = \nabla\Phi_N. \tag{60}$$

16

with

$$\tilde{\mu} \equiv (\Xi + k/4\pi\mu)^{-1}. \tag{61}$$

Consider the case $\mu \ll 1$ for which $\mu\big(k\ell^2(|\boldsymbol{\nabla}\phi|)^2\big) \approx (k/3)^{1/2}\ell|\boldsymbol{\nabla}\phi|$ (see Sec. III E). Eliminating $\boldsymbol{\nabla}\Phi_N$ between Eqs. (59) and (60) and defining

$$\mathfrak{a}_0 \equiv \frac{(3k)^{1/2}}{4\pi\Xi\ell} \tag{62}$$

we obtain a quadratic equation for μ with positive root

$$\mu = (k/8\pi\Xi)\big(-1 + \sqrt{1 + 4|\boldsymbol{\nabla}\Phi|/\mathfrak{a}_0}\,\big) \tag{63}$$

This is obviously valid only when $|\boldsymbol{\nabla}\Phi| \ll (4\pi/k)^2\mathfrak{a}_0$ since otherwise μ is not small. From Eq. (61) we now deduce the MOND function

$$\tilde{\mu} = \frac{1}{\Xi}\frac{-1 + \sqrt{1 + 4|\boldsymbol{\nabla}\Phi|/\mathfrak{a}_0}}{1 + \sqrt{1 + 4|\boldsymbol{\nabla}\Phi|/\mathfrak{a}_0}} \tag{64}$$

For $|\boldsymbol{\nabla}\Phi| \ll \mathfrak{a}_0$ (which is consistent with the above restriction since $k \ll 1$) this equation gives to lowest order in K and ϕ_c

$$\tilde{\mu} \approx |\boldsymbol{\nabla}\Phi|/\mathfrak{a}_0. \tag{65}$$

Thus if we identify our $\mathfrak{a}_0$ with Milgrom's constant, Eq. (60) with this $\tilde{\mu}$ coincides with the MOND formula (1) in the extreme low acceleration regime. Therefore, TeVeS recovers MOND's successes in regard to low surface brightness disk galaxies, dwarf spheroidal galaxies, and the outer regions of spiral galaxies. For all these the low acceleration limit of Eq. (1) is known to summarize the phenomenology correctly.

Now suppose $|\boldsymbol{\nabla}\Phi|$ varies from an order below $\mathfrak{a}_0$ up to a couple of orders above it. This respects the condition $|\boldsymbol{\nabla}\Phi| \ll (4\pi/k)^2\mathfrak{a}_0$. Then Eq. (64) shows $\tilde{\mu}$ to grow monotonically from about 0.1 to 0.9. Then Eq. (60) is essentially formula (1) in the intermediate MOND regime. This regime is relevant for the disks of massive spiral galaxies well outside the central bulges but not quite in their outer reaches. It is known that the precise form of $\tilde{\mu}$ makes little difference for the task of predicting detailed rotation curves from surface photometry.

We see that TeVeS reproduces the MOND paradigm encapsulated in Eq. (1) for not too large values of $|\boldsymbol{\nabla}\Phi|/\mathfrak{a}_0$. What happens for very large $|\boldsymbol{\nabla}\Phi|/\mathfrak{a}_0$?

C. The Newtonian limit: spherical symmetry

According to our choice of $y(\mu)$, Eq. (50), the limit $\mu \to 1$ corresponds to $y \to \infty$, that is to say $|\boldsymbol{\nabla}\phi| \to \infty$. By Eqs. (59)–(61) we simultaneously have $|\boldsymbol{\nabla}\Phi| \to \infty$ and $\tilde{\mu} \to (\Xi + k/4\pi)^{-1}$. Defining the Newtonian gravitational constant by

$$G_N = (\Xi + k/4\pi)G, \tag{66}$$

we see from Eq. (60) that $\boldsymbol{\nabla}\Phi$ is obtained from $\boldsymbol{\nabla}\Phi_N$ by just replacing $G \to G_N$ in it. In other words, in the nonrelativistic and arbitrarily large $|\boldsymbol{\nabla}\Phi|$ regime, TeVeS is equivalent to Newtonian gravity, but with a "renormalized" value of the gravitational constant. Now Ξ is really a surrogate of $C = (1 - K/2)^{-1}e^{-2\phi_c}$; hence for $K < 2$, G_N is positive. As mentioned, we here assume $K \ll 1$.

But how close are dynamics to Newtonian for *large but finite* $|\boldsymbol{\nabla}\Phi|/\mathfrak{a}_0$? Expanding the r.h.s. of Eq. (50) in the neighborhood of $\mu = 1$ gives

$$y = \frac{3/4}{1 - \mu} + \mathcal{O}(1 - \mu). \tag{67}$$

We also have by Eqs. (59) and (60) that $y \equiv k\ell^2|\boldsymbol{\nabla}\phi|^2 \approx (k^3\ell^2/16\pi^2)|\boldsymbol{\nabla}\Phi|^2$ where we have dropped corrections of higher order in $(k/4\pi)$. Dropping the $\mathcal{O}(1 - \mu)$ term in $y(\mu)$ and eliminating ℓ in favor of $\mathfrak{a}_0$ (with $\Xi = 1$) we get

$$\mu \approx 1 - \frac{64\pi^4}{k^4}\frac{\mathfrak{a}_0{}^2}{|\boldsymbol{\nabla}\Phi|^2} \tag{68}$$

Thus to trust the approximation $\mu \approx 1$ we must have $|\boldsymbol{\nabla}\Phi|/\mathfrak{a}_0 \gg 8\pi^2 k^{-2}$. Using Eqs. (68) and (61) we obtain, again after dropping higher order terms in k, that

$$\tilde{\mu} \approx \frac{G}{G_N}\left(1 - \frac{16\pi^3}{k^3}\frac{\mathfrak{a}_0{}^2}{|\boldsymbol{\nabla}\Phi|^2}\right). \tag{69}$$

Here the factor (G/G_N) just reflects the mentioned "renormalization" of the gravitational constant; it is the next factor which interests us as a measure of departures from strict Newtonian behavior. For example, if $k = 0.03$ there is a 5.3×10^{-9} fractional enhancement of the sun's Newtonian field at Earth's orbit where $|\boldsymbol{\nabla}\Phi| = 0.59\,\mathrm{cm}\,\mathrm{s}^{-2}$. This is probably unobservable today. At Saturn's orbit where $|\boldsymbol{\nabla}\Phi| = 0.0065\,\mathrm{cm}\,\mathrm{s}^{-2}$ the fractional correction is 4.3×10^{-5}, corresponding to an excess acceleration $2.8 \times 10^{-7}\,\mathrm{cm}\,\mathrm{s}^{-2}$ (at this point μ departs from unity by only 0.018 so that Eq. (68) is still reliable). Although this departure from Newtonian predictions seems serious, it should be remembered that navigational data from the Pioneer 10 and 11 spacecrafts seem to disclose a constant acceleration in excess of Newtonian of about $8 \times 10^{-8}\,\mathrm{cm}\,\mathrm{s}^{-2}$ between Uranus' orbit and the trans-Plutonian region [63]. It is, however, unclear whether the correction in Eq. (69), sensitive as it is to the choice of F, has anything to do with the "Pioneer anomaly".

D. Nonspherical systems

We now consider *generically* asymmetric systems. Since any system has a region where μ differs from unity and is variable, Eq. (59) is not the general solution of Eq. (53) and must be replaced by

$$\boldsymbol{\nabla}\phi = (k/4\pi\mu)(\boldsymbol{\nabla}\Phi_N + \boldsymbol{\nabla}\times\boldsymbol{h}), \tag{70}$$

where $\boldsymbol{h}$ is some regular vector field which is determined up to a gradient by the condition that the curl of the r.h.s. of Eq. (70) vanish.

The freedom inherent in $\boldsymbol{h}$ allows it to be made divergenceless. Then by Gauss' theorem $\boldsymbol{h}$ must fall off faster than $1/r^2$ and $\boldsymbol{\nabla}\times\boldsymbol{h}$ faster than $1/r^3$ at large distances. On physical grounds $|\boldsymbol{\nabla}\times\boldsymbol{h}|$ is expected to be of the same order as $|\boldsymbol{\nabla}\Phi_N|$ well inside the matter. But since the latter quantity falls off as $1/r^2$ well outside the matter, the curl term in Eq. (70) must rapidly become negligible well outside the system. We thus expect the discussion in Sec. IV B to apply well outside any nonspherical galaxy just as it applies anywhere inside a spherical one. The interior and near exterior of such a galaxy, where $\boldsymbol{\nabla}\times\boldsymbol{h}$ is still important, must be treated by numerical methods which would be no different than those developed by Milgrom within the old AQUAL theory [51].

Needless to say, an asymmetric system so dense that the Newtonian regime (μ approximately constant) obtains in its interior, e.g. an oblate globular cluster like ω Centauri, can be described everywhere without an $\boldsymbol{h}$. For in the interior $\boldsymbol{h}$ is not needed since even in its absence the curl of the r.h.s. of Eq. (70) vanishes (approximately). And μ begins to differ substantially from unity only well outside the system where we know from our previous argument that any $\boldsymbol{h}$ is becoming negligible. Hence both Newtonian and MOND regimes of the system may be described as in Secs. IV B and IV C.

In summary, we see that the extragalactic predictions of the MOND equation (1) are recovered from TeVeS; at the same time TeVeS hints at non-Newtonian behavior in the reaches of the solar system, though the effect is sensitive to the choice of F in the theory.

V. THE POST-NEWTONIAN CORRECTIONS

The upshot of the discussion at the end of Sec. III C 2 is that in the solar system (regarded as a static system—with rotation neglected—embedded in a FRW cosmological background), $\tilde{G}_{\alpha\beta} = 8\pi G\tilde{T}_{\alpha\beta} + \mathcal{O}(k) + \mathcal{O}(K)$. Here we compute the consequent $\mathcal{O}(k) + \mathcal{O}(K)$ corrections to the Schwarzschild metric

$$g_{\alpha\beta}\,dx^\alpha dx^\beta = -\frac{(1 - Gm/2\varrho)^2}{(1 + Gm/2\varrho)^2}dt^2 + (1 + Gm/2\varrho)^4[d\varrho^2 + \varrho^2(d\theta^2 + \sin^2\theta\,d\varphi^2)] \tag{71}$$

that describes the exterior of a spherically mass m, and determine the post-Newtonian parameters of TeVeS which we compare with those of GR.

Rather than just extending the Newtonian limit calculation of Sec. IV C, we start from scratch. First we write the spherically symmetric and static metric of the sun (inside and outside it) as

$$g_{\alpha\beta}\,dx^\alpha dx^\beta = -e^\nu dt^2 + e^\varsigma[d\varrho^2 + \varrho^2(d\theta^2 + \sin^2\theta\,d\varphi^2)] \tag{72}$$

with $\nu = \nu(\varrho)$ and $\varsigma = \varsigma(\varrho)$. Just as for metric (71), outside the sun these functions should admit the expansions (α_i and β_i are dimensionless constants)

$$e^\nu = 1 - r_g/\varrho + \alpha_2(r_g/\varrho)^2 + \cdots \tag{73}$$

$$e^\varsigma = 1 + \beta_1 r_g/\varrho + \beta_2(r_g/\varrho)^2 + \cdots, \tag{74}$$

where r_g is a lengthscale to be determined (see Appendix D). The magnitude of the coefficient of the r_g/ϱ term in Eq. (73) has been absorbed into r_g; its sign must be negative, as shown, because gravity is attractive. From the fact that TeVeS approaches GR for small k and K, we may infer that r_g is close to $2G$ times the system's Newtonian mass. This is made precise below.

Taking $\phi = \phi(\varrho)$ and $\tilde{T}_{\alpha\beta}$ from Eq. (40), we may write the scalar equation (42) as

$$\varrho^{-2} e^{-(\nu+3\varsigma)/2} [\mu e^{(\nu+\varsigma)/2} \varrho^2 \phi']' = kG e^{-2\phi}(\tilde{\rho} + 3\tilde{p}). \tag{75}$$

Here $'$ stands for $d/d\varrho$. The first integral of Eq. (75) is

$$\phi' = \frac{kG e^{-(\nu+\varsigma)/2}}{\mu \varrho^2} \int_0^\varrho (\tilde{\rho} + 3\tilde{p}) e^{\nu/2 + 3\varsigma/2 - 2\phi} \varrho^2 d\varrho, \tag{76}$$

where the integration constant has been chosen so that ϕ is regular at $\varrho = 0$.

Supposing the matter's boundary is at $\varrho = R$, we define the (positive) "scalar mass"

$$m_s \equiv 4\pi \int_0^R (\tilde{\rho} + 3\tilde{p}) e^{\nu/2 + 3\varsigma/2 - 2\phi} \varrho^2 d\varrho. \tag{77}$$

Because for a nonrelativistic fluid $\tilde{p} \ll \tilde{\rho}$, m_s must be close to the Newtonian mass. In fact, as shown in Appendix D, m_s and an appropriately defined gravitational mass m_g differ only by a fraction of $\mathcal{O}(Gm_g/R)$ which amounts to 10^{-5} for the inner solar system. For $\varrho > R$ we may expand ϕ' as

$$\phi' = \frac{kG m_s}{4\pi\mu} \left[\frac{1}{\rho^2} + \frac{(1-\beta_1) r_g}{2\varrho^3} + \mathcal{O}(\varrho^{-4}) \right]. \tag{78}$$

It is obvious from this that ϕ *decreases inward*. Its asymptotic value, as will be explained in Sec. VII, is positive and of $\mathcal{O}(k)$. The decrement in ϕ down to "radius" ϱ is, according to Eq. (76), or its integral Eq. (92) below, of $\mathcal{O}(kGm_s/4\pi\varrho)$. In any weakly gravitating system, $Gm_s/\varrho \ll 1$ and for strongly gravitating systems like a neutron star, Gm_s/ϱ is still well below unity (black holes require a special discussion which we defer to another occasion). Thus ϕ remains positive and small throughout space for all systems, and for the solar system in particular. This will have repercussions for the causality question examined in Sec. VIII.

Since we are not here interested in purely MOND corrections, we shall take $\mu = 1$ in Eq. (78) as well as in the terms in $\tau_{\alpha\beta}$, Eq. (32), which explicitly involve ϕ derivatives. The μ in the F term of $\tau_{\alpha\beta}$ is not so easily disposed of because with our choice of F, and indeed with any viable one, F must be singular at $\mu = 1$. If neglecting the F term in $8\pi G\tau_{\alpha\beta}$ can be justified, then using Eq. (78) we may compute from Eq. (32) that for $\varrho > R$

$$8\pi G\tau_{tt} = 8\pi G\tau_{\varrho\varrho} = \frac{kG^2 m_s^2}{4\pi\varrho^4} + \mathcal{O}(\varrho^{-5}). \tag{79}$$

Now by the approximation (68) the ratio of the F term in $8\pi G\tau_{\alpha\beta}$ to these last terms is

$$\frac{8\pi^2 \mu^2 |F(\mu)| \varrho^4}{k^3 \ell^2 G^2 m_s^2} = \frac{128\pi^4 \mathfrak{a}_0^2 \mu^2 |F(\mu)|}{3k^4 |\boldsymbol{\nabla}\Phi_N|^2} \approx \frac{2}{3}(1-\mu)|F(\mu)| \tag{80}$$

which numerically does not exceed 0.04 for $\mu > 0.99$. This justifies Eq. (79) in any region where MOND effects are totally negligible. However, as pointed out in Sec. IV C, at Saturn's orbit μ already departs from unity by two percent. In such cases the contribution of the F term to $\tau_{\alpha\beta}$ must be taken into account, and its post-Newtonian effects compared with the MOND departure from strict Newtonian behavior calculated in Sec. IV C. Here we shall only be concerned with inner solar system dynamics where μ is very close to unity. Because τ_{tt} is dominated by the derivative terms, the energy density contributed by the scalar fields is evidently positive.

Clearly in our situation (see Sec. III C 2)

$$\mathfrak{U}^\alpha = \{e^{-\nu/2}, 0, 0, 0\}. \tag{81}$$

Using this in Eqs. (33) and (38) we find for $\varrho > R$ that

$$\lambda = \frac{K(2 + \beta_1 - 4\alpha_2)r_g{}^2}{4\varrho^4} + \mathcal{O}(\varrho^{-5}) \tag{82}$$

$$\Theta_{tt} = \frac{K(-2\beta_1 - 3 + 8\alpha_2)r_g{}^2}{8\varrho^4} + \mathcal{O}(\varrho^{-5}) \tag{83}$$

$$\Theta_{\varrho\varrho} = -\frac{Kr_g{}^2}{8\varrho^4} + \mathcal{O}(\varrho^{-5}) \tag{84}$$

With this we now turn to Einstein's equations (31) for all ϱ. By virtue of $\mathfrak{U}^\alpha$'s form here, the tt and $\varrho\varrho$ components simplify to

$$-e^{\nu-\varsigma}\left(\varsigma'' + \tfrac{1}{4}\varsigma'^2 + 2\varsigma'/\varrho\right) = 8\pi G\left[(2e^{-4\phi} - 1)\tilde{T}_{tt} + \tau_{tt}\right] + \Theta_{tt} \tag{85}$$

$$\tfrac{1}{4}\varsigma'^2 + \tfrac{1}{2}\varsigma'\nu' + (\varsigma' + \nu')/\varrho = 8\pi G\left[\tilde{T}_{\varrho\varrho} + \tau_{\varrho\varrho}\right] + \Theta_{\varrho\varrho} \tag{86}$$

First we solve these for $\varrho > R$ where $\tilde{T}_{\alpha\beta} = 0$. From Eqs. (73) and (74) it follows that

$$\nu' = r_g/\varrho^2 + (1 - 2\alpha_2)r_g{}^2/\varrho^3 + \cdots \tag{87}$$

$$\varsigma' = -\beta_1 r_g/\varrho^2 + (\beta_1{}^2 - 2\beta_2)r_g{}^2/\varrho^3 + \cdots \tag{88}$$

Substituting these together with Eqs. (73), (74), (79) and (83) in Eqs. (85)-(86), matching coefficients of like powers of $1/\varrho$, and solving the three resulting algebraic conditions gives to lowest order in k and K

$$\beta_1 = 1 \tag{89}$$

$$\alpha_2 = \frac{1}{2} \tag{90}$$

$$\beta_2 = \frac{3}{8} + \frac{1}{16}K - \frac{kG^2 m_s{}^2}{8\pi r_g{}^2} \tag{91}$$

Using these results we show in Appendix D that $r_g = 2Gm_g[1 + \mathcal{O}(kGm_g/R) + \mathcal{O}(KGm_g/R)]$ with m_g, the gravitational mass, defined by Eq. (D4). The relative correction here is $\ll 10^{-5}$ for the inner solar system. We also remark that with the values (89)-(91) the energy density contributed by Θ_{tt} is positive (see Eq. (83)).

For solar system tests of $TeVeS$ we must know the physical metric $\tilde{g}_{\mu\nu}$. According to Eqs. (22) and (81), $\tilde{g}_{tt} = -e^{2\phi+\nu}$, $\tilde{g}_{\varrho\varrho} = \tilde{g}_{\theta\theta}/\varrho^2 = g_{\varphi\varphi}/\varrho^2 \sin^2\theta = e^{-2\phi+\varsigma}$, so we need ϕ. Integration of Eq. (78) in light of Eq. (89) gives

$$\phi(\varrho) = \phi_c - \frac{kGm_s}{4\pi\varrho} + \mathcal{O}(\varrho^{-3}), \tag{92}$$

whereupon

$$e^{\pm 2\phi} = e^{\pm 2\phi_c}\left(1 \mp \frac{kGm_s}{2\pi\varrho} + \frac{k^2G^2 m_s^2}{8\pi^2\varrho^2} + \mathcal{O}(\varrho^{-3})\right). \tag{93}$$

The integration constant ϕ_c is evidently the cosmological value of ϕ at the epoch in question. This value changes slowly over solar system timescales, so we can ignore its drift for most purposes. Thus by taking the advantage of the isotropic form of the metric (72), and rescaling the t and ϱ coordinates appropriately, we absorb the factors $e^{2\phi_c}$ and $e^{-2\phi_c}$ that would otherwise appear in $\tilde{g}_{\mu\nu}$ so that it can asymptote to Minkowskian form as expected. With this precaution one can calculate as if ϕ_c vanished. It must be stressed that this strategy works at a particular cosmological era.

Accordingly

$$\tilde{g}_{tt} = -1 + 2G_N m \varrho^{-1} - 2\beta G_N{}^2 m^2 \varrho^{-2} + \mathcal{O}(\varrho^{-3}) \tag{94}$$

$$\tilde{g}_{\varrho\varrho} = 1 + 2\gamma G_N m \varrho^{-1} + \mathcal{O}(\varrho^{-2}) \tag{95}$$

$$G_N m \equiv r_g/2 + (kGm_s/4\pi) \tag{96}$$

$$\beta = 1 \tag{97}$$

$$\gamma = 1 \tag{98}$$

As previously, G_N is defined by Eq. (66). Recalling the relations between r_g, m_g and m_s (Appendix D), we find that $m = m_g[1 + \mathcal{O}(kGm_g/R)] + \mathcal{O}(KGm_g/R)]$, i.e., in the inner solar system m and m_g differ fractionally by $\ll 10^{-5}$. Setting $r_g = 2Gm_g = 2Gm$ gives the second form of β. Our results for β and γ are consistent with those obtained by Eiling and Jacobson [62] for the relevant case of the Jacobson-Mattingly theory.

The β and γ are the standard post-Newtonian coefficients measurable by the classical tests of gravity theory [54]. They are both unity in TeVeS, exactly as in GR (for β this was first noticed by Giannios). Consequently the classical tests (perihelion precession, light deflection and radar time delay) cannot distinguish between the two theories with present experimental precision.

The β and γ are *not* the only PPN coefficients. Future work should look at those coefficients having to do with preferred frame effects, as well as at the Nortvedt effect, which should not be null in TeVeS.

VI.   GRAVITATIONAL LENSING IN TeVeS

In Sec. V we touched upon gravitational lensing in the Newtonian regime. Here we show that in the low acceleration regime, TeVeS predicts gravitational lensing of the correct magnitude to explain the observations of intergalactic lensing without any dark matter. First by following the essentially exact method of Ref. 44, we show this for a spherically symmetric structure; in nature many elliptical galaxies and galaxy clusters are well modelled as spherically symmetric. We then use linearized theory to give a short proof of the same result for asymmetric systems. Our discussion refers to lensing of both rays that pass through the system and those that skirt it, and is thus a generalization of the implicit result about light deflection in Sec. V in more than one way.

A.   Spherically symmetric systems

We adopt the Einstein metric (72); the physical metric is obtained by replacing $e^\nu \to e^{\nu+2\phi}$ and $e^\varsigma \to e^{\varsigma-2\phi}$ in it. Consider a light ray which propagates in the equatorial plane of the metric (which may, of course, be chosen to suit any light ray). The 4-velocity $\dot{x}^\alpha$ of the ray (derivative taken with respect to some suitable parameter) must satisfy

$$- e^{\nu+2\phi}\,\dot{t}^2 + e^{\varsigma-2\phi}(\dot{\varrho}^2 + \varrho^2\dot{\varphi}^2) = 0. \tag{99}$$

From the metric's stationarity follows the conservation law $e^{\nu+2\phi}\dot{t} = E$ where E is a constant characteristic of the ray. From spherical symmetry it follows that $e^{\varsigma-2\phi}\varrho^2\dot{\varphi} = L$ where L is another constant property of the ray. Let us write $\dot{\varrho} = (d\varrho/d\varphi)\dot{\varphi}$. Now eliminating $\dot{t}$ and $\dot{\varphi}$ from Eq. (99) in favor of E and L, and dividing by E^2 yields

$$- e^{-\nu-2\phi} + (b/\varrho)^2 e^{-\varsigma+2\phi}[\varrho^{-2}(d\varrho/d\varphi)^2 + 1] = 0, \tag{100}$$

where $b \equiv L/E$. By going to infinity where the metric factors approach unity one sees that b is just the ray's impact parameter with respect to the matter distribution's center at $\varrho = 0$. This last equation has the quadrature

$$\varphi = \int^\varrho \left[e^{\varsigma-\nu-4\varphi}\left(\frac{\varrho}{b}\right)^2 - 1 \right]^{-1/2} \frac{d\varrho}{\varrho}. \tag{101}$$

Were the physical metric exactly flat, this relation would describe a line with φ varying from 0 to π as ϱ decreased from infinity to its value ϱ_{turn} at the turning point, and then returned to infinity. Hence the deflection of the ray due to gravity is

$$\Delta\varphi = 2\int_{\varrho_{turn}}^\infty \left[e^{\varsigma-\nu-4\varphi}\left(\frac{\varrho}{b}\right)^2 - 1 \right]^{-1/2} \frac{d\varrho}{\varrho} - \pi. \tag{102}$$

This last integral is difficult. So let us take advantage of the weakness of extragalactic fields which allow us to assume that ν, ς and ϕ are all small compared to unity. Then the above result is closely approximated by

$$\Delta\varphi = -4\frac{\partial}{\partial\alpha} \int_{\varrho_{turn}}^\infty \left[(1+\varsigma-\nu-4\varphi)\left(\frac{\varrho}{b}\right)^2 - \alpha \right]^{1/2} \frac{d\varrho}{\varrho}\bigg|_{\alpha=1} - \pi. \tag{103}$$

The rewriting in terms of an α derivative allows us to Taylor expand the radical in the small quantity $\varsigma - \nu - 4\varphi$ without incurring a divergence of the integral at its lower limit. The zeroth order of the expansion yields a well known integral which cancels the π. Thus, to first order in small quantities

$$\Delta\varphi = -\frac{2}{b}\frac{\partial}{\partial\alpha} \int_{b\sqrt{\alpha}}^\infty \frac{(\varsigma-\nu-4\phi)\varrho\,d\varrho}{(\varrho^2-\alpha b^2)^{1/2}}\bigg|_{\alpha=1}. \tag{104}$$

At this point it pays to integrate by parts:

$$\Delta\varphi = -\frac{2}{b}\frac{\partial}{\partial\alpha}\left[\lim_{\varrho\to\infty}(\varsigma - \nu - 4\phi)(\varrho^2 - \alpha b^2)^{1/2} - \int_{b\sqrt{\alpha}}^{\infty}(\varsigma' - \nu' - 4\phi')(\varrho^2 - \alpha b^2)^{1/2}d\varrho\right]\Bigg|_{\alpha=1} \tag{105}$$

Since ν, ς and ϕ all decrease asymptotically as ϱ^{-1}, the integrated term, being α independent, contributes nothing. Carrying out the α derivative, and introducing the usual Cartesian x coordinate along the initial ray by $x \equiv \pm(\varrho^2 - b^2)^{1/2}$, we have

$$\Delta\varphi = \frac{b}{2}\int_{-\infty}^{\infty}\frac{\nu' - \varsigma' + 4\phi'}{\varrho}\,dx. \tag{106}$$

A factor $1/2$ appears because we have included the integral in Eq. (105) twice, once with ϱ decreasing to, and once with ϱ increasing from b. The integral is now performed over an infinite straight line following the original ray.

The difference between GR with dark matter and TeVeS in this respect is that with dark matter one would have $\phi = 0$ and would compute ν and ς from Einstein's equations including dark matter as source, whereas in TeVeS one has a nontrivial ϕ, and computes ν and ς on the basis of the visible matter alone.

We may simplify the above result by means of Einstein's equation (86). We shall neglect the ς'^2 and $\varsigma'\nu'$ terms because they are of second order, and thus smaller than ν'/ϱ by a factor $G \cdot \text{mass}/\varrho$ which amounts to v^2, with v the typical orbital velocity in the system. Using the residual terms we eliminate ς' from Eq. (106):

$$\Delta\varphi = b\int_{-\infty}^{\infty}\frac{\nu' + 2\phi'}{\varrho}\,dx - 4\pi Gb\int_{-\infty}^{\infty}\left(\tilde{T}_{\varrho\varrho} + \tau_{\varrho\varrho} + \Theta_{\varrho\varrho}/8\pi G\right)dx. \tag{107}$$

Now by Sec. IV A, $\nu = 2V + \mathcal{O}(V^2)$ and $\Phi = V + \phi$. Hence with fractional corrections of $\mathcal{O}(V^2)$,

$$\Delta\varphi = 2b\int_{-\infty}^{\infty}\frac{\Phi'}{\varrho}\,dx - 4\pi Gb\int_{-\infty}^{\infty}\left(\tilde{T}_{\varrho\varrho} + \tau_{\varrho\varrho} + \Theta_{\varrho\varrho}/8\pi G\right)dx. \tag{108}$$

The first integral here depends exclusively on the potential Φ which determines nonrelativistic motion. That is, the observed stellar or galactic dynamics will uniquely fix this part of $\Delta\varphi$. For this reason the first term makes the same predictions for lensing by nonrelativistic systems in TeVeS as in GR (where $\Phi = \Phi_N$, the last calculated assuming dark matter). We next show that for nonrelativistic systems the second integral is negligible.

In astrophysical matter the radial pressure $\tilde{T}_{\varrho\varrho}$ is of order $\tilde{\rho}$ times the local squared random velocity of the matter particles (stars, gas clouds, galaxies). Thus $\int \tilde{T}_{\varrho\varrho}\,dx = \langle v^2\rangle\int\tilde{\rho}\,dx$ with $\langle v^2\rangle$ a suitably averaged v^2. But by Poisson's equation $4\pi G\tilde{\rho} = \boldsymbol{\nabla}\cdot\boldsymbol{\nabla}\Phi_N \sim \Phi_N'/\varrho = \tilde{\mu}\Phi'/\varrho$ where we have also used Eq. (60). Thus the term with the integral over $\tilde{T}_{\varrho\varrho}$ is smaller than the first term in Eq. (108) by a factor of $\mathcal{O}(\tilde{\mu}\langle v^2\rangle)$. In GR (for which effectively $\tilde{\mu} = 1$) this factor is no larger than 10^{-5} for all extragalactic systems which have a missing mass problem; in TeVeS it is even smaller because typically $\tilde{\mu} \ll 1$ for such systems.

Turning now to $\tau_{\varrho\varrho}$ we recall from Sec. IV A that in the quasistatic situation in question, the F part is dominated by the term quadratic in ϕ derivatives. Using Eqs. (59)-(60) we work out that $4\pi G\tau_{\varrho\varrho} \approx (k\tilde{\mu}/8\pi\mu)\Phi'\Phi_N'$. Evidently $\Phi' \sim \Phi/\varrho$, and since $\Phi = \mathcal{O}(v^2)$ and $(k\tilde{\mu}/8\pi\mu) < \frac{1}{2}$, the contribution of $\tau_{\varrho\varrho}$ to the second term of Eq. (108) is no larger than that coming from $\tilde{T}_{\varrho\varrho}$.

Finally we note that the λ term in $\Theta_{\varrho\varrho}$ vanishes in a quasistatic situation because then $\mathfrak{U}_\alpha \approx -(1 + \Phi_N)\delta_{\alpha t}$. And from this last formula we estimate $|\Theta_{\varrho\varrho}| \approx \frac{1}{2}K(\Phi_N')^2 \sim K\tilde{\mu}^2|\Phi\Phi'|/\varrho$. Since $\tilde{\mu} < 1$ and by Sec. V we must take $K < 10^{-2}$, it is clear that the contribution of $\Theta_{\varrho\varrho}$ is much smaller than that coming from $\tilde{T}_{\varrho\varrho}$. From all the above the light ray deflection in TeVeS is

$$\Delta\varphi = 2b[1 + \mathcal{O}(\tilde{\mu}v^2)]\int_{-\infty}^{\infty}\frac{\Phi'}{\varrho}dx. \tag{109}$$

In GR with dark matter the same formula is valid with $\mathcal{O}(\tilde{\mu}v^2)$ replaced by $\mathcal{O}(v^2)$. Since these corrections are beyond foreseeable accuracy of extragalactic astronomy, it is clear that for given dynamics (given Φ), both theories predict identical lensing. We shall elaborate on this statement shortly.

B. Asymmetric systems

We now turn to systems with no particular spatial symmetry. The weakness of the gravitational potentials typical of nonrelativistic systems entitles us to use linearized theory [64] in which the metric is viewed as a perturbed Lorentz metric:

$$g_{\alpha\beta} = \eta_{\alpha\beta} + \bar{h}_{\alpha\beta} - \frac{1}{2}\eta_{\alpha\beta}\eta^{\gamma\delta}\bar{h}_{\gamma\delta} \tag{110}$$

with $|\bar{h}_{\alpha\beta}| \ll 1$. By small coordinate transformations one enforces the gauge conditions $\eta^{\beta\delta}\bar{h}_{\gamma\delta,\beta} = 0$; as a consequence to first order in the $\bar{h}$ fields

$$G_{\alpha\beta} = -\frac{1}{2}\eta^{\gamma\delta}\partial_{\gamma}\partial_{\delta}\,\bar{h}_{\alpha\beta}, \tag{111}$$

so that Einstein's equations take the form of wave equations in flat spacetime with the r.h.s. of Eq. (31) as sources. Of course there are motions and changes in galaxies and clusters of galaxies, but the associated changes in the metric are mostly very slow. Thus we confine ourselves to quasistationary situations where we can drop time derivatives (but not yet the g_{ti} components since galaxies do rotate). This tells us that

$$G_{tt} = -\frac{1}{2}\nabla^2\,\bar{h}_{tt} = 8\pi G\Big[\tilde{T}_{tt} + 2(1 - e^{-4\phi})\mathfrak{U}^{\mu}\tilde{T}_{\mu t}\mathfrak{U}_t + \tau_{tt}\Big] + \Theta_{tt}. \tag{112}$$

The various parts of the source here were explored in Sec. IV A; from that discussion it follows that

$$\bar{h}_{tt} = -4V = -4\Xi\Phi_N. \tag{113}$$

In regard to the spatio-temporal source components of Eq. (31), we observe that the $\tilde{T}_{it}$ is an $\mathcal{O}(v)$ below $\tilde{T}_{tt}$ (momentum density is velocity times mass density). Further, the dominant contributions to τ_{ti} are $\bar{h}_{ti}$ multiplied by $\sigma^2\eta^{jk}\phi_{,j}\phi_{,k}$ and by $(G/\ell^2)\sigma^4 F$. Of these the first dominates (see Sec. IV A), and it is small on the scale of $\tilde{\rho}$ both because it is of second order (c.f. Sec. V), and because $|\bar{h}_{ti}| \ll 1$. We can guess that $\mathfrak{U}_i$ is at most of order $\bar{h}_{ti}$ (it would vanish in a truly static situation), and since by Eq. (56) λ is below $8\pi G\tilde{\rho}$ by factors of $\mathcal{O}(K)$ and $\mathcal{O}(\phi_c)$, we see that the $\lambda\mathfrak{U}_t\mathfrak{U}_i$ term contribution to Θ_{ti} is small compared to $8\pi G\tilde{\rho}$. Similarly, the $Kg^{\mu\nu}\mathfrak{U}_{[\mu,t]}\mathfrak{U}_{[\nu,i]}$ contribution to Θ_{ti}, being of second order in $V_{,i}$ and first order in $\bar{h}_{ti}$, or first order in $V_{,i}$ and first order in $\bar{h}_{ti,j}$ (aside of carrying the small coefficient K), must be very small. We conclude that the source of the spatio-temporal Einstein equation can be neglected, so that to the accuracy of Eq. (113), $\bar{h}_{ti} \approx 0$.

Things are similar for the spatial-spatial components. We have already remarked that $\tilde{T}_{ij}$ is an $\mathcal{O}(v^2)$ below $\tilde{T}_{tt}$. The τ_{ij} consists of a term quadratic in $\phi_{,i}$ and one with a F factor which has been argued to be smaller. Hence τ_{ij} is small. Again the $Kg^{\mu\nu}\mathfrak{U}_{[\mu,i]}\mathfrak{U}_{[\nu,j]}$ contributions to Θ_{ij} are quadratic in $V_{,i}$ and suppressed by the K coefficient, so they are also small. And the λ, which we remarked above to be small, is multiplied by two factors $\bar{h}_{ti}$, and so is also small. So by the same logic as above we neglect the sources of the spatial-spatial components $\bar{h}_{ij}$ and conclude that $\bar{h}_{ij} \approx 0$. Substituting all these results in Eq. (110) we obtain

$$g_{\alpha\beta} = (1 - 2V)\eta_{\alpha\beta} - 4V\delta_{\alpha t}\delta_{\beta t}. \tag{114}$$

The absence of g_{ti} in this approximation makes the situation truly static (rather than just stationary); hence $\mathfrak{U}^{\alpha} = \delta_t^{\alpha}$. Calculating the physical metric from Eq. (22) with $e^{\pm 2\phi} \approx 1 \pm 2\phi$ we have

$$\tilde{g}_{\alpha\beta} = (1 - 2V - 2\phi)\eta_{\alpha\beta} - 4(V + \phi)\delta_{\alpha t}\delta_{\beta t} \tag{115}$$

which is equivalent to

$$\tilde{g}_{\alpha\beta}dx^{\alpha}dx^{\beta} = -(1 + 2\Phi)dt^2 + (1 - 2\Phi)\delta_{ij}dx^i dx^j \tag{116}$$

with $\Phi = V + \phi$ as in Sec. IV A.

Metric (116) has the same form as the GR metric for weak gravity [64]. Thus in TeVeS just as in GR the same potential governs dynamics and gravitational lensing. This accords with the conclusion of Sec. VI A for the spherically symmetry case. What does this mean in practice ? In GR Φ's role is played by the Newtonian potential due to the visible matter together with the putative dark matter; in TeVeS Φ is the sum of the scalar field and the renormalized Newtonian potential generated by the visible matter alone. These two prescriptions for Φ need not agree *a priori*, but as we argued in Sec. IV B, nonrelativistic dynamics in TeVeS are approximately of MOND form, and MOND's predictions have been found to agree with much of galaxy dynamics phenomenology. We thus expect TeVeS's predictions for gravitational lensing by galaxies and some clusters of galaxies to be as good as those of dark halo models within GR. But, of course, the early MOND formula (1), and TeVeS with our choice (51) for $F(\mu)$ both claim that asymptotically the potential Φ of an isolated galaxy grows logarithmically with distance indefinitely. Dark halo models do not. So TeVeS for a specific choice of F is in principle falsifiable. Dark matter is less falsifiable because of the essentially unlimited choice of halo models and choices of their free parameters. One should also remember that gravitational lensing affords the opportunity to map the Φ to greater distances than can dynamics; for unlike the latter, lensing can be measured outside the gas or galaxy distribution. Using this Φ both GR and TeVeS would predict the same dynamics for stars or galaxies, while disagreeing on the implied distribution of mass.

VII. COSMOLOGICAL EVOLUTION OF ϕ

A. Persistence of cosmological expansion

This section (where we write ϕ rather than ϕ_c) shows that for a range of initial conditions, FRW cosmological models with flat spaces in TeVeS expand forever, have $0 \leq \phi \ll 1$ throughout, and their law of expansion is very similar to that in GR. The second point is crucial for our discussion of causality in Sec. VIII.

First using Eq. (22) we transform metric (43) to the physical metric

$$\tilde{g}_{\alpha\beta} \, dx^\alpha dx^\beta = -d\tilde{t}^2 + \tilde{a}(\tilde{t})^2 [d\chi^2 + f(\chi)^2 (d\theta^2 + \sin^2\theta \, d\varphi^2)], \tag{117}$$

$$d\tilde{t} = e^\phi dt; \qquad \tilde{a} = e^{-\phi} a. \tag{118}$$

In what follows we take the initial moment, conventionally written as $\tilde{t} = 0$, at the end of the quantum era with $\tilde{a}(0)$ a very small scale; furthermore we take the zero of t to coincide with $\tilde{t} = 0$. For illustration we assume the initial conditions $\dot{\phi}(0) = 0$ (an overdot always denotes $\partial/\partial t$) and $0 < \phi_0 \equiv \phi(0) \ll 1$. Hence a also starts off from a very small scale, a_0, and can only increase initially.

We now show that the spatially flat $(f(\chi) \equiv \chi)$ FRW models in TeVeS persist and cannot recollapse, i.e. $\tilde{a}$ has no finite maximum. As in Sec. III C 1 we have $\mathfrak{U}^\alpha = \delta_t{}^\alpha$ which causes $\mathfrak{U}^{[\alpha;\beta]}$ to vanish. As a consequence $\Theta_{\alpha\beta} = -\lambda\delta_\alpha^t \delta_\beta^t$ with λ given by Eq. (46). Since $\phi = \phi(t)$, Eq. (32) gives $\tau_{tt} = 2\sigma^2\dot{\phi}^2 + G(4\ell^2)^{-1}\sigma^4 F(\mu)$. As mentioned in Sec. III C 1, $\mathfrak{U}^\beta \tilde{T}_{\alpha\beta} = -\tilde{\rho}e^{2\phi}\mathfrak{U}_\alpha$. Using $g^{\alpha\beta}\mathfrak{U}_\alpha \mathfrak{U}_\beta = -1$ gives us $\tilde{T}_{tt} + (1 - e^{-4\phi})\mathfrak{U}^\alpha \tilde{T}_{\alpha(t}\mathfrak{U}_{t)} = (2e^{-4\phi} - 1)\tilde{\rho}e^{2\phi}$. Substituting all the above in the tt component of Eq. (31), we get the following analog of Friedmann's equation:

$$\begin{aligned}
\frac{\dot{a}^2}{a^2} &= \frac{8\pi G}{3}\tilde{\rho}e^{-2\phi} + \frac{8\pi G\sigma^2\dot{\phi}^2}{3} + \frac{2\pi}{3k^2\ell^2}\mu^2 F(\mu) \\
&= \frac{8\pi G}{3}\tilde{\rho}e^{-2\phi} + \frac{4\pi}{3k^2\ell^2}\left[-\mu y(\mu) + \frac{1}{2}\mu^2 F(\mu)\right]
\end{aligned} \tag{119}$$

With the choice (50) for $y(\mu)$ we have $\mu > 0$, $y(\mu) < 0$ and $F > 0$ in the cosmological domain. Thus the scalar fields contribute positive energy density and the r.h.s. of Eq. (119) is positive definite ($\tilde{\rho} < 0$ is physically unacceptable). It follows that $\dot{a}$ cannot vanish for any t, so that by our earlier remark it must always be positive. Now the relations (118) imply that

$$d\tilde{a}/d\tilde{t} = e^{-2\phi}(\dot{a} - a\dot{\phi}). \tag{120}$$

We shall show in the sequel that although $\dot{\phi}$ can be positive, it is always the case that $|\dot{\phi}| \ll \dot{a}/a$. As a consequence $d\tilde{a}/d\tilde{t}$ is always strictly positive: in TeVeS a FRW model with flat spaces cannot recollapse.

The fact that $\dot{\phi}$ is given by an integral over time [see Eq. (45)] means that in a cosmological phase transition, where $\tilde{\rho}$ may change suddenly, $\dot{\phi}$ (and of course ϕ) will nevertheless evolve continuously in time. It follows that F will also evolve continuously in time [see Eq. (36)]. A consequence of Eq. (119) is that any jump in $\tilde{\rho}$ will be reflected in a similar jump in $(\dot{a}/a)^2$ or in the square of the Hubble function $\tilde{H} \equiv \tilde{a}^{-1} \, d\tilde{a}/d\tilde{t}$.

B. The proto-radiation era

Contemporary cosmology regards the inflationary era as preceded by a brief radiation dominated era, the proto-radiation era, in which the physical scale factor $\tilde{a}$ expands by just a few orders following the quantum gravity regime. As in any radiation dominated regime, here the equation of state is $\tilde{\rho} = 3\tilde{p}$ with both $\tilde{p}$ and $\tilde{\rho}$ varying as $\tilde{a}^{-4}$. It follows from Eq. (45) that throughout the era

$$\mu\dot{\phi} = -\frac{k}{a^3}\int_0^t G\tilde{\rho}e^{-2\phi}a^3 dt, \tag{121}$$

Because in the cosmological regime $\mu > 2$, we have $\dot{\phi} < 0$ throughout this era. Thus as promised $d\tilde{a}/d\tilde{t}$ in Eq. (120) is positive. Using the constancy of $(G\tilde{\rho})^{1/2}a^2 e^{-2\phi}$ we can now write

$$\mu\dot{\phi} = -\frac{k(G\tilde{\rho})^{1/2}e^{-2\phi}}{a}\int_0^t (G\tilde{\rho})^{1/2}a \, dt. \tag{122}$$

Tentatively assuming that $|\phi| \ll 1$ throughout the era we may, according to Eq. (119), bound both instances of $(G\tilde{\rho})^{1/2}$ from above by $(3/8\pi)^{1/2}\dot{a}/a$. The consequent integral is then trivial, and since a_0 is essentially zero we get

$$\mu|\dot{\phi}| < (3k/8\pi)(\dot{a}/a). \tag{123}$$

Thus $|\dot{\phi}| < (3k/16\pi)(\dot{a}/a)$; since $k \ll 1$, we have by Eq. (120) that $d\tilde{a}/d\tilde{t} \approx \dot{a}$.

We can now show that the cosmological evolution during the proto-radiation era is very similar to that within GR. For the choice (51) both F and F' are positive in the cosmological domain (see Fig. 2). It follows from Eq. (36) that $\mu^2 F < -\mu y$ (recall that $y < 0$), so the last term on the r.h.s. of the Friedmann equation is less than half the second. Next we use $y = -2k\ell^2\dot{\phi}^2$ to infer from Eq. (123) that

$$\frac{4\pi}{3k^2\ell^2}\mu|y| < \frac{3k}{8\pi\mu}\left(\frac{\dot{a}}{a}\right)^2 \tag{124}$$

But this means that the scalar field contributions to the Friedmann equation are small compared to its l.h.s. Specifically, to within a fractional correction of $\mathcal{O}(k/16)$ (actually smaller than this because μ will turn out to be large), the relation between $\tilde{H}$ and $\tilde{\rho}$ is the same as in GR.

The fact that the scalar field contributions to the Friedmann equation are small compared to its l.h.s. also means that inequality (123) is nearly saturated, as must be its kin (124). Then

$$\mu^2|y(\mu)| \approx \frac{1}{6}(3k/4\pi)^4(\dot{a}/a)^2\,\mathfrak{a}_0^{-2}. \tag{125}$$

But $a/\dot{a}$ is a very short scale (in standard cosmological models $\tilde{H}^{-1} \sim 10^{-35}\,\mathrm{s}$ in the proto-radiation era) while $\mathfrak{a}_0^{-1} \sim 3 \times 10^{18}\,\mathrm{s}$. Thus $\mu^2 y(\mu) \gg 1$. Since by Eq. (50) this is possible only for $\mu \gg 1$, we can sharpen our earlier conclusion from Eq. (123): $|\dot{\phi}| \ll (3k/8\pi)\dot{a}/a$. Now it is even clearer that a and $\tilde{a}$ (as well as t and $\tilde{t}$) are essentially equal, so that the expansion in this era proceeds just as in GR. Further, integrating this last inequality gives

$$|\phi_{pr} - \phi_0| \ll (3k/8\pi)\ln(a_{pr}/a_0), \tag{126}$$

where the subscript "pr" stands for the end of the proto-radiation era. Since this era spans just a few e-foldings of the scale a, the logarithm here is of order unity. Hence ϕ is almost frozen at its initial value ϕ_0, provided this last is not extremely small. By choosing as initial condition $0 < \phi_0 \ll 1$, as we proposed, but avoiding extremely small ϕ_0, we get $0 < \phi \ll 1$ throughout the proto-radiation era, as assumed earlier. Thus our assumption was consistent.

C. The inflationary era

The equation of state during inflation is $\tilde{p} = -\tilde{\rho} = \mathrm{const}$. Then (45) tells us that

$$\mu\dot{\phi} = \frac{k}{a^3}\int_{t_{pr}}^{t} G\tilde{\rho}e^{-2\phi}a^3 dt + \mu_{pr}\dot{\phi}_{pr}\left(\frac{a_{pr}}{a}\right)^3. \tag{127}$$

The integration constant prefacing the last term is fixed by the condition that μ and $\dot{\phi}$ be continuous through the proto-radiation inflation divide. It is clear that after rapid expansion has suppressed the last (negative) term here, $\dot{\phi}$ becomes positive. Because $\tilde{\rho}$ is constant, we may pull a factor $(G\tilde{\rho})^{1/2}$ out of the integral. Then by Eq. (119) and assuming everywhere that $e^{-\phi} \approx 1$ (which we verify below), we have $(G\tilde{\rho})^{1/2}e^{-2\phi} < (3/8\pi)^{1/2}\dot{a}/a$ both in and outside the integral. Thus

$$\mu\dot{\phi} \;<\; \frac{3k\dot{a}}{8\pi a^4}\int_{t_{pr}}^{t} a^2\dot{a}\,dt + \mu_{pr}\dot{\phi}_{pr}\left(\frac{a_{pr}}{a}\right)^3 \tag{128}$$

$$= \frac{k\dot{a}}{8\pi a}\left(1 - \frac{a_{pr}^3}{a^3}\right) - \frac{3k}{8\pi}\left(\frac{\dot{a}}{a}\right)_{pr}\left(\frac{a_{pr}}{a}\right)^3. \tag{129}$$

where we have used Eq. (123) as an equality as the end of the proto-radiation era. Thus during inflation

$$-(3k/8\pi)(\dot{a}/a)_{pr} < \mu\dot{\phi} < (k/8\pi)(\dot{a}/a). \tag{130}$$

The l.h.s. here comes from the last term in Eq. (127) in light of inequality (123). In the passage from the proto-radiation era, which involves a phase transition, $\tilde{\rho}$ can change by a factor of order unity, but then settles down to a constant. Thus by Eq. (119) $\dot{a}/a$ remains *at least* of the same order of magnitude as $(\dot{a}/a)_{pr}$. Hence inequality (130) translates into one of the same form as (123) but valid during inflation. As in Sec. VII B, this tells us that $d\tilde{a}/d\tilde{t} \approx \dot{a}$ also during inflation. And the argument following inequality (123) can now be repeated to show that the $-\mu y$ and $\mu^2 F$ terms in Friedmann's equation amount to relative corrections of $\mathcal{O}(k/16)$ (actually smaller), so that inflation in TeVeS proceeds very much like in GR.

Repeating the argument leading to Eq. (129) in light of this last conclusion and the added realization that the a^{-3} terms disappear very rapidly, we conclude that during the $\dot{\phi} > 0$ part of inflation, inequality (123) is very nearly saturated. One can then rederive Eq. (125) as in Sec. VII B. Because the inflation timescale is again very short compared to $\mathfrak{a}_0^{-1}$, the argument yielding Eq. (126) can be repeated with slight modifications to show that during inflation $\mu \gg 1$, and consequently

$$\phi_i - \phi_{pr} \ll (3k/8\pi)\ln(a_i/a_{pr}),\qquad(131)$$

where a subscript "i" stands for the end of inflation. Thus, although in standard models inflation can span up to 70 e-foldings of a, the r.h.s. of this inequality is very small compared to unity. We conclude that inflation manages to raise ϕ above its value at the end of the proto-radiation era by a very small fraction of unity. This justifies our replacement of $e^{-\phi}$ by unity in deriving Eq. (129).

In what follows we shall denote by $\tilde{H}_i$, μ_i, ϕ_i and $\dot{\phi}_i$ the values of the Hubble parameter, $\mu(-2k\ell^2\dot{\phi}^2)$, ϕ and $\dot{\phi}$, respectively, at the end of inflation, $t = t_i$, where $a = a_i$.

D.   The radiation era

In the ensuing radiation era the equation of state switches back to $3\tilde{p} = \tilde{\rho}$ with both $\tilde{p}$ and $\tilde{\rho}$ varying as $\tilde{a}^{-4}$. Thus the integral in Eq. (45) is

$$\mu\dot{\phi} = -\frac{k}{a^3}\int_{t_i}^{t} G\tilde{\rho}e^{-2\phi}a^3 dt + \mu_i\dot{\phi}_i\left(\frac{a_i}{a}\right)^3,\qquad(132)$$

with the integration constant $\mu_i\dot{\phi}_i$ set so $\mu\dot{\phi}$ at the radiation's era outset equals that at inflation's end. Although initially $\dot{\phi} > 0$, clearly the integral will eventually dominate the last term making $\dot{\phi}$ negative thereafter.

Now according to Eq. (129), $\mu_i\dot{\phi}_i < (k/8\pi)(\dot{a}/a)_i$. Due to the approximate continuity of $(\dot{a}/a)$ across the inflation-radiation eras divide [which itself follows from the approximate continuity of $\tilde{\rho}$ and Eq. (119)], and from the fact that $(\dot{a}/a)$ falls off no faster than $(a_i/a)^2$ in the radiation era, Eq. (132) gives

$$\mu\dot{\phi} < (k/8\pi)(\dot{a}/a)_i\,(a_i/a)^3 < (k/8\pi)(\dot{a}/a).\qquad(133)$$

On the other hand, from $\tilde{\rho}\tilde{a}^4 = \text{const.}$ we can move a factor $(G\tilde{\rho})^{1/2}a^2 e^{-2\phi}$ out of the integral in Eq. (132). Using again $(G\tilde{\rho})^{1/2} < (3/8\pi)^{1/2}\dot{a}/a$ from Eq. (119) (if we assume provisionally that $e^{-\phi} \approx 1$) both in and outside the integral, we have

$$\mu\dot{\phi} > -\left(\frac{3k\dot{a}}{8\pi a^2}\right)\int_{t_i}^{t}(\dot{a}/a)a\,dt + \mu_i\dot{\phi}_i\left(\frac{a_i}{a}\right)^3.\qquad(134)$$

The integral is $a(t) - a_i$. Hence

$$\mu\dot{\phi} > (-3k/8\pi)(1 - a_i/a)(\dot{a}/a) + \mu_i\dot{\phi}_i(a_i/a)^3 > -(3k/8\pi)(\dot{a}/a)\qquad(135)$$

In view of Eqs. (133) and (135), inequality (123) is again valid here. Because $\mu > 2$ we get again from Eq. (120) that $d\tilde{a}/d\tilde{a} \approx \dot{a}/a$. We may now reproduce inequality (124) and show as in Sec. VII B that to within a fractional correction of $\mathcal{O}(k/16)$, the relation between $\tilde{H}$ and $\tilde{\rho}$ is the same as in GR.

Because of this last result, Eq. (133) and the rapid decay of a_i/a in Eq. (135), we may conclude that when $\dot{\phi} < 0$, inequality (123) is nearly saturated. We may then rederive Eq. (125) as before. Now in conventional cosmology at redshift z during the radiation era $\tilde{H} \sim 3 \times 10^{-20}(1+z)^2\,\text{s}^{-1}$, which by previous inference closely approximates $\dot{a}/a$ in our model. We thus obtain $\mu^2|y(\mu)| \approx 5 \times 10^{-6}k^4(1+z)^4$. Taking $k \sim 0.03$ on the basis of Sec. IV C we see that at the end of the radiation era ($z \approx 10^4$), $\mu^2|y(\mu)| \approx 4 \times 10^4$ which corresponds to $\mu \approx 10$. For earlier times $\mu \propto (1+z)^{4/5}$ so that it rises to 10^{19} at the beginning of the era at $z \approx 10^{27}$. Going back to inequality (123) we see that in the last three e-foldings of the era $\phi(t) - \phi_i > -8 \times 10^{-4}$ with the previous 50 e-foldings contributing an even smaller decrease. Our assumption that $e^{-\phi} \approx 1$ was evidently justified if ϕ_0 is taken small compared to unity, yet sufficiently positive to keep $\phi(t)$ positive throughout the era.

We shall denote by μ_r, ϕ_r and $\dot{\phi}_r$ the values of $\mu(-2k\ell^2\dot{\phi}^2)$, ϕ and $\dot{\phi}$, respectively, at the end of the radiation era, $t = t_r$ where $a = a_r$.

E.   The matter era

In the matter era $\tilde{p} \approx 0$ and $\tilde{\rho}$ varies as $\tilde{a}^{-3}$. Integrating Eq. (45) gives, c.f. Eq. (132)

$$\mu\dot{\phi} = \frac{-k}{2a^3}\int_{t_r}^{t} G\tilde{\rho}e^{-2\phi}a^3 dt + \mu_r\dot{\phi}_r\left(\frac{a_r}{a}\right)^3. \tag{136}$$

It is clear that $\dot{\phi}$ continues to be negative throughout the matter era. Using $\tilde{\rho}a^3 e^{-3\phi} = \text{const.}$ and setting henceforth $e^{\phi} = 1$ (whose consistency will be checked below), we explicitly perform the integral in Eq. (136) from t_r to t:

$$\mu\dot{\phi} = -\frac{1}{2} kG\tilde{\rho}(t - t_r) + \mu_r\dot{\phi}_r(a_r/a)^3. \tag{137}$$

Integrating the inequality $(G\tilde{\rho}a^3)^{1/2} < (3/8\pi)^{1/2}a^{1/2}\dot{a}$ coming from Eq. (119) we get $(G\tilde{\rho})^{1/2}(t-t_r) < (2/3)(3/8\pi)^{1/2}$. Both together give $G\tilde{\rho}(t-t_r) < (\dot{a}/4\pi a)$, which when substituted in Eq. (137) finally gives

$$\mu\dot{\phi} > (-k/8\pi)(\dot{a}/a) + \mu_r\dot{\phi}_r(a_r/a)^3. \tag{138}$$

Now according to Eq. (135) $\mu_r\dot{\phi}_r > (-3k/8\pi)(\dot{a}/a)_r$. Thus at the beginning of the matter era, where $a = a_r$, the lower bound on the second term on the r.h.s. of inequality (138) maybe as much as three times larger in magnitude than the first term, yet it decays as a^{-3} while the first term cannot do so faster than $a^{-3/2}$ [see Friedmann's equation (119)]. Hence within about one e-folding of a, the first term comes to dominate the r.h.s., and over most of the matter era

$$\mu|\dot{\phi}| < (k/8\pi)(\dot{a}/a). \tag{139}$$

From this follows a tighter version of bound (124) which again demonstrates that the scalar field terms in Einstein's equations are rather negligible. The fact that (139) may be exceeded by a factor of a few early in the matter era is no reason to exclude that epoch from the just mentioned conclusion: the rather large μ at the end of the radiation era ($\mu \sim 10$)—and a bit beyond—acts to suppress that factor. Using by now well worn logic we conclude that in the matter era as well, the relation between $\tilde{H}$ and $\tilde{\rho}$ is almost the same as in GR.

Integrating inequality (139) with the use of $\mu > 2$ (the first e-folding's relatively larger contribution is suppressed by the larger μ which holds sway then), we get

$$\phi(t) - \phi_r > -(k/16\pi)\ln(a/a_r). \tag{140}$$

Because the matter era thus far has spanned nine e-foldings, ϕ has decreased by less than 0.0054 during this era.

Note that we have not addressed the cosmological matter problem. In TeVe S the expansion is driven by just $\tilde{\rho}$, the visible matter's density, whereas the observations require that the source of Friedmann's equation which falls off like $\tilde{a}^{-3}$ should be larger by a factor of perhaps 6. There are at least two possible avenues for dealing with this embarrassment. First, we have stuck to a particular $F(\mu)$; possibly a more realistic $F(\mu)$ would change late cosmological evolution enough to resolve the problem. Second, we have insisted on ϕ being small. This is a consistent solution as we have shown, but it is perhaps not the unique solution. Plainly nonegligible values of ϕ can affect the Friedmann equation significantly.

F.   The accelerating expansion

Lately data from distant supernovae indicate that in recent times ($z < 0.5$) the cosmological expansion has began to accelerate, namely, that $d\tilde{H}/d\tilde{t} > -\tilde{H}^2$. The data are best interpreted in GR by accepting the existence a positive cosmological constant $\Lambda \approx 2\tilde{H}^2_{today}$ [65]. One can incorporate such accelerating epoch in the TeVe S Einstein equations (31) by adding to $\mu^2 F(\mu)$—purely phenomenologically—a constant (μ-independent) term of magnitude $\approx \Lambda k^2\ell^2/2\pi$. Such constant part, which corresponds to the integration constant involved in solving Eq. (36), leaves $y(\mu)$ unchanged, merely shifting the curve for $F(\mu)$ in Fig. 2 up. Furthermore, according to Eq. (62) and the empirical connection $\mathfrak{a}_0 \sim H_0$ [7], the added constant is $\sim 3k^3/16\pi^2$, that is very small. It cannot thus affect the discussion in earlier sections, and in particular F continues to make a positive contribution to the energy both in static systems, and in cosmology.

The appearance of the cosmological constant in F has almost no effect on the value of ϕ. To see why note that Λ does not directly affect the scalar equation (42), but only the Friedmann equation (119). Hence Eq. (137) is still valid. As the expansion accelerates, a begins to grow exponentially with t. Both terms on the r.h.s. of Eq. (137) thus fall off drastically, and ϕ becomes "stuck" at the value it had soon after the onset of acceleration. Consolidating the results of Secs. VII B-VII E with our conclusion we see that the range of initial conditions $0.007 < \phi_0 \ll 1$ insures that $\phi > 0$ and $e^{\phi} \approx 1$ throughout cosmological evolution.

VIII.   CAUSALITY IN TeVeS

TeVeS's predecessors, AQUAL and PCG, permitted superluminal propagation of scalar waves on a static background. In the case of PCG with a convex potential this occurs hand in hand with an instability of the background, so it is unclear if true causality violation occurs. How does TeVeS handle causality issues ?

The question is complicated here by the existence of two metrics, $g_{\alpha\beta}$ and $\tilde{g}_{\alpha\beta}$, whose null cones do not coincide (except where $\phi = 0$). Which of the two cones is the relevant one for causal considerations ? We shall take the view that since common rods and clocks are material systems with negligible self-gravity, the coordinates to which the Lorentz transformations of special relativity refer are those of local orthonormal frames of the physical metric $\tilde{g}_{\alpha\beta}$ and not of $g_{\alpha\beta}$. It is by ascertaining that in no such physical Lorentz frame can physical signals travel back in time that we shall certify the causal behavior of TeVeS. Now Lorentz transformations involve a parameter, the critical speed "c". We shall identify this with the speed of electromagnetic disturbances so that, as customary, the speed of light is the same in all Lorentz frames. Since we have built special relativity into TeVeS by insisting that all nongravitational physics equations (including Maxwell's equation) take their standard form when written with $\tilde{g}_{\alpha\beta}$, this procedure is consistent. In fact, all signals associated with particles of all sorts are subluminal or travel at light's speed with respect to $\tilde{g}_{\alpha\beta}$.

There remains the question of whether gravitational perturbations (tensor, vector or scalar) can ever exit $\tilde{g}_{\alpha\beta}$'s null cone. The analysis given below is quite different for tensor and vector perturbations on the one hand, and scalar perturbations on the other. One point in common, however, is that causality is guaranteed only in spacetime regions for which $\phi > 0$. As shown in Sec. VII, there is gamut of reasonable cosmological models for which ϕ is indeed positive throughout the expansion.

A.   Propagation of tensor and vector disturbances is causal

The characteristics of both Einstein's equations (31) and the vector equation (38) lie on the null cone of $g_{\alpha\beta}$ because all terms in them with two derivatives are the usual ones in Einstein's and gauge field's equations. Accordingly, we do not expect metric and vector perturbations to travel outside the null cone of the Einstein metric $g_{\alpha\beta}$. However, the interesting question is rather what is the speed of a wave of this class in terms of the physical metric $\tilde{g}_{\alpha\beta}$?

In the eikonal approximation the wavevector κ_α of metric perturbations, that is the 4-gradient of the characteristic function, will satisfy $g^{\alpha\beta}\kappa_\alpha\kappa_\beta = 0$. Hence Eq. (23) gives

$$\tilde{g}^{\alpha\beta}\kappa_\alpha\kappa_\beta - 2(\mathfrak{U}^\alpha\kappa_\alpha)^2 \sinh(2\phi) = 0. \tag{141}$$

We consider a generic situation where $\mathfrak{U}^\alpha$ may have both temporal and spatial components. The normalization (20) implies by Eq. (22) that $\tilde{g}_{\alpha\beta}\mathfrak{U}^\alpha\mathfrak{U}^\beta = -e^{2\phi}$. Thus in an appropriately oriented local Lorentz frame, $\mathfrak{L}$, of the metric $\tilde{g}_{\alpha\beta}$ we may parametrize $\mathfrak{U}^\alpha$ by

$$\mathfrak{U}^\alpha = e^\phi(1 - V^2)^{-1/2}\{1, -V, 0, 0\} \tag{142}$$

with $-1 < V < 1$. This V is actually the ordinary velocity (measured by the physical metric) of $\mathfrak{L}$ w.r.t. the privileged frame in which the matter is at rest (whether in cosmology or in a local static configuration), namely that in which $\mathfrak{U}^\alpha = \{e^\phi, 0, 0, 0\}$. This is evident by considering a Lorentz transformation from the matter rest frame to the coordinates appropriate to frame $\mathfrak{L}$.

In view of the above, Eq. (141) reduces to

$$0 = A\omega^2 + 2B\kappa_\|\omega + D\kappa_\|^2 - (1 - V^2)\kappa_\perp^2 \tag{143}$$

$$A \equiv e^{4\phi} - V^2 \tag{144}$$

$$B \equiv V(e^{4\phi} - 1) \tag{145}$$

$$D \equiv -1 + V^2 e^{4\phi} \tag{146}$$

with $\omega = -\kappa_t$ and $\kappa_\|$ and $\kappa_\perp$ the spatial components of κ_α collinear and normal to $\mathfrak{U}_i$ (the space part of $\mathfrak{U}_\alpha$), respectively. For arbitrary V (143) is an anisotropic inhomogeneous dispersion relation (ω depends on position through ϕ as well as on direction of the wavevector). However, in the rest frame of the matter ($V = 0$), it is isotropic (though still position dependent through ϕ) with group (or phase) speed equal to

$$v_0 = e^{-2\phi}. \tag{147}$$

The condition for tensor and vector perturbations not to propagate superluminally ($v_0 \leq 1$ as judged in the physical metric) is thus that $\phi > 0$, which as we saw, is satisfied in a wide range of cosmological models (see Sec. VII) as well as quasistatic systems embedded in them (Sec. V). Normally this conclusion could be carried over to all Lorentz frames without further calculations. But because TeVeS admits a locally privileged frame, that in which $\mathfrak{U}^\alpha = e^\phi\{1,0,0,0\}$, we investigate this conclusion in more detail for any $V^2 < 1$.

Solving Eq. (143) for ω gives

$$\omega = (-B\boldsymbol{\kappa}_\| \pm S)A^{-1}, \tag{148}$$

$$S \equiv (C\boldsymbol{\kappa}_\|^2 + A(1-V^2)\boldsymbol{\kappa}_\perp^2)^{1/2}, \tag{149}$$

$$C \equiv B^2 - AD = (1-V^2)^2 e^{4\phi}. \tag{150}$$

The condition $\phi > 0$ makes A here strictly positive. It is possible for the above expression for ω to change sign, so for given $\boldsymbol{\kappa}$ we must agree to always choose the branch of the square root that makes ω positive (negative ω with opposite sign $\boldsymbol{\kappa}$ is the same mode, of course). In what follows we call the modes with upper (lower) signs of the square root $+$ ($-$) modes. For the components of *group* velocity collinear and orthogonal to $\mathfrak{U}_i$, respectively, we derive

$$\mathbf{v}_\| = \partial\omega/\partial\boldsymbol{\kappa}_\| = (-B \pm CS^{-1}\boldsymbol{\kappa}_\|)A^{-1}, \tag{151}$$

$$\mathbf{v}_\perp = \partial\omega/\partial\boldsymbol{\kappa}_\perp = \pm(1-V^2)S^{-1}\boldsymbol{\kappa}_\perp. \tag{152}$$

Since these expressions are homogeneous of degree zero in $\boldsymbol{\kappa}$, there is no dispersion, but for $V \neq 0$ the propagation is anisotropic. For small ϕ one has analytically

$$v = 1 - 2(1 \pm V\cos\vartheta)^2 (1-V^2)^{-1}\phi + \mathcal{O}(\phi^2) \tag{153}$$

where $v \equiv (\mathbf{v}_\|^2 + |\mathbf{v}_\perp|^2)^{1/2}$ and ϑ is the angle between $\boldsymbol{\kappa}$ and $\mathfrak{U}_i$. Thus for moderate V the group speed v is subluminal, but obviously formula (153) becomes unreliable for V close to unity.

For arbitrary V it is profitable, as remarked by Milgrom, to write v in terms of ω. In fact a straightforward calculation gives

$$1 - v^2 = S^{-2}C(\boldsymbol{\kappa}_\|^2 + \boldsymbol{\kappa}_\perp^2 - \omega^2), \tag{154}$$

from which it is clear that v can become superluminal only if the (isotropic) phase speed $\omega(\boldsymbol{\kappa}_\|^2 + \boldsymbol{\kappa}_\perp^2)^{-1/2}$ does the same simultaneously. Since the latter was found subluminal at $V = 0$, we have only to ask if there is some $V < 1$ for which $\omega = (\boldsymbol{\kappa}_\|^2 + \boldsymbol{\kappa}_\perp^2)^{1/2}$; we might then suspect there is superluminal propagation for larger V. Suppose we substitute this last value of ω in Eq. (143) together with those of A, B and D. Collecting terms one can put the condition for the transition to superluminality in the form

$$(e^{4\phi} - 1)\left(V\boldsymbol{\kappa}_\| + \sqrt{\boldsymbol{\kappa}_\|^2 + \boldsymbol{\kappa}_\perp^2}\right)^2 = 0. \tag{155}$$

As we saw in Sec. VII, for a broad class of cosmological models $\phi > 0$ throughout the expansion, and as Sec. V testifies, variation of ϕ in the vicinity of localized masses embedded in such a cosmology is far short of what is required to turn the sign of ϕ. It is thus clear that even in the case $\boldsymbol{\kappa}_\| < 0$, condition (155) cannot be satisfied for $V < 1$. Hence superluminal propagation of vector and tensor perturbations is forbidden.

How does v vary with V? When $\boldsymbol{\kappa}_\perp \neq 0$, we find numerically the following behavior. For the $+$ mode with $\boldsymbol{\kappa}_\| \leq 0$, $\mathbf{v}_\| < 0$ for all V, and after experiencing a shallow maximum at modest V, v reaches a minimum at V very near unity, which is the deeper and farther from $V = 1$ the larger $|\boldsymbol{\kappa}_\perp/\boldsymbol{\kappa}_\||$. As V grows further, v rises and approaches unity for $V \to 1$. If $\boldsymbol{\kappa}_\| > 0$, $\mathbf{v}_\|$ starts positive for small V but eventually turns negative at a rather large V which grows with $|\boldsymbol{\kappa}_\perp/\boldsymbol{\kappa}_\||$. As V grows further, v reaches a minimum, which gets shallower with growing $|\boldsymbol{\kappa}_\perp/\boldsymbol{\kappa}_\||$, and then begins to rise. At a critical V the positive $\boldsymbol{\kappa}_\|$ $+$ mode terminates. However, the $-$ mode with *negative* $\boldsymbol{\kappa}_\|$ takes over onward from the critical V; it features $\mathbf{v}_\| < 0$, and for it v rises with V and approaches unity as $V \to 1$. The $-$ mode with $\boldsymbol{\kappa}_\| > 0$ is always unphysical.

For $\boldsymbol{\kappa}_\perp = 0$ and $\boldsymbol{\kappa}_\| < 0$ the $+$ mode has $\mathbf{v}_\| < 0$ throughout, and v rises monotonically with V approaching unity as $V \to 1$. For $\boldsymbol{\kappa}_\| > 0$ that same mode has $\mathbf{v}_\| > 0$ and v decreasing with increasing V up to a $V = V_c \approx e^{-2\phi}$ at which point both $\mathbf{v}_\|$ and v vanish. The terminated sequence is continued by the $-$ mode with $\boldsymbol{\kappa}_\| < 0$ for which $\mathbf{v}_\| < 0$ and v rises monotonically with V from zero at $V = V_c$ and approaches unity as $V \to 1$.

B.   Propagation of scalar perturbations is also causal

The terms with second derivatives in the scalar equation (37) have a nonstandard form reminiscent of those in relativistic AQUAL (see Appendix A). Do scalar perturbations propagate across $\tilde{g}_{\alpha\beta}$'s null cone, that is do they travel faster than electromagnetic waves ? We now show that the answer is negative. In the scalar equation (37) in free space we break ϕ into background and perturbation $\phi = \phi_B + \delta\phi$, but ignore perturbations of $g_{\alpha\beta}$ and $\mathfrak{U}_\alpha$. For convenience we shall call ϕ_B simply ϕ. To first order in $\delta\phi$ we get [c.f. Eqs. (A2)-(A4)]

$$0 = \left(h^{\alpha\beta} + 2\xi H^\alpha H^\beta\right)\delta\phi_{;\alpha\beta} + \cdots \tag{156}$$

$$H^\alpha \equiv \left(h^{\mu\nu}\phi_{,\mu}\phi_{,\nu}\right)^{-1/2}h^{\alpha\beta}\phi_{,\beta} \tag{157}$$

$$\xi \equiv d\ln\mu(y)/d\ln y \tag{158}$$

where the ellipsis denotes terms with $\delta\phi$ differentiated only once. We have *temporarily* assumed that H^α is spacelike. Using Eq. (23) we reexpress (156) in terms of the physical metric:

$$[e^{-2\phi}\tilde{g}^{\alpha\beta} - (2 - e^{-4\phi})\mathfrak{U}^\alpha\mathfrak{U}^\beta + 2\xi H^\alpha H^\beta]\delta\phi_{;\alpha\beta} + \cdots = 0 \tag{159}$$

1.   Quasistatic background

For a quasistatic background, e.g. a quiescent galaxy, H^α is indeed a purely space vector in coordinates that reflect the time symmetry. By (157) H^α is normalized to unity w.r.t. metric $g_{\alpha\beta}$ and to $e^{-2\phi}$ w.r.t. $\tilde{g}_{\alpha\beta}$. In a local Lorentz frame of $\tilde{g}_{\alpha\beta}$ at rest w.r.t. to those coordinates and appropriately oriented, a generic H^α will have the form $e^{-\phi}\{0, s, 0, \sqrt{1-s^2}\}$, with s the cosine of the angle between H_i and the positive x axis. Then in a Lorentz frame moving w.r.t. the first one at velocity V in the positive x direction

$$H^\alpha = e^{-\phi}(1 - V^2)^{-1/2}\{-Vs, s, 0, \sqrt{(1-s^2)(1-V^2)}\} \tag{160}$$

In this same frame $\mathfrak{U}^\alpha$ is given by Eq. (142).

In the eikonal approximation (c.f. Appendix A) one replaces in a Lorentz frame $\delta\phi_{;\alpha\beta} \mapsto -\kappa_\alpha\kappa_\beta\delta\phi$ and drops first derivatives. Again interpreting $-\kappa_t$ as ω this gives a generalization of (143), namely

$$0 = \hat{A}\omega^2 + 2(B_\|\boldsymbol{\kappa}_\| + B_\perp\boldsymbol{\kappa}_\perp)\omega + \hat{D}\boldsymbol{\kappa}_\|^2 - (1-V^2)(\kappa_\llcorner^2 + E\kappa_\perp^2) + 2B_\perp V^{-1}\boldsymbol{\kappa}_\|\boldsymbol{\kappa}_\perp \tag{161}$$

$$\hat{A} \equiv 2e^{4\phi} - (1 + 2\xi s^2)V^2 \tag{162}$$

$$B_\| \equiv V(2e^{4\phi} - 1 - 2\xi s^2) \tag{163}$$

$$B_\perp \equiv -2V\xi s\sqrt{(1-s^2)(1-V^2)} \tag{164}$$

$$\hat{D} \equiv 2V^2 e^{4\phi} - (1 + 2\xi s^2) \tag{165}$$

$$E \equiv 1 + 2\xi(1 - s^2), \tag{166}$$

where $\boldsymbol{\kappa}_\|$ is the component in the x direction, $\boldsymbol{\kappa}_\perp$ is that in a direction orthogonal to x in the plane spanning the x axis and H_i, and $\boldsymbol{\kappa}_\llcorner$ is the component orthogonal to that plane (we use vector symbols for components to keep with previous notation).

For $V = 0$ (rest frame of matter) there is nothing to distinguish the x axis from H_i's direction, so without restricting generality we may set $s = 1$ and speak jointly of $\boldsymbol{\kappa}_\perp$ and $\boldsymbol{\kappa}_\llcorner$ as a *vector* $\boldsymbol{\kappa}_\perp$. Then the group speed $v = |\partial\omega/\partial\boldsymbol{\kappa}|^{1/2}$ turns out to be

$$v_0 = \frac{e^{-2\phi}}{\sqrt{2}}\left[\frac{(1+2\xi)^2\boldsymbol{\kappa}_\|^2 + \boldsymbol{\kappa}_\perp^2}{(1+2\xi)\boldsymbol{\kappa}_\|^2 + \boldsymbol{\kappa}_\perp^2}\right]^{1/2}. \tag{167}$$

From Sec. III E we compute the logarithmic slope

$$\xi(\mu) = (\mu - 1)(\mu - 2)/(3\mu^2 - 6\mu + 4) \tag{168}$$

whose graph is shown in Fig. 3. In particular, $\xi \leq \frac{1}{2}$ in a quasistatic region. In the deep MOND regime $\mu(y) \approx \sqrt{y/3}$ so $\xi \approx \frac{1}{2}$, while in the high acceleration limit $\mu(y) \approx 1$ so $\xi \approx 0$. Consequently, in the deep MOND regime, $v_0 \leq e^{-2\phi}$ with equality for $\boldsymbol{\kappa}_\perp = 0$. In the Newtonian regime $v_0 = 2^{-1/2}e^{-2\phi}$ for all $\boldsymbol{\kappa}$. Finally, in the intermediate regime

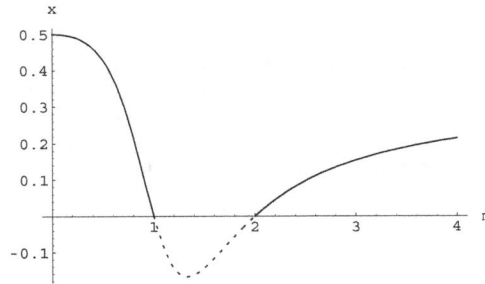

FIG. 3: **The logarithmic slope $\xi(\mu)$ as relevant for quasistationary systems, $0 < \mu < 1$, and for cosmology, $2 < \mu < \infty$.**

$2^{-1/2}e^{-2\phi} \le v_0 \le (1+2\xi)^{1/2}2^{-1/2}e^{-2\phi}$, with lower and upper equality for $\kappa_\parallel = 0$ and $\kappa_\perp = 0$, respectively. Summarizing, scalar waves propagate subluminally in the frame in which the matter is at rest, provided, of course, $\phi > 0$.

Since the vector $\mathfrak{U}^\alpha$ defines a privileged Lorentz frame, the form of the wave equation (159) is different in different frames. Thus we must check explicitly that the subluminal propagation of scalar waves remains valid in all Lorentz frames. Since the analytic expressions for general κ are cumbersome, we have done so numerically for small positive ϕ. For small V the group speed starts at the value (167). If $\kappa_\parallel < 0$, v for the $+$ mode rises with increasing V approaching unity as $V \to 1$. By contrast, if $\kappa_\parallel > 0$, v at first decreases with increasing V only to reach a minimum which can be quite narrow and deep for $\kappa_\parallel/|\kappa|$ near unity. Beyond the minimum is a critical V past which the $+$ mode with positive $\kappa_\parallel$ is no longer possible. It is replaced by the $-$ mode with opposite sign of $\kappa_\parallel$, whose v rises as V rises beyond the critical V, approaching unity for $V \to 1$.

In summary, provided $\phi > 0$ as guaranteed (see Sec. V) for the vicinity of masses embedded in the cosmologies studied in Sec. VII, no case of superluminal propagation is observed for scalar perturbations on a quasistatic background.

2. Cosmological background

Consider now propagation of scalar perturbations in FRW cosmology. Here $\mathfrak{U}^\alpha$ remains pointed in the time direction, and takes the form (142) in a local Lorentz frame of the physical metric which moves w.r.t. the matter at velocity V in the x direction. Since H^α is now timelike, one must change the sign of the argument of the square root in definition (157). Definition (158) then requires a switch in sign of the ξ term in Eq. (156). We may evidently write $\phi_{,\alpha} = \zeta \mathfrak{U}_\alpha$ (with ζ spacetime dependent). It follows from definition (142) that $H^\alpha = \sqrt{2}\,\mathfrak{U}^\alpha$ independent of ζ. Using all this in the modified wave equation (159), we obtain in the said Lorentz frame, after an eikonal approximation, a dispersion relation of the form (143) with the coefficients A, B and C modified according to the rule $e^{4\phi} \to (2+4\xi)e^{4\phi}$. Thus in the frame $\mathcal{L}$ where the matter is at rest ($V = 0$) we now find the isotropic group speed, c.f. Eq. (147),

$$v_0 = (2 + 4\xi)^{-1/2}e^{-2\phi}, \tag{169}$$

so that according to Fig. 3, for $\phi > 0$, v_0 never exceeds $1/\sqrt{2}$.

For $V > 0$ we use the analysis leading to Eqs. (154)-(155) with the substitution $e^{4\phi} \to (2+4\xi)e^{4\phi}$ to conclude that the passage to superluminality is forbidden. Numerical plots disclose a behavior of $v(V)$ very similar to the one for tensor waves. For $+$ type modes with $\kappa_\parallel < 0$, v grows monotonically approaching unity for $V \to 1$. For $\kappa_\parallel > 0$ modes there is a minimum of v at some high V, the narrower and deeper the larger $\kappa_\parallel/|\kappa|$. A mode of this type exists only up to a critical V beyond the minimum, and is thereafter taken over by the $-$ type mode whose $\kappa_\parallel$ is of opposite sign, and for which v approaches unity as $V \to 1$.

C. Caveats

Summing up, propagation of weak perturbations of the tensor, vector or scalar gravitational fields of TeVe$\,$S is always subluminal with respect to the physical metric. We have checked this in detail only for waves propagating on pure cosmological backgrounds or on quasistatic backgrounds. Furthermore, the analysis looked at perturbations of one field while keeping the others "frozen" at their background values. A more advanced analysis would have examined propagation of joint tensor-vector-scalar modes. This said, no mechanism is evident for the formation of causal loops. This under the condition $\phi > 0$ which, as we have seen, is widely obeyed in flat-space cosmological models and quasistatic systems embedded therein.

Acknowledgments

I thank Mordehai Milgrom, Robert Sanders, Arthur Kosowsky for a number of critical comments and for pointing out algebraic errors as well as easier ways to do things. Some results in Sec. V have been corrected for an error noticed by Dimitrios Giannios.

APPENDIX A: ACAUSALITY IN RELATIVISTIC AQUAL

This comes about because the wave equation for free propagation of ψ deriving from the $\mathcal{L}_\psi$ in Eq. (6) (covariant derivatives are w.r.t. $g_{\alpha\beta}$),

$$[\tilde{f}'(L^2 g^{\mu\nu}\psi_{,\mu}\,\psi_{,\nu})g^{\alpha\beta}\psi_{,\beta}]_{;\alpha} = 0, \tag{A1}$$

leads to the following linear equation for propagation of small perturbations $\delta\psi$ on the background $\{g^{\alpha\beta}, \psi_B\}$:

$$0 = \left(g^{\alpha\beta} + 2\xi X^\alpha X^\beta\right)\delta\psi_{;\alpha\beta} + \cdots \tag{A2}$$

$$X^\alpha \equiv (g^{\mu\nu}\psi_{B,\mu}\psi_{B,\nu})^{-1/2}g^{\alpha\beta}\psi_{B,\beta} \tag{A3}$$

$$\xi \equiv d\ln\tilde{f}'(y)/d\ln y \tag{A4}$$

In Eq. (A2) the ellipsis stands for terms where $\delta\psi$ is differentiated only once.

For a static background X^α is a unit purely space vector $\mathbf{X}$. In an appropriately oriented Cartesian coordinate system in a local Lorentz frame, it will point in the x direction. In such frame Eq. (A2) takes the form

$$0 = -\delta\psi_{,tt} + (1 + 2\xi)\delta\psi_{,xx} + \delta\psi_{,yy} + \delta\psi_{,zz} + \cdots \tag{A5}$$

In the eikonal approximation appropriate for short wavelengths, one sets $\psi = Ae^{i\varphi}$ and neglects terms with derivatives of A or of $k_\alpha \equiv \varphi_{,\alpha}$. Then Eq. (A5) gives

$$\omega = -k_t = [(1 + 2\xi)k_x{}^2 + k_y{}^2 + k_z{}^2]^{1/2} \tag{A6}$$

The group speed $v_g = |\partial\omega/\partial\mathbf{k}|^{1/2}$ turns out to be

$$v_g = \left[\frac{(1 + 2\xi)^2 k_x{}^2 + k_y{}^2 + k_z{}^2}{(1 + 2\xi)k_x{}^2 + k_y{}^2 + k_z{}^2}\right]^{1/2}. \tag{A7}$$

In the deep MOND regime $[\tilde{f}(y) = \frac{2}{3}y^{3/2}]$, $2\xi = 1$ while in the high acceleration limit $[\tilde{f}(y) \approx y]$, $\xi \approx 0$. Thus whatever the choice of $\tilde{f}$, $0 < \xi < 1$ over some range of y (acceleration). There $v_g > 1$ if $\mathbf{k}$ is not exactly orthogonal to $\mathbf{X}$ (distances and times measured w.r.t. metric $g_{\alpha\beta}$). On the other hand, light waves travel on light cones of $\tilde{g}_{\alpha\beta}$ while metric waves do so on null cones of $g_{\alpha\beta}$. The two metrics are conformally related so their null cones coincide: light and metric waves travel with unit speed. Thus most ψ waves are superluminal, in violation of the causality principle [see Sec. II B].

APPENDIX B: PROBLEMS FOR PCG IN SOLAR SYSTEM

The permissible ranges of η and ε are strongly constrained by the solar system. It can be shown [4] that the $1/r$ force in Eq. (17) causes the Kepler "constant" of planetary orbits with periods P and semimajor axes $\tilde{a}$ to vary slightly with $\tilde{a}$:

$$4\pi^2\tilde{a}^3/P^2 = GM_\odot(1 + \mathfrak{a}_0\tilde{a}/\kappa\eta). \tag{B1}$$

Assuming $M_\odot \ll M_c$, we get $\kappa = \frac{1}{2}$ so that as we pass from planet to planet, the "constant" varies by a fraction $\sim 2 \times 10^{-15}/\eta$. The inner planet periods P are known to better than one part in 10^8. Thus $\eta > 2 \times 10^{-7}$.

A stronger constraint comes the perihelia precessions of the planets. The anomalous force in Eq. (17) generates an extra precession [4] which in Mercury's case (excentricity 0.206 and $\tilde{a} = 6 \times 10^{12}$ cm) amounts to $3 \times 10^{-8}\eta^{-1}$ arcsec/century. With $\eta = 2 \times 10^{-7}$ this already amounts to 0.35% of the Einstein precession, which is measured to about that accuracy. Trying to assume $M_\odot > M_c$ just aggravates the problem. And we are not at liberty to raise η further because for fixed $\mathfrak{a}_0$, M_c scales as η^2. Thus, for example, with $\eta = 3 \times 10^{-7}$, the MOND limit of PCG would not apply to galaxies with $M < 8 \times 10^9$, a range including many dwarf spirals with missing mass problems ! Hence the perihelion precession marginally rules out PCG with a sextic potential.

APPENDIX C: RELATION BETWEEN DETERMINANTS g AND $\tilde{g}$

From Eqs. (22-23) it follows that

$$\tilde{g}^{\mu\nu} g_{\nu\alpha} = e^{2\phi}\left[\delta^\mu{}_\alpha + (1 - e^{-4\phi})\mathfrak{U}^\mu \mathfrak{U}_\alpha\right] \tag{C1}$$

Viewing this as multiplication of two matrices, we take the determinant:

$$\tilde{g}^{-1} g = e^{8\phi}\mathrm{Det}\,\mathcal{K}(\phi, \mathcal{U}); \qquad \mathcal{K}(\phi, \mathcal{U}) \equiv \mathcal{I} + (1 - e^{-4\phi})\mathcal{U} \tag{C2}$$

where $\mathcal{I}$ is the unit matrix whose components are $\delta^\mu{}_\alpha$ while $\mathcal{U}$ is a matrix with components $\mathfrak{U}^\mu \mathfrak{U}_\alpha$. Now both $\tilde{g}$ and g are scalar densities, so that their ratio must be a true scalar. Hence $\mathrm{Det}\,\mathcal{K}(\phi, \mathcal{U})$ is a scalar.

In a local Lorentz frame in which the unit timelike vector $\mathfrak{U}^\alpha$ has components $\{1, 0, 0, 0\}$, $\mathcal{U}$'s only nonvanishing component is $\mathcal{U}^0{}_0 = -1$. Therefore, $\mathrm{Det}\,\mathcal{K} = [1 - (1 - e^{-4\phi})] \times 1 \times 1 \times 1 = e^{-4\phi}$. Substituting this in Eq. (C2) we recover Eq. (28).

APPENDIX D: RELATIONS BETWEEN m_s, m_g AND r_g

To determine r_g one must delve into the region $\varrho < R$. Assuming that the ideal fluid modeling the matter is at rest in the global coordinates, we may write its 4-velocity as $\tilde{u}_\alpha = e^\phi \mathfrak{U}_\alpha = -e^{\phi+\nu/2}\delta_\alpha{}^t$ (see Sec.III C). Let us return to Eq. (85), substitute $\tilde{T}_{tt}$ from Eq. (40) and reorganize the left hand side to obtain

$$\varrho^{-2} e^{\nu - 5\varsigma/4}(\varrho^2 \varsigma' e^{\varsigma/4})' = -8\pi G \mathfrak{P} \tag{D1}$$

$$\mathfrak{P} \equiv \tilde{\rho} e^\nu (2e^{-2\phi} - e^{2\phi}) + \tau_{tt} + \Theta_{tt}/8\pi G \tag{D2}$$

Integration gives for $\varrho > R$

$$\varsigma' e^{\varsigma/4} = -\frac{2Gm_g}{\varrho^2} - \frac{1}{\varrho^2}\int_R^\varrho (8\pi G\tau_{tt} + \Theta_{tt})\, e^{5\varsigma/4-\nu}\varrho^2 d\varrho \tag{D3}$$

$$m_g \equiv 4\pi \int_0^R \mathfrak{P}\, e^{5\varsigma/4-\nu}\varrho^2 d\varrho, \tag{D4}$$

where the integral in Eq. (D3) does not contain $\tilde{\rho}$ since it extends only outside the fluid.

How much does the "gravitational mass" m_g differ from the scalar mass m_s ? For a star the volume integral of $\tilde{p}$ is of order the random kinetic energy, which by the Newtonian virial theorem is of order of the gravitational energy $\sim Gm_g/R$. According to Eqs. (73), (74) and (92) this is also the order of the fractional correction to m_s or to m_g coming from the metric factors and e^ϕ. We have not worked out τ_{tt} or Θ_{tt} in the interior, but from Eqs (79) and (83) we may estimate that the τ_{tt} and $\Theta_{tt}/8\pi G$ terms contribute to m_g terms of $\mathcal{O}(kGm_s{}^2/R)$ and $\mathcal{O}(Kr_g{}^2/GR)$, respectively. Because we assume small k and K, these last two terms are obviously subdominant contributions. We may conclude that m_g and m_s differ by a fraction of order Gm_g/R which is 10^{-5} for the solar system.

Let us now calculate $\varsigma' e^{\varsigma/4}$ at $\varrho = R$ using Eq. (74), (89) and (91) and equate the result to $-2Gm_g/R^2$ as stipulated by Eq. (D3):

$$r_g + \frac{Kr_g{}^2}{8R} - \frac{kG^2 m_s{}^2}{4\pi R} + \mathcal{O}(r_g{}^3/R^2) = 2Gm_g \tag{D5}$$

33

For the sun $r_g/R \sim Gm_s/R \sim 10^{-5}$; we see that $r_g \approx 2Gm_g$ with fractional accuracy much better than 10^{-5}.

---

[1] J. Oort, *Bull. Astron. Soc. Neth.* **6**, 249 (1932); **15**, 45 (1960).
[2] F. Zwicky, *Helv. Phys. Acta* **6**, 110 (1933); see also S. Smith, Astrophys. Journ. **83**, 23 (1936).
[3] *Dark Matter in the Universe*, G. R. Knapp and J. Kormendy, eds. (Reidel, Dordrecht 1987).
[4] J. D. Bekenstein, in *Second Canadian Conference on General Relativity and Relativistic Astrophysics*, A. Coley, C. Dyer and T. Tupper, eds. (World Scientific, Singapore 1988), p. 68.
[5] *The Birth and Early Evolution of our Universe: Proceedings of Nobel Symposium #79, Gräftåvallen, Östersund, Sweden*, Physica Scripta **T36** (1991).
[6] J. Ellis, Ref. 5; M. Turner, Ref. 5; C. Munoz, ArXiv hep-ph/0309346.
[7] M. Milgrom, Astrophys. Journ. **270**, 365 (1983).
[8] M. Milgrom, Astrophys. Journ. **270**, 371 (1983).
[9] M. Milgrom, Astrophys. Journ. **270**, 384 (1983).
[10] R. Berendzen, R. Hart, and D. Seeley, *Man Discovers the Galaxies* (Columbia University Press, New York 1987).
[11] J. Jeans, Month. Not. Roy. Astron. Soc. **84**, 60 (1923).
[12] F. Zwicky, *Morphological Astronomy* (Springer, Berlin 1957).
[13] A. Finzi, Month. Not. Roy. Astron. Soc. **127**, 21 (1963).
[14] R. B. Tully and J. R. Fisher, Astron. Astrophys. **54**, 661 (1977).
[15] M. Aaronson and J. Mould, Astrophys. Journ. **265**, 1 (1983); M. Aaronson, G. Bothun, J. Mould, J. Huchra, R. A. Schommer and N. E. Cornell, Astrophys. Journ. **302**, 536 (1986); W. L. Freedman, Astrophys. Journ. **355**, L35 (1990), R. H Sanders and M. A. W. Verheijen, Astrophys. Journ. **503**, 97 (1998).
[16] J. P. Ostriker, P. J. E. Peebles and A. Yahil, Astrophys. Journ. **193**, L1 (1974).
[17] A. Kalnajs in *The Internal Kinematics and Dynamics of Galaxies*, E. Athanassoula, ed. (Reidel, Dordrecht 1983), p. 87.
[18] J. N. Bahcall and S. Cassertano, Astrophys. Journ. **293**, L7 (1985).
[19] T. S. van Albada and R. Sancisi, Phil. Trans. Roy. Soc. London A **320**, 447 (1986).
[20] P. J. Quinn, J. K. Salmon and W. H. Zurek, Nature **322**, 329 (1986); J. Silk in *A Unified View of the Micro- and Macro-Cosmos*, A. de Rujula, D. V. Nanopoulos and P. A. Shaver, eds. (World Scientific, Singapore 1987).
[21] G. Blumenthal, S. M. Faber, R. A. Flores, and J. R. Primack, Astyrophys. Journ. **301**, 27 (1986); J. Barnes, in *Nearly Normal Galaxies*, S. M. Faber, ed. (Springer, New York 1987), p. 154; B. S. Ryden and J. E. Gunn, Astrophys. Journ. **318**, 15 (1987).
[22] R. H. Sanders, Astron. Astrophys. Rev. **2**, 1 (1990).
[23] M. Jobin and C. Carignan, Astron. Journ. **100**, 648 (1990); C. Carignan, R. Sancisi and T. S. van Albada, Astron. Journ. **95**, 37, (1988); C. Carignan and S. Beaulieu, Astrophys. Journ. **347**, 192, (1989).
[24] S. S. McGaugh and W. J. G. de Blok, Astrophys. Journ. **508**, 132 (1998).
[25] S. M. Kent, Astron. Journ. **93**, 816 (1987).
[26] M. Milgrom, Astrophys. Journ. **333**, 689 (1988).
[27] K. G. Begeman, A. H. Broeils and R. H. Sanders, Month. Not. Roy. Astron. Soc. **249**, 523 (1991).
[28] G. Gentile, P. Salucci, U. Klein, D. Vergani and P. Kalberla, ArXiv astro-ph/0403154.
[29] M. Milgrom and E. Braun, Astrophys. Journ. **334**, 130 (1988).
[30] S. S. McGaugh and E. de Blok, Astrophys. Journ. **499**, 66 (1998).
[31] Begum, A., and Chengalur, J. N. , Astron. and Astrophys. **413**, 525 (2004).
[32] R. H. Sanders and S. S. McGaugh, Ann. Rev. Astron. Astrophys. **40**, 263 (2002).
[33] S. S. McGaugh, ArXiv astro-ph/0403610.
[34] A. Aguirre, *Proceedings of the IAU Symposium 220 "Dark matter in galaxies"*, eds. S. Ryder, D. J. Pisano, M. Walker and K. Freeman (2003).
[35] R.H. Sanders, Mon. Not. R. Ast. Soc. 342, 901 (2003).
[36] M. Milgrom, **287**, 571, 1984; R. H. Sanders, Astron. Astrophys. **284**, (1994); M. Milgrom and R. H. Sanders, Astrophys. J. **599**, L25 (2003); R. Scarpa, G. Marconi and R. Gilmozzi, Astron. and Astrophys. **405**, L15 (2003).
[37] J. E. Felten, Astrophys. Journ. **286**, 3 (1984).
[38] R. H. Sanders, Mon. Not. Roy. Astron. Soc. **296**, 1009 (1998) .
[39] J. D. Bekenstein and M. Milgrom, Astrophys. Journ. **286**, 7 (1984).
[40] J. D. Bekenstein, Physics Letters B**202**, 497 (1988).
[41] R. H. Sanders, Month. Not. Roy. Ast. Soc. **235**, 105 (1988).
[42] J. D. Bekenstein in *Developments in General Relativity, Astrophysics and Quantum Theory*, F. I. Cooperstock, L. P. Horwitz and J. Rosen, eds. (IOP Publishing, Bristol 1990), p. 156.
[43] J. D. Bekenstein, in *Proceedings of the Sixth Marcel Grossman Meeting on General Relativity*, H. Sato and T. Nakamura, eds. (World Scientific, Singapore 1992), p. 905.
[44] J. D. Bekenstein and R. H. Sanders, Astrophys. Journ. **429**, 480 (1994).
[45] V V. Zhytnikov and J. M. Nester, Phys. Rev. Lett. **73**, 2950 (1994).
[46] A. Edery, Phys. Rev. Lett. **83**, 3990- (1999) and rebuttal J. D. Bekenstein, M. Milgrom and R. H. Sanders, Phys. Rev.

Lett. **85**, 1346 (2000).

[47] M. E. Soussa and R. P. Woodard, Class. Quant. Grav. **20**, 2737 (2003).

[48] M. E. Soussa and R. P. Woodard, Phys. Lett. B **578**, 253 (2004).

[49] M. E. Soussa, ArXiv astro-ph/0310531.

[50] R. H. Sanders, Astrophys. Journ. **480**, 492 (1997).

[51] M. Milgrom, Astrophys. Journ. **302**, 617 (1986).

[52] R. H. Sanders, Month. Not. Roy. Astron. Soc. **223**, 559 (1986).

[53] R. Romatka, Dissertation (University of Munich, 1992).

[54] C. Will, *Theory and Experiment in Gravitational Physics* (Cambridge University, Cambridge 1986).

[55] R. F. C. Vessot and M. A. Levine, Phys. Rev. Letters **45**, 2081 (1980).

[56] L. Schiff, Am. Journ. Phys. **28**, 340 (1960).

[57] E. Witten, Commun. Math. Phys. **80**, 381 (1981).

[58] J. D. Bekenstein, Int. Journ. Theor. Phys. **13**, 317 (1975).

[59] J. E. Tohline, in *Internal Kinematics and Dynamics of Galaxies*, A. Athanassoula, ed., (Reidel, Dordrecht 1982), p.205.

[60] J. R. Kuhn and L. Kruglyak, Astrophys. Journ. **313**, 1 (1987).

[61] T. Jacobson and D. Mattingly, Phys. Rev. **D**64, 024028 (2001).

[62] C. Eiling and T. Jacobson, Phys. Rev. **D**69, 064005 (2004).

[63] M. M. Nieto and S. G. Turyshev, ArXiv gr-qc/0308017.

[64] C. Misner, K. S. Thorne and J. A. Wheeler, *Gravitation* (Freeman, San Francisco 1973).

[65] A. G. Riess, L. G. Strolger et. al, ArXiv astro-ph/0402512.

[66] T. D. Lee and C. N. Yang, Phys. Rev. **98**, 1501 (1955).

[67] R. H. Dicke, *The Theoretical Significance of Experimental Relativity* (Gordon and Breach, New York 1964).

LIST OF PUBLICATIONS

Research Papers

1. Bekenstein JD, Krieger, J. (1969) Stark Effect in Hydrogenic Atoms *Phys. Rev.* **188**: 130–139.

2. Bekenstein JD, Krieger J. (1970) Stark Effect in Hydrogen Atoms for Nonuniform Fields *J. Math. Phys.* **11**: 2721–2727.

3. Bekenstein JD. (1971) Hydrostatic Equilibrium and Gravitational Collapse of Relativistic Charged Fluid Balls. *Phys. Rev.* **D4**: 2185–2190.

4. Bekenstein JD. (1972) Transcendence of the Law of Baryon Number Conservation in Black Hole Physics. *Phys. Rev. Letters* **28**: 452–455.

5. Bekenstein JD. (1972) Nonexistence of Baryon Number for Static Black Holes. *Phys. Rev.* **D5**: 1239–1246.

6. Bekenstein JD. (1972) Nonexistence of Baryon Number for Black Holes: II. *Phys. Rev.* **D5**: 2403–2412.

7. Bekenstein JD. (1972) Black Holes and the Second Law. *Lettere al Nuovo Cimento* **4**: 737–740.

8. Bekenstein JD. (1973) Extraction of Energy and Charge From Black Holes. *Phys. Rev.* **D7**: 949–953.

9. Bekenstein JD. (1973) Black Holes and Entropy. *Phys. Rev.* **D7**: 2333–2346.

10. Bekenstein JD. (1973) Gravitational Radiation Recoil and Runaway Black Holes. *Astrophys. J.* **183**: 657–664.

11. Bekenstein JD. (1974) Exact Solutions of Einstein-Conformal Scalar Equations. *Ann. Phys. (NY)* **82**: 535–547.

12. Bekenstein JD. (1974) Generalized Second Law of Thermodynamics in Black Hole Physics. *Phys. Rev.* **D9**: 3292–3300.

13. Bekenstein JD, Bowers R. (1974) Do OB Runaways Have Collapsed Companions? *Astrophys. J.* **190**: 653–659.

14. Bekenstein JD. (1974) The Quantum Mass Spectrum of the Kerr Black Hole. *Lettere al Nuovo Cimento* **11**: 467–470.

15. Bekenstein JD. (1975) Nonsingular General Relativistic Cosmologies. *Phys. Rev.* **D11**: 2072–2075.

16. Bekenstein JD. (1975) Black Holes with Scalar Charge. *Ann. Phys. (NY)* **91**: 75–82.

17. Bekenstein JD. (1975) Positivness of Mass and the Strong Energy Condition. *Int. J. Theor. Phys.* **13**: 317–321.

18. Bekenstein JD. (1975) Statistical Black Hole Thermodynamics. *Phys. Rev.* **D12**: 3077–3085.

19. Bekenstein JD. (1976) Supernovae in Binaries and the Possible Collapsed Nature of the Companion of HD 108. *Astrophys. J.* **210**: 544–548.

20. Bekenstein JD, Meisels A. (1977) Einstein A and B Coefficients for a Black Hole. *Phys. Rev.* **D15**: 2775–2781.

21. Bekenstein JD. (1977) Are Particle Rest Masses Variable?: Theory and Constraints from Solar System Experiments. *Phys. Rev.* **D15**: 1458–1468.

22. Bekenstein JD, Oron, E. (1978) New Conservation Laws in General Relativistic Magnetohydrodynamics. *Phys. Rev.* **D18**: 1809–1819.

23. Bekenstein JD, Rosenkrantz, M. (1978) Color Variability of Quasars. *Astrophys. J.* **224**: 812–815.

24. Bekenstein JD, Meisels A. (1978) General Relativity Without General Relativity. *Phys. Rev.* **D18**: 4378–4386.

25. Bekenstein JD, Oron, E. (1979) Interior Magnetohydrodynamic Structure of a Rotating Relativistic Star. *Phys. Rev.* **D19**: 2827–2837.

26. Bekenstein JD, Meisels, A. (1980) Is the Cosmological Singularity Compulsory? *Astrophys. J.* **237**: 342–348.

27. Bekenstein JD, Meisels, A. (1980) Conformal Invariance, Microscopic Physics and the Nature of Gravitation. *Phys. Rev.* **D22**: 1313–1324.

28. Bekenstein JD. (1981) Universal Bound on the Entropy to Energy Ratio for Bounded Systems. *Phys. Rev.* **D23**: 287–298.

29. Bekenstein JD, Parker LE. (1981) Path Integral Evaluation of Feynman Propagator in Curved Spacetime. *Phys. Rev.* **D23**: 2850–2869.

30. Bekenstein JD. (1981) The Energy Cost of Information Transfer. *Phys. Rev. Letters* **46**: 623–626.

31. Bekenstein JD. (1982) Black Holes and Everyday Physics. *General Relativity and Gravitation* **14**: 355–359.

32. Bekenstein JD. (1982) The Fine Structure Constant: Is it Really Constant? *Phys. Rev.* **D25**: 1527–1539.

33. Bekenstein JD. (1982) Specific Entropy and the Sign of the Energy. *Phys. Rev.* **D26**: 950–953.

34. Bekenstein JD. (1983) Entropy Bounds and the Second Law for Black Holes. *Phys. Rev.* **D27**: 2262–2270.

35. Bekenstein JD, Milgrom M. (1984) Does the Missing Mass Problem Signal the Breakdown of Newtonian Gravity ? *Astrophys. J.* **286**: 7–14.

36. Bekenstein JD. (1984) Entropy Content and Information Flow in Systems with Limited Energy. *Phys. Rev.* D**30**: 1669–1679.

37. Bekenstein JD, Eichler, D. (1985) Electrodynamic Confinement of Axisymmetric Flows. *Astrophys. J.* **298**: 493–501.

38. Bekenstein JD. (1986) Gravitation and Spontaneous Symmetry Breaking. *Found. Phys.* **16**: 409–422.

39. Bekenstein JD, Guendelman EI (1987) Symmetry Breaking Induced by Charge Density and the Entropy of Interacting Fields. *Phys. Rev.* D**35**: 716–731.

40. Bekenstein JD. (1987) Helicity Conservation Laws for Fluids and Plasmas. *Astrophys. J.* **319**: 207–214.

41. Bekenstein JD. (1988) Phase Coupled Gravitation and Gauge Fields. *Physics Letters* B**202**: 497–500.

42. Bekenstein JD. (1988) Communication and Energy. *Phys. Rev.* A**37**: 3437–3449.

43. Bekenstein JD, Schiffer, M. (1989) Proof of the Quantum Bound on Specific Entropy for Free Fields. *Phys. Rev.* D**39**: 1109–1115.

44. Bekenstein JD. (1989) Is the Cosmological Singularity Thermodynamically Consistent? *International J. of Theoretical Physics* **28**: 967–981.

45. Bekenstein JD, Maoz, E. (1990) On the Prevalence of *r* Density Profiles in Extragalactic Systems. *Astrophys. J.* **353**: 59–65.

46. Bekenstein JD, Zamir, R. (1990) Dynamical Friction in Binary Systems. *Astrophys. J.* **359**: 427–437.

47. Bekenstein JD, Schiffer, M. (1990) Do Zero Modes Contribute to the Entropy? *Phys. Rev.* D**42**: 3598–3599.

48. Bekenstein JD, Maoz, E. (1992) Dynamical Friction from Fluctuations in Stellar Dynamical Systems. *Astrophys. J.* **390**: 79–87.

49. Bekenstein JD. (1992) Chiral Cosmic Strings. *Phys. Rev.* D**45**: 2794–2801.

50. Bekenstein JD. (1992) Conservation Law for Linked Cosmic String Loops. *Phys. Letters* B**282**: 44–49.

51. Bekenstein JD. (1993) The Relation Between Physical and Gravitational Geometry. *Phys. Rev.* D**48**: 3641–3647.

52. Bekenstein JD. (1993) How Fast Does Information Leak Out from a Black Hole? *Phys. Rev. Letters* **70**: 3680–3683.

53. Bekenstein JD. (1994) Entropy Bounds and Black Hole Remnants. *Phys. Rev.* **D49**: 1912–1921.

54. Bekenstein JD, Schiffer, M. (1994) Universality in Grey Body Radiance: Extending Kirchhoff's Law to the Statistics of Quanta. *Phys. Rev. Letters* **72**: 2512–2515.

55. Bekenstein JD, Sanders RH. (1994) Gravitational Lenses and Unconventional Gravity. *Astrophys. J.* **429**: 480–490.

56. Bekenstein JD, Rosenzweig, C. (1994) Stability of the Black Hole Horizon and the Landau Ghost. *Phys. Rev.* **D50**: 7239–7243.

57. Bekenstein JD. (1995) Novel 'No Scalar Hair' Theorem for Black Holes. *Phys. Rev.* **D51**: R6608–R6611.

58. Bekenstein JD, Mukhanov VF. (1995) Spectroscopy of the Quantum Black Hole.) *Phys. Letters* **B360**: 7–12.

59. Bekenstein JD, Mayo AE. (1996) No Hair for Spherical Black Holes: Charged and Nonminimally Coupled Scalar Field with Self–interaction. *Phys. Rev.* **D54**: 5059–5069.

60. Bekenstein JD, Schiffer, M. (1998) The Many Faces of Superradiance. *Phys. Rev.* **D58**: 64014.

61. Bekenstein JD. (1999) Non-Archimedean Character of Quantum Buoyancy and the Generalized Second Law. *Phys. Rev.* **D60**: 124010.

62. Bekenstein JD, Mayo AE. (2000) Black Hole Polarization and New Entropy Bounds. *Phys. Rev.* **D61**: 024022.

63. Bekenstein JD. (2000) Holographic Bound and Second Law of Thermodynamics. *Physics Letters* **B481**: 339–345.

64. Bekenstein JD, Oron, A. (2000) Conservation of Circulation in Magnetohydrodynamics. *Phys. Rev.* **E62**: 5594–5602.

65. Bekenstein JD. (2001) The Limits of Information. *Stud. Hist. Phil. Mod. Phys.* **32**(4): 511–524.

66. Bekenstein JD, Oron, A. (2001) Extended Kelvin Theorem in Magnetohydrodynamics. *Found. Phys.* **31**: 895–907.

67. Bekenstein JD, Mayo AE. (2001) Black Holes are One-Dimensional. *Gen. Relat. Grav.* **33**: 2095–2099.

68. Bekenstein JD, Gour G. (2002) Building Blocks of a Black Hole. *Phys. Rev.* **D66**: 024005.

69. Bekenstein JD. (2002) Fine-Structure Constant Variability, Equivalence Principle and Cosmology. *Phys. Rev.* **D66**: 123514.

70. Bekenstein JD. (2004) Are there Hyperentropic Objects? *Phys. Rev.* **D70**: R121502.

71. Bekenstein JD. (2004) Relativistic Gravitation Theory for the Modified Newtonian Dynamics Paradigm. *Phys. Rev.* **D70**: 083509; Erratum (2005) *Phys. Rev.* D **71**: 069901(E).

72. Bekenstein JD. (2005) How Does the Entropy/Information Bound Work? *Found. Phys.* **35**: 1805–1823.

73. Bekenstein JD, Magueijo, J. (2006) Modified Newtonian Dynamics Habitats Within the Solar System. *Phys. Rev.* **D73**: 103513.

74. Bekenstein JD, Betschart, G. (2006) Perfect Magnetohydrodynamics as a Field Theory. *Phys. Rev.* **D74**: 083009.

75. Bekenstein JD, Sagi E. (2008) Black Holes in the TeVeS Theory of Gravity and their Thermodynamics. *Phys. Rev.* **D77**: 024010.

76. Bekenstein JD, Fouxon I. Betschart G. (2008) The Bound on Viscosity and the Generalized Second Law of Thermodynamics *Phys. Rev.* **D77**: 024016.

77. Bekenstein JD, Sagi E. (2008) Do Newton's G and Milgrom's a-30 Vary with Cosmological Epoch? *Phys. Rev.* **D77**: 103512.

78. Bekenstein JD, Eling C. (2009) Challenging the Generalized Second Law. *Phys. Rev.* **D79**: 024019.

79. Bekenstein JD, Schiffer M. (2009) Varying fine structure 'constant' and charged black holes. *Phys. Rev.* **D80**: 123508.

80. Bekenstein JD, Sanders RH. (2012) TeVeS/MOND is in Harmony with Gravitational Redshifts in Galaxy Clusters. *Mon. Not. Roy. Astron. Soc. Letters* **421**: L59–L61.

81. Bekenstein JD. (2012) Is a Tabletop Search for Planck Scale Signals Feasible? . *Phys. Rev.* **D86**: 12404.

82. Bekenstein JD. (2013) If Vacuum Energy can be Negative, Why is Mass Always Positive? Uses of the Subdominant Trace Energy Condition. *Phys. Rev.* **D88**: 125005.

Reviews

1. Bekenstein JD. (1980) Black Hole Thermodynamics. *Physics Today* **33**(1): 24–31.

2. Bekenstein JD. (1984) Black Hole Fluctuations. In: Christensen S (ed.), *Quantum Theory of Gravity: Essays in honor of Bryce DeWitt.* Adam Hilger, Bristol, pp. 148–159.

3. Bekenstein JD. (1986) The Fine Structure Constant: From Eddington's Time to Our Own. In: Ullmann-Margalit E. (ed.), *The Prism of Science.* Reidel, Dordrecht, pp. 209–224.

4. Bekenstein JD. (1990) Non-Newtonian Gravitation and Causality. In: Cooperstock FI, Horwitz LP, Rosen R. (eds.), *Developments in General Relativity, Astrophysics and Quantum Theory: A Jubilee in Honour of Nathan Rosen.* IOP Publishing, Bristol pp. 155–174.

5. Bekenstein JD, Schiffer M. (1990) Quantum Limitations on the Storage and Transmission of Information. *Int. J. Mod. Phys.* C1: 355–422.

6. Bekenstein JD. (1992) Gravitational Theories. In: Maran SP (ed.), *The Astronomy and Astrophysics Encyclopedia.* Van Nostrand-Rheinhold, New York, pp. 296–299.

7. Bekenstein JD. (1998) Disturbing the Black Hole: Festschrift for C. V. Vishveshwara. In: Iyer B, Bhawal B, (eds.), *Black Holes. Gravitation and the Universe.* Kluwer, Dordrecht, pp. 87–101.

8. Bekenstein JD. (2000) Black Holes: Classical Properties, Thermodynamics and Heuristic Quantization. In: Novello M. (ed.), *IX Brazilian School on Cosmology and Gravitation.* Atlantisciences, Paris, pp. 1–85.

9. Bekenstein JD. (2001) Limitations on Quantum Information from Black Hole Physics. In: Biesiada M, (ed.), *The XXV International School of Theoretical Physics, Particles and Astrophysics, Acta Phys. Pol.* B32: 3555–3570.

10. Bekenstein JD. (2002) Quantum Information and Quantum Black Holes. In: Bergmann PG, de Sabbata V, (ed.), *Advances in the Interplay Between Quantum and Gravity Physics.* Kluwer, Dordrecht, pp. 1–26.

11. Bekenstein JD. (2003) Information in the Holographic Universe. *Scientific American* **289**(2): pp. 58–65.

12. Bekenstein JD. (2003) Preparing the State of a Black Hole. In: Salim JM, Bergliaffa SEP, Oliveira LA, De Lorenci VA, (eds.), *Inquiring the Universe.* Frontier Group, pp. 17–32.

13. Bekenstein JD. (2004) Black Holes and Information Theory. *Contemporary Physics* **45**: 31–43.

14. Bekenstein JD. (2005) Black Holes: Physics and Astrophysics. In: Shapiro MM, Stanev T, Wefel JP, (eds.), *Neutrinos and Explosive Events in the Universe.* Springer, Dordrecht, pp. 149–173.

15. Bekenstein JD, Sanders RH. (2006) A Primer to Relativistic MOND Theory. In: Mamon G, Combes F, Deffayet C, Fort B, (eds.), *Mass Profiles and Cosmological Structures.* EAS Publication Series **20**, EDP Sciences, Les Ulis, pp. 225–230.

16. Bekenstein JD. (2006) The Modified Newtonian Dynamics—MOND and its Implications for New Physics. *Cont. Phys.* **47**: 387–403.

17. Bekenstein JD, Magueijo J. (2007) Testing Strong MOND Behavior in the Solar System. *J. Mod. Phys.* D**16**: 2035–2053.

18. Bekenstein JD. (2010) Modified Gravity as an Alternative to Dark Matter. In: Bertone G. (ed.), *Particle Dark Matter: Observations, Models and Searches.* Cambridge Univ. Press, Cambridge, pp. 95–114.

Papers in Conference Proceedings

1. Bekenstein JD. (1981) Gravitation, the Quantum and Statistical Physics. In: Y. Ne'eman Y. (ed.) *To Fulfill a Vision: Jerusalem Einstein Centennial Symposium on Gauge Theories and Unification.* Addison Wesley, Cambridge, Mass, pp. 42–62.

2. Bekenstein JD. (1986) Gravitation and the Origin of Large Structures in the Universe. In: *Absolute Values and the New Cultural Revolution: XIV International Conference on the Unity of Science.* International Cultural Foundation Press, New York, pp. 69–98.

3. Bekenstein JD, Milgrom M. (1987) The Modified Newtonian Dynamics as an Alternative to Hidden Matter. In: Knapp. G, Kormendy J, (eds.), *Dark Matter in the Universe.* Reidel, Dordrecht, pp. 319–333.

4. Bekenstein JD. (1988) The Missing Light Puzzle: A Hint About Gravity? In: Coley A, Dyer C, and Tupper B, (eds.), *Second Canadian Conference on General Relativity and Relativistic Astrophysics.* World Sci. Publ., Singapore, pp. 68–104.

5. Bekenstein JD. (1992) New Gravitational Theories as Alternatives to Dark Matter. In: Sato H, Nakamura, T, (eds.), *Proceedings of the Sixth Marcel Grossman Meeting on General Relativity.* World Sci. Publ., Singapore, pp. 905–924.

6. Bekenstein JD. (1996) Do We Understand Black Hole Entropy? In: Ruffini R, Keiser M. *Proceedings of the Seventh Marcel Grossmann Meeting on General Relativity.* World Sci. Publ., River Edge, NJ, pp. 39–58.

7. Bekenstein JD. (1997) 'No Hair': Twenty–five Years After. In: Dremin IM, Semikhatov AM. (eds.), *Proceedings of the Second International Andrei D. Sakharov Conference in Physics.* World Sci. Publ., Singapore, p. 216.

8. Bekenstein JD. (1997) Black Holes as Atoms: Classical Hair and Quantum Levels in the Light of General Relativity. In: da Silva AJ. *et al.* (eds.), *XVII Brazilian National Meeting on Particles and Fields,* Brazilian Physics Society, pp. 59–69.

9. Bekenstein JD. (1999) Quantum Black Holes as Atoms. In: Piran T, Ruffini R, (eds.), *Proceedings of the Eight Marcel Grossmann Meeting on General Relativity.* World Sci. Publ., Singapore, pp. 92–111.

10. Bekenstein JD. (2002) The Case for Discrete Energy Levels of a Black Hole. In: Duff MJ, Liu JT, (eds.), *2001 — A Spacetime Odyssey: Proceedings of the Inaugural Conference of the Michigan Center for Theoretical Physics.* World Sci. Publ., Singapore pp. 21–31.

11. Bekenstein JD. (2002) Holographic Bound from Second Law. In: Gurzadyan VG, Jantzen R, Ruffini R, (eds.), *Ninth Marcel Grossmann Meeting on General Relativity.* World Sci. Publ., Singapore, pp. 553–559.

12. Bekenstein JD. (2009) Relativistic MOND as an Alternative to the Dark Matter Paradigm. In: Tserruya I, Gal A, Ashery D, (eds.), *Proceedings of the Eighteenth Particles and Nuclei International Conference.* Elsevier, Amsterdam, pp. 937–942.

13. Bekenstein JD. (2011) Tensor-Vector-Scalar Modified Gravity: From Small Scale to Cosmology. In: *Proceedings of the Theo Murphy Meeting: Testing General Relativity with Cosmology. Phil. Trans. Roy. Soc.* A**369**: 5003–5017.

14. Bekenstein, JD. (2014) Can Quantum Gravity be Exposed in the Laboratory?: A Tabletop Experiment to Reveal the Quantum Foam. In: Scardigli, F. (ed.), *Proceedings of the Workshop "Horizons of Quantum Physics: From Foundations to Quantum Enabled Technologies"* To appear in *Found. Phys.*

Papers Reprinted in Collections

1. Bekenstein JD. (1980) Black Holes and the Second Law. Reprinted from *Lettere al Nuovo Cimento* In: Sato H, and Tomita K. (eds.), *Gravitational Collapse and Black Holes.* Physics Society of Japan, Tokyo. pp. 737–740.

2. Bekenstein JD. (1984) Black Hole Thermodynamics. Reprinted from *Physics Today* In: Cameron A, (ed.), *Astrophysics Today.* American Institute of Physics, New York, pp. 188–195.

3. Bekenstein JD. (1988) Gravitation and Spontaneous Symmetry Breaking. Reprinted from *Found. Phys.* In: Zurek WH, van der Merwe A, Miller WA, (eds.), *Between Quantum and Cosmos: Studies and Essays in Honor of John Archibald Wheeler.* Princeton University Press, Princeton, pp. 145–158.

4. Bekenstein JD. (1990) Gravitation and the Origin of Large Structures in the Universe. Reprinted from *Absolute Values and the New Cultural Revolution* In: Alonso M, (ed.), *Organization and Change in Complex Systems.* Paragon House, New York, pp. 1–23.

5. Bekenstein JD, Mukhanov VF (1998) Quantum Gravity and Hawking Radiation. Reprinted from *Phys. Letters* **B** In: Berezin VA, and Rubakov VA, Semikoz DV, (eds.) *Sixth Moscow Quantum Gravity Seminar.* World Sci. Publ., Singapore, p. 141.

6. Bekenstein JD. (2002) The Case for Discrete Energy Levels of a Black Hole. Reprinted from *2001: A Spacetime Oddysey* In: *Int. J. Mod. Phys.* A**17**, suppl. pp. 21–31.

7. Bekenstein JD. (2006) Information in the Holographic Universe. Reprinted from *Scientific American* In: *Scientific American, The Frontiers of Physics* **15**(3): pp. 74–81.

8. Bekenstein JD. (2007) Information in the Holographic Universe. Reprinted from *Scientific American* In: *Scientific American Reports* **17**(1): pp. 66–73.

9. Bekenstein JD. (2009) Relativistic MOND as an Alternative to the Dark Matter Paradigm. Reprinted from *Proceedings of the Eighteenth Particles and Nuclei International Conference* In: *Nucl. Phys.* A**827**: pp. 555–560.

Books

1. Bekenstein JD. (2001) *Buchi neri, Comunicazione, Energia.* Di Renzo Editore, Rome.

2. Bekenstein JD. (2006) *Of Gravity, Black holes and Information.* Di Renzo Editore, Rome.

3. Bekenstein JD, Mechoulam R, (eds.) (2012) *Albert Einstein Memorial Lectures* Israel Academy of Sciences and Humanities, Jerusalem and World Sci. Publ., Singapore.

Citation Classics

1. Bekenstein JD. (1973) Black Holes and Entropy. *Phys. Rev.* **D7**: 2333.

2. Bekenstein JD. (1974) Generalized Second Law of Thermodynamics in Black Hole Physics. *Phys. Rev.* **D9**: 3292.

3. Bekenstein JD. (1972) Black Holes and the Second Law. *Lettere al Nuovo Cimento* **4**: 737.

4. Bekenstein JD. (2004) Relativistic Gravitation Theory for the Modified Newtonian Dynamics Paradigm. *Phys. Rev.* **D70**: 083509

Commentaries

1. Bekenstein JD. (1979) Astronomical Consequences and Tests of Relativistic Theories of Variable Rest Masses. *Comm. Astrophys.* **8**: 89.

2. Bekenstein JD. (1981) Black-Hole Physics. *Physics Today* **34**(1): 69.

3. Bekenstein JD, Milgrom M, Sanders RH. (1995) Not the Only Game. *New Scientist* **145**: 49.

4. Bekenstein JD, Milgrom M, Sanders RH. (2000). Comment on "The Bright Side of Dark Matter". *Phys. Rev. Letters* **85**: 1346.

5. Bekenstein JD. (2013) What's Next — Discard Relativity. *New Scientist* **217**: 42.

Contributions On-Line

1. Bekenstein JD. (2008) Bekenstein bound. *Scholarpedia* **3** (10): 7374.

2. Bekenstein JD. (2008) Bekenstein-Hawking entropy. *Scholarpedia* **3** (10): 7375.

3. Bekenstein JD. (2004) Entropy/ Information Bounds and Gravitation. Lecture at *School and Workshop on Quantum Entanglement, Decoherence, Information, and Geometrical Phases in Complex Systems — Center for Theoretical Physics, Trieste.* November 1–12. http://agenda.ictp.trieste.it/smr.php? 1587

4. Bekenstein JD. (2005) Modified Gravity vs Dark Matter: Relativistic Theory for MOND. *Proc. Sci.* **jhw28**: 012.

Publications in Hebrew

1. Bekenstein JD. (1990) Time, Order and the Universe. Lecture in *Machshavoth* (IBM — Israel J. for Science), Dec.

2. Bekenstein JD. (1994) The Universe Through the Eyes of Man and the Satellite. Lecture in *Hauniversita* (Hebrew University's magazine) #9, p.12.

3. Bekenstein JD. (1995) General relativity. In: *Hebrew Encyclopedia* suppl. 3: 448-454.

4. Bekenstein JD. (1996) Nathan Rosen, great artist of theoretical physics. (Remarks on the scientific legacy of Prof. Rosen.) Israel Academy of Sciences, Jerusalem.

5. Bekenstein JD. (1999) Limits of information and the black hole. Prizewinning article in *Galileo* (Israeli popular science magazine) #35, pp. 38-45.

6. Bekenstein JD. (2004) Information in the Holographic Universe. Translation of *Scientific American* article. *Scientific American Israel* 2(9): 36.

7. Bekenstein JD. (2005) Black Holes and Information in a Holographic Universe. *PhysicaPlus* (Israel's popular physics magazine) #5.

Juan Ignacio Cirac

JUAN IGNACIO CIRAC

Birthplace

Born   Manresa, Spain, 11 October 1965

Education

1988   Licenciado (graduate) in Theoretical Physics, Universidad Complutense de Madrid

1991   Ph.D. in Physics, Universidad Complutense de Madrid

Research and Professional Experience

1991–1996   "Profesor Titular de Universidad", Departamento de Física Aplicada, Universidad de Castilla-La Mancha, Spain

1993–1994   Research Associate, Joint Institute for Laboratory Astrophysics, University of Colorado, USA

1996–2001   Professor, Institut für Theoretische Physik, Leopold Franzens Universität Innsbruck, Austria

2001–   Director of the Theory Division, Max Planck Institute of Quantum Optics, Garching, Germany

2002–   "Honorarprofessor", Technical University of Munich (Department of Physics), Germany

Membership of Societies

2003   Corresponding member of the Austrian Academy of Sciences

2002   Corresponding member of the Spanish Academy of Sciences

2002   Fellow of the American Physical Society

Honours

2005   "Doctor Honoris Causa", Universidad Castilla-La Mancha

2007   "Doctor Honoris Causa", Universidad Politecnica de Catalunya (Barcelona)

2015   "Doctor Honoris Causa", Universitat de València, Spain

2015   "Doctor Honoris Causa", Universitat Politècnica de València, Spain

Awards

Premio Nacional a Investigadores Noveles of the Royal Physical Society of Spain (1992), Felix Kuschenitz Preis of the Austrian Academy of Sciences (2001), Medal of the Royal Physical Society of Spain (2002), Quantum Electronics Prize of the European Physical Society (2005), Prince of Asturias Award for Technical and Scientific Research (2006), 6th International Quantum Communication Award (2006), National "Blas Cabrera" Prize for Physical, Material and Earth Sciences (2007), "Premios de las artes y de la ciencia" — Castellano-Manchegos del Mundo, Junta Castilla-La Mancha (2009), Carl Zeiss-Research Award (2009), BBVA Foundation

Frontiers of Knowledge Award (2009), Medal of Honour, Universidad Complutense de Madrid (2009), Benjamin Franklin Medal (2010), Premi Nacional de Pensament i Cultura Científica (2010), Gran Cruz de la Orden del Dos de Mayo de la Comunidad de Madrid (2011), Medal of Honour Niels Bohr Institute (2013), Wolf Prize in Physics (2013), Hamburg Prize for Theoretical Physics (2015).

COMMENTARY

I was born in Manresa, a small city near Barcelona (Spain) in 1965. My parents were high-school teachers there, teaching mathematics and Greek. At the age of three years, I moved to Barcelona and then, at ten, to Madrid. I was always very interested in science and engineering, read many science fiction books (among them, the whole Isaac Isamov's Foundation series), and I also liked technical drawing. I practiced many sports, like tennis, soccer, basketball, or taekwon-do.

When I finished high-school I enrolled in the school of Aeronautic Engineering, in Madrid. However, after a week of lectures I realized that I was not really convinced that I may want to devote my life to the field, but preferred to study something deeper and that required a thorough understanding. I thus enrolled in Physics, at the University of Madrid. My favorite subjects were those related to mathematics, which were taught at a very high level. However, in my third year I encountered for the first time a subject that changed my life: Quantum Physics. It combined everything I liked. On the one hand, it required some advanced mathematics in order to study it. It had also close connections with Philosophy, as some of its counterintuitive aspects had implications in that area. Besides that, and apart from describing what we are constituted from, it had led to the most relevant applications in the last century, like electronics, computing, lasers, etc. So, I decided to graduate in theoretical physics, where quantum physics was at the center. At the same time, I enrolled in an Engineering school, and studied three years Civil Engineering.

After I finished my studies, I went to do the military service for a year. After six months, I started my Ph.D. in the Optics Department of the University Complutense of Madrid, receiving the support of a generous grant from the Spanish Minister of Education and my parents. The subject of my Thesis was "Quantum Optics", which also combined some of the things I was interested in most: on the one hand, quantum physics, and on the other, a more applied approach to optics. The grant I enjoyed also gave me the opportunity to spend three months abroad. I decided to visit Prof. Zoller, at that time in Innsbruck, something that affected my life in an irreversible way. During that visit, I decided I wanted to become a scientist, and to work on quantum optics.

Right after my Ph.D. I was very lucky: I obtained a kind of tenure-track position (Profesor Titular de Universidad) at the University of Castilla-La Mancha, where I spent the next five years. At the same time, I traveled every year, for a total of about 18 months, to the University of Colorado in Boulder. There I worked with Peter Zoller as a postdoc, and started my interest in cold atoms and trapped ions physics. Together with him (and other collaborators, among them Rainer Blatt) we wrote several papers on how to cool atoms, how to prepare certain quantum states, and predicted several phenomena. With Peter Zoller and Maciej Lewenstein, we wrote some papers on Bose–Einstein condensation or cavity QED.

In summer of 1994, the ICAP conference took place in Boulder. They announced an invited talk by Artur Ekert, a young scientist from Oxford, about a very exciting subject: quantum computation. At that time, Peter Shor had written a paper (not published yet) showing that a quantum computer could solve the factorization problem in a much more efficient way than a classical one. Artur Ekert gave a wonderful talk explaining that and many other things, and finished his talk stating that it would be great if somebody figures out how to build a quantum computer. Peter Zoller and myself were sitting next to each other, and both of us thought right away that this should be possible with the systems we were studying. In fact, after few months we came up with an idea on how to do that using cold trapped ions. Fortunately enough, Christopher Monroe, David Wineland and their group experimentally demonstrated our ideas few months later, which started several activities worldwide to build small prototypes of quantum computers.

In 1996, I joined the Institute of Theoretical Physics at the University of Innsbruck, where Peter Zoller and Rainer Blatt had recently moved. Anton Zeilinger was also there, carrying out pioneering experiments (like teleportation). Thus, that place was a paradise for somebody interested in quantum physics or quantum information. From 1996–2001 I collaborated very closely with Peter Zoller, and we (together with students, visitors, and external collaborators) developed many concepts. Among them, the idea that sufficiently cold atoms in optical lattices would be described by the Hubbard-Model, a paradigm of quantum many-body physics in condensed matter. After few years, the group composed by Immanuel Bloch, Tilman Esslinger, Theodor Hänsch and others realized experimentally our proposal. This was, for us, a great accomplishment as it opened up a new area of research in atomic physics, with a lot of things to learn and understand. It also connected us with scientists of completely different areas, from whom we could learn and collaborate. During that period, until 2001, I also worked on Quantum Information Theory, with the goal of understanding, classifying and quantifying entanglement, which is the "fuel" in most applications in quantum information. That was a more abstract area of research, but I enjoyed it very much since I always had some passion for mathematics.

In the year 2001, I joined the Max-Planck Institute of Quantum Optics (MPQ) in Garching. This was another very important step in my career. I had to leave Innsbruck, where I had so much enjoyed working together with Peter Zoller. But it also opened up new challenges, and the possibility of continuing those collaborations as Innsbruck is not very far from Garching. There I continued working on quantum optics, cold atoms, and quantum information, making and analyzing proposals for experiments, to build information processing devices, or devising ways of performing quantum simulations with different systems. I also started working on a new area of research, together with Frank Verstraete (a postdoc in my group at that time), namely tensor networks. Together with many students and collaborators, we have been trying to express and solve many-body problems using a novel language. This is a very long-term project, but I believe that it will become relevant in other areas of physics soon.

And in 2013, the Wolf prize for Peter Zoller and myself arrived! I felt really privileged with such a great honor, and I have to thank many people for that. First of all, all my collaborators: I have had enormous luck to work with a bunch of brilliant and talented young scientists, who have taught me a lot. Second, my family and especially my parents, who have not only given me their unconditional support, but have always made everything possible so that I would obtain the best possible education.

Long-distance quantum communication with atomic ensembles and linear optics

L.-M. Duan*†, **M. D. Lukin**‡, **J. I. Cirac*** & **P. Zoller***

* *Institut für Theoretische Physik, Universität Innsbruck, A-6020 Innsbruck, Austria*
† *Laboratory of Quantum Communication and Computation, USTC, Hefei 230026, China*
‡ *Physics Department and ITAMP, Harvard University, Cambridge, Massachusetts 02138, USA*

Quantum communication holds promise for absolutely secure transmission of secret messages and the faithful transfer of unknown quantum states. Photonic channels appear to be very attractive for the physical implementation of quantum communication. However, owing to losses and decoherence in the channel, the communication fidelity decreases exponentially with the channel length. Here we describe a scheme that allows the implementation of robust quantum communication over long lossy channels. The scheme involves laser manipulation of atomic ensembles, beam splitters, and single-photon detectors with moderate efficiencies, and is therefore compatible with current experimental technology. We show that the communication efficiency scales polynomially with the channel length, and hence the scheme should be operable over very long distances.

The goal of quantum communication is to transmit quantum states between distant sites. Such transmission has potential application in the secret transfer of classical messages by means of quantum cryptography[1], and is also an essential element in the construction of quantum networks. The basic problem of quantum communication is to generate nearly perfect entangled states between distant sites. Such states can be used, for example, to implement secure quantum cryptography using the Ekert protocol[1], and to faithfully transfer quantum states via quantum teleportation[2]. All realistic schemes for quantum communication are at present based on the use of photonic channels. However, the degree of entanglement generated between two distant sites normally decreases exponentially with the length of the connecting channel, because of optical absorption and other channel noise. To regain a high degree of entanglement, purification schemes can be used[3], but this does not fully solve the long-distance communication problem. Because of the exponential decay of the entanglement in the channel, an exponentially large number of partially entangled states are needed to obtain one highly entangled state, which means that for a sufficiently long distance the task becomes nearly impossible.

To overcome the difficulty associated with the exponential fidelity decay, the concept of quantum repeaters can be used[4]. In principle, this allows the overall communication fidelity to be made very close to unity, with the communication time growing only polynomially with transmission distance. In analogy to fault-tolerant quantum computing[5,6], the proposed quantum repeater is a cascaded entanglement-purification protocol for communication systems. The basic idea is to divide the transmission channel into many segments, with the length of each segment comparable to the channel attenuation length. First, entanglement is generated and purified for each segment; the purified entanglement is then extended to a greater length by connecting two adjacent segments through entanglement swapping[2,7]. After this swapping, the overall entanglement is decreased, and has to be purified again. The rounds of entanglement swapping and purification can be continued until nearly perfect entangled states are created between two distant sites.

To implement the quantum repeater protocol, we need to generate entanglement between distant quantum bits (qubits), store them for a sufficiently long time and perform local collective operations on several of these qubits. Quantum memory is essential, because all purification protocols are probabilistic. When entanglement purification is performed for each segment of the channel, quantum memory can be used to keep the segment state if the purification succeeds, and to repeat the purification for the

segments only where the previous attempt fails. This is essential for ensuring polynomial scaling in the communication efficiency, because if there were no available memory, the purifications for all the segments would need to succeed at the same time; the probability of such an event decreases exponentially with channel length. The requirement of quantum memory implies that we need to store the local qubits in atomic internal states instead of photonic states, as it is difficult to store photons for a reasonably long time. With atoms as the local information carriers, it seems to be very hard to implement quantum repeaters: normally, one needs to achieve the strong coupling between atoms and photons by using high-finesse cavities for atomic entanglement generation, purification, and swapping[8,9], which, in spite of recent experimental advances[10-12], remains a very challenging technology.

Here we propose a different scheme, which realizes quantum repeaters and long-distance quantum communication with simple physical set-ups. The scheme is a combination of three significant advances in entanglement generation, connection, and applications, with each of the steps having built-in entanglement purification and resilience to realistic noise. The scheme for fault-tolerant entanglement generation originates from earlier proposals to entangle single atoms through single-photon interference at photodetectors[13,14]. But the present approach involves collective excitations in atomic ensembles rather than in single particles, which allows simpler realization and greatly improved generation efficiency. This is due to collectively enhanced coupling to light, which has been recently investigated both theoretically[15-19] and experimentally[20-22]. The entanglement connection is achieved through simple linear optical operations, and is inherently robust against realistic imperfections. Different schemes with linear optics have been proposed recently for quantum computation[23] and purification[24]. Finally, the resulting state of ensembles after the entanglement connection finds direct applications in realizing entanglement-based quantum communication protocols, such as quantum teleportation, cryptography, and Bell inequality detection. In all of these applications, the mixed entanglement is purified automatically to nearly perfect entanglement. As a combination of these three advances, our scheme circumvents the realistic noise and imperfections, and provides a feasible method of long-distance high-fidelity quantum communication. The required overhead in communication time increases with distance only polynomially.

Entanglement generation

The basic element of our system is a cloud of N_a identical atoms with the relevant level structure shown in Fig. 1. A pair of metastable lower states $|g\rangle$ and $|s\rangle$ can correspond to—for example—hyperfine or Zeeman sublevels of the electronic ground state of alkali-metal atoms. Long lifetimes for the relevant coherence have been observed both in a room-temperature dilute atomic gas (see, for example, ref. 21) and in a sample of cold trapped atoms (see, for example, refs 20, 22). To facilitate enhanced coupling to light, the atomic medium is preferably optically thick along one direction. This can be achieved either by working with a pencil-shaped atomic sample[20–22] or by placing the sample in a low-finesse ring cavity[17,25] (see Supplementary Information).

All the atoms are initially prepared in the ground state $|g\rangle$. A sample is illuminated by a short, off-resonant laser pulse that induces Raman transitions into the states $|s\rangle$. We are particularly interested in the forward-scattered Stokes light that is co-propagating with the laser. Such scattering events are uniquely correlated with the excitation of the symmetric collective atomic mode S (refs 15–22) given by $S \equiv (1/\sqrt{N_a})\Sigma_i |g\rangle_i \langle s|$, where the summation is taken over all the atoms. In particular, an emission of the single Stokes photon in a forward direction results in the state of atomic

ensemble given by $S^\dagger|0_a\rangle$, where the ensemble ground state $|0_a\rangle \equiv \otimes_i |g\rangle_i$.

We assume that the light–atom interaction time t_Δ is short, so that the mean photon number in the forward-scattered Stokes pulse is much smaller than 1. We can define an effective single-mode bosonic operator a for this Stokes pulse with the corresponding vacuum state denoted by $|0_p\rangle$. The whole state of the atomic collective mode and the forward-scattering Stokes mode can now be written in the following form (see Supplementary Information for details)

$$|\phi\rangle = |0_a\rangle|0_p\rangle + \sqrt{p_c}S^\dagger|0_a\rangle|0_p\rangle + o(p_c) \qquad (1)$$

where p_c is the small excitation probability, and $o(p_c)$ represents the terms with more excitations whose probabilities are equal to or smaller than p_c^2. Before proceeding, we note that a fraction of light is emitted in other directions owing to spontaneous emissions. But whenever N_a is large, the contribution from the spontaneous emissions to the population in the symmetric collective mode is small[15–22]. As a result, we have a large signal-to-noise ratio for the processes involving the collective mode, which greatly enhances the efficiency of the present scheme (see Box 1 and Supplementary Information).

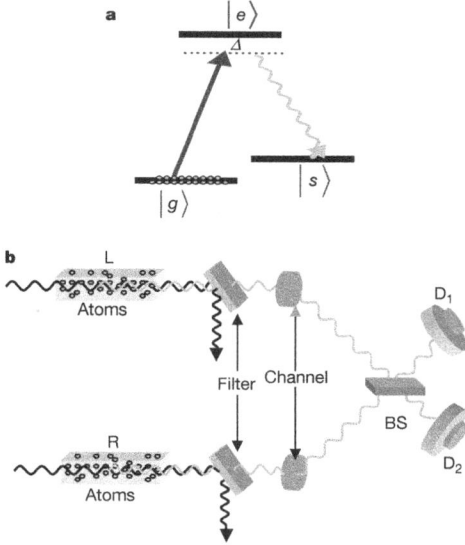

Figure 1 Set-up for entanglement generation. **a**, The relevant level structure of the atoms in the ensemble, with $|g\rangle$, the ground state, $|s\rangle$, the metastable state for storing a qubit, and $|e\rangle$, the excited state. The transition $|g\rangle \rightarrow |e\rangle$ is coupled by the classical laser (the pumping light) with the Rabi frequency Ω, and the forward-scattered Stokes light comes from the transition $|e\rangle \rightarrow |s\rangle$, which has a different polarization and frequency to the pumping light. For convenience, we assume off-resonant coupling with a large detuning Δ. **b**, Schematic set-up for generating entanglement between the two atomic ensembles L and R. The two ensembles are pencil-shaped, and illuminated by the synchronized classical pumping pulses. The forward-scattered Stokes pulses are collected and coupled to optical channels (such as fibres) after the filters, which are polarization- and frequency-selective to filter the pumping light. The pulses after the transmission channels interfere at a 50%-50% beam splitter BS, with the outputs detected respectively by two single-photon detectors D_1 and D_2. If there is a click in D_1 or D_2, the process is finished and we successfully generate entanglement between the ensembles L and R. Otherwise, we first apply a repumping pulse (to the transition $|s\rangle \rightarrow |e\rangle$) to the ensembles L and R, to set the state of the ensembles back to the ground state $|0_a\rangle_L \otimes |0_a\rangle_R$, then the same classical laser pulses as the first round are applied to the transition $|g\rangle \rightarrow |e\rangle$ and we detect again the forward-scattered Stokes pulses after the beam splitter. This process is repeated until finally we have a click in the D_1 or the D_2 detector.

Box 1

Collective enhancement

Long-lived excitations in atomic ensembles can be viewed as waves of excited spins. We are here particularly interested in the symmetric spin wave mode S. For a simple demonstration of collective enhancement, we assume that the atoms are placed in a low-finesse ring cavity[25], with a relevant cavity mode corresponding to forward-scattered Stokes radiation. The cavity-free case corresponds to the limit where the finesse tends to 1 (ref. 17). The interaction between the forward-scattered light mode and the atoms is described by the hamiltonian

$$H = \hbar\left(\sqrt{N_a}\,\Omega g_c/\Delta\right)S^\dagger b^\dagger + \text{h.c.}$$

where h.c. is the hermitian conjugation, $b^\dagger$ is the creation operator for cavity photons, Ω is the laser Rabi frequency, and g_c is the atom–field coupling constant. In addition to coherent evolution, the photonic field mode can leak out of the cavity at a rate κ, whereas atomic coherence is dephased by spontaneous photon scattering into random directions that occurs at a rate $\gamma_s' = \Omega^2/\Delta^2\gamma_s$ for each atom, with γ_s being the natural linewidth of the electronic excited state. We emphasize that in the absence of superradiant effects, spontaneous emission events are independent for each atom.

In the bad-cavity limit, we can adiabatically eliminate the cavity mode, and the resulting dynamics for the collective atomic mode is described by the Heisenberg–Langevin equation (see Supplementary Information for details)

$$\dot{S}^\dagger = \frac{(\kappa' - \gamma_s')}{2}S^\dagger - \sqrt{\kappa'}b_{in}(t) + \text{noise}$$

where $\kappa' = 4|\Omega|^2 g_c^2 N_a/(\Delta^2 \kappa)$, b_{in} is a vacuum field leading into the cavity, and the last term represents the fluctuating noise field corresponding to spontaneous emission. We note that the nature of the dynamics is determined by the ratio between the build-up of coherence due to forward-scattered photons κ' and coherence decay due to spontaneous emission γ_s'. The signal-to-noise ratio is therefore given by $R = \kappa'/\gamma_s' \equiv 4N_a g_c^2/(\kappa\gamma_s)$, which is large when a many-atom ensemble is used. In the cavity-free case, this expression corresponds to the optical depth (density-length product) of the sample. The result should be compared with the signal-to-noise ratio in the single-atom case $N_a = 1$, where to obtain $R > 1$ a high-Q microcavity is required[10–12]. The collective enhancement takes place because the coherent forward scattering involves only one collective atomic mode S, whereas the spontaneous emissions distribute excitation over all atomic modes. Therefore only a small fraction of spontaneous emission events influences the symmetric mode S, which results in a large signal-to-noise ratio.

We now show how to use this set-up to generate entanglement between two distant ensembles L (left) and R (right) using the configuration shown in Fig. 1. Here two laser pulses excite both ensembles simultaneously, and the whole system is described by the state $|\phi\rangle_L \otimes |\phi\rangle_R$, where $|\phi\rangle_L$ and $|\phi\rangle_R$ are given by equation (1) with all the operators and states distinguished by the subscript L or R. The forward-scattered Stokes light from both ensembles is combined at the beam splitter, and a photodetector click in either D_1 or D_2 measures the combined radiation from two samples, $a_+^\dagger a_+$ or $a_-^\dagger a_-$ with $a_\pm = (a_L \pm e^{i\varphi} a_R)/\sqrt{2}$. Here, φ denotes an unknown difference of the phase shifts in the left and the right side channels. We can also assume that φ has an imaginary part to account for the possible asymmetry of the set-up, which will also be corrected automatically in our scheme. But the set-up asymmetry can be easily made very small, and for simplicity of expressions we assume that φ is real in the following. Conditional on the detector click, we should apply a_+ or a_- to the whole state $|\phi\rangle_L \otimes |\phi\rangle_R$, and the projected state of the ensembles L and R is nearly maximally entangled, with the form (neglecting the high-order terms $o(p_c)$):

$$|\Psi_\varphi\rangle_{LR}^\pm = (S_L^\dagger \pm e^{i\varphi} S_R^\dagger)/\sqrt{2}|0_a\rangle_L |0_a\rangle_R \qquad (2)$$

The probability of getting a click is given by p_c for each round, so we need to repeat the process about $1/p_c$ times for a successful entanglement preparation, and the average preparation time is given by $T_0 \approx t_\Delta/p_c$. The states $|\Psi_\varphi\rangle_{LR}^+$ and $|\Psi_\varphi\rangle_{LR}^-$ can be transformed to each other by a simple local phase shift. Without loss of generality, we assume in the following that we generate the entangled state $|\Psi_\varphi\rangle_{LR}^+$.

As will be shown below, the presence of noise modifies the projected state of the ensembles to

$$\rho_{LR}(c_0,\varphi) = \frac{1}{c_0+1}(c_0|0_a 0_a\rangle_{LR}\langle 0_a 0_a| + |\Psi_\varphi\rangle_{LR}^+\langle\Psi_\varphi|) \qquad (3)$$

where the 'vacuum' coefficient c_0 is determined by the dark count rates of the photon detectors. It will be seen below that any state in the form of equation (3) will be purified automatically to a maximally entangled state in the entanglement-based communication schemes. We therefore call this state an effective maximally entangled (EME) state, with the vacuum coefficient c_0 determining the purification efficiency.

Entanglement connection through swapping
After the successful generation of entanglement within the

attenuation length, we want to extend the quantum communication distance. This is done through entanglement swapping with the configuration shown in Fig. 2. Suppose that we start with two pairs of entangled ensembles described by the state $\rho_{LI_1} \otimes \rho_{I_2R}$, where ρ_{LI_1} and ρ_{I_2R} are given by equation (3). In the ideal case, the set-up shown in Fig. 2 measures the quantities corresponding to operators $S_\pm^\dagger S_\pm$ with $S_\pm = (S_{I_1} \pm S_{I_2})/\sqrt{2}$. If the measurement is successful (that is, one of the detectors registers one photon), we will prepare the ensembles L and R into another EME state. The new φ-parameter is given by $\varphi_1 + \varphi_2$, where φ_1 and φ_2 denote the old φ-parameters for the two segment EME states. As will be seen below, even in the presence of realistic noise and imperfections, an EME state is still created after a detector click. The noise only influences the success probability of getting a click and the new vacuum coefficient in the EME state. In general, we can express the success probability p_1 and the new vacuum coefficient c_1 as $p_1 = f_1(c_0)$ and $c_1 = f_2(c_0)$, where the functions f_1 and f_2 depend on the particular noise properties.

The above method for connecting entanglement can be cascaded to arbitrarily extend the communication distance. For the ith ($i = 1, 2, ..., n$) entanglement connection, we first prepare in parallel two pairs of ensembles in the EME states with the same vacuum coefficient c_{i-1} and the same communication length L_{i-1}, and then perform entanglement swapping as shown in Fig. 2, which now succeeds with a probability $p_i = f_1(c_{i-1})$. After a successful detector click, the communication length is extended to $L_i = 2L_{i-1}$, and the vacuum coefficient in the connected EME state becomes $c_i = f_2(c_{i-1})$. As the ith entanglement connection needs to be repeated on average $1/p_i$ times, the total time needed to establish an EME state over the distance $L_n = 2^n L_0$ is given by $T_n = T_0\Pi_{i=1}^n(1/p_i)$, where L_0 denotes the distance of each segment in the entanglement generation.

Entanglement-based communication schemes
After an EME state has been established between two distant sites, we would like to use it in communication protocols, such as quantum teleportation, cryptography, and Bell inequality detection. It is not obvious that the EME state of equation (3), which is entangled in the Fock basis, is useful for these tasks, as in the Fock basis it is experimentally hard to do certain single-bit operations. We will now show how the EME states can be used to realize all these protocols with simple experimental configurations.

Quantum cryptography and Bell inequality detection are achieved with the set-up shown by Fig. 3a. The state of the two

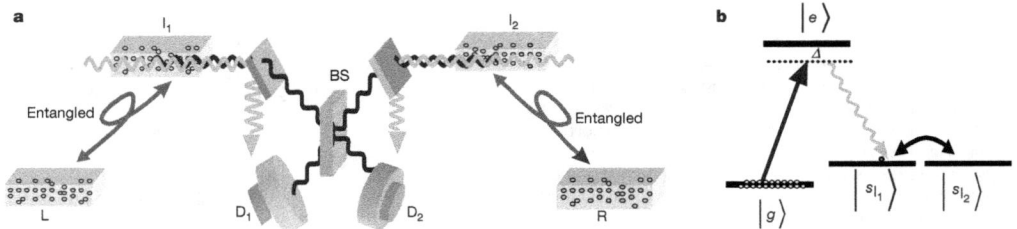

Figure 2 Set-up for entanglement connection. **a,** Illustrative set-up for the entanglement swapping. We have two pairs of ensembles—L and I_1, and I_2 and R—distributed at three sites L, I and R. Each of the ensemble-pairs L and I_1, and I_2 and R is prepared in an EME state in the form of equation (3). The stored atomic excitations of two nearby ensembles I_1 and I_2 are converted simultaneously into light. This is achieved by applying a retrieval pulses of suitable polarization that is near-resonant with the atomic transition $|s\rangle \rightarrow |e\rangle$, which causes coherent conversion of atomic excitations into photons that have a different polarization and frequency to the retrieval pulse[18,21,22]. The efficiency of this transfer can be very close to unity even at a single quantum level owing to collective enhancement[18,21,22]. After the transfer, the stimulated optical excitations interfere at a

50%-50% beam splitter, and then detected by the single-photon detectors D_1 and D_2. If either D_1 or D_2 clicks, the protocol is successful and an EME state in the form of equation (3) is established between the ensembles L and R with a doubled communication distance. Otherwise, the process fails, and we need to repeat the previous entanglement generation and swapping until finally we have a click in D_1 or D_2, that is, until the protocol finally succeeds. **b,** The two intermediate ensembles I_1 and I_2 can also be replaced by one ensemble but with two metastable states I_1 and I_2 to store the two different collective modes. The 50%-50% beam splitter operation can be simply realized by a $\pi/2$ pulse applied to the two metastable states before the collective atomic excitations are transferred to the optical excitations.

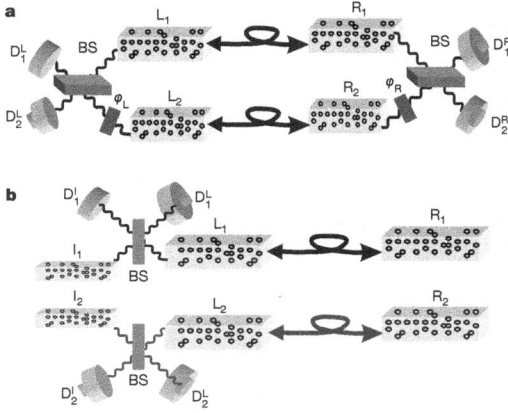

Figure 3 Set-up for entanglement-based communication schemes. **a**, Schematic set-up for the realization of quantum cryptography and Bell inequality detection. Two pairs of ensembles L_1, R_1 and L_2, R_2 (or two pairs of metastable states as shown in **b**) have been prepared in the EME states. The collective atomic excitations on each side are transferred to the optical excitations, which, respectively after a relative phase shift φ_L or φ_R and a 50%-50% beam splitter, are detected by the single-photon detectors D_1^L, D_2^L and D_1^R, D_2^R. We look at the four possible coincidences of D_1^R, D_2^R with D_1^L, D_2^L, which are functions of the phase difference $\varphi_L - \varphi_R$. Depending on the choice of φ_L and φ_R, this set-up can realize both the quantum cryptography and the Bell inequality detection. **b**, Schematic set-up for probabilistic quantum teleportation of the atomic 'polarization' state. Similarly, two pairs of ensembles L_1, R_1 and L_2, R_2 are prepared in the EME states. We want to teleport an atomic 'polarization' state $(d_0 S_{I_1}^\dagger + d_1 S_{I_2}^\dagger)|0_a 0_a\rangle_{I_1 I_2}$ with unknown coefficients d_0, d_1 from the left to the right side, where $S_{I_1}^\dagger$, $S_{I_2}^\dagger$ denote the collective atomic operators for the two ensembles I_1 and I_2 (or two metastable states in the same ensemble). The collective atomic excitations in the ensembles I_1, L_1 and I_2, L_2 are transferred to the optical excitations, which, after a 50%-50% beam splitter, are detected by the single-photon detectors D_1^I, D_1^L and D_2^I, D_2^L. If, and only if, there is one click in D_1^I, D_1^L, and one click in D_2^I or D_2^L, the protocol is successful. When the protocol succeeds, the collective excitation in the ensembles R_1 and R_2, if appearing, would be found in the same 'polarization' state $(d_0 S_{R_1}^\dagger + d_1 S_{R_2}^\dagger)|0_a 0_a\rangle_{R_1 R_2}$ up to a local π-phase rotation.

pairs of ensembles is expressed as $\rho_{L_1 R_1} \otimes \rho_{L_2 R_2}$, where $\rho_{L_i R_i}$ ($i = 1, 2$) denote the same EME state with the vacuum coefficient c_n if we have done entanglement connection n times. The φ-parameters in $\rho_{L_1 R_1}$ and $\rho_{L_2 R_2}$ are the same, provided that the two states are established over the same stationary channels. We register only the coincidences of the two-side detectors, so the protocol is successful only if there is a click on each side. Under this condition, the vacuum components in the EME states, together with the state components $S_{L_1}^\dagger S_{L_2}^\dagger|\mathrm{vac}\rangle$ and $S_{R_1}^\dagger S_{R_2}^\dagger|\mathrm{vac}\rangle$, where $|\mathrm{vac}\rangle$ denotes the ensemble state $|0_a 0_a 0_a 0_a\rangle_{L_1 L_2 R_1 R_2}$, have no contributions to the experimental results. So, for the measurement scheme shown by Fig. 3, the ensemble state $\rho_{L_1 R_1} \otimes \rho_{L_2 R_2}$ is effectively equivalent to the following 'polarization' maximally entangled (PME) state (the terminology of 'polarization' comes from an analogy to the optical case):

$$|\Psi\rangle_{\mathrm{PME}} = (S_{L_1}^\dagger S_{R_2}^\dagger + S_{L_2}^\dagger S_{R_1}^\dagger)/\sqrt{2}|\mathrm{vac}\rangle \quad (4)$$

The success probability for the projection from $\rho_{L_1 R_1} \otimes \rho_{L_2 R_2}$ to $|\Psi\rangle_{\mathrm{PME}}$ (that is, the probability of getting a click on each side) is given by $p_a = 1/[2(c_n + 1)^2]$. We can also check that in Fig. 3, the phase shift φ_Λ ($\Lambda = L$ or R) together with the corresponding beam-splitter operation are equivalent to a single-bit rotation in the basis $\{|0\rangle_\Lambda \equiv S_{\Lambda_1}^\dagger|0_a 0_a\rangle_{\Lambda_1 \Lambda_2}, |1\rangle_\Lambda \equiv S_{\Lambda_2}^\dagger|0_a 0_a\rangle_{\Lambda_1 \Lambda_2}\}$ with the rotation angle $\theta = \varphi_\Lambda/2$. Now it is clear how to do quantum cryptography and Bell inequality detection, as we have the PME state and we can

perform the desired single-bit rotations in the corresponding basis. For instance, to distribute a quantum key between the two remote sides, we simply choose φ_Λ randomly from the set $\{0, \pi/2\}$ with an equal probability, and keep the measurement results (to be 0 if D_1^Λ clicks, and 1 if D_2^Λ clicks) on both sides as the shared secret key if the two sides become aware that they have chosen the same phase shift after the public declaration of φ_Λ. This is exactly the Ekert scheme[1], and its absolute security follows directly from the proofs in refs 26 and 27. For the Bell inequality detection, we infer the correlations $E(\varphi_L, \varphi_R) \equiv P_{D_1^L D_1^R} + P_{D_2^L D_2^R} - P_{D_1^L D_2^R} - P_{D_2^L D_1^R} = \cos(\varphi_L - \varphi_R)$ from the measurement of the coincidences $P_{D_1^L D_1^R}$ and so on. For the set-up shown in Fig. 3a, we would have $|E(0, \pi/4) + E(\pi/2, \pi/4) + E(\pi/2, 3\pi/4) - E(0, 3\pi/4)| = 2\sqrt{2}$, whereas for any local hidden variable theories, the CHSH inequality[28] implies that this value should be below 2.

We can also use the established long-distance EME states for faithful transfer of unknown quantum states through quantum teleportation, with the set-up shown in Fig. 3b. As described in the figure legend, this set-up is used to teleport the polarization state of the collective atomic excitation in a probabilistic fashion. That is, even if the protocol succeeds—that is, two of the detectors register the counts on the left-hand side—an excitation is not necessarily present in the right (target) ensembles because the product of the EME states $\rho_{L_1 R_1} \otimes \rho_{L_2 R_2}$ contains vacuum components. However, if a collective excitation appears from the right-hand side, its 'polarization' state is exactly the same as the one input from the left side. So, as in the experiment of ref. 29, such a probabilistic teleportation needs posterior confirmation of the presence of the excitation; but if the presence is confirmed, the teleportation fidelity of its polarization state is nearly perfect. The success probability for the teleportation is also given by $p_a = 1/[2(c_n + 1)^2]$, which determines the average number of repetitions needed for a final successful teleportation.

Noise and built-in entanglement purification

We now discuss noise and imperfections in our schemes for entanglement generation, connection, and applications. In particular, we show that each step contains built-in entanglement purification, which makes the whole scheme resilient to realistic noise and imperfections.

In the entanglement generation, the dominant noise is due to photon loss, which includes contributions from channel attenuation, spontaneous emissions in the atomic ensembles (which result in the population of the collective atomic mode with the accompanying photon going in other directions), coupling inefficiency of the Stokes light into and out of the channel, and inefficiency of the single-photon detectors. The loss probability is denoted by $1 - \eta_p$ with the overall efficiency $\eta_p = \eta_p' e^{-L_0/L_{\mathrm{att}}}$, where we have separated the channel attenuation $e^{-L_0/L_{\mathrm{att}}}$ (L_{att} is the channel attenuation length) from other noise contributions η_p', with η_p' independent of the communication distance L_0. The photon loss decreases the success probability for getting a detector click from p_c to $\eta_p p_c$, but it has no influence on the resulting EME state. Owing to this noise, the entanglement preparation time should be replaced by $T_0 \approx t_\Delta/(\eta_p p_c)$. The second source of noise comes from the dark counts of the single-photon detectors. The dark count gives a detector click, but without population of the collective atomic mode, so it contributes to the vacuum coefficient in the EME state. If the dark count comes up with a probability p_{dc} for the time interval t_Δ, the vacuum coefficient is given by $c_0 = p_{dc}/(\eta_p p_c)$, which is typically much smaller than 1 as the Raman transition rate is much larger than the dark count rate. The final source of noise, which influences the fidelity of getting the EME state, is caused by the event in which more than one atom is excited to the collective mode S whereas there is only one click in D_1 or D_2. The conditional probability for that event is given by p_c, so we can estimate the fidelity imperfection $\Delta F_0 \equiv 1 - F_0$ for the entanglement generation by:

$$\Delta F_0 \approx p_c \qquad (5)$$

Note that by decreasing the excitation probability p_c, we can make the fidelity imperfection closer and closer to zero—with the price of a longer entanglement preparation time T_0. This is the basic idea of the entanglement purification. So, in this scheme, the confirmation of the click from the single-photon detector generates and purifies entanglement at the same time.

In the entanglement swapping, the dominant noise is still due to the losses, which include contributions from detector inefficiency, the inefficiency of the excitation transfer from the collective atomic mode to the optical mode[21,22], and the small decay of atomic excitation during storage[20-22]. Note that by introducing the detector inefficiency, we have automatically taken into account the imperfection that the detectors cannot distinguish between one and two photons. With all these losses, the overall efficiency in entanglement swapping is denoted by η_s. The loss in entanglement swapping gives contributions to the vacuum coefficient in the connected EME state, as in the presence of loss a single detector click might result from two collective excitations in the ensembles I_1 and I_2, and in this case, the collective modes in the ensembles L and R have to be in a vacuum state. After taking into account the realistic noise, we can specify the success probability and the new vacuum coefficient for the ith entanglement connection by the recursion relations $p_i \equiv f_1(c_{i-1}) = \eta_s(1 - \{\eta_s/[2(c_{i-1} + 1)]\})/(c_{i-1} + 1)$ and $c_i \equiv f_2(c_{i-1}) = 2c_{i-1} + 1 - \eta_s$. The coefficient c_0 for the entanglement preparation is typically much smaller than $1 - \eta_s$, so we have $c_i \approx (2^i - 1)(1 - \eta_s) = (L_i/L_0 - 1)(1 - \eta_s)$, where L_i denotes the communication distance after i times entanglement connection. With the expression for the c_i, we can evaluate the probability p_i and the communication time T_n for establishing an EME state over the distance $L_n = 2^n L_0$. After the entanglement connection, the fidelity of the EME state also decreases, and after n times connection, the overall fidelity imperfection $\Delta F_n \approx 2^n \Delta F_0 \approx (L_n/L_0)\Delta F_0$. We need to make ΔF_n small by decreasing the excitation probability p_c in equation (5).

We note that our entanglement connection scheme also has a built-in entanglement-purification function. This can be understood as follows: each time we connect entanglement, the imperfections of the set-up decrease the entanglement fraction $1/(c_i + 1)$ in the EME state. However, this fraction decays only linearly with distance (the number of segments), which is in contrast to the exponential decay of entanglement for connection schemes without entanglement purification. The reason for the slow decay is that for each time of entanglement connection, we need to repeat the protocol until there is a detector click, and the confirmation of a click removes part of the added vacuum noise, as a larger vacuum component in the EME state results in more repetitions. The built-in entanglement purification in the connection scheme is essential for the polynomial scaling law of the communication efficiency.

As in the entanglement generation and connection schemes, our entanglement application schemes also have built-in entanglement purification, which makes them resilient to realistic noise. First, we have seen that the vacuum components in the EME states are removed from the confirmation of the detector clicks, and thus have no influence on the fidelity of all the application schemes. Second, if the single-photon detectors and the atom-to-light excitation transitions in the application schemes are imperfect, with the overall efficiency denoted by η_a, we can show that these imperfections only influence the efficiency of getting detector clicks—with the success probability replaced by $p_a = \eta_a/[2(c_n + 1)^2]$—and have no effect on communication fidelity. Last, we have seen that the phase shifts in the stationary channels and the small asymmetry of the stationary set-up are removed automatically when we project the EME state to the PME state, and thus have no influence on the communication fidelity.

Noise not correctable by our scheme includes the detector dark count in the entanglement connection, the non-stationary channel noise and set-up asymmetries. The fidelity imperfection resulting from the dark count increases linearly with the number of segments L_n/L_0, and the imperfections from the non-stationary channel noise and set-up asymmetries increase by the random-walk law $\sqrt{L_n/L_0}$. For each time of entanglement connection, the dark count probability is about 10^{-5} if we make a typical choice that the collective emission rate is about 10 MHz and the dark count rate is 10^2 Hz. So this noise is negligible, even if we have communicated over a long distance (10^3 times the channel attenuation length L_{att}, for instance). The non-stationary channel noise and set-up asymmetries can also be safely neglected for such a distance. For instance, it is relatively easy to control the non-stationary asymmetries in local laser operations to values below 10^{-4} with the use of accurate polarization techniques[30] for Zeeman sublevels (as in Fig. 2b).

Scaling of the communication efficiency

We have shown that each of our entanglement generation, connection, and application schemes has built-in entanglement purification, and as a result of this property, we can fix the communication fidelity to be nearly perfect, and at the same time require the communication time to increase only polynomially with distance. Assume that we want to communicate over a distance $L = L_n = 2^n L_0$. By fixing the overall fidelity imperfection to be a desired small value ΔF_n, the entanglement preparation time becomes $T_0 \approx t_\Delta/(\eta_p \Delta F_0) \approx (L_n/L_0)t_\Delta/(\eta_p \Delta F_n)$. For effective generation of the PME state of equation (4), the total communication time $T_{tot} \approx T_n/p_a$ with $T_n = T_0 \Pi_{i=1}^n (1/p_i)$. So the total communication time scales with distance by the law

$$T_{tot} \approx 2(L/L_0)^2/(\eta_p p_a \Delta F_n \Pi_{i=1}^n p_i) \qquad (6)$$

where the success probabilities p_i, p_a for the ith entanglement connection and for the entanglement application have been specified above.

Equation (6) confirms that the communication time T_{tot} increases with distance L only polynomially. We show this explicitly by taking two limiting cases. In the first case, the inefficiency $1 - \eta_s$ for entanglement swapping is assumed to be negligibly small. We can deduce from equation (6) that in this case the communication time $T_{tot} \approx T_{con}(L/L_0)^2 e^{L_0/L_{att}}$, with the constant $T_{con} \equiv 2t_\Delta/(\eta_p' \eta_a \Delta F_n)$ being independent of the segment length and the total distance L_0 and L. The communication time T_{tot} increases with L quadratically. In the second case, we assume that the inefficiency $1 - \eta_s$ is fairly large. The communication time in this case is approximated by $T_{tot} \approx T_{con}(L/L_0)^{[\log_2(L/L_0)+1]/2 + \log_2(1/\eta_s - 1) + 2} e^{L_0/L_{att}}$, which increases with L still polynomially (or, more accurately, sub-exponentially, but this makes no difference in practice as the factor $\log_2(L/L_0)$ is well bounded from above for any reasonably long distance). If T_{tot} increases with L/L_0 by the mth power law $(L/L_0)^m$, there is an optimal choice of segment length ($L_0 = mL_{att}$) to minimize the time T_{tot}. As a simple estimation of the improvement in communication efficiency, we assume that the total distance L is about $100L_{att}$; for a choice of the parameter $\eta_s \approx 2/3$, the communication time $T_{tot}/T_{con} \approx 10^6$ with the optimal segment length $L_0 \approx 5.7L_{att}$. This is a notable improvement over the direct communication case, where the communication time T_{tot} for getting a PME state increases with distance L by the exponential law $T_{tot} \approx T_{con}e^{L/L_{att}}$. For the same distance $L \approx 100L_{att}$, we need $T_{tot}/T_{con} \approx 10^{43}$ for direct communication, which means that for this example the present scheme is 10^{37} times more efficient.

Outlook

We have presented a scheme for implementation of quantum repeaters and long-distance quantum communication. The proposed technique allows the generation and connection of entanglement, and its use in quantum teleportation, cryptography, and tests of Bell inequalities. All of the elements of the present scheme are within reach of current experimental technology, and all have the important property of built-in entanglement purification—which makes them resilient to realistic noise. As a result, the overhead required to implement the present scheme, such as the communication time, scales polynomially with the channel length. This is in marked contrast to direct communication, where an exponential overhead is required. Such efficient scaling, combined with the relative simplicity of the experimental set-up, opens up realistic prospects for quantum communication over long distances. □

Received 16 May; accepted 12 September 2001.

1. Ekert, A. Quantum cryptography based on Bell's theorem. *Phys. Rev. Lett.* **67**, 661–663 (1991).
2. Bennett, C. H. *et al.* Teleporting an unknown quantum state via dual classical and Einstein-Podolsky-Rosen channels. *Phys. Rev. Lett.* **73**, 3081–3084 (1993).
3. Bennett, C. H. *et al.* Purification of noisy entanglement and faithful teleportation via noisy channels. *Phys. Rev. Lett.* **76**, 722–725 (1991).
4. Briegel, H.-J., Duer, W., Cirac, J. I. & Zoller, P. Quantum repeaters: The role of imperfect local operations in quantum communication. *Phys. Rev. Lett.* **81**, 5932–5935 (1998).
5. Knill, E., Laflamme, R. & Zurek, W. H. Resilient quantum computation. *Science* **279**, 342–345 (1998).
6. Preskill, J. Reliable quantum computers. *Proc. R. Soc. Lond. A* **454**, 385–410 (1998).
7. Zukowski, M., Zeilinger, A., Horne, M. A. & Ekert, A. "Event-ready-detectors" Bell experiment via entanglement swapping. *Phys. Rev. Lett.* **71**, 4287–4290 (1993).
8. Cirac, J. I., Zoller, P., Kimble, H. J. & Mabuchi, H. Quantum state transfer and entanglement distribution among distant nodes in a quantum network. *Phys. Rev. Lett.* **78**, 3221–3224 (1997).
9. Enk, S. J., Cirac, J. I. & Zoller, P. Photonic channels for quantum communication. *Science* **279**, 205–207 (1998).
10. Ye, J., Vernooy, D. W. & Kimble, H. J. Trapping of single atoms in cavity QED. *Phys. Rev. Lett.* **83**, 4987–4990 (1999).
11. Hood, C. J. *et al.* The atom-cavity microscope: Single atoms bound in orbit by single photons. *Science* **287**, 1447–1453 (2000).
12. Pinkse, P. W. H., Fischer, T., Maunz, T. P. & Rempe, G. Trapping an atom with single photons. *Nature* **404**, 365–368 (2000).
13. Cabrillo, C., Cirac, J. I., G-Fernandez, P. & Zoller, P. Creation of entangled states of distant atoms by interference. *Phys. Rev. A* **59**, 1025–1033 (1999).
14. Bose, S., Knight, P. L., Plenio, M. B. & Vedral, V. Proposal for teleportation of an atomic state via cavity decay. *Phys. Rev. Lett.* **83**, 5158–5161 (1999).
15. Raymer, M. G., Walmsley, I. A., Mostowski, J. & Sobolewska, B. Quantum theory of spatial and temporal coherence properties of stimulated Raman scattering. *Phys. Rev. A* **32**, 332–344 (1985).
16. Kuzmich, A., Mölmer, K. & Polzik, E. S. Spin squeezing in an ensemble of atoms illuminated with squeezed light. *Phys. Rev. Lett.* **79**, 481 (1998).
17. Kuzmich, A., Bigelow, N. P. & Mandel, L. Atomic quantum non-demolition measurements and squeezing. *Europhys. Lett. A* **42**, 481–486 (1998).
18. Lukin, M. D., Yelin, S. F. & Fleischhauer, M. Entanglement of atomic ensembles by trapping correlated photon states. *Phys. Rev. lett.* **84**, 4232–4235 (2000).
19. Duan, L. M., Cirac, J. I., Zoller, P. & Polzik, E. S. Quantum communication between atomic ensembles using coherent light. *Phys. Rev. Lett.* **85**, 5643–5646 (2000).
20. Hald, J., Sorensen, J. L., Schori, C. & Polzik, E. S. Spin squeezed state: A macroscopic entangled ensemble created by light. *Phys. Rev. Lett.* **83**, 1319–1322 (1999).
21. Phillips, D. F. *et al.* Storage of light in atomic vapor. *Phys. Rev. Lett.* **86**, 783–786 (2001).
22. Liu, C., Dutton, Z., Behroozi, C. H. & Hau, L. V. Observation of coherent optical information storage in an atomic medium using halted light pulses. *Nature* **409**, 490–493 (2001).
23. Knill, E., Laflamme, R. & Milburn, G. J. A scheme for efficient quantum computation with linear optics. *Nature* **409**, 46–52 (2001).
24. Pan, J. W., Simon, C., Brukner, C. & Zeilinger, A. Feasible entanglement purification for quantum communication. *Nature* **410**, 1067–1070 (2001).
25. Roch, J.-F. *et al.* Quantum nondemolition measurements using cold trapped atoms. *Phys. Rev. Lett.* **78**, 634–637 (1997).
26. Lo, H. K. & Chau, H. F. Unconditional security of quantum key distribution over arbitrarily long distances. *Science* **283**, 2050–2056 (1999).
27. Shor, P. W. & Preskill, J. Simple proof of security of the BB84 quantum key distribution protocol. *Phys. Rev. Lett.* **85**, 441–444 (2000).
28. Clauser, J. F., Horne, M. A., Shimony, A. & Holt, R. A. Proposed experiment to test local hidden-variable theories. *Phys. Rev. Lett.* **23**, 880–884 (1969).
29. Bouwmeester, D. *et al.* Experimental quantum teleportation. *Nature* **390**, 575–579 (1997).
30. Budker, D., Yashuk, V. & Zolotorev, M. Nonlinear magneto-optic effects with ultranarrow width. *Phys. Rev. Lett.* **81**, 5788–5791 (1998).

Supplementary Information accompanies the paper on *Nature*'s website (http://www.nature.com).

Acknowledgements

This work was supported by the Austrian Science Foundation, the Europe Union project EQUIP, the ESF, the European TMR network Quantum Information, and the NSF through a grant to the ITAMP and ITR program. L.-M.D. was also supported by the Chinese Science Foundation.

Correspondence and requests for materials should be addressed to J.I.C. (e-mail: ignacio.cirac@uibk.ac.at).

Quantum state transfer and entanglement distribution among distant nodes in a quantum network

J. I. Cirac[1,2], P. Zoller[1,2], H. J. Kimble[1,3], and H. Mabuchi[1,3]

[1] Institute for Theoretical Physics, University of California at Santa Barbara, Santa Barbara, CA 93106-4030
[2] Institut für Theoretische Physik
Universität Innsbruck, Technikerstraße 25, A-6020 Innsbruck, Austria
[3] Norman Bridge Laboratory of Physics 12-33, California Institute of Technology, Pasadena CA 91125
(November 15, 1996)

We propose a scheme to utilize photons for ideal quantum transmission between atoms located at *spatially-separated* nodes of a quantum network. The transmission protocol employs special laser pulses which excite an atom inside an optical cavity at the sending node so that its state is mapped into a *time-symmetric* photon wavepacket that will enter a cavity at the receiving node and be absorbed by an atom there *with unit probability*. Implementation of our scheme would enable reliable transfer or sharing of entanglement among spatially distant atoms.

We consider a quantum network consisting of spatially separated nodes connected by quantum communication channels. Each node is a quantum system that stores quantum information in quantum bits and processes this information locally using quantum gates [1]. Exchange of information between the nodes of the network is accomplished via quantum channels. A physical implementation of such a network could consist *e.g.* of clusters of trapped atoms or ions representing the nodes, with optical fibers or similar photon "conduits" providing the quantum channels. Atoms and ions are particularly well suited for storing qubits in long-lived internal states, and recently proposed schemes for performing quantum gates between trapped atoms or ions provides an attractive method for local processing within an atom/ion node [2–4]. On the other hand, photons clearly represent the best qubit-carrier for fast and reliable communication over long distances [5,6], since fast and internal-state-preserving transportation of atoms or ions seems to be technically intractable.

To date, no process has actually been identified for using photons (or any other means) to achieve efficient *quantum transmission* between spatially distant atoms. In this letter we outline a scheme to implement this basic building block of communication in a distributed quantum network. Our scheme allows quantum transmission with (in principle) unit efficiency between distant atoms 1 and 2 (see Fig. 1). The possibility of combing local quantum processing with quantum transmission between the nodes of the network opens the possibility for a variety of novel applications ranging from entangled-state cryptography [7], teleportation [8] and purification [9], and is interesting from the perspective of distributed quantum computation [10].

The basic idea of our scheme is to utilize strong coupling between a high–Q optical cavity and the atoms [5] forming a given node of the quantum network. By applying laser beams, one first transfers the internal state of an atom at the first node to the optical state of the cavity mode. The generated photons leak out of the cavity, propagate as a wavepacket along the transmission line, and enter an optical cavity at the second node. Finally, the optical state of the second cavity is transferred to the internal state of an atom. Multiple-qubit transmissions can be achieved by sequentially addressing pairs of atoms (one at each node), as entanglements between arbitrarily located atoms are preserved by the state-mapping process.

The distinguishing feature of our protocol is that by controlling the atom-cavity interaction, one can absolutely avoid the reflection of the wavepackets from the second cavity, effectively switching off the dominant loss channel that would be responsible for decoherence in the communication process. For a physical picture of how this can be accomplished, let us consider that a photon leaks out of an optical cavity and propagates away as a wavepacket. Imagine that we were able to "time reverse" this wavepacket and send it back into the cavity; then this would restore the original (unknown) superposition state of the atom, provided we would also reverse the timing of the laser pulses. If, on the other hand, we are able to drive the atom in a transmitting cavity in such a way that the outgoing pulse were already symmetric in time, the wavepacket entering a receiving cavity would "mimic" this time reversed process, thus "restoring" the state of the first atom in the second one.

The simplest possible configuration of quantum transmission between two nodes consists of two two-level atoms 1 and 2 which are strongly coupled to their respective cavity modes (see Fig. 1). The Hamiltonian describing the interaction of each atom with the corresponding cavity mode is ($\hbar = 1$):

$$\hat{H}_i = \omega_c \hat{a}_i^\dagger \hat{a}_i + \omega_0 |r\rangle_{i\,i}\langle r| + g(|r\rangle_{i\,i}\langle g| \hat{a}_i + h.c.)$$
$$+ \frac{1}{2}\Omega_i(t)\left[e^{-i[\omega_L t + \phi_i(t)]}|r\rangle_{i\,i}\langle e| + h.c\right] \quad (i = 1, 2). \quad (1)$$

Here, $\hat{a}_i$ is the destruction operator for cavity mode i with frequency ω_c, $|g\rangle, |r\rangle$, and $|e\rangle$ form a three–level system of excitation frequency ω_0 (Fig. 1), and the qubit is stored

1

in a superposition of the two degenerate ground states. The states $|e\rangle$ and $|g\rangle$ are coupled by a Raman transition [3,4,11], where a laser of frequency ω_L excites the atom from $|e\rangle$ to $|r\rangle$ with a time dependent Rabi frequency $\Omega_i(t)$ and phase $\phi_i(t)$, followed by a transition $|r\rangle \rightarrow |e\rangle$ which is accompanied by emission of a photon into the corresponding cavity mode, with coupling constant g. In order to suppress spontaneous emission from the excited state during the Raman process, we assume that the laser is strongly detuned from the atomic transition $|\Delta| \gg \Omega_{1,2}(t), g, |\phi_{1,2}|$ (with $\Delta = \omega_L - \omega_0$). In such a case, one can eliminate adiabatically the excited states $|r\rangle_i$. The new Hamiltonian for the dynamics of the two ground states becomes, in a rotating frame for the cavity modes at the laser frequency,

$$\hat{H}_i = -\delta \hat{a}_i^\dagger \hat{a}_i + \frac{g^2}{\Delta} \hat{a}_i^\dagger \hat{a}_i |g\rangle_i {}_i\langle g| + \delta\omega_i(t)|e\rangle_i {}_i\langle e|$$
$$-ig_i(t)\left[e^{i\phi_i(t)}|e\rangle_i {}_i\langle g|\hat{a}_i - \text{h.c.}\right]. \quad (i = 1, 2) \quad (2)$$

The first term involves the Raman detuning $\delta = \omega_L - \omega_c$. The next two terms are AC–Stark shifts of the ground states $|g\rangle$ and $|e\rangle$ due to the cavity mode and laser field, respectively, with $\delta\omega_i(t) = \Omega_i(t)^2/(4\Delta)$. The last term is the familiar Jaynes–Cummings interaction, with an effective coupling constant $g_i(t) = g\Omega_i(t)/(2\Delta)$ [12].

Our goal is to select the time-dependent Rabi frequencies and laser phases [13] to accomplish the *ideal quantum transmission*

$$(|g\rangle_1|\chi_g\rangle + |e\rangle_1|\chi_e\rangle)|g\rangle_2 \otimes |0\rangle_1|0\rangle_2|vac\rangle$$
$$\rightarrow |g\rangle_1 (|g\rangle_2|\chi_g\rangle + |e\rangle_2|\chi_e\rangle) \otimes |0\rangle_1|0\rangle_2|vac\rangle, \quad (3)$$

where $|\chi_{g,e}\rangle$ are unnormalized states of other "spectator" atoms in the network. In (3), $|0\rangle_i$ and $|vac\rangle$ represent the vacuum state of the cavity modes and the free electromagnetic modes connecting the cavities. Transmission will occur by photon exchange via these modes.

In a quantum stochastic description employing the input–output formalism the cavity mode operators obey the quantum Langevin equations [14]:

$$\frac{d\hat{a}_i}{dt} = -i[\hat{a}_i, \hat{H}_i(t)] - \kappa \hat{a}_i - \sqrt{2\kappa}\,\hat{a}_{\text{in}}^{(i)}(t) \quad (i = 1, 2) \quad (4)$$

The first term on the RHS of this equation gives the systematic evolution due to the interaction with the atom, while the last two terms correspond to photon transmission through the mirror with loss rate κ, and (white) quantum noise of the vacuum field incident on the cavity i, respectively. The output of each cavity is given by the equation [14]:

$$\hat{a}_{\text{out}}^{(i)}(t) = \hat{a}_{\text{in}}^{(i)}(t) + \sqrt{2\kappa}\,\hat{a}_i(t) \quad (5)$$

which expresses the outgoing field at the mirror as a sum of the incident field plus the field radiated from the cavity. The output field of the first cavity constitutes the input

for the second cavity with an appropriate time delay, i.e., $\hat{a}_{\text{in}}^{(2)}(t) = \hat{a}_{\text{out}}^{(1)}(t - \tau)$, where τ is a constant related to retardation in the propagation between the mirrors. The output field of the second cavity is, therefore,

$$\hat{a}_{\text{out}}^{(2)}(t) = \hat{a}_{\text{in}}^{(1)}(t - \tau) + \sqrt{2\kappa}[\hat{a}_1(t - \tau) + \hat{a}_2(t)]. \quad (6)$$

Introducing this relation in Eqs. (4) we obtain

$$\frac{d\hat{a}_1}{dt} = -i[\hat{a}_1, \hat{H}_1(t)] - \kappa \hat{a}_1 - \sqrt{2\kappa}\,\hat{a}_{\text{in}}^{(1)}(t) \quad (7a)$$

$$\frac{d\hat{a}_2}{dt} = -i[\hat{a}_2, \hat{H}_2(t)] - \kappa \hat{a}_2 - 2\kappa \hat{a}_1(t - \tau)$$
$$-\sqrt{2\kappa}\,\hat{a}_{\text{in}}^{(1)}(t - \tau) \quad (7b)$$

Note that the first equation is decoupled from the second one, i.e. we consider here only a unidirectional coupling between the cavities (see Fig. 1) [15]. The present model is a particular example of a cascaded quantum system and can be described within the formalism developed by Gardiner and Carmichael [16,17]. We can eliminate the time delay τ in these equations by defining "time delayed" operators for the first system (atom + cavity), e.g. $\tilde{a}(t) \equiv \hat{a}(t - \tau)$, *etc.*; in a similar way we redefine the Rabi frequency $\tilde{\Omega}_1(t) = \Omega_1(t - \tau)$, and phase $\tilde{\phi}_1(t) = \phi_1(t - \tau)$. In the following we will assume that we have performed these transformations, and for simplicity of notation we will omit the tilde. This amounts to setting $\tau \rightarrow 0$ in all these equations. Equations (7) have to be solved with the corresponding equations for the atomic operators and with the condition that the field incident on the first cavity is in the vacuum state, i.e. $\hat{a}_{\text{in}}^{(1)}(t)|\Psi_0\rangle = 0\ \forall t$.

In the present context, it is convenient to reformulate the above problem in the language of quantum trajectories [17,18]. Let us consider a fictitious experiment where the output field of the second cavity is continuously monitored by a photodetector (see Fig. 1). The evolution of the quantum system under continuous observation, conditional to observing a particular trajectory of counts, can be described by a pure state wavefunction $|\Psi_c(t)\rangle$ in the system Hilbert space (where the radiation modes outside the cavity have been eliminated). During the time intervals when no count is detected, this wavefunction evolves according to a Schrödinger equation with non-hermitian effective Hamiltonian

$$\hat{H}_{\text{eff}}(t) = \hat{H}_1(t) + \hat{H}_2(t) - i\kappa \left(\hat{a}_1^\dagger \hat{a}_1 + \hat{a}_2^\dagger \hat{a}_2 + 2\hat{a}_2^\dagger \hat{a}_1\right). \quad (8)$$

The detection of a count at time t_r is associated with a quantum jump according to $|\Psi_c(t_r + dt)\rangle \propto \hat{c}|\Psi_c(t_r)\rangle$, where $\hat{c} = \hat{a}_1 + \hat{a}_2$ [16,18]. The probability density for a jump (detector click) to occur during the time interval from t to $t + dt$ is $\langle \Psi_c(t)|\hat{c}^\dagger \hat{c}|\Psi_c(t)\rangle dt$ [16,18].

We wish to design the laser pulses in both cavities in such a way that ideal quantum transmission condition

(3) is satisfied. A necessary condition for the time evolution is that a quantum jump (detector click, see Fig. 1) never occurs, i.e. $\hat{c}|\Psi_c(t)\rangle = 0 \,\forall t$, and thus the effective Hamiltonian will become a hermitian operator. In other words, the system will remain in a *dark* state of the cascaded quantum system. Physically, this means that the wavepacket is not reflected from the second cavity. We expand the state of the system as

$$|\Psi_c(t)\rangle = |\chi_g\rangle|gg\rangle|00\rangle$$
$$+ |\chi_e\rangle\Big[\alpha_1(t)e^{-i\phi_1(t)}|eg\rangle|00\rangle + \alpha_2(t)e^{-i\phi_2(t)}|ge\rangle|00\rangle$$
$$+ \beta_1(t)|gg\rangle|10\rangle + \beta_2(t)|gg\rangle|01\rangle\Big]. \quad (9)$$

Ideal quantum transmission (3) will occur for

$$\alpha_1(-\infty) = \alpha_2(+\infty) = 1, \ \phi_1(-\infty) = \phi_2(+\infty) = 0. \quad (10)$$

The first term on the RHS of (9) does not change under the time evolution generated by H_{eff}. Defining symmetric and antisymmetric coefficients $\beta_{1,2} = (\beta_s \mp \beta_a)/\sqrt{2}$, we find the following *evolution equations*

$$\dot{\alpha}_1(t) = g_1(t)\beta_a(t)/\sqrt{2}, \quad (11a)$$
$$\dot{\alpha}_2(t) = -g_2(t)\beta_a(t)/\sqrt{2}, \quad (11b)$$
$$\dot{\beta}_a(t) = -g_1(t)\alpha_1(t)/\sqrt{2} + g_2(t)\alpha_2(t)/\sqrt{2}. \quad (11c)$$

where we have chosen the laser frequencies $\omega_L + \dot{\phi}_{1,2}(t)$ so that $\delta = g^2/\Delta$ and

$$\dot{\phi}_{1,2}(t) = \delta\omega_i(t) \quad (12)$$

in order to compensate the AC–stark shifts; thus Eq. (11) are decoupled from the phases. The *dark state condition* implies $\beta_s(t) = 0$, and therefore

$$\dot{\beta}_s(t) = g_1(t)\alpha_1(t)/\sqrt{2} + g_2(t)\alpha_2(t)/\sqrt{2} + \kappa\beta_a(t) \equiv 0, \quad (13)$$

as well as the normalization condition

$$|\alpha_1(t)|^2 + |\alpha_2(t)|^2 + |\beta_a(t)|^2 = 1. \quad (14)$$

We note that the coefficients $\alpha_{1,2}(t)$ and $\beta_s(t)$ are real.

The mathematical problem is now to find pulse shapes $\Omega_{1,2}(t) \propto g_{1,2}(t)$ such that the conditions (10,11,13) are fulfilled. In general this is a difficult problem, as imposing conditions (10,13) on the solutions of the differential equations (11) give functional relations for the pulse shape whose solution are not obvious. We shall construct a class of solutions guided by the physical expectation that the time evolution in the second cavity should reverse the time evolution in the first one. Thus, we look for solutions satisfying the *symmetric pulse condition*

$$g_2(t) = g_1(-t) \quad (\forall t). \quad (15)$$

This implies $\alpha_1(t) = \alpha_2(-t)$, and $\beta_a(t) = \beta_a(-t)$. The latter relation leads to a symmetric shape of the photon wavepacket propagating between the cavities.

Suppose that we specify a pulse shape $\Omega_1(t) \propto g_1(t)$ for the second half of the pulse in the first cavity ($t \geq 0$) [19]. We wish to determine the first half $\Omega_1(-t) \propto g_1(-t)$ (for $t > 0$), such that the conditions for ideal transmission (3) are satisfied. From (13,10) we have

$$g_1(-t) = -\frac{\sqrt{2}\kappa\beta_a(t) + g_1(t)\alpha_1(t)}{\alpha_2(t)}, \quad (t > 0). \quad (16)$$

Thus, the pulse shape is completely determined provided we know the system evolution for $t \geq 0$. However, a difficulty arises when we try to find this evolution, since it depends on the yet unknown $g_2(t) = g_1(-t)$ for $t > 0$ [see Eqs.(11)]. In order to circumvent this problem, we use (13) to eliminate this dependence in Eqs. (11a,11c). This gives

$$\dot{\alpha}_1(t) = g_1(t)\beta_a(t)/\sqrt{2}, \quad (17a)$$
$$\dot{\beta}_a(t) = -\kappa\beta_a(t) - \sqrt{2}g_1(t)\alpha_1(t) \quad (17b)$$

for $t \geq 0$. These equations have to be integrated with the initial conditions

$$\alpha_1(0) = \left[\frac{2\kappa^2}{g_1(0)^2 + \kappa^2}\right]^{\frac{1}{2}} \quad (18a)$$
$$\beta_a(0) = \left[1 - 2\alpha_1(0)^2\right]^{\frac{1}{2}} \quad (18b)$$

which follow immediately from $\alpha_1(0) = \alpha_2(0)$, and (14,13) at $t = 0$. Given the solution of Eqs. (17), we can determine $\alpha_2(t)$ from the normalization (14). In this way, the problem is solved since all the quantities appearing on the RHS of Eq. (16) are known for $t \geq 0$. It is straightforward to find analytical expressions for the pulse shapes, for example by specifying $\Omega_1(t) = \text{const}$ for $t > 0$. This is the pulse that will be considered in the following discussion.

As an illustration, we have numerically integrated the full time–dependent Schrödinger equation with the effective Hamiltonian (8). The results are displayed in Fig. 2(a). We have used a pulse shape calculated using the above procedure, with $g_1(t) = 2\delta\omega_1(t) = \kappa \equiv \text{const}$ for $t > 0$ [see Fig. 2(b)]. As the figure shows, the quantum transmission is ideal.

In practice there will be several sources of imperfections. First, there is the possibility of spontaneous emission from the excited state during the Raman pulses. Its effects can be accounted for in the effective Hamiltonian (8) by the replacement $\Delta \to \Delta + i\Gamma/2$, where Γ is the decay rate from level $|r\rangle$. If we denote by τ ($\approx \max(1/\kappa, 1/g_{1,2})$) the effective pulse duration, the probability for a spontaneous emission is of the order of $\Gamma(\Omega_{1,2}^2 + 4g^2)/(8\Delta^2)\tau \ll 1$. For $g_1 \approx \kappa$ this probability scales like $1/\Delta$, so that the effects of spontaneous emission are suppressed for sufficiently large detunings.

A second source of decoherence will be losses in the mirror and during propagation. They can be taken into account by adding a term $-i\kappa'(\hat{a}_1^\dagger \hat{a}_1 + \hat{a}_2^\dagger \hat{a}_2)$ in H_{eff} (8), where κ' is the additional loss rate. Typically, we expect $\kappa' \ll \kappa$. Nevertheless, one can overcome the effects of photon losses by error correction [20]. We have included these imperfections in our numerical simulations. Fig. 3 shows the probability of a faithful transmission $\mathcal{F}$ as a function of κ'/κ for different values of Γ/Δ for the same parameters and pulse shapes as in Fig. 2.

In conclusion, we have proposed for the first time a protocol to accomplish ideal quantum transmission between two nodes of a quantum network. Our scheme has been tailored to a potential network implementation in which trapped atoms or ions constitute the nodes, and photon transmission lines provide communication channels between them. Extensions of the present scheme will be presented elsewhere [10], including error correction and new quantum gates in cavity quantum electrodynamics.

We thank the members of the ITP program *Quantum Computers and Quantum Coherence* for discussions. This work was supported in part by the Österreichischer Fonds zur Förderung der wissenschaftlichen Forschung, by NSF PHY94-07194 and PHY-93-13668, by DARPA/ARO through the QUIC program, and by the ONR.

[1] See, for example, D. P. DiVincenzo, Science **270**, 255 (1995).

[2] J. I. Cirac and P. Zoller, Phys. Rev. Lett. **74**, 4091 (1995).

[3] T. Pellizzari *et al*, Phys. Rev. Lett. **75** , 3788 (1995).

[4] C. Monroe *et al.*, Phys. Rev. Lett. **75**, 4714 (1995).

[5] Q. Turchette *et al.*, Phys. Rev. Lett. **75**, 4710 (1995); for experimental CQED in the microwave regime see, for example, M. Brune *et al..*, Phys. Rev. Lett. in press.

[6] K. Mattle *at al.*, Phys. Rev. Lett. **76**, 4656 (1996).

[7] C. H. Bennett, Phys. Today **24** (October 1995); A. K. Ekert, Phys. Rev. Lett. **67**, 661 (1991).

[8] C. H. Bennet *et al*, Phys. Rev. Lett. **70**, 1895 (1993).

[9] C. H. Bennet *et al.*, Phys. Rev. Lett. **76**, 722 (1996); D. Deutsch *et al.*, Phys. Rev. Lett. **77** , 2818 (1996); N. Gisin, Phys. Lett. A **210** 151 (1996),

[10] J. I. Cirac *et al.*, in preparation.

[11] C.K. Law and J.H. Eberly, Phys. Rev. Lett. **76**, 1055 (1996).

[12] We ignore for the moment the small effects produced by spontaneous emission during the Raman process. Its effects will be studied in the context of Fig. 3.

[13] One could also modulate the cavity transmission, but this is technically more difficult.

[14] C.W. Gardiner, *Quantum Noise* (Springer–Verlag, Berlin, 1991).

[15] In a perfect realization of the present scheme no light field will be reflected from the second mirror, and therefore the assumption of unidirectional propagation is not needed.

[16] C.W. Gardiner, Phys. Rev. Lett.**70**, 2269 (1993).

[17] H.J. Carmichael, Phys. Rev. Lett. **70**, 2273 (1993).

[18] For a review see P. Zoller and C. W. Gardiner in *Quantum Fluctuations*, Les Houches, ed. E. Giacobino *et al.* (Elsevier, NY, in press).

[19] $\Omega_1(t)$ has to be such that $\alpha_1(\infty) = 0$. This is fulfilled if $\Omega_1(\infty) > 0$, which also guarantees that the denominator in (16) does not vanish for $t > 0$.

[20] P.W. Shor, Phys. Rev. A **52**, R2493 (1995); A.M. Steane, Phys. Rev. Lett. **77**, 793 (1996); J. I. Cirac, T. Pellizzari and P. Zoller, Science **273**, 1207 (1996); P. Shor, *Fault–tolerant quantum computation*, quant–ph/9605011; D. DiVincenzo and P.W. Shor, Phys. Rev. Lett. **77**, 3260 (1996).

FIG. 1. Schematic representation of unidirectional quantum transmission between two atoms in optical cavities connected by a quantized transmission line (see text for explanation).

FIG. 2. Populations $\alpha_{1,2}(t)^2$ and $\beta_a(t)^2$ for the ideal transmission pulse $g_1(t) = g_2(-t)$ given in the inset, specified by $g_1(t \geq 0) = 2\delta\omega_1(t \geq 0) = \kappa = \text{const}$.

FIG. 3. Fidelity of a faithful transmission $\mathcal{F}$ including the effects of a mirror losses and spontaneous emission as a function of κ'/κ for $\Gamma/\Delta = 0, 0.01, 0.05$ (solid, dashed and dot-dashed lines, respectively). The other parameters are as in Fig. 2.

LIST OF PUBLICATIONS

1. J. I. Cirac and L. L. Sánchez-Soto, Analytic approximation to the interaction of a two-level atom with squeezed light, *Phys. Rev. A* **40**, 3743 (1989).

2. J. I. Cirac and L. L. Sánchez-Soto, Population trapping in the Jaynes-Cummings model via phase coupling, *Phys. Rev. A* **42**, 2851 (1990).

3. J. I. Cirac and L. L. Sánchez-Soto, Collective resonance fluorescence in a strongly squeezed vacuum, *Optics Comm.* **77**, 26 (1990).

4. J. I. Cirac and L. L. Sánchez-Soto, Trapping in the multiphoton Jaynes-Cummings model, *Optics Comm.* **80**, 67 (1990).

5. J. I. Cirac and L. L. Sánchez-Soto, Suppression of spontaneous emission by squeezed light in a cavity, *Phys. Rev. A* **44**,1948 (1991).

6. J. I. Cirac and L. L. Sánchez-Soto, Population trapping in two-level models: Spectral and statistical properties, *Phys.Rev. A* **44**, 3317 (1991).

7. J. I. Cirac, H. Ritsch and P. Zoller, Two-level system interacting with a finite-bandwidth thermal cavity mode, *Phys. Rev. A* **44**, 4541 (1991).

8. M. A. M. Marte, J. I. Cirac and P. Zoller, Deflection of atoms by circularly polarized light beam in triple Laue configuration, *J. of Mod. Opt.* **38**, 2265 (1991).

9. J. I. Cirac, R. Blatt, P. Zoller and W. D. Phillips, Laser cooling of trapped ions in a standing wave, *Phys. Rev. A* **46**, 2668 (1992).

10. J. I. Cirac, Interaction of a two-level atom with a cavity mode in the bad cavity limit, *Phys. Rev. A* **46**, 4354 (1992).

11. J. I. Cirac and P. Zoller, Laser cooling of a trapped ion in a squeezed vacuum, *Phys. Rev. A* **47**, 2191 (1993).

12. J. I. Cirac, R. Blatt, A. S. Parkins and P. Zoller, Preparation of Fock states by observation of quantum jumps, *Phys. Rev. Lett.* **70**, 762 (1993).

13. J. I. Cirac, A. S. Parkins, R. Blatt and P. Zoller, Dark squeezed states of the motion of a trapped ion, *Phys. Rev. Lett.* **70**, 556 (1993).

14. J. I. Cirac, A. S. Parkins, R. Blatt and P. Zoller, Cooling of a trapped ion coupled strongly to a quantized cavity, *Optics Comm.* **97**, 353 (1993).

15. J. I. Cirac, R. Blatt, A. S. Parkins and P. Zoller, Laser cooling of a trapped ion with polarization gradients, *Phys. Rev. A* **48**, 1434 (1993).

16. J. I. Cirac, R. Blatt, A. S. Parkins and P. Zoller, Spectrum of resonance fluorescence from a trapped ion, *Phys. Rev. A* **48**, 2169 (1993).

17. H.-R. Xia, J. I. Cirac, S. Swartz, B. Kohler, D. S. Elliot, J. L. Hall and P. Zoller, Phase-shifts and intensity dependence in frequency modulation spectroscopy, *J. of the Opt. Soc. of Am. B* **11**, 721 (1993).

18. J. I. Cirac, L. J. Garay, R. Blatt, A. S. Parkins and P. Zoller, Laser cooling a trapped ion: Effects of the micromotion, *Phys. Rev. A* **49**, 421 (1994).

19. J. I. Cirac, R. Blatt, A. S. Parkins and P. Zoller, Quantum collapse and revivals from a single trapped ion, *Phys. Rev. A* **49**, 1202 (1994).

20. I. Marzoli, J. I. Cirac, R. Blatt and P. Zoller, Laser cooling of trapped ions: Designing two-level atoms for sideband cooling, *Phys. Rev. A* **49**, 2771 (1994).

21. J. I. Cirac, R. Blatt and P. Zoller, Non-classical states of motion in a 3-dimensional trap by adiabatic passage, *Phys. Rev. A* (RC) **49**, R3174 (1994).

22. R. Taieb, R. Dum, J. I. Cirac, P. Marteand P. Zoller, Cooling and localization of atoms in laser induced potentials, *Phys. Rev. A* **49**, 4876 (1994).

23. J. I. Cirac, M. Lewenstein and P. Zoller, Quantum statistical properties of a laser cooled ideal gas, *Phys. Rev. Lett.* **72**, 2977 (1994).

24. J. I. Cirac, A. Schenzle and P. Zoller, Inhibition of quantum tunneling in by observing laser scattered light, *Europhys. Lett.* **27**, 123 (1994).

25. J. I. Cirac and P. Zoller, Preparation of macroscopic superpositions in many atom systems, *Phys. Rev. A* (RC) **50**, R2799 (1994).

26. J. I. Cirac, M. Lewenstein and P. Zoller, Quantum dynamics of a laser cooled ideal gas, *Phys. Rev. A* **50**, 3409 (1994).

27. J. I. Cirac, M. Lewenstein and P. Zoller, Generalized Bose–Einstein condensation and multistability of a laser cooled ideal gas, *Phys. Rev. A* **51**, 2899 (1994).

28. J. I. Cirac and A. S. Parkins, Schemes for atomic state teleportation, *Phys. Rev. A* (RC) **50**, R4441 (1994).

29. J. I. Cirac, M. Lewenstein and P. Zoller, Laser cooling of a trapped atom in a cavity: Bad cavity limit, *Phys. Rev. A* **51**, 1650 (1995).

30. R. Blatt, J. I. Cirac, A. S. Parkins and P. Zoller, Quantum motion of trapped ions, *Physica Scripta* **T59**, 294 (1995).

31. J. I. Cirac and P. Zoller, Quantum computations with cold trapped ions, *Phys. Rev. Lett.* **74**, 4091 (1995).

32. J. I. Cirac, M. Lewenstein and P. Zoller, Master equation for sympathetic cooling of trapped particles, *Phys. Rev. A* **51**, 4617 (1995).

33. R. Chacón and J. I. Cirac, Chaotic and regular behavior of a trapped ion interactiong with a laser field, *Phys. Rev. A* **51**, 4900 (1995).

34. R. Blatt, J. I. Cirac and P. Zoller, Trapping states with cold ions, *Phys. Rev. A* **52**, 518 (1995).

35. J. I. Cirac and M. Lewenstein, Cooling of atoms in external fields, *Phys. Rev. A* **52**, 4737 (1995).

36. T. Pellizari, S. Gardiner, J. I. Cirac and P. Zoller, Decoherence in a continuously monitored quantum computer based on cavity QED, *Phys. Rev. Lett.* **75**, 3788 (1995).

37. J. F. Poyatos, R. Walser, J. I. Cirac, R. Blatt and P. Zoller, Motion tomography of a trapped ion, *Phys. Rev. A* (RC) **53**, R1966 (1995).

38. J. I. Cirac and M. Lewenstein, Populationg a Bose condensate in the boson-accumulation regime, *Phys. Rev. A* **53**, 2466 (1995).

39. A. Vogt, J. I. Cirac and P. Zoller, Collective laser cooling of two trapped ions, *Phys. Rev. A* **53**, 950 (1996).

40. J. F. Poyatos, J. I. Cirac, R. Blatt and P. Zoller, Trapped ions in the strong excitation regime: Ion interferometry and non-classical states, *Phys. Rev. A* **54**, 1532 (1996).

41. J. I. Cirac, A. S. Parkins, R. Blatt and P. Zoller, Quantum motion of trapped ions, *Adv. At. and Mol. Phys.* **37**, 237 (1996).

42. M. Holland, K. Barnett, C. Gardiner, J. I. Cirac and P. Zoller, Theory of an atom laser, *Phys. Rev. A* (RC) **54**, R1757 (1996).

43. M. Naraschewski, H. Wallis, A. Schenzle, J. I. Cirac and P. Zoller, Interferences of Bose condensates, *Phys. Rev. A* **54**, 2185 (1996).

44. J. F. Poyatos, J. I. Cirac and P. Zoller, Quantum reservoir engineering with laser cooled trapped ions, *Phys. Rev. Lett.* **77**, 4728 (1996).

45. J. I. Cirac, T. Pellizari and P. Zoller, Enforcing coherent evolution in dissipative systems, *Science* **273**, 1207 (1996).

46. J. I. Cirac, C. Gardiner, M. Naraschewski and P. Zoller, Continuous observation of interference fringes from Bose condensates, *Phys. Rev. A* (RC) **54**, R3714 (1996).

47. J. I. Cirac, M. Lewenstein and P. Zoller, Collective laser cooling of trapped atoms, *Europh. Lett.* **35**, 647 (1996).

48. R. Walser, J. I. Cirac and P. Zoller, Magnetic tomography of a cavity state, *Phys. Rev. Lett.* **77**, 2658 (1996).

49. S. Gardiner, J. I. Cirac and P. Zoller, Non-classical states and measurement of general observables of a trapped ion, *Phys. Rev. A* **55**, 1683 (1996).

50. V. M. Perez, H. Michinel, J. I. Cirac, M. Lewenstein and P. Zoller, Low energy excitations of a Bose–Einstein-condensate: A variational approach, *Phys. Rev. Lett.* **77**, 5320 (1996).

51. J. F. Poyatos, J. I. Cirac and P. Zoller, Complete characterization of a quantum process: The two-bit quantum gate, *Phys. Rev. Lett.* **78**, 390 (1997).

52. J. I. Cirac, P. Zoller, H. J. Kimble and H. Mabuchi, Quantum state transfer and entanglement distribution among distant nodes in a quantum network, *Phys. Rev. Lett.* **78**, 3221 (1997).

53. G. Moriggi, J. I. Cirac, M. Lewenstein and P. Zoller, Laser cooling beyond the Lamb-Dicke limit, *Europhys. Lett.* **39**, (1997).

54. S. J. van Enk, J. I. Cirac and P. Zoller, Ideal quantum communication over noisy channels: A quantum optical implementation, *Phys. Rev. Lett.* **78**, 4293 (1997).

55. J. I. Cirac and N. Gisin, Coherent eavesdropping strategies for the 4 state quantum cryptography protocol, *Phys. Lett. A* **229**, 1 (1997).

56. V. Pérez-García, H. Michinel, J. I. Cirac, M. Lewenstein and P. Zoller, Dynamics of Bose–Einstein-condensates: Variational solutions of the Gross-Pitaevskii equation, *Phys. Rev. A* **56**, 1424 (1997).

57. S. van Enk, J. I. Cirac, P. Zoller, H. J. Kimble and H. Mabuchi, Quantum state transfer in a quantum network: A quantum optical implementation, *J. Mod. Opt.* **44** 1727 (1997).

58. Th. Busch, J. I. Cirac, V. M. Pérez-García and P. Zoller, Stability and collective excitations of a two-component Bose-condensed gas: A moment approach, *Phys. Rev. A* **56** 2978 (1997).

59. S. F. Huelga, C. Macchiavello, T. Pellizzari, A. K. Ekert, M. B. Plenio and J. I. Cirac, On the improvement of frequency standards with quantum entanglement, *Phys. Rev. Lett.* **79**, 3865 (1997).

60. S. van Enk, J. I. Cirac and P. Zoller, Purifying two bit quantum gates and joint measurements in cavity QED, *Phys. Rev. Lett.* **79**, 5178 (1997).

61. S. Gardiner, J. I. Cirac and P. Zoller, Quantum chaos in an ion trap: The delta-kicked harmonic oscillator, *Phys. Rev. Lett.* **79**, 4790 (1997).

62. N. Lütkenhaus, J. I. Cirac and P. Zoller, Mimicking a squeezed bath interaction: Quantum reservoir engineering with atoms, *Phys. Rev A.* **57**, 548 (1998).

63. J. I. Cirac, M. Lewenstein, K. Mølmer and P. Zoller, Quantum superposition states of Bose–Einstein-condensates, *Phys. Rev. A* **57**, 1208 (1998).

64. S. van Enk, J. I. Cirac and P. Zoller, Photonic channels for quantum communication, *Science* **279**, 205 (1998).

65. R. Dum, J. I. Cirac, M. Lewenstein and P. Zoller, Creation of dark solitons and vortices in Bose–Einstein-condensates, *Phys. Rev. Lett.* **80**, 2972 (1998).

66. G. Morigi, J.I. Cirac, K. Ellinger and P. Zoller, Laser cooling of single trapped atoms to the ground state: A dark state in position space, *Phys. Rev. A* **52**, 2909 (1998).

67. Y. Castin, J. I. Cirac and M. Lewenstein, Reabsorption of light by trapped atoms, *Phys. Rev. Lett.* **80**, 5305 (1998).

68. K. M. Gheri, C. Saavedra, P. Törm, J. I. Cirac and P. Zoller, Entanglement engineering of one-photon wavepackets using a single-atom source, *Phys. Rev. A* **58**, R2627 (1998).

69. J. F. Poyatos, J. I. Cirac and P. Zoller, Quantum gates with "hot" trapped ions, *Phys. Rev. Lett.* **81**, 1322 (1998).

70. D. Jacksh, C. Bruder, J. I. Cirac, C. Gardiner and P. Zoller, Cold bosonic atoms in optical lattices, *Phys. Rev. Lett.* **81**, 3108 (1998).

71. H.-J. Briegel, W. Dür, J. I. Cirac and P. Zoller, Quantum repeaters: the role of imperfect local operations in quantum communication, *Phys. Rev. Lett.* **81**, 5932 (1998).

72. H.-J. Briegel, W. Dür, S. J. Van Enk, J. I. Cirac and P. Zoller, Quantum communication and the creation of maximally entangled pairs of atoms over a noisy channel, *Phil. Tran. Roy. Soc. Lond. A* **356**, 1841 (1998).

73. T. Busch, J. R. Anglin, J. I. Cirac and P. Zoller, Inhibition of spontaneous emission in Fermi gases, *Europhys. Lett.* **44**, 1 (1998).

74. S. van Enk, J. I. Cirac, P. Zoller, J. H. Kimble and H. Mabuchi, Transmission of quantum information in a quantum network: A quantum optical implementation, *Fortschritte der Physik-Progress of Physics* **46**, 689 (1998).

75. J. F. Poyatos, J. I. Cirac and P. Zoller, Characterization of decoherence processes in quantum computation, *Optics Express* **2**, 372 (1998).

76. J. I. Cirac, S. J. van Enk, P. Zoller, H. J. Kimble and H. Mabuchi, Quantum communications in a quantum network, *Physica Scripta* **T76**, 223 (1998).

77. W. Dür, H.-J. Briegel, J. I. Cirac and P. Zoller, Quantum repeaters based on entanglement purification, *Phys. Rev. A* **59**, 169 (1999).

78. G. Giedke, H.-J. Briegel, J. I. Cirac and P. Zoller, Lower bounds for entanglement purification, *Phys. Rev. A* **59**, 2651 (1999).

79. C. Cabrillo, J. I. Cirac, P. Garcia-Fernandez and P. Zoller, Creation of entangled states of distant atoms by interference, *Phys. Rev. A* **59**, 1025, quant-ph/9810013 (1999).

80. D. Jaksch, H.-J. Briegel, J. I. Cirac, C. W. Gardiner and P. Zoller, Entanglement of atoms via cold controlled collisions, *Phys. Rev. Lett.* **82**, 1975 (1999).

81. J. I. Cirac, A. K. Ekert, S. F. Huelga and C. Macchiavello, Distributed quantum computation over noisy channels, *Phys. Rev. A* **59**, 4249 (1999).

82. G. Morigi, J. Eschner, J.I. Cirac and P. Zoller, Laser cooling of two trapped ions: Sideband cooling beyond the Lamb-Dicke limit, *Phys. Rev. A* **59**, 3797 (1999).

83. J. I. Cirac, A. K. Ekert and C. Macchiavello, Optimal purification of a single qubit, *Phys. Rev. Lett.* **82**, 4344 (1999).

84. S. J. van Enk, H. J. Kimble, J. I. Cirac and P. Zoller, Quantum Communication with dark photons, *Phys. Rev. A* **59**, 2659 (1999).

85. W. Dür, J. I. Cirac and R. Tarrach, Separability and distillability of multiparticle quantum systems, *Phys. Rev. Lett.* **83**, 3562 (1999).

86. W. Dür and J. I. Cirac, Multiparticle teleportation, *J. Mod. Optics* **27**, 247 (2000).

87. T. Calarco, E.A. Hinds, D. Jaksch, J. Schmiedmayer, J. I. Cirac, P. Zoller, Quantum gates with neutral atoms: Controlling collisional interactions in time dependent traps, *Phys. Rev. A* **61**, quant-ph/9905013 (2000).

88. J. I. Cirac and P. Zoller, A scalable quantum computer with ions in an array of microtraps, *Nature* **404**, 579 (2000).

89. W. Dür, J. I. Cirac, M. Lewenstein and D. Bruss, Distillability and partial transposition in bipartite systems, *Phys. Rev. A* **61**, 062313, quant-ph/9910022 (2000).

90. C. Saavedra, K. M. Gheri, P. Törma, J. I. Cirac and P. Zoller, Controlled source of entangled photonic qubits, *Phys. Rev. A* **61**, 062311 (2000).

91. J. J. Garcia-Ripoll, J. I. Cirac, J. Anglin, V. Perez-Garcia and P. Zoller, Spin monopoles with Bose–Einstein condenstates, *Phys. Rev. A* **61**, 053609, quant-ph/9811340 (2000).

92. B. Kraus, J. I. Cirac, S. Karnas and M. Lewenstein, Separability in 2 X N composite quantum systems, *Phys. Rev. A* **61**, 062302, quant-ph/9912010 (2000).

93. W. Dür and J. I. Cirac, Classification of multi-qubit mixed states: Separability and distillability properties, *Phys. Rev. A* **61**, 042314, quant-ph/9911044 (2000).

94. Lu-Ming Duan, G. Giedke, J. I. Cirac and P. Zoller, Inseparability criterion for continuous variable systems, *Phys. Rev. Lett.* **84**, 2722, quant-ph/9908056 (2000).

95. Lu-Ming Duan, G. Giedke, J. I. Cirac and P. Zoller, Entanglement purification of Gaussian continuous variable quantum states, *Phys. Rev. Lett.* **84**, 4002, quant-ph/9912017 (2000).

96. H.-J. Briegel, T. Calarco, D. Jaksch, J. I. Cirac and P. Zoller, Quantum computing with neutral atoms, *J. Mod. Optics* **47**, 415, quant-ph/9904010 (2000).

97. W. Dür and J. I. Cirac, Activating bound entanglement in multi-particle systems, *Phys. Rev. A* **62**, 22302, quant-ph/0002028 (2000).

98. S. A. Gardiner, D. Jaksch, R. Dum, J. I. Cirac and P. Zoller, Nonlinear matter wave dynamics with a "chaotic" potential, *Phys. Rev. A* **62**, 23612, quant-ph/9912092 (2000).

99. P. Horodecki, M. Lewenstein, G. Vidal and I. Cirac, Operational criterion and constructive checks for the separability of low rank density matrices, *Phys. Rev. A* **62**, 32310, quant-ph/0002089 (2000).

100. L.-M. Duan, G. Giedke, J. I. Cirac and P. Zoller, Physical implementation for entanglement purification of Gaussian continuous variable quantum states, *Phys. Rev. A* **62**, 32304, quant-ph/0003116 (2000).

101. J. F. Poyatos, J. I. Cirac and P. Zoller, From classical to quantum computers. Quantum computations with trapped ions, *Physica Scripta* **T86**, 72 (2000).

102. G. Vidal, W. Dür and J. I. Cirac, Reversible combination of inequivalent kinds of multipartite entanglement, *Phys. Rev. Lett.* **85**, 658, quant-ph/0004009 (2000).

103. D. Jaksch, J. I. Cirac, P. Zoller, S. L. Rolston, R. Cote and M. D. Lukin, Fast quantum gates for neutral atoms, *Phys. Rev. Lett.* **85**, 2208, quant-ph/0004038 (2000).

104. M. Lewenstein, B. Kraus, J. I. Cirac and P. Horodecki, Optimization of entanglement witnesses, *Phys. Rev. A* **62**, 52310, quant-ph/0005014 (2000).

105. M. Lewenstein, J. I. Cirac and L. Santos, Cooling of a small sample of Bose atoms with accidental degeneracy, *J. Phys. B* **33**, 4107, quant-ph/0005097 (2000).

106. M. Lewenstein, D. Bruss, J. I. Cirac, B. Kraus, M. Kus, J. Samsonowicz, A. Sanpera and R. Tarrach, Separability and distillability in composite quantum systems — A primer, *J. Mod. Opt.* **77**, 2481, quant-ph/0006064 (2000).

107. L. Santos, Z. Idziaszek, J. I. Cirac and M. Lewenstein, Laser induced condensation of trapped bosonic gases, *J. Phys. B* **33**, 4143, quant-ph/0005107 (2000).

108. L.-M. Duan, A. Sørensen, J. I. Cirac and P. Zoller, Squeezing and entanglement of atomic beams, *Phys. Rev. Lett.* **85**, 3991, quant-ph/0007048 (2000).

109. L. J. Garay, J. R. Anglin, J. I. Cirac and P. Zoller, Black holes in Bose–Einstein condensates, *Phys. Rev. Lett.* **85**, 4643, gr-qc/0002015 (2000).

110. Lu-Ming Duan, J. I. Cirac, P. Zoller and E. S. Polzik, Quantum communication between atomic ensembles using coherent light, *Phys. Rev. Lett.* **85**, 5643, quant-ph/0003111 (2000).

111. W. Dür, G. Vidal and J. I. Cirac, Three qubits can be entangled in two inequivalent ways, *Phys. Rev. A* **62**, 62314, quant-ph/0005115 (2000).

112. A. Sørensen, L.-M. Duan, J. I. Cirac and P. Zoller, Many-particle entanglement with Bose–Einstein-condensates, *Nature* **409**, 63, quant-ph/0006111 (2001).

113. C. Menotti, J. R. Anglin, J. I. Cirac and P. Zoller, Dynamic splitting of a Bose–Einstein-condensate, *Phys. Rev. A* **63**, 023601, cond-mat/0005302 (2001).

114. L. J. Garay, J. R. Anglin, J. I. Cirac and P. Zoller, Sonic black holes in dilute Bose–Einstein-condensates, *Phys. Rev. A* 63, 023611, gr-qc/0005131 (2001).

115. J. I. Cirac, W. Dür, B. Kraus and M. Lewenstein, Entangling operations and their implementation using a small amount of entanglement, *Phys. Rev. Lett.* **86**, 544, quant-ph/0007057 (2001).

116. M. Lewenstein, B. Kraus, P. Horodecki and J. I. Cirac, Characterization of separable states and entanglement witnesses, *Phys. Rev. A* **63**, 044304, quant-ph/0005112 (2001).

117. B. Paredes, P. Fedichev, J. I. Cirac, and P. Zoller, 1/2-Anyons in small atomic Bose–Einstein condensates, *Phys. Rev. Lett.* **87**, 010402, cond-mat/0103251 (2001).

118. G. Vidal and J. I. Cirac, Irreversibility in asymptotic manipulations of entanglement, *Phys. Rev. Lett.* **86**, 5803, quant-ph/0102036 (2001).

119. D. Jaksch, S. A. Gardiner, K. Schulze, J. I. Cirac and P. Zoller, Uniting Bose–Einstein condensates in optical resonators, *Phys. Rev. Lett.* **86**, 4733, cond-mat/0101057 (2001).

120. L.-M. Duan, J. I. Cirac and P. Zoller, Geometric manipulation of trapped ions for quantum computation, *Science* **292**, 1695, quant-ph/0111086 (2001).

121. W. Dür and J. I. Cirac, Non-local Operations: Purification, storage, compression, tomography, and probabilistic implementation, *Phys. Rev. A* **64**, 012317, quant-ph/0012148 (2001).

122. B. Kraus and J. I. Cirac, Optimal creation of entanglement using a two-qubit gate, *Phys. Rev. A* **63**, 062309, quant-ph/0011050 (2001).

123. T. Calarco, J.I. Cirac and P. Zoller, Entangling ions in arrays of microscopic traps, *Phys. Rev. A* **63**, 062304, quant-ph/0010105 (2001).

124. M. D. Lukin, M. Fleischhauer, R.Cote, L. M. Duan, D. Jaksch, J. I. Cirac and P. Zoller, Dipole blockade and quantum information processing in mesoscopic atomic ensembles, *Phys. Rev. Lett.* **87**, 037901, quant-ph/0011028 (2001).

125. J. Schliemann, J. I. Cirac, M. Ku's, M. Lewenstein and D. Loss, Quantum correlations in two-fermion systems, *Phys. Rev. A* **64**, 022303, quant-ph/0012094 (2001).

126. W. Dür, G. Vidal and J. I. Cirac, Visible compression of commuting mixed states, *Phys. Rev. A* **64**, 022308, quant-ph/0101111 (2001).

127. G. Giedke, B. Kraus, J. I. Cirac and M. Lewenstein, Separability properties of three-mode Gaussian states, *Phys. Rev. A* **64**, 052303, quant-ph/0103137 (2001).

128. G. Giedke, B. Kraus, M. Lewenstein and J. I. Cirac, Entanglement Criterion for all bipartite Gaussian States, *Phys. Rev. Lett.* **87**, 167904, quant-ph/0104050 (2001).

129. W. Dür, G. Vidal, J. I. Cirac, N. Linden and S. Popescu, Entanglement capabilities of non-local Hamiltonians, *Phys. Rev. Lett.* **87**, 137901, quant-ph/0006034 (2001).

130. W. Dür and J. I. Cirac, Multiparticle entanglement and its experimental detection, *J. Phys. A* **34**, 6837, quant-ph/0011025 (2001).

131. G. Giedke, L.-M. Duan, J. I. Cirac and P. Zoller, Distillability Criterion for all bipartite Gaussian States, *Quant. Inf. Comp.* **1**, 79, quant-ph/0104072 (2001).

132. G. Giedke, B. Kraus, L.-M. Duan, P. Zoller, M. Lewenstein and J. I. Cirac, Separability and distillability of Gaussian states — The complete story, *Fort. Phys.* **49**, Issue 10–11, 973 (2001).

133. L.-M. Duan, M. Lukin, J. I. Cirac and P. Zoller, Long-distance quantum communication with atomic ensembles and linear optics, *Nature* **414**, 413, quant-ph/0105105 (2001).

134. G. Vidal, L. Masanes and J.I. Cirac, Storing quantum dynamics in quantum states: Stochastic programmable gate for $U(1)$ operations, *Phys. Rev. Lett.* **88**, 047905, quant-ph/0102037 (2002).

135. G. Vidal and J. I. Cirac, Irreversibility in asymptotic manipulations of a distillable entangled state, *Phys. Rev. A* **65**, 012323, quant-ph/0107051 (2002).

136. L.-M. Duan, J. I. Cirac and P. Zoller, Quantum entanglement in spinor Bose–Einstein condensates, *Phys. Rev. A* **65**, 033619, quant-ph/0107055 (2002).

137. D. Jaksch, J. I. Cirac and P. Zoller, Dynamically turning off interactions in a two component condensate, *Phys. Rev. A* **65**, 033625, cond-mat/0110494 (2002).

138. G. Vidal and J. I. Cirac, Catalysis in non-local quantum operations, *Phys. Rev. Lett.* **88**, 167903, quant-ph/0108077 (2002).

139. B. Kraus, M. Lewenstein and J. I. Cirac, Characterization of distillable and activable states using entanglement witnesses, *Phys. Rev. A* **65,** 042327, quant-ph/0110174 (2002).

140. D. Jaksch, V. Venturi, J. I. Cirac, C. J. Williams and P. Zoller, Creation of a molecular condensate by dynamically melting a Mott-insulator, *Phys. Rev. Lett.* **89**, 040402, cond-mat/0204137 (2002).

141. G. Vidal, W. Dür and J. I. Cirac, Entanglement cost of bipartite mixed states, *Phys. Rev. Lett.* **89**, 027901, quant-ph/0112131 (2002).

142. G. Vidal, K. Hammerer and J. I. Cirac, Interaction cost of non-local gates, *Phys. Rev. Lett.* **88**, 237902, quant-ph/0112168 (2002).

143. W. Dür, G. Vidal and J. I. Cirac, Optimal conversion of non-local unitary operations, *Phys. Rev. Lett.* **89**, 057901, quant-ph/0112124 (2002).

144. C. H. Bennett, J. I. Cirac, M. S. Leifer, D. W. Leung, N. Linden, S. Popescu and G. Vidal, Optimal simulation of two-qubit Hamiltonians using general local operations, *Phys. Rev. A* **66**, 012305, quant-ph/0107035 (2002).

145. W. Dür and J. I. Cirac, Equivalence classes of non-local unitary operations, *Phys. Quant. Inf. Comp.* **2**, 240, quant-ph/0201112 (2002).

146. B. Paredes, P. Zoller and J. I. Cirac, Fermionizing a small gas of ultracold bosons, *Phys. Rev. A* **66**, 033609, cond-mat/0203061 (2002).

147. W. Hofstetter, J. I. Cirac, P. Zoller, E. Demler and M. D. Lukin, High-temperature superfluidity of fermionic atoms in optical lattices, *Phys. Rev. Lett.* **89**, 220407, cond-mat/0204237 (2002).

148. G. Giedke and J. I. Cirac, The characterization of Gaussian operations and distillation of Gaussian states, *Phys. Rev. A* **66**, 032316, quant-ph/0204085 (2002).

149. A. Recati, T. Calarco, P. Zanardi, J. I. Cirac and P. Zoller, Holonomic quantum computation with neutral atoms, *Phys. Rev. A* **66**, 032309, quant-ph/0204030 (2002).

150. L.-M. Duan, J. I. Cirac and P. Zoller, Three-dimensional theory for interaction between atomic ensembles and free-space light, *Phys. Rev. A* **66**, 023838, quant-ph/0205005 (2002).

151. G. Vidal and J. I. Cirac, Optimal simulation of nonlocal Hamiltonians using local operations and classical communication, *Phys. Rev. A* **66**, 022315, quant-ph/0108076 (2002).

152. W. Dür, C. Simon and J. I. Cirac, On the effective size of certain "Schrödinger cat" like states, *Phys. Rev. Lett.* **89**, 210402, quant-ph/0205099 (2002).

153. D. Bruss, J. I. Cirac, P. Horodecki, F. Hulpke, B. Kraus, M. Lewenstein and A. Sanpera, Reflections upon separability and distillability, *J. Mod. Opt.* **49**, 1399, quant-ph/0110081 (2002).

154. K. Hammerer, G. Vidal and J. I. Cirac, Characterization of non-local gates, *Phys. Rev. A* **66**, 062321, quant-ph/0205100 (2002).

155. E. Jané, G. Vidal, W. Dür, P. Zoller and J. I. Cirac, Simulation of quantum dynamics with quantum optical systems, *Quant. Inf. Comp.* **3**, 15, quant-ph/0207011 (2003).

156. A. Acin, G. Vidal and J. I. Cirac, On the structure of a reversible entanglement generating set for three-partite states, *Quant. Inf. Comp.* **3**, 55, quant-ph/0202056 (2003).

157. A. Micheli, D. Jaksch, J. I. Cirac and P. Zoller, Many particle entanglement in two-component Bose–Einstein condensates, *Phys. Rev. A* **67**, 013607, cond-mat/0205369 (2003).

158. B. Paredes and J. I. Cirac, From Cooper pairs to Luttinger liquids with bosonic atoms in optical lattices, *Phys. Rev. Lett.* **90**, 150402, cond-mat/0207040 (2003).

159. B. Kraus, K. Hammerer, G. Giedke and J. I. Cirac, Entanglement generation and Hamiltonian simulation in continuous-variable system, *Phys. Rev. A* **67**, 042314, quant-ph/0210136 (2003).

160. C. Schön and J. I. Cirac, Trapping atoms in the vacuum field of a cavity, *Phys. Rev. A* **67**, 043813, quant-ph/0212068 (2003).

161. J. J. Garcia-Ripoll and J. I. Cirac, Quantum computation with unknown parameters, *Phys. Rev. Lett.* **90**, 127902, quant-ph/0208143 (2003).

162. F. Verstraete and J. I. Cirac, Quantum-nonlocality in the presence of superselection rules and data hiding protocols, *Phys. Rev. Lett.* **91**, 010404, quant-ph/0302039 (2003).

163. T. S. Cubitt, F. Verstraete and J. I. Cirac, Separable states can be used to distribute entanglement, *Phys. Rev. Lett.* **91**, 037902, quant-ph/0302168 (2003).

164. G. Giedke, M. M. Wolf, O. Krueger, R. F. Werner and J. I. Cirac, Entanglement of formation for symmetric Gaussian states, *Phys. Rev. Lett.* **91**, 107901, quant-ph/0304042 (2003).

165. G. Giedke, J. Eisert, J. I. Cirac and M. Plenio, Entanglement transformations of pure Gaussian states, *Quant. Inf. and Comp.* **3**, 211, quant-ph/03010 (2003).

166. P. Rabl, A. J. Daley, P. O. Fedichev, J. I. Cirac and P. Zoller, Defect-suppressed atomic crystals in an optical lattice, *Phys. Rev. Lett.* **91**, 110403, cond-mat/0304026 (2003).

167. J. J. Garcia-Ripoll, P. Zoller and J. I. Cirac, Speed optimized two-qubit gates with laser coherent control techniques for ion trap quantum computing, *Phys. Rev. Lett.* **91**, 157901, quant-ph/0306006 (2003).

168. G. Toth, C. Simon and J. I. Cirac, Entanglement detection based on interference and particle counting, *Phys. Rev. A* **68**, 062310, quant-ph/0306086 (2003).

169. J. J. García-Ripoll and J. I. Cirac, Spin dynamics for Bosons in an optical lattice, *New J. of Phys.* **5**, 76.1 (2003).

170. J. I. Cirac and P. Zoller, How to manipulate cold atoms, *Science* **301**, 176 (2003).

171. J. J. Garcia-Ripoll and J. I. Cirac, Quantum computation with cold bosonic atoms in an optical lattice, *Phil. Trans. R. Soc. Lond. A* **361**, 1537 (2003).

172. B. Kraus and J. I. Cirac, Discrete entanglement distribution with squeezed light, *Phys. Rev. Lett.* **92**, 013602, quant-ph/0307158 (2004).

173. F. Verstraete, M. Popp and J. I. Cirac, Entanglement versus correlations in spin systems, *Phys. Rev. Lett.* **92**, 027901, quant-ph/0307009 (2004).

174. F. Verstraete, M. A. Martin-Delgado and J. I. Cirac, Diverging entanglement length in gapped quantum spin systems, *Phys. Rev. Lett.* **92**, 087201, quant-ph/0311087 (2004).

175. M. Wolf, F. Verstraete and J. I. Cirac, Entanglement frustration for Gaussian states on symmetric graphs, *Phys. Rev. Lett.* **92**, 087903, quant-ph/0307060 (2004).

176. N. Schuch, F. Verstraete and J. I. Cirac, Non-local resources in the presence of superselection rules, *Phys. Rev. Lett.* **92**, 087904, quant-ph/0310124 (2004).

177. A. Acin, J. I. Cirac and Ll. Massanes, Multipartite bound information exists and can be activated, *Phys. Rev. Lett.* **92**, 107903, quant-ph/0311064 (2004).

178. V. Murg and J. I. Cirac, Adiabatic time evolution in spin-systems, *Phys. Rev. A* **69**, 042320, quant-ph/0309026 (2004).

179. J. J. Garcia-Ripoll, J. I. Cirac, P. Zoller, C. Kollath, U. Schollwöck and J. von Delft, Variational ansatz for the superfluid mott-insulator transition in optical lattices, *Opt. Express* **12**, 42–54, cond-mat/0306162 (2004).

180. M. Wolf, G. Giedke, O. Krüger, R.-F. Werner and J. I. Cirac, Gaussian entanglement of formation, *Phys. Rev. A* **69**, 052320, quant-ph/0306177 (2004).

181. E. Moreno, F. J. Garcia-Vidal, D. Erni, J. I. Cirac and L. Martin-Moreno, Theory of plasmon-assisted transmission of entangled photons, *Phys. Rev. Lett.* **92**, 236801, quant-ph/0308075 (2004).

182. D. Porras and J. I. Cirac, Effective quantum spin systems with ion traps, *Phys. Rev. Lett.* **92**, 207901, quant-ph/0401102 (2004).

183. B. Paredes, A. Widera, V. Murg, O. Mandel, S. Fölling, I. Cirac, G. Shlyapnikov, T. W. Hänsch and I. Bloch, Tonks-Girardeau gas in an optical lattice, *Nature* **429**, 277 (2004).

184. L. Santos, M. A. Baranov, J. I. Cirac, H.-U. Everts, H. Fehrmann and M. Lewenstein, Atomic quantum gases in Kagomé lattices, *Phys. Rev. Lett.* **93**, 030601, cond-mat/0401502 (2004).

185. W. Dür, J. I. Cirac and P. Horodecki, Non-additivity of quantum capacity for multiparty communication channels, *Phys. Rev. Lett.* **93**, 020503, quant-ph/0403068 (2004).

186. N. Schuch, F. Verstraete and J. I. Cirac, Quantum entanglement theory in the presence of superselection rules, *Phys. Rev. A* **70**, 042310, quant-ph/0404079 (2004).

187. J. I. Cirac and P. Zoller, New frontiers in quantum information with atoms and ions, *Phys. Today* **57**, 38 (2004).

188. F. Verstraete, J. J. Garcia-Ripoll and J. I. Cirac, Matrix product density operators: Simulation of finite-T and dissipative systems, *Phys. Rev. Lett.* **93**, 207204, cond-mat/0406426 (2004).

189. K. Hammerer, K. Moelmer, E. S. Polzik and J. I. Cirac, Light-matter quantum interface, *Phys. Rev. A* **70**, 044304, quant-ph/0312156 (2004).

190. F. Verstraete, D. Porras and J. I. Cirac, DMRG and periodic boundary conditions: a quantum information perspective, *Phys. Rev. Lett.* **93**, 227205, cond-mat/0404706 (2004).

191. M. Popp, B. Paredes and J. I. Cirac, Adiabatic path to fractional quantum hall states of a few bosonic atoms, *Phys. Rev. A* **70**, 053612, cond-mat/04050195 (2004).

192. J. J. Garcia-Ripoll, M. A. Martin-Delgado and J. I. Cirac, Implementation of spin Hamiltonians in optical lattices, *Phys. Rev. Lett.* **93**, 250405, cond-mat/0404566 (2004).

193. B. Julsgaard, J. Sherson, J. I. Cirac, J. Fiurasek and E. S. Polzik, Experimental demonstration of quantum memory for light, *Nature* **432**, 482, quant-ph/0410072 (2004).

194. K. Vollbrecht, E. Solano and J. I. Cirac, Ensemble quantum computation with atoms in periodic potentials, *Phys. Rev. Lett.* **93**, 220502, quant-ph/0405014 (2004).

195. F. Verstraete and J. I. Cirac, Valence-bond states for quantum computation, *Phys. Rev. A* **70**, 060302, quant-ph/0311130 (2004).

196. D. Porras and J. I. Cirac, Bose–Einstein condensation and strong-correlation behavior of phonons in ion traps, *Phys. Rev. Lett.* **93**, 263602, quant-ph/0409015 (2004).

197. B. Paredes, C. Tejedor and J. I. Cirac, Fermionic atoms in optical superlattices, *Phys. Rev. A* **71**, 063608, cond-mat/0306497 (2005).

198. J. J. Garcia-Ripoll, P. Zoller and J. I. Cirac, Coherent control of trapped ions using off-resonant lasers, *Phys. Rev. A* **71**, 062309, quant-ph/0411103 (2005).

199. M. Popp, F. Verstraete, M. A. Martin-Delgado and J. I. Cirac, Localizable Entanglement, *Phys. Rev. A* **71**, 042306, quant-ph/0411123 (2005).

200. J. Korbicz, J. I. Cirac, J. Wehr and M. Lewenstein, Hilbert's 17th problem and the quantumness of states, *Phys. Rev. Lett.* **94**, 153601, quant-ph/0408029 (2005).

201. F. Verstraete, J. I. Cirac, J. I. Latorre, E. Rico and M. M. Wolf, Renormalization group transformations on quantum states, *Phys. Rev. Lett.* **94**, 140601, quant-ph/0410227 (2005).

202. K. Hammerer, M. M. Wolf, E. S. Polzik and J. I. Cirac, Quantum benchmark for storage and transmission of coherent states, *Phys. Rev. Lett.* **94**, 150503, quant-ph/0409109 (2005).

203. A. Retzker, J. I. Cirac and B. Reznik, Detection of vacuum entanglement in a linear ion trap, *Phys. Rev. Lett.* **94**, 050504, quant-ph/0408059 (2005).

204. M. Navascues, J. Bael, J. I. Cirac, M. Lewenstein, A. Sanpera and A. Acin, Quantum key distillation from Gaussian states by Gaussian operations, *Phys. Rev. Lett.* **94**, 010502, quant ph/0405047 (2005).

205. V. Murg, F. Verstraete and J. I. Cirac, Efficient evaluation of partition functions of frustrated and inhomogeneous spin systems, *Phys. Rev. Lett.* **95**, 057206, cond-mat/0501493 (2005).

206. T. Cubitt, F. Verstraete and J. I. Cirac, Entanglement flow in multipartite systems, *Phys. Rev. A*, **71**, 052308, quant-ph/0404179 (2005).

207. F. Verstraete and J. I. Cirac, Mapping local Hamiltonians of fermions to local Hamiltonians of spins, *J. Stat. Mech.*, P09012, cond-mat/0508353 (2005).

208. C. Schön, E. Solano, F. Verstraete, J. I. Cirac and M. Wolf, Sequential generation of entangled multi-qubit states, *Phys. Rev. Lett.* **95**, 110503, quant-ph/0501096 (2005).

209. J. Korbicz, J. I. Cirac and M. Lewenstein, Spin squeezing inequalities and entanglement of N qubit states, *Phys. Rev. Lett.* **95**, 120502, quant-ph/0504005 (2005).

210. B. Paredes, F. Verstraete and J. I. Cirac, Exploiting quantum parallelism to simulate quantum random many-body systems, *Phys. Rev. Lett.* **95**, 140501, cond-mat/0505288 (2005).

211. E. Moreno, A.I. Fernández-Domínguez, J. I. Cirac, F. J. García-Vidal and L. Martín-Moreno, Resonant transmission of cold atoms through subwavelength apertures, *Phys. Rev. Lett.* **95**, 170406 (2005).

212. K. Hammerer, E. S. Polzik and J. I. Cirac, Teleportation and spin squeezing utilizing multimode entanglement of light with atoms, *Phys. Rev. A* **72**, 052313, quant-ph/0511174 (2005).

213. W. Dür, M. Hein, J. I. Cirac and H.-J. Briegel, Standard forms of noisy quantum operations via depolarization, *Phys. Rev. A* **72**, 052326, quant-ph/0507134 (2005).

214. X.-L. Deng, D. Porras and J. I. Cirac, Effective spin quantum phases in systems of trapped ions *Phys, Rev. A* **72**, 063407, quant-ph/0509197 (2005).

215. K. Hammerer, E. S. Polzik and J. I. Cirac, High fidelity teleportation between light and atoms *Phys. Rev. A* **74**, 064301, quant-ph/0608133 (2007).

216. F. Verstraete, M. M. Wolf, D. Pérez-García and J. I. Cirac, Projected entangled states: Properties and applications, *Int. J. Mod. Phys. B* **20**, 5142-5153 (2006).

217. N. Schuch, J. I. Cirac and M. Wolf, Quantum states on harmonic lattices, *Commun. in Math. Phys.* **267**, 65–92, quant-ph/0509166 (2006).

218. M. Popp, J.-J. Garcia-Ripoll, K. G. H. Vollbrecht and J. I. Cirac, Cooling toolbox for atoms in optical lattices, *New J. Phys.* **8**, 164, cond-mat/0605198 (2006).

219. J. Sherson, H. Krauter, R. K. Olsson, B. Julsgaard, K. Hammerer, J. I. Cirac and E. Polzik, Quantum teleportation between light and matter, *Nature* **443**, 557-560, quant-ph/0605095 (2006).

220. M. Wolf, G. Ortiz, F. Verstraete and J. I. Cirac, Quantum phase transitions in matrix product systems, *Phys. Rev. Lett.* **97**, 110403, cond-mat/051218 (2006).

221. M. Popp, K. G. H. Vollbrecht and J. I. Cirac, Ensemble quantum computation and algorithmic cooling in optical lattices, *Fortschr. Phys.* **54**, No. 8–10, 686–701 (2006).

222. D. Porras and J. I. Cirac, Quantum manipulation of trapped ions in two dimensional Coulomb crystals, *Phys. Rev. Lett.* **96**, 250501, quant-ph/0601148 (2006).

223. M. Popp, J. J. Garcia-Ripoll, K. G. H. Vollbrecht and J. I. Cirac, Ground state cooling of atoms in optical lattices, *Phys. Rev. A* **74**, 013622, cond-mat/0603859 (2006).

224. M. Popp, F. Verstraete, M. A. Martin-Delgado and J. I. Cirac, Numerical computation of localizable entanglement in spin chains, *Appl. Phys. B* **82**, (2006).

225. Ch. A. Muschik, K. Hammerer, E.S. Polzik and J. I. Cirac, Efficient quantum memory and entanglement between light and an atomic ensemble using magnetic fields, *Phys. Rev. A* **73**, 062329, quant-ph/0512226. (2006).

226. F. Verstraete, M.Wolf, D. Pérez-García and J. I. Cirac, Criticality, the area law, and the computational power of projected entangled pair states, *Phys. Rev. Lett.* **96**, 220601, quant-ph/0601075 (2006).

227. D. Porras, F. Verstraete and J. I. Cirac, Renormalization algorithm for the calculation of spectra of interacting quantum systems, *Phys. Rev. B* **73**, 014410, cond-mat/0504717 (2006).

228. K. G. H. Vollbrecht and J. I. Cirac, Reversible universal quantum computation within translation-invariant systems, *Phys. Rev. A* **73**, 012324, quant-ph/0502143 (2006).

229. B. Kraus, W. Tittel, N. Gisin, M. Nilsson, S. Kroll and J. I. Cirac, Quantum memory for non-stationary light fields based on controlled reversible inhomogeneous broadening, *Phys. Rev. A* **73**, 020602, quant-ph/0502184 (2006).

230. M. Wolf, G. Giedke and J. I. Cirac, Extremality of Gaussian quantum states, *Phys. Rev. Lett.* **96**, 080502, quant-ph/0509154 (2006).

231. A. S. Parkins, E. Solano and J. I. Cirac, Unconditional two-mode squeezing of separated atomic ensembles, *Phys. Rev. Lett.* **96**, 053602, quant-ph/0510173 (2006).

232. F. Verstraete and J. I. Cirac, Matrix product states represent ground states faithfully, *Phys. Rev. B* **73**, 094423, cond-mat/0505140 (2006).

233. N. Schuch, M. M. Wolf and J. I. Cirac, Optimal squeezing and entanglement from noisy Gaussian operations, *Phys. Rev. Lett.* **96**, 023004, quant-ph/0505145 (2006).

234. M. Popp, K. G. H. Vollbrecht and J. I. Cirac, Ensemble quantum computation and algorithmic cooling, *Elements of Quantum Information*, **99** (2007).

235. B. Horstmann, J. I. Cirac and T. Roscilde, Dynamics of localization phenomena for hardcore bosons in optical lattices, *Phys. Rev. A* **76**, 043625, arXiv:0706.0823 (2007).

236. M. C. Bañuls, J. I. Cirac and M. M. Wolf, Entanglement in fermionic system, *Phys. Rev. A* **76**, 022311, arXiv:0705.1103 (2007).

237. T. Roscilde and J. I. Cirac, Quantum emulsion: A glassy phase of bosonic mixtures in optical lattices, *Phys. Rev. Lett.* **98**, 190402, cond-mat/0612100 (2007).

238. H. Christ, J. I. Cirac and G. Giedke, Quantum description of nuclear spin cooling in a quantum dot, *Phys. Rev. B* **75**, 155324, cond-mat/0611438 (2007).

239. K. G. H. Vollbrecht and J. I. Cirac, Delocalized entanglement of atoms in optical lattices *Phys. Rev. Lett.* **98**, 190502, quant-ph/0611132 (2007).

240. B. Paredes, T. Keilmann and J. I. Cirac, Pfaffian-like ground state for 3-body-hard-core bosons in 1D lattices, *Phys. Rev. A* **75**, 053611, cond-mat/0608012 (2007).

241. V. Murg, F. Verstraete and J. I. Cirac, Variational study of hard-core boson in a 2-D optical lattice using projected entangled pair states (PEPS), *Phys. Rev. A* **75**, 033605, cond-mat/0611522 (2007).

242. N. Schuch, M. M. Wolf, F. Verstraete and J. I. Cirac, Computational complexity of projected entangled pair states, *Phys. Rev. Lett.* **98**, 140506, quant-ph/0611050 (2007).

243. J. I. Korsbakken, K. B. Whaley, J. DuBois and J. I. Cirac, A measurement-based measure of the size of macroscopic quantum superpositions, *Phys. Rev. A* **75**, 042106, quant-ph/0611121 (2007).

244. C. Schoen, K. Hammerer, M. M. Wolf, J. I. Cirac and E. Solano, Sequential generation of matrix-product states in cavity QED, *Phys. Rev. A* **75**, 032311, quant-ph/0612101 (2007).

245. A. Acín, J. I. Cirac and M. Lewenstein, Entanglement percolation in quantum networks, *Nature Physics* Vol. 3, No. 4, quant-ph/0612167 (2007).

246. C. V. Kraus, M. M. Wolf and J. I. Cirac, Quantum simulations under translational symmetry *Phys. Rev. A* **75**, 022303, quant-ph/0607094 (2007).

247. L. Lamata, J. J. García-Ripoll and J. I. Cirac, How much entanglement can be generated between two atoms by detecting photons? *Phys. Rev. Lett.* **98**, 010502, quant-ph/0608158 (2007).

248. D. Perez-Garcia, F. Verstraete, M. M. Wolf and J.I . Cirac, Matrix product state representations, *Quant. Inf. And Comp.* **7**, 401, quant-ph/0608197 (2007).

249. D. Porras and J. I. Cirac, Collective generation of quantum states of light by entangled atoms *Phys. Rev. A* **78**, 053816, 0808.2732 (2008).

250. I. de Vega, D. Porras and J. I. Cirac, Matter-wave emission in optical lattices: Single particle and collective effects, *Phys. Rev. Lett.* **101**, 260404, 0807.1901 (2008).

251. S. Perseguers, L. Jiang, N. Schuch, F. Verstraete, M. D. Lukin, J. I. Cirac and K. G. H. Vollbrecht, One-shot entanglement generation over long distances in noisy quantum networks, *Phys. Rev. A* **78**, 062324, 0807.0208 (2008).

252. M. Aguado, G. K. Brennen, F. Verstraete and J. I. Cirac, Creation, manipulation, and detection of anyons in optical lattices, *Phys. Rev. Lett.* **101**, 260501, 0802.3163 (2008).

253. J. Jordan, R. Orus, G. Vidal, F. Verstraete and J. I. Cirac, Classical simulation of infinite-size quantum lattice systems in two spatial dimensions, *Phys. Rev. Lett.* **101**, 250602 (2008), cond-mat/0703788

254. N. Syassen, D. M. Bauer, M. Lettner, T. Volz, D. Dietze, J. J. García-Ripoll, J. I. Cirac, G. Rempe and S. Dürr, Strong dissipation inhibits losses and induces correlations in cold molecular gases, *Science* **320**, 1329–1331, 0806.4310 (2008).

255. A. Retzker, J. I. Cirac, M. B. Plenio and B. Reznik, Methods for detecting acceleration radiation in a Bose–Einstein condensate, *Phys. Rev. Lett.* **101**, 110402, 0709.2425 (2008).

256. M. M. Wolf, J. Eisert, T. S. Cubitt and J. I. Cirac, Assessing non-Markovian dynamics, *Phys. Rev. Lett.* **101**, 150402, 0711.3172 (2008).

257. H. Christ, J. I. Cirac and G. Giedke, Entanglement generation via a comletely mixed nuclear spin bath, *Phys. Rev. B* **78**, 125314, 0710.4120 (2008).

258. N. Schuch, M. M. Wolf, K. G. H. Vollbrecht and J. I. Cirac, On entropy growth and the hardness of simulation time evolution, *New J. Phys.* **10**, 033032, 12, 0801.2078 (2008).

259. R. Schmied, T. Roscilde, V. Murg, D. Porras and J. I. Cirac, Quantum phases of trapped ions in an optical lattice, *New J. Phys.* **10**, 045017, 18, 0712.4073 (2008).

260. D. Perez-García, F. Verstraete, J. I. Cirac and M. M. Wolf, PEPS as unique ground states of local Hamiltonians, *Quant. Inf and Comp.* **8**, 650, 0707.2260 (2008).

261. N. Schuch, J. I. Cirac and F. Verstraete, The computational difficulty of finding MPS ground states, *Phys. Rev. Lett.* **100**, 250501, 0802.3351 (2008).

262. D. Perez-Garcia, M. M. Wolf, M. Sanz, F. Verstraete and J. I. Cirac, String order and symmetries in quantum spin lattices, *Phys. Rev. Lett.* **100**, 167202, 0802.0447 (2008).

263. M. C. Bañuls, D. Pérez-García, M. M. Wolf, F. Verstraete and J. I. Cirac, Sequentially generated states for the study of two dimensional systems, *Phys. Rev. A.* **77**, 052306, 0802.2472 (2008).

264. D. Porras, F. Marquardt, J. von Delft and J. I. Cirac, Mesoscopic spin-boson models of trapped ions, *Phys. Rev. A* **78**, 010101, 0710.5145 (2008).

265. I. de Vega, I. Cirac and D. Porras, Detection of spin correlations in optical lattices by light scattering, *Phys. Rev. Lett. A* **77**, 051804, 0807.1901 (2008).

266. T. S. Cubitt and J. I. Cirac, Engineering correlation and entanglement dynamics in spin systems, *Phys. Rev. Lett.* **100**, 180406, quant-ph/0701053 (2008).

267. M. M. Wolf and J. I. Cirac, Dividing quantum channels, *Com. Math. Phys.* **279**, 147, math-ph/0611057 (2008).

268. N. Schuch, M. M. Wolf, F. Verstraete and J. I. Cirac, Simulation of quantum many-body systems with strings of operators and Monte Carlo tensor constractions, *Phys. Rev. Lett.* **100**, 040501, 0708.1567 (2008).

269. K. G. H. Vollbrecht and J. I. Cirac, Quantum simulators, continuous-time automata, and translationally invariant systems, *Phys. Rev. Lett.* **100**, 010501, 0704.3432 (2008).

270. N. Schuch, M. M. Wolf, F. Verstraete and J. I. Cirac, Entropy scaling and simulability by matrix product states, *Phys. Rev. Lett.* **100**, 030504, 0705.0292 (2008).

271. S. Perseguers, J. Wehr, A. Acin, M. Lewenstein and J. I. Cirac, Entanglement distribution in pure-state quantum networks, *Phys. Rev. A* **77**, 022308, 0708.1025 (2008).

272. Ch. A. Muschik, I. de Vega, D. Porras and J. I. Cirac, Quantum processing photonic states in optical lattices, *Phys. Rev. Lett.* **100**, 063601, quant-ph/0611093 (2008).

273. M. M. Wolf, F. Verstraete, M. B. Hastings and J. I. Cirac, Area laws in quantum systems: Mutual information and correlations, *Phys. Rev. Lett.* **100**, 070502, 0704.3906 (2008).

274. X.-L. Deng, D. Porras and J. I. Cirac, Quantum phases of interacting phonons in ion traps, *Phys. Rev. A* **77**, 033403, quant-ph/0703178 (2008).

275. F. Verstraete, V. Murg and J. I. Cirac, Matrix product states, projected entangled pair states, and variational renormalization group methods for quantum spin systems, *Advances in Physics*, Vol. 57, No. 2, pp. 143–224, (March 2008).

276. C. V. Kraus, M. M. Wolf, J. I. Cirac and G. Giedke, Pairing in fermionic systems: A quantum information perspective, *Phys. Rev. A* **79**, 012306, 0810.4772 (2009).

277. J. J. García-Ripoll, S. Dürr, N. Syassen, D. M. Bauer, M. Lettner, G. Rempe and J. I. Cirac, Dissipation induced tonks-girardeau gas in an optical lattice, *New J. Phys.* **11**, 013053, 0809.3679 (2009).

278. F. Verstraete, J. I. Cirac and J. I. Latorre, Quantum circuits for strongly correlated quantum systems, *Phys. Rev. A* **79**, 032316, 0804.1888 (2009).

279. S. Dürr, J. J. García-Ripoll, N. Syassen, D. M. Bauer, M. Lettner, J. I. Cirac and G. Rempe, Lieb-Liniger model of a dissipation-induced tonks-girardeau gas, *Phys. Rev. A* **79**, 023614, 0809.3696 (2009).

280. R. Renner and J. I. Cirac, A de finetti representation theorem for infinite dimensional quantum systems and applications to quantum cryptography, *Phys. Rev. Lett.* **102**, 110504, 0809.2243 (2009).

281. K. G. H. Vollbrecht and J. I. Cirac, Quantum simulations based on measurements and feedback control, *Phys. Rev. A* **79**, 042305, 0811.1844 (2009).

282. M. Sanz, M. M. Wolf, D. Perez-Garcia and J. I. Cirac, Matrix product states: Symmetries and two-body Hamiltonians, *Phys. Rev. A* **79**, 042308, 0901.2223 (2009).

283. V. Murg, F. Verstraete and J. I. Cirac, Exploring frustrated spin-systems using projected entangled pair states (PEPS), *Phys Rev. B* **79**, 195119, 0901.2019 (2009).

284. M. C. Bañuls, M. B. Hastings, F. Verstraete and J. I. Cirac, Matrix product states for dynamical simulation of infinite chains, *Phys. Rev. Lett.* **102**, 240603, 0904.1926 (2009).

285. T. Keilmann, J. I. Cirac and T. Roscilde, Dynamical creation of a supersolid in asymmetric mixtures of bosons, *Phys. Rev. Lett.* **102**, 255304, 0906.1110 (2009).

286. G. K. Brennen, M. Aguado and J. I. Cirac, Simulations of quantum double models, *New J. Physics* **11**, No 5, 0901.1345 (2009).

287. F. Pastawski, A. Kay, N. Schuch and J. I. Cirac, How long can a quantum memory withstand depolarizing noise? *Phys. Rev. Lett.* **103**, 080501, 0904.4861 (2009).

288. F. Verstraete, M. M. Wolf and J. I. Cirac, Quantum computation, quantum state engineering, and quantum phase transitions driven by dissipation, *Nature Physics* **5**, 633-636, 0803.1447 (20 July 2009).

289. M. C. Bañuls, J. I. Cirac and M. M. Wolf, Entanglement in systems of indistinguishable fermions, *Journal of Phys, Conf. Ser.* **171**, 012032, 012032 (2009).

290. F. Mezzacapo, N. Schuch, M. Boninsegni and J. I. Cirac, Ground-state properties of quantum many-body systems: Entangled-plaquette states and variational Monte Carlo, *New J. Physics* **11**, No. 8, 0905.3898 (2009).

291. A. Weichselbaum, F. Verstraete, U. Schollwöck, J. I. Cirac and J. von Delft, Variational matrix-product-state approach to quantum impurity models, *Phys. Rev. B* **80**, 165117, 165117 (2009).

292. J. I. Cirac and F. Verstraete, Renormalization and tensor product states in spin chains and lattices, *J. Phys. A: Math. Theor.* **42**, 504004, 0910.1130 (2009).

293. D. Perez-Garcia, M. Sanz, C. E. Gonzalez-Guillen, M. M. Wolf and J. I. Cirac, Characterizing symmetries in a projected entangled pair state, *New J. Phys.* **12**, 025010, 0908.1674 (2010).

294. V. Murg, J. I. Cirac, B. Pirvu and F. Verstraete, Matrix product operator representations, *New J. Phys.* **12**, 025012, 0804.3976 (2010).

295. O. Romero-Isart, M. L. Juan, R. Quidant and J. I. Cirac, Towards quantum superposition of living organisms, *New J. Phys.* **12**, 033015, 0909.1469 (2010).

296. H. Schwager, J. I. Cirac and G. Giedke, Quantum interface between light and nuclear spins in quantum dots, *Phys. Rev. B* **81**, 045309, 0810.4488 (2010).

297. L. Lamata, D. Porras, J. I. Cirac, J. Goldman and G. Gabrielse, Towards electron-electron entanglement in Penning traps, *Phys. Rev. A* **81**, 022301, 0905.0644 (2010).

298. M. Roncaglia, M. Rizzi and J. I. Cirac, Pfaffian state generation by strong three-body dissipation *Phays. Rev. Lett.* **104**, 096803, 0905.1247 (2010).

299. J. I. Cirac and G. Sierra, Infinite matrix product states, conformal field theory and the Haldane-Shastry model, *Phys. Rev. B* **81**, 104431, 0911.3029 (2010).

300. E. Kessler, S. Yelin, M. D. Lukin, J. I. Cirac and G. Giedke, Optical superradiance from nuclear spin environment of single photon emitters, *Phys. Rev. Lett.* **104**, 143601, 1002.1244 (2010).

301. S. Perseguers, M. Lewenstein, A. Acín and J. I. Cirac, Quantum random networks, *Nature Physics* **6**, 539, 0907.3283 (2010).

302. Ch. V. Kraus, N. Schuch, F. Verstraete and J. I. Cirac, Fermionic projected entangled pair states, *Phys. Rev. A* **81**, 052338, 0904.4667 (2010).

303. F. Verstraete and J. I. Cirac, Continuous matrix product states for quantum fields, *Phys. Rev. Lett.* **104**, 190405, 1002.1824 (2010).

304. A. Sfondrini, J. Cerrillo, N. Schuch and J. I. Cirac, Simulating two- and three-dimensional frustrated quantum systems with string-bond states, *Phys. Rev. B* **81**, 214426, 0908.4036 (2010).

305. B. Horstmann, B. Reznik, S. Fagnocchi and J. I. Cirac, Hawking radiation from an acoustic black hole on an ion ring, *Phys Rev. Lett.* **104**, 250403, 0904.4801 (2010).

306. P. Silvi, V. Giovannetti, S. Montangero, M. Rizzi, J. I. Cirac and R. Fazio, Homogeneous binary trees as ground states of quantum critical Hamiltonians, *Phys. Rev A* **81**, 062335, 0912.0466 (2010).

307. H. Schwager, J. I. Cirac and G. Giedke, Interfacing nuclear spins in quantum dots to a cavity or traveling-wave fields, *New J. Phys.* **12**, 043026, 0903.1727 (2010).

308. F. Pastawski, A. Kay, N. Schuch and J. I. Cirac, Limitations of passive protection of quantum information, *Quantum Information & Computation* **10**, 7–8, 0580–0618, 0911.3843 (2010).

309. Ph. Hauke, T. Roscilde, V. Murg, J. I. Cirac and R. Schmied, Modified spin-wave theory with ordering vector optimization: Frustrated bosons on the spatially anisotropic triangular lattice, *New J. Phys* **12**, 053036, 0912.5213 (2010).

310. N. Schuch and J. I. Cirac, Matrix product state and mean-field solutions for one-dimensional systems can be found efficiently, *Phys. Rev. A* **82**, 012314, 0910.4264 (2010).

311. J. I. Cirac, P. Maraner and J. K. Pachos, Cold atom simulation of interacting relativistic quantum field theories, *Phys. Rev. Lett.* **105**, 190403, 1006.2975 (2010).

312. L. Mazza, M. Rizzi, M. Lewenstein and J. I. Cirac, Emerging bosons with three-body interactions from spin-1 atoms in optical lattices, *Phys. Rev. A* **82**, 043629, 1007.2344 (2010).

313. F. Mezzacapo and J. I. Cirac, Ground-state properties of the spin-1/2 antiferromagnetic Heisenberg model on the triangular lattice: A variational study based on entangled-plaquette states, *NJP* **12**, 103039 (8pp), 1006.4480 (2010).

314. N. Schuch, I. Cirac and D. Perez-Garcia, PEPS as ground states: Degeneracy and topology *Annals of Physics*, Vol. 325, Iss 10, pp. 2153–2192, 1001.3807 (2010).

315. M. Sanz, D. Pérez-García, M. M. Wolf and J. I. Cirac, A quantum version of Wielandt's inequality information theory, *IEEE Transaction* Vol. 56, Iss. 9, pp. 4668–4673, 0909.5347 (2010).

316. Ch. V. Kraus and J. I. Cirac, Generalized hartree-fock theory for interacting fermions in lattices: Numerical methods, *NJP* **12**, 113001 (31pp),1005.5284 (2010).

317. J. Haegeman, J. I. Cirac, T. J. Osborne, H. Verschelde and F. Verstraete, Applying the variational principle to (1+1))-dimensional quantum field theories, *Phys. Rev. Lett.* **105**, 251601, 251601 (2010).

318. M. Issler, E. Kessler, G. Giedke, S. Yelin, J. I. Cirac, M. Lukin and A. Imamoglu, Nuclear spin cooling using overhauser field selective coherent populations trapping, *Phys. Rev. Lett.* **105**, 267202, 267202 (2010).

319. P. Hauke, F. M. Cucchietti, A. Müller-Hermes, M. C. Bañuls, J. I. Cirac and M. Lewenstein, Complete devil's staircase and crystal — Superfluid transitions in a dipolar XXZ spin chain: A trapped ion quantum simulation, *New J. Physics* **12**, 113037, 113037 (2010).

320. O. Romero-Isart, A. C. Pflanzer, M. L. Juan, R. Quidant, N. Kiesel, M. Aspelmeyer and J. I. Cirac, Optically levitating dielectrics in the quantum regime: Theory and protocols, *Phys. Rev. A* **83**, 013803,013803 (2011).

321. F. Pastawski, L. Clemente and J. I. Cirac, Quantum memories based on engineered dissipation *Phys. Rev. A* **83**, 012304, 012304 (2011).

322. M. C. Bañuls, J. I. Cirac and M. B. Hastings, Strong and weak thermalization of infinite nonintegrable quantum systems, *Phys. Rev. Lett.* **106**, 050405, 050405 (2011).

323. A. E. B. Nielsen, G. Sierra and J. I. Cirac, Violation of the area law and long-range correlations in infinite-dimensional-matrix product states, *Phys. Rev. A* **83**, 053807, 1103.2205 (2011).

324. Ch. A. Muschik, E. S. Polzik and J. I. Cirac, Dissipatively driven entanglement of two macroscopic atomic ensembles, *Phys. Rev. A* **83**, 052312, 1007.2209 (2011).

325. L. Jiang, T. Kitagawa, J. Alicea, A. R. Akhmerov, D. Pekker, G. Refael, J. I. Cirac, E. Demler, M. D. Lukin and P. Zoller, Majorana fermions in equilibrium and in driven cold-atom quantum wires, *Phys. Rev. Lett.* **106**, 220402, 1102.5367 (2011).

326. J. I. Cirac, D. Poilblanc, N. Schuch and F. Verstraete, Entanglement spectrum and boundary theories with projected entangled-pair states, *Phys. Rev. B* **83**, 245134, 1103.3427, (2011).

327. O. Romero-Isart, A. C. Pflanzer, F. Blaser, R. Kaltenbaek, N. Kiesel, M. Aspelmeyer and J. I. Cirac, Large quantum superpositions and interference of massive nanometer-sized objects, *Phys. Rev. Lett.* **107**, 020405, 1103.4081 (2011).

328. L. Lamata, D. R. Leibrandt, I. L. Chuang, J. I. Cirac, M. D. Lukin, V. Vuletic and S. F. Yelin, Ion crystal transducer for strong coupling between single ions and single photons, *Phys. Rev.Lett.* **107**, 030501, 1102.4141 (2011).

329. P. Hauke, T. Roscilde, V. Murg, J. I. Cirac and R. Schmied, Modified spin-wave theory with ordering vector optimization II: Spatialy anisotropic triangular lattice, *New J. Phys.* **12**, 053036, 1012.4491 (2011).

330. K. Jensen, W. Wasilewski, H. Krauter, T. Fernholz, B. M. Nielsen, J. M. Petersen, J. J. Renema, M. V. Balabas, M. Owari, M. B. Plenio, A. Serafini, M. M. Wolf, Ch. A. Muschik, J. I. Cirac, J. H. Müller and E. S. Polzik, Quantum memory, entanglement and sensing with room temperature atoms, *J. Phys. Conf. Ser.* **264**, 012022 (2011).

331. C. Navarrete–Benlloch, I. de Vega, D. Porras and J. I. Cirac, Simulating quantum-optical phenomena with cold atoms in optical lattices, *New J. Phys.* **13**, 023024, 1010.1730 (2011).

332. B. Horstmann, R. Schützhold, B. Reznik, S. Fagnocchi and J. I. Cirac, Hawking radiation on an ion ring in the quantum regime, *New J. Phys.* **13**, 045008, 1008.3494 (2011).

333. J. Haegeman, J. I. Cirac, T. J. Osborne, I. Pizorn, H. Verschelde and F. Verstraete, Time dependent variational principle for quantum lattices, *Phys. Rev. Lett.*, 1103.0936 (2011).

334. H. Krauter, Ch. A. Muschik, K. Jensen, W. Wasilewski, J. M. Petersen, J. I. Cirac and E. S. Polzik, Entanglement generated by dissipation and steady state entanglement of two macroscopic objects, *Phys. Rev. Lett.* **107**, 080503 (2011).

335. N. Schuch, D. Perez-Garcia and J. I. Cirac, Classifying quantum phases using MPS and PEPS, *Phys Rev. B* **84**, 165139, 1010.3732 (2011).

336. K. G. H. Vollbrecht, Ch. A. Muschik and J. I. Cirac, Entanglement distillation by dissipation and continuous quantum repeaters, *Phys. Rev. Lett.* **107**, 120502, 1011.4115 (2011).

337. M. Lubasch, V. Murg, U. Schneider, J. I. Cirac and M. C. Bañuls, Adiabatic preparation of a Heisenberg antiferromagnet using an optical superlattice, *Phys. Rev. Lett.* **107**, 165301, 1106.1628 (2011).

338. A. E. B. Nielsen, J. I. Cirac and G. Sierra, Quantum spin Hamiltonians for the SU(2)k WZW model, *J. Stat. Mech.*, P11014, 1109.5470 (2011).

339. N. Y. Yao, L. Jiang, A. V. Gorshkov, P. C. Maurer, G. Giedke, J. I. Cirac and M. D. Lukin, Scalable architecture for a room temperature solid-state quantum information processor, *Nature Communications* **3**, article number: 800, ncomms1788 (2012).

340. J. Haegeman, B. Pirvu, D. J. Weir, J. I. Cirac, T. J. Osborne, H. Verschelde and F. Verstraete, A variational matrix product ansatz for dispersion relations, *Phys. Rev. B* **84**, 100408 (2012).

341. J. I. Cirac and P. Zoller, Goals and opportunities in quantum simulation, *Nature Physics* **8**, 264–266, nphys2275 (2012).

342. P. C. Maurer, G. Kucsko, C. Latta, L. Jiang, N. Y. Yao, S. D. Bennett, F. Pastawski, D. Hunger, N. Chisholm, M. Markham, D. J. Twitchen, J. I. Cirac and M. D. Lukin, Room-temperature quantum bit memory exceeding one second, *Science* **336** (6068), 1283–1286, 1220513 (2012).

343. A. E. B. Nielsen, J. I. Cirac and G. Sierra, Laughlin spin liquid states on lattices obtained from conformal field theory, *Phys. Rev. Lett.* **108**, 257206 (2012).

344. Ch. A. Muschik, H. Krauter, K. Jensen, J. M. Petersen, J. I. Cirac and E. S. Polzik, Robust entanglement generation by reservoir engineering, *J. Phys. B: At. Mol. Opt. Phys.* **45** 124021 (2012).

345. A. C. Pflanzer, O. Romero-Isart and J. I. Cirac, Master equation approach to optomechanics with arbitrary dielectrics, *Phys. Rev. A* **86**, 013802 (2012).

346. A. Mueller-Hermes, J. I. Cirac and M. C. Banuls, Tensor network techniques for the computation of dynamical observables in one-dimensional quantum spin systems, *New Journal of Physics*, Vol. 14, 075003 (2012).

347. D. Poilblanc, N. Schuch, D. Pérez-García and J. I. Cirac, Topological and entanglement properties of resonating valence bond wavefunctions, *Phys. Rev. B* **86**, 014404 (2012).

348. E. M. Kessler, G. Giedke, A. Imamoglu, S. F. Yelin, M. D. Lukin and J. I. Cirac, Dissipative phase transition in central spin systems, *Phys. Rev. A* **86**, 012116 (2012).

349. J. Haegeman, D. Pérez-García, J. I. Cirac and N. Schuch, Order parameter for symmetry-protected phases in one dimension, *Phys. Rev. Lett.* **109**, 050402 (2012).

350. M. J. A. Schuetz, E. M. Kessler, J. I. Cirac and G. Giedke, Superradiance-like electron transport through a quantum dot, *Phys. Rev. B* **86**, 085322 (2012).

351. N. Schuch, D. Poilblanc, J. I. Cirac and D. Pérez-García, Resonating valence bond states in the PEPS formalism, *Phys. Rev. B* **86**, 115108 (2012).

352. E. Zohar, J. I. Cirac and B. Reznik, Simulating compact quantum electrodynamics with ultracold atoms: Probing confinement and nonperturbative effects, *Phys. Rev. Lett.* **109**, 125302 (2012).

353. F. Pastawski, N. Y. Yao, L. Jiang, M. D. Lukin and J. I. Cirac, Unforgeable noise-tolerant quantum tokens, *PNAS* Vol. 109 (No. 40), 16079–16082, 1203552109 (2012).

354. O. Romero-Isart, L. Clemente, C. Navau, A. Sanchez and J. I. Cirac, Quantum magnetomechanics with levitating superconducting microspheres, *Phys. Rev. Lett.* **109**, 147205 (2012).

355. M. Gullans, T. G. Tiecke, D. E. Chang, J. Feist, J. D. Thompson, J. I. Cirac, P. Zoller and M. D. Lukin, Nanoplasmonic lattices for ultracold atoms, *Phys. Rev. Lett.* **109**, 235309 (2012).

356. T. B. Wahl, D. Pérez-García and J. I. Cirac, Matrix product states with long-range localizable entanglement, *Phys. Rev. A* **86**, 062314 (2012).

357. V. M. Stojanović, T. Shi, C. Bruder and J. I. Cirac, Quantum simulation of small-polaron formation with trapped ions, *Phys. Rev. Lett.* **109**, 250501 (2012).

358. C. Fernández-González, N. Schuch, M. M. Wolf, J. I. Cirac and D. Pérez-García, Gapless Hamiltonians for the toric code using the PEPS formalism, *Phys. Rev. Lett.* **109**, 260401 (2012).

359. T. Shi and J. I. Cirac, Topological phenomena in trapped ion systems, *Phys. Rev. A* **87**, 013606 (2013).

360. B. Horstmann, J. I. Cirac and G. Giedke, Noise-driven dynamics and phase transitions in fermionic systems, *Phys. Rev. A* **87**, 012108 (2013).

361. A. Cadarso, M. Sanz, M. M. Wolf, J. I. Cirac and D. Pérez-García, Entanglement, fractional magnetization, and long-range interactions, *Phys. Rev. B* **87**, 035114 (2013).

362. E. Zohar, J. I. Cirac and B. Reznik, Simulating (2+1)-dimensional lattice QED with dynamical matter using ultracold atoms, *Phys. Rev. Lett.* **110**, 055302 (2013).

363. H. Schwager, J. I. Cirac and G. Giedke, Dissipative spin chains: Implementation with cold atoms and steady-state properties, *Phys. Rev. A* **87**, 022110 (2013).

364. N. Y. Yao, C. R. Laumann, A. V. Gorshkov, H. Weimer, L. Jiang, J. I. Cirac, P. Zoller and M. D. Lukin, Topologically protected quantum state transfer in a chiral spin liquid, *Nature Communications*, 4, 1585, ncomms2531 (2013).

365. D. E. Chang, J. I. Cirac and H. J. Kimble, Self-organization of atoms along a nanophotonic waveguide, *Phys. Rev. Lett.* **110**, 113606 (2013).

366. E. Zohar, J. I. Cirac and B. Reznik, Cold-atom quantum simulator for SU(2) Yang-Mills lattice gauge theory, *Phys. Rev. Lett.* **110**, 125304 (2013).

367. C. A. Muschik, K. Hammerer, E. S. Polzik and J. I. Cirac, Quantum teleportation of dynamics and effective interactions between remote systems, *Phys. Rev. Lett.* **111**, 020501 (2013).

368. J. Haegeman, J. I. Cirac, T. J. Osborne and F. Verstraete, Calculus of continuous matrix product states, *Phys. Rev. B* **88**, 085118 (2013).

369. N. Schuch, D. Poilblanc, J. I. Cirac and D. Pérez-García, Topological order in the projected entangled-pair states formalism: Transfer operator and boundary Hamiltonians, *Phys. Rev. Lett.* **111**, 090501 (2013).

370. E. Zohar, J. I. Cirac and B. Reznik, Quantum simulations of gauge theories with ultracold atoms: Local gauge invariance from angular momentum conservation, *Phys. Rev. A* **88**, 023617 (2013).

371. C. V. Kraus, M. Lewenstein and J. I. Cirac, Ground states of fermionic lattice Hamiltonians with permutation symmetry, *Phys. Rev. A* **88**, 022335 (2013).

372. A. C. Pflanzer, O. Romero-Isart and J. I. Cirac, Optomechanics assisted by a qubit: From dissipative state preparation to many-partite systems, *Phys. Rev. A* **88**, 033804 (2013).

373. J. I. Cirac, S. Michalakis, D. Perez-Garcia and N. Schuch, Robustness in projected entangled pair states, *Phys. Rev. B.* **88**, 115108 (2013).

374. O. Romero-Isart, C. Navau, A. Sanchez, P. Zoller and J. I. Cirac, Superconducting vortex lattices for ultracold atoms, *Phys. Rev. Lett.* **111**, 145304 (2013).

375. D. Poilblanc, N. Schuch and J. I. Cirac, Field-induced superfluids and Bose liquids in projected entangled pair states, *Phys. Rev. B* **88**, 144414 (2013).

376. A. E. B. Nielsen, G. Sierra and J. I. Cirac, Local models of fractional quantum Hall states in lattices and physical implementation, *Nature Communications*, **4**, 2864, ncomms3864 (4 Nov 2013).

377. M. C. Bañuls, K. Cichy, J. I. Cirac and K. Jansen, The mass spectrum of the Schwinger model with matrix product states, *Journal of High Energy Physics* **11**, 158 (2013).

378. L. Mazza, M. Rizzi, M. D. Lukin and J. I. Cirac, Robustness of quantum memories based on Majorana zero modes, *Phys. Rev. B* **88**, 205142 (26 Nov 2013).

379. T. B. Wahl, H.-H. Tu, N. Schuch and J. I. Cirac, Projected entangled-pair states can describe chiral topological states, *Phys. Rev. Lett.* **111**, 236805 (6 Dec 2013.)

380. M. J. A. Schuetz, E. M. Kessler, L. M. K. Vandersypen, J. I. Cirac and G. Giedke, Steady-state entanglement in the nuclear spin dynamics of a double quantum dot, *Phys. Rev. Lett.* **111**, 246802 (9 Dec 2013).

381. G. De las Cuevas, N. Schuch, D. Pérez-García and J. I. Cirac, Purifications of multipartite states: Limitations and constructive methods, *New J. Phys.* **15**, 123021 (2013).

382. H. Saito, M. C. Bañuls, K. Cichy, J. I. Cirac and K. Jansen, The temperature dependence of the chiral condensate in the Schwinger model with Matrix Product States, *PoS(LATTICE2014)*302.

383. S. Yang, L. Lehman, D. Poilblanc, K. Van Acoleyen, F. Verstraete, J. I. Cirac and N. Schuch, Edge Theories in Projected Entangled Pair State Models, *Phys. Rev. Lett.* **112**, 036402 (2014).

384. M. Lubasch, J. I. Cirac and M.C. Bañuls, Unifying projected entangled pair state contractions, *New J. Phys.* **16**, 033014 (2014).

385. H.-H. Tu, A. E. B. Nielsen, J. I. Cirac and G. Sierra, Lattice Laughlin states of bosons and fermions at filling fractions 1/q, *New J. Phys.* **16**, 033025 (2014).

386. M. J. A. Schuetz, E. M. Kessler, L. M. K. Vandersypen, J. I. Cirac and G. Giedke, Nuclear spin dynamics in double quantum dots: Multistability, dynamical polarization, criticality, and entanglement, *Phys. Rev. B* **89**, 195310 (2014).

387. C. Navau, J. Prat-Camps, O. Romero-Isart, J. I. Cirac and A. Sanchez, Long-Distance Transfer and Routing of Static Magnetic Fields, *Phys. Rev. Lett.* **112**, 253901 (2014).

388. D. Poilblanc, P. Corboz, N. Schuch and J. I. Cirac, Resonating-valence-bond superconductors with fermionic projected entangled pair states, *Phys. Rev. B* **89**, 241106 (2014).

389. A. E. B. Nielsen, G. Sierra and J. I. Cirac, Optical-lattice implementation scheme of a bosonic topological model with fermionic atoms, *Phys. Rev. A* **90**, 013606 (2014).

390. M. Lubasch, J. I. Cirac and M. C. Bañuls, Algorithms for finite projected entangled pair states, *Phys. Rev. B* **90**, 064425 (2014).

391. I. Glasser, J. I. Cirac, G. Sierra and A. E. B. Nielsen, Construction of spin models displaying quantum criticality from quantum field theory, *Nuclear Physics B* **886**, 63–74 (2014).

392. T. B. Wahl, S. T. Hassler, Hong-Hao Tu, J. I. Cirac and N. Schuch, Symmetries and boundary theories for chiral projected entangled pair states, *Phys. Rev. B* **90**, 115133 (2014).

393. S. Kühn, J. I. Cirac and M. C. Bañuls, Quantum simulation of the Schwinger model: A study of feasibility, *Phys. Rev. A* **90**, 042305 (2014).

394. C. Fernández-González, N. Schuch, M. M. Wolf, J. I. Cirac and D. Pérez-García, Frustration free gapless Hamiltonians for Matrix Product States, *Commun. Math. Phys.* **333**, 299–333 (2015).

395. A. Molnar, N. Schuch, F. Verstraete and J. I. Cirac, Approximating Gibbs states of local Hamiltonians efficiently with PEPS, *Phys. Rev. B* **91**, 045138 (2015).

396. J. Haegeman, K. Van Acoleyen, N. Schuch, J. I. Cirac and F. Verstraete, Gauging quantum states: from global to local symmetries in many-body systems, *Phys. Rev. X* **5**, 011024 (2015).

397. S. Yang, T. B. Wahl, H.-H. Tu, N. Schuch and J. I. Cirac, Chiral projected entangled-pair state with topological order, *Phys. Rev. Lett.* **114**, 106803 (2015).

398. A. González-Tudela, C.-L. Hung, D. E. Chang, J. I. Cirac and H. J. Kimble, Subwavelength vacuum lattices and atom-atom interactions in photonic crystals, *Nature Photonics* **9**, 320–325 (2015).

399. J. Cui, J. I. Cirac and M. C. Bañuls, Variational matrix product operators for the steady state of dissipative quantum systems, *Phys. Rev. Lett.* **114**, 220601 (2015).

400. D. Poilblanc, J. I. Cirac and N. Schuch, Chiral topological spin liquids with projected entangled pair states, *Phys. Rev. B* **91**, 224431 (2015).

401. H. Kim, M. C. Bañuls, J. I. Cirac, M. B. Hastings and D. A. Huse, Slowest local operators in quantum spin chains, *Phys. Rev. E* **92**, 012128 (2015).

402. S. Kühn, E. Zohar, J. I. Cirac and M. C. Bañuls, Non-Abelian string breaking phenomena with Matrix Product States, *JHEP* **1507** (2015) 130.

403. I. Glasser, J. I. Cirac, G. Sierra and A. E. B. Nielsen, Exact parent Hamiltonians of bosonic and fermionic Moore-Read states on lattices and local models, *New J. Phys.* **17**, 082001 (2015).

404. M. C. Bañuls, K. Cichy, J. I. Cirac, K. Jansen and H. Saito, Thermal evolution of the Schwinger model with Matrix Product Operators, *Phys. Rev. D* **92**, 034519 (2015).

405. M. J. A. Schuetz, E. M. Kessler, G. Giedke, L. M. K. Vandersypen, M. D. Lukin and J. I. Cirac, Universal Quantum Transducers based on Surface Acoustic Waves, *Phys. Rev. X* **5**, 031031 (2015).

406. A. González-Tudela, V. Paulisch, D. E. Chang, H. J. Kimble and J. I. Cirac, Deterministic Generation of Arbitrary Photonic States Assisted by Dissipation, *Phys. Rev. Lett.* **115**, 163603 (2015).

407. T. Caneva, M. T. Manzoni, T. Shi, J. S. Douglas, J. I. Cirac and D. E. Chang, Quantum dynamics of propagating photons with strong interactions: A generalized input-output formalism, *New J. Phys.* **17** (2015) 113001.

408. T. Shi, D. E. Chang and J. I. Cirac, Multiphoton scattering theory and generalized master equations, *Phys. Rev. A* **92**, 053834 (2015).

409. E. Zohar, M. Burrello, T. Wahl and J. I. Cirac, Fermionic Projected Entangled Pair States and Local U(1) Gauge Theories, *Annals of Phys.* **363**, 385–439 (2015).

410. B. Herwerth, G. Sierra, H.-H. Tu, J. I. Cirac and A. E. B. Nielsen, Edge states for the Kalmeyer-Laughlin wave function, *Phys. Rev. B* **92**, 245111 (2015).

411. E. Zohar, J. I. Cirac and B. Reznik, Quantum simulations of lattice gauge theories using ultracold atoms in optical lattices, *Rep. Prog. Phys.* (79) 014401 (2016).

412. Y. Ge, A. Molnar and J. I. Cirac, Rapid adiabatic preparation of injective PEPS and Gibbs states, *Phys. Rev. Lett* **116**, 080503 (2016).

Peter Zoller

PETER ZOLLER

Birthplace

Born   Innsbruck, Austria, 16 September 1952

Education

1977   Ph.D. University of Innsbruck

Research and Professional Experience

1978–1979   Max Kade Fellow, University of Southern California, Los Angeles
1977–1990   Assistant Professor, University of Innsbruck; tenure since 1981
1981–82, 88 JILA Visiting Fellow, University of Colorado, Boulder, Colorado
1991–1994   Professor of Physics with tenure, University of Colorado and JILA Fellow
1994–   Professor of Physics with tenure, University of Innsbruck, Austria
1994–   JILA Adjoint Fellow, Boulder, Colorado
2003–   Research Director, Institute for Quantum Optics and Quantum Information of the Austria Academy of Sciences

Membership of Societies

1993   American Physical Society (Fellow)
2001   Austrian Academy of Sciences
2008   Royal Netherlands Academy of Arts and Sciences
2008   Royal Spanish Academy of Sciences
2009   US National Academy of Sciences (Foreign Associate)
2010   German National Academy of Sciences (Leopoldina)
2013   Academy of Europe (Academia Europea)

Honours

2012   Honorary doctorate, Free University of Amsterdam

Awards

Ludwig Boltzmann Prize, 1983; Wittgenstein Award Austrian Science Foundation), 1998; Max Born Award (Optical Society of America), 1998; Schrödinger Prize of the Austrian Academy of Sciences, 1998; Senior Humboldt Award, 2000; Max Planck Medal, 2005; Niels Bohr/UNESCO Gold Medal, 2005; 6th International Quantum Communication Award, 2006; Dirac Medal, 2006; BBVA Foundation Frontiers of Knowledge Award, in the Basic Sciences category (with Ignacio Cirac), 2008; Benjamin Franklin Medal in Physics (with Juan Ignacio Cirac and David Wineland), 2010; Blaise Pascal Medal in Physics, 2011; Blaise-Pascal-Medal of the European Academy of Sciences, 2011; Hamburg Prize for Theoretical Physics, 2012; ERC Synergy Grant, 2012; David Ben Gurion Medal, 2013; Wolf Prize in Physics (with Juan Ignacio Cirac), 2013; Herbert Walther Award from the OSA, 2016.

COMMENTARY

I was born in Innsbruck, Austria, in 1952 into a family of teachers, where academic education had a high value. Both my father and my mother had a doctorate in economics. My father became the founding director of a college for teacher education, and my parents published together textbooks on social studies and accounting. The family on my mother's side had a long tradition as high-school teachers in Latin and ancient Greek, but I was the first one who went into natural sciences. My interest in science was triggered by discovering the science section of the public library, reading books on electronics, becoming a radio amateur at an early age, and becoming more and more intrigued by books on physics.

At the end of high-school it was clear to me that I wanted to go into natural sciences. I first tended towards medicine and biology, but then I decided to study physics at the University of Innsbruck — physics as a subject, which combines fundamental questions with potential applications. It was the lectures on theoretical physics, deriving and understanding nontrivial phenomena in physics from a few basic principles, which was the eye opening experience in studying physics. I had good lecturers in general physics courses, but I mainly taught myself reading books according to my interests, in particular, books on quantum physics, exploiting an — today definitely no longer existing — university system, which gave ultimate freedom in learning with remarkably few formal requirements. Only in retrospect it became clear to me that this — often naïve, and random — self-teaching in isolation, and abundance of time to study, made me learn and think about problems away from the main stream. My PhD advisor Fritz Ehlotzky directed me towards quantum optics and atom-laser interactions, and I still remember discovering Roy Glauber's lecture notes on quantum optics from the Les Houches school 1964: I had found what I wanted to do.

In the meantime, I had met Peter Lambropoulos from the University of Southern California, being intrigued by and having studied in detail his papers on multiphoton processes. He invited me to join his group as a Max Kade postdoctoral fellow, and it was a revelation being introduced and exposed to problems of current interest in atomic physics, like interactions with intense stochastic light fields — and provided the basis for submission of a Habilitation to the University of Innsbruck, and a permanent position as docent. In early 1980, Dan Walls invited me for the first time to New Zealand. This was the golden era of New Zealand theoretical quantum optics with Dan Walls as inspiration and driving force, and Crispin Gardiner as the master mind. I was introduced not only to quantum optics problems, but — more importantly — to the New Zealand way on how to "do" quantum optics and "think" quantum noise. For me this had a profound influence, and provided much of the basis ten years later for connecting quantum optics and quantum information; and it was the starting point of a life long collaboration and friendship with C. W. Gardiner, to me a father figure of modern techniques in quantum optics. Today, I am proud that I am also a coauthor of books on quantum noise and quantum optics, which he initiated.

In the 1980's, I also developed a close relationship with JILA, first by spending twice a year in Boulder as a Visiting Fellow, and eventually joining JILA in the 1990's as Fellow and Professor at the University of Colorado. JILA was not only a world institution in laser science, for me it provided for the first time an intense interaction with experiment; and experimentalists in atomic, molecular and optical physics, including S. Smith, J. Cooper, J. Hall, and C. Wieman; and D. Wineland at NIST Boulder and W.D. Phillips at NIST Gaithersburg; and close collaboration with young experimental postdocs like R. Blatt and W. Ertmer, and much later E. Cornell. I remember being immensely impressed by witnessing experimental progress, which at the end led from laser cooling to quantum degenerate atomic gases, and trapping of single ions. But I was equally impressed by the experimentalists' insightful way of developing physical pictures as a basis of understanding, feeling challenged and inspired as a theorist to consolidate this with my theoretical language. As JILA Fellow I had the opportunity to form my own theory group, parallel to C. Greene's atomic theory effort. With the incredibly talented and creative J. Ignacio Cirac joining, an excellent group of students, and frequent visitors R. Blatt, C. W. Gardiner, and M. Lewenstein, there was a remarkable critical mass on the theory side and room for novel ideas, inspired by the broader JILA surrounding pointing towards ultimate quantum control in atomic many-body systems, and the challenges provided by emerging field of quantum information.

Proposals like quantum computing with trapped ions, or later, quantum reservoir engineering, superfluid — Mott insulator transition of cold atoms in optical lattices as quantum simulation, and quantum state transfer between atoms in cavities, or the quantum repeater with atomic ensembles for long distance quantum communication were written in the following years. In the mean time I had accepted an offer to go back to Innsbruck as a chair professor, and Ignacio followed to join the Innsbruck faculty as a Professor, allowing us to run a joint group, and experimental quantum optics in Innsbruck gaining high visibility with Rainer Blatt and Anton Zeilinger as chair professors. For my colleagues and me, but also for my physics community in general, these were golden years, with atomic physics and quantum optics providing the unique experimental and theoretical environment for the interdisciplinary connections to quantum information and a novel condensed matter physics. The success of atomic, molecular and optical physics is, of course, very much connected with experimental achievements, which led from first demonstrations of quantum gates with ions by D. Wineland, and by R. Blatt to the small scale ion trap quantum computers of today. In quantum degenerate gases experiments by I. Bloch (MPQ) and collaborators, and others, led the way to quantum simulation with bosonic and fermionic atoms and molecules in optical lattices. New theoretical ideas emerged like Rydberg atoms and Rydberg blockade physics, in close collaboration with M. Lukin (Harvard), which in the mean time is a very active experimental field. And quantum communication with both CQED setups and with atomic ensembles were realized among others by my experimental friends H.J. Kimble (Caltech) and G. Rempe (MPQ).

During the last twenty years, we have seen Innsbruck quantum physics develop into an internationally recognized center. This includes the foundation of the Institute for Quantum Optics and Quantum Information of the Austrian Academy of Sciences established in 2003 in Innsbruck and Vienna. While I. Cirac left to become director of the Max-Planck Institute for Quantum Optics in Garching, and Anton Zeilinger took a chair in Vienna, we have been able to attract a next generation of top experimentalists and theorists in quantum physics, including R. Grimm, H. Briegel and F. Ferlaino.

What I find remarkable, is that quantum optics and atomic physics has — with all its success and competitiveness — remained a very friendly community where colleagues are also personal friends. Today I find myself a proud member of an ERC synergy grant with I. Bloch (MPQ), J. Dalibard (College de France) and E. Altman (Weizmann). What I find truly rewarding is that I was able to share prizes, like the Wolf Prize 2013, with my friend and colleague I. Cirac, recognizing our work as joint effort. My career as a scientist would not have been possible without the continuous support and understanding by my wife Johanna, and my family.

VOLUME 74, NUMBER 20   PHYSICAL REVIEW LETTERS   15 MAY 1995

Quantum Computations with Cold Trapped Ions

J. I. Cirac and P. Zoller*

Institut für Theoretische Physik, Universität Innsbruck, Technikerstrasse 25, A-6020 Innsbruck, Austria
(Received 30 November 1994)

A quantum computer can be implemented with cold ions confined in a linear trap and interacting with laser beams. Quantum gates involving any pair, triplet, or subset of ions can be realized by coupling the ions through the collective quantized motion. In this system decoherence is negligible, and the measurement (readout of the quantum register) can be carried out with a high efficiency.

PACS numbers: 89.80.+h, 03.65.Bz, 12.20.Fv, 32.80.Pj

A quantum computer (QC) obeys the laws of quantum mechanics, and its unique feature is that it can follow a superposition of computation paths simultaneously and produce a final state depending on the interference of these paths [1]. Recent results in quantum complexity theory and the development of algorithms indicate that quantum computers can solve some problems efficiently which are considered intractable on classical Turing machines. An example is the factorization of large composite numbers into primes [2], a problem which is the basis of the security of many classical key cryptosystems.

The task of designing a QC is equivalent to finding a physical implementation of quantum gates between quantum bits (or qubits), where a qubit refers to a two-state system $\{|0\rangle, |1\rangle\}$ [3]. It has been shown [4] that any operation can be decomposed into controlled-NOT gates between two qubits and rotations on a single qubit, where a controlled-NOT is defined by $\hat{C}_{12} : |\epsilon_1\rangle|\epsilon_2\rangle \rightarrow |\epsilon_1\rangle|\epsilon_1 \oplus \epsilon_2\rangle$ with $\epsilon_{1,2} = 0, 1$, and $\oplus$ denoting addition modulo 2. While there exist promising proposals to demonstrate the basic principle of gates in cavity QED [4], the experimental steps necessary to realize even a controlled-NOT gate indicate that extended networks would be exceedingly difficult to build. On the other hand, a number of interactions have been proposed for the construction of quantum computers [1,5], but so far no explicit physical system has been shown to serve as a realistic model. The main obstacle for a practical realization is the existence of decoherence processes due to the interaction of the system (the QC) with the environment [6].

In this Letter we show that a set of N cold ions interacting with laser light and moving in a linear trap [7] provides a realistic physical system to implement a quantum computer. The distinctive features of this system are (i) it allows the implementation of n-bit quantum gates between any set of (not necessarily neighboring) ions, (ii) decoherence can be made negligible during the whole computation, and (iii) the final readout can be performed with unit efficiency.

The basic elements of the computer (i.e., the qubits) are the ions themselves. The two states of the nth qubit are identified with two of the internal states of the corresponding ion; for example, a ground state $|g\rangle_n \equiv |0\rangle_n$ and an excited state $|e\rangle_n \equiv |1\rangle_n$. The state of the QC is a macroscopic superposition

$$|\psi\rangle = \sum_{x=0}^{2^N-1} c_x |x\rangle \equiv \sum_{\underline{x}=\{0,1\}^N} c_{\underline{x}} |\underline{x}\rangle$$

of quantum registers $|\underline{x}\rangle = |x_{N-1}\rangle_{N-1} \cdots |x_0\rangle_0$ with $x = \sum_{n=0}^{N-1} x_n 2^n$ the binary decomposition of x. In this system independent manipulation of each individual qubit is accomplished by directing different laser beams to each of the ions. The quantum controlled-NOT, and, more generally, the (controlled)^n-NOT gate between n arbitrary ions in the trap can be implemented by exciting the collective quantized motion of the ions with lasers [8]. The coupling of the motion of the ions is provided by the Coulomb repulsion which is much stronger than any other interaction for typical separations between the ions of a few optical wavelengths. Decoherence in an ion trap is due to spontaneous decay of the internal atomic states and damping of the motion of the ion. Application of stored ions in ultrahigh precision spectroscopy and time and frequency standards [9,10] shows that this decoherence time can be extremely long, much longer than the time required to perform many operations in a QC. Spontaneous emission is suppressed using metastable transitions [10]. Collisions with background atoms can be avoided at sufficiently low pressures for very long times, and other couplings that affect the moving charges can be made sufficiently small [9]. Furthermore, the final readout of the quantum register (state measurement of the individual qubits) at the end of the computation can be accomplished using the quantum jumps technique with unit efficiency [11].

The situation we have in mind is depicted in Fig. 1. N ions are confined in a linear trap, and interact with different laser beams [Fig. 1(a)] in standing wave configurations [12]. The confinement of the motion along X, Y, and Z directions can be described by an (anisotropic) harmonic potential of frequencies $\nu_x \ll \nu_y, \nu_z$. Nonharmonic traps can also be used leading to very similar results. The ions have been previously laser cooled in all three dimensions so that they undergo very small oscillations around the equilibrium position. In this case, the

VOLUME 74, NUMBER 20   PHYSICAL REVIEW LETTERS   15 MAY 1995

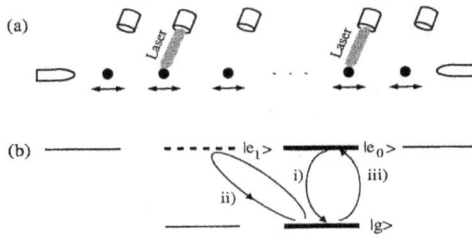

FIG. 1. (a) N ions in a linear trap interacting with N different laser beams; (b) atomic level scheme.

motion of the ions is described in terms of normal modes. Furthermore, we will assume that sideband cooling has left all the normal modes in their corresponding (quantum) ground states [13]. For this to be possible, one has to assume that the Lamb-Dicke limit (LDL) holds for all the modes [10]. This implies that their frequency is larger than the photon recoil frequency corresponding to the transition used for laser cooling. For example, for the $S_{1/2} \rightarrow D_{5/2}$ dipole-forbidden transition of a barium ion, this requires $\nu_{x,y,z} \gg 3$ kHz; in typical situations $\nu_{y,z} \gg \nu_x \sim 2\pi \times 50$ kHz [7]. The minimum frequency is that of the center-of-mass (CM) mode moving in the X direction, and coincides with ν_x. The next frequency is $\sqrt{3}\nu_x$, and all the others are larger. A remarkable feature of this system is that the frequency spacing is independent of the number of ions N in the trap.

Figure 1(b) shows a typical level scheme for an alkaline earth ion, corresponding to an electric dipole-forbidden transition [10]. The two-level system that we choose as the qubit is marked with thicker lines ($|g\rangle$ and $|e_0\rangle$). The other levels do not disturb the computation process. On the contrary, some of them are needed for implementing quantum gates, as we will show below.

When a laser beam acts on one of the ions, it induces transitions between its internal ground and excited levels and can change the state of the collective normal modes. However, in the LDL and for sufficiently low intensities, the laser beam will only cause transitions that modify the state of one of the modes. For example, with a laser frequency so that the detuning equals minus the CM mode frequency ($\delta_n = -\nu_x$), one excites the CM mode exclusively. This is so since the frequencies of the different normal modes are well separated in the excitation spectrum. This fact allows one to control the interactions between the ions through the CM motion, by selecting appropriately the frequency of the lasers.

Let $\hat{H}_0$ be the Hamiltonian for the system in the absence of any laser field. Now, consider that the laser acting on the nth ion is turned on. Obviously, this laser will leave the internal state of all the other ions unaffected. The laser frequency is chosen such that $\delta_n = -\nu_x$ and

the equilibrium position of the ion coincides with the node of the laser standing wave [12]. The Hamiltonian describing this situation in an interaction picture defined by the operator $\exp(-i\hat{H}_0 t)$ is ($\hbar = 1$)

$$\hat{H}_{n,q} = \frac{\eta}{\sqrt{N}} \frac{\Omega}{2} \left[|e_q\rangle_n \langle g| a e^{-i\phi} + |g\rangle_n \langle e_q| a^\dagger e^{i\phi} \right]. \quad (1)$$

Here $a^\dagger$ and a are the creation and annihilation operators of CM phonons, respectively, Ω is the Rabi frequency, ϕ is the laser phase, and $\eta = [\hbar k_\theta^2/(2M\nu_x)]^{1/2}$ is the LDL parameter [$k_\theta = k\cos(\theta)$, with k the laser wave vector and θ the angle between the X axis and the direction of propagation of the laser]. The subscript $q = 0, 1$ refers to the transition excited by the laser, which depends on the laser polarization [see Fig. 1(b)]. Equation (1) can be derived as a generalization of the single ion Hamiltonian for the case of a linear trap [14]. Physically, the factor $\sqrt{N}$ appears since the effective mass of the CM motion is NM, and the amplitude of the mode scales like $1/\sqrt{NM}$ (Mössbauer effect). A careful analysis shows that the model Hamiltonian (1) is valid for $(\Omega/2\nu_x)^2\eta^2 \ll 1$. Note that in the LDL $\eta \ll 1$. On the other hand, corrections to this Hamiltonian can be made arbitrarily small for sufficiently low laser intensities.

If the laser beam is on for a certain time $t = k\pi/(\Omega\eta/\sqrt{N})$ (i.e., using a $k\pi$ pulse), the evolution of the system will be described by the unitary operator

$$\hat{U}_n^{k,q}(\phi) = \exp\left[-ik\frac{\pi}{2} \left(|e_q\rangle_n \langle g| a e^{-i\phi} + \text{H.c.} \right) \right]. \quad (2)$$

It is easy to prove that this transformation keeps the state $|g\rangle_n|0\rangle$ unaltered, whereas

$$|g\rangle_n|1\rangle \longrightarrow \cos(k\pi/2)|g\rangle_n|1\rangle - ie^{i\phi}\sin(k\pi/2)|e_q\rangle_n|0\rangle,$$
$$|e\rangle_n|0\rangle \longrightarrow \cos(k\pi/2)|e_q\rangle_n|0\rangle - ie^{-i\phi}\sin(k\pi/2)|g\rangle_n|1\rangle,$$

where $|0\rangle$ ($|1\rangle$) denotes a state of the CM mode with no (one) phonon.

Let us now show how a two-bit gate can be performed using this interaction. We consider the following three-step process [see Fig. 1(b)]. (i) A π laser pulse with polarization $q = 0$ and $\phi = 0$ excites the mth ion. The evolution corresponding to this step is given by $\hat{U}_m^{1,0} \equiv \hat{U}_m^{1,0}(0)$. (ii) The laser directed on the nth ion is then turned on for a time of a 2π pulse with polarization $q = 1$ and $\phi = 0$. The corresponding evolution operator $\hat{U}_n^{2,1}$ changes the sign of the state $|g\rangle_n|1\rangle$ (without affecting the others) via a rotation through the auxiliary state $|e_1\rangle_n|0\rangle$. (iii) Same as (i). Thus the unitary operation for the whole process is $\hat{U}_{m,n} \equiv \hat{U}_m^{1,0}\hat{U}_n^{2,1}\hat{U}_m^{1,0}$ which is represented diagrammatically as follows:

VOLUME 74, NUMBER 20   PHYSICAL REVIEW LETTERS   15 MAY 1995

$$
\begin{array}{llll}
& \hat{U}_m^{1,0} & \hat{U}_n^{2,1} & \hat{U}_m^{1,0} \\
|g\rangle_m|g\rangle_n|0\rangle & \longrightarrow \ |g\rangle_m|g\rangle_n|0\rangle & \longrightarrow \ |g\rangle_m|g\rangle_n|0\rangle & \longrightarrow \ |g\rangle_m|g\rangle_n|0\rangle, \\
|g\rangle_m|e_0\rangle_n|0\rangle & \longrightarrow \ |g\rangle_m|e_0\rangle_n|0\rangle & \longrightarrow \ |g\rangle_m|e_0\rangle_n|0\rangle & \longrightarrow \ |g\rangle_m|e_0\rangle_n|0\rangle, \\
|e_0\rangle_m|g\rangle_n|0\rangle & \longrightarrow \ -i|g\rangle_m|g\rangle_n|1\rangle & \longrightarrow \ i|g\rangle_m|g\rangle_n|1\rangle & \longrightarrow \ |e_0\rangle_m|g\rangle_n|0\rangle, \\
|e_0\rangle_m|e_0\rangle_n|0\rangle & \longrightarrow \ -i|g\rangle_m|e_0\rangle_n|1\rangle & \longrightarrow \ -i|g\rangle_m|e_0\rangle_n|1\rangle & \longrightarrow \ -|e_0\rangle_m|e_0\rangle_n|0\rangle.
\end{array}
\tag{3}
$$

The effect of this interaction is to change the sign of the state only when both ions are initially excited. Note that the state of the CM mode is restored to the vacuum state $|0\rangle$ after the process. Equation (3) is equivalent to a controlled-NOT gate. To show this, let us denote by $|\pm\rangle = (|g\rangle \pm |e_0\rangle)/\sqrt{2}$. Then, this procedure can be summarized as $|g\rangle_m|\pm\rangle_n \to |g\rangle_m|\pm\rangle_n$ and $|e_0\rangle_m|\pm\rangle_n \to |e_0\rangle_m|\mp\rangle_n$. With an appropriate individual (one-bit) rotation applied to the nth ion this procedure then yields the controlled-NOT. These individual rotations acting on a single ion (without modifying the CM motion) can be performed using a laser frequency on resonance with the internal transition ($\delta_n = 0$), polarization $q = 0$, and with the equilibrium position of the ion coinciding with the antinode of the laser standing wave. In this case, the Hamiltonian is

$$
\hat{H}_n = (\Omega/2)[|e_0\rangle_n\langle g|e^{-i\phi} + |g\rangle_n\langle e_0|e^{i\phi}].
\tag{4}
$$

For an interaction time $t = k\pi/\Omega$ (i.e., using a $k\pi$ pulse), this process is described by the following unitary evolution operator:

$$
\hat{V}_n^k(\phi) = \exp\left[-ik\frac{\pi}{2}(|e_0\rangle_n\langle g|e^{-i\phi} + \text{H.c.})\right],
\tag{5}
$$

so that

$$
|g\rangle_n \longrightarrow \cos(k\pi/2)|g\rangle_n - ie^{i\phi}\sin(k\pi/2)|e_0\rangle_n,
$$
$$
|e_0\rangle_n \longrightarrow \cos(k\pi/2)|e_0\rangle_n - ie^{-i\phi}\sin(k\pi/2)|g\rangle_n.
$$

Thus the complete controlled-NOT gate for the states $|\epsilon_m\rangle|\epsilon_n\rangle$ ($\epsilon_{m,n} = g, e_0$) is given by $\hat{C}_{mn} = \hat{V}_n^{1/2}(\frac{\pi}{2})\hat{U}_{m,n}\hat{V}_n^{1/2}(-\frac{\pi}{2})$ [15].

Nonlocal three-bit gates can be implemented in a similar way between ions n, m, and l. The process takes place in five steps: (j) Same as (i); (jj) same as (ii), but with a π pulse; (jjj) same as step (ii) but with ion l; (jv) same as (jj); (v) same as (j). The corresponding unitary operation for this process is $\hat{U}_m^{1,0}\hat{U}_n^{1,1}\hat{U}_l^{2,1}\hat{U}_n^{1,1}\hat{U}_m^{1,0}$. This procedure only changes the sign of the state if all three ions were initially excited. One can easily generalize this procedure to the case of many ions. For example, a (control)^p-NOT gate acting on ions $n_1, n_2, \ldots, n_q$ corresponds to the unitary evolution

$$
\hat{V}_p^{\frac{1}{2}}\left(\frac{\pi}{2}\right)\hat{U}_{n_1}^{1,0}\left[\prod_{j=2}^{p-1}\hat{U}_{n_j}^{1,1}\right]\hat{U}_{n_p}^{2,1}\left[\prod_{j=p-1}^{2}\hat{U}_{n_j}^{1,1}\right]\hat{U}_{n_1}^{1,0}\hat{V}_p^{\frac{1}{2}}\left(-\frac{\pi}{2}\right).
$$

Using similar ideas with different laser phases and interaction times one can implement other n-bit gates [8].

In summary, the two key elements behind the above implementation of quantum gates are as follows. First, non-local entanglement between individual qubits is achieved by transferring the internal atomic coherence to and from the CM motion shared by all the ions ($\hat{U}_n^{k=1,q=0}$). Second, as an intermediate step we "hide atomic amplitudes" corresponding to the qubits in a third internal atomic level $|e_1\rangle$ ($\hat{U}_n^{k=1,q=1}$), and induce 2π rotations via this state to selectively change the sign of atomic amplitudes ($\hat{U}_n^{k=2,q=1}$). We note that no population is left in these auxiliary atomic and CM levels after the complete gate operation. Any population left in these states is an indication of an imprecise realization. This could be used to implement an error detection scheme by probing the population of these intermediate states, for example, with a laser inducing fluorescence after each gate operation [16].

The core of Shor's factorization scheme [2] is the high efficiency of a QC to find the period r of a given function by doing a discrete Fourier transform (FT) on a periodic state vector of the form $|\Psi\rangle \propto \sum_l |lr + k\rangle$. Here k is an integer number and $l = 0, \ldots, [(2^N - k)/r]$ with $[\ldots]$ the integer part. The FT is defined by the operation

$$
\hat{FT}|x\rangle = 1/\sqrt{2^N}\sum_y e^{2\pi ixy/2^N}|y\rangle
$$

on the quantum registers. This FT can be decomposed into a sequence of one- and two-bit operations [17,18]. The probability to measure the state $|y\rangle$ of the quantum register is then $P_y = |\langle y|\hat{FT}|\Psi\rangle|^2$. Shor has shown that this measurement gives with high probability an outcome that allows one to calculate r.

To show the capabilities of an ion trap as a QC, and to analyze how experimental uncertainties may affect the final results [6], we have simulated the above scheme on a (digital) computer. Figure 2 shows a comparison

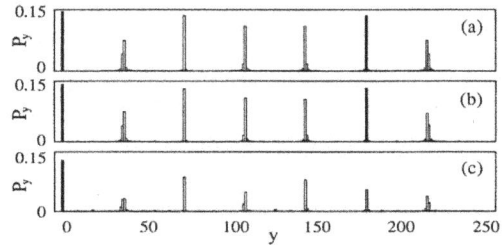

FIG. 2. Probability distribution P_y after FT (see text): (a) exact, (b) ion trap simulation, (c) simulation with 5% errors.

between the exact results [Fig. 2(a)] for P_y and the ion trap simulation [Figs. 2(b) and 2(c)] for a state with $k = 4$, $r = 7$, and eight ions. The existence of peaks in this spectrum (separated by $\simeq 2^N/r = 256/7$) allows one to determine the period r. Similar to Ref. [17] one can show that this probability distribution P_y can be obtained from the physical process corresponding to the sequence of operations $\hat{V}_0 \hat{W}_0 \hat{W}_1 \hat{V}_1 \cdots \hat{W}_{N-2} \hat{V}_{N-1}$. Here $\hat{W}_n = \hat{W}_n^{N-1} \hat{W}_n^{N-2} \cdots \hat{W}_n^{n+1}$ is a sequence of two-bit operations $\hat{W}_n^m = \hat{U}_m^{1,0}[\pi(1 - 2^{(n-m)})] \hat{U}_n^{1,1}[\pi(1 - 2^{(n-m)})] \hat{U}_n^{1,1} \hat{U}_m^{1,0}$ $(n < m)$, and $\hat{V}_n \equiv \hat{V}_n^{1/2}(-\pi/2)$ is a one-ion rotation [see (2) and (5)]. The specific form of the pulse sequence can be directly deduced from the definition of the operators W, and requires two- and one-bit gates between the ions. The simulation has been performed with the full Hamiltonian (to all orders in the Lamb-Dicke expansion) for $N = 8$ Ba^+ ions in a trap with $\nu_x = 2\pi \times 50$ kHz. The Rabi frequencies have been chosen as follows: $\Omega = 2\pi \times 1.5$ kHz for resonant excitations (at the antinode) and $\Omega = 2\pi \times 15$ kHz for off-resonant excitations (at the node). The rest of the parameters correspond to those of the Ba^+ ions. As shown in Fig. 2(b), with these realistic parameters, the result is nearly indistinguishable from the exact one. From our numerical simulations we could see that this result can even be improved by increasing the trap frequency (or decreasing the Rabi frequencies), in agreement with a perturbation theory analysis for the terms neglected in (1) and (4). Note that the total time required for the whole operation is about 35 ms, much smaller than the decoherence time due to spontaneous emission (the lifetime of the metastable state of Ba^+ is about 45 s, so that the decoherence time is $\simeq 6$ s). To analyze how experimental uncertainties affect the final results we have carried out numerical simulations assuming a 5% error in *all* the interaction times involved in the operation, 1 kHz of error in *all* the laser detunings, and a 5% $\pi/2$ error in all the angles in the problem (situation of the standing waves with respect to the position of the ions, and phases of the lasers). Figure 2(c) shows that even with all these errors the peaks in the distribution are still maintained, and the system of ions is remarkably robust to perform quantum computations.

Apart from one- and two-bit operations, (5) and (2), one can also prepare the most general entangled state of N ions [9,19]. For example, the maximal entangled state

$$|\Psi\rangle = 1/\sqrt{2}(|-\rangle_{N-1} \cdots |-\rangle_1$$
$$\times |-\rangle_0 - |+\rangle_{N-1} \cdots |+\rangle_1 |+\rangle_0)$$

can be obtained starting from $|g\rangle_{N-1}|g\rangle_{N-2} \cdots |g\rangle_0$ (as obtained after sideband cooling), by using the operations $\hat{V}_0 \hat{U}_{N-1,0} \cdots \hat{U}_{2,0} \hat{U}_{1,0} \hat{V}_{N-1} \cdots \hat{V}_1 \hat{V}_0$ [18].

In summary, linear ion traps are well suited to implement a QC. This is due to the negligible decoherence in these systems [9], as well as the possibility to manipulate the internal and CM degrees of freedom with external fields,

and to perform efficient state measurements. We have shown how to implement n-bit gates between n arbitrary ions, and have illustrated the performance of such a system with a numerical simulation. We believe that the present system provides a realistic implementation of a QC which can be built with present or planned technology.

We thank R. Blatt, A. Ekert, M. Lewenstein, and D. Wineland for helpful comments. This work was supported by the Austrian Science Foundation.

*Permanent address: Departamento de Fisica, Universidad de Castilla-La Mancha, 13071 Ciudad Real, Spain.

[1] For a review, see A. Ekert, in Proc. ICAP '94. edited by S. Smith, C. Wieman, and D. Wineland (to be published).

[2] P. W. Shor, in *Proceedings of the 35th Annual Symposium on the Foundations of Computer Science, Los Alamitos, CA* (IEEE Computer Society Press, New York, 1994), p. 124.

[3] D. Deutsch, Proc. R. Soc. London A **425**, 73 (1989).

[4] T. Sleator and H. Weinfurter, Phys. Rev. Lett. **74**, 4087 (1995).

[5] K. Obermayer, W. G. Teich, and G. Mahler, Phys. Rev. B **37**, 8096 (1988); S. Lloyd, Science **261**, 1569 (1993); D. P. Di Vincenzo, Phys. Rev. A **50**, 1015 (1995).

[6] R. Landauer, Proc. R. Soc. London A (to be published); W. G. Unruh (to be publishd).

[7] M. G. Raizen *et al.*, Phys. Rev. A **45**, 6493 (1992); H. Walther, Adv. At. Mol. Opt. Phys. **32**, 379 (1994).

[8] Although a (controlled)^n-NOT can be decomposed into a finite number of controlled-NOT gates plus one-bit rotations, this may require many operations. [H. Weinfurter (private communication)] Thus a direct implementation of the (controlled)^n-NOT gate may be interesting from a practical point of view.

[9] D. J. Wineland *et al.*, Phys. Rev. A **50**, 67 (1994); **46**, R6797 (1992).

[10] R. Blatt, in Proc. ICAP '94, Ref. [1].

[11] W. Nagourney *et al.*, Phys. Rev. Lett. **56**, 2797 (1986); J. C. Bergquist *et al.*, *ibid.* **56**, 1699 (1986); Th. Sauter *et al.*, *ibid.* **56**, 1696 (1986).

[12] A similar scheme can be used with traveling wave configurations. However, the standing wave minimizes the effects of unwanted transitions; see Ref. [14].

[13] F. Diedrich *et al.*, Phys. Rev. Lett. **62**, 403 (1989); here only the CM has to be cooled to the ground state.

[14] J. I. Cirac *et al.*, Phys. Rev. Lett. **70**, 762 (1993).

[15] The two-bit gate (3) (instead of the controlled-NOT) together with single bit rotations are sufficient to generate arbitrary unitary operations.

[16] Nonobservation of fluorescence corresponds to a projection of the state vector on $|g\rangle$, $|e_0\rangle$. This might be the basis of a partial error correction scheme.

[17] D. Coppersmith, IBM Research Report No. RC19642, 1994.

[18] FT and the preparation of general entangled states could be performed more efficiently using general n-bit gates (instead of a sequence of two-bit gates).

[19] D. M. Greenberger *et al.*, Am. J. Phys. **58**, 1131 (1990); see also N. D. Mermin, *ibid.* **58**, 8 (1990).

VOLUME 81, NUMBER 15   PHYSICAL REVIEW LETTERS   12 OCTOBER 1998

Cold Bosonic Atoms in Optical Lattices

D. Jaksch,[1,2] C. Bruder,[1,3] J. I. Cirac,[1,2] C. W. Gardiner,[1,4] and P. Zoller[1,2]

[1]*Institute for Theoretical Physics, University of Santa Barbara, Santa Barbara, California 93106-4030*
[2]*Institut für Theoretische Physik, Universität Innsbruck, A-6020 Innsbruck, Austria*
[3]*Institut für Theoretische Festkörperphysik, Universität Karlsruhe, D-76128 Karlsruhe, Germany*
[4]*School of Chemical and Physical Sciences, Victoria University, Wellington, New Zealand*
(Received 26 May 1998)

The dynamics of an ultracold dilute gas of bosonic atoms in an optical lattice can be described by a Bose-Hubbard model where the system parameters are controlled by laser light. We study the continuous (zero temperature) quantum phase transition from the superfluid to the Mott insulator phase induced by varying the depth of the optical potential, where the Mott insulator phase corresponds to a commensurate filling of the lattice ("optical crystal"). Examples for formation of Mott structures in optical lattices with a superimposed harmonic trap and in optical superlattices are presented. [S0031-9007(98)07267-6]

PACS numbers: 32.80.Pj, 03.75.Fi, 71.35.Lk

Optical lattices—arrays of microscopic potentials induced by the ac Stark effect of interfering laser beams—can be used to confine cold atoms [1–7]. The quantized motion of such atoms is described by the vibrational motion within an individual well and the tunneling between neighboring wells, leading to a spectrum describable as a band structure [3]. Near-resonant optical lattices, where dissipation associated with optical pumping produces cooling, have given filling factors of about one atom per ten lattice sites [1,6]. Higher filling factors will require lower temperatures, and hence will also require minimization of the optical dissipation. This can be achieved in a far-detuned optical lattice (especially with blue detuning), where photon scattering times of many minutes have been demonstrated [2]. Thus the lattice then behaves as a conservative potential, which could be loaded with a Bose condensed atomic vapor [8,9], for which present densities would correspond to tens of atoms per lattice site.

In this Letter we will study the dynamics of ultracold bosonic atoms loaded in an optical lattice. We will show that the dynamics of the bosonic atoms on the optical lattices realizes a Bose-Hubbard model (BHM) [10–16], describing the hopping of bosonic atoms between the lowest vibrational states of the optical lattice sites, the unique feature being the full control of the system's parameters by the laser parameters and configurations.

The BHM predicts phase transition from a superfluid (SF) phase to a Mott insulator (MI) at low temperatures and with increasing ratio of the on site interaction U (due to repulsion of atoms) to the tunneling matrix element J [10]. In the case of optical lattices this ratio can be varied by changing the laser intensity: with increasing depth of the optical potential the atomic wave function becomes more and more localized and the on site interaction increases, while at the same time the tunneling matrix element is reduced. In the MI phase the density (occupation number per site) is pinned at integer $n = 1, 2, \ldots$, corresponding to a commensurate filling of

the lattice, and thus represents an *optical crystal* with diagonal long range order with the period imposed by the laser light. The nature of the MI phase is reflected in the existence of a finite gap U in the excitation spectrum.

Our starting point is the Hamilton operator for bosonic atoms in an external trapping potential

$$H = \int d^3x \, \psi^\dagger(\mathbf{x}) \left(-\frac{\hbar^2}{2m} \nabla^2 + V_0(\mathbf{x}) + V_T(\mathbf{x}) \right) \psi(\mathbf{x})$$
$$+ \frac{1}{2} \frac{4\pi a_s \hbar^2}{m} \int d^3x \, \psi^\dagger(\mathbf{x}) \psi^\dagger(\mathbf{x}) \psi(\mathbf{x}) \psi(\mathbf{x}), \quad (1)$$

with $\psi(\mathbf{x})$ a boson field operator for atoms in a given internal atomic state, $V_0(\mathbf{x})$ is the optical lattice potential, and $V_T(\mathbf{x})$ describes an additional (slowly varying) external trapping potential, e.g., a magnetic trap (see Fig. 1a). In the simplest case, the optical lattice potential has the form $V_0(\mathbf{x}) = \sum_{j=1}^{3} V_{j0} \sin^2(k x_j)$ with wave vectors $k = 2\pi/\lambda$ and λ the wavelength of the laser light, corresponding to a lattice period $a = \lambda/2$. V_0 is proportional to the dynamic atomic polarizability times the laser intensity. The interaction potential between the

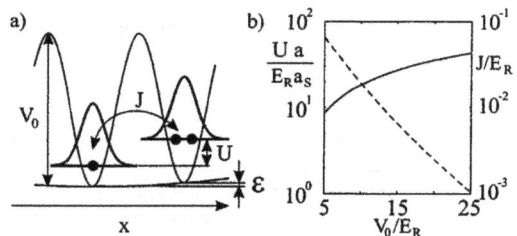

FIG. 1. (a) Realization of the BHM in an optical lattice (see text). The offset of the bottoms of the wells indicates a trapping potential V_T. (b) Plot of the scaled on site interaction U/E_R multiplied by a/a_s ($\gg 1$) (solid line; axis on left-hand side of graph) and J/E_R (dashed line; axis on right-hand side of graph) as a function of $V_0/E_R \equiv V_{x,y,z0}/E_R$ (3D lattice).

VOLUME 81, NUMBER 15   PHYSICAL REVIEW LETTERS   12 OCTOBER 1998

atoms is approximated by a short-range pseudopotential with a_s the s-wave scattering length and m the mass of the atoms. For single atoms the energy eigenstates are Bloch wave functions, and an appropriate superposition of Bloch states yields a set of Wannier functions which are well localized on the individual lattice sites. We assume the energies involved in the system dynamics to be small compared to excitation energies to the second band. Expanding the field operators in the Wannier basis and keeping only the lowest vibrational states, $\psi(\mathbf{x}) = \sum_i b_i w(\mathbf{x} - \mathbf{x}_i)$, Eq. (1) reduces to the Bose-Hubbard Hamiltonian

$$H = -J \sum_{\langle i,j \rangle} b_i^\dagger b_j + \sum_i \epsilon_i \hat{n}_i + \frac{1}{2} U \sum_i \hat{n}_i (\hat{n}_i - 1),$$
(2)

where the operators $\hat{n}_i = b_i^\dagger b_i$ count the number of bosonic atoms at lattice site i; the annihilation and creation operators b_i and $b_i^\dagger$ obey the canonical commutation relations $[b_i, b_j^\dagger] = \delta_{ij}$. The parameters $U = 4\pi a_s \hbar^2 \int d^3 x |w(\mathbf{x})|^4 / m$ correspond to the strength of the on site repulsion of two atoms on the lattice site i, $J = \int d^3 x\, w^*(\mathbf{x} - \mathbf{x}_i) [-\frac{\hbar^2}{2m} \nabla^2 + V_0(\mathbf{x})] w(\mathbf{x} - \mathbf{x}_j)$ is the hopping matrix element between adjacent sites i, j, and $\epsilon_i = \int d^3 x\, V_T(\mathbf{x}) |w(\mathbf{x} - \mathbf{x}_i)|^2 \approx V_T(\mathbf{x}_i)$ describes an energy offset of each lattice site.

For a given optical potential J and U are readily evaluated numerically. For the optical potential given above the Wannier functions can be written as products $w(\mathbf{x}) = w(x)w(y)w(z)$ which can be determined from a one-dimensional band structure calculation. Figure 1b shows U and J as a function of V_0 in units of the recoil energy $E_R = \hbar^2 k^2 / 2m$. Both the next-nearest neighbor amplitudes and the nearest-neighbor repulsion are typically 2 orders of magnitude smaller and can thus be neglected. Qualitative insight into the dependence of these parameters is obtained in a harmonic approximation expanding around the minima of the potential wells. The oscillation frequencies in the wells are $\nu_j = \sqrt{4 E_R V_{j0}} / \hbar$ which gives the separation to the first excited Bloch band. The oscillator ground state wave function of size $a_{j0} = \sqrt{\hbar / m \nu_j}$ allows us to obtain an estimate for the on site interaction $U = 2\hbar \bar{\nu}(a_s / \bar{a}_0) / \sqrt{2\pi}$ with the bar indicating geometric means. Consistency of our model requires $a_s \ll a_{j0} \ll \lambda/2$ and $\Delta E_i = \frac{1}{2} U n_i(n_i - 1) \ll \hbar \nu_j$. The first set of inequalities follows from the pseudopotential approximation and our requirement of a (large) energy separation from the first excited band. The second inequality expresses the requirement that the on site interaction associated with the presence of n_i particles at site i, which in our model is calculated in perturbation theory, must be much smaller than the excitation energy to the next band. These inequalities are readily satisfied in practice.

According to mean-field theory (MFT) in the homogeneous case [10,11] (see also [14]) the critical value of

the MI-SF transition for the phase $n = 1$ is at the critical value $U/zJ \approx 5.8$ with $z = 2d$ the number of nearest neighbors. According to Fig. 1b this parameter regime is accessible by varying V_0 in the regime of a few tens of recoil energies. As an example, for sodium [9] we have $E_R / \hbar = 2\pi \times 8.9$ kHz for a red detuned laser with $\lambda = 985$ nm, and the critical values for the first MI phase in 1D, 2D, and 3D are given by $V_{x0} = 10.8$, $V_{x,y0} = 14.4$, and $V_{x,y,z0} = 16.5 E_R$, and we assumed in 1D $V_{y,z0} = 25 E_R$ for the y and z directions in order to suppress tunneling in these other dimensions, and $V_{z0} = 25 E_R$ for 2D. For $V_0 = 15$ we have $U = 0.15$ and $J = 0.07$ in units of E_R. For a blue detuning [9] $\lambda = 514$ nm we find $E_R / \hbar = 2\pi \times 32$ kHz and the corresponding values are $V_{x0} = 8.4$, $V_{x,y0} = 11.9$, and $V_0 = 14.1$; $U = 0.2$, $J = 0.02$ for $V_0 = 10$ in units of E_R. For $V_0 \approx 10 E_R$ the single particle density at the center of the optical potential wells will be of the order of $1/a_0^3 \approx 10^{15}$ cm^{-3}. Thus we must discuss the role of collisions between ground state atoms (in the presence of a laser field) as a loss and decoherence mechanism [17]. This question is directly related to the problem of collisional loss of Bose-Einstein condensates in optical traps as studied in [9]. We emphasize that in the Mott phase with a single particle per site ($n = 1$) two and more particle loss channels are absent. For a MI phase with $n = 2$ there will be two particle losses: if we take as an order of magnitude the numbers published in Ref. [18] we estimate the corresponding lifetime to be >10 s. For $n = 3$ the lifetime due to three atom losses [18] will be of the order of $1/10$ s.

We have performed mean-field calculations for 1D and 2D configurations, as well as an exact diagonalization of the BH Hamiltonian in 1D to illustrate the formation of the Mott insulator phase in optical lattices, in particular, for the inhomogeneous case. Our mean-field calculations are based on a Gutzwiller ansatz for the ground state wave function $|\Psi_{\mathrm{MF}}\rangle = \prod_i |\phi_i\rangle$ with $|\phi_i\rangle = \sum_{n=0}^\infty f_n^{(i)} |n\rangle_i$, where $|n\rangle_i$ denotes the Fock state with n atoms at site i [11]. We minimize the expectation value of the Hamiltonian,

$$\langle \Psi_{\mathrm{MF}} | H | \Psi_{\mathrm{MF}} \rangle - \mu \langle \Psi_{\mathrm{MF}} | \sum_i \hat{n}_i | \Psi_{\mathrm{MF}} \rangle \to \min,$$
(3)

with respect to the coefficients $f_n^{(i)}$. The Lagrange multiplier μ enforces a given mean particle number $N = \sum_i \langle n_i \rangle$. This corresponds to a calculation in the grand canonical ensemble with chemical potential μ at temperature $T = 0$. A MI phase is indicated by solutions in the form of single Fock states, $|\phi_i\rangle \to |n_i\rangle_i$. A signature of a MI phase is integer occupation number (density) $\rho_i = \langle \hat{n}_i \rangle$ and fluctuations, $\sigma_i^2 = (\langle \hat{n}_i^2 \rangle - \langle \hat{n}_i \rangle^2) / \langle \hat{n}_i \rangle \to 0$. Solutions in the form of superposition of Fock states result in a mean-field $\phi_i = \langle b_i \rangle \neq 0$, indicating the presence of a SF component. The angular brackets indicate an average in the mean-field state. In the homogeneous case ($\epsilon_i = 0$) the phase diagram in the

VOLUME 81, NUMBER 15   PHYSICAL REVIEW LETTERS   12 OCTOBER 1998

J-μ plane consists of a series of lobes [10]. Inside the lobes (i.e., for J small in comparison with the on site repulsion energy U) the system is a Mott phase; outside it is superfluid.

In Fig. 2a we plot the density $\rho(x, y)$ and the superfluid component $|\phi(x, y)|^2$ in an optical lattice with a superimposed isotropic harmonic potential at the lattice points $(x/a, y/a) = (i, j)$ $(i, j = 0, \pm 1, \dots,)$. Figure 2a shows a MI phase with two atoms per site at the center of the trap ($\rho = 2$) surrounded by a Mott phase with a single atom ($\rho = 1$) and superfluid rings between the MI phases. For smaller values of the chemical potential only a single Mott phase would exist at the trap center. Qualitatively, this behavior is readily understood on the basis of the phase diagram in the homogeneous case [10] if we note that the offset $\epsilon_i = V_T(\mathbf{x}_i)$ leads to an effective local chemical potential $\mu - \epsilon_i$.

By use of interfering laser beams at different angles [4], one can produce a *superlattice*, in which the offset of the optical potential is modulated periodically in space on a scale larger than the lattice period. Figures 2b and 2c show the density $\rho(x, y)$ and the scaled density fluctuations $\sigma(x, y)$ of Mott structures formed in a superlattice.

With increasing μ we first find a Mott structure at the bottom of the superlattice potential, until the atoms are no longer confined to a particular well of the superlattice but form bridges connecting the superlattice wells.

In general, specific Mott structures can be designed by an appropriate choice of the laser configurations. An experimental signature to detect the Mott state is observation of reduced density-density fluctuations [see $\sigma(x, y)$ in Fig. 2]. This can be monitored directly in light scattering. Alternatively, the MI phase can be detected spectroscopically by observing the gapped particle-hole excitations.

In 1D and for systems with few atoms per superlattice well we expect fluctuations to be important, and the application of MFT becomes questionable. On the other hand, in this limit it is straightforward to diagonalize the Bose-Hubbard Hamiltonian exactly. Figure 3 is a plot of the density and the number fluctuations for the exact ground state for $N = 5$ atoms as a function of V_{x0}. With increasing V_{x0} the density shows a clear transition to the MI phase $\rho = 1$, even for this very small sample. The number fluctuations are suppressed in the MI phase but remain finite. The phase transition (which according to MFT in the homogeneous limit is expected for $V_0 = 7.4 E_R$) is smeared out, and fluctuations are strongly suppressed only for larger values of V_{x0}. Qualitatively, the mean-field theory for the inhomogeneous case agrees well with the exact calculations, even for these small systems. Figure 3 can be viewed as an adiabatic transfer into the MI phase as the laser intensity is varied slowly as a function of time.

The atomic level scheme of Fig. 1 allows only one adjustable parameter, the depth of the optical potential V_0. To adjust the tunneling matrix element J independently of the on site interaction U we can employ atomic configurations with two internal ground state levels $|g_1\rangle$ and $|g_2\rangle$, which are connected by an off-resonant Raman transition (Fig. 4a).

We assume that the two internal states move in optical potentials which are shifted relative to each other by $\lambda/4$, as is the case when they have polarizabilities of opposite sign. Expanding the bosonic field operators for the two

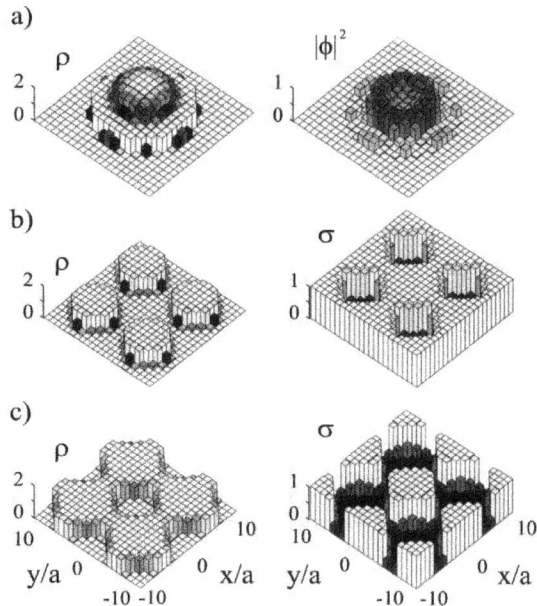

FIG. 2. (a) MI and SF phases in an optical potential and harmonic trap in 2D. Parameters: $U = 35J$, $V_T(x, y) = J(x^2 + y^2)/a^2$, and $\mu = 50J$. Density $\rho(x, y)$ (left plot) and superfluid density $|\phi(x, y)|^2$ (right plot). (b) Superlattice in 2D. Density $\rho(x, y)$ (left plot) and fluctuations $\sigma(x, y)$ (right plot). Parameters: $U = 45J$, $V_T(x, y) = 30J [\sin^2(\pi x/11a) + \sin^2(\pi y/11a)]$, and $\mu = 25J$. (c) Same as (b) with $\mu = 35J$. Four superlattice wells are shown.

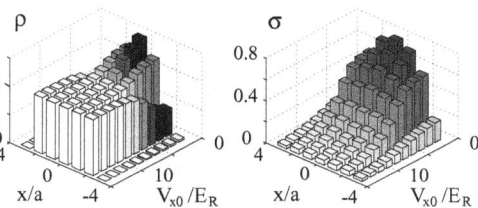

FIG. 3. Density ρ and fluctuations σ for the exact ground state in 1D for $N = 5$ atoms in a harmonic well as a function of V_{x0}/E_R for seven lattice cells. The parameters are $a_s/a = 1.1 \ 10^{-2}$ (corresponding to Na and $\lambda = 514$ nm, $V_{y0} = V_{z0} = 40 E_R$) and $V_T(x) = 0.06 \ E_R(x/a)^2$.

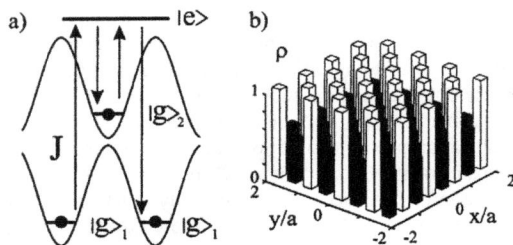

FIG. 4. (a) Atomic level scheme (see text). (b) Checkerboard pattern with a MI phase on one sublattice and a SF on the other obtained in MFT for the two species BHM, with parameters: $\mu = 25J$, $U_{aa} = U_{bb} = 45J$, $U_{ab} = 0$, $\delta = -25J$, and $\epsilon_i = 0$.

internal states we obtain a two-species Bose-Hubbard Hamiltonian

$$H = -\left(J \sum_{\langle i,j \rangle} a_i^\dagger b_j + \text{H.c.} \right) + \sum_i \epsilon_i a_i^\dagger a_i$$
$$+ \sum_j (\epsilon_j - \delta) b_j^\dagger b_j + \frac{U_{aa}}{2} \sum_i a_i^{\dagger 2} a_i^2$$
$$+ U_{ab} \sum_{\langle i,j \rangle} a_i^\dagger a_i b_j^\dagger b_j + \frac{U_{bb}}{2} \sum_j b_j^{\dagger 2} b_j^2, \quad (4)$$

with a_i and b_i bosonic destruction operators referring to atoms in the internal states $|g_1\rangle$ and $|g_2\rangle$, respectively. The first term in the Hamiltonian describes the Raman induced hopping between adjacent cells with coupling $J = \frac{1}{2} \int d^3\mathbf{x} \, w_a(\mathbf{x})^* \Omega_{\text{eff}}(\mathbf{x}) w_b(\mathbf{x} - \lambda/4)$, where Ω_{eff} is the effective two-photon Rabi frequency (including a possible phase). Direct tunneling has been neglected. The second and third terms contain offsets due to a trapping potential, and, in addition, a Raman detuning term $-\delta$ for atoms in the state $|g_2\rangle$. The second and third lines contain on site interactions of atoms a and b described by U_{aa} and U_{bb}, and a nearest-neighbor interaction U_{ab} whose value depends on the overlap of the Wannier functions between a and b. A Raman detuning δ shifts the chemical potential of species b relative to a. We can adjust the value of this detuning to generate checkerboard patterns, e.g., a MI phase of species a and a Mott phase of species b can coexist with different site occupation numbers. As an example, Fig. 4b plots the density $\rho(x, y)$ for a specific 2D homogeneous situation where a MI phase $|g_1\rangle$ coexists with a superfluid component in $|g_2\rangle$.

While the present discussion has emphasized periodic (ordered) Bose systems, adding a further optical potential with incommensurate lattice spacing allows the realization of a (pseudo)random potential [5] which leads to the study of disordered Bose systems and appearance of a Bose glass phase [10,15]. A study of the growth and fluctuations of the MI phase due to coupling to a finite temperature particle reservoir based on a master equation

treatment [19] will be presented elsewhere. The ability to manipulate both the lattice and the system parameters in our realization of a Bose-Hubbard model brings a new aspect to condensed matter physics: models and simplifying assumptions may be systematically investigated using the experimental techniques of quantum optics.

The authors thank the members of the BEC98 program at ITP UCSB for discussions. The work was supported in part by the NSF, the Austrian Science Foundation, EU TMR networks, and the Marsden Contract No. PVT-603.

[1] G. Raithel *et al.*, Phys. Rev. Lett. **78**, 630 (1997); T. Müller-Seydlitz *et al.*, *ibid.* **78**, 1038 (1997); S. E. Hamann *et al.*, *ibid.* **80**, 4149 (1998), and references therein.

[2] S. Friebel *et al.*, Phys. Rev. A **57**, R20 (1998).

[3] M. Raizen, C. Salomon, and Q. Niu, Phys. Today **50**, No. 7, 30 (1997).

[4] L. Guidoni and P. Verkerk, Phys. Rev. A **57**, R1501 (1998), and references therein.

[5] L. Guidoni *et al.*, Phys. Rev. Lett. **79**, 3363 (1997).

[6] K. I. Petsas, A. B. Coates, and G. Grynberg, Phys. Rev. A **50**, 5173 (1994).

[7] I. H. Deutsch and P. S. Jessen, Phys. Rev. A **57**, 1972 (1998).

[8] M. Anderson *et al.*, Science **269**, 198 (1995); K. B. Davis *et al.*, Phys. Rev. Lett. **75**, 3969 (1995); C. C. Bradley *et al.*, Phys. Rev. Lett. **75**, 1687 (1995); see the BEC homepage http://amo.phy.gasou.edu/bec.html

[9] D. M. Stamper-Kurn *et al.*, Phys. Rev. Lett. **80**, 2027 (1998); S. Inouye *et al.*, Nature (London) **392**, 151 (1998).

[10] M. P. A. Fisher *et al.*, Phys. Rev. B **40**, 546 (1989).

[11] W. Krauth, M. Caffarel, and J.-P. Bouchard, Phys. Rev. B **45**, 3137 (1992); K. Sheshadri *et al.*, Europhys. Lett. **22**, 257 (1993).

[12] A. P. Kampf and G. T. Zimanyi, Phys. Rev. B **47**, 279 (1993).

[13] C. Bruder, R. Fazio, and G. Schön, Phys. Rev. B **47**, 342 (1993); A. van Otterlo *et al.*, Phys. Rev. B **52**, 16 176 (1995).

[14] J. K. Freericks and H. Monien, Europhys. Lett. **26**, 545 (1994).

[15] W. Krauth, N. Trivedi, and D. Ceperley, Phys. Rev. Lett. **67**, 2307 (1991); N. V. Prokofev, B. V. Svistunov, and I. S. Tupitsyn, Phys. Lett. A **238**, 253 (1998).

[16] For other experimental realizations such as granular superconductors, Josephson junction arrays, and helium films, see references cited in [10,13].

[17] For a discussion of various aspects of inelastic processes in the presence of light, see P. O. Fedichev *et al.*, Phys. Rev. Lett. **77**, 2913 (1996); K.-A. Suominen *et al.*, Phys. Rev. A **57**, 3724 (1998).

[18] For decay rates of a Rb condensate in a magnetic trap, see E. A. Burt *et al.*, Phys. Rev. Lett. **79**, 337 (1997).

[19] C. W. Gardiner and P. Zoller, Phys. Rev. A **58**, 536 (1998).

LIST OF PUBLICATIONS

1. P. Zoller, Fokker-Planck equation treatment of atomic relaxation and resonance fluorescence in phase-modulated laser light, *J. Phys. B: At. Mol. Opt. Phys.* **10**, L321–L324 (1977) (ID: 377769)

2. P. Zoller and F. Ehlotzky, Resonance fluorescence in modulated laser fields, *J. Phys. B: At. Mol. Opt. Phys.* **10**, 3023–3032 (1977) (ID: 377764)

3. P. Zoller, Atomic relaxation and resonance fluorescence in intensity and phase-fluctuating laser light, *J. Phys. B: At. Mol. Opt. Phys.* **11**, 2825–2832 (1978) (ID: 377755)

4. P. Zoller and F. Ehlotzky, Resonance fluorescence in phase-frequency modulated laser fields, *Zeitschrift für Physik A* 285, 245–247 (1978). (ID: 377744)

5. P. Zoller and P. Lambropoulos, Non-Lorentzian laser lineshapes in intense field-atom interaction *J. Phys. B: At. Mol. Opt. Phys.* **12**, L547–L551 (1979) (ID: 377691)

6. A. Georges, P. Lambropoulos and P. Zoller, Saturation and stark splitting of resonant transitions in strong chaotic fields of arbitrary bandwidth, *Phys. Rev. Lett.* **42**, 1609–1613 (1979). (ID: 377687)

7. P. Zoller, ac Stark splitting in double optical resonance and resonance fluorescence by a nonmonochromatic chaotic field, *Phys. Rev. A* **20**, 1019–1031 (1979). (ID: 377692)

8. P. Zoller, Resonant multiphoton ionization by finite-bandwidth chaotic fields, *Phys. Rev. A* **20**, 1295 (1979). (ID: 377476)

9. P. Zoller, Saturation of two-level atoms in chaotic fields, *Phys. Rev. A* **20**, 2420–2423 (1979). (ID: 377475)

10. G. Alber and P. Zoller, Light statistical dependence of saturated two-photon transitions, *J. Phys. B: At. Mol. Opt. Phys.* **13**, 4567–4576 (1980) (ID: 377690)

11. S. N. Dixit, A. Georges, P. Lambropoulos and P. Zoller, Comments on the short-time behaviour of multiphoton ionisation, *J. Phys. B: At. Mol. Opt. Phys.* **13**, L157–L158 (1980) (ID: 377689)

12. C. Leubner and P. Zoller, Gauge invariant interpretation of multiphoton transition probabilities *J. Phys. B: At. Mol. Opt. Phys.* **13**, 3613–3617 (1980) (ID: 377686)

13. P. Zoller and P. Lambropoulos, Laser temporal coherence effects in two-photon resonant three-photon ionisation, *J. Phys. B: At. Mol. Opt. Phys.* **13**, 69–83 (1980) (ID: 377472)

14. P. Zoller, Laser photon correlation effects in electron scattering, *J. Phys. B: At. Mol. Opt. Phys.* **13**, L249–L252 (1980) (ID: 377471)

15. S. N. Dixit, P. Zoller and P. Lambropoulos, Non-Lorentzian laser line shapes and the reversed peak asymmetry in double optical resonance, *Phys. Rev. A* **21**, 1289–1296 (1980). (ID: 377688)

16. D. Wals, P. Zoller and M. Steyn-Ross, Optical bistability from three-level atoms, URL (ID: 377454)

17. S. N. Dixit, P. Lambropoulos and P. Zoller, Spin polarization of electrons in two-photon resonant three-photon ionization, *Phys. Rev. A* **24**, 318–325 (1981). (ID: 377470)

18. P. Lambropoulos and P. Zoller, Autoionizing states in strong laser fields, *Phys. Rev. A* **24**, 379–397 (1981). (ID: 377465)

19. D. Walls, C. Kunasz, P. Drummond and P. Zoller, Bifurcations and multistability in two-photon processes, *Phys. Rev. A* **24**, 627–630 (1981). (ID: 377462)

20. P. Zoller, G. Alber and R. Salvador, ac Stark splitting in intense stochastic driving fields with Gaussian statistics and non-Lorentzian line shape, *Phys. Rev. A* **24**, 398–410 (1981). (ID: 377422)

21. D. Walls and P. Zoller, Reduced quantum fluctuations in resonance fluorescence, *Phys. Rev. A* **47**, 709–711 (1981). (ID: 377456)

22. D. Walls and P. Zoller, Enhanced sensitivity of a gravitational wave detector, *Phys. Lett. A* **85**, 118–120 (1981). (ID: 377459)

23. M. Lewenstein, P. Zoller and J. Mostowski, Path integration method applied to (N-1)-resonant N-photon ionisation, *J. Phys. B: At. Mol. Opt. Phys.* **16**, 563–568 (1983) (ID: 377379)

24. P. Zoller, Stark shifts and resonant multiphoton ionisation in multimode laser fields, *J. Phys. B: At. Mol. Opt. Phys.* **15**, 2911 (1983) (ID: 377128)

25. C. Parigger, P. Zoller and D. Walls, Effect of Stark shift on two photon optical tristability **44**, 213–218 (1983). (ID: 377120)

26. G. Alber and P. Zoller, Harmonic generation and multiphoton ionization near an autoionizing resonance, *Phys. Rev. A* **27**, 1373–1388 (1983). (ID: 377390)

27. G. Alber and P. Zoller, Spin polarization by selective laser-induced interference, *Phys. Rev. A* **27**, 1713–1716 (1983). (ID: 377380)

28. E. Matthias, P. Zoller, D. Elliott, N. Piltch, S. Smith and G. Leuchs, Influence of configuration mixing in intermediate states on resonant multiphoton ionization, *Phys. Rev. Lett.* **50**, 1914–1917 (1983). (ID: 377377)

29. P. Zoller and J. Cooper, Nonlinear noise fields and strongly driven atomic transitions, URL (ID: 377127)

30. J. Cooper and P. Zoller, Radiative transfer equations in broad-band, time-varying fields, *The Astrophysical Journal* **277**, 813 (1984) (ID: 377081)

31. M. Helm and P. Zoller, On the model dependence of laser temporal coherence effects in multiphoton transitions, **49**, 324–328 (1984). (ID: 377075)

32. G. Alber and P. Zoller, Structure of autoionizing Rydberg series in strong laser fields: A multichannel-quantum-defect-theory approach, *Phys. Rev. A* **29**, 2290–2293 (1984). (ID: 377126)

33. H. Klar, P. Zoller and M. Fedorov, Laser-induced collective binding in two-electron systems *Phys. Rev. A* **30**, 658–660 (1984). (ID: 377073)

34. M. Collett, D. Walls and P. Zoller, Spectrum of squeezing in resonance fluorescence, **52**, 145–149 (1984). (ID: 377119)

35. F. Trombetta, G. Ferrante, K. Wodkiewicz and P. Zoller, Field correlation effects in laser-assisted electron scattering: The phase diffusion model, *J. Phys. B: At. Mol. Opt. Phys.* **18**, 2915–2930 (1985) (ID: 377072)

36. M. J. Fink and P. Zoller, One- and two-photon detachment of negative hydrogen ions: A hyperspherical adiabatic approach, *J. Phys. B: At. Mol. Opt. Phys.* **18**, L373–L377 (1985) (ID: 377066)

37. R. Blatt, P. Zoller, G. Holzmüller and I. Siemers, Brownian motion of a parametric oscillator: A model for ion confinement in radio frequency traps, *Zeitschrift für Physik D; Atoms, Molecules and Clusters* **4**, 121–126 (1986). (ID: 376916)

38. G. Alber, H. Ritsch and P. Zoller, Generation and detection of Rydberg wave packets by short laser pulses, *Phys. Rev. A* 34, 1058–1064 (1986). (ID: 377061)

39. R. Blatt, W. Ertmer, P. Zoller and J. Hall, Atomic-beam cooling: A simulation approach, *Phys. Rev. A* **34**, 3022–3033 (1986). (ID: 377060)

40. W. Henle and P. Zoller, Multichannel quantum defect parametrisation of resonant multiphoton ionisation, *J. Phys. B: At. Mol. Opt. Phys.* **20**, 4007–4025 (1987) (ID: 376899)

41. P. Zoller, M. Marte and D. Walls, Quantum jumps in atomic systems, *Phys. Rev. A* **35**, 198–207 (1987). (ID: 376897)

42. M. Hamilton, K. Arnett, S. Smith, D. Elliott, M. Dziemballa and P. Zoller, Saturation of an optical transition by a phase-diffusing laser field, *Phys. Rev. A* **36**, 178–188 (1987). (ID: 376901)

43. W. Henle, H. Ritsch and P. Zoller, Rydberg wave packets in many-electron atoms excited by short laser pulses, *Phys. Rev. A* **36**, 683–692 (1987). (ID: 376900)

44. A. Giusti-Suzor and P. Zoller, Rydberg electrons in laser fields: A finite-range-interaction problem, *Phys. Rev. A* **36**, 5178–5188 (1987). (ID: 376903)

45. G. Alber and P. Zoller, Near-threshold excitation of Rydberg series by strong laser fields, *Phys. Rev. A* **37**, 377–389 (1988). (ID: 376893)

46. H. Ritsch and P. Zoller, Atomic transitions in finite-bandwidth squeezed light, *Phys. Rev. Lett.* **61**, 1097–1100 (1988). (ID: 376814)

47. R. Blatt and P. Zoller, Quantum jumps in atomic systems, *European Journal of Physics* **9**, 250–256 (1988). (ID: 376844)

48. H. Ritsch and P. Zoller, Systems driven by colored squeezed noise: The atomic absorption spectrum, *Phys. Rev. A* **38**, 4657–4668 (1988). (ID: 376803)

49. T. Haslwanter, H. Ritsch, J. Cooper and P. Zoller, Laser-noise-induced population fluctuations in two- and three-level systems, *Phys. Rev. A* **38**, 5652–5659 (1988). (ID: 376816)

50. G. Alber, T. Haslwanter and P. Zoller, One-photon resonant two-photon excitation of Rydberg series close to threshold, URL (ID: 376892)

51. M. J. Fink and P. Zoller, Quantum-defect parametrization of perturbative two-photon ionization cross sections, *Phys. Rev. A* **39**, 2933–2947 (1989). (ID: 376800)

52. H. Gratl, G. Alber and P. Zoller, Near-threshold behaviour of multiphoton ionisation probabilities, *J. Phys. B: At. Mol. Opt. Phys.* **22**, L547–L551 (1989). (ID: 376799)

53. M. Marte and P. Zoller, Lasers with sub-Poissonian pump, *Phys. Rev. A* **40**, 5774–5782 (1989). (ID: 376796)

54. P. Zoller, Sub-Poissonian laser output due to optical pumping by squeezed light, *Quant. Opt.* **2**, 229–235 (1990) (ID: 376545)

55. M. H. Anderson, R. Jones, J. Cooper, S. Smith, D. Elliott, H. Ritsch and P. Zoller, Observation of population fluctuations in two-level atoms driven by a phase diffusing field, *Phys. Rev. Lett.* **64**, 1346–1349 (1990). (ID: 376550)

56. H. Ritsch, P. Zoller and J. Cooper, Power spectra and variance of laser-noise-induced population fluctuations in two-level atoms, *Phys. Rev. A* **41**, 2653–2667 (1990). (ID: 376539)

57. M. H. Anderson, R. Jones, J. Cooper, S. Smith, D. Elliott, H. Ritsch and P. Zoller, Variance and spectra of fluorescence-intensity fluctuations from two-level atoms in a phase-diffusing field, *Phys. Rev. A* **42**, 6690–6703 (1990). (ID: 376546)

58. G. Alber and P. Zoller, Laser induced excitation of electronic Rydberg wave packets, *Contemporary Physics* **32**, 185 (1991) (ID: 376525)

59. K. Ellinger, H. Gratl and P. Zoller, New aspects in laser excitation of Rydberg wavepackets, *Physica Scripta* **T34**, 60 (1991) (ID: 376491)

60. K. M. Gheri, M. Marte and P. Zoller, Atomic absorption in cross-correlated time-delayed stochastic laser fields, URL (ID: 376281)

61. G. Alber and P. Zoller, Laser excitation of electronic wave packets in rydberg atoms, *Physics Reports* **199**(5), 231–280 (1991). (ID: 376488)

62. P. Marte and P. Zoller, Hydrogen in intense laser fields: Radiative close-coupling equations and quantum-defect parametrization, *Phys. Rev. A* **43**, 1512–1522 (1991). (ID: 376269)

63. H. Ritsch, P. Zoller, Gardiner, Crispin and D. Walls, Sub-Poissonian laser light by dynamic pump-noise suppression, *Phys. Rev. A* **44**, 3361–3364 (1991). (ID: 376267)

64. J. I. Cirac, H. Ritsch, P. Zoller, Two-level system interacting with a finite-bandwidth thermal cavity mode, *Phys. Rev. A* **44**, 4541–4551 (1991). (ID: 376271)

65. P. Marte, P. Zoller and J. Hall, Coherent atomic mirrors and beam splitters by adiabatic passage in multilevel systems, *Phys. Rev. A* **44**, R4118–R4121 (1991). (ID: 376268)

66. M. Marte, J. I. Cirac and P. Zoller, Deflection of atoms by circularly polarized light beams in triple laue configuration, URL (ID: 376278)

67. R. Graham, M. Schlautmann and P. Zoller, Dynamical localization of atomic-beam deflection by a modulated standing light wave, *Mod. Phys. Lett. A* **45**, R19–R22 (1992). (ID: 375870)

68. H. Ritsch and P. Zoller, Dynamic quantum-noise reduction in multilevel-laser systems, *Phys. Rev. A* **45**, 1881–1892 (1992). (ID: 375590)

69. R. Walser, H. Ritsch, P. Zoller and J. Cooper Laser-noise-induced population fluctuations in two-level systems: Complex and real Gaussian driving fields, *Phys. Rev. A* **45**, 468–476 (1992). (ID: 375428)

70. R. Dumhart, P. Zoller and H. Ritsch, Monte Carlo simulation of the atomic master equation for spontaneous emission, *Phys. Rev. A* **45**, 4879–4887 (1992). (ID: 375891)

71. R. Graham, D. Walls and P. Zoller, Emission from atoms in linear superpositions of center-of-mass wave packets, *Phys. Rev. A* **45**, 5018–5030 (1992). (ID: 375865)

72. M. Marte and P. Zoller, Localization of atoms in light fields: Optical molasses, adiabatic compression and squeezing, **54/5**, 477–485 (1992). (ID: 375719)

73. A. S. Parkins, P. Zoller, •+-•- laser-cooling configuration with broadband laser fields: Instability at zero velocity, *Phys. Rev. A* **45**, R6161–R6164 (1992). (ID: 375596)

74. A. S. Parkins and P. Zoller, Laser cooling of atoms with broadband real Gaussian laser fields, *Phys. Rev. A* **45**, 6522–6538 (1992). (ID: 375592)

75. J. I. Cirac, R. Blatt, P. Zoller and W. D. Phillips, Laser cooling of trapped ions in a standing wave *Phys. Rev. A* **46**, 2668–2681 (1992). (ID: 375895)

76. R. Dumhart, A. S. Parkins, P. Zoller, Gardiner, Crispin, Monte Carlo simulation of master equations in quantum optics for vacuum, thermal, and squeezed reservoirs, *Phys. Rev. A* **46**, 4382–4396 (1992). (ID: 375893)

77. Gardiner, Crispin, A. S. Parkins and P. Zoller, Wave-function quantum stochastic differential equations and quantum-jump simulation methods, *Phys. Rev. A* **46**, 4363–4381 (1992). (ID: 375890)

78. A. S. Parkins, P. Zoller and H. J. Carmichael, Spectral linewidth narrowing in a strongly coupled atom-cavity system via squeezed-light excitation of a "vacuum" Rabi resonance, *Phys. Rev. A* **48**, 758–763 (1993). (ID: 375410)

79. J. I. Cirac, A. S. Parkins, R. Blatt and P. Zoller, "Dark" squeezed states of the motion of a trapped ion, *Phys. Rev. Lett.* **70**, 556–559 (1993). (ID: 375418)

80. P. Marte, R. Dumhart, R. Taïeb and P. Zoller, Resonance fluorescence from quantized one-dimensional molasses, *Phys. Rev. A* **47**, 1378–1390 (1993). (ID: 375414)

81. J. I. Cirac, R. Blatt, A. S. Parkins and P. Zoller, Preparation of Fock states by observation of quantum jumps in an ion trap, *Phys. Rev. Lett.* **70**, 762–765 (1993). (ID: 375422)

82. J. I. Cirac and P. Zoller, Laser cooling of trapped ions in a squeezed vacuum, *Phys. Rev. A* **47**, 2191–2195 (1993). (ID: 375417)

83. J. I. Cirac, A. S. Parkins, R. Blatt and P. Zoller, Cooling of a trapped ion coupled strongly to a quantized cavity mode, *Opt. Com.* **97**, 353–359 (1993). (ID: 375419)

84. M. H. Anderson, G. Vemuri, J. Cooper, P. Zoller and S. Smith, Experimental study of absorption and gain by two-level atoms in a time-delayed non-Markovian optical field, *Phys. Rev. A* **47**, 3202–3209 (1993). (ID: 375423)

85. R. Taïeb, P. Marte, R. Dumhart and P. Zoller, Spectrum of resonance fluorescence and cooling dynamics in quantized one-dimensional molasses: Effects of laser configuration, *Phys. Rev. A* **47**, 4986–4993 (1993). (ID: 375409)

86. H. R. Xia, J. I. Cirac, S. Swartz, B. Kohler, D. Elliott, J. Hall and P. Zoller, Phase shifts and intensity dependence in frequency-modulation spectroscopy, URL (ID: 375387)

87. J. I. Cirac, R. Blatt, A. S. Parkins and P. Zoller, Laser cooling of trapped ions with polarization gradients, *Phys. Rev. A* **48**, 1434–1445 (1993). (ID: 375420)

88. J. I. Cirac, R. Blatt, A. S. Parkins and P. Zoller, Spectrum of resonance fluorescence from a single trapped ion, *Phys. Rev. A* **48**, 2169–2181 (1993). (ID: 375421)

89. A. S. Parkins, P. Marte, P. Zoller and H. J. Kimble Synthesis of arbitrary quantum states via adiabatic transfer of Zeeman coherence, *Phys. Rev. Lett.* **71**, 3095–3098 (1993). (ID: 375413)

90. M. Lewenstein, J. I. Cirac and P. Zoller, Quantum dynamics of a laser-cooled ideal gas, *Phys. Rev. A* **50**, 3409–3422 (1994) (ID: 305615)

91. J. I. Cirac, L. J. Garay, R. Blatt, A. S. Parkins and P. Zoller, Laser cooling of trapped ions: The influence of micromotion, *Phys. Rev. A* **49**, 421–432 (1994). (ID: 375185)

92. J. I. Cirac, R. Blatt, A. S. Parkins and P. Zoller, Quantum collapse and revival in the motion of a single trapped ion, *Phys. Rev. A* **49**, 1202–1207 (1994). (ID: 375187)

93. L. S. Goldner, C. Gerz, S. L. Rolston, C. I. Westbrook, W. D. Phillips, P. Marte and P. Zoller, Momentum transfer in laser-cooled cesium by adiabatic passage in a light field, *Phys. Rev. Lett.* **97**, 997–1000 (1994). (ID: 375179)

94. I. Marzoli, J. I. Cirac, R. Blatt and P. Zoller, Laser cooling of trapped three-level ions: Designing two-level systems for sideband cooling, *Phys. Rev. A* **49**, 2771–2779 (1994). (ID: 375175)

95. J. I. Cirac, R. Blatt and P. Zoller, Nonclassical states of motion in a three-dimensional ion trap by adiabatic passage, *Phys. Rev. A* **49**, R3174–R3177 (1994). (ID: 375186)

96. K. Ellinger, J. Cooper and P. Zoller, Light-pressure force in N-atom systems, *Phys. Rev. A* **49**, 3909–3933 (1994). (ID: 375180)

97. J. I. Cirac, M. Lewenstein and P. Zoller, Quantum statistics of a laser cooled ideal gas, *Phys. Rev. Lett.* **72**, 2977–2980 (1994). (ID: 375183)

98. P. Marte, R. Dumhart, R. Taïeb, P. Zoller, M. S. Shahriar and M. Prentiss, Polarization-gradient-assisted subrecoil cooling: Quantum calculations in one dimension, *Phys. Rev. A* **49**, 4826–4836 (1994). (ID: 375174)

99. R. Taïeb, R. Dumhart, J. I. Cirac, P. Marte and P. Zoller, Cooling and localization of atoms in laser-induced potential wells, *Phys. Rev. A* **49**, 4876–4887 (1994). (ID: 375172)

100. R. Walser and P. Zoller, Laser-noise-induced polarization fluctuations as a spectroscopic tool, *Phys. Rev. A* **49**, 5067–5077 (1994). (ID: 375170)

101. S. Marksteiner, C. M. Savage, P. Zoller and S. L. Rolston, Coherent atomic waveguides from hollow optical fibers: Quantized atomic motion, *Phys. Rev. A* **50**, 2680–2690 (1994). (ID: 375173)

102. J. I. Cirac and P. Zoller, Preparation of macroscopic superpositions in many-atom systems, *Phys. Rev. A* 50, R2799–R2802 (1994). (ID: 375182)

103. R. Walser, J. Cooper and P. Zoller, Saturated absorption spectroscopy using diode-laser phase noise, *Phys. Lett. A* **50**, 4303–4309 (1994). (ID: 375171)

104. R. Dumhart, P. Marte, T. Pellizzari and P. Zoller, Laser cooling to a single quantum state in a trap, *Phys. Rev. Lett.* 73, 2829–2832 (1994). (ID: 375181)

105. R. Blatt, J. I. Cirac, A. S. Parkins and P. Zoller, Quantum motion of trapped ions, *Physica Scripta* **T59**, 294 (1995) (ID: 305661)

106. R. Blatt, J. I. Cirac and P. Zoller, Trapping states of motion with cold ions, *Phys. Rev. A* **52**, 518–524 (1995) (ID: 305623)

107. J. I. Cirac, M. Lewenstein and P. Zoller, Laser cooling a trapped atom in a cavity: Bad-cavity limit *Phys. Rev. A* **51**, 1650–1655 (1995) (ID: 305621)

108. J. I. Cirac, M. Lewenstein and P. Zoller, Generalized Bose–Einstein distributions and multi-stability of a laser-cooled gas, *Phys. Rev. A* **51**, 2899–2907 (1995) (ID: 305618)

109. J. I. Cirac and P. Zoller, Quantum computations with cold trapped ions, *Phys. Rev. Lett.* **74**, 4091–4094 (1995) (ID: 305614)

110. M. Lewenstein, J. I. Cirac and P. Zoller, Master equation for sympathetic cooling of trapped particles, *Phys. Rev. A* **51**, 4617–4627 (1995) (ID: 305613)

111. S. Marksteiner, R. Walser, P. Marte and P. Zoller, Localization of atoms in light fields: Optical molasses, adiabatic compression and squeezing, *Appl. Phys. B Las. Opt.* **60**, 145–153 (1995) (ID: 303542)

112. A. S. Parkins, P. Marte, P. Zoller, O. Carnal and H. J. Kimble, Quantum-state mapping between multilevel atoms and cavity light fields, *Phys. Rev. A* **51**, 1578–1596 (1995) (ID: 303497)

113. T. Pellizzari, S. Gardiner, J. I. Cirac and P. Zoller, Decoherence, continuous observation, and quantum computing: A cavity qed model, *Phys. Rev. Lett.* **75**, 3788–3791 (1995) (ID: 303489)

114. T. Pellizzari, P. Marte and P. Zoller, Laser cooling to a single quantum state in a trap: One-dimensional results, *Phys. Rev. A* **52**, 4709–4718 (1995) (ID: 303487)

115. L. You, J. Cooper and P. Zoller, Quantum-classical correspondences for atomic operators: A positive presentation approach, URL (ID: 303476)

116. J. I. Cirac, A. S. Parkins, R. Blatt and P. Zoller, Non-classical states of motion in ion traps, *Adv. Atom. Molec. and Opt. Physics* **37**, 238–296 (1996) (ID: 619592)

117. J. F. Poyatos, R. Walser, J. I. Cirac, R. Blatt and P. Zoller, Motion tomography of a single trapped ion, *Phys. Rev. A* **53**, R1966 (1996) (ID: 344250)

118. J. F. Poyatos, J. I. Cirac, R. Blatt and P. Zoller, Trapped ions in the strong excitation regime: Ion interferometry and non-classical states, *Phys. Rev. A* **54**, 1532 (1996) (ID: 344249)

119. R. Blatt and P. Zoller, Quantengatter für Quantenrechner, *Physikalische Blätter* **52**, 205 (1996) (ID: 344248)

120. V. M. Perez-Garcia, H. Michinel, J. I. Cirac, M. Lewenstein and P. Zoller, Low energy excitations of a Bose–Einstein condensate: A time-dependent variational analysis, *Phys. Rev. Lett.* **77**, 005320 (1996) (ID: 305704)

121. J. F. Poyatos, J. I. Cirac and P. Zoller, Quantum reservoir engineering with laser cooled trapped ions, *Phys. Rev. Lett.* **77**, 004728 (1996) (ID: 305700)

122. J. F. Poyatos, R. Walser, J. I. Cirac and P. Zoller, Motion tomography of a single trapped ion, *Phys. Rev. A* **53**, R1966–R1969 (1996) (ID: 305687)

123. J. I. Cirac, Gardiner, Crispin, M. Naraschewski and P. Zoller, Continuous observation of interference fringes from Bose condensates, *Phys. Rev. A* **54**, R3714–R3717 (1996) (ID: 305685)

124. R. Walser, J. I. Cirac and P. Zoller, Magnetic tomography of a cavity state, *Phys. Rev. Lett.* **77**, 002658 (1996) (ID: 305684)

125. M. Naraschewski, H. Wallis, A. Schenzle, J. I. Cirac and P. Zoller, Interference of Bose condensates, *Phys. Rev. A* **54**, 2185–2196 (1996) (ID: 305682)

126. H. Stecher, H. Ritsch, P. Zoller, F. Sander, T. Esslinger and T. W. Hänsch, All-optical gray lattice for atoms, *Phys. Rev. A* **55**, 545 (1996). (ID: 367816)

127. J. F. Poyatos, J. I. Cirac and P. Zoller, Complete characterization of a quantum process: The two-bit quantum gate, *Phys. Rev. Lett.* **78**, 390 (1997). (ID: 367817)

128. S. Gardiner, J. I. Cirac and P. Zoller, Nonclassical states and measurement of general motional observables of a trapped ion, *Phys. Rev. A* **55**, 1683 (1997). (ID: 367815)

129. Gardiner, Crispin and P. Zoller, Quantum kinetic theory: A quantum kinetic master equation for condensation of a weakly interacting Bose gas without a trapping potential, *Phys. Rev. A* **55**, 2902 (1997). (ID: 367811)

130. J. I. Cirac, P. Zoller, H. J. Kimble and H. Mabuchi, Quantum state transfer and entanglement distribution among distant nodes in a quantum network *Phys. Rev. Lett.* 78, 3221 (1997). (ID: 367814)

131. G. Morigi, J. I. Cirac, M. Lewenstein and P. Zoller, Ground-state laser cooling beyond the Lamb-Dicke limit, **39**, 13 (1997). (ID: 367813)

132. S. van Enk, J. I. Cirac and P. Zoller, Ideal quantum communication over noisy channels: A quantum optical implementation, *Phys. Rev. Lett.* **78**, 4293 (1997). (ID: 367812)

133. D. Jaksch, Gardiner, Crispin and P. Zoller, Quantum kinetic theory. II. Simulation of the quantum Boltzmann master equation, *Phys. Rev. A* **56**, 575 (1997). (ID: 367810)

134. V. M. Perez-Garcia, H. Michinel, J. I. Cirac, M. Lewenstein and P. Zoller, Dynamics of Bose–Einstein condensates: Variational solutions of the Gross-Pitaevskii equations, *Phys. Rev. A* **56**, 1424 (1997). (ID: 367809)

135. Gardiner, Crispin, P. Zoller, R. J. Ballagh and M. J. Davis, Kinetics of Bose–Einstein condensation in a trap, *Phys. Rev. Lett.* **79**, 1793 (1997). (ID: 367808)

136. J. I. Cirac, P. Zoller, H. J. Kimble and H. Mabuchi, Quantum state transfer in a quantum network: A quantum-optical implementation, *J. Mod. Opt.* **44**, 1727 (1997). (ID: 367805)

137. T. Busch, J. I. Cirac, V. M. Perez-Garcia and P. Zoller, Stability and collective excitations of a two-component Bose-Einstein condensed gas: A moment approach, *Phys. Rev. A* **56**, 2978 (1997). (ID: 367807)

138. S. Gardiner, J. I. Cirac and P. Zoller, Quantum chaos in an ion trap: The delta-kicked harmonic oscillator, *Phys. Rev. Lett.* 79, 4790 (1997). (ID: 367800)

139. S. van Enk, J. I. Cirac and P. Zoller, Purifying two-bit quantum gates and joint measurements in cavity QED, *Phys. Rev. Lett.* **79**, 5178 (1997). (ID: 367806)

140. P. Zoller, H. J. Kimble and H. Mabuchi, Transmission of quantum information in a quantum network: A quantum optical implementation URL (ID: 367803)

141. J. F. Poyatos, J. I. Cirac and P. Zoller, Characterization of decoherence processes in quantum computation, *Optics Express* **2**, 372 (1998) (ID: 367802)

142. J. I. Cirac and P. Zoller, Quantenkommunikation und Quantencomputing, URL (ID: 367793)

143. K. M. Gheri, K. Ellinger, T. Pellizzari and P. Zoller, Photon-wavepackets as flying quantum bits, URL (ID: 367773)

144. J. I. Cirac, P. Zoller, H. J. Kimble and H. Mabuchi, Quantum communication in a quantum network, URL (ID: 367768)

145. N. Lütkenhaus, J. I. Cirac and P. Zoller, Mimicking a squeezed-bath interaction: Quantum-reservoir engineering with atoms **57**, 548 (1997). (ID: 367777)

146. P. Zoller, Photonic channels for quantum communication, *Science* **279**, 205 (1998). (ID: 367794)

147. J. I. Cirac, M. Lewenstein, K. Moelmer and P. Zoller, Quantum superposition states of Bose–Einstein condensates, *Phys. Rev. A* **57**, 1208 (1998). (ID: 367775)

148. G. Morigi, J. I. Cirac, K. Ellinger and P. Zoller, Laser cooling of trapped atoms to the ground state: A dark state in position space, *Phys. Rev. A* **57**, 2909 (1998). (ID: 367774)

149. R. Dumhart, J. I. Cirac, M. Lewenstein and P. Zoller, Creation of dark solitons and vortices in Bose–Einstein condensates, *Phys. Rev. Lett.* **80**, 2972 (1998). (ID: 367776)

150. Gardiner, Crispin and P. Zoller, Quantum kinetic theory. III. Quantum kinetic master equation for strongly condensed trapped systems, *Phys. Rev. A* **58**, 536 (1998). (ID: 367769)

151. D. Jaksch, Gardiner, Crispin, K. M. Gheri and P. Zoller, Quantum kinetic theory. IV. Intensity and amplitude fluctuations of a Bose–Einstein condensate at finite temperature including trap loss, *Phys. Rev. A* **58**, 1450 (1998). (ID: 367772)

152. J. F. Poyatos, J. I. Cirac and P. Zoller, Quantum gates with "Hot" trapped ions, *Phys. Rev. Lett.* **81**, 1322 (1998). (ID: 367770)

153. T. Busch, J. R. Anglin, J. I. Cirac and P. Zoller, Inhibition of spontaneous emission in Fermi gases, **44**, 1 (1998). (ID: 367757)

154. H. J. Briegel, W. Dür, J. I. Cirac and P. Zoller, Quantum communication and the creation of maximally entangled pairs of atoms over a noisy channel, *Phil. Trans. R. Soc. Lond. A* **356**, 1841 (1998). (ID: 367767)

155. K. M. Gheri, C. Saavedra, P. Törmä, J. I. Cirac and P. Zoller, Entanglement engineering of one-photon wave packets using a single-atom source, *Phys. Rev. A* **58**, R2627 (1998). (ID: 367766)

156. D. Jaksch, C. Bruder, J. I. Cirac, Gardiner, Crispin and P. Zoller, Cold bosonic atoms in optical lattices, *Phys. Rev. Lett.* **81**, 3108 (1998). (ID: 367764)

157. Gardiner, Crispin, M. D. Lee, R. J. Ballagh, M. J. Davis, P. Zoller, Quantum kinetic theory of condensate growth: Comparison of experiment and theory, *Phys. Rev. Lett.* **81**, 5266 (1998). (ID: 367762)

158. H. J. Briegel, W. Dür, J. I. Cirac and P. Zoller, Quantum repeaters: The role of imperfect local operations in quantum communication, *Phys. Rev. Lett.* **81**, 5932 (1998). (ID: 367760)

159. K. M. Gheri, P. Törmä and P. Zoller, Quantum state engineering with photonic qubits, URL (ID: 352420)

160. W. Dür, H. J. Briegel, J. I. Cirac and P. Zoller, Quantum repeaters based on entanglement purification, *Phys. Rev. A* **59**, 000169 (1998). (ID: 352427)

161. J. I. Cirac and P. Zoller, Engineering entangled states of trapped ions, *Physics World* **01** (1998). (ID: 352424)

162. C. Cabrillo, J. I. Cirac, P. Garcia-Fernandez and P. Zoller, Creation of entangled states of distant atoms by interference, *Phys. Rev. A* **59**, 001025 (1999). (ID: 352426)

163. D. Jaksch, H. J. Briegel, J. I. Cirac, Gardiner, Crispin and P. Zoller, Entanglement of atoms via cold controlled collisions, *Phys. Rev. Lett.* **82**, 001975 (1999). (ID: 352425)

164. G. Giedke, H. J. Briegel, J. I. Cirac and P. Zoller, Lower bounds for attainable fidelities in entanglement purification, *Phys. Rev. A* **59**, 002641 (1999). (ID: 352423)

165. H. J. Kimble, J. I. Cirac and P. Zoller, Quantum communication with dark photons, *Phys. Rev. A* **59**, 002659 (1999). (ID: 352421)

166. G. Morigi, J. Eschner, J. I. Cirac and P. Zoller, Laser cooling of two trapped ions: Sideband cooling beyond the Lamb-Dicke limit, *Phys. Rev. A* **59**, 003793 (1999). (ID: 352422)

167. J. F. Poyatos, J. I. Cirac and P. Zoller, Schemes of quantum computations with trapped ions, URL (ID: 337598)

168. D. Jaksch, T. Calarco and P. Zoller, Auf dem Weg zum universellen Quantencomputer, URL (ID: 337597)

169. T. Calarco, E. A. Hinds, D. Jaksch, J. Schmiedmayer, J. I. Cirac and P. Zoller, Quantum gates with neutral atoms: Controlling collisional interactions in time-dependent traps, *Phys. Rev. A* **61**, 022304 (2000). (ID: 337613)

170. Gardiner, Crispin and P. Zoller, Quantum kinetic theory. V. Quantum kinetic master equation for mutual interaction of condensate and noncondensate, *Phys. Rev. A* **61**, 033601 (2000). (ID: 337612)

171. H. J. Briegel, T. Calarco, D. Jaksch, J. I. Cirac and P. Zoller, Quantum computing with neutral atoms *J. Mod. Opt.* **47**, 415 (2000). (ID: 337608)

172. L. Duan, G. Giedke, J. I. Cirac and P. Zoller, Inseparability criterion for continuous variable systems, *Phys. Rev. Lett.* **84**, 2722 (2000). (ID: 337606)

173. P. Zoller, Quantum optics: Tricks with a single photon, *Nature* **404**, 340 (2000). (ID: 337610)

174. J. I. Cirac and P. Zoller, A scalable quantum computer with ions in an array of microtraps, *Nature* **404**, 579 (2000). (ID: 337611)

175. J. J. García-Ripoll, J. I. Cirac, J. R. Anglin, V. M. Perez-Garcia and P. Zoller, Spin monopoles with Bose–Einstein condensates, *Phys. Rev. A* **61**, 053609 (2000). (ID: 337609)

176. L. Duan, G. Giedke, J. I. Cirac and P. Zoller, Entanglement purification of Gaussian continuous variable quantum states, *Phys. Rev. Lett.* **84**, 4002 (2000). (ID: 337607)

177. C. Saavedra, K. M. Gheri, P. Törmä, J. I. Cirac and P. Zoller, Controlled source of entangled photonic qubits, *Phys. Rev. A* **61**, 062311 (2000). (ID: 337605)

178. P. Törmä and P. Zoller, Laser probing of atomic cooper pairs, *Phys. Rev. Lett.* **85**, 487 (2000). (ID: 337603)

179. S. Gardiner, D. Jaksch, R. Dumhart, J. I. Cirac and P. Zoller, Nonlinear matter wave dynamics with a chaotic potential, *Phys. Rev. A* **62**, 023612 (2000). (ID: 337604)

180. L. Duan, G. Giedke, J. I. Cirac and P. Zoller, Physical implementation for entanglement purification of Gaussian continuous-variable quantum states, *Phys. Rev. A* **62**, 32304 (2000). (ID: 337600)

181. L. Santos, G. V. Shlyapnikov, P. Zoller and M. Lewenstein, Bose–Einstein condensation in trapped dipolar gases, *Phys. Rev. Lett.* **85**, 1791 (2000). (ID: 337602)

182. D. Jaksch, J. I. Cirac, P. Zoller, S. L. Rolston, R. Cote and M. Lukin, Fast quantum gates for neutral atoms, *Phys. Rev. Lett.* **85**, 2208 (2000). (ID: 337601)

183. T. Calarco, H. J. Briegel, D. Jaksch, J. I. Cirac and P. Zoller, Entangling neutral atoms for quantum information processing, URL (ID: 337599)

184. T. Calarco, H. J. Briegel, D. Jaksch, J. I. Cirac and P. Zoller, Quantum computing with trapped particles in microscopic potentials, *Fortschr. Phys.* **48**, 945 (2000). (ID: 337593)

185. L. Duan, A. Sorensen, J. I. Cirac and P. Zoller, Squeezing and entanglement of atomic beams, *Phys. Rev. Lett.* **85**, 3991 (2000). (ID: 337596)

186. L. Duan, G. Giedke, J. I. Cirac and P. Zoller, Continuous variable entanglement purification and its physical implementation, URL (ID: 337615)

187. L. J. Garay, J. R. Anglin, J. I. Cirac and P. Zoller, Sonic analog of gravitational black holes in Bose–Einstein condensates, *Phys. Rev. Lett.* **85**, 4643 (2000). (ID: 337595)

188. L. Duan, J. I. Cirac, P. Zoller and E. Polzik Quantum communication between atomic ensembles using coherent light, URL (ID: 337594)

189. C. Menotti, J. R. Anglin, J. I. Cirac and P. Zoller, Dynamic splitting of a Bose–Einstein condensate, *Phys. Rev. A* **63**, 023601 (2001). (ID: 337590)

190. A. Sorensen, L. Duan, J. I. Cirac and P. Zoller, Many-particle entanglement with Bose–Einstein condensates, *Nature* 409, **63** (2001). (ID: 337588)

191. L. J. Garay, J. R. Anglin, J. I. Cirac and P. Zoller, Sonic black holes in dilute Bose–Einstein condensates, *Phys. Rev. A* **63**, 023611 (2001). (ID: 337591)

192. S. Gardiner, K. M. Gheri and P. Zoller, Cavity-assisted quasiparticle damping in a Bose–Einstein condensate, *Phys. Rev. A* **63**, 051603(R) (2001). (ID: 337585)

193. T. Calarco, J. I. Cirac and P. Zoller, Entangling ions in arrays of microscopic traps, *Phys. Rev. A* **63**, 062304 (2001). (ID: 337587)

194. D. Jaksch, S. Gardiner, K. Schulze, J. I. Cirac and P. Zoller, Uniting Bose–Einstein condensates in optical resonators, *Phys. Rev. Lett.* **86**, 004733 (2001). (ID: 337582)

195. L. Duan, J. I. Cirac and P. Zoller, Geometric manipulation of trapped ions for quantum computation, *Science* **292**, 1695 (2001). (ID: 337578)

196. M. Lukin, M. Fleischhauer, R. Cote, L. Duan, D. Jaksch, J. I. Cirac and P. Zoller, Dipole blockade and quantum information processing in mesoscopic atomic ensembles, *Phys. Rev. Lett.* **87**, 037901 (2001). (ID: 337584)

197. B. Paredes, P. O. Fedichev, J. I. Cirac and P. Zoller, (1/2)-Anyons in small atomic Bose–Einstein condensates, *Phys. Rev. Lett.* **87**, 010402 (2001). (ID: 337583)

198. G. M. Bruun, P. Törmä, M. Rodriguez and P. Zoller, Laser probing of Cooper-paired trapped atoms, *Phys. Rev. A* **64**, 033609 (2001). (ID: 337586)

199. G. Giedke, L. Duan, J. I. Cirac and P. Zoller, Distillability criterion for all bipartite Gaussian states, URL (ID: 337581)

200. G. Giedke, B. Kraus, L. Duan, P. Zoller, M. Lewenstein and J. I. Cirac, Separability and distillability of bipartite Gaussian states — The complete story, *Fortschr. Phys.* **49**, 973 (2001). (ID: 337579)

201. L. Duan, M. Lukin, J. I. Cirac and P. Zoller, Long-distance quantum communication with atomic ensembles and linear optics, *Nature* **414**, 413 (2001). (ID: 337580)

202. L. Duan, J. I. Cirac and P. Zoller, Quantum entanglement in spinor Bose–Einstein condensates, *Phys. Rev. A* **65**, 033619 (2002). (ID: 337577)

203. D. Jaksch, J. I. Cirac and P. Zoller, Dynamically turning off interactions in a two-component condensate, *Phys. Rev. A* **65**, 033625 (2002). (ID: 337576)

204. R. M. Cavalcanti, P. Giacconi, G. Pupillo and R. Soldati, Bose–Einstein condensation in the presence of a uniform field and a pointlike impurity, *Phys. Rev. A* **65**, 053606 (2002). (ID: 340631)

205. P. Zoller, Making it with molecules, *Nature* **417**, 493 (2002). (ID: 337575)

206. A. Carollo and G. M. Palma, The role of auxiliary photons in state discrimination with linear optical devices, *J. Mod. Opt.* **49**, 1147 (2002). (ID: 340514)

207. D. Jaksch, V. Venturi, J. I. Cirac, C. Williams and P. Zoller, Creation of a molecular condensate by dynamically melting a mott insulator, *Phys. Rev. Lett.* **89**, 040402 (2002). (ID: 337573)

208. T. Calarco, D. Jaksch, J. I. Cirac and P. Zoller, Controlling dynamical phases in quantum optics, *J. Opt. B: Quantum Semiclass. Opt.* **4**, S430 (2002). (ID: 337572)

209. U. Dorner and P. Zoller, Laser-driven atoms in half-cavities, *Phys. Rev. A* **66**, 023816 (2002). (ID: 337571)

210. L. Duan, J. I. Cirac and P. Zoller, Three-dimensional theory for interaction between atomic ensembles and free-space light, *Phys. Rev. A* **66**, 023818 (2002). (ID: 337569)

211. A. Recati, T. Calarco, P. Zanardi, J. I. Cirac and P. Zoller, Holonomic quantum computation with neutral atoms, *Phys. Rev. A* **66**, 032309 (2002). (ID: 337570)

212. B. Paredes, P. Zoller and J. I. Cirac, Fermionizing a small gas of ultracold bosons, *Phys. Rev. A* **66**, 033609 (2002). (ID: 337568)

213. I. Fuentes-Guridi, A. Carollo, S. Bose and V. Vedral, Vacuum induced spin-1/2 Berry's phase, *Phys. Rev. Lett.* **89**, 220404 (2002). (ID: 340513)

214. W. Hofstetter, J. I. Cirac, P. Zoller, E. Demler and M. Lukin, High-temperature superfluidity of fermionic atoms in optical lattices, *Phys. Rev. Lett.* **89**, 220407 (2002). (ID: 337567)

215. A. Micheli, D. Jaksch, J. I. Cirac and P. Zoller, Many-particle entanglement in two-component Bose-Einstein condensates, *Phys. Rev. A* **67**, 013607 (2003) (ID: 336666)

216. A. Recati, P. O. Fedichev, W. Zwerger and P. Zoller, Spin-charge separation in ultracold quantum gases, *Phys. Rev. Lett.* **90**, 020401 (2003) (ID: 336665)

217. B. Damski, L. Santos, E. Tiemann, M. Lewenstein, S. Kotochigova, P. S. Julienne and P. Zoller, Creation of a dipolar superfluid in optical lattices, *Phys. Rev. Lett.* **90**, 110401 (2003) (ID: 336664)

218. E. Pazy, E. Biolatti, T. Calarco, I. D'Amico, P. Zanardi, F. Rossi and P. Zoller, Spin-based optical quantum computation via Pauli blocking in semiconductor quantum dots **62**, 175 (2003) (ID: 336663)

219. E. Jané, G. Vidal, W. Dür, P. Zoller and J. I. Cirac Simulation of quantum dynamics with quantum optical systems, URL (ID: 336662)

220. D. Jaksch and P. Zoller, Creation of effective magnetic fields in optical lattices: The Hofstadter butterfly for cold neutral atoms, *New J. Phys.* 5, **56** (2003) (ID: 336661)

221. J. I. Cirac and P. Zoller, How to manipulate cold atoms, *Science* **301**, 176 (2003) (ID: 336660)

222. E. Pazy, T. Calarco, I. D'Amico, P. Zanardi, F. Rossi and P. Zoller, Implementation of an all-optical spin-based quantum computer, *Physica Status Solidi B* **238**, 411 (2003) (ID: 336659)

223. B. Damski, J. Zakrzewski, L. Santos, P. Zoller and M. Lewenstein, Atomic Bose and Anderson glasses in optical lattices, *Phys. Rev. Lett.* **91**, 080403 (2003) (ID: 336658)

224. U. Dorner, P. O. Fedichev, D. Jaksch, M. Lewenstein and P. Zoller, Entangling strings of neutral atoms in 1D atomic pipeline structures, *Phys. Rev. Lett.* **91**, 073601 (2003) (ID: 336657)

225. A. Imamoglu, E. Knill, L. Tian and P. Zoller, Optical pumping of quantum-dot nuclear spins, *Phys. Rev. Lett.* **91**, 017402 (2003) (ID: 336656)

226. T. Calarco, A. Datta, P. O. Fedichev, E. Pazy and P. Zoller, Spin-based all-optical quantum computation with quantum dots: Understanding and suppressing decoherence, *Phys. Rev. A* **68**, 012310 (2003) (ID: 336655)

227. J. J. García-Ripoll, P. Zoller, J. I. Cirac Speed Optimized Two-Qubit Gates with Laser Coherent Control Techniques for Ion Trap Quantum Computing *Phys. Rev. Lett.* 91, 157901 (2003) (ID: 336653)

228. L. Tian and P. Zoller, Quantum computing with atomic Josephson junction arrays, *Phys. Rev. A* **68**, 042321 (2003) (ID: 336652)

230. J. I. Cirac, L. Duan, D. Jaksch and P. Zoller, Quantum information processing with quantum optics, *Ann. Henri Poincare* **4**, 661 (2003) (ID: 336650)

229. A. Recati, P. O. Fedichev, W. Zwerger and P. Zoller, Fermi 1D quantum gas: Luttinger liquid approach and spin-charge separation, *J. Opt. B: Quantum Semiclass. Opt.* **5**, 55 (2003) (ID: 336651)

231. P. Rabl, A. J. Daley, P. O. Fedichev, J. I. Cirac and P. Zoller, Defect-suppressed atomic crystals in an optical lattice, *Phys. Rev. Lett.* **91**, 110403 (2003) (ID: 335014)

232. E. Pazy, T. Calarco, I. D'Amico, P. Zanardi, F. Rossi and P. Zoller, All optical spin-based quantum information processing, *J Supercond Nov Magn* **16/2**, 383–385 (2003). (ID: 375698)

233. A. Carollo, I. Fuentes-Guridi, M. F. Santos and V. Vedral, Geometric phase in open systems, *Phys. Rev. Lett.* **90**, 160402 (2003). (ID: 340500)

234. L. Jacak, P. Machnikowski, J. Krasnyj and P. Zoller, Coherent and incoherent phonon processes in artificial atoms optical physics, *The European Physical Journal D* **22/3**, 319–331 (2003). (ID: 375605)

235. S. Bose, A. Carollo, I. Fuentes-Guridi, M. F. Santos and V. Vedral Vacuum induced Berry phase: Theory and experimental proposal, *J. Mod. Opt.* **50**, 1175 (2003). (ID: 340512)

236. A. Carollo, M. F. Santos and V. Vedral, Berry's phase in cavity QED: Proposal for observing an effect of field quantization, *Phys. Rev. A* **67**, 063804 (2003). (ID: 340503)

237. P. O. Fedichev, U. R. Fischer and A. Recati, Zero-temperature damping of Bose–Einstein condensate oscillations by vortex-antivortex pair creation, *Phys. Rev. A* **68**, 011602 (2003). (ID: 325337)

238. P. O. Fedichev and U. R. Fischer, Gibbons-Hawking effect in the sonic de sitter space-time of an expanding Bose–Einstein-condensed gas, *Phys. Rev. Lett.* **91**, 240407 (2003). (ID: 325333)

239. G. Pupillo, E. Tiesinga and C. Williams, Effects of inhomogeneity on the spectrum of the Mott-insulator state, *Phys. Rev. A* **68**, 063604 (2003). (ID: 340629)

240. I. Wilson-Rae, P. Zoller and A. Imamoglu, Laser cooling of a nanomechanical resonator mode to its quantum ground state, *Phys. Rev. Lett.* **92**, 075507 (2004) (ID: 336649)

241. A. Griessner, D. Jaksch and P. Zoller, Cavity-assisted nondestructive laser cooling of atomic qubits, *J. Phys. B: At. Mol. Opt. Phys.* **37**, 1419 (2004) (ID: 335134)

242. P. Rabl, A. Shnirman and P. Zoller, Generation of squeezed states of nanomechanical resonators by reservoir engineering, *Phys. Rev. B* **70**, 205304 (2004) (ID: 335015)

243. E. Pazy, T. Calarco and P. Zoller, Spin state readout by quantum jump technique: For the purpose of quantum computing IEEE transactions on nanotechnology, **3**, 10–16 (2004) (ID: 314773)

244. A. J. Daley, C. Kollath, U. Schollwöck and G. Vidal, Time-dependent density-matrix renormalization-group using adaptive effective Hilbert spaces, *J. Stat. Mech.* **04**, P04005 (2004) (ID: 314644)

245. I. Martin, A. Shnirman, L. Tian and P. Zoller, Ground-state cooling of mechanical resonators, *Phys. Rev. B* **69**, 125339 (2004) (ID: 314642)

246. J. J. García-Ripoll, J. I. Cirac, P. Zoller, C. Kollath, U. Schollwöck and J. von Delft, Variational ansatz for the superfluid Mott-insulator transition in optical lattices, arXiv:cond-mat/0306162 (ID: 314640)

247. L. Tian, P. Rabl, R. Blatt and P. Zoller, Interfacing quantum-optical and solid-state qubits, *Phys. Rev. Lett.* **92**, 247902 (2004) (ID: 314633)

248. T. Calarco, U. Dorner, P. S. Julienne, C. Williams and P. Zoller, Quantum computations with atoms in optical lattices: Marker qubits and molecular interactions, *Phys. Rev. A* **70**, 012306 (2004) (ID: 314632)

249. H. Büchler, P. Zoller and W. Zwerger, Spectroscopy of superfluid pairing in atomic Fermi gases, *Phys. Rev. Lett.* **93**, 080401 (2004) (ID: 314630)

250. W. V. Liu, F. Wilczek and P. Zoller, Spin-dependent Hubbard model and a quantum phase transition in cold atoms, *Phys. Rev. A* **70**, 033603 (2004) (ID: 314622)

251. A. Micheli, A. J. Daley, D. Jaksch and P. Zoller, Single atom transistor in a 1D optical lattice, *Phys. Rev. Lett.* **93**, 140408 (2004) (ID: 314620)

252. P. Zoller and L. Tian, Coupled ion-nanomechanical systems, *Phys. Rev. Lett.* **93**, 266403 (2004) (ID: 314555)

253. B. Paredes, P. Zoller, J. I. Cirac and J. J. García-Ripoll, Strong correlation effects and quantum information theory of low dimensional atomic gases, *J. Phys. IV* **116**, 135–168 (2004) (ID: 314540)

254. P. O. Fedichev and U. R. Fischer, "Cosmological" quasiparticle production in harmonically trapped superfluid gases, *Phys. Rev. A* **69**, 033602 (2004) (ID: 314392)

255. U. R. Fischer, A. Recati and P. O. Fedichev, Vortex liquids and vortex quantum Hall states in trapped rotating Bose gases, *J. Phys. B: At. Mol. Opt. Phys.* **37**, 301–310 (2004) (ID: 314366)

256. A. Carollo, I. Fuentes-Guridi, M. F. Santos and V. Vedral, Spin-1/2 geometric phase driven by decohering quantum fields, *Phys. Rev. Lett.* **92**, 020402 (2004). (ID: 340498)

257. A. J. Daley, P. O. Fedichev and P. Zoller, Single-atom cooling by superfluid immersion: A nondestructive method for qubits, *Phys. Rev. A* **69**, 022306 (2004). (ID: 324602)

258. P. O. Fedichev, M. J. Bijlsma and P. Zoller, Extended molecules and geometric scattering resonances in optical lattices, *Phys. Rev. Lett.* **92**, 080401 (2004). (ID: 324597)

259. P. Domokos and A. Vukics, Anomalous doppler-effect and polariton-mediated cooling of two-level atoms, *Phys. Rev. Lett.* **92**, 103601 (2004). (ID: 324599)

260. J. I. Cirac and P. Zoller, New frontiers in quantum information with atoms and ions, URL (ID: 324583)

261. R. Serra, A. Carollo, M. F. Santos and V. Vedral, Anyons and transmutation of statistics via a vacuum-induced Berry phase, *Phys. Rev. A* **70**, 044102 (2004). (ID: 340493)

262. U. Gavish, B. Yurke and Y. Imry, Generalized constraints on quantum amplification, *Phys. Rev. Lett.* **93**, 250601 (2004). (ID: 325935)

263. G. Pupillo, A. M. Rey, G. Brennen, C. Williams and C. W. Clark, Scalable quantum computation in systems with Bose–Hubbard dynamics, *J. Mod. Opt.* **51**, 2395 (2004). (ID: 340628)

264. J. Martikainen and P. Törmä, Quasi two-dimensional superfluid Fermi gases, *Phys. Rev. Lett.* **95**, 170407 (2005) (ID: 334530)

265. V. Ahufinger, L. Sanchez-Palencia, A. Kantian, A. Sanpera and M. Lewenstein, Disordered ultracold atomic gases in optical lattices: A case study of Fermi-Bose mixtures, *Phys. Rev. A* **72**, 063616 (2005) (ID: 332459)

266. L. Tian, R. Blatt and P. Zoller, Scalable ion trap quantum computing without moving ions, *Eur. Phys. J. D* **32**, 201–208 (2005) (ID: 308239)

267. D. Jaksch and P. Zoller, The cold atom Hubbard toolbox, *Annals of Physics* **315**, 52–79 (2005) (ID: 308231)

268. H. Büchler, M. Hermele, S. D. Huber, M. P. Fisher and P. Zoller, Atomic quantum simulator for lattice gauge theories and ring exchange models, arXiv:cond-mat/0503254 (ID: 302730)

269. J. Taylor, H. Engel, W. Dür, A. Yacoby, C. M. Marcus, P. Zoller and M. Lukin, Fault-tolerant architecture for quantum computation using electrically controlled semiconductor spins, *Nature Physics* and, 177–183 (2005). (ID: 308201)

270. A. Recati, P. O. Fedichev, W. Zwerger, J. von Delft and P. Zoller, Atomic quantum dots coupled to a reservoir of a superfluid Bose–Einstein condensate, *Phys. Rev. Lett.* **94**, 040404 (2005). (ID: 308229)

272. J. J. García-Ripoll, P. Zoller and J. I. Cirac, Quantum information processing with cold atoms and trapped ions, *J. Phys. B: At. Mol. Opt. Phys.* **38**, 567–578 (2005). (ID: 308228)

271. A. Micheli and P. Zoller, A single atom mirror for 1D atomic lattice gases, *Phys. Rev. A* **73**, 043613 (2005). (ID: 313116)

273. G. Brennen, G. Pupillo, A. M. Rey, C. W. Clark and C. Williams, Scalable register initialization for quantum computing in an optical lattice, *J. Phys. B: At. Mol. Opt. Phys.* **38**, 1687 (2005). (ID: 340627)

274. J. J. García-Ripoll, P. Zoller and J. I. Cirac, Coherent control of trapped ions using off-resonant lasers, *Phys. Rev. A* **71**, 062309 (2005). (ID: 308224)

275. K. Osterloh, M. Baig, L. Santos, P. Zoller and M. Lewenstein, Cold atoms in non-abelian gauge potentials: From the hofstadter "Moth" to lattice gauge theory, *Phys. Rev. Lett.* **95**, 010403 (2005). (ID: 308223)

276. U. Gavish and Y. Castin, Matter-wave localization in disordered cold atom lattices, *Phys. Rev. Lett.* **95**, 020401 (2005). (ID: 325911)

277. A. Carollo, The quantum trajectory approach to geometric phase for open systems, *Mod. Phys. Lett. A* **20**, 1635 (2005). (ID: 340489)

278. S. Powell, S. Sachdev and H. Büchler, Depletion of the Bose–Einstein condensate in Bose–Fermi mixtures, *Phys. Rev. B* **72**, 024534 (2005). (ID: 325091)

279. A. M. Rey, P. B. Blakie, G. Pupillo, C. Williams and C. W. Clark, Bragg spectroscopy of ultracold atoms loaded in an optical lattice, *Phys. Rev. A* **72**, 023407 (2005). (ID: 340626)

280. V. Steixner, P. Rabl and P. Zoller, Quantum feedback cooling of a single trapped ion in front of a mirror, *Phys. Rev. A* **72**, 043826 (2005). (ID: 308208)

281. W. H. Zurek, U. Dorner and P. Zoller, Dynamics of a quantum phase transition, *Phys. Rev. Lett.* **95**, 105701 (2005). (ID: 325892)

282. U. Dorner, T. Calarco, P. Zoller, A. Browaeys and P. Grangiere, Quantum logic via optimal control in holographic dipole traps, *J. Opt. B: Quantum Semiclass. Opt.* **7**, 341–346 (2005). (ID: 308221)

283. A. M. Rey, G. Pupillo, C. W. Clark and C. Williams, Ultracold atoms confined in an optical lattice plus parabolic potential: A closed-form approach, *Phys. Rev. A* **72**, 033616 (2005). (ID: 340625)

284. A. Griessner, A. J. Daley, D. Jaksch and P. Zoller, Fault-tolerant dissipative preparation of atomic quantum registers with fermions, *Phys. Rev. A* **72**, 032332 (2005). (ID: 308207)

285. J. I. Cirac and P. Zoller, Qubits, gatter und register, URL (ID: 332271)

286. A. Carollo and J. K. Pachos, Geometric phases and criticality in spin-chain systems, *Phys. Rev. Lett.* **95**, 157203 (2005). (ID: 340488)

287. P. Rabl, V. Steixner and P. Zoller, Quantum-limited velocity readout and quantum feedback cooling of a trapped ion via electromagnetically induced transparency, *Phys. Rev. A* **72**, 043823 (2005). (ID: 308209)

288. A. J. Daley, S. Clark, D. Jaksch and P. Zoller, Numerical analysis of coherent many-body currents in a single atom transistor, *Phys. Rev. A* **72**, 043618 (2005). (ID: 308211)

289. G. Paraoanu, Running-phase state in a Josephson washboard potential, *Review of Scientific Instruments* (Online) **72**, 134528 (2005). (ID: 308206)

290. P. Zoller, T. Beth, D. Binosi, R. Blatt, H. J. Briegel, D. Bruss, T. Calarco, J. I. Cirac, D. Deutsch, J. Eisert, A. Ekert, C. Fabre, N. Gisin, P. Grangiere, M. Grassl, S. Haroche, A. Imamoglu, A. Karlson, J. Kempe, L. Louwenhofen, S. Kröll, G. Leuchs, Quantum information processing and communication, *Eur. Phys. J. D* **36/2**, 203–228 (2005). (ID: 375863)

291. J. Kinnunen and P. Törmä, Beyond linear response spectroscopy of ultracold Fermi gases, *Phys. Rev. Lett.* **96**, 070402 (2006) (ID: 334754)

292. J. Gea-Banacloche, A. M. Rey, G. Pupillo, C. Williams and C. W. Clark, Mean-field treatment of the damping of the oscillations of a one-dimensional Bose gas in an optical lattice, *Phys. Rev. A* **73**, 013605 (2006). (ID: 340624)

293. A. Carollo, M. F. Santos and V. Vedral, Coherent evolution via reservoir driven holonomy, *Phys. Rev. Lett.* **96**, 020403 (2006). (ID: 340484)

294. G. Pupillo, C. Williams and N. Prokofev, Effects of finite temperature on the Mott-insulator state, *Phys. Rev. A* **73**, 013408 (2006). (ID: 340623)

295. P. Bushev, D. Rotter, A. Wilson, F. Dubin, C. Becher, J. Eschner, R. Blatt, V. Steixner, P. Rabl and P. Zoller, Feedback cooling of a single trapped ion, *Phys. Rev. Lett.* **96**, 043003 (2006). (ID: 332288)

296. A. S. Parkins, E. Solano and J. I. Cirac, Unconditional two-mode squeezing of separated atomic ensembles, *Phys. Rev. Lett.* **96**, 053602 (2006). (ID: 456305)

297. A. M. Rey, G. Pupillo and J. V. Porto, The role of interactions, tunneling, and harmonic confinement on the adiabatic loading of bosons in an optical lattice, *Phys. Rev. A* **73**, 023608 (2006). (ID: 340620)

298. A. M. Rey, G. Pupillo and J. V. Porto, The role of interactions, tunneling and harmonic confinement on the adiabatic loading of bosons in an optical lattice, *Phys. Rev. A* **73**, 023608 (2006). (ID: 325112)

299. U. Gavish, B. Yurke and Y. Imry, Quantum noise minimization in transistor amplifiers, *Phys. Rev. Lett.* **96**, 133602 (2006). (ID: 371234)

300. A. Carollo, A. Lozinski, G. M. Palma, M. F. Santos and V. Vedral, Geometric phase induced by a cyclically evolving squeezed vacuum reservoir, arXiv:quant-ph/0507101 (ID: 340619)

301. A. Micheli, G. Brennen and P. Zoller, A toolbox for lattice spin models with polar molecules, *Nature Physics* **2**, 341 (2006). (ID: 313100)

302. C. A. Muschik, K. Hammerer, E. Polzik and J. I. Cirac, Efficient quantum memory and entanglement between light and an atomic ensemble using magnetic fields, *Phys. Rev. A* **73**, 062329 (2006). (ID: 433791)

303. S. Trebst, U. Schollwöck, M. Troyer and P. Zoller, d-Wave resonating valence bond states of fermionic atoms in optical lattices, *Phys. Rev. Lett.* **96**, 250402 (2006). (ID: 372594)

304. G. Brennen, S. S. Bullock and D. P. O'Leary, Efficient circuits for exact-universal computation with qudits, *Quantum Information and Computation* **6**, 436 (2006). (ID: 371389)

305. G. Pupillo, A. M. Rey and G. G. Batrouni, Bragg spectroscopy of trapped one-dimensional strongly interacting bosons in optical lattices: Probing the cake structure, *Phys. Rev. A* **74**, 13601 (2006). (ID: 423709)

306. P. Rabl, D. DeMille, J. Doyle, M. Lukin, R. J. Schoelkopf and P. Zoller, Hybrid quantum processors: Molecular ensembles as quantum memory for solid state circuits, *Phys. Rev. Lett.* **97**, 033003 (2006). (ID: 353814)

307. M. Cozzini, T. Calarco, A. Recati and P. Zoller, Fast Rydberg gates without dipole blockade via quantum control, arXiv:quant-ph/0511118 (ID: 313117)

308. A. André, D. DeMille, J. Doyle, M. Lukin, S. E. Maxwell, P. Rabl, R. J. Schoelkopf and P. Zoller, A coherent all-electrical interface between polar molecules and mesoscopic superconducting resonators, *Nature Physics* **2**, 636 (2006). (ID: 363758)

309. G. Pupillo, A. M. Rey, C. Williams and C. W. Clark, Extended fermionization of 1-D bosons in optical lattices, *New J. Phys.* **8**, 161 (2006). (ID: 340635)

310. F. Plastina, G. Liberti and A. Carollo, Scaling of Berry's phase close to the Dicke quantum phase transition **76**, 182 (2006). (ID: 439140)

311. D. P. O'Leary, G. Brennen and S. S. Bullock, Parallelism for quantum computation with qudits, *Phys. Rev. A* **74**, 032334 (2006). (ID: 388511)

312. J. F. Sherson, H. Krauter, R. K. Olsson, B. Julsgaard, K. Hammerer, J. I. Cirac and E. Polzik, Quantum teleportation between light and matter, *Nature* **443**, 557 (2006). (ID: 433800)

313. A. Carollo and G. M. Palma, Observable geometric phase induced by a cyclically evolving dissipative process, *Laser Phys.* **16**, 1595 (2006). (ID: 439135)

314. G. Paraoanu, Microwave-induced coupling of superconducting qubits, *Phys. Rev. B* **74**, 140504 (2006). (ID: 430047)

315. G. Paraoanu, Interaction-free measurements with superconducting qubits, *Phys. Rev. Lett.* **97**, 180406 (2006). (ID: 430045)

316. K. Hammerer, E. Polzik and J. I. Cirac, High-fidelity teleportation between light and atoms *Phys. Rev. A* **74**, 064301 (2006). (ID: 433792)

317. A. Griessner, A. J. Daley, S. Clark, D. Jaksch and P. Zoller, Dark state cooling of atoms by superfluid immersion, *Phys. Rev. Lett.* **97**, 220403 (2006). (ID: 375452)

318. S. Saeidian, I. Lesanovsky and P. Schmelcher, Negative energy resonances of bosons in a magnetic quadrupole trap, **74**, 065402 (2006). (ID: 504084)

319. M. Titov, M. Müller and W. Belzig, Interaction-induced renormalization of Andreev reflection, *Phys. Rev. Lett.* **97**, 237006 (2006). (ID: 452688)

320. J. K. Pachos and A. Carollo, Geometric phases and criticality in spin systems, *Phil. Trans. R. Soc. Lond. A* **364**, 3463 (2006). (ID: 439139)

321. A. Griessner, A. J. Daley, S. Clark, D. Jaksch and P. Zoller, Dissipative dynamics of atomic Hubbard models coupled to a phonon bath: Dark state cooling of atoms within a Bloch band of an optical lattice, *New J. Phys.* **9**, 44 (2007) (ID: 428626)

322. K. Jähne, B. Yurke and U. Gavish, High-fidelity transfer of an arbitrary quantum state between harmonic oscillators, *Phys. Rev. A* **75**, 010301(R) (2007). (ID: 433635)

323. S. Tewari, S. D. Sarma, C. Nayak, C. Zhang and P. Zoller, Quantum computation using vortices and Majorana zero modes of a px+ipy superfluid of fermionic cold atoms, *Phys. Rev. Lett.* **98**, 010506 (2007). (ID: 436036)

324. I. Lesanovsky and W. v. Klitzing, Spontaneous emergence of angular momentum Josephson oscillations in coupled annular Bose–Einstein condensates, *Phys. Rev. Lett.* **98**, 050401 (2007). (ID: 504070)

325. H. Büchler, E. Demler, M. Lukin, A. Micheli, N. Prokofev, G. Pupillo and P. Zoller, Strongly correlated 2D quantum phases with cold polar molecules: Controlling the shape of the interaction potential, *Phys. Rev. Lett.* **98**, 060404 (2007). (ID: 376275)

326. S. D. Huber, E. Altman, H. Büchler and G. Blatter, Dynamical properties of ultracold bosons in an optical lattice, *Phys. Rev. B* **75**, 085106 (2007). (ID: 444465)

327. U. Schmidt, I. Lesanovsky and P. Schmelcher, Ultracold Rydberg atoms in a magneto-electric trap, *J. Phys. B: At. Mol. Opt. Phys.* **40**, 1003 (2007). (ID: 504061)

328. C. Schön, K. Hammerer, M. M. Wolf, J. I. Cirac and E. Solano, Sequential generation of matrix-product states in cavity QED, *Phys. Rev. A* **75**, 032311 (2007). (ID: 442644)

329. S. Diehl and C. Wetterich, Functional integral for ultracold zonic atoms, *Nucl. Phys. B* **770**, 206 (2007). (ID: 470566)

330. S. S. Bullock and G. Brennen, Qudit surface codes and gauge theory with finite cyclic groups, *J. Phys. A: Math. Gen.* **40**, 3481 (2007). (ID: 388513)

332. G. Brennen, A. Micheli and P. Zoller, Designing spin-1 lattice models using polar molecules, *New J. Phys.* **9**, 138 (2007). (ID: 430363)

331. G. Brennen and J. K. Pachos, Why should anyone care about computing with anyons, *Proc. R. Soc. A* **464**, 24 (2007). (ID: 504046)

333. S. Hofferberth, B. Fischer, T. Schumm, J. Schmiedmayer and I. Lesanovsky, Ultracold atoms in radio-frequency dressed potentials beyond the rotating-wave approximation, *Phys. Rev. A* **76**, 013401 (2007). (ID: 504074)

334. H. Büchler, A. Micheli and P. Zoller, Three-body interactions with cold polar molecules, *Nature Physics* B, 726 (2007). (ID: 462909)

335. S. Diehl, H. Gies, J. M. Pawlowski and C. Wetterich, Flow equations for the BCS-BEC crossover, *Phys. Rev. A* **76**, 021602(R) (2007). (ID: 470567)

336. I. Lesanovsky and W. v. Klitzing, Time-averaged adiabatic potentials: Versatile traps and waveguides for ultracold quantum gases, *Phys. Rev. Lett.* **99**, 083001 (2007). (ID: 504082)

337. S. Middelkamp, I. Lesanovsky and P. Schmelcher, Spectral properties of a Rydberg atom immersed in a Bose–Einstein condensate, *Phys. Rev. A* **76**, 022507 (2007). (ID: 504083)

338. S. Saeidian, I. Lesanovsky and P. Schmelcher, Atomic hyperfine resonances in a magnetic quadrupole field, *Phys. Rev. A* **76**, 023424 (2007). (ID: 582502)

339. M. Mayle, B. Hezel, I. Lesanovsky and P. Schmelcher, One-dimensional Rydberg gas in a magneto-electric trap, *Phys. Rev. Lett.* 113004 (2007). (ID: 504085)

340. W. Yi, G. Lin and L. Duan, Signal of Bose–Einstein condensation in an optical lattice at finite temperature, *Phys. Rev. A* **76**, 031602 (R) (2007). (ID: 554783)

341. Z. Idziaszek, T. Calarco and P. Zoller, Controlled collisions of a single atom and ion guided by movable trapping potentials, *Phys. Rev. A* **76**, 033409 (2007). (ID: 470101)

342. S. Hofferberth, I. Lesanovsky, B. Fischer, T. Schumm and J. Schmiedmayer, Non-equilibrium coherence dynamics in one-dimensional Bose gases, *Nature* **449**, 324 (2007). (ID: 504086)

343. A. Micheli, G. Pupillo, H. Büchler and P. Zoller, Cold polar molecules in 2D traps: Tailoring interactions with external fields for novel quantum phases, *Phys. Rev. A* **76**, 043604 (2007). (ID: 460560)

344. P. Rabl and P. Zoller, Molecular dipolar crystals as high fidelity quantum memory for hybrid quantum computing, *Phys. Rev. A* **76**, 042308 (2007). (ID: 494007)

345. M. Byrd and G. Brennen, General depolarized pure states: Identification and properties, *Phys. Lett. A* **372**, 1770 (2007). (ID: 504039)

346. A. Kantian, A. J. Daley, P. Törmä and P. Zoller, Atomic lattice excitons: From condensates to crystals, *New J. Phys.* **9**, 407 (2007). (ID: 514263)

347. B. Hezel, I. Lesanovsky and P. Schmelcher, Ultracold Rydberg atoms in a ioffe-pritchard trap, *Phys. Rev. A* **76**, 053417 (2007). (ID: 504087)

348. S. Diehl, H. Gies, J. M. Pawlowski and C. Wetterich, Renormalisation flow and universality for ultracold fermionic atoms, *Phys. Rev. A* **76**, 053627 (2007). (ID: 470570)

349. T. Calarco, P. Grangier, A. Wallraff and P. Zoller, Quantum leaps in small steps, *Nature Physics* **4**, 2 (2008). (ID: 602500)

350. S. Morrison and A. S. Parkins, Dynamical quantum phase transitions in the dissipative lipkin-meshkov-glick model with proposed realization in optical cavity QED, *Phys. Rev. Lett.* **100**, 040403 (2008). (ID: 558300)

351. G. Pupillo, A. Griessner, A. Micheli, M. Ortner, Wang, Daw-Wei and P. Zoller, Cold atoms and molecules in self-assembled dipolar lattices, *Phys. Rev. Lett.* **100**, 050402 (2008). (ID: 519056)

352. C. Genes, D. Vitali, P. Tombesi, S. Gigan and M. Aspelmeyer, Ground-state cooling of a micromechanical oscillator: Comparing cold damping and cavity-assisted cooling schemes, *Phys. Rev. A* **77**, 033804 (2008). (ID: 651385)

353. A. V. Gorshkov, L. Jiang, M. Greiner, P. Zoller and M. Lukin, Coherent quantum optical control with subwavelength resolution, *Phys. Rev. Lett.* **100**, 093005 (2008). (ID: 519563)

354. A. J. Daley, P. Zoller and B. Trauzettel, Andreev-like reflections with cold atoms, *Phys. Rev. Lett.* **100**, 110404 (2008). (ID: 530306)

355. S. Morrison and A. S. Parkins, Collective spin systems in dispersive optical cavity QED: Quantum phase transitions and entanglement, *Phys. Rev. A* **77**, 043810 (2008). (ID: 539919)

356. G. Brennen and J. K. Pachos, Vortex lattice locking in rotating two-component Bose–Einstein condensates, *New J. Phys.* **10**, 10 (2008). (ID: 509317)

357. S. Hofferberth, I. Lesanovsky, T. Schumm, A. Imambekov, V. Gritsev, E. Demler and J. Schmiedmayer, Probing quantum and thermal noise in an interacting many-body system, *Nature Physics* **4**, 489 (2008). (ID: 582503)

358. L. Jiang, G. Brennen, A. V. Gorshkov, K. Hammerer, M. Hafezi, E. Demler, M. Lukin and P. Zoller, Anyonic interferometry and protected memories in atomic spin lattices, *Nature Physics* **4**, 482 (2008). (ID: 538104)

359. B. Kraus, H. Büchler, S. Diehl, A. Kantian, A. Micheli and P. Zoller, Preparation of entangled states by quantum markov processes, *Phys. Rev. A* **78**, 042307 (2008). (ID: 576234)

360. M. Baranov, Theoretical progress in many-body physics with ultracold dipolar gases, *Physics Reports* **464**, 71 (2008). (ID: 610896)

361. M. Baranov, H. Fehrmann and M. Lewenstein, Wigner crystallization in rapidly rotating 2D dipolar Fermi gases, *Phys. Rev. Lett.* **100**, 200402 (2008). (ID: 593996)

362. M. Wallquist, P. Rabl, M. Lukin and P. Zoller, Theory of cavity-assisted microwave cooling of polar molecules, *New J. Phys.* **10**, 063005 (2008). (ID: 567588)

363. L. Tornberg, M. Wallquist, G. Johansson, V. S. Shumeiko and G. Wendin, Implementation of the three-qubit phase-flip error correction code with superconducting qubits, *Phys. Rev. B* **77**, 214528 (2008). (ID: 575605)

364. W. Yi, A. J. Daley, G. Pupillo and P. Zoller, State-dependent, addressable subwavelength lattices with cold atoms, *New J. Phys.* **10**, 073015 (2008). (ID: 582500)

365. X. Lu, C. Wu, A. Micheli and G. Pupillo, Structure and melting behavior of classical bilayer crystals of dipoles, *Phys. Rev. B* **78**, 024108 (2008). (ID: 596533)

366. S. Morrison, A. Kantian, A. J. Daley, H. Katzgraber, M. Lewenstein, H. Büchler and P. Zoller, Physical replicas and the Bose-glass in cold atomic gases, *New J. Phys.* **10**, 073032 (2008). (ID: 582497)

367. A. V. Gorshkov, P. Rabl, G. Pupillo, A. Micheli, P. Zoller, M. Lukin and H. Büchler, Suppression of inelastic collisions between polar molecules with a repulsive shield, *Phys. Rev. Lett.* **101**, 073201 (2008). (ID: 582039)

368. S. Diehl, H. C. Krahl and M. Scherer, Three-body scattering from nonperturbative flow equations, *Phys. Rev. C* **78**, 034001 (2008). (ID: 547706)

369. S. Diehl, A. Micheli, A. Kantian, B. Kraus, H. Büchler and P. Zoller, Quantum states and phases in driven open quantum systems with cold atoms, *Nature Physics* **4**, 878 (2008). (ID: 576233)

370. M. Müller, L. Liang, I. Lesanovsky and P. Zoller, Trapped Rydberg ions: From spin chains to fast quantum gates, *New J. Phys.* **10**, 093009 (2008). (ID: 520204)

371. M. Baranov, C. Lobo and G. V. Shlyapnikov, Superfluid pairing between fermions with unequal masses, *Phys. Rev. A* **78**, 033620 (2008). (ID: 593993)

372. K. Jähne, K. Hammerer and M. Wallquist, Ground state cooling of a nanomechanical resonator via a Cooper pair box qubit, *New J. Phys.* **10**, 095019 (2008). (ID: 619409)

373. A. J. Daley, M. M. Boyd, J. Ye and P. Zoller, Quantum computing with alkaline-earth-metal atoms, *Phys. Rev. Lett.* **101**, 170504 (2008). (ID: 606889)

374. S. Middelkamp, I. Lesanovsky and P. Schmelcher, Interaction-induced trapping and pulsed emission of a magnetically insensitive Bose–Einstein Condensate **84**, 40011 (2008). (ID: 635567)

375. S. Floerchinger, M. Scherer, S. Diehl and C. Wetterich, Particle-hole fluctuations in the BCS-BEC crossover, *Phys. Rev. B* **78**, 174528 (2008). (ID: 606176)

376. B. Capogrosso-Sansone, S. Wessel, H. Büchler, P. Zoller and G. Pupillo, Phase diagram of one-dimensional hard-core bosons with three-body interactions, *Phys. Rev. B* **79**, 020503 (R) (2009). (ID: 603828)

377. K. Hammerer, M. Aspelmeyer, E. Polzik and P. Zoller, Establishing Einstein-Poldosky-Rosen channels between nanomechanics and atomic ensembles, *Phys. Rev. Lett.* **102**, 020501 (2009). (ID: 637866)

378. I. Lesanovsky, M. Müller and P. Zoller, Trap assisted creation of giant molecules and Rydberg-mediated coherent charge transfer in a Penning trap, *Phys. Rev. A* **79**, 010701(R) (2009). (ID: 620459)

379. A. J. Daley, J. Taylor, S. Diehl, M. Baranov and P. Zoller, Atomic three-body loss as a dynamical three-body interaction, *Phys. Rev. Lett.* **102**, 040402 (2009). (ID: 627388)

380. S. Montangero, R. Fazio, P. Zoller and G. Pupillo, Dipole oscillations of confined lattice bosons in one dimension, *Phys. Rev. A* **79**, 041602(R) (2009). (ID: 620456)

381. H. Lignier, A. Zenesini, D. Ciampini, O. Morsch, E. Arimondo, S. Montangero, G. Pupillo and R. Fazio, Trap-modulation spectroscopy of the Mott-insulator transition in optical lattices, *Phys. Rev. A* **79**, 041601(R) (2009). (ID: 641974)

382. A. V. Gorshkov, A. M. Rey, A. J. Daley, M. M. Boyd, J. Ye, P. Zoller, and M. Lukin, Alkaline-earth atoms as few-qubit quantum registers, *Phys. Rev. Lett.* **102**, 110503 (2009). (ID: 644448)

383. C. Schwenke, T. Calarco, P. Zoller and C. Koch, Collective Rydberg excitations of an atomic gas confined in a ring lattice, *Phys. Rev. A* **79**, 043419 (2009). (ID: 650348)

384. M. Müller, I. Lesanovsky, H. Weimer, H. Büchler and P. Zoller, Mesoscopic Rydberg gate based on electromagnetically induced transparency, *Phys. Rev. Lett.* **102**, 170502 (2009). (ID: 628410)

385. M. Ortner, A. Micheli, G. Pupillo and P. Zoller, Quantum simulations of extended Hubbard models with dipolar crystals, *New J. Phys.* **11**, 055045 (2009). (ID: 665436)

386. H. Venzl, A. J. Daley, F. Mintert and A. Buchleitner, Simulability and regularity of complex quantum systems, *Phys. Rev. E* **79**, 056223 (2009). (ID: 627395)

387. K. Jähne, C. Genes, K. Hammerer, M. Wallquist, E. Polzik and P. Zoller, Cavity-assisted squeezing of a mechanical oscillator, *Phys. Rev. A* **79**, 063819 (2009). (ID: 679748)

388. K. Hammerer, M. Wallquist, C. Genes, M. Ludwig, F. Marquardt, P. Treutlein, P. Zoller, J. Ye and H. J. Kimble, Strong coupling of a mechanical oscillator and a single atom, *Phys. Rev. Lett.* **103**, 063005 (2009). (ID: 680115)

389. S. Gröblacher, K. Hammerer, M. Vanner and M. Aspelmeyer, Observation of strong coupling between a micromechanical resonator and an optical cavity field, *Nature* **460**, 724 (2009). (ID: 669662)

390. K. Stannigel, M. König, J. Niegemann and K. Busch, Discontinuous Galerkin time-domain computations of metallic nanostructures, *Opt. Express* **17**, 14934 (2009). (ID: 702959)

391. Y. Han, Y. Chan, W. Yi, A. J. Daley, S. Diehl, P. Zoller and L. Duan, Stabilization of the p-wave superfluid state in an optical lattice, *Phys. Rev. Lett.* **103**, 070404 (2009). (ID: 707070)

392. D. Chang, J. D. Thompson, H. Park, V. Vuletic, P. Zoller, A. Zibrov and M. Lukin, Trapping and manipulation of isolated atoms using nanoscale plasmonic structures, *Phys. Rev. Lett.* **103**, 123004 (2009). (ID: 686216)

393. S. Chang, Radio frequency response of the strongly interacting Fermi gases at finite temperatures, *Phys. Rev. A* **80**, 033623 (2009). (ID: 685305)

394. A. Kantian, M. Dalmonte, S. Diehl, W. Hofstetter, P. Zoller and A. J. Daley, Atomic color superfluid via three-body loss, *Phys. Rev. Lett.* **103**, 240401 (2009). (ID: 707092)

395. P. Rabl, C. Genes, K. Hammerer and M. Aspelmeyer, Phase-noise induced limitations on cooling and coherent evolution in optomechanical systems, *DFG — Physik, Math. u. Geowissen.* **80**, 063819 (2009). (ID: 666215)

396. P. Zoller, M. Wallquist, K. Hammerer, P. Rabl and M. Lukin, Hybrid quantum devices and quantum engineering, *Physica Scripta*, 014001 (2009). (ID: 716806)

397. D. Chang, C. A. Regal, S. B. Papp, D. J. Wilson, J. Ye, O. Painter, H. J. Kimble and P. Zoller, Cavity optomechanics using an optically levitated nanosphere, URL (ID: 708587)

398. B. Olmos, M. Müller and I. Lesanovsky, Thermalization of a strongly interacting 1D Rydberg lattice gas, *New J. Phys.* **12**, 013024 (2010). (ID: 717001)

399. P. Amann, M. Cordin, C. Braun, B. A. Lechner, A. Menzel, E. Bertel, C. Franchini, R. Zucca, J. Redinger, M. Baranov and S. Diehl, Electronically driven phase transitions in a quasi-one-dimensional adsorbate system, *Eur. Phys. J. B* (2010). (ID: 717176)

400. M. Wallquist, K. Hammerer, P. Zoller, C. Genes, M. Ludwig, F. Marquardt, P. Treutlein, J. Ye and H. Kimble, Single-atom cavity QED and optomicromechanics, *Phys. Rev. A* **81**, 023816 (2010). (ID: 716968)

401. B. Capogrosso-Sansone, C. Trefzger, M. Lewenstein, P. Zoller and G. Pupillo, Quantum phases of cold polar molecules in 2D optical lattices, *Phys. Rev. Lett.* **104**, 125301 (2010). (ID: 686428)

402. M. Baranov, A. Micheli, S. Ronen and P. Zoller, Bilayer superfluidity of fermionic polar molecules: Many body effects, *Phys. Rev. A* **83**, 043602 (2010). (ID: 717481)

403. J. Schachenmayer, G. Pupillo and A. J. Daley, Time-dependent currents of 1D bosons in an optical lattice, *New J. Phys.* **12**, 025014 (2010). (ID: 707111)

404. A. V. Gorshkov, M. Hermele, V. Gurarie, C. Xu, P. S. Julienne, J. Ye, P. Zoller, E. Demler, M. Lukin and A. M. Rey, Two-orbital SU(N) magnetism with ultracold alkaline-earth atoms, *Nature Physics* **6**, 289 (2010). (ID: 681277)

405. S. Ronen and J. L. Bohn, Zero sound in dipolar Fermi gases, *Phys. Rev. A*, 033601 (2010). (ID: 717480)

406. K. Hammerer, A. Sorensen and E. Polzik, Quantum interface between light and atomic ensembles, *Rev. Mod. Phys.* **82**, 1041 (2010). (ID: 637865)

407. H. Weimer, M. Müller, I. Lesanovsky, P. Zoller and H. Büchler, A Rydberg quantum simulator, *Nature Physics* **6**, 382 (2010). (ID: 695077)

408. A. Glätzle, K. Hammerer, A. J. Daley, R. Blatt and P. Zoller, A single trapped atom in front of an oscillating mirror, *Opt. Com.* **283**, 758 (2010). (ID: 717009)

409. Y. Deng, B. Zhao, R. Wei, S. Chen, Z. Chen and Pan, Jian-Wei, Light pulse in \Lamdba-type cold-atom gases, *Phys. Rev. A* **81**, 043403 (2010). (ID: 717242)

410. S. Diehl, M. Baranov, A. J. Daley and P. Zoller, Observability of quantum criticality and a continuous supersolid in atomic gases, *Phys. Rev. Lett.* **104**, 165301 (2010). (ID: 716766)

411. A. Kantian, A. J. Daley and P. Zoller, An eta-condensate of fermionic atom pairs via adiabatic state preparation, *Phys. Rev. Lett.* **104**, (2010). (ID: 716970)

412. B. Zhao, K. Hammerer, M. Müller and P. Zoller, Efficient quantum repeater based on deterministic Rydberg gates, *Phys. Rev. A* **81**, 052329 (2010). (ID: 717186)

413. M. Aspelmeyer, S. Gröblacher, K. Hammerer and N. Kiesel, Quantum optomechanics — Throwing a glance, *JOSA B* **27**, 189 (2010). (ID: 717340)

414. P. Rabl, S. J. Kolkowitz, F. H. Koppens, J. Harris, P. Zoller and M. Lukin, A quantum spin transducer based on nanoelectromechanical resonator arrays, *Nature Physics*, 602 (2010). (ID: 702633)

415. G. Pupillo, A. Micheli, M. Boninsegni, I. Lesanovsky and P. Zoller, Strongly correlated gases of Rydberg-dressed atoms: quantum and classical dynamics, *Phys. Rev. Lett.* **104**, 223002 (2010). (ID: 716985)

416. S. Diehl, S. Floerchinger, H. Gies, J. M. Pawlowski and C. Wetterich, Functional renormalization group approach to the BCS-BEC crossover, *Annals of Physics*, 615 (2010). (ID: 694286)

417. M. Gustavsson, E. Haller, M. J. Mark, J. G. Danzl, R. Hart, A. J. Daley and H. C. Nägerl, Interference of interacting matter waves, *New J. Phys.* **12**, 12 (2010). (ID: 717260)

418. S. Diehl, A. Tomadin, A. Micheli, R. Fazio and P. Zoller, Dynamical phase transitions and instabilities in open atomic many-body systems, *Phys. Rev. Lett.* **105**, 015702 (2010). (ID: 717159)

419. E. Haller, R. Hart, M. J. Mark, J. Danzl, L. Reichsöllner, M. Gustavsson, M. Dalmonte, G. Pupillo and H. C. Nägerl, Pinning quantum phase transition for a Luttinger liquid of strongly interacting bosons, *Nature* **466**, 597 (2010). (ID: 717204)

420. M. Ludwig, K. Hammerer and F. Marquardt, Entanglement of mechanical oscillators coupled to a nonequilibrium environment, *Phys. Rev. A* **82**, 012333 (2010). (ID: 716971)

421. S. Nimmrichter, K. Hammerer, P. Asenbaum, H. Ritsch and M. Arndt, Master equation for the motion of a polarizable particle in a multimode cavity, *New J. Phys.* **12**, 083003 (2010). (ID: 717185)

422. Z. Idziaszek, T. Calarco and P. Zoller, Ion-assisted ground-state cooling of a trapped polar molecule, *Phys. Rev. A* **83**, 053413 (2010). (ID: 717300)

423. A. Micheli, Z. Idziaszek, G. Pupillo, M. Baranov, P. Zoller and P. S. Julienne, Universal rates for reactive ultracold polar molecules in reduced dimensions, *Phys. Rev. Lett.* **105**, 073202 (2010). (ID: 717207)

424. S. Diehl, M. Baranov, A. J. Daley and P. Zoller, Quantum field theory for the three-body constrained lattice Bose gas — Part II: Application to the many-body problem, *Phys. Rev. B* **82**, 064510 (2010). (ID: 716840)

425. S. Diehl, M. Baranov, A. J. Daley and P. Zoller, Quantum field theory for the three-body constrained lattice Bose gas — Part I: Formal developments, *Phys. Rev. B* **82**, 064509 (2010). (ID: 716839)

426. K. Hammerer, K. Stannigel, C. Genes, P. Zoller, P. Treutlein, S. Camerer, D. Hunger and T. W. Hänsch, Optical lattices with micromechanical mirrors, *Phys. Rev. A* **82**, 021803(R) (2010). (ID: 717187)

427. F. Cinti, P. Jain, M. Boninsegni, A. Micheli, P. Zoller and G. Pupillo, Supersolid droplet crystal in a dipole-blockaded gas, *Phys. Rev. Lett.* **105**, 135301 (2010). (ID: 717214)

428. M. Dalmonte, G. Pupillo and P. Zoller, One-dimensional quantum liquids with power-law interactions: The luttinger staircase, *Phys. Rev. Lett.* **105**, 140401 (2010). (ID: 717312)

429. M. Cordin, B. A. Lechner, P. Amann, A. Menzel, E. Bertel, C. Franchini, R. Zucca, J. Redinger, M. Baranov and S. Diehl, Phase transitions driven by competing interactions in low-dimensional systems, *Europhysics Letters* **92**, 26004 (2010). (ID: 717348)

430. S. Diehl, W. Yi, A. J. Daley and P. Zoller, Dissipation-induced d-wave pairing of fermionic atoms in an optical lattice, *Phys. Rev. Lett.* **105**, 227001 (2010). (ID: 717276)

431. K. Stannigel, P. Rabl, A. Sorensen, P. Zoller and M. Lukin, Optomechanical transducers for long-distance quantum communication, *Phys. Rev. Lett.* **105**, 220501 (2010). (ID: 717347)

432. H. Pichler, A. J. Daley and P. Zoller, Nonequilibrium dynamics of bosonic atoms in optical lattices: Decoherence of many-body states due to spontaneous emission, *Phys. Rev. A* **82**, 063605 (2010). (ID: 717301)

433. G. Hétet, L. Slodicka, A. Glätzle, M. Hennrich and R. Blatt, QED with a spherical mirror, *Phys. Rev. A*, 063812 (2010). (ID: 717398)

434. J. Schachenmayer, I. Lesanovsky, A. Micheli and A. J. Daley, Dynamical crystal creation with polar molecules or Rydberg atoms in optical lattices, *New J. Phys.* **12**, 103044 (2010). (ID: 717357)

435. A. Tomadin, S. Diehl and P. Zoller, Nonequilibrium phase diagram of a driven and dissipative many-body system, *Phys. Rev. A* **83**, 013611 (2011). (ID: 717765)

436. J. T. Barreiro, M. Müller, P. Schindler, D. Nigg, T. Monz, M. Chwalla, M. Hennrich, C. F. Roos, P. Zoller and R. Blatt, An open-system quantum simulator with trapped ions, *Nature* **470**, 486 (2011). (ID: 717617)

437. I. Titvinidze, A. Privitera, S. Chang, S. Diehl, M. Baranov, A. J. Daley and W. Hofstetter, Magnetism and domain formation in SU(3)-symmetric multi-species Fermi mixtures, *New J. Phys.* **13**, 035013 (2011). (ID: 717395)

438. D. Marcos, C. Emary, T. Brandes and R. Aguado, Non-markovian effects in the quantum noise of interacting nanostructures, *Phys. Rev. B*, 125426 (2011). (ID: 717657)

439. M. Dalmonte, M. Di Dio, L. Barbiero and F. Ortolani, Homogeneous and inhomogeneous magnetic phases of constrained dipolar bosons, *Phys. Rev. B* **83**, 155110 (2011). (ID: 717917)

440. F. Schmidt-Kaler, T. Feldker, D. Kolbe, J. Walz, M. Müller, P. Zoller, W. Li and I. Lesanovsky, Rydberg excitation of trapped cold ions: A detailed case study, *New J. Phys.* **13**, 075014 (2011). (ID: 717667)

441. J. Schachenmayer, A. J. Daley and P. Zoller, Atomic matter-wave revivals with definite atom number in an optical lattice, *Phys. Rev. A* **83**, 043614 (2011). (ID: 717393)

442. L. Jiang, T. Kitagawa, J. Alicea, A. Akhmerov, D. Pekker, G. Refael, I. J. Cirac, E. Demler, M. Lukin and P. Zoller, Majorana fermions in equilibrium and in driven cold-atom quantum wires, *Phys. Rev. Lett.* **106**, 220402 (2011). (ID: 717687)

443. R. Wei, B. Zhao, Y. Deng, Y. Chen and J. Pan, Deterministic spin-wave interferometer based on the Rydberg blockade, *Phys. Rev. A* **83**, 063623 (2011). (ID: 717820)

444. J. Song and F. Zhou, Anomalous dimers in quantum mixtures near broad resonances: Pauli blocking, Fermi surface dynamics, and implications, **84**, 13601 (2011). (ID: 717884)

445. M. Vanner, I. Pikovski, G. Cole, M. S. Kim, Č. Brukner, K. Hammerer, G. Milburn and M. Aspelmeyer, Pulsed quantum optomechanics, *PNAS* **108**, 16182 (2011). (ID: 717339)

446. A. J. Daley, J. Ye and P. Zoller, State-dependent lattices for quantum computing with alkaline-earth-metal atoms, *Eur. Phys. J. D* **65**, 207 (2011). (ID: 717608)

447. P. Rabl, The photon blockade effect in optomechanical systems, *Phys. Rev. Lett.* **107**, 063601 (2011). (ID: 717658)

448. A. Privitera, I. Titvinidze, S. Chang, S. Diehl, A. J. Daley and W. Hofstetter, Loss-induced phase separation and pairing for 3-species atomic lattice fermions, *Phys. Rev. A* **84**, 021601 (2011). (ID: 717349)

449. M. Müller, K. Hammerer, Y. Zhou, C. F. Roos and P. Zoller, Simulating open quantum systems: From many-body interactions to stabilizer pumping, *New J. Phys.* **13**, 085007 (2011). (ID: 717750)

450. M. Dalmonte, E. Ercolessi and L. Taddia, Estimating quasi-long-range order via Rényi entropies, *Phys. Rev. B* **84**, 085110 (2011). (ID: 719239)

451. A. Safavi-Naini, P. Rabl, P. Weck and H. Sadeghpour, Microscopic model of electric-field-noise heating in ion traps, *Phys. Rev. A* **84**, 023412 (2011). (ID: 717725)

452. B. P. Lanyon, C. Hempel, D. Nigg, M. Müller, R. Gerritsma, F. Zähringer, P. Schindler, J. T. Barreiro, M. Rambach, G. Kirchmair, M. Hennrich, P. Zoller, R. Blatt and C. F. Roos, Universal digital quantum simulation with trapped ions, *Science* **334**, 57 (2011). (ID: 717768)

453. H. Zhang, X. M. Jin, J. Yang, H. N. Dai, S. J. Yang, T. M. Zhao, J. Rui, Y. He, X. Jiang, F. Yang, G. S. Pan, Z. S. Yuan, Y. Deng, Z. B. Chen, X. H. Bao, S. Chen, B. Zhao and J. Pan, Preparation and storage of frequency-uncorrelated entangled photons from cavity-enhanced spontaneous parametric downconversion, *Nature Photonics*, 628 (2011). (ID: 717980)

454. S. Diehl, E. Rico Ortega, M. Baranov and P. Zoller, Topology by dissipation in atomic quantum wires, *Nature Physics* **7**, 971 (2011). (ID: 717686)

455. T. Grass, M. Baranov and M. Lewenstein, Robustness of fractional quantum Hall states with dipolar atoms in artificial gauge fields, *Phys. Rev. A* **84**, 043605 (2011). (ID: 717780)

456. M. Dalmonte, P. Zoller and G. Pupillo, Trimer liquids and crystals of polar molecules in coupled wires, *Phys. Rev. Lett.* **107**, 163202 (2011). (ID: 717651)

457. M. Ortner, Y. Zhou, P. Rabl and P. Zoller, Quantum information processing in self-assembled crystals of cold polar molecules, *Quant. Inf. Proc.* **10**, 793 (2011). (ID: 717726)

458. R. Sandner, M. Müller, A. J. Daley and P. Zoller, Spatial Pauli blocking of spontaneous emission in optical lattices, *Phys. Rev. A* **84**, 043825 (2011). (ID: 717732)

459. K. Stannigel, P. Rabl, A. Sorensen, M. Lukin and P. Zoller, Optomechanical transducers for quantum information processing, *Phys. Rev. A* **84**, 042341 (2011). (ID: 717703)

460. M. Gibertini, A. Tomadin, F. Guinea, M. Katsnelson and M. Polini, Electron-hole puddles in the absence of charged impurities, *Phys. Rev. B* **85**, 201405(R) (2011). (ID: 717856)

461. Y. Zhou, M. Ortner and P. Rabl, Long-range and frustrated spin-spin interactions in crystals of cold polar molecules, *Phys. Rev. A* **84**, 052332 (2011). (ID: 717773)

462. E. Haller, M. Rabi, M. J. Mark, J. Danzl, R. Hart, K. Lauber, G. Pupillo and H. C. Nägerl, Three-body correlation functions and recombination rates for bosons in three and one dimensions, *Phys. Rev. Lett.* **107**, 230404 (2011). (ID: 717742)

463. J. Braun, S. Diehl and M. Scherer, Finite-size and particle-number effects in an ultracold Fermi gas at unitarity, *Phys. Rev. A* **84**, 063616 (2011). (ID: 717771)

464. L. Sieberer and M. Baranov, Collective modes, stability, and superfluid transition of a quasi-two-dimensional dipolar Fermi gas, *Phys. Rev. A* **84**, 063633 (2011). (ID: 717842)

465. T. Roscilde, C. Degli Esposti Boschi and M. Dalmonte, Pairing, crystallization and string correlations of mass-imbalanced atomic mixtures in one-dimensional optical lattices, *Europhysics Letters* **97**, (2012). (ID: 717893)

466. E. Berg, M. S. Rudner and S. A. Kivelson, Electronic liquid crystalline phases in a spin-orbit coupled two-dimensional electron gas, *Phys. Rev. B* **85**, 11 (2012). (ID: 717859)

467. S. J. Kolkowitz, A. Jayich, Q. Unterreithmeier, C. H. Bennett, P. Rabl, J. Harris and M. Lukin, Coherent sensing of a mechanical resonator with a single-spin qubit, *Science* **335**, 1603 (2012). (ID: 718079)

468. D. Borzov, M. S. Mashayekhi, S. Zhang, J. Song and F. Zhou, Three-dimensional Bose gas near a Feshbach resonance, *Phys. Rev. A* **85**, 023620 (2012). (ID: 717883)

469. K. Lakomy, R. Nath and L. Santos, Spontaneous crystallization and filamentation of solitons in dipolar condensates, *Phys. Rev. A* **85**, 033618 (2012). (ID: 718012)

470. M. Hafezi and P. Rabl, Optomechanically induced non-reciprocity in microring resonators, *Optics Express* **20**, 7672 (2012). (ID: 717802)

471. I. Cirac and P. Zoller, Goals and opportunities in quantum simulation. *Nature Physics* **8**, 264 (2012). (ID: 718063)

472. M. Dalmonte, E. Ercolessi and L. Taddia, Critical properties and Rényi entropies of the spin-3/2 XXZ chain, *Phys. Rev. B* **85**, 165112 (2012). (ID: 717902)

473. D. Marcos, A. Tomadin, S. Diehl and P. Rabl, Photon condensation in circuit QED by engineered dissipation, *New J. Phys.* **14**, 055005 (2012). (ID: 717877)

474. W. Yi, S. Diehl, A. J. Daley and P. Zoller, Driven-dissipative many-body pairing states for cold fermionic atoms in an optical lattice, *New J. Phys.* **14**, 055002 (2012). (ID: 717858)

475. B. Zhao, A. Glätzle, G. Pupillo and P. Zoller, Atomic Rydberg reservoirs for polar molecules *Phys. Rev. Lett.* **108**, 193007 (2012). (ID: 717841)

476. H. N. Dai, H. Zhang, S. J. Yang, T. M. Zhao, J. Rui, Y. Deng, L. Li, N. L. Liu, S. Chen, X. H. Bao, X. Jin, B. Zhao and J. Pan, Holographic storage of biphoton entanglement, *Phys. Rev. Lett.* **108**, 21501 (2012). (ID: 718080)

477. M. Dalmonte, K. Dieckmann, T. Roscilde, C. Hartl, A. E. Feiguin, U. Schollwöck and F. Heidrich-Meisner, Dimer, trimer and FFLO liquids in mass- and spin-imbalanced trapped binary mixtures in one dimension, *Phys. Rev. A* **85**, 063608 (2012). (ID: 718030)

478. K. Stannigel, P. Rabl and P. Zoller, Driven-dissipative preparation of entangled states in cascaded quantum-optical networks, *New J. Phys.* **14**, 063014 (2012). (ID: 717830)

479. N. Henkel, F. Cinti, P. Jain, G. Pupillo and T. Pohl, Supersolid vortex crystals in Rydberg-dressed Bose-Einstein condensates, *Phys. Rev. Lett.* **108**, 265301 (2012). (ID: 717975)

480. I. Boettcher, J. M. Pawlowski and S. Diehl, Ultracold atoms and the functional renormalization group, *Nucl. Phys. B* **228**, 63 (2012). (ID: 718053)

481. K. Lakomy, R. Nath and L. Santos, Soliton molecules in dipolar Bose–Einstein condensates, *Phys. Rev. A* **86**, 013610 (2012). (ID: 718074)

482. K. Stannigel, P. Komar, S. Habraken, S. D. Bennett, M. Lukin, P. Zoller and P. Rabl, Optomechanical quantum information processing with photons and phonons, *Phys. Rev. Lett.* **109**, 013603 (2012). (ID: 717990)

483. A. J. Daley, H. Pichler, J. Schachenmayer and P. Zoller, Measuring entanglement growth in quench dynamics of bosons in an optical lattice, *Phys. Rev. Lett.* **109**, 020505 (2012). (ID: 718070)

484. M. Baranov, M. Dalmonte, G. Pupillo and P. Zoller, Condensed matter theory of dipolar quantum gases, *Chem. Rev.*, (2012). (ID: 718140)

485. K. Lakomy, R. Nath and L. Santos, Faraday patterns in coupled one-dimensional dipolar condensates, *Phys. Rev. A* **86**, 023620 (2012). (ID: 718139)

486. A. Tomadin, S. Diehl, M. Lukin, P. Rabl and P. Zoller, Reservoir engineering and dynamical phase transitions in optomechanical arrays, *Phys. Rev. A* **86**, 033821 (2012). (ID: 718117)

487. Y. N. Obukhov, T. Ramos and G. Rubilar, Relativistic Lagrangian model of a nematic liquid crystal interacting with an electromagnetic field, *Phys. Rev. E* **86**, 031703 (2012). (ID: 718239)

488. C. Bardyn, M. Baranov, E. Rico Ortega, A. Imamoglu, P. Zoller and S. Diehl, Majorana modes in driven-dissipative atomic superfluids with zero chern number, *Phys. Rev. Lett.* **109**, 130402 (2012). (ID: 717881)

489. A. Glätzle, R. Nath, B. Zhao, G. Pupillo and P. Zoller, Driven-dissipative dynamics of a strongly interacting Rydberg gas, *Phys. Rev. A* **86**, 043403 (2012). (ID: 718134)

490. M. Cordin, P. Amann, A. Menzel, E. Bertel, M. Baranov, S. Diehl, J. Redinger and C. Franchini, Cleavage surface of the BaFe2-xCoxAs2 and FeySe1-xTex superconductors: A combined STM plus LEED study, *Phys. Rev. B* **86**, 167401 (2012). (ID: 718263)

491. D. Banerjee, M. Dalmonte, M. Müller, E. Rico Ortega, P. Stebler, U. Wiese and P. Zoller, Atomic quantum simulation of dynamical gauge fields coupled to fermionic matter: From string breaking to evolution after a quench, *Phys. Rev. Lett.* **109**, 175302 (2012). (ID: 718094)

492. S. Habraken, K. Stannigel, M. Lukin, P. Zoller and P. Rabl, Continuous mode cooling and phonon routers for phononic quantum networks, *New J. Phys.* **24**, 115004 (2012). (ID: 718101)

493. H. Pichler, J. Schachenmayer, J. Simon, P. Zoller and A. J. Daley, Noise- and disorder-resilient optical lattices, *Phys. Rev. A* **86**, 051605 (2012). (ID: 718093)

494. C. Kraus, S. Diehl, P. Zoller and M. Baranov, Preparing and probing atomic Majorana fermions and topological order in optical lattices, *New J. Phys.* **14**, 113036 (2012). (ID: 717882)

495. M. Gullans, T. Tiecke, S. Chang, J. Feist, J. D. Thompson, I. J. Cirac, P. Zoller and M. Lukin, Nanoplasmonic lattices for ultracold atoms, *Phys. Rev. Lett.* **109**, 235309 (2012). (ID: 718190)

496. S. D. Bennett, S. J. Kolkowitz, Q. Unterreithmeier, P. Rabl, A. Jayich, J. Harris and M. Lukin, Measuring mechanical motion with a single spin, *New J. Phys.* **14**, 125004 (2012). (ID: 718104)

497. C. Kraus and T. J. Osborne, Time-dependent variational principle for dissipative dynamics **86**, 062115 (2012). (ID: 718228)

498. N. Y. Yao, C. R. Laumann, A. V. Gorshkov, S. D. Bennett, E. Demler, P. Zoller and M. Lukin, Topological flat bands from dipolar spin systems, *Phys. Rev. Lett.* **109**, 266804 (2012). (ID: 718171)

499. M. Heyl and M. Vojta, Dynamical crossover from weak to infinite randomness in the one-dimensional transverse-field Ising model at criticality, arXiv:1310.6226 (ID: 718704)

500. E. Altman, J. Toner, L. Sieberer, S. Diehl and L. Chen, Two-dimensional superfluidity in driven systems requires strong anisotropy, arXiv:1311.0876 (ID: 718667)

501. P. Hauke, Quantum disorder in the spatially completely anisotropic triangular lattice, *Phys. Rev. B* **87**, 014415 (2013). (ID: 718452)

502. P. Komar, S. D. Bennett, K. Stannigel, S. Habraken, P. Rabl, P. Zoller and M. Lukin, Single-photon nonlinearities in two-mode optomechanics, *Phys. Rev. A* **87**, 013839 (2013). (ID: 718245)

503. F. Cinti, T. Macri, W. Lechner, G. Pupillo and T. Pohl, Defect-induced supersolidity with soft-core Bosons, *Nature Communications* **5**, (2013). (ID: 718489)

504. I. Boettcher, S. Diehl, J. M. Pawlowski and C. Wetterich, Tan contact and universal high momentum behavior of the fermion propagator in the BCS-BEC crossover, *Phys. Rev. A* **87**, 023606 (2013). (ID: 718237)

505. T. Ramos, V. Sudhir, K. Stannigel, P. Zoller and T. J. Kippenberg, Nonlinear quantum optomechanics via individual intrinsic two-level defects, *Phys. Rev. Lett.* **110**, 193602 (2013). (ID: 718426)

506. P. Hauke, R. J. Sewell, M. W. Mitchell and M. Lewenstein, Quantum control of spin correlations in ultracold lattice gases, *Phys. Rev. A* **87**, 021601 (2013). (ID: 718451)

507. B. Vogell, K. Stannigel, P. Zoller, K. Hammerer, M. T. Rakher, M. Korppi, A. Jöckel and P. Treutlein, Cavity-enhanced long-distance coupling of an atomic ensemble to a micromechanical membrane, *Phys. Rev. A* **87**, 023816 (2013). (ID: 718342)

508. M. S. Rudner and L. S. Levitov, Self-sustaining dynamical nuclear polarization oscillations in quantum dots, *Phys. Rev. Lett.* **110**, 086601 (2013). (ID: 718429)

509. E. Dalla Torre, S. Diehl, M. Lukin, S. Sachdev and P. Strack, Keldysh approach for non-equilibrium phase transitions in quantum optics: Dicke model in optical cavities, *Phys. Rev. A* **87**, 023831 (2013). (ID: 718248)

510. H. Pichler, J. Schachenmayer, A. J. Daley and P. Zoller, Heating dynamics of bosonic atoms in a noisy optical lattice, *Phys. Rev. A* **87**, 033606 (2013). (ID: 718344)

511. M. Dalmonte, E. Ercolessi, M. Mattioli, F. Ortolani and D. Vodola, Magnetic properties of commensurate Bose-Bose mixtures in one-dimensional optical lattices, *Eur. Phys. J. Special Topics* **217**, 13 (2013). (ID: 718338)

512. N. Y. Yao, C. R. Laumann, A. V. Gorshkov, H. Weimer, L. Jiang, I. J. Cirac, P. Zoller and M. Lukin, Topologically protected quantum state transfer in a chiral spin liquid, *Nature Communications* **4**, 1585 (2013). (ID: 717921)

513. Y. Su, D. Bimberg, A. Knorr and A. Carmele, Collective light emission revisited: Reservoir induced coherence, *Phys. Rev. Lett.* **110**, 113604 (2013). (ID: 718512)

514. D. Banerjee, M. Bögli, M. Dalmonte, E. Rico Ortega, P. Stebler, U. Wiese, P. Zoller, Atomic quantum simulation of U(N) and SU(N) non-abelian lattice gauge theories, *Phys. Rev. Lett.* **110**, 125303 (2013). (ID: 718423)

515. W. Lechner, S. Habraken, N. Kiesel, M. Aspelmeyer and P. Zoller, Cavity optomechanics of levitated nano-dumbbells: Non-equilibrium phases and self-assembly, *Phys. Rev. Lett.* **110**, 143604 (2013). (ID: 718335)

516. N. Goldman, J. Dalibard, A. Dauphin, F. Gerbier, M. Lewenstein, P. Zoller and I. Spielman, Direct imaging of topological edges states with cold atoms, *PNAS*, 6 (2013). (ID: 718336)

517. S. Bennett, N. Y. Yao, J. Otterbach, P. Zoller, P. Rabl and M. Lukin, Phonon-induced spin-spin interactions in diamond nanostructures: Application to spin squeezing, *Phys. Rev. Lett.* **110**, 156402 (2013). (ID: 718345)

518. A. V. Gorshkov, R. Nath and T. Pohl, Dissipative many-body quantum optics in Rydberg media, *Phys. Rev. Lett.* **110**, 153601 (2013). (ID: 718296)

519. S. Habraken, W. Lechner and P. Zoller, Resonances in dissipative optomechanics with nanoparticles: Sorting, speed rectification and transverse cooling, *Phys. Rev. A* **87**, 053808 (2013). (ID: 718463)

520. W. Li, A. Glätzle, R. Nath and I. Lesanovsky, Parallel execution of quantum gates in a long linear ion chain via Rydberg mode shaping, *Phys. Rev. A* **87**, 052304 (2013). (ID: 718339)

521. L. Sieberer, S. Huber, E. Altman and S. Diehl, Dynamical critical phenomena in driven-dissipative systems, *Phys. Rev. Lett.* **110**, 195301 (2013). (ID: 718369)

522. P. Schindler, M. Müller, D. Nigg, J. T. Barreiro, E. A. Martínez, M. Hennrich, T. Monz, S. Diehl, P. Zoller and R. Blatt, Quantum simulation of open-system dynamical maps with trapped ions, *Nature Physics* **9**, 361 (2013). (ID: 718325)

523. H. Pichler, L. Bonnes, A. J. Daley, A. Läuchli and P. Zoller, Thermal versus entanglement entropy: A measurement protocol for fermionic atoms with a quantum gas microscope, *New J. Phys.* **15**, 063003 (2013). (ID: 718531)

524. Y. Hu and Z. Liang, Dimensional crossover and dimensional effects in quasi-two-dimensional Bose gases, *Mod. Phys. Lett. B* **27**, 1330010 (2013). (ID: 718542)

525. M. Buchhold, P. Strack, S. Sachdev and S. Diehl, Dicke quantum spin and photon glass in optical cavities: Non-equilibrium theory and experimental signatures, *Phys. Rev. A* **87**, 063622 (2013). (ID: 718508)

526. T. S. Theuerholz, A. Carmele, M. Richter and A. Knorr, Influence of förster interaction on light emission statistics in hybrid systems, *Phys. Rev. B* **87**, 245313 (2013). (ID: 718546)

527. C. Bardyn, M. Baranov, C. Kraus, E. Rico Ortega, A. Imamoglu, P. Zoller and S. Diehl, Topology by dissipation, *New J. Phys.* **15**, 085001 (2013). (ID: 718440)

528. C. Kraus, M. Lewenstein and I. J. Cirac, Ground states of fermionic lattice hamiltonians with permutation symmetry, *Phys. Rev. A* **88**, 022335 (2013). (ID: 718543)

529. J. Struck, M. Weinberg, C. Ölschläger, P. Windpassinger, J. Simonet, K. Sengstock, R. Höppner, P. Hauke, A. Eckardt, M. Lewenstein and L. Mathey, Engineering Ising-XY spin models in a triangular lattice via tunable artificial gauge fields, *Nature Physics* **9**, 738 (2013). (ID: 718516)

530. D. Marcos, P. Rabl, E. Rico Ortega and P. Zoller, Superconducting circuits for quantum simulation of dynamical gauge fields, *Phys. Rev. Lett.* **111**, 110504 (2013). (ID: 718537)

531. L. Bonnes, H. Pichler and A. Läuchli, Entropy perspective on the thermal crossover in a fermionic Hubbard chainr in a fermionic Hubbard chain, *Phys. Rev. B* **88**, 155103 (2013). (ID: 718498)

532. O. Romero-Isart, C. Navau, A. Sanchez, P. Zoller and J. I. Cirac, Superconducting vortex lattices for ultracold atoms, *Phys. Rev. Lett.* **111**, 145304 (2013). (ID: 718430)

533. M. Mattioli, M. Dalmonte, W. Lechner and G. Pupillo, Cluster Luttinger liquids of Rydberg-dressed gases in optical lattices, *Phys. Rev. Lett.* **111**, 165302 (2013). (ID: 718487)

534. A. Carmele, A. Knorr and F. Milde, Stabilization of photon collapse and revival dynamics by a non-Markovian phonon bath, *New J. Phys.* **15**, 105024 (2013). (ID: 718646)

535. C. Kraus, M. Dalmonte, M. Baranov, A. Läuchli and P. Zoller, Majorana edge states in two atomic wires coupled by pair-hopping, *Phys. Rev. Lett.* **111**, 173004 (2013). (ID: 718420)

536. W. Lechner and P. Zoller, From classical to quantum glasses with ultracold polar molecules, *Phys. Rev. Lett.* **111**, 185306 (2013). (ID: 718563)

537. C. Kraus, P. Zoller and M. Baranov, Braiding of atomic Majorana fermions in wire networks and implementation of the deutsch-joursza algorithm, *Phys. Rev. Lett.* **111**, 203001 (2013). (ID: 718425)

538. P. Hauke and L. Tagliacozzo, Spread of correlations in long-range interacting systems, *Phys. Rev. Lett.* **111**, 207202 (2013). (ID: 718515)

539. C. Kraus and T. J. Osborne, Generalized hartree fock theory for dispersion relations of interacting fermionic lattice systems, *Phys. Rev. B* **88**, 195126 (2013). (ID: 718229)

540. M. Maik, P. Hauke, O. Dutta, J. Zakrzewski and M. Lewenstein, Density dependent tunneling in the extended Bose–Hubbard model, *New J. Phys.* **15**, 113041 (2013). (ID: 718545)

541. P. Hauke, D. Marcos, M. Dalmonte and P. Zoller, Quantum simulation of a lattice Schwinger model in a chain of trapped ions, *Phys. Rev. X* **3**, 041018 (2013). (ID: 718532)

542. W. Lechner, D. Polster, G. Maret, P. Keim and C. Dellago, Self-organized defect strings in two-dimensional crystals, *Phys. Rev. E* **88**, 060402 (2013). (ID: 718696)

543. M. Buchhold and S. Diehl, Non-equilibrium universality in the heating dynamics of interacting luttinger liquids, arXiv:1404.3740 (ID: 718898)

544. P. Hauke and M. Heyl, Many-body localization and quantum ergodicity in disordered long-range Ising models, arXiv:1410.1491 (ID: 719375)

545. L. Chen, Z. Liang, Y. Hu and Z. Zhang, Quench dynamics of three-dimensional disordered Bose gases: Condensation, superfluidity and fingerprint of dynamical Bose glass, arXiv:1411.7540 (ID: 719089)

546. A. M. Rey, A. V. Gorshkov, C. Kraus, M. J. Martin, M. Bishof, M. D. Swallows, X. Zhang, C. Benko, J. Ye, N. D. Lemke and A. D. Ludlow, Probing many-body interactions in an optical lattice clock, *Annals of Physics* **340**, 311 (2014). (ID: 718691)

547. R. C. Caballar, S. Diehl, H. Mäkelä, M. K. Oberthaler and G. Watanabe, Dissipative preparation of phase- and number-squeezed states with ultracold atoms, *Phys. Rev. A* **89**, 013620 (2014). (ID: 718618)

548. C. Laflamme, M. Baranov, P. Zoller and C. Kraus, Hybrid topological quantum computation with Majorana fermions: A cold atom setup, *Phys. Rev. A* **89**, 022319 (2014). (ID: 718700)

549. M. Punk, D. Chowdhury and S. Sachdev, Topological excitations and the dynamic structure factor of spin liquids on the kagome lattice, *Nature Physics*, (2014). (ID: 718869)

550. M. Heyl, Dynamical quantum phase transitions in systems with broken-symmetry phases, *Phys. Rev. Lett.* **113**, 205701 (2014). (ID: 718903)

551. K. Stannigel, P. Hauke, D. Marcos, M. Hafezi, S. Diehl, M. Dalmonte and P. Zoller, Constrained dynamics via the Zeno effect in quantum simulation: Implementing non-Abelian lattice gauge theories with cold atoms, *Phys. Rev. Lett.* **112**, 120406 (2014). (ID: 718579)

552. E. Mani, W. Lechner, W. Kegel and P. Bolhuis, Equilibrium and non-equilibrium cluster phases in colloids with competing interactions, *Soft Matter*, (2014). (ID: 718922)

553. J. Budich, B. Trauzettel and P. Michetti, Time reversal symmetric topological exciton condensate in bilayer HgTe quantum wells, *Phys. Rev. Lett.* **112**, 146405 (2014). (ID: 718934)

554. U. C. Tauber and S. Diehl, Perturbative field-theoretical renormalization group approach to driven-dissipative Bose–Einstein criticality, *Phys. Rev. X* **4**, 021010 (2014). (ID: 718875)

555. A. Glätzle, M. Dalmonte, R. Nath, I. Rousochatzakis, R. Mössner and P. Zoller, Quantum spin ice and dimer models with Rydberg atoms, *Phys. Rev. X* **4**, 041037 (2014). (ID: 718901)

556. S. A. Hartnoll, R. Mahajan, M. Punk and S. Sachdev, Transport near the Ising-nematic quantum critical point of metals in two dimensions, *Phys. Rev. B* **89**, 155130 (2014). (ID: 718825)

557. S. Walter and J. Budich, Teleportation-induced entanglement of two nanomechanical oscillators coupled to a topological superconductor, *Phys. Rev. B* **89**, 155431 (2014). (ID: 718933)

558. L. Sieberer, S. Huber, E. Altman and S. Diehl, Non-equilibrium functional renormalization for driven-dissipative Bose–Einstein condensation, *Phys. Rev. B* **89**, 134310 (2014). (ID: 718617)

559. J. Budich, J. Eisert and E. Bergholtz, Topological insulators with arbitrarily tunable entanglement, *Phys. Rev. B* **89**, 195120 (2014). (ID: 718932)

560. E. Rico Ortega, T. Pichler, M. Dalmonte, P. Zoller and S. Montangero, Tensor networks for lattice gauge theories and atomic quantum simulation, *Phys. Rev. Lett.* **112**, 201601 (2014). (ID: 718702)

561. A. Carmele, B. Vogell, K. Stannigel and P. Zoller, Opto-nanomechanics strongly coupled to a Rydberg superatom: Coherent vs. incoherent dynamics, *New J. Phys.* **16**, 063042 (2014). (ID: 718701)

562. W. Lechner, H. Büchler and P. Zoller, Role of quantum fluctuations in the hexatic phase of cold polar molecules, *Phys. Rev. Lett.* **112**, 255301 (2014). (ID: 718725)

563. S. Hein, F. Schulze, A. Carmele and A. Knorr, Optical feedback-enhanced photon entanglement from a Biexciton cascade, *Phys. Rev. Lett.* **113**, 027401 (2014). (ID: 718956)

564. L. Zou, D. Marcos, S. Diehl, S. Putz, J. Schmiedmayer, J. Majer and P. Rabl, Implementation of the Dicke lattice model in hybrid quantum system arrays, *Phys. Rev. Lett.* **113**, 023603 (2014). (ID: 718925)

565. P. Jurcevic, B. P. Lanyon, P. Hauke, C. Hempel, P. Zoller, R. Blatt and C. F. Roos, Quasiparticle engineering and entanglement propagation in a quantum many-body system, *Nature* **511**, 202 (2014). (ID: 718717)

566. P. Hauke, M. Lewenstein and A. Eckardt, Tomography of band-insulators from quench dynamics, *Phys. Rev. Lett.* **113**, 045303 (2014). (ID: 718873)

567. A. Bühler, N. Lang, C. Kraus, G. Möller, S. Huber and H. Büchler, Majorana modes and p-wave superfluids for fermionic atoms in optical lattices, *Nature Communications* **5**, 4504 (2014). (ID: 718871)

568. X. Zhang, M. Bishof, S. L. Bromley, C. Kraus, M. Safronova, P. Zoller, A. M. Rey and J. Ye, Spectroscopic observation of SU(N)-symmetric interactions in Sr orbital magnetism, *Science* 1254978, (2014). (ID: 718864)

569. J. Budich, J. Eisert, E. Bergholtz, S. Diehl and P. Zoller, Search for localized Wannier functions of topological band structures via compressed sensing, *Phys. Rev. B* **90**, 115110 (2014). (ID: 718935)

570. D. Marcos, P. Widmer, E. Rico Ortega, M. Hafezi, P. Rabl, U. Wiese and P. Zoller, Two-dimensional lattice gauge theories with superconducting quantum circuits, *Annals of Physics* **351**, 634 (2014). (ID: 718973)

571. Z. Zhou, Z. Cai, C. Wu and Y. Wang, Quantum Monte Carlo simulation of thermodynamic properties of SU(2N) ultracold fermions in optical lattices, *Phys. Rev. B* **90**, 235139 (2014). (ID: 719110)

572. M. Heyl and M. Vojta, Dynamics of symmetry breaking during quantum real-time evolution in a minimal model system, *Phys. Rev. Lett.* **113**, 180601 (2014). (ID: 718726)

573. M. Di Dio, L. Barbiero, A. Recati and M. Dalmonte, Spontaneous Peierls dimerization and emergent bond order in one-dimensional dipolar gases, *Phys. Rev. A* **90**, 063608 (2014). (ID: 718497)

574. T. Ramos, H. Pichler, A. J. Daley and P. Zoller, Quantum spin-dimers from chiral dissipation in cold atom chains, *Phys. Rev. Lett.* **113**, 237203 (2014). (ID: 718977)

575. M. Punk, Numerical optimisation using flow equations, *Phys. Rev. E* **90**, 063307 (2014). (ID: 718924)

576. Z. Cai, U. Schollwöck and L. Pollet, Identifying a bath-induced Bose liquid in interacting Spin-Boson models, *Phys. Rev. Lett.* **113**, 260403 (2014). (ID: 719111)

577. J. Budich and M. Heyl, Dynamical topological order parameters far from equilibrium, arXiv:1504.05599 (ID: 719244)

578. M. Heyl, Dynamical quantum phase transitions: Scaling and universality, arXiv:1505.02352 (ID: 719245)

579. T. Pichler, M. Dalmonte, E. Rico Ortega, P. Zoller and S. Montangero, Real-time dynamics in U(1) lattice gauge theories with tensor networks, arXiv:1505.04440 (ID: 719259)

580. M. Łącki, H. Pichler, A. Sterdyniak, A. Lyras, V. Lembessis, O. Al-Dossary, J. Budich and P. Zoller, Quantum hall physics with cold atoms in cylindrical optical lattices, arXiv:1507.00030 (ID: 719282)

581. S. Baier, M. J. Mark, D. Petter, K. Aikawa, L. Chomaz, Z. Cai, M. Baranov, P. Zoller and F. Ferlaino, Extended Bose–Hubbard models with ultracold magnetic atoms, arXiv:1507.03500 (ID: 719290)

582. C. Laflamme, W. Evans, M. Dalmonte, U. Gerber, H. Mejia-Diaz, W. Bietenholz, U. Wiese and P. Zoller, CP(N-1) quantum field theories with alkaline-earth atoms in optical lattices, arXiv:1507.06788 (ID: 719312)

583. J. Smith, A. Lee, P. Richerme, B. Neyenhuis, P. W. Hess, P. Hauke, M. Heyl, D. A. Huse and C. Monroe, Many-body localization in a quantum simulator with programmable random disorder, arXiv:1508.07026 (ID: 719341)

584. P. Hauke, M. Heyl, L. Tagliacozzo and P. Zoller, Measuring multipartite entanglement via dynamic susceptibilities, arXiv:1509.01739v1 (ID: 719333)

585. H. Pichler and P. Zoller, Photonic quantum circuits with time delays, arXiv:1510.04646 (ID: 719362)

586. C. Laflamme, W. Evans, M. Dalmonte, U. Gerber, H. Mejia-Diaz, W. Bietenholz, U. Wiese and P. Zoller, Proposal for the quantum simulation of the CP(2) model on optical lattices, arXiv:1510.08492 (ID: 719376)

587. N. Trautmann and P. Hauke, Quantum simulation of the dynamical Casimir effect with trapped ions arXiv:1512.00990 (ID: 719415)

588. B. Vermersch, M. Punk, A. Glätzle, C. Gross and P. Zoller, Dynamical preparation of laser-excited anisotropic Rydberg crystals in 2D optical lattices, *New J. Phys.* **17**, 013008 (2015). (ID: 718969)

589. M. Dalmonte, S. Mirzaei, P. R. Muppalla, D. Marcos, P. Zoller and G. Kirchmair, Dipolar spin models with arrays of superconducting qubits, *Phys. Rev. B* **92**, 174507 (2015). (ID: 719116)

590. T. Ramos, G. Rubilar and Y. N. Obukhov, First principles analysis of the Abraham-Minkowski controversy for the momentum of light in general linear media, *J. Opt.* **17**, 025611 (2015). (ID: 718649)

591. B. Vermersch, A. Glätzle and P. Zoller, Magic distances in the blockade mechanism of Rydberg p and d states, *Phys. Rev. A* **91**, 023411 (2015). (ID: 719085)

592. W. Li, L. Chen, Z. Chen, Y. Hu, Z. Zhang and Z. Liang, Probing flat band of optically-trapped spin-orbital coupled Bose gases using Bragg spectroscopy, *Phys. Rev. A* **91**, 023629 (2015). (ID: 719066)

593. W. Lechner, D. Polster, G. Maret, C. Dellago and P. Keim, Entropy and kinetics of point-defects in two-dimensional dipolar crystals, *Phys. Rev. E* **91**, 032304 (2015). (ID: 719182)

594. Z. Xu, X. Li, P. Zoller and W. Liu, Spontaneous quantum Hall effect in an atomic spinor Bose–Fermi mixture, *Phys. Rev. Lett.* **114**, 125303 (2015). (ID: 718957)

595. H. Pichler, T. Ramos, A. J. Daley and P. Zoller, Quantum optics of chiral spin networks, *Phys. Rev. A* **91**, 042116 (2015). (ID: 719040)

596. J. Budich, P. Zoller and S. Diehl, Dissipative preparation of chern insulators, *Phys. Rev. A* **91**, 042117 (2015). (ID: 719006)

597. B. Vogell, T. Kampschulte, M. Rakher, A. Faber, P. Treutlein, K. Hammerer and P. Zoller, Long distance coupling of a quantum mechanical oscillator to the internal states of an atomic ensemble, *New J. Phys.* **17**, 043044 (2015). (ID: 719098)

598. A. Glätzle, M. Dalmonte, R. Nath, C. Gross, I. Bloch and P. Zoller, Designing frustrated quantum magnets with laser-dressed Rydberg atoms, *Phys. Rev. Lett.* **114**, 173002 (2015). (ID: 719027)

599. C. Leitold, W. Lechner and C. Dellago, A string reaction coordinate for the folding of a polymer chain, *J. Phys. Con. Mat.* **27**, 194126 (2015). (ID: 719184)

600. M. Dalmonte, J. Carrasquilla, L. Taddia, E. Ercolessi and M. Rigol, Gap scaling at Berezinskii-Kosterlitz-Thouless quantum critical points in one-dimensional Hubbard and Heisenberg models, *Phys. Rev. B* **91**, 165136 (2015). (ID: 719100)

601. J. Budich and S. Diehl, Topology of density matrices, *Phys. Rev. B* **91**, 165140 (2015). (ID: 719279)

602. P. Hauke, L. Bonnes, M. Heyl and W. Lechner, Probing entanglement in adiabatic quantum optimization with trapped ions, *Frontiers in Physics* **3**, (2015). (ID: 719090)

603. A. Amaricci, J. Budich, M. Capone, B. Trauzettel and G. Sangiovanni, First order character and observable signatures of topological quantum phase transitions, *Phys. Rev. Lett.* **114**, 185701 (2015). (ID: 719102)

604. P. Jurcevic, P. Hauke, C. Maier, C. Hempel, B. P. Lanyon, R. Blatt and C. F. Roos, Spectroscopy of interacting quasiparticles in trapped ions, *Phys. Rev. Lett.* **115**, 100501 (2015). (ID: 719241)

605. H. Tercas, S. I. Ribeiro and J. T. Mendonca, Phonon-polaritons in Bose–Einstein condensates induced by Casimir-Polder interaction with graphene, *J. Phys. Con. Mat.* **27**, (2015). (ID: 719045)

606. O. Dutta, M. Gajda, P. Hauke, M. Lewenstein, D. Lühmann, T. Sowinski and J. Zakrzewski, Non-standard Hubbard models in optical lattices, *Rep. Prog. Phys.* **78**, 066001 (2015). (ID: 718986)

607. T. Grass, C. A. Muschik, A. Celi, R. Chhajlany and M. Lewenstein, Synthetic magnetic fluxes and topological order in one-dimensional spin systems, *Phys. Rev. A* **91**, 063612 (2015). (ID: 719101)

608. M. Punk, A. Allais and S. Sachdev, A quantum dimer model for the pseudogap metal, *PNAS* **112**, 9552 (2015). (ID: 719170)

609. R. Nath, M. Dalmonte, A. Glätzle, P. Zoller, F. Schmidt-Kaler and R. Gerritsma, Hexagonal plaquette spin-spin interactions and quantum magnetism in a two-dimensional ion crystal, *New J. Phys.* **17**, 065018 (2015). (ID: 719216)

610. M. Dalmonte, W. Lechner, Z. Cai, M. Mattioli, A. Läuchli and G. Pupillo, Cluster luttinger liquids and emergent supersymmetric conformal critical points in the one-dimensional soft-shoulder Hubbard model, *Phys. Rev. B* **92**, 045106 (2015). (ID: 719173)

611. P. Hauke, Viewpoint: Zooming in on entanglement, *Physics* **8**, 66 (2015). (ID: 719296)

612. W. Lechner and P. Zoller, Spatial patterns of Rydberg excitations from logarithmic pair interactions, *Phys. Rev. Lett.* **115**, 125301 (2015). (ID: 719185)

613. M. Mancini, G. Pagano, G. Cappellini, L. Livi, M. Rider, J. Catani, C. Sias, P. Zoller, M. Inguscio, M. Dalmonte and L. Fallani, Observation of chiral edge states with neutral fermions in synthetic Hall ribbons, *Science* **349**, 1510 (2015). (ID: 719172)

614. P. Hauke and M. Heyl, Many-body localization and quantum ergodicity in disordered long-range Ising models, *Phys. Rev. B* **92**, 134204 (2015). (ID: 719030)

615. Y. Hu, Z. Cai, M. Baranov and P. Zoller, Majoranas in noisy Kitaev wires, *Phys. Rev. B* **92**, 165118 (2015). (ID: 719277)

616. W. Lechner, P. Hauke and P. Zoller, A quantum annealing architecture with all-to-all connectivity from local interactions, *Science Advances* **1**, e1500838 (2015). (ID: 719363)

617. A. Carmele, M. Heyl, C. Kraus and M. Dalmonte, Stretched exponential decay of Majorana edge modes in many-body localized Kitaev chains under dissipation, *Phys. Rev. B* **92**, 195107 (2015). (ID: 719319)

618. R. Diaz-Mendez, F. Mezzacapo, F. Cinti, W. Lechner and G. Pupillo, Quantum and classical glasses of ultrasoft particles in two dimensions, *Phys. Rev. E* **92**, 52307 (2015). (ID: 718827)

619. M. Mattioli, A. Glätzle and W. Lechner, From classical to quantum non-equilibrium dynamics of Rydberg excitations in optical lattices, *New J. Phys.* **17**, 113039 (2015). (ID: 719270)

620. Y. Hu and M. Baranov, Effects of fluctuations on Majorana edge modes in atomic topological Kitaev wire with molecular reservoir, *Phys. Rev. A* **92**, 53615 (2015). (ID: 719088)

621. W. Lechner, F. Cinti and G. Pupillo, Tunable defect interactions in classical and quantum solids *Phys. Rev. A* **92**, 53625 (2015). (ID: 719183)

622. J. Budich, C. Laflamme, F. Tschirsich, S. Montangero and P. Zoller, Synthetic helical liquids with ultracold atoms in optical lattices, *Phys. Rev. B* **92**, 245121 (2015). (ID: 719243)

623. L. Chen, Z. Liang, Y. Hu and Z. Zhang, Non-equilibrium disordered Bose gases: Condensation, superfluidity and dynamical Bose glass, *J. Phys. B: At. Mol. Opt. Phys.* **49**, (2015). (ID: 719455)

www.ingramcontent.com/pod-product-compliance
Lightning Source LLC
Chambersburg PA
CBHW081206220326
41598CB00037B/6688

* 9 7 8 9 8 1 3 1 4 1 0 2 5 *